The elements

Name	Symbol	Atomic number	Molar mass (g mol⁻¹)
Actinium	Ac	89	227
Aluminium (aluminum)	Al	13	26.98
Americium	Am	95	243
Antimony	Sb	51	121.76
Argon	Ar	18	39.95
Arsenic	As	33	74.92
Astatine	At	85	210
Barium	Ba	56	137.33
Berkelium	Bk	97	247
Beryllium	Be	4	9.01
Bismuth	Bi	83	208.98
Bohrium	Bh	107	264
Boron	B	5	10.81
Bromine	Br	35	79.90
Cadmium	Cd	48	112.41
Caesium (cesium)	Cs	55	132.91
Calcium	Ca	20	40.08
Californium	Cf	98	251
Carbon	C	6	12.01
Cerium	Ce	58	140.12
Chlorine	Cl	17	35.45
Chromium	Cr	24	52.00
Cobalt	Co	27	58.93
Copernicum	?	112	?
Copper	Cu	29	63.55
Curium	Cm	96	247
Darmstadtium	Ds	110	271
Dubnium	Db	105	262
Dysprosium	Dy	66	162.50
Einsteinium	Es	99	252
Erbium	Er	68	167.27
Europium	Eu	63	151.96
Fermium	Fm	100	257
Fluorine	F	9	19.00
Francium	Fr	87	223
Gadolinium	Gd	64	157.25
Gallium	Ga	31	69.72
Germanium	Ge	32	72.64
Gold	Au	79	196.97
Hafnium	Hf	72	178.49
Hassium	Hs	108	269
Helium	He	2	4.00
Holmium	Ho	67	164.93
Hydrogen	H	1	1.008
Indium	In	49	114.82
Iodine	I	53	126.90
Iridium	Ir	77	192.22
Iron	Fe	26	55.84
Krypton	Kr	36	83.80
Lanthanum	La	57	138.91
Lawrencium	Lr	103	262
Lead	Pb	82	207.2
Lithium	Li	3	6.94
Lutetium	Lu	71	174.97
Magnesium	Mg	12	24.31
Manganese	Mn	25	54.94

Name	Symbol	Atomic number	Molar mass (g mol⁻¹)
Meitnerium	Mt	109	268
Mendelevium	Md	101	258
Mercury	Hg	80	200.59
Molybdenun	Mo	42	95.94
Neodymium	Nd	60	144.24
Neon	Ne	10	20.18
Neptunium	Np	93	237
Nickel	Ni	28	58.69
Niobium	Nb	41	92.91
Nitrogen	N	7	14.01
Nobelium	No	102	259
Osmium	Os	76	190.23
Oxygen	O	8	16.00
Palladium	Pd	46	106.42
Phosphorus	P	15	30.97
Platinum	Pt	78	195.08
Plutonium	Pu	94	244
Polonium	Po	84	209
Potassium	K	19	39.10
Praseodymium	Pr	59	140.91
Promethium	Pm	61	145
Protactinium	Pa	91	231.04
Radium	Ra	88	226
Radon	Rn	86	222
Rhenium	Re	75	186.21
Rhodium	Rh	45	102.91
Roentgenium	Rg	111	272
Rubidium	Rb	37	85.47
Ruthenium	Ru	44	101.07
Rutherfordium	Rf	104	261
Samarium	Sm	62	150.36
Scandium	Sc	21	44.96
Seaborgium	Sg	106	266
Selenium	Se	34	78.96
Silicon	Si	14	28.09
Silver	Ag	47	107.87
Sodium	Na	11	22.99
Strontium	Sr	38	87.62
Sulfur	S	16	32.06
Tantalum	Ta	73	180.95
Technetium	Tc	43	98
Tellurium	Te	52	127.60
Terbium	Tb	65	158.93
Thallium	Tl	81	204.38
Thorium	Th	90	232.04
Thulium	Tm	69	168.93
Tin	Sn	50	118.71
Titanium	Ti	22	47.87
Tungsten	W	74	183.84
Uranium	U	92	238.03
Vanadium	V	23	50.94
Xenon	Xe	54	131.29
Ytterbium	Yb	70	173.04
Yttrium	Y	39	88.91
Zinc	Zn	30	65.41
Zirconium	Zr	40	91.22

Shriver & Atkins'
Inorganic Chemistry

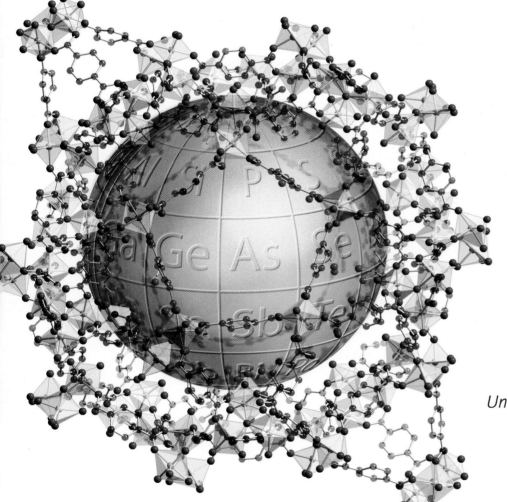

Shriver & Atkins'
Inorganic Chemistry

Fifth Edition

Peter Atkins
University of Oxford

Tina Overton
University of Hull

Jonathan Rourke
University of Warwick

Mark Weller
University of Southampton

Fraser Armstrong
University of Oxford

Michael Hagerman
Union College, New York

W. H. Freeman and Company
New York

Shriver and Atkins' Inorganic Chemistry, Fifth Edition

ISBN-13: 978-1-4292-1820-7
ISBN-10: 1-4292-1820-7

Published in Great Britain by Oxford University Press

This edition has been authorized by Oxford University Press for sale in the
United States and Canada only and not for export therefrom.

First printing

W. H. Freeman and Company
41 Madison Avenue, New York, NY 10010
www.whfreeman.com

Preface

Our aim in the fifth edition of *Shriver and Atkins' Inorganic Chemistry* is to provide a comprehensive and contemporary introduction to the diverse and fascinating discipline of inorganic chemistry. Inorganic chemistry deals with the properties of all of the elements in the periodic table. These elements range from highly reactive metals, such as sodium, to noble metals, such as gold. The nonmetals include solids, liquids, and gases, and range from the aggressive oxidizing agent fluorine to unreactive gases such as helium. Although this variety and diversity are features of any study of inorganic chemistry, there are underlying patterns and trends which enrich and enhance our understanding of the discipline. These trends in reactivity, structure, and properties of the elements and their compounds provide an insight into the landscape of the periodic table and provide a foundation on which to build understanding.

Inorganic compounds vary from ionic solids, which can be described by simple applications of classical electrostatics, to covalent compounds and metals, which are best described by models that have their origin in quantum mechanics. We can rationalize and interpret the properties of most inorganic compounds by using qualitative models that are based on quantum mechanics, such as atomic orbitals and their use to form molecular orbitals. The text builds on similar qualitative bonding models that should already be familiar from introductory chemistry courses. Although qualitative models of bonding and reactivity clarify and systematize the subject, inorganic chemistry is essentially an experimental subject. New areas of inorganic chemistry are constantly being explored and new and often unusual inorganic compounds are constantly being synthesized and identified. These new inorganic syntheses continue to enrich the field with compounds that give us new perspectives on structure, bonding, and reactivity.

Inorganic chemistry has considerable impact on our everyday lives and on other scientific disciplines. The chemical industry is strongly dependent on it. Inorganic chemistry is essential to the formulation and improvement of modern materials such as catalysts, semiconductors, optical devices, superconductors, and advanced ceramic materials. The environmental and biological impact of inorganic chemistry is also huge. Current topics in industrial, biological, and environmental chemistry are mentioned throughout the book and are developed more thoroughly in later chapters.

In this new edition we have refined the presentation, organization, and visual representation. All of the book has been revised, much has been rewritten and there is some completely new material. We have written with the student in mind, and we have added new pedagogical features and have enhanced others.

The topics in Part 1, *Foundations*, have been revised to make them more accessible to the reader with more qualitative explanation accompanying the more mathematical treatments.

Part 2, *The elements and their compounds*, has been reorganized. The section starts with a new chapter which draws together periodic trends and cross references forward to the descriptive chapters. The remaining chapters start with hydrogen and proceed across the periodic table from the s-block metals, across the p block, and finishing with the d- and f-block elements. Most of these chapters have been reorganized into two sections: *Essentials* describes the essential chemistry of the elements and the *Detail* provides a more thorough account. The chemical properties of each group of elements and their compounds are enriched with descriptions of current applications. The patterns and trends that emerge are rationalized by drawing on the principles introduced in Part 1.

Part 3, *Frontiers*, takes the reader to the edge of knowledge in several areas of current research. These chapters explore specialized subjects that are of importance to industry, materials, and biology, and include catalysis, nanomaterials, and bioinorganic chemistry.

All the illustrations and the marginal structures—nearly 1500 in all—have been redrawn and are presented in full colour. We have used colour systematically rather than just for decoration, and have ensured that it serves a pedagogical purpose.

We are confident that this text will serve the undergraduate chemist well. It provides the theoretical building blocks with which to build knowledge and understanding of inorganic chemistry. It should help to rationalize the sometimes bewildering diversity of descriptive chemistry. It also takes the student to the forefront of the discipline and should therefore complement many courses taken in the later stages of a programme.

Peter Atkins
Tina Overton
Jonathan Rourke
Mark Weller
Fraser Armstrong
Mike Hagerman

March 2009

Acknowledgements

We have taken care to ensure that the text is free of errors. This is difficult in a rapidly changing field, where today's knowledge is soon replaced by tomorrow's. We would particularly like to thank Jennifer Armstrong, University of Southampton; Sandra Dann, University of Loughborough; Rob Deeth, University of Warwick; Martin Jones, Jennifer Creen, and Russ Egdell, University of Oxford, for their guidance and advice.

Many of the figures in Chapter 27 were produced using PyMOL software; for more information see DeLano, W.L. The PyMOL Molecular Graphics System (2002), De Lano Scientific, San Carlos, CA, USA.

We acknowledge and thank all those colleagues who so willingly gave their time and expertise to a careful reading of a variety of draft chapters.

Rolf Berger, University of Uppsala, Sweden
Harry Bitter, University of Utrecht, The Netherlands
Richard Blair, University of Central Florida
Andrew Bond, University of Southern Denmark, Denmark
Darren Bradshaw, University of Liverpool
Paul Brandt, North Central College
Karen Brewer, Hamilton College
George Britovsek, Imperial College, London
Scott Bunge, Kent State University
David Cardin, University of Reading
Claire Carmalt, University College London
Carl Carrano, San Diego State University
Neil Champness, University of Nottingham
Ferman Chavez, Oakland University
Ann Chippindale, University of Reading
Karl Coleman, University of Durham
Simon Collison, University of Nottingham
Bill Connick, University of Cincinnati
Stephen Daff, University of Edinburgh
Sandra Dann, University of Loughborough
Nancy Dervisi, University of Cardiff
Richard Douthwaite, University of York
Simon Duckett, University of York
A.W. Ehlers, Free University of Amsterdam, The Netherlands
Anders Eriksson, University of Uppsala, Sweden
Andrew Fogg, University of Liverpool
Margaret Geselbracht, Reed College
Gregory Grant, University of Tennessee
Yurii Gun'ko, Trinity College Dublin
Simon Hall, University of Bristol
Justin Hargreaves, University of Glasgow

Richard Henderson, University of Newcastle
Eva Hervia, University of Strathclyde
Brendan Howlin, University of Surrey
Songping Huang, Kent State University
Carl Hultman, Gannon University
Stephanie Hurst, Northern Arizona University
Jon Iggo, University of Liverpool
S. Jackson, University of Glasgow
Michael Jensen, Ohio University
Pavel Karen, University of Oslo, Norway
Terry Kee, University of Leeds
Paul King, Birbeck, University of London
Rachael Kipp, Suffolk University
Caroline Kirk, University of Loughborough
Lars Kloo, KTH Royal Institute of Technology, Sweden
Randolph Kohn, University of Bath
Simon Lancaster, University of East Anglia
Paul Lickiss, Imperial College, London
Sven Lindin, University of Stockholm, Sweden
Paul Loeffler, Sam Houston State University
Paul Low, University of Durham
Astrid Lund Ramstrad, University of Bergen, Norway
Jason Lynam, University of York
Joel Mague, Tulane University
Francis Mair, University of Manchester
Mikhail Maliarik, University of Uppsala, Sweden
David E. Marx, University of Scranton
Katrina Miranda, University of Arizona
Grace Morgan, University College Dublin
Ebbe Nordlander, University of Lund, Sweden
Lars Öhrström, Chalmers (Goteborg), Sweden

Ivan Parkin, University College London
Dan Price, University of Glasgow
T. B. Rauchfuss, University of Illinois
Jan Reedijk, University of Leiden, The Netherlands
David Richens, St Andrews University
Denise Rooney, National University of Ireland, Maynooth
Graham Saunders, Queens University Belfast
Ian Shannon, University of Birmingham
P. Shiv Halasyamani, University of Houston
Stephen Skinner, Imperial College, London
Bob Slade, University of Surrey
Peter Slater, University of Surrey
LeGrande Slaughter, Oklahoma State University

Martin B. Smith, University of Loughborough
Sheila Smith, University of Michigan
Jake Soper, Georgia Institute of Technology
Jonathan Steed, University of Durham
Gunnar Svensson, University of Stockholm, Sweden
Andrei Verdernikov, University of Maryland
Ramon Vilar, Imperial College, London
Keith Walters, Northern Kentucky University
Robert Wang, Salem State College
David Weatherburn, University of Victoria, Wellington
Paul Wilson, University of Bath
Jingdong Zhang, Denmark Technical University

About the book

Inorganic chemistry is an extensive subject that at first sight can seem daunting. We have made every effort to help by organizing the information in this textbook systematically, and by including numerous features that are designed to make learning inorganic chemistry more effective and more enjoyable. Whether you work through the book chronologically or dip in at an appropriate point in your studies, this text will engage you and help you to develop a deeper understanding of the subject. We have also provided further electronic resources in the accompanying Book Companion Site. The following paragraphs explain the features of the text and website in more detail.

Organizing the information

Key points

The key points act as a summary of the main take-home message(s) of the section that follows. They will alert you to the principal ideas being introduced.

> #### 2.1 The octet rule
>
> **Key point:** Atoms share electron pairs until they have acquired an octet of valence electrons.
>
> Lewis found that he could account for the existence of a wide range of molecules by proposing the **octet rule**:

Context boxes

The numerous context boxes illustrate the diversity of inorganic chemistry and its applications to advanced materials, industrial processes, environmental chemistry, and everyday life, and are set out distinctly from the text itself.

> **BOX 11.1 Lithium batteries**
>
> The very negative standard potential and low molar mass of lithium make it an ideal anode material for batteries. These batteries have high specific energy (energy production divided by the mass of the battery) because lithium metal and compounds containing lithium are relatively light in comparison with some other materials used in batteries, such as lead and zinc. Lithium batteries are common, but there are many types based on different lithium compounds and reactions.
>
> The lithium rechargeable battery, used in portable computers and phones, mainly uses $Li_{1-x}CoO_2$ ($x < 1$) as the cathode with a lithium/graphite anode,
>
> the redox reaction in a similar way to the cobalt. The latest generation of electric cars uses lithium battery technology rather than lead-acid cells.
>
> Another popular lithium battery uses thionyl chloride, $SOCl_2$. This system produces a light, high-voltage cell with a stable energy output. The overall reaction in the battery is
>
> $$2\,Li(s) + 3\,SOCl_2(l) \rightarrow LiCl(s) + S(s) + SO_2(l)$$
>
> The battery requires no additional solvent as both $SOCl_2$ and SO_2 are liquids at the internal battery pressure. This battery is not rechargeable as

Further reading

Each chapter lists sources where more information can be found. We have tried to ensure that these sources are easily available and have indicated the type of information each one provided.

> **FURTHER READING**
>
> P. Atkins and J. de Paula, *Physical chemistry*. Oxford University Press and W.H. Freeman & Co (2010). An account of the generation and use of character tables without too much mathematical background.
>
> For more rigorous introductions, see: J.S. Ogden, *Introduction to molecular symmetry*. Oxford University Press (2001).
>
> P. Atkins and R. Friedman, *Molecular quantum mechanics*. Oxford University Press (2005).

Resource section

At the back of the book is a collection of resources, including an extensive data section and information relating to group theory and spectroscopy.

> ## Resource section 1
> ## Selected ionic radii
>
> Ionic radii are given (in picometres, pm) for the most common oxidation states and coordination geometries. The coordination number is given in parentheses. All d-block species are low-spin unless labelled with[†], in which case values for high-spin are quoted. Most data are taken from R.D. Shannon, *Acta Cryst.*, 1976, **A32**, 751, where values for

Problem solving

EXAMPLE 6.1 Identifying symmetry elements

Identify the symmetry elements in the eclipsed and staggered conformations of an ethane molecule.

Answer We need to identify the rotations, reflections, and inversions that leave the molecule apparently unchanged. Don't forget that the identity is a symmetry operation. By inspection of the molecular models, we see that the eclipsed conformation of a CH_3CH_3 molecule (1) has the elements E, C_3, C_2, σ_h, σ_v, and S_3. The staggered conformation (2) has the elements E, C_3, σ_d, i, and S_6.

Self-test 6.1 Sketch the S_4 axis of an NH_4^+ ion. How many of these axes does the ion possess?

EXERCISES

6.1 Draw sketches to identify the following symmetry elements: (a) a C_3 axis and a σ_v plane in the NH_3 molecule, (b) a C_4 axis and a σ_h plane in the square-planar $[PtCl_4]^{2-}$ ion.

6.2 Which of the following molecules and ions has (a) a centre of inversion, (b) an S_4 axis: (i) CO_2, (ii) C_2H_2, (iii) BF_3, (iv) SO_4^{2-}?

6.3 Determine the symmetry elements and assign the point group of (a) NH_2Cl, (b) CO_3^{2-}, (c) SiF_4, (d) HCN, (e) $SiFClBrI$, (f) BF_4^-.

6.4 How many planes of symmetry does a benzene molecule possess? What chloro-substituted benzene of formula $C_6H_nCl_{6-n}$ has exactly four planes of symmetry?

6.5 Determine the symmetry elements of objects with the same shape as the boundary surface of (a) an s orbital, (b) a p orbital, (c) a d_{xy} orbital, (d) a d_{z^2} orbital.

6.6 (a) Determine the symmetry group of an SO_3^{2-} ion. (b) What is the maximum degeneracy of a molecular orbital in this ion? (c) If the sulfur orbitals are 3s and 3p, which of them can contribute to molecular orbitals of this maximum degeneracy?

6.7 (a) Determine the point group of the PF_5 molecule. (Use VSEPR, if necessary, to assign geometry.) (b) What is the maximum degeneracy of its molecular orbitals? (c) Which P3p orbitals contribute to a molecular orbital of this degeneracy?

220, 213, and 83 cm^{-1}. Detailed analysis of the 369 and 295 cm^{-1} bands show them to arise from totally symmetric modes. Show that the Raman spectrum is consistent with a trigonal-bipyramidal geometry.

6.9 How many vibrational modes does an SO_3 molecule have (a) in the plane of the nuclei, (b) perpendicular to the molecular plane?

6.10 What are the symmetry species of the vibrations of (a) SF_6, (b) BF_3 that are both IR and Raman active?

6.11 What are the symmetry species of the vibrational modes of a C_{2v} molecule that are neither IR nor Raman active?

6.12 The $[AuCl_4]^-$ ion has D_{4h} symmetry. Determine the representations Γ of all $3N$ displacements and reduce it to obtain the symmetry species of the irreducible representations.

6.13 How could IR and Raman spectroscopy be used to distinguish between: (a) planar and pyramidal forms of PF_3, (b) planar and 90°-twisted forms of B_2F_4 (D_{2h} and D_{2d}, respectively).

6.14 (a) Take the four hydrogen 1s orbitals of CH_4 and determine how they transform under T_d. (b) Confirm that it is possible to reduce this representation to $A_1 + T_2$. (c) With which atomic orbitals on C would it be possible to form MOs with H1s SALCs of symmetry $A_1 + T_2$?

6.15 Consider CH_4. Use the projection operator method to construct the SALCs of $A_1 + T_2$ symmetry that derive from the four H1s orbitals.

PROBLEMS

6.1 Consider a molecule IF_3O_2 (with I as the central atom). How many isomers are possible? Assign point group designations to each isomer.

6.2 (a) Determine the point group of the most symmetric planar conformation of $B(OH)_3$ and the most symmetric nonplanar

conformation of $B(OH)_3$. Assume that the B−O−H bond angles are 109.5° in all conformations. (b) Sketch a conformation of $B(OH)_3$ that is chiral, once again keeping all three B−O−H bond angles equal to 109.5°.

Examples and Self-tests

We have provided numerous *Worked examples* throughout the text. Each one illustrates an important aspect of the topic under discussion or provides practice with calculations and problems.

Each *Example* is followed by a *Self-test*, where the answer is provided as a check that the method has been mastered. Think of *Self-tests* as in-chapter exercises designed to help you monitor your progress.

Exercises

There are many brief *Exercises* at the end of each chapter. Answers are found in the *Answers* section and fully worked answers are available in the separate *Solutions manual*. The *Exercises* can be used to check your understanding and gain experience and practice in tasks such as balancing equations, predicting and drawing structures, and manipulating data.

Problems

The *Problems* are more demanding in content and style than the *Exercises* and are often based on a research paper or other additional source of information. Problems generally require a discursive response and there may not be a single correct answer. They may be used as essay type questions or for classroom discussion.

New Molecular Modelling Problems

Over the past two decades computational chemistry has evolved from a highly specialized tool, available to relatively few researchers, into a powerful and practical alternative to experimentation, accessible to all chemists. The driving force behind this evolution is the remarkable progress in computer technology. Calculations that previously required hours or days on giant mainframe computers may now be completed in a fraction of time on a personal computer. It is natural and necessary that computational chemistry finds its way into the undergraduate chemistry curriculum. This requires a hands-on approach, just as teaching experimental chemistry requires a laboratory.

With this edition we have the addition of new molecular modelling problems for almost every chapter, which can be found on the text's companion web site. The problems were written to be performed using the popular *Spartan Student*™ software. With purchase of this text, students can purchase Wavefunction's *Spartan Student*™ at a significant discount from www.wavefun.com/cart/spartaned.html using the code WHFICHEM. While the problems are written to be performed using *Spartan Student*™ they can be completed using any electronic structure program that allows Hartree-Fock, density functional, and MP2 calculations.

About the Book Companion Site

The Book Companion Site which accompanies this book provides teaching and learning resources to augment the printed book. It is free of charge, and provides additional material for download, much of which can be incorporated into a virtual learning environment.

You can access the Book Companion Site by visiting
www.whfreeman.com/ichem5e

Please note that instructor resources are available only to registered adopters of the textbook. To register, simply visit **www.whfreeman.com/ichem5e** and follow the appropriate links. You will be given the opportunity to select your own username and password, which will be activated once your adoption has been verified.

Student resources are openly available to all, without registration.

Instructor resources

Artwork
An instructor may wish to use the figures from this text in a lecture. Almost all the figures are available in PowerPoint® format and can be used for lectures without charge (but not for commercial purposes without specific permission).

Tables of data
All the tables of data that appear in the chapter text are available and may be used under the same conditions as the figures.

New Molecular Modelling Problems
With this edition we have the addition of new molecular modelling problems for almost every chapter, which can be found on the text's companion web site. The problems were written to be performed using the popular *Spartan Student*™ software. With purchase of this text, students can purchase Wavefunction's *Spartan Student*™ at a significant discount from www.wavefun.com/cart/spartaned.html using the code WHFICHEM. While the problems are written to be performed using *Spartan Student*™ they can be completed using any electronic structure program that allows Hartree-Fock, density functional, and MP2 calculations.

Student resources

3D rotatable molecular structures
Nearly all the numbered molecular structures featured in the book are available in a three-dimensional, viewable, rotatable form along with many of the crystal structures and bioinorganic molecules. These have been produced in collaboration with Dr Karl Harrison, University of Oxford.

Group theory tables
Comprehensive group theory tables are available for downloading.

Videos of chemical reactions
Video clips showing demonstrations of inorganic chemistry reactions are available for viewing.

Solutions manual

As with the previous edition, Michael Hagerman, Christopher Schnabel, and Kandalam Ramanujachary have produced the solutions manual to accompany this book. A *Solution Manual* (978-142-925255-3) provides completed solutions to most end of chapter Exercises and Self-tests.

Spartan Student discount

With purchase of this text, students can purchase Wavefunction's *Spartan Student*™ at a significant discount at www.wavefun.com/cart/spartaned.html using the code WHFICHEM.

Answers to Self-tests and Exercises

Please visit the Book Companion Site at www.whfreeman.com/ichem5e/ to download a PDF document containing answers to the end-of-chapter exercises in this book.

Summary of contents

Contents

Glossary of chemical abbreviations

Ac	acetyl, CH_3CO
acac	acetylacetonato
aq	aqueous solution species
bpy	2,2'-bipyridine
cod	1,5-cyclooctadiene
cot	cyclooctatetraene
Cy	cyclohexyl
Cp	cyclopentadienyl
Cp*	pentamethylcyclopentadienyl
cyclam	tetraazacyclotetradecane
dien	diethylenetriamine
DMSO	dimethylsulfoxide
DMF	dimethylformamide
η	hapticity
edta	ethylenediaminetetraacetato
en	ethylenediamine (1,2-diaminoethane)
Et	ethyl
gly	glycinato
Hal	halide
^{i}Pr	isopropyl
KCP	$K_2Pt(CN)_4Br_{0.3} \cdot 3H_2O$
L	a ligand
μ	signifies a bridging ligand
M	a metal
Me	methyl
mes	mesityl, 2,4,6-trimethylphenyl
Ox	an oxidized species
ox	oxalato
Ph	phenyl
phen	phenanthroline
py	pyridine
Sol	solvent, or a solvent molecule
soln	nonaqueous solution species
^{t}Bu	tertiary butyl
THF	tetrahydrofuran
TMEDA	N, N,N',N'-tetramethylethylenediamine
trien	2,2',2''-triaminotriethylene
X	generally halogen, also a leaving group or an anion
Y	an entering group

PART 1
Foundations

The eight chapters in this part of the book lay the foundations of inorganic chemistry. The first three chapters develop an understanding of the structures of atoms, molecules, and solids. Chapter 1 introduces the structure of atoms in terms of quantum theory and describes important periodic trends in their properties. Chapter 2 develops molecular structure in terms of increasingly sophisticated models of covalent bonding. Chapter 3 describes ionic bonding and the structures and properties of a range of typical ionic solids. The next two chapters focus on two major types of reactions. Chapter 4 introduces the definitions of acids and bases, and uses their properties to systematize many inorganic reactions. Chapter 5 describes oxidation and reduction, and demonstrates how electrochemical data can be used to predict and explain the outcomes of redox reactions. Chapter 6 shows how a systematic consideration of the symmetry of molecules can be used to discuss the bonding and structure of molecules and help interpret the techniques described in Chapter 8. Chapter 7 describes the coordination compounds of the elements. We discuss bonding, structure, and reactions of complexes, and see how symmetry considerations can provide useful insight into this important class of compounds. Chapter 8 provides a toolbox for inorganic chemistry: it describes a wide range of the instrumental techniques that are used to identify and determine the structures of compounds.

Atomic structure

1

This chapter lays the foundations for the explanation of the trends in the physical and chemical properties of all inorganic compounds. To understand the behaviour of molecules and solids we need to understand atoms: our study of inorganic chemistry must therefore begin with a review of their structures and properties. We begin with discussion of the origin of matter in the solar system and then consider the development of our understanding of atomic structure and the behaviour of electrons in atoms. We introduce quantum theory qualitatively and use the results to rationalize properties such as atomic radii, ionization energy, electron affinity, and electronegativity. An understanding of these properties allows us to begin to rationalize the diverse chemical properties of the more than 110 elements known today.

The observation that the universe is expanding has led to the current view that about 15 billion years ago the currently visible universe was concentrated into a point-like region that exploded in an event called the **Big Bang**. With initial temperatures immediately after the Big Bang of about 10^9 K, the fundamental particles produced in the explosion had too much kinetic energy to bind together in the forms we know today. However, the universe cooled as it expanded, the particles moved more slowly, and they soon began to adhere together under the influence of a variety of forces. In particular, the **strong force**, a short-range but powerful attractive force between nucleons (protons and neutrons), bound these particles together into nuclei. As the temperature fell still further, the **electromagnetic force**, a relatively weak but long-range force between electric charges, bound electrons to nuclei to form atoms, and the universe acquired the potential for complex chemistry and the existence of life.

Table 1.1 summarizes the properties of the only subatomic particles that we need to consider in chemistry. All the known elements—by 2008, 112 had been confirmed and several more are candidates for confirmation—that are formed from these subatomic particles are distinguished by their **atomic number**, Z, the number of protons in the nucleus of an atom of the element. Many elements have a number of **isotopes**, which are atoms with the same atomic number but different atomic masses. These isotopes are distinguished by the **mass**

Table 1.1 Subatomic particles of relevance to chemistry

Practice	Symbol	Mass/m_u*	Mass number	Charge/$e^\dagger$	Spin
Electron	e^-	5.486×10^{-4}	0	-1	$\frac{1}{2}$
Proton	p	1.0073	1	$+1$	$\frac{1}{2}$
Neutron	n	1.0087	1	0	$\frac{1}{2}$
Photon	γ	0	0	0	1
Neutrino	ν	c. 0	0	0	$\frac{1}{2}$
Positron	e^+	5.486×10^{-4}	0	$+1$	$\frac{1}{2}$
α particle	α	[$^4_2\text{He}^{2+}$ nucleus]	4	$+2$	0
β particle	β	[e^- ejected from nucleus]	0	-1	$\frac{1}{2}$
γ photon	γ	[electromagnetic radiation from nucleus]	0	0	1

* Masses are expressed relative to the atomic mass constant, $m_u = 1.6605 \times 10^{-27}$ kg.
$\dagger$ The elementary charge is $e = 1.602 \times 10^{-19}$ C.

number, A, which is the total number of protons and neutrons in the nucleus. The mass number is also sometimes termed more appropriately the *nucleon number*. Hydrogen, for instance, has three isotopes. In each case $Z = 1$, indicating that the nucleus contains one proton. The most abundant isotope has $A = 1$, denoted ^1H, its nucleus consisting of a single proton. Far less abundant (only 1 atom in 6000) is deuterium, with $A = 2$. This mass number indicates that, in addition to a proton, the nucleus contains one neutron. The formal designation of deuterium is ^2H, but it is commonly denoted D. The third, short-lived, radioactive isotope of hydrogen is tritium, ^3H or T. Its nucleus consists of one proton and two neutrons. In certain cases it is helpful to display the atomic number of the element as a left suffix; so the three isotopes of hydrogen would then be denoted $^1_1H, {}^2_1H,$ and ^3_1H.

The origin of the elements

About two hours after the start of the universe, the temperature had fallen so much that most of the matter was in the form of H atoms (89 per cent) and He atoms (11 per cent). In one sense, not much has happened since then for, as Fig. 1.1 shows, hydrogen and helium

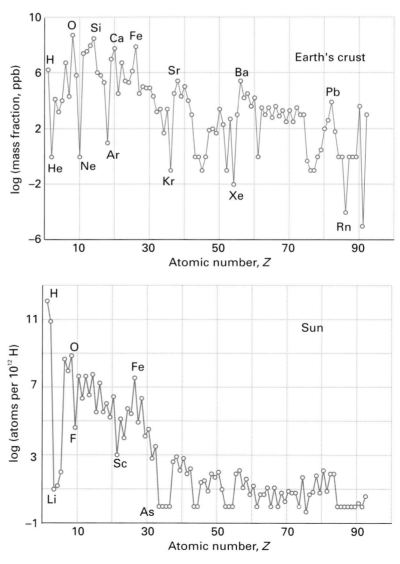

Figure 1.1 The abundances of the elements in the Earth's crust and the Sun. Elements with odd Z are less stable than their neighbours with even Z.

remain overwhelmingly the most abundant elements in the universe. However, nuclear reactions have formed a wide assortment of other elements and have immeasurably enriched the variety of matter in the universe, and thus given rise to the whole area of chemistry.

1.1 The nucleosynthesis of light elements

Key points: The light elements were formed by nuclear reactions in stars formed from primeval hydrogen and helium; total mass number and overall charge are conserved in nuclear reactions; a large binding energy signifies a stable nucleus.

The earliest stars resulted from the gravitational condensation of clouds of H and He atoms. The compression of these clouds under the influence of gravity gave rise to high temperatures and densities within them, and fusion reactions began as nuclei merged together. The earliest nuclear reactions are closely related to those now being studied in connection with the development of controlled nuclear fusion.

Energy is released when light nuclei fuse together to give elements of higher atomic number. For example, the nuclear reaction in which an α particle (a ^{4}He nucleus with two protons and two neutrons) fuses with a carbon-12 nucleus to give an oxygen-16 nucleus and a γ-ray photon (γ) is

$$^{12}_{6}\text{C} + ^{4}_{2}\alpha \rightarrow ^{16}_{8}\text{O} + \gamma$$

This reaction releases 7.2 MeV of energy.[1] Nuclear reactions are very much more energetic than normal chemical reactions because the strong force is much stronger than the electromagnetic force that binds electrons to nuclei. Whereas a typical chemical reaction might release about 10^3 kJ mol^{-1}, a nuclear reaction typically releases a million times more energy, about 10^9 kJ mol^{-1}. In this nuclear equation, the **nuclide**, a nucleus of specific atomic number Z and mass number A, is designated $^{A}_{Z}\text{E}$, where E is the chemical symbol of the element. Note that, in a balanced nuclear equation, the sum of the mass numbers of the reactants is equal to the sum of the mass numbers of the products (12 + 4 = 16). The atomic numbers sum similarly (6 + 2 = 8) provided an electron, e$^-$, when it appears as a β particle, is denoted $^{0}_{-1}\text{e}$ and a positron, e$^+$, is denoted $^{0}_{1}\text{e}$. A positron is a positively charged version of an electron: it has zero mass number (but not zero mass) and a single positive charge. When it is emitted, the mass number of the nuclide is unchanged but the atomic number decreases by 1 because the nucleus has lost one positive charge. Its emission is equivalent to the conversion of a proton in the nucleus into a neutron: $^{1}_{1}\text{p} \rightarrow ^{1}_{0}\text{n} + \text{e}^+ + \nu$. A neutrino, ν (nu), is electrically neutral and has a very small (possibly zero) mass.

Elements up to $Z = 26$ were formed inside stars. Such elements are the products of the nuclear fusion reactions referred to as 'nuclear burning'. The burning reactions, which should not be confused with chemical combustion, involved H and He nuclei and a complicated fusion cycle catalysed by C nuclei. (The stars that formed in the earliest stages of the evolution of the cosmos lacked C nuclei and used noncatalysed H-burning reactions.) Some of the most important nuclear reactions in the cycle are

Proton (p) capture by carbon-12: $\qquad ^{12}_{6}\text{C} + ^{1}_{1}\text{p} \rightarrow ^{13}_{7}\text{N} + \gamma$

Positron decay accompanied by neutrino (ν) emission: $\qquad ^{13}_{7}\text{N} \rightarrow ^{13}_{6}\text{C} + \text{e}^+ + \nu$

Proton capture by carbon-13: $\qquad ^{13}_{6}\text{C} + ^{1}_{1}\text{p} \rightarrow ^{14}_{7}\text{N} + \gamma$

Proton capture by nitrogen-14: $\qquad ^{14}_{7}\text{N} + ^{1}_{1}\text{p} \rightarrow ^{15}_{8}\text{O} + \gamma$

Positron decay, accompanied by neutrino emission: $\qquad ^{15}_{8}\text{O} \rightarrow ^{15}_{7}\text{N} + \text{e}^+ + \nu$

Proton capture by nitrogen-15: $\qquad ^{15}_{7}\text{N} + ^{1}_{1}\text{p} \rightarrow ^{12}_{6}\text{C} + ^{4}_{2}\alpha$

The net result of this sequence of nuclear reactions is the conversion of four protons (four ^{1}H nuclei) into an α particle (a ^{4}He nucleus):

$$4\,^{1}_{1}\text{p} \rightarrow ^{4}_{2}\alpha + 2\,\text{e}^+ + 2\nu + 3\gamma$$

---

[1] An electronvolt (1 eV) is the energy required to move an electron through a potential difference of 1 V. It follows that 1 eV = 1.602×10^{-19} J, which is equivalent to 96.48 kJ mol^{-1}; 1 MeV = 10^6 eV.

The reactions in the sequence are rapid at temperatures between 5 and 10 MK (where 1 MK = 10^6 K). Here we have another contrast between chemical and nuclear reactions, because chemical reactions take place at temperatures a hundred thousand times lower. Moderately energetic collisions between species can result in chemical change, but only highly vigorous collisions can provide the energy required to bring about most nuclear processes.

Heavier elements are produced in significant quantities when hydrogen burning is complete and the collapse of the star's core raises the density there to 10^8 kg m^{-3} (about 10^5 times the density of water) and the temperature to 100 MK. Under these extreme conditions, helium burning becomes viable. The low abundance of beryllium in the present-day universe is consistent with the observation that $_4^8$Be formed by collisions between α particles goes on to react with more α particles to produce the more stable carbon nuclide $_6^{12}$C:

$$_4^8\text{Be} + _2^4\alpha \rightarrow _6^{12}\text{C} + \gamma$$

Thus, the helium-burning stage of stellar evolution does not result in the formation of Be as a stable end product; for similar reasons, low concentrations of Li and B are also formed. The nuclear reactions leading to these three elements are still uncertain, but they may result from the fragmentation of C, N, and O nuclei by collisions with high-energy particles.

Elements can also be produced by nuclear reactions such as neutron (n) capture accompanied by proton emission:

$$_7^{14}\text{N} + _0^1\text{n} \rightarrow _6^{14}\text{C} + _1^1\text{p}$$

This reaction still continues in our atmosphere as a result of the impact of cosmic rays and contributes to the steady-state concentration of radioactive carbon-14 on Earth.

The high abundance of iron and nickel in the universe is consistent with these elements having the most stable of all nuclei. This stability is expressed in terms of the **binding energy**, which represents the difference in energy between the nucleus itself and the same numbers of individual protons and neutrons. This binding energy is often presented in terms of a difference in mass between the nucleus and its individual protons and neutrons because, according to Einstein's theory of relativity, mass and energy are related by $E = mc^2$, where c is the speed of light. Therefore, if the mass of a nucleus differs from the total mass of its components by $\Delta m = m_{\text{nucleons}} - m_{\text{nucleus}}$, then its binding energy is $E_{\text{bind}} = (\Delta m)c^2$. The binding energy of ^{56}Fe, for example, is the difference in energy between the ^{56}Fe nucleus and 26 protons and 30 neutrons. A positive binding energy corresponds to a nucleus that has a lower, more favourable, energy (and lower mass) than its constituent nucleons (Box 1.1).

Figure 1.2 shows the binding energy per nucleon, E_{bind}/A (obtained by dividing the total binding energy by the number of nucleons), for all the elements. Iron and nickel occur at the maximum of the curve, showing that their nucleons are bound more strongly than in any other nuclide. Harder to see from the graph is an alternation of binding energies as the atomic number varies from even to odd, with even-Z nuclides slightly more stable than their odd-Z neighbours. There is a corresponding alternation in cosmic abundances, with nuclides of even atomic number being marginally more abundant than those of odd atomic number. This stability of even-Z nuclides is attributed to the lowering of energy by pairing nucleons in the nucleus.

1.2 The nucleosynthesis of heavy elements

Key point: Heavier nuclides are formed by processes that include neutron capture and subsequent β decay.

Nuclei close to iron are the most stable and heavier elements are produced by a variety of processes that require energy. These processes include the capture of free neutrons, which are not present in the earliest stages of stellar evolution but are produced later in reactions such as

$$_{10}^{23}\text{Na} + _2^4\alpha \rightarrow _{12}^{26}\text{Mg} + _0^1\text{n}$$

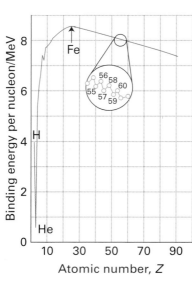

Figure 1.2 Nuclear binding energies. The greater the binding energy, the more stable is the nucleus. Note the alternation in stability shown in the inset.

BOX 1.1 Nuclear fusion and nuclear fission

If two nuclei with mass numbers lower than 56 merge to produce a new nucleus with a larger nuclear binding energy, the excess energy is released. This process is called **fusion**. For example, two neon-20 nuclei may fuse to give a calcium-40 nucleus:

$$2\,^{20}_{10}\text{Ne} \rightarrow\ ^{40}_{20}\text{Ca}$$

The value of E_{bind}/A for Ne is approximately 8.0 MeV. Therefore, the total binding energy of the species on the left-hand side of the equation is $2 \times 20 \times 8.0$ MeV = 320 MeV. The value of E_{bind}/A for Ca is close to 8.6 MeV and so the total energy of the species on the right-hand side is 40×8.6 MeV = 344 MeV. The difference in the binding energies of the products and reactants is therefore 24 MeV.

For nuclei with $A > 56$, binding energy can be released when they split into lighter products with higher values of E_{bind}/A. This process is called **fission**. For example, uranium-236 can undergo fission into (among many other modes) xenon-140 and strontium-93 nuclei:

$$^{236}_{92}\text{U} \rightarrow\ ^{140}_{54}\text{Xe} +\,^{93}_{38}\text{Sr} +\,^{1}_{0}\text{n}$$

The values of E_{bind}/A for ^{236}U, ^{140}Xe, and ^{93}Sr nuclei are 7.6, 8.4, and 8.7 MeV, respectively. Therefore, the energy released in this reaction is $(140 \times 8.4) + (93 \times 8.7) - (236 \times 7.6)$ MeV = 191.5 MeV for the fission of each ^{236}U nucleus.

Fission can also be induced by bombarding heavy elements with neutrons:

$$^{235}_{92}\text{U} +\,^{1}_{0}\text{n} \rightarrow\ \text{fission products} + \text{neutrons}$$

The kinetic energy of fission products from ^{235}U is about 165 MeV, that of the neutrons is about 5 MeV, and the γ-rays produced have an energy of about 7 MeV. The fission products are themselves radioactive and decay by β-, γ-, and X-radiation, releasing about 23 MeV. In a nuclear fission reactor the neutrons that are not consumed by fission are captured with the release of about 10 MeV. The energy produced is reduced by about 10 MeV, which escapes from the reactor as radiation, and about 1 MeV which remains as undecayed fission products in the spent fuel. Therefore, the total energy produced for one fission event is about 200 MeV, or 32 pJ. It follows that about 1 W of reactor heat (where 1 W = 1 J s^{-1}) corresponds to about 3.1×10^{10} fission events per second. A nuclear reactor producing 3 GW has an electrical output of approximately 1 GW and corresponds to the fission of 3 kg of ^{235}U per day.

The use of nuclear power is controversial in large part on account of the risks associated with the highly radioactive, long-lived spent fuel. The declining stocks of fossil fuels, however, make nuclear power very attractive as it is estimated that stocks of uranium could last for about 100 years. The cost of uranium ores is currently very low and one small pellet of uranium oxide generates as much energy as three barrels of oil or 1 tonne of coal. The use of nuclear power would also drastically reduce the rate of emission of greenhouse gases. The environmental drawback with nuclear power is the storage and disposal of radioactive waste and the public's continued nervousness about possible nuclear accidents and misuse in pursuit of political ambitions.

Under conditions of intense neutron flux, as in a supernova (one type of stellar explosion), a given nucleus may capture a succession of neutrons and become a progressively heavier isotope. However, there comes a point at which the nucleus will eject an electron from the nucleus as a β particle (a high-velocity electron, e$^-$). Because β decay leaves the mass number of the nuclide unchanged but increases its atomic number by 1 (the nuclear charge increases by 1 unit when an electron is ejected), a new element is formed. An example is

Neutron capture: $\quad^{98}_{42}\text{Mo} +\,^{1}_{0}\text{n} \rightarrow\ ^{99}_{42}\text{Mo} + \gamma$

Followed by β decay accompanied by neutrino emission: $\quad^{99}_{42}\text{Mo} \rightarrow\ ^{99}_{43}\text{Tc} + \text{e}^- + \nu$

The **daughter nuclide**, the product of a nuclear reaction ($^{99}_{33}\text{Tc}$, an isotope of technetium, in this example), can absorb another neutron, and the process can continue, gradually building up the heavier elements (Box 1.2).

BOX 1.2 Technetium—the first synthetic element

A synthetic element is one that does not occur naturally on Earth but that can be artificially generated by nuclear reactions. The first synthetic element was technetium (Tc, $Z = 43$), named from the Greek word for 'artificial'. Its discovery—more precisely, its preparation—filled a gap in the periodic table and its properties matched those predicted by Mendeleev. The longest-lived isotope of technetium (^{98}Tc) has a half-life of 4.2 million years so any produced when the Earth was formed has long since decayed. Technetium is produced in red giant stars.

The most widely used isotope of technetium is ^{99m}Tc, where the 'm' indicates a metastable isotope. Technetium-99m emits high-energy γ-rays but has a relatively short half-life of 6.01 hours. These properties make the isotope particularly attractive for use *in vivo* as the γ-ray energy is sufficient for it to be detected outside the body and its half-life means

that most of it will have decayed within 24 hours. Consequently, ^{99m}Tc is widely used in nuclear medicine, for example in radiopharmaceuticals for imaging and functional studies of the brain, bones, blood, lungs, liver, heart, thyroid gland, and kidneys. Technetium-99m is generated through nuclear fission in nuclear power plants but a more useful laboratory source of the isotope is a technetium generator, which uses the decay of ^{99}Mo to ^{99m}Tc. The half-life of ^{99}Mo is 66 hours, which makes it more convenient for transport and storage than ^{99m}Tc itself. Most commercial generators are based on ^{99}Mo in the form of the molybdate ion, MoO_4^{2-}, adsorbed on Al_2O_3. The $^{99}\text{MoO}_4^{2-}$ ion decays to the pertechnetate ion, $^{99m}\text{TcO}_4^-$, which is less tightly bound to the alumina. Sterile saline solution is washed through a column of the immobilized ^{99}Mo and the ^{99m}Tc solution is collected.

EXAMPLE 1.1 **Balancing equations for nuclear reactions**

Synthesis of heavy elements occurs in the neutron-capture reactions believed to take place in the interior of cool 'red giant' stars. One such reaction is the conversion of $^{68}_{30}$Zn to $^{69}_{31}$Ga by neutron capture to form $^{69}_{30}$Zn, which then undergoes β decay. Write balanced nuclear equations for this process.

Answer We use the fact that the sum of the mass numbers and the sum of the atomic numbers on each side of the equation must be the same. Neutron capture increases the mass number of a nuclide by 1 but leaves the atomic number (and hence the identity of the element) unchanged:

$$^{68}_{30}\text{Zn} + {}^{1}_{0}\text{n} \rightarrow {}^{69}_{30}\text{Zn} + \gamma$$

The excess energy is carried away as a photon. The loss of an electron from the nucleus by β decay leaves the mass number unchanged but increases the atomic number by 1. Because zinc has atomic number 30, the daughter nuclide has $Z = 31$, corresponding to gallium. Therefore, the nuclear reaction is

$$^{69}_{30}\text{Zn} \rightarrow {}^{69}_{31}\text{Ga} + \text{e}^-$$

In fact, a neutrino is also emitted, but this cannot be inferred from the data as a neutrino is effectively massless and electrically neutral.

Self-test 1.1 Write the balanced nuclear equation for neutron capture by $^{80}_{35}$Br.

The structures of hydrogenic atoms

The organization of the periodic table is a direct consequence of periodic variations in the electronic structure of atoms. Initially, we consider hydrogen-like or **hydrogenic atoms,** which have only one electron and so are free of the complicating effects of electron–electron repulsions. Hydrogenic atoms include ions such as He⁺ and C⁵⁺ (found in stellar interiors) as well as the hydrogen atom itself. Then we use the concepts these atoms introduce to build up an approximate description of the structures of **many-electron atoms** (or *polyelectron atoms*), which are atoms with more than one electron.

1.3 Spectroscopic information

Key points: Spectroscopic observations on hydrogen atoms suggest that an electron can occupy only certain energy levels and that the emission of discrete frequencies of electromagnetic radiation occurs when an electron makes a transition between these levels.

Electromagnetic radiation is emitted when an electric discharge is passed through hydrogen gas. When passed through a prism or diffraction grating, this radiation is found to consist of a series of components: one in the ultraviolet region, one in the visible region, and several in the infrared region of the electromagnetic spectrum (Fig. 1.3; Box 1.3).

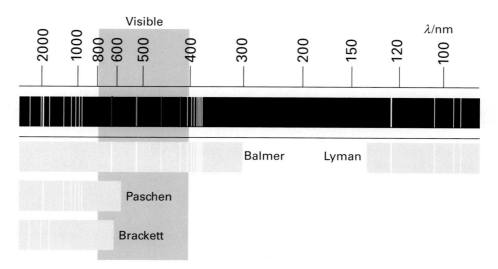

Figure 1.3 The spectrum of atomic hydrogen and its analysis into series.

The emission of light when atoms are excited is put to good use in lighting streets in many parts of the world. The widely used yellow street lamps are based on the emission of light from excited sodium atoms.

Low pressure sodium (LPS) lamps consist of a glass tube coated with indium tin oxide (ITO), a solid solution of In_2O_3 with typically 10 per cent by mass SnO_2. The indium tin oxide reflects the infrared radiation and transmits the visible light. Two inner glass tubes hold solid sodium and a small amount of neon and argon, the same mixture as found in neon

lights. When the lamp is turned on, the neon and argon emit a red glow and heat the sodium metal. The sodium rapidly starts to vaporize and the electrical discharge excites the atoms and they re-emit the energy as yellow light from the transition $3p \rightarrow 3s$. One advantage of sodium lamps over other types of street lighting is that their light output does not diminish with age. They do, however, use more energy towards the end of their life, which may make them less attractive from environmental and economic perspectives.

The nineteenth-century spectroscopist Johann Rydberg found that all the wavelengths (λ, lambda) can be described by the expression

$$\frac{1}{\lambda} = R\left(\frac{1}{n_1^2} - \frac{1}{n_2^2}\right) \tag{1.1}$$

where R is the **Rydberg constant**, an empirical constant with the value 1.097×10^7 m^{-1}. The n are integers, with $n_1 = 1, 2, \ldots$ and $n_2 = n_1 + 1, n_1 + 2, \ldots$. The series with $n_1 = 1$ is called the *Lyman series* and lies in the ultraviolet. The series with $n_1 = 2$ lies in the visible region and is called the *Balmer series*. The infrared series include the *Paschen series* ($n_1 = 3$) and the *Brackett series* ($n_1 = 4$).

The structure of the spectrum is explained if it is supposed that the emission of radiation takes place when an electron makes a transition from a state of energy $-hcR/n_2^2$ to a state of energy $-hcR/n_1^2$ and that the difference, which is equal to $hcR(1/n_1^2 - 1/n_2^2)$, is carried away as a photon of energy hc/λ. By equating these two energies, and cancelling hc, we obtain eqn 1.1.

The question these observations raise is why the energy of the electron in the atom is limited to the values $-hcR/n^2$ and why R has the value observed. An initial attempt to explain these features was made by Niels Bohr in 1913 using an early form of quantum theory in which he supposed that the electron could exist in only certain circular orbits. Although he obtained the correct value of R, his model was later shown to be untenable as it conflicted with the version of quantum theory developed by Erwin Schrödinger and Werner Heisenberg in 1926.

1.4 Some principles of quantum mechanics

Key points: Electrons can behave as particles or as waves; solution of the Schrödinger equation gives wavefunctions, which describe the location and properties of electrons in atoms. The probability of finding an electron at a given location is proportional to the square of the wavefunction. Wavefunctions generally have regions of positive and negative amplitude, and may undergo constructive or destructive interference with one another.

In 1924, Louis de Broglie suggested that because electromagnetic radiation could be considered to consist of particles called photons yet at the same time exhibit wave-like properties, such as interference and diffraction, then the same might be true of electrons. This dual nature is called **wave–particle duality**. An immediate consequence of duality is that it is impossible to know the linear momentum (the product of mass and velocity) and the location of an electron (and any particle) simultaneously. This restriction is the content of Heisenberg's **uncertainty principle**, that the product of the uncertainty in momentum and the uncertainty in position cannot be less than a quantity of the order of Planck's constant (specifically, $\frac{1}{2}\hbar$, where $\hbar = h/2\pi$).

Schrödinger formulated an equation that took account of wave–particle duality and accounted for the motion of electrons in atoms. To do so, he introduced the **wavefunction**, ψ (psi), a mathematical function of the position coordinates x, y, and z which describes the behaviour of an electron. The **Schrödinger equation**, of which the wavefunction is a solution, for an electron free to move in one dimension is

$$\underbrace{-\frac{\hbar^2}{2m_e}\frac{d^2\psi}{dx^2}}_{\text{Kinetic energy contribution}} + \underbrace{V(x)\psi(x)}_{\text{Potential energy contribution}} = \underbrace{E\psi(x)}_{\text{Total energy contribution}} \tag{1.2}$$

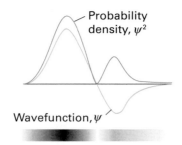

Figure 1.4 The Born interpretation of the wavefunction is that its square is a probability density. There is zero probability density at a node.

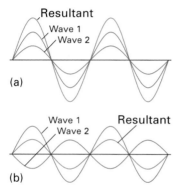

Figure 1.5 Wavefunctions interfere where they spread into the same region of space. (a) If they have the same sign in a region, they interfere constructively and the total wavefunction has an enhanced amplitude in the region. (b) If the wavefunctions have opposite signs, then they interfere destructively, and the resulting superposition has a reduced amplitude.

where m_e is the mass of an electron, V is the potential energy of the electron, and E is its total energy. The Schrödinger equation is a second-order differential equation that can be solved exactly for a number of simple systems (such as a hydrogen atom) and can be solved numerically for many more complex systems (such as many-electron atoms and molecules). However, we shall need only qualitative aspects of its solutions. The generalization of eqn 1.2 to three dimensions is straightforward, but we do not need its explicit form.

One crucial feature of eqn 1.2 and its analogues in three dimensions is that physically acceptable solutions exist only for certain values of E. Therefore, the **quantization** of energy, the fact that an electron can possess only certain discrete energies in an atom, follows naturally from the Schrödinger equation, in addition to the imposition of certain requirements ('boundary conditions') that restrict the number of acceptable solutions.

A wavefunction contains all the dynamical information possible about the electron, including where it is and what it is doing. Specifically, the probability of finding an electron at a given location is proportional to the square of the wavefunction at that point, ψ^2. According to this interpretation, there is a high probability of finding the electron where ψ^2 is large, and the electron will not be found where ψ^2 is zero (Fig. 1.4). The quantity ψ^2 is called the **probability density** of the electron. It is a 'density' in the sense that the product of ψ^2 and the infinitesimal volume element $d\tau = dxdydz$ (where τ is tau) is proportional to the probability of finding the electron in that volume. The probability is *equal* to $\psi^2 d\tau$ if the wavefunction is 'normalized'. A normalized wavefunction is one that is scaled so that the total probability of finding the electron somewhere is 1.

Like other waves, wavefunctions in general have regions of positive and negative amplitude, or sign. The sign of the wavefunction is of crucial importance when two wavefunctions spread into the same region of space and interact. Then a positive region of one wavefunction may add to a positive region of the other wavefunction to give a region of enhanced amplitude. This enhancement is called **constructive interference** (Fig. 1.5a). It means that, where the two wavefunctions spread into the same region of space, such as occurs when two atoms are close together, there may be a significantly enhanced probability of finding the electrons in that region. Conversely, a positive region of one wavefunction may be cancelled by a negative region of the second wavefunction (Fig. 1.5b). This **destructive interference** between wavefunctions reduces the probability that an electron will be found in that region. As we shall see, the interference of wavefunctions is of great importance in the explanation of chemical bonding. To help keep track of the relative signs of different regions of a wavefunction in illustrations, we label regions of opposite sign with dark and light shading (sometimes white in the place of light shading).

1.5 Atomic orbitals

The wavefunction of an electron in an atom is called an **atomic orbital**. Chemists use hydrogenic atomic orbitals to develop models that are central to the interpretation of inorganic chemistry, and we shall spend some time describing their shapes and significance.

(a) Hydrogenic energy levels

Key points: The energy of the bound electron is determined by n, the principal quantum number; in addition, l specifies the magnitude of the orbital angular momentum and m_l specifies the orientation of that angular momentum.

Each of the wavefunctions obtained by solving the Schrödinger equation for a hydrogenic atom is uniquely labelled by a set of three integers called **quantum numbers**. These quantum numbers are designated n, l, and m_l: n is called the **principal quantum number**, l is the **orbital angular momentum quantum number** (formerly the 'azimuthal quantum number'), and m_l is called the **magnetic quantum number**. Each quantum number specifies a physical property of the electron: n specifies the energy, l labels the magnitude of the orbital angular momentum, and m_l labels the orientation of that angular momentum. The value of n also indicates the size of the orbital, with high n, high-energy orbitals more diffuse than low n compact, tightly bound, low-energy orbitals. The value of l also indicates the angular shape of the orbital, with the number of lobes increasing as l increases. The value of m_l also indicates the orientation of these lobes.

The allowed energies are specified by the principal quantum number, n. For a hydrogenic atom of atomic number Z, they are given by

$$E_n = -\frac{hcRZ^2}{n^2} \tag{1.3}$$

with $n = 1, 2, 3, \ldots$ and

$$R = \frac{m_e e^4}{8h^3 c \varepsilon_0^2} \tag{1.4}$$

(The fundamental constants in this expression are given inside the back cover.) The calculated numerical value of R is 1.097×10^7 m^{-1}, in excellent agreement with the empirical value determined spectroscopically. For future reference, the value of hcR corresponds to 13.6 eV. The zero of energy (at $n = \infty$) corresponds to the electron and nucleus being widely separated and stationary. Positive values of the energy correspond to unbound states of the electron in which it may travel with any velocity and hence possess any energy. The energies given by eqn 1.3 are all negative, signifying that the energy of the electron in a bound state is lower than a widely separated stationary electron and nucleus. Finally, because the energy is proportional to $1/n^2$, the energy levels converge as the energy increases (becomes less negative, Fig. 1.6).

The value of l specifies the magnitude of the orbital angular momentum through $\{l(l + 1)\}^{1/2}\hbar$, with $l = 0, 1, 2, \ldots$. We can think of l as indicating the rate at which the electron circulates around the nucleus. As we shall see shortly, the third quantum number m_l specifies the orientation of this momentum, for instance whether the circulation is clockwise or anticlockwise.

(b) Shells, subshells, and orbitals

Key points: All orbitals with a given value of n belong to the same shell, all orbitals of a given shell with the same value of l belong to the same subshell, and individual orbitals are distinguished by the value of m_l.

In a hydrogenic atom, all orbitals with the same value of n have the same energy and are said to be **degenerate**. The principal quantum number therefore defines a series of **shells** of the atom, or sets of orbitals with the same value of n and hence with the same energy and approximately the same radial extent. Shells with $n = 1, 2, 3 \ldots$ are commonly referred to as K, L, M, $\ldots$ shells.

The orbitals belonging to each shell are classified into **subshells** distinguished by a quantum number l. For a given value of n, the quantum number l can have the values $l = 0, 1, \ldots, n - 1$, giving n different values in all. For example, the shell with $n = 1$ consists of just one subshell with $l = 0$, the shell with $n = 2$ consists of two subshells, one with $l = 0$ and the other with $l = 1$, the shell with $n = 3$ consists of three subshells, with values of l of 0, 1, and 2. It is common practice to refer to each subshell by a letter:

Value of l	0	1	2	3	4	$\ldots$
Subshell designation	s	p	d	f	g	$\ldots$

For most purposes in chemistry we need consider only s, p, d, and f subshells.

A subshell with quantum number l consists of $2l + 1$ individual orbitals. These orbitals are distinguished by the **magnetic quantum number,** m_l, which can have the $2l + 1$ integer values from $+l$ down to $-l$. This quantum number specifies the component of orbital angular momentum around an arbitrary axis (commonly designated z) passing through the nucleus. So, for example, a d subshell of an atom ($l = 2$) consists of five individual atomic orbitals that are distinguished by the values $m_l = +2, +1, 0, -1, -2$.

A note on good practice Write the sign of m_l, even when it is positive. Thus, we write $m_l = +2$, not $m_l = 2$.

The practical conclusion for chemistry from these remarks is that there is only one orbital in an s subshell ($l = 0$), the one with $m_l = 0$: this orbital is called an **s orbital**. There are three orbitals in a p subshell ($l = 1$), with quantum numbers $m_l = +1, 0, -1$; they are called **p orbitals**. The five orbitals of a d subshell ($l = 2$) are called **d orbitals**, and so on (Fig. 1.7).

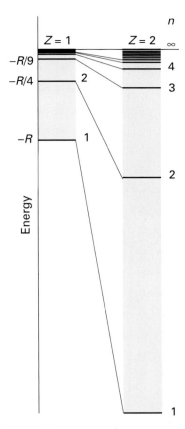

Figure 1.6 The quantized energy levels of an H atom ($Z = 1$) and an He$^+$ ion ($Z = 2$). The energy levels of a hydrogenic atom are proportional to Z^2.

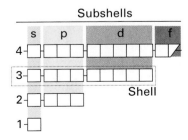

Figure 1.7 The classification of orbitals into subshells (same value of l) and shells (same value of n).

EXAMPLE 1.2 Identifying orbitals from quantum numbers

Which set of orbitals is defined by $n = 4$ and $l = 1$? How many orbitals are there in this set?

Answer We need to remember that the principal quantum number n identifies the shell and that the orbital quantum number l identifies the subshell. The subshell with $l = 1$ consists of p orbitals. The allowed values of $m_l = l, l - 1,\ldots, -l$ give the number of orbitals of that type. In this case, $m_l = +1, 0$, and -1. There are therefore three 4p orbitals.

Self-test 1.2 Which set of orbitals is defined by the quantum numbers $n = 3$ and $l = 2$? How many orbitals are there in this set?

(c) Electron spin

Key points: The intrinsic spin angular momentum of an electron is defined by the two quantum numbers s and m_s. Four quantum numbers are needed to define the state of an electron in a hydrogenic atom.

In addition to the three quantum numbers required to specify the spatial distribution of an electron in a hydrogenic atom, two more quantum numbers are needed to define the state of an electron. These additional quantum numbers relate to the intrinsic angular momentum of an electron, its **spin**. This evocative name suggests that an electron can be regarded as having an angular momentum arising from a spinning motion, rather like the daily rotation of a planet as it travels in its annual orbit around the sun. However, spin is a quantum mechanical property and this analogy must be viewed with great caution.

Spin is described by two quantum numbers, s and m_s. The former is the analogue of l for orbital motion but it is restricted to the single, unchangeable value $s = \frac{1}{2}$. The magnitude of the spin angular momentum is given by the expression $\{s(s + 1)\}^{1/2}\hbar$, so for an electron this magnitude is fixed at $\frac{1}{2}\sqrt{3}\hbar$ for any electron. The second quantum number, the **spin magnetic quantum number**, m_s, may take only two values, $+\frac{1}{2}$ (anticlockwise spin, imagined from above) and $-\frac{1}{2}$ (clockwise spin). The two states are often represented by the two arrows ↑ ('spin-up', $m_s = +\frac{1}{2}$) and ↓ ('spin-down', $m_s = -\frac{1}{2}$) or by the Greek letters α and β, respectively.

Because the spin state of an electron must be specified if the state of the atom is to be specified fully, it is common to say that the state of an electron in a hydrogenic atom is characterized by four quantum numbers, namely n, l, m_l, and m_s (the fifth quantum number, s, is fixed at $\frac{1}{2}$).

(d) Nodes

Key point: Regions where wavefunctions pass through zero are called nodes.

Inorganic chemists generally find it adequate to use visual representations of atomic orbitals rather than mathematical expressions. However, we need to be aware of the mathematical expressions that underlie these representations.

Because the potential energy of an electron in the field of a nucleus is spherically symmetric (it is proportional to Z/r and independent of orientation relative to the nucleus), the orbitals are best expressed in terms of the spherical polar coordinates defined in Fig. 1.8. In these coordinates, the orbitals all have the form

$$\psi_{nlm_l} = \underbrace{R_{nl}(r)}_{\text{Variation with } radius} \times \underbrace{Y_{lm_l}(\theta,\phi)}_{\text{Variation with angle}} \tag{1.5}$$

This expression expresses the simple idea that a hydrogenic orbital can be written as the product of a function $R(r)$ of the radius and a function $Y(\theta,\phi)$ of the angular coordinates. The positions where either component of the wavefunction passes through zero are called **nodes**. Consequently, there are two types of nodes. **Radial nodes** occur where the radial component of the wavefunction passes through zero and **angular nodes** occur where the angular component of the wavefunction passes through zero. The numbers of both types of node increase with increasing energy and are related to the quantum numbers n and l.

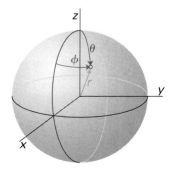

Figure 1.8 Spherical polar coordinates: r is the radius, θ (theta) the colatitude, and ϕ (phi) the azimuth.

(e) The radial variation of atomic orbitals

Key point: An s orbital has nonzero amplitude at the nucleus; all other orbitals (those with $l > 0$) vanish at the nucleus.

Figures 1.9 and 1.10 show the radial variation of some atomic orbitals. A 1s orbital, the wavefunction with $n = 1$, $l = 0$, and $m_l = 0$, decays exponentially with distance from the nucleus and never passes through zero. All orbitals decay exponentially at sufficiently great distances from the nucleus and this distance increases as n increases. Some orbitals oscillate through zero close to the nucleus and thus have one or more radial nodes before beginning their final exponential decay. As the principal quantum number of an electron increases, it is found further away from the nucleus and its energy increases.

An orbital with quantum numbers n and l in general has $n - l - 1$ radial nodes. This oscillation is evident in the 2s orbital, the orbital with $n = 2$, $l = 0$, and $m_l = 0$, which passes through zero once and hence has one radial node. A 3s orbital passes through zero twice and so has two radial nodes. A 2p orbital (one of the three orbitals with $n = 2$ and $l = 1$) has no radial nodes because its radial wavefunction does not pass through zero anywhere. However, a 2p orbital, like *all* orbitals other than s orbitals, is zero at the nucleus. For any series of the same type of orbital, the first occurrence has no radial nodes, the second has one radial node, and so on.

Although an electron in an s orbital may be found at the nucleus, an electron in any other type of orbital will not be found there. We shall soon see that this apparently minor detail, which is a consequence of the absence of orbital angular momentum when $l = 0$, is one of the key concepts for understanding chemistry.

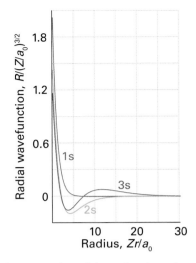

Figure 1.9 The radial wavefunctions of the 1s, 2s, and 3s hydrogenic orbitals. Note that the number of radial nodes is 0, 1, and 2, respectively. Each orbital has a nonzero amplitude at the nucleus (at $r = 0$).

EXAMPLE 1.3 Predicting numbers of radial nodes

How many radial nodes do 3p, 3d, and 4f orbitals have?

Answer We need to make use of the fact that the number of radial nodes is given by the expression $n - l - 1$ and use it to find the values of n and l. The 3p orbitals have $n = 3$ and $l = 1$ and the number of radial nodes will be $n - l - 1 = 1$. The 3d orbitals have $n = 3$ and $l = 2$. Therefore, the number of radial nodes will $n - l - 1 = 0$. The 4f orbitals have $n = 4$ and $l = 3$ and the number of radial nodes will be $n - l - 1 = 0$. The 3d and 4f orbitals are the first occurrence of the d and f orbitals so this also indicates that they will have no radial nodes.

Self-test 1.3 How many radial nodes does a 5s orbital have?

(f) The radial distribution function

Key point: A radial distribution function gives the probability that an electron will be found at a given distance from the nucleus, regardless of the direction.

The Coulombic (electrostatic) force that binds the electron is centred on the nucleus, so it is often of interest to know the probability of finding an electron at a given distance from the nucleus, regardless of its direction. This information enables us to judge how tightly the electron is bound. The total probability of finding the electron in a spherical shell of radius r and thickness dr is the integral of $\psi^2 d\tau$ over all angles. This result is written $P(r) dr$, where $P(r)$ is called the **radial distribution function**. In general,

$$P(r) = r^2 R(r)^2 \tag{1.6}$$

(For s orbitals, this expression is the same as $P = 4\pi r^2 \psi^2$.) If we know the value of P at some radius r, then we can state the probability of finding the electron somewhere in a shell of thickness dr at that radius simply by multiplying P by dr. In general, a radial distribution function for an orbital in a shell of principal quantum number n has $n - 1$ peaks, the outermost peak being the highest.

Because the wavefunction of a 1s orbital decreases exponentially with distance from the nucleus and the factor r^2 in eqn 1.6 increases, the radial distribution function of a 1s orbital goes through a maximum (Fig. 1.11). Therefore, there is a distance at which the electron is most likely to be found. In general, this most probable distance decreases as the

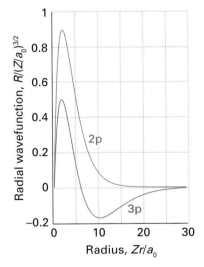

Figure 1.10 The radial wavefunctions of the 2p and 3p hydrogenic orbitals. Note that the number of radial nodes is 0 and 1, respectively. Each orbital has zero amplitude at the nucleus (at $r = 0$).

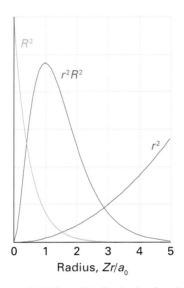

Figure 1.11 The radial distribution function of a hydrogenic 1s orbital. The product of $4\pi r^2$ (which increases as r increases) and ψ^2 (which decreases exponentially) passes through a maximum at $r = a_0/Z$.

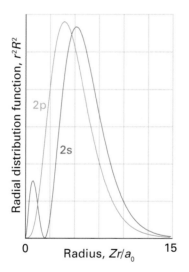

Figure 1.12 The radial distribution functions of hydrogenic orbitals. Although the 2p orbital is *on average* closer to the nucleus (note where its maximum lies), the 2s orbital has a high probability of being close to the nucleus on account of the inner maximum.

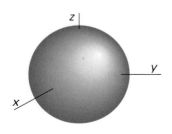

Figure 1.13 The spherical boundary surface of an s orbital.

nuclear charge increases (because the electron is attracted more strongly to the nucleus), and specifically

$$r_{max} = \frac{a_0}{Z} \tag{1.7}$$

where a_0 is the **Bohr radius**, $a_0 = \varepsilon_0 \hbar^2/\pi m_e e^2$, a quantity that appeared in Bohr's formulation of his model of the atom; its numerical value is 52.9 pm. The most probable distance increases as n increases because the higher the energy, the more likely it is that the electron will be found far from the nucleus.

EXAMPLE 1.4 Interpreting radial distribution functions

Figure 1.12 shows the radial distribution functions for 2s and 2p hydrogenic orbitals. Which orbital gives the electron a greater probability of close approach to the nucleus?

Answer By examining Fig. 1.12 we can see that the radial distribution function of a 2p orbital approaches zero near the nucleus faster than a 2s electron does. This difference is a consequence of the fact that a 2p orbital has zero amplitude at the nucleus on account of its orbital angular momentum. Thus, the 2s electron has a greater probability of close approach to the nucleus.

Self-test 1.4 Which orbital, 3p or 3d, gives an electron a greater probability of being found close to the nucleus?

(g) The angular variation of atomic orbitals

Key points: The boundary surface of an orbital indicates the region of space within which the electron is most likely to be found; orbitals with the quantum number *l* have *l* nodal planes.

The angular wavefunction expresses the variation of angle around the nucleus and this describes the orbital's angular shape. An s orbital has the same amplitude at a given distance from the nucleus whatever the angular coordinates of the point of interest: that is, an s orbital is spherically symmetrical. The orbital is normally represented by a spherical surface with the nucleus at its centre. The surface is called the **boundary surface** of the orbital, and defines the region of space within which there is a high (typically 90 per cent) probability of finding the electron. The planes on which the angular wavefunction passes through zero are called **angular nodes** or **nodal planes**. An electron will not be found anywhere on a nodal plane. A nodal plane cuts through the nucleus and separates the regions of positive and negative sign of the wavefunction.

In general, an orbital with the quantum number l has l nodal planes. An s orbital, with $l = 0$, has no nodal planes and the boundary surface of the orbital is spherical (Fig. 1.13).

All orbitals with $l > 0$ have amplitudes that vary with angle. In the most common graphical representation, the boundary surfaces of the three p orbitals of a given shell are identical apart from the fact that their axes lie parallel to each of the three different Cartesian axes centred on the nucleus, and each one possesses a nodal plane passing through the nucleus (Fig. 1.14). This representation is the origin of the labels p_x, p_y, and p_z, which are

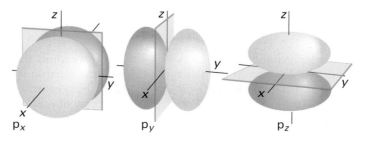

Figure 1.14 The boundary surfaces of p orbitals. Each orbital has one nodal plane running through the nucleus. For example, the nodal plane of the p orbital is the *xy*-plane. The lightly shaded lobe has a positive amplitude, the more darkly shaded one is negative.

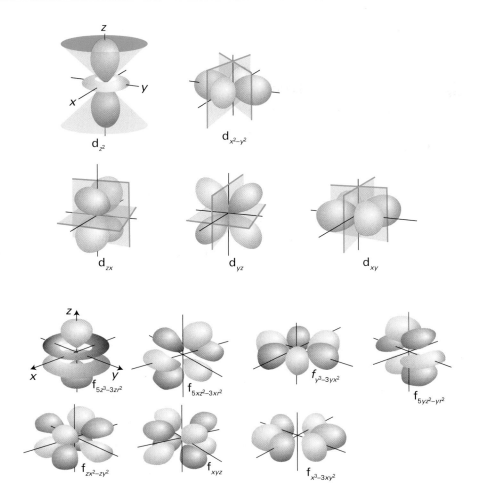

Figure 1.15 One representation of the boundary surfaces of the d orbitals. Four of the orbitals have two perpendicular nodal planes that intersect in a line passing through the nucleus. In the d_{z^2} orbital, the nodal surface forms two cones that meet at the nucleus.

Figure 1.16 One representation of the boundary surfaces of the f orbitals. Other representations (with different shapes) are also sometimes encountered.

alternatives to the use of m_l to label the individual orbitals. Each p orbital, with $l = 1$, has a single nodal plane.

The boundary surfaces and labels we use for the d and f orbitals are shown in Figs 1.15 and 1.16, respectively. The d_{z^2} orbital looks different from the remaining d orbitals. There are in fact six possible combinations of double dumb-bell shaped orbitals around three axes: three with lobes between the axes, as in d_{xy}, d_{yz}, and d_{zx}, and three with lobes along the axis. One of these orbitals is $d_{x^2-y^2}$. The d_{z^2} orbital can be thought of as the superposition of two contributions, one with lobes along the z- and x-axes and the other with lobes along the z- and y-axes. Note that a d orbital with ($l = 2$) has two nodal planes that intersect at the nucleus; a typical f orbital ($l = 3$) has three nodal planes.

Many-electron atoms

As we have remarked, a 'many-electron atom' is an atom with more than one electron, so even He, with two electrons, is technically a many-electron atom. The exact solution of the Schrödinger equation for an atom with N electrons would be a function of the $3N$ coordinates of all the electrons. There is no hope of finding exact formulas for such complicated functions; however, it is straightforward to perform numerical computations by using widely available software to obtain precise energies and probability densities. This software can also generate graphical representations of the resulting orbitals that can assist in the interpretation of the properties of the atom. For most of inorganic chemistry we rely on the **orbital approximation**, in which each electron occupies an atomic orbital that resembles those found in hydrogenic atoms. When we say that an electron 'occupies' an atomic orbital, we mean that it is described by the corresponding wavefunction.

1.6 Penetration and shielding

Key points: The ground-state electron configuration is a specification of the orbital occupation of an atom in its lowest energy state. The exclusion principle forbids more than two electrons to occupy a single orbital. The nuclear charge experienced by an electron is reduced by shielding by other electrons. Trends in effective nuclear charge can be used to rationalize the trends in many properties. As a result of the combined effects of penetration and shielding, the order of energy levels in a shell of a many-electron atom is s < p < d < f.

It is quite easy to account for the electronic structure of the helium atom in its **ground state**, its state of lowest energy. According to the orbital approximation, we suppose that both electrons occupy an atomic orbital that has the same spherical shape as a hydrogenic 1s orbital. However, the orbital will be more compact because, as the nuclear charge of helium is greater than that of hydrogen, the electrons are drawn in towards the nucleus more closely than is the one electron of an H atom. The ground-state **configuration** of an atom is a statement of the orbitals its electrons occupy in the ground state. For helium, with two electrons in the 1s orbital, the ground-state configuration is denoted $1s^2$ (read 'one s two').

As soon as we come to the next atom in the periodic table, lithium ($Z = 3$), we encounter several major new features. The configuration $1s^3$ is forbidden by a fundamental feature of nature known as the **Pauli exclusion principle:**

No more than two electrons may occupy a single orbital and, if two do occupy a single orbital, then their spins must be paired.

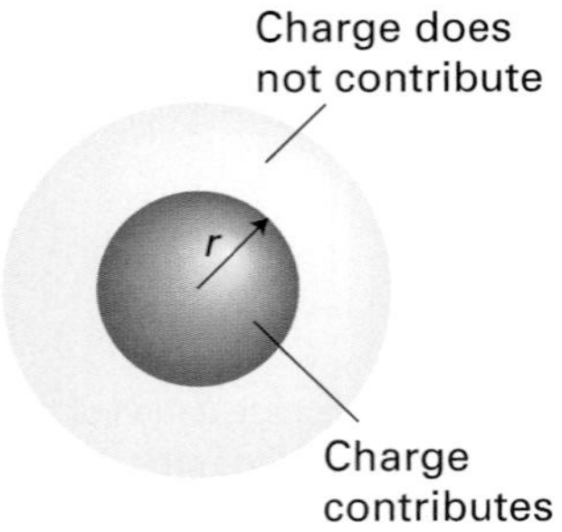

By 'paired' we mean that one electron spin must be ↑ and the other ↓; the pair is denoted ↑↓. Another way of expressing the principle is to note that, because an electron in an atom is described by four variable quantum numbers, n, l, m_l, and m_s, no two electrons can have the same four quantum numbers. The Pauli principle was introduced originally to account for the absence of certain transitions in the spectrum of atomic helium.

Because the configuration $1s^3$ is forbidden by the Pauli exclusion principle, the third electron must occupy an orbital of the next higher shell, the shell with $n = 2$. The question that now arises is whether the third electron occupies a 2s orbital or one of the three 2p orbitals. To answer this question, we need to examine the energies of the two subshells and the effect of the other electrons in the atom. Although 2s and 2p orbitals have the same energy in a hydrogenic atom, spectroscopic data and calculations show that this is not the case in a many-electron atom.

In the orbital approximation we treat the repulsion between electrons in an approximate manner by supposing that the electronic charge is distributed spherically around the nucleus. Then each electron moves in the attractive field of the nucleus and experiences an average repulsive charge from the other electrons. According to classical electrostatics, the field that arises from a spherical distribution of charge is equivalent to the field generated by a single point charge at the centre of the distribution (Fig. 1.17). This negative charge reduces the actual charge of the nucleus, Ze, to $Z_{eff}e$, where Z_{eff} (more precisely, $Z_{eff}e$) is called the **effective nuclear charge**. This effective nuclear charge depends on the values of n and l of the electron of interest because electrons in different shells and subshells approach the nucleus to different extents. The reduction of the true nuclear charge to the effective nuclear charge by the other electrons is called **shielding**. The effective nuclear charge is sometimes expressed in terms of the true nuclear charge and an empirical **shielding constant**, σ, by writing $Z_{eff} = Z - \sigma$. The shielding constant can be determined by fitting hydrogenic orbitals to those computed numerically.

The closer to the nucleus that an electron can approach, the closer is the value of Z_{eff} to Z itself because the electron is repelled less by the other electrons present in the atom. With this point in mind, consider a 2s electron in the Li atom. There is a nonzero probability that the 2s electron can be found inside the 1s shell and experience the full nuclear charge (Fig. 1.18). The presence of an electron inside shells of other electrons is called **penetration**. A 2p electron does not penetrate so effectively through the **core**, the filled inner shells of electrons, because its wavefunction goes to zero at the nucleus. As a consequence, it is more fully shielded from the nucleus by the core electrons. We can conclude that a 2s electron has a lower energy (is bound more tightly) than a 2p electron, and therefore that the 2s orbital will be occupied before the 2p orbitals, giving a ground-state electron configuration for Li of $1s^2 2s^1$. This configuration is commonly denoted $[He]2s^1$, where [He] denotes the atom's helium-like $1s^2$ core.

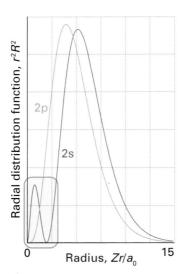

Figure 1.17 The electron at the r radius experiences a repulsion from the total charge within the sphere of radius r; charge outside that radius has no net effect.

Figure 1.18 The penetration of a 2s electron through the inner core is greater than that of a 2p electron because the latter vanishes at the nucleus. Therefore, the 2s electrons are less shielded than the 2p electrons.

Table 1.2 Effective nuclear charge, Z_{eff}

	H							He
Z	1							2
1s	1.00							1.69
	Li	Be	B	C	N	O	F	Ne
Z	3	4	5	6	7	8	9	10
1s	2.69	3.68	4.68	5.67	6.66	7.66	8.65	9.64
2s	1.28	1.91	2.58	3.22	3.85	4.49	5.13	5.76
2p			2.42	3.14	3.83	4.45	5.10	5.76
	Na	Mg	Al	Si	P	S	Cl	Ar
Z	11	12	13	14	15	16	17	18
1s	10.63	11.61	12.59	13.57	14.56	15.54	16.52	17.51
2s	6.57	7.39	8.21	9.02	9.82	10.63	11.43	12.23
2p	6.80	7.83	8.96	9.94	10.96	11.98	12.99	14.01
3s	2.51	3.31	4.12	4.90	5.64	6.37	7.07	7.76
3p			4.07	4.29	4.89	5.48	6.12	6.76

The pattern of energies in lithium, with 2s lower than 2p, and in general ns lower than np, is a general feature of many-electron atoms. This pattern can be seen from Table 1.2, which gives the values of Z_{eff} for a number of valence-shell atomic orbitals in the ground-state electron configuration of atoms. The typical trend in effective nuclear charge is an increase across a period, for in most cases the increase in nuclear charge in successive groups is not cancelled by the additional electron. The values in the table also confirm that an s electron in the outermost shell of the atom is generally less shielded than a p electron of that shell. So, for example, $Z_{eff} = 5.13$ for a 2s electron in an F atom, whereas for a 2p electron $Z_{eff} = 5.10$, a lower value. Similarly, the effective nuclear charge is larger for an electron in an np orbital than for one in an nd orbital.

As a result of penetration and shielding, the order of energies in many-electron atoms is typically ns $< n$p $< n$d $< n$f because, in a given shell, s orbitals are the most penetrating and f orbitals are the least penetrating. The overall effect of penetration and shielding is depicted in the energy-level diagram for a neutral atom shown in Fig. 1.19.

Figure 1.20 summarizes the energies of the orbitals through the periodic table. The effects are quite subtle, and the order of the orbitals depends strongly on the numbers of

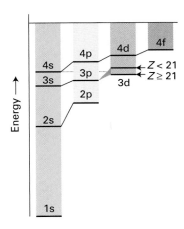

Figure 1.19 A schematic diagram of the energy levels of a many-electron atom with $Z < 21$ (as far as calcium). There is a change in order for $Z \geqslant 21$ (from scandium onwards). This is the diagram that justifies the building-up principle, with up to two electrons being allowed to occupy each orbital.

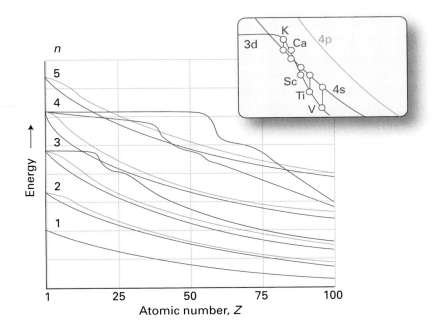

Figure 1.20 A more detailed portrayal of the energy levels of many-electron atoms in the periodic table. The inset shows a magnified view of the order near $Z = 20$, where the 3d series of elements begins.

electrons present in the atom and may change on ionization. For example, the effects of penetration are very pronounced for 4s electrons in K and Ca, and in these atoms the 4s orbitals lie lower in energy than the 3d orbitals. However, from Sc through Zn, the 3d orbitals in the neutral atoms lie close to but lower than the 4s orbitals. In atoms from Ga ($Z = 31$) onwards, the 3d orbitals lie well below the 4s orbital in energy, and the outermost electrons are unambiguously those of the 4s and 4p subshells.

1.7 The building-up principle

The ground-state electron configurations of many-electron atoms are determined experimentally by spectroscopy and are summarized in *Resource section 2*. To account for them, we need to consider both the effects of penetration and shielding on the energies of the orbitals and the role of the Pauli exclusion principle. The **building-up principle** (which is also known as the *Aufbau principle* and is described below) is a procedure that leads to plausible ground-state configurations. It is not infallible, but it is an excellent starting point for the discussion. Moreover, as we shall see, it provides a theoretical framework for understanding the structure and implications of the periodic table.

(a) Ground-state electron configurations

Key points: The order of occupation of atomic orbitals follows the order 1s, 2s, 2p, 3s, 3p, 4s, 3d, 4p, ... Degenerate orbitals are occupied singly before being doubly occupied; certain modifications of the order of occupation occur for d and f orbitals.

According to the building-up principle, orbitals of neutral atoms are treated as being occupied in the order determined in part by the principal quantum number and in part by penetration and shielding:

Order of occupation: 1s 2s 3s 3p 4s 3d 4p...

Each orbital can accommodate up to two electrons. Thus, the three orbitals in a p subshell can accommodate a total of six electrons and the five orbitals in a d subshell can accommodate up to ten electrons. The ground-state configurations of the first five elements are therefore expected to be

H	He	Li	Be	B
$1s^1$	$1s^2$	$1s^2 2s^1$	$1s^2 2s^2$	$1s^2 2s^2 2p^1$

This order agrees with experiment. When more than one orbital of the same energy is available for occupation, such as when the 2p orbitals begin to be filled in B and C, we adopt **Hund's rule**:

When more than one orbital has the same energy, electrons occupy separate orbitals and do so with parallel spins ($\uparrow\uparrow$).

The occupation of separate orbitals of the same value of l (such as a p_x orbital and a p_y orbital) can be understood in terms of the weaker repulsive interactions that exist between electrons occupying different regions of space (electrons in different orbitals) than between those occupying the same region of space (electrons in the same orbital). The requirement of parallel spins for electrons that do occupy different orbitals is a consequence of a quantum mechanical effect called **spin correlation**, the tendency for two electrons with parallel spins to stay apart from one another and hence to repel each other less. One consequence of this effect is that half-filled shells of electrons with parallel spins are particularly stable. For example, the ground state of the chromium atom is $4s^1 3d^5$ rather than $4s^2 3d^4$. Further examples of the effect of spin correlation will be seen later in this chapter.

It is arbitrary which of the p orbitals of a subshell is occupied first because they are degenerate, but it is common to adopt the alphabetical order p_x, p_y, p_z. It then follows from the building-up principle that the ground-state configuration of C is $1s^2 2s^2 2p_x^1 2p_y^1$ or, more simply, $1s^2 2s^2 2p^2$. If we recognize the helium-like core ($1s^2$), an even briefer notation is $[He]2s^2 2p^2$, and we can think of the electronic valence structure of the atom as consisting of two paired 2s electrons and two parallel 2p electrons surrounding a closed helium-like core. The electron configurations of the remaining elements in the period are similarly

C	N	O	F	Ne
$[He]2s^22p^2$	$[He]2s^22p^3$	$[He]2s^22p^4$	$[He]2s^22p^5$	$[He]2s^22p^6$

The $2s^22p^6$ configuration of neon is another example of a **closed shell**, a shell with its full complement of electrons. The configuration $1s^22s^22p^6$ is denoted [Ne] when it occurs as a core.

EXAMPLE 1.5 **Accounting for trends in effective nuclear charge**

The increase in Z_{eff} between C and N is 0.69 whereas the increase between N and O is only 0.62. Suggest a reason why the increase in Z_{eff} for a 2p electron is smaller between N and O than between C and N given the configurations of the atoms listed above.

Answer We need to identify the general trend and then think about an additional effect that might modify it. In this case, we expect to see an increase in effective nuclear charge across a period. However, on going from C to N, the additional electron occupies an empty 2p orbital whereas on going from N to O, the additional electron must occupy a 2p orbital that is already occupied by one electron. It therefore experiences stronger electron–electron repulsion, and the increase in Z_{eff} is not as great.

Self-test 1.5 Account for the larger increase in effective nuclear charge for a 2p electron on going from B to C compared with a 2s electron on going from Li to Be.

The ground-state configuration of Na is obtained by adding one more electron to a neon-like core, and is $[Ne]3s^1$, showing that it consists of a single electron outside a completely filled $1s^22s^22p^6$ core. Now a similar sequence of filling subshells begins again, with the 3s and 3p orbitals complete at argon, with configuration $[Ne]3s^23p^6$, which can be denoted [Ar]. Because the 3d orbitals are so much higher in energy, this configuration is effectively closed. Moreover, the 4s orbital is next in line for occupation, so the configuration of K is analogous to that of Na, with a single electron outside a noble-gas core: specifically, it is $[Ar]4s^1$. The next electron, for Ca, also enters the 4s orbital, giving $[Ar]4s^2$, which is the analogue of Mg. However, in the next element, Sc, the added electron occupies a 3d orbital, and filling of the d orbitals begins.

(b) Exceptions

The energy levels in Figs 1.19 and 1.20 are for individual atomic orbitals and do not fully take into account repulsion between electrons. For elements with an incompletely filled d subshell, the determination of actual ground states by spectroscopy and calculation shows that it is advantageous to occupy orbitals predicted to be *higher* in energy (the 4s orbitals). The explanation for this order is that the occupation of orbitals of higher energy can result in a reduction in the repulsions between electrons that would occur if the lower-energy 3d orbitals were occupied. It is essential when assessing the total energy of the electrons to consider all contributions to the energy of a configuration, not merely the one-electron orbital energies. Spectroscopic data show that the ground-state configurations of these atoms are mostly of the form $3d^n4s^2$, with the 4s orbitals fully occupied despite individual 3d orbitals being lower in energy.

An additional feature, another consequence of spin correlation, is that in some cases a lower total energy may be obtained by forming a half-filled or filled d subshell, even though that may mean moving an s electron into the d subshell. Therefore, as a half-filled d shell is approached the ground-state configuration is likely to be d^5s^1 and not d^4s^2 (as for Cr). As a full d subshell is approached the configuration is likely to be $d^{10}s^1$ rather than d^9s^2 (as for Cu) or $d^{10}s^0$ rather than d^8s^2 (as for Pd). A similar effect occurs where f orbitals are being occupied, and a d electron may be moved into the f subshell so as to achieve an f^7 or an f^{14} configuration, with a net lowering of energy. For instance, the ground-state electron configuration of Gd is $[Xe]4f^75d^16s^2$ and not $[Xe]4f^86s^2$.

For cations and complexes of the d-block elements the removal of electrons reduces the complicating effects of electron–electron repulsions and the 3d orbital energies fall well below that of the 4s orbitals. Consequently, all d-block cations and complexes have d^n configurations and no electrons in the outermost s orbitals. For example, the configuration of Fe is $[Ar]3d^64s^2$ whereas that of $[Fe(CO)_5]$ is $[Ar]3d^8$ and Fe^{2+} is $[Ar]3d^6$. For the purposes of chemistry, the electron configurations of the d-block ions are more important than those of the neutral atoms. In later chapters (starting in Chapter 19), we shall see the great significance of the configurations of the d-metal ions, for the subtle modulations of their energies provide the basis for the explanations of important properties of their compounds.

EXAMPLE 1.6 Deriving an electron configuration

Predict the ground-state electron configurations of (a) Ti and (b) Ti^{3+}.

Answer We need to use the building-up principle and Hund's rule to populate atomic orbitals with electrons. (a) For the neutral atom, for which $Z = 22$, we must add 22 electrons in the order specified above, with no more than two electrons in any one orbital. This procedure results in the configuration $[Ar]4s^2 3d^2$, with the two 3d electrons in different orbitals with parallel spins. However, because the 3d orbitals lie below the 4s orbitals for elements beyond Ca, it is appropriate to reverse the order in which they are written. The configuration is therefore reported as $[Ar]3d^2 4s^2$. (b) The cation has 19 electrons. We should fill the orbitals in the order specified above remembering, however, that the cation will have a d^n configuration and no electrons in the s orbital. The configuration of Ti^{3+} is therefore $[Ar]3d^1$.

Self-test 1.6 Predict the ground-state electron configurations of Ni and Ni^{2+}.

1.8 The classification of the elements

Key points: The elements are broadly divided into metals, nonmetals, and metalloids according to their physical and chemical properties; the organization of elements into the form resembling the modern periodic table is accredited to Mendeleev.

A useful broad division of elements is into **metals** and **nonmetals**. Metallic elements (such as iron and copper) are typically lustrous, malleable, ductile, electrically conducting solids at about room temperature. Nonmetals are often gases (oxygen), liquids (bromine), or solids that do not conduct electricity appreciably (sulfur). The chemical implications of this classification should already be clear from introductory chemistry:

1. Metallic elements combine with nonmetallic elements to give compounds that are typically hard, nonvolatile solids (for example sodium chloride).
2. When combined with each other, the nonmetals often form volatile molecular compounds (for example phosphorus trichloride).
3. When metals combine (or simply mix together) they produce alloys that have most of the physical characteristics of metals (for example brass from copper and zinc).

Some elements have properties that make it difficult to classify them as metals or nonmetals. These elements are called **metalloids**. Examples of metalloids are silicon, germanium, arsenic, and tellurium.

A note on good practice You will sometimes see metalloids referred to as 'semi-metals'. This name is best avoided because a semi-metal has a well defined and quite distinct meaning in physics (see Section 3.19).

(a) The periodic table

A more detailed classification of the elements is the one devised by Dmitri Mendeleev in 1869; this scheme is familiar to every chemist as the **periodic table**. Mendeleev arranged the known elements in order of increasing atomic weight (molar mass). This arrangement resulted in families of elements with similar chemical properties, which he arranged into the groups of the periodic table. For example, the fact that C, Si, Ge, and Sn all form hydrides of the general formula EH_4 suggests that they belong to the same group. That N, P, As, and Sb all form hydrides with the general formula EH_3 suggests that they belong to a different group. Other compounds of these elements show family similarities, as in the formulas CF_4 and SiF_4 in the first group, and NF_3 and PF_3 in the second.

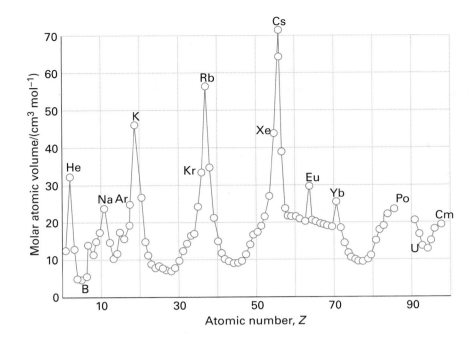

Figure 1.21 The periodic variation of molar volume with atomic number.

Mendeleev concentrated on the chemical properties of the elements. At about the same time Lothar Meyer in Germany was investigating their physical properties, and found that similar values repeated periodically with increasing molar mass. Figure 1.21 shows a classic example, where the molar volume of the element (its volume per mole of atoms) at 1 bar and 298 K is plotted against atomic number.

Mendeleev provided a spectacular demonstration of the usefulness of the periodic table by predicting the general chemical properties, such as the numbers of bonds they form, of unknown elements corresponding to gaps in his original periodic table. (He also predicted elements that we now know cannot exist and denied the presence of elements that we now know do exist, but that is overshadowed by his positive achievement and has been quietly forgotten.) The same process of inference from periodic trends is still used by inorganic chemists to rationalize trends in the physical and chemical properties of compounds and to suggest the synthesis of previously unknown compounds. For instance, by recognizing that carbon and silicon are in the same family, the existence of alkenes $R_2C=CR_2$ suggests that $R_2Si=SiR_2$ ought to exist too. Compounds with silicon–silicon double bonds (disila-ethenes) do indeed exist, but it was not until 1981 that chemists succeeded in isolating one. The periodic trends in the properties of the elements are explored further in Chapter 9.

(b) The format of the periodic table

Key points: The blocks of the periodic table reflect the identity of the orbitals that are occupied last in the building-up process. The period number is the principal quantum number of the valence shell. The group number is related to the number of valence electrons.

The layout of the periodic table reflects the electronic structure of the atoms of the elements (Fig. 1.22). We can now see, for instance, that a **block** of the table indicates the type of subshell currently being occupied according to the building-up principle. Each **period**, or row, of the table corresponds to the completion of the s and p subshells of a given shell. The period number is the value of the principal quantum number n of the shell which according to the building-up principle is currently being occupied in the main groups of the table. For example, Period 2 corresponds to the $n = 2$ shell and the filling of the 2s and 2p subshells.

The group numbers, G, are closely related to the number of electrons in the **valence shell**, the outermost shell of the atom. In the '1–18' numbering system recommended by IUPAC:

Block:	s	p	d
Number of electrons in valence shell:	G	$G-10$	G

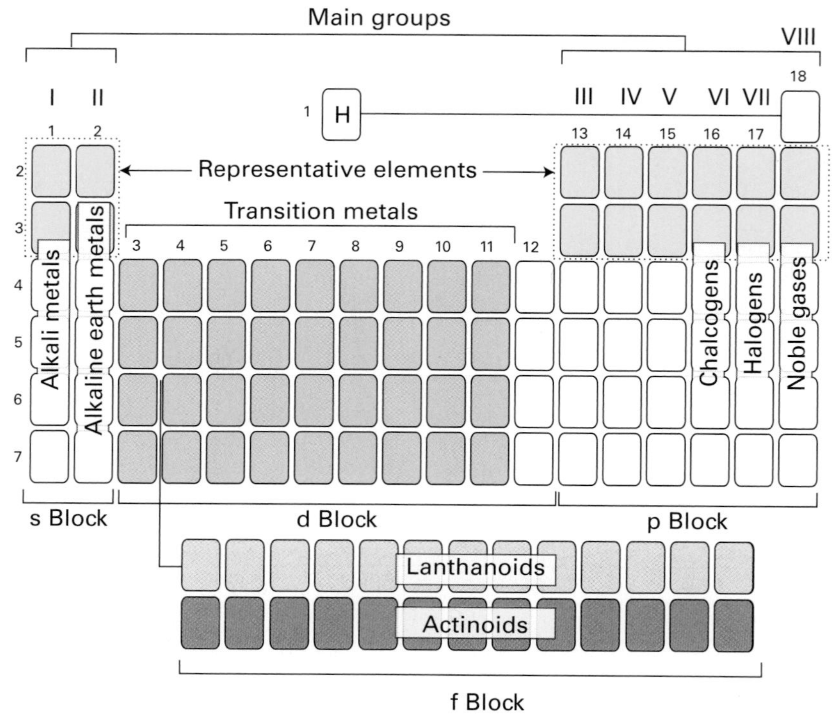

Figure 1.22 The general structure of the periodic table. Compare this template with the complete table inside the front cover for the identities of the elements that belong to each block.

For the purpose of this expression, the 'valence shell' of a d-block element consists of the ns and $(n-1)d$ orbitals, so a Sc atom has three valence electrons (two 4s and one 3d electron). The number of valence electrons for the p-block element Se (Group 16) is $16 - 10 = 6$, which corresponds to the configuration s^2p^4.

EXAMPLE 1.7 Placing elements within the periodic table.

State to which period, group, and block of the periodic table the element with the electron configuration $1s^2 2s^2 2p^6 3s^2 3p^4$ belongs. Identify the element.

Answer We need to remember that the period number is given by the principal quantum number, n, that the group number can be found from the number of valence electrons, and that the identity of the block is given by the type of orbital last occupied according to the building-up principle. The valence electrons have $n = 3$, therefore the element is in Period 3 of the periodic table. The six valence electrons identify the element as a member of Group 16. The electron added last is a p electron, so the element is in the p block. The element is sulfur.

Self-test 1.7 State to which period, group, and block of the periodic table the element with the electron configuration $1s^2 2s^2 2p^6 3s^2 3p^6 4s^2$ belongs. Identify the element.

1.9 Atomic properties

Certain characteristic properties of atoms, particularly their radii and the energies associated with the removal and addition of electrons, show regular periodic variations with atomic number. These atomic properties are of considerable importance for understanding the chemical properties of the elements and are discussed further in Chapter 9. A knowledge of these trends enables chemists to rationalize observations and predict likely chemical and structural behaviour without having to refer to tabulated data for each element.

(a) Atomic and ionic radii

Key points: Atomic radii increase down a group and, within the s and p blocks, decrease from left to right across a period. The lanthanide contraction results in a decrease in atomic radius for elements following the f block. All monatomic anions are larger than their parent atoms and all monatomic cations are smaller.

One of the most useful atomic characteristics of an element is the size of its atoms and ions. As we shall see in later chapters, geometrical considerations are central to explaining the structures of many solids and individual molecules. In addition, the average distance of electrons from the nucleus of an atom correlates with the energy needed to remove it in the process of forming a cation.

An atom does not have a precise radius because far from the nucleus the electron density falls off only exponentially (but sharply). However, we can expect atoms with numerous electrons to be larger, in some sense, than atoms that have only a few electrons. Such considerations have led chemists to propose a variety of definitions of atomic radius on the basis of empirical considerations.

The **metallic radius** of a metallic element is defined as half the experimentally determined distance between the centres of nearest-neighbour atoms in the solid (Fig. 1.23a, but see Section 3.7 for a refinement of this definition). The **covalent radius** of a nonmetallic element is similarly defined as half the internuclear distance between neighbouring atoms of the same element in a molecule (Fig. 1.23b). We shall refer to metallic and covalent radii jointly as **atomic radii** (Table 1.3). The periodic trends in metallic and covalent radii can be seen from the data in the table and are illustrated in Fig. 1.24. As will be familiar from introductory chemistry, atoms may be linked by single, double, and triple bonds, with multiple bonds shorter than single bonds between the same two elements. The **ionic radius** (Fig. 1.23c) of an element is related to the distance between the centres of neighbouring cations and anions in an ionic compound. An arbitrary decision has to be taken on how to apportion the cation–anion distance between the two ions. There have been many suggestions: in one common scheme, the radius of the O^{2-} ion is taken to be 140 pm (Table 1.4; see Section 3.7 for a refinement of this definition). For example, the ionic radius of Mg^{2+} is obtained by subtracting 140 pm from the internuclear distance between adjacent Mg^{2+} and O^{2-} ions in solid MgO.

The data in Table 1.3 show that *atomic radii increase down a group*, and that they *decrease from left to right across a period*. These trends are readily interpreted in terms of the electronic structure of the atoms. On descending a group, the valence electrons are found in orbitals of successively higher principal quantum number. The atoms within the group have a greater number of completed shells of electrons in successive periods and hence their radii increase down the group. Across a period, the valence electrons enter orbitals of the same shell; however, the increase in effective nuclear charge across the period draws in the electrons and results in progressively more compact atoms. The general increase in radius down a group and decrease across a period should be remembered as they correlate well with trends in many chemical properties.

Period 6 shows an interesting and important modification to these otherwise general trends. We see from Fig. 1.24 that the metallic radii in the third row of the d block are very similar to those in the second row, and not significantly larger as might be expected given their considerably greater numbers of electrons. For example, the atomic radii of Mo ($Z = 42$) and W ($Z = 74$) are 140 and 141 pm, respectively, despite the latter having many more electrons. The reduction of radius below that expected on the basis of a simple extrapolation down the group is called the **lanthanide contraction**. The name points to the

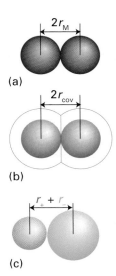

Figure 1.23 A representation of (a) metallic radius, (b) covalent radius, and (c) ionic radius.

Table 1.3 Atomic radii, r/pm*

Li	Be	Sc	Ti	V	Cr	Mn	Fe	Co	Ni	Cu	Zn	B	C	N	O	F
157	112											88	77	74	73	71
Na	Mg											Al	Si	P	S	Cl
191	160											143	118	110	104	99
K	Ca	Sc	Ti	V	Cr	Mn	Fe	Co	Ni	Cu	Zn	Ga	Ge	As	Se	Br
235	197	164	147	135	129	137	126	125	125	128	137	140	122	122	117	114
Rb	Sr	Y	Zr	Nb	Mo	Tc	Ru	Rh	Pd	Ag	Cd	In	Sn	Sb	Te	I
250	215	182	160	147	140	135	134	134	137	144	152	150	140	141	135	133
Cs	Ba	La	Hf	Ta	W	Re	Os	Ir	Pt	Au	Hg	Tl	Pb	Bi		
272	224	188	159	147	141	137	135	136	139	144	155	155	154	152		

* The values refer to coordination number 12 for metallic radii (see Section 3.2).

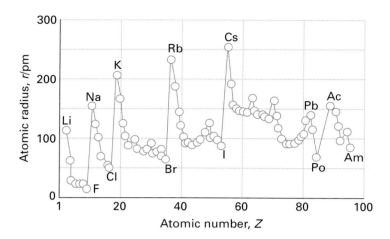

Figure 1.24 The variation of atomic radii through the periodic table. Note the contraction of radii following the lanthanoids in Period 6. Metallic radii have been used for the metallic elements and covalent radii have been used for the nonmetallic elements.

origin of the effect. The elements in the third row of the d block (Period 6) are preceded by the elements of the first row of the f block, the lanthanoids, in which the 4f orbitals are being occupied. These orbitals have poor shielding properties and so the valence electrons experience more attraction from the nuclear charge than might be expected. The repulsions between electrons being added on crossing the f block fail to compensate for the increasing nuclear charge, so Z_{eff} increases from left to right across a period. The dominating effect of the latter is to draw in all the electrons and hence to result in a more compact atom. A similar contraction is found in the elements that follow the d block for the same reasons. For example, although there is a substantial increase in atomic radius between C and Si (77 and 118 pm, respectively), the atomic radius of Ge (122 pm) is only slightly greater than that of Al.

Relativistic effects, especially the increase in mass as particles approach the speed of light, have an important role to play on the elements in and following Period 6 but are rather subtle. Electrons in s and p orbitals, which approach closely to the highly charged nucleus and experience strong accelerations, contract whereas electrons in the less penetrating d and f orbitals expand. One consequence of the latter expansion is that d and f electrons become less effective at shielding other electrons, and the outermost s electrons contract further. For light elements, relativistic effects can be neglected but for the heavier elements with high atomic numbers they become significant and can result in an approximately 20 per cent reduction in the size of the atom.

Another general feature apparent from Table 1.4 is that all monatomic anions are larger than their parent atoms and all monatomic cations are smaller than their parent atoms (in some cases markedly so). The increase in radius of an atom on anion formation is a result of the greater electron–electron repulsions that occur when an additional electron is added to form an anion. There is also an associated decrease in the value of Z_{eff}. The smaller radius of a cation compared with its parent atom is a consequence not only of the reduction in electron–electron repulsions that follow electron loss but also of the fact that cation formation typically results in the loss of the valence electrons and an increase in Z_{eff}. That loss often leaves behind only the much more compact closed shells of electrons. Once these gross differences are taken into account, the variation in ionic radii through the periodic table mirrors that of the atoms.

Although small variations in atomic radii may seem of little importance, in fact atomic radius plays a central role in the chemical properties of the elements. Small changes can have profound consequences, as we shall see in Chapter 9.

(b) Ionization energy

Key points: First ionization energies are lowest at the lower left of the periodic table (near caesium) and greatest near the upper right (near helium). Successive ionizations of a species require higher energies.

The ease with which an electron can be removed from an atom is measured by its **ionization energy**, I, the minimum energy needed to remove an electron from a gas-phase atom:

$$A(g) \rightarrow A^+(g) + e^-(g) \quad I = E(A^+, g) - E(A, g) \tag{1.8}$$

Table 1.4 Ionic radii, r/pm^*

Li⁺	Be²⁺	B³⁺				N³⁻	O²⁻	F⁻
59(4)	27(4)	11(4)				146	135(2)	128(2)
76(6)							138(4)	131(4)
							140(6)	133(6)
							142(8)	
Na⁺	Mg²⁺	Al³⁺				P³⁻	S²⁻	Cl⁻
99(4)	49(4)	39(4)				212	184(6)	181(6)
102(6)	72(6)	53(6)						
132(8)	103(8)							
K⁺	Ca²⁺	Ga³⁺				As³⁻	Se²⁻	Br⁻
138(6)	100(6)	62(6)				222	198(6)	196(6)
151(8)	112(8)							
159(10)	123(10)							
160(12)	134(12)							
Rb⁺	Sr²⁺	In³⁺	Sn²⁺	Sn⁴⁺			Te²⁻	I⁻
148(6)	118(6)	80(6)	83(6)	69(6)			221(6)	220(6)
160(8)	125(8)	92(8)	93(8)					
173(12)	144(12)							
Cs⁺	Ba²⁺	Tl³⁺						
167(6)	135(6)	89(6)						
174(8)	142(8)	Tl⁺						
188(12)	175(12)	150(6)						

*Numbers in parentheses are the coordination number of the ion. For more values, see *Resource section 1*.

The **first ionization energy**, I_1, is the energy required to remove the least tightly bound electron from the neutral atom, the **second ionization energy**, I_2, is the energy required to remove the least tightly bound electron from the resulting cation, and so on. Ionization energies are conveniently expressed in electronvolts (eV), but are easily converted into kilojoules per mole by using $1\text{ eV} = 96.485\text{ kJ mol}^{-1}$. The ionization energy of the H atom is 13.6 eV, so to remove an electron from an H atom is equivalent to dragging the electron through a potential difference of 13.6 V.

In thermodynamic calculations it is often more appropriate to use the **ionization enthalpy**, the standard enthalpy of the process in eqn 1.8, typically at 298 K. The molar ionization enthalpy is larger by $\frac{5}{2}RT$ than the ionization energy. This difference stems from the change from $T = 0$ (assumed implicitly for I) to the temperature T (typically 298 K) to which the enthalpy value refers, and the replacement of 1 mol of gas particles by 2 mol of gaseous ions plus electrons. However, because RT is only 2.5 kJ mol⁻¹ (corresponding to 0.026 eV) at room temperature and ionization energies are of the order of $10^2 - 10^3$ kJ mol⁻¹ (1−10 eV), the difference between ionization energy and enthalpy can often be ignored.

To a large extent, the first ionization energy of an element is determined by the energy of the highest occupied orbital of its ground-state atom. First ionization energies vary systematically through the periodic table (Table 1.5), being smallest at the lower left (near Cs) and greatest near the upper right (near He). The variation follows the pattern of effective nuclear charge, and (as Z_{eff} itself shows) there are some subtle modulations arising from the effect of electron–electron repulsions within the same subshell. A useful approximation is that for an electron from a shell with principal quantum number n

$$I \propto \frac{Z_{\text{eff}}^2}{n^2}$$

Ionization energies also correlate strongly with atomic radii, and elements that have small atomic radii generally have high ionization energies. The explanation of the correlation is

Table 1.5 First, second, and third (and some fourth) ionization energies of the elements, $I/(\text{kJ mol}^{-1})$

H							He
1312							2373
							5259
Li	Be	B	C	N	O	F	Ne
513	899	801	1086	1402	1314	1681	2080
7297	1757	2426	2352	2855	3386	3375	3952
11809	14844	3660	4619	4577	5300	6050	6122
		25018					
Na	Mg	Al	Si	P	S	Cl	Ar
495	737	577	786	1011	1000	1251	1520
4562	1476	1816	1577	1903	2251	2296	2665
6911	7732	2744	3231	2911	3361	3826	3928
		11574					
K	Ca	Ga	Ge	As	Se	Br	Kr
419	589	579	762	947	941	1139	1351
3051	1145	1979	1537	1798	2044	2103	3314
4410	4910	2963	3302	2734	2974	3500	3565
Rb	Sr	In	Sn	Sb	Te	I	Xe
403	549	558	708	834	869	1008	1170
2632	1064	1821	1412	1794	1795	1846	2045
3900	4210	2704	2943	2443	2698	3197	3097
Cs	Ba	Tl	Pb	Bi	Po	At	Rn
375	502	590	716	704	812	926	1036
2420	965	1971	1450	1610	1800	1600	
3400	3619	2878	3080	2466	2700	2900	

that in a small atom an electron is close to the nucleus and experiences a strong Coulombic attraction, making it difficult to remove. Therefore, as the atomic radius increases down a group, the ionization energy decreases and the decrease in radius across a period is accompanied by a gradual increase in ionization energy.

Some deviation from this general trend in ionization energy can be explained quite readily. An example is the observation that the first ionization energy of boron is smaller than that of beryllium, despite the former's higher nuclear charge. This anomaly is readily explained by noting that, on going to boron, the outermost electron occupies a 2p orbital and hence is less strongly bound than if it had occupied a 2s orbital. As a result, the value of I_1 decreases from Be to B. The decrease between N and O has a slightly different explanation. The configurations of the two atoms are

$$\text{N } [\text{He}]2s^2 2p_x^1 2p_y^1 2p_z^1 \quad \text{O } [\text{He}]2s^2 2p_x^2 2p_y^1 2p_z^1$$

We see that, in an O atom, two electrons are present in a single 2p orbital. They repel each other strongly, and this strong repulsion offsets the greater nuclear charge. Another contribution to the difference is the lower energy of the O^+ ion on account of its having a $2s^2 2p^3$ configuration: as we have seen, a half-filled subshell has a relatively low energy (Fig. 1.25). Additionally, the half-filled shell of p orbitals of nitrogen is a particularly stable configuration.

When considering F and Ne on the right of Period 2, the last electrons enter orbitals that are already half full, and continue the trend from O towards higher ionization energy. The higher values of the ionization energies of these two elements reflect the high value of Z_{eff}. The value of I_1 falls back sharply from Ne to Na as the outermost electron occupies the next shell with an increased principal quantum number and is therefore further from the nucleus.

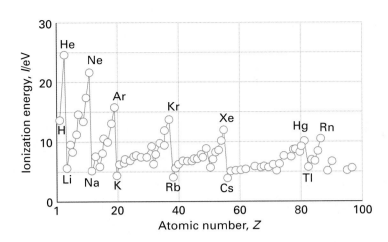

Figure 1.25 The periodic variation of first ionization energies.

EXAMPLE 1.8 Accounting for a variation in ionization energy

Account for the decrease in first ionization energy between phosphorus and sulfur.

Answer We approach this question by considering the ground-state configurations of the two atoms:

P $[Ne]3s^2 3p_x^1 3p_y^1 3p_z^1$ S $[Ne]3s^2 3p_x^2 3p_y^1 3p_z^1$

As in the analogous case of N and O, in the ground state of S, two electrons are present in a single 3p orbital. They are so close together that they repel each other strongly, and this increased repulsion offsets the effect of the greater nuclear charge of S compared with P. As in the difference between N and O, the half-filled subshell of S^+ also contributes to the lowering of energy of the ion and hence to the smaller ionization energy.

Self-test 1.8 Account for the decrease in first ionization energy between fluorine and chlorine.

Another important pattern is that successive ionizations of an element require increasingly higher energies (Fig. 1.26). Thus, the second ionization energy of an element E (the energy needed to remove an electron from the cation E^+) is higher than its first ionization energy, and its third ionization energy (the energy needed to remove an electron from E^{2+}) is higher still. The explanation is that the higher the positive charge of a species, the greater the electrostatic attraction experienced by the electron being removed. Moreover, when an electron is removed, Z_{eff} increases and the atom contracts. It is then even more difficult to remove an electron from this smaller, more compact, cation. The difference in ionization energy is greatly magnified when the electron is removed from a closed shell of the atom (as is the case for the second ionization energy of Li and any of its congeners) because the electron must then be extracted from a compact orbital in which it interacts strongly with the nucleus. The first ionization energy of Li, for instance, is 513 kJ mol^{-1}, but its second ionization energy is 7297 kJ mol^{-1}, more than ten times greater.

The pattern of successive ionization energies down a group is far from simple. Figure 1.26 shows the first, second, and third ionization energies of the members of Group 13. Although they lie in the expected order $I_1 < I_2 < I_3$, there is no simple trend. The lesson to be drawn is that whenever an argument hangs on trends in small differences in ionization energies, it is always best to refer to actual numerical values rather than to guess a likely outcome.

(c) Electron affinity

Key point: Electron affinities are highest for elements near fluorine in the periodic table.

The **electron-gain enthalpy**, $\Delta_{eg}H^{\ominus}$, is the change in standard molar enthalpy when a gaseous atom gains an electron:

$$A(g) + e^-(g) \rightarrow A^-(g)$$

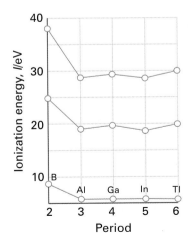

Figure 1.26 The first, second, and third ionization energies of the elements of Group 13. Successive ionization energies increase, but there is no clear pattern of ionization energies down the group.

EXAMPLE 1.9 Accounting for values of successive energies of ionization

Rationalize the following values for successive ionization energies of boron, where $\Delta_{ion}H(N)$ is the Nth enthalpy of ionization:

N	1	2	3	4	5
$\Delta_{ion}H(N)/(\text{kJ mol}^{-1})$	807	2433	3666	25033	32834

Answer When considering trends in ionization energy, a sensible starting point is the electron configurations of the atoms. The electron configuration of B is $1s^22s^22p^1$. The first ionization energy corresponds to removal of the electron in the 2p orbital. This electron is shielded from nuclear charge by the core and the full 2s orbital. The second value corresponds to removal of a 2s electron from the B$^+$ cation. This electron is more difficult to remove on account of the increased effective nuclear charge. The effective nuclear charge increases further on removal of this electron, resulting in an increase between $\Delta_{ion}H(2)$ and $\Delta_{ion}H(3)$. There is a large increase between $\Delta_{ion}H(3)$ and $\Delta_{ion}H(4)$ because the 1s shell lies at very low energy as it experiences almost the full nuclear charge and also has $n = 1$. The final electron to be removed experiences no shielding of nuclear charge so $\Delta_{ion}H(5)$ is very high, and is given by $hcRZ^2$ with $Z = 5$, corresponding to $(13.6\text{ eV}) \times 25 = 340\text{ eV}$ (32.8 MJ mol^{-1}).

Self-test 1.9 Study the values listed below of the first five ionization energies of an element and deduce to which group of the periodic table the element belongs. Give your reasoning.

N	1	2	3	4	5
$\Delta_{ion}H(N)/(\text{kJ mol}^{-1})$	1093	2359	4627	6229	37838

Electron gain may be either exothermic or endothermic. Although the electron-gain enthalpy is the thermodynamically appropriate term, much of inorganic chemistry is discussed in terms of a closely related property, the **electron affinity**, E_a, of an element (Table 1.6), which is the difference in energy between the gaseous atoms and the gaseous ions at $T = 0$.

$$E_a = E(A,g) - E(A^-,g) \tag{1.9}$$

Although the precise relation is $\Delta_{eg}H^\ominus = -E_a - \frac{5}{2}RT$, the contribution $\frac{5}{2}RT$ is commonly ignored. A positive electron affinity indicates that the ion A$^-$ has a lower, more negative energy than the neutral atom, A. The second electron-gain enthalpy, the enthalpy change for the attachment of a second electron to an initially neutral atom, is invariably positive because the electron repulsion outweighs the nuclear attraction.

The electron affinity of an element is largely determined by the energy of the *lowest unfilled* (or half-filled) orbital of the ground-state atom. This orbital is one of the two **frontier orbitals** of an atom, the other one being the *highest filled* atomic orbital. The frontier orbitals are the sites of many of the changes in electron distributions when bonds form,

Table 1.6 First electron affinities of the main-group elements, $E_a/(\text{kJ mol}^{-1})$*

H							He
72							−48
Li	Be	B	C	N	O	F	Ne
60	≤ 0	27	122	−8	141	328	−116
					−780		
Na	Mg	Al	Si	P	S	Cl	Ar
53	≤ 0	43	134	72	200	349	−96
					−492		
K	Ca	Ga	Ge	As	Se	Br	Kr
48	2	29	116	78	195	325	−96
Rb	Sr	In	Sn	Sb	Te	I	Xe
47	5	29	116	103	190	295	−77

* The first values refer to the formation of the ion X$^-$ from the neutral atom; the second value to the formation of X^{2-} from X$^-$.

EXAMPLE 1.10 Accounting for the variation in electron affinity

Account for the large decrease in electron affinity between Li and Be despite the increase in nuclear charge.

Answer When considering trends in electron affinities, as in the case of ionization energies, a sensible starting point is the electron configurations of the atoms. The electron configurations of Li and Be are [He]2s^1 and [He]2s^2, respectively. The additional electron enters the 2s orbital of Li but it enters the 2p orbital of Be, and hence is much less tightly bound. In fact, the nuclear charge is so well shielded in Be that electron gain is endothermic.

Self-test 1.10 Account for the decrease in electron affinity between C and N.

---

and we shall see more of their importance as the text progresses. An element has a high electron affinity if the additional electron can enter a shell where it experiences a strong effective nuclear charge. This is the case for elements towards the top right of the periodic table, as we have already explained. Therefore, elements close to fluorine (specifically O and Cl, but not the noble gases) can be expected to have the highest electron affinities as their Z_{eff} is large and it is possible to add electrons to the valence shell. Nitrogen has very low electron affinity because there is a high electron repulsion when the incoming electron enters an orbital that is already half full.

A note on good practice Be alert to the fact that some people use the terms 'electron affinity' and 'electron-gain enthalpy' interchangeably. In such cases, a positive electron affinity could indicate that A$^-$ has a higher energy than A.

(d) Electronegativity

Key points: The electronegativity of an element is the power of an atom of the element to attract electrons when it is part of a compound; there is a general increase in electronegativity across a period and a general decrease down a group.

The **electronegativity**, χ (chi), of an element is the power of an atom of the element to attract electrons to itself when it is part of a compound. If an atom has a strong tendency to acquire electrons, it is said to be highly electronegative (like the elements close to fluorine). Electronegativity is a very useful concept in chemistry and has numerous applications, which include a rationalization of bond energies and the types of reactions that substances undergo and the prediction of the polarities of bonds and molecules (Chapter 2).

Periodic trends in electronegativity can be related to the size of the atoms and electron configuration. If an atom is small and has an almost closed shell of electrons, then it is more likely to attract an electron to itself than a large atom with few valence electrons. Consequently, the electronegativities of the elements typically increase left to right across a period and decrease down a group.

Quantitative measures of electronegativity have been defined in many different ways. Linus Pauling's original formulation (which results in the values denoted χ_P in Table 1.7) draws on concepts relating to the energetics of bond formation, which will be dealt with in Chapter 2.[2] A definition more in the spirit of this chapter, in the sense that it is based on the properties of individual atoms, was proposed by Robert Mulliken. He observed that, if an atom has a high ionization energy, I, and a high electron affinity, E_a, then it will be likely to acquire rather than lose electrons when it is part of a compound, and hence be classified as highly electronegative. Conversely, if its ionization energy and electron affinity are both low, then the atom will tend to lose electrons rather than gain them, and hence be classified as electropositive. These observations motivate the definition of the **Mulliken electronegativity**, χ_M, as the average value of the ionization energy and the electron affinity of the element (both expressed in electronvolts):

$$\chi_M = \tfrac{1}{2}(I + E_a) \tag{1.10}$$

[2] Pauling values of electronegativity are used throughout the following chapters.

Table 1.7 Pauling χ_P, Mulliken, χ_M, and Allred–Rochow, χ_{AR}, electronegativities

H							He
2.20							5.5
3.06							
2.20							
Li	Be	B	C	N	O	F	Ne
0.98	1.57	2.04	2.55	3.04	3.44	3.98	
1.28	1.99	1.83	2.67	3.08	3.22	4.43	4.60
0.97	1.47	2.01	2.50	3.07	3.50	4.10	5.10
Na	Mg	Al	Si	P	S	Cl	Ar
0.93	1.31	1.61	1.90	2.19	2.58	3.16	
1.21	1.63	1.37	2.03	2.39	2.65	3.54	3.36
1.01	1.23	1.47	1.74	2.06	2.44	2.83	3.30
K	Ca	Ga	Ge	As	Se	Br	Kr
0.82	1.00	1.81	2.01	2.18	2.55	2.96	3.0
1.03	1.30	1.34	1.95	2.26	2.51	3.24	2.98
0.91	1.04	1.82	2.02	2.20	2.48	2.74	3.10
Rb	Sr	In	Sn	Sb	Te	I	Xe
0.82	0.95	1.78	1.96	2.05	2.10	2.66	2.6
0.99	1.21	1.30	1.83	2.06	2.34	2.88	2.59
0.89	0.99	1.49	1.72	1.82	2.01	2.21	2.40
Cs	Ba	Tl	Pb	Bi			
0.79	0.89	2.04	2.33	2.02			
0.70	0.90	1.80	1.90	1.90			
0.86	0.97	1.44	1.55	1.67			

The hidden complication in the apparently simple definition of the Mulliken electronegativity is that the ionization energy and electron affinity in the definition relate to the **valence state**, the electron configuration the atom is supposed to have when it is part of a molecule. Hence, some calculation is required because the ionization energy and electron affinity to be used in calculating χ_M are mixtures of values for various actual spectroscopically observable states of the atom. We need not go into the calculation, but the resulting values given in Table 1.7 may be compared with the Pauling values (Fig. 1.27). The two

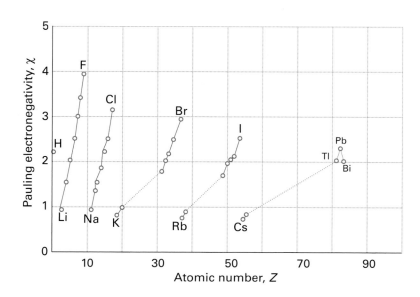

Figure 1.27 The periodic variation of Pauling electronegativities.

scales give similar values and show the same trends. One reasonably reliable conversion between the two is

$$\chi_P = 1.35\chi_M^{1/2} - 1.37 \qquad (1.11)$$

Because the elements near F (other than the noble gases) have high ionization energies and appreciable electron affinities, these elements have the highest Mulliken electronegativities. Because χ_M depends on atomic energy levels—and in particular on the location of the highest filled and lowest empty orbitals—the electronegativity of an element is high if the two frontier orbitals of its atoms are low in energy.

Various alternative 'atomic' definitions of electronegativity have been proposed. A widely used scale, suggested by A.L. Allred and E. Rochow, is based on the view that electronegativity is determined by the electric field at the surface of an atom. As we have seen, an electron in an atom experiences an effective nuclear charge Z_{eff}. The Coulombic potential at the surface of such an atom is proportional to Z_{eff}/r, and the electric field there is proportional to Z_{eff}/r^2. In the **Allred–Rochow definition** of electronegativity, χ_{AR} is assumed to be proportional to this field, with r taken to be the covalent radius of the atom:

$$\chi_{AR} = 0.744 + \frac{35.90Z_{eff}}{(r/\text{pm})^2} \qquad (1.12)$$

The numerical constants have been chosen to give values comparable to Pauling electronegativities. According to the Allred–Rochow definition, elements with high electronegativity are those with high effective nuclear charge and the small covalent radius: such elements lie close to F. The Allred–Rochow values parallel closely those of the Pauling electronegativities and are useful for discussing the electron distributions in compounds.

(e) Polarizability

Key points: A polarizable atom or ion is one with orbitals that lie close in energy; large, heavy atoms and ions tend to be highly polarizable.

The **polarizability**, α, of an atom is its ability to be distorted by an electric field (such as that of a neighbouring ion). An atom or ion (most commonly, an anion) is highly **polarizable** if its electron distribution can be distorted readily, which is the case if unfilled atomic orbitals lie close to the highest-energy filled orbitals. That is, the polarizability is likely to be high if the separation of the frontier orbitals is small and the polarizability will be low if the separation of the frontier orbitals is large (Fig. 1.28). Closely separated frontier orbitals are typically found for large, heavy atoms and ions, such as the atoms and ions of the heavier alkali metals and the heavier halogens, so these atoms and ions are the most polarizable. Small, light atoms, such as the atoms and ions near fluorine, typically have widely spaced energy levels, so these atoms and ions are least polarizable. Species that effectively distort the electron distribution of a neighbouring atom or anion are described as having **polarizing ability**.

We shall see the consequences of polarizability when considering the nature of bonding in Section 2.2, but it is appropriate to anticipate here that extensive polarization leads to covalency. **Fajan's rules** summarize the factors that affect polarization:

- Small, highly charged cations have polarizing ability.
- Large, highly charged anions are easily polarized.
- Cations that do not have a noble-gas electron configuration are easily polarized.

The last rule is particularly important for the d-block elements.

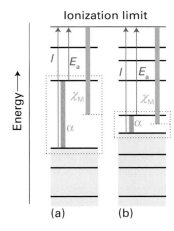

Figure 1.28 The interpretation of the electronegativity and polarizability of an element in terms of the energies of the frontier orbitals (the highest filled and lowest unfilled atomic orbitals). (a) Low electronegativity and polarizability; (b) high electronegativity and polarizability.

EXAMPLE 1.11 Identifying polarizable species

Which would be the more polarizable, an F^- ion or an I^- ion?

Answer We can make use of the fact that polarizable anions are typically large and highly charged. An F^- ion is small and singly charged. An I^- ion has the same charge but is large. Therefore, an I^- ion is likely to be the more polarizable.

Self-test 1.11 Which would be more polarizing, Na^+ or Cs^+?

FURTHER READING

M. Laing, The different periodic tables of Dmitrii Mendeleev. *J. Chem. Educ.*, 2008, **85**, 63.

M.W. Cronyn, The proper place for hydrogen in the periodic table. *J. Chem. Educ.* 2003, **80**, 947.

P.A. Cox, *Introduction to quantum theory and atomic structure*. Oxford University Press (1996). An introduction to the subject.

P. Atkins and J. de Paula, *Physical chemistry*. Oxford University Press and W.H. Freeman & Co. (2010). Chapters 7 and 8 give an account of quantum theory and atomic structure.

J. Emsley, *Nature's building blocks*. Oxford University Press (2003). An interesting guide to the elements.

D.M.P. Mingos, *Essential trends in inorganic chemistry*. Oxford University Press (1998). Includes a detailed discussion of the important horizontal, vertical, and diagonal trends in the properties of the atoms.

P.A. Cox, *The elements: their origin, abundance, and distribution*. Oxford University Press (1989). Examines the origin of the elements, the factors controlling their widely differing abundances, and their distributions in the Earth, the solar system, and the universe.

N.G. Connelly, T. Danhus, R.M. Hartshoin, and A.T Hutton, *Nomenclature of inorganic chemistry. Recommendations 2005*. Royal Society of Chemistry (2005). This book outlines the conventions for the periodic table and inorganic substances. It is known colloquially as the 'Red Book' on account of its distinctive red cover.

EXERCISES

1.1 Write balanced equations for the following nuclear reactions (show emission of excess energy as a photon of electromagnetic radiation, γ): (a) $^{14}N + {}^4He$ to produce ^{17}O, (b) $^{12}C + p$ to produce ^{13}N, (c) $^{14}N + n$ to produce ^3H and ^{12}C. (The last reaction produces a steady-state concentration of radioactive ^3H in the upper atmosphere.)

1.2 Balance the following nuclear reaction:

$$^{246}_{96}Cm + {}^{12}_{6}C \rightarrow ? + {}^1_0n$$

1.3 In general, ionization energies increase across a period from left to right. Explain why the second ionization energy of Cr is higher, not lower, than that of Mn.

1.4 One possible source of neutrons for the neutron-capture processes mentioned in the text is the reaction of ^{22}Ne with α particles to produce ^{25}Mg and neutrons. Write the balanced equation for the nuclear reaction.

1.5 ^9Be undergoes α decay to produce ^{12}C and neutrons. Write a balanced equation for this reaction.

1.6 The natural abundances of adjacent elements in the periodic table usually differ by a factor of 10 or more. Explain this phenomenon.

1.7 Explain how you would determine, using data you would look up in tables, whether or not the nuclear reaction in Exercise 1.2 corresponds to a release of energy.

1.8 What is the ratio of the energy of a ground-state He^+ ion to that of a Be^{3+} ion?

1.9 The ionization energy of H is 13.6 eV. What is the difference in energy between the $n = 1$ and $n = 6$ levels?

1.10 Calculate the wavenumber ($\tilde{\nu} = 1/\lambda$) and wavelength of the first transition in the visible region of the atomic spectrum of hydrogen.

1.11 Show that the following four lines in the Lyman series can be predicted from equation 1.1: 91.127, 97.202, 102.52, and 121.57 nm.

1.12 What is the relation of the possible angular momentum quantum numbers to the principal quantum number?

1.13 How many orbitals are there in a shell of principal quantum number n? (Hint: begin with $n = 1, 2,$ and 3 and see if you can recognize the pattern.)

1.14 Complete the following table:

n	l	m_l	Orbital designation	Number of orbitals
2			2p	
3	2			
			4s	
4		$+3, +2, \ldots, -3$		

1.15 What are the values of the n, l, and m_l quantum numbers that describe the 5f orbitals?

1.16 Use the data in Table 1.2 to calculate the screening constants for the outermost electron in the elements Li to F. Comment on the values you obtain.

1.17 Consider the process of shielding in atoms, using Be as an example. What is being shielded? What is it shielded from? What is doing the shielding?

1.18 Use sketches of 2s and 2p orbitals to distinguish between (a) the radial wavefunction and (b) the radial distribution function.

1.19 Compare the first ionization energy of Ca with that of Zn. Explain the difference in terms of the balance between shielding with increasing numbers of d electrons and the effect of increasing nuclear charge.

1.20 Compare the first ionization energies of Sr, Ba, and Ra. Relate the irregularity to the lanthanide contraction.

1.21 The second ionization energies of some Period 4 elements are

Ca	Sc	Ti	V	Cr	Mn
1145	1235	1310	1365	1592	1509 kJ mol^{-1}

Identify the orbital from which ionization occurs and account for the trend in values.

1.22 Give the ground-state electron configurations of (a) C, (b) F, (c) Ca, (d) Ga^{3+}, (e) Bi, (f) Pb^{2+}.

1.23 Give the ground-state electron configurations of (a) Sc, (b) V^{3+}, (c) Mn^{2+}, (d) Cr^{2+}, (e) Co^{3+}, (f) Cr^{6+}, (g) Cu, (h) Gd^{3+}.

1.24 Give the ground-state electron configurations of (a) W, (b) Rh^{3+}, (c) Eu^{3+}, (d) Eu^{2+}, (e) V^{5+}, (f) Mo^{4+}.

1.25 Identify the elements that have the ground-state electron configurations: (a) [Ne]$3s^2 3p^4$, (b) [Kr]$5s^2$, (c) [Ar]$4s^2 3d^3$, (d) [Kr]$5s^2 4d^5$, (e) [Kr]$5s^2 4d^{10} 5p^1$, (f) [Xe]$6s^2 4f^6$.

1.26 Without consulting reference material, draw the form of the periodic table with the numbers of the groups and the periods and identify the s, p, and d blocks. Identify as many elements as you can. (As you progress through your study of inorganic chemistry, you should learn the positions of all the s-, p-, and d-block elements and associate their positions in the periodic table with their chemical properties.)

1.27 Account for the trends across Period 3 in (a) ionization energy, (b) electron affinity, (c) electronegativity.

1.28 Account for the fact that the two Group 5 elements niobium (Period 5) and tantalum (Period 6) have the same atomic radii.

1.29 Identify the frontier orbitals of a Be atom in its ground state.

1.30 Use the data in Tables 1.6 and 1.7 to test Mulliken's proposition that electronegativity values are proportional to $I + E_a$.

PROBLEMS

1.1 Show that an atom with the configuration $ns^2 np^6$ is spherically symmetrical. Is the same true of an atom with the configuration $ns^2 np^3$?

1.2 According to the Born interpretation, the probability of finding an electron in a volume element $d\tau$ is proportional to $\psi^2 d\tau$. (a) What is the most probable location of an electron in an H atom in its ground state? (b) What is its most probable distance from the nucleus, and why is this different? (c) What is the most probable distance of a 2s electron from the nucleus?

1.3 The ionization energies of rubidium and silver are 4.18 and 7.57 eV, respectively. Calculate the ionization energies of an H atom with its electron in the same orbitals as in these two atoms and account for the differences in values.

1.4 When 58.4 nm radiation from a helium discharge lamp is directed on a sample of krypton, electrons are ejected with a velocity of 1.59 $\times 10^6$ m s^{-1}. The same radiation ejects electrons from Rb atoms with a velocity of 2.45 $\times 10^6$ m s^{-1}. What are the ionization energies (in electronvolts, eV) of the two elements?

1.5 Survey the early and modern proposals for the construction of the periodic table. You should consider attempts to arrange the elements on helices and cones as well as the more practical two-dimensional surfaces. What, in your judgement, are the advantages and disadvantages of the various arrangements?

1.6 The decision about which elements should be identified as belonging to the f block has been a matter of some controversy.

A view has been expressed by W.B. Jensen (*J. Chem. Educ.*, 1982, 59, 635). Summarize the controversy and Jensen's arguments. An alternative view has been expressed by L. Lavalle (*J. Chem. Educ.*, 2008, 85, 1482). Summarize the controversy and the arguments.

1.7 Draw pictures of the two d orbitals in the *xy*-plane as flat projections in the plane of the paper. Label each drawing with the appropriate mathematical function, and include a labelled pair of Cartesian coordinate axes. Label the orbital lobes correctly with $+$ and $-$ signs.

1.8 During 1999 several papers appeared in the scientific literature claiming that d orbitals of Cu_2O had been observed experimentally. In his paper 'Have orbitals really been observed?' (*J. Chem. Educ.*, 2000, 77, 1494), Eric Scerri reviews these claims and discusses whether orbitals can be observed physically. Summarize his arguments briefly.

1.9 At various times the following two sequences have been proposed for the elements to be included in Group 3: (a) Sc, Y, La, Ac, (b) Sc, Y, Lu, Lr. Because ionic radii strongly influence the chemical properties of the metallic elements, it might be thought that ionic radii could be used as one criterion for the periodic arrangement of the elements. Use this criterion to describe which of these sequences is preferred.

1.10 In the paper 'Ionization energies of atoms and atomic ions' (P.F. Lang and B.C. Smith, *J. Chem. Educ.*, 2003, 80, 938) the authors discuss the apparent irregularities in the first and second ionization energies of d- and f-block elements. Describe how these inconsistencies are rationalized.

2

Molecular structure and bonding

The interpretation of structures and reactions in inorganic chemistry is often based on semiquantitative models. In this chapter we examine the development of models of molecular structure in terms of the concepts of valence bond and molecular orbital theory. In addition, we review methods for predicting the shapes of molecules. This chapter introduces concepts that will be used throughout the text to explain the structures and reactions of a wide variety of species. The chapter also illustrates the importance of the interplay between qualitative models, experiment, and calculation.

Lewis structures

Lewis proposed that a **covalent bond** is formed when two neighbouring atoms share an electron pair. A **single bond**, a shared electron pair (A:B), is denoted A—B; likewise, a **double bond**, two shared electron pairs (A::B), is denoted A=B, and a **triple bond**, three shared pairs of electrons (A:::B), is denoted A≡B. An unshared pair of valence electrons on an atom (A:) is called a **lone pair**. Although lone pairs do not contribute directly to the bonding, they do influence the shape of the molecule and play an important role in its properties.

2.1 The octet rule

Key point: Atoms share electron pairs until they have acquired an octet of valence electrons.

Lewis found that he could account for the existence of a wide range of molecules by proposing the **octet rule**:

> *Each atom shares electrons with neighbouring atoms to achieve a total of eight valence electrons (an 'octet').*

As we saw in Section 1.8, a closed-shell, noble-gas configuration is achieved when eight electrons occupy the s and p subshells of the valence shell. One exception is the hydrogen atom, which fills its valence shell, the 1s orbital, with two electrons (a 'duplet').

The octet rule provides a simple way of constructing a **Lewis structure**, a diagram that shows the pattern of bonds and lone pairs in a molecule. In most cases we can construct a Lewis structure in three steps.

1. Decide on the number of electrons that are to be included in the structure by adding together the numbers of all the valence electrons provided by the atoms.

Each atom provides all its valence electrons (thus, H provides one electron and O, with the configuration $[\text{He}]2s^22p^4$, provides six). Each negative charge on an ion corresponds to an additional electron; each positive charge corresponds to one electron less.

2. Write the chemical symbols of the atoms in the arrangement that shows which atoms are bonded together.

In most cases we know the arrangement or can make an informed guess. The less electronegative element is usually the central atom of a molecule, as in CO_2 and SO_4^{2-}, but there are many well-known exceptions (H_2O and NH_3 among them).

3. Distribute the electrons in pairs so that there is one pair of electrons forming a single bond between each pair of atoms bonded together, and then supply electron pairs (to form lone pairs or multiple bonds) until each atom has an octet.

Each bonding pair (:) is then represented by a single line (−). The net charge of a polyatomic ion is supposed to be possessed by the ion as a whole, not by a particular individual atom.

EXAMPLE 2.1 **Writing a Lewis structure**

Write a Lewis structure for the BF_4^- ion.

Answer We need to consider the total number of electrons supplied and how they are shared to complete an octet around each atom. The atoms supply $3 + (4 \times 7) = 31$ valence electrons; the single negative charge of the ion reflects the presence of an additional electron. We must therefore accommodate 32 electrons in 16 pairs around the five atoms. One solution is (**1**). The negative charge is ascribed to the ion as a whole, not to a particular individual atom.

Self-test 2.1 Write a Lewis structure for the PCl_3 molecule.

Table 2.1 gives examples of Lewis structures of some common molecules and ions. Except in simple cases, a Lewis structure does not portray the shape of the species, but only the pattern of bonds and lone pairs: it shows the number of the links, not the geometry of the molecule. For example the BF_4^- ion is actually tetrahedral (**2**), not planar, and PF_3 is trigonal pyramidal (**3**).

2.2 Resonance

Key points: Resonance between Lewis structures lowers the calculated energy of the molecule and distributes the bonding character of electrons over the molecule; Lewis structures with similar energies provide the greatest resonance stabilization.

A single Lewis structure is often an inadequate description of the molecule. The Lewis structure of O_3 (**4**), for instance, suggests incorrectly that one O−O bond is different from the other, whereas in fact they have identical lengths (128 pm) intermediate between those of typical single O−O and double O=O bonds (148 pm and 121 pm, respectively). This deficiency of the Lewis description is overcome by introducing the concept of **resonance**,

1 BF_4^-

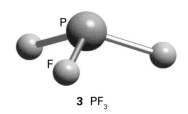

2 BF_4^-

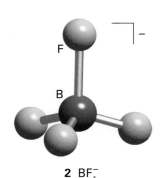

3 PF_3

Table 2.1 Lewis structures of some simple molecules*

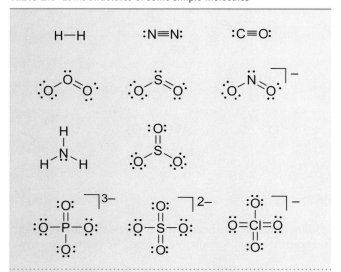

* Only representative resonance structures are given. Shapes are indicated only for diatomic and triatomic molecules.

4 O_3

in which the actual structure of the molecule is taken to be a superposition, or average, of all the feasible Lewis structures corresponding to a given atomic arrangement.

Resonance is indicated by a double-headed arrow, as in

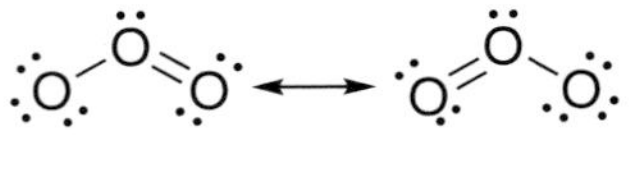

At this stage we are not indicating the shape of the molecule. Resonance should be pictured as a *blending* of structures, not a flickering alternation between them. In quantum mechanical terms, the electron distribution of each structure is represented by a wavefunction, ψ, and the actual wavefunction, ψ, of the molecule is the superposition of the individual wavefunctions for each contributing structure:[1]

$$\psi = \psi(O-O=O) + \psi(O=O-O)$$

The overall wavefunction is written as a superposition with equal contributions from both structures because the two structures have identical energies. The *blended* structure of two or more Lewis structures is called a **resonance hybrid**. Note that resonance occurs between structures that differ only in the allocation of electrons; resonance does not occur between structures in which the atoms themselves lie in different positions. For instance, there is no resonance between the structures SOO and OSO.

Resonance has two main effects:

1. Resonance averages the bond characteristics over the molecule.
2. The energy of a resonance hybrid structure is lower than that of any single contributing structure.

The energy of the O_3 resonance hybrid, for instance, is lower than that of either individual structure alone. Resonance is most important when there are several structures of identical energy that can be written to describe the molecule, as for O_3. In such cases, all the structures of the same energy contribute equally to the overall structure.

Structures with different energies may also contribute to an overall resonance hybrid but, in general, the greater the energy difference between two Lewis structures, the smaller the contribution of the higher energy structure. The BF_3 molecule, for instance, could be regarded as a resonance hybrid of the structures shown in (**5**), but the first structure dominates even though the octet is incomplete. Consequently, BF_3 is regarded *primarily* as having that structure with a small admixture of double-bond character. In contrast, for the NO_3^- ion (**6**), the last three structures dominate, and we treat the ion as having partial double-bond character.

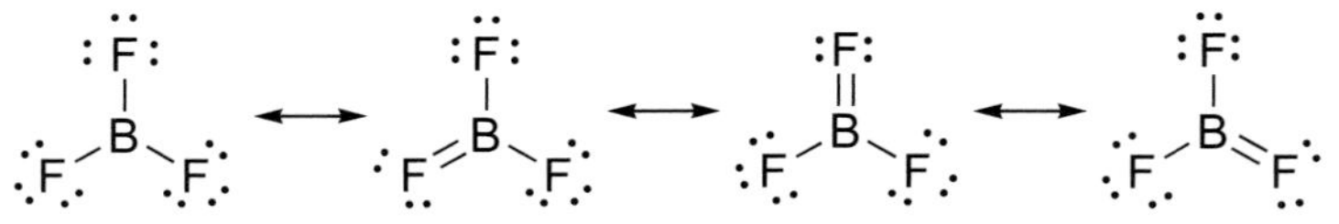

5 BF₃

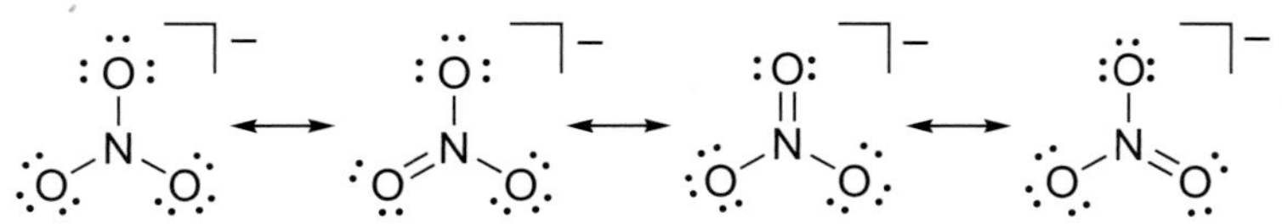

6 NO₃⁻

2.3 The VSEPR model

There is no simple method for predicting the numerical value of bond angles even in simple molecules, except where the shape is governed by symmetry. However, the **valence**

---

[1]This wavefunction is not normalized (Section 1.5). We shall often omit normalization constants from linear combinations in order to clarify their structure. The wavefunctions themselves are formulated in the valence bond theory, which is described later.

shell electron pair repulsion (VSEPR) model of molecular shape, which is based on some simple ideas about electrostatic repulsion and the presence or absence of lone pairs, is surprisingly useful.

(a) The basic shapes

Key points: In the VSEPR model, regions of enhanced electron density take up positions as far apart as possible, and the shape of the molecule is identified by referring to the locations of the atoms in the resulting structure.

The primary assumption of the VSEPR model is that regions of enhanced electron density, by which we mean bonding pairs, lone pairs, or the concentrations of electrons associated with multiple bonds, take up positions as far apart as possible so that the repulsions between them are minimized. For instance, four such regions of electron density will lie at the corners of a regular tetrahedron, five will lie at the corners of a trigonal bipyramid, and so on (Table 2.2).

Although the arrangement of regions of electron density, both bonding regions and regions associated with lone pairs, governs the shape of the molecule, the *name* of the shape is determined by the arrangement of *atoms*, not the arrangement of the regions of electron density (Table 2.3). For instance, the NH_3 molecule has four electron pairs that are disposed tetrahedrally, but as one of them is a lone pair the molecule itself is classified as trigonal pyramidal. One apex of the pyramid is occupied by the lone pair. Similarly, H_2O has a tetrahedral arrangement of its electron pairs but, as two of the pairs are lone pairs, the molecule is classified as angular (or 'bent').

To apply the VSEPR model systematically, we first write down the Lewis structure for the molecule or ion and identify the central atom. Next, we count the number of atoms and lone pairs carried by that atom because each atom (whether it is singly or multiply bonded to the central atom) and each lone pair counts as one region of high electron density. To achieve lowest energy, these regions take up positions as far apart as possible, so we identify the basic shape they adopt by referring to Table 2.2. Finally, we note which locations correspond to atoms and identify the shape of the molecule from Table 2.3. Thus, a PCl_5 molecule, with five single bonds and therefore five regions of electron density around the central atom, is predicted (and found) to be trigonal bipyramidal (**7**).

EXAMPLE 2.2 **Using the VSEPR model to predict shapes**

Predict the shape of (a) a BF_3 molecule, (b) an SO_3^{2-} ion, and (c) a PCl_4^+ ion.

Answer We begin by drawing the Lewis structure of each species and then consider the number of bonding and lone pairs of electrons and how they are arranged around the central atom. (a) The Lewis structure of BF_3 is shown in (**5**). To the central B atom there are attached three F atoms but no lone pairs. The basic arrangement of three regions of electron density is trigonal planar. Because each location carries an F atom, the shape of the molecule is also trigonal planar (**8**). (b) Two Lewis structures for SO_3^{2-} are shown in (**9**): they are representative of a variety of structures that contribute to the overall resonance structure. In each case there are three atoms attached to the central S atom and one lone pair, corresponding to four regions of electron density. The basic arrangement of these regions is tetrahedral. Three of the locations correspond to atoms, so the shape of the ion is trigonal pyramidal (**10**). Note that the shape deduced in this way is independent of which resonance structure is being considered. (c) Phosphorus has five valence electrons. Four of these electrons are used to form bonds to the four Cl atoms. One electron is removed to give the $+1$ charge on the ion, so all the electrons supplied by the P atom are used in bonding and there is no lone pair. Four regions adopt a tetrahedral arrangement and, as each one is associated with a Cl atom, the ion is tetrahedral (**11**).

Self-test 2.2 Predict the shape of (a) an H_2S molecule, (b) an XeO_4 molecule.

The VSEPR model is highly successful, but sometimes runs into difficulty when there is more than one basic shape of similar energy. For example, with five electron-dense regions around the central atom, a square-pyramidal arrangement is only slightly higher in energy than a trigonal-bipyramidal arrangement, and there are several examples of the former (**12**). Similarly, the basic shapes for seven electron-dense regions are less readily predicted than others, partly because so many different conformations correspond to similar energies. However, in the p block, seven-coordination is dominated by pentagonal-bipyramidal structures. For example, IF_7 is pentagonal bipyramidal and XeF_5^-, with five bonds and two

---

Table 2.2 The basic arrangement of regions of electron density according to the VSEPR model

Number of electron regions	Arrangement
2	Linear
3	Trigonal planar
4	Tetrahedral
5	Trigonal bipyramidal
6	Octahedral

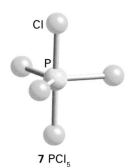

7 PCl_5

8 BF_3

$$:\ddot{O}-\underset{\underset{\ddot{\:}}{\overset{|}{\ddot{O}}}}{S}-\ddot{O}:\Big]^{2-} \longleftrightarrow :\ddot{O}-\underset{\underset{\ddot{\:}}{\overset{\|}{O}}}{S}-\ddot{O}:\Big]^{2-}$$

9 SO_3^{2-}

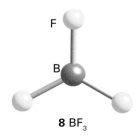

10 SO_3^{2-}

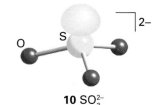

11 PCl_4^+

Table 2.3 The description of molecular shapes

Shape	Examples
Linear	HCN, CO_2
Angular (bent)	H_2O, O_3, NO_2^-
Trigonal planar	BF_3, SO_3, NO_3^-, CO_3^{2-}
Trigonal pyramidal	NH_3, SO_3^{2-}
Tetrahedral	CH_4, SO_4^{2-}
Square planar	XeF_4
Square pyramidal	$Sb(Ph)_5$
Trigonal bipyramidal	$PCl_5(g)$, SOF_4^*
Octahedral	SF_6, PCl_6^-, $IO(OH)_5^*$

*Approximate shape.

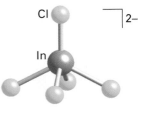

12 $[InCl_5]^{2-}$

lone pairs, is pentagonal planar. Lone pairs are stereochemically less influential when they belong to heavy p-block elements. The SeF_6^{2-} and $TeCl_6^{2-}$ ions, for instance, are octahedral despite the presence of a lone pair on the Se and Te atoms. Lone pairs that do not influence the molecular geometry are said to be **stereochemically inert** and are usually in the non-directional s orbitals.

(b) Modifications of the basic shapes

Key point: Lone pairs repel other pairs more strongly than bonding pairs do.

Once the basic shape of a molecule has been identified, adjustments are made by taking into account the differences in electrostatic repulsion between bonding regions and lone pairs. These repulsions are assumed to lie in the order

lone pair/lone pair > lone pair/bonding region > bonding region/bonding region

In elementary accounts, the greater repelling effect of a lone pair is explained by supposing that the lone pair is on average closer to the nucleus than a bonding pair and therefore repels other electron pairs more strongly. However, the true origin of the difference is obscure. An additional detail about this order of repulsions is that, given the choice between an axial and an equatorial site for a lone pair in a trigonal-bipyramidal array, the lone pair occupies the equatorial site. Whereas in the equatorial site the lone pair is repelled by the two bonding pairs at 90° (Fig. 2.1), in the axial position the lone pair is repelled by three bonding pairs at 90°. In an octahedral basic shape, a single lone pair can occupy any position but a second lone pair will occupy the position directly *trans* (opposite) to the first, which results in a square-planar structure.

In a molecule with two adjacent bonding pairs and one or more lone pairs, the bond angle is decreased relative to that expected when all pairs are bonding. Thus, the HNH angle in NH_3 is reduced from the tetrahedral angle (109.5°) of the underlying basic shape to a smaller value. This decrease is consistent with the observed HNH angle of 107°. Similarly, the HOH angle in H_2O is decreased from the tetrahedral value as the two lone pairs move apart. This decrease is in agreement with the observed HOH bond angle of 104.5°. A deficiency of the VSEPR model, however, is that it cannot be used to predict the actual bond angle adopted by the molecule.[2]

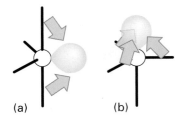

(a)   (b)

Fig. 2.1 In the VSEPR model a lone pair in (a) the equatorial position of a trigonal-bipyramidal arrangement interacts strongly with two bonding pairs, but in (b) an axial position it interacts strongly with three bonding pairs. The former arrangement is generally lower in energy.

EXAMPLE 2.3 Accounting for the effect of lone pairs on molecular shape

Predict the shape of an SF_4 molecule.

Answer We begin by drawing the Lewis structure of the molecule and identify the number of bonding and lone pairs of electrons; then we identify the shape of the molecule and finally consider any modifications

---

[2] There are also problems with hydrides and fluorides. See *Further reading*.

13 SF$_4$

due to the presence of lone pairs. The Lewis structure of SF$_4$ is shown in (**13**). The central S atom has four F atoms attached to it and one lone pair. The basic shape adopted by these five regions is trigonal bipyramidal. The potential energy is least if the lone pair occupies an equatorial site to give a molecular shape that resembles a see-saw, with the axial bonds forming the 'plank' of the see-saw and the equatorial bonds the 'pivot'. The S–F bonds then bend away from the lone pair (**14**).

Self-test 2.3 Predict the shape of an XeF$_2$ molecule.

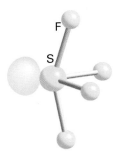

14 SF$_4$

Valence bond theory

The **valence bond theory** (VB theory) of bonding was the first quantum mechanical theory of bonding to be developed. Valence bond theory considers the interaction of atomic orbitals on separate atoms as they are brought together to form a molecule. Although the computational techniques involved have been largely superseded by molecular orbital theory, much of the language and some of the concepts of VB theory still remain and are used throughout chemistry.

2.4 The hydrogen molecule

Key points: In valence bond theory, the wavefunction of an electron pair is formed by superimposing the wavefunctions for the separated fragments of the molecule; a molecular potential energy curve shows the variation of the molecular energy with internuclear separation.

The two-electron wavefunction for two widely separated H atoms is $\psi = \chi_A(1)\chi_B(2)$, where χ_A and χ_B are H1s orbitals on atoms A and B. (Although χ, chi, is also used for electronegativity, the context makes it unlikely that the two usages will be confused: χ is commonly used to denote an atomic orbital in computational chemistry.) When the atoms are close, it is not possible to know whether it is electron 1 that is on A or electron 2. An equally valid description is therefore $\psi = \chi_A(2)\chi_B(1)$, in which electron 2 is on A and electron 1 is on B. When two outcomes are equally probable, quantum mechanics instructs us to describe the true state of the system as a superposition of the wavefunctions for each possibility, so a better description of the molecule than either wavefunction alone is the linear combination of the two possibilities.

$$\psi = \chi_A(1)\chi_B(2) + \chi_A(2)\chi_B(1) \tag{2.1}$$

This function is the (unnormalized) VB wavefunction for an H–H bond. The formation of the bond can be pictured as being due to the high probability that the two electrons will be found between the two nuclei and hence will bind them together (Fig. 2.2). More formally, the wave pattern represented by the term $\chi_A(1)\chi_B(2)$ interferes constructively with the wave pattern represented by the contribution $\chi_A(2)\chi_B(1)$ and there is an enhancement in the amplitude of the wavefunction in the internuclear region. For technical reasons stemming from the Pauli principle, only electrons with paired spins can be described by a wavefunction of the type written in eqn 2.1, so only paired electrons can contribute to a bond in VB theory. We say, therefore, that a VB wavefunction is formed by **spin pairing** of the electrons in the two contributing atomic orbitals. The electron distribution described by the wavefunction in eqn 2.1 is called a σ **bond**. As shown in Fig. 2.2, a σ bond has cylindrical symmetry around the internuclear axis, and the electrons in it have zero orbital angular momentum about that axis.

The **molecular potential energy curve** for H$_2$, a graph showing the variation of the energy of the molecule with internuclear separation, is calculated by changing the internuclear separation R and evaluating the energy at each selected separation (Fig. 2.3). The energy is found to fall below that of two separated H atoms as the two atoms are brought within bonding distance and each electron becomes free to migrate to the other atom. However, the resulting lowering of energy is counteracted by an increase in energy from the Coulombic (electrostatic) repulsion between the two positively charged nuclei. This positive contribution to the energy becomes large as R becomes small. Consequently, the total potential energy curve passes through a minimum and then climbs

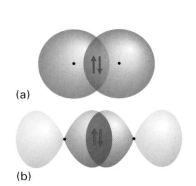

(a)

(b)

Fig. 2.2 The formation of a σ bond from (a) s orbital overlap, (b) p orbital overlap. A σ bond has cylindrical symmetry around the internuclear axis.

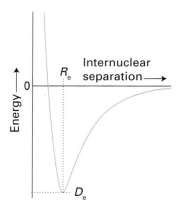

Fig. 2.3 A molecular potential energy curve showing how the total energy of a molecule varies as the internuclear separation is changed.

Fig. 2.4 The formation of a π bond.

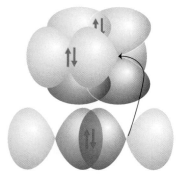

Fig. 2.5 The VB description of N_2. Two electrons form a σ bond and another two pairs form two π bonds. In linear molecules, where the x- and y-axes are not specified, the electron density of π bonds is cylindrically symmetrical around the internuclear axis.

to a strongly positive value at small internuclear separations. The depth of the minimum of the curve is denoted D_e. The deeper the minimum, the more strongly the atoms are bonded together. The steepness of the well shows how rapidly the energy of the molecule rises as the bond is stretched or compressed. The steepness of the curve, an indication of the *stiffness* of the bond, therefore governs the vibrational frequency of the molecule (Section 8.4).

2.5 Homonuclear diatomic molecules

Key point: Electrons in atomic orbitals of the same symmetry but on neighbouring atoms are paired to form σ and π bonds.

A similar description can be applied to more complex molecules, and we begin by considering **homonuclear diatomic molecules**, diatomic molecules in which both atoms belong to the same element (dinitrogen, N_2, is an example). To construct the VB description of N_2, we consider the valence electron configuration of each atom, which from Section 1.8 we know to be $2s^2 2p_z^1 2p_y^1 2p_x^1$. It is conventional to take the z-axis to be the internuclear axis, so we can imagine each atom as having a $2p_z$ orbital pointing towards a $2p_z$ orbital on the other atom, with the $2p_x$ and $2p_y$ orbitals perpendicular to the axis. A σ bond is then formed by spin pairing between the two electrons in the opposing $2p_z$ orbitals. Its spatial wavefunction is still given by eqn 2.1, but now χ_A and χ_B stand for the two $2p_z$ orbitals. A simple way of identifying a σ bond is to envisage rotation of the bond around the internuclear axis: if the wavefunction remains unchanged, the bond is classified as σ.

The remaining 2p orbitals cannot merge to give σ bonds as they do not have cylindrical symmetry around the internuclear axis. Instead, the orbitals merge to form two π **bonds**. A π bond arises from the spin pairing of electrons in two p orbitals that approach side by side (Fig. 2.4). The bond is so-called because, viewed along the internuclear axis, it resembles a pair of electrons in a p orbital. More precisely, an electron in a π bond has one unit of orbital angular momentum about the internuclear axis. A simple way of identifying a π bond is to envisage rotation of the bond through 180° around the internuclear axis. If the signs (as indicated by the shading) of the lobes of the orbital are interchanged, then the bond is classified as π.

There are two π bonds in N_2, one formed by spin pairing in two neighbouring $2p_x$ orbitals and the other by spin pairing in two neighbouring $2p_y$ orbitals. The overall bonding pattern in N_2 is therefore a σ bond plus two π bonds (Fig. 2.5), which is consistent with the structure $N\equiv N$. Analysis of the total electron density in a triple bond shows that it has cylindrical symmetry around the internuclear axis, with the four electrons in the two π bonds forming a ring of electron density around the central σ bond.

2.6 Polyatomic molecules

Key points: Each σ bond in a polyatomic molecule is formed by the spin pairing of electrons in any neighbouring atomic orbitals with cylindrical symmetry about the relevant internuclear axis; π bonds are formed by pairing electrons that occupy neighbouring atomic orbitals of the appropriate symmetry.

To introduce polyatomic molecules we consider the VB description of H_2O. The valence electron configuration of a hydrogen atom is $1s^1$ and that of an O atom is $2s^2 2p_z^2 2p_y^1 2p_x^1$. The two unpaired electrons in the O2p orbitals can each pair with an electron in an H1s orbital, and each combination results in the formation of a σ bond (each bond has cylindrical symmetry about the respective O−H internuclear axis). Because the $2p_y$ and $2p_z$ orbitals lie at 90° to each other, the two σ bonds also lie at 90° to each other (Fig. 2.6). We can predict, therefore, that H_2O should be an angular molecule, which it is. However, the theory predicts a bond angle of 90° whereas the actual bond angle is 104.5°. Similarly, to predict the structure of an ammonia molecule, NH_3, we start by noting that the valence electron configuration of an N atom given previously suggests that three H atoms can form bonds by spin pairing with the electrons in the three half-filled 2p orbitals. The latter are perpendicular to each other, so we predict a trigonal-pyramidal molecule with a bond angle of 90°. An NH_3 molecule is indeed trigonal pyramidal, but the experimental bond angle is 107°.

Another deficiency of the VB theory presented so far is its inability to account for the tetravalence of carbon, its ability to form four bonds. The ground-state configuration of

C is $2s^22p_z^12p_y^1$, which suggests that a C atom should be capable of forming only two bonds, not four. Clearly, something is missing from the VB approach.

These two deficiencies—the failure to account for bond angles and the valence of carbon—are overcome by introducing two new features, *promotion* and *hybridization*.

(a) Promotion

Key point: Promotion of electrons may occur if the outcome is to achieve more or stronger bonds and a lower overall energy.

Promotion is the excitation of an electron to an orbital of higher energy in the course of bond formation. Although electron promotion requires an investment of energy, that investment is worthwhile if the energy can be more than recovered from the greater strength or number of bonds that it allows to be formed. Promotion is not a 'real' process in which an atom somehow becomes excited and then forms bonds: it is a contribution to the overall energy change that occurs when bonds form.

In carbon, for example, the promotion of a 2s electron to a 2p orbital can be thought of as leading to the configuration $2s^12p_z^12p_y^12p_x^1$, with four unpaired electrons in separate orbitals. These electrons may pair with four electrons in orbitals provided by four other atoms, such as four H1s orbitals if the molecule is CH_4, and hence form four σ bonds. Although energy was required to promote the electron, it is more than recovered by the atom's ability to form four bonds in place of the two bonds of the unpromoted atom. Promotion, and the formation of four bonds, is a characteristic feature of carbon and of its congeners in Group 14 (Chapter 14) because the promotion energy is quite small: the promoted electron leaves a doubly occupied *ns* orbital and enters a vacant *np* orbital, hence significantly relieving the electron–electron repulsion it experiences in the ground state. This promotion on an electron becomes energetically less favourable as the group is descended and divalent compounds are common for tin and lead (Section 9.5).

(b) Hypervalence

Key point: Hypervalence and octet expansion occur for elements following Period 2.

The elements of Period 2, Li through Ne, obey the octet rule quite well, but elements of later periods show deviations from it. For example, the bonding in PCl_5 requires the P atom to have 10 electrons in its valence shell, one pair for each P–Cl bond (**15**). Similarly, in SF_6 the S atom must have 12 electrons if each F atom is to be bound to the central S atom by an electron pair (**16**). Species of this kind, which in terms of Lewis structures demand the presence of more than an octet of electrons around at least one atom, are called **hypervalent**.

The traditional explanation of hypervalence invokes the availability of low-lying unfilled d orbitals, which can accommodate the additional electrons. According to this explanation, a P atom can accommodate more than eight electrons if it uses its vacant 3d orbitals. In PCl_5, with its five pairs of bonding electrons, at least one 3d orbital must be used in addition to the four 3s and 3p orbitals of the valence shell. The rarity of hypervalence in Period 2 is then ascribed to the absence of 2d orbitals. However, the real reason for the rarity of hypervalence in Period 2 may be the geometrical difficulty of packing more than four atoms around a small central atom and may in fact have little to do with the availability of d orbitals. The molecular orbital theory of bonding, which is described later in this chapter, describes the bonding in hypervalent compounds without invoking participation of d orbitals.

(c) Hybridization

Key points: Hybrid orbitals are formed when atomic orbitals on the same atom interfere; specific hybridization schemes correspond to each local molecular geometry.

The description of the bonding in AB_4 molecules of Group 14 is still incomplete because it appears to imply the presence of three σ bonds of one type (formed from χ_B and χ_{A2p} orbitals) and a fourth σ bond of a distinctly different character (formed from χ_B and χ_{A2s}), whereas all the experimental evidence (bond lengths and strengths) points to the equivalence of all four A–B bonds, as in CH_4, for example.

This problem is overcome by realizing that the electron density distribution in the promoted atom is equivalent to the electron density in which each electron occupies a **hybrid**

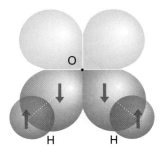

Fig. 2.6 The VB description of H_2O. There are two σ bonds formed by pairing electrons in O2p and H1s orbitals. This model predicts a bond angle of 90°.

15 PCl_5

16 SF_6

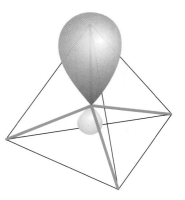

Fig. 2.7 One of the four equivalent sp^3 hybrid orbitals. Each one points towards a different vertex of a regular tetrahedron.

orbital formed by interference, or 'mixing', between the A2s and the A2p orbitals. The origin of the hybridization can be appreciated by thinking of the four atomic orbitals, which are waves centred on a nucleus, as being like ripples spreading from a single point on the surface of a lake: the waves interfere destructively and constructively in different regions, and give rise to four new shapes.

The specific linear combinations that give rise to four equivalent hybrid orbitals are

$$h_1 = s + p_x + p_y + p_z \quad h_2 = s - p_x - p_y + p_z$$
$$h_3 = s - p_x + p_y - p_z \quad h_4 = s + p_x - p_y - p_z \tag{2.2}$$

As a result of the interference between the component orbitals, each hybrid orbital consists of a large lobe pointing in the direction of one corner of a regular tetrahedron and a smaller lobe pointing in the opposite direction (Fig. 2.7). The angle between the axes of the hybrid orbitals is the tetrahedral angle, 109.47°. Because each hybrid is built from one s orbital and three p orbitals, it is called an **sp^3 hybrid orbital**.

It is now easy to see how the VB description of a CH_4 molecule is consistent with a tetrahedral shape with four equivalent C−H bonds. Each hybrid orbital of the promoted carbon atom contains a single unpaired electron; an electron in χ_{H1s} can pair with each one, giving rise to a σ bond pointing in a tetrahedral direction. Because each sp^3 hybrid orbital has the same composition, all four σ bonds are identical apart from their orientation in space.

A further feature of hybridization is that a hybrid orbital has pronounced directional character, in the sense that it has enhanced amplitude in the internuclear region. This directional character arises from the constructive interference between the s orbital and the positive lobes of the p orbitals. As a result of the enhanced amplitude in the internuclear region, the bond strength is greater than for an s or p orbital alone. This increased bond strength is another factor that helps to repay the promotion energy.

Hybrid orbitals of different compositions are used to match different molecular geometries and to provide a basis for their VB description. For example, sp^2 hybridization is used to reproduce the electron distribution needed for trigonal-planar species, such as on B in BF_3 and N in NO_3^-, and sp hybridization reproduces a linear distribution. Table 2.4 gives the hybrids needed to match the geometries of a variety of electron distributions.

Table 2.4 Some hybridization schemes

Coordination number	Arrangement	Composition
2	Linear	sp, pd, sd
	Angular	sd
3	Trigonal planar	sp^2, p^2d
	Unsymmetrical planar	spd
	Trigonal pyramidal	pd^2
4	Tetrahedral	sp^3, sd^3
	Irregular tetrahedral	spd^2, p^3d, pd^3
	Square planar	p^2d^2, sp^2d
5	Trigonal bipyramidal	sp^3d, spd^3
	Tetragonal pyramidal	sp^2d^2, sd^4, pd^4, p^3d^2
	Pentagonal planar	p^2d^3
6	Octahedral	sp^3d^2
	Trigonal prismatic	spd^4, pd^5
	Trigonal antiprismatic	p^3d^3

Molecular orbital theory

We have seen that VB theory provides a reasonable description of bonding in simple molecules. However, it does not handle polyatomic molecules very elegantly. **Molecular orbital theory** (MO theory) is a more sophisticated model of bonding that can be applied equally successfully to simple and complex molecules. In MO theory, we generalize the *atomic* orbital description of atoms in a very natural way to a **molecular orbital** description of molecules in which electrons spread over *all* the atoms in a molecule and bind them all together. In the spirit of this chapter, we continue to treat the concepts qualitatively and to give a sense of how inorganic chemists discuss the electronic structures of molecules by using MO theory. Almost all qualitative discussions and calculations on inorganic molecules and ions are now carried out within the framework of MO theory.

2.7 An introduction to the theory

We begin by considering homonuclear diatomic molecules and diatomic ions formed by two atoms of the same element. The concepts these species introduce are readily extended to heteronuclear diatomic molecules formed between two atoms or ions of different elements. They are also easily extended to polyatomic molecules and solids composed of huge numbers of atoms and ions. In parts of this section we shall include molecular fragments in the discussion, such as the SF diatomic group in the SF_6 molecule or the OO diatomic group in H_2O_2 as similar concepts also apply to pairs of atoms bound together as parts of larger molecules.

(a) The approximations of the theory

Key points: Molecular orbitals are constructed as linear combinations of atomic orbitals; there is a high probability of finding electrons in atomic orbitals that have large coefficients in the linear combination; each molecular orbital can be occupied by up to two electrons.

As in the description of the electronic structures of atoms, we set out by making the **orbital approximation**, in which we assume that the wavefunction, ψ, of the N_e electrons in the molecule can be written as a product of one-electron wavefunctions: $\psi = \psi(1)\psi(2) \ldots \psi(N_e)$. The interpretation of this expression is that electron 1 is described by the wavefunction $\psi(1)$, electron 2 by the wavefunction $\psi(2)$, and so on. These one-electron wavefunctions are the **molecular orbitals** of the theory. As for atoms, the square of a one-electron wavefunction gives the probability distribution for that electron in the molecule: an electron in a molecular orbital is likely to be found where the orbital has a large amplitude, and will not be found at all at any of its nodes.

The next approximation is motivated by noting that, when an electron is close to the nucleus of one atom, its wavefunction closely resembles an atomic orbital of that atom. For instance, when an electron is close to the nucleus of an H atom in a molecule, its wavefunction is like a 1s orbital of that atom. Therefore, we may suspect that we can construct a reasonable first approximation to the molecular orbital by superimposing atomic orbitals contributed by each atom. This modelling of a molecular orbital in terms of contributing atomic orbitals is called the **linear combination of atomic orbitals** (LCAO) approximation. A 'linear combination' is a sum with various weighting coefficients. In simple terms, we combine the atomic orbitals of contributing atoms to give molecular orbitals that extend over the entire molecule.

In the most elementary form of MO theory, only the valence shell atomic orbitals are used to form molecular orbitals. Thus, the molecular orbitals of H_2 are approximated by using two hydrogen 1s orbitals, one from each atom:

$$\psi = c_A\chi_A + c_B\chi_B \tag{2.3}$$

In this case the **basis set**, the atomic orbitals χ from which the molecular orbital is built, consists of two H1s orbitals, one on atom A and the other on atom B. The principle is exactly the same for more complex molecules. For example, the basis set for the methane molecule consists of the 2s and 2p orbitals on carbon and four 1s orbitals on the hydrogen atoms. The coefficients c in the linear combination show the extent to which each atomic orbital contributes to the molecular orbital: the greater the value of c, the greater the contribution of that atomic orbital to the molecular orbital. To interpret the coefficients in eqn 2.3 we note that c_A^2 is the probability that the electron will be found in the orbital χ_A and c_B^2 is the probability that the electron will be found in the orbital χ_B. The fact that both atomic orbitals contribute to the molecular orbital implies that there is interference between them where their amplitudes are nonzero, with the probability distribution being given by

$$\psi^2 = c_A^2\chi_A^2 + 2c_Ac_B\chi_A\chi_B + c_B^2\chi_B^2 \tag{2.4}$$

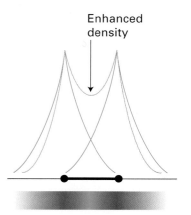

Enhanced
density

Fig. 2.8 The enhancement of electron density in the internuclear region arising from the constructive interference between the atomic orbitals on neighbouring atoms.

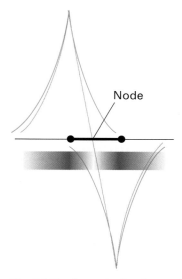

Node

Fig. 2.9 The destructive interference that arises if the overlapping orbitals have opposite signs. This interference leads to a nodal surface in an antibonding molecular orbital.

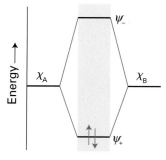

Fig. 2.10 The molecular orbital energy level diagram for H_2 and analogous molecules.

The term $2c_A c_B \chi_A \chi_B$ represents the contribution to the probability density arising from this interference.

Because H_2 is a homonuclear diatomic molecule, its electrons are equally likely to be found near each nucleus, so the linear combination that gives the lowest energy will have equal contributions from each 1s orbital ($c_A^2 = c_B^2$), leaving open the possibility that $c_A = +c_B$ or $c_A = -c_B$. Thus, ignoring normalization, the two molecular orbitals are

$$\psi_\pm = \chi_A \pm \chi_B \tag{2.5}$$

The relative signs of coefficients in LCAOs play a very important role in determining the energies of the orbitals. As we shall see, they determine whether atomic orbitals interfere constructively or destructively where they spread into the same region and hence lead to an accumulation or a reduction of electron density in those regions.

Two more preliminary points should be noted. We see from this discussion that *two* molecular orbitals may be constructed from *two* atomic orbitals. In due course, we shall see the importance of the general point that N molecular orbitals can be constructed from a basis set of N atomic orbitals. For example, if we use all four valence orbitals on each O atom in O_2, then from the total of eight atomic orbitals we can construct eight molecular orbitals. In addition, as in atoms, the Pauli exclusion principle implies that each molecular orbital may be occupied by up to two electrons; if two electrons are present, then their spins must be paired. Thus, in a diatomic molecule constructed from two Period 2 atoms and in which there are eight molecular orbitals available for occupation, up to 16 electrons may be accommodated before all the molecular orbitals are full. The same rules that are used for filling atomic orbitals with electrons (the building-up principle and Hund's rule, Section 1.8) apply to filling molecular orbitals with electrons.

The general pattern of the energies of molecular orbitals formed from N atomic orbitals is that one molecular orbital lies below that of the parent atomic energy levels, one lies higher in energy than they do, and the remainder are distributed between these two extremes.

(b) Bonding and antibonding orbitals

Key points: A bonding orbital arises from the constructive interference of neighbouring atomic orbitals; an antibonding orbital arises from their destructive interference, as indicated by a node between the atoms.

The orbital ψ_+ is an example of a **bonding orbital**. It is so-called because the energy of the molecule is lowered relative to that of the separated atoms if this orbital is occupied by electrons. The bonding character of ψ_+ is ascribed to the constructive interference between the two atomic orbitals and the resulting enhanced amplitude between the two nuclei (Fig. 2.8). An electron that occupies ψ_+ has an enhanced probability of being found in the internuclear region and can interact strongly with both nuclei. Hence orbital overlap, the spreading of one orbital into the region occupied by another, leading to enhanced probability of electrons being found in the internuclear region, is taken to be the origin of the strength of bonds.

The orbital ψ_- is an example of an **antibonding orbital**. It is so-called because, if it is occupied, the energy of the molecule is higher than for the two separated atoms. The greater energy of an electron in this orbital arises from the destructive interference between the two atomic orbitals, which cancels their amplitudes and gives rise to a nodal plane between the two nuclei (Fig. 2.9). Electrons that occupy ψ_- are largely excluded from the internuclear region and are forced to occupy energetically less favourable locations. It is generally true that the energy of a molecular orbital in a polyatomic molecule is higher the more internuclear nodes it has. The increase in energy reflects an increasingly complete exclusion of electrons from the regions between nuclei. Note that an antibonding orbital is slightly more antibonding than its partner bonding orbital is bonding: the asymmetry arises partly from the details of the electron distribution and partly from the fact that internuclear repulsion pushes the entire energy level diagram upwards.

The energies of the two molecular orbitals in H_2 are depicted in Fig. 2.10, which is an example of a **molecular orbital energy level diagram**, a diagram depicting the relative energies of molecular orbitals. The two electrons occupy the lower energy molecular orbital. An indication of the size of the energy gap between the two molecular orbitals is the observation of a spectroscopic absorption in H_2 at 11.4 eV (in the ultraviolet at 109 nm), which can be ascribed to the transition of an electron from the bonding orbital to the antibonding orbital. The dissociation energy of H_2 is 4.5 eV (434 kJ mol^{-1}), which gives an indication of the location of the bonding orbital relative to the separated atoms.

The Pauli exclusion principle limits to two the number of electrons that can occupy any molecular orbital and requires that those two electrons be paired ($\uparrow\downarrow$). The exclusion principle is the origin of the importance of the pairing of the electrons in bond formation in MO theory just as it is in VB theory: in the context of MO theory, two is the maximum number of electrons that can occupy an orbital that contributes to the stability of the molecule. The H_2 molecule, for example, has a lower energy than that of the separated atoms because two electrons can occupy the orbital ψ_+ and both can contribute to the lowering of its energy (as shown in Fig. 2.10). A weaker bond can be expected if only one electron is present in a bonding orbital, but nevertheless H_2^+ is known as a transient gas-phase ion; its dissociation energy is 2.6 eV (250.9 kJ mol^{-1}). Three electrons (as in H_2^-) are less effective than two electrons because the third electron must occupy the antibonding orbital ψ_- and hence destabilize the molecule. With four electrons, the antibonding effect of two electrons in ψ_- overcomes the bonding effect of two electrons in ψ_+. There is then no net bonding. It follows that a four-electron molecule with only 1s orbitals available for bond formation, such as He_2, is not expected to be stable relative to dissociation into its atoms.

So far, we have discussed interactions of atomic orbitals that give rise to molecular orbitals that are lower in energy (bonding) and higher in energy (antibonding) than the separated atoms. In addition, it is possible to generate a molecular orbital that has the same energy as the initial atomic orbitals. In this case, occupation of this orbital neither stabilizes nor destabilizes the molecule and so it is described as a **nonbonding orbital**. Typically, a nonbonding orbital is a molecular orbital that consists of a single orbital on one atom, perhaps because there is no atomic orbital of the correct symmetry for it to overlap on a neighbouring atom.

2.8 Homonuclear diatomic molecules

Although the structures of diatomic molecules can be calculated effortlessly by using commercial software packages, the validity of any such calculations must, at some point, be confirmed by experimental data. Moreover, elucidation of molecular structure can often be achieved by drawing on experimental information. One of the most direct portrayals of electronic structure is obtained from ultraviolet photoelectron spectroscopy (UPS, Section 8.8) in which electrons are ejected from the orbitals they occupy in molecules and their energies determined. Because the peaks in a photoelectron spectrum correspond to the various kinetic energies of photoelectrons ejected from different orbitals of the molecule, the spectrum gives a vivid portrayal of the molecular orbital energy levels of a molecule (Fig. 2.11).

(a) The orbitals

Key points: Molecular orbitals are classified as σ, π, or δ according to their rotational symmetry about the internuclear axis, and (in centrosymmetric species) as g or u according to their symmetry with respect to inversion.

Our task is to see how MO theory can account for the features revealed by photoelectron spectroscopy and the other techniques, principally absorption spectroscopy, that are used to study diatomic molecules. We are concerned predominantly with outer-shell valence orbitals, rather than core orbitals. As with H_2, the starting point in the theoretical discussion is the **minimal basis set**, the smallest set of atomic orbitals from which useful molecular orbitals can be built. In Period 2 diatomic molecules, the minimal basis set consists of the one valence s orbital and three valence p orbitals on each atom, giving eight atomic orbitals in all. We shall now see how the minimal basis set of eight valence shell atomic orbitals (four from each atom, one s and three p) is used to construct eight molecular orbitals. Then we shall use the Pauli principle to predict the ground-state electron configurations of the molecules.

The energies of the atomic orbitals that form the basis set are shown on either side of the molecular orbital diagram in Fig. 2.12. We form σ **orbitals** by allowing overlap between atomic orbitals that have cylindrical symmetry around the internuclear axis, which (as remarked earlier) is conventionally labelled z. The notation σ signifies that the orbital has cylindrical symmetry; atomic orbitals that can form σ orbitals include the 2s and 2p$_z$ orbitals on the two atoms (Fig. 2.13). From these four orbitals (the 2s and the 2p$_z$ orbitals

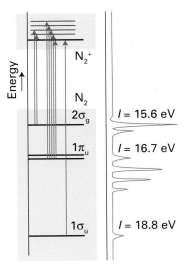

Fig. 2.11 The UV photoelectron spectrum of N_2. The fine structure in the spectrum arises from excitation of vibrations in the cation formed by photoejection of an electron.

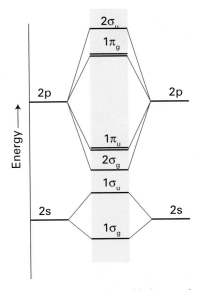

Fig. 2.12 The molecular orbital energy level diagram for the later Period 2 homonuclear diatomic molecules. This diagram should be used for O_2 and F_2.

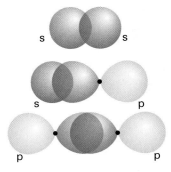

Fig. 2.13 A σ orbital can be formed in several ways, including s,s overlap, s,p overlap, and p,p overlap, with the p orbitals directed along the internuclear axis.

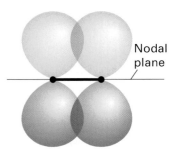

Fig. 2.14 Two p orbitals can overlap to form a π orbital. The orbital has a nodal plane passing through the internuclear axis, shown here from the side.

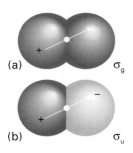

Fig 2.15 (a) Bonding and (b) antibonding σ interactions with the arrow indicating the inversion.

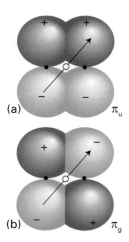

Fig 2.16 (a) Bonding and (b) antibonding π interactions with the arrow indicating the inversions.

on atom A and the corresponding orbitals on atom B) with cylindrical symmetry we can construct four σ molecular orbitals, two of which arise predominantly from interaction of the 2s orbitals, and two from interaction of the $2p_z$ orbitals. These molecular orbitals are labelled $1\sigma_g$, $1\sigma_u$, $2\sigma_g$, and $2\sigma_u$, respectively. Their energies resemble those shown in Fig. 2.12 but it is difficult to predict the precise locations of the central two orbitals. Interaction between a 2s on one atom and a $2p_z$ orbital on the other atom is possible if their relative energies are similar.

The remaining two 2p orbitals on each atom, which have a nodal plane containing the z-axis, overlap to give π orbitals (Fig. 2.14). Bonding and antibonding π orbitals can be formed from the mutual overlap of the two $2p_x$ orbitals, and also from the mutual overlap of the two $2p_y$ orbitals. This pattern of overlap gives rise to the two pairs of doubly degenerate energy levels (two energy levels of the same energy) shown in Fig. 2.12 and labelled $1\pi_u$ and $1\pi_g$.

For homonuclear diatomics, it is sometimes convenient (particularly for spectroscopic discussions) to signify the symmetry of the molecular orbitals with respect to their behaviour under inversion through the centre of the molecule. The operation of **inversion** consists of starting at an arbitrary point in the molecule, travelling in a straight line to the centre of the molecule, and then continuing an equal distance out on the other side of the centre. This procedure is indicated by the arrows in Figs 2.15 and 2.16. The orbital is designated g (for *gerade*, even) if it is identical under inversion, and u (for *ungerade*, odd) if it changes sign. Thus, a bonding σ orbital is g and an antibonding σ orbital is u (Fig. 2.15). On the other hand, a bonding π orbital is u and an antibonding π orbital is g (Fig. 2.16). Note that the σ_g orbitals are numbered separately from the σ_u orbitals, and similarly for the π orbitals.

The procedure can be summarized as follows:

1. From a basis set of four atomic orbitals on each atom, eight molecular orbitals are constructed.

2. Four of these eight molecular orbitals are σ orbitals and four are π orbitals.

3. The four σ orbitals span a range of energies, one being strongly bonding and another strongly antibonding; the remaining two lie between these extremes.

4. The four π orbitals form one doubly degenerate pair of bonding orbitals and one doubly degenerate pair of antibonding orbitals.

To establish the actual location of the energy levels, it is necessary to use electronic absorption spectroscopy, photoelectron spectroscopy, or detailed computation.

Photoelectron spectroscopy and detailed computation (the numerical solution of the Schrödinger equation for the molecules) enable us to build the orbital energy schemes shown in Fig. 2.17. As we see there, from Li_2 to N_2 the arrangement of orbitals is that shown in Fig. 2.18, whereas for O_2 and F_2 the order of the σ and π orbitals is reversed and the array is that shown in Fig. 2.12. The reversal of order can be traced to the increasing separation of the 2s and 2p orbitals that occurs on going to the right across Period 2. A general principle of quantum mechanics is that the mixing of wavefunctions is strongest if their energies are similar; mixing is not important if their energies differ by more than about 1 eV. When the s,p energy separation is small, each σ molecular orbital is a mixture of s and p character on each atom. As the s and p energy separation increases, the molecular orbitals become more purely s-like and p-like.

When considering species containing two neighbouring d-block atoms, as in Hg_2^{2+} and $[Cl_4ReReCl_4]^{2-}$, we should also allow for the possibility of forming bonds from d orbitals. A d_{z^2} orbital has cylindrical symmetry with respect to the internuclear (z) axis, and hence can contribute to the σ orbitals that are formed from s and p_z orbitals. The d_{yz} and d_{zx} orbitals both look like p orbitals when viewed along the internuclear axis, and hence can contribute to the π orbitals formed from p_x and p_y. The new feature is the role of $d_{x^2-y^2}$ and d_{xy}, which have no counterpart in the orbitals discussed up to now. These two orbitals can overlap with matching orbitals on the other atom to give rise to doubly degenerate pairs of bonding and antibonding δ **orbitals** (Fig. 2.19). As we shall see in Chapter 19, δ orbitals are important for the discussion of bonds between d-metal atoms, in d-metal complexes, and in organometallic compounds.

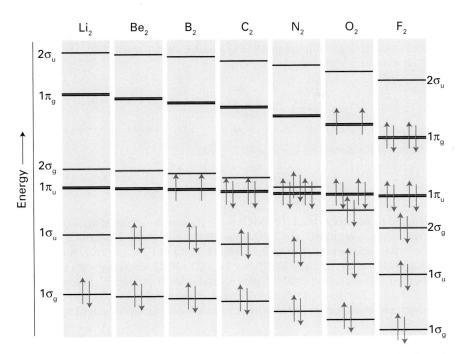

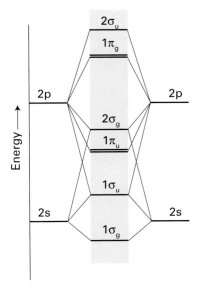

Fig. 2.18 The molecular orbital energy level diagram for Period 2 homonuclear diatomic molecules from Li_2 to N_2.

Fig. 2.17 The variation of orbital energies for Period 2 homonuclear diatomic molecules from Li_2 to F_2.

(b) The building-up principle for molecules

Key points: The building-up principle is used to predict the ground-state electron configurations by accommodating electrons in the array of molecular orbitals summarized in Fig. 2.12 or Fig. 2.18 and recognizing the constraints of the Pauli principle.

We use the building-up principle in conjunction with the molecular orbital energy level diagram in the same way as for atoms. The order of occupation of the orbitals is the order of increasing energy as depicted in Fig. 2.12 or Fig. 2.18. Each orbital can accommodate up to two electrons. If more than one orbital is available for occupation (because they happen to have identical energies, as in the case of pairs of π orbitals), then the orbitals are occupied separately. In that case, the electrons in the half-filled orbitals adopt parallel spins ($\uparrow\uparrow$), just as is required by Hund's rule for atoms (Section 1.7a). With very few exceptions, these rules lead to the actual ground-state configuration of the Period 2 diatomic molecules. For example, the electron configuration of N_2, with 10 valence electrons, is

$$N_2: 1\sigma_g^2 1\sigma_u^2 1\pi_u^4 2\sigma_g^2$$

Molecular orbital configurations are written like those for atoms: the orbitals are listed in order of increasing energy, and the number of electrons in each one is indicated by a superscript. Note that π^4 is shorthand for the occupation of two different π orbitals.

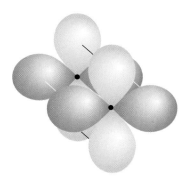

Fig. 2.19 The formation of δ orbitals by d-orbital overlap. The orbital has two mutually perpendicular nodal planes that intersect along the internuclear axis.

EXAMPLE 2.4 Predicting the electron configurations of diatomic molecules

Predict the ground-state electron configurations of the oxygen molecule, O_2, the superoxide ion, O_2^-, and the peroxide ion, O_2^{2-}.

Answer We need to determine the number of valence electrons and then populate the molecular orbitals with them in accord with the building-up principle. An O_2 molecule has 12 valence electrons. The first ten electrons recreate the N_2 configuration except for the reversal of the order of the $1\pi_u$ and $2\sigma_g$ orbitals (see Fig. 2.17). Next in line for occupation are the doubly degenerate $1\pi_g$ orbitals. The last two electrons enter these orbitals separately and have parallel spins. The configuration is therefore

$$O_2: 1\sigma_g^2 1\sigma_u^2 2\sigma_g^2 1\pi_u^4 1\pi_g^2$$

The **highest occupied molecular orbital** (HOMO) is the molecular orbital that, according to the building-up principle, is occupied last. The **lowest unoccupied molecular orbital** (LUMO) is the next higher molecular orbital. In Fig. 2.17, the HOMO of F_2 is $1\pi_g$ and its LUMO is $2\sigma_u$; for N_2 the HOMO is $2\sigma_g$ and the LUMO is $1\pi_g$. We shall increasingly see that these **frontier orbitals**, the LUMO and the HOMO, play special roles in the interpretation of structural and kinetic studies. The term SOMO, denoting a **singly occupied molecular orbital**, is sometimes encountered and is of crucial importance for the properties of radical species.

2.9 Heteronuclear diatomic molecules

The molecular orbitals of heteronuclear diatomic molecules differ from those of homonuclear diatomic molecules in having unequal contributions from each atomic orbital. Each molecular orbital has the form

$$\psi = c_A\chi_A + c_B\chi_B + \ldots \tag{2.6}$$

The unwritten orbitals include all the other orbitals of the correct symmetry for forming σ or π bonds but which typically make a smaller contribution than the two valence shell orbitals we are considering. In contrast to orbitals for homonuclear species, the coefficients c_A and c_B are not necessarily equal in magnitude. If $c_A^2 > c_B^2$, the orbital is composed principally of χ_A and an electron that occupies the molecular orbital is more likely to be found near atom A than atom B. The opposite is true for a molecular orbital in which $c_A^2 < c_B^2$. In heteronuclear diatomic molecules, the more electronegative element makes the larger contribution to bonding orbitals and the less electronegative element makes the greater contribution to the antibonding orbitals.

(a) Heteronuclear molecular orbitals

Key points: Heteronuclear diatomic molecules are polar; bonding electrons tend to be found on the more electronegative atom and antibonding electrons on the less electronegative atom.

The greater contribution to a bonding molecular orbital normally comes from the more electronegative atom: the bonding electrons are then likely to be found close to that atom and hence be in an energetically favourable location. The extreme case of a polar covalent bond, a covalent bond formed by an electron pair that is unequally shared by the two atoms, is an ionic bond. In an ionic bond, one atom gains complete control over the electron pair. The less electronegative atom normally contributes more to an antibonding orbital (Fig. 2.20), that is antibonding electrons are more likely to be found in an energetically unfavourable location, close to the less electronegative atom.

A second difference between homonuclear and heteronuclear diatomic molecules stems from the energy mismatch in the latter between the two sets of atomic orbitals. We have already remarked that two wavefunctions interact less strongly as their energies diverge. This dependence on energy separation implies that the lowering of energy as a result of the overlap of atomic orbitals on different atoms in a heteronuclear molecule is less pronounced than in a homonuclear molecule, in which the orbitals have the same energies. However, we cannot necessarily conclude that A–B bonds are weaker than A–A bonds because other factors (including orbital size and closeness of approach) are also important. The heteronuclear CO molecule, for example, which is isoelectronic with its homonuclear counterpart N_2, has an even higher bond enthalpy (1070 kJ mol^{-1}) than N_2 (946 kJ mol^{-1}).

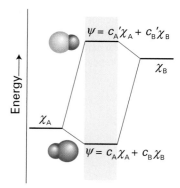

Fig. 2.20 The molecular orbital energy level diagram arising from interaction of two atomic orbitals with different energies. The lower molecular orbital is primarily composed of the lower energy atomic orbital, and vice versa. The shift in energies of the two levels is less than if the atomic orbitals had the same energy.

(b) Hydrogen fluoride

Key points: In hydrogen fluoride the bonding orbital is more concentrated on the F atom and the antibonding orbital is more concentrated on the H atom.

As an illustration of these general points, consider a simple heteronuclear diatomic molecule, HF. The five valence orbitals available for molecular orbital formation are the 1s orbital of H and the 2s and 2p orbitals of F; there are $1 + 7 = 8$ valence electrons to accommodate in the five molecular orbitals that can be constructed from the five basis orbitals.

The σ orbitals of HF can be constructed by allowing an H1s orbital to overlap the F2s and F2p$_z$ orbitals (z being the internuclear axis). These three atomic orbitals combine to give three σ molecular orbitals of the form $\psi = c_1\chi_{H1s} + c_2\chi_{F2s} + c_3\chi_{F2p}$. This procedure leaves the F2p$_x$ and F2p$_y$ orbitals unaffected as they have π symmetry and there is no valence H orbital of that symmetry. These π orbitals are therefore examples of the nonbonding orbitals mentioned earlier, and are molecular orbitals confined to a single atom. Note that, because there is no centre of inversion in a heteronuclear diatomic molecule, we do not use the g,u classification for its molecular orbitals.

Figure 2.21 shows the resulting energy level diagram. The 1σ bonding orbital is predominantly F2s in character as the energy difference between it and the H1s orbital is large. It is, therefore, confined mainly to the F atom and essentially nonbonding. The 2σ orbital is more bonding than the 1σ orbital and has both H1s and F2p character. The 3σ orbital is antibonding, and principally H1s in character: the 1s orbital has a relatively high energy (compared with the fluorine orbitals) and hence contributes predominantly to the high-energy antibonding molecular orbital.

Two of the eight valence electrons enter the 2σ orbital, forming a bond between the two atoms. Six more enter the 1σ and 1π orbitals; these two orbitals are largely nonbonding and confined mainly to the F atom. This is consistent with the conventional model of three lone pairs on the fluorine atom. All the electrons are now accommodated, so the configuration of the molecule is $1\sigma^2 2\sigma^2 1\pi^4$. One important feature to note is that all the electrons occupy orbitals that are predominantly on the F atom. It follows that we can expect the HF molecule to be polar, with a partial negative charge on the F atom, which is found experimentally.

(c) Carbon monoxide

Key points: The HOMO of a carbon monoxide molecule is an almost nonbonding σ orbital largely localized on C; the LUMO is an antibonding π orbital.

The molecular orbital energy level diagram for carbon monoxide is a somewhat more complicated example than HF because both atoms have 2s and 2p orbitals that can participate in the formation of σ and π orbitals. The energy level diagram is shown in Fig. 2.22. The ground-state configuration is

CO: $1\sigma^2 2\sigma^2 1\pi^4 3\sigma^2$

The 1σ orbital is localized mostly on the O atom and essentially nonbonding. The 2σ orbital is bonding. The 1π orbitals constitute the doubly degenerate pair of bonding π orbitals, with mainly C2p orbital character. The HOMO in CO is 3σ, which is predominantly C2p$_z$ in character, largely nonbonding, and located on the C atom. The LUMO is the doubly degenerate pair of antibonding π orbitals, with mainly C2p orbital character (Fig. 2.23). This combination of frontier orbitals—a full σ orbital largely localized on C and a pair of empty π orbitals—is one reason why metal carbonyls are such a characteristic feature of the d metals: in d-metal carbonyls, the HOMO lone pair orbital of CO participates in the formation of a σ bond and the LUMO antibonding π orbital participates in the formation of π bonds to the metal atom (Chapter 22).

Although the difference in electronegativity between C and O is large, the experimental value of the electric dipole moment of the CO molecule (0.1 D) is small. Moreover, the negative end of the dipole is on the C atom despite that being the less electronegative atom. This odd situation stems from the fact that the lone pairs and bonding pairs have a complex distribution. It is wrong to conclude that, because the bonding electrons are mainly on the O atom, O is the negative end of the dipole, as this ignores the balancing effect of the lone pair on the C atom. The inference of polarity from electronegativity is particularly unreliable when antibonding orbitals are occupied.

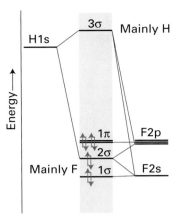

Fig. 2.21 The molecular orbital energy level diagram for HF. The relative positions of the atomic orbitals reflect the ionization energies of the atoms.

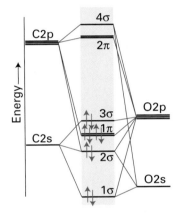

Fig. 2.22 The molecular orbital energy level diagram for CO.

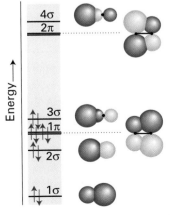

Fig. 2.23 A schematic illustration of the molecular orbitals of CO, with the size of the atomic orbital indicating the magnitude of its contribution to the molecular orbital.

ICl

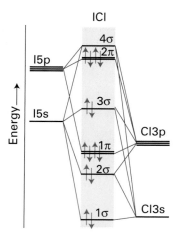

Fig. 2.24 A schematic illustration of the energies of the molecular orbitals of ICl.

2.10 Bond properties

We have already seen the origin of the importance of the electron pair: two electrons is the maximum number that can occupy a bonding orbital and hence contribute to a chemical bond. We now extend this concept by introducing the concept of 'bond order'.

(a) Bond order

Key points: The bond order assesses the net number of bonds between two atoms in the molecular orbital formalism; the greater the bond order between a given pair of atoms, the greater the bond strength.

The **bond order**, b, identifies a shared electron pair as counting as a 'bond' and an electron pair in an antibonding orbital as an 'antibond' between two atoms. More precisely, the bond order is defined as

$$b = \tfrac{1}{2}(n - n^*) \tag{2.7}$$

where n is the number of electrons in bonding orbitals and n^* is the number in antibonding orbitals. Nonbonding electrons are ignored when calculating bond order.

■ **A brief illustration.** Difluorine, F_2, has the configuration $1\sigma_g^2 1\sigma_u^2 2\sigma_g^2 1\pi_u^4 1\pi_g^4$ and, because $1\sigma_g$, $1\pi_u$, and $2\sigma_g$ orbitals are bonding but $1\sigma_u$ and $1\pi_g$ are antibonding, $b = \tfrac{1}{2}(2 + 2 + 4 - 2 - 4) = 1$. The bond order of F_2 is 1, which is consistent with the structure F—F and the conventional description of the molecule as having a single bond. Dinitrogen, N_2, has the configuration $1\sigma_g^2 1\sigma_u^2 1\pi_u^4 2\sigma_g^2$ and $b = \tfrac{1}{2}(2 + 4 + 2 - 2) = 3$. A bond order of 3 corresponds to a triply bonded molecule, which is in line with the structure N≡N. The high bond order is reflected in the high bond enthalpy of the molecule (946 kJ mol^{-1}), one of the highest for any molecule. ■

Isoelectronic molecules and ions have the same bond order, so F_2 and O_2^{2-} both have bond order 1. The bond order of the CO molecule, like that of the isoelectronic molecule N_2, is 3, in accord with the analogous structure C≡O. However, this method of assessing bonding is primitive, especially for heteronuclear species. For instance, inspection of the computed molecular orbitals suggests that 1σ and 3σ are best regarded as nonbonding orbitals largely localized on O and C, and hence should really be disregarded in the calculation of b. The resulting bond order is unchanged by this modification. The lesson is that the definition of bond order provides a useful indication of the multiplicity of the bond, but any interpretation of contributions to b needs to be made in the light of guidance from the composition of computed orbitals.

The definition of bond order allows for the possibility that an orbital is only singly occupied. The bond order in O_2^-, for example, is 1.5 because three electrons occupy the $1\pi_g$ antibonding orbitals. Electron loss from N_2 leads to the formation of the transient species N_2^+ in which the bond order is reduced from 3 to 2.5. This reduction in bond order is accompanied by a corresponding decrease in bond strength (from 946 to 855 kJ mol^{-1}) and increase in the bond length from 109 pm for N_2 to 112 pm for N_2^+.

EXAMPLE 2.6 Determining bond order

Determine the bond order of the oxygen molecule, O_2, the superoxide ion, O_2^-, and the peroxide ion, O_2^{2-}.

Answer We must determine the number of valence electrons, use them to populate the molecular orbitals, and then use eqn 2.7 to calculate b. The species O_2, O_2^- and O_2^{2-} have 12, 13, and 14 valence electrons, respectively. Their configurations are

O_2: $1\sigma_g^2 1\sigma_u^2 2\sigma_g^2 1\pi_u^4 1\pi_g^2$

O_2^-: $1\sigma_g^2 1\sigma_u^2 2\sigma_g^2 1\pi_u^4 1\pi_g^3$

O_2^{2-}: $1\sigma_g^2 1\sigma_u^2 2\sigma_g^2 1\pi_u^4 1\pi_g^4$

The $1\sigma_g$, $1\pi_u$, and $2\sigma_g$ orbitals are bonding and the $1\sigma_u$ and $1\pi_g$ orbitals are antibonding. Therefore, the bond orders are

O_2: $b = \frac{1}{2}(2 + 2 - 2 + 4 - 2) = 2$

O_2^-: $b = \frac{1}{2}(2 + 2 - 2 + 4 - 3) = 1.5$

O_2^{2-}: $b = \frac{1}{2}(2 + 2 - 2 + 4 - 4) = 1$

Self-test 2.6 Predict the bond order of the carbide anion, C_2^{2-}.

(b) Bond correlations

Key point: For a given pair of elements, bond strength increases and bond length decreases as bond order increases.

The strengths and lengths of bonds correlate quite well with each other and with the bond order. For a given pair of atoms:

Bond enthalpy increases as bond order increases.
Bond length decreases as bond order increases.

These trends are illustrated in Figs 2.25 and 2.26. The strength of the dependence varies with the elements. In Period 2 the correlation is relatively weak for CC bonds, with the result that a C=C double bond is less than twice as strong as a C–C single bond. This difference has profound consequences in organic chemistry, particularly for the reactions of unsaturated compounds. It implies, for example, that it is energetically favourable (but slow in the absence of a catalyst) for ethene and ethyne to polymerize: in this process, C–C single bonds form at the expense of the appropriate numbers of multiple bonds.

Familiarity with carbon's properties, however, must not be extrapolated without caution to the bonds between other elements. An N=N double bond (409 kJ mol^{-1}) is more than twice as strong as an N–N single bond (163 kJ mol^{-1}), and an N≡N triple bond (946 kJ mol^{-1}) is more than five times as strong. It is on account of this trend that NN multiply bonded compounds are stable relative to polymers or three-dimensional compounds having only single bonds. The same is not true of phosphorus, where the P–P, P=P, and P≡P bond enthalpies are 200, 310, and 490 kJ mol^{-1}, respectively. For phosphorus, single bonds are stable relative to the matching number of multiple bonds. Thus, phosphorus exists in a variety of solid forms in which P–P single bonds are present, including the tetrahedral P_4 molecules of white phosphorus. Diphosphorus molecules, P_2, are transient species generated at high temperatures and low pressures.

The two correlations with bond order taken together imply that, for a given pair of elements:

Bond enthalpy increases as bond length decreases.

This correlation is illustrated in Fig. 2.27: it is a useful feature to bear in mind when considering the stabilities of molecules because bond lengths may be readily available from independent sources.

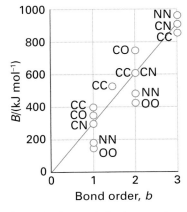

Fig. 2.25 The correlation between bond strength and bond order.

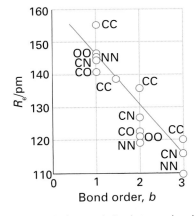

Fig. 2.26 The correlation between bond length and bond order.

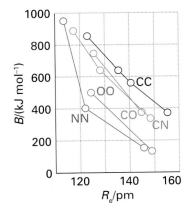

Fig. 2.27 The correlation between bond length and bond strength.

EXAMPLE 2.7 Predicting correlations between bond order, bond length, and bond strength

Use the bond orders of the oxygen molecule, O_2, the superoxide ion, O_2^-, and the peroxide ion, O_2^{2-}, calculated in Example 2.6 to predict the relative bond lengths and strengths of the species.

Answer We need to remember that bond enthalpy increases as bond order increases. The bond orders of O_2, O_2^-, and O_2^{2-} are 2, 1.5, and 1, respectively. Therefore, we expect the bond enthalpies to increase in the order $O_2^{2-} < O_2^- < O_2$. Bond length decreases as the bond enthalpy increases, so bond length should follow the opposite trend: $O_2^{2-} > O_2^- > O_2$. These predictions are supported by the gas phase bond enthalpies of O—O bonds (146 kJ mol^{-1}) and O=O bonds (496 kJ mol^{-1}) and the associated bond lengths of 132 and 121 pm, respectively.

Self-test 2.7 Predict the order of bond enthalpies and bond lengths for C—N, C=N, and C≡N bonds.

2.11 Polyatomic molecules

Molecular orbital theory can be used to discuss in a uniform manner the electronic structures of triatomic molecules, finite groups of atoms, and the almost infinite arrays of atoms in solids. In each case the molecular orbitals resemble those of diatomic molecules, the only important difference being that the orbitals are built from a more extensive basis set of atomic orbitals. As remarked earlier, a key point to bear in mind is that from N atomic orbitals it is possible to construct N molecular orbitals.

We saw in Section 2.8 that the general structure of molecular orbital energy level diagrams can be derived by grouping the orbitals into different sets, the σ and π orbitals, according to their shapes. The same procedure is used in the discussion of the molecular orbitals of polyatomic molecules. However, because their shapes are more complex than diatomic molecules, we need a more powerful approach. The discussion of polyatomic molecules will therefore be carried out in two stages. In this chapter we use intuitive ideas about molecular shape to construct molecular orbitals. In Chapter 6 we discuss the shapes of molecules and the use of their symmetry characteristics to construct molecular orbitals and account for other properties. That chapter rationalizes the procedures presented here.

The photoelectron spectrum of NH_3 (Fig. 2.28) indicates some of the features that a theory of the structure of polyatomic molecules must elucidate. The spectrum shows two bands. The one with the lower ionization energy (in the region of 11 eV) has considerable vibrational structure. This structure indicates that the orbital from which the electron is ejected plays a considerable role in the determination of the molecule's shape. The broad band in the region of 16 eV arises from electrons that are bound more tightly.

(a) Polyatomic molecular orbitals

Key points: Molecular orbitals are formed from linear combinations of atomic orbitals of the same symmetry; their energies can be determined experimentally from gas-phase photoelectron spectra and interpreted in terms of the pattern of orbital overlap.

The features that have been introduced in connection with diatomic molecules are present in all polyatomic molecules. In each case, we write the molecular orbital of a given symmetry (such as the σ orbitals of a linear molecule) as a sum of *all* the atomic orbitals that can overlap to form orbitals of that symmetry:

$$\psi = \sum_i c_i \chi_i \tag{2.8}$$

In this linear combination, the χ_i are atomic orbitals (usually the valence orbitals of each atom in the molecule) and the index i runs over all the atomic orbitals that have the appropriate symmetry. From N atomic orbitals we can construct N molecular orbitals. Then:

1. The greater the number of nodes in a molecular orbital, the greater the antibonding character and the higher the orbital energy.

2. Orbitals constructed from lower energy atomic orbitals lie lower in energy (so atomic s orbitals typically produce lower energy molecular orbitals than atomic p orbitals of the same shell).

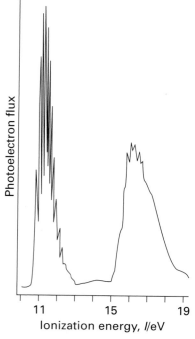

Fig. 2.28 The UV photoelectron spectrum of NH_3, obtained using He 21 eV radiation.

Photoelectron flux

Ionization energy, I/eV

3. Interactions between nonnearest-neighbour atoms are weakly bonding (lower the energy slightly) if the orbital lobes on these atoms have the same sign (and interfere constructively). They are weakly antibonding if the signs are opposite (and interfere destructively).

■ **A brief illustration.** To account for the features in the photoelectron spectrum of NH_3, we need to build molecular orbitals that will accommodate the eight valence electrons in the molecule. Each molecular orbital is the combination of seven atomic orbitals: the three H1s orbitals, the N2s orbital, and the three N2p orbitals. It is possible to construct seven molecular orbitals from these seven atomic orbitals (Fig. 2.29). ■

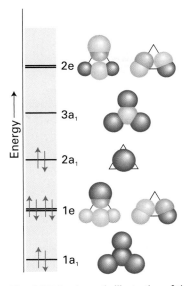

Fig. 2.29 A schematic illustration of the molecular orbitals of NH_3 with the size of the atomic orbital indicating the magnitude of its contribution to the molecular orbital. The view is along the z-axis.

It is not always strictly appropriate to use the notation σ and π in polyatomic molecules because these labels apply to a linear molecule. However, it is often convenient to continue to use the notation when concentrating on the *local* form of an orbital, its shape relative to the internuclear axis between two neighbouring atoms (this is an example of how the language of valence bond theory survives in MO theory). The correct procedure for labelling orbitals in polyatomic molecules according to their symmetry is described in Chapter 6. For our present purposes all we need know of this more appropriate procedure is the following:

a, b denote a nondegenerate orbital
e denotes a doubly degenerate orbital (two orbitals of the same energy)
t denotes a triply degenerate orbital (three orbitals of the same energy).

Subscripts and superscripts are sometimes added to these letters, as in a_1, b'', e_g, and t_2 because it is sometimes necessary to distinguish different a, b, e, and t orbitals according to a more detailed analysis of their symmetries.

The formal rules for the construction of the orbitals are described in Chapter 6, but it is possible to obtain a sense of their origin by imagining viewing the NH_3 molecule along its threefold axis (designated z). The $N2p_z$ and N2s orbitals both have cylindrical symmetry about that axis. If the three H1s orbitals are superimposed with the same sign relative to each other (that is, so that all have the same size and tint in the diagram, Fig. 2.29), then they match this cylindrical symmetry. It follows that we can form molecular orbitals of the form

$$\psi = c_1 \chi_{N2s} + c_2 \chi_{Np_z} + c_3 \{ \chi_{H1sA} + \chi_{H1sB} + \chi_{H1sC} \} \tag{2.9}$$

From these *three* basis orbitals (the specific combination of H1s orbitals counts as a single 'symmetry adapted' basis orbital), it is possible to construct three molecular orbitals (with different values of the coefficients c). The orbital with no nodes between the N and H atoms is the lowest in energy, that with a node between all the NH neighbours is the highest in energy, and the third orbital lies between the two. The three orbitals are nondegenerate and are labelled $1a_1$, $2a_1$, and $3a_1$ in order of increasing energy.

The $N2p_x$ and $N2p_y$ orbitals have π symmetry with respect to the z-axis, and can be used to form orbitals with combinations of the H1s orbitals that have a matching symmetry. For example, one such superposition will have the form

$$\psi = c_1 \chi_{N2p_x} + c_2 \{ \chi_{H1sA} + \chi_{H1sB} \} \tag{2.10}$$

As can be seen from Fig. 2.29, the signs of the H1s orbital combination match those of the $N2p_x$ orbital. The N2s orbital cannot contribute to this superposition, so only *two* combinations can be formed, one without a node between the N and H orbitals and the other with a node. The two orbitals differ in energy, the former being lower. A similar combination of orbitals can be formed with the $N2p_y$ orbital, and it turns out (by the symmetry arguments that we use in Chapter 6) that the two orbitals are degenerate with the two we have just described. The combinations are examples of e orbitals (because they form doubly degenerate pairs), and are labelled 1e and 2e in order of increasing energy.

The general form of the molecular orbital energy level diagram is shown in Fig. 2.30. The actual location of the orbitals (particularly the relative positions of the a and the e sets), can be found only by detailed computation or by identifying the orbitals responsible for the photoelectron spectrum. We have indicated the probable assignment of the 11 eV and 16 eV peaks, which fixes the locations of two of the occupied orbitals. The third occupied orbital is out of range of the 21 eV radiation used to obtain the spectrum.

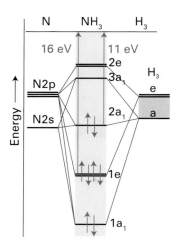

Fig. 2.30 The molecular orbital energy level diagram for NH_3 when the molecule has the observed bond angle (107°) and bond length.

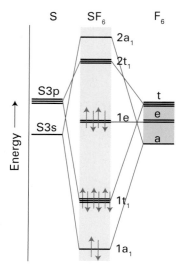

Fig. 2.31 A schematic molecular orbital energy level diagram for SF_6.

The photoelectron spectrum is consistent with the need to accommodate eight electrons in the orbitals. The electrons enter the molecular orbitals in increasing order of energy, starting with the orbital of lowest energy, and taking note of the requirement of the exclusion principle that no more than two electrons can occupy any one orbital. The first two electrons enter $1a_1$ and fill it. The next four enter the doubly degenerate $1e$ orbitals and fill them. The last two enter the $2a_1$ orbital, which calculations show is almost nonbonding and localized on the N atom. The resulting overall ground-state electron configuration is therefore $1a_1^2 1e^4 2a_1^2$. No antibonding orbitals are occupied, so the molecule has a lower energy than the separated atoms. The conventional description of NH_3 as a molecule with a lone pair is also mirrored in the configuration: the HOMO is $2a_1$, which is largely confined to the N atom and makes only a small contribution to the bonding. We saw in Section 2.3 that lone pair electrons play a considerable role in determining the shapes of molecules. The extensive vibrational structure in the 11 eV band of the photoelectron spectrum is consistent with this observation, as photoejection of a $2a_1$ electron removes the effectiveness of the lone pair and the shape of the ionized molecule is considerably different from that of NH_3 itself. Photoionization therefore results in extensive vibrational structure in the spectrum.

(b) Hypervalence in the context of molecular orbitals

Key point: The delocalization of molecular orbitals means that an electron pair can contribute to the bonding of more than two atoms.

In Section 2.3 we used valence bond theory to explain hypervalence by using d orbitals to allow the valence shell of an atom to accommodate more than eight electrons. Molecular orbital theory explains it rather more elegantly.

We consider SF_6, which has six S—F bonds and hence 12 electrons involved in forming bonds and is therefore hypervalent. The simple basis set of atomic orbitals that are used to construct the molecular orbitals consists of the valence shell s and p orbitals of the S atom and one p orbital of each of the six F atoms and pointing towards the S atom. We use the F2p orbitals rather than the F2s orbitals because they match the S orbitals more closely in energy. From these ten atomic orbitals it is possible to construct ten molecular orbitals. Calculations indicate that four of the orbitals are bonding and four are antibonding; the two remaining orbitals are nonbonding (Fig. 2.31).

There are 12 electrons to accommodate. The first two enter $1a_1$ and the next six enter $1t_1$. The remaining four fill the nonbonding pair of orbitals, resulting in the configuration $1a_1^2 1t_1^6 1e^4$. As we see, none of the antibonding orbitals ($2a_1$ and $2t_1$) is occupied. Molecular orbital theory, therefore, accounts for the formation of SF_6, with four bonding orbitals and two nonbonding orbitals occupied and does not need to invoke S3d orbitals and octet expansion. This does not mean that d orbitals cannot participate in the bonding, but it does show that they are not *necessary* for bonding six F atoms to the central S atom. The limitation of valence bond theory is the assumption that each atomic orbital on the central atom can participate in the formation of only one bond. Molecular orbital theory takes hypervalence into its stride by having available plenty of orbitals, not all of which are antibonding. Therefore, the question of when hypervalence can occur appears to depend on factors other than d-orbital availability, such as the ability of small atoms to pack around a large atom.

(c) Localization

Key points: Localized and delocalized descriptions of bonds are mathematically equivalent, but one description may be more suitable for a particular property, as summarized in Table 2.5.

A striking feature of the VB approach to chemical bonding is its accord with chemical instinct, as it identifies something that can be called 'an A—B bond'. Both OH bonds in H_2O, for instance, are treated as localized, equivalent structures because each one consists of an electron pair shared between O and H. This feature appears to be absent from MO theory because molecular orbitals are delocalized and the electrons that occupy them bind all the atoms together, not just a specific pair of neighbouring atoms. The concept of an A—B bond as existing independently of other bonds in the molecule, and of being transferable from one molecule to another, seems to have been lost. However, we shall now show that the molecular orbital description is mathematically almost equivalent to a localized

Table 2.5 A general indication of the properties for which localized and delocalized descriptions are appropriate

Localized appropriate	Delocalized appropriate
Bond strengths	Electronic spectra
Force constants	Photoionization
Bond lengths	Electron attachment
Brønsted acidity*	Magnetism
VSEPR description of molecular geometry	Walsh description of molecular geometry
	Standard potentials[†]

* Chapter 4.
[†] Chapter 5.

description of the overall electron distribution. The demonstration hinges on the fact that linear combinations of molecular orbitals can be formed that result in the same overall electron distribution, but the individual orbitals are distinctly different.

Consider the H_2O molecule. The two occupied bonding orbitals of the delocalized description, $1a_1$ and $1b_2$, are shown in Fig. 2.32. If we form the sum $1a_1 + 1b_2$, the negative half of $1b_2$ cancels half the $1a_1$ orbital almost completely, leaving a localized orbital between O and the other H. Likewise, when we form the difference $1a_1 - 1b_2$, the other half of the $1a_1$ orbital is cancelled almost completely, so leaving a localized orbital between the other pair of atoms. Therefore, by taking sums and differences of delocalized orbitals, localized orbitals are created (and vice versa). Because these are two equivalent ways of describing the same overall electron population, one description cannot be said to be better than the other.

Table 2.5 suggests when it is appropriate to select a delocalized description or a localized description. In general, a delocalized description is needed for dealing with global properties of the entire molecule. Such properties include electronic spectra (UV and visible transitions, Section 8.3), photoionization spectra, ionization and electron attachment energies (Section 1.9), and reduction potentials (Section 5.1). In contrast, a localized description is most appropriate for dealing with properties of a fragment of a total molecule. Such properties include bond strength, bond length, bond force constant, and some aspects of reactions (such as acid–base character): in these aspects the localized description is more appropriate because it focuses attention on the distribution of electrons in and around a particular bond.

(d) Localized bonds and hybridization

Key point: Hybrid atomic orbitals are sometimes used in the discussion of localized molecular orbitals.

The localized molecular orbital description of bonding can be taken a stage further by invoking the concept of hybridization. Strictly speaking, hybridization belongs to VB theory, but it is commonly invoked in simple qualitative descriptions of molecular orbitals.

We have seen that in general a molecular orbital is constructed from all atomic orbitals of the appropriate symmetry. However, it is sometimes convenient to form a mixture of orbitals on one atom (the O atom in H_2O, for instance), and then to use these hybrid orbitals to construct localized molecular orbitals. In H_2O, for instance, each OH bond can be regarded as formed by the overlap of an H1s orbital and a hybrid orbital composed of O2s and O2p orbitals (Fig. 2.33).

We have already seen that the mixing of s and p orbitals on a given atom results in hybrid orbitals that have a definite direction in space, as in the formation of tetrahedral hybrids. Once the hybrid orbitals have been selected, a localized molecular orbital description can be constructed. For example, four bonds in CF_4 can be formed by building bonding and antibonding localized orbitals by overlap of each hybrid and one F2p orbital directed towards it. Similarly, to describe the electron distribution of BF_3, we could consider each localized BF σ orbital as formed by the overlap of an sp² hybrid with an F2p orbital. A localized orbital description of a PCl_5 molecule would be in terms of five PCl σ bonds formed by overlap of each of the five trigonal-bipyramidal sp³d hybrid orbitals with a 2p

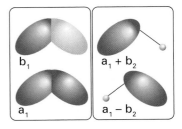

Fig. 2.32 The two occupied $1a_1$ and $1b_2$ orbitals of the H_2O molecule and their sum $1a_1 + 1b_2$ and difference $1a_1 - 1b_2$. In each case we form an almost fully delocalized orbital between a pair of atoms.

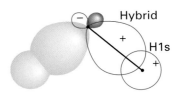

Fig. 2.33 The formation of localized O–H orbitals in H_2O by the overlap of hybrid orbitals on the O atom and H1s orbitals. The hybrid orbitals are a close approximation to the sp³ hybrids shown in Fig. 2.6.

orbital of a Cl atom. Similarly, where we wanted to form six localized orbitals in a regular octahedral arrangement (for example, in SF_6), we would need two d orbitals: the resulting six sp^3d^2 hybrids point in the required directions.

(e) Electron deficiency

Key point: The existence of electron-deficient species is explained by the delocalization of the bonding influence of electrons over several atoms.

The VB model of bonding fails to account for the existence of **electron-deficient compounds**, which are compounds for which, according to Lewis's approach, there are not enough electrons to form the required number of bonds. This point can be illustrated most easily with diborane, B_2H_6 (**17**). There are only 12 valence electrons but, according to Lewis's approach, at least eight electron pairs are needed to bind eight atoms together.

The formation of molecular orbitals by combining several atomic orbitals accounts effortlessly for the existence of these compounds. The eight atoms of this molecule contribute a total of 14 valence orbitals (three p and one s orbital from each B atom, making eight, and one s orbital each from the six H atoms). These 14 atomic orbitals can be used to construct 14 molecular orbitals. About seven of these molecular orbitals will be bonding or nonbonding, which is more than enough to accommodate the 12 valence electrons provided by the atoms.

The bonding can be best understood if we consider that the MOs produced are associated with either the terminal BH fragments or with the bridging BHB fragments. The localized MOs associated with the terminal BH bonds are constructed simply from atomic orbitals on two atoms (the H1s and a $B2s2p^n$ hybrid). The molecular orbitals associated with the two BHB fragments are linear combinations of the $B2s2p^n$ hybrids on each of the two B atoms and an H1s orbital of the H atom lying between them (Fig. 2.34). Three molecular orbitals are formed from these three atomic orbitals: one is bonding, one nonbonding, and the third is antibonding. The bonding orbital can accommodate two electrons and hold the BHB fragment together. The same remark applies to the second BHB fragment, and the two occupied 'bridging' bonding molecular orbitals hold the molecule together. Thus, overall, 12 electrons account for the stability of the molecule because their influence is spread over more than six pairs of atoms.

Electron deficiency is well developed not only in boron (where it was first clearly recognized) but also in carbocations and a variety of other classes of compounds that we encounter later in the text.

2.12 Molecular shape in terms of molecular orbitals

Key point: In the Walsh model, the shape of a molecule is predicted on the basis of the occupation of molecular orbitals that, in a correlation diagram, show a strong dependence on bond angle.

In MO theory, the electrons responsible for bonding are delocalized over the entire molecule. Current *ab initio* and semiempirical molecular orbital calculations, which are easily carried out with software, are able to predict the shapes of even quite complicated molecules with high reliability. Nevertheless, there is still a need to understand the qualitative factors that contribute to the shape of a molecule within the framework of MO theory.

Figure 2.35 shows the **Walsh diagram** for an XH_2 molecule. A Walsh diagram is a special case of a **correlation diagram**, a diagram that shows how one set of orbitals evolves into another as a parameter (such as a bond angle) is changed; we shall meet other examples later. A Walsh diagram adopts a simple pictorial approach to the task of analysing molecular shape in terms of delocalized molecular orbitals and was devised by A.D. Walsh in a classic series of papers published in 1953. Such diagrams play an important role in understanding the shapes, spectra, and reactions of polyatomic molecules. The XH_2 diagram has been constructed by considering how the composition and energy of each molecular orbital changes as the bond angle is varied from 90° to 180°.

The molecular orbitals are constructed from the 2s, $2p_z$, $2p_y$, and $2p_x$ atomic orbitals on X and the two H1s atomic orbitals. It is convenient to consider the possible combinations of H1s orbitals before forming the molecular orbitals with X. The linear combinations of

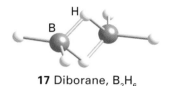

17 Diborane, B_2H_6

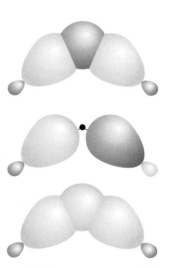

Fig. 2.34 The molecular orbital formed between two B atoms and one H atom lying between them, as in B_2H_6. Two electrons occupy the bonding combination and hold all three atoms together.

H1s orbitals ϕ_+ and ϕ_- are illustrated in Fig. 2.36. 'Symmetry adapted' combinations such as these will figure extensively in later discussions. The molecular orbitals to consider in the angular molecule are[3]

$$\psi_{a_1} = c_1 \chi_{2s} + c_2 \chi_{2p_z} + c_3 \phi_+$$

$$\psi_{b_1} = \chi_{2p_x} \tag{2.11}$$

$$\psi_{b_2} = c_4 \chi_{2p_y} + c_5 \phi_-$$

There are three a_1 orbitals and two b_2 orbitals; the lowest energy orbitals of each type are shown on the left of Fig. 2.37. In the linear molecule, the molecular orbitals are

$$\psi_{\sigma g} = c_1 \chi_{2s} + c_2 \phi_+$$

$$\psi_{\pi u} = \chi_{2p_y} \text{ and } \chi_{2p_z} \tag{2.12}$$

$$\psi_{\sigma u} = c_3 \chi_{2p_y} + c_4 \phi_-$$

These orbitals are shown on the right in Fig. 2.37.

The lowest energy molecular orbital in 90° H_2X is the one labelled $1a_1$, which is built from the overlap of the $X2p_z$ orbital with the ϕ_+ combination of H1s orbitals. As the bond angle changes to 180° the energy of this orbital decreases (Fig. 2.35) in part because the H−H overlap decreases and in part because the reduced involvement of the $X2p_z$ orbital decreases the overlap with ϕ_+ (Fig. 2.37). The energy of the $1b_2$ orbital is lowered because the H1s orbitals move into a better position for overlap with the $X2p_y$ orbital. The weakly antibonding H−H contribution is also decreased. The biggest change occurs for the $2a_1$ orbital. It has considerable X2s character in the 90° molecule, but the corresponding molecular orbital in the 180° molecule has pure $X2p_z$ orbital character. Hence, it shows a steep rise in energy as the bond angle increases. The $1b_1$ orbital is a nonbonding X2p orbital perpendicular to the molecular plane in the 90° molecule and remains nonbonding in the linear molecule. Hence, its energy barely changes with angle.

Each of the curves plotted on Fig. 2.35 will have a maximum or minimum on the 180° axis. The two lowest energy curves have a minimum on the 180° line. Therefore, we would expect XH_2 molecules having four valence electrons to be linear. XH_2 molecules with more than four valence electrons are expected to be angular because at least one electron is in the nonbonding $2a_1$ orbital. If the molecule is close to linear then the order of filling the molecular orbitals will be in the order of their increasing energy, which is $1a_1 < 2a_1 < 1b_2 < 1b_1$. The simplest XH_2 molecule in Period 2 is the transient gas-phase BeH_2 molecule (BeH_2 normally exists as a polymeric solid, with four-coordinate Be atoms). There are four valence electrons in a BeH_2 molecule, which occupy the lowest two molecular orbitals. If the lowest energy is achieved with the molecule angular, then that will be its shape. We can decide whether or not the molecule is likely to be angular by accommodating the electrons in the two lowest-energy orbitals corresponding to an arbitrary bond angle in Fig. 2.35. We then note that the HOMO decreases in energy on going to the right of the diagram (in the direction of increasing bond angle) and that the lowest total energy is obtained when the molecule is linear. Hence, BeH_2 is predicted to be linear and to have the configuration $1\sigma^2 2\sigma^2$. In CH_2, which has two more electrons than BeH_2, three of the molecular orbitals must be occupied. In this case, the lowest energy is achieved if the molecule is angular and has configuration $1a_1^2 2a_1^1 1b_2^2$.

In general, any XH_2 molecule with from five to eight valence electrons is predicted to be angular. The observed bond angles are

BeH_2	BH_2	CH_2	NH_2	OH_2
180°	131°	136°	103°	105°

These experimental observations are qualitatively in line with Walsh's approach, but for quantitative predictions we have to turn to detailed molecular orbital calculations.

---

[3] We continue to use the letters a and b to label nondegenerate orbitals, and will explain their full significance in Chapter 6.

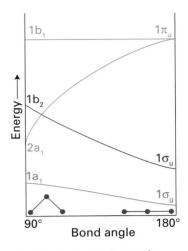

Fig. 2.35 The Walsh diagram for XH_2 molecules. Only the bonding and nonbonding orbitals are shown.

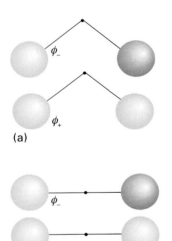

Fig. 2.36 The combination of H1s orbitals that are used to construct molecular orbitals in (a) angular and (b) linear XH_2.

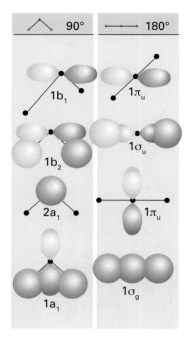

90° **180°**

$1b_1$

$1\pi_u$

$1b_2$

$1\sigma_u$

$2a_1$

$1\pi_u$

$1a_1$

$1\sigma_g$

Fig. 2.37 The composition of the molecular orbitals of an XH_2 molecule at the two extremes of the correlation diagram shown in Fig. 2.35.

EXAMPLE 2.8 Using a Walsh diagram to predict a shape

Predict the shape of an H_2O molecule on the basis of a Walsh diagram for an XH_2 molecule.

Answer We need to choose an angle between 90 and 180°, refer to the appropriate Walsh diagram, fill the molecular orbitals with the valence electrons, and assess whether the resulting configuration implies a linear or an angular shape. In this case we choose an intermediate bond angle along the horizontal axis of the XH_2 diagram in Fig. 2.37 and accommodate eight electrons. The resulting configuration is $1a_1^2 2a_1^2 1b_2^2 1b_1^2$. The $2a_1$ orbital is occupied, so we expect the nonlinear molecule to have a lower energy than the linear molecule.

Self-test 2.8 Is any XH_2 molecule, in which X denotes an atom of a Period 3 element, expected to be linear? If so, which?

Walsh applied his approach to molecules other than compounds of hydrogen, but the correlation diagrams soon become very complicated. His approach represents a valuable model because it traces the influences on molecular shapes of the occupation of orbitals spreading over the entire molecule. Correlation diagrams like those introduced by Walsh are frequently encountered in contemporary discussions of the shapes of complex molecules, and we shall see a number of examples in later chapters. They illustrate how inorganic chemists can sometimes identify and weigh competing influences by considering two extreme cases (such as linear and 90° XH_2 molecules), and then rationalize the fact that the state of a molecule is a compromise intermediate between the two extremes.

Structure and bond properties

Certain properties of bonds are approximately the same in different compounds of the elements. Thus, if we know the strength of an O−H bond in H_2O, then with some confidence we can use the same value for the O−H bond in CH_3OH. At this stage we confine our attention to two of the most important characteristics of a bond: its length and its strength. We also extend our understanding of bonds to predict the shapes of simple inorganic molecules.

2.13 Bond length

Key points: The equilibrium bond length in a molecule is the separation of the centres of the two bonded atoms; covalent radii vary through the periodic table in much the same way as metallic and ionic radii.

The **equilibrium bond length** in a molecule is the distance between the centres of the two bonded atoms. A wealth of useful and accurate information about bond lengths is available in the literature, most of it obtained by X-ray diffraction on solids (Section 8.1). Equilibrium bond lengths of molecules in the gas phase are usually determined by infrared or microwave spectroscopy, or more directly by electron diffraction. Some typical values are given in Table 2.6.

To a reasonable first approximation, equilibrium bond lengths can be partitioned into contributions from each atom of the bonded pair. The contribution of an atom to a covalent bond is called the **covalent radius** of the element (**18**). We can use the covalent radii in Table 2.7 to predict, for example, that the length of a P−N bond is 110 pm + 74 pm = 184 pm; experimentally, this bond length is close to 180 pm in a number of compounds. Experimental bond lengths should be used whenever possible, but covalent radii are useful for making cautious estimates when experimental data are not available.

Covalent radii vary through the periodic table in much the same way as metallic and ionic radii (Section 1.9a), for the same reasons, and are smallest close to F. Covalent radii are approximately equal to the separation of nuclei when the cores of the two atoms are in contact: the valence electrons draw the two atoms together until the repulsion between the cores starts to dominate. A covalent radius expresses the closeness of approach of *bonded* atoms; the closeness of approach of *nonbonded* atoms in neighbouring molecules that are in contact is expressed in terms of the **van der Waals radius** of the element, which is the internuclear separation when the *valence* shells of the two atoms are in nonbonding contact (**19**). van der Waals radii are of paramount importance for understanding the packing of molecular compounds in crystals, the conformations adopted by small but flexible molecules, and the shapes of biological macromolecules (Chapter 27).

Table 2.6 Equilibrium bond lengths, R_e/pm

H_2^+	106
H_2	74
HF	92
HCl	127
HBr	141
HI	160
N_2	109
O_2	121
F_2	144
Cl_2	199
I_2	267

2.14 Bond strength

Key points: The strength of a bond is measured by its dissociation enthalpy; mean bond enthalpies are used to make estimates of reaction enthalpies.

A convenient thermodynamic measure of the strength of an AB bond is the **bond dissociation enthalpy**, $\Delta H^{\ominus}(A-B)$, the standard reaction enthalpy for the process

$$AB(g) \rightarrow A(g) + B(g)$$

The **mean bond enthalpy**, B, is the average bond dissociation enthalpy taken over a series of A−B bonds in different molecules (Table 2.8).

Mean bond enthalpies can be used to estimate reaction enthalpies. However, thermodynamic data on actual species should be used whenever possible in preference to mean values because the latter can be misleading. For instance, the Si−Si bond enthalpy ranges from 226 kJ mol^{-1} in Si_2H_6 to 322 kJ mol^{-1} in $Si_2(CH_3)_6$. The values in Table 2.8 are best considered as data of last resort: they may be used to make rough estimates of reaction enthalpies when enthalpies of formation or actual bond enthalpies are unavailable.

EXAMPLE 2.9 Making estimates using mean bond enthalpies

Estimate the reaction enthalpy for the production of $SF_6(g)$ from $SF_4(g)$ given that the mean bond enthalpies of F_2, SF_4, and SF_6 are 158, 343, and 327 kJ mol^{-1}, respectively, at 25°C.

Answer We make use of the fact that the enthalpy of a reaction is equal to the difference between the sum of the bond enthalpies for broken bonds and the sum of the enthalpies of the bonds that are formed. The reaction is

$$SF_4(g) + F_2(g) \rightarrow SF_6(g)$$

In this reaction, 1 mol F−F bonds and 4 mol S−F bonds (in SF_4) must be broken, corresponding to an enthalpy change of 158 kJ + (4 × 343 kJ)= +1530 kJ. This enthalpy change is positive because energy will be used in breaking bonds. Then 6 mol S−F bonds (in SF_6) must be formed, corresponding to an enthalpy change of 6 × (−327 kJ) = −1962 kJ. This enthalpy change is negative because energy is released when the bonds are formed. The net enthalpy change is therefore

$$\Delta H^{\ominus} = +1530 \text{ kJ} - 1962 \text{ kJ} = -432 \text{ kJ}$$

Hence, the reaction is strongly exothermic. The experimental value for the reaction is −434 kJ, which is in excellent agreement with the estimated value.

Self-test 2.9 Estimate the enthalpy of formation of H_2S from S_8 (a cyclic molecule) and H_2.

2.15 Electronegativity and bond enthalpy

Key points: The Pauling scale of electronegativity is useful for estimating bond enthalpies and for assessing the polarities of bonds.

The concept of electronegativity was introduced in Section 1.9d, where it was defined as the power of an atom of the element to attract electrons to itself when it is part of a compound. The greater the difference in electronegativity between two elements A and B, the greater the ionic character of the A−B bond.

Linus Pauling's original formulation of electronegativity drew on concepts relating to the energetics of bond formation. For example, in the formation of AB from the diatomic A_2 and B_2 molecules,

$$A_2(g) + B_2(g) \rightarrow 2\,AB(g)$$

He argued that the excess energy, ΔE, of the A−B bond over the average energy of A−A and B−B bonds can be attributed to the presence of ionic contributions to the covalent bonding. He defined the difference in electronegativity as

$$|\chi_P(A) - \chi_P(B)| = 0.102(\Delta E/\text{kJ mol}^{-1})^{1/2} \tag{2.13a}$$

where

$$\Delta E = B(A-B) - \tfrac{1}{2}\{B(A-A) + B(B-B)\} \tag{2.13b}$$

Table 2.7 Covalent radii, r_{cov}/pm*

H			
37			
C	**N**	**O**	**F**
77 (1)	74 (1)	66 (1)	64
67 (2)	65 (2)	57 (2)	
60 (3)	54 (3)		
70 (a)			
Si	**P**	**S**	**Cl**
118	110	104 (1)	99
		95 (2)	
Ge	**As**	**Se**	**Br**
122	121	117	114
	Sb	**Te**	**I**
	141	137	133

* Values are for single bonds except where otherwise stated (in parentheses); (a) denotes aromatic.

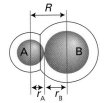

18 Covalent radius

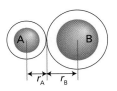

19 van der Waals radius

Table 2.8 Mean bond enthalpies, $B/(\text{kJ mol}^{-1})$*

	H	C	N	O	F	Cl	Br	I	S	P	Si
H	436										
C	412	348 (1)									
		612 (2)									
		837 (3)									
		518 (a)									
N	388	305 (1)	163 (1)								
		613 (2)	409 (2)								
		890 (3)	946 (3)								
O	463	360 (1)	157	146 (1)							
		743 (2)		497 (2)							
F	565	484	270	185	155						
Cl	431	338	200	203	254	242					
Br	366	276				219	193				
I	299	238				210	178	151			
S	338	259	464	523	343	250	212		264		
P	322 (1)									201	
										480 (3)	
Si	318			466							226

* Values are for single bonds except where otherwise stated (in parentheses); (a) denotes aromatic.

with $B(\text{A}-\text{B})$ the mean A−B bond enthalpy. Thus, if the A−B bond enthalpy is significantly greater than the average of the nonpolar A−A and B−B bonds, then it is presumed that there is a substantial ionic contribution to the wavefunction and hence a large difference in electronegativity between the two atoms. Pauling electronegativities increase with increasing oxidation number of the element and the values in Table 1.7 are for the most common oxidation state.

Pauling electronegativities are useful for estimating the enthalpies of bonds between elements of different electronegativity and to make qualitative assessments of the polarities of bonds. Binary compounds in which the difference in electronegativity between the two elements is greater than about 1.7 can generally be regarded as being predominantly ionic. However, this crude distinction was refined by Anton van Arkel and Jan Ketelaar in the 1940s, when they drew a triangle with vertices representing ionic, covalent, and metallic bonding. The **Ketelaar triangle** (more appropriately, the *van Arkel−Ketelaar triangle*) has been elaborated by Gordon Sproul, who constructed a triangle based on the difference in electronegativities ($\Delta\chi$) of the elements in a binary compound and their average electronegativity (χ_{mean}) (Fig. 2.38). The Ketelaar triangle is used extensively in Chapter 3, where we shall see how this basic concept can be used to classify a wide range of compounds of different kinds.

Ionic bonding is characterized by a large difference in electronegativity. Because a large difference indicates that the electronegativity of one element is high and that of the other is low, the average electronegativity must be intermediate in value. The compound CsF, for instance, with $\Delta\chi = 3.19$ and $\chi_{\text{mean}} = 2.38$, lies at the 'ionic' apex of the triangle. Covalent bonding is characterized by a small difference in electronegativities. Such compounds lie at the base of the triangle. Binary compounds that are predominantly covalently bonded are typically formed between nonmetals, which commonly have high electronegativities. It follows that the covalent region of the triangle is the lower, right-hand corner. This corner of the triangle is occupied by F_2, which has $\Delta\chi = 0$ and $\chi_{\text{mean}} = 3.98$ (the maximum value of any Pauling electronegativity). Metallic bonding is also characterized by a small electronegativity difference, and also lies towards the base of the triangle. In metallic bonding,

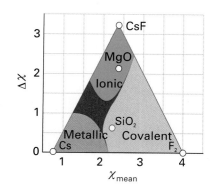

Fig. 2.38 A Ketelaar triangle, showing how a plot of average electronegativity against electronegativity difference can be used to classify the bond type for binary compounds.

however, electronegativities are low, the average values are therefore also low, and consequently metallic bonding occupies the lower, left-hand corner of the triangle. The outer corner is occupied by Cs, which has $\Delta\chi = 0$ and $\chi_{mean} = 0.79$ (the lowest value of Pauling electronegativity). The advantage of using a Ketelaar triangle over simple electronegativity difference is that it allows us to distinguish between covalent and metallic bonding, which are both indicated by a small electronegativity difference.

■ **A brief illustration.** For MgO, $\Delta\chi = 3.44 - 1.31 = 2.13$ and $\chi_{mean} = 2.38$. These values place MgO in the ionic region of the triangle. By contrast, for SiO_2, $\Delta\chi = 2.58 - 1.90 = 0.68$ and $\chi_{mean} = 2.24$. These values place SiO_2 lower on the triangle compared to MgO and in the covalent bonding region. ■

2.16 Oxidation states

Key point: Oxidation numbers are assigned by applying the rules in Table 2.9.

The **oxidation number**, N_{ox},[4] is a parameter obtained by exaggerating the *ionic* character of a bond. It can be regarded as the charge that an atom would have if the more electronegative atom in a bond acquired the two electrons of the bond completely. The **oxidation state** is the physical state of the element corresponding to its oxidation number. Thus, an atom may *be assigned* an oxidation number and be *in* the corresponding oxidation state.[5] The alkali metals are the most electropositive elements in the periodic table, so we can assume they will always be present as M^+ and are assigned an oxidation number of $+1$. Because oxygen's electronegativity is exceeded only by that of F, we can regard it as O^{2-} in combination with any element other than F, and hence it is ascribed an oxidation number of -2. Likewise, the exaggerated ionic structure of NO_3^- is $N^{5+}(O^{2-})_3$, so the oxidation number of nitrogen in this compound is $+5$, which is denoted either N(V) or N(+5). These conventions may be used even if the oxidation number is negative, so oxygen has oxidation number -2, denoted O(-2) or more rarely O($-$II), in most of its compounds.

Table 2.9 The determination of oxidation number*

	Oxidation number
1. The sum of the oxidation numbers of all the atoms in the species is equal to its total charge	
2. For atoms in their elemental form	0
3. For atoms of Group 1	$+1$
For atoms of Group 2	$+2$
For atoms of Group 13 (except B)	$+3(EX_3)$, $+1(EX)$
For atoms of Group 14 (except C, Si)	$+4(EX_4)$, $+2(EX_2)$
4. For hydrogen	$+1$ in combination with nonmetals
	-1 in combination with metals
5. For fluorine	-1 in all its compounds
6. For oxygen	-2 unless combined with F
	-1 in peroxides (O_2^{2-})
	$-\frac{1}{2}$ in superoxides (O_2^-)
	$-\frac{1}{3}$ in ozonides (O_3^-)
7. Halogens	-1 in most compounds, unless the other elements include oxygen or more electronegative halogens

* To determine an oxidation number, work through the rules in the order given. Stop as soon as the oxidation number has been assigned. These rules are not exhaustive, but they are applicable to a wide range of common compounds.

[4] There is no formally agreed symbol for oxidation number.

[5] In practice, inorganic chemists use the terms 'oxidation number' and 'oxidation state' interchangeably, but in this text we shall preserve the distinction.

In practice, oxidation numbers are assigned by applying a set of simple rules (Table 2.9). These rules reflect the consequences of electronegativity for the 'exaggerated ionic' structures of compounds and match the increase in the degree of oxidation that we would expect as the number of oxygen atoms in a compound increases (as in going from NO to NO_3^-). This aspect of oxidation number is taken further in Chapter 5. Many elements, for example nitrogen, the halogens, and the d-block elements, can exist in a variety of oxidation states (Table 2.9).

EXAMPLE 2.10 Assigning an oxidation number to an element

What is the oxidation number of (a) S in hydrogen sulfide, H_2S, (b) Mn in the permanganate ion, MnO_4^-?

Answer We need to work through the steps set out in Table 2.9 in the order given. (a) The overall charge of the species is 0; so $2N_{ox}(H) + N_{ox}(S) = 0$. Because $N_{ox}(H) = +1$ in combination with a nonmetal, it follows that $N_{ox}(S) = -2$. (b) The sum of the oxidation numbers of all the atoms is -1, so $N_{ox}(Mn) + 4N_{ox}(O) = -1$. Because $N_{ox}(O) = -2$, it follows that $N_{ox}(Mn) = -1 -4(-2) = +7$. That is, MnO_4^- is a compound of Mn(VII). Its formal name is tetraoxomanganese(VII) ion.

Self-test 2.10 What is the oxidation number of (a) O in O_2^+, (b) P in PO_4^{3-}?

FURTHER READING

R.J. Gillespie and I. Hargittai, *The VSEPR model of molecular geometry*. Prentice Hall (1992). An excellent introduction to modern attitudes to VSEPR theory.

R.J. Gillespie and P.L.A. Popelier, *Chemical bonding and molecular geometry: from Lewis to electron densities*. Oxford University Press (2001). A comprehensive survey of modern theories of chemical bonding and geometry.

M.J. Winter, *Chemical bonding*. Oxford University Press (1994). This short text introduces some concepts of chemical bonding in a descriptive and non-mathematical way.

T. Albright, *Orbital interactions in chemistry*. Wiley, New York (2005). This text covers the application of molecular orbital theory to organic, organometallic, inorganic, and solid-state chemistry.

D.M.P. Mingos, *Essential trends in inorganic chemistry*. Oxford University Press (1998). An overview of inorganic chemistry from the perspective of structure and bonding.

I.D. Brown, *The chemical bond in inorganic chemistry*. Oxford University Press (2006).

K. Bansal, *Molecular structure and orbital theory*. Campus Books International (2000).

J.N. Murrell, S.F.A. Kettle, and J.M. Tedder, *The chemical bond*. Wiley, New York (1985).

T. Albright and J.K. Burdett, *Problems in molecular orbital theory*. Oxford University Press (1993).

EXERCISES

2.1 What shapes would you expect for the species (a) H_2S, (b) BF_4^-, (c) NH_4^+?

2.2 What shapes would you expect for the species (a) SO_3, (b) SO_3^{2-}, (c) IF_5?

2.3 What shapes would you expect for the species (a) ClF_3, (b) ICl_4^-, (c) I_3^-?

2.4 In which of the species ICl_6^- and SF_4 is the bond angle closest to that predicted by the VSEPR model?

2.5 Solid phosphorus pentachoride is an ionic solid composed of PCl_4^- cations and PCl_6^- anions, but the vapour is molecular. What are the shapes of the ions in the solid?

2.6 Use the covalent radii in Table 2.7 to calculate the bond lengths in (a) CCl_4 (177 pm), (b) $SiCl_4$ (201 pm), (c) $GeCl_4$ (210 pm). (The values in parentheses are experimental bond lengths and are included for comparison.)

2.7 Given that $B(Si=O) = 640$ kJ mol^{-1}, show that bond enthalpy considerations predict that silicon–oxygen compounds are likely to contain networks of tetrahedra with Si–O single bonds and not discrete molecules with Si=O double bonds.

2.8 The common forms of nitrogen and phosphorus are $N_2(g)$ and $P_4(s)$, respectively. Account for the difference in terms of the single and multiple bond enthalpies.

2.9 Use the data in Table 2.8 to calculate the standard enthalpy of the reaction $2H_2(g) + O_2(g) \rightarrow 2H_2O(g)$. The experimental value is -484 kJ mol^{-1}. Account for the difference between the estimated and experimental values.

2.10 Predict the standard enthalpies of the reactions

(a) $S_2^{2-}(g) + \frac{1}{4}S_8(g) \rightarrow S_4^{2-}(g)$

(b) $O_2^{2-}(g) + O_2(g) \rightarrow O_4^{2-}(g)$

by using mean bond enthalpy data. Assume that the unknown species O_4^{2-} is a singly bonded chain analogue of S_4^{2-}.

2.11 Four elements arbitrarily labelled A, B, C, and D have electronegativities 3.8, 3.3, 2.8, and 1.3, respectively. Place the compounds AB, AD, BD, and AC in order of increasing covalent character.

2.12 Use the Ketelaar triangle in Fig. 2.38 and the electronegativity values in Table 1.7 to predict what type of bonding is likely to dominate in (a) BCl_3, (b) KCl, (c) BeO.

2.13 Predict the hybridization of orbitals required in (a) BCl_3, (b) NH_4^+, (c) SF_4, (d) XeF_4.

2.14 Use molecular orbital diagrams to determine the number of unpaired electrons in (a) O_2^-, (b) O_2^+, (c) BN, (d) NO_2.

2.15 Use Fig. 2.17 to write the electron configurations of (a) Be_2, (b) B_2, (c) C_2^-, (d) F_2^+ and sketch the form of the HOMO in each case.

2.16 When acetylene (ethyne) is passed through a solution of copper(I) chloride a red precipitate of copper acetylide, CuC_2, is formed. This is a common test for the presence of acetylene. Describe the bonding in the C_2^{2-} ion in terms of molecular orbital theory and compare the bond order to that of C_2.

2.17 Assume that the MO diagram of IBr is analogous to that of ICl (Fig. 2.24). (a) What basis set of atomic orbital would be used to generate the IBr molecular orbitals? (b) Calculate the bond order of IBr. (c) Comment on the relative stabilities and bond orders of IBr and IBr_2.

2.18 Determine the bond orders of (a) S_2, (b) Cl_2, and (c) NO_2 from their molecular orbital configurations and compare the values with the bond orders determined from Lewis structures. (NO has orbitals like those of O_2.)

2.19 What are the expected changes in bond order and bond distance that accompany the following ionization processes?

(a) $O_2 \rightarrow O_2^+ + e^-$ (b) $N_2 + e^- \rightarrow N_2^-$ (c) $NO \rightarrow NO^+ + e^-$

2.20 (a) How many independent linear combinations are possible for four 1s orbitals? (b) Draw pictures of the linear combinations of H1s orbitals for a hypothetical linear H_4 molecule. (c) From a consideration of the number of nonbonding and antibonding interactions, arrange these molecular orbitals in order of increasing energy.

2.21 (a) Construct the form of each molecular orbital in linear $[HHeH]^{2+}$ using 1s basis atomic orbitals on each atom and considering successive nodal surfaces. (b) Arrange the MOs in increasing energy. (c) Indicate the electron population of the MOs. (d) Should $[HHeH]^{2+}$ be stable in isolation or in solution? Explain your reasoning.

2.22 (a) Based on the MO discussion of NH_3 in the text, find the average NH bond order in NH_3 by calculating the net number of bonds and dividing by the number of NH groups.

2.23 From the relative atomic orbital and molecular orbital energies depicted in Fig. 2.31, describe the character as mainly F or mainly S for the frontier orbitals e (the HOMO) and 2t (the LUMO) in SF_6. Explain your reasoning.

2.24 Classify the hypothetical species (a) square H_4^{2+}, (b) angular O_3^{2-} as electron precise or electron deficient. Explain your answer and decide whether either of them is likely to exist.

PROBLEMS

2.1 Use the concepts from Chapter 1, particularly the effects of penetration and shielding on the radial wavefunction, to account for the variation of single bond covalent radii with position in the periodic table.

2.2 In valence bond theory, hypervalence is usually explained in terms of d-orbital participation in bonding. In the paper 'On the role of orbital hybridisation' (*J. Chem. Educ.* 2007, **84**, 783) the author argues that this is not the case. Give a concise summary of the method used and the author's reasoning.

2.3 Develop an argument based on bond enthalpies for the importance of SiO bonds in substances common in the Earth's crust in preference to SiSi or SiH bonds. How and why does the behaviour of silicon differ from that of carbon?

2.4 The van Arkel−Ketelaar triangle has been in use since the 1940s. A quantitative treatment of the triangle was carried out by Gordon Sproul in 1994 (*J. Phys. Chem.*, 1994, **98**, 6699). How many scales of electronegativity and how many compounds did Sproul investigate? What criteria were used to select compounds for the study? Which two electronegativity scales were found to give the best separation between areas of the triangle? What were the theoretical bases of these two scales?

2.5 When an He atom absorbs a photon to form the excited configuration $1s^1 2s^1$ (here called He*) a weak bond forms with another He atom to give the diatomic molecule HeHe*. Construct a molecular orbital description of the bonding in this species.

2.6 In their article 'Some observation on molecular orbital theory' (J.F. Harrison and D. Lawson, *J. Chem. Educ.*, 2005, **82**, 1205) the authors discuss several limitations of the theory. What are these

limitations? Sketch the MO diagram for Li_2 given in the paper. Why do you think this version does not appear in textbooks? Use the data given in the paper to construct MO diagrams for B_2 and C_2. Do these versions differ from that in Fig. 2.17 in this textbook? Discuss any variations.

2.7 Construct an approximate molecular orbital energy diagram for a hypothetical planar form of NH_3. You may refer to *Resource section* 4 to determine the form of the appropriate orbitals on the central N atom and on the triangle of H_3 atoms. From a consideration of the atomic energy levels, place the N and H_3 orbitals on either side of a molecular orbital energy-level diagram. Then use your judgement about the effect of bonding and antibonding interactions and energies of the parent orbitals to construct the molecular orbital energy levels in the centre of your diagram and draw lines indicating the contributions of the atomic orbitals to each molecular orbital. Ionization energies are $I(H1s) = 13.6$ eV, $I(N2s) = 26.0$ eV, and $I(N2p) = 13.4$ eV.

2.8 (a) Use a molecular orbital program or input and output from software supplied by your instructor to construct a molecular orbital energy level diagram to correlate the MO (from the output) and AO (from the input) energies and indicate the occupancy of the MOs (in the manner of Fig. 2.17) for one of the following molecules: HF (bond length 92 pm), HCl (127 pm), or CS (153 pm). (b) Use the output to sketch the form of the occupied orbitals, showing signs of the AO lobes by shading and their amplitudes by means of size of the orbital.

2.9 Use software to perform an MO calculation on H_3 by using the H energy given in Problem 2.7 and H−H distances from NH_3 (N−H length 102 pm, HNH bond angle 107°) and then carry out the same type of calculation for NH₃. Use energy data for N2s and N2p orbitals from Problem 2.7. From the output plot the molecular orbital energy levels with proper symmetry labels and correlate them with the N orbitals and H₃ orbitals of the appropriate symmetries. Compare the results of this calculation with the qualitative description in Problem 2.7.

2.10 Assign the lines in the UV photoelectron spectrum of CO shown in Fig. 2.39 and predict the appearance of the UV photoelectron spectrum of the SO molecule (see Section 8.3).

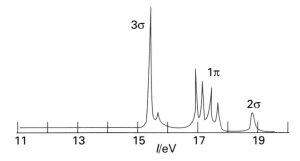

Fig. 2.39 The UV photoelectron spectrum of CO obtained using 21 eV radiation.

The structures of simple solids

3

An understanding of the chemistry of compounds in the solid state is central to the study of many important inorganic materials, such as alloys, simple metal salts, inorganic pigments, nanomaterials, zeolites, and high-temperature superconductors. This chapter surveys the structures adopted by atoms and ions in simple solids and explores why one arrangement may be preferred to another. We begin with the simplest model, in which atoms are represented by hard spheres and the structure of the solid is the outcome of stacking these spheres densely together. This 'close-packed' arrangement provides a good description of many metals and alloys and is a useful starting point for the discussion of numerous ionic solids. These simple solid structures can then be considered as building blocks for the construction of more complex inorganic materials. Introduction of partial covalent character into the bonding influences the choice of structure and thus trends in the adopted structural type correlate with the electronegativities of the constituent atoms. The chapter also describes some of the energy considerations that can be used to rationalize the trends in structure and reactivity. These arguments also systematize the discussion of the thermal stabilities and solubilities of ionic solids formed by the elements of Groups 1 and 2. Finally the electronic structures of materials are discussed in terms of an extension of molecular orbital theory to the almost infinite arrays of atoms found in solids. The classification of inorganic solids as conductors, semiconductors, and insulators is described in terms of this theory.

The majority of inorganic compounds exist as solids and comprise ordered arrays of atoms, ions, or molecules. Some of the simplest solids are the metals, the structures of which can be described in terms of regular, space-filling arrangements of the metal atoms. These metal centres interact through **metallic bonding,** a type of bonding that can be described in two ways. One view is that bonding occurs in metals when each atom loses one or more electrons to a common 'sea'. The strength of the bonding results from the combined attractions between all these freely moving electrons and the resulting cations. An alternative view is that metals are effectively enormous molecules with a multitude of atomic orbitals that overlap to produce molecular orbitals extending throughout the sample.

Metallic bonding is characteristic of elements with low ionization energies, such as those on the left of the periodic table, through the d block, and into part of the p block close to the d block. Most of the elements are metals, but metallic bonding also occurs in many other solids, especially compounds of the d-metals such as their oxides and sulfides. Compounds such as the lustrous-red rhenium oxide ReO_3 and 'fool's gold' (iron pyrites, FeS_2), illustrate the occurrence of metallic bonding in compounds.

The familiar properties of a metal stem from the characteristics of its bonding and in particular the delocalization of electrons throughout the solid. Thus, metals are malleable (easily deformed by the application of pressure) and ductile (able to be drawn into a wire) because the electrons can adjust rapidly to relocation of the metal atom nuclei and there is no directionality in the bonding. They are lustrous because the electrons can respond almost freely to an incident wave of electromagnetic radiation and reflect it.

In **ionic bonding** ions of different elements are held together in rigid, symmetrical arrays as a result of the attraction between their opposite charges. Ionic bonding also depends on electron loss and gain, so it is found typically in compounds of metals with electronegative elements. However, there are plenty of exceptions: not all compounds of metals are ionic and some compounds of nonmetals (such as ammonium nitrate) contain features of ionic bonding as well as covalent interactions. There are also materials that exhibit features of both ionic and metallic bonding.

Ionic and metallic bonding are nondirectional, so structures where these types of bonding occur are most easily understood in terms of space-filling models that maximize, for example, the number and strength of the electrostatic interactions between the ions. The regular arrays of atoms, ions, or molecules in solids that produce these structures are best represented in terms of the repeating units that are produced as a result of the efficient methods of filling space.

The description of the structures of solids

The arrangement of atoms or ions in simple solid structures can often be represented by different arrangements of hard spheres. The spheres used to describe metallic solids represent neutral atoms because each cation is still surrounded by its full complement of electrons. The spheres used to describe ionic solids represent the cations and anions because there has been a substantial transfer of electrons from one type of atom to the other.

3.1 Unit cells and the description of crystal structures

A crystal of an element or compound can be regarded as constructed from regularly repeating structural elements, which may be atoms, molecules, or ions. The 'crystal lattice' is the pattern formed by the points and used to represent the positions of these repeating structural elements.

(a) Lattices and unit cells

Key points: The lattice defines a network of identical points that has the translational symmetry of a structure. A unit cell is a subdivision of a crystal that, when stacked together without rotation or reflection, reproduces the crystal.

A **lattice** is a three-dimensional, infinite array of points, the **lattice points**, each of which is surrounded in an identical way by neighbouring points, and which defines the basic repeating structure of the crystal. In some cases the structural unit may be centred on the lattice point, but that is not necessary. The **crystal structure** itself is obtained by associating one or more identical structural units (such as molecules or ions) with each lattice point.

A **unit cell** of the crystal is an imaginary parallel-sided region (a 'parallelepiped') from which the entire crystal can be built up by purely translational displacements;[1] unit cells so generated fit perfectly together with no space excluded. Unit cells may be chosen in a variety of ways but it is generally preferable to choose the smallest cell that exhibits the greatest symmetry. Thus, in the two-dimensional pattern in Fig. 3.1, a variety of unit cells may be chosen, each of which repeats the contents of the box under translational displacements. Two possible choices of repeating unit are shown but (b) would be preferred to (a) because it is smaller. The relationship between the lattice parameters in three dimensions as a result of the symmetry of the structure gives rise to the seven **crystal systems** (Table 3.1 and Fig. 3.2). All ordered structures adopted by compounds belong to one of these crystal systems; most of those described in this chapter, which deals with simple compositions and stoichiometries, belong to the higher symmetry cubic and hexagonal systems. The angles (α, β, γ) and lengths (a, b, c) used to define the size and shape of a unit cell are the **unit cell parameters** (the 'lattice parameters'); the angle between a and b is denoted γ, that between b and c is α, and that between a and c is β; see the triclinic unit cell in Fig. 3.2.

A **primitive** unit cell (denoted by the symbol P) has just one lattice point in the unit cell (Fig. 3.3) and the translational symmetry present is just that on the repeating unit cell. More complex lattice types are **body-centred** (I, from the German word *innenzentriet*, referring to the lattice point at the unit cell centre) and **face-centred** (F) with two and four lattice points in each unit cell, respectively, and additional translational symmetry beyond that of

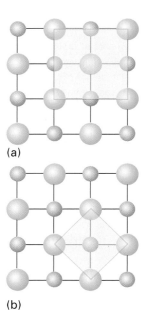

(a)

(b)

Fig. 3.1 A two-dimensional solid and two choices of a unit cell. The entire crystal is reproduced by translational displacements of either unit cell, but (b) is generally preferred to (a) because it is smaller.

---

[1] A translation exists where it is possible to move an original figure or motif in a defined direction by a certain distance to produce an exact image. In this case a unit cell reproduces itself exactly by translation parallel to a unit cell edge by a distance equal to the unit cell parameter.

Table 3.1 The seven crystal systems

System	Relationships between lattice parameters	Unit cell defined by	Essential symmetries
Triclinic	$a \neq b \neq c$ $\quad \alpha \neq \beta \neq \gamma \neq 90°$	$a\,b\,c\,\alpha\,\beta\,\gamma$	None
Monoclinic	$a \neq b \neq c$ $\quad \alpha \neq \gamma \neq 90°$ $\beta = 90°$	$a\,b\,c\,\alpha\,\gamma$	One twofold rotation axis and/or a mirror plane
Orthorhombic	$a \neq b \neq c$ $\quad \alpha = \beta = \gamma = 90°$	$a\,b\,c$	Three perpendicular twofold axes and/or mirror planes
Rhombohedral	$a = b = c$ $\quad \alpha = \beta = \gamma \neq 90°$	α	One threefold rotation axis
Tetragonal	$a = b \neq c$ $\quad \alpha = \beta = \gamma = 90°$	$a\,c$	One fourfold rotation axis
Hexagonal	$a = b \neq c$ $\quad \alpha = \beta = 90°$ $\gamma = 120°$	$a\,c$	One sixfold rotation axis
Cubic	$a = b = c$ $\quad \alpha = \beta = \gamma = 90°$	a	Four threefold rotation axes tetrahedrally arranged

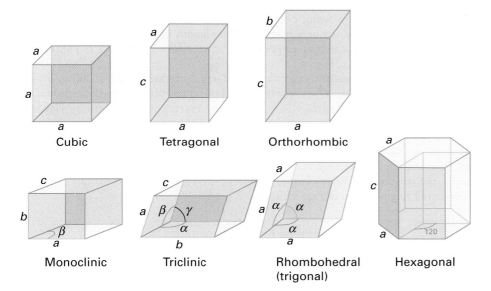

Fig. 3.2 The seven crystal systems.

Fig. 3.3 Lattice points describing the translational symmetry of a primitive cubic unit cell.

the unit cell (Figs 3.4 and 3.5). The additional translational symmetry in the **body-centred cubic** (bcc) lattice, equivalent to the displacement $(+\frac{1}{2}, +\frac{1}{2}, +\frac{1}{2})$ from the unit cell origin at $(0,0,0)$, produces a lattice point at the unit cell centre; note that the surroundings of each lattice point are identical, consisting of eight other lattice points at the corners of a cube. Centred lattices are sometimes preferred to primitive (although it is always possible to use a primitive lattice for any structure) for with them the essential structural symmetry of the cell is more apparent.

We use the following rules to work out the number of lattice points in a three-dimensional unit cell. The same process can be used to count the number of atoms, ions, or molecules that the unit cell contains (Section 3.9).

1. A lattice point in the body of, that is fully inside, a cell belongs entirely to that cell and counts as 1.
2. A lattice point on a face is shared by two cells and contributes $\frac{1}{2}$ to the cell.
3. A lattice point on an edge is shared by four cells and hence contributes $\frac{1}{4}$.
4. A lattice point at a corner is shared by eight cells that share the corner, and so contributes $\frac{1}{8}$.

Thus, for the face-centred cubic lattice depicted in Fig. 3.5 the total number of lattice points in the unit cell is $(8 \times \frac{1}{8}) + (6 \times \frac{1}{2}) = 4$. For the body-centred cubic lattice depicted in Fig. 3.4, the number of lattice points is $(1 \times 1) + (8 \times \frac{1}{8}) = 2$.

Fig. 3.4 Lattice points describing the translational symmetry of a body-centred cubic unit cell.

Fig. 3.5 Lattice points describing the translational symmetry of a face-centred cubic unit cell.

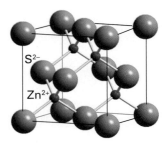

Fig 3.6 The cubic ZnS structure.

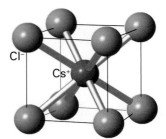

Fig 3.7 The cubic CsCl structure.

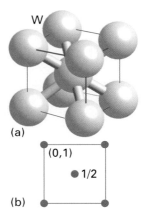

Fig. 3.8 (a) The structure of metallic tungsten and (b) its projection representation.

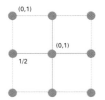

Fig. 3.9 The projection representation of an fcc unit cell.

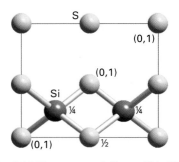

Fig. 3.10 The structure of silicon sulfide (SiS_2).

EXAMPLE 3.1 **Identifying lattice types**

Determine the translational symmetry present in the structure of cubic ZnS (Fig. 3.6) and identify the lattice type to which this structure belongs.

Answer We need to identify the displacements that, when applied to the entire cell, results in every atom arriving at an equivalent location (same atom type with the same coordination environment). In this case, the displacements $(0, +\frac{1}{2}, +\frac{1}{2})$, $(+\frac{1}{2}, +\frac{1}{2}, 0)$, and $(+\frac{1}{2}, 0, -\frac{1}{2})$ have this effect. For example starting at the Zn^{2+} ion on the bottom left-hand corner of the unit cell, which is surrounded by four S^{2-} ions at the corners of a tetrahedron, and applying the translation $(+\frac{1}{2}, 0, -\frac{1}{2})$ we arrive at the Zn^{2+} ion at the lower right-hand corner, which has an identical coordination to sulfur. These translations correspond to those of the face-centred lattice, so the lattice type is F.

Self-test 3.1 Determine the lattice type of CsCl (Fig. 3.7).

(b) Fractional atomic coordinates and projections

Key point: Structures may be drawn in projection, with atom positions denoted by fractional coordinates.

The position of an atom in a unit cell is normally described in terms of **fractional coordinates**, coordinates expressed as a fraction of the length of a side of the unit cell. Thus, the position of an atom located at xa parallel to a, yb parallel to b, and zc parallel to c is denoted (x,y,z), with $0 \leq x, y, z \geq 1$.

Three-dimensional representations of complex structures are often difficult to draw and to interpret in two dimensions.[2] A clearer method of representing three-dimensional structures on a two-dimensional surface is to draw the structure in projection by viewing the unit cell down one direction, typically one of the axes of the unit cell. The positions of the atoms relative to the projection plane are denoted by the fractional coordinate above the base plane and written next to the symbol defining the atom in the projection. If two atoms lie above each other, then both fractional coordinates are noted in parentheses. For example, the structure of body-centred tungsten, shown in three dimensions in Fig. 3.8a, is represented in projection in Fig. 3.8b.

EXAMPLE 3.2 **Drawing a three-dimensional representation in projection**

Convert the face-centred cubic lattice shown in Fig. 3.5 into a projection diagram.

Answer We need to identify the locations of the lattice points by viewing the cell from a position perpendicular to one of its faces. The faces of the cubic unit cell are square, so the projection diagram viewed from directly above the unit cell is a square. There is a lattice point at each corner of the unit cell, so the points at the corners of the square projection are labelled (0,1). There is a lattice point on each vertical face, which projects to points at fractional coordinate $\frac{1}{2}$ on each edge of the projection square. There is a lattice point on the lower and on the upper horizontal face of the unit cell, which projects to two points at the centre of the square at 0 and 1, respectively, so we place a final point in the centre of a square and label it (0,1). The resulting projection is shown in Fig. 3.9.

Self-test 3.2 Convert the projection diagram of the unit cell of the SiS_2 structure shown in Fig. 3.10 into a three-dimensional representation.

3.2 The close packing of spheres

Key points: The close packing of identical spheres can result in a variety of polytypes, of which hexagonal and cubic close-packed structures are the most common.

Many metallic and ionic solids can be regarded as constructed from entities, such as atoms and ions, represented as hard spheres. If there is no directional covalent bonding, these spheres are free to pack together as closely as geometry allows and hence adopt a **close-packed structure**, a structure in which there is least unfilled space. The **coordination number** (CN) of a sphere

[2] Nearly all the structures in this text are available as rotatable, three-dimensional versions in the *Online Resource Centre* for this text.

in a close-packed arrangement (the 'number of nearest neighbours') is 12, the greatest number that geometry allows.[3] When directional bonding is important, the resulting structures are no longer close-packed and the coordination number is less than 12.

Consider first a single layer of identical spheres (Fig. 3.11). The greatest number of immediate neighbours is 6 and there is only one way of constructing this close-packed layer.[4] A second close-packed layer of spheres is formed by placing spheres in the dips between the spheres of the first layer. (Note that only half the dips in the original layer are occupied, as there is insufficient space to place spheres into all the dips.) The third close-packed layer can be laid in either of two ways and hence can give rise to either of two **polytypes**, or structures that are the same in two dimensions (in this case, in the planes) but different in the third. Later we shall see that many different polytypes can be formed, but those described here are two very important special cases.

In one polytype, the spheres of the third layer lie directly above the spheres of the first. This ABAB... pattern of layers, where A denotes layers that have spheres directly above each other and likewise for B, gives a structure with a hexagonal unit cell and hence is said to be **hexagonally close-packed** (hcp, Figs 3.12a and 3.13). In the second polytype, the spheres of the third layer are placed above the gaps in the first layer. The second layer covers half the holes in the first layer and the third layer lies above the remaining holes. This arrangement results in an ABCABC... pattern, where C denotes a layer that has spheres not directly above spheres of the A or the B layer positions (but they will be directly above another C type layer). This pattern corresponds to a structure with a cubic unit cell and hence it is termed **cubic close-packed** (ccp, Figs 3.12b and 3.14). Because each ccp unit cell has a sphere at one corner and one at the centre of each face, a ccp unit cell is sometimes referred to as **face-centred cubic** (fcc).

A note on good practice The descriptions ccp and fcc are often used interchangeably, although strictly ccp refers only to a close-packed arrangement whereas fcc refers to the lattice type of the common representation of ccp. Throughout this text the term ccp will be used to describe this close-packing arrangement. It will be drawn as the cubic unit cell, with the fcc lattice type, as this representation is easiest to visualize.

The unoccupied space in a close-packed structure amounts to 26 per cent of the total volume (see Example 3.3). However, this unoccupied space is not empty in a real solid because electron density of an atom does not end as abruptly as the hard-sphere model suggests. The type and distribution of holes are important because many structures, including those of some alloys and many ionic compounds, can be regarded as formed from an expanded close-packed arrangement in which additional atoms or ions occupy all or some of the holes.

EXAMPLE 3.3 Calculating the unoccupied space in a close-packed array

Calculate the percentage of unoccupied space in a close-packed arrangement of identical spheres.

Answer Because the space occupied by hard spheres is the same in the ccp and hcp arrays, we can choose the geometrically simpler structure, ccp, for the calculation. Consider Fig. 3.15. The spheres of radius r are in contact across the face of the cube and so the length of this diagonal is $r + 2r + r = 4r$. The side of such a cell is $8^{1/2}r$ from Pythagoras' theorem (the square of the length of the diagonal $(4r)^2$ equals the sum of the squares of the two sides of length a, so $2 \times a^2 = (4r)^2$ giving $a = 8^{1/2}r$), so the cell volume is $(8^{1/2}r)^3 = 8^{3/2}r^3$. The unit cell contains $\frac{1}{8}$ of a sphere at each corner (for $8 \times \frac{1}{8} = 1$ in all) and half a sphere on each face (for $6 \times \frac{1}{2} = 3$ in all), for a total of 4. Because the volume of each sphere is $\frac{4}{3}\pi r^3$, the total volume occupied by the spheres themselves is $4 \times \frac{4}{3}\pi r^3 = \frac{16}{3}\pi r^3$. The occupied fraction is therefore $(\frac{16}{3}\pi r^3)/(8^{3/2}r^3) = \frac{16}{3}\pi/8^{3/2}$, which evaluates to 0.740. The unoccupied fraction is therefore 0.260, corresponding to 26.0 per cent.

Self-test 3.3 Calculate the fraction of space occupied by identical spheres in a primitive cubic unit cell.

---

[3] That this arrangement, where each sphere has 12 nearest-neighbours, is the highest possible density of packing spheres was conjectured by Johannes Kepler in 1611; the proof was found only in 1998.

[4] A good way of showing this yourself is to get a number of identical coins and push them together on a flat surface; the most efficient arrangement for covering the area is with six coins around each coin. This simple modelling approach can be extended to three dimensions by using any collection of identical spherical objects such as balls, oranges, or marbles.

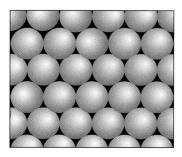

Fig. 3.11 A close-packed layer of hard spheres.

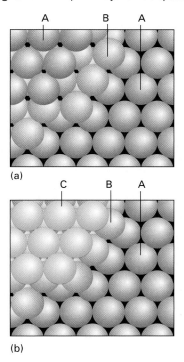

(a)

(b)

Fig. 3.12 The formation of two close-packed polytypes. (a) The third layer reproduces the first to give an ABA structure. (b) The third layer lies above the gaps in the first layer, giving an ABC structure. The different colours identify the different layers of identical spheres.

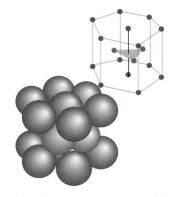

Fig. 3.13 The hexagonal close-packed (hcp) unit cell of the ABAB... polytype. The colours of the spheres correspond to the layers in Fig. 3.12a.

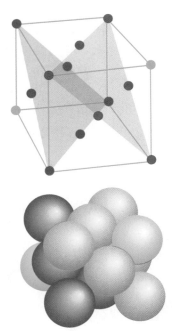

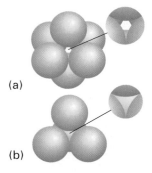

Fig. 3.14 The cubic close-packed (fcc) unit cell of the ABC... polytype. The colours of the spheres correspond to the layers in Fig 3.12b.

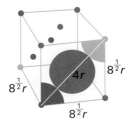

Fig. 3.15 The dimensions involved in the calculation of the packing fraction in a close-packed arrangement of identical spheres of radius r.

(a)

(b)

Fig. 3.17 (a) An octahedral hole and (b) a tetrahedral hole formed in an arrangement of close-packed spheres.

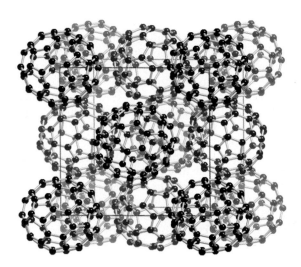

Fig. 3.16 The structure of solid C_{60} showing the packing of C_{60} polyhedra on an fcc unit cell.

The ccp and hcp arrangements are the most efficient simple ways of filling space with identical spheres. They differ only in the stacking sequence of the close-packed layers and other, more complex, close-packed layer sequences may be formed by locating successive planes in different positions relative to their neighbours (Section 3.4). Any collection of identical atoms, such as those in the simple picture of an elemental metal, or of approximately spherical molecules, is likely to adopt one of these close-packed structures unless there are additional energetic reasons—specifically covalent interactions—for adopting an alternative arrangement. Indeed, many metals adopt such close-packed structures (Section 3.4), as do the solid forms of the noble gases (which are ccp). Almost spherical molecules, such as C_{60}, in the solid state also adopt the ccp arrangement (Fig. 3.16), and so do many small molecules that rotate around their centres and thus appear spherical, such as H_2, F_2, and one form of solid oxygen, O_2.

3.3 Holes in close-packed structures

Key points: The structures of many solids can be discussed in terms of close-packed arrangements of one atom type in which the tetrahedral or octahedral holes are occupied by other atoms or ions. The ratio of spheres to octahedral holes to tetrahedral holes in a close-packed structure is 1:1:2.

The feature of a close-packed structure that enables us to extend the concept to describe structures more complicated than elemental metals is the existence of two types of **hole**, or unoccupied space between the spheres. An **octahedral hole** lies between two triangles of spheres on adjoining layers (Fig. 3.17). For a crystal consisting of N spheres in a close-packed structure, there are N octahedral holes. The distribution of these holes in an hcp unit cell is shown in Fig. 3.18a and those in a ccp unit cell Fig. 3.18b. This illustration also shows that the hole has local octahedral symmetry in the sense that it is surrounded by six nearest-neighbour spheres with their centres at the corners of an octahedron. If each hard sphere has radius r, and if the close-packed spheres are to remain in contact, then each octahedral hole can accommodate a hard sphere representing another type of atom with a radius no larger than $0.414r$.

EXAMPLE 3.4 Calculating the size of an octahedral hole

Calculate the maximum radius of a sphere that may be accommodated in an octahedral hole in a close-packed solid composed of spheres of radius r.

Answer The structure of a hole, with the top spheres removed, is shown in Fig. 3.19a. If the radius of a sphere is r and that of the hole is r_h, it follows from Pythagoras' theorem that $(r + r_h)^2 + (r + r_h)^2 = (2r)^2$ and therefore that $(r + r_h)^2 = 2r^2$, which implies that $r + r_h = 2^{1/2}r$. That is, $r_h = (2^{1/2} - 1)r$, which evaluates to $0.414r$. Note that this is the permitted maximum size subject to keeping the close-packed spheres in contact; if the spheres are allowed to separate slightly while maintaining their relative positions, then the hole can accommodate a larger sphere.

Self-test 3.4 Show that the maximum radius of a sphere that can fit into a tetrahedral hole (see below) is $r_h = 0.225r$; base your calculation on Fig. 3.19b.

A **tetrahedral hole**, T, Figs 3.17b and 3.20, is formed by a planar triangle of touching spheres capped by a single sphere lying in the dip between them. The tetrahedral holes in any close-packed solid can be divided into two sets: in one the apex of the tetrahedron is directed up (T) and in the other the apex points down (T'). In an arrangement of N close-packed spheres there are N tetrahedral holes of each set and $2N$ tetrahedral holes in all. In a close-packed structure of spheres of radius r, a tetrahedral hole can accommodate another hard sphere of radius no greater than $0.225r$ (see *Self-test* 3.4). The location of tetrahedral holes, and the four nearest-neighbour spheres for one hole, in the hcp arrangement is shown in Fig. 3.20a and for a ccp arrangement in Fig. 3.20b. Individual tetrahedral holes in ccp and hcp structures are identical (because they are properties of two neighbouring close-packed layers) but in the hcp arrangement neighbouring T and T' holes share a common tetrahedral face and are so close together that they are never occupied simultaneously.

Where two types of sphere of different radius pack together (for instance, when cations and anions stack together), the larger spheres (normally the anions) can form a close-packed array and the smaller spheres occupy the octahedral or tetrahedral holes. Thus simple ionic structures can be described in terms of the occupation of holes in close-packed arrays (Section 3.9).

The structures of metals and alloys

X-ray diffraction studies (Section 8.1) reveal that many metallic elements have close-packed structures, indicating that the bonds between the atoms have little directional covalent character (Table 3.2, Fig. 3.21). One consequence of this close-packing is that metals often have high densities because the most mass is packed into the smallest volume. Indeed, the elements deep in the d block, near iridium and osmium, include the densest solids known under normal conditions of temperature and pressure. Osmium has the highest density of all the elements at 22.61 g cm^{-3} and the density of tungsten, 19.25 g cm^{-3}, which is almost twice that of lead (11.3 g cm^{-3}), results in it being used as weighting material in fishing equipment and as ballast in high performance cars.

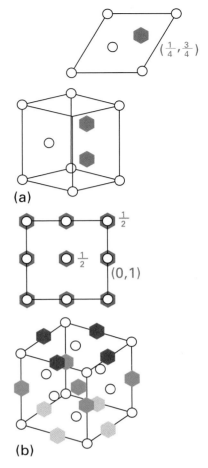

Fig. 3.18 (a) The location (represented by a hexagon) of the two octahedral holes in the hcp unit cell and (b) the locations (represented by hexagons) of the octahedral holes in the ccp unit cell.

EXAMPLE 3.5 **Calculating the density of a substance from a structure**

Calculate the density of gold, with a cubic close-packed array of atoms of molar mass $M = 196.97$ g mol^{-1} and a cubic lattice parameter $a = 409$ pm.

Answer Density is an intensive property; therefore the density of the unit cell is the same as the density of any macroscopic sample. We represent the ccp arrangement as a face-centred lattice with a sphere at each lattice point; there are four spheres associated with the unit cell. The mass of each atom is M/N_A, where N_A is Avogadro's constant, and the total mass of the unit cell is $4M/N_A$. The volume of the cubic unit cell is a^3. The mass density of the cell is $\rho = 4M/N_A a^3$. At this point we insert the data:

$$\rho = \frac{4 \times (196.97 \times 10^{-3}\, \text{kg mol}^{-1})}{(6.022 \times 10^{23}\, \text{mol}^{-1}) \times (409 \times 10^{-12}\, \text{m})^3} = 1.91 \times 10^4 \ \text{kg m}^{-3}$$

That is, the density of the unit cell, and therefore of the bulk metal, is 19.1 g cm^{-3}. The experimental value is 19.2 g cm^{-3}, in good agreement with this calculated value.

Self-test 3.5 Calculate the lattice parameter of silver assuming that it has the same structure as elemental gold but a density of 10.5 g cm^{-3}.

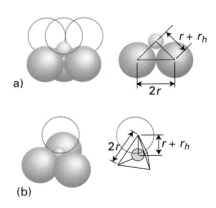

Fig. 3.19 The distances used to calculate the size of (a) an octahedral hole and (b) a tetrahedral hole.

A note on good practice It is always best to proceed symbolically with a calculation for as long as possible: that reduces the risk of numerical error and gives an expression that can be used in other circumstances.

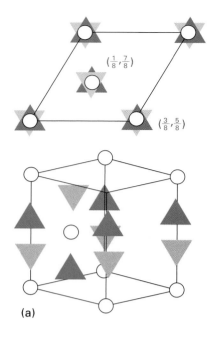

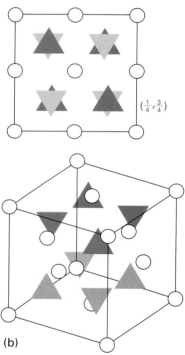

$(\frac{1}{8},\frac{7}{8})$

$(\frac{3}{8},\frac{5}{8})$

(a)

$(\frac{1}{4},\frac{3}{4})$

(b)

Fig. 3.20 (a) The locations (represented by triangles) of the tetrahedral holes in the hcp unit cell and (b) the locations of the tetrahedral holes in the ccp unit cell.

3.4 Polytypism

Key point: Polytypes involving complex stacking arrangements of close-packed layers occur for some metals.

Which of the common close-packed polytypes, hcp or ccp, a metal adopts depends on the details of the electronic structure of its atoms, the extent of interaction between second-nearest-neighbours, and the potential for some directional character in the bonding. Indeed, a close-packed structure need not be either of the common ABAB…or ABCABC… polytypes. An infinite range of close-packed polytypes can in fact occur, as the layers may stack in a more complex repetition of A, B, and C layers or even in some permissible random sequence. The stacking cannot be a completely random choice of A, B, and C sequences, however, because adjacent layers cannot have exactly the same sphere positions; for instance, AA, BB, and CC cannot occur because spheres in one layer must occupy dips in the adjacent layer.

Cobalt is an example of a metal that displays this more complex polytypism. Above 500°C, cobalt is ccp but it undergoes a transition when cooled. The structure that results is a nearly randomly stacked set (for instance, ABACBABABC…) of close-packed layers of Co atoms. In some samples of cobalt the polytypism is not random, as the sequence of planes of atoms repeats after several hundred layers. The long-range repeat may be a consequence of a spiral growth of the crystal that requires several hundred turns before a stacking pattern is repeated.

3.5 Nonclose-packed structures

Key points: A common nonclose-packed metal structure is body-centred cubic; a primitive cubic structure is occasionally encountered. Metals that have structures more complex than those described so far can sometimes be regarded as slightly distorted versions of simple structures.

Table 3.2 The crystal structures adopted by metals under normal conditions

Crystal structure	Element
Hexagonal close-packed (hcp)	Be, Ca, Co, Mg, Ti, Zn
Cubic close-packed (ccp)	Ag, Al, Au, Cd, Cu, Ni, Pb, Pt
Body-centred cubic (bcc)	Ba, Cr, Fe, W, alkali metals
Primitive cubic (cubic-P)	Po

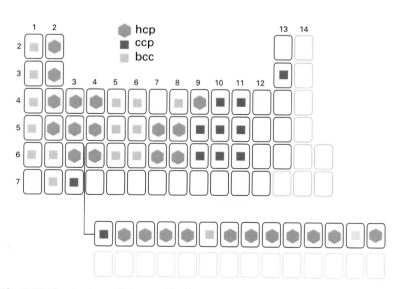

Fig. 3.21 The structures of the metallic elements at room temperature. Elements with more complex structures are left blank.

Not all elemental metals have structure based on close-packing and some other packing patterns use space nearly as efficiently. Even metals that are close-packed may undergo a phase transition to a less closely packed structure when they are heated and their atoms undergo large-amplitude vibrations.

One commonly adopted arrangement has the translational symmetry of the body-centred cubic lattice and is known as the **body-centred cubic** structure (cubic-I or bcc) in which a sphere is at the centre of a cube with spheres at each corner (Fig. 3.22a). Metals with this structure have a coordination number of 8 because the central atom is in contact with the atoms at the corners of the unit cell. Although a bcc structure is less closely packed than the ccp and hcp structures (for which the coordination number is 12), the difference is not very great because the central atom has six second-nearest neighbours, at the centres of the adjacent unit cells, only 15 per cent further away. This arrangement leaves 32 per cent of the space unfilled compared with 26 per cent in the close-packed structures (see Example 3.3). A bcc structure is adopted by 15 of the elements under standard conditions, including all the alkali metals and the metals in Groups 5 and 6. Accordingly, this simple arrangement of atoms is sometimes referred to as the 'tungsten type'.

The least common metallic structure is the **primitive cubic** (cubic-P) structure (Fig. 3.23), in which spheres are located at the lattice points of a primitive cubic lattice, taken as the corners of the cube. The coordination number of a cubic-P structure is 6. One form of polonium (α-Po) is the only example of this structure among the elements under normal conditions. Solid mercury (α-Hg), however, has a closely related structure: it is obtained from the cubic-P arrangement by stretching the cube along one of its body diagonals (Fig. 3.24a); a second form of solid mercury (β-Hg) has a structure based on the bcc arrangement but compressed along one cell direction (Fig. 3.24b). Although antimony and bismuth normally have structures based on layers of atoms, both convert to a cubic-P structure under pressure and then to close-packed structures at even higher pressures.

Metals that have structures more complex than those described so far can sometimes be regarded, like solid mercury, as having slightly distorted versions of simple structures. Zinc and cadmium, for instance, have almost hcp structures, but the planes of close-packed atoms are separated by a slightly greater distance than in perfect hcp. This difference suggests stronger bonding between the close-packed atoms in the plane than between the planes: the bonding draws these atoms together and, in doing so, squeezes out the atoms of the neighbouring layers.

3.6 Polymorphism of metals

Key points: Polymorphism is a common consequence of the low directionality of metallic bonding. At high temperatures a bcc structure is common for metals that are close-packed at low temperatures on account of the increased amplitude of atomic vibrations.

The low directionality of the bonds that metal atoms may form accounts for the wide occurrence of **polymorphism,** the ability to adopt different crystal forms under different conditions of pressure and temperature. It is often, but not universally, found that the most closely packed phases are thermodynamically favoured at low temperatures and that the less closely packed structures are favoured at high temperatures. Similarly, the application of high pressure leads to structures with higher packing densities, such as ccp and hcp.

The polymorphs of metals are generally labelled α, β, γ,... with increasing temperature. Some metals revert to a low-temperature form at higher temperatures. Iron, for example, shows several solid–solid phase transitions; α-Fe, which is bcc, occurs up to 906°C, γ-Fe, which is ccp, occurs up to 1401°C, and then α-Fe occurs again up to the melting point at 1530°C. The hcp polymorph, β-Fe, is formed at high pressures and was to believed to be the form that exists at the Earth's core, but recent studies indicate that a bcc polymorph is more likely (Box 3.1).

The bcc structure is common at high temperatures for metals that are close-packed at low temperatures because the increased amplitude of atomic vibrations in the hotter solid results in a less close-packed structure. For many metals (among them Ca, Ti, and Mn) the transition temperature is above room temperature; for others (among them Li and Na), the transition temperature is below room temperature. It is also found empirically that a bcc structure is favoured by metals with a small number of valence electrons per orbital.

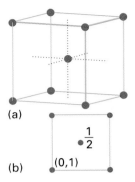

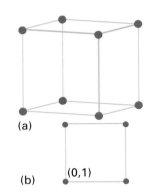

Fig. 3.22 (a) A bcc structure unit cell and (b) its projection representation.

Fig. 3.23 (a) A primitive cubic unit cell and (b) its projection representation.

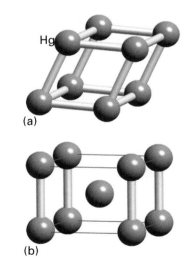

Fig 3.24 The structures of (a) α-mercury and (b) β-mercury that are closely related to the unit cells with primitive cubic and body-centred cubic lattices, respectively.

BOX 3.1   Metals under pressure

The Earth has an innermost core about 1200 km in diameter that consists of solid iron and is responsible for generating the planet's powerful magnetic field. The pressure at the centre of the Earth has been calculated to be around 370 GPa (about 3.7 million atm) at a temperature of 5000–6500°C. The polymorph of iron that exists under these conditions has been much debated with information from theoretical calculations and measurements using seismology. The current thinking is that the iron core consists of the body-centred cubic polymorph. It has been proposed that this exists either as a giant crystal or a large number of oriented crystals such that the long diagonal of the bcc unit cell aligns along the Earth's axis of rotation (Fig. B3.1, left).

The study of the structures and polymorphism of an element and compounds under high pressure conditions goes beyond the study of the Earth's core. Hydrogen, when subjected to pressures similar to those at the Earth's core, is predicted to become a metallic solid, similar to the alkali

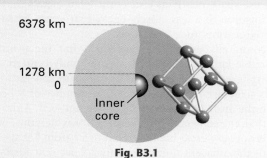

Fig. B3.1

metals, and the cores of planets such as Jupiter have been predicted to contain hydrogen in this form. When pressures of over 55 GPa are applied to iodine the I_2 molecules dissociate and adopt the simple face-centred cubic structure; the element becomes metallic and is a superconductor below 1.2 K.

3.7 Atomic radii of metals

Key point: The Goldschmidt correction converts atomic radii of metals to the value they would have in a close-packed structure with 12-fold coordination.

An informal definition of the atomic radius of a metallic element was given in Section 1.9 as half the distance between the centres of adjacent atoms in the solid. However, it is found that this distance generally increases with the coordination number of the lattice. The same atom in structures with different coordination numbers may therefore appear to have different radii, and an atom of an element with coordination number 12 appears bigger than one with coordination number 8. In an extensive study of internuclear separations in a wide variety of polymorphic elements and alloys, V. Goldschmidt found that the average relative radii are related as shown in Table 3.3.

It is desirable to put all elements on the same footing when comparing trends in their characteristics; that is when comparing the intrinsic properties of their atoms rather than the properties that stem from their environment. Therefore, it is common to adjust the empirical internuclear separation to the value that would be expected if the element were in fact close-packed (with coordination number 12).

Table 3.3 The variation of radius with coordination number

Coordination number	Relative radius
12	1
8	0.97
6	0.96
4	0.88

■ **A brief illustration.** The empirical atomic radius of Na is 185 pm, but that is for the bcc structure in which the coordination number is 8. To adjust to 12-coordination we multiply this radius by $1/0.97 = 1.03$ and obtain 191 pm as the radius that a Na atom would have if it were in a close-packed structure. ■

Goldschmidt radii of the elements were in fact the ones listed in Table 1.4 as 'metallic radii' and used in the discussion of the periodicity of atomic radius (Section 1.9). The essential features of that discussion to bear in mind now, with 'atomic radius' interpreted as Goldschmidt-corrected metallic radius in the case of metallic elements, are that metallic radii generally increase down a group and decrease from left to right across a period. As remarked in Section 1.9, trends in atomic radii reveal the presence of the lanthanide contraction in Period 6, with atomic radii of the elements that follow the lanthanoids found to be smaller than simple extrapolation from earlier periods would suggest. As also remarked there, this contraction can be traced to the poor shielding effect of f electrons. A similar contraction occurs across each row of the d block.

EXAMPLE 3.6   Calculating a metallic radius

The cubic unit cell parameter, a, of polonium (α-Po) is 335 pm. Use the Goldschmidt correction to calculate a metallic radius for this element.

Answer We need to infer the radius of the atoms from the dimensions of the unit cell and the coordination number, and then apply a correction to coordination number 12. Because the Po atoms of radius r are in contact along the unit cell edges, the length of the primitive cubic unit cell is $2r$. Thus, the metallic radius of 6-coordinate Po is $a/2$ with $a = 335$ pm. The conversion factor from 6-fold to 12-fold coordination from Table 3.3 ($1/0.960$) gives the metallic radius of Po as $\frac{1}{2} \times 335 \text{ pm} \times 1/0.960 = 174$ pm.

Self-test 3.6 Predict the lattice parameter for Po when it adopts a bcc structure.

3.8 Alloys

An **alloy** is a blend of metallic elements prepared by mixing the molten components and then cooling the mixture to produce a metallic solid. Alloys may be homogeneous solid solutions, in which the atoms of one metal are distributed randomly among the atoms of the other, or they may be compounds with a definite composition and internal structure. Alloys typically form from two electropositive metals, so they are likely to be located towards the bottom left-hand corner of a Ketelaar triangle (Fig. 3.25).

Solid solutions are classified as either 'substitutional' or 'interstitial'. A **substitutional solid solution** is a solid solution in which atoms of the solute metal occupy some of the locations of the solvent metal atoms (Fig. 3.26a). An **interstitial solid solution** is a solid solution in which the solute atoms occupy the interstices (the holes) between the solvent atoms (Fig. 3.26b). However, this distinction is not particularly fundamental because interstitial atoms often lie in a definite array (Fig. 3.26c), and hence can be regarded as a substitutional version of another structure. Some of the classic examples of alloys are brass (up to 38 atom per cent Zn in Cu), bronze (a metal other than Zn or Ni in Cu; casting bronze, for instance, is 10 atom per cent Sn and 5 atom per cent Pb), and stainless steel (over 12 atom per cent Cr in Fe).

(a) Substitutional solid solutions

Key point: A substitutional solid solution involves the replacement of one type of metal atom in a structure by another.

Substitutional solid solutions are generally formed if three criteria are fulfilled:

1. The atomic radii of the elements are within about 15 per cent of each other.
2. The crystal structures of the two pure metals are the same; this similarity indicates that the directional forces between the two types of atom are compatible with each other.
3. The electropositive characters of the two components are similar; otherwise compound formation, where electrons are transferred between species, would be more likely.

Thus, although sodium and potassium are chemically similar and have bcc structures, the atomic radius of Na (191 pm) is 19 per cent smaller than that of K (235 pm), and the two metals do not form a solid solution. Copper and nickel, however, two neighbours late in the d block, have similar electropositive character, similar crystal structures (both ccp), and similar atomic radii (Ni 125 pm, Cu 128 pm, only 2.3 per cent different), and form a continuous series of solid solutions, ranging from pure nickel to pure copper. Zinc, copper's other neighbour in Period 4, has a similar atomic radius (137 pm, 7 per cent larger), but it is hcp, not ccp. In this instance, zinc and copper are partially miscible and form solid solutions known as 'α-brasses' of composition $Cu_{1-x}Zn_x$ with $0 < x < 0.38$ and the same structural type as pure copper.

(b) Interstitial solid solutions of nonmetals

Key point: In an interstitial solid solution, additional small atoms occupy holes within the lattice of the original metal structure.

Interstitial solid solutions are often formed between metals and small atoms (such as boron, carbon, and nitrogen) that can inhabit the interstices in the structure. The small atoms enter the host solid with preservation of the crystal structure of the original metal and without the transfer of electrons and formation of ionic species. There is either a simple whole-number ratio of metal and interstitial atoms (as in tungsten carbide, WC) or the small atoms are distributed randomly in the available spaces or holes in the structure between the packed atoms. The former substances are true compounds and the latter can be considered as interstitial solid solutions or, on account of the variation in the atomic ratio of the two elements, nonstoichiometric compounds (Section 3.17).

Considerations of size can help to decide where the formation of an interstitial solid solution is likely to occur. Thus, the largest solute atom that can enter a close-packed solid without distorting the structure appreciably is one that just fits an octahedral hole, which as we have seen has radius $0.414r$. For small atoms such as B, C, or N the atomic radii of the possible host metal atom structures include those of the d-metals such as Fe, Co, and Ni. One important class of materials of this type consists of carbon steels in which C atoms occupy some of the octahedral holes in the Fe bcc lattice. Carbon steels typically contain

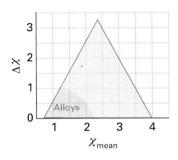

Fig. 3.25 The approximate locations of alloys in a Ketelaar triangle.

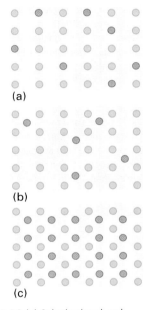

Fig. 3.26 (a) Substitutional and (b) interstitial alloys. (c) In some cases, an interstitial alloy may be regarded as a substitutional alloy derived from another lattice.

BOX 3.2 Steels

Steels are alloys of iron, carbon, and other elements. They are classified as mild and medium-or high-carbon steels according to the percentage of carbon they contain. Mild steels contain up to 0.25 atom per cent C, medium-carbon steels contain 0.25 to 0.45 atom per cent, and high-carbon steels contain 0.45 to 1.50 atom per cent. The addition of other metals to these carbon steels can have a major effect on the structure, properties, and therefore applications of the steel. Examples of metals added to carbon steels, so forming 'stainless steels', are listed in the table. Stainless steels are also classified by their crystalline structures, which are controlled by factors such as the rate of cooling following their formation in a furnace and the type of added metal. Thus pure iron adopts different polymorphs (Section 3.6) depending on temperature and some of these high-temperature structures can be stabilized at room temperature in steels or by quenching (cooling very rapidly).

Stainless steels with the *austenite* structure comprise over 70 per cent of total stainless steel production. Austenite is a solid solution of carbon and iron that exists in steel above 723°C and is ccp iron with about 2 per cent of the octahedral holes filled with carbon. As it cools, it breaks down into other materials including *ferrite* and *martensite* as the solubility of carbon in the iron drops to below 1 atom per cent. The rate of cooling determines the relative proportions of these materials and therefore the mechanical properties (for example, hardness and tensile strength) of the steel. The addition of certain other metals, such as Mn, Ni, and Cr, can allow the austenitic structure to survive cooling to room temperature. These steels contain a maximum of 0.15 atom per cent C and typically 10−20 atom per cent Cr plus Ni or Mn as a substitutional solid solution; they can retain an austenitic structure at all temperatures from the cryogenic region to the melting point of the alloy.

A typical composition, 18 atom per cent Cr and 8 atom per cent Ni, is known as *18/8 stainless*.

Ferrite is α-Fe with only a very small level of carbon, less than 0.1 atom per cent, with a bcc iron crystal structure. Ferritic stainless steels are highly corrosion resistant, but far less durable than austenitic grades. They contain between 10.5 and 27 atom per cent Cr and some Mo, Al, or W. Martensitic stainless steels are not as corrosion resistant as the other two classes, but are strong and tough as well as highly machineable, and can be hardened by heat treatment. They contain 11.5 to 18 atom per cent Cr and 1−2 atom per cent C which is trapped in the iron structure as a result of quenching compositions with the austenite structure type. The martensitic crystal structure is closely related to that of ferrite but the unit cell is tetragonal rather than cubic.

Metal	Atom percentage added	Effect on properties
Copper	0.2−1.5	Improves atmospheric corrosion resistance
Nickel	0.1−1	Benefits surface quality
Niobium	0.02	Increases tensile strength and yield point
Nitrogen	0.003−0.012	Improves strength
Manganese	0.2−1.6	Improves strength
Vanadium	Up to 0.12	Increases strength

between 0.2 and 1.6 atom per cent C. With increasing carbon content they become harder and stronger but less malleable (Box 3.2).

(c) Intermetallic compounds

Key point: Intermetallic compounds are alloys in which the structure adopted is different from the structures of either component metal.

There are materials formed between two metals that are best regarded as actual compounds despite the similarity of their electropositive nature. For instance, when some liquid mixtures of metals are cooled, they form phases with definite structures that are often unrelated to the parent structure. These phases are called **intermetallic compounds**. They include β-brass (CuZn) and compounds of composition $MgZn_2$, Cu_3Au, NaTl, and Na_5Zn_{21}. Note that some of these intermetallic compounds contain a very electropositive metal in combination with a less electropositive metal (for example, Na and Zn), and in a Ketelaar triangle lie above the true alloys (Fig. 3.27). Such combinations are called **Zintl phases**. These compounds are not fully ionic (although they are often brittle) and have some metallic properties, including lustre. A classic example of a Zintl phase is KGe with the structure shown in Fig. 3.28.

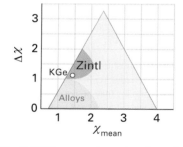

Fig. 3.27 The approximate locations of Zintl phases in a Ketelaar triangle. The point marks the location of one exemplar, KGe.

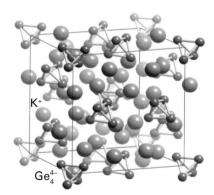

Fig. 3.28 The structure of the Zintl phase KGe showing the Ge_4^{4-} tetrahedral units and interspersed K^+ ions.

EXAMPLE 3.7 Composition, lattice type and unit cell content of iron and its alloys

What are the lattice types and unit cell contents of (a) iron metal (Fig. 3.29a) and (b) the iron/chromium alloy, FeCr (Fig. 3.29b)?

Answer We need to identify the translation symmetry of the unit cell and to count the net numbers of atoms present. (a) The iron structure consists of Fe atoms distributed over the sites at the centre and corners of a cubic unit cell with eightfold coordination to the nearest neighbours. All the occupied sites are equivalent, so the structure has the translational symmetry of a bcc lattice. The structure type is the bcc structure. The Fe atom at the centre counts 1 and the eight Fe atoms at the cell corners count $8 \times \frac{1}{8} = 1$, so there are two Fe atoms in the unit cell. (b) For FeCr, the atom at the centre of the unit cell (Cr) is different from the one on the corner (Fe) and thus the translational symmetry present is that of the entire unit cell (not half unit cell displacements characteristic of a bcc structure), so the lattice type is primitive, P. There is one Cr atom and $8 \times \frac{1}{8} = 1$ Fe atom in the unit cell, in accord with the stoichiometry FeCr.

Self-test 3.7 What are the stoichiometry and lattice type of the iron/chromium alloy shown in Fig. 3.29c ?

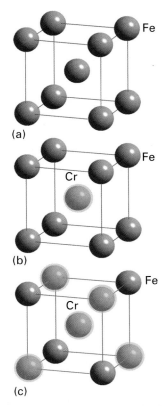

(a)

(b)

(c)

Fig. 3.29 The structures of (a) iron, (b) FeCr, and (c) as Fe, Cr alloy, see Self-test 3.7.

Ionic solids

Key points: The ionic model treats a solid as an assembly of oppositely charged spheres that interact by nondirectional electrostatic forces; if the thermodynamic properties of the solid calculated on this model agree with experiment, then the compound is normally considered to be ionic.

Ionic solids, such as NaCl and KNO_3, are often recognized by their brittleness because the electrons made available by cation formation are localized on a neighbouring anion instead of contributing to an adaptable, mobile electron sea. Ionic solids also commonly have high melting points and most are soluble in polar solvents, particularly water. However, there are exceptions: CaF_2, for example, is a high-melting ionic solid but it is insoluble in water. Ammonium nitrate, NH_4NO_3, is ionic in terms of its interactions between the ammonium and nitrate ions, but melts at 170°C. Binary ionic materials are typical of elements with large electronegativity differences, typically $\Delta\chi > 3$, and such compounds are therefore likely to be found at the top corner of a Ketelaar triangle (Fig. 3.27).

The classification of a solid as ionic is based on comparison of its properties with those of the **ionic model**, which treats the solid as an assembly of oppositely charged, hard spheres that interact primarily by nondirectional electrostatic forces (Coulombic forces) and repulsions between complete shells in contact. If the thermodynamic properties of the solid calculated on this model agree with experiment, then the solid may be ionic. However, it should be noted that many examples of coincidental agreement with the ionic model are known, so numerical agreement alone does not imply ionic bonding. The nondirectional nature of electrostatic interactions between ions in an ionic solid contrast with those present in a covalent solid, where the symmetries of the atomic orbitals play a strong role in determining the geometry of the structure. However, the assumption that ions can be treated as perfectly hard spheres (of fixed radius for a particular ion type) that have no directionality in their bonding, is far from true for real ions. For example, with halide anions some directionality might be expected in their bonding that results from the orientations of their p orbitals, and large ions, such as Cs^+ and I^-, are easily polarizable so do not behave as hard spheres. Even so, the ionic model is a useful starting point for describing many simple structures.

We start by describing some common ionic structures in terms of the packing of hard spheres of different sizes and opposite charges. After that, we see how to rationalize the structures in terms of the energetics of crystal formation. The structures described have been obtained by using X-ray diffraction (Section 8.1), and were among the first substances to be examined in this way.

3.9 Characteristic structures of ionic solids

The ionic structures described in this section are prototypes of a wide range of solids. For instance, although the rock-salt structure takes its name from a mineral form of NaCl, it is characteristic of numerous other solids (Table 3.4). Many of the structures can be regarded as derived from arrays in which the larger of the ions, usually the anions, stack together in ccp or hcp patterns and the smaller counter-ions (usually the cations) occupy the octahedral or tetrahedral holes in the lattice (Table 3.5). Throughout the following discussion, it will be helpful to refer back to Figs 3.18 and 3.20 to see how the structure being described is related to the hole patterns shown there. The close-packed layers usually need to expand to accommodate the counter-ions but this expansion is often a minor perturbation of the anion arrangement, which will still be referred to as ccp and hcp. This expansion avoids some of the strong repulsion between the identically charged ions and also allows larger species to be inserted into the holes between larger ions. Overall, examining the opportunities for hole-filling in a close-packed array of the larger ion type provides an excellent starting point for the descriptions of many simple ionic structures.

Table 3.4 The crystal structures of compounds

Crystal structure	Example*
Antifluorite	K_2O, K_2S, Li_2O, Na_2O, Na_2Se, Na_2S
Caesium chloride	**CsCl**, TlI, CsAu, CsCN, CuZn , NbO
Fluorite	**CaF_2**,UO_2, HgF_2, LaH_2, PbO_2
Nickel arsenide	**NiAs**, NiS, FeS, PtSn, CoS
Perovskite	**$CaTiO_3$** (distorted), $SrTiO_3$, $PbZrO_3$, $LaFeO_3$, $LiSrH_3$, $KMnF_3$
Rock salt	**NaCl**, KBr, RbI, AgCl, AgBr, MgO, CaO, TiO, FeO, NiO, SnAs, UC, ScN
Rutile	**TiO_2**, MnO_2, SnO_2,WO_2, MgF_2, NiF_2
Sphalerite (zinc blende)	**ZnS**, CuCl, CdS, HgS, GaP, InAs
Spinel	**$MgAl_2O_4$**, $ZnFe_2O_4$, $ZnCr_2S_4$
Wurtzite	**ZnS**, ZnO, BeO, MnS, AgI, AlN, SiC, NH_4F

*The substance in bold type is the one that gives its name to the structure.

Table 3.5 The relation of structure to the filling of holes

Close-packing type	Hole filling	Structure type (exemplar)
Cubic (ccp)	All octahedral	Rock salt (NaCl)
	All tetrahedral	Fluorite (CaF_2)
	Half octahedral	$CdCl_2$
	Half tetrahedral	Sphalerite (ZnS)
Hexagonal (hcp)	All octahedral	Nickel arsenide (NiAs); with some distortion from perfect hcp CdI_2
	Half octahedral	Rutile (TiO_2); with some distortion from perfect hcp
	All tetrahedral	No structure exists: tetrahedral holes share faces
	Half tetrahedral	Wurtzite (ZnS)

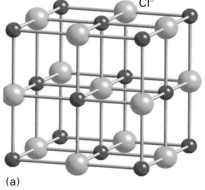

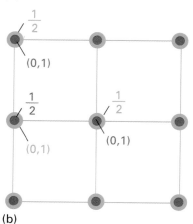

(a)

(b)

Fig. 3.30 (a) The rock-salt structure and (b) its projection representation. Note the relation of this structure to the fcc structure in Fig. 3.18 with an atom in each octahedral hole.

(a) Binary phases, AX$_n$

Key points: Important structures that can be expressed in terms of the occupation of holes include the rock-salt, caesium-chloride, sphalerite, fluorite, wurtzite, nickel-arsenide, and rutile structures.

The simplest ionic compounds contain just one type of cation (A) and one type of anion (X) present in various ratios covering compositions such as AX and AX_2. Several different structures may exist for each of these compositions, depending on the relative sizes of the cations and anions and which holes are filled and to what degree in the close-packed array (Table 3.5). We start by considering compositions AX with equal numbers of cations and anions and then consider AX_2, the other commonly found stoichiometry.

The **rock-salt structure** is based on a ccp array of bulky anions with cations in all the octahedral holes (Fig. 3.30). Alternatively, it can be viewed as a structure in which the anions occupy all the octahedral holes in a ccp array of cations. As the number of octahedral holes in a close-packed array is equal to the number of ions forming the array (the X ions), then filling them all with A ions yields the stoichiometry AX. Because each ion is surrounded by an octahedron of six counter-ions, the coordination number of each type of ion is 6 and the structure is said to have **(6,6)-coordination**. In this notation, the first number in parentheses is the coordination number of the cation, and the second number is the coordination number of the anion. The rock-salt structure can still be described as having a face-centred cubic lattice after this hole filling because the translational symmetry demanded by this lattice type is preserved when all the octahedral sites are occupied.

To visualize the local environment of an ion in the rock-salt structure, we should note that the six nearest neighbours of the central ion of the cell shown in Fig. 3.30 lie at the centres of the faces of the cell and form an octahedron around the central ion. All six neighbours have a charge opposite to that of the central ion. The 12 second-nearest neighbours of the central ion, those next further away, are at the centres of the edges of the cell, and all have the same charge as the central ion. The eight third-nearest neighbours are at the corners of the unit cell, and have a charge opposite to that of the central ion. We can use the rules described in Section 3.1 to determine the composition of the unit cell, the number of atoms or ions of each type present.

■ **A brief illustration.** In the unit cell shown in Fig. 3.30, there are the equivalent of $8 \times \frac{1}{8} + 6 \times \frac{1}{2} = 4\,Na^+$ ions and $12 \times \frac{1}{4} + 1 = 4\,Cl^-$ ions. Hence, each unit cell contains four NaCl formula units. The number of formula units present in the unit cell is commonly denoted Z, so in this case $Z = 4$. ■

The rock-salt arrangement is not just formed for simple monatomic species such as M^+ and X^- but also for many 1:1 compounds in which the ions are complex units such as $[Co(NH_3)_6][TlCl_6]$. The structure of this compound can be considered as an array of close-packed octahedral $[TlCl_6]^{3-}$ ions with $[Co(NH_3)_6]^{3+}$ ions in all the octahedral holes. Similarly, compounds such as CaC_2, CsO_2, KCN, and FeS_2 all adopt structures closely related to the rock-salt structure with alternating cations and complex anions (C_2^{2-}, O_2^-, CN^-, and S_2^{2-}, respectively), although the orientation of these linear diatomic species can lead to elongation of the unit cell and elimination of the cubic symmetry (Fig. 3.31). Further compositional flexibility, but retaining a rock-salt type of structure, can come from having more than one cation or anion type while maintaining the overall 1:1 ratio between opposite ions of opposite charge, as in $LiNiO_2$, which is equivalent to $(Li_{1/2}Ni_{1/2})O$.

Much less common than the rock-salt structure for compounds of stoichiometry AX is the **caesium-chloride structure** (Fig. 3.32), which is possessed by CsCl, CsBr, and CsI, as well as some other compounds formed of ions of similar radii to these, including TlI (see Table 3.4). The caesium-chloride structure has a cubic unit cell with each corner occupied by an anion and a cation occupying the 'cubic hole' at the cell centre (or vice versa); as a result, $Z = 1$. An alternative view of this structure is as two interlocking primitive cubic cells, one of Cs^+ and the other of Cl^-. The coordination number of both types of ion is 8, so the structure is described as having (8,8)-coordination. The radii are so similar that this energetically highly favourable coordination is feasible, with numerous counter-ions adjacent to a given ion. Note that NH_4Cl also forms this structure despite the relatively small size of the NH_4^+ ion because the cation can form hydrogen bonds with four of the Cl^- ions at the corners of the cube (Fig. 3.33). Many 1:1 alloys, such as AlFe and CuZn, have a caesium-chloride arrangement of the two metal atom types.

The **sphalerite structure** (Fig. 3.34), which is also known as the **zinc-blende structure**, takes its name from a mineral form of ZnS. Like the rock-salt structure it is based on an expanded ccp anion arrangement but now the cations occupy one type of tetrahedral hole, one half the tetrahedral holes present in a close-packed structure. Each ion is surrounded by four neighbours and so the structure has (4,4)-coordination and $Z = 4$.

■ **A brief illustration.** To count the ions in the unit cell shown in the sphalerite structure shown in Fig. 3.34, we draw up the following table:

Location (share)	Number of cations	Number of anions	Contribution
Body (1)	4×1	0	4
Face ($\frac{1}{2}$)	0	$6 \times \frac{1}{2}$	3
Edge ($\frac{1}{4}$)	0	0	0
Vertex ($\frac{1}{8}$)	0	$8 \times \frac{1}{8}$	1
Total:	4	4	8

There are four cations and four anions in the unit cell. This ratio is consistent with the chemical formula ZnS, with $Z = 4$. ■

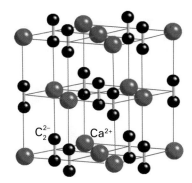

Fig. 3.31 The structure of CaC_2 is based on the rock-salt structure but is elongated in the direction parallel to the axes of the C_2^{2-} ions.

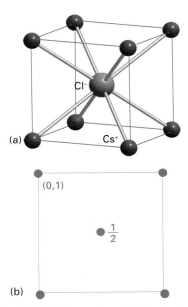

Fig. 3.32 (a) The caesium-chloride structure. The corner lattice points, which are shared by eight neighbouring cells, are surrounded by eight nearest-neighbour lattice points. The anion occupies a cubic hole in a primitive cubic lattice. (b) Its projection.

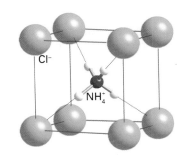

Fig. 3.33 The structure of ammonium chloride, NH_4Cl, reflects the ability of the tetrahedral NH_4^+ ion to form hydrogen bonds to the tetrahedral array of Cl^- ions around it.

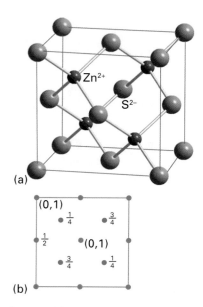

(a)

(b)

Fig. 3.34 (a) The sphalerite (zinc-blende) structure and (b) its projection representation. Note its relation to the ccp lattice in Fig. 3.18a with half the tetrahedral holes occupied by Zn^{2+} ions.

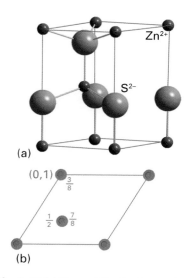

(a)

(b)

Fig. 3.35 (a) The wurtzite structure and (b) its projection representation.

The **wurtzite structure** (Fig. 3.35) takes its name from another polymorph of zinc sulfide. It differs from the sphalerite structure in being derived from an expanded hcp anion array rather than a ccp array, but as in sphalerite the cations occupy half the tetrahedral holes; that is just one of the two types (either T or T′ as discussed in Section 3.3). This structure, which has (4,4)-coordination, is adopted by ZnO, AgI, and one polymorph of SiC, as well as several other compounds (Table 3.4). The local symmetries of the cations and anions are identical with respect to their nearest neighbours in wurtzite and sphalerite but differ at the second-nearest neighbours.

The **nickel-arsenide structure** (NiAs, Fig. 3.36) is also based on an expanded, distorted hcp anion array, but the Ni atoms now occupy the octahedral holes and each As atom lies at the centre of a trigonal prism of Ni atoms. This structure is adopted by NiS, FeS, and a number of other sulfides. The nickel-arsenide structure is typical of MX compounds that contain polarizable ions and are formed from elements with smaller electronegativity differences than elements that, as ions, adopt the rock-salt structure. Compounds that form this structure type lie in the 'polarized ionic salt area' of a Ketelaar triangle (Fig. 3.37). There is also potential for some degree of metal–metal bonding between metal atoms in adjacent layers and this structure type (or distorted forms of it) is also common for a large number of alloys based on d- and p-block elements.

A common AX_2 structural type is the **fluorite structure**, which takes its name from its exemplar, the naturally occurring mineral fluorite, CaF_2. In fluorite, the Ca^{2+} ions lie in an expanded ccp array and the F^- ions occupy all the tetrahedral holes (Fig. 3.38). In this description it is the cations that are close-packed because the F^- anions are small. The lattice has (8,4)-coordination, which is consistent with there being twice as many anions as cations. The anions in their tetrahedral holes have four nearest neighbours and the cation site is surrounded by a cubic array of eight anions. An alternative description of the structure is that the anions form a nonclose-packed, primitive cubic lattice and the cations occupy half the cubic holes in this lattice. Note the relation of this structure to the caesium-chloride structure, in which all the cubic holes are occupied.

The **antifluorite** structure is the inverse of the fluorite structure in the sense that the locations of cations and anions are reversed. The latter structure is shown by some alkali metal oxides, including Li_2O. In it, the cations (which are twice as numerous as the anions) occupy all the tetrahedral holes of a ccp array of anions. The coordination is (4,8) rather than the (8,4) of fluorite itself.

The **rutile structure** (Fig. 3.39) takes its name from rutile, a mineral form of titanium(IV) oxide, TiO_2. The structure can also be considered an example of hole filling in an hcp anion arrangement, but now the cations occupy only half the octahedral holes and there is considerable buckling of the close-packed anion layers. This arrangement results in a structure that reflects the strong tendency of a Ti^{4+} ion to acquire octahedral coordination. Each Ti^{4+} atom is surrounded by six O atoms and each O atom is surrounded by three Ti^{4+} ions; hence the rutile structure has (6,3)-coordination. The principal ore of tin, cassiterite SnO_2, has the rutile structure, as do a number of metal difluorides (Table 3.4).

In the **cadmium-iodide structure** (as in CdI_2, Fig. 3.40), the octahedral holes between every other pair of hcp layers of I^- ions (that is half of the total number of octahedral

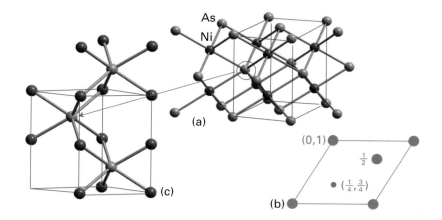

As
Ni

(a)

(c)

(b)

Fig. 3.36 (a) The nickel-arsenide structure, (b) the projection representation of the unit cell, and (c) the trigonal prismatic coordination around As.

holes) are filled by Cd^{2+} ions. The CdI_2 structure is often referred to as a 'layer-structure' as the repeating layers of atoms perpendicular to the close-packed layers form the sequence I–Cd–I···I–Cd–I···I–Cd–I with weak van der Waals interactions between the iodine atoms in adjacent layers. The structure has (6,3)- coordination, being octahedral for the cation and trigonal pyramidal for the anion. The structure type is found commonly for many d-metal halides and chalcogenides (for example, $FeBr_2$, MnI_2, ZrS_2, and $NiTe_2$).

The **cadmium-chloride structure** (as in $CdCl_2$, Fig. 3.41) is analogous to the CdI_2 structure but with a ccp arrangement of anions; half the octahedral sites between alternate anion layers are occupied. This layer structure has identical coordination numbers (6,3) and geometries for the ions to those found for the CdI_2 structure-type, although it is preferred for a number of d-metal dichlorides, such as $MnCl_2$ and $NiCl_2$.

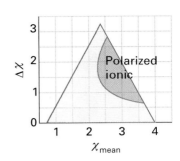

Fig. 3.37 The location of polarized ionic salts in a Ketelaar triangle.

EXAMPLE 3.8 Determining the stoichiometry of a hole-filled structure

Identify the stoichiometries of the following structures based on hole filling using a cation, A, in close-packed arrays of anions, X. (a) An hcp array in which one-third of the octahedral sites are filled. (b) A ccp array in which all the tetrahedral and all the octahedral sites are filled.

Answer We need to be aware that in an array of N close-packed spheres there are $2N$ tetrahedral holes and N octahedral holes (Section 3.3). Therefore, filling all the octahedral holes in a closed-packed array of anions X with cations A would produce a structure in which cations and anions were in the ratio 1:1, corresponding to the stoichiometry AX. (a) As only one-third of the holes are occupied, the A:X ratio is $\frac{1}{3}$:1, corresponding to the stoichiometry AX_3. An example of this type of structure is BiI_3. (b) The total number of A species is $2N + N$ with NX species. The A:X ratio is therefore 3:1, corresponding to the stoichiometry A_3X. An example of this type of structure is Li_3Bi.

Self-test 3.8 Determine the stoichiometry of an hcp array with two-thirds of the octahedral sites occupied.

(b) Ternary phases $A_aB_bX_n$

Key point: The perovskite and spinel structures are adopted by many compounds with the stoichiometries ABO_3 and AB_2O_4, respectively.

Structural possibilities increase very rapidly once the compositional complexity is increased to three ionic species. Unlike binary compounds, it is difficult to predict the most likely structure type based on the ion sizes and preferred coordination numbers. This section describes two important structures formed by ternary oxides; the O^{2-} ion is the most common anion, so oxide chemistry is central to a significant part of solid-state chemistry.

The mineral perovskite, $CaTiO_3$, is the structural prototype of many ABX_3 solids (Table 3.4), particularly oxides. In its ideal form, the **perovskite structure** is cubic with each A cation surrounded by 12 X anions and each B cation surrounded by six X anions (Fig. 3.42). In fact, the perovskite structure may also be described as a close-packed array of A cations and O^{2-} anions (arranged such that each A cation is surrounded by 12 O^{2-} anions from the original close-packed layers) with B cations in all the octahedral holes that are formed from six of the O^{2-} ions, giving $B_{n/4}[AO_3]_{n/4}$, which is equivalent to ABO_3.

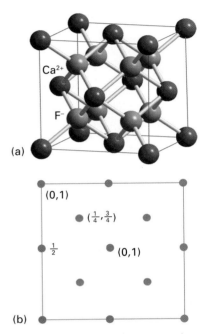

Fig. 3.38 (a) The fluorite structure and (b) its projection representation. This structure has a ccp array of cations and all the tetrahedral holes are occupied by anions.

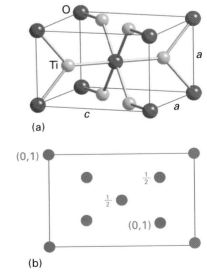

Fig. 3.39 (a) The rutile structure and (b) its projection representation. Rutile itself is one polymorph of TiO_2.

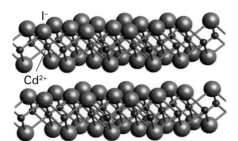

Fig. 3.40 The CdI_2 structure (left).

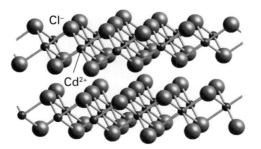

Fig. 3.41 The $CdCl_2$ structure (right).

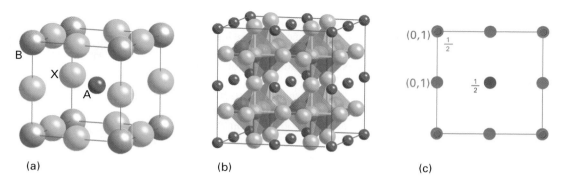

Fig. 3.42 (a) The perovskite structure, ABX_3, (b) a display that emphasizes the octahedral shape of the B^- sites, and (c) its projection representation.

In oxides, X = O and the sum of the charges on the A and B ions must be +6. That sum can be achieved in several ways ($A^{2+}B^{4+}$ and $A^{3+}B^{3+}$ among them), including the possibility of mixed oxides of formula $A(B_{0.5}B'_{0.5})O_3$, as in $La(Ni_{0.5}Ir_{0.5})O_3$. The A-type cation in perovskites is therefore usually a large ion (of radius greater than 110 pm) of lower charge, such as Ba^{2+} or La^{3+}, and the B cation is a small ion (of radius less than 100 pm) of higher charge, such as Ti^{4+}, Nb^{5+}, or Fe^{3+}.

The perovskite structure is closely related to the materials that show interesting electrical properties, such as piezoelectricity, ferroelectricity, and high-temperature superconductivity (Section 24.8).

EXAMPLE 3.9 Determining coordination numbers

Demonstrate that the coordination number of the Ti^{4+} ion in the perovskite $CaTiO_3$ is 6.

Answer We need to imagine eight of the unit cells shown in Fig. 3.42 stacked together with a Ti atom shared by them all. A local fragment of the structure is shown in Fig. 3.43; it shows that there are six O^{2-} ions around the central Ti^{4+} ion, so the coordination number of Ti in perovskite is 6. An alternative way of viewing the perovskite structure is as BO_6 octahedra sharing all vertices in three orthogonal directions with the A cations at the centres of the cubes so formed (Fig. 3.42b).

Self-test 3.9 What is the coordination number of a Ti^{4+} site and the O^{2-} site in rutile?

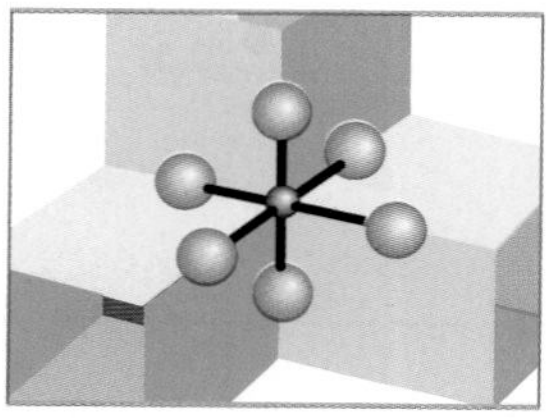

Fig. 3.43 The local coordination environment of a Ti atom in perovskite.

Spinel itself is $MgAl_2O_4$, and oxide spinels, in general, have the formula AB_2O_4. The **spinel structure** consists of a ccp array of O^{2-} ions in which the A cations occupy one-eighth of the tetrahedral holes and the B cations occupy half the octahedral holes (Fig. 3.44). Spinels are sometimes denoted $A[B_2]O_4$, the square brackets denoting the cation type (normally the smaller, higher charged ion of A and B) that occupies the octahedral holes. So, for example,

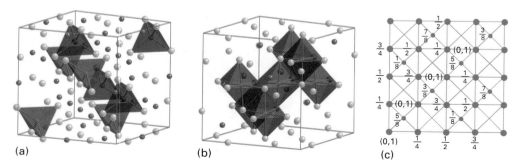

Fig. 3.44 (a) The spinel structure showing the tetrahedral oxygen environment around the B cations, (b) showing the octahedral oxygen environment around the A cations, and (c) its projection representation with only the cation locations specified.

$ZnAl_2O_4$ can be written $Zn[Al_2]O_4$ to show that all the Al^{3+} cations occupy octahedral sites. Examples of compounds that have spinel structures include many ternary oxides with the stoichiometry AB_2O_4 that contain a 3d-series metal, such as $NiCr_2O_4$ and $ZnFe_2O_4$, and some simple binary d-block oxides, such as Fe_3O_4, Co_3O_4, and Mn_3O_4; note that in these structures A and B are the same element but in different oxidation states, as in $Fe^{2+}[Fe^{3+}]_2O_4$. There are also a number of compositions termed **inverse spinels**, in which the cation distribution is $B[AB]O_4$ and in which the more abundant cation is distributed over both tetrahedral and octahedral sites. Spinels and inverse spinels are discussed again in Sections 20.1 and 24.8.

EXAMPLE 3.10 **Predicting possible ternary phases**

What ternary oxides with the perovskite or spinel structures might it be possible to synthesize that contain the cations Ti^{4+}, Zn^{2+}, In^{3+}, and Pb^{2+}? Use the ionic radii given in *Resource section 1*.

Answer We need to consider whether the sizes of the ions permit the occurrence of the two structures. We can predict that $ZnTiO_3$ does not exist as a perovskite as the Zn^{2+} ion is too small for the A-type site; likewise, $PbIn_2O_4$ does not adopt the spinel structure as the Pb^{2+} cation is too large for the tetrahedral sites. We conclude that the permitted structures are $PbTiO_3$ (perovskite), $TiZn_2O_4$ (spinel), and $ZnIn_2O_4$ (spinel).

Self-test 3.10 What additional oxide perovskite composition(s) might be obtained if La^{3+} is added to this list of cations?

3.10 The rationalization of structures

The thermodynamic stabilities and structures of ionic solids can be treated very simply using the ionic model. However, a model of a solid in terms of charged spheres interacting electrostatically is crude and we should expect significant departures from its predictions because many solids are more covalent than ionic. Even conventional 'good' ionic solids, such as the alkali metal halides, have some covalent character. Nevertheless, the ionic model provides an attractively simple and effective scheme for correlating many properties.

(a) Ionic radii

Key point: The sizes of ions, ionic radii, generally increase down a group, decrease across a period, increase with coordination number, and decrease with increasing charge number.

A difficulty that confronts us at the outset is the meaning of the term 'ionic radius'. As remarked in Section 1.9, it is necessary to apportion the single internuclear separation of nearest-neighbour ions between the two different species (for example, an Na^+ ion and a Cl^- ion in contact). The most direct way to solve the problem is to make an assumption about the radius of one ion, and then to use that value to compile a set of self-consistent values for all other ions. The O^{2-} ion has the advantage of being found in combination with a wide range of elements. It is also reasonably unpolarizable, so its size does not vary much as the identity of the accompanying cation is changed. In a number of compilations, therefore, the values are based on $r(O^{2-}) = 140$ pm. However, this value is by no means sacrosanct: a set of values compiled by Goldschmidt was based on $r(O^{2-}) = 132$ pm and other values use the F^- ion as the basis.

For certain purposes (such as for predicting the sizes of unit cells) ionic radii can be helpful, but they are reliable only if they are all based on the same fundamental choice (such as the value 140 pm for O^{2-}). If values of ionic radii are used from different sources, it is essential to verify that they are based on the same convention. An additional complication that was first noted by Goldschmidt is that, as we have already seen for metals, ionic radii increase with coordination number (Fig. 3.45). Hence, when comparing ionic radii, we should compare like with like, and use values for a single coordination number (typically 6).

The problems of the early workers have been resolved only partly by developments in X-ray diffraction (Section 8.1). It is now possible to measure the electron density between two neighbouring ions and identify the minimum as the boundary between them. However, as can be seen from Fig. 3.46, the electron density passes through a very broad minimum, and its exact location may be very sensitive to experimental uncertainties and to the

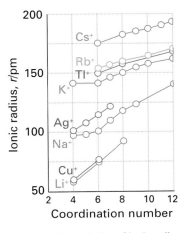

Fig. 3.45 The variation of ionic radius with coordination number.

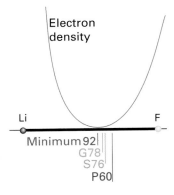

Fig. 3.46 The variation in electron density along the Li–F axis in LiF. The point P denotes the Pauling radii of the ions, G the original (1927) Goldschmidt radii, and S the Shannon radii.

identities of the two neighbours. That being so, it is still probably more useful to express the sizes of ions in a self-consistent manner than to seek calculated values of individual radii in certain combinations. After all, we are interested in compounds where we always have interactions between pairs of ions so we are consistent in the method of determination and application of ionic radii. Very extensive lists of self-consistent values that have been compiled by analysing X-ray data on thousands of compounds, particularly oxides and fluorides, exist and some are given in Table 1.4 and *Resource section 1*.

The general trends for ionic radii are the same as for atomic radii. Thus:

1. Ionic radii increase down a group. (The lanthanide contraction, Section 1.9, restricts the increase between the 4d- and 5d-series metal ions.)
2. The radii of ions of the same charge decrease across a period.
3. If an element can form cations with different charge numbers, then for a given coordination number its ionic radius decreases with increasing charge number.
4. Because a positive charge indicates a reduced number of electrons, and hence a more dominant nuclear attraction, cations are smaller than anions for elements with similar atomic numbers.
5. When an ion can occur in environments with different coordination numbers, the observed radius, as measured by considering the average distances to the nearest neighbours, increases as the coordination number increases. This increase reflects the fact that the repulsions between the surrounding ions are reduced if they move apart, so leaving more room for the central ion.

(b) The radius ratio

Key point: The radius ratio indicates the likely coordination numbers of the ions in a binary compound.

A parameter that figures widely in the literature of inorganic chemistry, particularly in introductory texts, is the 'radius ratio', γ (gamma), of the ions. The **radius ratio** is the ratio of the radius of the smaller ion (r_{small}) to that of the larger (r_{large}):

$$\gamma = \frac{r_{small}}{r_{large}} \tag{3.1}$$

In most cases, r_{small} is the cation radius and r_{large} is the anion radius. The minimum radius ratio that can support a given coordination number is then calculated by considering the geometrical problem of packing together spheres of different sizes (Table 3.6). It is argued that, if the radius ratio falls below the minimum given, then ions of opposite charge will not be in contact and ions of like charge will touch. According to a simple electrostatic argument, a lower coordination number, in which the contact of oppositely charged ions is restored, then becomes favourable. Another way of looking at this argument is that as the radius of the M^+ ion increases, more anions can pack around it, so giving a larger number of favourable Coulombic interactions. In this respect, compare CsCl, and its (8,8)-coordination, with NaCl, and its (6,6)-coordination.

We can use our previous calculations of hole size (Example 3.4), to put these ideas on a firmer footing. A cation of radius between $0.225r$ and $0.414r$ can occupy a tetrahedral hole in a close-packed or slightly expanded close-packed array of anions of radius r.

Table 3.6 The correlation of structural type with radius ratio

Radius ratio (γ)	CN for 1:1 and 1:2 stoichiometries	Binary AB structure type	Binary AB_2 structure type
1	12	None known	None known
0.732−1	8:8 and 8:4	CsCl	CaF_2
0.414−0.732	6:6 and 6:3	NaCl (ccp), NiAs (hcp)	TiO_2
0.225−0.414	4:4	ZnS (ccp and hcp)	

CN denotes coordination number.

However, once the radius of a cation reaches $0.414r$, the anions are forced so far apart that octahedral coordination becomes possible and most favourable. Note that $0.225r$ represents the size of the smallest ion that will fit in a tetrahedral hole and that cations between $0.225r$ and $0.414r$ will push the anions apart. However, the coordination number cannot increase to 6 with good contacts between cation and anions until the radius goes above $0.414r$. Similar arguments apply for the tetrahedral holes that can be filled by ions with sizes up to $0.225r$.

These concepts of ion packing based on radius ratios can often be used to predict which structure is most likely for any particular choice of cation and anion (Table 3.6). In practice, the radius ratio is most reliable when the cation coordination number is 8, and less reliable with six- and four-coordinate cations because directional covalent bonding becomes more important for these lower coordination numbers.

■ **A brief illustration.** To predict the crystal structure of TlCl we note that the ionic radii are $r(Tl^+) = 159$ pm and $r(Cl^-) = 181$ pm, giving $\gamma = 0.88$. We can therefore predict that TlCl is likely to adopt a caesium-chloride structure with (8,8)-coordination. That is the structure found in practice. ■

The ionic radii used in these calculations are those obtained by consideration of structures under normal conditions. At high pressures, different structures may be preferred, especially those with higher coordination numbers and greater density. Thus many simple compounds transform between the simple (4,4)-, (6,6)-, and (8,8)-coordination structures under pressure. Examples of this behaviour include most of the lighter alkali metal halides, which change from a (6,6)-coordinate rock-salt structure to an (8,8)-coordinate caesium-chloride structure at 5 kbar (the rubidium halides) or 10–20 kbar (the sodium and potassium halides). The ability to predict the structures of compounds under pressure is important for understanding the behaviour of ionic compounds under such conditions. Calcium oxide, for instance, is predicted to transform from the rock-salt to the caesium-chloride structure at around 600 kbar, the pressure in the Earth's lower mantle.

Similar arguments involving the relative ionic radii of cations and anions and their preferred coordination numbers (that is, preferences for octahedral, tetrahedral, or cubic geometries) can be applied throughout structural solid-state chemistry and aid the prediction of which ions might be incorporated into a particular structure type. For more complex stoichiometries, such as the ternary compounds with the perovskite and spinel structure types, the ability to predict which combinations of cations and anions will yield a specific structure type has proved very useful. One example is that for the high-temperature superconducting cuprates (Section 24.8), the design of a particular structure feature, such as Cu^{2+} in octahedral coordination to oxygen, can be achieved using ionic radii considerations.

(c) Structure maps

Key point: A structure map is a representation of the variation in crystal structure with the character of the bonding.

Even though the use of radius ratios is not totally reliable, it is still possible to rationalize structures by collecting enough information empirically and looking for patterns. This approach has motivated the compilation of 'structure maps'. A **structure map** is an empirically compiled map that depicts the dependence of crystal structure on the electronegativity difference between the elements present and the average principal quantum number of the valence shells of the two atoms.[5] As such, a structure map can be regarded as an extension of the ideas introduced in Chapter 2 in relation to Ketelaar's triangle. As we have seen, binary ionic salts are formed for large differences in electronegativity $\Delta\chi$, but as this difference is reduced, polarized ionic salts and more covalently bonded networks become preferred. Now we can focus on this region of the triangle and explore how small changes in electronegativity and polarizability affect the choice of ion arrangement.

The ionic character of a bond increases with $\Delta\chi$, so moving from left to right along the horizontal axis of a structure map correlates with an increase in ionic character in

[5] Structure maps were introduced by E. Mooser and W.B. Pearson, *Acta Cryst.*, 1959, **12**, 1015.

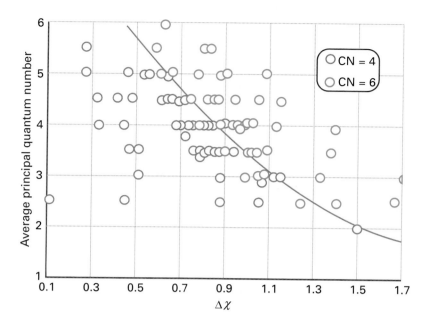

Fig. 3.47 A structure map for compounds of formula MX. A point is defined by the electronegativity difference ($\Delta\chi$) between M and X and their average principal quantum number n. The location in the map indicates the coordination number expected for that pair of properties. (Based on E. Mooser and W.B. Pearson, *Acta Crystallogr.*, 1959, **12**, 1015.)

the bonding. The principal quantum number is an indication of the radius of an ion, so moving up the vertical axis corresponds to an increase in the average radius of the ions. Because atomic energy levels also become closer as the atom expands, the polarizability of the atom increases too (Section 1.9e). Consequently, the vertical axis of a structure map corresponds to increasing size and polarizability of the bonded atoms. Figure 3.47 is an example of a structure map for MX compounds. We see that the structures we have been discussing for MX compounds fall in distinct regions of the map. Elements with large $\Delta\chi$ have (6,6)-coordination, such as is found in the rock-salt structure; elements with small $\Delta\chi$ (and hence where there is the expectation of covalence) have a lower coordination number. In terms of a structure map representation, GaN is in a more covalent region of Fig. 3.47 than ZnO because $\Delta\chi$ is appreciably smaller.

■ **A brief illustration.** To predict the type of crystal structure that should be expected for magnesium sulfide, MgS, we note that the electronegativities of magnesium and sulfur are 1.3 and 2.6 respectively, so $\Delta\chi = 1.3$. The average principal quantum number is 3 (both elements are in Period 3). The point $\Delta\chi = 1.3$, $n = 3$ lies just in the sixfold coordination region of the structure map in Fig. 3.47. This location is consistent with the observed rock-salt structure of MgS. ■

The energetics of ionic bonding

A compound tends to adopt the crystal structure that corresponds to the lowest Gibbs energy. Therefore, if for the process

$$M^+(g) + X^-(g) \rightarrow MX(s)$$

the change in standard reaction Gibbs energy, $\Delta_r G^\ominus$, is more negative for the formation of a structure A rather than B, then the transition from B to A is spontaneous under the prevailing conditions, and we can expect the solid to be found with structure A.

The process of solid formation from the gas of ions is so exothermic that at and near room temperature the contribution of the entropy to the change in Gibbs energy (as in $\Delta G^\ominus = \Delta H^\ominus - T\Delta S^\ominus$) may be neglected; this neglect is rigorously true at $T = 0$. Hence, discussions of the thermodynamic properties of solids normally focus, initially at least, on changes in enthalpy. That being so, we look for the structure that is formed most exothermically and identify it as the thermodynamically most stable form. Some typical values of lattice enthalpies are given in Table 3.7 for a number of simple ionic compounds.

Table 3.7 Lattice enthalpies of some simple inorganic solids

Compound	Structure type	$\Delta H_L^{exp}/(\text{kJ mol}^{-1})$	Compound	Structure type	$\Delta H_L^{exp}/(\text{kJ mol}^{-1})$
LiF	Rock salt	1030	SrCl₂	Fluorite	2125
LiI	Rock salt	757	LiH	Rock salt	858
NaF	Rock salt	923	NaH	Rock salt	782
NaCl	Rock salt	786	KH	Rock salt	699
NaBr	Rock salt	747	RbH	Rock salt	674
NaI	Rock salt	704	CsH	Rock salt	648
KCl	Rock salt	719	BeO	Wurtzite	4293
KI	Rock salt	659	MgO	Rock salt	3795
CsF	Rock salt	744	CaO	Rock salt	3414
CsCl	Caesium chloride	657	SrO	Rock salt	3217
CsBr	Caesium chloride	632	BaO	Rocksalt	3029
CsI	Caesium chloride	600	Li₂O	Antifluorite	2799
MgF₂	Rutile	2922	TiO₂	Rutile	12150
CaF₂	Fluorite	2597	CeO₂	Fluorite	9627

3.11 Lattice enthalpy and the Born–Haber cycle

Key points: Lattice enthalpies are determined from enthalpy data by using a Born-Haber cycle; the most stable crystal structure of the compound is commonly the structure with the greatest lattice enthalpy under the prevailing conditions.

The **lattice enthalpy**, $\Delta H_L^{\ominus}$, is the standard molar enthalpy change accompanying the formation of a gas of ions from the solid:

$$MX(s) \rightarrow M^+(g) + X^-(g) \qquad \Delta H_L^{\ominus}$$

A note on good practice The definition of lattice enthalpy as an endothermic (positive) term corresponding to the break up of the lattice is correct but contrary to many school and college texts where it is defined with respect to lattice formation (and listed as a negative quantity).

Because lattice disruption is always endothermic, lattice enthalpies are always positive and their positive signs are normally omitted from their numerical values. As remarked above, if entropy considerations are neglected, then the most stable crystal structure of the compound is the structure with the greatest lattice enthalpy under the prevailing conditions.

Lattice enthalpies are determined from enthalpy data by using a **Born–Haber cycle**, a closed path of steps that includes lattice formation as one stage, such as that shown in Fig. 3.48. The standard enthalpy of decomposition of a compound into its elements in their reference states (their most stable states under the prevailing conditions) is the negative of its standard enthalpy of formation, $\Delta_f H^{\ominus}$:

$$M(s) + X(s, l, g) \rightarrow MX(s) \qquad \Delta_f H^{\ominus}$$

Likewise, the standard enthalpy of lattice formation from the gaseous ions is the negative of the lattice enthalpy as specified above. For a solid element, the standard enthalpy of atomization, $\Delta_{atom} H^{\ominus}$, is the standard enthalpy of sublimation, as in the process

$$M(s) \rightarrow M(g) \qquad \Delta_{atom} H^{\ominus}$$

For a gaseous element, the standard enthalpy of atomization is the standard enthalpy of dissociation, $\Delta_{dis} H^{\ominus}$, as in

$$X_2(g) \rightarrow 2X(g) \qquad \Delta_{dis} H^{\ominus}$$

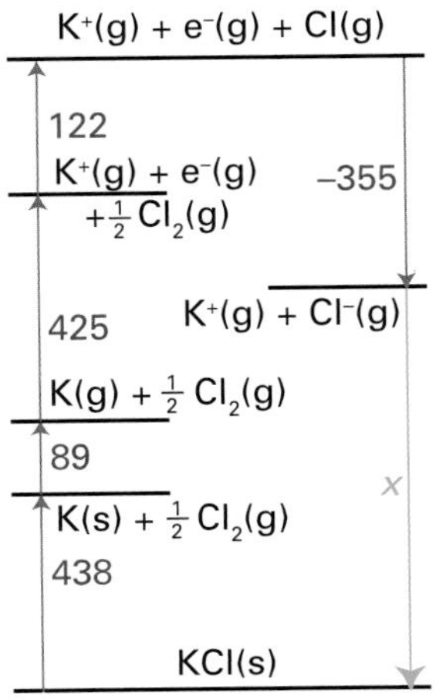

Fig. 3.48 The Born–Haber cycle for KCl. The lattice enthalpy is equal to $-x$. All numerical values are in kilojoules per mole.

The standard enthalpy of formation of ions from their neutral atoms is the enthalpy of ionization (for the formation of cations, $\Delta_{ion}H^{\ominus}$) and the electron-gain enthalpy (for anions, $\Delta_{eg}H^{\ominus}$):

$$M(g) \rightarrow M^+(g) + e^-(g) \qquad \Delta_{ion}H^{\ominus}$$

$$X(g) + e^-(g) \rightarrow X^-(g) \qquad \Delta_{eg}H^{\ominus}$$

The value of the lattice enthalpy—the only unknown in a well-chosen cycle—is found from the requirement that the sum of the enthalpy changes round a complete cycle is zero (because enthalpy is a state property).[6] The value of the lattice enthalpy obtained from a Born–Haber cycle depends on the accuracy of all the measurements being combined, and as a result there can be significant variations, typically ± 10 kJ mol^{-1}, in tabulated values.

	$\Delta H^{\ominus}$/(kJ mol^{-1})
Sublimation of K(s)	+89
Ionization of K(g)	+425
Dissociation of Cl$_2$(g)	+244
Electron gain by Cl(g)	−355
Formation of KCl(s)	−438

	$\Delta H^{\ominus}$/(kJ mol^{-1})
Sublimation of Mg(s)	+148
Ionization of Mg(g) to Mg^{2+}(g)	+2187
Vaporization of Br$_2$(l)	+31
Dissociation of Br$_2$(g)	+193
Electron gain by Br(g)	−331
Formation of MgBr$_2$(s)	−524

EXAMPLE 3.11   Using a Born-Haber cycle to determine a lattice enthalpy

Calculate the lattice enthalpy of KCl(s) using a Born-Haber cycle and the information in the margin.

Answer The required cycle is shown in Fig. 3.48. The sum of the enthalpy changes around the cycle is zero, so

$$\Delta H_L^{\ominus}/(\text{kJ mol}^{-1}) = 438 + 425 + 89 + 244/2 - 355 = 719$$

Note that the calculation becomes more obvious by drawing an energy level diagram showing the signs of the various steps of the cycle; all lattice enthalpies are positive. Also as only one Cl atom from Cl$_2$(g) is required to produce KCl, half the dissociation energy of Cl$_2$, $\frac{1}{2} \times 244$ kJ mol^{-1}, is used in the calculation.

Self-test 3.11 Calculate the lattice enthalpy of magnesium bromide from the data shown in the margin.

3.12 The calculation of lattice enthalpies

Once the lattice enthalpy is known, it can be used to judge the character of the bonding in the solid. If the value calculated on the assumption that the lattice consists of ions interacting electrostatically is in good agreement with the measured value, then it may be appropriate to adopt a largely ionic model of the compound. A discrepancy indicates a degree of covalence. As mentioned earlier, it is important to remember that numerical coincidences can be misleading in this assessment.

(a) The Born–Mayer equation

Key points: The Born-Mayer equation is used to estimate lattice enthalpy for an ionic lattice. The Madelung constant reflects the effect of the geometry of the lattice on the strength of the net Coulombic interaction.

To calculate the lattice enthalpy of a supposedly ionic solid we need to take into account several contributions, including the Coulombic attractions and repulsions between the ions and the repulsive interactions that occur when the electron densities of the ions overlap. This calculation yields the **Born–Mayer equation** for the lattice enthalpy at $T = 0$:

$$\Delta H_L^{\ominus} = \frac{N_A |z_A z_B| e^2}{4\pi\varepsilon_0 d}\left(1 - \frac{d^*}{d}\right)A \tag{3.2}$$

where $d = r_+ + r_-$ is the distance between centres of neighbouring cations and anions, and hence a measure of the 'scale' of the unit cell (for the derivation, see *Further information* 3.1). In this expression N_A is Avogadro's constant, z_A and z_B the charge numbers of the cation and anion, e the fundamental charge, ε_0 the vacuum permittivity, and d^* a constant (typically 34.5 pm) used to represent the repulsion between ions at short range. The quantity A is called the **Madelung constant**, and depends on the structure (specifically, on the relative distribution of ions, Table 3.8). The Born–Mayer equation in fact gives the lattice energy as distinct from the lattice enthalpy, but the two are identical at $T = 0$ and the difference may be disregarded in practice at normal temperatures.

---

[6] Note that when the lattice enthalpy is known from calculation, a Born–Haber cycle may be used to determine the value of another elusive quantity, the electron-gain enthalpy (and hence the electron affinity).

■ **A brief illustration.** To estimate the lattice enthalpy of sodium chloride, we use $z(Na^+) = +1$, $z(Cl^-) = -1$, from Table 3.8, $A = 1.748$, and from Table 1.4 $d = r_{Na^+} + r_{Cl^-} = 283$ pm; hence (using fundamental constants from inside the back cover):

$$\Delta H_L^{\ominus} = \frac{(6.022 \times 10^{23}\ \text{mol}^{-1}) \times |(+1) \times (-1)| \times (1.602 \times 10^{-19}\ \text{C})^2}{4\pi \times (8.854 \times 10^{-12}\ \text{J}^{-1}\ \text{C}^2\ \text{m}^{-1}) \times (2.82 \times 10^{-10}\ \text{m})} \times \left(1 - \frac{34.5\ \text{pm}}{283\ \text{pm}}\right) \times 1.748$$
$$= 7.56 \times 10^5\ \text{J mol}^{-1}$$

or 756 kJ mol^{-1}. This value compares reasonably well with the experimental value from the Born–Haber cycle, 788 kJ mol^{-1}. ■

Table 3.8 Madelung constants

Structural type	A
Caesium chloride	1.763
Fluorite	2.519
Rock salt	1.748
Rutile	2.408
Sphalerite	1.638
Wurtzite	1.641

The form of the Born–Mayer equation for lattice enthalpies allows us to account for their dependence on the charges and radii of the ions in the solid. Thus, the heart of the equation is

$$\Delta H_L^{\ominus} \propto \frac{|z_A z_B|}{d}$$

Therefore, a large value of d results in a low lattice enthalpy, whereas high ionic charges result in a high lattice enthalpy. This dependence is seen in some of the values given in Table 3.7. For the alkali metal halides, the lattice enthalpies decrease from LiF to LiI and from LiF to CsF as the halide and alkali metal ion radii increase, respectively. We also note that the lattice enthalpy of MgO ($|z_A z_B| = 4$) is almost four times that of NaCl ($|z_A z_B| = 1$) due to the increased charges on the ions for a similar value of d, noting that the Madelung constant is the same.

The Madelung constant typically increases with coordination number. For instance, $A = 1.748$ for the (6,6)-coordinate rock-salt structure but $A = 1.763$ for the (8,8)-coordinate caesium-chloride structure and 1.638 for the (4,4)-coordinate sphalerite structure. This dependence reflects the fact that a large contribution comes from nearest neighbours, and such neighbours are more numerous when the coordination number is large. However, a high coordination number does not necessarily mean that the interactions are stronger in the caesium-chloride structure because the potential energy also depends on the scale of the lattice. Thus, d may be so large in lattices with ions big enough to adopt eightfold-coordination that the separation of the ions reverses the effect of the small increase in the Madelung constant and results in a smaller lattice enthalpy.

(b) Other contributions to lattice enthalpies

Key point: Nonelectrostatic contributions to the lattice enthalpy include van der Waals interactions, particularly the dispersion interaction.

Another contribution to the lattice enthalpy is the **van der Waals interaction** between the ions and molecules, the weak intermolecular interactions that are responsible for the formation of condensed phases of electrically neutral species. An important and sometimes dominant contribution of this kind is the **dispersion interaction** (the 'London interaction'). The dispersion interaction arises from the transient fluctuations in electron density (and, consequently, instantaneous electric dipole moment) on one molecule driving a fluctuation in electron density (and dipole moment) on a neighbouring molecule, and the attractive interaction between these two instantaneous electric dipoles. The molar potential energy of this interaction, V, is expected to vary as

$$V = -\frac{N_A C}{d^6} \tag{3.3}$$

The constant C depends on the substance. For ions of low polarizability, this contribution is only about 1 per cent of the electrostatic contribution and is ignored in elementary lattice enthalpy calculations of ionic solids. However, for highly polarizable ions, such as Tl^+ and I^-, such terms can make significant contributions of several per cent. Thus, the dispersion interaction for compounds such as LiF and CsBr is estimated to contribute 16 kJ mol^{-1} and 50 kJ mol^{-1}, respectively.

3.13 Comparison of experimental and theoretical values

Key points: For compounds formed from elements with $\Delta\chi > 2$, the ionic model is generally valid and lattice enthalpy values derived using the Born–Mayer equation and Born–Haber cycles are similar. For structures formed with small electronegativity differences and polarizable ions there may be additional, nonionic contributions to the bonding.

The agreement between the experimental lattice enthalpy and the value calculated using the ionic model of the solid (in practice, from the Born–Mayer equation) provides a measure of the extent to which the solid is ionic. Table 3.9 lists some calculated and measured lattice enthalpies together with electronegativity differences. The ionic model is reasonably valid if $\Delta\chi > 2$, but the bonding becomes increasingly covalent if $\Delta\chi < 2$. However, it should be remembered that the electronegativity criterion ignores the role of polarizability of the ions. Thus, the alkali metal halides give fairly good agreement with the ionic model, the best with the least polarizable halide ions (F^-) formed from the highly electronegative F atom, and the worst with the highly polarizable halide ions (I^-) formed from the less electronegative I atom. This trend is also seen in the lattice enthalpy data for the silver halides in Table 3.9. The discrepancy between experimental and theoretical values is largest for the iodide, which indicates major deficiencies in the ionic model for this compound. Overall the agreement is much poorer with Ag than with Li as the electronegativity of silver ($\chi = 1.93$) is much higher than that of lithium ($\chi = 0.98$) and significant covalent character in the bonding would be expected.

It is not always clear whether it is the electronegativity of the atoms or the polarizability of the resultant ions that should be used as a criterion. The worst agreement with the ionic model is for polarizable-cation/polarizable-anion combinations that are substantially covalent. Here again, though, the difference between the electronegativities of the parent elements is small and it is not clear whether electronegativity or polarizability provides the better criterion.

EXAMPLE 3.12 Using the Born–Mayer equation to decide the theoretical stability of unknown compounds

Decide whether solid ArCl is likely to exist.

Answer The answer hinges on whether the enthalpy of formation of ArCl is significantly positive or negative: if it is significantly positive (endothermic), the compound is unlikely to be stable (there are, of course, exceptions). Consideration of a Born–Haber cycle for the synthesis of ArCl would show two unknowns, the enthalpy of formation of ArCl and its lattice enthalpy. We can estimate the lattice enthalpy of a purely ionic ArCl by using the Born–Mayer equation assuming the radius of Ar^+ to be midway between that of Na^+ and K^+. That is, the lattice enthalpy is somewhere between the values for NaCl and KCl at about 745 kJ mol^{-1}. So, taking the enthalpy of dissociation of $\frac{1}{2}$ Cl$_2$ to form Cl as 122 kJ mol^{-1}, the ionization enthalpy of Ar as 1524 kJ mol^{-1}, and the electron affinity of Cl as 356 kJ mol^{-1} gives $\Delta_fH^{\ominus}(ArCl, s) = 1524 - 745 - 356 + 122$ kJ mol^{-1} = +545 kJ mol^{-1}. That is, the compound is predicted to be very unstable with respect to its elements, mainly because the large ionization enthalpy of Ar is not compensated by the lattice enthalpy.

Self-test 3.12 Predict whether CsCl$_2$ with the fluorite structure is likely to exist.

Table 3.9 Comparison of experimental and theoretical lattice enthalpies for rock-salt structures

	$\Delta H_L^{calc}/(\text{kJ mol}^{-1})$	$\Delta H_L^{exp}/(\text{kJ mol}^{-1})$	$(\Delta H_L^{exp} - \Delta H_L^{calc})/(\text{kJ mol}^{-1})$
LiF	1029	1030	1
LiCl	834	853	19
LiBr	788	807	19
LiI	730	757	27
AgF	920	953	33
AgCl	832	903	71
AgBr	815	895	80
AgI	777	882	105

Calculations like that in Example 3.12 were used to predict the stability of the first noble gas compounds. The ionic compound $O_2^+PtF_6^-$ had been obtained from the reaction of oxygen with PtF_6. Consideration of the ionization energies of O_2 (1176 kJ mol^{-1}) and Xe (1169 kJ mol^{-1}) showed them to be almost identical and the sizes of Xe^+ and O_2^+ would be expected to be similar, implying similar lattice enthalpies for their compounds. Hence once O_2 had been found to react with platinum hexafluoride, it could be predicted that Xe should too, as indeed it does, to give an ionic compound which is believed to contain XeF^+ and PtF_6^- ions. Similar calculations may be used to predict the stability, or otherwise, of a wide variety of compounds, for example the stability of alkaline earth monohalides, such as MgCl. Calculations based on Born–Mayer lattice enthalpies and Born–Haber cycles show that such a compound would be expected to disproportionate into Mg and $MgCl_2$. Note that calculations of this type provide only an estimate of the enthalpies of formation of ionic compounds and, hence, some idea of the thermodynamic stability of a compound. It may still be possible to isolate a thermodynamically unstable compound if its decomposition is very slow. Indeed, a compound containing Mg(I) was reported in 2007 (Section 12.13).

3.14 The Kapustinskii equation

Key point: The Kapustinskii equation is used to estimate lattice enthalpies of ionic compounds and to give a measure of the thermochemical radii of the constituent ions.

A.F. Kapustinskii observed that, if the Madelung constants for a number of structures are divided by the number of ions per formula unit, N_{ion}, then approximately the same value is obtained for them all. He also noted that the value so obtained increases with the coordination number. Therefore, because ionic radius also increases with coordination number, the variation in $A/N_{ion}d$ from one structure to another can be expected to be fairly small. This observation led Kapustinskii to propose that there exists a hypothetical rock-salt structure that is energetically equivalent to the true structure of any ionic solid and therefore that the lattice enthalpy can be calculated by using the rock-salt Madelung constant and the appropriate ionic radii for (6,6)-coordination. The resulting expression is called the **Kapustinskii equation**:

$$\Delta H_L^\ominus = \frac{N_{ion} \left| z_A z_B \right|}{d} \left(1 - \frac{d^*}{d} \right) \kappa \tag{3.4}$$

In this equation $\kappa = 1.21 \times 10^5$ kJ pm mol^{-1}.

The Kapustinskii equation can be used to ascribe numerical values to the 'radii' of non-spherical molecular ions, as their values can be adjusted until the calculated value of the lattice enthalpy matches that obtained experimentally from the Born–Haber cycle. The self-consistent parameters obtained in this way are called **thermochemical radii** (Table 3.10). They may be used to estimate lattice enthalpies, and hence enthalpies of formation, of a wide range of compounds without needing to know the structure, assuming that the bonds are essentially ionic.

■ **A brief illustration.** To estimate the lattice enthalpy of potassium nitrate, KNO_3 we need the number of ions per formula unit ($N_{ion} = 2$), their charge numbers, $z(K^+) = +1$, $z(NO_3^-) = -1$, and the sum of their thermochemical radii, 138 pm $+ 189$ pm $= 327$ pm. Then, with $d^* = 34.5$ pm,

$$\Delta H_L^\ominus = \frac{2 \left| (+1)(-1) \right|}{327 \text{ pm}} \times \left(1 - \frac{34.5 \text{ pm}}{327 \text{ pm}} \right) \times \left(1.21 \times 10^5 \text{ kJ pm mol}^{-1} \right)$$
$$= 622 \text{ kJ mol}^{-1}$$

■

3.15 Consequences of lattice enthalpies

The Born–Mayer equation shows that, for a given lattice type (a given value of A), the lattice enthalpy increases with increasing ion charge numbers (as $|z_A z_B|$). The lattice enthalpy also increases as the ions come closer together and the scale of the lattice decreases. Energies that vary as the **electrostatic parameter**, ξ (xi),

$$\xi = \frac{\left| z_A z_B \right|}{d} \tag{3.5}$$

Table 3.10 The thermochemical radii of ions, r/pm

Main-group elements

BeF_4^{2-}	BF_4^-	CO_3^{2-}	NO_3^-	OH^-	
245	228	185	189	140	
		CN^-	NO_3^-	O_2^{2-}	
		182	155	180	
			PO_4^{3-}	SO_4^{2-}	ClO_4^-
			238	230	236
			AsO_4^{3-}	SeO_4^{2-}	
			248	243	
			SbO_4^{3-}	TeO_4^{2-}	IO_4^-
			260	254	249
					IO_3^-
					182

Complex ions				*d-Metal oxoanions*
$[TiCl_6]^{2-}$	$[IrCl_6]^{2-}$	$[SiF_6]^{2-}$	$[GeCl_6]^{2-}$	CrO_4^{2-} MnO_4^-
248	254	194	243	230 240
$[TiBr_6]^{2-}$	$[PtCl_6]^{2-}$	$[GeF_6]^{2-}$	$[SnCl_6]^{2-}$	MoO_4^{2-}
261	259	201	247	254
$[ZrCl_6]^{2-}$			$[PbCl_6]^{2-}$	
247			248	

Source: A.F. Kaputinskii, *Q. Rev. Chem. Soc.*, 1956, **10**, 283.

(which is often written more succinctly as $\xi = z^2/d$) are widely adopted in inorganic chemistry as indicative that an ionic model is appropriate. In this section we consider three consequences of lattice enthalpy and its relation to the electrostatic parameter.

(a) Thermal stabilities of ionic solids

Key point: Lattice enthalpies may be used to explain the chemical properties of many ionic solids, including their thermal decomposition.

The particular aspect we consider here is the temperature needed to bring about thermal decomposition of carbonates (although the arguments can easily be extended to many inorganic solids):

$$MCO_3(s) \rightarrow MO\ (s) + CO_2(g)$$

Magnesium carbonate, for instance, decomposes when heated to about 300°C, whereas calcium carbonate decomposes only if the temperature is raised to over 800°C. The decomposition temperatures of thermally unstable compounds (such as carbonates) increase with cation radius (Table 3.11). In general, large cations stabilize large anions (and vice versa).

Table 3.11 Decomposition data for carbonates*

	$MgCO_3$	$CaCO_3$	$SrCO_3$	$BaCO_3$
$\Delta G^{\ominus}/(\text{kJ mol}^{-1})$	+48.3	+130.4	+183.8	+218.1
$\Delta H^{\ominus}/(\text{kJ mol}^{-1})$	+100.6	+178.3	+234.6	+269.3
$\Delta S^{\ominus}/(\text{kJ mol}^{-1})$	+175.0	+160.6	+171.0	+172.1
$\theta_{\text{decomp}}/°\text{C}$	300	840	1100	1300

*Data are for the reaction $MCO_3(s) \rightarrow MO(s) + CO_2(g)$ at 298 K. θ is the temperature required to reach $p(CO_2) = 1$ bar and has been estimated from the thermodynamic data at 298 K.

The stabilizing influence of a large cation on an unstable anion can be explained in terms of trends in lattice enthalpies. First, we note that the decomposition temperatures of solid inorganic compounds can be discussed in terms of their Gibbs energies of decomposition into specified products. The standard Gibbs energy for the decomposition of a solid, $\Delta G^{\ominus} = \Delta H^{\ominus} - T\Delta S^{\ominus}$, becomes negative when the second term on the right exceeds the first, which is when the temperature exceeds

$$T = \frac{\Delta H^{\ominus}}{\Delta S^{\ominus}} \qquad (3.6)$$

In many cases it is sufficient to consider only trends in the reaction enthalpy, as the reaction entropy is essentially independent of M because it is dominated by the formation of gaseous CO_2. The standard enthalpy of decomposition of the solid is then given by

$$\Delta H^{\ominus} \approx \Delta_{\text{decomp}} H^{\ominus} + \Delta H_L^{\ominus}(MCO_3, s) - \Delta H_L^{\ominus}(MO, s)$$

where $\Delta_{\text{decomp}} H^{\ominus}$ is the standard enthalpy of decomposition of CO_3^{2-} in the gas phase (Fig. 3.49):

$$CO_3^{2-}(g) \rightarrow O^{2-}(g) + CO_2(g)$$

Because $\Delta_{\text{decomp}} H^{\ominus}$ is large and positive, the overall reaction enthalpy is positive (decomposition is endothermic), but it is less strongly positive if the lattice enthalpy of the oxide is markedly greater than that of the carbonate because then $\Delta H_L^{\ominus}(MCO_3, s) - \Delta H_L^{\ominus}(MO, s)$ is negative. It follows that the decomposition temperature will be low for oxides that have relatively high lattice enthalpies compared with their parent carbonates. The compounds for which this is true are composed of small, highly charged cations, such as Mg^{2+}, which explains why a small cation increases the lattice enthalpy of an oxide more than that of a carbonate.

Figure 3.50 illustrates why a small cation has a more significant influence on the change in the lattice enthalpy as the cation size is varied. The *change* in separation is relatively small when the parent compound has a large cation initially. As the illustration shows in an exaggerated way, when the cation is very big, the change in size of the anion barely affects the scale of the lattice. Therefore, with a given unstable polyatomic anion, the lattice enthalpy difference is more significant and favourable to decomposition when the cation is small than when it is large.

The difference in lattice enthalpy between MO and MCO_3 is magnified by a larger charge on the cation as $\Delta H_L^{\ominus} \propto |z_A z_B|/d$. As a result, thermal decomposition of a carbonate will occur at lower temperatures if it contains a higher charged cation. One consequence of this dependence on cation charge is that alkaline earth carbonates (M^{2+}) decompose at lower temperatures than the corresponding alkali metal carbonates (M^+).

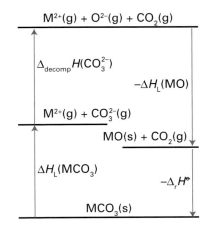

Fig 3.49 A thermodynamic cycle showing the enthalpy changes involved in the decomposition of a solid carbonate MCO_3.

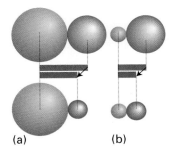

Fig. 3.50 A greatly exaggerated representation of the change in lattice parameter d for cations of different radii. (a) When the anion changes size (as when CO_3^{2-} decomposes into O^{2-} and CO_2, for instance) and the cation is large, the lattice parameter changes by a small amount. (b) If the cation is small, however, the relative change in lattice parameter is large and decomposition is thermodynamically more favourable.

EXAMPLE 3.13 Assessing the dependence of stability on ionic radius

Present an argument to account for the fact that, when they burn in oxygen, lithium forms the oxide Li_2O but sodium forms the peroxide Na_2O_2.

Answer We need to consider the role of the relative sizes of cations and ions in determining the stability of a compound. Because the small Li^+ ion results in Li_2O having a more favourable lattice enthalpy (in comparison with M_2O_2) than Na_2O, the decomposition reaction $M_2O_2(s) \rightarrow M_2O(s) + \frac{1}{2} O_2(g)$ is thermodynamically more favourable for Li_2O_2 than for Na_2O_2.

Self-test 3.13 Predict the order of decomposition temperatures of alkaline earth metal sulfates in the reaction $MSO_4(s) \rightarrow MO(s) + SO_3(g)$.

The use of a large cation to stabilize a large anion that is otherwise susceptible to decomposition, forming a smaller anionic species, is widely used by inorganic chemists to prepare compounds that are otherwise thermodynamically unstable. For example, the interhalogen anions, such as ICl_4^-, are obtained by the oxidation of I^- ions by Cl_2 but are susceptible to decomposition to iodine monochloride and Cl^-:

$$MI(s) + 2Cl_2(g) \rightarrow MICl_4(s) \rightarrow MCl(s) + ICl(g) + Cl_2(g)$$

To disfavour the decomposition, a large cation is used to reduce the lattice enthalpy difference between $MICl_4$ and MCl/MI. The larger alkali metal cations such as K^+, Rb^+, and Cs^+ can be used in some cases, but it is even better to use a really bulky alkylammonium ion, such as $N^tBu_4^+$ (with $^tBu = C(CH_3)_3$).

(b) The stabilities of oxidation states

Key point: The relative stabilities of different oxidation states in solids can often be predicted from considerations of lattice enthalpies.

A similar argument can be used to account for the general observation that high metal oxidation states are stabilized by small anions. In particular, F has a greater ability than the other halogens to stabilize the high oxidation states of metals. Thus, the only known halides of Ag(II), Co(III), and Mn(IV) are the fluorides. Another sign of the decrease in stability of the heavier halides of metals in high oxidation states is that the iodides of Cu(II) and Fe(III) decompose on standing at room temperature (to CuI and FeI_2). Oxygen is also a very effective species for stabilizing the highest oxidation states of elements because of the high charge and small size of the O^{2-} ion.

To explain these observations consider the reaction

$$MX(s) + \tfrac{1}{2} X_2(g) \rightarrow MX_2(s)$$

where X is a halogen. The aim is to show why this reaction is most strongly spontaneous for X = F. If we ignore entropy contributions, we must show that the reaction is most exothermic for fluorine.

One contribution to the reaction enthalpy is the conversion of $\tfrac{1}{2} X_2$ to X^-. Despite F having a lower electron affinity than Cl, this step is more exothermic for X = F than for X = Cl because the bond enthalpy of F_2 is lower than that of Cl_2. The lattice enthalpies, however, play the major role. In the conversion of MX to MX_2, the charge number of the cation increases from +1 to +2, so the lattice enthalpy increases. As the radius of the anion increases, however, this difference in the two lattice enthalpies diminishes, and the exothermic contribution to the overall reaction decreases too. Hence, both the lattice enthalpy and the X^- formation enthalpy lead to a less exothermic reaction as the halogen changes from F to I. Provided entropy factors are similar, which is plausible, we expect an increase in thermodynamic stability of MX relative to MX_2 on going from X = F to X = I down Group 17. Thus many iodides do not exist for metals in their higher oxidation states and compounds such as $Cu^{2+}(I^-)_2$, $Tl^{3+}(I^-)_3$, and VI_5 are unknown, whereas the corresponding fluorides CuF_2, TlF_3, and VF_5 are easily obtained. In effect, the high-oxidation-state metal oxidizes I^- ions to I_2, leading to formation of a lower metal oxidation state such as Cu(I), Tl(I), and V(III) in the iodides of these metals.

(c) Solubility

Key point: The solubilities of salts in water can be rationalized by considering lattice and hydration enthalpies.

Lattice enthalpies play a role in solubilities, as the dissolution involves breaking up the lattice, but the trend is much more difficult to analyse than for decomposition reactions. One rule that is reasonably well obeyed is that *compounds that contain ions with widely different radii are soluble in water*. Conversely, the least water-soluble salts are those of ions with similar radii. That is, in general, *difference in size favours solubility in water*. It is found empirically that an ionic compound MX tends to be very soluble when the radius of M^+ is smaller than that of X^- by about 80 pm.

Two familiar series of compounds illustrate these trends. In gravimetric analysis, Ba^{2+} is used to precipitate SO_4^{2-}, and the solubilities of the Group 2 sulfates decrease from $MgSO_4$ to $BaSO_4$. In contrast, the solubility of the Group 2 hydroxides increases down the group: $Mg(OH)_2$ is the sparingly soluble 'milk of magnesia' but $Ba(OH)_2$ can be used as a soluble hydroxide for preparation of solutions of OH^-. The first case shows that a large anion requires a large cation for precipitation. The second case shows that a small anion requires a small cation for precipitation.

Before attempting to rationalize the observations, we should note that the solubility of an ionic compound depends on the standard reaction Gibbs energy for

$$MX(s) \rightarrow M^+(aq) + X^-(aq)$$

In this process, the interactions responsible for the lattice enthalpy of MX are replaced by hydration (and by solvation in general) of the ions. However, the exact balance of enthalpy and entropy effects is delicate and difficult to assess, particularly because the entropy change also depends on the degree of order of the solvent molecules that is brought about by the presence of the dissolved solute. The data in Fig. 3.51 suggest that enthalpy considerations are important in some cases at least, as the graph shows that there is a correlation between the enthalpy of solution of a salt and the difference in hydration enthalpies of the two ions. If the cation has a larger hydration enthalpy than its anion partner (reflecting the difference in their sizes) or vice versa, then the dissolution of the salt is exothermic (reflecting the favourable solubility equilibrium).

The variation in enthalpy can be explained using the ionic model. The lattice enthalpy is inversely proportional to the distance between the centres of the ions:

$$\Delta H_L^{\ominus} \propto \frac{1}{r_+ + r_-}$$

However, the hydration enthalpy, with each ion being hydrated individually, is the sum of individual ion contributions:

$$\Delta_{hyd}H \propto \frac{1}{r_+} + \frac{1}{r_-}$$

If the radius of one ion is small, the term in the hydration enthalpy for that ion will be large. However, in the expression for the lattice enthalpy one small ion cannot make the denominator of the expression small by itself. Thus, one small ion can result in a large hydration enthalpy but not necessarily lead to a high lattice enthalpy, so ion size asymmetry can result in exothermic dissolution. If both ions are small, then both the lattice enthalpy and the hydration enthalpy may be large, and dissolution might not be very exothermic.

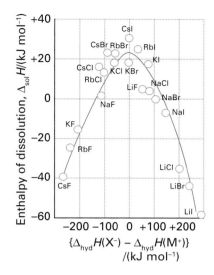

Fig. 3.51 The correlation between enthalpies of solution of halides and the differences between the hydration enthalpies of the ions. Dissolution is most exothermic when the difference is large.

EXAMPLE 3.14 **Accounting for trends in the solubility of s-block compounds**

What is the trend in the solubilities of the Group 2 metal carbonates?

Answer We need to consider the role of the relative sizes of cations and anions. The CO_3^{2-} anion has a large radius and has the same magnitude of charge as the cations M^{2+} of the Group 2 elements. The least soluble carbonate of the group is predicted to be that of the largest cation, Ra^{2+}. The most soluble is expected to be the carbonate of the smallest cation, Mg^{2+}. (Beryllium has too much covalent character in its bonding for it to be included in this analysis.) Although magnesium carbonate is more soluble than radium carbonate, it is still only sparingly soluble: its solubility constant (its solubility product, K_{sp}) is only 3×10^{-8}.

Self-test 3.14 Which can be expected to be more soluble in water, $NaClO_4$ or $KClO_4$?

Defects and nonstoichiometry

Key points: Defects, vacant sites, and misplaced atoms are a feature of all solids as their formation is thermodynamically favourable.

All solids contain **defects**, or imperfections of structure or composition. Defects are important because they influence properties such as mechanical strength, electrical conductivity, and chemical reactivity. We need to consider both **intrinsic defects**, which are defects that occur in the pure substance, and **extrinsic defects**, which stem from the presence of impurities. It is also common to distinguish **point defects**, which occur at single sites, from **extended defects**, which are ordered in one, two, and three dimensions. Point defects are random errors in a periodic lattice, such as the absence of an atom at its usual site or the presence of an atom at a site that is not normally occupied. Extended defects involve various irregularities in the stacking of the planes of atoms.

3.16 The origins and types of defects

Solids contain defects because they introduce disorder into an otherwise perfect structure and hence increase its entropy. The Gibbs energy, $G = H - TS$, of a solid with defects has contributions from the enthalpy and the entropy of the sample. The formation of defects is normally endothermic because, as the lattice is disrupted, the enthalpy of the solid rises. However, the term $-TS$ becomes more negative as defects are formed because they introduce disorder into the lattice and the entropy rises. Provided $T > 0$, therefore, the Gibbs energy will have a minimum at a nonzero concentration of defects and their formation will be spontaneous (Fig. 3.52a). Moreover, as the temperature is raised, the minimum in G shifts to higher defect concentrations (Fig. 3.52b), so solids have a greater number of defects as their melting points are approached.

(a) Intrinsic point defects

Key points: Schottky defects are site vacancies, formed in cation/anion pairs, and Frenkel defects are displaced, interstitial atoms; the structure of a solid influences the type of defect that occurs, with Frenkel defects forming in solids with lower coordination numbers and more covalency and Schottky defects in more ionic materials.

The solid-state physicists W. Schottky and J. Frenkel identified two specific types of point defect. A **Schottky defect** (Fig. 3.53) is a vacancy in an otherwise perfect arrangement of atoms or ions in a structure. That is, it is a point defect in which an atom or ion is missing from its normal site in the structure. The overall stoichiometry of a solid is not affected by the presence of Schottky defects because, to ensure charge balance, the defects occur in pairs in a compound of stoichiometry MX and there are equal numbers of vacancies at cation and anion sites. In solids of different composition, for example MX_2, the defects must occur with balanced charges, so two anion vacancies must be created for each cation lost. Schottky defects occur at low concentrations in purely ionic solids, such as NaCl; they occur most commonly in structures with high coordination numbers, such as close-packed metals, where the enthalpy penalty of reducing the average coordination number of the remaining atoms (from 12 to 11, for instance) is relatively low.

A **Frenkel defect** (Fig. 3.54) is a point defect in which an atom or ion has been displaced onto an interstitial site. For example, in silver chloride, which has the rock-salt structure, a small number of Ag^+ ions reside in tetrahedral sites (**1**). The stoichiometry of the compound is unchanged when a Frenkel defect forms and it is possible to have Frenkel defects involving either one (M or X displaced) or both (some M and some X interstitials) of the ion types in a binary compound, MX. Thus the Frenkel defects that occur in, for example, PbF_2 involve the displacement of a small number of F^- ions from their normal sites in the fluorite structure, on the tetrahedral holes in the close-packed Pb^{2+} ion array, to sites that

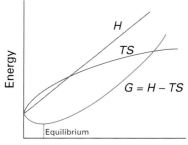

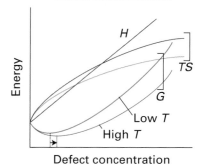

Fig. 3.52 (a) The variation of the enthalpy and entropy of a crystal as the number of defects increases. The resulting Gibbs energy $G = H - TS$ has a minimum at a nonzero concentration, and hence defect formation is spontaneous. (b) As the temperature is increased, the minimum in the Gibbs energy moves to higher defect concentrations, so more defects are present at equilibrium at higher temperatures than at low temperatures.

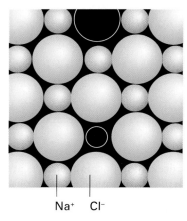

Fig. 3.53 A Schottky defect is the absence of ions on normally occupied sites; for charge neutrality there must be equal numbers of cation and anion vacancies in a 1:1 compound.

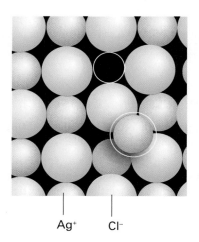

Fig. 3.54 A Frenkel defect forms when an ion moves to an interstitial site.

correspond to the octahedral holes. A useful generalization is that Frenkel defects are most often encountered in structures such as wurtzite and sphalerite in which coordination numbers are low (6 or less) and the more open structure provides sites that can accommodate the interstitial atoms. This is not to say that Frenkel defects are exclusive to such structures; as we have seen, the (8,4)-coordination fluorite structure can accommodate such interstitials although some local repositioning of adjacent anions is required to allow for the presence of the displaced anion.

The concentration of Schottky defects varies considerably from one type of compound to the next. The concentration of vacancies is very low in the alkali metal halides, being of the order of $10^6\,\text{cm}^{-3}$ at 130°C. That concentration corresponds to about one defect per 10^{14} formula units. Conversely, some d-metal oxides, sulfides, and hydrides have very high concentrations of vacancies. An extreme example is the high-temperature form of TiO, which has vacancies on both the cation and anion sites at a concentration corresponding to about one defect per 10 formula units.

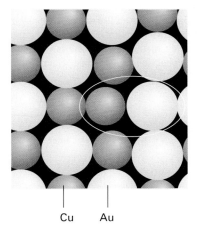

1 Interstitial Ag^+

EXAMPLE 3.15 Predicting defect types

What type of intrinsic defect would you expect to find in (a) MgO and (b) CdTe?

Answer The type of defect formed depends on factors such as the coordination numbers and the level of covalency in the bonding with high coordination numbers and ionic bonding favouring Schottky defects and low coordination numbers and partial covalency in the bonding favouring Frenkel defects. (a) MgO has the rock-salt structure and the ionic bonding in this compound generally favours Schottky defects. (b) CdTe adopts the wurtzite structure with (4,4)-coordination, favouring Frenkel defects.

Self-test 3.15 Predict the most likely type of intrinsic defects for CsF.

Schottky and Frenkel defects are only two of the many possible types of defect. Another type is an **atom-interchange** or **anti-site defect**, which consists of an interchanged pair of atoms. This type of defect is common in metal alloys with exchange of neutral atoms. It is expected to be very unfavourable for binary ionic compounds on account of the introduction of strongly repulsive interactions between neighbouring similarly charged ions. For example, a copper/gold alloy of exact overall composition CuAu has extensive disorder at high temperatures, with a significant fraction of Cu and Au atoms interchanged (Fig. 3.55). The interchange of similarly charged species on different sites in ternary and compositionally more complex compounds is common; thus in spinels (Section 24.8) the partial swapping of the metal ions between tetrahedral and octahedral sites is often observed.

Both Schottky and Frenkel defects are stoichiometry defects in that they do not change the overall composition of the material because the vacancies occur in charge-balanced pairs (Schottky) or each interstitial is derived from one displaced atom or ion (Frenkel, a vacancy-interstitial pair). Similar types of defects, vacancies, and interstitials occur in many inorganic materials and may be balanced by changes in the oxidation number of one component in the system rather than by their creation as charge-balanced pairs. This behaviour, as seen for example in $La_2CuO_{4.1}$ with extra interstitial O^{2-} ions, is discussed more fully in Section 24.8.

Fig. 3.55 Atom exchange can also give rise to a point defect as in CuAu.

Cu   Au

(b) Extrinsic point defects

Key point: Extrinsic defects are defects introduced into a solid as a result of doping with an impurity atom.

Extrinsic defects, those resulting from the presence of impurities, are inevitable because perfect purity is unattainable in practice in crystals of any significant size. Such behaviour is commonly seen in naturally occurring minerals. Thus, the incorporation of low levels of Cr into the Al_2O_3 structure produces the gemstone ruby, whereas replacement of some Al by Fe and Ti results in the blue gemstone sapphire (Box 3.3). The dopant species normally has a similar atomic or ionic radius to the species which it replaces. Thus Cr^{3+} in ruby and Fe^{3+} in sapphire have similar ionic radii to Al^{3+}. Impurities can also be introduced intentionally by doping one material with another; an example is the introduction of As into Si to modify the latter's semiconducting properties. Synthetic equivalents of ruby and sapphire can also be synthesized easily in the laboratory by incorporating small levels of Cr, Fe, and Ti into the Al_2O_3 structure.

BOX 3.3 Defects and gemstones

Defects and dopant ions are responsible for the colours of many gemstones. Whereas aluminium oxide (Al_2O_3), silica (SiO_2), and fluorite (CaF_2) in their pure forms are colourless, brightly coloured materials may be produced by substituting in small levels of dopant ions or producing vacant sites that trap electrons. The impurities and defects are often found in naturally occurring minerals on account of the geological and environmental conditions under which they are formed. For example, d-metal ions are often present in the solutions from which the gemstones grew and the presence of ionizing radiation from radioactive species in the natural environment generates electrons that become trapped in their structure.

The most common origin of colour in a gemstone is a d-metal ion dopant (see table). Thus ruby is Al_2O_3 containing around $0.2-1$ atom per cent Cr^{3+} ions in place of the Al^{3+} ions and its red colour results from the absorption of green light in the visible spectrum as a result of the excitation of Cr3d electrons (Section 20.4). The same ion is responsible for the green of emeralds; the different colour reflects a different local coordination environment of the dopant. The host structure is beryl, beryllium aluminium silicate, $Be_3Al_2(SiO_3)_6$, and the Cr^{3+} ion is surrounded by six silicate ions, rather than the six O^{2-} ions in ruby, producing absorption at a different energy. Other d-metal ions are responsible for the colours of other gemstones. Iron(II) produces the red of garnets and the yellow-green of peridots. Manganese(II) is responsible for the pink colour of tourmaline.

In ruby and emerald the colour is caused by excitation of electrons on a single dopant d-metal ion, Cr^{3+}. When more than one dopant species, which may be of different type or oxidation state, is present it is possible to transfer an electron between them. One example of this behaviour is sapphire. Sapphire, like ruby, is alumina but in this gemstone some adjacent pairs of Al^{3+} ions are replaced by Fe^{2+} and Ti^{4+} pairs. This material absorbs visible radiation in the yellow as an electron is transferred from Fe^{2+} to Ti^{4+}, so producing a brilliant blue colour (the complementary colour of yellow).

In other gemstones and minerals, colour is a result of doping a host structure with a species that has a different charge from the ion that it replaces or by the presence of a vacancy (Schottky-type defect). In both cases a colour-centre or F-centre (F from the German word *farbe* for colour) is formed. As the charge at an F-centre is different from that of a normally occupied site in the same structure, it can easily supply an electron to or receive an electron from another ion. This electron can then be excited by absorbing visible light, so producing colour. For instance, in purple fluorite, CaF_2, an F-centre is formed from a vacancy on a normally occupied F^- ion site. This site then traps an electron, generated by exposure of the mineral to ionizing radiation in the natural environment. Excitation of the electron, which acts like a particle in a box, absorbs visible light in the wavelength range $530-600$ nm, producing the violet/purple colours of this mineral.

In amethyst, the purple derivative of quartz, SiO_2, some Si^{4+} ions are substituted by Fe^{3+} ions. This replacement leaves a hole (one missing electron) and excitation of this hole, by ionizing radiation for instance, traps it by forming Fe^{4+} or O^- in the quartz matrix. Further excitation of the electrons in this material now occurs by the absorption of visible light at 540 nm, producing the observed purple colour. If an amethyst crystal is heated to 450°C the hole is freed from its trap. The colour of the crystal reverts to that typical of iron-doped silica and is a characteristic of the yellow semi-precious gemstone citrine. If citrine is irradiated the trapped-hole is regenerated and the original colour restored.

Colour centres can also be produced by nuclear transformations. An example of such a transformation is the β-decay of ^{14}C in a diamond. This decay produces a ^{14}N atom, with an additional valence electron, embedded in the diamond structure. The electron energy levels associated with these N atoms allow absorption in the visible region of the spectrum and produce the colouration of blue and yellow diamonds.

Table B3.1 Gemstones and the origin of their colours

Mineral or gemstone	Colour	Parent formula	Dopant or defect responsible for the colour
Ruby	Red	Al_2O_3	Cr^{3+} replacing Al^{3+} in octahedral sites
Emerald	Green	$Be_3Al_2(SiO_3)_6$	Cr^{3+} replacing Al^{3+} in octahedral sites
Tourmaline	Green or pink	$Na_3Li_3Al_6(BO_3)_3(SiO_3)_6F_4$	Cr^{3+} or Mn^{2+} replacing Li^+ and Al^{3+} in octahedral sites respectively
Garnet	Red	$Mg_3Al_2(SiO_4)_3$	Fe^{2+} replacing Mg^{2+} in 8-coordinate sites
Peridot	Yellow-green	Mg_2SiO_4	Fe^{2+} replacing Mg^{2+} in 6-coordinate sites
Sapphire	Blue	Al_2O_3	Electron transfer between Fe^{2+} and Ti^{4+} replacing Al^{3+} in adjacent octahedral sites
Diamond	Colourless, pale blue or yellow	C	Colour centres from N
Amethyst	Purple	SiO_2	Colour centre based on Fe^{3+}/Fe^{4+}
Fluorite	Purple	CaF_2	Colour centre based on trapped electron

When the dopant species is introduced into the host the latter's structure remains essentially unchanged. If attempts are made to introduce high levels of the dopant species a new structure incorporating all the elements present often forms or the dopant species is not incorporated. This behaviour usually limits the level of extrinsic point defects to low levels.

The composition of ruby is typically $(Al_{0.998}Cr_{0.002})_2O_3$, with 0.2 per cent of metal sites as extrinsic Cr^{3+} dopant ions. Some solids may tolerate much higher levels of defects (Section 3.17a).

Dopants often modify the electronic structure of the solid. Thus, when an As atom replaces an Si atom, the additional electron from each As atom enters the conduction band. In the more ionic substance ZrO_2, the introduction of Ca^{2+} impurities in place of Zr^{4+} ions is accompanied by the formation of an O^{2-} ion vacancy to maintain charge neutrality (Fig. 3.56).

Another example of an extrinsic point defect is a **colour centre**, a generic term for defects responsible for modifications to the IR, visible, and UV absorption characteristics of solids that have been irradiated or exposed to chemical treatment. One type of colour centre is produced by heating an alkali metal halide crystal in the vapour of the alkali metal and gives a material with a colour characteristic of the system: NaCl becomes orange, KCl violet, and KBr blue-green. The process results in the introduction of an alkali metal cation at a normal cation site and the associated electron from the metal atom occupies a halide ion vacancy. A colour centre consisting of an electron in a halide ion vacancy is called an **F-centre** (Fig. 3.57).[5] The colour results from the excitation of the electron in the localized environment of its surrounding ions. An alternative method of producing F-centres involves exposing a material to an X-ray beam that ionizes electrons into anion vacancies. F-centres and extrinsic defects are important in producing colour in gemstones, Box 3.3.

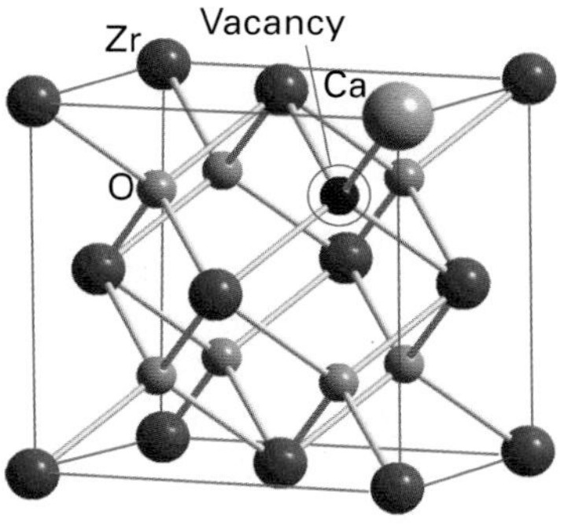

Fig. 3.56 Introduction of a Ca^{2+} ion into the ZrO_2 lattice produces a vacancy on the O^{2-} sublattice. This substitution helps to stabilize the cubic fluorite structure for ZrO_2.

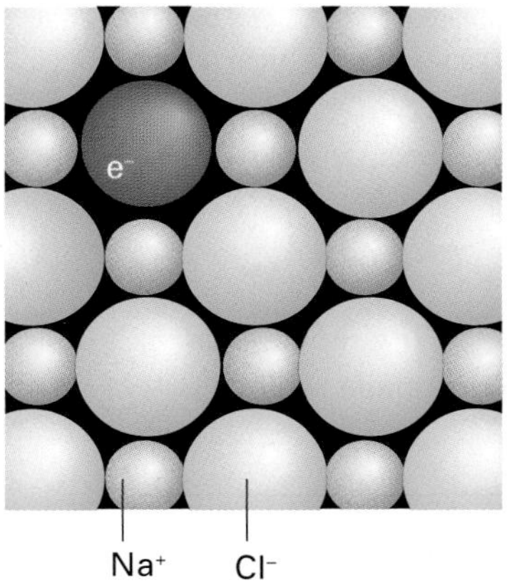

EXAMPLE 3.16 Predicting possible dopant ions

What ions might substitute for Al^{3+} in beryl, $Be_3Al_2(SiO_3)_6$, forming extrinsic defects?

Answer We need to identify ions of similar charge and size. Ionic radii are listed in *Resource section* 1. Triply charged cations with ionic radii similar to Al^{3+} ($r = 53$ pm) should prove to be suitable dopant ions. Candidates could be Fe^{3+} ($r = 55$ pm), Mn^{3+} ($r = 65$ pm), and Cr^{3+} ($r = 62$ pm). Indeed when the extrinsic defect is Cr^{3+} the material is a bright green beryl, the gemstone emerald. For Mn^{3+} the material is a red or pink beryl and for Fe^{3+} it is the yellow beryl heliodor.

Self-test 3.16 What elements other than As might be used to form extrinsic defects in silicon?

3.17 Nonstoichiometric compounds and solid solutions

The statement that the stoichiometry of a compound is fixed by its chemical formula is not always true for solids: differences in the composition of unit cells can occur throughout a solid, perhaps because there are defects at one or more atom sites, interstitial atoms are present, or substitutions have occurred at one position.

(a) Nonstoichiometry

Key point: Deviations from ideal stoichiometry are common in the solid-state compounds of the d-, f-, and later p-block elements.

A **nonstoichiometric compound** is a substance that exhibits variable composition but retains the same structure type. For example, at 1000°C the composition of 'iron monoxide', which is sometimes referred to as wüstite, $Fe_{1-x}O$, varies from $Fe_{0.89}O$ to $Fe_{0.96}O$. Gradual changes in the size of the unit cell occur as the composition is varied, but all the features of the rock-salt structure are retained throughout this composition range. The fact that the lattice parameter of the compound varies smoothly with composition is a defining criterion of a nonstoichiometric compound because a discontinuity in the value of the lattice parameter indicates the formation of a new crystal phase. Moreover, the thermodynamic properties of nonstoichiometric compounds also vary continuously as the composition changes. For example, as the partial pressure of oxygen above a metal oxide is varied, both the lattice parameter and the equilibrium composition of the oxide change continuously (Figs 3.58 and 3.59). The gradual change in the lattice parameter of a solid as a function of its composition is known as **Vegard's rule**.

Table 3.12 lists some representative nonstoichiometric hydrides, oxides, and sulfides. Note that as the formation of a nonstoichiometric compound requires overall changes in

Fig. 3.57 An F-centre is an electron that occupies an anion vacancy. The energy levels of the electron resemble those of a particle in a three-dimensional square well.

---

[5] The name comes from the German word for colour centre, *Farbenzentrum*.

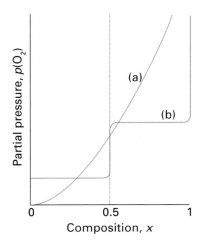

Fig. 3.58 Schematic representation of the variation of the partial pressure of oxygen with composition at constant pressure for (a) a nonstoichiometric oxide MO_{1+x} and (b) a stoichiometric pair of metal oxides MO and MO_2. The x-axis is the atom ratio in MO_{1+x}.

Table 3.12   Representative composition ranges* of nonstoichiometric binary hydrides, oxides, and sufides

d block			f block			
Hydrides						
TiH_x	1−2				Fluorite type	Hexagonal
ZrH_x	1.5−1.6		GdH_x		1.9−2.3	2.85−3.0
HfH_x	1.7−1.8		ErH_x		1.95−2.31	2.82−3.0
NbH_x	0.64−1.0		LuH_x		1.85−2.23	1.74−3.0
Oxides						
	Rock-salt type	Rutile type				
TiO_x	0.7−1.25	1.9−2.0				
VO_x	0.9−1.20	1.8−2.0				
NbO_x	0.9−1.04					
Sulfides						
ZrS_x	0.9−1.0					
YS_x	0.9−1.0					

* Expressed as the range of values x may take.

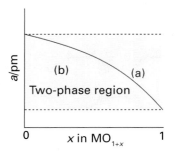

Fig. 3.59 Schematic representation of the variation of one lattice parameter with composition for (a) a nonstoichiometric oxide MO_{1+x} and (b) a stoichiometric pair of metal oxides MO and MO_2 with no intermediate stoichiometric phases (which would produce a two-phase mixture for $0 < x < 1$, each phase in the mixture having the lattice parameter of the end member.

composition, it also requires at least one element to exist in more than one oxidation state. Thus in wüstite, $Fe_{1−x}O$, as x increases some iron(II) must be oxidized to iron(III) in the structure. Hence deviations from stoichiometry are usual only for d- and f-block elements, which commonly adopt two or more oxidation states, and for some heavy p-block metals that have two accessible oxidation states.

(b) Solid solutions

Key point: A solid solution occurs where there is a continuous variation in compound stoichiometry without a change in structural type.

Because many substances adopt the same structural type, it is often energetically feasible to replace one type of atom or ion with another. Such behaviour is seen in many simple metal alloys such as those discussed in Section 3.8. Thus zinc/copper brasses exist for the complete range of compositions $Cu_{1−x}Zn_x$ with $0 < x < 0.38$, where Cu atoms in the structure are gradually replaced by Zn atoms. This replacement occurs randomly throughout the solid, and individual unit cells contain an arbitrary number of Cu and Zn atoms (but such that the sum of their contents gives the overall brass stoichiometry).

Another good example is the perovskite structure adopted by many compounds of stoichiometry ABX_3 (Section 3.9), in which the composition can be varied continuously by varying the ions that occupy the A, B, and X sites. For instance, both $LaFeO_3$ and $SrFeO_3$ adopt the perovskite structure and we can consider a perovskite crystal that has, randomly distributed, half $SrFeO_3$ unit cells (with Sr on the A-type cation site) and half $LaFeO_3$ unit cells (with La on the A-site). The overall compound stoichiometry is $LaSrFe_2O_6$, which is better written $(La_{0.5}Sr_{0.5})FeO_3$, to reflect the normal ABO_3 perovskite stoichiometry. Other proportions of these unit cells are possible and the series of compounds $La_{1−x}Sr_xFeO_3$ for $0 \leq x \leq 1$ can be prepared. This system is called a **solid solution** because all the phases formed as x is varied have the same perovskite structure. A solid solution occurs when there is a single structural type for a range of compositions and there is a smooth variation in lattice parameter over that range.

Solid solutions occur most frequently for d-metal compounds because the change in one component might require a change in the oxidation state of another component to preserve the charge balance. Thus, as x increases in $La_{1−x}Sr_xFeO_3$ and La(III) is replaced by Sr(II), the oxidation state of iron must change from Fe(III) to Fe(IV). This change can occur through a gradual replacement of one exact oxidation state, here Fe(III), by another, Fe(IV), on a proportion of the cation sites within the structure. Alternatively, if the material is metallic and has delocalized electrons, then the change can be accommodated by altering the number of electrons in a conduction band, which corresponds to the delocalization of the change in oxidation state rather than its identification with individual atoms. Some other solid solutions include the high-temperature superconductors of composition $La_{2-x}Ba_xCuO_4$ ($0 \le x \le 0.4$), which are superconducting for $0.12 \le x \le 0.25$, and the spinels $Mn_{1-x}Fe_{2+x}O_4$ ($0 \le x \le 1$). It is also possible to combine solid-solution behaviour on a cation site with nonstoichiometry caused by defects on a different ion site. An example is the system $La_{1-x}Sr_xFeO_{3-y}$, with $0 \le x \le 1.0$ and $0.0 \le y \le 0.5$, which has vacancies on the O^{2-} ion sites.

The electronic structures of solids

The previous sections have introduced concepts associated with the structures and energetics of ionic solids in which it was necessary to consider almost infinite arrays of ions and the interactions between them. Similarly, an understanding of the electronic structures of solids, and the derived properties such as electric conductivity, magnetism, and many optical effects, needs to consider the interactions of electrons with each other and extended arrays of ions. One simple approach is to regard a solid as a single huge molecule and to extend the ideas of molecular orbital theory introduced in Chapter 2 to very large numbers of orbitals. Similar concepts are used in later chapters to understand other key properties of large three-dimensional arrays of electronically interacting centres such as ferromagnetism, superconductivity, and the colours of solids.

3.18 The conductivities of inorganic solids

Key points: A metallic conductor is a substance with an electric conductivity that decreases with increasing temperature; a semiconductor is a substance with an electric conductivity that increases with increasing temperature.

The molecular orbital theory of small molecules can be extended to account for the properties of solids, which are aggregations of an almost infinite number of atoms. This approach is strikingly successful for the description of metals; it can be used to explain their characteristic lustre, their good electrical and thermal conductivity, and their malleability. All these properties arise from the ability of the atoms to contribute electrons to a common 'sea'. The lustre and electrical conductivities stem from the mobility of these electrons in response to either the oscillating electric field of an incident ray of light or to a potential difference. The high thermal conductivity is also a consequence of electron mobility because an electron can collide with a vibrating atom, pick up its energy, and transfer it to another atom elsewhere in the solid. The ease with which metals can be mechanically deformed is another aspect of electron mobility because the electron sea can quickly readjust to a deformation of the solid and continue to bind the atoms together.

Electronic conduction is also a characteristic of semiconductors. The criterion for distinguishing between a metallic conductor and a semiconductor is the temperature dependence of the electric conductivity (Fig. 3.60):

A **metallic conductor** is a substance with an electric conductivity that *decreases* with increasing temperature.
A **semiconductor** is a substance with an electric conductivity that increases with increasing temperature.

It is also generally the case (but not the criterion for distinguishing them) that the conductivities of metals at room temperature are higher than those of semiconductors. Typical values are given in Fig. 3.60. A solid **insulator** is a substance with a very low electrical conductivity. However, when that conductivity can be measured, it is found to increase with temperature, like that of a semiconductor. For some purposes, therefore, it is possible to disregard the classification 'insulator' and to treat all solids as either metals or semiconductors. **Superconductors** are a special class of materials that have zero electrical resistance below a critical temperature.

3.19 Bands formed from overlapping atomic orbitals

The central idea underlying the description of the electronic structure of solids is that the valence electrons supplied by the atoms spread through the entire structure. This concept is expressed more formally by making a simple extension of MO theory in which the solid

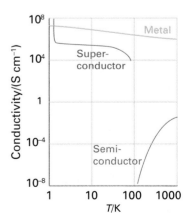

Fig. 3.60 The variation of the electrical conductivity of a substance with temperature is the basis of the classification of the substance as a metallic conductor, a semiconductor, or a superconductor.

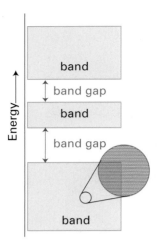

Fig. 3.61 The electronic structure of a solid is characterized by a series of bands of orbitals separated by gaps at energies where orbitals do not occur.

is treated like an indefinitely large molecule. In solid-state physics, this approach is called the **tight-binding approximation**. The description in terms of delocalized electrons can also be used to describe nonmetallic solids. We therefore begin by showing how metals are described in terms of molecular orbitals. Then we go on to show that the same principles can be applied, but with a different outcome, to ionic and molecular solids.

(a) Band formation by orbital overlap

Key point: The overlap of atomic orbitals in solids gives rise to bands of energy levels separated by energy gaps.

The overlap of a large number of atomic orbitals in a solid leads to a large number of molecular orbitals that are closely spaced in energy and so form an almost continuous **band** of energy levels (Fig. 3.61). Bands are separated by **band gaps**, which are values of the energy for which there is no molecular orbital.

The formation of bands can be understood by considering a line of atoms, and supposing that each atom has an s orbital that overlaps the s orbitals on its immediate neighbours (Fig. 3.62). When the line consists of only two atoms, there is a bonding and an antibonding molecular orbital. When a third atom joins them, there are three molecular orbitals. The central orbital of the set is nonbonding and the outer two are at low energy and high energy, respectively. As more atoms are added, each one contributes an atomic orbital, and hence one more molecular orbital is formed. When there are N atoms in the line, there are N molecular orbitals. The orbital of lowest energy has no nodes between neighbouring atoms. The orbital of highest energy has a node between every pair of neighbours. The remaining orbitals have successively 1, 2,...internuclear nodes and a corresponding range of energies between the two extremes.

The total width of the band, which remains finite even as N approaches infinity (as shown in Fig. 3.63), depends on the strength of the interaction between neighbouring atoms. The greater the strength of interaction (in broad terms, the greater the degree of overlap between neighbours), the greater the energy separation of the non-node orbital and the all-node orbital. However, whatever the number of atomic orbitals used to form the molecular orbitals, there is only a finite spread of orbital energies (as depicted in Fig. 3.63). It follows that the separation in energy between neighbouring orbitals must approach zero as N approaches infinity, otherwise the range of orbital energies could not be finite. That is, a band consists of a countable number but near-continuum of energy levels.

The band just described is built from s orbitals and is called an **s band**. If there are p orbitals available, a **p band** can be constructed from their overlap as shown in Fig. 3.64.

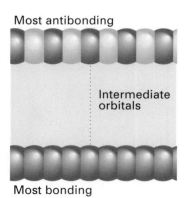

Fig. 3.62 A band can be thought of as formed by bringing up atoms successively to form a line of atoms. N atomic orbitals give rise to N molecular orbitals.

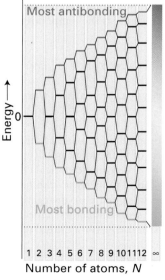

Fig. 3.63 The energies of the orbitals that are formed when N atoms are brought up to form a one-dimensional array.

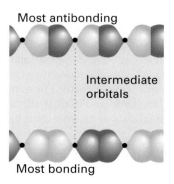

Fig. 3.64 An example of a p band in a one-dimensional solid.

Because p orbitals lie higher in energy than s orbitals of the same valence shell, there is often an energy gap between the s band and the p band (Fig. 3.65). However, if the bands span a wide range of energy and the atomic s and p energies are similar (as is often the case), then the two bands overlap. The d **band** is similarly constructed from the overlap of d orbitals. The formation of bands is not restricted to one type of atomic orbital and bands may be formed in compounds by combinations of different orbital types, for example the d orbitals of a metal atom may overlap the p orbitals of neighbouring O atoms.

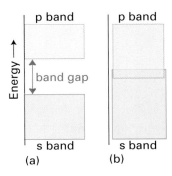

Fig. 3.65 (a) The s and p bands of a solid and the gap between them. Whether or not there is in fact a gap depends on the separation of the s and p orbitals of the atoms and the strength of the interaction between them in the solid. (b) If the interaction is strong, the bands are wide and may overlap.

EXAMPLE 3.17 Identifying orbital overlap

Decide whether any d orbitals on titanium in TiO (with the rock-salt structure) can overlap to form a band.

Answer We need to decide whether there are d orbitals on neighbouring metal atoms that can overlap with one another. Figure 3.66 shows one face of the rock-salt structure with the d_{xy} orbital drawn in on each of the Ti atoms. The lobes of these orbitals point directly towards each other and will overlap to give a band. In a similar fashion the d_{zx} and d_{yz} orbitals overlap in the directions perpendicular to the xz and yz faces.

Self-test 3.17 Which d orbitals can overlap in a metal having a primitive structure?

(b) The Fermi level

Key point: The Fermi level is the highest occupied energy level in a solid at $T = 0$.

At $T = 0$, electrons occupy the individual molecular orbitals of the bands in accordance with the building-up principle. If each atom supplies one s electron, then at $T = 0$ the lowest $\frac{1}{2}N$ orbitals are occupied. The highest occupied orbital at $T = 0$ is called the **Fermi level**; it lies near the centre of the band (Fig. 3.67). When the band is not completely full, the electrons close to the Fermi level can easily be promoted to nearby empty levels. As a result, they are mobile and can move relatively freely through the solid, and the substance is an electrical conductor.

The solid is in fact a *metallic* conductor. We have seen that the criterion of metallic conduction is the decrease of electrical conductivity with increasing temperature. This behaviour is the opposite of what we might expect if the conductivity were governed by thermal promotion of electrons above the Fermi level. The competing effect can be identified once we recognize that the ability of an electron to travel smoothly through the solid in a conduction band depends on the uniformity of the arrangement of the atoms. An atom vibrating vigorously at a site is equivalent to an impurity that disrupts the orderliness of the orbitals. This decrease in uniformity reduces the ability of the electron to travel from one edge of the solid to the other, so the conductivity of the solid is less than at $T = 0$. If we think of the electron as moving through the solid, then we would say that it was 'scattered' by the atomic vibration. This carrier scattering increases with increasing temperature as the lattice vibrations increase, and the increase accounts for the observed inverse temperature dependence of the conductivity of metals.

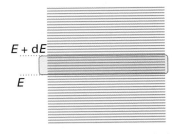

Fig. 3.66 One face of the TiO rock-salt structure showing how orbital overlap can occur for the d_{xy}, d_{yz}, and d_{zx} orbitals.

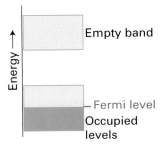

Fig. 3.67 If each of the N atoms supplies one s electron, then at $T = 0$ the lower $\frac{1}{2}N$ orbitals are occupied and the Fermi level lies near the centre of the band.

(c) Densities of states

Key point: The density of states is not uniform across a band: in most cases, the states are densest close to the centre of the band.

The number of energy levels in an energy range divided by the width of the range is called the **density of states**, ρ (Fig. 3.68). The density of states is not uniform across a band because the energy levels are packed together more closely at some energies than at others. This variation is apparent even in one dimension, for—compared with its edges—the centre of the band is relatively sparse in orbitals (as can be seen in Fig. 3.63). In three dimensions, the variation of density of states is more like that shown in Fig. 3.69, with the greatest density of states near the centre of the band and the lowest density at the edges. The reason for this behaviour can be traced to the number of ways of producing a particular linear combination of atomic orbitals. There is only one way of forming a fully bonding molecular orbital (the lower edge of the band) and only one way of forming a fully antibonding orbital (the upper edge). However, there are many ways (in a three-dimensional array of atoms) of forming a molecular orbital with an energy corresponding to the interior of a band.

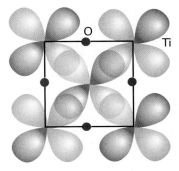

Fig. 3.68 The density of states is the number of energy levels in an infinitesimal range of energies between E and $E + dE$.

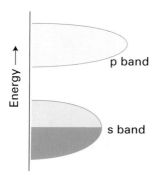

Fig. 3.69 Typical densities of states for two bands in a three-dimensional metal.

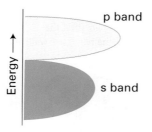

Fig. 3.70 The densities of states in a semimetal.

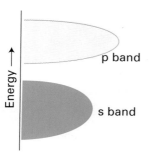

Fig. 3.71 The structure of a typical insulator: there is a significant gap between the filled and empty bands.

Table 3.13   Some typical band gaps at 298 K

Material	E_g/eV
Carbon (diamond)	5.47
Silicon carbide	3.00
Silicon	1.11
Germaniun	0.66
Gallium arsenide	1.35
Indium arsenide	0.36

The density of states is zero in the band gap itself—there is no energy level in the gap. In certain special cases, however, a full band and an empty band might coincide in energy but with a zero density of states at their conjunction (Fig. 3.70). Solids with this band structure are called **semimetals**. One important example is graphite, which is a semimetal in directions parallel to the sheets of carbon atoms.

A note on good practice This use of the term 'semimetal' should be distinguished from its other use as a synonym for metalloid. In this text we avoid the latter usage.

(d) Insulators

Key point: A solid insulator is a semiconductor with a large band gap.

A solid is an insulator if enough electrons are present to fill a band completely and there is a considerable energy gap before an empty orbital becomes available (Fig. 3.71). In a sodium chloride crystal, for instance, the $N\,Cl^-$ ions are nearly in contact and their 3s and three 3p valence orbitals overlap to form a narrow band consisting of $4N$ levels. The Na^+ ions are also nearly in contact and also form a band. The electronegativity of chlorine is so much greater than that of sodium that the chlorine band lies well below the sodium band, and the band gap is about 7 eV. A total of $8N$ electrons are to be accommodated (seven from each Cl atom, one from each Na atom). These $8N$ electrons enter the lower chlorine band, fill it, and leave the sodium band empty. Because the energy of thermal motion available at room temperature is $kT \approx 0.03$ eV (k is Boltzmann's constant), very few electrons have enough energy to occupy the orbitals of the sodium band.

In an insulator the band of highest energy that contains electrons (at $T = 0$) is normally termed the **valence band**. There next higher band (which is empty at $T = 0$) is called the **conduction band**. In NaCl the band derived from the Cl orbitals is the valence band and the band derived from the Na orbitals is the conduction band.

We normally think of an ionic or molecular solid as consisting of discrete ions or molecules. According to the picture just described, however, they can be regarded as having a band structure. The two pictures can be reconciled because it is possible to show that a full band is equivalent to a sum of localized electron densities. In sodium chloride, for example, a full band built from Cl orbitals is equivalent to a collection of discrete Cl^- ions.

3.20 Semiconduction

The characteristic physical property of a semiconductor is that its electrical conductivity increases with increasing temperature. At room temperature, the conductivities of semiconductors are typically intermediate between those of metals and insulators. The dividing line between insulators and semiconductors is a matter of the size of the band gap (Table 3.13); the conductivity itself is an unreliable criterion because, as the temperature is increased, a given substance may have in succession a low, intermediate, and high conductivity. The values of the band gap and conductivity that are taken as indicating semiconduction rather than insulation depend on the application being considered.

(a) Intrinsic semiconductors

Key point: The band gap in a semiconductor controls the temperature dependence of the conductivity through an Arrhenius-like expression.

In an **intrinsic semiconductor**, the band gap is so small that the energy of thermal motion results in some electrons from the valence band populating the empty upper band (Fig. 3.72). This occupation of the conduction band introduces **positive holes**, equivalent to an absence of electrons, into the lower band, and as a result the solid is conducting because both the holes and the promoted electrons can move. A semiconductor at room temperature generally has a much lower conductivity than a metallic conductor because only very few electrons and holes can act as charge carriers. The strong, increasing temperature dependence of the conductivity follows from the exponential Boltzmann-like temperature dependence of the electron population in the upper band.

It follows from the exponential form of the population of the conduction band that the conductivity of a semiconductor should show an Arrhenius-like temperature dependence of the form

$$\sigma = \sigma_0 e^{-E_g/2kT} \tag{3.7}$$

where E_g is the width of the band gap. That is, the conductivity of a semiconductor can be expected to be Arrhenius-like with an activation energy equal to half the band gap, $E_a \approx \frac{1}{2}E_g$. This is found to be the case in practice.

(b) Extrinsic semiconductors

Key points: p-Type semiconductors are solids doped with atoms that remove electrons from the valence band; n-type semiconductors are solids doped with atoms that supply electrons to the conduction band.

An **extrinsic semiconductor** is a substance that is a semiconductor on account of the presence of intentionally added impurities. The number of electron carriers can be increased if atoms with more electrons than the parent element can be introduced by the process called **doping**. Remarkably low levels of dopant concentration are needed—only about one atom per 10^9 of the host material—so it is essential to achieve very high purity of the parent element initially.

If arsenic atoms ($[Ar]4s^2 4p^3$) are introduced into a silicon crystal ($[Ne]3s^2 3p^2$), one additional electron will be available for each dopant atom that is substituted. Note that the doping is *substitutional* in the sense that the dopant atom takes the place of an Si atom in the silicon structure. If the donor atoms, the As atoms, are far apart from each other, their electrons will be localized and the donor band will be very narrow (Fig. 3.73a). Moreover, the foreign atom levels will lie at higher energy than the valence electrons of the host structure and the filled dopant band is commonly near the empty conduction band. For $T > 0$, some of its electrons will be thermally promoted into the empty conduction band. In other words, thermal excitation will lead to the transfer of an electron from an As atom into the empty orbitals on a neighbouring Si atom. From there it will be able to migrate through the structure in the band formed by Si–Si overlap. This process gives rise to **n-type semiconductivity**, the 'n' indicating that the charge carriers are negatively charged (that is, electrons).

An alternative substitutional procedure is to dope the silicon with atoms of an element with fewer valence electrons on each atom, such as gallium ($[Ar]4s^2 4p^1$). A dopant atom of this kind effectively introduces holes into the solid. More formally, the dopant atoms form a very narrow, empty **acceptor band** that lies above the full Si band (Fig. 3.73b). At $T = 0$ the acceptor band is empty but at higher temperatures it can accept thermally excited electrons from the Si valence band. By doing so, it introduces holes into the latter and hence allows the remaining electrons in the band to be mobile. Because the charge carriers are now effectively positive holes in the lower band, this type of semiconductivity is called **p-type semiconductivity**. Semiconductor materials are essential components of all modern electronic circuits and some devices based on them are described in Box 3.4.

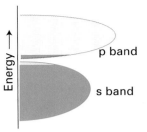

Fig. 3.72 In an intrinsic semiconductor, the band gap is so small that the Fermi distribution results in the population of some orbitals in the upper band.

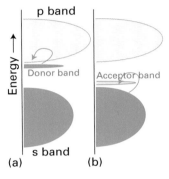

Fig. 3.73 The band structure in (a) an n-type semiconductor and (b) a p-type semiconductor.

BOX 3.4 Applications of semiconductors

Semiconductors have many applications because their properties can be easily modified by the addition of impurities to produce, for example, n- and p-type semiconductors. Furthermore, their electrical conductivities can be controlled by application of an electric field, by exposure to light, by pressure, and by heat; as a result, they can be used in many sensor devices.

Diodes and photodiodes

When the junction of a p-type and an n-type semiconductor is under 'reverse bias' (that is, with the p-side at a lower electric potential), the flow of current is very small, but it is high when the junction is under 'forward bias' (with the p-side at a higher electric potential). The exposure of a semiconductor to light can generate electron-hole pairs, which increases its conductivity through the increased number of free carriers (electrons or holes). Diodes that use this phenomenon are known as *photodiodes*. Compound semiconductor diodes can also be used to generate light, as in light-emitting diodes and laser diodes (Section 24.28).

Transistors

Bipolar junction transistors (BJT) are formed from two p-n junctions, in either an npn or a pnp configuration, with a narrow central region termed the base. The other regions, and their associated terminals, are known as the *emitter* and the *collector*. A small potential difference applied across the base and the emitter junction changes the properties of the base–collector junction so that it can conduct current even though it is reverse biased. Thus a transistor allows a current to be controlled by a small change in potential difference and is consequently used in amplifiers. Because the current flowing through a BJT is dependent on temperature they can be used as temperature sensors. Another type of transistor, the *field effect transistor* (FET) operates on the principle that semiconductor conductivity can be increased or decreased by the presence of an electric field. The electric field increases the number of charge carriers, thereby changing its conductivity. These FETs are used in both digital and analogue circuits to amplify or switch electronic signals.

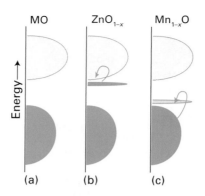

Fig 3.74 The band structure in (a) a stoichiometric oxide, (b) an anion-deficient oxide, and (c) an anion-excess oxide.

Several d-metal oxides, including ZnO and Fe_2O_3, are n-type semiconductors. In their case, the property is due to small variations in stoichiometry and a small deficit of O atoms. The electrons that should be in localized O atomic orbitals (giving a very narrow oxide band, essentially localized individual O^{2-} ions) occupy a previously empty conduction band formed by the metal orbitals (Fig. 3.74). The electrical conductivity decreases after the solids have been heated in oxygen and cooled slowly back to room temperature because the deficit of O atoms is partly replaced and, as the atoms are added, electrons are withdrawn from the conduction band to form oxide ions. However, when measured at high temperatures the conductivity of ZnO increases as further oxygen is lost from the structure, so increasing the number of electrons in the conduction band.

p-Type semiconduction is observed for some low oxidation number d-metal chalcogenides and halides, including Cu_2O, FeO, FeS, and CuI. In these compounds, the loss of electrons can occur through a process equivalent to the oxidation of some of the metal atoms, with the result that holes appear in the predominantly metal band. The conductivity increases when these compounds are heated in oxygen (or sulfur and halogen sources for FeS and CuI, respectively) because more holes are formed in the metal band as oxidation progresses. n-Type semiconductivity, however, tends to occur for oxides of metals in higher oxidation states, as the metal can be reduced to a lower oxidation state by occupation of a conduction band formed from the metal orbitals. Thus typical n-type semiconductors include Fe_2O_3, MnO_2, and CuO. By contrast, p-type semiconductivity occurs when the metal is in a low oxidation state, such as MnO and Cr_2O_3.

EXAMPLE 3.18 Predicting extrinsic semiconducting properties

Which of the oxides WO_3, MgO, and CdO are likely to show p- or n-type extrinsic semiconductivity?

Answer The type of semiconductivity depends on the defect levels that are likely to be introduced which is, in turn, determined by whether the metal present can be easily oxidized or reduced. If the metal can easily be oxidized (which may be the case if it has a low oxidation number), then n-type semiconductivity is expected. On the other hand, if the metal can easily be reduced (which may be the case if it has a high oxidation number), then p-type semiconductivity is expected. Thus, WO_3, with tungsten present in the high oxidation state W(VI), is readily reduced and accepts electrons from the O^{2-} ions, which escape as elemental oxygen. The excess electrons enter a band formed from the W d orbitals, resulting in n-type semiconductivity. Similarly, CdO, like ZnO, readily loses oxygen and is predicted to be an n-type semiconductor. In contrast, Mg^{2+} ions are neither easily oxidized nor reduced, therefore MgO does not lose or gain even small quantities of oxygen and is an insulator.

Self-test 3.18 Predict p- or n-type extrinsic semiconductivity for V_2O_5 and CoO.

Further information

3.1 **The Born–Mayer equation**

Consider a one-dimensional line of alternating cations A and anions B of charges $+e$ and $-e$ separated by a distance d. The Coulomb potential energy of a single cation is the sum of its interactions with all the other ions:

$$V = \frac{e^2}{4\pi\varepsilon_0}\left(-\frac{2}{d} + \frac{2}{2d} - \frac{2}{3d} + \cdots\right) = -\frac{2e^2}{4\pi\varepsilon_0 d}\left(1 - \frac{1}{2} + \frac{1}{3} - \cdots\right)$$

The sum of the series in parentheses is ln 2, so for this arrangement of ions

$$V = -\frac{2e^2 \ln 2}{4\pi\varepsilon_0 d}$$

The total molar contribution of all the ions is this potential energy multiplied by Avogadro's constant N_A (to convert to a molar value) and divided by 2 (to avoid counting each interaction twice):

$$V = \frac{N_A e^2}{4\pi\varepsilon_0 d} A$$

The factor $A = \ln 2$ is an example of a Madelung constant, a constant that represents the geometrical distribution of the ions (here, a straight line of constant separation). Two- and three-dimensional arrays of ions may be treated similarly, and give the values of A listed in Table 3.8.

The total molar potential energy includes the repulsive interaction between the ions. We can model that by a short-range exponential function of the form Be^{-d/d^*}, with d^* a constant that defines the range of the repulsive interaction and B a constant that defines its magnitude. The total molar potential energy of interaction is therefore

$$V = -\frac{N_A e^2}{4\pi\varepsilon_0 d} A + Be^{-d/d^*}$$

This potential energy passes through a minimum when $dV/dd = 0$, which occurs at

$$\frac{dV}{dd} = \frac{N_A e^2}{4\pi\varepsilon_0 d^2} A - \frac{B}{d^*} e^{-d/d^*} = 0$$

It follows that, at the minimum,

$$Be^{-d^*/d} = \frac{N_A e^2 d^*}{4\pi\varepsilon_0 d^2} A$$

This relation can be substituted into the expression for V, to give

$$V = -\frac{N_A e^2}{4\pi\varepsilon_0 d}\left(1 - \frac{d^*}{d}\right) A$$

On identifying $-V$ with the lattice enthalpy (more precisely, with the lattice energy at $T = 0$), we obtain the Born–Mayer equation (eqn 3.2) for the special case of singly charged ions. The generalization to other charge types is straightforward.

If a different expression for the repulsive interaction between the ions is used then this expression will be modified. One alternative is to use an expression such as $1/r^n$ with a large n, typically $6 \leq n \leq 12$, which then gives rise to a slightly different expression for V known as the **Born–Landé equation:**

$$V = -\frac{N_A e^2}{4\pi\varepsilon_0 d}\left(1 - \frac{1}{n}\right) A$$

The semiempirical Born–Mayer expression, with $d^* = 34.5$ pm determined from the best agreement with experimental data, is generally preferred to the Born–Landé equation.

FURTHER READING

R.D. Shannon in *Encyclopaedia of inorganic chemistry* (ed. R.B. King). Wiley, New York (2005). A survey of ionic radii and their determination.

A.F. Wells, *Structural inorganic chemistry*. Oxford University Press (1985). The standard reference book, which surveys the structures of a huge number of inorganic solids.

J.K. Burdett, *Chemical bonding in solids*. Oxford University Press (1995). Further details of the electronic structures of solids.

Some introductory texts on solid-state inorganic chemistry are:

U. Müller, *Inorganic structural chemistry*. Wiley, New York (1993).

A.R. West, *Basic solid state chemistry*. Wiley, New York (1999).

S.E. Dann, *Reactions and characterization of solids*. Royal Society of Chemistry, Cambridge (2000).

L.E. Smart and E.A. Moore, *Solid state chemistry: an introduction*. Taylor and Francis, CRC Press (2005).

P.A. Cox, *The electronic structure and chemistry of solids*. Oxford University Press (1987).

Two very useful texts on the application of thermodynamic arguments to inorganic chemistry are:

W.E. Dasent, *Inorganic energetics*. Cambridge University Press (1982).

D.A. Johnson, *Some thermodynamic aspects of inorganic chemistry*. Cambridge University Press (1982).

EXERCISES

3.1 What are the relationships between the unit cell parameters in the orthorhombic crystal system?

3.2 What are the fractional coordinates of the lattice points shown in the face-centred cubic unit cell (Fig. 3.5)? Confirm by counting lattice points and their contributions to the cubic unit cell that a face-centred, F, lattice contains four lattice points in the unit cell and a body-centred one two lattice points.

3.3 Which of the following schemes for the repeating pattern of close-packed planes are not ways of generating close-packed lattices? (a) ABCABC … , (b) ABAC … , (c) ABBA … , (d) ABCBC … , (e) ABABC … , (f) ABCCB …

3.4 Determine the formula of a compound produced by filling $\frac{1}{4}$ of the tetrahedral holes with cations X in a hexagonal close-packed array of anions A.

3.5 Potassium reacts with C_{60} (Fig. 3.16) to give a compound in which all the octahedral and tetrahedral holes are filled by potassium ions. Derive a stoichiometry for this compound.

3.6 Calculate a value for the atomic radius of the Cs atom with a coordination number of 12 given that the empirical atomic radius of Cs in the metal with a bcc structure is 272 pm.

3.7 Metallic sodium adopts a bcc structure with density 970 kg m⁻³. What is the length of the edge of the unit cell?

3.8 An alloy of copper and gold has the structure shown in Fig. 3.75. Calculate the composition of this unit cell. What is the lattice type of this structure? Given that 24 carat gold is pure gold, what carat gold does this alloy represent?

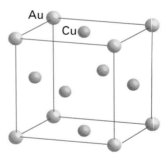

Fig 3.75 The structure of Cu_3Au.

3.9 Using Ketelaar's triangle would you classify Sr_2Ga ($\chi(Sr) = 0.95$; $\chi(Ga) = 1.81$) as an alloy or a Zintl phase?

3.10 Depending on temperature, RbCl can exist in either the rock-salt or caesium-chloride structure. (a) What is the coordination number of the cation and anion in each of these structures? (b) In which of these structures will Rb have the larger apparent radius?

3.11 Consider the structure of caesium chloride. How many Cs^+ ions occupy second-nearest-neighbour locations of a Cs^+ ion?

3.12 Describe the coordination around the anions in the perovskite structure in terms of coordination to the A- and B-type cations.

3.13 Use radius-ratio rules and the ionic radii given in *Resource section* 1 to predict structures of (a) PuO_2, (b) FrI, (c) BeO, (d) InN.

3.14 Based on the variation of ionic radii down a group, what structure would you predict for FrBr?

3.15 What are the most significant terms in the Born–Haber cycle for the formation of Ca_3N_2?

3.16 By considering the parameters that change in the Born–Mayer expression estimate lattice enthalpies for MgO and AlN given that MgO and AlN both adopt the rock-salt structure with similar lattice parameters and $\Delta H_L^{\ominus}\left(NaCl\right) = 786$ kJ mol⁻¹.

3.17 Use the Kapustinskii equation and the ionic and thermochemical radii given in *Resource section* 1 and Table 3.10, and $r(Bk^{4+}) = 96$ pm to calculate lattice enthalpies of (a) BkO_2, (b) K_2SiF_6, and (c) $LiClO_4$.

3.18 Which member of each pair is likely to be more soluble in water: (a) $SrSO_4$ or $MgSO_4$, (b) NaF or $NaBF_4$?

3.19 On the basis of the factors that contribute to lattice enthalpies place LiF, CaO, RbCl, AlN, NiO, and CsI, all of which adopt the rock-salt structure, in order of increasing lattice energy.

3.20 Recommend a specific cation for the quantitative precipitation of carbonate ion in water. Justify your recommendations.

3.21 Predict what type of intrinsic defect is most likely to occur in (a) Ca_3N_2, (b) HgS.

3.22 Explain why the number of defects in a solid increases as it is heated.

3.23 By considering which dopant ions produce the blue of sapphires provide an explanation for the origin of the colour in the blue form of beryl known as aquamarine.

3.24 For which of the following compounds might nonstoichiometry be found: magnesium oxide, vanadium carbide, manganese oxide?

3.25 Would VO or NiO be expected to show metallic properties?

3.26 Describe the difference between a semiconductor and a semimetal.

3.27 Classify the following as to whether they are likely to show n- or p-type semiconductivity: Ag_2S, VO_2, CuBr.

PROBLEMS

3.1 Draw a cubic unit cell (a) in projection (showing the fractional heights of the atoms) and (b) as a three-dimensional representation that has atoms at the following positions: Ti at $(\frac{1}{2}, \frac{1}{2}, \frac{1}{2})$, O at $(\frac{1}{2}, \frac{1}{2}, 0)$, $(0, \frac{1}{2}, \frac{1}{2})$, and $(\frac{1}{2}, 0, \frac{1}{2})$, and Ba at $(0,0,0)$. Remember that a cubic unit cell with an atom on the cell face, edge, or corner will have equivalent atoms displaced by the unit cell repeat in any direction. Of what structural type is the cell?

3.2 Draw a tetragonal unit cell and mark on it a set of points that would define (a) a face-centred lattice and (b) a body-centred lattice.

Demonstrate, by considering two adjacent unit cells, that a tetragonal face-centred lattice of dimensions a and c can always be redrawn as a body-centred tetragonal lattice with dimensions $a/2^{1/2}$ and c.

3.3 Draw one layer of close-packed spheres. On this layer mark the positions of the centres of the B layer atoms using the symbol ⊗ and, with the symbol ○, mark the positions of the centres of the C layer atoms of an fcc lattice.

3.4 In the structure of MoS_2, the S atoms are arranged in close-packed layers that repeat themselves in the sequence AAA … The Mo atoms occupy holes with coordination number 6. Show that each Mo atom is surrounded by a trigonal prism of S atoms.

3.5 The ReO_3 structure is cubic with an Re atom at each corner of the unit cell and one O atom on each unit cell edge midway between the Re atoms. Sketch this unit cell and determine (a) the coordination numbers of the ions and (b) the identity of the structure type that would be generated if a cation were inserted in the centre of each ReO_3 unit cell.

3.6 Consider the structure of rock salt. (a) How many Na^+ ions occupy second-nearest-neighbour locations of an Na^+ ion? (b) Pick out the closest-packed plane of Cl^- ions. (*Hint:* This hexagonal plane is perpendicular to a threefold axis.)

3.7 Imagine the construction of an MX_2 structure from the CsCl structure by removal of half the Cs^+ ions to leave tetrahedral coordination around each Cl^- ion. Identify this MX_2 structure.

3.8 Obtain formulae (MX_n or M_nX) for the following structures derived from hole filling in close-packed arrays with (a) half the octahedral holes filled, (b) one-quarter of the tetrahedral holes filled, and (c) two-thirds of the octahedral holes filled. What are the average coordination numbers of M and X in (a) and (b)?

3.9 Given the following data for the length of a side of the unit cell for compounds that crystallize in the rock-salt structure, determine the cation radii: MgSe (545 pm), CaSe (591 pm), SrSe (623 pm), BaSe (662 pm). (*Hint:* To determine the radius of Se^{2-}, assume that the Se^{2-} ions are in contact in MgSe.)

3.10 Use the structure map in Fig. 3.47 to predict the coordination numbers of the cations and anions in (a) LiF, (b) RbBr, (c) SrS, (d) BeO. The observed coordination numbers are (6,6) for LiF, RbBr, and SrS and (4,4) for BeO. Propose a possible reason for the discrepancies.

3.11 Describe how the structures of the following can be described in terms of simple structure types of Table 3.4 but with complex ions K_2PtCl_6, $[Ni(H_2O)_6].[SiF_6]$, CsCN.

3.12 The structure of calcite $CaCO_3$ is shown in Fig. 3.76. Describe how this structure is related to that of NaCl.

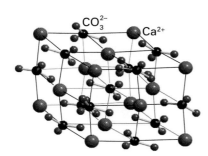

Fig 3.76 The structure of $CaCO_3$.

3.13 Using the accepted ionic radius of the ammonium ion, NH_4^+, NH_4Br is predicted to have a rock-salt structure with (6,6)-coordination. However, at room temperature NH_4Br has a caesium-chloride structure. Explain this observation.

3.14 (a) Calculate the enthalpy of formation of the hypothetical compound KF_2 assuming a CaF_2 structure. Use the Born–Mayer equation to obtain the lattice enthalpy and estimate the radius of K^{2+} by extrapolation of trends in Table 1.4 and *Resource section 1*. Ionization enthalpies and electron gain enthalpies are given in

Tables 1.5 and 1.6. (b) What factor prevents the formation of this compound despite the favourable lattice enthalpy?

3.15 The common oxidation number for an alkaline earth metal is +2. Using the Born–Mayer equation and a Born–Haber cycle, show that CaCl is an exothermic compound. Use a suitable analogy to estimate an ionic radius for Ca^+. The sublimation enthalpy of Ca(s) is 176 kJ mol^{-1}. Show that an explanation for the nonexistence of CaCl can be found in the enthalpy change for the reaction $2\,CaCl(s) \rightarrow Ca(s) + CaCl_2(s)$.

3.16 The Coulombic attraction of nearest-neighbour cations and anions accounts for the bulk of the lattice enthalpy of an ionic compound. With this fact in mind, estimate the order of increasing lattice enthalpy of (a) MgO, (b) NaCl, (c) AlN, all of which crystallize in the rock-salt structure. Give your reasoning.

3.17 There are two common polymorphs of zinc sulfide: cubic and hexagonal. Based on the analysis of Madelung constants alone, predict which polymorph should be more stable. Assume that the Zn–S distances in the two polymorphs are identical.

3.18 (a) Explain why lattice energy calculations based on the Born–Mayer equation reproduce the experimentally determined values to within 1 per cent for LiCl but only 10 per cent for AgCl given that both compounds have the rock-salt structure. (b) Identify a pair of compounds containing M^{2+} ions that might be expected to show similar behaviour.

3.19 Which of the following pairs of isostructural compounds are likely to undergo thermal decomposition at lower temperature? Give your reasoning. (a) $MgCO_3$ and $CaCO_3$ (decomposition products $MO + CO_2$). (b) CsI_3 and $N(CH_3)_4I_3$ (both compounds contain I_3^-; decomposition products $MI + I_2$; the radius of $N(CH_3)_4^+$ is much greater than that of Cs^+).

3.20 The Kapustinskii equation shows that lattice enthalpies are inversely proportional to the ion separations. Later work has shown that further simplification of the Kapustinskii equation allows lattice enthalpies to be estimated from the molecular (formula) unit volume (the unit cell volume divided by the number of formula units, Z, it contains) or the mass density (see, for example, H.D.B. Jenkins and D. Tudela, *J. Chem. Educ.*, 2003, **80**, 1482). How would you expect the lattice enthalpy to vary as a function of (a) the molecular unit volume and (b) the mass density? Given the following unit cell volumes (all in cubic angstrom, Å^3; 1 Å = 10^{-10} m) for the alkaline earth carbonates MCO_3 and oxides, predict the observed decomposition behaviour of the carbonates.

MgCO$_3$	CaCO$_3$	SrCO$_3$	BaCO$_3$
47	61	64	76
MgO	CaO	SrO	BaO
19	28	34	42

3.21 By considering the rock-salt structure and the distances and charges around one central ion show that the first six terms of the Na^+ Madelung series are

$$\frac{6}{\sqrt{1}} - \frac{12}{\sqrt{2}} + \frac{8}{\sqrt{3}} - \frac{6}{\sqrt{4}} + \frac{24}{\sqrt{5}} - \frac{24}{\sqrt{6}}$$

Discuss methods for showing this series converges to 1.748 by reference to R.P. Grosso, J.T. Fermann, and W.J. Vining, *J. Chem. Educ.*, 2001, **78**, 1198.

3.22 Explain why higher levels of defects are found in solids at high temperatures and close to their melting points. How would pressure affect the equilibrium number of defects in a solid?

3.23 By considering the effect on the lattice energies of incorporating large numbers of defects and the resultant changes in oxidation numbers of the ions making up the structure, predict which of the following systems should show nonstoichiometry over a large range of x: $Zn_{1+x}O$, $Fe_{1-x}O$, UO_{2+x}.

3.24 Graphite is a semimetal with a band structure of the type shown in Fig. 3.70. Reaction of graphite with potassium produces C_8K while reaction with bromine yields C_8Br. Assuming the graphite sheets remain intact and potassium and bromine enter the graphite structure as K^+ and Br^- ions respectively, discuss whether you would expect the compounds C_8K and C_8Br to exhibit metallic, semimetallic, semiconducting, or insulating properties.

Acids and bases

4

This chapter focuses on the wide variety of species that are classified as acids and bases. The acids and bases described in the first part of the chapter take part in proton transfer reactions. Proton transfer equilibria can be discussed quantitatively in terms of acidity constants, which are a measure of the tendency for species to donate protons. In the second part of the chapter, we broaden the definition of acids and bases to include reactions that involve electron-pair sharing between a donor and an acceptor. This broadening enables us to extend our discussion of acids and bases to species that do not contain protons and to nonaqueous media. Because of the greater diversity of these species, a single scale of strength is not appropriate. Therefore, we describe two approaches: in one, acids and bases are classified as 'hard' or 'soft'; in the other, thermochemical data are used to obtain a set of parameters characteristic of each species.

The original distinction between acids and bases was based, hazardously, on criteria of taste and feel: acids were sour and bases felt soapy. A deeper chemical understanding of their properties emerged from Arrhenius's (1884) conception of an acid as a compound that produced hydrogen ions in water. The modern definitions that we consider in this chapter are based on a broader range of chemical reactions. The definition due to Brønsted and Lowry focuses on proton transfer, and that due to Lewis is based on the interaction of electron pair acceptor and electron pair donor molecules and ions.

Acid–base reactions are common, although we do not always immediately recognize them as such, especially if they involve more subtle definitions of what it is to be an acid or base. For instance, production of acid rain begins with a very simple reaction between sulfur dioxide and water:

$$SO_2(g) + H_2O(l) \rightarrow HOSO_2^-(aq) + H^+(aq)$$

This will turn out to be a type of acid–base reaction. Saponification is the process used in soapmaking:

$$NaOH(aq) + RCOOR'(aq) \rightarrow NaRCO_2(aq) + R'OH(aq)$$

This too is a type of acid–base reaction. There are many such reactions, and in due course we shall see why they should be regarded as reactions between acids and bases.

Brønsted acidity

Key points: A Brønsted acid is a proton donor and a Brønsted base is a proton acceptor. A proton has no separate existence in chemistry and it is always associated with other species. A simple representation of a hydrogen ion in water is as the hydronium ion, H_3O^+.

Johannes Brønsted in Denmark and Thomas Lowry in England proposed (in 1923) that the essential feature of an acid–base reaction is the transfer of a hydrogen ion, H^+, from one species to another. In the context of this definition, a hydrogen ion is often referred to as a proton. They suggested that any substance that acts as a proton donor should be classified as an acid, and any substance that acts as a proton acceptor should be classified as a base. Substances that act in this way are now called 'Brønsted acids' and 'Brønsted bases', respectively:

A **Brønsted acid** is a proton donor.
A **Brønsted base** is a proton acceptor.

The definitions make no reference to the environment in which proton transfer occurs, so they apply to proton transfer behaviour in any solvent and even in no solvent at all.

An example of a Brønsted acid is hydrogen fluoride, HF, which can donate a proton to another molecule, such as H_2O, when it dissolves in water:

$$HF(g) + H_2O(l) \rightarrow H_3O^+(aq) + F^-(aq)$$

An example of a Brønsted base is ammonia, NH_3, which can accept a proton from a proton donor:

$$H_2O(l) + NH_3(aq) \rightarrow NH_4^+(aq) + OH^-(aq)$$

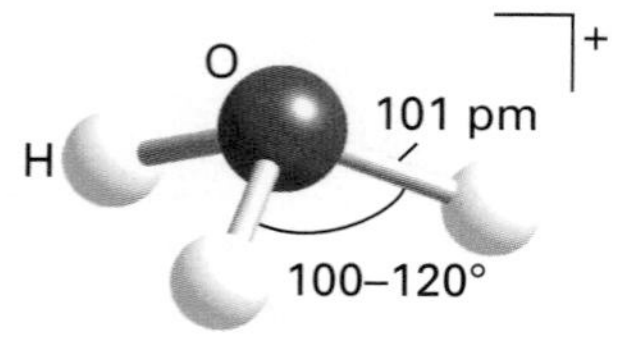

As these two examples show, water is an example of an **amphiprotic** substance, a substance that can act as both a Brønsted acid and a Brønsted base.

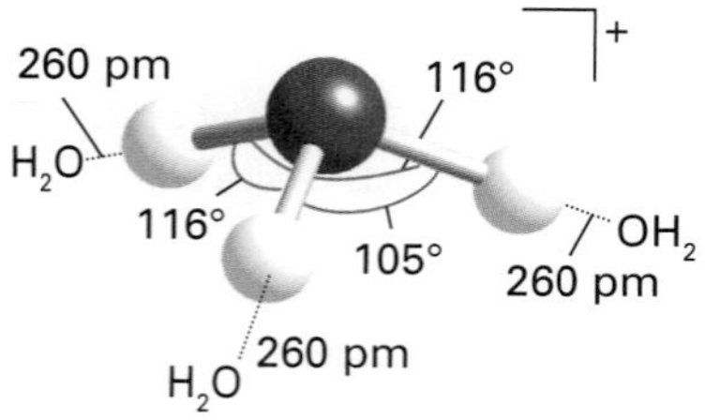

When an acid donates a proton to a water molecule, the latter is converted into a *hydronium ion*, H_3O^+ (**1**; the dimensions are taken from the crystal structure of $H_3O^+ClO_4^-$). However, the entity H_3O^+ is almost certainly an oversimplified description of the proton in water, for it participates in extensive hydrogen bonding, and a better representation is $H_9O_4^+$ (**2**). Gas-phase studies of water clusters using mass spectrometry suggest that a cage of H_2O molecules can condense around one H_3O^+ ion in a regular pentagonal dodecahedral arrangement, resulting in the formation of the species $H^+(H_2O)_{21}$. As these structures indicate, the most appropriate description of a proton in water varies according to the environment and the experiment under consideration; for simplicity, we shall use the representation H_3O^+ throughout.

1

2 $H_9O_4^+$

4.1 Proton transfer equilibria in water

Proton transfer between acids and bases is fast in both directions, so the dynamic equilibria

$$HF(aq) + H_2O(l) \rightleftharpoons H_3O^+(aq) + F^-(aq)$$

$$H_2O(l) + NH_3(aq) \rightleftharpoons NH_4^+(aq) + OH^-(aq)$$

give a more complete description of the behaviour of the acid HF and the base NH_3 in water than the forward reaction alone. The central feature of Brønsted acid–base chemistry in aqueous solution is that of rapid attainment of equilibrium in the proton transfer reaction, and we concentrate on this aspect.

(a) Conjugate acids and bases

Key points: When a species donates a proton, it becomes the conjugate base; when a species gains a proton, it becomes the conjugate acid. Conjugate acids and bases are in equilibrium in solution.

The form of the two forward and reverse reactions given above, both of which depend on the transfer of a proton from an acid to a base, is expressed by writing the general Brønsted equilibrium as

$$Acid_1 + Base_2 \rightleftharpoons Acid_2 + Base_1$$

The species $Base_1$ is called the **conjugate base** of $Acid_1$, and $Acid_2$ is the **conjugate acid** of $Base_2$. The conjugate base of an acid is the species that is left after a proton is lost. The conjugate acid of a base is the species formed when a proton is gained. Thus, F^- is the conjugate base of HF and H_3O^+ is the conjugate acid of H_2O. There is no *fundamental* distinction between an acid and a conjugate acid or a base and a conjugate base: a conjugate acid is just another acid and a conjugate base is just another base.

EXAMPLE 4.1 Identifying acids and bases

Identify the Brønsted acid and its conjugate base in the following reactions:

(a) $HSO_4^-(aq) + OH^-(aq) \rightarrow H_2O(l) + SO_4^{2-}(aq)$

(b) $PO_4^{3-}(aq) + H_2O(l) \rightarrow HPO_4^{2-}(aq) + OH^-$

Answer We need to identify the species that loses a proton and its conjugate partner. (a) The hydrogensulfate ion, HSO_4^-, transfers a proton to hydroxide; it is therefore the acid and the SO_4^{2-} ion produced is its conjugate base. (b) The H_2O molecule transfers a proton to the phosphate ion acting as a base; thus H_2O is the acid and the OH^- ion is its conjugate base.

Self-test 4.1 Identify the acid, base, conjugate acid, and conjugate base in the following reactions:
(a) $HNO_3(aq) + H_2O(l) \rightarrow H_3O^+(aq) + NO_3^-(aq)$ (b) $CO_3^{2-}(aq) + H_2O(l) \rightarrow HCO_3^-(aq) + OH^-$
(c) $NH_3(aq) + H_2S(aq) \rightarrow NH_4^+(aq) + HS^-(aq)$

(b) The strengths of Brønsted acids

Key points: The strength of a Brønsted acid is measured by its acidity constant, and the strength of a Brønsted base is measured by its basicity constant; the stronger the base, the weaker is its conjugate acid.

Throughout this discussion, we shall need the concept of pH, which we assume to be familiar from introductory chemistry:

$$pH = -\log [H_3O^+], \text{ and hence } [H_3O^+] = 10^{-pH} \tag{4.1}$$

The strength of a Brønsted acid, such as HF, in aqueous solution is expressed by its **acidity constant** (or 'acid ionization constant'), K_a:

$$HF(aq) + H_2O(l) \rightleftharpoons H_3O^+(aq) + F^-(aq) \qquad K_a = \frac{[H_3O^+][F^-]}{[HF]}$$

More generally:

$$HX(aq) + H_2O(l) \rightleftharpoons H_3O^+(aq) + X^-(aq) \qquad K_a = \frac{[H_3O^+][X^-]}{[HX]} \tag{4.2}$$

In this definition, $[X^-]$ denotes the numerical value of the molar concentration of the species X^- (so, if the molar concentration of HF molecules is 0.001 mol dm^{-3}, then $[HF] = 0.001$). A value $K_a \ll 1$ implies that $[HX]$ is large with respect to $[X^-]$, and so proton retention by the acid is favoured. The experimental value of K_a for hydrogen fluoride in water is 3.5×10^{-4}, indicating that under normal conditions only a very small fraction of HF molecules are deprotonated in water. The actual fraction deprotonated can be calculated as a function of acid concentration from the numerical value of K_a.

A note on good practice In precise work, K_a is expressed in terms of the activity of X, $a(X)$, its effective thermodynamic concentration. The acidity constant is based on the assumption that the solutions are sufficiently dilute for it to be permissible to write $a(H_2O) = 1$.

EXAMPLE 4.2 Calculating acidity constants

The pH of 0.145 M $CH_3COOH(aq)$ is 2.80. Calculate K_a of ethanoic acid.

Answer To calculate K_a we need to calculate the concentrations of H_3O^+, $CH_3CO_2^-$, and CH_3COOH in the solution. The concentration of H_3O^+ is obtained from the pH by writing $[H_3O^+] = 10^{-pH}$, so in a solution of pH = 2.80, the molar concentration of H_3O^+ is 1.6×10^{-3} mol dm^{-3}. Each deprotonation event produces one H_3O^+ ion and one $CH_3CO_2^-$ ion, so the concentration of $CH_3CO_2^-$ is the same as that of the H_3O^+ ions (provided the autoprotolysis of water can be neglected). The molar concentration of the remaining acid is $0.145 - 0.0016$ mol dm^{-3} = 0.143 mol dm^{-3}. Therefore

$$K_a = \frac{(1.6 \times 10^{-3})^2}{0.143} = 1.8 \times 10^{-5}$$

This value corresponds to pK_a = 4.75.

Self-test 4.2 For hydrofluoric acid $K_a = 3.5 \times 10^{-4}$. Calculate the pH of 0.10 M HF(aq).

The proton transfer equilibrium characteristic of a base, such as NH_3, in water can also be expressed in terms of an equilibrium constant, the **basicity constant**, K_b:

$$NH_3(aq) + H_2O(l) \rightleftharpoons NH_4^+(aq) + OH^-(aq) \qquad K_b = \frac{[NH_4^+][OH^-]}{[NH_3]}$$

More generally:

$$B(aq) + H_2O(l) \rightleftharpoons HB^+(aq) + OH^-(aq) \qquad K_b = \frac{[HB^+][OH^-]}{[B]} \qquad (4.3)$$

If $K_b \ll 1$, then $[HB^+] \ll [B]$ at typical concentrations of B and only a small fraction of B molecules are protonated. Therefore, the base is a weak proton acceptor and its conjugate acid is present in low concentration in solution. The experimental value of K_b for ammonia in water is 1.8×10^{-5}, indicating that under normal conditions, only a very small fraction of NH_3 molecules are protonated in water. As for the acid calculation, the actual fraction of base protonated can be calculated from the numerical value of K_b.

Because water is amphiprotic, a proton transfer equilibrium exists even in the absence of added acids or bases. The proton transfer from one water molecule to another is called **autoprotolysis** (or 'autoionization'). Proton transfer in water is very fast because it involves the interchange of weak hydrogen bonds between neighbouring molecules (Section 10.6). The extent of autoprotolysis and the composition of the solution at equilibrium is described by the **autoprotolysis constant** (or 'autoionization constant') of water:

$$2H_2O(l) \rightleftharpoons H_3O^+(aq) + OH^-(aq) \qquad K_w = [H_3O^+][OH^-]$$

The experimental value of K_w is 1.00×10^{-14} at 25°C, indicating that only a very tiny fraction of water molecules are present as ions in pure water. Indeed, we know that because the pH of pure water is 7.00, and $[H_3O^+] = [OH^-]$, then $[H_3O^+] = 1.0 \times 10^{-7}$ mol dm^{-3}. Tap and bottled water have a pH slightly lower than 7 due to the dissolved carbon dioxide.

An important role for the autoprotolysis constant of a solvent is that it enables us to express the strength of a base in terms of the strength of its conjugate acid. Thus, the value of K_b for the ammonia equilibrium in which NH_3 acts as a base is related to the value of K_a for the equilibrium

$$NH_4^+(aq) + H_2O(l) \rightleftharpoons H_3O^+(aq) + NH_3(aq)$$

in which its conjugate acid acts as an acid by

$$K_a K_b = K_w \qquad (4.4)$$

This relation may be verified by multiplying together the expressions for the acidity constant of NH_4^+ and the basicity constant of NH_3. The implication of eqn 4.4 is that the larger the value of K_b, the smaller the value of K_a. That is, the stronger the base, the weaker is its conjugate acid. A further implication of eqn 4.4 is that the strengths of bases may be reported in terms of the acidity constants of their conjugate acids.

■ **A brief illustration.** The K_b of ammonia in water is 1.8×10^{-5}. It follows that K_a of the conjugate acid NH_4^+ is

$$K_a = \frac{K_w}{K_b} = \frac{1 \times 10^{-14}}{1.8 \times 10^{-5}} = 5.6 \times 10^{-10} . \blacksquare$$

Because, like molar concentrations, acidity constants span many orders of magnitude, it is convenient to report them as their common logarithms (logarithms to the base 10) like pH by using

$$pK = -\log K \qquad (4.5)$$

where K may be any of the constants we have introduced. At 25°C, for instance, $pK_w = 14.00$. It follows from this definition and the relation in eqn 4.4 that

$$pK_a + pK_b = pK_w \qquad (4.6)$$

A similar expression applies to the strengths of conjugate acids and bases in any solvent, with pK_w replaced by the appropriate autoprotolysis constant of the solvent, pK_{sol}.

(c) Strong and weak acids and bases

Key points: An acid or base is classified as either weak or strong depending on the size of its acidity constant.

Table 4.1 lists the acidity constants of some common acids and conjugate acids of some bases in water. A substance is classified as a **strong acid** if the proton transfer equilibrium lies strongly in favour of donation of a proton to the solvent. Thus, a substance with $pK_a < 0$ (corresponding to $K_a > 1$ and usually to $K_a \gg 1$) is a strong acid. Such acids are commonly regarded as being fully deprotonated in solution (but it must never be forgotten that that is only an approximation). For example, hydrochloric acid is regarded as a solution of H_3O^+ and Cl^- ions, and a negligible concentration of HCl molecules. A substance with $pK_a > 0$ (corresponding to $K_a < 1$) is classified as a **weak acid**; for such species, the proton transfer equilibrium lies in favour of nonionized acid. Hydrogen fluoride is a weak acid in water, and hydrofluoric acid consists of hydronium ions, fluoride ions, and a high proportion of HF molecules.

A **strong base** reacts with water to become almost fully protonated. An example is the oxide ion, O^{2-}, which is immediately converted into OH^- ions in water. A **weak base** is only partially protonated in water. An example is NH_3, which dissolves in water to give only a small proportion of NH_4^+ ions. The conjugate base of any strong acid is a weak base because it is thermodynamically unfavourable for such a base to accept a proton.

(d) Polyprotic acids

Key points: A polyprotic acid loses protons in succession, and successive deprotonations are progressively less favourable; a distribution diagram summarizes how the fraction of each species present depends on the pH of the solution.

A **polyprotic acid** is a substance that can donate more than one proton. An example is hydrogen sulfide, H_2S, a diprotic acid. For a diprotic acid, there are two successive proton donations and two acidity constants:

$$H_2S(aq) + H_2O(l) \rightleftharpoons HS^-(aq) + H_3O^+(aq) \qquad K_{a1} = \frac{[H_3O^+][HS^-]}{[H_2S]}$$

$$HS^-(aq) + H_2O(l) \rightleftharpoons S^{2-}(aq) + H_3O^+(aq) \qquad K_{a2} = \frac{[H_3O^+][S^{2-}]}{[HS^-]}$$

Table 4.1 Acidity constants for species in aqueous solution at 25°C

Acid	HA	A^-	K_a	pK_a	Acid	HA	A^-	K_a	pK_a
Hydriodic	HI	I^-	10^{11}	−11	Ethanoic	CH_3COOH	$CH_3CO_2^-$	1.74×10^{-5}	4.76
Perchloric	$HClO_4$	ClO_4^-	10^{10}	−10	Pyridinium ion	$HC_5H_5N^+$	C_5H_5N	5.6×10^{-6}	5.25
Hydrobromic	HBr	Br^-	10^9	−9	Carbonic	H_2CO_3	HCO_3^-	4.3×10^{-7}	6.37
Hydrochloric	HCl	Cl^-	10^7	−7	Hydrogen sulfide	H_2S	HS^-	9.1×10^{-8}	7.04
Sulfuric	H_2SO_4	HSO_4^-	10^2	−2	Boric acid*	$B(OH)_3$	$B(OH)_4^-$	7.2×10^{-10}	9.14
Nitric	HNO_3	NO_3^-	10^2	−2	Ammonium ion	NH_4^+	NH_3	5.6×10^{-10}	9.25
Hydronium ion	H_3O^+	H_2O	1	0.0	Hydrocyanic	HCN	CN^-	4.9×10^{-10}	9.31
Chloric	$HClO_3$	ClO_3^-	10^{-1}	1	Hydrogencarbonate ion	HCO_3^-	CO_3^{2-}	4.8×10^{-11}	10.32
Sulfurous	H_2SO_3	HSO_3^-	1.5×10^{-2}	1.81	Hydrogenarsenate ion	$HAsO_4^{2-}$	AsO_4^{3-}	3.0×10^{-12}	11.53
Hydrogensulfate ion	HSO_4^-	SO_4^{2-}	1.2×10^{-2}	1.92	Hydrogensulfide ion	HS^-	S^{2-}	1.1×10^{-19}	19
Phosphoric	H_3PO_4	$H_2PO_4^-$	7.5×10^{-3}	2.12	Hydrogenphosphate ion	HPO_4^{2-}	PO_4^{3-}	2.2×10^{-13}	12.67
Hydrofluoric	HF	F^-	3.5×10^{-4}	3.45	Dihydrogenphosphate ion	$H_2PO_4^-$	HPO_4^{2-}	6.2×10^{-8}	7.21
Formic	HCOOH	HCO_2^-	1.8×10^{-4}	3.75					

* The proton transfer equilibrium is $B(OH)_3(aq) + 2H_2O(l) \rightleftharpoons H_3O^+(aq) + B(OH)_4^-(aq)$.

From Table 4.1, $K_{a1} = 9.1 \times 10^{-8}$ ($pK_{a1} = 7.04$) and $K_{a2} \approx 1.1 \times 10^{-19}$ ($pK_{a2} = 19$). The second acidity constant, K_{a2}, is almost always smaller than K_{a1} (and hence pK_{a2} is generally larger than pK_{a1}). The decrease in K_a is consistent with an electrostatic model of the acid in which, in the second deprotonation, a proton must separate from a centre with one more negative charge than in the first deprotonation. Because additional electrostatic work must be done to remove the positively charged proton, the deprotonation is less favourable.

EXAMPLE 4.3 Calculating the concentration of ions in polyprotic acids

Calculate the concentration of carbonate ions in 0.10 M $H_2CO_3(aq)$. K_{a1} is given in Table 4.1; $K_{a2} = 4.6 \times 10^{-11}$.

Answer We need to consider the equilibria for the successive deprotonation steps with their acidity constants:

$$H_2CO_3(aq) + H_2O(l) \rightleftharpoons HCO_3^-(aq) + H_3O^+(aq) \qquad K_{a1} = \frac{[H_3O^+][HCO_3^-]}{[H_2CO_3]}$$

$$HCO_3^-(aq) + H_2O(l) \rightleftharpoons CO_3^{2-}(aq) + H_3O^+(aq) \qquad K_{a2} = \frac{[H_3O^+][CO_3^{2-}]}{[HCO_3^-]}$$

We suppose that the second deprotonation is so slight that it has no effect on the value of $[H_3O^+]$ arising from the first deprotonation, in which case we can write $[H_3O^+] = [HCO_3^-]$ in K_{a2}. These two terms therefore cancel in the expression for K_{a2}, which results in

$$K_{a2} = [CO_3^{2-}]$$

independent of the initial concentration of the acid. It follows that the concentration of carbonate ions in the solution is 4.6×10^{-11} mol dm^{-3}.

Self-test 4.3 Calculate the pH of 0.20 M $H_2C_4H_4O_6(aq)$ (tartaric acid), given $K_{a1} = 1.0 \times 10^{-3}$ and $K_{a2} = 4.6 \times 10^{-5}$.

The clearest representation of the concentrations of the species that are formed in the successive proton transfer equilibria of polyprotic acids is a **distribution diagram**, a diagram showing the fraction of solute present as a specified species X, $f(X)$, plotted against the pH. Consider, for instance, the triprotic acid H_3PO_4, which releases three protons in succession to give $H_2PO_4^-$, HPO_4^{2-}, and PO_4^{3-}. The fraction of solute present as intact H_3PO_4 molecules is

$$f(H_3PO_4) = \frac{[H_3PO_4]}{[H_3PO_4] + [H_2PO_4^-] + [HPO_4^{2-}] + [PO_4^{3-}]} \qquad (4.7)$$

The concentration of each solute at a given pH can be calculated from the pK_a values.[1] Figure 4.1 shows the fraction of all four solute species as a function of pH and hence summarizes the relative importance of each acid and its conjugate base at each pH. Conversely, the diagram indicates the pH of the solution that contains a particular fraction of the species. We see, for instance, that if pH < pK_{a1}, corresponding to high hydronium ion concentrations, then the dominant species is the fully protonated H_3PO_4 molecule. However, if pH > pK_{a3}, corresponding to low hydronium ion concentrations, then the dominant species is the fully deprotonated PO_4^{3-} ion. The intermediate species are dominant when pH values lie between the relevant pK_as.

(e) Factors governing the strengths of acids and bases

Key points: Proton affinity is the negative of the gas phase proton-gain enthalpy. The proton affinities of p-block conjugate bases decrease to the right along a period and down a group. Solution proton affinities for binary acids are lower than gas phase proton affinities. Small highly charged ions are stabilized in polar solvents.

A quantitative understanding of the relative acidities of X–H protons can be obtained by considering the enthalpy changes accompanying proton transfer. We shall consider gas-phase proton transfer reactions first and then consider the effects of the solvent.

---

[1] For the calculations involved, see P. Atkins and L. Jones, *Chemical principles*. W.H. Freeman & Co. (2010).

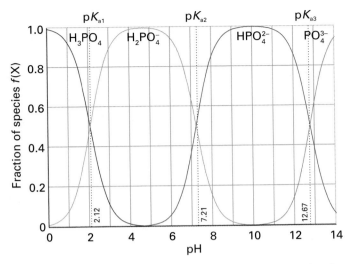

Figure 4.1 The distribution diagram for the various forms of the triprotic acid phosphoric acid in water, as a function of pH.

The simplest reaction of a proton is its attachment to a base, A^- (which, although denoted here as a negatively charged species, could be a neutral molecule, such as NH_3), in the gas phase:

$$A^-(g) + H^+(g) \rightarrow HA(g)$$

The standard enthalpy of this reaction is the **proton-gain enthalpy**, $\Delta_{pg}H^\ominus$. The negative of this quantity is often reported as the **proton affinity**, A_p (Table 4.2). When $\Delta_{pg}H^\ominus$ is large and negative, corresponding to an exothermic proton attachment, the proton affinity is high, indicating strongly basic character in the gas phase. If the proton gain enthalpy is only slightly negative, then the proton affinity is low, indicating a weaker basic (or more acidic) character.

The proton affinities of the conjugate bases of p-block binary acids HA decrease to the right along a period and down a group, indicating an increase in gas-phase acidity. Thus, HF is a stronger acid than H_2O and HI is the strongest acid of the hydrogen halides. In other words, the order of proton affinities of their conjugate bases is $I^- < OH^- < F^-$. These trends can be explained by using a thermodynamic cycle such as that shown in Fig. 4.2, in which proton gain can be thought of as the outcome of three steps:

Electron loss from A^-: $A^-(g) \rightarrow A(g) + e^-(g)$ $\qquad\qquad -\Delta_{eg}H^\ominus(A) = A_e(A)$

(the reverse of electron gain by A)

Table 4.2 Gas phase and solution proton affinities*

Conjugate acid	Base	A_p/kJ mol^{-1}	A'_p/kJ mol^{-1}
HF	F^-	1553	1150
HCl	Cl^-	1393	1090
HBr	Br^-	1353	1079
HI	I^-	1314	1068
H_2O	OH^-	1643	1188
HCN	CN^-	1476	1183
H_3O^+	H_2O	723	1130
NH_4^+	NH_3	865	1182

* A_p is the gas phase proton affinity, A'_p is the effective proton affinity for the base in water.

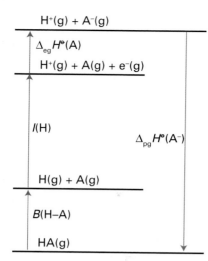

Figure 4.2 Thermodynamic cycle for a proton gain reaction.

Electron gain by H^+: $H^+(g) + e^-(g) \rightarrow H(g)$ $-\Delta_i H^{\ominus}(H) = -I(H)$

(the reverse of the ionization of H)

Combination of H and A: $H(g) + A(g) \rightarrow HA(g)$ $-B(H-A)$

(the reverse of H–A bond dissociation)

The proton-gain enthalpy of the conjugate base A^- is the sum of these enthalpy changes:

Overall: $H^+(g) + A^-(g) \rightarrow HA(g)$ $\Delta_{pg} H^{\ominus}(A^-) = A_e(A) - I(H) - B(H-A)$

Therefore, the proton affinity of A^- is

$$A_p(A^-) = B(H-A) + I(H) - A_e(A) \tag{4.8}$$

The dominant factor in the variation in proton affinity across a period is the trend in electron affinity of A, which increases from left to right and hence lowers the proton affinity of A^-. Thus, because the proton affinity of A^- decreases, the gas-phase acidity of HA *increases* across a period as the electron affinity of A increases. Because increasing electron affinity correlates with increasing electronegativity (Section 1.9), the gas-phase acidity of HA also increases as the electronegativity of A increases. The dominant factor when descending a group is the decrease in the H–A bond dissociation enthalpy, which lowers the proton affinity of A^- and therefore results in an increase in the gas-phase acid strength of HA. The overall result of these effects is a decrease in gas-phase proton affinity of A^-, and therefore an increase in the gas-phase acidity of HA, from the top left to bottom right of the p block. On this basis we see that HI is a much stronger acid than CH_4.

The correlation we have described is modified when a solvent (typically water) is present. The gas-phase process $A^-(g) + H^+(g) \rightarrow AH(g)$ becomes

$A^-(aq) + H^+(aq) \rightarrow HA(aq)$

and the negative of the accompanying proton-gain enthalpy is called the **effective proton affinity**, A_p', of $A^-(aq)$.

If the species A^- denotes H_2O itself, the effective proton affinity of H_2O is the enthalpy change accompanying the process

$H_2O(l) + H^+(aq) \rightarrow H_3O^+(aq)$

The energy released as water molecules are attached to a proton in the gas phase, the process

$n\,H_2O(g) + H^+(g) \rightarrow H^+(H_2O)_n(g)$

can be measured by mass spectrometry and used to assess the energy change for the hydration process in solution. It is found that the energy released passes through a maximum value of 1130 kJ mol^{-1} as n increases, and this value is taken to be the effective proton affinity of H_2O in bulk water. The effective proton affinity of the ion OH^- in water is simply the negative of the enthalpy of the reaction

$OH^-(g) + H^+(aq) \rightarrow H_2O(l)$

which can be measured by conventional means (such as the temperature dependence of its equilibrium constant, K_w). The value found is 1188 kJ mol^{-1}.

The reaction

$HA(aq) + H_2O(l) \rightarrow H_3O^+(aq) + A^-(aq)$

is exothermic if the effective proton affinity of $A^-(aq)$ is lower than that of $H_2O(l)$ (less than 1130 kJ mol^{-1}) and—provided entropy changes are negligible and enthalpy changes are a guide to spontaneity—will give up protons to the water and be strongly acidic. Likewise, the reaction

$A^-(aq) + H_2O(l) \rightarrow HA(aq) + OH^-(aq)$

is exothermic if the effective proton affinity of $A^-(aq)$ is higher than that of $OH^-(aq)$ (1188 kJ mol^{-1}). Provided enthalpy changes are a guide to spontaneity, $A^-(aq)$ will accept protons and will act as a strong base.

■ **A brief illustration.** The effective proton affinity of I⁻ in water is 1068 kJ mol⁻¹ compared to 1314 kJ mol⁻¹ in the gas phase, showing that the I⁻ ion is stabilized by hydration. The effective proton affinity is also smaller than the effective proton affinity of water (1130 kJ mol⁻¹), which is consistent with the fact that HI is a strong acid in water. All the halide ions except F⁻ have effective proton affinities smaller than that of water, which is consistent with all the hydrogen halides except HF being strong acids in water. ■

The effects of solvation can be rationalized in terms of an electrostatic model in which the solvent is treated as a continuous dielectric medium. The solvation of a gas-phase ion is always strongly exothermic. The magnitude of the enthalpy of solvation $\Delta_{solv}H^{\ominus}$ (the enthalpy of hydration in water, $\Delta_{hyd}H^{\ominus}$) depends on the radius of the ions, the relative permittivity of the solvent, and the possibility of specific bonding (especially hydrogen bonding) between the ions and the solvent.

When considering the gas phase we assume that entropy contributions for the proton transfer process are small and so $\Delta G^{\ominus} \approx \Delta H^{\ominus}$. In solution, entropy effects cannot be ignored and we must use $\Delta G^{\ominus}$. The Gibbs energy of solvation of an ion can be identified as the energy involved in transferring the anion from a vacuum into a solvent of relative permittivity ε_r. The **Born equation** can be derived using this model:[2]

$$\Delta_{solv}G^{\ominus} = -\frac{N_A z^2 e^2}{8\pi\varepsilon_0 r}\left(1 - \frac{1}{\varepsilon_r}\right) \tag{4.9}$$

where z is the charge number of the ion, r is its effective radius, which includes part of the radii of solvent molecules, N_A is Avogadro's constant, ε_0 is the vacuum permittivity, and ε_r is the relative permittivity (the dielectric constant). Because $z^2/r = \xi$, the electrostatic parameter of the ion (Section 3.15), this expression can be written

$$\Delta_{solv}G^{\ominus} = -\frac{N_A e^2 \xi}{8\pi\varepsilon_0}\left(1 - \frac{1}{\varepsilon_r}\right) \tag{4.10}$$

The Gibbs energy of solvation is proportional to ξ, so small, highly charged ions are stabilized in polar solvents (Fig. 4.3). The Born equation also shows that the larger the relative permittivity the more negative the value of $\Delta_{solv}G^{\ominus}$. This stabilization is particularly important for water, for which $\varepsilon_r = 80$ (and the term in parentheses is close to 1), compared with nonpolar solvents for which ε_r may be as low as 2 (and the term in parentheses is close to 0.5).

Because $\Delta_{solv}G^{\ominus}$ is the change in molar Gibbs energy when an ion is transferred from the gas phase into aqueous solution, a large, negative value of $\Delta_{solv}G^{\ominus}$ favours the formation of ions in solution compared with the gas phase (Fig. 4.3). The interaction of the charged ion with the polar solvent molecules stabilizes the conjugate base A⁻ relative to the parent acid HA, and as a result the acidity of HA is enhanced by the polar solvent. On the other hand, the effective proton affinity of a neutral base B is higher than in the gas phase because the conjugate acid HB⁺ is stabilized by solvation. Because cationic acids, such as NH_4^+, are stabilized by solvation, their effective proton affinity is higher than in the gas phase and their acidity is lowered by a polar solvent.

The Born equation ascribes stabilization to Coulombic interactions. However, hydrogen bonding is an important factor in protic solvents such as water and leads to the formation of hydrogen-bonded clusters around some solutes. As a result, water has a greater stabilizing effect on small, highly charged ions than the Born equation predicts. This stabilizing effect is particularly great for F⁻, OH⁻, and Cl⁻, with their high charge densities and for which water acts as a hydrogen-bond donor. Because water has lone pairs of electrons on O, it can also be a hydrogen-bond acceptor. Acidic ions such as NH_4^+ are stabilized by hydrogen bonding and consequently have a lower acidity than predicted by the Born equation.

4.2 Solvent levelling

Key point: A solvent with a large autoprotolysis constant can be used to discriminate between a wide range of acid and base strengths.

An acid that is weak in water may appear strong in a solvent that is a more effective proton acceptor, and vice versa. Indeed, in sufficiently basic solvents (such as liquid ammonia), it

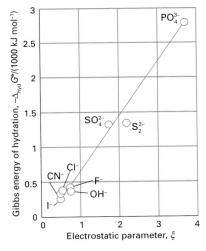

Figure 4.3 The correlation between $\Delta_{solv}G^{\ominus}$ and the electrostatic parameter ξ of selected anions. To obtain ξ as a small dimensionless number we have used $\xi = 100z^2/(r/\mathrm{pm})$.

---

[2] For the derivation of the Born equation see P. Atkins and J. de Paula, *Physical chemistry*, Oxford University Press and W.H. Freeman & Co. (2010).

may not be possible to discriminate between their strengths because all of them will be fully deprotonated. Similarly, bases that are weak in water may appear strong in a more strongly proton-donating solvent (such as anhydrous acetic acid). It may not be possible to arrange a series of bases according to strength, for all of them will be effectively fully protonated in acidic solvents. We shall now see that the autoprotolysis constant of a solvent plays a crucial role in determining the range of acid or base strengths that can be distinguished for species dissolved in it.

Any acid stronger than H_3O^+ in water donates a proton to H_2O and forms H_3O^+. Consequently, no acid significantly stronger than H_3O^+ can remain protonated in water. No experiment conducted in water can tell us which of HBr and HI is the stronger acid because both transfer their protons essentially completely to give H_3O^+. In effect, solutions of the strong acids HX and HY behave as though they are solutions of H_3O^+ ions regardless of whether HX is intrinsically stronger than HY. Water is therefore said to have a **levelling effect** that brings all stronger acids down to the acidity of H_3O^+. The strengths of such acids can be distinguished by using a less basic solvent. For instance, although HBr and HI have indistinguishable acid strengths in water, in acetic acid HBr and HI behave as weak acids and their strengths can be distinguished: in this way it is found that HI is a stronger proton donor than HBr.

The levelling effect can be expressed in terms of the pK_a of the acid. An acid such as HCN dissolved in a solvent, HSol, is classified as strong if $pK_a < 0$, where K_a is the acidity constant of the acid in the solvent Sol:

$$HCN(sol) + HSol(l) \rightleftharpoons H_2Sol^+(sol) + CN^-(sol) \qquad K_a = \frac{[H_2Sol^+][CN^-]}{[HCN]}$$

That is, all acids with $pK_a < 0$ (corresponding to $K_a > 1$) display the acidity of H_2Sol^+ when they are dissolved in the solvent HSol.

An analogous effect can be found for bases in water. Any base that is strong enough to undergo complete protonation by water produces an OH^- ion for each molecule of base added. The solution behaves as though it contains OH^- ions. Therefore, we cannot distinguish the proton-accepting power of such bases, and we say that they are levelled to a common strength. Indeed, the OH^- ion is the strongest base that can exist in water because any species that is a stronger proton acceptor immediately forms OH^- ions by proton transfer from water. For this reason, we cannot study NH_2^- or CH_3^- in water by dissolving alkali metal amides or methides because both anions generate OH^- ions and are fully protonated to NH_3 and CH_4:

$$KNH_2(s) + H_2O(l) \rightarrow K^+(aq) + OH^-(aq) + NH_3(aq)$$

$$Li_4(CH_3)_4(s) + 4 H_2O(l) \rightarrow 4 Li^+(aq) + 4 OH^-(aq) + 4 CH_4(g)$$

The base levelling effect can be expressed in terms of the pK_b of the base. A base dissolved in HSol is classified as strong if $pK_b < 0$, where K_b is the basicity constant of the base in HSol:

$$NH_3(sol) + HSol(l) \rightleftharpoons NH_4^+(sol) + Sol^-(sol) \qquad K_b = \frac{[NH_4^+][Sol^-]}{[NH_3]}$$

That is, all bases with $pK_b < 0$ (corresponding to $K_b > 1$) display the basicity of Sol^- in the solvent HSol. Now, because $pK_a + pK_b = pK_{sol}$, this criterion for levelling may be expressed as follows: all bases with $pK_a > pK_{sol}$ give a negative value for pK_b and behave like Sol^- in the solvent HSol.

It follows from this discussion of acids and bases in a common solvent HSol that, because any acid is levelled if $pK_a < 0$ in HSol and any base is levelled if $pK_a > pK_{sol}$ in the same solvent, then the window of strengths that are not levelled in the solvent is from $pK_a = 0$ to pK_{sol}. For water, $pK_w = 14$. For liquid ammonia, the autoprotolysis equilibrium is

$$2 NH_3(l) \rightleftharpoons NH_4^+(sol) + NH_2^-(sol) \qquad pK_{am} = 33$$

It follows from these figures that acids and bases are discriminated much less in water than they are in ammonia. The discrimination windows of a number of solvents are shown in Fig. 4.4. The window for dimethylsulfoxide (DMSO, $(CH_3)_2SO$) is wide because $pK_{DMSO} = 37$.

The strengths of aqua acids typically increase with increasing positive charge of the central metal ion and with decreasing ionic radius. This variation can be rationalized to some extent in terms of an ionic model, in which the metal cation is represented by a sphere of radius r_+ carrying z positive charges. Because protons are more easily removed from the vicinity of cations of high charge and small radius, the model predicts that the acidity should increase with increasing z and with decreasing r_+.

The validity of the ionic model of acid strengths can be judged from Fig. 4.5. Aqua ions of elements that form ionic solids (principally those from the s block) have pK_a values that are quite well described by the ionic model. Several d-block ions (such as Fe^{2+} and Cr^{3+}) lie reasonably near the same straight line, but many ions (particularly those with low pK_a, corresponding to high acid strength) deviate markedly from it. This deviation indicates that the metal ions repel the departing proton more strongly than is predicted by the ionic model. This enhanced repulsion can be rationalized by supposing that the positive charge of the cation is not confined to the central ion but is delocalized over the ligands and hence is closer to the departing proton. The delocalization is equivalent to attributing covalence to the element–oxygen bond. Indeed, the correlation is worst for ions that are disposed to form covalent bonds.

For the later d- and the p-block metal ions (such as Cu^{2+} and Sn^{2+}, respectively), the strengths of the aqua acids are much greater than the ionic model predicts. For these species, covalent bonding is more important than ionic bonding and the ionic model is unrealistic. The overlap between metal orbitals and the orbitals of an oxygen ligand increases from left to right across a period. It also increases down a group, so aqua ions of heavier d-block metals tend to be stronger acids.

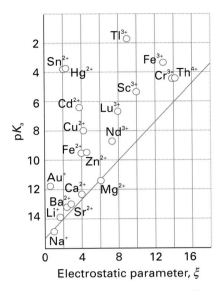

Figure 4.5 The correlation between acidity constant and the electrostatic parameter ξ of aqua ions.

EXAMPLE 4.6 Accounting for trends in aqua acid strength

Account for the trend in acidity $[Fe(OH_2)_6]^{2+} < [Fe(OH_2)_6]^{3+} < [Al(OH_2)_6]^{3+} \approx [Hg(OH_2)]^{2+}$.

Answer We need to consider the charge density on the metal centre and its effect on the ease with which the H_2O ligands can be deprotonated. The weakest acid is the Fe^{2+} complex on account of its relatively large ionic radius and low charge. The increase of charge to $+3$ increases the acid strength. The greater acidity of Al^{3+} can be explained by its smaller radius. The anomalous ion in the series is the Hg^{2+} complex. This complex reflects the failure of an ionic model because in the complex there is a large transfer of positive charge to oxygen as a result of covalent bonding.

Self-test 4.6 Arrange $[Na(OH_2)_6]^+$, $[Sc(OH_2)_6]^{3+}$, $[Mn(OH_2)_6]^{2+}$, and $[Ni(OH_2)_6]^{2+}$ in order of increasing acidity.

4.5 Simple oxoacids

The simplest oxoacids are the **mononuclear acids**, which contain one atom of the parent element. They include H_2CO_3, HNO_3, H_3PO_4, and H_2SO_4.[3] These oxoacids are formed by the electronegative elements at the upper right of the periodic table and by other elements in high oxidation states (Table 4.3). One interesting feature in the table is the occurrence of planar H_2CO_3 and HNO_3 molecules but not their analogues in later periods. As we saw in Chapter 2, π bonding is more important among the Period 2 elements, so their atoms are more likely to be constrained to lie in a plane.

(a) Substituted oxoacids

Key points: Substituted oxoacids have strengths that may be rationalized in terms of the electron-withdrawing power of the substitutent; in a few cases, a nonacidic H atom is attached directly to the central atom of an oxoacid.

One or more –OH groups of an oxoacid may be replaced by other groups to give a series of substituted oxoacids, which include fluorosulfuric acid, $O_2SF(OH)$, and aminosulfuric acid, $O_2S(NH_2)OH$ (**6**). Because fluorine is highly electronegative, it withdraws electrons

---

[3] These acids are more helpfully written as $(HO)_2CO$, $HONO_2$, $(HO)_3PO$, and $(HO)_2SO_2$, and boric acid written as $B(OH)_3$ rather than H_3BO_3. In this text we use both forms of notation depending upon the properties being explained.

Table 4.3   The structure and pK_a values of oxoacids*

$p = 0$	$p = 1$	$p = 2$	$p = 3$
HO—Cl 7.2	HO—C(=O)—OH 3.6	O=N(=O)—OH −1.4	
HO—Si(OH)₃ (OH, OH, OH) 10	HO—P(=O)(OH)₂ 2.1, 7.4, 12.7 O=Cl—OH 2.0	O=S(=O)(OH)₂ −1.9	O=Cl(=O)(=O)—OH −10
HO, HO—Te(—OH)(OH)—OH 7.8, 11.2	HO, HO—I(=O)(—OH)(OH) 1.6, 7.0 HO—P(=O)(OH)H 1.8, 6.6	O=Cl(OH)(=O) −1.0	
HO—B(OH)—OH 9.1*	HO—As(=O)(OH)(OH) 2.3, 6.9, 11.5 O=Se(OH)(OH) 2.6, 8.0		

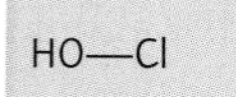

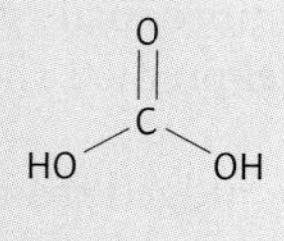

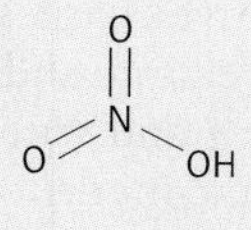

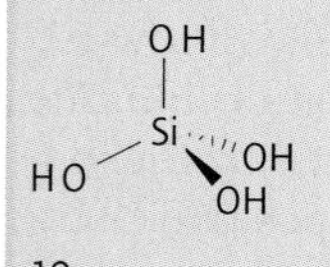

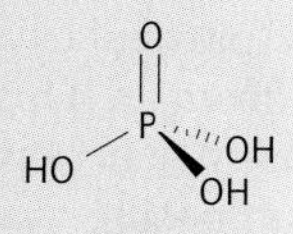

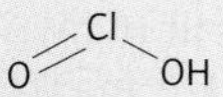

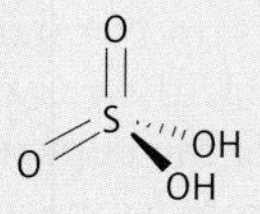

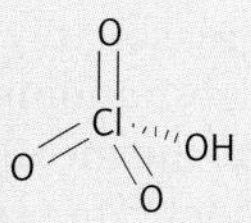

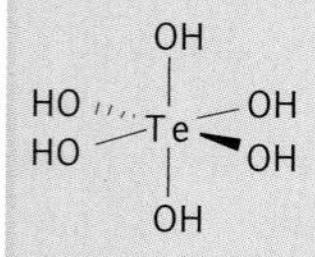

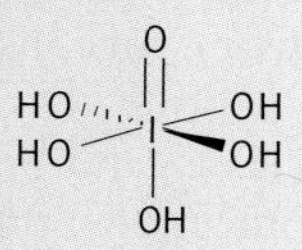

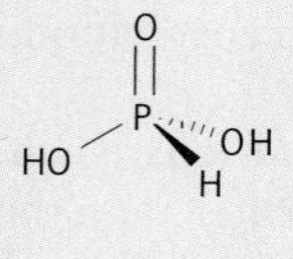

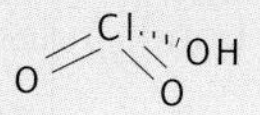

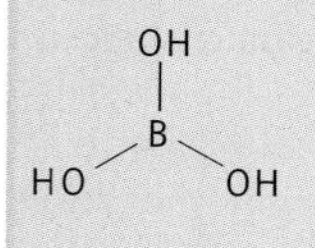

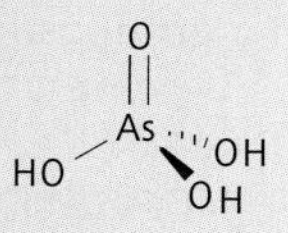

* p is the number of nonprotonated O atoms.
† Boric acid is a special case; see Section 13.5.

from the central S atom and confers on S a higher effective positive charge. As a result, the substituted acid is stronger than $O_2S(OH)_2$. Another electron acceptor substituent is $-CF_3$, as in the strong acid trifluoromethylsulfonic acid, CF_3SO_3H (that is, $O_2S(CF_3)(OH)$). By contrast, the $-NH_2$ group, which has lone pair electrons, can donate electron density to S by π bonding. This transfer of charge reduces the positive charge of the central atom and weakens the acid.

A trap for the unwary is that not all oxoacids follow the familiar structural pattern of a central atom surrounded by OH and O groups. Occasionally an H atom is attached directly to the central atom, as in phosphonic (phosphorous) acid, H_3PO_3. Phosphonic acid is in fact only a *di*protic acid, as the substitution of two OH groups leaves a P–H bond (**7**) and consequently a nonacidic proton. This structure is consistent with NMR and vibrational spectra, and the structural formula is $OPH(OH)_2$. The nonacidity of the H–P bond reflects the much lower electron-withdrawing ability of the central P atom compared to O (Section 4.1e). Substitution for an oxo group (as distinct from a hydroxyl group) is another example of a structural change that can occur. An important example is the thiosulfate ion, $S_2O_3^{2-}$ (**8**), in which an S atom replaces an O atom of a sulfate ion.

A note on good practice The structures of oxoacids are drawn with double bonds to oxo groups, $=O$. This representation indicates the connectivity of the O atom to the central atom but in reality resonance lowers the calculated energy of the molecule and distributes the bonding character of the electrons over the molecule.

(b) Pauling's rules

Key point: The strengths of a series of oxoacids containing a specific central atom with a variable number of oxo and hydroxyl groups are summarized by Pauling's rules.

For a series of mononuclear oxoacids of an element E, the strength of the acids increases with increasing number of O atoms. This trend can be explained qualitatively

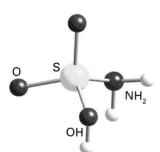

6 $O_2S(NH_2)OH$

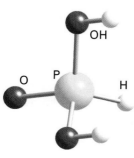

7 $OPH(OH)_2$, H_3PO_3

by considering the electron-withdrawing properties of oxygen. The O atoms withdraw electrons, so making each O–H bond weaker. Consequently, protons are more readily released. In general, for any series of oxoacids, the one with the most O atoms is the strongest. For example, the acid strengths of the oxoacids of chlorine decrease in the order $HOCl_4 > HClO_3 > HClO_2 > HClO$. Similarly, H_2SO_4 is stronger than H_2SO_3 and HNO_3 is stronger than HNO_2.

Another important factor is the degree to which differing numbers of terminal oxo groups stabilize the deprotonated (conjugate) base by resonance. For example, the conjugate base of H_2SO_4, the HSO_4^- anion, can be described as a resonance hybrid of three contributions (**9**), whereas the conjugate base of H_2SO_3, the HSO_3^- anion, has only two resonance contributions (**10**). Consequently, H_2SO_4 is a stronger acid than H_2SO_3.

The trends can be systematized semiquantitatively by using two empirical rules devised by Linus Pauling, where p is the number of oxo groups and q is the number of hydroxyl groups:

1. For the oxoacid $O_pE(OH)_q$, $pK_a \approx 8 - 5p$.
2. The successive pK_a values of polyprotic acids (those with $q > 1$), increase by 5 units for each successive proton transfer.

Rule 1 predicts that neutral hydroxoacids with $p = 0$ have $pK_a \approx 8$, acids with one oxo group have $pK_a \approx 3$, and acids with two oxo groups have $pK_a \approx -2$. For example, sulfuric acid, $O_2S(OH)_2$, has $p = 2$ and $q = 2$, and $pK_{a1} \approx -2$ (signifying a strong acid). Similarly, pK_{a2} is predicted to be $+3$, although comparison with the experimental value of 1.9 reminds us that these rules are only approximations.

The success of Pauling's rules may be gauged by inspection of Table 4.3, in which acids are grouped according to p. The variation in strengths down a group is not large, and the complicated, and perhaps cancelling, effects of changing structures allow the rules to work moderately well. The more important variation across the periodic table from left to right and the effect of change of oxidation number are taken into account by the number of oxo groups. In Group 15, the oxidation number $+5$ requires one oxo group (as in $OP(OH)_3$) whereas in Group 16 the oxidation number $+6$ requires two (as in $O_2S(OH)_2$).

(c) Structural anomalies

Key point: In certain cases, notably H_2CO_3 and H_2SO_3, a simple molecular formula misrepresents the composition of aqueous solutions of nonmetal oxides.

An interesting use of Pauling's rules is to detect structural anomalies. For example, carbonic acid, $OC(OH)_2$, is commonly reported as having $pK_{a1} = 6.4$, but the rules predict $pK_{a1} = 3$. The anomalously low acidity indicated by the experimental value is the result of treating the concentration of dissolved CO_2 as if it were all H_2CO_3. However, in the equilibrium

$$CO_2(aq) + H_2O(l) \rightleftharpoons OC(OH)_2(aq)$$

only about 1 per cent of the dissolved CO_2 is present as $OC(OH)_2$, so the actual concentration of acid is much less than the concentration of dissolved CO_2. When this difference is taken into account, the true pK_{a1} of H_2CO_3 is about 3.6, as Pauling's rules predict.

The experimental value $pK_{a1} = 1.8$ reported for sulfurous acid, H_2SO_3, suggests another anomaly, in this case acting in the opposite way. In fact, spectroscopic studies have failed to detect the molecule $OS(OH)_2$ in solution, and the equilibrium constant for

$$SO_2(aq) + H_2O(l) \rightleftharpoons H_2SO_3(aq)$$

is less than 10^{-9}. The equilibria of dissolved SO_2 are complex, and a simple analysis is inappropriate. The ions that have been detected include HSO_3^- and $S_2O_5^{2-}$, and there is evidence for an SH bond in the solid salts of the hydrogensulfite ion.

This discussion of the composition of aqueous solutions of CO_2 and SO_2 calls attention to the important point that not all nonmetal oxides react fully with water to form acids. Carbon monoxide is another example: although it is formally the anhydride of methanoic acid, HCOOH, carbon monoxide does not in fact react with water at room temperature to give the acid. The same is true of some metal oxides: OsO_4, for example, can exist as dissolved neutral molecules.

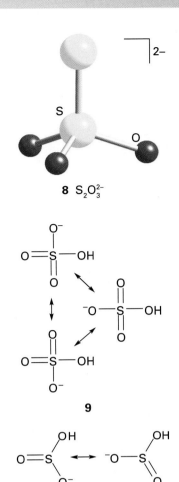

8 $S_2O_3^{2-}$

9

10

EXAMPLE 4.7 Using Pauling's rules

Identify the structural formulas that are consistent with the following pK_a values: H_3PO_4, 2.12; H_3PO_3, 1.80; H_3PO_2, 2.0.

Answer We can use Pauling's rules to use the pK_a values to predict the number of oxo groups. All three values are in the range that Pauling's first rule associates with one oxo group. This observation suggests the formulas $(HO)_3P{=}O$, $(HO)_2HP{=}O$, and $(HO)H_2P{=}O$. The second and the third formulas are derived from the first by replacement of $-OH$ by H bound to P (as in structure **7**).

Self-test 4.7 Predict the pK_a values of (a) H_3PO_4, (b) $H_2PO_4^-$, (c) HPO_4^{2-}.

4.6 Anhydrous oxides

We have treated oxoacids as being derived by deprotonation of their parent aqua acids. It is also useful to take the opposite viewpoint and to consider aqua acids and oxoacids as being derived by hydration of the oxides of the central atom. This approach emphasizes the acid and base properties of oxides and their correlation with the location of the element in the periodic table.

(a) Acidic and basic oxides

Key points: Metallic elements typically form basic oxides; nonmetallic elements typically form acidic oxides.

An **acidic oxide** is an oxide that, on dissolution in water, binds an H_2O molecule and releases a proton to the surrounding solvent:

$$CO_2(g) + H_2O(l) \rightleftharpoons OC(OH)_2(aq)$$

$$OC(OH)_2(aq) + H_2O(l) \rightleftharpoons H_3O^+(aq) + O_2COH^-(aq)$$

An equivalent interpretation is that an acidic oxide is an oxide that reacts with an aqueous base (an alkali):

$$CO_2(g) + OH^-(aq) \rightarrow O_2C(OH)^-(aq)$$

A **basic oxide** is an oxide to which a proton is transferred when it dissolves in water:

$$BaO(s) + H_2O(l) \rightarrow Ba^{2+}(aq) + 2\,OH^-(aq)$$

The equivalent interpretation in this case is that a basic oxide is an oxide that reacts with an acid:

$$BaO(s) + 2\,H_3O^+(aq) \rightarrow Ba^{2+}(aq) + 3\,H_2O(l)$$

Because acidic and basic oxide character often correlates with other chemical properties, a wide range of properties can be predicted from a knowledge of the character of oxides. In a number of cases the correlations follow from the basic oxides being largely ionic and of acidic oxides being largely covalent. For instance, an element that forms an acidic oxide is likely to form volatile, covalent halides. By contrast, an element that forms a basic oxide is likely to form solid, ionic halides. In short, the acidic or basic character of an oxide is a chemical indication of whether an element should be regarded as a metal or a nonmetal. Generally, metals form basic oxides and nonmetals form acidic oxides.

(b) Amphoterism

Key points: The frontier between metals and nonmetals in the periodic table is characterized by the formation of amphoteric oxides; amphoterism also varies with the oxidation state of the element.

An **amphoteric oxide** is an oxide that reacts with both acids and bases.[4] Thus, aluminium oxide reacts with acids and alkalis:

$$Al_2O_3(s) + 6\,H_3O^+(aq) + 3\,H_2O(l) \rightarrow 2\,[Al(OH_2)_6]^{3+}(aq)$$

$$Al_2O_3(s) + 2\,OH^-(aq) + 3\,H_2O(l) \rightarrow 2\,[Al(OH)_4]^-(aq)$$

---

[4] The word 'amphoteric' is derived from the Greek word for 'both'.

Amphoterism is observed for the lighter elements of Groups 2 and 13, as in BeO, Al_2O_3, and Ga_2O_3. It is also observed for some of the d-block elements in high oxidation states, such as MoO_3 and V_2O_5, in which the central atom is very electron withdrawing, and some of the heavier elements of Groups 14 and 15, such as SnO_2 and Sb_2O_5.

Figure 4.6 shows the location of elements that in their characteristic group oxidation states have amphoteric oxides. They lie on the frontier between acidic and basic oxides, and hence serve as an important guide to the metallic or nonmetallic character of an element. The onset of amphoterism correlates with a significant degree of covalent character in the bonds formed by the elements, either because the metal ion is strongly polarizing (as for Be) or because the metal ion is polarized by the O atom attached to it (as for Sb).

An important issue in the d block is the oxidation number necessary for amphoterism. Figure 4.7 shows the oxidation number for which an element in the first row of the block has an amphoteric oxide. We see that on the left of the block, from titanium to manganese and perhaps iron, oxidation state +4 is amphoteric (with higher values on the border of acidic and lower values of the border of basic). On the right of the block, amphoterism occurs at lower oxidation numbers: the oxidation states +3 for cobalt and nickel and +2 for copper and zinc are fully amphoteric. There is no simple way of predicting the onset of amphoterism. However, it presumably reflects the ability of the metal cation to polarize the oxide ions that surround it—that is, to introduce covalence into the metal–oxygen bond. The degree of covalence typically increases with the oxidation number of the metal as the increasingly positively charged cation becomes more strongly polarizing (Section 1.9e).

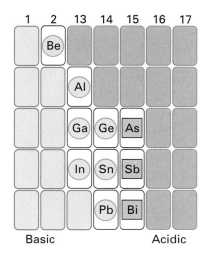

Figure 4.6 The location of elements having amphoteric oxides. The circled elements have amphoteric oxides in all oxidation states. The elements in boxes have acidic oxides in the highest oxidation state and amphoteric oxides in lower oxidation states.

EXAMPLE 4.8 Using oxide acidity in qualitative analysis

In the traditional scheme of qualitative analysis, a solution of metal ions is oxidized and then aqueous ammonia is added to raise the pH. The ions Fe^{3+}, Ce^{3+}, Al^{3+}, Cr^{3+}, and V^{3+} precipitate as hydrous oxides. The addition of H_2O_2 and NaOH redissolves the aluminium, chromium, and vanadium oxides. Discuss these steps in terms of the acidities of oxides.

Answer When the oxidation number of the metal is +3, all the metal oxides are sufficiently basic to be insoluble in a solution with pH ≈ 10. Aluminium(III) oxide is amphoteric and redissolves in alkaline solution to give aluminate ions, $[Al(OH)_4]^-$. Vanadium(III) and chromium(III) oxides are oxidized by H_2O_2 to give vanadate ions, $[VO_4]^{3-}$, and chromate ions, $[CrO_4]^{2-}$, which are the anions derived from the acidic oxides V_2O_5 and CrO_3, respectively.

Self-test 4.8 If Ti(IV) ions were present in the sample, how would they behave?

4.7 Polyoxo compound formation

Key points: Acids containing the OH group condense to form polyoxoanions; polycation formation from simple aqua cations occurs with the loss of H_2O. Oxoanions form polymers as the pH is lowered whereas aqua ions form polymers as the pH is raised.

As the pH of a solution is increased, the aqua ions of metals that have basic or amphoteric oxides generally undergo polymerization and precipitation. Because the precipitation occurs quantitatively at a pH characteristic of each metal, one application of this behaviour is the separation of metal ions,.

With the exception of Be^{2+} (which is amphoteric), the elements of Groups 1 and 2 have no important solution species beyond the aqua ions $M^+(aq)$ and $M^{2+}(aq)$. By contrast, the solution chemistry of the elements becomes very rich as the amphoteric region of the periodic table is approached. The two most common examples are polymers formed by Fe(III) and Al(III), both of which are abundant in the Earth's crust. In acidic solutions, both form octahedral hexaaqua ions, $[Al(OH_2)_6]^{3+}$ and $[Fe(OH_2)_6]^{3+}$. In solutions of pH > 4, both precipitate as gelatinous hydrous oxides:

$$[Fe(OH_2)_6]^{3+}(aq) + (3 + n)\, H_2O(l) \rightarrow Fe(OH)_3 \cdot nH_2O(s) + 3\, H_3O^+(aq)$$

$$[Al(OH_2)_6]^{3+}(aq) + (3 + n)\, H_2O(l) \rightarrow Al(OH)_3 \cdot nH_2O(s) + 3\, H_3O^+(aq)$$

The precipitated polymers, which are often of colloidal dimensions (between 1 nm and 1 μm), slowly crystallize to stable mineral forms. The extensive network structure of

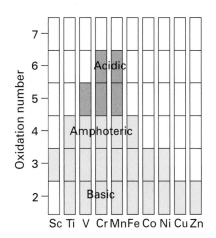

Figure 4.7 The oxidation numbers for which elements in the first row of the d block have amphoteric oxides. Predominantly acidic oxides are shown shaded pink, predominantly basic oxides are shaded blue.

aluminium polymers, which are neatly packed in three dimensions, contrasts with the linear polymers of their iron analogues.

Polyoxoanion formation from oxoanions occurs by protonation of an O atom and its departure as H_2O:

$$2\,[CrO_4]^{2-}(aq) + 2\,H_3O^+(aq) \rightarrow [O_3CrOCrO_3]^{2-}(aq) + 3\,H_2O(l)$$

The importance of polyoxo anions can be judged by the fact that they account for most of the mass of oxygen in the Earth's crust, as they include almost all silicate minerals. They also include the phosphate polymers (such as ATP (**11**)) used for energy transfer in living cells.

The formation of polyoxoanions is important for early d-block ions, particularly V(V), Mo(VI), W(VI), and (to a lesser extent) Nb(V), Ta(V), and Cr(VI); see Section 19.8. They are formed when base is added to aqueous solutions of the ions or oxides in high oxidation states. Polyoxoanions are also formed by some nonmetals, but their structures are different from those of their d-metal analogues. The common species in solution are rings and chains. The silicates are very important examples of polymeric oxoanions, and we discuss them in detail in Chapter 14. One example of a polysilicate mineral is $MgSiO_3$, which contains an infinite chain of SiO_3^{2-} units. In this section we illustrate some features of polyoxoanions using phosphates as examples.

The simplest condensation reaction, starting with the orthophosphate ion, PO_4^{3-}, is

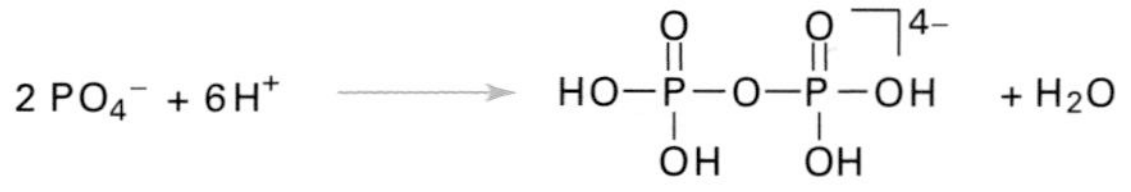

$$2\,PO_4^- + 6H^+ \longrightarrow HO-\overset{\overset{\displaystyle O}{\|}}{\underset{\underset{\displaystyle OH}{|}}{P}}-O-\overset{\overset{\displaystyle O}{\|}}{\underset{\underset{\displaystyle OH}{|}}{P}}-OH \Bigg]^{4-} + H_2O$$

The elimination of water consumes protons and decreases the average charge number of each P atom to −2. If each phosphate group is represented as a tetrahedron with the O atoms located at the corners, the diphosphate ion, $P_2O_7^{4-}$ (**12**), can be drawn as (**13**). Phosphoric acid can be prepared by hydrolysis of the solid phosphorus(V) oxide, P_4O_{10}. An initial step using a limited amount of water produces a metaphosphate ion with the formula $P_4O_{12}^{4-}$ (**14**). This reaction is only the simplest among many, and the separation of products from the hydrolysis of phosphorus(V) oxide by chromatography reveals the presence of chain species with from one to nine P atoms. Higher polymers are also present and can be removed from the column only by hydrolysis. Figure 4.8 is a schematic representation of a two-dimensional paper chromatogram: the upper spot sequence corresponds to linear polymers and the lower sequence corresponds to rings. Chain polymers of formula P_n with $n = 10$ to 50 can be isolated as mixed amorphous glasses analogous to those formed by silicates (Section 14.15).

The polyphosphates are biologically important. At physiological pH (close to 7.4), the P–O–P entity is unstable with respect to hydrolysis. Consequently, its hydrolysis can serve as a mechanism for providing the energy to drive a reaction (the Gibbs energy). Similarly, the formation of the P–O–P bond is a means of storing Gibbs energy. The key to energy

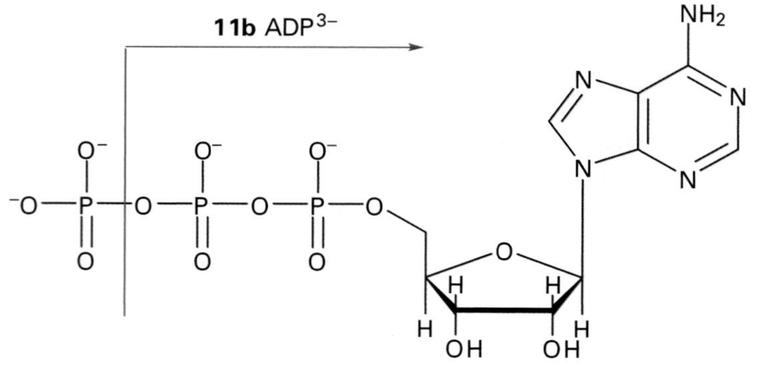

11b ADP^{3-}

11a ATP^{4-}

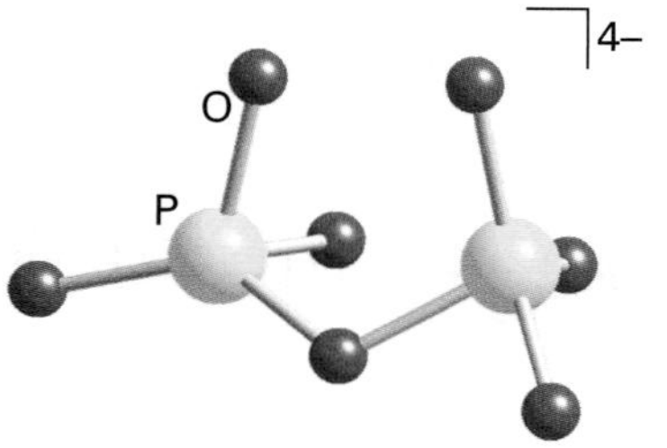

12 $P_2O_7^{4-}$

exchange in metabolism is the hydrolysis of adenosine triphosphate, ATP (**11a**), to adenosine diphosphate, ADP (**11b**):

$$ATP^{4-} + 2\,H_2O \rightarrow ADP^{3-} + HPO_4^{2-} + H_3O^+ \quad \Delta_r G^{\ominus} = -41 \text{ kJ mol}^{-1} \text{ at pH} = 7.4$$

Energy flow in metabolism depends on the subtle construction of pathways to make ATP from ADP. The energy is used metabolically by pathways that have evolved to exploit the delivery of a thermodynamic driving force resulting from the hydrolysis of ATP.

4.8 Nonaqueous solvents

Not all proton transfer reactions take place in aqueous media. Nonaqueous solvents can be selected for reactions of molecules that are readily hydrolyzed, to avoid levelling by water, or to enhance the solubility of a solute. Nonaqueous solvents are often selected on the basis of their liquid range and relative permittivity. Some physical properties of some common nonaqueous solvents are given in Table 4.4. The solvent system definition of acids and bases applies to both protic and aprotic nonaqueous solvents.

(a) Liquid ammonia

Key points: Liquid ammonia is a useful nonaqueous solvent. Many reactions in liquid ammonia are analogous to those in water.

Liquid ammonia is widely used as a nonaqueous solvent. It boils at −33°C at 1 atm and, despite a somewhat lower relative permittivity ($\varepsilon_r = 22$) than that of water, it is a good solvent for inorganic compounds such as ammonium salts, nitrates, cyanides, and thiocyanides, and organic compounds such as amines, alcohols, and esters. It closely resembles the aqueous system as can be seen from the autoionization

$$2\,NH_3(l) \rightleftharpoons NH_4^+(sol) + NH_2^-(sol)$$

Solutes that increase the concentration of NH_4^+, the solvated proton, are acids. Solutes that decrease the concentration of NH_4^+ or increase the concentration of NH_2^- are defined as bases. Thus, ammonium salts are acids in liquid ammonia and amines are bases.

Liquid ammonia is a more basic solvent than water and enhances the acidity of many compounds that are weak acids in water. For example, acetic acid is almost completely ionized in liquid ammonia:

$$CH_3COOH(sol) + NH_3(l) \rightarrow NH_4^+(sol) + CH_3CO_2^-(aq)$$

Many reactions in liquid ammonia are analogous to those in water. The following acid–base neutralization can be carried out:

$$NH_4Cl(sol) + NaNH_2(sol) \rightarrow NaCl(sol) + 2NH_3(l)$$

Liquid ammonia is a very good solvent for alkali and alkali earth metals, with the exception of beryllium. The alkali metals are particularly soluble and 336 g of caesium can be dissolved in 100 g of liquid ammonia at −50°C. The metals can be recovered by evaporating the ammonia. These solutions are very conducting and are blue when dilute and bronze when concentrated. Electron paramagnetic resonance spectra (Section 8.6) show that the

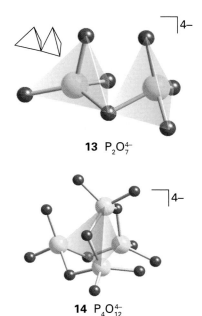

13 $P_2O_7^{4-}$

14 $P_4O_{12}^{4-}$

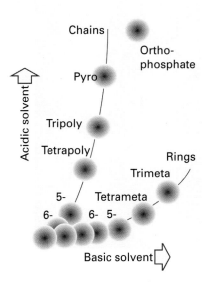

Figure 4.8 A representation of a two-dimensional paper chromatogram of a complex mixture of phosphates formed by condensation reactions. The sample spot was placed at the lower left corner. Basic solvent separation was used first, followed by acidic solvent perpendicular to the basic one. This separates open chains from rings. The upper spot sequence corresponds to linear polymers and the lower sequence corresponds to rings.

Table 4.4 Physical properties of some nonaqueous solvents

Solvent	Melting point/°C	Boiling point/°C	Relative permittivity
Liquid ammonia	−77.7	−33.5	23.9 (at −33°C)
Glacial acetic acid	16.7	117.9	6.15
Sulfuric acid	10.4	290 (decomposes)	100
Hydrogen fluoride	−83.4	19.5	80
Ethanol	−114.5	78.3	24.55
Dinitrogen tetroxide	−11.2	21.1	2.42
Bromine trifluoride	8.8	125.8	107
Dimethyl sulfoxide (DMSO)	18.5	189	46.45

solutions contain unpaired electrons. The blue colour typical of the solutions is the outcome of a very broad optical absorption band in the near IR with a maximum near 1500 nm. The metal is ionized in ammonia solution to give 'solvated electrons':

$$Na(s) \xrightarrow{\text{NH}_3(l)} Na^+(sol) + e^-(sol)$$

The blue solutions survive for long times at low temperature but decompose slowly to give hydrogen and sodium amide, $NaNH_2$. The exploitation of the blue solutions to produce compounds called 'electrides' is discussed in Section 11.13.

(b) Hydrogen fluoride

Key point: Hydrogen fluoride is a reactive toxic solvent that is highly acidic.

Liquid hydrogen fluoride (bp 19.5°C) is an acidic solvent with a relative permittivity (ε_r = 84 at 0°C) comparable to that of water (ε_r = 78 at 25°C). It is a good solvent for ionic substances. However, as it is both highly reactive and toxic, it presents handling problems, including its ability to etch glass. In practice, liquid hydrogen fluoride is usually contained in polytetrafluoroethylene and polychlorotrifluoroethylene vessels. Hydrogen fluoride is particularly hazardous because it penetrates tissue rapidly and interferes with nerve function. Consequently, burns may go undetected and treatment may be delayed. It can also etch bone and reacts with calcium in the blood.

Liquid hydrogen fluoride is a highly acidic solvent as it has a high autoprotolysis constant and produces solvated protons very readily (Section 4.1(b)):

$$3\,HF(l) \rightarrow H_2F^+(sol) + HF_2^-(sol)$$

Although the conjugate base of HF is formally F⁻, the ability of HF to form a strong hydrogen bond to F⁻ means that the conjugate base is better regarded as the bifluoride ion, HF_2^-. Only very strong acids are able to donate protons and function as acids in HF, for example fluorosulfonic acid:

$$HSO_3F(sol) + HF(l) \rightleftharpoons H_2F^+(sol) + SO_3F^-(sol)$$

Organic compounds such as acids, alcohols, ethers, and ketones can accept a proton and act as bases in HF(l). Other bases increase the concentration of HF_2^- to produce basic solutions:

$$CH_3COOH(l) + 2\,HF(l) \rightleftharpoons CH_3C(OH)_2^+(sol) + HF_2^-(sol)$$

In this reaction acetic acid, an acid in water, is acting as a base.

Many fluorides are soluble in liquid HF as a result of the formation of the HF_2^- ion; for example

$$LiF(s) + HF(l) \rightarrow Li^+(sol) + HF_2^-(sol)$$

(c) Anhydrous sulfuric acid

Key point: The autoionization of anhydrous sulfuric acid is complex, with several competing side reactions.

Anhydrous sulfuric acid is an acidic solvent. It has a high relative permittivity and is viscous because of extensive hydrogen bonding (Section 10.6). Despite this association the solvent is appreciably autoionized at room temperature. The major autoionization is

$$2\,H_2SO_4(l) \rightleftharpoons H_3SO_4^+(sol) + HSO_4^-(sol)$$

However, there are secondary autoionizations and other equilibria, such as

$$H_2SO_4(l) \rightleftharpoons H_2O(sol) + SO_3(sol)$$

$$H_2O(sol) + H_2SO_4(l) \rightleftharpoons H_3O^+(sol) + HSO_4^-(sol)$$

$$SO_3(sol) + H_2SO_4(l) \rightleftharpoons H_2S_2O_7(sol)$$

$$H_2S_2O_7(sol) + H_2SO_4(l) \rightleftharpoons H_3SO_4^+(sol) + HS_2O_7^-(sol)$$

The high viscosity and high level of association through hydrogen bonding would usually lead to low ion mobilities. However, the mobilities of $H_3SO_4^+$ and HSO_4^- are comparable to those of H_3O^+ and OH^- in water, indicating that similar proton transfer mechanisms are taking place. The main species taking part are $H_3SO_4^+$ and HSO_4^-:

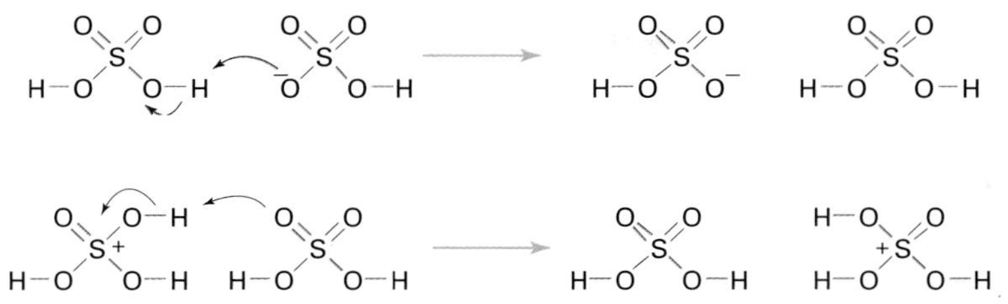

Most strong oxo acids accept a proton in anhydrous sulfuric acid and are thus bases:

$$H_3PO_4(sol) + H_2SO_4(l) \rightleftharpoons H_4PO_4^+(sol) + HSO_4^-(sol)$$

An important reaction is that of nitric acid with sulfuric acid to generate the nitronium ion, NO_2^+, which is the active species in aromatic nitration reactions:

$$HNO_3(sol) + 2\,H_2SO_4(l) \rightleftharpoons NO_2^+(sol) + H_3O^+(sol) + 2\,HSO_4^-(sol)$$

Some acids that are very strong in water act as weak acids in anhydrous sulfuric acids, for example perchloric acid, $HClO_4$, and fluorosulfuric acid, $HFSO_3$.

(d) Dinitrogen tetroxide

Key point: Dinitrogen tetroxide autoionizes by two reactions. The preferred route can be enhanced by addition of electron-pair donors or acceptors.

Dinitrogen tetroxide, N_2O_4, has a narrow liquid range with a freezing point at $-11.2°C$ and boiling point of $21.2°C$. Two autoionization reactions occur:

$$N_2O_4(l) \rightleftharpoons NO^+(sol) + NO_3^-(sol)$$
$$N_2O_4(l) \rightleftharpoons NO_2^+(sol) + NO_2^-(sol)$$

The first autoionization is enhanced by addition of electron pair donors (which in the next section we see to be 'Lewis bases'), such as diethyl ether:

$$N_2O_4(l) + :X \rightleftharpoons XNO^+(sol) + NO_3^-(sol)$$

Electron pair acceptors ('Lewis acids'; see the next section) such as BF_3 enhance the second autoionization reaction:

$$N_2O_4(l) + BF_3(sol) \rightleftharpoons NO_2^+(sol) + F_3BNO_2^-(sol)$$

Dinitrogen tetroxide has a low relative permittivity and is not a very useful solvent for inorganic compounds. It is, however, a good solvent for many esters, carboxylic acids, halides, and organic nitro compounds.

Lewis acidity

Key points: A Lewis acid is an electron pair acceptor; a Lewis base is an electron pair donor.

The Brønsted–Lowry theory of acids and bases focuses on the transfer of a proton between species. The solvent system generalizes the Brønsted–Lowry theory to include the transfer of cationic and anionic species other than protons. Whereas both definitions are more general than any that preceded them, they still fail to take into account reactions between substances that show similar features but in which no proton or other charged species is transferred. This deficiency was remedied by a more general theory of acidity introduced by G.N. Lewis in the same year as Brønsted and Lowry introduced theirs (1923). Lewis's approach became influential only in the 1930s.

A **Lewis acid** is a substance that acts as an electron pair acceptor. A **Lewis base** is a substance that acts as an electron pair donor. We denote a Lewis acid by A and a Lewis base by :B, often omitting any other lone pairs that may be present. The fundamental reaction of Lewis acids and bases is the formation of a **complex** (or adduct), A–B, in which A and :B bond together by sharing the electron pair supplied by the base.

A note on good practice The terms 'Lewis acid' and 'Lewis base' are used in discussions of the equilibrium properties of reactions. In the context of reaction rates, an electron pair donor is called a nucleophile and an electron acceptor is called an electrophile.

4.9 Examples of Lewis acids and bases

Key points: Brønsted acids and bases exhibit Lewis acidity and basicity; the Lewis definition can be applied to aprotic systems.

A proton is a Lewis acid because it can attach to an electron pair, as in the formation of NH_4^+ from NH_3. It follows that any Brønsted acid, as it provides protons, exhibits Lewis acidity too. Note that the Brønsted acid HA is the complex formed by the Lewis acid H^+ with the Lewis base A^-. We say that a Brønsted acid *exhibits* Lewis acidity rather than that a Brønsted acid *is* a Lewis acid. All Brønsted bases are Lewis bases because a proton acceptor is also an electron pair donor: an NH_3 molecule, for instance, is a Lewis base as well as a Brønsted base. Therefore, the whole of the material presented in the preceding sections of this chapter can be regarded as a special case of Lewis's approach. However, because the proton is not essential to the definition of a Lewis acid or base, a wider range of substances can be classified as acids and bases in the Lewis scheme than can be classified in the Brønsted scheme.

   We meet many examples of Lewis acids later, but we should be alert to the following possibilities:

1. A molecule with an incomplete octet of valence electrons can complete its octet by accepting an electron pair.
   A prime example is $B(CH_3)_3$, which can accept the lone pair of NH_3 and other donors:

   Hence, $B(CH_3)_3$ is a Lewis acid.

2. A metal cation can accept an electron pair supplied by the base in a coordination compound. This aspect of Lewis acids and bases is treated at length in Chapters 7 and 20. An example is the hydration of Co^{2+}, in which the lone pairs of H_2O (acting as a Lewis base) donate to the central cation to give $[Co(OH_2)_6]^{2+}$. The Co^{2+} cation is therefore the Lewis acid.

3. A molecule or ion with a complete octet may be able to rearrange its valence electrons and accept an additional electron pair.
   For example, CO_2 acts as a Lewis acid when it forms HCO_3^- by accepting an electron pair from an O atom in an OH^- ion:

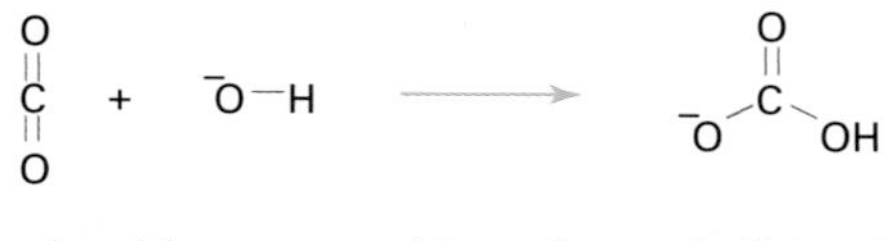

4. A molecule or ion may be able to expand its valence shell (or simply be large enough) to accept another electron pair. An example is the formation of the complex $[SiF_6]^{2-}$ when two F^- ions (the Lewis bases) bond to SiF_4 (the acid).

   This type of Lewis acidity is common for the halides of the heavier p-block elements, such as SiX_4, AsX_3, and PX_5 (with X a halogen).

EXAMPLE 4.9 Identifying Lewis acids and bases

Identify the Lewis acids and bases in the reactions (a) $BrF_3 + F^- \rightarrow BrF_4^-$, (b) $KH + H_2O \rightarrow KOH + H_2$.

Answer We need to identify the electron pair acceptor (the acid) and the electron pair donor (the base). (a) The acid BrF_3 accepts a pair of electrons from the base F^-. Therefore BrF_3 is a Lewis acid and F^- is a Lewis base. (b) The ionic hydride complex KH provides H^-, which displaces H^+ from water to give H_2 and OH^-. The net reaction is

$$H^- + H_2O \rightarrow H_2 + OH^-$$

If we think of this reaction as

$$H^- + H^+{:}OH^- \rightarrow HH + {:}OH^-$$

we see that H^- provides a lone pair and is therefore a Lewis base. It reacts with H_2O to drive out OH^-, another Lewis base.

Self-test 4.9 Identify the acids and bases in the reactions (a) $FeCl_3 + Cl^- \rightarrow FeCl_4^-$, (b) $I^- + I_2 \rightarrow I_3^-$.

4.10 Group characteristics of Lewis acids

An understanding of the trends in Lewis acidity and basicity enables us to predict the outcome of many reactions of the s- and p-block elements.

(a) Lewis acids and bases of the s-block elements

Key point: Alkali metal ions act as Lewis acids with water, forming hydrated ions.

The existence of hydrated alkali metal ions in water can be regarded as an aspect of their Lewis acid character, with H_2O the Lewis base. Alkali metal ions do not act as Lewis bases but their fluorides act as a source of the Lewis base F^- and form fluoride complexes with Lewis acids, such as SF_4:

$$CsF + SF_4 \rightarrow Cs^+[SF_5]^-$$

The Be atom in beryllium dihalides acts as a Lewis acid by forming a polymeric chain structure in the solid state (**15**). In this structure, a σ bond is formed when a lone pair of electrons of a halide ion, acting as a Lewis base, is donated into an empty sp^3 hybrid orbital on the Be atom. The Lewis acidity of beryllium chloride is also demonstrated by the formation of adducts such as $BeCl_4^{2-}$ (**16**).

(b) Group 13 Lewis acids

Key points: The ability of boron trihalides to act as Lewis acids generally increases in the order $BF_3 < BCl_3 < BBr_3$; aluminium halides are dimeric in the gas phase and are used as catalysts in solution.

The planar molecules BX_3 and AlX_3 have incomplete octets, and the vacant p orbital perpendicular to the plane (**17**) can accept a lone pair from a Lewis base:

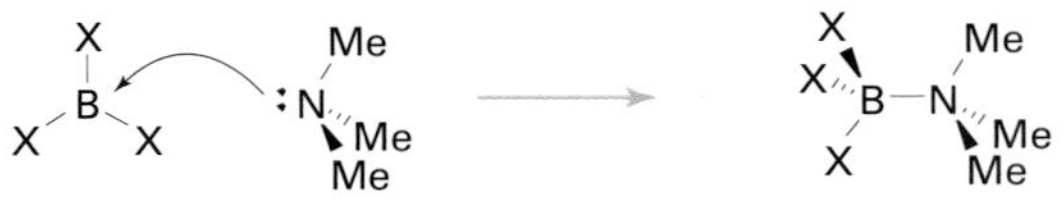

The acid molecule becomes pyramidal as the complex is formed and the B–X bonds bend away from their new neighbours.

The order of thermodynamic stability of complexes of $:N(CH_3)_3$ with BX_3 is $BF_3 < BCl_3 < BBr_3$. This order is opposite to that expected on the basis of the relative electronegativities of the halogens: an electronegativity argument would suggest that F, the most electronegative halogen, ought to leave the B atom in BF_3 most electron deficient and hence able to form the strongest bond to the incoming base. The currently accepted explanation is that the halogen atoms in the BX_3 molecule can form π bonds with the empty B2p orbital (**18**), and that these π bonds must be disrupted to make the acceptor orbital available for complex formation. The π bond also favours the planar structure of the molecule, a structure that must be converted into tetrahedral in the adduct. The small F atom forms the

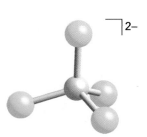

15 $BeHal_2$

16 $BeCl_4^{2-}$

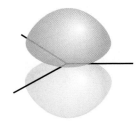

17 AlX_3 and BX_3

strongest π bonds with the B2p orbital: recall that p–p π bonding is strongest for Period 2 elements, largely on account of the small atomic radii of these elements and the significant overlap of their compact 2p orbitals (Section 2.5). Thus, the BF_3 molecule has the strongest π bond to be broken when the amine forms an N–B bond.

Boron trifluoride is widely used as an industrial catalyst. Its role is to extract bases bound to carbon and hence to generate carbocations:

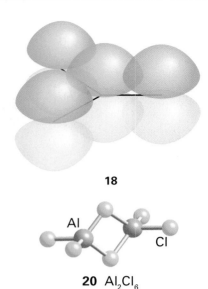

18

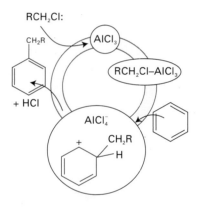

20 Al_2Cl_6

Boron trifluoride is a gas at room temperature and pressure, but it dissolves in diethyl ether to give a solution that is convenient to use. This dissolution is also an aspect of Lewis acid character because, as BF_3 dissolves, it forms a complex with the :O atom of a solvent molecule.

Aluminium halides are dimers in the gas phase; aluminium chloride, for example, has molecular formula Al_2Cl_6 in the vapour state (**19**). Each Al atom acts as an acid towards a Cl atom initially belonging to the other Al atom. Aluminium chloride is widely used as a Lewis acid catalyst for organic reactions. The classic examples are Friedel–Crafts alkylation (the attachment of R^+ to an aromatic ring) and acylation (the attachment of RCO) during which $AlCl_4^-$ is formed. The catalytic cycle is shown in Fig. 4.9.

(c) Group 14 Lewis acids

Key points: Group 14 elements other than carbon exhibit hypervalence and act as Lewis acids by becoming five- or six-coordinate; tin(II) chloride is both a Lewis acid and a Lewis base.

Unlike carbon, a Si atom can expand its valence shell (or is simply large enough) to become hypervalent. For example, a five-coordinate trigonal bipyramidal structure is possible (**20**). A representative Lewis acid–base reaction is that of SiF_4 with two F^- ions:

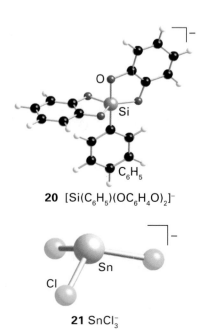

Figure 4.9 The catalytic cycle for the Friedel-Crafts alkylation reaction.

Germanium and tin fluorides can react similarly. Because the Lewis base F^-, aided by a proton, can displace O^{2-} from silicates, hydrofluoric acid is corrosive towards glass (SiO_2). The trend in acidity for SiX_4, which follows the order $SiI_4 < SiBr_4 < SiCl_4 < SiF_4$, correlates with the increase in the electron-withdrawing power of the halogen from I to F and is the reverse of that for BX_3.

Tin(II) chloride is both a Lewis acid and a Lewis base. As an acid, $SnCl_2$ combines with Cl^- to form $SnCl_3^-$ (**21**). This complex retains a lone pair, and it is sometimes more revealing to write its formula as $:SnCl_3^-$. It acts as a base to give metal–metal bonds, as in the complex $(CO)_5Mn–SnCl_3$ (**22**). Compounds containing metal–metal bonds are currently the focus of much attention in inorganic chemistry, as we see later in the text (Section 19.11). Tin(IV) halides are Lewis acids. They react with halide ions to form SnX_6^{2-}:

$$SnCl_4 + 2\,Cl^- \rightarrow SnCl_6^{2-}$$

The strength of the Lewis acidity follows the order $SnF_4 > SnCl_4 > SnBr_4 > SnI_4$.

20 $[Si(C_6H_5)(OC_6H_4O)_2]^-$

21 $SnCl_3^-$

EXAMPLE 4.10 Predicting the relative Lewis basicity of compounds

Rationalize the following relative Lewis basicities: (a) $(H_3Si)_2O < (H_3C)_2O$; (b) $(H_3Si)_3N < (H_3C)_3N$.

Answer Nonmetallic elements in Period 3 and later can expand their valence shells by delocalization of the O or N lone pairs to create multiple bonds (O and N are thus acting as π-electron donors). The silyl ether and silyl amine are therefore the weaker Lewis bases in each pair.

Self-test 4.10 Given that π bonding between Si and the lone pairs of N is important, what difference in structure between $(H_3Si)_3N$ and $(H_3C)_3N$ do you expect?

(d) Group 15 Lewis acids

Key points: Oxides and halides of the heavier Group 15 elements act as Lewis acids.

Phosphorus pentafluoride is a strong Lewis acid and forms complexes with ethers and amines. The heavier elements of the nitrogen group (Group 15) form some of the most important Lewis acids, SbF_5 being one of the most widely studied compounds. The reaction with HF produces a **superacid** (Section 4.15)

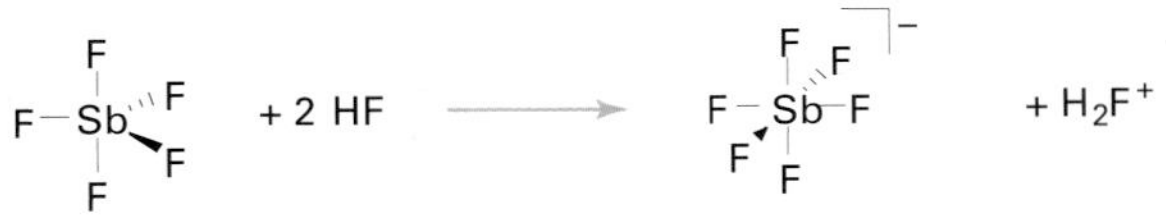

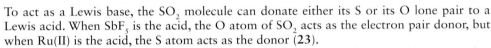

(e) Group 16 Lewis acids

Key points: Sulfur dioxide can act as a Lewis acid by the formation of a complex; to act as a Lewis base, the SO_2 molecule can donate either its S or its O lone pair to a Lewis acid.

Sulfur dioxide is both a Lewis acid and a Lewis base. Its Lewis acidity is illustrated by the formation of a complex with a trialkylamine acting as a Lewis base:

To act as a Lewis base, the SO_2 molecule can donate either its S or its O lone pair to a Lewis acid. When SbF_5 is the acid, the O atom of SO_2 acts as the electron pair donor, but when Ru(II) is the acid, the S atom acts as the donor (**23**).

Sulfur trioxide is a strong Lewis acid and a very weak (O donor) Lewis base. Its acidity is illustrated by the reaction

A classic aspect of the acidity of SO_3 is its highly exothermic reaction with water in the formation of sulfuric acid. The resulting problem of having to remove large quantities of heat from the reactor used for the commercial production of sulfuric acid is alleviated by exploiting the Lewis acidity of sulfur trioxide further to carry out the hydration by a two-stage process. Before dilution, sulfur trioxide is dissolved in sulfuric acid to form the mixture known as *oleum*. This reaction is an example of Lewis acid–base complex formation:

The resulting $H_2S_2O_7$ can then be hydrolyzed in a less exothermic reaction:

$$H_2S_2O_7 + H_2O \rightarrow 2 H_2SO_4$$

(f) Halogens as Lewis acids

Key point: Bromine and iodine molecules act as mild Lewis acids.

Lewis acidity is expressed in an interesting and subtle way by Br_2 and I_2, which are both strongly coloured. The strong visible absorption spectra of Br_2 and I_2 arise from transitions to low-lying unfilled antibonding orbitals. The colours of the species therefore suggest that the empty orbitals may be low enough in energy to serve as acceptor orbitals in

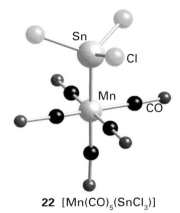

22 $[Mn(CO)_5(SnCl_3)]$

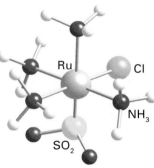

23 $[RuCl(NH_3)_4(SO_2)]^+$

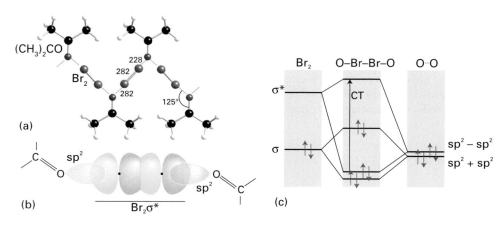

Figure 4.10 The interaction of Br_2 with the carbonyl group of propanone. (a) The structure of $(CH_3)_2COBr_2$ shown by X-ray diffraction. (b) The orbital overlap responsible for the complex formation. (c) A partial molecular orbital energy level diagram for the σ and σ^* orbitals of Br_2 with the appropriate combinations of the sp^2 orbitals on the two O atoms. The charge transfer transition is labelled CT.

Lewis acid–base complex formation.[5] Iodine is violet in the solid and gas phases, and in nondonor solvents such as trichloromethane. In water, propanone (acetone), or ethanol, all of which are Lewis bases, iodine is brown. The colour changes because a solvent–solute complex is formed from the lone pair of donor molecule O atoms and a low-lying σ^* orbital of the dihalogen.

The interaction of Br_2 with the carbonyl group of propanone is shown in Fig. 4.10. The illustration also shows the transition responsible for the new absorption band observed when a complex is formed. The orbital from which the electron originates in the transition is predominantly the lone pair orbital of the base (the ketone). The orbital to which the transition occurs is predominantly the LUMO of the acid (the dihalogen). Thus, to a first approximation, the transition transfers an electron from the base to the acid and is therefore called a **charge-transfer transition.**

The triiodide ion, I_3^-, is an example of a complex between a halogen acid (I_2) and a halide base (I^-). One of the applications of its formation is to render molecular iodine soluble in water so that it can be used as a titration reagent:

$$I_2(s) + I^-(aq) \rightleftharpoons I_3^-(aq) \qquad K = 725$$

The triiodide ion is one example of a large class of polyhalide ions (Section 17.8).

Reactions and properties of Lewis acids and bases

Reactions of Lewis acids and bases are widespread in chemistry, the chemical industry, and biology. For example, cement is made by grinding together limestone ($CaCO_3$) and a source of aluminosilicates, such as clay, shale, or sand, which are then heated to 1500°C in a rotary cement kiln. The limestone is heated and decomposes to lime (CaO), which reacts with the silicates to form molten calcium silicates of varying compositions such as Ca_2SiO_4, Ca_3SiO_5, and $Ca_3Al_2O_6$.

$$2\,CaO(s) + SiO_2(s) \rightarrow Ca_2SiO_4(s)$$

In industry carbon dioxide is removed from flue gas in order to reduce atmospheric emissions and to supply the demands of the soft drinks industry. This is achieved by using liquid amine scrubbers.

$$2\,RNH_2(aq) + CO_2(g) + H_2O(l) \rightarrow (RNH_3)_2CO_3(aq)$$

---

[5] The terms *donor–acceptor complex* and *charge-transfer complex* were at one time used to denote these complexes. However, the distinction between these complexes and the more familiar Lewis acid–base complexes is arbitrary and in the current literature the terms are used more or less interchangeably.

The toxicity of carbon monoxide to animals is an example of a Lewis acid–base reaction. Normally, oxygen forms a bond to the Fe(II) atom of haemoglobin and does so reversibly. Carbon monoxide is a much better Lewis acid than O_2 and forms a strong, almost irreversible, bond to the iron(II) site of haemoglobin:

$$Hb-Fe^{II} + CO \rightarrow Hb-Fe^{II}CO$$

All reactions between d-block metal atoms or ions to form coordination compounds (Chapter 7) are examples of reactions between a Lewis acid and a Lewis base:

$$Ni^{2+}(aq) + 6\,NH_3 \rightarrow [Ni(NH_3)_6]^{2+}$$

Friedel–Crafts alkylations and acylations are widely used in synthetic organic chemistry. They require a strong Lewis acid catalyst such as $AlCl_3$ or $FeCl_3$.

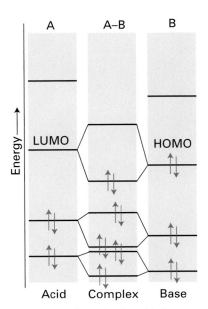

The first step is the reaction between the Lewis acid and the alkyl halide:

$$RCl + AlCl_3 \rightarrow R^+ + [AlCl_4]^-$$

4.11 The fundamental types of reaction

Lewis acids and bases undergo a variety of characteristic reactions. The simplest Lewis acid–base reaction in the gas phase or noncoordinating solvents is **complex formation**:

$$A + :B \rightarrow A-B$$

Two examples are

Both reactions involve Lewis acids and bases that are independently stable in the gas phase or in solvents that do not form complexes with them. Consequently, the individual species (as well as the complexes) may be studied experimentally.

Figure 4.11 shows the interaction of orbitals responsible for bonding in Lewis complexes. The exothermic character of the formation of the complex stems from the fact that the newly formed bonding orbital is populated by the two electrons supplied by the base whereas the newly formed antibonding orbital is left unoccupied. As a result, there is a net lowering of energy when the bond forms.

(a) Displacement reactions

Key point: In a displacement reaction, an acid or base drives out another acid or base from a Lewis complex.

A **displacement** of one Lewis base by another is a reaction of the form

$$B-A + :B' \rightarrow B: + A-B'$$

An example is

Figure 4.11 The molecular orbital representation of the orbital interactions responsible for formation of a complex between Lewis acid A and Lewis base :B.

All Brønsted proton transfer reactions are of this type, as in

$$HS^-(aq) + H_2O(l) \rightarrow S^{2-}(aq) + H_3O^+(aq)$$

In this reaction, the Lewis base H_2O displaces the Lewis base S^{2-} from its complex with the acid H^+. Displacement of one acid by another,

$$A' + B{-}A \rightarrow A'{-}B + A$$

is also possible, as in the reaction

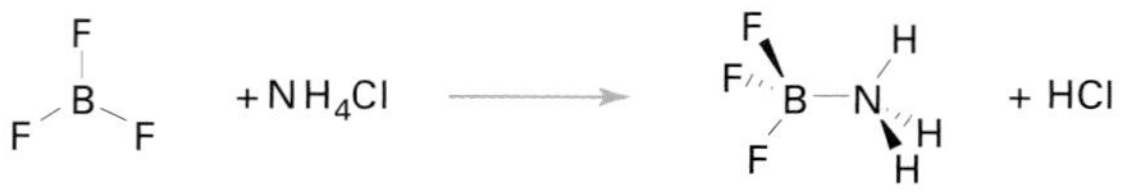

In the context of d-metal complexes, a displacement reaction in which one ligand is driven out of the complex and is replaced by another is generally called a **substitution reaction** (Section 21.1).

(b) Metathesis reactions

Key point: A metathesis reaction is a displacement reaction assisted by the formation of another complex.

A **metathesis reaction** (or 'double displacement reaction') is an interchange of partners:[6]

$$A{-}B + A'{-}B' \rightarrow A{-}B' + A'{-}B$$

The displacement of the base :B by :B' is assisted by the extraction of :B by the acid A'. An example is the reaction

Here the base Br^- displaces I^-, and the extraction is assisted by the formation of the less soluble AgI.

4.12 Hard and soft acids and bases

The proton (H^+) was the key electron pair acceptor in the discussion of Brønsted acid and base strengths. When considering Lewis acids and bases we must allow for a greater variety of acceptors and hence more factors that influence the interactions between electron pair donors and acceptors in general.

(a) The classification of acids and bases

Key points: Hard and soft acids and bases are identified empirically by the trends in stabilities of the complexes that they form: hard acids tend to bind to hard bases and soft acids tend to bind to soft bases.

It proves helpful when considering the interactions of Lewis acids and bases containing elements drawn from throughout the periodic table to consider at least two main classes of substance. The classification of substances as 'hard' and 'soft' acids and bases was introduced by R.G. Pearson; it is a generalization—and a more evocative renaming—of the distinction between two types of behaviour that were originally named simply 'class *a*' and 'class *b*' respectively, by S. Ahrland, J. Chatt, and N.R. Davies.

The two classes are identified empirically by the opposite order of strengths (as measured by the equilibrium constant, K_f, for the formation of the complex) with which they form complexes with halide ion bases:

- Hard acids bond in the order: $I^- < Br^- < Cl^- < F^-$.
- Soft acids bond in the order: $F^- < Cl^- < Br^- < I^-$.

---

[6] The name metathesis comes from the Greek word for exchange.

Figure 4.12 shows the trends in K_f for complex formation with a variety of halide ion bases. The equilibrium constants increase steeply from F^- to I^- when the acid is Hg^{2+}, indicating that Hg^{2+} is a soft acid. The trend is less steep but in the same direction for Pb^{2+}, which indicates that this ion is a borderline soft acid. The trend is in the opposite direction for Zn^{2+}, so this ion is a borderline hard acid. The steep downward slope for Al^{3+} indicates that it is a hard acid. A useful rule of thumb is that small cations, which are not easily polarized, are hard and form complexes with small anions. Large cations are more polarizable and are soft.

For Al^{3+}, the binding strength increases as the electrostatic parameter ($\xi = z^2/r$) of the anion increases, which is consistent with an ionic model of the bonding. For Hg^{2+}, the binding strength increases with increasing polarizability of the anion. These two correlations suggest that hard acid cations form complexes in which simple Coulombic, or ionic, interactions are dominant, and that soft acid cations form more complexes in which covalent bonding is important.

A similar classification can be applied to neutral molecular acids and bases. For example, the Lewis acid phenol forms a more stable complex by hydrogen bonding to $(C_2H_5)_2O$: than to $(C_2H_5)_2S$:. This behaviour is analogous to the preference of Al^{3+} for F^- over Cl^-. By contrast, the Lewis acid I_2 forms a more stable complex with $(C_2H_5)_2S$:. We can conclude that phenol is hard whereas I_2 is soft.

In general, acids are identified as hard or soft by the thermodynamic stability of the complexes they form, as set out for the halide ions above and for other species as follows:

- Hard acids bond in the order: $R_3P \ll R_3N, R_2S \ll R_2O$.
- Soft acids bond in the order: $R_2O \ll R_2S, R_3N \ll R_3P$.

Bases can also be defined as soft or hard. Bases such as halides and oxoanions are classified as hard because ionic bonding will be predominant in most of the complexes they form. Many soft bases bond through a carbon atom, such as CO or CN^-. In addition to donating electron density to the metal through a σ interaction, these small multiply bonded ligands are able to accept electron density through the low-lying empty π orbitals (the LUMO) present on the base (See Chapter 2). The bonding is, consequently, predominantly covalent in character. As these soft bases are able to accept electron density into π orbitals they are known as π **acids**. The nature of this bonding will be explored in Chapter 20.

It follows from the definition of hardness that:

- Hard acids tend to bind to hard bases.
- Soft acids tend to bind to soft bases.

When species are analysed with these rules in mind, it is possible to identify the classification summarized in Table 4.5.

(b) Interpretation of hardness

Key points: Hard acid–base interactions are predominantly electrostatic; soft acid–base interactions are predominantly covalent.

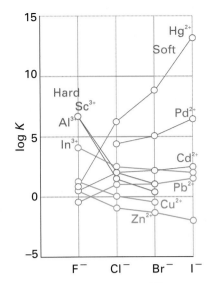

Figure 4.12 The trends in stability constants for complex formation with a variety of halide ion bases. Hard ions are indicated by the blue lines, soft ions by the red line. Borderline hard or borderline soft ions are indicated by green lines.

Table 4.5 The classification of Lewis acids and bases*

Hard	Borderline	Soft
Acids		
H^+, Li^+, Na^+, K^+	Fe^{2+}, Co^{2+}, Ni^{2+}	Cu^+, Au^+, Ag^+, Tl^+, Hg_2^{2+}
Be^{2+}, Mg^{2+}, Ca^{2+}	Cu^{2+}, Zn^{2+}, Pb^{2+}	Pd^{2+}, Cd^{2+}, Pt^{2+}, Hg^{2+}
Cr^{2+}, Cr^{3+}, Al^{3+}	SO_2, BBr_3	BH_3
SO_3, BF_3		
Bases		
F^-, OH^-, H_2O, NH_3	NO_2^-, SO_3^{2-}, Br^-	H^-, R^-, $\underline{C}N^-$, CO, I^-
CO_3^{2-}, NO_3^-, O^{2-}	N_3^-, N_2	$\underline{S}CN^-$, R_3P, C_6H_5
SO_4^{2-}, PO_4^{3-}, ClO_4^-	C_5H_5N, $SC\underline{N}^-$	R_2S

* The underlined element is the site of attachment to which the classification refers.

The bonding between hard acids and bases can be described approximately in terms of ionic or dipole–dipole interactions. Soft acids and bases are more polarizable than hard acids and bases, so the acid–base interaction has a more pronounced covalent character.

Although the type of bond formation is a major reason for the distinction between the two classes, there are other contributions to the Gibbs energy of complex formation and hence to the equilibrium constant. Important factors are:

1) Competition with the solvent in reactions in solution.

2) The rearrangement of the substituents of the acid and base that may be necessary to permit formation of the complex.

3) Steric repulsion between substituents on the acid and the base.

Any of these contributions can have a marked effect on the outcome of a reaction.

It is important to note that although we associate soft acid/soft base interactions with covalent bonding, the bond itself may be surprisingly weak. This point is illustrated by reactions involving Hg^{2+}, a representative soft acid. The metathesis reaction

$$BeI_2 + HgF_2 \rightarrow BeF_2 + HgI_2$$

is exothermic, as predicted by the hard–soft rule. The bond dissociation energies (in kilojoules per mole) measured for these molecules in the gas phase are.

Be–F	632	Hg–F	268
Be–I	289	Hg–I	145

Therefore, it is not the large Hg–I bond energy that ensures that the reaction is exothermic but the especially strong bond between Be and F, which is an example of a hard–hard interaction. In fact an Hg atom forms only weak bonds to any other atom. In aqueous solution, the reason why Hg^{2+} forms a much more stable complex with iodide ions compared to chloride ions is the much more favourable hydration energy of Cl^-.

(c) Chemical consequences of hardness

Key points: Hard-hard and soft-soft interactions help to systematize complex formation but must be considered in the light of other possible influences on bonding.

The concepts of hardness and softness help to rationalize a great deal of inorganic chemistry. For instance, they are useful for choosing preparative conditions and predicting the directions of reactions, and they help to rationalize the outcome of metathesis reactions. However, the concepts must always be used with due regard for other factors that may affect the outcome of reactions. This deeper understanding of chemical reactions will grow in the course of the rest of the book. For the time being we shall limit the discussion to a few straightforward examples.

The classification of molecules and ions as hard or soft acids and bases helps to clarify the terrestrial distribution of the elements described in Chapter 1. Hard cations such as Li^+, Mg^{2+}, Ti^{3+}, and Cr^{3+} are found in association with the hard base O^{2-}. The soft cations Cd^{2+}, Pb^{2+}, Sb^{2+}, and Bi^{2+} are found in association with the soft anions S^{2-}, Se^{2-}, and Te^{2-}. The consequences of this correlation are discussed in more detail in Section 9.3.

Polyatomic anions may contain two or more donor atoms differing in their hard–soft character. For example, the SCN^- ion is a base that comprises both the harder N atom and the softer S atom. The ion binds to the hard Si atom through N. However, with a soft acid, such as a metal ion in a low oxidation state, the ion bonds through S. Platinum(II), for example, forms Pt–SCN in the complex $[Pt(SCN)_4]^{2-}$.

4.13 Thermodynamic acidity parameters

Key points: The standard enthalpies of complex formation are reproduced by the E and C parameters of the Drago-Wayland equation that reflect, in part, the ionic and covalent contributions to the bond in the complex.

An important alternative to the hard–soft classification of acids and bases makes use of an approach in which electronic and structural rearrangement, and steric effects are incorporated into a small set of parameters. The standard reaction enthalpies of complex formation

$$A(g) + B(g) \rightarrow A\text{–}B(g) \qquad \Delta_r H^{\ominus}(A\text{–}B)$$

can be reproduced by the Drago–Wayland equation:

$$-\Delta_r H^{\ominus}(A\text{–}B)/\text{kJ mol}^{-1} = E_A E_B + C_A C_B \qquad (4.11)$$

The parameters E and C were introduced with the idea that they represent 'electrostatic' and 'covalent' factors, respectively, but in fact they must accommodate all factors except solvation. The compounds for which the parameters are listed in Table 4.6 satisfy the equation with an error of less than ±3 kJ mol^{-1}, as do a much larger number of examples in the original papers.

■ **A brief illustration.** From Table 4.6 we find $E = 20.2$ and $C = 3.31$ for BF$_3$ and $E = 2.78$ and $C = 7.08$ for NH$_3$. The Drago–Wayland equation gives $\Delta_r H^{\ominus} = -[(20.2 \times 2.78) + (3.31 \times 7.08)] = -79.59$ kJ mol^{-1}, indicating an exothermic reaction for the formation of the adduct NH$_3$BF$_3$. ■

The Drago–Wayland equation is semiempirical but very successful and useful. In addition to providing estimates of the enthalpies of complex formation for over 1500 complexes, these enthalpies can be combined to calculate the enthalpies of displacement and metathesis reactions. Moreover, the equation is useful for reactions of acids and bases in nonpolar, noncoordinating solvents as well as for reactions in the gas phase. The major limitation is that the equation is restricted to substances that can conveniently be studied in the gas phase or in noncoordinating solvents; hence, in the main it is limited to neutral molecules.

4.14 Solvents as acids and bases

The solvent system definition of acids and bases allows solutes to be defined as acids and bases by considering the autoionization products of the solvent. Most solvents are also either electron pair acceptors or donors and hence are either Lewis acids or bases. The chemical consequences of solvent acidity and basicity are considerable, as they help to account for the differences between reactions in aqueous and nonaqueous media. It follows that a displacement reaction often occurs when a solute dissolves in a solvent, and that the subsequent reactions of the solution are also usually either displacements or metatheses. For example, when antimony pentafluoride dissolves in bromine trifluoride, the following displacement reaction occurs:

$$SbF_5(s) + BrF_3(l) \rightarrow BrF_2^+(sol) + SbF_6^-(sol)$$

In the reaction, the strong Lewis acid SbF$_5$ abstracts F$^-$ from BrF$_3$. A more familiar example of the solvent as participant in a reaction is in Brønsted theory. In this theory, the acid (H$^+$) is always regarded as complexed with the solvent, as in H$_3$O$^+$ if the solvent is water, and reactions are treated as the transfer of the acid, the proton, from a basic solvent molecule to another base. Only the saturated hydrocarbons among common solvents lack significant Lewis acid or base character.

(a) Basic solvents

Key points: Basic solvents are common; they may form complexes with the solute and participate in displacement reactions.

Solvents with Lewis base character are common. Most of the well-known polar solvents, including water, alcohols, ethers, amines, dimethylsulfoxide (DMSO, (CH$_3$)$_2$SO), dimethylformamide (DMF, (CH$_3$)$_2$NCHO), and acetonitrile (CH$_3$CN), are hard Lewis bases. Dimethylsulfoxide is an interesting example of a solvent that is hard on account of its O donor atom and soft on account of its S donor atom. Reactions of acids and bases in these solvents are generally displacements:

Table 4.6 Drago–Wayland parameters for some acids and bases*

	E	C
Acids		
Antimony pentachloride	15.1	10.5
Boron trifluoride	20.2	3.31
Iodine	2.05	2.05
Iodine monochloride	10.4	1.70
Phenol	8.86	0.90
Sulfur dioxide	1.88	1.65
Trichloromethane	6.18	0.32
Trimethylboron	12.6	3.48
Bases		
Acetone	2.02	4.67
Ammonia	2.78	7.08
Benzene	0.57	1.21
Dimethylsulfide	0.70	15.26
Dimethylsulfoxide	2.76	5.83
Methylamine	2.66	12.00
p-Dioxane	2.23	4.87
Pyridine	2.39	13.10
Trimethylphosphine	17.2	13.40

* E and C parameters are often reported to give ΔH in kcal mol^{-1}; we have multiplied both by $\sqrt{(4.184)}$ to obtain ΔH in kJ mol^{-1}.

EXAMPLE 4.11 Accounting for properties in terms of the Lewis basicity of solvents

Silver perchlorate, $AgClO_4$, is significantly more soluble in benzene than in alkane solvents. Account for this observation in terms of Lewis acid–base properties.

Answer We need to consider how the solvent interacts with the solute. The π electrons of benzene, a soft base, are available for complex formation with the empty orbitals of the cation Ag^+, a soft acid. The Ag^+ ion is thus solvated favourably by benzene. The species $[Ag-C_6H_6]^+$ is the complex of the acid Ag^+ with π electrons of the weak base benzene.

Self-test 4.11 Boron trifluoride, BF_3, a hard acid, is often used in the laboratory as a solution in diethyl ether, $(C_2H_5)_2O$:, a hard base. Draw the structure of the complex that results from the dissolution of $BF_3(g)$ in $(C_2H_5)_2O(l)$.

(b) Acidic and neutral solvents

Key points: Hydrogen bond formation is an example of Lewis complex formation; other solvents may also show Lewis acid character.

Hydrogen bonding (Section 10.6) can be regarded as an example of complex formation. The 'reaction' is between A–H (the Lewis acid) and :B (the Lewis base) and gives the complex conventionally denoted A–H … B. Hence, many solutes that form hydrogen bonds with a solvent can be regarded as dissolving because of complex formation. A consequence of this view is that an acidic solvent molecule is displaced when proton transfer occurs:

Liquid sulfur dioxide is a good soft acidic solvent for dissolving the soft base benzene. Unsaturated hydrocarbons may act as acids or bases by using their π or π* orbitals as frontier orbitals. Alkanes with electronegative substituents, such as haloalkanes (for example, $CHCl_3$), are significantly acidic at the hydrogen atom. Saturated fluorocarbon solvents lack Lewis acid and base properties.

Applications of acid–base chemistry

The Brønsted and Lewis definitions of acids and bases do not have to be considered separately from each other. In fact, many applications of acid–base chemistry utilize both Lewis and Brønsted acids or bases simultaneously.

4.15 Superacids and superbases

Key point: Superacids are more efficient proton donors than anhydrous sulfuric acid. Superbases are more efficient proton acceptors than the hydroxide ion.

A **superacid** is a substance that is a more efficient proton donor than pure H_2SO_4. Superacids are typically viscous, corrosive liquids and can be up to 10^{18} times more acidic than H_2SO_4 itself. They are formed when a powerful Lewis acid is dissolved in a powerful Brønsted acid. The most common superacids are formed when SbF_5 is dissolved in fluorosulfonic acid, HSO_3F, or anhydrous HF. An equimolar mixture of SbF_5 and HSO_3F is known as 'magic acid', so named because of its ability to dissolve candle wax. The enhanced acidity is due to the formation of a solvated proton, which is a better proton donor than the acid:

$$SbF_5(l) + 2\,HSO_3F(l) \rightarrow H_2SO_3F^+(sol) + SbF_5SO_3F^-(sol)$$

An even stronger superacid is formed when SbF_5 is added to anhydrous HF:

$$SbF_5(l) + 2\,HF(l) \rightarrow H_2F^+(sol) + SbF_6^-(sol)$$

Other pentafluorides also form superacids in HSO_3F and HF and the acidity of these compounds decreases in the order $SbF_5 > AsF_5 > TaF_5 > NbF_5 > PF_5$.

Superacids are known that can protonate almost any organic compound. In the 1960s, George Olah and his colleagues found that carbonium ions were stabilized when hydrocarbons were dissolved in superacids.[7] In inorganic chemistry, superacids have been used to observe a wide variety of reactive cations such as S_8^{2+}, $H_3O_2^+$, Xe_2^+, and HCO^+, some of which have been isolated for structural characterization.

A **superbase** is a compound that is a more efficient proton acceptor than the OH^- ion, the strongest base that can exist in aqueous solution. Superbases react with water to produce the OH^- ion. Inorganic superbases are usually salts of Group 1 or Group 2 cations with small, highly charged anions. The highly charged anions are attracted to acid solvents such as water and ammonia. For example, lithium nitride, Li_3N, reacwts violently with water:

$$Li_3N(s) + 3\,H_2O(l) \rightarrow 3\,LiOH(aq) + NH_3(g)$$

The nitride anion is a stronger base than the hydride ion and deprotonates hydrogen:

$$Li_3N(s) + 2\,H_2(g) \rightarrow LiNH_2(s) + 2\,LiH(s)$$

Lithium nitride is a possible hydrogen storage material as this reaction is reversible at 270°C (Box 10.4).

Sodium hydride is a superbase that is used in organic chemistry to deprotonate carboxylic acids, alcohols, phenols, and thiols. Calcium hydride reacts with water to liberate hydrogen:

$$CaH_2(s) + 2\,H_2O(l) \rightarrow Ca(OH)_2(s) + 2\,H_2(g)$$

Calcium hydride is used as a dessicant, to inflate weather balloons, and as a laboratory source of pure hydrogen.

4.16 Heterogeneous acid–base reactions

Key point: The surfaces of many catalytic materials and minerals have Brønsted and Lewis acid sites.

Some of the most important reactions involving the Lewis and Brønsted acidity of inorganic compounds occur at solid surfaces. For example, **surface acids**, which are solids with a high surface area and Lewis acid sites, are used as catalysts in the petrochemical industry for the interconversion of hydrocarbons. The surfaces of many materials that are important in the chemistry of soil and natural waters also have Brønsted and Lewis acid sites.

Silica surfaces do not readily produce Lewis acid sites because –OH groups remain tenaciously attached at the surface of SiO_2 derivatives; as a result, Brønsted acidity is dominant. The Brønsted acidity of silica surfaces themselves is only moderate (and comparable to that of acetic acid). However, as already remarked, aluminosilicates display strong Brønsted acidity. When surface OH groups are removed by heat treatment, the aluminosilicate surface possesses strong Lewis acid sites. The best-known class of aluminosilicates is the zeolites (Section 14.15), which are widely used as environmentally benign heterogeneous catalysts (Chapter 26). The catalytic activity of zeolites arises from their acidic nature and they are known as **solid acids**. Other solid acids include supported heteropoly acids and acidic clays. Some reactions occurring at these catalysts are very sensitive to the presence of Brønsted or Lewis acid sites. For example, toluene can be subjected to Friedel–Crafts alkylation over a bentonite clay catalyst:

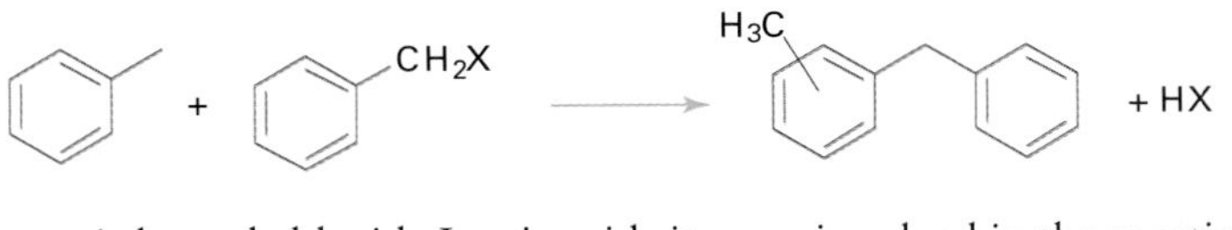

When the reagent is benzyl chloride Lewis acid sites are involved in the reaction, and when the reagent is benzyl alcohol Brønsted sites are involved.

---

[7] Carbocations could not be studied before Olah's experiments, and he won the 1994 Nobel Prize for Chemistry for this work.

Surface reactions carried out using the Brønsted acid sites of silica gels are used to prepare thin coatings of a wide variety of organic groups using surface modification reactions
such as

Thus, silica gel surfaces can be modified to have affinities for specific classes of molecules.
This procedure greatly expands the range of stationary phases that can be used for chromatography. The surface –OH groups on glass can be modified similarly, and glassware
treated in this manner is sometimes used in the laboratory when proton-sensitive compounds are being studied.

Solid acids are finding new applications in green chemistry. Traditional industrial processes generate large volumes of hazardous waste during the final stages of the process
when the product is separated from the reagents and byproducts. Solid catalysts are easily
separated from liquid products and reactions can often operate under milder conditions
and give greater selectivity.

FURTHER READING

W. Stumm and J.J. Morgan, *Aquatic chemistry: chemical equilibria and
rates in natural waters*. Wiley, New York (1995). The classic text on
the chemistry of natural waters.

N. Corcoran, *Chemistry in non-aqueous solvents*. Kluwer Academic
Publishers, Dordrecht, The Netherlands (2003). A comprehensive
account.

J. Chipperfield, *Non-aqueous solvents*. Oxford University Press (1999).
A readable introduction to the topic.

J. Burgess, *Ions in solution*. Ellis Horwood, Chichester (1988). A
readable account of solvation with an introduction to acidity and
polymerization.

J. Burgess, *Ions in solution: basic principles of chemical interactions*.
Ellis Horwood, Chichester (1999).

G.A. Olah, G.K. Prakash, and J. Sommer, *Superacids*. Wiley, New York
(1985).

G.A. Olah, 'My search for carbocations and their role in chemistry',
Nobel lectures in chemistry 1991–1995, ed. B.G. Malmstrom. World
Scientific Publishing, Singapore (1996).

R.J. Gillespie and J. Laing, Superacid solutions in hydrogen fluoride.
J. Am. Chem. Soc., 1988, **110**, 6053.

E.S. Stoyanov, K.-C Kim, and C.A. Reed, A strong acid that does not
protonate water. *J. Phys. Chem. A.*, 2004, **108**, 9310.

EXERCISES

4.1 Sketch an outline of the s and p blocks of the periodic table and
indicate on it the elements that form (a) strongly acidic oxides,
(b) strongly basic oxides, and (c) show the regions for which
amphoterism is common.

4.2 Identify the conjugate bases corresponding to the following acids:
$[Co(NH_3)_5(OH_2)]^{3+}$, HSO_4^-, CH_3OH, $H_2PO_4^-$, $Si(OH)_4$, HS^-.

4.3 Identify the conjugate acids of the bases C_5H_5N (pyridine), HPO_4^{2-},
O^{2-}, CH_3COOH, $[Co(CO)_4]^-$, CN^-.

4.4 Calculate the equilibrium concentration of H_3O^+ in a 0.10 M solution
of butanoic acid ($K_a = 1.86 \times 10^{-5}$). What is the pH of this solution?

4.5 The K_a of ethanoic acid, CH_3COOH, in water is 1.8×10^{-5}.
Calculate K_b of the conjugate base, $CH_3CO_2^-$.

4.6 The value of K_b for pyridine, C_5H_5N, is 1.8×10^{-9}. Calculate K_a for
the conjugate acid, $C_5H_5NH^+$.

4.7 The effective proton affinity of F^- in water is 1150 kJ mol^{-1}.
Predict whether it will behave as an acid or a base in water.

4.8 Draw the structures of chloric acid and chlorous acid, and predict
their pK_a values using Pauling's rules.

4.9 Aided by Fig. 4.2 (taking solvent levelling into account), identify
which bases from the following lists are (a) too strong to be studied
experimentally, (b) too weak to be studied experimentally, or (c) of
directly measurable base strength. (i) CO_3^{2-}, O^{2-}, ClO_4^-, and NO_3^- in
water; (ii) HSO_4^-, NO_3^-, ClO_4^- in H_2SO_4.

4.10 The aqueous solution pK_a values for HOCN, H_2NCN, and
CH_3CN are approximately 4, 10.5, and 20 (estimated), respectively.
Explain the trend in these cyano derivatives of binary acids and
compare them with H_2O, NH_3, and CH_4. Is the CN group electron
donating or withdrawing?

4.11 The pK_a value of $HAsO_4^{2-}$ is 11.6. Is this value consistent with Pauling's rules?

4.12 Use Pauling's rules to place the following acids in order of increasing acid strength: HNO_2, H_2SO_4, $HBrO_3$, and $HClO_4$ in a nonlevelling solvent.

4.13 Draw the structures and indicate the charges of the tetraoxoanions of X = Si, P, S, and Cl. Summarize and account for the trends in the pK_a values of their conjugate acids.

4.14 Which member of the following pairs is the stronger acid? Give reasons for your choice. (a) $[Fe(OH_2)_6]^{3+}$ or $[Fe(OH_2)_6]^{2+}$, (b) $[Al(OH_2)_6]^{3+}$ or $[Ga(OH_2)_6]^{3+}$, (c) $Si(OH)_4$ or $Ge(OH)_4$, (d) $HClO_3$ or $HClO_4$, (e) H_2CrO_4 or $HMnO_4$, (f) H_3PO_4 or H_2SO_4.

4.15 Arrange the oxides Al_2O_3, B_2O_3, BaO, CO_2, Cl_2O_7, SO_3 in order from the most acidic through amphoteric to the most basic.

4.16 Arrange the acids HSO_4^-, H_3O^+, H_4SiO_4, CH_3GeH_3, NH_3, HSO_3F in order of increasing acid strength.

4.17 The ions Na^+ and Ag^+ have similar radii. Which aqua ion is the stronger acid? Why?

4.18 Which of the elements Al, As, Cu, Mo, Si, B, Ti form oxide polyanions and which form oxide polycations?

4.19 When a pair of aqua cations forms an M–O–M bridge with the elimination of water, what is the general rule for the change in charge per M atom on the ion?

4.20 Write a balanced equation for the formation of $P_2O_7^{4-}$ from PO_4^{3-}. Write a balanced equation for the condensation of $[Fe(OH_2)_6]^{3+}$ to give $[(H_2O)_4Fe(OH)_2Fe(OH_2)_4]^{4+}$.

4.21 Write balanced equations for the main reaction occurring when (a) H_3PO_4 and Na_2HPO_4 and (b) CO_2 and $CaCO_3$ are mixed in aqueous media.

4.22 Hydrogen fluoride acts as an acid in anhydrous sulfuric acid and as a base in liquid ammonia. Give the equations for both reactions.

4.23 Explain why hydrogen selenide is a stronger acid than hydrogen sulfide.

4.24 Sketch the p block of the periodic table. Identify as many elements as you can that act as Lewis acids in one of their lower oxidation states and give the formula of a representative Lewis acid for each element.

4.25 For each of the following processes identify the acids and bases involved and characterize the process as complex formation or acid–base displacement. Identify the species that exhibit Brønsted acidity as well as Lewis acidity.

  (a) $SO_3 + H_2O \rightarrow HSO_4^- + H^+$
  (b) $CH_3[B_{12}] + Hg^{2+} \rightarrow [B_{12}]^+ + CH_3Hg^+$; $[B_{12}]$ designates the coporphyrin, vitamin B_{12}.
  (c) $KCl + SnCl_2 \rightarrow K^+ + [SnCl_3]^-$
  (d) $AsF_3(g) + SbF_5(l) \rightarrow [AsF_2]^+[SbF_6]^-(s)$
  (e) Ethanol dissolves in pyridine to produce a nonconducting solution.

4.26 Select the compound on each line with the named characteristic and state the reason for your choice.

(a) Strongest Lewis acid:
  BF_3 BCl_3 BBr_3
  $BeCl_2$ BCl_3
  $B(n\text{-}Bu)_3$ $B(t\text{-}Bu)_3$

(b) More basic towards $B(CH_3)_3$
  Me_3N Et_3N
  $2\text{-}CH_3C_5H_4N$ $4\text{-}CH_3C_5H_4N$

4.27 Using hard–soft concepts, which of the following reactions are predicated to have an equilibrium constant greater than 1? Unless otherwise stated, assume gas-phase or hydrocarbon solution and 25°C.

  (a) $R_3PBBr_3 + R_3NBF_3 \rightleftharpoons R_3PRBF_3 + R_3NBBr_3$
  (b) $SO_2 + (C_6H_5)_3P{:}HOC(CH_3)_3 \rightleftharpoons (C_6H_5)_3PSO_2 + HOC(CH_3)_3$
  (c) $CH_3HgI + HCl \rightleftharpoons CH_3HgCl + HI$
  (d) $[AgCl_2]^{2-}(aq) + 2\,CN^-(aq) \rightleftharpoons [Ag(CN)_2]^-(aq) + 2\,Cl^-(aq)$

4.28 The molecule $(CH_3)_2N–PF_2$ has two basic atoms, P and N. One is bound to B in a complex with BH_3, the other to B in a complex with BF_3. Decide which is which and state the reason for your decision.

4.29 The enthalpies of reaction of trimethylboron with NH_3CH_3, NH_2, $(CH_3)_2NH$, and $(CH_3)_3N$ are –58, –74, –81, and –74 kJ mol^{-1}, respectively. Why is trimethylamine out of line?

4.30 With the aid of the table of E and C values (Table 4.6), discuss the relative basicity in (a) acetone and dimethylsulfoxide, (b) dimethylsulfide and dimethylsulfoxide. Comment on a possible ambiguity for dimethylsulfoxide.

4.31 Give the equation for the dissolution of SiO_2 glass by HF and interpret the reaction in terms of Lewis and Brønsted acid–base concepts.

4.32 Aluminium sulfide, Al_2S_3, gives off a foul odour characteristic of hydrogen sulfide when it becomes damp. Write a balanced chemical equation for the reaction and discuss it in terms of acid–base concepts.

4.33 Describe the solvent properties that would (a) favour displacement of Cl$^-$ by I$^-$ from an acid centre, (b) favour basicity of R_3As over R_3N, (c) favour acidity of Ag^+ over Al^{3+}, (d) promote the reaction $2\,FeCl_3 + ZnCl_2 \rightarrow Zn^{2+} + 2\,[FeCl_4]^-$. In each case, suggest a specific solvent that might be suitable.

4.34 The Lewis acid $AlCl_3$ catalysis of the acylation of benzene was described in Section 4.10b. Propose a mechanism for a similar reaction catalysed by an alumina surface.

4.35 Use acid–base concepts to comment on the fact that the only important ore of mercury is cinnabar, HgS, whereas zinc occurs in nature as sulfides, silicates, carbonates, and oxides.

4.36 Write balanced Brønsted acid–base equations for the dissolution of the following compounds in liquid hydrogen fluoride: (a) CH_3CH_2OH, (b) NH_3, (c) C_6H_5COOH.

4.37 Is the dissolution of silicates in HF a Lewis acid–base reaction, a Brønsted acid–base reaction, or both?

4.38 The f-block elements are found as M(III) lithophiles in silicate minerals. What does this indicate about their hardness?

4.39 Use the data in Table 4.6 to calculate the enthalpy change for the reaction of iodine with phenol.

PROBLEMS

4.1 In analytical chemistry a standard procedure for improving the detection of the stoichiometric point in titrations of weak bases with strong acids is to use acetic acid as a solvent. Explain the basis of this approach.

4.2 In the gas phase, the base strength of amines increases regularly along the series $NH_3 < CH_3NH_2 < (CH_3)_2NH < (CH_3)_3N$. Consider the role of steric effects and the electron-donating ability of CH_3 in determining this order. In aqueous solution, the order is reversed. What solvation effect is likely to be responsible for this change?

4.3 The hydroxoacid $Si(OH)_4$ is weaker than H_2CO_3. Write balanced equations to show how dissolving a solid, M_2SiO_4, can lead to a reduction in the pressure of CO_2 over an aqueous solution. Explain why silicates in ocean sediments might limit the increase of CO_2 in the atmosphere.

4.4 The precipitation of $Fe(OH)_3$ discussed in the chapter is used to clarify waste waters because the gelatinous hydrous oxide is very efficient at coprecipitating some contaminants and entrapping others. The solubility constant of $Fe(OH)_3$ is $K_s = [Fe^{3+}][OH^-]^3 \approx 1.0 \times 10^{-38}$. As the autoprotolysis constant of water links $[H_3O^+]$ to $[OH^-]$ by $K_w = [H_3O^+][OH^-] = 1.0 \times 10^{-14}$, we can rewrite the solubility constant by substitution as $[Fe^{3+}]/[H^+]^3 = 1.0 \times 10^4$. (a) Balance the chemical equation for the precipitation of $Fe(OH)_3$ when iron(III) nitrate is added to water. (b) If 6.6 kg of $Fe(NO_3)_3 \cdot 9H_2O$ is added to 100 dm^3 of water, what is the final pH of the solution and the molar concentration of Fe^{3+}, neglecting other forms of dissolved Fe(III)? Give formulas for two Fe(III) species that have been neglected in this calculation.

4.5 The frequency of the symmetrical M–O stretching vibration of the octahedral aqua ions $[M(OH_2)_6]^{2+}$ increases along the series, $Ca^{2+} < Mn^{2+} < Ni^{2+}$. How does this trend relate to acidity?

4.6 An electrically conducting solution is produced when $AlCl_3$ is dissolved in the basic polar solvent CH_3CN. Give formulas for the most probable conducting species and describe their formation using Lewis acid–base concepts.

4.7 The complex anion $[FeCl_4]^-$ is yellow whereas $[Fe_2Cl_6]$ is reddish. Dissolution of 0.1 mol $FeCl_3(s)$ in 1 dm^3 of either $POCl_3$ or $PO(OR)_3$ produces a reddish solution that turns yellow on dilution. Titration of red solutions in $POCl_3$ with Et_4NCl solutions leads to a sharp colour change (from red to yellow) at a 1:1 mole ratio of $FeCl_3/Et_4NCl$. Vibrational spectra suggest that oxochloride solvents form adducts with typical Lewis acids by coordination of oxygen. Compare the following two sets of reactions as possible explanations of the observations.

(a) $Fe_2Cl_6 + 2\,POCl_3 \rightleftharpoons 2\,[FeCl_4]^- + 2\,[POCl_2]^+$
   $POCl_3^+ + Et_4NCl \rightleftharpoons Et_4N^+ + POCl_3$

(b) $Fe_2Cl_6 + 4\,POCl_3 \rightleftharpoons [FeCl_2(OPCl_3)_4]^+ + [FeCl_4]^-$

Both sets of equilibria are shifted to products by dilution.

4.8 In the traditional scheme for the separation of metal ions from solution that is the basis of qualitative analysis, ions of Au, As, Sb, and Sn precipitate as sulfides but redissolve on addition of excess ammonium polysulfide. By contrast, ions of Cu, Pb, Hg, Bi, and Cd precipitate as sulfides but do not redissolve. In the language of this chapter, the first group is amphoteric for reactions involving SH^- in place of OH^-. The second group is less acidic. Locate the amphoteric boundary in the periodic table for sulfides implied by this information. Compare this boundary with the amphoteric boundary for hydrous

oxides in Fig. 4.6. Does this analysis agree with describing S^{2-} as a softer base than O^{2-}?

4.9 The compounds SO_2 and $SOCl_2$ can undergo an exchange of radioactively labelled sulfur. The exchange is catalysed by Cl^- and $SbCl_5$. Suggest mechanisms for these two exchange reactions with the first step being the formation of an appropriate complex.

4.10 In the reaction of *t*-butyl bromide with $Ba(NCS)_2$, the product is 91 per cent S-bound *t*-Bu-SCN. However, if $Ba(NCS)_2$ is impregnated into solid CaF_2, the yield is higher and the product is 99 per cent *t*-Bu-NCS. Discuss the effect of alkaline earth metal salt support on the hardness of the ambident nucleophile SCN^-. (See T. Kimura, M. Fujita, and T. Ando, *J. Chem Soc., Chem. Commun.*, 1990, 1213.)

4.11 Pyridine forms a stronger Lewis acid–base complex with SO_3 than with SO_2. However, pyridine forms a weaker complex with SF_6 than with SF_4. Explain the difference.

4.12 Predict whether the equilibrium constants for the following reactions should be greater than 1 or less than 1:

(a) $CdI_2(s) + CaF_2(s) \rightleftharpoons CdF_2(s) + CaI_2(s)$

(b) $[CuI_4]^{2-}(aq) + [CuCl_4]^{3-}(aq) \rightleftharpoons [CuCl_4]^{2-}(aq) + [CuI_4]^{3-}(aq)$

(c) $NH_2^-(aq) + H_2O(l) \rightleftharpoons NH_3(aq) + OH^-(aq)$

4.13 For parts (a), (b), and (c), state which of the two solutions has the lower pH:

(a) 0.1 M $Fe(ClO_4)_2(aq)$ or 0.1 M $Fe(ClO_4)_3(aq)$

(b) 0.1 M $Ca(NO_3)_2(aq)$ or 0.1 M $Mg(NO_3)_2(aq)$

(c) 0.1 M $Hg(NO_3)_2(aq)$ or 0.1 M $Zn(NO_3)_2(aq)$

4.14 A paper by Gillespie and Liang entitled 'Superacid solutions in hydrogen fluoride' (*J. Am. Chem. Soc.*, 1988, **110**, 6053) discusses the acidity of various solutions of inorganic compounds in HF.

(a) Give the order of acid strength of the pentafluorides determined during the investigation. (b) Give the equations for the reactions of SbF_5 and AsF_5 with HF. (c) SbF_5 forms a dimer, $Sb_2F_{11}^-$, in HF. Give the equation for the equilibrium between the monomeric and the dimeric species.

4.15 Why are strongly acidic solvents (e.g. SbF_5/HSO_3F) used in the preparation of cations such as I_2^+ and Se_8^{2+}, whereas strongly basic solvents are needed to stabilize anionic species such as S_4^{2-} and Pb_9^{4-}?

4.16 In their paper 'The strengths of the hydrohalic acids' (*J. Chem. Educ.*, 2001, **78**, 116), R. Schmid and A. Miah discuss the validity of literature values of the pK_as for HF, HCl, HBr, and HI. (a) On what basis have the literature values been estimated? (b) To what is the low acid strength of HF relative to HCl usually attributed? (c) What reason do the authors suggest for the high acid strength of HCl?

4.17 Superacids are well known. Superbases also exist and are usually based on hydrides of Group 1 and Group 2 elements. Write an account of the chemistry of superbases.

Oxidation and reduction

5

Oxidation is the removal of electrons from a species; reduction is the addition of electrons. Almost all elements and their compounds can undergo oxidation and reduction reactions and the element is said to exhibit one or more different oxidation states. In this chapter we present examples of this 'redox' chemistry and develop concepts for understanding why oxidation and reduction reactions occur, considering mainly their thermodynamic aspects. We discuss the procedures for analysing redox reactions in solution and see that the electrode potentials of electrochemically active species provide data that are useful for determining and understanding the stability of species and solubility of salts. We describe procedures for displaying trends in the stabilities of various oxidation states, including the influence of pH. Next, we describe the applications of this information to environmental chemistry, chemical analysis, and inorganic synthesis. The discussion concludes with a thermodynamic examination of the conditions needed for some major industrial oxidation and reduction processes, particularly the extraction of metals from their ores.

A large class of reactions of inorganic compounds can be regarded as occurring by the transfer of electrons from one species to another. Electron gain is called **reduction** and electron loss is called **oxidation**; the joint process is called a **redox reaction**. The species that supplies electrons is the **reducing agent** (or 'reductant') and the species that removes electrons is the **oxidizing agent** (or 'oxidant'). Many redox reactions release a great deal of energy and they are exploited in combustion or battery technologies.

Many redox reactions occur between reactants in the same physical state. Some examples are:

in gases:

$$2\,NO(g) + O_2(g) \rightarrow 2\,NO_2(g)$$

$$2\,C_4H_{10}(g) + 13\,O_2(g) \rightarrow 8\,CO_2(g) + 10\,H_2O(g)$$

in solution:

$$Fe^{3+}(aq) + Cr^{2+}(aq) \rightarrow Fe^{2+}(aq) + Cr^{3+}(aq)$$

$$3\,CH_3CH_2OH(aq) + 2\,CrO_4^{2-}(aq) + 10\,H^+(aq) \rightarrow$$
$$3\,CH_3CHO(aq) + 2\,Cr^{3+}(aq) + 8\,H_2O(l)$$

in biological systems:

$$'Mn_4'(V,IV,IV,IV) + 2\,H_2O(l) \rightarrow$$
$$'Mn_4'(IV,III,III,III) + 4\,H^+(aq) + O_2(g)$$

in solids:

$$LiCoO_2(s) + C(s) \rightarrow Li@C(s) + CoO_2(s)$$

The biological example refers to the production of O_2 from water by an 'Mn$_4$' cofactor contained in one of the photosynthetic complexes of plants (Section 27.10). In the solid-state example the symbol Li@C(s) indicates that a Li^+ ion has penetrated between the graphene sheets of graphite to form an **intercalation compound**. The reaction takes place in a lithium-ion battery during charging and its reverse takes place during discharge. Redox reactions can also occur at interfaces (phase boundaries), such as a gas/solid or a solid/liquid interface. Examples include the dissolution of a metal and reactions occurring at an electrode.

Because of the diversity of redox reactions it is often convenient to analyse them by applying a set of formal rules expressed in terms of oxidation numbers (Section 2.1) and not to think in terms of actual electron transfers. Oxidation then corresponds to an increase in the oxidation number of an element and reduction corresponds to a decrease in its oxidation number. If no element in a reaction undergoes a change in oxidation number, then the reaction is not redox. We shall adopt this approach when we judge it appropriate.

■ **A brief illustration.** The simplest redox reactions involve the formation of cations and anions from the elements. Examples include the oxidation of lithium to Li^+ ions when it burns in air to form Li_2O and the reduction of chlorine to Cl^- when it reacts with calcium to form $CaCl_2$. For the Group 1 and 2 elements the only oxidation numbers commonly encountered are those of the element (0) and of the ions, $+1$ and $+2$, respectively. However, many of the other elements form compounds in more than one oxidation state. Thus lead is commonly found in its compounds as Pb(II), as in PbO, and as Pb(IV), as in PbO_2. ■

The ability to exhibit multiple oxidation numbers is seen at its fullest in d-metal compounds, particularly in Groups 6, 7, and 8; osmium, for instance, forms compounds that span oxidation numbers between -2, as in $[Os(CO)_4]^{2-}$, and $+8$, as in OsO_4. Because the oxidation state of an element is often reflected in the properties of its compounds, the ability to express the tendency of an element to form a compound in a particular oxidation state quantitatively is very useful in inorganic chemistry.

Reduction potentials

Because electrons are transferred between species in redox reactions, electrochemical methods (using pairs of electrodes to measure electron transfer reactions under controlled thermodynamic conditions) are of major importance and lead to the construction of tables of 'standard potentials'. The tendency of an electron to migrate from one species to another is expressed in terms of the differences between their standard potentials.

5.1 Redox half-reactions

Key point: A redox reaction can be expressed as the difference of two reduction half-reactions.

It is convenient to think of a redox reaction as the combination of two conceptual **half-reactions** in which the electron loss (oxidation) and gain (reduction) are displayed explicitly. In a reduction half-reaction, a substance gains electrons, as in

$$2\,H^+(aq) + 2\,e^- \rightarrow H_2(g)$$

In an oxidation half-reaction, a substance loses electrons, as in

$$Zn(s) \rightarrow Zn^{2+}(aq) + 2\,e^-$$

Electrons are not ascribed a state in the equation of a half-reaction: they are 'in transit'. The oxidized and reduced species in a half-reaction constitute a **redox couple**. A couple is written with the oxidized species before the reduced, as in H^+/H_2 and Zn^{2+}/Zn, and typically the phases are not shown.

For reasons that will become clear, it is useful to represent oxidation half-reactions by the corresponding reduction half-reaction. To do so, we simply reverse the equation for the oxidation half-reaction. Thus, the reduction half-reaction associated with the oxidation of zinc is written

$$Zn^{2+}(aq) + 2\,e^- \rightarrow Zn(s)$$

A redox reaction in which zinc is oxidized by hydrogen ions,

$$Zn(s) + 2\,H^+(aq) \rightarrow Zn^{2+}(aq) + H_2(g)$$

is then written as the *difference* of the two reduction half-reactions. In some cases it may be necessary to multiply each half-reaction by a factor to ensure that the numbers of electrons released and used match.

EXAMPLE 5.1 Combining half-reactions

Write a balanced equation for the oxidation of Fe^{2+} by permanganate ions (MnO_4^-) in acid solution.

Answer Balancing redox reactions often requires additional attention to detail because species other than products and reactants, such as electrons and hydrogen ions, often need to be considered. A systematic approach is as follows:

1. Write the unbalanced half-reactions for the two species as reductions.
2. Balance the elements other than hydrogen.
3. Balance O atoms by adding H_2O to the other side of the arrow.
4. If the solution is acidic, balance the H atoms by adding H^+; if the solution is basic, balance the H atoms by adding OH^- to one side and H_2O to the other.
5. Balance the charge by adding e^-.
6. Multiply each half-reaction by a factor to ensure that the numbers of e^- match.
7. Subtract one half-reaction from the other and cancel redundant terms.

The half-reaction for the reduction of Fe^{3+} is straightforward as it involves only the balance of charge:

$$Fe^{3+}(aq) + e^- \rightarrow Fe^{2+}(aq)$$

The unbalanced half-reaction for the reduction of MnO_4^- is

$$MnO_4^-(aq) \rightarrow Mn^{2+}(aq)$$

Balance the O with H_2O:

$$MnO_4^-(aq) \rightarrow Mn^{2+}(aq) + 4H_2O(l)$$

Balance the H with H^+(aq):

$$MnO_4^-(aq) + 8H^+(aq) \rightarrow Mn^{2+}(aq) + 4H_2O(l)$$

Balance the charge with e^-:

$$MnO_4^-(aq) + 8H^+(aq) + 5e^- \rightarrow Mn^{2+}(aq) + 4H_2O(l)$$

To balance the number of electrons in the two half-reactions the first is multiplied by 5 and the second by 2 to give $10e^-$ in each case. Then subtracting the iron half-reaction from the permanganate half-reaction and rearranging so that all stoichiometric coefficients are positive gives

$$MnO_4^-(aq) + 8H^+(aq) + 5Fe^{2+}(aq) \rightarrow Mn^{2+}(aq) + 5Fe^{3+}(aq) + 4H_2O(l)$$

Self-test 5.1 Use reduction half-reactions to write a balanced equation for the oxidation of zinc metal by permanganate ions in acid solution.

5.2 Standard potentials and spontaneity

Key point: A reaction is thermodynamically favourable (spontaneous) in the sense $K > 1$, if $E^\ominus > 0$, where $E^\ominus$ is the difference of the standard potentials corresponding to the half-reactions into which the overall reaction may be divided.

Thermodynamic arguments can be used to identify which reactions are spontaneous (that is, have a natural tendency to occur). The thermodynamic criterion of spontaneity is that, at constant temperature and pressure, the reaction Gibbs energy change, $\Delta_r G$, is negative. It is usually sufficient to consider the standard reaction Gibbs energy, $\Delta_r G^\ominus$, which is related to the equilibrium constant through

$$\Delta_r G^\ominus = -RT \ln K \tag{5.1}$$

A negative value of $\Delta_r G^\ominus$ corresponds to $K > 1$ and therefore to a 'favourable' reaction in the sense that the products dominate the reactants at equilibrium. It is important to realize, however, that $\Delta_r G$ depends on the composition and that all reactions are spontaneous (that is, have $\Delta_r G < 0$) under appropriate conditions.

Because the overall chemical equation is the difference of two reduction half-reactions, the standard Gibbs energy of the overall reaction is the difference of the standard Gibbs energies of the two half-reactions. However, because reduction half-reactions always occur in pairs in any actual chemical reaction, only the difference in their standard

Gibbs energies has any significance. Therefore, we can choose one half-reaction to have $\Delta_r G^\ominus = 0$, and report all other values relative to it. By convention, the specially chosen half-reaction is the reduction of hydrogen ions:

$$H^+(aq) + e^- \rightarrow \tfrac{1}{2}H_2(g) \quad \Delta_r G^\ominus = 0$$

at all temperatures.

■ **A brief illustration.** The standard Gibbs energy for the reduction of Zn^{2+} ions is found by determining experimentally that

$$Zn^{2+}(aq) + H_2(g) \rightarrow Zn(s) + 2H^+(aq) \quad \Delta_r G^\ominus = +147 \text{ kJ mol}^{-1}$$

Then, because the H^+ reduction half-reaction makes zero contribution to the reaction Gibbs energy (according to our convention), it follows that

$$Zn^{2+}(aq) + 2e^- \rightarrow Zn(s) \quad \Delta_r G^\ominus = +147 \text{ kJ mol}^{-1} \quad ■$$

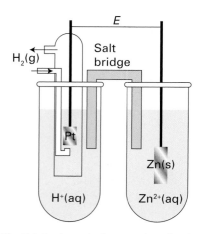

Fig. 5.1 A schematic diagram of a galvanic cell. The standard potential, $E_{cell}^\ominus$, is the potential difference when the cell is not generating current and all the substances are in their standard states.

Standard reaction Gibbs energies may be measured by setting up a **galvanic cell,** an electrochemical cell in which a chemical reaction is used to generate an electric current, in which the reaction driving the electric current through the external circuit is the reaction of interest (Fig. 5.1). The potential difference between its electrodes is then measured. The **cathode** is the electrode at which reduction occurs and the **anode** is the site of oxidation. In practice, we must ensure that the cell is acting reversibly in a thermodynamic sense, which means that the potential difference must be measured with no current flowing. If desired, the measured potential difference can be converted to a reaction Gibbs energy by using $\Delta_r G = -vFE$, where v is the stoichiometric coefficient of the electrons transferred when the half-reactions are combined and F is Faraday's constant ($F = 96.48 \text{ kC mol}^{-1}$). Tabulated values, normally for standard conditions, are usually kept in the units in which they were measured, namely volts (V).

■ **A brief comment.** Standard conditions are all substances at 1 bar and unit activity. For reactions involving H^+ ions, standard conditions correspond to pH = 0, approximately 1 M acid. Pure solids and liquids have unit activity. Although we use v (nu) for the stoichiometric coefficient of the electron, electrochemical equations in inorganic chemistry are also commonly written with n in its place; we use v to emphasize that it is a dimensionless number, not an amount in moles. ■

The potential that corresponds to the $\Delta_r G^\ominus$ of a half-reaction is written $E^\ominus$, with

$$\Delta_r G^\ominus = -vFE^\ominus \tag{5.2}$$

The potential $E^\ominus$ is called the **standard potential** (or 'standard reduction potential', to emphasize that, by convention, the half-reaction is a reduction and written with the oxidized species and electrons on the left). Because $\Delta_r G^\ominus$ for the reduction of H^+ is arbitrarily set at zero, the standard potential of the H^+/H_2 couple is also zero at all temperatures:

$$H^+(aq) + e^- \rightarrow \tfrac{1}{2}H_2(g) \quad E^\ominus(H^+, H_2) = 0$$

■ **A brief illustration.** For the Zn^{2+}/Zn couple, for which $v = 2$, it follows from the measured value of $\Delta_r G^\ominus$ that at 25°C:

$$Zn^{2+}(aq) + 2e^- \rightarrow Zn(s) \quad E^\ominus(Zn^{2+}, Zn) = -0.76 \text{ V} \quad ■$$

Because the standard reaction Gibbs energy is the difference of the $\Delta_r G^\ominus$ values for the two contributing half-reactions, $E_{cell}^\ominus$ for an overall reaction is also the difference of the two standard potentials of the reduction half-reactions into which the overall reaction can be divided. Thus, from the half-reactions given above it follows that the difference is

$$2H^+(aq) + Zn(s) \rightarrow Zn^{2+}(aq) + H_2(g) \quad E_{cell}^\ominus = +0.76V$$

Note that the $E^\ominus$ values for couples (and their half-reactions) are called standard potentials and that their difference is denoted $E_{cell}^\ominus$ and called the **standard cell potential.** The consequence of the negative sign in eqn 5.2 is that a reaction is favourable (in the sense $K > 1$) if the corresponding standard cell potential is positive. Because $E^\ominus > 0$ for the reaction in the illustration ($E^\ominus = +0.76$ V), we know that zinc has a thermodynamic tendency to reduce H^+ ions under standard conditions (aqueous, pH = 0, and Zn^{2+} at unit activity); that is zinc metal dissolves in acids. The same is true for any metal that has a couple with a negative standard potential.

A note on good practice The cell potential used to be called (and in practice is still widely called) the electromotive force (emf). However, a potential is not a force, and IUPAC favours the name 'cell potential'.

EXAMPLE 5.2 **Calculating a standard cell potential**

Use the following standard potentials to calculate the standard potential of a copper−zinc cell.

$$Cu^{2+}(aq) + 2e^- \rightarrow Cu(s) \quad E^{\ominus}(Cu^{2+}, Cu) = +0.34\,V$$
$$Zn^{2+}(aq) + 2e^- \rightarrow Zn(s) \quad E^{\ominus}(Zn^{2+}, Zn) = -0.76\,V$$

Answer For this calculation we note from the standard potentials that Cu^{2+} is the more oxidizing species (the couple with the higher potential), and will be reduced by the species with the lower potential (Zn in this case). The spontaneous reaction is therefore $Cu^{2+}(aq) + Zn(s) \rightarrow Zn^{2+}(aq) + Cu(s)$, and the cell potential is the difference of the two half-reactions (copper−zinc),

$$E_{cell}^{\ominus} = E^{\ominus}(Cu^{2+}, Cu) - E^{\ominus}(Zn^{2+}, Zn)$$
$$= +0.34\,V - (-0.76\,V) = +1.10\,V$$

The cell will produce a potential difference of 1.1 V (under standard conditions).

Self-test 5.2 Is copper metal expected to dissolve in dilute hydrochloric acid? Is copper metal expected to be oxidized by dilute hydrochloric acid?

Combustion is a familiar type of redox reaction, and the energy that is released can be exploited in heat engines. A **fuel cell** converts a chemical fuel directly into electrical power (Box 5.1).

5.3 Trends in standard potentials

Key point: The atomization and ionization of a metal and the hydration enthalpy of its ions all contribute to the value of the standard potential.

The factors that contribute to the standard potential of the couple M^+/M can be identified by consideration of a thermodynamic cycle and the corresponding changes in Gibbs energy that contribute to the overall reaction

$$M^+(aq) + \tfrac{1}{2}H_2(g) \rightarrow H^+(aq) + M(s)$$

The thermodynamic cycle shown in Fig. 5.2 has been simplified by ignoring the reaction entropy, which is largely independent of the identity of M. The entropy contribution $T\Delta S^{\ominus}$ lies in the region of -20 to -40 kJ mol^{-1}, which is small in comparison with the reaction enthalpy, the difference between the standard enthalpies of formation of $H^+(aq)$ and $M^+(aq)$. In this analysis we use the absolute values of the enthalpies of formation of M^+ and H^+, not the values based on the convention $\Delta_f H^{\ominus}$ (H^+, aq) $= 0$. Thus, we use $\Delta_f H^{\ominus}$ (H^+, aq) $= +445$ kJ mol^{-1}, which is obtained by considering the formation of an H atom from $\tfrac{1}{2}H_2(g)$ ($+218$ kJ mol^{-1}), ionization to $H^+(g)$ ($+1312$ kJ mol^{-1}), and hydration of $H^+(g)$ (approximately -1085 kJ mol^{-1}).

The analysis of the cell potential into its thermodynamic contributions allows us to account for trends in the standard potentials. For instance, the variation of standard potential down Group 1 seems contrary to expectation based on electronegativities insofar as Cs^+/Cs ($\chi = 0.79$, $E^{\ominus} = -2.94\,V$) has a less negative standard potential than Li^+/Li ($\chi = 2.20$, $E^{\ominus} = -3.04\,V$) despite Li having a higher electronegativity than Cs. Lithium has a higher enthalpy of sublimation and ionization energy than Cs, and in isolation this difference would imply a less negative standard potential as formation of the ion is less favourable. However, Li^+ has a large negative enthalpy of hydration, which results from its small size (its ionic radius is 90 pm) compared with Cs^+ (181 pm) and its consequent strong electrostatic interaction with water molecules. Overall, the favourable enthalpy of hydration of Li^+ outweighs terms relating to the formation of $Li^+(g)$ and gives rise to a more negative standard potential. The relatively less negative standard potential for Na^+/Na ($-2.71\,V$) in comparison with the rest of Group 1 (close to $-2.9\,V$) is a result of a combination of a fairly high sublimation enthalpy and moderate hydration enthalpy (Table 5.1).

The value of $E^{\ominus}$ (Na^+, Na) $= -2.71\,V$ may also be compared with that for $E^{\ominus}$ (Ag^+, Ag) $= +0.80\,V$. The (6-coordinate) ionic radii of these ions ($r_{Na^+} = 102$ pm and $r_{Ag^+} = 115$ pm)

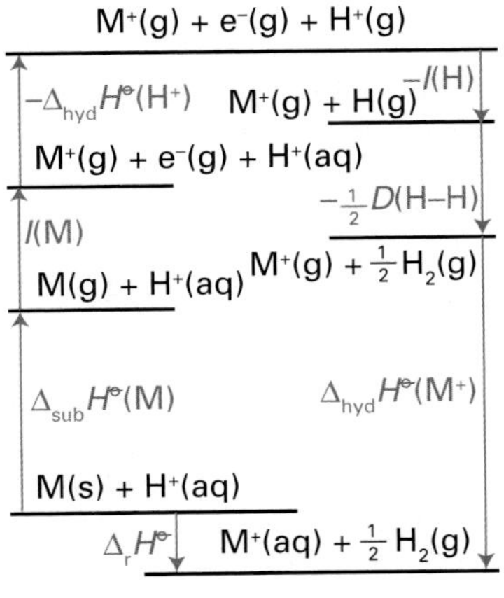

Fig. 5.2 A thermodynamic cycle showing the properties that contribute to the standard potential of a metal couple. Endothermic processes are drawn with upward-pointing arrows and exothermic contributions with downward pointing arrows.

BOX 5.1 Fuel cells

A *fuel cell* converts a chemical fuel, such as hydrogen (used for larger power requirements) or methanol (a convenient fuel for small applications), directly into electrical power, using O_2 or air as the oxidant. As power sources, fuel cells offer several advantages over rechargeable batteries or combustion engines, and their use is steadily increasing. Compared to batteries, which have to be replaced or recharged over a significant period of time, a fuel cell operates as long as fuel is supplied. Furthermore, a fuel cell does not contain large amounts of environmental contaminants such as Ni and Cd, although relatively small amounts of Pt and other metals are required as electrocatalysts. The operation of a fuel cell is more efficient than combustion devices, with near-quantitative conversion of fuel to H_2O and (for methanol) CO_2. Fuel cells are also much less polluting because nitrogen oxides are not produced at the relatively low temperatures that are used. Because an individual cell potential is less than about 1 V, fuel cells are connected in series known as 'stacks' in order to produce a useful voltage.

Important classes of hydrogen fuel cell are the *proton-exchange membrane fuel cell* (PEMFC), the *alkaline fuel cell* (AFC), and the *solid oxide fuel cell* (SOFC), which differ in their mode of electrode reactions, chemical charge transfer, and operational temperature. Details are included in the table.

Fuel cell	Reaction at anode	Electrolyte	Transfer ion	Reaction at cathode	Temp. range/°C	Pressure/atm	Efficiency/%
PEMFC	$H_2 \rightarrow 2H^+ + 2e^-$	H^+-conducting polymer (PEM)	H^+	$2H^+ + \frac{1}{2}O_2 + 2e^- \rightarrow H_2O$	80−100	1−8	35−40
AFC	$H_2 \rightarrow 2H^+ + 2e^-$	Aqueous alkali	OH^-	$H_2O + \frac{1}{2}O_2 + 2e^- \rightarrow 2OH^-$	80−250	1−10	50−60
SOFC	$H_2 + O^{2-} \rightarrow H_2O + 2e^-$	Solid oxide	O^{2-}	$\frac{1}{2}O_2 + 2e^- \rightarrow O^{2-}$	800−1000	1	50−55
DMFC	$CH_3OH + H_2O \rightarrow CO_2 + 6H^+ + 6e^-$	H^+-conducting polymer	H^+	$2H^+ + \frac{1}{2}O_2 + 2e^- \rightarrow H_2O$	0−40	1	20−40

The basic principles of fuel cells are illustrated by a PEMFC (Fig. B5.1), which operates at modest temperatures (80−100°C) and is suitable as an on-board power supply for road vehicles. At the anode, a continuous supply of H_2 is oxidized and the resulting H^+ ions, the chemical charge carriers, pass through a membrane to the cathode, at which O_2 is reduced to H_2O. This process produces a flow of electrons from anode to cathode (the current) that is directed through the load (typically an electric motor). The anode (the site of H_2 oxidation) and the cathode (the site of O_2 reduction) are both loaded with a Pt catalyst to obtain efficient electrochemical conversions of fuel and oxidant. The major factor limiting the efficiency of PEMFC and other fuel cells is the sluggish reduction of O_2 at the cathode, which involves expenditure of a few tenths of a volt (the 'overpotential') just to drive this reaction at a practical rate. The operating voltage is usually about 0.7 V. The membrane is composed of an H^+-conducting polymer, sodium perfluorosulfonate (invented by Du Pont and known commercially as Nafion®).

An AFC is more efficient than a PEMFC because the reduction of O_2 at the Pt cathode is much easier under alkaline conditions. Hence the operating voltage is typically greater than about 0.8 V. The membrane of the PEMFC is now replaced by a pumped flow of hot aqueous alkali between the two electrodes. Alkaline fuel cells were used to provide power for the pioneering Apollo spacecraft moon missions.

An SOFC operates at much higher temperatures (800−1100°C) and is used to provide electricity and heating in buildings (in the arrangement called combined heat and power, CHP). The cathode is typically a complex metal oxide based on $LaCoO_3$, such as $La_{(1-x)}Sr_xMn_{(1-y)}Co_yO_3$, whereas the anode is typically NiO mixed with RuO_2 and a lanthanoid oxide such as $Ce_{(1-x)}Gd_xO_{1.95}$. The chemical charge is carried by a ceramic oxide such as ZrO_2 doped with yttrium, which allows conduction by O^{2-} ion transfer at high temperatures (Section 23.4). The high operating temperature relaxes the requirement for such an efficient catalyst as Pt.

Methanol is used as a fuel in either of two ways. One exploits methanol as an 'H_2 carrier', because the reforming reaction (see Chapter 10) is used to generate H_2 which is then supplied *in situ* to a normal hydrogen fuel cell as mentioned above. This indirect method avoids the need to store H_2 under pressure. The other is the direct methanol fuel cell (DMFC) which incorporates anode and cathode each loaded with Pt or a Pt alloy, and a PEM. The methanol is supplied

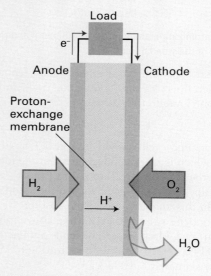

Fig. B5.1 A schematic diagram of a proton-exchange membrane (PEM) fuel cell. The anode and cathode are loaded with a catalyst (Pt) to convert fuel (H_2) and oxidant (O_2) into H^+ and H_2O, respectively. The membrane (usually a material called Nafion®) allows the H^+ ions produced at the anode to be transferred to the cathode.

to the anode as an aqueous solution (at 1 mol dm^{-3}). The DMFC is particularly suitable for small low-power devices such as mobile phones and portable electronic processors and it provides a promising alternative to the Li-ion battery. The principal disadvantage of the DMFC is its relatively low efficiency. This inefficiency arises from two factors that lower the operating voltage: the sluggish kinetics at the anode (oxidation of CH_3OH to CO_2 and H_2O) in addition to the poor cathode kinetics already mentioned, and transfer of methanol across the membrane to the cathode ('crossover'), which occurs because methanol permeates the hydrophilic PEM easily. A 50/50 Pt/Ru mixture supported on carbon is used as the anode catalyst to improve the rate of methanol oxidation.

Table 5.1 Thermodynamic contributions to $E^\ominus$ for a selection of metals at 298 K

	Li	Na	Cs	Ag
$\Delta_{sub}H^\ominus/(\text{kJ mol}^{-1})$	$+161$	$+109$	$+79$	$+284$
$I/(\text{kJ mol}^{-1})$	526	502	382	735
$\Delta_{hyd}H^\ominus/(\text{kJ mol}^{-1})$	-520	-406	-264	-468
$\Delta_f H^\ominus(M^+,\text{aq})/(\text{kJ mol}^{-1})$	$+167$	$+206$	$+197$	$+551$
$\Delta_r H^\ominus/(\text{kJ mol}^{-1})$	$+278$	$+240$	$+248$	-106
$T\Delta_r S^\ominus/(\text{kJ mol}^{-1})$	-16	-22	-34	-29
$\Delta_r G^\ominus/(\text{kJ mol}^{-1})$	$+294$	$+262$	$+282$	-77
$E^\ominus/V$	-3.04	-2.71	-2.92	$+0.80$

$\Delta_f H^\ominus(H^+,\text{aq}) = +455\,\text{kJ mol}^{-1}$.

Table 5.2 Selected standard potentials at 298 K; further values are included in *Resource section* 3

Couple	$E^\ominus/V$
$F_2(g) + 2e^- \rightarrow 2F^-(aq)$	$+2.87$
$Ce^{4+}(aq) + e^- \rightarrow Ce^{3+}(aq)$	$+1.76$
$MnO_4^-(aq) + 8H^+(aq) + 5e^- \rightarrow Mn^{2+}(aq) + 4H_2O(l)$	$+1.51$
$Cl_2(g) + 2e^- \rightarrow 2Cl^-(aq)$	$+1.36$
$O_2(g) + 4H^+(aq) + 4e^- \rightarrow 2H_2O(l)$	$+1.23$
$[IrCl_6]^{2-}(aq) + e^- \rightarrow [IrCl_6]^{3-}(aq)$	$+0.87$
$Fe^{3+}(aq) + e^- \rightarrow Fe^{2+}(aq)$	$+0.77$
$[PtCl_4]^{2-}(aq) + 2e^- \rightarrow Pt(s) + 4Cl^-(aq)$	$+0.60$
$I_3^-(aq) + 2e^- \rightarrow 3I^-(aq)$	$+0.54$
$[Fe(CN)_6]^{3-}(aq) + e^- \rightarrow [Fe(CN)_6]^{4-}(aq)$	$+0.36$
$AgCl(s) + e^- \rightarrow Ag(s) + Cl^-(aq)$	$+0.22$
$2H^+(aq) + 2e^- \rightarrow H_2(g)$	0
$AgI(s) + e^- \rightarrow Ag(s) + I^-(aq)$	-0.15
$Zn^{2+}(aq) + 2e^- \rightarrow Zn(s)$	-0.76
$Al^{3+}(aq) + 3e^- \rightarrow Al(s)$	-1.68
$Ca^{2+}(aq) + 2e^- \rightarrow Ca(s)$	-2.84
$Li^+(aq) + e^- \rightarrow Li(s)$	-3.04

are similar, and consequently their ionic hydration enthalpies are also similar. However, the much higher enthalpy of sublimation of silver, and particularly its high ionization energy, which is due to the poor screening by the 4d electrons, results in a positive standard potential. This difference is reflected in the very different behaviour of the metals when treated with a dilute acid: sodium reacts and dissolves explosively, producing hydrogen, whereas silver is unreactive. Similar arguments can be used to explain many of the trends observed in the standard potentials given in Table 5.2. For example, the positive potentials characteristic of the noble metals result in large part from their very high sublimation enthalpies.

5.4 The electrochemical series

Key points: The oxidized member of a couple is a strong oxidizing agent if $E^\ominus$ is positive and large; the reduced member is a strong reducing agent if $E^\ominus$ is negative and large.

A negative standard potential ($E^\ominus < 0$) signifies a couple in which the reduced species (the Zn in Zn^{2+}/Zn) is a reducing agent for H^+ ions under standard conditions in aqueous solution. That is, if $E^\ominus$ (Ox, Red) < 0, then the substance 'Red' is a strong enough reducing agent to reduce H^+ ions (in the sense that $K > 1$ for the reaction). A short compilation of $E^{\ominus}$ values at 25°C is given in Table 5.2. The list is arranged in the order of the **electrochemical series**:

Ox/Red couple with strongly positive $E^{\ominus}$ [Ox is strongly oxidizing]

$\vdots$

Ox/Red couple with strongly negative $E^{\ominus}$ [Red is strongly reducing]

An important feature of the electrochemical series is that the reduced member of a couple has a thermodynamic tendency to reduce the oxidized member of any couple that lies above it in the series. Note that the classification refers only to the thermodynamic aspect of the reaction—its spontaneity under standard conditions and the value of K, not its rate. Thus even reactions that are found to be thermodynamically favourable from the electrochemical series may not progress, or progress only extremely slowly, if the kinetics of the process are unfavourable.

EXAMPLE 5.3   **Using the electrochemical series**

Among the couples in Table 5.2 is the permanganate ion, MnO_4^-, the common analytical reagent used in redox titrations of iron. Which of the ions Fe^{2+}, Cl^-, and Ce^{3+} can permanganate oxidize in acidic solution?

Answer We need to note that a reagent that is capable of reducing MnO_4^- ions must be the reduced form of a redox couple having a more negative standard potential than the couple MnO_4^-/Mn^{2+}. The standard potential of the couple MnO_4^-/Mn^{2+} in acidic solution is $+1.51$ V. The standard potentials of Fe^{3+}/Fe^{2+}, Cl_2/Cl^-, and Ce^{4+}/Ce^{3+} are $+0.77$, $+1.36$, and $+1.76$ V, respectively. It follows that MnO_4^- ions are sufficiently strong oxidizing agents in acidic solution (pH = 0) to oxidize Fe^{2+} and Cl^-, which have less positive standard potentials. Permanganate ions cannot oxidize Ce^{3+}, which has a more positive standard potential. It should be noted that the presence of other ions in the solution can modify the potentials and the conclusions (Section 5.10); this variation with conditions is particularly important in the case of H^+ ions, and the influence of pH is discussed in Section 5.6. The ability of MnO_4^- ions to oxidize Cl^- means that HCl cannot be used to acidify redox reactions involving permanganate but H_2SO_4 is used instead.

Self-test 5.3 Another common analytical oxidizing agent is an acidic solution of dichromate ions, $Cr_2O_7^{2-}$, for which $E^{\ominus}(Cr_2O_7^{2-}, Cr^{3+}) = +1.38$ V. Is the solution useful for a redox titration of Fe^{2+} to Fe^{3+}? Could there be a side reaction when Cl^- is present?

5.5 The Nernst equation

Key point: The cell potential at an arbitrary composition of the reaction mixture is given by the Nernst equation.

To judge the tendency of a reaction to run in a particular direction at an arbitrary composition, we need to know the sign and value of $\Delta_r G$ at that composition. For this information, we use the thermodynamic result that

$$\Delta_r G = \Delta_r G^{\ominus} + RT \ln Q \tag{5.3a}$$

where Q is the reaction quotient[1]

$$a\,Ox_A + b\,Red_B \rightarrow a'\,Red_A + b'\,Ox_B \qquad Q = \frac{[Red_A]^{a'}[Ox_B]^{b'}}{[Ox_A]^a[Red_B]^b} \tag{5.3b}$$

The reaction quotient has the same form as the equilibrium constant K but the concentrations refer to an arbitrary stage of the reaction; at equilibrium, $Q = K$. When evaluating Q and K, the quantities in square brackets are to be interpreted as the numerical values of the molar concentrations. Both Q and K are therefore dimensionless quantities. The reaction is spontaneous at an arbitrary stage if $\Delta_r G < 0$. This criterion can be expressed

---

[1] For reactions involving gas-phase species, the molar concentrations of the latter are replaced by partial pressures relative to $p^{\ominus} = 1$ bar.

in terms of the potential of the corresponding cell by substituting $E_{cell} = -\Delta_r G/\nu F$ and $E_{cell}^{\ominus} = -\Delta_r G^{\ominus}/\nu F$ into eqn 5.3a, which gives the **Nernst equation**:

$$E_{cell} = E_{cell}^{\ominus} - \frac{RT}{\nu F}\ln Q \qquad (5.4)$$

A reaction is spontaneous if, under the prevailing conditions, $E_{cell} > 0$, for then $\Delta_r G < 0$. At equilibrium $E_{cell} = 0$ and $Q = K$, so eqn 5.4 implies the following very important relationship between the standard potential of a cell and the equilibrium constant of the cell reaction at a temperature T:

$$\ln K = \frac{\nu F E_{cell}^{\ominus}}{RT} \qquad (5.5)$$

Table 5.3 lists the values of K that correspond to cell potentials in the range -2 to $+2$ V, with $\nu = 1$ and at 25°C. The table shows that, although electrochemical data are often compressed into the range -2 to $+2$ V, this narrow range corresponds to 68 orders of magnitude in the value of the equilibrium constant for $\nu = 1$.

If we regard the cell potential E_{cell} as the difference of two reduction potentials, just as $E_{cell}^{\ominus}$ is the difference of two *standard* reduction potentials, then the potential of each couple, E, that contributes to the cell reaction can be written like eqn 5.4,

$$E = E^{\ominus} - \frac{RT}{\nu F}\ln Q \qquad (5.6a)$$

but with

$$a\,Ox + \nu\,e^- \rightarrow a'\,Red \qquad Q = \frac{[Red]^{a'}}{[Ox]^{a}} \qquad (5.6b)$$

By convention, the electrons do not appear in the expression for Q.

The temperature dependence of a standard cell potential (eqns 5.6a and 5.6b) provides a straightforward way to determine the standard entropy of many redox reactions. From eqn 5.2, we can write

$$-\nu F E_{cell}^{\ominus} = \Delta_r G^{\ominus} = \Delta_r H^{\ominus} - T\,\Delta_r S^{\ominus} \qquad (5.7a)$$

Then, if we suppose that $\Delta_r H^{\ominus}$ and $\Delta_r S^{\ominus}$ are independent of temperature over the small range usually of interest, it follows that

$$-\nu F E_{cell}^{\ominus}(T_2) - \left\{-\nu F E_{cell}^{\ominus}(T_1)\right\} = -\left(T_2 - T_1\right)\Delta_r S^{\ominus}$$

and therefore that

$$\Delta_r S^{\ominus} = \frac{\nu F\left\{E_{cell}^{\ominus}(T_2) - E_{cell}^{\ominus}(T_1)\right\}}{T_2 - T_1} \qquad (5.7b)$$

In other words, $\Delta_r S^{\ominus}$ is proportional to the slope of a graph of a plot of the standard cell potential against temperature.

The standard reaction entropy $\Delta_r S^{\ominus}$ often reflects the change in solvation accompanying a redox reaction: for each half-cell reaction a positive entropy contribution is expected when the corresponding reduction results in a decrease in electric charge (solvent molecules are less tightly bound and more disordered). Conversely, a negative contribution is expected when there is an increase in charge. As discussed in Section 5.3, entropy contributions to standard potentials are usually very similar when comparing redox couples involving the same change in charge.

Table 5.3 The relationship between K and $E^{\ominus}$

$E^{\ominus}/V$	K
$+2$	10^{34}
$+1$	10^{17}
0	1
-1	10^{-17}
-2	10^{-34}

EXAMPLE 5.4 The potential generated by a fuel cell

Calculate the cell potential (measured using a load of such high resistance that negligible current flows) produced by a fuel cell in which the overall reaction is $2\,H_2(g) + O_2(g) \rightarrow 2\,H_2O(l)$ with H_2 and O_2 each at 1 bar and 25°C. (Note that in a working PEM fuel cell the temperature is usually $80-100$°C to improve performance.)

Answer We note that under zero-current conditions, the cell potential is given by the difference of standard potentials of the two redox couples. For the reaction as stated, we write

Right : $O_2(g) + 4H^+(aq) + 4e^- \rightarrow 2H_2O(l)$ $E^{\ominus} = +1.23\,V$
Left : $2\,H^+(aq) + 2\,e^- \rightarrow H_2(g)$ $E^{\ominus} = 0$
Overall$(Right - Left)$: $2H_2(g) + O_2(g) \rightarrow 2H_2O(l)$

The standard potential of the cell is therefore

$$E_{cell}^{\ominus} = (+1.23\,V) - 0 = +1.23\,V$$

The reaction is spontaneous as written, and the right-hand electrode is the cathode (the site of reduction).

Self-test 5.4 What potential difference would be produced in a fuel cell operating with oxygen and hydrogen with both gases at 5.0 bar?

Redox stability

When assessing the thermodynamic stability of a species in solution, we must bear in mind all possible reactants: the solvent, other solutes, the species itself, and dissolved oxygen. In the following discussion, we focus on the types of reaction that result from the thermodynamic instability of a solute. We also comment briefly on kinetic factors, but the trends they show are generally less systematic than those shown by stabilities.

5.6 The influence of pH

Key point: Many redox reactions in aqueous solution involve transfer of H^+ as well as electrons and the electrode potential therefore depends on the pH.

For many reactions in aqueous solution the electrode potential varies with pH because reduced species of a redox couple are usually much stronger Brønsted bases than the oxidized species. For a redox couple in which there is transfer of v_e electrons and v_H protons, it follows from eqn 5.6b that

$$Ox + v_e e^- + v_{H^+} \rightarrow RedH_{v_{H^+}} \qquad Q = \frac{[RedH_{v_{H^+}}]}{[Ox][H^+]^{v_{H^+}}}$$

and

$$E = E^{\ominus} - \frac{RT}{v_e F} \ln \frac{[RedH_{v_{H^+}}]}{[Ox][H^+]^{v_{H^+}}} = E^{\ominus} - \frac{RT}{v_e F} \ln \frac{[RedH_{v_{H^+}}]}{[Ox]} + \frac{v_{H^+} RT}{v_e F} \ln[H^+]$$

(We have used $\ln x = \ln 10 \log x$.) If the concentrations of Red and Ox are combined with $E^{\ominus}$ we define E' as

$$E' = E^{\ominus} - \frac{RT}{v_e F} \ln \frac{[RedH_{v_{H^+}}]}{[Ox]}$$

and if we use $\ln[H^+] = \ln 10 \log[H^+]$ with $pH = -\log[H^+]$, the potential of the electrode can be written

$$E = E' - \frac{v_{H^+} RT \ln 10}{v_e F} pH \qquad\qquad (5.8a)$$

At 25°C,

$$E = E' - \frac{(0.059\,V)v_{H^+}}{v_e} pH \qquad\qquad (5.8b)$$

That is, the potential decreases (becoming more negative) as the pH increases and the solution becomes more basic.

■ **A brief illustration.** The half-reaction for the perchlorate/chlorate (ClO_4^-/ClO_3^-) couple is

$$ClO_4^-(aq) + 2\,H^+(aq) + 2\,e^- \rightarrow ClO_3^-(aq) + H_2O(l)$$

Therefore whereas at pH = 0, $E^{\ominus} = +1.201$ V, at pH = 7 the reduction potential for the ClO_4^-/ClO_3^- couple is $1.201 - (2/2)(7 \times 0.059)$ V $= +0.788$ V. The perchlorate anion is a stronger oxidant under acid conditions. ■

A note on good practice Always include the sign of a reduction potential, even when it is positive.

Standard potentials in neutral solution (pH = 7) are denoted $E_W^{\ominus}$. These potentials are particularly useful in biochemical discussions because cell fluids are buffered near pH = 7. The condition pH = 7 (with unit activity for the other electroactive species present) corresponds to the so-called **biological standard state**; in biochemical contexts they are sometimes denoted either $E^{\oplus}$ or E_{m7}, the 'm7' denoting the 'midpoint' potential at pH = 7.

■ **A brief illustration.** To determine the reduction potential of the H^+/H_2 couple at pH = 7.0, the other species being present in their standard states, we note that $E' = E^{\ominus}(H^+, H_2) = 0$. The reduction half-reaction is $2\,H^+(aq) + 2\,e^- \rightarrow H_2(g)$, so $\nu_e = 2$ and $\nu_H = 2$. The biological standard potential is therefore

$$E^{\oplus} = 0 - (2/2)(7 \times 0.059)\,V = -0.41\,V\ ■$$

5.7 Reactions with water

Water may act as an oxidizing agent, when it is reduced to H_2:

$$H_2O(l) + e^- \rightarrow \tfrac{1}{2}H_2(g) + OH^-(aq)$$

For the equivalent reduction of hydronium ions in water at any pH (and partial pressure of H_2 of 1 bar) we have seen that the Nernst equation gives

$$H^+(aq) + e^- \rightarrow \tfrac{1}{2}H_2(g) \qquad E = -0.059\,V \times pH \tag{5.9}$$

This is the reaction that chemists typically have in mind when they refer to 'the reduction of water'. Water may also act as a reducing agent, when it is oxidized to O_2:

$$2H_2O(l) \rightarrow O_2(g) + 4\,H^+(aq) + 4\,e^-$$

When the partial pressure of O_2 is 1 bar, the Nernst equation for the $O_2, 2H^+/2H_2O$ half-reaction becomes

$$E = 1.23\,V - (0.059\,V \times pH) \tag{5.10}$$

because $\nu_{H^+}/\nu_e = 4/4 = 1$. Both H^+ and O_2 therefore have the same pH dependence for their reduction half-reactions. The variation of these two potentials with pH is shown in Fig. 5.3.

(a) Oxidation by water

Key point: For metals with large, negative standard potentials, reaction with aqueous acids leads to the production of H_2 unless a passivating oxide layer is formed.

The reaction of a metal with water or aqueous acid is in fact the oxidation of the metal by water or hydrogen ions because the overall reaction is one of the following processes (and their analogues for more highly charged metal ions):

$$M(s) + H_2O(l) \rightarrow M^+(aq) + \tfrac{1}{2}H_2(g) + OH^-(aq)$$
$$M(s) + H^+(aq) \rightarrow M^+(aq) + \tfrac{1}{2}H_2(g)$$

These reactions are thermodynamically favourable when M is an s-block metal, a 3d-series metal from Group 3 to at least Group 8 or 9 and beyond (Ti, V, Cr, Mn, Ni), or a lanthanoid. An example from Group 3 is

$$2\,Sc(s) + 6\,H^+(aq) \rightarrow 2\,Sc^{3+}(aq) + 3\,H_2(g)$$

When the standard potential for the reduction of a metal ion to the metal is negative, the metal should undergo oxidation in 1 M acid with the evolution of hydrogen.

Although the reactions of magnesium and aluminium with moist air are spontaneous, both metals can be used for years in the presence of water and oxygen. They survive because they

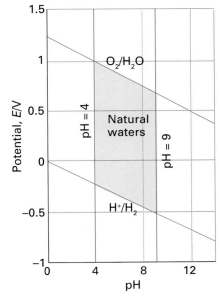

Fig. 5.3 The variation of the reduction potentials of water with pH. The sloping lines defining the upper and lower limits of thermodynamic water stability are the potentials for the O_2/H_2O and H^+/H_2 couples, respectively. The central zone represents the stability range of natural waters.

are **passivated**, or protected against reaction, by an impervious film of oxide. Magnesium oxide and aluminium oxide both form a protective skin on the parent metal beneath. A similar passivation occurs with iron, copper, and zinc. The process of 'anodizing' a metal, in which the metal is made an anode in an electrolytic cell, is one in which partial oxidation produces a smooth, hard passivating film on its surface. Anodizing is especially effective for the protection of aluminium by the formation of an inert, cohesive, and impenetrable Al_2O_3 layer.

Production of H_2 by electrolysis or photolysis of water is widely viewed as one of the renewable energy solutions for the future and is discussed in more detail in Chapter 10.

(b) Reduction by water

Key point: Water can act as a reducing agent, that is be oxidized by other species.

The strongly positive potential of the O_2, H^+/H_2O couple (eqn 5.10) shows that acidified water is a poor reducing agent except towards strong oxidizing agents. An example of the latter is $Co^{3+}(aq)$, for which $E^{\ominus}(Co^{3+}, Co^{2+}) = +1.92$ V. It is reduced by water with the evolution of O_2 and Co^{3+} does not survive in aqueous solution:

$$4\,Co^{3+}(aq) + 2\,H_2O(l) \rightarrow 4\,Co^{2+}(aq) + O_2(g) + 4\,H^+(aq) \qquad E_{cell}^{\ominus} = +0.69\text{ V}$$

Because H^+ ions are produced in the reaction, lower acidity (higher pH) favours the oxidation; lowering the concentration of H^+ ions encourages the formation of the products.

Only a few oxidizing agents (Ag^{2+} is another example) can oxidize water rapidly enough to give appreciable rates of O_2 evolution. Standard potentials greater than $+1.23$ V occur for several redox couples that are regularly used in aqueous solution, including Ce^{4+}/Ce^{3+} ($E^{\ominus} = +1.76$ V), the acidified dichromate ion couple $Cr_2O_7^{2-}/Cr^{3+}$ ($E^{\ominus} = +1.38$ V), and the acidified permanganate couple MnO_4^-/Mn^{2+} ($E^{\ominus} = +1.51$ V). The origin of the barrier to reaction is a kinetic one, stemming from the need to transfer four electrons and to form an $O-O$ bond.

Given that the rates of redox reactions are often controlled by the slow rate at which an $O-O$ bond can be formed, it remains a challenge for inorganic chemists to find good catalysts for O_2 evolution. The importance of this process is not due to any economic demand for O_2 but because of the desire to generate H_2 (a 'green' fuel) from water by electrolysis or photolysis. Some progress has been made using Co, Ru, and Ir complexes. Existing catalysts include the relatively poorly understood coatings that are used in the anodes of cells for the commercial electrolysis of water. They also include the enzyme system found in the O_2 evolution apparatus of the plant photosynthetic centre. This system is based on a special cofactor containing four Mn atoms and one Ca atom (Section 27.10). Although Nature is elegant and efficient, it is also complex, and the photosynthetic process is only slowly being elucidated by biochemists and bioinorganic chemists.

(c) The stability field of water

Key point: The stability field of water shows the region of pH and reduction potential where couples are neither oxidized by nor reduce hydrogen ions.

A reducing agent that can reduce water to H_2 rapidly, or an oxidizing agent that can oxidize water to O_2 rapidly, cannot survive in aqueous solution. The **stability field** of water, which is shown in Fig. 5.3, is the range of values of potential and pH for which water is thermodynamically stable towards both oxidation and reduction.

The upper and lower boundaries of the stability field are identified by finding the dependence of E on pH for the relevant half-reactions. As we have seen above, both oxidation (to O_2) and reduction of water have the same pH dependence (a slope of -0.059 V when E is plotted against pH at 25°C) and the stability field is confined within the boundaries of a pair of parallel lines of that slope. Any species with a potential more negative than that given in eqn 5.9 can reduce water (specifically, can reduce H^+) with the production of H_2; hence the lower line defines the low-potential boundary of the stability field. Similarly, any species with a potential more positive than that given in eqn 5.10 can liberate O_2 from water and the upper line gives the high-potential boundary. Couples that are thermodynamically unstable in water lie outside (above or below) the limits defined by the sloping lines in Fig. 5.3: species that are oxidized by water have potentials lying below the H_2 production line and species that are reduced by water have potentials lying above the O_2 production line.

The stability field in 'natural' water is represented by the addition of two vertical lines at pH = 4 and pH = 9, which mark the limits on pH that are commonly found in lakes and streams. A diagram like that shown in the illustration is known as a **Pourbaix diagram** and is widely used in environmental chemistry, as we shall see in Section 5.14.

5.8 Oxidation by atmospheric oxygen

Key point: The oxygen present in air and dissolved in water can oxidize metals and metal ions in solution.

The possibility of reaction between the solutes and dissolved O_2 must be considered when a solution is contained in an open beaker or is otherwise exposed to air. As an example, consider an aqueous solution containing Fe^{2+} in contact with an inert atmosphere such as N_2. Because $E^{\ominus}$ (Fe^{3+}, Fe^{2+}) = +0.77 V, which lies within the stability field of water, we expect Fe^{2+} to survive in water. Moreover, we can also infer that the oxidation of metallic iron by H^+(aq) should not proceed beyond Fe(II) because further oxidation to Fe(III) is unfavourable (by 0.77 V) under standard conditions. However, the picture changes considerably in the presence of O_2. In nature, many elements are found as oxidized species, either as soluble oxoanions such as SO_4^{2-}, NO_3^-, and MoO_4^{2-} or as ores such as Fe_2O_3. In fact, Fe(III) is the most common form of iron in the Earth's crust, and most iron in sediments that have been deposited from aqueous environments is present as Fe(III). The reaction

$$4\,Fe^{2+}(aq) + O_2(g) + 4\,H^+(aq) \rightarrow 4\,Fe^{3+}(aq) + 2\,H_2O(l)$$

is the difference of the following two half-reactions:

$$O_2(g) + 4\,H^+(aq) + 4\,e^- \rightarrow 2\,H_2O \qquad E^{\ominus} = +1.23\ V$$

$$Fe^{3+}(aq) + e^- \rightarrow Fe^{2+}(aq) \qquad E^{\ominus} = +0.77\ V$$

which implies that $E_{cell}^{\ominus}$ = +0.46 V at pH = 0. The oxidation of Fe^{2+}(aq) by O_2 is therefore spontaneous (in the sense $K > 1$) at pH = 0 and also at higher pH, although Fe(III) aqua species are hydrolysed and are precipitated as 'rust' (Section 5.13).

EXAMPLE 5.5 Judging the importance of atmospheric oxidation

The oxidation of copper roofs to a green substance (typically 'basic copper carbonate') is an example of atmospheric oxidation in a damp environment. Estimate the potential for oxidation of copper by oxygen in acid-to-neutral aqueous solution. Cu^{2+}(aq) is not deprotonated between pH = 0 and 7, so we may assume no hydrogen ions are involved in the half-reaction.

Answer We need to consider the reaction between Cu metal and atmospheric O_2 in terms of the two relevant reduction half-reactions:

$$O_2(g) + 4\,H^+(aq) + 4\,e^- \rightarrow 2\,H_2O \qquad E = +1.23\ V - (0.059\ V) \times pH$$

$$Cu^{2+}(aq) + 2\,e^- \rightarrow Cu(s) \qquad E^{\ominus} = +0.34\ V$$

The difference is

$$E_{cell} = 0.89\ V - (0.059\ V) \times pH$$

Therefore, E_{cell} = +0.89 V at pH = 0 and +0.48 V at pH = 7, so atmospheric oxidation by the reaction

$$2\,Cu(s) + O_2(g) + 4\,H^+(aq) \rightarrow 2\,Cu^{2+}(aq) + 2\,H_2O(l)$$

has $K > 1$ in both neutral and acid environments. Nevertheless, copper roofs do last for more than a few minutes: their familiar green surface is a passive layer of an almost impenetrable hydrated copper(II) carbonate, sulfate, or, near the sea, chloride. These compounds are formed from oxidation in the presence of atmospheric CO_2, SO_2, or salt water and the anion is also involved in the redox chemistry.

Self-test 5.5 The standard potential for the conversion of sulfate ions, SO_4^{2-}, to SO_2(aq) by the reaction SO_4^{2-}(aq) + 4 H^+(aq) + 2 $e^- \rightarrow SO_2$(aq) + 2 H_2O(l) is +0.16 V. What is the thermodynamically expected fate of SO_2 emitted into fog or clouds?

5.9 Disproportionation and comproportionation

Key point: Standard potentials can be used to define the inherent stability and instability of different oxidation states in terms of disproportionation and comproportionation.

Because $E^{\ominus}(Cu^+, Cu) = +0.52$ V and $E^{\ominus}(Cu^{2+}, Cu^+) = +0.16$ V, and both potentials lie within the stability field of water, Cu^+ ions neither oxidize nor reduce water. Nevertheless, Cu(I) is not stable in aqueous solution because it can undergo **disproportionation**, a redox reaction in which the oxidation number of an element is simultaneously raised and lowered. In other words, the element undergoing disproportionation serves as its own oxidizing and reducing agent:

$$2\,Cu^+(aq) \rightarrow Cu^{2+}(aq) + Cu(s)$$

This reaction is the difference of the following two half-reactions:

$$Cu^+(aq) + e^- \rightarrow Cu(s) \qquad\qquad E^{\ominus} = +0.52\text{ V}$$
$$Cu^{2+}(aq) + e^- \rightarrow Cu^+(aq) \qquad\qquad E^{\ominus} = +0.16\text{ V}$$

Because $E_{cell}^{\ominus} = 0.52$ V $- 0.16$ V $= +0.36$ V for the disproportionation reaction, $K = 1.3 \times 10^6$ at 298 K, so the reaction is highly favourable. Hypochlorous acid also undergoes disproportionation:

$$5\,HClO(aq) \rightarrow 2\,Cl_2(g) + ClO_3^-(aq) + 2\,H_2O(l) + H^+(aq)$$

This redox reaction is the difference of the following two half-reactions:

$$4\,HClO(aq) + 4\,H^+(aq) + 4\,e^- \rightarrow \quad 2\,Cl_2(g) + 4\,H_2O(l) \qquad E^{\ominus} = +1.63\,V$$
$$ClO_3^-(aq) + 5\,H^+(aq) + 4\,e^- \rightarrow \quad HClO(aq) + 2\,H_2O(l) \qquad E^{\ominus} = +1.43\,V$$

So overall $E_{cell}^{\ominus} = 1.63$ V $- 1.43$ V $= +0.20$ V, and $K = 3 \times 10^{13}$ at 298 K.

EXAMPLE 5.6 Assessing the likelihood of disproportionation

Show that Mn(VI) is unstable with respect to disproportionation into Mn(VII) and Mn(II) in acidic aqueous solution.

Answer To answer this question we need to consider the two half-reactions, one an oxidation, the other a reduction, that involve the species Mn(VI). The overall reaction (noting, from Pauling's rules, Section 4.5, that the Mn(VI) oxoanion MnO_4^{2-} should be protonated at pH $= 0$)

$$5\,HMnO_4^-(aq) + 3\,H^+(aq) \rightarrow 4\,MnO_4^-(aq) + Mn^{2+}(aq) + 4\,H_2O(l)$$

is the difference of the following two half-reactions

$$HMnO_4^-(aq) + 7\,H^+(aq) + 4\,e^- \rightarrow Mn^{2+}(aq) + 4\,H_2O(l) \qquad E^{\ominus} = +1.63\,V$$
$$4\,MnO_4^-(aq) + 4\,H^+(aq) + 4\,e^- \rightarrow 4\,HMnO_4^-(aq) \qquad E^{\ominus} = +0.90\,V$$

The difference of the standard potentials is $+0.73$ V, so the disproportionation is essentially complete ($K = 10^{50}$ at 298 K). A practical consequence of the disproportionation is that high concentrations of Mn(VI) ions cannot be obtained in acidic solution; they can, however, be obtained in basic solution, as we see in Section 5.12.

Self-test 5.6 The standard potentials for the couples Fe^{2+}/Fe and Fe^{3+}/Fe^{2+} are -0.41 V and $+0.77$ V, respectively. Should we expect Fe^{2+} to disproportionate in aqueous solution?

In **comproportionation**, the reverse of disproportionation, two species with the same element in different oxidation states form a product in which the element is in an intermediate oxidation state. An example is

$$Ag^{2+}(aq) + Ag(s) \rightarrow 2\,Ag^+(aq) \qquad E_{cell}^{\ominus} = +1.18\,V$$

The large positive potential indicates that Ag(II) and Ag(0) are completely converted to Ag(I) in aqueous solution ($K = 1 \times 10^{20}$ at 298 K).

5.10 The influence of complexation

Key points: The formation of a more thermodynamically stable complex when the metal is in the higher oxidation state of a couple favours oxidation and makes the standard potential more negative; the formation of a more stable complex when the metal is in the lower oxidation state of the couple favours reduction and the standard potential becomes more positive.

The formation of metal complexes (see Chapter 7) affects standard potentials because the ability of a complex (ML) formed by coordination of a ligand (L) to accept or release an electron differs from that of the corresponding aqua ion (M).

$$M^{\nu+}(aq) + e^- \rightarrow M^{(\nu-1)+}(aq) \qquad E^{\ominus}(M)$$

$$ML^{\nu+}(aq) + e^- \rightarrow ML^{(\nu-1)+}(aq) \qquad E^{\ominus}(ML)$$

The change in standard potential for the ML redox couple relative to that of M reflects the degree to which the ligand L coordinates more strongly to the oxidized or reduced form of M. The change in standard potential is analysed by considering the thermodynamic cycle shown in Fig. 5.4. Because the sum of reaction Gibbs energies round the cycle is zero, we can write

$$-FE^{\ominus}(M) - RT \ln K^{ox} + FE^{\ominus}(ML) + RT \ln K^{red} = 0 \qquad (5.11)$$

where K^{ox} and K^{red} are equilibrium constants for L binding to $M^{\nu+}$ and $M^{(\nu-1)+}$, respectively (of the form $K = [ML]/[M][L]$), and we have used $\Delta_r G^{\ominus} = -RT \ln K$ in each case. This expression rearranges to

$$E^{\ominus}(M) - E^{\ominus}(ML) = \frac{RT}{F} \ln \frac{K^{ox}}{K^{red}} \qquad (5.12a)$$

At 25°C and with $\ln x = \ln 10 \log x$

$$E^{\ominus}(M) - E^{\ominus}(ML) = (0.059\ \mathrm{V}) \log \frac{K^{ox}}{K^{red}} \qquad (5.12b)$$

Thus, every ten-fold increase in the equilibrium constant for ligand binding to $M^{\nu+}$ compared to $M^{(\nu-1)+}$ decreases the reduction potential by 0.059 V.

■ **A brief illustration.** The standard potential for the half-reaction $[Fe(CN)_6]^{3-}(aq) + e^- \rightarrow [Fe(CN)_6]^{4-}(aq)$ is 0.36 V; that is, 0.41 V more negative than that of the aqua redox couple $[Fe(OH_2)_6]^{3+}(aq) + e^- \rightarrow [Fe(OH_2)_6]^{2+}(aq)$. This equates to CN^- having a 10^7-fold greater affinity (in the sense $K^{ox} \approx 10^7 K^{red}$) for Fe(III) compared to Fe(II). ■

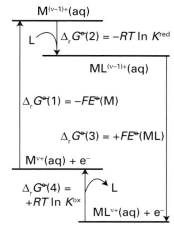

Fig. 5.4 Thermodynamic cycle showing how the standard potential of the couple M^+/M is altered by the presence of a ligand L.

Figure labels:
$M^{(\nu-1)+}(aq)$
$L \quad \Delta_r G^{\ominus}(2) = -RT \ln K^{red}$
$ML^{(\nu-1)+}(aq)$
$\Delta_r G^{\ominus}(1) = -FE^{\ominus}(M)$
$\Delta_r G^{\ominus}(3) = +FE^{\ominus}(ML)$
$M^{\nu+}(aq) + e^-$
$\Delta_r G^{\ominus}(4) = +RT \ln K^{ox} \quad L$
$ML^{\nu+}(aq) + e^-$

EXAMPLE 5.7 Interpreting potential data to identify bonding trends in complexes

Ruthenium is located immediately below iron in the periodic table. The following reduction potentials have been measured for species of Ru in aqueous solution. What do these values suggest when compared to their Fe counterparts?

$$[Ru(OH_2)_6]^{3+} + e^- \rightarrow [Ru(OH_2)_6]^{2+} \qquad E^{\ominus} = +0.25\ \mathrm{V}$$
$$[Ru(CN)_6]^{3-} + e^- \rightarrow [Ru(CN)_6]^{4-} \qquad E^{\ominus} = +0.85\ \mathrm{V}$$

Answer We can answer this question by noting that if complexation by a certain ligand causes the reduction potential of a metal ion to shift in a *positive* direction, then the new ligand must be stabilizing the reduced metal ion. In this case we see that CN^- stabilizes Ru(II) with respect to Ru(III). This behaviour is in stark contrast to the behaviour of Fe (see the preceding *brief illustration*) where we noted that CN^- stabilizes Fe(III), a result more in keeping with Fe–CN bonds being more ionic. The contrasting effects for species having identical charges suggest that the bonding between CN^- and Ru(II) is particularly strong.

Self-test 5.7 The ligand bpy (1) forms complexes with Fe(III) and Fe(II). The standard potential of the $[Fe(bpy)_3]^{3+}/[Fe(bpy)_3]^{2+}$ couple is +1.02 V. Does bpy bind preferentially to Fe(III) or Fe(II)?

1 2,2′-bipyridine (bpy)

5.11 The relation between solubility and standard potentials

Key points: The standard cell potential can be used to determine the solubility product.

The solubility of sparingly soluble compounds is expressed by an equilibrium constant known as the **solubility product**, K_{sp}. The approach is analogous to that introduced above for relating complexation equilibria to standard potentials. For metal ions $M^{\nu+}$ forming a precipitate MX_ν with anions X^-,

$$M^{\nu+}(aq) + \nu X^-(aq) \rightleftharpoons MX_\nu(s) \quad K_{sp} = [M^{\nu+}][X^-]^\nu \tag{5.13}$$

To generate the overall (non-redox) solubility reaction we use the difference of the two reduction half-reactions

$$M^{\nu+}(aq) + \nu e^- \rightarrow M(s) \qquad\qquad E^{\ominus}(M^{\nu+}/M)$$

$$MX_\nu(s) + \nu e^- \rightarrow M(s) + \nu X^-(aq) \qquad\qquad E^{\ominus}(MX_\nu / M,X^-)$$

From which it follows that

$$\ln K_{sp} = \frac{\nu F\{E^{\ominus}(MX/M,X^-) - E^{\ominus}(M^{\nu+}/M)\}}{RT} \tag{5.14}$$

EXAMPLE 5.8 Determining a solubility product from standard potentials

The possibility of plutonium waste leaking from nuclear facilities is a serious environmental problem. Calculate the solubility product of $Pu(OH)_4$ based on the following potentials measured in acid or basic solution. Hence, comment on the consequences of Pu(IV) waste leaking into environments of low pH as compared to high pH.

$$Pu^{4+}(aq) + 4e^- \rightarrow Pu(s) \qquad\qquad E^{\ominus} = -1.28 \text{ V}$$

$$Pu(OH)_4(s) + 4e^- \rightarrow Pu(s) + 4OH^-(aq) \qquad\qquad E = -2.06 \text{ V at pH} = 14$$

Answer We need to consider a thermodynamic cycle that combines the changes in Gibbs energy for the electrode reactions at pH = 0 and 14 using the potentials given, and the standard Gibbs energy for the reaction $Pu^{4+}(aq)$ with $OH^-(aq)$. The solubility product for $Pu(OH)_4$ is $K_{sp} = [Pu^{4+}][OH^-]^4$, so the corresponding Gibbs energy term is $-RT \ln K_{sp}$. For the thermodynamic cycle $\Delta G = 0$, so we obtain

$$-RT \ln K_{sp} = 4FE^{\ominus}(Pu^{4+}/Pu) - 4FE^{\ominus}(Pu(OH)_4/Pu)$$

and therefore

$$\ln K_{sp} = \frac{4F\{(-2.06 \text{ V}) - (-1.28 \text{ V})\}}{RT}$$

It follows that $K_{sp} = 1.7 \times 10^{-53}$.

Self-test 5.8 Given that the standard potential for the Ag^+/Ag couple is $+0.80$ V, calculate the potential of the $AgCl/Ag,Cl^-$ couple under conditions of $[Cl^-] = 1.0$ mol dm^{-3}, given that $K_{sp} = 1.77 \times 10^{-10}$.

The diagrammatic presentation of potential data

There are several useful diagrammatic summaries of the relative stabilities of different oxidation states in aqueous solution. Latimer diagrams are useful for summarizing quantitative data for individual elements. Frost diagrams are useful for the qualitative portrayal of the relative and inherent stabilities of oxidation states of a range of elements. We use them frequently in this context in the following chapters to convey the sense of trends in the redox properties of the members of a group.

5.12 Latimer diagrams

In a **Latimer diagram** (also known as a *reduction potential diagram*) for an element, the numerical value of the standard potential (in volts) is written over a horizontal line (or arrow) connecting species with the element in different oxidation states. The most highly oxidized form of the element is on the left, and in species to the right the element is in successively lower oxidation states.

A Latimer diagram summarizes a great deal of information in a compact form and (as we explain) shows the relationships between the various species in a particularly clear manner.

(a) Construction

Key points: In a Latimer diagram, oxidation numbers decrease from left to right and the numerical values of $E^\ominus$ in volts are written above the line joining the species involved in the couple.

The Latimer diagram for chlorine in acidic solution, for instance, is

$$\underset{+7}{ClO_4^-} \xrightarrow{+1.20} \underset{+5}{ClO_3^-} \xrightarrow{+1.18} \underset{+3}{HClO_2} \xrightarrow{+1.65} \underset{+1}{HClO} \xrightarrow{+1.67} \underset{0}{Cl_2} \xrightarrow{+1.36} \underset{-1}{Cl^-}$$

As in this example, oxidation numbers are sometimes written under (or over) the species. Conversion of a Latimer diagram to a half-reaction often involves balancing elements by including the predominant species present in acidic aqueous solution (H^+ and H_2O). The procedure for balancing redox equations was shown in Section 5.1. The standard state for this couple includes the condition that pH = 0. For example: the notation

$$HClO \xrightarrow{+1.67} Cl_2$$

denotes

$$2\,HClO(aq) + 2\,H^+(aq) + 2\,e^- \rightarrow Cl_2(g) + 2\,H_2O(l) \qquad E^\ominus = +1.67\,V$$

Similarly,

$$ClO_4^- \xrightarrow{+1.20} ClO_3^-$$

denotes

$$ClO_4^-(aq) + 2\,H^+(aq) + 2\,e^- \rightarrow ClO_3^-(aq) + H_2O(l) \qquad E^\ominus = +1.20\,V$$

Note that both of these half-reactions involve hydrogen ions, and therefore the potentials depend on pH.

In basic aqueous solution (corresponding to pOH = 0 and therefore pH = 14), the Latimer diagram for chlorine is

$$\underset{+7}{ClO_4^-} \xrightarrow{+0.37} \underset{+5}{ClO_3^-} \xrightarrow{+0.30} \underset{+3}{ClO_2^-} \xrightarrow{+0.68} \underset{+1}{ClO^-} -\xrightarrow{+0.42} \underset{0}{Cl_2} \xrightarrow{+1.36} \underset{-1}{Cl^-}$$

Note that the value for the Cl_2/Cl^- couple is the same as in acidic solution because its half-reaction does not involve the transfer of protons.

(b) Nonadjacent species

Key points: The standard potential of a couple that is the combination of two other couples is obtained by combining the standard Gibbs energies, not the standard potentials, of the half-reactions.

The Latimer diagram given above includes the standard potential for two nonadjacent species (the couple ClO^-/Cl^-). This information is redundant in the sense that it can be inferred from the data on adjacent species, but it is often included for commonly used couples as a convenience. To derive the standard potential of a nonadjacent couple when it is not listed explicitly we cannot in general just add their standard potentials but must make use of eqn 5.2 ($\Delta_r G^\ominus = -\nu F E^\ominus$) and the fact that the overall $\Delta_r G^\ominus$ for two successive steps a and b is the sum of the individual values:

$$\Delta_r G^\ominus (a + b) = \Delta_r G^\ominus (a) + \Delta_r G^\ominus (b)$$

To find the standard potential of the composite process, we convert the individual $E^\ominus$ values to $\Delta_r G^\ominus$ by multiplication by the relevant factor $-\nu F$, add them together, and then convert the sum back to $E^\ominus$ for the nonadjacent couple by division by $-\nu F$ for the overall electron transfer:

$$-\nu F E^\ominus (a + b) = -\nu(a) F E^\ominus(a) - \nu(b) F E^\ominus(b)$$

Because the factors $-F$ cancel and $\nu = \nu(a) + \nu(b)$, the net result is

$$E^\ominus(a + b) = \frac{\nu(a) E^\ominus(a) + \nu(b) E^\ominus(b)}{\nu(a) + \nu(b)} \qquad (5.15)$$

■ **A brief example.** To use the Latimer diagram to calculate the value of $E^{\ominus}$ for the ClO_2^-/Cl_2 couple in basic aqueous solution we note the following two standard potentials:

$$ClO_2^-(aq) + 2H^+(aq) + 2e^- \rightarrow ClO^-(aq) + H_2O(l) \qquad E^{\ominus}(a) = +0.68\,V$$

$$ClO^-(aq) + e^- \rightarrow \tfrac{1}{2}Cl_2(aq) \qquad E^{\ominus}(b) = +0.42\,V$$

Their sum,

$$ClO_2^-(aq) + 2H^+(aq) + 3e^- \rightarrow \tfrac{1}{2}Cl_2(g) + H_2O(l)$$

is the half-reaction for the couple we require. We see that $v(a) = 2$ and $v(b) = 1$. It follows from eqn 5.15 that the standard potential of the ClO/Cl^- couple is

$$E^{\ominus} = \frac{(2)(0.68\,V) + (1)(0.42\,V)}{3} = +0.59\,V \quad ■$$

(c) Disproportionation

Key point: A species has a tendency to disproportionate into its two neighbours if the potential on the right of the species in a Latimer diagram is higher than that on the left.

Consider the disproportionation

$$2\,M^+(aq) \rightarrow M(s) + M^{2+}(aq)$$

This reaction has $K > 1$ if $E^{\ominus} > 0$. To analyse this criterion in terms of a Latimer diagram, we express the overall reaction as the difference of two half-reactions:

$$M^+(aq) + e^- \rightarrow M(s) \qquad E^{\ominus}(R)$$

$$M^{2+}(aq) + e^- \rightarrow M^+(aq) \qquad E^{\ominus}(L)$$

The designations L and R refer to the relative positions, left and right respectively, of the couples in a Latimer diagram (recall that the more highly oxidized species lies to the left). The standard potential for the overall reaction is $E^{\ominus} = E^{\ominus}(R) - E^{\ominus}(L)$, which is positive if $E^{\ominus}(R) > E^{\ominus}(L)$. We can conclude that a species is inherently unstable (that is, it has a tendency to disproportionate into its two neighbours) if the potential on the right of the species is higher than the potential on the left.

EXAMPLE 5.9 **Identifying a tendency to disproportionate**

A part of the Latimer diagram for oxygen is

$$O_2 \xrightarrow{+0.70} H_2O_2 \xrightarrow{+1.76} H_2O$$

Does hydrogen peroxide have a tendency to disproportionate in acid solution?

Answer We can approach this question by reasoning that if H_2O_2 is a stronger oxidant than O_2, then it should react with itself to produce O_2 by oxidation and $2\,H_2O$ by reduction. The potential to the right of H_2O_2 is higher than that to its left, so we anticipate that H_2O_2 should disproportionate into its two neighbours under acid conditions. From the two half-reactions

$$2\,H^+(aq) + 2\,e^- + H_2O_2(aq) \rightarrow 2\,H_2O \qquad E^{\ominus} = +1.76\,V$$

$$O_2 + 2\,H^+(aq) + 2\,e^- \rightarrow H_2O_2(aq) \qquad E^{\ominus} = +0.70\,V$$

we conclude that for the overall reaction

$$2\,H_2O_2(aq) \rightarrow 2\,H_2O(l) + O_2(g) \qquad E^{\ominus} = +1.06\,V$$

and is spontaneous (in the sense $K > 1$).

Self-test 5.9 Use the following Latimer diagram (acid solution) to discuss whether (a) Pu(IV) disproportionates to Pu(III) and Pu(V) in aqueous solution; (b) Pu(V) disproportionates to Pu(VI) and Pu(IV).

$$PuO_2^{2+} \underset{+6}{\xrightarrow{+1.02}} PuO_2^+ \underset{+5}{\xrightarrow{+1.04}} Pu^{4+} \underset{+4}{\xrightarrow{+1.01}} Pu^{3+}_{} \;\; {\scriptstyle +3}$$

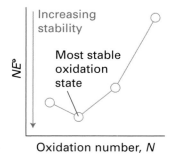

Fig. 5.5 Oxidation state stability as viewed in a Frost diagram.

5.13 **Frost diagrams**

A **Frost diagram** (also known as an *oxidation state diagram*) of an element X is a plot of $NE^{\ominus}$ for the couple $X(N)/X(0)$ against the oxidation number, N, of the element. The general form of a Frost diagram is given in Fig. 5.5. Frost diagrams depict whether a particular species

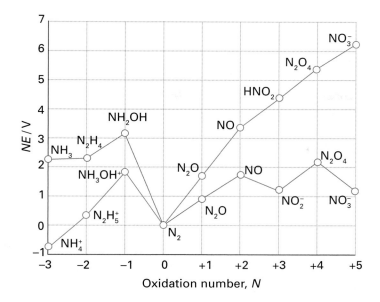

Fig. 5.6 The Frost diagram for nitrogen: the steeper the slope of a line, the higher the standard potential for the couple. The red line refers to standard (acid) conditions (pH = 0), the blue line refers to pH = 14. Note that because HNO_3 is a strong acid, even at pH = 0 it is present as its conjugate base NO_3^-.

X(N) is a good oxidizing agent or reducing agent. They also provide an important guide for identifying the oxidation states of an element that are inherently stable or unstable.

(a) Gibbs energies of formation for different oxidation states

Key point: A Frost diagram shows how the Gibbs energies of formation of different oxidation states of an element vary with oxidation number. The most stable oxidation state of an element corresponds to the species that lies lowest in its Frost diagram.

For a half-reaction in which a species X with oxidation number N is converted to its elemental form, the reduction half-reaction is written

$$X(N) + Ne^- \rightarrow X(0)$$

Because $NE^{\ominus}$ is proportional to the standard reaction Gibbs energy for the conversion of the species X(N) to the element (explicitly, $NE^{\ominus} = -\Delta_r G^{\ominus}/F$, where $\Delta_r G^{\ominus}$ is the standard reaction Gibbs energy for the half-reaction given above), a Frost diagram can also be regarded as a plot of standard reaction Gibbs energy against oxidation number. Consequently, the most stable states of an element in aqueous solution correspond to species that lie lowest in its Frost diagram. The example given in Fig. 5.6 shows data for nitrogen species formed in aqueous solution at pH = 0 and pH = 14. Only $NH_4^+(aq)$ is exergonic ($\Delta_f G^{\ominus} < 0$); all other species are endergonic ($\Delta_f G^{\ominus} > 0$). The diagram shows that the higher oxides and oxoacids are highly endergonic in acid solution but relatively stabilized in basic solution. The opposite is generally true for species with $N < 0$, except that hydroxylamine is particularly unstable regardless of pH.

EXAMPLE 5.10 Constructing a Frost diagram

Construct a Frost diagram for oxygen from the Latimer diagram in Example 5.9.

Answer We begin by placing the element in its zero oxidation state (O_2) at the origin for the $NE^{\ominus}$ and N axes. For the reduction of O_2 to H_2O_2 (for which $N = -1$), $E^{\ominus} = +0.70$ V, so $NE^{\ominus} = -0.70$ V. Because the oxidation number of O in H_2O is -2 and $E^{\ominus}$ for the O_2/H_2O couple is $+1.23$ V, $NE^{\ominus}$ at $N = -2$ $E^{\ominus} = -2.46$ V. These results are plotted in Fig. 5.7.

Self-test 5.10 Construct a Frost diagram from the Latimer diagram for Tl:

$$Tl^{3+} \xrightarrow{\ +1.25\ } Tl^+ \xrightarrow{\ -0.34\ } Tl$$

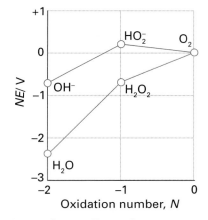

Fig. 5.7 The Frost diagram for oxygen in acidic solution (red line, pH = 0) and alkaline solution (blue line, pH = 14).

(b) Construction and interpretation

Key point: Frost diagrams are conveniently constructed by using electrode potential data. They may be used to gauge the inherent stabilities of different oxidation states of an element and to decide whether particular species are good oxidizing or reducing agents.

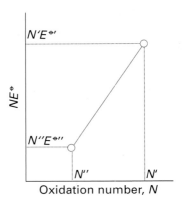

Fig. 5.8 The general structure of a region of a Frost diagram used to establish the relationship between the slope of a line and the standard potential of the corresponding couple.

To interpret the qualitative information contained in a Frost diagram it will be useful to keep the following features in mind.

1. The slope of the line joining any two points in a Frost diagram is equal to the standard potential of the couple formed by the two species that the points represent (Fig. 5.8).

It follows that the steeper the line joining two points (left to right) in a Frost diagram, the higher the standard potential of the corresponding couple (Fig. 5.9a).

■ **A brief illustration.** Refer to the oxygen diagram in Fig. 5.7. At the point corresponding to $N = -1$ (for H_2O_2), $(-1) \times E^{\ominus} = -0.70$ V, and at $N = -2$ (for H_2O), $(-2) \times E^{\ominus} = -2.46$ V. The difference of the two values is -1.76 V. The change in oxidation number of oxygen on going from H_2O_2 to H_2O is -1. Therefore, the slope of the line is $(-1.76 \text{ V})/(-1) = +1.76$ V, in accord with the value for the H_2O_2/H_2O couple in the Latimer diagram. ■

2. The oxidizing agent in the couple with the more positive slope (the more positive $E^{\ominus}$) is liable to undergo reduction (Fig. 5.9b).

3. The reducing agent of the couple with the less positive slope (the most negative $E^{\ominus}$) is liable to undergo oxidation (Fig. 5.9b).

For example, the steep slope connecting NO_3^- to lower oxidation numbers in Fig. 5.6 shows that nitrate is a good oxidizing agent under standard conditions.

We saw in the discussion of Latimer diagrams that a species is liable to undergo disproportionation if the potential for its reduction from $X(N)$ to $X(N - 1)$ is greater than its potential for oxidation from $X(N)$ to $X(N + 1)$. The same criterion can be expressed in terms of a Frost diagram (Fig. 5.9c):

4. A species in a Frost diagram is unstable with respect to disproportionation if its point lies above the line connecting the two adjacent species (on a convex curve).

When this criterion is satisfied, the standard potential for the couple to the left of the species is greater than that for the species on the right. A specific example is NH_2OH; as can be seen in Fig. 5.6, this compound is unstable with respect to disproportionation into NH_3 and N_2. The origin of this rule is illustrated in Fig. 5.9d, where we show geometrically that the reaction Gibbs energy of a species with intermediate oxidation number lies above the average value for the two species on either side. As a result, there is a tendency for the intermediate species to disproportionate into the two other species.

The criterion for comproportionation to be spontaneous can be stated analogously (Fig. 5.9e):

5. Two species will tend to comproportionate into an intermediate species that lies below the straight line joining the terminal species (on a concave curve).

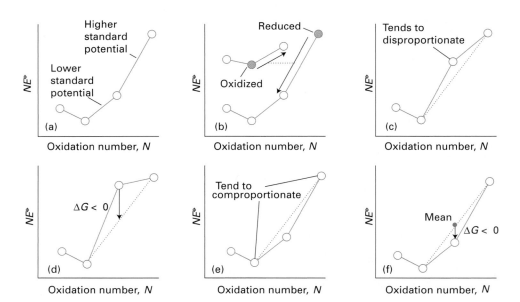

Fig. 5.9 The interpretation of a Frost diagram to gauge (a) reduction potential, (b) tendency towards oxidation and reduction, (c, d) disproportionation, and (e, f) comproportionation.

A substance that lies below the line connecting its neighbours in a Frost diagram is inherently more stable than they are because their average molar Gibbs energy is higher (Fig. 5.9f) and hence comproportionation is thermodynamically favourable. The nitrogen in NH_4NO_3, for instance, has two ions with oxidation numbers -3 (NH_4^+) and $+5$ (NO_3^-). Because N_2O lies below the line joining NH_4^+ to NO_3^-, their comproportionation is spontaneous:

$$NH_4^+(aq) + NO_3^-(aq) \rightarrow N_2O(g) + 2H_2O(l)$$

However, although the reaction is expected to be spontaneous on thermodynamic grounds under standard conditions, the reaction is kinetically inhibited in solution and does not ordinarily occur. The corresponding reaction

$$NH_4NO_3(s) \rightarrow N_2O(g) + 2H_2O(g)$$

in the solid state is both thermodynamically spontaneous ($\Delta_r G^\ominus = -168$ kJ mol^{-1}) and, once initiated by a detonation, explosively fast. Indeed, ammonium nitrate is often used in place of dynamite for blasting rocks.

Frost diagrams can equally well be constructed for other conditions. The potentials at pH = 14 are denoted $E_B^\ominus$ and the blue line in Fig. 5.6 is a '*basic* Frost diagram' for nitrogen. The important difference from the behaviour in acidic solution is the stabilization of NO_2^- against disproportionation: its point in the basic Frost diagram no longer lies above the line connecting its neighbours. The practical outcome is that metal nitrites are stable in neutral and basic solutions and can be isolated, whereas HNO_2 cannot (although solutions of HNO_2 have some short-term stability as their decomposition is kinetically slow). In some cases there are marked differences between strongly acidic and basic solutions, as for the phosphorus oxoanions. This example illustrates an important general point about oxoanions: when their reduction requires removal of oxygen, the reaction consumes H^+ ions, and all oxoanions are stronger oxidizing agents in acidic than in basic solution.

EXAMPLE 5.11 **Using a Frost diagram to judge the thermodynamic stability of ions in solution**

Figure 5.10 shows the Frost diagram for manganese. Comment on the stability of Mn^{3+} in acidic aqueous solution.

Answer We approach this question by inspecting how the $NE^\ominus$ value for Mn^{3+} ($N = +3$) compares with the values for species on either side ($N < +3$, $N > +3$)). Because Mn^{3+} lies *above* the line joining Mn^{2+} to MnO_2, it should disproportionate into these two species. The chemical reaction is

$$2Mn^{3+}(aq) + 2H_2O(l) \rightarrow Mn^{2+}(aq) + MnO_2(s) + 4H^+(aq)$$

Self-test 5.13 What is the oxidation number of Mn in the product when MnO_4^- is used as an oxidizing agent in aqueous acid?

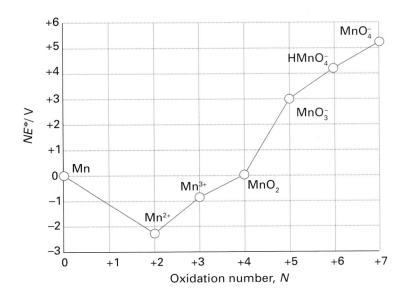

Fig. 5.10 The Frost diagram for manganese in acidic solution (pH = 0). Note that because $HMnO_3$ and $HMnO_4$ are strong acids, even at pH = 0 they are present as their conjugate bases.

Modified Frost diagrams summarize potential data under specified conditions of pH; their interpretation is the same as for pH = 0, but oxoanions often display markedly different thermodynamic stabilities; all oxoanions are stronger oxidizing agents in acidic than in basic solution.

EXAMPLE 5.12 Application of Frost diagrams at different pH

Potassium nitrite is stable in basic solution but, when the solution is acidified, a gas is evolved that turns brown on exposure to air. What is the reaction?

Answer To answer this we use the Frost diagram (Fig. 5.6) to compare the inherent stabilities of N(III) in acid and basic solutions. The point representing the NO_2^- ion in basic solution lies below the line joining NO to NO_3^-; the ion therefore is not liable to disproportionation. On acidification, the HNO_2 point rises and the straightness of the line through NO, HNO_2, and N_2O_4 (dimeric NO_2) implies that all three species are present at equilibrium. The brown gas is NO_2 formed from the reaction of NO evolved from the solution with air. In solution, the species of oxidation number +2 (NO) tends to disproportionate. However, the escape of NO from the solution prevents its disproportionation to N_2O and HNO_2.

Self-test 5.12 By reference to Fig. 5.6, compare the strength of NO_3^- as an oxidizing agent in acidic and basic solution.

5.14 Pourbaix diagrams

Key points: A Pourbaix diagram is a map of the conditions of potential and pH under which species are stable in water. A horizontal line separates species related by electron transfer only, a vertical line separates species related by proton transfer only, and sloped lines separate species related by both electron and proton transfer.

A **Pourbaix diagram** (also known as an *E–pH diagram*) indicates the conditions of pH and potential under which a species is thermodynamically stable. The diagrams were introduced by Marcel Pourbaix in 1938 as a convenient way of discussing the chemical properties of species in natural waters and they are particularly useful in environmental and corrosion science.

Iron is essential for almost all life forms and the problem of its uptake from the environment is discussed further in Chapter 27. Figure 5.11 is a simplified Pourbaix diagram for iron, omitting such low concentration species as oxygen-bridged Fe(III) dimers. This diagram is useful for the discussion of iron species in natural waters (see Section 5.15) because the total iron concentration is low; at high concentrations complex multinuclear iron species can form. We can see how the diagram has been constructed by considering some of the reactions involved.

The reduction half-reaction

$$Fe^{3+}(aq) + e^- \rightarrow Fe^{2+}(aq) \qquad E^{\ominus} = +0.77\ V$$

does not involve H^+ ions, so its potential is independent of pH and hence corresponds to a horizontal line on the diagram. If the environment contains a couple with a potential above this line (a more positive, oxidizing couple), then the oxidized species, Fe^{3+}, will be the major species. Hence, the horizontal line towards the top left of the diagram is a boundary that separates the regions where Fe^{3+} and Fe^{2+} dominate.

Another reaction to consider is

$$Fe^{3+}(aq) + 3H_2O(l) \rightarrow Fe(OH)_3(s) + 3H^+(aq)$$

This reaction is not a redox reaction (there is no change in oxidation number of any element), so it is insensitive to the electric potential in its environment and therefore is represented by a vertical line on the diagram. However, this boundary does depend on pH, with $Fe^{3+}(aq)$ favoured by low pH and $Fe(OH)_3(s)$ favoured by high pH. We adopt the convention that Fe^{3+} is the dominant species in the solution if its concentration exceeds 10 µmol dm^{-3} (a typical freshwater value). The equilibrium concentration of Fe^{3+} varies with pH, and the vertical boundary at pH = 3 represents the pH at which Fe^{3+} becomes dominant according to this definition. In general, a vertical line in a Pourbaix diagram does not involve a redox reaction but signifies a pH-dependent change of state of either the oxidized or reduced form.

As the pH is increased, the Pourbaix diagram includes reactions such as

$$Fe(OH)_3(s) + 3H^+(aq) + e^- \rightarrow Fe^{2+}(aq) + 3H_2O(l)$$

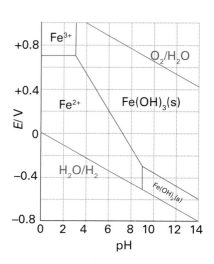

Fig. 5.11 A simplified Pourbaix diagram for some important naturally occurring aqua-species of iron.

(for which the slope of potential against pH, according to eqn 5.8b, is $v_H/v_e = -3(0.059\ V)$) and eventually $Fe^{2+}(aq)$ is also precipitated as $Fe(OH)_2$. Inclusion of the metal dissolution couple ($Fe^{2+}/Fe(s)$) would complete construction of the Pourbaix diagram for well-known aqua species of iron.

5.15 Natural waters

Key point: Pourbaix diagrams show that iron can exist in solution as Fe^{2+} under acidic reducing conditions such as in contact with soils rich in organic matter.

The chemistry of natural waters can be rationalized by using Pourbaix diagrams of the kind we have just constructed. Thus, where fresh water is in contact with the atmosphere, it is saturated with O_2, and many species may be oxidized by this powerful oxidizing agent. More fully reduced forms are found in the absence of oxygen, especially where there is organic matter to act as a reducing agent. The major acid system that controls the pH of the medium is $CO_2/H_2CO_3/HCO_3^-/CO_3^{2-}$, where atmospheric CO_2 provides the acid and dissolved carbonate minerals provide the base. Biological activity is also important because respiration releases CO_2. This acidic oxide lowers the pH and hence makes the potential more positive. The reverse process, photosynthesis, consumes CO_2, thus raising the pH and making the potential more negative. The condition of typical natural waters—their pH and the potentials of the redox couples they contain—is summarized in Fig. 5.12.

From Fig. 5.11 we see that Fe^{3+} can exist in water if the environment is oxidizing; hence, where O_2 is plentiful and the pH is low (below 3), iron will be present as Fe^{3+}. Because few natural waters are so acidic, $Fe^{3+}(aq)$ is very unlikely to be found in the environment. The iron in insoluble Fe_2O_3 or insoluble hydrated forms such as $FeO(OH)$ can enter solution as Fe^{2+} if it is reduced, which occurs when the condition of the water lies below the sloping boundary in the diagram. We should observe that, as the pH rises, Fe^{2+} can form only if there are strong reducing couples present, and its formation is very unlikely in oxygen-rich water. Figure 5.11 shows that iron will be reduced and dissolved in the form of Fe^{2+} in both bog waters and organic-rich waterlogged soils (at pH near 4.5 in both cases and with corresponding E values near $+0.03\ V$ and $-0.1\ V$, respectively).

It is instructive to analyse a Pourbaix diagram in conjunction with an understanding of the physical processes that occur in water. As an example, consider a lake where the temperature gradient, cool at the bottom and warmer above, tends to prevent vertical mixing. At the surface, the water is fully oxygenated and the iron must be present in particles of the insoluble $FeO(OH)$; these particles tend to settle. At greater depth, the O_2 content is low. If the organic content or other sources of reducing agents are sufficient, the oxide will be reduced and iron will dissolve as Fe^{2+}. The Fe(II) ions will then diffuse towards the surface where they encounter O_2 and are oxidized to insoluble $FeO(OH)$ again.

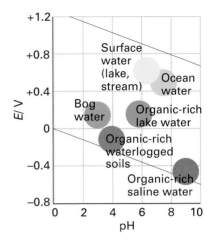

Fig. 5.12 The stability field of water showing regions typical of various natural waters.

> **EXAMPLE 5.13 Using a Pourbaix diagram**
>
> Figure 5.13 is part of a Pourbaix diagram for manganese. Identify the environment in which the solid MnO_2 or its corresponding hydrous oxides are important. Is Mn(III) formed under any conditions?
>
> **Answer** We approach this problem by locating the zone of stability for MnO_2 on the Pourbaix diagram and inspecting its position relative to the boundary between O_2 and H_2O. Manganese dioxide is the thermodynamically favoured state in well-oxygenated water under all pH conditions with the exception of strong acid (pH < 1). Under mildly reducing conditions, in waters having neutral-to-acidic pH, the stable species is $Mn^{2+}(aq)$. Manganese(III) species are stabilized only in oxygenated waters at higher pH.
>
> **Self-test 5.13** Use Fig. 5.11 to evaluate the possibility of finding $Fe(OH)_3(s)$ in a waterlogged soil.

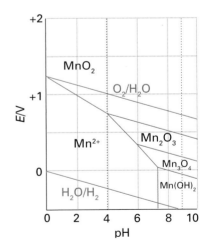

Fig. 5.13 A section of the Pourbaix diagram for manganese. The broken black vertical lines represent the normal pH range in natural waters.

Chemical extraction of the elements

The original definition of 'oxidation' was a reaction in which an element reacts with oxygen and is converted to an oxide. 'Reduction' originally meant the reverse reaction, in which an oxide of a metal is converted to the metal. Although both terms have been generalized and expressed in terms of electron transfer and changes in oxidation state, these special cases are still the basis of a major part of chemical industry and laboratory chemistry. In the following sections we discuss the extraction of the elements in terms of changing their oxidation number from its value in a naturally occurring compound to zero (corresponding to the element).

5.16 Chemical reduction

Only a few metals, such as gold, occur in nature as their elements. Most metals are found as their oxides, such as Fe_2O_3, or as ternary compounds, such as $FeTiO_3$. Sulfides are also common, particularly in mineral veins where deposition occurred under water-free and oxygen-poor conditions. Slowly, prehistoric humans learned how to transform ores to produce metals for making tools and weapons. Copper could be extracted from its ores by aerial oxidation at temperatures attainable in the primitive hearths that became available about 6000 years ago.

$$2\,Cu_2S(s) + 3\,O_2(g) \rightarrow 2\,Cu_2O(s) + 2\,SO_2(g)$$

$$2\,Cu_2O(s) + Cu_2S(s) \rightarrow 6\,Cu(s) + SO_2(g)$$

It was not until nearly 3000 years ago that higher temperatures could be reached and less readily reduced elements, such as iron, could be extracted, leading to the Iron Age. These elements were produced by heating the ore to its molten state with a reducing agent such as carbon. This process is known as *smelting*. Carbon remained the dominant reducing agent until the end of the nineteenth century, and metals that needed higher temperatures for their production remained unavailable even though their ores were reasonably abundant.

The availability of electric power expanded the scope of carbon reduction because electric furnaces can reach much higher temperatures than carbon-combustion furnaces, such as the blast furnace. Thus, magnesium is a metal of the twentieth century because one of its modes of recovery, the *Pidgeon process*, involves the very high temperature, electrothermal reduction of the oxide by carbon:

$$MgO(s) + C(s) \xrightarrow{\Delta} Mg(l) + CO(g)$$

Note that the carbon is oxidized only to carbon monoxide, the product favoured thermodynamically at the very high reaction temperatures used.

The technological breakthrough in the nineteenth century that resulted in the conversion of aluminium from a rarity into a major construction metal was the introduction of electrolysis, the driving of a nonspontaneous reaction (including the reduction of ores) by the passage of an electric current.

(a) Thermodynamic aspects

Key points: An Ellingham diagram summarizes the temperature dependence of the standard Gibbs energies of formation of metal oxides and is used to identify the temperature at which reduction by carbon or carbon monoxide becomes spontaneous.

As we have seen, the standard reaction Gibbs energy, $\Delta_r G^\ominus$, is related to the equilibrium constant, K, through $\Delta_r G^\ominus = -RT \ln K$, and a negative value of $\Delta_r G^\ominus$ corresponds to $K > 1$. It should be noted that equilibrium is rarely attained in commercial processes as many such systems involve dynamic stages where, for example, reactants and products are in contact only for short times. Furthermore, even a process at equilibrium for which $K < 1$ can be viable if the product (particularly a gas) is swept out of the reaction chamber and the reaction continues to chase the ever-vanishing equilibrium composition. In principle, we also need to consider rates when judging whether a reaction is feasible in practice, but reactions are often fast at high temperature and thermodynamically favourable reactions are likely to occur. A fluid phase (typically a gas or solvent) is usually required to facilitate what would otherwise be a sluggish reaction between coarse particles.

To achieve a negative $\Delta_r G^\ominus$ for the reduction of a metal oxide with carbon or carbon monoxide, one of the following reactions

(a) $C(s) + \frac{1}{2}O_2(g) \rightarrow CO(g)$ $\qquad \Delta_r G^\ominus(C,CO)$

(b) $\frac{1}{2}C(s) + \frac{1}{2}O_2(g) \rightarrow \frac{1}{2}CO_2(g)$ $\qquad \Delta_r G^\ominus(C,CO_2)$

(c) $CO(g) + \frac{1}{2}O_2(g) \rightarrow CO_2(g)$ $\qquad \Delta_r G^\ominus(CO,CO_2)$

must have a more negative $\Delta_r G^{\ominus}$ than a reaction of the form

(d) $x\,\mathrm{M(s\ or\ l)} + \tfrac{1}{2}O_2(g) \to M_xO(s)$ $\qquad \Delta_r G^{\ominus}(M, M_xO)$

under the same reaction conditions. If that is so, then one of the reactions

(a − d) $M_xO(s) + C(s) \to x\,\mathrm{M(s\ or\ l)} + CO(g)$ $\qquad \Delta_r G^{\ominus}(C, CO) - \Delta_r G^{\ominus}(M, M_xO)$

(b − d) $M_xO(s) + \tfrac{1}{2}C(s) \to x\,\mathrm{M(s\ or\ l)} + \tfrac{1}{2}CO_2(g)$ $\qquad \Delta_r G^{\ominus}(C, CO_2) - \Delta_r G^{\ominus}(M, M_xO)$

(c − d) $M_xO(s) + CO(g) \to x\,\mathrm{M(s\ or\ l)} + CO_2(g)$ $\qquad \Delta_r G^{\ominus}(CO, CO_2) - \Delta_r G^{\ominus}(M, M_xO)$

will have a negative standard reaction Gibbs energy, and therefore have $K > 1$. The procedure followed here is similar to that adopted with half-reactions in aqueous solution (Section 5.1), but now all the reactions are written as oxidations with $\tfrac{1}{2}O_2$ in place of e^-, and the overall reaction is the difference of reactions with matching numbers of oxygen atoms. The relevant information is commonly summarized in an **Ellingham diagram** (Fig. 5.14), which is a graph of $\Delta_r G^{\ominus}$ against temperature.

We can understand the appearance of an Ellingham diagram by noting that $\Delta_r G^{\ominus} = \Delta_r H^{\ominus} - T\Delta_r S^{\ominus}$ and using the fact that the enthalpy and entropy of reaction are, to a reasonable approximation, independent of temperature. That being so, the slope of a line in an Ellingham diagram should therefore be equal to $-\Delta_r S^{\ominus}$ for the relevant reaction. Because the standard molar entropies of gases are much larger than those of solids, the reaction entropy of (a), in which there is a net formation of gas (because 1 mol CO replaces $\tfrac{1}{2}$ mol O_2), is positive, and its line therefore has a negative slope. The standard reaction entropy of (b) is close to zero as there is no net change in the amount of gas, so its line is horizontal. Reaction (c) has a negative reaction entropy because $\tfrac{3}{2}$ mol of gas molecules is replaced by 1 mol CO_2; hence the line in the diagram has a positive slope. The standard reaction entropy of (d), in which there is a net consumption of gas, is negative, and hence the plot has a positive slope (Fig. 5.15). The kinks in the lines, where the slope of the metal oxidation line changes, are where the metal undergoes a phase change, particularly melting, and the reaction entropy changes accordingly. At temperatures for which the C/CO line (a) lies above the metal oxide line (d), $\Delta_r G^{\ominus}(M, M_xO)$ is more negative than $\Delta_r G^{\ominus}(C, CO)$. At these temperatures, $\Delta_r G^{\ominus}(C, CO) - \Delta_r G^{\ominus}(M, M_xO)$ is positive, so the reaction (a − d) has $K < 1$. However, for temperatures for which the C/CO line lies below the metal oxide line, the reduction of the metal oxide by carbon has $K > 1$. Similar remarks apply to the temperatures at which the other two carbon oxidation lines (b) and (c) lie above or below the metal oxide lines. In summary:

- For temperatures at which the C/CO line lies below the metal oxide line, carbon can be used to reduce the metal oxide and itself is oxidized to carbon monoxide.

- For temperatures at which the C/CO$_2$ line lies below the metal oxide line, carbon can be used to achieve the reduction, but is oxidized to carbon dioxide.

- For temperatures at which the CO/CO$_2$ line lies below the metal oxide line, carbon monoxide can reduce the metal oxide to the metal and is oxidized to carbon dioxide.

Figure 5.16 shows an Ellingham diagram for a selection of common metals. In principle, production of all the metals shown in the diagram, even magnesium and calcium, could be accomplished by **pyrometallurgy**, heating with a reducing agent. However, there are severe practical limitations. Efforts to produce aluminium by pyrometallurgy (most notably in Japan, where electricity is expensive) were frustrated by the volatility of Al_2O_3 at the very high temperatures required. A difficulty of a different kind is encountered in the pyrometallurgical extraction of titanium, where titanium carbide, TiC, is formed instead of the metal. In practice, pyrometallurgical extraction of metals is confined principally to magnesium, iron, cobalt, nickel, zinc, and a variety of ferroalloys (alloys with iron).

EXAMPLE 5.14 **Using an Ellingham diagram**

What is the lowest temperature at which ZnO can be reduced to zinc metal by carbon? What is the overall reaction at this temperature?

Answer To answer this question we examine the Ellingham diagram in Fig. 5.16 and estimate the temperature at which the ZnO line crosses the C, CO line. The C, CO line lies below the ZnO line at approximately

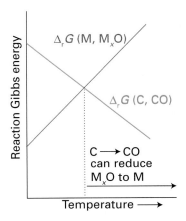

Fig. 5.14 The variation of the standard reaction Gibbs energies for the formation of a metal oxide and carbon monoxide with temperature. The formation of carbon monoxide from carbon can reduce the metal oxide to the metal at temperatures higher than the point of intersection of the two lines. More specifically, at the intersection the equilibrium constant changes from $K < 1$ to $K > 1$. This type of display is an example of an Ellingham diagram.

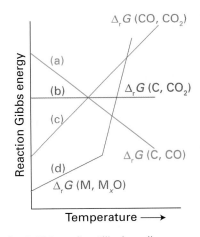

Fig. 5.15 Part of an Ellingham diagram showing the standard Gibbs energy for the formation of a metal oxide and the three-carbon oxidation Gibbs energies. The slopes of the lines are determined largely by whether or not there is net gas formation or consumption in the reaction. A phase change generally results in a kink in the graph (because the entropy of the substance changes).

1200°C; above this temperature reduction of the metal oxide is spontaneous. The contributing reactions are reaction (a) and the reverse of

$$Zn(g) + \tfrac{1}{2}O_2(g) \rightarrow ZnO(s)$$

so the overall reaction is the difference, or

$$C(s) + ZnO(s) \rightarrow CO(g) + Zn(g)$$

The physical state of zinc is given as a gas because the element boils at 907°C (the corresponding inflection in the ZnO line in the Ellingham diagram can be seen in Fig. 5.16).

Self-test 5.14 What is the minimum temperature for reduction of MgO by carbon?

Similar principles apply to reductions using other reducing agents. For instance, an Ellingham diagram can be used to explore whether a metal M′ can be used to reduce the oxide of another metal M. In this case, we note from the diagram whether at a temperature of interest the M′/M′O line lies below the M/MO line, as M′ is now taking the place of C. When

$$\Delta_r G^{\ominus} = \Delta_r G^{\ominus}(M', M'O) - \Delta_r G^{\ominus}(M, MO)$$

is negative, where the Gibbs energies refer to the reactions

(a) $M'(s \text{ or } l) + \tfrac{1}{2}O_2(g) \rightarrow M'O(s)$ $\Delta_r G^{\ominus}(M', M'O)$

(b) $M(s \text{ or } l) + \tfrac{1}{2}O_2(g) \rightarrow MO(s)$ $\Delta_r G^{\ominus}(M, MO)$

the reaction

(a − b) $MO(s) + M'(s \text{ or } l) \rightarrow M(s \text{ or } l) + M'O(s)$

and its analogues for MO_2, and so on) is feasible (in the sense $K > 1$). For example, because in Fig. 5.16 the line for MgO lies below the line for SiO_2 at temperatures below 2400°C, magnesium may be used to reduce SiO_2 below that temperature. This reaction has in fact been used to produce low-grade silicon as discussed in the following section.

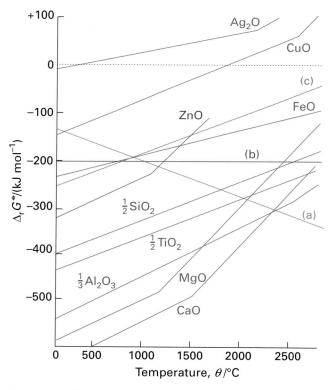

Fig. 5.16 An Ellingham diagram for the reduction of metal oxides.

(b) Survey of processes

Key points: A blast furnace produces the conditions required to reduce iron oxides with carbon; electrolysis may be used to bring about a nonspontaneous reduction as required for the extraction of aluminium from its oxide.

Industrial processes for achieving the reductive extraction of metals show a greater variety than the thermodynamic analysis might suggest. An important factor is that the ore and carbon are both solids, and a reaction between two solids is rarely fast. Most processes exploit gas/solid or liquid/solid heterogeneous reactions. Current industrial processes are varied in the strategies they adopt to ensure economical rates, exploit materials, and avoid environmental problems. We can explore these strategies by considering three important examples that reflect low, moderate, and extreme difficulty of reduction.

The least difficult reductions include those of copper ores. Roasting and smelting are still widely used in the pyrometallurgical extraction of copper. However, some recent techniques seek to avoid the major environmental problems caused by the production of the large quantity of SO_2, released to the atmosphere, that accompanies roasting. One promising development is the **hydrometallurgical extraction** of copper, the extraction of a metal by reduction of aqueous solutions of its ions, using H_2 or scrap iron as the reducing agent. In this process, Cu^{2+} ions, leached from low-grade ores by acid or bacterial action, are reduced by hydrogen in the reaction or by a similar reduction using iron. This process is less harmful to the environment provided the acid by-product is used or neutralized locally rather than contributing to acidic atmospheric pollutants. It also allows economic exploitation of lower grade ores.

$$Cu^{2+}(aq) + H_2(g) \rightarrow Cu(s) + 2H^+(aq)$$

That extraction of iron is of intermediate difficulty is shown by the fact that the Iron Age followed the Bronze Age. In economic terms, iron ore reduction is the most important application of carbon pyrometallurgy. In a blast furnace (Fig. 5.17), which is still the major source of the element, the mixture of iron ores (Fe_2O_3, Fe_3O_4), coke (C), and limestone ($CaCO_3$) is heated with a blast of hot air. Combustion of coke in this blast raises the temperature to 2000°C, and the carbon burns to carbon monoxide in the lower part of the furnace. The supply of Fe_2O_3 from the top of the furnace meets the hot CO rising from below. The iron(III) oxide is reduced, first to Fe_3O_4 and then to FeO at 500−700°C, and the CO is oxidized to CO_2. The final reduction to iron, from FeO, by carbon monoxide occurs between 1000 and 1200°C in the central region of the furnace. Thus overall

$$Fe_2O_3(s) + 3CO(g) \rightarrow 2Fe(l) + 3CO_2(g)$$

The function of the lime, CaO, formed by the thermal decomposition of calcium carbonate is to combine with the silicates present in the ore to form a molten layer of calcium silicates (slag) in the hottest (lowest) part of the furnace. Slag is less dense than iron and can be drained away. The iron formed melts at about 400°C below the melting point of the pure metal on account of the dissolved carbon it contains. The impure iron, the densest phase, settles to the bottom and is drawn off to solidify into 'pig iron', in which the carbon content is high (about 4 per cent by mass). The manufacture of steel is then a series of reactions in which the carbon content is reduced and other metals are used to form alloys with the iron (see Box 3.1).

More difficult than the extraction of either copper or iron is the extraction of silicon from its oxide: indeed, silicon is very much an element of the twentieth century. Silicon of 96 to 99 per cent purity is prepared by reduction of quartzite or sand (SiO_2) with high purity coke. The Ellingham diagram (Fig. 5.16) shows that the reduction is feasible only at temperatures in excess of about 1700°C. This high temperature is achieved in an electric arc furnace in the presence of excess silica (to prevent the accumulation of SiC):

$$SiO_2(l) + 2C(s) \xrightarrow{1700°C} Si(l) + 2CO(g)$$

$$2SiC(s) + SiO_2(l) \rightarrow 3Si(l) + 2CO(g)$$

Very pure silicon (for semiconductors) is made by converting crude silicon to volatile compounds, such as $SiCl_4$. These compounds are purified by exhaustive fractional distillation and then reduced to silicon with pure hydrogen. The resulting semiconductor-grade silicon is melted and large single crystals are pulled slowly from the cooled surface of a melt: this procedure is called the *Czochralski process*.

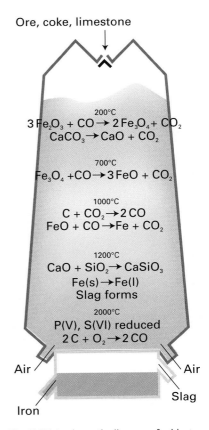

Ore, coke, limestone

200°C
$3Fe_2O_3 + CO \rightarrow 2Fe_3O_4 + CO_2$
$CaCO_3 \rightarrow CaO + CO_2$

700°C
$Fe_3O_4 + CO \rightarrow 3FeO + CO_2$

1000°C
$C + CO_2 \rightarrow 2CO$
$FeO + CO \rightarrow Fe + CO_2$

1200°C
$CaO + SiO_2 \rightarrow CaSiO_3$
$Fe(s) \rightarrow Fe(l)$
Slag forms

2000°C
P(V), S(VI) reduced
$2C + O_2 \rightarrow 2CO$

Air   Air

Iron

Slag

Fig. 5.17 A schematic diagram of a blast furnace showing the typical composition and temperature profile.

As we have remarked, the Ellingham diagram shows that the direct reduction of Al_2O_3 with carbon becomes feasible only above 2400°C, which makes it uneconomically expensive and wasteful in terms of any fossil fuels used to heat the system. However, the reduction can be brought about **electrolytically** (Section 5.18).

5.17 Chemical oxidation

Key point: Elements obtained by chemical oxidation include the heavier halogens, sulfur, and (in the course of their purification) certain noble metals.

As oxygen is available from fractional distillation of air, chemical methods for its production are not necessary. Sulfur is an interesting mixed case. Elemental sulfur is either mined or produced by oxidation of the H_2S that is removed from 'sour' natural gas and crude oil. The oxidation is accomplished by the **Claus process**, which consists of two stages. In the first, some hydrogen sulfide is oxidized to sulfur dioxide:

$$2\,H_2S(g) + 3\,O_2(g) \rightarrow 2\,SO_2(g) + 2\,H_2O(l)$$

In the second stage, this sulfur dioxide is allowed to react in the presence of a catalyst with more hydrogen sulfide:

$$2\,H_2S(g) + SO_2(g) \xrightarrow{\text{Oxide catalyst, 300°C}} 3\,S(s) + 2\,H_2O(l)$$

The catalyst is typically Fe_2O_3 or Al_2O_3. The Claus process is environmentally benign; otherwise it would be necessary to burn the toxic hydrogen sulfide to polluting sulfur dioxide.

The only important metals extracted in a process using an oxidation stage are the ones that occur in native form (that is, as the element). Gold is an example because it is difficult to separate the granules of metal in low-grade ores by simple 'panning'. The dissolution of gold depends on oxidation, which is favoured by complexation with CN^- ions:

$$Au(s) + 2\,CN^-(aq) \rightarrow [Au(CN)_2]^-(aq)\ +\ e^-$$

This complex is then reduced to the metal by reaction with another reactive metal, such as zinc:

$$2[Au(CN)_2]^-(aq) + Zn(s) \rightarrow 2\,Au(s) + [Zn(CN)_4]^{2-}(aq)$$

However, because of the toxicity of cyanide, alternative methods of extracting gold have been used, involving processing by bacteria.

The lighter, strongly oxidizing halogens are extracted electrochemically, as described in Section 5.18. The more readily oxidizable halogens, Br_2 and I_2, are obtained by chemical oxidation of the aqueous halides with chlorine. For example,

$$2\,NaBr(aq) + Cl_2(g) \rightarrow 2\,NaCl(aq) + Br_2(l)$$

5.18 Electrochemical extraction

Key points: Elements obtained by electrochemical reduction include aluminium; those obtained by electrochemical oxidation include chlorine.

The extraction of metals from ores electrochemically is confined mainly to the more electropositive elements, as discussed in the case of aluminium later in this section. For other metals produced in bulk quantities, such as iron and copper, the more energy-efficient and cleaner routes used by industry in practice, using chemical methods of reduction, were described in Section 5.16b. In some specialist cases, electrochemical reduction is used to isolate small quantities of platinum group metals. So, for example, treatment of spent catalytic converters with acids under oxidizing conditions produces a solution containing complexes of Pt(II) and other platinum group metals, which can then be reduced electrochemically. The metals are deposited at the cathode with an overall 80 per cent efficient extraction from the ceramic catalytic converter.

As we saw in Section 5.16, an Ellingham diagram shows that the reduction of Al_2O_3 with carbon becomes feasible only above 2400°C, which is uneconomical. However, the reduction can be brought about **electrolytically**, and all modern production uses the electrochemical *Hall−Héroult process*, which was invented in 1886 independently by Charles Hall and Paul Héroult. The process requires pure aluminium hydroxide that is extracted from aluminium ores using the *Bayer process*. In this process the bauxite ore used as a source of aluminium is a mixture of the acidic oxide SiO_2 and amphoteric oxides and hydroxides, such as Al_2O_3, $AlOOH$, and Fe_2O_3. The Al_2O_3 is dissolved in hot aqueous sodium hydroxide, which separates the aluminium from much of the less soluble Fe_2O_3, although silicates are also rendered soluble in these strongly basic conditions. Cooling the sodium aluminate solution results in the precipitation of $Al(OH)_3$, leaving the silicates in solution. For the final stage, in the Hall–Héroult process, the aluminium hydroxide is dissolved in molten cryolite (Na_3AlF_6) and the melt is reduced **electrolytically** at a steel cathode with graphite **anodes**. The latter participate in the electrochemical reaction by reacting with the evolved oxygen atoms so that the overall process is

$$2\,Al_2O_3(s) + 3\,C(s) \rightarrow 4\,Al(s) + 3\,CO_2(g)$$

As the power consumption of a typical plant is huge, aluminium is often produced where electricity is cheap (for example, from hydroelectric sources in Canada) and not where bauxite is mined (in Jamaica, for example).

The lighter halogens are the most important elements extracted by electrochemical oxidation. The standard reaction Gibbs energy for the oxidation of Cl^- ions in water

$$2\,Cl^-(aq) + 2\,H_2O(l) \rightarrow 2\,OH^-(aq) + H_2(g) + Cl_2(g) \quad \Delta_r G^\ominus = +422 \ \text{kJ mol}^{-1}$$

is strongly positive, which suggests that electrolysis is required. The minimum potential difference that can achieve the oxidation of Cl^- is about 2.2 V (from $\Delta_r G^\ominus = -\nu F E^\ominus$ and $\nu = 2$)

It may appear that there is a problem with the competing reaction

$$2\,H_2O(l) \rightarrow 2\,H_2(g) + O_2(g) \quad \Delta_r G^\ominus = +414 \ \text{kJ mol}^{-1}$$

which can be driven forwards by a potential difference of only 1.2 V (in this reaction, $\nu = 4$). However, the rate of oxidation of water is very slow at potentials at which it first becomes favourable thermodynamically. This slowness is expressed by saying that the reduction requires a high **overpotential**, η (eta), the potential that must be applied in addition to the equilibrium value before a significant rate of reaction is achieved. Consequently, the electrolysis of brine produces Cl_2, H_2, and aqueous $NaOH$, but not much O_2.

Oxygen, not fluorine, is produced if aqueous solutions of fluorides are electrolysed. Therefore, F_2 is prepared by the electrolysis of an anhydrous mixture of potassium fluoride and hydrogen fluoride, an ionic conductor that is molten above 72°C.

FURTHER READING

P. Zanello, *Inorganic electrochemistry: theory, practice and applications*. Royal Society of Chemistry (2003). An introduction to electrochemical investigations.

A.J. Bard, M. Stratmann, F. Scholtz, and C.J. Pickett, *Encyclopedia of Electrochemistry: Inorganic Chemistry*, Vol. 7b. Wiley (2006).

J.-M. Savéant, *Elements of molecular and biomolecular electrochemistry: an electrochemical approach to electron-transfer chemistry*. Wiley (2006).

R.M. Dell and D.A.J. Rand, *Understanding batteries*. Royal Society of Chemistry (2001).

A.J. Bard, R. Parsons, and R. Jordan, *Standard potentials in aqueous solution*. M. Dekker, New York (1985). A collection of cell potential data with discussion.

I. Barin, *Thermochemical data of pure substances*, Vols 1 and 2. VCH, Weinheim (1989). A comprehensive source of thermodynamic data for inorganic substances.

J. Emsley, *The elements*. Oxford University Press (1998). Excellent source of data on the elements, including standard potentials.

A.G. Howard, *Aquatic environmental chemistry*. Oxford University Press (1998). Discussion of the compositions of freshwater and marine systems explaining the effects of oxidation and reduction processes.

M. Pourbaix, *Atlas of electrochemical equilibria in aqueous solution*. Pergamon Press, Oxford (1966). The original and still good source of Pourbaix diagrams.

W. Stumm and J.J. Morgan, *Aquatic chemistry*. Wiley, New York (1996). A standard reference on natural water chemistry.

C. Spiegel, *Design and building of fuel cells*. McGraw Hill (2007).

J. Larminie and A. Dicks, *Fuel cell systems explained*. Wiley (2003).

EXERCISES

5.1 Assign oxidation numbers for each of the elements participating in the following reactions.

$$2NO(g) + O_2(g) \rightarrow 2NO_2(g)$$

$$2Mn^{3+}(aq) + 2H_2O \rightarrow MnO_2 + Mn^{2+} + 4H^+(aq)$$

$$LiCoO_2(s) + C(s) \rightarrow Li^+@C(s) + CoO_2(s)$$

$$Ca(s) + H_2(g) \rightarrow CaH_2(s)$$

5.2 Use data from *Resource section* 3 to suggest chemical reagents that would be suitable for carrying out the following transformations and write balanced equations for the reactions: (a) oxidation of HCl to chlorine gas, (b) reduction of Cr(III)(aq) to Cr(II)(aq), (c) reduction of Ag$^+$ to Ag(s), (d) reduction of I$_2$ to I$^-$.

5.3 Use standard potential data from *Resource section* 3 as a guide to write balanced equations for the reactions that each of the following species might undergo in aerated aqueous acid. If the species is stable, write 'no reaction'. (a) Cr^{2+}, (b) Fe^{2+}, (c) Cl$^-$, (d) HClO, (e) Zn(s).

5.4 Use the information in *Resource section* 3 to write balanced equations for the reactions, including disproportionations, that can be expected for each of the following species in aerated acidic aqueous solution: (a) Fe^{2+}, (b) Ru^{2+}, (c) HClO$_2$, (d) Br$_2$.

5.5 Explain why the standard potentials for the half-cell reactions

$$[Ru(NH_3)_6]^{3+}(aq) + e^- \rightarrow [Ru(NH_3)_6]^{2+}(aq)$$

$$[Fe(CN)_6]^{3-}(aq) + e^- \rightarrow [Fe(CN)_6]^{4-}(aq)$$

vary with temperature in opposite directions.

5.6 Balance the following redox reaction in acid solution: MnO$_4^-$ + H$_2$SO$_3$ → Mn^{2+} + HSO$_4^-$. Predict the qualitative pH dependence on the net potential for this reaction (that is, increases, decreases, remains the same).

5.7 Write the Nernst equation for

(a) the reduction of O$_2$(g): O$_2$(g) + 4H$^+$(aq) + 4e$^-$ → 2H$_2$O(l)

(b) the reduction of Fe$_2$O$_3$(s): Fe$_2$O$_3$(s) + 6H$^+$(aq) + 6e$^-$ → 2Fe(s) + 3H$_2$O(l)

In each case express the formula in terms of pH. What is the potential for the reduction of O$_2$ at pH = 7 and p(O$_2$) = 0.20 bar (the partial pressure of oxygen in air)?

5.8 Answer the following questions using the Frost diagram in Fig. 5.18. (a) What are the consequences of dissolving Cl$_2$ in aqueous basic solution? (b) What are the consequences of dissolving Cl$_2$ in aqueous acid? (c) Is the failure of HClO$_3$ to disproportionate in aqueous solution a thermodynamic or a kinetic phenomenon?

5.9 Use standard potentials as a guide to write equations for the main net reaction that you would predict in the following experiments: (a) N$_2$O is bubbled into aqueous NaOH solution, (b) zinc metal is added to aqueous sodium triiodide, (c) I$_2$ is added to excess aqueous HClO$_3$.

5.10 Adding NaOH to an aqueous solution containing Ni^{2+} results in precipitation of Ni(OH)$_2$. The standard potential for the Ni^{2+}/Ni

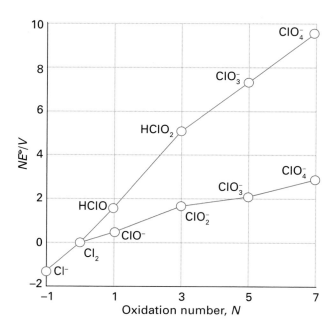

Fig. 5.18 A Frost diagram for chlorine. The red line refers to acid conditions (pH = 0) and the blue line to pH = 14. Note that because HClO$_3$ and HClO$_4$ are strong acids, even at pH= 0 they are present as their conjugate bases.

couple is +0.26 V and the solubility product K_{sp} = [Ni^{2+}][OH$^-$]^2 = 1.5 × 10^{-16}. Calculate the electrode potential at pH = 14.

5.11 Characterize the condition of acidity or basicity that would most favour the following transformations in aqueous solution: (a) Mn^{2+} → MnO$_4^-$, (b) ClO$_4^-$ → ClO$_3^-$, (c) H$_2$O$_2$ → O$_2$, (d) I$_2$ → 2I$^-$.

5.12 Use the Latimer diagram for chlorine to determine the potential for reduction of ClO$_4^-$ to Cl$_2$. Write a balanced equation for this half-reaction.

5.13 Calculate the equilibrium constant of the reaction Au$^+$(aq) + 2CN$^-$(aq) → [Au(CN)$_2$]$^-$(aq) from the standard potentials

$$Au^+(aq) + e^- \rightarrow Au(s) \qquad\qquad E^\ominus = +1.69V$$

$$[Au(CN)_2]^-(aq) + e^- \rightarrow Au(s) + 2CN^-(aq) \qquad E^\ominus = -0.6V$$

5.14 Use Fig. 5.12 to find the approximate potential of an aerated lake at pH = 6. With this information and Latimer diagrams from *Resource section* 3, predict the species at equilibrium for the elements (a) iron, (b) manganese, (c) sulfur.

5.15 Using the following Latimer diagram, which shows the standard potentials for sulfur species in acid solution (pH = 0), construct a Frost diagram and calculate the standard potential for the HSO$_4^-$/S$_8$(s) couple.

$$S_2O_8^{2-} \xrightarrow{+1.96} HSO_4^- \xrightarrow{+0.16} H_2SO_3 \xrightarrow{+0.40}$$

$$S_2O_3^{2-} \xrightarrow{+0.60} S \xrightarrow{+0.14} H_2S$$

5.16 Using the following aqueous acid solution reduction potentials $E^\ominus$(Pd^{2+}, Pd) = +0.915 V and $E^\ominus$([PdCl$_4$]$^{2-}$, Pd) = +0.50 V, calculate the equilibrium constant for the reaction Pd^{2+}(aq) + 4Cl$^-$(aq) ⇌ [PdCl$_4$]$^{2-}$(aq) in 1M HCl(aq).

5.17 Calculate the reduction potential at 25°C for the conversion of MnO_4^- to $MnO_2(s)$ in aqueous solution at pH = 9.00 and 1 M $MnO_4^-(aq)$ given that $E^{\ominus}(MnO_4^-, MnO_2) = +1.69$ V.

5.18 Draw a Frost diagram for mercury in acid solution, given the following Latimer diagram:

$$Hg^{2+} \xrightarrow{+0.911} Hg_2^{2+} \xrightarrow{+0.796} Hg$$

Comment on the tendency of any of the species to act as an oxidizing agent, a reducing agent, or to undergo disproportionation.

5.19 From the following Latimer diagram, calculate the value of $E^{\ominus}$ for the reaction $2\,HO_2(aq) \rightarrow O_2(g) + H_2O_2(aq)$.

$$O_2 \xrightarrow{-0.125} HO_2 \xrightarrow{+1.510} H_2O_2$$

Comment on the thermodynamic tendency of HO_2 to undergo disproportionation.

5.20 Explain why water with high concentrations of dissolved carbon dioxide and open to atmospheric oxygen is very corrosive towards iron.

5.21 The species Fe^{2+} and H_2S are important at the bottom of a lake where O_2 is scarce. If the pH = 6, what is the maximum value of E characterizing the environment?

5.22 The ligand EDTA forms stable complexes with hard acid centres. How will complexation with EDTA affect the reduction of M^{2+} to the metal in the 3d-series?

5.23 In Fig. 5.11, which of the boundaries depend on the choice of Fe^{2+} concentration as 10^{-5} mol dm^{-3}?

5.24 Consult the Ellingham diagram in Fig. 5.16 and determine if there are any conditions under which aluminium might be expected to reduce MgO. Comment on these conditions.

PROBLEMS

5.1 Use standard potential data to suggest why permanganate (MnO_4^-) is not a suitable oxidizing agent for the quantitative estimation of Fe^{2+} in the presence of HCl but becomes so if sufficient Mn^{2+} and phosphate ion are added to the solution. (*Hint:* Phosphate forms complexes with Fe^{3+}, thereby stabilizing it.)

5.2 Many of the tabulated data for standard potentials have been determined from thermochemical data rather than direct electrochemical measurements of cell potentials. Carry out a calculation to illustrate this approach for the half-reaction $Sc_2O_3(s) + 3\,H_2O(l) + 6\,e^- \rightarrow 2\,Sc(s) + 6\,OH^-(aq)$.

	$Sc^{3+}(aq)$	$OH^-(aq)$	$H_2O(l)$	$Sc_2O_3(s)$	$Sc(s)$
$\Delta_f H^{\ominus}/(kJ\ mol^{-1})$	−614.2	−230.0	−285.8	−1908.7	0
$S_m^{\ominus}/(J\ K^{-1}\ mol^{-1})$	−255.2	−10.75	69.91	77.0	34.76

$S_m^{\ominus}$ is the standard molar entropy

5.3 The reduction potential of an ion such as OH^- can be strongly influenced by the solvent. (a) From the review article by D.T. Sawyer and J.L. Roberts, *Acc. Chem. Res.*, 1988, **21**, 469, describe the magnitude of the change in the potential of the OH/OH^- couple on changing the solvent from water to acetonitrile, CH_3CN. (b) Suggest a qualitative interpretation of the difference in solvation of the OH^- ion in these two solvents.

5.4 Given the following standard potentials in basic solution

$$CrO_4^{2-}(aq) + 4\,H_2O(l) + 3\,e^- \rightarrow Cr(OH)_3(s) + 5\,OH^-(aq)$$
$$E^{\ominus} = -0.11\,V$$

$$[Cu(NH_3)_2]^+(aq) + e^- \rightarrow Cu(s) + 2\,NH_3(aq)$$
$$E^{\ominus} = -0.10\,V$$

and assuming that a reversible reaction can be established on a suitable catalyst, calculate $\Delta_r G^{\ominus}$, and K for the reductions of (a) CrO_4^{2-} and (b) $[Cu(NH_3)_2]^+$ in basic solution. Comment on why $\Delta_r G^{\ominus}$ and K are so different between the two cases despite the values of $E^{\ominus}$ being so similar.

5.5 Explain the significance of reduction potentials in inorganic chemistry, highlighting their applications in investigations of stability, solubility, and reactivity in water.

5.6 Using *Resource sections* 1–3, and data for atomization of the elements $\Delta_f H^{\ominus} = 397$ kJ mol^{-1} (Cr) and 664 kJ mol^{-1} (Mo), construct thermodynamic cycles for the reactions of Cr or Mo with dilute acids and thus consider the importance of metallic bonding in determining the standard reduction potentials for formation of cations from metals.

5.7 Discuss how the equilibrium $Cu^{2+}(aq) + Cu(s) \rightleftharpoons 2\,Cu^+(aq)$ can be shifted by complexation with chloride ions (see J. Malyyszko and M. Kaczor, *J. Chem. Educ.*, 2003, **80**, 1048).

5.8 In their article 'Variability of the cell potential of a given chemical reaction' (*J. Chem. Educ.*, 2004, **81**, 84), L.H. Berka and I. Fishtik conclude that $E^{\ominus}$ for a chemical reaction is not a state function because the half-reactions are arbitrarily chosen and may contain different numbers of transferred electrons. Discuss this objection.

5.9 The standard potentials for phosphorus species in aqueous solution at pH = 0 and pH = 14 are represented by the following Latimer diagrams.

pH = 0

$$H_3PO_4 \xrightarrow{-0.276} H_3PO_3 \xrightarrow{-0.499} H_3PO_2$$
$$\xrightarrow{-0.508} P \xrightarrow{-0.063} PH_3$$

pH = 14

$$PO_4^{3-} \xrightarrow{-1.12} HPO_3^{2-} \xrightarrow{-1.57} H_2PO_2^-$$
$$\xrightarrow{-2.05} P \xrightarrow{-0.89} PH_3$$

(a) Account for the difference in reduction potentials between pH 0 and pH 14.

(b) Construct a single Frost diagram showing both sets of data.

(c) Phosphine (PH_3) can be prepared by heating phosphorus with aqueous alkali. Discuss the reactions that are feasible and estimate their equilibrium constants.

5.10 Construct an Ellingham diagram for the thermal splitting of water ($H_2O(g) \rightarrow H_2(g) + \frac{1}{2}O_2(g)$) by using $\Delta_r H^{\ominus} = +260$ kJ mol^{-1}; $\Delta_r S^{\ominus} = +60$ J K^{-1} mol^{-1} (you may assume that $\Delta H^{\ominus}$ is independent of temperature). Hence, calculate the temperature at which H_2 may be obtained by spontaneous decomposition of water (*Chem. Rev.* 2007, **107**, 4048). Comment on the feasibility of producing H_2 by this means.

5.11 Enterobactin (Ent) is a special ligand secreted by some bacteria to sequester Fe from the environment (Fe is an essential nutrient for almost all living species, see Chapter 27). The equilibrium constant for formation of $[Fe(III)(Ent)]$ ($K = 10^{52}$) is at least 40 orders of magnitude higher than the equilibrium constant for the corresponding Fe(II) complex. Determine the feasibility of releasing Fe from $[Fe(III)(Ent)]$ by its reduction to Fe(II) under conditions of neutral pH, noting that the strongest common reducing agent normally available to bacteria is H_2.

5.12 Standard potentials at 25°C for indium and thallium in aqueous solution (pH = 0) are given below.

$In^{3+}(aq) + 3e^- \rightarrow In(s)$	$E^{\ominus} = -0.338$ V
$In^{+}(aq) + e^- \rightarrow In(s)$	$E^{\ominus} = -0.126$ V
$Tl^{3+}(aq) + 3e^- \rightarrow Tl(s)$	$E^{\ominus} = +0.72$ V
$Tl^{+}(aq) + e^- \rightarrow Tl(s)$	$E^{\ominus} = -0.336$ V

Use the data to construct a Frost diagram for the two elements and discuss the relative stabilities of the species.

Molecular symmetry

6

Symmetry governs the bonding and hence the physical and spectroscopic properties of molecules. In this chapter we explore some of the consequences of molecular symmetry and introduce the systematic arguments of group theory. We shall see that symmetry considerations are essential for constructing molecular orbitals and analysing molecular vibrations. They also enable us to extract information about molecular and electronic structure from spectroscopic data.

The systematic treatment of symmetry makes use of a branch of mathematics called **group theory**. Group theory is a rich and powerful subject, but we shall confine our use of it at this stage to the classification of molecules in terms of their symmetry properties, the construction of molecular orbitals, and the analysis of molecular vibrations and the selection rules that govern their excitation. We shall also see that it is possible to draw some general conclusions about the properties of molecules without doing any calculations at all.

An introduction to symmetry analysis

That some molecules are 'more symmetrical' than others is intuitively obvious. Our aim though, is to define the symmetries of individual molecules precisely, not just intuitively, and to provide a scheme for specifying and reporting these symmetries. It will become clear in later chapters that symmetry analysis is one of the most pervasive techniques in inorganic chemistry.

6.1 Symmetry operations, elements and point groups

Key points: Symmetry operations are actions that leave the molecule apparently unchanged; each symmetry operation is associated with a symmetry element. The point group of a molecule is identified by noting its symmetry elements and comparing these elements with the elements that define each group.

A fundamental concept of the chemical application of group theory is the **symmetry operation,** an action, such as rotation through a certain angle, that leaves the molecule apparently unchanged. An example is the rotation of an H_2O molecule by $180°$ around the bisector of the HOH angle (Fig. 6.1). Associated with each symmetry operation there is a **symmetry element**, a point, line, or plane with respect to which the symmetry operation is performed. Table 6.1 lists the most important symmetry operations and their corresponding elements. All these operations leave at least one point unchanged (the centre of the molecule), and hence they are referred to as the operations of **point-group symmetry**.

The **identity operation**, E, consists of doing nothing to the molecule. Every molecule has at least this operation and some have only this operation, so we need it if we are to classify all molecules according to their symmetry.

The rotation of an H_2O molecule by $180°$ around a line bisecting the HOH angle (as in Fig. 6.1) is a symmetry operation, denoted C_2. In general, an **n-fold rotation** is a symmetry operation if the molecule appears unchanged after rotation by $360°/n$. The corresponding symmetry element is a line, an **n-fold rotation axis**, C_n, about which the rotation is performed. There is only one rotation operation associated with a C_2 axis (as in H_2O) because clockwise and anticlockwise rotations by $180°$ are identical. The trigonal-pyramidal NH_3

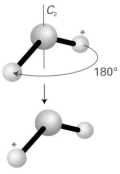

Figure 6.1 An H_2O molecule may be rotated through any angle about the bisector of the HOH bond angle, but only a rotation of $180°$ (the C_2 operation) leaves it apparently unchanged.

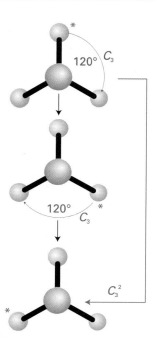

Figure 6.2 A threefold rotation and the corresponding C_3 axis in NH_3. There are two rotations associated with this axis, one through 120° (C_3) and one through 240° (C_3^2).

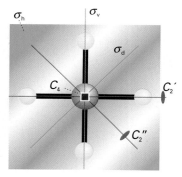

Figure 6.3 Some of the symmetry elements of a square-planar molecule such as XeF_4.

Table 6.1 Symmetry operations and symmetry elements

Symmetry operation	Symmetry element	Symbol
Identity	'whole of space'	E
Rotation by 360°/n	n-fold symmetry axis	C_n
Reflection	mirror plane	σ
Inversion	centre of inversion	i
Rotation by 360°/n followed by reflection in a plane perpendicular to the rotation axis	n-fold axis of improper rotation*	S_n

*Note the equivalences $S_1 = \sigma$ and $S_2 = i$.

molecule has a threefold rotation axis, denoted C_3, but there are now two operations associated with this axis, one a clockwise rotation by 120° and the other an anticlockwise rotation by 120° (Fig. 6.2). The two operations are denoted C_3 and C_3^2 (because two successive clockwise rotations by 120° are equivalent to an anticlockwise rotation by 120°), respectively.

The square-planar molecule XeF_4 has a fourfold C_4 axis, but in addition it also has two pairs of twofold rotation axes that are perpendicular to the C_4 axis: one pair (C_2') passes through each *trans*-FXeF unit and the other pair (C_2'') passes through the bisectors of the FXeF angles (Fig. 6.3). By convention, the highest order rotational axis, which is called the **principal axis**, defines the z-axis (and is typically drawn vertically).

The reflection of an H_2O molecule in either of the two planes shown in Fig. 6.4 is a symmetry operation; the corresponding symmetry element, the plane of the mirror, is a **mirror plane**, σ. The H_2O molecule has two mirror planes that intersect at the bisector of the HOH angle. Because the planes are 'vertical', in the sense of containing the rotational (z) axis of the molecule, they are labelled with a subscript v, as in σ_v and σ_v'. The XeF_4 molecule in Fig. 6.3 has a mirror plane σ_h in the plane of the molecule. The subscript h signifies that the plane is 'horizontal' in the sense that the vertical principal rotational axis of the molecule is perpendicular to it. This molecule also has two more sets of two mirror planes that intersect the fourfold axis. The symmetry elements (and the associated operations) are denoted σ_v for the planes that pass through the F atoms and σ_d for the planes that bisect the angle between the F atoms. The d denotes 'dihedral' and signifies that the plane bisects the angle between two C_2' axes (the FXeF axes).

To understand the **inversion operation**, i, we need to imagine that each atom is projected in a straight line through a single point located at the centre of the molecule and then out to an equal distance on the other side (Fig. 6.5). In an octahedral molecule such as SF_6, with the point at the centre of the molecule, diametrically opposite pairs of atoms at the corners of the octahedron are interchanged. The symmetry element, the point through

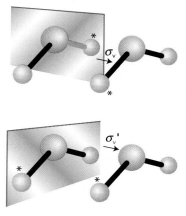

Figure 6.4 The two vertical mirror planes σ_v and σ_v' in H_2O and the corresponding operations. Both planes cut through the C_2 axis.

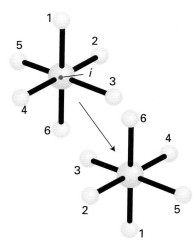

Figure 6.5 The inversion operation and the centre of inversion i in SF_6.

which the projections are made, is called the **centre of inversion**, *i*. For SF_6, the centre of inversion lies at the nucleus of the S atom. Likewise, the molecule CO_2 has an inversion centre at the C nucleus. However, there need not be an atom at the centre of inversion: an N_2 molecule has a centre of inversion midway between the two nitrogen nuclei. An H_2O molecule does not possess a centre of inversion. No tetrahedral molecule has a centre of inversion. Although an inversion and a twofold rotation may sometimes achieve the same effect, that is not the case in general and the two operations must be distinguished (Fig. 6.6).

An **improper rotation** consists of a rotation of the molecule through a certain angle around an axis followed by a reflection in the plane perpendicular to that axis (Fig. 6.7). The illustration shows a fourfold improper rotation of a CH_4 molecule. In this case, the operation consists of a 90° (that is, 360°/4) rotation about an axis bisecting two HCH bond angles, followed by a reflection through a plane perpendicular to the rotation axis. Neither the 90° (C_4) operation nor the reflection alone is a symmetry operation for CH_4 but their overall effect is a symmetry operation. A fourfold improper rotation is denoted S_4. The symmetry element, the **improper-rotation axis**, S_n (S_4 in the example), is the corresponding combination of an *n*-fold rotational axis and a perpendicular mirror plane.

An S_1 axis, a rotation through 360° followed by a reflection in the perpendicular plane, is equivalent to a reflection alone, so S_1 and σ_h are the same; the symbol σ_h is generally used rather than S_1. Similarly, an S_2 axis, a rotation through 180° followed by a reflection in the perpendicular plane, is equivalent to an inversion, *i* (Fig. 6.8); the symbol *i* is employed rather than S_2.

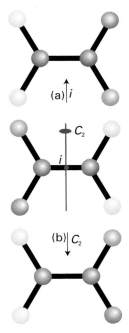

Figure 6.6 Care must be taken not to confuse (a) an inversion operation with (b) a twofold rotation. Although the two operations may sometimes appear to have the same effect, that is not the case in general.

EXAMPLE 6.1 Identifying symmetry elements

Identify the symmetry elements in the eclipsed and staggered conformations of an ethane molecule.

Answer We need to identify the rotations, reflections, and inversions that leave the molecule apparently unchanged. Don't forget that the identity is a symmetry operation. By inspection of the molecular models, we see that the eclipsed conformation of a CH_3CH_3 molecule (**1**) has the elements E, C_3, C_2, σ_h, σ_v, and S_3. The staggered conformation (**2**) has the elements E, C_3, σ_d, *i*, and S_6.

Self-test 6.1 Sketch the S_4 axis of an NH_4^+ ion. How many of these axes does the ion possess?

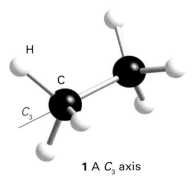

1 A C_3 axis

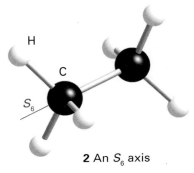

2 An S_6 axis

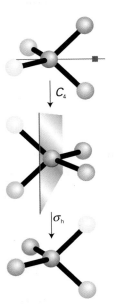

Figure 6.7 A fourfold axis of improper rotation S_4 in the CH_4 molecule.

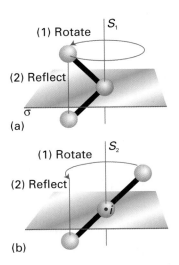

Figure 6.8 (a) An S_1 axis is equivalent to a mirror plane and (b) an S_2 axis is equivalent to a centre of inversion.

The assignment of a molecule to its **point group** consists of two steps:

1. Identify the symmetry elements of the molecule.
2. Refer to Table 6.2.

Table 6.2 The composition of some common groups

Point group	Symmetry elements	Shape	Examples
C_1	E	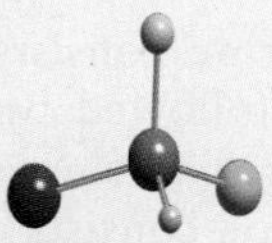	SiHClBrF
C_2	E, C_2		H_2O_2
C_s	E, σ	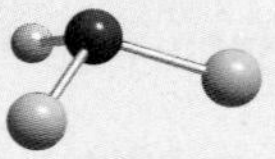	NHF_2
C_{2v}	$E, C_2, \sigma_v, \sigma_v'$		SO_2Cl_2, H_2O
C_{3v}	$E, 2C_3, 3\sigma_v$		$NH_3, PCl_3, POCl_3$
$C_{\infty v}$	$E, C_2, 2C_\varphi, \infty\sigma_v$		OCS, CO, HCl
D_{2h}	$E, 3C_2, i, 3\sigma$	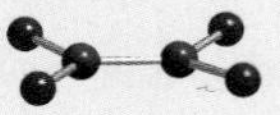	N_2O_4, B_2H_6
D_{3h}	$E, 2C_3, 3C_2, \sigma_h, 2S_3, 3\sigma_v$		BF_3, PCl_5
D_{4h}	$E, 2C_4, C_2, 2C_2', 2C_2'', i, 2S_4, \sigma_h, 2\sigma_v, 2\sigma_d$		$XeF_4,$ trans-$[MA_4B_2]$
$D_{\infty h}$	$E, \infty C_2', 2C_\varphi, i, \infty\sigma_v, 2S_\varphi$		CO_2, H_2, C_2H_2
T_d	$E, 8C_3, 3C_2, 6S_4, 6\sigma_d$	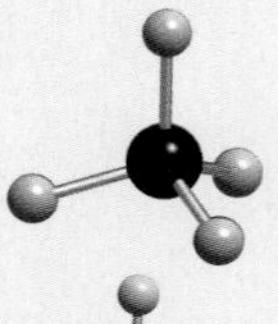	$CH_4, SiCl_4$
O_h	$E, 8C_3, 6C_2, 6C_4, 3C_2, i, 6S_4, 8S_6, 3\sigma_h, 6\sigma_d$		SF_6

In practice, the shapes in the table give a very good clue to the identity of the group to which the molecule belongs, at least in simple cases. The decision tree in Fig. 6.9 can also be used to assign most common point groups systematically by answering the questions at each decision point. The name of the point group is normally its **Schoenflies symbol**, such as C_{2v} for a water molecule.

EXAMPLE 6.2 **Identifying the point group of a molecule**

To what point groups do H_2O and XeF_4 belong?

Answer We need to work through Fig. 6.9. (a) The symmetry elements of H_2O are shown in Fig. 6.10. H_2O possesses the identity (E), a twofold rotation axis (C_2), and two vertical mirror planes (σ_v and σ_v'). The set of elements ($E, C_2, \sigma_v, \sigma_v'$) corresponds to the group C_{2v}. (b) The symmetry elements of XeF$_4$ are shown in Fig. 6.3. XeF$_4$ possesses the identity (E), a fourfold axis (C_4), two pairs of twofold rotation axes that are perpendicular to the principal C_4 axis, a horizontal reflection plane σ_h in the plane of the paper, and two sets of two vertical reflection planes, σ_v and σ_d. This set of elements identifies the point group as D_{4h}.

Self-test 6.2 Identify the point groups of (a) BF$_3$, a trigonal-planar molecule, and (b) the tetrahedral SO$_4^{2-}$ ion.

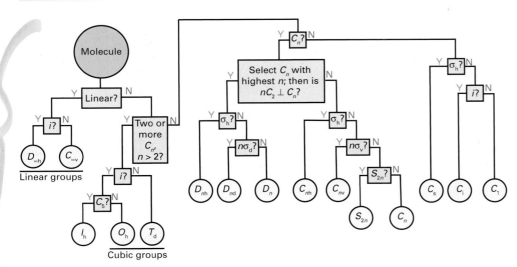

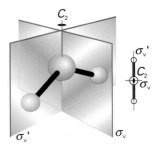

Figure 6.10 The symmetry elements of H$_2$O. The diagram on the right is the view from above and summarizes the diagram on the left.

Figure 6.9 The decision tree for identifying a molecular point group. The symbols of each point refer to the symmetry elements.

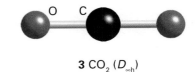

3 CO$_2$ ($D_{\infty h}$)

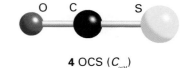

4 OCS ($C_{\infty v}$)

It is very useful to be able to recognize immediately the point groups of some common molecules. Linear molecules with a centre of symmetry, such as H$_2$, CO$_2$ (**3**), and HC≡CH belong to $D_{\infty h}$. A molecule that is linear but has no centre of symmetry, such as HCl or OCS (**4**) belongs to $C_{\infty v}$. Tetrahedral (T_d) and octahedral (O_h) molecules have more than one principal axis of symmetry (Fig. 6.11): a tetrahedral CH$_4$ molecule, for instance, has four C_3 axes, one along each CH bond. The O_h and T_d point groups are known as **cubic groups** because they are closely related to the symmetry of a cube. A closely related group, the **icosahedral group**, I_h, characteristic of the icosahedron, has 12 fivefold axes (Fig. 6.12). The icosahedral group is important for boron compounds (Section 13.11) and the C$_{60}$ fullerene molecule (Section 14.6).

The distribution of molecules among the various point groups is very uneven. Some of the most common groups for molecules are the low-symmetry groups C_1 and C_s. There are many examples of polar molecules in groups C_{2v} (such as SO$_2$) and C_{3v} (such as NH$_3$). There are many linear molecules, which belong to the groups $C_{\infty v}$ (HCl, OCS) and $D_{\infty h}$ (Cl$_2$ and CO$_2$), and a number of planar-trigonal molecules, D_{3h} (such as BF$_3$, **5**), trigonal-bipyramidal molecules (such as PCl$_5$, **6**), which are D_{3h}, and square-planar molecules, D_{4h} (**7**). So-called 'octahedral' molecules with two identical substituents opposite each other, as in (**8**), are also D_{4h}. The last example shows that the point-group classification of a molecule is more precise than the casual use of the terms 'octahedral' or 'tetrahedral' that indicate molecular geometry. For instance, a molecule may be called octahedral (that is, it has octahedral geometry) even if it has six different groups attached to the central atom. However, the 'octahedral molecule' belongs to the octahedral point group O_h only if all six groups and the lengths of their bonds to the central atom are identical and all angles are 90°.

6.2 **Character tables**

Key point: The systematic analysis of the symmetry properties of molecules is carried out using character tables.

We have seen how the symmetry properties of a molecule define its point group and how that point group is labelled by its Schoenflies symbol. Associated with each point group is

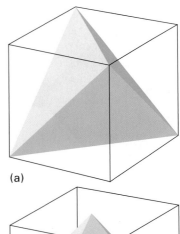

(a)

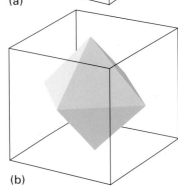

(b)

Figure 6.11 Shapes having cubic symmetry. (a) The tetrahedron, point group T_d. (b) The octahedron, point group O_h.

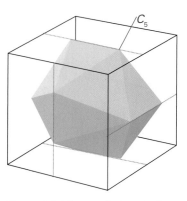

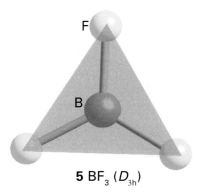

5 BF_3 (D_{3h})

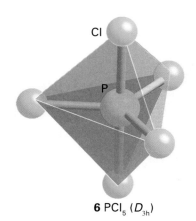

6 PCl_5 (D_{3h})

Figure 6.12 The regular icosahedron, point group I_h, and its relation to a cube.

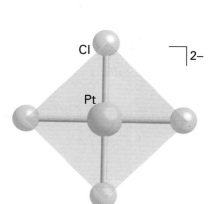

7 $[PtCl_4]^{2-}$ (D_{4h})

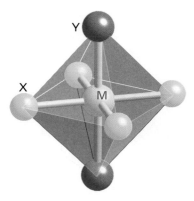

8 *trans*-$[MX_4Y_2]$ (D_{4h})

a **character table**. A character table displays all the symmetry elements of the point group together with a description, as we explain below, of how various objects or mathematical functions transform under the corresponding symmetry operations. A character table is complete: every possible object or mathematical function relating to the molecule belonging to a particular point group must transform like one of the rows in the character table of that point group.

The structure of a typical character table is shown in Table 6.3. The entries in the main part of the table are called **characters**, χ (chi). Each character shows how an object or mathematical function, such as an atomic orbital, is affected by the corresponding symmetry operation of the group. Thus:

Character	Significance
1	the orbital is unchanged
–1	the orbital changes sign
0	the orbital undergoes a more complicated change

For instance, the rotation of a p_z orbital about the z axis leaves it apparently unchanged (hence its character is 1); a reflection of a p_z orbital in the xy plane changes its sign (character –1). In some character tables, numbers such as 2 and 3 appear as characters: this feature is explained later.

The **class** of an operation is a specific grouping of symmetry operations of the same geometrical type: the two (clockwise and anticlockwise) threefold rotations about an axis form one class, reflections in a mirror plane form another, and so on. The number of members of each class is shown in the heading of each column of the table, as in $2C_3$, denoting that there are two members of the class of threefold rotations. All operations of the same class have the same character.

Each row of characters corresponds to a particular **irreducible representation** of the group. An irreducible representation has a technical meaning in group theory but, broadly speaking, it is a fundamental type of symmetry in the group (like the symmetries represented

Table 6.3 The components of a character table

Name of point group*	Symmetry operations R arranged by class (E, C_n, etc.)	Functions	Further functions	Order of group, h
Symmetry species (Γ)	Characters (χ)	Translations and components of dipole moments (x, y, z), of relevance to IR activity; rotations	Quadratic functions such as z^2, xy, etc., of relevance to Raman activity	

* Schoenflies symbol.

by σ and π orbitals for linear molecules). The label in the first column is the **symmetry species** (essentially, a label, like σ and π) of that irreducible representation. The two columns on the right contain examples of functions that exhibit the characteristics of each symmetry species. One column contains functions defined by a single axis, such as translations or p orbitals (x,y,z) or rotations (R_x,R_y,R_z), and the other column contains quadratic functions such as d orbitals $(xy, \text{etc.})$. Character tables for a selection of common point groups are given in *Resource section* 4.

EXAMPLE 6.3 Identifying the symmetry species of orbitals

Identify the symmetry species of the oxygen valence-shell atomic orbitals in an H_2O molecule, which has C_{2v} symmetry.

Answer The symmetry elements of the H_2O molecule are shown in Fig. 6.10 and the character table for C_{2v} is given in Table 6.4. We need to see how the orbitals behave under these symmetry operations. An s orbital on the O atom is unchanged by all four operations, so its characters are (1,1,1,1) and thus it has symmetry species A_1. Likewise, the $2p_z$ orbital on the O atom is unchanged by all operations of the point group and is thus totally symmetric under C_{2v}: it therefore has symmetry species A_1. The character of the $O2p_x$ orbital under C_2 is –1, which means simply that it changes sign under a twofold rotation. A p_x orbital also changes sign (and therefore has character –1) when reflected in the yz-plane (σ_v'), but is unchanged (character 1) when reflected in the xz-plane (σ_v). It follows that the characters of an $O2p_x$ orbital are (1,–1,1,–1) and therefore that its symmetry species is B_1. The character of the $O2p_y$ orbital under C_2 is –1, as it is when reflected in the xz-plane (σ_v). The $O2p_y$ is unchanged (character 1) when reflected in the yz-plane (σ_v'). It follows that the characters of an $O2p_y$ orbital are (1,–1,–1,1) and therefore that its symmetry species is B_2.

Self-test 6.3 Identify the symmetry species of all five d orbitals of the central Xe atom in XeF_4 (D_{4h}, Fig. 6.3).

Table 6.4 The C_{2v} character table

C_{2v}	E	C_2	σ_v	σ_v'	$h = 4$	
A_1	1	1	1	1	z	x^2, y^2, z^2
A_2	1	1	–1	–1	R_z	
B_1	1	–1	1	–1	x, R_y	xy
B_2	1	–1	–1	1	y, R_x	zx, yz

The letter A used to label a symmetry species in the group C_{2v} means that the function to which it refers is symmetric with respect to rotation about the twofold axis (that is, its character is 1). The label B indicates that the function changes sign under that rotation (the character is –1). The subscript 1 on A_1 means that the function to which it refers is also symmetric with respect to reflection in the principal vertical plane (for H_2O this is the plane that contains all three atoms). A subscript 2 is used to denote that the function changes sign under this reflection.

Now consider the slightly more complex example of NH_3, which belongs to the point group C_{3v} (Table 6.5). An NH_3 molecule has higher symmetry than H_2O. This higher symmetry is apparent by noting the **order**, h, of the group, the total number of symmetry operations that can be carried out. For H_2O, $h = 4$ and for NH_3, $h = 6$. For highly symmetric molecules, h is large; for example $h = 48$ for the point group O_h.

Inspection of the NH_3 molecule (Fig. 6.13) shows that whereas the $N2p_z$ orbital is unique (it has A_1 symmetry), the $N2p_x$ and $N2p_y$ orbitals both belong to the symmetry representation E. In other words, the $N2p_x$ and $N2p_y$ orbitals have the same symmetry characteristics, are degenerate, and must be treated together.

The characters in the column headed by the identity operation E give the degeneracy of the orbitals:

Symmetry label	Degeneracy
A, B	1
E	2
T	3

Table 6.5 The C_{3v} character table

C_{3v}	E	$2C_3$	$3\sigma_v$	$h = 6$	
A_1	1	1	1	z	z^2
A_2	1	1	–1	R_z	
E	2	–1	0	$(x, y) \, (R_x, R_y)$	$(zx, yz) \, (x^2 - y^2, xy)$

Figure 6.13 The nitrogen $2p_z$ orbital in ammonia is symmetric under all operations of the C_{3v} point group and therefore has A_1 symmetry. The $2p_x$ and $2p_y$ orbitals behave identically under all operations (they cannot be distinguished) and are given the symmetry label E.

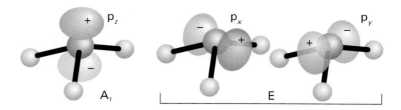

Be careful to distinguish the italic E for the operation and the roman E for the label: all operations are italic and all labels are roman.

Degenerate irreducible representations also contain zero values for some operations because the character is the sum of the characters for the two or more orbitals of the set, and if one orbital changes sign but the other does not, then the total character is 0. For example, the reflection through the vertical mirror plane containing the y-axis in NH_3 results in no change of the p_y orbital, but an inversion of the p_x orbital.

EXAMPLE 6.4 Determining degeneracy

Can there be triply degenerate orbitals in BF_3?

Answer To decide if there can be triply degenerate orbitals in BF_3 we note that the point group of the molecule is D_{3h}. Reference to the character table for this group (*Resource section* 4) shows that, because no character exceeds 2 in the column headed E, the maximum degeneracy is 2. Therefore, none of its orbitals can be triply degenerate.

Self-test 6.4 The SF_6 molecule is octahedral. What is the maximum possible degree of degeneracy of its orbitals?

Applications of symmetry

Important applications of symmetry in inorganic chemistry include the construction and labelling of molecular orbitals and the interpretation of spectroscopic data to determine structure. However, there are several simpler applications, one being to use group theory to decide whether a molecule is polar or chiral. In many cases the answer may be obvious and we do not need to use group theory. However, that is not always the case and the following examples illustrate the approach that can be adopted when the result is not obvious.

There are two aspects of symmetry. Some properties require a knowledge only of the point group to which a molecule belongs. These properties include its polarity and chirality. Other properties require us to know the detailed structure of the character table. These properties include the classification of molecular vibrations and the identification of their IR and Raman activity. We illustrate both types of application in this section.

6.3 Polar molecules

Key point: A molecule cannot be polar if it belongs to any group that includes a centre of inversion, any of the groups D and their derivatives, the cubic groups (T, O), the icosahedral group (I), and their modifications.

A **polar molecule** is a molecule that has a permanent electric dipole moment. A molecule cannot be polar if it has a centre of inversion. Inversion implies that a molecule has matching charge distributions at all diametrically opposite points about a centre, which rules out a dipole moment. For the same reason, a dipole moment cannot lie perpendicular to any mirror plane or axis of rotation that the molecule may possess. For example, a mirror plane demands identical atoms on either side of the plane, so there can be no dipole moment across the plane. Similarly, a symmetry axis implies the presence of identical atoms at points related by the corresponding rotation, which rules out a dipole moment perpendicular to the axis.

In summary:

1. A molecule cannot be polar if it has a centre of inversion.

2. A molecule cannot have an electric dipole moment perpendicular to any mirror plane.

3. A molecule cannot have an electric dipole moment perpendicular to any axis of rotation.

EXAMPLE 6.5 Judging whether or not a molecule can be polar

The ruthenocene molecule (**9**) is a pentagonal prism with the Ru atom sandwiched between two C_5H_5 rings. Can it be polar?

Answer We should decide whether the point group is *D* or cubic because in neither case can it have a permanent electric dipole. Reference to Fig. 6.9 shows that a pentagonal prism belongs to the point group D_{5h}. Therefore, the molecule must be nonpolar.

Self-test 6.5 A conformation of the ferrocene molecule that lies 4 kJ mol^{-1} above the lowest energy configuration is a pentagonal antiprism (**10**). Is it polar?

6.4 Chiral molecules

Key point: A molecule cannot be chiral if it possesses an improper rotation axis (S_n).

A **chiral molecule** (from the Greek word for 'hand') is a molecule that cannot be superimposed on its own mirror image. An actual hand is chiral in the sense that the mirror image of a left hand is a right hand, and the two hands cannot be superimposed. A chiral molecule and its mirror image partner are called **enantiomers** (from the Greek word for 'both parts'). Chiral molecules that do not interconvert rapidly between enantiomeric forms are **optically active** in the sense that they can rotate the plane of polarized light. Enantiomeric pairs of molecules rotate the plane of polarization of light by equal amounts in opposite directions.

A molecule with an improper rotation axis, S_n, cannot be chiral. A mirror plane is an S_1 axis of improper rotation and a centre of inversion is equivalent to an S_2 axis; therefore, molecules with either a mirror plane or a centre of inversion have axes of improper rotation and cannot be chiral. Groups in which S_n is present include D_{nh}, D_{nd}, and some of the cubic groups (specifically, T_d and O_h). Therefore, molecules such as CH_4 and $Ni(CO)_4$ that belong to the group T_d are not chiral. That a 'tetrahedral' carbon atom leads to optical activity (as in CHClFBr) should serve as another reminder that group theory is stricter in its terminology than casual conversation. Thus CHClFBr (**11**) belongs to the group C_1, not to the group T_d; it has tetrahedral geometry but not tetrahedral symmetry.

When judging chirality, it is important to be alert for axes of improper rotation that might not be immediately apparent. Molecules with neither a centre of inversion nor a mirror plane (and hence with no S_1 or S_2 axes) are usually chiral, but it is important to verify that a higher-order improper-rotation axis is not also present. For instance, the quaternary ammonium ion (**12**) has neither a mirror plane (S_1) nor an inversion centre (S_2), but it does have an S_4 axis and so it is not chiral.

EXAMPLE 6.6 Judging whether or not a molecule is chiral

The complex [Mn(acac)$_3$], where acac denotes the acetylacetonato ligand (CH$_3$COCHCOCH$_3^-$), has the structure shown as (**13**). Is it chiral?

Answer We begin by identifying the point group in order to judge whether it contains an improper-rotation axis either explicitly or in a disguised form. The chart in Fig. 6.9 shows that the ion belongs to the point group D_3, which consists of the elements (E, C_3, $3C_2$) and hence does not contain an S_n axis either explicitly or in a disguised form. The complex ion is chiral and hence, because it is long-lived, optically active.

Self-test 6.6 Is the conformation of H$_2$O$_2$ shown in (**14**) chiral? The molecule can usually rotate freely about the O–O bond: comment on the possibility of observing optically active H$_2$O$_2$.

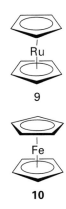

9

10

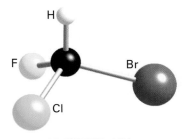

11 CHClFBr (C_1)

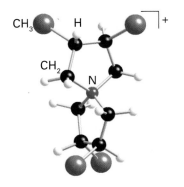

12 [N(CH$_2$CH(CH$_3$)CH(CH$_3$)CH$_2$)$_2$]$^+$

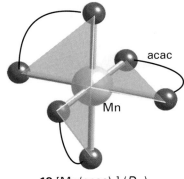

13 [Mn(acac)$_3$] (D_{3d})

14 H_2O_2

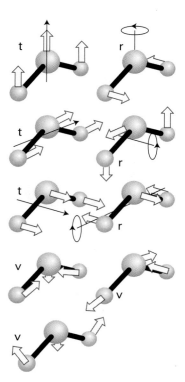

Figure 6.14 An illustration of the counting procedure for displacements of the atoms in a nonlinear molecule.

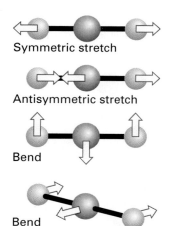

Symmetric stretch

Antisymmetric stretch

Bend

Bend

Figure 6.15 The stretches and bends of a CO_2 molecule.

6.5 Molecular vibrations

Key points: If a molecule has a centre of inversion, none of its modes can be both IR and Raman active; a vibrational mode is IR active if it has the same symmetry as a component of the electric dipole vector; a vibrational mode is Raman active if it has the same symmetry as a component of the molecular polarizability.

A knowledge of the symmetry of a molecule can assist and greatly simplify the analysis of infrared (IR) and Raman spectra (Chapter 8). It is convenient to consider two aspects of symmetry. One is the information that can be obtained directly by knowing to which point group a molecule as a whole belongs. The other is the additional information that comes from knowing the symmetry species of each normal mode. All we need to know at this stage is that the absorption of infrared radiation can occur when a vibration results in a change in the electric dipole moment of a molecule; a Raman transition can occur when the polarizability of a molecule changes during a vibration.

For a molecule of N atoms there are $3N$ displacements to consider as the atoms move. For a nonlinear molecule, three of these displacements correspond to translational motion of the molecule as a whole, and three correspond to an overall rotation, leaving $3N - 6$ vibrational modes. There is no rotation around the axis if the molecule is linear, so the molecule has only two rotational degrees of freedom instead of three, leaving $3N - 5$ vibrational displacements.

(a) The exclusion rule

The three-atom nonlinear molecule H_2O has $3 \times 3 - 6 = 3$ vibrational modes (Fig. 6.14). It should be intuitively obvious (and can be confirmed by group theory) that all three vibrational displacements lead to a change in the dipole moment. It follows that all three modes of this C_{2v} molecule are IR active. It is much more difficult to judge intuitively whether or not a mode is Raman active because it is hard to know whether a particular distortion of a molecule results in a change of polarizability (although modes that result in a swelling of the molecule such as the symmetric stretch of CO_2 are good prospects). This difficulty is partly overcome by the **exclusion rule**, which is sometimes helpful:

If a molecule has a centre of inversion, none of its modes can be both IR and Raman active. (A mode may be inactive in both.)

EXAMPLE 6.7 Using the exclusion rule

There are four vibration modes of the linear triatomic CO_2 molecule (Fig. 6.15). Which are IR or Raman active?

Answer To establish whether or not a stretch is IR active, we need to consider its effect on the dipole moment of the molecule. If we consider the symmetric stretch ν_1 we can see it leaves the electric dipole moment unchanged at zero and so it is IR inactive: it may therefore be Raman active (and is). In contrast, for the antisymmetric stretch, ν_3, the C atom moves opposite to that of the two O atoms: as a result, the electric dipole moment changes from zero in the course of the vibration and the mode is IR active. Because the CO_2 molecule has a centre of inversion, it follows from the exclusion rule that this mode cannot be Raman active. Both bending modes cause a departure of the dipole moment from zero and are therefore IR active. It follows from the exclusion rule that the two bending modes are Raman inactive.

Self-test 6.7 The bending mode of N_2O is active in the IR. Can it also be Raman active?

(b) Information from the symmetries of normal modes

So far, we have remarked that it is often intuitively obvious whether a vibrational mode gives rise to a changing electric dipole and is therefore IR active. When intuition is unreliable, perhaps because the molecule is complex or the mode of vibration is difficult to visualize, a symmetry analysis can be used instead. We shall illustrate the procedure by considering the two square-planar palladium species (**15**) and (**16**). The Pt analogues of these species and the distinction between them are of considerable social and practical significance because the *cis* isomer is used as a chemotherapeutic agent against certain cancers (whereas the *trans* isomer is therapeutically inactive (Section 27.18).

First, we note that the *cis* isomer (**15**) has C_{2v} symmetry, whereas the *trans* isomer (**16**) is D_{2h}. Both species have bands in the Pd–Cl stretching region between 200 and 400 cm^{-1}. We know immediately from the exclusion rule that the two modes of the *trans* isomer (which has a centre of symmetry) cannot be active in both IR and Raman. However, to decide which modes are IR active and which are Raman active we consider the characters of the modes themselves. It follows from the symmetry properties of dipole moments and polarizabilities (which we do not verify here) that:

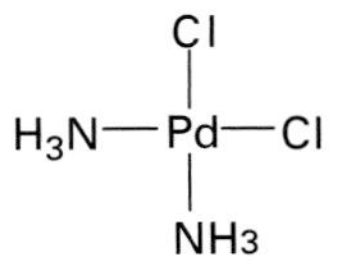

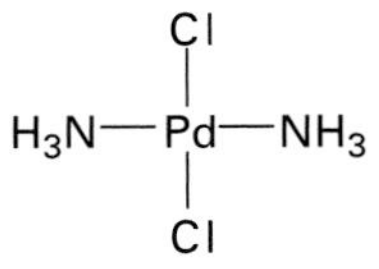

> The symmetry species of the vibration must be the same as that of x, y, or z in the character table for the vibration to be IR active and the same as that of a quadratic function, such as xy or x^2, for it to be Raman active.

Our first task, therefore, is to classify the normal modes according to their symmetry species, and then to identify which of these modes have the same symmetry species as x, etc. and xy, etc. by referring to the final columns of the character table of the molecular point group.

Figure 6.16 shows the symmetric (left) and antisymmetric (right) stretches of the Pd–Cl bonds for each isomer, where the NH_3 group is treated as a single mass point. To classify them according to their symmetry species in their respective point groups we use an approach similar to the symmetry analysis of molecular orbitals we will use for determining SALCs (Section 6.10).

Consider the *cis* isomer and its point group C_{2v} (Table 6.4). For the symmetric stretch, we see that the pair of displacement vectors representing the vibration is apparently unchanged by each operation of the group. For example, the twofold rotation simply interchanges two equivalent displacement vectors. It follows that the character of each operation is 1:

E	C_2	σ_v	σ_v'
1	1	1	1

The symmetry of this vibration is therefore A_1. For the antisymmetric stretch, the identity E leaves the displacement vectors unchanged and the same is true of σ_v', which lies in the plane containing the two Cl atoms. However, both C_2 and σ_v interchange the two oppositely directed displacement vectors, and so convert the overall displacement into -1 times itself. The characters are therefore

E	C_2	σ_v	σ_v'
1	-1	-1	1

The C_{2v} character table identifies the symmetry species of this mode as B_2. A similar analysis of the *trans* isomer, but using the D_{2h} group, results in the labels A_g and B_{2u} for the symmetric and antisymmetric Pd–Cl stretches, respectively.

EXAMPLE 6.8 Identifying the symmetry species of vibrational displacements

The *trans* isomer in Fig. 6.16 has D_{2h} symmetry. Verify that the symmetry species of the antisymmetric Pd-Cl stretches is B_{2u}.

Answer We need to start by considering the effect of the various elements of the group on the displacement vectors of the Cl^- ligands. The elements of D_{2h} are E, $C_2(x)$, $C_2(y)$, $C_2(z)$, i, $\sigma(xy)$, $\sigma(yz)$, and $\sigma(zx)$. Of these, E, $C_2(y)$, $\sigma(xy)$, and $\sigma(yz)$ leave the displacement vectors unchanged and so have characters 1. The remaining operations reverse the directions of the vectors, so giving characters of -1:

E	$C_2(x)$	$C_2(y)$	$C_2(z)$	i	$\sigma(xy)$	$\sigma(yz)$	$\sigma(zx)$
1	-1	1	-1	-1	1	1	-1

We now compare this set of characters with the D_{2h} character table and establish that the symmetry species is B_{2u}.

Self-test 6.8 Confirm that the symmetry species of the symmetric mode of the Pd-Cl stretches in the *trans* isomer is A_g.

As we have remarked, a vibrational mode is IR active if it has the same symmetry species as the displacements x, y, or z. In C_{2v}, z is A_1 and y is B_2. Both A_1 and B_2 vibrations of the *cis* isomer are therefore IR active. In D_{2h}, x, y, and z are B_{3u}, B_{2u}, and B_{1u}, respectively, and only vibrations with these symmetries can be IR active. The antisymmetric Pd–Cl stretch of the *trans* isomer has symmetry B_{2u} and is IR active. The symmetric A_g mode of the *trans* isomer is not IR active.

15

16

Figure 6.16 The Pd–Cl stretching modes of *cis* and *trans* forms of [Pd(Cl)$_2$(NH$_3$)$_2$]. The motion of the Pd atom (which preserves the centre of mass of the molecule) is not shown.

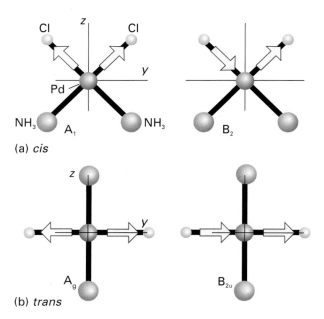

(a) *cis*

(b) *trans*

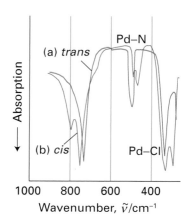

Figure 6.17 The IR spectra of *cis* (red) and *trans* (blue) forms of [Pd(Cl)$_2$(NH$_3$)$_2$]. (R. Layton, D.W. Sink, and J.R. Durig, *J. Inorg. Nucl. Chem.*, 1966, **28**, 1965.)

To determine the Raman activity, we note that in C_{2v} the quadratic forms xy, etc. transform as A$_1$, A$_2$, B$_1$, and B$_2$ and therefore in the *cis* isomer the modes of symmetry A$_1$, A$_2$, B$_1$, and B$_2$ are Raman active. In D_{2h}, however, only A$_g$, B$_{1g}$, B$_{2g}$, and B$_{3g}$ are Raman active.

The experimental distinction between the *cis* and *trans* isomers now emerges. In the Pd–Cl stretching region, the *cis* (C_{2v}) isomer has two bands in both the Raman and IR spectra. By contrast, the *trans* (D_{2h}) isomer has one band at a different frequency in each spectrum. The IR spectra of the two isomers are shown in Fig. 6.17.

(c) The assignment of molecular symmetry from vibrational spectra

An important application of vibrational spectra is the identification of molecular symmetry and hence shape and structure. An especially important example arises in metal carbonyls in which CO molecules are bound to a metal atom. Vibrational spectra are especially useful because the CO stretch is responsible for very strong characteristic absorptions between 1850 and 2200 cm^{-1} (Section 22.5).

The set of characters obtained by considering the symmetries of the displacements of atoms are often found not to correspond to any of the rows in the character table. However, because the character table is a complete summary of the symmetry properties of an object, the characters that have been determined must correspond to a sum of two or more of the rows in the table. In such cases we say that the displacements span a **reducible representation**. Our task is to find the **irreducible representations** that they span. To do so, we identify the rows in the character table that must be added together to reproduce the set of characters that we have obtained. This process is called **reducing a representation**. In some cases the reduction is obvious; in others it may be carried out systematically by using a procedure explained in Section 6.9.

EXAMPLE 6.9 **Reducing a representation**

One of the first metal carbonyls to be characterized was the tetrahedral (T_d) molecule Ni(CO)$_4$. The vibrational modes of the molecule that arise from stretching motions of the CO groups are four combinations of the four CO displacement vectors. Which modes are IR or Raman active? The CO displacements of Ni(CO)$_4$ are shown in Fig. 6.18.

Answer We need to consider the motion of the four CO displacement vectors and then consult the character table for T_d (Table 6.6). Under the E operation all four vectors remain unchanged, under a C_3 only one remains the same, under both C_2 and S_4 none of the vectors remains unchanged, and under σ_d two remain the same. The characters are therefore:

E	$8C_3$	$3C_2$	$6S_4$	$6\sigma_d$
4	1	0	0	2

This set of characters does not correspond to any one symmetry species. However, it does correspond to the sum of the characters of symmetry species A_1 and T_2:

	E	$8C_3$	$3C_2$	$6S_4$	$6\sigma_d$
A_1	1	1	1	1	1
T_2	3	0	-1	-1	1
$A_1 + T_2$	4	1	0	0	2

It follows that the CO displacement vectors transform as $A_1 + T_2$. By consulting the character table for T_d, we see that the combination labelled A_1 transforms like $x^2 + y^2 + z^2$, indicating that it is Raman active but not IR active. By contrast, x, y, and z and the products xy, yz, and zx transform as T_2, so the T_2 modes are both Raman and IR active. Consequently, a tetrahedral carbonyl molecule is recognized by one IR band and two Raman bands in the CO stretching region.

Self-test 6.9 Show that the four CO displacements in the square-planar (D_{4h}) $[Pt(CO)_4]^{2+}$ cation transform as $A_{1g} + B_{1g} + E_u$. How many bands would you expect in the IR and Raman spectra for the $[Pt(CO)_4]^{2+}$ cation?

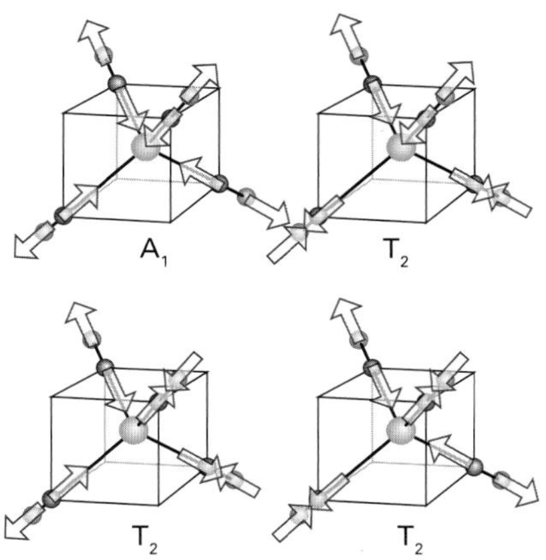

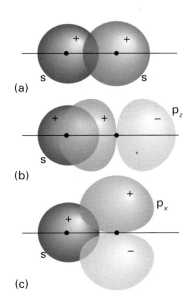

Figure 6.18 The modes of $Ni(CO)_4$ that correspond to the stretching of CO bonds.

Table 6.6 The T_d character table

T_d	E	$8C_3$	$3C_2$	$6S_4$	$6\sigma_d$	$h = 24$	
A_1	1	1	1	1	1		$x^2 + y^2 + z^2$
A_2	1	1	1	-1	-1		
E	2	-1	2	0	0		$(2x^2 - y^2 - z^2, x^2 - y^2)$
T_1	3	0	-1	1	-1	(R_x, R_y, R_z)	
T_2	3	0	-1	-1	1	(x, y, z)	(xy, yz, zx)

The symmetries of molecular orbitals

We shall now see in more detail the significance of the labels used for molecular orbitals introduced in Sections 2.7 and 2.8 and gain more insight into their construction. At this stage the discussion will continue to be informal and pictorial, our aim being to give an introduction to group theory but not the details of the calculations involved. The specific objective here is to show how to identify the symmetry label of a molecular orbital from a drawing like those in *Resource section* 5 and, conversely, to appreciate the significance of a symmetry label. The arguments later in the book are all based on simply 'reading' molecular orbital diagrams qualitatively.

6.6 Symmetry-adapted linear combinations

Key point: Symmetry-adapted linear combinations of orbitals are combinations of atomic orbitals that conform to the symmetry of a molecule and are used to construct molecular orbitals of a given symmetry species.

A fundamental principle of the MO theory of diatomic molecules (Section 2.8) is that molecular orbitals are constructed from atomic orbitals of the same symmetry. Thus, in a diatomic molecule, an s orbital may have nonzero overlap with another s orbital or with a p_z orbital on the second atom (where z is the internuclear direction, Fig. 6.19), but not with a p_x or p_y orbital. Formally, whereas the p_z orbital of the second atom has the same rotational symmetry as the s orbital of the first atom and the same symmetry with respect to reflection in a mirror plane containing the internuclear axis, the p_x and p_y orbitals do not. The restriction that σ, π, or δ bonds can be formed from atomic orbitals of the same symmetry species stems from the requirement that all components of the molecular orbital must behave identically under transformation if they are to have nonzero overlap.

Exactly the same principle applies in polyatomic molecules, where the symmetry considerations may be more complex and require us to use the systematic procedures provided by group theory. The general procedure is to group atomic orbitals, such as the three H1s orbitals of NH_3, together to form combinations of a particular symmetry and then to build molecular orbitals by allowing combinations of the same symmetry on different atoms to overlap, such as an N2s orbital and the appropriate combination of the three H1s orbitals.

Figure 6.19 An s orbital can overlap (a) an s or (b) a p_z orbital on a second atom with constructive interference. (c) An s orbital has zero net overlap with a p_x or p_y orbital because the constructive interference between the parts of the atomic orbitals with the same sign exactly matches the destructive interference between the parts with opposite signs.

Specific combinations of atomic orbitals that are used to build molecular orbitals of a given symmetry are called **symmetry-adapted linear combinations** (SALCs). A collection of commonly encountered SALCs of orbitals is shown in *Resource section 5*; it is usually simple to identify the symmetry of a combination of orbitals by comparing it with the diagrams provided there.

EXAMPLE 6.10 Identifying the symmetry species of a SALC

Identify the symmetry species of the SALCs that may be constructed from the H1s orbitals of NH_3.

Answer We start by establishing how the set of H1s orbitals transform under the operations of the appropriate symmetry group of the molecule. An NH_3 molecule has symmetry C_{3v} and the three H1s orbitals of all remain unchanged under the identity operation E. None of the H1s orbitals remains unchanged under a C_3 rotation, and only one remains unchanged under a vertical reflection σ_v. As a set they therefore span a representation with the characters

E	$2C_3$	$3\sigma_v$
3	0	1

We now need to reduce this set of characters, and by inspection we can see they correspond to $A_1 + E$. It follows that the three H1s orbitals contribute two SALCs, one with A_1 symmetry and the other with E symmetry. The SALC with E symmetry has two members of the same energy. In more complicated examples the reduction might not be obvious and we use the systematic procedure discussed in Section 6.10.

Self-test 6.10 What is the symmetry label of the SALC $\phi = \psi_{A1s} + \psi_{B1s} + \psi_{C1s} + \psi_{D1s}$ in CH_4, where ψ_{J1s} is an H1s orbital on atom J?

The generation of SALCs of a given symmetry is a task for group theory, as we explain in Section 6.10. However, they often have an intuitively obvious form. For instance, the fully symmetrical A_1 SALC of the H1s orbitals of NH_3 (Fig. 6.20) is

$$\phi_1 = \psi_{A1s} + \psi_{B1s} + \psi_{C1s}$$

To verify that this SALC is indeed of symmetry A_1 we note that it remains unchanged under the identity E, each C_3 rotation, and any of the three vertical reflections, so its characters are $(1,1,1)$ and hence it spans the fully symmetrical irreducible representation of C_{3v}. The E SALCs are less obvious, but as we shall see are

$$\phi_2 = 2\psi_{A1s} - \psi_{B1s} - \psi_{C1s}$$

$$\phi_3 = \psi_{B1s} - \psi_{C1s}$$

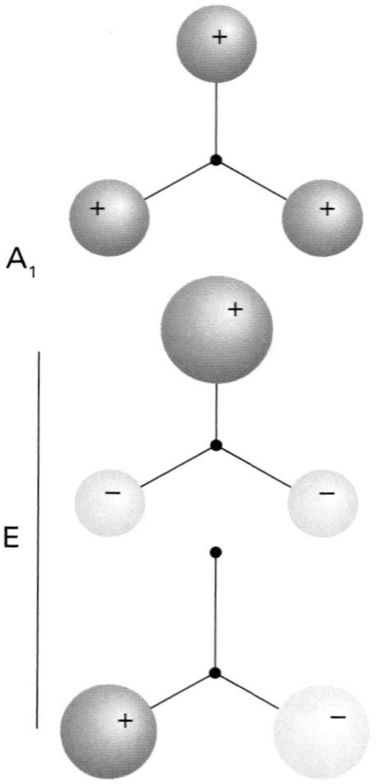

Figure 6.20 The (a) A_1 and (b) E symmetry-adapted linear combinations of H1s orbitals in NH_3.

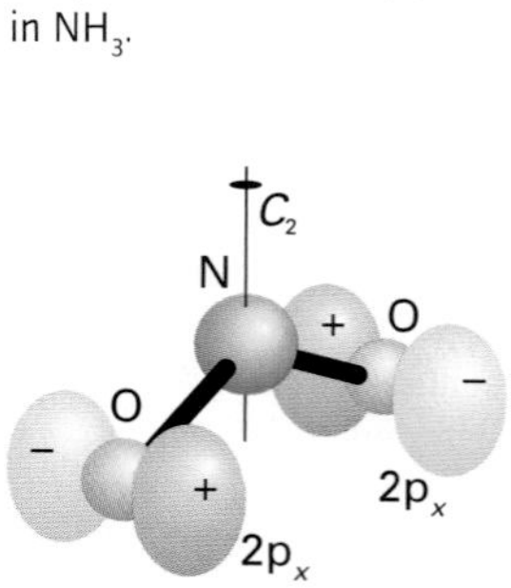

Figure 6.21 The combination of $O2p_x$ orbitals referred to in Example 6.11.

EXAMPLE 6.11 Identifying the symmetry species of SALCs

Identify the symmetry species of the SALC $\phi = \psi'_O - \psi''_O$ in the C_{2v} molecule NO_2, where ψ'_O is a $2p_x$ orbital on one O atom and ψ''_O is a $2p_x$ orbital on the other O atom.

Answer To establish the symmetry species of a SALC we need to see how it transforms under the symmetry operations of the group. A picture of the SALC is shown in Fig. 6.21, and we can see that under C_2, ϕ changes into itself, implying a character of 1. Under σ_v, both atomic orbitals change sign, so ϕ is transformed into $-\phi$, implying a character of -1. The SALC also changes sign under σ_v', so the character for this operation is also -1. The characters are therefore

E	C_2	σ_v	σ_v'
1	1	-1	-1

Inspection of the character table for C_{2v} shows that these characters correspond to symmetry species A_2.

Self-test 6.11 Identify the symmetry species of the combination $\phi = \psi_{A1s} - \psi_{B1s} + \psi_{C1s} - \psi_{D1s}$ for a square-planar (D_{4h}) array of H atoms A, B, C, D.

6.7 The construction of molecular orbitals

Key point: Molecular orbitals are constructed from SALCs and atomic orbitals of the same symmetry species.

We have seen that the SALC ϕ_1 of H1s orbitals in NH_3 has A_1 symmetry. The N2s and N2p_z also have A_1 symmetry in this molecule, so all three can contribute to the same molecular orbitals. The symmetry species of these molecular orbitals will be A_1, like their components, and they are called **a_1 orbitals**. Note that the labels for molecular orbitals are lowercase versions of the symmetry species of the orbital. Three such molecular orbitals are possible, each of the form

$$\psi = c_1\psi_{N2s} + c_2\psi_{N2p_z} + c_3\phi_1$$

with c_i coefficients that are found by computational methods. They are labelled $1a_1$, $2a_1$, and $3a_1$ in order of increasing energy (the order of increasing number of internuclear nodes), and correspond to bonding, nonbonding, and antibonding combinations (Fig. 6.22).

We have also seen (and can confirm by referring to *Resource section 5*) that in a C_{3v} molecule the SALCs ϕ_2 and ϕ_3 of the H1s orbitals have E symmetry. The C_{3v} character table shows that the same is true of the N2p_x and N2p_y orbitals (Fig. 6.23). It follows that ϕ_2 and ϕ_3 can combine with these two N2p orbitals to give doubly degenerate bonding and antibonding orbitals of the form

$$\psi = c_4\psi_{N2p_x} + c_5\phi_2, \text{ and } c_6\psi_{N2p_y} + c_7\phi_3$$

These molecular orbitals have E symmetry and are therefore called **e orbitals**. The pair of lower energy, denoted 1e, are bonding and the upper pair, 2e, are antibonding.

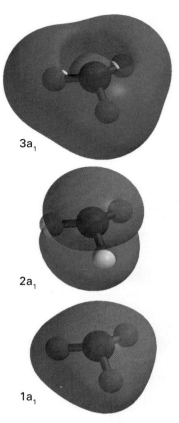

3a₁

2a₁

1a₁

Figure 6.22 The three a_1 molecular orbitals of NH_3 as computed by molecular modelling software.

EXAMPLE 6.12 Constructing molecular orbitals from SALCs

The two SALCs of H1s orbitals in the C_{2v} molecule H_2O are $\phi_1 = \psi_{A1s} + \psi_{B1s}$ (**17**) and $\phi_2 = \psi_{A1s} - \psi_{B1s}$ (**18**). Which oxygen orbitals can be used to form molecular orbitals with them?

Answer We start by establishing how the SALCs transform under the symmetry operations of the group (C_{2v}). Under E neither SALC changes sign, so their characters are 1. Under C_2, ϕ_1 does not change sign but ϕ_2 does; their characters are therefore 1 and –1, respectively. Under σ_v the combination ϕ_1 does not change sign but ϕ_2 does change sign, so their characters are again +1 and –1, respectively. Under the reflection σ_v' neither SALC changes sign, so their characters are 1. The characters are therefore

	E	C_2	σ_v	σ_v'
ϕ_1	1	1	1	1
ϕ_2	1	–1	–1	1

We now consult the character table and identify their symmetry labels as A_1 and B_2, respectively. The same conclusion could have been obtained more directly by referring to *Resource section* 5. According to the entries on the right of the character table, the O2s and O2p_z orbitals also have A_1 symmetry; O2p_y has B_2 symmetry. The linear combinations that can be formed are therefore

$$\psi = c_1\psi_{O2s} + c_2\psi_{O2p_z} + c_3\phi_1$$

$$\psi = c_4\psi_{O2p_y} + c_5\phi_2$$

The three a_1 orbitals are bonding, intermediate, and antibonding in character according to the relative signs of the coefficients c_1, c_2, and c_3. Similarly, depending on the relative signs of the coefficients c_4 and c_5, one of the two b_2 orbitals is bonding and the other is antibonding.

Self-test 6.12 The four SALCs built from Cl3s orbitals in the square planar (D_{4h}) $[PtCl_4]^{2-}$ anion have symmetry species A_{1g}, B_{1g}, and E_u. Which Pt atomic orbitals can combine with which of these SALCs?

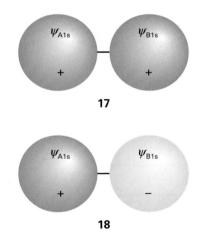

ψ_{A1s} ψ_{B1s}

+   +

17

ψ_{A1s} ψ_{B1s}

+   −

18

A symmetry analysis has nothing to say about the energies of orbitals other than to identify degeneracies. To calculate the energies, and even to arrange the orbitals in order, it is necessary to use quantum mechanics; to assess them experimentally it is necessary to use techniques such as photoelectron spectroscopy. In simple cases, however, we can use the general rules set out in Section 2.8 to judge the relative energies of the orbitals. For example, in NH_3, the $1a_1$ orbital, containing the low-lying N2s orbital, can be expected to lie lowest in energy and its antibonding partner, $3a_1$, will probably lie highest, with the nonbonding $2a_1$ approximately half-way between. The 1e bonding orbital is next higher in energy after $1a_1$, and the 2e correspondingly lower in energy than the $3a_1$ orbital. This qualitative analysis leads to the energy level scheme shown in Fig. 6.24. These days, there is no difficulty in using one of the widely available software packages to calculate the energies of

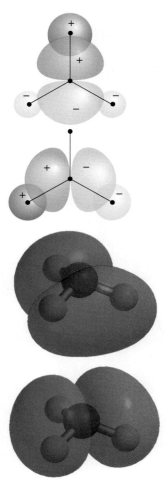

Figure 6.23 The two bonding e orbitals of NH_3 as schematic diagrams and as computed by molecular modelling software.

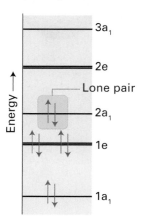

Figure 6.24 A schematic molecular orbital energy level diagram for NH_3 and an indication of its ground-state electron configuration.

the orbitals directly by either an *ab initio* or a semi-empirical procedure; the energies given in Fig. 6.24 have in fact been calculated in this way. Nevertheless, the ease of achieving computed values should not be seen as a reason for disregarding the understanding of the energy level order that comes from investigating the structures of the orbitals.

The general procedure for constructing a molecular orbital scheme for a reasonably simple molecule can now be summarized as follows:

1. Assign a point group to the molecule.
2. Look up the shapes of the SALCs in *Resource section 5*.
3. Arrange the SALCs of each molecular fragment in increasing order of energy, first noting whether they stem from s, p, or d orbitals (and put them in the order s < p < d), and then their number of internuclear nodes.
4. Combine SALCs of the same symmetry type from the two fragments, and from N SALCs form N molecular orbitals.
5. Estimate the relative energies of the molecular orbitals from considerations of overlap and relative energies of the parent orbitals, and draw the levels on a molecular orbital energy level diagram (showing the origin of the orbitals).
6. Confirm, correct, and revise this qualitative order by carrying out a molecular orbital calculation by using appropriate software.

6.8 The vibrational analogy

Key point: The shapes of SALCs are analogous to stretching displacements.

One of the great strengths of group theory is that it enables disparate phenomena to be treated analogously. We have already seen how symmetry arguments can be applied to molecular vibrations, so it should come as no surprise that SALCs have analogies in the normal modes of molecules. In fact, the illustrations of SALCs in the Resource section can be interpreted as contributions to the normal vibrational modes of molecules. The following example illustrates how this is done.

EXAMPLE 6.13 Predicting the IR and Raman bands of an octahedral molecule

Consider an AB_6 molecule, such as SF_6, that belongs to the O_h point group. Sketch the normal modes of A–B stretches and comment on their activities in IR or Raman spectroscopy.

Answer We argue by analogy with the shapes of SALCs and identify the SALCs that can be constructed from s orbitals in an octahedral arrangement (*Resource section 4*). These orbitals are the analogues of the stretching displacements of the A–B bonds and the signs represent their relative phases. They have the symmetry species A_{1g}, E_g, and T_{1u}. The resulting linear combinations of stretches are illustrated in Fig. 6.25. The A_{1g} (totally symmetric) and E_g modes are Raman active and the T_{1u} mode is IR active.

Self-test 6.13 Consider only bands due to stretching vibrations and predict how the IR and Raman spectra of SF_5Cl differ from those of SF_6.

Representations

We now move on to a more quantitative treatment and introduce two topics that are important for applying symmetry arguments in the treatment of molecular orbitals and spectroscopy systematically.

6.9 The reduction of a representation

Key point: A reducible representation can be resolved into its constituent irreducible representations by using the reduction formula.

We have seen that the three H1s orbitals of NH_3 give rise to—the technical term is 'span'—two irreducible representations in C_{3v}, one of symmetry species A_1 and the other

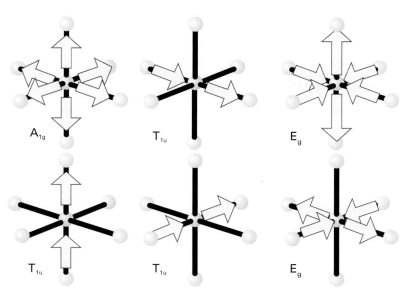

Figure 6.25 The A_{1g} and T_{1u} M–L stretching modes of an octahedral ML_6 complex. The motion of the central metal atom M, which preserves the centre of mass of the molecule, is not shown (it is stationary in both the A_{1g} and E_g modes).

of symmetry species E. Here we present a systematic way for arriving at the identification of the symmetry species spanned by a set of orbitals or atom displacements.

The fact the three H1s orbitals of NH_3 span two particular irreducible representations is expressed formally by writing $\Gamma = A_1 + E$ where Γ (uppercase gamma) denotes the symmetry species of the reducible representation. In general, we write

$$\Gamma = c_1\Gamma_1 + c_2\Gamma_2 + \cdots \quad (6.1a)$$

where the Γ_i denote the various symmetry species of the group and the c_i tell us how many times each symmetry species appears in the reduction. A very deep theorem from group theory (see *Further reading*) provides an explicit formula for calculating the coefficients c_i in terms of the characters χ_i of the irreducible representation Γ_i and the corresponding characters χ of the original reducible representation Γ:

$$c_i = \frac{1}{h}\sum_C g(C)\chi_i(R)\chi(R) \quad (6.1b)$$

Here h is the order of the point group (the number of symmetry elements; it is given in the top row of the character table) and the sum is over each class C of the group with $g(C)$ the number of elements in that class. How this expression is used is illustrated by the following example.

EXAMPLE 6.14 Using the reduction formula

Consider the molecule *cis*-[$PdCl_2(NH_3)_2$], which, if we ignore the hydrogen atoms, belongs to the point group C_{2v}. What are the symmetry species spanned by the displacements of the atoms?

Answer To analyse this problem we consider the 15 displacements of the five nonhydrogen atoms (Fig. 6.26) and obtain the characters of what will turn out to be a reducible representation Γ by examining what happens when we apply the symmetry operations of the group. Then we use eqn 6.1b to identify the symmetry species of the irreducible representations into which that reducible representation can be reduced. To identify the characters of Γ we note that each displacement that moves to a new location under a particular symmetry operation contributes 0 to the character of that operation; those that remain the same contribute 1; those that are reversed contribute –1. Thus, because all 15 displacements remain unmoved under the identity, $\chi(E) = 15$. A C_2 rotation leaves only the z displacement on Pd unchanged (contributing 1) but reverses the x and y displacements on Pd (contributing –2), so

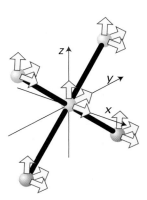

Figure 6.26 The atomic displacements in *cis*-[$PdCl_2(NH_3)_3$] with the H atoms ignored.

$\chi(C_2) = -1$. Under the reflection σ_v the z and x displacements on Pd are unchanged (contributing 2) and the y displacement on Pd is reversed (contributing -1), so $\chi(\sigma_v) = 1$. Finally, for any reflection in a vertical plane passing through the plane of the atoms, the five z displacements on these atoms remain the same (contributing 5), so do the five y displacements (another 5), but the five x displacements are reversed (contributing -5); therefore $\chi(\sigma_v') = 5$. The characters of Γ are therefore

E	C_2	σ_v	σ_v'
15	-1	1	5

Now we use eqn 6.1b, noting that $h = 4$ for this group, and noting that $g(C) = 1$ for all C. To find how many times the symmetry species A_1 appears in the reducible representation, we write

$$c_1 = \tfrac{1}{4}\{1 \times 15 + 1 \times (-1) + 1 \times 1 + 1 \times 5\} = 5$$

By repeating this procedure for the other species, we find

$$\Gamma = 5A_1 + 2A_2 + 3B_1 + 5B_2$$

For C_{2v}, the translations of the entire molecule span $A_1 + B_1 + B_2$ (as given by the functions x, y, and z in the final column in the character table) and the rotations span $A_2 + B_1 + B_2$ (as given by the functions R_x, R_y, and R_z in the final column in the character table). By subtracting these symmetry species from the ones we have just found, we can conclude that the vibrations of the molecule span $4A_1 + A_2 + B_1 + 3B_2$.

Self-test 6.14 Determine the symmetries of all the vibration modes of $[PdCl_4]^{2-}$, a D_{4h} molecule.

Many of the modes of cis-$[PdCl_2(NH_3)_2]$ found in Example 6.14 are complex motions that are not easy to visualize: they include Pd–N stretches and various buckling motions of the plane. However, even without being able to visualize them easily, we can infer at once that the A_1, B_1, and B_2 modes are IR active (because the functions x, y, and z, and hence the components of the electric dipole, span these symmetry species) and all the modes are Raman active (because the quadratic forms span all four species).

6.10 Projection operators

Key point: A projection operator is used to generate SALCs from a basis of orbitals.

To generate an unnormalized SALC of a particular symmetry species from an arbitrary set of basis atomic orbitals, we select any one of the set and form the following sum:

$$\phi = \sum_R \chi_i(R)R\psi \tag{6.2}$$

where $\chi_i(R)$ is the character of the operation R for the symmetry species of the SALC we want to generate. Once again, the best way to illustrate the use of this expression is with an example.

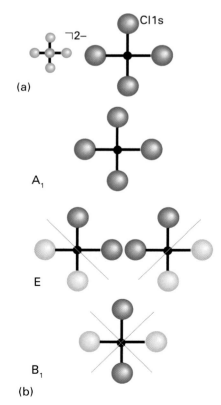

Figure 6.27 (a) The Cl orbital basis used to construct SALCs in $[PtCl_4]^{2-}$, and (b) the SALCs constructed for $[PtCl_4]^{2-}$.

EXAMPLE 6.15 Generating a SALC

Generate the SALC of Clσ orbitals for $[PtCl_4]^{2-}$. The basis orbitals are denoted ψ_1, ψ_2, ψ_3, and ψ_4 and are shown in Fig. 6.27a.

Answer To implement eqn 6.2 we start with one of the basis orbitals and subject it to all the symmetry operations of the D_{4h} point group, writing down the basis function $R\psi$ into which it is transformed. For example, the operation C_4 moves ψ_1 into the position occupied by ψ_2, C_2 moves it to ψ_3 and C_4^3 moves it to ψ_4. Continuing for all operations we obtain

Operation R:	E	C_4	C_4^3	C_2	C_2'	C_2'	C_2''	C_2''	i	S_4	S_4^3	σ_h	σ_v	σ_v	σ_d	σ_d
$R\psi_1$	ψ_1	ψ_2	ψ_4	ψ_3	ψ_1	ψ_3	ψ_2	ψ_4	ψ_3	ψ_2	ψ_4	ψ_1	ψ_1	ψ_3	ψ_2	ψ_4

We now add together all the new basis functions, and for each class of operation we multiply by the character $\chi_i(R)$ for the irreducible representation we are interested in. Thus, for A_{1g} (as all characters are 1) we obtain $4\psi_1 + 4\psi_2 + 4\psi_3 + 4\psi_4$. The (unnormalized) SALC is therefore

$$\phi(A_{1g}) = \psi_1 + \psi_2 + \psi_3 + \psi_4$$

As we continue down the character table using the various symmetry species, the SALCs emerge as follows:

$$\phi(B_{1g}) = \psi_1 - \psi_2 + \psi_3 - \psi_4$$

$$\phi(E_u) = \psi_1 - \psi_3$$

Under all other irreducible representations the projection operators vanish (thus no SALCs exist of those symmetries). We then continue by using ψ_2 as our basis function, whereupon we obtain the same SALCs except for

$$\phi(B_{1g}) = \psi_2 - \psi_1 + \psi_4 - \psi_3$$

$$\phi(E_u) = \psi_2 - \psi_4$$

Completing the process with ψ_3 and ψ_4 gives similar SALCs (only the signs of some of the component orbitals change). The forms of the SALCs are therefore $A_{1g} + B_{1g} + E_u$ (Fig. 6.27b).

Self-test 6.15 Use projection operators in SF_6 to determine the SALCs for σ bonding in an octahedral complex. (Use the O point group rather than O_h.)

FURTHER READING

P. Atkins and J. de Paula, *Physical chemistry*. Oxford University Press and W.H. Freeman & Co (2010). An account of the generation and use of character tables without too much mathematical background.

For more rigorous introductions, see: J.S. Ogden, *Introduction to molecular symmetry*. Oxford University Press (2001).

P. Atkins and R. Friedman, *Molecular quantum mechanics*. Oxford University Press (2005).

EXERCISES

6.1 Draw sketches to identify the following symmetry elements: (a) a C_3 axis and a σ_v plane in the NH_3 molecule, (b) a C_4 axis and a σ_h plane in the square-planar $[PtCl_4]^{2-}$ ion.

6.2 Which of the following molecules and ions has (a) a centre of inversion, (b) an S_4 axis: (i) CO_2, (ii) C_2H_2, (iii) BF_3, (iv) SO_4^{2-}?

6.3 Determine the symmetry elements and assign the point group of (a) NH_2Cl, (b) CO_3^{2-}, (c) SiF_4, (d) HCN, (e) $SiFClBrI$, (f) BF_4^-.

6.4 How many planes of symmetry does a benzene molecule possess? What chloro-substituted benzene of formula $C_6H_nCl_{6-n}$ has exactly four planes of symmetry?

6.5 Determine the symmetry elements of objects with the same shape as the boundary surface of (a) an s orbital, (b) a p orbital, (c) a d_{xy} orbital, (d) a d_{z^2} orbital.

6.6 (a) Determine the symmetry group of an SO_3^{2-} ion. (b) What is the maximum degeneracy of a molecular orbital in this ion? (c) If the sulfur orbitals are 3s and 3p, which of them can contribute to molecular orbitals of this maximum degeneracy?

6.7 (a) Determine the point group of the PF_5 molecule. (Use VSEPR, if necessary, to assign geometry.) (b) What is the maximum degeneracy of its molecular orbitals? (c) Which P3p orbitals contribute to a molecular orbital of this degeneracy?

6.8 Reaction of $AsCl_3$ with Cl_2 at low temperature yields a product, believed to be $AsCl_5$, which shows Raman bands at 437, 369, 295,

220, 213, and 83 cm^{-1}. Detailed analysis of the 369 and 295 cm^{-1} bands show them to arise from totally symmetric modes. Show that the Raman spectrum is consistent with a trigonal-bipyramidal geometry.

6.9 How many vibrational modes does an SO_3 molecule have (a) in the plane of the nuclei, (b) perpendicular to the molecular plane?

6.10 What are the symmetry species of the vibrations of (a) SF_6, (b) BF_3 that are both IR and Raman active?

6.11 What are the symmetry species of the vibrational modes of a C_{6v} molecule that are neither IR nor Raman active?

6.12 The $[AuCl_4]^-$ ion has D_{4h} symmetry. Determine the representations Γ of all $3N$ displacements and reduce it to obtain the symmetry species of the irreducible representations.

6.13 How could IR and Raman spectroscopy be used to distinguish between: (a) planar and pyramidal forms of PF_3, (b) planar and 90°-twisted forms of B_2F_4 (D_{2h} and D_{2d}, respectively).

6.14 (a) Take the four hydrogen 1s orbitals of CH_4 and determine how they transform under T_d. (b) Confirm that it is possible to reduce this representation to $A_1 + T_2$. (c) With which atomic orbitals on C would it be possible to form MOs with H1s SALCs of symmetry $A_1 + T_2$?

6.15 Consider CH_4. Use the projection operator method to construct the SALCs of $A_1 + T_2$ symmetry that derive from the four H1s orbitals.

6.16 Use the projection operator method to determine the SALCs required for formation of σ bonds in (a) BF_3, (b) PF_5.

PROBLEMS

6.1 Consider a molecule IF_3O_2 (with I as the central atom). How many isomers are possible? Assign point group designations to each isomer.

6.2 (a) Determine the point group of the most symmetric planar conformation of $B(OH)_3$ and the most symmetric nonplanar

conformation of $B(OH)_3$. Assume that the B−O−H bond angles are 109.5° in all conformations. (b) Sketch a conformation of $B(OH)_3$ that is chiral, once again keeping all three B−O−H bond angles equal to 109.5°.

6.3 How many isomers are there for 'octahedral' molecules with the formula MA_3B_3, where A and B are monoatomic ligands? What is the point group of each isomer? Are any of the isomers chiral? Repeat this exercise for molecules with the formula $MA_2B_2C_2$.

6.4 Group theory is often used by chemists as an aid in the interpretation of infrared spectra. For example, there are four NH bonds in NH_4^+ and four stretching modes are possible: each one is a linear combination of the four stretching modes, and each one has a characteristic symmetry. There is the possibility that several vibrational modes occur at the same frequency, and hence are degenerate. A quick glance at the character table will tell if degeneracy is possible. (a) In the case of the tetrahedral NH_4^+ ion, is it necessary to consider the possibility of degeneracies? (b) Are degeneracies possible in any of the vibrational modes of $NH_2D_2^+$?

6.5 Determine whether the number of IR and Raman active stretching modes could be used to determine uniquely whether a sample of gas is BF_3, NF_3, or ClF_3.

6.6 Figure 6.28 shows the electronic energy levels of CH_3^+. What is the point group used for this illustration? Identify the following contributions:

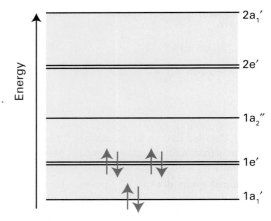

Figure 6.28

(a) H1s to a_1'

(b) C2s and C2p to a_1'

(c) H1s to e'

(d) C2s and C2p to e_1

(e) C2s and C2p to a_2''

(f) H1s to a_2''

Now add two H1s orbitals on the z-axis (above and below the plane), modify the linear combinations of each symmetry type accordingly, and construct a new A_2'' linear combination. Are there bonding and nonbonding (or only weakly antibonding orbitals) that can accommodate ten electrons and allow carbon to become hypervalent?

6.7 Consider the p orbitals on the four Cl atoms of tetrahedral $[CoCl_4]^-$, with one p orbital on each Cl pointing directly at the central metal atom. (a) Confirm that the four p orbitals which point at the metal transform in an identical manner to the four s orbitals on the Cl atoms. How might these p orbitals contribute to the bonding of the complex? (b) Take the remaining eight p orbitals and determine how they transform. Reduce the representation you derive to determine the symmetry of the SALCs these orbitals contribute to. Which metal orbitals can these SALCs bond with? (c) Generate the SALCs referred to in (b).

6.8 Consider all 12 of the p orbitals on the four Cl atoms of a square planar complex like $[PtCl_4]^{2-}$. (a) Determine how these p orbitals transform under D_{4h} and reduce the representation. (b) Which metal orbitals can these SALCs bond with? (c) Which SALCs and which metal orbitals contribute to σ bonds? (d) Which SALCs and which metal orbitals contribute to the in-plane π bonds? (e) Which SALCs and which metal orbitals contribute to the out of plane π bonds?

6.9 Take an octahedral complex and construct all the σ- and π-bonding SALCs.

An introduction to coordination compounds

7

Metal complexes, in which a single central metal atom or ion is surrounded by several ligands, play an important role in inorganic chemistry, especially for elements of the d block. In this chapter, we introduce the common structural arrangements for ligands around a central metal atom and the isomeric forms that are possible.

In the context of metal coordination chemistry, the term **complex** means a central metal atom or ion surrounded by a set of ligands. A **ligand** is an ion or molecule that can have an independent existence. Two examples of complexes are $[Co(NH_3)_6]^{3+}$, in which the Co^{3+} ion is surrounded by six NH_3 ligands, and $[Na(OH_2)_6]^+$, in which the Na^+ ion is surrounded by six H_2O ligands. We shall use the term **coordination compound** to mean a neutral complex or an ionic compound in which at least one of the ions is a complex. Thus, $Ni(CO)_4$ (**1**) and $[Co(NH_3)_6]Cl_3$ (**2**) are both coordination compounds. A complex is a combination of a Lewis acid (the central metal atom) with a number of Lewis bases (the ligands). The atom in the Lewis base ligand that forms the bond to the central atom is called the **donor atom** because it donates the electrons used in bond formation. Thus, N is the donor atom when NH_3 acts as a ligand, and O is the donor atom when H_2O acts as a ligand. The metal atom or ion, the Lewis acid in the complex, is the **acceptor atom**. All metals, from all blocks of the periodic table, form complexes.

The principal features of the geometrical structures of metal complexes were identified in the late nineteenth and early twentieth centuries by Alfred Werner, whose training was in organic stereochemistry. Werner combined the interpretation of optical and geometrical isomerism, patterns of reactions, and conductance data in work that remains a model of how to use physical and chemical evidence effectively and imaginatively. The striking colours of many d- and f-metal coordination compounds, which reflect their electronic structures, were a mystery to Werner. This characteristic was clarified only when the description of electronic structure in terms of orbitals was applied to the problem in the period from 1930 to 1960. We look at the electronic structure of d-metal complexes in Chapter 20 and f-metal complexes in Chapter 23.

The geometrical structures of metal complexes can now be determined in many more ways than Werner had at his disposal. When single crystals of a compound can be grown, X-ray diffraction (Section 8.1) gives precise shapes, bond distances, and angles. Nuclear magnetic resonance (Section 8.5) can be used to study complexes with lifetimes longer than microseconds. Very short-lived complexes, those with lifetimes comparable to diffusional encounters in solution (a few nanoseconds), can be studied by vibrational and electronic spectroscopy. It is possible to infer the geometries of complexes with long lifetimes in solution (such as the classic complexes of Co(III), Cr(III), and Pt(II) and many organometallic compounds) by analysing patterns of reactions and isomerism. This method was originally exploited by Werner, and it still teaches us much about the synthetic chemistry of the compounds as well as helping to establish their structures.

The language of coordination chemistry

Key points: In an inner-sphere complex, the ligands are attached directly to a central metal ion; outer-sphere complexes occur where cation and anion associate in solution.

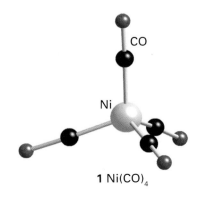

1 $Ni(CO)_4$

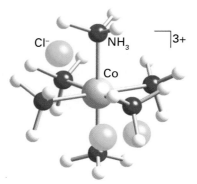

2 $[Co(NH_3)_6]Cl_3$

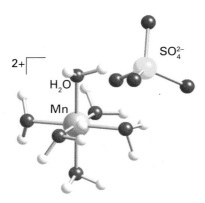

3 $[Mn(OH_2)_6]SO_4$

In what we normally understand as a complex, more precisely an **inner-sphere complex**, the ligands are attached directly to the central metal atom or ion. These ligands form the **primary coordination sphere** of the complex and their number is called the **coordination number** of the central metal atom. As in solids, a wide range of coordination numbers can occur, and the origin of the structural richness and chemical diversity of complexes is the ability of the coordination number to range up to 12.

Although we shall concentrate on inner-sphere complexes throughout this chapter, we should keep in mind that complex cations can associate electrostatically with anionic ligands (and, by other weak interactions, with solvent molecules) without displacement of the ligands already present. The product of this association is called an **outer-sphere complex**. With $[Mn(OH_2)_6]^{2+}$ and SO_4^{2-} ions, for instance, the equilibrium concentration of the outer-sphere complex $\{[Mn(OH_2)_6]^{2+}SO_4^{2-}\}$ (**3**) can, depending on the concentration, exceed that of the inner-sphere complex $[Mn(OH_2)_5SO_4]$ in which the ligand SO_4^{2-} is directly attached to the metal ion. It is worth remembering that most methods of measuring complex formation equilibria do not distinguish outer-sphere from inner-sphere complex formation but simply detect the sum of all bound ligands. Outer-sphere complexation should be suspected whenever the metal and ligands have opposite charges.

A large number of molecules and ions can behave as ligands, and a large number of metal ions form complexes. We now introduce some representative ligands and consider the basics of naming complexes.

7.1 Representative ligands

Key points: Polydentate ligands can form chelates; a bidentate ligand with a small bite angle can result in distortions from standard structures.

Table 7.1 gives the names and formulas of a number of common simple ligands and Table 7.2 gives the common prefixes used. Some of these ligands have only a single donor pair of electrons and will have only one point of attachment to the metal: such ligands are classified as **monodentate** (from the Latin meaning 'one-toothed'). Ligands that have more than one point of attachment are classified as **polydentate**. Ligands that specifically have two points of attachment are known as **bidentate**, those with three, **tridentate**, and so on.

Table 7.1 Typical ligands and their names

Name	Formula	Abbreviation	Donor atoms	Number of donors
Acetylacetonato	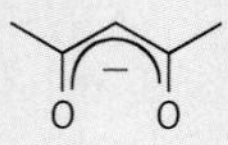	acac⁻	O	2
Ammine	NH_3		N	1
Aqua	H_2O		O	1
2,2-Bipyridine	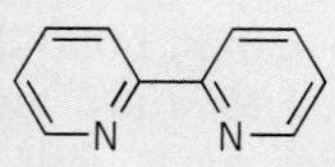	bpy	N	2
Bromido	Br^-		Br	1
Carbanato	CO_3^{2-}		O	1 or 2
Carbonyl	CO		C	1
Chlorido	Cl^-		Cl	1
1,4,7,10,13,16-Hexaoxacyclooctadecane	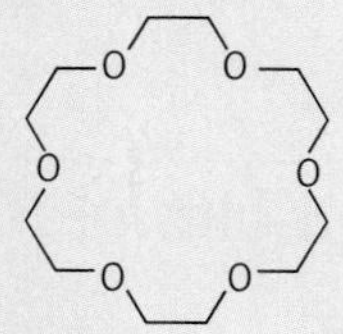	18-crown-6	O	6

Table 7.1 (*Continued*)

Name	Formula	Abbreviation	Donor atoms	Number of donors
4,7,13,16,21-Pentaoxa-1, 10-diaza-bicyclo [8.8.5]tricosane	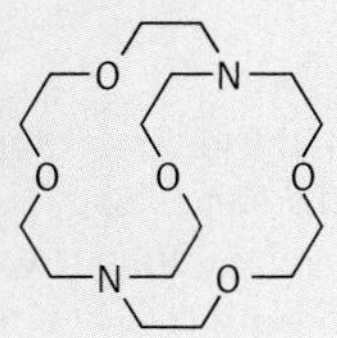	2.2.1 crypt	N, O	2N, 5O
Cyanido	CN^-		C	1
Diethylenetriamine	$NH(CH_2CH_2NH_2)_2$	dien	N	3
Bis(diphenylphosphino)ethane	 Ph_2P PPh_2	dppe	P	2
Bis(diphenylphosphino)methane	 Ph_2P PPh_2	dppm	P	2
Cyclopentadienyl	$C_5H_5^-$	Cp^-	C	5
Ethylenediamine (1,2-diaminoethane)	$NH_2CH_2CH_2NH_2$	en	N	2
Ethylenediaminetetraacetato	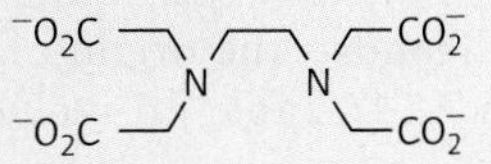	$edta^{4-}$	N, O	2N, 4O
Fluorido	F^-		F	1
Glycinato	$NH_2CH_2CO_2^-$	gly	N, O	1N, 1O
Hydrido	H^-		H	1
Hydroxido	OH^-		O	1
Iodido	I^-		I	1
Nitrato	NO_3^-		O	1 or 2
Nitrito$-\kappa O$	NO_2^-		O	1
Nitrito$-\kappa N$	NO_2^-		N	1
Oxido	O^{2-}		O	1
Oxalato	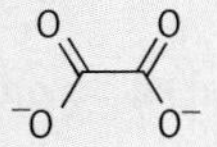	ox	O	2
Pyridine	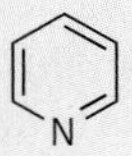	py	N	1
Sulfido	S^{2-}		S	1
Tetraazacyclotetradecane	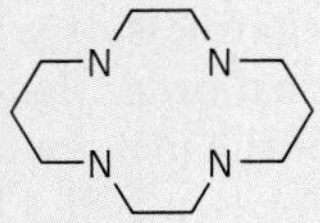	cyclam	N	4
Thiocyanato$-\kappa N$	NCS^-		N	1
Thiocyanato$-\kappa S$	SCN^-		S	1
Thiolato	RS^-		S	1
Triaminotriethylamine	$N(CH_2CH_2NH_2)_3$	tren	N	4
Tricyclohexylphosphine	$P(C_6H_{11})_3$	PCy_3	P	1
Trimethylphosphine	$P(CH_3)_3$	PMe_3	P	1
Triphenylphosphine	$P(C_6H_5)_3$	PPh_3	P	1

Table 7.2 Prefixes used for naming complexes

Prefix	Meaning
mono-	1
di-, bis-	2
tri-,tris-	3
tetra-, tetrakis-	4
penta-	5
hexa-	6
hepta-	7
octa-	8
nona-	9
deca-	10
undeca-	11
dodeca-	12

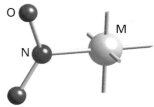

4 Nitrito-κ*N* ligand

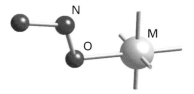

5 Nitrito-κ*O* ligand

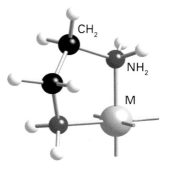

6 Ethylenediamine ligand (en)

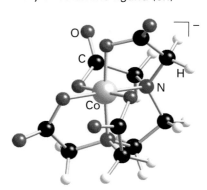

7 [Co(edta)]⁻

Ambidentate ligands have more than one different potential donor atom. An example is the thiocyanate ion (NCS^-), which can attach to a metal atom either by the N atom, to give thiocyanato-κ*N* complexes, or by the S atom, to give thiocyanato-κ*S* complexes. Another example of an ambidentate ligand is NO_2^-: as $M-NO_2^-$ (**4**) the ligand is nitrito-κ*N* and as $M-ONO^-$ (**5**) it is nitrito-κ*O*.

A note on good practice The 'κ terminology' in which the letter κ (kappa) is used to indicate the atom of ligation has only recently been introduced, and the old names isothiocyanato, indicating attachment by the N atom, or thiocyanato, indicating attachment by the S atom, are still widely encountered. Similarly, the old names nitro, indicating attachment by the N atom, or nitrito, indicating attachment by the O atom, are also still widely encountered.

Polydentate ligands can produce a **chelate** (from the Greek for 'claw'), a complex in which a ligand forms a ring that includes the metal atom. An example is the bidentate ligand ethylenediamine (1,2-diaminoethane, en, $NH_2CH_2CH_2NH_2$), which forms a five-membered ring when both N atoms attach to the same metal atom (**6**). It is important to note that normal chelating ligands will attach to the metal only at two adjacent coordination sites, in a *cis* fashion. The hexadentate ligand ethylene-diaminetetraacetic acid, as its anion ($edta^{4-}$), can attach at six points (at two N atoms and four O atoms) and can form an elaborate complex with five five-membered rings (**7**). This ligand is used to trap metal ions, such as Ca^{2+} ions, in 'hard' water. Complexes of chelating ligands often have additional stability over those of non-chelating ligands—the origin of this so called **chelate effect** is discussed later in this chapter (Section 7.14). Table 7.1 includes some of the most common chelating ligands.

In a chelate formed from a saturated organic ligand, such as ethylenediamine, a five-membered ring can fold into a conformation that preserves the tetrahedral angles within the ligand and yet still achieve an L−M−L angle of 90°, the angle typical of octahedral complexes. Six-membered rings may be favoured sterically or by electron delocalization through their π orbitals. The bidentate β-diketones, for example, coordinate as the anions of their enols in six-membered ring structures (**8**). An important example is the acetylacetonato anion ($acac^-$, **9**). Because biochemically important amino acids can form five- or six-membered rings, they also chelate readily. The degree of strain in a chelating ligand is often expressed in terms of the **bite angle**, the L−M−L angle in the chelate ring (**10**).

7.2 Nomenclature

Key points: The cation and anion of a complex are named according to a set of rules; cations are named first and ligands are named in alphabetical order.

Detailed guidance on nomenclature is beyond the scope of this book and only a general introduction is given here. In fact, the names of complexes often become so cumbersome that inorganic chemists often prefer to give the formula rather than spell out the entire name.

For compounds that consist of one or more ions, the cation is named first followed by the anion (as for simple ionic compounds), regardless of which ion is complex. Complex ions are named with their ligands in alphabetical order (ignoring any numerical prefixes). The ligand names are followed by the name of the metal with either its oxidation number in parentheses, as in hexaamminecobalt(III) for $[Co(NH_3)_6]^{3+}$, or with the overall charge on the complex specified in parentheses, as in hexaamminecobalt(3+). The suffix -ate is added to the name of the metal (sometimes in its Latin form) if the complex is an anion, as in the name hexacyanoferrate(II) for $[Fe(CN)_6]^{4-}$.

The number of a particular type of ligand in a complex is indicated by the prefixes mono-, di-, tri-, and tetra-. The same prefixes are used to state the number of metal atoms if more than one is present in a complex, as in octachloridodirhenate(III), $[Re_2Cl_8]^{2-}$ (**11**). Where confusion with the names of ligands is likely, perhaps because the name already includes a prefix, as with ethylenediamine, the alternative prefixes bis-, tris-, and tetrakis- are used, with the ligand name in parentheses. For example, dichlorido- is unambiguous but tris(ethylenediamine) shows more clearly that there are three ethylenediamine ligands, as in tris(ethylenediamine)cobalt(II), $[Co(en)_3]^{2+}$. Ligands that bridge two metal centres are denoted by a prefix μ (mu) added to the name of the relevant ligand, as in

H_3C ⌒ CH_3
O⟍M⟋O
8

H_3C ⌒ CH_3
O O
9

μ-oxido-bis(pentamminecobalt(III)) (**12**). If the number of centres bridged is greater than two, a subscript is used to indicate the number; for instance a hydride ligand bridging three metal atoms is denoted μ_3-H.

> **A note on good practice** The letter κ is also used to indicate the number of points of attachment: thus a bidentate ethylenediamine ligand bound through both N atoms is indicated as $\kappa^2 N$. The letter η (eta) is used to indicate bonding modes of certain organometallic ligands (Section 22.4).

Square brackets are used to indicate which groups are bound to a metal atom, and should be used whether the complex is charged or not; however, in casual usage, neutral complexes and oxoanions are often written without brackets, as in $Ni(CO)_4$ for tetracarbonylnickel(0)[1] and MnO_4^- for tetraoxidomanganate(VII) ('permanganate'). The metal symbol is given first, then the ligands in alphabetical order (the earlier rule that anionic ligands precede neutral ligands has been superseded), as in $[Co(Cl)_2(NH_3)_4]^+$ for tetraamminedichloridocobalt(III). This order is sometimes varied to clarify which ligand is involved in a reaction. Polyatomic ligand formulas are sometimes written in an unfamiliar sequence (as for OH_2 in $[Fe(OH_2)_6]^{2+}$ for hexaaquairon(II)) to place the donor atom adjacent to the metal atom and so help to make the structure of the complex clear. The donor atom of an ambidentate ligand is sometimes indicated by underlining it, for example $[Fe(OH_2)_5(\underline{N}CS)]^{2+}$. Note that, somewhat confusingly, the ligands in the formula are in alphabetical order of binding element, and thus the formula and name of the complex may differ in the order in which the ligands appear.

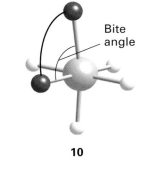

10

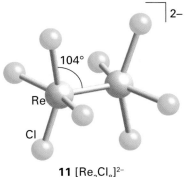

11 $[Re_2Cl_8]^{2-}$

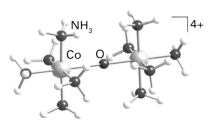

12 $[(H_3N)_5CoOCo(NH_3)_5]^{4+}$

> **EXAMPLE 7.1** Naming complexes
>
> Name the complexes (a) $[Pt(Cl)_2(NH_3)_4]^{2+}$; (b) $[Ni(CO)_3(py)]$; (c) $[Cr(edta)]^-$; (d) $[Co(Cl)_2(en)_2]^+$; (e) $[Rh(CO)_2I_2]^-$.
>
> **Answer** To name a complex, we start by working out the oxidation number of the central metal atom and then add the names of the ligands in alphabetical order. (a) The complex has two anionic ligands (Cl⁻), four neutral ligands (NH₃) and an overall charge of +2; hence the oxidation number of platinum must be +4. According to the alphabetical order rules, the name of the complex is tetraamminedichloridoplatinum(IV). (b) The ligands CO and py (pyridine) are neutral, so the oxidation number of nickel must be 0. It follows that the name of the complex is tricarbonylpyridinenickel(0). (c) This complex contains the hexadentate edta⁴⁻ ion as the sole ligand. The four negative charges of the ligand result in a complex with a single negative charge if the central metal ion is Cr³⁺. The complex is therefore ethylenediaminetetraacetatochromate(III). (d) This complex contains two anionic chloride ligands and two neutral en ligands. The overall charge of +1 must be the result of the cobalt having oxidation number +3. The complex is therefore dichloridobis(ethylenediamine)cobalt(III). (e) This complex contains two anionic I⁻ (iodido) ligands and two neutral CO ligands. The overall charge of −1 must be the result of the rhodium having oxidation number +1. The complex is therefore dicarbonyldiiododorhodate(I).
>
> **Self-test 7.1** Write the formulas of the following complexes: (a) diaquadichlorido-platinum(II); (b) diamminetetra(thiocyanato-κN)chromate(III); (c) tris(ethylenediamine)rhodium(III); (d) bromidopentacarbonylmanganese(I); (e) chloridotris(triphenylphosphine)rhodium(I).

Constitution and geometry

Key points: The number of ligands in a complex depends on the size of the metal atom, the identity of the ligands, and the electronic interactions.

The coordination number of a metal atom or ion is not always evident from the composition of the solid, as solvent molecules and species that are potentially ligands may simply fill spaces within the structure and not have any direct bonds to the metal ion. For example, X-ray diffraction shows that $CoCl_2\cdot6H_2O$ contains the neutral complex $[Co(Cl)_2(OH_2)_4]$ and two uncoordinated H_2O molecules occupying well-defined positions in the crystal. Such additional solvent molecules are called **solvent of crystallization**.

---

[1] When assigning oxidation numbers in carbonyl complexes, CO is ascribed a net oxidation number of 0.

Three factors govern the coordination number of a complex:

1. The size of the central atom or ion.

2. The steric interactions between the ligands.

3. Electronic interactions between the central atom or ion and the ligands.

In general, the large radii of atoms and ions lower down the periodic table favour higher coordination numbers. For similar steric reasons, bulky ligands often result in low coordination numbers, especially if the ligands are also charged (when unfavourable electrostatic interactions also come into play). High coordination numbers are also most common on the left of a period, where the ions have larger radii. They are especially common when the metal ion has only a few electrons because a small number of valence electrons means that the metal ion can accept more electrons from Lewis bases; one example is $[Mo(CN)_8]^{4-}$. Lower coordination numbers are found on the right of the d block, particularly if the ions are rich in electrons; an example is $[PtCl_4]^{2-}$. Such atoms are less able to accept electrons from any Lewis bases that are potential ligands. Low coordination numbers occur if the ligands can form multiple bonds with the central metal, as in MnO_4^- and CrO_4^{2-}, as now the electrons provided by each ligand tend to exclude the attachment of more ligands. We consider these coordination number preferences in more detail in Chapter 20.

7.3 Low coordination numbers

Key points: Two-coordinate complexes are found for Cu^+ and Ag^+; these complexes often accommodate more ligands if they are available. Complexes may have coordination numbers higher than their empirical formulas suggest.

The best known complexes of metals with coordination number 2 that are formed in solution under ordinary laboratory conditions are linear species of the Group 11 ions. Linear two-coordinate complexes with two identical symmetric ligands have $D_{\infty h}$ symmetry. The complex $[AgCl_2]^-$, which is responsible for the dissolution of solid silver chloride in aqueous solutions containing excess Cl^- ions, is one example, dimethyl mercury, $Me-Hg-Me$, is another. A series of linear Au(I) complexes of formula LAuX, where X is a halogen and L is a neutral Lewis base such as a substituted phosphine, R_3P, or thioether, R_2S, are also known. Two-coordinate complexes often gain additional ligands to form three- or four-coordinate complexes.

A formula that suggests a certain coordination number in a solid compound might conceal a polymeric chain with a higher coordination number. For example, CuCN appears to have coordination number 1, but it in fact exists as linear $-Cu-CN-Cu-CN-$ chains in which the coordination number of copper is 2.

Three-coordination is rare among metal complexes, but is found with bulky ligands such as tricyclohexylphosphine, as in $[Pt(PCy_3)_3]$ (**13**), where Cy denotes cyclohexyl ($-C_6H_{11}$), with its trigonal arrangement of the ligands. MX_3 compounds, where X is a halogen, are usually chains or networks with a higher coordination number and shared ligands. Three-coordinate complexes with three identical symmetric ligands normally have D_{3h} symmetry.

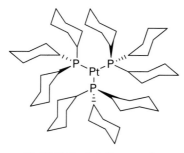

13 $[Pt(PCy_3)_3]$, Cy= *cyclo-* C_6H_{11}

7.4 Intermediate coordination numbers

Complexes of metal ions with the intermediate coordination numbers four, five, and six are the most important class of complex. They include the vast majority of complexes that exist in solution and almost all the biologically important complexes.

(a) Four-coordination

Key points: Tetrahedral complexes are favoured over higher coordinate complexes if the central atom is small or the ligands large; square-planar complexes are typically observed for metals with d^8 configurations.

Four-coordination is found in an enormous number of compounds. Tetrahedral complexes of approximately T_d symmetry (**14**) are favoured over higher coordination numbers when

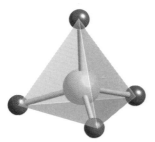

14 Tetrahedral complex (T_d)

the central atom is small and the ligands are large (such as Cl^-, Br^-, and I^-) because then ligand–ligand repulsions override the energy advantage of forming more metal–ligand bonds. Four-coordinate s- and p-block complexes with no lone pair on the central atom, such as $[BeCl_4]^{2-}$, $[BF_4]^-$, and $[SnCl_4]$, are almost always tetrahedral, and tetrahedral complexes are common for oxoanions of metal atoms on the left of the d block in high oxidation states, such as $[MoO_4]^{2-}$. Examples of tetrahedral complexes from Groups 5–11 are: $[VO_4]^{3-}$, $[CrO_4]^{2-}$, $[MnO_4]^-$, $[FeCl_4]^{2-}$, $[CoCl_4]^{2-}$, $[NiBr_4]^{2-}$, and $[CuBr_4]^{2-}$.

Another type of four-coordinate complex is also found: those where the four ligands surround the central metal in a square-planar arrangement (**15**). Complexes of this type were originally identified because they can lead to different isomers when the complex has the formula MX_2L_2. We discuss this isomerism in Section 7.7. Square-planar complexes with four identical symmetric ligands have D_{4h} symmetry.

Square-planar complexes are rarely found for s- and p-block complexes but are abundant for d^8 complexes of the elements belonging to the 4d- and 5d-series metals such as Rh^+, Ir^+, Pd^{2+}, Pt^{2+}, and Au^{3+}, which are almost invariably square planar. For 3d-metals with d^8 configurations (for example, Ni^{2+}), square-planar geometry is favoured by ligands that can form π bonds by accepting electrons from the metal atom, as in $[Ni(CN)_4]^{2-}$. Examples of square-planar complexes from Groups 9, 10, and 11 are $[RhCl(PPh_3)_3]$, *trans*-$[Ir(CO)Cl(PMe_3)_2]$, $[Ni(CN)_4]^{2-}$, $[PdCl_4]^{2-}$, $[Pt(NH_3)_4]^{2+}$, and $[AuCl_4]^-$. Square-planar geometry can also be forced on a central atom by complexation with a ligand that contains a rigid ring of four donor atoms, much as in the formation of a porphyrin complex (**16**). Section 20.1f gives a detailed explanation of the factors that help to stabilize square-planar complexes.

(b) Five-coordination

Key points: In the absence of polydentate ligands that enforce the geometry, the energies of the various geometries of five-coordinate complexes differ little from one another and such complexes are often fluxional.

Five-coordinate complexes, which are less common than four- or six-coordinate complexes, are normally either square pyramidal or trigonal bipyramidal. A square-pyramidal complex would have C_{4v} symmetry if all the ligands were identical and the trigonal-bipyramidal complex would have D_{3h} symmetry with identical ligands. Distortions from these ideal geometries are common, and structures are known at all points between the two ideal geometries. A trigonal-bipyramidal shape minimizes ligand–ligand repulsions, but steric constraints on ligands that can bond through more than one site to a metal atom can favour a square-pyramidal structure. For instance, square-pyramidal five-coordination is found among the biologically important porphyrins, where the ligand ring enforces a square-planar structure and a fifth ligand attaches above the plane. Structure (**17**) shows part of the active centre of myoglobin, the oxygen transport protein; the location of the Fe atom above the plane of the ring is important to its function (Section 27.7). In some cases, five-coordination is induced by a polydentate ligand containing a donor atom that can bind to an axial location of a trigonal bipyramid, with its remaining donor atoms reaching down to the three equatorial positions (**18**). Ligands that force a trigonal-bipyramidal structure in this fashion are called **tripodal**.

(c) Six-coordination

Key point: The overwhelming majority of six-coordinate complexes are octahedral or have shapes that are small distortions of octahedral.

Six-coordination is the most common arrangement for metal complexes and is found in s-, p-, d-, and f-metal coordination compounds. Almost all six-coordinate complexes are octahedral (**19**), at least if we consider the ligands as represented by structureless points. A regular octahedral (O_h) arrangement of ligands is highly symmetric (Fig. 7.1). It is especially important, not only because it is found for many complexes of formula ML_6 but also because it is the starting point for discussions of complexes of lower symmetry, such as those shown in Fig. 7.2. The simplest deviation from O_h symmetry is tetragonal (D_{4h}), and occurs when two ligands along one axis differ from the other four; these two ligands, which are *trans* to each other, might be closer in than the other four or, more commonly, further away. For the d^9 configuration (particularly for Cu^{2+} complexes), a tetragonal distortion may occur even when all ligands are identical because of an inherent effect known

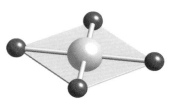

15 Square-planar complex (D_{4h})

16

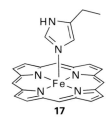

17

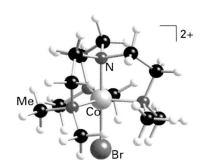

18 $[CoBrN(CH_2CH_2NMe_2)_3]^{2+}$

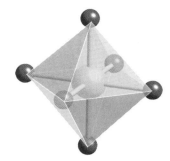

19 Octahedral complex (O_h)

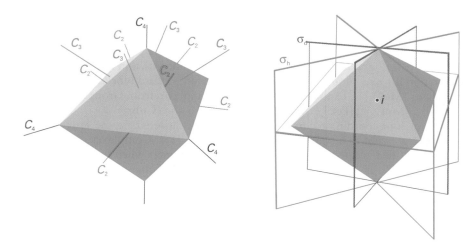

Figure 7.1 The highly symmetric octahedral arrangement of six ligands around a central metal atom and the corresponding symmetry elements of an octahedron. Note that not all the σ_d are shown.

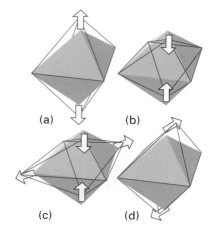

Figure 7.2 Distortions of a regular octahedron: (a) and (b) tetragonal distortions, (c) rhombic distortion, (d) trigonal distortion.

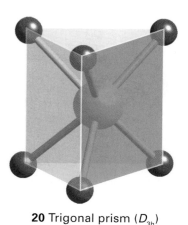

20 Trigonal prism (D_{3h})

Figure 7.3 A chelating ligand that permits only a small bite angle can distort an octahedral complex into trigonal-prismatic geometry.

as the Jahn−Teller distortion (Section 20.1g). Rhombic (D_{2h}) distortions, in which a *trans* pair of ligands are close in and another *trans* pair are further out, can occur. Trigonal (D_{3d}) distortions occur when two opposite faces of the octahedron move away and give rise to a large family of structures that are intermediate between regular octahedral and trigonal prismatic (**20**); such structures are sometimes referred to as rhombohedral.

Trigonal-prismatic (D_{3h}) complexes are rare, but have been found in solid MoS_2 and WS_2; the trigonal prism is also the shape of several complexes of formula $[M(S_2C_2R_2)_3]$ (**21**). Trigonal-prismatic d^0 complexes such as $[Zr(CH_3)_6]^{2-}$ have also been isolated. Such structures require either very small σ-**donor ligands**, ligands that bind by forming a σ bond to the central atom, or favourable ligand−ligand interactions that can constrain the complex into a trigonal prism; such ligand−ligand interactions are often provided by ligands that contain sulfur atoms, which can form long, weak covalent bonds to each other. A chelating ligand that permits only a small bite angle can cause distortion from octahedral towards trigonal-prismatic geometry in six-coordinate complexes (Fig. 7.3).

7.5 Higher coordination numbers

Key points: Larger atoms and ions, particularly those of the f block, tend to form complexes with high coordination numbers; nine-coordination is particularly important in the f block.

Seven-coordination is encountered for a few 3d complexes and many more 4d and 5d complexes, where the larger central atom can accommodate more than six ligands. Seven-coordination resembles five-coordination in the similarity in energy of its various geometries. These limiting 'ideal' geometries include the pentagonal bipyramid (**22**), a capped octahedron (**23**), and a capped trigonal prism (**24**); in each of the latter two, the seventh capping ligand occupies one face. There are a number of intermediate structures, and interconversion between them is often rapid at room temperature. Examples include $[Mo(CNR)_7]^{2+}$, $[ZrF_7]^{3-}$, $[TaCl_4(PR_3)_3]$, and $[ReCl_6O]^{2-}$ from the d block and $[UO_2(OH_2)_5]^{2+}$ from the f block. A method to force seven- rather than six-coordination on the lighter elements is to synthesize a ring of five donor atoms (**25**) that then occupy the equatorial positions, leaving the axial positions free to accommodate two more ligands.

Stereochemical non-rigidity is also shown in eight-coordination; such complexes may be square antiprismatic (**26**) in one crystal but dodecahedral (**27**) in another. Two examples of

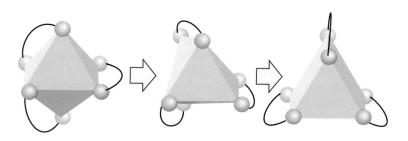

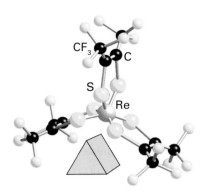

21 [Re(S(CF$_3$)C=C(CF$_3$)S)$_3$]

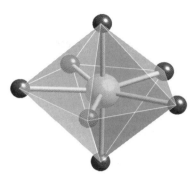

22 Pentagonal bipyramid (D_{5h})

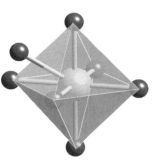

23 Capped octahedron

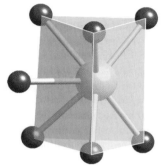

24 Capped trigonal prism

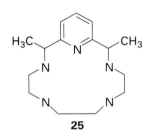

25

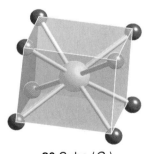

26 Square antiprism (D_{4d})

complexes with these geometries are shown as (**28**) and (**29**), respectively. Cubic geometry (**30**) is rare.

Nine-coordination is important in the structures of f-block elements; their relatively large ions can act as host to a large number of ligands. A simple example of a nine-coordinate lanthanoid complex is [Nd(OH$_2$)$_9$]$^{3+}$. More complex examples arise with the MCl$_3$ solids, with M ranging from La to Gd, where a coordination number of nine is achieved through metal−halide−metal bridges (Section 23.5). An example of nine-coordination in the d block is [ReH$_9$]$^{2-}$ (**31**), which has small enough ligands for this coordination number to be feasible; the geometry can be thought of as a tricapped trigonal prism.

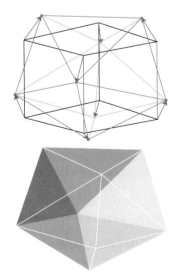

27 Dodecahedron; triangulated decahedron (D_{2d})

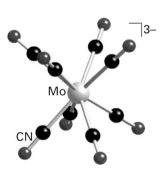

28 [Mo(CN)$_8$]$^{3-}$ (D_4)

29 [Zr(ox)$_4$]$^{4-}$

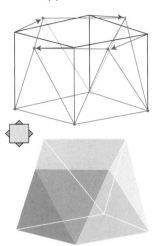

30 Cube (O_h)

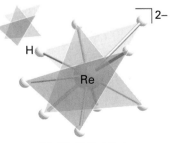

31 [ReH$_9$]$^{2-}$ (D_{3h})

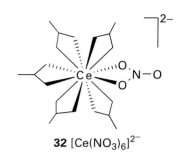

32 $[Ce(NO_3)_6]^{2-}$

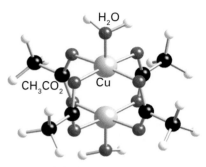

33 $[(H_2O)Cu(\mu\text{-}CH_3CO_2)_4Cu(OH_2)]$

34 $[Fe_4S_4(SR)_4]^{2-}$

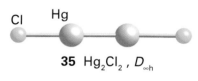

35 Hg_2Cl_2 , $D_{\infty h}$

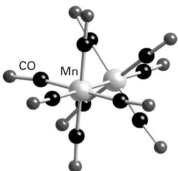

36 $[(CO)_5MnMn(CO)_5]$

Ten- and twelve-coordination are encountered in complexes of the f-block M^{3+} ions. Examples include $[Ce(NO_3)_6]^{2-}$ (**32**), which is formed in the reaction of Ce(IV) salts with nitric acid. Each NO_3^- ligand is bonded to the metal atom by two O atoms. An example of a ten-coordinate complex is $[Th(ox)_4(OH_2)_2]^{4-}$, in which each oxalate ion ligand (ox, $C_2O_4^{2-}$) provides two O donor atoms. These high coordination numbers are rare with s-, p-, and d- block ions.

7.6 Polymetallic complexes

Key point: Polymetallic complexes are classified as metal clusters if they contain M—M bonds or as cage complexes if they contain ligand-bridged metal atoms.

Polymetallic complexes are complexes that contain more than one metal atom. In some cases, the metal atoms are held together by bridging ligands; in others there are direct metal—metal bonds; in yet others there are both types of link. The term **metal cluster** is usually reserved for polymetallic complexes in which there are direct metal—metal bonds that form triangular or larger closed structures. This rigorous definition, however, would exclude linear M—M compounds, and is normally relaxed. We shall consider any M—M bonded system to be a cluster. When no metal—metal bond is present, polymetallic complexes are referred to as **cage complexes** (or 'cage compounds').[2] Polymetallic complexes are known for metals in every coordination number and geometry.

Cage complexes may be formed with a wide variety of anionic ligands. For example, two Cu^{2+} ions can be held together with acetate-ion bridges (**33**). Structure (**34**) is an example of a cubic structure formed from four Fe atoms bridged by RS^- ligands. This type of structure is of great biological importance as it is involved in a number of biochemical redox reactions (Section 27.8). With the advent of modern structural techniques, such as automated X-ray diffractometers and multinuclear NMR, many polymetallic clusters containing metal—metal bonds have been discovered and have given rise to an active area of research. A simple example is the mercury(I) cation Hg_2^{2+}, and complexes derived from it, such as $[Hg_2(Cl)_2]$ (**35**), which is commonly written simply Hg_2Cl_2. A metal cluster containing nine CO ligands and two Mn atoms is shown as (**36**).

Isomerism and chirality

Key points: A molecular formula may not be sufficient to identify a coordination compound: linkage, ionization, hydrate, and coordination isomerism are all possible for coordination compounds.

A molecular formula often does not give enough information to identify a compound unambiguously. We have already noted that the existence of ambidentate ligands gives rise to the possibility of **linkage isomerism**, in which the same ligand may link through different atoms. This type of isomerism accounts for the red and yellow isomers of the formula $[Co(NH_3)_5(NO_2)]^{2+}$. The red compound has a nitrito-κO Co—O link (5); the yellow isomer, which forms from the unstable red form over time, has a nitro nitrito-κN Co—N link (4). We will consider three further types of isomerism briefly before looking at geometric and optical isomerism in more depth. **Ionization isomerism** occurs when a ligand and a counterion in one compound exchange places. An example is $[PtCl_2(NH_3)_4]Br_2$ and $[PtBr_2(NH_3)_4]Cl_2$. If the compounds are soluble, the two isomers exist as different ionic species in solution (in this example, with free Br^- and Cl^- ions, respectively). Very similar to ionization isomerism is **hydrate isomerism**, which arises when one of the ligands is water, for example there are three differently coloured hydration isomers of a compound with molecular formula $CrCl_3 \cdot 6H_2O$: the violet $[Cr(OH_2)_6]Cl_3$, the pale green $[CrCl(OH_2)_5]Cl_2 \cdot H_2O$, and the dark green $[CrCl_2(OH_2)_4]Cl \cdot 2H_2O$. **Coordination isomerism** arises when there are different complex ions that can form from the same molecular formula, as in $[Co(NH_3)_6][Cr(CN)_6]$ and $[Cr(NH_3)_6][Co(CN)_6]$.

---

[2] The term 'cage compound' has a variety of meanings in inorganic chemistry and it is important to keep them distinct. For example, another use of the term is as a synonym of a clathrate compound (an inclusion compound, in which a species is trapped in a cage formed by molecules of another species).

Once we have established which ligands bind to which metals, and through which donor atoms, we can consider how to arrange these ligands in space. The three-dimensional character of metal complexes can result in a multitude of possible arrangements of the ligands. We now explore these varieties of isomerism by considering the permutations of ligand arrangement for each of the common complex geometries: this type of isomerism is known as **geometric isomerism**.

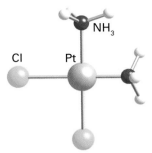

37 *cis*-[PtCl$_2$(NH$_3$)$_2$]

EXAMPLE 7.2 **Isomerism in metal complexes**

What types of isomerism are possible for complexes with the following molecular formulas: (a) [Pt(PEt$_3$)$_3$SCN]$^+$, (b) CoBr(NH$_3$)$_5$SO$_4$, (c) FeCl$_2$.6H$_2$O?

Answer (a) The complex contains the ambidentate thiocyanate ligand, SCN$^-$, which can bind through either the S or the N atom to give rise to two linkage isomers: [Pt(SCN)(PEt$_3$)$_3$]$^+$ and [Pt(NCS)(PEt$_3$)$_3$]$^+$. (b) With an octahedral geometry and five coordinated ammonia ligands, it is possible to have two ionization isomers: [Co(NH$_3$)$_5$SO$_4$]Br and [CoBr(NH$_3$)$_5$]SO$_4$. (c) Hydrate isomerism occurs as complexes of formula [Fe(OH$_2$)$_6$]Cl$_2$, [FeCl(OH$_2$)$_5$]Cl.H$_2$O, and [FeCl$_2$(OH$_2$)$_4$].2H$_2$O are possible.

Self-test 7.2 Two types of isomerism are possible for the six-coordinate complex Cr(NO$_2$)$_2$.6H$_2$O. Identify all isomers.

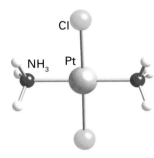

38 *trans*-[PtCl$_2$(NH$_3$)$_2$]

7.7 Square-planar complexes

Key point: The only simple isomers of square-planar complexes are *cis*−*trans* isomers.

Werner studied a series of four-coordinate Pt(II) complexes formed by the reactions of PtCl$_2$ with NH$_3$ and HCl. For a complex of formula MX$_2$L$_2$, only one isomer is expected if the species is tetrahedral, but two isomers are expected if the species is square planar, (**37**) and (**38**). Because Werner was able to isolate two nonelectrolytes of formula [PtCl$_2$(NH$_3$)$_2$], he concluded that they could not be tetrahedral and were, in fact, square planar. The complex with like ligands on adjacent corners of the square is called a *cis* isomer (**37**, point group C$_{2v}$) and the complex with like ligands opposite is the *trans* isomer (**38**, D$_{2h}$). Geometric isomerism is far from being of only academic interest: platinum complexes are used in cancer chemotherapy, and it is found that only *cis*-Pt(II) complexes can bind to the bases of DNA for long enough to be effective.

In the simple case of two sets of two different monodentate ligands, as in [MA$_2$B$_2$], there is only the case of *cis/trans* isomerism to consider, (**39**) and (**40**). With three different ligands, as in [MA$_2$BC], the locations of the two A ligands also allow us to distinguish the geometric isomers as *cis* and *trans*, (**41**) and (**42**). When there are four different ligands, as in [MABCD], there are three different isomers and we have to specify the geometry more explicitly, as in (**43**), (**44**), and (**45**). Bidentate ligands with different endgroups, as in [M(AB)$_2$], can also give rise to geometrical isomers that can be classified as *cis* (**46**) and *trans* (**47**).

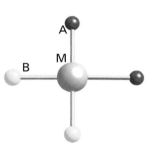

39 *cis*-[MA$_2$B$_2$]

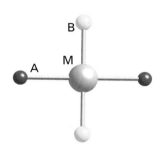

40 *trans*-[MA$_2$B$_2$]

EXAMPLE 7.3 **Identifying isomers from chemical evidence**

Use the reactions indicated in Fig. 7.4 to show how the *cis* and *trans* geometries of a pair of platinum complexes may be assigned.

Answer The *cis* diamminedichlorido isomer reacts with Ag$_2$O to lose Cl$^-$, and the product adds one oxalato dianion (C$_2$O$_4^{2-}$) at adjacent positions. The *trans* isomer loses Cl$^-$, but the product cannot displace the two OH$^-$ ligands with only one C$_2$O$_4^{2-}$ anion. A reasonable explanation is that the C$_2$O$_4^{2-}$ anion cannot reach across the square plane to bridge two *trans* positions. This conclusion is supported by X-ray crystallography.

Self-test 7.3 The two square-planar isomers of [PtBrCl(PR$_3$)$_2$] (where PR$_3$ is a trialkylphosphine) have different ^{31}P-NMR spectra (Fig. 7.5). For the sake of this exercise, we ignore coupling to ^{195}Pt ($I = \frac{1}{2}$ at 33 per cent abundance). One isomer (A) shows a single ^{31}P resonance; the other (B) shows two ^{31}P resonances, each of which is split into a doublet by the second ^{31}P nucleus. Which isomer is *cis* and which is *trans*?

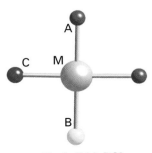

41 *cis*-[MA$_2$BC]

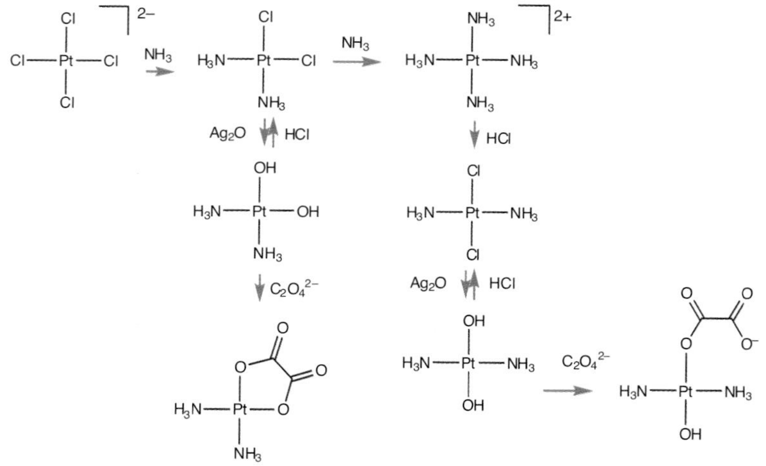

Figure 7.4 The preparation of *cis-* and *trans*-diamminedichloridoplatinum(II) and a chemical method for distinguishing the isomers.

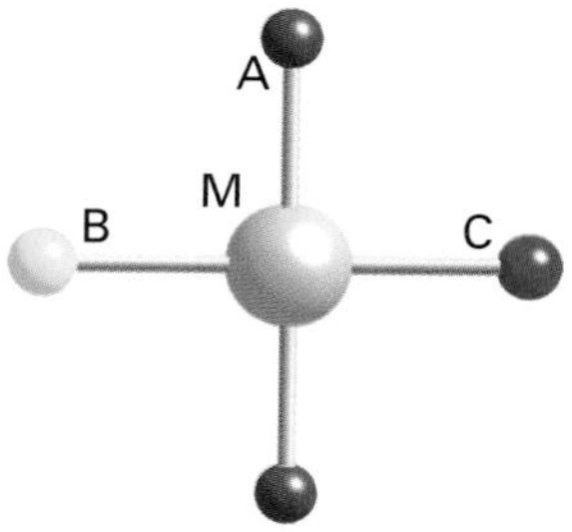

42 *trans*-[MA₂BC]

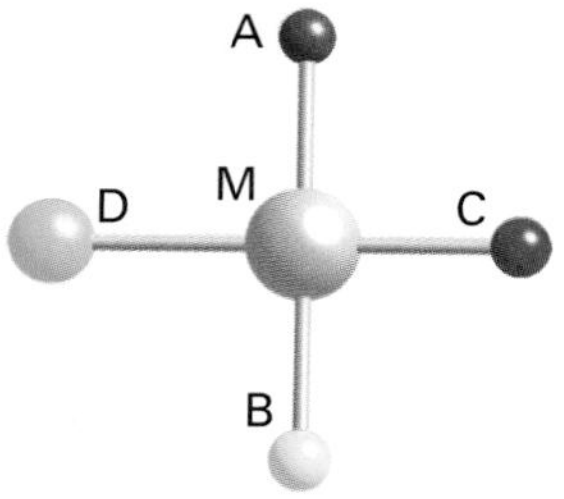

43 [MABCD], A *trans* to B

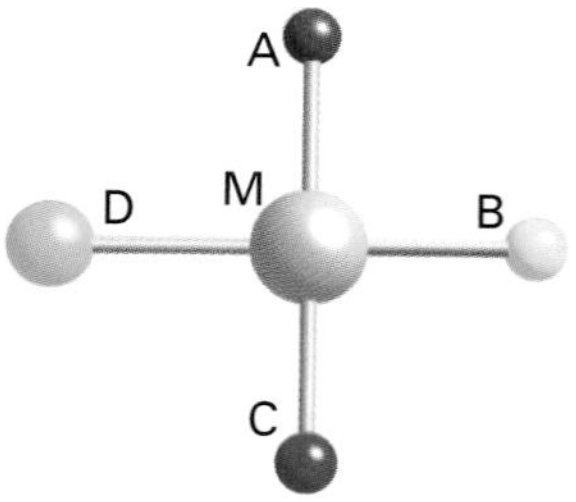

44 [MABCD] A *trans* to C

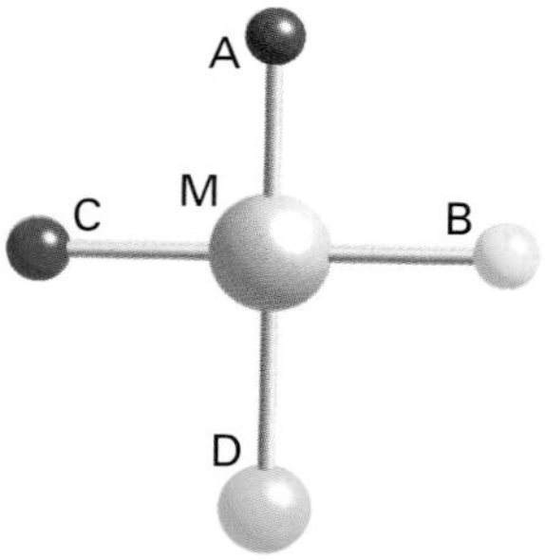

45 [MABCD] A *trans* to D

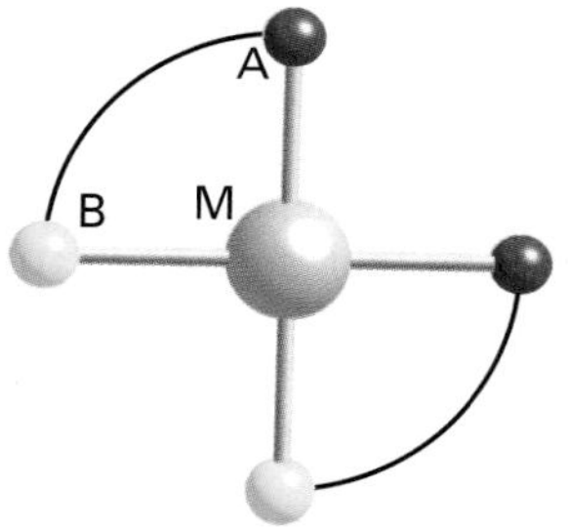

46 *cis*-[M(AB)₂]

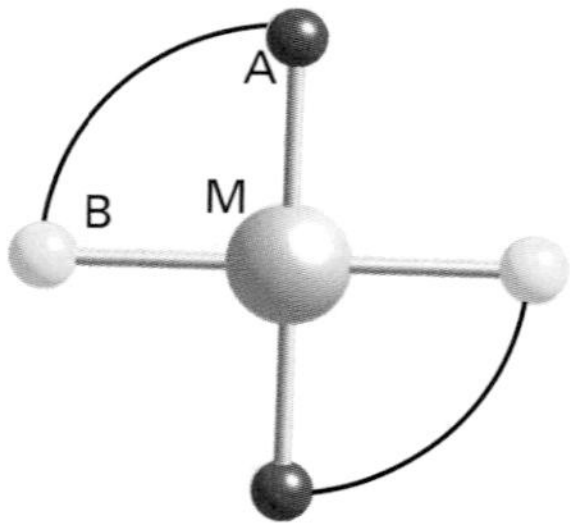

47 *trans*-[M(AB)₂]

7.8 Tetrahedral complexes

Key point: The only simple isomers of tetrahedral complexes are optical isomers.

The only isomers of tetrahedral complexes normally encountered are those where either all four ligands are different or where there are two unsymmetrical bidentate chelating ligands. In both cases, (**48**) and (**49**), the molecules are **chiral**, not superimposable on their mirror image (Section 6.4). Two mirror-image isomers jointly make up an **enantiomeric pair**. The existence of a pair of chiral complexes that are each other's mirror image (like a right hand and a left hand), and that have lifetimes that are long enough for them to be separable, is called **optical isomerism**. Optical isomers are so-called because they are **optically active**, in the sense that one enantiomer rotates the plane of polarized light in one direction and the other rotates it through an equal angle in the opposite direction.

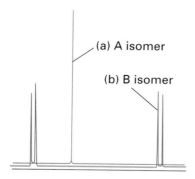

(a) A isomer

(b) B isomer

Figure 7.5 Idealized ³¹P-NMR spectra of two isomers of [PtBrCl(PR₃)₂]. The fine structure due to Pt is not shown.

7.9 Trigonal-bipyramidal and square-pyramidal complexes

Key points: Five-coordinate complexes are not stereochemically rigid; two chemically distinct coordination sites exist within both trigonal-bipyramidal and square-pyramidal complexes.

The energies of the various geometries of five-coordinate complexes often differ little from one another. The delicacy of this balance is underlined by the fact that $[Ni(CN)_5]^{3-}$ can exist as both square-pyramidal (**50**) and trigonal-bipyramidal (**51**) conformations in the same crystal. In solution, trigonal-bipyramidal complexes with monodentate ligands are often highly fluxional (that is, able to twist into different shapes), so a ligand that is axial at one moment becomes equatorial at the next moment: the conversion from one stereochemistry to another may occur by a **Berry pseudorotation** (Fig. 7.6). Thus, although isomers of five-coordinate complexes do exist, they are commonly not separable. It is important to be aware that both trigonal-bipyramidal and square-pyramidal complexes have two chemically distinct sites: axial (a) and equatorial (e) for the trigonal bipyramid (**52**) and axial (a) and basal (b) for the square pyramid (**53**). Certain ligands will have preferences for the different sites because of their steric and electronic requirements, but we do not go into detail here.

7.10 Octahedral complexes

There are huge numbers of complexes with nominally octahedral geometry, where in this context the nominal structure 'ML_6' is taken to mean a central metal atom surrounded by six ligands, not all of which are necessarily the same.

(a) Geometrical isomerism

Key points: *Cis* and *trans* isomers exist for octahedral complexes of formula $[MA_4B_2]$, and *mer* and *fac* isomers are possible for complexes of formula $[MA_3B_3]$. More complicated ligand sets lead to further isomers.

Whereas there is only one way of arranging the ligands in octahedral complexes of general formula $[MA_6]$ or $[MA_5B]$, the two B ligands of an $[MA_4B_2]$ complex may be placed on adjacent octahedral positions to give a *cis* isomer (**54**) or on diametrically opposite positions to give a *trans* isomer (**55**). Provided we treat the ligands as structureless points, the *trans* isomer has D_{4h} symmetry and the *cis* isomer has C_{2v} symmetry.

There are two ways of arranging the ligands in $[MA_3B_3]$ complexes. In one isomer, three A ligands lie in one plane and three B ligands lie in a perpendicular plane (**56**). This complex is designated the *mer* isomer (for meridional) because each set of ligands can be regarded as lying on a meridian of a sphere. In the second isomer, all three A (and B) ligands are adjacent and occupy the corners of one triangular face of the octahedron (**57**); this complex is designated the *fac* isomer (for facial). Provided we treat the ligands as structureless points, the *mer* isomer has C_{2v} symmetry and the *fac* isomer has C_{3v} symmetry.

For a complex of composition $[MA_2B_2C_2]$, there are five different geometrical isomers: an all-*trans* isomer (**58**); three different isomers where one pair of ligands are *trans* with the other two *cis*, as in (**59**), (**60**), and (**61**); and an enantiomeric pair of all-*cis* isomers (**62**). More complicated compositions, such as $[MA_2B_2CD]$ or $[MA_3B_2C]$, result in more extensive geometrical isomerism. For instance the rhodium compound $[RhH(C{\equiv}CR)_2(PMe_3)_3]$ exists as three different isomers: *fac* (**63**), *mer-trans* (**64**), and *mer-cis* (**65**). Although octahedral complexes are normally stereochemically rigid, isomerization reactions do sometimes occur (Section 21.9).

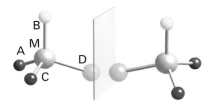

48 [MABCD] enantiomers

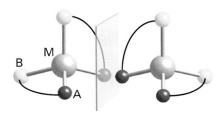

49 [M(AB)$_2$] enantiomers

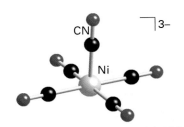

50 [Ni(CN)$_5$]$^{3-}$, square pyramidal (C_{4v})

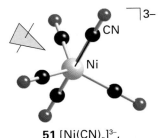

51 [Ni(CN)$_5$]$^{3-}$,
trigonal bipyramidal (D_{3h})

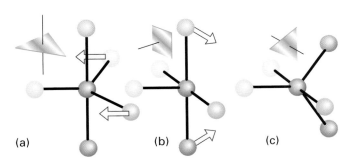

(a)   (b)   (c)

Figure 7.6 A Berry pseudorotation in which (a) a trigonal-bipyramidal Fe(CO)$_5$ complex distorts into (b) a square-pyramidal isomer and then (c) becomes trigonal bipyramidal again, but with the two initially axial ligands now equatorial.

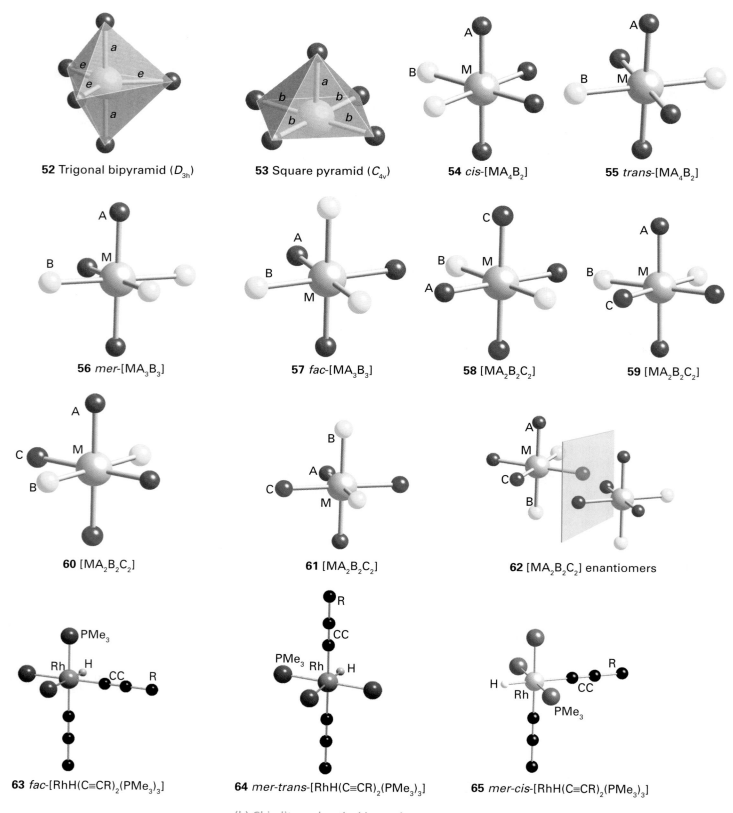

52 Trigonal bipyramid (D_{3h})

53 Square pyramid (C_{4v})

54 *cis*-[MA$_4$B$_2$]

55 *trans*-[MA$_4$B$_2$]

56 *mer*-[MA$_3$B$_3$]

57 *fac*-[MA$_3$B$_3$]

58 [MA$_2$B$_2$C$_2$]

59 [MA$_2$B$_2$C$_2$]

60 [MA$_2$B$_2$C$_2$]

61 [MA$_2$B$_2$C$_2$]

62 [MA$_2$B$_2$C$_2$] enantiomers

63 *fac*-[RhH(C≡CR)$_2$(PMe$_3$)$_3$]

64 *mer-trans*-[RhH(C≡CR)$_2$(PMe$_3$)$_3$]

65 *mer-cis*-[RhH(C≡CR)$_2$(PMe$_3$)$_3$]

(b) Chirality and optical isomerism

Key points: A number of ligand arrangements at an octahedral centre give rise to chiral compounds; isomers are designated Δ or Λ depending on their configuration.

In addition to the many examples of geometrical isomerism shown by octahedral compounds, many are also chiral. A very simple example is [Mn(acac)$_3$] (**66**), where three bidentate

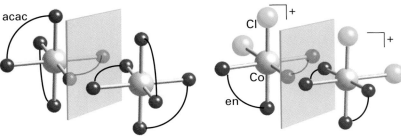

66 [Mn(acac)$_3$] enantiomers   **67** *cis*-[CoCl$_2$(en)$_2$]$^+$ enantiomers

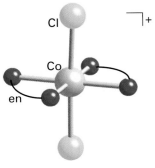

68 *trans*-[CoCl$_2$(en)$_2$]$^+$

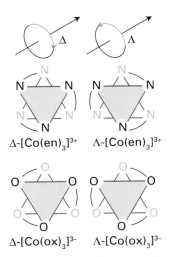

Figure 7.7 Absolute configurations of M(L−L)$_3$ complexes. Δ is used to indicate clockwise rotation of the helix and Λ to indicate anticlockwise rotation.

acetylacetonato (acac) ligands result in the existence of enantiomers. One way of looking at the optical isomers that arise in complexes of this nature is to imagine looking down one of the threefold axes and see the ligand arrangement as a propellor or screw thread.

Chirality can also exist for complexes of formula [MA$_2$B$_2$C$_2$] when the ligands of each pair are *cis* to each other (**62**). In fact, many examples of optical isomerism are known for octahedral complexes with both monodentate and polydentate ligands, and we must always be alert to the possibility of optical isomerism.

As a further example of optical isomerism, consider the products of the reaction of cobalt(III) chloride and ethylenediamine in a 1:2 mole ratio. The product includes a pair of dichlorido complexes, one of which is violet (**67**) and the other green (**68**); they are, respectively, the *cis* and *trans* isomers of dichloridobis(ethylenediamine)cobalt(III), [CoCl$_2$(en)$_2$]$^+$. As can be seen from their structures, the *cis* isomer cannot be superimposed on its mirror image. It is therefore chiral and hence (because the complexes are long-lived) optically active. The *trans* isomer has a mirror plane and can be superimposed on its mirror image; it is achiral and optically inactive.

The absolute configuration of a chiral octahedral complex is described by imagining a view along a threefold rotation axis of the regular octahedron and noting the handedness of the helix formed by the ligands (Fig. 7.7). Clockwise rotation of the helix is then designated Δ (delta) whereas the anticlockwise rotation is designated Λ (lambda). The designation of the absolute configuration must be distinguished from the experimentally determined direction in which an isomer rotates polarized light: some Λ compounds rotate in one direction, others rotate in the opposite direction, and the direction may change with wavelength. The isomer that rotates the plane of polarization clockwise (when viewed into the oncoming beam) at a specified wavelength is designated the *d*-isomer, or the (+)-isomer; the one rotating the plane anticlockwise is designated the *l*-isomer, or the (−)-isomer. Box 7.1 describes how the specific isomers of a complex might be synthesized and Box 7.2 describes how enantiomers of metal complexes may be separated.

Complexes with coordination numbers of greater than six have the potential for a great number of isomers, both geometrical and optical. As these complexes are often stereochemically nonrigid, the isomers are usually not separable and we do not consider them further.

BOX 7.1 The synthesis of specific isomers

The synthesis of specific isomers often requires subtle changes in synthetic conditions. For example, the most stable Co(II) complex in ammoniacal solutions of Co(II) salts, [Co(NH$_3$)$_6$]$^{2+}$, is only slowly oxidized. As a result, a variety of complexes containing other ligands as well as NH$_3$ can be prepared by bubbling air through a solution containing ammonia and a Co(II) salt. Starting with ammonium carbonate yields [Co(CO$_3$)(NH$_3$)$_4$]$^+$, in which CO$_3^{2-}$ is a bidentate ligand that occupies two adjacent coordination positions. The complex *cis*-[CoL$_2$(NH$_3$)$_4$] can be prepared by displacement of the CO$_3^{2-}$ ligand in acidic solution. When concentrated hydrochloric acid is used, the violet *cis*-[CoCl$_2$(NH$_3$)$_4$]Cl compound (**B1**) can be isolated:

$$[Co(CO_3)(NH_3)_4]^+(aq) + 2\,H^+(aq) + 3\,Cl^-(aq) \rightarrow$$
$$cis\text{-}[CoCl_2(NH_3)_4]Cl(s) + H_2CO_3(aq)$$

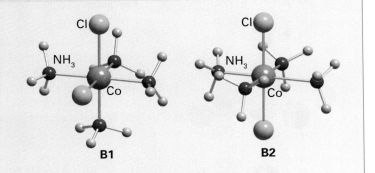

B1   **B2**

By contrast, reaction of [Co(NH$_3$)$_6$]$^{3+}$ directly with a mixture of HCl and H$_2$SO$_4$ in air gives the bright green *trans*-[CoCl$_2$(NH$_3$)$_4$]Cl isomer (**B2**).

BOX 7.2 The resolution of enantiomers

Optical activity is the only physical manifestation of chirality for a compound with a single chiral centre. However, as soon as more than one chiral centre is present, other physical properties, such as solubility and melting points, are affected because they depend on the strengths of intermolecular forces, which are different between different isomers (just as there are different forces between a given nut and bolts with left- and right-handed threads). One method of separating a pair of enantiomers into the individual isomers is therefore to prepare *diastereomers*. As far as we need be concerned, diastereomers are isomeric compounds that contain two chiral centres, one being of the same absolute configuration in both components and the other being enantiomeric between the two components. An example of diastereomers is provided by the two salts of an enantiomeric pair of cations, A, with an optically pure anion, B, and hence of composition [Δ-A][Δ-B] and [Λ-A][Δ-B]. Because diastereomers differ in physical properties (such as solubility), they are separable by conventional techniques.

A classical chiral resolution procedure begins with the isolation of a naturally optically active species from a biochemical source (many naturally occurring compounds are chiral). A convenient compound is *d*-tartaric acid (**B3**), a carboxylic acid obtained from grapes. This molecule is a chelating ligand for complexation of antimony, so a convenient resolving agent is the potassium salt of the singly charged antimony *d*-tartrate anion. This anion is used for the resolution of [Co(en)$_2$(NO$_2$)$_2$]$^+$ as follows:

The enantionmeric mixture of the cobalt(III) complex is dissolved in warm water and a solution of potassium antimony *d*-tartrate is added. The mixture

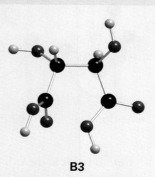

B3

is cooled immediately to induce crystallization. The less soluble diastereomer {*l*-[Co(en)$_2$(NO$_2$)$_2$]}{*d*-[SbOC$_4$H$_4$O$_6$]} separates as fine yellow crystals. The filtrate is reserved for isolation of the *d*-enantiomer. The solid diastereomer is ground with water and sodium iodide. The sparingly soluble compound *l*-[Co(en)$_2$(NO$_2$)$_2$]I separates, leaving sodium antimony tartrate in the solution. The *d*-isomer is obtained from the filtrate by precipitation of the bromide salt.

Further reading

A. von Zelewsky, *Stereochemistry of coordination compounds*. Wiley, Chichester (1996).

W.L. Jolly, *The synthesis and characterization of inorganic compounds*. Waveland Press, Prospect Heights (1991).

EXAMPLE 7.4 Identifying types of isomerism

When the four-coordinate square-planar complex [IrCl(PMe$_3$)$_3$] (where PMe$_3$ is trimethylphosphine) reacts with Cl$_2$, two six-coordinate products of formula [Ir(Cl)$_3$(PMe$_3$)$_3$] are formed. ^{31}P-NMR spectra indicate one P environment in one of these isomers and two in the other. What isomers are possible?

Answer Because the complexes have the formula [MA$_3$B$_3$], we expect meridional and facial isomers. Structures (**69**) and (**70**) show the arrangement of the three Cl$^-$ ions in the *fac* and *mer* isomers, respectively. All P atoms are equivalent in the *fac* isomer and two environments exist in the *mer* isomer.

Self-test 7.4 When the anion of the amino acid glycine, H$_2$NCH$_2$CO$_2^-$ (gly$^-$), reacts with cobalt(III) oxide, both the N and an O atom of gly$^-$ coordinate and two Co(III) nonelectrolyte *mer* and *fac* isomers of [Co(gly)$_3$] are formed. Sketch the two isomers.

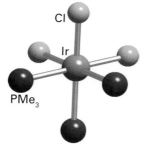

69 *fac*-[IrCl$_3$(PMe$_3$)$_3$]

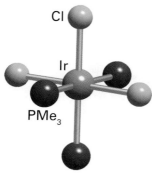

70 *mer*-[IrCl$_3$(PMe$_3$)$_3$]

7.11 Ligand chirality

Key point: Coordination to a metal can stop a ligand inverting and hence lock it into a chiral configuration.

In certain cases, achiral ligands can become chiral on coordination to a metal, leading to a complex that is chiral. Usually the nonchiral ligand contains a donor that rapidly inverts as a free ligand, but becomes locked in one configuration on coordination. An example is MeNHCH$_2$CH$_2$NHMe, where the two N atoms become chiral centres on coordination to a metal atom. For a square-planar complex, this imposed chirality results in four isomers, one pair of chiral enantiomers (**71**) and two complexes that are not chiral (**72**) and (**73**).

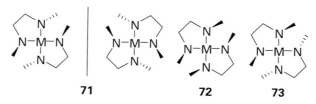

71   **72**   **73**

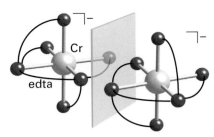

74 [Cr(edta)]⁻ enantiomers

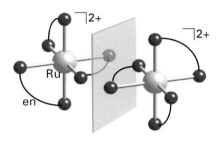

75 [Ru(en)₃]²⁺ enantiomers

EXAMPLE 7.5 Recognizing chirality

Which of the complexes (a) [Cr(edta)]⁻, (b) [Ru(en)₃]²⁺, (c) [Pt(dien)Cl]⁺ are chiral?

Answer If a complex has either a mirror plane or centre of inversion, it cannot be chiral. If we look at the schematic complexes in (**74**), (**75**), and (**76**), we can see that neither (**74**) nor (**75**) has a mirror plane or a centre of inversion; so both are chiral (they also have no higher S_n axis). Conversely, (**76**) has a plane of symmetry and hence is achiral. (Although the CH₂ groups in a dien ligand are not in the mirror plane, they oscillate rapidly above and below it.)

Self-test 7.5 Which of the complexes (a) *cis*-[Cr(Cl)₂(ox)₂]³⁻, (b) *trans*-[Cr(Cl)₂(ox)₂]³⁻, (c) *cis*-[RhH(CO)(PR₃)₂] are chiral?

The thermodynamics of complex formation

When assessing chemical reactions we need to consider both thermodynamic and kinetic aspects because although a reaction may be thermodynamically feasible, there might be kinetic constraints.

7.12 Formation constants

Key points: A formation constant expresses the interaction strength of a ligand relative to the interaction strength of the solvent molecules (usually H₂O) as a ligand; a stepwise formation constant is the formation constant for each individual solvent replacement in the synthesis of the complex; an overall formation constant is the product of the stepwise formation constants.

Consider the reaction of Fe(III) with SCN⁻ to give [Fe(SCN)(OH₂)₅]²⁺, a red complex used to detect either iron(III) or the thiocyanate ion:

$$[Fe(OH_2)_6]^{3+}(aq) + SCN^-(aq) \rightleftharpoons [Fe(SCN)(OH_2)_5]^{2+}(aq) + H_2O(l)$$

$$K_f = \frac{[Fe(SCN)(OH_2)_5^{2+}]}{[Fe(OH_2)_6^{3+}][SCN^-]}$$

The equilibrium constant, K_f, of this reaction is called the **formation constant** of the complex. The concentration of solvent (normally H₂O) does not appear in the expression because it is taken to be constant in dilute solution and ascribed unit activity. The value of K_f indicates the strength of binding of the ligand relative to H₂O: if K_f is large, the incoming ligand binds more strongly than the solvent, H₂O; if K_f is small, the incoming ligand binds more weakly than H₂O. Because the values of K_f can vary over a huge range (Table 7.3), they are often expressed as their logarithms, log K_f.

A note on good practice In expressions for equilibrium constants and rate equations, we omit the brackets that are part of the chemical formula of the complex; the surviving square brackets denote molar concentration of a species (with the units mol dm⁻³ removed).

The discussion of stabilities is more involved when more than one ligand may be replaced. For instance, in the reaction of [Ni(OH₂)₆]²⁺ to give [Ni(NH₃)₆]²⁺,

$$[Ni(OH_2)_6]^{2+}(aq) + 6NH_3(aq) \rightarrow [Ni(NH_3)_6]^{2+}(aq) + 6H_2O(l)$$

there are at least six steps, even if *cis–trans* isomerization is ignored. For the general case of the complex ML$_n$, for which the overall reaction is M + nL → ML$_n$, the **stepwise formation constants** are

$$M + L \rightleftharpoons ML \qquad K_{f1} = \frac{[ML]}{[M][L]}$$

$$ML + L \rightleftharpoons ML_2 \qquad K_{f2} = \frac{[ML_2]}{[ML][L]}$$

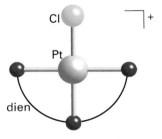

76 [PtCl(dien)]⁺

Table 7.3 Formation constants for the ction $[M(H_2O)_n]^{m+} + L \rightleftharpoons [M(L)(OH_2)_{n-1}]^{m+} + H_2O$

Ion	Ligand	K_f	log K_f	Ion	Ligand	K_f	log K_f
Mg^{2+}	NH_3	1.7	0.23	Pd^{2+}	Cl^-	1.25×10^5	5.1
Ca^{2+}	NH_3	0.64	−0.2	Na^+	SCN^-	1.2×10^4	4.08
Ni^{2+}	NH_3	525	2.72	Cr^{3+}	SCN^-	1.2×10^3	3.08
Cu^+	NH_3	8.50×10^5	5.93	Fe^{3+}	SCN^-	234	2.37
Cu^{2+}	NH_3	2.0×10^4	4.31	Co^{2+}	SCN^-	11.5	1.06
Hg^{2+}	NH_3	6.3×10^8	8.8	Fe^{2+}	pyridine	5.13	0.71
Rb^+	Cl^-	0.17	−0.77	Zn^{2+}	pyridine	8.91	0.95
Mg^{2+}	Cl^-	4.17	0.62	Cu^{2+}	pyridine	331	2.52
Cr^{3+}	Cl^-	7.24	0.86	Ag^+	pyridine	93	1.97
Co^{2+}	Cl^-	4.90	0.69				

and so on, and in general

$$ML_{n-1} + L \rightleftharpoons ML_n \qquad\qquad K_{fn} = \frac{[ML_n]}{[ML_{n-1}][L]}$$

These stepwise constants are the ones to consider when seeking to understand the relationships between structure and reactivity.

When we want to calculate the concentration of the final product (the complex ML_n) we use the **overall formation constant, β_n:**

$$M + nL \rightleftharpoons ML_n \qquad\qquad \beta_n = \frac{[ML_n]}{[M][L]^n}$$

As may be verified by multiplying together the individual stepwise constants, the overall formation constant is the product of the stepwise constants:

$$\beta_n = K_{f1}K_{f2}...K_{fn}$$

The inverse of each K_f, the **dissociation constant, K_d,** is also sometimes useful, and is often preferred when we are interested in the concentration of ligand that is required to give a certain concentration of complex:

$$ML \rightleftharpoons M + L \qquad\qquad K_{d1} = \frac{[M][L]}{[ML]} = \frac{1}{K_{f1}}$$

For a 1:1 reaction, like the one above, when half the metal ions are complexed and half are not, so that $[M] = [ML]$, then $K_{d1} = [L]$. In practice, if initially $[L] >> [M]$, so that there is an insignificant change in the concentration of L when M is added and undergoes complexation, K_d is the ligand concentration required to obtain 50 per cent complexation.

Because K_d has the same form as K_a for acids, with L taking the place of H^+, its use facilitates comparisons between metal complexes and Brønsted acids. The values of K_d and K_a can be tabulated together if the proton is considered to be simply another cation. For instance, HF can be considered as the complex formed from the Lewis acid H^+ with the Lewis base F^- playing the role of a ligand.

7.13 Trends in successive formation constants

Key points: Stepwise formation constants typically lie in the order $K_{fn} > K_{fn+1}$, as expected statistically; deviations from this order indicate a major change in structure.

The magnitude of the formation constant is a direct reflection of the sign and magnitude of the standard Gibbs energy of formation (because $\Delta_r G^\ominus = -RT \ln K_f$). It is commonly observed that stepwise formation constants lie in the order $K_{f1} > K_{f2} > ... > K_{fn}$. This general trend can be explained quite simply by considering the decrease in the number of the ligand and H_2O molecules available for replacement in the formation step, as in

$$[M(OH_2)_5L](aq) + L(aq) \rightleftharpoons [M(OH_2)_4L_2](aq) + H_2O(l)$$

Table 7.4 Formation constants of Ni(II) ammines, $[Ni(NH_3)_n(OH_2)_{6-n}]^{2+}$

n	K_f	$\log K_f$	K_n/K_{n-1} Experimental	Statistical*
1	525	2.72		
2	148	2.17	0.28	0.42
3	45.7	1.66	0.31	0.53
4	13.2	1.12	0.29	0.56
5	4.7	0.63	0.35	0.53
6	1.1	0.04	0.23	0.42

* Based on ratios of numbers of ligands available for replacement, with the reaction enthalpy assumed constant.

compared with

$$[M(OH_2)_4L_2](aq) + L(aq) \rightleftharpoons [M(OH_2)_3L_3](aq) + H_2O(l)$$

The decrease in the stepwise formation constants reflects the diminishing statistical factor as successive ligands are replaced, coupled with the fact that an increase in the number of bound ligands increases the likelihood of the reverse reaction. That such a simple explanation is more or less correct is illustrated by data for the successive complexes in the series from $[Ni(OH_2)_6]^{2+}$ to $[Ni(NH_3)_6]^{2+}$ (Table 7.4). The reaction enthalpies for the six successive steps are known to vary by less than 2 kJ mol^{-1}.

A reversal of the relation $K_{fn} > K_{fn+1}$ is usually an indication of a major change in the electronic structure of the complex as more ligands are added. An example is the observation that the tris(bipyridine) complex of Fe(II), $[Fe(bpy)_3]^{2+}$, is strikingly stable compared with the bis complex, $[Fe(bpy)_2(OH_2)_2]^{2+}$. This observation can be correlated with the change in electronic configuration from a high-spin (weak-field) $t_{2g}^4 e_g^2$ configuration in the bis complex (note the presence of weak-field H_2O ligands) to a low-spin (strong-field) t_{2g}^6 configuration in the tris complex, where there is a considerable increase in the LFSE (see Sections 20.1 and 20.2).

$$[Fe(OH_2)_6]^{2+}(aq) + bpy(aq) \rightleftharpoons [Fe(bpy)(OH_2)_4]^{2+}(aq) + 2H_2O(l) \qquad \log K_{f1} = 4.2$$

$$[Fe(bpy)(OH_2)_4]^{2+}(aq) + bpy(aq) \rightleftharpoons [Fe(bpy)_2(OH_2)_2]^{2+}(aq) + 2H_2O(l) \qquad \log K_{f2} = 3.7$$

$$[Fe(bpy)_2(OH_2)_2]^{2+}(aq) + bpy(aq) \rightleftharpoons [Fe(bpy)_3]^{2+}(aq) + 2H_2O(l) \qquad \log K_{f3} = 9.3$$

A contrasting example is the halogeno complexes of Hg(II), where K_{f3} is anomalously low compared with K_{f2}:

$$[Hg(OH_2)_6]^{2+}(aq) + Cl^-(aq) \rightleftharpoons [HgCl(OH_2)_5]^+(aq) + H_2O(l) \qquad \log K_{f1} = 6.74$$

$$[HgCl(OH_2)_5]^+(aq) + Cl^-(aq) \rightleftharpoons [HgCl_2(OH_2)_4](aq) + H_2O(l) \qquad \log K_{f2} = 6.48$$

$$[HgCl_2(OH_2)_4](aq) + Cl^-(aq) \rightleftharpoons [HgCl_3(OH_2)]^-(aq) + 3H_2O(l) \qquad \log K_{f3} = 0.95$$

The decrease between the second and third values is too large to be explained statistically and suggests a major change in the nature of the complex, such as the onset of four-coordination:

EXAMPLE 7.6 Interpreting irregular successive formation constants

The successive formation constants for complexes of cadmium with Br$^-$ are $K_{f1} = 36.3$, $K_{f2} = 3.47$, $K_{f3} = 1.15$, $K_{f4} = 2.34$. Suggest an explanation of why $K_{f4} > K_{f3}$.

Answer The anomaly suggests a structural change, so we need to consider what it might be. Aqua complexes are usually six-coordinate whereas halogeno complexes of M^{2+} ions are commonly tetrahedral. The reaction of the complex with three Br^- groups to add the fourth is

$$[CdBr_3(OH_2)_3]^-(aq) + Br^-(aq) \rightarrow [CdBr_4]^{2-}(aq) + 3\,H_2O(l)$$

This step is favoured by the release of three H_2O molecules from the relatively restricted coordination sphere environment. The result is an increase in K_f.

Self-test 7.6 Assuming the displacement of a water by a ligand were so favoured that the back reaction could be ignored, calculate all the stepwise formation constants you would expect in the formation of $[ML_6]^{2+}$ from $[M(OH_2)_6]^{2+}$, and the overall formation constant, given that $K_{f1} = 1 \times 10^5$.

7.14 The chelate and macrocyclic effects

Key points: The chelate and macrocyclic effects are the greater stability of complexes containing co-ordinated polydentate ligands compared with a complex containing the equivalent number of analogous monodentate ligands; the chelate effect is largely an entropic effect; the macrocyclic effect has an additional enthalpic contribution.

When K_{f1} for the formation of a complex with a bidentate chelate ligand, such as ethylenediamine (en), is compared with the value of β_2 for the corresponding bis(ammine) complex, it is found that the former is generally larger:

$$[Cd(OH_2)_6]^{2+}(aq) + en(aq) \rightleftharpoons [Cd(en)(OH_2)_4]^{2+}(aq) + 2\,H_2O(l)$$

$$\log K_{f1} = 5.84 \qquad \Delta_r H^{\ominus} = -29.4\ kJ\ mol^{-1} \quad \Delta_r S^{\ominus} = +13.0\ J\ K^{-1}\ mol^{-1}$$

$$[Cd(OH_2)_6]^{2+}(aq) + 2\,NH_3(aq) \rightleftharpoons [Cd(NH_3)_2(OH_2)_4]^{2+}(aq) + 2\,H_2O(l)$$

$$\log \beta_2 = 4.95 \qquad \Delta_r H^{\ominus} = -29.8\ kJ\ mol^{-1} \quad \Delta_r S^{\ominus} = -5.2\ J\ K^{-1}\ mol^{-1}$$

Two similar Cd−N bonds are formed in each case, yet the formation of the chelate-containing complex is distinctly more favourable. This greater stability of chelated complexes compared with their nonchelated analogues is called the **chelate effect**.

The chelate effect can be traced primarily to differences in reaction entropy between chelated and nonchelated complexes in dilute solutions. The chelation reaction results in an increase in the number of independent molecules in solution. By contrast, the nonchelating reaction produces no net change (compare the two chemical equations above). The former therefore has the more positive reaction entropy and hence is the more favourable process. The reaction entropies measured in dilute solution support this interpretation.

The entropy advantage of chelation extends beyond bidentate ligands, and applies, in principle, to any polydentate ligand. In fact, the greater the number of donor sites the multidentate ligand has, the greater is the entropic advantage of displacing monodentate ligands. Macrocyclic ligands, where multiple donor atoms are held in a cyclic array, such as crown ethers or phthalocyanin (**77**), give complexes of even greater stability than might otherwise be expected. This so-called **macrocyclic effect** is thought to be a combination of the entropic effect seen in the chelate effect, together with an additional energetic contribution that comes from the preorganized nature of the ligating groups (that is, no additional strains are introduced to the ligand on coordination).

The chelate and macrocyclic effects are of great practical importance. The majority of reagents used in complexometric titrations in analytical chemistry are polydentate chelates like edta^{4-}, and most biochemical metal binding sites are chelating or macrocyclic ligands. A formation constant as high as 10^{12} to 10^{25} is generally a sign that the chelate or macrocyclic effect is in operation.

In addition to the thermodynamic rationalization for the chelate effect we have described, there is an additional role in the chelate effect for kinetics. Once one ligating group of a polydentate ligand has bound to a metal ion, it becomes more likely that its other

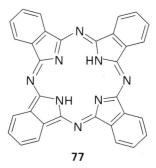

77

ligating groups will bind, as they are now constrained to be in close proximity to the metal ion; thus chelate complexes are favoured kinetically too.

7.15 Steric effects and electron delocalization

Key point: The stability of chelate complexes of d metals involving diimine ligands is a result of the chelate effect in conjunction with the ability of the ligands to act as π acceptors as well as σ donors.

Steric effects have an important influence on formation constants. They are particularly important in chelate formation because ring completion may be difficult geometrically. Chelate rings with five members are generally very stable because their bond angles are near ideal in the sense of there being no ring strain. Six-membered rings are reasonably stable and may be favoured if their formation results in electron delocalization. Three-, four-, and seven-membered (and larger) chelate rings are found only rarely because they normally result in distortions of bond angles and unfavourable steric interactions.

Complexes containing chelating ligands with delocalized electronic structures may be stabilized by electronic effects in addition to the entropy advantages of chelation. For example, diimine ligands (**78**), such as bipyridine (**79**) and phenanthroline (**80**), are constrained to form five-membered rings with the metal atom. The great stability of their complexes with d metals is probably a result of their ability to act as π acceptors as well as σ donors and to form π bonds by overlap of the full metal d orbitals and the empty ring π* orbitals (Section 20.2). This bond formation is favoured by electron population in the metal t_{2g} orbitals, which allows the metal atom to act as a π donor and transfer electron density to the ligand rings. An example is the complex $[Ru(bpy)_3]^{2+}$ (**81**). In some cases the chelate ring that forms can have appreciable aromatic character, which stabilizes the chelate ring even more.

Box 7.3 describes how complicated chelating and macrocyclic ligands might be synthesized.

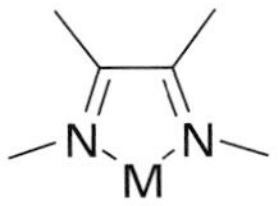

78

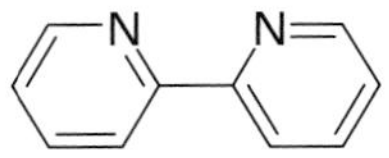

79

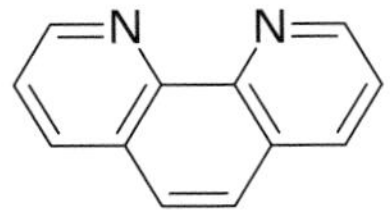

80

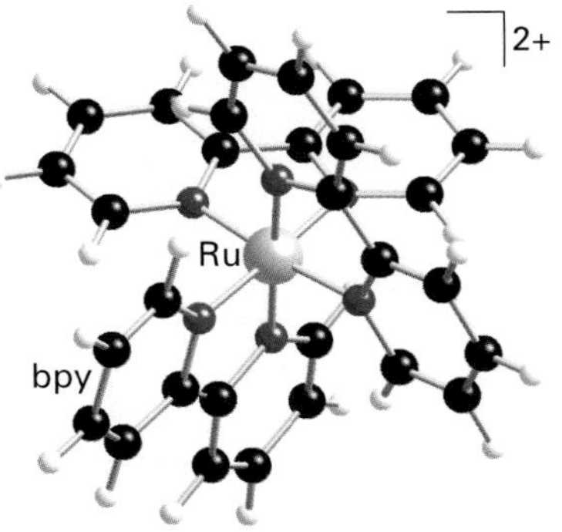

81 $[Ru(bpy)_3]^{2+}$

BOX 7.3 Making rings and knots

A metal ion such as Ni(II) can be used to assemble a group of ligands that then undergo a reaction among themselves to form a *macrocyclic ligand*, a cyclic molecule with several donor atoms. A simple example is

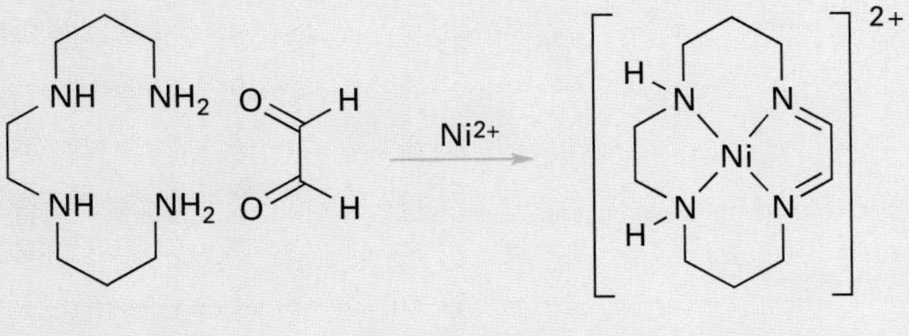

This phenomenon, which is called the *template effect*, can be applied to produce a surprising variety of macrocyclic ligands. The reaction shown above is an example of a *condensation reaction*, a reaction in which a bond is formed between two molecules, and a small molecule (in this case H_2O) is eliminated. If the metal ion had not been present, the condensation reaction of the component ligands would have been an ill-defined polymeric mixture, not a macrocycle. Once the macrocycle has been formed, it is normally stable on its own, and the metal ion may be removed to leave a multidentate ligand that can be used to complex other metal ions.

A wide variety of macrocyclic ligands can be synthesized by the template approach. Two more complicated ligands are shown below.

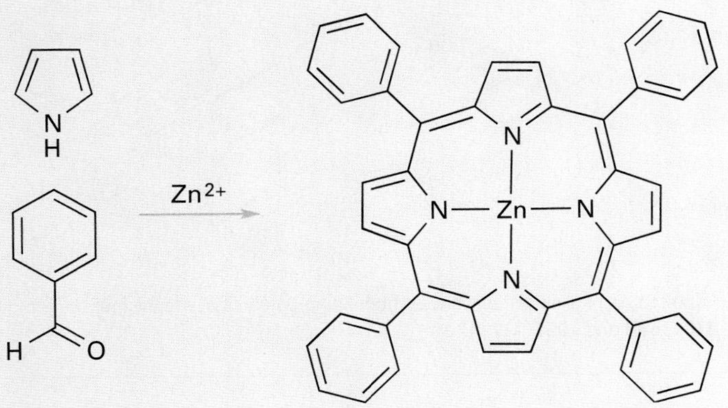

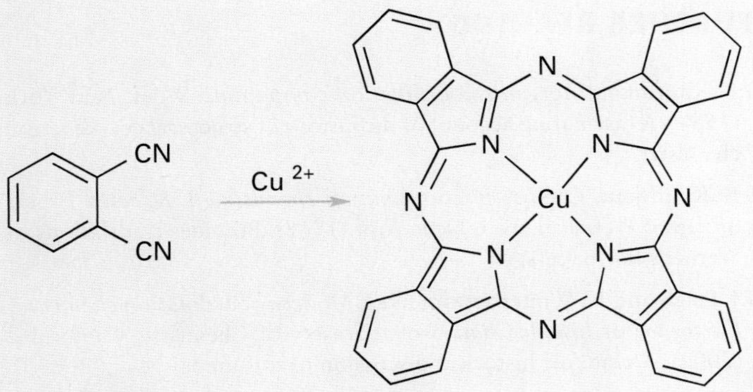

(Continued)

BOX 7.3 (*Continued*)

The origin of the template effect may be either kinetic or thermodynamic. For example, the condensation may stem either from the increase in the rate of the reaction between coordinated ligands (on account of their proximity or electronic effects) or from the added stability of the chelated ring product.

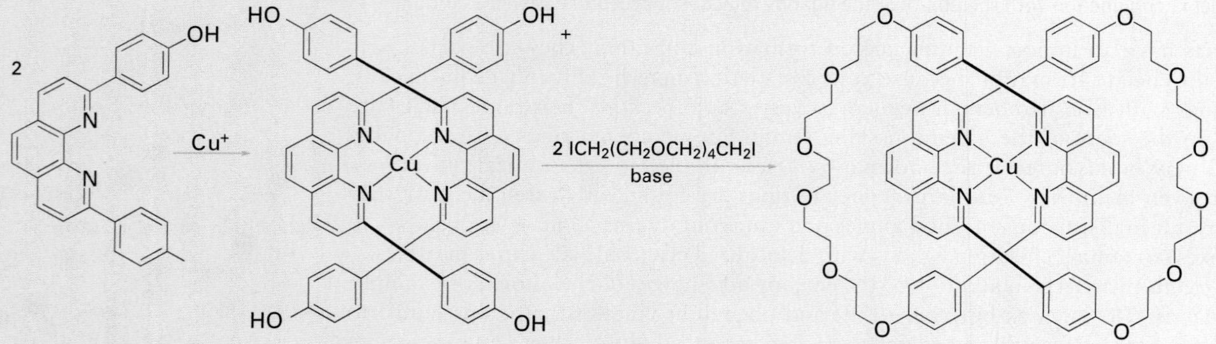

Here, two bipyridine-based ligands are coordinated to a copper ion, and then the ends of each ligand are joined by a flexible linkage. The metal ion can then be removed to give a *catenand* (*catenane* ligand), which can be used to complex other metal ions.

More complicated template syntheses can be used to construct topologically complex molecules, such as the chain-like *catenanes*, molecules that consist of interlinked rings. An example of the synthesis of a catenane containing two rings is shown below.

Even more complicated systems, equivalent to knots and links,[1] can be constructed with multiple metals. The following synthesis gives rise to a single molecular strand tied in a trefoil knot:

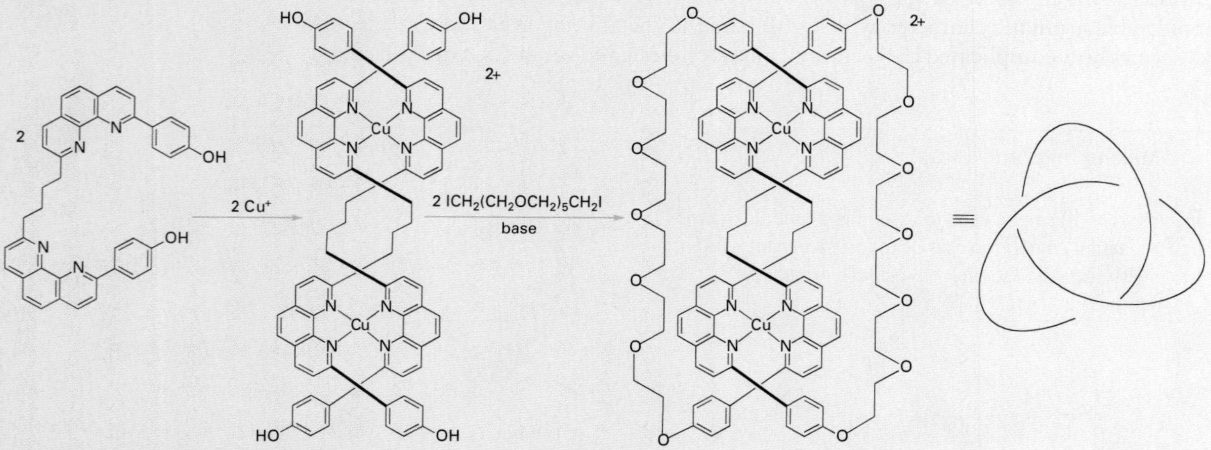

[1] Knotted and linked systems are far from being purely of academic interest and many proteins exist in these forms: see C. Liang and K. Mislow, *J. Am. Chem. Soc.*, 1994, **116**, 3588 and 1995, **117**, 4201.

FURTHER READING

G.B. Kauffman, *Inorganic coordination compounds*. Wiley, New York (1981). A fascinating account of the history of structural coordination chemistry.

G.B. Kauffman, *Classics in coordination chemistry: I. Selected papers of Alfred Werner*. Dover, New York (1968). Provides translations of Werner's key papers.

G.J. Leigh and N. Winterbottom (ed.), *Modern coordination chemistry: the legacy of Joseph Chatt*. Royal Society of Chemistry, Cambridge (2002). A readable historical discussion of this area.

A. von Zelewsky, *Stereochemistry of coordination compounds*. Wiley, Chichester (1996). A readable book that covers chirality in detail.

J.A. McCleverty and T.J. Meyer (ed.), *Comprehensive coordination chemistry II*. Elsevier (2004).

N.G. Connelly, T. Damhus, R.M. Hartshorn, and A.T. Hutton, *Nomenclature of inorganic chemistry, IUPAC recommendations 2005*. Royal Society of Chemistry, Cambridge (2005). Also known as 'The IUPAC red book', the definitive guide to naming inorganic compounds.

R.A. Marusak, K. Doan, and S.D. Cummings, *Integrated approach to coordination chemistry—an inorganic laboratory guide*. Wiley (2007). This unusual textbook describes the concepts of coordination chemistry and illustrates these concepts through well-explained experimental projects.

J.-M. Lehn (ed.), *Transition metals in supramolecular chemistry*, Volume 5 of *Perspectives in Supramolecular Chemistry*. Wiley (2007). Inspiring accounts of new developments and applications in coordination chemistry.

EXERCISES

7.1 Name and draw structures of the following complexes: (a) $[Ni(CO)_4]$, (b) $[Ni(CN)_4]^{2-}$, (c) $[CoCl_4]^{2-}$, (d) $[Mn(NH_3)_6]^{2+}$.

7.2 Give formulas for (a) chloridopentaamminecobalt(III) chloride, (b) hexaaquairon(3+) nitrate, (c) *cis*-dichloridobis(ethylenediamine)-ruthenium(II), (d) μ-hydroxidobis(penta-amminechromium(III)) chloride.

7.3 Name the octahedral complex ions (a) *cis*-$[CrCl_2(NH_3)_4]^+$, (b) *trans*-$[Cr(NH_3)_2(NCS)_4]^-$, (c) $[Co(C_2O_4)(en)_2]^+$.

7.4 (a) Sketch the two structures that describe most four-coordinate complexes. (b) In which structure are isomers possible for complexes of formula MA_2B_2?

7.5 Sketch the two structures that describe most five-coordinate complexes. Label the two different sites in each structure.

7.6 (a) Sketch the two structures that describe most six-coordinate complexes. (b) Which one of these is rare?

7.7 Explain the meaning of the terms *monodentate*, *bidentate*, and *tetradentate*.

7.8 What type of isomerism can arise with ambidentate ligands? Give an example.

7.9 Which of the following molecules could act as bidentate ligands? Which could act as chelating ligands?

(a) $P(OPh)_3$ (b) Me_2P PMe_2 (c) [structure] (d) [structure]

7.10 Draw the structures of representative complexes that contain the ligands (a) en, (b) ox^{2-}, (c) phen, (d) $edta^{4-}$.

7.11 The two compounds $[RuBr(NH_3)_5]Cl$ and $[RuCl(NH_3)_5]Br$ are what types of isomers?

7.12 For which of the following tetrahedral complexes are isomers possible? Draw all the isomers. $[CoBr_2Cl_2]^-$, $[CoBrCl_2(OH_2)]$, $[CoBrClI(OH_2)]$.

7.13 For which of the following square-planar complexes are isomers possible? Draw all the isomers. $[Pt(NH_3)_2(ox)]$, $[PdBrCl(PEt_3)_2]$, $[IrH(CO)(PR_3)_2]$, $[Pd(gly)_2]$.

7.14 For which of the following octahedral complexes are isomers possible? Draw all the isomers. $[FeCl(OH_2)_5]^{2+}$, $[Ir(Cl)_3(PEt_3)_3]$, $[Ru(bpy)_3]^{2+}$, $[Co(Cl)_2(en)(NH_3)_2]^+$, $[W(CO)_4(py)_2]$.

7.15 Ignoring optical isomers, how many isomers are possible for an octahedral complex of general formula $[MA_2BCDE]$? How many isomers are possible (include optical isomers)?

7.16 Which of the following complexes are chiral? (a) $[Cr(ox)_3]^{3-}$, (b) *cis*-$[Pt(Cl)_2(en)]$, (c) *cis*-$[Rh(Cl)_2(NH_3)_4]^+$, (d) $[Ru(bpy)_3]^{2+}$, (e) *fac*-$[Co(NO_2)_3(dien)]$, (f) *mer*-$[Co(NO_2)_3(dien)]$. Draw the enantiomers of the complexes identified as chiral and identify the plane of symmetry in the structures of the achiral complexes.

7.17 Which isomer is the following tris(acac) complex?

7.18 Draw both Λ and Δ isomers of the $[Ru(en)_3]^{2+}$ cation.

7.19 The stepwise formation constants for complexes of NH_3 with $[Cu(OH_2)_6]^{2+}(aq)$ are $\log K_{f1} = 4.15$, $\log K_{f2} = 3.50$, $\log K_{f3} = 2.89$, $\log K_{f4} = 2.13$, and $\log K_{f5} = -0.52$. Suggest a reason why K_{f5} is so different.

7.20 The stepwise formation constants for complexes of $NH_2CH_2CH_2NH_2$ (en) with $[Cu(OH_2)_6]^{2+}(aq)$ are $\log K_{f1} = 10.72$ and $\log K_{f2} = 9.31$. Compare these values with those of ammonia given in Exercise 7.19 and suggest why they are different.

PROBLEMS

7.1 The compound Na_2IrCl_6 reacts with triphenylphosphine in diethyleneglycol in an atmosphere of CO to give *trans*-$[IrCl(CO)(PPh_3)_2]$, known as 'Vaska's compound'. Excess CO gives a five-coordinate species and treatment with $NaBH_4$ in ethanol gives $[IrH(CO)_2(PPh_3)_2]$. Derive a formal name for 'Vaska's compound'. Draw and name all isomers of the two five-coordinate complexes.

7.2 A pink solid has the formula $CoCl_3.5NH_3.H_2O$. A solution of this salt is also pink and rapidly gives three 3 mol AgCl on titration with silver nitrate solution. When the pink solid is heated, it loses 1 mol H_2O to give a purple solid with the same ratio of $NH_3:Cl:Co$. The purple solid, on dissolution and titration with $AgNO_3$, releases two of its chlorides rapidly. Deduce the structures of the two octahedral complexes and draw and name them.

7.3 The hydrated chromium chloride that is available commercially has the overall composition $CrCl_3.6H_2O$. On boiling a solution, it becomes violet and has a molar electrical conductivity similar to that of $[Co(NH_3)_6]Cl_3$. By contrast, $CrCl_3.5H_2O$ is green and has a lower molar conductivity in solution. If a dilute acidified solution of the green complex is allowed to stand for several hours, it turns violet. Interpret these observations with structural diagrams.

7.4 The complex first denoted β-$[PtCl_2(NH_3)_2]$ was identified as the *trans* isomer. (The *cis* isomer was denoted α.) It reacts slowly with solid Ag_2O to produce $[Pt(NH_3)_2(OH_2)_2]^{2+}$. This complex does not react with ethylenediamine to give a chelated complex. Name and draw the structure of the diaqua complex. A third isomer of composition $PtCl_2.2NH_3$ is an insoluble solid that, when ground with $AgNO_3$, gives a mixture containing $[Pt(NH_3)_4](NO_3)_2$ and a new solid phase of composition $Ag_2[PtCl_4]$. Give the structures and names of each of the three Pt(II) compounds.

7.5 Phosphane (phosphine) and arsane (arsine) analogues of $[PtCl_2(NH_3)_2]$ were prepared in 1934 by Jensen. He reported zero dipole moments for the β isomers, where the designation β represents the product of a synthetic route analogous to that of the ammines. Give the structures of the complexes.

7.6 Air oxidation of Co(II) carbonate and aqueous ammonium chloride gives a pink chloride salt with a ratio of $4NH_3:Co$. On addition of HCl to a solution of this salt, a gas is rapidly evolved and the solution slowly turns violet on heating. Complete evaporation of the violet solution yields $CoCl_3.4NH_3$. When this is heated in concentrated HCl, a green salt can be isolated with composition $CoCl_3.4NH_3.HCl$. Write balanced equations for all the transformations occurring after the air oxidation. Give as much information as possible concerning the isomerism occurring and the basis of your reasoning. Is it helpful to know that the form of $[Co(Cl)_2(en)_2]^+$ that is resolvable into enantiomers is violet?

7.7 When cobalt(II) salts are oxidized by air in a solution containing ammonia and sodium nitrite, a yellow solid, $[Co(NO_2)_3(NH_3)_3]$, can be isolated. In solution it is nonconducting; treatment with HCl gives a complex that, after a series of further reactions, can be identified as $trans$-$[Co(Cl)_2(NH_3)_3(OH_2)]^+$. It requires an entirely different route to prepare cis-$[Co(Cl)_2(NH_3)_3(OH_2)]^+$. Is the yellow substance fac or mer? What assumption must you make to arrive at a conclusion?

7.8 The reaction of $[ZrCl_4(dppe)]$ (dppe is a bidentate phosphine ligand) with $Mg(CH_3)_2$ gives $[Zr(CH_3)_4(dppe)]$. NMR spectra indicate that all methyl groups are equivalent. Draw octahedral and trigonal prism structures for the complex and show how the conclusion from NMR supports the trigonal prism assignment. (P.M. Morse and G.S. Girolami, *J. Am. Chem. Soc.*, 1989, **111**, 4114.)

7.9 The resolving agent d-$cis[Co(NO_2)_2(en)_2]Br$ can be converted to the soluble nitrate by grinding in water with $AgNO_3$. Outline the use of this species for resolving a racemic mixture of the d and l enantiomers of $K[Co(edta)]$. (The l-$[Co(edta)]^-$ enantiomer forms the less soluble diastereomer. See F.P. Dwyer and F.L. Garvan, *Inorg. Synth.*, 1965, **6**, 192.)

7.10 Show how the coordination of two $MeHNCH_2CH_2NH_2$ ligands to a metal atom in a square-planar complex results in not only *cis* and *trans* but also optical isomers. Identify the mirror planes in the isomers that are not chiral.

7.11 The equilibrium constants for the successive reactions of ethylenediamine with Co^{2+}, Ni^{2+}, and Cu^{2+} are as follows.

$$[M(OH_2)_6]^{2+} + en \rightleftharpoons [M(en)(OH_2)_4]^{2+} + 2H_2O \qquad K_1$$
$$[M(en)(OH_2)_4]^{2+} + en \rightleftharpoons [M(en)_2(OH_2)_2]^{2+} + 2H_2O \qquad K_2$$
$$[M(en)_2(OH_2)_2]^{2+} + en \rightleftharpoons [M(en)_3]^{2+} + 2H_2O \qquad K_3$$

Ion	log K_1	log K_2	log K_3
Co^{2+}	5.89	4.83	3.10
Ni^{2+}	7.52	6.28	4.26
Cu^{2+}	10.72	9.31	−1.0

Discuss whether these data support the generalizations in the text about successive formation constants. How do you account for the very low value of K_3 for Cu^{2+}?

7.12 How may the aromatic character of a chelate ring provide additional stabilization of a complex? See A. Crispini and M. Ghedini, *J. Chem. Soc., Dalton Trans.* 1997, 75.

7.13 Use the internet to find out what rotaxanes are. Discuss how coordination chemistry might be used to synthesize such molecules.

Physical techniques in inorganic chemistry

8

All the structures of the molecules and materials to be covered in this book have been determined by applying one or more kinds of physical technique. The techniques and instruments available vary greatly in complexity and cost, as well as in their suitability for meeting particular challenges. All the methods produce data that help to determine a compound's structure, its composition, or its properties. Many of the physical techniques used in contemporary inorganic research rely on the interaction of electromagnetic radiation with matter and there is hardly a section of the electromagnetic spectrum that is not used. In this chapter, we introduce the most important physical techniques that are used to investigate the atomic and electronic structures of inorganic compounds and study their reactions.

Diffraction methods

Diffraction techniques, particularly those using X-rays, are the most important methods available to the inorganic chemist for the determination of structures. X-ray diffraction has been used to determine the structures a quarter of a million different substances, including tens of thousands of purely inorganic compounds and many organometallic compounds. The method is used to determine the positions of the atoms and ions that make up a solid compound and hence provides a description of structures in terms of features such as bond lengths, bond angles, and the relative positions of ions and molecules in a unit cell. This structural information has been interpreted in terms of atomic and ionic radii, which then allow chemists to predict structure and explain trends in many properties. Diffraction methods are nondestructive in the sense that the sample remains unchanged and may be analysed further by using a different technique.

8.1 X-ray diffraction

Key points: The scattering of radiation with wavelengths of about 100 pm from crystals gives rise to diffraction; the interpretation of the diffraction patterns gives quantitative structural information and in many cases the complete molecular or ionic structure.

Diffraction is the interference between waves that occurs as a result of an object in their path. X-rays are scattered elastically (with no change in energy) by the electrons in atoms, and diffraction can occur for a periodic array of scattering centres separated by distances similar to the wavelength of the radiation (about 100 pm), such as exist in a crystal. If we think of scattering as equivalent to reflection from two adjacent parallel planes of atoms separated by a distance d (Fig. 8.1), then the angle at which constructive interference occurs (to produce a diffraction intensity maximum) between waves of wavelength λ is given by **Bragg's equation**:

$$2d \sin \theta = n\lambda \tag{8.1}$$

where n is an integer. Thus an X-ray beam impinging on a crystalline compound with an ordered array of atoms will produce a set of diffraction maxima, termed a **diffraction pattern**, with each maximum, or **reflection**, occurring at an angle θ corresponding to a different separation of planes of atoms, d, in the crystal.

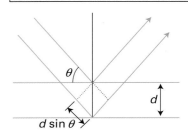

Figure 8.1 Bragg's equation is derived by treating layers of atoms as reflecting planes. X-rays interfere constructively when the additional path length $2d \sin \theta$ is equal to an integral multiple of the wavelength λ.

An atom or ion scatters X-rays in proportion to the number of electrons it possesses and the intensities of the measured diffraction maxima are proportional to the square of that number. Thus the diffraction pattern produced is characteristic of the positions and types (in terms of their number of electrons) of atom present in the crystalline compound and the measurement of X-ray diffraction angles and intensities provides structural information. Because of its dependence on the number of electrons, X-ray diffraction is particularly sensitive to any electron-rich atoms in a compound. Thus, X-ray diffraction by $NaNO_3$ displays all three nearly isoelectronic atoms similarly, but for $Pb(OH)_2$ the scattering and structural information is dominated by the Pb atom.

There are two principal X-ray techniques: the **powder method**, in which the materials being studied are in polycrystalline form, and **single-crystal diffraction**, in which the sample is a single crystal of dimensions of several tens of micrometres or larger.

(a) Powder X-ray diffraction

Key point: Powder X-ray diffraction is used mainly for phase identification and the determination of lattice parameters and lattice type.

A powdered (polycrystalline) sample contains an enormous number of very small crystallites, typically 0.1 to 10 μm in dimension and orientated at random. An X-ray beam striking a polycrystalline sample is scattered in all directions; at some angles, those given by Bragg's equation, constructive interference occurs. As a result, each set of planes of atoms with lattice spacing d gives rise to a cone of diffraction intensity. Each cone consists of a set of closely spaced diffracted rays, each one of which represents diffraction from a single crystallite within the powder sample (Fig. 8.2). With a very large number of crystallites these rays merge together to form the diffraction cone. A **powder diffractometer** (Fig. 8.3a) uses an electronic detector to measure the angles of the diffracted beams. Scanning the detector around the sample along the circumference of a circle cuts through the diffraction cones at the various diffraction maxima and the intensity of the X-rays detected is recorded as a function of the detector angle (Fig. 8.3b).

The number and positions of the reflections depend on the cell parameters, crystal system, lattice type, and wavelength used to collect the data; the peak intensities depend on the types of atoms present and their positions. Nearly all crystalline solids have a unique powder X-ray diffraction pattern in terms of the angles of the reflections and their intensities. In mixtures of compounds, each crystalline phase present contributes to the powder diffraction pattern its own unique set of reflection angles and intensities. Typically, the method is sensitive enough to detect a small level (5 to 10 per cent by mass) of a particular crystalline component in a mixture.

The effectiveness of powder X-ray diffraction has led to it becoming the major technique for the characterization of polycrystalline inorganic materials (Table 8.1). Many of the powder diffraction data sets collected from inorganic, organometallic, and organic compounds have been compiled into a database by the Joint Committee on Powder Diffraction Standards (JCPDS). This database, which contains over 50 000 unique powder X-ray diffraction patterns, can be used like a fingerprint library to identify an unknown material from its powder pattern alone. Powder X-ray diffraction is used routinely in the investigation of phase formation and changes in structures of solids. The synthesis of a metal oxide can be verified by collecting a powder diffraction pattern and demonstrating that the data are consistent with a single pure phase of that material. Indeed, the progress of a chemical reaction is often monitored by observing the formation of the product phase at the expense of the reactants.

Basic crystallographic information, such as lattice parameters, can normally be extracted easily from powder X-ray diffraction data, usually with high precision. The presence or absence of certain reflections in the diffraction pattern permits the determination of the lattice type. In recent years the technique of fitting the intensities of the peaks in the diffraction pattern has become a popular method of extracting structural information such as atomic positions. The analysis, which is known as the **Rietveld method**, involves fitting a calculated diffraction pattern to the experimental trace. The technique is not as powerful as the single-crystal methods, for it gives less accurate atomic positions, but has the advantage of not requiring the growth of a single crystal.

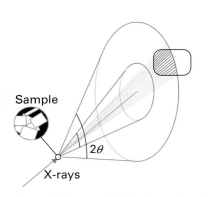

Figure 8.2 A cone of diffraction that results from X-ray scattering by a powdered sample. The cone consists of thousands of individual diffraction spots from individual crystallites that merge together.

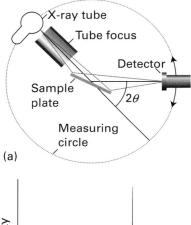

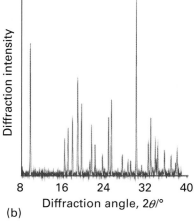

Figure 8.3 (a) Schematic diagram of a powder diffractometer operating in reflection mode in which the X-ray scattering occurs from a sample mounted as a flat plate. For weakly absorbing compounds the samples may be mounted in a capillary and the diffraction data collected in transmission mode. (b) The form of a typical powder diffraction pattern showing a series of reflections as a function of angle.

Table 8.1 Application of powder X-ray diffraction

Application	Typical use and information extracted
Identification of unknown materials	Rapid identification of most crystalline phases
Determination of sample purity	Monitoring the progress of a chemical reaction occurring in the solid state
Determination and refinement of lattice parameters	Phase identification and monitoring structure as a function of composition
Investigation of phase diagrams/ new materials	Mapping out composition and structure
Determination of crystallite size/stress	Particle size measurement and uses in metallurgy
Structure refinement	Extraction of crystallographic data from a known structure type
Ab initio structure determination	Structure determination (often at high precision) is possible in some cases without initial knowledge of the crystal structure
Phase changes/expansion coefficients	Studies as a function of temperature (cooling or heating typically in the range 100–1200 K). Observation of structural transitions

EXAMPLE 8.1 Using powder X-ray diffraction

Titanium dioxide exists as several polymorphs, the most common of which are anatase, rutile, and brookite. The experimental diffraction angles for the six strongest reflections collected from each of these different polymorphs are summarized in the table in the margin. The powder X-ray diffraction pattern collected using 154 pm X-radiation from a sample of white paint, known to contain TiO_2 in one or more of these polymorphic forms, showed the diffraction pattern in Fig. 8.4. Identify the TiO_2 polymorphs present.

Answer We need to identify the polymorph that has a diffraction pattern that matches the one observed. The lines closely match those of rutile (strongest reflections) and anatase (a few weak reflections), so the paint contains these phases with rutile as the major TiO_2 phase.

Self-test 8.1 Chromium(IV) oxide also adopts the rutile structure. By consideration of Bragg's equation and the ionic radii of Ti^{4+} and Cr^{4+} (*Resource section* 1) predict the main features of the CrO_2 powder X-ray diffraction pattern.

Rutile	Anatase	Brookite
27.50	25.36	19.34
36.15	37.01	25.36
39.28	37.85	25.71
41.32	38.64	30.83
44.14	48.15	32.85
54.44	53.97	34.90

(b) Single-crystal X-ray diffraction

Key point: The analysis of the diffraction patterns obtained from single crystals allows the full determination of the structure.

Analysis of the diffraction data obtained from single crystals is the most important method of obtaining the structures of inorganic solids. Provided a compound can be grown as a

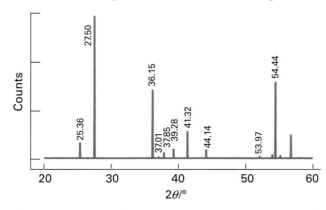

Figure 8.4 A powder diffraction pattern obtained from a mixture of TiO_2 polymorphs (see Example 8.1).

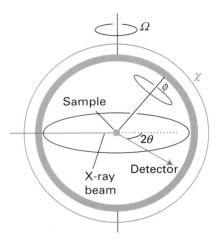

Figure 8.5 The layout of a four-circle diffractometer. A computer controls the location of the detector as the four angles are changed systematically.

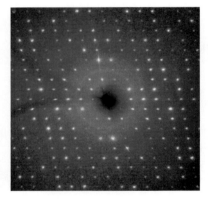

Figure 8.6 Part of a single-crystal X-ray diffraction pattern. Individual spots arise by diffraction of X-rays scattered from different planes of atoms within the crystal.

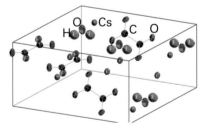

Figure 8.7 An ORTEP diagram of caesium oxalate monohydrate, $Cs_2C_2O_4 \cdot H_2O$. The ellipsoids correspond to a 90 per cent probability of locating the atoms.

crystal of sufficient size and quality, the data provide definitive information about molecular and extended lattice structures.

The collection of diffraction data from a single crystal is normally carried out by using a four-circle or area-detector diffractometer (Fig. 8.5). A **four-circle diffractometer** uses a scintillation detector to measure the diffracted X-ray beam intensity as a function of the angles shown in the illustration. An **area-detector diffractometer** uses an image plate that is sensitive to X-rays and so can measure a large number of diffraction maxima simultaneously; many new systems use this technology because the data can typically be collected in just a few hours (Fig. 8.6).

Analysis of the diffraction data from single crystals is formally a complex process involving the locations and intensities of many thousands of reflections, but with increasing advances in computation power a skilled crystallographer can complete the structure determination of a small inorganic molecule in under an hour. Single-crystal X-ray diffraction can be used to determine the structures of the vast majority of inorganic compounds when they can be obtained as crystals with dimensions of about $50 \times 50 \times 50$ μm or larger. Positions for most atoms, including C, N, O, and metals, in most inorganic compounds can be determined with sufficient accuracy that bond lengths can be defined to within a fraction of a picometre. As an example, the S—S bond length in monoclinic sulfur has been reported as 204.7 ± 0.3 pm.

A note on good (or at least conventional) practice Crystallographers still generally use the ångström (1 Å $= 10^{-10}$ m $= 10^{-8}$ cm $= 10^{2}$ pm) as a unit of measurement. This unit is convenient because bond lengths typically lie between 1 and 3 Å. The S—S bond length in monoclinic sulfur would be reported as 2.047 ± 0.003 Å.

The positions of H atoms can be determined for inorganic compounds that contain only light atoms (Z less than about 18, Ar), but their locations in many inorganic compounds that also contain heavy atoms, such as the 4d- and 5d-series elements, can be difficult or impossible. The problem lies with the small number of electrons on an H atom (just 1), which is often reduced even further when H forms bonds to other atoms. Other techniques, such as neutron diffraction (Section 8.2), can often be applied to determine the positions of H in inorganic compounds.

Molecular structures obtained by the analysis of single-crystal X-ray diffraction data are often represented in ORTEP diagrams (Fig. 8.7; the acronym stands for Oak Ridge Thermal Ellipsoid Program). In an ORTEP diagram an ellipsoid is used to represent the volume within which the atomic nucleus most probably lies, taking into account its thermal motion. The size of the ellipsoid increases with temperature and, as a result, so does the imprecision of the bond lengths extracted from the data.

(c) X-ray diffraction at synchrotron sources

Key point: High-intensity X-ray beams generated by synchrotron sources allow the structures of very complex molecules to be determined.

Much more intense X-ray beams than are available from laboratory sources can be obtained by using **synchrotron radiation**. Synchrotron radiation is produced by electrons circulating close to the speed of light in a storage ring and is typically several orders of magnitude more intense than laboratory sources. Because of their size, synchrotron X-ray sources are normally national or international facilities. Diffraction equipment located at such an X-ray source permits the study of much smaller samples and crystals as small as $10 \times 10 \times 10$ μm can be used. Furthermore, data collection can be undertaken much more rapidly and more complex structures, such as those of enzymes, can be determined more easily.

8.2 Neutron diffraction

Key point: The scattering of neutrons by crystals yields diffraction data that give additional information on structure, particularly the positions of light atoms.

Diffraction occurs from crystals for any particle with a velocity such that its associated wavelength (through the de Broglie relation, $\lambda = h/mv$) is comparable to the separations of the atoms or ions in the crystal. Neutrons and electrons travelling at suitable velocities have wavelengths of the order of 100–200 pm and thus undergo diffraction by crystalline inorganic compounds.

Neutron beams of the appropriate wavelength are generated by 'moderating' (slowing down) neutrons generated in nuclear reactors or through a process known as **spallation**, in which neutrons are chipped off the nuclei of heavy elements by accelerated beams of protons. The instrumentation used for collecting data and analysing single-crystal or powder neutron diffraction patterns is often similar to that used for X-ray diffraction. The scale is much larger, however, because neutron beam fluxes are much lower than laboratory X-ray sources. Furthermore whereas many chemistry laboratories have X-ray diffraction equipment for structure characterization, neutron diffraction can be undertaken only at a few specialist sources worldwide. The investigation of an inorganic compound with this technique is therefore much less routine and its application is essentially limited to systems where X-ray diffraction fails.

The advantages of neutron diffraction stem from the fact that neutrons are scattered by nuclei rather than by the surrounding electrons. As a result, neutrons are sensitive to structural parameters that often complement those for X-rays. In particular, the scattering is not dominated by the heavy elements, which can be a problem with X-ray diffraction for most inorganic compounds. For example, locating the position of a light element such as H and Li in a material that also contains Pb can be impossible with X-ray diffraction, as almost all the electron density is associated with the Pb atoms. With neutrons, in contrast, the scattering from light atoms is often similar to that of heavy elements, so the light atoms contribute significantly to the intensities in the diffraction pattern. Thus neutron diffraction is frequently used in conjunction with X-ray diffraction techniques to define an inorganic structure more accurately in terms of atoms such as H, Li, and O when they are in the presence of heavier, electron-rich metal atoms. Typical applications include studies of the complex metal oxides, such as the high-temperature superconductors (where accurate oxide ion positions are required in the presence of metals such as Ba and Tl) and systems where H atom positions are of interest.

Another use for neutron diffraction is to distinguish nearly isoelectronic species. In X-ray scattering, pairs of neighbouring elements in a period of the periodic table, such as O and N or Cl and S, are nearly isoelectronic and scatter X-rays to about the same extent, therefore they are hard to tell apart in a crystal structure that contains them both. However, the atoms of these pairs do scatter neutrons to very different extents, N 50 per cent more strongly than O, and Cl about four times better than S, so the identification of the atoms is much easier than by X-ray diffraction.

Absorption spectroscopy

The majority of physical techniques used to investigate inorganic compounds involve the absorption and sometimes the re-emission of electromagnetic radiation. The frequency of the radiation absorbed provides useful information on the energy levels of an inorganic compound and the intensity of the absorption can often be used to provide quantitative analytical information. Absorption spectroscopy techniques are normally nondestructive as after the measurement the sample can be recovered for further analysis.

The spectrum of electromagnetic radiation used in chemistry ranges from the short wavelengths associated with γ- and X-rays (about 1 nm), to radiowaves with wavelengths of several metres (Fig. 8.8). This spectrum covers the full range of atomic and molecular energies associated with characteristic phenomena such as ionization, vibration, rotation, and nuclear reorientation. Thus, X- and ultraviolet (UV) radiation can be used to determine the electronic structures of atoms and molecules and infrared (IR) radiation can be used to examine their vibrational behaviour. Radiofrequency (RF) radiation, in nuclear magnetic resonance (NMR), can be used to explore the energies associated with reorientations of the nucleus in a magnetic field, and those energies are sensitive to the chemical environment of the nucleus. In general, absorption

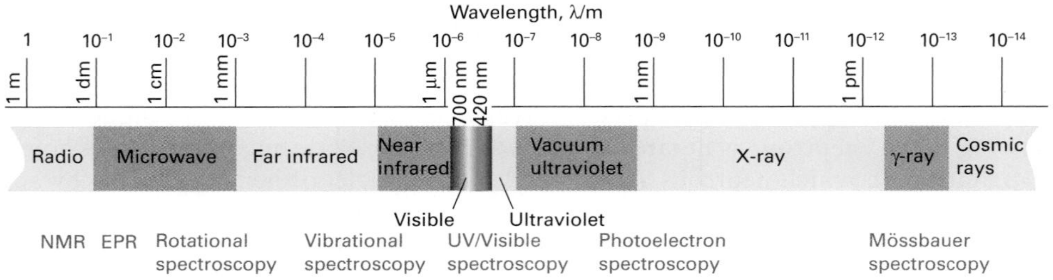

Wavelength, λ/m

| 1 | 10^{-1} | 10^{-2} | 10^{-3} | 10^{-4} | 10^{-5} | 10^{-6} | 10^{-7} | 10^{-8} | 10^{-9} | 10^{-10} | 10^{-11} | 10^{-12} | 10^{-13} | 10^{-14} |

1 m 1 dm 1 cm 1 mm 1 μm 700 nm 420 nm 1 nm 1 pm

| Radio | Microwave | Far infrared | Near infrared | Vacuum ultraviolet | X-ray | γ-ray | Cosmic rays |

Visible   Ultraviolet

NMR EPR Rotational spectroscopy Vibrational spectroscopy UV/Visible spectroscopy Photoelectron spectroscopy Mössbauer spectroscopy

Figure 8.8 The electromagnetic spectrum with wavelengths and techniques that make use of the different regions.

spectroscopic methods make use of the absorption of electromagnetic radiation by a molecule or material at a characteristic frequency corresponding to the energy of a transition between the relevant energy levels. The intensity is related to the probability of the transition, which in turn is determined in part by symmetry rules, such as those described in Chapter 6 for vibrational spectroscopy.

The various spectroscopic techniques involving electromagnetic radiation have different associated timescales. This variation can influence the structural information that is extracted. When a photon interacts with an atom or molecule we need to consider factors such as the lifetime of any excited state and a how a molecule may change during that interval. Table 8.2 summarizes the timescales associated with various spectroscopic techniques discussed in this section. Thus IR spectroscopy takes a much faster snapshot of the molecular structure than NMR, for a molecule may have time to reorientate or change shape in a nanosecond. The temperature at which data are collected should also be taken into account as molecular reorientation rates increase with increasing temperature.

Table 8.2 Typical timescales of some common characterization methods

X-ray diffraction	10^{-18} s
Mössbauer	10^{-18} s
Electronic spectroscopy UV–visible	10^{-15} s
Vibrational spectroscopy IR/Raman	10^{-12} s
NMR	c.10^{-3}–10^{-6} s
EPR	10^{-6} s

■ **A brief illustration.** Iron pentacarbonyl, $Fe(CO)_5$, illustrates why consideration of such timescales is important when analysing the spectra to obtain structural information. Infrared spectroscopy suggests that $Fe(CO)_5$ has D_{3h} symmetry with distinct axial and equatorial carbonyl groups whereas NMR suggests that all the carbonyl groups are equivalent. ■

8.3 Ultraviolet–visible spectroscopy

Key points: The energies and intensities of electronic transitions provide information on electronic structure and chemical environment; changes in spectral properties are used to monitor the progress of reactions.

Ultraviolet–visible spectroscopy (UV–visible spectroscopy) is the observation of the absorption of electromagnetic radiation in the UV and visible regions of the spectrum. It is sometimes known as **electronic spectroscopy** because the energy is used to excite electrons to higher energy levels. UV–visible spectroscopy is among the most widely used techniques for studying inorganic compounds and their reactions, and most laboratories possess a UV–visible spectrophotometer (Fig. 8.9). This section describes only basic principles; they are elaborated in later chapters, particularly Chapter 20.

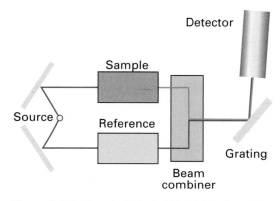

Figure 8.9 The layout of a typical UV–visible absorption spectrometer.

(a) Measuring a spectrum

The sample for a UV–visible spectrum determination is usually a solution but may also be a gas or a solid. A gas or liquid is contained in a cell (a 'cuvette') constructed of an optically transparent material such as glass or, for UV spectra at wavelengths below 320 nm, pure silica. Usually, the beam of incident radiation is split into two, one passing through the sample and the other passing through a cell that is identical except for the absence of the sample. The emerging beams are compared at the detector (a photodiode) and the absorption is obtained as a function of wavelength. Conventional spectrometers sweep the wavelength of the incident beam by changing the angle of a diffraction grating, but it is now more common for the entire spectrum to be recorded at once using a diode array detector. For solid samples, the intensity of UV–visible radiation reflected from the sample is more easily measured than that transmitted through a solid and an absorption spectrum is obtained by subtraction of the reflected intensity from the intensity of the incident radiation (Fig. 8.10).

The intensity of absorption is measured as the **absorbance**, A, defined as

$$A = \log_{10}\left(\frac{I_0}{I}\right) \tag{8.2}$$

where I_0 is the incident intensity and I is the measured intensity after passing through the sample. The detector is the limiting factor for strongly absorbing species because the measurement of low photon flux is unreliable.

■ **A brief illustration.** A sample that attenuates the light intensity by 10 per cent (so $I_0/I = 100/90$) has an absorbance of 0.05, one that attenuates it by 90 per cent (so $I_0/I = 100/10$) has an absorbance of 1.0, and one that attenuates it by 99 per cent (so $I_0/I = 100/1$) an absorbance of 2.0, and so on. ■

The empirical **Beer–Lambert law** is used to relate the absorbance to the molar concentration [J] of the absorbing species J and optical pathlength L:

$$A = \varepsilon[\text{J}]L \tag{8.3}$$

where ε (epsilon) is the **molar absorption coefficient** (still commonly referred to as the 'extinction coefficient' and sometimes the 'molar absorptivity'). Values of ε range from above 10^5 dm^3 mol^{-1} cm^{-1} for fully allowed transitions, for example an electron transferring from the 3d to 4p energy levels in an atom ($\Delta l = 1$), to less than 1 dm^3 mol^{-1} cm^{-1} for 'forbidden' transitions, such as those with $\Delta l = 0$. Selection rules also exist for transitions between molecular orbitals, although in complex molecules they are frequently broken (Chapter 20). For small molar absorption coefficients, the absorbing species may be difficult to observe unless the concentration or pathlength is increased accordingly.

Figure 8.11 shows a typical solution UV–visible spectrum obtained from a d-metal compound, in this case Ti(III), which has a d^1 configuration. From the wavelength of the radiation absorbed, the energy levels of the compound, including the effect of the ligand environment on d-metal atoms, can be inferred. The type of transition involved can often be inferred from the value of ε. The proportionality between absorbance and concentration provides a way to measure properties that depend on concentration, such as equilibrium compositions and the rates of reaction.

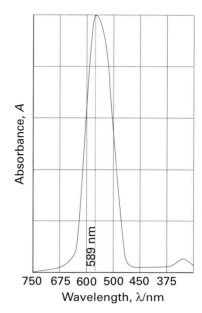

Figure 8.10 The UV–visible absorption spectrum of the solid ultramarine blue Na$_7$[SiAlO$_4$]$_6$(S$_3$).

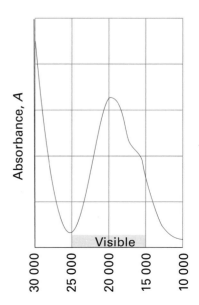

Figure 8.11 The UV–visible spectrum of [Ti(OH$_2$)$_6$]$^{3+}$(aq). Absorbance is given as a function of wavenumber.

EXAMPLE 8.2 Relating UV–visible spectra and colour

Figure 8.12 shows the UV–visible absorption spectra of PbCrO$_4$ and TiO$_2$. What colour would you expect PbCrO$_4$ to be?

Answer We need to be aware that the removal of light of a particular wavelength from incident white light results in the remaining light being perceived as having its complementary colour. Complementary colours are diametrically opposite each other on an artist's colour wheel (Fig. 8.13). The only absorption from

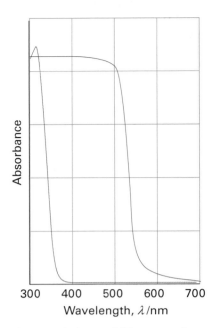

Figure 8.12 The UV–visible spectra of $PbCrO_4$ (red) and TiO_2 (blue). Absorbance is given as a function of wavelength.

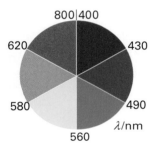

Figure 8.13 An artist's colour wheel: complementary colours lie opposite each other across a diameter.

$PbCrO_4$ in the visible region is in the blue region of the spectrum. The rest of the light is scattered into the eye, which, according to Fig. 8.13, perceives the complementary colour yellow.

Self-test 8.2 Explain why TiO_2 is widely used in sunscreens to protect against harmful UVA radiation (UV radiation with wavelengths in the range 320–360 nm).

(b) Spectroscopic monitoring of titrations and kinetics

When the emphasis is on **measurement of the intensities rather than the energies of transitions**, the spectroscopic investigation is usually called **spectrophotometry**. Provided at least one of the species involved has a suitable absorption band, it is usually straightforward to carry out a 'spectrophotometric titration' in which the extent of reaction is monitored by measuring the concentrations of the components present in the mixture. The measurement of UV–visible absorption spectra of species in solution also provides a method for monitoring the progress of reactions and determining rate constants.

Techniques that use UV–visible spectral monitoring range from those measuring reactions with half-lives of picoseconds (photochemically initiated by an ultrafast laser pulse) to the monitoring of slow reactions with half-lives of hours and even days. The **stopped-flow technique** (Fig. 8.14) is commonly used to study reactions with half-lives of between 5 ms and 10 s and which can be initiated by mixing. Two solutions, each containing one of the reactants, are mixed rapidly by a pneumatic impulse, then the flowing, reacting solution is brought to an abrupt stop by filling a 'stop-syringe' chamber, and triggering the monitoring of absorbance. The reaction can be monitored at a single wavelength or successive spectra can be measured very rapidly by using a diode array detector.

The spectral changes incurred during a titration or the course of a reaction also provide information about the number of species that form during its progress. An important case is the appearance of one or more **isosbestic points**, wavelengths at which two species have equal values for their molar absorption coefficients (Fig. 8.15; the name comes from the Greek for 'equal extinguishing'). The detection of isosbestic points in a titration or during the course of a reaction is evidence for there being only two dominant species (reactant and product) in the solution.

8.4 Infrared and Raman spectroscopy

Key points: Infrared and Raman spectroscopy are often complementary in that a particular type of vibration may be observed in one method but not the other; the information is used in many ways, ranging from structural determination to the investigation of reaction kinetics.

Vibrational spectroscopy is used to characterize compounds in terms of the strength, stiffness, and number of bonds that are present. It is also used to detect the presence of known compounds (fingerprinting), to monitor changes in the concentration of a species during a reaction, to determine the components of an unknown compound (such as the presence of CO ligands), to determine a likely structure for a compound, and to measure properties of bonds (their force constants).

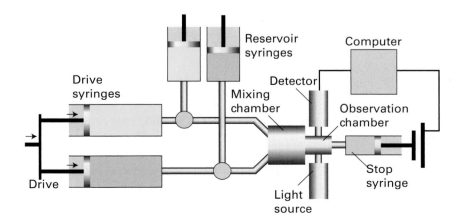

Figure 8.14 The structure of a stopped-flow instrument for studying fast reactions in solution.

(a) The energies of molecular vibrations

A bond in a molecule behaves like a spring: stretching it through a distance x produces a restoring force F. For small displacements, the restoring force is proportional to the displacement and $F = -kx$, where k is the **force constant** of the bond: the stiffer the bond, the greater the force constant. Such a system is known as a harmonic oscillator, and solution of the Schrödinger equation gives the energies

$$E_v = (v + \tfrac{1}{2}) \hbar \omega \tag{8.4a}$$

where $\omega = (k/\mu)^{1/2}$, $v = 0, 1, 2, \ldots$, and μ is the **effective mass** of the oscillator. For a diatomic molecule composed of atoms of masses m_A and m_B,

$$\mu = \frac{m_A m_B}{m_A + m_B} \tag{8.4b}$$

This effective mass is different for isotopologues (molecules composed of different isotopes of an element), which in turn leads to changes in E_v. If $m_A \gg m_B$, then $\mu \approx m_B$ and only atom B moves appreciably during the vibration: in this case the vibrational energy levels are determined largely by m_B, the mass of the lighter atom. The frequency ω is therefore high when the force constant is large (a stiff bond) and the effective mass of the oscillator is low (only light atoms are moved during the vibration). Vibrational energies are usually expressed in terms of the wavenumber $\tilde{v} = \omega/2\pi c$; typical values of $\tilde{v}$ lie in the range 300–3800 cm^{-1} (Table 8.3).

> *A note on good practice* You will commonly see μ referred to as the 'reduced mass', for the same term appears in the separation of internal motion from translational motion. However, in polyatomic molecules each vibrational mode corresponds to the motion of different quantities of mass that depends on the individual masses in a much more complicated way, and the 'effective mass' is the more general term for use when discussing vibrational modes.

A molecule consisting of N atoms can vibrate in $3N - 6$ different, independent ways if it is nonlinear and $3N - 5$ different ways if it is linear. These different, independent vibrations are called **normal modes**. For instance, a CO_2 molecule has four normal modes of vibration (as was shown in Fig. 6.15), two corresponding to stretching the bonds and two corresponding to bending the molecule in two perpendicular planes. Bending modes typically occur at lower frequencies than stretching modes and their effective masses, and therefore their frequencies, depend on the masses of the atoms in a complicated way that reflects the extents to which the various atoms move in each mode. The modes are labelled v_1, v_2, etc. and are sometimes given evocative names, such as 'symmetric stretch' and 'antisymmetric stretch'. Only normal modes that correspond to a changing electric dipole moment can absorb infrared radiation so only these modes are **IR active** and contribute to an IR spectrum. A normal mode is **Raman active** if it corresponds to a change in polarizability. As we saw in Chapter 6, group theory is a powerful tool for predicting the IR and Raman activities of molecular vibrations.

The lowest level ($v = 0$) of any normal mode corresponds to $E_0 = \tfrac{1}{2}\hbar\omega$, the so-called **zero-point energy**, which is the lowest vibrational energy that a bond can possess. In addition to the fundamental transitions with $\Delta v = +1$, vibrational spectra may also show bands arising from double quanta ($\Delta v = +2$) at $2\tilde{v}$ known as **overtones**, and **combinations** of two different vibrational modes (for example, $v_1 + v_2$). These special transitions may be helpful as they often arise even when the fundamental transition is not allowed by the selection rules.

(b) The techniques

The IR vibrational spectrum of a compound is obtained by exposing the sample to infrared radiation and recording the variation of the absorbance with frequency, wavenumber, or wavelength.

> *A note on good practice* An absorption is often stated as occurring at, say, 'a frequency of 1000 wavenumbers'. This usage is doubly wrong. First, 'wavenumber' is the name of a physical observable related to frequency v by $\tilde{v} = v / c$. Second, wavenumber is not a unit: the dimensions of wavenumber are 1/length and it is commonly reported in inverse centimetres (cm^{-1}).

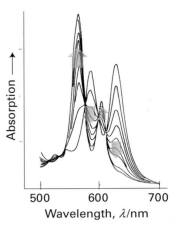

Figure 8.15 Isosbestic points observed in the absorption spectral changes during reaction of HgTPP (TPP = tetraphenylporphyrin) with Zn^{2+} in which Zn replaces Hg in the macrocycle. The initial and final spectra are those of the reactant and product, which indicates that free TPP does not reach a detectable concentration during the reaction. (Adapted from C. Grant and P. Hambright, *J. Am. Chem. Soc.* 1969, **91**, 4195.)

Table 8.3 Characteristic fundamental stretching wavenumbers of some common molecular species as free ions or coordinated to metal centres

Species	Range/cm^{-1}
OH	3400–3600
NH	3200–3400
CH	2900–3200
BH	2600–2800
CN$^-$	2000–2200
CO (terminal)	1900–2100
CO (bridging)	1800–1900
$\backslash$C = 0	1600–1760
NO	1675–1870
O_2^-	920–1120
O_2^{2-}	800–900
Si—O	900–1100
Metal—Cl	250–500
Metal—metal bonds	120–400

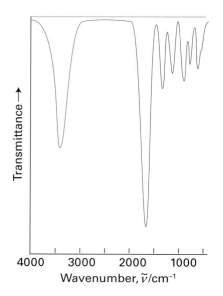

Figure 8.16 The IR spectrum of nickel acetate tetrahydrate showing characteristic absorptions due to water and the carbonyl group (OH stretch at 3600 cm^{-1} and $\rangle$C = O stretch at c.1700 cm^{-1}).

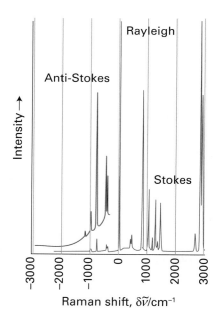

Figure 8.17 A typical Raman spectrum showing Rayleigh scattering (scattering of the laser light with no change in wavelength) and Stokes and anti-Stokes lines.

In early spectrometers, the transmission was measured as the frequency of the radiation was swept between two limits. Now the spectrum is extracted from an interferogram by **Fourier transformation**, which converts information in the time domain (based on the interference of waves travelling along paths of different lengths) to the frequency domain. The sample must be contained in a material that does not absorb IR radiation, which means that glass cannot be used and aqueous solutions are unsuitable unless the spectral bands of interest occur at frequencies not absorbed by water. Optical windows are typically constructed from CsI. Traditional procedures of sample preparation include KBr pellets (where the sample is diluted with dried KBr and then pressed into a translucent disc) and paraffin mulls (where the sample is produced as a suspension that is then placed as a droplet between the optical windows). These methods are still widely used, although becoming more popular are total internal reflectance devices in which the sample is simply placed in position. A typical range of an IR spectrum is between 4000 and 250 cm^{-1}, which corresponds to a wavelength of electromagnetic radiation of between 2.5 and 40 μm; this range covers many important vibrational modes or inorganic bonds. Figure 8.16 shows a typical spectrum.

In **Raman spectroscopy** the sample is exposed to intense laser radiation in the visible region of the spectrum. Most of the photons are scattered elastically (with no change of frequency) but some are scattered inelastically, having given up some of their energy to excite vibrations. The latter photons have frequencies different from that of the incident radiation (ν_0) by amounts equivalent to vibrational frequencies (ν_i) of the molecule. An advantage of Raman spectroscopy over IR spectroscopy is that aqueous solutions can be used, but a disadvantage is that linewidths are usually much greater. Conventional Raman spectroscopy involves the photon causing a transition to a 'virtual' excited state which then collapses back to a real lower state, emitting the detected photon in the process. The technique is not very sensitive but great enhancement is achieved if the species under investigation is coloured and the excitation laser is tuned to a real electronic transition. The latter technique is known as **resonance Raman spectroscopy** and is particularly valuable for studying the environment of d-metal atoms in enzymes because only vibrations close to the electronic chromophore (the group primarily responsible for the electronic excitation) are excited and the many thousands of bonds in the rest of the molecule are silent.

Raman spectra may be obtained over a similar range to IR spectra (200–4000 cm^{-1}) and Fig. 8.17 shows a typical spectrum. Note that energy can be transferred to the sample, leading to **Stokes lines**, lines in the spectrum at energies lower (at a lower wavenumber and greater wavelength) than the excitation energy, or transferred from the sample to the photon, leading to **anti-Stokes lines**, which appear at energies higher than the excitation energy. Raman spectroscopy is often complementary to IR spectroscopy as the two techniques probe vibrational modes with different activities: one mode might correspond to a change in dipole moment (IR active and observed in the IR spectrum) and another to a change in polarizability (a change in the electron distribution in a molecule caused by an applied electric field) and be seen in the Raman spectrum. We saw in Sections 6.5 and 6.6 that for group-theoretical reasons no mode can be both IR and Raman active in a molecule with a centre of inversion (the exclusion rule).

(c) Applications of IR and Raman spectroscopy

One important application of vibration spectroscopy is in the determination of the shape of an inorganic molecule. A five-coordinate structure, AX$_5$, for instance, may adopt a square-pyramidal (C_{4v}) or trigonal-bipyramidal (D_{3h}) geometry. A normal mode analysis of these geometries of the kind explained in Section 6.6 reveals that a trigonal-bipyramidal AX$_5$ molecule has five stretching modes (of symmetry species $2A' + A_2'' + E'$, the last is a pair of doubly degenerate vibrations) of which three are IR active ($A_2'' + E'$, corresponding to two absorption bands) and four are Raman active ($2A' + E'$, corresponding to three bands when the degeneracy of the E' modes are taken into account). A similar analysis of square-pyramidal geometry shows it has four IR active stretching modes ($2A_1 + E$, three bands) and five Raman active stretching modes ($2A_1 + B_1 + E$, four bands). The spectra of BrF$_5$ show three IR and four Raman Br−F stretching bands, showing that this molecule is square-pyramidal, just as expected from VSEPR theory (Section 2.3).

Figure 8.18 shows the IR and Raman spectra obtained from XeF_4 over the same range of wavenumbers. Does XeF_4 adopt a square-planar or tetrahedral geometry?

Answer An AB_4 molecule may be tetrahedral (T_d), square-planar (D_{4h}), square-pyramidal (C_{4v}), or see-saw (C_{2v}). The spectra have no absorption energy in common and therefore it is likely that the molecule has a centre of symmetry. Of the symmetry groups available, only square-planar geometry, D_{4h}, has a centre of symmetry.

Self-test 8.3 Use VSEPR theory to predict a molecular shape for XeF_2 and hence determine the total number of vibrational modes expected to be observed in its IR and Raman spectra. Would any of these absorptions occur at the same frequency in both the Raman and IR spectra?

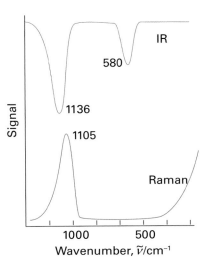

Figure 8.18 The IR and Raman spectra of XeF_4.

A major use of IR and Raman spectroscopy is the study of numerous compounds of the d block that contain carbonyl ligands, which give rise to intense vibrational absorption bands in a region where few other molecules produce absorptions. Free CO absorbs at 2143 cm^{-1}, but when coordinated to a metal atom in a compound the stretching frequency (and correspondingly the wavenumber) is lowered by an amount that depends on the extent to which electron density is transferred into the 2π antibonding orbital (the LUMO) by back donation from the metal atom (Section 22.5). The CO stretching absorption also allows terminal and bridging ligands to be distinguished, with bridging ligands occurring at lower frequencies. Isotopically labelled compounds show a shift in the absorption bands (by about 40 cm^{-1} to lower wavenumbers when ^{13}C replaces ^{12}C in a CO group) through a change in effective mass (eqn 8.4b), and this effect can be used to assign spectra and probe reaction mechanisms involving this ligand. The speed of data acquisition possible with Fourier-transform IR (FTIR) spectroscopy has meant that it can be incorporated into rapid kinetic techniques, including ultrafast laser photolysis and stopped-flow methods.

Raman and IR spectroscopy are excellent methods for studying molecules that are formed and trapped in inert matrices, the technique known as **matrix isolation**. The principle of matrix isolation is that highly unstable species that would not normally exist can be generated in an inert matrix such as solid xenon.

Resonance techniques

Several techniques of structural investigation depend on bringing energy level separations into resonance with electromagnetic radiation, with the separations in some cases controlled by the application of a magnetic field. Two of these techniques involve **magnetic resonance**: in one, the energy levels are those of magnetic nuclei (nuclei with non-zero spin, $I > 0$); in the other, they are the energy levels of unpaired electrons.

8.5 Nuclear magnetic resonance

Key points: Nuclear magnetic resonance is suitable for studying compounds containing elements with magnetic nuclei, especially hydrogen. The technique gives information on molecular structure, including chemical environment, connectivity, and internuclear separations. It also probes molecular dynamics and is an important tool for investigating rearrangement reactions occurring on a millisecond timescale.

Nuclear magnetic resonance (NMR) is the most powerful and widely used spectroscopic method for the determination of molecular structures in solution and pure liquids. In many cases it provides information about shape and symmetry with greater certainty than is possible with other spectroscopic techniques, such as IR and Raman spectroscopy. It also provides information about the rate and nature of the interchange of ligands in fluxional molecules and can be used to follow reactions, in many cases providing exquisite mechanistic detail. The technique has been used to obtain the structures of protein molecules with molar masses of up to 30 kg mol^{-1} (corresponding to a molecular mass of 30 kDa) and complements the more static descriptions obtained with X-ray single-crystal diffraction. However, unlike X-ray diffraction, NMR studies of molecules in solution generally cannot provide precise bond distances and angles, although it can provide some information on internuclear separations. It is a nondestructive technique because the sample can be recovered from solution after the resonance spectrum has been collected.

The sensitivity of NMR depends on several factors, including the abundance of the isotope and the size of its nuclear magnetic moment. For example, ^1H, with 99.98 per cent natural abundance and a large magnetic moment, is easier to observe than ^{13}C, which has a smaller magnetic moment and only 1.1 per cent natural abundance. With modern multinuclear NMR techniques it is particularly easy to observe spectra for ^1H, ^{19}F, and ^{31}P, and useful spectra can also be obtained for many other elements. Table 8.4 lists a selection of nuclei and their sensitivities. A common limitation for exotic nuclei is the presence of a nuclear quadrupole moment, a nonuniform distribution of electric charge (which is present for all nuclei with nuclear spin quantum number $I > \frac{1}{2}$), which broadens signals and degrades spectra. Nuclei with even atomic numbers and even mass numbers (such as ^{12}C and ^{16}O) have zero spin and are invisible in NMR.

(a) Observation of the spectrum

A nucleus of spin I can take up $2I + 1$ orientations relative to the direction of an applied magnetic field. Each orientation has a different energy (Fig. 8.19), with the lowest level the most highly populated. The energy separation of the two states $m_I = +\frac{1}{2}$ and $m_I = -\frac{1}{2}$ of a spin-$\frac{1}{2}$ nucleus (such as ^1H or ^{13}C) is

$$\Delta E = \hbar \gamma B_0 \tag{8.5}$$

where B_0 is the magnitude of the applied magnetic field (more precisely, the magnetic induction in tesla, $1\ T = 1\ kg\ s^{-2}\ A^{-1}$) and γ is the magnetogyric ratio of the nucleus, that is the ratio of its magnetic moment to its spin angular momentum. With modern superconducting magnets producing a field of 5–20 T, resonance is achieved with electromagnetic radiation in the range 200–900 MHz. Because the difference in energy of the $m_I = +\frac{1}{2}$ and $m_I = -\frac{1}{2}$ states in the applied magnetic field is small, the population of the lower level is only marginally greater than that of the higher level. Consequently,

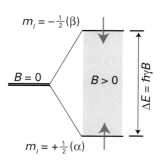

Figure 8.19 When a nucleus with spin $I > 0$ is in a magnetic field, its $2I + 1$ orientations (designated m_I) have different energies. This diagram shows the energy levels of a nucleus with $I = \frac{1}{2}$ (as for ^1H, ^{13}C, ^{31}P).

Table 8.4 Nuclear spin characteristics of common nuclei

Nucleus	Natural abundance/%	Sensitivity*	Spin	NMR frequency/MHz†
^1H	99.98	5680	$\frac{1}{2}$	100.000
^2H	0.015	0.00821	1	15.351
^7Li	92.58	1540	$\frac{3}{2}$	38.863
^{11}B	80.42	754	$\frac{3}{2}$	32.072
^{13}C	1.11	1.00	$\frac{1}{2}$	25.145
^{15}N	0.37	0.0219	$\frac{1}{2}$	10.137
^{17}O	0.037	0.0611	$\frac{3}{2}$	13.556
^{19}F	100	4730	$\frac{1}{2}$	94.094
^{23}Na	100	525	$\frac{3}{2}$	26.452
^{29}Si	4.7	2.09	$\frac{1}{2}$	19.867
^{31}P	100	377	$\frac{1}{2}$	40.481
^{89}Y	100	0.668	$\frac{1}{2}$	4.900
^{103}Rh	100	0.177	$\frac{1}{2}$	3.185
^{109}Ag	48.18	0.276	$\frac{1}{2}$	4.654
^{119}Sn	8.58	28.7	$\frac{1}{2}$	37.272
^{183}W	14.4	0.0589	$\frac{1}{2}$	4.166
^{195}Pt	33.8	19.1	$\frac{1}{2}$	21.462
^{199}Hg	16.84	5.42	$\frac{1}{2}$	17.911

* Sensitivity is relative to $^{13}C = 1$ and is the product of the relative sensitivity of the isotope and the natural abundance.
† At 2.349 T (a '100 MHz spectrometer').

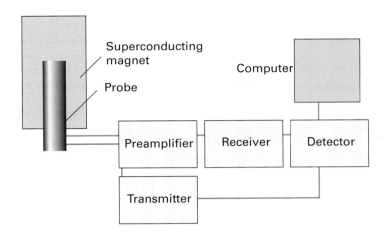

Figure 8.20 The layout of a typical NMR spectrometer. The link between transmitter and detector is arranged so that only low-frequency signals are processed.

the sensitivity of NMR is low but can be increased by using a stronger magnetic field, which increases the energy difference and therefore the population difference and the signal intensity.

Spectra were originally obtained in a continuous wave (CW) mode in which the sample is subjected to a constant radiofrequency and the resonances encountered as the field is increased are recorded as the spectrum, or the field is held constant and the radiofrequency is swept. In contemporary spectrometers, the energy separations are identified by exciting nuclei in the sample with a sequence of radiofrequency pulses and then observing the return of the nuclear magnetization back to equilibrium. Fourier transformation then converts the time-domain data to the frequency domain with peaks at frequencies corresponding to transitions between the different nuclear energy levels. Figure 8.20 shows the experimental arrangement of an NMR spectrometer.

(b) Chemical shifts

The frequency of an NMR transition depends on the local magnetic field experienced by the nucleus and is expressed in terms of the **chemical shift**, δ, the difference between the resonance frequency of nuclei (ν) in the sample and that of a reference compound (ν°):

$$\delta = \frac{\nu - \nu^\circ}{\nu^\circ} \times 10^6 \tag{8.6}$$

> **A note on good practice** The chemical shift δ is dimensionless. However, common practice is to report is as 'parts per million' (ppm) in acknowledgement of the factor of 10^6 in the definition. This practice is unnecessary.

A common standard for ^1H, ^{13}C, or ^{29}Si spectra is tetramethylsilane $Si(CH_3)_4$ (TMS). When $\delta < 0$ the nucleus is said to be **shielded** (with a resonance that is said to occur to 'high field') relative to the standard; $\delta > 0$ corresponds to a nucleus that is **deshielded** (with a resonance that is said to occur to 'low field') with respect to the reference. An H atom bound to a closed-shell, low-oxidation-state, d-block element from Groups 6 to 10 (such as $[HCo(CO)_4]$) is generally found to be highly shielded whereas in an oxoacid (such as H_2SO_4) it is deshielded. From these examples it might be supposed that the higher the electron density around a nucleus, the greater its shielding. However, as several factors contribute to the shielding, a simple physical interpretation of chemical shifts in terms of electron density is generally not possible.

The chemical shifts of ^1H and other nuclei in various chemical environments are tabulated, so empirical correlations can often be used to identify compounds or the element to which the resonant nucleus is bound. For example the H chemical shift in CH_4 is only 0.1 because the H nuclei are in an environment similar to that in tetramethylsilane, but the H chemical shift in GeH_4 is $\delta = 3.1$ (Fig. 8.21). Chemical shifts are different for the same element in inequivalent positions within a molecule. For instance, in ClF_3 the chemical shift of the unique ^{19}F nucleus is separated by $\Delta\delta = 120$ from that of the other two F nuclei (Fig. 8.22).

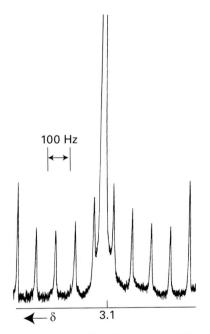

Figure 8.21 The ^1H-NMR spectrum of GeH_4.

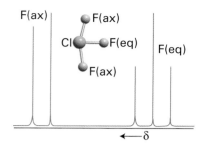

Figure 8.22 The ^{19}F-NMR spectrum of ClF$_3$.

(c) Spin–spin coupling

Structural assignment is often helped by the observation of the **spin–spin coupling**, which gives rise to multiplets in the spectrum due to interactions between nuclear spins. Spin–spin coupling arises when the orientation of the spin of a nearby nucleus affects the energy of another nucleus and causes small changes in the location of the latter's resonance. The strength of spin–spin coupling, which is reported as the **spin–spin coupling constant**, J (in hertz, Hz), decreases rapidly with distance through chemical bonds, and in many cases is greatest when the two atoms are directly bonded to each other. In **first-order spectra**, which are being considered here, the coupling constant is equal to the separation of adjacent lines in a multiplet. As can be seen in Fig. 8.21, $J(^1\text{H}-^{73}\text{Ge}) \approx 100$ Hz. The resonances of chemically equivalent nuclei do not display the effects of their mutual spin–spin coupling. Thus, a single ^{1}H signal is observed for the CH$_3$I molecule even though there is coupling between the H nuclei.

A multiplet of $2I + 1$ lines is obtained when a spin-$\frac{1}{2}$ nucleus (or a set of symmetry-related spin-$\frac{1}{2}$ nuclei) is coupled to a nucleus of spin I. In the spectrum of GeH$_4$ shown in Fig. 8.21, the single central line arises from the four equivalent H nuclei in GeH$_4$ molecules that contain Ge isotopes with $I = 0$. This central line is flanked by ten evenly spaced but less intense lines that arise from a small fraction of GeH$_4$ that contains ^{73}Ge, for which $I = \frac{9}{2}$, the four ^{1}H nuclei are coupled to the ^{73}Ge nucleus to yield a ten-line multiplet ($2 \times \frac{9}{2} + 1 = 10$).

The coupling of the nuclear spins of different elements is called **heteronuclear coupling**; the Ge−H coupling just discussed is an example. **Homonuclear coupling** between nuclei of the same element is detectable when the nuclei are in chemically inequivalent locations.

■ **A brief illustration.** The ^{19}F NMR spectrum of ClF$_3$ is shown in Fig. 8.22. The signal ascribed to the two axial F nuclei (each with $I = \frac{1}{2}$) is split into a doublet by the single equatorial ^{19}F nucleus, and the latter is split into a triplet by the two axial ^{19}F nuclei (^{19}F is in 100 per cent abundance). Thus, the pattern of ^{19}F resonances readily distinguishes this unsymmetrical structure from trigonal-planar and trigonal-pyramidal structures, both of which would have equivalent F nuclei and hence a single ^{19}F resonance. ■

The sizes of ^{1}H−^{1}H homonuclear coupling constants in organic molecules are typically 18 Hz or less. By contrast, ^{1}H−X heteronuclear coupling constants can be several hundred hertz. Homonuclear and heteronuclear coupling between nuclei other than ^{1}H can lead to coupling constants of many kilohertz. The sizes of coupling constants are often related to the geometry of a molecule by noting empirical trends. In square-planar Pt(II) complexes, J(Pt−P) is sensitive to the group *trans* to a phosphine ligand and the value of J(Pt−P) increases in the following order of *trans* ligands:

$$R^- < H^- < PR_3 < NH_3 < Br^- < Cl^-$$

For example, *cis*-[PtCl$_2$(PEt$_3$)$_2$], where Cl$^-$ is *trans* to P, has J(Pt–P) = 3.5 kHz, whereas *trans*-[PtCl$_2$(PEt$_3$)$_2$], with P *trans* to P, has J(Pt−P) = 2.4 kHz. These systematic variations allow *cis* and *trans* isomers to be distinguished quite readily. The rationalization for the variation in the sizes of the coupling constants above stems from the fact that a ligand that exerts a large *trans* influence (Section 21.4) substantially weakens the bond *trans* to itself, causing a reduction in the NMR coupling between the nuclei.

(d) Intensities

The integrated intensity of a signal arising from a group of chemically equivalent nuclei is proportional to the number of nuclei in the group. Provided sufficient time is allowed during spectrum acquisition for the full relaxation of the observed nucleus, integrated intensities can be used to aid spectral assignment for most nuclei. However, for nuclei with low sensitivities (such as ^{13}C), allowing sufficient time for full relaxation may not be realistic, so quantitative information from signal intensities is difficult to obtain. For instance, in the spectrum of ClF$_3$ the integrated ^{19}F intensities are in the ratio 2:1 (for the doublet and triplet, respectively). This pattern is consistent with the structure and the splitting pattern because it indicates the presence of two equivalent F nuclei and one inequivalent F nucleus. Thus, the pattern of ^{19}F resonances distinguishes this less symmetrical structure from a trigonal-planar structure, D_{3h}, which would have a single resonance with all F environments equivalent.

The relative intensities of the $2N + 1$ lines in a multiplet that arises from coupling to N equivalent spin-$\frac{1}{2}$ nuclei are given by Pascal's triangle (**1**); thus three equivalent protons give a 1:3:3:1 quartet. Groups of nuclei with higher spin quantum numbers give different patterns. The ^{1}H-NMR spectrum of HD, for instance, consists of three lines of equal intensity as a result of coupling to the ^{2}H nucleus ($I = 1$, with $2I + 1 = 3$ orientations).

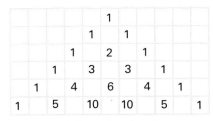

1 Pascal's triangle

EXAMPLE 8.4 Interpreting an NMR spectrum

Explain why the ^{19}F-NMR spectrum of SF$_4$ consists of two 1:2:1 triplets of equal intensity.

Answer We need to recall that an SF$_4$ molecule (**2**) has a distorted tetrahedral or 'see-saw' structure based on a trigonal-bipyramidal arrangement of electron pairs with a lone pair occupying one of the equatorial sites. The two axial F nuclei are chemically different from the two equatorial F nuclei and give two signals of equal intensity. The signals are in fact 1:2:1 triplets, as each ^{19}F nucleus couples to the two chemically distinct ^{19}F nuclei.

Self-test 8.4 (a) Explain why the ^{77}Se-NMR spectrum of SeF$_4$ consists of a triplet of triplets. (b) Use the isotope information in Table 8.4 to explain why the ^{1}H resonance of the hydrido (H$^-$) ligand in *cis*-[Rh(CO)H(PMe$_3$)$_2$] consists of eight lines of equal intensity (^{77}Se $I = \frac{1}{2}$).

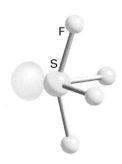

2 SF$_4$

(e) Fluxionality

The timescale of NMR is slow in the sense that structures can be resolved provided their lifetime is not less than a few milliseconds. For example, Fe(CO)$_5$ shows just one ^{13}C resonance, indicating that, on the NMR timescale, all five CO groups are equivalent. However, the IR spectrum (of timescale about 1 ps) shows distinct axial and equatorial CO groups, and by implication a trigonal-bipyramidal structure. The observed ^{13}C-NMR spectrum of Fe(CO)$_5$ is the weighted average of these separate resonances.

Because the temperature at which NMR spectra are recorded can easily be changed, samples can often be cooled down to a temperature at which the rate of interconversion becomes slow enough for separate resonances to be observed. Figure 8.23, for instance, shows the idealized ^{31}P-NMR spectra of [RhMe(PMe$_3$)$_4$] at room temperature and at $-80°$C. At low temperatures the spectrum consists of a doublet of doublets of relative intensity 3 near $\delta = -24$, which arises from the equatorial P atoms (coupled to ^{103}Rh and the single axial ^{31}P), and a quartet of doublets of intensity 1, derived from the equatorial P atom (coupled to ^{103}Rh and the three equatorial ^{31}P atoms). At room temperature the scrambling of the PMe$_3$ groups makes them all equivalent and a doublet is observed (from coupling to ^{103}Rh).

Careful control allows determination of the temperature at which the spectrum changes from the high-temperature form to the low-temperature form ('the coalescence temperature') and thence to determine the barrier to interconversion.

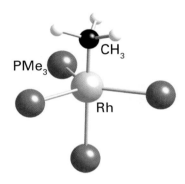

3 [RhMe(PMe$_3$)$_4$]

(f) Solid-state NMR

The NMR spectra of solids rarely show the high resolution that can be obtained from solution NMR. This difference is mainly due to anisotropic interactions such as dipolar magnetic couplings between nuclei, which are averaged in solution due to molecular tumbling, and

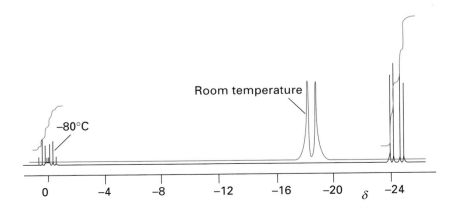

Figure 8.23 The ^{31}P-NMR spectra of [RhMe(PMe$_3$)$_4$] (**3**) at room temperature and at $-80°$C.

long-range magnetic interactions arising from the fixed atomic positions. These effects mean that in the solid state chemically equivalent nuclei might be in different magnetic environments and so have different resonance frequencies. A typical result of these additional couplings is to produce very broad resonances, often more than 10 kHz wide.

To average out anisotropic interactions, samples are spun at very high speeds (typically 10–25 kHz) at the 'magic angle' (54.7°) with respect to the field axis. At this angle the interaction of parallel magnetic dipoles and quadrupole interactions, which typically vary as $1 - 3 \cos^2 \theta$, is zero. This so-called **magic-angle spinning** (MAS) reduces the effect of anisotropy substantially but often leaves signals that are still significantly broadened compared to those from solution. The broadening of signals can sometimes be so great that signal widths are comparable to the chemical shift range for some nuclei. This broadening is a particular problem for ^{1}H, which typically has a chemical shift range of $\Delta\delta = 10$. Broad signals are less of a problem for nuclei such as ^{195}Pt, for which the range in chemical shifts is $\Delta\delta = 16\ 000$, although this large range can be reflected in large anisotropic linewidths. Quadrupolar nuclei (those with $I > \frac{1}{2}$) present additional problems as the peak position becomes field dependent and can no longer be identified with the chemical shift.

Despite these difficulties, developments in the technique have made possible the observation of high-resolution NMR spectra for solids and are of far-reaching importance in many areas of chemistry. An example is the use of ^{29}Si-MAS-NMR to determine the environments of Si atoms in natural and synthetic aluminosilicates such as zeolites. The techniques of homonuclear and heteronuclear 'decoupling' enhance the resolution of spectra and the use of multiple pulse sequences has allowed the observation of spectra with some difficult samples. The high-resolution technique CPMAS-NMR, a combination of MAS with **cross-polarization** (CP), usually with heteronuclear decoupling, has been used to study many compounds containing ^{13}C, ^{31}P, and ^{29}Si. The technique is also used to study molecular compounds in the solid state. For example, the ^{13}C-CPMAS spectrum of $[Fe_2(C_8H_8)(CO)_5]$ at $-160°C$ indicates that all C atoms in the C_8 ring are equivalent on the timescale of the experiment. The interpretation of this observation is that the molecule is fluxional even in the solid state.

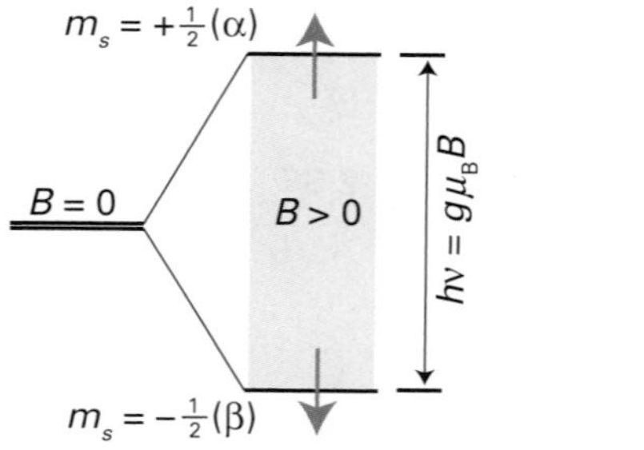

8.6 Electron paramagnetic resonance

Key points: Electron paramagnetic resonance spectroscopy is used to study compounds possessing unpaired electrons, particularly those containing a d-block element; it is often the technique of choice for identifying and studying metals such as Fe and Cu at the active sites of metalloenzymes.

Electron paramagnetic resonance (EPR; or electron spin resonance, ESR) spectroscopy, the observation of resonant absorption by unpaired electrons in a magnetic field, is a technique for studying paramagnetic species such as organic and main-group radicals. Its principal importance in inorganic chemistry is for characterizing compounds containing the d- and f-block elements.

The simplest case is for a species having one unpaired electron ($s = \frac{1}{2}$): by analogy with NMR, the application of an external magnetic field B_0 produces a difference in energy between the $m_s = +\frac{1}{2}$ and $m_s = -\frac{1}{2}$ states of the electron, with

$$\Delta E = g\mu_B B_0 \tag{8.7}$$

where μ_B is the Bohr magneton and g is a numerical factor known simply as the **g value** (Fig. 8.24). The conventional method of recording EPR spectra is to use a continuous wave (CW) spectrometer (Fig. 8.25), in which the sample is irradiated with a constant microwave frequency and the applied magnetic field is varied. The resonance frequency for most spectrometers is approximately 9 GHz, and the instrument is then known as an 'X-band spectrometer'. The magnetic field in an X-band spectrometer is about 0.3 T. Laboratories specializing in EPR spectroscopy often have a range of instruments, each operating at different fields. Thus an S-band (resonant frequency 3 GHz) spectrometer and particularly those operating at high fields, Q-band (35 GHz) and W-band (95 GHz) spectrometers, are used to complement the information gained with an X-band spectrometer. The use of high fields improves resolution and can simplify spectra by reducing interactions between paramagnetic centres.

Pulsed EPR spectrometers are becoming commercially available and offering new opportunities analogous to the way that pulsed Fourier-transform techniques have revolutionized

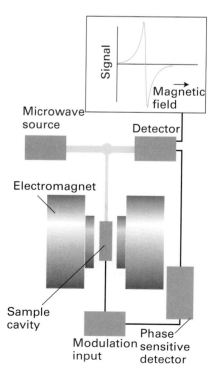

Figure 8.24 When an unpaired electron is in a magnetic field, its two orientations (α, $m_s = +\frac{1}{2}$ and β, $m_s = -\frac{1}{2}$) have different energies. Resonance is achieved when the energy separation matches the energy of the incident microwave photons.

Figure 8.25 The layout of a typical continuous wave EPR spectrometer.

NMR. Pulsed EPR techniques provide time resolution, making it possible to measure the dynamic properties of paramagnetic systems.

(a) The g value

For a free electron $g = 2.0023$ but in compounds this value is altered by spin–orbit coupling which changes the local magnetic field experienced by the electron. For many species, particularly d-metal complexes, g values may be highly anisotropic so that the resonance condition depends on the angle that the paramagnetic species makes to the applied field (Fig. 8.26). The illustration shows the EPR spectra of frozen solutions or 'glasses', expected for isotropic (all three g values the same along perpendicular axes), axial (two the same), and rhombic (all three different) spin systems.

The sample (which is usually contained in a quartz tube) comprises the paramagnetic species in dilute form, either in the solid state (doped crystals or powders) or in solution. Relaxation is so efficient for d-metal ions that the spectra are often too broad to detect; consequently, liquid nitrogen and sometimes liquid helium are used to cool the sample. Frozen solutions behave as amorphous powders, so resonances are observed at all the g values of the compound, analogous to powder X-ray diffraction. More detailed studies can be made with oriented single crystals. Provided relaxation is slow, EPR spectra can also be observed at room temperature in liquids.

Spectra can also be obtained for systems having more than one unpaired electron, such as triplet states, but the theoretical background is much more complicated. Whereas species having an odd number of electrons are usually detectable, it can be difficult to observe spectra for systems having an even number of electrons. Table 8.5 shows the suitability of common paramagnetic species for EPR detection.

(b) Hyperfine coupling

The **hyperfine structure** of an EPR spectrum, the multiplet structure of the resonance lines, is due to the coupling of the electron spin to any magnetic nuclei present. A nucleus with spin I splits an EPR line into $2I + 1$ lines of the same intensity (Fig. 8.27). A distinction is sometimes made between the hyperfine structure due to coupling to the nucleus of the atom on which the unpaired electron is primarily located and the 'superhyperfine coupling', the coupling to ligand nuclei. Superhyperfine timescoupling to ligand nuclei is used to measure the extent of electron delocalization and covalence in metal complexes (Fig. 8.28).

Table 8.5 EPR detectability of common d-metal ions

Usually easy to study		Usually difficult to study or diamagnetic	
Species	S	Species	S
Ti(III)	$\frac{1}{2}$	Ti(II)	1
Cr(III)	$\frac{3}{2}$	Ti(IV)	0
V(IV)	$\frac{1}{2}$	Cr(II)	2
Fe(III)	$\frac{1}{2}, \frac{5}{2}$	V(III)	1
Co(II)	$\frac{3}{2}, \frac{1}{2}$	V(V)	0
Ni(III)	$\frac{3}{2}, \frac{1}{2}$	Fe(II)	2, 1, 0
Ni(I)	$\frac{1}{2}$	Co(III)	0
Cu(II)	$\frac{1}{2}$	Co(I)	0
Mo(V)	$\frac{1}{2}$	Ni(II)	1
W(V)	$\frac{1}{2}$	Cu(I)	0
		Mo(VI)	0
		Mo(IV)	1, 0
		W(VI)	0

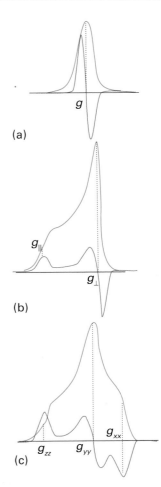

Figure 8.26 The forms of EPR powder (frozen solution) spectra expected for different types of g value anisotropy. The green line is the absorption and the red line the first derivative of the absorption (its slope). For technical reasons related to the detection technique, the first derivative is normally observed in EPR spectrometers.

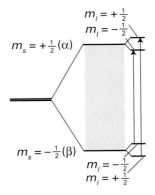

Figure 8.27 When a magnetic nucleus is present, its $2I + 1$ orientations give rise to a local magnetic field that splits each spin state of an electron into $2I + 1$ levels. The allowed transitions ($\Delta m_s = +1$, $\Delta m_I = 0$) give rise to the hyperfine structure of an EPR spectrum.

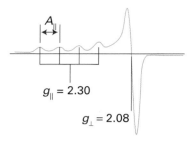

$g_{\parallel} = 2.30$

$g_{\perp} = 2.08$

Figure 8.28 The EPR spectrum of Cu^{2+} (d^9, one unpaired electron) in frozen aqueous solution. The tetragonally distorted Cu^{2+} ion shows an axially symmetrical spectrum in which hyperfine coupling to Cu ($I = \frac{3}{2}$, four hyperfine lines) is clearly evident for the $g_{\parallel}$ component.

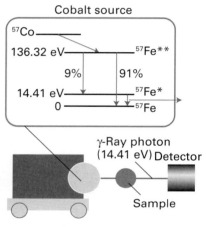

Figure 8.29 The layout of a Mössbauer spectrometer. The speed of the carriage is adjusted until the Doppler shifted frequency of the emitted γ-ray matches the corresponding nuclear transition in the sample. The inset shows the nuclear transitions responsible for the emission of the γ-ray.

EXAMPLE 8.5 Interpreting superhyperfine coupling

Use the data provided in Table 8.4 to suggest how the EPR spectrum of a Co(II) complex might change when a single OH^- ligand is replaced by an F^- ligand.

Answer We need to note that ^{19}F (100 per cent abundance) has nuclear spin $I = \frac{1}{2}$ whereas ^{16}O (close to 100 per cent) has $I = 0$. Consequently, any part of the EPR spectrum might be split into two lines.

Self-test 8.5 Predict how you might show that an EPR signal that is characteristic of a new material arises from tungsten sites.

8.7 Mössbauer spectroscopy

Key point: Mössbauer spectroscopy is based on the resonant absorption of γ-radiation by nuclei and exploits the fact that nuclear energies are sensitive to the electronic and magnetic environment.

The **Mössbauer effect** makes use of recoilless absorption and emission of γ-radiation by a nucleus. To understand what is involved, consider a radioactive ^{57}Co nucleus that decays by electron capture to produce an excited state of ^{57}Fe, denoted $^{57}Fe^{**}$ (Fig. 8.29). This nuclide decays to another excited state, denoted $^{57}Fe^*$, that lies 14.41 eV above the ground state and which emits a γ-ray of energy 14.41 eV as it decays. If the emitting nucleus (the source) is pinned down in a rigid lattice the nucleus does not recoil and the radiation produced is highly monochromatic.

If a sample containing ^{57}Fe (which occurs with 2 per cent natural abundance) is placed close to the source, the monochromatic γ-ray emitted by $^{57}Fe^*$ can be expected to be absorbed resonantly. However, because changes in the exact electronic and magnetic environment affect the nuclear energy levels to a small degree, resonant absorption occurs only if the environment of the sample ^{57}Fe is chemically identical to that of the emitting $^{57}Fe^*$ nucleus. It might seem that the energy of the γ-ray cannot easily be varied, but the Doppler effect can be exploited for moving the transmitting nucleus at a speed v relative to the sample and a frequency shift is induced of magnitude $\Delta \nu = (v/c)\nu_{\gamma}$. This shift is sufficient even for velocities of a few millimetres per second, to match the absorption frequency to the transmitted frequency. A **Mössbauer spectrum** is a portrayal of the resonant absorption peaks that occur as the velocity of the source is changed.

The Mössbauer spectrum of a sample containing iron in a single chemical environment can be expected to consist of a single line due to absorption of radiation at the energy required (ΔE) to excite the nucleus from its ground state to the excited state. The difference between ΔE of the sample and that of metallic ^{57}Fe is called the **isomer shift** and is expressed in terms of the velocity that is required to achieve resonance by the Doppler shift. The value of ΔE depends on the magnitude of the electron density at the nucleus and although this effect is primarily due to s electrons (because their wavefunctions are nonzero at the nucleus), shielding effects cause ΔE to be sensitive to the number of p and d electrons too. As a result, different oxidation states, such as Fe(II), Fe(III), and Fe(IV), can be distinguished as well as ionic and covalent bonding.

The element most suited for study by Mössbauer spectroscopy is iron. This element has great importance and is common in minerals, oxides, alloys, and biological samples, for which other many other characterisation techniques, for example NMR, are less effective. Mössbauer spectroscopy is also used to study some other nuclei, including ^{119}Sn, ^{129}I, and ^{197}Au which have suitable nuclear energy levels and γ-emission half-lives to get useful spectra. Because a $^{57}Fe^*$ nucleus has $I = \frac{3}{2}$, it possesses an electric quadrupole moment (a nonspherical distribution of electric charge) which interacts with electric field gradients. As a result, and providing the electronic environment of the nucleus is not isotropic, the Mössbauer spectrum splits into two lines of separation ΔE_Q (Fig. 8.30). The splitting is a good guide as to the state of Fe in proteins and minerals as it depends on the oxidation state and the distribution of d-electron density. Magnetic fields produced by a large magnet or internally in some ferromagnetic materials also cause changes in the energies of the various spin orientation states of an $^{57}Fe^*$ nucleus. As a result, the Mössbauer spectrum splits into six lines.

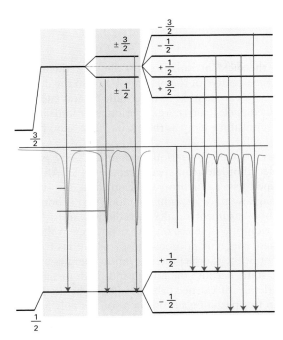

Figure 8.30 The effects of an electric field gradient and a magnetic field on the energy levels involved in the Mössbauer technique for Fe samples. The derived spectra show, from left to right, the origins of the isomer shift, quadrupole coupling, and magnetic hyperfine coupling (but no quadrupole splitting). The spectrum of (a) $K_4Fe(CN)_6 \cdot 3H_2O$, octahedral low spin d^6, is representative of a highly symmetric environment and shows a single peak with an isomer shift. The spectrum of (b) $FeSO_4 \cdot 7H_2O$, d^6 in a nonsymmetric environment, shows quadrupolar splitting.

EXAMPLE 8.6 Interpreting a Mössbauer spectrum

The isomer shifts of Fe(II) compounds relative to metallic iron, Fe(0), are generally in the range $+1$ to $+1.5$ mm s^{-1} whereas isomer shifts for Fe(III) compounds lie in the range $+0.2$ to $+0.5$ mm s^{-1}. Explain these values in terms of the electronic configurations of Fe(0), Fe(II), and Fe(III).

Answer The outermost electron configurations of Fe(0), Fe(II), and Fe(III) are $4s^2 3d^6$, $3d^6$, and $3d^5$, respectively. The s-electron density at the nucleus is reduced in Fe(II) compared with Fe(0), producing a large positive isomer shift. When a 3d electron is removed to produce Fe(III) from Fe(II), there is a small increase in s-electron density at the nucleus (as the 3d electrons partly screen the nucleus from the inner s electrons) and the isomer shift becomes less positive.

Self-test 8.6 Predict a likely isomer shift for iron in Sr_2FeO_4.

Ionization-based techniques

Ionization-based techniques measure the energies of products, electrons, or molecular fragments generated when a sample is ionized by bombardment with high-energy radiation or particles.

8.8 Photoelectron spectroscopy

Key point: Photoelectron spectroscopy is used to determine the energies and order of orbitals in molecules and solids by analysing the kinetic energies of photoejected electrons.

The basis of **photoelectron spectroscopy** (PES) is the measurement of the kinetic energies of electrons (photoelectrons) emitted by ionization of a sample that is irradiated with high-energy monochromatic radiation (Fig. 8.31). It follows from the conservation of energy

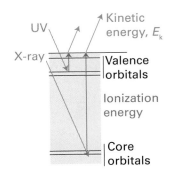

Figure 8.31 In photoelectron spectroscopy, high-energy electromagnetic radiation (UV for the ejection of valence electrons, X-ray for core electrons) expels an electron from its orbital, and the kinetic energy of the photoelectron is equal to the difference between the photon energy and the ionization energy of the electron.

that the kinetic energy of the ejected photoelectrons, E_k, is related to their ionization energies, E_i, from their orbitals by the relation

$$E_k = h\nu - E_i \tag{8.8}$$

where ν is the frequency of the incident radiation. **Koopmans' theorem** states that the ionization energy is equal to the negative of the orbital energy, so the determination of the kinetic energies of photoelectrons can be used to determine orbital energies. The theorem assumes that the energy involved in electron reorganization after ionization is offset by the increase in electron–electron repulsion energy as the orbital contracts. This approximation is usually taken as reasonably valid.

There are two major types of photoionization technique, **X-ray photoelectron spectroscopy** (XPS) and **ultraviolet photoelectron spectroscopy** (UPS). Although much more intense sources can be obtained using synchrotron beam lines, the standard laboratory source for XPS is usually a magnesium or aluminium anode that is bombarded by a high-energy electron beam. This bombardment results in radiation at 1.254 and 1.486 keV, respectively, due to the transition of a 2p electron into a vacancy in the 1s orbital caused by ejection of an electron. These energetic photons cause ionizations from core orbitals in other elements that are present in the sample; the ionization energies are characteristic of the element and its oxidation state. Because the linewidth is high (usually 1–2 eV), XPS is not suitable for probing fine details of valence orbitals but can be used to study the band structures of solids. The mean free path of electrons in a solid is only about 1 nm, so XPS is suitable for surface elemental analysis, and in this application it is commonly known as **electron spectroscopy for chemical analysis** (ESCA).

The source for UPS is typically a helium discharge lamp that emits He(I) radiation (21.22 eV) or He(II) radiation (40.8 eV). Linewidths are much smaller than in XPS, so the resolution is far greater. The technique is used to study valence-shell energy levels and the vibrational fine structure often provides important information on the bonding or antibonding character of the orbitals from which electrons are ejected (Fig. 8.32). When the electron is removed from a nonbonding orbital the product is formed in its vibrational ground state, and a narrow line is observed. However, when the electron is removed from a bonding or antibonding orbital the resulting ion is formed in several different vibrational states and extensive fine structure is observed. Bonding and antibonding orbitals can be distinguished by determining whether the vibrational frequencies in the resulting ion are higher or lower than for the original molecule.

Another useful aid is the comparison of photoelectron intensities for a sample irradiated with He(I) and He(II). The higher energy source preferentially ejects electrons from d or f orbitals, allowing these contributions to be distinguished from s and p orbitals, for which He(I) causes higher intensities. The origin of this effect lies in differences in absorption cross-sections (see *Further reading*).

8.9 X-ray absorption spectroscopy

Key point: X-ray absorption spectra can be used to determine the oxidation state of an element in a compound and to investigate its local environment.

As mentioned in the previous section, the intense X-ray radiation from synchrotron sources may be used to eject electrons from the cores of elements present in a compound. **X-ray absorption spectra** (XAS) are obtained by varying the photon energy across a range of energies at which electrons in the various atoms present in a compound can be excited and ionized (typically between 0.1 and 100 keV). The characteristic absorption energies correspond to the binding energies of different inner-shell electrons of the various elements present. Thus the frequency of an X-ray beam may be swept across an absorption edge of a selected element and information on the oxidation state and neighbourhood of this chosen chemical element obtained.

Figure 8.33 shows a typical X-ray absorption spectrum. Each region of the spectrum can provide different useful information on the chemical environment of the element under investigation:

1. Just prior to the absorption edge is the 'pre-edge' where core electrons are excited to higher empty orbitals but not ejected. This 'pre-edge structure' can provide information on the energies of excited electronic states and also on the local symmetry of the atom.

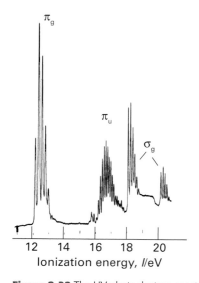

Figure 8.32 The UV photoelectron spectra of O_2. Loss of an electron from $2\sigma_g$ (see the MO energy-level diagram, Fig. 2.12) gives rise to two bands because the unpaired electron that remains can be parallel or antiparallel to the two unpaired electrons in the $1\pi_g$ orbitals.

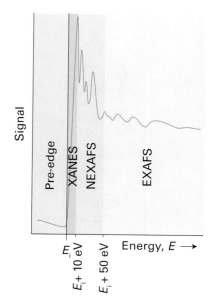

Figure 8.33 A typical X-ray absorption edge spectrum defining the various regions discussed in the text.

2. In the edge region, where the photon energy, E is between E_i and $E_i + 10$ eV, where E_i is the ionization energy, the 'X-ray absorption near-edge structure' (XANES) is observed. Information that can be extracted from the XANES region includes oxidation state and the coordination environment, including any subtle geometrical distortions. The near-edge structure can also be used as a 'fingerprint', as it is characteristic of a specific environment and valence state. The presence and quantity of a compound in a mixture may be determined from analysis of this region of the spectrum.

3. The 'near-edge X-ray absorption fine structure' (NEXAFS) region lies between $E_i + 10$ eV and $E_i + 50$ eV. It has particular application to chemisorbed molecules on surfaces because it is possible to infer information about the orientation of the adsorbed molecule.

4. The 'extended X-ray absorption fine structure' (EXAFS) region lies at energies greater than $E_i + 50$ eV. The photoelectrons ejected from a particular atom by absorption of X-ray photons with energies in this region may be backscattered by any adjacent atoms. This effect can result in an interference pattern that is detected as periodic variations in intensity at energies just above the absorption edge. In EXAFS these variations are analysed to reveal the nature (in terms of their electron density) and number of nearby atoms and the distance between the absorbing atom and the scattering atom. An advantage of this method is that it can provide bond lengths in amorphous samples and for species in solutions.

8.10 Mass spectrometry

Key point: Mass spectrometry is a technique for determining the mass of a molecule and of its fragments.

Mass spectrometry measures the mass-to-charge ratio of gaseous ions. The ions can be either positively or negatively charged, and it is normally trivial to infer the actual charge on an ion and hence the mass of a species. It is a destructive analytical technique because the sample cannot be recovered for further analysis.

The precision of measurement of the mass of the ions varies according to the use being made of the spectrometer (Fig. 8.34). If all that is required is a crude measure of the mass, for instance to within $\pm m_u$ (where m_u is the atomic mass constant, 1.66054×10^{-27} kg), then the resolution of the mass spectrometer need be of the order of only 1 part in 10^4. In contrast, to determine the mass of individual atoms so that the mass defect can be determined, the precision must approach 1 part in 10^{10}. With a mass spectrometer of this precision, molecules of nominally the same mass such as $^{12}C^{16}O$ (of mass $27.9949m_u$) can be distinguished from $^{14}N_2$ (of mass $28.0061m_u$) and the elemental and isotopic composition of ions of nominal mass less than $1000m_u$ may be determined unambiguously.

(a) Ionization and detection methods

The major practical challenge with mass spectrometry is the conversion of a sample into gaseous ions. Typically, less than a milligram of compound is used. Many different experimental arrangements have been devised to produce gas-phase ions but all suffer from a tendency to fragment the compound of interest. **Electron impact ionization** (EI) relies on bombarding a sample with high-energy electrons to cause both vaporization and ionization. A disadvantage is that EI tends to induce considerable decomposition in larger molecules. **Fast atom bombardment** (FAB) is similar to EI, but bombardment of the sample with

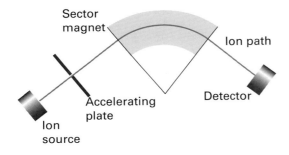

Figure 8.34 A magnetic sector mass spectrometer. The molecular fragments are deflected according to their mass-to-charge ratio, allowing for separation at the detector.

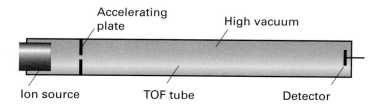

Figure 8.35 A time-of-flight (TOF) mass spectrometer. The molecular fragments are accelerated to different speeds by the potential difference and arrive at different times at the detector.

fast neutral atoms is used to vaporize and ionize the sample; it induces less fragmentation than EI. **Matrix-assisted laser desorption/ionization** (MALDI) is similar to EI, but a short laser pulse is used to the same effect; this technique is particularly effective with polymeric samples. In **electrospray ionization** (ESI), charged droplets of solution are sprayed into a vacuum chamber where solvent evaporation results in generation of individually charged ions; ESI mass spectrometry is becoming more widely used and is often the method of choice for ionic compounds in solution.

The traditional method of ion separation relies on the acceleration of ions with an electric field and then using a magnetic field to deflect the moving ions: ions with a lower mass-to-charge ratio are deflected more than heavier ions. As the magnetic field is changed, ions with different mass-to-charge ratio are directed on to the detector (Fig. 8.34). In a **time-of-flight** (TOF) mass spectrometer, the ions from a sample are accelerated by an electric field for a fixed time and then allowed to fly freely (Fig. 8.35). Because the force on all the ions of the same charge is the same, the lighter ions are accelerated to higher speeds than the heavier ions and strike a detector sooner. In an **ion cyclotron resonance** (ICR) mass spectrometer (often denoted FTICR, for Fourier transform-ICR) ions are collected in a small cyclotron cell inside a strong magnetic field. The ions circle round in the magnetic field, effectively behaving as an electric current. Because an accelerated current generates electromagnetic radiation, the signal generated by the ions can be detected and used to establish their mass-to-charge ratio.

Mass spectrometry is most widely used in organic chemistry but is also very useful for the analysis of inorganic compounds. However, many inorganic compounds, such as those with ionic structures or covalently bonded networks (for example SiO_2), are not volatile and do not fragment into molecular ion units (even with the MALDI technique) so cannot be analysed by this method. Conversely, the weaker bonding in some inorganic coordination compounds means that they fragment much more easily than organic compounds in the mass spectrometer.

(b) Interpretation

Figure 8.36 shows a typical mass spectrum. To interpret a spectrum, it is helpful to detect a peak corresponding to the singly charged, intact molecular ion. Sometimes a peak occurs at half the molecular mass and is then ascribed to a doubly charged ion. Peaks from multiply charged ions are usually easy to identify because the separation between the peaks

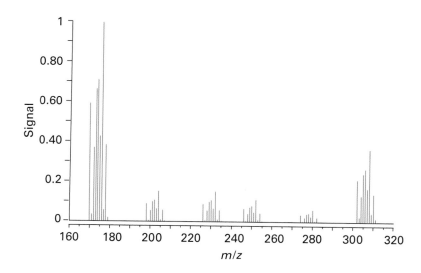

Figure 8.36 The mass spectrum of $[Mo(\eta^6\text{-}C_6H_6)(CO)_2PMe_3]$.

from the different isotopomers is no longer m_u but fractions of that mass. For instance, in a doubly charged ion, isotopic peaks are $\frac{1}{2}m_u$ apart, in a triply charged ion they are $\frac{1}{3}m_u$ apart, and so on.

In addition to indicating the mass of the molecule or ion that is being studied (and hence its molar mass), a mass spectrum also provides information about fragmentation pathways of molecules. This information can be used to confirm structural assignments. For example, complex ions often lose ligands and peaks are observed that correspond to the complete ion less one or more ligands.

Multiple peaks are observed when an element is present as a number of isotopes (for instance, chlorine is 75.5 per cent ^{35}Cl and 24.5 per cent ^{37}Cl). Thus, for a molecule containing chlorine, the mass spectrum will show two peaks $2m_u$ apart in an intensity ratio of about 3:1. Different patterns of peaks are obtained for elements with a more complex isotopic composition and can be used to identify the presence of an element in compounds of unknown composition. An Hg atom, for instance, has six isotopes in significant abundance (Fig. 8.37). The actual proportion of isotopes of an element varies according to its geographic source, and this subtle aspect is easily identified with high-resolution mass spectrometers. Thus, the precise determination of the proportions of isotopes can be used to determine the source of a sample.

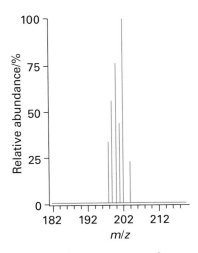

Figure 8.37 The mass spectrum of a sample containing mercury showing the isotopic composition of the atoms.

EXAMPLE 8.7 **Interpreting a mass spectrum**

Figure 8.36 shows part of the mass spectrum of $[Mo(\eta^6\text{-}C_6H_6)(CO)_2PMe_3]$. Assign the main peaks.

Answer The complex has an average molecular mass of $306m_u$, but a simple molecular ion is not seen because Mo has a large number of isotopes. Ten peaks centred on $306m_u$ are detected. The most abundant isotope of Mo is ^{98}Mo (24 per cent) and the ion that contains this isotope has the highest intensity of the peaks of the molecular ion. In addition to the peaks representing the molecular ion, peaks at $M^+ - 28$, $M^+ - 56$, $M^+ - 76$, $M^+ - 104$, and $M^+ - 132$ are seen. These peaks represent the loss of one CO, two CO, PMe_3, $PMe_3 + CO$, and $PMe_3 + 2CO$ ligands, respectively, from the parent compound.

Self-test 8.7 Explain why the mass spectrum of $ClBr_3$ (Fig. 8.38) consists of five peaks separated by $2m_u$.

Chemical analysis

A classic application of physical techniques is to the determination of the elemental composition of compounds. The techniques now available are highly sophisticated and in many cases can be automated to achieve rapid, reliable results. In this section we include thermal techniques that can be used to follow the phase changes of substances without change of composition, as well as processes that result in changes of composition. In each of these techniques the compound is usually destroyed during the analysis.

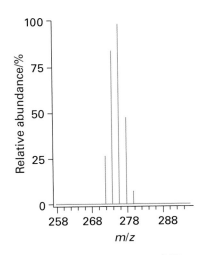

Figure 8.38 The mass spectrum of $ClBr_3$.

8.11 Atomic absorption spectroscopy

Key point: Almost every metallic element can be determined quantitatively by using the spectral absorption characteristics of atoms.

The principles of atomic absorption spectroscopy are similar to those of UV–visible spectroscopy except that the absorbing species are free atoms or ions. Unlike molecules, atoms and ions do not have rotational or vibrational energy levels and the only transitions that occur are between electronic energy levels. Consequently, atomic absorption spectra consist of sharply defined lines rather than the broad bands typical of molecular spectroscopy.

Figure 8.39 shows the basic components of an atomic absorption spectrophotometer. The gaseous sample is exposed to radiation of a specific wavelength from a 'hollow cathode' lamp, which consists of a cathode constructed of a particular element and a tungsten anode in a sealed tube filled with neon. If a particular element is present in the sample, the radiation emitted by the lamp for that element is reduced in intensity because it stimulates absorption. By determining the level of absorption relative to standard materials a quantitative measurement can be made of the amount of the element. A different lamp is required for each element that is to be analysed.

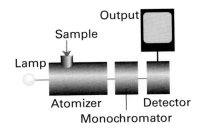

Figure 8.39 The layout of a typical atomic absorption spectrophotometer.

The major differences in instrumentation arise from the different methods used to convert the analyte (the substance being analysed) to free, unbound atoms or ions. In **flame atomization** the analyte solution is mixed with the fuel in a 'nebulizer', which creates an aerosol. The aerosol enters the burner where it passes into a fuel–oxidant flame. Typical fuel–oxidant mixtures are acetylene–air, which produces flame temperatures of up to 2500 K, and acetylene–nitrous oxide, which generates temperatures of up to 3000 K. A common type of 'electrothermal atomizer' is the graphite furnace. The temperatures reached in the furnace are comparable to those attained in a flame atomizer but detection limits can be 1000 times better. The increased sensitivity is due to the ability to generate atoms quickly and keep them in the optical path for longer. Another advantage of the graphite furnace is that solid samples may be used. Because the ionization process may produce spectral lines from other components of the analyte, a monochromator is placed after the atomizer to isolate the desired wavelength for passage to the detector.

Almost every metallic element can be analysed using atomic absorption spectroscopy, although not all with high sensitivity or a usefully low detection limit. For example, the detection limit for Cd in a flame ionizer is 1 part per billion (1 ppb = 1 in 10^9) whereas that for Hg is only 500 ppb. Limits of detection using a graphite furnace can be as low as 1 part in 10^{15}. Direct determination is possible for any element for which hollow cathode lamp sources are available. Other species can be determined by indirect procedures. For example, PO_4^{3-} reacts with MoO_4^{2-} in acid conditions to form $H_3PMo_{12}O_{40}$, which can be extracted into an organic solvent and analysed for molybdenum. To analyse for a particular element, a set of calibration standards is prepared in a similar matrix to the sample, and the standards and the sample are analysed under the same conditions.

8.12 CHN analysis

Key point: The carbon, hydrogen, nitrogen, oxygen, and sulfur content of a sample can be determined by high-temperature decomposition.

Instruments are available that allow automated analysis of C, H, N, O, and S. Figure 8.40 shows the arrangement for an instrument that analyses for C, H, and N, sometimes referred to as **CHN analysis**. The sample is heated to 900°C in oxygen and a mixture of carbon dioxide, carbon monoxide, water, nitrogen, and nitrogen oxides is produced. A stream of helium sweeps the products into a tube furnace at 750°C, where copper reduces nitrogen oxides to nitrogen and removes oxygen. Copper oxide converts carbon monoxide to carbon dioxide. The resulting mixture is analysed by passing it through a series of three thermal conductivity detectors. The first detector measures hydrogen and then water is removed in a trap. At the second detector the carbon is measured, and carbon dioxide is removed in a second trap. The remaining nitrogen is measured at the third detector. The data obtained from this technique are reported as mass percentage C, H, and N.

Oxygen may be analysed if the reaction tube is replaced with a quartz tube filled with carbon that has been coated with catalytic platinum. When the gaseous products are swept through this tube, the oxygen is converted to carbon monoxide, which is then converted to carbon dioxide by passage over hot copper oxide. The rest of the procedure is the same as described above. Sulfur can be measured if the sample is oxidized in a tube filled with copper oxide. Water is removed by trapping in a cool tube and the sulfur dioxide is determined at what is normally the hydrogen detector.

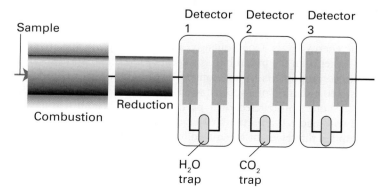

Figure 8.40 The layout of the apparatus used for CHN analysis.

EXAMPLE 8.8 Interpreting CHN analytical data

A CHN analysis of a compound of iron gave the following mass percentages for the elements present C: 64.54, N: 0, and H: 5.42, with the residual mass being iron. Determine the empirical formula of the compound.

Answer The molar masses of C, H, and Fe are 12.01, 1.008, and 55.85 g mol^{-1}, respectively. The mass of each element in exactly 100 g of the sample is 64.54 g C, 5.42 g H, and (the difference from 100 g) 30.04 g Fe. The amounts present are therefore

$$n(C) = \frac{64.54 \text{ g}}{12.01 \text{ g mol}^{-1}} = 5.37 \text{ mol}$$

$$n(H) = \frac{5.42 \text{ g}}{1.008 \text{ g mol}^{-1}} = 5.38 \text{ mol}$$

$$n(Fe) = \frac{30.04 \text{ g}}{55.85 \text{ g mol}^{-1}} = 0.538 \text{ mol}$$

These amounts are in the ratio 5.37:5.38:0.538 ≈ 10:10:1. The empirical formula of the compound is therefore C$_{10}$H$_{10}$Fe.

Self-test 8.8 Why are the percentages of hydrogen in 5d-series compounds as determined by CHN methods less accurate than those determined for the analogous 3d-series compounds?

8.13 X-ray fluorescence elemental analysis

Key points: Qualitative and quantitative information on the elements present in a compound may be obtained by exciting and analysing X-ray emission spectra.

As discussed in Section 8.9, ionization of core electrons can occur when a material is exposed to short-wavelength X-rays. When an electron is ejected in this way, an electron from a higher energy orbital can take its place and the difference in energy is released in the form of a photon, which is also typically in the X-ray region with an energy characteristic of the atoms present. This fluorescent radiation can be analysed either by energy-dispersive analysis or by wavelength-dispersive analysis. By matching the peaks in the spectrum with the characteristic values of the elements it is possible to identify the presence of a particular element. This is the basis of the X-ray fluorescence (XRF) technique. The intensity of the characteristic radiation is also directly related to the amount of each element in the material. Once the instrument is calibrated with appropriate standards it can be used to determine quantitatively most elements with Z > 8 (oxygen). Figure 8.41 shows a typical XRF energy-dispersive spectrum.

A technique similar to XRF is used in electron microscopes where the method is known as energy-dispersive analysis of X-rays (EDAX) or energy-dispersive spectroscopy (EDS). Here the X-rays generated by bombardment of a sample by energetic electrons result in the ejection of core electrons and X-ray emission occurs as the outer electrons fall into the vacancies in the core levels. These X-rays are characteristic of the elements present and their intensities are representative of the amounts present. The spectrum can be analysed to determine qualitatively and quantitatively the presence and amount of most elements (generally those with Z > 8) in the material.

8.14 Thermal analysis

Key points: Thermal methods include thermogravimetric analysis, differential thermal analysis, and differential scanning calorimetry.

Thermal analysis is the analysis of a change in a property of a sample induced by heating. The sample is usually a solid and the changes that occur include melting, phase transition, sublimation, and decomposition.

The analysis of the change in the mass of a sample on heating is known as **thermogravimetric analysis** (TGA). The measurements are carried out using a *thermobalance*,

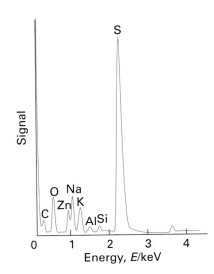

Figure 8.41 An XRF spectrum obtained from a metal silicate sample showing the presence of various elements by their characteristic X-ray emission lines.

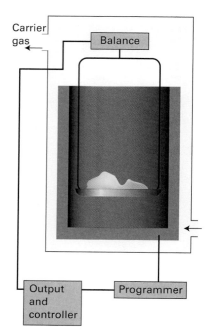

Figure 8.42 A thermogravimetric analyser: the mass of the sample is monitored as the temperature is raised.

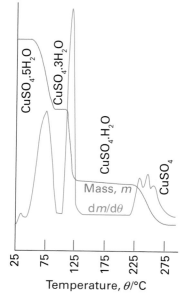

Figure 8.43 The thermogravimetric curve obtained for $CuSO_4.5H_2O$ as the temperature is raised from 20°C to 500°C. The red line is the mass of the sample and the green line is its first derivative (the slope of the red line).

which consists of an electronic microbalance, a temperature-programmable furnace, and a controller, which enables the sample to be simultaneously heated and weighed (Fig. 8.42). The sample is weighed into a sample holder and then suspended from the balance within the furnace. The temperature of the furnace is usually increased linearly, but more complex heating schemes, isothermal heating (heating that maintains constant temperature at a phase transition), and cooling protocols can also be used. The balance and furnace are situated within an enclosed system so that the atmosphere can be controlled. That atmosphere may be inert or reactive, depending on the nature of the investigation, and can be static or flowing. A flowing atmosphere has the advantage of carrying away any volatile or corrosive species and prevents the condensation of reaction products. In addition, any species produced can be fed into a mass spectrometer for identification.

Thermogravimetric analysis is most useful for desorption, decomposition, dehydration, and oxidation processes. For example, the thermogravimetric curve for $CuSO_4.5H_2O$ from room temperature to 300°C shows three stepwise mass losses (Fig. 8.43), corresponding to the three stages in the dehydration to form first $CuSO_4.3H_2O$, then $CuSO_4.H_2O$, and finally $CuSO_4$.

The most widely used thermal method of analysis is **differential thermal analysis** (DTA). In this technique the temperature of the sample is compared to that of a reference material while they are both subjected to the same heating procedure. In a DTA instrument, the sample and reference are placed in low thermal conductivity sample holders that are then located within cavities in a block in the furnace. Common reference samples for the analysis of inorganic compounds are alumina, Al_2O_3, and carborundum, SiC. The temperature of the furnace is increased linearly and the difference in temperature between the sample and the reference is plotted against the furnace temperature. If an endothermic event takes place within the sample, the temperature of the sample lags behind that of the reference and a minimum is observed on the DTA curve. If an exothermal event takes place, the temperature of the sample rises above that of the reference and a maximum is observed on the curve. The area under the endotherm or exotherm (the resulting curve in each case) is related to the enthalpy change accompanying the thermal event.

A technique closely related to DTA is **differential scanning calorimetry** (DSC). In DSC, the sample and the reference are maintained at the same temperature throughout the heating procedure by using separate power supplies to the sample and reference holders. Any difference between the power supplied to the sample and reference is recorded against the furnace temperature. Thermal events appear as deviations from the DSC baseline as either endotherms or exotherms, depending on whether more or less power has to be supplied to the sample relative to the reference. In DSC, endothermic reactions are usually represented as positive deviations from the baseline, corresponding to increased power supplied to the sample. Exothermic events are represented as negative deviations from the baseline.

The information obtained from DTA and DSC is very similar. The former can be used up to higher temperatures although the quantitative data, such as the enthalpy of a phase change, obtained from DSC are more reliable. Both DTA and DSC are used for 'fingerprint' comparison of the results obtained from a sample with those of a reference material. Information about the temperatures and enthalpy changes of transitions, such as a change in structure or melting, can be extracted.

EXAMPLE 8.9 Interpreting thermal analysis data

When a sample of bismuth nitrate hydrate, $Bi(NO_3)_3.nH_2O$, of mass 100 mg was heated to 500°C and dryness, the loss in mass observed was 18.56 mg. Determine n.

Answer We need to make a stoichiometric analysis of the decomposition, $Bi(NO_3)_3.nH_2O \rightarrow Bi(NO_3)_3 + nH_2O$ to determine the value of n. The molar mass of $Bi(NO_3)_3.nH_2O$ is $395.01 + 18.02n$ g mol^{-1}, so the initial amount of $Bi(NO_3)_3.nH_2O$ present is $(100$ mg$)/(395.01 + 18.02n$ g mol$^{-1})$. As each formula unit of $Bi(NO_3)_3.nH_2O$ contains n mol H_2O, the amount of H_2O present in the solid is n times this amount, or $n(100$ mg$)/(395.01 + 18.02n$ g mol$^{-1}) = 100n/(395.01 + 18.02n)$ mmol. The mass loss is 18.56 mg. This loss is entirely due to the loss of water, so the amount of H_2O lost is $(18.56$ mg$)/(18.02$ g mol$^{-1}) = 1.030$ mmol. We equate this amount to the amount of H_2O in the solid initially:

$$\frac{100n}{395.01 + 18.02n} = 1.030$$

(The units mmol have cancelled.) It follows that $n = 5$, and that the solid is $Bi(NO_3)_3.5H_2O$.

Self-test 8.9 Reduction of a 10.000 mg sample of an oxide of tin in hydrogen at 600°C resulted in the formation of 7.673 mg of tin metal. Determine the stoichiometry of the tin oxide.

Magnetometry

Key point: Magnetometry is used to determine the characteristic response of a sample to an applied magnetic field.

The classic way of monitoring the magnetic properties of a sample is to measure the attraction into or repulsion out of an inhomogeneous magnetic field by monitoring the change in the apparent weight of a sample when the magnetic field is applied by using a **Gouy balance** (Fig. 8.44a). The sample is hung on one side of a balance by a fine thread so that one end of it lies in the field of a powerful electromagnet and the other end of the sample is in just the Earth's magnetic field. The sample is weighed with the electromagnet field applied and then with it turned off. From the change in apparent weight, the force acting on the sample as a result of the application of the field can be determined and from this, with a knowledge of various instrumental constants, sample volume and the molar mass, the molar susceptibility. The effective magnetic moment of a d-metal ion present in a material may be deduced from the magnetic susceptibility and used to infer the number of unpaired electrons and the spin state (Chapter 20). In a **Faraday balance** a magnetic field gradient is generated between two curved magnets; this technique yields precise susceptibility measurements and also allows collection of magnetization data as a function of the magnitude and direction of the applied field.

The more modern **vibrating sample magnetometer** (VSM, Fig. 8.44b) measures the magnetic properties of a material by using a modified version of a Gouy balance. The sample is placed in a uniform magnetic field, which induces a net magnetization. As the sample is vibrated, an electrical signal is induced in suitably placed pick-up coils. The signal has the same frequency of vibration and its amplitude is proportional to the induced magnetization. The vibrating sample may be cooled or heated, allowing a study of magnetic properties as a function of temperature.

Measurements of magnetic properties are now made more routinely using a **superconducting quantum interference device** (SQUID, Fig. 8.45). A SQUID makes use of the quantization of magnetic flux and the property of current loops in superconductors that are part of the circuit. The current that flows in the loop in a magnetic field is determined by the value of the magnetic flux and hence the magnetic susceptibility of the sample.

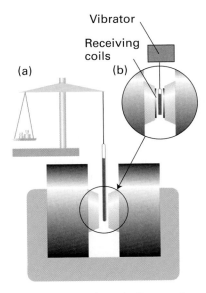

Figure 8.44 A schematic diagram of (a) a Gouy balance and (b, insert) a schematic diagram of a vibrating sample magnetometer modification.

Electrochemical techniques

Key point: Cyclic voltammetry measures the electrical currents due to reduction and oxidation of electroactive species in solution.

In **cyclic voltammetry** the current flowing between two electrodes immersed in a solution is measured as the potential difference is changed cyclically. It provides direct information on reduction potentials and the stabilities of different products of oxidation or reduction. The technique gives rapid qualitative insight into the redox properties of an electroactive compound and reliable quantitative information on its thermodynamic and kinetic properties. The 'working electrode' at which the electrochemical reaction of interest occurs is usually constructed from platinum, silver, gold, or graphite. The reference electrode is normally a silver/silver-chloride electrode and the counter electrode is normally platinum.

To understand what is involved, consider the redox couple $[Fe(CN)_6]^{3-}/[Fe(CN)_6]^{4-}$, where, initially, only the reduced form (the Fe(II) complex) is present (Fig. 8.46). The concentration of electroactive species is usually quite low (less than 0.001 M) and the solution contains a relatively high concentration of inert 'supporting' electrolyte (at concentrations greater than about 0.1 M) to provide conductivity. The potential difference is applied between the working electrode and the reference electrode and is scanned back and forth between two limits, tracing out a triangular waveform.

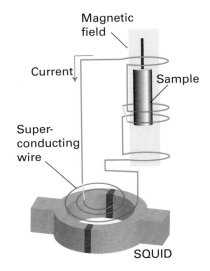

Figure 8.45 The magnetic susceptibility of a sample is measured by using a SQUID: the sample, which is exposed to a magnetic field, is moved through the loops in small increments and the potential difference across the SQUID is monitored.

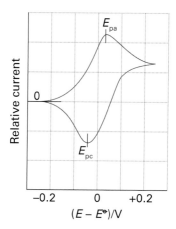

Figure 8.46 The cyclic voltammogram for an electroactive species present in solution as the reduced form and displaying a reversible one-electron reaction at an electrode. The peak potentials E_{pa} and E_{pc} for oxidation and reduction, respectively, are separated by 0.06 V. The reduction potential is the mean of E_{pa} and E_{pc}.

No current flows while the potential is low. As it approaches the reduction potential of the Fe(III)/Fe(II) couple, the Fe(II) is oxidized at the working electrode and a current starts to flow. This current rises to a peak then decreases steadily because Fe(II) becomes depleted close to the electrode (the solution is unstirred) and must be supplemented by species diffusing from increasingly distant regions of the solution. Once the upper potential limit is reached, the potential sweep is reversed. Initially, Fe(II) diffusing to the electrode continues to be oxidized but eventually the potential difference becomes sufficiently negative to reduce the Fe(III) that had been formed; the current reaches a peak then decreases gradually to zero as the lower potential limit is reached.

The average of the two peak potentials is a good approximation to the reduction potential E under the prevailing conditions (E is not in general the standard potential because nonstandard conditions are usually adopted). In the ideal case, the oxidation and reduction peaks are similar in magnitude and separated by a small potential increment, which is usually $(59 \text{ mV})/\nu_e$ at 25°C, where ν_e is the number of electrons transferred during the reaction at the electrode. This case is an example of a reversible redox reaction, with electron transfer at the electrode sufficiently fast that equilibrium is always maintained throughout the potential sweep. In such a case the current is usually limited by diffusion of electroactive species to the electrode.

Slow kinetic processes at the electrode result in a large separation of reduction and oxidation peaks that increases with increasing scan rate. This separation arises because an overpotential (Section 5.18, effectively a driving force) is required to overcome barriers to electron transfer in each direction. Moreover, the peak due to a reduction or an oxidation in the initial part of the cycle process is often not matched by a corresponding peak in the reverse direction. This absence occurs because the species that is initially generated undergoes a further chemical reaction during the cycle and produces either a species with a different reduction potential or one that is not electroactive within the range of potential scanned. Inorganic chemists often refer to this behaviour as 'irreversible'.

An electrochemical reaction followed by a chemical reaction is known as an **EC process**. By analogy, a **CE process** is a reaction in which the species able to undergo the electrochemical (E) reaction must first be generated by a chemical reaction. Thus, for a molecule that is suspected of decomposing upon oxidation, it may be possible to observe the initial unstable species formed by the E process provided the scan rate is sufficiently fast to re-reduce it before it undergoes further reaction. Consequently, by varying the scan rate, the kinetics of the chemical reaction can be determined.

Computational techniques

Key point: Computational procedures use either *ab initio* methods or parametrized semi-empirical methods to calculate the properties of molecules and solids. Graphical techniques can be used to display the results.

Computation has proved to be one of the most important techniques in chemistry. **Computer modelling** is the use of numerical models for exploring the structures and properties of individual molecules and materials. The methods used range from rigorous, and therefore computationally very time-consuming, treatments, known as *ab initio* methods, based on the numerical solution of the Schrödinger equation for the system, to the more rapid and necessarily less detailed 'semi-empirical techniques', which use approximate or 'effective functions' to describe the forces between particles.

There are two principal approaches to solving the Schrödinger equation numerically for many-electron polyatomic molecules. In the more fundamental *ab initio* methods, an attempt is made to calculate structures from first principles, using only the atomic numbers of the atoms present and their general arrangement in space. Such an approach is intrinsically reliable but computationally very demanding. For complex problems involving molecules and materials with numerous atoms, such methods are so computationally time-consuming that alternative methods involving experimental data are needed. In these **semi-empirical methods**, integrals that occur in the formal solution of the Schrödinger equation are set equal to parameters that have been chosen to lead to the best fit to experimental quantities, such as enthalpies of formation. Semi-empirical methods are applicable to a wide range of molecules with an almost limitless number of atoms, and are widely popular.

Both types of procedure typically adopt a **self-consistent field** (SCF) procedure, in which an initial guess about the composition of the linear combinations of atomic orbitals (LCAO) used to model molecular orbitals is successively refined until the composition and the corresponding energy remains unchanged in a cycle of calculation. The most common type of *ab initio* calculation is based on the **Hartree–Fock method** in which the primary approximation is applied to the electron–electron repulsion. Various methods of correcting for the explicit electron–electron repulsion, referred to as the **correlation problem**, are the Møller–Plesset perturbation theory (MPn, where n is the order of correction), the generalized valence bond (GVB) method, multi-configurations self-consistent field (MCSCF), configuration interaction (CI), and coupled cluster theory (CC).

A currently popular alternative to the *ab initio* method is **density functional theory** (DFT), in which the total energy is expressed in terms of the total electron density $\rho = |\psi|^2$ rather than the wavefunction ψ itself. When the Schrödinger equation is expressed in terms of ρ, it becomes a set of equations called the **Kohn–Sham equations**, which are solved iteratively starting from an initial estimate and continuing until they are self-consistent. The advantage of the DFT approach is that it is less demanding computationally, requires less computer time, and—in some cases, particularly d-metal complexes—gives better agreement with experimental values than is obtained from other procedures.

Semi-empirical methods are set up in the same general way as Hartree–Fock calculations but within this framework certain pieces of information, such as integrals representing the interaction between two electrons, are approximated by importing empirical data or simply ignored. To soften the effect of these approximations, parameters representing other integrals are adjusted so as to give the best agreement with experimental data. Semi-empirical calculations are much faster than the *ab initio* calculations but the quality of results is very dependent on using a reasonable set of experimental parameters that can be transferred from structure to structure. Thus semi-empirical calculations have been very successful in organic chemistry with just a few types of element and molecular geometries. Semi-empirical methods have also been devised specifically for the description of inorganic species.

The raw output of a molecular structure calculation is a list of the coefficients of the atomic orbitals in each molecular orbital and the energies of these orbitals. The graphical representation of a molecular orbital uses stylized shapes to represent the basis set and then scales their size to indicate the value of the coefficient in the LCAO. Different signs of the wavefunctions are typically represented by different colours. The total electron density at any point (the sum of the squares of the wavefunctions evaluated at that point) is commonly represented by an **isodensity surface**, a surface of constant total electron density (Fig. 8.47). An important aspect of a molecule other than its geometrical shape is the distribution of charge over its surface. A common procedure begins with calculation of the net electric potential at each point on an isodensity surface by subtracting the potential due to the electron density at that point from the potential due to the nuclei. The result is an **electrostatic potential surface** (an 'elpot surface') in which net positive potential is shown in one colour and net negative potential is shown in another, with intermediate gradations of colour.

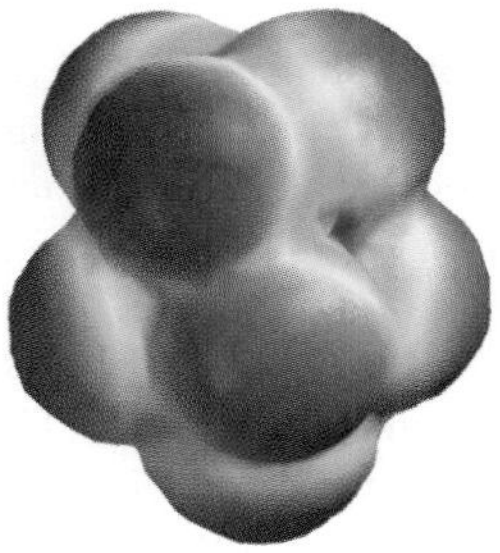

Computer modelling is applied to solids as well as to individual molecules and is useful for predicting the behaviour of a material, for example for indicating which crystal structure of a compound is energetically most favourable, for predicting phase changes, for calculating thermal expansion coefficients, identifying preferred sites for dopant ions, and calculating a diffusion pathway through a lattice. However, it is worth noting that very few aspects of inorganic chemistry can be computed exactly. Although modelling computation can give a very useful insight into materials chemistry, it is not yet at a stage where it can be used reliably to predict the exact structure or properties of any complex compound.

Figure 8.47 The output of computations of the electronic structure of a molecule is conveyed in a variety of ways. Here we show the electric potential surface of SF_5CF_3, a molecule that has been found to act as a very powerful greenhouse gas but of uncertain origin in the atmosphere. Red areas indicate regions of negative potential and green regions of positive potential.

FURTHER READING

Although this chapter has introduced many of the methods used by chemists to characterize inorganic compounds, it is not exhaustive. Other techniques used for investigating the structures and properties of solids and solutions include electron microscopy (Chapter 25) and inelastic neutron scattering to name two. The following references are a source of information on these techniques and a greater depth of coverage on the major techniques that have been introduced here.

A.K. Brisdon, *Inorganic spectroscopic methods*. Oxford Science Publications (1998).

R.P. Wayne, *Chemical instrumentation*. Oxford Science Publications (1994).

D.A. Skoog, F.J. Holler, and T.A. Nieman, *Principles of instrumental analysis*. Brooks Cole (1997).

R.S. Drago, *Physical methods for chemists*. Saunders (1992).

S.K. Chatterjee, *X-ray diffraction: Its theory and applications*. Prentice-Hall of India (2004).

B.D. Cullity and S.R. Stock, *Elements of X-ray diffraction*. Prentice Hall (2003).

B. Henderson and G. F. Imbusch, *Optical spectroscopy of inorganic solids (Monographs on the Physics & Chemistry of Materials)*. Oxford University Press (2006).

E.I. Solomon and A.B.P. Lever, *Inorganic electronic structure and spectroscopy: Methodology: Volume 1*. Wiley (2006).

E.I. Solomon and A.B.P. Lever, *Inorganic electronic structure and spectroscopy: applications and case studies: Volume 2*. Wiley (2006).

J.S. Ogden, *Introduction to molecular symmetry*. Oxford University Press (2001).

F. Siebert and P. Hildebrandt, *Vibrational spectroscopy in life science*. Wiley VCH (2007).

J.R. Ferraro and K. Nakamoto, *Introductory Raman spectroscopy*. Academic Press (1994).

K. Nakamoto, *Infrared and Raman spectra of inorganic and coordination compounds*. Wiley–Interscience (1997).

J.K.M. Saunders and B.K. Hunter, *Modern NMR spectroscopy, a guide for chemists*. Oxford University Press (1993).

J.A. Iggo, *NMR spectroscopy in inorganic chemistry*. Oxford University Press (1999).

J.W. Akitt and B.E. Mann, *NMR and chemistry*. Stanley Thornes, Cheltenham (2000).

K.J.D. MacKenzie and M.E. Smith, *Multinuclear solid-state nuclear magnetic resonance of inorganic materials*. Pergamon (2004).

D.P.E. Dickson and F.J. Berry, *Mössbauer spectroscopy*. Cambridge University Press; (2005).

M.E. Brown, *Introduction to thermal analysis*. Kluwer Academic Press (2001).

P.J. Haines, *Principles of thermal analysis and calorimetry*. Royal Society of Chemistry (2002).

A.J. Bard and L.R. Faulkner, *Electrochemical methods: fundamentals and applicatons*. 2nd edn. Wiley (2001).

O. Kahn, *Molecular magnetism*. VCH, New York (1993).

EXERCISES

8.1 How might you determine what crystalline components are present in a natural mineral sample?

8.2 The reaction of sodium carbonate, boron oxide, and silicon dioxide gives a borosilicate glass. Explain why the powder diffraction pattern of this product shows no diffraction maxima.

8.3 The minimum size of crystal that can typically be studied using a laboratory single crystal diffractometer is $50 \times 50 \times 50$ μm. The X-ray flux from a synchrotron source is expected to be 10^6 times the intensity of a laboratory source. Calculate the minimum size of a cubic crystal that could be studied on a diffractometer by using this source. A neutron flux is 10^3 times weaker. Calculate the minimum size of crystal that could be studied by single crystal neutron diffraction.

8.4 Calculate the wavelength associated with a neutron moving at 2.20 km s^{-1}. Is this wavelength suitable for diffraction studies ($m_n = 1.675 \times 10^{-27}$ kg)?

8.5 Suggest reasons for the order of stretching frequencies observed with diatomic species: $CN^- > CO > NO$.

8.6 Use the data in Table 8.3 to estimate the O−O stretching wavenumber expected for a compound believed to contain the oxygenyl species O_2^+. Would you expect to observe this stretching vibration in (i) the IR spectrum or (ii) the Raman spectrum?

8.7 Figure 8.48 shows the UV photoelectron spectrum of NH_3. Explain why the band at approximately 11 eV shows such a long and sharply resolved progression.

8.8 Explain why a Raman band assigned to the symmetric N−C stretching mode in $N(CH_3)_3$ shows a shift to lower frequency when ^{14}N is substituted by ^{15}N but no such shift is observed for the N−Si symmetric stretch in $N(SiH_3)_3$.

8.9 Explain why the ^{13}C-NMR spectrum of $Co_2(CO)_9$ shows only a single peak at room temperature.

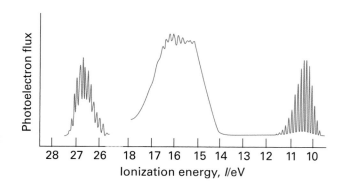

Figure 8.48

8.10 Predict the form of the ^{19}F-NMR and the ^{77}Se-NMR spectra of $^{77}SeF_4$. For ^{77}Se, $I = \frac{1}{2}$.

8.11 Explain the observation that the ^{19}F-NMR spectrum of XeF_5^- consists of a central peak symmetrically flanked by two peaks, each of which is roughly one-sixth of the intensity of the central peak.

8.12 Determine the g values of the EPR spectrum shown in Fig. 8.49, measured for a frozen sample using a microwave frequency of 9.43 GHz.

8.13 Which technique is sensitive to the slowest processes, NMR or EPR?

8.14 For a paramagnetic compound of a d-metal compound having one unpaired electron, outline the main difference you would expect to see between an EPR spectrum measured in aqueous solution at room temperature and that recorded for a frozen solution.

8.15 Predict a value for the isomer shift for iron in $BaFe^{(VI)}O_4$.

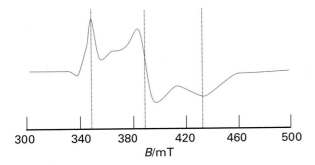
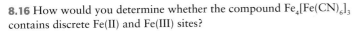

Figure 8.49

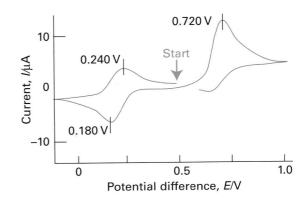

Figure 8.50

8.16 How would you determine whether the compound $Fe_4[Fe(CN)_6]_3$ contains discrete Fe(II) and Fe(III) sites?

8.17 Suggest a reason why no resolved quadrupole splitting is observed in the ^{121}Sb ($I = {}^5/_2$) Mössbauer spectrum of solid SbF_5.

8.18 Explain why, even though the average atomic mass of silver is $107.9m_u$, no peak at $108m_u$ is observed in the mass spectrum of pure silver. What effect does this absence have on the mass spectra of silver compounds?

8.19 What peaks should you expect in the mass spectrum of $[Mo(C_6H_6)(CO)_3]$?

8.20 Interpret the cyclic voltammogram shown in Fig. 8.50, which has been recorded for an Fe(III) complex in aqueous solution.

8.21 Thermogravimetric analysis of a zeolite of composition $CaAl_2Si_6O_{16}.nH_2O$ shows a mass loss of 20 per cent on heating to dryness. Determine n.

8.22 A cobalt(II) salt was dissolved in water and reacted with excess acetylacetone (1,2-pentanedione, $CH_3COCHCOCH_3$) and hydrogen peroxide. A green solid was formed that gave the following results for elemental analysis: C, 50.4 per cent; H, 6.2 per cent; Co, 16.5 per cent (all by mass). Determine the ratio of cobalt to acetylacetonate ion in the product.

PROBLEMS

8.1 Discuss the importance of X-ray crystallography in inorganic chemistry. See for example: The history of molecular structure determination viewed through the Nobel Prizes. W.P. Jensen, G.J. Palenik, and I.-H. Suh, *J. Chem. Educ.*, 2003, **80**, 753.

8.2 Discuss why the length of an O−H bond obtained from X-ray diffraction experiments averages 85 pm whereas that obtained in neutron diffraction experiments averages 96 pm. Would you expect to see similar effects with C−H bond lengths measured by these techniques?

8.3 How is single crystal neutron diffraction of use in inorganic chemistry?

8.4 Refer to Fig. 8.51. Discuss the likely reason for the shifts in λ_{max} for spectra of I_2 recorded in different solvent media: (a) heptane, (b) benzene, (c) diethylether/CCl_4, (d) pyridine/heptane, (e) triethylamine/heptane.

8.5 Discuss how you would carry out the following analyses: (a) calcium levels in breakfast cereal, (b) mercury in shell fish, (c) the geometry of BrF_5, (d) number of organic ligands in a d-metal complex, (e) water of crystallization in an inorganic salt.

8.6 Write a review of the applications of thermal methods of analysis in inorganic chemistry.

8.7 Discuss the challenges involved in carrying out each of the following scenarios: (a) identification of inorganic pigments in a valuable painting, (b) analysis of evolved inorganic gases from a volcano, (c) determination of pollutants at the sea bed.

8.8 How may a consideration of ^{31}P chemical shifts and ^1H coupling constants be used to distinguish the isomers of the octahedral complex $[Rh(CCR)_2H(PMe_3)_3]$? (See J.P. Rourke, G. Stringer, D.S. Yufit, J.A.K. Howard, and T.B. Marder, *Organometallics*, 2002, **21**, 429.)

8.9 Ionization techniques in mass spectrometry frequently induce fragmentation and other undesirable reactions. However, under some circumstances these reactions can be helpful. Using fullerenes as an example, discuss this phenomenon. (See M.M. Boorum, Y.V. Vasil'ev, T. Drewello, and L.T. Scott, *Science*, 2001, **294**, 828.)

8.10 The iron content of a dietary supplement tablet was determined using atomic absorption spectrometry. A tablet (0.4878 g) was ground to a fine powder and 0.1123 g was dissolved in dilute sulfuric acid and transferred to a 50 cm³ volumetric flask. A 10 cm³ sample of this solution was taken and made up to 100 cm³ in another volumetric flask. A series of standards was prepared that contained 1.00, 3.00, 5.00, 7.00, and 10.0 ppm iron. The absorptions of

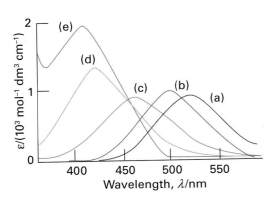

Figure 8.51

the standards and the sample solution were measured at the iron absorption wavelength:

Concentration/ppm	Absorbance
1.00	0.095
3.00	0.265
5.00	0.450
7.00	0.632
10.0	0.910
Sample	0.545

Calculate the mass of iron in the tablet.

8.11 A sample of water from a reservoir was analysed for copper content. The sample was filtered and diluted tenfold with deionized water. A set of standards was prepared with copper concentrations between 100 and 500 ppm. The standards and the sample were aspirated into an atomic absorption spectrometer and the absorbance was measured at the copper absorption wavelength. The results obtained are given below. Calculate the concentration of copper in the reservoir.

Concentration/ppm	Absorbance
100	0.152
200	0.388
300	0.590
400	0.718
500	0.865
Sample	0.751

8.12 A sample from an effluent stream was analysed for phosphate levels. Dilute hydrochloric acid and excess sodium molybdate were added to a 50 cm^3 sample of the effluent. The molybdophosphoric acid, $H_3PMo_{12}O_{40}$, that was formed was extracted into two 10 cm^3 portions of an organic solvent. A molybdenum standard with a concentration of 10 ppm was prepared in the same solvent. The combined extract and the standard were aspirated into an atomic absorption spectrometer that was set up to measure molybdenum. The extracts gave an absorbance of 0.573 and the standard gave an absorbance of 0.222. Calculate the concentration of phosphate in the effluent.

PART 2
The elements and their compounds

This part of the book describes the physical and chemical properties of the elements as they are set out in the periodic table. This 'descriptive chemistry' of the elements reveals a rich tapestry of patterns and trends, many of which can be rationalized and explained by application of the concepts developed in Part 1.

The first chapter of this part, Chapter 9, summarizes the trends and patterns in the context of the periodic table and the principles described in Part 1. The trends described in this chapter are illustrated throughout the following chapters. Chapter 10 deals with the chemistry of the unique element hydrogen. The following eight chapters (Chapters 11 − 18) proceed systematically across the main groups of the periodic table. The elements of these groups demonstrate the diversity, intricacy, and fascinating nature of inorganic chemistry.

The chemical properties of the d-block elements are so diverse and extensive that the next four chapters of the part are devoted to them. Chapter 19 reviews the descriptive chemistry of the three series of d-block elements. Chapter 20 describes how electronic structure affects the chemical and physical properties of the d-metal complexes and Chapter 21 describes their reactions in solution. Chapter 22 deals with the industrially vital d-metal organometallic compounds.

Our tour of the periodic table draws to a close in Chapter 23 with a description of the remarkably uniform f-block elements.

Periodic trends

9

The periodic table provides an organizing principle that coordinates and rationalizes the diverse physical and chemical properties of the elements. Periodicity is the regular manner in which the physical and chemical properties of the elements vary with atomic number. This chapter reviews the material in Chapter 1 and summarizes these variations in a manner that should be kept in mind throughout the chapters of this part of the text.

Although the chemical properties of the elements can seem bewilderingly diverse, the periodic table helps to show that they vary reasonably systematically with atomic number. Once these trends and patterns are recognized and understood, much of the detailed properties of the elements no longer seem like a random collection of unrelated facts and reactions. In this chapter we summarize some of the trends in the physical and chemical properties of the elements and interpret them in terms of the underlying principles presented in Chapter 1.

Periodic properties of the elements

The general structure of the modern periodic table was discussed in Section 1.8. Almost all trends in the properties of the elements can be traced to the electronic configuration of the atoms and atomic radii, and their variation with atomic number.

9.1 Valence electron configurations

The valence electron configuration of the ground state of an atom of an element can be inferred from its group number. For example, in Group 1 all the elements have an ns^1 valence configuration, where n is the period number. As we saw in Chapter 1, the valence electron configurations vary with group number as follows:

1	2	13	14	15	16	17	18
ns^1	ns^2	ns^2np^1	ns^2np^2	ns^2np^3	ns^2np^4	ns^2np^5	ns^2np^6

Electron configurations in the d block are slightly less systematic, but involve the filling of the $(n-1)$d orbitals. In Period 4 they are as follows:

3	4	5	6	7	8	9	10	11	12
$3d^14s^2$	$3d^24s^2$	$3d^34s^2$	$3d^54s^1$	$3d^54s^2$	$3d^64s^2$	$3d^74s^2$	$3d^84s^2$	$3d^{10}4s^1$	$3d^{10}4s^2$

Note the manner in which half-filled and full d subshells are favoured.

9.2 Atomic parameters

Although this part of the text deals with the *chemical* properties of the elements and their compounds, we need to keep in mind that these chemical properties spring from the *physical* characteristics of atoms. As we saw in Chapter 1, these physical characteristics—the radii of atoms and ions, and the energy changes associated with the formation of ions—vary periodically. Here we review these variations.

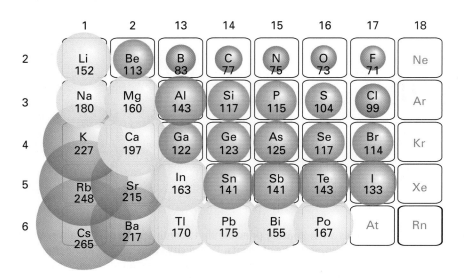

Figure 9.1 The variation of atomic radii (in picometres) through the main groups of the periodic table.

(a) Atomic radii

Key points: Atomic radii increase down a group and, within the s and p blocks, decrease from left to right across a period.

As we saw in Section 1.9, atomic radii increase down a group and decrease from left to right across a period. Across a period, as a result of the joint effects of penetration and shielding, there is an increase in effective nuclear charge. This increase draws in the electrons and results in a smaller atom. On descending a group, electrons occupy successive shells outside a completed core and the radii increase (Fig. 9.1).

Atomic radii in the 5d series of the d block are very similar to their congeners in the 4d series even though the atoms have a greater number of electrons. For example, the radii of Mo and W in Group 6 are 140 and 141 pm, respectively. This reduction of radius below that expected, the *lanthanide contraction*, is due to the presence of 4f electrons in the intervening lanthanoids: the poor shielding properties of f electrons results in a higher effective nuclear charge than expected on the basis of a simple extrapolation from other atoms. A similar contraction is found in the elements that follow the d block. For example, although there is a substantial increase in atomic radius between C and Si (77 and 118 pm, respectively), the atomic radius of Ge (122 pm) is only slightly greater than that of Si.

(b) Ionization energies and electron affinities

Key point: Ionization energy increases across a period and decreases down a group. Electron affinities are highest for elements near fluorine, particularly the halogens.

We need to be aware of the energies needed to form cations and anions of the elements. Ionization energies are relevant to the formation of cations; electron affinities are relevant to the formation of anions.

The ionization energy of an element is the energy required to remove an electron from a gas-phase atom (Section 1.9). Ionization energies correlate strongly with atomic radii, and elements that have small atomic radii generally have high ionization energies. Therefore, as the atomic radius increases down a group, the ionization energy decreases. Likewise, the decrease in radius across a period is accompanied by an increase in ionization energy (Fig. 9.2). As discussed in Section 1.9, there are variations in this trend: in particular, high ionization energies occur when electrons are removed from half-filled or full shells or subshells. Thus the first ionization energy of nitrogen ([He]$2s^2 2p^3$) is 1402 kJ mol^{-1}, which is higher than the value for oxygen ([He]$2s^2 2p^4$, 1314 kJ mol^{-1}). Similarly, the ionization energy of phosphorus (1011 kJ mol^{-1}) is higher than that of sulfur (1000 kJ mol^{-1}).

The energies of electrons in hydrogenic atoms are proportional to Z^2/n^2. A first approximation to the energies of electrons in many-electron atoms is that they are proportional to Z_{eff}^2/n^2, where Z_{eff} is the effective nuclear charge (Section 1.6), although this proportionality should not be taken too seriously. Plots of first ionization energies against Z_{eff}^2/n^2 for the outermost electrons for the elements Li to Ne ($n = 2$) and Hf to Hg ($n = 6$) are shown

Cross Reference: Section 19.2

Cross Reference: Table 14.1

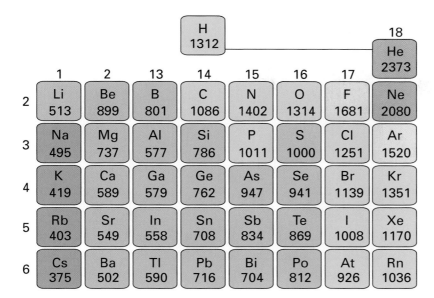

Figure 9.2 The variation of first ionization energy (in kilojoules per mole) through the main groups of the periodic table.

in Figs 9.3 and 9.4, respectively. The graphs confirm that the proportionality is broadly followed, especially at high values of n when the outermost electron experiences an interaction with an almost point-like core.

The electron affinity plays a role in assessing the energy required to form an anion. As we saw in Section 1.9c, an element has a high electron affinity if the additional electron can enter a shell where it experiences a strong effective nuclear charge. Therefore, elements close to F (other than the noble gases) have the highest electron affinities as Z_{eff} is then large. The addition of an electron to a singly charged ion (as in the formation of O^{2-} from O^{-}) is invariably negative (that is, the process is endothermic) because it takes energy to push an electron on to a negatively charged species. However, that does not mean it cannot happen, for it is important to assess the overall consequences of ion formation, and it is often the case that the interaction between highly charged ions in the solid overcomes the additional energy needed to form them. When assessing the energetics of compound formation, it is essential to think globally and not to rule out an overall process simply because an individual step is endothermic.

(c) Electronegativity

Key points: Electronegativity increases across a period and decreases down a group.

We saw in Section 1.9 that the electronegativity, χ, is the power of an atom of the element to attract electrons to itself when it is part of a compound. As we saw in Section 1.9b, trends in electronegativity can be correlated with trends in atomic radii. This correlation is most readily understood in terms of the Mulliken definition of electronegativity as the mean of the ionization energy and electron affinity of an element. If an atom has a high ionization energy (so it is unlikely to give up electrons) and a high electron affinity (so there are energetic advantages in its gaining electrons), then it is more likely to attract an electron to itself. Consequently, the electronegativities of the elements, tracking the trends in ionization energies and electron affinities, which in turn track atomic radii, typically increase left to right across a period and decrease down a group. However, the Pauling values of electronegativity are commonly used (Fig. 9.5).

There are some exceptions to this general trend as can be seen from the following electronegativities:

Al	Si
1.61	1.90
Ga	Ge
1.81	2.01
In	Sn
1.78	1.96

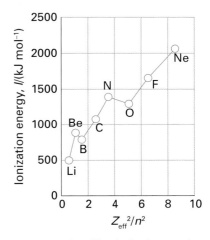

Figure 9.3 Plot of first ionization energies against Z_{eff}^2/n^2 for the outermost electrons for the elements lithium to neon ($n = 2$).

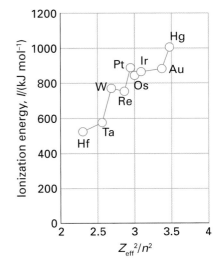

Figure 9.4 Plot of first ionization energies against Z_{eff}^2/n^2 for the outermost electrons for the elements hafnium to mercury ($n = 6$).

	1	2	13	14	15	16	17	18
					H 2.20			He
2	Li 0.98	Be 1.57	B 2.04	C 2.55	N 3.04	O 3.44	F 3.98	Ne
3	Na 0.93	Mg 1.31	Al 1.61	Si 1.90	P 2.19	S 2.58	Cl 3.16	Ar
4	K 0.82	Ca 1.00	Ga 1.81	Ge 2.01	As 2.18	Se 2.55	Br 2.96	Kr
5	Rb 0.82	Sr 0.95	In 1.78	Sn 1.96	Sb 2.05	Te 2.10	I 2.66	Xe
6	Cs 0.79	Ba 0.89	Tl 2.04	Pb 2.33	Bi 2.02	Po 2.0	At	Rn

Figure 9.5 The variation of Pauling electronegativity through the main groups of the periodic table.

Cross Reference: Sections 13.1, 14.5, and 15.11b

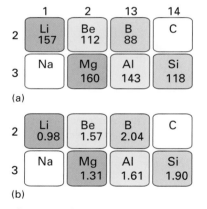

(a)

	1	2	13	14
2	Li 157	Be 112	B 88	C
3	Na	Mg 160	Al 143	Si 118

(b)

	1	2	13	14
2	Li 0.98	Be 1.57	B 2.04	C
3	Na	Mg 1.31	Al 1.61	Si 1.90

Figure 9.6 The diagonal relationship between (a) atomic radii (in picometres) and (b) Pauling electronegativity in Periods 2 and 3.

Cross Reference: Section 11.3

Cross Reference: Section 12.3

Cross Reference: Section 13.1

This departure from a smooth decrease down the group is called the **alternation effect** and is due to the intervention of the 3d subshell earlier in Period 4. The alternation effect also appears in a more chemically direct manner, where it summarizes (but does not explain) the nonexistence of various compounds in Groups 13 to 15, as in the following examples from Group 15, where the grey denotes unknown compounds ($AsCl_5$ is unstable above $-50°C$):

$$NF_5 \qquad NCl_5 \qquad NBr_5$$
$$PF_5 \qquad PCl_5 \qquad PBr_5$$
$$AsF_5 \qquad AsCl_5 \qquad AsBr_5$$
$$SbF_5 \qquad SbCl_5 \qquad SbBr_5$$
$$BiF_5 \qquad BiCl_5 \qquad BiBr_5$$

Although electronic factors such as electronegativity no doubt play a role in these examples, steric effects are important too, especially for N.

The term 'electropositive', which describes an element's ability to lose electrons, needs to be used with caution. In some applications it means that the element has a low electronegativity; in others it means that the redox couple M^{n+}/M has a strongly negative standard potential (so that M is a reducing metal). We use the term only in the latter sense in this book.

(d) Diagonal relationships

Key points: The atomic radius, and hence some chemical properties, of some Period 2 elements is similar to that of the element to their lower right in the periodic table.

Most of the trends in the chemical properties of the elements within the periodic table are best discussed in terms of vertical trends within groups or horizontal trends across periods. The element at the head of each group also commonly possesses a diagonal relationship with the element to its lower right. Diagonal relationships arise because the atomic radii, charge densities, electronegativities, and hence many of the chemical properties, of the two elements are similar (Fig. 9.6). The most striking diagonal relationships are those between Li and Mg. For example, whereas the Group 1 elements form compounds that are essentially ionic in nature, Li and Mg salts have some degree of covalent character in their bonding. There is a strong diagonal relationship between Be and Al: both elements form covalent hydrides, halides, and oxides; the analogous compounds of Group 2 are predominantly ionic. The diagonal relationship between B and Si is illustrated by the fact that both elements form flammable, gaseous hydrides whereas aluminium hydride is a solid.

(e) Enthalpies of atomization

Key point: The enthalpy of atomization increases with increasing number of valence electrons.

The enthalpy of atomization of an element, $\Delta_a H^{\ominus}$, is a measure of the energy required to form gaseous atoms. For solids, the enthalpy of atomization is the enthalpy change

Table 9.1 Enthalpies of atomization, $\Delta_a H^\ominus/(\text{kJ mol}^{-1})$

Li	Be											B	C	N	O	F
161	321											590	715	473	248	79
Na	Mg											Al	Si	P	S	Cl
109	150											314	439	315	223	121
K	Ca	Sc	Ti	V	Cr	Mn	Fe	Co	Ni	Cu	Zn	Ga	Ge	As	Se	Br
90	193	340	469	515	398	279	418	427	431	339	130	289	377	290	202	112
Rb	Sr	Y	Zr	Nb	Mo	Tc	Ru	Rh	Pd	Ag	Cd	In	Sn	Sb	Te	I
86	164	431	611	724	651	648	640	556	390	289	113	244	301	254	199	107
Cs	Ba	La	Hf	Ta	W	Re	Os	Ir	Pt	Au	Hg	Tl	Pb	Bi	Po	
79	176	427	669	774	844	791	782	665	565	369	61	186	196	208	144	

associated with the atomization of the solid; for molecular species, it is the enthalpy of dissociation of the molecules. As can be seen in Table 9.1, enthalpies of atomization first increase and then decrease across Periods 2 and 3, reaching a maximum at C in Period 2 and Si in Period 3. The values decrease between C and N, and Si and P: even though N and P each have five valence electrons, two of these electrons form a lone pair and only three are involved in bonding. A similar effect is seen between N and O, where O has six valence electrons of which four form lone pairs and only two are involved in bonding. These trends are shown in Fig. 9.7.

The enthalpies of atomization of the d-block elements are higher than those of the s- and p-block elements, in line with their greater number of valence electrons and consequently stronger bonding. The values reach a maximum at Groups 5 and 6 (Fig. 9.8), where there is a maximum number of unpaired electrons available to form bonds. The middle of each row shows an irregularity due to spin correlation (Section 1.7a), which favours a half-filled d shell for the free atom. This effect is particularly evident for the 3d series, in which Cr ($3d^5 4s^1$) and Mn ($3d^5 4s^2$) have significantly lower atomization energies than expected from a simple consideration of their number of valence electrons.

The enthalpy of atomization decreases down a group in the s and p blocks but increases down a group in the d block. Thus s and p orbitals become less effective at forming bonds as the period number increases, whereas d orbitals become more effective. These trends are attributed to the expansion of p orbitals on descending a group from optimal for overlap to too diffuse for extensive overlap and, in contrast, d orbitals expanding in size from too contracted to optimal for overlap. The same trends can be seen in the melting points (Table 9.2) of the elements, where a greater number of valence electrons leads to greater binding energy and a higher melting temperature.

9.3 Occurrence

Key point: Hard—hard and soft—soft interactions help to systematize the terrestrial distribution of the elements.

Some elements occur in their elemental state in nature, for example the gases nitrogen and oxygen, the nonmetal sulfur, and the metals silver and gold. Most elements however, occur naturally in ores as compounds with other elements.

The concept of hardness and softness (Section 4.12) helps to rationalize a great deal of inorganic chemistry, including the type of compound that the element forms in Nature. Thus, soft acids tend to bond to soft bases and hard acids tend to bond to hard bases. These tendencies explain certain aspects of the **Goldschmidt classification** of the elements into four types (Fig 9.9), a scheme widely used in geochemistry:

Lithophiles are found primarily in the Earth's crust (the lithosphere) in silicate minerals, and include Li, Mg, Ti, Al, and Cr (as their cations). These cations are hard, and are found in association with the hard base O^{2-}.

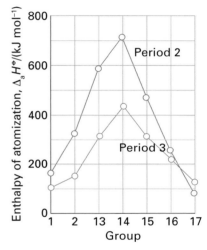

Figure 9.7 Variation of the enthalpy of atomization in the s- and p-block elements.

Cross Reference: Section 19.3b

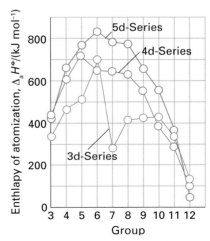

Figure 9.8 Variation of enthalpy of atomization in the d-block elements.

Table 9.2 Normal melting points of the elements, $\theta_{mp}/°C$

Li	Be	Sc	Ti	V	Cr	Mn	Fe	Co	Ni	Cu	Zn	B	C	N	O	F
180	1280											2300	3730	−210	−218	−220
Na	Mg											Al	Si	P	S	Cl
97.8	650											660	1410		113	−110
K	Ca	Sc	Ti	V	Cr	Mn	Fe	Co	Ni	Cu	Zn	Ga	Ge	As	Se	Br
63.7	850	1540	1675	1900	1890	1240	1535	1492	1453	1083	420	29.8	937		217	−7.2
Rb	Sr	Y	Zr	Nb	Mo	Tc	Ru	Rh	Pd	Ag	Cd	In	Sn	Sb	Te	I
38.9	768	1500	1850	2470	2610	2200	2500	1970	1550	961	321	2000	232	630	450	114
Cs	Ba	La	Hf	Ta	W	Re	Os	Ir	Pt	Au	Hg	Tl	Pb	Bi	Po	
28.7	714	920	2220	3000	3410	3180	3000	2440	1769	1063	13.6	304	327	271	254	

Chalcophiles are often found in combination with sulfide (and selenide and telluride) minerals, and include Cd, Pb, Sb, and Bi. These elements (as their cations) are soft, and are found in association with the soft base S^{2-} (or Se^{2-} and Te^{2-}). Zinc cations are borderline hard, but softer than Al^{3+} and Cr^{3+}, and Zn is also often found as its sulfide. **Siderophiles** are intermediate in terms of hardness and softness and show an affinity for both oxygen and sulfur. They occur mainly in their elemental state and include Pt, Pd, Ru, Rh, and Os.

Atmophiles are gases such as H, N, and Group 18 elements (the noble gases).

EXAMPLE 9.1 Explaining the Goldschmidt classification

The common ores of Ni and Cu are sulfides. By contrast Al is obtained from the oxide and Ca from the carbonate. Can these observations be explained in terms of hardness?

Answer We need to assess whether the hard–hard and soft–soft rule applies. From Table 4.5 we know that O^{2-} and CO_3^{2-} are hard bases; S^{2-} is a soft base. The table also shows that the cations Ni^{2+} and Cu^{2+} are considerably softer acids than Al^{3+} or Ca^{2+}. Hence the hard–hard and soft–soft rule accounts for the sorting observed.

Self-test 9.1 Of the metals Cd, Rb, Cr, Pb, Sr, and Pd, which might be expected to be found in aluminosilicate minerals and which in sulfides?

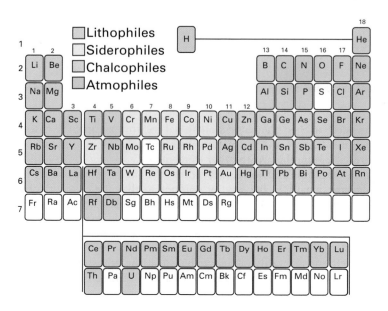

Figure 9.9 The Goldschmidt classification of the elements.

9.4 Metallic character

Key point: The metallic character of the elements decreases across a period and increases down a group. Many elements in the p block exist as allotropes.

The chemical properties of the metallic elements can be considered as arising from the ability of the elements to lose electrons to form the electron sea that binds together the cations and accounts for metallic bonding (Section 3.19). Consequently, elements with low ionization energies are likely to be metals and those with high ionization energies are likely to be nonmetals. Thus, as ionization energies decrease down a group the elements become more metallic, and as the ionization energies increase across a row the elements become less metallic (Fig. 9.10). These trends can also be directly related to the trends in atomic radii as large atoms typically have low ionization energies and are more metallic in character. This trend is most noticeable within Groups 13 to 16, where the elements at the head of the group are nonmetals and those at the foot of the group are metals. Within this general trend there are allotropic variations in the sense that some elements exist as both metals and nonmetals. An example is Group 15: N and P are nonmetals, As exists as nonmetal, metalloid, and metallic allotropes, and Sb and Bi are metals. Elements in the p block typically form several allotropes (Table 9.3).

Cross Reference: Sections 13.1, 14.1, 15.1, and 16.1

Cross Reference: Section 15.1

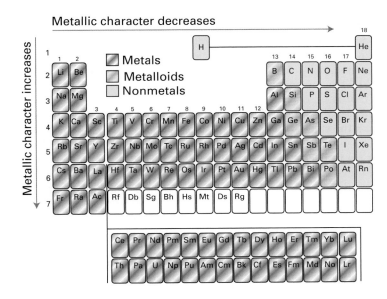

Figure 9.10 The variation of metallic character through the periodic table.

Table 9.3 Some allotropes of the p-block elements

C Diamond, graphite, amorphous, fullerenes		**O** Dioxygen, ozone
	P White, red, black	**S** Many catenated rings, chains, amorphous
	As Yellow, metallic/grey, black	**Se** Red (α, β, γ), grey, black
Sn Grey, white	**Sb** Blue, yellow, black	
	Bi Amorphous, crystalline	

9.5 Oxidation states

Key points: The group oxidation number can be predicted from the electron configuration of an element. The inert pair effect leads to an increasing stability of an oxidation state that is 2 less than the group oxidation number for the heavier elements. d-Block elements exhibit a variety of oxidation states.

The trends in stable oxidation states within the periodic table can be understood to some extent by considering electron configurations. Related factors such as ionization energies and spin correlation also play a role. A complete or half-full valence shell imparts greater stability than a partially filled shell. Therefore, there is a tendency for atoms to gain or lose electrons until they acquire that configuration.

A noble-gas configuration is achieved in the s and p blocks when eight electrons occupy the s and p subshells of the valence shell. In Groups 1, 2, and 13 the loss of electrons to leave the inner complete shell can be achieved with a relatively small input of energy. Thus, for these elements the oxidation numbers typical of the groups are +1, +2, and +3, respectively. From Group 14 to Group 17 it becomes increasingly energetically favourable—provided we consider the overall contributions to the energy, such as the interaction between oppositely charged ions—for the atoms to accept electrons in order to complete the valence shell. Consequently, the group oxidation numbers are −4, −3, −2, −1 with more electronegative elements. Group 18 elements already have a complete octet of electrons and are neither readily oxidized nor reduced.

Cross Reference: Sections 13.14, 14.8, and 15.13

The heavier elements of the p block also form compounds with the element with an oxidation number 2 less than the group oxidation number. The relative stability of an oxidation state in which the oxidation number is 2 less than the group oxidation number is an example of the *inert pair effect* and it is a recurring theme within the p block. For example, in Group 13 whereas the group oxidation number is +3, the +1 oxidation state increases in stability down the group. In fact, the most common oxidation state of thallium is Tl(I). There is no simple explanation for this effect: it is often ascribed to the large energy that is needed to remove the ns^2 electrons after the np^1 electron has been removed. However, the sum of the first three ionization energies of Tl (5438 kJ mol^{-1}) is not higher than the value for Ga (5521 kJ mol^{-1}) and only slightly higher than the value for In (5083 kJ mol^{-1}). Another contribution to the effect may be the low M−X bond enthalpies for the heavier p-block elements and the decreasing lattice energy as the atomic radii increase down a group.

Cross Reference: Section 19.3a

Cross Reference: Section 19.4

The group oxidation number is achieved in the d block only for Groups 3 to 8 and even then highly oxidizing F or O are required to bring it out. For Group 7 and 8, only O can produce anions or neutral oxides with an element with oxidation numbers +7 and +8, as in the permanganate anion, MnO_4^- and osmium tetraoxide, OsO_4. The range of observed oxidation states is shown in Table 9.4. As can be seen, up to Mn, all the 3d and 4s electrons can participate in bonding and the maximum oxidation state corresponds to the group number. Once the d^5 electron configuration is exceeded the tendency for the d electrons to participate in bonding decreases and high oxidation states are not observed. Similar trends exist for the 4d and 5d series. However, the stability of high oxidation states increases down a group in Groups 4 to 10 as the atomic radius increases. The relative stabilities of oxidation states of the 4d and 5d series members of each group are similar as their

Table 9.4 The range of observed positive oxidation states for 3d-series elements

Sc	Ti	V	Cr	Mn	Fe	Co	Ni	Cu	Zn
d^1s^2	d^2s^2	d^3s^2	d^5s^1	d^5s^2	d^6s^2	d^7s^2	d^8s^2	$d^{10}s^1$	$d^{10}s^2$
		+1					+1		
+2	+2	+2	+2	+2	+2	+2	+2	+2	+2
+3	+3	+3	+3	+3	+3	+3	+3	+3	
	+4	+4	+4	+4	+4	+4	+4		
		+5	+5	+5	+5	+5			
			+6	+6	+6				
				+7					

atomic radii are very similar (due to the lanthanide contraction). As already remarked, half-filled shells of electrons with parallel spins are particularly stable due to spin correlation (Section 1.7a). This additional stability has important consequences for the chemistry of the d-block elements with exactly half-filled shells. Manganese has the configuration $3d^5 4s^2$; as a result, Mn(II) is particularly stable and Mn(III) compounds are uncommon. The importance of spin correlation diminishes as the orbitals become larger, for example Tc and Re, which are the 4d and 5d counterparts of Mn, do not form M(II) compounds. d-Block elements also form stable compounds with the metal in its zero oxidation state. These complexes are typically stabilized by a ligand that acts as a π-acid.

Cross Reference: Section 22.12

Periodic characteristics of compounds

The number and type of bonds that elements form depend to a large extent on the relative strength of the bonds and the relative sizes of the atoms.

9.6 Coordination numbers

Key points: Low coordination numbers generally dominate for small atoms; high coordination numbers are possible as a group is descended.

The coordination number of an atom in a compound depends very much on the relative sizes of the central atom and the surrounding atoms. In the p block, low coordination numbers are most common for the compounds of Period 2 elements but higher coordination numbers are observed as each group is descended and the radius of the central atom increases. For example, in Group 15 N forms three-coordinate molecules such as NCl_3 and four-coordinate ions such as NH_4^+, whereas its congener P forms molecules with coordination numbers of 3 and 5, as in PCl_3 and PCl_5, and six-coordinate ionic species such as PCl_6^-. This higher coordination number in Period 3 elements is an example of hypervalence and is sometimes attributed to the participation of d orbitals in bonding. However, as remarked in Section 2.3b, it is more likely to be due to the possibility of arranging a greater number of atoms or molecules around the larger central atom. In the d block the 4d and 5d series of elements tend to exhibit higher coordination numbers than the 3d series elements due to their larger radii. For example, in Group 3 Sc forms the 6-coordinate ScF_6^{3-} ion whereas La forms the fluoride-bridged 9-coordinate LaF_9^{6-} ion. Very high coordination numbers have been observed for some large atoms, for example Th forms the 10-coordinate $[Th(C_2O_4)_4(OH_2)_2]^{4-}$ ion.

Cross Reference: Section 15.1

Cross Reference: Section 19.7

9.7 Bond enthalpy trends

Key points: Bond enthalpies E—H for p-block elements decrease down a group whereas in the d block they increase. For an atom E that has no lone pairs, the E—X bond enthalpy decreases down the group; for an atom that has lone pairs, it typically increases between Periods 2 and 3, and then decreases down the group.

As explained in Section 2.10, the mean bond enthalpy is the average bond dissociation enthalpy taken over a series of compounds. In the p block, E—H bonds get weaker on descending a group whereas in the d block they get stronger. These trends are attributed to the same orbital contraction and expansion effects as affect atomization enthalpies (Section 9.2f).

Mean E—X bond enthalpies for atoms that have lone pairs typically decrease down a group. However, the bond enthalpy for an element at the head of a group in Period 2 is anomalous and smaller than that of an element in Period 3:

	$B/(\text{kJ mol}^{-1})$		$B/(\text{kJ mol}^{-1})$
N—N	163	N—Cl	200
P—P	201	P—Cl	319
As—As	180	As—Cl	317

The relative weakness of single bonds between atoms of Period 2 elements is often ascribed to the proximity of the lone pairs on neighbouring atoms and the repulsion between them. For a p-block element E that has no lone pairs, the E—X bond enthalpy decreases down a group:

	$B/(\text{kJ mol}^{-1})$		$B/(\text{kJ mol}^{-1})$
C—C	348	C—Cl	338
Si—Si	226	Si—Cl	391
Ge—Ge	188	Ge—Cl	342

Smaller atoms form stronger bonds because the shared electrons are closer to each of the atomic nuclei. The strength of the Si—Cl bond is attributed to the fact that the atomic orbitals of the two elements have similar energies and efficient overlap. High values are also sometimes attributed to a contribution from π-bonding involving d orbitals.

EXAMPLE 9.2 Using bond enthalpies to rationalize structures

Explain why elemental sulfur forms rings or chains with S—S single bonds, whereas oxygen exists as diatomic molecules.

Answer We need to consider the relative magnitude of the bond enthalpies for the single and double bonds:

$B/(\text{kJ mol}^{-1})$		$B/(\text{kJ mol}^{-1})$	
O—O	142	O=O	498
S—S	263	S=S	431

Because an O=O bond is more than three times as strong as an O—O bond, there is a much stronger tendency for oxygen to form O=O bonds than O—O bonds, as in dioxygen, O_2. An S=S bond is less than twice as strong as an S—S bond, so the tendency to form S=S bonds is not as strong as in oxygen and the formation of S—S bonds is more likely.

Self-test 9.2 Why does sulfur form catenated polysulfides of formula $[S-S-S]^{2-}$ and $[S-S-S-S-S]^{2-}$ whereas polyoxygen anions beyond O_3^- are unknown?

One application of bond enthalpy arguments concerns the existence of subvalent compounds, compounds in which fewer bonds are formed than valence rules suggest, such as PH_2. Although this compound is thermodynamically stable with respect to dissociation into the constituent atoms, it is unstable with respect to the disproportionation:

$$3\,PH_2(g) \rightarrow 2\,PH_3(g) + \tfrac{1}{4}P_4(s)$$

The origin of the spontaneity of this reaction is the strength of the P—P bonds in molecular phosphorus, P_4. There are the same number (six) of P—H bonds in the reactants as there are in the products, but the reactants have no P—P bonds.

In the d block, bond enthalpies generally increase down a group, a trend that is the opposite of the general trend for the p block:

	$B/(\text{kJ mol}^{-1})$		$B/(\text{kJ mol}^{-1})$
Co—H	226	Fe—C	390
Rh—H	247	Ru—C	528
		Os—C	598

As we saw in Section 9.3e, d orbitals appear to become more effective at forming bonds down a group as they expand in size from contracted to optimal for overlap, and unpaired electrons become much more common.

9.8 Anomalies

Key point: The first member of each group within the p block shows differences from the rest of the group that are attributed to smaller atomic radii and a lack of low-lying d orbitals. The 3d metals form compounds with lower coordination numbers and oxidation states than the 4d and 5d elements. The oxidation state Ln(III) dominates in compounds of the lanthanoids, whereas the actinoids exhibit a variety of oxidation states.

The chemical properties of the first member of each group in the p block are significantly different from its congeners. These anomalies are attributable to the small atomic radius and its correlates, high ionization energies, high electronegativities, and low coordination numbers. For example, in Group 14 carbon forms an enormous number of catenated hydrocarbons with strong C−C bonds. Carbon also forms strong multiple bonds in the alkenes and alkynes. This tendency to catenation is much reduced for its congeners and the longest silane formed contains just four Si atoms. Nitrogen shows distinct differences from phosphorus and the rest of Group 15. Thus, nitrogen commonly exhibits a coordination number of 3, as in NF_3, and a coordination number of 4 in species such as NH_4^+ and NF_4^+ whereas phosphorus can form 3- and 5-coordinate compounds, such as PF_3 and PF_5, and 6-coordinate species such as PF_6^-.

Cross Reference: Section 14.7a

Cross Reference: Section 14.7b

Cross Reference: Section 15.1

The extent of hydrogen bonding is much greater in the compounds of the first member of each group. For example, the boiling point of ammonia is −33°C, which is higher than that of the other Group 15 hydrides. Likewise, water and hydrogen fluoride are liquids at room temperature whereas H_2S and HCl are gases.

Cross Reference: Section 15.10

Cross Reference: Sections 16.8a and 17.8

The striking differences between the Period 2 elements and their congeners is also reflected, but less markedly, in the d block. Thus, the properties of the 3d-series metals differ from those of the 4d and 5d series. The lower oxidation states are more stable in the 3d series, with the stability of higher oxidation states increasing down each group. For example, the most stable oxidation state of chromium is Cr(III) whereas it is M(VI) for Mo and W. The degree of covalence and the coordination numbers increase between the compounds of 3d and the 4d and 5d elements. For instance, the dihalides of the 3d-series elements are mostly ionic solids, as in CrF_2, whereas the 4d- and 5d-series elements form higher halides such as MoF_6 and WF_6, which are liquids at room temperature. These differences are attributable to the smaller ionic radii of the 3d-series elements and the fact that the radii of the 4d- and 5d-series elements are quite similar (due to lanthanide contraction).

Cross Reference: Section 19.4

Cross Reference: Section 19.7

Cross Reference: Section 23.3

The anomalies of the heading element are also found in the f block. The elements Ce through to Lu, generically Ln, are all highly electropositive, with the Ln^{3+}/Ln standard potential lying between those of Li and Mg. The elements favour the oxidation state Ln(III) with a uniformity that is unprecedented in the periodic table. Other properties of the elements vary significantly. For example, the radii of Ln^{3+} ions contract steadily across the series. This decrease is attributed in part to the increase in Z_{eff} as electrons are added to the 4f subshell, but relativistic effects also make a substantial contribution (Section 1.9a). This decrease in radius leads to an increase in the hydration enthalpy across the series. The reduction potentials of the lanthanoids are all similar, with values ranging from −1.99 V for Eu^{3+}/Eu to −2.38 V for La^{3+}/La (an honorary member of the f block).

The elements from thorium (Th, Z = 90) to lawrencium (Lr, Z = 103) have ground-state electron configurations that involve the filling of the 5f subshell, and in this sense are analogues of the lanthanoids. However, the actinoids do not exhibit the chemical uniformity of the lanthanoids and occur in a rich variety of oxidation states. The f orbitals of the actinoids extend beyond the Rn core and participate in bonding. Like the lanthanoids, the actinoids have large atomic and ionic radii and as a result often have high coordination numbers. For example, U in solid UCl_4 is 8-coordinate and in solid UBr_4 it is 7-coordinate in a pentagonal-bipyramidal array. Solid-state structures with coordination numbers up to 12 have been observed (Section 7.5).

Cross Reference: Section 23.9

In addition to the differences between the first member of each group and its congeners there are also similarities between elements with atomic numbers Z and Z + 8; for the d block, the similarities are between elements with atomic numbers Z and Z + 22. For example, Al (Z = 13) shows similarities to Sc (Z = 21). The similarities are understandable in terms of the electron configurations, for Al in Group 13 and Sc in Group 3 both have three valence electrons. Their atomic radii are fairly similar: Al is 143 pm and Sc is 160 pm and the standard potential of Al^{3+}/Al (−1.66 V) is closer to that of Sc^{3+}/Sc (−1.88 V) than that of Ga^{3+}/Ga (−0.53 V). Similarities also exist between the following pairs:

Z	14	15	16	17
	Si	P	S	Cl
Z + 8	22	23	24	25
	Ti	V	Cr	Mn

For example, S and Cr form anions of the type SO_4^{2-}, $S_2O_7^{2-}$, CrO_4^{2-}, and $Cr_2O_7^{2-}$, and both Cl and Mn form the oxidizing peroxoanions, ClO_4^- and MnO_4^-. These similarities are observed when the elements are in their highest oxidation states and the d-block element has a d^0 configuration. For d-block elements the similarity is between the Z and Z+22 elements.

EXAMPLE 9.3 **Predicting the chemical properties of a Z + 8 element**

The perchlorate ion, ClO_4^-, is a very powerful oxidizing agent and its compounds can detonate on contact or with heat. Predict whether the analogous compound of the $Z + 8$ element would be a suitable replacement for a perchlorate compound in a reaction.

Answer We need to identify the $Z + 8$ element. Because the atomic number of Cl is 17, the $Z + 8$ element is Mn ($Z = 25$). The compound of Mn analogous to ClO_4^- is the permanganate ion, MnO_4^-, which is in fact an oxidizing agent, but less hazardous. Permanganate is likely to be a suitable replacement for perchlorate.

Self-test 9.3 Xenon is very unreactive but forms a few compounds with oxygen and fluorine such as XeO_4. Predict the shape of XeO_4 and identify the $Z + 22$ compound with the same structure.

9.9 Binary compounds

The simple binary compounds of the elements exhibit interesting trends in their structure and properties. Hydrogen, oxygen, and the halogens form compounds with most elements and the hydrides, oxides, and halides are reviewed here to give some insight into trends in bonding and properties.

(a) Hydrides of the elements

Key point: The hydrides of the elements are classified as molecular, saline, or metallic.

Cross Reference: Sections 13.6, 14.7, 15.10, 16.8, and 17.2

Hydrogen reacts with most elements to form hydrides that can be described as molecular, saline, or metallic, although some cannot be easily classified and are termed intermediate (Fig. 9.11). Molecular compounds of hydrogen are common for the nonmetallic, electronegative elements of Groups 13 to 17; some examples are B_2H_6, CH_4, NH_3, H_2O, and HF. These covalent hydrides are gases, with the exception of water (due to extensive hydrogen bonding). The saline hydrides are formed by the electropositive elements of Group 1 and Group 2 (with the exception of Be). The saline hydrides are ionic solids with high melting points. Nonstoichiometric metallic hydrides are formed by all the d-block metals of Groups 3, 4, and 5, and by the f-block elements.

Cross Reference: Section 11.6 and 12.6

Cross Reference: Section X.X

(b) Oxides of the elements

Key point: Metals form basic oxides and nonmetals form acidic oxides. The elements form normal oxides, peroxides, superoxides, suboxides, and nonstoichiometric oxides.

The high reactivity of oxygen and its high electronegativity leads to a large number of binary oxygen compounds, many of which bring out high oxidation states in the second element. The range of possible oxides is shown in Table 9.5.

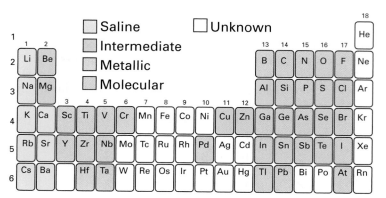

Figure 9.11 Classification of the binary hydrides of the s-, p-, and d-block elements.

Table 9.5 Possible oxides of the elements

1	2	3	4	5	6	7	8	9	10	11	12	13	14	15	16	17	18
H_2O H_2O_2																	
Li_2O	BeO											B_2O_3 (network solids glasses)	CO, CO_2, C_3O_2	N_2O, NO, N_2O_3, NO_2, N_2O_4, N_2O_5	O_2, O_3	OF_2, O_2F_2	
Na_2O, Na_2O_2	MgO, MgO_2											Al_2O_3	SiO_2 (glasses minerals)	P_4O_6, P_4O_{10}	SO_2, SO_3	Cl_2O, Cl_2O_3, ClO_2, Cl_2O_4, Cl_2O_6, Cl_2O_7	
K_2O, K_2O_2, KO_2, KO_3	CaO, CaO_2	Sc_2O_3	TiO, TiO_2, Ti_2O_3	VO, VO_2, V_2O_3, V_2O_5, V_3O_5	CrO_2, CrO_3, Cr_2O_3, Cr_3O_4	MnO, MnO_2, Mn_2O_3, Mn_2O_7, Mn_3O_4	FeO, Fe_2O_3, Fe_3O_4	CoO, Co_3O_4	NiO, Ni_2O_3	CuO, Cu_2O	ZnO	Ga_2O_3	GeO, GeO_2	As_2O_3, As_2O_5	SeO_2, SeO_3	Br_2O, Br_2O_3, BrO_2	
Rb_2O, Rb_2O_2, RbO_2, RbO_3, Rb_9O_2	SrO, SrO_2	Y_2O_3	ZrO_2	NbO, NbO_2, Nb_2O_5	MoO, MoO_2, MoO_3, Mo_2O_3, Mo_2O_5	TcO_2, Tc_2O_7	RuO_2, RuO_3	RhO_2, Rh_2O_3	PdO, PdO_2	AgO, Ag_2O	CdO	In_2O_3	SnO, SnO_2	Sb_2O_3, Sb_2O_5	TeO_2, TeO_3	I_2O_4, I_2O_4, I_2O_5, I_4O_9	XeO_3, XeO_4
Cs_2O, Cs_2O_2, CsO_2, CsO_3	BaO, BaO_2	La_2O_3	HfO_2	TaO, TaO_2, Ta_2O_3, Ta_2O_5	WO_2, WO_3	ReO_2, ReO_3, Re_2O_3, Re_2O_7	OsO_2, OsO_4	IrO_2, Ir_2O_3	PtO, PtO_2, PtO_3	Au_2O_3	HgO, Hg_2O	Tl_2O, Tl_2O_2	PbO, PbO_2, Pb_3O_4	Bi_2O_3, Bi_2O_5			

Metals typically form basic oxides. The electropositive metal forms a cation readily and the oxide anion abstracts a proton from water. For example, OH^- ions are produced when barium oxide reacts with water:

$$BaO(s) + H_2O(l) \rightarrow Ba^{2+}(aq) + 2\,OH^-(aq)$$

Nonmetals form acidic oxides. The electronegative element pulls in electrons from coordinated H_2O molecules, liberating H^+. For example, sulfur trioxide reacts with water to produce hydronium ions (represented here as simply $H^+(aq)$):

$$SO_3(g) + H_2O(l) \rightarrow 2\,H^+(aq) + SO_4^{2-}(aq)$$

The acidic nature of the oxides increases across a row and decreases down a group for a given oxidation state (Fig. 9.12). In Group 13 the element at the head of the group, B, is a nonmetal and forms the acidic oxide B_2O_3. At the bottom of the group the metallic character has increased and the inert pair effect has reduced the stable oxidation state from $+3$ to $+1$ and the oxide of thallium is the basic Tl_2O.

(c) Halides of the elements

Key point: The s-block halides are predominantly ionic and the p-block halides are predominantly covalent. In the d block, low oxidation state halides tend to be ionic and high oxidation state halides tend to be covalent.

The halogens form compounds with most elements, but not always directly. The range of chlorides that are formed is illustrated in Table 9.6.

With the exception of Li and Be, the s-block halides are ionic and the p-block fluorides are predominantly covalent. Fluorine and Cl bring out the group oxidation number in most of the elements as well as an oxidation number 2 lower, as expected from the inert pair effect.

Cross Reference: Section 11.8 and 12.8

Cross Reference: Sections 15.13, 16.12, and 17.2

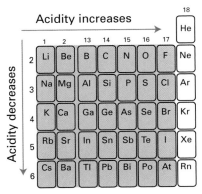

Figure 9.12 The general variation of acidic nature of the oxides of the elements through the periodic table.

Table 9.6 Simple chlorides of the elements

1	2	3	4	5	6	7	8	9	10	11	12	13	14	15	16	17
HCl																
LiCl	BeCl$_2$											BCl$_3$	CCl$_4$	NCl$_3$	OCl$_2$	ClF, ClF$_3$, ClF$_5$
NaCl	MgCl$_2$											AlCl3	SiCl$_4$	PCl$_3$, PCl$_5$	S$_2$Cl$_2$, SCl$_2$	Cl$_2$
KCl	CaCl$_2$	ScCl$_3$	TiCl$_2$, TiCl$_3$, TiCl$_4$	VCl$_2$, VCl$_3$, VCl$_4$	CrCl$_2$, CrCl$_3$, CrCl$_4$	MnCl$_2$, MnCl$_3$	FeCl$_2$, FeCl$_3$	CoCl$_2$, CoCl$_3$	NiCl$_2$	CuCl$_2$, CuCl$_3$	ZnCl$_2$	GaCl$_3$	GeCl$_4$	AsCl$_3$, AsCl$_5$	SeCl$_4$	BrCl
RbCl	SrCl$_2$	YCl$_3$	ZrCl$_2$, ZrCl$_4$	NbCl$_3$, NbCl$_4$, NbCl$_5$	MoCl$_2$, MoCl$_3$, MoCl$_4$, MoCl$_5$, MoCl$_6$	TcCl$_4$, MoCl$_6$	RuCl$_2$, RuCl$_3$	RhCl$_3$	PdCl$_2$	AgCl	CdCl$_2$	InCl, InCl$_2$, InCl$_3$	SnCl$_2$, SnCl$_4$	SbCl$_3$, SbCl$_5$	TeCl$_4$	ICl, ICl$_3$, I$_2$Cl$_6$
CsCl	BaCl$_2$	LaCl$_3$	HfCl$_4$	TaCl$_3$, TaCl$_4$, TaCl$_5$	WCl$_2$, WCl$_4$, WCl$_6$	ReCl$_4$, ReCl$_5$, ReCl$_6$	OsCl$_4$, OsCl$_5$, OsCl$_6$	IrCl$_2$, IrCl$_3$, IrCl4	PtCl$_3$, PtCl$_4$	AuCl	HgCl$_2$, Hg2Cl2	TlCl, TlCl$_2$, TlCl$_3$	PbCl$_2$, PbCl$_4$	BiCl$_3$, BiCl$_5$		At

Exceptions are N, O, and S, where only lower oxidation state halides are formed. The d-block elements form halides with a range of oxidation states. The higher oxidation state halides are formed with F and Cl. The lower oxidation state halides are ionic solids. The higher oxidation state chlorides formed predominantly by the 4d- and 5d-series elements are covalent and there is an increased tendency to form cluster compounds with metal–metal bonds.

Cross Reference: Section 19.7

9.10 Wider aspects of periodicity

The differences in chemical properties of elements and compounds result from a complex interplay of periodic trends. In this final section we illustrate how these trends compensate, conflict with, and enhance each other.

(a) Ionic chlorides

Key point: For ionic compounds, trends in lattice enthalpies, ionization energies, and enthalpies of atomization have significant effects on the enthalpy of formation of ionic halides.

Cross Reference: Section 12.7 and 11.7

As can be seen from Table 9.7, the values of $\Delta_f H^\ominus$ for the Group 1 halides are reasonably constant on descending a group. The ionization energy and enthalpy of atomization both become less positive down the group as the atomic radii increase but these trends are largely offset by changes in the lattice enthalpy (Section 3.11). The values of $\Delta_f H^\ominus$ for Group 2 halides are up to twice the values for the Group 1 halides. The ionization energy and enthalpy of atomization both become more positive between Groups 1 and 2 as the radii decrease. However, the more significant factor is the very large increase in lattice enthalpy as the radii decrease and the charges on the ions increase from +1 and −1 to +2 and 2 × −1. As a result, Group 2 halides are more stable than those of Group 1.

(b) Covalent halides

Key point: Bond enthalpy and entropy effects are the most important factors in determining whether or not Group 16 halides exist.

Table 9.7 Standard enthalpies of formation of Group 1 and 2 chlorides, $\Delta_f H^\ominus/(\text{kJ mol}^{-1})$

LiCl	−409	BeCl$_2$	−512
NaCl	−411	MgCl$_2$	−642
KCl	−436	CaCl$_2$	−795
RbCl	−431	SrCl$_2$	−828
CsCl	−433	BaCl$_2$	−860

Compounds formed between sulfur and the halogens can provide some insight into the factors that influence the values of $\Delta_f H^{\ominus}$ for covalent halides. Sulfur forms several different compounds with F, most of which are gases. Sulfur hexafluoride, SF_6, sulfur difluoride, SF_2, and sulfur dichloride, SCl_2, exist whereas SCl_6 is not known. The values of $\Delta_f H^{\ominus}$ calculated from bond enthalpy data are:

Cross Reference: Section 16.9

	SF_2	SF_6	SCl_2	SCl_6
$\Delta_f H^{\ominus}/(\text{kJ mol}^{-1})$	-298	-1220	-49	-74

Thus, although the formation of SCl_6 is more exothermic than SCl_2, other factors ensure that SCl_6 cannot be prepared under standard conditions. The explanation can be found by considering the bond enthalpies of sulfur−halogen bonds:

	F−SF	F−SF$_5$	Cl−SCl
$B/(\text{kJ mol}^{-1})$	367	329	271

There is a decrease in bond enthalpy between SF_2 and SF_6 due possibly to steric crowding around the S atom and repulsion between crowded F atoms. A similar decrease would be expected between SCl_2 and SCl_6. This weak bond is one factor in the nonexistence of SCl_6. In addition, there would be a higher entropic cost when a SCl_6 molecule is formed from three dihalogen molecules compared to just one when SCl_2 is formed. In contrast, compounds containing the PCl_6^- ion are known. Bonding between P and Cl should be stronger than that between S and Cl because P is less electronegative than S.

EXAMPLE 9.4 Assessing the factors affecting the formation of a compound

Estimate $\Delta_f H^{\ominus}(SH_6, g)$ by assuming that $B(H-S)$ is the same for $H-SH_5$ as for $H-SH$ (375 kJ mol^{-1}). The value of $B(H-H)$ is 436 kJ mol^{-1} and $B(S-S)$ is 263 kJ mol^{-1}. Suggest a way of reconciling your answer with observation.

Answer We estimate $\Delta_f H^{\ominus}(SH_6, g)$ from the difference between the enthalpies of bonds broken and bonds formed in the reaction

$$\tfrac{1}{8}S_8(s) + 3H_2(g) \rightarrow SH_6(g)$$

The enthalpy change accompanying bond breaking is 263 kJ mol^{-1} + 3(436 kJ mol^{-1}) = 1571 kJ mol^{-1}. The enthalpy change accompanying bond making is $-6 \times$ (375 kJ mol^{-1}) = -2250 kJ mol^{-1}. Therefore,

$$\Delta_f H^{\ominus}(SH_6, g) = 1571 \text{ kJ mol}^{-1} - 2250 \text{ kJ mol}^{-1} = -679 \text{ kJ mol}^{-1}$$

indicating that the compound is exothermic and, on the basis of this calculation, can be expected to exist. However, SH_6 does not exist. The reason must lie in the S−H bond being much weaker than that used in the calculation. A further contribution to the Gibbs energy of formation is the unfavourable entropy change due to forming the molecule from three H_2 molecules instead of only one when SH_2 is formed.

Self-test 9.4 Comment on the following $\Delta_f H^{\ominus}$ values (kJ mol^{-1}):

S(g)	Se(g)	Te(g)	SF$_4$	SeF$_4$	TeF$_4$	SF$_6$	SeF$_6$	TeF$_6$
+223	+202	+199	−762	−850	−1036	−1220	−1030	−1319

(c) Oxides in the d block

Key point: The d-block oxides deviate from the ionic model as metal−metal bonding becomes more important.

The contrasting enthalpies of formation of various metal oxides of formula MO (Table 9.8) give some interesting insights into different aspects of periodicity. The high exothermocity of the Group 2 oxides is a result of the relatively low ionization energy and low enthalpy of atomization of the s-block metals. The experimental lattice enthalpies are very close to those calculated from the Kapustinskii equation, indicating that the compounds conform well to the ionic model. Values of $\Delta_f H^{\ominus}(MO)$ for the 3d-series elements become less negative across

Cross Reference: Section 12.8

Table 9.8 Some thermodynamic data (in kJ mol^{-1}) for d-metal oxides, MO

	$\Delta_{\text{ion}(1+2)}H^{\ominus}$	$\Delta_a H^{\ominus}$		$\Delta_f H^{\ominus}$	$\Delta H_L^{\ominus}$ (calc*)	$\Delta H_L^{\ominus}$ (exp)
Ca	1735	177	CaO	−636	3464	3390
V	2064	514	VO	−431	3728	4037
Ni	2490	430	NiO	−240	3860	4436
Nb	2046	726	NbO	−406	3760	4206

* As calculated from the Kapustinskii equation.

Cross Reference: Section 19.8

the period. There are opposing trends across this series because, although the ionization energy increases, the atomization enthalpy decreases. The experimental lattice enthalpies of the oxides on the right of the series deviate from those calculated from the Kapustinskii equation, indicating that the ionic model is no longer adequate. Although the ionization energy of a 4d-series element is lower than that of its 3d-series congener, its atomization enthalpy is much higher, reflecting the stronger metallic bonding due to better overlap between 4d orbitals than between 3d orbitals.

EXAMPLE 9.5 Predicting the thermal stabilities of d-block oxides

Compare the stabilities of V_2O_5 and Nb_2O_5 towards thermal decomposition in the reaction

$$M_2O_5(s) \rightarrow 2\,MO(s) + \tfrac{3}{2}O_2(g)$$

Use the data in Table 9.8 and the enthalpies of formation of Nb_2O_5 and V_2O_5 of −1901 and −1552 kJ mol^{-1}, respectively.

Answer We need to consider the reaction enthalpy for each oxide. The enthalpies of reaction can be calculated from the difference between the enthalpies of formation of products and those of the reactants:

For Nb_2O_5: $\Delta_r H^{\ominus} = 2(−406\text{ kJ mol}^{-1}) − (−1901\text{ kJ mol}^{-1}) = +1089\text{ kJ mol}^{-1}$

For V_2O_5: $\Delta_r H^{\ominus} = 2(−431\text{ kJ mol}^{-1}) − (−1552\text{ kJ mol}^{-1}) = +690\text{ kJ mol}^{-1}$

The enthalpy of the reaction for V_2O_5 is less endothermic than that of Nb_2O_5. Therefore, V_2O_5 is thermally more stable.

Self-test 9.5 Given that $\Delta_f H^{\ominus}(P_4O_{10}, s) = −3012$ kJ mol^{-1}, what further data would be useful when drawing comparisons with the value for V_2O_5?

FURTHER READING

P. Enghag, *Encyclopedia of the elements*. John Wiley & Sons (2004).

D.M.P. Mingos, *Essential trends in inorganic chemistry*. Oxford University Press (1998). An overview of inorganic chemistry from the perspective of structure and bonding.

N.C. Norman, *Periodicity and the s- and p-block elements*. Oxford University Press (1997). Includes coverage of essential trends and features of s-block chemistry.

E.R. Scerri, *The periodic table: Its story and its significance*. Oxford University Press (2007).

EXERCISES

9.1 Give the expected maximum stable oxidation state for (a) Ba, (b) As, (c) P, (d) Cl.

9.2 With the exception of one member of the group, the elements form saline hydrides. They form oxides and peroxides and all the carbides react with water to liberate a hydrocarbon. Identify this group of elements.

9.3 The elements vary from metals through metalloids to nonmetals. They form halides in oxidation states +5 and +3 and the hydrides are all toxic gases. Identify this group of elements.

9.4 Draw a Born−Haber cycle for the formation of the hypothetical compound $NaCl_2$. State which thermochemical step is responsible for the fact that $NaCl_2$ does not exist.

9.5 Predict how the inert pair effect would manifest itself beyond Group 15 and compare your predictions with the chemical properties of the elements involved.

9.6 Summarize the relationship between ionic radii, ionization energy, and metallic character.

9.7 Give the names of the ores from which (a) Mg, (b) Al, and (c) Pb are extracted.

9.8 Identify the $Z + 8$ element for P. Briefly summarize any similarities between the elements.

9.9 Use the following data to calculate average values of $B(Se−F)$ in SeF_4 and SeF_6. Comment on your answers in view of the corresponding values for $B(S-F)$ in SF_4 (+340 kJ mol^{-1}) and SF_6 (+329 kJ mol^{-1}): $\Delta_a H^\ominus$ (Se) = +227 kJ mol^{-1}, $\Delta_a H^\ominus$(F) = +159 kJ mol^{-1}, $\Delta_f H^\ominus$ l(SeF$_6$, g) = −1030 kJ mol^{-1}, $\Delta_f H^\ominus$l(SeF$_4$, g) = −850 kJ mol^{-1}.

PROBLEMS

9.1 In the paper 'What and how physics contributes to understanding the periodic law' (*Foundations of Chemistry*, 2001, 3, 145) V. Ostrovsky describes the philosophical and methodological approaches taken by physicists to explain periodicity. Compare and contrast the approaches of physics and chemistry to explaining chemical periodicity.

9.2 P. Christiansen *et al.* describe 'Relativistic effects in chemical systems' in their 1985 paper (*Ann. Rev. Phys. Chem.*, 2001, **36**, 407).

How did they define relativistic effects? Briefly summarize the most important consequences of relativistic effects in chemistry.

9.3 Many models of the periodic table have been proposed since the version devised by Mendeleev. Review the more recent versions and discuss the theoretical basis for each one.

10 Hydrogen

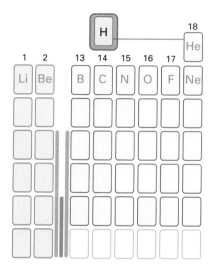

Hydrogen has a very rich chemistry despite its simple atomic structure. In this chapter we discuss reactions of hydrogen species that are particularly interesting in terms of fundamental chemistry as well as important applications, including energy. We describe how H_2 is produced in the laboratory, on an industrial scale from fossil fuels, and by using renewable resources. We summarize the properties of binary compounds that range from volatile, molecular compounds to salt-like and metallic solids. We shall see that many of the properties of these compounds are understandable in terms of their ability to supply H^- or H^+ ions, and that it is often possible to anticipate which one is likely to dominate. We explain how H-bonding stabilizes the structures of H_2O and DNA and how H_2 can behave as a ligand to d-block metal atoms and ions.

PART A: THE ESSENTIALS

Hydrogen is the most abundant element in the universe and the tenth most abundant by mass on Earth, where it is found in the oceans, minerals, and in all forms of life. The partial depletion of elemental hydrogen from Earth reflects its volatility during formation of the planet. The stable form of elemental hydrogen under normal conditions is dihydrogen, H_2, which occurs at trace levels in the Earth's lower atmosphere (0.5 ppm) and is essentially the only component of the extremely thin outer atmosphere. Dihydrogen has many uses (Fig. 10.1). It is produced naturally, as a product of fermentation and as a byproduct of ammonia biosynthesis (Box 10.1). It is often cited as the 'fuel of the future' on account of its availability from fully renewable resources (water and sunlight) and its clean and highly exothermic reaction with O_2, but the volatility and low energy density of H_2 pose challenges for storage.

10.1 The element

The hydrogen atom, with ground-state configuration $1s^1$, has only one electron so it might be thought that the element's chemical properties will be limited, but this is far from the case. Hydrogen has richly varied chemical properties and forms compounds with nearly every other element. It ranges in character from being a strong Lewis base (the hydride ion, H^-) to being a strong Lewis acid (as the hydrogen cation, H^+, the proton; Section 4.1). Under certain circumstances H atoms can form bonds to more than one other atom simultaneously. The 'hydrogen bond' formed when an H atom bridges two electronegative atoms is fundamental to life: it is because of hydrogen bonding that water occurs as a liquid rather than a gas, and proteins and nucleic acids fold into the complex, highly organized three-dimensional structures that determine their functions.

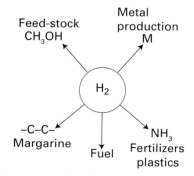

Figure 10.1 The major uses of H_2.

BOX 10.1 The biological hydrogen cycle

Hydrogen is cycled by microbial organisms using metalloenzymes (Section 27.14). Although H_2 is present only to the extent of about 0.5 ppm at the Earth's surface, its levels are hundreds of times higher in anaerobic environments, such as wetland soils and sediments at the bottoms of deep lakes and hot springs. Hydrogen is produced in these O_2-free zones as a waste product by strict anaerobes (fermentative bacteria) which break down organic material (biomass) using H^+ as an oxidant acting as a terminal electron acceptor. It is also produced by thermophilic organisms that derive their carbon and energy entirely from CO, and by nitrogen fixing bacteria, which yield H_2 as a byproduct of ammonia formation. Other microorganisms, many of them aerobic, use H_2 instead as a 'food' (a fuel) and are responsible for the formation of the familiar gases CH_4 (methanogens) and H_2S (*Desulfovibrio*) as well as nitrite and other products. Figure B10.1 summarizes some of the overall processes occurring in a freshwater environment.

In animals, including humans, the anaerobic environment of the large intestine is host to bacteria that form H_2 by the breakdown of carbohydrates. The mucus layer of the mouse intestine has been found to contain H_2 at levels above 0.04 mmol dm^{-3}, equivalent to a partial atmosphere of 5 per cent H_2. In turn, this H_2 is utilized by methanogens, such as those found in ruminating mammals, to produce CH_4 and by other bacteria, including dangerous pathogens such as those of the *Salmonella* genus and *Helicobacter pylori*, which is responsible for gastric ulcers. High levels of H_2 in the breath have been used to diagnose conditions related to carbohydrate-intolerance and these levels may reach >70 ppm following lactose ingestion by lactose-intolerant patients.

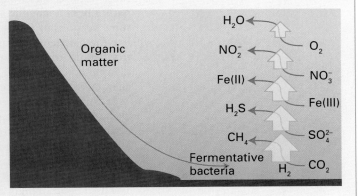

Figure B10.1 Some of the processes that contribute to the biological hydrogen cycle in a freshwater environment.

Industrial production of H_2 by microorganisms (*biohydrogen*) is an important area for research and development. There are two different approaches, each of which uses renewable energy. The first of these approaches is to use anaerobic organisms to ferment biomass from sources ranging from cultivated biomass (including seaweed) to domestic waste. The second involves manipulation of photosynthetic organisms such as green algae and cyanobacteria to produce H_2 as well as biomass. In either case H_2 can be extracted continuously by gas filters without the interruptions that would be required were harvesting required.

(a) The atom and its ions

Key points: The proton, H^+, is always found in combination with a Lewis base and is highly polarizing; the hydride ion, H^-, is highly polarizable.

There are three isotopes of hydrogen: hydrogen itself (^1H), deuterium (D, ^2H), and tritium (T, ^3H); tritium is radioactive. The lightest isotope, ^1H (very occasionally called protium), is by far the most abundant. Deuterium has variable natural abundance with an average value of about 16 atoms in 100 000. Tritium occurs to the extent of only 1 in 10^{21}. The different names and symbols for the three isotopes reflect the significant differences in their masses and the chemical properties that stem from mass, such as the rates of diffusion and bond-cleavage reactions. The nuclear spin of ^1H ($I = \frac{1}{2}$) is exploited in NMR spectroscopy (Section 8.5) for identifying hydrogen-containing molecules and determining their structures.

The free hydrogen cation (H^+, the proton) has a very high charge/radius ratio and it is not surprising to find that it is a very strong Lewis acid. In the gas phase it readily attaches to other molecules and atoms; it even attaches to He to form HeH^+. In the condensed phase, H^+ is always found in combination with a Lewis base, and its ability to transfer between Lewis bases gives it the special role in chemistry explored in detail in Chapter 4. The molecular cations H_2^+ and H_3^+ have only a transitory existence in the gas phase and are unknown in solution. In contrast to H^+, which is highly polarizing; the hydride ion, H^-, is highly polarizable because two electrons are bound by just one proton. The radius of H^- varies considerably depending on the atom to which it is attached. The lack of core electrons to scatter X-rays means that bond distances and angles involving a H atom in a compound are difficult to measure by X-ray diffraction: for this reason, neutron diffraction is used when it is crucial to determine the precise positions of H atoms.

(b) Properties and reactions

Key points: Hydrogen has unique atomic properties that place it in a special position in the periodic table. Dihydrogen is an inert molecule and its reactions require a catalyst or initiation by radicals.

Hydrogen's unique properties distinguish it from all other elements in the periodic table. It is often placed at the head of Group 1 because, like the alkali metals, it has only one electron in its valence shell. That position, however, does not truly reflect the chemical or physical properties of the element. In particular, its ionization energy is far higher than those of the other Group 1 elements, so hydrogen is not a metal, although it may be found naturally in a metallic state where extreme pressures exist, such as the core of Jupiter. In some versions of the periodic table hydrogen is placed at the head of Group 17 because, like the halogens, it requires only one electron to complete its valence shell. But the electron affinity of hydrogen is far lower than that of any of the elements of Group 17 and the discrete hydride ion, H⁻, is encountered only in certain compounds. To reflect its unique characteristics, we place H in its own special position at the head of the entire table.

Because H_2 has so few electrons, the intermolecular forces between H_2 molecules are weak, and at 1 atm the gas condenses to a liquid only when cooled to 20 K. If an electric discharge is passed through H_2 gas at low pressure, the molecules dissociate, ionize, and recombine, forming a plasma containing, in addition to H_2, spectroscopically observable amounts of H, H^+, H_2^+, and H_3^+.

The H_2 molecule has a high bond enthalpy (436 kJ mol⁻¹) and a short bond length (74 pm). The high bond strength results in H_2 being an inert molecule. However, it is still much easier to dissociate H_2 homolytically than heterolytically,[1] because in heterolytic dissociation energy is required to separate opposite charges:

$$H_2(g) \rightarrow H(g) + H(g) \qquad \Delta_r H^{\ominus} = +436 \text{ kJ mol}^{-1}$$

$$H_2(g) \rightarrow H^+(g) + H^-(g) \qquad \Delta_r H^{\ominus} = +1675 \text{ kJ mol}^{-1}$$

Because of its inherently inert nature, reactions of H_2 do not occur readily unless a special activation pathway has been provided. These pathways include homolytic or heterolytic dissociation catalysed by molecules or active surfaces, and initiation of radical chain reactions. The explosive reaction of H_2 with O_2, for instance

$$2 H_2(g) + O_2(g) \rightarrow 2 H_2O(g) \qquad \Delta_r H^{\ominus} = -242 \text{ kJ mol}^{-1}$$

proceeds by a complex radical chain mechanism. Hydrogen is an excellent fuel for large rockets on account of its high specific enthalpy (the standard enthalpy of combustion divided by the mass), which is approximately three times that of a typical hydrocarbon (Box 10.2).

In addition to reactions that result in dissociation of the H−H bond, H_2 6n also react reversibly without cleavage to form dihydrogen d-metal complexes (Section 10.6 and Section 22.7).

10.2 Simple compounds

The nature of the bonding in binary compounds of hydrogen (EH_n) is largely rationalized by noting that an H atom has a high ionization energy (1312 kJ mol⁻¹) and a low but positive electron affinity (73 kJ mol⁻¹). Although binary hydrogen compounds are often known as 'hydrides', very few actually contain a discrete H⁻ anion. The (Pauling) electronegativity of 2.2 (Section 1.9) is intermediate in value, so it is normally assigned the oxidation number −1 when in combination with metals (as in NaH and AlH_3) and +1 when in combination with nonmetals (as in H_2O and HCl).

(a) Classification of binary compounds

Key points: Compounds formed between hydrogen and other elements vary in their nature and stability. In combination with metals, hydrogen is often regarded as a hydride; hydrogen compounds with elements of similar electronegativity have low polarity.

The binary compounds of hydrogen fall into three classes, although there is a range of structural types, and some elements form compounds with hydrogen that do not fall strictly into any one category:

1. **Molecular hydrides** exist as individual, discrete molecules; they are usually formed with p-block elements of similar or higher electronegativity than H. Their E−H bonds are best regarded as covalent.

Familiar examples of molecular hydrides include methane, CH_4 (**1**), ammonia, NH_3 (**2**), and water, H_2O (**3**).

2. **Saline hydrides**, also known as *ionic hydrides*, are formed with the most electropositive elements.

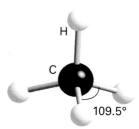

1 Methane, CH_4

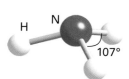

2 Ammonia, NH_3

3 Water, H_2O

---

[1] In homolytic dissociation the bond breaks symmetrically to give one product. In heterolytic dissociation the bond breaks unsymmetrically to give two different products.

BOX 10.2 Dihydrogen as a fuel

The use of H_2 as a fuel (an energy carrier) has been investigated seriously since the 1970s when oil prices first rose dramatically; interest has increased greatly in more recent times owing to environmental pressures on further use of fossil fuels. Hydrogen is clean burning, nontoxic, and its production from fully renewable resources is slowly but inevitably replacing its production from fossil carbon feedstocks. Table B10.1 compares the performance data for H_2 and other energy carriers, including hydrocarbon fuels and a lithium ion battery. Among all fuels, H_2 has the highest specific enthalpy (its standard enthalpy of combustion divided by its mass) but has a very low energy density (its standard enthalpy of combustion divided by its volume).

It is obvious that H_2 is an excellent fuel for vehicles provided the problems of on-board containment are solved (see Box 10.4). In addition to its choice as a rocket fuel (due to its high specific enthalpy), H_2 can be used in conventional internal combustion engines with little if any modification

Table B10.1 Specific enthalpies and energy densities of common energy carriers (1 MJ = 0.278 kWh)

Fuel	Specific enthalpy /(MJ kg^{-1})	Energy density /(MJ dm^{-3})
Liquid H$_2$*	120	8.5
H$_2$ at 200 bar*	120	1.9
Liquid natural gas	50	20.2
Natural gas at 200 bar	50	8.3
Petrol (gasoline)	46	34.2
Diesel*	45	38.2
Coal	30	27.4
Ethanol*	27	22.0
Methanol	20	15.8
Wood*	15	14.4
Lithium battery*	2.0	6.1

(Li$_{1-x}$CoO$_2$, see Box 11.1)

....

* Denotes an energy carrier that is easily derived or recharged from renewable resources.

to their design or specifications. However, the most important way of utilizing H_2 in a vehicle is to react it in a fuel cell to produce electricity directly (Box 5.1). The efficient and reliable power output of H_2 fuel cells makes it viable to produce H_2 'on board' by steam reforming of methanol, a transportable and energy-dense fuel. (Direct methanol fuel cells, discussed in Box 5.1, produce less power than H_2 fuel cells and are therefore less attractive for vehicles.) An automotive steam reformer (the principle of which is shown in Fig. B10.2) mixes methanol vapour with H_2O (steam) and with O_2 (air) to produce H_2 by the following reactions

$$CH_3OH(g) + H_2O(g) \xrightarrow{Cu/ZnO} CO_2(g) + 3H_2(g) \quad \Delta_rH^\ominus = +49 \text{ kJ mol}^{-1}$$

$$CH_3OH(g) + \tfrac{1}{2}O_2(g) \xrightarrow{Pd} CO_2(g) + 2H_2(g) \quad \Delta_rH^\ominus = -155 \text{ kJ mol}^{-1}$$

The reactions, which occur over the temperature range 200–350°C, are controlled to ensure that the heat produced by the exothermic oxidation reaction just offsets that required for the reaction with steam and the vaporization of all components. Excessive heat results in production of CO, which poisons the Pt catalyst of the PEM fuel cell. The CO_2 and H_2 products are separated with a Pd membrane.

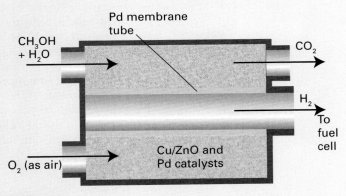

Figure B10.2 Schematic cross-sectional view of an on-board methanol reformer.

Saline hydrides, such as LiH and CaH$_2$, are nonvolatile, electrically nonconducting, crystalline solids although only those in Group 1 and the heavier elements of Group 2 should be regarded as hydride 'salts' containing H$^-$ ions.

3. Metallic hydrides are non-stoichiometric, electrically conducting solids with a metallic lustre.

Metallic hydrides are formed with many d- and f-block elements (Table 3.12). The H atoms are often regarded as occupying interstitial sites within the metal structure although this occupation rarely occurs without expansion or phase change. Figure 10.2 (which reproduces Fig. 9.11) summarizes this classification and the distribution of the different classes through the periodic table. It also identifies 'intermediate' hydrides that do not fall strictly into any of these categories, and elements for which binary hydrides have not been characterized.

In addition to binary compounds, hydrogen is found in complex anions of some p-block elements, examples being the BH$_4^-$ ion (tetrahydridoborate, also known in older texts as 'borohydride') in NaBH$_4$ or the AlH$_4^-$ ion (tetrahydridoaluminate, also known in older texts as 'aluminiumhydride') in LiAlH$_4$.

(b) Thermodynamic considerations

Key points: In the s and p blocks, strengths of E—H bonds decrease down each group. In the d block, strengths of E—H bonds increase down each group.

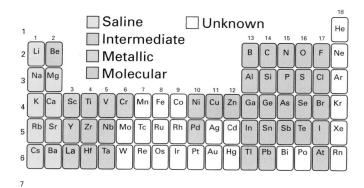

Figure 10.2 Classification of the binary hydrogen compounds of the s-, p-, and d-block elements. Although some d-block elements such as iron and ruthenium do not form binary hydrides they do form metal complexes containing the hydride ligand.

The standard Gibbs energies of formation of the hydrogen compounds of s- and p-block elements reveal a regular variation in stability (Table 10.1). With the possible exception of BeH_2 (for which good data are not available) all the s-block hydrides are exergonic ($\Delta_f G^\ominus < 0$) and therefore thermodynamically stable with respect to their elements at room temperature. The trend is erratic in Group 13, in that only AlH_3 is exergonic at room temperature. In all the other groups of the p block, the simple hydrogen compounds of the first members of the groups (CH_4, NH_3, H_2O, and HF) are exergonic but the analogous compounds of their congeners become progressively less stable down the group, a trend that is illustrated by decreasing E−H bond energies (Fig. 10.3). The heavier hydrides become more stable on going from Group 14 across to the halogens. For example, SnH_4 is highly endergonic ($\Delta_f G^\ominus > 0$) whereas HI is barely so.

These thermodynamic trends can be traced to the variation in atomic properties. The H−H bond is the strongest single homonuclear bond known (apart from D−D or T−T bonds) and in order for a compound to be exergonic and stable with respect to its elements, it needs to have E−H bonds that are even stronger than H−H. For molecular hydrides of the p-block elements, bonding is strongest with the Period 2 elements and becomes progressively weaker down each group. The weak bonds formed by the heavier p-block elements are due to the poor overlap between the relatively compact H1s orbital and the more diffuse s and p orbitals of their atoms. Although d-block elements do not form binary molecular compounds, many complexes contain one or more hydride ligands. Metal−hydrogen bond strengths in the d block increase down a group because the 3d orbitals are too contracted to overlap well with the H1s orbital and better overlap is afforded by 4d and 5d orbitals.

(c) Reactions of binary compounds

Key point: The reactions of binary compounds of hydrogen fall into three classes depending on the polarity of the E−H bond.

In compounds where E and H have similar electronegativities, cleavage of the E−H bond tends to be homolytic, producing, initially, an H atom and a radical, each of which can go on to combine with other available radicals.

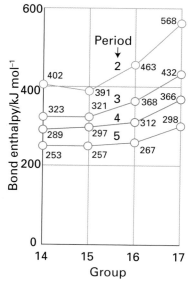

Figure 10.3 Average bond energies (kJ mol⁻¹) for binary molecular hydrides of p-block elements.

Table 10.1 Standard Gibbs energy of formation, $\Delta_f G^\ominus /(\text{kJ mol}^{-1})$ of binary s- and p-block hydrogen compounds at 25°C. Values in parentheses are estimates.

Period	Group						
	1	2	13	14	15	16	17
2	LiH(s)	BeH₂(s)	B₂H₆(g)	CH₄(g)	NH₃(g)	H₂O(l)	HF(g)
	−68.4	(+20)	+86.7	−50.7	−16.5	−237.1	−273.2
3	NaH(s)	MgH₂(s)	AlH₃(s)	SiH₄	PH₃(g)	H₂S(g)	HCl(g)
	−33.5	−35.9	(+91)	+56.9	+13.4	−33.6	−95.3
4	KH(s)	CaH₂(s)	Ga₂H₆(s)	GeH₄(g)	AsH₃(g)	H₂Se(g)	HBr(g)
	(−36)	−147.2	>0	+113.4	+68.9	+15.9	−53.5
5	RbH(s)	SrH₂(s)		SnH₄(g)	SbH₃(g)	H₂Te(g)	HI(g)
	(−30)	(−141)		+188.3	+147.8	>0	+1.7
6	CsH(s)	BaH₂(s)					
	(−32)	(−140)					

For $\chi(E) \approx \chi(H)$: $\qquad$ $E-H \rightarrow E\cdot + H\cdot$

Common examples of homonuclear cleavage include the thermolysis and combustion of hydrocarbons.

In compounds where E is more electronegative than H, heterolytic cleavage occurs, releasing a proton.

For $\chi(E) > \chi(H)$: $\qquad$ $E-H \rightarrow E^- + H^+$

The compound behaves as a Brønsted acid and is able to transfer H^+ to a base. In such compounds the H atom is termed **protonic**. Heterolytic bond cleavage also occurs in compounds where E is less electronegative than H, including saline hydrides

For $\chi(E) < \chi(H)$: $\qquad$ $E-H \rightarrow E^+ + H^-$

In this case the H atom is **hydridic** and an H^- ion is transferred to a Lewis acid. The reducing agents $NaBH_4$ and $LiAlH_4$ used in organic synthesis are examples of hydride-transfer reagents.

PART B: **THE DETAIL**

In this part of the chapter we present a more detailed discussion of the chemical properties of hydrogen, identifying and interpreting trends. We explain how dihydrogen is prepared on a small scale in the laboratory and how it is produced industrially from fossil fuels, then we outline methods for its production from water using renewable energy. We describe the reactions that dihydrogen undergoes with other elements and classify the different types of compounds formed. Finally we present the strategies for synthesizing various hydrogen-containing compounds.

10.3 Nuclear properties

Key point: The three hydrogen isotopes H, D, and T have large differences in their atomic masses and different nuclear spins, which give rise to easily observed changes in IR, Raman, and NMR spectra of molecules containing these isotopes.

Neither ^{1}H nor ^{2}H (deuterium, D) is radioactive, but ^{3}H (tritium, T) decays by the loss of a β particle to yield a rare but stable isotope of helium:

$$^3_1H \rightarrow {}^3_2He + \beta^-$$

The half-life for this decay is 12.4 years. Tritium's abundance of 1 in 10^{21} hydrogen atoms in surface water reflects a steady state between its production by bombardment of cosmic rays on the upper atmosphere and its loss by radioactive decay. Tritium can be synthesized by neutron bombardment of ^{6}Li or ^{7}Li.

$$^1_0n + {}^6_3Li \rightarrow {}^3_1H + {}^4_2He + 4.78 \text{ MeV}$$

$$^1_0n + {}^7_3Li \rightarrow {}^3_1H + {}^4_2He + {}^1_0n - 2.87 \text{ MeV}$$

Continuous production of tritium from lithium is a key step in the projected future generation of energy from nuclear fusion rather than nuclear fission. In a fusion reactor tritium and deuterium are heated to over 100 MK to give a plasma in which the nuclei react to produce ^{4}He and a neutron.

$$^2_1H + {}^3_1H \rightarrow {}^4_2He + {}^1_0n + 17.6 \text{ MeV}$$
$$(\Delta H = -1698 \text{ MJ mol}^{-1})$$

The neutron is used to bombard a lithium blanket that has been enriched in ^{6}Li to generate further tritium. This process carries far fewer environmental risks than fission of ^{235}U and is essentially renewable: of the two primary fuels required, deuterium is readily available from water, although the natural abundance of lithium is low (Chapter 11).

The physical and chemical properties of **isotopologues**, isotopically substituted molecules, are usually very similar, but not when D is substituted for H, as the mass of the substituted atom is doubled. Table 10.2 shows that the differences in boiling points and bond enthalpies are easily measurable for H_2 and D_2. The difference in boiling point between H_2O and D_2O reflects the greater strength of the $O\cdots D-O$ hydrogen bond (Section 10.6) compared with that of the $O\cdots H-O$ bond because the zero-point energy (Section 8.4) of the former is lower. The compound D_2O is known as 'heavy water' and is used as a moderator in the nuclear power industry; it slows down emitted neutrons and increases the rate of induced fission.

> *A note on good practice* An *isotopologue* is a molecular entity that differs only in isotopic composition. An *isotopomer* is an isomer having the same number of each isotopic atom but differing in their positions.

Reaction rates are often measurably different for processes in which $E-H$ and $E-D$ bonds, where E is another element, are broken, made, or rearranged. The detection of this **kinetic isotope effect** can often help to support a proposed reaction mechanism. Kinetic isotope effects are frequently observed when an H atom is transferred from one atom to another in an activated complex. For example, the electrochemical reduction of $H^+(aq)$ to $H_2(g)$ occurs with a substantial isotope effect, with H_2 being liberated much more rapidly. A practical consequence of the difference in rates of formation of H_2 and D_2 is that D_2O may be concentrated electrolytically, thus facilitating separation of

Table 10.2 The effect of deuteration on physical properties

	H_2	D_2	H_2O	D_2O
Normal boiling point/°C	−252.8	−249.7	100.0	101.4
Mean bond enthalpy/(kJ mol⁻¹)	436.0	443.3	463.5	470.9

the two isotopes: the pure D_2O that accumulates is then used to produce pure HD (by reaction with $LiAlH_4$) or D_2 (by electrolysis). In general, reactions involving D_2O occur more slowly than those involving H_2O and, not surprisingly, D_2O and D-substituted foods ingested in large quantities are poisonous for higher organisms.

Because the frequencies of molecular vibrations depend on the masses of atoms, they are strongly influenced by substitution of D for H. The heavier isotope results in the lower frequency (Section 8.4). The isotope effect can be exploited by observing the IR spectra of isotopologues to determine whether a particular infrared absorption involves significant motion of a hydrogen atom in the molecule.

The distinct properties of the isotopes make them useful as **tracers**. The involvement of H and D through a series of reactions can be followed by infrared (IR, Section 8.4) and mass spectrometry (Section 8.10) as well as by NMR spectroscopy (Section 8.5). Tritium can be detected by its radioactivity, which can be a more sensitive probe than spectroscopy.

Another important property of the hydrogen nucleus is its spin. The nucleus of hydrogen, a proton, has $I = \frac{1}{2}$; the nuclear spins of D and T are 1 and $\frac{1}{2}$, respectively. As explained in Chapter 8, proton NMR detects the presence of H nuclei in a compound and is a powerful method for determining structures of molecules, even proteins with molecular masses as high as 20 kDa. Figure 10.4 shows some typical ^{1}H-NMR chemical shifts for some compounds of p- and d-block elements. Hydrogen atoms bonded to electronegative elements (protonic H atoms) display deshielded NMR signals (at more positive values of chemical shift), whereas hydrogens coordinated to metal ions with incomplete d subshells typically display more negative chemical shifts.

Molecular hydrogen, H_2, exists in two forms that differ in the relative orientations of the two nuclear spins: in *ortho*-hydrogen the spins are parallel ($I = 1$), in *para*-hydrogen the spins are antiparallel ($I = 0$). At $T = 0$ hydrogen is 100 per cent *para*. As the temperature is raised, the proportion of the *ortho* form in a mixture at equilibrium increases until at room temperature there is approximately 75 per cent *ortho* and 25 per cent *para*. Most physical properties of the two forms are the same, but the melting and boiling points of *para*-hydrogen are about 0.1°C lower than those of normal hydrogen, and the thermal conductivity of *para*-hydrogen is about 50 per cent greater than that of the *ortho* form. The heat capacities also differ.

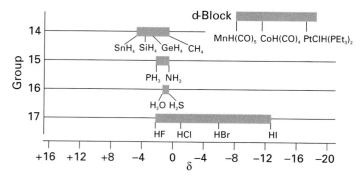

Figure 10.4 Typical ^{1}H-NMR chemical shifts. The tinted boxes show families of elements together.

10.4 Production of dihydrogen

Hydrogen is important both as a raw material for the chemical industry and, increasingly, as a fuel. Although it is not present in significant quantities in the Earth's atmosphere or in underground gas deposits, there is a high biological turnover because various microorganisms use H^+ as an oxidant or H_2 as a fuel (Box 10.1). Industrially, most H_2 is produced from natural gas by using steam reforming (in the USA, about 95 per cent is produced in this way). Increasingly, H_2 is being produced by other methods, notably coal gasification (with carbon dioxide capture, Box 14.4) and thermally assisted electrolysis. In 2007, world production of H_2 was approximately 50 Mt. Most H_2 is used close to its site of production for the synthesis of ammonia (the Haber process), hydrogenation of unsaturated fats and large-scale manufacture of organic chemicals. In the future, H_2 may be produced from entirely renewable sources such as water and capturing the energy of sunlight.

(a) Small-scale preparation

Key points: In the laboratory, H_2 is easily produced by the reactions of electropositive elements with aqueous acid or alkali, or by hydrolysis of saline hydrides. It is also produced by electrolysis.

There are many straightforward procedures for preparing small quantities of pure H_2. In the laboratory, H_2 is produced by reaction of Al or Si with hot alkali solution.

$$2\,Al(s) + 2\,OH^-(aq) + 6\,H_2O(l) \rightarrow 2\,Al(OH)_4^-(aq) + 3\,H_2(g)$$

$$Si(s) + 2\,OH^-(aq) + H_2O(l) \rightarrow SiO_3^{2-}(aq) + 2\,H_2(g)$$

or, at room temperature, by reaction of Zn with mineral acids:

$$Zn(s) + 2\,H_3O^+(aq) \rightarrow Zn^{2+}(aq) + H_2(g) + 2\,H_2O(l)$$

The reaction of metal hydrides with water provides a convenient way to obtain small amounts of H_2 outside the laboratory. Calcium hydride is particularly suited for this purpose as it is commercially available and inexpensive and it reacts with H_2O at room temperature:

$$CaH_2(s) + 2\,H_2O(l) \rightarrow Ca(OH)_2(s) + 2\,H_2(g)$$

Calcium hydride is used as a portable H_2 generator and has important applications in remote places, such as filling meteorological balloons.

Pure H_2 is also produced in small amounts using a simple electrolysis cell; electrolysis of heavy water is also a convenient way to prepare pure D_2.

(b) Production from fossil sources

Key point: Most H_2 for industry is produced by high-temperature reaction of H_2O with CH_4 or a similar reaction with coke.

Hydrogen is produced in huge quantities to satisfy the needs of industry, in fact production is often integrated directly (without transport) into chemical processes that require H_2 as a feedstock. The main commercial process for the production of H_2 is currently *steam reforming*, the catalysed reaction of H_2O (as steam) and hydrocarbons (typically methane from natural gas) at high temperatures:

$$CH_4(g) + H_2O(g) \rightarrow CO(g) + 3\,H_2(g) \quad \Delta_r H^\ominus = +206.2 \text{ kJ mol}^{-1}$$

Increasingly, coal or coke is used. This reaction, which occurs at 1000°C, is

$$C(s) + H_2O(g) \rightarrow CO(g) + H_2(g) \qquad \Delta_r H^{\ominus} = +131.4 \text{ kJ mol}^{-1}$$

The mixture of CO and H_2 is known as *water gas* and further reaction with water (the water gas shift reaction) produces more H_2:

$$CO(g) + H_2O(g) \rightarrow CO_2(g) + H_2(g) \qquad \Delta_r H^{\ominus} = -41.2 \text{ kJ mol}^{-1}$$

Overall, coal gasification (and hydrocarbon reforming) result in production of CO_2 and H_2.

$$C(s) + 2H_2O(g) \rightarrow CO_2(g) + 2H_2(g) \qquad \Delta_r H^{\ominus} = +90.2 \text{ kJ mol}^{-1}$$

By implementing a system for capturing CO_2 from the mixture (Box 14.4), it is possible to use fossil fuels and minimize release of the greenhouse gas CO_2 into the atmosphere. However, this process is not a renewable route for H_2 production as it is based on the use of fossil fuels. Dihydrogen for immediate consumption by on-board fuel cells in vehicles is produced from methanol by using an automotive steam reformer (Box 10.2).

(c) Production from renewable sources

Key points: Production of H_2 by electrolysis of water is costly and viable only in areas where electricity is cheap or if it is a byproduct of an economically important process. Environmental pressures are driving technologies to produce H_2 more efficiently from surplus or renewable energy, including solar and biological sources.

Electrolysis is used to produce H_2 that is free from contaminants.

$$H_2O(l) \rightarrow H_2(g) + \tfrac{1}{2}O_2(g) \quad E_{cell}^{\ominus} = -1.23 \text{ V}, \; \Delta_r G^{\ominus} = +237 \text{ kJ mol}^{-1}$$

To drive this reaction, a large overpotential is required to offset the sluggish electrode kinetics, particularly for the production of O_2. The best catalysts are based on platinum, but it is too expensive to justify its use in large-scale plants. As a consequence, electrolysis of water is economical and environmentally benign only if the electrical power stems from cheap, renewable resources or if it is surplus to demand. These conditions are found in countries that have plenty of hydroelectric or nuclear energy. There is also scope for off-shore wind farms that can employ resources far from the electricity grids and are far from the population areas that are otherwise necessary for conventional power generation. Electrolysis is carried out using hundreds of cells arranged in series, each operating at 2 V with iron or nickel electrodes and aqueous NaOH (or an ion-selective membrane) as electrolyte (Fig. 10.5). Temperatures of 80−85°C are used to increase the electrolytic current and to lower the overpotential. The most important electrolytic H_2 production method is the *chlor-alkali process* (Box 11.2), in which H_2 is produced as a byproduct of NaOH manufacture. In this process the other gaseous product is Cl_2, which requires a lower overpotential than O_2.

As yet, however, only about 0.1 per cent of the global H_2 demand is produced by electrolysis, including that produced in the chlor-alkali process. This percentage could be improved by the development of cheap and efficient electrocatalysts to reduce the economically wasteful overpotential.

Hydrogen can be produced by fermentation, using anaerobic bacteria that use cultivated biomass or biological waste as their energy source (Box 10.1). Research is also under way to establish how best to produce H_2 by exploiting solar energy directly either through physical methods (solar-powered thermolysis or photo-

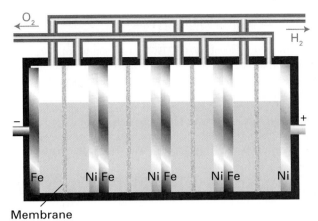

Figure 10.5 An industrial electrolysis cell for H_2 production using Ni anodes and Fe cathodes connected in series.

electron-chemistry) or biological processes (modified photosynthesis). The physical methods under investigation are outlined in Box 10.3. Biological production could take place in 'hydrogen farms' by nurturing photosynthetic microorganisms that have been modified to produce H_2 as well as organic molecules.

10.5 Reactions of dihydrogen

Key points: Molecular hydrogen is activated by homolytic or heterolytic dissociation on a metal or metal oxide surface or by coordination to a d-block metal. Reactions of hydrogen with O_2 and halogens involve a radical chain mechanism.

Although H_2 is quite an inert molecule, it reacts very rapidly under special conditions. Conditions for activating H_2 include:

1. Homolytic dissociation into H atoms, induced by adsorption at certain metal surfaces:

$$H_2 \; + \; -Pt-Pt-Pt-Pt- \; \rightleftharpoons \; \begin{array}{c} H \;\; H \\ | \;\; | \\ -Pt-Pt-Pt-Pt- \end{array}$$

2. Heterolytic dissociation into H^+ and H^- ions induced by adsorption on a heteroatom surface, such as a metal oxide:

$$H_2 \; + \; -Zn-O-Zn-O- \; \rightleftharpoons \; \begin{array}{c} H^+ \;\; H^- \\ | \;\; | \\ -Zn-O-Zn-O- \end{array}$$

3. Initiation of a radical chain reaction:

$$H_2 \xrightarrow{\; X \;} XH\cdot + H\cdot \xrightarrow{\; O_2 \;} HOO\cdot$$

$$\xrightarrow{\; XH\cdot \;} 2OH\cdot \xrightarrow{\; H_2 \;} H_2O + H\cdot \; ... \text{ etc.}$$

(a) Homolytic dissociation

High temperatures are required to dissociate H_2 into atoms. An important example of homolytic dissociation at normal temperatures is the reaction of H_2 at finely divided Pt or Ni metal (Section 26.11).

BOX 10.3 Hydrogen from solar energy

The Earth receives about 100 000 TW from the Sun, which is approximately 7000 times greater than the present global rate of energy consumption (15 TW). Solar energy is already harnessed in several familiar ways, such as wind turbines, photosynthesis (biomass), and photovoltaic cells but ultimately the use of solar energy to generate H_2 from water (water splitting) provides the greatest opportunity to end the world's dependence on fossil fuels and help curb global climate change. Two technologies under development are high-temperature solar H_2 production and solar photoelectrochemical H_2 production.

The so-called 'sunbelt' regions, which include Australia, southern Europe, the Sahara desert, and southwestern states of the USA, receive about 1 kW m^{-2} of solar power. These regions are suitable sites for high-temperature solar H_2 production using solar concentrating systems that reflect and focus solar radiation onto a receiver furnace, producing temperatures in excess of 1500°C. The intense heat, which is also available in the mantel surrounding a nuclear reactor, can be used to drive a turbine for generating electricity or to split water into H_2 and O_2, so producing a fuel.

Direct, single-step thermolysis of water requires temperatures in excess of 4000°C, which is well above the threshold readily attainable in a solar concentrator or compatible with containment materials and engineering. By using a multi-step process, however, it is possible to produce H_2 at much lower temperatures. Many systems are under investigation and development, the simplest of which are two-stage processes involving metal oxides, such as the sequence

$$Fe_3O_4(s) \rightarrow 3\,FeO(s) + \tfrac{1}{2}O_2(g) \qquad \Delta_f H^{\ominus} = +319.5 \text{ kJ mol}^{-1}$$

$$H_2O(l) + 3\,FeO(s) \rightarrow Fe_3O_4(s) + H_2(g) \qquad \Delta_f H^{\ominus} = -33.6 \text{ kJ mol}^{-1}$$

Dihydrogen production by this route still requires temperatures in excess of 2200°C. Water splitting at lower temperatures has been achieved with hybrid processes that combine thermochemical and electrochemical reactions, such as

$$2\,Cu(s) + 2\,HCl(g) \xrightarrow{425°C} H_2(g) + 2\,CuCl(s)$$

$$4\,CuCl(s) \xrightarrow{electrolysis} 2\,Cu(s) + 2\,CuCl_2(s)$$

$$2\,CuCl_2(s) + H_2O \xrightarrow{325°C} Cu_2OCl_2(s) + 2\,HCl(g)$$

$$Cu_2OCl_2(s) \xrightarrow{550°C} 2\,CuCl(s) + \tfrac{1}{2}O_2(g)$$

Solar photoelectrochemical H_2 production operates on a principle similar to that used by plants for photosynthesis. To split water electrochemically a cell potential greater than 1.23 V is required, which in principle could be provided by light with a wavelength below 1000 nm. The principle of a photoelectrochemical water-splitting system based on light-sensitive particles is shown in Fig. B10.3, which represents water splitting in terms of two separate half-reactions. The essentials are (a) a mechanism for generating an excited electronic state by photon capture, (b) efficient transfer of electrons between the site of excitation and catalytic centres, (c) sites for the H_2 production half-reaction, and (d) sites for the O_2 production half-reaction. The excitation is produced at a photosensitive centre P, an organic molecule or metal complex attached to the particle. The sites for H_2 and O_2 production must be catalysts in order for these reactions to occur on timescales that are short compared to the lifetimes of the excited and oxidized states of P. For H_2, the catalyst can be Pt, although much cheaper alternatives are necessary in order to achieve a feasible industrial-scale system. The major challenge for photoelectrochemical water splitting is to achieve rapid and efficient production of O_2, and there are intense efforts to find substances that mimic the Mn catalyst used in plant photosynthesis (Section 27.10).

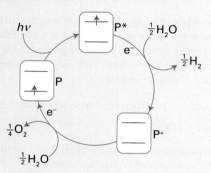

Figure B10.3 Schematic cycle for photoelectrochemical production of H_2 and O_2 by catalysts attached to particles of a conducting material such as doped TiO_2, to which a photosensitiser P is also attached. Excitation by visible light produces P*, a powerful reducing agent for producing H_2. Electron transfer from P* produces P$^+$, a powerful oxidizing agent for producing O_2.

This reaction, in which H_2 is dissociatively chemisorbed as H atoms, is used to catalyse the hydrogenation of alkenes and the reduction of aldehydes to alcohols. Platinum is also used as the electrocatalyst for H_2 oxidation in proton-exchange membrane fuel cells (Box 5.1). The facile chemisorption of H_2 at Pt anodes results in a minimal overpotential for H_2 oxidation and optimal performance.

Another example of homolytic cleavage involves the initial coordination of molecular H_2 as an η^2-H_2 species in discrete metal complexes, which is described briefly in Section 10.6d and in more detail in Section 22.7. Dihydrogen complexes provide examples of species intermediate between molecular H_2 and a dihydrido complex. No dihydrogen complexes are known for the early d-block (Groups 3, 4, and 5), f-block, or p-block metals. If the metal is sufficiently electron rich, back donation of d electrons into the $1\sigma_u$ orbital splits the H–H bond, resulting in the formation of a *cis*-dihydrido complex in which the formal oxidation number of the metal has increased by 2:

$$M^{n+} + H_2 \longrightarrow \overset{H-H}{\underset{M^{n+}}{|}} \longrightarrow \overset{H^-\ \ H^-}{\underset{M^{(n+2)+}}{\diagdown}}$$

(b) Heterolytic dissociation

Heterolytic dissociation of H_2 depends on a metal ion (for hydride coordination) and a Brønsted base being in close proximity. Reaction of H_2 with a ZnO surface appears to produce a Zn(II)-bound hydride and an O-bound proton. This reaction is involved in the production of methanol by catalytic hydrogenation of carbon monoxide over $Cu/ZnO/Al_2O_3$.

$$CO(g) + 2\,H_2(g) \rightarrow CH_3OH(g)$$

Another example in which H_2 is dissociated into a hydride and a proton is during its oxidation at the active site of metalloenzymes known as hydrogenases (Section 27.14).

(c) Radical chain reactions

Radical chain mechanisms account for the thermally or photochemically initiated reactions between H_2 and the halogens in which atoms are generated that act as radical chain carriers in the propagation reaction:

Initiation, by heat or light: $Br_2 \rightarrow Br\cdot + Br\cdot$

Propagation: $Br\cdot + H_2 \rightarrow HBr + H\cdot$

$H\cdot + Br_2 \rightarrow HBr + Br\cdot$

The activation energy for radical attack is low because a new bond is formed as one bond is lost, so once initiated the formation and consumption of radicals is self-sustaining and the production of HBr is very rapid. Chain termination occurs when the radicals recombine:

Termination: $H\cdot + H\cdot \rightarrow H_2$

$Br\cdot + Br\cdot \rightarrow Br_2$

$H\cdot + Br\cdot \rightarrow HBr$

Termination becomes more important towards the end of the reaction when the concentrations of H_2 and Br_2 are low.

The highly exothermic reaction of H_2 with O_2 also occurs by a radical chain mechanism. Certain mixtures explode violently when detonated:

$$2 H_2(g) + O_2(g) \rightarrow 2 H_2O(g) \qquad \Delta_r H^{\ominus} = -242 \text{ kJ mol}^{-1}$$

10.6 Compounds of hydrogen

Hydrogen forms compounds with most of the elements. These compounds are classified into molecular hydrides, saline hydrides (salts of the hydride anion), metallic hydrides (interstitial compounds of d-block elements), and discrete complexes of d-block elements in which hydride or dihydrogen are ligands.

(a) Molecular hydrides

Molecular hydrides are formed with p-block elements and Be. The bonding is covalent but variations in bond polarity (depending on the electronegativity of the atoms to which hydrogen is attached) result in a range of reaction types in which hydrogen is formally transferred as H^+, H^-, or $H\cdot$.

(i) Nomenclature and classification

Key point: Molecular compounds of hydrogen are classified as electron-rich, electron-precise, or electron-deficient. Electron-deficient hydrides provide some of the most intriguing examples of molecular structure and bonding as their simplest units tend to associate via bridging hydrogen atoms to form dimers and higher polymers.

The systematic names of the molecular hydrogen compounds are formed from the name of the element and the suffix -ane, as in phosphane for PH_3. The more traditional names, however, such as phosphine and hydrogen sulfide (H_2S, sulfane) are still widely used (Table 10.3). The common names ammonia and water are universally used rather than their systematic names azane and oxidane. Molecular compounds of hydrogen are divided further into three subcategories:

Electron-precise, in which all valence electrons of the central atom are engaged in bonds.

Electron-rich, in which there are more electron pairs on the central atom than are needed for bond formation (that is, there are lone pairs on the central atom).

Electron-deficient, in which there are too few electrons to be able to write a Lewis structure for the molecule.

Table 10.3 Some common molecular hydrogen compounds

Group	Formula	Traditional name	IUPAC name
13	B_2H_6	Diborane	Diborane(6)
14	CH_4	Methane	Methane
	SiH_4	Silane	Silane
	GeH_4	Germane	Germane
	SnH_4	Stannane	Stannane
15	NH_3	Ammonia	Azane
	PH_3	Phosphine	Phosphane
	AsH_3	Arsine	Arsane
	SbH_3	Stibine	Stibane
16	H_2O	Water	Oxidane
	H_2S	Hydrogen sulfide	Sulfane
	H_2Se	Hydrogen selenide	Sellane
	H_2Te	Hydrogen telluride	Tellane

Electron-precise molecular hydrogen compounds include hydrocarbons such as methane and ethane, and their heavier analogues silane, SiH_4, and germane, GeH_4 (Chapter 14). All these molecules are characterized by the presence of two-centre, two-electron bonds ($2c,2e$ bonds) and the absence of lone pairs on the central atom. Electron-rich compounds are formed by the elements in Groups 15 to 17. Important examples include ammonia, water, and the hydrogen halides. Electron-deficient hydrogen compounds are common for boron and aluminium. The analogous simple hydride of boron, BH_3, is not found. instead it occurs as a dimer, B_2H_6 (diborane, **4**) in which the two B atoms are bridged by a pair of H atoms in two three-centre, two-electron bonds ($3c,2e$ bonds).

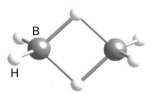

4 Diborane, B_2H_6

The shapes of the molecules of the electron-precise and electron-rich compounds can all be predicted by the VSEPR rules (Section 2.3). Thus, CH_4 is tetrahedral (**1**), NH_3 is trigonal pyramidal (**2**), H_2O is angular (**3**), and HF is (necessarily) linear.

Electron-deficient compounds provide some of the most interesting and unusual examples of structure and bonding. A Lewis structure for diborane, B_2H_6, would require at least 14 valence electrons to bind the eight atoms together, but the molecule has only 12 valence electrons. The simple explanation of its structure is the presence of BHB three-centre, two-electron bonds ($3c,2e$; Section 2.11e) acting as bridges between the two B atoms, so two electrons contribute to binding three atoms. These bridging B−H bonds are longer and weaker than the terminal B−H bonds. Another way of viewing this structure is that each BH_3

moiety is a strong Lewis acid and gains the share of an electron pair from a B—H bond in the other BH_3 moiety. Being so small the H atoms pose little or no steric hindrance to dimer formation. The structures of boron hydrides are described more fully in Chapter 13.

As expected, aluminium shows related behaviour that is modified by the larger atomic radius of this Period 3 element. The compound AlH_3 does not exist as a monomer but forms a polymer in which each relatively large Al atom is surrounded octahedrally by six H atoms. Beryllium, unlike its congeners, exhibits a diagonal relationship with Al and also forms a polymeric covalent hydride BeH_2. Although BH_3 and AlH_3 do not exist as monomers they do form important complex anions in combination with the hydride anion. The common reagents sodium tetrahydridoborate ($NaBH_4$) and lithium tetrahydridoaluminate ($LiAlH_4$) are examples of adduct formation between BH_3 or AlH_3, each a Lewis acid, and the Lewis base H^-.

(ii) Reactions of molecular hydrides

Key points: Homolytic dissociation of an E—H bond to produce a radical E· and hydrogen atom H occurs most readily for the hydrides of the heavy p-block elements. Hydrogen attached to an electronegative element has protic character and the compound is typically a Brønsted acid. Hydrogen attached to an electropositive element can be transferred to an acceptor as a hydride ion.

As was summarized briefly in Section 10.2 of Part A, the reactions of binary molecular hydrides are discussed in terms of their ability to undergo homolytic dissociation and, when the dissociation is heterolytic, in terms of their protic or hydridic character.

Homolytic dissociation occurs readily for the hydrogen compounds of some p-block elements, especially the heavier elements. For example, the use of a radical initiator greatly facilitates the reaction of trialkylstannanes, R_3SnH, with haloalkanes, RX, as a result of the formation of $R_3Sn\cdot$ radicals:

$$R_3SnH + R'X \rightarrow R'H + R_3SnX$$

The tendency towards radical reactions increases towards the heavier elements in each group, and Sn—H compounds are in general more prone to radical reactions than are Si—H compounds. The ease of homolytic E—H bond cleavage correlates with the decrease in E—H bond strength down a group.

The order of reactivity for haloalkanes with trialkylstannanes is

$$RF < RCl < RBr < RI$$

Thus fluoroalkanes do not react with R_3SnH, chloroalkanes require heat, photolysis, or chemical radical initiators, and bromoalkanes and iodoalkanes react spontaneously at room temperature. This trend indicates that the initiation step is halogen abstraction.

Thermal decomposition reactions of molecular hydrides yielding H_2 and the element occur by homolytic dissociation. Decomposition temperatures usually correlate with E—H bond energies and inversely with enthalpies of formation. For example, AsH_3 (As—H bond enthalpy 297 kJ mol^{-1}), which is an endothermic hydride, decomposes quantitatively at $250-300°C$:

$$AsH_3(g) \rightarrow As(s) + \tfrac{3}{2}H_2(g) \qquad \Delta_r H^\ominus = -66.4 \text{ kJ mol}^{-1}$$

In contrast, water (O—H bond enthalpy 464 kJ mol^{-1}), which is a highly exothermic hydride, is only 4 per cent dissociated into H_2 and O_2 at 2200°C (Box 10.3):

$$H_2O(g) \rightarrow \tfrac{1}{2}O_2(g) + H_2(g) \qquad \Delta_r H^\ominus = +242 \text{ kJ mol}^{-1}$$

Direct thermolysis of water is therefore not a practical solution for H_2 production.

As we saw in Section 10.2, compounds reacting by proton donation are said to show protic behaviour: in other words, they are Brønsted acids. We saw in Section 4.1 that Brønsted acid strength increases from left to right across a period in the p block (in the order of increasing electron affinity) and down a group (in the order of decreasing bond energy). One striking example of this trend is the increase in acidity from CH_4 to HF and then from HF to HI. Binary hydrogen compounds of elements on the right of the periodic table typically undergo these reactions.

Molecules in which hydrogen is bound to a more electropositive element can act as hydride ion donors. Important examples are the complex hydrido anions such as BH_4^- and AlH_4^-, which are used to hydrogenate compounds containing a multiple bond.

$$AlH_4^- + 4RCHO \rightarrow Al(OCH_2R)_4^- \xrightarrow{4H_2O} Al(OH)_4^- + 4RCH_2OH$$

EXAMPLE 10.1 Determining which hydrogen atoms in a molecule are the most acidic

Phosphorous acid, H_3PO_3, is a diprotic acid and is more helpfully written as $OP(H)(OH)_2$. Explain why the H atom bound to P is much less protonic than the two H atoms bound to O.

Answer We approach this problem by adapting the principles used to explain the Brønsted acidity of simple molecules. In Section 4.1 we saw that the Brønsted acidity of an acid EH depends upon the E—H bond enthalpy and electron affinity of B. In $OP(H)(OH)_2$ the P—H bond (bond enthalpy in PH_3 = 321 kJ mol^{-1}) is considerably weaker than an O—H bond (bond enthalpy in H_2O = 464 kJ mol^{-1}) and on this basis we would expect a P—H bond to be more protonic. But the determining factor is that O is much more electronegative than P and therefore better able to accommodate the negative charge left by the departing H^+. Formic acid, $HCO(OH)$, is another example of a molecule containing H atoms with very different protonic character.

Self-test 10.1 Which of the following, CH_4, SiH_4, or GeH_4, would you expect to be (a) the strongest Brønsted acid and (b) the strongest hydride donor?

(iii) Hydrogen bonding

Key points: Compounds and functionalities containing H atoms attached to electronegative elements with at least one lone pair often associate through hydrogen bonds.

An E—H bond between an electronegative element E and hydrogen is highly polar, $^{\delta-}E—H^{\delta+}$, and the partially positively charged H atom can interact with a lone pair on the E atom of another molecule, forming a bridge that is known as a **hydrogen bond**. Striking evidence for hydrogen bonding is provided by the trends in normal boiling points, which are unusually high for the strongly hydrogen-bonded molecules water (in which there are O—H···O bonds), ammonia (containing N—H···N bonds), and hydrogen fluoride (containing F—H···F bonds, Fig. 10.6). The relatively low boiling points of PH_3, H_2S, HCl, and the heavier p-block molecular hydrides indicate that these molecules do not form strong hydrogen bonds. Although hydrogen bonds are usually much weaker than conventional bonds (Table 10.4), their collective action is responsible for stabilizing complex structures such as the open network structure of ice (Fig. 10.7). Collective hydrogen bonding interactions play a large part in maintaining the structure of protein molecules (Section 27.2). They are also responsible for the recognition between specific DNA bases, adenine/thymine and guanine/cytosine, that underlies gene replication (Fig. 10.8) Similarly, solid HF consists of chain structures that survive partially even in the vapour (5).

Table 10.4 Comparison of hydrogen bond enthalpies with the corresponding E—H covalent bond enthalpies (kJ mol^{-1})

	Hydrogen bond			Covalent bond
HS—H···SH$_2$	7		S—H	363
H$_2$N—H···NH$_3$	17		N—H	386
HO—H···OH$_2$	22		O—H	464
F—H···FH	29		F—H	574
HO—H···Cl$^-$	55		Cl—H	428
F—H···F$^-$	163			

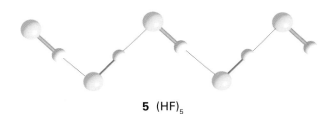

5 (HF)$_5$

Hydrogen bonding may be symmetrical or unsymmetrical. In unsymmetrical hydrogen bonding, the H atom is not midway between the two nuclei, even when the heavier linked atoms are identical. For example, the [ClHCl]$^-$ ion is linear but the H atom is not midway between the Cl atoms (Fig. 10.9). By contrast, in the bifluoride ion, [FHF]$^-$, the H atom lies midway between the F atoms; the F—F separation (226 pm) is significantly less than twice the van der Waals radius of the F atom (2 × 135 pm).

Hydrogen bonding is readily detected by the shift to lower frequency and broadening of E—H stretching bands in infrared spectra (Fig. 10.10) and by unusual proton chemical shifts in ^{1}H-NMR. The structures of hydrogen-bonded complexes have been observed in the gas phase by microwave spectroscopy. The lone-pair orientation of electron-rich compounds implied by VSEPR theory (Section 2.3) shows good agreement with the HF orientation (Fig. 10.11). For example, HF is oriented along the threefold axis of NH_3, collinear with HCN, out of the H_2O plane in its complex with H_2O, and off the HF axis in the HF dimer. X-ray single crystal structure determinations often show the same patterns as, for example, in the structure of ice and in

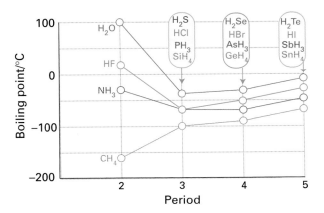

Figure 10.6 Normal boiling points of p-block binary hydrogen compounds.

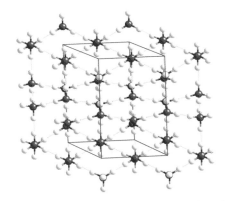

Figure 10.7 The structure of ice. The structure shows all possible atom positions, but only half are actually occupied.

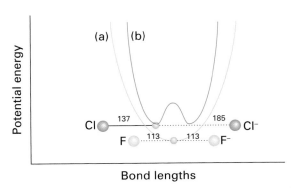

Figure 10.9 The variation of the potential energy with the position of the proton between two atoms in a hydrogen bond. (a) The single minimum potential characteristic of a strong hydrogen bond. (b) The double minimum potential characteristic of a weak hydrogen bond.

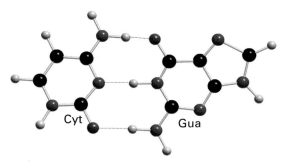

Figure 10.8 Base pairing in DNA. Cytosine recognizes guanine through the formation of three hydrogen bonds.

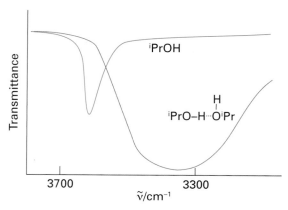

Figure 10.10 Infrared spectra of 2-propanol. In the upper curve, 2-propanol is present as unassociated molecules in dilute solution. In the lower curve, the pure alcohol is associated through hydrogen bonds. The association lowers the frequency and broadens the O—H stretching absorption band. (from N.B. Colthrup, L.H. Daly, and S.E. Wiberley, *Introduction to infrared and Raman spectroscopy.* Academic Press, New York (1975).)

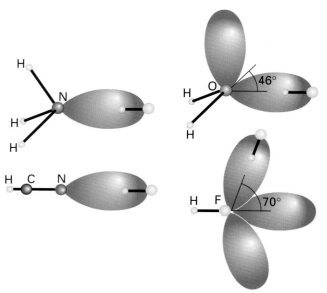

Figure 10.11 The orientation of lone pairs as indicated by VSEPR theory compared with the orientation of HF in the gas-phase hydrogen-bonded complex. The HF molecule is oriented along the threefold axis of NH_3, collinear with HCN, out of the H_2O plane in its complex with H_2O, and off the HF axis in the HF dimer.

solid HF, but packing forces in solids may have a strong influence on the orientation of the relatively weak hydrogen bond.

One of the most interesting manifestations of hydrogen bonding is the structure of ice. There are at least ten different phases of ice but only one is stable under ordinary conditions. The familiar low-pressure phase of ice, ice-I, crystallizes in a hexagonal unit cell with each O atom surrounded tetrahedrally by four others (as shown in Fig. 10.7). These O atoms are held together by hydrogen bonds with O—H⋯O and O⋯H—O bonds largely randomly distributed through the solid. The resulting structure is quite open, which accounts for the density of ice being lower than that of water. When ice melts, the network of hydrogen bonds partially collapses.

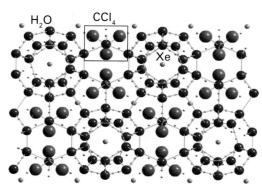

Figure 10.12 The cages of water molecules in clathrate hydrates; in this case $Xe_4(CCl_4)_8(H_2O)_{68}$.

Water can also form **clathrate hydrates**, consisting of hydrogen-bonded cages of water molecules surrounding foreign molecules or ions. One example is the clathrate hydrate of composition $Xe_4(CCl_4)_8(H_2O)_{68}$ (Fig. 10.12). In this structure, the cages with O atoms defining their corners consist of 14-faced and 12-faced polyhedra in the ratio 3:2. These O atoms are held together by hydrogen bonds and guest molecules occupy the interiors of the polyhedra. Aside from their interesting structures, which illustrate the organization that can be enforced by hydrogen bonding, clathrate hydrates are often used as models for the way in which water appears to become organized around nonpolar groups, such as those in proteins. Methane clathrate hydrates occur in the Earth at high pressures, and it is estimated that huge quantities of CH_4 are trapped in these formations (see Box 14.2).

Some ionic compounds form clathrate hydrates in which the anion is incorporated into the framework by hydrogen bonding. This type of clathrate is particularly common with the very strong hydrogen bond acceptors F^- and OH^-. One such example is $N(CH_3)_4F\cdot4H_2O$ (**6**).

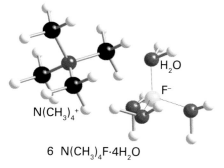

6 $N(CH_3)_4F\cdot4H_2O$

(b) Saline hydrides

Key points: Hydrogen compounds of the most electropositive metals may be regarded as ionic hydrides; they liberate H_2 in contact with Brønsted acids and transfer H^- to electrophiles. As direct hydride donors they react with halide compounds to form anionic hydride complexes.

The saline hydrides are ionic solids containing discrete H^- ions and are analogous to corresponding halide salts. The ionic radius of H^- varies from 126 pm in LiH to 154 pm in CsH. This wide variability reflects the poor control that the single charge of the proton has on its two surrounding electrons and the resulting high compressibility and polarizability of H^-. Hydrides of Group 1 and 2 elements, with the exception of Be, are ionic compounds. All Group 1 hydrides adopt the rock-salt structure. With the exception of MgH_2, which has the rutile structure, the Group 2 hydrides adopt the $PbCl_2$ structure (Table 10.5).

Table 10.5 Structures of s-block hydrides

Compound	Crystal structure
LiH, NaH, RbH, CsH	Rock salt
MgH_2	Rutile
CaH_2, SrH_2, BaH_2	Distorted $PbCl_2$

The saline hydrides are insoluble in common nonaqueous solvents but they do dissolve in molten alkali halides. Electrolysis of these molten-salt solutions produces hydrogen gas at the anode (the site of oxidation):

$$2\,H^-(melt) \rightarrow H_2(g) + 2\,e^-$$

This reaction provides chemical evidence for the existence of H^- ions. The reaction of saline hydrides with water, as in

$$NaH(s) + H_2O(l) \rightarrow NaOH(aq) + H_2(g)$$

is dangerously vigorous.

Alkali metal hydrides are convenient reagents for making other hydride compounds because they are direct providers of H^- ions for the following synthetically useful reactions:

1. Metathesis with a halide, such as the reaction of finely divided lithium hydride with silicon tetrachloride in dry diethyl ether (et):

$$4\,LiH(s) + SiCl_4(et) \rightarrow 4\,LiCl(s) + SiH_4(g)$$

2. Addition to a Lewis acid, for example reaction with a trialkylboron compound yields a hydride complex that is a useful reducing agent and source of hydride ions in organic solvents:

$$NaH(s) + B(C_2H_5)_3(et) \rightarrow Na[HB(C_2H_5)_3](et)$$

where 'et' denotes solution in diethyl ether

3. Reaction with a proton source, to produce H_2:

$$NaH(s) + CH_3OH(et) \rightarrow NaOCH_3(s) + H_2(g)$$

The absence of convenient solvents limits the use of saline hydrides as reagents, but this problem is partially overcome by the availability of commercial dispersions of finely divided NaH in oil. Even more finely divided and reactive alkali metal hydrides can be prepared from the metal alkyl and hydrogen.

Saline hydrides are pyrophoric; indeed, finely divided sodium hydride can ignite simply if it is left exposed to humid air. Such fires are difficult to extinguish because even carbon dioxide is reduced when it comes into contact with hot metal hydrides (water, of course, forms even more flammable hydrogen); they may, however, be blanketed with an inert solid, such as sand.

Magnesium dihydride, MgH_2, is under investigation as a hydrogen storage medium for transport purposes, where lightness is important (Box 10.4). The amount of H atoms in a given volume of MgH_2 is about 50 per cent higher than in the same volume of liquid H_2. Calcium dihydride is used as a portable H_2 generator.

(c) Metallic hydrides

Key points: No stable binary metal hydrides are known for the metals in Groups 7 to 9; metallic hydrides have metallic conductivity and in many the hydrogen is very mobile.

Many of the d- and f-block elements react with H_2 to produce metallic hydrides. Most of these compounds (and the hydrides of alloys) have a metallic lustre and are electrically conducting (hence their name). They are less dense than the parent metal and are brittle. Most metallic hydrides have variable composition (they are nonstoichiometric). For example, at $550°C$ zirconium hydride exists over a composition range from $ZrH_{1.30}$ to $ZrH_{1.75}$; it has the fluorite structure (Fig. 3.38) with a variable number of anion sites unoccupied. The variable stoichiometry and metallic conductivity of these hydrides can be understood in terms of a model in which the band of delocalized orbitals responsible for the conductivity accommodates the electrons supplied by arriving H atoms. In this model, the H atoms as well as the metal atoms take up equilibrium positions in the electron sea. The conductivities of metallic hydrides typically vary with hydrogen content, and this variation can be correlated with the extent to which the conduction band is filled or emptied as hydrogen is added or removed. Thus, whereas CeH_{2-x} is a metallic conductor, CeH_3 (which has a full conduction band) is an insulator and is more like a saline hydride.

Metallic hydrides are formed by all the d-block elements of Groups 3, 4, and 5 and by almost all f-block elements (Fig. 10.13). However, the only hydride in Group 6 is CrH, and no hydrides are known for the unalloyed metals of Groups 7, 8, and 9. The region of the periodic table covered by Groups 7 through to 9 is sometimes referred to as the **hydride gap** because few, if any, stable binary metal−hydrogen compounds are formed by these elements. However, these metals are important as hydrogenation catalysts because they can activate hydrogen. Claims have been made, however, that hydrogen dissolves in iron at very high pressure and that iron hydride is abundant at the centre of the Earth.

The Group 10 metals, especially Ni and Pt, are often used as hydrogenation catalysts in which surface hydride formation is thought to be involved (Chapter 26). However, somewhat surprisingly, at moderate pressures only Pd forms a stable bulk phase; its composition is PdH_x, with $x < 1$. Nickel forms hydride phases at very high pressures but Pt does not form any at all. Apparently the Pt−H bond enthalpy is sufficiently great to disrupt the H−H bond but not strong enough to offset the loss of Pt−Pt bonding, which would occur on formation of a bulk platinum hydride. In agreement with this interpretation, the enthalpies of sublimation, which reflect M−M bond enthalpies, increase in the order Pd ($378\ kJ\ mol^{-1}$) < Ni ($430\ kJ\ mol^{-1}$) < Pt ($565\ kJ\ mol^{-1}$). The M−H bond enthalpy is a crucial factor in the design of metal hydride batteries, which are described in Box 10.5.

Another striking property of many metallic hydrides is the high mobility of hydrogen within the material at slightly elevated temperatures. This mobility is used in the ultrapurification of H_2 by diffusion through a palladium/silver alloy tube (Fig. 10.14).

BOX 10.4 The quest for reversible H₂ storage materials

The need to develop practical systems for on-board hydrogen storage is considered to be a major obstacle to the future use of H₂ as an energy carrier in vehicles. The problem is only partly resolved by compression and liquefaction. Compression to high-pressure gaseous H₂ at 200 bar (energy density 0.53 kWh dm⁻³) then refrigeration to form liquid H₂ (energy density 2.37 kWh dm⁻³) requires considerable energy and containment costs, and these are particularly prohibitive for small private vehicles for which space and cost are also of prime concern. The challenge, therefore, is to identify and develop materials that can store H₂ in a fully reversible manner at high rates under reasonable temperatures and pressure conditions. One such material is LaNi₅H₆, which stores H₂ reversibly with a density of 2 per cent by mass, but for transport these materials also have to be lightweight. Materials under investigation include hydrides, borohydrides, and amides of the lightest metals. Examples of these compounds and their H₂ storage densities (as a percentage of hydrogen by mass) are MgH₂ (8 per cent), LiBH₄ (20 per cent), LiNH₂ (10 per cent), and Al(BH₄)₃ (17 per cent). The last is a liquid that melts at −65°C.

Some of the principles are illustrated by the LiNH₂ system, for which reversible hydrogen storage takes place in two reactions:

$$Li_3N(s) + H_2(g) \xrightleftharpoons[\text{Vacuum, >320°C}]{\text{3 bar H}_2, \text{ 210°C}} Li_2NH(s) + LiH(s)$$

$$(\Delta_r H^\ominus = +148 \text{ kJ mol}^{-1} \text{ at 298 K})$$

$$Li_2NH(s) + H_2(g) \xrightleftharpoons[\text{Vacuum, <200°C}]{\text{3 bar H}_2, \text{ 255°C}} LiNH_2(s) + LiH(s)$$

$$(\Delta_r H^\ominus = +45 \text{ kJ mol}^{-1} \text{ at 298 K})$$

The second reaction is the thermodynamically more accessible.

Comparisons of the structures of Li₂NH and LiNH₂ suggest how the kinetics of absorption and desorption depend on ion mobility. The structure

of Li₂NH (antifluorite) is closely related to that of LiNH₂ (defect antifluorite structure with half the Li sites occupied in an ordered manner). The small Li⁺ ions can migrate within such a structure by a hopping mechanism involving transitory defect sites, so allowing H₂ uptake by protonation of NH²⁻ with corresponding expansion of the unit cell (Fig. B10.4) and coupled formation of a LiH phase. A problem that must be overcome with amides and other complex hydrides is their tendency to decompose to undesirable products, such as NH₃.

Liquid organic nitrogen-containing heterocyclic compounds that can take up H₂ reversibly are also under investigation. Compounds based on imidazole are easily handled and the discharged (dehydrogenated) product could simply be exchanged for hydrogenated fuel at a filling station. Another direction for reversible H₂ storage is as molecular clathrates, such as H₂(H₂O)₂. The H₂ molecule is held by weak van der Waals forces and is released by decreasing the pressure or increasing the temperature.

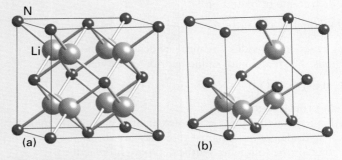

Figure B10.4 Structure relationship between Li₂NH and LiNH₂ which may facilitate transport of Li⁺ ions and reversible uptake of H₂.

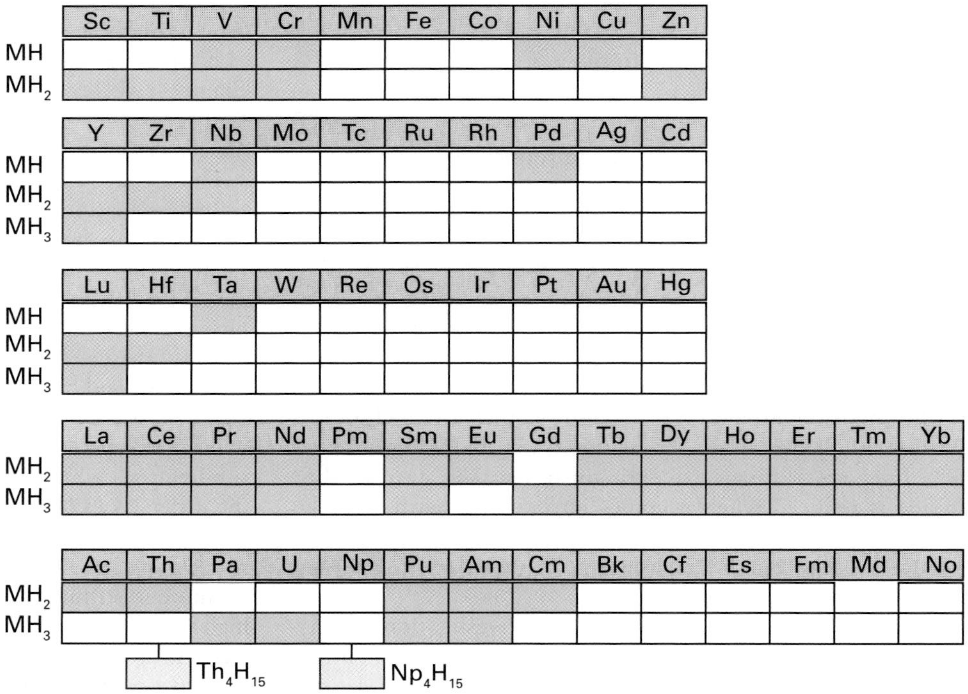

	Sc	Ti	V	Cr	Mn	Fe	Co	Ni	Cu	Zn
MH			░	░				░		
MH₂	░	░	░	░						░

	Y	Zr	Nb	Mo	Tc	Ru	Rh	Pd	Ag	Cd
MH			░					░		
MH₂	░	░	░							
MH₃	░									

	Lu	Hf	Ta	W	Re	Os	Ir	Pt	Au	Hg
MH			░							
MH₂	░	░								
MH₃	░									

	La	Ce	Pr	Nd	Pm	Sm	Eu	Gd	Tb	Dy	Ho	Er	Tm	Yb
MH₂	░	░	░	░	░	░		░	░	░	░	░	░	
MH₃	░	░	░	░	░	░	░	░	░	░	░	░	░	░

	Ac	Th	Pa	U	Np	Pu	Am	Cm	Bk	Cf	Es	Fm	Md	No
MH₂	░	░	░		░	░	░	░						
MH₃	░		░	░										

[ ] Th₄H₁₅   [ ] Np₄H₁₅

Figure 10.13 Hydrides formed by d- and f-block elements. The formulae are limiting stoichiometries based on the structure type.

BOX 10.5 Metal–hydride batteries

A nickel metal–hydride battery is a type of rechargeable battery similar to the widely used nickel–cadmium (NiCad) battery. The main advantages of metal–hydride over NiCad batteries are that they are more easily recycled and do not contain the very toxic element Cd. However, nickel metal–hydride batteries have a high self-discharge rate of approximately 30 per cent per month. This rate is higher than that of NiCad batteries, which is around 20 per cent per month. Despite this, nickel metal–hydride batteries are being investigated as possible power sources for electric vehicles. In contrast to vehicles powered by the internal combustion engine, electric vehicles are emission-free (if the generation of electricity elsewhere is ignored). In addition, the energy efficiency of generating electricity for vehicles is almost twice that of the internal combustion engine. Electric power also reduces society's reliance on oil and increases the opportunities for using renewable energy and also using coal and gas in such a way that the CO_2 can be captured (Box 14.4)

The attractive properties of nickel metal–hydride batteries include high power, long life, a wide range of operating temperatures, short recharging times, and sealed, maintenance-free operation. The cathode is made from a mixed metal alloy at which the metal hydrides are formed reversibly. The anode is made from nickel hydroxide. The electrolyte is a basic solution of 30 per cent by mass KOH. During charging the electrode reactions are

At the cathode: $M + H_2O + e^- \rightarrow M–H + OH^-$

At the anode: $Ni(OH)_2 + OH^- \rightarrow Ni(O)OH + H_2O + e^-$

There is no net change in the electrolyte concentration over the charge-discharge cycle.

The strength of the M—H bond in the metal hydride is crucial to the operation of the battery. The ideal bond enthalpy falls in the range 25–50 kJ mol^{-1}. If the bond enthalpy is too low, the hydrogen does not react with the alloy and H_2 is evolved instead. If the bond enthalpy is too high, the reaction is not reversible. Other factors influence the choice of metal. For example, the alloy must not react with KOH solution, must be resistant to oxidation and corrosion, and must tolerate overcharge (during which O_2 is generated at the Ni(O)OH electrode) and overdischarge (during which H_2 is generated at the $Ni(OH)_2$ electrode). To satisfy these diverse requirements the alloys have disordered structures and use metals that would not be suitable if used alone, including Li, Mg, Al, Ca, V, Cr, Mn, Fe, Cu, and Zr. The number of H atoms per metal atom can be increased by using Mg, Ti, V, Zr, and Nb, and the M—H bond enthalpy can be adjusted by using V, Mn, and Zr. The charge and discharge reactions are catalysed by Al, Mn, Co, Fe, and Ni, and the corrosion resistance is improved by using Cr, Mo, and W. This wide range of properties allows nickel metal–hydride battery performance to be optimized for different applications.

The high mobility of the hydrogen they contain and their variable composition make the metallic hydrides potential hydrogen storage media. On cooling from red heat, palladium absorbs up to 900 times its own volume of H_2, which is given off again on heating. As a result, palladium is sometimes referred to as a 'hydrogen sponge'. The intermetallic compound LaNi$_5$ forms a hydride phase with a limiting composition LaNi$_5$H$_6$, and at this composition it contains a greater density of hydrogen than liquid H_2. A less expensive system with the composition FeTiH$_x$ ($x < 1.95$) is now commercially available for low-pressure hydrogen storage and it has been tested as a fuel source in vehicle trials.

EXAMPLE 10.2 Correlating the classification and properties of hydrogen compounds

Classify the compounds PH$_3$, CsH, and B$_2$H$_6$ and discuss their probable physical properties. For the molecular compounds specify their subclassification (electron-deficient, electron-precise, or electron-rich).

Answer We need to consider the group to which the element E belongs. The compound CsH is a compound of a Group 1 element, and so it is expected to be a saline hydride, typical of the s-block metals. It is an electrical insulator with the rock-salt structure. As with the hydrogen compounds of other p-block elements, the hydrides PH$_3$ and B$_2$H$_6$ are molecular with low molar masses and high volatilities. They are in fact gases under normal conditions. The Lewis structure indicates that PH$_3$ has a lone pair on the phosphorus atom and that it is therefore an electron-rich molecular compound. On the other hand, diborane, B$_2$H$_6$, is an electron-deficient compound.

Self-test 10.2 Give balanced equations (or NR, for no reaction) for (a) Ca + H$_2$, (b) NH$_3$ + BF$_3$, (c) LiOH + H$_2$.

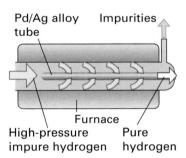

Figure 10.14 Schematic diagram of a hydrogen purifier. Because of a pressure differential and the mobility of H atoms in palladium, hydrogen diffuses through the palladium-silver alloy as H atoms but impurities do not.

(d) Hydrido and dihydrogen complexes

Key points: A large number of d-block metals complexes are known in which the dihydrogen molecule or hydride anion are ligands. These complexes play important roles in catalysis and hydrogen activation.

The H atom and the H_2 molecule play an important role in organometallic chemistry, particularly catalysis involving hydrogenation of alkenes and carbonyl groups (see Sections 22.7 and 26.4). An individual, bound H atom is usually regarded as a H$^-$ (hydrido) ligand: H$^-$ is highly polarizable and behaves as a soft, two-electron σ-donor (Section 22.3). There are a very large number of complexes of d-block and f-block elements containing one or more hydride ligands; these complexes include those of elements in the 'hydride gap' that do not form metallic hydrides. Hydrido complexes can be synthesized by many routes, such as the reaction of a metal ion or complex with a suitable hydrogen source (water) and (usually) a reducing agent.

$$Rh^{3+}(aq) \xrightarrow{\ Zn(s)\,/\,NH_3(aq)\ } [RhH(NH_3)_5]^{2+}$$

$$[FeI_2(CO)_4] \xrightarrow{\ NaBH_4\,(thf)\ } [Fe(H)_2(CO)_4]$$

As with binary compounds of the main group, the coordinated H ligand can be protonic or hydridic depending on the electron-withdrawing or electron-donating character of the metal atom, which in turn depends on the nature of the other ligands. Electron-withdrawing CO ligands in the coordination shell render the H atom protonic, and compounds such as $Co(H)(CO)_4$ are strong Brønsted acids. As with some main-group hydrides, an H atom can also occupy a bridging position (in a 3c,2e bond) between two metal atoms, usually in conjunction with a metal–metal bond. The complex $[(\mu\text{-}H)W_2(CO)_{10}]^-$ (7) provides a rare example of an H atom bridging two metal atoms that are not otherwise bonded.

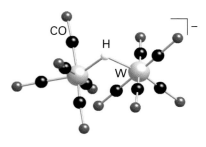

7 $[(\mu\text{-}H)W_2(CO)_{10}]^-$

A **homoleptic complex** is a complex that contains only one type of ligand. Examples of homoleptic hydrido metal complexes are provided by Fe, Rh, and Tc. The dark green compound Mg_2FeH_6, which contains the octahedral $[FeH_6]^{4-}$ complex anion, is obtained by reacting the elements together under pressure. The complex anion $[ReH_9]^{2-}$ (8) is formed by reducing perrhenate, ReO_4^-, with K or Na in ethanol. In the solid state the H atoms form a tricapped trigonal prism around the Re, which is formally in the oxidation state +7. The TcH_9^{2-} complex has the same structure.

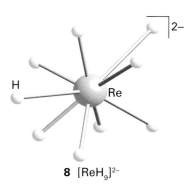

8 $[ReH_9]^{2-}$

The H_2 molecule can also coordinate, intact, using the $1\sigma_g$ orbital to donate an electron pair and the $1\sigma_u$ orbital to accept an electron pair back from the metal, in what is known as **π-back donation** or **synergic bonding** (Section 22.7). If the metal is electron-rich and in a sufficiently low oxidation state the π-back donation results in homolytic cleavage of the H–H bond and the two H atoms are formally reduced to H^- with concomitant oxidation of the metal. This process is known as **oxidative**

addition and is discussed in more detail in Section 22.22. Oxidative addition of H_2 is exemplified with the so-called 'Vaska's compound' $[IrClCO(PPh_3)_2]$ (9). In the product (10), the two H atoms are regarded as hydrido (H^-) ligands and the formal oxidation number of Ir has increased by 2. Many d-block complexes have been isolated that contain a relatively stable, intact H_2 ligand. The first such compound to be identified was $[W(CO)_3(H_2)(P^iPr_3)_2]$ (11), where ^iPr denotes isopropyl, $CH(CH_3)_2$.

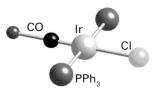

9 $[IrCl(CO)(PPh_3)_2]$, Ph = C_6H_5

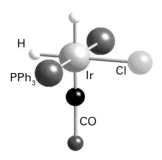

10 $[IrCl(H)_2CO(PPh_3)_2]$

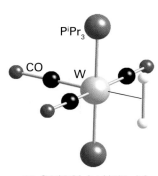

11 $[W(CO)_3(H_2)(P^iPr_3)_2]$

Both H atoms and H_2 molecules may be coordinated as ligands to the same metal atom. The complex $[Ru(H)_2(H_2)_2(PCyp_3)_2]$ (12) contains six H atoms in the inner coordination sphere: the two kinds of ligand, H^- and H_2, can be distinguished by neutron diffraction.

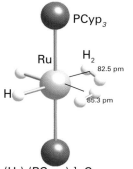

12 $[Ru(H)_2(H_2)_2(PCyp_3)_2]$, Cyp = *cyclo*-$C_5H_9$

10.7 General methods for synthesis

Key points: The general routes to binary hydrogen compounds are direct reaction of H_2 and the element, protonation of nonmetal anions, and metathesis between a hydride source and a halide or pseudohalide.

A negative Gibbs energy of formation is a clue that the direct combination of hydrogen and an element may be the preferred synthetic route for a hydrogen compound. When a compound is thermodynamically unstable with respect to its elements, an indirect synthetic route from other compounds can often be found, but each step in the indirect route must be thermodynamically favourable.

There are three common methods for synthesizing binary hydrogen compounds:

1. Direct combination of the elements (hydrogenolysis):

$$2\,E + H_2(g) \rightarrow 2\,EH$$

2. Protonation of a Brønsted basic anion:

$$E^- + H_2O(l) \rightarrow EH + OH^-(aq)$$

3. Reaction of an ionic hydride or hydride donor with a halide (metathesis):

$$EH + EX \rightarrow E^+X^- + EH$$

In such general equations, the symbol E can also denote an element with higher valence, with corresponding changes of detail in the formulas and stoichiometric coefficients.

Direct combination is used commercially for the synthesis of compounds that have negative Gibbs energies of formation, including NH_3 and the hydrides of lithium, sodium, and calcium. However, in some cases high pressure, high temperature, and a catalyst are necessary to overcome unfavourable kinetic barriers. The high temperature used for the lithium reaction is an example: it melts the metal and hence helps to break up the surface layer of hydride that would otherwise passivate it. This inconvenience is avoided in many laboratory preparations by adopting one of the alternative synthesis routes, which may also be used to prepare compounds with positive Gibbs energies of formation.

An example of protonation of a Brønsted base, such as a nitride ion, is

$$Li_3N(s) + 3\,H_2O(l) \rightarrow 3\,LiOH(aq) + NH_3(g)$$

Lithium nitride is too expensive for the reaction to be suitable for the commercial production of ammonia, but it is very useful in the laboratory for the preparation of ND_3 (by using D_2O in place of H_2O). The success of the reaction depends on the Brønsted acid being a better proton donor than the conjugate acid of the N^{3-} anion (NH_3 in this case). Water is a sufficiently strong acid to protonate the very strong base N^{3-}, but a stronger acid, such as H_2SO_4, is required to protonate the weak base Cl^-:

$$NaCl(s) + H_2SO_4(l) \rightarrow NaHSO_4(s) + HCl(g)$$

An example of synthesis by metathesis is the preparation of silane

$$LiAlH_4(s) + SiCl_4(l) \rightarrow LiAlCl_4(s) + SiH_4(l)$$

This reaction involves (at least formally) the exchange of Cl^- ions for H^- ions in the coordination sphere of the Si atom. Hydrides of the more electropositive elements (LiH, NaH, and the AlH_4^- anion) are the most active H^- sources. The favoured sources are often the AlH_4^- and BH_4^- ions in salts such as $LiAlH_4$ and $NaBH_4$, which are soluble in ether solvents that solvate the alkali metal ion. Of these two anion complexes, AlH_4^- is much the stronger hydride donor.

EXAMPLE 10.3 Using hydrogen compounds in synthesis

Suggest a procedure for synthesizing lithium tetraethoxyaluminate, $Li[Al(OEt)_4]$, from $LiAlH_4$ and reagents and solvents of your choice.

Answer We need to note that AlH_4^- is a H^- donor. Because H^- is an even stronger Brønsted base than ethoxide ($CH_3CH_2O^- = EtO^-$) it should react with ethanol to produce H_2 and yield EtO^-, which will thus replace H^-. The reaction of the slightly acidic compound ethanol with the strongly hydridic lithium tetrahydridoaluminate should yield the desired alkoxide and hydrogen. The reaction might be carried out by dissolving $LiAlH_4$ in tetrahydrofuran and dropping ethanol into this solution slowly:

$$LiAlH_4(thf) + 4\,C_2H_5OH(l) \rightarrow Li[Al(OEt)_4](thf) + 4\,H_2(g)$$

This type of reaction should be carried out slowly under a stream of inert gas (N_2 or Ar) to dilute the H_2, which is explosively flammable.

Self-test 10.3 Suggest a way of making triethylmethylstannane, $MeEt_3Sn$, from triethylstannane, Et_3SnH, and a reagent of your choice.

FURTHER READING

T.I. Sigfusson, Pathways to hydrogen as an energy carrier. *Phil. Trans. Royal. Society. A.*, 2007, **365**, 1025.

B. Sørensen, *Hydrogen and fuel cells*. Elsevier Academic Press (2005).

W. Grochala and P.P. Edwards. Thermal decomposition of the non-interstitial hydrides for the storage and production of hydrogen. *Chem. Rev.*, 2004, **104**, 1283.

G.A. Jeffrey, *An introduction to hydrogen bonding*. Oxford University Press (1997).

G.A. Jeffrey, *Hydrogen bonds in biological systems*. Oxford University Press (1994).

R.B. King, *Inorganic chemistry of the main group elements*. Wiley (1994).

J.S. Rigden, *Hydrogen: the essential element*. Harvard University Press (2002).

P. Enghag, *Encyclopedia of the elements*. John Wiley & Sons (2004).

P. Ball, *H₂O: A biography of water*. Phoenix (2004). An entertaining look at the chemistry and physics of water.

G.W. Crabtree, M.S. Dresselhaus, and M.V. Buchanan. The hydrogen economy. *Physics Today*, 2004, **57**, 39.

W. Lubitz and W. Tumas (eds.), Hydrogen. *Chemical Reviews* (100th Thematic Edition), 2007, **107**.

S.-I. Orimo, Y. Nakamori, J.R. Eliseo, A. Züttel, and C.M. Jensen, Complex hydrides for hydrogen storage. *Chem. Rev.*, 2007, **107**, 4111.

R.H. Crabtree, Hydrogen storage in liquid organic heterocycles. *Energy and Env. Sci.*, 2008, **1**, 134.

T. Kodama and N. Gokon, Thermochemical cycles for high-temperature solar hydrogen production. *Chem. Rev.*, 2007, **107**, 4048.

N.S. Lewis and D.G. Nocera, Powering the planet: Chemical challenges in solar energy utilization. *Proc. Nat. Acad. Sci. USA*, 2006, **103**, 157.

A. Kudo and Y. Miseki, Heterogeneous photocatalyst materials for water splitting, *Chem. Soc. Rev.*, 2009, **38**, 253.

EXERCISES

10.1 It has been suggested that hydrogen could be placed in Group 1, Group 14, or Group 17 of the periodic table. Give arguments for and against each of these positions.

10.2 Explain the relatively low reactivity of hydrogen.

10.3 Assign oxidation numbers to the elements in (a) H_2S, (b) KH, (c) $[ReH_9]^{2-}$, (d) H_2SO_4, (e) $H_2PO(OH)$.

10.4 Write balanced chemical equations for three major industrial preparations of hydrogen gas. Propose two different reactions that would be convenient for the preparation of hydrogen in the laboratory.

10.5 Preferably without consulting reference material, construct the periodic table, identify the elements, and (a) indicate positions of salt-like, metallic, and molecular hydrides, (b) add arrows to indicate trends in $\Delta_f G^\oplus$ for the hydrogen compounds of the p-block elements, and (c) identify the areas where the molecular hydrides are electron-deficient, electron-precise, and electron-rich.

10.6 Describe the expected physical properties of water in the absence of hydrogen bonding.

10.7 Which hydrogen bond would you expect to be stronger, $S-H \cdots O$ or $O-H \cdots S$? Why?

10.8 Name and classify the following hydrogen compounds: (a) BaH_2, (b) SiH_4, (c) NH_3, (d) AsH_3, (e) $PdH_{0.9}$, (f) HI.

10.9 Identify the compounds from Exercise 10.8 that provide the most pronounced example of the following chemical characteristics and give a balanced equation that illustrates each of the characteristics: (a) hydridic character, (b) Brønsted acidity, (c) variable composition, (d) Lewis basicity.

10.10 Divide the compounds in Exercise 10.8 into those that are solids, liquids, or gases at room temperature and pressure. Which of the solids are likely to be good electrical conductors?

10.11 Use Lewis structures and VSEPR theory to predict the shapes of H_2Se, P_2H_4, and H_3O^+ and to assign point groups. Assume a skew structure for P_2H_4.

10.12 Identify the reaction that is most likely to give the highest proportion of HD and give your reasoning: (a) $H_2 + D_2$ equilibrated over a platinum surface, (b) $D_2O + NaH$, (c) electrolysis of HDO.

10.13 Identify the compound in the following list that is most likely to undergo radical reactions with alkyl halides and describe the reason for your choice: H_2O, NH_3, $(CH_3)_3SiH$, $(CH_3)_3SnH$.

10.14 Arrange H_2O, H_2S, and H_2Se in order of (a) increasing acidity and (b) increasing basicity towards a hard acid such as the proton.

10.15 Describe the three different common methods for the synthesis of binary hydrogen compounds and illustrate each one with a balanced chemical equation.

10.16 What is the trend in hydridic character of BH_4^-, AlH_4^-, and GaH_4^-? Which is the strongest reducing agent? Give the equations for the reaction of GaH_4^- with excess 1 M HCl(aq).

10.17 Describe the important physical differences and a chemical difference between each of the hydrogen compounds of the p-block elements in Period 2 with their counterparts in Period 3.

10.18 Stibane, SbH_3 ($\Delta_f H^\oplus$ = +145.1 kJ mol^{-1}) decomposes above $-45°C$). Assess the difficulty in preparing a sample of BiH_3 ($\Delta_f H^\oplus$ = +277.8 kJ mol^{-1}) and suggest a method for its preparation.

10.19 What type of substance is formed by the interaction of water and krypton at low temperatures and elevated krypton pressure? Describe the structure in general terms.

10.20 Sketch the approximate potential energy surfaces for the hydrogen bond between H_2O and the Cl^- ion, and contrast this with the potential energy surface for the hydrogen bond in $[FHF]^-$.

10.21 Dihydrogen is a familar reducing agent, but it is also an oxidizing agent. Explain this statement, giving examples.

PROBLEMS

10.1 What is the expected infrared stretching wavenumber of gaseous $^3H^{35}Cl$ given that the corresponding value for $^1H^{35}Cl$ is 2991 cm^{-1}?

10.2 Consult Chapter 8 and then sketch the qualitative splitting pattern and relative intensities within each set for the ^1H- and ^{31}P-NMR spectra of PH_3.

10.3 In his paper 'The proper place for hydrogen' (*J. Chem. Educ.*, 2003, 80, 947), M.W. Cronyn argues that hydrogen should be placed at the head of Group 14 immediately above carbon. Summarize his reasoning.

10.4 (a) Sketch a qualitative molecular orbital energy level diagram for the HeH^+ molecule ion and indicate the correlation of the molecular orbital levels with the atomic energy levels. The ionization energy of H is 13.6 eV and the first ionization energy of He is 24.6 eV. (b) Estimate the relative contribution of H1s and He1s orbitals to the bonding orbital and predict the location of the partial positive charge of the polar molecule. (c) Why do you suppose that HeH^+ is unstable on contact with common solvents and surfaces?

10.5 Spectroscopic evidence has been obtained for the existence of $[Ir(C_5H_5)(H_3)(PR_3)]^+$, a complex in which one ligand is formally H_3^+. Devise a plausible molecular orbital scheme for the bonding in the complex, assuming that an angular H_3 unit occupies one coordination

site and interacts with the e_g and t_{2g} orbitals of the metal. An alternative formulation of the structure of the complex, however, is as a trihydro species with very large coupling constants (see *J. Am. Chem. Soc.*, 1991, 113, 6074 and the references therein, and especially *J. Am. Chem. Soc.*, 1990, 112, 909 and 920). Review the evidence for this alternative formulation.

10.6 Correct the faulty statements in the following description of hydrogen compounds. 'Hydrogen, the lightest element, forms thermodynamically stable compounds with all of the nonmetals and most metals. The isotopes of hydrogen have mass numbers of 1, 2, and 3, and the isotope of mass number 2 is radioactive. The structures of the hydrides of the Group 1 and 2 elements are typical of ionic compounds because the H^- ion is compact and has a well-defined radius. The structures of the hydrogen compounds of the nonmetals are adequately described by VSEPR theory. The compound $NaBH_4$ is a versatile reagent because it has greater hydridic character than the simple Group 1 hydrides such as NaH. Heavy element hydrides such as the tin hydrides frequently undergo radical reactions, in part because of the low E−H bond energy. The boron hydrides are called electron-deficient compounds because they are easily reduced by hydrogen.'

The Group 1 elements

11

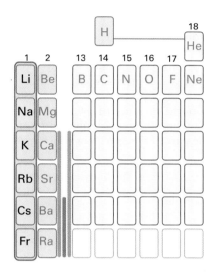

All the Group 1 elements are metallic but, unlike most metals, they have low densities and are very reactive. In this chapter we provide a survey of the chemistry of these elements, highlighting the similarities and trends in properties and commenting on the slightly anomalous behaviour of lithium. We then provide a detailed review of the chemistry of the alkali metals, discussing the occurrence of the elements in the natural environment and how they are extracted and used. The chapter also interprets trends in the properties of the simple binary compounds in terms of the ionic model and the nature of complexes and organometallic compounds of the elements.

PART A: **THE ESSENTIALS**

The Group 1 elements, the **alkali metals**, are lithium, sodium, potassium, rubidium, caesium (cesium), and francium. We shall not discuss francium, which exists naturally only in minute quantities and is highly radioactive. All the elements are metals and form simple ionic compounds, most of which are soluble in water. The elements form a limited number of complexes and organometallic compounds. In this first section of the chapter we summarize the key features of the chemistry of the Group 1 elements.

11.1 The elements

Key point: The trends in the properties of the Group 1 metals and their compounds can be explained in terms of variations in their atomic radii and ionization energies.

Sodium and potassium have high natural abundances, occurring widely as salts such as the chlorides. Lithium is relatively rare, occurring mainly in the mineral spodumene $LiAlSi_2O_6$. Rubidium and caesium are rarer still but occur in reasonable concentration in some minerals such as the zeolite *pollucite*, $Cs_2Al_2Si_4O_{12}.nH_2O$. Sodium and lithium metals are extracted by the electrolysis of molten metal chloride. Potassium is obtained by reacting KCl with sodium metal and rubidium and caesium by reaction of the metal chloride with calcium or barium.

All the Group 1 elements are metals with valence electron configuration ns^1. They conduct electricity and heat, are soft, and have low melting points that decrease down the group. Their softness and low melting points stem from the fact that their metallic bonding is weak because each atom contributes only one electron to the molecular orbital band (Section 3.19). This softness is particularly evident for Cs, which melts at only 29°C. Liquid sodium and a sodium/potassium mixture have been used as the coolant in some

Table 11.1 Selected properties of the Group 1 elements

	Li	Na	K	Rb	Cs
Metallic radius/pm	152	186	231	244	262
Ionic radius/pm (coordination number)	59(4)	102(6)	138(6)	148(6)	174(8)
Ionization energy/(kJ mol⁻¹)	519	494	418	402	376
Standard potential/V	−3.04	−2.71	−2.94	−2.92	−3.03
Density/(g cm⁻³)	0.53	0.97	0.86	1.53	1.90
Melting point/°C	180	98	64	39	29
$\Delta_{hyd}H^{\ominus}/(kJ\,mol^{-1})$	−519	−406	−322	−301	−276
$\Delta_{sub}H^{\ominus}/(kJ\,mol^{-1})$	161	109	90	86	79

nuclear power plants due to their excellent thermal conductivities. All the elements adopt a body-centred cubic structure (Section 3.5) and because that structure-type is not close-packed and their atomic radii are large, they have low densities. The metals readily form alloys among themselves, for example NaK, and with many other metals, such as sodium/mercury amalgam. Table 11.1 summarizes some important properties.

Flame tests are commonly used for the identification of the presence of the alkali metals and their compounds. Electronic transitions occur within the metal atoms and ions formed in the flames with energies that fall in the visible part of the spectrum which give a characteristic colour to the flame:

Li	Na	K	Rb	Cs
crimson	yellow	red to violet	violet	blue

The intensity of the emission spectrum obtained from an alkali metal salt solution can be measured with a flame photometer to provide a quantitative measurement of the element's concentration in the solution.

The chemical properties of the Group 1 elements correlate with the trend in their atomic radii (Fig. 11.1). The increase in atomic radius from Li to Cs leads to a decrease in first ionization energy down the group because the valence shell is increasingly distant from the nucleus (Fig. 11.2; Section 1.9). Because their first ionization energies are all low, the metals are reactive and form M^+ ions increasingly readily down the group. Their reaction with water

$$2\,M(s) + 2\,H_2O(l) \rightarrow 2\,MOH(aq) + H_2(g)$$

illustrates this trend:

Li	Na	K	Rb	Cs
gently	vigorously	vigorously with ignition	explosively	explosively

A part of the reason for the explosive character of the reaction of Rb and Cs with water is that both metals are denser than water, sink below the surface, and the sudden ignition of the hydrogen scatters the water violently.

The thermodynamic tendency to form M^+ in aqueous conditions is confirmed by the standard potentials of the couples M^+/M, which are all large and negative (Table 11.1), indicating that the metals are readily oxidized. The surprising uniformity of the standard potentials of the alkali metals can be explained by studying the thermodynamic cycle for the reduction half-reaction (Fig. 11.3). The enthalpies of sublimation and ionization both decrease down the group (making oxidation more favourable); however, this trend is counteracted by a smaller enthalpy of hydration as the radii of the ions increase (making oxidation less favourable).

All the elements must be stored under a hydrocarbon oil to prevent reaction with atmospheric oxygen, although Li, Na, and K can be handled in air for short periods; Rb and Cs must be handled under an inert atmosphere at all times.

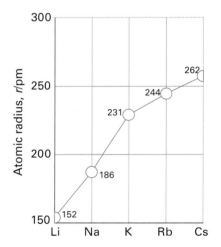

Figure 11.1 The variation in atomic radius of the elements of Group 1.

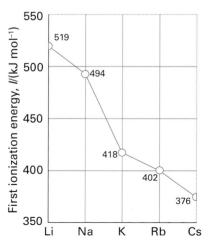

Figure 11.2 The variation in first ionization energy of the elements of Group 1.

11.2 Simple compounds

Key point: The binary compounds of the alkali metals contain the cations of the elements and exhibit predominantly ionic bonding.

Group 1 elements form ionic (saline) hydrides with the rock-salt structure; the anion present is the hydride ion, H^-. These hydrides were discussed in some detail in Section 10.6b. All the Group 1 elements form halides, MX. They can be obtained by direct combination of the elements or more normally from solutions, for example reaction of the metal hydroxide or carbonate with hydrohalic acid (HX, X = F, Cl, Br, Cl). The halides occur widely, for example a litre of sea water contains about 35 g of NaCl. Most of the halides have the (6,6)-coordinate rock-salt structure (Fig. 11.4), but CsCl, CsBr, and CsI have the (8,8)-coordinate caesium-chloride structure (Fig. 11.5; Section 3.9).

The Group 1 elements react vigorously with oxygen. Only Li reacts directly with oxygen to give a simple oxide, Li_2O. Sodium reacts with oxygen to give the peroxide, Na_2O_2, which contains the peroxide ion, O_2^{2-}, and the other Group 1 elements form the superoxides, which contain the paramagnetic superoxide ion, O_2^-. All the hydroxides are white, translucent, deliquescent solids. They absorb water from the atmosphere in an exothermic reaction. Lithium hydroxide, LiOH, forms the stable hydrate, $LiOH.8H_2O$. The solubility of the hydroxides makes them a ready source of OH^- ions in the laboratory and in industry. The metals react with sulfur to form compounds with the formula M_2S_x, where x lies in the range 1 to 6. The simple sulfides Na_2S and K_2S have the antifluorite structure whereas the polysulfides, with $n > 2$, contain S_n^{2-} chains. Lithium readily forms a nitride, Li_3N, when it is heated in nitrogen (or more slowly at room temperature), but the other alkali metals do not react directly.

Only Li reacts directly with carbon to form a carbide of the stoichiometry Li_2C_2 and containing the dicarbide (acetylide) anion, C_2^{2-}. Similar carbides are formed by the other alkali metals by heating them in ethyne. Potassium, Rb, and Cs react with graphite to form intercalation compounds such as C_8K (Section 14.5). In combination with the p-block metals (from Groups 13 to 15) the alkali metals are strongly reducing and often form Zintl phases; the latter contain the alkali metal cation together with a reduced, complex anionic species, such as Ge_4^{4-}.

All the common salts of the Group 1 metals are soluble in water, although most of their solid salts are anhydrous. There are a few exceptions for the smaller Li and Na ions, for example $LiX.3H_2O$ for X = Cl, Br, I, and $LiOH.8H_2O$. Lithium iodide is deliquescent, absorbing water rapidly from the air to form $LiI.3H_2O$ and then a solution.

Sodium dissolves in liquid ammonia without evolving hydrogen, producing, at low concentrations, deep blue solutions that contain solvated electrons. These solutions survive for long periods at and below the normal boiling point of ammonia ($-33°C$) and in the absence of air. Concentrated metal–ammonia solutions have a metallic bronze colour and have electrical conductance close to that of a solid metal, about 10^7 S m^{-1}. It is possible to isolate the alkalide anions, M^-, from solutions of the metals in amines, which are formed by disproportionation of the element into M^+ and M^-.

The Group 1 element ions are hard Lewis acids (Section 4.12) and complexes are formed mainly with small, hard donors such as O or N atoms. Their hardness decreases down the group with increasing ionic radius and there is evidence for more covalent character in their bonding, for example complexes where Cs coordinates to P and S. Interactions with monodentate ligands are weak and the hydrated species, $M(OH_2)_n^+$, readily exchange H_2O ligands with the surrounding solvent. Chelating ligands such as the hexadentate ethylenediaminetetraacetate ion, $(O_2C)_2NCH_2CH_2N(CO_2)_2^{4-}$, have much higher formation constants. Macrocycles and crown ethers can form strong complexes with the Group 1 elements provided their ions have the correct radius to fit into the ligand coordination environment (Section 7.14).

The lighter Group 1 elements form organometallic compounds which are highly reactive, hydrolysed by water, liberating hydrogen, and are pyrophoric (spontaneously igniting) in air. Protic (proton-donating) organic compounds are reduced by the elements to form ionic organometallic compounds, for example cyclopentadiene reacts with Na metal in the solvent THF to form $Na^+[C_5H_5]^-$. Lithium alkyls and aryls are by far the most important Group 1 organometallic compounds. They are thermally stable, soluble in organic and nonpolar solvents such as THF, and widely used as a source of nucleophilic alkyl or aryl groups in organic synthesis.

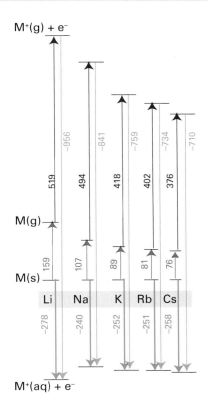

Figure 11.3 Thermochemical cycle (standard enthalpy changes in kJ mol^{-1}) for the oxidation half-reaction M(s) → M$^+$(aq) + e$^-$.

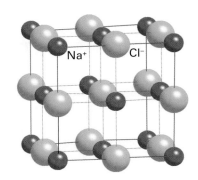

Figure 11.4 The rock-salt structure adopted by the majority of Group I metal halides.

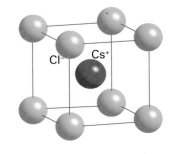

Figure 11.5 The CsCl structure adopted by CsCl, CsBr, and CsI under normal conditions.

11.3 The atypical properties of lithium

Key point: The chemical properties of Li are anomalous due to its small ionic radius and tendency to exhibit covalent bonding.

As we saw in Chapter 9, most of the trends in the chemical properties of the elements in the periodic table are best discussed in terms of vertical trends within groups or horizontal trends across periods. However, the lightest member of a group, in this case Li, often displays properties that are markedly different from that of its congeners. This difference can often be expressed as a diagonal relationship with the element to its lower right in the periodic table (Section 9.2b). In the case of Group 1 and Li the following differences can be noted:

1. Lithium can exhibit a high degree of covalent character in its bonding. This covalent character is due to the high polarizing power of the Li^+ ion associated with high charge density (Section 1.9).
2. Lithium forms the normal oxide when burnt in oxygen whereas other Group 1 elements form peroxides or superoxides.
3. Lithium is the only alkali metal to form a nitride, Li_3N, when heated in nitrogen and a carbide, Li_2C_2, when heated with graphite.
4. Some lithium salts such as the carbonate, phosphate, and fluoride have very low solubilities in water. Other lithium salts crystallize as hydrates or are hygroscopic.
5. Lithium forms many stable organometallic compounds.
6. Lithium nitrate decomposes directly to the oxide whereas the other alkali metals initially form nitrites, MNO_2.
7. Lithium hydride is stable to heating to 900°C whereas the other hydrides decompose above 400°C.

Lithium's very low molar mass, which also makes it the least dense metal (0.53 g cm^{-3}), leads to applications where low weight is important. Examples include rechargeable batteries ($LiCoO_2$, $LiFePO_4$, LiC_6) and systems for hydrogen storage such as lithium metal hydrides, lithium borohydrides, and lithium amides and imides (Box 10.4).

PART B: **THE DETAIL**

In this section we present a more detailed discussion of the chemistry of the elements of Group 1, interpreting some of the observed properties in thermodynamic terms. As the bonding in the compounds formed by these elements is usually ionic, we apply the concepts of the ionic model.

11.4 Occurrence and extraction

Key point: The Group 1 elements can be extracted by electrolysis.

The name lithium comes from the Greek *lithos* for stone. The natural abundance of lithium is low, the most abundant minerals being spodumene, $LiAlSi_2O_6$, from which lithium is most commonly extracted, and lepidolite, which has the approximate formula $K_2Li_3Al_4Si_7O_{21}(F,OH)_3$. Spodumene is first converted to LiCl and then electrolysed to produce lithium metal.

Sodium occurs as the mineral rock salt (NaCl) and in salt lakes and seawater. Sodium chloride makes up 2.6 per cent by mass of the biosphere with the oceans containing 4×10^{19} kg of the salt. Sodium chloride occurs naturally as the mineral rock salt, the deposits of ancient dried saline lakes. Many of these deposits are underground and are mined in the conventional way. Alternatively, water may be pumped underground to dissolve the rock salt, which is then pumped out as saturated brine solution.

The metal is extracted by *Down's process*, the electrolysis of molten sodium chloride:

$$2\,NaCl(l) \rightarrow 2\,Na(l) + Cl_2(g)$$

The sodium chloride is kept molten at 600°C, a temperature considerably below its melting point of 808°C, by the addition of calcium chloride. A high potential difference, typically between 4 and 8 V, is applied between a carbon anode and an iron cathode immersed in the molten salt. The electrolysis liberates liquid sodium metal at the cathode, which rises to the surface of the cell where it is collected under an inert atmosphere. This process is also used for the industrial production of chlorine, which is generated at the anode.

Potassium occurs naturally as potash (K_2CO_3) and carnallite ($KClMgCl_2.6H_2O$). Natural potassium contains 0.012 per cent of the radioactive isotope ^{40}K, which undergoes β decay, with a half-life of 1.25 Ga, to ^{40}Ca and electron capture to ^{40}Ar. The ratio of ^{40}K and ^{40}Ar can be used for dating rocks, specifically the time when the rock solidified, at which point it traps any ^{40}Ar formed. In principle, potassium could be extracted electrolytically but the high reactivity of the element makes this far too hazardous. Instead, molten sodium and molten potassium chloride are heated together and potassium and sodium chloride are formed.

$$Na(l) + KCl(l) \rightleftharpoons NaCl(l) + K(g)$$

At the temperature of operation, potassium is a vapour and removing it from the system drives the equilibrium to the right.

Rubidium (from the Latin *rubidus* for deep red) and caesium (*caesius* for sky blue) were discovered by Robert Bunsen in 1861 and named from the colour their salts impart to a flame. Both elements occur as minor constituents of the mineral lepidolite, from which they are obtained as byproducts of the extraction of lithium. Prolonged treatment of lepidolite with sulfuric acid forms the alums of the alkali metals, $M_2SO_4.Al_2(SO_4)_3.nH_2O$. The alums are separated by multiple fractional crystallizations and then converted to the hydroxide by reaction with $Ba(OH)_2$ and then to the chloride by ion exchange. The metals are obtained from the molten chloride by reduction with calcium or barium.

$$2\,RbCl(l) + Ca(s) \rightarrow CaCl_2(s) + 2\,Rb(s)$$

Caesium also occurs as the mineral pollucite, $Cs_4Al_4Si_9O_{26}.H_2O$. The element is extracted from the mineral by leaching with sulfuric acid to form the alum $Cs_2SO_4Al_2(SO_4)_3.24H_2O$, which is then converted to the sulfate by roasting with carbon. The chloride is formed by ion exchange and is then reduced with calcium or barium as described above. Caesium metal can also be obtained by the electrolysis of molten CsCN.

11.5 Uses of the elements and their compounds

Key points: Common uses of lithium are related to its low density; the most widely used compounds of Group 1 are sodium chloride and sodium hydroxide.

The applications of lithium metal are in large part due to its low atomic mass and consequently low density. It is used in alloys where weight is of a premium concern, such as aircraft parts; Al containing around 2 per cent Li has a mass density 6 per cent lower than that of pure Al and is used, for example, in parts of aircraft wings to reduce the overall weight and thereby improve fuel consumption. Similar lithium-containing alloys have been used in aerospace applications such as the booster tanks of the space shuttle.

The low molar mass of lithium (6.94 g mol^{-1}), only 3.3 per cent that of lead, coupled with the strongly negative standard potential of the Li$^+$/Li couple (Table 11.1), make lithium batteries an attractive alternative to lead–acid batteries (Box 11.1). Lithium carbonate is widely used to treat bipolar conditions (manic depression) and lithium stearate is a widely used lubricant in the automotive industry. The high polarizing power of Li$^+$ means that some complex oxides, such as $LiNbO_3$ and lithium tantalate, show important nonlinear optical and acousto-optical effects, and they are widely used in mobile communication devices.

Sodium and potassium are essential for physiological function (Section 27.3) and a major use of NaCl is in flavouring food. Sodium is used in the extraction of rarer metals, such as Ti from titanium(IV) chloride. The major uses of NaCl are road de-icing and production of NaOH for the chloralkali industry (Box 11.2). However, due to concerns with the effects on the environment of distributing large amounts of rock salt as a de-icing agent, alternative materials using less NaCl are being sought. For example, a mixture of sodium chloride and molasses has been employed. Sodium hydroxide is in the ten most important industrial chemicals in terms of annual tonnage produced. Other common applications of sodium and its compounds include the use of the metal in some kinds of street lamps, which produce a distinctive yellow glow when an electrical discharge is passed through sodium vapour (Box 1.3), table salt, baking soda, and caustic soda (NaOH). Sodium salts and compounds with ion-exchangable Na$^+$ are also widely used in water softening equipment (Box 11.3).

Potassium hydroxide is used in soap manufacture to make 'soft' liquid soaps. Potassium chloride and sulfate are used as fertilizers; the nitrate and chlorate are used in fireworks. Potassium bromide has been used as an antaphrodisiac (a compound that reduces libido). Potassium cyanide is used in the metal extraction and plating industries to obtain or aid in the deposition of copper, silver, and gold.

Rubidium and caesium are often used in the same applications, and one element may often be substituted for the other. The

---

BOX 11.1 Lithium batteries

The very negative standard potential and low molar mass of lithium make it an ideal anode material for batteries. These batteries have relatively high specific energy (energy production divided by the mass of the battery) because lithium metal and compounds containing lithium are light in comparison with some other materials used in batteries, such as lead and zinc. Lithium batteries are common, but there are many types based on different lithium compounds and reactions.

The lithium rechargeable battery, used in portable computers and phones, mainly uses $Li_{1-x}CoO_2$ ($x < 1$) as the cathode with a lithium/graphite anode, LiC_6. Lithium ions are produced at the anode during the battery discharge. To maintain charge balance, Co(IV) is reduced to Co(III) in the form of $LiCoO_2$ at the cathode. The reactions occurring during battery discharge are

Cathode $\quad Li_{1-x}CoO_2(s) + x\,Li^+(sol) + x\,e^- \rightarrow LiCoO_2(s)$

Anode $\quad C_6Li \rightarrow 6\,C(graphite) + Li^+(sol) + e^-$

The battery is rechargeable because both the cathode and the anode can act as host for the Li$^+$ ions, which can move back and forth between them when charging and discharging. There are many other lithium batteries using different electrode materials, mainly d-metal compounds that take part in the redox reaction in a similar way to the cobalt. The latest generation of electric cars uses lithium battery technology rather than lead–acid cells.

Another popular lithium battery uses thionyl chloride, $SOCl_2$. This system produces a light, high-voltage cell with a stable energy output. The overall reaction in the battery is

$$4\,Li(s) + 2\,SOCl_2(l) \rightarrow 4LiCl(s) + S(s) + SO_2(l)$$

The battery requires no additional solvent as both $SOCl_2$ and SO_2 are liquids at the internal battery pressure. This battery is not rechargeable as both sulfur and LiCl are precipitated. It is used in military applications and in spacecraft. Another battery system is based on the reduction of SO_2:

$$2\,Li(s) + 2\,SO_2(l) \rightarrow Li_2S_2O_4(s)$$

The system is also not rechargeable as solid $Li_2S_2O_4$ deposits on the cathode. This battery uses acetonitrile (CH_3CN) as a cosolvent and the handling of this compound and the SO_2 present safety hazards. The batteries are hermetically sealed and not available to the general public. They are used in military communications and automated external defibrillators that are used to restore normal heart rhythm.

Rechargeable batteries are discussed further in Chapter 24.

BOX 11.2 The chloralkali industry

The chloralkali industry has its roots in the industrial revolution when large quantities of alkali were required for the manufacture of soap, paper, and textiles. Today, sodium hydroxide is one of the top ten most important inorganic chemicals in terms of quantity produced and continues to be important in the manufacture of other inorganic chemicals and in the pulp and paper industries. Chlorine and dihydrogen are the gaseous products. Chlorine is very important industrially and is used in the manufacture of PVC, in the extraction of titanium, and in the pulp and paper industries.

The industrial process is based on the electrolysis of aqueous sodium chloride. Water is reduced to hydrogen gas and hydroxide ions at the cathode, and chloride ions are oxidized to chlorine gas at the anode:

$$2\,H_2O(l) + 2\,e^- \rightarrow H_2(g) + 2\,OH^-$$

$$2\,Cl^-(aq) \rightarrow Cl_2(g) + 2\,e^-$$

There are three different types of cells that are used for the electrolysis. In a *diaphragm cell* there is a diaphragm which prevents the OH^- ions produced at the cathode from coming into contact with the Cl_2 gas produced at the anode. This diaphragm used to be made of asbestos, but it is now made of a polytetrafluoroethylene mesh. During the electrolysis, the solution at the cathode is removed continuously and evaporated

in order to crystallize the sodium chloride impurities. The final sodium hydroxide typically contains approximately 1 per cent by mass NaCl.

A *membrane cell* functions like the diaphragm cell except that the anode and cathode solutions are separated by a microporous polymer membrane that is permeable only to Na^+ ions. The sodium hydroxide solution produced using this cell typically contains approximately 50 ppm Cl^-. The disadvantage with this method is that the membrane is very expensive and it can become clogged by trace impurities.

The *mercury cell* uses liquid mercury as the cathode. Chlorine gas is produced at the anode but sodium metal is produced at the cathode:

$$Na^+(aq) + e^- \rightarrow Na(Hg)$$

The sodium–mercury amalgam is reacted with water on a graphite surface:

$$2\,Na(Hg) + 2\,H_2O(l) \rightarrow 2\,NaOH(aq) + H_2(g)$$

The sodium hydroxide solution produced by this route is very pure and the mercury cell is the preferred source of high-quality, solid sodium hydroxide. Unfortunately, the process is accompanied by discharge of mercury into the environment by a number of routes. Consequently, the chloralkali industry is under pressure to move away from the use of mercury electrodes.

BOX 11.3 Sodium ion-exchange materials

Hard water contains high levels of Ca^{2+} and Mg^{2+} ions, which precipitate from the water on heating (as limescale, which is largely $CaCO_3$) and inhibit the formation of a lather with soap or detergent, so reducing their effectiveness. Domestic water-softeners contain zeolites, or ion-exchange resins, which contain Na^+ ions that exchange with the Ca^{2+} and Mg^{2+} ions. Zeolites are often also a component of washing detergents, where they perform the same role.

Zeolites are microporous aluminosilicates that contain weakly held cations and water molecules within their cavities (Section 14.15). They are often available naturally or are synthesized as their sodium-containing form and denoted 'Na-zeolite'. The ion-exchange reaction that occurs when hard water is exposed to Na-zeolite is

$$2\,Na\text{-zeolite(s)} + Ca^{2+}(aq) \rightarrow Ca\text{-zeolite(s)} + 2\,Na^+(aq)$$

with the result that Ca^{2+} ions are removed from the solution to the solid phase. The softened water contains mainly Na^+ ions as the dissolved cation species. As sodium carbonate and the sodium salts of soap and detergent

molecules are very soluble, the softened water is more effective for washing and avoids problems such as the deposition of limescale on the heating elements of kettles, dishwashers, and washing machines.

The reverse reaction, which regenerates the ion-exchanger Na-zeolite, can be performed in a water softener by exposing the exhausted zeolite to high concentrations of Na^+ ions (for example, sodium chloride solution):

$$Ca\text{-zeolite(s)} + 2\,Na^+(aq) \rightarrow 2\,Na\text{-zeolite(s)} + Ca^{2+}(aq)$$

In the case of zeolites added to detergents, the Ca-zeolite that is produced exists as a finely divided solid that is flushed away. This procedure is environmentally benign because the effluent contains calcium, silicon, aluminium, oxygen, and water—the same as many natural minerals.

Instead of zeolites some water softeners contain resins, which are porous polymeric organic compounds formed from crosslinked polystyrene decorated with functional groups such as carboxylates and sulfonates. The charge on these anionic groups is balanced by Na^+ ions on the surface of the resin, which are readily exchanged with Ca^{2+} and Mg^{2+}.

market for these elements is small and highly specialized. Applications include glass for fibre-optics in the telecommunications industry, night vision equipment, and photoelectric cells. The 'caesium clock' (atomic clock) is used for the international standard measure of time and for the definition of the second and the metre. Caesium salts are also used as high-density drilling fluids: the high density of the solutions arises from the high atomic mass of Cs.

11.6 Hydrides

Key point: The hydrides of the Group 1 elements are ionic and contain the H^- ion.

The Group 1 elements react with hydrogen to form ionic (saline) hydrides with the rock-salt structure; the anion present

is the hydride ion, H^-. These hydrides were discussed in detail in Section 10.6.

The hydrides react violently with water:

$$NaH(s) + H_2O(l) \rightarrow NaOH(aq) + H_2(g)$$

Finely divided sodium hydride can ignite if it is left exposed to humid air. Such fires are difficult to extinguish because even carbon dioxide is reduced when it comes into contact with hot metal hydrides. Hydrides are useful as non-nucleophilic bases and reductants:

$$NaH(s) + NH_3(l) \rightarrow NaNH_2(am) + H_2(g)$$

where am denotes a solution in ammonia.

11.7 Halides

Key point: On descending the group, the enthalpy of formation becomes less negative for the fluorides but more negative for the chlorides, bromides, and iodides.

All the Group 1 elements form halides, MX, by direct combination of the elements. Most of the halides have the (6,6)-coordinate rock-salt structure, but CsCl, CsBr, and CsI have the (8,8)-coordinate caesium-chloride structure (Section 3.9). The simple radius-ratio arguments presented in Section 3.10 may be used to help rationalize this choice of structure. Table 11.2 summarizes the radius ratio (γ) for the various alkali metal halides. As we saw in Section 3.10, a rock-salt structure with (6,6) coordination is expected for ratios between 0.414 and 0.732, and the caesium-chloride structure is expected for larger values; a (4,4)-coordination zinc-sulfide structure is expected for values below 0.414. However, the lattice energies of the CsCl and rock-salt arrangements differ by only a small percentage and factors such as polarization (Section 3.12) help to stabilize the rock-salt structure over the caesium-chloride structure for most of the alkali metal halides. At 445°C the caesium chloride structure changes to the rock-salt structure, and on cooling below room temperature RbCl converts to the caesium-chloride structure.

Table 11.2 Radius ratio, γ, for the alkali metal halides*

	F	Cl	Br	I
Li	0.57	0.42	0.39	0.35
Na	0.77	0.56	0.52	0.46
K	0.96	0.76	0.70	0.63
Rb	0.90	0.82	0.76	0.67
Cs	0.80	0.92	0.85	0.76

* Based on ionic radii for 6-coordination. Values in italic type are compounds that adopt the rock-salt structure type.

fluorides on descending the group but become more negative for the chlorides, bromides, and iodides (Fig. 11.6). These trends can be rationalized by considering a Born−Haber cycle for formation of the halides from the elements (Fig. 11.7). These calculations treat the bonding in alkali metal compounds as purely ionic; however, for the heavier metal ions there is increasing contribution from covalence as the ions become larger, more polarizable, and less hard.

As we saw in Section 3.11, the requirement that the sum of enthalpy changes round a Born−Haber cycle be zero implies that the enthalpy of formation of a compound is

$$\Delta_f H^\ominus = \Delta_{sub} H^\ominus + \Delta_{ion} H^\ominus + \Delta_{dis} H^\ominus + \Delta_{eg} H^\ominus - \Delta H_L \qquad (11.1)$$

EXAMPLE 11.1 Investigating a pressure-induced phase transition using powder X-ray diffraction

The powder X-ray diffraction data collected from RbI at standard conditions show that the lattice type is face-centred cubic and the lattice parameter is 734 pm. Application of 4 kbar causes the powder X-ray diffraction pattern to change, showing that the lattice type becomes primitive with a lattice parameter of 446 pm. Interpret these data given that the ionic radii for Rb^+ and I^- in sixfold (eightfold in parentheses) coordination are 148 (160) pm and 220 (232) pm, respectively.

Answer On the basis of radius-ratio rules and $\gamma = 0.67$, the predicted structure for rubidium iodide is rock salt with (6,6) coordination (Section 3.10b). The lattice type of this structure is face-centred cubic, in agreement with the diffraction data. It is also possible to predict the lattice parameter for this structure by using the ionic radii. In the rock-salt structure the lattice parameter is equivalent to the overall Rb−I−Rb distance (Fig. 11.4), which is twice the sum of the anion and cation ionic radii. Thus, the predicted lattice parameter is 736 pm, in good agreement with the experimental value. Structures under pressure have a thermodynamic tendency to rearrange to denser arrangements. The radius ratio for rubidium iodide lies close to 0.732, above which the caesium-chloride structure is expected. Therefore, under pressure, a phase transformation occurs in RbI. The lattice type of the CsCl structure type is primitive (Fig. 11.5), in agreement with the X-ray diffraction data, and a lattice parameter for this structure type can be calculated by using the ionic radii of Rb^+ and I^- in eightfold coordination (160 and 232 pm, respectively) as 453 pm (from $2(r_+ + r_-)/3^{1/2}$). This result closely matches the experimental data: the diffraction data are collected at 4 kbar, so the lattice parameter is slightly smaller than the calculated value, which is based on ionic radii evaluated at 1 bar.

Self-test 11.1 Predict the differences in the experimental powder X-ray diffraction pattern of CsCl collected at room temperature and at 600°C.

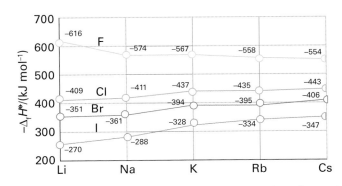

Figure 11.6 The standard enthalpies of formation of the halides of Group 1 elements at 298 K.

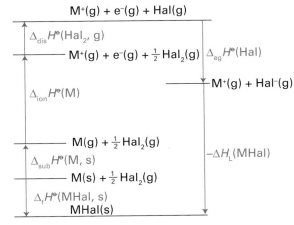

Figure 11.7 The Born−Haber cycle for the formation of the Group 1 halides. The sum of the enthalpy changes round the cycle is zero.

The enthalpies of formation of all the halides are large and negative, becoming less negative from the fluoride to iodide for each element. The enthalpies of formation become less negative for the

Table 11.3 Selected data for the discussion of the stabilities of the Group 1 halides

	F	Cl	Br	I
Ionic radius, r/pm	133	181	196	220
$\frac{1}{2}\Delta_{dis}H^{\ominus}$ / $\left(kJ\ mol^{-1}\right)$	79	121	112	107
$\Delta_{eg}H^{\ominus}$ / $\left(kJ\ mol^{-1}\right)$	−328	−349	−325	−295
$\left(\frac{1}{2}\Delta_{dis}H^{\ominus} + \Delta_{eg}H^{\ominus}\right)$ / $\left(kJ\ mol^{-1}\right)$	−249	−228	−213	−188

The first two terms on the right in this expression are constant for a series of halides of a given element. The next two terms vary from fluoride to iodide and, as can be seen from the data in Table 11.3, their sum becomes less negative from F to I. The final term is the lattice enthalpy, which (from the Born−Mayer equation, Section 3.12a) we know to be inversely proportional to the sum of the ionic radii. As the radius of the anion increases from F$^-$ to I$^-$, the lattice enthalpy becomes smaller. Consequently, $\Delta_f H^{\ominus}$ becomes less negative.

If we consider the formation of a series of Group 1 halides, the terms $\Delta_{dis}H^{\ominus}$ and $\Delta_{eg}H^{\ominus}$ are constant. The terms $\Delta_{sub}H^{\ominus}$ and $\Delta_{ion}H^{\ominus}$ vary with the metal and, as can be seen from Table 11.1, the sum of their values decreases down the group. The lattice enthalpy also decreases as the radius of the cation increases down a group. The trend in the enthalpy of formation depends on the relative difference between these values, that is $\left(\Delta_{sub}H^{\ominus} + \Delta_{ion}H^{\ominus}\right) - \Delta H_L^{\ominus}$. For the chlorides, bromides, and iodides, the variation in values of the sum $\Delta_{sub}H^{\ominus} + \Delta_{ion}H^{\ominus}$ is greater than the variation in $\Delta H_L^{\ominus}$ and the enthalpies of formation become more negative down the group. However, for the fluorides, the small ionic radius of fluorine ensures that the differences in $\Delta H_L^{\ominus}$ are greater than those in $\Delta_{sub}H^{\ominus} + \Delta_{ion}H^{\ominus}$ and the enthalpies of formation become less negative down the group.

EXAMPLE 11.2 **Calculating enthalpies of formation**

Use data from Tables 11.1 and 11.3 to calculate the enthalpies of formation of NaF(s) and NaCl(s) and comment on the values obtained.

Answer The lattice energies of the compounds can be calculated by using the Kapustinskii equation (eqn 3.4), which gives 879 kJ mol^{-1} for NaF and 751 kJ mol^{-1} for NaCl. Then, from eqn 11.1,

$\Delta_f H^{\ominus}$ (NaF) $= 109 + 494 + 79 - 328 - 879$ kJ mol^{-1} $= -525$ kJ mol^{-1}

$\Delta_f H^{\ominus}$ (NaCl) $= 109 + 494 + 121 - 349 - 751$ kJ mol^{-1} $= -376$ kJ mol^{-1}

The enthalpy of formation for NaF is the more negative and therefore the fluoride is expected to be more stable than the chloride. In this case, the most important term in the expression for $\Delta_f H^{\ominus}$ is the lattice enthalpy, $\Delta H_L^{\ominus}$, which is larger for NaF because of the smaller size of the anion.

Self-test 11.2 Calculate the enthalpies of formation of LiF(s) and NaF(s) and comment on the values obtained.

The halides are all soluble in water with the exception of LiF, which is only sparingly soluble. This low solubility of LiF can be traced to the fact that the high lattice enthalpy due to the small ionic radii is not offset by the enthalpy of hydration.

11.8 Oxides and related compounds

Key points: Only Li forms a normal oxide on direct reaction with oxygen; Na forms the peroxide and the heavier elements form the superoxides.

As mentioned previously, all the Group 1 elements react vigorously with oxygen. Only Li reacts directly with excess oxygen to give just the oxide, Li$_2$O, which adopts the antifluorite structure (Section 3.9).

$$4\,Li(s) + O_2(g) \rightarrow 2\,Li_2O(s)$$

Sodium reacts with oxygen to give the peroxide, Na$_2$O$_2$, which contains the peroxide ion, O$_2^{2-}$:

$$2\,Na(s) + O_2(g) \rightarrow Na_2O_2(s)$$

The other Group 1 elements form the superoxides:

$$K(s) + O_2(g) \rightarrow KO_2(s)$$

These compounds contain the paramagnetic superoxide ion, O$_2^-$ and adopt a CaC$_2$ type structure that is based on the rock-salt structure (Fig. 3.31, see Fig. 11.8 for an alternative view).

All the varieties of oxides are basic and react with water to give the OH$^-$ ion by extraction of H$^+$ from H$_2$O in a Lewis acid−base reaction:

$$Li_2O(s) + H_2O(l) \rightarrow 2\,Li^+(aq) + 2\,OH^-(aq)$$
$$Na_2O_2(s) + 2\,H_2O(l) \rightarrow 2\,Na^+(aq) + 2\,OH^-(aq) + H_2O_2(aq)$$
$$2\,KO_2(s) + 2\,H_2O(l) \rightarrow 2\,K^+(aq) + 2\,OH^-(aq) + H_2O_2(aq) + O_2(g)$$

The oxide and the peroxide react by proton transfer from H$_2$O. The initial formation of 'hydrogen superoxide, HO$_2$' by proton transfer to the superoxide ion is followed immediately by the disproportionation of HO$_2$ into O$_2$ and H$_2$O$_2$.

The normal oxides of Na, K, Rb, and Cs can be prepared by heating the metal with a limited amount of oxygen or by thermal decomposition of the peroxide or superoxide:

$$Na_2O_2(s) \rightarrow Na_2O(s) + \tfrac{1}{2}O_2(g)$$

The oxides Na$_2$O, K$_2$O, and Rb$_2$O adopt the antifluorite structure. The stability of the peroxides and superoxides to this decomposition increases down the group, Li$_2$O$_2$ being the least stable and Cs$_2$O$_2$ the most. Sodium peroxide is widely used as an oxidizing agent as it provides a ready source of oxygen on warming.

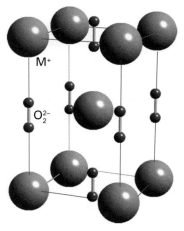

Figure 11.8 The structure of the superoxides MO$_2$ of the Group 1 elements.

The tendency of the peroxide or superoxide to decompose to the oxide can be explained by examining the lattice enthalpies of the compounds. As remarked earlier, the lattice enthalpy is inversely proportional to the sum of the ionic radii. Consequently, as the O^{2-} ion is smaller than either O_2^{2-} or O_2^{-}, the lattice enthalpy of any oxide is larger than that of the corresponding peroxide or superoxide. On descending the group, the radii of the cations increase and the lattice enthalpies of both the oxide and peroxide (or superoxide) decrease. Therefore, the difference between the two lattice enthalpies decreases and the tendency to decompose also decreases.

Potassium superoxide, KO_2, absorbs carbon dioxide and liberates oxygen. This reaction is exploited to purify air in applications such as submarines and breathing apparatus.

$$4\,KO_2(s) + 2\,CO_2(g) \rightarrow 2\,K_2CO_3(s) + 3\,O_2(g)$$
$$K_2CO_3(s) + CO_2(g) + H_2O(g) \rightarrow 2\,KHCO_3(s)$$

For applications in aerospace, lithium peroxide is often used instead of KO_2 to reduce weight.

Ozonides, compounds that contain the ozonide ion, O_3^{-}, exist for all the Group 1 elements. The ozonides of K, Rb, and Cs are obtained by heating the peroxide or superoxides with ozone. Sodium and lithium ozonides may be prepared by ion exchange of CsO_3 in liquid ammonia. These compounds are very unstable and explode violently:

$$2\,KO_3(s) \rightarrow 2\,KO_2(s) + O_2(g)$$

Partial oxidation of Rb and Cs yields suboxides of various compositions. Special conditions are needed to form these compounds, in which the elements occur with average oxidation numbers lower than +1. These compounds are formed only when air, water, and other oxidizing agents are rigorously excluded. For example, a series of metal-rich oxides are formed by the reaction of Rb or Cs with a limited supply of oxygen. These compounds are dark, highly reactive metallic conductors with formulas such as Rb_6O, Rb_9O_2, Cs_4O, and Cs_7O. A clue to the nature of these compounds is that Rb_9O_2 consists of O atoms surrounded by octahedra of six Rb atoms, with two neighbouring octahedra sharing faces (Fig. 11.9). These compounds were some of the earliest metal cluster compounds to be synthesized and characterized, although many other systems, for example the Zintl phases (Section 11.15), have now been found. The metallic conduction of the compounds suggests that the valence electrons are delocalized beyond the individual Rb_9O_2 clusters.

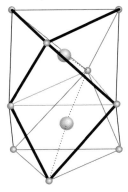

Figure 11.9 The structure of Rb_9O_2. Each O atom is surrounded by an octahedron of Rb atoms, and neighbouring octahedra share triangular faces.

EXAMPLE 11.3 Predicting the stabilities of peroxides

The ionic radii of the O^{2-} and O_2^{2-} ions are 126 and 180 pm, respectively. Use this information to confirm that there is a decreased tendency for the peroxide to decompose on descending the group.

Answer To assess stabilities we need to compare lattice enthalpies. To do so, we use the data in Table 11.1 and the Kapustinskii equation (eqn 3.4) to calculate the difference in lattice enthalpy between Na_2O and Na_2O_2 and then for Rb_2O and Rb_2O_2. We need to remember that the peroxide formula is $(M^+)_2(O_2^{2-})$ so in the Kapustinskii equation the number of ions is 3 and their charge numbers are +1 and −2. Insertion of the data gives the following values:

Na_2O	Na_2O_2	Rb_2O	Rb_2O_2	
2702	2260	2316	1980	$kJ\ mol^{-1}$

The difference between the values for Na_2O and Na_2O_2 is 442 kJ mol^{-1}, and the difference for Rb_2O and Rb_2O_2 is 336 kJ mol^{-1}. These results show that the difference between the values decreases down the group, suggesting that there is a lower thermodynamic tendency for the peroxide to form the oxide (providing entropy considerations are similar).

Self-test 11.3 All the Group 1 ozonides are unstable. Predict whether the instability increases or decreases down the group.

11.9 Sulfides, selenides, and tellurides

Key point: The Group 1 elements form simple sulfides, M_2S, and polysulfides in combination with sulfur.

All the alkali metals form a simple sulfide of stoichiometry M_2S; those of the smaller ions, Li^+ to K^+, adopt the antifluorite structure with simple S^{2-} ions. The polysulfides, M_2S_n, with n ranging from 2 to 6, are also known for the heavier alkali metals where the softer acids, M^+, stabilize the soft bases S_n^{2-}. For $n \geq 3$, the structures contain polysulfide anions as zig-zag chains separated by the alkali metal cations (Fig 11.10). The sodium/sulfur battery is being studied as a possible stationary energy storage system for use in combination with wind farms and solar energy plants (see Box 11.4). Selenium and tellurium react with the alkali metals to form selenides such as K_2Se, and tellurides, respectively; polyselenides, K_2Se_5, and polytellurides, Cs_2Te_5 are also known.

11.10 Hydroxides

Key point: All Group 1 hydroxides are soluble in water and absorb water and carbon dioxide from the atmosphere.

All the hydroxides of Group 1 elements are white, translucent, deliquescent solids. They absorb water and carbon dioxide from the atmosphere in an exothermic reaction. Lithium hydroxide, LiOH, forms the stable hydrate, $LiOH.8H_2O$. The solubility of the hydroxides makes them a ready source of OH^- ions in the laboratory and in industry. Potassium hydroxide, KOH, is soluble in ethanol and this 'ethanolic KOH' is a useful reagent in organic synthesis. Alkali metal hydroxide solutions rapidly absorb carbon dioxide from the air

$$2\,MOH(aq) + CO_2(g) \rightarrow M_2CO_3(aq) + H_2O(l)$$

and solutions left open to the air rapidly become contaminated with carbonate. For this reason the concentration of an MOH

The **sodium-sulfur battery** uses the power generated by the reaction of sodium with sulfur. The battery has a high energy density, a good efficiency of charge and discharge (90 per cent), a long cycle life, and is fabricated using inexpensive materials. Molten sodium metal forms the anode and is separated from the cathode (steel in contact with sulfur absorbed into a porous carbon) by a β-alumina solid electrolyte. Sodium β-alumina is an ionic conductor, but a poor electrical conductor, so avoiding self-discharge of the battery. When the battery is discharging Na gives up an electron to the external circuit and the resulting Na⁺ ions migrate through the sodium β-alumina to the sulfur container. At the cathode, electrons from the external circuit react with sulfur to form sodium polysulfides, S_n^{2-}. The overall battery discharge process is

$$2\,Na(l) + 4\,S(l) \rightarrow Na_2S_4(l) \qquad E_{cell} \approx 2.1\ V$$

During charging, the reverse process takes place and small heat losses in the system keep it at the operating temperature of 300 to 350°C. On account of the high temperature of their operation and the highly corrosive nature of the battery components, such cells are primarily suitable for large-scale, static applications rather than transport. Sodium-sulfur batteries therefore offer an energy-storage system that could be used in association with renewable energy plants that only operate at certain periods. Examples include wind farms, wave energy, and solar generation plants. At a wind farm the battery would store energy during times of high wind but low power demand and the stored energy would then be discharged from the batteries during peak load periods.

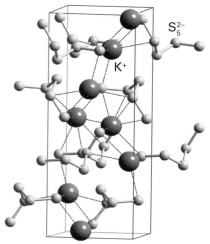

Figure 11.10 The structure of K_2S_5.

solution to be used in qualitative volumetric analysis should be checked prior to its use. Concentrated MOH solutions also react slowly at room temperature (more rapidly when heated) with silicate glass to produce sodium silicate, so reactions involving their use at high temperatures should be undertaken in inert plastic laboratory-ware.

Sodium hydroxide is produced by the chloralkali industry (Box 11.2) and is used as a reagent in the organic chemical industry and in the preparation of other inorganic chemicals. It is also used in the papermaking industry and by the food industry to break down proteins. For example, olives are soaked in sodium hydroxide solution to make the skins soft enough to be edible and the Norwegian delicacy *lutefisk* has a jelly-like consistency produced by the dissolution of the proteins from dried cod. Domestic applications are based on the action of NaOH on grease and it is used extensively in oven and drain cleaners. In some 'foaming' drain cleaners it is mixed with aluminium powder. The aluminium reacts with the aqueous hydroxide ions to liberate hydrogen gas.

11.11   Compounds of oxoacids

The Group 1 elements form salts with most oxoacids. The most industrially important Group 1 salts of oxoacids are sodium carbonate, commonly called *soda ash*, and sodium hydrogen carbonate, commonly known as *sodium bicarbonate*.

(a) Carbonates

Key point: The Group 1 carbonates are soluble and decompose to the oxide when heated strongly.

The Group 1 elements form the only soluble carbonates (with the exception of the NH_4^+ ion), although lithium carbonate is only sparingly soluble.

Sodium carbonate has been produced by the *Solvay process* for many years. The overall reaction, which uses the commonly available feedstocks of NaCl and $CaCO_3$, can be represented by the equilibrium

$$2\,NaCl(aq) + CaCO_3(s) \rightleftharpoons Na_2CO_3(s) + CaCl_2(aq)$$

However, the equilibrium lies to the left on account of the high lattice energy of $CaCO_3$, and the actual process uses a complex stepwise route involving ammonia. Calcium oxide, produced by the thermal decomposition of calcium carbonate, is reacted with ammonium chloride to generate ammonia

$$2\,NH_4Cl(s) + CaO(s) \rightarrow 2\,NH_3(g) + CaCl_2(s) + H_2O(l)$$

Ammonia and carbon dioxide (from the thermal decomposition of $CaCO_3$ and $NaHCO_3$) are passed into a saturated sodium chloride solution to form a solution of NH_4^+, Na^+, Cl^-, and HCO_3^- ions:

$$NaCl(aq) + CO_2(g) + NH_3(g) + H_2O(l) \rightarrow$$
$$Na^+(aq) + HCO_3^-(aq) + NH_4^+(aq) + Cl^-(aq)$$

When cooled to below 15°C, $NaHCO_3$ precipitates from the solution, is filtered off, and is then heated to produce the desired Na_2CO_3, with evolution of CO_2. The residual NH_4Cl is isolated and reused in the initial reaction stage with $CaCO_3$. The process is energy intensive and produces large amounts of $CaCl_2$ as a byproduct. These problems mean that Na_2CO_3 is mined wherever sources of the mineral *trona*, sodium sesquicarbonate ($Na_3(CO_3)(HCO_3).2H_2O$), exist.

The main uses of sodium carbonate are in glass manufacture, where it is heated with silica to form sodium silicate, $Na_2O.xSiO_2$, and as a water softener, where it removes Ca^{2+} ions as the calcium carbonate, the 'scale' formed in kettles in hard water areas.

Potassium carbonate is produced by treating KOH with carbon dioxide and is used in glass and ceramics manufacture.

Lithium carbonate decomposes when heated above 650°C:

$$Li_2CO_3(s) \xrightarrow{\Delta} Li_2O(s) + CO_2(g)$$

The carbonates of the heavier elements only decompose significantly when heated above 800°C. This stabilizing influence of a large cation on a large anion can be explained in terms of trends in lattice energies and was discussed in Section 3.15.

EXAMPLE 11.4 **Predicting the thermal stabilities of carbonates**

Justify the remark that the thermal stabilities of carbonates increase down Group 1.

Answer Once again, we need to focus on lattice enthalpies. To identify a trend we can use the Kapustinskii equation (eqn 3.4) to estimate the difference between the lattice enthalpies of Na_2CO_3 and Na_2O and then Cs_2CO_3 and Cs_2O. Ionic radii are given in Table 11.1; the ionic and thermochemical radii of the oxide and carbonate ions are 126 pm and 185 pm, respectively. Substitution of the data into eqn 3.4 then gives the following values:

	Na_2CO_3	Na_2O	Rb_2CO_3	Rb_2O
ΔH_L / (kJ mol^{-1})	2246	2732	1954	2316
Difference/(kJ mol^{-1})		486		362

This calculation shows that the differences between the lattice enthalpies of the carbonate and the oxide decrease down the group, suggesting that there is a lower thermodynamic tendency for the carbonate to form the oxide on descending the group (entropy effects being supposed similar). The decomposition temperature also increases: sodium carbonate begins to decompose above 800°C whereas Rb_2CO_3 requires heating to near 1000°C

Self-test 11.4 Sketch a thermochemical cycle for the decomposition of a Group 1 carbonate into the oxide and carbon dioxide.

(b) Hydrogencarbonates

Key points: Sodium hydrogencarbonate is less soluble than sodium carbonate and liberates CO_2 when heated.

Sodium hydrogencarbonate (sodium bicarbonate) is less soluble than sodium carbonate in water and can be prepared by bubbling carbon dioxide through a saturated solution of the carbonate.

$$Na_2CO_3(aq) + CO_2(g) + H_2O(l) \rightarrow 2NaHCO_3(s)$$

The reverse of this reaction occurs when the hydrogencarbonate is heated:

$$2NaHCO_3(s) \rightarrow Na_2CO_3(s) + CO_2(g) + H_2O(l)$$

This reaction provides the basis for the use of sodium hydrogencarbonate as a fire extinguisher. The powdered salt smothers the flames and decomposes in the heat to liberate carbon dioxide and water, which themselves act as extinguishers. This reaction is also the basis for the use of sodium hydrogencarbonate in baking, when the carbon dioxide and water vapour released during the baking process cause the dough to rise. A more effective raising agent is baking powder, in which sodium hydrogencarbonate is mixed with calcium dihydrogenphosphate:

$$2NaHCO_3(s) + Ca(H_2PO_4)_2(s) \rightarrow Na_2HPO_4(s) + CaHPO_4(s) + 2CO_2(g) + 2H_2O(l)$$

Potassium hydrogencarbonate is used as a buffer in wine production and in water treatment. It is also used as a buffer in low pH liquid detergents, as an additive in soft drinks, and as an antacid to combat indigestion.

(c) Other oxosalts

Key point: The nitrates of Group 1 elements are used as fertilizers and explosives.

Sodium sulfate, Na_2SO_4, is very soluble and readily forms hydrates. The major commercial source of sodium sulfate is as a byproduct of the production of hydrochloric acid from sodium chloride:

$$NaCl(aq) + H_2SO_4(aq) \rightarrow Na_2SO_4(aq) + 2HCl(aq)$$

It is also obtained as a byproduct of several other industrial processes, including flue-gas desulfurization and the manufacture of rayon. The principal use of sodium sulfate is in processing wood pulp for making the tough brown paper used in packaging and cardboard. During the process, the sodium sulfate is reduced to sodium sulfite, which dissolves the lignin in the wood. (The lignin is recovered from the pulp and used as an adhesive and binder.) It is also used in glass manufacture, in detergents, and as a mild laxative.

Sodium nitrate, $NaNO_3$, is deliquescent and is used in making other nitrates, fertilizers, and explosives. Potassium nitrate, KNO_3, occurs naturally as the mineral saltpetre. It is slightly soluble in cold water and very soluble in hot water. It has been used extensively in the manufacture of gunpowder since about the 12th century and is used in explosives, fireworks, matches, and fertilizers.

EXAMPLE 11.5 **Applying thermogravimetric analysis to study the decomposition of alkali metal nitrates**

When heated above 600°C a sample of lithium nitrate, $LiNO_3$, of mass 100.0 mg loses 71.76 per cent of its mass in a single stage whereas potassium nitrate heated to the same temperature loses mass in two stages, with respective total mass losses of 15.82 per cent (at 350°C) and 53.42 per cent (above 450°C) of the original sample. Determine the compositions of the various products formed in the decompositions of potassium and lithium nitrates.

Answer We need to consider the changes in molar mass that the data represent and then identify the corresponding empirical formulas. The molar mass of $LiNO_3$ is 68.95 g mol^{-1}, so 100.0 mg corresponds to (100.0 mg /68.95 g mol^{-1}) = 1.450 mmol $LiNO_3$. Because 1 mol $LiNO_3$ produces 1 mol of solid lithium-containing decomposition product X, 1.450 mmol $LiNO_3$ produces 1.450 mmol X. However, we know that the mass of X produced is 28.24 mg. Therefore, its molar mass is (28.24 mg)/(1.450 mmol) = 19.48 g mol^{-1}. This molar mass corresponds to the empirical formula $LiO_{0.5}$ (or Li_2O) and results from the loss of $NO_2(g)$ and $O_2(g)$ with the overall equation for the decomposition of lithium nitrate as

$$LiNO_3(s) \rightarrow \tfrac{1}{2}Li_2O(s) + NO_2(g) + \tfrac{1}{4}O_2(g)$$

Similar calculations for KNO_3 show the initial mass loss corresponds to the formation of KNO_2 (potassium nitrite) at 350°C and K_2O at 450°C with the sequential reactions

$$KNO_3(s) \rightarrow KNO_2(s) + \tfrac{1}{2}O_2(g)$$

$$2\,KNO_2(s) \rightarrow K_2O\,(s) + 2\,NO(g) + \tfrac{1}{2}O_2(g)$$

The decompositions of lithium and potassium nitrates to their oxides proceed by different routes, which is another example of the atypical behaviour of lithium for this group. The larger alkali metal cations stabilize the NO_2^- ion against its immediate decomposition to oxide. A similar difference in decomposition route and temperature occurs for lithium carbonate, which is the only alkali metal carbonate to decompose readily on heating.

Self-test 11.5 Use similar arguments to rationalize the different decomposition temperatures of the two alkali metal nitrates to the final product.

11.12 Nitrides and carbides

Key point: Only Li forms a nitride and a carbide by direct reaction with nitrogen and carbon, respectively.

Although Li is the least reactive of the Group 1 metals, it is the only one that (like Mg) forms a nitride (which is normally red) by direct reaction with nitrogen:

$$6\,Li(s) + N_2(g) \rightarrow 2\,Li_3N(s)$$

The structure of lithium nitride (Figure 11.11) consists of sheets of composition Li_2N, containing 6-coordinate Li^+ ions that are separated by other Li^+ ions. The Li^+ ions in solid lithium nitride are highly mobile, as there are vacant sites in the structures on to which lithium ions can hop, and it is therefore classified as a 'fast ion conductor'. It is being studied as a solid electrolyte and as a possible anode material for use in rechargeable batteries.

Lithium nitride also shows potential as a hydrogen storage material (Box 10.4). It stores up to 11.5 per cent by mass of hydrogen when exposed to hydrogen gas at elevated temperatures and pressures. The Li_3N reacts with hydrogen to form $LiNH_2$ and LiH in a reversible reaction:

$$Li_3N(s) + 2\,H_2(g) \rightleftharpoons LiNH_2(s) + 2\,LiH(s)$$

When heated to 1700°C the $LiNH_2$ and LiH react together to form Li_3N and liberate hydrogen.

Sodium nitride has recently been synthesized by deposition of Na and N atoms on a cooled sapphire surface at liquid nitrogen temperatures. Its structure is analogous to the ReO_3 structure type (Section 24.8), with N^{3-} replacing Re(VI) and Na^+ replacing O^{2-}. The other Group 1 elements do not form nitrides, although the azides, which contain the N_3^- ion, can be obtained by the reaction

$$2\,NaNH_2(s) + N_2O(g) \rightarrow NaN_3(s) + NaOH(s) + NH_3(g)$$

Lithium reacts directly with carbon at high temperatures to form a carbide of the stoichiometry Li_2C_2, which contains the dicarbide (acetylide) anion C_2^{2-}. The other alkali metals do not form carbides by direct reaction of the elements although ionic compounds of the stoichiometry M_2C_2 are obtained by heating the metal in ethyne. Potassium, Rb, and Cs react with graphite at low temperatures to form intercalation compounds such as C_8K (Section 14.13). Lithium may be inserted into graphite electrochemically to produce LiC_6, which has an important role in some rechargeable lithium battery systems (Box 11.1). The alkali metals Na to Cs also react with fullerene, C_{60}, to form fullerides such as Na_2C_{60}, Cs_3C_{60}, and K_6C_{60}, which contain the alkali metal cation and a fulleride anion, C_{60}^{n-}. The structure of

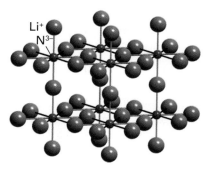

Figure 11.11 The structure of Li_3N.

K_3C_{60} is described in Section 14.6 and contains K^+ ions in all the octahedral and tetrahedral holes of a close-packed array of C_{60}^{3-} anions; this material becomes superconducting below 30 K.

EXAMPLE 11.6 Applying NMR to study Group 1 compounds

All the Group 1 elements have quadrupolar nuclei, for instance $I(^{23}Na) = \tfrac{3}{2}$ and $I(^{133}Cs) = \tfrac{7}{2}$. However, NMR spectra, including solid-state MAS-NMR spectra (Section 8.5), can be obtained for such nuclei, particularly if they are in high symmetry environments. The ^{23}Na-NMR spectrum of the fulleride Na_3C_{60} (which is obtained by reacting sodium metal with the fullerene C_{60}) shows two resonances at 170 K which coalesce when the spectrum is collected above room temperature. Interpret this information and describe how the structure of Na_3C_{60} is related to that of solid C_{60}.

Answer The two resonances in the low temperature spectrum indicate that the compound contains two different environments for Na. We know that C_{60} adopts a structure formed by cubic close-packing of the C_{60} molecules (Section 3.9). In the reaction with sodium metal, the C_{60} molecules are reduced to anions; the small Na^+ cations can occupy all the available tetrahedral and octahedral holes of a slightly expanded, but still close-packed, array of C_{60}^{3-} anions. Each type of hole corresponds to one of the environments detected by NMR.

Self-test 11.6 Predict the ^7Li-NMR spectrum of Li_3N at high and low temperatures assuming that a high-resolution spectrum can be obtained for this nucleus.

11.13 Solubility and hydration

Key points: There is wide variation in the solubility of the common salts; only Li and Na form hydrated salts.

All the common salts of the Group 1 elements are soluble in water. The solubilities cover a wide range of values, some of the most soluble being those for which there is the greatest difference between the radii of the cation and anion. Thus, the solubilities of the Li halides increase from the fluoride to the bromide whereas for Cs the trend is reversed. The explanation for these trends was discussed in Section 3.15.

Not all alkali metal salts occur as their hydrates. The lattice enthalpies of hydrated salts are lower than for the anhydrous salt because the radius of the cation is effectively increased by the hydration sphere and is further from its surrounding anions. The hydrated salt will be favoured if this decrease in lattice enthalpy is offset by the hydration enthalpy. The hydration enthalpy depends on the ion−dipole interaction between the cation and the polar water molecule. This interaction is greatest when the cation has high charge density. The Group 1 metal cations have low charge density on account of their large radii and low charge. Consequently, most of their salts are anhydrous. There are a few exceptions for the smaller Li^+ and Na^+ ions, for example $LiOH \cdot 8H_2O$ and $Na_2SO_4 \cdot 10H_2O$ (Glauber's salt).

11.14 Solutions in liquid ammonia

Key points: Sodium dissolves in liquid ammonia to give a solution that is blue when dilute and bronze when concentrated.

Sodium dissolves in pure, anhydrous liquid ammonia (without hydrogen evolution) to give solutions that are deep blue when dilute. The colour of these **metal–ammonia solutions** originates from the tail of a strong absorption band that peaks in the near infrared.[1] The dissolution of sodium in liquid ammonia to give a very dilute solution is represented by the equation

$$Na(s) \rightarrow Na^+(am) + e^-(am)$$

These solutions survive for long periods at the temperature of boiling ammonia ($-33°C$) and in the absence of air. However, they are only metastable and their decomposition is catalysed by some d-block compounds:

$$Na^+(am) + e^-(am) + NH_3(l) \rightarrow NaNH_2(am) + \tfrac{1}{2}H_2(g)$$

Concentrated metal–ammonia solutions have a metallic bronze colour and have electrical conductance close to that of a metal. These solutions have been described as 'expanded metals' in which $e^-(am)$ associates with the ammoniated cation. This description is supported by the fact that, in saturated solutions, the ammonia-to-metal ratio is between 5 and 10, which corresponds to a reasonable coordination number for the metal.

The blue metal–ammonia solutions are excellent reducing agents. For example, the Ni(I) complex $[Ni_2(CN)_6]^{4-}$, in which nickel is in an unusually low oxidation state, Ni(I), may be prepared by the reduction of Ni(II) with potassium in liquid ammonia:

$$2K_2[Ni(CN)_4] + 2K^+(am) + 2e^-(am) \rightarrow$$
$$K_4[Ni_2(CN)_6](am) + 2KCN$$

The reaction is performed in the absence of air in a vessel cooled to the boiling point of ammonia. Other reactions of M(am) as a strong reducing reagent include the formation of graphitic intercalates (Section 14.5), fullerides (Section 14.6), and Zintl phases (Section 11.15) by, for example, the following reactions

$$8C(graphite) + K^+(am) + e^-(am) \rightarrow [K(am)]^+[C_8]^-(s)$$

$$C_{60}(s) + 3Rb^+(am) + 3e^-(am) \rightarrow [Rb(am)]_3C_{60} \xrightarrow{\Delta} Rb_3C_{60}(s)$$

The alkali metals also dissolve in ethers and alkylamines to give solutions with absorption spectra that depend on the identity of the metal. The dependence on the metal suggests that the spectrum is associated with charge transfer from an *alkalide ion*, M^- (such as a sodide ion, Na^-), to the solvent. When ethylenediamine (1,2-diaminoethane, en) is used as a solvent the dissolution equation is written as

$$2Na(s) \rightarrow Na^+(en) + Na^-(en)$$

[1] Other electropositive metals with low enthalpies of sublimation, including Ca and Eu, dissolve in liquid ammonia to give solutions with a blue colour that is independent of the metal.

Further evidence for the occurrence of alkalide ions is the diamagnetism associated with the species assigned as M^-, which would have the spin-paired ns^2 valence-electron configuration. Another observation in agreement with this interpretation is that, when sodium/potassium alloy is dissolved, the metal-dependent band is the same as for solutions of Na itself.

$$NaK(l) \rightarrow K^+(en) + Na^-(en)$$

11.15 Zintl phases containing alkali metals

Key point: The alkali metals reduce the Group 13 to 16 metals to produce Zintl phases containing polymeric anions.

Zintl phases are formed when a Group I element is combined with a p-block metal from Groups 13 to 16. Alkali metal solutions in liquid ammonia are strong reducing agents and react with the metal to form such phases. Alternatively, Zintl phases can be obtained by direct reaction of the Group 1 element and the p-block element at high temperatures. Group 1 Zintl phases are ionic compounds in which electrons are transferred from the alkali metal atom to a cluster of the p-block atoms to form a polyanion; these compounds are normally diamagnetic, semiconducting or poor conductors, and brittle.

With the Group 14 elements (E), compounds of stoichiometry M_4E_4, which contain a tetrahedral E_4^{4-} anion, and M_4E_9, for example Cs_4Ge_9 containing the monocapped square antiprismatic Ge_9^{4-} anion, can be obtained (Fig. 11.12). For Group 13, compounds such as Rb_2In_3 (containing In_6 octahedra) and KGa (with Ga_8 polyhedral anions) are known. The compound Cs_5Bi_4 contains tetrameric chains of stoichiometry Bi_4^{4-}. Even more exotic Zintl phases obtained with Group 1 elements include the fullerene-type structures of $Na_{96}In_{91}M_2$ and $Na_{172}In_{192}M_2$, M = Ni, Pd, Pt (Fig. 11.13).

11.16 Coordination compounds

Key point: The Group 1 elements form stable complexes with polydentate ligands.

The Group 1 ions, particularly Li^+ to K^+, are hard Lewis acids (Section 4.12). Therefore, most of the complexes they form arise from Coulombic interactions with small, hard donors, such as those possessing O or N atoms. Monodentate ligands are only

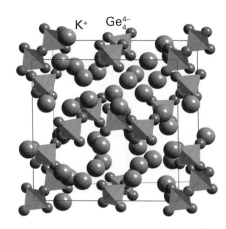

Figure 11.12 The structure of K_4Ge_4.

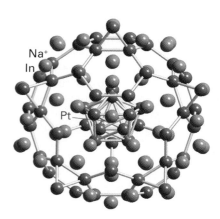

Figure 11.13 Part of the structure of $Na_{172}In_{192}Pt_2$ showing the complex fulleride-like networks formed by the In atoms around Na ions.

1 18-crown-6

2 2.2.1-crypt

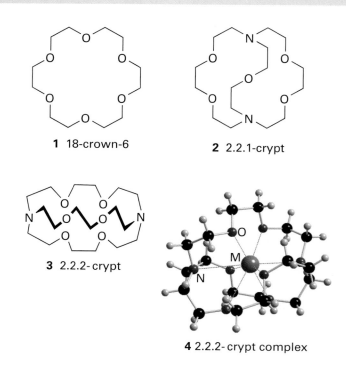

3 2.2.2- crypt

4 2.2.2- crypt complex

weakly bound on account of the weak Coulombic interactions and lack of significant covalent bonding by these ions. However, a number of factors (such as the formation of peroxides and ozonides, rather than oxides, by the heavier alkali metals and the insolubility of their perchlorates) indicate that the metals become less hard as the group is descended.

The H_2O ligands in $M(OH_2)_n^+$ species readily exchange with the surrounding H_2O molecules of the solvent, although this is slowest for the very hard Li^+ ion and faster for the increasingly less hard Rb^+ and Cs^+ ions. Chelating ligands such as the ethylenediaminetetraacetate ion, $(O_2C)_2NCH_2CH_2N(CO_2)_2^{4-}$, have much higher formation constants, particularly with the larger alkali metal cations. Macrocycles and related ligands form the most stable complexes. Crown ethers such as 18-crown-6 (**1**) form complexes with alkali metal ions that are reasonably stable in nonaqueous solution. Bicyclic cryptand ligands, such as 2.2.1-crypt (**2**) and 2.2.2-crypt (**3**), form complexes with alkali metals that are even more stable, and they can survive even in aqueous solution (**4**). These ligands are selective for a particular metal ion, the dominant factor being the fit between the cation and the cavity in the ligand that accommodates it (Fig. 11.14).

Another example of this fit between cation and ligand cavity is thought to be responsible for the transport of Na^+ and K^+ ions across cell membranes (Section 27.3). The ions cross the hydrophobic cell membrane by means of embedded protein molecules

that contain cavities lined with donor atoms. The donor atoms are arranged to form a cavity, the size of which determines whether Na^+ or K^+ is bound. Such *ion channels* modulate the Na^+/K^+ concentration differential across the cell membrane that is essential for particular functions of the cell. The naturally occurring molecule valinomycin (**5**) is an antibiotic that selectively coordinates K^+: the resulting hydrophobic 1:1 complex transports K^+ through a bacterial cell membrane, depolarizing the ion differential and resulting in cell death.

The complexation of sodium with a cryptand can be used to prepare solid sodides, such as $[Na(2.2.2)]^+Na^-$, where (2.2.2) denotes the cryptand ligand. X-ray structure determination reveals the presence of $[Na(2.2.2)]^+$ and Na^- ions, with the latter located in a cavity of the crystal with an apparent radius larger than that of I^-. The precise nature of the products of this reaction varies with the ratio of sodium to cryptand. It is also possible to crystallize solids containing solvated electrons, the so-called **electrides**, and to obtain their X-ray crystal structures. Figure 11.15, for example, shows the inferred position of the

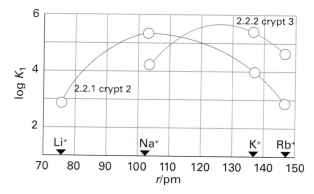

Figure 11.14 The formation constants of complexes of Group 1 metals with cryptand ligands plotted against cation size. Note that the smaller 2.2.1 crypt favours complex formation with Na^+ and the larger 2.2.2 crypt favours K^+.

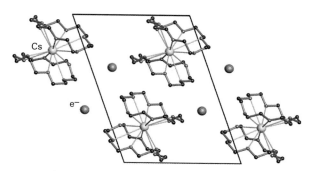

Figure 11.15 The crystal structure of $[Cs(18\text{-crown-6})_2]^+e^-$. The pink spheres mark the sites of highest electron density and so indicate the locations of the 'anion' e^-. (From S.B. Dawes, D.L. Ward, R.H. Huang, and J.J. Dye, *J. Am. Chem. Soc.*, 1986, **108**, 3534.)

5 Valinomycin

maxima of the electron density in such a solid. The preparation of sodides and other alkalides demonstrates the powerful influence of solvents and complexing agents on the chemical properties of metals. A further example of these influences is the ability of crown ethers to produce reactive Cl^- ions in organic solvents. When an aqueous solution of NaCl is shaken in a separating funnel with a solution of 18-crown-6 in organic solvent, the Na^+ ions cross into the organic phase, drawing Cl^- ions with them. The poorly solvated Cl^- ions are highly reactive.

11.17 Organometallic compounds

Key point: The organometallic compounds of the Group 1 elements react rapidly with water and are pyrophoric.

Group 1 elements form a number of organometallic compounds that are unstable in the presence of water and are pyrophoric in air. They are prepared in organic solvents such as tetrahydrofuran (THF). Protic (proton-donating) organic compounds form ionic organometallic compounds with the Group 1 metals. For example, cyclopentadiene reacts with sodium metal in THF:

$$Na(s) + C_5H_6(l) \rightarrow Na^+[C_5H_5]^- (sol) + \tfrac{1}{2}H_2(g)$$

The resulting cyclopentadienide anion is an important intermediate in the synthesis of d-block organometallic compounds (Chapter 22). Lithium, sodium, and potassium form intensely coloured compounds with aromatic species. The oxidation of the metal results in transfer of an electron to the aromatic system to produce a **radical anion**, an anion that possesses an unpaired electron:

$$Na + \text{(naphthalene)} \longrightarrow Na^+[C_{10}H_8]^-$$

Sodium and potassium alkyls are colourless solids that are insoluble in organic solvents and, when stable, have fairly high melting temperatures. They are produced by a **transmetallation reaction**, which involves breaking a metal−carbon bond and forming a metal−carbon bond to a different metal. Alkylmercury compounds are often the starting materials in these reactions. For example, methylsodium is produced in the reaction between sodium metal and dimethylmercury in a hydrocarbon solvent:

$$Hg(CH_3)_2 + 2Na \rightarrow 2NaCH_3 + Hg$$

Organolithiums are by far the most important Group 1 organometallic compounds. They are liquids or low melting solids, are the most thermally stable of the entire group, and are soluble in organic and nonpolar solvents such as THF. They can be synthesized from an alkyl halide and lithium metal or by reacting the organic species with butyllithium, $Li(C_4H_9)$, commonly abbreviated to BuLi.

$$BuCl + 2Li \rightarrow BuLi + LiCl$$
$$BuLi + C_6H_6 \rightarrow Li(C_6H_5) + C_4H_{10}$$

A feature of many main-group organometallic compounds is the presence of bridging alkyl groups. When ethers are the solvent, methyl lithium exists as $Li_4(CH_3)_4$, with a tetrahedron of Li atoms and bridging CH_3 groups (6). In hydrocarbon solvents, $Li_6(CH_3)_6$ (7) is formed; its structure is based on an octahedral arrangement of Li atoms. Other alkyllithiums adopt similar structures except when the alkyl groups become very bulky, as in the case of t-butyl, $-C(CH_3)_3$, when tetramers are the largest species formed. Many of these alkyllithiums are electron-deficient compounds and contain the 3c,2e bonds characteristic of such compounds (Section 2.11).

Organolithium compounds are very important in organic synthesis, the most important reactions being those in which they act as nucleophiles and attack, for example, a carbonyl group:

$$R^- \quad \diagdown C=O \longrightarrow R-C-O^-$$

Organolithium compounds are also used to convert p-block halides to organoelement compounds, as we see in later chapters. For example, boron trichloride reacts with butyllithium in THF to give an organoboron compound:

$$BCl_3 + 3BuLi \rightarrow Bu_3B + 3LiCl$$

The driving force for this and many other reactions of s- and p-block organometallic compounds is the formation of the insoluble halide compound of the less electronegative metal.

Alkyllithiums are important industrially in the stereospecific polymerization of alkenes to form synthetic rubber. Butyllithium is used as an initiator in solution polymerization to produce a wide range of elastomers and polymers. Organolithium compounds are also used in the synthesis of a range of pharmaceuticals, including vitamins A and D, analgesics, antihistamines, antidepressants, and anticoagulants. Alkyllithiums can be used in the synthesis of other organometallic compounds. For example, they can be used

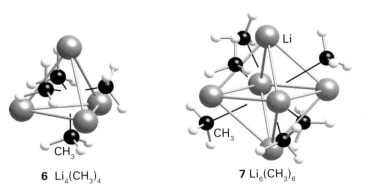

6 $Li_4(CH_3)_4$ **7** $Li_6(CH_3)_6$

to introduce alkyl groups into d-metal organometallic compounds (Section 22.8):

$$(C_5H_5)_2MoCl_2 + 2CH_3Li \rightarrow (C_5H_5)_2Mo(CH_3)_2 + 2LiCl$$

The reactivity and solubility of alkyllithium is enhanced by adding a chelating ligand such as tetramethylethylenediamine, TMEDA (8), which breaks up any tetramers to give complexes such as [BuLi(TMEDA)]$_2$.

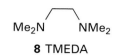

8 TMEDA

FURTHER READING

R.B. King, *Inorganic chemistry of the main group elements*. John Wiley & Sons (1994).

P. Enghag, *Encyclopedia of the elements*. John Wiley & Sons (2004).

D.M.P. Mingos, *Essential trends in inorganic chemistry*. Oxford University Press (1998). A survey of inorganic chemistry from the perspective of structure and bonding.

V.K. Grigorovich, *The metallic bond and the structure of metals*. Nova Science Publishers (1989).

N.C. Norman, *Periodicity and the s- and p-block elements*. Oxford University Press (1997). Includes coverage of essential trends and features of s-block chemistry.

A. Sapse and P.V. Schleyer (ed.), Lithium chemistry: a theoretical and experimental overview. John Wiley & Sons (1995).

EXERCISES

11.1 Why are Group 1 elements (a) strong reducing agents, (b) poor complexing agents?

11.2 Use the data given in Tables 11.1 and 11.3 to calculate the enthalpies of formation for the Group 1 fluorides and chlorides. Plot the data and comment on the trends observed.

11.3 Give equations for the synthesis of (a) C_2H_5Li, (b) C_2H_5Na.

11.4 Which of the following pairs are most likely to form the desired compound? Describe the periodic trend and the physical basis for your answer in each case. (a) acetate ion or edta^{4-} ion with Cs$^+$, (b) Li$^+$ or K$^+$, to form a complex with crypt 2.2.2.

11.5 Identify compounds A, B, C, and D in the following array of reactions when (i) M is Li and (ii) M is Cs.

$$A \xleftarrow{H_2O} M \xrightarrow{O_2} B \xrightarrow{\Delta} C$$
$$\downarrow NH_3$$
$$D$$

11.6 Account for the fact that LiF and CsI have low solubility in water whereas LiI and CsF are very soluble.

11.7 Explain why LiH has greater thermal stability than the other Group 1 hydrides whereas Li_2CO_3 decomposes at a lower temperature than the other Group 1 carbonates.

11.8 Draw the structures of NaCl and CsCl, and give the coordination number of the metal in each case. Explain why the compounds adopt different structures.

11.9 Explain how the nature of the alkyl group affects the structure of lithium alkyls.

11.10 Predict the products of the following reactions:

(a) $CH_3Br + Li \rightarrow$
(b) $MgCl_2 + LiC_2H_5 \rightarrow$
(c) $C_2H_5Li + C_6H_6 \rightarrow$

PROBLEMS

11.1 Describe the origin of the diagonal relationship between Li and Mg.

11.2 Identify the incorrect statement or statements in the following description and provide corrections and an explanation of the trend. (a) Sodium dissolves in ammonia and amines to produce the sodium cation and solvated electrons or the sodide ion. (b) Sodium dissolved in liquid ammonia will not react with NH_4^+ because of strong hydrogen bonding with the solvent.

11.3 Z. Jedlinski and M. Sokol describe the solubility of alkali metals in nonaqueous supramolecular systems (*Pure Appl. Chem.*, 1995, **67**, 587). They dissolved the metals in THF containing crown ethers or cryptands. Sketch the structure of the 18-crown-6 ligand. Give the equations proposed for the dissolution process. Outline the two methods used to prepare the alkali metal solutions. What factors affect the stability of the solutions?

11.4 Alkali metal halides can be extracted from aqueous solution by solid-phase ditopic salt receptors. (See J.M. Mahoney, A.M. Beatty, and B.D. Smith, *Inorg. Chem.*, 2004, **43**, 7617.) (a) What is a ditopic receptor? (b) What is the order of selectivity to extraction of the alkali metal ions in aqueous solution? (c) What is the order of selectivity to extraction from the solid phase? (d) Explain the observed order of selectivity.

11.5 The molecular geometries of crown ether derivatives play an important role in capturing and transporting alkali metal ions. K. Okano and co-workers (see K. Okano, H. Tsukube, and K. Hori, *Tetrahedron*, 2004, **60**, 10877) studied stable conformations of 12-crown-O3N and its Li$^+$ complex in aqueous and acetonitrile solutions. (a) Which three programs did the authors use in their study and what did each program calculate? (b) Which Li$^+$ complex was found to be most stable in (i) aqueous and (ii) acetonitrile solutions?

The Group 2 elements

12

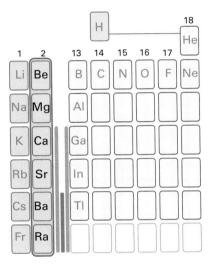

In this chapter we look at the occurrence and isolation of the Group 2 elements and study the chemical properties of their simple compounds, complexes, and organometallic compounds. Throughout the chapter we draw comparisons with the elements of Group 1 and show how the chemical properties of beryllium differ from those of the other Group 2 elements. We see how the insolubility of some of the calcium compounds in particular leads to the existence of many inorganic minerals that provide the raw materials for the infrastructure of our built environment and provide the building blocks for the compounds from which many rigid biological structures are formed.

The elements calcium, strontium, barium, and radium are known as the **alkaline earth metals,** but the term is often applied to the whole of Group 2. All the elements are silvery white metals and the bonding in their compounds is normally described in terms of the ionic model (Section 3.9). Some aspects of the chemical properties of beryllium are more like those of a metalloid with a degree of covalence in its bonding. The elements are denser, harder, and less reactive than the elements of Group 1 but are still more reactive than many typical metals. The lighter elements beryllium and magnesium form a number of complexes and organometallic compounds.

PART A: THE ESSENTIALS

In this first section of the chapter we summarize the key features of the chemistry of the Group 2 elements.

12.1 The elements

Key point: The most important factors influencing the chemical properties of the Group 2 elements are their ionization energies and ionic radii.

Beryllium occurs naturally as the semiprecious mineral beryl, $Be_3Al_2(SiO_3)_6$. Magnesium is the eighth most abundant element in the Earth's crust and the third most abundant element dissolved in seawater; it is commercially extracted from seawater and the mineral dolomite, $CaCO_3.MgCO_3$. Calcium is the fifth most abundant element in the Earth's crust but only the seventh most common in seawater due to the low solubility of $CaCO_3$; it occurs widely in its carbonate as limestone, marble, and chalk, and it is a major component of biominerals, such as shells and coral. Calcium, strontium, and barium are all extracted by electrolysis of their molten chlorides. Radium can be extracted from uranium-bearing minerals although all its isotopes are radioactive.

Table 12.1 Selected properties of the Group 2 elements

	Be	Mg	Ca	Sr	Ba	Ra
Metallic radius/pm	112	150	197	215	217	220
Ionic radius, $r(M^{2+})$/pm (coordination number)	27(4)	72(6)	100(6)	126(8)	142(8)	170(12)
First ionization energy, I/(kJ mol^{-1})	900	736	590	548	502	510
$E^{\ominus}(M^{2+}, M)$/V	-1.85	-2.38	-2.87	-2.89	-2.90	-2.92
Density, ρ/(g cm^{-3})	1.85	1.74	1.54	2.62	3.51	5.00
Melting point/°C	1280	650	850	768	714	700
$\Delta_{hyd}H^{\ominus}$/(kJ mol^{-1})	-2500	-1920	-1650	-1480	-1360	–
$\Delta_{sub}H^{\ominus}$/(kJ mol^{-1})	321	150	193	164	176	130

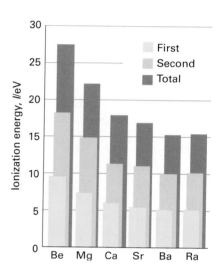

Figure 12.1 The variation of first, second, and total (first plus second) ionization energies in Group 2.

The greater mechanical hardness and higher melting points of the Group 2 compared with the Group 1 elements indicates an increase in the strength of metallic bonding on going from Group 1 to Group 2, which can be attributed to the increased number of electrons available (Section 3.19). The atomic radii of the Group 2 elements are smaller than those of Group 1. This reduction between the groups is responsible for their higher densities and ionization energies (Table 12.1). The ionization energies of the elements decrease down the group as the radius increases (Fig. 12.1), and the elements become more reactive and more electropositive as it becomes easier to form the +2 ions. This decrease in ionization energy is reflected in the trend in standard potentials for the M^{2+}/M couples, which become more negative down the group. Thus, whereas calcium, strontium, barium, and radium react readily with cold water, magnesium reacts only with hot water:

$$M(s) + 2\,H_2O(l) \rightarrow M(OH)_2(aq) + H_2(g)$$

All the elements occur as hexagonal closed-packed structures with the exception of barium and radium, which adopt the more open body-centred cubic structure. The density decreases from Be to Mg to Ca (in contrast to the lighter Group 1 elements) as a result of very strong metallic bonding in the Group 2 elements, which leads to short metal–metal distances in the lighter elements (225 pm in beryllium, for instance) and as a result small unit cells. Beryllium is inert in air as its surface is passivated by the formation of a thin layer of BeO. Magnesium and calcium metals also tarnish in air with the formation of an oxide layer, but will burn completely to their oxides and nitrides when heated. Strontium and barium, especially in powdered forms, ignite in air and are stored under hydrocarbon oils.

As with the Group 1 elements (Section 11.1), flame tests are commonly used for the identification of the presence of the heavier Group 2 elements and their compounds:

Ca	Sr	Ba	Ra
orange-red	crimson	yellowish-green	deep red

Compounds of the Group 2 elements are used to colour fireworks.

12.2 Simple compounds

Key point: The binary compounds of the Group 2 metals contain the cations of the elements and exhibit predominantly ionic bonding.

All the elements occur as M(II) in their simple compounds, which is consistent with their ns^2 valence-electron configuration. Apart from Be, their compounds are predominantly ionic.

With the exception of Be, the Group 2 elements form ionic (saline) hydrides; the anion present is the hydride ion, H$^-$. In contrast, beryllium hydride adopts a three-dimensional network of linked BeH$_4$ tetrahedra. Magnesium hydride, MgH$_2$, loses hydrogen when heated above 250°C and is being studied as a hydrogen-storage material. The hydrides react with water to produce hydrogen gas.

All the elements form halides, MX$_2$, by direct combination of the elements. The halides of the elements other than Be, however, are normally formed from solution, such as by reaction of the metal hydroxide or carbonate with a hydrohalic acid (HX(aq), X = Cl, Br, I) followed by dehydration of the resulting hydrous salt. The fluorides of the larger cations (from Ca to Ba) adopt the (8,4)-coordinate fluorite structure (Fig. 12.2), but MgF_2 crystallizes with a rutile structure. The beryllium halides form covalently bonded networks of edge- or corner-linked tetrahedra.

Beryllium oxide, BeO, is a white, insoluble solid with the wurtzite structure with (4,4)-coordination, as expected for the small Be^{2+} ion; the oxides of the other Group 2 elements all adopt the rock-salt structure with (6,6)-coordination. Magnesium oxide is insoluble but reacts slowly with water to form $Mg(OH)_2$; likewise CaO reacts with water to form the partially soluble $Ca(OH)_2$. The oxides of Sr and Ba, SrO and BaO, dissolve in water to form the strongly basic hydroxide solutions:

$$BaO(s) + H_2O(l) \rightarrow Ba^{2+}(aq) + 2\,OH^-(aq)$$

Magnesium hydroxide, $Mg(OH)_2$, is basic but only very sparingly soluble; beryllium hydroxide, $Be(OH)_2$, is amphoteric and in strongly basic solutions it forms the tetrahydroxyberyllate ion, $Be(OH)_4^{2-}$:

$$Be(OH)_2(s) + 2\,OH^-(aq) \rightarrow Be(OH)_4^{2-}(aq)$$

The sulfides can be prepared by direct reaction of the elements and adopt the rock-salt structure for all except Be, which has a sphalerite structure, Fig. 3.34. Beryllium carbide, Be_2C, has an antifluorite structure formally containing Be^{2+} and C^{4-} ions. The carbides of the other members of the group have the formula MC_2 and contain the dicarbide (acetylide) anion, C_2^{2-}; they react with water to generate ethyne, C_2H_2. The elements Mg–Ra react directly with nitrogen when heated to produce the nitrides M_3N_2, which react with water to produce ammonia.

With the exception of the fluorides, the salts of singly charged anions are usually soluble in water, although beryllium salts—once again on account of the highly polarizing nature of the Be^{2+} ion—often hydrolyse in aqueous solutions with the formation of $[Be(OH_2)_3(OH)]^+$ and H_3O^+. The radium halides are the least soluble of the group halides: this property is used to extract radium using fractional crystallization. In general, the salts of the Group 2 elements are generally much less soluble in water than those of Group 1 on account of the higher lattice enthalpies of structures containing doubly charged cations, especially when they are in combination with highly charged anions: the carbonates, sulfates, and phosphates are insoluble or only sparingly soluble.

The carbonates and sulfates of the Group 2 elements have important roles in natural water systems, rock formation, and as materials for forming hard structures. The carbonates and sulfates are insoluble, as a result of the high lattice energy of structure formed from 2+ and 2− ions. The solubility of calcium carbonate increases if CO_2 is dissolved in the water, as in rainwater, due to the formation of HCO_3^- with its lower charge. 'Temporary hardness' of water is caused by the presence of magnesium and calcium hydrogencarbonates; the cations are precipitated as carbonates on boiling solutions containing the hydrogencarbonates. Calcium carbonate is widely used by living organisms in the construction of hard structural biomaterials such as shells, bones, and teeth (Chapter 27). When heated, an alkaline earth carbonate decomposes to the oxide, although for Sr and Ba this decomposition process requires temperatures above 800°C. Calcium sulfate is used widely in the construction industry (plaster) and occurs naturally as gypsum, which is the dihydrate, $CaSO_4.2H_2O$.

The Group 2 cations form complexes with charged polydentate ligands, such as the analytically important ethylenediaminetetraacetate ion (edta, see Table 7.1) and crown and crypt ligands. The most important macrocyclic complexes are the chlorophylls, which are porphyrin complexes of Mg and are involved in photosynthesis (Section 27.10d).

Beryllium forms an extensive series of organometallic compounds. Alkyl- and arylmagnesium halides are very well known as Grignard reagents and are widely used in synthetic organic chemistry, where they behave as a source of alkyl and aryl anions.

12.3 The anomalous properties of beryllium

The small size of Be^{2+} (ionic radius 27 pm) and its consequent high charge density and polarizing power results in the compounds of Be being largely covalent; the ion is a strong Lewis acid. The coordination number most commonly observed for this small atom is 4

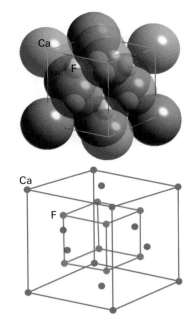

Figure 12.2 The fluorite structure adopted by CaF_2, SrF_2, BaF_2, and $SrCl_2$.

and the local geometry tetrahedral. Beryllium's larger congeners typically have coordination numbers of 6 or more. Some consequences of these properties are:

- A significant covalent contribution to the bonding in compounds such as the beryllium halides $BeCl_2$, $BeBr_2$, and BeI_2 and the hydride, BeH_2.
- A greater tendency to form complexes, with the formation of molecular compounds such as $Be_4O(O_2CCH_3)_6$.
- Hydrolysis (deprotonation) of beryllium salts in aqueous solution, forming species such as $[Be(OH_2)_3OH]^+$ and acidic solutions. Hydrated beryllium salts tend to decompose by hydrolysis reactions, where beryllium oxo- or hydroxo salts are formed, rather than by the simple loss of water.
- The oxide and other chalcogenides of Be adopt structures with the more directional (4,4)-coordination structures.
- Beryllium forms many stable organometallic compounds, including methylberyllium ($Be(CH_3)_2$), ethylberyllium, t-butylberyllium, and beryllocene ($(C_5H_5)_2Be$).

Another important general feature of Be is its strong diagonal relationship with Al (Section 9.2d):

- Both Be and Al form covalent hydrides and halides; the analogous compounds of the other Group 2 elements are predominantly ionic.
- The oxides of Be and Al are amphoteric whereas the oxides of the rest of the Group 2 elements are basic.
- In the presence of excess OH^- ions, Be and Al form $[Be(OH)_4]^{2-}$ and $[Al(OH)_4]^-$, respectively; no equivalent chemistry is observed for Mg.
- Both elements form structures based on linked tetrahedra: Be forms structures built from $[BeO_4]^{n-}$ and $[BeX_4]^{n-}$ tetrahedra (X = halide) and Al forms numerous aluminates and aluminosilicates containing the $[AlO_4]^{n-}$ unit.
- Both elements form carbides that contain the C^{4-} ion and produce methane on reaction with water; the other Group 2 carbides contain the C_2^{2-} ion and produce ethyne on reaction with water.
- The alkyl compounds of Be and Al are electron-deficient compounds that contain $M-C-M$ bridges.

There are also analogies between the chemical properties of Be and Zn. For example, Zn also dissolves in strong bases, to produce zincates, and structures containing linked $[ZnO_4]^{n-}$ tetrahedra are common.

PART B: **THE DETAIL**

In this section we present a more detailed discussion of the chemistry of the elements of Group 2 and their compounds. Because the bonding in the compounds formed by these elements is usually ionic (as always, bearing in mind the individuality of Be) we can usually interpret their properties in terms of the ionic model.

12.4 Occurrence and extraction

Key points: Magnesium is the only Group 2 element extracted on an industrial scale; magnesium, calcium, strontium, and barium can be extracted from the molten chloride.

Beryllium occurs naturally as the semiprecious mineral beryl, $Be_3Al_2(SiO_3)_6$, from which its name is taken. Beryl is the basis of the gemstone emerald, in which a small fraction of Al^{3+} is replaced by Cr^{3+}. Beryllium is extracted by heating beryl with sodium hexafluorosilicate, Na_2SiF_6, to produce BeF_2, which is then reduced to the element by magnesium.

Magnesium is the eighth most abundant element in the Earth's crust. It occurs naturally in a number of minerals such as dolomite, $CaCO_3.MgCO_3$, and magnesite, $MgCO_3$, and is the third most abundant element dissolved in seawater (after Na and Cl), from which it is commercially extracted. A litre of seawater contains more than 1 g of magnesium ions. The extraction from seawater relies on the fact that magnesium hydroxide is less soluble than calcium hydroxide because the solubility of the salts of mononegative anions increases down the group (Section 12.11). Either CaO (quicklime) or $Ca(OH)_2$ (slaked lime) is added to seawater and $Mg(OH)_2$ precipitates. The hydroxide is converted to the chloride by treatment with hydrochloric acid:

$$CaO(s) + H_2O(l) \rightarrow Ca^{2+}(aq) + 2OH^-(aq)$$
$$Mg^{2+}(aq) + 2OH^-(aq) \rightarrow Mg(OH)_2(s)$$
$$Mg(OH)_2(s) + 2HCl(aq) \rightarrow MgCl_2(aq) + 2H_2O(l)$$

The magnesium is then extracted by electrolysis of molten magnesium chloride:

Cathode: $Mg^{2+}(l) + 2e^- \rightarrow Mg(s)$ Anode: $2Cl^-(l) \rightarrow Cl_2(g) + 2e^-$

Magnesium is also extracted from dolomite. The dolomite is heated in air to give magnesium and calcium oxides. This mixture is heated with ferrosilicon (FeSi), which forms calcium silicate, Ca_2SiO_4, iron, and magnesium. Magnesium is a liquid at the high operating temperatures used in the process and can be removed by distillation.

A major problem in the production of magnesium is its high reactivity towards water and moist air. Nitrogen, which is commonly used to provide an inert atmosphere for the production of many other reactive metals, cannot be used for magnesium because it reacts to form the nitride, Mg_3N_2. Sulfur hexafluoride or sulfur dioxide added to dry air are used as alternatives to nitrogen as these gases inhibit the formation of MgO. Argon may also be used. Although magnesium is very reactive towards oxygen and water, the metal can be handled safely on account of the presence of an inert passivating oxide film on its surface.

Calcium is the fifth most abundant element in the Earth's crust and occurs widely as limestone, $CaCO_3$. The name 'calcium' comes from the Latin *calx*, which means lime. Calcium concentrations in seawater are lower than those of magnesium due to the lower solubility of $CaCO_3$ compared with $MgCO_3$ and the greater use of calcium by marine organisms. The element is a major component of biominerals such as bone, shells, and teeth, and is central to cell signalling processes such as hormonal or electrical activation of enzymes in higher organisms (Section 27.4). The average adult human contains approximately 1 kg of calcium. Calcium binds strongly to oxalate ions to form insoluble $Ca(C_2O_4)$; kidney stones are formed when this reaction occurs in the kidneys.

Calcium is extracted by electrolysis of the molten chloride, which is itself obtained as a byproduct of the Solvay process for the production of sodium carbonate (Section 11.11). Calcium tarnishes in air and ignites on heating with the formation of calcium oxide and nitride. Strontium is named after the Scottish village of Strontian where the strontium-containing ore was first found. It is extracted by electrolysis of molten $SrCl_2$ or by reduction of SrO with Al:

$$6SrO(s) + 2Al(s) \rightarrow 3Sr(s) + Sr_3Al_2O_6(s)$$

The metal reacts vigorously with water and as a finely divided powder ignites in air; initially the product is SrO but once burning the nitride, Sr_3N_2, is also formed. Barium is extracted by electrolysis of the molten chloride or by reduction of BaO with Al. It reacts very vigorously with water and ignites very readily in air.

All the isotopes of radium are radioactive. They undergo α, β, and γ decay with half-lives that vary from 42 minutes to 1599 years. Radium was discovered by Pierre and Marie Curie in 1898 after painstaking extraction from the uranium-bearing mineral pitchblende. Pitchblende is a complex mineral containing many elements: it contains approximately 1 g of Ra in 10 t of ore and the Curies took three years to isolate 0.1 g of $RaCl_2$.

12.5 Uses of the elements and their compounds

Key points: Magnesium and its compounds have major applications in pyrotechnics, alloys, and common medicines; calcium compounds are widely used in the construction industry; magnesium and calcium are very important for biological function.

Beryllium is unreactive in air on account of a passivating layer of an inert oxide film on its surface, which makes it very resistant to corrosion. This inertness, combined with the fact that it is one of the lightest metals, results in its use in alloys to make precision instruments, aircraft, and missiles. It is highly transparent to X-rays due to its low atomic number (and thus electron count) and is used for X-ray tube windows. Beryllium is also used as a moderator for nuclear reactions (where it slows down fast-moving neutrons through inelastic collisions) because the beryllium nucleus is a very weak absorber of neutrons and the metal has a high melting point.

Most of the applications of elemental magnesium are based on the formation of light alloys, especially with aluminium, that are widely used in construction in applications where weight is an issue, such as aircraft. A magnesium–aluminium alloy was previously used in warships but was discovered to be highly flammable when subjected to missile attack. Some of the uses of magnesium are based on the fact that the metal burns in air with an intense white flame, and so it is used in fireworks and flares.

As beryllium oxide is extremely toxic and carcinogenic by inhalation and soluble beryllium salts are mildly poisonous, the industrial applications of beryllium compounds are limited; BeO is used as an insulator in high-power electrical devices where high thermal conductivity is also necessary. Various applications of magnesium compounds include 'Milk of Magnesia', $Mg(OH)_2$, which is a common remedy for indigestion, and 'Epsom Salts', $MgSO_4 \cdot 7H_2O$, which is used for a variety of health treatments, including as a treatment for constipation, a purgative, and a soak for sprains and bruises. Magnesium oxide, MgO, is used as a refractory lining for furnaces. Organomagnesium compounds are widely used in organic synthesis as Grignard reagents (Section 12.13).

The compounds of calcium are much more useful than the element itself. Calcium oxide (as lime or quicklime) is a major component of mortar and cement (Box 12.1). It is also used in steelmaking and papermaking. Calcium sulfate dihydrate, $CaSO_4 \cdot 2H_2O$ is widely used in building materials, such as plasterboard, and anhydrous $CaSO_4$ is a common drying agent. Calcium carbonate is used in the Solvay process (Section 11.11) for the production of sodium carbonate (except in the USA, where sodium carbonate is mined as trona) and as the raw material for production of CaO. Calcium fluoride is insoluble and transparent over a wide range of wavelengths. It is used to make cells and windows for infrared and ultraviolet spectrometers.

Strontium is used in pyrotechnics (Box 12.2), phosphors, and in glasses for the now rapidly declining market for colour television tubes. Barium compounds, taking advantage of the large number of electrons of each Ba^{2+} ion, are very effective at absorbing X-rays: they are used as 'barium meals' and 'barium enemas' to investigate the intestinal tract. Barium is highly toxic, so the insoluble sulfate is used in this application. Barium carbonate is used in glassmaking and as a flux to aid the flow of glazes and enamels. It is also used as rat poison. The sulfide has been used as a depilatory, to remove unwanted body hair. Barium sulfate is pure white, with no absorption in the visible region of the electromagnetic spectrum, and it is used as a reference standard in UV-visible spectroscopy.

Soon after its discovery, radium was used to treat malignant tumours; its compounds are still used as precursors for radon used in similar applications. Luminous radium paint was once

BOX 12.1   Cement and concrete

Cement is made by grinding together limestone and a source of aluminosilicates, such as clay, shale, or sand, and then heating the mixture to 1500°C in a rotary cement kiln. The first important reaction to occur in the lower temperature portion of the kiln (900°C) is the calcining of limestone (the process of heating to a high temperature to oxidize or decompose a substance and convert it to a powder), when calcium carbonate (limestone) decomposes to calcium oxide (lime) and carbon dioxide is driven off. At higher temperatures the calcium oxide reacts with the aluminosilicates and silicates to form molten Ca_2SiO_4, Ca_3SiO_5, and $Ca_3Al_2O_6$. The relative proportions of these compounds determine the properties of the final cement. As the compounds cool, they solidify into a form called *clinker*. The clinker is ground to a fine powder and a small amount of calcium sulfate (gypsum) is added to form Portland cement.

Concrete is produced by mixing cement with sand, gravel, or crushed stone and water. Often small amounts of additives are added to achieve particular properties. For example, flow and dispersion are improved by adding polymeric materials such as phenolic resins, and resistance to frost damage is improved by adding surfactants. When the water is added to the cement, complex hydration reactions occur that produce hydrates such as $Ca_3Si_2O_7.H_2O$, $Ca_3Si_2O_7.3H_2O$, and $Ca(OH)_2$:

$$2Ca_2SiO_4(s) + 2H_2O(l) \rightarrow Ca_3Si_2O_7.H_2O(s) + Ca(OH)_2(aq)$$
$$2Ca_2SiO_4(s) + 4H_2O(l) \rightarrow Ca_3Si_2O_7.3H_2O(s) + Ca(OH)_2(aq)$$

These hydrates form a gel or slurry that coats the surfaces of the sand or aggregates and fills the voids to form the solid concrete. The properties of concrete are determined by the relative proportions of calcium silicates and calcium aluminosilicates in the cement used, the additives, and the amount of water, which determines the degree of hydration.

The raw materials for cement manufacture often contain traces of sodium and potassium sulfates, and sodium and potassium hydroxides are formed during the hydration process. These hydroxides are responsible for the cracking, swelling, and distortion of many ageing concrete structures. The hydroxides take part in a complex series of reactions with the aggregate material to form an alkali silicate gel. This gel is hygroscopic and expands as it absorbs water, producing stress in the concrete, which leads to cracking and deformation. The susceptibility of concrete to this 'alkali silicate reaction' is now monitored by calculating the total alkali levels in the concrete produced and strategies are in place to minimize its effects. For example, adding 'fly ash' to the mix, which is a waste product from coal-fired power stations, can reduce the problem.

BOX 12.2   Fireworks and flares

Fireworks use exothermic reactions to produce heat, light, and sound. Common oxidants are nitrates and perchlorates, which decompose when heated to liberate oxygen. Common fuels are carbon, sulfur, powdered aluminium or magnesium, and organic materials such as poly(vinyl chloride) (PVC), starch, and gums. The most common constituent of fireworks is gunpowder or black powder, a mixture of potassium nitrate, sulfur, and charcoal, and thus both an oxidant and a fuel. Special effects, such as colours, flashes, smoke, and noises, are provided by additives to the firework mixture. The Group 2 elements are used in fireworks to provide colour.

Barium compounds are added to fireworks to produce green flames. The species responsible for the colour is $BaCl^+$, which is produced when Ba^{2+} ions combine with Cl^- ions. The Cl^- ions are produced during decomposition of the perchlorate oxidant or during combustion of the PVC fuel:

$$KClO_4(s) \rightarrow KCl(s) + 2O_2(g)$$
$$KCl(s) \rightarrow K^+(g) + Cl^-(g)$$
$$Ba^{2+}(g) + Cl^-(g) \rightarrow BaCl^+(g)$$

Barium chlorate, $Ba(ClO_3)_2$, has been used instead of $KClO_4$ and a barium compound but is too unstable to shock and friction. Similarly, strontium nitrate and carbonate are used to produce a red colour on formation of $SrCl^+$. Strontium chlorate and perchlorate are effective at producing the red colour but are too unstable to shock and friction for routine use.

Distress flares also use strontium compounds. Strontium nitrate is mixed with sawdust, waxes, sulfur, and $KClO_4$ and packed into a waterproof tube. When ignited, the flares burn with an intense red flame for up to 30 minutes.

As well as being used as a fuel, powdered magnesium is added to fireworks and flares to maximize light output. As well as the magnesium producing an intense white light, illumination is increased by the incandescence of high-temperature MgO particles that are produced in the oxidation reaction.

widely used on clock and watch faces but has been replaced by less hazardous compounds.

Magnesium and calcium are of great biological importance. Magnesium is a component of chlorophyll but also it is coordinated by many other biologically important ligands, including ATP (adenosine triphosphate, Section 27.2b). It is essential for human health, being responsible for the activity of many enzymes. The recommended adult human dose is approximately 0.3 g per day and the average adult contains about 25 g of magnesium. The bioinorganic chemistry of calcium is discussed in detail in Chapter 27.

12.6 Hydrides

Key point: All the Group 2 elements form saline hydrides with the exception of beryllium, which formvs a polymeric covalent compound.

Like the Group 1 elements, the Group 2 elements, with the exception of Be, form ionic, saline hydrides that contain the H− ion. They can be prepared by direct reaction between the metal and hydrogen. Beryllium hydride is covalent and must be prepared from alkylberyllium (Section 12.13). It has a network structure

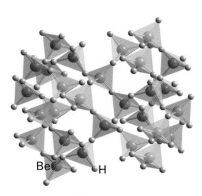

Figure 12.3 The structure of BeH$_2$.

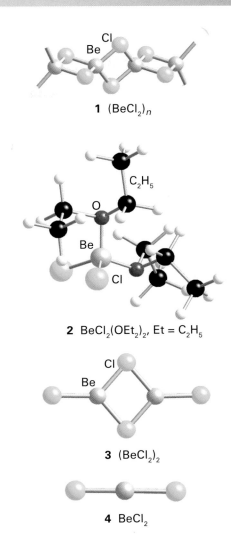

1 (BeCl$_2$)$_n$

2 BeCl$_2$(OEt$_2$)$_2$, Et = C$_2$H$_5$

3 (BeCl$_2$)$_2$

4 BeCl$_2$

with bridging H atoms (Fig. 12.3); the long-held view that it is a linear chain is incorrect.

The ionic hydrides of the heavier elements react violently with water to produce hydrogen:

$$MgH_2(s) + 2 H_2O(l) \rightarrow Mg(OH)_2(s) + 2 H_2(g)$$

This reaction is not as violent as that for the Group 1 elements and can be used as a source of hydrogen in fuel cells. For hydrogen storage, a reversible reaction involving uptake of hydrogen near room temperature is needed. Magnesium hydride loses hydrogen on heating above 250°C so the process

$$Mg(s) + H_2(g) \rightarrow MgH_2(s)$$

is reversible and the the low molar mass of Mg (24.3 g mol^{-1}) makes MgH$_2$ potentially an excellent hydrogen storage material. Attempts to reduce the decomposition temperature to closer to room temperature involve making complex magnesium hydrides doped with other metals and fabricating nanoparticulate forms (Box 10.4).

12.7 Halides

Key points: The halides of beryllium are covalent; all the fluorides, except BeF$_2$, are insoluble in water; all other halides are soluble.

All the beryllium halides are covalent. Beryllium fluoride is prepared from the thermal decomposition of (NH$_4$)$_2$BeF$_4$ and is a glassy solid that exists in several temperature-dependent phases similar to those of SiO$_2$ (Section 14.10). It is soluble in water, forming the hydrate [Be(OH$_2$)$_4$]$^{2+}$. Beryllium chloride, BeCl$_2$, can be made from the oxide:

$$BeO(s) + C(s) + Cl_2(g) \rightarrow BeCl_2(s) + CO(g)$$

The chloride, as well as BeBr$_2$ and BeI$_2$, can also be prepared from direct reaction of the elements at elevated temperatures.

The structure of solid BeCl$_2$ is a polymeric chain (**1**). The local structure is an almost regular tetrahedron around the Be atom and the bonding can be considered to be based on sp^3 hybridization. In BeCl$_2$ the chloride ion has sufficient electron density for 2c,2e covalent bonding to take place. Beryllium chloride is a Lewis acid, readily forming adducts with electron-pair donors such as diethyl ether (**2**). In the vapour phase the compound tends to form a dimer based on sp^2 hybridization (**3**), and when the temperature is above 900°C linear monomers are formed, indicating sp hybridization (**4**).

The anhydrous halides of magnesium are prepared by the direct combination of the elements as preparation from aqueous solutions yields the hydrates, which are partially hydrolysed on heating. The anhydrous halides of the heavier Group 2 elements can be prepared by dehydration of the hydrates. All the fluorides except BeF$_2$ are sparingly soluble, although the solubility increases slightly down the group. As the radius of the cation increases from Be to Ba, the cation coordination number increases from 4 to 8, with CaF$_2$, SrF$_2$, and BaF$_2$ adopting the fluorite structure (Section 3.9). The other halides of Group 2 form layer structures, reflecting the increasing polarizability of the halide ions. Magnesium chloride adopts the cadmium-chloride layered structure in which the layers are arranged so that the Cl$^-$ ions are cubic close packed (Fig. 12.4). Both MgI$_2$ and CaI$_2$ adopt the closely related cadmium-iodide structure in which the layers of I$^-$ ions are hexagonal close packed.

The most important fluoride of the group is CaF$_2$. Its mineral form, fluorite or fluorspar, is the only large-scale source of fluorine. Anhydrous hydrogen fluoride is prepared by the action of concentrated sulfuric acid on fluorspar:

$$CaF_2(s) + H_2SO_4(l) \rightarrow CaSO_4(s) + 2 HF(l)$$

All the chlorides are deliquescent and form hydrates. They have lower melting points than the fluorides. Magnesium chloride is the most important chloride for industry. It is extracted from

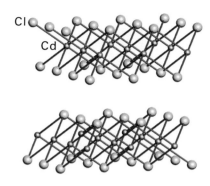

Figure 12.4 The cadmium-chloride structure adopted by $MgCl_2$.

seawater and then used in the production of magnesium metal. Calcium chloride is also of great importance and is produced on a massive scale industrially. Its hygroscopic character leads to its widespread use as a laboratory drying agent. It is also used to de-ice roads, where it is more effective than NaCl for two reasons. First, the dissolution is very exothermic:

$$CaCl_2(s) \rightarrow Ca^{2+}(aq) + 2Cl^-(aq) \quad \Delta_{sol}H^{\ominus} = -82kJ\,mol^{-1}$$

The heat generated helps to melt the ice. Second, the minimum freezing mixture of $CaCl_2$ in water has a freezing point of $-55°C$ compared to $-18°C$ for that of NaCl in water. The exothermic dissolution also leads to another application in instant heating packs and self-heating drink containers. Concentrated solutions of $CaCl_2$ have a very sticky consistency and they are sprayed on to unfinished roads to minimize dust production.

Radium has the least soluble halides as a result of the low hydration enthalpy of the large Ra^{2+} ion. This property is used to separate Ra^{2+} from other Group 2 metals by fractional crystallization.

EXAMPLE 12.1 **Predicting the nature of the halides**

Use the data in Table 1.7 and Fig. 2.38 to predict whether CaF_2 is predominantly ionic or covalent.

Answer One approach is to identify the electronegativities of the two elements in the compound and then to refer to the Ketalaar triangle to judge the type of bonding present. The Pauling electronegativity values of Ca and F are 1.00 and 3.98, respectively. The average electronegativity is therefore 2.49 and the difference is 2.98. These values on the Ketelaar triangle in Fig. 2.2 indicate that CaF_2 should be ionic.

Self-test 12.1 Predict whether (a) $BeCl_2$ and (b) $BaCl_2$ are predominantly ionic or covalent.

12.8 Oxides, sulfides, and hydroxides

The Group 2 elements react with O_2 to form the oxides. All the elements except Be also form unstable peroxides. The oxides of Mg to Ra react with water to form the basic hydroxides; BeO and $Be(OH)_2$ are amphoteric.

(a) Oxides, peroxides, and complex oxides

Key points: All the Group 2 elements form normal oxides with oxygen except Ba, which forms the peroxide; all the peroxides decompose to the oxides, their stabilities increasing down the group.

Beryllium oxide is obtained by ignition of the metal in oxygen. It is a white, insoluble solid with the wurtzite structure (Section 3.9). Its high melting point (2570°C), low reactivity, and excellent thermal conductivity, the highest of any oxide, lead to its use as a refractory material. It is highly toxic on inhalation, leading to chronic beryllosis, a disease of the lungs, and cancer. This problem is exacerbated by its low density (3 g cm^{-3}) as dust particles remain airborne for long periods, but BeO is safe for many applications when used as a sintered monolith.

The oxides of the other Group 2 elements can be obtained by direct combination of the elements (except Ba, which forms the peroxide) but they are more commonly obtained by decomposition of the carbonates:

$$MCO_3(s) \xrightarrow{\Delta} MO(s) + CO_2(g)$$

The oxides of the elements from Mg to Ba all adopt the rock-salt structure (Section 3.9). Their melting points decrease down the group as the lattice enthalpies decrease with increasing cation radius. Magnesium oxide is a high-melting-point solid (as is BeO) and is used as a refractory lining in industrial furnaces. Like BeO, MgO has a high thermal conductivity coupled with a low electrical conductivity. This combination of properties leads to its use as an electrically insulating material around the heating elements of domestic appliances and in electrical cables.

Calcium oxide (as lime or quicklime) is used in large quantities in the steel industry to remove P, Si, and S. When heated, CaO is thermoluminescent and emits a bright white light (hence 'limelight'). Calcium oxide is also used as a water softener to remove hardness by reacting with soluble carbonates and hydrogencarbonates to form the insoluble $CaCO_3$. It reacts with water to form $Ca(OH)_2$, which is sometimes known as *slaked lime* and is used to neutralize acidic soils.

The peroxides of Mg, Ca, Sr, and Ba are prepared by a variety of routes; only SrO_2 and BaO_2 can be made by direct reaction of the elements, albeit at elevated temperatures and pressures. All the peroxides are strong oxidizing agents and decompose to the oxide:

$$MO_2(s) \rightarrow MO(s) + \tfrac{1}{2}O_2(g)$$

The thermal stability of the peroxides increases down the group as the radius of the cation increases. This trend is explained by considering the lattice enthalpies of the peroxide and the oxide, and their dependence on the relative radii of the cations and anions. As O^{2-} is smaller than O_2^{2-}, the lattice enthalpy of the oxide is greater than that of the corresponding peroxide. The difference between the two lattice enthalpies decreases down the group as both values become smaller with increasing cation radius, therefore the tendency to decompose decreases. Magnesium peroxide, MgO_2, is consequently the least stable peroxide and is used as an *in situ* source of oxygen in a range of applications, including bioremediation to clean up polluted waterways.

EXAMPLE 12.2 **Explaining the thermal stabilities of peroxides**

Estimate the difference between the lattice enthalpies of the peroxide and oxide of Mg and Ba, and comment on the values obtained.

Answer The Kapustinskii equation (eqn 3.4) can be used to estimate lattice enthalpies, using the ionic radii in Table 12.1 and the radii of the oxide and peroxide ions (126 and 180 pm, respectively). Remembering that peroxide is a single anion, O_2^{2-}, substitution of the values gives:

	MgO	MgO_2		BaO	BaO_2
$\Delta_L H/(\text{kJ mol}^{-1})$	4037	3315		3147	2684
Difference/(kJ mol^{-1})		722			463

This calculation confirms that the difference between the lattice enthalpies of the oxide and peroxide decreases down the group.

Self-test 12.2 Calculate the lattice enthalpies for CaO and CaO_2 and check that the above trend is confirmed.

The elements also form a number of complex oxides such as perovskite, $SrTiO_3$, and spinel, $MgAl_2O_4$ (Section 3.9). The range of ionic radii available, from Mg^{2+} (72 pm for coordination number 6) to Ba^{2+} (142 pm and larger for coordination number 8 and higher) means that complex oxides can be synthesized that contain these cations in a large variety of different structure types. Important examples of such complex oxides include the ferroelectric perovskite, $BaTiO_3$, the phosphor $SrAl_2O_4$:Eu, and many of the high temperature superconductors, for example $YBa_2Cu_3O_7$ and $Bi_2Sr_2CaCu_2O_8$ (Section 24.8). The coordination number preferences of the cations are important in solid state chemistry where they can be used to control the structures of many complex oxides. If a doubly charged ion is required to occupy the A site of a perovskite structure (Section 3.9), with a coordination number of 12, then Sr^{2+} or Ba^{2+} is normally selected (as in $SrTiO_3$). On the other hand, for the spinel structure, general formula AB_2O_4, with 6-coordinate B-type sites, Mg^{2+} is a good choice (as in $GeMg_2O_4$).

■ **A brief illustration.** The control of the local coordination number by the Group 2 ions is illustrated by considering the structure of the superconducting phase $Tl_2Ba_2Ca_2Cu_3O_{10}$; the larger Ba^{2+} cations demand a higher coordination number to O than does Ca^{2+} so the former are found exclusively on the 9-coordinate sites and the latter on 8-coordinate sites. It would be impossible to synthesize $Tl_2Ca_2Ca_2Cu_3O_{10}$ with Ca^{2+} replacing Ba^{2+} on the higher coordinate site as that location would be thermodynamically unfavourable. ■

(b) Sulfides

Key point: The sulfides mostly adopt the rock-salt structure and have applications as phosphors.

Beryllium sulfide adopts a zinc-blende structure whereas the sulfides of the heavier elements all crystallize with the rock-salt structure. Barium sulfide, produced by reducing the naturally occurring barytes, $BaSO_4$, with coke,

$$BaSO_4(s) + 2\,C(s) \rightarrow BaS(s) + 2\,CO_2(g)$$

displays strong phosphoresence and was the first synthetic phosphor. A mixed calcium/strontium sulfide doped with bismuth is a long-lifetime phosphor and has been used in glow-in-the-dark pigments.

EXAMPLE 12.3 Predicting and identifying the structure type of a Group 2 chalcogenide

Analysis of the powder X-ray diffraction pattern obtained from CaSe showed that the lattice type was face-centred with lattice parameter 592 pm. Predict a structure type for this compound by using the ionic radii

for Ca^{2+} and Se^{2-} as 100 and 184 pm, respectively. Does your prediction agree with observation?

Answer First, we need to be aware that a binary compound such as CaSe is likely to adopt one of the simple AX structure types described in Section 3.9; then we can use the radius-ratio rule to guide us to likely structural type. Two of the AX structure types have face-centred lattice types, namely rock-salt and zinc-blende. The radius ratio is (100 pm)/(194 pm) = 0.52, which according to Table 3.6 suggests that the rock-salt structure is preferred. The lattice parameter for the rock-salt structure is calculated by considering the unit cell in Fig 3.30 as it is equal to the length of the side of the unit cell. As can be seen from the figure, this length is twice the sum $r(Ca^{2+}) + r(Se^{2-})$. Thus, from the ionic radii the lattice parameter is predicted to be $2 \times (100 + 194)$ pm = 588 pm, in good agreement with the X-ray diffraction value which, therefore, demonstrates that CaSe adopts the rock-salt structure.

Self-test 12.3 Use ionic radii to predict a structure type of BeSe.

(c) Hydroxides

Key point: The solubility of the hydroxides increases down the group.

All the hydroxides are formed by reaction of the oxides with water. Beryllium hydroxide, $Be(OH)_2$, is amphoteric. The hydroxides become apparently more basic down the group because their solubility increases from $Mg(OH)_2$ to $Ba(OH)_2$. Magnesium hydroxide, $Mg(OH)_2$, is sparingly soluble and forms a mildly basic solution because a saturated solution contains a low concentration of the OH^- ions. Calcium hydroxide, $Ca(OH)_2$, is more soluble than $Mg(OH)_2$, so a saturated solution has a higher concentration of OH^- ions and the solution is described as moderately basic. A saturated solution of $Ca(OH)_2$ is called *limewater* and is used to test for the presence of CO_2. If CO_2 is bubbled through the limewater a white precipitate of $CaCO_3$ is formed, which then disappears on further reaction with CO_2 to form the hydrogencarbonate ion:

$$Ca(OH)_2(aq) + CO_2(g) \rightarrow CaCO_3(s) + H_2O(l)$$
$$CaCO_3(s) + CO_2(g) + H_2O(l) \rightarrow Ca^{2+}(aq) + 2\,HCO_3^-(aq)$$

Barium hydroxide, $Ba(OH)_2$, is soluble and aqueous solutions are described as strongly basic.

■ **A brief illustration.** The molar solubility of $Mg(OH)_2$ is 1.54×10^{-4} mol dm^{-3}. Therefore the concentration of OH^- ions in the saturated solution is 3.08×10^{-4} mol dm^{-3}. We know from Section 4.1b that $K_w = 1.0 \times 10^{-14}$, so (ignoring deviations from ideality) $[H_3O^+] = K_w/[OH^-] = 3.25 \times 10^{-11}$ mol dm^{-3}. Provided we can identify activities with molar concentrations, this concentration corresponds to pH = 10.5. ■

12.9 Nitrides and carbides

Key point: The Group 2 nitrides and carbides react with water to produce ammonia and either methane or ethyne, respectively.

All the elements form nitrides of composition M_3N_2 which react with water to form ammonia and the metal hydroxide.

$$M_3N_2(s) + 6\,H_2O(l) \rightarrow 3\,M(OH)_2(s,aq) + 2\,NH_3(g)$$

Magnesium burns in nitrogen to form greenish-yellow Mg_3N_2, which has been used as a catalyst for preparing cubic BN (Section 13.9). Calcium nitride, Ca_3N_2, reacts with hydrogen gas at

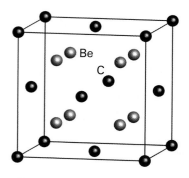

Figure 12.5 The antifluorite structure adopted by Be₂C.

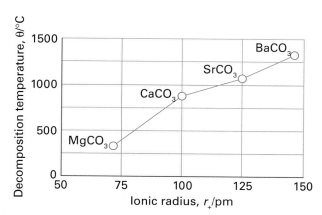

Figure 12.6 The variation of the decomposition temperature of the Group 2 carbonates with ionic radius.

400°C to produce CaNH and CaH₂. Beryllium nitride melts at 2200°C and is used as a refractory material.

All the elements also form carbides. Beryllium carbide, Be₂C, formally contains the methide ion, C⁴⁻, although some covalency would be expected in the bonding in this compound; it is a crystalline solid with the antifluorite structure (Fig. 12.5). The carbides of Mg, Ca, Sr, and Ba have the formula MC₂ and contain the dicarbide (acetylide) anion, C_2^{2-}. The carbides of Ca, Sr, and Ba are prepared by heating the oxide or carbonate with carbon in a furnace at 2000°C:

$$MO(s) + 3C(s) \rightarrow MC_2(s) + CO(g)$$
$$MCO_3(s) + 4C(s) \rightarrow MC_2(s) + 3CO(g)$$

All the carbides react with water to produce the hydrocarbon corresponding to the carbon ion present: beryllium carbide produces methane, whereas the other elements produce ethyne (acetylene):

$$Be_2C(s) + 4H_2O(l) \rightarrow 2Be(OH)_2(s) + CH_4(g)$$
$$CaC_2(s) + 2H_2O(l) \rightarrow Ca(OH)_2(s) + C_2H_2(g)$$

Methane and ethyne are flammable and the latter burns with a bright light from the incandescent carbon particles that form in the flame. When this reaction was discovered in the late nineteenth century, calcium carbide found widespread use in vehicle lights, enabling safe night-time driving for the first time, and in miners' lamps.

12.10 Salts of oxoacids

The most important oxo compounds of the Group 2 elements are the carbonates, hydrogencarbonates, and sulfates.

(a) Carbonates and hydrogencarbonates

Key points: All the carbonates are sparingly soluble in water, with the exception of BeCO₃; the carbonates decompose to the oxide on heating, most readily high in the group. The hydrogencarbonates are more soluble than the carbonates.

Beryllium carbonate exists as a soluble hydrate that is susceptible to hydrolysis due to the high charge density on the Be²⁺ ion and its polarization of the O–H bond of a hydrating H₂O molecule:

$$[Be(OH_2)_4]^{2+}(aq) + H_2O(l) \rightarrow [Be(OH_2)_3(OH)]^+(aq) + H_3O^+(aq)$$

This equation is a greatly simplified version of the actual processes that occur; in fact several reactions occur and hydroxo-bridged

species are present. The carbonates of the other elements are all sparingly soluble and are decomposed to the oxide on heating:

$$MCO_3(s) \xrightarrow{\Delta} MO(s) + CO_2(g)$$

The temperature at which this decomposition occurs increases from 350°C for Mg to 1360°C for Ba (Fig. 12.6). The Group 2 carbonates have similar thermal stability to the Group 1 carbonates. As discussed in Section 3.15, these trends can be explained in terms of trends in lattice enthalpies and hence, more fundamentally, in terms of trends in ionic radii.

Calcium carbonate is the most important oxo compound of the elements. It occurs widely in nature as limestone, chalk, marble, dolomite (with magnesium), and as coral, pearl, and seashell. Calcium carbonate crystallizes in a variety of polymorphs. The most common forms are calcite, aragonite, and vaterite (Fig. 12.7, Box 12.3). Calcium carbonate is an important biomineral and a major constituent of bones and shells (Box 12.4). It also has widespread use in construction and road building, and is used as an antacid, an abrasive in toothpaste, in chewing gum, and as a health supplement where it is taken to maintain bone density. Powdered limestone is known as *agricultural lime* and is used to neutralize acidic soil:

$$CaCO_3(s) + 2H^+(aq) \rightarrow Ca^{2+}(aq) + CO_2(g) + H_2O(l)$$

Calcium carbonate is sparingly soluble in water but its solubility is increased if CO₂ is dissolved in the water, as in rainwater. Thus, caverns have been eroded in limestone rock by the reaction to form the more soluble hydrogencarbonate:

$$CaCO_3(s) + H_2O(l) + CO_2(g) \rightarrow Ca^{2+}(aq) + 2HCO_3^-(aq)$$

This reaction is reversible and over time calcium carbonate stalactites and stalagmites are formed.

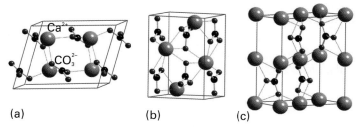

Figure 12.7 The structures of the (a) calcite, (b) aragonite and (c) vaterite polymorphs of CaCO₃.

BOX 12.3 The polymorphs of calcium carbonate

Calcium carbonate occurs as vast deposits of sedimentary rocks that are formed from the fossilized remains of marine creatures. The most common and stable form is the hexagonal *calcite*, which has been identified in numerous different crystalline forms. Calcite comprises about 4 per cent by mass of the Earth's crust and is formed in many different geological environments. Calcite can form rocks of considerable mass and constitutes a significant part of all three major rock classification types: igneous, sedimentary, and metamorphic. Calcite is a major component in the igneous rock called carbonatite and forms the major portion of many hydrothermal veins. Limestone is the sedimentary form of calcite.

Limestone metamorphoses to marble from the heat and pressure of metamorphic events, which increase the density of the rock and destroy any texture. Pure white marble is the result of metamorphism of very pure limestone. The characteristic swirls and veins of many coloured marble varieties are usually due to various mineral impurities such as clay, silt, sand, and iron oxides. *Iceland spar* is a form of transparent, colourless calcite originally found in Iceland. It exhibits birefringence. Birefringence, or double refraction, is the division of a ray of light into two rays when it passes through certain types of material, depending on the polarization of the light. This behaviour is explained by assigning two different refractive indices to the material for different polarizations. As the two beams exit the crystal they are bent into two different angles of refraction.

Aragonite is a less abundant polymorph of calcite. Aragonite is orthorhombic with three crystal forms. It is less stable than calcite and converts into it at 400°C and, given enough time, aragonite in the environment will convert to calcite. Most bivalve animals, such as oysters, clams, mussels, and corals, secrete aragonite and their shells and pearls are composed of mostly aragonite. The pearlization and iridescent colours in seashells such as abalone are due to several layers of aragonite (Box 25.2). Other natural sources of aragonite include hot springs and cavities in volcanic rocks.

Synthetic aragonite can be formed and is used as a filler in the paper industry, where its fine texture, whiteness, and absorbent properties add quality to the product. Powdered limestone is heated in a kiln to form CaO, lime. The lime is then slurried with water to form *milk of lime*. Carbon dioxide is then bubbled through the slurry until aragonite is formed.

$$CaCO_3(s) \xrightarrow{\Delta} CaO(s) + CO_2(g)$$
$$CaO(s) + H_2O(l) \rightarrow Ca(OH)_2(s)$$
$$Ca(OH)_2(s) + CO_2(g) \rightarrow CaCO_3(s) + H_2O(l)$$

The conditions, such as temperature and flow rate of carbon dioxide, determine the final particle size distribution and crystal type.

Vaterite is an even rarer polymorph of calcium carbonate, also adopting a hexagonal unit cell. It is more soluble than calcite or aragonite and converts slowly to these forms in contact with water. It is deposited from some mineral springs in cold climates; gallstones are often composed of this form of $CaCO_3$. *Travertine* is a white, naturally occurring, very hard form of calcium carbonate. It is deposited from the water of hot mineral springs or streams containing CaO.

BOX 12.4 Calcium carbonate biominerals and minerals

Biomineralization involves the production by organisms of inorganic solids and over 50 biominerals have been identified of which calcium carbonate is one of the most common. Nature is adept at controlling mineralization processes to produce single crystals and polycrystalline and amorphous structures with remarkable morphologies and mechanical properties (Chapter 27). The biomineral often fulfills a structural role in the organism, for example as teeth bones or shells. Many shells are composed mostly of calcium carbonate with only up to about 2 per cent protein.

Most of the calcium carbonate mineral deposits found today were formed by sea creatures that manufactured shells and skeletons of calcium carbonate. When these animals died, their shells settled on the sea floor and were compressed to form limestone and chalk. The White Cliffs of Dover in England are chalk made from the shells of microscopic sea creatures called *Foraminifera* that lived about 136 Ma ago.

In pearls and mother-of-pearl, very small $CaCO_3$ crystals are deposited to form smooth, hard, lustrous layers of alternating layers of calcite and aragonite. The different structural forms of the calcium carbonate layers make the shell very strong as they have different preferred directions of fracture and the interleaved layers resist breaking under pressure. The thickness of the layers is similar to the wavelength of light, which produces the interference effects and the observed pearlescence.

Many attempts are being made to replicate these complicated crystal growth patterns of $CaCO_3$ in the laboratory. Crystals can sometimes be grown around a template, which is then removed to produce a porous structure, or the addition of certain chemicals can cause different shapes of crystals to grow from a particular solution. Similarly the growth of bone, which is mainly calcium hydroxyapatite $Ca_5(PO_4)_3(OH)_2$, is being investigated, including the development of synthetic bone materials.

■ **A brief illustration.** In calcite and aragonite the CO_3^{2-} anions are surrounded by Ca^{2+} ions to produce distinct local environments. As a result, the vibrational modes of CO_3^{2-} observed in their infrared spectra occur at slightly different frequencies. Moreover, due to differences in the local symmetry, certain vibrational modes are detected for each polymorph, as summarized in the table. The spectrum obtained from a piece of snail shell is shown in Fig. 12.8. It shows a strong feature at around 1080 cm⁻¹ that is characteristic of the aragonite form of calcium carbonate. ■

Wavenumber/cm⁻¹	
Calcite	Aragonite
714	698
876	857
	1080
1420 broad	1480 broad
1800	1785

Because HCO_3^- has a lower charge than CO_3^{2-}, Group 2 hydrogencarbonates do not precipitate from solutions containing these ions. Temporary hardness of water (hardness which

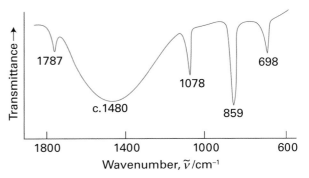

Figure 12.8 The IR spectrum of a piece of snail shell.

is removed by boiling the water) is caused by the presence of magnesium and calcium hydrogencarbonates in solution. These ions are precipitated as carbonate on boiling, when the equilibrium in the following reaction is moved to the right

$$Ca(HCO_3)_2(aq) \rightleftharpoons CaCO_3(s) + CO_2(g) + H_2O(l)$$

Temporary hardness may also be removed by adding $Ca(OH)_2$, which also precipitates the carbonate:

$$Ca(HCO_3)_2(aq) + Ca(OH)_2(aq) \rightleftharpoons 2\,CaCO_3(s) + 2\,H_2O(l)$$

If temporary hard water is not treated to remove the Ca^{2+} and Mg^{2+} ions then these ions can react with soap (sodium stearate, $NaC_{17}H_{35}CO_2$) or detergent molecules to form an insoluble precipitate, scum, which reduces the effectiveness of the detergent

$$2\,NaC_{17}H_{35}CO_2(aq) + Ca^{2+}(aq) \rightarrow Ca(C_{17}H_{35}CO_2)_2(s) + 2\,Na^+(aq)$$

(b) Sulfates and nitrates

Key points: The most important sulfate is calcium sulfate, which occurs naturally as gypsum and alabaster.

'Permanent hardness' (so-called because the hardness is not removed by boiling) is caused by magnesium and calcium sulfates. In this case the water is softened by passing through an ion-exchange resin, which replaces the Mg^{2+} and Ca^{2+} ions with Na^+ ions.

Calcium sulfate is the most important of the Group 2 sulfates. It occurs naturally as gypsum, which is the dihydrate, $CaSO_4.2H_2O$ (Fig. 12.9). Alabaster is a dense, fine-grained form of the same composition that resembles marble and can be carved into sculptures. When the dihydrate is heated above 150°C it loses water to form the hemihydrate, $CaSO_4.\frac{1}{2}H_2O$, which is also known as *plaster of Paris* as it was first mined in the Montmartre district of the city. When mixed with water, plaster of Paris expands as it forms the dihydrate, so providing a strong structure that is used to make casts for broken limbs. Gypsum is mined and used as a building material. One application is in fire-proof wallboard. In the event of fire the dihydrate will dehydrate to form the hemihydrate and release water vapour:

$$2\,CaSO_4.2H_2O(s) \rightarrow 2\,CaSO_4.\frac{1}{2}H_2O(s) + \frac{3}{2}H_2O(g)$$

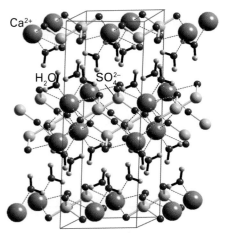

Figure 12.9 The structure of $CaSO_4.2H_2O$ highlighting the water molecules that are mainly lost on heating above 150°C.

The reaction is endothermic ($\Delta_r H^{\ominus} = +117$ kJ mol^{-1}) so it absorbs heat from the fire. In addition, the water produced absorbs heat and evaporates, and then the gaseous water provides an inert barrier, reducing the supply of oxygen to the fire.

The insolubility of $BaSO_4$ coupled to the strong X-ray absorbing properties of barium, due to its high atomic number (56), lead to its use in X-ray imaging of the digestive tract. The white pigment lithopone is a mixture of $BaSO_4$ and ZnS and has excellent chemical stability, being inert to attack by sulfides, unlike lead white, $PbCO_3$. Barium sulfate is a component of many drilling muds, which clean and cool the drill bit and then carry the drilled rock away.

Hydrated nitrates, such as $Ca(NO_3)_2.4H_2O$, can be obtained by treating the oxides, hydroxides, and carbonates with nitric acid and crystallizing the salt from the resulting aqueous solution. For Mg to Ba the anhydrous salts are easily obtained by thermal dehydration. Heating hydrated beryllium nitrate $Be(NO_3)_2.4H_2O$ results in its decomposition and the evolution of NO_2. Anhydrous beryllium nitrate, $Be(NO_3)_2$, can be obtained by dissolving $BeCl_2$ in N_2O_4 and heating the resultant solvate $Be(NO_3)_2.2N_2O_4$ gently to drive off NO_2. Heating $Be(NO_3)_2$ further produces the basic nitrate $Be_4O(NO_3)_6$, which contains a central Be_4O tetrahedral unit with edge-bridging nitrate groups with a structure similar to that of the basic acetate (5).

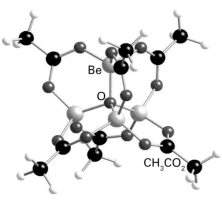

5 $Be_4O(O_2CCH_3)_6$

12.11 Solubility, hydration, and beryllates

Key points: The large negative hydration enthalpies of the salts of mononegative ions ensure that they are soluble. For salts of dinegative ions, the lattice enthalpies are more influential and the salts are insoluble.

Compounds of the Group 2 elements are generally much less soluble in water than those of Group 1 elements even though the hydration enthalpies are more negative:

	Na^+	K^+	Mg^{2+}	Ca^{2+}
$\Delta_{hyd}H^{\ominus}/(kJ\,mol^{-1})$	−406	−322	−1920	−1650

With the exception of the fluorides, the salts of singly charged anions are usually soluble in water and those of doubly charged anions (such as oxides) are usually only sparingly soluble. For the latter, such as the carbonates and the sulfates, the high lattice enthalpy arising from the high charge of the anion is the deciding factor, outweighing the influence of the enthalpy of hydration. This insolubility is responsible for the enormous deposits of the magnesium- and calcium-containing minerals, such as limestone, gypsum, and dolomite, that are widely exploited in the construction industry. With the exception of BeF_2, the fluorides are all insoluble in water because the small size of F^- leads to a high lattice enthalpy. Beryllium fluoride has a very high hydration enthalpy due to the high charge density on the small cation: in its case, hydration enthalpy rather than lattice enthalpy is the dominant factor.

$$BeF_2(s) + 4H_2O(l) \rightarrow [Be(OH_2)_4]^{2+}(aq) + 2F^-(aq)$$

$$\Delta_r H^{\ominus} = -250 \text{ kJ mol}^{-1}$$

The H_2O molecules directly coordinated to Be^{2+} are very strongly held in aqueous solution, exchanging only very slowly with free water. The hydrated Be^{2+} cation, $[Be(OH_2)_4]^{2+}$, acts as an acid in water.

$$[Be(OH_2)_4]^{2+}(aq) + H_2O(l) \rightarrow [Be(OH_2)_3(OH)]^+(aq) + H_3O^+(aq)$$

This reaction can be traced to the high polarizing power of the small, doubly charged cation. Solutions of the hydrated salts of the heavier elements are neutral.

These trends in solubility and hydrolysis can also be explained in hard/soft acid–base terms. As the group is descended the ions become less hard; the fluorides and hydroxides (small, hard anions) are insoluble for the harder Be^{2+} and Mg^{2+} but more soluble for Ba^{2+}. Be^{2+} is most strongly hydrated by the hard O atom in H_2O and the relatively soft Ba^{2+} much more weakly.

EXAMPLE 12.4 Assessing the factors affecting solubility

Estimate the lattice enthalpies of $MgCl_2$ and $MgCO_3$. Comment on the likely implications for their solubilities.

Answer Once again we can use the data in Table 12.1 and the Kapustinskii equation (eqn 3.4); we also need to know that the ionic radii of Cl^- and CO_3^{2-} are 167 and 185 pm, respectively (*Resource section* 1). Substitution of the data gives the lattice enthalpies of $MgCO_3$ and $MgCl_2$ as 3260 and 2478 kJ mol^{-1}, respectively. As the value for $MgCO_3$ is larger, it is more likely to offset the hydration enthalpy and the solubility will be lower than that of $MgCl_2$.

Self-test 12.4 Calculate the lattice enthalpy of MgF_2 and comment on how it will affect the solubility compared to $MgCl_2$.

The amphoteric nature of Be leading to the formation of $[Be(OH)_4]^{2-}$ under strongly basic conditions means that the element forms an extensive series of beryllates that are built from BeO_4 tetrahedra. The beryl family of minerals, $Be_3Al_2(SiO_3)_6$, which includes emerald, aquamarine, and morganite, contains this unit as do a number of other complex beryllates such as phenakite, Be_2SiO_4, and the zeolite nabesite, $Na_2BeSi_4O_{10}\cdot4H_2O$. The compound $BeAl_2O_4$ occurs naturally as the mineral chrysoberyl, which in one form, alexandrite, is doped with chromium and changes in colour from green in daylight to purple under incandescent lighting.

12.12 Coordination compounds

Key points: Only beryllium forms coordination compounds with simple ligands such as the halides; the most stable complexes are formed with polydentate chelating ligands such as edta.

Compounds of Be show properties consistent with a greater covalent character than those of its congeners, and some of its complexes with ordinary ligands are stable. The complexes are usually tetrahedral, although the coordination number of Be can fall to 3 or 2 if the ligands are bulky. The most nonlabile complexes are formed with halide or chelating O-donor ligands, such as oxalate, alkoxides, and diketonates. For example, basic beryllium acetate (beryllium oxoethanoate, $Be_4O(O_2CCH_3)_6$) consists of a central O atom surrounded by a tetrahedron of four Be atoms, which in turn are bridged by ethanoate ions (**5**). It can be prepared by the reaction of ethanoic (acetic) acid with beryllium carbonate:

$$4BeCO_3(s) + 6CH_3COOH(l) \rightarrow$$
$$4CO_2(g) + 3H_2O(l) + Be_4O(O_2CCH_3)_6(s)$$

Basic beryllium acetate is a colourless, sublimable, molecular compound; it is soluble in chloroform, from which it can be recrystallized.

Group 2 cations form complexes with crown and crypt ligands. The least labile of these complexes are formed with the larger Sr^{2+} and Ba^{2+} cations. All the complexes are more stable than those of the smaller Group 1 cations. The most stable complexes are formed with charged polydentate ligands, such as the analytically important ethylenediaminetetraacetate ion (edta). The formation constants of edta complexes lie in the order $Ca^{2+} > Mg^{2+} > Sr^{2+} > Ba^{2+}$. In the solid state, the structure of the Mg^{2+} edta complex is seven-coordinate (**6**), with H_2O at one coordination site. The Ca^{2+} complex is either seven- or eight-coordinate, depending on the counter ion, with one or two H_2O molecules serving as ligands.

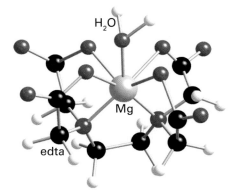

6 $[Mg(edta)(OH_2)]^{2-}$

Many complexes of Ca^{2+} and Mg^{2+} occur naturally. The most important macrocyclic complexes are the chlorophylls (**7**), which are porphyrin complexes of Mg and are central to photosynthesis

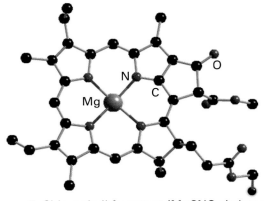

7 Chlorophyll fragment (MgCNO skeleton)

(Section 27.10d). Magnesium is involved in phosphate transfer and carbohydrate metabolism. Calcium is a component of biominerals and is also coordinated by proteins, notably those involved in cell signalling and muscle action (Section 27.4).

12.13 Organometallic compounds

Key points: Alkylberyllium compounds polymerize in the solid phase; Grignard reagents are some of the most important main-group organometallic compounds.

Organometallic compounds of Be are pyrophoric in air and unstable in water. Methylberyllium can be prepared by transmetallation from methylmercury in a hydrocarbon solvent:

$$Hg(CH_3)_2(sol) + Be(s) \rightarrow Be(CH_3)_2(sol) + Hg(l)$$

Another synthetic route is by halogen exchange or metathesis reactions in which a beryllium halide reacts with an alkyllithium compound. The products are the lithium halide and an alkylberyllium compound. In this way, the halogen and organic groups are transferred between the two metal atoms. The driving force for this and similar reactions is the formation of the halide of the more electropositive metal.

$$2\,n\text{-}BuLi(sol) + BeCl_2(sol) \rightarrow (n\text{-}Bu)_2Be(sol) + 2\,LiCl(s)$$

Grignard reagents in ether can also be used in the synthesis of organoberyllium compounds:

$$2\,RMgCl(sol) + BeCl_2(sol) \rightarrow R_2Be(sol) + 2\,MgCl_2(s)$$

Methylberyllium, $Be(CH_3)_2$, is predominantly a monomer in the vapour phase and in hydrocarbon solvents, where it adopts a linear structure, as expected from the VSEPR model. In the solid it forms polymeric chains in which the bridging CH_3 groups form $3c,2e$ bridging bonds (Section 2.11e, **8**). Bulkier alkyl groups lead to a lower degree of polymerization; ethylberyllium (**9**) is a dimer and *t*-butylberyllium (**10**) is a monomer.

An interesting organoberyllium compound is beryllocene, $(C_5H_5)_2Be$, which, although the formula suggests an analogy with ferrocene (Section 22.19), in fact has a different structure in the crystalline state, with the Be atom positioned directly above the

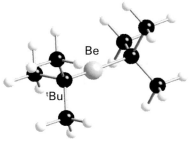

10 Be^tBu_2, $^tBu = (CH_3)_3C$

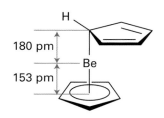

11 $BeCp_2$, $Cp = C_5H_5$

centre of one cyclopentadienyl ring and below a single C atom on the other ring (**11**). However, the low-temperature (−135°C) NMR spectrum of this compound in solution suggests that the two rings are equivalent, indicating that even at this low temperature the Be atom and the C_5H_5 rings are rearranging rapidly.

Alkyl- and arylmagnesium halides are very well known as **Grignard reagents** and are widely used in synthetic organic chemistry where they behave as a source of R⁻. They are prepared from magnesium metal and an organohalide. As the surface of magnesium is covered by a passivating oxide film, it has to be activated before the reaction can proceed. A trace of iodine is usually added to the reactants, forming magnesium iodide; this compound is soluble in the solvent used and dissolves to expose an activated magnesium surface. Alternatively a highly active, finely powdered form of magnesium can be generated by reducing $MgCl_2$ with potassium in THF. The reaction to produce the Grignard reagents is carried out in ether or tetrahydrofuran:

$$Mg(s) + RBr(sol) \rightarrow RMgBr(sol)$$

The structures of Grignard reagents are far from simple. The metal atom has a coordination number of 2 only in solution and when the alkyl group is bulky. Otherwise, it is solvated with a tetrahedral arrangement of solvent molecules around the Mg atom (**12**). In addition, complex equilibria in solution, known as **Schlenk equilibria**, lead to the presence of several species, the

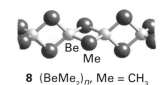

8 $(BeMe_2)_n$, Me = CH_3

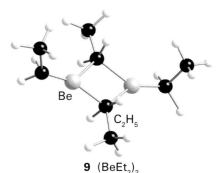

9 $(BeEt_2)_2$

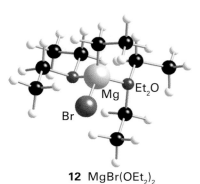

12 $MgBr(OEt_2)_2$

exact nature of which depend on temperature, concentration, and solvent. For example R_2Mg, $RMgX$, and MgX_2 have all been detected:

$$2\,RMgX(sol) \rightleftharpoons R_2Mg(sol) + MgX_2(sol)$$

Grignard reagents are widely used in the synthesis of organometallic compounds of other metals, as in the formation of alkylberyllium compounds mentioned above. They are also widely used in organic synthesis. One reaction is **organomagnesiation**, which involves addition of the Grignard reagent to an unsaturated bond:

$$R^1MgX\,(sol) + R^2R^3C{=}CR^4R^5(sol) \rightarrow$$
$$R^1R^2R^3CCR^4R^5MgX\,(sol)$$

Grignard reagents undergo side reactions such as **Wurtz coupling** to form a carbon–carbon bond:

$$R^1MgX\,(sol) + R^2X\,(sol) \rightarrow R^1R^2\,(sol) + MgX_2(sol)$$

The organometallic compounds of Ca, Sr, and Ba are generally ionic and very unstable. They all form analogues of Grignard reagents by direct interaction of the finely divided metal with organohalide.

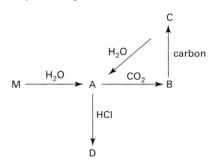

13 LMgMgL, L = [Ar(NC)(N^{i}Pr$_2$)NAr]$^-$, ^{i}Pr = (CH$_3$)$_2$CH

Finally, although the members of the group occur almost exclusively in oxidation state +2 in their compounds, the reduction of Mg(II) to Mg(I) by potassium has been achieved in the synthesis of the compounds LMg−MgL where L is [Ar(NC)(NPr$^i_2$)N(Ar)]$^-$ and Ar = 2,6-diisopropylphenyl. The structures contain a central Mg−Mg bond of length 285 pm (**13**), which is shorter than the Mg−Mg distance in magnesium metal (320 pm).

FURTHER READING

R.B. King, *Inorganic chemistry of the main group elements*. John Wiley & Sons (1994).

P. Enghag, *Encyclopedia of the elements*. John Wiley & Sons, Ltd (2004).

D.M.P. Mingos, *Essential trends in inorganic chemistry*. Oxford University Press (1998). A survey of inorganic chemistry from the perspective of structure and bonding.

N.C. Norman, *Periodicity and the s- and p-block elements*. Oxford University Press (1997). Includes coverage of essential trends and features of s-block chemistry.

J.A.H. Oates, *Lime and limestone: chemistry and technology, production and uses*. John Wiley & Sons (1998).

EXERCISES

12.1 Explain why compounds of beryllium are mainly covalent whereas those of the other Group 2 elements are predominantly ionic.

12.2 Why are the properties of beryllium more similar to aluminium and zinc than to magnesium?

12.3 Identify the compounds A, B, C, and D for M = Ba.

12.4 Why does beryllium fluoride form a glass when cooled from a melt?

12.5 Why is magnesium hydroxide a much more effective antacid than calcium or barium hydroxide?

12.6 Explain why Group 1 hydroxides are much more corrosive to metals than Group 2 hydroxides.

12.7 Which of the salts $MgSeO_4$ or $BaSeO_4$ would be expected to be more soluble in water?

12.8 Which Group 2 salts are used as drying agents and why?

12.9 How do Group 2 salts give rise to scaling from hard water?

12.10 Predict structures for BeTe and BaTe.

12.11 Use the data in Table 1.7 and the Ketelaar triangle in Fig. 2.38 to predict the nature of the bonding in $BeBr_2$, $MgBr_2$, and $BaBr_2$.

12.12 The two Grignard compounds C_2H_5MgBr and 2,4,6-$(CH_3)_3C_6H_2MgBr$ dissolve in THF. What differences would be expected in the structures of the species formed in these solutions?

12.13 Predict the products of the following reactions:

(a) $MgCl_2 + LiC_2H_5 \rightarrow$
(b) $Mg + (C_2H_5)_2Hg \rightarrow$
(c) $Mg + C_2H_5HgCl \rightarrow$

PROBLEMS

12.1 Marble and limestone buildings are eroded by contact with acid rain. Define the term 'acid rain' and discuss the origins of the acidity. Describe the processes by which the marble and limestone are attacked. List the compounds that are used as scrubbers in power plants to minimize emissions implicated in acid rain and describe how they are effective.

12.2 In their paper 'Noncovalent interaction of chemical bonding between alkaline earth cations and benzene?' (*Chem. Phys. Lett.*, 2001, **349**, 113), X.J. Tan and co-workers carried out theoretical calculations of complexes formed between beryllium, magnesium, and calcium ions and benzene. To which orbital interactions were the binding of the alkali metal to benzene attributed? How was the $C-C$ bond length in benzene affected by this interaction? Place the $M-C$ bonds in order of increasing bond enthalpy. How did the strength of the bonds compare to those formed between Group 1 elements and benzene? Sketch the geometry of the metal–benzene complexes.

12.3 Discuss beryllium fluoride glasses, comparing how the chemistry of BeF_2 is analogous to that of SiO_2.

12.4 P.C. Junk and J.W. Steed (*J. Chem. Soc., Dalton. Trans.*, 1999, 407) prepared crown ether complexes from the nitrates of Mg, Ca, Sr, and Ba. Outline the general procedure that was used for the syntheses. Sketch the structures of the two crown ethers used. Comment on the structures of the complexes and how they change with different cations.

12.5 By considering the thermodynamic factors that affect the dissolution of a crystalline solid in water (Section 3.15c), discuss why the solubility of the Group 2 halides decreases in the order Mg to Ra.

12.6 The aluminosilicates form a large group of minerals based on linked AlO_4 and SiO_4 tetrahedra which includes many clay and zeolite minerals whose compositions may be written $M^{x+}[Al_xSi_{1-x}]O_4]^{x-}$. Discuss the occurrence of the BeO_4 unit in natural minerals. To what extent has it proved possible to produce beryllophosphates based on linked BeO_4 and PO_4 tetrahedral units that are structural and compositional analogues of $M^{x+}[Al_xSi_{1-x}]O_4]^{x-}$ and $M^{x+}[Be_xP_{1-x}]O_4]^{x-}$?

12.7 An experiment was carried out to determine the hardness of domestic water. A few drops of pH = 10 buffer were added to a 100 cm³ sample of the water. This sample was titrated against 0.01 M edta(aq) using Eriochrome Black T indicator and gave a titre of 33.8 cm³. Under these circumstances both Mg^{2+} and Ca^{2+} ions react with edta. A second 100 cm³ sample was titrated against the edta solution after 5.0 cm³ of 0.1 M NaOH(aq) and a few drops of murexide indicator had been added. Under these conditions only the Ca^{2+} ions react with the edta and a titre of 27.5 cm³ was obtained. Determine the hardness of the water sample in terms of the concentration of the Mg^{2+} and of the Ca^{2+} ions.

12.8 The synthesis of a Mg(I) compound has been reported (*Science*, 2007, **318**, 1754). Describe how this synthesis was carried out and how this unusual Group 2 oxidation state was stabilized.

The Group 13 elements

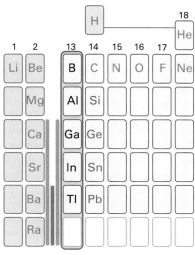

13

There are some clear trends in chemical properties of the Group 13 elements, such as oxidation number and amphoteric character, that we shall see repeated in the other groups of the p block. In this chapter we look at the occurrence and isolation of each element in Group 13 and consider the chemical properties of the elements and their simple compounds, coordination compounds, and organometallic compounds. In addition, we introduce the extensive range of boron clusters.

The elements of Group 13, boron, aluminium, gallium, indium, and thallium, have diverse physical and chemical properties. The first member of the group, boron, is essentially nonmetallic whereas the properties of the heavier members of the group are distinctly metallic. Aluminium is the most important element commercially and is produced on a massive scale for a wide range of applications. Boron forms a large number of cluster compounds involving hydrogen, metals, and carbon. Gallium and indium in alloys and compounds have important electronic and optical properties.

PART A: **THE ESSENTIALS**

In this section we discuss the essential features of the chemistry of the Group 13 elements.

13.1 The elements

Key points: Boron is the only nonmetal in the group. Aluminium is the most abundant Group 13 element.

The elements of Group 13 show a wide variation in abundance in crustal rocks, the oceans, and the atmosphere. Aluminium is abundant but the low cosmic and terrestrial abundance of boron, like that of lithium and beryllium, reflects how the light elements are sidestepped in nucleosynthesis (Section 1.1). The low abundance of heavier members of the group is in keeping with the progressive decrease in nuclear stability of the elements that follow iron. Boron occurs naturally as *borax*, $Na_2B_4O_5(OH)_4.8H_2O$, and *kernite*, $Na_2B_4O_5(OH)_4.2H_2O$, from which the impure element is obtained. Aluminium occurs in numerous clays and aluminosilicate minerals but the commercially most important mineral is *bauxite*, a complex mixture of hydrated aluminium hydroxide and aluminium oxide, from which it is extracted on an immense scale. Gallium oxide occurs as an impurity in bauxite and is normally recovered as a byproduct of the manufacture of aluminium. Indium and thallium occur in trace amounts in many minerals.

Whereas the elements of the s and d blocks are all metallic, the elements of the p block range from nonmetals, through metalloids, to metals. This variety results in a diversity of chemical properties and some distinctive trends (Section 9.4). There is an increase in metallic

Table 13.1 Selected properties of the elements

	B	Al	Ga	In	Tl
Covalent radius/pm	80	125	125	150	155
Metallic radius/pm		143	141	166	171
Ionic radius, $r(M^{3+})$/pm	27	53	62	94	98
Melting point/°C	2300	660	30	157	304
Boiling point/°C	3930	2470	2400	2000	1460
First ionization energy, I_1/(kJ mol^{-1})	799	577	577	556	590
Second ionization energy, I_2/(kJ mol^{-1})	2427	1817	1979	1821	1971
Third ionization energy, I_3/(kJ mol^{-1})	3660	2745	2963	2704	2878
Electron affinity, E_a/(kJ mol^{-1})	26.7	42.5	28.9	28.9	
Pauling electronegativity	2.0	1.6	1.8	1.8	2.0
$E^{\ominus}(M^{3+},M)$ / V	−0.89	−1.68	−0.53	−0.34	+1.26*

* For coordination number 6.

character from B to Tl: B is a nonmetal, Al is essentially metallic, although it is often classed as a metalloid on account of its amphoteric character, and Ga, In, and Tl are metals. Associated with this trend is a variation from predominantly covalent to ionic bonding in the compounds of the elements that can be rationalized in terms of the increase in atomic radius and related decrease in ionization energy down the group (Table 13.1). Because the ionization energies of the heavier elements are low, the metals form cations increasingly readily down the group. In contrast to the expected trend in electronegativity (Section 1.9), Ga exhibits the alternation effect (Section 9.2c), being more electronegative than Al.

As we discussed in Section 9.8 the first member of each group differs from its congeners on account of its small atomic radius. This difference is particularly evident in Group 13, where the chemical properties of B are distinct from those of the rest of the group. However, B does have a pronounced diagonal relationship with Si in Group 14:

1. Boron and silicon form acidic oxides, B_2O_3 and SiO_2; aluminium forms an amphoteric oxide.

2. Boron and silicon form many polymeric oxide structures and glasses.

3. Boron and silicon form flammable, gaseous hydrides; aluminium hydride is a solid.

The valence electron configuration of the Group 13 elements is ns^2np^1 and, as this configuration suggests, all the elements adopt the +3 oxidation state in their compounds. However, the heavier elements of the group also form compounds with the metal in the +1 oxidation state and this state increases in stability down the group. In fact, the most common oxidation state of Tl is Tl(I). This trend is particularly evident within the halides and is a consequence of the inert-pair effect (Section 9.5). The consequences of the inert-pair effect are evident in Group 13. Thallium(I) is intensely poisonous because its ionic radius is very similar to that of potassium and sodium ions: it enters cells and disrupts the mechanisms of potassium and sodium transport (Section 27.3).

Boron exists in several allotropes. Amorphous B is a brown powder but the hard and refractory crystalline B forms shiny black crystals. The three solid phases for which crystal structures are available contain the icosahedral (20-faced) B_{12} unit as a building block (Fig. 13.1). This icosahedral unit is a recurring motif in boron chemistry and we shall meet it again in the structures of metal borides and boron hydrides. The icosahedral unit is also found in some intermetallic compounds of other Group 13 elements, such as Al_5CuLi_3, $RbGa_7$, and K_3Ga_{13}. Boron is inert, and under normal conditions finely divided B is attacked only by F_2 and HNO_3.

Even though Al is an electropositive metal it is very inert on account of the presence of a passivating surface oxide film. If this film is removed then Al is rapidly oxidized by air. Aluminium has a high reflectance, which is maintained in the powdered form, making it a useful component of silver-coloured paints. It is a good thermal and electrical conductor.

Gallium is brittle at low temperatures but liquefies at 30°C. Its low melting point is attributed to its crystal structure, in which each Ga atom has only one nearest neighbour

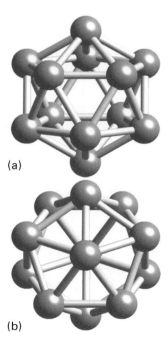

(a)

(b)

Figure 13.1 A view of the B_{12} icosahedron in a-rhombohedral boron (a) along and (b) perpendicular to the threefold axis of the crystal. The individual icosahedra are linked by 3c,2e bonds.

and six next-nearest neighbours: thus, the Ga atoms tend to form Ga−Ga pairs. Gallium has the widest liquid range (30−2420°C) of any element with the exception of Hg and Cs. Unlike Hg, Ga wets glass and skin, making it more difficult to handle. Gallium readily forms alloys with other metals and diffuses into their lattices, making them brittle. Indium forms a distorted ccp lattice and Tl is hexagonal close packed.

13.2 Compounds

Key point: All of the elements form hydrides, oxides, and halides in the $+3$ oxidation state. The $+1$ oxidation state becomes more stable down the group and is the most stable oxidation state for compounds of thallium.

A most striking feature of the lighter Group 13 elements is their ns^2np^1 electron configuration, which contributes up to a maximum of six electrons in the valence shell when three covalent bonds are formed by electron sharing. As a result, many of their compounds have an incomplete octet and act as Lewis acids, being able to complete their octet by accepting a pair of electrons from a donor. Moreover, as is typical of an element at the head of its group, the chemical properties of B and its compounds are strikingly different from those of its congeners.

> *A note on good practice* Be careful to distinguish electron deficiency from the possession of an incomplete octet. The former refers to the lack of sufficient electrons to be able to account for the connections between atoms as normal covalent bonds; the latter is the possession of less than eight electrons in a valence shell.

The binary hydrogen compounds of B are called boranes. The simplest member of the series, diborane, B_2H_6 (**1**), is electron deficient and its structure is commonly described in terms of $2c,2e$ and $3c,2e$ bonds (Section 2.11): bridging $3c,2e$ bonds are a recurring theme in borane chemistry. All the boron hydrides burn with a characteristic green flame and several of them ignite explosively on contact with air. Alkali metal tetrahydridoboranates, $NaBH_4$ and $LiBH_4$, are very useful in the laboratory as general reducing agents and as precursors for most boron−hydrogen compounds.

Boron trihalides consist of trigonal-planar BX_3 molecules. Unlike the halides of the other elements in the group, they are monomeric in the gas, liquid, and solid states. Boron trifluoride and boron trichloride are gases, the tribromide is a volatile liquid, and the triiodide is a solid (Table 13.2). This trend in volatility is consistent with the increase in strength of dispersion forces with the number of electrons in the molecules. Boron trihalides have an incomplete octet and are Lewis acids. The order of Lewis acidity is $BF_3 < BCl_3 \leq BBr_3$ and contrary to the order of electronegativity of the attached halogens (Section 4.8). The electron deficiency is partially removed by X−B π bonding between the halogen atoms and the B atom, giving rise to the partial occupation of the vacant p orbital on the B atom by electrons donated by the halogen atoms (Fig. 13.2). The trend in Lewis acidity stems from more efficient X−B π bonding for the lighter, smaller halogens, the F−B bond being one of the strongest single bonds known.

The most important oxide of B, B_2O_3, is prepared by dehydration of boric acid.

$$4\,B(OH)_3(s) \xrightarrow{\Delta} 2\,B_2O_3(s) + 6\,H_2O(l)$$

The vitreous form of the oxide consists of a network of partially ordered trigonal BO_3 units. Crystalline B_2O_3 consists of an ordered network of BO_3 units joined through O atoms. Metal oxides dissolve in molten B_2O_3 to give coloured glasses. Boron oxide and

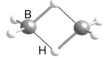

1 Diborane, B_2H_6

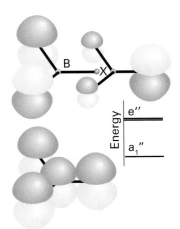

Figure 13.2 The bonding π orbitals of boron trihalide are largely localized on the electronegative halogen atoms, but overlap with a p orbital of boron is significant in the a_1'' orbital.

Table 13.2 Properties of the boron trihalides

	BF$_3$	BCl$_3$	BBr$_3$	BI$_3$
Melting point/°C	−127	−107	−46	50
Boiling point/°C	−100	13	91	210
Bond length/pm	130	175	187	210
$\Delta_f G^{\ominus}$/(kJ mol^{-1})	−1112	−339	−232	+21

silica are the main constituents of borosilicate glass, which, because of the low thermal expansivity of the glass due to the strong B—O bonds, is used to make heat-resistant laboratory glassware.

There are many molecular compounds that contain BN bonds and many of them are analogous to carbon compounds. The similarities between compounds containing BN and CC units can be explained by the fact that these units are isoelectronic. The simplest compound of B and N, boron nitride, BN, is easily synthesized by heating boron oxide with a nitrogen compound (Box 13.1):

$$B_2O_3(l) + 2NH_3(g) \xrightarrow{1200°C} 2BN(s) + 3H_2O(g)$$

The structure of one form of boron nitride consists of planar sheets of atoms like those in graphite (Section 14.5) and some of the physical properties of BN are similar to those of graphite. For example, both graphite and BN have a slippery feel and are used as lubricants. However, BN is a white, nonconducting solid, not a black, metallic conductor. Apart from layered boron nitride, the best-known unsaturated compound of B and N is borazine, $B_3N_3H_6$ (**2**), which is isoelectronic and isostructural with benzene and, like benzene, is a colourless liquid (b.p. 55°C).

The elements Al, Ga, In, and Tl are metals with many similarities in their chemical properties. Like B, they form electron-deficient compounds that act as Lewis acids. Aluminium forms alloys with many other metals and produces light, corrosion-resistant materials. When Al is alloyed with Ga, the Ga prevents the formation of the tightly held passivating oxide film on the Al. When the alloy is dropped in water the Al reacts with the water, forming aluminium oxide and liberating hydrogen. This reaction has been suggested as a solution to the problem of storing hydrogen for use in fuel cells. However, the energy cost of recycling the aluminium oxide and gallium mixture to metallic Al and Ga makes this application unlikely.

Aluminium hydride, AlH_3, is a solid that is best regarded as saline, like the hydrides of the s-block metals. Unlike CaH_2 and NaH, which are more readily available commercially, AlH_3 has few applications in the laboratory. The alkylaluminium hydrides, such as $Al_2(C_2H_5)_4H_2$, are well-known molecular compounds and contain Al—H—Al $3c,2e$ bonds (Section 2.11).

All the elements form trihalides with the metal in its +3 oxidation state. However, as we expect from the inert-pair effect (Section 9.5), the +1 oxidation state becomes more common on descending the group and Tl forms stable monohalides. Because the F^- ion is so small, the trifluorides are mechanically hard ionic solids that have much higher melting points and sublimation enthalpies than the other halides. Their high lattice enthalpies also result in them having very limited solubility in most solvents, and they do not act as Lewis acids to simple donor molecules. The heavier trihalides of Al, Ga, and In are soluble in a wide variety of polar solvents and are excellent Lewis acids. The trigonal planar MX_3 monomer occurs only at elevated temperatures in the gas phase. Otherwise, the trihalides exist as M_2X_6 dimers in the vapour phase and in solution. The volatile solids are dimeric. An exception is $AlCl_3$, which has a six-coordinate layer structure in the solid phase and converts to four-coordinate molecular dimers at its melting point. The dimers contain co-ordinate M—X bonds in which a lone pair on X belonging to one AlX_3 unit completes the octet of M belonging to the second MX_3 unit (**3**). This arrangement results in a tetrahedral arrangement of X atoms around each M atom. In contrast to the other elements in the group, Tl(I) is the most stable oxidation state of the halides.

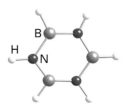

2 Borazine, $B_3N_3H_6$

3 Al_2Cl_6

BOX 13.1 Applications of boron nitride

Hexagonal boron nitride was first developed to meet the needs of the aerospace industry. It is stable in oxygen and is not attacked by steam below 900°C. It is a good thermal insulator, has low thermal expansion, and is resistant to thermal shock. These applications have led to its use in industry to make high-temperature crucibles. The powder is used as a mould release and thermal insulator. Boron nitride nanotubes have been formed by depositing boron and nitrogen on a tungsten surface under high vacuum. These nanotubes could be suitable for high-temperature conditions under which carbon nanotubes would burn.

The softness and sheen of powdered boron nitride has led to its widest application, in the cosmetics and personal care industries. It is nontoxic and presents no known hazard, and is added to many products up to around 10 per cent. It adds a pearlescent sheen to products such as nail polishes and lipsticks, and is added to foundations to hide wrinkles. Its light-reflective properties scatter the light, making wrinkles less noticeable.

The most stable form of Al_2O_3, α-alumina, is a very hard, refractory, and amphoteric material. Dehydration of aluminium hydroxide at temperatures below 900°C leads to the formation of γ-alumina, which is a metastable polycrystalline form with a defect spinel structure (Section 3.9b) and a very high surface area. The α and γ forms of Ga_2O_3 have the same structures as their Al analogues. Indium and thallium form In_2O_3 and Tl_2O_3. Thallium also forms the Tl(I) oxide and peroxide, Tl_2O and Tl_2O_2, respectively.

The most important oxosalts of Group 13 are the *alums*, $MAl(SO_4)_2.12H_2O$, where M is a univalent cation such as Na^+, K^+, Rb^+, Cs^+, Tl^+, or NH_4^+. Gallium and In can also form analogous series of salts of this type but B and Tl do not: a B atom is too small and a Tl atom is too large. The alums can be thought of as double salts containing the hydrated trivalent cation $[Al(OH_2)_6]^{3+}$. The remaining water molecules form hydrogen bonds between the cations and sulfate ions. The mineral *alum*, $KAl(SO_4)_2.12H_2O$, from which aluminium takes its name, is the only common, water-soluble, aluminium-bearing mineral. It has been used since ancient times as a mordant to fix dyes to textiles. The term 'alum' is used widely to describe other compounds with the general formula $M^IM'^{III}(SO_4)_2.12H_2O$, where M' is often a d metal, such as Fe in 'ferric alum', $KFe(SO_4)_2.12H_2O$.

13.3 Boron clusters

Key point: Boron forms an extensive range of polymeric, cage-like compounds which include the borohydrides, metallaboranes, and the carboranes.

In addition to the simple hydrides, B forms several series of neutral and anionic polymeric cage-like boron−hydrogen compounds. Borohydrides are formed with up to 12 B atoms and fall into three classes called *closo*, *nido*, and *arachno*.

The borohydrides with the formula $[B_nH_n]^{2-}$ have a ***closo*** structure, a name derived from the Greek for 'cage'. This series of anions is known for $n = 5$ to 12, and examples include the trigonal-bipyramidal $[B_5H_5]^{2-}$ ion (**4**), the octahedral $[B_6H_6]^{2-}$ ion (**5**), and the icosahedral $[B_{12}H_{12}]^{2-}$ ion (**6**). When boron clusters have the formula B_nH_{n+4} they adopt the ***nido*** structure, a name derived from the Latin for 'nest'. An example is B_5H_9 (**7**). Clusters of formula B_nH_{n+6} have an ***arachno*** structure, from the Greek for 'spider' (as they resemble untidy spiders' webs). One example is pentaborane(11) (B_5H_{11}, **8**).

Boron forms many metal-containing clusters called the **metallaboranes**. In some cases the metal is attached to a borohydride ion through hydrogen bridges. A more common and generally more robust group of metallaboranes have direct M−B bonds.

Closely related to the polyhedral boranes and borohydrides are the **carboranes** (more formally, the *carbaboranes*), a large family of clusters that contain both B and C atoms. An analogue of $B_6H_6^{2-}$ (**9**) is the neutral carborane $B_4C_2H_6$ (**10**).

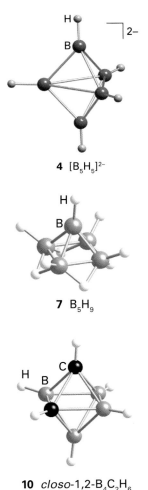

4 $[B_5H_5]^{2-}$

7 B_5H_9

10 *closo*-1,2-$B_4C_2H_6$

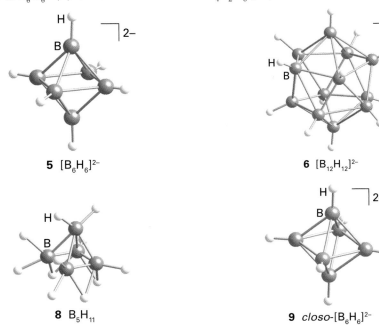

5 $[B_6H_6]^{2-}$

6 $[B_{12}H_{12}]^{2-}$

8 B_5H_{11}

9 *closo*-$[B_6H_6]^{2-}$

PART B: **THE DETAIL**

In this section we present a more detailed discussion of the chemistry of the elements of Group 13, interpreting some of the observed properties in terms of the trends from nonmetallic to metallic character down the group and the impact of the incomplete octet and associated Lewis acidity on their properties. The properties of boron are dealt with separately.

13.4 Occurrence and recovery

Key points: Aluminium is highly abundant; thallium and indium are the least abundant of the Group 13 elements.

Boron exists in several hard and refractory allotropes. The three solid phases for which crystal structures are available contain the icosahedral (20-faced) B_{12} unit as a building block (Fig. 13.1). This icosahedral unit is a recurring motif in boron chemistry and we shall meet it again in the structures of metal borides and boron hydrides. The icosahedral unit is also found in some intermetallic compounds and Zintl phases (Section 3.8c) of other Group 13 elements, such as Al_5CuLi_3, $RbGa_7$, and K_3Ga_{13}.

Boron occurs naturally as *borax*, $Na_2B_4O_5(OH)_4.8H_2O$, and *kernite*, $Na_2B_4O_5(OH)_4.2H_2O$, from which the impure element is obtained. The borax is converted to boric acid, $B(OH)_3$, and then to boron oxide, B_2O_3. The oxide is reduced with magnesium and washed with alkali and then hydrofluoric acid. Pure B is produced by reduction of BBr_3 vapour with H_2:

$$2\,BBr_3(g) + 3\,H_2(g) \rightarrow 2\,B(s) + 6\,HBr(g)$$

Aluminium is the most abundant metallic element in the crustal regions of the Earth and makes up approximately 8 per cent by mass of crustal rocks. It occurs in numerous clays and aluminosilicate minerals but the commercially most important mineral is *bauxite*, a complex mixture of hydrated aluminium hydroxide and aluminium oxide, from which it is extracted by the Hall–Héroult process on an immense scale (Section 5.18). In this process the bauxite is dissolved in molten cryolite, Na_3AlF_6, the mixture is electrolysed and aluminium is deposited at the cathode. The process is very expensive but this expense is offset by the scale of production, the availability of the raw material, and the use of hydroelectric power. The oxide, alumina, occurs naturally as ruby, sapphire, corundum, and emery.

Gallium oxide occurs as an impurity in bauxite and is normally recovered as a byproduct of the manufacture of aluminium. The process results in the concentration of gallium in the residues from which it is extracted by electrolysis. Indium is produced as a byproduct of the extraction of lead and zinc, and is isolated by electrolysis. Thallium compounds are found in flue dust, which is dissolved in dilute sulfuric acid; hydrochloric acid is then added to precipitate thallium(I) chloride and the metal is extracted by electrolysis.

13.5 Uses of the elements and their compounds

Key points: The most useful compound of boron is borax; the most commercially important element is aluminium.

The main use of B is in borosilicate glasses. Borax has many domestic uses, for example as a water softener, cleaner, and mild pesticide. Boric acid, $B(OH)_3$, is used as a mild antiseptic. Amorphous

brown boron is used in pyrotechnics to impart a bright green colour. Boron is an essential micronutrient in plants. Lightweight, strong boron filaments are used in composite materials for the aerospace industry and in sports equipment. Many compounds of B are superhard materials, having hardness approaching that of diamond. Cubic boron nitride is synthesized at high pressures, which makes it expensive. Rhenium diboride does not require high pressures so production is relatively cheap, but Re is an expensive metal. The material known as 'heterodiamond', sometimes labelled BCN, is formed from diamond and boron nitride by explosive shock synthesis. These compounds are used as substitutes for diamond in cutting tools and blades. Sodium perborate, $NaBO_3$, is used as a chlorine-free bleach in laundry products, cleaning materials, and tooth whitener. It is less aggressive to textiles than chlorine bleaches and active at low temperatures when mixed with an activator such as tetraacetylethylenediamine, which is commonly abbreviated to TAED. Boranes used to be popuar as fuels for rockets but were found to be too pyrophoric to be handled safely. Boranes are being investigated as possible hydrogen storage materials with the hydrogen stored as the ammonia-borane complex $NH_3:BH_3$ (Section 10.4 and 13.3).

Aluminium is the most widely used nonferrous metal. The technological uses of aluminium metal exploit its lightness, resistance to corrosion, and the fact that it is easily recycled. It is used in cans, foils, utensils, in construction, and in aircraft alloys (Box 13.2). Many Al compounds are used as mordants, in water and sewage treatment, in paper production, as food additives, and for waterproofing textiles. Aluminium chloride and chlorohydride are used in antiperspirants and the hydroxide is used as an antacid. Sodium tetrahydridoaluminate, $NaAlH_4$, doped with TiF_3 is used as a hydrogen storage material.

Because the melting point of Ga (30°C) is just above room temperature, it is used in high-temperature thermometers. Gallium and In form a low-melting-point alloy that is used as the safety device in sprinkler systems. Both elements are deposited on glass surfaces to form corrosion-resistant mirrors, and In_2O_3 doped with Sn is used as a transparent, conducting coating for electronic displays and as a heat-reflective coating for light bulbs. Gallium nitride is used in blue laser diodes and is the basis of Blu-ray technology. It is insensitive to ionizing radiation and is used in solar cells in satellites. Gallium arsenide is a semiconductor and used in integrated circuits, light-emitting diodes and solar cells. Thallium compounds were once used to treat ringworm and as a rat and ant poison. However, this application has been banned because of their very high toxicity, which arises from the transport of Tl^+ ions across cell membranes together with K^+ ions (Section 26.3). Thallium is absorbed more efficiently by tumour cells and has been used in nuclear medicine as an imaging agent.

13.6 Simple hydrides of boron

The simplest hydride of boron is gaseous diborane, B_2H_6. Higher boranes exist and can be liquids such as B_5H_9 and solids such as $B_{10}H_{14}$. The boranes are cleaved by Lewis bases.

(a) Boranes

Key points: Diborane can be synthesized by metathesis between a boron halide and a hydride source; many of the higher boranes can be prepared

---

BOX 13.2 Aluminium alloys: lightening the load

The chemical and physical properties of aluminium make it the most widely used nonferrous metal. It is light, has high electrical and thermal conductivity, high reflectance, and is easily machined. These properties are enhanced by the impervious oxide layer on the surface, which makes it resistant to corrosion. However, aluminium is soft and lacks strength, and alloying the metal with small amounts (typically less than 2 per cent by mass) of other elements gives a much more desirable weight to strength ratio.

The properties of the alloys depend on the chemical composition and processing involved. In work-hardening alloys strength is achieved by the amount of 'cold work' applied to the alloy, for example by rolling. In heat-treatable or precipitation alloys the strength and properties are achieved by heat treatments of varying complexity. Binary, tertiary, quaternary, and more complex alloys are used commercially. An Al/Mn alloy is the oldest and most widely used. It almost always contains traces of Si and Fe, which give added strength and hardness. Al/Mg alloys have good ductility and corrosion resistance but are less strong. Al/Si alloys have good fluidity and are used in welding wires. Al/Li alloys have very low densities and high

elasticity. Adding Cu to this alloy increases the density but also increases the strength and corrosion resistance.

The greatest use of aluminium alloys worldwide is in building and construction, where they are used to make walls, roofs, gutters, window frames, and doors. Their strength, weight, and weather resistance make them particularly appropriate for these applications. In the USA and Europe, packaging has replaced construction as the major consumer of aluminium alloys as they are impermeable to gas, flexible, and recyclable. They are used to make drink cans, easy-open lids, aerosol cans, foil−paper−plastic laminates, and sheets and foils. A growing application is in transport, where weight and fuel consumption are serious issues. An Al/Cu/Mg alloy is used in aircraft and an Al/Cu alloy is used in satellites and space vehicles. The use of aluminium alloys in cars is increasingly prevalent and they are used to make body panels, engine blocks, wheels, bumpers, radiators, and trims. Around the home aluminium alloys are used to make cooking utensils, refrigerator bodies, air conditioning units, toys, tools, and collapsible toothpaste and cosmetic tubes.

---

by the partial pyrolysis of diborane; all the boron hydrides are flammable, sometimes explosively, and many of them are susceptible to hydrolysis.

Diborane, B_2H_6, can be prepared in the laboratory by metathesis of a boron halide with either $LiAlH_4$ or $LiBH_4$ in ether:

$$3\,LiBH_4\,(et) + 4\,BF_3\,(et) \rightarrow \quad 3\,LiBF_4\,(et) + 2\,B_2H_6\,(g)$$

That this reaction is a metathesis (an exchange of partners) can be seen by writing it in the simplified form

$$\tfrac{3}{4}\,BH_4^-(et) + BF_3(et) \rightarrow \tfrac{3}{4}\,BF_4^-(et) + BH_3(g)$$

Both $LiBH_4$ and $LiAlH_4$, like LiH, are good reagents for the transfer of H^-, but they are generally preferred over LiH and NaH because they are soluble in ethers. The synthesis is carried out with the strict exclusion of air (typically in a vacuum line because diborane ignites on contact with air). Diborane decomposes very slowly at room temperature, forming higher boron hydrides and a nonvolatile and insoluble yellow solid that consist of $B_{10}H_{14}$ and the polymeric species, BH_n.

The compounds fall into two classes. One class has the formula B_nH_{n+4} and the other, which is richer in hydrogen and less stable, has the formula B_nH_{n+6}. Examples include pentaborane (11), B_5H_{11} (8), tetraborane(10), B_4H_{10} (11), and pentaborane(9), B_5H_9 (7). The nomenclature, in which the number of B atoms is specified by a prefix and the number of H atoms is given in parentheses, should be noted. Thus, the systematic name for diborane is diborane(6); however, as there is no diborane(8), the simpler term 'diborane' is almost always used.

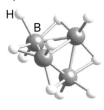

11 B_4H_{10}

All the boranes are colourless and diamagnetic. They range from gases (B_2H_6 and B_4H_8), through volatile liquids (B_5H_9 and

B_6H_{10} hydrides), to the sublimable solid $B_{10}H_{14}$. All the boron hydrides are flammable, and several of the lighter ones, including diborane, react spontaneously with air, often with explosive violence and a green flash (an emission from an excited state of the reaction intermediate BO). The final product of the reaction is the hydrated oxide:

$$B_2H_6\,(g) + 3\,O_2\,(g) \rightarrow \quad 2\,B(OH)_3\,(s)$$

The lighter boranes are readily hydrolysed by water:

$$B_2H_6\,(g) + 6\,H_2O\,(l) \rightarrow \quad 2\,B(OH)_3\,(aq) + 6\,H_2\,(g)$$

As described below, B_2H_6 is a Lewis acid, and the mechanism of this hydrolysis reaction involves coordination of H_2O acting as a Lewis base. Molecular hydrogen then forms as a result of the combination of the partially positively charged H atom on O with the partially negatively charged H atom on B.

(b) Lewis acidity

Key points: Soft and bulky Lewis bases cleave diborane symmetrically; more compact and hard Lewis bases cleave the hydrogen bridge unsymmetrically; although it reacts with many hard Lewis bases, diborane is best regarded as a soft Lewis acid.

As implied by the mechanism of hydrolysis, diborane and many other light boron hydrides act as Lewis acids and are cleaved by reaction with Lewis bases. Two different cleavage patterns have been observed, namely symmetric cleavage and unsymmetric cleavage. In **symmetric cleavage**, B_2H_6 is broken symmetrically into two BH_3 fragments, each of which forms a complex with a Lewis base:

Many complexes of this kind exist. They are interesting partly because they are isoelectronic with hydrocarbons. For instance, the product of this reaction is isoelectric with 2,2-dimethyl-propane, $C(CH_3)_4$. Stability trends indicate that BH_3 is a soft Lewis acid, as illustrated by the reaction

$$H_3B{-}N(CH_3)_3 + F_3B{-}S(CH_3)_2 \rightarrow H_3B{-}S(CH_3)_2 + F_3B{-}N(CH_3)_3$$

in which BH_3 transfers to the soft S donor atom and the harder Lewis acid, BF_3, combines with the hard N donor atom.

The direct reaction of diborane and ammonia results in **unsymmetrical cleavage**, which is cleavage leading to an ionic product:

Unsymmetrical cleavage of this kind is generally observed when diborane and a few other boron hydrides react with strong, sterically uncrowded bases at low temperatures. The steric repulsion is such that only two small ligands can attack one B atom in the course of the reaction.

EXAMPLE 13.1 Using NMR to identify reaction products

Explain how ^{11}B-NMR could be used to determine whether cleavage of diborane with an NMR-inactive Lewis base is symmetrical or unsymmetrical (Section 8.5).

Answer We need to identify the possible products of the two reactions and then decide how the features of their NMR spectra will differ. Symmetrical cleavage of B_2H_6 with L yields $BH_3L + BH_3L$ and unsymmetrical cleavage yields $BH_2L_2^+$ and BH_4^-. In the former, ^{11}B is coupled to three equivalent ^{1}H nuclei and we would therefore observe a quartet in the NMR spectrum. In unsymmetrical cleavage the first product has ^{11}B coupled to two equivalent ^{1}H nuclei, which would produce a triplet. The second product has ^{11}B coupled to four equivalent nuclei, which would produce a quintuplet.

Self-test 13.1 ^{11}B nuclei have $I = \frac{3}{2}$. Predict the number of lines and their relative intensities in the ^{1}H-NMR spectrum of BH_4^-.

(c) Hydroboration

Key points: Hydroboration, the reaction of diborane with alkenes in ether solvent, produces organoboranes that are useful intermediates in synthetic organic chemistry.

An important component of a synthetic chemist's repertoire of reactions is hydroboration, the addition of HB across a multiple bond:

$$H_3B{-}OR_2 + H_2C{=}CH_2 \xrightarrow{\Delta,\ \text{ether}} CH_3CH_2BH_2 + R_2O$$

From the viewpoint of an organic chemist, the C–B bond in the primary product of hydroboration is an intermediate stage in the stereospecific formation of C–H or C–OH bonds, to which it can be converted. From the viewpoint of the inorganic chemist, the reaction is a convenient method for the preparation of a wide variety of organoboranes. The hydroboration reaction is one of

a class of reactions in which EH adds across the multiple bond; hydrosilylation (Section 14.7b) is another important example.

(d) The tetrahydridoborate ion

Key point: The tetrahydridoborate ion is a useful intermediate for the preparation of metal hydride complexes and borane adducts.

Diborane reacts with alkali metal hydrides to produce salts containing the tetrahydridoborate ion, BH_4^-. Because of the sensitivity of diborane and LiH to water and oxygen, the synthesis must be carried out in the absence of air and in a nonaqueous solvent such as the short-chain polyether $CH_3OCH_2CH_2OCH_3$ (denoted here 'polyet'):

$$B_2H_6\,(\text{polyet}) + 2\,\text{LiH}\,(\text{polyet}) \rightarrow 2\,\text{LiBH}_4\,(\text{polyet})$$

We can view this reaction as another example of the Lewis acidity of BH_3 towards the strong Lewis basicity of H^-. The BH_4^- ion is isoelectronic with CH_4 and NH_4^+, and the three species show the following variation in chemical properties as the electronegativity of the central atom increases:

$$BH_4^- \qquad CH_4 \qquad NH_4^+$$
Character: hydridic   –   protic

where 'protic' denotes Brønsted acid (proton donating) character; CH_4 is neither acidic nor basic under the conditions prevailing in aqueous solution.

Alkali metal tetrahydridoborates are very useful laboratory and commercial reagents. They are often used as a mild source of H^- ions, as general reducing agents, and as precursors for most boron–hydrogen compounds, and they are utilized as hydrogen storage materials (Box 13.3 and 10.4). Most of these reactions are carried out in polar nonaqueous solvents. The preparation of diborane, mentioned previously,

$$3\,\text{LiEH}_4 + 4\,BF_3 \rightarrow 2\,B_2H_6 + 3\,\text{LiEF}_4 \qquad (E = B, Al)$$

is one example and $NaBH_4$ in tetrahydrofuran (THF) is used to reduce aldehydes and ketones to alcohols. Although BH_4^- is thermodynamically unstable with respect to hydrolysis, the reaction is very slow at high pH and some synthetic applications have been devised in water. For example, germane (GeH_4) can be prepared by dissolving GeO_2 and KBH_4 in aqueous potassium hydroxide and then acidifying the solution:

$$HGeO_3^-\,(\text{aq}) + BH_4^-\,(\text{aq}) + 2\,H^+\,(\text{aq}) \rightarrow GeH_4\,(g) + B(OH)_3\,(\text{aq})$$

Aqueous BH_4^- also can serve as a simple reducing agent, as in the reduction of aqua ions such as Ni^{2+} or Cu^{2+} to the metal or metal boride. With halogen complexes of 4d and 5d elements that also have stabilizing ligands, such as phosphines, tetrahydroborate ions can be used to introduce a hydride ligand by a metathesis reaction in a nonaqueous solvent:

$$RuCl_2(PPh_3)_3 + NaBH_4 + PPh_3 \xrightarrow{\Delta\ \text{benzene/ethanol}} RuH_2(PPh_3)_4 + \text{other products}$$

It is probable that many of these metathesis reactions proceed through a transient BH_4^- complex. Indeed, hydridoborate complexes are known, especially with highly electropositive metals: they include $Al(BH_4)_3$ (**12**), which contains a diborane-like double hydride bridge, and $Zr(BH_4)_4$ (**13**), in which triple hydride

BOX 13.3 Group 13 elements in hydrogen storage

Hydrogen fuel cells are seen as an alternative to carbon-based fuel and are starting to find applications in mobile technologies and motor vehicles. Efficient fuel cells demand an efficient source of hydrogen and many methodologies for storing hydrogen have been investigated. These include using high pressure and porous materials but others focus on chemical compounds that generate H_2 on heating or on reaction with water. The boron and aluminium hydrides come into this last category. Attractive compounds have a high mass percentage hydrogen content. The values for $LiBH_4$, $NaBH_4$, $LiAlH_4$, and AlH_3 are approximately 18, 11, 11, and 10 mass per cent, respectively.

Sodium tetrahydridoborate, $NaBH_4$, reacts with water to generate hydrogen gas in an exothermic reaction.

$$NaBH_4(aq) + 4H_2O(l) \rightarrow 4H_2(g) + NaB(OH)_4(aq) \quad \Delta_r H^{\ominus} = -300 \text{ kJ mol}^{-1}$$

The reaction requires a nickel or platinum catalyst and rapidly produces moist hydrogen for the engine or fuel cell. The $NaBH_4$ is used as a 30 mass per cent solution in water and the fuel is thus a nonvolatile, nonflammable liquid at atmospheric pressure. There are no side reactions or volatile by-products and the borate product can be recycled.

Ammonia borane, BH_3NH_3, with a hydrogen content of 21 mass per cent, has also been investigated for hydrogen generation. It was investigated as a rocket fuel in the 1950s but the studies were abandoned. Ammonia borane decomposes to liberate hydrogen when heated to 500°C. The residue is boron nitride, which cannot be easily recycled. Recent studies have investigated the hydrogen-storage potential of the ammonia complex of magnesium borohydride $Mg(BH_4)_2 \cdot 2NH_3$. The complex contains 16 mass per cent hydrogen, which is released when a solution of the complex flows over a ruthenium catalyst. The complex begins to decompose at 150°C, with a maximum hydrogen release rate at 205°C, making it competitive with ammonia borane BH_3NH_3 as a hydrogen-storage material.

---

bridges are present. We see from these examples that many compounds can be described in terms of $3c,2e$ bonds.

EXAMPLE 13.2 Predicting the reactions of boron–hydrogen compounds

By means of a chemical equation, indicate the products resulting from the interaction of equal amounts of $[HN(CH_3)_3]Cl$ with $LiBH_4$ in tetrahydrofuran (THF).

Answer We should expect LiCl, with its high lattice enthalpy, to be a likely product. If this is the case we shall be left with BH_4^- and $[HN(CH_3)_3]^+$. The interaction of the hydridic BH_4^- ion with the protic $[HN(CH_3)_3]^+$ ion will evolve hydrogen to produce trimethylamine and BH_3. In the absence of other Lewis bases, the BH_3 molecule would coordinate to THF; however, the stronger Lewis base trimethylamine is produced in the initial reactions, so the overall reaction will be

$$[HN(CH_3)_3]Cl + LiBH_4 \rightarrow H_2 + H_3BN(CH_3)_3 + LiCl$$

The B atom in $H_3BN(CH_3)_3$ has four groups attached tetrahedrally.

Self-test 13.2 Write an equation for the reaction of B_2H_6 with propene in ether solvent and a 1:1 stoichiometry and another equation for its reaction with ammonium chloride in THF with the same stoichiometry.

13.7 Boron trihalides

Key points: Boron trihalides are useful Lewis acids, with BCl_3 stronger than BF_3, and important electrophiles for the formation of boron–element bonds; subhalides with B–B bonds, such as B_2Cl_4, are also known.

All the boron trihalides except BI_3 may be prepared by direct reaction between the elements. However, the preferred method for BF_3 is the reaction of B_2O_3 with CaF_2 in H_2SO_4. This reaction is driven in part by production of HF from the reaction of the H_2SO_4 with CaF_2 and the stability of the $CaSO_4$:

$$B_2O_3(s) + 3CaF_2(s) + 6H_2SO_4(l) \rightarrow$$
$$2BF_3(g) + 3[H_3O][HSO_4](soln) + 3CaSO_4(s)$$

All the boron trihalides form simple Lewis complexes with suitable bases, as in the reaction

$$BF_3(g) + :NH_3(g) \rightarrow F_3B\text{—}NH_3(s)$$

However, boron chlorides, bromides, and iodides are susceptible to protolysis by mild proton sources such as water, alcohols, and even amines. As shown in Fig. 13.3, this reaction, together with metathesis reactions, is very useful in preparative chemistry. An example is the rapid hydrolysis of BCl_3 to give boric acid, $B(OH)_3$:

$$BCl_3(g) + 3H_2O(l) \rightarrow B(OH)_3(aq) + 3HCl(aq)$$

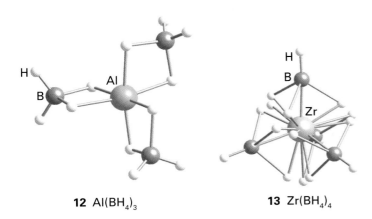

12 $Al(BH_4)_3$ **13** $Zr(BH_4)_4$

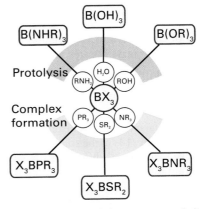

Figure 13.3 The reactions of boron–halogen compounds (X = halogen).

It is probable that a first step in this reaction is the formation of the complex Cl_3B—OH_2, which then eliminates HCl and reacts further with water.

EXAMPLE 13.3 **Predicting the products of reactions of the boron trihalides**

Predict the likely products of the following reactions and write the balanced chemical equations: (a) BF_3 and excess NaF in acidic aqueous solution, (b) BCl_3 and excess NaCl in acidic aqueous solution, (c) BBr_3 and excess $NH(CH_3)_2$ in a hydrocarbon solvent.

Answer We need to consider whether the B—X bond is susceptible to hydrolysis. (a) The F^- ion is a chemically hard and fairly strong base; BF_3 is a hard and strong Lewis acid with a high affinity for the F^- ion. Hence, the reaction should result in a complex:

$$BF_3(g) + F^-(aq) \rightarrow BF_4^-(aq)$$

Excess F^- and acid prevent the formation of hydrolysis products such as BF_3OH^-, which are formed at high pH. (b) Unlike B—F bonds, which are very strong and only mildly susceptible to hydrolysis, the other boron—halogen bonds are hydrolysed vigorously by water. We can anticipate that BCl_3 will undergo hydrolysis rather than coordinate to aqueous Cl^-:

$$BCl_3(g) + 3 H_2O(l) \rightarrow B(OH)_3(aq) + 3 HCl(aq)$$

(c) Boron tribromide will undergo protolysis with formation of a B—N bond:

$$BBr_3(g) + 6 NH(CH_3)_2 \rightarrow B(N(CH_3)_2)_3 + 3 [NH_2(CH_3)_2]Br$$

In this reaction the HBr produced by the protolysis protonates excess dimethylamine.

Self-test 13.3 Write and justify balanced equations for plausible reactions between (a) BCl_3 and ethanol, (b) BCl_3 and pyridine in hydrocarbon solution, (c) BBr_3 and $F_3BN(CH_3)_3$.

The tetrafluoridoborate anion, BF_4^-, which is mentioned in Example 13.3, is used in preparative chemistry when a relatively large noncoordinating anion is needed. The tetrahalidoborate anions BCl_4^- and BBr_4^- can be prepared in nonaqueous solvents. However, because of the ease with which B—Cl and B—Br bonds undergo solvolysis, they are stable in neither water nor alcohols.

Boron halides are the starting point for the synthesis of many boron—carbon and boron—pseudohalogen compounds (Section 17.3).[1] Examples include the formation of alkylboron and arylboron compounds, such as trimethylboron, by the reaction of boron trifluoride with a methyl Grignard reagent in ether solution:

$$BF_3 + 3 CH_3MgI \rightarrow B(CH_3)_3 + magnesium\ halides$$

When an excess of the Grignard (or organolithium) reagent is present, tetraalkyl or tetraaryl borates are formed:

$$BF_3 + Li_4(CH_3)_4 \rightarrow Li[B(CH_3)_4] + 3 LiF$$

Boron halides containing B—B bonds have been prepared. The best known of these compounds have the formula B_2X_4, with X = F, Cl, and Br, and the tetrahedral cluster compound B_4Cl_4. The B_2Cl_4 molecules are planar (**14**) in the solid state but staggered (**15**) in the gas. This conformational difference suggests that rotation about the B—B bond is quite easy, as is expected for a single bond.

[1]Pseudohalogens are species that resemble the halogens in their chemical properties. Cyanogen, $(CN)_2$, is a pseudohalogen and the cyanide ion, CN^-, is a pseudohalide.

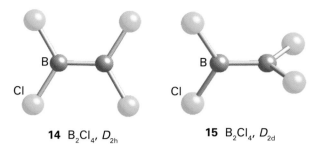

14 B_2Cl_4, D_{2h} **15** B_2Cl_4, D_{2d}

One route to B_2Cl_4 is to pass an electric discharge through BCl_3 gas in the presence of a Cl atom scavenger, such as mercury vapour. Spectroscopic data indicate that BCl is produced by electron impact on BCl_3:

$$BCl_3(g) \xrightarrow{\text{electron impact}} BCl(g) + 2 Cl(g)$$

The Cl atoms are scavenged by mercury vapour and removed as $Hg_2Cl_2(s)$, and the BCl fragment is thought to combine with BCl_3 to yield B_2Cl_4. Metathesis reactions can be used to make B_2X_4 derivatives from B_2Cl_4. The thermal stability of these derivatives increases with increasing tendency of the X group to form a π bond with B:

$$B_2Cl_4 < B_2F_4 < B_2(OR)_4 \ll B_2(NR_2)_4$$

It was thought for a long time that X groups with lone pairs were essential for the existence of B_2X_4 compounds, but diboron compounds with alkyl or aryl groups have been prepared. Compounds that survive at room temperature can be obtained when the groups are bulky, as in $B_2(^tBu)_4$.

A secondary product in the synthesis of B_2Cl_4 is B_4Cl_4, a pale yellow solid composed of molecules with the four B atoms forming a tetrahedron (**16**). Like B_2Cl_4, B_4Cl_4 does not have a formula analogous to those of the boranes (such as B_2H_6) discussed below. This difference may lie in the tendency of halogens to form π bonds with boron by the donation of lone electron pairs on the halide into the otherwise vacant p orbital on B, as in Fig. 13.2 (Section 4.10b).

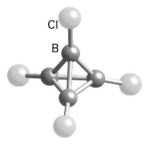

16 B_4Cl_4, T_d

13.8 Boron—oxygen compounds

Key point: Boron forms B_2O_3, polyborates, and borosilicate glasses.

Boric acid, $B(OH)_3$, is a very weak Brønsted acid in aqueous solution. However, the equilibria are more complicated than the simple Brønsted proton transfer reactions characteristic of the later p-block oxoacids. Boric acid is in fact primarily a weak Lewis acid, and the complex it forms with H_2O, $H_2OB(OH)_3$, is the actual source of protons:

$$B(OH)_3(aq) + 2 H_2O(l) \rightleftharpoons H_3O^+(aq) + [B(OH)_4]^-(aq) \quad pK_a = 9.2$$

As is typical of many of the lighter elements of the p block, there is a tendency for the anion to polymerize by condensation, with the loss of H_2O. Thus, in concentrated neutral or basic solution, equilibria such as

$$3\,B(OH)_3(aq) \rightleftharpoons [B_3O_3(OH)_4]^-(aq) + H^+(aq) + 2\,H_2O(l)$$

$$pK_a = 0.85$$

occur to yield polynuclear anions (**17**).

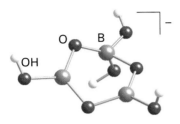

17 $[B_3O_3(OH)_4]^-$

The reaction of boric acid with an alcohol in the presence of sulfuric acid leads to the formation of simple borate esters, which are compounds of the form $B(OR)_3$:

$$B(OH)_3 + 3\,CH_3OH \xrightarrow{H_2SO_4} B(OCH_3)_3 + 3\,H_2O$$

Borate esters are much weaker Lewis acids than the boron trihalides, presumably because the O atom acts as an intramolecular π donor, like the F atom in BF_3 (Section 4.10b), and donates electron density to the p orbital of the B atom. Hence, judging from Lewis acidity, an O atom is more effective than an F atom as a π donor towards B. 1,2-Diols have a particularly strong tendency to form borate esters on account of the chelate effect (Section 20.1), and produce a cyclic borate ester (**18**).

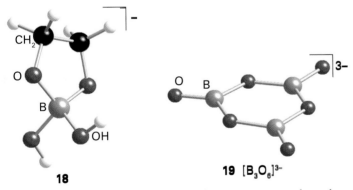

18

19 $[B_3O_6]^{3-}$

As with silicates and aluminates, there are many polynuclear borates, and both cyclic and chain species are known. An example is the cyclic polyborate anion $B_3O_6^{3-}$ (**19**). A notable feature of borate formation is the possibility of both three-coordinate B atoms, as in (**19**), and four-coordinate B atoms, as in $[B(OH)_4]^-$. Polyborates form by sharing one O atom with a neighbouring B atom, as in (**19**); structures in which two adjacent B atoms share two or three O atoms are unknown.

Boron oxide, B_2O_3, is acidic and is prepared by dehydration of boric acid:

$$2\,B(OH)_3(s) \xrightarrow{\Delta} B_2O_3(s) + 3\,H_2O(g)$$

The rapid cooling of molten B_2O_3 or metal borates often leads to the formation of borate glasses. Although these glasses themselves have little technological significance, the fusion of sodium borate with silica leads to the formation of borosilicate glasses (such as Pyrex). Borosilicate glasses are resistant to thermal shock and can be heated over a flame or other source of direct heat.

Sodium perborate is used as a bleach in laundry powders, automatic dishwasher powders, and whitening toothpastes. Although the formula is often given as $NaBO_3.4H_2O$, the compound contains the peroxide anion, O_2^{2-}, and is more accurately described as $Na_2[B_2(O_2)_2(OH)_4].6H_2O$. The compound is preferred to hydrogen peroxide in many applications because it is more stable and liberates oxygen only at elevated temperatures.

13.9 Compounds of boron with nitrogen

Key points: Compounds containing BN, which is isoelectronic with CC, include the ethane analogue ammonia borane H_3NBH_3, the benzene analogue $H_3N_3B_3H_3$, and BN analogues of graphite and diamond.

The thermodynamically stable phase of boron nitride, BN, consists of planar sheets of atoms like those in graphite (Section 14.5). The planar sheets of alternating B and N atoms consist of edge-shared hexagons and, as in graphite, the B—N distance within the sheet (145 pm) is much shorter than the distance between the sheets (333 pm, Fig. 13.4). The difference between the structures of graphite and boron nitride, however, lies in the register of the atoms of neighbouring sheets: in BN, the hexagonal rings are stacked directly over each other, with B and N atoms alternating in successive layers; in graphite, the hexagons are staggered. Molecular orbital calculations suggest that the stacking in BN stems from a partial positive charge on B and a partial negative charge on N. This charge distribution is consistent with the electronegativity difference of the two elements ($\chi_P(B) = 2.04$, $\chi_P(N) = 3.04$).

As with impure graphite, layered boron nitride is a slippery material that is used as a lubricant. Unlike graphite, however, it is a colourless electrical insulator, as there is a large energy gap between the filled and vacant π bands. The size of the band gap is consistent with its high electrical resistivity and lack of absorption in the visible spectrum. In keeping with this large band gap, BN forms a much smaller number of intercalation compounds than graphite (Section 14.5). In contrast to graphite, layered boron nitride is stable in air up to 1000°C, making it a useful refractory material.

Layered boron nitride changes into a denser cubic phase at high pressures and temperatures (60 kbar and 2000°C, Fig. 13.5).

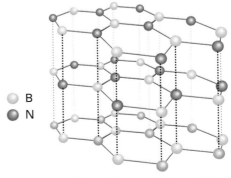

B
N

Figure 13.4 The structure of layered hexagonal boron nitride. Note that the rings are in register between layers.

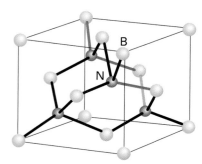

Figure 13.5 The sphalerite structure of cubic boron nitride.

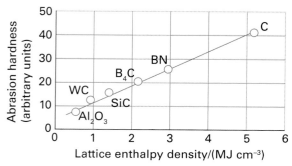

Figure 13.6 The correlation of hardness with lattice enthalpy density (the lattice enthalpy divided by the molar volume of the substance). The point for carbon represents diamond; that for boron nitride represents the diamond-like sphalerite structure.

This phase is a hard crystalline analogue of diamond but, as it has a lower lattice enthalpy, it has a slightly lower mechanical hardness (Fig. 13.6). Cubic boron nitride is manufactured and used as an abrasive for certain high-temperature applications in which diamond cannot be used because it forms carbides with the material being ground.

The fact that BN and CC are isoelectronic suggests that there might be analogies between these compounds and hydrocarbons. Many **amine-boranes**, the boron−nitrogen analogues of saturated hydrocarbons, can be synthesized by reaction between a nitrogen Lewis base and a boron Lewis acid:

$$\tfrac{1}{2}B_2H_6 + N(CH_3)_3 \rightarrow H_3BN(CH_3)_3$$

However, although amine-boranes are isoelectronic with hydrocarbons, their properties are significantly different, in large part due to the difference in electronegativities of B and N. For example, whereas ammoniaborane, H_3NBH_3, is a solid at room temperature with a vapour pressure of a few pascals, its analogue ethane, H_3CCH_3, is a gas that condenses at $-89°C$. This difference can be traced to the difference in polarity of the two molecules: ethane is nonpolar, whereas ammoniaborane has a large dipole moment of 5.2 D (**20**).

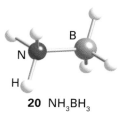

20 NH_3BH_3

Several BN analogues of the amino acids have been prepared, including ammoniacarboxyborane, H_3NBH_2COOH, the analogue of propionic acid, CH_3CH_2COOH. These compounds display significant physiological activity, including tumour inhibition and reduction of serum cholesterol.

The simplest unsaturated boron−nitrogen compound is aminoborane, H_2NBH_2, which is isoelectronic with ethene. It has only a transient existence in the gas phase because it readily forms cyclic ring compounds such as a cyclohexane analogue (**21**). However, the aminoboranes do survive as monomers when the double bond is shielded from reaction by bulky alkyl groups on the N atom and by Cl atoms on the B atom (**22**). For instance, monomeric aminoboranes can be synthesized readily by the reaction of a dialkylamine and a boron halide:

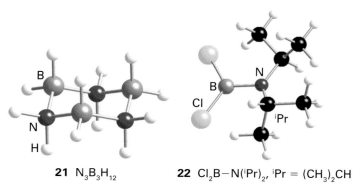

The reaction also occurs with xylyl (2,4,6-trimethylphenyl) groups in place of isopropyl groups.

21 $N_3B_3H_{12}$ **22** $Cl_2B-N(^iPr)_2$, $^iPr = (CH_3)_2CH$

Apart from layered boron nitride, the best known unsaturated compound of boron and nitrogen is borazine, $B_3N_3H_6$ (**23**), which is isoelectronic and isostructural with benzene. Borazine was first prepared by Alfred Stock in 1926 by the reaction between diborane and ammonia. Since then, many symmetrically trisubstituted derivatives have been made by procedures that depend on the protolysis of BCl bonds of BCl_3 by an ammonium salt (**24**):

$$3NH_4Cl + 3BCl_3 \xrightarrow{\Delta} \text{(borazine ring)} + 9HCl$$

The use of an alkylammonium chloride yields *N*-alkyl substituted *B*-trichloroborazines.

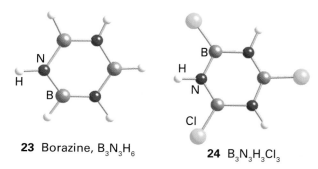

23 Borazine, $B_3N_3H_6$ **24** $B_3N_3H_3Cl_3$

Despite their structural resemblance, there is little chemical resemblance between borazine and benzene. Once again, the difference in the electronegativities of boron and nitrogen is influential, and BCl bonds in trichloroborazine are much more labile than the CCl bonds in chlorobenzene. In the borazine compound, the π electrons are concentrated on the N atoms and there is a partial positive charge on the B atoms that leaves them open to electrophilic attack. A sign of the difference is that the reaction of a chloroborazine with a Grignard reagent or hydride source results in the substitution of Cl by alkyl, aryl, or hydride groups. Another example of the difference is the ready addition of HCl to borazine to produce a trichlorocyclohexane analogue (**25**):

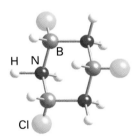

The electrophile, H^+, in this reaction attaches to the partially negative N atom and the nucleophile Cl^- attaches to the partially positive B atom.

25 $B_3N_3H_9Cl_3$

EXAMPLE 13.4 Preparing borazine derivatives

Give balanced chemical equations for the synthesis of borazine starting with NH_4Cl and other reagents of your choice.

Answer As we have just seen, the first step will be the protolysis of the B—Cl bond in BCl_3 by the ammonium ion. Therefore, reaction of NH_4Cl with BCl_3 will yield

$$3\,NH_4Cl + 3\,BCl_3 \rightarrow H_3N_3B_3Cl_3 + 9\,HCl$$

The Cl atoms in *B*-trichloroborazine can then be displaced by hydride ions from reagents such as $LiBH_4$, to yield borazine:

$$3\,LiBH_4 + H_3N_3B_3Cl_3 \xrightarrow{THF} H_3B_3N_3H_3 + 3\,LiCl + 3\,THF \cdot BH_3$$

Self-test 13.4 Suggest a reaction or series of reactions for the preparation of *N,N′,N″*-trimethyl-*B,B′,B″*-trimethylborazine starting with methylamine and boron trichloride.

13.10 Metal borides

Key point: Metal borides include boron anions as isolated B atoms, linked *closo*-boron polyhedra, and hexagonal boron networks.

The direct reaction of elemental boron and a metal at high temperatures provides a useful route to many metal borides. An example is the reaction of Ca and some other highly electropositive metals with B to produce a phase of composition MB_6:

$$Ca(l) + 6\,B(s) \rightarrow CaB_6(s)$$

Metal borides are found with a wide range of compositions, as B can occur in numerous types of structure, including isolated B atoms, chains, planar and puckered nets, and clusters. The simplest metal borides are metal-rich compounds that contain isolated B^{3-} ions. The most common examples of these compounds have the formula M_2B, where M may be one of the middle to late 3d metals (Mn to Ni) in low oxidation states. Another important class of metal borides contains planar or puckered hexagonal nets that have the composition MB_2 (Fig. 13.7). These compounds are formed primarily by electropositive metals, including magnesium, aluminium, the early d metals (from Sc to Mn, for instance, in Period 4), and uranium (Box 13.4).

The boron-rich borides, typically MB_6 and MB_{12}, where M is an electropositive metal, are of even greater structural interest. In them, the B atoms link to form an intricate network of interconnecting cages. In MB_6 compounds (which are formed by the electropositive s-block metals, such as Na, K, Ca, Sr, and Ba, and f-block metals), the B_6 octahedra are linked by their vertices to form a cubic framework (Fig. 13.8).

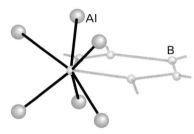

Figure 13.7 The AlB_2 structure. To give a clear picture of the hexagonal layer B atoms outside the unit cell are displayed.

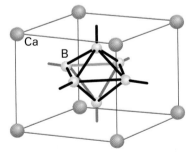

Figure 13.8 The CaB_6 structure. Note that the B_6 octahedra are connected by a bond between vertices of adjacent B_6 octahedra. The crystal is a simple cube analogue of CsCl. Thus eight Ca atoms surround the central B_6 octahedron.

26 B_{12} cuboctahedron

Magnesium diboride, MgB_2, is a cheap compound that has been known in the laboratory for over 50 years. In 2001 this simple compound was found to have superconducting properties (Secrtion 24.7e). Jun Akimitsu and his co-workers discovered by chance that MgB_2 loses its electrical resistance when cooled. At the time, they were characterizing materials used to enhance the performance of known high-temperature superconductors. The discovery led to a major flurry of research on this new superconductor around the world.

In bulk materials the transition temperature of MgB_2 is 38 K and is exceeded only by the much more complicated perovskite cuprate structures (Section 24.7). The discovery of this latest superconductor was even more serendipitous than the discovery of the cuprate superconductors by Bednorz and Müller in 1986, for which they shared the Nobel Prize a year later. Many of the first measurements were made using MgB_2 powder straight from the bottle. High-quality MgB_2 can be synthesized by heating fine

boron and magnesium powders together at around 950°C under pressure. Thin films, wires, and tapes have since been formed that have potential for applications in superconducting magnets, microwave communications, and power applications.

Magnesium diboride has a simple structure in which the B atoms are arranged in graphite-like planes with alternating layers of Mg atoms. The Mg atoms donate their two valence electrons to the network of B atoms. Varying the number of electrons donated to the boron conduction bands can dramatically affect the transition temperature. The transition temperature of the compound falls if some of the Mg atoms are replaced by Al and increases when doped with Cu. The transition temperature, T_c, of MgB_2 is approximately 15 K higher than theory predicts. This difference has been explained in terms of vibrations in the lattice that allow two electrons to form a Cooper pair, which then travels resistance-free through the material.

The linked B_6 clusters bear a charge of -1, -2, or -3 depending on the cation with which they are associated. In the MB_{12} compounds the B-atom networks are based on linked cuboctahedra (**26**) rather than the more familiar icosahedron. This type of compound is formed by some of the heavier electropositive metals, particularly those of the f block.

13.11 Higher boranes and borohydrides

Key point: The bonding in boron hydrides and polyhedral borohydride ions can be approximated by conventional 2c,2e bonds together with 3c,2e bonds.

In this section we describe the structures and properties of the cage-like boranes and borohydrides, which include Stock's series B_nH_{n+4} and B_nH_{n+6} as well as the more recently discovered $B_nH_n^{2-}$ closed polyhedra. The borohydrides have been studied for many years as an interesting class of compounds and have found applications only recently (Box 13.5).

Boron cluster compounds are best considered from the standpoint of fully delocalized molecular orbitals containing electrons that contribute to the stability of the entire molecule. However, it is sometimes fruitful to identify groups of three atoms and to regard them as bonded together by versions of the 3c,2e bonds of the kind that occur in diborane itself (**1**). In the more complex boranes, the three centres of the 3c,2e bonds may be BHB bridge bonds, but they may also be bonds in which three B atoms lie at the corners of an equilateral triangle with their sp^3 hybrid orbitals overlapping at its centre (**27**). To reduce the complexity of

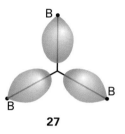

27

the structural diagrams, the illustrations that follow will not in general indicate the 3c,2e bonds in the structures.

(a) Wade's rules

Key points: Wade's rules can be used to predict the structures of polyhedral borohydrides; boron hydride structures include simple polyhedral *closo* compounds and the progressively more open *nido* and *arachno* structures.

A correlation between the number of electrons (counted in a specific way), the formula, and the shape of the molecule was established by Kenneth Wade in the 1970s. These so-called **Wade's rules** apply to a class of polyhedra called **deltahedra** (because they are made up of triangular faces resembling Greek deltas, Δ) and can be used in two ways. For molecular and anionic boranes, they enable us to predict the general shape of the molecule or anion from its formula. However, because the rules are also expressed in terms of the number of electrons, we can extend them to analogous species in which there are atoms other than boron, such as carboranes and other p-block clusters.

A promising new form of radiotherapy for brain tumours involves the irradiation of boron compounds with low-energy neutrons. Boron neutron capture therapy (BNCT) involves injecting the patient with a ^{10}B-labelled boron compound which preferentially binds to tumour cells. When irradiated with neutrons the ^{10}B undergoes nuclear fission and produces a helium nucleus (an alpha particle) and $^7Li^+$ nucleus and liberates approximately 2.4 MeV of energy:

$$^{10}_{5}B + ^{1}_{0}n \rightarrow ^{4}_{2}He + ^{7}_{3}Li^+$$

The most promising boron-containing compounds for this application have been polyhedral borohydrides, $Na_2B_{12}H_{11}SH$ and $Na_2B_{12}H_{12}$, but the factor limiting the progress is the amount of boron that can be introduced into a tumour cell. A breakthrough may come with the recent development of boron carbide nanoparticles. The nanoparticles are introduced into a sample of the patient's own T-cells, which are then injected back into the patient where they travel to the tumour and deliver the nanoparticles. The nanoparticles have also been coated with a peptide that improves cell uptake and labelled with a fluorescent dye that enables the nanoparticles to be tracked within the body.

Here we concentrate on the boron clusters, where knowing the formula is sufficient for predicting the shape. However, so that we can cope with other clusters we shall show how to count the framework electrons too.

The building block from which the deltahedron is constructed is assumed to be one BH group (**28**) that contributes two electrons. The electrons in the B−H bond are ignored in the counting procedure, but all others are included whether or not it is obvious that they help to hold the skeleton together. By the 'skeleton' is meant the framework of the cluster with each BH group counted as a unit. If a B atom happens to carry two H atoms, only one of the B−H bonds is treated as a unit. The second B–H bond lies within the same spherical surface as the B atoms and is included in the skeleton electron count. For instance, in B_5H_{11}, one of the B atoms has two 'terminal' H atoms but only one BH entity is treated as a unit, the other pair of electrons being treated as part of the skeleton and hence referred to as 'skeletal electrons'. A BH group makes two electrons available to the skeleton (the B atom provides three electrons and the H atom provides one but, of these four, two are used for the B−H bond).

■ **A brief illustration.** To count the number of skeletal electrons in B_4H_{10} (**11**) we consider the number of BH units and the number of H atoms. There are four BH units, which contribute $4 \times 2 = 8$ electrons, and the six additional H atoms, which contribute a further six electrons, giving 14 in all. The resulting seven pairs are distributed as shown in (**29**): two are used for the additional terminal B−H bonds, four are used for the four BHB bridges, and one is used for the central B−B bond. ■

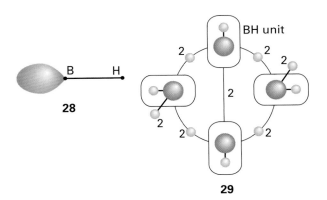

28

BH unit

29

According to Wade's rules (Table 13.3), species of formula $[B_nH_n]^{2-}$ and $n + 1$ pairs of skeletal electrons have a *closo* structure, with a B atom at each corner of a closed deltahedron and no B−H−B bonds. This series of anions is known for $n = 5$ to 12, and examples include the trigonal-bipyramidal $[B_5H_5]^{2-}$ ion, the octahedral $[B_6H_6]^{2-}$ ion, and the icosahedral $[B_{12}H_{12}]^{2-}$ ion. The *closo*-borohydrides and their carborane analogues (Section 13.11) are typically thermally stable and moderately unreactive.

Boron clusters of formula B_nH_{n+4} and $n + 2$ pairs of skeletal electrons have the *nido* structure. They can be regarded as derived from a *closo*-borane that has lost one vertex but have B−H−B bonds as well as B−B bonds. An example is B_5H_9. In general, the thermal stability of the *nido*-boranes is intermediate between that of *closo*- and *arachno*-boranes.

Clusters of formula B_nH_{n+6} and $n+3$ skeletal electron pairs have an *arachno* structure. They can be regarded as *closo*-borane

Table 13.3 Classification and electron count of boron hydrides

Type	Formula* pairs	Skeletal electron	Examples
Closo	$[B_nH_n]^{2-}$	$n + 1$	$[B_5H_5]^{2-}$ to $[B_{12}H_{12}]^{2-}$
Nido	B_nH_{n+4}	$n + 2$	B_2H_6, B_5H_9, B_6H_{10}
Arachno	B_nH_{n+6}	$n + 3$	B_4H_{10}, B_5H_{11}
Hypho†	B_nH_{n+8}	$n + 4$	None‡

* In some cases, protons can be removed; thus $[B_5H_8]^-$ arises from the deprotonation of B_5H_9.
†The name comes from the Greek for 'net'.
‡Some derivatives are known.

polyhedra less two vertices (and must have BHB bonds). One example of an *arachno*-borane is pentaborane(11), (B_5H_{11}). As with most *arachno*-boranes, pentaborane(11) is thermally unstable at room temperature and is highly reactive.

> **EXAMPLE 13.5 Using Wade's rules**
>
> Infer the structure of $[B_6H_6]^{2-}$ from its formula and from its electron count.
>
> **Answer** We should note that the formula $[B_6H_6]^{2-}$ belongs to a class of borohydrides having the formula $[B_nH_n]^{2-}$, which is characteristic of a *closo* species. Alternatively, we can count the number of skeletal electron pairs and from that deduce the structural type. Assuming one B−H bond per B atom, there are six BH units to take into account and therefore 12 skeletal electrons plus two from the overall charge of -2: $6 \times 2 + 2 = 14$, or seven electron pairs which is $n + 1$ with $n = 6$. This number is characteristic of *closo* clusters. The closed polyhedron must contain triangular faces and six vertices; therefore an octahedral structure is indicated.
>
> **Self-test 13.5** How many framework electron pairs are present in B_4H_{10} and to what structural category does it belong? Sketch its structure.

(b) The origin of Wade's rules

Key points: The molecular orbitals in a *closo*-borane can be constructed from BH units, each of which contributes one radial atomic orbital pointing towards the centre of the cluster and two perpendicular p orbitals that are tangential to the polyhedron.

Wade's rules have been justified by molecular orbital calculations. We shall indicate the kind of reasoning involved by considering the first of them (the $n + 1$ rule). In particular, we shall show that $[B_6H_6]^{2-}$ has a low energy if it has an octahedral *closo* structure, as predicted by the rules.

A B−H bond uses one electron and one orbital of the B atom, leaving three orbitals and two electrons for the skeletal bonding. One of these orbitals, which is called a **radial orbital**, can be considered to be a boron sp hybrid pointing towards the interior of the fragment (as in **28**). The remaining two boron p orbitals, the **tangential orbitals,** are perpendicular to the radial orbital (**30**). The shapes of the 18 symmetry-adapted linear combinations of these 18 orbitals in an octahedral B_6H_6 cluster can be inferred from the drawings in *Resource section* 4, and we show the ones with net bonding character in Fig. 13.9.

The lowest energy molecular orbital is totally symmetric (a_{1g}) and arises from in-phase contributions from all the radial orbitals.

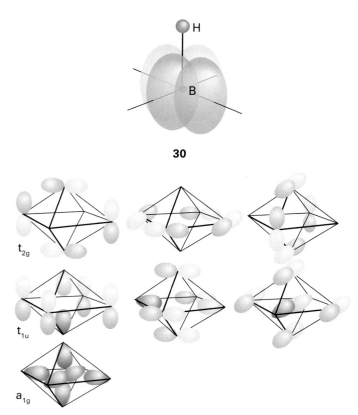

30

Figure 13.9 Radial and tangential bonding molecular orbitals for $[B_6H_6]^{2-}$. The relative energies are $a_{1g} < t_{1u} < t_{2g}$.

Calculations show that the next higher orbitals are the t_{1u} orbitals, each of which is a combination of four tangential and two radial orbitals. Above these three degenerate orbitals lie another three t_{2g} orbitals, which are tangential in character, giving seven bonding orbitals in all. Hence, there are seven orbitals with net bonding character delocalized over the skeleton, and they are separated by a considerable gap from the remaining 11 largely antibonding orbitals (Fig. 13.10).

There are seven electron pairs to accommodate, one pair from each of the six B atoms and one pair from the overall charge (-2). These seven pairs can all enter and fill the seven bonding skeleton orbitals, and hence give rise to a stable structure, in accord with the $n + 1$ rule. Note that the unknown neutral

octahedral B_6H_6 molecule would have too few electrons to fill the t_{2g} bonding orbitals. Similar arguments can be used for all *closo* structures.

(c) Structural correlations

Key points: Conceptually, the *closo*, *nido*, and *arachno* structures are related by the successive removal of a BH fragment and addition of H or electrons.

A very useful structural correlation between *closo*, *nido*, and *arachno* species is based on the observation that clusters with the same numbers of skeletal electrons are related by removal of successive BH groups and the addition of the appropriate numbers of electrons and H atoms. This conceptual process provides a good way to think about the structures of the various boron clusters but does not represent how they are interconverted chemically.

The idea is amplified in Fig. 13.11, where the removal of a BH unit and two electrons and the addition of four H atoms converts the octahedral *closo*-$[B_6H_6]^{2-}$ anion to the square-pyramidal *nido*-B_5H_9 borane. A similar process (removal of a BH unit and addition of two H atoms) converts *nido*-B_5H_9 into a butterfly-like *arachno*-B_4H_{10} borane. Each of these three boranes has 14 skeletal electrons but, as the number of skeletal electrons per B atom increases, the structure becomes more open. A more systematic correlation of this type is indicated for many different boranes in Fig. 13.12.

(d) Synthesis of higher boranes and borohydrides

Key point: Pyrolysis followed by rapid quenching provides one method of converting small boranes to larger boranes.

As discovered by Stock and perfected by many subsequent workers, the controlled pyrolysis of B_2H_6 in the gas phase provides a route to most of the higher boranes and borohydrides, including B_4H_{10}, B_5H_9, and $B_{10}H_{14}$. A key first step in the proposed mechanism is the dissociation of B_2H_6 and the condensation of the resulting BH_3 fragment with borane fragments. For example, the mechanism of the formation of tetraborane(10) by the pyrolysis of diborane appears to be

$$B_2H_6 \rightarrow BH_3 + BH_3$$
$$B_2H_6 + BH_3 \rightarrow B_3H_7 + H_2$$
$$BH_3 + B_3H_7 \rightarrow B_4H_{10}$$

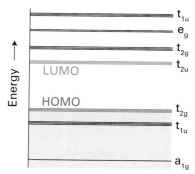

Figure 13.10 Schematic molecular orbital energy levels of the B-atom skeleton of $[B_6H_6]^{2-}$. The form of the bonding orbitals is shown in Fig. 13.9.

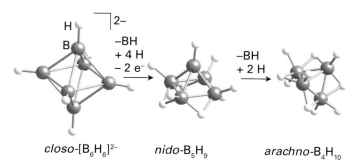

closo-$[B_6H_6]^{2-}$ *nido*-B_5H_9 *arachno*-B_4H_{10}

Figure 13.11 Structural correlations between a B_6 *closo* octahedral structure, a B_5 *nido* square pyramid, and a B_4 *arachno* butterfly.

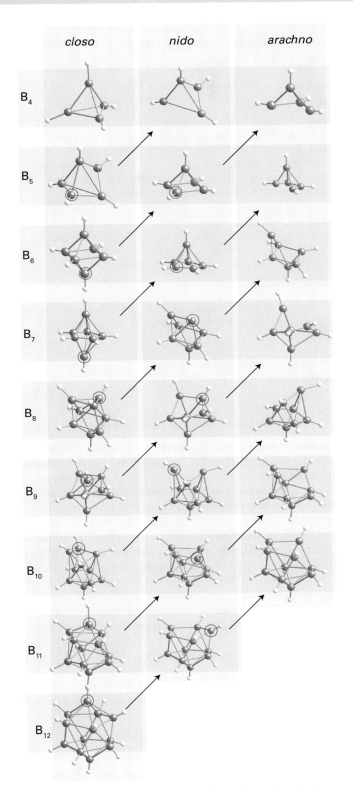

Figure 13.12 Structural relations between *closo*, *nido*, and *arachno* boranes and heteroatomic boranes. Diagonal lines connect species that have the same number of skeletal electrons. Hydrogen atoms beyond those in the B—H framework and charges have been omitted. The encircled atom shows which one is removed to produce the structure to its upper right. (Based on R.W. Rudolph, *Acc. Chem. Res.*, 1976, **9**, 446.)

The synthesis of tetraborane(10), B_4H_{10}, is particularly difficult because it is highly unstable, in keeping with the instability of the B_nH_{n+6} (*arachno*) series. To improve the yield, the product that emerges from the hot reactor is immediately quenched on a cold surface. Pyrolytic syntheses to form species belonging to the more stable B_nH_{n+4} (*nido*) series proceed in higher yield, without the need for a rapid quench. Thus B_5H_9 and $B_{10}H_{14}$ are readily prepared by the pyrolysis reaction. More recently, these brute-force methods of pyrolysis have given way to more specific methods that are described below.

(e) Characteristic reactions of boranes and borohydrides

Key points: Characteristic reactions of boranes are cleavage of a BH_2 group from diborane and tetraborane by NH_3, deprotonation of large boron hydrides by bases, reaction of a boron hydride with a borohydride ion to produce a larger borohydride anion, and Friedel—Crafts-type substitution of an alkyl group for hydrogen in pentaborane and some larger boron hydrides.

The characteristic reactions of boron clusters with a Lewis base range from cleavage of BH_n from the cluster to deprotonation of the cluster, cluster enlargement, and abstraction of one or more protons.

Lewis base cleavage reactions have already been introduced in Section 13.6 in connection with diborane. With the robust higher borane B_4H_{10}, cleavage may break some BHB bonds, leading to partial fragmentation of the cluster:

Deprotonation, rather than cleavage, occurs readily with the large borane $B_{10}H_{14}$:

$$B_{10}H_{14} + N(CH_3)_3 \rightarrow [HN(CH_3)_3]^+[B_{10}H_{13}]^-$$

The structure of the product anion indicates that deprotonation occurs from a $3c,2e$ BHB bridge, leaving the electron count on the boron cluster unchanged. This deprotonation of a BHB $3c,2e$ bond to yield a $2c,2e$ bond occurs without major disruption of the bonding:

The Brønsted acidity of boron hydrides increases approximately with size:

$$B_4H_{10} < B_5H_9 < B_{10}H_{14}$$

This trend correlates with the greater delocalization of charge in the larger clusters, in much the same way that delocalization accounts for the greater acidity of phenol than methanol. The variation in acidity is illustrated by the observation that, as shown above, the weak base trimethylamine deprotonates decaborane(10), but the much stronger base methyllithium is required to deprotonate B_5H_9:

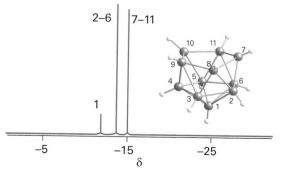

$$+ CH_4$$

Hydridic character is most characteristic of small anionic borohydrides. As an illustration, whereas BH_4^- readily surrenders an H^- ion in the reaction

$$BH_4^- + H^+ \rightarrow \tfrac{1}{2}B_2H_6 + H_2$$

the $[B_{10}H_{10}]^{2-}$ ion survives even in strongly acidic solution. Indeed, the hydronium salt $(H_3O)_2B_{10}H_{10}$ can even be crystallized.

The cluster-building reaction between a borane and a borohydride provides a convenient route to higher borohydride ions:

$$5K[B_9H_{14}] + 2B_5H_9 \xrightarrow{\text{polyether, }85°C} 5K[B_{11}H_{14}] + 9H_2$$

Similar reactions are used to prepare other borohydrides, such as $B_{10}H_{10}^{2-}$. This type of reaction has been used to synthesize a wide range of polynuclear borohydrides. ^{11}B-NMR spectroscopy reveals that the boron skeleton in $[B_{11}H_{14}]^-$ consists of an icosahedron with a missing vertex (Fig. 13.13).

The **electrophilic displacement** of H^+ provides a route to alkylated and halogenated species. As with Friedel–Crafts reactions, the electrophilic displacement of H is catalysed by a Lewis acid, such as aluminium chloride, and the substitution generally occurs on the closed portion of the boron clusters:

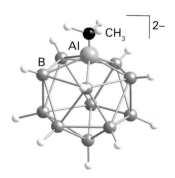

$$+ HCl$$

Figure 13.13 The proton-decoupled ^{11}B-NMR spectrum of $[B_{11}H_{14}]^-$. The *nido* structure (a truncated icosahedron) is indicated by the 1:5:5 pattern.

EXAMPLE 13.6 Proposing a structure for a boron-cluster reaction product

Propose a structure for the product of the reaction of $B_{10}H_{14}$ with $LiBH_4$ in a refluxing polyether, $CH_3OC_2H_4OCH_3$ (which boils at 162°C).

Answer The prediction of the probable outcome for the reactions of a boron cluster is difficult because several products are often plausible and the actual outcome can be sensitive to the conditions of the reaction. In the present case we note that an acidic borane, $B_{10}H_{14}$, is brought into contact with the hydridic anion BH_4^- under rather vigorous conditions. Therefore, we might expect the evolution of hydrogen:

$$B_{10}H_{14} + Li[BH_4] \xrightarrow{\text{ether, }R_2O} Li[B_{10}H_{13}] + R_2OBH_3 + H_2$$

This set of products suggests the further possibility of condensation of the neutral BH_3 complex with $[B_{10}H_{13}]^{2-}$ to yield a larger borohydride. That is in fact the observed outcome under these conditions:

$$Li[B_{10}H_{13}] + R_2OBH_3 \rightarrow Li[B_{11}H_{14}] + H_2 + R_2O$$

It turns out that, in the presence of excess $LiBH_4$, the cluster building continues to give the very stable icosahedral $[B_{12}H_{12}]^{2-}$ anion:

$$Li[\textit{nido-} B_{11}H_{14}] + Li[BH_4] \rightarrow Li_2[\textit{closo-} B_{12}H_{12}] + 3H_2$$

Self-test 13.6 Propose a plausible product for the reaction between $Li[B_{10}H_{13}]$ and $Al_2(CH_3)_6$.

13.12 Metallaboranes and carboranes

Key points: Main-group and d-block metals may be incorporated into boron hydrides through BHM bridges or more robust B–M bonds. When CH is introduced in place of BH in a polyhedral boron hydride, the charge of the resulting carboranes is one unit more positive; carborane anions are useful precursors of boron-containing organometallic compounds.

Metallaboranes are metal-containing B clusters. In some cases the metal is appended to a borohydride ion through hydrogen bridges. A more common and generally more robust group of metallaboranes have direct metal–boron bonds. An example of a main-group metallaborane with an icosahedral framework is *closo*-$[B_{11}H_{11}AlCH_3]^{2-}$ (**31**). It is prepared by interaction of the acidic hydrogens in $Na_2[B_{11}H_{13}]$ with trimethylaluminium:

$$2[B_{11}H_{13}]^{2-} + Al_2(CH_3)_6 \xrightarrow{\Delta} 2[B_{11}H_{11}AlCH_3]^{2-} + 4CH_4$$

When B_5H_9 is heated with $Fe(CO)_5$, a metallated analogue of pentaborane is formed (**32**). Generally, boranes are quite reactive

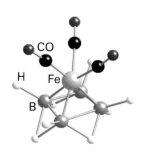

31 *closo*-$[B_{11}H_{11}AlCH_3]^{2-}$

32 $[Fe(CO)_3B_4H_8]$

to metal reagents and attack can occur at several points on the polyhedral cage. Therefore, reactions produce complex mixtures of metallaboranes from which individual species can be isolated.

Closely related to the polyhedral boranes and borohydrides are the **carboranes** (more formally, the *carbaboranes*), a large family of clusters that contain both B and C atoms. Now we begin to see the full generality of Wade's electron counting rules, as BH^- is isolectronic and isolobal with CH (**33**), and we can expect the polyhedral borohydrides and carboranes to be related.

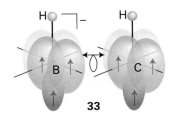

33

One entry into the interesting and diverse world of carboranes is the conversion of decaborane(14) to *closo*-1,2-$B_{10}C_2H_{12}$ (**34**). The first reaction in this preparation is the displacement of an H_2 molecule from decaborane by a thioether:

$$B_{10}H_{14} + 2SEt_2 \rightarrow B_{10}H_{12}(SEt_2)_2 + H_2$$

The loss of two H atoms in this reaction is compensated by the donation of electron pairs by the added thioethers, so the electron count is unchanged. The product of the reaction is then converted to the carborane by the addition of an alkyne:

$$B_{10}H_{12}(SEt_2)_2 + C_2H_2 \rightarrow B_{10}C_2H_{12} + 2SEt_2 + H_2$$

The four π electrons of ethyne displace two thioether molecules (two two-electron donors) and an H_2 molecule (which leaves with two additional electrons). The net loss of two electrons correlates with the change in structure from a *nido* starting material to the *closo* product. The C atoms are in adjacent (1,2) positions, reflecting their origin from ethyne. This *closo*-carborane survives in air and can be heated without decomposition. At 500°C in an inert atmosphere it undergoes isomerization into 1,7-$B_{10}C_2H_{12}$ (**35**), which in turn isomerizes at 700°C to the 1,12-isomer (**36**).

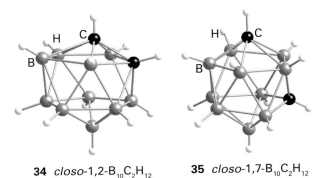

34 *closo*-1,2-$B_{10}C_2H_{12}$ **35** *closo*-1,7-$B_{10}C_2H_{12}$

The H atoms attached to carbon in *closo*-$B_{10}C_2H_{12}$ are very mildly acidic, so it is possible to lithiate these compounds with butyllithium:

$$B_{10}C_2H_{12} + 2LiC_4H_9 \rightarrow B_{10}C_2H_{10}Li_2 + 2C_4H_{10}$$

These dilithiocarboranes are good nucleophiles and undergo many of the reactions characteristic of organolithium reagents

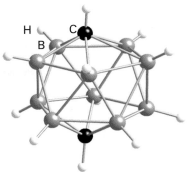

36 *closo*-1,12-$B_{10}C_2H_{12}$

(Section 11.17). Thus, a wide range of carborane derivatives can be synthesized. For example, reaction with CO_2 gives a carborane dicarboxylic acid:

$$B_{10}C_2H_{10}Li_2 \xrightarrow{(1)2CO_2;(2)2H_2O} B_{10}C_2H_{10}(COOH)_2$$

Similarly, I_2 leads to the diiodocarborane and NOCl yields $B_{10}C_2H_{10}(NO)_2$.

Although 1,2-$B_{10}C_2H_{12}$ is very stable, the cluster can be partially fragmented in strong base, and then deprotonated with NaH to yield *nido*-$[B_9C_2H_{11}]^{2-}$:

$$2B_{10}C_2H_{12} + 2EtO^- + 4EtOH \rightarrow 2[B_9C_2H_{11}]^- + 2B(OEt)_3 + 3H_2$$

$$Na[B_9C_2H_{12}] + NaH \rightarrow Na_2[B_9C_2H_{11}] + H_2$$

The importance of these reactions is that *nido*-$[B_9C_2H_{11}]^{2-}$ (Fig. 13.14a) is an excellent ligand. In this role it mimics the cyclopentadienyl ligand ($[C_5H_5]^-$; Fig. 13.14b), which is widely used in organometallic chemistry:

$$2Na_2[B_9C_2H_{11}] + FeCl_2 \xrightarrow{THF} 2NaCl + Na_2[Fe(B_9C_2H_{11})_2]$$

$$2Na[C_5H_5] + FeCl_2 \xrightarrow{THF} 2NaCl + Fe(C_5H_5)_2$$

Although we shall not go into the details of their synthesis, a wide range of metal-coordinated carboranes can be synthesized. A notable feature is the ease of formation of multi-decker sandwich compounds containing carborane ligands, (**37**) and (**38**). The highly negative $[B_3C_2H_5]^{4-}$ ligand has a much greater tendency to form stacked sandwich compounds than the less negative and therefore poorer donor $[C_5H_5]^-$.

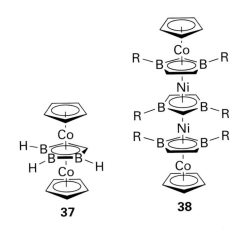

37   **38**

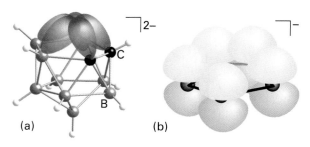

Figure 13.14 The isolobal relation between (a) $[B_9C_2H_{11}]^{2-}$ and (b) $[C_5H_5]^-$. The H atoms have been omitted for clarity.

EXAMPLE 13.7 Planning the synthesis of a carborane derivative

Give balanced chemical equations for the synthesis of $1,2\text{-}B_{10}C_2H_{10}(Si(CH_3)_3)_2$ starting with decaborane(10) and other reagents of your choice.

Answer We need to note that the H atoms attached to carbon in *closo*-$B_{10}C_2H_{12}$ are very mildly acidic, so we can lithiate these compounds with butyllithium. We first prepare $1,2\text{-}B_{10}C_2H_{12}$ from decaborane:

$$B_{10}H_{14} + 2\,SR_2 \rightarrow B_{10}H_{12}(SR_2)_2 + H_2$$
$$B_{10}H_{12}(SR_2)_2 + C_2H_2 \rightarrow B_{10}C_2H_{12} + 2\,SR_2 + H_2$$

The product is then lithiated by lithium alkyl, where the alkyl carbanion abstracts the slightly acidic hydrogen atoms from $B_{10}C_2H_{50}$, replacing them with Li^+:

$$B_{10}C_2H_{12} + 2\,LiC_4H_9 \rightarrow B_{10}C_2H_{10}Li_2 + 2\,C_4H_{10}$$

The resulting carborane is then used in a nucleophilic displacement on $Si(CH_3)_3Cl$ to yield the desired product:

$$B_{10}C_2H_{10}Li_2 + 2\,Si(CH_3)_3Cl \rightarrow B_{10}C_2H_{10}(Si(CH_3)_3)_2 + 2\,LiCl$$

Self-test 13.7 Propose a synthesis for the polymer precursor $1,7\text{-}B_{10}C_2H_{10}(Si(CH_3)_2Cl)_2$ from $1,2\text{-}B_{10}C_2H_{12}$ and other reagents of your choice.

13.13 The hydrides of aluminium and gallium

Key points: $LiAlH_4$ and $LiGaH_4$ are useful precursors of MH_3L_2 complexes; $LiAlH_4$ is also used as a source of H^- ions in the preparation of metalloid hydrides, such as SiH_4. The alkyl aluminium hydrides are used to couple alkenes.

Aluminium hydride, AlH_3, is a polymeric solid. The alkylaluminium hydrides, such as $Al_2(C_2H_5)_4H_2$, are well-known molecular compounds and contain $Al-H-Al$ $3c,2e$ bonds (Section 2.11). Hydrides of this kind are used to couple alkenes, the initial step being the addition of the AlH entity across the C=C double bond, as in hydroboration (Section 13.6c). Pure Ga_2H_6 has been prepared only relatively recently but derivatives have been known for some time. The hydrides of indium and thallium are very unstable.

The metathesis of the halides with LiH leads to lithium tetrahydridoaluminate, $LiAlH_4$, or the analogous $LiGaH_4$:

$$4\,LiH + ECl_3 \xrightarrow{\Delta,\,ether} LiEH_4 + 3\,LiCl \quad (E = Al, Ga)$$

The direct reaction of Li, Al, and H_2 leads to the formation of either $LiAlH_4$ or Li_3AlH_6, depending on the conditions of the

reaction. Their formal analogy with halide complexes such as $AlCl_4^-$ and AlF_6^{3-} should be noted.

The AlH_4^- and GaH_4^- ions are tetrahedral, and are much more hydridic than BH_4^-. Their hydridic character is consistent with the higher electronegativity of B compared with Al and Ga and the fact that BH_4^- is more covalent than AlH_4^- and GaH_4^-. For example, $NaAlH_4$ reacts violently with water but, as we saw earlier, basic aqueous solutions of $NaBH_4$ are useful in synthetic chemistry. They are also much stronger reducing agents; $LiAlH_4$ is commercially available and widely used as a strong hydride source and as a reducing agent.

With the halides of many nonmetallic elements, AlH_4^- serves as a hydride source in metathesis reactions, such as the reaction of lithium tetrahydridoaluminate with silicon tetrachloride in tetrahydrofuran solution to produce silane:

$$LiAlH_4 + SiCl_4 \xrightarrow{THF} LiAlCl_4 + SiH_4$$

The general rule in this important type of reaction is that H^- migrates from the element of lower electronegativity (Al in the example) to the element of greater electronegativity (Si).

Under conditions of controlled protolysis, both AlH_4^- and GaH_4^- lead to complexes of aluminium or gallium hydride:

$$LiEH_4 + [(CH_3)_3NH]Cl \rightarrow (CH_3)_3N\!-\!EH_3 + LiCl + H_2$$
$$(E = Al, Ga)$$

In striking contrast to BH_3 complexes, these complexes will add a second molecule of base to form five-coordinate complexes of aluminium or gallium hydride:

$$(CH_3)_3N\!-\!EH_3 + N(CH_3)_3 \rightarrow ((CH_3)N)_2EH_3 \quad (E = Al, Ga)$$

This behaviour is consistent with the trend for Period 3 and heavier p-block elements to form five- and six-coordinate hypervalent compounds (Section 2.6b).

13.14 Trihalides of aluminium, gallium, indium, and thallium

Key points: Aluminium, gallium, and indium all favour the $+3$ oxidation state, and their trihalides are Lewis acids. Thallium trihalides are less stable than those of its cogeners.

Although direct reaction of Al, Ga, or In with a halogen yields a halide, these electropositive metals also react with HCl or HBr gas, and the latter is usually a more convenient route:

$$2\,Al(s) + 6\,HCl(g) \xrightarrow{100°C} 2\,AlCl_3(s) + 3\,H_2(g)$$

AlF_3 and GaF_3 form salts of the type Na_3AlF_6 (cryolite) and Na_3GaF_6, which contain octahedral $[MF_6]^{3-}$ complex ions. Cryolite occurs naturally and molten synthetic cryolite is used as a solvent for bauxite in the industrial extraction of aluminium.

The Lewis acidities of the trihalides reflect the relative chemical hardness of the Group 13 elements. Thus, towards a hard Lewis base (such as ethyl acetate, which is hard because of its O donor atoms), the Lewis acidities of the halides weaken as the softness of the acceptor element increases, so the Lewis acidities fall in the order:

$$BCl_3 > AlCl_3 > GaCl_3$$

By contrast, towards a soft Lewis base (such as dimethylsulfane, Me_2S, which is soft because of its S atom), the Lewis acidities strengthen as the softness of the acceptor element increases:

$$GaX_3 > AlX_3 > BX_3 \quad (X = Cl \text{ or } Br)$$

Aluminium trichloride is a useful starting material for the synthesis of other Al compounds:

$$AlCl_3(sol) + 3\,LiR(sol) \rightarrow AlR_3(sol) + 3\,LiCl(s)$$

This reaction is an example of **transmetallation**, which is important in the preparation of main-group organometallic compounds. In transmetallation reactions, the halide formed is that of the more electronegative element and the high lattice enthalpy of that compound can be regarded as the 'driving force' of the reaction (as we saw in Example 13.7). The main industrial application of $AlCl_3$ is as a Friedel−Crafts catalyst in organic synthesis.

The thallium trihalides are much less stable than those of its lighter congeners. A trap for the unwary is that thallium triiodide is a compound of Tl(I) rather than Tl(III), as it contains the I_3^- ion, not I^-. This is confirmed by considering the standard potentials, which indicate that Tl(III) is rapidly reduced to Tl(I) by iodide:

$$Tl^{3+}(aq) + 2e^- \rightarrow Tl^+(aq) \quad E^\ominus = +1.25\,V$$
$$I_3^-(aq) + 2e^- \rightarrow 3I^-(aq) \quad E^\ominus = +0.55\,V$$

However, in excess iodide Tl(III) is stabilized by the formation of a complex:

$$TlI_3(s) + I^-(aq) \rightarrow [TlI_4]^-(aq)$$

In keeping with the general tendency towards higher coordination numbers for the larger atoms of later p-block elements, halides of Al and its heavier congeners may take on more than one Lewis base:

$$AlCl_3 + N(CH_3)_3 \rightarrow Cl_3AlN(CH_3)_3$$
$$Cl_3AlN(CH_3)_3 + N(CH_3)_3 \rightarrow Cl_3Al(N(CH_3)_3)_2$$

13.15 Low-oxidation-state halides of aluminium, gallium, indium, and thallium

Key point: The +1 oxidation state becomes progressively more stable from aluminium to thallium.

All the AlX compounds, GaF, and InF are unstable, gaseous species that disproportionate in the solid phase:

$$3\,AlX(s) \rightarrow 2\,Al(s) + AlX_3(s)$$

The other monohalides of Ga, In, and Tl are more stable. Gallium monohalides are formed by reacting GaX_3 with the metal in a 1:2 ratio:

$$GaX_3(s) + 2\,Ga(s) \rightarrow 3\,GaX(s) \quad (X = Cl, Br, or\ I)$$

The stability increases from the chloride to the iodide. The stability of the +1 oxidation state is increased by the formation of complexes such as $Ga[AlX_4]$. The apparently divalent $GaCl_2$ can be prepared by heating GaX_3 with gallium metal in a 2:1 ratio:

$$2\,GaX_3(s) + Ga(s) \xrightarrow{\Delta} 3\,GaX_2(s) \quad (X = Cl, Br, or\ I)$$

The formula $GaCl_2$ is deceiving, as this solid and most other apparently divalent salts do not contain Ga(II); instead they are mixed-oxidation-state compounds containing Ga(I) and Ga(III). Mixed-oxidation-state halogen compounds are also known for the heavier metals, such as $InCl_2$ and $TlBr_2$. The presence of M^{3+} ions is indicated by the existence of MX_4^- complexes in these salts with short M−X distances, and the presence of M^+ ions is indicated by longer and less regular separation from the halide ions. There is in fact only a fine line between the formation of a mixed-oxidation-state ionic compound and the formation of a compound that contains M−M bonds. For example, mixing $GaCl_2$ with a solution of $[N(CH_3)_4]Cl$ in a nonaqueous solvent yields the compound $[N(CH_3)_4]_2[Cl_3Ga−GaCl_3]$, in which the anion has an ethane-like structure with a Ga−Ga bond.

Indium monohalides are prepared by direct interaction of the elements or by heating the metal with HgX_2. The stability increases from the chloride to the iodide and is enhanced by the formation of complexes such as $In[AlX_4]$. Gallium(I) and In(I) halides both disproportionate when dissolved in water:

$$3\,MX(s) \rightarrow 2\,M(s) + M^{3+}(aq) + 3\,X^-(aq)$$
$$(M = Ga, In;\ X = Cl, Br, I)$$

Thallium(I) is stable with respect to disproportionation in water because Tl^{3+} is difficult to achieve. Thallium(I) halides are prepared by the action of HX on an acidified solution of a soluble Tl(I) salt. Thallium(I) fluoride has a distorted rock-salt structure whereas TlCl and TlBr have the caesium-chloride structure (Section 3.9). Yellow TlI has an orthorhombic layer structure in which the inert pair manifests itself structurally but, when pressure is applied, is converted to red TlI with a caesium-chloride structure. Thallium(I) iodide is used in photomultiplier tubes to detect ionizing radiation.

Other low-oxidation-state halides of indium and thallium are known: TlX_2 is actually $Tl^I[Tl_3^{III}X_4]$ and Tl_2X_3 is $Tl_3^I[Tl^{III}X_6]$ and In_4Br_6 is $In_6^I[I^{III}Br_6]_2$.

EXAMPLE 13.8 Proposing reactions of Group 13 halides

Propose chemical equations (or indicate no reaction) for reactions between (a) $AlCl_3$ and $(C_2H_5)_3NGaCl_3$ in toluene, (b) $(C_2H_5)_3NGaCl_3$ and GaF_3 in toluene, (c) TlCl and NaI in water.

Answer (a) We need to note that the trichlorides are excellent Lewis acids and that Al(III) is a stronger and harder Lewis acid than Ga(III). Therefore, the following reaction can be expected:

$$AlCl_3 + (C_2H_5)_3NGaCl_3 \rightarrow (C_2H_5)_3NAlCl_3 + GaCl_3$$

(b) In this case we need to note that the fluorides are ionic, so GaF_3 has a very high lattice enthalpy and is not a good Lewis acid. There is no reaction. (c) Now we note that Tl(I) is a chemically borderline soft Lewis acid, so it combines with the softer I^- ion rather than Cl^-:

$$TlCl(s) + NaI(aq) \rightarrow TlI(s) + NaCl(aq)$$

Like silver halides, Tl(I) halides have low solubility in water, so the reaction will probably proceed very slowly.

Self-test 13.8 Propose, with reasons, the chemical equation (or indicate no reaction) for reactions between (a) $(CH_3)_2SAlCl_3$ and $GaBr_3$, (b) TlCl and formaldehyde (HCHO) in acidic aqueous solution. (*Hint:* Formaldehyde is easily oxidized to CO_2 and H^+.)

13.16 Oxo compounds of aluminium, gallium, indium, and thallium

Key points: Aluminium and gallium form α and β forms of the oxide in which the elements are in their $+3$ oxidation state; thallium forms an oxide, in which it is in its $+1$ oxidation state, and a peroxide.

The most stable form of Al_2O_3, α-alumina, is a very hard and refractory material. In its mineral form it is known as *corundum* and as a gemstone it is *sapphire* or *ruby*, depending on the metal ion impurities. The blue of sapphire arises from a charge-transfer transition from Fe^{2+} to Ti^{4+} ion impurities (Section 20.5). Ruby is α-alumina in which a small fraction of the Al^{3+} ions are replaced by Cr^{3+}. The structure of α-alumina and gallia, Ga_2O_3, consists of an hcp array of O^{2-} ions with the metal ions occupying two-thirds of the octahedral holes in an ordered array.

Dehydration of aluminium hydroxide at temperatures below 900°C leads to the formation of γ-alumina, which is a metastable polycrystalline form with a defect spinel structure (Section 3.9b) and a very high surface area. Partly because of its surface acid and base sites, this material is used as a solid phase in chromatography and as a heterogeneous catalyst and catalyst support (Section 26.10)

The α and γ forms of Ga_2O_3 have the same structures as their Al analogues. The metastable form is β-Ga_2O_3, which has a ccp structure with Ga(III) in distorted octahedral and tetrahedral sites. Half the Ga(III) ions are therefore four-coordinate despite its large radius (compared to Al(III)). This coordination may be due to the effect of the filled $3d^{10}$ shell of electrons, as remarked previously. Indium and Tl form In_2O_3 and Tl_2O_3, respectively. Thallium also forms the Tl(I) oxide and peroxide, Tl_2O and Tl_2O_2.

Indium tin oxide (ITO) is In_2O_3 doped with 10 per cent by mass SnO_2 to form an n-type semiconductor. The material is transparent in the visible region, and is electrically conducting. It is deposited in thin films on surfaces by a variety of methods such as physical vapour deposition and ion beam sputtering. The main uses of these films are as transparent conducting coatings for liquid crystal and plasma displays, touch panels, solar cells, and organic light-emitting diodes. They are also used as infrared reflecting mirrors and as an anti-reflective coating on binoculars, telescopes, and spectacles. A coating of ITO on aircraft renders them invisible to radar. The melting point of ITO is 1900°C, which makes ITO thin film strain gauges very useful in harsh environments such as jet engines and gas turbines.

13.17 Sulfides of gallium, indium, and thallium

Key point: Gallium, indium, and thallium form many sulfides with a wide range of structures.

The only sulfide of Al is Al_2S_3, which is prepared by direct reaction of the elements at elevated temperatures:

$$2\,Al(s) + 3\,S(s) \xrightarrow{\Delta} Al_2S_3(s)$$

It is rapidly hydrolysed in aqueous solution:

$$Al_2S_3(s) + 6\,H_2O(l) \rightarrow 2\,Al(OH)_3(s) + 3\,H_2S(g)$$

Table 13.4 Selected sulfides of gallium, indium, and thallium

Sulfide	Structure
GaS	Layer structure with Ga—Ga bonds
α-Ga_2S_3	Defect wurtzite structure (hexagonal)
γ-Ga_2S_3	Defect sphalerite structure (cubic)
InS	Layer structure with In—In bonds
β-In_2S_3	Defect spinel (as γ-Al_2O_3)
TlS	Chains of edge-shared $Tl^{III}S_4$ tetrahedra
Tl_4S_3	Chains of $[Tl^{III}S_4]$ and $Tl^I[Tl^{III}S_3]$ tetrahedra

Aluminium sulfide exists in α, β, and γ forms. The structures of the α and β forms are based on the wurtzite structure (Section 3.9): in α-Al_2S_3 the S^{2-} ions are hcp and the Al^{3+} ions occupy two-thirds of the tetrahedral sites in an ordered fashion; in β-Al_2S_3 the Al^{3+} ions occupy two-thirds of the tetrahedral sites randomly. The γ form adopts the same structure as γ-Al_2O_3.

The sulfides of Ga, In, and Tl are more numerous and varied than those of Al and adopt many different structural types. Some examples are given in Table 13.4. Many of the sulfides are semiconductors, photoconductors, or light emitters and are used in electronic devices.

13.18 Compounds with Group 15 elements

Key point: Aluminium, gallium, and indium react with phosphorus, arsenic, and antimony to form materials that act as semiconductors.

The compounds formed between Group 13 and Group 15 elements (the nitrogen group) are important commercially and technologically as they are isoelectronic with Si and Ge and act as semiconductors (Sections 14.1 and 24.19). The nitrides adopt the wurtzite structure and the phosphides, arsenides, and stibnides all adopt the zinc blende (sphalerite) structure (Section 3.9). All the Group 13/15 (still commonly 'Group III/V') binary compounds can be prepared by direct reaction of the elements at high temperature and pressure.

$$Ga(s) + As(s) \rightarrow GaAs(s)$$

The most widely used Group 13/15 semiconductor is gallium arsenide, GaAs, which is used to make devices such as integrated circuits, light-emitting diodes, and laser diodes. Its band gap is similar to that of Si and larger than those of other Group 13/15 compounds (Table 13.5). Gallium arsenide is superior to Si for such applications because it has higher electron mobility, allowing it to function at frequencies in excess of 250 GHz. Gallium arsenide devices also generate less electronic noise than silicon devices.

Table 13.5 Band gap at 298 K

	E_g/eV
GaAs	1.35
GaSb	0.67
InAs	0.36
InSb	0.16
Si	1.11

One disadvantage of Group 13/15 semiconductors is that the compounds decompose in moist air and must be kept under an inert atmosphere, usually nitrogen, or be completely encapsulated.

13.19 Zintl phases

Key point: Group 13 elements form Zintl phases with Group 1 and Group 2 elements which are poor conductors and diamagnetic.

The Group 13 elements form Zintl phases (Section 3.8c) with Group 1 or Group 2 metals. Zintl phases are compounds of two metals which are brittle, diamagnetic, and poor conductors. They are therefore quite different from alloys. Zintl phases are formed between a very electropositive Group 1 or 2 element and a moderately electronegative p-block metal or metalloid. They are ionic, with electrons being transferred from the Group 1 or 2 metal to the more electronegative element. The anion, referred to as the 'Zintl ion', has a complete octet of valence electrons and is polymeric; the cations are located within the anionic lattice. The structure of NaTl consists of the polymeric anion in a covalent diamond structure with Na^+ ions fitted into the anionic lattice. In Na_2Tl the polymeric anion is tetrahedral Tl_4^{8-}. The Zintl anions can be isolated by reacting with salts containing the tetraalkylammonium ion, which substitutes for the Group 1 or 2 metal ion, or by encapsulation within a cryptand. Some compounds appear to be Zintl phases but are conducting and paramagnetic. For example, K_8In_{11} contains the In_{11}^{8-} anion, which has one delocalized electron per formula unit.

13.20 Organometallic compounds

The most important organometallic compounds of the Group 13 elements are those of B and Al. Organoboron compounds are commonly treated as organometallic compounds even though B is not a metal.

(a) Organoboron compounds

Key points: Organoboron compounds are electron deficient and act as Lewis acids; tetraphenylborate is an important ion.

Organoboranes of the type BR_3 can be prepared by hydroboration of an alkene with diborane.

$$B_2H_6 + 6CH_2=CH_2 \rightarrow 2B(CH_2CH_3)_3$$

Alternatively, they can be produced from a Grignard reagent (Section 12.13):

$$(C_2H_5)_2O:BF_3 + 3RMgX \rightarrow BR_3 + 3MgXF + (C_2H_5)_2O$$

Alkylboranes are not hydrolysed but are pyrophoric. The aryl species are more stable. They are all monomeric and planar. Like other B compounds, the organoboron species are electron deficient and consequently act as Lewis acids and form adducts easily.

An important anion is the tetraphenylborate ion, $[B(C_6H_5)_4]^-$, more commonly written BPh_4^-, analogous to the tetrahydridoboranate ion, BH_4^- (Section 13.6). The sodium salt can be obtained by a simple addition reaction:

$$BPh_3 + NaPh \rightarrow Na^+[BPh_4]^-$$

The Na salt is soluble in water but the salts of most large, monopositive ions are insoluble. Consequently, the anion is useful as a precipitating agent and can be used in gravimetric analysis.

(b) Organoaluminium compounds

Key points: Methyl- and ethylaluminium are dimers; bulky alkyl groups result in monomeric species.

Alkylaluminium compounds can be prepared on a laboratory scale by transmetallation of a mercury compound:

$$2Al + 3Hg(CH_3)_2 \rightarrow Al_2(CH_3)_6 + 3Hg$$

Trimethylaluminium is prepared commercially by the reaction of Al metal with chloromethane to give $Al_2Cl_2(CH_3)_4$. This intermediate is then reduced with Na and the $Al_2(CH_3)_6$ (**39**) is removed by fractional distillation.

39 $Al_2(CH_3)_6$

Alkylaluminium dimers are similar in structure to the analogous dimeric halides (Section 13.6) but the bonding is different. In the halides, the bridging $Al-Cl-Al$ bonds are $2c,2e$ bonds, that is each $Al-Cl$ bond involves an electron pair. In the alkylaluminium dimers the $Al-C-Al$ bonds are longer than the terminal $Al-C$ bonds, which suggests that they are $3c,2e$ bonds, with one bonding pair shared across the $Al-C-Al$ unit, somewhat analogous to the bonding in diborane, B_2H_6 (Section 13.6).

Triethylaluminium and higher alkyl compounds are prepared from the metal, an appropriate alkene, and hydrogen gas at elevated temperatures and pressures.

$$2Al + 3H_2 + 6CH_2=CH_2 \xrightarrow{60-110°C, 10-20MPa} Al_2(CH_2CH_3)_6$$

This route is relatively cost effective and, as a result, alkylaluminium compounds have found many commercial applications. Triethylaluminium, often written as the monomer $Al(C_2H_5)_3$, is an organometallic complex of Al of major industrial importance. It is used in the Ziegler–Natta polymerization catalyst (Section 26.16).

Steric factors have a powerful effect on the structures of alkylaluminiums. Where dimers are formed, the long weak bridging bonds are easily broken. This tendency increases with the bulkiness of the ligand. So, for example, triphenylaluminium is a dimer but trimesitylaluminium (where trimesityl is $2,4,6-(CH_3)_3C_6H_2-$) is a monomer.

FURTHER READING

R.B. King, *Inorganic chemistry of the main group elements*. John Wiley & Sons (1994).

D.M.P. Mingos, *Essential trends in inorganic chemistry*. Oxford University Press (1998). A survey of inorganic chemistry from the perspective of structure and bonding.

N.C. Norman, *Periodicity and the s- and p-block elements*. Oxford University Press (1997). Includes coverage of essential trends and features of s-block chemistry.

R.B. King (ed.), *Encyclopedia of inorganic chemistry*. John Wiley & Sons (2005).

C.E. Housecroft, *Boranes and metalloboranes*. Ellis Horwood, Chichester (2005). An introduction to borane chemistry.

EXERCISES

13.1 Give a balanced chemical equation and conditions for the recovery of boron.

13.2 Describe the bonding in (a) BF_3, (b) $AlCl_3$, (c) B_2H_6.

13.3 Arrange the following in order of increasing Lewis acidity: BF_3, BCl_3, $AlCl_3$. In the light of this order, write balanced chemical reactions (or no reaction) for (a) $BF_3N(CH_3)_3 + BCl_3 \rightarrow$, (b) $BH_3CO + BBr_3 \rightarrow$.

13.4 Thallium tribromide (1.11 g) reacts quantitatively with 0.257 g of NaBr to form a product A. Deduce the formula of A. Identify the cation and anion.

13.5 Identify compounds A, B, and C.

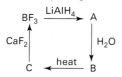

13.6 Does B_2H_6 survive in air? If not, write the equation for the reaction.

13.7 Predict how many different boron environments would be present in the proton-decoupled ^{11}B-NMR of a) B_5H_{11}, b) B_4H_{10}.

13.8 Predict the products from the hydroboration of (a) $(CH_3)_2C=CH_2$, (b) $CH\equiv CH$.

13.9 Diborane has been used as a rocket propellant. Calculate the energy released from 1.00 kg of diborane given the following values of $\Delta_f H^{\ominus}$/kJ mol^{-1}: $B_2H_6 = 31$, $H_2O = -242$, $B_2O_3 = -1264$. The combustion reaction is $B_2H_6(g) + 3 O_2(g) \rightarrow 3 H_2O(g) + B_2O_3(s)$. What would be the problem with diborane as a fuel?

13.10 Using BCl_3 as a starting material and other reagents of your choice, devise a synthesis for the Lewis acid chelating agent, $F_2B-C_2H_4-BF_2$.

13.11 Given $NaBH_4$, a hydrocarbon of your choice, and appropriate ancillary reagents and solvents, give formulas and conditions for the synthesis of (a) $B(C_2H_5)_3$, (b) Et_3NBH_3.

13.12 Draw the B_{12} unit that is a common motif of boron structures; take a viewpoint along a C_2 axis.

13.13 Which boron hydride would you expect to be more thermally stable, B_6H_{10} or B_6H_{12}? Give a generalization by which the thermal stability of a borane can be judged.

13.14 How many skeletal electrons are present in B_5H_9?

13.15 (a) Give a balanced chemical equation (including the state of each reactant and product) for the air oxidation of pentaborane(9). (b) Describe the probable disadvantages, other than cost, for the use of pentaborane as a fuel for an internal combustion engine.

13.16 (a) From its formula, classify $B_{10}H_{14}$ as *closo*, *nido*, or *arachno*. (b) Use Wade's rules to determine the number of framework electron pairs for decaborane(14). (c) Verify by detailed accounting of valence electrons that the number of cluster valence electrons of $B_{10}H_{14}$ is the same as that determined in (b).

13.17 Starting with $B_{10}H_{14}$ and other reagents of your choice, give the equations for the synthesis of $[Fe(nido-B_9C_2H_{11})_2]^{2-}$, and sketch the structure of this species.

13.18 (a) What are the similarities and differences in structure of layered BN and graphite (Section 14.5)? (b) Contrast their reactivity with Na and Br_2. (c) Suggest a rationalization for the differences in structure and reactivity.

13.19 Devise a synthesis for the borazines (a) $Ph_3N_3B_3Cl_3$ and (b) $Me_3N_3B_3H_3$, starting with BCl_3 and other reagents of your choice. Draw the structures of the products.

13.20 Give the structures and names of B_4H_{10}, B_5H_9, and 1,2-$B_{10}C_2H_{12}$.

13.21 Arrange the following boron hydrides in order of increasing Brønsted acidity, and draw a structure for the probable structure of the deprotonated form of one of them: B_2H_6, $B_{10}H_{14}$, B_5H_9.

PROBLEMS

13.1 Borane exists as the molecule B_2H_6 and trimethylborane exists as a monomer $B(CH_3)_3$. In addition, the molecular formulas of the compounds of intermediate compositions are observed to be $B_2H_5(CH_3)$, $B_2H_4(CH_3)_2$, $B_2H_3(CH_3)_3$, and $B_2H_2(CH_3)_4$. Based on these facts, describe the probable structures and bonding in the latter series.

13.2 ^{11}B-NMR is an excellent spectroscopic tool for inferring the structures of boron compounds. With ^{11}B-^{11}B coupling ignored, it is possible to determine the number of attached H atoms by the multiplicity of a resonance: BH gives a doublet, BH_2 a triplet, and BH_3 a quartet. B atoms on the closed side of *nido* and *arachno* clusters are

generally more shielded than those on the open face. Assuming no B—B or B—H—B coupling, predict the general pattern of the ^{11}B-NMR spectra of (a) BH_3CO and (b) $[B_{12}H_{12}]^{2-}$.

13.3 Identify the incorrect statements in the following description of Group 13 chemistry and provide corrections along with explanations of the principle or chemical generalization that applies. (a) All the elements in Group 13 are nonmetals. (b) The increase in chemical hardness on going down the group is illustrated by greater oxophilicity and fluorophilicity for the heavier elements. (c) The Lewis acidity increases for BX_3 from X = F to Br and this may be explained by stronger Br—B π bonding. (d) *Arachno*-boron hydrides have a $2(n + 3)$ skeletal electron count and they are more stable than *nido*-boron hydrides. (e) In a series of *nido*-boron hydrides acidity increases with increasing size. (f) Layered boron nitride is similar in structure to graphite and because it has a small separation between HOMO and LUMO, it is a good electrical conductor.

13.4 Use suitable molecular orbital software to calculate the wavefunctions and energy levels for *closo*-$[B_6H_6]^{2-}$. From that output, draw a molecular orbital energy diagram for the orbitals primarily involved in B—B bonding and sketch the form of the orbitals. How do these orbitals compare qualitatively with the qualitative description for this anion in this chapter? Is B—H bonding neatly separated from the B—B bonding in the computed wavefunctions?

13.5 In his paper 'Covalent and ionic molecules: why are BeF_2 and AlF_3 high melting point solids whereas BF_3 and SiF_4 are gases?' (*J. Chem. Educ.*, 1998, **75**, 923), R.J. Gillespie makes a case for the classification of the bonding in BF_3 and SiF_4 as predominantly ionic. Summarize his arguments and describe how these differ from the conventional view of bonding in gaseous molecules.

13.6 Nanotubes of C and BN have been synthesized by C. Colliex et al. (*Science*, 1997, **278**, 653). (a) What are the advantageous properties of these nanotubes over carbon analogues? (b) Outline the method used for the preparation of these compounds. (c) What was the main structural feature of the nanotubes and how could this be exploited in applications?

13.7 M. Montiverde discusses 'Pressure dependence of the superconducting temperature of MgB_2' (*Science*, 2001, **292**, 75). (a) Describe the theoretical bases of the two theories that have been postulated to explain the superconductivity of MgB_2. (b) How does the T_c of MgB_2 vary with pressure? What insight does this provide into the superconductivity?

13.8 Use the references in Z.W. Pan, Z.R. Dai, and Z.L. Wang (*Science*, 2001, **291**, 1947) as a starting point to write a review of wire-like nanomaterials of Group 13 elements. Indicate how In_2O_3 nanobelts were prepared and give the dimensions of a typical nanobelt.

13.9 In their paper 'New structural motifs in metallaborane chemistry: synthesis, characterization, and solid-state structures of $(Cp*W)_3(\mu\text{-H})B_8H_8$, $(Cp*W)_2B_7H_9$, and $(Cp*Re)_2B_7H_7$ ($Cp* = \eta^5\text{-}C_5Me_5$)' (*Organometallics*, 1999, **18**, 853). A.S. Weller, M. Shang, and T.P. Fehlner discuss the synthesis and characterization of some novel boron-rich metalloboranes. Sketch and explain the ^{11}B- and ^{1}H-NMR spectra of $(Cp*W)_2B_7H_9$.

14 The Group 14 elements

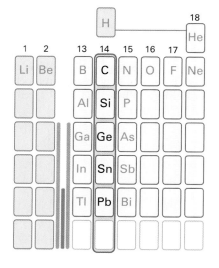

The elements of Group 14 are arguably the most important of all, carbon providing the basis for life on Earth and silicon being vital for the physical structure of the natural environment in the form of crustal rocks. The elements of this group exhibit great diversity in their properties, ranging as they do from the nonmetallic carbon to the well-known metals tin and lead. All the elements form binary compounds with other elements. In addition, silicon forms a diverse range of network solids. Many of the organocompounds of the Group 14 elements are commercially important.

The elements of Group 14, carbon, silicon, germanium, tin, and lead, show considerable diversity in their chemical and physical properties. Carbon, of course, is the building block of life and central to organic chemistry. In this chapter, our focus with carbon is on its *inorganic* chemistry. Silicon is widely distributed in the natural environment, and tin and lead find widespread applications in industry and manufacturing.

PART A: **THE ESSENTIALS**

The elements of Group 14 (the carbon group) are of fundamental importance in industry and nature. We discuss carbon in many contexts throughout this text, including organometallic compounds in Chapter 22 and catalysis in Chapter 26. The focus of this section is on the essential aspects of the chemistry of Group 14.

14.1 The elements

Key points: The lightest elements of the group are nonmetals; tin and lead are metals. All the elements except lead exist as several allotropes.

The lightest members of the group, carbon and silicon, are nonmetals, germanium is a metalloid, and tin and lead are metals. This increase in metallic properties on descending a group is a striking feature of the p block and can be understood in terms of the increasing atomic radius and associated decrease in ionization energy down the group (Table 14.1). Because the ionization energies of the heavier elements are low, the metals form cations increasingly readily down the group.

As the valence configuration ns^2np^2 suggests, the +4 oxidation state is dominant in the compounds of the elements. The major exception is lead, for which the most common oxidation state is +2, two less than the group maximum. The relative stability of the low oxidation state is an example of the inert-pair effect (Section 9.5), which is such a striking feature of the heaviest p-block elements.

Table 14.1 Selected properties of the Group 14 elements

	C	Si	Ge	Sn	Pb
Melting point/°C	3730 (graphite sublimes)	1410	937	232	327
Atomic radius/pm	77	117	122	140	154
Ionic radius, $r(M^{n+})$/pm			73 (+2)	93	119 (+2)
			53 (+4)	69 (+4)	78 (+4)
First ionization energy, I/(kJ mol^{-1})	1090	786	762	707	716
Pauling electronegativity	2.5	1.9	2.0	1.9	2.3
Electron affinity, E_a/(kJ mol^{-1})	154	134	116	107	35
$E^{\ominus}(M^{4+}, M^{2+})$ / V				+0.15	+1.69
$E^{\ominus}(M^{2+}, M)$ / V				−0.14	−0.13

The electronegativities of carbon and silicon are similar to that of hydrogen and they form many covalent hydrogen and alkyl compounds. Carbon and silicon are strong **oxophiles** and **fluorophiles**, in the sense that they have high affinities for the hard anions O^{2-} and F^-, respectively (Section 4.12). Their oxophilic character is evident in the existence of an extensive series of oxoanions, the carbonates and silicates. In contrast, Pb^{2+} forms more stable compounds with soft anions, such as I^- and S^{2-}, than with hard anions, and is therefore classified as chemically soft.

Two almost pure forms of carbon, *diamond* and *graphite*, are mined. There are many less pure forms, such as *coke*, which is made by the pyrolysis of coal, and *lamp black*, which is the product of incomplete combustion of hydrocarbons. Silicon occurs widely distributed in the natural environment and makes up 26 per cent by mass of the Earth's crust. It occurs as sand, quartz, amethyst, agate, and opal, and is also found in asbestos, feldspar, clays, and micas. Germanium is low in abundance and occurs naturally in the ore *germanite*, $Cu_{13}Fe_2Ge_2S_{16}$, in zinc ores and in coal. Tin occurs as the mineral cassiterite, SnO_2, and lead occurs as *galena*, PbS.

Diamond and graphite, the two common crystalline forms of elemental carbon, are strikingly different. Diamond is effectively an electrical insulator; graphite is a good conductor. Diamond is the hardest known natural substance and hence the ultimate abrasive; impure (partially oxidized) graphite is slippery and frequently used as a lubricant. The origin of these widely different physical properties can be traced to the very different structures and bonding in the two allotropes.

In diamond, each C atom forms single bonds of length 154 pm with four adjacent C atoms at the corners of a regular tetrahedron (Fig. 14.1); the result is a rigid, covalent, three-dimensional framework. Graphite consists of stacks of planar graphene layers within which each C atom has three nearest neighbours at 142 pm (Fig. 14.2). The σ bonds between neighbours within the sheets are formed from the overlap of sp^2 hybrid orbitals, and the remaining perpendicular p orbitals overlap to form π bonds that are delocalized over the plane. The ready cleavage of graphite parallel to the planes of atoms (which is largely due to the presence of impurities) accounts for its slipperiness. Diamond can be cleaved, but this ancient craft requires considerable expertise as the forces in the crystal are more symmetrical.

Diamond and graphite are not the only allotropes of carbon. The fullerenes (known informally as 'buckyballs') were discovered in the 1980s and have given rise to a new field within the inorganic chemistry of carbon.

All the elements of the group except lead have at least one solid phase with a diamond structure (Fig. 14.1). The cubic phase of tin, which is called *grey tin* or *α-tin* (α-Sn), is not stable at room temperature. It converts to the more stable, common phase, *white tin* or β-*tin* (β-Sn) in which an Sn atom has six nearest neighbours in a highly distorted octahedral array. When white tin is cooled to 13.2°C, it converts to grey tin. The effects of this transformation were first recognized on organ pipes in medieval European cathedrals, where it was believed to be due to the Devil's work. Legend has it that Napoleon's armies

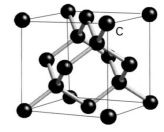

Figure 14.1 The cubic diamond structure.

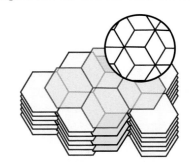

Figure 14.2 The structure of graphite. The rings are in register in alternate planes, not adjacent planes.

were defeated in Russia because, as the temperature fell, the white tin buttons on the soldiers' uniforms were converted to grey tin, which then crumbled away.

The gap between the valence and conduction bands (Section 3.19) decreases steadily from diamond, which is classed as a wide-band-gap semiconductor but commonly regarded as an insulator, to tin, which behaves like a metal above its transition temperature.

Elemental carbon in the form of coal or coke is used as a fuel and reducing agent in the recovery of metals from their ores. Graphite is used as a lubricant and in pencils, and diamond is used in industrial cutting tools. The band gap and consequent semiconductivity of silicon leads to its many applications in integrated circuits, computer chips, solar cells, and other electronic solid-state devices. Silica (SiO_2) is the major raw material used to make glass. Germanium was the first widely used material for the construction of transistors because it was easier to purify than silicon and, having a smaller band gap than silicon (0.72 eV for Ge, 1.11 eV for Si), is a better intrinsic semiconductor.

Tin is resistant to corrosion and is used to plate steel for use in tin cans. Bronze is an alloy of tin and copper that typically contains less than 12 per cent by mass of tin; bronze with higher tin content is used to make bells. Solder is an alloy of tin and lead, and has been in use since Roman times. Window glass or float glass is made by floating molten glass on the surface of molten tin. The 'tin side' of window glass can be seen as a haze of tin(IV) oxide when viewed with ultraviolet radiation. Trialkyl and triaryltin compounds are in widespread use as fungicides and biocides.

The softness and malleability of lead has resulted in its use in plumbing, although this application is now illegal in many countries due to concerns over lead poisoning. Its low melting point contributes to its use in solder and its high density (11.34 g cm^{-3}) leads to its use in ammunition and as shielding from ionizing radiation. Lead oxide is added to glass to raise its refractive index and form 'lead' or 'crystal' glass.

14.2 Simple compounds

Key points: All the Group 14 elements form simple binary compounds with hydrogen, oxygen, the halogens, and nitrogen. Carbon and silicon also form carbides and silicides with metals.

All the Group 14 elements form tetravalent hydrides, EH_4. In addition, carbon and silicon form series of catenated molecular hydrides. Carbon forms an enormous range of hydrocarbon compounds that are best regarded from the viewpoint of organic chemistry.

Carbon forms a series of simple hydrocarbons, the alkanes, with the general formula C_nH_{2n+2}. The stability of the long-chain, catenated hydrocarbons is due to the high C–C and C–H bond enthalpies (Table 14.2; Section 9.7). Carbon also forms strong multiple bonds in the unsaturated alkenes and alkynes (Table 14.2). The strength of the C–C bond and the ability to form multiple bonds are largely responsible for the diversity and stability of carbon compounds.

The data in Table 14.2 illustrate how the E–E bond enthalpy decreases on descending the group. As a result, the tendency to catenation decreases from C to Pb. Silicon forms a series of compounds analogous to the alkanes, the *silanes*, but the longest chain contains just seven Si atoms, as heptasilane, Si_7H_{16}. The silanes, with their greater number of electrons and stronger intermolecular forces, are less volatile than their hydrocarbon analogues. Thus, whereas propane, C_3H_8, is a gas under normal conditions, its silicon analogue trisilane,

Table 14.2 Selected mean bond enthalpies, $B(X-Y)/(\text{kJ mol}^{-1})$

C–H	412	Si–H	318	Ge–H	288	Sn–H	250	Pb–H	<157
C–O	360	Si–O	466	Ge–O	350				
C=O	743	Si=O	642						
C–C	348	Si–Si	326	Ge–Ge	186	Sn–Sn	150	Pb–Pb	87
C=C	612								
C≡C	837								
C–F	486	Si–F	584	Ge–F	466				
C–Cl	322	Si–Cl	390	Ge–Cl	344	Sn–Cl	320	Pb–Cl	301

Si_3H_8, is a liquid that boils at 53°C. The decreasing stability of the hydrides on going down the group severely limits the accessible chemical properties of stannanes and plumbane.

The tetrahalomethanes, the simplest halocarbons, vary from the highly stable and volatile CF_4 to the thermally unstable solid CI_4. The full range of tetrahalides is known for silicon and germanium; all of them are volatile molecular compounds. Germanium shows signs of an inert-pair effect (Section 9.5) in that it also forms nonvolatile dihalides. Evidence of the inert-pair effect becomes more prominent in the chemistry of tin and lead as the +2 oxidation state becomes increasingly stable.

The two familiar oxides of carbon are CO and CO_2. Among the less familiar oxides is carbon suboxide, O=C=C=C=O. Physical data on all three compounds are summarized in Table 14.3. It should be noted that the bond in CO is short and strong (bond enthalpy 1076 kJ mol^{-1}) and its force constant is high. These features are in accord with its possession of a triple bond, as in the Lewis structure :C≡O:. Carbon dioxide, CO_2, shows a number of significant differences from carbon monoxide. The bonds are longer and the stretching force constants smaller in CO_2 than in CO, which is consistent with the bonds being double rather than triple.

The high affinity of silicon for oxygen accounts for the existence of a vast array of silicate minerals and synthetic silicon−oxygen compounds, which are important in mineralogy, industrial processing, and the laboratory. Aside from rare high-temperature phases, the structures of silicates are confined to tetrahedral four-coordinate Si. Thus, orthosilicate is $[SiO_4]^{4-}$ (**1**), disilicate is $[O_3SiOSiO_3]^{6-}$ (**2**). Silica and many silicates crystallize slowly. Amorphous solids known as **glasses** can be obtained instead of crystals by cooling the melt at an appropriate rate. In some respects these glasses resemble liquids. As with liquids, their structures are ordered over distances of only a few interatomic spacings (such as within a single SiO_4 tetrahedron). Unlike liquids, however, their viscosities are very high, and for most practical purposes they behave like solids.

Germanium(IV) oxide, GeO_2, resembles silica. Germanium(II) oxide, GeO, disproportionates readily to Ge and GeO_2. Tin(II) oxide, SnO, exists as blue-black and red polymorphs. Both forms are readily oxidized to SnO_2 when heated in air. Lead forms the brown lead(IV) oxide, PbO_2, red and yellow forms of lead(II) oxide, PbO, and a mixed oxide, Pb_3O_4, which contains Pb(IV) and Pb(II) and is known as 'red lead'. The inert-pair effect (Section 9.5) is evident once again in the stability of Pb(II) oxide relative to the Pb(IV) oxide.

Carbon forms hydrogen cyanide, HCN, ionic cyanides containing the CN$^-$ ion, and the gas cyanogen $(CN)_2$. They are all extremely toxic. The direct reaction of silicon and nitrogen gas at high temperatures produces silicon nitride, Si_3N_4. This substance is very hard and inert, and is used in high-temperature ceramic materials.

Carbon forms numerous binary carbides with metals and metalloids. Group 1 and 2 metals form ionic saline carbides, d-block metals form metallic carbides and boron and silicon form covalent solids. Silicon carbide, SiC, is widely used as the abrasive *carborundum*.

14.3 Extended silicon−oxygen compounds

Key point: As well as forming simple binary compounds with oxygen, silicon forms a wide range of extended network solids that find a range of applications in industry.

Aluminosilicates are formed when Al atoms replace some of the Si atoms in a silicate and occur naturally as clays, minerals, and rocks. Zeolite aluminosilicates are widely used as molecular sieves, microporous catalysts, and catalyst support materials. Because Al occurs as Al(III), its presence in place of Si(IV) in an aluminosilicate renders the overall charge

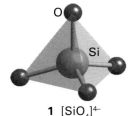

2 $[Si_2O_7]^{6-}$

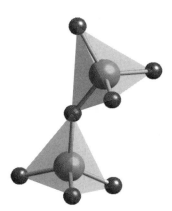

1 $[SiO_4]^{4-}$

Table 14.3 Properties of some oxides of carbon

Oxide	m.p./°C	b.p./°C	$\tilde{v}$ (CO)/cm^{-1}		k(CO)/(N m^{-1})	Bond length/pm	
						CC	CO
CO	−199	−192	2145		1860	113	
CO_2		−78	2449	1318	1550	116	
OCCCO	−111	7	2200	2290		128	116

more negative by one unit. An additional cation, such as H^+, Na^+, or $\frac{1}{2}Ca^{2+}$, is therefore required for each Al atom that replaces an Si atom. These additional cations have a profound effect on the properties of the materials.

Many important minerals are varieties of layered aluminosilicates that also contain metals such as lithium, magnesium, and iron: they include clays, talc, and various micas. An example of a simple layered aluminosilicate is the mineral *kaolinite*, $Al_2(OH)_4Si_2O_5$, which is used commercially as china clay and in some medical applications. It has long been used in diarrhea remedies and a more recent application uses kaolinite nanoparticle-impregnated bandages to stop bleeding, as the mineral triggers blood clotting.

In the mineral *talc*, $Mg_3(OH)_2Si_4O_{10}$, Mg^{2+} and OH^- ions are sandwiched between layers of $Si_4O_{10}^{4-}$ anions. The arrangement is electrically neutral, and as a result talc readily cleaves between the layers and accounts for talc's familiar slippery feel. Muscovite mica, $KAl_2(OH)_2Si_3AlO_{10}$, has charged layers because one Al(III) atom substitutes for one Si(IV) atom and the resulting negative charge is compensated by a K^+ ion that lies between the repeating layers. Because of this electrostatic cohesion, muscovite is not soft like talc but it is readily cleaved into sheets. There are many minerals based on a three-dimensional aluminosilicate framework. The *feldspars*, for instance, are the most important class of rock-forming minerals.

The **molecular sieves** are crystalline microporous aluminosilicates having open structures with apertures of molecular dimensions. The name 'molecular sieve' is prompted by the observation that these materials absorb only molecules that are smaller than the aperture dimensions and so can be used to separate molecules of different sizes. A subclass of molecular sieves, the *zeolites*,[1] have an aluminosilicate framework with cations (typically from Groups 1 or 2) trapped inside tunnels or cages (Fig. 14.3). In addition to their function as molecular sieves, zeolites are used as ion-exchange resins as they can exchange their ions for those in a surrounding solution. Zeolites are also used for shape-selective heterogeneous catalysis (Chapter 26).

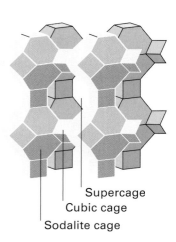

Supercage
Cubic cage
Sodalite cage

Figure 14.3 Framework representation of a type-A zeolite. Note the sodalite cages (truncated octahedral), the small cubic cages, and the central supercage.

PART B: **THE DETAIL**

In this section we discuss the detailed chemistry of the Group 14 elements, interpreting the reasons for the decreasing tendency to form catenated compounds and increasing metallic character down the group.

14.4 **Occurrence and recovery**

Key points: Elemental carbon is mined as graphite and diamond; elemental silicon is recovered from SiO_2 by carbon-arc reduction; the much less abundant germanium is found in zinc ores.

Carbon occurs as diamond and graphite, and in several forms of low crystallinity. The 1996 Nobel Prize in Chemistry was awarded to Richard Smalley, Robert Curl, and Harold Kroto for their discovery of a new allotrope of carbon, C_{60}, named buckminsterfullerene after the geodesic domes designed by the architect Buckminster Fuller (see Section 14.6). Carbon occurs as carbon dioxide in the atmosphere and dissolved in natural waters and as the insoluble carbonates of calcium and magnesium.

Elemental silicon is produced from silica, SiO_2, by high-temperature reduction with carbon in an electric arc furnace:

$$SiO_2(s) + 2\,C(s) \rightarrow Si(s) + 2\,CO(g)$$

Germanium is low in abundance and generally not concentrated in nature. It is obtained by the reduction of GeO_2 with carbon monoxide or hydrogen (Section 5.16). Tin is produced by the reduction of the mineral *cassiterite*, SnO_2, with coke in an electric furnace. Lead is obtained from its sulfide ores, which are converted to oxide and reduced by carbon in a blast furnace.

14.5 **Diamond and graphite**

Key points: Graphite consists of stacked two-dimensional carbon sheets; oxidizing agents or reducing agents may be intercalated between these sheets with concomitant electron transfer.

Diamond has the highest known thermal conductivity because its structure (shown in Fig. 14.1) distributes thermal motion in three dimensions very efficiently. The measurement of thermal conductivity is used to identify fake diamonds. Because of its durability, clarity, and high refractive index, diamond is one of the most highly prized gemstones.

The ready cleavage of graphite parallel to the planes of atoms (as shown in Fig. 14.2) is largely due to the presence of impurities and accounts for its slipperiness. These graphene planes are widely separated from each other (at 335 pm), indicating that

---

[1] The name zeolite is derived from the Greek for 'boiling stone'. Geologists found that certain rocks seemed to boil when subjected to the flame of a blowpipe.

BOX 14.1 Synthetic diamonds

There has been interest in the synthesis of diamonds since Antoine Lavoisier discovered that they were composed entirely of carbon. The first reported synthesis was in 1880 by J.B. Henney. He claimed that he had made diamond from hydrocarbons, bone oil, and lithium but his results were never reproduced.

Diamonds were synthesized in 1955 after many failed attempts, using graphite and a d metal heated to 1500−2000 K and subjected to 7 GPa. The graphite and metal must both be molten for diamond to be produced, so the temperature of synthesis depends on the melting point of the metal. The d metal (typically nickel) dissolves the graphite and the less soluble diamond phase crystallizes from it. The size, shape, and colour of the diamonds depend upon the conditions. Low-temperature synthesis produces dark, impure crystals. High-temperature synthesis produces paler, purer crystals. Common impurities are species that can be accommodated into the diamond lattice with minimum distortion. The diamonds are often contaminated with graphite or the metal catalyst. For example, the lattice dimensions of nickel are similar to those of diamond and crystallites of nickel may be included in the diamond lattice.

The diamond crystals can be grown by seeding with small diamond crystals but the new growth is often uneven with gaps and inclusions. Better quality diamonds are formed when the source of carbon is diamond and the seed crystals are in a cooler part of the apparatus. The difference in the solubility with the change in temperature causes the carbon to crystallize in a slow, controlled way, giving high-quality diamonds. Diamonds up to 1 carat (200 mg) may take up to a week to crystallize in this way.

Diamonds can be synthesized directly from graphite without a metal catalyst if the temperature and pressure are high enough. The shock synthesis method (the *Du Pont method*) exposes graphite to the intense pressure generated by a charge of high explosive. The graphite reaches a temperature of 1000 K and a pressure of 30 GPa for a few milliseconds and some of it is converted to diamond. The *static pressure method* heats graphite in high-pressure equipment by the discharge from a capacitor. Polycrystalline lumps of diamond are formed at 3300−4500 K and 13 GPa. Hydrocarbons may also be used as the carbon source in this method. Aromatic compounds such as naphthalene and anthracene produce graphite but aliphatic compounds such as paraffin wax and camphor produce diamond.

Because the high-pressure synthesis of diamond is costly and cumbersome, a low-pressure process would be highly attractive. It has in fact been known for a long time that microscopic diamond crystals mixed with graphite can be formed by depositing C atoms on a hot surface in the absence of air. The C atoms are produced by the pyrolysis of methane, and the atomic hydrogen also produced in the pyrolysis plays an important role in favouring diamond over graphite. One property of the atomic hydrogen is that it reacts more rapidly with the graphite than with diamond to produce volatile hydrocarbons, so the unwanted graphite is swept away. Although the process is not fully perfected, synthetic diamond films are already finding applications ranging from the hardening of surfaces subjected to wear, such as cutting tools and drills, to the construction of electronic devices. For example, boron-doped diamond films are very conducting and are used as electrodes in electrochemistry.

A promising new method of synthesis that is environmentally more friendly and cheaper than any of the high-temperature and high-pressure methods uses silicon carbide. Carbon is extracted as diamond under Cl_2 and H_2 gases close to 1 atm and the relatively low temperature of 1300 K.

---

there are weaker forces between them. These forces are sometimes, but not very appropriately, called 'van der Waals forces' (because in the common impure form of graphite, graphitic oxide, they are weak, like intermolecular forces), and consequently the region between the planes is called the **van der Waals gap**. Unlike diamond, graphite is soft and black with a slightly metallic lustre; it is neither durable nor particularly attractive.

The conversion of diamond to graphite at room temperature and pressure is spontaneous ($\Delta_{trs}G^{\ominus} = -2.90$ kJ mol^{-1}) but does not occur at an observable rate under ordinary conditions: diamonds older than the solar system have been isolated from meteorites. Diamond is the denser phase (3.51 g cm^{-3} instead of 2.26 g cm^{-3}), so it is favoured by high pressures, and large quantities of diamond abrasive are manufactured commercially by a d-metal-catalysed high-temperature, high-pressure process (Box 14.1). Thin films of boron-doped diamond are piezoresistive (their electrical resistance changes when pressure is applied) and are deposited on silica surfaces for use as high-temperature pressure sensors.

The electrical conductivity and many of the chemical properties of graphite are closely related to the structure of its delocalized π bonds. Its electrical conductivity perpendicular to the planes is low (5 S cm^{-1} at 25°C) and increases with increasing temperature, signifying that graphite is a semiconductor in that direction. The electrical conductivity is much higher parallel to the planes (30 kS cm^{-1} at 25°C) but decreases as the temperature is raised, indicating that graphite behaves as a metal, more precisely a semimetal,[2] in that direction. This effect is most striking in pyrolytic graphite, which is manufactured by the decomposition of a hydrocarbon gas at high temperature in a vacuum furnace. The resulting graphite is of very high purity with desirable mechanical, thermal, and electrical properties. Pyrolytic graphite is used in ion beam grids, thermal insulators, rocket nozzles, heater elements, and as an electrode material.

Graphite can act as either an electron donor or an electron acceptor towards atoms and ions that penetrate between its sheets and give rise to an **intercalation compound**. Thus, K atoms reduce graphite by donating their valence electron to the empty orbitals of the π^* band and the resulting K$^+$ ions penetrate between the layers (Section 14.4). The electrons added to the band are mobile, and therefore alkali metal graphite intercalates have high electrical conductivity. The stoichiometry of the compound depends on the quantity of alkali metal and the reaction conditions. The different stoichiometries are associated with an interesting series of structures, where the alkali metal ion may insert between neighbouring layers of C atoms, every other layer, and so on in a process known as *staging* (Fig. 14.4).

An example of an oxidation of graphite by removal of electrons from the π band is the formation of **graphite bisulfates** by heating graphite with a mixture of sulfuric and nitric acids. In this reaction, electrons are removed from the π band, and HSO$_4^-$ ions penetrate between the sheets to give substances of approximate formula $(C_{24})^+HSO_4^-$. In this oxidative intercalation reaction, the removal of electrons from the full π band leads to a higher conductivity than that of pure graphite. This process is analogous to the formation of p-type silicon by electron-accepting dopants (Section 3.18). When graphite bisulfates are treated with water, the layers are disrupted. When the water

---

[2] A semimetal (Section 3.19) is a substance in which two neighbouring bands have zero density of states at their edge but zero band gap between them.

Graphite   KC$_8$ KC$_{36}$

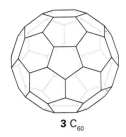

Figure 14.4 Potassium graphite compounds showing two types of alternation of intercalated atoms.

is subsequently removed at high temperatures, a highly flexible form of graphite is formed; this *graphite tape* is used to make sealing gaskets, valves, and brake linings.

The halogens show an alternation effect in their tendency to form intercalation compounds with graphite. Graphite reacts with fluorine to produce 'graphite fluoride', a nonstoichiometric species with formula $(CF)_n$ ($0.59 < n < 1$). This compound is black when n is low in its range and colourless when n approaches 1. It is used as a lubricant in high vacuum applications and as the cathode in lithium batteries. At elevated temperatures the products of the reaction also include C_2F and C_4F. Chlorine reacts slowly with graphite to form C_8Cl and iodine does not react at all. By contrast, bromine intercalates readily to give C_8Br, $C_{16}Br$, and $C_{20}Br$ in another example of staging.

14.6 Other forms of carbon

Carbon also exists in several less crystalline forms, as well as the fullerenes and related compounds.

(a) Carbon clusters

Key point: Fullerenes are formed when an electric arc is discharged between carbon electrodes in an inert atmosphere.

Metal and nonmetal cluster compounds have been known for decades, but the discovery of the soccer-ball shaped C_{60} cluster in the 1980s created great excitement in the scientific community and in the popular press. Much of this interest undoubtedly stemmed from the fact that carbon is a common element and there had seemed little likelihood that new molecular carbon structures would be found.

When an electric arc is struck between carbon electrodes in an inert atmosphere, a large quantity of soot is formed together with significant quantities of C_{60} and much smaller quantities of related **fullerenes**, such as C_{70}, C_{76}, and C_{84}. The fullerenes can be dissolved in a hydrocarbon or halogenated hydrocarbon and separated by chromatography on an alumina column. The structure of C_{60} has been determined by X-ray crystallography on the solid at low temperature and electron diffraction in the gas phase. The molecule consists of five- and six-membered carbon rings, and the overall symmetry is icosahedral in the gas phase (**3**).

Fullerenes can be reduced to form [60]fulleride salts, C_{60}^{n-} ($n = 1$ to 12). Fullerides of alkali metals are solids having compositions such as K_3C_{60}. The structure of K_3C_{60} consists of a

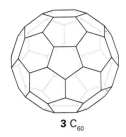

3 C_{60}

face-centred cubic array of C_{60} ions in which K$^+$ ions occupy the one octahedral and two tetrahedral sites available to each C_{60}^{3-} ion (Fig. 14.5). The compound is a metallic conductor at room temperature and a superconductor below 18 K. Other superconducting salts include Rb_2CsC_{60}, which has a superconducting transition temperature (T_c) of 33 K, and Cs_3C_{60}, with $T_c = 40$ K. The conductivity of E_3C_{60} compounds can be explained by considering that the conduction electrons are donated to the C_{60} molecules and are mobile because of overlapping C_{60} molecular orbitals (Section 24.21).

Some of the most interesting consequences of fullerene research have been the identification of *carbon nanotubes*. Carbon nanotubes consist of one or more concentric cylindrical tubes formed of graphene sheets in which the ends may be closed by fullerene-like caps containing six five-membered rings of atoms. The preparation of nanotubes has stimulated much research and the compounds could ultimately find a wide range of practical applications such as hydrogen storage and catalysis. They are covered more fully in Chapter 25.

Graphene has very high electrical conductivity, opacity, and strength and is being investigated for use in electronic devices, batteries, and for gas storage. It is currently very expensive and it is very difficult to produce perfect layers that give the best electronic performance. The method that produces the cleanest graphene surface is exfoliation, where the surface is mechanically ripped from a graphite crystal. This can be achieved quite simply by using sticky tape, but separating the useful thin flakes from the graphite debris is time-consuming. A simple and inexpensive route to graphene could revolutionize its use. A step in this direction is a new method that has been developed which can produce gram

Figure 14.5 The structure of K_3C_{60}. The full cell is face-centred cubic. (The structure of solid C_{60} itself is shown in Fig. 3.16.)

quantities by a simple chemical process: sodium metal is heated with ethanol for three days before being heated rapidly. The resulting solid consists of fused graphene sheets which are then washed and dried to separate them. The quality of the sheets is not as good as those produced by exfoliation, but this may be the first step to large-scale production of perfect graphene flakes.

(b) Fullerene–metal complexes

Key point: The polyhedral fullerenes undergo reversible multielectron reduction and form complexes with d-metal organometallic compounds and with OsO_4.

Reasonably efficient methods of synthesizing the fullerenes have been developed, and their redox and coordination chemistry have been extensively investigated. In keeping with the formation of alkali metal fullerides, C_{60} undergoes five electrochemically reversible electron transfer steps in nonaqueous solvents (Fig. 14.6). These observations suggest that the fullerenes ought to serve as either electrophiles or nucleophiles when paired with the appropriate metal. One illustration of this ability is the attack of electron-rich Pt(0) phosphine complexes on C_{60}, yielding compounds such as (**4**), in which the Pt atom spans a pair of C atoms in the fullerene molecule (section 22.9). This reaction is analogous to the coordination of double bonds to Pt–phosphine complexes. Although analogy with η^6-benzenechromium complexes (Section 22.19) suggests that a metal atom might coordinate to a sixfold face of C_{60}, that such hexahapto complexes do not in fact form is attributed to the radial arrangement of the C2pπ orbitals (**5**), which results in them having a poor overlap with d orbitals of a metal atom centred above a sixfold face of the molecule.

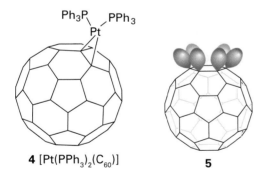

4 $[Pt(PPh_3)_2(C_{60})]$ **5**

In contrast to the poor interaction of a fullerene sixfold face with a single metal atom, a larger array of metal atoms, the triruthenium cluster $Ru_3(CO)_{12}$, reacts to form a $Ru_3(CO)_9$ cap on a sixfold face of C_{60}. In the process, three CO ligands are

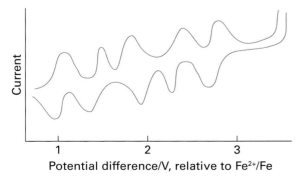

Figure 14.6 The cyclic voltammogram of C_{60} in toluene at 25°C.

displaced (**6**). The relatively large triangle of three metal atoms provides a favourable geometry for overlap with the radially orientated C2pπ orbitals.

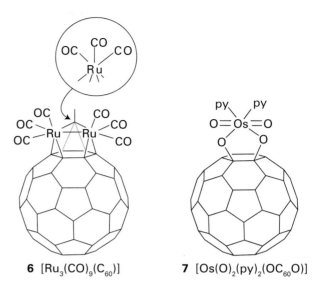

6 $[Ru_3(CO)_9(C_{60})]$ **7** $[Os(O)_2(py)_2(OC_{60}O)]$

The chemical properties of C_{60} are not limited to its interaction with electron-rich metal complexes. Reaction with a strong electrophile and oxidant, OsO_4 in pyridine, yields an oxo bridge complex analogous to the adducts of OsO_4 with alkenes (**7**).

In addition to complexes formed with the metal atom outside the fullerene cage, **endohedral fullerenes** are formed in which one or more atoms are accommodated inside the C_{60} shell. Such complexes are denoted M@C_{60}, indicating that the M atom is inside the C_{60} cage. Small inert gas atoms and molecules may be driven inside the cage at high temperatures (>600°C) and pressures (>2000 atm) to give, for example, $H_3@C_{60}$. Alternatively, the carbon cage can be formed around the endohedral atom by using a metal-doped carbon rod in an electric arc. Larger shells, such as La@C_{82} and $La_3@C_{106}$, are often formed.

(c) Partially crystalline carbon

Key points: Amorphous and partially crystalline carbon in the form of small particles are used on a large scale as adsorbents and as strengthening agents for rubber; carbon fibres impart strength to polymeric materials.

There are many forms of carbon that have a low degree of crystallinity. These partially crystalline materials have considerable commercial importance; they include *carbon black*, *activated carbon*, and *carbon fibres*. Because single crystals suitable for complete X-ray analyses of these materials are not available, their structures are uncertain. However, what information there is suggests that their structures are similar to that of graphite, but the degree of crystallinity and shapes of the particles differ.

'Carbon black' is a very finely divided form of carbon. It is prepared (on a scale that exceeds 8 Mt annually) by the combustion of hydrocarbons under oxygen-deficient conditions. Planar stacks, like those of graphite, and multilayer balls, reminiscent of the fullerenes, have both been proposed for its structure (Fig. 14.7). Carbon black is used on a huge scale as a pigment, in printer's ink (as on this page), and as a filler for rubber goods, including tyres, where it greatly improves the strength and wear resistance of the rubber and helps to protect it from degradation by sunlight.

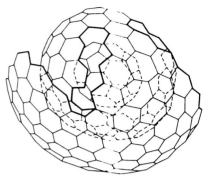

Figure 14.7 A proposed structure for a soot particle resulting from imperfect closure of a curved C-atom network. Graphite-like structures have also been proposed.

'Activated carbon' is prepared from the controlled pyrolysis of organic material, including coconut shells. It has a high surface area (in some cases exceeding 1000 m² g⁻¹), that arises from the small particle size. It is therefore a very efficient adsorbent for molecules, including organic pollutants from drinking water, noxious gases from the air, and impurities from reaction mixtures. There is evidence that the parts of the surface defined by the edges of the hexagonal sheets are covered with oxida-

8

tion products, including carboxyl and hydroxyl groups (**8**). This structure may account for some of its surface activity.

Carbon fibres are made by the controlled pyrolysis of asphalt fibres or synthetic fibres and are incorporated into a variety of high-strength plastic products, such as tennis rackets and aircraft components. Their structure bears a resemblance to that of graphite, but in place of the extended sheets the layers consist of ribbons parallel to the axis of the fibre. The strong in-plane bonds (which resemble those in graphite) give the fibre its very high tensile strength.

EXAMPLE 14.1 Comparing bonding in diamond and boron

Each B atom in elemental boron is bonded to five other B atoms but each C atom in diamond is bonded to four nearest neighbours. Suggest an explanation of this difference.

Answer We need to consider the valence electrons on each atom and the orbitals available for bond formation. The B and C atoms both have four orbitals available for bonding (one s and three p). However, a C atom has four valence electrons, one for each orbital, and it can therefore use all its electrons and orbitals in forming 2c,2e bonds with the four neighbouring C atoms. By contrast, B has only three electrons, therefore to use all four orbitals it forms 3c,2e bonds. The formation of these three-centre bonds brings another B atom into binding distance.

Self-test 14.1 Describe how the electronic structure of graphite is altered when it reacts with (a) potassium, (b) bromine.

14.7 Hydrides

The Group 14 elements form tetravalent hydrides, EH_4, with hydrogen, and carbon and silicon form catenated molecular hydrides.

(a) Hydrocarbons

Key point: The stability of catenated hydrocarbons can be attributed to high C–C and C–H bond enthalpies.

Methane, CH_4, an odourless, flammable gas, is the simplest hydrocarbon. It is found in large natural underground deposits from which it is extracted as natural gas and used as domestic and industrial fuel:

$$CH_4(g) + 2O_2(g) \rightarrow CO_2(g) + 2H_2O(g) \quad \Delta_{comb}H^{\ominus} = -882\,kJ\,mol^{-1}$$

Apart from this combustion reaction, methane is not very reactive. It is not hydrolysed by water (Box 14.2) and reacts with halogens only when exposed to ultraviolet radiation:

$$CH_4(g) + Cl_2(g) \xrightarrow{h\nu} CH_3Cl(g) + HCl(g)$$

The alkanes up to butane, C_4H_{10} (b.p. −0.5°C) are gases, those containing from 5 to 17 carbon atoms are liquids, and the heavier hydrocarbons are solids.

(b) Silanes

Key points: Silane is a reducing agent; with a platinum complex as a catalyst it undergoes hydrosilylation and forms $Si(OR)_4$ with alcohols.

Silane, SiH_4, is prepared commercially by the reduction of SiO_2 with Al under a high pressure of hydrogen in a molten salt mixture of NaCl and $AlCl_3$. An idealized equation for this reaction is

$$6H_2(g) + 3SiO_2(s) + 4Al(s) \rightarrow 3SiH_4(g) + 2Al_2O_3(s)$$

The silanes are much more reactive than the alkanes and their stability decreases with increasing chain length. Silane itself, SiH_4, is spontaneously flammable in air, reacts violently with halogens, and is hydrolysed on contact with water. This increased reactivity compared to hydrocarbons is attributed to the large atomic radius of Si, which leaves it open to attack by nucleophiles, the greater polarity of the Si–H bond, and the availability of low-lying d orbitals, which may facilitate the formation of adducts. Silane is a reducing agent in aqueous solution. For example, when silane is bubbled through an oxygen-free aqueous solution containing Fe^{3+}, it reduces the iron to Fe^{2+}.

Bonds between silicon and hydrogen are not readily hydrolysed in neutral water, but the reaction is rapid in strong acid or in the presence of traces of base. Similarly, alcoholysis is accelerated by catalytic amounts of alkoxide:

$$SiH_4 + 4ROH \xrightarrow{\Delta, OR^-} Si(OR)_4 + 4H_2$$

Kinetic studies indicate that the reaction proceeds through a structure in which OR^- attacks the Si atom while H_2 is being formed via a kind of H···H hydrogen bond between hydridic and protic H atoms.

BOX 14.2 Methane clathrates — fossil fuels from the ocean floor

Methane clathrates are crystalline solids formed at low temperatures when ice crystallizes around CH_4 molecules. Clathrates are also referred to as *methane hydrates* or *natural gas hydrates* and their formation has caused major problems in the past by clogging gas pipelines in cold climates. The hydrates may contain other small, gaseous molecules such as ethane and propene. Several different clathrate structures are known. The unit cell of the most common one, known as Structure I, contains 46 H_2O molecules and up to eight CH_4 molecules. Recently, clathrates have received attention as a possible energy source because 1 m^3 of clathrate liberates up to 164 m^3 of methane gas.

Clathrates have been found under sediments on the ocean floors. They are thought to form by migration of methane from beneath the ocean floor along geological faults, followed by crystallization on contact with cold seawater. The methane in clathrates is also generated by bacterial degradation of organic matter in low-oxygen environments at the ocean floor. Where sedimentation rates and organic carbon levels are high, the water in the pores of the sediment is low in oxygen, and methane is produced by anaerobic bacteria. Below the zone of solid clathrates, large volumes of methane may occur as bubbles of free gas in the sediment. Methane hydrates are stable at low temperatures and high pressures. Because of these conditions and the need for relatively large amounts of organic matter for bacterial methanogenesis, clathrates are mainly restricted to high latitudes and along continental margins in the oceans. On the continental margins the supply of organic material is high enough to generate enough methane and water temperatures are close to freezing. In polar regions the gas hydrates are commonly linked to the occurrence of permafrost. The permafrost reservoir of methane has been estimated at about 400 Gt of carbon in the Arctic but no estimates have been made of possible Antarctic reservoirs. The oceanic reservoir has been estimated to be about 10−11 Tt of carbon.

In recent years, many governments have become very interested in the possible use of methane hydrates as fossil fuels. The realization that huge reservoirs of methane hydrates occur on the ocean floor and in permafrost regions has led to exploration and investigation of how to use hydrates as an energy source. The USSR tried unsuccessfully to recover gas hydrates from permafrost reservoirs in the 1960s and 1970s. Not enough is known about how clathrate deposits occur in ocean sediments to be able to plan for their recovery and drilling has been carried out in very few places.

The potential recovery of methane from clathrates is not without serious implications. As methane is a greenhouse gas, the discharge of large amounts of it into the atmosphere would increase global warming. Methane levels in the atmosphere were lower during glacial periods than during interglacial periods. Disturbances could destabilize sea-floor methane hydrates, triggering submarine landslides and huge releases of methane.

EXAMPLE 14.2 Investigating the formation of catenated species

Use the bond enthalpy data in Table 14.2 and the additional data given below to calculate the standard enthalpy of formation of $C_2H_6(g)$ and Si_2H_6 (g).

$\Delta_{vap}H^{\ominus}$ (C, graphite) = 715 kJ mol^{-1} $\Delta_{atom}H^{\ominus}$ (Si, s) = 439 kJ mol^{-1}

B(H−H) = 436 kJ mol^{-1}

Answer The enthalpy of formation of a compound can be calculated as the difference in energy between the bonds broken and the bonds formed in the formation reaction. Therefore, the relevant equations for the formation of $C_2H_6(g)$ and $Si_2H_6(g)$ are

2 C(graphite) + 3 $H_2(g)$ → $C_2H_6(g)$

2 Si(s) + 3 $H_2(g)$ → $Si_2H_6(g)$

It follows that

$\Delta_f H^{\ominus}(C_2H_6, g)$ = [2(715) + 3(436)] − [348 + 6(412)] kJ mol^{-1} = −82 kJ mol^{-1}

and

$\Delta_f H^{\ominus}(Si_2H_6, g)$ = [2(439) + 3(436)] − [326 + 6(318)] kJ mol^{-1} = −48 kJ mol^{-1}

The more negative value for ethane is due, to a large extent, to the greater C−H bond enthalpy compared to that for Si−H.

Self-test 14.2 Use the bond enthalpy data in Table 14.2 and above to calculate the standard enthalpy of formation of CH_4 and SiH_4.

The silicon analogue of hydroboration (Section 13.6c) is **hydrosilylation**, the addition of SiH across the multiple bonds of alkenes and alkynes. This reaction, which is used in both industrial and laboratory syntheses, can be carried out under conditions (300°C or ultraviolet irradiation) that produce a radical intermediate. In practice, it is usually performed under far milder conditions by using a platinum complex as catalyst:

$$CH_2{=}CH_2 + SiH_4 \xrightarrow{\Delta, H_2PtCl_6, isopropanol} CH_3CH_2SiH_3$$

The current view is that this reaction proceeds through an intermediate in which both the alkene and silane are attached to the Pt atom.

Silane is used in the production of semiconductor devices such as solar cells and in the hydrosilylation of alkenes; it is prepared commercially by the high-pressure reaction of hydrogen, silicon dioxide, and aluminium.

(c) Germane, stannane, and plumbane

Key point: Thermal stability decreases from germane to stannane and plumbane.

Germane (GeH_4) and stannane (SnH_4) can be synthesized by the reaction of the appropriate tetrachloride with $LiAlH_4$ in tetrahydrofuran solution. Plumbane (PbH_4) has been synthesized in trace amounts by the protolysis of a magnesium/lead alloy but it is extremely unstable. The stability of the tetrahydrides varies in the order $SiH_4 < GeH_4 > SnH_4 > PbH_4$, which is an example of the alternation effect (Section 9.2c). The presence of alkyl or aryl groups stabilizes the hydrides of all three elements. For example, trimethylplumbane, $(CH_3)_3PbH$, begins to decompose at −30°C, but it can survive for several hours at room temperature.

14.8 Compounds with halogens

Silicon, germanium, and tin react with all the halogens to form tetrahalides. Carbon reacts only with fluorine and lead forms stable dihalides.

(a) Halides of carbon

Key points: Nucleophiles displace halogens in carbon−halogen bonds; organometallic nucleophiles produce new M−C bonds; mixtures of polyhalocarbons and alkali metals are explosion hazards.

Carbon tetrafluoride is a colourless gas, CCl_4 is a dense liquid, CBr_4 is a pale yellow solid, and CI_4 is a red solid. The stabilities of the tetrahalomethanes decrease from CF_4 to CI_4 (Table 14.4). These tetrahalomethanes and analogous partially halogenated alkanes provide a route to a wide variety of derivatives, mainly by nucleophilic displacement of one or more halogen atoms. Some useful and interesting reactions from an inorganic perspective are outlined in Fig. 14.8. Note in particular the metal−carbon bond-forming reactions, which take place either by complete displacement of halogen or by oxidative addition.

The rates of nucleophilic displacement increase greatly from fluorine to iodine, and lie in the order F << Cl < Br < I. All tetrahalomethanes are thermodynamically unstable with respect to hydrolysis:

$$CX_4 (l \text{ or } g) + 2H_2O(l) \rightarrow CO_2(g) + 4HX(aq)$$

However, the reaction for C−F bonds is very slow, and fluorocarbon polymers such as poly(tetrafluoroethene) are highly resistant to attack by water.

Tetrahalomethanes can be reduced by strong reducing agents, such as alkali metals. For example, the reaction of carbon tetrachloride with sodium is highly exoergic:

$$CCl_4(l) + 4Na(s) \rightarrow 4NaCl(s) + C(s) \quad \Delta_r G^{\ominus} = -249\,kJ\,mol^{-1}$$

This reaction can occur with explosive violence with CCl_4 and other polyhalocarbons, so alkali metals such as sodium should never be used to dry them. Analogous reactions occur on the surface of poly(tetrafluoroethene) when it is exposed to alkali metals or strongly reducing organometallic compounds. Fluorocarbons, together with other fluorine-containing molecules, exhibit many interesting properties, such as high volatility and strong electron-withdrawing character (Chapter 17).

Carbon tetrachloride used to be used widely as a laboratory solvent and as a dry-cleaning fluid, refrigerant, and in fire extinguishers. Its use has declined steeply since the 1980s because it has been identified as a greenhouse gas and a carcinogen.

The **carbonyl halides** (Table 14.5) are planar molecules and useful chemical intermediates. The simplest of these compounds, $OCCl_2$, phosgene (**9**), is a highly toxic gas. It is prepared on a large scale by the reaction of chlorine with carbon monoxide:

$$CO(g) + Cl_2(g) \xrightarrow{200°C, charcoal} OCCl_2(g)$$

The utility of phosgene lies in the ease of nucleophilic displacement of Cl to produce carbonyl compounds and isocyanates (Fig. 14.9). The fact that hydrolysis leads to CO_2 rather than carbonic acid, $(HO)_2CO$, can be traced to the stability of the double bonds in CO_2.

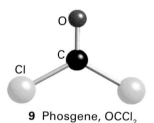

9 Phosgene, $OCCl_2$

Table 14.5 Properties of carbonyl halides

	COF_2	$COCl_2$	$COBr_2$
Melting point/°C	−114	−128	
Boiling point/°C	−83	8	65
$\Delta_f G^{\ominus}/(kJ\,mol^{-1})$	−619	−205	−111

Table 14.4 Properties of tetrahalomethanes

	CF_4	CCl_4	CBr_4	CI_4
Melting point/°C	−187	−23	90	171 dec
Boiling point/°C	−128	77	190	Sub
$\Delta_f G^{\ominus}/(kJ\,mol^{-1})$	−879	−65	+48	>0

dec, decomposes; *sub*, sublimes.

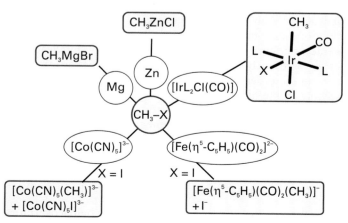

Figure 14.8 Some characteristic reactions of carbon−halogen bonds (X = halogen).

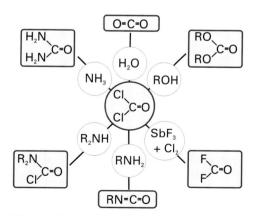

Figure 14.9 Characteristic reactions of phosgene, Cl_2CO.

(b) Compounds of silicon and germanium with halogens

Key point: Because silicon can form hypervalent intermediates states whereas carbon cannot, substitution reactions of silicon halides occur more readily than those of carbon halides.

Among the silicon tetrahalides, the most important is the tetrachloride, which is prepared by direct reaction of the elements or by chlorination of silica in the presence of carbon:

$$Si(s) + 2\,Cl_2(g) \rightarrow SiCl_4(l)$$

$$SiO_2(s) + 2\,Cl_2(g) + 2\,C(s) \xrightarrow{\Delta} SiCl_4(l) + 2\,CO(g)$$

Silicon and germanium halides are mild Lewis acids and add one or two ligands to yield five- or six-coordinate complexes:

$$SiF_4(g) + 2\,F^-(aq) \rightarrow SiF_6^{2-}(aq)$$
$$GeCl_4(l) + N\equiv CCH_3(l) \rightarrow Cl_4GeN\equiv CCH_3(s)$$

Hydrolysis of the Si and Ge tetrahalides is fast, and can be represented schematically as

$$EX_4 + 2\,H_2O \rightarrow EX_4(OH_2)_2 \rightarrow EO_2 + 4\,HX$$
$$(E = Si\ or\ Ge, X = halogen)$$

The corresponding carbon tetrahalides are kinetically more resistant to hydrolysis because of the lack of access to the sterically shrouded C atom to form an intermediate aqua complex.

The substitution reactions of halosilanes have been studied extensively. The reactions are more facile than for their carbon analogues because a Si atom can readily expand its coordination sphere to accommodate the incoming nucleophile. The stereochemistry of these substitution reactions indicates that a five-coordinate intermediate is formed with the most electronegative substituents adopting the axial position. Moreover, substituents leave from the axial position. The H^- ion is a poor leaving group, and alkyl groups are even poorer:

Note that in these examples, the R^4 substituent replaces H with retention of configuration.

(c) Tin and lead halides

Key points: Tin forms dihalides and tetrahalides; for lead, only the dihalides are stable.

Aqueous and nonaqueous solutions of tin(II) salts are useful mild reducing agents, but they must be stored under an inert atmosphere because air oxidation is spontaneous and rapid:

$$Sn^{2+}(aq) + \tfrac{1}{2}O_2(g) + 2\,H^+(aq) \rightarrow Sn^{4+}(aq) + H_2O(l)$$
$$E^{\ominus} = +1.08V$$

Tin dihalides and tetrahalides are both well known. The tetrachloride, tetrabromide, and tetraiodide are molecular compounds, but the tetrafluoride is an ionic solid formed from close

packing of SnF_6 octahedra. Lead tetrafluoride can be considered as an ionic solid but, as a manifestation of the inert-pair effect, $PbCl_4$ is an unstable, covalent, yellow oil that decomposes into $PbCl_2$ and Cl_2 at room temperature. Lead tetrabromide and tetraiodide are unknown, so the dihalides dominate the halogen compounds of lead. The arrangement of halogen atoms around the central metal atom in the dihalides of tin and lead often deviates from simple tetrahedral or octahedral coordination and is attributed to the presence of a stereochemically active lone pair. The tendency to achieve the distorted structure is more pronounced with the small F^- ion, and less distorted structures are observed with larger halides.

Both Sn(IV) and Sn(II) form a variety of complexes. Thus, $SnCl_4$ forms complex ions such as $SnCl_5^-$ and $SnCl_6^{2-}$ in acidic solution. In nonaqueous solution, a variety of donors interact with the moderately strong Lewis acid $SnCl_4$ to form complexes such as cis-$SnCl_4(OPMe_3)_2$. In aqueous and nonaqueous solutions Sn(II) forms trihalo complexes, such as $SnCl_3^-$, where the pyramidal structure indicates the presence of a stereochemically active lone pair (**10**). The $SnCl_3^-$ ion can act as a soft donor to d-metal ions. One unusual example of this ability is the red cluster compound $Pt_3Sn_8Cl_{20}$, which is trigonal bipyramidal (**11**).

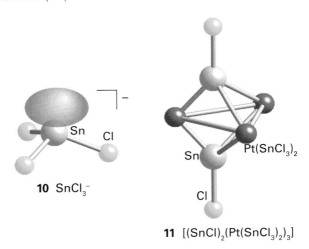

10 $SnCl_3^-$

11 $[(SnCl)_2(Pt(SnCl_3)_2)_3]$

14.9 Compounds of carbon with oxygen and sulfur

Key points: Carbon monoxide is a key reducing agent in the production of iron and a common ligand in d-metal chemistry; carbon dioxide is much less important as a ligand and is the acid anhydride of carbonic acid; the sulfur compounds CS and CS_2 have similar structures to their oxygen analogues.

Carbon forms CO, CO_2, and the suboxide, $O=C=C=C=O$ (Table 14.3). The uses of CO include the reduction of metal oxides in a blast furnace (Section 5.16) and the shift reaction (Section 10.4) for the production of H_2:

$$CO(g) + H_2O(g) \rightleftharpoons CO_2(g) + H_2(g)$$

In Chapter 26, where we deal with catalysis, we describe the conversion of carbon monoxide to acetic acid and aldehydes. The CO molecule has very low Brønsted basicity and negligible Lewis acidity towards neutral electron-pair donors. Despite its weak Lewis acidity, however, CO is attacked by strong Lewis bases at high pressure and somewhat elevated temperatures. Thus, the reaction with OH^- ions yields the formate ion, HCO_2^-:

$$CO(g) + OH^-(s) \rightarrow HCO_2^-(s)$$

Similarly, the reaction with methoxide ions (CH_3O^-) yields the acetate ion, $CH_3CO_2^-$.

Carbon monoxide is an excellent ligand towards d-metal atoms in low oxidation states (Section 22.5). Its well-known toxicity is an example of this behaviour: it binds to the Fe atom in haemoglobin, so excluding the attachment of O_2, and the victim suffocates. An interesting point is that H_3BCO can be prepared from B_2H_6 and CO at high pressures in a rare example of the coordination of CO to a simple Lewis acid. A complex of similar stability is not formed by BF_3; this observation is consistent with the classification of BH_3 as a soft acid and BF_3 as a hard acid.

Carbon dioxide is only a very weak Lewis acid. For example, only a small fraction of molecules are complexed with water to form H_2CO_3 in acidic aqueous solution but, at higher pH, OH^- coordinates to the C atom, so forming the hydrogencarbonate (bicarbonate) ion, HCO_3^-. This reaction is very slow; yet the attainment of rapid equilibrium between CO_2 and HCO_3^- is so important to life that it is catalysed by a Zn-containing enzyme carbon dioxide hydratase (carbonic anhydrase, Section 27.9a). The enzyme accelerates the reaction by a factor of about 10^9.

Carbon dioxide is one of several polyatomic molecules that are implicated in the **greenhouse effect**. In this effect, a polyatomic molecule in the atmosphere permits the passage of visible light but, because of its vibrational infrared absorptions, it blocks the immediate radiation of heat from the Earth. There is strong evidence for a significant increase in atmospheric CO_2 since the industrialization of society. In the past, nature has managed to stabilize the concentration of atmospheric CO_2, in part by precipitation of calcium carbonate in the deep oceans, but it seems that the rate of diffusion of CO_2 into the deep waters is too slow to compensate for the increased influx of CO_2 into the atmosphere (Box 14.3). There is convincing evidence for increasing concentrations of the greenhouse gases CO_2, CH_4, N_2O, and chlorofluorocarbons, and it is clear that they are having an impact on global temperatures. One method of slowing the rate of increase in atmospheric CO_2 is *carbon dioxide sequestration* (Box 14.4), in which CO_2 is captured from industrial flues by reaction with amines. It is then released and liquefied by compression and pumped underground—often back into gas or oil wells in order to drive out further oil or gas.

From an economic perspective, an important reaction is CO_2 with ammonia to yield ammonium carbonate, $(NH_4)_2CO_3$, which at elevated temperatures is converted directly to urea, $CO(NH_2)_2$, a fertilizer, a feed supplement for cattle, and a chemical intermediate. Another important use of CO_2 is in the soft drinks industry, where it dissolves under pressure to give a pleasant acidic taste of carbonic acid, H_2CO_3, and comes out of solution in the form of bubbles when the pressure is released. In organic chemistry, a common synthetic reaction is that between CO_2 and carbanion reagents to produce carboxylic acids. In the crucial biological process known as the *Calvin cycle*, CO_2 is 'fixed' (to the extent of 100 Gt per year) into organic molecules by reaction with the electron-rich C=C double bond of a pentose enolate ligand coordinated to a Mg^{2+} ion in the enzyme known as 'Rubisco' (Chapter 27.9b).

BOX 14.3 The carbon cycle

The carbon cycle is of particular interest because of the increased levels of carbon dioxide in the Earth's atmosphere and its potential to cause climate change through an enhanced greenhouse effect. On a global scale, the biological carbon cycle cannot be discussed without also considering the oxygen cycle (Box 16.1). The intimate relation between the two cycles is shown in Fig. B14.1.

Photosynthesis involves reduction of CO_2 to organic compounds and oxidation of H_2O to O_2 (Chapter 27). Oxygenic photosynthesis is present in the chloroplasts of higher plants, in a variety of algae and in cyanobacteria. In effect, oxygenic photosynthesis produces O_2 as a side product of H_2O oxidation, the principal aim being to scavenge the H atoms. When this process first occurred, the evolved O_2 would have been a toxin, producing reactive oxygen species capable of destroying most contemporary biomolecules.

Oxygen was not present when the Earth first cooled and liquid water first became available, and CO_2 was the principal atmospheric gas. Early organisms used photosynthesis or *chemolithotropy* (inorganic reactions) to produce the energy needed to reduce carbon dioxide or hydrogencarbonate ions to the organic molecules needed for cellular function. The electron donated in this reduction process did not come from water, the dominant electron donor of the past 2 Ga (1 Ga = 10^9 years). The first photosynthetic organisms on the early Earth used far simpler forms of photosynthesis. Some of these processes persist in modern bacteria, which use molecules such as H_2S, S_8, thiosulfate, H_2, and organic acids to reduce CO_2. As these molecules are in limited supply (compared to water), non O_2 evolving photosynthesis is capable of reducing only a small fraction of CO_2. However, once oxygenic photosynthesis evolved, with H_2O serving as the electron donor, planetary biomass could be produced and sustained at levels two to three orders of magnitude larger than previously.

The mass balance of the biological cycle in Fig. B14.1 is not quantitatively complete. Whereas there is input of CO_2 from the eruption of volcanoes and consumption of CO_2 in the weathering of silicate solids, as far as oxygen and organic carbon are concerned, there is no purely geochemical source. Therefore, for the cycle to be truly complete, no O_2 would ever accumulate:

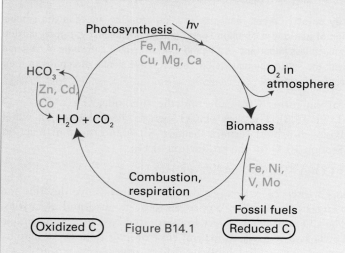

Figure B14.1

BOX 14.3 Continued

all the O_2 produced on the left side of the cycle by photosynthesis would be consumed on the right side by respiration and combustion. However, with each pass around the cycle, some of the reduced carbon biomass is buried in sediments, mostly land plants and algae in shallow marine basins and lakes. This small amount of buried biomass gradually becomes unavailable for oxidation, and some of it is transformed into hydrocarbon fossil fuels. Over geological time scales this buried reduced organic matter accumulates and is converted into the coal, shale, oil, and natural gas that constitute our fossil fuel reserves.

As this reduced carbon is buried, it is no longer available to be oxidized by O_2, which begins to accumulate in the atmosphere. Thus, over hundreds of millions of years, the process that created our fossil fuel reserves also formed the O_2 of the atmosphere and helped to decrease the initially high level of CO_2. The global accumulation of O_2 was slow on the early Earth on account of the vast amounts of iron(II) present in the oceans. This iron was oxidized by the O_2, yielding banded iron(III) formations. Once the iron(II) and reduced sulfur were consumed, O_2 began to accumulate in the atmosphere, achieving roughly modern levels about 1 Ga ago.

Currently, we are extracting and burning fossil fuels on a geologically very short time scale, thereby disturbing the relationship between oxygen and carbon. The combustion reactions are obviously the major factor, but some oil or gas reaches the surface through natural and human activities. The oil

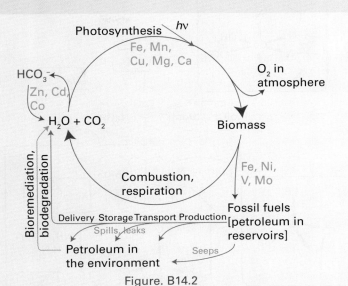

Figure. B14.2

or gas that is not burned can be biodegraded to produce CO_2 and complete the carbon cycle shown in Fig. B14.2. The biodegradation is carried out by aerobic organisms that, almost exclusively, use iron-dependent enzymes.

BOX 14.4 Reducing atmospheric CO_2 levels

The increasing use of fossil fuels since the industrial revolution has lead to an increase in atmospheric levels of CO_2 which may be contributing to the greenhouse effect and the associated climate change. There are several ways in which increases in the levels of atmospheric CO_2 could be minimized. Our reliance on carbon-based fossil fuels could be reduced by using energy more efficiently and reducing our consumption. Alternatively, our use of low-carbon fuels such as nuclear and renewable energies could be increased.

Another way of managing atmospheric carbon dioxide levels is by carbon dioxide sequestration. Carbon dioxide sequestration refers to the removal of carbon dioxide from the atmosphere and its long-term storage underground. One of the major sources of atmospheric CO_2 is coal and gas-fired power stations. A typical new 1 GW coal-fired power station produces around 6 Mt of CO_2 annually. Adding CO_2 capture scrubbers to remove CO_2 from flue gas can significantly reduce these emissions. The process uses aqueous solutions of various amines to remove CO_2 (and H_2S) from gases. The CO_2 reacts with the amines to form solid ammonium carbamate, $NH_4(NH_2CO_2)$. One of the problems with this process is that the aqueous phase evaporates

in the gas stream. However, new nonvolatile CO_2 capture materials have been developed. One approach is to produce ionic liquids with an attached amine group. These low-temperature molten ionic salts react reversibly with CO_2, do not need water to function, and can be recycled.

The principal chemical properties of CO_2 are summarized in Fig. 14.10. These properties are based on its mild Lewis acidity towards hard donors, as in the formation of CO_3^{2-} ions in strongly basic solution. The ability of CO_2 to form limestone and thereby to moderate the concentration of atmospheric CO_2 has been mentioned above. Similarly, carbonato complexes of metals can be formed in which CO_3^{2-} is a ligand (**12**). These complexes are often useful intermediates because CO_3^{2-} can be displaced in acidic solution to yield complexes that are otherwise difficult to prepare:

$$[Co(NH_3)_5(CO_3)]^+ + 2\,HF \rightarrow [Co(NH_3)_5F]^{2+} + CO_2 + H_2O + F^-$$

Carbonato complexes can be either monodentate, as shown above, or bidentate; in the latter case the CO_3^{2-} ion has a small bite angle.

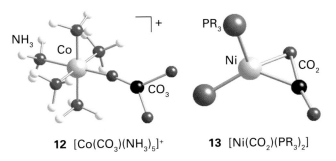

12 $[Co(CO_3)(NH_3)_5]^+$ **13** $[Ni(CO_2)(PR_3)_2]$

Metal complexes of CO_2 are known (**13**), but they are rare and far less important than the metal carbonyls. In its interaction with a low-oxidation state, electron-rich metal centre, the

neutral CO_2 molecule acts as a Lewis acid and the bonding is dominated by electron donation from the metal atom into an antibonding π orbital of CO_2.

An important use of superfluid CO_2 (that is, highly compressed carbon dioxide but above its critical temperature) is as a solvent. Applications range from decaffeination of coffee beans to its use in chemical synthesis in place of conventional solvents as an important part of the strategy for implementing the procedures of 'green chemistry'.

The sulfur analogues of carbon monoxide and carbon dioxide, CS and CS_2, are known. The former is an unstable transient molecule and the latter is endoergic ($\Delta_f G^\oplus = +65$ kJ mol^{-1}). Some complexes of CS (**14**) and CS_2 (**15**) exist,

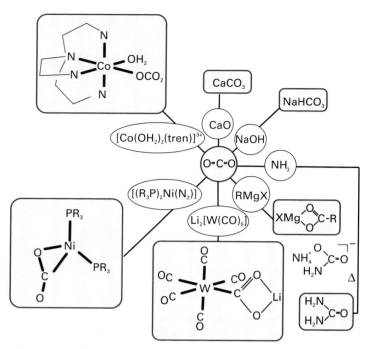

Figure 14.10 Characteristic reactions of carbon dioxide.

and their structures are similar to those formed by CO and CO_2. In basic aqueous solution, CS_2 undergoes hydrolysis and yields a mixture of carbonate ions, CO_3^{2-}, and trithiocarbonate ions, CS_3^{2-}.

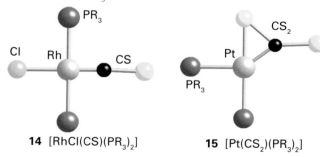

14 [RhCl(CS)(PR$_3$)$_2$]   **15** [Pt(CS$_2$)(PR$_3$)$_2$]

EXAMPLE 14.3 Proposing a synthesis that uses the reactions of carbon monoxide

Propose a synthesis of $CH_3{}^{13}CO_2^-$ that uses ^{13}CO, a primary starting material for many carbon-13 labelled compounds.

Answer We should bear in mind that CO_2 is readily attacked by strong nucleophiles such as $LiCH_3$ to produce acetate ions. Therefore, an appropriate procedure would be to oxidize ^{13}CO to $^{13}CO_2$ and then to react the latter with $LiCH_3$. A strong oxidizing agent, such as solid MnO_2, can be used in the first step to avoid the problem of excess O_2 in the direct oxidation.

$$^{13}CO(g) + 2MnO_2(s) \xrightarrow{\Delta} {}^{13}CO_2(g) + Mn_2O_3(s)$$
$$4\,{}^{13}CO_2(g) + Li_4(CH_3)_4\,(et) \rightarrow 4\,Li[CH_3\,{}^{13}CO_2](et)$$

where et denotes solution in ether. (Another method involves the reaction of $[Rh(I)(CO)_2]^-$ with ^{13}CO. The basis of this reaction is discussed in Chapter 26.)

Self-test 14.3 Propose a synthesis of $D^{13}CO_2^-$ starting from ^{13}CO.

14.10 Simple compounds of silicon with oxygen

Key point: The Si−O−Si link is present in silica, a wide range of metal silicate minerals, and silicone polymers.

The complicated silicate structures are often easier to comprehend if the tetrahedral SiO_4 unit from which they are built is drawn as a tetrahedron with the Si atom at the centre and O atoms at the vertices. The representation is often cut to the bone by drawing the SiO_4 unit as a simple tetrahedron with the atoms omitted. Each terminal O atom contributes -1 to the charge of the SiO_4 unit, but each shared O atom contributes 0. Thus, orthosilicate is $[SiO_4]^{4-}$ (**1**), disilicate is $[O_3SiOSiO_3]^{6-}$ (**2**), and the SiO_2 unit of silica has no net charge because all the O atoms are shared.

With these principles of charge balance in mind, it should be clear that an endless single-stranded chain or a ring of SiO_4 units, which has two shared O atoms for each Si atom, will have the formula and charge $[(SiO_3)^{2-}]_n$. An example of a compound containing such a cyclic metasilicate ion is the mineral *beryl*, $Be_3Al_2Si_6O_{18}$, which contains the $[Si_6O_{18}]^{12-}$ ion (**16**). Beryl is a major source of beryllium. The gemstone emerald is beryl in which Cr^{3+} ions are substituted for some Al^{3+} ions. A chain metasilicate (**17**) is present in the mineral *jadeite*, $NaAl(SiO_3)_2$, one of two different minerals sold as jade, the green colour arising from traces of iron impurities. In addition to other configurations for the single chain, there are double-chain silicates, which include the family of minerals known commercially as asbestos (Box 14.5).

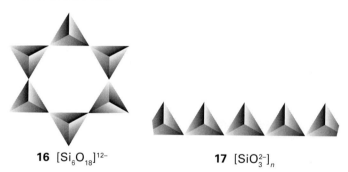

16 [Si$_6$O$_{18}$]$^{12-}$ **17** [SiO$_3^{2-}$]$_n$

■ **A brief illustration.** The cyclic silicate anion $[Si_3O_9]^{n-}$ is a six-membered ring with alternating Si and O atoms and six terminal O atoms, two on each Si atom. Because each terminal O atom contributes -1 to the charge, the overall charge is -6. From another perspective, the conventional oxidation numbers of silicon and oxygen, $+4$ and -2, respectively, also indicate a charge of -6 for the anion. ■

The composition of silicate glasses has a strong influence on their physical properties. For example, fused quartz (amorphous SiO_2) softens at about 1300°C, borosilicate glass (which contains boron oxide, Section 13.8) softens at about 800°C, and soda-lime glass softens at even lower temperatures. The variation in softening point can be understood by appreciating that the Si−O−Si links in silicate glasses form the framework that imparts rigidity. When basic oxides such as Na_2O and CaO are incorporated (as in soda-lime glass), they react with the SiO_2 melt and convert Si−O−Si links into terminal SiO groups and hence lower its softening temperature.

Very different properties are found for the $-Si-O-Si-$ backbone of silicone polymers, which are described later in this chapter.

14.11 Oxides of germanium, tin, and lead

Key point: The +2 oxide becomes more stable on going down the group from Ge to Pb.

Germanium(II) oxide, GeO, is a reducing agent and disproportionates to Ge and GeO_2. The structure of germanium(IV) oxide, GeO_2, is based on tetrahedral four-coordinate GeO_4 units. It also exists in a six-coordinate crystalline form with a rutile-like structure and in a vitreous form that resembles fused silica. Germanium analogues of silicates and aluminosilicates are also known (Section 14.15).

In the blue—black form of SnO the Sn(II) ions are four-coordinate (Fig. 14.11), but the O^{2-} ions around the Sn(II) lie in a square to one side with the lone pair on Sn pointing away from the square. This structure can be rationalized by the presence of a stereochemically active lone pair on the Sn atom and can be described as a fluorite structure (Section 3.9) with alternate layers of anions missing. The red form of SnO has a similar structure and can be converted to the blue—black form by heat, pressure, and treatment with alkali.

When heated in the absence of air, SnO disproportionates into Sn and SnO_2. The latter occurs naturally as the mineral *cassiterite* and has a rutile structure (Section 3.9). It has low solubility in glasses and glazes, and is used in large quantities as an opacifier and pigment carrier in ceramic glazes to make them less transparent.

The oxides of lead are very interesting from both fundamental and technological standpoints. The red form of PbO has the same structure as blue—black SnO with a stereochemically active lone pair (Fig. 14.11). Lead also forms mixed oxidation state oxides. The best known is 'red lead', Pb_3O_4, which contains Pb(IV) in an octahedral environment and Pb(II) in an irregular six-coordinate environment. The assignment of different oxidation numbers to the lead in these two sites is based on the shorter PbO distances for the atom identified as Pb(IV). The maroon form of lead(IV) oxide, PbO_2, crystallizes in the rutile structure. This oxide is a component of the cathode of a lead—acid battery (Box 14.6).

14.12 Compounds with nitrogen

Key points: The cyanide ion, CN^-, forms complexes with many d-metal ions; its coordination to the active sites of enzymes such as cytochrome *c* oxidase accounts for its high toxicity.

Hydrogen cyanide, HCN, is produced in large amounts by the high-temperature catalytic partial oxidation of methane and ammonia, and is used as an intermediate in the synthesis of many common polymers, such as poly(methyl methacrylate) and poly(acrylonitrile). It is highly volatile (b.p. 26°C) and, like the CN^- ion, highly poisonous. In some respects the toxicity of the CN^- ion is similar to that of the isoelectronic CO molecule because both form complexes with iron porphyrin molecules. However, whereas CO attaches to the Fe in haemoglobin and causes oxygen starvation, CN^- targets the Fe in the active site of cytochrome *c* oxidase (the enzyme in mitochondria that reduces oxygen to water), which results in a rapid and catastrophic collapse of energy production.

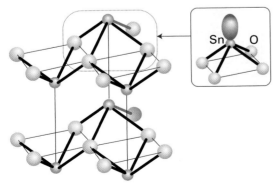

Figure 14.11 The structure of blue—black SnO showing parallel layers of square-based pyramidal SnO_4 units.

BOX 14.6 The lead–acid battery

The chemistry of the lead–acid battery is noteworthy because, as well as being the most successful rechargeable battery, it illustrates the role of both kinetics and thermodynamics in the operation of cells.

In its fully charged state, the active material on the cathode is PbO_2 and at the anode it is lead metal; the electrolyte is dilute sulfuric acid. One feature of this arrangement is that the lead-containing reactants and products at both electrodes are insoluble. When the cell is producing current, the reaction at the cathode is the reduction of Pb(IV) as PbO_2 to Pb(II), which in the presence of sulfuric acid is deposited on the electrode as insoluble $PbSO_4$:

$$PbO_2(s) + HSO_4^-(aq) + 3H^+(aq) + 2e^- \rightarrow PbSO_4(s) + 2H_2O(l)$$

At the anode, lead is oxidized to Pb(II), which is also deposited as the sulfate:

$$Pb(s) + SO_4^{2-}(aq) \rightarrow PbSO_4 + 2e^-$$

The overall reaction is

$$PbO_2(s) + 2HSO_4^-(aq) + 2H^+(aq) + Pb(s) \rightarrow 2PbSO_4(s) + 2H_2O(l)$$

The potential difference of about 2 V is remarkably high for a cell in which an aqueous electrolyte is used, and exceeds by far the potential for the oxidation of water to O_2, which is 1.23 V. The success of the battery hinges on the high overpotentials (and hence low rates) of oxidation of H_2O on PbO_2 and of reduction of H_2O on lead.

Unlike the neutral ligand CO, the negatively charged CN^- ion is a strong Brønsted base ($pK_a = 9.4$) and a much poorer Lewis acid π acceptor. The CO ligand can form complexes with metals in a zero oxidation state as it can remove electron density through the π system. However, the coordination chemistry of CN^- is more often associated with metal ions in positive oxidation states, as with Fe^{2+} in the hexacyanoferrate(II) complex, $[Fe(CN)_6]^{4-}$, as there will be less electron density on the metal ion.

The toxic, flammable gas cyanogen, $(CN)_2$ (**18**), is known as a **pseudohalogen** because of its similarity to a halogen. It dissociates to give ·CN radicals and forms interpseudohalogen compounds, such as FCN and ClCN. Similarly, CN^- is an example of a **pseudohalide ion** (Section 17.7).

18 Cyanogen, $(CN)_2$

The direct reaction of Si and N_2 at high temperatures produces silicon nitride, Si_3N_4. This substance is very hard and inert, and is used in high-temperature ceramic materials. Current industrial research projects focus on the use of suitable organosilicon–nitrogen compounds that might undergo pyrolysis to yield silicon nitride fibres and other shapes. When SiO_2 is heated with carbon, CO is evolved and silicon carbide, SiC, forms. This very hard material is widely used as the abrasive *carborundum*.

Trisilylamine, $(H_3Si)_3N$, the silicon analogue of trimethylamine, has very low Lewis basicity. It has a planar structure, or is fluxional with a very low barrier to inversion. The low basicity and planar structure have traditionally been attributed to d-orbital participation in bonding, allowing sp^2 hybridization around the N atom and delocalization of the lone pair through π bonding. However, quantum mechanical calculations indicate that whereas d orbitals play a role in delocalization, they are not responsible for the planar structure. Because the electronegativity of Si is lower than that of C, the Si–N bond is more polar than the C–N bond. This difference leads to long-range electrostatic repulsion between the silyl groups in trisilylamine and hence a planar structure.

14.13 Carbides

The numerous binary compounds of carbon with metals and metalloids, the *carbides*, are classified as follows:

Saline carbides, which are largely ionic solids; they are formed by the elements of Groups 1 and 2 and by aluminium.

Metallic carbides, which have a metallic conductivity and lustre; they are formed by the d-block elements.

Metalloid carbides, which are hard covalent solids formed by boron and silicon.

Figure 14.12 summarizes the distribution of the different types in the periodic table; it also includes binary molecular compounds

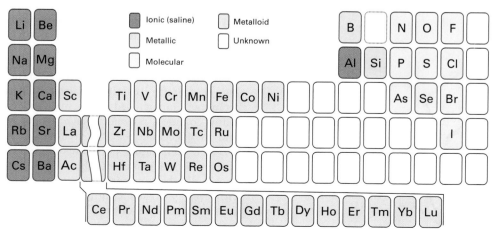

Figure 14.12 The distribution of carbides in the periodic table. Molecular compounds of carbon are included for completeness, but are not carbides.

of carbon with electronegative elements, which are not normally regarded as carbides. This classification is very useful for correlating chemical and physical properties, but (as so often in inorganic chemistry) the borderlines are sometimes indistinct.

(a) Saline carbides

Key points: Metal—carbon compounds of highly electropositive metals are saline; nonmetal carbides are mechanically hard and are semiconductors.

Saline carbides of the Group 1 and 2 metals may be divided into three subcategories: **graphite intercalation compounds**, such as KC_8, **dicarbides** (or 'acetylides'), which contain the C_2^{2-} anion, and **methides**, which formally contain the C^{4-} anion.

Graphite intercalation compounds are formed by the Group 1 metals (Section 11.12). They are formed by a redox process, and specifically by the reaction of graphite with alkali metal vapour or with metal—ammonia solution. For example, contact between graphite and potassium vapour in a sealed tube at 300°C leads to the formation of KC_8 in which the alkali metal ions lie in an ordered array between the graphite sheets (Fig. 14.13). A series of alkali metal—graphite intercalation compounds can be prepared with different metal:carbon ratios, including KC_8 and KC_{16}.

The dicarbides are formed by a broad range of electropositive metals, including those from Groups 1 and 2 (Section 11.11) and the lanthanoids. The C_2^{2-} ion has a very short CC distance in some dicarbides (for example, 119 pm in CaC_2), which is consistent with it being a triply bonded $[C\equiv C]^{2-}$ ion isoelectronic

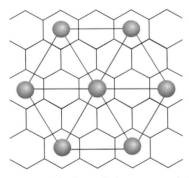

Figure 14.13 In KC_8, a graphite intercalation compound, the potassium atoms lie in a symmetrical array between the sheets. (See Fig. 14.4 for a view parallel to the sheets.)

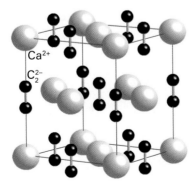

Figure 14.14 The calcium carbide structure. Note that this structure bears a similarity to the rock-salt structure. Because C_2^{2-} is not spherical, the cell is elongated along one axis. This crystal is therefore tetragonal rather than cubic.

with $[C\equiv N]^-$ and $N\equiv N$. Some dicarbides have a structure related to rock salt, but replacement of the spherical Cl^- ion by the elongated $[C\equiv C]^{2-}$ ion leads to an elongation of the crystal along one axis, and a resulting tetragonal symmetry (Fig. 14.14). The CC bond is significantly longer in the lanthanoid dicarbides, which suggests that for them the simple triply bonded structure is not a good approximation.

Carbides such as Be_2C and Al_4C_3 are borderline between saline and metalloid, and the isolated C ion is only formally C^{4-}. The existence of directional bonding to the C atom in carbides (as distinct from the nondirectional character expected of purely ionic bonding) is indicated by the crystal structures of methides, which are not those expected for the simple packing of spherical ions.

The principal synthetic routes to the saline carbides and acetylides of Groups 1 and 2 are very straightforward:
Direct reaction of the elements at high temperatures:

$$Ca(l) + 2C(s) \xrightarrow{>2000°C} CaC_2(s)$$

The formation of graphite intercalation compounds is another example of a direct reaction, but is carried out at much lower temperatures. The intercalation reaction is more facile because no CC covalent bonds are broken when an ion slips between the graphite layers.
Reaction of a metal oxide and carbon at a high temperature:

$$CaO(s) + 3C(s) \xrightarrow{2000°C} CaC_2(s) + CO(g)$$

Crude calcium carbide is prepared in electric arc furnaces by this method. The carbon serves both as a reducing agent to remove the oxygen and as a source of carbon to form the carbide.
Reaction of ethyne (acetylene) with a metal—ammonia solution:

$$2Na(am) + C_2H_2(g) \rightarrow Na_2C_2(s) + H_2(g)$$

This reaction occurs under mild conditions and leaves the carbon—carbon bonds of the starting material intact. As the ethyne molecule is a very weak Brønsted acid ($pK_a = 25$), the reaction can be regarded as a redox reaction between a highly active metal and a weak acid to yield H_2 (with H^+ the oxidizing agent) and the metal dicarbide.

The saline carbides have high electron density on the C atom, so they are readily oxidized and protonated. For example, calcium carbide reacts with the weak acid water to produce ethyne:

$$CaC_2(s) + 2H_2O(l) \rightarrow Ca(OH)_2(s) + HC\equiv CH(g)$$

This reaction is readily understood as the transfer of a proton from a Brønsted acid (H_2O) to the conjugate base (C_2^{2-}) of a weaker acid ($HC\equiv CH$). Similarly, the controlled hydrolysis or oxidation of the graphite intercalation compound KC_8 restores the graphite and produces a hydroxide or oxide of the metal:

$$2KC_8(s) + 2H_2O(g) \rightarrow 16C(graphite) + 2KOH(s) + H_2(g)$$

(b) Metallic carbides

Key point: d-Metal carbides are often hard materials with the carbon atom octahedrally surrounded by metal atoms.

The d metals provide the largest class of carbides. Examples are Co_6Mo_6C and Fe_3Mo_3C. They are sometimes referred to as **interstitial carbides** because it was long thought that the structures were the same as those of the metals and that they were formed by the insertion of C atoms in octahedral holes. In fact, the structure of the metal and the metal carbide often differ. For example, tungsten metal has a body-centred structure whereas tungsten carbide (WC) is hexagonal close packed. The name 'interstitial carbide' gives the erroneous impression that the metallic carbides are not legitimate compounds. In fact the hardness and other properties of metallic carbides demonstrate that strong metal−carbon bonding is present in them. Some of these carbides are economically and technologically useful materials. Tungsten carbide (WC), for example, is used for cutting tools and high-pressure apparatus such as that used to produce diamond. Cementite, Fe_3C, is a major constituent of steel and cast iron.

Metallic carbides of composition MC have an fcc or hcp arrangement of metal atoms with the C atoms in the octahedral holes. The fcc arrangement results in a rock-salt structure. The C atoms in carbides of composition M_2C occupy only half the octahedral holes between the close-packed metal atoms. A C atom in an octahedral hole is formally **hypercoordinate** (that is, has an untypically high coordination number) because it is surrounded by six metal atoms. However, the bonding can be expressed in terms of delocalized molecular orbitals formed from the C2s and C2p orbitals and the d orbitals (and perhaps other valence orbitals) of the surrounding metal atoms.

It has been found empirically that the formation of simple compounds in which the C atom resides in an octahedral hole of a close-packed structure occurs when $r_C/r_M < 0.59$, where r_C is the covalent radius of C and r_M is the metallic radius of M. This relationship also applies to metal compounds containing nitrogen or oxygen.

14.14 Silicides

Key points: Silicon−metal compounds (silicides) contain isolated Si, tetrahedral Si_4 units, or hexagonal nets of Si atoms.

Silicon, like its neighbours boron and carbon, forms a wide variety of binary compounds with metals. Some of these **silicides** contain isolated Si atoms. The structure of ferrosilicon, Fe_3Si, for instance, which plays an important role in steel manufacture, can be viewed as an fcc array of Fe atoms with some atoms replaced by Si. Compounds such as K_4Si_4 contain isolated tetrahedral cluster anions $[Si_4]^{4-}$ that are isoelectronic with P_4. Many of the f-block elements form compounds with the formula MSi_2 that have the hexagonal layers that adopt the AlB_2 structure shown in Fig. 13.8.

14.15 Extended silicon−oxygen compounds

As well as forming simple binary compounds with oxygen, silicon forms a wide range of extended network solids that find a range of applications in industry. Aluminosilicates occur naturally as clays, minerals, and rocks. Zeolite aluminosilicates are widely used as molecular sieves, catalysts, and catalyst support materials. These compounds are discussed further in Chapters 24 and 26.

(a) Aluminosilicates

Key point: Aluminium may replace silicon in a silicate framework to form an aluminosilicate. The brittle layered aluminosilicates are the primary constituents of clay and some common minerals.

Even greater structural diversity than that displayed by the silicates themselves is possible when Al atoms replace some of the Si atoms. The resulting aluminosilicates are largely responsible for the rich variety of the mineral world. We have already seen that in γ-alumina, Al^{3+} ions are present in both octahedral and tetrahedral holes (Section 3.3). This versatility carries over into the aluminosilicates, where Al may substitute for Si in tetrahedral sites, enter an octahedral environment external to the silicate framework, or, more rarely, occur with other coordination numbers. Because aluminium occurs as Al(III), its presence in place of Si(IV) in an aluminosilicate renders the overall charge negative by one unit. An additional cation, such as H^+, Na^+, or half as many Ca^{2+}, is therefore required for each Al atom that replaces an Si atom. As we shall see, these additional cations have a profound effect on the properties of the materials.

Many important minerals are varieties of layered aluminosilicates that also contain metals such as Li, Mg, and Fe: they include clays, talc, and various micas. In one class of layered aluminosilicate the repeating unit consists of a silicate layer with the structure shown in Fig. 14.15. An example of a simple aluminosilicate of this type (simple, that is, in the sense of there being no additional

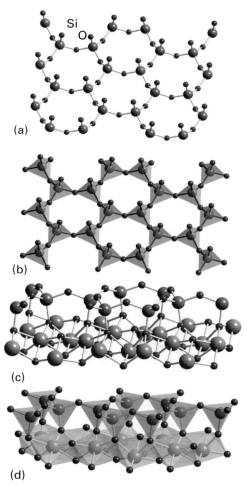

Figure 14.15 (a) A net of SiO_4 tetrahedra and (b) its tetrahedral representations. (c) Edge view of the above net and (d) its polyhedral representation. The structures (c) and (d) represent a double layer from the mineral chrysotile, for which M is Mg. When M is Al^{3+} and the anions in the bottom layers are replaced by an OH^- group this structure is close to that of the 1:1 clay mineral kaolinite.

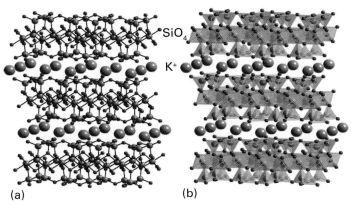

Figure 14.16 (a) The structure of 2:1 clay minerals such as muscovite mica $KAl_2(OH)_2Si_3AlO_{10}$, in which K^+ resides between the charged layers (exchangeable cation sites), Si^{4+} resides in sites of coordination number 4, and Al^{3+} in sites of coordination number 6. (b) The polyhedral representation. In talc, Mg^{2+} ions occupy the octahedral sites and O atoms on the top and bottom are replaced by OH groups and the K^+ sites are vacant.

elements) is the mineral *kaolinite*, $Al_2(OH)_4Si_2O_5$, which is used commercially as *china clay*. The electrically neutral layers are held together by rather weak hydrogen bonds, so the mineral readily cleaves and incorporates water between the layers.

A larger class of aluminosilicates has Al^{3+} ions sandwiched between silicate layers (Fig. 14.16). One such mineral is *pyrophyllite*, $Al_2(OH)_2Si_4O_{10}$. The mineral *talc*, $Mg_3(OH)_2Si_4O_{10}$, is obtained when three Mg^{2+} ions replace two Al^{3+} ions in the octahedral sites. As remarked earlier, in talc (and in pyrophyllite) the repeating layers are neutral, and as a result talc readily cleaves between them. Muscovite *mica*, $KAl_2(OH)_2Si_3AlO_{10}$, has charged layers because one Al(III) atom substitutes for one Si(IV) atom in the pyrophyllite structure. The resulting negative charge is compensated by a K^+ ion that lies between the repeating layers and results in greater hardness.

There are many minerals based on a three-dimensional aluminosilicate framework. The *feldspars*, for instance, which are the most important class of rock-forming minerals (and contribute to granite), belong to this class. The aluminosilicate frameworks of feldspars are built up by sharing all vertices of SiO_4 or AlO_4 tetrahedra. The cavities in this three-dimensional network accommodate ions such as K^+ and Ba^{2+}. Two examples are the feldspars *orthoclase*, $KAlSi_3O_8$, and *albite*, $NaAlSi_3O_8$.

(b) Microporous solids

Key point: Zeolite aluminosilicates have large open cavities or channels giving rise to useful properties such as ion exchange and molecular absorption.

The molecular sieves are crystalline aluminosilicates having open structures with apertures of molecular dimensions. These 'microporous' substances, which include the zeolites in which cations (typically from Groups 1 or 2) are trapped in an aluminosilicate framework, represent a major triumph of solid-state chemistry, for their synthesis and our understanding of their properties combine challenging determinations of structures, imaginative synthetic chemistry, and important practical applications. The cages are defined by the crystal structure, so they are highly regular and of precise size. Consequently, molecular sieves capture molecules with greater selectivity than high surface area solids such as silica gel or activated carbon, where molecules may be caught in irregular voids between the small particles.

Zeolites are used for shape-selective heterogeneous catalysis. For example, the molecular sieve ZSM-5 is used to synthesize 1,2-dimethylbenzene (*o*-xylene) for use as an octane booster in gasoline. The other xylenes are not produced because the catalytic process is controlled by the size and shape of the zeolite cages and tunnels. This and other applications are summarized in Table 14.6 and discussed in Chapters 24 and 26.

Synthetic procedures have added to the many naturally occurring zeolite varieties and have produced zeolites that have specific cage sizes and specific chemical properties within the cages. These synthetic zeolites are sometimes made at atmospheric pressure, but more often they are produced in a high-pressure autoclave. Their open structures seems to form around hydrated cations or other large cations such as NR_4^+ ions introduced into the reaction mixture. For example, a synthesis may be performed by heating colloidal silica to $100-200°C$ in an autoclave with an aqueous solution of tetrapropylammonium hydroxide. The microcrystalline product, which has the typical composition $[N(C_3H_7)_4]OH(SiO_2)_{48}$, is converted into the zeolite by burning away the C, H, and N of the quaternary ammonium cation at $500°C$ in air. Aluminosilicate zeolites are made by including high surface area alumina in the starting materials.

A wide range of zeolites has been prepared with varying cage and bottleneck sizes (Table 14.7). Their structures are based on approximately tetrahedral MO_4 units, which in the great majority of cases are SiO_4 and AlO_4. Because the structures involve many such tetrahedral units, it is common practice to abandon the polyhedral representation in favour of one that emphasizes the position of the Si and Al atoms. In this scheme, the Si or Al atom lies at the intersection of four line segments and the O atom bridge lies on the line segment (Fig. 14.17). This **framework representation** has the advantage of giving a clear impression of the shapes of the cages and channels in the zeolite. Some examples are illustrated in Fig. 14.18.

The important zeolites have structures that are based on the 'sodalite cage' (Fig 14.3), a truncated octahedron formed by slicing off each vertex of an octahedron (**19**). The truncation leaves a square face in the place of each vertex and the triangular faces of the octahedron are transformed into regular hexagons. The substance known as 'zeolite type A' is based on sodalite cages that are joined by O bridges between the square faces. Eight such sodalite cages are linked in a cubic pattern with a large central cavity called an **α cage**. The α cages share octagonal faces,

Table 14.6 Some uses of zeolites

Function	Application
Ion exchange	Water softeners in detergents
Absorption of molecules	Selective gas separation Gas chromatography
Solid acid	Cracking high molar mass hydrocarbons for fuel and petrochemical intermediates Shape-selective alkylation and isomerization of aromatics for petroleum and polymer intermediates

Table 14.7 Composition and properties of some molecular sieves

Molecular sieve	Composition	Bottleneck diameter/pm	Chemical properties
A	$Na_{12}[(AlO_2)_{12}(SiO_2)_{12}].xH_2O$	400	Absorbs small molecules; ion exchanger, hydrophilic
X	$Na_{86}[(AlO_2)_{86}(SiO_2)_{106}].xH_2O$	800	Absorbs medium-sized molecules; ion exchanger, hydrophilic
Chabazite	$Ca_2[(AlO_2)_4(SiO_2)_8].xH_2O$	400–500	Absorbs small molecules; ion exchanger, hydrophilic; acid catalyst
ZSM-5	$Na_3[(AlO_2)_3(SiO_2)_{93}].xH_2O$	550	Moderately hydrophilic
ALPO-5	$AlPO_4.xH_2O$	800	Moderately hydrophobic
Silicalite	SiO_2	600	Hydrophobic

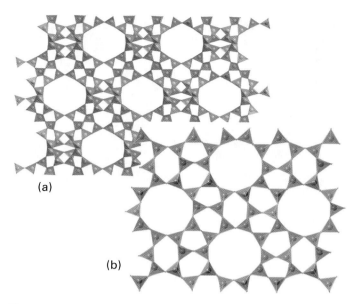

Figure 14.18 Two zeolite framework structures: (a) Zeolite-X and (b) ZSM-5. In each case just the SiO_4 tetrahedra that form the framework are shown; nonframework atoms such as charge-balancing cations and water molecules are omitted.

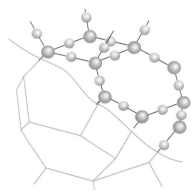

Figure 14.17 Framework representation of a truncated octahedron (truncation perpendicular to the fourfold axes of the octahedron) and the relationship of Si and O atoms to the framework. Note that a Si atom is at each vertex of the truncated octahedron and an O atom is approximately along each edge.

with an open diameter of 420 pm. Thus H_2O or other small molecules can fill them and diffuse through octagonal faces. However, these faces are too small to permit the entrance of molecules with van der Waals diameters larger than 420 pm.

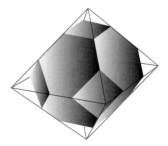

19 Truncated octahedron

■ **A brief illustration.** To identify the fourfold and sixfold axes in the truncated octahedral polyhedron used to describe the sodalite cage we note that there is one fourfold axis running through each pair of opposite square faces for a total of three fourfold axes. Similarly, a set of four sixfold axes runs through opposite sixfold faces. ■

The charge on the aluminosilicate zeolite framework is neutralized by cations lying within the cages. In the type-A zeolite, Na^+ ions are present and the formula is $Na_{12}(AlO_2)_{12}(SiO_2)_{12}.xH_2O$. Numerous other ions, including d-block cations and NH_4^+, can be introduced by ion exchange with aqueous solutions. Zeolites are therefore used for water softening and as a component of laundry detergent to remove the di- and tri-positive ions that decrease the effectiveness of the surfactant. Zeolites have in part replaced polyphosphates because the latter, which are plant nutrients, find their way into natural waters and stimulate the growth of algae.

In addition to the control of properties by selecting a zeolite with the appropriate cage and bottleneck size, the zeolite can be chosen for its affinity for polar or nonpolar molecules according to its polarity (Table 14.7). The aluminosilicate zeolites, which always contain charge-compensating ions, have high affinities for polar molecules such as H_2O and NH_3. By contrast, the nearly pure silica molecular sieves bear no net electric charge and are nonpolar to the point of being mildly hydrophobic. Another group of hydrophobic zeolites is based on the aluminium phosphate frameworks; $AlPO_4$ is isoelectronic with Si_2O_4 and the framework is similarly uncharged.

One interesting aspect of zeolite chemistry is that large molecules can be synthesized from smaller molecules inside the zeolite cage. The result is like a ship in a bottle because once assembled the molecule is too big to escape. For example, Na^+ ions in a Y-type zeolite may be replaced by Fe^{2+} ions (by ion exchange). The resulting $Fe^{2+}-Y$ zeolite is heated with phthalonitrile, which diffuses into the zeolite and condenses around the Fe^{2+} ion to form iron phthalocyanine (**20**), which remains imprisoned in the cage.

20

14.16 Organosilicon compounds

Key points: Methylchlorosilanes are important starting materials for the manufacture of silicone polymers; the properties of silicone polymers are determined by the degree of cross-linking and may be liquids, gels, or resins.

All silicon tetraalkyls and tetraaryls are monomeric with a tetrahedral Si centre. The C−Si bond is strong and the compounds are fairly stable. They can be prepared in a variety of ways, examples of which are shown below.

$$SiCl_4 + 4\,RLi \rightarrow SiR_4 + 4\,LiCl$$
$$SiCl_4 + LiR \rightarrow RSiCl_3 + LiCl$$

The *Rochow process* provides a cost-effective industrial route to methylchlorosilane, which is an important starting material in the manufacture of silicones:

$$n\,MeCl + Si/Cu \rightarrow Me_nSiCl_{4-n}$$

These methylchlorosilanes, Me_nSiCl_{4-n}, where n = 1–3, can be hydrolysed to form silicones or polysiloxanes:

$$Me_3SiCl + H_2O \rightarrow Me_3SiOH + HCl$$
$$2\,Me_3SiOH \rightarrow Me_3SiOSiMe_3 + H_2O$$

The reaction yields oligomers that contain the tetrahedral silicon group and oxygen atoms that form Si−O−Si bridges. Hydrolysis of Me_2SiCl_2 produces chains or rings and the hydrolysis of $MeSiCl_3$ produces a cross-linked polymer (Fig. 14.19). It is interesting to note that most silicon polymers are based on a Si−O−Si backbone, whereas carbon polymers are generally based on a C−C backbone, reflecting the strengths of the Si−O and C−C bonds (Table 14.2).

Silicone polymers have a range of structures and uses. Their properties depend on the degree of polymerization and cross-linking, which are influenced by the choice and mix of reactants, and the use of dehydrating agents such as sulfuric acid and elevated temperatures. The liquid silicones are more stable than hydrocarbon oils. Moreover, unlike hydrocarbons, their viscosity changes only slightly with temperature. Thus silicones are used as lubricants and wherever inert fluids are needed, for example in hydraulic braking systems. Silicones are very hydrophobic

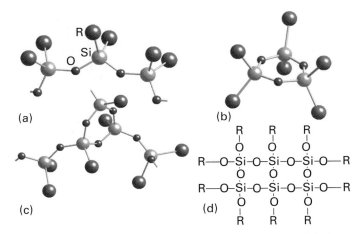

Figure 14.19 The structure of (a) a chain, (b) a ring, and (c) cross-linked silicone; (d) the chemical formula of a fragment.

and are used in water-repellent sprays for shoes and other items. The lower molar mass silicones are essential in personal-care products such as shampoos, conditioners, shaving foams, hair gels, and toothpastes, and impart to them a 'silky' feel. At the other end of the spectrum of delicacy, silicone greases, oils, and resins are used as sealants, lubricants, varnishes, waterproofing, synthetic rubbers, and hydraulic fluids.

14.17 Organometallic compounds

Key points: Tin and lead form tetravalent organo compounds; organotin compounds are used as fungicides and pesticides.

Many organometallic compounds of Group 14 are of great commercial importance, although the use of lead is declining owing to its toxicity (see below). Organotin compounds are used to stabilize poly(vinyl chloride) (PVC) as antifouling agents on ships, as wood preservatives, and as pesticides. Generally, organometallic compounds of the group are tetravalent and have low polarity bonds. Their stability decreases from silicon to lead.

Organotin compounds differ from organosilicon and organogermanium compounds in several ways. There is a greater occurrence of the +2 oxidation state, a greater range of coordination numbers, and halide bridges are often present. Most organotin compounds are colourless liquids or solids that are stable to air and water. The structures of R_4Sn compounds are all similar, with a tetrahedral tin atom (**21**).

The halide derivatives, R_3SnX, often contain Sn−X−Sn bridges and form chain structures. The presence of bulky R

21 SnR_4

groups may affect the shape. For example, in $(SnFMe_3)_n$ (**22**), the Sn−F−Sn backbone is in a zig-zag arrangement, in Ph_3SnF the chain has straightened, and $(Me_3SiC)Ph_2SnF$ is a monomer. The haloalkyls are more reactive than the tetraalkyls and are useful in the synthesis of tetraalkyl derivatives.

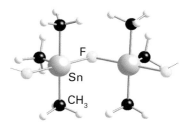

22 $(SnFMe_3)_n$, Me = CH_3

Alkyltin compounds may be prepared in a variety of ways, including by using a Grignard reagent and by metathesis:

$$SnCl_4 + 4\,RMgBr \rightarrow SnR_4 + 4\,MgBrCl$$
$$3\,SnCl_4 + 2\,Al_2R_6 \rightarrow 3\,SnR_4 + 2\,Al_2Cl_6$$

Organotin compounds have the widest range of uses of all main-group organometallic compounds and their annual worldwide industrial production exceeds 50 kt. Their major application is in the stabilization of PVC plastics. Without the additive, halogenated polymers are rapidly degraded by heat, light, and atmospheric oxygen to give discoloured, brittle products. The tin stabilizers scavenge labile Cl^- ions that initiate the loss of HCl, the first step in the degradation process. Organotin compounds also have a wide range of applications relating to their biocidal

effects. They are used as fungicides, algaecides, wood preservative, and antifouling agents. However, their widespread use on boats to prevent fouling and attachment of barnacles has caused environmental concerns as high levels of organotin compounds kill some species of marine life and affect the growth and reproduction of others. Many nations now restrict the use of organotin compounds to vessels over 25 m long.

Tetraethyl lead used to be made on a huge scale as an anti-knock agent in petrol. However, concerns about the levels of lead in the environment has led to it being phased out (Box 14.7). Alkyllead compounds, R_4Pb, can be made in the laboratory by using a Grignard reagent or an organolithium compound:

$$2\,PbCl_2 + 4\,RLi \rightarrow R_4Pb + 4\,LiCl + Pb$$
$$2\,PbCl_2 + 4\,RMgBr \rightarrow R_4Pb + Pb + 4\,MgBrCl$$

They are all monomeric molecules with tetrahedral geometry around the Pb atom. The halide derivatives may contain bridging halide atoms to form chains. Monomers are favoured by more bulky organic substituents. For example, $Pb(CH_3)_3Cl$ exists as a chain structure with bridging Cl atoms (**23**) whereas the mesityl derivative $Pb(Me_3C_6H_2)_3Cl$ is a monomer.

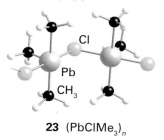

23 $(PbClMe_3)_n$

BOX 14.7 Lead in the environment

Of all the toxic elements in the environment, lead is perhaps the most pervasive. It poisons thousands of people yearly, most of them children in urban areas. The key aspect of lead in the body is that Pb^{2+} can be absorbed through the intestine and is stored in bones because Pb^{2+}, having a similar ionic radius to Ca^{2+}, can substitute for Ca^{2+} in hydroxyapatite, $Ca_5(PO_4)_3(OH)$. Bone is continuously formed and resorbed, allowing Pb^{2+} to circulate in the blood and reach critical targets in the blood-forming and nerve tissues. Several epidemiological studies have associated blood lead, down to quite low levels, with impairments in growth, hearing, and mental development. Drinking water can be a significant source of lead exposure because of lead in pipes or in the solder used in fitting connections. On contact with oxygenated water, metallic lead can be oxidized and solubilized:

$$2\,Pb + O_2 + 4\,H^+ \rightarrow 2\,Pb^{2+} + 2\,H_2O$$

Because the reaction consumes protons, the dissolution rate depends on the pH and is greatest in areas of soft water (low in Ca^{2+} and Mg^{2+}), where there has been little neutralization of rainwater's natural acidity. Districts with soft water sometimes add phosphate to form a protective lead phosphate precipitate that limits further dissolution of lead.

$$3\,Pb^{2+} + 2\,PO_4^{3-} \rightarrow Pb_3(PO_4)_2$$

Modern pipes and fittings are lead free, but older lead-based stock is still in use. The major other exposure route is lead-laced dust, which is deposited on food crops or directly ingested, particularly by children, who play in the dust and sometimes eat it. The two major dust sources are leaded paint and leaded fuel. Lead has been added to fuel in the form of tetraethyl or tetramethyl lead

since the 1920s to suppress 'knocking', the tendency of the fuel−air mixture to pre-ignite when compressed in an engine's cylinder. The lead inhibits pre-ignition by catalysing recombination of radicals. To prevent lead from depositing in the cylinder, dichloroethene or dibromoethene is also added, so that during combustion the lead is volatilized as PbX_2 (X = Cl or Br). Once in the atmosphere, the PbX_2 condenses into fine particles. Although these particles can circulate widely, most of them settle out in the dust near roadways; lead levels in urban dust correlate strongly with traffic congestion.

Lead additives started to be phased out in the developed world in the early 1970s, when catalytic converters were introduced for control of emissions because lead particles in the exhaust gas deactivate the catalytic surfaces. However, there are many areas of Africa, Asia, and South America where use of leaded fuels continues, and is even increasing as traffic increases. This is a public health issue because it has become clear that blood lead levels in the population correlate strongly with the use of leaded fuel.

The remaining problem involves lead in paint. Lead salts are brightly coloured and have been widely used in pigments and paint bases: $PbCrO_4$ is yellow, Pb_3O_4 is red, and $Pb_3(OH)_2CO_3$ is white. As the paint wears away, the lead compound is dispersed in dust. Dust and paint chips are particularly hazardous indoors; lead was banned from indoor paint in 1927 throughout much of Europe but only in 1971 in the USA. Some older housing, particularly in inner cities, still has leaded paint on interior walls. Leaded paints have also been widely used on building exteriors, and weathering can result in high lead levels in the adjacent soils, where children often play. In the USA, about one-fifth of children living in inner-city homes built before 1946 are estimated to have elevated blood lead levels.

FURTHER READING

M.A. Pitt and D.W. Johnson, Main group supramolecular chemistry, *Chem. Soc. Rev.*, 2007, **36**, 1441.

A. Schnepf, Metalliod Group 14 cluster compounds: an introduction and perspectives on this novel group of cluster compounds, *Chem. Soc. Rev.*, 2007, **36**, 745.

H. Berke, The invention of blue and purple pigments in ancient times, *Chem. Soc. Rev.*, 2007, **36**, 15. An interesting account of the uses of silicate pigments.

R.B. King, *Inorganic chemistry of the main group elements*. John Wiley & Sons (1994).

D.M.P. Mingos, *Essential trends in inorganic chemistry*. Oxford University Press (1998). A survey of inorganic chemistry from the perspective of structure and bonding.

N.C. Norman, *Periodicity and the s- and p-block elements*. Oxford University Press (1997). Includes coverage of essential trends and features of p-block chemistry.

R.B. King (ed.), *Encyclopedia of inorganic chemistry*. John Wiley & Sons (2005).

P.R. Birkett, A round-up of fullerene chemistry. *Educ. Chem.*, 1999, **36**, 24. A readable survey of fullerene chemistry.

J. Baggot, *Perfect symmetry: the accidental discovery of buckminsterfullerene*. Oxford University Press (1994). A general account of the story of the discovery of the fullerenes.

P.J.F. Harris, *Carbon nanotubes and related structures*. Cambridge University Press (2002).

EXERCISES

14.1 Silicon forms the chlorofluorides $SiCl_3F$, $SiCl_2F_2$, and $SiClF_3$. Sketch the structures of these molecules.

14.2 Explain why CH_4 burns in air whereas CF_4 does not. The enthalpy of combustion of CH_4 is -888 kJ mol^{-1} and the C−H and C−F bond enthalpies are -412 and -486 kJ mol^{-1} respectively.

14.3 SiF_4 reacts with $(CH_3)_4NF$ to form $[(CH_3)_4N][SiF_5]$. (a) Use the VSEPR rules to determine the shape of the cation and anion in the product; (b) Account for the fact that the ^{19}F NMR spectrum shows two fluorine environments.

14.4 Draw the structure and determine the charge on the cyclic anion $[Si_4O_{12}]^{n-}$.

14.5 Predict the appearance of the ^{119}Sn-NMR spectrum of $Sn(CH_3)_4$.

14.6 Predict the appearance of the ^1H-NMR spectrum of $Sn(CH_3)_4$.

14.7 Use the data in Table 14.2 and the additional bond enthalpy data given here to calculate the enthalpy of hydrolysis of CCl_4 and CBr_4. Bond enthalpies/(kJ mol^{-1}): O−H = 463, H−Cl = 431, H−Br = 366.

14.8 Identify the compounds A to F:

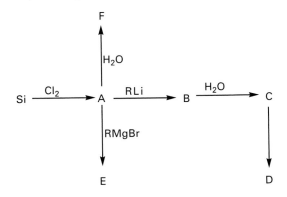

14.9 (a) Summarize the trends in relative stabilities of the oxidation states of the elements of Group 14, and indicate the elements that display the inert-pair effect. (b) With this information in mind, write balanced chemical reactions or NR (for no reaction) for the following combinations, and explain how the answer fits the trends.

(i) $Sn^{2+}(aq) + PbO_2(s)$ (excess) → (air excluded)

(ii) $Sn^{2+}(aq) + O_2(air)$ →

14.10 Use data from *Resource section* 3 to determine the standard potential for each of the reactions in Exercise 14.9(b). In each case, comment on the agreement or disagreement with the qualitative assessment you gave for the reactions.

14.11 Give balanced chemical equations and conditions for the recovery of silicon and germanium from their ores.

14.12 (a) Describe the trend in band gap energy, E_g, for the elements carbon (diamond) to tin (grey). (b) Does the electrical conductivity of silicon increase or decrease when its temperature is changed from 20°C to 40°C?

14.13 Preferably without consulting reference material, draw a periodic table and indicate the elements that form saline, metallic, and metalloid carbides.

14.14 Describe the preparation, structure and classification of (a) KC_8, (b) CaC_2, (c) K_3C_{60}.

14.15 Write balanced chemical equations for the reactions of K_2CO_3 with HCl(aq) and of Na_4SiO_4 with aqueous acid.

14.16 Describe in general terms the nature of the $[SiO_3]_n^{2n-}$ ion in jadeite and the silica-alumina framework in kaolinite.

14.17 (a) How many bridging O atoms are in the framework of a single sodalite cage? (b) Describe the (supercage) polyhedron at the centre of the Zeolite A structure in Fig. 14.3.

PROBLEMS

14.1 Correct any inaccuracies in the following descriptions of Group 14 chemistry. (a) None of the elements in this group is a metal. (b) At very high pressures, diamond is a thermodynamically stable phase of carbon. (c) Both CO_2 and CS_2 are weak Lewis acids and the hardness increases from CO_2 to CS_2. (d) Zeolites are layered materials exclusively composed of aluminosilicates. (e) The reaction of calcium carbide with water yields ethyne and this product reflects the presence of a highly basic C_2^{2-} ion in calcium carbide.

14.2 The lightest p-block elements often display different physical and chemical properties from the heavier members. Discuss the similarities and differences by comparison of:

(a) The structures and electrical properties of carbon and silicon.

(b) The physical properties and structures of the oxides of carbon and silicon.

(c) The Lewis acid–base properties of the tetrahalides of carbon and silicon.

14.3 Describe the physical properties of pyrophilite and muscovite mica and explain how these properties arise from the composition and structures of these closely related aluminosilicates.

14.4 There are major commercial applications for semicrystalline and amorphous solids, many of which are formed by Group 14 elements or their compounds. List four different examples of amorphous or partially crystalline solids described in this chapter and briefly state their useful properties.

14.5 The layered silicate compound $CaAl_2Si_2O_8$ contains a double aluminosilicate layer, with both Si and Al in four-coordinate sites. Sketch an edge-on view of a reasonable structure for the double layer, involving only vertex sharing between the SiO_4 and AlO_4 units. Discuss the likely sites occupied by Ca^{2+} in relation to the silica–alumina double layer.

14.6 Discuss the solid-state chemistry of silicon with reference to silicon dioxide, mica, asbestos, and silicate glasses.

14.7 One of your friends is studying English and is taking a course in science fiction. A common theme in the sources is silicon-based life forms. Your friend wonders why silicon should be chosen and why all life is carbon based. Prepare a short article that presents arguments for and against silicon-based life.

14.8 Karl Marx remarked in *Das Kapital* that '*If we could succeed, at a small expenditure of labour, in converting carbon into diamonds, their value might fall below that of bricks*'. Review current methods of synthesizing diamonds and discuss why these developments have not resulted in a large decrease in the value of diamonds.

14.9 The combination of mesoporosity with semiconduction would produce materials with interesting properties. A synthesis of such a material was discussed in 'Hexagonal mesoporous germanium' by Gerasimo *et al.* (*Science*, 2006, **313**, 5788). Summarize the expected advantages of such a material and describe how the mesoporous germanium was synthesized.

14.10 In 'An atomic seesaw switch formed by tilted asymmetric Sn-Ge dimers on a Ge (001) surface' Tomatsu *et al.* describe the synthesis and operation of a molecular switch. Describe how such molecular switches operate and summarize their current and potential applications. (*Science*, 2007, **315**, 1696).

The Group 15 elements

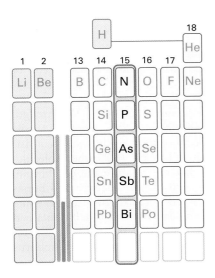

15

The chemical properties of the Group 15 elements are very diverse. Although the simple trends that we observed for Groups 13 and 14 are still apparent, they are complicated by the fact that the Group 15 elements exhibit a wide range of oxidation states and form many complex compounds with oxygen. Nitrogen makes up a large proportion of the atmosphere and is widely distributed in the biosphere. Phosphorus is essential for both plant and animal life. In stark contrast, arsenic is a well-known poison.

The Group 15 elements—nitrogen, phosphorus, arsenic, antimony, and bismuth—are some of the most important elements for life, geology, and industry. They range from gaseous nitrogen to metallic bismuth. The members of this group, the 'nitrogen group', are sometimes referred to collectively as the **pnictogens** (from the Greek for to stifle, a property of nitrogen). This name is neither widely used nor officially sanctioned. As in the rest of the p block, the element at the head of Group 15, nitrogen, differs significantly from its congeners. Its coordination number is generally lower and it is the only member of the group to exist as a gaseous, diatomic molecule under normal conditions.

PART A: **THE ESSENTIALS**

The properties of the Group 15 elements are diverse and more difficult to rationalize in terms of atomic radii and electron configuration than the p-block elements encountered so far. The usual trends of increasing metallic character down a group and stability of low oxidation states at the foot of the group are still evident but they are complicated by the wide range of oxidation states available.

15.1 The elements

Key points: Nitrogen is a gas; the heavier elements are all solids that exist in several allotropic forms.

All the members of the group other than N are solids under normal conditions. However, the trend to increasing metallic character down the group is not clear-cut because the electrical conductivities of the heavier elements actually decrease from As to Bi (Table 15.1). The normal increase in conductivity down a group reflects the closer spacing of the atomic energy levels in heavier elements and hence a smaller separation of the valence and conduction bands (Section 3.17). The opposite trend in conductivity in this group suggests that there must be a more pronounced molecular character in the solid state. Indeed, the structures of solid As, Sb, and Bi have three nearest-neighbour atoms and three more at significantly larger distances. The ratio of these long and short interactions decreases down the group, indicating the onset of a

Table 15.1 Selected properties of the Group 15 elements

	N	P	As	Sb	Bi
Melting point/°C	−210	44 (white)	613	630	271
		590 (red)	(sublimes)		
Atomic radius/pm	74	110	121	141	170
First ionization energy/(kJ mol^{-1})	1402	1011	947	833	704
Electrical conductivity/(10^6 S m^{-1})		10	3.33	2.50	0.77
Pauling electronegativity	3.0	2.2	2.2	2.0	2.0
Electron affinity/(kJ mol^{-1})	−8	72	78	103	91
B(E−H)/(kJ mol^{-1})	390	322	275		

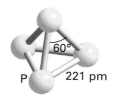

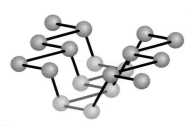

1 P_4, T_d

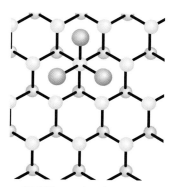

Figure 15.1 One of the puckered layers of black phosphorus. Note the trigonal pyramidal coordination of the atoms.

Figure 15.2 The puckered network structure of bismuth. Each Bi atom has three nearest neighbours; the dark shaded atoms indicate next-nearest neighbours from an adjacent puckered sheet. Lightest atoms are furthest from the viewer.

polymeric, molecular structure. The band structure of Bi suggests a low density of conduction electrons and holes, and it is best classified as a metalloid rather than as a semiconductor or a true metal.

The solid elements of Group 15 exist as a number of allotropes. Like the gaseous N_2 molecule, P_2 has a formal triple bond and a short bond length (189 pm). The strength of π bonds formed by Period 3 elements is weak relative to those of Period 2, so the allotrope P_2 is much less favoured than N_2. *White phosphorus* is a waxy solid consisting of tetrahedral P_4 molecules (**1**). Despite the small PPP angle (60°), the molecules persist in the vapour up to about 800°C, but above that temperature the equilibrium concentration of P_2 becomes appreciable. White phosphorus is very reactive and bursts into flame in air to yield P_4O_{10}. *Red phosphorus* can be obtained by heating white phosphorus at 300°C in an inert atmosphere for several days. It is normally obtained as an amorphous solid, but crystalline materials can be prepared that have very complex three-dimensional network structures. Unlike white phosphorus, red phosphorus does not ignite readily in air. When phosphorus is heated under high pressure, a series of phases of *black phosphorus* are formed, the thermodynamically most stable form. One of the phases consists of puckered layers composed of pyramidal three-coordinate P atoms (Fig. 15.1). In contrast to the usual practice of choosing the most stable phase of an element as the reference phase for thermodynamic calculations, white phosphorus is adopted because it is more accessible and better characterized than the other forms.

Arsenic exists in two solid forms, *yellow arsenic* and *metallic arsenic*. Yellow arsenic and gaseous arsenic both consist of tetrahedral As_4 molecules. The most stable structure at room temperature of metallic As, and of Sb and Bi, is built from puckered hexagonal layers in which each atom has three nearest neighbours. The layers stack in a way that gives three more distant neighbours in the adjacent net, as described above (Fig. 15.2).

Bismuth has recently been found to be radioactive, decaying by α emission with a half-life of 1.9×10^{19} years, which is much longer than the current age of the universe.

Nitrogen is readily available as dinitrogen, N_2, as it makes up 78 per cent by mass of the atmosphere. The principal raw material for the production of elemental phosphorus is phosphate rock, the insoluble, crushed, and compacted remains of ancient organisms, which consists primarily of the minerals fluorapatite, $Ca_5(PO_4)_3F$, and hydroxyapatite, $Ca_5(PO_4)_3OH$. The chemically softer elements As, Sb, and Bi are often found in sulfide ores. Arsenic is found naturally in the ores realgar, As_4S_4, orpiment, As_2S_3, arsenolite, As_2O_3, and arsenopyrite, FeAsS. Antimony occurs naturally as the minerals stibnite, Sb_2S_3, and ullmanite, NiSbS.

15.2 Simple compounds

Key point: The Group 15 elements form binary compounds on direct interaction with many elements. Nitrogen achieves oxidation number +5 only with oxygen and fluorine. Oxidation state +5 is common for phosphorus, arsenic, and antimony but rare for bismuth, for which the +3 state is the more stable.

The wide variety of possible oxidation states of the Group 15 elements can be understood to a large extent by considering the valence electron configuration of the elements, which is ns^2np^3. This configuration suggests that the highest oxidation state should be +5, as is indeed the case. According to the inert-pair effect (Section 9.5), we should also expect the $+3$ oxidation state to be more stable for Bi, as is in fact observed.

Nitrogen has a very high electronegativity (exceeded by only O, F, and Cl) and in many compounds, for example the nitrides, which contain the N^{3-} ion, and ammonia, NH_3, nitrogen is in a negative oxidation state. Nitrogen achieves positive oxidation states only in compounds with the more electronegative elements O and F. Nitrogen does achieve the group oxidation state ($+5$), but only under much stronger oxidizing conditions than are necessary to achieve this state for the other elements of the group.

The distinctive nature of nitrogen is due in large part to its high electronegativity, its small atomic radius, and the absence of accessible d orbitals. Thus, N seldom has coordination numbers greater than 4 in simple molecular compounds, but the heavier elements frequently reach coordination numbers of 5 and 6, as in PCl_5 and AsF_6^-.

Nitrogen forms binary compounds, the nitrides, with almost all the elements. The nitrides are classified as saline, covalent, and interstitial. Nitrogen also forms the azides, which contain the N_3^- ion, in which the average oxidation number of nitrogen is $-\frac{1}{3}$. Like N, P forms compounds with almost every element in the periodic table. There are many varieties of phosphides, with formulas ranging from M_4P to MP_{15}. The P atoms may be arranged in rings, chains, or cages, for example P_7^{3-} (**2**), P_8^{2-} (**3**), and P_{11}^{3-} (**4**). The arsenides and antimonides of the Group 13 elements In and Ga are semiconductors.

All the elements form simple hydrides. Ammonia NH_3 (**5**) is a pungent gas that is toxic at high levels of exposure. Ammonia is an excellent solvent for the Group 1 metals, for instance it is possible to dissolve 330 g of Cs in 100 g of liquid ammonia at $-50°C$. These highly coloured, electrically conducting solutions contain solvated electrons (Section 11.14). The chemical properties of ammonium salts are similar to those of the Group 1 ions, especially K^+ and Rb^+. Ammonium salts decompose on heating and ammonium nitrate is a component of some explosives. It is also widely used as a fertilizer. Nitrogen also forms the colourless liquid hydrazine, N_2H_4. The other hydrides of Group 15 are phosphine (formally phosphane, PH_3), arsine (arsane, AsH_3), and stibine (stibane, SbH_3), which are all poisonous gases.

> ***A note on good practice*** Although phosphane is the correct formal name of phosphine, the latter name is widely used and we adopt it here. However, we shall use the formal names arsane and stibane for these less common compounds and their derivatives.

Halogen compounds of P, As, and Sb are numerous and important in synthetic chemistry. Trihalides are known for all the Group 15 elements. However, whereas pentafluorides are known for all members of the group from P to Bi, pentachlorides are known only for P, As, and Sb, and the pentabromide is known only for P. Nitrogen does not reach its group oxidation state ($+5$) in neutral binary halogen compounds, but it does achieve it in NF_4^+. Presumably, an N atom is too small for NF_5 to be sterically feasible. The difficulty of oxidizing Bi(III) to Bi(V) by chlorine or bromine is an example of the inert-pair effect (Section 9.5). Bismuth pentafluoride, BiF_5, exists, but $BiCl_5$ and $BiBr_5$ do not.

Nitrogen forms many oxides and oxoanions, which are treated separately in Section 15.3. Phosphorus, As, Sb, and Bi form oxides and oxoanions in a range of oxidation states from $+5$ to $+1$. The most common oxidation state is $+5$ but the $+3$ state becomes increasingly more important for bismuth.

The complete combustion of phosphorus yields phosphorus(V) oxide, P_4O_{10}. Each P_4O_{10} molecule has a cage structure in which a tetrahedron of P atoms is held together by bridging O atoms, and each P atom has a terminal O atom (**6**). Combustion in a limited supply of oxygen results in the formation of phosphorus(III) oxide, P_4O_6; this molecule has the same O-bridged framework as P_4O_{10}, but lacks the terminal O atoms (**7**). Arsenic, Sb, and Bi form As_2O_3, Sb_2O_3, and Bi_2O_3.

15.3 Oxides and oxanions of nitrogen

Key points: The nitrate ion is a strong but slow oxidizing agent. The intermediate oxidation states of nitrogen are often susceptible to disproportionation. Dinitrogen oxide is unreactive.

Nitrogen forms oxo compounds and oxoanions in all oxidation states from $+5$ to $+1$. Nitrogen is in the oxidation state $+5$ in nitric acid, HNO_3, which is a major industrial

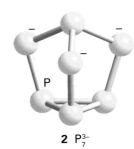

2 P_7^{3-}

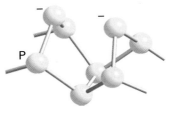

3 $(P_8^{2-})_n$

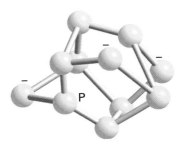

4 P_{11}^{3-}

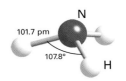

101.7 pm

107.8°

5 Ammonia, NH_3, C_{3v}

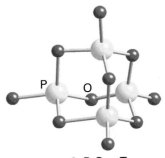

6 P_4O_{10}, T_d

7 P_4O_6, T_d

chemical used in the production of fertilizers, explosives, and a wide variety of nitrogen-containing chemicals. The nitrate ion, NO_3^-, is a moderately strong oxidizing agent. When concentrated nitric acid is mixed with concentrated hydrochloric acid, the orange, fuming *aqua regia* is formed, which is one of the few reagents that are able to dissolve platinum and gold. The anhydride of nitric acid is N_2O_5. It is a crystalline solid of composition $[NO_2^+][NO_3^-]$.

Nitrogen(IV) oxide, which is commonly called nitrogen dioxide, exists as an equilibrium mixture of the brown NO_2 radical and its colourless dimer, N_2O_4 (dinitrogen tetroxide). The equilibrium constant for the dimerization

$$N_2O_4(g) \rightleftharpoons 2\,NO_2(g)$$

is $K = 0.115$ at 25°C.

In nitrous acid, HNO_2, nitrogen is present as N(III). Nitrous acid is a strong oxidizing agent. Dinitrogen trioxide, N_2O_3, the anhydride of nitrous acid, is a blue solid that melts above -100°C to give a blue liquid that dissociates into NO and NO_2.

Nitrogen(II) oxide, more commonly nitric oxide, NO, is an odd-electron molecule. However, unlike NO_2 it does not form a stable dimer in the gas phase because the odd electron is distributed almost equally over both atoms and not, as in NO_2, largely confined to the N atom. Until the late 1980s, no beneficial biological roles were known for NO. However, since then it has been found that NO is generated *in vivo*, and that it performs functions such as the reduction of blood pressure, neurotransmission, and the destruction of microbes. Thousands of scientific papers have been published on the physiological functions of NO but our fundamental knowledge of its biochemistry is still quite meagre.

The average oxidation number of nitrogen in dinitrogen oxide, N_2O (specifically, NNO), which is commonly called nitrous oxide, is +1. N_2O is a colourless, unreactive gas. One sign of this inertness is that N_2O has been used as the propellant gas for instant whipping cream. Similarly, N_2O was used for many years as a mild anaesthetic; however, this practice has been discontinued because of undesirable physiological side-effects, particularly mild hysteria, indicated by its common name of *laughing gas*. It is still used as a 50:50 mixture with oxygen as an analgesic in childbirth and in clinical procedures, such as wound suturing.

PART B: **THE DETAIL**

In this section we review the detailed chemistry of the Group 15 elements. We shall see the wide variety of oxidation states achieved by the elements, particularly nitrogen and phosphorus.

15.4 Occurrence and recovery

Key points: Nitrogen is recovered by distillation from liquid air; it is used as an inert gas and in the production of ammonia. Elemental phosphorus is recovered from the minerals fluorapatite and hydroxyapatite by carbon arc reduction; the resulting white phosphorus is a molecular solid, P_4. Treatment of apatite with sulfuric acid yields phosphoric acid, which is converted to fertilizers and other chemicals.

Nitrogen is obtained on a massive scale by the distillation of liquid air. Liquid nitrogen is a very convenient way of storing and handling N_2 in the laboratory. Membrane materials that are more permeable to O_2 than to N_2 are used in laboratory-scale separations from air at room temperature (Fig. 15.3).

Phosphorus was first isolated by Hennig Brandt in 1669. Brandt, misinterpreting their colour, was trying to extract gold from urine and sand, and instead extracted a white solid, which glowed in the dark. This element was called phosphorus after the Greek for 'light bearer'. Today, phosphorus is produced by the action of concentrated sulfuric acid on the mineral fluorapatite

to generate phosphoric acid from which elemental phosphorous is subsequently extracted:

$$Ca_5(PO_4)_3F(s) + 5\,H_2SO_4(l) \rightarrow 3\,H_3PO_4(l) + 5\,CaSO_4(s) + HF(g)$$

The potential pollutant, hydrogen fluoride, from the fluoride component of the rock is scavenged by reaction with silicates to yield the less reactive SiF_6^{2-} complex ion.

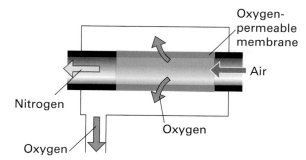

Figure 15.3 Schematic diagram of a membrane separator for nitrogen and oxygen.

The product of the treatment of phosphate rock with acid contains d-metal contaminants that are difficult to remove completely, so its use is largely confined to fertilizers and metal treatment. Most pure phosphoric acid and phosphorus compounds are still produced from the element because it can be purified by sublimation. The production of elemental phosphorus starts with crude calcium phosphate (as calcined phosphate rock), which is reduced with carbon in an electric arc furnace. Silica is added (as sand) to produce a slag of calcium silicate:

$$2\,Ca_3(PO_4)_2(s) + 6\,SiO_2(s) + 10\,C(s) \xrightarrow{1500°C}$$
$$6\,CaSiO_3(l) + 10\,CO(g) + P_4(g)$$

The slag is molten at these high temperatures and so can easily be removed from the furnace. The phosphorus vaporizes and is condensed to the solid, which is stored under water to protect it from reaction with air. Most phosphorus produced in this way is burned to form P_4O_{10}, which is then hydrated to yield pure phosphoric acid.

Arsenic is usually extracted from the flue dust of copper and lead smelters (Box 15.1). However, it is also obtained by heating the ores in the absence of oxygen:

$$FeAsS(s) \xrightarrow{700°C} FeS(s) + As(g)$$

Antimony is extracted by heating the ore stibnite with iron, which produces the metal and iron sulfide:

$$Sb_2S_3(s) + 3\,Fe(s) \rightarrow 2\,Sb(s) + 3\,FeS(s)$$

Bismuth occurs as bismite, Bi_2O_3, and bismuthinite, Bi_2S_3. It is produced as a byproduct of the extraction of copper, tin, lead, and zinc.

15.5 Uses

Key points: Nitrogen is essential for the industrial production of ammonia and nitric acid; the major use of phosphorus is in the manufacture of fertilizers.

The major nonchemical use of nitrogen gas is as an inert atmosphere in metal processing, petroleum refining, and food processing. Nitrogen gas is used to provide an inert atmosphere in the laboratory, and liquid nitrogen (b.p. $-196°C$, 77 K) is a convenient

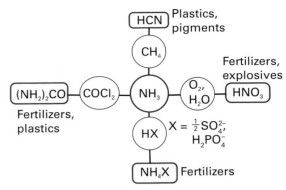

Figure 15.4 The industrial uses of ammonia.

refrigerant in both industry and the laboratory. The major industrial use of nitrogen is in the production of ammonia by the Haber process (Section 15.6) and conversion to nitric acid by the Ostwald process (Section 15.13). Ammonia provides a route to a wide range of nitrogen compounds, which include fertilizers, plastics, and explosives (Fig. 15.4). Nitrogen plays a crucial role in biology as it is a constituent of amino acids, nucleic acids, and proteins, and the nitrogen cycle is one of the most important processes in the ecosystem (Box 15.2 and Section 27.13).

Phosphorus is used in pyrotechnics, smoke bombs, steel making, and alloys. Red phosphorus mixed with sand is used as the striking strip on matchboxes. Sodium phosphate is used as a cleaning agent, a water softener, and to prevent scaling in boilers and pipes. Condensed phosphates are added to detergents as builders that enhance the detergency by softening the water by forming complexes with metal ions. In the natural environment, phosphorus is usually present as phosphate ions. Phosphorus (together with N and K) is an essential plant nutrient. However, as a result of the low solubility of many metal phosphates it is often depleted in soil, and hence hydrogenphosphates are important components of balanced fertilizers. Approximately 85 per cent of the phosphoric acid produced goes into fertilizer manufacture. Phosphorus is also an important constituent of bones and teeth (which are predominantly calcium phosphate),

---

BOX 15.1 Arsenic in the environment

The environmental toxicity of arsenic is a problem of groundwater contamination. The worst occurrence of arsenic pollution is in Bangladesh and the neighbouring Indian province of West Bengal, where hundreds of thousands of people have been diagnosed with arsenicosis. Three major rivers drain into this region, bringing iron-laden sediments from the mountains. The fertile delta is heavily farmed, and organic matter leaches into the shallow aquifer, creating reducing conditions. Arsenic levels are correlated with iron levels in groundwater and the arsenic is thought to be released on the dissolution of iron oxides and hydroxides from the ores.

Paradoxically, the problem grew out of a scheme sponsored by the United Nations, starting in the 1960s, to provide clean drinking water (replacing contaminated surface water) by sinking inexpensive tube wells into the aquifer. These wells did in fact improve health greatly by reducing the incidence of water-borne diseases, but the high arsenic content went unrecognized for many years. The tube wells are typically 20–100 m deep. Ground water closer to the surface has not had time to develop high

concentrations of arsenic and below 100 m the sediment has depleted in arsenic over time. As many as half of the four million tube wells exceed the Bangladesh arsenic standard of 50 ppb (the World Health Organization's guideline is 10 ppb), whereas levels routinely exceed 500 ppb in the more contaminated areas. There are several schemes to treat the well water to remove arsenic, and new wells could be dug into deeper, uncontaminated aquifers. The World Bank is coordinating a mitigation plan, but the massive effort could take years.

Arsenicosis develops over a period of up to 20 years. The first symptoms are keratoses of the skin, which develop into cancers; the liver and kidneys also deteriorate. The early stage is reversible if arsenic ingestion is discontinued, but once cancers develop, effective treatment becomes more difficult. The biochemistry of these effects is uncertain. Arsenate is reduced to As(III) complexes in the body, which probably act by binding sulfhydryl groups. A plausible link to cancer is suggested by the laboratory finding that low levels of arsenic inhibit hormone receptors that turn on cancer suppressor genes.

BOX 15.2 The nitrogen cycle

Most of the molecules used by biological systems contain nitrogen, including proteins, nucleic acids, chlorophyll, various enzymes and vitamins, and many other cellular constituents. In all these compounds, nitrogen is in its reduced form with an oxidation number of -3. The principal reservoir of nitrogen is N_2 in the atmosphere. Therefore, a major challenge to biology (and technology) involves the reduction of N_2 for incorporation into essential nitrogen compounds.

The nitrogen cycle is shown in Fig. B15.1. The cycle can be viewed as a set of enzymatically catalysed redox reactions that lead to an accessible supply of reduced nitrogen compounds. Microorganisms are almost entirely responsible for the interconversion of inorganic forms of nitrogen. The enzymes that catalyse these conversions have Fe, Mo, and Cu at their active site. Enzymes of the nitrogen cycle are discussed in Section 27.13.

Although N_2 is the most abundant constituent of the Earth's atmosphere, its usefulness is limited by its unreactivity. Although small amounts of N_2 react with O_2 under extreme conditions (for instance, in lightning discharges and high-temperature combustions), such processes provide only a small percentage of the nitrogen requirements of the biosphere. The remainder comes from the process of nitrogen fixation. The biological reduction of nitrogen to ammonia is carried out exclusively by prokaryotes, including both bacteria and archaea. The enzyme system for nitrogen

fixation functions anaerobically and O_2 rapidly and irreversibly destroys the enzyme. Nevertheless, nitrogen fixation also occurs in aerobic bacteria. In addition, important symbioses and associations exist between higher plants and nitrogen-fixing bacteria (such as rhizobia). In these symbioses, bacteria live within controlled environments in the plant, such as root nodules, that have low O_2 levels. The plant provides the bacterium with reduced carbon compounds from photosynthesis, while the bacterium provides fixed nitrogen to the plant.

A reduction potential of about -0.28 V is required for biological nitrogen fixation. Reduced ferredoxins or flavoproteins with reduction potentials of -0.4 to -0.5 V are readily available in biological systems (Chapter 27). Whereas these potentials indicate that nitrogen fixation is thermodynamically feasible, this is not the case kinetically. The kinetic barrier to N_2 reduction apparently arises from the need to form bound intermediates in the conversion of N_2 to ammonia. Organisms invest metabolic energy from adenosine triphosphate (ATP) hydrolysis, for which $\Delta_r G^{\ominus} \approx -31$ kJ mol^{-1}, for conversion to adenosine diphosphate (ADP) and inorganic phosphate (P_i) to produce the key intermediates in the N_2 fixation process. The reduction of N_2 consumes 16 molecules of ATP for each molecule of N_2 reduced. Given the opportunity, most organisms capable of nitrogen fixation use available fixed nitrogen sources (ammonia, nitrate, or nitrite) and repress the synthesis of the elaborate nitrogen fixation system.

Once nitrogen is reduced, organisms incorporate the nitrogen into organic molecules, where it enters the biosynthetic pathways of the cell. When organisms die and biomass decays, organonitrogen compounds decompose and release nitrogen to the environment in the form of NH_3 or NH_4^+, depending on the conditions.

The growing human population and its dependence on synthetic fertilizers have had enormous impact on the nitrogen cycle. Ammonia synthesis is carried out by the Haber–Bosch process (Section 26.12), which augments the total fixed nitrogen available to life on Earth. Between a third and a half of all nitrogen fixed occurs through technological and agricultural, rather than natural, means. In addition to ammonia itself, nitrate salts are produced industrially from ammonia for use in fertilizers. Both ammonia and nitrates enter the nitrogen cycle as fertilizer, which increases all segments of the natural cycle. The natural reservoirs are inadequate sinks for the excess input. Under such conditions, nitrate or nitrite may accumulate as an undesirable component of ground water or produce eutrophication in lakes, wetlands, river deltas, and coastal areas.

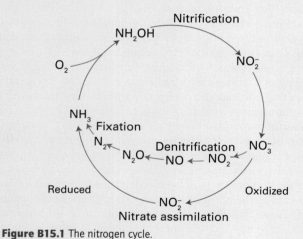

Figure B15.1 The nitrogen cycle.

cell membranes (phosphate esters of fatty acids), and nucleic acids, including DNA, RNA, and adenosine triphosphate (ATP), the energy-transfer unit of living organisms. The phosphines, PX_3, are widely used ligands (Section 7.1).

Arsenic is used as a dopant in solid-state devices such as integrated circuits and lasers. Although As is a well-known poison it is also an essential trace element in chickens, rats, goats, and pigs, and arsenic deficiency leads to restricted growth (Box 15.3).

BOX 15.3 Arsenicals

'Arsenicals' is the term used to describe chemicals containing arsenic. Arsenic and its compounds are intensely toxic and all the applications of arsenicals are based on this broad spectrum toxicity.

Inorganic arsenicals in the form of the mineral realgar and arsenolite were used in ancient times to treat ulcers, skin diseases, and leprosy. In the early 1900s an organoarsenic compound was found to be an effective treatment for syphilis and led to a rapid increase in research in this area. This initial treatment has now been replaced by penicillin, but organoarsenic compounds are still used today to treat trypanosiosis, or sleeping sickness, which is caused by a parasite in the blood. Arsenoamide, $C_{11}H_{12}AsNO_5S_2$, is used in veterinary medicines to treat heartworm in dogs.

Arsenilic acid, $C_6H_8AsNO_3$, and sodium arsenilate, $NaAsC_6H_8$, are used as antimicrobial agents in animal and poultry feed to prevent the growth of moulds. Another powerful antimicrobial agent is 10,10'-oxybisphenoxarsine (OBPA), which is used extensively in the manufacture of plastics.

Arsenicals are also used as insecticides and herbicides. Methylarsonic acid, CH_3AsO_3, is used to control weeds in cotton and turf crops. It is water soluble but is strongly absorbed into the soil and is not easily leached out. The first arsenic-containing insecticide was *Paris green*, $Cu(CH_3CO_2)_2.3Cu(AsO_2)_2$, which was manufactured in 1865 for the treatment of the Colorado potato beetle. Sodium arsenite, $NaAsO_2$, is used in poison baits to control grasshoppers and as a dip to prevent parasites in livestock.

Arsenic can be detected by the Marsh test in which the arsenic compound is reacted with zinc powder and sulfuric acid to liberate AsH_3:

$$As_2O_3 + 6\,Zn + 6\,H_2SO_4 \rightarrow 2\,AsH_3 + 6\,ZnSO_4 + 3\,H_2O$$

Ignition of AsH_3 produces arsenic, which can be observed as a black powder. The development of this test effectively made arsenic poisoning detectable for the first time.

Antimony is used in semiconductor technologies to produce infrared detectors and light-emitting diodes. It is used in alloys, where it leads to stronger and harder products. Antimony oxide is used to increase the activity of chlorinated hydrocarbon flame retardants, where it enhances the release of halogenated radicals.

In keeping with the general trend down the p block, the +3 oxidation state becomes more favourable relative to +5 on going down the group from P to Bi. Consequently, Bi(V) compounds are useful oxidizing agents. The other major uses of Bi compounds are in medicine. Bismuth subsalicylate, $HOC_6H_4CO_2BiO$, is used in conjunction with antibiotics and as a treatment for peptic ulcers. Bismuth(III)oxide is used in haemorrhoid creams.

EXAMPLE 15.1 **Examining the electronic structure and chemistry of P_4**

Draw the Lewis structure of P_4, and discuss its possible role as a ligand.

Answer We use the rules described in Section 2.1 to develop the Lewis structure. There is a total of $4 \times 5 = 20$ valence electrons. If each P atom forms a bond to each of the other three P atoms then 12 electrons will be accounted for, leaving eight electrons, or one lone pair on each P atom (**8**). This structure, together with the fact that the electronegativity of P is moderate ($\chi_P = 2.06$), suggests that P_4 might be a moderately good donor ligand. Indeed, though rare, P_4 complexes are known.

8 P_4

Self-test 15.1 Consider the Lewis structure of a segment of the structure of bismuth shown in Fig. 15.2. Is this puckered structure consistent with the VSEPR model?

15.6 Nitrogen activation

Key points: The commercial Haber process requires high temperatures and pressures to yield ammonia, which is a major ingredient in fertilizers and an important chemical intermediate.

Nitrogen occurs in many compounds, but N_2 itself, with a triple bond between the two atoms, is strikingly unreactive. A few strong reducing agents can transfer electrons to the N_2 molecule at room temperature, leading to scission of the N−N bond, but usually the reaction needs extreme conditions. The prime example of this reaction is the slow reaction of lithium metal at room temperature, which yields Li_3N. Similarly, when Mg (the diagonal neighbor of Li) burns in air it forms the nitride as well as the oxide.

The slowness of the reactions of N_2 appears to be the result of several factors. One is the strength of the N−N triple bond

and hence the high activation energy required for breaking it. (The strength of this bond also accounts for the lack of nitrogen allotropes.) Another factor is the relatively large size of the HOMO−LUMO gap in N_2 (Section 2.8b), which makes the molecule resistant to simple electron-transfer redox processes. A third factor is the low polarizability of N_2, which does not encourage the formation of the highly polar transition states that are often involved in electrophilic and nucleophilic displacement reactions.

Cheap methods of nitrogen activation, its conversion into useful compounds, are highly desirable because they would have a profound effect on the economy, particularly in poorer agricultural economies. In the *Haber process* for the production of ammonia, H_2 and N_2 are combined at high temperatures and pressures over an Fe catalyst, as we discuss in detail in Section 15.10. Much of the recent research aimed at achieving more economical ways of activating N_2 has been inspired by the way in which bacteria carry out the transformation at room temperature. Catalytic conversion of nitrogen to NH_4^+ involves the metalloenzyme nitrogenase, which occurs in nitrogen-fixing bacteria such as those found in the root nodules of legumes. The mechanism by which nitrogenase carries out this reaction, at an active site containing Fe, Mo, and S, is the topic of considerable research. In this connection, dinitrogen complexes of metals were discovered in 1965, at about the same time that it was realized that nitrogenase contains Mo (Section 27.13). These developments led to optimism that efficient homogeneous catalysts might be developed in which metal ions would coordinate to N_2 and promote its reduction. Many N_2 complexes have in fact been prepared, and in some cases the preparation is as simple as bubbling N_2 through an aqueous solution of a complex:

$$[Ru(NH_3)_5(OH_2)]^{2+}(aq) + N_2(g) \rightarrow$$
$$[Ru(NH_3)_5(N_2)]^{2+}(aq) + H_2O(l)$$

As with the isoelectronic CO molecule, end-on bonding is typical of N_2 when it acts as a ligand (**9**, Section 22.17). The N−N bond length in the Ru(II) complex is only slightly altered from that in the free molecule. However, when N_2 is coordinated to a more strongly reducing metal centre, this bond is considerably lengthened by back-donation of electron density into the π^* orbitals of N_2.

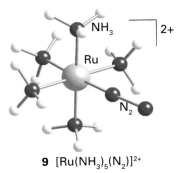

9 $[Ru(NH_3)_5(N_2)]^{2+}$

A recent advance has been the direct reduction of N_2 to ammonia at room temperature and atmospheric pressure with a molybdenum catalyst that contains a tetradentate triamidoamine ligand known as $HIPTN_3N$. Nitrogen coordinates to the Mo centre and is converted to NH_3 on addition of a proton source and a reducing agent.

15.7 Nitrides and azides

Nitrogen forms simple binary compounds with other elements; they are classified as nitrides or azides.

(a) Nitrides

Key point: Nitrides are classified as saline, covalent, or interstitial.

The nitrides of metals can be prepared by direct interaction of the element with nitrogen or ammonia or by thermal decomposition of an amide:

$$6\,Li(s) + N_2(g) \rightarrow 2\,Li_3N(s)$$
$$3\,Ca(s) + 2\,NH_3(l) \rightarrow Ca_3N_2(s) + 3\,H_2(g)$$
$$3\,Zn(NH_2)_2(s) \rightarrow Zn_3N_2(s) + 4\,NH_3(g)$$

The compounds of N with H, O, and the halogens are treated separately.

The **saline nitrides** can be regarded as containing the nitride ion, N^{3-}. However, the high charge of this ion means that it is highly polarizing (Section 1.9e) and saline nitrides are likely to have considerable covalent character. Saline nitrides occur for lithium, Li_3N, and the Group 2 elements, M_3N_2.

The **covalent nitrides**, in which the E−N bond is covalent, possess a wide range of properties depending on the element to which N is bonded. Some examples of covalent nitrides are boron nitride, BN, cyanogen, $(CN)_2$, phosphorus nitride, P_3N_5, tetrasulfur tetranitride, S_4N_4, and disulfur dinitride, S_2N_2. These compounds are discussed in the context of the other elements.

The largest category of nitrides consists of the **interstitial nitrides** of the d-block elements with formula MN, M_2N, or M_4N. The N atom occupies some or all of the octahedral sites within the cubic or hexagonal close-packed lattice of metal atoms. The compounds are hard and inert, with a metallic lustre and conductivity. They are widely used as refractory materials and find applications as crucibles, high-temperature reaction vessels, and thermocouple sheaths.

The nitride ion, N^{3-}, is often found as a ligand in d-metal complexes. Its high negative charge, small size, and ability to serve as a good π-donor as well as a σ-donor, means that it can stabilize metals in high oxidation states. The short coordinate bond between the ion and the metal atom is often represented as M≡N. An example is the complex $[Os(N)(NH_3)_5]^{2+}$ (**10**).

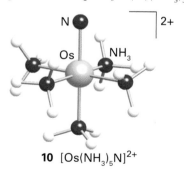

10 $[Os(NH_3)_5N]^{2+}$

(b) Azides

Key points: Azides are toxic and unstable; they are used as detonators in explosives.

Azides, in which nitrogen is present as N_3^-, may be synthesized by the oxidation of sodium amide with either NO_3^- ions or N_2O at elevated temperatures:

$$3\,NH_2^- + NO_3^- \xrightarrow{175°C} N_3^- + 3\,OH^- + NH_3$$
$$2\,NH_2^- + N_2O \xrightarrow{190°C} N_3^- + OH^- + NH_3$$

The average oxidation number of N in the azide ion is $-\tfrac{1}{3}$. The ion is isoelectronic with both dinitrogen oxide, N_2O, and CO_2 and, like these two molecules, is linear. It is a reasonably strong Brønsted base, the pK_a of its conjugate acid, hydrazoic acid, HN_3, being 4.77. It is also a good ligand towards d-block ions. However, heavy-metal complexes or salts, such as $Pb(N_3)_2$ and $Hg(N_3)_2$, are shock-sensitive detonators and decompose to produce the metal and nitrogen:

$$Pb(N_3)_2(s) \rightarrow Pb(s) + 3\,N_2(g)$$

Ionic azides such as NaN_3 are thermodynamically unstable but kinetically inert; they can be handled at room temperature. Sodium azide is toxic and is used as a chemical preservative and in pest control. When alkali metal azides are heated or detonated by impact they explode, liberating N_2; this reaction is used in the inflation of air bags in cars, in which the heating of the azide is electrical.

■ **A brief illustration.** A typical airbag contains approximately 50 g of NaN_3. To estimate the volume of nitrogen produced when the azide is detonated at room temperature and pressure (20°C and 1.0 atm) we need to consider the amount (in moles) of N_2 molecules produced in the decomposition reaction $2\,NaN_3(s) \rightarrow 2\,Na(s) + 3\,N_2(g)$. Because 50 g of NaN_3 contains 0.77 mol NaN_3, it liberates 1.2 mol N_2. This amount occupies 26 dm³ at 20°C and 1.0 atm. As the airbag is restricted in volume, the pressure of nitrogen in the airbag will be high, so providing protection to the driver. ■

Compounds containing the polynitrogen cation, N_5^+ (**11**), have been synthesized from species containing N_3^- and N_2F^+ ions. For example, N_5AsF_6 is prepared from N_2FAsF_6 and HN_3 in anhydrous HF solvent:

$$N_2FAsF_6(sol) + HN_3(sol) \rightarrow N_5AsF_6(sol) + HF(l)$$

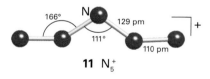

11 N_5^+

The compound is a white solid that decomposes explosively above 250°C. It is a powerful oxidizing agent and ignites organic material even at low temperatures. Salts of dipositive anions can be prepared from metathesis with the salts of monopositive anions in anhydrous HF:

$$2\,N_5SbF_6(sol) + Cs_2SnF_6(sol) \rightarrow (N_5)_2SnF_6(sol) + 2\,CsSbF_6(s)$$

The product is a white solid that is friction-sensitive and decomposes to N_5SnF_6 above 250°C. This product, N_5SnF_6, is stable up to 500°C.

15.8 Phosphides

Key point: Phosphides may be metal-rich or phosphorus-rich.

The compounds of phosphorus with hydrogen, oxygen, and the halogens are discussed separately. The phosphides of other elements can be prepared by heating the appropriate element with red phosphorus in an inert atmosphere:

$$n\,M + m\,P \rightarrow M_n P_m$$

There are many varieties of phosphides, with formulas ranging from M_4P to MP_{15}. They include metal-rich phosphides, in which M:P > 1, monophosphides, in which M:P = 1, and phosphorus-rich phosphides, in which M:P < 1. Metal-rich phosphides are usually very inert, hard, brittle refractory materials and resemble the parent metal in having high electrical and thermal conductivities. The structures have a trigonal prismatic arrangement of six, seven, eight, or nine metal ions around a P atom (**12**). Monophosphides adopt a variety of structures depending on the relative size of the other atom. For example, AlP adopts the zinc-blende structure, SnP adopts the rock-salt structure, and VP adopts the nickel-arsenide structure (Section 3.9). Phosphorus-rich phosphides have lower melting points and are less stable than metal-rich phosphides and monophosphides. They are semiconductors rather than conductors.

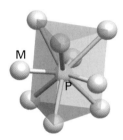

12

15.9 Arsenides, antimonides, and bismuthides

Key point: Indium and gallium arsenides and antimonides are semiconductors.

The compounds formed between metals and arsenic, antimony, and bismuth can be prepared by direct reaction of the elements:

$$Ni(s) + As(s) \rightarrow NiAs(s)$$

The arsenides and antimonides of the Group 13 elements In and Ga are semiconductors. Gallium arsenide (GaAs) is the more important and is used to make devices such as integrated circuits, light-emitting diodes, and laser diodes. Its band gap is similar to that of silicon and larger than those of other Group 13/15 semiconductors (Table 13.5; Section 24.19). Gallium arsenide is superior to Si for such applications because it has higher electron mobility and the devices produce less electronic noise. Silicon still has major advantages over GaAs in the sense that silicon is cheap and the wafers are stronger than those of GaAs, so processing is easier. Gallium arsenide integrated circuits are commonly used in mobile phones, satellite communications, and some radar systems.

15.10 Hydrides

All the Group 15 elements form binary compounds with hydrogen. All the EH_3 hydrides are toxic. Nitrogen also forms a catenated hydride, hydrazine, N_2H_4.

(a) Ammonia

Key points: Ammonia is produced by the Haber process; it is used to manufacture fertilizers and many other useful nitrogen-containing chemicals.

Ammonia is produced in huge quantities worldwide for use as a fertilizer and as a primary source of nitrogen in the production of many chemicals. As already mentioned, the Haber process is used for the entire global production. In this process, N_2 and H_2 combine directly at high temperature (450°C) and pressure (100 atm) over a promoted Fe catalyst:

$$N_2(g) + 3H_2(g) \rightarrow 2NH_3(g)$$

The promoters (compounds that enhance the catalyst's activity) include SiO_2, MgO, and other oxides (Section 26.12). The high temperature and catalyst are required to overcome the kinetic inertness of N_2, and the high pressure is needed to overcome the thermodynamic effect of an unfavourable equilibrium constant at the operating temperature.

So novel and great were the chemical and engineering problems arising from the then (early twentieth century) uncharted area of large-scale high-pressure technology, that two Nobel Prizes were awarded in connection with the process. One went to Fritz Haber (in 1918), who developed the chemical process. The other went to Carl Bosch (in 1931), the chemical engineer who designed the first plants to realize Haber's process. The **Haber−Bosch process** has had a major impact on civilization because ammonia is the primary source of most nitrogen-containing compounds, including fertilizers and most commercially important compounds of nitrogen. Before the development of the process the main sources of nitrogen for fertilizers were guano (bird droppings) and saltpetre, which had to be mined and transported from South America. In the early twentieth century there were predictions of widespread starvation across Europe, predictions that were never realized because of the widespread availability of nitrogen-based fertilizers.

The boiling point of ammonia is −33°C, which is higher than that of the hydrides of the other elements in the group and indicates the influence of extensive hydrogen bonding. Liquid ammonia is a useful nonaqueous solvent for solutes such as alcohols, amines, ammonium salts, amides, and cyanides. Reactions in liquid ammonia closely resemble those in aqueous solution, as indicated by the following autoprotolysis equilibria:

$$2H_2O(l) \rightleftharpoons H_3O^+(aq) + OH^-(aq) \qquad pK_w = 14.00 \text{ at } 25°C$$

$$2NH_3(l) \rightleftharpoons NH_4^+(am) + NH_2^-(am) \qquad pK_{am} = 34.00 \text{ at } -33°C$$

Many of the reactions are analogous to those carried out in water. For example, simple acid−base neutralization reactions can be carried out:

$$NH_4Cl(am) + NaNH_2(am) \rightarrow NaCl(am) + 2NH_3(l)$$

Ammonia is a water-soluble weak base:

$$NH_3(aq) + H_2O(l) \rightleftharpoons NH_4^+(aq) + OH^-(aq) \qquad pK_b = 4.75$$

The chemical properties of ammonium salts are very similar to those of Group 1 salts, especially of K^+ and Rb^+. They are soluble in water and solutions of the salts of strong acids, such as NH_4Cl, are acidic due to the equilibrium:

$$NH_4^+(aq) + H_2O(l) \rightleftharpoons H_3O^+(aq) + NH_3(aq) \qquad pK_a = 9.25$$

Ammonium salts decompose readily on heating and for many salts, such as the halides, carbonate, and sulfates, ammonia is evolved:

$$NH_4Cl(s) \rightarrow NH_3(g) + HCl(g)$$
$$(NH_4)_2SO_4(s) \rightarrow 2NH_3(g) + H_2SO_4(l)$$

When the anion is oxidizing, as in the case of NO_3^-, ClO_4^-, and $Cr_2O_7^{2-}$, the NH_4^+ is oxidized to N_2 or N_2O:

$$NH_4NO_3(s) \rightarrow N_2O(g) + 2H_2O(g)$$

When ammonium nitrate is heated strongly or detonated, the decomposition of 2 mol $NH_4NO_3(s)$ in the reaction

$$2NH_4NO_3(s) \rightarrow 2N_2(g) + O_2(g) + 4H_2O(g)$$

produces 7 mol of gaseous molecules, corresponding to an increase in volume from about approximately 200 cm^3 to about 140 dm^3, a factor of 700. This features leads to the use of ammonium nitrate as an explosive and nitrate fertilizers are often mixed with materials such as calcium carbonate or ammonium sulfate to make them more stable. Ammonium sulfate and the ammonium hydrogenphosphates, $NH_4H_2PO_4$ and $(NH_4)_2HPO_4$, are also used as fertilizers because phosphate is a plant nutrient. Ammonium perchlorate is used as the oxidizing agent in solid-fuel rocket propellants.

(b) Hydrazine and hydroxylamine

Key point: Hydrazine is a weaker base than ammonia and forms two series of salts.

Hydrazine, N_2H_4, is a fuming, colourless liquid with an odour like that of ammonia. It has a liquid range similar to that of water ($2 - 114$°C), indicating the presence of hydrogen bonding. In the liquid phase hydrazine adopts a *gauche* conformation around the N−N bond (**13**).

13 Hydrazine, N_2H_4

Hydrazine is manufactured by the *Raschig process*, in which ammonia and sodium hypochlorite react in dilute aqueous solution. The reaction proceeds through several steps, which can be simplified to

$$NH_3(aq) + NaOCl(aq) \rightarrow NH_2Cl(aq) + NaOH(aq)$$
$$2NH_3(aq) + NH_2Cl(aq) \rightarrow N_2H_4(aq) + NH_4Cl(aq)$$

There is a competing side reaction that is catalysed by d-metal ions:

$$N_2H_4(aq) + 2NH_2Cl(aq) \rightarrow N_2(g) + 2NH_4Cl(aq)$$

Gelatine is added to the reaction mixture to trap the d-metal ions by forming a complex with them. The dilute aqueous solution of hydrazine so produced is converted to a concentrated solution of hydrazine hydrate, $N_2H_4.H_2O$, by distillation. This product is often preferred commercially as it is cheaper than hydrazine and has a wider liquid range. Hydrazine is produced by distillation of the hydrate in the presence of a drying agent such as solid NaOH or KOH.

Hydrazine is a weaker base than ammonia:

$$N_2H_4(aq) + H_2O(l) \rightleftharpoons N_2H_5^+(aq) + OH^-(aq) \quad pK_{b1} = 7.93$$

$$N_2H_5^+(aq) + H_2O(l) \rightleftharpoons N_2H_6^{2+}(aq) + OH^-(aq) \quad pK_{b2} = 15.05$$

It reacts with acids HX to form two series of salts, N_2H_5X and $N_2H_6X_2$.

The major use of hydrazine and its methyl derivatives, CH_3NHNH_2 and $(CH_3)_2NNH_2$, is as a rocket fuel. Hydrazine is also used as a foam-blowing agent and as a treatment in boiler water to scavenge dissolved oxygen and prevent oxidation of pipes. Both N_2H_4 and $N_2H_5^+$ are reducing agents and are used in the recovery of precious metals.

EXAMPLE 15.2 Evaluating rocket fuels

Hydrazine, N_2H_4, and dimethylhydrazine, $N_2H_2(CH_3)_2$, are used as rocket fuels. Given the following data, suggest which would be the more efficient fuel thermochemically.

	$\Delta_fH^{\ominus}$/kJ mol^{-1}
$N_2H_4(l)$	+50.6
$N_2H_2(CH_3)_2(l)$	+42.0
$CO_2(g)$	−394
$H_2O(g)$	−242

Answer We must evaluate which combustion reactions release most heat by calculating their (standard) enthalpies of combustion. The combustion reactions are

$$N_2H_4(l) + O_2(g) \rightarrow N_2(g) + 2H_2O(g)$$

$$N_2H_2(CH_3)_2(l) + O_2(g) \rightarrow N_2(g) + 4H_2O(g) + 2CO_2(g)$$

The enthalpy of reaction (in this case, combustion) is calculated from

$$\Delta_cH^{\ominus} = \sum_{products} \Delta_fH^{\ominus} - \sum_{reactants} \Delta_fH^{\ominus}$$

We find −535 kJ mol^{-1} for N_2H_4 and −1798 kJ mol^{-1} for $N_2H_2(CH_3)_2$. An important factor for selecting rocket fuels is the specific enthalpy (the enthalpy of combustion divided by the mass of fuel), which for these fuels has values −16.7 and −29.9 kJ g^{-1}, respectively, indicating that $N_2H_2(CH_3)_2$ is the better fuel even when mass is significant.

Self-test 15.2 Refined hydrocarbons and liquid hydrogen are also used as rocket fuel. What are the advantages of dimethylhydrazine over these fuels?

Hydroxylamine, NH_2OH (**14**), is a colourless, hygroscopic solid with a low melting point (32°C). It is usually available as one of its salts or in aqueous solution. It is a weaker base than either ammonia or hydrazine:

$$NH_2OH(aq) + H_2O(l) \rightleftharpoons NH_3OH^+(aq) + OH^-(aq) \quad pK_b = 8.18$$

Anhydrous hydroxylamine can be prepared by adding sodium butoxide, NaC_4H_9O (NaOBu), to a solution of hydroxylamine hydrochloride in 1-butanol. The NaCl produced is filtered off and the hydroxylamine precipitated by the addition of ether.

$$[NH_3OH]Cl(sol) + NaOBu \rightarrow NH_2OH(sol) + NaCl(s) + BuOH(l)$$

The major commercial use of hydroxylamine is in the synthesis of caprolactam, which is an intermediate in the manufacture of nylon.

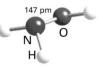

14 Hydroxylamine, NH_2OH

(c) Phosphine, arsane, and stibane

Key points: Unlike liquid ammonia, liquid phosphine, arsane, and stibane do not associate through hydrogen bonding; their much more stable alkyl and aryl analogues are useful soft ligands.

In contrast to the commanding role that ammonia plays in nitrogen chemistry, the highly poisonous hydrides of the heavier nonmetallic elements of Group 15 (particularly phosphine, PH_3, and arsane, AsH_3) are of minor importance in the chemistry of their respective elements. Both phosphine and arsane are used in the semiconductor industry to dope Si or to prepare other semiconductor compounds, such as GaAs, by chemical vapour deposition. These thermal decomposition reactions reflect the positive Gibbs energy of formation of these hydrides.

The commercial synthesis of PH_3 uses the disproportionation of white phosphorus in basic solution:

$$P_4(s) + 3OH^-(aq) + 3H_2O(l) \rightarrow PH_3(g) + 3H_2PO_2^-(aq)$$

Arsane and stibane may be prepared by the protolysis of compounds that contain an electropositive metal in combination with arsenic or antimony:

$$Zn_3E_2(s) + 6H_3O^+(aq) \rightarrow 2EH_3(g) + 3Zn^{2+}(aq) + 6H_2O(l)$$
$$(E = As, Sb)$$

Phosphine and arsane are poisonous gases that readily ignite in air, but the much more stable organic derivatives PR_3 and AsR_3 (R = alkyl or aryl groups) are widely used as ligands in metal coordination chemistry. In contrast with the hard donor properties of ammonia and alkylamine ligands, the organophosphines and organoarsanes, such as $P(C_2H_5)_3$ and $As(C_6H_5)_3$, are soft ligands and are therefore often incorporated into metal complexes having central metal atoms in low oxidation states. The stability of these complexes correlates with the soft acceptor nature of the metals in low oxidation states, and the stability of soft-donor−soft-acceptor combinations (Section 4.12).

All the Group 15 hydrides are pyramidal, but the bond angle decreases down the group:

$$NH_3, 107.8° \qquad PH_3, 93.6° \qquad AsH_3, 91.8° \qquad SbH_3, 91.3°$$

The large change in bond angle has been attributed to a decrease in the extent of sp^3 hybridization from NH_3 to SbH_3 but is more likely to be due to steric effects. The E−H bonding pairs of electrons will repel each other. This repulsion is greatest when the central element, E, is small, as in NH_3, and the H atoms will be as far away from each other as possible in a near-tetrahedral arrangement. As the size of the central atom is increased down the series, the repulsion between bonding pairs decreases and the bond angle is close to 90°.

It is evident from the boiling points plotted in Fig. 10.6 that PH_3, AsH_3, and SbH_3 are subject to little, if any, hydrogen bonding with themselves; however, PH_3 and AsH_3 can be protonated by strong acids, such as HI, to form phosphonium and arsonium ions, PH_4^+ and AsH_4^+, respectively.

15.11 Halides

All the elements form a trihalide with at least one halogen. Phosphorus, arsenic, and antimony form stable pentahalides.

(a) Nitrogen halides

Key point: Except for NF_3, nitrogen trihalides have limited stability and nitrogen triiodide is dangerously explosive.

Nitrogen trifluoride, NF_3, is the only exergonic binary halogen compound of nitrogen. This pyramidal molecule is not very reactive. Thus, unlike NH_3, it is not a Lewis base because the strongly electronegative F atoms make the lone pair of electrons unavailable: whereas the polarity of the N−H bond in NH_3 is $^{\delta-}N-H^{\delta+}$, that of the N−F bond in NF_3 is $^{\delta+}N-F^{\delta-}$. Nitrogen trifluoride can be converted into the N(V) species NF_4^+ by the reaction

$$NF_3(l) + 2F_2(g) + SbF_5(l) \rightarrow [NF_4^+][SbF_6^-](sol)$$

Nitrogen trichloride, NCl_3, is a highly endergonic, explosive, yellow oil. It is prepared commercially by the electrolysis of an aqueous solution of ammonium chloride and was once used as an oxidizing bleach for flour. The electronegativities of nitrogen and chlorine are the same and the N−Cl bond is almost nonpolar. Nitrogen tribromide, NBr_3, is an explosive, deep red oil. Nitrogen triiodide, NI_3, is highly explosive. Nitrogen is more electronegative than both bromine and iodine, and so the N−X bonds are polar in the sense $^{\delta-}N-X^{\delta+}$ and, formally, the oxidation numbers are −3 for the N and +1 for each halogen.

(b) Halides of the heavy elements

Key points: Whereas the halides of nitrogen have limited stability, their heavier congeners form an extensive series of compounds; the trihalides and pentahalides are useful starting materials for the synthesis of derivatives by metathetical replacement of the halide.

The trihalides and pentahalides of Group 15 elements other than nitrogen are used extensively in synthetic chemistry and their simple empirical formulas conceal an interesting and varied structural chemistry.

The trihalides range from gases and volatile liquids, such as PF_3 (b.p. −102°C) and AsF_3 (b.p. 63°C), to solids, such as BiF_3 (m.p. 649°C). A common method of preparation is direct reaction of the element and halogen. For phosphorus, the trifluoride is prepared by metathesis of the trichloride and a fluoride:

$$2PCl_3(l) + 3ZnF_2(s) \rightarrow 2PF_3(g) + 3ZnCl_2(s)$$

The trichlorides PCl_3, $AsCl_3$, and $SbCl_3$ are useful starting materials for the preparation of a variety of alkyl, aryl, alkoxy, and amino derivatives because they are susceptible to protolysis and metathesis:

$$ECl_3(sol) + 3EtOH(l) \rightarrow E(OEt)_3(sol) + 3HCl(sol) \quad (E = P, As, Sb)$$
$$ECl_3(sol) + 6Me_2NH(sol) \rightarrow E(NMe_2)_3(sol) + 3[Me_2NH_2]Cl(sol) \quad (E = P, As, Sb)$$

Phosphorus trifluoride, PF_3, is an interesting ligand because in some respects it resembles CO. Like CO, it is a weak σ donor but a strong π acceptor, and complexes of PF_3 exist that are the analogues of carbonyls, such as $Ni(PF_3)_4$, the analogue of $Ni(CO)_4$ (Section 22.18). The π-acceptor character is attributed to a P−F antibonding LUMO, which has mainly P p-orbital character. The trihalides also act as mild Lewis acids towards Lewis bases such as trialkylamines and halides. Many halide complexes have been isolated, such as the simple mononuclear species $AsCl_4^-$ (**15**) and SbF_5^{2-} (**16**). More complex dinuclear and polynuclear anions linked by halide bridges, such as the polymeric chain ($[BiBr_3]^{2-})_n$ in which Bi(I) is surrounded by a distorted octahedron of Br atoms, are also known.

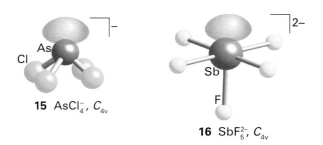

15 $AsCl_4^-$, C_{4v}

16 SbF_5^{2-}, C_{4v}

The pentahalides vary from gases, such as PF_5 (b.p. $-85°C$) and AsF_5 (b.p. $-53°C$), to solids, such as PCl_5 (sublimes at $162°C$) and BiF_5 (m.p. $154°C$). The five-coordinate gas-phase molecules are trigonal bipyramidal. In contrast to PF_5 and AsF_5, SbF_5 is a highly viscous liquid in which the molecules are associated through F-atom bridges. In solid SbF_5 these bridges result in a cyclic tetramer (**17**), which reflects the tendency of Sb(V) to achieve a coordination number of 6. A related phenomenon occurs with PCl_5, which in the solid state exists as $[PCl_4^+][PCl_6^-]$. In this case, the ionic contribution to the lattice enthalpy provides the driving force for the transfer of a Cl^- ion from one PCl_5 molecule to another. Another contributing factor may be the more efficient packing of the PCl_4 and PCl_6 units compared to the less efficient stacking of PCl_5 units. The pentafluorides of P, As, Sb, and Bi are strong Lewis acids (Section 4.10). SbF_5 is a very strong Lewis acid; it is much stronger, for example, than the aluminium halides. When SbF_5 or AsF_5 is added to anhydrous HF, a *superacid* is formed. (see Section 4.15):

$$SbF_5(l) + 2HF(l) \rightarrow H_2F^+(sol) + SbF_6^-(sol)$$

Of the pentachlorides, PCl_5 and $SbCl_5$ are stable whereas $AsCl_5$ is very unstable. This difference is a manifestation of the alternation effect (Section 9.2c). The instability of $AsCl_5$ is attributed to the increased effective nuclear charge arising from the poor shielding of the 3d electrons, which leads to a 'd-block contraction' and a lowering of the energy of the 4s orbitals in As. Consequently, it is more difficult to promote a 4s electron to form $AsCl_5$.

17 $(SbF_5)_4$

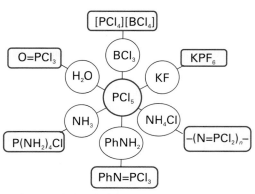

Figure 15.5 The uses of phosphorus pentachloride.

The pentahalides of P and Sb are very useful in syntheses. Phosphorus pentachloride, PCl_5, is widely used in the laboratory and in industry as a starting material and some of its characteristic reactions are shown in Fig. 15.5. Note, for example, that reaction of PCl_5 with Lewis acids yields PCl_4^+ salts, and simple Lewis bases like F^- give six-coordinate complexes such as PF_6^-. Compounds containing the NH_2 group lead to the formation of PN bonds, and the interaction of PCl_5 with either H_2O or P_4O_{10} yields $O=PCl_3$.

15.12 Oxohalides

Key points: Nitrosyl and nitryl halides are useful halogenating agents; phosphoryl halides are important industrially in the synthesis of organophosphorus derivatives.

Nitrogen forms all the nitrosyl halides, NOX, and the nitryl halides, NO_2X. The nitrosyl halides and NO_2F are prepared by direct interaction of the halogen with NO or NO_2, respectively:

$$2NO(g) + Cl_2(g) \rightarrow 2NOCl(g)$$
$$2NO_2(g) + F_2(g) \rightarrow 2NO_2F(g)$$

They are all reactive gases and the oxofluorides and oxochlorides are useful fluorinating and chlorinating agents.

Phosphorus readily forms the phosphoryl halides $POCl_3$ and $POBr_3$ by the reaction of the trihalides PX_3 with O_2 at room temperature. The fluorine and iodine analogues are prepared by the reaction of $POCl_3$ with a metal fluoride or iodide:

$$POCl_3(l) + 3NaF(s) \rightarrow POF_3(g) + 3NaCl(s)$$

All the molecules are tetrahedral and contain a P=O bond. POF_3 is gaseous, $POCl_3$ is a colourless liquid, $POBr_3$ is a brown solid, and POI_3 is a violet solid. They are all readily hydrolysed, fume in air, and form adducts with Lewis acids. They provide a route to the synthesis of organophosphorus compounds, which are manufactured on a large scale for use as plasticizers, oil additives, pesticides, and surfactants. For example, reaction with alcohols and phenols gives $(RO)_3PO$ and Grignard reagents (Section 12.13) yield R_nPOCl_{3-n}:

$$3ROH(l) + POCl_3(l) \rightarrow (RO)_3PO(sol) + 3HCl(sol)$$
$$nRMgBr(sol) + POCl_3(sol) \rightarrow R_nPOCl_{3-n}(sol) + nMgBrCl(s)$$

15.13 Oxides and oxoanions of nitrogen

Key point: Reactions of nitrogen–oxygen compounds that liberate or consume N_2 are generally very slow at normal temperatures and pH = 7.

We can infer the redox properties of the compounds of the elements in Group 15 in acidic aqueous solution from the Frost diagram in Fig. 15.6. The steepness of the slopes of the lines on the far right of the diagram show the thermodynamic tendency for reduction of the +5 oxidation states of the elements. They show, for instance, that Bi_2O_5 is potentially a very strong oxidizing agent, which is consistent with the inert-pair effect and the tendency of Bi(V) to form Bi(III). The next strongest oxidizing agent is NO_3^-. Both As(V) and Sb(V) are milder oxidizing agents, and P(V), in the form of phosphoric acid, is a very weak oxidant.

The redox properties of nitrogen are important because of its widespread occurrence in the atmosphere, the biosphere, industry, and the laboratory. Nitrogen chemistry is quite complex, partly because of the large number of accessible oxidation states but also because reactions that are thermodynamically favourable are often slow or have rates that depend crucially on the identity of the reactants. As the N_2 molecule is kinetically inert, redox reactions that consume N_2 are slow. Moreover, the formation of N_2 is often slow and may be sidestepped in aqueous solution (Fig. 15.7). As with several other p-block elements, the barriers to reaction of high oxidation state oxoanions, such as NO_3^-, are greater than for low oxidation state oxoanions, such as NO_2^-. We should also remember that low pH enhances the oxidizing power of oxoanions (Section 5.6). Low pH also often accelerates their oxidizing reactions by protonation, and this step is thought to facilitate subsequent NO bond breaking.

Table 15.2 summarizes some of the properties of the nitrogen oxides, and Table 15.3 does the same for the nitrogen oxoanions. Both tables will help us to navigate through the details of their properties.

(a) Nitrogen(V) oxides and oxoanions

Key points: The nitrate ion is a strong but slow oxidizing agent at room temperature; strong acid and heating accelerate the reaction.

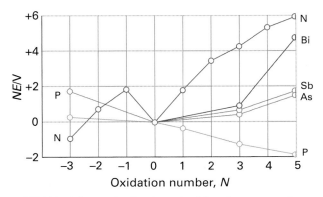

Figure 15.6 Frost diagram for the elements of the nitrogen group in acidic solution. The species with oxidation number −3 are NH_3, PH_3, and AsH_3, and those with oxidation numbers −2 and −1 are N_2H_4 and NH_2OH, respectively. The positive oxidation states refer to the most stable oxo or hydroxo species in acidic solution, and may be oxides, oxoacids, or oxoanions.

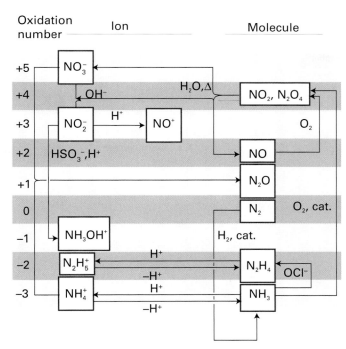

Figure 15.7 The interconversion of important nitrogen species.

The most common source of N(V) is nitric acid, HNO_3, which is a major industrial chemical used in the production of fertilizers, explosives, and a wide variety of nitrogen-containing chemicals. It is produced by modern versions of the *Ostwald process*, which make use of an indirect route from N_2 to the highly oxidized compound HNO_3 via the fully reduced compound NH_3. Thus, after nitrogen has been reduced to the −3 state as NH_3 by the Haber process, it is oxidized to the +4 state:

$$4\,NH_3(g) + 7\,O_2(g) \rightarrow 6\,H_2O(g) + 4\,NO_2(g)$$
$$\Delta_r G^\ominus = -308.0\,kJ\,(mol\,NO_2)^{-1}$$

The NO_2 then undergoes disproportionation into N(II) and N(V) in water at elevated temperatures:

$$3\,NO_2(aq) + H_2O(l) \rightarrow 2\,HNO_3(aq) + NO(g)$$
$$\Delta_r G^\ominus = -5.0\,kJ\,(mol\,HNO_3)^{-1}$$

All the steps are thermodynamically favourable. The byproduct NO is oxidized with O_2 to NO_2 and recirculated. Such an indirect route is used because the direct oxidation of N_2 to NO_2 is thermodynamically unfavourable, with $\Delta_r G^\ominus(NO_2, g) = +51$ kJ mol^{-1}. In part, this endergonic character is due to the great strength of the N≡N bond.

Standard potential data imply that the NO_3^- ion is a moderately strong oxidizing agent. However, its reactions are generally slow in dilute acid solution. Because protonation of an O atom promotes NO bond breaking, concentrated HNO_3 (in which NO_3^- is protonated) undergoes more rapid reactions than the dilute acid (in which HNO_3 is fully deprotonated). It is also a thermodynamically more potent oxidizing agent at low pH. A sign of this oxidizing character is the yellow colour of the concentrated acid, which indicates its instability with respect to decomposition into NO_2:

$$4\,HNO_3(aq) \rightarrow 4\,NO_2(aq) + O_2(g) + 2\,H_2O(l)$$

Table 15.2 Oxides of nitrogen

Oxidation number	Formula	Name	Structure (gas phase)	Comments
+1	N_2O	Nitrous oxide (dinitrogen oxide)	 119 pm	Colourless gas, not very reactive
+2	NO	Nitric oxide (nitrogen monoxide)	 115 pm	Colourless, reactive, paramagnetic gas
+3	N_2O_3	Dinitrogen trioxide	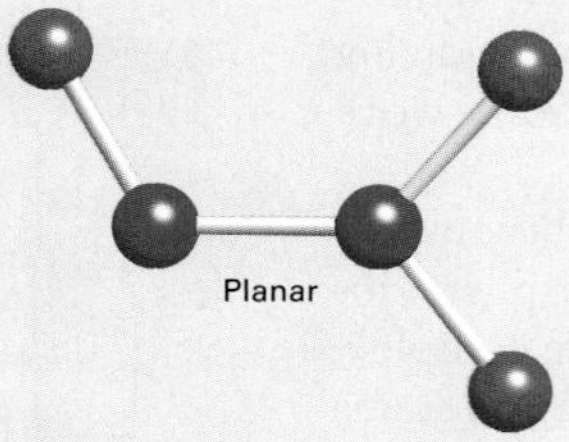 Planar	Blue solid (m.p. 101°C); dissociates into NO and NO_2 in the gas phase
+4	NO_2	Nitrogen dioxide	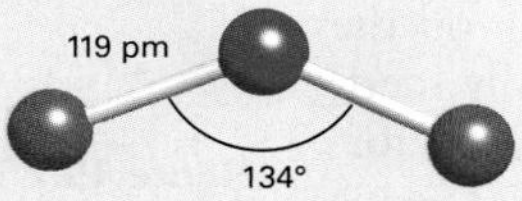 119 pm, 134°	Brown, reactive, paramagnetic gas
+4	N_2O_4	Dinitrogen tetroxide	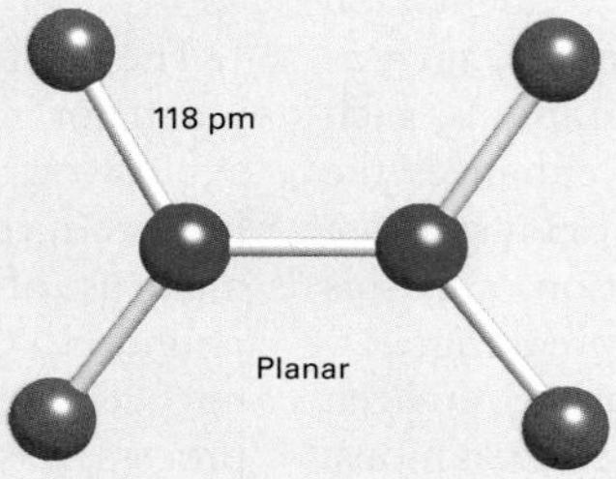 118 pm Planar	Colourless liquid (m.p. −11°C); in equilibrium with NO_2 in the gas phase
+5	N_2O_5	Dinitrogen pentoxide (dinitrogen pentoxide)	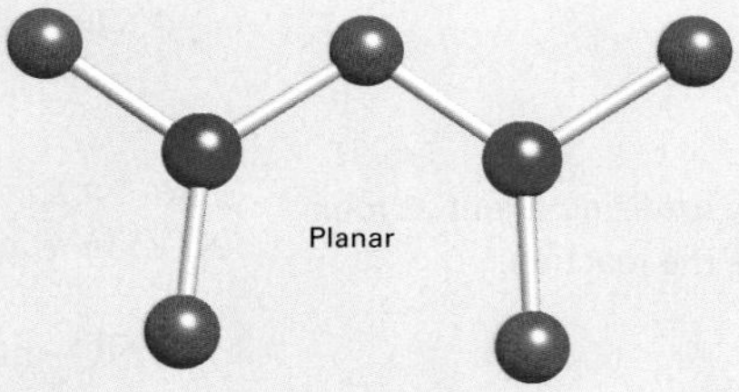 Planar	Colourless, ionic solid $[NO_2][NO_3]$ (m.p. 32°C); unstable

This decomposition is accelerated by light and heat.

The reduction of NO_3^- ions rarely yields a single product, as so many lower oxidation states of nitrogen are available. For example, a strong reducing agent such as zinc can reduce a substantial proportion of dilute HNO_3 as far as oxidation state −3:

$$HNO_3(aq) + 4\,Zn(s) + 9\,H^+(aq) \rightarrow NH_4^+(aq) + 3\,H_2O(l) + 4\,Zn^{2+}(aq)$$

A weaker reducing agent, such as copper, proceeds only as far as oxidation state +4 in the concentrated acid:

$$2\,HNO_3(aq) + Cu(s) + 2\,H^+(aq) \rightarrow 2\,NO_2(g) + Cu^{2+}(aq) + 2\,H_2O(l)$$

With the dilute acid, the +2 oxidation state is favoured, and NO is formed:

$$2\,NO_3^-(aq) + 3\,Cu(s) + 8\,H^+(aq) \rightarrow 2\,NO(g) + 3\,Cu^{2+}(aq) + 4\,H_2O(l)$$

Aqua regia is a mixture of concentrated nitric acid and concentrated hydrochloric acid, which is yellow due to the presence of the decomposition products NOCl and Cl_2. It loses its potency as these volatile products are formed:

$$HNO_3(aq) + 3\,HCl(aq) \rightarrow NOCl(g) + Cl_2(g) + 2\,H_2O(l)$$

Aqua regia is Latin for 'royal water' and was so called by alchemists because of its ability to dissolve the noble metals gold and platinum. Gold will dissolve to a very small extent in concentrated nitric acid. In *aqua regia* the Cl^- ions present react immediately

Table 15.3 Nitrogen-oxygen ions

Oxidation number	Formula	Common name	Structure	Comments
+1	$N_2O_2^{2-}$	Hyponitrite		Usually acts as a reducing agent
+3	NO_2^-	Nitrite	124 pm, 115°	Weak base; acts as an oxidizing agent and a reducing agent
+3	NO^+	Nitrosonium (nitrosyl cation)		Oxidizing agent and Lewis acid; π-acceptor ligand
+5	NO_3^-	Nitrate	122 pm	Very weak base; an oxidizing agent
+5	NO_2^+	Nitronium (nitryl cation)	115 pm	Oxidizing agent, nitrating agent, Lewis acid

with the Au^{3+} ions formed to produce $[AuCl_4]^-$ and thereby remove Au^{3+} from the product side of the oxidation reaction:

$$Au(s) + NO_3^-(aq) + 4\,Cl^-(aq) + 4\,H^+(aq) \rightarrow [AuCl_4]^-(aq) + NO(g) + 2\,H_2O(l)$$

The anhydride of nitric acid is N_2O_5. It is a crystalline solid with the more accurate formula $[NO_2^+][NO_3^-]$ and can be prepared by dehydration of nitric acid with P_4O_{10}:

$$4\,HNO_3(l) + P_4O_{10}(s) \rightarrow 2\,N_2O_5(s) + 4\,HPO_3(l)$$

The solid sublimes at 320°C and the gaseous molecules dissociate to give NO_2 and O_2. The compound is a strong oxidizing agent and can be used for the synthesis of anhydrous nitrates:

$$N_2O_5(s) + Na(s) \rightarrow NaNO_3(s) + NO_2(g)$$

EXAMPLE 15.3 Correlating trends in the stabilities of N(V) and Bi(V)

Compounds of N(V) and Bi(V) are stronger oxidizing agents than the +5 oxidation states of the three intervening elements. Correlate this observation with trends in the periodic table.

Answer We need to consider some of the periodic trends discussed earlier in Chapter 9. The light p-block elements are more electronegative than the elements immediately below them in the periodic table; accordingly, these light elements are generally less easily oxidized themselves and are thus good oxidizing agents. Nitrogen is generally a good oxidizing agent in its positive oxidation states. Bismuth is much less electronegative, but favours the +3 oxidation state in preference to the +5 state on account of the inert-pair effect.

Self-test 15.3 From trends in the periodic table, decide whether phosphorus or sulfur is likely to be the stronger oxidizing agent.

(b) Nitrogen(IV) and nitrogen(III) oxides and oxoanions

Key point: The intermediate oxidation states of nitrogen are often susceptible to disproportionation.

Nitrogen(IV) oxide exists as an equilibrium mixture of the brown NO_2 radical and its colourless dimer, N_2O_4 (dinitrogen tetroxide):

$$N_2O_4(g) \rightleftharpoons 2\,NO_2(g) \quad K = 0.115 \text{ at } 25°C$$

This readiness to dissociate is consistent with the N−N bond in N_2O_4 (**18**) being long and weak and arises because the molecular orbital occupied by the unpaired electron is spread almost equally over all three atoms in NO_2 rather than being concentrated on the N atom. This structure is in contrast to the isoelectronic oxalate ion, $C_2O_4^{2-}$, where the C−C bond is stronger because in CO_2^- the electron is more concentrated on the C atom.

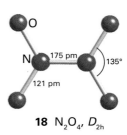

18 N_2O_4, D_{2h}

Nitrogen(IV) oxide is a poisonous oxidizing agent that is present in low concentrations in the atmosphere, especially in photochemical smog. In basic aqueous solution it disproportionates into N(III) and N(V), forming NO_2^- and NO_3^- ions:

$$2\,NO_2(aq) + 2\,OH^-(aq) \rightarrow NO_2^-(aq) + NO_3^-(aq) + H_2O(l)$$

In acidic solution (as in the Ostwald process) the reaction product is N(II) in place of N(III) because nitrous acid itself readily disproportionates:

$$3\,HNO_2(aq) \rightarrow NO_3^-(aq) + 2\,NO(g) + H_3O^+(aq)$$
$$E^\ominus = +0.05\,V, K = 50$$

Nitrous acid, HNO_2, is a strong oxidizing agent:

$$HNO_2(aq) + H^+(aq) + e^- \rightarrow NO(g) + H_2O(l) \quad E^\ominus = +1.00\,V$$

and its reactions as an oxidizing agent are often more rapid than its disproportionation.

The rate at which nitrous acid oxidizes is increased by acid as a result of its conversion to the nitrosonium ion, NO^+:

$$HNO_2(aq) + H^+(aq) \rightarrow H_2NO_2^+(aq) \rightarrow NO^+(aq) + H_2O(l)$$

The nitrosonium ion is a strong Lewis acid and forms complexes rapidly with anions and other Lewis bases. The resulting species may not themselves be susceptible to oxidation (as in the case of SO_4^{2-} and F^- ions, which form $[O_3SONO]^-$ (**19**) and ONF (**20**), respectively). Thus there is good experimental evidence that the reaction of HNO_2 with I^- ions leads to the rapid formation of INO:

$$I^-(aq) + NO^+(aq) \rightarrow INO(aq)$$

followed by the rate-determining second-order reaction between two INO molecules:

$$2\,INO(aq) \rightarrow I_2(aq) + 2\,NO(g)$$

Nitrosonium salts containing poorly coordinating anions, such as $[NO][BF_4]$, are useful reagents in the laboratory as facile oxidizing agents and as a source of NO^+.

Dinitrogen trioxide, N_2O_3, the anhydride of nitrous acid, is a blue solid that melts above $-100°C$ to give a blue liquid that dissociates to give NO and NO_2:

$$N_2O_3(l) \rightarrow NO(g) + NO_2(g)$$

The brown colour of NO_2 means that the liquid becomes progressively more green as the dissociation proceeds.

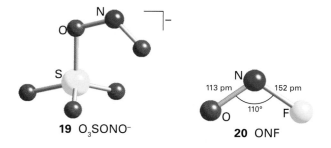

19 O_3SONO^-

20 ONF

(c) Nitrogen(II) oxide

Key points: Nitric oxide is a strong π-acceptor ligand, and a troublesome pollutant in urban atmospheres; the molecule acts as a neurotransmitter.

Nitrogen(II) oxide reacts with O_2 to generate NO_2, but in the gas phase the rate law is second-order in NO because a transient dimer, $(NO)_2$, is produced that subsequently collides with an O_2 molecule. Because the reaction is second-order, atmospheric NO (which is produced in low concentrations by coal-fired power plants and by internal combustion engines) is slow to convert to NO_2.

Because NO is endergonic, it should be possible to find a catalyst to convert the pollutant NO to the natural atmospheric gases N_2 and O_2 at its source in exhausts. It is known that Cu^+ in a zeolite catalyses the decomposition of NO, and a reasonable understanding of the mechanism has been developed; however, this system is not yet sufficiently robust for a practical automobile exhaust catalytic converter.

(d) Low oxidation state nitrogen–oxygen compounds

Key points: Dinitrogen oxide is unreactive, for kinetic reasons; the iso-electronic azide ion, N_3^-, forms many metal complexes.

Dinitrogen oxide, N_2O is a colourless, unreactive gas and is produced by the comproportionation of molten ammonium nitrate. Care must be taken to avoid an explosion in this reaction, in which the cation is oxidized by the anion:

$$NH_4NO_3(l) \xrightarrow{250°C} N_2O(g) + 2\,H_2O(g)$$

Standard potential data suggest that N_2O should be a strong oxidizing agent in acidic and basic solutions:

$$N_2O(g) + 2\,H^+(aq) + 2\,e^- \rightarrow N_2(g) + H_2O(l)$$
$$E^\ominus = +1.77\,V \text{ at } pH = 0$$
$$N_2O(g) + H_2O(l) + 2\,e^- \rightarrow N_2(g) + 2\,OH^-(aq)$$
$$E^\ominus = +0.94\,V \text{ at } pH = 14$$

However, kinetic considerations are paramount, and the gas is unreactive towards many reagents at room temperature.

EXAMPLE 15.4 Comparing the redox properties of nitrogen oxoanions and oxo compounds

Compare (a) NO_3^- and NO_2^- as oxidizing agents, (b) NO_2, NO, and N_2O with respect to their ease of oxidation in air, (c) N_2H_4 and H_2NOH as reducing agents.

Answer We need to refer to the Frost diagram for nitrogen, which is included in Fig. 15.6, and use the interpretation described in Section 5.13. (a) Both NO_3^- and NO_2^- ions are strong oxidizing agents. The reactions of the former are often sluggish but are generally faster in acidic solution. The reactions of NO_2^- ions are generally faster and become even faster in acidic solution, where the NO^+ is a common identifiable intermediate. (b) NO_2 is stable with respect to oxidation in air. N_2O and NO are thermodynamically susceptible to oxidation. However, the reaction of N_2O with oxygen is slow, and at low NO concentrations the reaction between NO and O_2 is slow because the rate law is second-order in NO. (c) Hydrazine and hydroxylamine are both good reducing agents. In basic solution hydrazine becomes a stronger reducing agent.

Self-test 15.4 Summarize the reactions that are used for the synthesis of hydrazine and hydroxylamine. Are these reactions best described as electron-transfer processes or nucleophilic displacements?

15.14 Oxides of phosphorus, arsenic, antimony, and bismuth

Key points: The oxides of phosphorus include P_4O_6 and P_4O_{10}, both of which are cage compounds with T_d symmetry; on progressing from arsenic to bismuth, the +5 oxidation state is more readily reduced to +3.

Phosphorus forms phosphorus(V) oxide, P_4O_{10}, and phosphorus(III) oxide, P_4O_6. It is also possible to isolate the intermediate compositions having one, two, or three O atoms terminally attached to the P atoms. Both principal oxides can be hydrated to yield the corresponding acids, the P(V) oxide giving phosphoric acid, H_3PO_4, and the P(III) oxide giving phosphonic acid, H_3PO_3. As remarked in Section 4.5, phosphonic acid has one H atom attached directly to the P atom; it is therefore a diprotic acid and better represented as $PHO(OH)_2$.

In contrast to the high stability of phosphorus(V) oxide, arsenic, antimony, and bismuth more readily form oxides with oxidation number $+3$, specifically As_2O_3, Sb_2O_3, and Bi_2O_3. In the gas phase, the arsenic(III) and antimony(III) oxides have the molecular formula E_4O_6, with the same tetrahedral structure as P_4O_6. Arsenic, Sb, and Bi do form oxides with oxidation state $+5$, but Bi(V) oxide is unstable and has not been structurally characterized. This is another example of the consequences of the inert-pair effect. Note the coordination number of 6 for Sb and 4 for the lighter elements P and As, with their smaller atoms.

15.15 Oxoanions of phosphorus, arsenic, antimony, and bismuth

Key points: Important oxoanions are the P(I) species hypophosphite, $H_2PO_2^-$, the P(III) species phosphite, HPO_3^{2-}, and the P(V) species phosphate, PO_4^{3-}. The existence of P−H bonds and the highly reducing character of the two lower oxidation states is notable. Phosphorus(V) also forms an extensive series of O-bridged polyphosphates. In contrast to N(V), P(V) species are not strongly oxidizing. As(V) is more easily reduced than P(V).

It can be seen from the Latimer diagram in Table 15.4 that elemental P and most of its compounds other than P(V) are strong reducing agents. White phosphorus disproportionates into phosphine, PH_3 (oxidation number -3), and hypophosphite ions (oxidation number $+1$) in basic solution:

$$P_4(s) + 3OH^-(aq) + 3H_2O(l) \rightarrow PH_3(g) + 3H_2PO_2^-(aq)$$

Table 15.5 lists some common P oxoanions (Box 15.4). The approximately tetrahedral environment of the P atom in their structures should be noted, as should the existence of P−H bonds in the hypophosphite and phosphite anions. The synthesis of various P(III) oxoacids and oxoanions, including HPO_3^{2-} and alkoxophosphanes, is conveniently performed by solvolysis of phosphorus(III) chloride under mild conditions, such as in cold tetrachloromethane solution:

$$PCl_3(l) + 3H_2O(l) \rightarrow \quad H_3PO_3(sol) + 3HCl(sol)$$
$$PCl_3(l) + 3ROH(sol) + 3N(CH_3)_3(sol) \rightarrow \quad P(OR)_3(sol) +$$
$$3[HN(CH_3)_3]Cl(sol)$$

Reductions with $H_2PO_2^-$ and HPO_3^{2-} are usually fast. One of the commercial applications of this lability is the use of $H_2PO_2^-$ to reduce $Ni^{2+}(aq)$ ions and so coat surfaces with metallic Ni in the process called 'electrodeless plating'.

$$Ni^{2+}(aq) + 2H_2PO_2^-(aq) + 2H_2O(l) \rightarrow Ni(s) + 2H_2PO_3^-(aq)$$
$$+H_2(g) + 2H^+(aq)$$

The Frost diagram for the elements shown in Fig. 15.6 reveals similar trends in aqueous solution, with oxidizing character following the order $PO_4^{3-} \approx AsO_4^{3-} < Sb(OH)_6^- \approx Bi(V)$. The thermodynamic tendency and kinetic ease of reducing AsO_4^{3-} is thought to be key to its toxicity towards animals. Thus, As(V) as AsO_4^{3-} readily mimics PO_4^{3-}, and so may be incorporated into

Table 15.4 Latimer diagrams for phosphorus

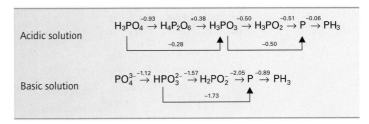

cells. There, unlike P, it is reduced to an As(III) species, which is thought to be the actual toxic agent. This toxicity may stem from the affinity of As(III) for sulfur-containing amino acids. The enzyme arsenite oxidase, which contains a Mo cofactor, is produced by certain bacteria and is used to reduce the toxicity of As(III) by converting it to As(V).

15.16 Condensed phosphates

Key point: Dehydration of phosphoric acid leads to the formation of chain or ring structures that may be based on many PO_4 units.

When phosphoric acid, H_3PO_4, is heated above 200°C, condensation occurs resulting in the formation of P−O−P bridges between two neighbouring PO_4^{3-} units. The extent of this condensation depends upon the temperature and duration of heating.

$$2H_3PO_4(l) \rightarrow H_4P_2O_7(l) + H_2O(g)$$
$$H_3PO_4(l) + H_4P_2O_7(l) \rightarrow H_5P_3O_{10}(l) + H_2O(g)$$

The simplest condensed phosphate is thus $H_4P_2O_7$. The most commercially important condensed phosphate is the sodium salt of the triacid, $Na_5P_3O_{10}$ (**21**). It is widely used in detergents for laundry and dishwashers, and in other cleaning products and in water treatment (Box 15.5). Polyphosphates are also used in various ceramics and as food additives. Triphosphates such as adenosine triphosphate (ATP) are of vital importance in living organisms (Section 26.8).

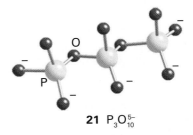

21 $P_3O_{10}^{5-}$

A range of condensed phosphates occurs with chain lengths ranging from those based on two PO_4 units to polyphosphates having chain lengths of several thousand units. Di-, tri-, tetra-, and pentapolyphosphates have been isolated but higher members of the series always contain mixtures. However, the average chain length can be determined by the usual methods used in polymer analysis or by titration. Just as the three successive acidity constants for phosphoric acid differ, so do the acidity constants for the two types of OH group of the polyphosphoric acids. The terminal OH groups, of which there are two per molecule, are weakly acidic. The remaining OH groups, of which there is one per P atom, are strongly acidic because they are situated

Table 15.5 Some phosphorus oxoanions

Oxidation number	Formula	Name	Structure	Comments
+1	$H_2PO_2^-$	Hypophosphite (dihydrodioxophosphate)	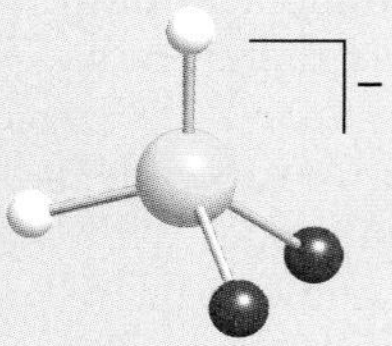	Facile reducing agent
+3	HPO_3^{2-}	Phosphite	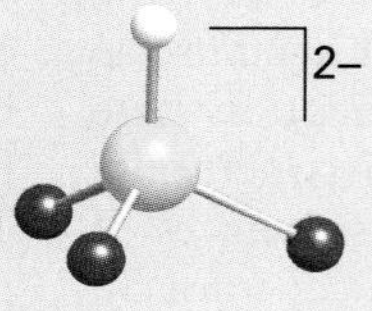	Facile reducing agent
+4	$P_2O_6^{4-}$	Hypophosphate	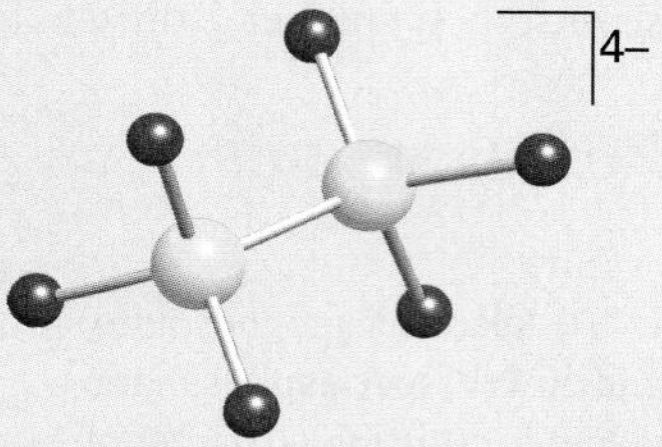	Basic
+5	PO_4^{3-}	Phosphate	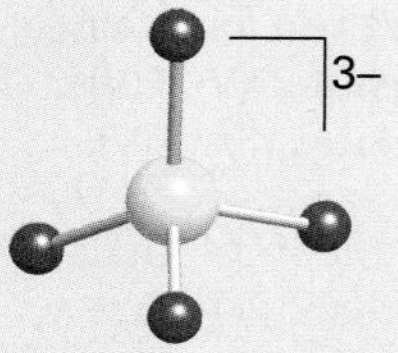	Strongly basic
+5	$P_2O_7^{4-}$	Diphosphate		Basic; longer chain

---

BOX 15.4 Phosphates and the food industry

Phosphorus in the form of phosphates is essential to life and phosphate fertilizers in forms such as bone, fish, and guano have been used since ancient times. The phosphate industry started in the mid-nineteenth century when sulfuric acid was used to decompose bones and phosphate minerals to make the phosphate more readily available. The development of more economical routes led to the diversification of the industrial applications of phosphoric acid and phosphate salts.

More than 90 per cent of world production of phosphoric acid is used to make fertilizers but there are several other applications. One of the most important is the food industry. A dilute solution of phosphoric acid is nontoxic and has an acidic taste. It is used extensively in beverages to give a tart taste, as a buffering agent in jams and jellies, and as a purifying agent in sugar refining.

The phosphates and hydrogenphosphates have many applications in the food industries. Sodium dihydrogenphosphate, NaH_2PO_4, is added to animal feeds as a dietary supplement. The disodium salt, Na_2HPO_4, is used as an emulsifier for processing cheese. It interacts with the protein casein and prevents separation of the fat and water. The potassium salts are more soluble and more expensive than the sodium salts. The dipotassium salt, K_2HPO_4, is used as an anti-coagulant in coffee creamer. It interacts with the protein and prevents coagulation by the coffee acids. Calcium dihydrogenphosphate monohydrate, $Ca(H_2PO_4)_2 \cdot H_2O$, is used as a raising agent in bread, cake mixes, and self-raising flour. Together with $NaHCO_3$ it produces CO_2 during the baking process but it also reacts with the protein in the flour to control the elasticity and viscosity of the dough or mixture. The largest use of calcium monohydrogenphosphate, $CaHPO_4 \cdot 2H_2O$, is as a dental polish in nonfluoride toothpaste. Calcium diphosphate, $Ca_2P_2O_7$, is used in fluoride toothpaste. Calcium phosphate, Ca_3PO_4, is added to sugar and salt to improve their flow.

BOX 15.5 Polyphosphates

The most widely used polyphosphate is sodium tripolyphosphate, $Na_5P_3O_{10}$. Its major use is as a 'builder' for synthetic detergents used in domestic laundry products, car shampoos, and industrial cleaners. Its role in these applications is to form stable complexes with the calcium and magnesium ions in hard water, effectively making them unavailable for precipitation, the process called 'sequestration'. It also acts as a buffer and prevents the flocculation of dirt and the redeposition of soil particles.

Food-grade sodium tripolyphosphate is used in the curing of hams and bacon. It interacts with the proteins and leads to good moisture retention during curing. It is also used to improve the quality of processed chicken and seafood products. The technical-grade product is used as a water softener, by sequestration as above, and in the paper pulping and textile industries, where it is used to help break down cellulose.

Potassium tripolyphosphate is more soluble and more expensive than the sodium analogue and is used in liquid detergents. For some applications an effective trade-off between solubility and cost can be achieved by using sodium potassium tripolyphosphate, $Na_3K_2P_3O_{10}$.

The use of polyphosphates in detergents has been implicated in excessive algae growth and eutrophication of some natural waters. This has led to restrictions in their use in many countries and the reduction in their use in home laundry products. However, phosphate is still widely used as an essential fertilizer and more of it enters rivers and lakes by run-off from farm land than from detergents.

opposite the strongly electron-withdrawing =O groups. The ratio of weakly to strongly acidic protons gives an indication of average chain length. The long-chain polyphosphates are viscous liquids or glasses.

EXAMPLE 15.5 Determining the chain length of a polyphosphoric acid by titration

A sample of a polyphosphoric acid was dissolved in water and titrated with dilute NaOH(aq). Two stoichiometric points were observed at 16.8 and 28.0 cm^3. Determine the chain length of the polyphosphate.

Answer We need to determine the ratio of the two different types of OH group. The strongly acidic OH groups are titrated by the first 16.8 cm^3. The two terminal OH groups are titrated by the remaining 28.0 − 16.8 cm^3 = 11.2 cm^3. Because the concentrations of analyte and titrant are such that each OH group requires 5.6 cm^3 of the titrant (because 11.2 cm^3 is used to titrate two such groups), we conclude that there are (16.8 cm^3)/(5.6 cm^3) = 3 strongly acidic OH groups per molecule. A molecule with two terminal OH groups and three further OH groups is a tripolyphosphate.

Self-test 15.5 When titrated against base a sample of polyphosphate gave end points at 30.4 and 45.6 cm^3. What is the chain length?

If NaH_2PO_4 is heated and the water vapour allowed to escape, then the tricyclo anion $P_3O_9^{3-}$ (**22**) is formed. If this reaction is carried out in a closed system, the product is 'Maddrell's salt', a crystalline material that contains long chains of PO_4 units. The tetracyclo anion (**23**) is formed when P_4O_{10} is treated with cold aqueous solutions of NaOH or $NaHCO_3$.

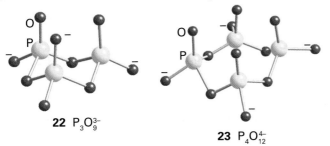

22 $P_3O_9^{3-}$

23 $P_4O_{12}^{4-}$

15.17 Phosphazenes

Key points: The range of PN compounds is extensive, and includes cyclic and polymeric phosphazenes, $(PX_2N)_n$; phosphazenes form highly flexible elastomers.

Many analogues of phosphorus–oxygen compounds exist in which the O atom is replaced by the isolobal NR or NH group, such as $P_4(NR)_6$ (**24**), the analogue of P_4O_6 (Section 22.20c). Other compounds exist in which OH or OR groups are replaced by the isolobal NH_2 or NR_2 groups. An example is $P(NMe_2)_3$, the analogue of $P(OMe)_3$. Another indication of the scope of PN chemistry, and a useful point to remember, is that PN is structurally equivalent to SiO. For example, various phosphazenes, which are chains and rings containing R_2PN units (**25**), are analogous to the siloxanes (Chapter 14) and their R_2SiO units (**26**).

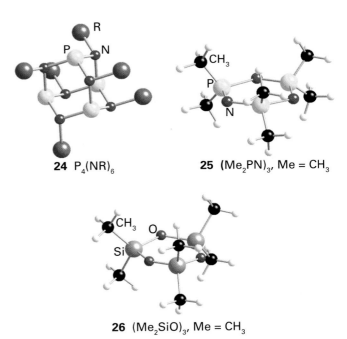

24 $P_4(NR)_6$

25 $(Me_2PN)_3$, Me = CH_3

26 $(Me_2SiO)_3$, Me = CH_3

The cyclic phosphazene dichlorides are good starting materials for the preparation of the more elaborate phosphazenes. They are easily synthesized:

$$n\,PCl_5 + n\,NH_4Cl \rightarrow (Cl_2PN)_n + 4n\,HCl \qquad n = 3 \text{ or } 4$$

A chlorocarbon solvent and temperatures near 130°C produce the cyclic trimer (**27**) and tetramer (**28**), and when the trimer is heated to about 290°C it changes to polyphosphazene

(Box 15.6). The Cl atoms in the trimer, tetramer, and polymer are readily displaced by other Lewis bases

$$(Cl_2PN)_n + 2nCF_3CF_2O^- \rightarrow [(CF_3CF_2O)_2PN]_n + 2nCl^-$$

Like silicone rubber, the polyphosphazenes remain rubbery at low temperatures because, like the isoelectronic SiOSi group, the molecules are helical and the PNP groups are highly flexible.

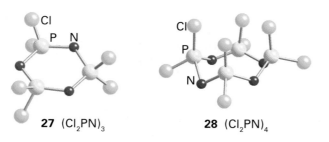

27 $(Cl_2PN)_3$ **28** $(Cl_2PN)_4$

A phosphorus−nitrogen compound that was particularly useful in the laboratory until its carcinogenic properties were recognized is the aprotic solvent hexamethylphosphoramide, $((CH_3)_2N)_3P=O$, which is sometimes designated HMPA. The large bis(triphenylphosphine)iminium cation, $[Ph_3P=N=PPh_3]^+$, which is commonly abbreviated as PPN$^+$, is very useful in forming salts of large anions. The salts of this cation are usually soluble in polar aprotic solvents such as HMPA, dimethylformamide, and even dichloromethane.

■ **A brief illustration.** To prepare $[NP(OCH_3)_2]_4$ from PCl_5, NH_4Cl, and NaOCH the cyclic chlorophosphazene is synthesized first:

$$4PCl_5 + 4NH_4Cl \xrightarrow{130°C} (Cl_2PN)_4 + 16HCl$$

Then, because Cl atoms are readily replaced by strong Lewis bases, such as alkoxides, the chlorophosphazene is used as follows:

$$(Cl_2PN)_4 + 8NaOCH_3 \rightarrow [(CH_3O)_2PN]_4 + 8NaCl ■$$

15.18 Organometallic compounds of arsenic, antimony, and bismuth

Oxidation states +3 and +5 are encountered in many of the organometallic compounds of arsenic, antimony, and bismuth. An example of a compound with an element in the +3 oxidation state is $As(CH_3)_3$ (**29**) and an example of the +5 state is $As(C_6H_5)_5$ (**30**). Organoarsenic compounds were once widely used to treat bacterial infections and as herbicides and fungi-

cides. However, because of their high toxicity they no longer have major commercial applications.

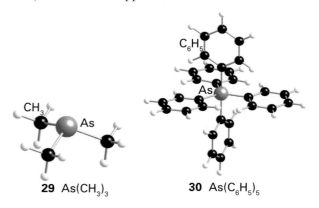

29 $As(CH_3)_3$ **30** $As(C_6H_5)_5$

(a) Oxidation state +3

Key points: The stability of the organometallic compounds decreases in the order As > Sb > Bi; the aryl compounds are more stable than the alkyl compounds.

Organometallic compounds of arsenic(III), antimony(III), and bismuth(III) can be prepared in an ether solvent by using a Grignard reagent, an organolithium compound, or an organohalide:

$$AsCl_3(et) + 3RMgCl(et) \rightarrow AsR_3(et) + 3MgCl_2(et)$$

$$2As(et) + 3RBr(et) \xrightarrow{Cu/\Delta} AsRBr_2(et) + AsR_2Br(et)$$

$$AsR_2Br(et) + R'Li(et) \rightarrow AsR_2R'(et) + LiBr(et)$$

The compounds are all readily oxidized but are stable to water. The M−C bond strength decreases for a given R group in the order As > Sb > Bi. Consequently, the stability of the compounds decreases in the same order. In addition, the aryl compounds, such as $(C_6H_5)_3As$, are generally more stable than the alkyl compounds. The halogen-substituted compounds R_nMX_{3-n} have been prepared and characterized.

All the compounds act as Lewis bases and form complexes with d-metal ions. The basicity decreases in the order As > Sb > Bi. Many complexes of alkyl- and arylarsanes have been prepared but fewer stibane complexes are known. A useful ligand, for example, is the bidentate compound known as diars (**31**). Because of their soft-donor character, many aryl- and alkylarsane complexes of the soft species Rh(I), Ir(I), Pd(II), and Pt(II) have been prepared. However, hardness criteria are only approximate, so we should not be surprised to see phosphine and arsane complexes of some metals in higher oxidation states.

For example, the unusual +4 oxidation state of palladium is stabilized by the diars ligand (**32**).

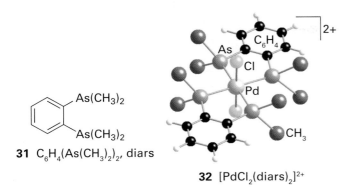

31 C$_6$H$_4$(As(CH$_3$)$_2$)$_2$, diars

32 [PdCl$_2$(diars)$_2$]$^{2+}$

The synthesis of diars provides a good illustration of some common reactions in the synthesis of organoarsenic compounds. The starting material is (CH$_3$)$_2$AsI. This compound is not conveniently prepared by metathesis reaction between AsI$_3$ and a Grignard or similar carbanion reagent because that reaction is not selective to partial substitution on the As atom when the organic group is compact. Instead, the compound can be prepared by the direct action of a haloalkane, CH$_3$I, on metallic arsenic:

$$4\,As(s) + 6\,CH_3I(l) \rightarrow 3\,(CH_3)_2AsI(sol) + AsI_3(sol)$$

In the next step, the action of sodium on (CH$_3$)$_2$AsI is used to produce [(CH$_3$)$_2$As]$^-$:

$$(CH_3)_2AsI(sol) + 2\,Na(sol) \rightarrow Na[(CH_3)_2As](sol) + NaI(s)$$

The resulting powerful nucleophile [(CH$_3$)$_2$As]$^-$ is then used to displace chlorine from 1,2-dichlorobenzene:

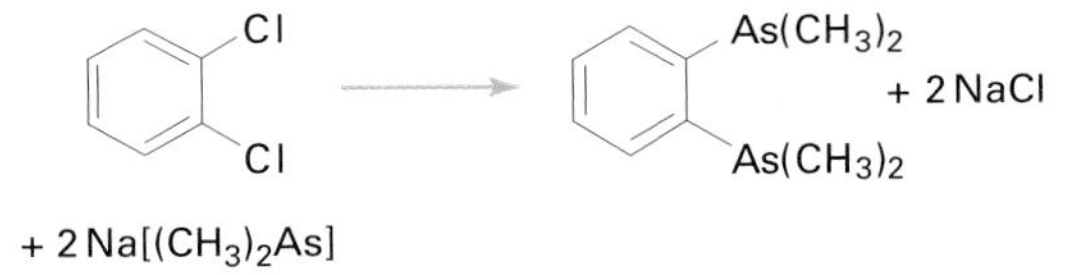

Polyarsane compounds, (RAs)$_n$, can be prepared in ether by reduction of a pentavalent organometallic compound, R$_5$As, or by treating an organohaloarsenic compound with Li:

$$n\,RAsX_2(et) + 2n\,Li(et) \rightarrow (RAs)_n(et) + 2n\,LiX(et)$$

The compound R$_2$AsAsR$_2$ is very reactive because the As—As bond is readily cleaved. It reacts with oxygen, sulfur, and species containing C=C bonds, and forms complexes with d-metal species in which the As—As bond may be cleaved or left intact:

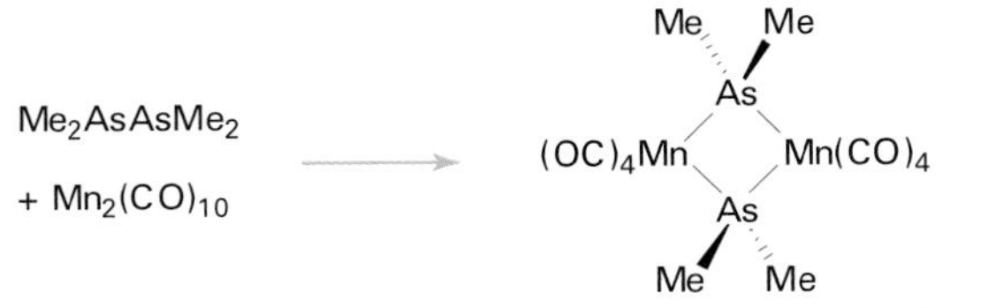

Polyarsanes of up to six units have been characterized. Polymethylarsane exists as a yellow, puckered cyclic pentamer (**33**) and as a purple–black ladder-like structure (**34**). The strength of the M—M bond decreases in the order As > Sb > Bi, so although arsenic forms catenated organometallic compounds, only R$_2$Bi—BiR$_2$ has been isolated.

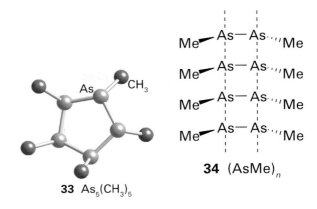

33 As$_5$(CH$_3$)$_5$

34 (AsMe)$_n$

As well as forming single M—C bonds, As, Sb, and Bi also form M=C bonds. A well-studied group of compounds are the arylometals in which a metal atom forms part of a heterocyclic six-membered benzene-like ring (**35**). Arsabenzene, C$_5$H$_5$As, is stable up to 200°C, stibabenzene, C$_5$H$_5$Sb, can be isolated but readily polymerizes, and bismabenzene, C$_5$H$_5$Bi, is very unstable. These compounds exhibit typical aromatic character although arsabenzene is 1000 times more reactive than benzene. A related group of compounds is arsole, stibole, and bismuthole, C$_4$H$_5$M, in which the metal atom forms part of a five-membered ring (**36**).

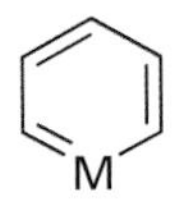

35 C$_5$H$_5$M

36 C$_4$H$_6$M

(b) Oxidation state +5

Key point: The tetraphenylarsonium ion is a starting material for the preparation of other As(V) organometallic compounds.

The trialkylarsanes act as nucleophiles towards haloalkanes to produce tetraalkylarsonium salts, which contain As(V):

$$As(CH_3)_3(sol) + CH_3Br(sol) \rightarrow [As(CH_3)_4]Br(sol)$$

This type of reaction cannot be used for the preparation of the tetraphenylarsonium ion, [AsPh$_4$]$^+$ because triphenylarsane is a much weaker nucleophile than trimethylarsane. Instead, a suitable synthetic reaction is:

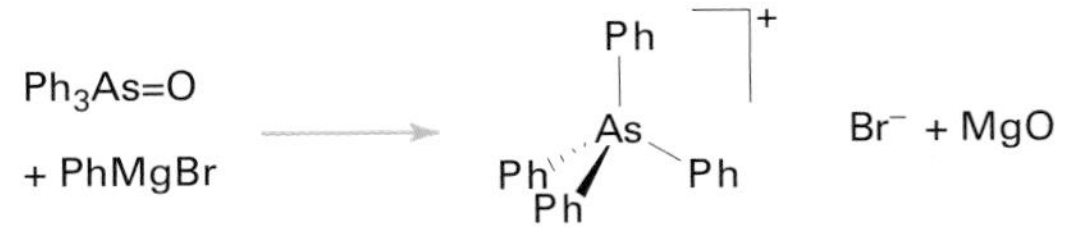

Ph$_3$As=O + PhMgBr $\longrightarrow$ [AsPh$_4$]$^+$ Br$^-$ + MgO

This reaction may look unfamiliar, but it is simply a metathesis in which the Ph$^-$ anion replaces the formal O^{2-} ion attached to the As atom, resulting in a compound in which the arsenic retains its +5 oxidation state. The formation of the highly exergonic compound MgO also contributes to the Gibbs energy of this reaction, and its formation drives the reaction forward.

The tetraphenylarsonium, tetraalkylammonium, and tetraphenylphosphonium cations are used in synthetic inorganic chemistry as bulky cations to stabilize bulky anions. The tetraphenylarsonium ion is also a starting material for the preparation of other As(V) organometallic compounds. For instance, the action of phenyllithium on a tetraphenylarsonium salt produces pentaphenylarsenic (**30**), a compound of As(V):

$$[AsPh_4]Br(sol) + LiPh(sol) \rightarrow AsPh_5(sol) + LiBr(s)$$

Pentaphenylarsenic, $AsPh_5$, is trigonal bipyramidal, as expected from VSEPR considerations. We have seen (Section 2.3) that a square-pyramidal structure is often close in energy to the trigonal-bipyramidal structure, and the antimony analogue, $SbPh_5$, is in fact square pyramidal (**37**). A similar reaction under carefully controlled conditions yields the unstable compound $As(CH_3)_5$.

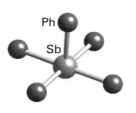

37 $SbPh_5$, Ph = C_6H_5

FURTHER READING

R.B. King, *Inorganic chemistry of the main group elements*. John Wiley & Sons (1994).

D.M.P. Mingos, *Essential trends in inorganic chemistry*. Oxford University Press (1998). A survey of inorganic chemistry from the perspective of structure and bonding.

R.B. King (ed.), *Encyclopedia of inorganic chemistry*. John Wiley & Sons (2005).

H.R. Allcock, *Chemistry and applications of polyphosphazenes*. John Wiley & Sons (2002).

J. Emsley, *The shocking history of phosphorus: a biography of the devil's element*. Pan (2001).

W.T. Frankenberger, *The environmental chemistry of arsenic*. Marcel Dekker (2001).

G.J. Leigh, *The world's greatest fix: a history of nitrogen and agriculture*. Oxford University Press (2004).

N.N. Greenwood and A. Earnshaw, *Chemistry of the elements*. Butterworth-Heinemann (1997).

EXERCISES

15.1 List the elements in Groups 15 and indicate the ones that are (a) diatomic gases, (b) nonmetals, (c) metalloids, (d) true metals. Indicate those elements that display the inert-pair effect.

15.2 (a) Give complete and balanced chemical equations for each step in the synthesis of H_3PO_4 from hydroxyapatite to yield (a) high-purity phosphoric acid and (b) fertilizer-grade phosphoric acid. (c) Account for the large difference in costs between these two methods.

15.3 Ammonia can be prepared by (a) the hydrolysis of Li_3N or (b) the high-temperature, high-pressure reduction of N_2 by H_2. Give balanced chemical equations for each method starting with N_2, Li, and H_2, as appropriate. (c) Account for the lower cost of the second method.

15.4 Show with an equation why aqueous solutions of NH_4NO_3 are acidic.

15.5 Carbon monoxide is a good ligand and is toxic. Why is the isoelectronic N_2 molecule not toxic?

15.6 Compare and contrast the formulas and stabilities of the oxidation states of the common nitrogen chlorides with the phosphorus chlorides.

15.7 Use the VSEPR model to predict the probable shapes of (a) PCl_4^+, (b) PCl_4^-, (c) $AsCl_5$.

15.8 Give balanced chemical equations for each of the following reactions: (a) oxidation of P_4 with excess oxygen, (b) reaction of the product from part (a) with excess water, (c) reaction of the product from part (b) with a solution of $CaCl_2$ and name the product.

15.9 Starting with $NH_3(g)$ and other reagents of your choice, give the chemical equations and conditions for the synthesis of (a) HNO_3, (b) NO_2^-, (c) NH_2OH, (d) N_3^-.

15.10 Write the balanced chemical equation corresponding to the standard enthalpy of formation of $P_4O_{10}(s)$. Specify the structure, physical state (s, l, or g), and allotrope of the reactants. Do either of the reactants differ from the usual practice of taking as reference state the most stable form of an element?

15.11 Without reference to the text, sketch the general form of the Frost diagrams for phosphorus (oxidation states 0 to +5) and bismuth (0 to +5) in acidic solution and discuss the relative stabilities of the +3 and +5 oxidation states of both elements.

15.12 Are reactions of NO_2^- as an oxidizing agent generally faster or slower when pH is lowered? Give a mechanistic explanation for the pH dependence of NO_2^- oxidations.

15.13 When equal volumes of nitric oxide (NO) and air are mixed at atmospheric pressure a rapid reaction occurs, to form NO_2 and N_2O_4. However, nitric oxide from an automobile exhaust, which is present in the parts per million concentration range, reacts slowly with air. Give an explanation for this observation in terms of the rate law and the probable mechanism.

15.14 Give balanced chemical equations for the reactions of the following reagents with PCl_5 and indicate the structures of the products: (a) water (1:1), (b) water in excess, (c) $AlCl_3$, (d) NH_4Cl.

15.15 Use data in *Resource section 3* to calculate the standard potential of the reaction of H_3PO_2 with Cu^{2+}. Are HPO_2^{2-} and $H_2PO_2^{2-}$ useful as oxidizing or reducing agents?

15.16 Identify the compounds A, B, C, and D.

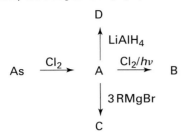

15.17 Sketch the two possible geometric isomers of the octahedral [AsF$_4$Cl$_2$]$^-$ and explain how they could be distinguished by ^{19}F-NMR.

15.18 Identify the nitrogen compounds A, B, C, D, and E.

15.19 Use the Latimer diagrams in *Resource section* 3 to determine which species of N and P disproportionate in acid conditions.

PROBLEMS

15.1 Explain how you could use ^{31}P-NMR to distinguish between PF$_3$ and POF$_3$.

15.2 The tetrahedral P$_4$ molecule may be described in terms of localized 2*c*,2*e* bonds. Determine the number of skeletal valence electrons and from this decide whether P$_4$ is *closo*, *nido*, or *arachno* (these terms are specified in Section 13.3). If it is not *closo*, determine the parent *closo* polyhedron from which the structure of P$_4$ could be formally derived by the removal of one or more vertices.

15.3 Describe sewage treatment methods that result in a decrease in phosphate levels in wastewater. Outline a laboratory method that could be used to monitor phosphate levels in water.

15.4 On account of their slow reactions at electrodes, the potentials of most redox reactions of nitrogen compounds cannot be measured in an electrochemical cell. Instead, the values must be determined from other thermodynamic data. Illustrate such a calculation by using $\Delta_f G^{\ominus}(NH_3, aq) = -26.5$ kJ mol^{-1} to calculate the standard potential of the N$_2$/NH$_3$ couple in basic aqueous solution.

15.5 A compound containing pentacoordinate nitrogen has been characterized (A. Frohmann, J. Riede, and H. Schmidbaur, *Nature*, 1990, **345**, 140). Describe (a) the synthesis, (b) the structure of the compound, (c) the bonding.

15.6 Two articles (A. Lykknes and L. Kvittingen, 'Arsenic: not so evil after all?', *J. Chem. Educ.*, 2003, **80**, 497 and J. Wang and C.M. Chien, 'Arsenic in drinking water—a global environmental problem',

J. Chem. Educ., 2004, **81**, 207) present opposing perspectives on the toxic nature of arsenic. Use these references to produce a critical assessment of the beneficial and detrimental effects of arsenic.

15.7 A paper published by N. Tokitoh *et al.* (*Science*, 1997, **277**, 78) gives an account of the synthesis and characterization of a stable bismuthene, containing Bi=Bi double bonds. Give the equations for the synthesis of the compound. Name and sketch the structure of the steric protecting group that was used. Why was the isolation of the product simple? What methods were used to determine the structure of the compound?

15.8 A paper published by Y. Zhang *et al.* (*Inorg. Chem.*, 2006, **45**, 10446) describes the synthesis of phosphazene cations as precursors for polyphosphazenes. Polyphosphazenes are prepared by ring opening polymerization of the cyclic (NPCl$_2$)$_3$ and the reaction is initiated by phosphazene cations. Discuss which Lewis acids were used to produce the cations and give the reaction scheme for the ring opening polymerization of (NPCl$_2$)$_3$.

15.9 In their paper 'Catalytic reduction of dinitrogen to ammonia at single molybdenum center' (*Science*, 2003, **301**, 5629) D. Yandulov and R. Schrock describe the catalytic conversion of nitrogen to ammonia at room temperature and atmospheric pressure. Discuss why this development could be important commercially. Review nonbiological methods of nitrogen activation.

16

The Group 16 elements

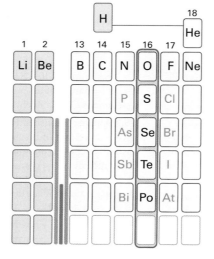

Group 16 contains two of the most important elements for life. Oxygen is most commonly found as water and in the atmosphere, both of which are essential for almost all life forms. Dioxygen is produced from water by photosynthesis and recycled in respiration by higher organisms. Sulfur is also essential to all life forms and even selenium is required in trace amounts. Sulfur and selenium exhibit a tendency to catenation and form rings and chains.

The Group 16 elements oxygen, sulfur, selenium, tellurium, and polonium are often called the **chalcogens**. The name derives from the Greek word for bronze, and refers to the association of sulfur and its congeners with copper. As in the rest of the p block, the element at the head of the group, oxygen, differs significantly from the other members of the group. The coordination numbers of its compounds are generally lower, and it is the only member of the group to exist as diatomic molecules under normal conditions.

PART A: **THE ESSENTIALS**

All the members of Group 16 other than oxygen are solids under normal conditions and, as we have seen previously, metallic character generally increases down the group. In this section we discuss the essential features of the chemistry of the Group 16 elements.

16.1 **The elements**

Key points: Oxygen is the most electronegative element in Group 16 and is the only gas; all the elements occur in several allotropic forms.

Oxygen, sulfur, and selenium are nonmetals, tellurium is a metalloid, and polonium is a metal. Allotropy and polymorphism are important features of Group 16 and sulfur occurs in more natural allotropes and polymorphs than any other element.

The group electron configuration of ns^2np^4 suggests a group maximum oxidation number of $+6$ (Table 16.1). Oxygen never achieves this maximum oxidation state, although the other elements do in some circumstances. The electron configuration also suggests that stability may be achieved with an oxidation number of -2, which is overwhelmingly common for O. The most remarkable feature of S is that it forms stable compounds with oxidation numbers between -2 and $+6$.

In addition to its distinctive physical properties, O is significantly different chemically from the other members of the group (Section 9.8). It is the second most electronegative element in the periodic table and significantly more electronegative than its congeners.

Table 16.1 Selected properties of the elements

	O	S	Se	Te	Po
Covalent radius/pm	74	104	117	137	140
Ionic radius/pm	140	184	198	221	
First ionization energy/(kJ mol^{-1})	1310	1000	941	870	812
Melting point/°C	-218	113 (α)	217	450	254
Boiling point/°C	-183	445	685	990	960
Pauling electronegativity	3.4	2.6	2.6	2.1	2.0
Electron affinity*/(kJ mol^{-1})	141	200	195	190	183
	-844	-532			

* The first value is for $X(g) + e^-(g) \rightarrow X^-(g)$, the second value is for $X^-(g) + e^-(g) \rightarrow X^{2-}(g)$.

This high electronegativity has an enormous influence on the chemical properties of the element. The small atomic radius of O and absence of accessible d orbitals also contribute to its distinctive chemical character. Thus, O seldom has a coordination number greater than 3 in simple molecular compounds, but its heavier congeners frequently reach coordination numbers of 5 and 6, as in SF$_6$.

Dioxygen (O$_2$) oxidizes many elements and reacts with many organic and inorganic compounds under suitable conditions. Only the noble gases He, Ne, and Ar do not form oxides directly. The oxides of the elements are discussed in each relevant chapter and will not be revisited here. Even though the O=O bond energy of $+494$ kJ mol^{-1} is high, many exothermic combustion reactions occur because the resulting E$-$O covalent bond enthalpies or MO$_n$ lattice enthalpies are also high. One of the most important reactions of dioxygen is the coordination to the oxygen-transport protein haemoglobin (Section 27.7b).

Oxygen is the most abundant element in the Earth's crust at 46 per cent by mass and is present in all silicate minerals. It comprises 86 per cent by mass of the oceans and 89 per cent of water. The average human is two-thirds oxygen by mass. Dioxygen, which is derived completely from the water-splitting action of photosynthetic organisms, makes up 21 per cent by mass of the atmosphere (Box 16.1). Oxygen is also the third most abundant element in the Sun, and the most abundant element on the surface of the Moon (46 per cent by mass). Oxygen also occurs as ozone, O$_3$, a highly reactive pungent

---

BOX 16.1 The oxygen atmosphere

In the course of the evolution of the Earth's atmosphere, the proliferation of oxygen-evolving photosynthesis eventually resulted in the presence of O$_2$ in the atmosphere at the present level of 21 per cent by volume. Photosynthesis also produced organic carbon compounds, which became the materials of cellular biomass (Box 14.3).

Oxygen was a toxic constituent of the atmosphere of the early Earth and led to the extinction of many species. Some species retreated to habitats deeper in the soil or waters, where anaerobic conditions remained and where their descendents remain today. Other organisms adapted differently and evolved to exploit this now abundant and powerful oxidant. These organisms are the *aerobes*, among which were our ancestors. The shift from an anaerobic atmosphere to an oxygenic atmosphere had a profound effect on the composition of the waters. Sulfur, which was present largely in the form of sulfide in anaerobic waters, was oxidized to sulfate. Metal ion concentrations also changed dramatically. Two of the metals most profoundly affected were molybdenum and iron.

In Earth's modern oceans, molybdenum is the most abundant d metal (at 0.01 ppm). However, before the oxygenation of the oceans and atmosphere, molybdenum was present as insoluble solids, mainly MoO$_2$ and MoS$_2$. Oxidation of these solids produced the molybdate ion:

$$2\,MoS_2(s) + 7\,O_2(g) + 2\,H_2O(l) \rightarrow 2\,MoO_4^{2-}(aq) + 4\,SO_2(g) + 4\,H^+(aq)$$

The highly soluble molybdate ion became available to aquatic organisms and is now transported into cells by methods that differ dramatically from those used to acquire the more widespread cationic d-metal species in the marine environment. In addition, competition between molybdate and sulfate, both tetrahedral dinegative ions, remains a challenge for modern organisms.

Iron suffered the opposite fate to molybdenum. In the ancient oceans, the element was present as Fe(II). Iron(II) hydroxide and sulfide are essentially soluble and so iron would have been readily available to aquatic organisms. However, on oxygenation of the atmosphere, the oxidation of Fe^{2+} to Fe^{3+} led to the precipitation of iron(III) hydroxides and oxides. Massive banded formations containing magnetite (Fe$_3$O$_4$) and hematite (Fe$_2$O$_3$) in Canada and Australia are testimony to the precipitation of the iron from the oceans between 2 and 3 Ga ago.

gas that is crucial to the protection of life on Earth, for it shields the surface from solar ultraviolet radiation.

Sulfur occurs as deposits of the native element, in meteorites, volcanoes, and hot springs, as the ores *galena*, PbS, and *barite*, $BaSO_4$, and as Epsom salts, $MgSO_4 \cdot 7H_2O$. It also occurs as H_2S in natural gas and as organosulfur compounds in crude oil. That sulfur can exist in a large number of allotropic forms can be explained by the ability of S atoms to catenate because of the high S−S bond energy of 265 kJ mol^{-1}, which is exceeded only by C−C (330 kJ mol^{-1}) and H−H (436 kJ mol^{-1}). All the crystalline forms of sulfur that can be isolated at room temperature consist of S_n rings.

The striking difference between O−O and S−S single bond energies has important consequences. The O−O bond enthalpy is 146 kJ mol^{-1} and peroxides are powerful oxidizing agents: in contrast, the S−S bond enthalpy of 265 kJ mol^{-1} is so high that it is used in biology to stabilize protein structure by forming permanent linkages (RS−SR) between cysteine residues on different protein strands and different regions of one strand.

The chemically soft elements Se and Te occur in metal sulfide ores and their principal source is the electrolytic refining of copper. Polonium occurs in 27 known isotopes and all are radioactive.

16.2 Simple compounds

Key point: The elements of Group 16 form simple binary compounds with hydrogen, halogens, oxygen, and metals.

The most important hydride of any element is that of oxygen, namely *water*. The properties and reactions of water and of reactions in water are of paramount importance to inorganic chemists and are discussed throughout this text.

Water is the only Group 16 hydride that is not a poisonous, malodorous gas. Its melting and boiling points (0°C and 100°C, respectively) are both very high compared to compounds of similar molecular mass and the analogous molecules in Group 16 (Table 16.2). This high boiling point is due to extensive hydrogen bonding between H and the highly electronegative oxygen, O−H···O (Section 10.6). Oxygen also forms hydrogen peroxide, H_2O_2, which is also a liquid (from 0°C to 150°C) on account of extensive hydrogen bonding.

Oxygen forms oxides with most metals, and peroxides and superoxides with Group 1 and 2 metals. When the oxidation number of the metal is lower than +4 the oxide is commonly ionic. When the oxidation number of the metal is +4 or greater the oxide is molecular. Sulfur forms sulfides, S^{2-}, and disulfides, S_2^{2-}, with metals. Selenium and tellurium form selenides and tellurides, Se^{2-} and Te^{2-}.

Sulfur, Se, Te, and Po have a rich halogen chemistry, and some of the most common halides are summarized in Table 16.3. Sulfur forms very unstable iodides, but the iodides of Te and Po are more robust, which is an example of a large anion stabilizing a large cation (Section 3.15a). Of the halogens, only F brings out the maximum group oxidation state of the chalcogen elements, but the low oxidation state fluorides of Se, Te, and Po are unstable with respect to disproportionation into the element and a higher oxidation

Table 16.2 Selected properties of the Group 16 hydrides

Property	H_2O	H_2S	H_2Se	H_2Te	H_2Po
Melting point/°C	0.0	−85.6	−65.7	−51	−36
Boiling point/°C	100.0	−60.3	−41.3	−4	37
$\Delta_f H^\ominus / (\text{kJ mol}^{-1})$	−285.6 (l)	−20.1	+73.0	+99.6	
Bond length/pm	96	134	146	169	
Bond angle/°	104.5	92.1	91	90	
Acidity constants					
pK_{a1}	14.00	6.89	3.89	2.64	
pK_{a2}		14.15	11	10.80	

Table 16.3 Some halides of sulfur, selenium, and tellurium

Oxidation number	Formula	Structure	Remarks
$+\frac{1}{2}$	Te_2X (X = Br,I)	Halide bridges	Silver-grey
$+1$	S_2F_2	Two isomers: 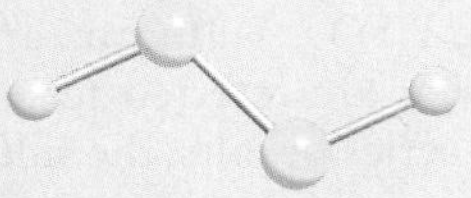 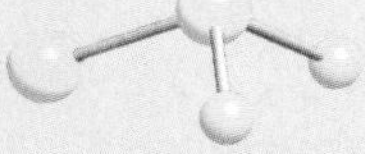	
	S_2Cl_2	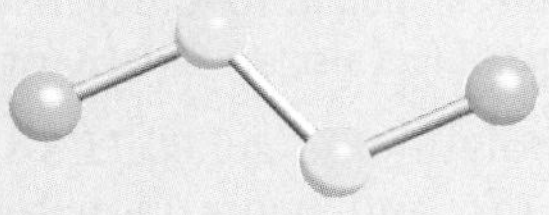	Reactive
$+2$	SCl_2	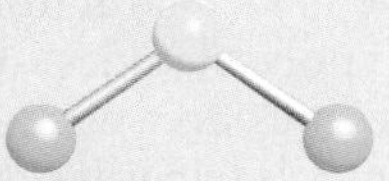	Reactive
$+4$	SF_4 SeX_4 (X = F, Cl, Br) TeX_4 (X = F, Cl, Br I)		Gas SeF_4 liquid TeF_4 solid
$+5$	S_2F_4 Se_2F_{10}	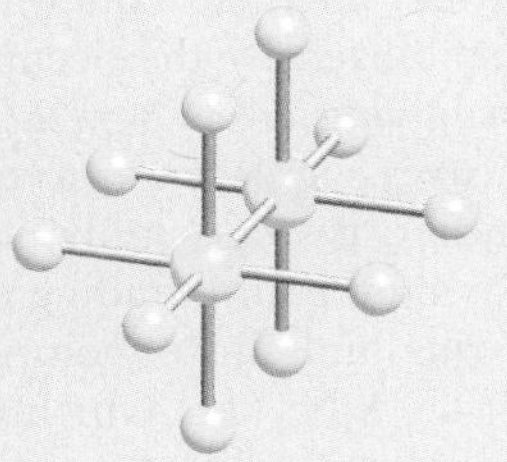	Reactive
$+6$	SF_6, SeF_6 TeF_6	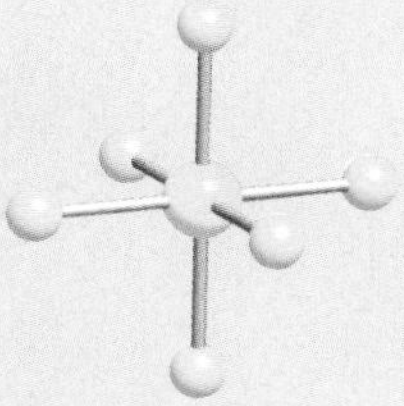	Colourless gases Liquid (b.p. 36°C)

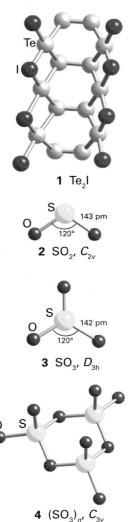

1 Te_2I

2 SO_2, C_{2v}
143 pm, 120°

3 SO_3, D_{3h}
142 pm, 120°

4 $(SO_3)_n$, C_{3v}

state fluoride. A series of catenated subhalides exist for the heavy members of the group. For example, Te_2I and Te_2Br consist of ribbons of edge-shared Te hexagons with halogen bridges (**1**).

The molecules of the two common oxides of S, SO_2 (b.p. $-10°C$) and SO_3 (b.p. 44.8°C), are angular (**2**) and trigonal planar (**3**), respectively, in the gas phase. In the solid, sulfur trioxide exists as cyclic trimers (**4**). Sulfur dioxide is a poisonous gas with a sharp, choking odour. The major use of SO_2 is in the manufacture of sulfuric acid, where it is first oxidized to SO_3. It is also used as a bleach, disinfectant, and food preservative. Sulfur trioxide, SO_3, is made on a huge scale by catalytic oxidation of SO_2. It is seldom isolated but immediately converted to sulfuric acid, H_2SO_4. Because sulfur trioxide is extremely corrosive, anhydrous SO_3 is seldom handled in the laboratory. It is available as oleum, or fuming sulfuric acid, $H_2S_2O_7$, which is a solution of $25-65$ per cent SO_3 by mass in concentrated sulfuric acid. Sulfur trioxide reacts with water to give H_2SO_4 in a vigorous

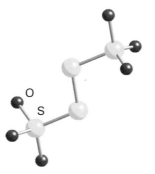

5 Tetrathionate ion, $S_4O_6^{2-}$

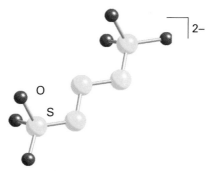

6 Pentathionate ion, $S_5O_6^{2-}$

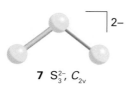

7 S_3^{2-}, C_{2v}

and very exothermic reaction. The reaction with metal oxides to produce sulfates is used to scrub undesirable SO_3 from effluent gases from industrial processes. Selenium, Te, and Po all form a dioxide and trioxide.

Sulfuric acid, H_2SO_4, is a dense viscous liquid. It is a strong acid (for the first deprotonation step), is a useful nonaqueous solvent, and exhibits extensive autoprotolysis (Section 4.1). Concentrated sulfuric acid extracts water from organic matter to leave a charred, carbonaceous residue. Sulfuric acid forms two series of salts, the sulfates, SO_4^{2-}, and the hydrogensulfates, HSO_4^-. Sulfurous acid, H_2SO_3, has never been isolated. Aqueous solutions of SO_2, which are referred to as 'sulfurous acid', are better considered as hydrates, $SO_2 \cdot nH_2O$. Two series of salts, the sulfites, SO_3^{2-}, and the hydrogensulfites, HSO_3^-, are known and are moderately strong reducing agents, becoming oxidized to sulfates, SO_4^{2-}, or dithionates, $S_2O_6^{2-}$.

16.3 Ring and cluster compounds

Key point: Ring and chain compounds of Group 16 elements are anionic or cationic. Neutral heteroatomic ring and chain compounds also are formed with other p-block elements.

Sulfur forms many polythionic acids, $H_2S_nO_6$, with up to six S atoms, such as the tetrathionate, $S_4O_6^{2-}$ (**5**), and pentathionate, $S_5O_6^{2-}$ (**6**), ions. Many polysulfides of electropositive elements have been characterized. They all contain the S_n^{2-} ions, where $n = 2-6$, as in (**7**). The smaller polyselenides and polytellurides resemble the polysulfides. The structures of the larger ones are more complex and depend to some extent on the nature of the cation. The polyselenides up to Se_9^{2-} are chains but larger molecules form rings such as Se_{11}^{2-}, which has a Se atom at the centre of two six-membered rings in a square-planar arrangement (**8**). The polytellurides may be bicyclic, as in Te_7^{2-} (**9**).

Many cationic chain, ring, and cluster compounds of the p-block elements have been prepared. The majority of them contain S, Se, or Te. The square-planar ions E_4^{2+} (E = S, Se, Te; **10**) are stabilized by the occupation of molecular orbitals. Each E atom has six valence electrons, giving $24 - 2 = 22$ electrons in all. There are two lone pairs on each E atom, which leaves six electrons to occupy the available molecular orbitals. Of those orbitals, one is bonding, two are nonbonding, and one is antibonding. The electrons occupy the first three, leaving the antibonding orbital unoccupied.

Neutral heteroatomic ring and cluster compounds of the p-block elements include the cyclic tetrasulfurtetranitride, S_4N_4 (**11**) which decomposes explosively. Disulfurdinitride, S_2N_2 (**12**) is even less stable but polymerizes to form a superconducting polymer, $(SN)_n$, that is stable up to 240°C.

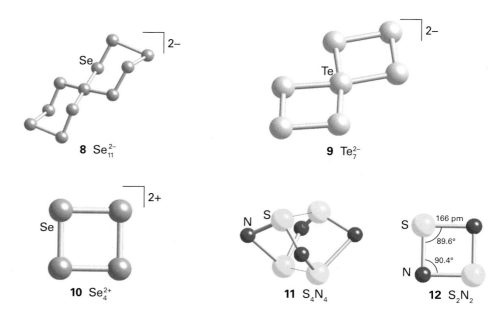

8 Se_{11}^{2-}

9 Te_7^{2-}

10 Se_4^{2+}

11 S_4N_4

12 S_2N_2

PART B: **THE DETAIL**

In this section we discuss the detailed chemistry of the Group 16 elements and observe the rich variation in the structures of the compounds they form.

16.4 Oxygen

Key points: Oxygen has two allotropes, dioxygen and ozone. Dioxygen has a triplet ground state and oxidizes hydrocarbons by a radical chain mechanism. Reaction with an excited state molecule can produce a fairly long-lived singlet state that can react as an electrophile. Ozone is an unstable and highly aggressive oxidizing agent.

Dioxygen is a biogenic gas (that is, one that has been produced by the action of organisms) and most of it is the result of photosynthesis, although some is produced by the action of ultraviolet radiation on water. It is colourless, odourless, and soluble in water to the extent of 3.08 cm³ per 100 cm³ water at 25°C and atmospheric pressure. This solubility falls to below 2.0 cm³ in seawater but is still sufficient to support aerobic marine life. The solubility of O_2 in organic solvents is approximately ten times greater than in water. The high solubility of O_2 makes it necessary to purge all solvents used in the synthesis of oxygen-sensitive compounds.

Oxygen is readily available as O_2 from the atmosphere and is obtained on a massive scale by the liquefaction and distillation of liquid air. The main commercial motivation is to recover O_2 for steel making, in which it reacts exothermically with coke (carbon) to produce carbon monoxide. The high temperature is necessary to achieve a fast reduction of iron oxides by CO and carbon (Section 5.16). Pure oxygen, rather than air, is advantageous in this process because energy is not wasted in heating the nitrogen. About 1 tonne (1 t = 10^3 kg) of oxygen is needed to make 1 tonne of steel. Oxygen is also required by industry in the production of the white pigment TiO_2 by the *chloride process*:

$$TiCl_4\,(l) + O_2\,(g) \rightarrow TiO_2\,(s) + 2\,Cl_2\,(g)$$

Oxygen is used in many oxidation processes, for example the production of oxirane (ethylene oxide) from ethene. Oxygen is also supplied on a large scale for sewage treatment, renewal of polluted waterways, paper-pulp bleaching, and as an artificial atmosphere in medical and submarine applications.

Liquid oxygen is very pale blue and boils at $-183°C$. Its colour arises from electronic transitions involving pairs of neighbouring molecules: one photon from the red−yellow−green region of the visible spectrum can raise two O_2 molecules to an excited state to form a molecular pair. Under high pressure the colour of solid oxygen changes from light blue to orange and then to red at approximately 10 GPa.

The molecular orbital description of O_2 implies the existence of a double bond; however, as we saw in Section 2.8, the outermost two electrons occupy different antibonding π orbitals with parallel spins; as a result, the molecule is paramagnetic (Fig. 16.1). The term symbol for the ground state is $^3\Sigma_g^-$, and henceforth the molecule will be denoted $O_2(^3\Sigma_g^-)$.

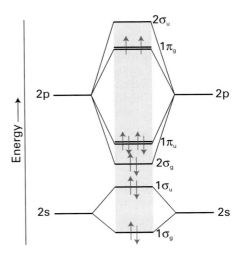

Figure 16.1 The molecular orbital diagram for O_2.

when it is appropriate to specify the spin state.[1] The singlet state $^1\Sigma_g^-$, with paired electrons in the same two π^* orbitals as in the ground state, is higher in energy by 1.61 eV (155 kJ mol⁻¹), and another singlet state $^1\Delta_g$ ('singlet delta'), with electrons paired in one π^* orbital, lies between these two terms at 0.95 eV (91.7 kJ mol⁻¹) above the ground state. Of the two singlet states, the latter has much the longer excited state lifetime (its return to the ground state is spin-forbidden) and $O_2(^1\Delta_g)$ survives long enough to participate in chemical reactions. When it is needed for reactions, $O_2(^1\Delta_g)$ can be generated in solution by energy transfer from a photoexcited molecule. Thus $[Ru(bpy)_3]^{2+}$ can be excited by absorption of blue light (452 nm) to give an electronically excited state, denoted $^*[Ru(bpy)_3]^{2+}$ (Section 20.7), and this state transfers energy to $O_2(^3\Sigma_g^-)$:

$$^*[Ru(bpy)_3]^{2+} + O_2\left(^3\Sigma_g^-\right) \rightarrow [Ru(bpy)_3]^{2+} + O_2\left(^1\Delta_g\right)$$

Another efficient way to generate $O_2(^1\Delta_g)$ is through the thermal decomposition of an ozonide:

In contrast to the radical character of many $O_2(^3\Sigma_g^-)$ reactions, $O_2(^1\Delta_g)$ reacts as an electrophile. This mode of reaction is feasible because $O_2(^1\Delta_g)$ has an empty π^* orbital, rather than two that are each occupied by a single electron. For example, $O_2(^1\Delta_g)$

---

[1] The symbols Σ, Π, and Δ are used for linear molecules such as dioxygen in place of the symbols S, P, and D used for atoms. The Greek letters represent the magnitude of the total orbital angular momentum around the internuclear axis.

adds across a diene, thus mimicking the Diels−Alder reaction of butadiene with an electrophilic alkene:

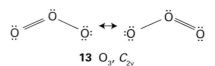

Singlet oxygen is implicated as one of the biologically hazardous products of photochemical smog.

The other allotrope of oxygen, *ozone*, O_3, boils at $-112°C$ and is an explosive and highly reactive endoergic blue gas $(\Delta_f G^\ominus = +163 \text{ kJ mol}^{-1})$. It decomposes into dioxygen

$$2O_3(g) \rightarrow 3O_2(g)$$

but this reaction is slow in the absence of a catalyst or ultraviolet radiation.

Ozone has a pungent odour; this property is reflected in its name, which is derived from the Greek *ozein*, to smell. The O_3 molecule is angular, in accord with the VSEPR model (**13**) and has bond angle 117°; it is diamagnetic. Gaseous ozone is blue, liquid ozone is blue−black, and solid ozone is violet−black. Ozone is produced from electrical discharges or ultraviolet radiation acting on O_2. This second method is used to produce low concentrations of ozone for the preservation of foodstuffs. The ability of O_3 to absorb strongly in the 220−290 nm region of the spectrum is vital in preventing the harmful ultraviolet rays of the Sun from reaching the Earth's surface (Box 17.3). Ozone reacts with unsaturated polymers, causing undesirable cross-linking and degradation.

13 O_3, C_{2v}

Reactions of ozone typically involve oxidation and transfer of an O atom. Ozone is very unstable in acidic solution and much more stable in basic conditions:

$$O_3(g) + 2H^+(aq) + 2e^- \rightarrow O_2(g) + H_2O(l) \qquad E^\ominus = +2.08 \text{ V}$$
$$O_3(g) + H_2O(l) + 2e^- \rightarrow O_2(g) + 2OH^-(aq) \qquad E^\ominus = +1.25 \text{ V}$$

Ozone is exceeded in oxidizing power only by F_2, atomic O, the OH radical, and perxenate ions (Section 18.5). Ozone forms ozonides with Group 1 and 2 elements (Sections 11.8 and 12.8). They are prepared by passing gaseous ozone over the powdered hydroxide, MOH or $M(OH)_2$, at temperatures below $-10°C$. The ozonides are red−brown solids that decompose on warming:

$$MO_3(s) \rightarrow MO_2(s) + \tfrac{1}{2}O_2(g)$$

The ozonide ion, O_3^-, is angular, like O_3, but with the slightly larger bond angle of 119.5°.

16.5 Reactivity of oxygen

Key point: The reactions of dioxygen are often thermodynamically favourable but sluggish.

Oxygen is by no means an inert molecule, yet many of its reactions are sluggish (a point first made in connection with overpotentials in Section 5.18). For example, a solution of Fe^{2+} is only slowly oxidized by air even though the reaction is thermodynamically favourable.

Several factors contribute to the appreciable activation energy of many reactions of O_2. One factor is that, with weak reducing agents, single-electron transfer to O_2 is mildly unfavourable thermodynamically:

$$O_2(g) + H^+(aq) + e^- \rightarrow HO_2(g) \qquad E^\ominus = -0.13 \text{ V at pH} = 0$$
$$O_2(g) + e^- \rightarrow O_2^-(aq) \qquad E^\ominus = -0.33 \text{ V at pH} = 14$$

A single-electron reducing reagent must exceed these potentials for the reaction to be thermodynamically viable and must exceed them to achieve a significant rate. Second, the ground state of O_2, with both π^* orbitals singly occupied, is neither an effective Lewis acid nor an effective Lewis base, and therefore has little tendency to undergo displacement reactions with p-block Lewis bases or acids. Finally, the high bond energy of O_2 (497 kJ mol^{-1}) results in a high activation energy for reactions that depend on its dissociation. Radical chain mechanisms can provide reaction paths that circumvent some of these activation barriers in combustion processes at elevated temperatures, and radical oxidations also occur in solution. In metalloenzymes (Section 27.10) O_2 coordinates to metals such as Fe and Cu, and the enzyme catalyses the four-electron reduction of O_2 to water.

16.6 Sulfur

Key points: Sulfur is extracted as the element from underground deposits. It has many allotropic and polymorphic forms, including a metastable polymer, but its most stable form is the cyclic S_8 molecule.

Sulfur can be extracted from deposits of the element by the *Frasch process*, in which underground deposits are forced to the surface using superheated water and steam, and compressed air. The extracted S is molten and is allowed to cool in large basins. The process is energy intensive and commercial success depends on access to cheap water and energy. Extraction from natural gas and crude oil by the *Claus process* has become increasingly important. In this process, H_2S is first oxidized in air at 1000−1400°C. This step produces some SO_2 that then reacts with the remaining H_2S at 200−350°C over a catalyst:

$$2H_2S(g) + SO_2(g) \rightarrow 3S(l) + 2H_2O(l)$$

Unlike O, S (and all the heavier members of the group) tends to form single bonds with itself rather than double bonds because of the poor π overlap of its orbitals, which are held apart by the bulky atomic cores of neighbouring atoms. As a result, it aggregates into larger molecules or extended structures and hence is a solid at room temperature. Sulfur vapour, which is formed at high temperatures, consists partially of paramagnetic disulfur molecules, S_2, that resemble O_2 in having a triplet ground state and a formal double bond.

The common yellow orthorhombic polymorph, α-S_8, consists of crown-like eight-membered rings (**14**) and all other forms of S eventually revert to this form. Orthorhombic α-sulfur is an electrical and thermal insulator. When it is heated to 93°C, the packing of the S_8 rings is modified and monoclinic β-S_8 forms. When molten sulfur that has been heated above 150°C is cooled slowly, monoclinic γ-sulfur is formed. This polymorph consists of S_8 rings like the α and β forms but the packing of the rings is more efficient, resulting in a higher density.

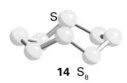

14 S_8

It is possible to synthesize and crystallize sulfur rings with from six to 20 S atoms (Table 16.4). An additional complexity is that some of these allotropes exist in several crystalline forms. For example, S_7 is known in four crystalline forms and S_{18} is known in two. Orthorhombic sulfur melts at 113°C; the yellow liquid darkens above 160°C and becomes more viscous as the sulfur rings break open and polymerize. The resulting helical S_n polymers (**15**) can be drawn from the melt and quenched to form metastable rubber-like materials that slowly revert to α-S_8 at room temperature. In the gas phase, S_2 and S_3 are observed. S_3 is a cherry red, angular molecule like ozone. The more stable species is the violet S_2 molecule that, like O_2, is doubly bonded with a bond dissociation energy of 421 kJ mol^{-1}.

Sulfur reacts directly with many elements at room or elevated temperatures. It ignites in F_2 to form SF_6, reacts rapidly with Cl_2 to form S_2Cl_2, and dissolves in Br_2 to give S_2Br_2, which readily dissociates. It does not react with liquid I_2, which can therefore be used as a low-temperature solvent for sulfur. Atomic sulfur, S, is extremely reactive, and triplet and singlet states are possible with different reactivities, as with O.

Table 16.4 Properties of selected sulfur allotropes and polymorphs

Allotrope	Melting point*/°C	Appearance
S_3	Gas	Cherry red
S_6	50d	Orange red
S_7	39d	Yellow
α-S_8	113	Yellow
β-S_8	119	Yellow
γ-S_8	107	Pale yellow
S_{10}	0d	Yellow green
S_{12}	148	Pale yellow
S_{18}	128	Lemon yellow
S_{20}	124	Pale yellow
S_∞	104	Yellow

* d, decomposes.

15 S_n

Most of the sulfur produced is used to manufacture sulfuric acid, H_2SO_4, which is one of the most important manufactured chemicals. Sulfuric acid has many uses, including the synthesis of fertilizers and in dilute aqueous solution as the electrolyte in lead−acid batteries (Box 14.6). Sulfur is a component of gunpowder (a mixture of potassium nitrate, KNO_3, carbon, and sulfur). It is also used in the vulcanization of natural rubber.

16.7 Selenium, tellurium, and polonium

Key points: Selenium and tellurium crystallize in helical chains; polonium crystallizes in a primitive cubic form.

Selenium can be extracted from the waste sludge from sulfuric acid plants. Selenium and tellurium can be extracted from copper sulfide ores, where they occur as the copper selenide or telluride. The extraction method depends on the other compounds or elements present. The first step usually involves oxidation in the presence of sodium carbonate:

$$Cu_2Se(aq) + Na_2CO_3(aq) + 2\,O_2(g) \rightarrow$$
$$2\,CuO(s) + Na_2SeO_3(aq) + CO_2(g)$$

The solution containing Na_2SeO_3 and Na_2TeO_3 is acidified with sulfuric acid. The Te precipitates out as the dioxide, leaving selenous acid, H_2SeO_3, in solution. Selenium is recovered by treatment with SO_2:

$$H_2SeO_3(aq) + 2\,SO_2(g) + H_2O(l) \rightarrow Se(s) + 2\,H_2SO_4(aq)$$

Tellurium is liberated by dissolving the TeO_2 in aqueous sodium hydroxide followed by electrolytic reduction:

$$TeO_2(s) + 2\,NaOH(aq) \rightarrow Na_2TeO_3(aq) + H_2O(l) \rightarrow$$
$$Te(s) + 2\,NaOH(aq) + O_2(g)$$

As with S, three polymorphs of Se exist that contain Se_8 rings and differ only in the packing of the rings to give α, β, and γ forms of red selenium. The most stable form at room temperature is metallic grey selenium, a crystalline material composed of helical chains. The common commercial form of the element is amorphous black selenium; it has a very complex structure comprising rings containing up to 1000 Se atoms. Another amorphous form of Se, obtained by deposition of the vapour, is used as the photoreceptor in the xerographic photocopying process. Selenium is an essential element for humans, but, as with many essential elements, there is only a narrow range of concentration between the minimum daily requirement and toxicity. An early indication of Se poisoning is a garlicky smell on the breath, which is due to methylated selenium.

Selenium exhibits both photovoltaic character, where light is converted directly into electricity, and photoconductive character. The photoconductivity of grey selenium arises from the ability of incident light to excite electrons across its reasonably small band gap (2.6 eV in the crystalline material, 1.8 eV in the amorphous material). These properties make Se useful in the production of photocells and exposure meters for photographic use, as well as solar cells. Selenium is also a p-type semiconductor (Section 3.20) and is used in electronic and solid-state applications. It is also used in photocopier toner and in the glass industry to make red glasses and enamels.

Tellurium crystallizes in a chain structure like that of grey selenium. Polonium crystallizes in a primitive cubic structure and a closely related higher temperature form above 36°C. We remarked in Section 3.5 that the primitive cubic structure represents inefficient packing of atoms, and Po is the only element that adopts this structure under normal conditions. Tellurium and Po are both highly toxic; the toxicity of Po is enhanced by its intense radioactivity. Mass for mass, it is about 2.5×10^{11} times as toxic as hydrocyanic acid. All 29 isotopes of polonium are radioactive. It has been found in tobacco as a contaminant and in uranium ores. It can be produced in small amounts (gram quantities) through irradiation of ^{209}Bi (atomic number 83) with neutrons, which gives ^{210}Po (atomic number 84):

$$^{209}_{83}\text{Bi} + {}^{1}_{1}\text{n} \rightarrow {}^{210}_{84}\text{Po} + \text{e}^-$$

Metallic Po can then be separated from the remaining Bi by fractional distillation or electrodeposited on to a metal surface.

Selenium, Te, and Po combine directly with most elements, although less readily than O or S. The occurrence of multiple bonds is lower than with O or S, as is the tendency towards catenation (compared with S) and the number of allotropes. The unexpected difficulty of oxidizing Se to Se(VI) (Fig. 16.2) is due to the lanthanide-like contraction (Section 9.2a) in radius following the 3d elements.

16.8 Hydrides

The impact of hydrogen bonding is seen clearly in the hydrides of the Group 16 elements. The hydrides of oxygen are water and hydrogen peroxide, which are both liquids. The hydrides of the heavier elements are all toxic, foul-smelling gases.

(a) Water

Key point: Hydrogen bonding in water results in a high boiling liquid and a highly structured arrangement in the solid, ice.

At least nine distinct forms of ice have been identified. At 0°C and atmospheric pressure, hexagonal ice I_h forms (Fig. 10.7) but between −120 and −140°C the cubic form, I_c, is produced. At very high pressures several higher density polymorphs are formed,

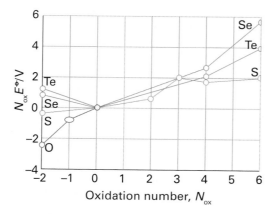

Figure. 16.2 Frost diagram for the elements of Group 16 in acidic solution. The species with oxidation number −2 are H_2E. For oxidation number −1 the compound is H_2O_2. The positive oxidation numbers refer to the oxo acids or oxoanions.

some of which are based on silica-like structures (Section 14.10).

Water is formed by the direct interaction of the elements:

$$H_2(g) + \tfrac{1}{2}O_2(g) \rightarrow H_2O(l) \qquad \Delta_f H^{\ominus}(H_2O,l) = -286\,\text{kJ mol}^{-1}$$

This reaction is very exothermic and provides the basis for the development of the hydrogen economy and hydrogen fuel cells (Fig. 5.1, Box 10.2, and Section 24.1).

Water is the most widely used solvent not only because it is so widely available but also because of its high relative permittivity (dielectric constant), wide liquid range, and—through a combination of its polar character and ability to form hydrogen bonds—solvating ability. Many anhydrous and hydrated compounds dissolve in water to give hydrated cations and anions. Some predominantly covalent compounds, such as ethanol and ethanoic (acetic) acid, are soluble in water or miscible with it because of hydrogen-bonded interactions with the solvent. Many other covalent compounds react with water in hydrolysis reactions; examples are discussed in the appropriate chapters. In addition to simple dissolution and hydrolysis reactions, the importance of aqueous solution chemistry can be seen in redox reactions (Chapter 5) and acid−base reactions (Chapter 4). Water also acts as a Lewis base ligand in metal complexes (Section 7.1). The deprotonated forms, OH^- and particularly the oxide ion O^{2-} are important ligands for stabilizing higher oxidation states, examples of which are found in the simple oxo-cations of early d-block elements, such as the vanadyl ion, VO^{2+}.

(b) Hydrogen peroxide

Key point: Hydrogen peroxide is susceptible to decomposition by disproportionation at elevated temperatures or in the presence of catalysts.

Hydrogen peroxide is a very pale blue, viscous liquid. It has a higher boiling point than water (150°C) and a greater density (1.445 g cm^{-3} at 25°C). It is miscible in water and is usually handled in aqueous solution. The Frost diagram for oxygen (Fig. 16.2) shows that H_2O_2 is a good oxidizing agent, but it is unstable with respect to disproportionation:

$$H_2O_2(l) \rightarrow H_2O(l) + \tfrac{1}{2}O_2(g) \qquad \Delta_r G^{\ominus} = -119\,\text{kJ mol}^{-1}$$

This reaction is slow but is explosive when catalysed by a metal surface or alkali dissolved from glass. For this reason, hydrogen peroxide and its solutions are stored in plastic bottles and a stabilizer is added. This reaction can be considered in terms of the reduction half-reactions

$$\tfrac{1}{2}H_2O_2(aq) + H^+(aq) + e^- \rightarrow H_2O(l) \qquad E^{\ominus} = +1.68\,\text{V}$$
$$H^+(aq) + \tfrac{1}{2}O_2(g) + e^- \rightarrow \tfrac{1}{2}H_2O_2(aq) \qquad E^{\ominus} = +0.70\,\text{V}$$

Any substance with a standard potential in the range $0.70−1.68$ V that has suitable binding sites will catalyse this reaction. As can be inferred from these standard potentials, hydrogen peroxide is a very powerful oxidizing agent in acid solution:

$$2\,Ce^{3+}(aq) + H_2O_2(aq) + 2\,H^+(aq) \rightarrow 2\,Ce^{4+}(aq) + 2\,H_2O(l)$$

However, in basic solution hydrogen peroxide can act as a reducing agent:

$$2\,Ce^{4+}(aq) + H_2O_2(aq) + 2\,OH^-(aq) \rightarrow$$
$$2\,Ce^{3+}(aq) + 2\,H_2O(l) + O_2(g)$$

The underlying reason for the oxidizing nature of hydrogen peroxide lies in the weakness of the O−O single bond (146 kJ mol^{-1}). Hydrogen peroxide reacts with d-metal ions such as Fe^{2+} to form the hydroxyl radical in the *Fenton reaction*:

$$Fe^{2+}(aq) + H_2O_2(aq) \rightarrow Fe^{3+}(aq) + OH^-(aq) + OH(aq)$$

The Fe^{3+} product can react with a second H_2O_2 to regenerate Fe^{2+}, so that the production of hydroxyl radical is catalytic. The hydroxyl radical is one of the strongest oxidizing agents known ($E^\ominus = +2.85$ V) and the reaction is used to oxidize organic matter. In living cells, its reaction with DNA has potentially lethal consequences.

EXAMPLE 16.1 Deciding whether an ion can catalyse H_2O_2 disproportionation

Is Fe^{3+} thermodynamically capable of catalysing the decomposition of H_2O_2?

Answer For Fe^{3+} to catalyse the decomposition of H_2O it needs to act in the reaction

$$2 Fe^{3+}(aq) + H_2O_2(aq) \rightarrow 2 Fe^{2+}(aq) + O_2(g) + 2 H^+(aq)$$

and then be regenerated in the reaction

$$2 Fe^{2+}(aq) + H_2O_2(aq) + 2 H^+(aq) \rightarrow 2 Fe^{3+}(aq) + 2 H_2O(l)$$

So the net reaction is simply the decomposition of H_2O_2:

$$2 H_2O_2(aq) \rightarrow O_2(g) + 2 H_2O(l)$$

The first reaction is the difference between the half-reactions

$$Fe^{3+}(aq) + e^- \rightarrow Fe^{2+}(aq) \qquad E^\ominus = +0.77\,V$$
$$O_2(g) + 2 H^+(aq) + 2 e^- \rightarrow H_2O_2(aq) \qquad E^\ominus = +0.70\,V$$

and therefore $E_{cell}^\ominus = +0.07$ V. This reaction is therefore spontaneous ($K > 1$). The second reaction is the difference of the half-reactions

$$H_2O_2(aq) + 2 H^+(aq) + 2 e^- \rightarrow 2 H_2O(l) \qquad E^\ominus = +1.76\,V$$
$$Fe^{3+}(aq) + e^- \rightarrow Fe^{2+}(aq) \qquad E^\ominus = +0.77\,V$$

and therefore $E_{cell}^\ominus = +0.99$ V and this reaction is also spontaneous ($K > 1$). Because both reactions are spontaneous (in the sense $K > 1$), catalytic decomposition is thermodynamically favoured. In fact, the rates also are high, so Fe^{3+} is a highly effective catalyst for the decomposition of H_2O_2, and in its manufacture great pains are taken to minimize contamination by iron.

Self-test 16.1 Determine whether the decomposition of H_2O_2 is spontaneous in the presence of either Br^- or Cl^-.

Hydrogen peroxide is a slightly stronger acid than water:

$$H_2O_2(aq) + H_2O(l) \rightleftharpoons H_3O^+(aq) + HO_2^-(aq) \qquad pK_a = 11.65$$

Deprotonation occurs in other basic solvents such as liquid ammonia, and NH_4OOH has been isolated and found to consist of NH_4^+ and HO_2^- ions. When solid NH_4OOH melts (at 25°C), the melt contains hydrogen-bonded NH_3 and H_2O_2 molecules.

The oxidizing ability of hydrogen peroxide and the harmless nature of its byproducts lead to its many applications. It is used in water treatment to oxidize pollutants, as a mild antiseptic, and as a bleach in the textile, paper, and hair-care industries (Box 16.2).

(c) Hydrides of sulfur, selenium, and tellurium

Key points: The extent of hydrogen bonding is much less for these hydrides than for water; all the hydrides are gases.

Hydrogen sulfide, H_2S, is toxic, its toxicity made more hazardous by the fact that it tends to anaesthetize the olfactory nerves, making intensity of smell a dangerously inaccurate guide to concentration. Hydrogen sulfide is produced by volcanoes and by some micro-organisms (Box 16.3). It is an impurity in natural gas and must be removed before the gas is used.

Pure H_2S can be prepared by direct combination of the elements above 600°C:

$$H_2(g) + S(l) \rightarrow H_2S(s)$$

Hydrogen sulfide is easily generated in the laboratory by trickling dilute hydrochloric or phosphoric acid on to FeS:

$$FeS(s) + 2 HCl(aq) \rightarrow H_2S(g) + FeCl_2(aq)$$

It can also be prepared by hydrolysis of aluminium sulfide, which is easily generated by ignition of a mixture of the elements:

$$2 Al(s) + 3 S(s) \rightarrow Al_2S_3(s)$$
$$Al_2S_3(s) + 3 H_2O(l) \rightarrow Al_2O_3(s) + 3 H_2S(g)$$

It is readily soluble in water, and is a weak acid:

$$H_2S(aq) + H_2O(l) \rightleftharpoons H_3O^+(aq) + HS^-(aq) \qquad pK_{a1} = 6.89$$
$$HS^-(aq) + H_2O(l) \rightleftharpoons H_3O^+(aq) + S^{2-}(aq) \qquad pK_{a2} = 14.15$$

Acidic solutions of H_2S are mild reducing agents and deposit elemental S on standing.

In a similar way, H_2Se can be made by direct combination of the elements, by the reaction of FeSe with hydrochloric acid, or by hydrolysis of Al_2Se_3:

$$H_2(g) + Se(s) \rightarrow H_2Se(g)$$
$$FeSe(s) + 2 HCl(aq) \rightarrow H_2Se(g) + FeCl_2(aq)$$
$$Al_2Se_3(s) + 3 H_2O(l) \rightarrow Al_2O_3(s) + 3 H_2Se(g)$$

On the other hand, H_2Te is made by hydrolysis of Al_2Te_3 or by the action of hydrochloric acid on Mg, Zn, or Al tellurides.

$$Al_2Te_3(s) + 6 H_2O(l) \rightarrow 3 H_2Te(g) + 2 Al(OH)_3(aq)$$
$$MgTe(s) + 2 HCl(aq) \rightarrow H_2Te(g) + MgCl_2(aq)$$

The solubilities of H_2Se and H_2Te in water are similar to that of H_2S. The acidity constants of the hydrides (which are protic) increase from H_2S to H_2Te (Table 16.2). Like their sulfur analogue, aqueous solutions of H_2Se and H_2Te are readily oxidized and deposit elemental selenium and tellurium on standing.

16.9 Halides

Key points: The halides of oxygen have limited stability but its heavier congeners form an extensive series of halogen compounds; typical formulas are EX_2, EX_4, and EX_6.

The oxidation number of O is -2 in all its compounds with the halogens other than F. Oxygen difluoride, OF_2, is the highest

BOX 16.2 Environmentally friendly bleach

Hydrogen peroxide is rapidly replacing chlorine and hypochlorite bleaches in industrial applications as it is environmentally benign, producing only water and oxygen.

The major consumers of hydrogen peroxide bleach are the paper, textile, and wood pulp industries. A growing market is in de-inking of recycled paper and in the manufacture of kraft paper (strong brown paper). Approximately 85 per cent of all cotton and wool is bleached with hydrogen peroxide. One of its advantages over chlorine-based bleaches is that it does not affect many modern dyes. It is also used to decolour oils and waxes.

Hydrogen peroxide is used to treat domestic and industrial effluent and sewage. It minimizes odours by preventing the production of H_2S by anaerobic reactions in sewers and pipes. It also acts as a source of oxygen in sewage sludge treatment plants. Other industrial uses of hydrogen peroxide are the epoxidation of soybean and linseed oils to produce plasticizers and stabilizers for the plastics industry and as a propellant for torpedoes and missiles. It has been speculated that the sinking of the Russian submarine *Kursk* in 2000 was due to an explosion involving the hydrogen peroxide used to fuel its torpedoes.

Hydrogen peroxide is increasingly being used as a green oxidant. It can be used in aqueous solution if generated *in situ* and the only byproduct is water.

BOX 16.3 The sulfur cycle

Sulfur is essential to all life forms through its presence in the amino acids cysteine and methionine, and in many key active site structures, including the inorganic sulfide in Fe$-$S proteins, and all molybdenum and tungsten enzymes. Moreover, many organisms obtain energy by the oxidation or reduction of inorganic sulfur compounds. The resultant transformations constitute the sulfur cycle.

The redox extremes of sulfur chemistry are demonstrated by sulfate, the most oxidized form, and by H_2S and its ionized forms, HS^- or S^{2-}, the most reduced forms. Many classes of organisms occupy ecological niches defined by the sulfur. The sulfate-reducing bacteria (SRBs), such as *Desulfovibrio*, reduce sulfate to sulfide under anaerobic conditions, oxidizing organic compounds in the process. Sulfide oxidizing organisms, such as *Thiobacilli*, are generally, but not always, aerobic and use O_2 to oxidize sulfide, polysulfide ions, elemental sulfur, or thiosulfate to sulfate. Figure B16.1 is an incomplete version of the sulfur cycle, highlighting some of the known participating molecules.

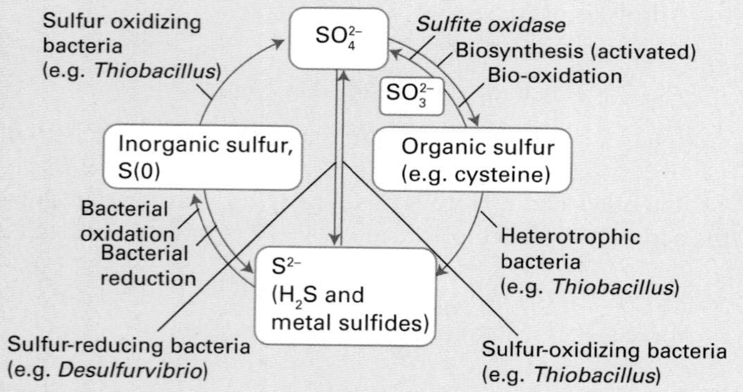

Figure B16.1 The sulfur cycle.

SRBs use sulfate as their electron acceptor and generate sulfide under anaerobic conditions. These anaerobic bacteria are found in environments where both SO_4^{2-} and reduced organic matter are found, for example in anoxic marine sediments and in the rumen of sheep and cattle. SRBs are important in sulfide ore formation, bio-corrosion, the souring of petroleum under anaerobic conditions, the Cu$-$Mo antagonism in ruminants, and many other physiological, ecological, and biogeochemical contexts.

The reduction of sulfate is carried out in two steps:

$$SO_4^{2-} + 8\ e^- + 10\ H^+ \rightarrow H_2S + 4\ H_2O$$

First, the relatively unreactive sulfate must be activated. This step is achieved through reaction with ATP to form adenosine phosphosulfate (APS) and pyrophosphate. The further hydrolysis of pyrophosphate $\left(\Delta_r H^{\ominus} = -30.5\ \text{kJ mol}^{-1}\right)$ ensures that the APS formation reaction goes to the right:

$$ATP + SO_4^{2-} \rightarrow APS + P_2O_7^{4-}$$

The enzyme APS reductase carries out the catalytic reduction of the sulfate intermediate to sulfite:

$$APS + 2\ e^- + H^+ \rightarrow AMP + HSO_3^-$$

Conversion of sulfite to sulfide is then catalysed by the enzyme sulfite reductase:

$$HSO_3^- + 6\ e^- + 7\ H^+ \rightarrow H_2S + 3\ H_2O$$

The oxidative part of the sulfur cycle is the province of bacteria that gain energy from various interconversions. Some *Thiobacilli* species can oxidize sulfide in ores, for example iron sulfides. The oxidation of sulfide to sulfate produces an acidic environment in which some species of *Thiobacilli* thrive and can alter the pH to produce acidic conditions favourable to their own metabolic processes. Acid mine drainage water can have a microbially produced pH as low as 1.5 and *Thiobacilli* are used commercially to mobilize metals from sulfide ores. For example, *Thiobacillus ferrooxidans* not only oxidizes the sulfur in iron sulfide deposits but also oxidizes the iron(II) to the soluble iron(III):

$$4\ FeS_2 + 15\ O_2 + 2\ H_2O \rightarrow 4\ Fe^{3+} + 8\ SO_4^{2-} + 4\ H^+$$

Thiobacilli live solely on inorganic materials: they use energy obtained from sulfide oxidation to drive all their cellular reactions, including the fixation of carbon from CO_2.

fluoride of oxygen and hence contains O in its highest oxidation state (+2).

The structures of the sulfur halides S_2F_2, SF_4, SF_6, and S_2F_{10} (Table 16.3) are all in line with the VSEPR model. Thus, SF_4 has ten valence electrons around the S atom, two of which form a lone pair in an equatorial position of a trigonal bipyramid. We have already mentioned the theoretical evidence that the molecular orbitals bonding the F atoms to the central atom in SF_6

primarily use the sulfur 4s and 4p orbitals, with the 3d orbitals playing a relatively unimportant role (Section 2.11a). The same seems to be true of SF_4 and S_2F_{10}.

Sulfur hexafluoride is a gas at room temperature. It is very unreactive and its inertness stems from the suppression, presumably by steric protection of the central S atom, of thermodynamically favourable reactions, such as the hydrolysis

$$SF_6(g) + 4H_2O(l) \rightarrow 6HF(aq) + H_2SO_4(aq)$$

The less sterically crowded SeF_6 molecule is easily hydrolysed and is generally more reactive than SF_6. Similarly, the sterically less hindered molecule SF_4 is reactive and undergoes rapid partial hydrolysis:

$$SF_4(g) + H_2O(l) \rightarrow OSF_2(aq) + 2HF(aq)$$

Both SF_4 and SeF_4 are selective fluorinating agents for the conversion of $-COOH$ into $-CF_3$ and $C=O$ and $P=O$ groups into CF_2 and PF_2 groups:

$$2R_2CO(l) + SF_4(g) \rightarrow 2R_2CF_2(sol) + SO_2(g)$$

Sulfur chlorides are commercially important. The reaction of molten S with Cl_2 yields the foul-smelling and toxic substance disulfur dichloride, S_2Cl_2, which is a yellow liquid at room temperature (b.p. 138°C). Disulfur dichloride and its further chlorination product sulfur dichloride, SCl_2, an unstable red liquid, are produced on a large scale for use in the vulcanization of rubber. In this process, S atom bridges are introduced between polymer chains so the rubber object can retain its shape.

16.10 Metal oxides

Key points: The oxides formed by metals include the basic oxides with high oxygen coordination number that are formed with most M^+ and M^{2+} ions. Oxides of metals in intermediate oxidation states often have more complex structures and are amphoteric. Metal peroxides and superoxides are formed between O_2 and alkali metals and alkaline earth metals. Terminal $E=O$ linkages and $E-O-E$ bridges are common with nonmetals and with metals in high oxidation states.

The O_2 molecule readily removes electrons from metals to form a variety of metal oxides containing the anions O^{2-} (oxide), O_2^- (superoxide), and O_2^{2-} (peroxide). Even though the existence of O^{2-} can be rationalized in terms of a closed-shell noble gas electron configuration, the formation of $O^{2-}(g)$ from $O_2(g)$ is highly endothermic, and the ion is stabilized in the solid state by its strong Coulombic interaction with the surrounding cations.

Alkali metals and alkaline earth metals often form peroxides or superoxides (Sections 11.8 and 12.8), but peroxides and superoxides of other metals are rare. Among the metals, only some of the noble metals do not form thermodynamically stable oxides. However, even where no bulk oxide phase is formed, an atomically clean metal surface (which can be prepared only in an ultrahigh vacuum) is quickly covered with a surface layer of oxide when it is exposed to traces of oxygen.

Structural trends in the metal oxides are not readily summarized but for oxides in which the metal has oxidation number +1, +2, or +3, the O^{2-} ion is generally in a site of high coordination number:

M(I): M_2O oxides often have a rutile or antifluorite structure (6,3)- and (8,4)-coordination, respectively.

M(II): MO oxides usually have the rock-salt structure (6,6)-coordination,

M(III): M_2O_3 oxides often have (6,4)-coordination.

At the other extreme, MO_4 compounds are molecular: the tetrahedral compound osmium tetroxide, OsO_4, is an example. The structures of the oxides with metals in high oxidation states and the oxides of nonmetallic elements often have multiple bond character. Deviations from these simple structures are common with p-block metals, where the less symmetric packing of O^{2-} ions around the metal can often be rationalized in terms of the existence of a stereochemically active lone pair, as in PbO (Section 14.11). Another common structural motif for nonmetals and some metals in high oxidation states is a bridging oxygen atom, as in $E-O-E$, in angular and linear structures.

16.11 Metal sulfides, selenides, tellurides, and polonides

Key point: Monatomic and polyatomic sulfide ions are known as discrete anions and as ligands.

Many metals occur naturally as their sulfide ores. The ores are roasted in air to form the oxide or the water-soluble sulfate, from which the metals are extracted. The sulfides can be prepared in the laboratory or industry by a number of routes; direct combination of the elements, reduction of a sulfate, or precipitation of an insoluble sulfide from solution by addition of H_2S:

$$Fe(s) + S(s) \rightarrow FeS(s)$$
$$MgSO_4(s) + 4C(s) \rightarrow MgS(s) + 4CO(g)$$
$$M^{2+}(aq) + H_2S(g) \rightarrow MS(s) + 2H^+(aq)$$

The solubilities of the metal sulfides vary enormously. The Group 1 and 2 sulfides are soluble, whereas the sulfides of the heavy elements of Group 11 and 12 are among the least soluble compounds known. The wide variation enables selective separation of metals to take place on the basis of the solubilities of the sulfides.

The Group 1 sulfides, M_2S, adopt the antifluorite structure (Section 3.9). The Group 2 elements and some of the f-block elements form monosulfides, MS, with a rock-salt structure. The first-row d-block metals form monosulfides with the NiAs structure, whereas the heavier elements have a greater tendency to covalence and adopt a zinc-blende structure. The d-metals form disulfides that have a layered structure or contain discrete S_2^{2-} ions. These compounds are discussed in Chapter 19.

The selenides and tellurides are the most common naturally occurring sources of the elements. Group 1 and 2 selenides, tellurides, and polonides are prepared by direct interaction of the elements in liquid ammonia. They are water-soluble solids that are rapidly oxidized in air to give the elements, with the exception of the polonides, for which they are among the most stable compounds of the element. The selenides and tellurides of Li, Na, and K adopt the antifluorite structure; those of the heavier elements of Group 1 adopt the rock-salt structure. Selenides, tellurides, and polonides of the d metals are also prepared by direct interaction of the elements and are nonstoichiometric.

Two examples are compounds of approximate stoichiometry Ti$_2$Se and Ti$_3$Se.

16.12 Oxides

Oxides of Group 16 elements are described in the relevant group chapters. In this section, we concentrate on compounds formed between oxygen and its congeners in Group 16.

(a) Sulfur oxides and oxohalides

Key points: Sulfur dioxide is a mild Lewis acid towards p-block bases, and OSCl$_2$ is a useful drying agent.

Sulfur dioxide and sulfur trioxide are both Lewis acids, with the S atom the acceptor site, but SO$_3$ is the much the stronger and harder acid. The high Lewis acidity of SO$_3$ accounts for its occurrence as a cyclic trimeric O-bridged solid at room temperature and pressure (**4**).

Sulfur dioxide is manufactured on a large scale by combustion of sulfur or H$_2$S or by roasting sulfide ores in air:

$$4\,FeS(s) + 7\,O_2(g) \rightarrow 4\,SO_2(g) + 2\,Fe_2O_3(s)$$

It is soluble in water and gives a solution commonly referred to as sulfurous acid, H$_2$SO$_3$, but which is in fact a complex mixture of numerous species (Box 16.4). Sulfur dioxide forms weak complexes with simple p-block Lewis bases. For example, although it does not form a stable complex with H$_2$O, it does form stable complexes with stronger Lewis bases, such as trimethylamine and F$^-$ ions. Sulfur dioxide is a useful solvent for acidic substances (see Box 16.5).

EXAMPLE 16.2 Deducing the structures and properties of SO$_2$ complexes

Suggest the probable structures of SO$_2$F$^-$ and (CH$_3$)$_3$NSO$_2$, and predict their reactions with OH$^-$.

Answer A good starting point for the discussion of shape is to draw a Lewis structure. The Lewis structure of SO$_2$ is shown in (**16**). We know that SO$_2$ can act as either a Lewis acid or a Lewis base, but in both cases SO$_2$ has acted as

---

a Lewis acid and formed a complex with the base, F$^-$ or (CH$_3$)$_3$N. Both the complexes still have a lone pair on S, and the resulting four electron pairs form a tetrahedron around the S atom, yielding the trigonal pyramidal complexes (**17**) and (**18**). As the OH$^-$ ion is a stronger Lewis base than either F$^-$ or N(CH$_3$)$_3$, it will form a complex with the SO$_2$ in preference to either of them. Therefore, exposure of either complex to OH$^-$ will yield the hydrogensulfite ion, HSO$_3^-$ which has been found to exist in two isomers, (**19**) and (**20**).

Self-test 16.2 Draw the Lewis structures and identify the point groups of (a) SO$_3$(g) and (b) SO$_3$F$^-$.

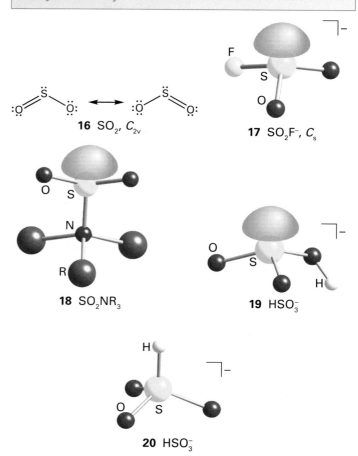

16 SO$_2$, C_{2v}

17 SO$_2$F$^-$, C_s

18 SO$_2$NR$_3$

19 HSO$_3^-$

20 HSO$_3^-$

BOX 16.4 Acid rain

The main components of acid rain are nitric and sulfuric acids produced by interaction of the oxides with hydroxyl radicals:

$$HO\cdot + NO_2 \rightarrow HNO_3$$
$$HO\cdot + SO_2 \rightarrow HSO_3\cdot$$
$$HSO_3\cdot + O_2 + H_2O \rightarrow H_2SO_4 + HO_2\cdot$$

The resulting hydroperoxyl radicals produce additional hydroxyl radicals:

$$HO_2\cdot + X \rightarrow XO + HO\cdot \qquad X = NO \text{ or } SO_2$$

Sulfuric and nitric acid molecules form hydrogen bonds and interact strongly with one another, with metal oxides and gases in the atmosphere, and with water, to form particles. These small particles are a major health threat in polluted air. Recent studies have convincingly associated increased concentrations of particulate matter in the size range of 2.5 μm or less with increased mortality from pulmonary, and especially heart, disease. These

particles are small enough to lodge deep in the lungs, and they can carry noxious chemicals on their surface.

In addition to their physiological effects, these particles affect the ecosystem because of the acids they contain. As the acidity of rainfall increases, the protons increasingly wash alkali metal (Na$^+$, K$^+$) and alkaline earth (Ca^{2+}, Mg^{2+}) ions from the soil, where they are held in ion exchange sites in clay and humus or in limestone. Depletion of these nutrients limits plant growth. The same chemistry also erodes marble statues and buildings. Granite-lined lakes (which have low buffer capacity) can be acidified, leading to the disappearance of fish and other aquatic life. Because emissions from combustion sources can travel long distances, acid rain is a regional problem, with large areas at risk, particularly downwind of coal-fired power plants, which have tall smokestacks to disperse the NO and SO$_2$ exhaust gases. These environmental and health effects combine to make NO and SO$_2$ a main focus of air pollution regulatory activity.

BOX 16.5 Synthesis in nonaqueous solvents

The nonaqueous solvents liquid ammonia, liquid sulfur dioxide, and sulfuric acid show an interesting set of contrasts as they range from a good Lewis base with a resistance to reduction (ammonia) to a strong Brønsted acid with resistance to oxidation (sulfuric acid).

Liquid ammonia (b.p. $-33°C$) and the solutions it forms may be handled in an open Dewar flask in an efficient fume hood (Fig. B16.2), in a vacuum line when the liquid is kept below its boiling point, or (with caution) in sealed heavy-walled glass tubes at room temperature (the vapour pressure of ammonia is about 10 atm at room temperature). Liquid ammonia solutions of the s-block metals are excellent reducing agents. One example of their application is in the preparation of a complex of Ni in its unusual $+1$ oxidation state:

$$2\,K(am) + 2\,[Ni(CN)_4]^{2-}(am) \rightarrow [Ni_2(CN)_6]^{4-}(am) + 2\,KCN(am)$$

Liquid sulfur dioxide (b.p. $-10°C$) can be handled like liquid ammonia. An example of its use is the reaction

$$NH_2OH(s) + SO_2(l) \rightarrow H_2NSO_2OH(s)$$

This reaction is performed in a sealed thick-walled borosilicate glass tube, which is sometimes called a Carius tube (Fig. B16.3). The hydroxylamine is introduced into the tube and the sulfur dioxide is condensed into the tube after it has been cooled to about $-45°C$. After the tube has been sealed, it is allowed to warm to room temperature behind a blast shield in a hood. The reaction is complete after several days at room temperature; the tube is cooled (to reduce the pressure) and broken open. The product is known as sulfamic acid and is used in the synthesis of compounds that taste sweet.

Sulfuric acid and solutions of SO_3 in sulfuric acid (fuming sulfuric acid, $H_2S_2O_7$) are used as oxidizing acidic media for the preparation of polychalcogen cations:

$$8\,Se(s) + 5\,H_2SO_4(l) \rightarrow Se_8^{2+}(sol) + 2\,H_3O^+(sol)$$
$$+\,4\,HSO_4^-(sol) + SO_2(g)$$

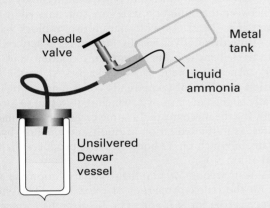

Figure B16.2 The transfer of liquid ammonia from a pressurized tank. (Based on W.L. Jolly, *The synthesis and characterization of inorganic compounds*. Waveland Press, Prospect Heights (1991).)

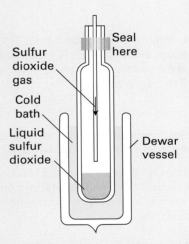

Figure B16.3 The condensation of sulfur dioxide into a Carius tube.

(b) Oxides of selenium and tellurium

Key points: Selenium and tellurium dioxides are polymorphic; selenium dioxide is thermodynamically less stable than SO_2 or TeO_2, and selenium trioxide, SeO_3, is thermodynamically less stable than SeO_2.

The dioxides of Se, Te, and Po can be prepared by direct reaction of the elements. Selenium dioxide is a white solid that sublimes at $315°C$. It has a polymeric structure in the solid state (**21**). It is thermodynamically less stable than SO_2 or TeO_2 and is reduced to selenium on reaction with NH_3, N_2H_4, or aqueous SO_2.

$$3\,SeO_2(s) + 4NH_3(l) \rightarrow 3\,Se(s) + 2\,N_2(g) + 6\,H_2O(l)$$

It is used an oxidizing agent in organic chemistry.

Tellurium dioxide occurs naturally as the mineral tellurite, $\beta\text{-}TeO_2$, which has a layer structure in which TeO_4 units form dimers (**22**). Synthetic $\alpha\text{-}TeO_2$ consists of similar TeO_4 units that share all vertices to form a three-dimensional rutile-like structure (**23**). Polonium dioxide exists as the yellow form with the fluorite structure and the red tetragonal form.

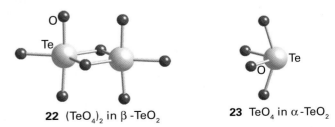

22 $(TeO_4)_2$ in $\beta\text{-}TeO_2$ **23** TeO_4 in $\alpha\text{-}TeO_2$

Selenium trioxide, unlike SO_3 or TeO_3, is thermodynamically less stable than the dioxide (Table 16.5). It is a white hygroscopic solid that sublimes at $100°C$ and decomposes at $165°C$. In the solid state the structure is based on Se_4O_{12} tetramers (**24**)

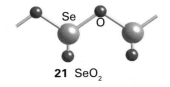

21 SeO_2

Table 16.5 Standard enthalpies of formation, $\Delta_f H^{\oplus}$/ (kJ mol^{-1}), of sulfur, selenium, and tellurium oxides

SO$_2$	−297	SO$_3$	−432
SeO$_2$	−230	SeO$_3$	−184
TeO$_2$	−325	TeO$_3$	−348

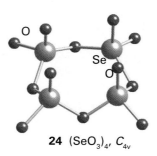

24 (SeO$_3$)$_4$, C_{4v}

but it is monomeric in the vapour phase. Tellurium trioxide exists as the yellow α-TeO$_3$, which is prepared by dehydration of Te(OH)$_6$, and the more stable β-TeO$_3$, which is made by heating α-TeO$_3$ or Te(OH)$_6$ in oxygen.

(c) Chalcogen oxohalides

Key points: The most important oxohalides are those of sulfur; selenium and tellurium oxofluorides are known and the 'teflate' ion is a useful ligand.

Many chalcogen oxohalides are known. The most important are the thionyl dihalides, OSX$_2$, and the sulfuryl dihalides, O$_2$SX$_2$. One laboratory application of thionyl dichloride is the dehydration of metal chlorides:

$$MgCl_2.6H_2O(s) + 6\,OSCl_2\,(l) \rightarrow$$
$$MgCl_2\,(s) + 6\,SO_2\,(g) + 12\,HCl(g)$$

The compound F$_5$TeOTeF$_5$ and its selenium analogue are known, and the OTeF$_5^-$ ion, which is known informally as 'teflate', is a bulky, electronegative anion. It is a well-established ligand for high oxidation state d-metal and main-group element complexes such as [Ti(OTeF$_5$)$_6$]$^{2-}$ (**25**), [Xe(OTeF$_5$)$_6$], and [M(C$_5$H$_5$)$_2$(OTeF$_5$)$_2$], where M = Ti, Zr, Hf, W, and Mo.

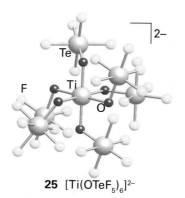

25 [Ti(OTeF$_5$)$_6$]$^{2-}$

16.13 Oxoacids of sulfur

Sulfur (like N and P) forms many oxoacids. These exist in aqueous solution or as the solid salts of the oxoanions (Table 16.6). Many of them are important in the laboratory and in industry.

(a) Redox properties of the oxoanions

Key points: The oxoanions of sulfur include the sulfite ion, SO$_3^{2-}$, which is a good reducing agent, the rather unreactive sulfate ion, SO$_4^{2-}$, and the strongly oxidizing peroxodisulfate ion, O$_3$SOOSO$_3^{2-}$. As with sulfur, the redox reactions of selenium and tellurium oxoanions are often slow.

Sulfur's common oxidation numbers are −2, 0, +2, +4, and +6, but there are also many S−S bonded species that are assigned odd and fractional average oxidation numbers. A simple example is the thiosulfate ion, S$_2$O$_3^{2-}$, in which the average oxidation number of S is +2, but in which the environments of the two S atoms are quite different. The thermodynamic relations between the oxidation states are summarized by the Frost diagram (Fig. 16.2). As with many other p-block oxoanions, many of the thermodynamically favourable reactions are slow when the element is in its maximum oxidation state (+6), as in SO$_4^{2-}$. Another kinetic factor is suggested by the fact that oxidation numbers of compounds containing a single S atom generally change in steps of 2, which requires an O atom transfer path for the mechanism. In some cases a radical mechanism operates, as in the oxidation of thiols and alcohols by peroxodisulfate, in which O−O bond cleavage produces the transient radical anion SO$_4^-$.

We saw in Section 5.6 that the pH of a solution has a marked effect on the redox properties of oxoanions. This strong dependence is true for SO$_2$ and SO$_3^{2-}$ because the former is easily reduced in acidic solution and is therefore an oxidizing agent, whereas the latter in basic solution is primarily a reducing agent:

$$SO_2\,(aq) + 4\,H^+(aq) + 4\,e^- \rightarrow S(s) + 2\,H_2O(l)$$
$$E^{\oplus} = +0.50\,V$$

$$SO_4^{2-}\,(aq) + H_2O(l) + 2\,e^- \rightarrow SO_3^{2-}\,(aq) + 2\,OH^-(aq)$$
$$E^{\oplus} = -0.94\,V$$

The principal species present in acidic solution is SO$_2$, not H$_2$SO$_3$, but in more basic solution HSO$_3^-$ exists in equilibrium with H−SO$_3^-$ and H−OSO$_2^-$. The oxidizing character of SO$_2$ accounts for its use as a mild disinfectant and preservative for foodstuffs, such as dried fruit and wine.

The peroxodisulfate ion, O$_3$SOOSO$_3^{2-}$, is a powerful and useful oxidizing agent:

$$E^{\oplus} = +2.01\,V$$

although this reactivity reflects the properties of O rather than S because it arises from the weakness of the O−O bond, as we have discussed above for hydrogen peroxide.

Table 16.6 Some sulfur oxoanions

Oxidation number	Formula	Name	Structure	Remarks
One S atom				
+4	SO_3^{2-}	Sulfite	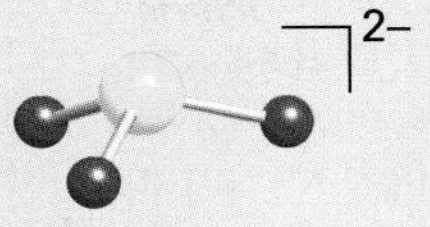	Basic, reducing agent
+6	SO_4^{2-}	Sulfate	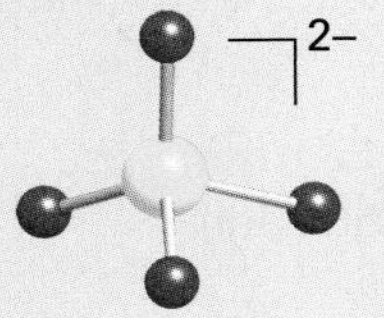	Weakly basic
Two S atoms				
+2	$S_2O_3^{2-}$	Thiosulfate	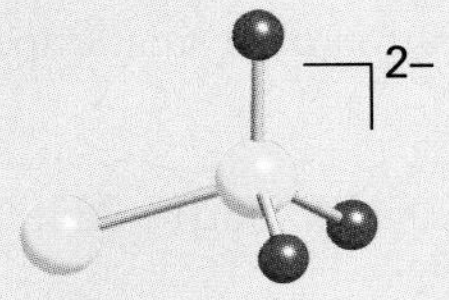	Moderately strong reducing agent
+3	$S_2O_4^{2-}$	Dithionite	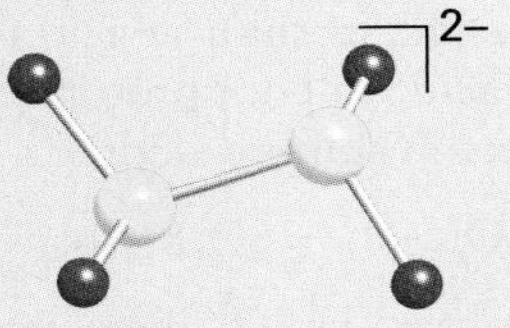	Strong reducing agent
+4	$S_2O_5^{2-}$	Disulfite	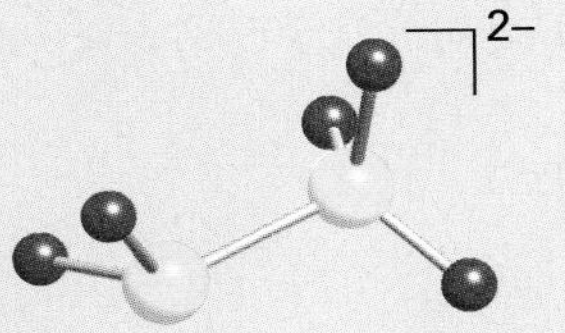	
+5	$S_2O_6^{2-}$	Dithionate	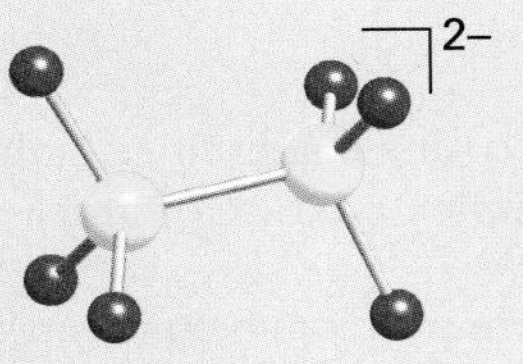	Resists oxidation and reduction
Polysulfur oxanions				
Variable	$S_nO_{2n+2}^{2-}$ $3 \leq n \leq 20$	$n = 3$, trithionate	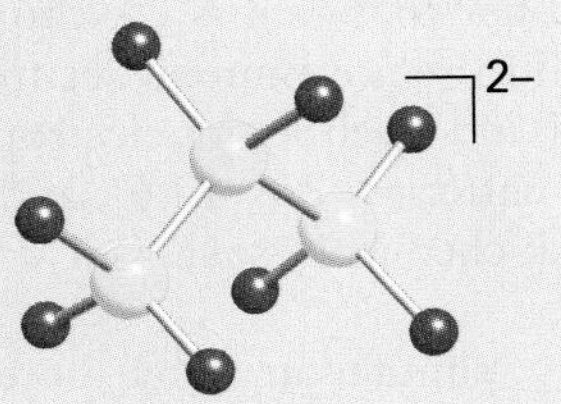	

The oxoanions of selenium and tellurium are a much less diverse and extensive group. Selenic acid is thermodynamically a strong oxidizing acid:

$$SeO_4^{2-}(aq) + 4H^+(aq) + 2e^- \rightarrow H_2SeO_3(aq) + H_2O(l)$$
$$E^\ominus = +1.15\,V$$

However, like SO_4^{2-} and in common with the behaviour of oxoanions of other elements in high oxidation states, the reduction of SeO_4^{2-} is generally slow. Telluric acid exists as $Te(OH)_6$ and also as $(HO)_2TeO_2$ in solution. Again, its reduction is thermodynamically favourable but kinetically sluggish.

(b) Sulfuric acid

Key points: Sulfuric acid is a strong acid; it is a useful nonaqueous solvent due to its extensive autoprotolysis.

Sulfuric acid is a dense viscous liquid. It dissolves in water in a highly exothermic reaction:

$$H_2SO_4(l) \rightarrow H_2SO_4(aq) \quad \Delta_rH^\ominus = -880\,kJ\,mol^{-1}$$

It is a strong Brønsted acid in water ($pK_{a1} = -2$) but not for its second deprotonation ($pK_{a2} = 1.92$). Anhydrous H_2SO_4 has a very high relative permittivity and high electrical conductivity consistent with extensive autoprotolysis:

$$2\,H_2SO_4(l) \rightleftharpoons H_3SO_4^+(sol) + HSO_4^-(sol) \qquad K = 2.7 \times 10^{-4}$$

The equilibrium constant for this autoprotolysis is greater than that of water by a factor of more than 10^{10}. This property leads to the use of sulfuric acid as a nonaqueous, protic solvent (Box 16.5).

Bases (proton acceptors) increase the concentration of HSO_4^- ions in anhydrous sulfuric acid: they include water and salts of weaker acids, such as nitrates:

$$H_2O(l) + H_2SO_4(sol) \rightarrow H_3O^+(sol) + HSO_4^-(sol)$$
$$NO_3^-(s) + H_2SO_4(l) \rightarrow HNO_3(sol) + HSO_4^-(sol)$$

Another example of this kind is the reaction of concentrated sulfuric acid with concentrated nitric acid to produce the nitronium ion, NO_2^+, which is responsible for the nitration of aromatic species:

$$HNO_3(aq) + 2\,H_2SO_4(aq) \rightarrow$$
$$NO_2^+(aq) + H_3O^+(aq) + 2\,HSO_4^-(aq)$$

The number of species that are acidic in sulfuric acid is much smaller than in water because the acid is a poor proton acceptor. For example, HSO_3F is a weak acid in sulfuric acid:

$$HSO_3F(sol) + H_2SO_4(l) \rightleftharpoons H_3SO_4^+(sol) + SO_3F^-(sol)$$

As well as undergoing autoprotolysis, H_2SO_4 dissociates into H_2O and SO_3, which react further with H_2SO_4 to give a number of products:

$$H_2O + H_2SO_4 \rightleftharpoons H_3O^+ + HSO_4^-$$

$$SO_3 + H_2SO_4 \rightleftharpoons H_2S_2O_7$$

$$H_2S_2O_7 + H_2SO_4 \rightleftharpoons H_3SO_4^+ + HS_2O_7^-$$

Consequently, rather than being a single substance, anhydrous sulfuric acid is made up of a complex mixture of at least seven characterized species.

Sulfuric acid is one of the most important chemicals produced on an industrial scale. Over 80 per cent of it is used to manufacture fertilizers. It is also used to remove impurities from petroleum, to 'pickle' (clean) iron and steel before electroplating, as the electrolyte in lead−acid batteries (Box 14.6), and in the manufacture of many other bulk chemicals, such as hydrochloric and nitric acids.

Concentrated sulfuric acid is manufactured by the *contact process*. The first stage is the oxidation of a sulfur compound to SO_2. Most plants use elemental sulfur but metal sulfides and H_2S are also used:

$$S(s) + O_2(g) \rightarrow SO_2(g)$$
$$4\,FeS(s) + 7\,O_2(g) \rightarrow 2\,Fe_2O_3 + 4\,SO_2(g)$$
$$2\,H_2S(g) + 3\,O_2(g) \rightarrow 2\,SO_2(g) + 2\,H_2O(g)$$

The second stage is the oxidation of SO_2 to SO_3. This reaction is carried out at high temperatures and pressures over a V_2O_5 catalyst supported on silica beads:

$$2\,SO_2(g) + O_2(g) \rightarrow 2\,SO_3(g)$$

The SO_3 is then passed into the bottom of a packed column and washed by running oleum, $H_2S_2O_7$, from the top of the column. The gas is then washed in a second column with 98 per cent by mass H_2SO_4. The SO_3 reacts with the 2 per cent of water to produce sulfuric acid, H_2SO_4:

$$SO_3(g) + H_2O(sol) \rightarrow H_2SO_4(l)$$

(c) Sulfurous acid and disulfurous acid

Key points: Sulfurous and disulfurous acids have never been isolated. However, salts of both acids exist; sulfites are moderately strong reducing agents and are used as bleaches; disulfites rapidly decompose in acidic conditions.

Although an aqueous solution of SO_2 is referred to as 'sulfurous acid', H_2SO_3 has never been isolated and the predominant species present are the hydrates $SO_2.nH_2O$. The first and second proton donation are therefore best represented as follows:

$$SO_2.nH_2O(aq) + 2\,H_2O(l) \rightleftharpoons$$
$$H_3O^+(aq) + HSO_3^-(aq) + n\,H_2O(l) \qquad pK_a = 1.79$$

$$HSO_3^-(aq) + H_2O(l) \rightleftharpoons H_3O^+(aq) + SO_3^{2-}(aq) \qquad pK_a = 7.00$$

Anhydrous sodium sulfite, Na_2SO_3, is produced on an industrial scale and used as a bleach in the pulp and paper industry, as a reducing agent in photography, and as an oxygen scavenger in boiler treatments.

Disulfurous acid, $H_2S_2O_5$ (**26**), does not exist in the free state but salts are readily obtained from a concentrated solution of hydrogensulfites:

$$2\,HSO_3^-(aq) \rightleftharpoons S_2O_5^{2-}(aq) + H_2O(l)$$

Acidic solutions of disulfites rapidly decompose to give HSO_3^- and SO_3^{2-}.

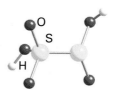

26 Disulfurous acid, $H_2S_2O_5$

(d) Thiosulfuric acid

Key points: Thiosulfuric acid decomposes but the salts are stable; thiosulfate ion is a moderately strong reducing agent.

Aqueous thiosulfuric acid, $H_2S_2O_3$ (**27**), decomposes rapidly in a complex process that produces a number of products, such as S, SO_2, H_2S, and H_2SO_4. The anhydrous acid is more stable and decomposes slowly to H_2S and SO_3. In contrast to the acid, the thiosulfate salts are stable and can be prepared by boiling the sulfites or hydrogensulfites with elemental sulfur or by oxidation of polysulfides:

$$8\,K_2SO_3(aq) + S_8(s) \rightarrow 8\,K_2S_2O_3(aq)$$
$$2\,CaS_2(s) + 3\,O_2(g) \rightarrow 2\,CaS_2O_3(s)$$

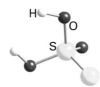

27 Thiosulfuric acid, $H_2S_2O_3$

The thiosulfate ion, $S_2O_3^{2-}$, is a moderately strong reducing agent:

$$\tfrac{1}{2}S_4O_6^{2-}(aq) + e^- \rightarrow S_2O_3^{2-}(aq) \qquad E^\oplus = +0.09\,V$$

The reaction with iodine is the basis of iodometric titrations in analytical chemistry:

$$\tfrac{1}{2}I_2(aq) + e^- \rightarrow I^-(aq) \qquad E^\oplus = +0.54\,V$$
$$2S_2O_3^{2-}(aq) + I_2(aq) \rightarrow S_4O_6^{2-}(aq) + 2I^-(aq)$$

Stronger oxidizing agents, such as chlorine, oxidize thiosulfate to sulfate, which has led to the use of thiosulfate to remove excess chlorine in the bleaching industries.

(e) Peroxosulfuric acids

Key point: Peroxodisulfate salts are strong oxidizing agents.

Peroxomonosulfuric acid, H_2SO_5 (**28**), is a crystalline solid that can be prepared by the reaction of H_2SO_4 with peroxodisulfates or as a byproduct of the synthesis of $H_2S_2O_8$ by electrolysis of H_2SO_4. The salts are unstable and decompose, producing H_2O_2. Peroxodisulfuric acid, $H_2S_2O_8$ (**29**), is also a crystalline solid. Its ammonium and potassium salts are prepared on an industrial scale by the oxidation of ammonia and potassium sulfates. They are strong oxidizing and bleaching agents.

$$\tfrac{1}{2}S_2O_8^{2-}(aq) + H^+(aq) + e^- \rightarrow HSO_4^-(aq) \qquad E^\oplus = +2.12\,V$$

When $K_2S_2O_8$ is heated, ozone and oxygen are evolved.

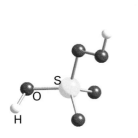

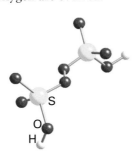

28 Peroxomonosulfuric acid, H_2SO_5

29 Peroxodisulfuric acid, $H_2S_2O_8$

(f) Dithionous and dithionic acids

Key point: Dithionite salts may disproportionate whereas dithionate salts may be oxidized or reduced.

Neither anhydrous dithionous acid, $H_2S_2O_4$ (**30**), nor dithionic acid, $H_2S_2O_6$ (**31**), can be isolated. However, the dithionous and dithionate salts are stable crystalline solids. Dithionites, $S_2O_4^{2-}$, can be prepared by the reduction of sulfites with zinc dust or

sodium amalgam. Sodium dithionite is an important reducing agent in biochemistry. Neutral and acidic solutions of dithionite disproportionate into HSO_3^- and $S_2O_3^{2-}$:

$$2S_2O_4^{2-}(aq) + H_2O(l) \rightarrow 2HSO_3^-(aq) + S_2O_3^{2-}(aq)$$

Dithionates, $S_2O_6^{2-}$, are prepared by oxidation of the corresponding sulfite. Strong oxidizing agents such as MnO_4^- oxidize dithionate to sulfate:

$$SO_4^{2-}(aq) + 2H^+(aq) + e^- \rightarrow \tfrac{1}{2}S_2O_6^{2-}(aq) + H_2O(l)$$
$$E^\oplus = -0.25\,V$$

Strong reducing agents such as sodium amalgam reduce it to SO_3^{2-}:

$$\tfrac{1}{2}S_2O_6^{2-}(aq) + 2H^+(aq) + e^- \rightarrow H_2SO_3(aq) \qquad E^\oplus = +0.57\,V$$

Neutral and acidic solutions of dithionate slowly decompose to SO_2 and SO_4^{2-}:

$$S_2O_6^{2-}(aq) \rightarrow SO_2(g) + SO_4^{2-}(aq)$$

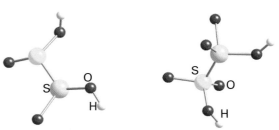

30 Dithionous acid, $H_2S_2O_4$ **31** Dithionic acid, $H_2S_2O_6$

(g) Polythionic acids

Key point: Polythionic acids can be prepared with up to six S atoms.

Many polythionic acids, $H_2S_nO_6$, were first identified by the study of 'Wackenroder's solution', which consists of H_2S in aqueous SO_2. Among those first characterized are the tetrathionate, $S_4O_6^{2-}$ (**5**), and pentathionate, $S_5O_6^{2-}$ (**6**), ions. More recently, a wide variety of preparative routes have been developed, many of which are complicated by numerous redox and catenation reactions. Typical examples are oxidation of thiosulfates with I_2 or H_2O_2, and reaction of polysulfanes, H_2S_n, with SO_3 to yield $H_2S_{n+2}O_6$, where $n = 2-6$:

$$H_2S_n(aq) + 2SO_3(aq) \rightarrow H_2S_{n+2}O_6(aq)$$

16.14 Polyanions of sulfur, selenium, and tellurium

Key points: Sulfur forms polyanions with up to six sulfur atoms; polyselenides form chains and rings, and polytellurides form chains and bicyclic structures.

Many polysulfides of electropositive elements have been characterized. They all contain the S_n^{2-} ions where $n = 2-6$, as in (**7**), (**32**), and (**33**). Typical examples are Na_2S_2, BaS_2, Na_4S_4, K_2S_4, and Cs_2S_6. They can be prepared by heating stoichiometric amounts of S and the element in a sealed tube.

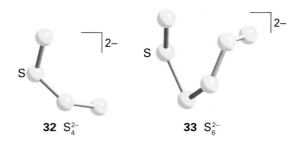

32 S_4^{2-} **33** S_6^{2-}

The polysulfides can also act as ligands. An example is $[Mo_2(S_2)_6]^{2-}$ (**34**), which is formed from ammonium polysulfide and MoO_4^{2-}; it contains side-bonded S_2^{2-} ligands. The larger polysulfides bond to metal atoms, forming chelate rings, as in $[WS(S_4)_2]^{2-}$ (**35**), which contains chelating S_4 ligands. The mineral iron pyrites, also known as 'Fools' gold', has the formula FeS_2 and consists of Fe^{2+} and discrete S_2^{2-} anions in a rock-salt structure.

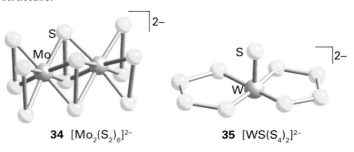

34 $[Mo_2(S_2)_6]^{2-}$ **35** $[WS(S_4)_2]^{2-}$

More polyselenides and polytellurides have been characterized than polysulfides. Structurally, the smaller polyanion solids resemble the polysulfides. The structures of the larger polyanion solids are more complex and depend to some extent on the nature of the cation. The polyselenides up to Se_9^{2-} are chains but larger molecules form rings such as Se_{11}^{2-}, which has a Se atom at the centre of two six-membered rings in a square-planar arrangement (**8**). The polytellurides are structurally more complex and there is a greater occurrence of bicyclic arrangements, such as Te_7^{2-} (**9**) and Te_8^{2-} (**36**). d-Metal complexes of larger polyselenides and polytellurides are known, such as $[Ti(Cp)_2Se_5]$ (**37**). It appears that in polysulfides, polyselenides, and polytellurides electron density is concentrated at the ends of an E_n^{2-} chain, which accounts for coordination through the terminal atoms, as shown in (**35**) and (**37**).

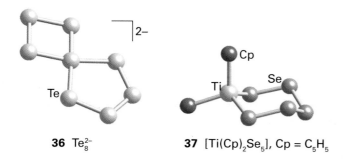

36 Te_8^{2-} **37** $[Ti(Cp)_2Se_5]$, $Cp = C_5H_5$

16.15 Polycations of sulfur, selenium, and tellurium

Key point: Polyatomic cations of S, Se, and Te can be produced by the action of mild oxidizing agents on the elements in strong acid media.

Many cationic chain, ring, and cluster compounds of the p-block elements have been prepared. The majority of them contain S, Se, or Te. Because these cations are oxidizing agents and Lewis acids, the preparative conditions are quite different from those used to synthesize the highly reducing polyanions. For example, S_8 is oxidized by AsF_5 in liquid sulfur dioxide to yield the S_8^{4+} ion:

$$S_8 + 3\,AsF_5 \xrightarrow{SO_2} [S_8][AsF_6]_2 + AsF_3$$

A solvent is used that is more acidic than the polycations, such as fluorosulfuric acid. Sulfur, Se, and Te each form ions of the type E_4^{2+}. For example, Se_4^{2+} is formed by oxidation of elemental Se by the strongly oxidizing peroxide compound FO_2SOOSO_2F:

$$4\,Se + S_2O_6F_2 \xrightarrow{HSO_3F} [Se_4][SO_3F]_2$$

The E_4^{2+} ions have square-planar (D_{4h}) structures (**10**). In the molecular orbital model of the bonding, the cations have a closed-shell configuration in which six electrons fill the a_{2u} and e_g orbitals, leaving the higher energy antibonding b_{2u} orbital vacant. In contrast, most of the larger ring systems can be understood in terms of localized $2c,2e$ bonds. For these larger rings, the removal of two electrons brings about the formation of an additional $2c,2e$ bond, thereby preserving the local electron count on each element. This change is readily seen for the oxidation of S_8 to S_8^{2+} (**38**). An X-ray single-crystal structure determination shows that the transannular bonds in S_8^{2+} are long compared with the other bonds. Long transannular bonds are common in these types of compounds.

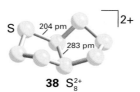

38 S_8^{2+}

16.16 Sulfur–nitrogen compounds

Key points: Neutral heteroatomic ring and cluster compounds of the p-block elements include P_4S_{10} and cyclic S_4N_4. Disulfurdinitride transforms into a polymer that is superconducting at very low temperatures.

Sulfur–nitrogen compounds have structures that can be related to the polycations discussed above. The oldest known, and easiest to prepare, is the pale yellow–orange tetrasulfurtetranitride, S_4N_4 (**11**), which is made by passing ammonia through a solution of SCl_2:

$$6\,SCl_2(l) + 16\,NH_3(g) \rightarrow S_4N_4(s) + \tfrac{1}{4}S_8(s) + 12\,NH_4Cl(sol)$$

Tetrasulfurtetranitride is endergonic ($\Delta_f G^\ominus = +536\ kJ\ mol^{-1}$) and may decompose explosively. The 'cradle-like' molecule is an eight-membered ring with the four N atoms in a plane and bridged by S atoms that project above and below the plane. The short S–S distance (258 pm) suggests that there is a weak interaction between pairs of S atoms. Lewis acids such as BF_3, SbF_5, and SO_3 form 1:1 complexes with one of the N atoms and in the process the S_4N_4 ring rearranges (**39**).

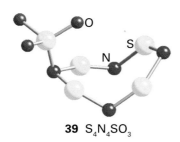

39 $S_4N_4SO_3$

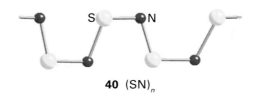

40 $(SN)_n$

Disulfurdinitride, S_2N_2 (**12**), is formed (together with Ag_2S and N_2) when S_4N_4 vapour is passed over hot silver wool. It is even more sensitive than its precursor and explodes above room temperature. When allowed to stand at 0°C for several days, disulfurdinitride transforms into a bronze-coloured zig-zag polymer of composition $(SN)_n$ (**40**), which is much more stable than its precursor, not exploding until 240°C. The compound exhibits metallic conductivity along the chain axis and becomes superconducting below 0.3 K. The discovery of this superconductivity was important because it was the first example of a superconductor that had no metal constituents. Halogenated derivatives have been synthesized that have even higher conductivity. For example, partial bromination of $(SN)_n$ produces blue−black single crystals of $(SNBr_{0.4})_n$, which has room temperature conductivity an order of magnitude greater than $(SN)_n$. Treatment of S_4N_4 with ICl, IBr, and I_2 produces highly conducting, nonstoichiometric polymers with conductivities greater by 16 orders of magnitude than $(SN)_n$.

The compound S_4N_2 can be prepared by heating S_4N_4 with sulfur in CS_2 at 120°C and increased pressure:

$$S_4N_4 + 4S \xrightarrow{CS_2/120°C} 2S_4N_2$$

It forms dark red, needle-like crystals that melt to a dark red liquid at 25°C. It decomposes explosively at 100°C.

FURTHER READING

J.S. Thayer, Relativistic effects and the chemistry of the heaviest main-group elements, *J. Chem. Ed.*, 2005, **82**, 1721. This paper explains the reasons for the different properties of the heaviest elements in each group compared to the lighter elements in terms of relativistic effects.

R.B. King, *Inorganic chemistry of the main group elements*. John Wiley & Sons (1994).

D.M.P. Mingos, *Essential trends in inorganic chemistry*. Oxford University Press (1998). A survey of inorganic chemistry from the perspective of structure and bonding.

R.B. King (ed.), *Encyclopedia of inorganic chemistry*. John Wiley & Sons (2005).

N. Saunders, *Oxygen and the elements of group 16*. Heinemann (2003).

P. Ball, *H₂O: a biography of water*. Phoenix (2004). An entertaining look at the chemistry and physics of water.

R. Steudel, *Elemental sulfur and sulfur-rich compounds*. Springer-Verlag (2003).

N.N. Greenwood and A. Earnshaw, *Chemistry of the elements*. Butterworth-Heinemann (1997).

EXERCISES

16.1 State whether the following oxides are acidic, basic, neutral, or amphoteric: CO_2, P_2O_5, SO_3, MgO, K_2O, Al_2O_3, CO.

16.2 (a) Use standard potentials (*Resource section 3*) to calculate the standard potential of the disproportionation of H_2O_2 in acid solution. (b) Is Cr^{2+} a likely catalyst for the disproportionation of H_2O_2? (c) Given the Latimer diagram

$$O_2 \xrightarrow{-0.13} HO_2^- \xrightarrow{+1.51} H_2O_2$$

in acidic solution, calculate $\Delta_r G^{\ominus}$ for the disproportionation of hydrogen superoxide (HO_2^-) into O_2 and H_2O_2, and compare the result with its value for the disproportionation of H_2O_2.

16.3 Which hydrogen bond would be stronger: $S—H\cdots O$ or $O—H\cdots S$?

16.4 Which of the solvents ethylenediamine (which is basic and reducing) or SO_2 (which is acidic and oxidizing) might not react with (a) Na_2S_4, (b) K_2Te_3?

16.5 Rank the following species from the strongest reducing agent to the strongest oxidizing agent: SO_4^{2-}, SO_3^{2-}, $O_3SO_2SO_3^{2-}$.

16.6 Predict which oxidation states of Mn will be reduced by sulfite ions in basic conditions.

16.7 (a) Give the formula for Te(VI) in acidic aqueous solution and contrast it with the formula for S(VI). (b) Offer a plausible explanation for this difference.

16.8 Use the standard potential data in *Resource section 3* to predict which oxoanions of sulfur will disproportionate in acidic conditions.

16.9 Use the standard potential data in *Resource section 3* to predict whether SeO_3^{2-} is more stable in acidic or basic solution.

16.10 Predict whether any of the following will be reduced by thiosulfate ions, $S_2O_3^{2-}$, in acidic conditions: VO^{2+}, Fe^{3+}, Cu^+, Co^{3+}.

16.11 SF$_4$ reacts with BF$_3$ to form [SF$_3$][BF$_4$]. Use VSEPR theory to predict the shapes of the cation and anion.

16.12 Tetramethylammonium fluoride (0.70 g) reacts with SF$_4$ (0.81 g) to form an ionic product. (a) Write a balanced equation for the reaction and (b) sketch the structure of the anion. (c) How many lines would be observed in the ^{19}F-NMR spectrum of the anion?

16.13 Identify the sulfur-containing compounds A, B, C, D, E, and F.

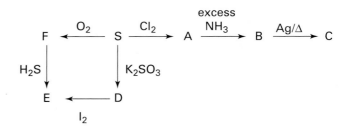

PROBLEMS

16.1 The bond lengths in O$_2$, O$_2^+$, and O$_2^{2-}$ are 121, 112, and 149 pm, respectively. Describe the bonding in these molecules in terms of molecular orbital theory and use this description to rationalize the differences in bond lengths.

16.2 Correct any inaccuracies in the following statements and, after correction, provide examples to illustrate each statement. (a) Elements in the middle of Group 16 are easier to oxidize to the group oxidation number than are the lightest and heaviest members. (b) In its ground state, O$_2$ is a triplet and it undergoes Diels–Alder electrophilic attack on dienes. (c) The diffusion of ozone from the stratosphere into the troposphere poses a major environmental problem.

16.3 Write a comparative account of the properties of sulfuric, selenic, and telluric acids.

16.4 In their paper 'Formation of tellurium nanotubes through concentration depletion at the surfaces of seeds' (*Advanced Materials*, 2002, **14**, 279) Mayers and Xia describe the synthesis of tellurium nanotubes. Describe how their method differs from those used to produce carbon nanotubes. What were the dimensions of tellurium nanotubes produced? What applications do the authors envisage for tellurium nanotubes?

16.5 In November 2006 the former KGB agent Alexander Litvinenko was found to have been poisoned by radioactive polonium-210. Write a review of the chemical and radiological properties of Po and discuss its toxicity.

16.6 A mechanistic study of reaction between chloramine and sulfite has been reported (B.S. Yiin, D.M. Walker, and D.W. Margerum, *Inorg. Chem.*, 1987, **26**, 3435). Summarize the observed rate law and the proposed mechanism. Accepting the proposed mechanism, why should SO$_2$(OH)$^-$ and HSO$_3^-$ display different rates of reaction? Explain why it was not possible to distinguish the reactivity of SO$_2$(OH)$^-$ from that of HSO$_3^-$.

16.7 Tetramethyltellurium, Te(CH$_3$)$_4$, was prepared in 1989 (R.W. Gedrige, D.C. Harris, K.R. Higa, and R.A. Nissan, *Organometallics*, 1989, **8**, 2817), and its synthesis was soon followed by the preparation of the hexamethyl compound (L. Ahmed and J.A. Morrison, *J. Am. Chem. Soc.*, 1990, **112**, 7411). Explain why these compounds are so unusual, give equations for their syntheses, and speculate on why these synthetic procedures are successful. In relation to the last point, speculate on why reaction of TeF$_4$ with methyllithium does not yield tetramethyltellurium.

16.8 The bonding in the square-planar Se$_4^{2+}$ ion is described in Section 16.15. Explore this proposition in more detail by carrying out computations, using software of your choice, on S$_4^{2+}$ with S–S bond distances of 200 pm (sulfur is recommended because its semiempirical parameters are more reliable than those of Se). From the output (a) draw the molecular orbital energy level diagram, (b) assign the symmetry of each level, and (c) sketch the highest energy molecular orbital. Is a closed-shell molecule predicted?

16.9 The nature of the sulfur cycle in ancient times has been investigated (J. Farquhar, H. Bao, and M. Thiemen, *Science*, 2000, **289**, 756). What three factors influence the modern-day cycle? When did the authors establish that a significant change in the cycle had occurred and how were the differences between the ancient and modern cycles explained?

16.10 H. Keppler has investigated the concentration of sulfur in volcano magma (*Science*, 1999, **284**, 1652). In what forms is sulfur erupted from volcanoes? What concentration of sulfur was found in the magma erupted from Mount Pinatubo in 1991? Discuss whether this concentration was expected and how any deviation from the expected value was explained.

The Group 17 elements

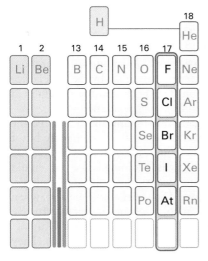

All the Group 17 elements are nonmetals. As with the elements in Groups 15 and 16, we shall see that the oxoanions of the halogens are oxidizing agents that often react by atom transfer. There is a useful correlation between the oxidation number of the central atom and the rates of redox reactions. A wide range of oxidation states is observed for most of the halogens. The reactions of the dihalogens are often fast.

The Group 17 elements, fluorine, chlorine, bromine, iodine, and astatine, are known as the **halogens** from the Greek for 'salt giver'. Fluorine and chlorine are poisonous gases, bromine is a toxic, volatile liquid, and iodine is a sublimable solid. They are among the most reactive nonmetallic elements. The chemical properties of the halogens are extensive and their compounds have been mentioned many times already and were reviewed in Section 9.9c. Therefore, in this chapter we highlight their systematic features, their compounds with oxygen, and discuss the interhalogens.

PART A: **THE ESSENTIALS**

As we discuss the elements of the penultimate group in the p block, we shall see that many of the systematic themes that were helpful for discussing preceding groups again prove useful. For instance, the VSEPR model can be used to predict the shapes of the wide range of molecules that the halogens form among themselves, with oxygen, and with xenon.

17.1 The elements

Key points: Except for fluorine and the highly radioactive astatine, the halogens exist with oxidation numbers ranging from −1 to +7; the small and highly electronegative fluorine atom is effective in oxidizing many elements to high oxidation states.

The atomic properties of the halogens are listed in Table 17.1; they all have the valence electron configuration ns^2np^5. The features to note include their high ionization energies and their high electronegativities and electron affinities. Their electron affinities are high because the incoming electron can occupy an orbital of an incomplete valence shell and experience a strong nuclear attraction: recall that Z_{eff} increases progressively across the period (Section 1.6).

We have seen when discussing the earlier groups in the p block that the element at the head of each group has properties that are distinct from those of its heavier congeners. The anomalies are much less striking for the halogens, and the most notable difference is that F has a lower electron affinity than Cl. Intuitively, this feature seems to be at odds with the high electronegativity of F, but it stems from the larger electron–electron repulsion in the

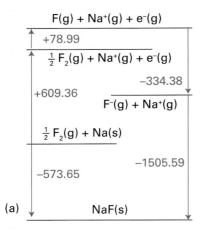

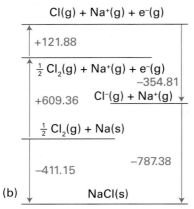

Figure 17.1 Thermochemical cycles for (a) sodium fluoride and (b) sodium chloride. All values are in kilojoules per mole (kJ mol^{-1}).

Table 17.1 Selected properties of the elements

	F	Cl	Br	I	At
Covalent radius/pm	71	99	114	133	140
Ionic radius/pm	131	181	196	220	
First ionization energy/(kJ mol^{-1})	1681	1251	1139	1008	926
Melting point/°C	−220	−101	−7.2	114	302
Boiling point/°C	−188	−34.7	58.8	184	
Pauling electronegativity	4.0	3.2	3.0	2.6	2.2
Electron affinity/(kJ mol^{-1})	328	349	325	295	270
$E^{\ominus}(X_2, X^-)/V$	+3.05	+1.36	+1.09	+0.54	

compact F atom as compared with the larger Cl atom. This electron–electron repulsion is also responsible for the weakness of the F−F bond in F$_2$. Despite this difference in electron affinity, the enthalpies of formation of metal fluorides are generally much greater than those of metal chlorides because the low electron affinity of F is more than offset by the high lattice enthalpies of ionic compounds containing the small F$^-$ ion (Fig. 17.1) and the strengths of bonds in covalent species (for example, the fluorides of metals in high oxidation states).

Fluorine is a pale yellow gas that reacts with most inorganic and organic molecules and the noble gases Kr, Xe, and Rn (Section 18.6). Consequently, it is very difficult to handle but it can be stored in steel or monel metal (a nickel/copper alloy) as these alloys form a passivating metal fluoride surface film. Chlorine is a green–yellow toxic gas. Bromine is the only liquid nonmetallic element at room temperature and pressure and is dark red, toxic, and volatile. Iodine is a purple–grey solid that sublimes to a violet vapour. The violet colour persists when it is dissolved in nonpolar solvents such as CCl$_4$. However, in polar solvents it dissolves to give red–brown solutions, indicating the presence of polyiodide ions such as I$_3^-$ (Section 17.10c).

The halogens are so reactive that they are found naturally only as compounds. They occur mainly as halides, but the most easily oxidized element, I, is also found as sodium or potassium iodate, KIO$_3$, in alkali metal nitrate deposits. Because many chlorides, bromides, and iodides are soluble, these anions occur in the oceans and in brines. The primary source of F is calcium fluoride, which has low solubility in water and is often found in sedimentary deposits (as fluorite, CaF$_2$). Chlorine occurs as sodium chloride in rock salt. Bromine is obtained by the displacement of the element from brine by chlorine. Iodine accumulates in seaweed, from which it can be extracted.

BOX 17.1 Preparation of anhydrous metal halides

Because anhydrous halides are common starting materials for inorganic syntheses, their preparation and the removal of water from impure commercial halides is of considerable practical importance. The principal synthetic methods involve the direct reaction of a metal and a halogen, and the reaction of halogen compounds with metals or metal oxides. An example of direct reaction is shown in Fig. B17.1, in which the gaseous halogen reacts with the metal in a heated tube. When the halide is volatile at the temperature of the reaction, it sublimes to the exit end of the tube.

It is sometimes advantageous to use a halogen compound for the synthesis of a metal halide. For example, the reaction of ZrO$_2$ with CCl$_4$ vapour in a heated tube is a good method for the preparation of ZrCl$_4$. A contribution to the Gibbs energy of reaction comes from the production of the C=O bond in phosgene, COCl$_2$:

$$ZrO_2(s) + 2CCl_4(g) \rightarrow ZrCl_4(s) + 2COCl_2(g)$$

Anhydrous halides can often be prepared by the removal of water from hydrated metal chlorides. Simply heating the sample in a dry gas stream is usually not satisfactory for metals with high charge-to-radius ratio because significant hydrolysis may lead to the oxide or oxohalide:

$$2CrCl_3 \cdot 6H_2O(s) \rightarrow Cr_2O_3(s) + 6HCl(g) + 9H_2O(g)$$

This hydrolysis can be suppressed by carrying out the dehydration in a heated tube in a stream of hydrogen halide, or by dehydration with thionyl chloride (b.p. 79°C), which produces the desired anhydrous chloride and volatile HCl and SO$_2$ byproducts:

$$FeCl_3 \cdot 6H_2O(s) + 6SOCl_2(l) \rightarrow FeCl_3(s) + 6SO_2(g) + 12HCl(g)$$

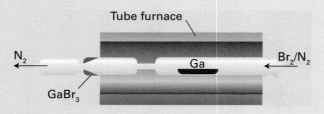

Fig. B17.1 A tube furnace used for the synthesis of anhydrous gallium bromide.

17.2 Simple compounds

Key points: All the halogens form hydrogen halides; HF is a liquid and HCl, HBr, and HI are gases. All the Group 17 elements form oxo compounds and oxoanions.

Because fluorine is the most electronegative of all elements, it is never found in a positive oxidation state (except in the transient gas-phase species F_2^+). With the possible exception of At, the other halogens occur with oxidation numbers ranging from -1 to $+7$. Compounds of Br(VII) are very unstable compared to those of Cl and I, this being yet another example of the alternation effect (Section 9.2c). The dearth of chemical information on At stems from its lack of any stable isotopes and the relatively short half-life (8.3 hours) of the most long-lived of its 33 known isotopes. Astatine solutions are intensely radioactive and can be studied only in high dilution. Astatine appears to exist as the anion At^- and as At(I) and At(III) oxoanions; no evidence for At(VII) has yet been obtained.

The high electronegativity of F leads to enhanced Brønsted acidity in compounds containing the element compared to nonfluorine analogues. A combination of high electronegativity and small atomic radius, which means that many atoms can pack round a central atom, results in F being able to stabilize high oxidation states of most elements, for example UF_6 and IF_7. (Oxygen, however, can often achieve more because it counts for -2 in the calculation of the oxidation number whereas F counts for only -1.) The metal fluorides in low oxidation states tend to be predominantly ionic whereas extensive covalent character is found in metal chlorides, bromides, and iodides.

The synthesis of fluorocarbon compounds, such as polytetrafluoroethene (PTFE), is of great technological importance because these compounds are useful in applications ranging from coatings for nonstick cookware and halogen-resistant laboratory vessels to the volatile fluorocarbons used as refrigerants in air conditioners and refrigerators. Fluorocarbon derivatives have also been the topic of considerable exploratory synthetic research because their derivatives often have unusual properties. Hydrofluorocarbons are used as replacements for chlorofluorocarbons as refrigerants and propellants. They are also used as anaesthetics, where they are less flammable than their nonfluorinated counterparts. Tetrafluorethane, CHF_2CHF_2, is an important solvent for the extraction of natural products such as vanilla and taxol, which is used in chemotherapy.

All the Group 17 elements form protic molecular hydrides, the *hydrogen halides*. Largely on account of its ability to participate in extensive hydrogen bonding, the properties of HF contrast starkly with those of the other hydrogen halides (Table 17.2). Because hydrogen bonding (Section 10.2) is extensive, HF is a volatile liquid whereas HCl, HBr, and HI are gases at room temperature. Hydrogen fluoride has a wide liquid range, a high relative permittivity, and high conductivity. All the hydrogen halides are Brønsted acids: aqueous HF is a weak acid ('hydrofluoric acid') whereas HCl, HBr, and HI are all essentially fully deprotonated in water (Table 17.2).

Although hydrofluoric acid is a weak acid it is one of the most toxic and corrosive substances known, being able to attack glass, metals, concrete, and organic matter. It is much more hazardous to handle than other acids because it is very readily absorbed through the skin, and even brief contact can lead to severe burning and necrosis of the skin and deep tissue, and damage to bone by decalcification by formation of CaF_2 from calcium phosphate.

Table 17.2 Selected properties of the hydrogen halides

	HF	HCl	HBr	HI
Melting point/°C	-84	-114	-89	-51
Boiling point/°C	20	-85	-67	-35
Relative permittivity	83.6 (at 0°C)	9.3 (at -95°C)	7.0 (at -85°C)	3.4 (at -50°C)
Electrical conductivity/ (S cm^{-1})	$c.\ 10^{-6}$ (at 0°C)	$c.\ 10^{-9}$ (at -85°C)	$c.\ 10^{-9}$ (at -85°C)	$c.\ 10^{-10}$ (at -50°C)
$\Delta_f G^{\ominus}$/(kJ mol^{-1})	-273.2	-95.3	-54.4	$+1.72$
Bond dissociation energy/(kJ mol^{-1})	567	431	366	298
pK_a	3.45	$c.\ -7$	$c.\ -9$	$c.\ -11$

Table 17.3 Selected oxides of chlorine

Oxidation number	+1	+3	+4		+6	+7
Formula	Cl_2O	Cl_2O_3	ClO_2	Cl_2O_4	Cl_2O_6	Cl_2O_7
Colour	brown–yellow	dark brown	yellow	pale yellow	dark red	colourless
State	gas	solid	gas	liquid	liquid	liquid

Many binary compounds of the halogens and oxygen are known, but most are unstable and not commonly encountered in the laboratory. We shall mention only a few of the most important.

Oxygen difluoride, OF_2, is the most stable oxide of F but decomposes above $-100°C$. Chlorine occurs with many different oxidation numbers in its oxides (Table 17.3). Some of these oxides are odd-electron species, including ClO_2, in which Cl has the unusual oxidation number +4, and Cl_2O_6, which exists as a mixed oxidation state ionic solid, $[ClO_2^+]$ $[ClO_4^-]$. All the chlorine oxides are endergonic ($\Delta_f G^\oplus > 0$) and unstable. They all explode when heated. There are fewer oxides of Br than Cl. The best characterized compounds are Br_2O, Br_2O_3, and BrO_2. The oxides of I are the most stable halogen oxides. The most important of these is I_2O_5. Both BrO and IO (which are odd-electron species) have been implicated in ozone depletion and both are produced naturally by volcanic activity.

All the Group 17 elements form oxoanions and oxoacids. The wide range of oxoanions and oxoacids of the halogens presents a challenge to those who devise systems of nomenclature. We shall use the common names, such as chlorate for ClO_3^-, rather than the systematic names, such as trioxidochlorate(V). Table 17.4 lists the oxoanions of Cl with common and systematic nomenclature. The strength of the acids can be predicted by using Pauling's rules (Section 4.5b). All the oxoanions are strong oxidizing agents.

■ **A brief illustration.** To use Pauling's rules we write perchloric acid, $HClO_4$, as $O_3Cl(OH)$. The rules then predict that $pK_a = 8 - 5p$, with $p = 3$. Thus the pK_a of perchloric acid is predicted to be -7 (corresponding to a strong acid). ■

17.3 The interhalogens

Key point: All the halogens form compounds with other members of the group.

The **interhalogens** are formed between Group 17 elements. The binary interhalogens are molecular compounds with formulas XY, XY_3, XY_5, and XY_7, where the heavier, less electronegative halogen X is the central atom. They also form ternary interhalogens of the type XY_2Z and XYZ_2, where Z is also a halogen atom. The interhalogens are of special importance as highly reactive intermediates and for providing useful insights into bonding.

Table 17.4 Halogen oxoanions

Oxidation number	Formula	Name*	Point group	Shape	Remarks
+1	ClO^-	Hypochlorite [monoxidochlorate(I)]	$C_{\infty v}$	Linear	Good oxidizing agent
+2	ClO_2^-	Chlorite [dioxidochlorate(III)]	C_{2v}	Angular	Strong oxidizing agent, disproportionates
+5	ClO_3^-	Chlorate [trioxidochlorate(V)]	C_{3v}	Pyramidal	Oxidizing agent
+7	ClO_4^-	Perchlorate [tetraoxidochlorate(VII)]	T_d	Tetrahedral	Oxidizing agent, very weak ligand

* IUPAC names in square brackets.

The diatomic interhalogens, XY, have been made for all combinations of the elements, but many of them do not survive for long. All the F interhalogen compounds are exergonic ($\Delta_f G^\ominus < 0$). The least labile interhalogen is ClF, but ICl and IBr can also be obtained in pure crystalline form. Their physical properties are intermediate between those of their component elements. For example, the deep red α-ICl (m.p. 27°C, b.p. 97°C) is intermediate between yellowish–green Cl_2 (m.p. −101°C, b.p. −35°C) and dark purple I_2 (m.p. 114°C, b.p. 184°C). Photoelectron spectra indicate that the molecular orbital energy levels in the mixed dihalogen molecules lie in the order $3\sigma^2 < 1\pi^4 < 2\pi^4$, which is the same as in the homonuclear dihalogen molecules (Fig. 17.2). An interesting historical note is that ICl was discovered before Br_2 in the early nineteenth century, and, when later the first samples of the dark red–brown Br_2 (m.p. −7°C, b.p. 59°C) were prepared, they were mistaken for ICl.

Most of the higher interhalogens are fluorides (Table 17.5). The only neutral interhalogen with the central atom in a +7 oxidation state is IF_7, but the cation ClF_6^+, a compound of Cl(VII), is known. The absence of a neutral ClF_7 reflects the destabilizing effect of nonbonding electron repulsions between F atoms (indeed, coordination numbers greater than six are not observed for other p-block central atoms in Period 3). The lack of BrF_7 might be rationalized in a similar way, but in addition we shall see later that bromine is reluctant to achieve its maximum oxidation state. This is another manifestation of the alternation effect (Section 9.2c). In this respect, it resembles some other Period 4 p-block elements, notably arsenic and selenium.

The shapes of interhalogen molecules (**1**), (**2**), and (**3**) are largely in accord with the VSEPR model (Section 2.3). For example, the XY_3 compounds (such as ClF_3) have five valence electron pairs around the X atom in a trigonal–bipyramidal arrangement. The Y atoms attach to the two axial pairs and one of the three equatorial pairs, and then the two axial bonding pairs move away from the two equatorial lone pairs. As a result, XY_3 molecules have a C_{2v} bent T shape. There are some discrepancies: for example, ICl_3 is a Cl-bridged dimer.

The Lewis structure of XF_5 has five bonding pairs and one lone pair on the central X atom and, as expected from the VSEPR model, XF_5 molecules are square pyramidal. As already mentioned, the only known XY_7 compound is IF_7, which is predicted to be pentagonal bipyramidal. The experimental evidence for its actual structure is inconclusive. As with other hypervalent molecules, the bonding in IF_7 can be explained without invoking d-orbital participation by adopting a molecular orbital model in which bonding and nonbonding orbitals are occupied but antibonding orbitals are not.

Polymeric interhalogens may also be formed and may be cationic or anionic. Examples of cationic polyhalides are I_3^+ (**4**) and I_5^+ (**5**). Anionic polyhalides are most numerous for iodine. The I_3^- ion is the most stable but others with the general formula $[(I_2)_n I^-]$ are formed. Other anionic polyhalides include Cl_3^- and BrF_4^-.

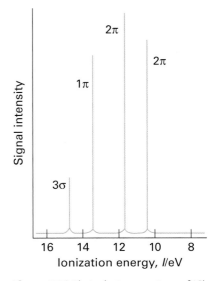

Figure 17.2 Photoelectron spectrum of ICl. The 2π levels give rise to two peaks because of spin-orbit interaction in the positive ion.

Table 17.5 Representative interhalogens

XY	XY$_3$	XY$_5$	XY$_7$
ClF	ClF$_3$	ClF$_5$	
BrF*	BrF$_3$	BrF$_5$	
IF	(IF$_3$)$_n$	IF$_5$	IF$_7$
BrCl			
ICl	I$_2$Cl$_6$		
IBr			

* very unstable

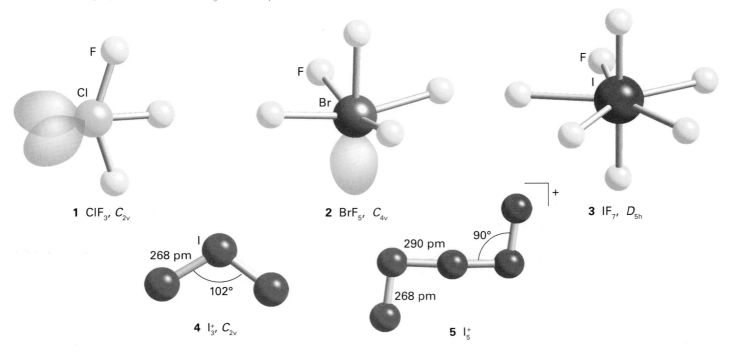

1 ClF$_3$, C_{2v}

2 BrF$_5$, C_{4v}

3 IF$_7$, D_{5h}

4 I$_3^+$, C_{2v}

268 pm

102°

5 I$_5^+$

290 pm

268 pm

90°

PART B: **THE DETAIL**

In this section we look in some detail at the chemistry of the halogens. Most elements form halides with the halogens and these have been dealt with in individual group chapters so our focus here is particularly on the interhalogens and the halogen oxides.

17.4 Occurrence, recovery, and uses

Key points: Fluorine, chlorine, and bromine are prepared by electrochemical oxidation of halide salts; chlorine is used to oxidize Br^- and I^- to the corresponding dihalogen.

All the dihalogens (except the radioactive At_2) are produced commercially on a large scale, with chlorine production by far the greatest, followed by fluorine. The principal method of production of the elements is by electrolysis of the halides (Section 5.18). The strongly positive standard potentials $E^{\ominus}(F_2,F^-) = +2.87$ V and $E^{\ominus}(Cl_2,Cl^-) = +1.36$ V indicate that the oxidation of F^- and Cl^- ions requires a strong oxidizing agent. Only electrolytic oxidation is commercially feasible. An aqueous electrolyte cannot be used for fluorine production because water is oxidized at a much lower potential ($+1.23$ V) and any fluorine produced would react rapidly with water. The isolation of elemental fluorine is achieved by electrolysis of a 1:2 mixture of molten KF and HF in a cell like that shown in Fig. 17.3. It is important to keep fluorine and the byproduct, hydrogen, separate because they react violently.

Most commercial chlorine is produced by the electrolysis of aqueous sodium chloride solution in a *chloralkali cell* (Fig. 17.4). The half-reactions are

Anode half-reaction: $2\,Cl^-(aq) \rightarrow Cl_2(g) + 2\,e^-$
Cathode half-reaction: $2\,H_2O(l) + 2\,e^- \rightarrow 2\,OH^-(aq) + H_2(g)$

The oxidation of water at the anode is suppressed by using an electrode material that has a higher overpotential for O_2 evolution than for Cl_2 evolution (Section 5.18). The best anode material seems to be RuO_2 (Section 19.8). This process is the basis of

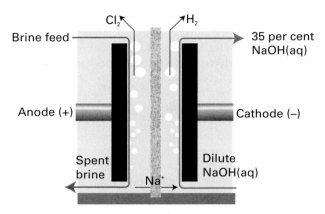

Figure 17.4 Schematic diagram of a chloralkali cell using a cation transport membrane, which has high permeability to Na^+ ions and low permeability to OH^- and Cl^- ions.

the chloralkali industry, which produces sodium hydroxide on a massive scale (see Box 11.2):

$$2\,NaCl(s) + 2\,H_2O(l) \rightarrow 2\,NaOH(aq) + H_2(g) + Cl_2(g)$$

Bromine is obtained by the chemical oxidation of Br^- ions in seawater. A similar process is used to recover iodine from certain natural brines that are rich in I^-. The more strongly oxidizing halogen, chlorine, is used as the oxidizing agent in both processes, and the resulting Br_2 and I_2 are driven from the solution in a stream of air:

$$Cl_2(g) + 2\,X^-(aq) \xrightarrow{\text{air}} 2\,Cl^-(aq) + X_2(g) \qquad (X = Br\ or\ I)$$

EXAMPLE 17.1   Analysing the recovery of Br_2 from brine

Show that from a thermodynamic standpoint bromide ions can be oxidized to Br_2 by Cl_2 and by O_2, and suggest a reason why O_2 is not used for this purpose.

Answer We need to consider the relevant standard potentials and recall that a redox couple can be driven in the direction of oxidation by a couple with a more positive standard potential. The two half-reactions we need to consider for oxidation by chlorine are

$Cl_2(g) + 2\,e^- \rightarrow 2\,Cl^-(aq)$ $E^{\ominus} = +1.358$ V
$Br_2(g) + 2\,e^- \rightarrow 2\,Br^-(aq)$ $E^{\ominus} = +1.087$ V

Because $E^{\ominus}(Cl_2,Cl^-) > E^{\ominus}(Br_2,Br^-)$, chlorine can be used to oxidize Br^- in the reaction

$Cl_2(g) + 2\,Br^-(aq) \rightarrow\ 2\,Cl^-(aq) + Br_2(g)$ $E_{cell}^{\ominus} = +0.271$ V

To encourage the formation of bromine, the Br_2 is removed in a steam–air mixture. Oxygen would be thermodynamically capable of carrying out this reaction in acidic solution:

$O_2(g) + 4\,H^+(aq) + 4\,e^- \rightarrow 2\,H_2O(l)$ $E^{\ominus} = +1.229$ V
$Br_2(l) + 2\,e^- \rightarrow 2\,Br^-$ $E^{\ominus} = +1.087$ V

Resulting in

$O_2(g) + 4\,Br^-(aq) + 4\,H^+(aq) \rightarrow 2\,H_2O(l) + 2\,Br_2(l)$ $E_{cell}^{\ominus} = +0.142$ V

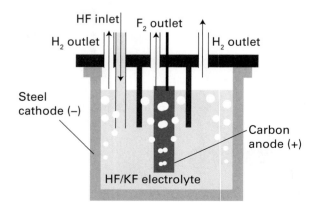

Figure 17.3 Schematic diagram of an electrolysis cell for the production of fluorine from KF dissolved in liquid HF.

but the reaction is not favourable at pH = 7, when $E_{cell}^{\ominus}$ = −0.15 V. Even though the reaction is thermodynamically favourable in acidic solution, it is doubtful that the rate would be adequate because an overpotential of about 0.6 V is associated with the reactions of O_2 (Section 5.18). Even if the oxidation by O_2 in acidic solution were kinetically favourable, the process would be unattractive because of the cost of acidifying large quantities of brine and then neutralizing the effluent.

Self-test 17.1 One source of iodine is sodium iodate, $NaIO_3$. Which of the reducing agents $SO_2(aq)$ or $Sn^{2+}(aq)$ would seem practical from the standpoints of thermodynamic feasibility and plausible judgements about cost? Standard potentials are given in *Resource section* 3.

All the Group 17 elements undergo thermal or photochemical dissociation in the gas phase to form radicals. These radicals take part in chain reactions, such as:

$$X_2 \xrightarrow{\Delta/h\nu} X + X\cdot$$
$$H_2 + X \rightarrow HX + H\cdot$$
$$H + X_2 \rightarrow HX + X\cdot$$

A reaction of this type between chlorine and methane is used in the industrial synthesis of chloroform, CH_3Cl, and dichloromethane, CH_2Cl_2.

Compounds of F are used throughout industry. Fluorine as F^- ions is added to some domestic water supplies and toothpaste to prevent tooth decay (Box 17.2). It is used as UF_6 in the nuclear power industry for the separation of the isotopes of uranium. Hydrogen fluoride is used to etch glass and as a nonaqueous solvent. Chlorine is widely used in industry to make chlorinated hydrocarbons and in applications in which a strong oxidizing agent is needed, including disinfectants and bleaches. These applications are in decline, however, because some organic Cl compounds are carcinogenic and chlorofluorocarbons (CFCs) are implicated in the destruction of ozone in the stratosphere (Box 17.3). Hydrofluorocarbons (HFCs) are now replacing CFCs in applications such as refrigeration and air conditioning. Organobromine compounds are used in synthetic organic chemistry: the C−Br bond is not as strong as the C−Cl bond and Br can be more readily displaced (and recycled). Iodine is an essential element and iodine deficiency is a cause of goitre, the enlargement of the thyroid gland. For this reason, small amounts of potassium iodide are added to table salt.

17.5 Molecular structure and properties

Key points: The F−F bond is weak relative to the Cl−Cl bond; bond strengths decrease down the group from chlorine.

Among the most striking physical properties of the halogens are their colours. In the vapour they range from the almost colourless F_2, through yellow–green Cl_2 and red–brown Br_2, to purple I_2. The progression of the maximum absorption to longer wavelengths reflects the decrease in the HOMO–LUMO gap on descending the group. In each case, the optical absorption spectrum arises primarily from transitions in which an electron is promoted from the highest filled $2\sigma_g$ and $1\pi_g$ orbitals into the vacant antibonding $2\sigma_u$ orbital (Fig. 17.5).

Except for F_2, the analysis of the UV absorption spectra gives precise values for the dihalogen bond dissociation energies (Fig. 17.6). It is found that bond strengths decrease down the group from Cl_2. The UV spectrum of F_2, however, is a broad

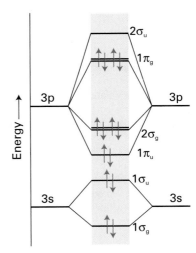

Figure 17.5 Schematic molecular orbital energy level diagram for Cl_2 (similarly Br_2 and I_2). For F_2 the order of the π_u and upper σ_g orbitals is reversed.

BOX 17.3 Chlorofluorocarbons and the ozone hole

The ozone layer extends from 10 km to 50 km above the Earth's surface and plays a crucial role in protecting us from the harmful effects of the Sun's ultraviolet rays by absorbing radiation at wavelengths below 300 nm, thereby attenuating the spectrum of sunlight at ground level.

Ozone is produced naturally by the action of UV radiation on O_2 in the upper atmosphere:

$$O_2 \xrightarrow{h\nu} O + O$$
$$O + O_2 \rightarrow O_3$$

When ozone absorbs an ultraviolet photon, it dissociates:

$$O_3 \xrightarrow{h\nu} O_2 + O$$

The resulting O atom can then remove ozone in the reaction

$$O_3 + O \rightarrow O_2 + O_2$$

These reactions constitute the principal steps of the *oxygen–ozone cycle* that maintains an equilibrium (but seasonally varying) concentration of ozone. If all the atmospheric O_3 were condensed into a single layer at 1 atm and 25°C, it would cover the Earth to a depth of about 3 mm.

The stratosphere also contains naturally occurring species, such as the hydroxyl radical and nitric oxide, that catalyse the destruction of ozone by reactions such as

$$X + O_3 \rightarrow XO + O_2$$
$$XO + O \rightarrow X + O_2$$

However, the main concern about the loss of ozone centres on Cl and Br atoms introduced artificially by industrial activities, which catalyse O_3 destruction very efficiently. Chlorine and bromine are carried into the stratosphere as part of organohalogen molecules, RHal, which release the halogen atoms when the C—Hal bond is fragmented by far-UV photons. The ozone-destroying potential of these molecules was pointed out in 1974 by Mario Molina and Sherwood Rowland, who won the 1995 Nobel Prize in chemistry (together with Paul Crutzen) for their work.

International action followed 13 years later (in the form of the 1987 Montreal Protocol), and was given added impetus by the discovery of the 'ozone hole' over Antarctica, which provided dramatic evidence of the vulnerability of atmospheric ozone. This hole surprised even the scientists working on the problem; its explanation required additional chemistry, involving the polar stratospheric clouds that form in winter. The ice crystals in these clouds adsorb molecules of chlorine or bromine nitrate, $ClONO_2$ or $BrONO_2$, which form when stratospheric ClO or BrO combine with NO_2. Once on the ice surface, these molecules react with water:

$$H_2O + XONO_2 \rightarrow HOX + HNO_3$$

where X = Cl or Br. They also react with co-adsorbed HCl or HBr (formed by attack of Cl^- and Br^- on the methane escaping from the troposphere):

$$HX + XONO_2 \rightarrow X_2 + HNO_3$$

The nitric acid, being very hygroscopic, enters the ice crystals and the HOX or X_2 molecules are released during the dark polar winter. When the sunlight strengthens in the spring, these molecules photolyse, releasing high concentrations of ozone-destroying radicals.

$$HOX \xrightarrow{h\nu} HO\cdot + \ X\cdot$$
$$X_2 \xrightarrow{h\nu} 2X\cdot$$

To threaten the ozone layer, organohalogen molecules must survive their migration from the Earth's surface. Those containing H atoms are mostly broken down in the troposphere (the lowest region of the atmosphere) by reaction with HO radicals. Even so, they may be a problem if released in sufficient amounts. There is currently a major controversy over the use of bromomethane, CH_3Br, as an agricultural fumigant. However, the greatest potential for ozone destruction rests with molecules that lack H atoms, the chlorofluorocarbons (CFCs), which have been used in many industrial applications, and their brominated analogues (the halons), which are used to extinguish fires. These compounds have no tropospheric sink and eventually reach the stratosphere unaltered. They are the main focus of the international regulatory regime worked out in 1987 (and amended in 1990 and 1992). Most CFCs and halons have been phased out of production, and their atmospheric concentrations are beginning to decline. The CFCs present an additional problem as they are also potent greenhouse gases.

continuum that lacks structure because absorption is accompanied by dissociation of the F_2 molecule. The lack of discrete absorption bands makes it difficult to estimate the dissociation energy spectroscopically, and thermochemical methods are complicated by the highly corrosive nature of this reactive halogen. When these problems were solved, the F—F bond enthalpy was found to be less than that of Br_2 and thus out of line with the trend in the group. However, the low F—F bond enthalpy is consistent with the low single-bond enthalpies of N—N, O—O, and various combinations of N, F, and O (Fig. 17.7). The simplest explanation (like the explanation of the low electron affinity of fluorine) is that the bond is weakened by the strong repulsions between nonbonding electrons in the small F_2 molecule. In molecular orbital terms, the molecule has numerous electrons in strongly antibonding orbitals.

Chlorine, bromine, and iodine all crystallize in lattices of the same symmetry (Fig. 17.8), so it is possible to make a detailed comparison of distances between bonded and nonbonded adjacent atoms (Table 17.6). The important conclusion is that nonbonded distances do not increase as rapidly as the bond lengths.

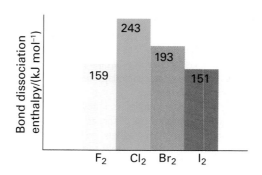

Figure 17.6 Bond dissociation enthalpies of the halogens in kilojoules per mole ($kJ\ mol^{-1}$).

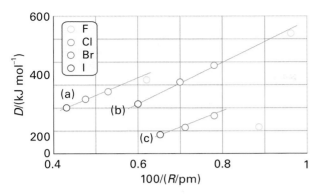

Figure 17.7 Dissociation enthalpies of (a) carbon–halogen, (b) hydrogen–halogen, and (c) halogen–halogen bonds plotted against the reciprocal of the bond length.

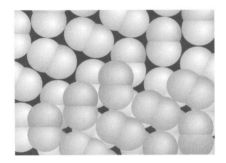

Figure 17.8 Solid chlorine, bromine, and iodine have similar structures. The closest nonbonded interactions are relatively less compressed in Cl_2 and Br_2 than in I_2.

Table 17.6 Bonding and shortest nonbonding distances for solid dihalogens

Element	Temperature/°C	Bond length/pm	Nonbonding distance/pm	Ratio
Cl_2	−160	198	332	1.68
Br_2	−106	227	332	1.46
I_2	−163	272	350	129

This observation suggests the presence of weak intermolecular bonding interactions that strengthen on going from Cl_2 to I_2. Solid iodine is a semiconductor and under high pressure exhibits metallic conductivity.

17.6 Reactivity trends

Key points: Fluorine is the most oxidizing halogen; the oxidizing power of the halogens decreases down the group.

Fluorine, F_2, is the most reactive nonmetal and is the strongest oxidizing agent among the halogens. The rapidity of many of its reactions with other elements may in part be due to a low kinetic barrier associated with the weak F–F bond. Despite the thermodynamic stability of most metal fluorides, fluorine can be handled in containers made from some metals, such as Ni, because a substantial number of them form a passive metal fluoride surface film on contact with fluorine gas. Fluorocarbon polymers, such as polytetrafluoroethene (PTFE), are also useful

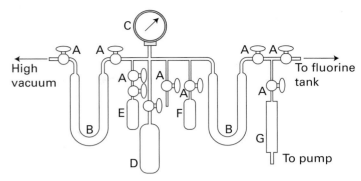

Figure 17.9 A typical metal vacuum system for handling fluorine and reactive fluorides. Nickel tubing is used throughout. (A) Monel valves, (B) nickel U-traps, (C) monel pressure gauge, (D) nickel container, (E) PTFE reaction tube, (F) nickel reaction vessel, and (G) nickel canister filled with soda lime to neutralize HF and react with F_2 and fluorine compounds.

materials for the construction of apparatus to contain fluorine and oxidizing fluorine compounds (Fig. 17.9). Few laboratories have the equipment and expertise for research involving elemental F_2.

The standard potentials for the halogens (Table 17.1) indicate that F_2 is a much stronger oxidizing agent than Cl_2. The decrease in oxidizing strength continues in more modest steps from Cl_2 through Br_2 to I_2. Although the half-reaction

$$\tfrac{1}{2}X_2(g) + e^- \rightarrow X^-(aq)$$

is favoured by a high electron affinity (which suggests that F should have a lower standard potential than Cl), the process is favoured by the low bond enthalpy of F_2 and by the highly exothermic hydration of the small F^- ion (Fig. 17.10). The net outcome of these three competing effects of size is that F is the most strongly oxidizing element of the group.

17.7 Pseudohalogens

Key points: Pseudohalogens and pseudohalides mimic halogens and halides, respectively; the pseudohalogens exist as dimers and form molecular compounds with nonmetals and ionic compounds with alkali metals.

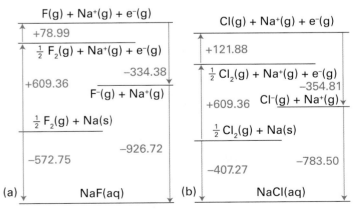

Figure 17.10 Thermochemical cycles for the enthalpy of formation of (a) aqueous sodium fluoride and (b) aqueous sodium chloride. The hydration is much more exothermic for F^- than for Cl^-. All values are in kilojoules per mole (kJ mol^{-1}).

Table 17.7 Pseudohalides, pseudohalogens, and corresponding acids

Pseudohalide	Pseudohalogen	$E^{\ominus}/V$	Acid	pK_a
CN⁻	NCCN	+0.27	HCN	9.2
Cyanide	Cyanogen		Hydrogen cyanide	
NCS⁻	NCSSCN	+0.77	HNCS	−1.9
Thiocyanate	Dithiocyanogen		Hydrogen thiocyanate	
NCO⁻			HNCO	3.5
Cyanate			Isocyanic acid	
CNO⁻			HCNO	3.66
Fulminate			Fulminic acid	
NNN⁻			HNNN	4.92
Azide			Hydrazoic acid	

A number of compounds have properties so similar to those of the halogens that they are called **pseudohalogens** (Table 17.7). For example, like the dihalogens, cyanogen, $(CN)_2$, undergoes thermal and photochemical dissociation in the gas phase; the resulting CN radicals are isolobal with halogen atoms and undergo similar reactions, such as a chain reaction with hydrogen:

$$NC{-}CN \xrightarrow{\text{heat or light}} 2\,CN\cdot$$
$$H_2 + CN\cdot \rightarrow HCN + H\cdot$$
$$H\cdot + NC{-}CN \rightarrow HCN + CN\cdot$$

Overall: $H_2 + C_2N_2 \rightarrow 2\,HCN$

Another similarity is the reduction of a pseudohalogen:

$$\tfrac{1}{2}(CN)_2(aq)(aq) + e^- \rightarrow CN^-(aq)$$

The anion formally derived from a pseudohalogen is called a *pseudohalide ion*. An example is the cyanide anion, CN⁻. Covalent pseudohalides similar to the covalent halides of the p-block elements are also common. They are often structurally similar to the corresponding covalent halides (compare (6) and (7)), and undergo similar metathesis reactions.

As with all analogies, the concepts of pseudohalogen and pseudohalide have many limitations. For example, pseudohalogen ions are not spherical, so the structures of their ionic compounds often differ: NaCl is fcc but NaCN is similar to CaC₂ (Section 11.12). The pseudohalogens are generally less

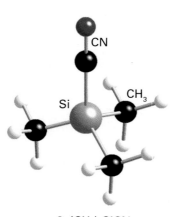

6 (CH₃)₃SiCN

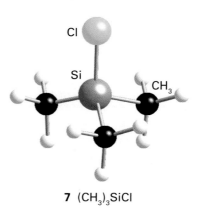

7 (CH₃)₃SiCl

electronegative than the lighter halogens and some pseudohalides have more versatile donor properties. The thiocyanate ion, SCN⁻, for instance, acts as an ambidentate ligand with a soft base site, S, and a hard base site, N (Section 4.15).

17.8 Special properties of fluorine compounds

Key points: Fluorine substituents promote volatility, increase the strengths of Lewis and Brønsted acids, and stabilize high oxidation states.

The boiling points in Table 17.8 demonstrate that molecular compounds of F tend to be highly volatile, in some cases even more volatile than the corresponding hydrogen compounds (compare, for example, PF₃, b.p. −101.5°C, and PH₃, b.p. −87.7°C) and in all cases much more volatile than the Cl analogues. The volatilities of the compounds are a result of variations in the strength of the dispersion interaction (the interaction between instantaneous transient electric dipole moments), which is strongest for highly polarizable molecules. The electrons in the small F atoms are gripped tightly by the nuclei, and consequently F compounds have low polarizabilities and hence weak dispersion interactions.

There are some opposite effects on volatility that can be traced to hydrogen bonding. The structure of solid HF is a planar zigzag chain polymer of F−H···F units. Although liquid HF has a lower density and viscosity than water, which suggests the absence of an extensive three-dimensional network of H bonds, in the gas phase HF forms H-bonded oligomers, $(HF)_n$ with n up to 5 or 6. As with H_2O and NH_3, the properties of HF, such as a wide liquid range, make it an excellent nonaqueous solvent.

Hydrogen fluoride undergoes autoprotolysis:

$$2\,HF(l) \rightleftharpoons H_2F^+(sol) + F^-(sol) \qquad pK_{auto} = 12.3$$

It is a much weaker acid (pK_a = 3.45 in water) than the other hydrogen halides. Although this difference is sometimes attributed to the formation of an ion pair ($H_3O^+F^-$), theoretical considerations show that its poor proton donor properties are a

Table 17.8 Normal boiling points (in °C) of compounds of fluorine and their analogues

F₂	−188.2	H₂	−252.8	Cl₂	−34.0
CF₄	−127.9	CH₄	−161.5	CCl₄	76.7
PF₃	−101.5	PH₃	−87.7	PCl₃	75.5

direct result of the very strong H−F bond. Carboxylic acids act as bases in anhydrous HF and are protonated:

$$HCOOH(l) + 2\,HF(l) \rightarrow \quad HC(OH)_2^+(sol) + HF_2^-(sol)$$

An important characteristic is the ability of an F atom in a compound to withdraw electrons from the other atoms present and, if the compound is a Brønsted acid, to enhance its acidity. An example of this effect is the increase by three orders of magnitude in the acidity of trifluoromethanesulfonic acid, $HOSO_2CF_3$ ($pK_a = 3.0$ in nitromethane), over that of methanesulfonic acid, $HOSO_2CH_3$ ($pK_a = 6.0$ in nitromethane). The presence of F atoms in a molecule also results—for the same reason—in an enhanced Lewis acidity. For example, we saw in Sections 4.15 and 15.11b that SbF_5 is one of the strongest Lewis acids of its type and much stronger than $SbCl_5$.

Some examples of high oxidation state compounds of F are IF_7, PtF_6, BiF_5, $KAgF_4$, UF_6, and ReF_7. Rhenium(VII) heptafluoride is the only example of a thermally stable metal heptafluoride and uranium(VI) hexafluoride is important in the separation of U isotopes in the preprocessing of nuclear fuels. All these compounds are examples of the highest oxidation state attainable for these elements, the rare oxidation state Ag(III) being perhaps the most notable. Another example is the stability of PbF_4, compared to all other Pb(IV) halides.

A related phenomenon is the tendency of fluorine to disfavour low oxidation states. Thus, solid copper(I) fluoride, CuF, is unstable but CuCl, CuBr, and CuI are stable with respect to disproportionation. Similar trends were discussed in Section 3.11 in terms of a simple ionic model in which the small size of the F^- ion in combination with a small, highly charged cation results in a high lattice enthalpy. As a result, there is a thermodynamic tendency for CuF to disproportionate and form copper metal and CuF_2 (because Cu^{2+} is doubly charged and its ionic radius is smaller than that of Cu^+ and it has a greater lattice enthalpy).

Compounds that accept F^- ions are Lewis acids and compounds that donate F^- are Lewis bases:

$$SbF_5(s) + HF(l) \rightarrow SbF_6^-(sol) + H^+(sol)$$
$$XeF_6(s) + HF(l) \rightarrow XeF_5^+(sol) + HF_2^-(sol)$$

Ionic fluorides dissolve in HF to give highly conducting solutions. The fact that chlorides, bromides, and iodides react with HF to give the corresponding fluoride and HX provides a preparative route to anhydrous fluorides:

$$TiCl_4(l) + 4\,HF(l) \rightarrow TiF_4(s) + 4\,HCl(g)$$

17.9 Structural features

Metal difluorides, MF_2, where M is a Group 2 or d-metal, generally adopt the CaF_2 or rutile structures and are described well by the ionic model. In contrast, whereas the Group 2 dichlorides, dibromides, and diiodides may be described by the ionic model, the d-metal analogues adopt the CdI_2 or $CdCl_2$ layer structures and their bonding is not described well by either the ionic or covalent models. Many metal trifluorides have three-dimensional ionic structures but the trichlorides, tribromides, and triiodides have layered structures. The compounds NbF_3 and FeF_3 (at high temperature) adopt the ReO_3 structure type (Section 3.6) and many other metal trifluorides (including AlF_3, ScF_3, and CoF_3) have a slightly distorted variant of this structure type.

As the oxidation number of the metal atom increases, the halides become more covalent. Thus all metal hexahalides such as MoF_6 and WCl_6, are molecular covalent compounds. For intermediate oxidation states (such as MF_4 and MF_5) the structures normally consist of linked MF_6 polyhedra. Titanium tetrafluoride has a structure based on columns of triangular Ti_3F_{15} units formed from three TiF_6 octahedra (**8**) whereas NbF_5 is built from four NbF_6 octahedra forming a square unit of composition Nb_4F_{20} (**9**).

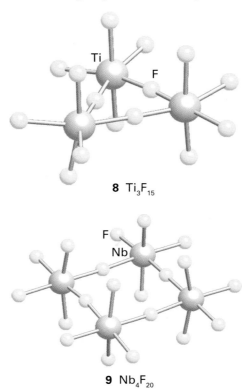

8 Ti_3F_{15}

9 Nb_4F_{20}

Although not as important in applications as complex oxides, complex solid fluorides and chlorides such as the ternary phases $MM'F_n$ and $MM'Cl_n$, and the quaternary compounds $MM'M''F_n$, have structures similar to their oxide counterparts. As F^- has an oxidation number of -1 compared to -2 for O^{2-}, the compositionally equivalent fluorides or chlorides generally contain d metals in lower oxidation states than the equivalent oxide. Thus, ternary fluorides of stoichiometry ABF_3, with, for example, A = K, Rb, and Cs and M a dipositive d-metal ion, adopt the perovskite structure (Section 3.9). One example is $KMnF_3$, which precipitates when potassium fluoride is added to Mn(II) solutions. Molten cryolite, Na_3AlF_6, is used to dissolve aluminium oxide in electrochemical extraction of aluminium. Its structure is related to that of perovskite (ABO_3) with Na in the A sites and a mixture of Na and Al in the B sites: the formula $Na(Al_{1/2}Na_{1/2})F_3$, which is equivalent to Na_3AlF_6. Mixed-anion compounds containing halides are also well characterized and include the superconducting cuprate $Sr_{2-x}Na_xCuO_2F_2$.

17.10 The interhalogens

The halogens form many compounds between themselves with formulas ranging from XY to XY_7 (Table 17.9). Their structures can usually be predicted accurately by the VSEPR rules and verified by techniques such as ^{19}F-NMR.

Table 17.9 Properties of interhalogens

XY	XY$_3$	XY$_5$	XY$_7$
ClF	ClF$_3$	ClF$_5$	
Colourless	Colourless	Colourless	
m.p. −156°C	m.p. −76°C	m.p. −103°C	
b.p. −100°C	b.p. 12°C	b.p. −13°C	
BrF*	BrF$_3$	BrF$_5$	
Light brown	Yellow	Colourless	
m.p. ≈240°C	m.p. 9°C	m.p. −61°C	
b.p. −20°C	b.p. 126°C	b.p. 41°C	
IF*	(IF$_3$)$_n$	IF$_5$	IF$_7$
	Yellow	Colourless	Colourless
	Dec. −28°C	m.p. 9°C	m.p. 6.5
		b.p. 105°C	(triple point)
			Subl. 5°C
BrCl*			
Red-brown			
m.p. ≈−66°C			
b.p. 5°C			
α-ICl, β-ICl	I$_2$Cl$_6$		
Ruby red solid, black liquid	Bright yellow		
m.p. 27°C, m.p. 14°C	m.p. 101°C (16 atm)		
b.p. 97−100°C			
IBr			
Black solid			
m.p. 41°C			
b.p.≈116°C			

* Very unstable. dec., decomposes; subl., sublimes.

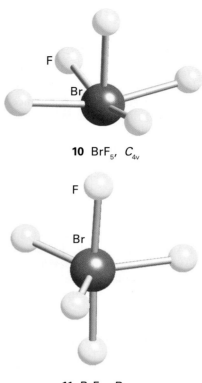

10 BrF$_5$, C_{4v}

11 BrF$_5$, D_{3h}

(a) Chemical properties

Key point: Fluorine–containing interhalogens are typically Lewis acids and strong oxidizing agents.

All the interhalogens are oxidizing agents. In general, the rates of oxidation of interhalogens do not bear a simple relation to their thermodynamic stabilities. As with all the known interhalogen fluorides, ClF$_3$ is an exergonic compound, so thermodynamically it is a weaker fluorinating agent than F$_2$ itself. However, the rate at which it fluorinates substances generally exceeds that of fluorine, so it is in fact an aggressive fluorinating agent towards many elements and compounds. The fluorides ClF$_3$ and BrF$_3$ are much more aggressive fluorinating agents than BrF$_5$, IF$_5$, and IF$_7$; iodine pentafluoride, for instance, is a convenient mild fluorinating agent that can be handled in glass apparatus. One use of ClF$_3$ as a fluorinating agent is in the formation of a passivating metal fluoride film on the inside of the nickel apparatus used in fluorine chemistry.

Both ClF$_3$ and BrF$_3$ react vigorously (often explosively) with organic matter, burn asbestos, and expel oxygen from many metal oxides:

$$2\,Co_3O_4(s) + 6\,ClF_3(g) \rightarrow 6\,CoF_3(s) + 3\,Cl_2(g) + 4\,O_2(g)$$

Bromine trifluoride autoionizes in the liquid state:

$$2\,BrF_3(l) \rightleftharpoons BrF_2^+(sol) + BrF_4^-(sol)$$

This Lewis acid-base behaviour is shown by its ability to dissolve a number of halide salts:

$$CsF(s) + BrF_3(l) \rightarrow Cs^+(sol) + BrF_4^-(sol)$$

Bromine trifluoride is a useful solvent for ionic reactions that must be carried out under highly oxidizing conditions. The Lewis acid character of BrF$_3$ is shared by other interhalogens, which react with alkali metal fluorides to produce anionic fluoride complexes.

EXAMPLE 17.2 Predicting the shape of an interhalogen molecule.

BrF$_5$ is a fluxional molecule that rapidly interconverts between a square pyramidal (**10**) and a trigonal-bipyramidal structure (**11**). If these two structures could be isolated explain how ^{19}F-NMR could be used to differentiate between them.

Answer We need to identify the number of different ^{19}F environments in each case and consider how each is coupled to other F environments. In the case of the square pyramid there are two different F environments of four and one F atoms, respectively. The ^{19}F-NMR would show a signal equivalent to four F atoms split into a doublet by coupling to the other one F atom. A second signal equivalent to one F atom would be split into a quintuplet by coupling. The trigonal-bipyramidal structure has two F environments equivalent to three and two F atoms, respectively. Their resonances would be split into a triplet and quartet by spin−spin coupling.

Self-test 17.2 Predict the ^{19}F-NMR pattern for IF$_7$.

(b) Cationic interhalogens

Key point: Cationic interhalogen compounds have structures in accord with the VSEPR model.

Under special strongly oxidizing conditions, such as in fuming sulfuric acid, I_2 is oxidized to the blue paramagnetic diiodinium cation, I_2^+. The dibrominium cation, Br_2^+, is also known. The bonds of these cations are shorter than those of the corresponding neutral dihalogens, which is the expected result for loss of an electron from a π^* orbital and the accompanying increase in bond order from 1 to 1.5 (see Fig. 17.5). Three higher polyhalogen cations, Br_3^+, I_3^+, and I_5^+, are known, and X-ray diffraction studies of the iodine species have established the structures shown in (4) and (5). The angular shape of I_3^+ is in line with the VSEPR model because the central I atom has two lone pairs of electrons.

Another class of polyhalogen cations of formula XF_n^+ is obtained when a strong Lewis acid, such as SbF_5, abstracts F^- from interhalogen fluorides:

$$ClF_3 + SbF_5 \rightarrow [ClF_2^+][SbF_6^-]$$

This formulation is idealized because X-ray diffraction of solid compounds that contain these cations indicates that the F^- abstraction from the cations is incomplete and that the anions remain weakly associated with the cations by fluorine bridges (12). Table 17.10 lists a variety of interhalogen cations that are prepared in a similar manner.

12 $(ClF_2)(SbF_6)_2$

(c) Polyhalides

Key points: Polyiodides, such as I_3^-, are formed by adding I_2 to I^-; they are stabilized by large cations. Some of the most stable polyhalogen anions contain fluorine as the substituent; their structures usually conform to the VSEPR model.

A deep brown colour develops when I_2 is added to a solution of I^- ions. This colour is characteristic of the polyiodides, which include triiodide ions, I_3^-, and pentaiodide ions, I_5^-. These polyiodides are Lewis acid-base complexes in which I^- and I_3^- act as the bases and I_2 acts as the acid (Fig. 17.11). The Lewis structure of I_3^- has three equatorial lone pairs on the central I atom and two axial bonding pairs in a trigonal-bipyramidal arrangement. This hypervalent Lewis structure is consistent with the observed linear structure of I_3^-, which is described in more detail below.

An I_3^- ion can interact with other I_2 molecules to yield larger mononegative polyiodides of composition $[(I_2)_n I^-]$. The I_3^- ion

Table 17.10 Representative interhalogen cations

Compound	Shape
ClF_2^+, BrF_2^+, ICl_2^+	Bent
ClF_4^+, BrF_4^+, IF_4^+	See-saw
ClF_6^+, BrF_6^+, IF_6^+	Octahedral

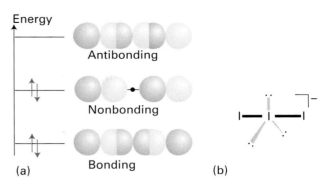

Figure 17.11 Some representations of the I_3^- polyiodide ion. (a) The σ interaction. (b) Lewis and VSEPR rationalization of the linear structure, where the five electron pairs are arranged around the central atom in a trigonal-pyramidal array.

is the most stable member of this series. In combination with a large cation, such as $[N(CH_3)_4]^+$, it is symmetrical and linear with a longer I–I bond than in I_2. However, the structure of the triiodide ion, like that of the polyiodides in general, is highly sensitive to the identity of the counterion. For example, Cs^+, which is smaller than the tetramethylammonium ion, distorts the I_3^- ion and produces one long and one short I–I bond (13). The ease with which the ion responds to its environment is a reflection of the weakness of bonds that just manage to hold the atoms together. An example of sensitivity to the cation is provided by NaI_3, which can be formed in aqueous solution but decomposes when the water is evaporated:

$$Na^+(aq) + I_3^-(aq) \xrightarrow{\text{remove water}} NaI(s) + I_2(s)$$

13 I_3^-

A more extreme example is $NI_3 \cdot NH_3$, which is a black powder formed when iodine crystals are added to concentrated ammonia solution. The free NI_3 can be prepared from reacting iodine monofluoride with boron nitride:

$$3\,IF(g) + BN(s) \rightarrow NI_3(s) + BF_3(g)$$

Nitrogen triiodide and the ammoniate are extremely unstable and detonate at the slightest touch or vibration:

$$2\,NI_3 \cdot NH_3(s) \rightarrow N_2(g) + 3\,I_2(s) + 2\,NH_3(g)$$

Although the formula of nitrogen triiodide is usually written NI_3, it would be more accurate to write I_3N, as the compound is thought to consist of I^+ and N^{3-} ions and its sensitivity to shock is due to the redox instability of these ions. This behaviour is also another example of the instability of large anions in combination with small cations, which, as we saw in Section 3.15, can be rationalized by the ionic model.

The existence and structures of the higher polyiodides are sensitive to the counterion for similar reasons, and large cations are necessary to stabilize them in the solid state. In fact, entirely different shapes are observed for polyiodide ions in combination with various large cations, as the structure of the anion is determined in large measure by the manner in which the ions

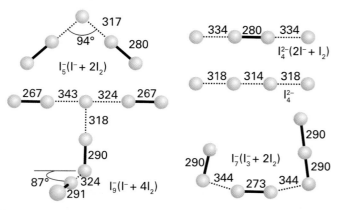

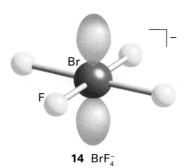

14 BrF_4^-

Figure 17.12 Some representative polyiodide structures and their approximate description in terms of I^-, I_3^-, and I_2 building blocks. Bond lengths and angles vary with the identity of the cation.

Table 17.11 Representative interhalogen anions

Compound	Shape
ClF_2^-, IF_2^-, ICl_2^-, IBr_2^-	Linear
ClF_4^-, BrF_4^-, IF_4^-, ICl_4^-	Square planar
ClF_6^-, BrF_6^-	Octahedral
IF_6^-	Trigonally distorted octahedron
IF_8^-	Square antiprism

pack together in the crystal. The bond lengths in a polyiodide ion often suggest that it can be regarded as a chain of associated I^-, I_2, I_3^-, and sometimes I_4^{2-} units (Fig. 17.12). Solids containing polyiodides exhibit electrical conductivity, which may arise either from the hopping of electrons (or holes) or by an ion relay along the polyiodide chain (Fig. 17.13).

Some dinegative polyiodides are known. They contain an even number of I atoms and their general formula is $[I^-(I_2)_nI^-]$. They have the same sensitivity to the cation as their mononegative counterparts.

Although polyhalide formation is most pronounced for iodine, other polyhalides are also known. They include Cl_3^-, Br_3^-, and BrI_2^-, which are known in solution and (in partnership with large cations) as solids too. Even F_3^- has been detected spectroscopically at low temperatures in an inert matrix. This technique, known as *matrix isolation*, makes use of the co-deposition of the reactants with a large excess of noble gas at very low temperatures (in the region of 4–14 K). The solid noble gas forms an inert matrix within which the F_3^- ion can sit in chemical isolation.

In addition to complex formation between dihalogens and halide ions, some interhalogens can act as Lewis acids towards halide ions. The reaction results in the formation of polyhalides that, in contrast to the chain-like polyiodides, are assembled around a central halogen acceptor atom in a high oxidation state. As mentioned earlier, for instance, BrF_3 reacts with CsF to form $CsBrF_4$, which contains the square-planar BrF_4^- anion (**14**). Many of these interhalogen anions have been synthesized (Table 17.11).

Their shapes generally agree with the VSEPR model, but there are some interesting exceptions. Two such exceptions are ClF_6^- and BrF_6^-, in which the central halogen has a lone pair of electrons, but the apparent structure is octahedral. The ion IF_6^- participates in an extended array through $I-F\cdots I$ interactions.

EXAMPLE 17.3 Proposing a bonding model for I^+ complexes

In some cases the interaction of I_2 with strong donor ligands leads to the formation of cationic complexes such as bis(pyridine)iodine(+1), [py—I—py]$^+$. Propose a bonding model for this linear complex from (a) the standpoint of the VSEPR model and (b) simple molecular orbital considerations.

Answer (a) The Lewis electron structure places 10 electrons around the central I^+ in [py—I—py]$^+$, six from the iodine cation and four from the lone pairs on the two pyridine ligands. According to the VSEPR model, these pairs should form a trigonal bipyramid. The lone pairs will occupy the equatorial positions and consequently the complex should be linear. (b) From a molecular orbital perspective, the orbitals of the N—I—N array can be pictured as being formed from an iodine 5p orbital and an orbital of σ symmetry from each of the two ligand atoms. Three orbitals can be constructed: 1σ (bonding), 2σ (nearly nonbonding), and 3σ (antibonding). There are four electrons to accommodate (two from each ligand atom; the iodine 5p orbital is empty). The resulting configuration is $1\sigma^2 2\sigma^2$, which is net bonding.

Self-test 17.3 From the perspective of structure and bonding, indicate several polyhalides that are analogous to [py—I—py]$^+$, and describe their bonding.

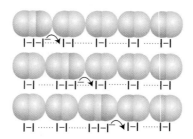

Figure 17.13 One possible mode of charge transport along a polyiodide chain is the shift of long and short bonds, resulting in the effective migration of an I^- ion along a chain. Three successive stages in the migration are shown. Note that the iodide ion from the I_3^- on the left is not the same one emerging on the right.

17.11 Halogen oxides

Key points: The only fluorine oxides are OF_2 and O_2F_2; chlorine oxides are known for Cl oxidation numbers of +1, +4, +6, and +7; the strong and facile oxidizing agent ClO_2 is the most commonly used halogen oxide.

Oxygen difluoride (FOF; m.p. −224°C, b.p. −145°C), the most stable binary compound of O and F, is prepared by passing fluorine through dilute aqueous hydroxide solution:

$$2\,F_2\,(g) + 2\,OH^-(aq) \rightarrow OF_2\,(g) + 2\,F^-(aq) + H_2O(l)$$

The pure difluoride survives in the gas phase above room temperature and does not react with glass. It is a strong fluorinating agent, but less so than fluorine itself. As suggested by the VSEPR model, the OF_2 molecule is angular.

Dioxygen difluoride (FOOF; m.p. $-154°C$, b.p. $-57°C$) can be synthesized by photolysis of a liquid mixture of the two elements. It is unstable in the liquid state and decomposes rapidly above $-100°C$, but can be transferred (with some decomposition) as a low-pressure gas in a metal vacuum line. Dioxygen difluoride is an even more aggressive fluorinating agent than ClF_3. For example, it oxidizes plutonium metal and its compounds to PuF_6, which is an intermediate in reprocessing of nuclear fuels, in a reaction that ClF_3 cannot accomplish:

$$Pu(s) + 3O_2F_2(g) \rightarrow PuF_6(g) + 3O_2(g)$$

Chlorine dioxide is the only halogen oxide produced on a large scale. The reaction used is the reduction of ClO_3^- with HCl or SO_2 in strongly acidic solution:

$$2ClO_3^-(aq) + SO_2(g) \xrightarrow{\text{acid}} 2ClO_2(g) + SO_4^{2-}(aq)$$

Because chlorine dioxide is a strongly endergonic compound ($\Delta_f G^{\ominus} = +121$ kJ mol^{-1}), it must be kept dilute to avoid explosive decomposition and is therefore used at the site of production. Its major uses are to bleach paper pulp and to disinfect sewage and drinking water. Some controversy surrounds these applications because the action of chlorine (or its product of hydrolysis, HClO) and chlorine dioxide on organic matter produces low concentrations of chlorocarbon compounds, some of which are potential carcinogens. However, the disinfection of water undoubtedly saves many more lives than the carcinogenic byproducts may take. Chlorine bleaches are being replaced by oxygen-based bleaches such as hydrogen peroxide (Box 16.1).

The most well known oxides of bromine are given below:

Oxidation number	+1	+3	+4
Formula	Br_2O	Br_2O_3	BrO_2
Colour	Dark brown	Orange	Pale yellow
State	Solid	Solid	Solid

The structure of BrO_2 has been found to be a mixed $Br(I)/Br(VII)$ oxide, $BrOBrO_3$. All the bromine oxides are thermally unstable above $-40°C$ and explode on heating.

The most stable halogen oxides are those formed by iodine. The most important of these is I_2O_5 (15), which is used to oxidize carbon monoxide quantitatively to carbon dioxide in the analysis of CO in blood and in air. The compound is a white, hygroscopic solid. It dissolves in water to give iodic acid, HIO_3.

The less stable iodine oxides I_2O_4 and I_4O_9 are both yellow solids that decompose on heating to give I_2O_5:

$$5I_2O_4(s) \rightarrow 4I_2O_5(s) + I_2(g)$$
$$4I_4O_9(s) \rightarrow 6I_2O_5(s) + 2I_2(g) + 3O_2(g)$$

17.12 Oxoacids and oxoanions

Key points: The halogen oxoanions are thermodynamically strong oxidizing agents; perchlorates of oxidizable cations are unstable.

The strengths of the oxoacids vary systematically with the number of O atoms on the central atom (Table 17.12; see Pauling's rules in Section 4.5b). Periodic acid, H_5IO_6, is the I(VII) analogue of perchloric acid. It is a weak acid ($pK_{a1} = 3.29$), which can be explained as soon as we note that its formula is $(HO)_5IO$ and that there is only one I=O group. The O atoms in the conjugate base $H_4IO_6^-$ are very labile on account of the rapid equilibration:

$$H_4IO_6^-(aq) \rightleftharpoons IO_4^-(aq) + 2H_2O(l) \qquad K = 40$$

IO_4^- is the dominant ion. The tendency to have an expanded coordination shell is shared by the oxoacids of the neighbouring Group 16 element tellurium, which in its maximum oxidation state forms the weak acid $Te(OH)_6$.

The halogen oxoanions, like many oxoanions, form metal complexes, including the metal perchlorates and periodates discussed here. In this connection we note that, because $HClO_4$ is a very strong acid and H_5IO_6 is a weak acid, it follows that ClO_4^- is a very weak base and $H_4IO_6^-$ is a relatively strong base.

In view of the low Brønsted basicity and single negative charge of the perchlorate ion, ClO_4^-, it is not surprising that it is a weak Lewis base with little tendency to form complexes with cations in aqueous solution. Therefore, metal perchlorates are often used to study the properties of hexaaqua ions in solution. The ClO_4^- ion is used as a weakly coordinating ion that can readily be displaced from a complex by other ligands, or as a medium-sized anion that might stabilize solid salts containing large cationic complexes with easily displaced ligands.

However, the ClO_4^- ion is a treacherous ally. Because it is a powerful oxidizing agent, solid compounds of perchlorate should be avoided whenever there are oxidizable ligands or ions present (which is commonly the case). In some cases the danger lies in wait, as the reactions of ClO_4^- are generally slow and it is possible to prepare many metastable perchlorate complexes or salts that may be handled with deceptive ease. However, once reaction has been initiated by mechanical action, heat, or static electricity, these compounds can detonate with disastrous consequences. Such explosions have injured chemists who may

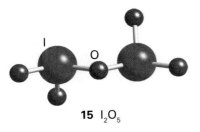

15 I_2O_5

Table 17.12 Acidities of chlorine oxoacids

Acid	p/q	pK_a
HOCl	0	7.53 (weak)
HOClO	1	2.00
HOClO$_2$	2	-1.2
HOClO$_3$	3	-10 (strong)

have handled a compound many times before it unexpectedly exploded. Some readily available and more docile weakly basic anions may be used in place of ClO_4^-; they include trifluoridomethanesulfonate $[SO_3CF_3]^-$, tetrafluoridoborate BF_4^-, and hexafluoridophosphate $[PF_6^-]$.

For many years it was believed that perbromate did not exist. However, it was prepared in 1968 by a radiochemical route based on the β-decay of ^{83}Se and chemical syntheses have now been devised. The perbromate ion is more oxidizing than any other oxohalide. The instability of perbromate compared to perchlorate and periodate is an example of the reluctance of post-3d elements to achieve their highest possible oxidation state and is a manifestation of the alternation effect (Section 9.2c).

In contrast to perchlorate, periodate is a rapid oxidizing agent and a stronger Lewis base. These properties lead to the use of periodate as an oxidizing agent and stabilizing ligand for metal ions in high oxidation states. Some of the high oxidation states it can be used to form are very unusual: they include Cu(III) in a salt containing the $[Cu(HIO_6)_2]^{5-}$ complex and Ni(IV) in an extended complex containing the $[Ni(IO_6)]^-$ unit. The periodate ligand is bidentate in these complexes, and in the last example it forms a bridge between Ni(IV) ions.

17.13 Thermodynamic aspects of oxoanion redox reactions

Key point: The oxoanions of halogens are strong oxidizing agents, especially in acidic solution.

The thermodynamic tendencies of the halogen oxoanions and oxoacids to participate in redox reactions have been extensively studied. As we shall see, we can summarize their behaviour with a Frost diagram that is quite easy to rationalize. It is a very different story with the rates of the reactions, which vary widely. Their mechanisms are only partly understood despite many years of investigation. Recent progress in the understanding of some of these mechanisms stems from advances

in techniques for fast reactions and interest in oscillating reactions (Box 17.4).

We saw in Section 5.13 that, if in a Frost diagram a species lies above the line joining its two neighbours of higher and lower oxidation numbers, then it is unstable with respect to disproportionation into them. From the Frost diagram for the halogen oxoanions and oxoacids in Fig. 17.14 we can see that many of the oxoanions in intermediate oxidation states are susceptible to disproportionation. Chlorous acid, $HClO_2$, for instance, lies above the line joining its two neighbours, and is liable to disproportionation:

$$2\,HClO_2(aq) \rightarrow ClO_3^-(aq) + HClO(aq) + H^+(aq)\quad E^{\ominus}_{cell} = +0.52\ V$$

Although BrO_2^- is well characterized, the corresponding I(III) species is so unstable that it does not exist in solution, except perhaps as a transient intermediate.

We also saw in Section 5.13 that the more positive the slope for the line from a lower to higher oxidation state species in a Frost diagram, the stronger the oxidizing power of the couple. A glance at Fig. 17.14 shows that all three Frost diagrams have steep positively sloping lines, which immediately shows that all the oxidation states except the lowest (Cl^-, Br^-, and I^-) are strongly oxidizing.

Finally, basic conditions decrease reduction potentials for oxoanions as compared with their conjugate acids (Section 5.5). This decrease is evident in the less steep slopes of the lines in the Frost diagrams for the oxoanions in basic solution. The numerical comparison for ClO_4^- ions in 1 M acid compared with 1 M base makes this clear:

At pH = 0: $ClO_4^-(aq) + 2\,H^+(aq) + 2\,e^- \rightarrow ClO_3^-(aq) + H_2O(l)$
$$E^{\ominus} = +1.20\ V$$
At pH = 14: $ClO_4^-(aq) + H_2O(l) + 2\,e^- \rightarrow ClO_3^-(aq) + 2\,OH^-(aq)$
$$E^{\ominus}_B = +0.37\ V$$

The reduction potentials show that perchlorate is thermodynamically a weaker oxidizing agent in basic solution than in acidic solution.

BOX 17.4 Oscillating reactions

Clock reactions and oscillating reactions are an active topic of research and provide fascinating lecture demonstrations. Most oscillating reactions are based on the reactions of halogen oxoanions, apparently because of the variety of oxidation states and their sensitivity to changes in pH.

In 1895, H. Landot discovered that a mixture of sulfite, iodate, and starch in acidic aqueous solution remains nearly colourless for an initial period and then suddenly switches to the dark purple of the I_2–starch complex. When the concentrations are properly adjusted, the reaction oscillates between nearly colourless and opaque blue. The reactions leading to this oscillation are the reduction of iodate to iodide by sulfite, where all reactants are colourless:

$$IO_3^-(aq) + 3\,SO_3^{2-}(aq) \rightarrow I^-(aq) + 3\,SO_4^{2-}(aq)$$

A comproportionation reaction between I^- and IO_3^- then produces I_2, which forms an intensely coloured complex with starch:

$$IO_3^-(aq) + 6\,H^+(aq) + 5\,I^-(aq) \rightarrow 3\,H_2O(l) + 3\,I_2(starch)$$

Under some conditions the I_2–starch complex is the final state, but adjustment of concentrations may lead to bleaching of the complex by sulfite reduction of iodine to the colourless $I^-(aq)$ ion:

$$3\,I_2(starch) + 3\,SO_3^{2-}(aq) + 3\,H_2O(l) \rightarrow 6\,I^-(aq) + 6\,H^+(aq) + 3\,SO_4^{2-}(aq)$$

The reaction may then oscillate between colourless and blue as the I_2/I^- ratio changes.

The detailed analysis of the kinetic conditions for oscillating reactions is pursued by chemists and chemical engineers. In the former case the challenge is to use kinetic data determined separately for the individual steps to model the observed oscillations with a view to testing the validity of the overall scheme. As oscillating reactions have been observed in commercial catalytic processes, the concern of the chemical engineer is to avoid large fluctuations or even chaotic reactions that might degrade the process. Oscillating reactions are of more than industrial interest, for they also maintain the rhythm of the heartbeat and their interruption can result in fibrillation and death.

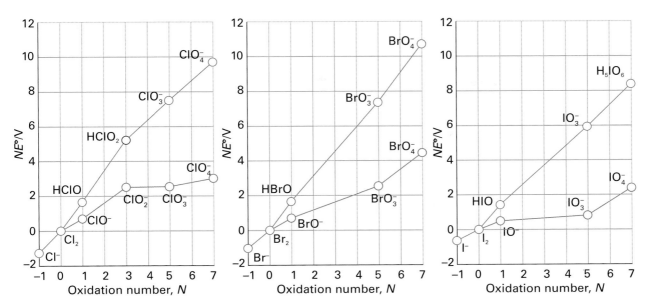

Figure 17.14 Frost diagrams for chlorine, bromine, and iodine in acidic solution (red line) and in basic solution (blue line).

17.14 Trends in rates of oxoanion redox reactions

Key points: Oxidation by halogen oxoanions is faster for the lower oxidation states; rates and thermodynamics of oxidation are both enhanced by an acidic medium.

Mechanistic studies show that the redox reactions of halogen oxoanions are complex. Nevertheless, despite this complexity, a few discernible patterns help to correlate the trends in rates of reaction. These correlations have practical value and give some clues about the mechanisms that may be involved.

The oxidation of many molecules and ions by halogen oxoanions becomes progressively faster as the oxidation number of the halogen decreases. Thus the rates observed are often in the order

$$ClO_4^- < ClO_3^- < ClO_2^- \approx ClO^- \approx Cl_2$$
$$BrO_4^- < BrO_3^- \approx BrO^- \approx Br_2$$
$$IO_4^- < IO_3^- < I_2$$

For example, aqueous solutions containing Fe^{2+} and ClO_4^- are stable for many months in the absence of dissolved oxygen, but an equilibrium mixture of aqueous HClO and Cl_2 rapidly oxidizes Fe^{2+}.

Oxoanions of the heavier halogens tend to react most rapidly, particularly for the elements in their highest oxidation states:

$$ClO_4^- < BrO_4^- < IO_4^-$$

As we have remarked, perchlorates in dilute aqueous solution are usually unreactive, but periodate oxidations are fast enough to be used for titrations. The mechanistic details are often complex, but the existence of both four- and six-coordinate periodate ions shows that the I atom in periodate is accessible to nucleophiles.

We have already seen that the thermodynamic tendency of oxoanions to act as oxidizing agents increases as the pH is lowered. It is found that their rates are increased too. Thus, kinetics and equilibria unite to bring about otherwise difficult oxidations. The oxidation of halides by BrO_3^- ions, for instance, is second-order in H^+:

$$\text{Rate} = k_r[BrO_3^-][X^-][H^+]^2$$

and so the rate increases as the pH is decreased. The acid is thought to protonate the oxo group in the oxoanion, so aiding oxygen–halogen bond scission. Another role of protonation is to increase the electrophilicity of the halogen. An example is HClO, where, as described below, the Cl atom may be viewed as an electrophile towards an incoming reducing agent (**16**). An illustration of the effect of acidity on rate is the use of a mixture of H_2SO_4 and $HClO_4$ in the final stages of the oxidation of organic matter in certain analytical procedures.

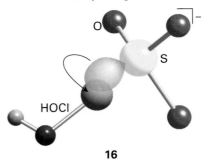

16

17.15 Redox properties of individual oxidation states

Key point: Dihalogen molecules disproportionate in aqueous solution. Hypochlorite is a facile oxidizing agent; hypohalite and halite ions undergo disproportionation. Chlorate ions undergo disproportionation in solution but bromates and iodates do not.

With the general redox properties of the halogens now outlined, we can consider the characteristic properties and reactions of specific oxidation states. Although we are dealing here with halogen oxides, it is convenient to mention, for the sake of completeness, the redox properties of halogen(0) species. Figure 17.15 summarizes some of the reactions that interconvert the oxoanions and oxoacids of chlorine in its various oxidation states. One point to note is the major role of disproportionation and electrochemical reactions in the scheme. For example, the figure includes the production of Cl_2 by the electrochemical oxidation of Cl^-, which was discussed in Section 17.4.

Oxidation
number

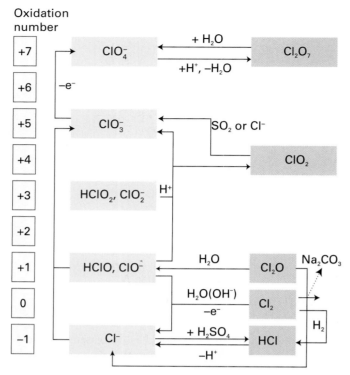

Figure 17.15 The interconversion of oxidation states of some important chlorine species.

Disproportionation is thermodynamically favourable for basic solutions of Cl_2, Br_2, and I_2. The equilibria in basic aqueous solution are:

$$X_2(aq) + 2\,OH^-(aq) \rightleftharpoons XO^-(aq) + X^-(aq) + H_2O(l)$$

$$K = \frac{[XO^-][X^-]}{[X_2][OH^-]^2}$$

with $K = 7.5 \times 10^{15}$ for $X = Cl$, 2×10^8 for Br, and 30 for I.

A note on good practice When writing the associated equilibrium expressions in aqueous solution, we conform to the usual convention that the activity of water is 1, so H_2O does not appear in the equilibrium constant.

Disproportionation is much less favourable in acidic solution, as would be expected from the fact that H^+ is a product of the reaction:

$$Cl_2(aq) + H_2O(l) \rightleftharpoons HClO(aq) + H^+(aq) + Cl^-(aq)$$
$$K = 3.9 \times 10^{-4}$$

Because the redox reactions of Cl_2 are often fast, Cl_2 in water is widely used as an inexpensive and powerful oxidizing agent. The equilibrium constants for the hydrolysis of Br_2 and I_2 in acid solution are smaller than for Cl_2, and both elements are unchanged when dissolved in slightly acidified water. Because F_2 is a much stronger oxidizing agent than the other halogens, it produces mainly O_2 and H_2O_2 when in contact with water. As a result, hypofluorous acid, HFO, was discovered long after the other hypohalous acids.

The aqueous Cl(I) species hypochlorous acid, HClO (the angular molecular species H–O–Cl), and hypochlorite ions, ClO^-, are facile oxidizing agents that are used as household bleach and disinfectant, and as laboratory oxidizing agents (Box 17.5). The ready access to the unobstructed, electrophilic Cl atom in HClO appears to be one feature that leads to the very fast redox reactions of this compound. These rates contrast with the much slower redox reactions of perchlorate ions, in which access to the Cl atom is blocked by the surrounding O atoms.

A note on good practice Hypohalous acids are widely denoted HOX to emphasize their structure. We have adopted the formula HXO to emphasize their relationship to the other oxoacids, HXO_n.

Hypohalite ions undergo disproportionation. For instance, ClO^- disproportionates into Cl^- and ClO_3^-:

$$3\,ClO^-(aq) \rightleftharpoons 2\,Cl^-(aq) + ClO_3^-(aq) \qquad K = 1.5 \times 10^{27}$$

This reaction (which is used for the commercial production of chlorates) is slow at or below room temperature for ClO^- but is much faster for BrO^-. It is so fast for IO^- that this ion has been detected only as a reaction intermediate.

Chlorite ions, ClO_2^-, and bromite ions, BrO_2^-, are both susceptible to disproportionation. However, the rate is strongly dependent on pH and ClO_2^- (and to a lesser extent BrO_2^-) can be handled in basic solution with only slow decomposition.

BOX 17.5 Chlorine-based bleaches

Substances that are used as bleaches are powerful oxidizing agents. As mentioned in Section 17.2, the oxidizing power of the halogen oxoanions increases as the oxidation number of the halogen decreases. It is not surprising then that the chlorine-based bleaches contain Cl in a low oxidation state.

Chlorine disproportionates in water to produce the oxidizing hypochlorite ion, ClO^-, and Cl^-. Solutions of up to 15 per cent by mass of sodium hypochlorite are used as industrial bleaches in the paper, textiles, and laundry industries and for disinfecting swimming pools. Household bleach is a more dilute (5 per cent) solution of NaClO. A 0.5 per cent aqueous solution of NaClO is used by dentists during root-canal work, where it is used kill pathogens and dissolve necrotic tissue.

Other hypochlorite salts are also used as oxidants. Calcium hypochlorite, $Ca(ClO)_2$, is used as a disinfectant in dairies, breweries, food processing,

and bottling plants. It is also used in domestic mildew removers. *Bleaching powder* is a mixture of $Ca(ClO)_2$ and $CaCl_2$ and is used for large-scale applications such as disinfecting seawater, reservoirs, and sewers. It is also used as a decontaminant in areas where chemical weapons, such as mustard gas, have been deployed.

Chlorine dioxide gas is widely used as a bleach in the wood pulp industry where it produces whiter and stronger paper than other bleaches because, unlike oxidizing bleaches such as chlorine, ozone, and hydrogen peroxide, it does not attack the cellulose and therefore preserves the mechanical strength of the pulp. Chlorine-based bleaches lead to the production of toxic chlorinated organic compounds. The most toxic polychlorinated phenols, such as dioxins, are mostly produced by ClO_2, but the levels can be drastically reduced by substituting some of the ClO_2 with Cl_2.

By contrast, chlorous acid, $HClO_2$, and bromous acid, $HBrO_2$, both disproportionate rapidly. Iodine(III) is even more elusive, and HIO_2 has been identified only as a transient species in aqueous solution.

The Frost diagram for Cl shown in Fig. 17.14 indicates that chlorate ions, ClO_3^-, are unstable with respect to disproportionation in both acidic and basic solution:

$$4\,ClO_3^-(aq) \rightleftharpoons 3\,ClO_4^-(aq) + Cl^-(aq) \quad \Delta_r G^\ominus = -24\,kJ\,mol^{-1}$$
$$K = 1.4 \times 10^{25}$$

Because $HClO_3$ is a strong acid, and this reaction is slow at both low and high pH, ClO_3^- ions can be handled readily in aqueous solution. Bromates and iodates are thermodynamically stable with respect to disproportionation.

Of the three XO_4^- ions, BrO_4^- is the strongest oxidizing agent. That perbromate is out of line with its adjacent halogen congeners fits a general pattern for anomalies in the chemistry of p-block elements of Period 4. However, the reduction of periodate in dilute acid is faster than that of perchlorate or perbromate and periodates are therefore used in analytical chemistry as oxidizing titrants and also in syntheses, such as the oxidative cleavage of diols:

■ **A brief illustration.** To confirm that perbromate is the most oxidizing perhalate ion we need to consider the slope of the lines joining the perhalate ions to their neighbours in the Frost diagram. The more positive the slope of the line, the stronger the oxidizing power of the couple. Inspection of the diagram reveals that the line joining the

BrO_4^-/BrO_3^- couple is the most positive. In fact the $E^\ominus$ values for the ClO_4^-/ClO_3^-, BrO_4^-/BrO_3^-, and IO_4^-/IO_3^- couples in acidic conditions are 1.201, 1.853, and 1.600 V, respectively, confirming that perbromate is the strongest oxidizing agent. ■

17.16 Fluorocarbons

Key point: Fluorocarbon molecules and polymers are resistant to oxidation.

Fluorocarbons find many useful applications (Box 17.6). The direct reaction of an aliphatic hydrocarbon with an oxidizing metal fluoride leads to the formation of strong C—F bonds ($456\,kJ\,mol^{-1}$) and produces HF as a byproduct:

$$RH(l) + 2\,CoF_3(s) \rightarrow RF(sol) + 2\,CoF_2(s) + HF(sol)$$
$$R = alkyl\ or\ aryl$$

When R is aryl, CoF_3 yields the cyclic saturated fluoride:

$$C_6H_6(l) + 18\,CoF_3(s) \rightarrow C_6F_{12}(l) + 18\,CoF_2(s) + 6\,HF(l)$$

The strongly oxidizing fluorinating agent used in these reactions, CoF_3, is regenerated by the reaction of CoF_2 with fluorine:

$$2\,CoF_2(s) + F_2(g) \rightarrow 2\,CoF_3(s)$$

Another important method of CF bond formation is halogen exchange by the reaction of a nonoxidizing fluoride, such as HF, with a chlorocarbon in the presence of a catalyst, such as SbF_3:

$$CCl_4(l) + HF(l) \rightarrow CCl_3F(l) + HCl(g)$$
$$CHCl_3(l) + 2\,HF(l) \rightarrow CHClF_2(l) + 2\,HCl(g)$$

These processes used to be performed on a large scale to produce the chlorofluorocarbons (CFCs) and hydrochlorofluorocarbons (HCFCs) that were used as refrigerant fluids, the propellant in spray cans, and in the blowing agent in plastic foam products. These applications have been banned in some countries and are being phased out worldwide because of the role of CFCs

---

BOX 17.6 PTFE: a high-performance polymer

Polytetrafluoroethene, PTFE, is a unique product in the plastics industry. It is chemically inert, thermally stable over a wide temperature range (-196 to $260°C$), is an excellent electrical insulator, and has a low coefficient of friction. It is a white solid that is manufactured by the polymerization of tetrafluoroethene:

$$n\,CF_2{=}CF_2 \rightarrow (CF_2CF_2)_n$$

PTFE is an expensive polymer because of the cost of synthesizing and purifying the monomer by a multistage process:

$$CH_4(g) + 3\,Cl_2(g) \rightarrow CHCl_3(g) + 3\,HCl(g)$$
$$CHCl_3(g) + 2\,HF(g) \rightarrow CHClF_2(g) + 2\,HCl(g)$$
$$2\,CHClF_2(g) \xrightarrow{\Delta} CF_2{=}CF_2(g) + 2\,HCl(g)$$

The hydrogen fluoride is generated by the action of sulfuric acid on fluorite:

$$CaF_2(s) + H_2SO_4(l) \rightarrow CaSO_4(s) + 2\,HF(l)$$

As the process employs HF and HCl, the reactors have to be lined with platinum. Many byproducts are produced, which leads to complex purification of the final product.

The tetrafluoroethene is polymerized in two ways: solution polymerization with vigorous agitation produces a resin known as *granular PTFE*; emulsion polymerization with a dispersing agent and gentle agitation produces small particles known as *dispersed PTFE*. The molten polymer does not flow, so the usual methods of processing cannot be used. Instead processes similar to those used for metals are applied. For example, the dispersed form can be cold extruded (which is a method used for processing lead).

The remarkable properties of PTFE arise from the protective sheath that the F atoms form around the carbon polymer backbone. The F atoms are just the right size to form a smooth sheath. This smooth sheath reduces the disruption of intermolecular forces at the surface, leading to a low coefficient of friction and the familiar nonstick properties. The polymer is used in a wide range of applications. Its low electrical conductivity leads to its use in electrical tapes, wires, and coaxial cable. Its mechanical properties make it an ideal material for seals, piston rings, and bearings. It is used as a packaging material, in hose lines, and as thread-sealant tape. Familiar applications are as the nonstick coating on cookware, and as the porous fabric Gore-Tex®.

and HCFCs in ozone depletion. They are being replaced by hydrofluorocarbons (HFCs) after investment by the chemical industry because, in contrast to the simple one-step synthesis of CFCs and HCFCs, HFC production is a complex multistage process. For example, the preferred route to CF_3CH_2F, which is one of the preferred CFC replacements is

$$CCl_2{=}CCl_2 \xrightarrow{HF+Cl_2} CClF_2CCl_2F \xrightarrow{isomerize} CF_3CCl_3$$
$$\xrightarrow{HF} CF_3CCl_2F \xrightarrow{H_2} CF_3CH_2F$$

When heated, chlorodifluoromethane is converted to the useful monomer, C_2F_4:

$$2\,CHClF_2 \xrightarrow{600\text{--}800^\circ C} C_2F_4 + 2\,HCl$$

The polymerization of tetrafluoroethene is carried out with a radical initiator:

$$n\,C_2F_4 \xrightarrow{ROO\cdot} (-CF_2{-}CF_2{-})_n$$

Polytetrafluoroethene (PTFE) is sold under many trade names, one of which is Teflon (DuPont). Its depolymerization at high temperatures is the most convenient method of preparing tetrafluoroethene in the laboratory:

$$(-CF_2{-}CF_2{-})_n \xrightarrow{600^\circ C} n\,C_2F_4$$

Although tetrafluoroethene is not highly toxic, a byproduct, 1,1,3,3,3-pentafluoro-2-trifluoromethyl-1-propene, is toxic and its presence dictates care in handling crude tetrafluoroethene.

FURTHER READING

M. Schnürch, M. Spina, A.F. Khan, M.D. Mihovilovic, and P. Stanetty, Halogen dance reactions—A review, *Chem. Soc. Rev.*, 2007, **36**, 1046.

S. Purser, P.R. Moore, S. Swallow, and V. Gouverneur, Fluorine in medicinal chemistry, *Chemical Society Reviews*, 2008, **37**, 2, 320.

A.G. Massey, *Main group chemistry*. Wiley (2000).

D.M.P. Mingos, *Essential trends in inorganic chemistry*. Oxford University Press (1998).

R.B. King (ed.), *Encyclopedia of inorganic chemistry*. Wiley (2005).

N.N. Greenwood and A. Earnshaw, *Chemistry of the elements*. Butterworth-Heinemann (1997).

P. Schmittinger, *Chlorine: principles and industrial practice*. Wiley–VCH (2000).

M. Howe-Grant, *Fluorine chemistry*. Wiley (1995).

EXERCISES

17.1 Preferably without consulting reference material, write out the halogens as they appear in the periodic table, and indicate the trends in (a) physical state (s, l, or g) at room temperature and pressure, (b) electronegativity, (c) hardness of the halide ion, (d) colour.

17.2 Describe how the halogens are recovered from their naturally occurring halides and rationalize the approach in terms of standard potentials. Give balanced chemical equations and conditions where appropriate.

17.3 Sketch a choralkali cell. Show the half-cell reactions and indicate the direction of diffusion of the ions. Give the chemical equation for the unwanted reaction that would occur if OH^- migrated through the membrane and into the anode compartment.

17.4 Sketch the form of the vacant σ^* orbital of a dihalogen molecule and describe its role in the Lewis acidity of the dihalogens.

17.5 Which dihalogens are thermodynamically capable of oxidizing H_2O to O_2?

17.6 Nitrogen trifluoride, NF_3, boils at $-129^\circ C$ and is a very weak Lewis base. By contrast, the lower molar mass compound NH_3 boils at $-33^\circ C$ and is well known as a Lewis base. (a) Describe the origins of this very large difference in volatility. (b) Describe the probable origins of the difference in basicity.

17.7 Based on the analogy between halogens and pseudohalogens write: (a) the balanced equation for the probable reaction of cyanogen, $(CN)_2$, with aqueous sodium hydroxide, (b) the equation for the probable reaction of excess thiocyanate with the oxidizing agent $MnO_2(s)$ in acidic aqueous solution, (c) a plausible structure for trimethylsilyl cyanide.

17.8 Given that 1.84 g of IF_3 reacts with 0.93 g of $[(CH_3)_4N]F$ to form a product X, (a) identify X, (b) use the VSEPR model to predict the shapes of IF_3 and the cation and anion in X, (c) predict how many ^{19}F-NMR signals would be observed in IF_3 and X.

17.9 Use the VSEPR model to predict the shapes of $SbCl_5$, $FClO_3$, and $[ClF_6]^+$.

17.10 Indicate the product of the reaction between ClF_5 and SbF_5.

17.11 Sketch all the isomers of the complexes MCl_4F_2 and MCl_3F_3. Indicate how many fluorine environments would be indicated in the ^{19}F-NMR spectrum of each isomer.

17.12 (a) Use the VSEPR model to predict the probable shapes of $[IF_6]^+$ and IF_7. (b) Give a plausible chemical equation for the preparation of $[IF_6][SbF_6]$.

17.13 Predict the shape of the doubly chlorine-bridged I_2Cl_6 molecule by using the VSEPR model and assign the point group.

17.14 Predict the structure and identify the point group of ClO_2F.

17.15 Predict whether each of the following solutes is likely to make liquid BrF_3 act as a Lewis acid or a Lewis base: (a) SbF_5, (b) SF_6, (c) CsF.

17.16 Predict the appearance of the ^{19}F-NMR spectrum of IF_5^+.

17.17 Predict whether each of the following compounds is likely to be dangerously explosive in contact with BrF_3 and explain your answer: (a) SbF_5, (b) CH_3OH, (c) F_2, (d) S_2Cl_2.

17.18 The formation of Br_3^- from a tetraalkylammonium bromide and Br_2 is only slightly exergonic. Write an equation (or NR for no reaction) for the interaction of $[NR_4][Br_3]$ with I_2 in CH_2Cl_2 solution and give your reasoning.

17.19 Explain why $CsI_3(s)$ is stable with respect to decomposition but $NaI_3(s)$ is not.

17.20 Write plausible Lewis structures for (a) ClO_2 and (b) I_2O_6 and predict their shapes and the associated point group.

17.21 (a) Give the formulas and the probable relative acidities of perbromic acid and periodic acid. (b) Which is the more stable?

17.22 (a) Describe the expected trend in the standard potential of an oxoanion in a solution with decreasing pH. (b) Demonstrate this phenomenon by calculating the reduction potential of ClO_4^- at pH = 7 and comparing it with the tabulated value at pH = 0.

17.23 With regard to the general influence of pH on the standard potentials of oxoanions, explain why the disproportionation of an oxoanion is often promoted by low pH.

17.24 Which oxidizing agent reacts more readily in dilute aqueous solution, perchloric acid or periodic acid? Give a mechanistic explanation for the difference.

17.25 (a) For which of the following anions is disproportionation thermodynamically favourable in acidic solution: ClO^-, ClO_2^-, ClO_3^-, and ClO_4^-? (If you do not know the properties of these ions, determine them from a table of standard potentials as in *Resources section* 3.) (b) For which of the favourable cases is the reaction very slow at room temperature?

17.26 Which of the following compounds present an explosion hazard? (a) NH_4ClO_4, (b) $Mg(ClO_4)_2$, (c) $NaClO_4$, (d) $[Fe(OH_2)_6][ClO_4]_2$. Explain your reasoning.

17.27 Use standard potentials to predict which of the following will be oxidized by ClO^- ions in acidic conditions: (a) Cr^{3+}, (b) V^{3+}, (c) Fe^{2+}, (d) Co^{2+}.

PROBLEMS

17.1 Many of the acids and salts corresponding to the positive oxidation numbers of the halogens are not listed in the catalogue of a major international chemical supplier: (a) $KClO_4$ and KIO_4 are available but $KBrO_4$ is not, (b) $KClO_3$, $KBrO_3$, and KIO_3 are all available, (c) $NaClO_2$ and $NaBrO_2 \cdot 3H_2O$ are available but salts of IO_2^- are not, (d) only ClO^- salts are available but the bromine and iodine analogues are not. Describe the probable reason for the missing salts of the oxoanions.

17.2 Identify the incorrect statements among the following descriptions and provide correct statements. (a) Oxidation of the halides is the only commercial method of preparing the halogens F_2 through I_2. (b) ClF_4^- and I_5^- are isolobal and isostructural. (c) Atom-transfer processes are common in the mechanisms of oxidations by the halogen oxoanions and an example is the O atom transfer in the oxidation of SO_3^{2-} by ClO^-. (d) Periodate appears to be a more facile oxidizing agent than perchlorate because the former can coordinate to the reducing agent at the I(VII) centre, whereas the Cl(VII) centre in perchlorate is inaccessible to reducing agents.

17.3 The reaction of I^- ions is often used to titrate ClO^-, giving deeply coloured I_3^- ions, along with Cl^- and H_2O. Although never proved, it was thought that the initial reaction proceeds by O atom transfer from Cl to I. However, it is now believed that the reaction proceeds by Cl atom transfer to give ICl as the intermediate (K. Kumar, R.A. Day,

and D.W. Margerum, *Inorg. Chem.*, 1986, **25**, 4344). Summarize the evidence for Cl atom transfer.

17.4 Given the bond lengths and angles in I_5^+ (5), describe the bonding in terms of two-centre and three-centre σ bonds and account for the structure in terms of the VSEPR model.

17.5 Until the work of K.O. Christe (*Inorg. Chem.*, 1986, **25**, 3721), F_2 could be prepared only electrochemically. Give chemical equations for Christe's preparation and summarize the reasoning behind it.

17.6 The use of templates to synthesize long-chain polyiodide ions has been described (A.J. Blake *et al.*, *Chem. Soc. Rev.*, 1998, **27**, 195). (a) According to the authors what is the longest polyiodide that has been characterized? (b) How does the nature of the cation influence the structure of the polyanion? (c) What was the templating agent used for the synthesis of I_7^- and I_{12}^-? (c) Which spectroscopic method was used for the characterization of the polyanions in this study?

17.7 Review published studies on the fluoridation of drinking water in your country. Summarize both the reasons for continuing fluoridation and the main concerns expressed by those opposed to it.

17.8 Write a review of the environmental problems associated with the use of chlorine-based bleaches in industry and suggest possible solutions.

18

The Group 18 elements

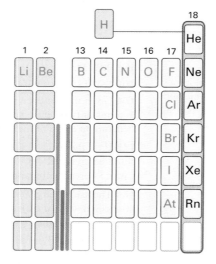

The final group in the p block contains six elements that are so unreactive that they form only a very limited range of compounds. The existence of the Group 18 elements was not suspected until late in the nineteenth century and their discovery led to the redrawing of the periodic table and played a key role in the development of bonding theories.

The Group 18 elements, helium, neon, argon, krypton, xenon, and radon, are all monatomic gases. They are the least reactive elements and have been given and then lost various collective names over the years as different aspects of their properties have been identified and then disproved. Thus they have been called the *rare gases* and the *inert gases*, and are currently called the **noble gases**. The first name is inappropriate because argon is by no means rare (it is substantially more abundant than CO_2 in the atmosphere). The second has been inappropriate since the discovery of compounds of xenon. The name 'noble gases' is now accepted because it gives the sense of low but significant reactivity.

PART A: **THE ESSENTIALS**

In this section we survey the limited chemistry of the noble gases and concentrate in particular on the well-characterized compounds of xenon.

18.1 **The elements**

Key point: Of the noble gases, only xenon forms a significant range of compounds with fluorine and oxygen.

All the Group 18 elements are very unreactive. Their unreactivity can be understood in terms of their atomic properties (Table 18.1) and in particular their ground-state valence electron configurations, ns^2np^6. The features to note include their high ionization energies and negative electron affinities. The first ionization energy is high because the effective nuclear charge is high at the far right of the period. The electron affinities are negative because an incoming electron needs to occupy an orbital belonging to a new shell.

Helium makes up 23 per cent by mass of the Universe and the Sun, and is the second most abundant element after hydrogen; it is rare in the atmosphere because its atoms travel fast enough to escape from the Earth. All the other noble gases occur in the atmosphere. The crustal abundances of argon (0.94 per cent by volume) and neon (1.5×10^{-3} per cent) make these two elements more plentiful than many familiar elements, such as arsenic and bismuth, in the Earth's crust (Fig. 18.1). Xenon and radon are the rarest elements of

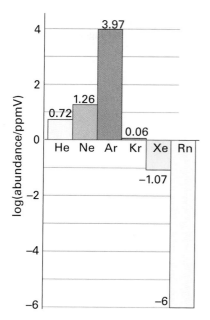

Figure 18.1 Abundances of the noble gases in the Earth's crust. The values are log (atmospheric parts per million by volume).

Table 18.1 Selected properties of the elements

	He	Ne	Ar	Kr	Xe	Rn
Covalent radius/pm	99	160	192	197	217	
Melting point/°C	−272	−249	−189	−157	−112	−71
Boiling point/°C	−269	−246	−186	−152	−108	−62
Electron affinity/(kJ mol⁻¹)	−48.2	−115.8	−96.5	−96.5	−77.2	
First ionization energy/(kJ mol⁻¹)	2373	2080	1520	1350	1170	1036

the group. Radon is a product of radioactive decay and, as its atomic number exceeds that of lead, is itself unstable; it accounts for around 50 per cent of background radiation.

18.2 Simple compounds

Key point: Xenon forms fluorides, oxides, and oxofluorides.

The most important oxidation numbers of Xe in its compounds are +2, +4, and +6. Compounds with Xe−F, Xe−O, Xe−N, Xe−H, Xe−C, and Xe−metal bonds are known and Xe can behave as a ligand. The chemical properties of xenon's lighter congener krypton are much more limited. The study of radon chemistry, like that of astatine, is inhibited by the high radioactivity of the element.

Xenon reacts directly with fluorine to produce XeF_2 (**1**), XeF_4 (**2**), and XeF_6 (**3**). Solid XeF_6 is more complex than its gas-phase structure and contains fluoride ion-bridged XeF_5^+ cations. The xenon fluorides are strong oxidizing agents and form complexes with F^- ions, such as XeF_5^-.

Xenon forms xenon trioxide, XeO_3 (**4**), and xenon tetraoxide, XeO_4 (**5**), which both decompose explosively. The white crystalline perxenates of several alkali metals have been prepared and contain the XeO_6^{4-} ion. Xenon also forms several oxofluorides. The oxofluoride $XeOF_2$ has a T-shape (**6**) and XeO_3F_2 is a trigonal bipyramid (**7**). The square-pyramidal molecule $XeOF_4$ (**8**) is remarkably similar to IF_5 in its physical and chemical properties.

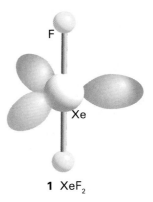

1 XeF_2

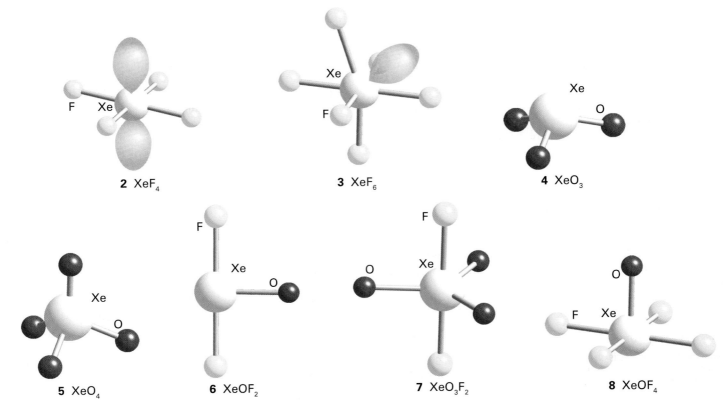

2 XeF_4

3 XeF_6

4 XeO_3

5 XeO_4

6 $XeOF_2$

7 XeO_3F_2

8 $XeOF_4$

Xenon forms a number of hydrides, such as HXeH, HXeOH, and HXeOXeH. Xenon, argon, and krypton form clathrates when frozen with water at high pressure (see Box 14.2). The noble gas atoms are guests within the three-dimensional ice structure and have the composition E.6H_2O. Clathrates provide a convenient way of handling the radioactive isotopes of krypton and xenon (Section 10.6a).

PART B: **THE DETAIL**

In this section we describe the detailed chemistry of the Group 18 elements. Although they are the least reactive of all elements, the noble gases, especially Xe, form a surprisingly wide range of compounds with hydrogen, oxygen, and the halogens.

18.3 **Occurrence and recovery**

Key points: The noble gases are monatomic; radon is radioactive.

Because they are so unreactive and are rarely concentrated in nature, the noble gases eluded recognition until the end of the nineteenth century. Indeed, Mendeleev did not make a place for them in his periodic table because the chemical regularities of the other elements, on which his table was based, did not suggest their existence. However, in 1868 a new spectral line observed in the spectrum of the Sun did not correspond to a known element. This line was eventually attributed to helium, and in due course the element itself and its congeners were found on Earth.

The noble gases were given names that reflect their curious nature; helium from the Greek *helios* for sun, neon from the Greek *neos* for new, argon from *argos* meaning inactive, krypton from *kryptos* meaning hidden, and xenon from *xenos* meaning strange. Radon was named after radium as it is a product of the radioactive decay of that element.

Helium atoms are too light to be retained by the Earth's gravitational field, so most of the He on Earth (5 parts per million by volume) is the product of α-emission in the decay of radioactive elements. High He concentrations of up to 7 per cent by mass are found in certain natural gas deposits (mainly in the USA and eastern Europe) from which it can be recovered by low-temperature distillation. Some He arrives from the Sun as a solar wind of α particles. Neon, Ar, Kr, and Xe are extracted from liquid air by low-temperature distillation.

The elements are all monatomic gases at room temperature. In the liquid phase they form low concentrations of dimers that are held together by dispersion forces. The low boiling points of the lighter noble gases (Table 18.1) follow from the weakness of these forces between the atoms and the absence of other forces. When helium (specifically ^{4}He, not the rarer isotope ^{3}He) is cooled below 2.178 K it undergoes a transformation into a second liquid phase known as **helium-II**. This phase is classed as a superfluid, as it flows without viscosity. Solid He is formed only under pressure.

18.4 **Uses**

Key points: Helium is used as an inert gas and as a light source in lasers and electric discharge lamps; liquid helium is a very low temperature refrigerant.

On account of its low density and nonflammability, He is used in balloons and lighter-than-air craft. Its very low boiling point leads to its wide use in cryogenics and as a very low temperature refrigerant; it is the coolant for superconducting magnets used for NMR spectroscopy and magnetic resonance imaging. It is also used as an inert atmosphere for growing crystals of semiconducting materials, such as Si. It is mixed with O_2 in a 4:1 ratio to provide an artificial atmosphere for divers, where its lower solubility than nitrogen minimizes the danger of causing the 'bends', or decompression sickness.

The most widespread use of Ar is to provide an inert atmosphere for the production of air-sensitive compounds and as an inert gas blanket to suppress the oxidation of metals when they are welded. Argon is also used as a cryogenic refrigerant and to fill the gap between the panes in sealed double-glazed windows as its low thermal conductivity reduces heat loss.

Xenon has anaesthetic properties but is not very widely used because it is approximately 2000 times more expensive than N_2O. Isotopes of Xe are also used in medical imaging (Box 18.1).

Radon, which is a product of nuclear power plants and of the radioactive decay of naturally occurring Th and U, is a health hazard because of the ionizing nuclear radiation it produces. It is usually a minor contributor to the background radiation arising from cosmic rays and terrestrial sources. However, in regions where the soil, underlying rocks, or building materials contain significant concentrations of U, excessive amounts of the gas have been found in buildings.

Because of their lack of chemical reactivity, the noble gases are extensively used in various light sources, including conventional sources (neon signs, fluorescent lamps, and xenon flash lamps) and lasers (helium–neon, argon-ion, and krypton-ion lasers). Argon is used as the inert atmosphere in incandescent light bulbs, where it reduces burning of the filament. In each case, an electric discharge through the gas ionizes some of the atoms and promotes both ions and neutral atoms into excited states that then emit electromagnetic radiation on return to a lower state.

18.5 **Synthesis and structure of xenon fluorides**

Key point: Xenon reacts with fluorine to form XeF_2, XeF_4, and XeF_6.

The reactivity of the noble gases has been investigated sporadically ever since their discovery, but all early attempts to coerce them into compound formation were unsuccessful. Until the 1960s the only known compounds were the unstable diatomic species such as He_2^+ and Ar_2^+, which were detected only spectroscopically. However, in March 1962, Neil Bartlett, then at the University of British Columbia, observed the reaction

Magnetic resonance imaging (MRI) is widely used in medicine to produce high-quality images of soft matter within the body. The technique relies on the protons of water molecules in tissue to provide the NMR signal, but protons are difficult to image in some parts of the body, particularly the lungs and the brain. However, other NMR-active nuclei can also be used for MRI, most notably ^{129}Xe, which is 26.4 per cent abundant with $I = \frac{1}{2}$. Because the extent of polarization of nuclear spin in ^{129}Xe is too low to give a good signal, it is enhanced by using low temperatures or increasing the applied magnetic field. Alternatively, the polarization can be enhanced by a factor of about 10^5 by spin-exchange with a polarized alkali metal. MRI images of the lungs can be obtained by inhaling this hyperpolarized ^{129}Xe into the lungs. The xenon is then transferred from the lungs to the blood and then on to other tissues. Consequently, images of the circulatory system, the brain, and other vital organs can be obtained.

^{129}Xe-NMR is also used in materials chemistry, where it is applied to examine the structures of mesoporous materials such as zeolites and ceramics, and to soft matter, such as polymer melts and elastomers. The technique is also useful for spectroscopic characterization of compounds of xenon. The fact that ^{129}Xe is only 26.4 per cent abundant does not affect the Xe spectra but does produce satellite peaks with relative intensities 13:74:13 in the spectra of other NMR nuclei, especially ^{19}F.

of a noble gas. Bartlett's report, and another from Rudolf Hoppe's group in the University of Munster a few weeks later, set off a flurry of activity throughout the world. Within a year, a series of xenon fluorides and oxo compounds had been synthesized and characterized. The field is somewhat limited, but compounds with bonds to nitrogen, carbon, and metals have been prepared.

Bartlett's motivation for studying xenon was based on the observations that PtF_6 can oxidize O_2, to give the solid O_2PtF_6, and that the ionization energy of xenon is similar to that of molecular oxygen. Indeed, reaction of xenon with PtF_6 did give a solid, but the reaction is complex and the complete formulation of the product (or products) remains unclear. The direct reaction of xenon and fluorine leads to a series of compounds with oxidation numbers +2 (XeF_2), +4 (XeF_4), and +6 (XeF_6).

The structures of XeF_2 and XeF_4 are well-established from diffraction and spectroscopic methods. Similar measurements on XeF_6 in the gas phase, however, led to the conclusion that this molecule is fluxional. Infrared spectra and electron diffraction on XeF_6 show that a distortion occurs about a threefold axis, suggesting that a triangular face of F atoms opens up to accommodate a lone pair of electrons, as in (3). One interpretation is that the fluxional process arises from the migration of the lone pair from one triangular face to another. Solid XeF_6 consists of F^--bridged XeF_5^+ units and in solution it forms Xe_4F_{24} tetramers. The gaseous and solid structures bear a molecular and electronic structural resemblance to the isoelectronic polyhalide anions I_3^- and ClF_4^- (Section 17.10c).

The xenon fluorides are synthesized by direct reaction of the elements, usually in a nickel reaction vessel that has been passivated by exposure to F_2 to form a thin protective NiF_2 coating. This treatment also removes surface oxide, which would react with the xenon fluorides. The synthetic conditions indicated in the following equations show that formation of the higher halides is favoured by a higher proportion of fluorine and higher total pressure:

$$Xe(g) + F_2(g) \xrightarrow{400°C, 1\,atm} XeF_2(g) \quad (Xe\,in\,excess)$$

$$Xe(g) + 2F_2(g) \xrightarrow{600°C, 6\,atm} XeF_4(g) \quad (Xe : F_2 = 1:5)$$

$$Xe(g) + 3F_2(g) \xrightarrow{300°C, 60\,atm} XeF_6(g) \quad (Xe : F_2 = 1:20)$$

A simple 'window-sill' synthesis is also possible. Xenon and fluorine are sealed in a glass bulb (rigorously dried to prevent the formation of HF and the attendant etching of the glass) and

the bulb is exposed to sunlight, whereupon beautiful crystals of XeF_2 slowly form in the bulb. It will be recalled that F_2 undergoes photodissociation (Chapter 17), and in this synthesis the photochemically generated F atoms react with Xe atoms.

18.6 Reactions of xenon fluorides

Key points: Xenon fluorides are strong oxidizing agents and form complexes with, F^- such as XeF_5^-, XeF_7^-, and XeF_8^{2-}; they are used in the preparation of compounds containing $Xe-O$ and $Xe-N$ bonds.

The reactions of the xenon fluorides are similar to those of the high oxidation state interhalogens (Section 17.10), and redox and metathesis reactions dominate. One important reaction of XeF_6 is metathesis with oxides:

$$XeF_6(s) + 3H_2O(l) \rightarrow XeO_3(aq) + 6HF(g)$$

$$2XeF_6(s) + 3SiO_2(s) \rightarrow 2XeO_3(s) + 3SiF_4(g)$$

Another striking chemical property of the xenon fluorides is their strong oxidizing power:

$$2XeF_2(s) + 2H_2O(1) \rightarrow 2Xe(g) + 4HF(g) + O_2(g)$$

$$XeF_4(s) + Pt(s) \rightarrow Xe(g) + PtF_4(s)$$

As with the interhalogens, the xenon fluorides react with strong Lewis acids to form xenon fluoride cations:

$$XeF_2(s) + SbF_5(l) \rightarrow [XeF]^+[SbF_6]^-(s)$$

These cations are associated with the counterion by F^- bridges.

Another similarity with the interhalogens is the reaction of XeF_4 with the Lewis base F^- in acetonitrile (cyanomethane, CH_3CN) solution to produce the XeF_5^- ion:

$$XeF_4 + [N(CH_3)_4]F \rightarrow [N(CH_3)_4]^+[XeF_5]^-$$

The XeF_5^- ion is pentagonal planar (9), and in the VSEPR model the two electron pairs on Xe occupy axial positions on opposite sides of the plane. Similarly, it has been known for many years that reaction of XeF_6 with an F^- source produces the XeF_7^- or XeF_8^{2-} ions depending on the proportion of fluoride. Only the shape of XeF_8^{2-} is known: it is a square antiprism (10), which is difficult to reconcile with the simple VSEPR model because this shape does not provide a site for the lone pair on Xe.

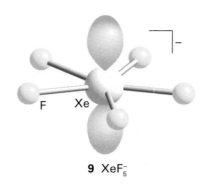

9 XeF_5^-

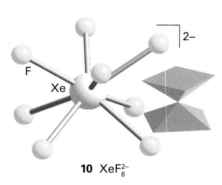

10 XeF_8^{2-}

The xenon fluorides are the gateway to the preparation of compounds of the noble gases with elements other than F and O. The reaction of nucleophiles with a xenon fluoride is one useful strategy for the synthesis of such bonds. For instance, the reaction

$$XeF_2 + HN(SO_2F)_2 \rightarrow FXeN(SO_2F)_2 + HF$$

is driven forward by the stability of the product HF and the energy of formation of the Xe–N bond (**11**). A strong Lewis acid such as AsF_5 can extract F^- from the product of this reaction to yield the cation $[XeN(SO_2F)_2]^+$. Another route to XeN bonds is the reaction of one of the fluorides with a strong Lewis acid:

$$XeF_2 + AsF_5 \rightarrow [XeF]^+[AsF_6]^-$$

followed by the introduction of a Lewis base, such as CH_3CN, to yield $[CH_3CNXe]^+ [AsF_6]^-$.

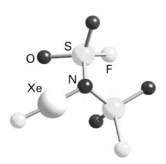

11 $FXeN(SO_2F)_2$

18.7 Xenon–oxygen compounds

Key point: The xenon oxides are unstable and highly explosive.

Xenon oxides are endergonic ($\Delta_f G^\circ > 0$) and cannot be prepared by direct interaction of the elements. The oxides and oxofluorides are prepared by the hydrolysis of xenon fluorides:

$$XeF_6(s) + 3H_2O(l) \rightarrow XeO_3(s) + 6HF(aq)$$
$$3XeF_4(s) + 6H_2O(l) \rightarrow XeO_3(s) + 2Xe(g) + \tfrac{3}{2}O_2(g) + 12HF(aq)$$
$$XeF_6(s) + H_2O(l) \rightarrow XeOF_4(s) + 2HF(aq)$$

The pyramidal xenon trioxide, XeO_3 (**4**) presents a serious hazard because this endergonic compound is highly explosive. It is a very strong oxidizing agent in acidic solution, with $E^\circ(XeO_3,Xe) = +2.10$ V. In basic aqueous solution the Xe(VI) oxoanion $HXeO_4^-$ slowly decomposes in a coupled disproportionation and water oxidation to yield a Xe(VIII) perxenate ion, XeO_6^{4-}, and xenon:

$$2HXeO_4^-(aq) + 2OH^-(aq) \rightarrow$$
$$XeO_6^{4-}(aq) + Xe(g) + O_2(g) + 2H_2O(l)$$

The perxenates of several alkali metal ions have been prepared by treatment of XeO_3 with ozone in basic conditions. These compounds are white crystalline solids with octahedral XeO_6^{4-} units (**12**). They are powerful oxidizing agents in acidic aqueous solution:

$$XeO_6^{4-}(aq) + 3H^+(aq) \rightarrow HXeO_4^-(aq) + \tfrac{1}{2}O_2(g) + H_2O(l)$$

Treating Ba_2XeO_6 with concentrated sulfuric acid produces the only other known oxide of xenon, XeO_4 (**5**), which is an explosively unstable gas.

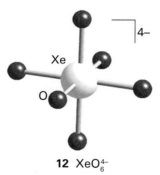

12 XeO_6^{4-}

■ **A brief illustration.** The structure of many xenon compounds can be successfully predicted by using the VSEPR model. The Lewis structure of the perxenate ion is shown in (**13**). With six electron pairs around the Xe atom, the VSEPR model predicts an octahedral arrangement of bonding electron pairs and an octahedral overall structure. ■

13 XeO_6^{4-}

Xenon forms the oxofluorides $XeOF_2$ (**6**), XeO_3F_2 (**7**), and $XeOF_4$ (**8**). When alkali metal fluorides are dissolved in $XeOF_4$, solvated fluoride ions are formed of composition $F^-\cdot 3XeOF_4$. Attempted removal of $XeOF_4$ from the solvate yields $XeOF_5^-$ (**14**), which is a pentagonal pyramid.

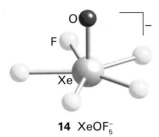

14 $XeOF_5^-$

■ **A brief illustration.** The structures of compounds of xenon can be probed using ^{129}Xe-NMR spectroscopy. For example, the ^{129}Xe-NMR spectrum of $XeOF_4$ (**8**) consists of a quintuplet of peaks. These peaks correspond to the single Xe environment, which is coupled to four equivalent ^{19}F atoms. ■

EXAMPLE 18.1 Describing the synthesis of a noble gas compound

Describe a procedure for the synthesis of potassium perxenate, starting with xenon and other reagents of your choice.

Answer We know that Xe—O compounds are endergonic, so they cannot be prepared by direct reaction of xenon and oxygen, so we need to look for an indirect method. As was described in the text, the hydrolysis of XeF_6 yields XeO_3, which undergoes disproportionation in basic solution to yield perxenate, XeO_6^{4-}. Thus XeF_6 could be synthesised by the reaction of xenon and excess F_2 at 300°C and 6 MPa in a stout nickel container. The resulting XeF_6 might then be converted to the perxenate in one step by exposing it to aqueous KOH solution. The resulting potassium perxenate (which turns out to be a hydrate) could then be crystallized.

Self-test 18.1 Write a balanced equation for the decomposition of xenate ions in basic solution for the production of perxenate ions, xenon, and oxygen.

18.8 Xenon insertion compounds

Key point: Xenon can insert into H—Y bonds.

A number of noble gas hydrides have been isolated at low temperatures with the general formula HEY, where E is the Group 18 element and Y is an electronegative element or fragment. The first species identified were HXeCl, HXeBr, HXeI, and HKrCl. More recently characterized species include HKrCN and $HXeC_3N$. They are all prepared by UV photolysis of the HY precursors in the solid noble gas at low temperatures. When xenon reacts in this way with water, both HXeOH (**15**) and the HXeO radical are produced. The latter can react further with another Xe atom and hydrogen to form HXeOXeH (**16**), which is the smallest molecule to contain two Xe atoms. Xenon atoms have been successfully inserted into H—C bonds of hydrocarbons by photolysis and annealing of a solid mixture of C_2H_2 and Xe. The noble gas hydrides HXeCCH (**17**) and HXeCCXeH (**18**) have been prepared in this way.

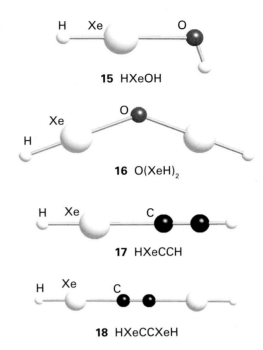

15 HXeOH

16 $O(XeH)_2$

17 HXeCCH

18 HXeCCXeH

These insertion reactions may hold the key to explaining the so-called 'missing xenon' phenomenon. Relative to the other noble gases the amount of xenon in the atmosphere is depleted by a factor of 20. One theory is that xenon in the Earth's interior can form stable compounds. The formation of these insertion compounds indicates that xenon compound formation under extreme conditions may indeed be possible. In fact, recent studies have indicated that Xe will exchange with Si in silicate minerals under high temperature and pressure.

18.9 Organoxenon compounds

Key point: Organoxenon compounds can be prepared by xenodeborylation of an organoboron compound.

The first compound containing Xe—C bonds was reported in 1989. Since then a wide variety of organoxenon compounds have been prepared. The most useful routes to organoxenon compounds are through the fluorides XeF_2 and XeF_4.

Organoxenon(II) salts can be prepared from organoboranes by **xenodeborylation**, the substitution of B by Xe. For example, tris(pentafluorophenyl)borane reacts with XeF_2 in dichloromethane to produce arylxenon(II) fluoroborates (**19**):

$$(C_6F_5)_3B + XeF_2 \xrightarrow{CH_2Cl_2} $$
$$[C_6F_5Xe]^+ + [(C_6F_5)_nBF_{4-n}]^- \quad n = 1, 2$$

When this reaction is carried out in anhydrous HF, all the C_6F_5 groups are transferred to the Xe:

$$(C_6F_5)_3B + 3XeF_2 \xrightarrow{HF} 3[C_6F_5Xe]^+ + [BF_4]^- + 2[F(HF)_n]^-$$

A general route that can be used to introduce other organic groups uses organodifluoroboranes, RBF_2:

$$RBF_2 + XeF_2 \xrightarrow{CH_2Cl_2} [RXe]^+ + [BF_4]^-$$

In addition to xenodeborylation, organoxenon(II) compounds (**20**) can be prepared from $C_6F_5SiMe_3$:

$$3C_6F_5SiMe_3 + 2XeF_2 \xrightarrow{CH_2Cl_2} Xe(C_6F_5)_2 + C_6F_5XeF + 3Me_3SiF$$

Organoxenon(II) compounds are thermally unstable and decompose above $-40°C$.

19 $C_6F_5Xe^+$

20 $Xe(C_6H_5)_2$

The first organoxenon(IV) compound was prepared by the reaction between XeF_4 and $C_6F_5BF_2$ in CH_2Cl_2:

$$C_6F_5BF_2 + XeF_4 \xrightarrow{CH_2Cl_2} [C_6F_5XeF_2][BF_4]$$

Organoxenon(IV) compounds are less thermally stable than the analogous Xe(II) compounds. In all Xe(II) and Xe(IV) salts the Xe atom is bonded to a C atom that is part of a π system. Extended π systems, such as in aryl groups, increase the stability of the Xe−C bond. This stability is further favoured by the presence of electron-withdrawing substituents (such as fluorine) on the aryl group.

18.10 Coordination compounds

Key points: Argon, xenon, and krypton form coordination compounds that are usually studied by matrix isolation; the stability of the complexes decreases in the order Xe > Kr > Ar.

Coordination compounds of noble gases have been known since the mid-1970s. The first stable noble-gas coordination compound to be synthesized was $[AuXe_4]^{2+}[SbF_{11}]^{2-}$, which contains a square-planar $[AuXe_4]^{2+}$ cation (**21**). The compound is made by reduction of AuF_3 by HF/SbF_5 in elemental xenon to yield dark red crystals that are stable up to $-78°C$. Alternatively, addition of Xe to a solution of Au^{2+} in HF/SbF_5 gives a dark red solution that is stable up to $-40°C$. This solution is stable at room temperature under a xenon pressure of 1 MPa (about 10 atm). During the reduction of Au^{3+} to Au^{2+} the extreme Brønsted acidity of HF/SbF_5 (Section 15.11) is essential and the overall reaction indicates the role of protons:

$$AuF_3 + 6Xe + 3H^+ \xrightarrow{HF/SbF_5} [AuXe_4]^{2+} + Xe_2^+ + 3HF$$

Green crystals of $[Xe_2]^+[Sb_4F_{21}]^-$ are also produced at $-60°C$. The Xe−Xe bond length is 309 pm and is the longest homonuclear bond known for a main-group element.

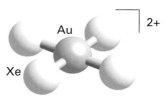

21 $[AuXe_4]^{2+}$

Many complexes of the noble gases are transient species that have been characterized by matrix isolation. The complex $Fe(CO)_4Xe$ is formed when $Fe(CO)_5$ is photolysed in solid xenon at 12 K. Similarly, $M(CO)_5E$ is formed when $M(CO)_6$ (M = Cr, Mo, or W) is photolysed in solid argon, krypton, or xenon at 20 K, where E = Ar, Kr, or Xe, respectively. An alternative method of synthesizing these complexes is to generate them in an argon, krypton, or xenon atmosphere. The Xe complex has also been isolated in liquid xenon. The stabilities of the complexes decrease in the order W > Mo ≈ Cr and Xe > Kr > Ar. The complexes are octahedral (**22**) and the bonding is thought to involve interactions between the p orbitals on the noble gas and orbitals on the equatorial CO groups. Thus, noble gases can be thought of as potential ligands, and indeed have been fully characterized as such (including by NMR; Section 22.16).

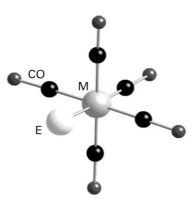

22 $[M(CO)_5E]$ E = Ar, Kr, Xe

When $[Rh(\eta^5\text{-}Cp)(CO)_2]$ or $[Rh(\eta^5\text{-}Cp^*)(CO)_2]$ is photolysed in supercritical xenon or krypton at room temperature, the complexes $[Rh(\eta^5\text{-}Cp)(CO)E]$ and $[Rh(\eta^5\text{-}Cp^*)(CO)E]$ are formed, where E = Xe or Kr. The $\eta^5\text{-}Cp^*$ complexes are less stable than the $\eta^5\text{-}Cp$ analogues and the Kr complexes are less stable than those of Xe.

18.11 Other compounds of noble gases

Key point: Krypton and radon fluorides are known but their chemical properties are much less extensive than those of xenon.

Radon has a lower ionization energy than Xe, so it can be expected to form compounds even more readily. Evidence exists for the formation of RnF_2, and cationic compounds, such as

[RnF$^+$][SbF$_6^-$], but detailed characterization is frustrated by their radioactivity. Krypton has a much higher ionization energy than Xe (Table 18.1) and its ability to form compounds is more limited. Krypton difluoride, KrF$_2$, is prepared by passing an electric discharge or ionizing radiation through a fluorine–krypton mixture at low temperatures ($-196°$C). As with XeF$_2$, the krypton compound is a colourless volatile solid and the molecule is linear. It is an endergonic and highly reactive compound that must be stored at low temperatures.

When monomeric HF is photolysed in solid argon and annealed to 18 K, HArF is formed. This compound is stable up to 27 K and contains the HAr$^+$ and F$^-$ ions. The related molecular ions, HHe$^+$, HNe$^+$, and HXe$^+$ have been observed by spectroscopy.

The heavier noble gases form clathrates. Argon, Kr, and Xe form clathrates with quinol (1,4-C$_6$H$_4$(OH)$_2$) with one gas atom to three quinol molecules. They also form clathrate hydrates with water in a ratio of one gas atom to 46 H$_2$O molecules. Helium and Ne are too small to form stable clathrates. Titan, Saturn's moon, has a dense atmosphere in which levels of Kr and Xe are depleted in comparison to Ar. It is believed that the Kr and Xe are trapped in clathrates whereas the smaller Ar atoms are trapped less effectively.

Endohedral fullerene complexes, in which the 'guest' atom or ion sits inside a fullerene cage (Section 14.6b), have been observed for C$_{60}^{n+}$ and C$_{70}^{n+}$ (n = 1, 2, or 3) with He and C$_{60}^+$ with Ne. Molecular orbital calculations indicate that Ar should be able to penetrate the C$_{60}^+$ cage, although this complex has not yet been observed.

Other than the fullerene complexes, those identified transiently in high-energy molecular beams, or van der Waals complexes in the gas phase, there are no known compounds of He. However, theoretical calculations predict HeBeO to be exergonic.

FURTHER READING

W. Grochala, Atypical compounds of gases which have been called 'noble', *Chem. Soc. Rev.*, 2007, **36**, 1632.

A.G. Massey, *Main group chemistry*. Wiley (2000).

D.M.P. Mingos, *Essential trends in inorganic chemistry*. Oxford University Press (1998).

M.S. Albert, G.D. Cates, B. Driehuys, W. Happer, B. Saam, C.S. Springer, and A. Wishnia, Biological magnetic resonance imaging using laser-polarized ^{129}Xe, *Nature*, **370**, 199–201 (1994).

R.B. King (ed.), *Encyclopedia of inorganic chemistry*. Wiley (2005).

M. Ozima and F.A. Podosec, *Noble gas geochemistry*. Cambridge University Press (2002).

P. Lazlo and G.J. Schrobilgen, *Angew. Chem., Int. Ed. Engl.*, 1988, **27**, 479. An enjoyable account of the early failure and final success in the quest for compounds of the noble gases.

J. Holloway, Twenty-five years of noble gas chemistry. *Chemistry in Britain*, 1987, 658. A good summary of the development of the field.

H. Frohn and V.V. Bardin, *Organometallics*, 2001, **20**, 4750. A readable review of the organo compounds of the noble gases.

EXERCISES

18.1 Explain why helium is present in low concentration in the atmosphere even though it is the second most abundant element in the universe.

18.2 Which of the noble gases would you choose as (a) the lowest temperature liquid refrigerant, (b) an electric discharge light source requiring a safe gas with the lowest ionization energy, (c) the least expensive inert atmosphere?

18.3 By means of balanced chemical equations and a statement of conditions, describe a suitable synthesis of (a) xenon difluoride, (b) xenon hexafluoride, (c) xenon trioxide.

18.4 Draw the Lewis structures of (a) XeOF$_4$, (b) XeO$_2$F$_2$, and (c) XeO$_6^{2-}$.

18.5 Give the formula and describe the structure of a noble gas species that is isostructural with (a) ICl$_4^-$, (b) IBr$_2^-$, (c) BrO$_3^-$, (d) ClF.

18.6 (a) Give a Lewis structure for XeF$_7^-$. (b) Speculate on its possible structures by using the VSEPR model and analogy with other xenon fluoride anions.

18.7 Use molecular orbital theory to calculate the bond order of the diatomic species E$_2^+$ with E = He and Ne.

18.8 Identify the xenon compounds A, B, C, D, and E.

$$D \xleftarrow{\text{H}_2\text{O}} C \xleftarrow{xs\ \text{F}_2} Xe \xrightarrow{\text{F}_2} A \xrightarrow{\text{MeBF}_2} B$$
$$\Big\downarrow 2\text{F}_2$$
$$E$$

18.9 Predict the appearance of the ^{129}Xe-NMR spectrum of XeOF$_3^+$.

18.10 Predict the appearance of the ^{19}F-NMR spectrum of XeOF$_4$.

PROBLEMS

18.1 The first compound containing an Xe–N bond was reported by R.D. LeBlond and K.K. DesMarteau (*J. Chem. Soc., Chem. Commun.*, 1974, 554). Summarize the method of synthesis and characterization. (The proposed structure was later confirmed in an X-ray crystal structure determination.)

18.2 (a) Use the references in the paper by O.S. Jina, X.Z. Sun, and M.W. George (*J. Chem. Soc., Dalton Trans.*, 2003, 1773) to produce a review of the use of matrix isolation in the characterization of organometallic noble gas complexes. (b) How did the methods used by these authors differ from the usual techniques of matrix isolation? (c) Place the complexes [MnCp(CO)$_2$Xe], [RhCp(CO)Xe], [MnCp(CO)$_2$Kr], [Mo(CO)$_5$Kr], and [W(CO)$_5$Kr] in order of increasing stability towards CO substitution.

18.3 In their paper 'Xenon as a complex ligand: the tetra xenon gold(II) cation in AuXe$_4^{2+}$(Sb$_2$F$_{11}^-$)$_2$', S. Seidel and K. Seppelt, *Science*, 2000, **290**, 117 describe the first synthesis of a stable noble gas coordination compound. Give details of the synthesis and characterization of the compound.

18.4 The synthesis and characterization of the $XeOF_5^-$ anion has been described by A. Ellern and K. Seppelt (*Angew. Chem., Int. Ed. Engl*, 1995, **34**, 1586). (a) Summarize the similarities between $XeOF_4$ and IF_5. (b) Give possible reasons for the differences in structure between $XeOF_5^-$ and IF_6^-. (c) Summarize how $XeOF_5^-$ was prepared.

18.5 In their paper 'Helium chemistry: theoretical predictions and experimental challenge', (*J. Am. Chem. Soc.*, 1987, **109**, 5917) W. Koch and co-workers used quantum mechanical calculations to demonstrate that helium can form strong bonds with carbon in cations. (a) Give the range of bond lengths calculated for these He—C cations. (b) What are the requirements for an element to form a strong bond with He? (c) To which branch of science do the authors suggest that this work would be particularly relevant?

18.6 In their paper 'Observation of superflow in solid helium' (*Science*, 2004, **305**, 5692, 1941) Kim and Chan described the observed superfluidity of solid helium. Define superfluidity, describe the experiment carried out by the authors to demonstrate the property, and summarize their explanation of superfluidity.

The d-block elements

19

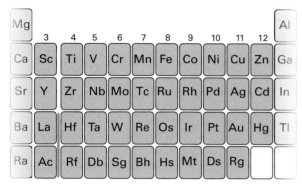

The metallic elements are the most numerous of the elements and of the metallic elements the d-block elements are the most important: their chemical properties are central to both industry and contemporary research. We shall see systematic trends across the block and trends in their chemical properties within each group. One of the most important trends is the variation in stability of the oxidation states because these stabilities are closely related to the ease of recovery of metals from their ores and to the ways in which the various metals and their compounds are handled in the laboratory. It will be seen that simple binary compounds, such as metal halides and oxides, often follow systematic trends; however, when the metal is in a low oxidation state, interesting variations such as metal–metal bonding may be encountered.

The two terms **d-block metal** and **transition metal** are often used interchangeably; however, they do not mean the same thing. The name transition metal originally derived from the fact that their chemical properties were transitional between those of the s and p blocks. Now, however, the IUPAC definition of a **transition element** is that it is an element that has an incomplete d subshell in either the neutral atom or its ions. Thus the Group 12 elements (Zn, Cd, Hg) are members of the d block but are not transition elements. In the following discussion, it will be convenient to refer to each row of the d block as a series, with the 3d series the first row of the block (Period 4), the 4d series the second row (Period 5), and so on. It will prove important to note the intrusion of the f block, the **inner transition elements**, the lanthanoids, into the 5d series. Elements towards the left of the d block are often referred to as *early* and those towards the right are referred to as *late*.

The elements

The chemical properties of the d-block elements (or d metals as we shall commonly call them) will occupy this and the next three chapters. It is therefore appropriate to begin with a survey of their occurrence and properties. The trends in their properties and the correlation with their electronic structures and therefore location in the periodic table should be kept in mind throughout these chapters.

19.1 Occurrence and recovery

Key point: Chemically soft members of the block occur as sulfide minerals and are partially oxidized to obtain the metal; the more electropositive 'hard' metals occur as oxides and are extracted by reduction.

The elements on the left of the 3d series occur in nature primarily as metal oxides or as metal cations in combination with oxoanions (Table 19.1). Of these elements, titanium ores are the most difficult to reduce, and the element is widely produced by heating TiO_2 with chlorine and carbon to produce $TiCl_4$, which is then reduced by molten magnesium at about 1000°C in an inert-gas atmosphere. The oxides of Cr, Mn, and Fe are reduced with carbon (Section 5.16), a much cheaper reagent. To the right of Fe in the 3d series, Co, Ni, Cu, and Zn occur mainly as sulfides and arsenides, which is consistent with the increasingly soft Lewis acid character of their dipositive ions. Sulfide ores are usually roasted in air either to the metal directly (for example, Ni) or to an oxide that is subsequently reduced (for example, Zn). Copper is used in large quantities for electrical conductors; electrolysis is used to refine crude copper to achieve the high purity needed for high electrical conductivity.

The difficulty of reducing the early 4d and 5d metals Mo and W is apparent from Table 19.1. It reflects the tendency of these elements to have stable high oxidation states, as discussed later. The platinum metals (Ru and Os, Rh and Ir, and Pd and Pt), which are found at the lower right of the d block, occur as sulfide and arsenide ores, usually in association with larger quantities of Cu, Ni, and Co. They are collected from the sludge that forms during the electrolytic refinement of copper and nickel. Gold (and to some extent silver) is found in its elemental form.

The role that certain d metals play in the environment is summarized in Box 19.1.

19.2 Physical properties

Key points: The properties of the d metals are largely derived from their electronic structure, with the strength of metallic bonding peaking at Group 6; the lanthanide contraction is responsible for some of the anomalous behaviour of the metals in the 5d series.

The d block of the periodic table contains the metals most important to modern society. It contains the immensely strong and light titanium, the major components of most steels (Fe, Cr, Mn, Mo), the highly electrically conducting copper, the malleable gold and platinum, and the very dense osmium and iridium. To a large extent these properties derive from the nature of the metallic bonding that binds the atoms together.

Table 19.1 Mineral sources and methods of recovery of some commercially important d metals

Metal	Principal minerals	Method of recovery	Note
Titanium	Ilmenite, $FeTiO_3$ Rutile, TiO_2	$TiO_2 + 2C + 2Cl_2 \rightarrow TiCl_4 + 2CO$ followed by reduction of $TiCl_4$ with Na or Mg	
Chromium	Chromite, $FeCr_2O_4$	$FeCr_2O_4 + 4C \rightarrow Fe + 2Cr + 4CO$	(a)
Molybdenum	Molybdenite, MoS_2	$2MoS_2 + 7O_2 \rightarrow 2MoO_3 + 4SO_2$ followed by either $MoO_3 + 2Fe \rightarrow Mo + Fe_2O_3$ or $MoO_3 + 3H_2 \rightarrow Mo + 3H_2O$	
Tungsten	Scheelite, $CaWO_4$ Wolframite, $FeMn(WO_4)_2$	$CaWO_4 + 2HCl \rightarrow WO_3 + CaCl_2 + H_2O$ followed by $WO_3 + 3H_2 \rightarrow W + 3H_2O$	
Manganese	Pyrolusite, MnO_2	$MnO_2 + 2C \rightarrow Mn + 2CO$	(b)
Iron	Haematite, Fe_2O_3 Magnetite, Fe_3O_4 Limonite, $FeO(OH)$	$Fe_2O_3 + 3CO \rightarrow 2Fe + 3CO_2$	
Cobalt	CoAsS Smaltite, $CoAs_2$ Linnaeite, Co_3S_4	Byproduct of copper and nickel production	
Nickel	Pentlandite, $(Fe,Ni)_6S_8$	$NiS + O_2 \rightarrow Ni + SO_2$	(c)
Copper	Chalcopyrite, $CuFeS_2$ Chalocite, Cu_2S	$2CuFeS_2 + 2SiO_2 + 5O_2 \rightarrow 2Cu + 2FeSiO_3 + 4SO_2$	

(a) The iron−chromium alloy is used directly for stainless steel.
(b) The reaction is carried out in a blast furnace with Fe_2O_3 to produce alloys.
(c) NiS is formed by smelting the ore and is then separated by physical processes. NiO is used in a blast furnace with iron oxides to produce steel. Nickel is purified by electrolysis or the Mond process via $Ni(CO)_4$, Box 22.1.

BOX 19.1 Toxic metals in the environment

The biosphere evolved in close association with all the elements in the periodic table, and has had a long time to adapt to them. Metals are released from rocks by weathering, and are processed by a variety of mechanisms, including biological ones. Indeed, many metals have been harnessed for essential biochemical functions (Chapter 27), and other biochemical systems have evolved to sequester metals and keep them out of harm's way. However, the natural biogeochemical cycles have been greatly perturbed by mining: circulating levels of most metals have increased dramatically since pre-industrial times.

Many metals and metalloids, including Be, Mn, Cr, Ni, Cd, Hg, Pb, Se, and As, are hazardous in occupational or environmental settings. Exposure to these elements varies widely, and depends on the pattern of their industrial use, and on their environmental chemistry. The metals of greatest environmental concern are heavy metals, such as mercury, which, being chemically very soft, bind tightly to thiol groups in proteins.

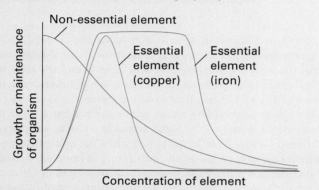

Figure B19.1 The idealized dose–response curves for a non-essential element and two essential elements.

The effect of a substance on a living organism is often plotted as a *dose–response curve*, such as those in Fig. B19.1. The shape of the dose–response curves for various metals varies widely, depending on physiological variables, and on the chemistry of the metals. For example, iron and copper are both essential elements but have very different dose–response curves: toxicity occurs at a much lower concentration for copper than for iron. It is reasonable to associate this toxicity with the higher affinity of copper ions for sulfur and nitrogen ligands, which makes copper more likely to interfere with crucial sites in proteins. Nevertheless, iron is harmful at higher doses because it can catalyse the production of oxygen radicals, and also because excess iron can stimulate the growth of bacteria.

The ability of elements to exist in alternative redox states can be of crucial importance to the hazard they present. For example, the solubilities of different oxidation states of a particular metal are often very different. Many metals are found as insoluble sulfides in the Earth's crust and are thus immobile. However, when they are disturbed and brought into contact with air, which oxidizes them, they can become very mobile. An example is mercury(II) sulfide, which is oxidized to the mobile mercury(II) sulfate on exposure to air.

Another effect of oxidation is exemplified by iron, which occurs widely as the mineral pyrite, FeS_2. Mining operations often lead to acid drainage due to the reaction

$$FeS_2 + \tfrac{15}{4}O_2 + \tfrac{7}{2}H_2O \rightarrow Fe(OH)_3 + 2SO_4^{2-} + 4H^+$$

In this process (which is generally catalysed by certain bacteria), S_2^{2-} is oxidized to sulfuric acid and Fe^{2+} is oxidized to Fe^{3+}, which hydrolyses to $Fe(OH)_3$, yielding additional H^+ ions. Streams emanating from iron mines, many of them abandoned, are thus often highly acidic.

Redox status can have opposite effects on solubility for different metals. Thus, whereas manganese is mobile under reducing conditions, other metals are mobilized under oxidizing conditions. For instance, Cr(III) is immobile, forming an insoluble oxide or kinetically inert complexes with soil constituents; in addition, there is no mechanism for transporting free Cr(III) ions across biological membranes. However, Cr(III) is solubilized upon oxidation to the chromate(VI) ion, CrO_4^{2-}. Chromate(VI) is highly toxic, and is implicated in cancer. The CrO_4^{2-} ion has the same shape and charge as SO_4^{2-}, and can be ferried across biological membranes by sulfate transport proteins. Once inside a cell, chromate is converted to Cr(III) complexes by endogenous reductants, such as ascorbate. In the process, reactive oxygen species are generated, which can damage DNA; in addition, the Cr(III) reduction product can complex and cross-link DNA. Cr(III) accumulates inside the chromate-exposed cells, as it does not have a mechanism for transport through biological membranes.

For other metals, toxicity is controlled by the interplay of membrane transport and intracellular binding with the redox state. For example, mercury salts are relatively harmless because membranes, which present a barrier to ionic species, are impermeable to Hg^{2+}. Likewise, metallic mercury is not absorbed through the gut, so it is not toxic when swallowed. However, mercury vapour is highly toxic (that is why all mercury spills must be rigorously cleaned up) because the neutral atoms readily pass through the lungs' membranes and also across the blood–brain barrier. Once in the brain, the Hg(0) is oxidized to Hg(II) by the vigorous oxidizing activity of brain cell mitochondria, and the Hg(II) binds tightly to crucial thiolate groups of neuronal proteins. Mercury is a powerful neurotoxin, but only if it gets inside nerve cells. Even more hazardous than mercury vapour are organomercury compounds, particularly methylmercury. Thus, CH_3Hg^+ is taken up through the gut because it is complexed by the stomach's chloride, forming CH_3HgCl, which, being electrically neutral, can pass through a membrane. Once inside cells, CH_3Hg^+ binds to thiolate groups and accumulates.

The environmental toxicity of mercury is associated almost entirely with eating fish. Methylmercury is produced by the action of sulfate-reducing bacteria on Hg^{2+} in sediments, and accumulates as little fish are eaten by bigger fish further up the aquatic food chain. Fish everywhere have some level of mercury present. Mercury levels can increase markedly if sediments are contaminated by additional mercury. The worst known case of environmental mercury poisoning occurred in the 1950s in the Japanese fishing village of Minamata. A polyvinyl chloride plant, using Hg^{2+} as a catalyst, discharged mercury-laden residues into the bay, where fish accumulated methylmercury to levels approaching 100 ppm. Thousands of people were poisoned by eating the fish, and a number of infants suffered mental disabilities and motor disturbance from exposure *in utero*. This disaster led to strict standards for fish consumption. Limited consumption is advised for fish at the top of the food chain, such as pike and bass in fresh waters, and swordfish and tuna in the oceans.

Regulatory action has been aimed at reducing mercury discharges and emissions, and industrial point sources have been largely controlled. Chloralkali plants, producing the large-volume industrial chemicals Cl_2 and NaOH by electrolysis of NaCl, were a major source as a mercury pool electrode was used to transfer metallic sodium to a separate hydroxide-generating compartment. However, this is now accomplished by separating the two electrode compartments with a cation-exchange membrane, which prevents migration of the anions. Combustion can vent mercury to the atmosphere if the fuel contains mercury compounds. Whereas municipal waste and hospital incinerators have been equipped with filters to reduce

(Continued)

BOX 19.1 *(Continued)*

mercury emissions to the air, coal contains small amounts of mercury minerals. Because of the huge quantities burned, coal is a major contributor to environmental mercury.

Mercury is a global problem because mercury vapour and volatile organomercurial compounds can travel long distances in the atmosphere. Eventually elemental mercury is oxidized and organomercurials are decomposed, both to Hg^{2+}, by reaction with ozone or by atmospheric hydroxyl or halogen radicals. The Hg^{2+} ions are solvated by water molecules and deposited in rainfall. Thus mercury deposition can occur far from the emission source, and is distributed fairly uniformly around the globe. For example, it is estimated that only a third of North American mercury emissions are in fact deposited in the USA, and this accounts for only half of the mercury deposition in the USA. Even the gold fields of

Brazil contribute because miners use mercury to extract the gold, which is recovered by heating the resultant amalgam to drive off the mercury. This practice is estimated to account for 2 per cent of global mercury emissions (half of South American emissions). The picture is further complicated by the fact that much of the deposited mercury is recirculated through processes that produce volatile compounds or mercury vapour. For example, much of the biomethylation activity of the sulfate-reducing bacteria produces dimethylmercury, $(CH_3)_2Hg$, which, being volatile, is vented to the atmosphere. Other bacteria have an enzyme (methylmercury lyase) that breaks the methyl−mercury bond of CH_3Hg^+, and another enzyme (methylmercury reductase) that reduces the resulting Hg(II) to Hg(0); this is a protective mechanism for the microorganisms, ridding them of mercury as volatile Hg(0).

Metallic bonding was covered in Chapter 3, which introduced the concept of band structure. Generally speaking, the same band structure is present for all the d-block metals and arises from the overlap of the $(n+1)$s orbitals to give an s band and of the nd orbitals to give a d band. The principal differences between the metals is the number of electrons available to occupy these bands: Ti ($3d^2 4s^2$) has four bonding electrons, V ($3d^3 4s^2$) five, Cr ($3d^5 4s^1$) six, and so on. The lower, net bonding region of the valence band is therefore progressively filled with electrons on going to the right across the block, which results in stronger bonding, until around Group 7 (at Mn, Tc, Re) when the electrons begin to populate the upper, net antibonding part of the band. This trend in bonding strength is reflected in the increase in melting point from the low-melting alkali metals (effectively only one bonding electron for each atom, resulting in melting points typically less than 100°C) up to Cr, and its decline thereafter to the low-melting Group 12 metals (mercury being a liquid at room temperature, Fig. 19.1 and Section 9.2). The strength of metallic bonding in tungsten is such that its melting point (3410°C) is exceeded by only one other element, carbon.

The radii of d-metal ions depend on the effective charge of the nucleus, and ionic radii generally decrease on moving to the right as the atomic number increases. The radius of the metal atoms in the solid element is determined by a combination of the strength of the metallic bonding and the size of the ions. Thus, the separations of the centres of the atoms in the solid generally follow a similar pattern to the melting points: they decrease to the middle of the d block, followed by an increase back up to Group 12, with the smallest separations occurring in and near Groups 7 and 8.

The atomic radii of the elements in the 5d series (Hf, Ta, W,...) are not much bigger than those of their 4d-series congeners (Zr, Nb, Mo,...). In fact, the atomic radius of Hf

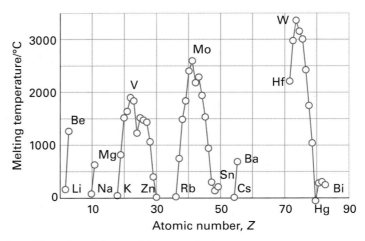

Figure 19.1 The melting points of the metals in Groups 1−15.

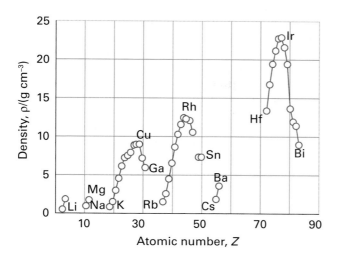

Figure 19.2 The densities of the elements in Groups 1 – 15.

is smaller than that of Zr even though it appears in a later period. To understand this anomaly, we need to consider the effect of the lanthanoids (the first row of the f block).

The intervention of the lanthanoid elements in Period 6 corresponds to the occupation of the poorly shielding 4f orbitals. Because the atomic number has increased by 32 between Zr in Period 5 and its congener Hf in Period 6 without a corresponding increase in shielding, the overall effect is that the atomic radii of the 5d-series elements are much smaller than expected. This reduction in radius is the *lanthanide contraction* introduced in Section 1.9a. The lanthanide contraction also affects the ionization energies of the 5d-series elements, making them higher than expected on the basis of a straightforward extrapolation. Some of the metals—specifically Au, Pt, Ir, and Os—have such high ionization energies that they are unreactive under normal conditions.

Atomic mass increases with atomic number, and the combination of this increase with the changes in the radii of the metal atoms in the metal lattice means that the mass densities of the elements reach a peak with Ir (density 22.65 g cm^{-3}). Figure 19.2 illustrates this trend.

Trends in chemical properties

Many of the d metals display a wide range of oxidation states, which leads to a rich and fascinating chemistry. They also form an extensive range of coordination compounds (Chapters 7, 20, and 21) and organometallic compounds (Chapter 22). Many of the trends discussed in Chapter 9 are applicable to the d metals; in particular we should recall that ionization energies generally decrease down a group and increase across a period.

19.3 Oxidation states across a series

The range of oxidation states of the d metals accounts for the interesting electronic properties of many solid compounds (Chapters 24 and 25), their ability to participate in catalysis (Chapter 26), and their subtle and interesting role in biochemical processes (Chapter 27). In this section we focus on trends in the stabilities of oxidation states within the d block.

(a) High oxidation states

Key point: The group oxidation state can be achieved by elements that lie towards the left of the d block but not by elements on the right. Oxygen is usually more effective than fluorine at bringing out the highest oxidation states because less crowding is involved.

The group oxidation state (in which the oxidation number is equal to the Group number) can be achieved by elements that lie towards the left of the d block but not by elements on the right (Section 9.5). For example, Sc, Y, and La in Group 3, with configurations nd^1($n+1$)s^2, are found in aqueous solution only in oxidation state +3, which corresponds to the loss of all their outermost electrons, and the majority of their complexes contain the elements in this state. The group oxidation state is never achieved after Group 8 (Fe, Ru, and Os). This limit on the maximum oxidation state correlates with the increase in ionization energy and hence noble character from left to right across each series in the d block.

The trend in thermodynamic stability of the group oxidation states of the 3d-series elements is illustrated in Fig. 19.3, which shows the Frost diagram for species in aqueous acidic solution. We see that the group oxidation states of Sc, Ti, and V fall in the lower part of the diagram. This location indicates that the element and any species in intermediate oxidation states are readily oxidized to the group oxidation state. By contrast, species in the group oxidation state for Cr and Mn (+6 and +7, respectively) lie in the upper part of the diagram. This location indicates that they are very susceptible to reduction. The Frost diagram shows that the group oxidation state is not achieved in Groups 8−12 of the 3d series (Fe, Co, Ni, Cu, and Zn), and also shows the oxidation states that are most stable under acid conditions; namely Ti^{3+}, V^{3+}, Cr^{3+}, Mn^{2+}, Fe^{2+}, Co^{2+}, and Ni^{2+}.

The binary compounds of the 3d-series elements with the halogens and with oxygen also illustrate the trend in stabilities of the group oxidation states. The earliest metals can achieve their group oxidation states in compounds with chlorine (for example, $ScCl_3$ and $TiCl_4$), but the more strongly oxidizing halogen fluorine is necessary to achieve the group oxidation state of V (Group 5) and Cr (Group 6), which form VF_5 and CrF_6, respectively. Beyond Group 6 in the 3d series, even fluorine cannot produce the group oxidation state, and MnF_7 and FeF_8 have not been prepared. Oxygen brings out the group oxidation state for many elements more readily than does fluorine because fewer O atoms than F atoms are needed to achieve the same oxidation number, thus decreasing steric crowding. For example, the group oxidation state of +7 for Mn is achieved in manganate(VII) salts, such as potassium permanganate, $KMnO_4$.

As can be inferred from the Frost diagram in Fig. 19.3, chromate(VI) CrO_4^{2-}, manganate(VII) MnO_4^-, and ferrate(VI) FeO_4^{2-} are strong oxidizing agents and become stronger from CrO_4^{2-} to FeO_4^{2-}. This trend is another illustration of the decreasing stability of the maximum attainable oxidation state for Groups 6, 7, and 8. Yet another example of the greater difficulty of oxidizing an element to the right of Cr to its group oxidation state is that the air oxidation of MnO_2 in molten potassium hydroxide does not take Mn to its group oxidation state but instead yields the deep green compound potassium manganate(VI), K_2MnO_4. The disproportion of MnO_4^{2-} in acidic aqueous solution yields manganese(IV) oxide (MnO_2) and the deep purple manganate(VII) ion MnO_4^-:

$$3\,MnO_4^{2-}(aq) + 4\,H^+(aq) \rightarrow 2\,MnO_4^-(aq) + MnO_2(s) + 2\,H_2O(l)$$

(b) Intermediate oxidation states in the 3d series

Key point: Oxidation state +3 is common to the left of the 3d series and +2 is common for metals from the middle to the right of the block.

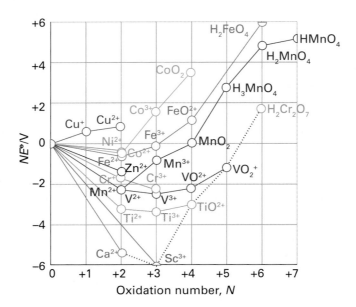

Figure 19.3 A Frost diagram for the first series of the d-block elements in acidic solution (pH = 0). The broken line connects species in their group oxidation states.

Most unipositive d-metal ions (M$^+$) disproportionate (to M and M^{2+}) because the bonds in the solid metal are so strong and, for the 3d metals, the +2 oxidation state is generally the lowest one it is necessary to consider in aqueous solution. Exceptions are mainly confined to metal–metal bonded and organometallic compounds, such as the metal carbonyls Ni(CO)$_4$ and Mo(CO)$_6$ discussed in Chapter 22, in which the metal is in oxidation state 0. Figure 19.4 shows the second and third ionization energies of the 3d metals, and we can see the expected increase across the period, in line with increasing nuclear charge. The anomalous values for manganese and iron are a result of the very stable d^5 configurations of the Mn^{2+} and Fe^{3+} ions.

The dipositive aqua ions, M^{2+}(aq) (specifically, the octahedral complexes [M(OH$_2$)$_6$]$^{2+}$), play an important role in the chemistry of the 3d-series metals. Many of these ions are coloured as a result of d–d transitions in the visible region of the spectrum (Section 20.4). For example, Cr^{2+}(aq) is blue, Fe^{2+}(aq) is green, Co^{2+}(aq) is pink, Ni^{2+}(aq) is green, and Cu^{2+}(aq) is blue.

The +3 oxidation state is common at the left of the period, and is the only oxidation state normally encountered for scandium. Careful inspection of the Frost diagrams in Fig. 19.3 gives the balance of stability between the +3 and +2 oxidation states. Titanium, vanadium, and chromium all form a wide range of compounds in oxidation state +3, and under normal conditions the +3 oxidation state is more stable than the +2 state. Manganese(II) is especially stable due to its half-filled d shell, and relatively few Mn(III) compounds are known. Beyond manganese, many Fe(III) complexes are known, but are often oxidizing. In acid solution, Co^{3+}(aq) is powerfully oxidizing and O$_2$ is evolved:

$$4\,Co^{3+}(aq) + 2\,H_2O(l) \rightarrow 4\,Co^{2+}(aq) + 4\,H^+(aq) + O_2 \qquad E_{cell}^{\ominus} = +0.58\ \text{V}$$

Species such as Co(III) are stabilized as oxido compounds (for instance, CoO(OH)), which are formed under basic conditions (Section 19.8), or when complexed by good donor ligands such as amines. Aqua ions of Ni^{3+} and Cu^{3+} have not been prepared.

In contrast, M(II) becomes increasingly common from left to right across the series. For example, among the early members of the 3d series, Sc^{2+}(aq) (Group 3) is unknown and Ti^{2+}(aq) (Group 4) is formed only upon bombarding solutions of Ti^{3+} with electrons in the technique known as **pulse radiolysis**. For Groups 5 and 6, V^{2+}(aq) and Cr^{2+}(aq) are thermodynamically unstable with respect to oxidation by H$^+$ ions:

$$2\,V^{2+}(aq) + 2\,H^+(aq) \rightarrow 2\,V^{3+}(aq) + H_2(g) \qquad E_{cell}^{\ominus} = +0.26\ \text{V}$$

Despite this instability, the slowness of the evolution of H$_2$ makes it possible to work with aqueous solutions of these dipositive ions in the absence of air, and as a result they are useful reducing agents. Beyond Cr (for Mn^{2+}, Fe^{2+}, Co^{2+}, Ni^{2+}, and Cu^{2+}), M(II) is stable with respect to reaction with water, and only Fe^{2+} is oxidized by air.

Water is incompatible with many metal ions because it can serve as an oxidizing agent. A wider range of M^{2+} ions is therefore known in solids than in aqueous solution. For example, TiCl$_2$ can be prepared by the reduction of TiCl$_4$ with hexamethyldisilane,

$$(CH_3)_3Si-Si(CH_3)_3(l) + TiCl_4(l) \rightarrow TiCl_2(s) + 2\,(CH_3)_3SiCl(l)$$

but it is oxidized by both water and air to give TiO$_2$. With the exception of ScCl$_2$, these dihalides are well described by structures containing isolated M^{2+} ions (Table 19.2). The fluorides have a rutile structure (Section 3.9) and most of the heavier halides are layered;

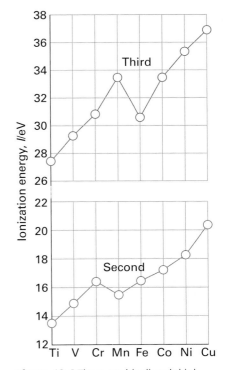

Figure 19.4 The second (red) and third (blue) ionization energies of the 3d metals.

Table 19.2 Structures of the d-metal dihalides*

	Ti	V	Cr	Mn	Fe	Co	Ni	Cu
F	R	R†	R†	R	R	R	R†	L
Cl	L	L	R†	L	L	L	L	L
Br	L	L	L	L	L	L	L	L
I	L	L	L	L	L	L	L	

* R = rutile, L = layered (CdI$_2$, CdCl$_2$ or related structures).

† The structure is distorted from ideal.

Adapted from A.F. Wells, *Structural inorganic chemistry*. Oxford University Press (1984).

in both cases the metal is in an octahedral site. This shift in structure is understandable because the rutile structure type is associated with ionic bonding and the layered structures are associated with more covalent bonding.

EXAMPLE 19.1 Judging trends in redox stability in the d block

On the basis of trends in the properties of the 3d-series elements, suggest possible M^{2+} aqua ions for use as reducing agents and write a balanced chemical equation for the reaction of one of these ions with O_2 in acidic solution.

Answer We need to identify an element that has an accessible, but oxidizable, M(II) oxidation state. The M(II) state is most stable for the late 3d-series elements with ions of the metals on the left of the series, such as $V^{2+}(aq)$ and $Cr^{2+}(aq)$ being too strongly reducing to be used easily in water. By contrast, the $Fe^{2+}(aq)$ ion is only weakly reducing and the $Co^{2+}(aq)$, $Ni^{2+}(aq)$, and $Cu^{2+}(aq)$ ions not reducing in water. The Latimer diagram for iron indicates that Fe^{3+} is the only accessible higher oxidation state of iron in acidic solution:

$$Fe^{3+} \xrightarrow{\;+0.77\;} Fe^{2+} \xrightarrow{\;-0.44\;} Fe$$

The chemical equation for the oxidation is then

$$4\,Fe^{2+}(aq) + O_2(g) + 4\,H^+(aq) \rightarrow 4\,Fe^{3+}(aq) + 2\,H_2O(l)$$

Self-test 19.1 Refer to the appropriate Latimer diagram in *Resource section* 3 and identify the oxidation state and formula of the species that is thermodynamically favoured when an acidic aqueous solution of V^{2+} is exposed to oxygen.

(c) Intermediate oxidation states in the 4d and 5d series

Key points: Complexes of M(II) with σ-donor ligands are common for the 3d-series metals but complexes of M(II) of the 4d- and 5d-series metals are less common; they generally contain π-acceptor ligands.

In contrast to the 3d-series, the 4d- and 5d-series metals only rarely form simple $M^{2+}(aq)$ ions. A few examples have been characterized, including $[Ru(OH_2)_6]^{2+}$, $[Pd(OH_2)_4]^{2+}$, and $[Pt(OH_2)_4]^{2+}$. However, the 4d- and 5d-series metals do form many M(II) complexes with ligands other than H_2O; they include the very stable d^6 octahedral complexes, such as (**1**), and the much rarer square-pyramidal d^6 complexes, such as (**2**), which form with bulky ligands. Palladium(II) and platinum(II) form many square-planar d^8 complexes, such as $[PtCl_4]^{2-}$. Much of the rationalization of this behaviour relates to the bonding of the ligands, and cannot be fully understood without reference to the concepts introduced in Chapter 20; however, examples of complexes are included here for completeness.

The Ru(II) complex in (**1**) is obtained by reducing $RuCl_3.3H_2O$ with zinc in the presence of ammonia; it is a useful starting material for the synthesis of a range of ruthenium(II) pentaammine complexes that have π-acceptor ligands, such as CO, as the sixth ligand:

$$[Ru(NH_3)_5(OH_2)]^{2+}(aq) + L(aq) \rightarrow [Ru(NH_3)_5L]^{2+}(aq) + H_2O(l)$$

$$(L = CO, N_2, N_2O)$$

These Ru and the related Os pentaammine species are strong π donors, and consequently the complexes they form with the π acceptors CO and N_2 are stable.

19.4 Oxidation states down a group

Key points: In Groups 4–10, the highest oxidation state of an element becomes more stable on descending a group, with the greatest change in stability occurring between the first two rows of the d block; ease of oxidation of the metal does not correlate with the highest available oxidation state.

The stability of an oxidation state of an element changes on descending a group. Figure 19.5 is the Frost diagram for the chromium group: note that Mo(VI) and W(VI) lie below Cr(VI) in $H_2Cr_2O_7$, indicating that the maximum oxidation state is more stable for Mo and W, with the greatest change in stability occurring between Cr and Mo. This pattern is repeated

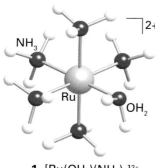

1 $[Ru(OH_2)(NH_3)_5]^{2+}$

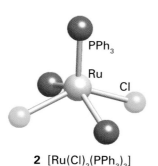

2 $[Ru(Cl)_2(PPh_3)_3]$

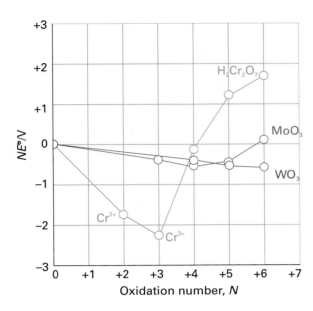

Figure 19.5 A Frost diagram for the chromium group in the d block (Group 6) in acidic solution (pH = 0).

across the d block: for Groups 4−10 the highest oxidation state of an element becomes more stable on descending a group, with the greatest change in stability occurring between the first two series of the block.

The increasing stability of high oxidation states for the heavier d metals can be seen in the formulas of their halides (Table 19.3), and the limiting formulas MnF_4, TcF_6, and ReF_7 show a greater ease of oxidizing the 4d- and 5d-series metals than the 3d-series. The hexa-fluorides of the heavier d metals (as in PtF_6) have been prepared from Group 6 through to Group 10 except for Pd. In keeping with the stability of high oxidation states for the heavier metals, WF_6 is not a significant oxidizing agent. However, the oxidizing character of the hexafluorides increases to the right, and PtF_6 is so potent that it can oxidize O_2 to O_2^+:

$$O_2(g) + PtF_6(s) \rightarrow (O_2)PtF_6(s)$$

Even Xe can be oxidized by PtF_6 (Section 18.5).

The ability to achieve the highest oxidation state does not correlate with the ease of oxidation of the bulk metal to an intermediate oxidation state. For example, although elemental iron is susceptible to oxidation by $H^+(aq)$ under standard conditions,

$$Fe(s) + 2H^+(aq) \rightarrow Fe^{2+}(aq) + H_2(g) \qquad E_{cell}^\ominus = +0.44 \text{ V}$$

no chemical oxidizing agent has been found that will take it to its group oxidation state of +8 in solution. By contrast, although the two heavier metals in Group 8 (Ru and Os) are not oxidized by H^+ ions in acidic aqueous solution:

$$Os(s) + 2H_2O(aq) \rightarrow OsO_2(s) + 2H_2(g) \qquad E_{cell}^\ominus = -0.65 \text{ V}$$

they can be oxidized by oxygen to the +8 state, as RuO_4 and OsO_4:

$$Os(s) + 2O_2(g) \rightarrow OsO_4(s)$$

Table 19.3 Highest oxidation states of the d-block binary halides*

Group							
4	5	6	7	8	9	10	11
TiI_4	VF_5	CrF_5†	MnF_4	$FeBr_3$	CoF_4	NiF_4	$CuBr_2$
ZrI_4	NbI_5	$MoCl_6$	$TcCl_6$	RuF_6	RhF_6	PdF_4	AgF_3
HfI_4	TaI_5	WBr_6	ReF_7	OsF_6	IrF_6	PtF_6	AuF_5

* The formulas show the least electronegative halide that brings out the highest oxidation state of the d metal.
† CrF_6 exists for several days at room temperature in a passivated Monel chamber.

We see from Fig. 19.5 that the Frost diagrams for the positive oxidation states of Mo and (especially) W are quite flat. This flatness indicates that neither element exhibits the marked tendency to form M(III) that is so characteristic of chromium. Mononuclear complexes of Mo and W in oxidation states +3, +4, +5, and +6 are common.

19.5 Structural trends

Key points: The 4d- and 5d-series elements often exhibit higher coordination numbers than their 3d-series congeners; compounds of d-metals in high oxidation states tend to have covalent structures.

We might expect the ionic radii of d-metal ions to follow the same pattern as atomic radii and decrease across each period. However, in addition to this general trend, there are some subtle effects on the size of the ions caused by the order in which d orbitals are occupied and which will be explained more fully in Chapter 20. Figure 19.6 shows the variation in radius of the M^{2+} ions for six-coordinate complexes of the 3d-series metals. To understand the two trends shown in the illustration we need to know that three of the 3d orbitals point between the ligands and that the remaining two point directly at them (this feature is explained more fully in Section 20.1). For the so-called 'low-spin complexes', in which electrons first individually occupy the three 3d orbitals that point between the ligands, there is a general decrease in radius across the series up to the d^6 ion Fe^{2+}. After Fe^{2+}, the additional electrons occupy the two d orbitals that point towards the ligands, which they repel slightly and thus result in an effective increase in radius. The trend for the so-called 'high-spin complexes', in which the electrons occupy all five 3d orbitals singly before pairing with any already present, is more complicated. Initially, at Ti^{2+} (d^2) and V^{2+} (d^3), the electrons occupy the three 3d orbitals that point between the ligands and the radius decreases. The next two electrons occupy the two orbitals that point at the ligands, and the radii increase accordingly. Then the sequence starts again at Fe^{2+} (d^6), as the additional electrons pair with those already present, first in the 'non-repelling' set of three orbitals and then, finally, in the two 'repelling' orbitals.

As might be anticipated from consideration of atomic and ionic radii, the 4d- and 5d-series elements often have higher coordination numbers than their smaller 3d-series congeners. Table 19.4 illustrates this trend for the fluorido and cyanido complexes of the early d metals. Note that with the small F^- ligand, these 3d-series metals tend to form six-coordinate complexes but that the larger 4d- and 5d-series metals in the same oxidation state tend to form seven-, eight-, and nine-coordinate complexes. The octacyanidomolybdate complex, $[Mo(CN)_8]^{3-}$, illustrates the tendency towards high coordination numbers with compact ligands. The same complex is readily reduced electrochemically or chemically without change of coordination number:

$$[Mo(CN)_8]^{3-}(aq) + e^- \rightarrow [Mo(CN)_8]^{4-}(aq) \qquad E^\ominus = +0.73 \text{ V}$$

Structural changes also result from changes in metal oxidation state. Compounds of d metals in low oxidation states often exist as ionic solids, whereas compounds of d metals in high oxidation states tend to take on covalent character: compare OsO_2, which is an ionic solid with the rutile structure, and OsO_4, which is a covalent molecular species (Section 19.8). We discussed the effect in Section 1.9e.

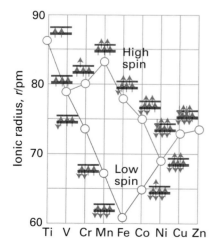

Figure 19.6 The ionic radii of the M^{2+} ions of the 3d metals. Where there are alternatives, red represents high-spin and blue low-spin complexes. In the orbital energy diagrams, the three lower levels are the d orbitals that point between ligands and the two upper levels are the d orbitals that point directly at them.

Table 19.4 Coordination numbers of some early d-block fluorido and cyanido complexes*

	Group 3	4	5
3d	$(NH_4)_3[ScF_6]$ (6)	$Na_2[TiF_6]$ (6)	$K[VF_6]$ (6); $K_2[V(CN)_7]\cdot 2H_2O$ (7)
4d	$Na_6[YF_9]$ (9)	$Na_3[ZrF_7]$ (7)	$K_2[NbF_7]$ (7); $K_4[Nb(CN)_8]$ (8)
5d	$Na_6[LaF_9]$ (9)	$Na_3[HfF_7]$ (7)	$K_3[TaF_8]$ (8)

* The number in parentheses is the coordination number of the d-metal atom in the complex anion enclosed in square brackets.

19.6 Noble character

Key point: Metals on the right of the d block tend to exist in low oxidation states and form compounds with soft ligands.

With the exception of Group 12, the metals at the lower right of the d block are resistant to oxidation. This resistance is largely due to strong intermetallic bonding and high ionization energies. It is most evident for Ag, Au, and the 4d- and 5d-series metals in Groups 8−10 (Fig. 19.7). The latter are referred to as the **platinum metals** because they occur together in platinum-bearing ores. In recognition of their traditional use, Cu, Ag, and Au are referred to as the **coinage metals**. Gold occurs as the metal; silver, gold, and the platinum metals are also recovered in the electrolytic refining of copper. The prices of the individual platinum metals vary widely because they are recovered together but their consumption is not proportional to their abundance. Rhodium is by far the most expensive metal in this group because it is widely used in industrial catalytic processes and in automotive catalytic converters (Box 26.1).

Copper, silver, and gold are not susceptible to oxidation by H^+ under standard conditions, and this noble character accounts for their use, together with platinum, in jewellery and ornaments. Aqua regia, a 3:1 mixture of concentrated hydrochloric and nitric acids, is an old but effective reagent for the oxidation of gold and platinum. Its function is twofold: the NO_3^- ions provide the oxidizing power and the Cl^- ions act as complexing agents. The overall reaction is

$$Au(s) + 4H^+(aq) + NO_3^-(aq) + 4Cl^-(aq) \rightarrow [AuCl_4]^-(aq) + NO(g) + 2H_2O(l)$$

The active species in solution are thought to be Cl_2 and NOCl, which are generated in the reaction

$$3HCl(aq) + HNO_3(aq) \rightarrow Cl_2(aq) + NOCl(aq) + 2H_2O(l)$$

Oxidation state preferences are erratic in Group 11. For Cu, the +1 and +2 states are most common, but for Ag +1 is typical, and for Au +1 and +3 are common. The simple aqua ions $Cu^+(aq)$ and $Au^+(aq)$ undergo disproportionation in aqueous solution:

$$2Cu^+(aq) \rightarrow Cu(s) + Cu^{2+}(aq)$$

$$3Au^+(aq) \rightarrow 2Au(s) + Au^{3+}(aq)$$

Complexes of Cu(I), Ag(I), and Au(I) are often linear. For example, $[H_3NAgNH_3]^+$ forms in aqueous solution and linear $[XAgX]^-$ complexes have been identified by X-ray crystallography. The currently preferred explanation for the tendency towards linear coordination is the similarity in energy of the outer nd and $(n+1)$s and $(n+1)$p orbitals, which permits the formation of collinear spd hybrids (Fig. 19.8).

The soft Lewis acid character of Cu^+, Ag^+, and Au^+ is illustrated by their affinity order, which is $I^- > Br^- > Cl^-$. Complex formation, as in the formation of $[Cu(NH_3)_2]^+$ and $[AuI_2]^-$, provides a means of stabilizing the +1 oxidation state of these metals in aqueous solution. Many tetrahedral complexes are also known for Cu(I), Ag(I), and Au(I) (Section 7.8).

Square-planar complexes are common for the platinum metals and gold in oxidation states that yield the d^8 electronic configuration, which include Rh(I), Ir(I), Pd(II), Pt(II), and Au(III) (Section 20.1f). An example is $[Pt(NH_3)_4]^{2+}$. Characteristic reactions for these

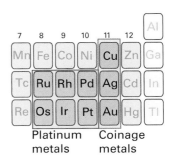

Figure 19.7 The location of the platinum and coinage metals in the periodic table.

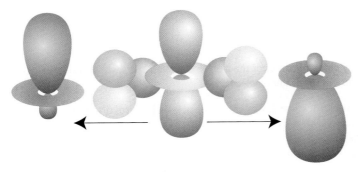

Figure 19.8 The hybridization of s, p_z, and d_{z^2} with the choice of phases shown here produces a pair of collinear orbitals that can be used to form strong σ bonds.

complexes are ligand substitution (Section 21.3) and, except for Au(III) complexes, oxidative addition (Section 22.22).

The noble character that these elements developed across the d block is suddenly lost at Group 12 (Zn, Cd, Hg), where the metals once again become susceptible to atmospheric oxidation. The greater ease of oxidation of the Group 12 metals is due to a reduction in the extent of intermetallic bonding and an abrupt lowering of d-orbital energies at the end of the d block, with the higher energy $(n+1)$s electrons participating in reactions.

Representative compounds

As a result of the convention of assigning a negative oxidation number to any nonmetal in combination with a metal, high formal oxidation states for d metals may be encountered in their compounds, such as Re(VII) in $[ReH_9]^{2-}$ and W(VI) in $W(CH_3)_6$. However, these compounds are not oxidizing agents in the usual sense, and are best discussed together with other organometallic compounds (Chapter 22). The discussion here is confined to compounds containing electronegative ligands such as the halogens, oxygen, nitrogen, and sulfur.

19.7 Metal halides

Key points: Binary halides of the d-block elements span all metals and most oxidation states; dihalides are typically ionic solids, with higher halides taking on covalent character.

Binary metal halides of the d-block elements occur for all the elements with nearly all oxidation states represented. As we should expect, the more strongly oxidizing halogens bring out the higher oxidation states, with the corollary that the low oxidation state binary halides are more stable as iodides and bromides.

Of all the groups, only the members of Group 11 (Cu, Ag, Au) have simple monohalides. For Cu, these salts are highly insoluble in water and dissolve only when complexed by other ligands (which include halides). Silver(I) halides are sparingly soluble and photosensitive, decomposing to the metal (a process that is exploited in photography). The only monohalide of Au that exists is the chloride; it is oxidized by water.

More common are the dihalides which, as noted in Section 19.3b (Table 19.2), are typically ionic solids that dissolve in water to give $M^{2+}(aq)$ ions; some of the earlier dihalides are reducing. Many of the intermediate halides exhibit metal−metal bonds and form clusters (Section 19.11).

Higher halides exist for most of the d block, and covalent character becomes more prevalent with high oxidation state, especially for the lower halogens. For instance, in Group 4, whereas TiF_4 is a solid with melting point 284°C, $TiCl_4$ melts at −24°C and boils at 136°C. In Group 6, not even the fluoride has ionic character and both MoF_6 and WF_6 are liquids at room temperature.

19.8 Metal oxides and oxido complexes

Because oxygen is readily available in an aqueous environment, in the atmosphere, and as the donor atom in many organic molecules, it is not surprising that oxides and oxygen-containing ligands play a major role in the chemistry of the metallic elements.

(a) Metal oxides

Key point: Many different oxides of the d-block elements exist, with a wide variety of structures, varying from ionic lattices to covalent molecules.

Many different oxides are known for the d-block elements, with a number of different structures. We have already noted the ability of oxygen to bring out the highest oxidation state for some elements, but oxides exist for some elements in very low oxidation states: in Cu_2O, copper is present as Cu(I). Monoxides are known for all of the 3d-series metals, except Cr. The monoxides have the rock-salt structure characteristic of ionic solids but their properties, which are discussed in more detail in Chapter 24, indicate significant deviations from the simple ionic $M^{2+}O^{2-}$ model. For example, TiO has metallic conductivity and FeO is always deficient in iron. The early d-block monoxides are strong reducing agents. Thus, TiO is easily oxidized by water or oxygen, and MnO is a convenient oxygen scavenger that is used in the laboratory to remove oxygen impurity in inert gases down to the parts-per-billion range.

As we have already noted, very high oxidation state oxides can show covalent structures. For example ruthenium tetroxide and osmium tetroxide are low melting, highly volatile, toxic, molecular compounds that are used as selective oxidizing agents. Indeed, osmium tetroxide is used as the standard reagent to oxidize alkenes to *cis*-diols:

(b) Mononuclear oxido complexes

Key points: The conversion of an aqua ligand to an oxido ligand is favoured by a high pH and by a high oxidation state of the central metal atom; in vanadium complexes in oxidation state $+4$ or $+5$, the site *trans* to the oxido ligand may be vacant or occupied by a weakly coordinating ligand.

Our focus in this section is on the oxido ligand (O^{2-}) and its ability to bring out the high oxidation states of the chemically hard metals on the left of the d block. Another important issue is the relation between oxido and aqua (H_2O) ligands resulting from proton transfer equilibria.

Elements in high oxidation states typically occur as oxoanions in aqueous solution, such as MnO_4^-, which contains Mn(VII), and CrO_4^{2-}, which contains Cr(VI). The existence of these oxoanions contrasts with the existence of simple aqua ions for the same metals in lower oxidation states, such as $[Mn(OH_2)_6]^{2+}$ for manganese(II) and $[Cr(OH_2)_6]^{3+}$ for chromium(III).

The formation of an oxido complex rather than an aqua complex is favoured by high pH because the OH^- ions in basic solution tend to remove protons from the aqua ligands. The occurrence of aqua complexes for low oxidation state metal cations is explained by noting the relatively small electron-withdrawing effect exerted by these cations on the O atoms of the OH_2 ligand. As a result, the aqua ligands are only weak proton donors. A metal ion in a high oxidation state, however, depletes the electron density on the O atoms attached to it and thereby increases the Brønsted acidity of H_2O and OH^- ligands.

The influence of pH on the stabilities of different oxidation states is best summarized by a Pourbaix diagram (Section 5.14). The example shown in Fig. 19.9 is for a complex (**3**) containing a tetradentate ligand L, which results in a similar coordination geometry for a wide range of conditions. The aqua complex *cis*-[RuLCl(OH₂)]⁺ in solution at pH = 2 is stable up to $+0.40$ V. Just above this potential, simple oxidation (that is, electron removal) occurs to give *cis*-[RuLCl(OH₂)]²⁺. Under even more strongly oxidizing conditions (at $+0.95$ V) and pH = 2, both oxidation and deprotonation occur and result in the formation of a Ru(IV) oxido species, *cis*-[RuLCl(O)]⁺. At an even higher potential (about $+1.4$ V), further oxidation yields *cis*-[RuLCl(O)]²⁺. As already remarked, the influence of more basic conditions is to deprotonate an OH_2 ligand and thus to favour hydroxido or oxido complexes. For example, at pH = 8, the Ru(III) species is deprotonated to the hydroxido complex *cis*-[RuLCl(OH)]⁺, which in turn is converted to the oxido-Ru(IV) complex at a lower potential than for the Ru(III)−Ru(IV) transformation at pH = 2.

Table 19.5 gives a list of simple complexes containing oxido ligands. Many complexes are known that contain the vanadium(IV) group, VO^{2+} (sometimes referred to as vanadyl), with V in its penultimate oxidation state ($+4$). Vanadyl complexes generally contain four additional ligands and are square pyramidal (**4**). Many of these d^1 complexes are blue (as a result of a d−d transition) and may take on a weakly bound sixth ligand *trans* to the oxido ligand. The vanadyl VO bond length (158 pm) in $[VO(acac)_2]$ is short compared with the four VO bond lengths to the acac ligand (197 pm). The short bond lengths in vanadyl complexes, together with high VO stretching wavenumbers (940−1000 cm⁻¹) provide strong evidence for VO multiple bonding in which lone pairs on the oxido ligand are donated to the central V atom. The V3d−O2p π-orbital overlap involved in a multiple bond is illustrated in Fig. 19.10; thus it can be seen that the O^{2-} ligand is not only a σ donor but also a π donor capable of making two π bonds and resulting in a bond order of up to three (though by convention such interactions are depicted as M=O, preserving electroneutrality). This strong

3 *cis*-[RuLCl(OH₂)]²⁺

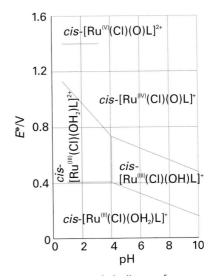

Figure 19.9 A Pourbaix diagram for *cis*-[RuLCl(OH₂)]²⁺ (**3**) and related species. (Adapted from C.-K. Li, W.-T. Wang, C.-M. Chi, K.-Y. Wong, R.-J. Wang, and T.C.W. Mak, *J. Chem. Soc., Dalton Trans.* 1991, 1909.)

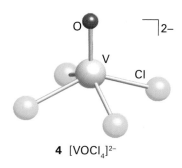

4 $[VOCl_4]^{2-}$

Table 19.5 Some common monoxido and dioxido complexes

Group	Element, configuration	Structure	Formula
5	V(IV), d^1	Square pyramidal	$[V(O)(acac)_2]$, $[V(O)Cl_4]^{2-}$
	V(V), d^0	*cis*-octahedral	$[V(O)_2(OH_2)_4]^+$
6	Mo(VI), d^0; W(VI), d^0	Tetrahedral	$[M(O)_2(Cl)_2]$
		cis-octahedral	$[Mo(O)_2(acac)_2]$
7, 8	Re(V), d^2; Os(VI), d^2	*trans*-octahedral	$[Re(O)_2(CN)_4]^{3-}$
			$[Os(O)_2(Cl)_2]$
			$[Os(O)_2(Cl)_4]^{2-}$

multiple bonding with the O atom appears to be responsible for the *trans* influence of the oxido ligand, which disfavours attachment of the ligand *trans* to O (Section 21.4).

Vanadium in its highest (and group) oxidation state, $+5$, forms an extensive series of oxido compounds, many of which are the polyoxo species discussed later. The simplest oxido complex $[V(O)_2(OH_2)_4]^+$ exists in the acidic solution formed when the sparingly soluble vanadium(V) oxide, V_2O_5, dissolves in water. This pale yellow complex has a *cis* geometry (**5**). Here again we see the *trans* influence of an oxido ligand: the *cis* geometry minimizes competition between the oxido ligands for π-bonding to the V atom. As shown in Fig. 19.11, the two O^{2-} ligands in the *trans* configuration can π-bond with the same two d orbitals; in the *cis* configuration the metal atom has only one d orbital in common with the π orbitals on the O atoms.

In contrast to the *cis* structure of the d^0 complex $[V(O)_2(OH_2)_4]^+$, many *trans*-dioxido complexes are known for the d^2 metal centres Os(VI) and Re(V) (**6**); Table 19.5 gives some examples. This geometry is probably favoured because the *trans* configuration leaves a vacant low-energy orbital that can be occupied by the two d electrons. According to this explanation, the avoidance of the destabilizing effect of the d^2 electrons more than offsets the disadvantage of the ligands having to compete for the same d orbital in the *trans*-oxido geometry.

(c) Polyoxometallates

Key points: In their highest oxidation states, metals in Groups 5 and 6 readily form polyoxometallates and heteropolyoxometallates; pH is crucial in determining which compounds form.

A **polyoxometallate** is an oxoanion containing more than one metal atom. The H_2O ligand created by protonation of an oxido ligand at low pH can be eliminated from the central metal atom and thus lead to the condensation of mononuclear oxometallates. A familiar example is the reaction of a basic chromate solution, which is yellow, with excess acid to form the oxido-bridged dichromate ion, which is orange:

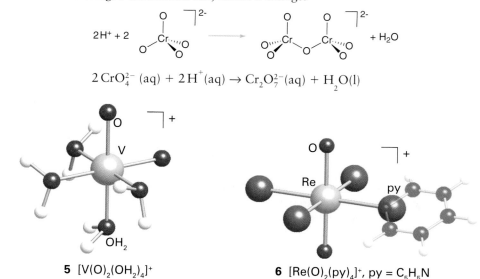

$$2\,CrO_4^{2-}\,(aq) + 2\,H^+(aq) \rightarrow Cr_2O_7^{2-}(aq) + H_2O(l)$$

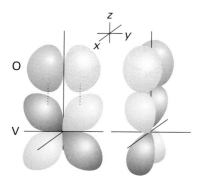

Figure 19.10 The d−p π-bonding between O and V in the vanadyl group VO^{2+}. It is convenient to think of the O as having oxidation number -2 and V as $+4$. Thus, both electron pairs for the two π bonds are donated from the O^{2-} ion into the metal d_{yz} and d_{zx} orbitals.

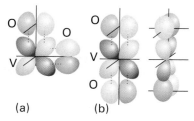

(a)   (b)

Figure 19.11 Comparison of the competition between *cis* and *trans* oxygens attached to vanadium. (a) In the *cis* configuration the metal has only one d orbital in common with the π orbitals on the O atoms. (b) In the *trans* configuration the metal atom has two d orbitals in common with the π orbitals on the two O atoms.

5 $[V(O)_2(OH_2)_4]^+$

6 $[Re(O)_2(py)_4]^+$, py $= C_5H_5N$

In highly acidic solution, oxido-bridged Cr(VI) species with longer chains are formed. The tendency for Cr(VI) to form polyoxo species is limited by the fact that the O tetrahedra link only through vertices: edge and face bridging would result in too close an approach of the metal atoms. By contrast, it is found that five- and six-coordinate metal oxido complexes, which are common with the larger 4d- and 5d-series metal atoms, can share oxido ligands between either vertices or edges. These structural possibilities lead to a richer variety of polyoxometallates than found with the 3d-series metals.

Chromium's neighbours in Groups 5 and 6 form six-coordinate polyoxo complexes (Fig. 19.12). In Group 5, the polyoxometallates are most numerous for vanadium, which forms many V(V) complexes and a few V(IV) or mixed oxidation state V(IV)−V(V) polyoxido complexes. Polyoxometallate formation is most pronounced in Groups 5 and 6 for V(V), Mo(VI), and W(VI).

It is often convenient to represent the structures of the polyoxometallate ions by polyhedra, with the metal atom understood to be in the centre and O atoms at the vertices. For example, the sharing of O atom vertices in the dichromate ion, $Cr_2O_7^{2-}$, may be depicted in either the traditional way (**7**) or in the polyhedral representation (**8**). Similarly, the important M_6O_{19} structure of $[Nb_6O_{19}]^{8-}$, $[Ta_6O_{19}]^{8-}$, $[Mo_6O_{19}]^{2-}$, and $[W_6O_{19}]^{2-}$ is depicted by the conventional or polyhedral structures shown in Fig. 19.13. The structures for this series of polyoxometallates contain terminal O atoms (those projecting out from a single metal atom) and two types of bridging O atoms: two-metal bridges, M−O−M, and one hypercoordinated O atom in the centre of the structure that is common to all six metal atoms. The structure consists of six MO_6 octahedra, each sharing an edge with four neighbours. The overall symmetry of the M_6O_{19} array is O_h. Another example of a polyoxometallate is $[W_{12}O_{40}(OH)_2]^{10-}$ (**9**). As shown by this formula, the polyoxoanion is partially protonated.

Polyoxometallate anions can be prepared by carefully adjusting pH and concentrations, for example polyoxomolybdates and polyoxotungstates are formed by acidification of solutions of the simple molybdate or tungstate:

$$6\,[MoO_4]^{2-}(aq) + 10\,H^+(aq) \rightarrow [Mo_6O_{19}]^{2-}(aq) + 5\,H_2O(l)$$

$$8\,[MoO_4]^{2-}(aq) + 12\,H^+(aq) \rightarrow [Mo_8O_{26}]^{4-}(aq) + 6\,H_2O(l)$$

Mixed metal polyoxometallates are also common, as in $MoV_9O_{28}^{5-}$, and there is a large class of heteropolyoxometallates, such as the molybdates and tungstates, that also incorporate P, As, and other heteroatoms. For example, $[PMo_{12}O_{40}]^{3-}$ contains a PO_4^{3-} tetrahedron that shares O atoms with surrounding octahedral MoO_6 groups (**10**). Many different heteroatoms can be incorporated into this structure, and the general formulation is $[X(+N)Mo_{12}O_{40}]^{(8-N)-}$ where $X(+N)$ represents the oxidation state of the heteroatom X, which may be As(V), Si(IV), Ge(IV), or Ti(IV). An even broader range of heteroatoms is observed with the analogous tungsten heteropolyoxoanions. Heteropolyoxomolybdates and tungstates can undergo one-electron reduction with no change in structure but with the formation of a deep blue colour. The colour seems to arise from the excitation of the added electron from Mo(V) or W(V) to an adjacent Mo(VI) or W(VI) site.

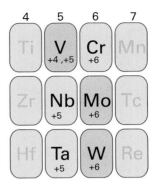

Figure 19.12 Elements in the d block that form polyoxometallates. The elements coloured yellow form the greatest variety of polyoxometallates.

7 $Cr_2O_7^{2-}$

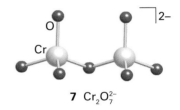

8 $Cr_2O_7^{2-}$

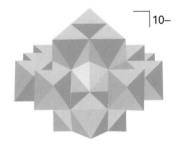

9 $[W_{12}O_{40}(OH)_2]^{10-}$

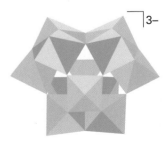

10 $[PMo_{12}O_{40}]^{3-}$

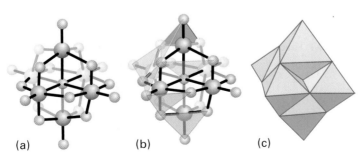

(a)   (b)   (c)

Figure 19.13 (a) Conventional and (c) polyhedral representation of the six edge-shared octahedra as found in $[M_6O_{19}]^{2-}$. The intermediate structure (b) shows the process of constructing the polyhedral representation in progress.

19.9 Metal sulfides and sulfide complexes

Sulfur is softer as a Lewis base and less electronegative than oxygen; it therefore has a broader range of oxidation states and a stronger affinity for the softer metals on the right of the d block. For example, zinc(II) sulfide is readily precipitated when $Zn^{2+}(aq)$ is added to an aqueous solution containing H_2S and NH_3 (to adjust the pH) whereas the harder Lewis acid $Sc^{3+}(aq)$ gives $Sc(OH)_3(s)$ instead. The covalent contribution to the lattice enthalpy of all the softer metal sulfides cannot be overcome in water, and in general the sulfides are only very sparingly soluble. Another striking feature of sulfur is its tendency to form catenated sulfide ions, Section 16.14. A broad set of compounds containing S_n^{2-} ions can be prepared, and the smaller polysulfides can serve as chelating ligands towards d-metal ions.

(a) Monosulfides

Key point: Monosulfides with the nickel-arsenide structure are formed by most 3d metals.

As with the d-metal monoxides, the monosulfides are most common in the 3d series (Table 19.6). In contrast to the monoxides, however, most of the monosulfides have the nickel-arsenide structure (Fig. 3.36). The different structures are consistent with the rock-salt structure of the monoxides being favoured by the ionic (harder) cation−anion combination and the nickel-arsenide structure being favoured by more covalent (softer) combinations. The nickel-arsenide structure is expected when there is appreciable metal−metal bonding, resulting in the shorter metal−metal distances found for the hexagonal packing. Monosulfides are formed by most 3d-series metal ions.

(b) Disulfides

Key points: The 4d- and 5d-series metals often form disulfides with alternating layers of metal ions and sulfide ions; binary disulfides of the early d metals often have a layered structure, whereas Fe^{2+} and many of the later d-metal disulfides contain discrete S_2^{2-} ions.

The disulfides of the d metals fall into two broad classes (Table 19.7). One class consists of layered compounds with either the CdI_2 or the MoS_2 structure; the other consists of compounds containing discrete S_2^{2-} groups, the pyrites and marcasite structures.

The layered disulfides are built from a sulfide layer, a metal layer, and then another sulfide layer (for example, Fig. 19.14). These sandwiches stack together in the crystal with sulfide layers in one slab adjacent to a sulfide layer in the next. Clearly, this crystal

Table 19.6 Structures of d-block MS compounds*

	Group						
	4	5	6	7	8	9	10
Nickel-arsenide structure (shaded)	Ti	V		Mn†	Fe	Co	Ni
Rock-salt structure (unshaded)	Zr	Nb					

* Metal monosulfides of Group 6 are not shown; some of the heavier metals have more complex structures.
† MnS has two polymorphs; one has a rock-salt structure, the other has a wurtzite structure.

Table 19.7 Structures of d-block MS$_2$ compounds*

	Group							
	4	5	6	7	8	9	10	11
Layered (shaded)	Ti			Mn	Fe	Co	Ni	Cu
Pyrite or marcasite	Zr	Nb	Mo		Ru	Rh		
(unshaded)	Hf	Ta	W	Re	Os	Ir	Pt	

* Metals not shown do not form disulfides or have disulfides with complex structures.
Adapted from A.F. Wells, *Structural inorganic chemistry*. Oxford University Press (1984).

structure is not consistent with a simple ionic model and its formation is a sign of covalence in the bonds between the soft sulfide ion and d-metal cations. The metal ion in these layered structures is surrounded by six S atoms. Its coordination environment is octahedral in some cases (such as PtS_2, which adopts the CdI_2 structure shown in Fig. 19.14) and trigonal prismatic in others (MoS_2). The layered MoS_2 structure is favoured by S—S bonding as indicated by short S—S distances within each of the MoS_2 slabs. The common occurrence of the trigonal-prismatic structure in many of these compounds is in striking contrast to the isolated metal complexes, where the octahedral arrangement of ligands is by far the most common. Some of the layered metal sulfides readily undergo intercalation reactions in which ions or molecules penetrate between adjacent sulfide layers (Section 24.10).

Compounds containing discrete S_2^{2-} ions adopt the pyrite or marcasite structure (Fig. 19.15). The stability of the formal S_2^{2-} ion in metal sulfides is much greater than that of the O_2^{2-} ion in peroxides, and there are many more metal sulfides in which the anion is S_2^{2-} than there are peroxides.

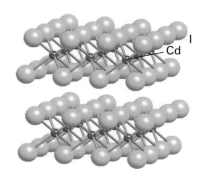

Figure 19.14 The CdI_2 structure adopted by many disulfides.

EXAMPLE 19.2 **Contrasting the structures of two different d-block disulfides**

Compare the structures of MoS_2 and FeS_2, and explain their existence in terms of the oxidation states of the metal ions.

Answer We need to decide whether the metals are likely to be present as M(IV), in which case the two S atoms would be present as S^{2-} ions, or whether the metal is present as M(II) with the S atoms present as the S—S bonded species S_2^{2-}. Because it is readily oxidized, S^{2-} will be found only with a metal ion in an oxidation state that is not easily reduced. As with many of the 4d- and 5d-series metals, Mo is easily oxidized to Mo(IV), therefore Mo(IV) can coexist with S^{2-}. Molybdenum(IV) sulfide (MoS_2) has the layered structure typical of metal disulfides. In contrast, Fe is not readily oxidized to Fe(IV). Therefore, Fe(IV) cannot coexist with S^{2-}. The compound is therefore likely to contain Fe(II) and S_2^{2-}. The mineral name for FeS_2 is pyrite; its common name, 'fool's gold', indicates its misleading colour.

Self-test 19.2 Suggest a use for molybdenum(IV) sulfide that makes use of its solid-state structure. Rationalize your suggestion.

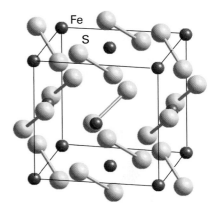

Figure 19.15 The structure of pyrite, FeS_2.

(c) Sulfide complexes

Key point: Chelating polysulfide ligands are common in metal—sulfur coordination compounds of the 4d- and 5d-series metals.

The coordination chemistry of sulfur is quite different from that of oxygen. Much of this difference is connected with the ability of sulfur to catenate (form chains), the availability of low-lying empty 3d orbitals that allow an S atom to act as a π acceptor, and the preference of sulfur for metal centres that are not highly oxidizing (Section 16.11).

Simple thiometallate complexes such as $[MoS_4]^{2-}$ can be synthesized easily by passing H_2S gas through a strongly basic aqueous solution of molybdate or tungstate ions:

$$[MoO_4]^{2-}(aq) + 4H_2S(g) \rightarrow [MoS_4]^{2-}(aq) + 4H_2O(l)$$

These tetrathiometallate anions are building blocks for the synthesis of complexes containing more metal atoms. For example, they will coordinate to many dipositive metal ions, such as Co^{2+} and Zn^{2+}:

$$Co^{2+}(aq) + 2[MoS_4]^{2-}(aq) \rightarrow [S_2MoS_2CoS_2MoS_2]^{2-}(aq)$$

The polysulfide ions, such as S_2^{2-} and S_3^{2-}, which are formed by addition of elemental sulfur to a solution of ammonium sulfide, can also act as ligands. An example is $[Mo_2(S_2)_6]^{2-}$ (**11**), which is formed from ammonium polysulfide and MoO_4^{2-}; it contains side-bonded S_2^{2-} ligands. The larger polysulfides bond to metal atoms forming chelate rings, as in $[MoS(S_4)_2]^{2-}$ (**12**), which contains chelating S_4^{2-} ligands.

Clusters of Fe and S atoms ('Fe—S clusters') are cofactors in many enzymes, including nitrogenase, which converts N_2 to NH_3 under ambient conditions (Chapter 15). Synthetic analogues such as the cubane shown in (**13**) have been prepared and characterized. The cubane contains Fe and S atoms on alternate corners, so each S atom bridges three Fe

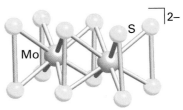

11 $[Mo_2(S_2)_6]^{2-}$

12 $[MoS(S_4)_2]^{2-}$

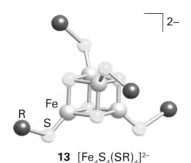

13 [Fe₄S₄(SR)₄]²⁻

atoms; each of the Fe atoms has a thiolate group, RS^-, occupying a terminal position. These compounds can be obtained from simple starting materials in the absence of air using a polar aprotic solvent such as dimethylsulfoxide:

$$4\,FeCl_3 + 4\,NaSH + 6\,RSH + 10\,NaOCH_3 \rightarrow$$
$$[Fe_4S_4(SR)_4]^{2-} + RS-SR + 14\,Na^+ + 12\,Cl^- + 10\,CH_3OH$$

The HS^- ion provides the sulfide ligands for the cage, the RS^- ion serves as both a ligand and a reducing agent, and the CH_3O^- ion acts as a base. Addition of the bulky cation from $(Et_4N)Cl$ results in the formation of black crystals of $(Et_4N)_2[Fe_4S_4(SR)_4]$. The observation that this reaction is successful, in good yield, with many different R groups indicates that the Fe_4S_4 cage is thermodynamically more stable than other possibilities. The cage undergoes reversible one-electron reduction to $[Fe_4S_4(SR)_4]^{3-}$, a reaction that is important in biological oxidation and reduction (Section 27.8).

19.10 Nitrido and alkylidyne complexes

Key points: Multiple M—L bonds are common with high oxidation state metals of the early d-metal series; these bonds weaken the bonds to the trans ligands.

The nitrido ligand ($N\equiv$) and alkylidyne ligand ($RC\equiv$, previously known as carbido) are present formally as N^{3-} and RC^{3-}. Thus, complexes containing them are usually of very high oxidation state. The highly reactive compound $(Me_3CO)_3W\equiv W(OCMe_3)_3$ cleaves the $C\equiv N$ bond in benzonitrile to give both a nitrido and an alkylidyne ligand:

$$(Me_3CO)_3W\equiv W(OCMe_3)_3 + PhC\equiv N \rightarrow (Me_3CO)_3W\equiv CPh + (Me_3CO)_3W\equiv N$$

The remarkable feature of this reaction is that the $C\equiv N$ bond is broken despite its considerable strength (890 kJ mol⁻¹), so the $W\equiv C$ and $W\equiv N$ bond enthalpies must be substantial.

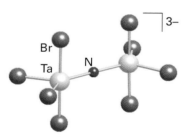

14 [M(N)Cl₄]⁻

15 [Br₄Ta(μ-N)TaBr₄]³⁻

Both nitrido and alkylidyne complexes are known to have short M—N and M—C bonds, which confirms the existence of multiple metal—ligand bonds. Like the O^{2-} group, these ligands are strong π donors and weaken the *trans* metal—ligand bond. The order of this influence is $RC\equiv > N\equiv > O=$; this weakening is attributed to competition for π donation into a common set of metal d orbitals.

Many nitrido complexes are known for the elements in Groups 5–8. They are most numerous for Mo and W in Group 6, Re in Group 7, and Ru and Os in Group 8. Square-pyramidal nitrido complexes of formula $[MNX_4]^-$ are known for M(VI), with M = Mo, Re, Ru, and Os, and X = F, Cl, Br, and (in some cases) I. These square-pyramidal structures (**14**) have short $M\equiv N$ bonds (157 to 166 pm). In a few cases, a ligand is attached in the sixth site, but the resulting bond is long and weak, like the oxido complexes discussed earlier. Examples of M=N=M species are known, such as $[Ta_2NBr_8]^{3-}$ (**15**), but nitrogen analogues of the polyoxometallates described earlier are unknown.

19.11 Metal—metal bonded compounds and clusters

The traditional view of the chemical properties of the d metals is from the perspective of ionic solids and complexes that contain a single central metal ion surrounded by a set of ligands. However, with the development of improved techniques for the determination of structure, it has been recognized that there are also many d-metal compounds that have metal—metal (M—M) bond distances comparable to or shorter than those in the elemental metal. These findings have stimulated additional X-ray structural investigations of compounds that might contain M—M bonds and have inspired research into syntheses designed to broaden the range of such **cluster compounds**. Examples of these metal (and nonmetal) cluster compounds are now known in every block of the periodic table (Fig. 19.16), but they are most numerous in the d block.

A rigorous definition of metal clusters restricts them to molecular complexes with metal—metal bonds that form triangular or larger structures. This definition, however, would exclude linear M—M compounds, and is normally relaxed. We shall consider any M—M bonded system to be a cluster.

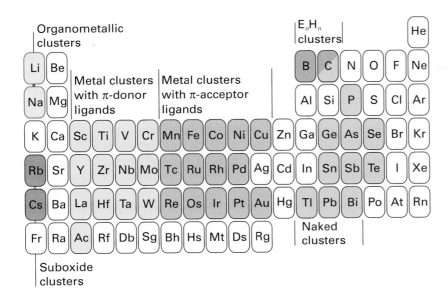

Figure 19.16 The major classes of cluster-forming elements. Note that carbon is in two classes (for example it occurs as C_8H_8 and as C_{60}). (Adapted from D.M.P. Mingos and D.J. Wales, *Introduction to cluster chemistry*. Prentice Hall, Englewood Cliffs (1990).)

(a) Metal–metal bonds

Key point: Metal–metal bonds with bond orders up to five are formed by many d metals in low oxidation states.

The first d-block metal–metal bonded species to be identified was the Hg_2^{2+} ion of mercury(I) compounds, as occurs in Hg_2Cl_2, and examples of metal–metal bonded compounds and clusters are now known for most of the d metals. Some of their common structural motifs are an ethane-like structure (**16**), an edge-shared bioctahedron (**17**), a face-shared bioctahedron (**18**), and the tetragonal prism of $[Re_2Cl_8]^{2-}$ (**19**).

If we consider the possible overlap between d orbitals on adjacent metal atoms, we can see that (Fig. 19.17):

- a σ bond between two metal atoms can arise from the overlap of a d_{z^2} orbital from each atom
- two π bonds can arise from the overlap of d_{zx} or d_{yz} orbitals
- two δ bonds can be formed from the overlap of two face-to-face d_{xy} or $d_{x^2-y^2}$ orbitals.

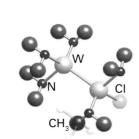

16 $[(Me_2N)_3WWCl(NMe_2)_2]$, Me = CH_3

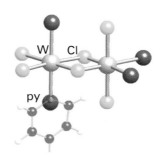

17 $[(Cl)_2(py)_2W(\mu\text{-}Cl)_2W(Cl)_2(py)_2]$

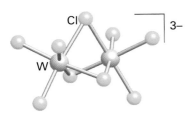

18 $[(Cl)_3W(\mu\text{-}Cl)_3W(Cl)_3]^{3-}$

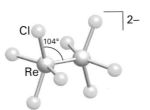

19 $[Re_2Cl_8]^{2-}$

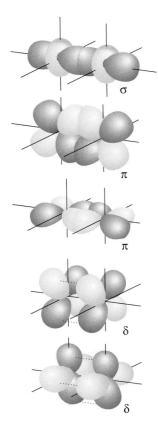

Figure 19.17 The origin of σ, π, and δ interactions between the d orbitals of two d-metal atoms situated along the z-axis. Only bonding combinations are shown.

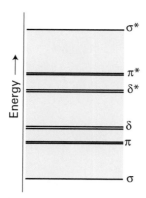

Figure 19.18 Approximate molecular orbital energy level scheme for M–M interactions.

20

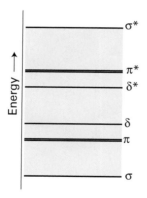

Figure 19.19 Approximate molecular orbital energy level scheme for the M–M interactions in a quadruply bonded system, where only the $d_{x^2-y^2}$ is utilized in bonding to the ligands.

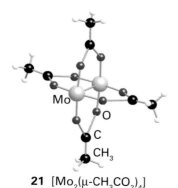

21 $[Mo_2(\mu\text{-}CH_3CO_2)_4]$

Thus a quintuple bond could result if all the bonding orbitals are occupied to give the electron configuration $\sigma^2\pi^4\delta^4$ (Fig.19.18).

Five d electrons would be needed from each metal for a quintuple bond, and this is exactly the case for the d^5 Cr(I) centres in (**20**). In this molecule, the two Cr atoms are separated by a very short distance of 183.5 pm, unsupported by any additional bridging ligand interactions; compare this distance with the separation of Cr atoms in the bulk metal, which is 258 pm. The Re compound (**19**) also lacks bridging ligand interactions, but here the two d^4 Re(III) centres can form only a quadruple Re–Re bond. The four d electrons from each Re atom in (**19**) result in the configuration $\sigma^2\pi^4\delta^2$, and it is believed that the $d_{x^2-y^2}$ orbital is involved in bonding to the Cl^- ligands. Evidence for quadruple bonding comes from the observation that $[Re_2Cl_8]^{2-}$ has an eclipsed array of Cl ligands, which is sterically unfavourable. It is argued that the δ bond, which is formed only when the d_{xy} orbitals are confacial, locks the complex in the eclipsed conformation.

Many other species with multiple metal–metal bonds, where the $d_{x^2-y^2}$ orbital is involved in bonding to ligand species, are known. In all these complexes, the molecular orbital diagram shown in Figure 19.19 becomes appropriate and the maximum possible bond order is 4. A well-known example is the quadruply bonded compound molybdenum(II) acetate (**21**), which is prepared by heating $Mo(CO)_6$ with acetic acid:

$$2\,Mo(CO)_6 + 4\,CH_3COOH \rightarrow Mo_2(O_2CCH_3)_4 + 2\,H_2 + 12\,CO$$

The dimolybdenum complex is an excellent starting material for the preparation of other Mo–Mo compounds. For example, the quadruply bonded chlorido complex is obtained when the acetato complex is treated with concentrated hydrochloric acid at below room temperature:

$$Mo_2(O_2CCH_3)_4(aq) + 4\,H^+(aq) + 8\,Cl^-(aq) \rightarrow [Mo_2Cl_8]^{4-}(aq) + 4\,CH_3COOH(aq)$$

As shown in Table 19.8, incomplete occupation of the bonding orbitals can result in a reduction of the formal bond order to 3.5 or to the triply bonded M≡M systems. These complexes are more numerous than the quadruply bonded complexes and, because δ bonds are weak, M≡M bond lengths are often similar to those of quadruply bonded systems. A decrease of bond order can also stem from the occupation of both the δ* orbitals and, once these are fully occupied, successive occupation of the two higher lying π* orbitals leads to further decrease in the bond order from 2.5 to 1.

As with carbon–carbon multiple bonds, metal–metal multiple bonds are centres of reaction. However, the variety of structures resulting from the reactions of metal–metal multiple bonded compounds is more diverse than for organic compounds. For example:

Cp(OC)$_2$Mo≡Mo(CO)$_2$Cp + HI ⟶ Cp(OC)$_2$Mo⟨H I⟩Mo(CO)$_2$Cp

In this reaction, HI adds across a triple bond but both the H and I bridge the metal atoms; the outcome is quite unlike the addition of HX to an alkyne, which results in a substituted alkene. The reaction product can be regarded as containing a $3c,2e$ MHM bridge and an iodide anion bonding by two conventional $2c,2e$ bonds, one to each Mo atom.

Larger metal clusters can be synthesized by addition to a metal–metal multiple bond. For example, $Pt(PPh_3)_4$ loses two triphenylphosphine ligands when it adds to the Mo≡Mo triple bond, resulting in a three-metal cluster:

Cp(OC)$_2$Mo≡Mo(CO)$_2$Cp + Pt(PPh$_3$)$_4$ ⟶ Cp(OC)$_2$Mo=Mo(CO)$_2$Cp with Pt(PPh$_3$)$_2$ bridge + 2PPh$_3$

(b) Clusters

Key points: Metal clusters may be formed; for early d-block elements they are favoured by good donor ligands, whereas for later d-block elements π-acceptor ligands are needed.

Table 19.8 Examples of metal—metal bonded tetragonal prismatic complexes[*]

Complex	Configuration	Bond order	M—M bond length/pm
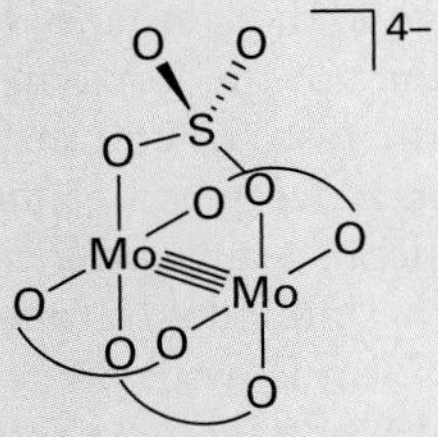 [Mo≡Mo] SO bridged complex, 4−	$\sigma^2\pi^4\delta^2$	4	211
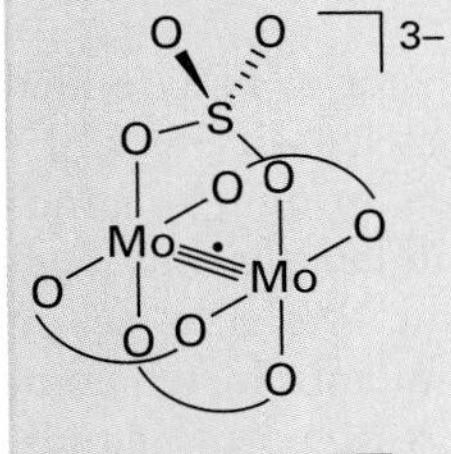 [Mo≡Mo] SO bridged complex, 3−	$\sigma^2\pi^4\delta^1$	3.5	217
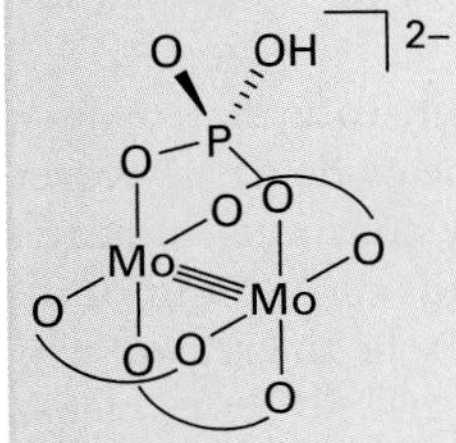 [Mo≡Mo] P,OH bridged complex, 2−	$\sigma^2\pi^4$	3	222
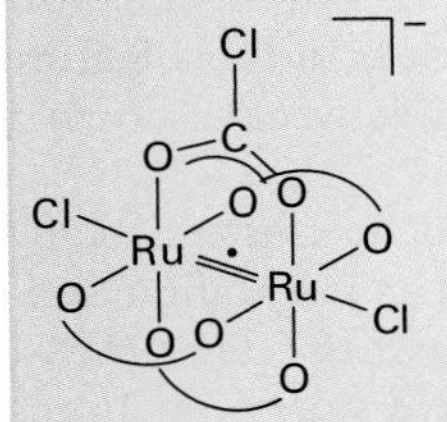 [Ru≡Ru] Cl, C bridged complex, 1−	$\sigma^2\pi^4\delta^2\delta*^1\pi*^2$	2.5	227
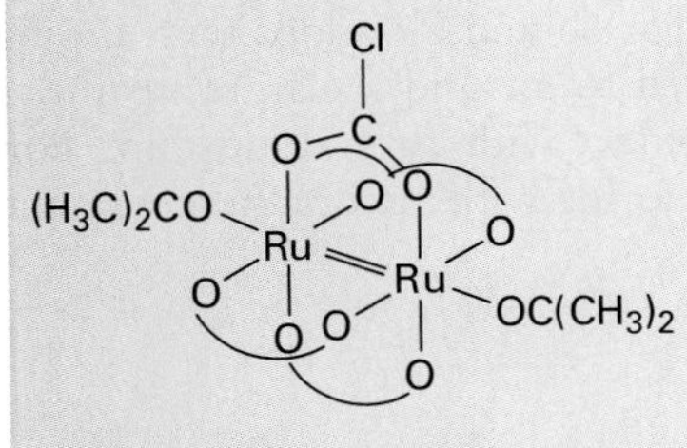 [Ru≡Ru] Cl, $(H_3C)_2CO$, $OC(CH_3)_2$ bridged complex	$\sigma^2\pi^4\delta^2\delta*^2\pi*^2$	2	238
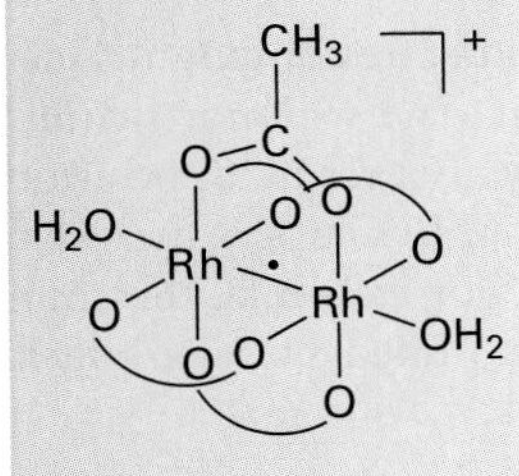 [Rh—Rh] CH_3, H_2O, OH_2 bridged complex, +	$\sigma^2\pi^4\delta^2\delta*^1\pi*^4$	1.5	232
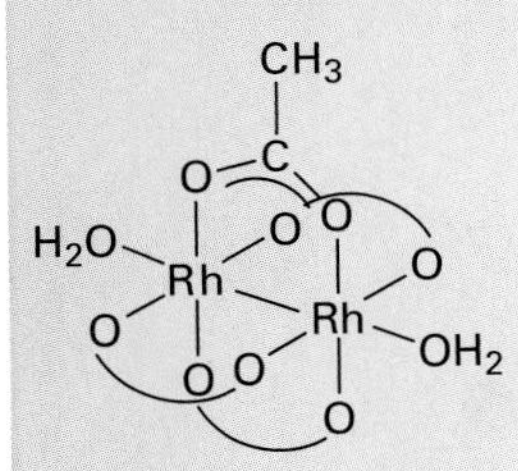 [Rh—Rh] CH_3, H_2O, OH_2 bridged complex	$\sigma^2\pi^4\delta^2\delta*^2\pi*^4$	1	239

[*] When multiple bridging ligands are present, only one is shown in detail.

Figure 19.16 showed how cluster compounds may be classified according to ligand type. Although this classification is not appropriate in detail for all cluster compounds, it provides information on the principal ligand types and some insight into the bonding. For example, alkyllithium compounds often have alkyl groups attached to a metal cluster by two-electron multicentre bonds. Clusters of the early d-block elements and the lanthanoids generally contain donor ligands such as Br^-. These ligands can fill some of the low-lying orbitals of these electron-poor metal atoms by σ- and π-electron donation. In contrast, electron-rich metal clusters on the right of the d block generally contain π-acceptor ligands such as CO, which remove some of the electron density from the metal. Many p-block elements do not require ligands to complete the valence shells of the individual atoms in clusters and can exist as **naked clusters**, which are clusters of the form E_n with no accompanying ligands. The ionic clusters Pb_5^{2-} and Sn_9^{4-} are two examples of naked clusters formed by metals.

Clusters may involve M—M bonds within a discrete molecular cluster or in extended solid-state compounds. When no bridging ligands are present in a cluster, as in $[Re_2Cl_8]^{2-}$ (**19**), the presence of a metal—metal bond is unambiguous. When bridging ligands are present, careful observations and measurement (typically of bond lengths and magnetic properties) are needed to identify direct metal—metal bonding. We shall see in Section 22.20 that there is an extensive range of organometallic metal cluster compounds of the middle-to-late d-block elements that are stabilized by π-acceptor ligands, particularly CO.

The bonding patterns in most metal cluster compounds are so intricate that metal—metal bond strengths cannot be determined with great precision. Some evidence, however, such as the stability of compounds and the magnitudes of M—M force constants, indicates that there is an increase in M—M bond strength down a group, perhaps on account of the greater spatial extension of d orbitals in heavier atoms. This trend may be the reason why there are so many more metal—metal bonded compounds for the 4d- and 5d-series metals than for their 3d-series counterparts. Figure 19.1 indicates that for the bulk metals, the metal—metal bonds in the d block are strongest in the 4d and 5d series, and this feature carries over into their compounds. By contrast, element—element bonds weaken down a group in the p block.

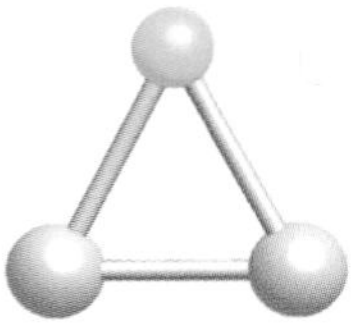

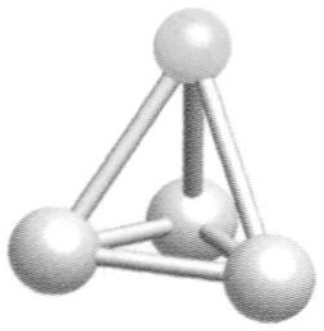

Not all the early d-block metal—metal bonded compounds are discrete clusters. Many extended metal—metal bonded compounds exist, such as the multiple-chain scandium subhalides, Sc_7Cl_{10} (Fig. 19.20) and Sc_5Cl_6, and layered compounds such as ZrCl (Fig. 19.21). In ZrCl, the Zr atoms are within bonding distance in two adjacent layers of metal atoms sandwiched between Cl^- layers. We have already seen that Sc and Zr adopt their group oxidation states (+3 and +4, respectively) when exposed to air and moisture, so these metal—metal bonded compounds are prepared out of contact with air and moisture. For example, ZrCl is prepared by reducing zirconium tetrachloride with zirconium metal in a sealed tantalum tube at high temperatures:

$$3\,Zr(s) + ZrCl_4(g) \xrightarrow{600-800°C} 4\,ZrCl(s)$$

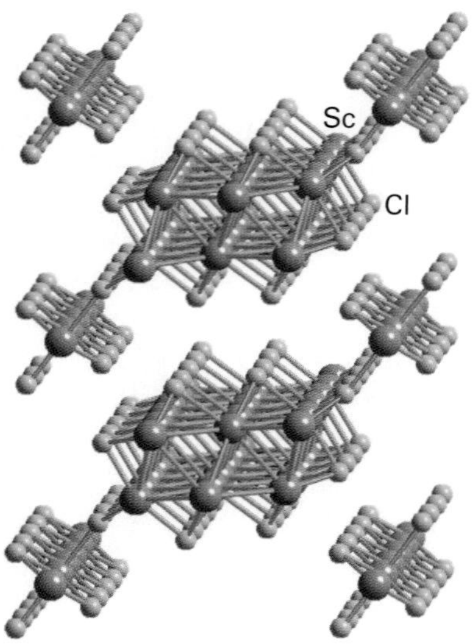

Discrete clusters—as distinct from the extended M—M bonded solid-state compounds just described—are often soluble, and can be manipulated in solution. The π-donor ligands in these clusters typically occur in one of several locations, namely, in a terminal position (**22**), bridging two metal atoms (**23**), or bridging three metal atoms (**24**). Clusters may also be linked by bridging halides or chalcogenides in the solid state. For example, the solid compound of formula $MoCl_2$ consists of octahedral Mo clusters linked by Cl bridges.

22

23

24

Sc

Cl

Figure 19.20 The structure of Sc_7Cl_{10}.

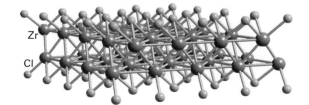

Zr

Cl

Figure 19.21 The structure of ZrCl consists of layers of metal atoms in graphite-like hexagonal nets.

The bridged structure of $MoCl_2$ is in sharp contrast to $CrCl_2$, which has discrete Cr atoms arranged in a distorted rutile structure. Samples of $MoCl_2$ may be prepared in a sealed glass tube by the reaction of $MoCl_5$ in a mixture of molten $NaAlCl_4$ and $AlCl_3$, together with aluminium metal as the reducing agent:

$$MoCl_5(s) + Al(s) \xrightarrow{NaAlCl_4/AlCl_3(l), 200°C} MoCl_2(s) + AlCl_3(l)$$

The compound can withstand oxidation under mild conditions. When treated with hydrochloric acid it forms the anionic cluster $[Mo_6Cl_{14}]^{2-}$. This cluster contains an octahedral array of Mo atoms with a Cl atom bridging each triangular face and a terminal Cl atom on each Mo vertex (**25**). These terminal Cl atoms can be replaced by other halogens, alkoxides, and phosphines. An analogous series of tungsten cluster compounds is known. Once formed, these Mo and W compounds can be handled in air and water at room temperature on account of their kinetic barriers to decomposition. The oxidation number of the metal is $+2$, so the metal valence electron count is $(6 - 2) \times 6 = 24$. One-electron oxidation and reduction of the clusters is possible.

Similar octahedral clusters are known for Nb and Ta in Group 5, and for Zr in Group 4. The cluster $[Nb_6Cl_{12}L_6]^{2+}$ and its Ta analogue have an octahedral framework with edge-bridging Cl atoms and six terminal ligands (**26**). The metal valence electron count in this case is 16, but these clusters can be oxidized in several steps. One example from Group 4 is $[Zr_6Cl_{18}C]^{4-}$, which has the same metal and Cl atom array together with a C atom in the centre of the octahedron.

Rhenium trichloride, which consists of Re_3Cl_9 clusters linked by weak halide bridges (**27**), provides the starting material for the preparation of a series of M_3 clusters that have been studied thoroughly. As with molybdenum dichloride, the intercluster bridges in the solid state can be broken by reaction with potential ligands. For example, treatment of Re_3Cl_9 with Cl^- ions produces the discrete complex $[Re_3Cl_{12}]^{3-}$. Neutral ligands, such as trialkylphosphines, can also occupy these coordination sites and result in clusters of the general formula $Re_3Cl_9L_3$.

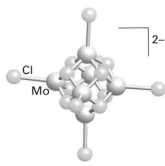

25 $[Mo_6Cl_{14}]^{2-}$

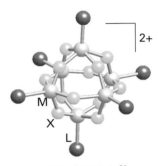

26 $[M_6X_{12}L_6]^{2+}$

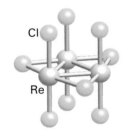

27 Re_3Cl_9

EXAMPLE 19.3 Metal−metal bonding and clusters

Suggest which interactions might be responsible for, and thus the bond order of, the metal−metal bond in the Hg_2^{2+} ion.

Answer We need to judge the types of bonds that can form from the available atomic orbitals on each metal atom and the bond order that results from their occupation. The oxidation state of mercury in Hg_2^{2+} is Hg(I) and therefore its electron configuration is $d^{10}s^1$. Although overlap of the d orbitals in the manner depicted in Fig. 19.17 is possible, the 20 d electrons from the two Hg ions would result in complete filling of both the bonding and the antibonding orbitals, with no effective bonding. Therefore the bonding must come from the overlap of s orbitals on each ion and the two remaining s electrons: σ bonding and antibonding orbitals can be constructed and as only the former is occupied, it results in a single Hg−Hg bond. Even if a degree of sd hybridization is invoked, the description still demands the involvement of s orbitals in the bonding and the bond order remains 1.

Self-test 19.3 Describe the probable structure of the compound formed when Re_3Cl_9 is dissolved in a solvent containing PPh_3.

FURTHER READING

D.M.P. Mingos, *Essential trends in inorganic chemistry*. Oxford University Press (1998). A survey of inorganic chemistry from the perspective of structure and bonding.

R.B. King (ed.), *Encyclopedia of inorganic chemistry*. Wiley (2005).

M.T. Pope, Polyoxoanions: synthesis and structure. In *Comprehensive coordination chemistry II*, Vol. 4 (ed. J.A. McCleverty and T.J. Meyer),

Chapter 10. Elsevier (2004). A discussion of metal oxides in solution, focusing on their tendency to form polyoxoanions.

M.H. Chisholm (ed.) *Early transition metal clusters with π-donor ligands*. VCH, Weinheim (1995).

EXERCISES

19.1 Without reference to a periodic table, sketch the first series of the d block, including the symbols of the elements. Indicate those elements for which the group oxidation number is common by C, those for which the group oxidation number can be reached but is a powerful oxidizing agent by O, and those for which the group oxidation number is not achieved by N.

19.2 Explain why the enthalpy of sublimation of Re(s) is significantly greater than that of Mn(s).

19.3 State the trend in the stability of the group oxidation state on descending a group of metallic elements in the d block. Illustrate the trend using standard potentials in acidic solution for Groups 5 and 6.

19.4 For each part, give balanced chemical equations or NR (for no reaction) and rationalize your answer in terms of trends in oxidation states.

(a) $Cr^{2+}(aq) + Fe^{3+}(aq) \rightarrow$

(b) $CrO_4^{2-}(aq) + MoO_2(s) \rightarrow$

(c) $MnO_4^-(aq) + Cr^{3+}(aq) \rightarrow$

19.5 (a) Which ion, $Ni^{2+}(aq)$ or $Mn^{2+}(aq)$, is more likely to form a sulfide in the presence of H_2S? (b) Rationalize your answer with the trends in hard and soft character across Period 4. (c) Give a balanced chemical equation for the reaction.

19.6 Preferably without reference to the text (a) write out the d block of the periodic table, (b) indicate the metals that form difluorides with the rutile or fluorite structures, and (c) indicate the region of the periodic table in which metal−metal bonded halide compounds are formed, giving one example.

19.7 Write a balanced chemical equation for the reaction that occurs when cis-$[RuLCl(OH_2)]^+$ (see Fig. 19.9) in acidic solution at +0.2 V is made strongly basic at the same potential. Write a balanced equation for each of the successive reactions when this same complex at pH = 6

and +0.2 V is exposed to progressively more oxidizing environments up to +1.0 V. Give other examples and a reason for the redox state of the metal centre affecting the extent of protonation of coordinated oxygen.

19.8 Give plausible balanced chemical reactions (or NR for no reaction) for the following combinations, and state the basis for your answer: (a) $MnO_4^-(aq)$ plus $Fe^{2+}(aq)$ in acidic solution, (b) the preparation of $[Mo_6O_{19}]^{2-}(aq)$ from $K_2MoO_4(s)$, (c) $ReCl_5(s)$ plus $KMnO_4(aq)$, (d) $MoCl_2(s)$ plus warm HBr(aq), (e) TiO(s) with HCl(aq) under an inert atmosphere, (f) Cd(s) added to $Hg^{2+}(aq)$.

19.9 Speculate on the structures of the following species and present bonding models to justify your answers: (a) $[Re(O)_2(py)_4]^+$, (b) $[V(O)_2(ox)_2]^{3-}$, (c) $[Mo(O)_2(CN)_4]^{4-}$, (d) $[VOCl_4]^{2-}$.

19.10 Which of the following are likely to have structures that are typical of (a) predominantly ionic, (b) significantly covalent, (c) metal−metal bonded compounds: NiI_2, $NbCl_4$, FeF_2, PtS, and WCl_2? Rationalize the differences and speculate on the structures.

19.11 Indicate the probable occupancy of σ, π, and δ bonding and antibonding orbitals, and the bond order for the following tetragonal prismatic complexes: (a) $[Mo_2(O_2CCH_3)_4]$, (b) $[Cr_2(O_2CC_2H_5)_4]$, (c) $[Cu_2(O_2CCH_3)_4]$.

19.12 Explain the differences in the following redox couples, measured at 25°C:

MnO_4^-/MnO_2	+1.69 V
TcO_4^-/TcO_2	+0.74 V
ReO_4^-/ReO_2	+0.51 V

19.13 Addition of sodium ethanoate to aqueous solutions of Cr(II) gives a red diamagnetic product. Draw the structure of the product, noting any features of interest.

19.14 Consider the two ruthenium complexes in Table 19.8. Using the bonding scheme depicted in Figure 19.19, confirm the bonding orders and electron configurations given in the table.

PROBLEMS

19.1 An amateur chemist claimed the existence of a new metallic element, grubium (Gr), which has the following characteristics. Metallic Gr reacts with 1 M $H^+(aq)$ in the absence of air to produce $Gr^{3+}(aq)$ and $H_2(g)$. In the absence of air $GrCl_2(s)$ dissolves in 1 M $H^+(aq)$ and very slowly yields $H_2(g)$ plus $Gr^{3+}(aq)$. When $Gr^{3+}(aq)$ is exposed to air $GrO^{2+}(aq)$ is produced. From this information, estimate the range of potentials for (a) the reduction of $Gr^{3+}(aq)$ to Gr(s), (b) the reduction of $Gr^{3+}(aq)$ to $GrCl_2(s)$, and (c) the reduction of $GrO^{2+}(aq)$ to $Gr^{3+}(aq)$. Suggest a known element that fits the description of grubium.

19.2 Compared with the p block, the variation in the properties of the elements of the d block across each period is rather modest. Provide evidence for this statement by reference to isostructural compounds. Speculate on reasons why this statement is true.

19.3 Think about the mechanism of electrical conduction and then suggest how electrical conductivity might vary across the d block.

19.4 Many metal salts can be vaporized to a small extent at high temperatures and their structures studied in the vapour phase by electron diffraction. Speculate on the structures that you might expect to find for the following gas-phase species and present your reasoning: (a) TaF_5, (b) MoF_6.

19.5 Discuss ways in which the triamidoamine ligands $[(RNCH_2CH_2)_3N]^{3-}$ (where R is a bulky substituent) can be used to stabilize nitrido, phosphido, and arsenido groups (see R.R. Schrock, *Acc. Chem. Res.*, 1997, **30**, 9).

19.6 Discuss how the presence of a quintuple bond in (20) was confirmed, and how a sextuple bond between two metal atoms might arise (see T. Nguyen, A.D. Sutton, M. Brynda, J.C. Fettinger, G.J. Long, and P.P. Power, *Science* 2005, **310**, 844).

d-Metal complexes: electronic structure and properties

20

d-Metal complexes play an important role in inorganic chemistry. In this chapter, we discuss the nature of ligand—metal bonding in terms of two theoretical models. We start with the simple but useful crystal-field theory, which is based on an electrostatic model of the bonding, and then progress to the more sophisticated ligand-field theory. Both theories invoke a parameter, the ligand-field splitting parameter, to correlate spectroscopic and magnetic properties. We then examine the electronic spectra of complexes and see how ligand-field theory allows us to interpret the energies and intensities of electronic transitions.

We now examine in detail the bonding, electronic structure, electronic spectra, and magnetic properties of the d-metal complexes introduced in Chapter 7. The striking colours of many d-metal complexes were a mystery to Werner when he elucidated their structures, and the origin of the colours was clarified only when the description of electronic structure in terms of orbitals was applied to the problem in the period from 1930 to 1960. Tetrahedral and octahedral complexes are the most important, and the discussion begins with them.

Electronic structure

There are two widely used models of the electronic structure of d-metal complexes. One ('crystal-field theory') emerged from an analysis of the spectra of d-metal ions in solids; the other ('ligand-field theory') arose from an application of molecular orbital theory. Crystal-field theory is more primitive, and strictly speaking it applies only to ions in crystals; however, it can be used to capture the essence of the electronic structure of complexes in a straightforward manner. Ligand-field theory builds on crystal-field theory: it gives a more complete description of the electronic structure of complexes and accounts for a wider range of properties.

20.1 Crystal-field theory

In **crystal-field theory**, a ligand lone pair is modelled as a point negative charge (or as the partial negative charge of an electric dipole) that repels electrons in the d orbitals of the central metal ion. The theory concentrates on the resulting splitting of the d orbitals into groups with different energies, and uses that splitting to rationalize and correlate the optical spectra, thermodynamic stability, and magnetic properties of complexes.

(a) Octahedral complexes

Key points: In the presence of an octahedral crystal field, d orbitals are split into a lower-energy triply degenerate set (t_{2g}) and a higher-energy doubly degenerate set (e_g) separated by an energy Δ_o; the ligand-field splitting parameter increases along a spectrochemical series of ligands and varies with the identity and charge of the metal atom.

In the model of an octahedral complex used in crystal-field theory, six point negative charges representing the ligands are placed in an octahedral array around the central metal ion. These charges (which we shall refer to as the 'ligands') interact strongly with the

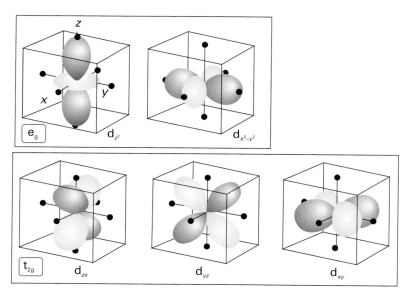

Figure 20.1 The orientation of the five d orbitals with respect to the ligands of an octahedral complex: the degenerate (a) e_g and (b) t_{2g} orbitals.

Spherical environment   Octahedral crystal field

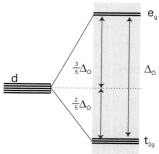

Figure 20.2 The energies of the d orbitals in an octahedral crystal field. Note that the mean energy remains unchanged relative to the energy of the d orbitals in a spherically symmetrical environment (such as in a free atom).

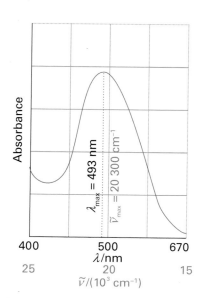

Figure 20.3 The optical absorption spectrum of $[Ti(OH_2)_6]^{3+}$.

central metal ion, and the stability of the complex stems in large part from this attractive interaction between opposite charges. However, there is a much smaller but very important secondary effect arising from the fact that electrons in different d orbitals interact with the ligands to different extents. Although this differential interaction is little more than about 10 per cent of the overall metal–ligand interaction energy, it has major consequences for the properties of the complex and is the principal focus of this section.

Electrons in d_{z^2} and $d_{x^2-y^2}$ orbitals (which are of symmetry type e_g in O_h; Section 6.1) are concentrated close to the ligands, along the axes, whereas electrons in d_{xy}, d_{yz}, and d_{zx} orbitals (which are of symmetry type t_{2g}) are concentrated in regions that lie between the ligands (Fig. 20.1). As a result, the former are repelled more strongly by the negative charge on the ligands than the latter and lie at a higher energy. Group theory shows that the two e_g orbitals have the same energy (although this is not readily apparent from drawings), and that the three t_{2g} orbitals also have the same energy. This simple model leads to an energy-level diagram in which the three degenerate t_{2g} orbitals lie below the two degenerate e_g orbitals (Fig. 20.2). The separation of the two sets of orbitals is called the **ligand-field splitting parameter**, Δ_O (where the subscript O signifies an octahedral crystal field).

A note on good practice In the context of crystal-field theory, the ligand-field splitting parameter should be called the *crystal-field splitting parameter*, but we use ligand-field splitting parameter to avoid a proliferation of names.

The energy level that corresponds to the hypothetical spherically symmetrical environment (in which the negative charge due to the ligands is evenly distributed over a sphere instead of being localized at six points) defines the **barycentre** of the array of levels, with the two e_g orbitals lying at $\frac{3}{5}\Delta_O$ above the barycentre and the three t_{2g} orbitals lying at $\frac{2}{5}\Delta_O$ below it. As in the representation of the configurations of atoms, a superscript is used to indicate the number of electrons in each set, for example t_{2g}^2.

The simplest property that can be interpreted by crystal-field theory is the absorption spectrum of a one-electron complex. Figure 20.3 shows the optical absorption spectrum of the d^1 hexaaquatitanium(III) ion, $[Ti(OH_2)_6]^{3+}$. Crystal-field theory assigns the first absorption maximum at 493 nm (20 300 cm^{-1}) to the transition $e_g \leftarrow t_{2g}$ and identifies 20 300 cm^{-1} with Δ_O for the complex. It is not so straightforward to obtain values of Δ_O for complexes with more than one d electron because the energy of a transition then depends not only on orbital energies but also on the electron–electron repulsion energies. This aspect is treated more fully in Section 20.4 and the results from the analyses described there have been used to obtain the values of Δ_O in Table 20.1.

Table 20.1 Ligand-field splitting parameters Δ_O of ML_6 complexes[*]

Ions		Ligands				
		Cl^-	H_2O	NH_3	en	CN^-
d^3	Cr^{3+}	13 700	17 400	21 500	21 900	26 600
d^5	Mn^{2+}	7500	8500		10 100	30 000
	Fe^{3+}	11 000	14 300			(35 000)
d^6	Fe^{2+}		10 400			(32 800)
	Co^{3+}		(20 700)	(22 900)	(23 200)	(34 800)
	Rh^{3+}	(20 400)	(27 000)	(34 000)	(34 600)	(45 500)
d^8	Ni^{2+}	7500	8500	10 800	11 500	

* Values are in cm^{-1}; entries in parentheses are for low-spin complexes.
Source: H.B. Gray, *Electrons and chemical bonding*. Benjamin, Menlo Park (1965).

A note on good practice The convention in spectroscopic notation is to indicate transitions as [upper state] ← [lower state].

The ligand-field splitting parameter, Δ_O, varies systematically with the identity of the ligand. For instance, in the series of complexes $[CoX(NH_3)_5]^{n+}$ with $X = I^-$, Br^-, Cl^-, H_2O, and NH_3, the colours range from purple (for $X = I^-$) through pink (for Cl^-) to yellow (with NH_3). This sequence indicates that the energy of the lowest energy electronic transition (and therefore Δ_O) increases as the ligands are varied along the series. The same order is followed regardless of the identity of the metal ion. Thus ligands can be arranged in a **spectrochemical series**, in which the members are arranged in order of increasing energy of transitions that occur when they are present in a complex:

$$I^- < Br^- < S^{2-} < \underline{S}CN^- < Cl^- < \underline{NO}_2^- < N^{3-} < F^- < OH^- < C_2O_4^{2-} < O^{2-} < H_2O$$
$$< \underline{N}CS^- < CH_3C{\equiv}N < py < NH_3 < en < bpy < phen < \underline{N}O_2^- < PPh_3 < \underline{C}N^- < CO$$

(The donor atom in an ambidentate ligand is underlined.) Thus, the series indicates that, for the same metal, the optical absorption of the cyano complex will occur at higher energy than that of the corresponding chlorido complex. A ligand that gives rise to a high-energy transition (such as CO) is referred to as a **strong-field ligand**, whereas one that gives rise to a low-energy transition (such as Br^-) is referred to as a **weak-field ligand**. Crystal-field theory alone cannot explain these strengths, but ligand-field theory can, as we shall see in Section 20.2.

The ligand-field strength also depends on the identity of the central metal ion, the order being approximately:

$$Mn^{2+} < Ni^{2+} < Co^{2+} < Fe^{2+} < V^{2+} < Fe^{3+} < Co^{3+} < Mo^{3+} < Rh^{3+} < Ru^{3+} < Pd^{4+}$$
$$< Ir^{3+} < Pt^{4+}$$

The value of Δ_O increases with increasing oxidation state of the central metal ion (compare the two entries for Fe and Co) and also increases down a group (compare, for instance, the locations of Co, Rh, and Ir). The variation with oxidation state reflects the smaller size of more highly charged ions and the consequently shorter metal–ligand distances and stronger interaction energies. The increase down a group reflects the larger size of the 4d and 5d orbitals compared with the compact 3d orbitals and the consequent stronger interactions with the ligands.

(b) Ligand-field stabilization energies

Key point: The ground-state configuration of a complex reflects the relative values of the ligand-field splitting parameter and the pairing energy. For $3d^n$ species with $n = 4–7$, high-spin and low-spin complexes occur in the weak-field and strong-field cases, respectively. Complexes of 4d- and 5d-series metals are typically low-spin.

Because the d orbitals in a complex do not all have the same energy, the ground-state electron configuration of a complex is no longer immediately obvious. To predict it, we use the

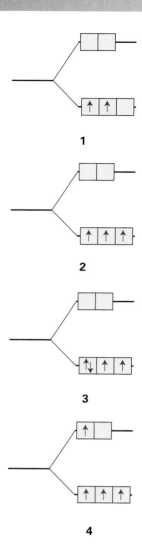

1

2

3

4

d-orbital energy level diagram shown in Fig. 20.2 as a basis for applying the building-up principle. That is, we identify the lowest energy configuration subject to the Pauli exclusion principle (a maximum of two electrons in an orbital) and (if more than one degenerate orbital is available) to the requirement that electrons first occupy separate orbitals and do so with parallel spins.

First, we consider complexes formed by the 3d-series elements. In an octahedral complex, the first three d electrons of a $3d^n$ complex occupy separate t_{2g} nonbonding orbitals, and do so with parallel spins. For example, the ions Ti^{2+} and V^{2+} have electron configurations $3d^2$ and $3d^3$, respectively. The d electrons occupy the lower t_{2g} orbitals as shown in (**1**) and (**2**), respectively. The energy of a t_{2g} orbital relative to the barycentre of an octahedral ion is $-0.4\Delta_O$ and the complexes are stabilized by $2 \times (0.4\Delta_O) = 0.8\Delta_O$ (for Ti^{2+}) and $3 \times (0.4\Delta_O) = 1.2\Delta_O$ (for V^{2+}). This additional stability, relative to the barycentre is called the **ligand-field stabilization energy (LFSE)**.

A note on good practice The term *crystal-field stabilization energy* (CFSE) is widely used in place of LFSE, but strictly speaking the term is appropriate only for ions in crystals.

The next electron needed for the $3d^4$ ion Cr^{2+} may enter one of the t_{2g} orbitals and pair with the electron already there (**3**). However, if it does so, it experiences a strong Coulombic repulsion, which is called the **pairing energy**, P. Alternatively, the electron may occupy one of the e_g orbitals (**4**). Although the pairing penalty is now avoided, the orbital energy is higher by Δ_O. In the first case (t_{2g}^4), there is a stabilisation of $1.6\Delta_O$, countered by the pairing energy of P, giving a net LFSE of $1.6\Delta_O - P$. In the second case ($t_{2g}^3 e_g^1$), the LFSE is $3 \times (0.4\Delta_O) - 0.6\Delta_O = 0.6\Delta_O$, as there is no pairing energy to consider. Which configuration is adopted depends on which of $1.60\Delta_O - P$ and $0.60\Delta_O$ is the larger.

If $\Delta_O < P$, which is called the **weak-field case**, a lower energy is achieved when the upper orbital is occupied to give the configuration $t_{2g}^3 e_g^1$. If $\Delta_O > P$, which is called the **strong-field case**, a lower energy is achieved by occupying only the lower orbitals despite the cost of the pairing energy. The resulting configuration is now t_{2g}^4. For example, $[Cr(OH_2)_6]^{2+}$ has the ground-state configuration $t_{2g}^3 e_g^1$ whereas $[Cr(CN)_6]^{4-}$, with relatively strong-field ligands (as indicated by the spectrochemical series), has the configuration t_{2g}^4. In the weak-field case all the electrons occupy different orbitals and have parallel spins. The resulting spin correlation effect (the tendency of electrons of the same spin to avoid each other) helps to offset the cost of occupying orbitals of higher energy.

The ground-state electron configurations of $3d^1$, $3d^2$, and $3d^3$ complexes are unambiguous because there is no competition between the additional stabilization achieved by occupying the t_{2g} orbitals and the pairing energy: the configurations are t_{2g}^1, t_{2g}^2, and t_{2g}^3, respectively, with each electron in a separate orbital. As remarked above, there are two possible configurations for $3d^4$ complexes; the same is true of $3d^n$ complexes in which $n = 5$, 6, or 7. In the strong-field case, the lower orbitals are occupied preferentially and in the weak-field case, electrons avoid the pairing energy by occupying the upper orbitals.

When alternative configurations are possible, the species with the smaller number of parallel electron spins is called a **low-spin complex**, and the species with the greater number of parallel electron spins is called a **high-spin complex**. As we have noted, an octahedral $3d^4$ complex is likely to be low-spin if the ligand field is strong but high-spin if the field is weak (Fig. 20.4); the same applies to $3d^5$, $3d^6$, and $3d^7$ complexes:

	Weak-field ligands		Strong-field ligands	
	Configuration	Unpaired electrons	Configuration	Unpaired electrons
$3d^4$	$t_{2g}^3 e_g^1$	4	t_{2g}^4	2
$3d^5$	$t_{2g}^3 e_g^2$	5	t_{2g}^5	1
$3d^6$	$t_{2g}^4 e_g^2$	4	t_{2g}^6	0
$3d^7$	$t_{2g}^5 e_g^2$	3	$t_{2g}^6 e_g^1$	1

The ground-state electron configurations of $3d^8$, $3d^9$, and $3d^{10}$ complexes are unambiguous and the configurations are $t_{2g}^6 e_g^2$, $t_{2g}^6 e_g^3$, and $t_{2g}^6 e_g^4$.

In general, the net energy of a $t_{2g}^x e_g^y$ configuration relative to the barycentre, without taking the pairing energy into account, is $(0.4x - 0.6y)\Delta_O$. Pairing energies need to be taken into account only for pairing that is additional to the pairing that occurs in a spherical field. Figure 20.5 shows the case of a d^6 ion. In both the free ion and the high-spin complex two electrons are paired, whereas in the low-spin case all six electrons occur as three pairs. Thus we do not need to consider the pairing energy in the high-spin case, as there is no additional pairing. There are two additional pairings in the low-spin case, so two pairing energy contributions must be taken into account. In general, high-spin complexes always have the same number of unpaired electrons as in a spherical field (free ion), and we therefore do not need to consider pairing energies for high-spin complexes. Table 20.2 lists the values for the LFSE of the various configurations of octahedral ions, with the appropriate pairing energies taken into account for the low-spin complexes. Remember that the LFSE is generally only a small fraction of the overall interaction between the metal atom and the ligands.

The strength of the crystal field (as measured by the value of Δ_O) and the spin-pairing energy (as measured by P) depend on the identity of both the metal and the ligand, so it is not possible to specify a universal point in the spectrochemical series at which a complex changes from high spin to low spin. For 3d-metal ions, low-spin complexes commonly occur for ligands that are high in the spectrochemical series (such as CN^-) and high-spin complexes are common for ligands that are low in the series (such as F^-). For octahedral d^n complexes with $n = 1-3$ and $8-10$ there is no ambiguity about the configuration (see Table 20.2), and the designations high-spin and low-spin are not used.

As we have seen, the values of Δ_O for complexes of 4d- and 5d-series metals are typically higher than for the 3d-series metals. Pairing energies for the 4d- and 5d-series metals tend to be lower than for the 3d-series metals because the orbitals are less compact and electron−electron repulsions correspondingly weaker. Consequently, complexes of these metals generally have electron configurations that are characteristic of strong crystal fields and typically have low spin. An example is the $4d^4$ complex $[RuCl_6]^{2-}$, which has a t_{2g}^4 configuration, which is typical of a strong crystal field despite Cl^- being low in the spectrochemical series. Likewise, $[Ru(ox)_3]^{3-}$ has the low-spin configuration t_{2g}^5 whereas $[Fe(ox)_3]^{3-}$ has the high-spin configuration $t_{2g}^3 e_g^2$.

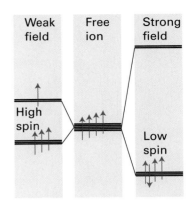

Figure 20.4 The effect of weak and strong ligand fields on the occupation of electrons for a d^4 complex. The former results in a high-spin configuration and the latter in a low-spin configuration.

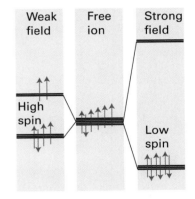

Figure 20.5 The effect of weak and strong ligand fields on the occupation of electrons for a d^6 complex. The former results in a high-spin configuration and the latter in a low-spin configuration.

Table 20.2 Ligand-field stabilization energies for octahedral complexes*

d^n	Example	N (high spin)	LFSE/Δ_O	N (low spin)	LFSE/Δ_O
d^0		0	0		
d^1	Ti^{3+}	1	0.4		
d^2	V^{3+}	2	0.8		
d^3	Cr^{3+}, V^{2+}	3	1.2		
d^4	Cr^{2+}, Mn^{3+}	4	0.6	2	$1.6 - P$
d^5	Mn^{2+}, Fe^{3+}	5	0	1	$2.0 - 2P$
d^6	Fe^{2+}, Co^{3+}	4	0.4	0	$2.4 - 2P$
d^7	Co^{2+}	3	0.8	1	$1.8 - P$
d^8	Ni^{2+}	2	1.2		
d^9	Cu^{2+}	1	0.6		
d^{10}	Cu^+, Zn^{2+}	0	0		

* N is the number of unpaired electrons.

EXAMPLE 20.1 Calculating the LFSE

Determine the LFSE for the following octahedral ions from first principles and confirm the value matches those in Table 20.2: (a) d^3, (b) high-spin d^5, (c) high-spin d^6, (d) low-spin d^6, (e) d^9.

Answer We need to consider the total orbital energy in each case and, when appropriate, the pairing energy. (a) A d^3 ion has configuration t_{2g}^3 (no pairing of electrons) and therefore LFSE $= 3 \times (0.4\Delta_O) = 1.2\Delta_O$. (b) A high spin d^5 ion has configuration $t_{2g}^3 e_g^2$ (no pairing of electrons) therefore LFSE $= 3 \times (0.4\Delta_O) - 2 \times (0.6\Delta_O) = 0$. (c) A high-spin d^6 ion has configuration $t_{2g}^4 e_g^2$ with the pairing of two electrons. However, since those two electrons would be paired in a spherical field there is no additional pairing energy to be concerned with. Therefore LFSE $= 4 \times (0.4\Delta_0) - 2 \times (0.6\Delta_0) = 0.4\Delta_0$. (d) A low-spin d^6 ion has configuration t_{2g}^6 with the pairing of three pairs of electrons. However, since one pair of electrons would be paired in a spherical field the additional pairing energy is $2P$. Therefore LFSE $= 6 \times (0.4\Delta_0) - 2P = 2.4\Delta_0 - 2P$. (e) A d^9 ion has configuration $t_{2g}^6 e_g^3$ with the pairing of four pairs of electrons. However, since all four pairs of electrons would be paired in a spherical field there is no additional pairing energy. Therefore LFSE $= 6 \times (0.4\Delta_0) - 3 \times (0.6\Delta_0) = 0.6\Delta_0$.

Self-test 20.1 What is the LFSE for both high- and low-spin d^7 configurations?

(c) Magnetic measurements

Key points: Magnetic measurements are used to determine the number of unpaired spins in a complex and hence to identify its ground-state configuration. A spin-only calculation may fail for low-spin d^5 and for high-spin $3d^6$ and $3d^7$ complexes.

The experimental distinction between high-spin and low-spin octahedral complexes is based on the determination of their magnetic properties. Compounds are classified as **diamagnetic** if they are repelled by a magnetic field and **paramagnetic** if they are attracted by a magnetic field. The two classes are distinguished experimentally by magnetometry (Chapter 8). The magnitude of the paramagnetism of a complex is commonly reported in terms of the magnetic dipole moment it possesses: the higher the magnetic dipole moment of the complex, the greater the paramagnetism of the sample.

In a free atom or ion, both the orbital and the spin angular momenta give rise to a magnetic moment and contribute to the paramagnetism. When the atom or ion is part of a complex, any orbital angular momentum is normally **quenched**, or suppressed, as a result of the interactions of the electrons with their nonspherical environment. However, if any electrons are unpaired the net electron spin angular momentum survives and gives rise to **spin-only paramagnetism**, which is characteristic of many d-metal complexes. The spin-only magnetic moment, μ, of a complex with total spin quantum number S is

$$\mu = 2\{S(S+1)\}^{\frac{1}{2}}\mu_B \tag{20.1}$$

where μ_B is the **Bohr magneton**, $\mu_B = e\hbar/2m_e$ with the value 9.274×10^{-24} J T^{-1}. Because $S = \frac{1}{2}N$, where N is the number of unpaired electrons, each with spin $s = \frac{1}{2}$,

$$\mu = \{N(N+2)\}^{\frac{1}{2}}\mu_B \tag{20.2}$$

A measurement of the magnetic moment of a d-block complex can usually be interpreted in terms of the number of unpaired electrons it contains, and hence the measurement can be used to distinguish between high-spin and low-spin complexes. For example, magnetic measurements on a d^6 complex easily distinguish between a high-spin $t_{2g}^4 e_g^2$ ($N = 4$, $S = 2$, $\mu = 4.90\mu_B$) configuration and a low-spin t_{2g}^6 ($N = 0$, $S = 0$, $\mu = 0$) configuration.

The spin-only magnetic moments for some electron configurations are listed in Table 20.3 and compared there with experimental values for a number of 3d complexes. For most 3d complexes (and some 4d complexes), experimental values lie reasonably close to spin-only predictions, so it becomes possible to identify correctly the number of unpaired electrons and assign the ground-state configuration. For instance, $[Fe(OH_2)_6]^{3+}$ is paramagnetic with a magnetic moment of $5.9\mu_B$. As shown in Table 20.3, this value is consistent with there being five unpaired electrons ($N = 5$ and $S = \frac{5}{2}$), which implies a high-spin $t_{2g}^3 e_g^2$ configuration.

Table 20.3 Calculated spin-only magnetic moments

Ion	Electron configuration	S	μ/μ_B Calculated	Experimental
Ti^{3+}	t_{2g}^1	$\frac{1}{2}$	1.73	1.7−1.8
V^{3+}	t_{2g}^2	1	2.83	2.7−2.9
Cr^{3+}	t_{2g}^3	$\frac{3}{2}$	3.87	3.8
Mn^{3+}	$t_{2g}^3 e_g^1$	2	4.90	4.8−4.9
Fe^{3+}	$t_{2g}^3 e_g^2$	$\frac{5}{2}$	5.92	5.9

The interpretation of magnetic measurements is sometimes less straightforward than this example might suggest. For example, the potassium salt of $[Fe(CN)_6]^{3-}$ has $\mu = 2.3\mu_B$, which is between the spin-only values for one and two unpaired electrons ($1.7\mu_B$ and $2.8\mu_B$, respectively). In this case, the spin-only assumption has failed because the orbital contribution to the magnetic moment is substantial.

For orbital angular momentum to contribute, and hence for the paramagnetism to differ significantly from the spin-only value, there must be one or more unfilled or half-filled orbitals similar in energy to the orbitals occupied by the unpaired spins and of the appropriate symmetry (one that is related to the occupied orbital by rotation round the direction of the applied field). If that is so, the applied magnetic field can force the electrons to circulate around the metal ion by using the low-lying orbitals and hence it generates orbital angular momentum and a corresponding orbital contribution to the total magnetic moment (Fig. 20.6). Departure from spin-only values is generally large for low-spin d^5 and for high-spin $3d^6$ and $3d^7$ complexes. It is also possible for the electronic state of the metal ion to change (for example with temperature), leading to a change from high-spin to low-spin and a change in the magnetic moment. Such complexes are referred to as **spin-crossover** complexes and are discussed in more detail, together with the effects of cooperative magnetism, in Sections 20.8 and 20.9.

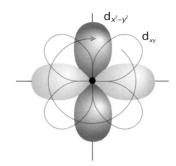

Figure 20.6 If there is a low-lying orbital of the correct symmetry, the applied field may induce the circulation of the electrons in a complex and hence generate orbital angular momentum. This diagram shows the way in which circulation may arise when the field is applied perpendicular to the xy-plane (perpendicular to this page).

EXAMPLE 20.2 Inferring an electron configuration from a magnetic moment

The magnetic moment of a certain octahedral Co(II) complex is $4.0\mu_B$. What is its d-electron configuration?

Answer We need to match the possible electron configurations of the complex with the observed magnetic dipole moment. A Co(II) complex is d^7. The two possible configurations are $t_{2g}^5 e_g^2$ (high spin, $N = 3$, $S = \frac{3}{2}$) with three unpaired electrons or $t_{2g}^6 e_g^1$ (low spin, $N = 1$, $S = \frac{1}{2}$) with one unpaired electron. The spin-only magnetic moments are $3.87\mu_B$ and $1.73\mu_B$, respectively (see Table 20.3). Therefore, the only consistent assignment is the high-spin configuration $t_{2g}^5 e_g^2$.

Self-test 20.2 The magnetic moment of the complex $[Mn(NCS)_6]^{4-}$ is $6.06\mu_B$. What is its electron configuration?

(d) Thermochemical correlations

Key point: The experimental variation in hydration enthalpies reflects a combination of the variation in radii of the ions (the linear trend) and the variation in LFSE (the saw-tooth variation).

The concept of ligand-field stabilization energy helps to explain the double-humped variation in the hydration enthalpies of the high-spin octahedral 3d-metal M^{2+} ions (Fig. 20.7). The nearly linear increase across a period shown by the filled circles represents the increasing strength of the bonding between H_2O ligands and the central metal ion as the ionic radii decrease from left to right across the period. The deviation of hydration enthalpies from a straight line reflects the variation in the ligand-field stabilization energies. As Table 20.2 shows, the LFSE increases from d^1 to d^3, decreases again to d^5, then rises to d^8. The filled circles in Fig. 20.7 were calculated by subtracting the high-spin LFSE from $\Delta_{hyd}H$ by using the spectroscopic values of Δ_O in Table 20.1. We see that the LFSE calculated from spectroscopic data accounts for the additional ligand binding energy for the complexes shown in the illustration.

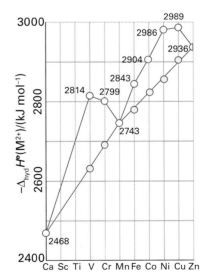

Figure 20.7 The hydration enthalpy of M^{2+} ions of the first row of the d block. The straight line shows the trend when the ligand-field stabilization energy has been subtracted from the observed values. Note the general trend to greater hydration enthalpy (more exothermic hydration) on crossing the period from left to right.

EXAMPLE 20.3 Using the LFSE to account for thermochemical properties

The oxides of formula MO, which all have octahedral coordination of the metal ions in a rock-salt structure, have the following lattice enthalpies:

CaO	TiO	VO	MnO
3460	3878	3913	3810 kJ mol^{-1}

Account for the trends in terms of the LFSE.

Answer We need to consider the simple trend that would be expected on the basis of trends in ionic radii and then deviations that can be traced to the LFSE. The general trend across the d block is the increase in lattice enthalpy from CaO (d^0) to MnO (d^5) as the ionic radii of the metals decrease (recall that lattice enthalpy is proportional to $1/(r_+ + r_-)$, Section 3.12). The Ca^{2+} ion has an LFSE of zero as it has no d electrons and the Mn^{2+} ion, being high spin (O^{2-} is a weak field ligand), also has an LFSE of zero. For a linear increase in lattice enthalpy from calcium to manganese oxides we would expect the lattice enthalpies to increase by $(3810 - 3460)/5$ kJ mol^{-1} from Ca^{2+} to Sc^{2+} to Ti^{2+} to V^{2+} to Mn^{2+}. We would therefore expect TiO and VO to have lattice enthalpies of 3600 and 3670 kJ mol^{-1}, respectively. In fact, TiO (d^2) has a lattice enthalpy of 3878 kJ mol^{-1} and we can ascribe this difference of 278 kJ mol^{-1} to an LFSE of $0.8\Delta_O$. Likewise the actual lattice enthalpy of 3913 kJ mol^{-1} for VO (d^3) is 243 kJ mol^{-1} greater than predicted, with this difference arising from an LFSE of $1.2\Delta_O$.

Self-test 20.3 Account for the variation in lattice enthalpy of the solid fluorides in which each metal ion is surrounded by an octahedral array of F$^-$ ions: MnF_2 (2780 kJ mol^{-1}), FeF_2 (2926 kJ mol^{-1}), CoF_2 (2976 kJ mol^{-1}), NiF_2 (3060 kJ mol^{-1}), and ZnF_2 (2985 kJ mol^{-1}).

(e) Tetrahedral complexes

Key points: In a tetrahedral complex, the e orbitals lie below the t_2 orbitals; only the high-spin case need be considered.

Four-coordinate tetrahedral complexes are second only in abundance to octahedral complexes for the 3d metals. The same kind of arguments based on crystal-field theory can be applied to these species as we used for octahedral complexes.

A tetrahedral crystal field splits d orbitals into two sets but with the two e orbitals (the $d_{x^2-y^2}$ and the d_{z^2}) lower in energy than the three t_2 orbitals (the d_{xy}, the d_{yz} and the d_{zx}) (Fig. 20.8).[1] The fact that the e orbitals lie below the t_2 orbitals can be understood from a consideration of the spatial arrangement of the orbitals: the e orbitals point between the positions of the ligands and their partial negative charges whereas the t_2 orbitals point more directly towards the ligands (Fig. 20.9). A second difference is that the ligand-field splitting parameter in a tetrahedral complex, Δ_T, is less than Δ_O, as should be expected for complexes with fewer ligands, none of which is oriented directly at the d orbitals (in fact, $\Delta_T \approx \frac{4}{9}\Delta_O$). The pairing energy is invariably more unfavourable than Δ_T, and normally only high-spin tetrahedral complexes are encountered.

Ligand-field stabilization energies can be calculated in exactly the same way as for octahedral complexes. Since tetrahedral complexes are always high-spin, there is never any

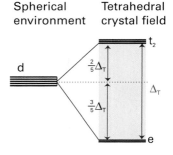

Figure 20.8 The orbital energy level diagram used in the application of the building-up principle in a crystal-field analysis of a tetrahedral complex.

Spherical environment / Tetrahedral crystal field

d → t_2 ($\frac{2}{5}\Delta_T$), Δ_T, e ($\frac{3}{5}\Delta_T$)

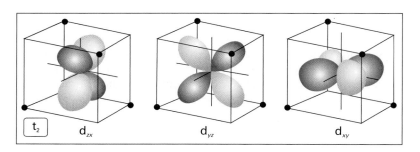

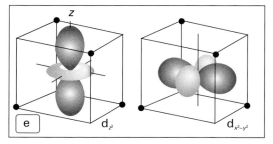

Figure 20.9 The effect of a tetrahedral crystal field on a set of d orbitals is to split them into two sets; the e pair (which point less directly at the ligands) lie lower in energy than the t_2 triplet.

---

[1] Because there is no centre of inversion in a tetrahedral complex, the orbital designation does not include the parity label g or u.

Table 20.4 Ligand-field stabilization energies for tetrahedral complexes*

d^n	Configuration	N	LFSE/Δ_T
d^0		0	0
d^1	e^1	1	0.6
d^2	e^2	2	1.2
d^3	$e^2t_2^1$	3	0.8
d^4	$e^2t_2^2$	4	0.4
d^5	$e^2t_2^3$	5	0
d^6	$e^3t_2^3$	4	0.6
d^7	$e^4t_2^3$	3	1.2
d^8	$e^4t_2^4$	2	0.8
d^9	$e^4t_2^5$	1	0.4
d^{10}	$e^4t_2^6$	0	0

*N is the number of unpaired electrons.

Table 20.5 Values of Δ_T for representative tetrahedral complexes

Complex	Δ_T/cm^{-1}
VCl_4	9010
$[CoCl_4]^{2-}$	3300
$[CoBr_4]^{2-}$	2900
$[CoI_4]^{2-}$	2700
$[Co(NCS)_4]^{2-}$	4700

need to consider the pairing energy in the LFSE and the only differences compared with octahedral complexes are the order of occupation (e before t_2) and the contribution of each orbital to the total energy ($\frac{3}{5}\Delta_T$ for an e orbital and $-\frac{2}{5}\Delta_T$ for a t_2 orbital). Table 20.4 lists the configurations of tetrahedral d^n complexes together with the calculated values of the LFSE and Table 20.5 lists some experimental values of Δ_T for a number of complexes.

(f) Square-planar complexes

Key point: A d^8 configuration, coupled with a strong ligand field, favours the formation of square-planar complexes. This tendency is enhanced with the 4d and 5d metals due to their larger size and the greater ease of electron pairing.

Although a tetrahedral arrangement of four ligands is the least sterically demanding arrangement, some complexes exist with four ligands in an apparently higher energy square-planar arrangement. A square-planar arrangement gives the d-orbital splitting shown in Fig. 20.10, with $d_{x^2-y^2}$ raised above all the others. This arrangement may become energetically favourable when there are eight d electrons and the crystal field is strong enough to favour the low-spin $d_{yz}^2 d_{zx}^2 d_{z^2}^2 d_{xy}^2$ configuration. In this configuration the electronic stabilization energy can more than compensate for any unfavourable steric interactions. Thus, many square-planar complexes are found for complexes of the large $4d^8$ and $5d^8$ Rh(I), Ir(I), Pt(II), Pd(II), and Au(III) ions, in which unfavourable steric constraints have less effect and there is a large ligand-field splitting associated with the 4d- and 5d-series metals. By contrast, small 3d-series metal complexes such as $[NiX_4]^{2-}$, with X a halogen, are generally tetrahedral because the ligand-field splitting parameter is generally quite small and will not compensate sufficiently for the unfavourable steric interactions. Only when the ligand is high in the spectrochemical series is the LFSE large enough to result in the formation of a square-planar complex, as, for example, with $[Ni(CN)_4]^{2-}$. We have already noted that pairing energies for the 4d- and 5d-series metals tend to be lower than for the 3d-series metals, and this difference provides a further factor that favours the formation of low-spin square-planar complexes with these metals.

The sum of the three distinct orbital splittings in Fig. 20.10 is denoted Δ_{SP}. Simple theory predicts that $\Delta_{SP} = 1.3\Delta_O$ for complexes of the same metal and ligands with the same M—L bond lengths.

(g) Tetragonally distorted complexes: the Jahn–Teller effect

Key points: A tetragonal distortion can be expected when the ground electronic configuration of a complex is orbitally degenerate; the complex will distort so as to remove the degeneracy and achieve a lower energy.

Six-coordinate d^9 complexes of copper(II) usually depart considerably from octahedral geometry and show pronounced tetragonal distortions (Fig. 20.11). High-spin d^4 (for instance,

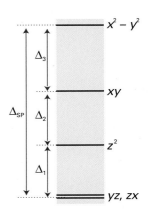

Figure 20.10 The orbital splitting parameters for a square-planar complex.

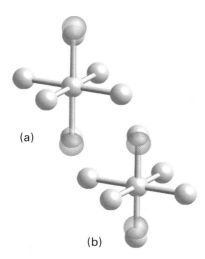

(a)

(b)

Figure 20.11 (a) A tetragonally distorted complex where two of the ligands have moved further away from the central ion. (b) A tetragonally distorted complex where two of the ligands have moved closer towards the central ion.

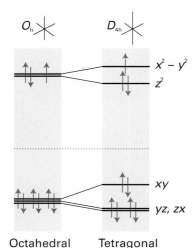

Figure 20.12 The effect of tetragonal distortions (compression along x and y and extension along z) on the energies of d orbitals. The electron occupation is for a d^9 complex.

Mn^{3+}) and low-spin d^7 six-coordinate complexes (for instance, Ni^{3+}) may show a similar distortion, but they are less common. These distortions are manifestations of the **Jahn–Teller effect**: if the ground electronic configuration of a nonlinear complex is orbitally degenerate, and asymmetrically filled, then the complex distorts so as to remove the degeneracy and achieve a lower energy.

The physical origin of the effect is quite easy to identify. Thus, a tetragonal distortion of a regular octahedron, corresponding to extension along the z-axis and compression on the x- and y-axes, lowers the energy of the $e_g(d_{z^2})$ orbital and increases the energy of the $e_g(d_{x^2-y^2})$ orbital (Fig. 20.12). Therefore, if one or three electrons occupy the e_g orbitals (as in high-spin d^4, low-spin d^7, and d^9 complexes) a tetragonal distortion may be energetically advantageous. For example, in a d^9 complex (with configuration that would be $t_{2g}^6 e_g^3$ in O_h), such a distortion leaves two electrons in the d_{z^2} orbital with a lower energy and one in the $d_{x^2-y^2}$ orbital with a higher energy. Similar distortions can occur with tetrahedral complexes.

The Jahn–Teller effect identifies an unstable geometry (a nonlinear complex with an orbitally degenerate ground state); it does not predict the preferred distortion. For instance, with an octahedral complex, instead of axial elongation and equatorial compression the degeneracy can also be removed by axial compression and equatorial elongation. Which distortion occurs in practice is a matter of energetics, not symmetry. However, because axial elongation weakens two bonds but equatorial elongation weakens four, axial elongation is more common than axial compression.

A Jahn–Teller distortion can hop from one orientation to another and give rise to the **dynamic Jahn–Teller effect**. For example, below 20 K the EPR spectrum of $[Cu(OH_2)_6]^{2+}$ shows a static distortion (more precisely, one that is effectively stationary on the timescale of the resonance experiment). However, above 20 K the distortion disappears because it hops more rapidly than the timescale of the EPR observation.

A Jahn–Teller effect is possible for other electron configurations (for an octahedral complex the d^1, d^2, low-spin d^4 and d^5, high-spin d^6, and d^7 configurations, for a tetrahedral complex the d^1, d^3, d^4, d^6, d^8, and d^9 configurations). However, as neither the t_{2g} orbitals in an octahedral complex nor any of the d orbitals in a tetrahedral complex point directly at the ligands, the effect is too small to induce a measurable distortion in the structure.

(h) Octahedral versus tetrahedral coordination

Key points: Consideration of the LFSE predicts that d^3 and d^8 ions strongly prefer an octahedral geometry over a tetrahedral one; for other configurations the preference is less pronounced, and LFSE has no bearing on the geometry of d^0, high-spin d^5, and d^{10} ions.

An octahedral complex has six M–L bonding interactions and, in the absence of significant steric and electronic effects, this arrangement will have a lower energy than a tetrahedral complex with just four M–L bonding interactions. We have already discussed the effects of steric bulk on a complex (Section 7.3), and have just seen the electronic reasons that favour a square-planar complex. We can now complete the discussion by considering the electronic effects that favour an octahedral complex over a tetrahedral one.

Figure 20.13 illustrates the variation of the LFSE for tetrahedral and high-spin octahedral complexes for all electronic configurations. It is apparent that, in terms of LFSE, octahedral geometries are strongly preferred over tetrahedral for d^3 and d^8 complexes: chromium(III) (d^3) and nickel(II) (d^8) do indeed show an exceptional preference for octahedral geometries. Similarly, d^4 and d^9 configurations show a preference for octahedral complexes (for example Mn(III) and Cu(II); note that the Jahn–Teller effect enhances this preference), whereas tetrahedral complexes of d^1, d^2, d^6, and d^7 ions will not be too disfavoured; thus V(II) (d^2) and Co(II) (d^7) form tetrahedral complexes (MX_4^{2-}) with halide ligands. The geometry of complexes of ions with d^0, d^5, and d^{10} configurations will not be affected by the number of d electrons, as there is no LFSE for these species.

Because the size of the d-orbital splitting, and hence the LFSE, depends on the ligand, it follows that a preference for octahedral coordination will be least pronounced for weak-field ligands. With strong-field ligands, low-spin complexes might be preferred and, although the situation is complicated by the pairing energy, the LFSE of a low-spin octahedral complex will be greater than that of a high-spin complex. There will thus be a correspondingly greater preference for octahedral over tetrahedral coordination when the octahedral complex is low-spin.

This preference for octahedral over tetrahedral coordination plays an important role in the solid state by influencing the structures that are adopted by d-metal compounds. This influence is demonstrated by the ways in which the different metal ions A and B in spinels (of formula AB_2O_4, Sections 3.9b and 24.7c) occupy the octahedral or tetrahedral sites. Thus, Co_3O_4 is a normal spinel because the low-spin d^6 Co(III) ion strongly favours octahedral coordination, resulting in $(Co^{2+})_t(2Co^{3+})_oO_4$, whereas Fe_3O_4 (magnetite) is an inverse spinel because Fe(II), but not Fe(III), can acquire greater LFSE by occupying an octahedral site. Thus magnetite is formulated as $(Fe^{3+})_t(Fe^{2+}Fe^{3+})_oO_4$.

(i) The Irving–Williams series

Key point: The Irving–Williams series summarizes the relative stabilities of complexes formed by M^{2+} ions, and reflects a combination of electrostatic effects and LFSE.

Figure 20.14 shows $\log K_f$ values (Section 7.12) for complexes of the octahedral M^{2+} ions of the 3d series. The variation in formation constants shown there is summarized by the Irving–Williams series:

$$Ba^{2+} < Sr^{2+} < Ca^{2+} < Mg^{2+} < Mn^{2+} < Fe^{2+} < Co^{2+} < Ni^{2+} < Cu^{2+} > Zn^{2+}$$

The order is relatively insensitive to the choice of ligands.

In general, the increase in stability correlates with ionic radius, which suggests that the Irving–Williams series reflects electrostatic effects. However, beyond Mn^{2+} there is a sharp increase in the value of K_f for d^6 Fe(II), d^7 Co(II), d^8 Ni(II), and d^9 Cu(II), with strong-field ligands. These ions experience an additional stabilization proportional to the ligand-field stabilization energies (Table 20.2). There is one important exception: the stability of Cu(II) complexes is greater than that of Ni(II) even though Cu(II) has an additional antibonding e_g electron. This anomaly is a consequence of the stabilizing influence of the Jahn–Teller effect, which results in strong binding of four of the ligands in the plane of the tetragonally distorted Cu(II) complex, and that stabilization enhances the value of K_f.

20.2 Ligand-field theory

Crystal-field theory provides a simple conceptual model that can be used to interpret magnetic, spectroscopic, and thermochemical data by using empirical values of Δ_O. However, the theory is defective because it treats ligands as point charges or dipoles and does not take into account the overlap of ligand and metal atom orbitals. One consequence of this oversimplification is that crystal-field theory cannot account for the ligand spectrochemical series. **Ligand-field theory,** which is an application of molecular orbital theory that concentrates on the d orbitals of the central metal atom, provides a more substantial framework for understanding the origins of Δ_O.

The strategy for describing the molecular orbitals of a d-metal complex follows procedures similar to those described in Chapter 2 for bonding in polyatomic molecules: the valence orbitals on the metal and ligand are used to form symmetry-adapted linear combinations (SALCs; Section 6.6), and then estimating the relative energies of the molecular orbitals by using empirical energy and overlap considerations. These relative energies can be verified and positioned more precisely by comparison with experimental data (particularly UV/visible absorption and photoelectron spectroscopy).

We shall first consider octahedral complexes, initially taking into account only the metal–ligand σ bonding. We then consider the effect of π bonding, and see that it is essential for understanding Δ_O (which is one reason why crystal-field theory cannot explain the spectrochemical series). Finally, we consider complexes with different symmetries, and see that similar arguments apply to them. Later in the chapter we shall see how information from optical spectroscopy is used to refine the discussion and provide quantitative data on the ligand-field splitting parameter and electron–electron repulsion energies.

(a) σ Bonding

Key point: In ligand-field theory, the building-up principle is used in conjunction with a molecular orbital energy level diagram constructed from metal atom orbitals and symmetry-adapted linear combinations of ligand orbitals.

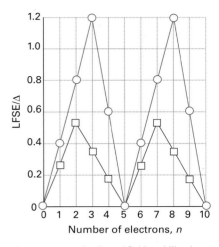

Figure 20.13 The ligand-field stabilization energy for d^n complexes in octahedral (high-spin, circles) and tetrahedral (squares) complexes. The LFSE is shown in terms of Δ_O, by applying the relationship $\Delta_T = \frac{4}{9}\Delta_O$.

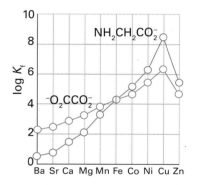

Figure 20.14 The variation of formation constants for the M^{2+} ions of the Irving–Williams series.

We begin by considering an octahedral complex in which each ligand (L) has a single valence orbital directed towards the central metal atom (M); each of these orbitals has local σ symmetry with respect to the M−L axis. Examples of such ligands include the NH_3 molecule and the F^- ion.

In an octahedral (O_h) environment, the orbitals of the central metal atom divide by symmetry into four sets (Fig. 20.15 and *Resource section* 4):

Metal orbital	Symmetry label	Degeneracy
s	a_{1g}	1
p_x, p_y, p_z	t_{1u}	3
$d_{x^2-y^2}, d_{z^2}$	e_g	2
d_{xy}, d_{yz}, d_{zx}	t_{2g}	3

Six symmetry-adapted linear combinations of the six ligand σ orbitals can also be formed. These combinations can be taken from *Resource section* 5 and are also shown in Fig. 20.15. One (unnormalized) SALC has symmetry a_{1g}:

$$a_{1g}:\ \sigma_1 + \sigma_2 + \sigma_3 + \sigma_4 + \sigma_5 + \sigma_6$$

where σ_i denotes a σ orbital on ligand i. There are three SALCs of symmetry t_{1u}:

$$t_{1u}:\ \sigma_1 - \sigma_3,\quad \sigma_2 - \sigma_4,\quad \sigma_5 - \sigma_6$$

and two SALCs of symmetry e_g:

$$e_g:\quad \sigma_1 - \sigma_2 + \sigma_3 - \sigma_4,\qquad 2\sigma_6 + 2\sigma_5 - \sigma_1 - \sigma_2 - \sigma_3 - \sigma_4$$

These six SALCs account for all the ligand orbitals of σ symmetry: there is no combination of ligand σ orbitals that has the symmetry of the metal t_{2g} orbitals, so the latter do not participate in σ bonding.[2]

Molecular orbitals are formed by combining SALCs and metal atomic orbitals of the same symmetry. For example, the (unnormalized) form of an a_{1g} molecular orbital is $c_M \psi_{Ms} + c_L \psi_{La1g}$, where ψ_{Ms} is the s orbital on the metal atom M and ψ_{La1g} is the ligand SALC of symmetry a_{1g}. The metal s orbital and ligand a_{1g} SALC overlap to give two molecular orbitals, one bonding and one antibonding. Similarly, the doubly degenerate metal e_g orbitals and the ligand e_g SALCs overlap to give four molecular orbitals (two degenerate bonding, two degenerate antibonding), and the triply degenerate metal t_{1u} orbitals and the three t_{1u} SALCs overlap to give six molecular orbitals (three degenerate bonding, three degenerate antibonding). There are therefore six bonding combinations in all and six antibonding combinations. The three triply degenerate metal t_{2g} orbitals remain nonbonding and fully localized on the metal atom. Calculations of the resulting energies (adjusted to agree with a variety of spectroscopic data of the kind to be discussed in Section 20.4) result in the molecular orbital energy level diagram shown in Fig. 20.16.

The greatest contribution to the molecular orbital of lowest energy is from atomic orbitals of lowest energy (Section 2.9). For NH_3, F^-, and most other ligands, the ligand σ orbitals are derived from atomic orbitals with energies that lie well below those of the metal d orbitals. As a result, the six bonding molecular orbitals of the complex are mainly ligand-orbital in character (that is, $c_L^2 > c_M^2$). These six bonding orbitals can accommodate the 12 electrons provided by the six ligand lone pairs. The electrons that we can regard as provided by the ligands are therefore largely confined to the ligands in the complex, just as the crystal-field theory presumes. However, because the coefficients c_M are nonzero, the bonding molecular orbitals do have some d-orbital character and the 'ligand electrons' are partly delocalized on to the central metal atom.

The total number of electrons to accommodate, in addition to those supplied by the ligands, now depends on the number of d electrons, n, supplied by the metal atom. These additional electrons enter the orbitals next in line for occupation, which are the nonbonding d orbitals (the t_{2g} orbitals) and the antibonding combination (the upper e_g orbitals) of the d orbitals and ligand orbitals. The t_{2g} orbitals are wholly confined (in the present approximation) to the metal atom and the antibonding e_g orbitals are largely metal atom in

---

[2] The normalization constants (with overlap neglected) are $N(a_{1g}) = (\tfrac{1}{6})^{\frac{1}{2}}$, $N(t_{1u}) = (\tfrac{1}{2})^{\frac{1}{2}}$ for all three orbitals, and $N(e_g) = \tfrac{1}{2}$ and $(\tfrac{1}{12})^{\frac{1}{2}}$, respectively.

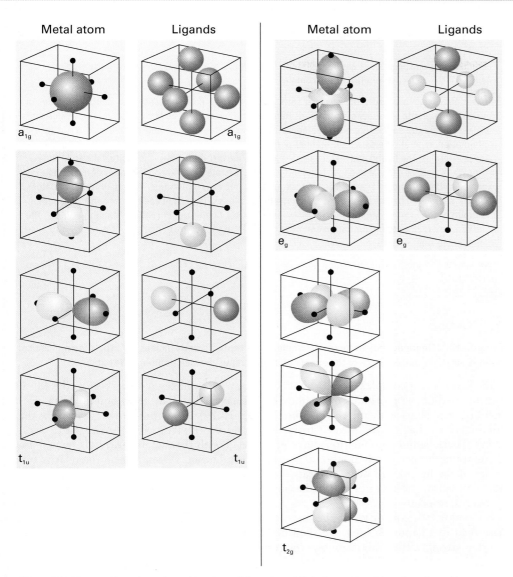

Metal atom   Ligands   Metal atom   Ligands   Metal   Complex   Ligands

Figure 20.16 Molecular orbital energy levels of a typical octahedral complex. The frontier orbitals are inside the tinted box.

Figure 20.15 Symmetry-adapted combinations of ligand σ orbitals (represented here by spheres) in an octahedral complex. For symmetry-adapted orbitals in other point groups, see *Resource section* 5.

character too, so the n electrons supplied by the central atom remain largely on that atom. The frontier orbitals of the complex are therefore the nonbonding entirely metal t_{2g} orbitals and the antibonding, mainly metal e_g orbitals. Thus, we have arrived at an arrangement that is qualitatively the same as in crystal-field theory. In the ligand-field approach the octahedral ligand-field splitting parameter, Δ_O, is the separation between the molecular orbitals largely, but not completely, confined to the metal atom, Fig. 20.16.

With the molecular orbital energy level diagram established, we use the building-up principle to construct the ground-state electron configuration of the complex. For a d^n complex, there are $12 + n$ electrons to accommodate. The six bonding molecular orbitals accommodate the 12 electrons supplied by the ligands. The remaining n electrons are accommodated in the nonbonding t_{2g} orbitals and the antibonding e_g orbitals. Now the story is essentially the same as for crystal-field theory, the types of complexes that are obtained (high-spin or low-spin, for instance) depending on the relative values of Δ_O and the pairing energy P. The principal difference from the crystal-field discussion is that ligand-field theory gives deeper insight into the origin of the ligand-field splitting, and we can begin to understand why some ligands are strong and others are weak. For instance, a good σ-donor ligand should result in strong metal–ligand overlap, hence a more strongly antibonding e_g set and consequently a larger value of Δ_O. However, before drawing further conclusions, we must go on to consider what crystal-field theory ignores completely, the role of π bonding.

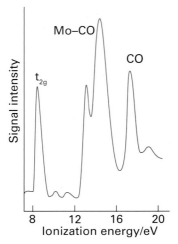

Figure 20.17 The He(II) (30.4 nm) photoelectron spectrum of Mo(CO)$_6$. With six electrons from Mo and twelve from :CO, the ground-state configuration of the complex is $a_{1g}^2 t_{1u}^6 e_g^4 t_{2g}^6$. (From B.R. Higginson, D.R. Lloyd, P. Burroughs, D.M. Gibson, and A.F. Orchard, *J. Chem. Soc., Faraday II*, 1973, 69, 1659.)

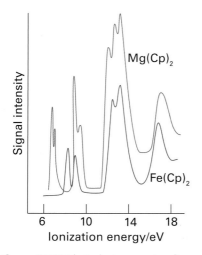

Figure 20.18 Photoelectron spectra of ferrocene and magnesocene.

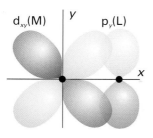

Figure 20.19 The π overlap that may occur between a ligand p orbital perpendicular to the M−L axis and a metal d_{xy} orbital.

EXAMPLE 20.4 Using a photoelectron spectrum to obtain information about a complex

The photoelectron spectrum of gas-phase [Mo(CO)$_6$] is shown in Fig. 20.17. Use the spectrum to infer the energies of the molecular orbitals of the complex.

Answer We need to identify the electron configuration of the complex, and then match the order of ionization energies to the order of the orbitals from which the electrons are likely to come. Twelve electrons are provided by the six CO ligands (treated as :CO); they enter the bonding orbitals and result in the configuration $a_{1g}^2 t_{1u}^6 e_g^4$. The oxidation number of molybdenum, Group 6, is 0, so Mo provides a further six valence electrons. The ligand and metal valence electrons are distributed over the orbitals shown in the box in Fig. 20.15 and, as CO is a strong-field ligand, the ground-state electron configuration of the complex is expected to be low-spin $a_{1g}^2 t_{1u}^6 e_g^4 t_{2g}^6$. The HOMOs are the three t$_{2g}$ orbitals largely confined to the Mo atom, and their energy can be identified by ascribing the peak of lowest ionization energy (close to 8 eV) to them. The group of ionization energies around 14 eV are probably due to the Mo−CO σ-bonding orbitals. The value of 14 eV is close to the ionization energy of CO itself, so the variety of peaks at that energy also arise from bonding orbitals in CO.

Self-test 20.4 Suggest an interpretation of the photoelectron spectra of [Fe(C$_5$H$_5$)$_2$] and [Mg(C$_5$H$_5$)$_2$] shown in Fig. 20.18.

(b) π Bonding

Key points: π-Donor ligands decrease Δ_O whereas π-acceptor ligands increase Δ_O; the spectrochemical series is largely a consequence of the effects of π bonding when such bonding is feasible.

If the ligands in a complex have orbitals with local π symmetry with respect to the M−L axis (as two of the p orbitals of a halide ligand have), they may form bonding and antibonding π orbitals with the metal orbitals (Fig. 20.19). For an octahedral complex the combinations that can be formed from the ligand π orbitals include SALCs of t$_{2g}$ symmetry. These ligand combinations have net overlap with the metal t$_{2g}$ orbitals, which are therefore no longer purely nonbonding on the metal atom. Depending on the relative energies of the ligand and metal orbitals, the energies of the now molecular t$_{2g}$ orbitals lie above or below the energies they had as nonbonding atomic orbitals, so Δ_O is decreased or increased, respectively.

To explore the role of π bonding in more detail, we need two of the general principles described in Chapter 2. First, we shall make use of the idea that, when atomic orbitals overlap strongly, they mix strongly: the resulting bonding molecular orbitals are significantly lower in energy and the antibonding molecular orbitals are significantly higher in energy than the atomic orbitals. Second, we note that atomic orbitals with similar energies interact strongly, whereas those of very different energies mix only slightly even if their overlap is large.

A **π-donor ligand** is a ligand that, before any bonding is considered, has filled orbitals of π symmetry around the M−L axis. Such ligands include Cl$^-$, Br$^-$, OH$^-$, O^{2-} and even H$_2$O. In Lewis acid−base terminology (Section 4.9), a π-donor ligand is a π **base**. The energies of the full π orbitals on the ligands will not normally be higher than their σ-donor orbitals (HOMO) and must therefore also be lower in energy than the metal d orbitals. Because the full π orbitals of π donor ligands lie lower in energy than the partially filled d orbitals of the metal, when they form molecular orbitals with the metal t$_{2g}$ orbitals, the bonding combination lies lower than the ligand orbitals and the antibonding combination lies above the energy of the d orbitals of the free metal atom (Fig. 20.20). The electrons supplied by the ligand π orbitals occupy and fill the bonding combinations, leaving the electrons originally in the d orbitals of the central metal atom to occupy the antibonding t$_{2g}$ orbitals. The net effect is that the previously nonbonding metal t$_{2g}$ orbitals become antibonding and hence are raised closer in energy to the antibonding e$_g$ orbitals. It follows that π-donor ligands *decrease* Δ_O.

A **π-acceptor ligand** is a ligand that has empty π orbitals that are available for occupation. In Lewis acid−base terminology, a π-acceptor ligand is a π **acid**. Typically, the π-acceptor orbitals are vacant antibonding orbitals on the ligand (usually the LUMO), as in CO and N$_2$, which are higher in energy than the metal d orbitals. The two π* orbitals of CO, for instance, have their largest amplitude on the C atom and have the correct symmetry for overlap with the metal t$_{2g}$ orbitals, so CO can act as a π-acceptor ligand

(Section 22.5). Phosphines (PR_3) are also able to accept π-electron density and also act as π acceptors (Section 22.6).

Because the π-acceptor orbitals on most ligands are higher in energy than the metal d orbitals, they form molecular orbitals in which the bonding t_{2g} combinations are largely of metal d-orbital character (Fig. 20.21). These bonding combinations lie lower in energy than the d orbitals themselves. The net result is that π-acceptors *increase* Δ_O.

We can now put the role of π bonding in perspective. The order of ligands in the spectrochemical series is partly that of the strengths with which they can participate in M–L σ bonding. For example, both CH_3^- and H^- are very high in the spectrochemical series because they are very strong σ donors. However, when π bonding is significant, it has a strong influence on Δ_O: π-donor ligands decrease Δ_O and π-acceptor ligands increase Δ_O. This effect is responsible for CO (a strong π acceptor) being high on the spectrochemical series and for OH^- (a strong π donor) being low in the series. The overall order of the spectrochemical series may be interpreted in broad terms as dominated by π effects (with a few important exceptions), and in general the series can be interpreted as follows:

$$— \text{increasing } \Delta_O \rightarrow$$
$$\pi \text{ donor} < \text{weak } \pi \text{ donor} < \text{no } \pi \text{ effects} < \pi \text{ acceptor}$$

Representative ligands that match these classes are

π donor	weak π donor	no π effects	π acceptor
I^-, Br^-, Cl^-, F^-	H_2O	NH_3	PR_3, CO

Notable examples of where the effect of σ bonding dominates include amines (NR_3), CH_3^-, and H^-, none of which has orbitals of π symmetry of an appropriate energy and thus are neither π-donor nor π-acceptor ligands. It is important to note that the classification of a ligand as strong-field or weak-field does not give any guide as to the strength of the M—L bond.

Electronic spectra

Now that we have considered the electronic structure of d-metal complexes, we are in a position to understand their electronic spectra and to use the data they provide to refine the discussion of structure. The magnitudes of ligand-field splittings are such that the energy of electronic transitions corresponds to an absorption of ultraviolet radiation and visible light. However, the presence of electron–electron repulsions within the metal orbitals means that the absorption frequencies are not in general a direct portrayal of the ligand-field splitting. The role of electron–electron repulsion was originally determined by the analysis of atoms and ions in the gas phase, and much of that information can be used in the analysis of the spectra of metal complexes provided we take into account the lower symmetry of a complex.

Keep in mind that the purpose of the following development is to find a way to extract the value of the ligand-field splitting parameter from the electronic absorption spectrum of a complex with more than one d electron, when electron–electron repulsions are important. First, we discuss the spectra of free atoms and see how to take electron–electron repulsions into account. Then we see what energy states atoms adopt when they are embedded in an octahedral ligand field. Finally, we see how to represent the energies of these states for various field strengths and electron–electron repulsion energies (in the Tanabe–Sugano diagrams of Section 20.4e), and how to use these diagrams to extract the value of the ligand-field splitting parameter.

20.3 Electronic spectra of atoms

Key point: Electron–electron repulsions result in multiple absorptions in the electronic spectrum.

Figure 20.22 sets the stage for our discussion by showing the electronic absorption spectrum of the d^3 complex $[Cr(NH_3)_6]^{3+}$ in aqueous solution. The band at lowest energy (longest wavelength) is very weak; later we shall see that it is an example of a 'spin-forbidden' transition. Next are two bands with intermediate intensities; these are 'spin-allowed' transitions between the t_{2g} and e_g orbitals of the complex, which are mainly derived from the

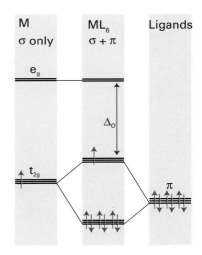

Figure 20.20 The effect of π bonding on the ligand-field splitting parameter. Ligands that act as π donors decrease Δ_O. Only the π orbitals of the ligand are shown.

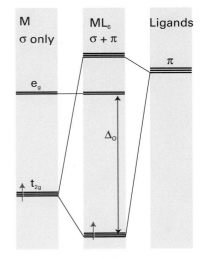

Figure 20.21 Ligands that act as π acceptors increase Δ_O. Only the π orbitals of the ligand are shown.

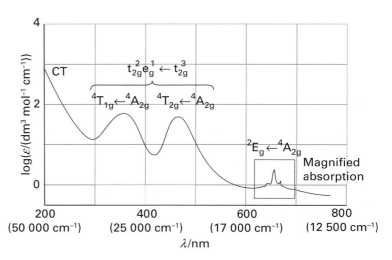

Figure 20.22 The spectrum of the d^3 complex $[Cr(NH_3)_6]^{3+}$, which illustrates the features studied in this section, and the assignments of the transitions as explained in the text.

metal d orbitals. The third feature in the spectrum is an intense charge-transfer band at short wavelength (labelled CT, denoting 'charge transfer'), of which only the low-energy tail is evident in the illustration.

One problem that immediately confronts us is why two absorptions can be ascribed to the apparently single transition $t_{2g}^2 e_g^1 \leftarrow t_{2g}^3$. This splitting of a single transition into two bands is in fact an outcome of the electron−electron repulsions mentioned above. To understand how it arises, and to extract the information it contains, we need to consider the spectra of free atoms and ions.

(a) Spectroscopic terms

Key points: Different microstates exist for the same electronic configuration; for light atoms, Russell−Saunders coupling is used to describe the terms, which are specified by symbols in which the value of L is indicated by one of the letters S, P, D,..., and the value of 2S + 1 is given as a left superscript.

In Chapter 1 we expressed the electronic structures of atoms by giving their electronic configurations, the designation of the number of electrons in each orbital (as in $1s^2 2s^1$ for Li). However, a configuration is an incomplete description of the arrangement of electrons in atoms. In the configuration $2p^2$, for instance, the two electrons might occupy orbitals with different orientations of their orbital angular momenta (that is, with different values of m_l from among the possibilities +1, 0, and −1 that are available when $l = 1$). Similarly, the designation $2p^2$ tells us nothing about the spin orientations of the two electrons $m_s = +\frac{1}{2}$ or $-\frac{1}{2}$. The atom may in fact have several different states of total orbital and spin angular momenta, each one corresponding to the occupation of orbitals with different values of m_l by electrons with different values of m_s. The different ways in which the electrons can occupy the orbitals specified in the configuration are called the **microstates** of the configuration. For example, one microstate of a $2p^2$ configuration is $(1^+,1^-)$; this notation signifies that both electrons occupy an orbital with $m_l = +1$ but do so with opposite spins, the superscript + indicating $m_s = +\frac{1}{2}$ and − indicating $m_s = -\frac{1}{2}$. Another microstate of the same configuration is $(-1^+,0^+)$. In this microstate, both electrons have $m_s = +\frac{1}{2}$ but one occupies the 2p orbital with $m_l = -1$ and the other occupies the orbital with $m_l = 0$.

The microstates of a given configuration have the same energy only if electron−electron repulsions on the atom are negligible. However, because atoms and most molecules are compact, interelectronic repulsions are strong and cannot always be ignored. As a result, microstates that correspond to different relative spatial distributions of electrons have different energies. If we group together the microstates that have the same energy when electron−electron repulsions are taken into account, we obtain the spectroscopically distinguishable energy levels called **terms**.

For light atoms and the 3d series, it turns out that the most important property of a microstate for helping us to decide its energy is the relative orientation of the spins of the electrons.

Next in importance is the relative orientation of the orbital angular momenta of the electrons. It follows that we can identify the terms of light atoms and put them in order of energy by sorting the microstates according to their total spin quantum number S (which is determined by the relative orientation of the individual spins) and then according to their total orbital angular momentum quantum number L (which is determined by the relative orientation of the individual orbital angular momenta of the electrons). For heavy atoms, such as those of the 4d and 5d series, the relative orientations of orbital momenta or of spin momenta are less important. In these atoms the spin and orbital angular momenta of individual electrons are strongly coupled together by **spin–orbit coupling**, so the relative orientation of the spin and orbital angular momenta of each electron is the most important feature for determining the energy. The terms of heavy atoms are therefore sorted on the basis of the values of the total angular momentum quantum number j for an electron in each microstate.

The process of combining electron angular momenta by summing first the spins, then the orbital momenta, and finally combining the two resultants is called **Russell–Saunders coupling**. This coupling scheme is used to identify the terms of light atoms (i.e. the 3d metals), and we consider it in detail here. The coupling scheme most appropriate to heavy atoms (that is, atoms of the 4d and 5d series of elements) is called *jj*-**coupling**, but we shall not consider it further.

Our first task is to identify the values of L and S that can arise from the orbital and spin angular momenta of individual electrons. Suppose we have two electrons with quantum numbers l_1, s_1 and l_2, s_2. Then, according to the **Clebsch–Gordan series**, the possible values of L and S are

$$L = l_1 + l_2, l_1 + l_2 - 1, \dots, |l_1 - l_2| \qquad S = s_1 + s_2, s_1 + s_2 - 1, \dots, |s_1 - s_2| \qquad (20.3)$$

(The modulus signs appear because neither L nor S can, by definition, be negative.) For example, an atom with configuration d^2 ($l_1 = 2, l_2 = 2$) can have the following values of L:

$$L = 2 + 2, 2 + 2 - 1, \dots, |2 - 2| = 4, 3, 2, 1, 0$$

The total spin (because $s_1 = \tfrac{1}{2}, s_2 = \tfrac{1}{2}$) can be

$$S = \tfrac{1}{2} + \tfrac{1}{2}, \tfrac{1}{2} + \tfrac{1}{2} - 1, \dots, |\tfrac{1}{2} - \tfrac{1}{2}| = 1, 0$$

To find the values of L and S for atoms with three electrons, we continue the process by combining l_3 with the value of L just obtained, and likewise for s_3.

Once L and S have been found, we can write down the allowed values of the quantum numbers M_L and M_S,

$$M_L = L, L - 1, \dots, -L \qquad M_S = S, S - 1, \dots, -S$$

These quantum numbers give the orientation of the angular momentum relative to an arbitrary axis: there are $2L + 1$ values of M_L for a given value of L and $2S + 1$ values of M_S for a given value of S. The values of M_L and M_S for a given microstate can be found very easily by adding together the values of m_l or m_s for the individual electrons. Therefore, if one electron has the quantum number m_{l1} and the other has m_{l2}, then

$$M_L = m_{l1} + m_{l2}$$

A similar expression applies to the total spin:

$$M_S = m_{s1} + m_{s2}$$

Thus, for example, $(0^+, -1^-)$ is a microstate with $M_L = 0 - 1 = -1$ and $M_S = \tfrac{1}{2} + (-\tfrac{1}{2}) = 0$ and may contribute to any term for which these two quantum numbers apply.

By analogy with the notation s, p, d, … for orbitals with $l = 0, 1, 2,\dots$, the total orbital angular momentum of an atomic term is denoted by the equivalent uppercase letter:

$$L = 0 \quad 1 \quad 2 \quad 3 \quad 4$$
$$ S \quad P \quad D \quad F \quad G \quad \text{then alphabetical (omitting J)}$$

The total spin is normally reported as the value of $2S + 1$, which is called the **multiplicity** of the term:

$$S = 0 \quad \tfrac{1}{2} \quad 1 \quad \tfrac{3}{2} \quad 2$$
$$2S + 1 = 1 \quad 2 \quad 3 \quad 4 \quad 5$$

The multiplicity is written as a left superscript on the letter representing the value of L, and the entire label of a term is called a **term symbol**. Thus, the term symbol ^3P denotes a term (a collection of nearly degenerate states) with $L = 1$ and $S = 1$, and is called a *triplet term*.

EXAMPLE 20.5 Deriving term symbols

Give the term symbols for an atom with the configurations (a) s^1, (b) p^1, and (c) s^1p^1.

Answer We need to use the Clebsch−Gordan series to couple any angular momenta, identify the letter for the term symbol from the table above, and then attach the multiplicity as a left superscript. (a) The single s electron has $l = 0$ and $s = \frac{1}{2}$. Because there is only one electron, $L = 0$ (an S term), $S = s = \frac{1}{2}$, and $2S + 1 = 2$ (a doublet term). The term symbol is therefore ^2S. (b) For a single p electron, $l = 1$ so $L = 1$ and the term is ^2P. (These terms arise in the spectrum of an alkali metal atom, such as Na.) (c) With one s and one p electron, $L = 0 + 1 = 1$, a P term. The electrons may be paired ($S = 0$) or parallel ($S = 1$). Hence both ^1P and ^3P terms are possible.

Self-test 20.5 What terms arise from a p^1d^1 configuration?

(b) The classification of microstates

Key point: The allowed terms of a configuration are found by identifying the values of L and S to which the microstates of an atom can contribute.

The Pauli principle restricts the microstates that can occur in a configuration and consequently affects the terms that can occur. For example, two electrons cannot both have the same spin and be in a d orbital with $m_l = +2$. Therefore, the microstate $(2^+,2^+)$ is forbidden and so are the values of L and S to which such a microstate might contribute. We shall illustrate how to determine what terms are allowed by considering a d^2 configuration, as the outcome will be useful in the discussion of the complexes encountered later in the chapter. An example of a species with a d^2 configuration is a Ti^{2+} ion.

We start the analysis by setting up a table of microstates of the d^2 configuration (Table 20.6); only the microstates allowed by the Pauli principle have been included. We then use a process of elimination to classify all the microstates. First, we note the largest value of M_L, which for a d^2 configuration is $+4$. This state must belong to a term with $L = 4$ (a G term). Table 20.6 shows that the only value of M_S that occurs for this term is $M_S = 0$, so the G term must be a singlet. Moreover, as there are nine values of M_L when $L = 4$, one of the microstates in each of the boxes in the column below $(2^+,2^-)$ must belong to this term.[3] We can therefore strike out one microstate from each row in the central column of Table 20.6, which leaves 36 microstates.

Table 20.6 Microstates of the d^2 configuration

M_L	-1	M_S 0	$+1$
$+4$		$(2^+,2^-)$	
$+3$	$(2^-,1^-)$	$(2^+,1^-)(2^-,1^+)$	$(2^+,1^+)$
$+2$	$(2^-,0^-)$	$(2^+,0^-)(2^-,0^+)(1^+,1^-)$	$(2^+,0^+)$
$+1$	$(2^-,-1^-)(1^-,0^-)$	$(2^+,-1^-)(2^-,-1^+)$ $(1^+,0^-)(1^-,0^+)$	$(2^+,-1^+)(1^+,0^+)$
0	$(1^-,-1^-)(2^-,-2^-)$	$(1^+,-1^-)(1^-,-1^+)$ $(2^+,-2^-)(2^-,-2^+)$ $(0^+,0^-)$	$(1^+,-1^+)(2^+,-2^+)$
-1 to -4^*			

* The lower half of the table is a reflection of the upper half.

---

[3] In fact, it is unlikely that one of the microstates itself will correspond to one of these states: in general, a state is a linear combination of microstates. However, as N linear combinations can be formed from N microstates, each time we cross off one microstate, we are taking one linear combination into account, so the bookkeeping is correct even though the detail may be wrong.

The next largest value is $M_L = +3$, which must stem from $L = 3$ and hence belong to an F term. That row contains one microstate in each column (that is, each box contains one unassigned combination for $M_S = -1$, 0, and $+1$), which signifies $S = 1$ and therefore a triplet term. Hence the microstates belong to ^{3}F. The same is true for one microstate in each of the rows down to $M_L = -3$, which accounts for a further $3 \times 7 = 21$ microstates. If we strike out one state in each of the 21 boxes, we are left with 15 to be assigned.

There is one unassigned microstate in the row with $M_L = +2$ (which must arise from $L = 2$) and the column under $M_S = 0$ ($S = 0$), which must therefore belong to a ^{1}D term. This term has five values of M_L, which removes one microstate from each row in the column headed $M_S = 0$ down to $M_L = -2$, leaving 10 microstates unassigned. Because these unassigned microstates include one with $M_L = +1$ and $M_S = +1$, nine of these microstates must belong to a ^{3}P term. There now remains only one microstate in the central box of the table, with $M_L = 0$ and $M_S = 0$. This microstate must be the one and only state of a ^{1}S term (which has $L = 0$ and $S = 0$).

At this point we can conclude that the terms of a 3d^2 configuration are ^{1}G, ^{3}F, ^{1}D, ^{3}P, and ^{1}S. These terms account for all 45 permitted states (see table in the margin).

Term	Number of states
^{1}G	$9 \times 1 = 9$
^{3}F	$7 \times 3 = 21$
^{1}D	$5 \times 1 = 5$
^{3}P	$3 \times 3 = 9$
^{1}S	$1 \times 1 = 1$
Total:	45

(c) The energies of the terms

Key point: Hund's rules indicate the likely ground term of a gas-phase atom or ion.

Once the values of L and S that can arise from a given configuration are known, it is possible to identify the term of lowest energy by using Hund's rules. The first of these empirical rules was introduced in Section 1.7, where it was expressed as 'the lowest energy configuration is achieved if the electron spins are parallel'. Because a high value of S stems from parallel electron spins, an alternative statement is

1. For a given configuration, the term with the greatest multiplicity lies lowest in energy.

The rule implies that a triplet term of a configuration (if one is permitted) has a lower energy than a singlet term of the same configuration. For the d^2 configuration, this rule predicts that the ground state will be either ^{3}F or ^{3}P.

By inspecting spectroscopic data, Hund also identified a second rule for the relative energies of the terms of a given multiplicity:

2. For a term of given multiplicity, the term with the greatest value of L lies lowest in energy.

The physical justification for this rule is that when L is high, the electrons can stay clear of one another and hence experience a lower repulsion. If L is low, the electrons are more likely to be closer to each other, and hence repel one another more strongly. The second rule implies that, of the two triplet terms of a d^2 configuration, the ^{3}F term is lower in energy than the ^{3}P term. It follows that the ground term of a d^2 species such as Ti^{2+} is expected to be ^{3}F.

The spin multiplicity rule is fairly reliable for predicting the ordering of terms, but the 'greatest L' rule is reliable only for predicting the ground term, the term of lowest energy; there is generally little correlation of L with the order of the higher terms. Thus, for d^2 the rules predict the order

$$^3F < \, ^3P < \, ^1G < \, ^1D < \, ^1S$$

but the order observed for Ti^{2+} from spectroscopy is

$$^3F < \, ^1D < \, ^3P < \, ^1G < \, ^1S$$

Normally, all we want to know is the identity of the ground term of an atom or ion. The procedure may then be simplified and summarized as follows:

1. Identify the microstate that has the highest value of M_S.

This step tells us the highest multiplicity of the configuration.

2. Identify the highest permitted value of M_L for that multiplicity.

This step tells us the highest value of L consistent with the highest multiplicity.

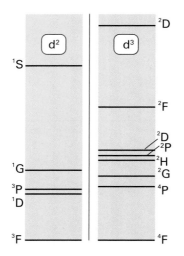

Figure 20.23 The relative energies of the terms arising from d² (left) and d³ (right) configurations of a free atom.

EXAMPLE 20.6 Identifying the ground term of a configuration

What is the ground term of the configurations (a) 3d⁵ of Mn^{2+} and (b) 3d³ of Cr^{3+}?

Answer First we need to identify the term with maximum multiplicity, as this will be the ground term. Then we need to identify the L value for any terms that have the maximum multiplicity, for the term with the highest L value will be the ground term. (a) Because the d⁵ configuration permits occupation of each d orbital singly with parallel spins, the maximum value of S is $\frac{5}{2}$, giving a multiplicity of $2 \times \frac{5}{2} + 1 = 6$, a sextet term. If each of the electrons is to have the same spin quantum number, all must occupy different orbitals and hence have different M_L values. Thus, the M_L values of the occupied orbitals will be $+2, +1, 0, -1$, and -2. This configuration is the only one possible for a sextet term. Because the sum of the M_L is 0, it follows that $L = 0$ and the term is ⁶S. (b) For the configuration d³, the maximum multiplicity corresponds to all three electrons having the same spin quantum number, so $S = \frac{3}{2}$. The multiplicity is therefore $2 \times \frac{3}{2} + 1 = 4$, a quartet. Again, the three M_L values must be different if the electrons are all parallel. There are several possible arrangements that give quartet terms, but the one that gives a maximum value of M_L has the three electrons with $M_L = +2, +1$ and 0, giving a total of $+3$, which must arise from a term with $L = 3$, an F term. Hence, the ground term of d³ is ⁴F.

Self-test 20.6 Identify the ground terms of (a) 2p² and (b) 3d⁹. (Hint: Because d⁹ is one electron short of a closed shell with $L = 0$ and $S = 0$, treat it on the same footing as a d¹ configuration.)

Figure 20.23 shows the relative energies of the terms for the d² and d³ configurations of free atoms. Later, we shall see how to extend these diagrams to include the effect of a crystal field (Section 20.4).

(d) Racah parameters

Key points: The Racah parameters summarize the effects of electron−electron repulsion on the energies of the terms that arise from a single configuration; the parameters are the quantitative expression of the ideas underlying Hund's rules and account for deviations from them.

Different terms of a configuration have different energies on account of the repulsion between electrons. To calculate the energies of the terms we must evaluate these electron−electron repulsion energies as complicated integrals over the orbitals occupied by the electrons. Mercifully, however, all the integrals for a given configuration can be collected together in three specific combinations and the repulsion energy of any term of a configuration can be expressed as a sum of these three quantities. The three combinations of integrals are called the **Racah parameters** and denoted A, B, and C. The parameter A corresponds to an average of the total interelectron repulsion, and B and C relate to the repulsion energies between individual d electrons. We do not even need to know the theoretical values of the parameters or the theoretical expressions for them because it is more reliable to use A, B, and C as empirical quantities obtained from gas-phase atomic spectroscopy.

Each term stemming from a given configuration has an energy that may be expressed as a linear combination of all three Racah parameters. For a d² configuration a detailed analysis shows that

$$E(^1S) = A + 14B + 7C \quad E(^1G) = A + 4B + 2C \quad E(^1D) = A - 3B + 2C$$

$$E(^3P) = A + 7B \quad\quad\quad E(^3F) = A - 8B$$

The values of A, B, and C can be determined by fitting these expressions to the observed energies of the terms. Note that A is common to all the terms (as remarked above it is the average of the total interelectron repulsion energy); therefore, if we are interested only in their relative energies, we do not need to know its value. All three Racah parameters are positive as they represent electron−electron repulsions. Therefore, provided $C > 5B$, the energies of the terms of the d² configuration lie in the order

$$^3F < {}^3P < {}^1D < {}^1G < {}^1S$$

This order is nearly the same as obtained by using Hund's rules. However, if $C < 5B$, the advantage of having an occupation of orbitals that corresponds to a high orbital angular momentum is greater than the advantage of having a high multiplicity, and the ³P term lies above ¹D (as is in fact the case for Ti^{2+}). Table 20.7 shows some experimental values of

Table 20.7 Racah parameters for some d-block ions*

	1+	2+	3+	4+
Ti		720 (3.7)		
V		765 (3.9)	860 (4.8)	
Cr		830 (4.1)	1030 (3.7)	1040 (4.1)
Mn		960 (3.5)	1130 (3.2)	
Fe		1060 (4.1)		
Co		1120 (3.9)		
Ni		1080 (4.5)		
Cu	1220 (4.0)	1240 (3.8)		

* The table gives the B parameter in cm^{-1} with the value of C/B in parentheses.

B and C. The values in parentheses indicate that $C \approx 4B$ so the ions listed there are in the region where Hund's rules are not reliable for predicting anything more than the ground term of a configuration.

The parameter C appears only in the expressions for the energies of states that differ in multiplicity from the ground state. Hence if, as is usual, we are interested only in the relative energies of terms of the same multiplicity as the ground state (that is excitation without a change in spin state), we do not need to know the value of C. The parameter B is of the most interest, and we return to factors that affect its value in Section 20.4f.

20.4 Electronic spectra of complexes

The preceding discussion related only to free atoms, and we will now expand our discussion to encompass complex ions. The spectrum of $[Cr(NH_3)_6]^{3+}$ in Fig. 20.22 has two central bands with intermediate intensities and with energies that differ on account of the electron–electron repulsions (as we explain soon). Because both the transitions are between orbitals that are predominantly metal d orbital in character, with a separation characterized by the strength of the ligand-field splitting parameter Δ_O, these two transitions are called **d–d transitions** or **ligand-field transitions**.

(a) Ligand-field transitions

Key point: Electron–electron repulsion splits ligand-field transitions into components with different energies.

According to the discussion in Section 20.1, we expect the octahedral d^3 complex $[Cr(NH_3)_6]^{3+}$ to have the ground-state configuration t_{2g}^3. The absorption near 25 000 cm^{-1} can be identified as arising from the excitation $t_{2g}^2 e_g^1 \leftarrow t_{2g}^3$ because the corresponding energy (close to 3 eV) is typical of ligand-field splittings in complexes.

Before we embark on a Racah-like analysis of the transition, it will be helpful to see qualitatively from the viewpoint of molecular orbital theory why the transition gives rise to two bands. First, note that a $d_{z^2} \leftarrow d_{xy}$ transition, which is one way of achieving $e_g \leftarrow t_{2g}$, promotes an electron from the xy-plane into the already electron-rich z-direction: that axis is electron-rich because both d_{yz} and d_{zx} are occupied (Fig. 20.24). However, a $d_{z^2} \leftarrow d_{zx}$ transition, which is another way of achieving $e_g \leftarrow t_{2g}$, merely relocates an electron that is already largely concentrated along the z-axis. In the former case, but not in the latter, there is a distinct increase in electron repulsion and, as a result, the two $e_g \leftarrow t_{2g}$ transitions lie at different energies. There are six possible $t_{2g}^2 e_g^1 \leftarrow t_{2g}^3$ transitions, and all resemble one or other of these two cases: three of them fall into one group and the other three fall into the second group.

(b) The spectroscopic terms

Key points: The terms of an octahedral complex are labelled by the symmetry species of the overall orbital state; a superscript prefix shows the multiplicity of the term.

The two bands we are discussing in Fig. 20.22 are labelled $^4T_{2g} \leftarrow {}^4A_{2g}$ (at 21 550 cm^{-1}) and $^4T_{1g} \leftarrow {}^4A_{2g}$ (at 28 500 cm^{-1}). The labels are **molecular term symbols** and serve a

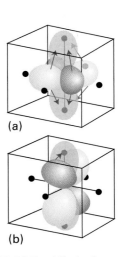

(a)

(b)

Figure 20.24 The shifts in electron density that accompany the two transitions discussed in the text. There is a considerable relocation of electron density towards the ligands on the z-axis in (a), but a much less substantial relocation in (b).

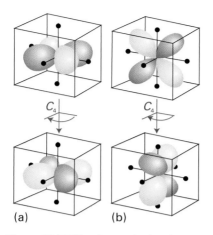

Figure 20.25 The changes in sign that occur under C_4 rotations about the z-axis: (a) a d_{xy} orbital is rotated into the negative of itself; (b) a d_{yz} orbital is rotated into a d_{zx} orbital.

Table 20.8 The correlation of spectroscopic terms for d electrons in O_h complexes

Atomic term	Number of states	Terms in O_h symmetry
S	1	A_{1g}
P	3	T_{1g}
D	5	$T_{2g} + E_g$
F	7	$T_{1g} + T_{2g} + A_{2g}$
G	9	$A_{1g} + E_g + T_{1g} + T_{2g}$

purpose similar to that of atomic term symbols. The left superscript denotes the multiplicity, so the superscript 4 denotes a quartet state with $S = \frac{3}{2}$, as expected when there are three unpaired electrons. The rest of the term symbol is the symmetry label of the overall electronic orbital state of the complex. For example, the nearly totally symmetrical ground state of a d^3 complex (with an electron in each of the three t_{2g} orbitals) is denoted A_{2g}. We say *nearly* totally symmetrical because close inspection of the behaviour of the three occupied t_{2g} orbitals shows that the C_3 rotation of the O_h point group transforms the product $t_{2g} \times t_{2g} \times t_{2g}$ into itself, which identifies the complex as an A symmetry species (see the character table in *Resource section* 4). Moreover, because each orbital has even parity (g), the overall parity is also g. However, each C_4 rotation transforms one t_{2g} orbital into the negative of itself and the other two t_{2g} orbitals into each other (Fig. 20.25), so overall there is a change of sign under this operation and its character is -1. The term is therefore A_{2g} rather than the totally symmetrical A_{1g} of a closed shell.

It is more difficult to establish that the term symbols that can arise from the quartet $t_{2g}^2 e_g^1$ excited configuration are $^4T_{2g}$ and $^4T_{1g}$, and we shall not consider this aspect here. The superscript 4 implies that the upper configuration continues to have the same number of unpaired spins as in the ground state, and the subscript g stems from the even parity of all the contributing orbitals.

(c) Correlating the terms

Key point: In the ligand field of an octahedral complex the free atom terms split and are then labelled by their symmetry species as enumerated in Table 20.8.

In a free atom, where all five d orbitals in a shell are degenerate, we needed to consider only the electron–electron repulsions to arrive at the relative ordering of the terms of a given d^n configuration. In a complex, the d orbitals are not all degenerate and it is necessary to take into account the difference in energy between the t_{2g} and e_g orbitals as well as the electron–electron repulsions.

Consider the simplest case of an atom or ion with a single valence electron. Because a totally symmetric orbital in one environment becomes a totally symmetric orbital in another environment, an s orbital in a free atom becomes an a_{1g} orbital in an octahedral field. We express the change by saying that the s orbital of the atom 'correlates' with the a_{1g} orbital of the complex. Similarly, the five d orbitals of a free atom correlate with the triply degenerate t_{2g} and doubly degenerate e_g sets in an octahedral complex.

Now consider a many-electron atom. In exactly the same way as for a single electron, the totally symmetrical overall S term of a many-electron atom correlates with the totally symmetrical A_{1g} term of an octahedral complex. Likewise, an atomic D term splits into a T_{2g} term and an E_g term in O_h symmetry. The same kind of analysis can be applied to other states, and Table 20.8 summarizes the correlations between free atom terms and terms in an octahedral complex.

EXAMPLE 20.7 Identifying correlations between terms

What terms in a complex with O_h symmetry correlate with the ^3P term of a free atom with a d^2 configuration?

Answer We argue by analogy: if we know how p orbitals correlate with orbitals in a complex, then we can use that information to express how the overall states correlate, simply by changing to uppercase letters. The three p orbitals of a free atom become the triply degenerate t_{1u} orbitals of an octahedral complex. Therefore, if we disregard parity for the moment, a P term of a many-electron atom becomes a T_1 term in the point group O_h. Because d orbitals have even parity, the term overall must be g, and specifically T_{1g}. The multiplicity is unchanged in the correlation, so the ^3P term becomes a $^3T_{1g}$ term.

Self-test 20.7 What terms in a d^2 complex of O_h symmetry correlate with the ^3F and ^1D terms of a free atom?

(d) The energies of the terms: weak- and strong-field limits

Key points: For a given metal ion, the energies of the individual terms respond differently to ligands of increasing field strength, and the correlation between free atom terms and terms of a complex can be displayed on an Orgel diagram.

Electron–electron repulsions are difficult to take into account, but the discussion is simplified by considering two extreme cases. In the weak-field limit the ligand field, as measured by Δ_O, is so weak that only electron−electron repulsions are important. As the Racah parameters B and C fully describe the interelectron repulsions, these are the only parameters we need at this limit. In the strong-field limit the ligand field is so strong that electron−electron repulsions can be ignored and the energies of the terms can be expressed solely in terms of Δ_O. Then, with the two extremes established, we can consider intermediate cases by drawing a correlation diagram between the two. We shall illustrate what is involved by considering two simple cases, namely, d^1 and d^2. Then we show how the same ideas are used to treat more complicated cases.

The only term arising from the d^1 configuration of a free atom is ^2D. In an octahedral complex the configuration is either t_{2g}^1, which gives rise to a $^2T_{2g}$ term, or e_g^1, which gives rise to a ^2E_g term. Because there is only one electron, there are no electron−electron repulsions to worry about, and the separation of the $^2T_{2g}$ and ^2E_g terms is the same as the separation of the t_{2g} and e_g orbitals, which is Δ_O. The correlation diagram for the d^1 configuration will therefore resemble that shown in Fig. 20.26.

We saw earlier that for a d^2 configuration the lowest energy term in the free atom is the triplet ^3F. We need consider only electronic transitions that start from the ground state, and, in this section, will discuss only those in which there is no change in spin. There is an additional triplet term (^3P); relative to the lower term (^3F), the energies of the terms are $E(^3F) = 0$ and $E(^3P) = 15B$. These two energies are marked on the left of Fig. 20.27. Now consider the very strong field limit. A d^2 atom has the configurations

$$t_{2g}^2 < t_{2g}^1 e_g^1 < e_g^2$$

In an octahedral field, these configurations have different energies; that is, as we noted earlier, the ^3F term splits into three terms. From the information in Fig. 20.2, we can write their energies as

$$E(t_{2g}^2) = 2(-\tfrac{2}{5}\Delta_O) = -0.8\Delta_O$$

$$E(t_{2g}^1 e_g^1) = (-\tfrac{2}{5} + \tfrac{3}{5})\Delta_O = +0.2\Delta_O$$

$$E(e_g^2) = 2(\tfrac{3}{5})\Delta_O = +1.2\Delta_O$$

Therefore, relative to the energy of the lowest term, their energies are

$$E(t_{2g}^2, T_{1g}) = 0 \qquad E(t_{2g}^1 e_g^1, T_{2g}) = \Delta_O \qquad E(e_g^2, A_{2g}) = 2\Delta_O$$

These energies are marked on the right in Fig. 20.27.

Our problem now is to account for the energies when neither the ligand-field nor the electron repulsion terms is dominant. To do so, we correlate the terms in the two extreme cases. The triplet t_{2g}^2 configuration gives rise to a $^3T_{1g}$ term, and this correlates with the ^3F term of the free atom. The remaining correlations can be established similarly, and we see that the $t_{2g}^1 e_g^1$ configuration gives rise to a T_{2g} term and that the e_g^2 configuration gives rise to a A_{2g} term; both terms correlate with the ^3F term of the free atom. Note that some terms, such as the $^3T_{1g}$ term that correlates with ^3P, are independent of the ligand-field strength. All the correlations are shown in Fig. 20.27, which is a simplified version of an **Orgel diagram**. An Orgel diagram can be constructed for any d-electron configuration, and several electronic configurations can be combined on the same diagram. Orgel diagrams are of considerable value for simple discussions of the electronic spectra of complexes; however, they consider only some of the possible transitions (the spin-allowed transitions, which is why we considered only the triplet terms) and cannot be used to extract a value for the ligand-field splitting parameter, Δ_O.

(e) Tanabe−Sugano diagrams

Key point: Tanabe−Sugano diagrams are correlation diagrams that depict the energies of electronic states of complexes as a function of the strength of the ligand field.

Diagrams showing the correlation of all terms can be constructed for any electron configuration and strength of ligand field. The most widely used versions are called **Tanabe−Sugano diagrams**, after the scientists who devised them. Figure 20.28 shows the diagram for d^2 and we can see splittings for all the atomic terms that split; thus the ^3F splits into three, the ^1D into two, and the ^1G into four. In these diagrams the term energies, E, are expressed as E/B and plotted against Δ_O/B, where B is the Racah parameter. The relative energies of the

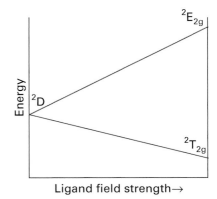

Figure 20.26 Correlation diagram for a free ion (left) and the strong-field terms (right) of a d^1 configuration.

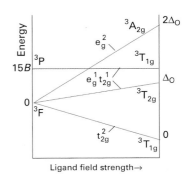

Figure 20.27 Correlation diagram for a free ion (left) and the strong-field terms (right) of a d^2 configuration.

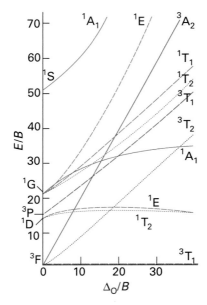

Figure 20.28 The Tanabe–Sugano diagram for the d^2 configuration. Note that the left-hand axis corresponds to Fig. 20.23 (left). A complete collection of diagrams for d^n configurations is given in *Resource section 6*. The parity subscript g has been omitted from the term symbols for clarity.

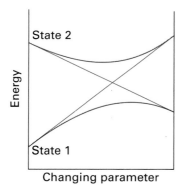

Figure 20.29 The noncrossing rule states that if two states of the same symmetry are likely to cross as a parameter is changed (as shown by the blue lines), they will in fact mix together and avoid the crossing (as shown by the brown lines).

terms arising from a given configuration are independent of A, and by choosing a value of C (typically setting $C \approx 4B$), terms of all energies can be plotted on the same diagrams. Some lines in Tanabe–Sugano diagrams are curved because of the mixing of terms of the same symmetry type. Terms of the same symmetry obey the **noncrossing rule**, which states that, if the increasing ligand field causes two weak-field terms of the same symmetry to approach, then they do not cross but bend apart from each other (Fig. 20.29). The effect of the noncrossing rule can be seen for the two ^1E terms, the two $^1T_2$ terms, and the two $^1A_1$ terms in Fig. 20.28.

Tanabe–Sugano diagrams for O_h complexes with configurations d^2 to d^8 are given in *Resource section 6*. The zero of energy in a Tanabe–Sugano diagram is always taken as that of the lowest term. Hence the lines in the diagrams have abrupt changes of slope when there is a change in the identity of the ground term brought about by the change from high-spin to low-spin with increasing field strength (see the diagram for d^4, for instance).

The purpose of the preceding discussion has been to find a way to extract the value of the ligand-field splitting parameter from the electronic absorption spectrum of a complex with more than one d electron, when electron–electron repulsions are important, such as that in Fig. 20.22. The strategy involves fitting the observed transitions to the correlation lines in a Tanabe–Sugano diagram and identifying the values of Δ_O and B where the observed transition energies match the pattern. This procedure is illustrated in Example 20.8.

As we shall see in Section 20.6, certain transitions are allowed and certain transitions are forbidden. In particular, those that correspond to a change of spin state are forbidden, whereas those that do not are allowed. In general, spin-allowed transitions will dominate the UV/visible absorption spectrum, thus we might only expect to see three transitions for a d^2 ion, specifically $^3T_{2g} \leftarrow {}^3T_{1g}$, $^3T_{1g} \leftarrow {}^3T_{1g}$, and $^3A_{2g} \leftarrow {}^3T_{1g}$. However, some complexes (for instance high-spin d^5 ions such as Mn^{2+}) do not have any spin-allowed transitions and none of the 11 possible transitions dominates.

> **EXAMPLE 20.8** Calculating Δ_O and B using a Tanabe–Sugano diagram
>
> Deduce the values of Δ_O and B for $[Cr(NH_3)_6]^{3+}$ from the spectrum in Fig. 20.22 and a Tanabe–Sugano diagram.
>
> ***Answer*** We need to identify the relevant Tanabe–Sugano diagram and then locate the position on the diagram where the observed ratio of transition energies (as wavenumbers) matches the theoretical ratio. The relevant diagram (for d^3) is shown in Fig. 20.30. We need concern ourselves only with the spin-allowed transitions, of which there are three for a d^3 ion (a $^4T_{2g} \leftarrow {}^4A_{2g}$ and two $^4T_{1g} \leftarrow {}^4A_{2g}$ transitions). We have seen that the spectrum in Fig. 20.22 exhibits two low-energy ligand-field transitions at 21 550 and 28 500 cm^{-1}, which correspond to the two lowest energy transitions ($^4T_{2g} \leftarrow {}^4A_{2g}$ and $^4T_{1g} \leftarrow {}^4A_{2g}$). The ratio of the energies of the transitions is 1.32, and the only point in Fig. 20.30 where this energy ratio is satisfied is on the far right. Hence, we can read off the value of $\Delta_O/B = 33.0$ from the location of this point. The tip of the arrow representing the lower energy transition lies at $32.8B$ vertically, so equating $32.8B$ and 21 550 cm^{-1} gives $B = 657\ cm^{-1}$ and therefore $\Delta_O = 21\ 700\ cm^{-1}$.
>
> ***Self-test 20.8*** Use the same Tanabe–Sugano diagram to predict the energy of the first two spin-allowed quartet bands in the spectrum of $[Cr(OH_2)_6]^{3+}$ for which $\Delta_O = 17\ 600\ cm^{-1}$ and $B = 700\ cm^{-1}$.

A Tanabe–Sugano diagram also provides some understanding of the widths of some absorption lines. Consider Fig. 20.28 and the $^3T_{2g} \leftarrow {}^3T_{1g}$ and $^3A_{2g} \leftarrow {}^3T_{1g}$ transitions. The line representing $^3T_{2g}$ is not parallel to that representing the ground state $^3T_{1g}$, and so any variation in the value of Δ_O (such as those caused by molecular vibration) results in a change in the energy of the transition and thus a broadening of the absorption band. The line representing $^3A_{2g}$ is even less parallel to the line representing $^3T_{1g}$; the energy of this transition will be affected even more by variations in Δ_O and will thus be even broader. By contrast, the line representing the lower $^1T_{2g}$ term is almost parallel to that representing $^3T_{1g}$ and thus the energy of this transition is largely unaffected by variations in Δ_O and the absorption is consequently very sharp (albeit weak, because it is forbidden).

(f) The nephelauxetic series

Key points: Electron–electron repulsions are lower in complexes than in free ions because of electron delocalization; the nephelauxetic parameter is a measure of the extent of d-electron delocalization on to the ligands of a complex; the softer the ligand, the smaller the nephelauxetic parameter.

In Example 20.8 we found that $B = 657$ cm^{-1} for $[Cr(NH_3)_6]^{3+}$, which is only 64 per cent of the value for a Cr^{3+} ion in the gas phase. This reduction is a general observation and indicates that electron repulsions are weaker in complexes than in the free atoms and ions. The weakening occurs because the occupied molecular orbitals are delocalized over the ligands and away from the metal. The delocalization increases the average separation of the electrons and hence reduces their mutual repulsion.

The reduction of B from its free ion value is normally reported in terms of the **nephelauxetic parameter, β:**[4]

$$\beta = B(\text{complex})/B(\text{free ion}) \qquad (20.4)$$

The values of β depend on the identity of the metal ion and the ligand, and a list of ligands ordered by the value of β gives the **nephelauxetic series:**

$$Br^- < CN^-, Cl^- < NH_3 < H_2O < F^-$$

A small value of β indicates a large measure of d-electron delocalization on to the ligands and hence a significant covalent character in the complex. Thus the series shows that a Br$^-$ ligand results in a greater reduction in electron repulsions in the ion than an F$^-$ ion, which is consistent with a greater covalent character in bromido complexes than in analogous fluorido complexes. As an example, compare $[NiF_6]^{4-}$, for which $B = 843$ cm^{-1}, with $[NiBr_4]^{2-}$, for which $B = 600$ cm^{-1}. Another way of expressing the trend represented by the nephelauxetic series is: *the softer the ligand, the smaller the nephelauxetic parameter.*

20.5 Charge-transfer bands

Key points: Charge-transfer bands arise from the movement of electrons between orbitals that are predominantly ligand in character and orbitals that are predominantly metal in character; such transitions are identified by their high intensity and the sensitivity of their energies to solvent polarity.

Another feature in the spectrum of $[Cr(NH_3)_6]^{3+}$ in Fig. 20.22 that remains to be explained is the very intense shoulder of an absorption that appears to have a maximum at well above 50 000 cm^{-1}. The high intensity suggests that this transition is not a simple ligand-field transition, but is consistent with a **charge-transfer transition (CT transition)**. In a CT transition, an electron migrates between orbitals that are predominantly ligand in character and orbitals that are predominantly metal in character. The transition is classified as a **ligand-to-metal charge-transfer transition (LMCT transition)** if the migration of the electron is from the ligand to the metal, and as a **metal-to-ligand charge-transfer transition (MLCT transition)** if the charge migration occurs in the opposite direction. An example of an MLCT transition is the one responsible for the red colour of tris(bipyridyl)iron(II), the complex used for the colorimetric analysis of Fe(II). In this case, an electron makes a transition from a d orbital of the central metal into a π^* orbital of the ligand. Figure 20.31 summarizes the transitions we classify as charge transfer.

Several lines of evidence are used to identify a band as due to a CT transition. The high intensity of the band, which is evident in Fig. 20.22, is one strong indication. Another indication is if such a band appears following the replacement of one ligand with another, as this implies that the band is strongly dependent on the ligand. The CT character is most often identified (and distinguished from $\pi^* \leftarrow \pi$ transitions on ligands) by demonstrating **solvatochromism**, the variation of the transition frequency with changes in solvent permittivity. Solvatochromism indicates that there is a large shift in electron density as a result of the transition, which is more consistent with a metal–ligand transition than a ligand–ligand or metal–metal transition.

Figure 20.32 shows another example of a CT transition in the visible and UV spectrum of $[CrCl(NH_3)_5]^{2+}$ (5). If we compare this spectrum with that of $[Cr(NH_3)_6]^{3+}$ in Fig. 20.22, then we can recognize the two ligand-field bands in the visible region. The replacement of one NH$_3$ ligand by a weaker field Cl$^-$ ligand moves the lowest energy ligand-field bands to lower energy than those of $[Cr(NH_3)_6]^{3+}$. Also, a shoulder appears on the high-energy side of one of the ligand-field bands, indicating an additional transition that is the result of the reduction in symmetry from O_h to C_{4v}. The major new feature in the spectrum is the strong absorption maximum in the ultraviolet, near 42 000 cm^{-1}. This band is at lower energy than the corresponding band in the spectrum of $[Cr(NH_3)_6]^{3+}$ and is due to an

[4] The name is from the Greek words for 'cloud expanding'.

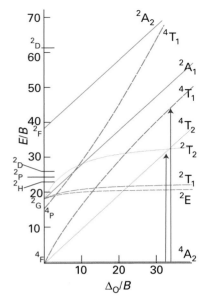

Figure 20.30 The Tanabe–Sugano diagram for the d^3 configuration. Note that the left-hand axis corresponds to Fig. 20.22 (left). A complete collection of diagrams for d^n configurations is given in *Resource section 6*. The parity subscript g has been omitted from the term symbols for clarity.

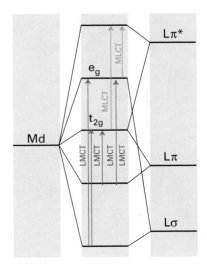

Figure 20.31 A summary of the charge-transfer transitions in an octahedral complex.

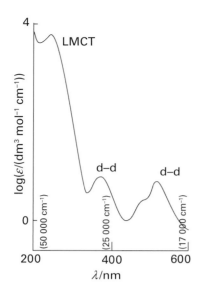

Figure 20.32 The absorption spectrum of [CrCl(NH₃)₅]²⁺ in water in the visible and ultraviolet regions. The peak corresponding to the transition $^2E \leftarrow {}^4A$ is not visible on this magnification.

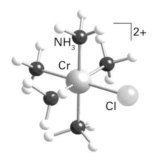

5 [CrCl(NH₃)₅]²⁺

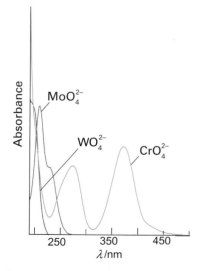

Figure 20.33 Optical absorption spectra of the ions CrO_4^{2-}, WO_4^{2-}, and MoO_4^{2-}. On descending the group, the absorption maximum moves to shorter wavelengths, indicating an increase in the energy of the LMCT band.

LMCT transition from the Cl^- ligand to the metal. The LMCT character of similar bands in $[CoX(NH_3)_5]^{2+}$ is confirmed by the decrease in energy in steps of about 8000 cm⁻¹ as X is varied from Cl to Br to I. In this LMCT transition, a lone-pair electron of the halide ligand is promoted into a predominantly metal orbital.

(a) LMCT transitions

Key points: Ligand-to-metal charge-transfer transitions are observed in the visible region of the spectrum when the metal is in a high oxidation state and ligands contain nonbonding electrons; the variation in the position of LMCT bands can be parameterized in terms of optical electronegativities.

Charge-transfer bands in the visible region of the spectrum (and hence contributing to the intense colours of many complexes) may occur if the ligands have lone pairs of relatively high energy (as in sulfur and selenium) or if the metal atom has low-lying empty orbitals.

The tetraoxidoanions of metals with high oxidation numbers (such as MnO_4^-) provide what are probably the most familiar examples of LMCT bands. In these, an O lone-pair electron is promoted into a low-lying empty metal e orbital. High metal oxidation numbers correspond to a low d-orbital population (many are formally d^0), so the acceptor level is available and low in energy. The trend in LMCT energies is:

Oxidation number

$$+7 \quad MnO_4^- < TcO_4^- < ReO_4^-$$

$$+6 \quad CrO_4^{2-} < MoO_4^{2-} < WO_4^{2-}$$

$$+5 \quad VO_4^{3-} < NbO_4^{3-} < TaO_4^{3-}$$

The UV/visible spectra of the tetraoxido anions of the Group 6 metals, CrO_4^{2-}, MoO_4^{2-}, and WO_4^{2-} are shown in Fig. 20.33. The energies of the transitions correlate with the order of the electrochemical series, with the lowest energy transitions taking place to the most easily reduced metal ions. This correlation is consistent with the transition being the transfer of an electron from the ligands to the metal ion, corresponding, in effect, to the reduction of the metal ion by the ligands. Polymeric and monomeric oxidoanions follow the same trends, with the oxidation state of the metal the determining factor. The similarity suggests that these LMCT transitions are localized processes that take place on discrete molecular fragments.

The variation in the position of LMCT bands can be expressed in terms of the **optical electronegativities** of the metal, χ_{metal}, and the ligands, χ_{ligand}. The wavenumber of the transition is then written as the difference between the two electronegativities:

$$\tilde{\nu} = \left| \chi_{ligand} - \chi_{metal} \right| \tilde{\nu}_o$$

Optical electronegativities have values comparable to Pauling electronegativities (Table 1.7) if we set $\tilde{\nu}_o = 3.0 \times 10^4$ cm⁻¹ (Table 20.9), and can be used in a similar fashion. If the LMCT transition terminates in an e_g orbital, Δ_O must be added to the energy predicted by this equation. Electron pairing energies must also be taken into account if the transition results in the population of an orbital that already contains an electron. The values for metals are different in complexes of different symmetry, and the ligand values are different if the transition originates from a π orbital rather than a σ orbital.

Table 20.9 Optical electronegativities

Metal	O_h	T_d	Ligand	π	σ
Cr(III)	1.8–1.9		F⁻	3.9	4.4
Co(III)*	2.3		Cl⁻	3.0	3.4
Ni(II)		2.0–2.1	Br⁻	2.8	3.3
Co(II)		1.8–1.9	I⁻	2.5	3.0
Rh(III)*	2.3		H₂O	3.5	
Mo(VI)	2.1		NH₃	3.3	

* Low-spin complexes.

(b) MLCT transitions

Key point: Charge-transfer transitions from metal to ligand are observed when the metal is in a low oxidation state and the ligands have low-lying acceptor orbitals.

Charge-transfer transitions from metal to ligand are most commonly observed in complexes with ligands that have low-lying π^* orbitals, especially aromatic ligands. The transition occurs at low energy and appears in the visible spectrum if the metal ion is in a low oxidation state, as its d orbitals are then relatively close in energy to the empty ligand orbitals.

The family of ligands most commonly involved in MLCT transitions are the diimines, which have two N donor atoms: two important examples are 2,2′-bipyridine (bpy, **6**) and 1,10-phenanthroline (phen, **7**). Complexes of diimines with strong MLCT bands include tris(diimine) species such as tris(2,2′-bipyridyl)ruthenium(II) (**8**), which is orange. A diimine ligand may also be easily substituted into a complex with other ligands that favour a low oxidation state. Two examples are $[W(CO)_4(phen)]$ and $[Fe(CO)_3(bpy)]$. However, the occurrence of MLCT transitions is by no means limited to diimine ligands. Another important ligand type that shows typical MLCT transitions is dithiolene, $S_2C_2R_2^{2-}$ (**9**). Resonance Raman spectroscopy (Section 8.4) is a powerful technique for the study of MLCT transitions.

The MLCT excitation of tris(2,2′-bipyridyl)ruthenium(II) has been the subject of intense research efforts because the excited state that results from the charge transfer has a lifetime of microseconds, and the complex is a versatile photochemical redox reagent. The photochemical behaviour of a number of related complexes has also been studied on account of their relatively long excited-state lifetimes.

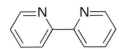

6 2,2′-Bipyridine (bpy)

7 1,10-Phenanthroline (phen)

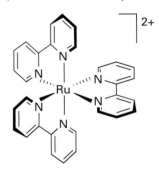

8 Tris(2,2′-bipyridyl)ruthenium(II)

9 Dithiolene

20.6 Selection rules and intensities

Key point: The strength of an electronic transition is determined by the transition dipole moment.

The contrast in intensity between typical charge-transfer bands and typical ligand-field bands raises the question of the factors that control the intensities of absorption bands. In an octahedral, nearly octahedral, or square-planar complex, the maximum molar absorption coefficient ε_{max} (which measures the strength of the absorption)[5] is typically less than or close to 100 $dm^3\ mol^{-1}\ cm^{-1}$ for ligand-field transitions. In tetrahedral complexes, which have no centre of symmetry, ε_{max} for ligand-field transitions might exceed 250 $dm^3\ mol^{-1}\ cm^{-1}$. By contrast, charge-transfer bands usually have an ε_{max} in the range 1000–50 000 $dm^3\ mol^{-1}\ cm^{-1}$.

To understand the intensities of transitions in complexes we have to explore the strength with which the complex couples with the electromagnetic field. Intense transitions indicate strong coupling; weak transitions indicate feeble coupling. The strength of coupling when an electron makes a transition from a state with wavefunction ψ_i to one with wavefunction ψ_f is measured by the **transition dipole moment**, which is defined as the integral

$$\boldsymbol{\mu}_{fi} = \int \psi_f^* \boldsymbol{\mu} \psi_i\, d\tau \qquad (20.5)$$

where $\boldsymbol{\mu}$ is the electric dipole moment operator, $-er$. The transition dipole moment can be regarded as a measure of the impulse that a transition imparts to the electromagnetic field: a large impulse corresponds to an intense transition; zero impulse corresponds to a forbidden transition. The intensity of a transition is proportional to the square of its transition dipole moment.

A spectroscopic **selection rule** is a statement about which transitions are allowed and which are forbidden. An **allowed transition** is a transition with a nonzero transition dipole moment, and hence nonzero intensity. A **forbidden transition** is a transition for which the transition dipole moment is calculated as zero. Formally forbidden transitions may occur in a spectrum if the assumptions on which the transition dipole moment were calculated are invalid, such as the complex having a lower symmetry than assumed.

---

[5] The molar absorption coefficient is the constant in the Beer–Lambert law for the transmittance $T = I_f/I_i$ when light passes through a length L of solution of molar concentration [X] and is attenuated from an intensity I_i to an intensity I_f: $\log T = -\varepsilon[X]L$ (the logarithm is a common logarithm, to the base 10). Its older but still widely used name is the 'extinction coefficient'.

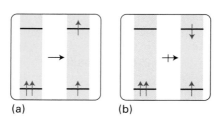

Figure 20.34 (a) A spin-allowed transition does not change the multiplicity. (b) A spin-forbidden transition results in a change in the multiplicity.

(a) Spin selection rules

Key points: Electronic transitions with a change of multiplicity are forbidden; intensities of spin-forbidden transitions are greater for 4d- or 5d-series metal complexes than for comparable 3d-series complexes.

The electromagnetic field of the incident radiation cannot change the relative orientations of the spins of the electrons in a complex. For example, an initially antiparallel pair of electrons cannot be converted to a parallel pair, so a singlet ($S = 0$) cannot undergo a transition to a triplet ($S = 1$). This restriction is summarized by the rule $\Delta S = 0$ for **spin-allowed transitions,** Fig. 20.34.

The coupling of spin and orbital angular momenta can relax the spin selection rule, but such **spin-forbidden,** $\Delta S \neq 0$, transitions are generally much weaker than spin-allowed transitions. The intensity of spin-forbidden bands increases as the atomic number increases because the strength of the spin−orbit coupling is greater for heavy atoms than for light atoms. The breakdown of the spin selection rule by spin−orbit coupling is often called the **heavy-atom effect.** In the 3d series, in which spin−orbit coupling is weak, spin-forbidden bands have ε_{max} less than about 1 dm^3 mol^{-1} cm^{-1}; however, spin-forbidden bands are a significant feature in the spectra of heavy d-metal complexes.

The very weak transition labelled $^2E_g \leftarrow {}^4A_{2g}$ in Fig. 20.22 is an example of a spin-forbidden transition. Some metal ions, such as the d^5 Mn^{2+} ion, have no spin-allowed transitions, and hence are only weakly coloured.

(b) The Laporte selection rule

Key points: Transitions between d orbitals are forbidden in octahedral complexes; asymmetric vibrations relax this restriction.

The **Laporte selection rule** states that *in a centrosymmetric molecule or ion, the only allowed transitions are those accompanied by a change in parity.* That is, transitions between g and u terms are permitted, but a g term cannot undergo a transition to another g term and a u term cannot undergo a transition to another u term:

$$g \leftrightarrow u \qquad g \nleftrightarrow g \qquad u \nleftrightarrow u$$

In many cases it is enough to note that in a centrosymmetric complex, if there is no change in quantum number l, then there can be no change in parity. Thus, s−s, p−p, d−d , and f−f transitions are forbidden. Since s and d orbitals are g, whereas p and f orbitals are u it follows that s−p, p−d, and d−f transitions are allowed whereas s−d and p−f transitions are forbidden.

A more formal treatment of the Laporte selection rule is based on the properties of the transition dipole moment, which is proportional to r. Because r changes sign under inversion (and is therefore u), the entire integral in eqn 20.5 also changes sign under inversion if ψ_i and ψ_f have the same parity because g × u × g = u and u × u × u = u. Therefore, because the value of an integral cannot depend on the choice of coordinates used to evaluate it,[6] it vanishes if ψ_i and ψ_f have the same parity. However, if they have opposite parity, the integral does not change sign under inversion of the coordinates because g × u × u = g and therefore need not vanish.

In a centrosymmetric complex, d−d ligand-field transitions are g ↔ g and are therefore forbidden. Their forbidden character accounts for the relative weakness of these transitions in octahedral complexes (which are centrosymmetric) compared with those in tetrahedral complexes, on which the Laporte rule is silent (they are noncentrosymmetric, and have no g or u as a subscript).

The question remains why d−d ligand-field transitions in octahedral complexes occur at all, even weakly. The Laporte selection rule may be relaxed in two ways. First, a complex may depart slightly from perfect centrosymmetry in its ground state, perhaps on account of the intrinsic asymmetry in the structure of polyatomic ligands or a distortion imposed by the environment of a complex packed into a crystal. Alternatively, the complex might undergo an asymmetrical vibration, which also destroys its centre of inversion. In either case, a Laporte-forbidden d−d ligand-field band tends to be much more intense than a spin-forbidden transition.

[6] An integral is an area, and areas are independent of the coordinates used for their evaluation.

Table 20.10 summarizes typical intensities of electronic transitions of complexes of the 3d-series elements. The width of spectroscopic absorption bands is due principally to the simultaneous excitation of vibration when the electron is promoted from one distribution to another. According to the **Franck–Condon principle**, the electronic transition takes place within a stationary nuclear framework. As a result, after the transition has occurred, the nuclei experience a new force field and the molecule begins to vibrate anew.

Table 20.10 Intensities of spectroscopic bands in 3d complexes

Band type	$\varepsilon_{max}/$ (dm^3 mol^{-1} cm^{-1})
Spin-forbidden	< 1
Laporte-forbidden d–d	20–100
Laporte-allowed d–d	c. 250
Symmetry-allowed (e.g. CT)	1000–50 000

EXAMPLE 20.9 Assigning a spectrum using selection rules

Assign the bands in the spectrum in Fig. 20.32 by considering their intensities.

Answer If we assume that the complex is approximately octahedral, examination of the Tanabe–Sugano diagram for a d^3 ion reveals that the ground term is $^4A_{2g}$. Transitions to the higher terms ^2E_g, $^2T_{1g}$, and $^2T_{2g}$ are spin-forbidden and will have $\varepsilon_{max} < 1$ dm^3 mol^{-1} cm^{-1}. Thus, very weak bands for these transitions are predicted, and will be difficult to distinguish. The next two higher terms of the same multiplicity are $^4T_{2g}$ and $^4T_{1g}$. These terms are reached by spin-allowed but Laporte-forbidden ligand-field transitions, and have $\varepsilon_{max} \approx 100$ dm^3 mol^{-1} cm^{-1}: these are the two bands at 360 and 510 nm. In the near UV, the band with $\varepsilon_{max} \approx 10\,000$ dm^3 mol^{-1} cm^{-1} corresponds to the LMCT transitions in which an electron from a chlorine π lone pair is promoted into a molecular orbital that is principally metal d orbital in character.

Self-test 20.9 The spectrum of [Cr(NCS)$_6$]$^{3-}$ has a very weak band near 16 000 cm^{-1}, a band at 17 700 cm^{-1} with $\varepsilon_{max} = 160$ dm^3 mol^{-1} cm^{-1}, a band at 23 800 cm^{-1} with $\varepsilon_{max} = 130$ dm^3 mol^{-1} cm^{-1}, and a very strong band at 32 400 cm^{-1}. Assign these transitions using the d^3 Tanabe–Sugano diagram and selection rule considerations. (*Hint:* NCS$^-$ has low-lying π^* orbitals.)

20.7 Luminescence

Key points: A luminescent complex is one that re-emits radiation after it has been electronically excited. Fluorescence occurs when there is no change in multiplicity, whereas phosphorescence occurs when an excited state undergoes intersystem crossing to a state of different multiplicity and then undergoes radiative decay.

A complex is **luminescent** if it emits radiation after it has been electronically excited by the absorption of radiation. Luminescence competes with nonradiative decay by thermal degradation of energy to the surroundings. Relatively fast radiative decay is not especially common at room temperature for d-metal complexes, so strongly luminescent systems are comparatively rare. Nevertheless, they do occur, and we can distinguish two types of process. Traditionally, rapidly decaying luminescence was called 'fluorescence' and luminescence that persists after the exciting illumination is extinguished was called 'phosphorescence'. However, because the lifetime criterion is not reliable, the modern definitions of the two kinds of luminescence are based on the distinctive mechanisms of the processes. **Fluorescence** is radiative decay from an excited state of the same multiplicity as the ground state. The transition is spin-allowed and is fast; fluorescence half-lives are a matter of nanoseconds. **Phosphorescence** is radiative decay from a state of different multiplicity from the ground state. It is a spin-forbidden process, and hence is often slow.

The initial excitation of a phosphorescent complex usually populates a state by a spin-allowed transition, so the mechanism of phosphorescence involves **intersystem crossing**, the nonradiative conversion of the initial excited state into another excited state of different multiplicity. This second state acts as an energy reservoir because radiative decay to the ground state is spin-forbidden. However, just as spin–orbit coupling allows the intersystem crossing to occur, it also breaks down the spin selection rule, so the radiative decay can occur. Radiative decay back to the ground state is slow, so a phosphorescent state of a d-metal complex may survive for microseconds or even longer.

An important example of phosphorescence is provided by ruby, which consists of a low concentration of Cr^{3+} ions in place of Al^{3+} in alumina. Each Cr^{3+} ion is surrounded octahedrally by six O^{2-} ions and the initial excitations are the spin-allowed processes

$$t_{2g}^2 e_g^1 \leftarrow t_{2g}^3: \qquad {}^4T_{2g} \leftarrow {}^4A_{2g} \text{ and } {}^4T_{1g} \leftarrow {}^4A_{2g}$$

These absorptions occur in the green and violet regions of the spectrum and are responsible for the red colour of the gem (Fig. 20.35). Intersystem crossing to a ^2E term of the t_{2g}^3 configuration occurs in a few picoseconds or less, and red 627 nm phosphorescence occurs

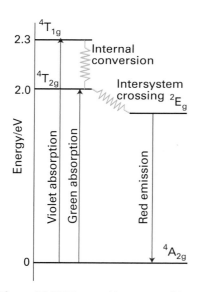

Figure 20.35 The transitions responsible for the absorption and luminescence of Cr^{3+} ions in ruby.

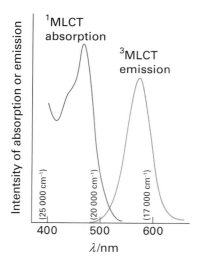

Figure 20.36 The absorption and phosphorescence spectra of $[Ru(bpy)_3]^{2+}$.

as this doublet decays back into the quartet ground state. This red emission adds to the red perceived by the subtraction of green and violet light from white light, and adds lustre to the gem's appearance. This effect was utilised in the first laser to be constructed (in 1960).

A similar $^2E \rightarrow {}^4A$ phosphorescence can be observed from a number of Cr(III) complexes in solution. The ^2E term arises from the t_{2g}^3 configuration, which is the same as the ground state, and thus the strength of the ligand field is not important. Hence the emission is always in the red (and close to the wavelength of ruby emission). If the ligands are rigid, as in $[Cr(bpy)_3]^{3+}$, the ^2E term may live for several microseconds in solution.

Another interesting example of a phosphorescent state is found in $[Ru(bpy)_3]^{2+}$. The excited singlet term produced by a spin-allowed MLCT transition of this d^6 complex undergoes intersystem crossing to the lower energy triplet term of the same configuration, $t_{2g}^5 \pi^{*1}$. Bright orange emission then occurs with a lifetime of about 1 μs (Fig. 20.36). The effects of other molecules (quenchers) on the lifetime of the emission may be used to monitor the rate of electron transfer from the excited state.

Magnetism

The diamagnetic and paramagnetic properties of complexes were introduced in Section 20.1c, but the discussion was restricted to magnetically dilute species, where the individual paramagnetic centres, the atoms with unpaired d-electrons, are separate from each other. We now consider two further aspects of magnetism, one where magnetic centres can interact with one another and one where the spin-state may change.

20.8 Cooperative magnetism

Key point: In solids, the spins on neighbouring metal centres may interact to produce magnetic behaviour, such as ferromagnetism and antiferromagnetism, that are representative of the whole solid.

In the solid state the individual magnetic centres are often close together and separated by only a single atom, typically O. In such arrays cooperative properties can arise from interactions between electron spins on different atoms.

The **magnetic susceptibility**, χ, of a material is a measure of how easy it is to align electron spins with the applied magnetic field in the sense that the induced magnetic moment is proportional to the applied field, with χ the constant of proportionality. A paramagnetic material has a positive susceptibility and a diamagnetic material has a negative susceptibility. Magnetic effects arising from cooperative phenomena can be very much larger than those arising from individual atoms and ions. The susceptibility and its variation with temperature are different for different types of magnetic materials and are summarized in Table 20.11 and Fig. 20.37.

The application of a magnetic field to a paramagnetic material results in the partial alignment of the spins parallel to the field. As a paramagnetic material is cooled, the disordering effect of thermal motion is reduced, more spins become aligned, and the magnetic susceptibility increases. In a **ferromagnetic substance**, which is one example of a cooperative magnetic property, the spins on different metal centres are coupled into a parallel alignment that is sustained over thousands of atoms to form a **magnetic domain** (Fig. 20.38). The net magnetic moment, and hence the magnetic susceptibility, may be

Table 20.11 Magnetic behaviour of materials

Magnetic behaviour	Typical value of χ	Variation of χ with temperature	Field dependence
Diamagnetism (no unpaired spins)	-8×10^{-6} for Cu	None	No
Paramagnetism	4×10^{-3} for $FeSO_4$	Decreases	No
Ferromagnetism	5×10^3 for Fe	Decreases	Yes
Antiferromagnetism	$0 - 10^{-2}$	Increases	(Yes)

very large because the magnetic moments of individual spins add to each other. Moreover, once established and with the temperature maintained below the **Curie temperature** (T_C), the magnetization persists after the applied field is removed because the spins are locked together. Ferromagnetism is exhibited by materials containing unpaired electrons in d or, more rarely, f orbitals that couple with unpaired electrons in similar orbitals on surrounding atoms. The key feature is that this interaction is strong enough to align spins but not so strong as to form covalent bonds, in which the electrons would be paired. At temperatures above T_C the disordering effect of thermal motion overcomes the ordering effect of the interaction and the material becomes paramagnetic (Fig. 20.37).

The magnetization, M, of a ferromagnet, its bulk magnetic moment, is not proportional to the applied field strength H. Instead, a 'hysteresis loop' is observed like that shown in Fig. 20.39. For **hard ferromagnets** the loop is broad and M remains large when the applied field has been reduced to zero. Hard ferromagnets are used for permanent magnets where the direction of the magnetization does not need to be reversed. A **soft ferromagnet** has a narrower hysteresis loop and is therefore much more responsive to the applied field. Soft ferromagnets are used in transformers, where they must respond to a rapidly oscillating field.

In an **antiferromagnetic material**, neighbouring spins are locked into an antiparallel alignment (Fig. 20.40). As a result, the collection of individual magnetic moments cancel and the sample has a low magnetic moment and magnetic susceptibility (tending, in fact, to zero). Antiferromagnetism is often observed when a paramagnetic material is cooled to a low temperature and is indicated by a sharp decrease in magnetic susceptibility at the **Néel temperature**, T_N (Fig. 24.37). Above T_N the magnetic susceptibility is that of a paramagnetic material, and decreases as the temperature is raised.

The spin coupling responsible for antiferromagnetism generally occurs through intervening ligands by a mechanism called **superexchange**. As indicated in Fig. 20.41, the spin on one metal atom induces a small spin polarization on an occupied orbital of a ligand, and this spin polarization results in an antiparallel alignment of the spin on the adjacent metal atom. This alternating ...↑↓↑↓... alignment of spins then propagates throughout the material. Many d-metal oxides exhibit antiferromagnetic behaviour that can be ascribed to a superexchange mechanism involving O atoms, for example MnO is antiferromagnetic below 122 K and Cr_2O_3 is antiferromagnetic below 310 K. Coupling of spins through intervening ligands is frequently observed in molecular complexes containing two ligand-bridged metal ions but it is weaker than with a simple O^{2-} link between metal sites and as a result the ordering temperatures are much lower, typically below 100 K.

In **ferrimagnetism**, a net magnetic ordering of ions with different individual magnetic moments is observed below the Curie temperature. These ions can order with opposed spins, as in antiferromagnetism, but because the individual spin moments are different, there is incomplete cancellation and the sample has a net overall moment. As with antiferro-magnetism, these interactions are generally transmitted through the ligands; an example is magnetite Fe_3O_4.

There are a large number of molecular systems where magnetic coupling is observed. Typical systems have two or more metal atoms bridged by ligands that mediate the coupling. Simple examples include copper acetate (**10**), which exists as a dimer with antiferromagnetic coupling between the two d⁹ centres. Many metalloenzymes (Sections 27.9 to 27.14) have multiple metal centres that show magnetic coupling.

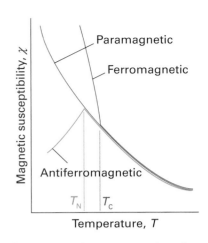

Figure 20.37 The temperature dependence of the susceptibilities of paramagnetic, ferromagnetic, and antiferromagnetic substances.

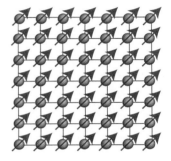

Figure 20.38 The parallel alignment of individual magnetic moments in a ferromagnetic material.

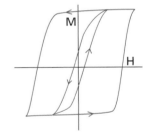

Figure 20.39 Magnetization curves for ferromagnetic materials. A hysteresis loop results because the magnetization of the sample with increasing field (→) is not retraced as the field is decreased (←). Blue line: hard ferromagnet; red line: soft ferromagnet.

Figure 20.40 The antiparallel arrangement of individual magnetic moments in an antiferromagnetic material.

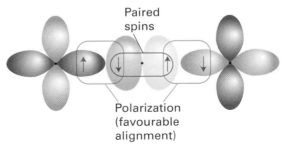

Figure 20.41 Antiferromagnetic coupling between two metal centres created by spin polarization of a bridging ligand.

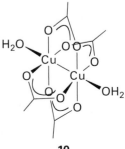

10

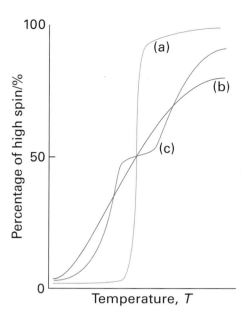

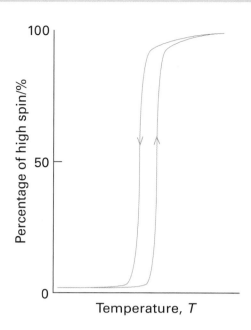

Figure 20.42 The change to high-spin might be (a) abrupt, (b) gradual, or (c) stepped.

Figure 20.43 A hysteresis loop that can occur with some spin crossover systems.

20.9 Spin crossover complexes

Key point: When the factors that determine the spin state of a d-metal centre are closely matched, complexes that change spin state in response to external stimuli are possible.

We have seen how a number of factors, such as oxidation state and ligand type, determine whether a complex is high- or low-spin. With some complexes, normally of the 3d-series metals, there is only a very small energy difference between the two states, leading to the possibility of **spin crossover** complexes. Such complexes change their spin state in response to an external stimulus (such as heat or pressure), which in turn leads to a change in their bulk magnetic properties. An example is the d^6 iron complex of two diphenylterpyridine ligands (**11**), which is low-spin ($S = 0$) below 300 K, but high-spin ($S = 2$) above 323 K. The transition from one spin state to another can be abrupt, gradual, or even stepped (Fig 20.42). In the solid state, a further feature of spin crossover complexes is the existence of cooperativity between the magnetic centres, which can lead to hysteresis like that shown in Fig. 20.43.

A preference normally exists for the low-spin state under high pressure and low temperature. This preference can be understood on the basis that the e_g orbitals, which are more extensively occupied in the high-spin state, have significant metal−ligand antibonding character. Thus the high-spin form occupies a larger volume, which is favoured by low pressure or high temperature.

Spin crossover complexes occur in many geological systems, are implicated in the binding of O_2 to haemoglobin, and have the potential to be exploited in both practical magnetic information storage and pressure-sensitive devices.

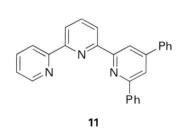

11

FURTHER READING

E.I. Solomon and A.B.P. Lever, *Inorganic electronic structure and spectroscopy*. Wiley, Chichester (2006). A thorough account of the material covered in this chapter, including a useful discussion of solvatochromism.

S.F.A. Kettle, *Physical inorganic chemistry: a co-ordination chemistry approach*. Oxford University Press (1998).

B.N. Figgis and M.A. Hitchman, *Ligand field theory and its applications*. Wiley, Chichester (2000).

E.U. Condon and G.H. Shortley, *The theory of atomic spectra*. Cambridge University Press (1935); revised as E.U. Condon and H. Odabaşi, *Atomic structure*. Cambridge University Press (1980). The standard reference text on atomic spectra.

A.F. Orchard, *Magnetochemistry*. Oxford University Press (2003). This book provides detailed, modern explanations, based on ligand-field theory, of the origins and interpretations of magnetic effects in complexes and materials.

EXERCISES

20.1 Determine the configuration (in the form $t_{2g}^x e_g^y$ or $e^x t_2^y$, as appropriate), the number of unpaired electrons, and the ligand-field stabilization energy in terms of Δ_O or Δ_T and P for each of the following complexes using the spectrochemical series to decide, where relevant, which are likely to be high-spin and which low-spin. (a) $[Co(NH_3)_6]^{3+}$, (b) $[Fe(OH_2)_6]^{2+}$, (c) $[Fe(CN)_6]^{3-}$, (d) $[Cr(NH_3)_6]^{3+}$, (e) $[W(CO)_6]$, (f) tetrahedral $[FeCl_4]^{2-}$, (g) tetrahedral $[Ni(CO)_4]$.

20.2 Both H^- and $P(C_6H_5)_3$ are ligands of similar field strength, high in the spectrochemical series. Recalling that phosphines act as π acceptors, is π-acceptor character required for strong-field behaviour? What orbital factors account for the strength of each ligand?

20.3 Estimate the spin-only contribution to the magnetic moment for each complex in Exercise 20.1.

20.4 Solutions of the complexes $[Co(NH_3)_6]^{2+}$, $[Co(OH_2)_6]^{2+}$ (both O_h), and $[CoCl_4]^{2-}$ are coloured. One is pink, another is yellow, and the third is blue. Considering the spectrochemical series and the relative magnitudes of Δ_T and Δ_O, assign each colour to one of the complexes.

20.5 For each of the following pairs of complexes, identify the one that has the larger LFSE:

(a) $[Cr(OH_2)_6]^{2+}$ or $[Mn(OH_2)_6]^{2+}$

(b) $[Mn(OH_2)_6]^{2+}$ or $[Fe(OH_2)_6]^{3+}$

(c) $[Fe(OH_2)_6]^{3+}$ or $[Fe(CN)_6]^{3-}$

(d) $[Fe(CN)_6]^{3-}$ or $[Ru(CN)_6]^{3-}$

(e) tetrahedral $[FeCl_4]^{2-}$ or tetrahedral $[CoCl_4]^{2-}$

20.6 Interpret the variation, including the overall trend across the 3d series, of the following values of oxide lattice enthalpies (in kJ mol^{-1}). All the compounds have the rock-salt structure: CaO (3460), TiO (3878), VO (3913), MnO (3810), FeO (3921), CoO (3988), NiO (4071).

20.7 A neutral macrocyclic ligand with four donor atoms produces a red diamagnetic low-spin d^8 complex of Ni(II) if the anion is the weakly coordinating perchlorate ion. When perchlorate is replaced by two thiocyanate ions, SCN^-, the complex turns violet and is high-spin with two unpaired electrons. Interpret the change in terms of structure.

20.8 Bearing in mind the Jahn–Teller theorem, predict the structure of $[Cr(OH_2)_6]^{2+}$.

20.9 The spectrum of d^1 $Ti^{3+}(aq)$ is attributed to a single electronic transition $e_g \leftarrow t_{2g}$. The band shown in Fig. 20.3 is not symmetrical and suggests that more than one state is involved. Suggest how to explain this observation using the Jahn–Teller theorem.

20.10 Write the Russell–Saunders term symbols for states with the angular momentum quantum numbers (L,S): (a) $(0,\frac{5}{2})$, (b) $(3,\frac{3}{2})$, (c) $(2,\frac{1}{2})$, (d) $(1,1)$.

20.11 Identify the ground term from each set of terms: (a) ^1P, ^3P, ^3F, ^1G, (b) ^3P, ^5D, ^3H, ^1I, ^1G, (c) ^6S, ^4P, ^4G, ^2I.

20.12 Give the Russell–Saunders terms of the configurations: (a) $4s^1$, (b) $3p^2$. Identify the ground term.

20.13 The gas-phase ion V^{3+} has a ^3F ground term. The ^1D and ^3P terms lie, respectively, 10 642 and 12 920 cm^{-1} above it. The energies of the terms are given in terms of Racah parameters as $E(^3F) = A - 8B$, $E(^3P) = A + 7B$, $E(^1D) = A - 3B + 2C$. Calculate the values of B and C for V^{3+}.

20.14 Write the d-orbital configurations and use the Tanabe–Sugano diagrams (*Resource section* 6) to identify the ground term of (a) low-spin $[Rh(NH_3)_6]^{3+}$, (b) $[Ti(OH_2)_6]^{3+}$, (c) high-spin $[Fe(OH_2)_6]^{3+}$.

20.15 Using the Tanabe–Sugano diagrams in *Resource section* 6, estimate Δ_O and B for (a) $[Ni(OH_2)_6]^{2+}$ (absorptions at 8500, 15 400, and 26 000 cm^{-1}) and (b) $[Ni(NH_3)_6]^{2+}$ (absorptions at 10 750, 17 500, and 28 200 cm^{-1}).

20.16 The spectrum of $[Co(NH_3)_6]^{3+}$ has a very weak band in the red and two moderate intensity bands in the visible to near-UV. How should these transitions be assigned?

20.17 Explain why $[FeF_6]^{3-}$ is colourless whereas $[CoF_6]^{3-}$ is coloured but exhibits only a single band in the visible region of the spectrum.

20.18 The Racah parameter B is 460 cm^{-1} in $[Co(CN)_6]^{3-}$ and 615 cm^{-1} in $[Co(NH_3)_6]^{3+}$. Consider the nature of bonding with the two ligands and explain the difference in nephelauxetic effect.

20.19 An approximately 'octahedral' complex of Co(III) with ammine and chloro ligands gives two bands with ε_{max} between 60 and 80 dm^3 mol^{-1} cm^{-1}, one weak peak with ε_{max} =2 dm^3 mol^{-1} cm^{-1}, and a strong band at higher energy with ε_{max} =2 × 10^4 dm^3 mol^{-1} cm^{-1}. What do you suggest for the origins of these transitions?

20.20 Ordinary bottle glass appears nearly colourless when viewed through the wall of the bottle but green when viewed from the end so that the light has a long path through the glass. The colour is associated with the presence of Fe^{3+} in the silicate matrix. Suggest which transitions are responsible for the colour.

20.21 Solutions of $[Cr(OH_2)_6]^{3+}$ ions are pale blue–green but the chromate ion, CrO_4^{2-}, is an intense yellow. Characterize the origins of the transitions and explain the relative intensities.

20.22 Classify the symmetry type of the d orbitals in a tetragonal C_{4v} symmetry complex, such as $[CoCl(NH_3)_5]^{2+}$, where the Cl lies on the z-axis. (a) Which orbitals will be displaced from their position in the octahedral molecular orbital diagram by π interactions with the lone pairs of the Cl$^-$ ligand? (b) Which orbital will move because the Cl$^-$ ligand is not as strong a σ base as NH$_3$? (c) Sketch the qualitative molecular orbital diagram for the C_{4v} complex.

20.23 Consider the molecular orbital diagram for a tetrahedral complex (based on Fig. 20.8) and the relevant d-orbital configuration and show that the purple colour of MnO_4^- ions cannot arise from a ligand-field transition. Given that the wavenumbers of the two transitions in MnO_4^- are 18 500 and 32 200 cm^{-1}, explain how to estimate Δ_T from an assignment of the two charge-transfer transitions, even though Δ_T cannot be observed directly.

20.24 The lowest energy band in the spectrum of $[Fe(OH_2)]^{3+}$ (in 1M HClO$_4$) occurs at lower energy than the equivalent transition in the spectrum of $[Mn(OH_2)]^{2+}$. Explain why this is.

PROBLEMS

20.1 In a fused magma liquid from which silicate minerals crystallize, the metal ions can be four-coordinate. In olivine crystals, the M(II) co-ordination sites are octahedral. Partition coefficients, which are defined as $K_p = [M(II)]_{olivine}/[M(II)]_{melt}$, follow the order Ni(II) > Co(II) > Fe(II) > Mn(II). Account for this in terms of ligand-field theory. (See I.M. Dale and P. Henderson, *24th Int. Geol. Congress, Sect.* 1972, **10**, 105.)

20.2 In Problem 7.11 we looked at the successive formation constants for ethylenediamine complexes of three different metals. Using the same data, discuss the effect of the metal on the formation constant. How might the Irving–Williams series provide insight into these formation constants?

20.3 By considering the splitting of the octahedral orbitals as the symmetry is lowered, draw the symmetry-adapted linear combinations and the molecular orbital energy level diagram for σ bonding in a *trans*-[ML$_4$X$_2$] complex. Assume that the ligand X is lower in the spectrochemical series than L.

20.4 By referring to *Resource section 5*, draw the appropriate symmetry-adapted linear combinations and the molecular orbital diagram for σ bonding in a square-planar complex. The point group is D_{4h}. Take note of the small overlap of the ligand with the d$_{z^2}$ orbital. What is the effect of π bonding?

20.5 Figures 20.12 and 20.10 show the relationship between the frontier orbitals of octahedral and square-planar complexes. Construct similar diagrams for two-coordinate linear complexes.

20.6 Consider a trigonal prismatic six-coordinate ML$_6$ complex with D_{3h} symmetry. Use the D_{3h} character table to divide the d orbitals of the metal into sets of defined symmetry type. Assume that the ligands are at the same angle relative to the *xy*-plane as in a tetrahedral complex.

20.7 In a trigonal bipyramidal complex ligand the axial and equatorial sites have different steric and electronic interactions with the central metal ion. Consider a range of some common ligands and decide which coordination site in a trigonal bipyramidal complex they would favour. (See A.R. Rossi and R. Hoffmann, *Inorg. Chem.*, 1975, **14**, 365.)

20.8 Vanadium(IV) species that have the V=O group have quite distinct spectra. What is the d-electron configuration of V(IV)? The most symmetrical of such complexes are VOL$_5$ with C_{4v} symmetry with the O atom on the *z*-axis. What are the symmetry species of the five d orbitals in VOL$_5$ complexes? How many d–d bands are expected in the spectra of these complexes? A band near 24 000 cm^{-1} in these complexes shows vibrational progressions of the V=O vibration, implicating an orbital involving V=O bonding. Which d–d transition is a candidate? (See C.J. Ballhausen and H.B. Gray, *Inorg. Chem.*, 1962, **1**, 111.)

20.9 The compound Ph$_3$Sn−Re(CO)$_3$(^{t}Bu-DAB) (^{t}Bu-DAB = ^{t}Bu−N=CH=N−^{t}Bu) has a metal-to-diimine ligand MLCT band as the lowest energy transition in its spectrum. Irradiation of this band gives an Re(CO)$_3$(^{t}Bu-DAB) radical as a photochemical product. The EPR spectrum shows extensive hyperfine splitting by ^{14}N and ^{1}H. The radical is assigned as a complex of the DAB radical anion with Re(I). Explain the argument (see D.J. Stufkens, *Coord. Chem. Rev.*, 1990, **104**, 39).

20.10 MLCT bands can be recognized by the fact that the energy is a sensitive function of the polarity of the solvent (because the excited state is more polar than the ground state). Two simplified molecular orbital diagrams are shown in Fig. 20.44. In (a) is a case with a ligand π level higher than the metal d orbital. In (b) is a case in which the metal d orbital and the ligand level are at the same energy. Which of the two MLCT bands should be more solvent sensitive? These two cases are realized by [W(CO)$_4$(phen)] and [W(CO)$_4$(^{i}Pr-DAB)], where DAB = 1,4-diaza-1,3-butadiene, respectively. (See P.C. Servas, H.K. van Dijk, T.L. Snoeck, D.J. Stufkens, and A. Oskam, *Inorg. Chem.*, 1985, **24**, 4494.) Comment on the CT character of the transition as a function of the extent of back-donation by the metal atom.

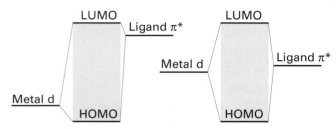

Figure 20.44 Representation of the orbitals involved in MLCT transitions for cases in which the energy of the ligand π* orbital varies with respect to the energy of the metal d orbital. See Problem 20.10.

20.11 Consider spin crossover complexes and identify the features that a complex would need for it to be used in (a) a practical pressure sensor and (b) a practical information storage device (see P. Gütlich, Y. Garcia, and H.A. Goodwin, *Chem. Soc. Rev.*, 2000, **29**, 419).

Coordination chemistry: reactions of complexes

21

We now look at the evidence and experiments that are used in the analysis of the reaction pathways of metal complexes and so develop a deeper understanding of their mechanisms. Because a mechanism is rarely known definitively, the nature of the evidence for it should always be kept in mind in order to recognize that there might be other consistent possibilities. In the first part of this chapter we consider ligand exchange reactions and describe how reaction mechanisms are classified. We consider the steps by which the reactions take place and the details of the formation of the transition state. These concepts are then used to describe the mechanisms of the redox reactions of complexes.

Coordination chemistry is not the sole preserve of d metals. Whereas Chapter 20 dealt exclusively with d metals, this chapter builds on the introduction to coordination chemistry in Chapter 7 and applies it to all metals regardless of the block to which they belong. However, there are special features of each block, and we shall point them out.

Ligand substitution reactions

The most fundamental reaction a complex can undergo is **ligand substitution**, a reaction in which one Lewis base displaces another from a Lewis acid:

$$Y + M-X \rightarrow M-Y + X$$

This class of reaction includes complex formation reactions, in which the **leaving group**, the displaced base X, is a solvent molecule and the **entering group**, the displacing base Y, is some other ligand. An example is the replacement of a water ligand by Cl^-:

$$[Co(OH_2)_6]^{2+}(aq) + Cl^-(aq) \rightarrow [CoCl(OH_2)_5]^+(aq) + H_2O(l)$$

The thermodynamic aspects of complex formation are discussed in Sections 7.12 to 7.15.

21.1 Rates of ligand substitution

Key points: The rates of substitution reactions span a very wide range and correlate with the structures of the complexes; complexes that react quickly are called labile, those that react slowly are called inert or nonlabile.

Rates of reaction are as important as equilibria in coordination chemistry. The numerous isomers of the ammines of Co(III) and Pt(II), which were so important to the development of the subject, could not have been isolated if ligand substitutions and interconversion of the isomers had been fast. But what determines whether one complex will survive for long periods whereas another will undergo rapid reaction?

The rate at which one complex converts into another is governed by the height of the activation energy barrier that lies between them. Thermodynamically unstable complexes that survive for long periods (by convention, at least a minute) are commonly called 'inert', but **nonlabile** is more appropriate and is the term we shall use. Complexes that undergo more rapid equilibration are called **labile**. An example of each type is the labile complex $[Ni(OH_2)_6]^{2+}$, which has a half-life of the order of milliseconds before the H_2O is replaced by another H_2O or a stronger base, and the nonlabile complex $[Co(NH_3)_5(OH_2)]^{3+}$, in which H_2O survives for several minutes as a ligand before it is replaced by a stronger base.

Figure 21.1 shows the characteristic lifetimes of the important aqua metal ion complexes. We see a range of lifetimes starting at about 1 ns, which is approximately the time it takes for a molecule to diffuse one molecular diameter in solution. At the other end of the scale are lifetimes in years. Even so, the illustration does not show the longest times that could be considered, which are comparable to geological eras.

We shall examine the lability of complexes in greater detail when we discuss the mechanism of reactions later in this section, but we can make two broad generalizations now. The first is that complexes of metals that have no additional factor to provide extra stability (for instance, the LFSE and chelate effects) are among the most labile. Any additional stability of a complex results in an increase in activation energy for a ligand replacement reaction and hence decreases the lability of the complex. A second generalization is that very small ions are often less labile because they have greater M−L bond strengths and it is sterically very difficult for incoming ligands to approach the metal atom closely.

Some further generalizations are as follows:

1. All complexes of s-block ions except the smallest (Be^{2+} and Mg^{2+}) are very labile.
2. Complexes of the M(III) ions of the f block are all very labile.
3. Complexes of the d^{10} ions (Zn^{2+}, Cd^{2+}, and Hg^{2+}) are normally very labile.
4. Across the 3d series, complexes of d-block M(II) ions are generally moderately labile, with distorted Cu(II) complexes among the most labile.
5. Complexes of d-block M(III) ions are distinctly less labile than d-block M(II) ions.
6. d-Metal complexes with d^3 and low-spin d^6 configurations (for example Cr(III), Fe(II), and Co(III)) are generally nonlabile as they have large LFSEs. Chelate complexes with the same configuration, such as $[Fe(phen)_3]^{2+}$, are particularly inert.
7. Nonlability is common among the complexes of the 4d and 5d series, which reflects the high LFSE and strength of the metal−ligand bonding.

Table 21.1 illustrates the range of timescales for a number of reactions.

The natures of the ligands in the complex also affect the rates of reactions. The identity of the incoming ligand has the greatest effect, and equilibrium constants of displacement reactions can be used to rank ligands in order of their strength as Lewis bases. However, a different order may be found if bases are ranked according to the rates at which they displace a ligand from the central metal ion. Therefore, for kinetic considerations, we replace the equilibrium concept of basicity by the kinetic concept of **nucleophilicity**, the rate of attack on a complex by a given Lewis base relative to the rate of attack by a reference Lewis base. The shift from equilibrium to kinetic considerations is emphasized by referring to ligand displacement as **nucleophilic substitution**.

Ligands other than the entering and leaving groups may play a significant role in controlling the rates of reactions; these ligands are referred to as **spectator ligands**.

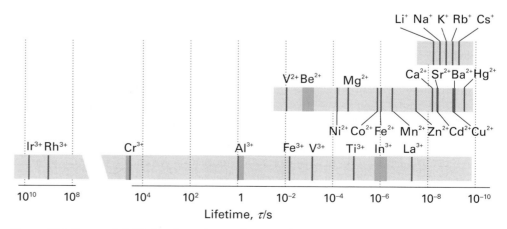

Figure 21.1 Characteristic lifetimes for exchange of water molecules in aqua complexes.

Table 21.1 Representative timescales of chemical and physical processes

Timescale*	Process	Example
10^8 s	Ligand exchange (inert complex)	$[Cr(OH_2)_6]^{3+} - H_2O$ (c. 6 days)
60 s	Ligand exchange (nonlabile complex)	$[V(OH_2)_6]^{3+} - H_2O$ (50 s)
1 ms	Ligand exchange (labile complex)	$[Pt(OH_2)_4]^{2+} - H_2O$ (0.4 ms)
1 μs	Intervalence charge transfer	$(H_3N)_5Ru^{II}-N\diagup\!\!\!\diagdown N- Ru^{III}(NH_3)$
		(0.5 μs)
1 ns	Ligand exchange (labile complex)	$[Ni(OH_2)_5(py)]^{2+} - H_2O$ (1 ns)
10 ps	Ligand association	$Cr(CO)_5 + THF$ (10 ps)
1 ps	Rotation time in liquid	CH_3CN (1 ps)
1 fs	Molecular vibration	$Sn-Cl$ stretch (300 fs)

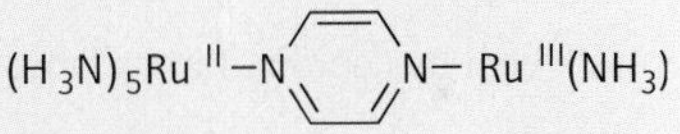

* Approximate time at room temperature.

For instance, it is observed for square-planar complexes that the ligand *trans* to the leaving group X has a great effect on the rate of substitution of X by the entering group Y.

21.2 The classification of mechanisms

The **mechanism** of a reaction is the sequence of elementary steps by which the reaction takes place. Once the mechanism has been identified, attention turns to the details of the activation process of the rate-determining step. In some cases the overall mechanism is not fully resolved, and the only information available is the rate-determining step.

(a) Association, dissociation, and interchange

Key points: The mechanism of a nucleophilic substitution reaction is the sequence of elementary steps by which the reaction takes place and is classified as associative, dissociative, or interchange; an associative mechanism is distinguished from an interchange mechanism by demonstrating that the intermediate has a relatively long life.

The first stage in the kinetic analysis of a reaction is to study how its rate changes as the concentrations of reactants are varied. This type of investigation leads to the identification of **rate laws,** the differential equations governing the rate of change of the concentrations of reactants and products. For example, the observation that the rate of formation of $[Ni(NH_3)(OH_2)_5]^{2+}$ from $[Ni(OH_2)_6]^{2+}$ is proportional to the concentration of both NH_3 and $[Ni(OH_2)_6]^{2+}$ implies that the reaction is first order in each of these two reactants, and that the overall rate law is:

$$rate = k_r[Ni(OH_2)_6^{2+}][NH_3] \tag{21.1}$$

A note on good practice In rate equations, as in expressions for equilibrium constants, we omit the brackets that are part of the chemical formula of the complex; the surviving brackets denote molar concentration. We denote rate constants by k_r to avoid possible confusion with Boltzmann's constant.

In simple sequential reaction schemes, the slowest elementary step of the reaction dominates the overall reaction rate and the overall rate law, and is called the **rate-determining step.** However, in general, all the steps in the reaction may contribute to the rate law and affect its rate. Therefore, in conjunction with stereochemical and isotopic labelling studies, the determination of the rate law is the route to the elucidation of the mechanism of the reaction.

Three main classes of reaction mechanism have been identified. A **dissociative mechanism,** denoted D, is a reaction sequence in which an intermediate of reduced coordination number is formed by the departure of the leaving group:

$$ML_nX \rightarrow ML_n + X$$

$$ML_n + Y \rightarrow ML_nY$$

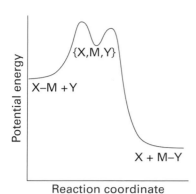

Figure 21.2 The typical form of the reaction profile of a reaction with a dissociative mechanism.

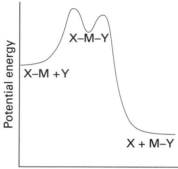

Figure 21.3 The typical form of the reaction profile of a reaction with an associative mechanism.

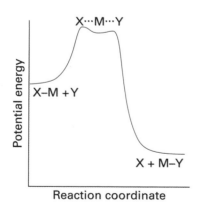

Figure 21.4 The typical form of the reaction profile of a reaction with an interchange mechanism.

Here ML_n (the metal atom and any spectator ligands) is a true intermediate that can, in principle, be detected (or even isolated). The typical form of the corresponding reaction profile is shown in Fig. 21.2.

■ **A brief illustration**. The substitution of hexacarbonyltungsten by phosphine takes place by dissociation of CO from the complex

$$W(CO)_6 \rightarrow W(CO)_5 + CO$$

followed by coordination of phosphine:

$$W(CO)_5 + PPh_3 \rightarrow W(CO)_5(PPh_3)$$

Under the conditions in which this reaction is usually performed in the laboratory, the intermediate $W(CO)_5$ is rapidly captured by the solvent, such as tetrahydrofuran, to form $[W(CO)_5(thf)]$. This complex in turn is converted to the phosphine product, presumably by a second dissociative process. ■

An **associative mechanism**, denoted *A*, involves a step in which an intermediate is formed with a higher coordination number than the original complex:

$$ML_nX + Y \rightarrow ML_nXY$$
$$ML_nXY \rightarrow ML_nY + X$$

Once again, the intermediate ML_nXY can, in principle at least, be detected. This mechanism plays a role in many reactions of square-planar Au(III), Pt(II), Pd(II), Ni(II), and Ir(I) d⁸ complexes. The typical form of the reaction profile is similar to that of the dissociative mechanism, and is shown in Fig. 21.3.

■ **A brief illustration**. The first step in the exchange of ¹⁴CN⁻ with the ligands in the square-planar complex $[Ni(CN)_4]^{2-}$ is the coordination of a ligand to the complex:

$$[Ni(CN)_4]^{2-} + {}^{14}CN^- \rightarrow [Ni(CN)_4({}^{14}CN)]^{3-}$$

A ligand is then discarded:

$$[Ni(CN)_4({}^{14}CN)]^{3-} \rightarrow [Ni(CN)_3({}^{14}CN)]^{2-} + CN^-$$

The radioactivity of carbon-14 provides a means of monitoring this reaction, and the intermediate $[Ni(CN)_5]^{3-}$ has been detected and isolated. ■

An **interchange mechanism**, denoted *I*, takes place in one step:

$$ML_nX + Y \rightarrow X \cdots ML_n \cdots Y \rightarrow ML_nY + X$$

The leaving and entering groups exchange in a single step by forming a transition state but not a true intermediate. The interchange mechanism is common for many reactions of six-coordinate complexes. The typical form of the reaction profile is shown in Fig. 21.4.

The distinction between the *A* and *I* mechanisms hinges on whether or not the intermediate persists long enough to be detectable. One type of evidence is the isolation of an intermediate in another related reaction or under different conditions. If an argument by extrapolation to the actual reaction conditions suggests that a moderately long-lived intermediate might exist during the reaction in question, then the *A* path is indicated. For example, the synthesis of the first trigonal-bipyramidal Pt(II) complex, $[Pt(SnCl_3)_5]^{3-}$, indicates that a five-coordinate platinum complex may be plausible in substitution reactions of square-planar Pt(II) ammine complexes. Similarly, the fact that $[Ni(CN)_5]^{3-}$ is observed spectroscopically in solution, and that it has been isolated in the crystalline state, provides support for the view that it is involved when CN⁻ exchanges with the square-planar tetracyanidonickelate(II) ion.

A second indication of the persistence of an intermediate is the observation of a stereochemical change, which implies that the intermediate has lived long enough to undergo rearrangement. *Cis* to *trans* isomerization is observed in the substitution reactions of certain square-planar phosphine Pt(II) complexes, which is in contrast to the retention of configuration usually observed. This difference implies that the trigonal-bipyramidal intermediate lives long enough for an exchange between the axial and equatorial ligand positions to occur.

Direct spectroscopic detection of the intermediate, and hence an indication of A rather than I, may be possible if a sufficient amount accumulates. Such direct evidence, however, requires an unusually stable intermediate with favourable spectroscopic characteristics.

(b) The rate-determining step

Key point: The rate-determining step is classified as associative or dissociative according to the dependence of its rate on the identity of the entering group.

Now we consider the rate-determining step of a reaction and the details of its formation. The step is called **associative** and denoted a if its rate depends strongly on the identity of the incoming group. Examples are found among reactions of the d^8 square-planar complexes of Pt(II), Pd(II), and Au(III), including

$$[PtCl(dien)]^+(aq) + I^-(aq) \rightarrow [PtI(dien)]^+(aq) + Cl^-(aq)$$

where dien is diethylenetriamine ($NH_2CH_2CH_2NHCH_2CH_2NH_2$). It is found, for instance, that use of I^- instead of Br^- increases the rate constant by an order of magnitude. Experimental observations on the substitution reactions of square-planar complexes support the view that the rate-determining step is associative.

The strong dependence of the rate-determining step on entering group Y indicates that the transition state must involve significant bonding to Y. A reaction with an associative mechanism (A) will be associatively activated (a) if the attachment of Y to the initial reactant ML_nX is the rate-determining step; such a reaction is designated A_a, and in this case the intermediate ML_nXY would not be detected. A reaction with a dissociative mechanism (D) is associatively activated (a) if the attachment of Y to the intermediate ML_n is the rate-determining step; such a reaction is designated D_a. Figure 21.5 shows the reaction profiles for associatively activated A and D mechanisms. For the reactions to proceed, it is necessary to have established a population of an encounter complex $X-M, Y$ in a pre-equilibrium step.

The rate-determining step is called **dissociative** and denoted d if its rate is largely independent of the identity of Y. This category includes some of the classic examples of ligand substitution in octahedral d-metal complexes, including

$$[Ni(OH_2)_6]^{2+}(aq) + NH_3(aq) \rightarrow [Ni(NH_3)(OH_2)_5]^{2+}(aq) + H_2O(l)$$

It is found that replacement of NH_3 by pyridine in this reaction changes the rate by at most a few per cent.

The weak dependence on Y of a dissociatively activated process indicates that the rate of formation of the transition state is determined largely by the rate at which the bond to the leaving group X can break. A reaction with an associative mechanism (A) will be dissociatively activated (d) provided the loss of X from the intermediate YML_nX is the rate-determining step; such a reaction is designated A_d. A reaction with a dissociative mechanism (D) is dissociatively activated (d) if the initial loss of X from the reactant ML_nX is the rate-determining step, such a reaction is designated D_d. In this case, the intermediate ML_n would not be detected. Figure 21.6 shows the reaction profiles for dissociatively activated A and D mechanisms.

A reaction that has an interchange mechanism (I) can be either associatively or dissociatively activated, and is designated either I_a or I_d, respectively. In an I_a mechanism, the rate of reaction depends on the rate at which the $M \cdots Y$ bond forms, whereas in an I_d reaction the rate of reaction depends on the rate at which the $M \cdots X$ bond breaks (Fig. 21.7).

The distinction between these possibilities may be summarized as follows, where ML_nX denotes the initial complex:

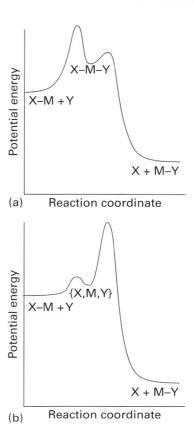

Figure 21.5 The typical form of the reaction profile of reactions with an associatively activated step: (a) associative mechanism, A_a; (b) a dissociative mechanism, D_a.

Mechanism:	A		I		D	
Activation:	a	d	a	d	a	d
Rate-determining step	Y attaching to ML_nX	Loss of X from YML_nX	Y attaching to ML_nX	Loss of X from YML_nX	Y attaching to ML_n	Loss of X from ML_nX
Detect intermediate?	no	ML_nXY detectable	no	no	ML_n detectable	no

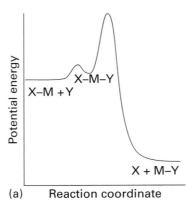

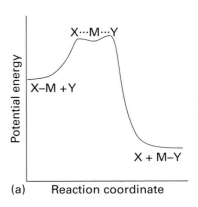

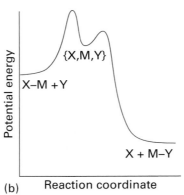

Figure 21.6 The typical form of the reaction profile of reactions with a dissociatively activated step: (a) an associative mechanism, A_d; (b) a dissociative mechanism, D_d.

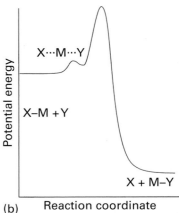

Figure 21.7 The typical form of the reaction profile of reactions with an interchange mechanism: (a) associatively activated, I_a; (b) dissociatively activated, I_d.

Ligand substitution in square-planar complexes

The mechanism of ligand exchange in square-planar Pt complexes has been studied extensively, largely because the reactions occur on a timescale that is very amenable to investigation. We might expect an associative mechanism of ligand exchange because square-planar complexes are sterically uncrowded—they can be considered as octahedral complexes with two ligands missing—but it is rarely that simple. The elucidation of the mechanism of the substitution of square-planar complexes is often complicated by the occurrence of alternative pathways. For instance, if a reaction such as

$$[PtCl(dien)]^+(aq) + I^-(aq) \rightarrow [PtI(dien)]^+(aq) + Cl^-(aq)$$

is first order in the complex and independent of the concentration of I^-, then the rate of reaction will be equal to $k_{r,1}[PtCl(dien)^+]$. However, if there is a pathway in which the rate law is first order in the complex *and* first order in the incoming group (that is, overall second order) then the rate would be given by $k_{r,2}[PtCl(dien)^+][I^-]$. If both reaction pathways occur at comparable rates, the rate law has the form

$$\text{rate} = (k_{r,1} + k_{r,2}[I^-])[PtCl(dien)^+] \tag{21.2}$$

A reaction like this is usually studied under the conditions $[I^-] \gg [\text{complex}]$ so that $[I^-]$ does not change significantly during the reaction. This simplifies the treatment of the data as $k_{r,1} + k_{r,2}[I^-]$ is effectively constant and the rate law is now pseudo-first order:

$$\text{rate} = k_{r,\text{obs}}[PtCl(dien)^+] \qquad k_{r,\text{obs}} = k_{r,1} + k_{r,2}[I^-] \tag{21.3}$$

A plot of the observed pseudo-first-order rate constant against $[I^-]$ gives $k_{r,2}$ as the slope and $k_{r,1}$ as the intercept.

In the following sections, we examine the factors that affect the second-order reaction and then consider the first-order process.

21.3 The nucleophilicity of the entering group

Key points: The nucleophilicity of an entering group is expressed in terms of the nucleophilicity parameter defined in terms of the substitution reactions of a specific square-planar platinum complex; the sensitivity of other platinum complexes to changes in the entering group is expressed in terms of the nucleophilic discrimination factor.

We start by considering the variation of the rate of the reaction as the entering group Y is varied. The reactivity of Y (for instance, I^- in the reaction above) can be expressed in terms of a **nucleophilicity parameter**, n_{Pt}:

$$n_{Pt} = \log \frac{k_{r,2}(Y)}{k_{r,2}^{o}} \qquad (21.4)$$

where $k_{r,2}(Y)$ is the second-order rate constant for the reaction

$$\textit{trans-}[PtCl_2(py)_2] + Y \rightarrow \textit{trans-}[PtClY(py)_2]^+ + Cl^-$$

and $k_{r,2}^{o}$ is the rate constant for the same reaction with the reference nucleophile methanol. The entering group is highly nucleophilic, or has a high nucleophilicity, if n_{Pt} is large.

Table 21.2 gives some values of n_{Pt}. One striking feature of the data is that, although the entering groups in the table are all quite simple, the rate constants span nearly nine orders of magnitude. Another feature is that the nucleophilicity of the entering group towards Pt appears to correlate with soft Lewis basicity (Section 4.12), with $Cl^- < I^-$, $O < S$, and $NH_3 < PR_3$.

The nucleophilicity parameter is defined in terms of the reaction rates of a specific platinum complex. When the complex itself is varied we find that the reaction rates show a range of different sensitivities towards changes in the entering group. To express this range of sensitivities we rearrange eqn 21.4 into

$$\log k_{r,2}(Y) = n_{Pt}(Y) + C \qquad (21.5)$$

where $C = \log k_{r,2}^{o}$. Now consider the analogous substitution reactions for the general complex $[PtL_3X]$:

$$[PtL_3X] + Y \rightarrow [PtL_3Y] + X$$

The relative rates of these reactions can be expressed in terms of the same nucleophilicity parameter n_{Pt} provided we replace eqn 21.5 by

$$\log k_{r,2}(Y) = S n_{Pt}(Y) + C \qquad (21.6)$$

The parameter S, which characterizes the sensitivity of the rate constant to the nucleophilicity parameter, is called the **nucleophilic discrimination factor**. We see that the straight line obtained by plotting $\log k_{r,2}(Y)$ against n_{Pt} for reactions of Y with $\textit{trans-}[PtCl_2(PEt_3)_2]$, the red circles in Fig. 21.8, is steeper than that for reactions with $\textit{cis-}[PtCl_2(en)]$, the blue squares in Fig. 21.8. Hence, S is larger for the former reaction, which indicates that the rate of the reaction is more sensitive to changes in the nucleophilicity of the entering group.

Some values of S are given in Table 21.3. Note that S is close to 1 in all cases, so all the complexes are quite sensitive to n_{Pt}. This sensitivity is what we expect for associatively activated reactions. Another feature to note is that larger values of S are found for complexes of platinum with softer base ligands.

EXAMPLE 21.1 Using the nucleophilicity parameter

The second-order rate constant for the reaction of I^- with $\textit{trans-}[Pt(CH_3)Cl(PEt_3)_2]$ in methanol at 30°C is 40 dm^3 mol^{-1} s^{-1}. The corresponding reaction with N_3^- has $k_2 = 7.0$ dm^3 mol^{-1} s^{-1}. Estimate S and C for the reaction given the n_{Pt} values of 5.42 and 3.58, respectively, for the two nucleophiles.

Table 21.2 A selection of n_{Pt} values for a range of nucleophiles

Nucleophile	Donor atom	n_{Pt}
CH_3OH	O	0
Cl^-	Cl	3.04
Br^-	Br	4.18
I^-	I	5.42
CN^-	C	7.14
SCN^-	S	5.75
N_3^-	N	3.58
C_6H_5SH	S	4.15
NH_3	N	3.07
$(C_6H_5)_3P$	P	8.93

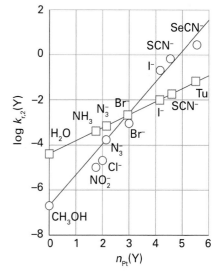

Figure 21.8 The slope of the straight line obtained by plotting $\log k_{r,2}(Y)$ against the nucleophilicity parameter $n_{Pt}(Y)$ for a series of ligands is a measure of the responsiveness of the complex to the nucleophilicity of the entering group (data from U. Belluso, L. Cattaini, F. Basolo, R.G. Pearson, and A. Turco; *J. Am. Chem. Soc.*, 1965, **87**, 241).

Table 21.3 Nucleophilic discrimination factors

	S
trans-[PtCl$_2$(PEt$_3$)$_2$]	1.43
trans-[PtCl$_2$(py)$_2$]	1.00
[PtCl$_2$(en)]	0.64
trans-[PtCl(dien)]$^+$	0.65

Answer To determine S and C, we need to use the two pieces of information to set up and solve two simultaneous equations based on eqn 21.6. Substituting the two values of n_{Pt} into eqn 21.6 gives

$$1.60 = 5.42S + C \quad \text{(for I}^-\text{)}$$

$$0.85 = 3.58S + C \quad \text{(for N}_3^-\text{)}$$

Solving these two simultaneous equations gives $S = 0.41$ and $C = -0.62$. The value of S is fairly small, showing that the discrimination of this complex among different nucleophiles is not great. This lack of sensitivity is related to the fairly large value of C, which corresponds to the rate constant being large and hence to the complex being reactive. It is commonly found that high reactivity correlates with low selectivity.

Self-test 21.1 Calculate the second-order rate constant for the reaction of the same complex with NO$_2^-$, for which $n_{Pt} = 3.22$.

21.4 The shape of the transition state

Careful studies of the variation of the reaction rates of square-planar complexes with changes in the composition of the reactant complex and the conditions of the reaction shed light on the general shape of the transition state. They also confirm that substitution almost invariably has an associative rate-determining stage; hence intermediates are rarely detected.

(a) The *trans* effect

Key point: A strong σ-donor ligand or π-acceptor ligand greatly accelerates substitution of a ligand that lies in the trans position.

The spectator ligands T that are *trans* to the leaving group in square-planar complexes influence the rate of substitution. This phenomenon is called the ***trans* effect**. It is generally accepted that the *trans* effect arises from two separate influences: one arising in the ground state and the other in the transition state itself.

The ***trans* influence** is the extent to which the ligand T weakens the bond *trans* to itself in the ground state of the complex. The *trans* influence correlates with the σ-donor ability of the ligand T because, broadly speaking, ligands *trans* to each other use the same orbitals on the metal for bonding. Thus if one ligand is a strong σ donor, then the ligand *trans* to it cannot donate electrons to the metal so well, and thus has a weaker interaction with the metal. The *trans* influence is assessed quantitatively by measuring bond lengths, stretching frequencies, and metal-to-ligand NMR coupling constants (Section 8.5). The **transition state effect** correlates with the π-acceptor ability of the ligand. Its origin is thought to be the increase in electron density on the metal atom arising due to the incoming ligand: any ligand that can accept this increased electron density will stabilize the transition state (**1**). The *trans* effect is the combination of both effects; it should be noted that the same factors contribute to a large ligand-field splitting. *Trans* effects are listed in Table 21.4 and follow the order:

For a T σ-donor: OH$^-$ < NH$_3$ < Cl$^-$ < Br$^-$ < CN$^-$, CH$_3^-$ < I$^-$ < SCN$^-$ < PR$_3$, H$^-$

For a T π-acceptor: Br$^-$ < I$^-$ < NCS$^-$ < NO$_2^-$ < CN$^-$ < CO, C$_2$H$_4$

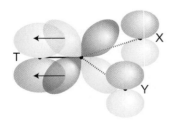

1

Table 21.4 The effect of the *trans* ligand in reactions of trans-[PtCl(PEt$_3$)$_2$L]

L	$k_{r,1}/\text{s}^{-1}$	$k_{r,2}/(\text{dm}^3\,\text{mol}^{-1}\,\text{s}^{-1})$
CH$_3^-$	1.7×10^{-4}	6.7×10^{-2}
C$_6$H$_5^-$	3.3×10^{-5}	1.6×10^{-2}
Cl$^-$	1.0×10^{-6}	4.0×10^{-4}
H$^-$	1.8×10^{-2}	4.2
PEt$_3$	1.7×10^{-2}	3.8

EXAMPLE 21.2 Using the *trans* effect synthetically

Use the *trans* effect series to suggest synthetic routes to *cis*- and *trans*-[PtCl$_2$(NH$_3$)$_2$] from [Pt(NH$_3$)$_4$]$^{2+}$ and [PtCl$_4$]$^{2-}$.

Answer If we consider the reaction of [Pt(NH$_3$)$_4$]$^{2+}$ with HCl we can see it leads to [PtCl(NH$_3$)$_3$]$^+$. Now, because the *trans* effect of Cl$^-$ is greater than that of NH$_3$, substitution reactions will occur preferentially *trans* to Cl$^-$, and further action of HCl gives *trans*-[PtCl$_2$(NH$_3$)$_2$]:

$$[\text{Pt(NH}_3)_4]^{2+} + \text{Cl}^- \rightarrow [\text{PtCl(NH}_3)_3]^+ \rightarrow \textit{trans-}[\text{PtCl}_2(\text{NH}_3)_2]$$

However, when the starting complex is $[PtCl_4]^{2-}$, reaction with NH_3 leads first to $[PtCl_3(NH_3)]^-$. A second step should substitute one of the two mutually *trans* Cl^- ligands with NH_3 to give *cis*-$[PtCl_2(NH_3)_2]$.

$$[PtCl_4]^{2-} + NH_3 \rightarrow [PtCl_3(NH_3)]^- \rightarrow \textit{cis-}[PtCl_2(NH_3)_2]$$

Self-test 21.2 Given the reactants PPh_3, NH_3, and $[PtCl_4]^{2-}$, propose efficient routes to both *cis*- and *trans*-$[PtCl_2(NH_3)(PPh_3)]$.

(b) Steric effects

Key point: Steric crowding at the reaction centre usually inhibits associative reactions and facilitates dissociative reactions.

Steric crowding at the reaction centre by bulky groups that can block the approach of attacking nucleophiles will inhibit associative reactions. The rate constants for the replacement of Cl^- by H_2O in *cis*-$[PtClL(PEt_3)_2]$ complexes at 25°C illustrate the point:

L =	pyridine	2-methylpyridine	2,6-dimethylpyridine
k_r/s^{-1}	8×10^{-2}	2.0×10^{-4}	1.0×10^{-6}

The methyl groups adjacent to the N donor atom greatly decrease the rate. In the 2-methylpyridine complex they block positions either above or below the plane. In the 2,6-dimethylpyridine complex they block positions both above and below the plane (**2**). Thus, along the series, the methyl groups increasingly hinder attack by H_2O.

The effect is smaller if L is *trans* to Cl^-. This difference is explained by the methyl groups then being further from the entering and leaving groups in the trigonal-bipyramidal transition state if the pyridine ligand is in the trigonal plane (**3**). Conversely, the decrease in coordination number that occurs in a dissociative reaction can relieve the steric overcrowding and thus increase the rate of the dissociative reaction.

(c) Stereochemistry

Key point: Substitution of a square-planar complex preserves the original geometry, which suggests a trigonal-pyramidal transition state.

Further insight into the nature of the transition state is obtained from the observation that substitution of a square-planar complex preserves the original geometry. That is, a *cis* complex gives a *cis* product and a *trans* complex gives a *trans* product. This behaviour is explained by the formation of an approximately trigonal-bipyramidal transition state with the entering, leaving, and *trans* groups in the trigonal plane (**4**).[1] Trigonal-bipyramidal intermediates of this type account for the relatively small influence that the two *cis* spectator ligands have on the rate of substitution, as their bonding orbitals will be largely unaffected by the course of the reaction.

The steric course of the reaction is shown in Fig. 21.9. We can expect a *cis* ligand to exchange places with the T ligand in the trigonal plane only if the intermediate lives long enough to be stereomobile. That is, it must be a long-lived associative (*A*) intermediate, with release of the ligand from the five-coordinate intermediate being the rate-determining step.

(d) Temperature and pressure dependence

Key point: Negative volumes and entropies of activation support the view that the rate-determining step of square-planar Pt(II) complexes is associative.

Another clue to the nature of the transition state comes from the entropies and volumes of activation for reactions of Pt(II) and Au(III) complexes (Table 21.5). The entropy of activation is obtained from the temperature dependence of the rate constant, and indicates the change in disorder (of reactants and solvent) when the transition state forms. Likewise, the volume of activation, which is obtained (with considerable difficulty) from the pressure dependence of the rate constant, is the change in volume that occurs on formation of the transition state.

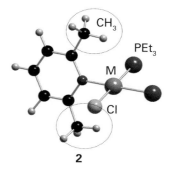

2

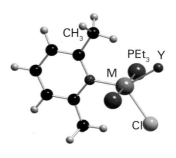

3

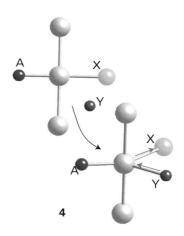

4

---

[1] Note that the stereochemistry is quite different from that of p-block central atoms, such as Si(IV) and P(V), where the leaving group departs from the more crowded axial position.

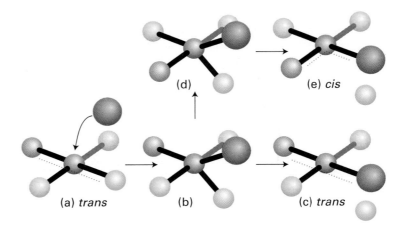

Figure 21.9 The stereochemistry of substitution in a square-planar complex. The normal path (resulting in retention) is from (a) to (c). However, if intermediate (b) is sufficiently long-lived, it can undergo pseudorotation to (d), which leads to isomer (e).

Table 21.5 Activation parameters for substitution in square-planar complexes (in methanol)

Reaction	$k_{r,1}$			$k_{r,2}$		
	$\Delta^{\ddagger}H$	$\Delta^{\ddagger}S$	$\Delta^{\ddagger}V$	$\Delta^{\ddagger}H$	$\Delta^{\ddagger}S$	$\Delta^{\ddagger}V$
trans-[PtCl(NO$_2$)(py)$_2$] + py				50	−100	−38
trans-[PtBrP$_2$(mes)] + SC(NH$_2$)$_2$	71	−84	−46	46	−138	−54
cis-[PtBrP$_2$(mes)] + I⁻	84	−59	−67	63	−121	−63
cis-[PtBrP$_2$(mes)] + SC(NH$_2$)$_2$	79	−71	−71	59	−121	−54
[AuCl(dien)]$^{2+}$ + Br⁻				54	−17	

[PtBrP$_2$(mes)] is [PtBr(PEt$_3$)$_2$(2,4,6-Me$_3$C$_6$H$_2$)].
Enthalpy in kilojoules per mole (kJ mol^{-1}); entropy in J K^{-1} mol^{-1} and volume in cm^3 mol^{-1}.

The limiting cases for the volume of activation in ligand substitution reactions correspond to the increase in molar volume of the outgoing ligand (for a dissociative reaction) and the decrease in molar volume of the incoming ligand (for an associative reaction).

The two striking aspects of the data in the table are the consistently strongly negative values of both quantities. The simplest explanation of the decrease in disorder and the decrease in volume is that the entering ligand is being incorporated into the transition state without release of the leaving group. That is, we can conclude that the rate-determining step is associative.

(e) The first-order pathway

Key point: The first-order contribution to the rate law is a pseudo-first-order process in which the solvent participates.

Having considered factors that affect the second-order pathway, we can now consider the first-order pathway for the substitution of square-planar complexes. The first issue we must address is the first-order pathway in the rate equation and decide whether $k_{r,1}$ in the rate law in eqn 21.2 and its generalization

$$\text{rate} = (k_{r,1} + k_{r,2}[Y])[PtL_4] \tag{21.7}$$

does indeed represent the operation of an entirely different reaction mechanism. It turns out that it does not, and $k_{r,1}$ represents an associative reaction involving the solvent. In this pathway the substitution of Cl⁻ by pyridine in methanol as solvent proceeds in two steps, with the first rate-determining:

$$[PtCl(dien)]^+ + CH_3OH \rightarrow [Pt(CH_3OH)(dien)]^{2+} + Cl^- \quad \text{(slow)}$$

$$[Pt(CH_3OH)(dien)]^{2+} + py \rightarrow [Pt(py)(dien)]^{2+} + CH_3OH \quad \text{(fast)}$$

The evidence for this two-step mechanism comes from a correlation of the rates of these reactions with the nucleophilicity parameters of the solvent molecules and the observation that reactions of entering groups with solvent complexes are rapid when compared to the step in which the solvent displaces a ligand. Thus, the substitution of ligands at a square-planar platinum complex results from two competing associative reactions.

Ligand substitution in octahedral complexes

Octahedral complexes occur for a wide variety of metals in a wide range of oxidation states and with a great diversity of bonding modes. We might therefore expect a wide variety of mechanisms of substitution; however, almost all octahedral complexes react by the interchange mechanism. The only real question is whether the rate-determining step is associative or dissociative. The analysis of rate laws for reactions that take place by such a mechanism helps to formulate the precise conditions for distinguishing these two possibilities and identifying the substitution as I_a (interchange with an associative rate-determining stage) or I_d (interchange with a dissociative rate-determining stage). The difference between the two classes of reaction hinges on whether the rate-determining step is the formation of the new $Y{\cdots}M$ bond or the breaking of the old $M{\cdots}X$ bond.

21.5 Rate laws and their interpretation

Rate laws provide an insight into the detailed mechanism of a reaction in the sense that any proposed mechanism must be consistent with the observed rate law. In the following section we see how the rate laws found experimentally for ligand substitution are interpreted.

(a) The Eigen–Wilkins mechanism

Key points: In the Eigen–Wilkins mechanism, an encounter complex is formed in a pre-equilibrium step and the encounter complex forms products in a subsequent rate-determining step.

As an example of a ligand substitution reaction, we consider

$$[Ni(OH_2)_6]^{2+} + NH_3 \rightarrow [Ni(NH_3)(OH_2)_5]^{2+} + H_2O$$

The first step in the **Eigen–Wilkins mechanism** is an encounter in which the complex ML_6, in this case $[Ni(OH_2)_6]^{2+}$, and the entering group Y, in this case NH_3, diffuse together and come into contact:

$$[Ni(OH_2)_6]^{2+} + NH_3 \rightarrow \{[Ni(OH_2)_6]^{2+}, NH_3\}$$

The two components of the encounter pair, the entity {A,B}, may also separate at a rate governed by their ability to migrate by diffusion through the solvent:

$$\{[Ni(OH_2)_6]^{2+}, NH_3\} \rightarrow [Ni(OH_2)_6]^{2+} + NH_3$$

Because in aqueous solution the lifetime of an encounter pair is approximately 1 ns, the formation of the pair can be treated as a pre-equilibrium in all reactions that take longer than a few nanoseconds. Consequently, we can express the concentrations in terms of a pre-equilibrium constant K_E:

$$ML_6 + Y \rightleftharpoons \{ML_6, Y\} \qquad K_E = \frac{[\{ML_6, Y\}]}{[ML_6][Y]}$$

The second step in the mechanism is the rate-determining reaction of the encounter complex to give products:

$$\{[Ni(OH_2)_6]^{2+}, NH_3\} \rightarrow [Ni(NH_3)(OH_2)_5]^{2+} + H_2O$$

and in general

$$\{ML_6, Y\} \rightarrow ML_5Y + L \qquad \text{rate} = k_r[\{ML_6, Y\}]$$

We cannot simply substitute $[\{ML_6, Y\}] = K_E[ML_6][Y]$ into this expression because the concentration of ML_6 must take into account the fact that some of it is present as the encounter pair; that is, $[M]_{tot} = [\{ML_6, Y\}] + [ML_6]$, the total concentration of the complex. It follows that

$$\text{rate} = \frac{k_r K_E [M]_{tot} [Y]}{1 + K_E[Y]} \tag{21.8}$$

It is rarely possible to conduct experiments over a range of concentrations wide enough to test eqn 21.8 exhaustively. However, at such low concentrations of the entering group that $K_E[Y] \ll 1$, the rate law reduces to

$$\text{rate} = k_{r,obs}[M]_{tot}[Y] \qquad k_{r,obs} = k_r K_E \qquad (21.9)$$

Because $k_{r,obs}$ can be measured and K_E can be either measured or estimated as we describe below, the rate constant k_r can be found from $k_{r,obs}/K_E$. The results for reactions of Ni(II) hexaaqua complexes with various nucleophiles are shown in Table 21.6. The very small variation in k_r indicates a model I_d reaction with very slight sensitivity to the nucleophilicity of the entering group.

When Y is a solvent molecule the encounter equilibrium is 'saturated' in the sense that, because the complex is always surrounded by solvent, a solvent molecule is always available to take the place of one that leaves the complex. In such a case $K_E[Y] \gg 1$ and $k_{r,obs} = k_r$. Thus, reactions with the solvent can be directly compared to reactions with other entering ligands without needing to estimate the value of K_E.

(b) The Fuoss–Eigen equation

Key point: The Fuoss–Eigen equation provides an estimate of the pre-equilibrium constant based on the strength of the Coulombic interaction between the reactants and their distance of closest approach.

The equilibrium constant K_E for the encounter pair can be estimated by using a simple equation proposed independently by R.M. Fuoss and M. Eigen. Both sought to take the complex size and charge into account, expecting larger, oppositely charged ions to meet more frequently than small ions of the same charge. Fuoss used an approach based on statistical thermodynamics and Eigen used one based on kinetics. Their result, which is called the **Fuoss–Eigen equation**, is

$$K_E = \tfrac{4}{3}\pi a^3 N_A e^{-V/kT} \qquad (21.10)$$

In this expression a is the distance of closest approach of ions of charge numbers z_1 and z_2 in a medium of permittivity ε, V is the Coulombic potential energy ($z_1 z_2 e^2/4\pi\varepsilon a$) of the ions at that distance, and N_A is Avogadro's constant. Although the value predicted by this equation depends strongly on the details of the charges and radii of the ions, typically it clearly favours the encounter if the reactants are large (so a is large) or oppositely charged (V negative).

> ■ **A brief illustration.** If one of the reactants is uncharged (as in the case of substitution by NH_3), then $V = 0$ and $K_E = \frac{4}{3}\pi a^3 N_A$. For an encounter distance of 200 pm for neutral species, we find
>
> $$K_E = \frac{4\pi}{3} \times \left(2.00 \times 10^{-10}\ \text{m}\right)^3 \times \left(6.022 \times 10^{23}\ \text{mol}^{-1}\right) = 2.02 \times 10^{-5}\ \text{m}^3\ \text{mol}^{-1}$$
>
> or $2.02 \times 10^{-2}\ \text{dm}^3\ \text{mol}^{-1}$. For two singly charged ions of opposite charge in water at 298 K (when $\varepsilon_r = 78$), other factors being equal, the value of K_E is increased by a factor of
>
> $$e^{-V/kT} = e^{e^2/4\pi\varepsilon\, akT} = 36 \quad ■$$

Table 21.6 Complex formation by the $[Ni(OH_2)_6]^{2+}$ ion

Ligand	$k_{r,obs}/(\text{dm}^3\ \text{mol}^{-1}\ \text{s}^{-1})$	$K_E/(\text{dm}^3\ \text{mol}^{-1})$	$(k_{r,obs}/K_E)/\text{s}^{-1}$
$CH_3CO_2^-$	1×10^5	3	3×10^4
F^-	8×10^5	1	8×10^3
HF	3×10^3	0.15	2×10^4
H_2O^*			3×10^3
NH_3	5×10^3	0.15	3×10^4
$[NH_2(CH_2)_2NH_3]^+$	4×10^2	0.02	2×10^4
SCN^-	6×10^3	1	6×10^3

* The solvent is always in encounter with the ion so that K_E is undefined and all rates are inherently first order.

21.6 The activation of octahedral complexes

Many studies of substitution in octahedral complexes support the view that the rate-determining step is dissociative, and we summarize these studies first. However, the reactions of octahedral complexes can acquire a distinct associative character in the case of large central ions (as in the 4d and 5d series) or where the d-electron population at the metal is low (the early members of the d block). More room for attack or lower π^* electron density appears to facilitate nucleophilic attack and hence permit association.

(a) Leaving-group effects

Key points: A large effect of the leaving group X is expected in I_d reactions; a linear relation is found between the logarithms of the rate constants and equilibrium constants.

We can expect the identity of the leaving group X to have a large effect in dissociatively activated reactions because their rates depend on the scission of the M$\cdots$X bond. When X is the only variable, as in the reaction

$$[CoX(NH_3)_5]^{2+} + H_2O \rightarrow [Co(NH_3)_5(OH_2)]^{3+} + X^-$$

it is found that the rate constant and equilibrium constant of the reaction are related by

$$\ln k_r = \ln K + c \tag{21.11}$$

This correlation is illustrated in Fig. 21.10. Because both logarithms are proportional to Gibbs energies ($\ln k_r$ is approximately proportional to the activation Gibbs energy, $\Delta^\ddagger G$, and $\ln K$ is proportional to the standard reaction Gibbs energy, $\Delta_r G^\ominus$), we can write the following **linear free energy relation (LFER)**:

$$\Delta^\ddagger G = p\Delta_r G^\ominus + b \tag{21.12}$$

with p and b constants (and $p \approx 1$).

The existence of an LFER of unit slope, as for the reaction of $[CoX(NH_3)_5]^{2+}$, shows that changing X has the same effect on $\Delta^\ddagger G$ for the conversion of Co$-$X to the transition state as it has on $\Delta_r G^\ominus$ for the complete elimination of X$^-$ (Fig. 21.11). This observation in turn suggests that in a reaction with an interchange mechanism and a dissociative rate-determining step (I_d), the leaving group (an anionic ligand) has already become a solvated ion in the transition state. An LFER with a slope of less than 1, indicating some associative character, is observed for the corresponding complexes of Rh(III). For Co(III), the reaction rates are in the order I$^-$>Br$^-$>Cl$^-$ whereas for Rh(III) they are reversed, and are in the order I$^-$ < Br$^-$ < Cl$^-$. This difference should be expected as the softer Rh(III) centre forms more stable complexes with I$^-$, compared with Br$^-$ and Cl$^-$, whereas the harder Co(III) centre forms more stable complexes with Cl$^-$.

(b) The effects of spectator ligands

Key points: In octahedral complexes, spectator ligands affect rates of substitution; the effect is related to the strength of the metal$-$ligand interaction, with stronger donor ligands increasing the reaction rate by stabilizing the transition state.

In Co(III), Cr(III), and related octahedral complexes, both *cis* and *trans* ligands affect rates of substitution in proportion to the strength of the bonds they form with the metal atom. For instance, hydrolysis reactions such as

$$[NiXL_5]^+ + H_2O \rightarrow [NiL_5(OH_2)] + X^-$$

are much faster when L is NH$_3$ than when it is H$_2$O. This difference can be explained on the grounds that, as NH$_3$ is a stronger σ donor than H$_2$O, it increases the electron density at the metal atom and hence facilitates the scission of the M$-$X bond and the formation of X$^-$. In the transition state, the stronger donor stabilizes the reduced coordination number.

(c) Steric effects

Key point: Steric crowding favours dissociative activation because formation of the transition state can relieve strain.

Steric effects on reactions with dissociative rate-determining steps can be illustrated by considering the rate of hydrolysis of the first Cl$^-$ ligand in two complexes of the type $[CoCl_2(bn)_2]^+$:

$$[CoCl_2(bn)_2]^+ + H_2O \rightarrow [CoCl(OH_2)(bn)_2]^{2+} + Cl^-$$

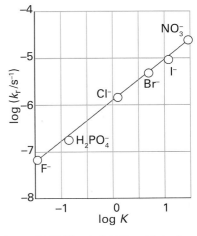

Figure 21.10 The straight line obtained when the logarithm of a rate constant is plotted against the logarithm of an equilibrium constant shows the existence of a linear free energy relation. The graph is for the reaction $[Co(NH_3)_5X]^{2+} + H_2O \rightarrow [Co(NH_3)_5(OH_2)]^{3+} + X^-$ with different leaving groups X.

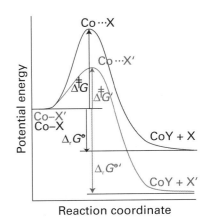

Figure 21.11 The existence of a linear free energy relation with unit slope shows that changing X has the same effect on $\Delta^\ddagger G$ for the conversion of M$-$X to the transition state as it has on $\Delta_r G^\ominus$ for the complete elimination of X$^-$. The reaction profile shows the effect of changing the leaving group from X to X$'$.

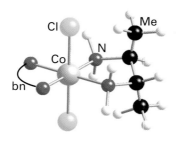

5 $[Co(Cl)_2(bn)_2]^+$

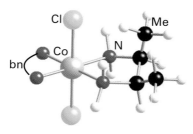

6 $[Co(Cl)_2(bn)_2]^+$

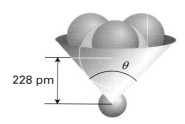

228 pm

θ

Figure 21.12 The determination of the ligand cone angles from space-filling molecular models of the ligand and an assumed M−P bond length of 228 pm.

The ligand bn is 2,3-butanediamine, and may be either chiral (**5**) or achiral (**6**). The important observation is that the complex formed with the chiral form of the ligand hydrolyses 30 times more slowly than the complex of the achiral form. The two ligands have very similar electronic effects, but the CH_3 groups are on the opposite sides of the chelate ring in (**5**) but adjacent and crowded in (**6**). The latter arrangement is more reactive because the strain is relieved in the dissociative transition state with its lowered coordination number. In general, steric crowding favours an I_d process because the five-coordinate transition state can relieve strain.

Quantitative treatments of the steric effects of ligands have been developed using molecular modelling computer software that takes into account van der Waals interactions. However, a more pictorial semiquantitative approach was introduced by C.A. Tolman. In this approach, the extent to which various ligands (especially phosphines) crowd each other is assessed by approximating the ligand by a cone with an angle determined from a space-filling model and, for phosphine ligands, an M−P bond length of 228 pm (Fig. 21.12 and Table 21.7).[2] The ligand CO is small in the sense of having a small cone angle; $P({}^tBu)_3$ is regarded as bulky because it has a large cone angle. Bulky ligands have considerable steric repulsion with each other when packed around a metal centre. They favour dissociative activation and inhibit associative activation.

As an illustration, the rate of the reaction of $[Ru(CO)_3(PR_3)(SiCl_3)_2]$ (**7**) with Y to give $[Ru(CO)_2Y(PR_3)(SiCl_3)_2]$ is independent of the identity of Y, which suggests that the rate-determining step is dissociative. Furthermore, it has been found that there is only a small variation in rate for the substituents Y with similar cone angles but significantly different values of pK_a. This observation supports the assignment of the rate changes to steric effects because changes in pK_a should correlate with changes in electron distributions in the ligands.

(d) Activation energetics

Key point: A significant loss of LFSE on going from the starting complex to the transition state results in nonlabile complexes.

One factor that has a strong bearing on the activation of complexes is the difference between ligand field stabilization energy (LFSE, Section 20.1) of the reacting complex and that of the transition state (the $LFSE^‡$). This difference is known as the **ligand field activation energy** (LFAE):

$$LFAE = LFSE^‡ - LFSE \qquad (21.13)$$

Table 21.8 gives calculated values of LFAE for the replacement of H_2O at a hexaaqua ion, assuming a square-pyramidal transition state (that is, a dissociatively activated reaction), and shows there is correlation between a large $\Delta^‡H$ and a large LFAE. Thus, we can begin to see why Ni^{2+} and V^{2+} complexes are not very labile: they have large activation energies which arise, in part, from a significant loss of LFSE on moving from an octahedral complex to a transition state.

Table 21.7 Tolman cone angles for various ligands

Ligand	$\theta/°$	Ligand	$\theta/°$
CH_3	90	$P(OC_6H_5)_3$	127
CO	95	PBu_3	130
Cl, Et	102	PEt_3	132
PF_3	104	$\eta^5\text{-}C_5H_5$ (Cp)	136
Br, Ph	105	PPh_3	145
I, $P(OCH_3)_3$	107	$\eta^5\text{-}C_5Me_5$ (Cp*)	165
PMe_3	118	$2,4\text{-}Me_2C_5H_3$	180
t-Butyl	126	$P(t\text{-}Bu)_3$	182

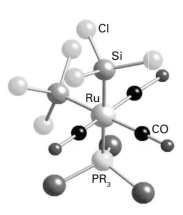

7 $[Ru(CO)_3(PR_3)(SiCl_3)_2]$

[2] Tolman's studies were on nickel complexes, so strictly speaking we are talking about a Ni−P distance of 228 pm.

Table 21.8 Activation parameters for the H_2O exchange reactions
$[M(OH_2)_6]^{2+} + H_2^{17}O \rightarrow [M(OH_2)_5(^{17}OH_2)]^{2+} + H_2O$

	$\Delta^{\ddagger}H/(kJ\ mol^{-1})$	$LFSE^*/\Delta_o$	$(LFSE)^{\ddagger}/\Delta_o^{\dagger}$	$LFAE/\Delta_o$	$\Delta^{\ddagger}V/(cm^3\ mol^{-1})$
$Ti^{2+}(d^2)$		0.8	0.91	−0.11	
$V^{2+}(d^3)$	68.6	1.2	1	0.2	−4.1
$Cr^{2+}(d^4, hs)$		0.6	0.91	−0.31	
$Mn^{2+}(d^5, hs)$	33.9	0	0	0	−5.4
$Fe^{2+}(d^6, hs)$	31.2	0.4	0.46	−0.06	+3.8
$Co^{2+}(d^7, hs)$	43.5	0.8	0.91	−0.11	+6.1
$Ni^{2+}(d^8)$	58.1	1.2	1	0.2	+7.2

* Octahedral. † Square pyramidal. hs, high spin.

(e) Associative activation

Key point: A negative volume of activation indicates association of the entering group into the transition state.

As we have seen, the activation volume reflects the change in compactness (including that of the surrounding solvent) when the transition state forms from the reactants. The last column in Table 21.8 gives $\Delta^{\ddagger}V$ for some H_2O ligand-exchange reactions. We see that $\Delta^{\ddagger}V$ becomes more positive, from $-4.1\ cm^3\ mol^{-1}$ for V^{2+} to $+7.2\ cm^3\ mol^{-1}$ for Ni^{2+}. Because a negative volume of activation can be interpreted as the result of shrinkage when an H_2O molecule becomes part of the transition state, we can infer that the activation has significant associative character.[3] The increase in $\Delta^{\ddagger}V$ follows the increase in the number of nonbonding d electrons from d^3 to d^8 across the 3d series. In the earlier part of the d block, associative reaction appears to be favoured by a low population of d electrons.

Negative volumes of activation are also observed in the 4d and 5d series, such as for Rh(III), and indicate an associative interaction of the entering group in the transition state of the reaction. Associative activation begins to dominate when the metal centre is more accessible to nucleophilic attack, either because it is large or has a low (nonbonding or π^*) d-electron population, and the mechanism shifts from I_d towards I_a. Table 21.9 shows some data for the formation of Br^-, Cl^-, and NCS^- complexes from $[Cr(NH_3)_5(OH_2)]^{3+}$ and $[Cr(OH_2)_6]^{3+}$. In contrast to the strong dependence of the hexaaqua complex, the penta-ammine complex shows only a weak dependence on the identity of the nucleophile. The two complexes probably mark a transition from I_d to I_a. In addition, the rate constants for the replacement of H_2O in $[Cr(OH_2)_6]^{3+}$ by Cl^-, Br^-, or NCS^- are smaller by a factor of about 10^4 than those for the analogous reactions of $[Cr(NH_3)_5(OH_2)]^{3+}$. This difference suggests that the NH_3 ligands, which are stronger σ donors than H_2O, promote dissociation of the sixth ligand more effectively. As we saw above, this behaviour is to be expected in dissociatively activated reactions.

Table 21.9 Kinetic parameters for anion attack on Cr(III)*

X	$k_r/$ $(10^{-8}\ dm^3\ mol^{-1}\ s^{-1})$	L = H_2O $\Delta^{\ddagger}H/$ $(kJ\ mol^{-1})$	$\Delta^{\ddagger}S/$ $(J\ K^{-1}\ mol^{-1})$	L = NH_3 $k_r/$ $(10^{-4}\ dm^3\ mol^{-1}\ s^{-1})$
Br^-	0.46	122	8	3.7
Cl^-	1.15	126	38	0.7
NCS^-	48.7	105	4	4.2

* Reaction is $[CrL_5(OH_2)]^{3+} + X^- \rightarrow [CrL_5X]^{2+} + H_2O$

[3] The limiting value for $\Delta^{\ddagger}V$ is approximately $\pm 18\ cm^3\ mol^{-1}$, the molar volume of water, with A reactions having negative and D reactions positive values.

EXAMPLE 21.3 Interpreting kinetic data in terms of a mechanism

The second-order rate constants for formation of $[VX(OH_2)_5]^+$ from $[V(OH_2)_6]^{2+}$ and X^- for $X^- = Cl^-$, NCS^-, and N_3^- are in the ratio 1:2:10. What do the data suggest about the rate-determining step for the substitution reaction?

Answer We need to consider the factors that might affect the rate of the reaction. All three ligands are singly charged anions of similar size, so we can expect the encounter equilibrium constants to be similar. Therefore, the second-order rate constants are proportional to first-order rate constants for substitution in the encounter complex. The second-order rate constant is equal to $K_E k_{r,2}$, where K_E is the pre-equilibrium constant and $k_{r,2}$ is the first-order rate constant for substitution of the encounter complex. The greater rate constants for NCS^- than for Cl^-, and especially the five-fold difference of NCS^- from its close structural analogue N_3^-, suggest some contribution from nucleophilic attack and an associative reaction. By contrast, there is no such systematic pattern for the same anions reacting with Ni(II), for which the reaction is believed to be dissociative.

Self-test 21.3 Use the data in Table 21.8 to estimate an appropriate value for K_E and calculate $k_{r,2}$ for the reactions of V(II) with Cl^- if the observed second-order rate constant is $1.2 \times 10^2 \ dm^3 \ mol^{-1} \ s^{-1}$.

21.7 Base hydrolysis

Key points: Octahedral substitution can be greatly accelerated by OH^- ions when ligands with acidic hydrogens are present as a result of the decrease in charge of the reactive species and the increased ability of the deprotonated ligand to stabilize the transition state.

Consider a substitution reaction in which the ligands possess acidic protons, such as

$$[CoCl(NH_3)_5]^{2+} + OH^- \rightarrow [Co(OH)(NH_3)_5]^{2+} + Cl^-$$

An extended series of studies has shown that whereas the rate law is overall second order, with rate = $k_r[CoCl(NH_3)_5^{2+}][OH^-]$, the mechanism is not a simple bimolecular attack by OH^- on the complex. For instance, whereas the replacement of Cl^- by OH^- is fast, the replacement of Cl^- by F^- is slow, even though F^- resembles OH^- more closely in terms of size and nucleophilicity. There is a considerable body of indirect evidence relating to the problem but one elegant experiment makes the essential point. This conclusive evidence comes from a study of the $^{18}O/^{16}O$ isotope distribution in the product $[Co(OH)(NH_3)_5]^{2+}$. It is known that the $^{18}O/^{16}O$ ratio differs between H_2O and OH^- at equilibrium, and this fact can be used to establish whether the incoming group is H_2O or OH^-. The $^{18}O/^{16}O$ isotope ratio in the cobalt product matches that for H_2O, not that for the OH^- ions, proving that it is an H_2O molecule that is the entering group.

The mechanism that takes these observations into account supposes that the role of OH^- is to act as a Brønsted base, not an entering group:

$$[CoCl(NH_3)_5]^{2+} + OH^- \rightarrow [CoCl(NH_2)(NH_3)_4]^+ + H_2O$$

$$[CoCl(NH_2)(NH_3)_4]^+ \rightarrow [Co(NH_2)(NH_3)_4]^{2+} + Cl^- \ (slow)$$

$$[Co(NH_2)(NH_3)_4]^{2+} + H_2O \rightarrow [Co(OH)(NH_3)_5]^{2+} \ (fast)$$

In the first step, an NH_3 ligand acts as a Brønsted acid, resulting in the formation of its conjugate base, the NH_2^- ion, as a ligand. Because the deprotonated form of the complex has a lower charge, it will be able to lose a Cl^- ion more readily than the protonated form, thus accelerating the reaction. In addition, according to the **conjugate base mechanism**, loss of a proton from an NH_3 ligand changes it from a pure σ-donor ligand to a strong σ and π donor (as NH_2^-) and so helps to stabilize the five-coordinate transition state, greatly accelerating the loss of the Cl^- ion (see the next *brief illustration*).

21.8 Stereochemistry

Key point: Reaction through a square-pyramidal intermediate results in retention of the original geometry but reaction through a trigonal-bipyramidal intermediate can lead to isomerization.

Classic examples of octahedral substitution stereochemistry are provided by Co(III) complexes. Table 21.10 shows some data for the hydrolysis of *cis*- and *trans*-$[CoAX(en)_2]^+$

Table 21.10 Stereochemical course of hydrolysis reactions of $[CoAX(en)_2]^+$

	A	X	Percentage *cis* in product
cis	OH^-	Cl^-	100
	Cl^-	Cl^-	100
	NCS^-	Cl^-	100
	Cl^-	Br^-	100
trans	NO_2^-	Cl^-	0
	NCS^-	Cl^-	50–70
	Cl^-	Cl^-	35
	OH^-	Cl^-	75

X is the leaving group.

(8) and (9), respectively, where X is the leaving group (either Cl⁻ or Br⁻) and A is OH⁻, NCS⁻, or Cl⁻. The stereochemical consequences of substitution of octahedral complexes are much more intricate than those of square-planar complexes. The *cis* complexes do not undergo isomerization when substitution occurs, whereas the *trans* forms show a tendency to isomerize in the order $A = NO_2^- < Cl^- < NCS^- < OH^-$.

The data can be understood in terms of an I_d mechanism and by recognizing that the five-coordinate metal centre in the transition state may resemble either of the two stable geometries for five-coordination, namely square-pyramidal or trigonal-bipyramidal. As can be seen from Fig. 21.13, reaction through the square-pyramidal complex results in retention of the original geometry but reaction through the trigonal-bipyramidal complex can lead to isomerization. The *cis* complex gives rise to a square-pyramidal intermediate but the *trans* isomer gives a trigonal-bipyramidal intermediate. For d metals, trigonal-bipyramidal complexes are favoured when the ligands in the equatorial positions are good π donors, and a good π-donor ligand *trans* to the leaving group Cl⁻ favours isomerization (10).

■ **A brief illustration.** Substitution of Co(III) complexes of the type [CoAX(en)₂]⁺ results in *trans* to *cis* isomerization, but only when the reaction is catalysed by a base. In a base hydrolysis reaction, one of the NH₂R groups of the en ligands loses a proton and becomes its conjugate base, :NHR⁻. The :NHR⁻ ligand group is a strong π-donor ligand and favours a trigonal bipyramid of the type shown in Fig. 21.13 and it may be attacked in the way shown there. If the direction of attack of the incoming ligands were random, we would expect 33 per cent *trans* and 67 per cent *cis* product. ■

21.9 Isomerization reactions

Key points: Isomerization of a complex can take place by mechanisms that involve substitution, bond cleavage, and reformation, or twisting.

Isomerization reactions are closely related to substitution reactions; indeed, a major pathway for isomerization is often via substitution. The square-planar Pt(II) and octahedral Co(III) complexes we have discussed can form five-coordinate trigonal-bipyramidal transition states. The interchange of the axial and equatorial ligands in a trigonal-bipyramidal complex can be pictured as occurring by a Berry pseudorotation through a square-pyramidal conformation (Section 7.4 and Fig. 21.14). As we have seen, when a trigonal-bipyramidal complex adds a ligand to produce a six-coordinate complex, a new direction of attack of the entering group can result in isomerization.

If a chelate ligand is present, isomerization can occur as a consequence of metal−ligand bond breaking, and substitution need not occur. An example is the exchange of the 'outer' CD₃ group with the 'inner' CH₃ group during the isomerization of a substituted tris(acetylacetonato)cobalt(III) complex, (11) → (12). An octahedral complex can also undergo isomerization by an intramolecular twist without loss of a ligand or breaking of a bond. There is evidence, for example, that racemization of [Ni(en)₃]²⁺ occurs by such an internal twist. Two possible paths are the **Bailar twist** and the **Ray−Dutt twist** (Fig. 21.15).

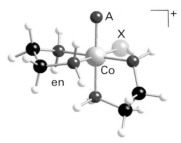

8 *cis*-[CoAX(en)₂]⁺

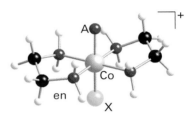

9 *trans*-[CoAX(en)₂]⁺

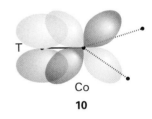

10

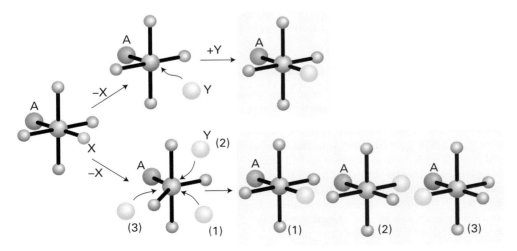

Figure 21.13 Reaction through a square-pyramidal complex (top path) results in retention of the original geometry, but reaction through a trigonal-bipyramidal complex (bottom path) can lead to isomerization.

Figure 21.14 The exchange of axial and equatorial ligands by a twist through a square-pyramidal conformation of the complex.

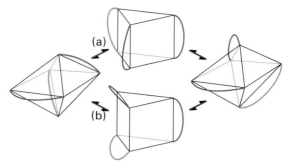

Figure 21.15 (a) The Bailar twist and (b) the Ray–Dutt twist by which an octahedral complex can undergo isomerization without losing a ligand or breaking a bond.

11

12

Redox reactions

As remarked in Chapter 5, redox reactions can occur by the direct transfer of electrons (as in some electrochemical cells and in many solution reactions) or by the transfer of atoms and ions, as in the transfer of O atoms in reactions of oxidoanions. Because redox reactions in solution involve both an oxidizing and a reducing agent, they are usually bimolecular in character. The exceptions are reactions in which one molecule has both oxidizing and reducing centres.

21.10 The classification of redox reactions

Key points: In an inner-sphere redox reaction a ligand is shared to form a transition state; in an outer-sphere redox reaction there is no bridging ligand between the reacting species.

In the 1950s, Henry Taube identified two mechanisms of redox reactions for metal complexes. One is the **inner-sphere mechanism**, which includes atom-transfer processes. In an inner-sphere mechanism, the coordination spheres of the reactants share a ligand transitorily and form a bridged transition state. The other is an **outer-sphere mechanism**, which includes many simple electron transfers. In an outer-sphere mechanism, the complexes come into contact without sharing a bridging ligand and the electron tunnels from one metal atom to the other.

The mechanisms of some redox reactions have been definitively assigned as inner- or outer-sphere. However, the mechanisms of a vast number of reactions are unknown because it is difficult to make unambiguous assignments when complexes are labile. Much of the study of well-defined examples is directed towards the identification of the parameters that differentiate the two paths with the aim of being able to making correct assignments in more difficult cases.

21.11 The inner-sphere mechanism

Key point: The rate-determining step of an inner-sphere redox reaction may be any one of the component processes, but a common one is electron transfer.

The inner-sphere mechanism was first confirmed for the reduction of the nonlabile complex $[CoCl(NH_3)_5]^{2+}$ by $Cr^{2+}(aq)$. The products of the reaction included both $Co^{2+}(aq)$ and $[CrCl(OH_2)_5]^{2+}$, and addition of $^{36}Cl^-$ to the solution did not lead to the incorporation of any of the isotope into the Cr(III) product. Furthermore, the reaction is much faster than reactions that remove Cl^- from nonlabile Co(III) or introduce Cl^- into the nonlabile $[Cr(OH_2)_6]^{3+}$ complex. These observations suggest that Cl has moved directly from the coordination sphere of one complex to that of the other during the reaction. The Cl^- attached to Co(III) can easily enter into the labile coordination sphere of $[Cr(OH_2)_6]^{2+}$ to produce a bridged intermediate (**13**).

Inner-sphere reactions, though involving more steps than outer-sphere reactions, can be fast. Figure 21.16 summarizes the steps necessary for such a reaction to occur.

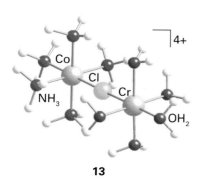

13

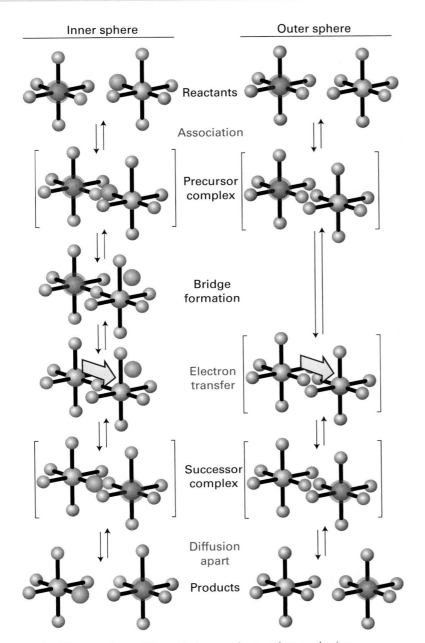

Figure 21.16 The different pathways followed by inner- and outer-sphere mechanisms.

The first two steps of an inner-sphere reaction are the formation of a precursor complex and the formation of the bridged binuclear intermediate. These two steps are identical to the first two steps in the Eigen−Wilkins mechanism (Section 21.5). The final steps are electron transfer through the bridging ligand to give the successor complex, followed by dissociation to give the products.

The rate-determining step of the overall reaction may be any one of these processes, but the most common one is the electron-transfer step. However, if both metal ions have a nonlabile electron configuration after electron transfer, then the break-up of the bridged complex is rate determining. An example is the reduction of $[RuCl(NH_3)_5]^{2+}$ by $[Cr(OH_2)_6]^{2+}$, in which the rate-determining step is the dissociation of the Cl-bridged complex $[Ru^{II}(NH_3)_5(\mu\text{-Cl})Cr^{III}(OH_2)_5]^{4+}$. Reactions in which the formation of the bridged complex is rate determining tend to have similar rate constants for a series of partners of a given species. For example, the oxidation of $V^{2+}(aq)$ has similar rate constants for a long series of Co(III) oxidants with different bridging ligands. The explanation is that the rate-determining step is the substitution of an H_2O molecule from the coordination sphere of V(II), which is quite slow (Table 21.8).

Table 21.11 Second-order rate constants for selected inner-sphere reactions with variable bridging ligands

Oxidant	Reductant	Bridging ligand	$k_r/(\text{dm}^3\,\text{mol}^{-1}\,\text{s}^{-1})$
$[\text{Co(NH}_3)_6]^{3+}$	$[\text{Cr(OH}_2)_6]^{2+}$		8×10^{-5}
$[\text{CoF(NH}_3)_5]^{2+}$	$[\text{Cr(OH}_2)_6]^{2+}$	F^-	2.5×10^5
$[\text{CoCl(NH}_3)_5]^{2+}$	$[\text{Cr(OH}_2)_6]^{2+}$	Cl^-	6.0×10^5
$[\text{CoI(NH}_3)_5]^{2+}$	$[\text{Cr(OH}_2)_6]^{2+}$	I^-	3.0×10^6
$[\text{Co(NCS)(NH}_3)_5]^{2+}$	$[\text{Cr(OH}_2)_6]^{2+}$	NCS^-	1.9×10^1
$[\text{Co(SCN)(NH}_3)_5]^{2+}$	$[\text{Cr(OH}_2)_6]^{2+}$	SCN^-	1.9×10^5
$[\text{Co(OH}_2)(\text{NH}_3)_5]^{2+}$	$[\text{Cr(OH}_2)_6]^{2+}$	H_2O	1.0×10^{-1}
$[\text{CrF(OH}_2)_5]^{2+}$	$[\text{Cr(OH}_2)_6]^{2+}$	F^-	7.4×10^{-3}

The numerous reactions in which electron transfer is rate determining do not display such simple regularities. Rates vary over a wide range as metal ions and bridging ligands are varied.[4] The data in Table 21.11 show some typical variations as bridging ligand, oxidizing metal, and reducing metal are changed.

All the reactions in Table 21.11 result in the change of oxidation number by ±1. Such reactions are still often called **one-equivalent processes**, the name reflecting the largely outmoded term 'chemical equivalent'. Similarly, reactions that result in the change of oxidation number by ±2 are often called **two-equivalent processes** and may resemble nucleophilic substitutions. This resemblance can be seen by considering the reaction

$$[\text{Pt}^{II}\text{Cl}_4]^{2-} + [\text{Pt}^{IV}\text{Cl}_6]^{2-} \rightarrow [\text{Pt}^{IV}\text{Cl}_6]^{2-} + [\text{Pt}^{II}\text{Cl}_4]^{2-}$$

which occurs through a Cl^- bridge (**14**). The reaction depends on the transfer of a Cl^- ion in the break-up of the successor complex.

There is no difficulty in assigning an inner-sphere mechanism when the reaction involves ligand transfer from an initially nonlabile reactant to a nonlabile product. With more labile complexes, inner-sphere reactions should always be suspected when ligand transfer occurs as well as electron transfer, and if good bridging groups such as Cl^-, Br^-, I^-, N_3^-, CN^-, SCN^-, pyrazine (**15**), 4,4'-bipyridine (**16**), and 4-dimethylaminopyridine (**17**) are present. Although all these ligands have lone pairs to form the bridge, this may not be an essential requirement. For instance, just as the carbon atom of a methyl group can act as a bridge between OH^- and I^- in the hydrolysis of iodomethane, so it can act as a bridge between Cr(II) and Co(III) in the reduction of methylcobalt species by Cr(II).

The oxidation of a metal centre by oxidoanions is also an example of an inner-sphere process. For example, in the oxidation of Mo(IV) by NO_3^- ions, an O atom of the nitrate ion binds to the Mo atom, facilitating the electron transfer from Mo to N, and then remains bound to the Mo(VI) product:

$$\text{Mo(IV)} + \text{NO}_3^- \rightarrow \text{Mo}-\text{O}-\text{NO}_2 \rightarrow \text{Mo}=\text{O} + \text{NO}_2^-$$

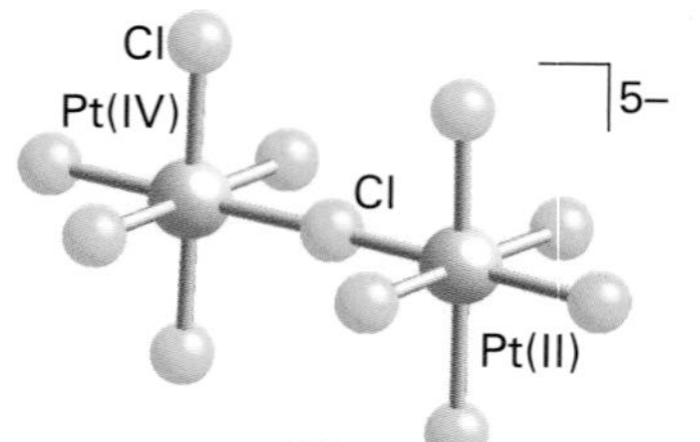

■ **A brief illustration.** The rate constant for the oxidation of the Ru(II) centre by the Co(III) centre in the bimetallic complex (**18**) is $1.0 \times 10^2\ \text{dm}^3\,\text{mol}^{-1}\,\text{s}^{-1}$, whereas the rate constant for complex (**19**) is $1.6 \times 10^{-2}\ \text{dm}^3\,\text{mol}^{-1}\,\text{s}^{-1}$. In both complexes there is a pyridine carboxylic acid group bridging the two metal centres. These groups are bound to both metal atoms and could facilitate an

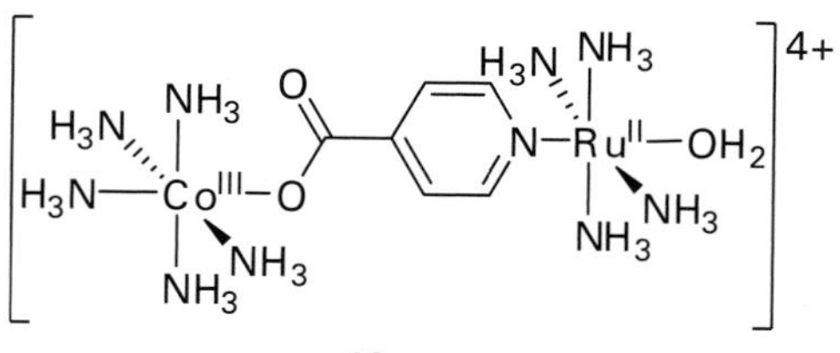

14

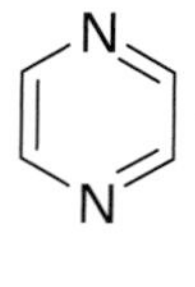

15 Pyrazine

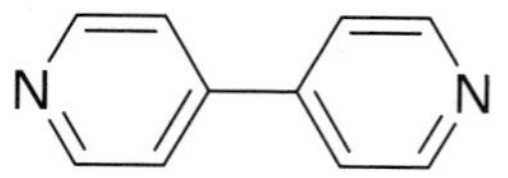

16 4,4'-Bipyridine

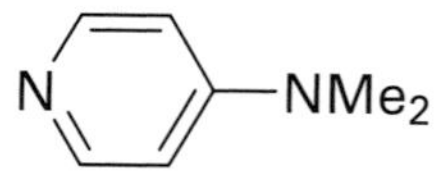

17 4-Dimethylaminopyridine

18

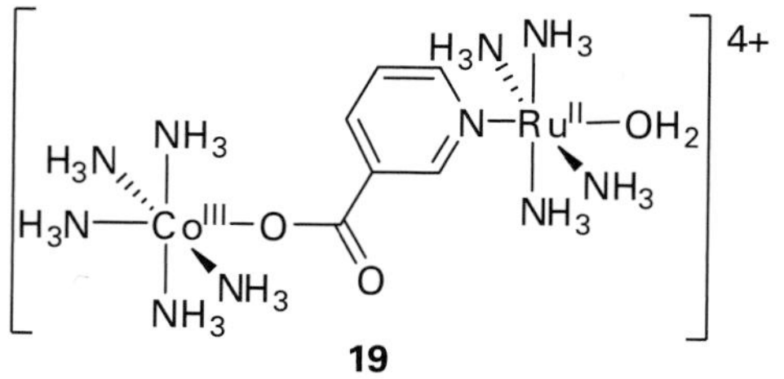

19

[4] Some bridged intermediates have been isolated with the electron clearly located on the bridge, but we shall not consider these here.

electron-transfer process through the bridge, suggesting an inner-sphere process. The fact that the rate constant changes between the two complexes, when the only substantive difference between them is in the substitution pattern of the pyridine ring, confirms that the bridge must be playing a role in the electron transfer process. ■

21.12 The outer-sphere mechanism

Key points: An outer-sphere redox reaction involves electron tunnelling between two reactants without any major disturbance of their covalent bonding or inner coordination spheres; the rate constant depends on the electronic and geometrical structures of the reacting species and on the Gibbs energy of reaction.

A conceptual starting point for understanding the principles of outer-sphere electron transfer is the deceptively simple reaction called **electron self-exchange**. A typical example is the exchange of an electron between $[Fe(OH_2)_6]^{3+}$ and $[Fe(OH_2)_6]^{2+}$ ions in water.

$$[Fe(OH_2)_6]^{3+} + [Fe(OH_2)_6]^{2+} \rightarrow [Fe(OH_2)_6]^{2+} + [Fe(OH_2)_6]^{3+}$$

Self-exchange reactions can be studied over a wide dynamic range with techniques ranging from isotopic labelling to NMR, with EPR being useful for even faster reactions. The rate constant of the Fe^{3+}/Fe^{2+} reaction is about 1 dm^3 mol^{-1} s^{-1} at 25°C.

To set up a scheme for the mechanism, we suppose that Fe^{3+} and Fe^{2+} come together to form a weak outer-sphere complex (Fig. 21.17). We need to consider, by assuming that the overlap of their respective acceptor and donor orbitals is sufficient to give a reasonable tunnelling probability,[5] how rapidly an electron transfers between the two metal ions. To explore this problem, we invoke the Franck–Condon principle introduced originally to account for the vibrational structure of electronic transitions in spectroscopy, which states that electronic transitions are so fast that they take place in a stationary nuclear framework. In Fig. 21.18 the nuclear motions associated with the 'reactant' Fe^{3+} and its 'conjugate product' Fe^{2+} are represented as displacements along a reaction coordinate. If $[Fe(OH_2)_6]^{3+}$ lies at its energy minimum, then an instantaneous electron transfer would give a compressed state of $[Fe(OH_2)_6]^{2+}$. Likewise, the removal of an electron from Fe^{2+} at its energy minimum would give an expanded state of $[Fe(OH_2)_6]^{3+}$. The only instant at which the electron can transfer within the precursor complex is when both $[Fe(OH_2)_6]^{3+}$ and $[Fe(OH_2)_6]^{2+}$ have achieved the same nuclear configuration by thermally induced fluctuations. That configuration corresponds to the point of intersection of the two curves, and the energy required to reach this position is the Gibbs energy of activation, $\Delta^\ddagger G$. If $[Fe(OH_2)_6]^{3+}$ and $[Fe(OH_2)_6]^{2+}$ differ in their nuclear configurations, $\Delta^\ddagger G$ is larger and electron exchange is slower. The difference in rates is expressed quantitatively by the **Marcus equation**

$$k_{ET} = v_N \kappa_e e^{-\Delta^\ddagger G/RT} \tag{21.14}$$

in which k_{ET} is the rate constant for electron transfer and $\Delta^\ddagger G$ is given by

$$\Delta^\ddagger G = \tfrac{1}{4}\lambda \left(1 + \frac{\Delta_r G^\ominus}{\lambda}\right)^2 \tag{21.15}$$

with $\Delta_r G^\ominus$ the standard reaction Gibbs energy (which is obtained from the difference in standard potentials of the redox partners) and λ the **reorganization energy**, the energy required to move the nuclei associated with the reactant to the positions they adopt in the product immediately before the transfer of the electron. This energy depends on the changes in metal–ligand bond lengths (the so-called *inner-sphere reorganization energy*) and alterations in solvent polarization, principally the orientation of the solvent molecules around the complex (the *outer-sphere reorganization energy*).

The pre-exponential factor in eqn 21.14 has two components, the **nuclear frequency factor** v_N and the **electronic factor** κ_e. The former is the frequency at which the two complexes, having already encountered each other in the solution, attain the transition state. The electronic factor gives the probability on a scale from 0 to 1 that an electron will transfer when the transition state is reached; its precise value depends on the extent of overlap of the donor and acceptor orbitals.

A small reorganization energy and a value of κ_e close to 1 corresponds to a redox couple capable of fast electron self-exchange. The first requirement is achieved if the transferred

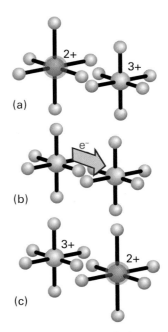

Figure 21.17 Electron transfer between two metal ions in a precursor complex is not productive until their coordination shells have reorganized to be of equal size. (a) Reactants; (b) reactant complexes having distorted into the same geometry; (c) products.

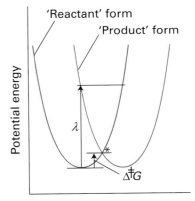

Figure 21.18 The potential energy curves for electron self-exchange. The nuclear motions of both the oxidized and reduced species (shown displaced along the reaction coordinate) and the surrounding solvent are represented by potential wells. Electron transfer to oxidized metal ion (left) occurs once fluctuations of its inner and outer coordination shell bring it to a point (denoted *) on its energy surface that coincides with the energy surface of its reduced state (right). This point is at the intersection of the two curves. The activation energy depends on the horizontal displacement of the two curves (representing the difference in sizes of the oxidized and reduced forms).

[5] Tunnelling refers to a process where, according to classical physics, the electrons do not have sufficient energy to overcome the barrier but penetrate into or through it.

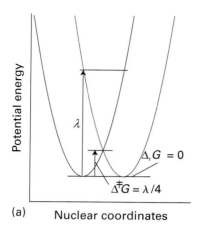

(a) Nuclear coordinates

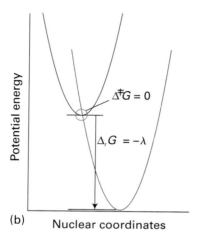

(b) Nuclear coordinates

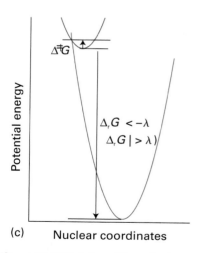

(c) Nuclear coordinates

Figure 21.19 Variation of the activation Gibbs energy ($\Delta^{\ddagger}G$) with reaction Gibbs energy ($\Delta_r G^{\ominus}$). (a) In a self-exchange reaction, $\Delta_r G^{\ominus} = 0$ and $\Delta^{\ddagger}G = \lambda/4$. (b) A reaction is 'activationless' when $\Delta_r G^{\ominus} = -\lambda$. (c) $\Delta^{\ddagger}G$ increases (the rate diminishes) as $\Delta_r G^{\ominus}$ becomes more negative beyond $\Delta_r G^{\ominus} = -\lambda$.

electron is removed from or added to a nonbonding orbital, as the change in metal–ligand bond length is then least. It is also likely if the metal ion is shielded from the solvent, in the sense of it being sterically difficult for solvent molecules to approach close to the metal ion, because the polarization of the solvent is normally a major component of the reorganization energy. Simple metal ions such as aqua species typically have λ well in excess of 1 eV, whereas buried redox centres in enzymes, which are very well shielded from the solvent, can have values as low as 0.25 eV. A value of κ_e close to 1 is achieved if there is good orbital overlap between the two components of the precursor complex.

For a self-exchange reaction, $\Delta_r G^{\ominus} = 0$ and therefore, from eqn 21.15, $\Delta^{\ddagger}G = \frac{1}{4}\lambda$ and the rate of electron transfer is controlled by the reorganization energy (Fig. 21.19a). To a considerable extent, the rates of self-exchange can be interpreted in terms of the types of orbitals involved in the transfer (Table 21.12). In the $[Cr(OH_2)_6]^{3+/2+}$ self-exchange reaction, an electron is transferred between antibonding σ^* orbitals, and the consequent extensive change in metal–ligand bond lengths results in a large inner-sphere reorganization energy and therefore a slow reaction. The $[Co(NH_3)_6]^{3+/2+}$ couple has an even greater reorganization energy because two electrons are moved into the σ^* orbital as rearrangement occurs and the reaction is even slower. With the other hexaaqua and hexaammine complexes in the table, the electron is transferred between weakly antibonding or nonbonding π orbitals, the inner-sphere reorganization is less extensive, and the reactions are faster. The bulky, hydrophobic chelating ligand bipyridyl acts as a solvent shield, thus decreasing the outer-sphere reorganization energy.

Bipyridyl and other π-acceptor ligands allow electrons in an orbital with π symmetry on the metal ion to delocalize on to the ligand. This delocalization effectively lowers the reorganization energy when the electron is transferred between π orbitals, as occurs with Fe and Ru, where the electron transfer is between t_{2g} orbitals (which, as explained in Section 20.2, can participate in π bonding), but not with Ni, where the electron transfer is between e_g orbitals. Delocalization can also increase the electronic factor.

Self-exchange reactions are helpful for pointing out the concepts that are involved in electron transfer, but chemically useful redox reactions occur between different species and involve net electron transfer. For the latter reactions, $\Delta_r G^{\ominus}$ is nonzero and contributes to the rate through eqns 21.14 and 21.15. Provided $|\Delta_r G^{\ominus}| \ll |\lambda|$ eqn 21.15 becomes

$$\Delta^{\dagger\dagger}G = \tfrac{1}{4}\lambda\left(1 + \frac{\Delta_r G}{\lambda}\right)^2 \approx \tfrac{1}{4}\lambda\left(1 + \frac{2\Delta_r G}{\lambda}\right) = \tfrac{1}{4}\left(\lambda + 2\Delta_r G\right)$$

and then according to eqn 21.14

$$k_{ET} \approx \nu_N \kappa_e e^{-(\lambda + 2\Delta_r G^{\ominus})/4RT}$$

Because $\lambda > 0$ and $\Delta_r G^{\ominus} < 0$ for thermodynamically feasible reactions, provided $|\Delta_r G^{\ominus}| \ll |\lambda|$ the rate constant increases exponentially as $\Delta_r G^{\ominus}$ becomes increasingly favourable (that is, more negative). However, as $|\Delta_r G^{\ominus}|$ becomes comparable to $|\lambda|$, this equation breaks down and we see that the reaction rate peaks before declining as $|\Delta_r G^{\ominus}| > |\lambda|$.

Equation 21.15 shows that $\Delta^{\ddagger}G = 0$ when $\Delta_r G^{\ominus} = -\lambda$. That is, the reaction becomes 'activationless' when the standard reaction Gibbs energy and the reorganization energy

Table 21.12 Correlations between rate constants for electron self-exchange reactions

Reaction	Electron configuration	Δd/pm	$k_{11}/$ (dm^3 mol^{-1} s^{-1})
$[Cr(OH_2)_6]^{3+/2+}$	$t_{2g}^3/t_{2g}^3 e_g^1$	20	1×10^{-5}
$[V(OH_2)_6]^{3+/2+}$	t_{2g}^2/t_{2g}^3	13	1×10^{-5}
$[Fe(OH_2)_6]^{3+/2+}$	$t_{2g}^3 e_g^2/t_{2g}^4 e_g^2$	13	1.1
$[Ru(OH_2)_6]^{3+/2+}$	t_{2g}^5/t_{2g}^6	9	20
$[Ru(NH_3)_6]^{3+/2+}$	t_{2g}^5/t_{2g}^6	4	9.6×10^2
$[Co(NH_3)_6]^{3+/2+}$	$t_{2g}^6/t_{2g}^5 e_g^2$	22	2×10^{-8}
$[Fe(bpy)_3]^{3+/2+}$	t_{2g}^5/t_{2g}^6	0	3×10^8
$[Ru(bpy)_3]^{3+/2+}$	t_{2g}^5/t_{2g}^6	0	4×10^8
$[Ni(bpy)_3]^{3+/2+}$	$t_{2g}^6 e_g^1/t_{2g}^6 e_g^2$	12	1.5×10^3

cancel (Fig. 21.19b). The activation energy now *increases* as $\Delta_r G^{\oplus}$ becomes more negative and the reaction rate decreases. This slowing of the reaction as the standard Gibbs energy of the reaction becomes more exergonic is called **inverted behaviour** (Fig. 21.19c). Inverted behaviour has important consequences, a notable one relating to the long-range electron transfer involved in photosynthesis. Photosystems are complex proteins containing light-excitable pigments such as chlorophyll and a chain of redox centres having low reorganization energies. In this chain, one highly exergonic recombination of the photoelectron with oxidized chlorophyll is sufficiently retarded (to 30 ns) to allow the electron to escape (in 200 ps) and proceed down the photosynthetic electron-transport chain, ultimately to produce reduced carbon compounds (Section 27.10). The theoretical dependence of the reaction rate of a reaction on the standard Gibbs energy is plotted in Fig. 21.20, and Fig. 21.21 shows the observed variation of reaction rate with $\Delta_r G^{\oplus}$ for the iridium complex (20). The results plotted in Fig. 21.21 represent the first unambiguous experimental observation of an inverted region for a synthetic complex.

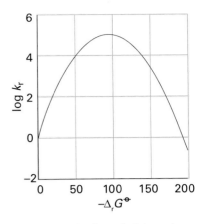

Figure 21.20 The theoretical dependence of the $\log_{10}$ of the reaction rate (in arbitrary units) on $\Delta_r G^{\oplus}$ for a redox reaction with $\lambda = 1.0$ eV (96.5 kJ mol^{-1}).

20

The Marcus equation can be used to predict the rate constants for outer-sphere electron transfer reactions between different species. Consider the electron-transfer reaction between an oxidant Ox_1 and reductant Red_2:

$$Ox_1 + Red_2 \rightarrow Red_1 + Ox_2$$

If we suppose that the reorganization energy for this reaction is the average of the values for the two self-exchange processes, we can write $\lambda_{12} = \frac{1}{2}(\lambda_{11} + \lambda_{22})$ and then manipulation of eqns 21.14 and 21.15 gives the **Marcus cross-relation**

$$k_{12} = (k_{11}k_{22}K_{12}f_{12})^{1/2} \tag{21.16}$$

in which k_{12} is the rate constant, K_{12} is the equilibrium constant obtained from $\Delta_r G^{\oplus}$ and k_{11} and k_{22} are the respective self-exchange rate constants for the two reaction partners. For reactions between simple ions in solution, when the standard Gibbs reaction energy is not too large, an LFER of the kind expressed by eqn 21.12 exists between $\Delta^{\ddagger}G$ and $\Delta_r G^{\oplus}$ and f_{12} can normally be set to 1. However, for reactions that are highly favourable thermodynamically (that is, $\Delta_r G^{\oplus}$ large and negative) the LFER breaks down. The term f_{12} takes into consideration the nonlinearity of the relationship between $\Delta^{\ddagger}G$ and $\Delta_r G^{\oplus}$ and is given by

$$\log f_{12} = \frac{(\log K_{12})^2}{4\log(k_{11}k_{22}/Z^2)} \tag{21.17}$$

where Z is the constant of proportionality between the encounter density in solution (in moles of encounters per cubic decimetre per second) and the molar concentrations of the reactants; it is often taken to be 10^{11} mol^{-1} dm^3 s^{-1}.

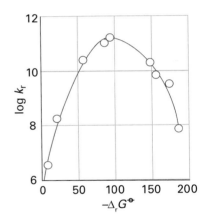

Figure 21.21 Plot of $\log_{10} k_r$ against $-\Delta_r G^{\oplus}$ for the iridium complex (18) in acetonitrile solution at room temperature. The rates are for electron transfer from the photoexcited Ir_2 unit to the outlying pyridinium functionalities. (Data from L.S. Fox, M. Kozik, J.R. Winkler, and H.B. Gray, *Science* 1990, **247**, 1069.)

■ **A brief illustration.** The rate constants for

$$[Co(bpy)_3]^{2+} + [Co(bpy)_3]^{3+} \xrightarrow{k_{11}} [Co(bpy)_3]^{3+} + [Co(bpy)_3]^{2+}$$
$$[Co(terpy)_2]^{2+} + [Co(terpy)_2]^{3+} \xrightarrow{k_{22}} [Co(terpy)_2]^{3+} + [Co(terpy)_2]^{2+}$$

(where bpy is bipyridyl and terpy is tripyridyl) are $k_{11} = 9.0$ dm^3 mol^{-1} s^{-1} and $k_{22} = 48$ dm^3 mol^{-1} s^{-1}, and $K_{12} = 3.57$. Then for the outer-sphere reduction of $[Co(bpy)_3]^{3+}$ by $[Co(terpy)_2]^{2+}$, eqn 21.16 with $f_{12} = 1$ (as noted above) gives

$$k_{12} = (9.0 \times 48 \times 3.57)^{1/2} \text{ dm}^3 \text{ mol}^{-1} \text{ s}^{-1} = 39 \text{ dm}^3 \text{ mol}^{-1} \text{ s}^{-1}$$

This result compares reasonably well with the experimental value, which is 64 dm^3 mol^{-1} s^{-1}. ■

Photochemical reactions

The absorption of a photon of ultraviolet radiation or visible light increases the energy of a complex by between 170 and 600 kJ mol^{-1}. Because these energies are larger than typical activation energies it should not be surprising that new reaction channels are opened. However, when the high energy of a photon is used to provide the energy of the primary forward reaction, the back reaction is almost always very favourable, and much of the design of efficient photochemical systems lies in trying to avoid the back reaction.

21.13 Prompt and delayed reactions

Key point: Reactions of electronically excited species are classified as prompt or delayed.

In some cases, the excited state formed after absorption of a photon dissociates almost immediately after it is formed. Examples include formation of the pentacarbonyl intermediates that initiate ligand substitution in metal carbonyl compounds:

$$Cr(CO)_6 \xrightarrow{h\nu} Cr(CO)_5 + CO$$

and the scission of Co−Cl bonds:

$$[Co^{III}Cl(NH_3)_5]^{2+} \xrightarrow{h\nu \, (\lambda = 350 \, nm)} [Co^{II}(NH_3)_5]^{2+} + Cl\cdot$$

Both processes occur in less than 10 ps and hence are called **prompt reactions**.

In the second reaction, the **quantum yield**, the amount of reaction per mole of photons absorbed, increases as the wavelength of the radiation is decreased (and the photon energy correspondingly increased, $E_{photon} = hc/\lambda$). The energy in excess of the bond energy is available to the newly formed fragments and increases the probability that they will escape from each other through the solution before they have an opportunity to recombine.

Some excited states have long lifetimes. They may be regarded as energetic isomers of the ground state that can participate in **delayed reactions**. The excited state of $[Ru^{II}(bpy)_3]^{2+}$ created by photon absorption in the metal-to-ligand charge-transfer band (Section 20.5) may be regarded as a Ru(III) cation complexed to a radical anion of the ligand. Its redox reactions can be explained by adding the excitation energy (expressed as a potential by using $-FE = \Delta_r G$ and equating $\Delta_r G$ to the molar excitation energy) to the ground-state reduction potential (Fig. 21.22).

21.14 d−d and charge-transfer reactions

Key point: A useful first approximation is to associate photosubstitution and photoisomerization with d−d transitions and photoredox reactions with charge-transfer transitions, but the rule is not absolute.

There are two main types of spectroscopically observable electron promotion in d-metal complexes, namely d−d transitions and charge-transfer transitions (Sections 20.4 and 20.5). A d−d transition corresponds to the essentially angular redistribution of electrons within a d shell. In octahedral complexes, this redistribution often corresponds to the occupation of M−L antibonding e_g orbitals. An example is the $^4T_{1g} \leftarrow {}^4A_{2g}$ ($t_{2g}^2 e_g^1 \leftarrow t_{2g}^3$) transition in $[Cr(NH_3)_6]^{3+}$. The occupation of the antibonding e_g orbital results in a quantum yield close to 1 (specifically 0.6) for the photosubstitution

$$[Cr(NH_3)_6]^{3+} + H_2O \xrightarrow{h\nu} [Cr(NH_3)_5(OH_2)]^{3+} + NH_3$$

This is a prompt reaction, as it occurs in less than 5 ps.

Charge-transfer transitions correspond to the *radial* redistribution of electron density. They correspond to the promotion of electrons into predominantly ligand orbitals if the transition is metal-to-ligand or into orbitals of predominantly metal character if the transition is ligand-to-metal. The former process corresponds to oxidation of the metal centre and the latter to its reduction. These excitations commonly initiate photoredox reactions of the kind already mentioned in connection with Co(III) and Ru(II).

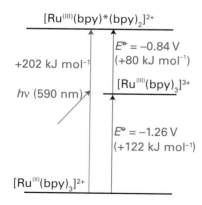

$[Ru^{(III)}(bpy)*(bpy)_2]^{2+}$

$E^{\ominus} = -0.84 \, V$
(+80 kJ mol^{-1})

+202 kJ mol^{-1}

$[Ru^{(III)}(bpy)_3]^{3+}$

$h\nu$ (590 nm)

$E^{\ominus} = -1.26 \, V$
(+122 kJ mol^{-1})

$[Ru^{(III)}(bpy)_3]^{2+}$

Figure 21.22 The photoexcitation of $[Ru^{II}(bpy)_3]^{2+}$ can be treated as if the excited state is a Ru(III) cation complexed to a radical anion of the ligand.

Although a useful first approximation is to associate photosubstitution and photo-isomerization with d−d transitions and photoredox with charge-transfer transitions, the rule is not absolute. For example, it is not uncommon for a charge-transfer transition to result in photosubstitution by an indirect path:

$$[Co^{III}Cl(NH_3)_5]^{2+} + H_2O \xrightarrow{h\nu} [Co^{II}(NH_3)_5(OH_2)]^{2+} + Cl\cdot$$

$$[Co^{II}(NH_3)_5(OH_2)]^{2+} + Cl\cdot \rightarrow [Co^{III}(NH_3)_5(OH_2)]^{3+} + Cl^-$$

In this case, the aqua complex formed after the homolytic fission of the Co−Cl bond is reoxidized by the Cl atom. The net result leaves the Co substituted. Conversely, some excited states show no differences in substitutional reactivity compared with the ground state: the long-lived excited ^2E state of $[Cr(bpy)_3]^{3+}$ results from a pure d−d transition and its lifetime of several microseconds allows the excess energy to enhance its redox reactions. The standard potential (+1.3 V), calculated by adding the excitation energy to the ground-state value, accounts for its function as a good oxidizing agent, in which it undergoes reduction to $[Cr(bpy)_3]^{2+}$.

21.15 Transitions in metal−metal bonded systems

Key points: Population of a metal−metal antibonding orbital can sometimes initiate photodissociation; such excited states have been shown to initiate multielectron redox photochemistry.

We might expect the $\delta^* \leftarrow \delta$ transition in metal−metal bonded systems to initiate photo-dissociation as it results in the population of an antibonding orbital of the metal−metal system. It is more interesting that such excited states have also been shown to initiate multi-electron redox photochemistry.

One of the best characterized systems is the dinuclear platinum complex $[Pt_2(\mu\text{-}P_2O_5H_2)_4]^{4-}$, called informally 'PtPOP' (**21**). There is no metal−metal bonding in the ground state of this Pt(II)−Pt(II) d^8−d^8 species. The HOMO−LUMO pattern indicates that excitation populates a bonding orbital between the two metal atoms (Fig. 21.23). The lowest-lying excited state has a lifetime of 9 μs and is a powerful reducing agent, reacting by both electron and halogen-atom transfer. The most interesting oxidation products are Pt(III)—Pt(III), which contain X^- ligands (where X is halogen or pseudohalogen) at both ends and a metal−metal single bond. Irradiation in the presence of $(Bu)_3SnH$ gives a dihy-drido product that can eliminate H_2.

Irradiation of the quadruply bonded dinuclear cluster $[Mo_2(O_2P(OC_6H_5)_2)_4]$ (**22**) at 500 nm in the presence of $ClCH_2CH_2Cl$ results in production of ethene and the addition of two Cl atoms to the two Mo atoms, with a two-electron oxidation. The reaction proceeds in one-electron steps, and requires a complex with the metal atoms shielded by sterically crowding ligands. If smaller ligands are present, the reaction that occurs instead is a photo-chemical oxidative addition of the organic molecule.

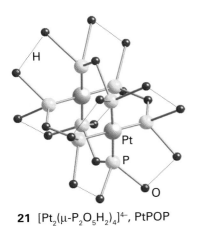

21 $[Pt_2(\mu\text{-}P_2O_5H_2)_4]^{4-}$, PtPOP

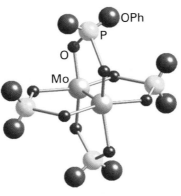

22 $[Mo_2(O_2P(OPh)_2)_4]$

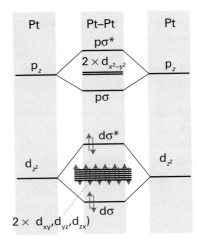

Figure 21.23 The dinuclear complex $[Pt_2(\mu\text{-}P_2O_5H_2)_4]^{4-}$ consists of two face-to-face square-planar complexes held together by a bridging pyrophosphito ligand. The metal p_z and d_{z^2} orbitals interact along the Pt−Pt axis. The other p and d orbitals are considered to be nonbonding. Photoexcitation results in an electron in the antibonding σ^* orbital moving into the bonding σ orbital.

FURTHER READING

G.J. Leigh and N. Winterbottom (ed.), *Modern coordination chemistry: the legacy of Joseph Chatt*. Royal Society of Chemistry, Cambridge (2002). A readable historical discussion of this area.

M.L. Tobe and J. Burgess, *Inorganic reaction mechanisms*. Longman, Harlow (1999).

R.G. Wilkins, *Kinetics and mechanism of reactions of transition metal complexes*. VCH, Weinheim (1991).

A special issue of *Coordination Chemistry Reviews* has been dedicated to the work of Henry Taube. See *Coord. Chem. Rev.*, 2005, **249**.

Two readable discussions of redox processes are to be found in Taube's 1983 Nobel Prize lecture reprinted in *Science*, 1984, **226**, 1028 and in Marcus's 1992 Nobel Prize lecture published in *Nobel Lectures: Chemistry 1991–1995*, World Scientific, Singapore (1997).

EXERCISES

21.1 The rate constants for the formation of $[CoX(NH_3)_5]^{2+}$ from $[Co(NH_3)_5OH_2]^{3+}$ for X = Cl^-, Br^-, N_3^-, and SCN^- differ by no more than a factor of two. What is the mechanism of the substitution?

21.2 If a substitution process is associative, why may it be difficult to characterize an aqua ion as labile or inert?

21.3 The reactions of $Ni(CO)_4$ in which phosphines or phosphites replace CO to give $Ni(CO)_3L$ all occur at the same rate regardless of which phosphine or phosphite is being used. Is the reaction *d* or *a*?

21.4 Write the rate law for formation of $[MnX(OH_2)_5]^+$ from the aqua ion and X^-. How would you undertake to determine if the reaction is *d* or *a*?

21.5 Octahedral complexes of metal centres with high oxidation numbers or of d metals of the second and third series are less labile than those of low oxidation number and d metals of the first series of the block. Account for this observation on the basis of a dissociative rate-determining step.

21.6 A Pt(II) complex of tetramethyldiethylenetriamine is attacked by Cl^- 10^5 times less rapidly than the diethylenetriamine analogue. Explain this observation in terms of an associative rate-determining step.

21.7 The rate of loss of chlorobenzene, PhCl, from $[W(CO)_4L(PhCl)]$ increases with increase in the cone angle of L. What does this observation suggest about the mechanism?

21.8 The pressure dependence of the replacement of chlorobenzene (PhCl) by piperidine in the complex $[W(CO)_4(PPh_3)(PhCl)]$ has been studied. The volume of activation is found to be $+11.3$ cm^3 mol^{-1}. What does this value suggest about the mechanism?

21.9 Does the fact that $[Ni(CN)_5]^{3-}$ can be isolated help to explain why substitution reactions of $[Ni(CN)_4]^{2-}$ are very rapid?

21.10 Reactions of $[Pt(Ph)_2(SMe_2)_2]$ with the bidentate ligand 1,10-phenanthroline (phen) give $[Pt(Ph)_2phen]$. There is a kinetic pathway with activation parameters $\Delta^\ddagger H = +101$ kJ mol^{-1} and $\Delta^\ddagger S = +42$ J K^{-1} mol^{-1}. Propose a mechanism.

21.11 Design two-step syntheses of *cis*- and *trans*-$[PtCl_2(NO_2)(NH_3)]^-$ starting from $[PtCl_4]^{2-}$.

21.12 How does each of the following modifications affect the rate of a square-planar complex substitution reaction? (a) Changing a *trans* ligand from H to Cl. (b) Changing the leaving group from Cl to I. (c) Adding a bulky substituent to a *cis* ligand. (d) Increasing the positive charge on the complex.

21.13 The rate of attack on Co(III) by an entering group Y is nearly independent of Y with the spectacular exception of the rapid reaction

with OH^-. Explain the anomaly. What is the implication of your explanation for the behaviour of a complex lacking Brønsted acidity on the ligands?

21.14 Predict the products of the following reactions:

(a) $[Pt(PR_3)_4]^{2+} + 2 Cl^-$

(b) $[PtCl_4]^{2-} + 2 PR_3$

(c) *cis*-$[Pt(NH_3)_2(py)_2]^{2+} + 2 Cl^-$

21.15 Put in order of increasing rate of substitution by H_2O the complexes (a) $[Co(NH_3)_6]^{3+}$, (b) $[Rh(NH_3)_6]^{3+}$, (c) $[Ir(NH_3)_6]^{3+}$, (d) $[Mn(OH_2)_6]^{2+}$, (e) $[Ni(OH_2)_6]^{2+}$.

21.16 State the effect on the rate of dissociatively activated reactions of Rh(III) complexes of (a) an increase in the overall charge on the complex, (b) changing the leaving group from NO_3^- to Cl^-, (c) changing the entering group from Cl^- to I^-, (d) changing the *cis* ligands from NH_3 to H_2O.

21.17 Write out the inner- and outer-sphere pathways for reduction of azidopentaamminecobalt(III) ion with $V^{2+}(aq)$. What experimental data might be used to distinguish between the two pathways?

21.18 The compound $[Fe(SCN)(OH_2)_5]^{2+}$ can be detected in the reaction of $[Co(NCS)(NH_3)_5]^{2+}$ with $Fe^{2+}(aq)$ to give $Fe^{3+}(aq)$ and $Co^{2+}(aq)$. What does this observation suggest about the mechanism?

21.19 Calculate the rate constants for electron transfer in the oxidation of $[V(OH_2)_6]^{2+}$ $(E^\ominus(V^{3+}/V^{2+}) = -0.255$ V) and the oxidants (a) $[Ru(NH_3)_6]^{3+}$ $(E^\ominus(Ru^{3+}/Ru^{2+}) = +0.07$ V), (b) $[Co(NH_3)_6]^{3+}$ $(E^\ominus(Co^{3+}/Co^{2+}) = +0.10$ V). Comment on the relative sizes of the rate constants.

21.20 Calculate the rate constants for electron transfer in the oxidation of $[Cr(OH_2)_6]^{2+}$ $(E^\ominus(Cr^{3+}/Cr^{2+}) = -0.41$ V) and each of the oxidants $[Ru(NH_3)_6]^{3+}$ $(E^\ominus(Ru^{3+}/Ru^{2+}) = +0.07$ V), $[Fe(OH_2)_6]^{3+}$ $(E^\ominus(Fe^{3+}/Fe^{2+}) = +0.77$ V) and $[Ru(bpy)_3]^{3+}$ $(E^\ominus(Ru^{3+}/Ru^{2+}) = +1.26$ V). Comment on the relative sizes of the rate constants.

21.21 The photochemical substitution of $[W(CO)_5(py)]$ (py = pyridine) with triphenylphosphine gives $W(CO)_5(P(C_6H_5)_3)$. In the presence of excess phosphine, the quantum yield is approximately 0.4. A flash photolysis study reveals a spectrum that can be assigned to the intermediate $W(CO)_5$. What product and quantum yield do you predict for substitution of $[W(CO)_5(py)]$ in the presence of excess triethylamine? Is this reaction expected to be initiated from the ligand field or MLCT excited state of the complex?

21.22 From the spectrum of $[CrCl(NH_3)_5]^{2+}$ shown in Fig. 20.32, propose a wavelength for photoinitiation of reduction of Cr(III) to Cr(II) accompanied by oxidation of a ligand.

PROBLEMS

21.1 Given the following mechanism for the formation of a chelate complex

$$[Ni(OH_2)_6]^{2+} + L-L \rightleftharpoons [Ni(OH_2)_6]^{2+},L-L \qquad K_E, \text{rapid}$$

$$[Ni(OH_2)_6]^{2+},L-L \rightleftharpoons [Ni(OH_2)_5L-L]^{2+} + H_2O \qquad k_a, k_a'$$

$$[Ni(OH_2)_5L-L]^{2+} \rightleftharpoons [Ni(OH_2)_4L-L]^{2+} + H_2O \qquad k_b, k_b'$$

derive the rate law for the formation of the chelate. Discuss the step that is different from that for two monodentate ligands. The formation of chelates with strongly bound ligands occurs at the rate of formation of the analogous monodentate complex but the formation of chelates of weakly bound ligands is often significantly slower. Assuming an I_d mechanism, explain this observation. (See R.G. Wilkins, *Acc. Chem. Res.*, 1970, **3**, 408.)

21.2 The complex $[PtH(PEt_3)_3]^+$ was studied in deuterated acetone in the presence of excess PEt_3. In the absence of excess ligand the ^{1}H-NMR spectrum in the hydride region exhibits a doublet of triplets. As excess PEt_3 ligand is added the hydride signal begins to change, the line shape depending on the ligand concentration. Suggest a mechanism to account for the effects of excess PEt_3.

21.3 Solutions of $[PtH_2(PMe_3)_2]$ exist as a mixture of *cis* and *trans* isomers. Addition of excess PMe_3 led to formation of $[PtH_2(PMe_3)_3]$ at a concentration that could be detected using NMR. This complex exchanged phosphine ligands rapidly with the *trans* isomer but not the *cis*. Propose a pathway. What are the implications for the *trans* effect of H versus PMe_3? (See D.L. Packett and W.G. Trogler, *Inorg. Chem.*, 1988, **27**, 1768.)

21.4 Figure 21.24 (which is based on J.B. Goddard and F. Basolo, *Inorg. Chem.*, 1968, **7**, 936) shows the observed first-order rate constants for the reaction of $[PdBrL]^+$ with various Y^- to give $[PdYL]^+$, where L is $Et_2NCH_2CH_2NHCH_2CH_2NEt_2$. Note the large slope for $S_2O_3^{2-}$ and zero slopes for $Y^- = N_3^-$, I^-, NO_2^-, and SCN^-. Propose a mechanism.

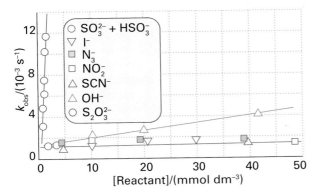

Figure 21.24 The data required for Problem 21.4.

21.5 The substitution reactions of the bridged dinuclear Rh(II) complex $[Rh_2(\mu\text{-}O_2CCH_3)_4XY]$ (**23**) have been studied by M.A.S. Aquino and D.H. Macartney (*Inorg. Chem.*, 1987, **26**, 2696). Reaction rates show little dependence on the choice of the entering group. The table below shows the dependence on the leaving group, X, and the ligand on the opposite Rh, (*trans*) Y, at 298K. What conclusions can you draw concerning the mechanism? Note that the complex has a d^7 configuration at each Rh and a single Rh–Rh bond.

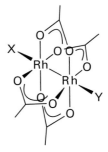

23

X	Y	$k_r/(\text{dm}^3 \text{ mol}^{-1} \text{ s}^{-1})$
H_2O	H_2O	$10^5 - 10^7$
CH_3OH	CH_3OH	2×10^6
CH_3CN	CH_3CN	1.1×10^5
PPh_3	PPh_3	1.5×10^5
CH_3CN	PR_3	$10^7 - 10^9$
PR_3	CH_3CN	$10^{-1} - 10^2$
N-donor	H_2O	$10^2 - 10^3$

21.6 The activation enthalpy for the reduction of *cis*-$[CoCl_2(en)_2]^+$ by $Cr^{2+}(aq)$ is -24 kJ mol^{-1}. Explain the negative value. (See R.C. Patel, R.E. Ball, J.F. Endicott, and R.G. Hughes, *Inorg. Chem.*, 1970, **9**, 23.)

21.7 The rate of reduction of $[Co(NH_3)_5(OH_2)]^{3+}$ by Cr(II) is seven orders of magnitude slower than reduction of its conjugate base, $[Co(NH_3)_5(OH)]^{2+}$, by Cr(II). For the corresponding reductions with $[Ru(NH_3)_6]^{2+}$, the two differ by less than a factor of 10. What do these observations suggest about mechanisms?

21.8 Consider the complexes (**18**) and (**19**) discussed in the *illustration* on page 526. Think about the potential routes of the electron transfer in the two complexes and suggest why there is such a difference in the rates of electron transfer between the two complexes.

21.9 Calculate the rate constants for outer-sphere reactions from the following data. Compare your results to the measured values in the last column.

Reaction	$k_{11}/(\text{dm}^3 \text{ mol}^{-1} \text{ s}^{-1})$	$k_{22}/(\text{dm}^3 \text{ mol}^{-1} \text{ s}^{-1})$	$E^{\ominus}/V$	$k_{obs}/(\text{dm}^3 \text{ mol}^{-1} \text{ s}^{-1})$
$Cr^{2+} + Fe^{2+}$	2×10^{-5}	4.0	$+1.18$	2.3×10^3
$[W(CN)_8]^{4-} + Ce(IV)$	$>4 \times 10^4$	4.4	$+0.90$	$>10^8$
$[Fe(CN)_6]^{4-} + MnO_4^-$	7.4×10^2	3×10^3	$+0.20$	1.7×10^5
$[Fe(phen)_3]^{2+} + Ce(IV)$	$>3 \times 10^7$	4.4	$+0.36$	1.4×10^5

21.10 In the presence of catalytic amounts of $[Pt(P_2O_5H_2)_4]^{4-}$ (**21**) and light, 2-propanol produces H_2 and acetone (E.L. Harley, A.E. Stiegman, A. Vlcek, Jr., and H.B. Gray, *J. Am. Chem. Soc.*, 1987, **109**, 5233; D.C. Smith and H.B. Gray, *Coord. Chem. Rev.*, 1990, **100**, 169). (a) Give the equation for the overall reaction. (b) Give a plausible molecular orbital scheme for the metal–metal bonding in this tetragonal-prismatic complex and indicate the nature of the excited state that is thought to be responsible for the photochemistry. (c) Indicate the metal complex intermediates and the evidence for their existence.

22 d-Metal organometallic chemistry

Organometallic chemistry is the chemistry of compounds containing metal−carbon bonds. Much of the basic organometallic chemistry of the s- and p-block metals was understood by the early part of the twentieth century and has been discussed in Chapters 11−16. The organometallic chemistry of the d and f blocks has been developed much more recently. Since the mid-1950s this field has grown into a thriving area that spans new types of reactions, unusual structures, and practical applications in organic synthesis and industrial catalysis. We discuss the organometallic chemistry of the d and f blocks separately, covering d metals in this chapter and f metals in the next. The widespread use of organometallic compounds in synthesis is covered in Chapter 26 (on catalysis).

A few d-block organometallic compounds were synthesized and partially characterized in the nineteenth century. The first of them (**1**), an ethene complex of platinum(II), was prepared by W.C. Zeise in 1827, with the first metal carbonyls, $[PtCl_2(CO)_2]$ and $[PtCl_2(CO)]_2$, being reported by P. Schützenberger in 1868. The next major discovery was tetracarbonylnickel (**2**), which was synthesized by L. Mond, C. Langer, and F. Quinke in 1890. Beginning in the 1930s, W. Hieber synthesized a wide variety of metal carbonyl cluster compounds, many of which are anionic, including $[Fe_4(CO)_{13}]^{2-}$ (**3**). It was clear from this work that metal carbonyl chemistry was potentially a very rich field. However, as the structures of these and other d- and f-block organometallic compounds are difficult or impossible to deduce by chemical means alone, fundamental advances had to await the development of X-ray diffraction for precise structural data on solid samples and of IR and NMR spectroscopy for structural information in solution. The discovery of the remarkably stable organometallic compound ferrocene, $Fe(C_5H_5)_2$ (**4**), occurred at a time (in 1951) when these techniques were becoming widely available. The 'sandwich' structure of ferrocene was soon correctly inferred from its IR spectrum and then determined in detail by X-ray crystallography.

The stability, structure, and bonding of ferrocene defied the classical Lewis description and therefore captured the imagination of chemists. This puzzle in turn set off a train of synthesizing, characterizing, and theorizing that led to the rapid development of d-block organometallic chemistry. Two highly productive research workers in the formative stage of the subject, Ernst-Otto Fischer in Munich and Geoffrey Wilkinson in London, were awarded the Nobel Prize in 1973 for their contributions. Similarly, f-block organometallic chemistry blossomed soon after the discovery in the late 1970s that the pentamethylcyclopentadienyl ligand, C_5Me_5, forms stable f-block compounds (**5**).

We adhere to the convention that an organometallic compound contains at least one metal–carbon (M−C) bond. Thus, compounds (**1**) to (**5**) clearly qualify as organometallic, whereas a complex such as $[Co(en)_3]^{3+}$, which contains carbon but has no M−C bonds, does not. Cyano complexes, such as hexacyanidoferrate(II) ions, do have M−C bonds, but as their properties are more akin to those of conventional coordination complexes they are generally not considered as organometallic. In contrast, complexes of the isoelectronic ligand CO are considered to be organometallic. The justification for this somewhat arbitrary distinction is that many metal carbonyls are significantly different from coordination complexes both chemically and physically.

In general, the distinctions between the two classes of compounds are clear: coordination complexes normally are charged, with variable d-electron count, and are soluble in water; organometallic compounds are often neutral, with fixed d-electron count, and are

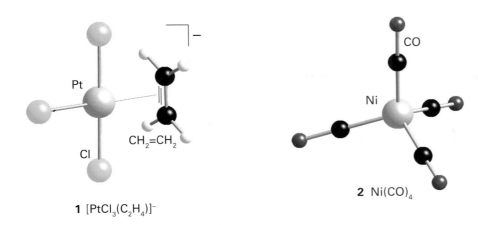

1 [PtCl$_3$(C$_2$H$_4$)]$^-$

2 Ni(CO)$_4$

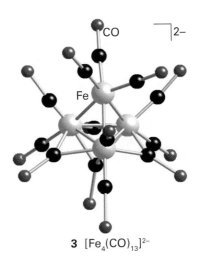

3 [Fe$_4$(CO)$_{13}$]$^{2-}$

soluble in organic solvents such as tetrahydrofuran. Most organometallic compounds have properties that are much closer to organic compounds than inorganic salts, with many of them having low melting points (some are liquid at room temperature).

Bonding

Although there are many organometallic compounds of the s and p blocks, the bonding in these compounds is often relatively simple and normally adequately described solely by σ bonds. The d metals, in contrast, form a large number of organometallic compounds with many different bonding modes. For instance, to describe fully the bonding of a cyclopentadienyl group to iron in ferrocene (and in general to any d metal), we need to invoke σ, π, and δ bonds.

Unlike coordination compounds, d-metal organometallic compounds normally have relatively few stable electron configurations and often have a total of 16 or 18 valence electrons around the metal atom. This restriction to a limited number of electronic configurations is due to the strength of the π (and δ, where appropriate) bonding interactions between the metal atom and the carbon-containing ligands.

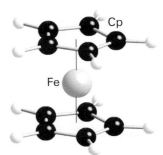

4 FeCp$_2$, Cp = C$_5$H$_5^-$

22.1 Stable electron configurations

We start by examining the bonding patterns so that we can appreciate the importance of π bonds and understand the origin of the restriction of the d-metal organometallic compounds to certain electron configurations.

(a) 18-Electron compounds

Key points: Six σ-bonding interactions are possible in an octahedral complex and, when π-acceptor ligands are present, bonding combinations can be made with the three orbitals of the t$_{2g}$ set, leading to nine bonding MOs, and space for a total of 18 electrons.

In the 1920s, N.V. Sidgwick recognized that the metal atom in a simple metal carbonyl, such as Ni(CO)$_4$, has the same valence electron count (18) as the noble gas that terminates the long period to which the metal belongs. Sidgwick coined the term 'inert gas rule' for this indication of stability, but it is now usually referred to as the **18-electron rule**.[1] It becomes readily apparent, however, that the 18-electron rule is not as uniformly obeyed for d-block organometallic compounds as the octet rule is obeyed for compounds of Period 2 elements, and we need to look more closely at the bonding to establish the reasons for the stability of both the compounds that have the 18-electron configurations and those that do not.

Figure 22.1 shows the energy levels that arise when a strong-field ligand such as carbon monoxide bonds to a d-metal atom (Section 20.2). Carbon monoxide is a strong-field ligand, even though it is a poor σ donor, because it can use its empty π* orbitals to act as

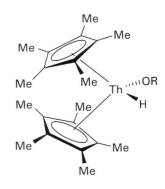

5 [Th(Cp*)$_2$H(OR)], Cp* = C$_5$(CH$_3$)$_5^-$

---

[1]The 18-electron rule is sometimes referred to as the *effective atomic number* or *EAN* rule.

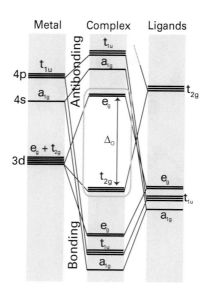

Figure 22.1 The energy levels of the d orbitals of an octahedral complex with strong-field ligands.

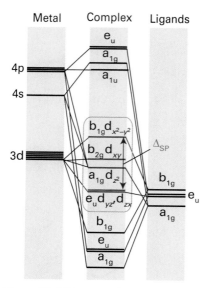

Figure 22.2 The energy levels of the molecular orbitals of a square-planar complex with strong-field ligands. The eight lowest MOs correspond to bonding interactions, with the higher MOs corresponding to antibonding interactions; the MOs are labelled with the d orbitals from which they are derived.

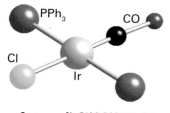

6 *trans*-[IrCl(CO)(PPh$_3$)$_2$]

a good π acceptor. In this picture of the bonding, the t_{2g} orbitals of the metal atom are no longer nonbonding, as they would be in the absence of π interactions, but are bonding. The energy level diagram shows six bonding MOs that result from the ligand–metal σ interactions, and three bonding MOs that result from π interactions. Thus up to 18 electrons can be accommodated in the nine bonding MOs. Compounds that have this configuration are remarkably stable, for instance the 18-electron Cr(CO)$_6$ is a colourless air-stable compound. An indication of the size of the HOMO-LUMO gap (Δ_O) can be gained from a consideration of its lack of colour, which results from a lack of any electronic transitions in the visible region of the spectrum, that is Δ_O is so large that such transitions are shifted to the UV.

The only way to accommodate more than 18 valence electrons in an octahedral complex with strong-field ligands is to use an antibonding orbital. As a result, such complexes are unstable, being particularly prone to electron loss and acting as reducing agents. Compounds with fewer than 18 electrons will not necessarily be very unstable, but such complexes will find it energetically favourable to acquire extra electrons by reaction and so populate their bonding MOs fully. As we shall see later, compounds with fewer than 18 electrons often occur as intermediates in reaction pathways.

The bonding characteristic of the carbonyl ligand is replicated with other ligands, which are often poor σ donors but good π acceptors. Hence, octahedral organometallic compounds are most stable when they have a total of 18 valence electrons around their central metal ion.

Similar arguments can be used to rationalize the stability of the 18-electron configuration for other geometries, such as tetrahedral and trigonal bipyramidal, although in practice relatively few tetrahedral organometallic compounds are known. The steric requirements of most ligands normally preclude coordination numbers of greater than six for d-metal organometallic compounds.

(b) 16-Electron square-planar compounds

Key point: With strong-field ligands, a square-planar complex has only eight bonding MOs, thus a 16-electron configuration is the most energetically favourable configuration.

One other geometry already discussed in the context of coordination chemistry is the square-planar arrangement of four ligands (Section 20.1), where we noted that it occurred only for strong-field ligands and a d^8 metal ion. Because organometallic ligands often produce a strong field, many square-planar organometallic compounds exist. Stable square-planar complexes are normally found with a total of 16 valence electrons, which results in the population of all the bonding and none of the antibonding MOs (Fig. 22.2).

The ligands in square-planar complexes can normally provide only two electrons each, for a total of eight electrons. Therefore, to reach 16 electrons, the metal ion must provide an additional eight electrons. As a result, organometallic compounds with 16 valence electrons are common only on the right of the d block, particularly in Groups 9 and 10 (Table 22.1). Examples of such complexes include [Ir(CO)Cl(PPh$_3$)$_2$] (**6**) and the anion of Zeise's salt, [Pt(C$_2$H$_4$)Cl$_3$]$^-$ (**1**). Square-planar 16-electron complexes are particularly common for the heavier elements in Groups 9 and 10, especially for Rh(I), Ir(I), Pd(II), and Pt(II), because the ligand-field splitting is large and the ligand-field stabilization energy of these complexes favours the square-planar configuration.

22.2 Electron count preference

Key point: On the left of the d block, steric requirements can mean that it is not possible to assemble enough ligands around a metal atom to achieve a total of 16 or 18 valence electrons.

The preference of a metal atom for one particular geometry and electron count is not normally so strong that other geometries do not occur. For instance, although the chemistry of

Table 22.1 Validity of the 16/18-electron rule for d-metal organometallic compounds

Usually fewer than 18 electrons			Usually 18 electrons			16 or 18 electrons	
Sc	Ti	V	Cr	Mn	Fe	Co	Ni
Y	Zr	Nb	Mo	Tc	Ru	Rh	Pd
La	Hf	Ta	W	Re	Os	Ir	Pt

both Pd(II) and Rh(I) is dominated by 16-electron square-planar complexes, the cyclopenta-dienyl palladium complex (**7**) and the rhodium complex (**8**) are both 18-electron compounds.

Steric factors may restrict the number of ligands that can bond to a metal atom and sta-bilize compounds with lower electron counts than might have been expected. For instance, the tricyclohexylphosphine ligands (**9**) in the trigonal Pt(0) compound Pt(PCy$_3$)$_3$ are so large that only three can be fitted around the metal atom, which consequently has only 16 valence electrons. Steric stabilization of the metal centre is partially a kinetic effect, in which the large groups protect the metal centre from reaction. Many complexes readily lose or gain ligands, forming other configurations transitorily during reactions—indeed, the accessibility of these other configurations is precisely why the organometallic chemistry of the d metals is so interesting.

Unusual electron configurations are common on the left of the d block, where the metal atoms have fewer electrons, and it is often not possible to crowd enough ligands around the atom to bring the electron count up to 16 or 18. For example, the simplest carbonyl in Group 5, [V(CO)$_6$], is a 17-electron complex. Other examples include [W(CH$_3$)$_6$], which has 12 valence electrons, and [Cr(η^5-Cp)(CO)$_2$(PPh$_3$)] with 17. The latter compound pro-vides another good example of the role of steric crowding. When the compact CO ligand is present in place of bulky triphenylphosphine (**10**), a dimeric compound with a long but definite Cr−Cr bond is observed in the solid state and in solution. The formation of the Cr−Cr bond in [Cr(η^5-Cp)(CO)$_3$]$_2$ raises the electron count on each metal to 18.

22.3 Electron counting and oxidation states

The dominance of 16- and 18-electron configurations in organometallic chemistry makes it imperative to be able to count the number of valence electrons on a central metal atom because knowing that number allows us to predict the stabilities of compounds and to suggest patterns of reactivity. Although the concept of 'oxidation state' for organometallic compounds is regarded by many as tenuous at best, the vast majority of the research com-munity uses it as a convenient shorthand for describing electron configurations. Oxida-tion states (and the corresponding oxidation number) help to systematize reactions such as oxidative addition (Section 22.22), and also bring out analogies between the chemical properties of organometallic and coordination complexes. Fortunately, the business of counting electrons and assigning oxidation numbers can be combined.

Two models are routinely used to count electrons, the so-called **neutral-ligand method** (sometimes called the *covalent method*) and the **donor-pair method** (sometimes known as the *ionic method*). We introduce both briefly—they give identical results for electron counting—but in what follows we use the donor-pair method as it may easily be used to assign oxidation numbers too.

(a) Neutral-ligand method

Key point: All ligands are treated as neutral and are categorized according to how many electrons they are considered to donate.

For the sake of counting electrons, each metal atom and ligand is treated as neutral. We in-clude in the count all valence electrons of the metal atom and all the electrons donated by the ligands. If the complex is charged, we simply add or subtract the appropriate number of electrons to the total. Ligands are defined as **L type** if they are neutral two-electron donors (like CO, PMe$_3$) and **X type** if, when they are considered to be neutral, they are one-electron radical donors (like halogen atoms, H, CH$_3$). For example, Fe(CO)$_5$ acquires 18 electrons from the eight valence electrons on the Fe atom and the 10 electrons donated by the five CO ligands. Some ligands are considered combinations of these types, for instance cyclopentadienyl is considered as a five-electron L$_2$X donor; see Table 22.2.

7

8 [Rh(Me)(PMe$_3$)$_4$], Me = CH$_3$

9 PCy$_3$, Cy = *cyclo*-C$_6$H$_{11}$

10 PPh$_3$, Ph = C$_6$H$_5$

EXAMPLE 22.1 Counting electrons using the neutral-ligand method

Do (a) [IrBr$_2$(CH$_3$)(CO)(PPh$_3$)$_2$] and (b) [Cr(η^5-C$_5$H$_5$)(η^6-C$_6$H$_6$)] obey the 18-electron rule?

Answer (a) We start with the Ir atom (Group 9) which has nine valence electrons, then add in the electrons from the two Br atoms and the CH$_3$ group (each is a one-electron donor) and finally add in the electrons from the CO and PPh$_3$ (both are two-electron donors). Thus, the number of valence electrons on the metal atom is $9 + (3 \times 1) + (3 \times 2) = 18$. (b) In a similar fashion, the Cr atom (Group 6) has six valence electrons, the η^5-C$_5$H$_5$ ligand donates five electrons, and the η^6-C$_6$H$_6$ ligand donates six, so the number of metal valence electrons is $6 + 5 + 6 = 17$. This complex does not obey the 18-electron rule and is not stable. A related but stable 18-electron compound is $[\text{Cr}(\eta^6\text{-C}_6\text{H}_6)_2]$.

Self-test 22.1 Is [Mo(CO)$_7$] likely to be stable?

The advantage of the neutral-ligand method is that, given the information in Table 22.2, it is trivial to establish the electron count. The disadvantage, however, is that the method

Table 22.2 Typical ligands and their electron counts

(a) Neutral-ligand method

Ligand	Formula	Designation	Electrons donated
Carbonyl	CO	L	2
Phosphine	PR$_3$	L	2
Hydride	H	X	1
Chloride	Cl	X	1
Dihydrogen	H$_2$	L	2
η^1-Alkyl, -alkenyl, -alkynyl, and -aryl groups	R	X	1
η^2-Alkene	CH$_2$=CH$_2$	L	2
η^2-Alkyne	RCCR	L	2
Dinitrogen	N$_2$	L	2
Butadiene	CH$_2$=CH−CH=CH$_2$	L$_2$	4
Benzene	C$_6$H$_6$	L$_3$	6
η^3-Allyl	CH$_2$CHCH$_2$	LX	3
η^5-Cyclopentadienyl	C$_5$H$_5$	L$_2$X	5

(b) Donor-pair method*

Ligand	Formula	Electrons donated
Carbonyl	CO	2
Phosphine	PR$_3$	2
Hydride	H$^-$	2
Chloride	Cl$^-$	2
Dihydrogen	H$_2$	2
η^1-Alkyl, -alkenyl, -alkynyl, and -aryl groups	R$^-$	2
η^2-Alkene	CH$_2$=CH$_2$	2
η^2-Alkyne	RCCR	2
Dinitrogen	N$_2$	2
Butadiene	CH$_2$=CH−CH=CH$_2$	4
Benzene	C$_6$H$_6$	6
η^3-Allyl	CH$_2$CHCH$_2^-$	4
η^5-Cyclopentadienyl	C$_5$H$_5^-$	6

* We use this method throughout this book.

overestimates the degree of covalence and thus underestimates the charge at the metal. Moreover, it becomes confusing to assign an oxidation number to a metal, and meaningful information on some ligands is lost.

(b) Donor-pair method

Key point: Ligands are considered to donate electrons in pairs, resulting in the need to treat some ligands as neutral and others as charged.

The donor-pair method requires a calculation of the oxidation number. The rules for calculating the oxidation number of an element in an organometallic compound are the same as for conventional coordination compounds. Neutral ligands, such as CO and phosphine, are considered to be two-electron donors and are formally assigned an oxidation number of 0. Ligands such as halides, H, and CH_3 are formally considered to take an electron from the metal atom, and are treated as Cl^-, H^-, and CH_3^- (and hence are assigned oxidation number -1); in this anionic state they are considered to be two-electron donors. The cyclopentadienyl ligand, C_5H_5 (Cp), is treated as $C_5H_5^-$ (it is assigned an oxidation number of -1); in this anionic state it is considered to be a six-electron donor. Then:

The *oxidation number* of the metal atom is the total charge of the complex minus the charges of any ligands.

The *number of electrons* the metal provides is its group number minus its oxidation number.

The *total electron count* is the sum of the number of electrons on the metal atom and the number of electrons provided by the ligands.

The main advantage of this method is that with a little practice both the electron count and the oxidation number may be determined in a straightforward manner. The main disadvantage is that it overestimates the charge on the metal atom and can suggest reactivity that might be incorrect (see Section 22.7 on hydrides). Table 22.2 lists the maximum number of electrons available for donation to a metal for most common ligands.

EXAMPLE 22.2 Assigning oxidation numbers and counting the electrons using the donor-pair method

Assign the oxidation number and count the valence electrons on the metal atom in (a) $[IrBr_2(CH_3)(CO)(PPh_3)_2]$, (b) $[Cr(\eta^5\text{-}C_5H_5)(\eta^6\text{-}C_6H_6)]$, and (c) $[Mn(CO)_5]^-$.

Answer (a) We treat the two Br groups and the CH_3 as three singly negatively charged two-electron donors and the CO and the two PPh_3 ligands as three two-electron donors, providing 12 electrons in all. Because the complex is neutral overall, the Group 9 Ir atom must have a charge of $+3$ (that is, have oxidation number $+3$) to balance the charge of the three anionic ligands, and thus contributes $9 - 3 = 6$ electrons. This analysis gives a total of 18 electrons for the Ir(III) complex. (b) We treat the $\eta^5\text{-}C_5H_5$ ligand as $C_5H_5^-$ and thus it donates six electrons, with the $\eta^6\text{-}C_6H_6$ ligand donating a further six. To maintain neutrality, the Group 6 Cr atom must have a charge of $+1$ (and an oxidation number of $+1$) and contributes $6 - 1 = 5$ electrons. The total number of metal electrons is $12 + 5 = 17$ for a Cr(I) complex. As noted before, this complex does not obey the 18-electron rule and is unlikely to be stable. (c) We treat each CO ligand as neutral and contributing two electrons, giving 10 electrons. The overall charge of the complex is -1; because all the ligands are neutral, we consider this charge to reside formally on the metal atom, giving it an oxidation number -1. The Group 7 Mn atom thus contributes $7 + 1$ electrons, giving a total of 18 for a Mn(-1) complex.

Self-test 22.2 What is the electron count for and oxidation number of platinum in the anion of Zeise's salt, $[Pt(CH_2=CH_2)Cl_3]^-$? Treat $CH_2=CH_2$ as a neutral two-electron donor.

22.4 Nomenclature

Key point: The naming of organometallic compounds is similar to the naming of coordination compounds, but certain ligands have multiple bonding modes, which is reported as the hapticity.

According to the recommended convention, we use the same system of nomenclature for organometallic compounds as set out for coordination complexes in Section 7.2.

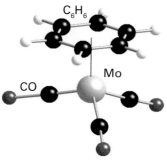

11 $[Mo(\eta^6\text{-}C_6H_6)(CO)_3]$

12 η^1-Cyclopentadienyl

13 η^3-Cyclopentadienyl

14 η^5-Cyclopentadienyl

Thus, ligands are listed in alphabetical order followed by the name of the metal, all of which is written as one word. The name of the metal should be followed by its oxidation number in parentheses. The nomenclature used in research journals, however, does not always obey these rules, and it is common to find the name of the metal buried in the middle of the name of the compound and the oxidation number omitted. For example, (**11**) is sometimes referred to as benzenemolybdenum-tricarbonyl, rather than the preferred name benzene(tricarbonyl)molybdenum(0).

The IUPAC recommendation for the formula of an organometallic compound is to write it in the same form as for a coordination complex: the symbol for the metal is written first, followed by the ligands, listed in alphabetical order based on their chemical symbol. We shall follow these conventions unless a different order of ligands helps to clarify a particular point.

Often a ligand with carbon donor atoms can exhibit multiple bonding modes—for instance, the cyclopentadienyl group can commonly bond to a d-metal atom in three different ways—thus, we need some additional nomenclature. Without going into the intimate details of the bonding of the various ligands (we do that later in this chapter), the extra information we need to describe a bonding mode is the number of points of attachment. This procedure gives rise to the notion of **hapticity**, the number of ligand atoms that are considered formally to be bonded to the metal atom. The hapticity is denoted η^n, where n is the number of atoms (and η is eta). For example, a CH_3 group attached by a single M—C bond is monohapto, η^1, and if the two C atoms of an ethene ligand are both within bonding distance of the metal, the ligand is dihapto, η^2. Thus, three cyclopentadienyl complexes might be described as having η^1 (**12**), η^3 (**13**), or η^5 (**14**) cyclopentadienyl groups.

Some ligands (including the simplest of them all, the hydride ligand, H^-) can bond to more than one metal atom in the same complex, and are then referred to as **bridging ligands**. We do not require any new concepts to understand bridging ligands other than those introduced in Section 2.11. Recall from Section 7.2 that the Greek letter μ (mu) is used to indicate how many atoms the ligand bridges. Thus a μ_2-CO is a carbonyl group that bridges two metal atoms and a μ_3-CO bridges three.

EXAMPLE 22.3 Naming organometallic compounds

Give the formal names of (a) ferrocene (**4**) and (b) $[RhMe(PMe_3)_4]$ (**8**).

Answer (a) Ferrocene contains two cyclopentadienyl groups that are both bound to the metal atom through all five carbon atoms, thus both groups are designated η^5. The full name for ferrocene is thus bis(η^5-cyclopentadienyl)iron(II). (b) The rhodium compound contains one formally anionic methyl group and four neutral trimethylphosphine ligands, therefore the formal name is methyltetrakis-(trimethylphosphine)rhodium(I).

Self-test 22.3 What is the formal name of $[Ir(Br)_2(CH_3)(CO)(PPh_3)_2]$?

Ligands

A large number of ligands are found in organometallic complexes, with many different bonding modes. Because the reactivity of the metal atom and the ligands is affected by the M—L bonding, it is important to look at each ligand in some detail.

22.5 Carbon monoxide

Key point: The 3σ orbital of CO serves as a very weak donor and the π^* orbitals act as acceptors.

Carbon monoxide is a very common ligand in organometallic chemistry, where it is known as the *carbonyl group*. Carbon monoxide is particularly good at stabilizing very low oxidation states, with many compounds (such as $Fe(CO)_5$) having the metal in its zero oxidation state. We described the molecular orbital structure of CO in Section 2.9, and it would be sensible to review that section.

A simple picture of the bonding of CO to a metal atom is to treat the lone pair on the carbon atom as a Lewis σ base (an electron-pair donor) and the empty CO antibonding orbital as a Lewis π acid (an electron-pair acceptor), which accepts π-electron density from the filled d orbitals on the metal atom. In this picture, the bonding can be considered to be made up of two parts: a σ bond from the ligand to the metal atom (**15**) and a π bond from the metal atom to the ligand (**16**). This type of π bonding is sometimes referred to as π **backbonding**.

Carbon monoxide is not appreciably nucleophilic, which suggests that σ bonding to a d-metal atom is weak. As many d-metal carbonyl compounds are very stable, we can also infer that π backbonding is strong and the stability of carbonyl complexes arises mainly from the π-acceptor properties of CO. Further evidence for this view comes from the observation that stable carbonyl complexes exist only for metals that have filled d orbitals of an energy suitable for donation to the CO antibonding orbital. For instance, elements in the s and p blocks do not form stable carbonyl complexes. However, the bonding of CO to a d metal atom is best regarded as a synergistic (that is, mutually enhancing) outcome of both σ and π bonding: the π backbonding from the metal to the CO increases the electron density on the CO, which in turn increases the ability of the CO to form a σ bond to the metal atom.

A more formal description of the bonding can be derived from the molecular orbital scheme for CO (Fig. 22.3), which shows that the HOMO has σ symmetry and is essentially a lobe that projects away from the C atom. When CO acts as a ligand, this 3σ orbital serves as a very weak donor to a metal atom, and forms a σ bond with the central metal atom. The LUMOs of CO are the π^* orbitals. These two orbitals play a crucial role because they can overlap with metal d orbitals that have local π symmetry (such as the t_{2g} orbitals in an O_h complex). The π interaction leads to the delocalization of electrons from filled d orbitals on the metal atom into the empty π^* orbitals on the CO ligands, so the ligand also acts as a π acceptor.

One important consequence of this bonding scheme is the effect on the strength of the CO triple bond: the stronger the metal–carbon bond becomes through pushing electron density from the metal atom into the π bond, the weaker the CO bond becomes, as this electron density enters a CO antibonding orbital. In the extreme case, when two electrons are fully donated by the metal atom, a formal metal–carbon double bond is formed; because the two electrons occupy a CO antibonding orbital, this donation results in a decrease in the bond order of the CO to 2.

In practice, the bonding is somewhere between M—C≡O, with no backbonding, and M=C=O, with complete backbonding. Infrared spectroscopy is a very convenient method of assessing the extent of π bonding; the CO stretch is clearly identifiable as it is both strong and normally clear of all other absorptions. In CO gas, the absorption for the triple bond is at 2143 cm^{-1}, whereas a typical metal carbonyl complex has a stretching mode in the range 2100–1700 cm^{-1} (Table 22.3). The number of IR absorptions that occur in a particular carbonyl compound is discussed in Section 22.18g.

Carbonyl stretching frequencies are often used to determine the order of acceptor or donor strengths for the other ligands present in a complex. The basis of the approach is that the CO stretching frequency is decreased when it serves as a π acceptor. However, as other π acceptors in the same complex compete for the d electrons of the metal atom, they cause the CO frequency to increase. This behaviour is opposite to that observed with donor ligands, which cause the CO stretching frequency to decrease as they supply electrons to the metal atom and hence, indirectly, to the CO π^* orbitals. Thus, strong σ-donor ligands attached to a metal carbonyl and a formal negative charge on a metal carbonyl anion both result in slightly greater CO bond lengths and significantly lower CO stretching frequencies.

Carbon monoxide is versatile as a ligand because, as well as the bonding mode we have described so far (often referred to as 'terminal'), it can bridge two (**17**) or three (**18**) metal atoms. Although the description of the bonding is now more complicated, the concepts of σ-donor and π-acceptor ligands remain useful. The CO stretching frequencies generally follow the order MCO > M_2CO > M_3CO, which suggests an increasing occupation of the π^* orbital as the CO molecule bonds to more metal atoms and more electron density from the metals enters the CO π^* orbitals. As a rule of thumb, carbonyls bridging two metal atoms typically have stretching bands in the range 1900–1750 cm^{-1}, and those that bridge three atoms have stretching bands in the range 1800–1600 cm^{-1} (Fig. 22.4). We consider carbon monoxide to be a two-electron neutral ligand when terminal or bridging (thus a carbonyl bridging two metals can be considered to give one electron to each).

A further bonding mode for CO that is sometimes observed is when the CO is terminally bound to one metal atom and the CO triple bond binds side-on to another metal (**19**).

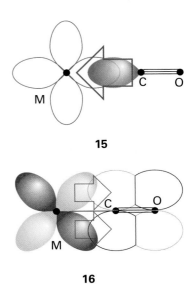

15

16

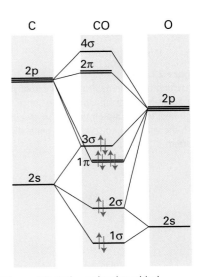

Figure 22.3 The molecular orbital scheme for CO shows that the HOMO has σ symmetry and is essentially a lobe that projects away from the C atom. The LUMO has π symmetry.

Table 22.3 The influence of coordination and charge on CO stretching bands

Compound	$\tilde{v} / cm^{-1}$
CO	2143
$[Mn(CO)_6]^+$	2090
$Cr(CO)_6$	2000
$[V(CO)_6]^-$	1860
$[Ti(CO)_6]^{2-}$	1750

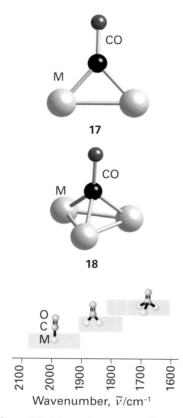

17

18

19

Figure 22.4 Approximate ranges for CO stretching bands in neutral metal carbonyls. Note that high wavenumbers (and hence high frequencies) are on the left, in keeping with the way infrared spectra are generally plotted.

This description is best considered as two separate bonding interactions, with the terminal interaction being the same as that described above and the side-on bonding being essentially identical to that of other side-on π donors such as alkynes and N_2, which are discussed later.

The synthesis, properties, and reactivities of compounds containing the carbonyl ligand are discussed in more detail in Section 22.18.

22.6 Phosphines

Key point: Phosphines bond to metals by a combination of σ donation from the P atom and π backbonding from the metal atom.

Although phosphine complexes are not organometallic because the ligands do not bond to the metal atom through a carbon atom, they are best discussed here as their bonding has many similarities to that of carbon monoxide.

Phosphine, PH_3 (formally, phosphane), is a reactive, noxious, poisonous, and flammable gas (Section 15.10). Like ammonia, phosphine can behave as a Lewis base and use its lone pair to donate electron density to a Lewis acid, and so act as a ligand. However, given the problems associated with handling phosphine, it is rarely used as a ligand. Substituted phosphines, on the other hand, such as trialkylphosphines (for example, PMe_3, PEt_3), or triarylphosphines (for example, PPh_3, **10**), or trialkyl- or triarylphosphites (for example, $P(OMe)_3$, $P(OPh)_3$, **20**) and a whole host of bridged multidentate di- and triphosphines (for example, $Ph_2PCH_2CH_2PPh_2$ = dppe, **21**) are easy to handle (indeed some are air-stable odourless solids with no appreciable toxicity) and are widely used as ligands; all are colloquially referred to as 'phosphines'.

Phosphines have a lone pair on the P atom that is appreciably basic and nucleophilic, and can serve as a σ donor. Phosphines also have empty orbitals on the P atom that can overlap with filled d orbitals on 3d-metal ions and behave as π acceptors (**22**). The bonding of phosphines to a d-metal atom, made up of a σ bond from the ligand to the metal and a π bond from the metal back to the ligand, is completely analogous to the bonding of CO to a d-metal atom. Quite which orbitals on the P atom behave as π acceptors has been the subject of considerable debate in the past with some groups claiming a role for unoccupied 3d orbitals on P, and some claiming a role for the P−R σ* orbitals; current consensus favours the σ* orbitals. In any case, each phosphine provides an additional two electrons to the valence electron count.

As we have remarked, a huge variety of phosphines are both possible and widely available, including chiral systems such as 2,2′-bis(diphenylphosphino)-1,1′-binaphthyl (BINAP, **23**), in which steric constraints result in compounds that can be resolved into diastereomers. Generally there are two properties of phosphine ligands that are considered important in discussions of the reactivity of their complexes: their steric bulk and their electron donating (and accepting) ability.

We described in Section 21.6 how the steric bulk of phosphines can be expressed in terms of the notional cone occupied by the bonded ligand, and Table 22.4 lists some of the derived cone angles. The bonding of phosphines to d-metal atoms is, as we have seen, a composite of σ bonding from the ligand to the metal atom, and π backbonding from the metal atom to the ligand. The σ-donating ability and π-acceptor ability of phosphines are inversely

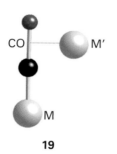

20 $P(OPh)_3$

21 $Ph_2PCH_2CH_2PPh_2$, dppe

22

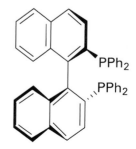

23 2,2′-bis(diphenylphosphino)-1,1′-binaphthyl, BINAP

correlated in the sense that electron-rich phosphines, such as PMe_3, are good σ donors and poor π acceptors, whereas electron-poor phosphines, such as PF_3, are poor σ donors and good π acceptors. Thus Lewis basicity can normally be used as a single scale to indicate their donor/acceptor ability. The generally accepted order of basicity of phosphines is:

$$PCy_3 > PEt_3 > PMe_3 > PPh_3 > P(OMe)_3 > P(OPh)_3 > PCl_3 > PF_3$$

and is easily understood in terms of the electronegativity of the substituents on the P atom. The basicity of a phosphine is not simply related to the strength of the M−P bond in a complex, for instance an electron-poor metal atom forms a stronger bond with an electron-rich (basic) phosphine, whereas an electron-rich metal atom will form a stronger bond with an electron-poor phosphine.

If there are carbonyl ligands present in a metal−phosphine complex, then the carbonyl stretching frequency can be used to assess the basicity of the phosphine ligand: this method allows us to conclude that PF_3 is a π acceptor comparable to CO.

The vast range of phosphines that are commonly used in organometallic chemistry is a testament to their versatility as ligands: judicious choice allows control over both steric and electronic properties of the metal atom in a complex. Phosphorus-31 (which occurs in 100 per cent natural abundance) is easy to observe by NMR and both the ^{31}P chemical shift and the coupling constant to the metal atom (where appropriate) give considerable insight into the bonding and reactivity of a complex. Like carbonyls, phosphines can bridge either two or three metal atoms, providing additional variety in bonding modes.

Table 22.4 Tolman cone angles (in degrees) for selected phosphines

PF_3	104
$P(OMe)_3$	107
PMe_3	118
PCl_3	125
$P(OPh)_3$	127
PEt_3	132
PPh_3	145
PCy_3	169
P^tBu_3	182
$P(o\text{-tolyl})_3$	193

EXAMPLE 22.4 Interpreting carbonyl stretching frequencies and phosphine complexes

(a) Which of the two isoelectronic compounds $Cr(CO)_6$ and $[V(CO)_6]^-$ will have the higher CO stretching frequency? (b) Which of the two chromium compounds $[Cr(CO)_5(PEt_3)]$ and $[Cr(CO)_5(PPh_3)]$ will have the lower CO stretching frequency? Which will have the shorter M−C bond?

Answer We need to think about whether backbonding to the CO ligands is enhanced or diminished: more backbonding results in a weaker carbon−oxygen bond. (a) The negative charge on the V complex will result in greater π backbonding to the CO π^* orbitals, compared to the Cr complex. This backbonding results in a weakening of the CO bond, with a corresponding decrease in stretching frequency. Thus the Cr complex has the higher CO stretching frequency. (b) PEt_3 is more basic than PPh_3 and thus the PEt_3 complex will have greater electron density on the metal atom than the PPh_3 complex. The greater electron density will result in greater backbonding and thus both a lower CO stretching frequency and a shorter M−C bond.

Self-test 22.4 Which of the two iron compounds $Fe(CO)_5$ and $[Fe(CO)_4(PEt_3)]$ will have the higher CO stretching frequency? Which will have the longer M−C bond?

22.7 Hydrides and dihydrogen complexes

Key points: The bonding of a hydrogen atom to a metal atom is a σ interaction, whereas the bonding of a dihydrogen ligand involves π backbonding.

A hydrogen atom directly bonded to a metal is commonly found in organometallic complexes and is referred to as a **hydride ligand**. The name 'hydride' can be misleading as it implies a H$^-$ ligand. Although the formulation H$^-$ might be appropriate for most hydrides, such as $[CoH(PMe_3)_4]$, some hydrides are appreciably acidic and behave as though they contain H$^+$, for example $[CoH(CO)_4]$ is an acid with $pK_a = 8.3$ (in acetonitrile). The acidity of organometallic carbonyls is described in Section 22.18. In the donor-pair method of electron counting, we consider the hydride ligand to contribute two electrons and to have a single negative charge (that is, to be H$^-$).

The bonding of a hydrogen atom to a metal atom is simple because the only orbital of appropriate energy for bonding on the hydrogen is H1s and the M−H bond can be considered as a σ interaction between the two atoms. Hydrides are readily identified by NMR spectroscopy as their chemical shift is rather unusual, typically occurring in the range $-50 < \delta < 0$. Infrared spectroscopy can also be useful in identifying metal hydrides as they normally have a stretching band in the range 2850−2250 cm^{-1}. X-ray diffraction, normally so valuable for identifying the structure of crystalline materials, is of little use in identifying hydrides because the diffraction is related to electron density, and the hydride ligand will have at most two electrons around it, compared with, for instance, 78 for a platinum.

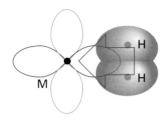

24

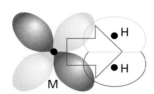

25

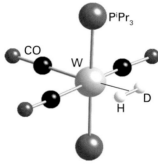

26 [W(HD)(P^{i}Pr$_3$)$_2$(CO)$_3$],
^{i}Pr = CH(CH$_3$)$_2$

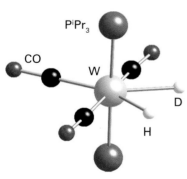

27 [W(H)(D)(P^{i}Pr$_3$)$_2$(CO)$_3$]

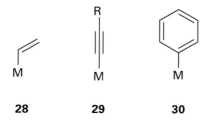

28   **29**   **30**

Neutron diffraction is of more use in locating hydride ligands, especially if the hydrogen atom is replaced by a deuterium atom, because deuterium has a large neutron scattering cross-section.

An M—H bond can sometimes be produced by protonation of an organometallic compound, such as neutral and anionic metal carbonyls (Section 22.18e). For example, ferrocene can be protonated in strong acid to produce an Fe—H bond:

$$Cp_2Fe + HBF_4 \rightarrow [Cp_2FeH]^+[BF_4]^-$$

Bridging hydrides exist, where an H atom bridges either two or three metal atoms: here the bonding can be treated in exactly the same way we considered bridging hydrides in diborane, B$_2$H$_6$ (Section 2.11).

Although the first organometallic metal hydride was reported in 1931, complexes of hydrogen gas, H$_2$, were identified only in 1984. In such compounds, the dihydrogen molecule, H$_2$, bonds side-on to the metal atom (in the older literature, such compounds were sometimes called *non-classical hydrides*). The bonding of dihydrogen to the metal atom is considered to be made up of two components: a σ donation of the two electrons in the H$_2$ bond to the metal atom (**24**) and a π backdonation from the metal to the σ* antibonding orbital of H$_2$ (**25**). This picture of the bonding raises a number of interesting issues. In particular, as the π backbonding from the metal atom increases, the strength of the H—H bond decreases and the structure tends to that of a dihydride:

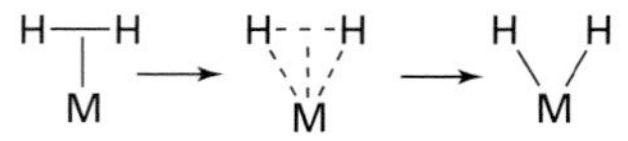

A dihydrogen molecule is treated as a neutral two-electron donor. Thus the transformation of the dihydrogen into two hydrides (each considered to have a single negative charge and to contribute two electrons) requires the formal charge on the metal atom to increase by two. That is, the metal is oxidized by two units and the dihydrogen is reduced. Although it might seem that this oxidation of the metal is just an anomaly thrown up by our method of counting electrons, two of the electrons on the metal atom have been used to backbond to the dihydrogen, and these two electrons are no longer available to the metal atom for further bonding. This transformation of the dihydrogen molecule to a dihydride is an example of *oxidative addition*, and is discussed more fully later in this chapter (Section 22.22).

It is now recognized that complexes exist with structures at all points between these two extremes, and in some cases an equilibrium can be identified between the two. Work by G. Kubas on tungsten complexes used the H—D coupling constant to show that it is possible to detect both the dihydrogen complex (**26**, $^1J_{HD}$ = 34 Hz) and the dihydride (**27**, $^2J_{HD}$ < 2 Hz), and to follow the conversion from one to the other. Certain microbes contain enzymes known as *hydrogenases* that use Fe and Ni at their catalytic centres to catalyse the rapid oxidation of H$_2$ and reduction of H$^+$, via intermediate metal–dihydrogen and hydride species (Section 27.14).

22.8 η^1-Alkyl, -alkenyl, -alkynyl, and -aryl ligands

Key point: The metal—ligand bonding of η^1-hydrocarbon ligands is a σ interaction.

Alkyl groups are often found as ligands in d-metal organometallic chemistry and their bonding presents no new features: it is best considered a simple covalent σ interaction between the metal atom and the carbon atom of the organic fragment. Alkyl groups with a hydrogen atom on a carbon atom adjacent to the one that bonds to the metal are prone to decompose by a process known as *β-hydrogen elimination* (Section 22.25) and hence those alkyl groups that cannot react in this fashion, such as methyl, benzyl, (CH$_2$C$_6$H$_5$), neopentyl (CH$_2$CMe$_3$), and trimethylsilylmethyl (CH$_2$SiMe$_3$), are more stable than those that can, such as ethyl.

Alkenyl (**28**), alkynyl (**29**), and aryl (**30**) groups can bond to a metal atom in a similar fashion, binding to the metal atom through a single carbon atom, and hence are described as monohapto (η^1). Although there is potential for each of these three groups to accept π electron density into antibonding orbitals, there is little evidence that this happens. For instance, even though an η^1-alkynyl group might be considered analogous to a CO group, the stretching frequency of the triple bond in alkynyl complexes changes little on attachment to a metal. Bridging alkyl and aryl groups also exist, and the bonding can be considered in the same way as we have considered other bridging ligands, with 3c,2e bonds.

Alkyl, alkenyl, alkynyl, and aryl groups are commonly introduced into organometallic complexes by the displacement of a halide at a metal centre with a lithium or Grignard reagent. For example:

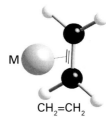

We consider alkyl, alkenyl, alkynyl, and aryl ligands to be two-electron donors with a single negative charge (for example, Me^-, Ph^-) in the donor-pair scheme of electron counting.

22.9 η²-Alkene and -alkyne ligands

Key point: The bonding of an alkene or an alkyne to a metal atom is best described as a σ interaction from the multiple bond to the metal atom, with a π backbonding interaction from the metal atom to the π* antibonding orbital on the alkene or alkyne.

Alkenes are routinely found bound to metal centres: the first organometallic compound isolated, Zeise's salt (**1**), was a complex of ethene. Alkenes normally bond side-on to a metal atom with both carbon atoms of the double bond equidistant from the metal with the other groups on the alkene approximately perpendicular to the plane of the metal atom and the two carbon atoms (**31**). In this arrangement, the electron density of the C=C π bond can be donated to an empty orbital on the metal atom to form a σ bond. In parallel with this interaction, a filled metal d orbital can donate electron density back to the empty π* orbitals of the alkene to form a π bond. This description is called the **Dewar−Chatt−Duncanson model** (Fig. 22.5) and η²-alkenes are considered to be two-electron neutral ligands.

Electron donor and acceptor character appear to be fairly evenly balanced in most ethene complexes of the d metals, but the degree of donation and backdonation can be altered by substituents on the metal atom and on the alkene. When the π backbonding from the metal atom increases, the strength of the C=C bond decreases as the electron density is located in the C=C antibonding orbital and the structure tends to that of a C−C singly bonded structure, a metallocyclopropane:

$$\overline{\underset{M}{|}} \longrightarrow \underset{M}{\nabla} \longrightarrow \underset{M}{\triangledown}$$

Dihaptoalkenes with only a small degree of electron donation from the metal have their substituents bent slightly away from the metal atom, and the C=C bond length is only slightly greater than in the free alkene (134 pm). When the degree of backdonation is greater, substituents on the alkene are bent away more from the metal atom and the C−C bond length approaches that characteristic of a single bond. Steric constraints can also force the other groups on the alkene to bend away from the metal atom.

Alkynes have two π bonds and hence the potential to be four-electron donors. When side-on to a single metal atom, the η²-carbon−carbon triple bond is best considered as a two-electron donor, with the π* orbitals accepting electron density from a metal atom in the same way as for alkenes. When strongly electron-withdrawing groups are attached to an alkyne, the ligand can become an excellent π acceptor and displace other ligands such as phosphines; the compound commonly known as dimethylacetylenedicarboxylate, $CH_3OCOC≡CCO_2CH_3$, is a good example.

Substituted alkynes can form very stable polymetallic complexes in which the alkyne can be regarded as a four-electron donor. An example is η²-diphenylethyne-(hexacarbonyl)dicobalt(0), in which we can view one π bond as donating to one of the Co atoms and the second π bond as overlapping with the other Co atom (**32**). In this example, the alkyl or aryl groups present on the alkyne impart stability by lowering the tendency towards secondary reactions of the coordinated ethyne, such as loss of the slightly acidic ethynic H atom to the metal atom.

22.10 Nonconjugated diene and polyene ligands

Key point: The bonding of nonconjugated alkenes to a metal atom is best described as independent multiple alkenes bonding to a metal centre.

Nonconjugated diene (−C=C−X−C=C−) and polyene ligands can also bond to metal atoms. It is simplest to consider them as linked alkenes, and hence they present no new bonding concepts. As with the chelate effect in coordination complexes (Section 7.14), the

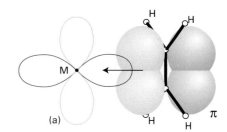

$CH_2=CH_2$

31

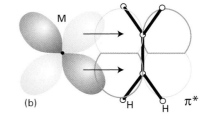

(a)

π

(b)

π*

Figure 22.5 The interaction of ethene with a metal atom. (a) Donation of electron density from the filled π molecular orbital of ethene to a vacant metal σ orbital. (b) Acceptance of electron density from a filled dπ orbital into the vacant π* orbital of ethene.

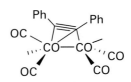

32 [Co₂(PhC≡CPh)(CO)₆]

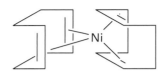

33 Ni(cod)$_2$

34 Cycloocta-1,5-diene, cod

resulting polyene complexes are usually more stable than the equivalent complex with individual ligands because the entropy of dissociation of the complex is much smaller than when the liberated ligands can move independently. For example, bis(η^4-cycloocta-1,5-diene)nickel(0) (**33**) is more stable than the corresponding complex containing four ethene ligands. Cycloocta-1,5-diene (**34**) is a fairly common ligand in organometallic chemistry, where it is referred to engagingly as 'cod', and is normally introduced into the metal coordination sphere by simple ligand displacement reactions. An example is:

$$\text{Cl—Pd(—NCPh)(—NCPh)—Cl} \xrightarrow{\text{cod}} \text{Cl—Pd—Cl(cod)} + 2\,\text{PhCN}$$

Metal–cod complexes are commonly used as starting materials because they often have intermediate stability. Many of them are sufficiently stable to be isolated and handled, but cod can be displaced by many other ligands. For example, if the highly toxic Ni(CO)$_4$ molecule is needed in a reaction, then it may be generated from Ni(cod)$_2$ directly in the reaction flask:

$$\text{Ni(cod)}_2(\text{soln}) + 4\,\text{CO(g)} \rightarrow \text{Ni(CO)}_4(\text{soln}) + 2\,\text{cod(soln)}$$

22.11 Butadiene, cyclobutadiene, and cyclooctatetraene

Key points: Some insight into the bonding of butadiene and cyclobutadiene can be gained by treating them as containing two alkene units, but a full understanding needs a consideration of the molecular orbitals. Cyclooctatetraene bonds in a number of different ways; the most common mode in d-metal chemistry is as an η^4-donor, analogous to butadiene.

The temptation with both butadiene and cyclobutadiene is to treat them like two isolated double bonds. However, a proper molecular orbital approach is necessary to understand the bonding fully because the ligand–metal atom interactions are different in the two cases.

Figure 22.6 shows the molecular orbitals for the π system in butadiene. The two occupied lower energy MOs can behave as donors to the metal, the lowest a σ donor and the next a π donor. The next higher unoccupied MO, the LUMO, can act as a π acceptor from the metal atom. Thus, attachment of a butadiene molecule to a metal atom results in population of an MO that is bonding between the two central C atoms (which are already nominally singly bonded) and antibonding between the nominally doubly bonded C atoms. The resulting modification of electron density results in a shortening of the central C—C bond and a lengthening of the double C—C bonds; in some complexes the central C—C bond is found to be even shorter than the other two C—C bonds. In theory, a δ-bonding interaction is possible between the d$_{xy}$ orbital of the metal atom and the most antibonding of the butadiene MOs, but there is no definitive evidence that it occurs. Butadiene is therefore considered to be a four-electron neutral ligand in electron counting schemes.

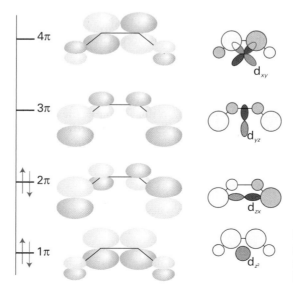

Figure 22.6 The molecular orbitals of the π system in butadiene; also shown are metal d orbitals of appropriate symmetry to form bonding interactions.

Cyclobutadiene is rectangular (D_{2h}) and unstable as a free molecule, as a result of bond angle constraints and its four-electron anti-aromatic configuration. However, stable complexes are known, including [Ru(η^4-C$_4$H$_4$)(CO)$_3$] (**35**). This species is one of many in which coordination to a metal atom stabilizes an otherwise unstable molecule.

As a result of a distortion from the square arrangement,[2] cyclobutadiene has an MO diagram similar to that of butadiene (Fig. 22.7). Population of the LUMO by backbonding now leads to greater bonding on the long sides of the rectangular cyclobutadiene molecule, tending towards a square (D_{4h}) arrangement. If the cyclobutadiene formally accepts two electrons from the metal atom, it will have six π electrons with three MOs occupied and no incentive to distort from square; two of the MOs are degenerate in this configuration. All cyclobutadiene–metal complexes are found to be square, suggesting that a six-electron aromatic configuration more accurately describes the bonding within the carbon ring than a four-electron configuration does. This view of the bonding has led some to regard cyclobutadiene complexes as complexes of the R$_4$C$_4^{2-}$ dianion (a six-electron donor), although for convenience most treat cyclobutadiene as a four-electron neutral ligand. Once again, a δ-bonding interaction is possible between the d$_{xy}$ orbital of the metal atom and the most antibonding of the butadiene MOs, but again there seems to be no definitive evidence that it occurs.

Because cyclobutadiene is unstable, the ligand must be generated in the presence of the metal to which it is to be coordinated. This synthesis can be accomplished in a variety of ways. One method is the dehalogenation of a halogenated cyclobutene:

Another procedure is the dimerization of a substituted ethyne:

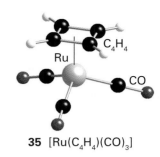

35 [Ru(C$_4$H$_4$)(CO)$_3$]

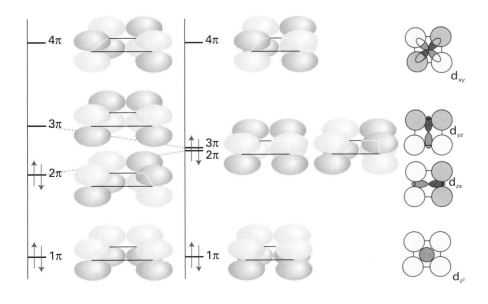

Figure 22.7 The molecular orbitals of the π system in cyclobutadiene; also shown are metal d orbitals of appropriate symmetry to form bonding interactions.

[2] We can regard this distortion as an organic example of the Jahn–Teller effect (Section 20.1g).

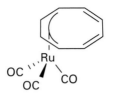

36 Cyclooctatetraene

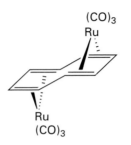

37 [Ru(η⁴-C₈H₈)(CO)₃]

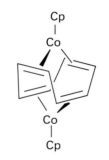

38 [(CO)₃Ru(C₈H₈)Ru(CO)₃]

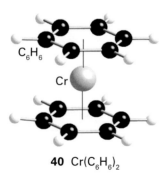

39 [CpCo(C₈H₈)CoCp]

40 Cr(C₆H₆)₂

Cyclooctatetraene (**36**) is a large ligand that is found in a wide variety of bonding arrangements. Like cyclobutadiene, it is anti-aromatic as the free molecule. Cyclooctatetraene can bond to a metal atom in an octahapto arrangement, in which it is planar and all C—C bond lengths are equal. In a similar fashion to cyclobutadiene, in this arrangement cyclooctatetraene is considered to extract two electrons and to become (formally) the aromatic dinegative ligand [C₈H₈]²⁻ (that is, a 10-electron donor). Cyclooctatetraenes bound in this fashion are rarely found in d-metal compounds and are normally found only with lanthanoids and actinoids (Section 23.8). For example, two such dinegative ligands are found in bis(η⁸-cyclooctatetraenyl)-uranium(IV), [U(η⁸-C₈H₈)₂], commonly referred to as uranocene.

The more usual bonding modes for cyclooctatetraenes in d-metal compounds are as puckered η⁴-C₈H₈ ligands such as (**37**), where the bonding part of the ligand can be treated as a butadiene. Bridging modes, (**38**) and (**39**), are also possible.

22.12 Benzene and other arenes

Key point: A consideration of the MOs of benzene leads to a picture of bonding of benzene to a metal atom that includes a significant δ backbonding interaction.

If benzene is considered to have three localized double bonds, each double bond can behave as a ligand and the molecule could behave as a tridentate η⁶-ligand. A compound such as bis(η⁶-benzene)chromium (**40**) could then be considered to be made up of six coordinated double bonds, each donating two electrons, bound to a d⁶ metal atom, giving a total of 18 valence electrons for the octahedral complex. Bis(η⁶-benzene)chromium does exist, and is remarkably stable: it can be handled in air and sublimes with no decomposition. Although this description of the bonding is a first step towards understanding its structure, the true picture needs a deeper consideration of the molecular orbitals involved.

In the molecular orbital picture of the π bonding in benzene there are three bonding and three antibonding orbitals. If we consider a single benzene molecule bonding to a single metal, and consider only the d orbitals, the strongest interaction is a σ interaction between the most strongly bonding a₁ benzene MO and the d_z² orbital of the metal atom; π bonds are possible between the two other bonding benzene MOs and the d_zx and d_yz orbitals. Backbonding from the metal atom to the benzene is possible as a δ interaction between the d_x²−y² and d_xy orbitals and the empty antibonding e₂ orbitals of benzene (Fig. 22.8). η⁶-Arenes are considered to be neutral ligands that donate six electrons and are normally considered to take up three coordination sites at a metal.

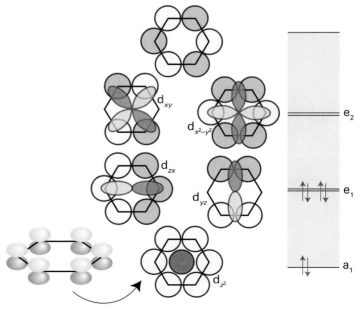

Figure 22.8 The molecular orbitals of the π system in benzene; also shown are metal d orbitals of appropriate symmetry to form bonding interactions.

Hexahapto (η^6) arene complexes are very easy to make, often simply by dissolving a compound that has three replaceable ligands in the arene and refluxing the solution:

One commonly invoked reaction intermediate of η^6-arene complexes is a 'slipping' to an η^4 complex, which donates only four electrons to the metal, and therefore allows a substitution reaction to proceed without an initial ligand loss:

η^2-Arenes are also known and are analogous to η^2-alkenes; they have an important role in the activation of arenes by metal complexes.

22.13 The allyl ligand

Key points: A consideration of the MOs of η^3-allyl complexes leads to a picture that gives two identical C−C bond lengths; because the type of bonding of the allyl ligand is so variable, η^3-allyl complexes are often highly reactive.

The allyl ligand, $CH_2=CH−CH_2^-$, can bind to a metal atom in either of two configurations. As an η^1-ligand (**41**) it should be considered just like an η^1-alkyl group (that is, as a two-electron donor with a single negative charge). However, the allyl ligand can also use its double bond as an additional two-electron donor and act as an η^3-ligand (**42**); in this arrangement it acts as a four-electron donor with a single negative charge. The η^3-allyl ligand can be thought of as a resonance between two forms (**43**), and because all evidence points towards a symmetrical structure, it is often depicted with a curved line representing all the bonding electrons (**44**).

As with benzene, a more detailed understanding of the bonding of an allyl group needs a consideration of the molecular orbitals of the organic fragment (Fig. 22.9), whereupon it becomes apparent why a symmetrical arrangement is the correct description of the η^3-bonding mode. The filled 1π orbital on the allyl group behaves as a σ donor (into the d_{z^2} orbital), the 2π orbital behaves as a π donor (into the d_{zx} orbital) and the 3π orbital behaves as a π acceptor (from the d_{yz} orbital). Thus the interactions of the metal atom with each of the terminal carbon atoms are identical and a symmetrical arrangement results.

The terminal substituents of an η^3-allyl group are bent slightly out of the plane of the three-carbon backbone and are either *syn* (**45**) or *anti* (**46**) relative to the central hydrogen. It is common to observe *anti* and *syn* group exchange, which in some cases is fast on an NMR timescale. A mechanism that involves the transformation η^3 to η^1 to η^3 is often invoked to explain this exchange.

Because of this flexibility in the bonding, η^3-allyl complexes are often highly reactive as transformation to the η^1-form allows them to bind readily to another ligand.

There are many synthetic routes to allyl complexes. One is the nucleophilic attack of an allyl Grignard reagent on a metal halide:

$$2\,C_3H_5MgBr + NiCl_2 \rightarrow [Ni(\eta^3\text{-}C_3H_5)_2] + 2\,MgBrCl$$

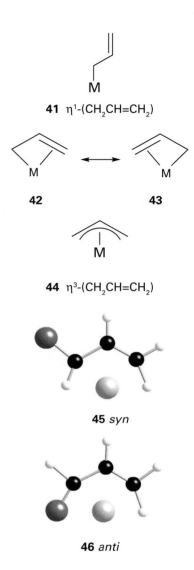

41 η^1-$(CH_2CH=CH_2)$

42   **43**

44 η^3-$(CH_2CH=CH_2)$

45 *syn*

46 *anti*

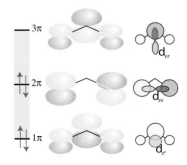

Figure 22.9 The molecular orbitals of the π system of the allyl$^-$ group; also shown are metal d orbitals of appropriate symmetry to form bonding interactions.

Nucleophilic attack on a haloalkane by a metal atom in a low oxidation state also yields allyl complexes:

$$[Mn(CO)_5]^- + CH_2=CHCH_2Cl \longrightarrow$$

+ Cl⁻   + CO

In complexes where the metal centre is not protonated directly, the protonation of a butadiene ligand can lead to an η³-allyl complex:

+ HCl $\longrightarrow$

22.14 Cyclopentadiene and cycloheptatriene

Key points: The common η⁵-bonding mode of a cyclopentadienyl ligand can be understood on the basis of both σ and π donation from the organic fragment to the metal in conjunction with δ backbonding; cycloheptatriene commonly forms either η⁶-complexes or η⁷-complexes of the aromatic tropilium cation $(C_7H_7)^+$.

Cyclopentadiene, C_5H_6, is a mildly acidic hydrocarbon that can be deprotonated to form the cyclopentadienyl anion, $C_5H_5^-$. The stability of the cyclopentadienyl anion can be understood when it is realized that the six electrons in its π system make it aromatic. The delocalization of these six electrons results in a ring structure with five equal bond lengths. As a ligand, the cyclopentadienyl group, commonly denoted Cp, has played a major role in the development of organometallic chemistry and continues to be the archetype of cyclic polyene ligands. We have already alluded to the role ferrocene (**4**) had in the development of organometallic chemistry. A huge number of metal cyclopentadienyl and substituted cyclopentadienyl compounds are known. Some compounds have $C_5H_5^-$ as a monohapto ligand (**12**), in which case it is treated like an η¹-alkyl group; others contain $C_5H_5^-$ as a trihapto ligand (**13**), in which case it is treated like an η³-allyl group. Usually, though, $C_5H_5^-$ is present as a pentahapto ligand, bound through all five carbons of its ring.

We treat the η⁵-$C_5H_5^-$ group as a six-electron donor. Formally, the electron donation to the metal now comes from the filled 1π (σ bonding) and 2π (π bonding) MOs (Fig. 22.10) with δ backbonding to the d_{xy} and $d_{x^2-y^2}$ orbitals on the metal atom. As we shall see in Section 22.19, coordinated Cp ligands behave as though they maintain their six-electron aromatic structure.

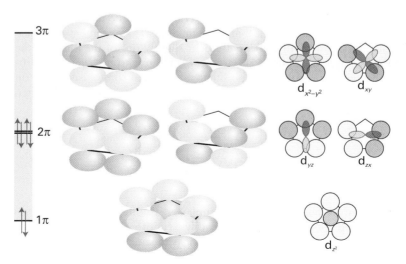

Figure 22.10 The molecular orbitals for the π systems of the cyclopentadienyl⁻ group; also shown are metal d orbitals of appropriate symmetry to form bonding interactions.

The electronic and steric properties of cyclopentadiene can easily be tuned: electron withdrawing and donating groups can be attached to the five-membered ring, and steric bulk can be enhanced by additional substitution. The pentamethylcyclopentadienyl ligand (Cp*) is commonly used to provide greater electron density on, and greater steric protection of, a metal atom. Chiral groups are often added to Cp groups so that complexes can be used in stereoselective reactions: the neo-menthyl group (**47**) is commonly used.

The synthesis, properties, and reactivities of compounds containing the cyclopentadienyl ligand are discussed in more detail in Section 22.19.

Cycloheptatriene, C_7H_8 (**48**), can form η^6-complexes such as (**49**), which may be treated as having three η^2-alkene molecules bound to the metal atom. Hydride abstraction from these complexes results in formation of η^7-complexes of the six-electron aromatic cation $C_7H_7^+$ (**50**), for example (**51**). In η^7-cycloheptatrienyl complexes, all carbon−carbon bond lengths are equal; bonding to the metal atom and backbonding from the metal are similar to those found in arene and cyclopentadienyl complexes.

22.15 Carbenes

Key points: Fischer- and Schrock-type carbene complexes are considered to have a metal−carbon double bond; N-heterocyclic carbenes are considered to have a metal−carbon single bond, together with π backbonding.

Carbene, CH_2, has only six electrons around its C atom and is consequently highly reactive. Other substituted carbenes exist and are substantially less reactive and can behave as ligands towards metals.

In principle, carbenes can exist in one of two electronic configurations: with a linear arrangement of the two groups bound to the carbon atom and the two remaining electrons unpaired in two p orbitals (**52**), or with the two groups bent, the two remaining electrons paired, and an empty p orbital (**53**). Carbenes with a linear arrangement of groups are referred to as 'triplet carbenes' (because the two unpaired electrons are unpaired and $S = 1$) and are favoured when sterically very bulky groups are attached to the carbene carbon. Carbenes with the bent arrangement are known as 'singlet carbenes' (the two electrons are paired and $S = 0$) and are the normal form for carbenes. The electron pair on the carbon atom of a singlet carbene is suitable to bond to a metal atom, resulting in a ligand-to-metal bond. The empty p orbital on the C atom can then accept electron density from the metal atom, thus stabilizing the electron-poor carbon atom (**54**). For historical reasons, carbenes bonded to metal atoms in this fashion are known as **Fischer carbenes** and are represented by a metal−carbon double bond. Fischer carbenes are electron deficient at the C atom and consequently are easily attacked by nucleophiles. When the backbonding to the C atom is very strong, the carbene can become electron rich and is thus prone to attack by electrophiles. Carbenes of this type are known as **Schrock carbenes**, after their discoverer. The term **alkylidene** technically refers only to carbenes with alkyl substituents (CR_2), but is sometimes used to mean both Fischer and Schrock carbenes.

More recently, a large number of derivatives of what are known as **N-heterocyclic carbenes** (NHCs) have been used as ligands. In most NHCs, two nitrogen atoms are adjacent to the carbene carbon atom, and if the lone pair on the nitrogen is considered to be largely p-orbital based, then strong π-donor interactions from the two nitrogen atoms can help to stabilize the carbene (**55**). Tying the carbene C atom and the two N atoms into a ring helps to stabilize the carbene, and five-membered rings are common (**56**). Additional stability can be achieved by a double bond in the ring, which provides an additional two electrons that may be considered part of a six-electron aromatic resonance structure (**57**).

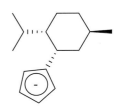

47 *neo*-Menthylcyclpentadienyl

48 Cycloheptatriene

49 $[Mo(\eta^6\text{-}C_7H_8)(CO)_3]$

50 $C_7H_7^+$

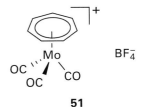

BF_4^-

51

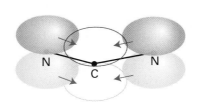

52

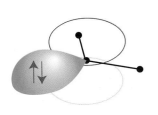

53

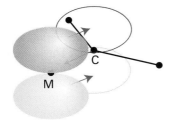

54

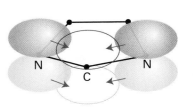

55

56

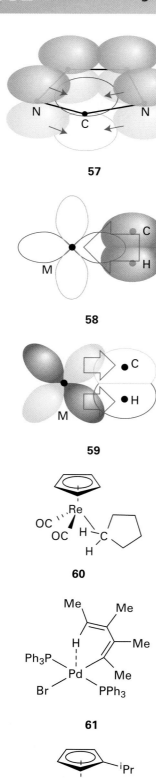

57

58

59

60

61

62

63

NHC ligands are considered to be two-electron σ donors and initial descriptions of the bonding suggested only minimal π backbonding from the metal atom. However, the current view is that there is actually significant π backbonding from the metal atom to the NHC.

22.16 Alkanes, agostic hydrogens, and noble gases

Key point: Alkanes can donate the electron density from C−H single bonds to a metal atom and, in the absence of other donors, even the electron density of a noble gas atom can enable it to behave as a ligand.

Highly reactive metal intermediates can be generated by photolysis and, in the absence of any other ligands, alkanes and noble gases have been observed to coordinate to the metal atom. Such species were first identified in the 1970s in solid methane and noble gas matrices, and were initially regarded as mere curiosities. However, both species have recently been fully characterized in solution and are now accepted as important intermediates in some reactions.

Alkanes are considered to donate electron density from a C−H σ bond to the metal atom (**58**), and accept π-electron density back from the metal atom into the corresponding σ* orbital (**59**), just like dihydrogen (Section 22.7). Although most alkane complexes are short lived, with the alkane being readily displaced, in 1998 the cyclopentane complex (**60**) was unambiguously identified in solution by NMR.

Interactions between the C−H bond of an already coordinated ligand and the metal atom have also been observed. These species are referred to as having **agostic** C−H interactions, from the Greek for 'to hold on to oneself', and are thought to have additional stability due to the chelate effect (Section 7.14). Many examples of compounds with agostic interactions are now known; an example is (**61**). Though weakly bonding, each C−H to metal atom interaction, whether it be agostic or not, is considered formally to donate two electrons to the metal.

Unlikely as it might seem, noble gas atoms can behave as ligands towards metal centres, and a number of complexes of Kr and Xe have been identified by IR spectroscopy, with the relatively long-lived Xe complex (**62**) characterized in solution by NMR in 2005. These complexes are stable only in the absence of better ligands (such as alkanes). The noble gas is formally considered to be neutral and donate two electrons.

22.17 Dinitrogen and nitrogen monoxide

Key points: The bonding of dinitrogen to a metal atom is weak but contains both a σ-donating and a π-accepting component; nitrogen monoxide can bind to a metal in two different ways—either bent or straight.

Neither dinitrogen, N_2, nor nitrogen monoxide, NO, is strictly an organometallic ligand, although they are sometimes found in organometallic compounds. Dinitrogen is a much sought after ligand as complexes can potentially take part in a catalytic reduction of nitrogen to more useful species. Dinitrogen can bind to metals in a number of different ways. The majority of complexes have a terminal monohapto link, η^1-N_2, in which the bonding can be considered to be like that of the isoelectronic CO ligand (**63**). Dinitrogen is both a weaker σ donor and a weaker π acceptor than CO, and hence is bound less strongly; in fact only good π-donor metal atoms bind N_2. Like CO, the N_2 ligand has a distinctive IR stretching band lying in the range $2150-1900$ cm^{-1}.

A dinitrogen molecule can participate in two bonding interactions and bridge two metal atoms (**64**). If the backdonation to the nitrogen is extensive in this kind of complex, it can formally be considered to have been reduced to a hydrazine (**65**). Occasionally, dinitrogen ligands are found bound in a dihapto (η^2) side-on fashion (**66**). In these complexes the ligand is best considered analogous to an η^2-alkyne. The side-on bonding mode seems to be particularly common in complexes of the f metals (Chapter 23).

Nitrogen monoxide (nitric oxide) is a radical with 11 valence electrons. When bound, NO is referred to as the nitrosyl ligand and can bond in one of two modes to d-metal atoms, in either a bent or a linear fashion. In the linear arrangement (**67**), the ligand is considered to be the NO+ cation. The NO+ cation is isoelectronic with CO, and the bonding can be considered in a similar fashion (a two-electron σ donor, with strong π-acceptor ability).

In the bent arrangement (**68**), NO is considered to behave as NO^-, again donating two electrons. In many complexes, NO can change its coordination mode; in effect, moving from the linear to the bent mode reduces the number of electrons on the metal by 2.

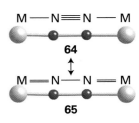

64

65

66

67

68

EXAMPLE 22.5 Counting electrons in complexes

Which of the following compounds have 18 electrons: (a) the agostic Pd compound shown as (**61**), (b) [Re(^iPr-Cp)(CO)(PF_3)Xe] (**62**)?

Answer (a) We consider both the η^1-alkenyl and the bromide ligand as two-electron singly negatively charged donors, with PPh_3 ligands being considered as two-electron neutral donors. Thus the compound is a complex of Pd(II), which provides eight further electrons. The total number of electrons, before we consider the agostic interaction, is therefore $(4 \times 2) + 8 = 16$. The agostic interaction can be considered to donate a further two electrons, leading to an 18-electron compound. (b) The ^iPr-Cp ligand is considered a six-electron donor with a single negative charge, the CO, PF_3, and Xe ligands are each considered to be two-electron neutral donors, implying that the Re must have oxidation number $+1$, so providing a further six electrons. The total electron count is therefore $6 + 2 + 2 + 2 + 6 = 18$.

Self-test 22.5 Show that both (a) [Mo(η^6-C_7H_8)(CO)_3] (**49**) and (b) [Mo(η^7-C_7H_7)(CO)_3]^+ (**51**) are 18-electron species.

Compounds

The preceding discussion of ligands and their bonding modes suggests that there are likely to be a large number of organometallic compounds that have either 16 or 18 valence electrons. A detailed discussion of all these compounds is well beyond the scope of this book. However, we shall examine a number of different classes of compounds because they provide insight into the structures and properties of many other compounds that can be derived from them. Thus, we shall now consider the structures, bonding, and reactions of metal carbonyls, which historically formed the foundation of much of d-block organometallic chemistry. Then we consider some sandwich compounds, before describing the structures and reactions of metal cluster compounds.

22.18 d-Block carbonyls

d-Block carbonyls have been studied extensively since the discovery of tetracarbonylnickel in 1890. Interest in carbonyl compounds has not waned, with many important industrial processes relying on carbonyl intermediates.

(a) Homoleptic carbonyls

Key points: The carbonyls of the Period 4 elements of Groups 6 to 10 obey the 18-electron rule; they have alternately one and two metal atoms and a decreasing number of CO ligands.

A **homoleptic complex** is a complex with only one kind of ligand. Simple homoleptic metal carbonyls can be prepared for most of the d metals, but those of Pd and Pt are so unstable that they exist only at low temperatures. No simple neutral metal carbonyls are known for Cu, Ag, and Au or for the members of Group 12. The metal carbonyls are useful synthetic precursors for other organometallic compounds and are used in organic syntheses and as industrial catalysts.

The 18-electron rule helps to systematize the formulas of metal carbonyls. As shown in Table 22.5, the carbonyls of the Period 4 elements of Groups 6 to 10 have alternately one and two metal atoms and a decreasing number of CO ligands. The binuclear carbonyls are formed by elements of the odd-numbered groups, which have an odd number of valence electrons and therefore dimerize by forming metal−metal (M−M) bonds (each M−M bond effectively increases the electron count on the metal by 1). The decrease in the number of CO ligands from left to right across a period matches the need for fewer CO ligands to achieve 18 valence electrons. The simple vanadium carbonyl $V(CO)_6$ is an exception as it only has 17 valence electrons and is too sterically crowded to dimerize; it is, however, readily reduced to the 18-electron $V(CO)_6^-$ anion.

Table 22.5 Formulas and electron count for some 3d-series carbonyls

Group	formula	Valence electrons		Structure
6	$Cr(CO)_6$	Cr	6	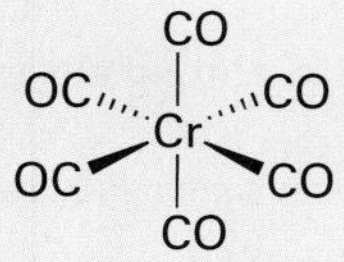
		6(CO)	12	
		Total	18	
7	$Mn_2(CO)_{10}$	Mn	7	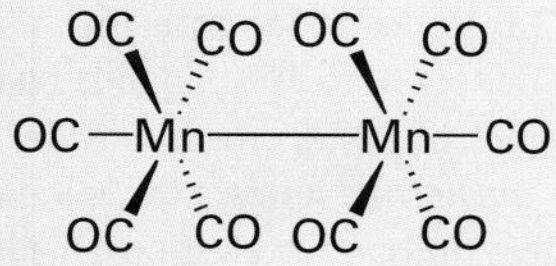
		5(CO)	10	
		M–M	1	
		Total	18	
8	$Fe(CO)_5$	Fe	8	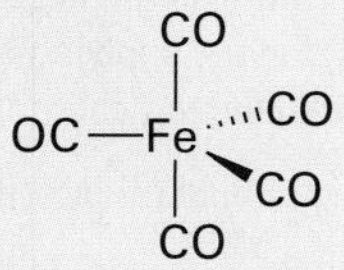
		5(CO)	10	
		Total	18	
9	$Co_2(CO)_8$	Co	9	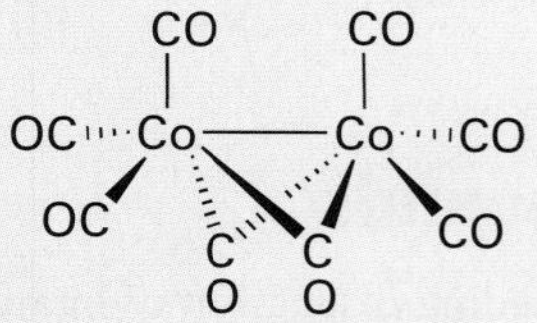
		4(CO)	8	
		M–M	1	
		Total	18	
8	$Ni(CO)_4$	Ni	10	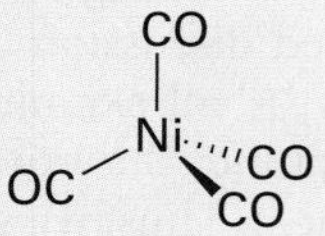
		4(CO)	8	
		Total	18	

Simple metal carbonyl molecules often have well-defined, simple, symmetrical shapes that correspond to the CO ligands taking up the most distant locations, like regions of enhanced electron density in the VSEPR model. Thus, the Group 6 hexacarbonyls are octahedral, pentacarbonyliron(0) is trigonal bipyramidal, tetracarbonylnickel(0) is tetrahedral, and decacarbonyldimanganese(0) consists of two square-pyramidal $Mn(CO)_5$ groups joined by a metal−metal bond. Bridging carbonyls are also found, for example one isomer of octacarbonyldicobalt(0) has its metal−metal bond bridged by two CO ligands.

(b) Synthesis of homoleptic carbonyls

Key points: Some metal carbonyls are formed by direct reaction, but of those that can be formed in this way most require high pressures and temperatures; metal carbonyls are commonly formed by reductive carbonylation.

The two principal methods for the synthesis of monometallic metal carbonyls are direct combination of carbon monoxide with a finely divided metal and the reduction of a metal salt in the presence of carbon monoxide under pressure. Many polymetallic carbonyls are synthesized from monometallic carbonyls.

In 1890 Mond, Langer, and Quinke discovered that the direct combination of nickel and carbon monoxide produced tetracarbonylnickel(0), $Ni(CO)_4$, a reaction that is used in the Mond process for purifying nickel (Box 22.1):

$$Ni(s) + 4\,CO(g) \xrightarrow{\text{50°C, 1 atm CO}} Ni(CO)_4(g)$$

Tetracarbonylnickel(0) is in fact the metal carbonyl that is most readily synthesized in this way, with other metal carbonyls, such as $Fe(CO)_5$, being formed more slowly. They are therefore synthesized at high pressures and temperatures (Fig. 22.11):

$$Fe(s) + 5\,CO(g) \xrightarrow{\text{200°C, 200 atm}} Fe(CO)_5(l)$$

$$2\,Co(s) + 8\,CO(g) \xrightarrow{\text{150°C, 35 atm}} Co_2(CO)_8(s)$$

Direct reaction is impractical for most of the remaining d metals, and **reductive carbonylation**, the reduction of a salt or metal complex in the presence of CO, is normally employed instead. Reducing agents vary from active metals such as aluminium and sodium, to alkyl-aluminium compounds, H_2, and CO itself:

$$CrCl_3(s) + Al(s) + 6\,CO(g) \xrightarrow{\text{AlCl}_3,\ \text{benzene}} AlCl_3(\text{soln}) + Cr(CO)_6(\text{soln})$$

$$3\,Ru(acac)_3(\text{soln}) + H_2(g) + 12\,CO(g) \xrightarrow{150°C,\ 200\ \text{atm},\ \text{CH}_3\text{OH}} Ru_3(CO)_{12}(\text{soln}) + \ldots$$

$$Re_2O_7(s) + 17\,CO(g) \xrightarrow{250°C,\ 350\ \text{atm}} Re_2(CO)_{10}(s) + 7\,CO_2(g)$$

(c) Properties of homoleptic carbonyls

Key points: All the mononuclear carbonyls are volatile; all the mononuclear and many of the polynuclear carbonyls are soluble in hydrocarbon solvents; polynuclear carbonyls are coloured.

Iron and nickel carbonyls are liquids at room temperature and pressure but all other common carbonyls are solids. All the mononuclear carbonyls are volatile; their vapour pressures at room temperature range from approximately 50 kPa for tetracarbonylnickel(0) to approximately 10 Pa for hexacarbonyltungsten(0). The high volatility of $Ni(CO)_4$, coupled with its extremely high toxicity, means that unusual care is required in its handling. Although the other carbonyls appear to be less toxic, they too must not be inhaled or allowed to touch the skin.

Because they are nonpolar, all the mononuclear and many of the polynuclear carbonyls are soluble in hydrocarbon solvents. The most striking exception among the common carbonyls is nonacarbonyldiiron(0), $[Fe_2(CO)_9]$, which has a very low vapour pressure and is insoluble in solvents with which it does not react. By contrast, $[Mn_2(CO)_{10}]$ and $[Co_2(CO)_8]$ are soluble in hydrocarbon solvents and sublime readily.

Most of the mononuclear carbonyls are colourless or lightly coloured. Polynuclear carbonyls are coloured, the intensity of the colour increasing with the number of metal atoms. For example, pentacarbonyliron(0) is a light straw-coloured liquid, nonacarbonyldiiron(0) forms golden-yellow flakes, and dodecacarbonyltriiron(0) is a deep green compound that looks black in the solid state. The colours of polynuclear carbonyls arise from electronic transitions between orbitals that are largely localized on the metal framework.

The principal reactions of the metal centre of simple metal carbonyls are substitution (Section 22.21), oxidation, reduction, and condensation into clusters (Section 22.20). In certain cases, the CO ligand itself is also subject to attack by nucleophiles or electrophiles.

(d) Oxidation and reduction of carbonyls

Key points: Most metal carbonyls can be reduced to metal carbonylates; some metal carbonyls disproportionate in the presence of a strongly basic ligand, producing the ligated cation and a carbonylate anion; metal carbonyls are susceptible to oxidation by air; metal−metal bonds undergo oxidative cleavage.

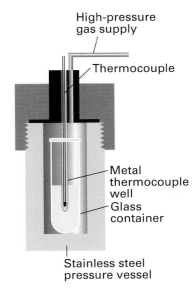

Figure 22.11 A high-pressure reaction vessel. The reaction mixture is in a glass container.

BOX 22.1 The Mond process

Ludwig Mond, Carl Langer, and Friederich Quinke discovered $Ni(CO)_4$ in the course of studying the corrosion of nickel valves in process gas containing CO in 1890. They were not able to characterize the new compound fully (calling it 'nickel-carbon-oxide') commenting, 'We have at present no suggestion to offer as to the constitution of this remarkable compound'. They were, however, able to assign the formula '$Ni(CO)_4$' to their new compound and were quick to apply their discovery to develop a new industrial process (the *Mond process*) for the purification of nickel. This process was such a success that it brought nickel all the way from Canada to Mond's factory in Wales.

The Mond process relies on the ease of synthesis of $Ni(CO)_4$. At a pressure of 1 atm of carbon monoxide, nickel metal will react at around 50°C to give $Ni(CO)_4$:

$$Ni + 4\,CO \rightarrow Ni(CO)_4$$

At this temperature $Ni(CO)_4$ (b.p. 34°C) is a gas and is easily separated from the residues and from the impure nickel. Tetracarbonylnickel decomposes to give pure nickel at about 220°C, liberating carbon monoxide, which can then be reused. Typically the impure nickel is generated from the reduction of nickel oxide ores with a mixture of hydrogen and carbon monoxide.

Mond, Langer, and Quinke also attempted the synthesis of analogous compounds with other metals, but were unable to isolate anything new. They did, however, succeed in extracting nickel contaminants from samples of cobalt, suggesting a way to purify cobalt too.

The centenary of the discovery of tetracarbonylnickel is celebrated in a special volume of *J. Organomet. Chem.*, 1990, **383**, which is devoted to metal carbonyl chemistry.

Most neutral metal carbonyl complexes can be reduced to an anionic form known as a **metal carbonylate**. In monometallic carbonyls, two-electron reduction is generally accompanied by loss of the two-electron donor CO ligand, thus preserving the electron count at 18:

$$2\,Na + Fe(CO)_5 \xrightarrow{\text{THF}} (Na^+)_2[Fe(CO)_4]^{2-} + CO$$

The metal carbonylate contains Fe with oxidation number -2, and it is rapidly oxidized by air. That much of the negative charge is delocalized over the CO ligands is confirmed by the observation of a low CO stretching band in the IR spectrum at about $1730\ cm^{-1}$. Polynuclear carbonyls, which obey the 18-electron rule through the formation of M—M bonds, are generally cleaved by strong reducing agents. The 18-electron rule is obeyed in the product and a mononegative mononuclear carbonylate results:

$$2\,Na + (OC)_5Mn-Mn(CO)_5 \xrightarrow{\text{THF}} 2\,Na[Mn(CO)_5]$$

Some metal carbonyls disproportionate in the presence of a strongly basic ligand, producing the ligated cation and a carbonylate. Much of the driving force for this reaction is the stability of the metal cation when it is surrounded by strongly basic ligands. Octacarbonyldicobalt(0) is highly susceptible to this type of reaction when exposed to a good Lewis base such as pyridine (py):

$$3\,[Co_2^{(0)}(CO)_8] + 12\,py \rightarrow 2\,[Co^{(+2)}(py)_6][Co^{(-1)}(CO)_4]_2 + 8\,CO$$

It is also possible for the CO ligand to be oxidized in the presence of the strongly basic ligand OH^-, the net outcome being the reduction of a metal centre:

$$3\,[Fe^{(0)}(CO)_5] + 4\,OH^- \rightarrow [Fe_3^{(-\frac{2}{3})}(CO)_{11}]^{2-} + CO_3^{2-} + 2\,H_2O + 3\,CO$$

Carbonyl compounds that have only 17 electrons are particularly prone to reduction to give 18-electron carbonylates.

Metal carbonyls are susceptible to oxidation by air. Although uncontrolled oxidation produces the metal oxide and CO or CO_2, of more interest in organometallic chemistry are the controlled reactions that give rise to organometallic halides. One of the simplest of these is the oxidative cleavage of an M—M bond:

$$[(OC)_5Mn^{(0)}-Mn^{(0)}(CO)_5] + Br_2 \rightarrow 2\,[Mn^{(+1)}Br(CO)_5]$$

In keeping with the loss of electron density from the metal when a halogen atom is attached, the CO stretching frequencies of the product are significantly higher than those of $[Mn_2(CO)_{10}]$.

(e) Metal carbonyl basicity

Key points: Most organometallic carbonyl compounds can be protonated at the metal centre; the acidity of the protonated form depends on the other ligands on the metal.

Many organometallic compounds can be protonated at the metal centre. Metal carbonylates provide many examples of this basicity:

$$[Mn(CO)_5]^- + H^+ \rightarrow [MnH(CO)_5]$$

The affinity of metal carbonylates for the proton varies widely (Table 22.6). It is observed that, the greater the electron density on the metal centre of the anion, the higher its Brønsted basicity and hence the lower the acidity of its conjugate acid (the metal carbonyl hydride).

As we noted in Section 22.7, d-block M—H complexes are commonly referred to as 'hydrides', which reflects the assignment of oxidation number -1 to an H atom attached to a metal atom. Nevertheless, most of the carbonyl hydrides of metals to the right of the d block are Brønsted acids. The Brønsted acidity of a metal carbonyl hydride is a reflection of the π-acceptor strength of a CO ligand, which stabilizes the conjugate base. Thus, $[CoH(CO)_4]$ is acidic whereas $[CoH(PMe_3)_4]$ is strongly hydridic. In striking contrast to p-block hydrogen compounds, the Brønsted acidity of d-block M—H compounds decreases on descending a group.

Neutral metal carbonyls (such as pentacarbonyliron, $[Fe(CO)_5]$) can be protonated in air-free concentrated acid; the Brønsted basicity of a metal atom with oxidation number zero is associated with the presence of nonbonding d electrons. Compounds having metal—metal bonds, such as clusters (Section 22.20), are even more easily protonated; here the

Table 22.6 Acidity constants of d-metal hydrides in acetonitrile at 25°C

Hydride	pK_a
$[CoH(CO)_4]$	8.3
$[CoH(CO)_3P(OPh)_3]$	11.3
$[Fe(H)_2(CO)_4]$	11.4
$[CrH(Cp)(CO)_3]$	13.3
$[MoH(Cp)(CO)_3]$	13.9
$[MnH(CO)_5]$	15.1
$[CoH(CO)_3PPh_3]$	15.4
$[WH(Cp)(CO)_3]$	16.1
$[MoH(Cp^*)(CO)_3]$	17.1
$[Ru(H)_2(CO)_4]$	18.7
$[FeH(Cp)(CO)_2]$	19.4
$[RuH(Cp)(CO)_2]$	20.2
$[Os(H)_2(CO)_4]$	20.8
$[ReH(CO)_5]$	21.1
$[FeH(Cp^*)(CO)_2]$	26.3
$[WH(Cp)(CO)_2PMe_3]$	26.6

Brønsted basicity is associated with the ready protonation of M—M bonds to produce a formal $3c,2e$ bond like that in diborane:

$$[Fe_3(CO)_{11}]^{2-} + H^+ \rightarrow [Fe_3H(CO)_{11}]^-$$

The M—H—M bridge is by far the most common bonding mode of hydrogen in clusters.

Metal basicity is turned to good use in the synthesis of a wide variety of organometallic compounds. For example, alkyl and acyl groups can be attached to metal atoms by the reaction of an alkyl or acyl halide with an anionic metal carbonyl:

$$[Mn(CO)_5]^- + CH_3I \rightarrow [Mn(CH_3)(CO)_5] + I^-$$

$$[Co(CO)_4]^- + CH_3COI \rightarrow [Co(COCH_3)(CO)_4] + I^-$$

A similar reaction with organometallic halides may be used to form M—M bonds:

$$[Mn(CO)_5]^- + [ReBr(CO)_5] \rightarrow [(OC)_5Mn-Re(CO)_5] + Br^-$$

(f) Reactions of the CO ligand

Key points: The C atom of CO is susceptible to attack by nucleophiles if it is attached to a metal atom that is electron poor; the O atom of CO is susceptible to attack by electrophiles in electron-rich carbonyls.

The C atom of CO is susceptible to attack by nucleophiles if it is attached to a metal atom that is not electron rich. Thus, terminal carbonyls with high CO stretching frequencies are liable to attack by nucleophiles. The d electrons in these neutral or cationic metal carbonyls are not extensively delocalized on to the carbonyl C atom and so that atom can be attacked by electron-rich reagents. For example, strong nucleophiles (such as methyllithium, Section 11.17) attack the CO in many neutral metal carbonyl compounds:

$$\tfrac{1}{4}[Li_4(CH_3)_4] + [Mo(CO)_6] \rightarrow Li[Mo(COCH_3)(CO)_5]$$

The resulting anionic acyl compound reacts with carbocation reagents to produce a stable and easily handled neutral product:

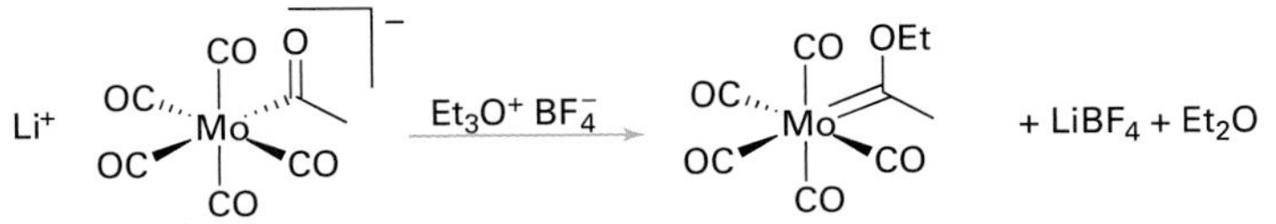

The product of this reaction, with a direct M=C bond, is a Fischer carbene (Section 22.15). The attack of a nucleophile on the C atom is also important for the mechanism of the hydroxide-induced dissociation of metal carbonyls:

$$[(OC)_nM(CO)] + OH^- \rightarrow [(OC)_nM(COOH)]^-$$

$$[(OC)_nM(COOH)]^- + 3\,OH^- \rightarrow [M(CO)_n]^{2-} + CO_3^{2-} + 2\,H_2O$$

In electron-rich metal carbonyls, considerable electron density is delocalized on the CO ligand. As a result, in some cases the O atom of a CO ligand is susceptible to attack by electrophiles. Once again, IR data provide an indication of when this type of reaction should be expected, as a low CO stretching frequency indicates significant backdonation to the CO ligand and hence appreciable electron density on the O atom. Thus a bridging carbonyl is particularly susceptible to attack at the O atom:

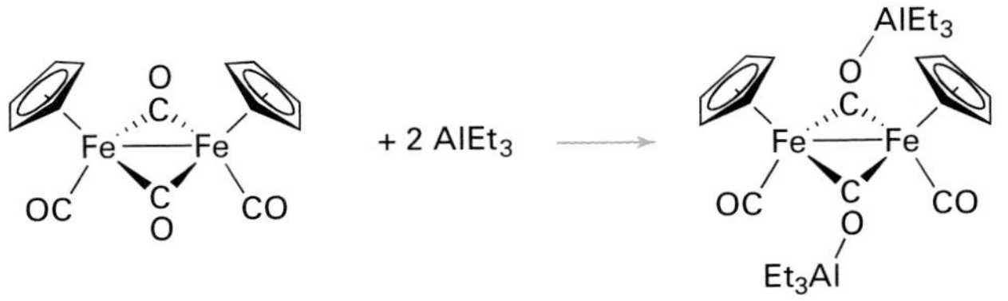

The attachment of an electrophile to the oxygen of a CO ligand, as in the structure on the right of this equation, promotes migratory insertion reactions (Section 22.24) and C—O cleavage reactions.

The ability of some alkyl-substituted metal carbonyls to undergo a migratory insertion reaction to give acyl ligands, —(CO)R, is discussed in detail in Section 22.24.

EXAMPLE 22.6 Converting CO to carbene and acyl ligands

Propose a set of reactions for the formation of $[W(C(OCH_3)Ph)(CO)_5]$ starting with hexacarbonyltungsten(0) and other reagents of your choice.

Answer We know that CO ligands in hexacarbonyltungsten(0) are susceptible to attack by nucleophiles, and therefore that the reaction with phenyllithium should give a C-phenyl intermediate:

$$W(CO)_6 + PhLi \longrightarrow$$

This anion can then react with a carbon electrophile to attach an alkyl group to the O atom of the CO ligand:

$$+ LiBF_4 + Me_2O$$

Self-test 22.6 Propose a synthesis for $[Mn(COCH_3)(CO)_4(PPh_3)]$ starting with $[Mn_2(CO)_{10}]$, PPh_3, Na, and CH_3I.

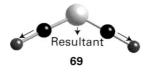

69

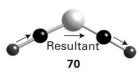

70

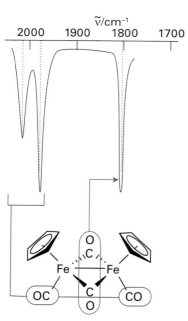

Figure 22.12 The infrared spectrum of $[Fe_2(Cp)_2(CO)_4]$. Note the two high-frequency terminal CO stretches and the lower frequency absorption of the bridging CO ligands. Although two bridging CO bands would be expected on account of the low symmetry of the complex, a single band is observed because the two bridging CO groups are nearly collinear.

(g) Spectroscopic properties of carbonyl compounds

Key points: The CO stretching frequency is decreased when it serves as a π acceptor; donor ligands cause the CO stretching frequency to decrease as they supply electrons to the metal; ^{13}C-NMR is of less use as many carbonyl compounds are fluxional on the NMR timescale.

Infrared and ^{13}C-NMR spectroscopy are widely used to determine the arrangement of atoms in metal carbonyl compounds, as separate signals are observed for inequivalent CO ligands. NMR spectra generally contain more detailed structural information than IR spectra provided the molecule is not fluxional (the timescales of the NMR and IR transition are different, Sections 8.4, 8.5). However, IR spectra are often simpler to obtain, and are particularly useful for following reactions. Most CO stretching bands occur in the range $2100-1700 \text{ cm}^{-1}$, a region that is generally free of bands due to organic groups. Both the range of CO stretching frequencies (see Fig. 22.4) and the number of CO bands (Table 22.7) are important for making structural inferences.

Group theory allows us to predict the number of active CO stretches in both the IR and Raman spectra (Section 6.5). If the CO ligands are not related by a centre of inversion or a threefold or higher axis of symmetry, a molecule with N CO ligands will have N CO stretching absorption bands. Thus a bent OC−M−CO group (with only a twofold symmetry axis) will have two infrared absorptions because both the symmetric (**69**) and antisymmetric (**70**) stretches cause the electric dipole moment to change and are IR active. Highly symmetrical molecules have fewer bands than CO ligands. Thus, in a linear OC−−M−CO group, only one IR band (corresponding to the out-of-phase stretching of the two CO ligands) is observed in the CO stretching region because the symmetrical stretch leaves the overall electric dipole moment unchanged. As shown in Fig. 22.12, the positions of the CO ligands in a metal carbonyl may be more symmetrical than the point group of the whole compound suggests and then fewer bands will be observed than are predicted on the basis of the overall point group. Raman spectroscopy can be very useful in assigning structures because the selection rules complement those of IR (Sections 6.5 and 8.4). Thus, for a linear OC−M−CO group, the symmetrical stretching of the two CO ligands is observed in the Raman spectrum.

As we noted in Section 22.5, infrared spectroscopy is also useful for distinguishing terminal CO (M−CO) from bridging CO (μ_2-CO) and face-bridging CO (μ_3-CO). It can also be used to determine the order of π-acceptor strengths for the other ligands present in a complex.

A motionally averaged NMR signal is observed when a molecule undergoes changes in structure more rapidly than the technique can resolve (Section 8.5). Although this phenomenon is quite common in the NMR spectra of organometallic compounds, it is not normally observed in their IR or Raman spectra. An example of this difference is $Fe(CO)_5$,

Table 22.7 Relation between the structure of a carbonyl complex and the number of CO stretching bands in its IR spectrum

Complex	Isomer	Structure	Point group	Number of bands*	Complex	Isomer	Structure	Point group	Number of bands*
$M(CO)_6$		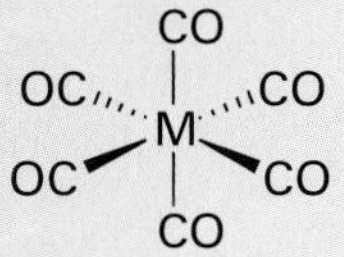 (octahedral, OC/CO)	O_h	1	$M(CO)_5$	ax	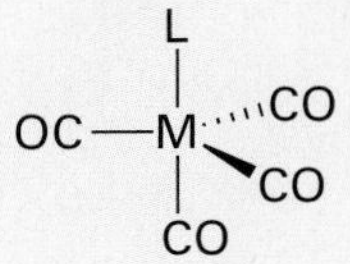 (trigonal bipyramidal)	C_{3v}	3§
$M(CO)_5L$		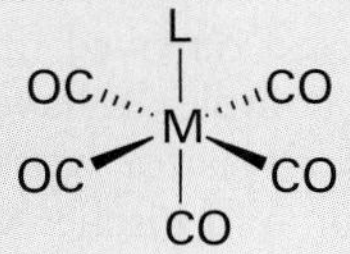 (L axial)	C_{4v}	3†	$M(CO)_4L$	eq	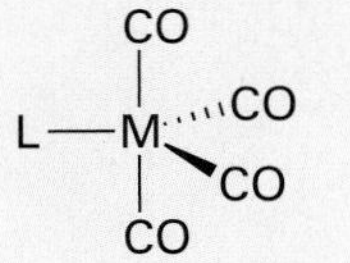 (L equatorial)	C_{2v}	4
$M(CO)_4L_2$	trans	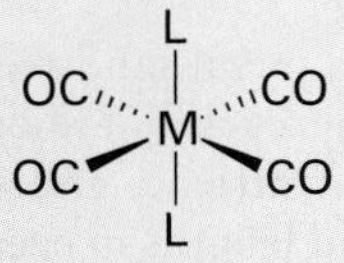 (L axial trans)	D_{4h}	1	$M(CO)_3L_2$	trans	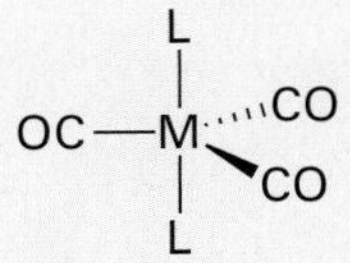 (L axial)	D_{3h}	1
$M(CO)_4L_2$	cis	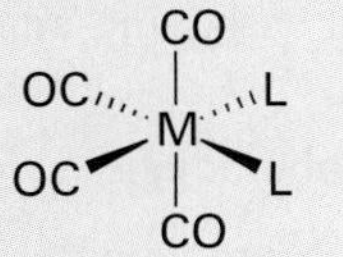 (cis)	C_{2v}	4‡	$M(CO)_3L_2$	cis	 (cis)	C_s	3
$M(CO)_3L_3$	mer	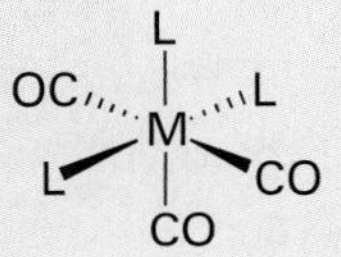 (mer)	C_{2v}	3‡	$M(CO)_4$		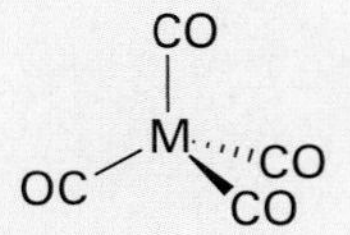 (tetrahedral)	T_d	1
$M(CO)_3L_3$	fac	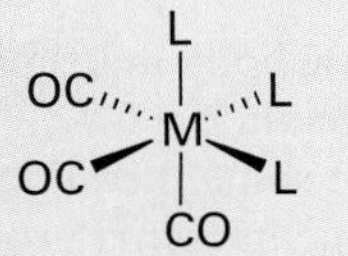 (fac)	C_{3v}	2	$M(CO)_3L_3$	fac	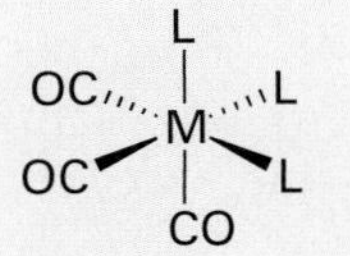 (fac)	C_{3v}	2
$M(CO)_5$		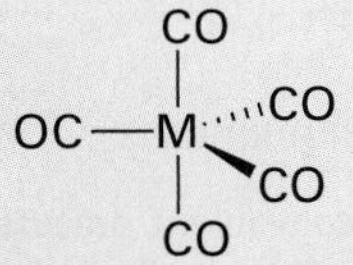 (trigonal bipyramidal)	D_{3h}	2					

* The number of IR bands expected in the CO stretching region is based on formal selection rules, and in some cases fewer bands are observed, as explained above.
† If the fourfold array of CO ligands lies in the same plane as the metal atom, two bands will be observed.
‡ If the *trans*-CO ligands are nearly collinear, one fewer band will be observed.
§ If the threefold array of CO ligands is nearly planar, only two bands will be observed..

for which the ^{13}C-NMR signal shows a single line at $\delta = 210$, whereas IR and Raman spectra are consistent with a trigonal-bipyramidal structure.

EXAMPLE 22.7 Determining the structure of a carbonyl from IR data

The complex $[Cr(CO)_4(PPh_3)_2]$ has one very strong IR absorption band at 1889 cm^{-1} and two other very weak bands in the CO stretching region. What is the probable structure of this compound? (The CO stretching frequencies are lower than in the corresponding hexacarbonyl because the phosphine ligands are better σ donors and poorer π acceptors than CO.)

Answer If we consider the possible isomers, we can see that a disubstituted hexacarbonyl may exist with either *cis* or *trans* configurations. In the *cis* isomer the four CO ligands are in a low-symmetry (C_{2v}) environment and therefore four IR bands should be observed, as indicated in Table 22.7. The *trans* isomer has a square-planar array of four CO ligands (D_{4h}), for which only one band in the CO stretching region is expected (Table 22.7). The *trans* CO arrangement is indicated by the data because it is reasonable to assume that the weak bands reflect a small departure from D_{4h} symmetry imposed by the PPh$_3$ ligands.

Self-test 22.7 The IR spectrum of $[Ni_2(\eta^5\text{-Cp})_2(CO)_2]$ has a pair of CO stretching bands at 1857 cm^{-1} (strong) and 1897 cm^{-1} (weak). Does this complex contain bridging or terminal CO ligands, or both? (Substitution of $\eta^5\text{-C}_5H_5$ ligands for CO ligands leads to small shifts in the CO stretching frequencies for a terminal CO ligand.)

22.19 Metallocenes

As we have already noted, cyclopentadienyl compounds are often remarkably stable and the discovery of (Cp)$_2$Fe, ferrocene, in 1951 sparked renewed interest in the whole field of d-block organometallic compounds. Many Cp complexes have two ring systems, with the metal sandwiched between the two rings, and it was for work on so-called 'sandwich compounds', formally **metallocenes**, that Wilkinson and Fischer were awarded the Nobel Prize in 1972.

In keeping with the picture of a metallocene as a metal sandwiched between two planar carbon rings, we can consider compounds of η^4-cyclobutadiene, η^5-cyclopentadienyl, η^6-arenes, η^7-cycloheptatrienyl (C$_7$H$_7^+$), η^8-cyclooctatriene, and even η^3-cyclopropenium to be metallocenes. As all the carbon−carbon bond lengths in each of these bound ligands are identical, it makes sense to treat each ligand as having an aromatic configuration; that is,

η^3-cyclopropenium$^+$ η^4-cyclobutadiene^{2-} η^8-cyclooctatetraenyl^{2-}

η^5-cyclopentadienyl$^-$

η^6- arenes

η^7-cycloheptatrienyl$^+$

π electrons: 2   6   10

This picture of the structure of metallocenes is not wholly in keeping with either system of electron counting, but is closer to the donor-pair method.

We touched on the structures and reactivities of some metallocenes when we discussed the mode of bonding of ligands such as the cyclopentadienyl group, and here we look at some further aspects of their bonding and reactivity.

(a) Synthesis and reactivity of cyclopentadienyl compounds

Key points: Deprotonation of cyclopentadiene gives a convenient precursor to many metal cyclopentadienyl compounds; bound cyclopentadienyl rings behave as aromatic compounds and will undergo Friedel−Crafts electrophilic reactions.

Sodium cyclopentadienide, NaCp, is a common starting material for the preparation of cyclopentadienyl compounds. It can conveniently be prepared by the action of metallic sodium on cyclopentadiene in tetrahydrofuran solution:

$$2\,Na + 2\,C_5H_6 \xrightarrow{\text{THF}} 2\,Na[C_5H_5] + H_2$$

Sodium cyclopentadienide can then be used to react with d-metal halides to produce metallocenes. Cyclopentadiene itself is acidic enough that potassium hydroxide will deprotonate it in solution and, for example, ferrocene can be prepared with a minimum of fuss:

$$2\,KOH + 2\,C_5H_6 + FeCl_2 \xrightarrow{\text{DMSO}} Fe(C_5H_5)_2 + 2\,H_2O + 2\,KCl$$

Because of their great stability, the 18-electron Group 8 compounds ferrocene, ruthenocene, and osmocene maintain their ligand−metal bonds under rather harsh conditions, and it is possible to carry out a variety of transformations on the cyclopentadienyl ligands. For example, they undergo reactions similar to those of simple aromatic hydrocarbons, such as Friedel−Crafts acylation:

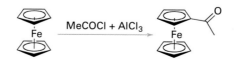

It also is possible to replace H on a C_5H_5 ring by Li:

As might be imagined, the lithiated product is an excellent starting material for the synthesis of a wide variety of ring-substituted products and in this respect resembles simple organo-lithium compounds (Section 11.17). Most Cp complexes of other metals undergo reactions similar to these two types where the five-membered ring behaves as an aromatic system.

(b) Bonding in bis(cyclopentadienyl)metal complexes

Key points: The MO picture of bonding in bis(Cp) metal complexes shows that the frontier orbitals are neither strongly bonding nor strongly antibonding; thus complexes that do not obey the 18-electron rule are possible.

We start by looking at ferrocene, where, although details of the bonding are not settled, the molecular orbital energy level diagram shown in Fig. 22.13 accounts for a number of experimental observations. This diagram refers to the eclipsed (D_{5h}) form of the complex which, in the gas phase, is about 4 kJ mol^{-1} lower in energy than the staggered conform-ation (Section 22.19c). We shall focus our attention on the frontier orbitals. As shown in Fig. 22.13, the e_1'' symmetry-adapted linear combinations of ligand orbitals have the same

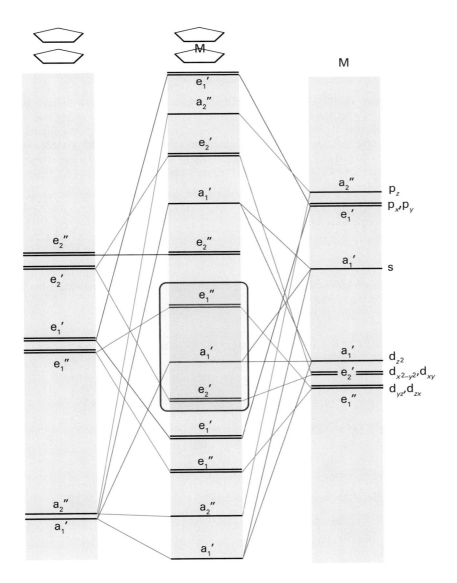

Figure 22.13 Molecular orbital energy diagram for a M(Cp)$_2$ with D_{5h} symmetry. The energies of the symmetry-adapted π orbitals of the C_5H_5 ligands are shown on the left, relevant d orbitals of the metal are on the right, and the resulting molecular orbital energies are in the centre. Eighteen electrons can be accommodated by filling the molecular orbitals up to and including the a_1' orbital in the box. The box denotes the orbitals typically regarded as frontier orbitals in these molecules.

symmetry as the d_{zx} and d_{yz} orbitals of the metal atom. The lower energy frontier orbital (a_1') is composed of d_{z^2} and the corresponding SALC of ligand orbitals. However, there is little interaction between the ligands and the metal orbitals because the ligand π orbitals happen to lie, by accident, in the conical nodal surface of d_{z^2} orbital of the metal atom. In ferrocene and the other 18-electron bis(cyclopentadienyl) complexes, the a_1' frontier orbital and all lower orbitals are full but the e_1'' frontier orbital and all higher orbitals are empty.

The frontier orbitals are neither strongly bonding nor strongly antibonding. This characteristic permits the possibility of the existence of bis(cyclopentadienyl) complexes that diverge from the 18-electron rule, such as the 17-electron complex $[Fe(\eta^5\text{-}Cp)_2]^+$ and the 20-electron complex $[Ni(\eta^5\text{-}Cp)_2]$. Deviations from the 18-electron rule, however, do lead to significant changes in M−C bond lengths that correlate fairly well with the molecular orbital scheme (Table 22.8). Similarly, the redox properties of the complexes can be understood in terms of the electronic structure.

We have already mentioned that ferrocene is fairly readily oxidized to the ferrocinium ion, $[Fe(\eta^5\text{-}Cp)_2]^+$. From an orbital viewpoint, this oxidation corresponds to the removal of an electron from the nonbonding a_1' orbital. The 19-electron complex $[Co(\eta^5\text{-}Cp)_2]$ is much more readily oxidized than ferrocene because the electron is lost from the antibonding e_1' orbital to give the 18-electron $[Co(\eta^5\text{-}Cp)_2]^+$ ion.

A useful comparison can be made with octahedral complexes. The e_1'' frontier orbital of a metallocene is the analogue of the e_g orbital in an octahedral complex, and the a_1' orbital plus the e_2'' pair of orbitals are analogous to the t_{2g} orbitals of an octahedral complex. This formal similarity extends to the existence of high- and low-spin bis(cyclopentadienyl) complexes.

EXAMPLE 22.8 Identifying metallocene electronic structure and stability

Refer to Fig. 22.13. Discuss the occupancy and nature of the HOMO in $[Co(\eta^5\text{-}Cp)_2]^+$ and the change in metal−ligand bonding relative to neutral cobaltocene.

Answer If we consider the $[Co(\eta^5\text{-}Cp)_2]^+$ ion we can see that it contains 18 valence electrons (six from Co(III), 12 from the two Cp ligands). If we then assume that the molecular orbital energy level diagram for ferrocene is applicable, the 18-electron count leads to double occupancy of the orbitals up to a_1'. The 19-electron cobaltocene molecule has an additional electron in the e_1'' orbital, which is antibonding with respect to the metal and ligands. Therefore, the metal−ligand bonds should be stronger and shorter in $[Co(\eta^5\text{-}Cp)_2]^+$ than in $[Co(\eta^5\text{-}Cp)_2]$. This conclusion is borne out by structural data.

Self-test 22.8 By using the same molecular orbital diagram, comment on whether the removal of an electron from $[Fe(\eta^5\text{-}Cp)_2]$ to produce $[Fe(\eta^5\text{-}Cp)_2]^+$ should produce a substantial change in M−C bond length relative to neutral ferrocene.

(c) Fluxional behaviour of metallocenes

Key point: Many metallocenes exhibit fluxionality and undergo internal rotation because the barrier to the interconversion of the various forms is low.

Table 22.8 Electronic configuration and M−C bond length in $[M(\eta^5\text{-}Cp)_2]$ complexes

Complex	Valence electrons	Electron configuration	M−C bond length/pm
$[V(\eta^5\text{-}Cp)_2]$	15	$e_2'^2 a_1'^1$	228
$[Cr(\eta^5\text{-}Cp)_2]$	16	$e_2'^3 a_1'^1$	217
$[Mn(\eta^5\text{-}Me\text{-}C_5H_4)_2]^*$	17	$e_2'^3 a_1'^2$	211
$[Fe(\eta^5\text{-}Cp)_2]$	18	$e_2'^4 a_1'^2$	206
$[Co(\eta^5\text{-}Cp)_2]$	19	$e_2'^4 a_1'^2 e_1''^1$	212
$[Ni(\eta^5\text{-}Cp)_2]$	20	$e_2'^4 a_1'^2 e_1''^2$	220

*Data are quoted for this complex because $[Mn(\eta^5\text{-}Cp)_2]$ has a high-spin configuration and hence an anomalously long M−C bond (238 pm).

One of the most remarkable aspects of many cyclic polyene complexes is their stereo-chemical nonrigidity (their fluxionality). For example, at room temperature the two rings in ferrocene rotate rapidly relative to each other as there is only a low staggered-eclipsed conversion barrier. This type of fluxional process is called **internal rotation**, and is similar to the process by which the two CH_3 groups rotate relative to each other in ethane. We have already noted how, in the gas phase, the eclipsed conformation of ferrocene is slightly more stable than the staggered one; however, the steric bulk of substituents on the metal-locene rings can change the barrier and can make staggered the preferred conformation. The rings of metallocenes are often drawn in a staggered conformation simply because there is then a little more space to illustrate substitutions.

Of greater interest is the stereochemical nonrigidity that is often seen when a conjugated cyclic polyene is attached to a metal atom through some, but not all, of its C atoms. In such complexes the metal–ligand bonding may hop around the ring; in the informal jargon of organometallic chemists this internal rotation is called 'ring whizzing'. A simple example is found in $[Ge(\eta^1\text{-}Cp)(CH_3)_3]$, in which the single site of attachment of the Ge atom to the cyclopentadiene ring hops around the ring in a series of **1,2-shifts**, a motion in which a C–M bond is replaced by a C–M bond to the next C atom around the ring; this motion is known as a 1,2-shift, because the bond starts on atom 1 and ends up on the adjacent atom 2 (Fig. 22.14). The great majority of fluxional conjugated polyene complexes that have been investigated migrate by 1,2-shifts, but it is not known whether these shifts are controlled by a principle of least motion or by some aspect of orbital symmetry.

Nuclear magnetic resonance provides the primary evidence for the existence and mecha-nism of these fluxional processes, as they occur on a timescale of 10^{-2} to 10^{-4} s, and can be studied by ^1H- and ^{13}C-NMR. The compound $[Ru(\eta^4\text{-}C_8H_8)(CO)_3]$ (71) provides a good illustration of the approach. At room temperature, its ^1H-NMR spectrum consists of a single, sharp line that could be interpreted as arising from a symmetrical $\eta^8\text{-}C_8H_8$ ligand. However, X-ray diffraction studies of single crystals show unambiguously that the ligand is tetrahapto. This conflict is resolved by ^1H-NMR spectra at lower temperatures because as the sample is cooled the signal broadens and then separates into four peaks. These peaks are expected for the four pairs of protons of an $\eta^4\text{-}C_8H_8$ ligand. The interpretation is that at room temperature the ring is 'whizzing' around the metal atom rapidly compared with the timescale of the NMR experiment, so an averaged signal is observed. At lower temperatures the motion of the ring is slower, and the distinct conformations exist long enough to be resolved. A detailed analysis of the line shape of the NMR spectra can be used to measure the activation energy of the migration.

(d) Bent metallocene complexes

Key point: The structures of bent sandwich compounds can be systematized in terms of a model in which three metal atom orbitals project towards the open face of the bent Cp_2M fragment.

In addition to the simple bis(cyclopentadienyl) and bis(arene) complexes with parallel rings, there are many related structures. In the jargon of this area, these species are referred to as 'bent sandwich compounds' (72), 'half-sandwich' or 'piano stool' compounds (73), and, inevitably, 'triple deckers' (74). Bent sandwich compounds play a major role in the

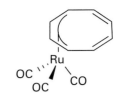

71 $[Ru(\eta^4\text{-}C_8H_8)(CO)_3]$

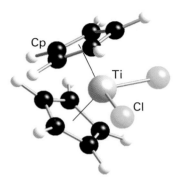

72 $[Ti(Cp)_2(Cl)_2]$

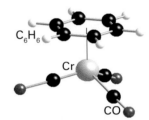

73 $[Cr(\eta^6\text{-}C_6H_6)(CO)_3]$

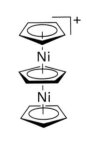

74 $[Ni_2(Cp)_3]^+$

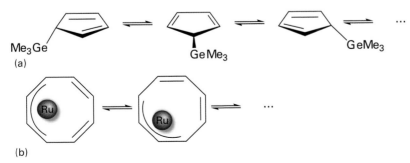

Figure 22.14 (a) The fluxional process in $[Ge(\eta^1\text{-}Cp)(Me)_3]$ occurs by a series of 1,2-shifts. (b) The fluxionality of $[Ru(\eta^4\text{-}C_8H_8)(CO)_3]$ can be described similarly. We need to imagine that the Ru atom is out of the plane of the page with the CO ligands omitted for clarity.

Figure 22.15 Bent sandwich compounds with their electron counts.

organometallic chemistry of the early and middle d-block elements, and examples include $[Ti(\eta^5\text{-}Cp)_2Cl_2]$, $[Re(\eta^5\text{-}Cp)_2Cl]$, $[W(\eta^5\text{-}Cp)_2(H)_2]$, and $[Nb(\eta^5\text{-}Cp)_2Cl_3]$.

As shown in Fig. 22.15, bent sandwich compounds occur with a variety of electron counts and stereochemistries. Their structures can be systematized in terms of a model in which three metal atom orbitals project out of the open face of the bent $M(Cp)_2$ fragment. According to this model, the metal atom often satisfies its electron deficiency when the electron count is less than 18 by interaction with lone pairs or agostic C−H groups on the ligands.

Box 22.2 describes some of the commercial uses of ferrocene.

22.20 Metal−metal bonding and metal clusters

Organic chemists must go to heroic efforts to synthesize their cage-like molecules, such as cubane(**75**). Bycontrast, one of the distinctive characteristics of inorganic chemistry is the large number of closed polyhedral molecules, such as the tetrahedral P_4 molecule (Section 15.1), the octahedral halide-bridged early d-block clusters (Section 19.11), the polyhedral carboranes (Section 13.12), and the organometallic cluster compounds that we discuss here. The structures of clusters often resemble the close-packed structures of the metal itself, and this similarity provides the main rationale for studying them—the idea that the chemical properties of the ligands of a cluster reflect the behaviour on a metal surface. In more recent years, electronic properties of clusters that depend on how large the cluster is have emerged, such as quantum dots and nanoparticles (Section 25.2).

(a) Structure of clusters

Key points: A cluster includes all compounds with metal−metal bonds that form triangular or larger cyclic structures.

A rigorous definition of **metal clusters** restricts them to molecular complexes with metal−metal bonds that form triangular or larger cyclic structures. This definition excludes linear M−M compounds and cage compounds, in which several metal atoms are held together exclusively by ligand bridges. However, this rigorous definition is normally relaxed, and

75 Cubane, C_8H_8

we shall consider any M−M bonded system as a cluster. The distinction between cage and cluster compounds can seem arbitrary as the presence of bridging ligands in a cluster such as (**76**) raises the possibility that the atoms are held together by M−L−M interactions rather than M−M bonds. Bond lengths are of some help in resolving this issue. If the M−M distance is much greater than twice the metallic radius, then it is reasonable to conclude that the M−M bond is either very weak or absent. However, if the metal atoms are within a reasonable bonding distance, the proportion of the bonding that is attributable to direct M−M interaction is ambiguous. For example, there has been much debate about the extent of Fe−Fe bonding in $[Fe_2(CO)_9]$ (**77**).

Metal−metal bond strengths in metal complexes cannot be determined with great precision, but a variety of pieces of evidence—such as the stability of compounds and M−M force constants—indicate that there is an increase in M−M bond strengths down a group in the d block. This trend contrasts with that in the p block, where element−element bonds are usually weaker for the heavier members of a group. As a consequence of this trend, metal−metal bonded systems are most numerous for the 4d- and 5d-series metals.

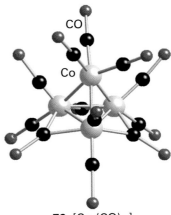

76 $[Co_4(CO)_{12}]$

(b) Electron counting in clusters

Key points: The 18-electron rule is suitable for identifying the correct number of electrons for clusters with fewer than six metal atoms; the Wade−Mingos−Lauher rules identify a correlation between the valence electron count and the structures of larger organometallic complexes.

Organometallic cluster compounds are rare for the early d metals and unknown for the f metals, but a large number of metal carbonyl clusters exist for the elements of Groups 6 to 10. The bonding in the smaller clusters can be readily explained in terms of local M−M and M−L electron pair bonding and the 18-electron rule.

If we take $Mn_2(CO)_{10}$ and $Os_3(CO)_{12}$ as examples we can arrive at a simple, yet illuminating, picture. In $Mn_2(CO)_{10}$ each Mn atom is considered to have 17 electrons (seven from Mn and 10 from the five CO ligands) before we take into account the Mn−Mn bond. This Mn−Mn bond consists of two electrons shared between the two metal atoms, and hence raises the electron count of each by 1, resulting in two 18-electron metal atoms, but with a total electron count of only 34, not 36. In $Os_3(CO)_{12}$, each $Os(CO)_4$ fragment has 16 electrons before metal−metal bonding is taken into consideration and each metal shares two further electrons with adjacent metals, so increasing the number of electrons around each metal to 18, but with a total of only 48 and not 54 electrons. The bonding electrons we are dealing with are referred to as the **cluster valence electrons** (CVEs), and it quickly becomes apparent that a cluster of x metal atoms with y metal−metal bonds needs $18x - 2y$ electrons. Octahedral M_6 and larger clusters do not conform to this pattern, and the polyhedral skeletal electron pair rules, known as Wade's rules (Section 13.11), have been refined by D.M.P. Mingos and J. Lauher to apply to metal clusters. These **Wade−Mingos−Lauher rules** are summarized in Table 22.9; they apply most reliably to metal clusters in Groups 6 to 9. In general, and as in the boron hydrides where similar considerations apply, more open structures (which have fewer metal−metal bonds) occur when there is a higher CVE count.

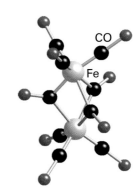

77 $[Fe_2(CO)_9]$

EXAMPLE 22.9 Correlating spectroscopic data, cluster valence electron count, and structure

The reaction of chloroform (trichloromethane) with $[Co_2(CO)_8]$ yields a compound of formula $[Co_3(CH)(CO)_9]$. Both NMR and IR data indicate the presence of only terminal CO ligands and the presence of a CH group. Propose a structure consistent with the spectra and the correlation of CVE with structure.

Answer We assume that the CH ligand is simply C-bonded: one C electron is used for the C−H bond, so three are available for bonding in the cluster. Electrons available for the cluster are then 27 for three Co atoms, 18 for nine CO ligands, and three for CH. The resulting total CVE of 48 indicates a triangular cluster (see Table 22.9). A structure consistent with this conclusion and the presence of only terminal CO ligands and a capping CH ligand is (**78**).

Self-test 22.9 The compound $[Fe_4(Cp)_4(CO)_4]$ is a dark-green solid. Its IR spectrum shows a single CO stretch at 1640 cm^{-1}. The ^1H NMR spectrum is a single line even at low temperatures. From this spectroscopic information and the CVE, propose a structure for $[Fe_4(Cp)_4(CO)_4]$.

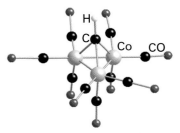

78 $[Co_3(CH)(CO)_9]$

Table 22.9 Correlation of cluster valence electron (CVE) count and structure

Number of metal atoms	Structure of metal framework		CVE count	Example
1	Single atom		18	$Ni(CO)_4$ (**2**)
2	Linear		34	$Mn_2(CO)_{10}$
3	Closed triangle	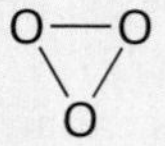	48	$[Co_3(CH)(CO)_9]$ (**78**)
4	Tetrahedron		60	$Co_4(CO)_{12}$ (**76**)
	Butterfly		62	$[Fe_4(CO)_{12}C]^{2-}$
	Square	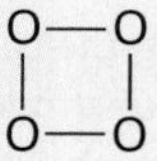	64	$Os_4(CO)_{16}$
5	Trigonal bipyramid	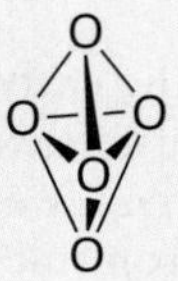	72	$Os_5(CO)_{16}$
	Square pyramid	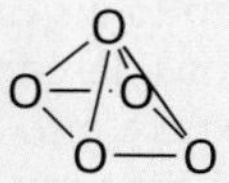	74	$Fe_5C(CO)_{15}$
6	Octahedron	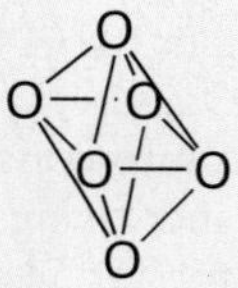	86	$Ru_6C(CO)_{17}$
	Trigonal prism	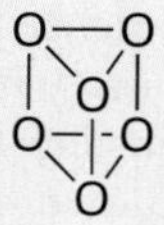	90	$[Rh_6C(CO)_{15}]^{2-}$

(c) Isolobal analogies

Key point: Structurally analogous fragments of molecules are described as isolobal; groups of isolobal molecular fragments are used to suggest patterns of bonding between seemingly unrelated fragments, allowing the rationalization of diverse structures.

We can identify analogies in the structures of apparently unrelated molecules. Thus, we may view $N(CH_3)_3$ as derived from NH_3 by substitution of a CH_3 fragment for each H atom. In current terminology, the structurally analogous fragments are said to be **isolobal**, and the relationship is expressed by the symbol $\overline{0}$. The origin of the name is the lobe-like shape of a hybrid orbital in a molecular fragment. Two fragments are isolobal if their highest energy orbitals have the same symmetry (such as the σ symmetry of the H1s and a Csp^3 hybrid orbital), similar energies, and the same electron occupation (one in each case in H1s and Csp^3). Table 22.10 lists some selected isolobal fragments and the first line shows isolobal fragments with a single frontier orbital. The recognition of this family permits us to anticipate by analogy with H–H that molecules such as $H_3C–CH_3$ and $(OC)_5Mn–CH_3$ can be formed. The second line of Table 22.10 lists some isolobal fragments with two frontier orbitals, and the third line lists some with three.

Isolobal analogies provide a good way to picture the incorporation of hetero atoms into a metal cluster. These analogies allow us to draw a parallel between $[Co_3(CH)(CO)_9]$ (**78**)

Table 22.10 Selected isolobal fragments

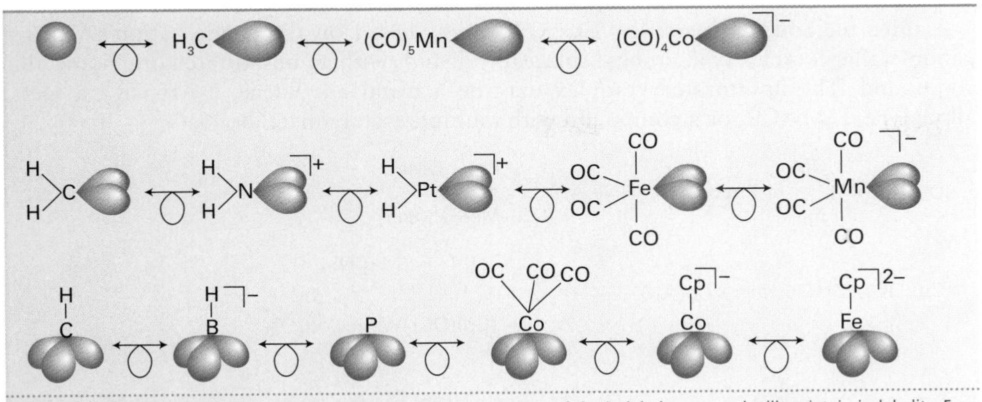

Note that electrons can be added to or subtracted from each member of the isolobal group and still maintain isolobality. For example, CH₃⁺ ⟷ Mn(CO)₃⁺ ⟷ Co(CO)₄

and [Co₄(CO)₁₂] (**76**), both of which can be regarded as triangular Co₃(CO)₉ fragments capped on one side either by Co(CO)₃ or by CH. A minor complication in this comparison is the occurrence of Co(CO)₂ groups together with bridging CO ligands in [Co₄(CO)₁₂] because bridging and terminal ligands often have similar energies. Further inspection of the isolobal fragments given in Table 22.10 shows that a P atom is isolobal with CH; accordingly, a cluster similar to [Co₃(CH)(CO)₉] (**78**) is known, but with a capping P atom. Similarly, the ligands CR₂ and Fe(CO)₄ are both capable of bonding to two metal atoms in a cluster; CH₃ and Mn(CO)₅ can bond to one metal atom.

As a final example of isolobal analogies, consider the mixed manganese and platinum complex (**79**): the three-membered {Mn,Pt,P} rings can be considered as the metallocyclopropane form of a coordinated double bond (**80**). Both halves of the Mn=P fragments are isolobal with CH₂ if we treat them as PR₂⁺ and Mn(CO)₄⁻; the complete fragment can then be treated as analogous to an ethene molecule. This treatment means that (**79**) can be considered analogous to bis-(ethene)carbonylplatinum(0) (**81**), a known simple 16-electron organometallic compound.

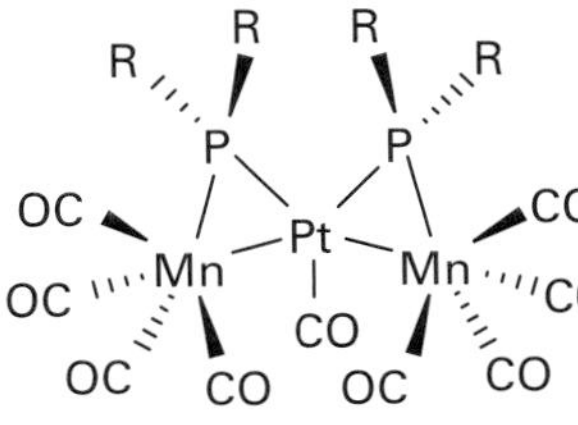

(d) Synthesis of clusters

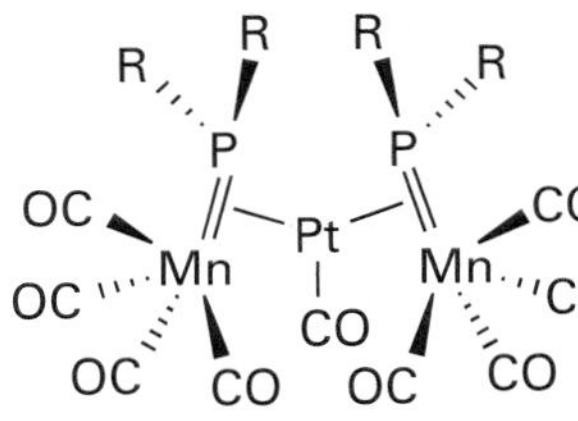

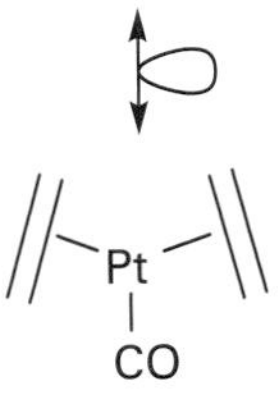

Key point: Three methods are commonly used to prepare metal clusters: thermal expulsion of CO from a metal carbonyl, the condensation of a carbonyl anion and a neutral organometallic complex, and the condensation of an organometallic complex with an unsaturated organometallic compound.

One of the oldest methods for the synthesis of metal clusters is the thermal expulsion of CO from a metal carbonyl. The pyrolytic formation of metal cluster compounds can be viewed from the standpoint of electron count: a decrease in valence electrons around the metal resulting from loss of CO is compensated by the formation of M−M bonds. One example is the synthesis of [Co₄(CO)₁₂] by heating [Co₂(CO)₈]:

$$2\,[Co_2(CO)_8] \rightarrow [Co_4(CO)_{12}] + 4\,CO$$

This reaction proceeds slowly at room temperature, so samples of octacarbonyldicobalt(0) are usually contaminated with dodecacarbonyltetracobalt(0).

A widely used and more controllable reaction is based on the condensation of a carbonyl anion and a neutral organometallic complex:

$$[Ni_5(CO)_{12}]^{2-} + [Ni(CO)_4] \rightarrow [Ni_6(CO)_{12}]^{2-} + 4\,CO$$

The Ni₅ complex has a CVE of 76 whereas the Ni₆ complex has a count of 86. The descriptive name **redox condensation** is often given to reactions of this type, which are very useful for the preparation of anionic metal carbonyl clusters. In this example, a trigonal-bipyramidal cluster containing Ni with formal oxidation number $-\frac{2}{5}$ and Ni(CO)₄ containing Ni(0) is converted into an octahedral cluster having Ni with oxidation number $-\frac{1}{3}$. The [Ni₅(CO)₁₂]²⁻ cluster, which has four electrons in excess of the 72 expected for a trigonal

79

80

81

bipyramid, illustrates a fairly common tendency for the Group 10 metal clusters to have an electron count in excess of that expected from the Wade−Mingos−Lauher rules.

A third method, pioneered by F.G.A. Stone, is based on the condensation of an organometallic complex containing displaceable ligands with an unsaturated organometallic compound. The unsaturated complex may be a metal alkylidene, $L_nM=CR_2$, a metal alkylidyne, $L_nM\equiv CR$, or a compound with multiple metal−metal bonds:

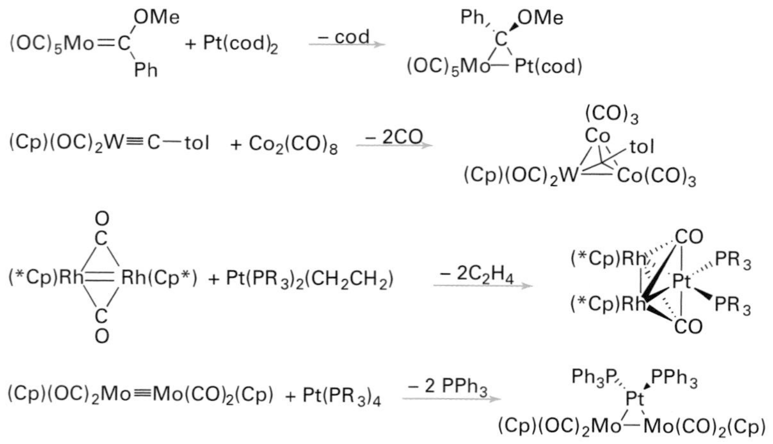

Reactions

The fact that most organometallic compounds can be made to react in a variety of ways is responsible for their use as catalysts. In the previous sections we looked at ligands and how to introduce them into a metal centre, and in this section we look at how ligands might react further or with each other. Implicit in the following discussion is the fact that coordinatively saturated complexes are less reactive than unsaturated ones.

22.21 Ligand substitution

Key points: The substitution of ligands in organometallic complexes is very similar to the substitution of ligands in coordination complexes, with the additional constraint that the valence electron count at the metal atom does not increase above 18; steric crowding of ligands increases the rate of dissociative processes and decreases the rate of associative processes.

Extensive studies of CO substitution reactions of simple carbonyl complexes have revealed systematic trends in mechanisms and rates, and much that has been established for these compounds is applicable to all organometallic complexes. The simple replacement of one ligand by another in organometallic complexes is very similar to that observed with coordination compounds, where reactions sometimes go by an associative, a dissociative, or an interchange pathway, with the reaction being either associatively or dissociatively activated (Section 21.2).

The simplest examples of substitution reactions involve the replacement of CO by another electron-pair donor, such as a phosphine. Studies of the rates at which trialkylphosphines and other ligands replace CO in $Ni(CO)_4$, $Fe(CO)_5$, and the hexacarbonyls of the chromium group show that they are relatively insensitive to the incoming group, indicating that a dissociatively activated mechanism is in operation. In some cases a solvated intermediate such as $[Cr(CO)_5(THF)]$ has been detected. This intermediate then combines with the entering group in a bimolecular process:

$$Cr(CO)_6 + sol \rightarrow Cr(CO)_5(sol) + CO$$

$$Cr(CO)_5(sol) + L \rightarrow Cr(CO)_5L + sol$$

A dissociatively activated substitution reaction would be expected with metal carbonyl complexes as associative activation would require reaction intermediates with more than 18 valence electrons, formation of which corresponds to populating high-energy antibonding MOs.

Whereas the loss of the first CO group from $Ni(CO)_4$ occurs easily, and substitution is fast at room temperature, the CO ligands are much more tightly bound in the Group 6 carbonyls, and loss of CO often needs to be promoted thermally or photochemically. For example, the substitution of CO by CH_3CN is carried out in refluxing acetonitrile, using a stream of nitrogen to sweep away the carbon monoxide and hence drive the reaction to completion. To achieve photolysis, mononuclear carbonyls (which do not absorb strongly in the visible region) are exposed to near-UV radiation in an apparatus like that shown in Fig. 22.16. As with the thermal process, there is strong evidence that the photoassisted substitution reaction leads to the formation of a labile intermediate complex with the solvent, which is then displaced by the entering group. Solvated intermediates in the photolysis of metal carbonyls have been detected, not only in polar solvents such as THF but also in every solvent that has been tried, even in alkanes and noble gases.[3]

The rates of substitution of ligands in 16-electron complexes are sensitive to the identity and concentration of the entering group, which indicates associative activation. For example, the reactions of $[Ir(CO)Cl(PPh_3)_2]$ with triethylphosphine are associatively activated:

$$[Ir(CO)Cl(PPh_3)_2] + PEt_3 \rightarrow [Ir(CO)Cl(PPh_3)_2(PEt_3)] \rightarrow [Ir(CO)Cl(PPh_3)(PEt_3)] + PPh_3$$

Sixteen-electron organometallic compounds appear to undergo associatively activated substitution reactions because the 18-electron activated complex is energetically more favourable than the 14-electron activated complex that would occur in dissociative activation.

As in the reactions of coordination complexes, we can expect steric crowding between ligands to accelerate dissociative processes and to decrease the rates of associative processes (Section 21.6). The extent to which various ligands crowd each other is approximated by the Tolman cone angle and we can see how it influences the equilibrium constant for ligand binding by examining the dissociation constants of $[Ni(PR_3)_4]$ complexes (Table 22.11). These complexes are slightly dissociated in solution if the phosphine ligands are compact, such as PMe_3, with a cone angle of 118°. However, a complex such as $[Ni(P^tBu_3)_4]$, where the cone angle is huge (182°), is highly dissociated. The ligand P^tBu_3 is so bulky that the 14-electron complex $[Pt(P^tBu_3)_2]$ can be identified.

The rate of CO substitution in six-coordinate metal carbonyls often decreases as more strongly basic ligands replace CO, and two or three alkylphosphine ligands often represent the limit of substitution. With bulky phosphine ligands, further substitution may be thermodynamically unfavourable on account of ligand crowding, but increased electron density on the metal centre, which arises when a π-acceptor ligand is replaced by a net donor ligand, appears to bind the remaining CO ligands more tightly and therefore reduce the rate of CO dissociative substitution. The explanation of the influence of σ-donor ligands on CO bonding is that the increased electron density contributed by the phosphine leads to stronger π backbonding to the remaining CO ligands and therefore strengthens the M−CO bond. This stronger M−C bond decreases the tendency of CO to leave the metal atom and therefore decreases the rate of dissociative substitution. It is also observed that the second carbonyl that is replaced is normally *cis* to the site of the first and that replacement of a third carbonyl results in a *fac* complex. The reason for this regiochemistry is that CO ligands have very high *trans* effects (Section 21.4).

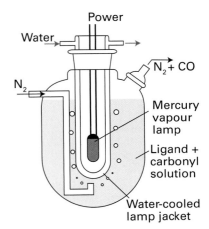

Figure 22.16 Apparatus for photochemical ligand substitution of metal carbonyls.

Table 22.11 Cone angles and dissociation constants for Some Ni complexes

L	$\theta/°$	K_d
PMe_3	118	$< 10^{-9}$
PEt_3	137	1.2×10^{-5}
$PMePh_2$	136	5.0×10^{-2}
PPh_3	145	Large
P^tBu_3	182	Large

Data are for $NiL_4 \rightleftharpoons NiL_3 + L$ in benzene at 25°C.

EXAMPLE 22.10 Preparing substituted metal carbonyls

Starting with MoO_3 as a source of Mo, and CO and PPh_3 as the ligand sources, plus other reagents of your choice, give equations and conditions for the synthesis of $[Mo(CO)_5PPh_3]$.

Answer Considering the materials available to us, a sensible procedure might be to synthesize $Mo(CO)_6$ first and then carry out a ligand substitution. Reductive carbonylation of MoO_3 can be performed using $Al(CH_2CH_3)_3$ as a reducing agent in the presence of carbon monoxide under pressure. The temperature and pressure required for this reaction are less than those for the direct combination of molybdenum and carbon monoxide.

$$MoO_3 + Al(CH_2CH_3)_3 + 6\,CO \xrightarrow{\text{50 atm, 150°C, heptane}} Mo(CO)_6 + \text{oxidation products of } Al(CH_2CH_3)_3$$

---

[3] Infrared spectroscopy has been used to demonstrate that, as quickly as can be measured (that is, in less than one picosecond), the photolysis of $W(CO)_6$ results in the formation of $W(CO)_5(sol)$.

82

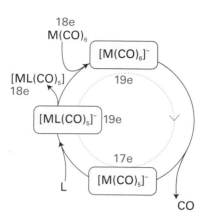

18e
M(CO)$_6$

[M(CO)$_6$]$^-$

[ML(CO)$_5$]
18e

19e

[ML(CO)$_5$]$^-$ 19e

17e

[M(CO)$_5$]$^-$

L

CO

Figure 22.17 Schematic diagram of an electron transfer catalysed CO substitution. After addition of a small amount of a reducing initiator, the cycle continues until the limiting reagent M(CO)$_6$ or L has been consumed.

83 [ReCp(CO)$_3$]

84

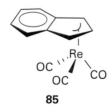

85

86 Fluorenyl$^-$

The subsequent substitution could be carried out photochemically by using the apparatus illustrated in Fig. 22.16:

$$Mo(CO)_6 + PPh_3 \xrightarrow{THF,\ h\nu} Mo(CO)_5PPh_3 + CO$$

The progress of the reaction can be followed by IR spectroscopy in the CO stretching region using small samples that are removed periodically from the reaction vessel.

Self-test 22.10 If the highly substituted complex [Mo(CO)$_3$L$_3$] is desired, which of the ligands PMe$_3$ or P(^{t}Bu)$_3$ would be preferred? Give reasons for your choice.

Although the generalizations above apply to a wide range of reactions, some exceptions are observed, especially if cyclopentadienyl or nitrosyl ligands are present. In these cases it is common to find evidence of associatively activated substitution even for 18-electron complexes. The common explanation is that NO may switch from being linear (as in **67**) to being angular (as in **68**), whereupon it donates two fewer electrons (Section 22.17). Similarly, the η^5-Cp six-electron donor can slip relative to the metal and become an η^3-Cp four-electron donor. In this case, the C$_5$H$_5^-$ ligand is regarded as having a three-carbon interaction with the metal while the remaining two electrons form a simple C=C bond that is not engaged with the metal (**82**), and the relatively electron-depleted central metal atom becomes susceptible to substitution:

$$[V(CO)_5(NO)] + PPh_3 \rightarrow [V(CO)_4(NO)(PPh_3)] + CO$$

$$[Re(\eta^5\text{-}Cp)(CO)_3] + PPh_3 \rightarrow [Re(\eta^3\text{-}Cp)(CO)_2(PPh_3)] + CO$$

It has been found that, for some metal carbonyls, the displacement of CO can be catalysed by electron-transfer processes that create anion or cation radicals. These radicals do not have 18 electrons, and a typical process of this type is illustrated in Fig. 22.17. As can be seen, the key feature is the lability of CO in the 19-electron anion radical compared to the metal carbonyl starting material. Similarly, the less common 19- and 17-electron metal compounds are labile with respect to substitution.

The substitution of ligands at a cluster is often not a straightforward process because fragmentation is common. Fragmentation occurs because the M—M bonds in a cluster are generally comparable in strength to the M—L bonds, and so the breaking of the M—M bonds provides a reaction pathway with a low activation energy. For example, dodecacarbonyltriiron(0) reacts with triphenylphosphine under mild conditions to yield simple mono- and disubstituted products as well as some cluster fragmentation products:

$$3\,[Fe_3(CO)_{12}] + 6\,PPh_3 \rightarrow [Fe_3(CO)_{11}PPh_3] + [Fe_3(CO)_{10}(PPh_3)_2] + [Fe(CO)_5]$$
$$+ [Fe(CO)_4PPh_3] + [Fe(CO)_3(PPh_3)_2] + 3\,CO$$

However, for somewhat longer reaction times or elevated temperatures, only monoiron cleavage products are obtained. Because the strength of M—M bonds increases down a group, substitution products of the heavier clusters, such as [Ru$_3$(CO)$_{10}$(PPh$_3$)$_2$] or [Os$_3$(CO)$_{10}$(PPh$_3$)$_2$], can be prepared without significant fragmentation into mononuclear complexes.

EXAMPLE 22.11 Assessing substitutional reactivity

Which of the compounds (**83**) or (**84**) will undergo substitution of a CO ligand for a phosphine more readily?

Answer If we consider the ligands present on the compounds, we can see that the 18-electron compound (**84**) contains the indenyl ligand, which can ring slip to a 16-electron η^3 bound form (**85**) more readily than the normal Cp compound (**83**), as the double bond that forms becomes part of a six-membered aromatic ring. This ring slipping provides a low-energy route to coordinative unsaturation and thus the indenyl compound reacts with an incoming ligand much more rapidly than the Cp compound.

Self-test 22.11 Assess the relative substitutional reactivities of indenyl and fluorenyl (**86**) compounds.

22.22 Oxidative addition and reductive elimination

Key points: Oxidative addition occurs when a molecule X—Y adds to a metal atom to form new M—X and M—Y bonds with cleavage of the X—Y bond; oxidative addition results in increasing the coordination number of the metal atom by 2 and increasing the oxidation number by 2; reductive elimination is the reverse of oxidative addition.

When we discussed the bonding of dihydrogen to a metal atom in Section 22.7, we noted that the oxidation number on the metal atom increased by 2 when the dihydrogen reacted to give a dihydride:

$$M(N_{ox}) + H_2 \rightarrow [M(N_{ox}+2)(H)_2]$$

The increase in oxidation number of the metal by 2 arises because dihydrogen is treated as a neutral ligand, whereas the hydride ligands are treated as H$^-$: thus the formation of two M—H bonds from a H$_2$ molecule corresponds to a formal increase in the charge on the metal by 2. Whilst it might seem that this oxidation of the metal is just an anomaly thrown up by our method of counting electrons, two of the electrons on the metal atom have been used to backbond to the dihydrogen, and these two electrons are no longer available to the metal for further bonding. This type of reaction is quite general and is known as **oxidative addition**. A large number of molecules add oxidatively to a metal atom, including the alkyl and aryl halides, dihydrogen, and simple hydrocarbons. In general, the addition of any molecule X—Y to a metal atom, where both X and Y are more electronegative than the metal, can be classed as oxidative addition. Thus the reaction of a metal with an acid such as HCl is an oxidative addition reaction. Oxidative addition reactions are not restricted to d-block metals: the reaction of magnesium to form Grignard reagents (Section 12.13) is an oxidative addition reaction.

Oxidative addition reactions result in two more ligands bound to the metal with an increase in the total electron count at the metal of 2. Thus oxidative addition reactions normally require a coordinatively unsaturated metal centre, and are particularly common for 16-electron square-planar metal complexes:

16e Pt(II)   18e Pt(IV)

The oxidative addition of hydrogen is a concerted reaction: dihydrogen coordinates to form a σ-bonded H$_2$ ligand, and then backbonding from the metal results in cleavage of the H—H bond and the formation of *cis* dihydrides:

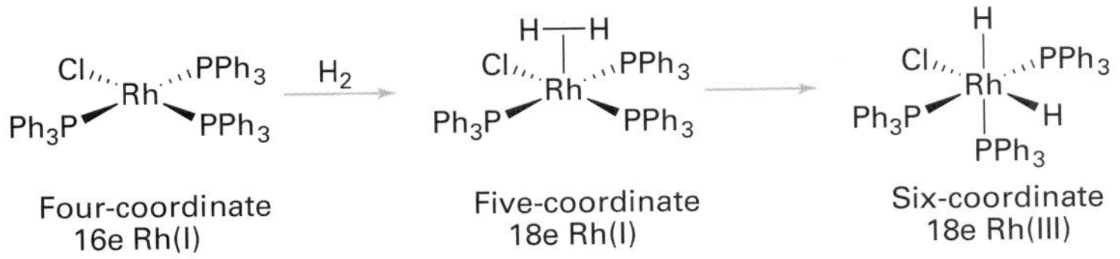

| Four-coordinate | Five-coordinate | Six-coordinate |
| 16e Rh(I) | 18e Rh(I) | 18e Rh(III) |

Other molecules, such as alkanes and aryl halides, are known to react in a concerted fashion, and in all these cases the two incoming ligands end up *cis* to each other.

Some oxidative addition reactions are not concerted and either go through radical intermediates or are best thought of as S$_N$2 displacement reactions. Radical oxidative addition reactions are rare and will not be discussed further here. In an S$_N$2 oxidative addition reaction, a lone pair on the metal attacks the X—Y molecule displacing Y$^-$, which subsequently bonds to the metal:

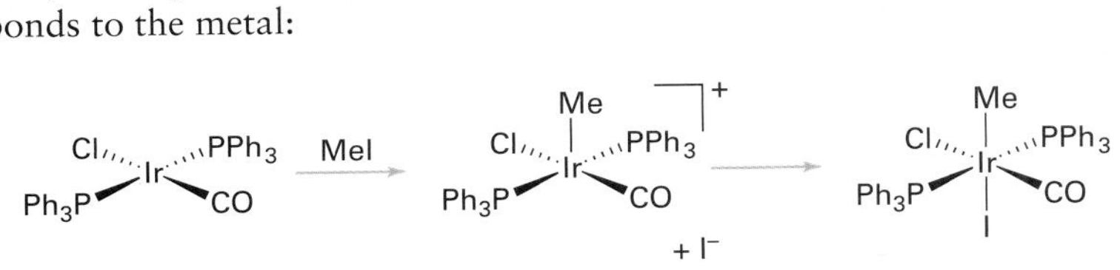

There are two stereochemical consequences of this reaction. First, the two incoming ligands need not end up *cis* to each other and, second, unlike the concerted reaction, any chirality at the X group is inverted. An S_N2-type oxidative addition is common for polar molecules such as alkyl halides.

The opposite of oxidative addition, where two ligands couple and eliminate from a metal centre, is known as **reductive elimination**:

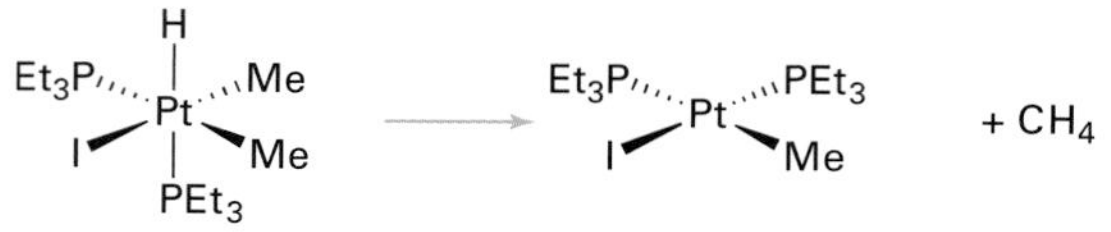

18e Pt(IV)   **16e Pt(II)**

Reductive elimination reactions require both eliminating fragments to be *cis* to each other, and are best thought of as the reverse of the concerted form of oxidative addition.

Oxidative addition and reductive elimination reactions are, in principle, reversible. However, in practice, one direction is normally thermodynamically favoured over the other. Oxidative addition and reductive elimination reactions play a major role in many catalytic processes (Chapter 26).

EXAMPLE 22.12 **Identifying oxidative addition and reductive elimination**

Show that the reaction

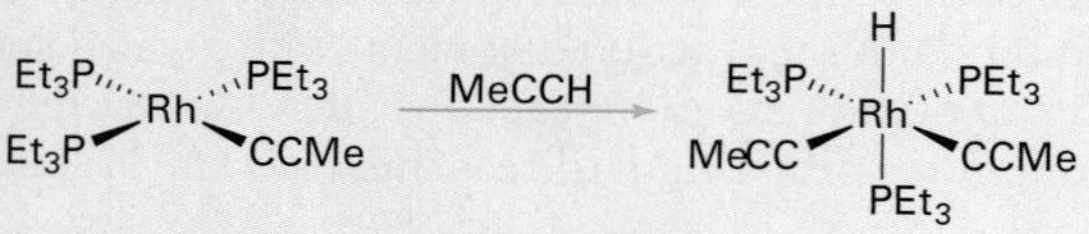

is an example of an oxidative addition reaction.

Answer In order to identify an oxidative addition reaction, we need to establish the valence electron counts and oxidation states of both the starting material and the product. The four-coordinate square-planar Rh starting material contains an η^1-alkynyl ligand as well as three neutral phosphine ligands; it is therefore a 16-electron Rh(I) species. The six-coordinate octahedral product contains two η^1-alkynyl ligands, a hydride ligand, and three neutral phosphine ligands; it is therefore an 18-electron Rh(III) species. The increase in both coordination number and oxidation number by 2 identifies it as an oxidative addition.

Self-test 22.12 Show that the reaction:

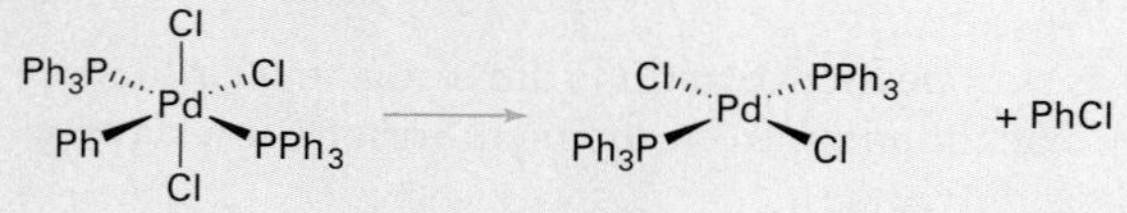

is an example of reductive elimination.

22.23 σ-Bond metathesis

Key point: A σ-bond metathesis reaction is a concerted process that sometimes occurs when oxidative addition cannot take place.

A reaction sequence that appears to be an oxidative addition followed by a reductive elimination may in fact be the exchange of two species by a process known as σ-**bond metathesis**. σ-Bond metathesis reactions are common for early d-metal complexes where there are not enough electrons on the metal atom for it to participate in oxidative addition. For instance, the 16-electron compound $[ZrHMe(Cp)_2]$ cannot react with H_2 to give a trihydride as all its electrons are involved in bonding to the existing ligands. A four-membered transition state is proposed in such cases, and a concerted bond-making and bond-breaking step results in elimination of methane:

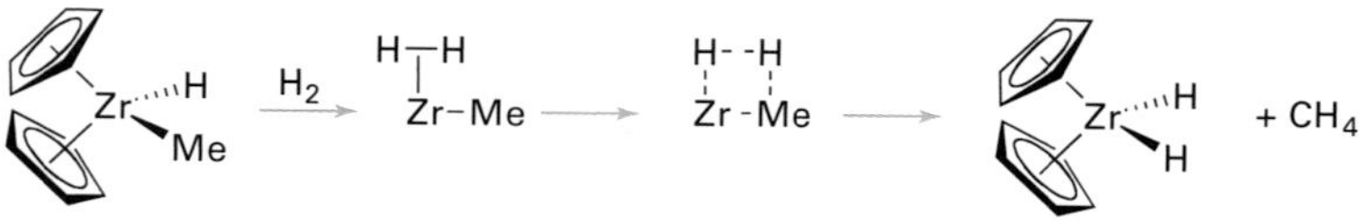

22.24 1,1-Migratory insertion reactions

Key point: 1,1-Migratory insertion reactions result from the migration of a species such as a hydride or alkyl group to an adjacent ligand such as carbonyl to give a metal complex with two fewer electrons on the metal atom.

A **1,1-migratory insertion reaction** is exemplified by reactions of the η^1-CO ligand, where the following change can take place:

The reaction is called a '1,1-reaction' because the X group that was one bond away from the metal atom ends up on an atom that is one bond away from the metal atom. Typically the X group is an alkyl or aryl species and then the product contains an acyl group. In principle, the reaction could proceed by a migration of the X group, or an insertion of the CO into the M−X bond. An uncertainty about the actual mechanism has led to the apparently contradictory name **migratory insertion**. Colloquially, however, the terms 'migratory insertion', 'migration', and 'insertion' are used interchangeably. The overall reaction results in a decrease in the number of electrons on the metal atom by 2, with no change in the oxidation state. It is therefore possible to induce 1,1-migratory insertion reactions by the addition of another species that can act as a ligand:

$$[Mn(CH_3)(CO)_5] + PPh_3 \rightarrow [Mn(CH_3CO)(CO)_4PPh_3]$$

The classic study of the migratory insertion of CO with $[CH_3Mn(CO)_5]$ illustrates a number of key features of reactions of this type.[4] First, in the reaction

$$[Mn(CH_3)(CO)_5] + {}^{13}CO \rightarrow [Mn(CH_3CO)(CO)_4({}^{13}CO)]$$

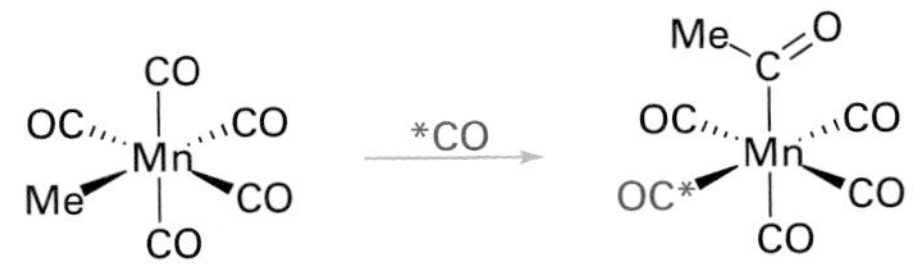

the product has only one labelled CO, and that group is *cis* to the newly formed acyl group. This stereochemistry demonstrates that the incoming CO group does not insert into the Mn−CH$_3$ bond, and that either the methyl group migrates to an adjacent CO ligand, or a CO ligand adjacent to the methyl group inserts into the Mn−CH$_3$ bond. Second, in the reverse reaction

$$cis\text{-}[Mn(CH_3CO)(CO)_4({}^{13}CO)] \rightarrow [MnCH_3(CO)_5] + CO$$

it is possible to distinguish between the migration of the methyl group and the insertion of the CO ligand: *cis*-$[Mn(CH_3CO)(CO)_4({}^{13}CO)]$ must lose a CO ligand *cis* to the acyl group in order for the reaction to proceed. The scheme below summarizes the potential reaction pathways. In one quarter of the instances, this ligand will be the labelled CO and there will be no significant information gained. In half the instances, a nonlabelled CO will be lost, leaving a vacant site *cis* to both the labelled CO ligand and the acyl group. In this case, either (a) migration of the methyl group back to the metal atom or (b) extrusion of CO will lead to the methyl group and the ^{13}CO ligand being *cis* to each other and no information is gained. However, in the remaining one quarter of the instances, the CO that is *trans* to the labelled CO ligand will be lost, and in this case it is possible to distinguish (c) CO ligand extrusion from (d) methyl group migration. If the methyl group migrates, it ends up *trans* to the labelled CO, whereas if the CO is extruded, the methyl group ends up *cis* to the CO.

---

[4] The Mn species are not fluxional and do not rearrange; if they did, we would not be able to draw the conclusions made here.

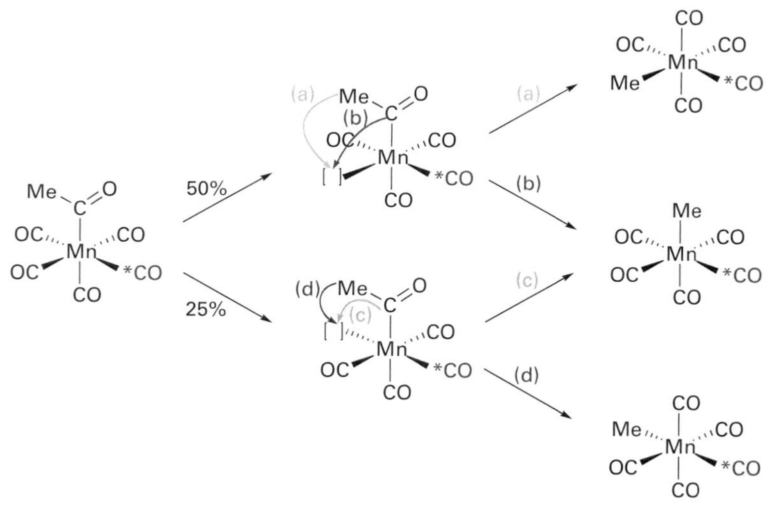

Because the product with CH₃ and ¹³CO *trans* to each other constitutes about 25 per cent of the product, we can conclude that the CH₃ group does indeed migrate. Application of the principle of microscopic reversibility[5] allows us to conclude that the forward reaction proceeds by methyl-group migration. All 1,1-migratory insertions are now thought to proceed by migration of the X group. An important consequence of this pathway is that the relative positions of the other groups on the migrating atom are left unchanged, so the stereochemistry at the X group is preserved.

22.25 1,2-Insertions and β-hydride elimination

Key points: 1,2-Insertion reactions are observed with η^2 ligands such as alkenes and result in the formation of an η^1 ligand with no change in oxidation state of the metal; β-hydride elimination is the reverse of 1,2-insertion.

1,2-Insertion reactions are commonly observed with η^2-ligands, such as alkenes and alkynes, and are exemplified by the reaction:

The reaction is a **1,2-insertion** because the X group that was one bond away from the metal atom ends up on an atom that is two bonds away from the metal. Typically, the X group is a hydride, alkyl, or aryl species, in which case the product contains a (substituted) alkyl group. Like 1,1-insertion reactions, the overall reaction results in a decrease in the number of electrons on the metal atom by 2, with no change in the oxidation state.

If, in the above reaction with X=H, another ethene molecule were to coordinate, the resultant ethyl group could migrate to give a butyl group:

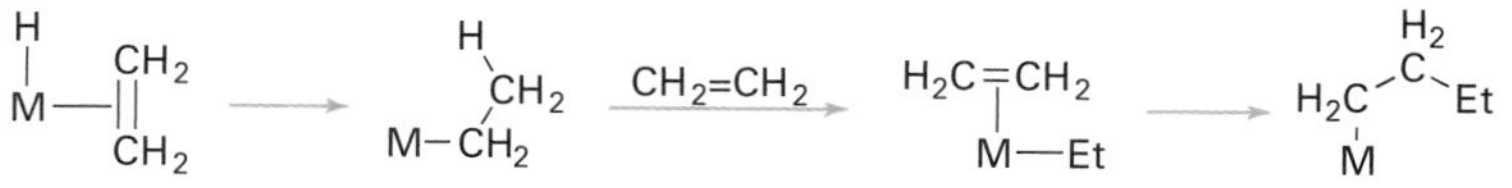

Repetition of this process gives polyethene. Catalytic reactions of this kind are of considerable industrial importance and are discussed in Chapter 26.

The reverse of 1,2-insertions can occur but this is rare except when X=H, when the reaction is known as β-**hydride elimination**:[6]

---

[5]The principle of microscopic reversibility states that both forward and reverse reactions proceed by the same mechanism.

[6]The reaction is known as 'β-hydride elimination' because the H atom that is eliminated is on the second carbon atom from the metal atom (the carbon atom bonded to the metal atom is the α carbon, the third one the γ carbon, etc.).

The experimental evidence shows that both 1,2-insertion and β-hydride elimination proceed through a syn intermediate:

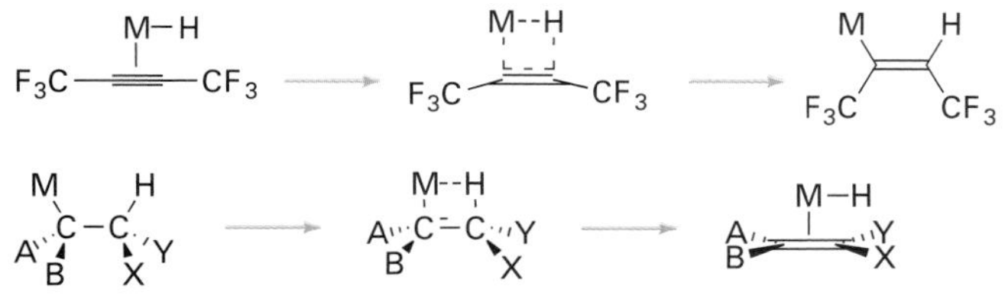

As noted in Section 22.8, a β-hydride elimination reaction can provide a facile route for decomposition of alkyl-containing compounds. The 1,2-insertion reaction, coupled with the β-hydride elimination, can also provide a low-energy route to alkene isomerization:

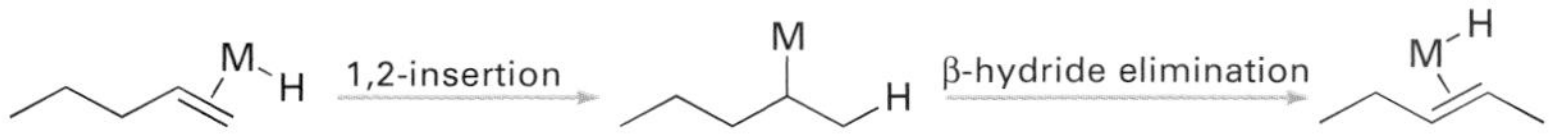

22.26 α-, γ-, and δ-Hydride eliminations and cyclometallations

Key point: Cyclometallation reactions, in which a metal inserts into a remote C—H bond, are equivalent to hydride elimination reactions.

α-Hydride eliminations are occasionally found for complexes that have no β-hydrogens and the reaction gives rise to a carbene that is often highly reactive:

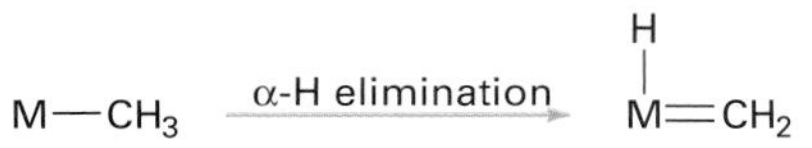

γ-Hydride and δ-hydride eliminations are more commonly observed. Because the product contains a **metallocycle**, a cyclic structure incorporating a metal atom, these reactions are normally described as **cyclometallation** reactions:

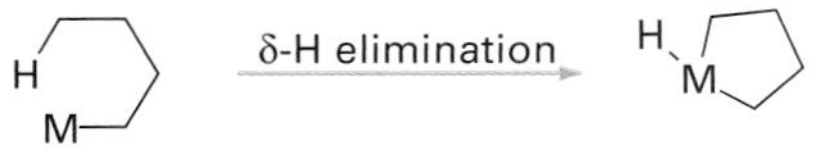

A cyclometallation reaction is often also thought of as the oxidative addition to an adjacent C—H bond. Both α- and β-hydride eliminations can also be considered as cyclometallation reactions. This identification is more obvious for the β-hydride elimination if we consider an alkene in its metallacyclopropane form:

EXAMPLE 22.13 Predicting the outcomes of insertion and elimination reactions

What product, including its stereochemistry, would you expect from the reaction between [MnMe(CO)$_5$] and PPh$_3$?

Answer If we consider the reaction between [MnMe(CO)$_5$] and PPh$_3$ we can see that it is unlikely to be the simple replacement of a carbonyl ligand by the phosphine as this reaction would require a strongly bound carbonyl ligand to dissociate. A more likely reaction would be the migration of the methyl group on to an adjacent CO ligand to give an acyl group, with the phosphine filling the vacated coordination site. This reaction has a low activation barrier, and the product would therefore be expected to be *cis*-[Mn(MeCO)(PPh$_3$)(CO)$_4$].

Self-test 22.13 Explain why [Pt(Et)(Cl)(PEt$_3$)$_2$] readily decomposes, whereas [Pt(Me)(Cl)(PEt$_3$)$_2$] does not.

FURTHER READING

R.H. Crabtree, *The organometallic chemistry of the transition metals.* Wiley, New York (2005). The best single volume book on the subject.

C. Elschenbroich, *Organometallics.* Wiley-VCH, Weinheim (2006).

R.H. Crabtree and D.M.P. Mingos (ed.) *Comprehensive organometallic chemistry III.* Elsevier, Oxford, 2006. The definitive reference work that builds on the two earlier editions.

See *J. Organomet. Chem.*, 1975, **100**, 273 for Wilkinson's personal account of the development of metallocene chemistry.

G.J. Kubas, *Chem. Rev.*, 2006, **107**, 4152. A historical perspective and a full account of the discovery of dihydrogen complexes.

W. Scherer and G.S. McGrady, *Angew. Chem., Int. Ed. Engl.*, 2004, **43**, 1782. A review of agostic interactions that gives a good historical perspective.

D.C. Grills and M.W. George, *Adv. Inorg. Chem.*, 2001, **52**, 113. A review of the topic of noble gas complexes.

D.M.P. Mingos and D.J. Wales, *Introduction to cluster chemistry.* Prentice Hall, Englewood Cliffs (1990); J.W. Lauher, *J. Am. Chem. Soc.*, 1978, **100**, 5305. Descriptions of some of the concepts of cluster bonding.

R. Hoffmann, *Angew. Chem., Int. Ed. Engl.*, 1982, **21**, 711. The application of isolobal analogies to metal cluster compounds (Hoffmann's Nobel Prize lecture).

EXERCISES

22.1 Name the species, draw the structures of, and give valence electron counts to the metal atoms in: (a) $Fe(CO)_5$, (b) $Mn_2(CO)_{10}$, (c) $V(CO)_6$, (d) $[Fe(CO)_4]^{2-}$, (e) $La(\eta^5-Cp^*)_3$, (f) $Fe(\eta^3-allyl)(CO)_3Cl$, (g) $Fe(CO)_4(PEt_3)$, (h) $Rh(Me)(CO)_2(PPh_3)$, (i) $Pd(Me)(Cl)(PPh_3)_2$, (j) $Co(\eta^5-C_5H_5)(\eta^4-C_4Ph_4)$, (k) $[Fe(\eta^5-C_5H_5)(CO)_2]^-$, (l) $Cr(\eta^6-C_6H_6)$ $(\eta^6-C_7H_8)$, (m) $Ta(\eta^5-C_5H_5)_2Cl_3$, (n) $Ni(\eta^5-C_5H_5)NO$. Do any of the complexes deviate from the 18-electron rule? If so, how is this reflected in their structure or chemical properties?

22.2 (a) Sketch an η^2 interaction of 1,4-butadiene with a metal atom and (b) do the same for an η^4 interaction.

22.3 What hapticities are possible for the interaction of each of the following ligands with a single d-block metal atom such as cobalt? (a) C_2H_4, (b) cyclopentadienyl, (c) C_6H_6, (d) cyclooctadiene, (e) cyclooctatetraene.

22.4 Draw plausible structures and give the electron count of (a) $Ni(\eta^3-C_3H_5)_2$, (b) η^4-cyclobutadiene-η^5-cyclopentadienylcobalt, (c) $Co(\eta^3-C_3H_5)(CO)_2$. If the electron count deviates from 18, is the deviation explicable in terms of periodic trends?

22.5 State the two common methods for the preparation of simple metal carbonyls and illustrate your answer with chemical equations. Is the selection of method based on thermodynamic or kinetic considerations?

22.6 Suggest a sequence of reactions for the preparation of $Fe(CO)_3(dppe)$, given iron metal, CO, dppe ($Ph_2PCH_2CH_2PPh_2$), and other reagents of your choice.

22.7 Suppose that you are given a series of metal tricarbonyl compounds having the respective symmetries C_{2v}, D_{3h}, and C_s. Without consulting reference material, which of these should display the greatest number of CO stretching bands in the IR spectrum? Check your answer and give the number of expected bands for each by consulting Table 22.7.

22.8 Provide plausible reasons for the differences in IR wavenumbers between each of the following pairs: (a) $Mo(CO)_3(PF_3)_3$ 2040, 1991 cm^{-1} versus $Mo(CO)_3(PMe_3)_3$ 1945, 1851 cm^{-1}, (b) $MnCp(CO)_3$ 2023, 1939 cm^{-1} versus $MnCp^*(CO)_3$ 2017, 1928 cm^{-1}.

22.9 The compound $Ni_3(C_5H_5)_3(CO)_2$ has a single CO stretching absorption at 1761 cm^{-1}. The IR data indicate that all C_5H_5 ligands are pentahapto and probably in identical environments. (a) On the basis of these data, propose a structure. (b) Does the electron count for each metal in your structure agree with the 18-electron rule? If not, is nickel in a region of the periodic table where deviations from the 18-electron rule are common?

22.10 Decide which of the two complexes (a) $W(CO)_6$ or (b) $Ir(CO)$ $Cl(PPh_3)_2$ should undergo the fastest exchange with ^{13}CO. Justify your answer.

22.11 Which metal carbonyl in each of (a) $[Fe(CO)_4]^{2-}$ or $[Co(CO)_4]^-$, (b) $[Mn(CO)_5]^-$ or $[Re(CO)_5]^-$ should be the most basic towards a proton? What are the trends on which your answer is based?

22.12 Using the 18-electron rule as a guide, indicate the probable number of carbonyl ligands in (a) $W(\eta^6-C_6H_6)(CO)_n$, (b) $Rh(\eta^5-C_5H_5)$ $(CO)_n$, and (c) $Ru_3(CO)_n$.

22.13 Propose two syntheses for $MnMe(CO)_5$, both starting with $Mn_2(CO)_{10}$, with one using Na and one using Br_2. You may use other reagents of your choice.

22.14 Give the probable structure of the product obtained when $Mo(CO)_6$ is allowed to react first with LiPh and then with the strong carbocation reagent, $CH_3OSO_2CF_3$.

22.15 $Na[W(\eta^5-C_5H_5)(CO)_3]$ reacts with 3-chloroprop-1-ene to give a solid, A, which has the molecular formula $W(C_3H_5)(C_5H_5)(CO)_3$. Compound A loses carbon monoxide on exposure to light and forms compound B, which has the formula $W(C_3H_5)(C_5H_5)(CO)_2$. Treating compound A with hydrogen chloride and then potassium hexafluorophosphate, $K^+PF_6^-$, results in the formation of a salt, C. Compound C has the molecular formula $[W(C_3H_6)(C_5H_5)(CO)_3]PF_6$. Use this information and the 18-electron rule to identify the compounds A, B, and C. Sketch a structure for each, paying particular attention to the hapticity of the hydrocarbon.

22.16 Treatment of $TiCl_4$ at low temperature with EtMgBr gives a compound that is unstable above $-70°C$. However, treatment of $TiCl_4$ at low temperature with MeLi or $LiCH_2SiMe_3$ gives compounds that are stable at room temperature. Rationalize these observations.

22.17 Suggest syntheses of (a) $[Mo(\eta^7-C_7H_7)(CO)_3]BF_4$ from $Mo(CO)_6$ and (b) $[Ir(COMe)(CO)(Cl)_2(PPh_3)_2]$ from $[Ir(CO)Cl(PPh_3)_2]$.

22.18 When $Fe(CO)_5$ is refluxed with cyclopentadiene compound **A** is formed which has the empirical formula $C_8H_6O_3Fe$ and a complicated ^1H NMR spectrum. Compound **A** readily loses CO to give compound **B** with two ^1H-NMR resonances, one at negative chemical shift (relative intensity one) and one at around 5 ppm (relative intensity five). Subsequent heating of **B** results in the loss of H_2 and the formation of compound **C**. Compound **C** has a single ^1H-NMR resonance and the empirical formula $C_7H_5O_2Fe$. Compounds **A, B,** and **C** all have 18 valence electrons: identify them and explain the observed spectroscopic data.

22.19 When $Mo(CO)_6$ is refluxed with cyclopentadiene compound **D** is formed which has the empirical formula $C_8H_5O_3Mo$ and an absorption in the IR spectrum at 1960 cm^{-1}. Compound **D** can be treated with bromine to yield **E** or with Na/Hg to give compound **F**. There are absorptions in the IR spectra of **E** and **F** at 2090 and 1860 cm^{-1}, respectively. Compounds **D**, **E**, and **F** all have 18 valence electrons: identify them and explain the observed spectroscopic data.

22.20 Which compound would you expect to be more stable, $Rh(\eta^5\text{-}C_5H_5)_2$ or $Ru(\eta^5\text{-}C_5H_5)_2$? Give a plausible explanation for the difference in terms of simple bonding concepts.

22.21 Give the equation for a workable reaction that will convert $Fe(\eta^5\text{-}C_5H_5)_2$ into (a) $Fe(\eta^5\text{-}C_5H_5)(\eta^5\text{-}C_5H_4COCH_3)$ and (b) $Fe(\eta^5\text{-}C_5H_5)(\eta^5\text{-}C_5H_4CO_2H)$.

22.22 Sketch the a_1' symmetry-adapted orbitals for the two eclipsed C_5H_5 ligands stacked together with D_{5h} symmetry. Identify the s, p, and d orbitals of a metal atom lying between the rings that may have nonzero overlap, and state how many a_1' molecular orbitals may be formed.

22.23 The compound $Ni(\eta^5\text{-}C_5H_5)_2$ readily adds one molecule of HF to yield $[Ni(\eta^5\text{-}C_5H_5)(\eta^4\text{-}C_5H_6)]^+$ whereas $Fe(\eta^5\text{-}C_5H_5)_2$ reacts with strong acid to yield $[Fe(\eta^5\text{-}C_5H_5)_2H]^+$. In the latter compound the H atom is attached to the Fe atom. Provide a reasonable explanation for this difference.

22.24 Write a plausible mechanism, giving your reasoning, for the reactions

(a) $[Mn(CO)_5(CF_2)]^+ + H_2O \rightarrow [Mn(CO)_6]^+ + 2\ HF$

(b) $Rh(CO)(C_2H_5)(PR_3)_2 \rightarrow RhH(CO)(PR_3)_2 + C_2H_4$

22.25 The mechanism of CO insertion is thought to be

$$RMn(CO)_5 \rightarrow (RCO)Mn(CO)_4$$

$$(RCO)Mn(CO)_4 + L \rightarrow (RCO)MnL(CO)_4$$

In the first step, the acyl intermediate is formed, leaving a vacant coordination position. In the second step, a ligand enters. At high ligand concentrations, the rate is independent of [L]. At this limit, what rate constant can be extracted from rate data?

22.26 (a) What cluster valence electron (CVE) count is characteristic of octahedral and trigonal prismatic complexes? (b) Can these CVE values be derived from the 18-electron rule? (c) Determine the probable geometry (octahedral or trigonal prismatic) of $[Fe_6(C)(CO)_{16}]^{2-}$ and $[Co_6(C)(CO)_{16}]^{2-}$. (The C atom in both cases resides in the centre of the cluster and can be considered to be a four-electron donor.)

22.27 Based on isolobal analogies, choose the groups that might replace the group in boldface in

(a) $Co_2(CO)_9\mathbf{CH} \rightarrow OCH_3, N(CH_3)_2,$ or $SiCH_3$

(b) $(OC)_5Mn\mathbf{Mn(CO)}_5 \rightarrow I, CH_2,$ or CCH_3

22.28 Ligand substitution reactions on metal clusters are often found to occur by associative mechanisms, and it is postulated that these occur by initial breaking of an M—M bond, thereby providing an open coordination site for the incoming ligand. If the proposed mechanism is applicable, which would you expect to undergo the fastest exchange with added ^{13}CO, $Co_4(CO)_{12}$ or $Ir_4(CO)_{12}$? Suggest an explanation.

PROBLEMS

22.1 Propose the structure of the product obtained by the reaction of $[Re(CO)(\eta^5\text{-}C_5H_5)(NO)(PPh_3)]^+$ with $Li[HBEt_3]$. The latter contains a strongly nucleophilic hydride. (For full details see: W. Tam, G. Y. Lin, W.K. Wong, W.A. Kiel, V. Wong, and J.A. Gladysz, *J. Am. Chem. Soc.*, 1982, **104**, 141.)

22.2 Treatment of $TiCl_4$ with 4 equivalents of NaCp gives a single organometallic compound (together with NaCl byproduct). At room temperature the ^1H-NMR spectrum shows a single sharp singlet; on cooling to $-40°C$ this singlet separates into two singlets of equal intensity; further cooling results in one of the singlets separating into three signals with intensity ratios 1:2:2. Explain these results.

22.3 When several CO ligands are present in a metal carbonyl, an indication of the individual bond strengths can be determined by means of force constants derived from the experimental IR frequencies. In $Cr(CO)_5(Ph_3P)$ the *cis*-CO ligands have the higher force constants, whereas in $Ph_3SnCo(CO)_4$ the force constants are higher for the *trans*-CO. Suggest why, and explain which carbonyl C atoms should be susceptible to nucleophilic attack in these two cases. (For details see D.J. Darensbourg and M.Y. Darensbourg, *Inorg. Chem.*, 1970, **9**, 1691.)

22.4 It is often possible to assign different resonance structures to an organometallic compound, for example the structure below can be thought of as existing as either the charge separated aromatic structure on the left, or the carbene form on the right.

Suggest ways of distinguishing the two forms. (For details see: G.W.V. Cave, A.J. Hallett, W. Errington, and J.P. Rourke, *Angew. Chem., Int. Ed. Engl.*, 1998, **37**, 3270 and C.P. Newman, G.J. Clarkson, N.W. Alcock, and J.P. Rourke, *Dalton Trans.*, 2006, 3321.)

22.5 Describe how NMR has been used to identify alkane complexes of d metals unambiguously. See S. Geftakis and G.E. Ball, *J. Am. Chem. Soc.*, 1998, **120**, 9953 and D.J. Lawes, S. Geftakis, and G.E. Ball, *J. Am. Chem. Soc.*, 2005, **127**, 4134.

22.6 The dinitrogen complex $Zr_2(\eta^5\text{-}Cp^*)_4(N_2)_3$ has been isolated and its structure determined by single-crystal X-ray diffraction. Each Zr atom is bonded to two Cp^* and one terminal N_2. The third N_2 bridges between the Zr atoms in a nearly linear ZrNNZr array. Before consulting the reference, write a plausible structure for this compound that accounts for the ^1H-NMR spectrum obtained on a sample held at $-7°C$. This spectrum shows two singlets, indicating that the Cp^*

rings are in two different environments. At somewhat above room temperature these rings become equivalent on the NMR timescale and ^{15}N-NMR indicates that N_2 exchange between the terminal ligands and dissolved N_2 is correlated with the process that interconverts the Cp* ligand sites. Propose a way in which this equilibration could interconvert the sites of the Cp* ligands. (For further details see J.M. Manriquez, D.R. McAlister, E. Rosenberg, H.M. Shiller, K.L. Willamson, S.I. Chan, and J.E. Bercaw, *J. Am. Chem. Soc.*, 1978, **100**, 3078.)

22.7 How might you unambiguously identify an organometallic complex containing a noble gas atom as a ligand? See G.E. Ball, T.A. Darwish, S. Geftakis, M.W. George, D.J. Lawes, P. Portius, and J.P. Rourke, *Proc. Natl Acad. Sci.*, 2005, **102**, 1853.

22.8 What conclusions can you draw from the bonding and reactivity of dihydrogen bound to a low oxidation state d-block metal that might be applicable to the bonding of an alkane to a metal?

What implications might this have for the oxidative addition of a hydrocarbon to a metal atom? (See, for example, R.H. Crabtree, *J. Organomet. Chem.*, 2004, **689**, 4083.)

22.9 Compare and contrast Fischer and Schrock carbenes. See E.O. Fischer, *Adv. Organomet. Chem.*, 1976, **14**, 1 and R.R. Schrock, *Acc. Chem. Res.*, 1984, **12**, 98.

22.10 Rearrangements of ligands so that differing numbers of carbons interact with the central metal atom are known as *haptotropic rearrangements*. Consider the (fluorenyl)(cyclo-pentadienyl)iron complex. What haptotropic isomers are possible? See E. Kirillov, S. Kahlal, T. Roisnel, T. Georgelin, J. Saillard, and J. Carpentier, *Organometallics* 2008, **27**, 387.

22.11 Suggest two plausible routes by which a carbonyl ligand in Mo(Cp)(CO)$_3$Me might exchange for a phosphine. Neither route should invoke the initial dissociation of a CO.

The f-block elements

23

There are two series of elements in the f block, each consisting of 14 elements, but relatively little diverse chemistry has been reported for such a large number of elements. For the 4f elements (the lanthanoids) this lack of diversity is normally ascribed to the buried nature of the 4f electrons. The chemical properties of the lanthanoids are similar to those of other electropositive metals. The applications of the lanthanoids derive mainly from the optical spectra of their ions. The fact that only two of the 5f elements (the actinoids) occur naturally, all other actinoids being synthetic and often intensely radioactive, has hindered their study, particularly for the elements with the highest atomic numbers.

The two series of elements in the f block derive from the filling of the seven 4f and 5f orbitals, respectively. This occupation of f orbitals from f^1 to f^{14} corresponds to the elements cerium (Ce) to lutetium (Lu) in Period 6 and from thorium (Th) to lawrencium (Lr) in Period 7; however, given the similarity of their chemical properties, the elements lanthanum (La) and actinium (Ac) are normally included in discussion of the f block, as we do here.[1] The 4f elements are collectively the **lanthanoids** (formerly and still commonly 'the lanthanides', but this terminology is potentially confusing as the suffix -ide generally refers to an anionic form of an element) and the 5f elements are the **actinoids** (still commonly the 'actinides'). The lanthanoids are sometimes referred to as the 'rare earth elements'. That name, however, is inappropriate because they are not particularly rare, except for promethium, which has no stable isotope. A general lanthanoid is represented by the symbol Ln and an actinoid by An. Because the names and symbols of the f-block elements are less familiar than those elsewhere in the periodic table, we give discreet reminders in the margins or in the text in appropriate places.

The chemical properties of the lanthanoids are quite different from those of the actinoids and we discuss them separately. As we shall see, there is a striking uniformity in the properties of the lanthanoids and greater diversity in the chemistry of the actinoids.

The elements

We begin our account of the f-block elements by considering the properties and the extraction of the elements.

23.1 Occurrence and recovery

Key points: The principal sources of the lanthanoids are phosphate minerals; the most important actinoid, uranium, is recovered from its oxide.

---

[1] The majority of periodic tables, including the one used here, place La in Group 3 and Lu at the end of the f block; however, both these elements have $ns^2(n-1)d^1$ electron configurations and never exhibit partially filled f orbitals in their compounds, so a reasonable alternative, which is adopted by some, is to place La at the start of the f block and Lu in Group 3.

Other than promethium (Pm), the lanthanoids are reasonably common in the Earth's crust; indeed, even the 'rarest' lanthanoid, thulium, has a crustal abundance that is similar to that of iodine. The principal mineral source for the early lanthanoids is *monazite*, $(Ln,Th)PO_4$, which contains mixtures of lanthanoids and thorium. Another phosphate mineral, *xenotime* (of similar composition, $LnPO_4$), is the principal source of the heavier lanthanoids. The common oxidation state for all the lanthanoids is Ln(III), which makes separation difficult, although cerium, which can be oxidized to Ce(IV), and europium, which can be reduced to Eu(II), are separable from the other lanthanoids by exploiting their redox chemistry. Separation of the remaining Ln^{3+} ions is accomplished on a large scale by multistep liquid−liquid extraction in which the ions are distributed between an aqueous phase and an organic phase containing complexing agents. Ion-exchange chromatography is used to separate the individual lanthanoid ions when high purity is required. Pure and mixed lanthanoid metals are prepared by the electrolysis of molten lanthanoid halides.

Beyond lead ($Z = 82$) no element has a stable isotope, but two of the actinoids, thorium (Th, $Z = 90$) and uranium (U, $Z = 92$), have isotopes that are sufficiently long-lived that significant quantities have persisted from their formation in the supernova that preceded the formation of the Sun (Section 1.2). Indeed the levels of Th and U in the Earth's crust exceed those of both Sn and I. Table 23.1 gives the half-lives of the most stable actinoid isotopes.

Thorium is extracted from either monazite or *thorite* (essentially $ThSiO_4$); the major ores for uranium, *uranite* and *pitchblende*, have the approximate formulas UO_2 and U_3O_8, respectively. The primary source of the other actinoids is synthesis by nuclear reactions; all of them are more radioactive than thorium and uranium.

23.2 Physical properties and applications

Key points: Lanthanoids are reactive metals that find small-scale uses in specialist applications; the primary uses of actinoids are restricted by their radioactivity.

The lanthanoids are soft white metals that have densities comparable to those of the 3d metals ($6-10$ g cm^{-3}). For metals, they are relatively poor conductors of both heat and electricity, with thermal and electrical conductivities 25 and 50 times less than that of copper, respectively. The light lanthanoid metals are highly reactive towards oxygen and are normally stored in sealed glass ampoules. The metals react with steam and dilute acids

Table 23.1 Half-lives of the most stable actinoid isotopes

Z	Element	Symbol	Mass number	$t_{1/2}$
89	Actinium	Ac	227	21.8 a
90	Thorium	Th	232	14.1 Ga
91	Protactinium	Pa	231	32.8 ka
92	Uranium	U	238	4.47 Ga
93	Neptunium	Np	237	2.14 Ma
94	Plutonium	Pu	244	81 Ma
95	Americium	Am	243	7.38 ka
96	Curium	Cm	247	16 Ma
97	Berkelium	Bk	247	1.38 ka
98	Californium	Cf	251	900 a
99	Einsteinium	Es	252	460 d
100	Fermium	Fm	257	100 d
101	Mendelevium	Md	258	55 d
102	Nobelium	No	259	1.0 h
103	Lawrencium	Lr	260	3 min

a = year, d= day, h = hour, min = minute

but are somewhat passivated by an oxide coating. The majority of the metals adopt the hexagonal cubic close-packed structure type, although cubic close-packed forms are also known for most of the elements, particularly under high pressure.

A mixture of the early lanthanoid metals, with mainly Ce and La and small amounts of Pr and Nd reflecting the composition of the monazite ore, is referred to in commerce as *mischmetal*. It is used in steel-making to remove impurities such as H, O, S, and As (by forming very stable lanthanoid oxysulfides), which reduce the mechanical strength and ductility of steel. Compounds of the lanthanoids find a wide range of applications, many of which depend on the optical properties associated with their f−f electronic transitions (Section 23.4): europium oxide and europium orthovanadate are used as red phosphors in displays and lighting, and neodymium (as Nd^{3+}), samarium (as Sm^{3+}), and holmium (as Ho^{3+}) are used in solid-state lasers.

The alloys $SmCo_5$ and Sm_2Co_{17} of samarium and cobalt have very high magnetic strengths, more than ten times that of iron and some magnetic iron oxides (Fe_3O_4). They also have excellent corrosion resistance and good stability at elevated temperatures. Neodymium iron boride, $Nd_2Fe_{14}B$, shows similar magnetic properties and is cheaper to produce, but because it is susceptible to corrosion the magnets are often coated with zinc, nickel, or epoxy resins. Applications of these strongly magnetic materials include headphones, microphones, magnetic switches, and components of particle beams (for guidance of the beam). Toy magnetic building sets have used them extensively.

The density of the actinoids increases from 10.1 g cm^{-3} for Ac to 20.4 g cm^{-3} for Np, before decreasing for the remainder of the series. Plutonium has at least six phases at atmospheric pressure with density differences of more than 20 per cent. Many of the physical and chemical properties of the actinoids are unknown because only miniscule quantities have ever been isolated; in addition to being radioactive, many actinoids are known to be poisonous and all are treated as hazardous in the laboratory. Their current primary use is in nuclear reactors and nuclear weapons.

Lanthanoid chemistry

The lanthanoids are all electropositive metals with a remarkable uniformity of chemical properties. The significant difference between two lanthanoids is often only their atomic or ionic radius, and the ability to choose a lanthanoid of a particular size often allows the 'tuning' of the properties of their compounds. For example, the magnetic and electronic properties of a material often depend on the exact separation of the atoms present and degree of overlap of various atomic orbitals. By choosing a lanthanoid of an appropriate atomic radius this separation can be controlled, with consequences for its electrical conductivity or magnetic transition temperature.

23.3 General trends

Key points: The lanthanoids are electropositive metals that commonly occur in their compounds as Ln(III); other oxidation states are stable only when an empty, half-filled, or full f subshell is produced.

The elements La through to Lu are all highly electropositive, with standard potentials of the Ln^{3+}/Ln couple similar to that of Mg^{2+}/Mg (Table 23.2). They favour the oxidation state Ln(III) with a uniformity that is unprecedented in the periodic table. Other properties of the elements vary significantly. For example, the radii of the Ln^{3+} ions contract steadily from 116 pm for La^{3+} to 98 pm for Lu^{3+}. The decrease in ionic radius is attributed in part to the increase in the effective atomic number, Z_{eff}, as electrons are added to the poorly shielding 4f subshell (Section 1.9), but detailed calculations indicate that subtle relativistic effects also make a substantial contribution.

Figure 23.1 shows that the decrease in ionic radius corresponds to two gentle curves intersecting at Gd^{3+} (f^7). The effect is similar to that in the d block and can be traced to the splitting of the f-orbital energies by the crystal field generated by the ligand environment (Fig. 23.2), but the splitting is much smaller than in the d block. On the left of the series, the lower energy f orbitals (which point away from the ligands) are occupied as the number of f electrons increases (f^1 to f^4). For later configurations, such as f^6 and f^7, the higher energy

Table 23.2 Element names, symbols, and selected properties of the lanthanoids

Z	Name	Symbol	Configuration (M^{3+})	$E^{\ominus}$ / V	$r(Ln^{3+})$/pm*	N_{ox}†
57	Lanthanum	La	[Xe]	−2.38	116	2(n), **3**, 4
58	Cerium	Ce	$[Xe]f^1$	−2.34	114	2(n), **3**, 4
59	Praseodymium	Pr	$[Xe]f^2$	−2.35	113	2(n), **3**, 4
60	Neodymium	Nd	$[Xe]f^3$	−2.32	111	2(n), **3**
61	Promethium	Pm	$[Xe]f^4$	−2.29	109	**3**
62	Samarium	Sm	$[Xe]f^5$	−2.30	108	2(n), **3**
63	Europium	Eu	$[Xe]f^6$	−1.99	107	2, **3**
64	Gadolinium	Gd	$[Xe]f^7$	−2.28	105	**3**
65	Terbium	Tb	$[Xe]f^8$	−2.31	104	**3**, 4
66	Dysprosium	Dy	$[Xe]f^9$	−2.29	103	2(n), **3**
67	Holmium	Ho	$[Xe]f^{10}$	−2.33	102	**3**
68	Erbium	Er	$[Xe]f^{11}$	−2.32	100	**3**
69	Thulium	Tm	$[Xe]f^{12}$	−2.32	99	2(n), **3**
70	Ytterbium	Yb	$[Xe]f^{13}$	−2.22	99	2, **3**
71	Lutetium	Lu	$[Xe]f^{14}$	−2.30	98	**3**

* Ionic radii for coordination number 8 from R.D. Shannon, *Acta Cryst.*, 1976, **A32**, 751.
† Oxidation numbers (N_{ox}) in bold type indicate the most stable states; (n) indicates that the state is stable only in nonaqueous conditions.

f orbitals are occupied, and as these point towards the ligands the increased electron–ligand repulsion results in a smaller than expected decrease in ionic radius.

The 18 per cent decrease in ionic radius from La^{3+} to Lu^{3+} leads to an increase in the hydration enthalpy across the series. The standard potentials of the lanthanoids are all very similar, with $E^{\ominus}(La^{3+}/La) = -2.38$ V similar to $E^{\ominus}(Lu^{3+}/Lu) = -2.30$ V at the other end of the series. This similarity reflects the balance of increasing hydration and atomization enthalpies as the atomic radius decreases.

The common occurrence of the lanthanoids as Ln(III) is normally ascribed to the fact that once the valence s and d electrons have been removed the f electrons are held tightly by the nucleus and do not extend beyond the xenon-like core of the atom. A further

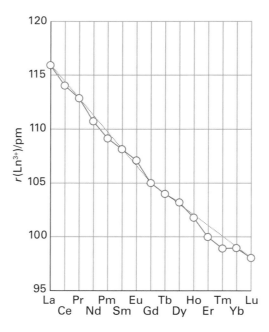

Figure 23.1 Variation of the ionic radius of Ln^{3+}.

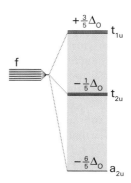

Figure 23.2 The splitting of f orbitals in an octahedral crystal field.

consequence of the burying of the f electrons is that a Ln^{3+} ion has no frontier orbitals with directional preference, so ligand-field stabilization plays only a small part in the properties of lanthanoid complexes. For a few complexes weak stabilization energies (of a few kilojoules per mole) are observed for configurations such as f^4, f^{10}, and f^{11} when compared with f^0, f^7, and f^{14}. This stabilization can be explained following similar arguments to those described in Section 20.1 but using f orbitals and acknowledging the much weaker effects of the ligand field.

Superimposed on the common occurrence of Ln^{3+} there are some atypical oxidation states that are most prevalent when the ion can attain the relatively more stable empty (f^0), half-filled (f^7), or filled (f^{14}) subshell (Table 23.2). Thus, Ce^{3+} (f^1) can be oxidized to Ce^{4+} (f^0) and the latter is a strong and useful oxidizing agent. The next most common of the atypical oxidation states is Eu^{2+} (f^7), and there are a number of stable Eu^{2+} compounds, including EuI_2, $EuSO_4$, and $EuCO_3$, and solutions of this ion are stable. The ions Sm^{2+} and Yb^{2+} also have an extensive chemistry but reduce water to hydrogen in aqueous solutions. Recently Dy(II), Nd(II), and Tm(II) have been produced in molecular complexes in solution; an example is $NdI_2(THF)_5$. Other reasonably stable oxidation states include Pr(IV) and Tb(IV), and the oxides of these elements formed in air, Pr_6O_{11} and Tb_4O_7, contain mixtures of Ln(III) and Ln(IV). Under very strongly oxidizing conditions Dy(IV) and Nd(IV) can be obtained.

A Ln^{3+} ion is a hard Lewis acid, as indicated by its preference for F^- and oxygen-containing ligands and its occurrence with PO_4^{3-} in minerals.

Dy	Dysprosium
Eu	Europium
Nd	Neodymium
Pr	Praseodymium
Sm	Samarium
Tb	Terbium
Tm	Thulium
Yb	Ytterbium

23.4 Electronic, optical, and magnetic properties

Much of the commercial and technological value of the lanthanoids stems from their optical and magnetic properties.

(a) Electronic absorption spectra

Key point: Lanthanoid ions typically display weak but sharp absorption spectra because the f orbitals overlap only weakly with the ligand orbitals.

Most lanthanoid ions are only weakly coloured because their absorptions in the visible region of the spectrum are commonly $f-f$ transitions which are symmetry forbidden (Table 23.3). The spectra of their complexes generally show much narrower and more distinct absorption bands than those of d-metal complexes. Both the narrowness of the spectral features and their insensitivity to the nature of coordinated ligands indicate that the f orbitals have a smaller radial extension than the filled 5s and 5p orbitals. As noted in Section 1.6, the 5s and 5p orbitals are expected to shield the 4f electrons from the ligands.

We shall not go into as complete an analysis of $f-f$ electronic transitions as we provided for $d-d$ electronic transitions in Chapter 20 as they can be very complex; for instance, there are 91 microstates of an f^2 configuration. However, the discussion is simplified somewhat by the fact that f orbitals are relatively deep inside the atom and overlap only weakly with ligand orbitals. Hence, as a first approximation, their electronic states (and therefore electronic spectra) can be discussed in the free-ion limit and the Russell−Saunders coupling scheme remains a reasonable approximation despite the elements having high atomic numbers.

EXAMPLE 23.1 Deriving the ground state term symbol of a lanthanoid ion

What is the ground state term symbol of Pr^{3+} (f^2)?

Answer The procedure for deriving the ground state term symbols, which have the general form $^{2S+1}\{L\}_J$, for d-block elements was summarized in Section 20.3 and we can proceed in a similar way. According to Hund's rules, the ground state will have the two electrons in different f orbitals; noting that $l = 3$ for an f orbital, the maximum value of $M_L = m_{l1} + m_{l2}$ is therefore $M_L = (+3) + (+2) = +5$, which must stem from a state with $L = 5$, an H term. The lower spin arrangement of two electrons in different orbitals is a triplet with $S = 1$, so the term will be ^3H. According to the Clebsch−Gordan series (Section 20.3a), the total angular momentum of a term with $L = 5$ and $S = 1$ will be $J = 6$, 5, or 4. According to Hund's rules, for a less than half-full shell the level with the lowest value of J lies lowest (in this case, $J = 4$), so we can expect its term symbol to be $^3H_4$.

Self-test 23.1 Derive the ground state of the Tm^{3+} ion

The large number of microstates for each electronic configuration means a correspondingly large number of terms and hence of possible transitions between them. As electrons in f orbitals interact only weakly with the ligands there is little coupling of the electronic transitions with molecular vibrations, with the consequence that the bands are narrow. As the terms are derived almost purely from f orbitals, and there is only very little d−f orbital mixing or mixing with ligand orbitals, the transitions are Laporte forbidden (Section 20.6). Hence, in contrast to the d metals, which normally show one or two broad bands of moderate intensity, the visible spectra of lanthanoids usually consist of a large number of sharp, low intensity peaks that are barely affected by changing the coordination environment of the central metal ion.

Figure 23.3a shows the relative energies associated with the Pr^{3+} ion, including the ground state term $^3H_4$; ground state terms for all the Ln^{3+} ions are given in Table 23.3. Figure 23.3b shows the experimental absorption spectrum of $Pr^{3+}(aq)$ from the near IR to the UV region. For this ion the absorptions occur mainly between 450 and 500 nm (blue), and at 580 nm (yellow) so the residual light that reaches the eye after reflection from a Pr^{3+} compound is mainly green and red, giving this ion its characteristic green colour.

In summary, the spectra of the lanthanoid ions are normally characterized by the following properties:

- Numerous absorptions due to the large number of microstates.
- Weak absorptions due to lack of orbital mixing. Molar absorption coefficients(ε) are typically $1-10$ dm^3 mol^{-1} cm^{-1} compared with d metals (close to 100 dm^3 mol^{-1} cm^{-1}).
- Sharp absorptions due to the weak interaction of the f orbitals with the ligand vibrations.
- Spectra that are to a large degree independent of the ligand type and coordination number.

A Nd^{3+} ion has a strong absorption at 580 nm due to the transition $^4I_{9/2} \leftarrow {}^4F_{3/2}$. This wavelength corresponds almost exactly to the main yellow emission from excited Na atoms (Section 11.1). As a result, Nd is incorporated into goggles used by glassblowers, where it reduces the glare produced by hot sodium silicate glasses. In some cases transitions between

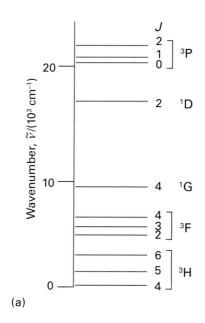

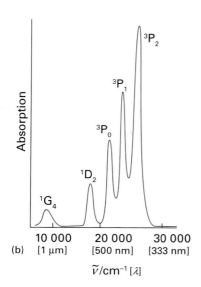

Figure 23.3 (a) Energy level diagram for the free Pr^{3+} (f^2) ion, ground state $^3H_4$. (b) The absorption spectrum of the $Pr^{3+}(aq)$ ion in the visible region and excited state terms for each band.

Table 23.3 Electron configurations, colours, and magnetic moments (at 298 K) of the Ln^{3+} ions

Element	Configuration	Colour in aqueous solution	Ground state	Magnetic moment	
				Theory μ/μ_B	Observed[§] μ/μ_B
Lanthanum	[Xe]]	Colourless	$^1S_0$	0	0
Cerium	[Xe]f^1	Colourless	$^2F_{5/2}$	2.54	2.46
Praseodymium	[Xe]f^2	Green-yellow	$^3H_4$	3.58	3.47−3.61
Neodymium	[Xe]f^3	Violet	$^4I_{9/2}$	3.68	3.44−3.65
Promethium	[Xe]f^4	−	$^5I_4$	2.83	−
Samarium	[Xe]f^5	Yellow	$^6H_{5/2}$	0.84 (1.55−1.65)[¶]	1.54−1.65
Europium	[Xe]f^6	Pink	$^7F_0$	0 (3.40−3.51)[¶]	3.32−3.54
Gadolinium	[Xe]f^7	Colourless	$^8S_{7/2}$	7.94	7.9−8.0
Terbium	[Xe]f^8	Pink	$^7F_6$	9.72	9.69−9.81
Dysprosium	[Xe]f^9	Yellow-green	$^6H_{15/2}$	10.63	10.0−10.6
Holmium	[Xe]f^{10}	Yellow	$^5I_8$	10.60	10.4−10.7
Erbium	[Xe]f^{11}	Lilac	$^4I_{15/2}$	9.59	9.4−9.5
Thulium	[Xe]f^{12}	Green	$^3H_6$	7.57	7.0−7.5
Ytterbium	[Xe]f^{13}	Colourless	$^2F_{7/2}$	4.54	4.0−4.5
Lutetium	[Xe]f^{14}	Colourless	$^1S_0$	0	0

[¶] The values in parentheses include expected contribution from terms other than the ground state.
[§] From compounds such as $Ln_2(SO_4)_3.8H_2O$ and $Ln(Cp)_2$.

4f and 5d orbitals can occur; as $\Delta l = 1$, these transitions are allowed and intense, but they are generally in the UV region of the spectrum. For example, the transition $4f^{10}5d^1 \leftarrow 4f^{11}$ in Er^{3+} occurs at about 150 nm.

(b) Emission spectra and fluorescence

Key point: Lanthanoid ions show strong emission spectra with applications in phosphors and lasers.

Some of the most important applications of the lanthanoids derive from their emission spectra produced after excitation (using energetic photons or electron beams) of the f electrons. These emission spectra show many features of the absorption spectra in that they consist of sharply defined frequencies characteristic of the lanthanoid cation and mainly independent of the ligand.

All the lanthanoid ions except La^{3+} (f^0) and Lu^{3+} (f^{14}) show luminescence, with Eu^{3+} (f^6) and Tb^{3+} (f^8) being particularly strong. In part, the strong luminescence is due to the large number of excited states that exist, which increases the probability of intersystem crossing (the formation of excited states of different multiplicity from the ground state, Section 20.7). The strength of the emission also stems from the excited electron interacting only weakly with its environment and so having a long nonradiative lifetime (milliseconds to nanoseconds). That is, an electronically excited lanthanoid species rarely loses energy by transfer into vibrational modes because f orbitals overlap only weakly with the ligand orbitals whereas this mechanism provides a pathway for such relaxation in many d-metal systems where there is stronger d-orbital-to-ligand-orbital overlap. These properties lead to applications of lanthanoid compounds in display systems, such as cathode-ray and plasma screens (Box 23.1). By stimulating the emission from the excited state, high intensity laser radiation can be obtained, as in Nd:YAG lasers.

BOX 23.1 Lanthanoid-based phosphors and lasers

Phosphors are fluorescent materials that normally convert high energy photons, typically in the UV region of the electromagnetic spectrum, into lower energy visible wavelengths. Whereas a number of fluorescent materials are based on d-metal systems (for example, Cu and Mn doped into zinc sulfide) for some emitted energies lanthanoid-containing materials offer the best performance. This is particularly so of the red phosphors needed in displays, such as plasma screens, and fluorescent lighting where the emission spectrum of Eu^{3+} occurs in a series of lines between 580 nm (orange) and 700 nm (red). Terbium, as Tb^{3+}, is also widely used in similar phosphors, producing emissions between 480 and 580 nm (green). The phosphor compounds, such as Eu^{3+} doped into YVO_4, are coated on the interiors of fluorescent light tubes and convert the UV radiation generated from the mercury discharge into visible light. By using a combination of phosphors fluorescing in different regions of the visible spectrum, white light is emitted by the tube (Fig. B23.1).

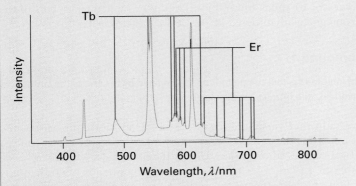

Figure B23.1 Spectrum of a fluorescent lamp showing lines due to lanthanoid cations in the phosphor.

Neodymium can be doped into the garnet structure of yttrium aluminium oxide (YAG), $Y_3Al_5O_{12}$, at levels of 1 per cent Nd for Y, to produce the material that is used in Nd:YAG lasers. The Nd^{3+} ions are strongly absorbing at wavelengths between 730−760 nm and 790−820 nm, such as the high intensity light produced by krypton-containing flashlamps. The ions are excited from their ground $^4I_{9/2}$ state to a number of excited states which then transfer into a relatively long-lived excited $^4F_{3/2}$ state.

Radiative decay of this state to $^4I_{11/2}$ (which lies just above the ground state $^4I_{9/2}$) is stimulated by a photon of the same frequency and results in laser radiation. So once a large number of ions exist in an excited state, they can all be stimulated to emit light simultaneously, so producing very intense radiation.

A Nd:YAG laser typically emits at 1.064 μm, in the near-infrared region of the electromagnetic spectrum in pulsed and continuous modes. There are weaker transitions near 0.940, 1.120, 1.320, and 1.440 μm. The high-intensity pulses may be frequency-doubled to generate laser light at 532 nm (in the visible region) or even higher harmonics at 355 and 266 nm. Applications of Nd:YAG lasers include removal of cataracts and unwanted hair, range finders, and for marking plastics and glass. In the last application, a white pigment that absorbs strongly in the infrared region, near the main Nd:YAG emission line of 1.064 μm, is added to a transparent plastic. When the laser is shone on the plastic the pigment absorbs the energy and heats up sufficiently to burn the plastic at the point exposed to the laser, which becomes indelibly marked. There are several YAG gain media with other lanthanoid dopants, for example Yb:YAG emits at either 1.030 μm (strongest line) or 1.050 μm and is often used where a thin disk of lasing material is needed, and Er:YAG lasers emit at 2.94 μm and are used in dentistry and for skin resurfacing.

Lanthanoids are widely used in other laser materials including glass fibres, where they are used as fibre lasers, and for the amplification of optical signals, In an erbium doped fibre amplifier (EDFA), a silica glass-fibre core is doped with Er^{3+}, and an optical signal passing along the glass fibre is amplified (converted to a higher intensity) when a pump laser is directed on to the fibre.

(c) Magnetic properties

Key point: The magnetic moments of lanthanoid compounds arise from both spin and orbital contributions.

The magnetic moment μ of many d-metal ions can be calculated by using the spin-only approximation because the strong ligand field quenches the orbital contribution (Section 20.1c). For the lanthanoids, where the spin−orbital coupling is strong, the orbital angular momentum contributes to the magnetic moment, and the ions behave like almost free atoms. Therefore, the magnetic moment must be expressed in terms of the total angular momentum quantum number J:

$$\mu = g_J \left\{ J(J+1) \right\}^{1/2} \mu_B$$

where the Landé g-factor is

$$g_J = 1 + \frac{S(S+1) - L(L+1) + J(J+1)}{2J(J+1)}$$

and μ_B is the Bohr magneton. Theoretical values of the magnetic moment of the ground states of the Ln^{3+} ions are summarized in Table 23.3; in general these values agree well with experimental data.

■ **A brief illustration.** As we have seen, the ground state term symbol for Pr^{3+} (f^2) is $^3H_4$ with $L = 5$, $S = 1$, and $J = 4$. It follows that

$$g_J = 1 + \frac{1(1+1) - 5(5+1) + 4(4+1)}{2 \times 4(4+1)} = 1 + \frac{2 - 30 + 20}{40} = \frac{4}{5}$$

Therefore

$$\mu = g_J \{ J(J+1) \}^{1/2} \mu_B = \tfrac{4}{5} \{ 4(4+1) \}^{1/2} \mu_B = 3.58 \mu_B \quad ■$$

The analysis in the *brief illustration* assumes that only one $^{2S+1}\{L\}_J$ level is occupied at the temperature of the experiment; for most lanthanoid ions this is a good assumption. For example, the first excited state of Ce^{3+} ($^5F_{5/2}$) is 1000 cm^{-1} above the ground state ($^2F_{3/2}$) and nearly unpopulated at room temperature when $kT \approx 200$ cm^{-1}. Small contributions from the higher energy terms result in the minor deviations of observed values from those based on the population of a single term. For Eu^{3+} and to a lesser extent Sm^{3+} the first excited state lies close to the ground state (for Eu^{3+}, $^7F_1$ lies only 300 cm^{-1} above the $^7F_0$ ground state) and is partly populated even at room temperature. Although the value of μ based on occupation of only the ground state is zero (because $J = 0$), the experimentally observed value is nonzero.

Long-range magnetic ordering effects, ferromagnetism, and antiferro-magnetism (Section 20.8), are observed in many lanthanoid compounds, although in general the coupling between metal atoms is much weaker than in the d-block compounds and as a result the magnetic ordering temperatures are much lower. In $BaTbO_3$ the magnetic moments of the Tb^{3+} ion become antiferromagnetically ordered below 36 K.

23.5 Binary ionic compounds

Key points: The structures of ionic lanthanoid compounds are determined by the size of the lanthanoid ion; binary oxides, halides, hydrides, and nitrides are all known.

Lanthanoid(III) ions have radii that vary between 116 and 98 pm; for comparison, the ionic radius of Fe^{3+} is 64 pm. Thus the volume occupied by a Ln^{3+} ion is typically four to five times that occupied by a typical 3d-metal ion. Unlike the 3d metals, which rarely exceed a coordination number of 6 (with 4 being common too), compounds of lanthanoids often have high coordination numbers, typically between 6 and 12, and a wide variety of coordination environments.

The binary lanthanoid(III) oxides, Ln_2O_3, have moderately complex structures with the coordination number of the Ln^{3+} ions being typically 7 (or a mixture of 6 and 7). Several

Ce	Cerium
Pr	Praseodymium
Tb	Terbium

related structure types termed A-, B-, C-Ln_2O_3 are known and many of the oxides are polymorphic with transitions between the structures occurring as the temperature is changed. The coordination geometries are determined by the radius of the lanthanoid ion, with the average cation coordination number in the structures decreasing with decreasing ionic radius, for example the La^{3+} ion in La_2O_3 has coordination number 7, whereas the Lu^{3+} ion in Lu_2O_3 has coordination number 6. In cases where Ln^{4+} ions can be obtained (for example with Ce, Pr, and Tb), the LnO_2 adopts the fluorite structure as expected from radius-ratio rules (Section 3.10).

Sulfides of stoichiometry LnS may be obtained by direct reaction of the elements at 1000°C and adopt the rock-salt structure. Phases of composition Ln_2S_3 can also be obtained by reaction of the lanthanoid trichloride with H_2S; they have been studied as replacements for the toxic CdS and CdSe as possible pigments due to their intense red-orange-yellow colours.

The lanthanoid(III) trihalides have complex structural characteristics as a result of the high coordination numbers for these large ions. For example, in LaF_3 the La^{3+} ion is in an irregular 11-coordinate environment and in $LaCl_3$ it is in a nine-coordinate, capped anti-square prismatic environment (Fig. 23.4). Towards the end of the series the trihalides of the smaller lanthanoids have different structure types with lower coordination numbers for the same halide, as expected in view of the decrease in ionic radius. In LnF_3, Ln has a nine-coordinate environment which can be considered as a capped anti-square prismatic environment distorted or a tricapped trigonal prism (**1**) and the compounds $LnCl_3$ have layer structures based on six-coordinate Ln in a cubic close-packed array of Cl^- ions. Cerium is the only lanthanoid to form a tetrahalide (CeF_4); it crystallizes with a structure formed from vertex-sharing CeF_8 polyhedra (Fig. 23.5).

All the lanthanoids form hydrides of stoichiometry LnH_2 which adopt the fluorite structure (Section 3.9a) based on cubic close-packed H^- ions with lanthanoid ions in all the tetrahedral holes. Cerium hydride may be further oxidized by hydrogen to form a series of nonstoichiometric phases of formula CeH_{2+x}, with additional H^- ions incorporated into the fluorite lattice. Some of the smaller lanthanoids (for instance Dy, Yb, and Lu) form stoichiometric trihydrides, LnH_3. Complex metal hydrides containing lanthanum, such as $LaNi_5H_6$, have been studied intensively as possible hydrogen-storage materials, Box 10.4 and Sections 24.14, 24.15, as they can be prepared by heating the alloy $LaNi_5$ in hydrogen gas under pressure; the reverse reaction occurs when $LaNi_5H_6$ is heated under ambient pressure conditions evolving H_2 gas.

Nitrides of composition LnN exist for all the lanthanoids and adopt the expected rock-salt structure with alternating Ln^{3+} and N^{3-} ions. Three different lanthanoid carbide stoichiometries are known: M_3C, M_2C_3, and MC_2. The M_3C phases form for the heavier lanthanoids and contain isolated carbon atoms and are hydrolysed by water to produce methane. The M_2C_3 phases form for the lighter lanthanoids La–Ho and contain the dicarbide anion C_2^{2-}, as is also found in CaC_2 (Section 3.9). The MC_2 phases show metallic properties and except for the elements that form stable divalent cations, such as Yb, they can be expressed as $Ln^{3+}(C_2^{2-},e^-)$ with the electron in a conduction band formed from Ln6s

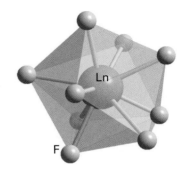

1 Ln coordination in LnF_3

Dy	Dysprosium
Yb	Ytterbium
Lu	Lutetium

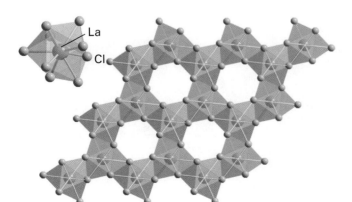

Figure 23.4 The structure of $LaCl_3$ shown as vertex-linked $LaCl_9$-capped antiprisms; a single unit is shown as inset.

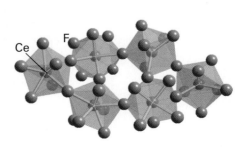

Figure 23.5 The structure of CeF_4 contains vertex-sharing CeF_8 antiprisms.

and Ln6p orbital overlap; they react with water to form ethyne and other hydrocarbons. The lanthanoid nickel borocarbides, $LnNi_2B_2C$, have structures containing alternating layers of the stoichiometries LnC and Ni_2B_2. These borocarbides are superconductors at low temperature: the transition temperature for $LuNi_2B_2C$, for instance, is 16 K.

23.6 Ternary and complex oxides

Key point: Lanthanoid ions are often found in perovskites and garnets, where the ability to change the size of the ion allows the properties of the materials to be modified.

The lanthanoids are a good source of stable, large, tripositive cations with a reasonable range of ionic radii. As a result, they can take one or more of the cation positions in ternary and more complex oxides. For example, perovskites of the type ABO_3 can readily be prepared with La on the A cation site; an example is $LaFeO_3$. Indeed, some distorted structure types are named after lanthanoids; an example is the structural type $GdFeO_3$ (Fig. 23.6), which has vertex-linked FeO_6 octahedra around the Gd^{3+} ion (as in the parent perovskite structure, Fig. 3.9) but the octahedra are tilted relative to each other. This tilting allows better coordination to the central Gd^{3+} ion. The ability to change the size of the B^{3+} ion in a series of compounds $LnBO_3$ allows the physical properties of the complex oxide to be modified in a controlled manner. For example, in the series of compounds $LnNiO_3$ for Ln = Pr to Eu, the insulating−metallic transition temperature T_{IM} (at which the properties change from that of a metal to that of an insulator, Section 3.19) increases with decreasing lanthanoid ionic radius:

	$PrNiO_3$	$NdNiO_3$	$EuNiO_3$
$r(Ln^{3+})$/pm	113	111	107
T_{IM}/K	135	200	480

As the perovskite unit cell is a structural building block often found in more complex oxide structures, lanthanoids are frequently used in such materials. Famous examples are the original high-temperature superconducting cuprate (HTSC) $La_{1.8}Ba_{0.2}CuO_4$ and the family of '123' complex oxides, $LnBa_2Cu_3O_7$, which become superconducting below 93 K. The best known of these HTSCs is the d-block (yttrium) compound $YBa_2Cu_3O_7$, but they are also found for all the lanthanoids (Section 24.8). Another system where choice of lanthanoid is crucial in obtaining the required property is the complex manganites $Ln_{1-x}Sr_xMnO_3$ that exhibit resistance effects strongly dependent on the applied magnetic field and temperature; the optimized properties are found when Ln = Pr.

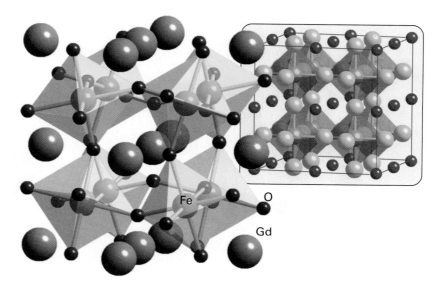

Figure 23.6 The $GdFeO_3$ structure type, with the FeO_6 octahedra outlined. The inset shows its relation to the ideal perovskite structure.

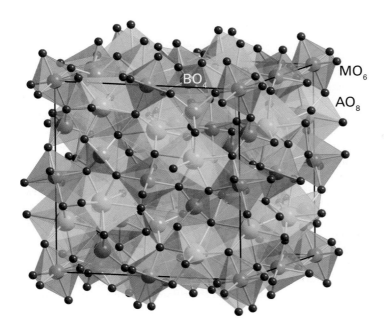

Figure 23.7 The garnet structure shown in the form of linked AO_8, BO_4, and MO_6 polyhedra. The eight-coordinate A sites often occupied by yttrium can be occupied by other lanthanoids.

The spinel structure (Fig. 3.44) has only small tetrahedral and octahedral holes in the close-packed O^{2-} ion array and cannot accommodate bulky lanthanoid ions. However, the garnet structure adopted by materials of stoichiometry $M_3M'_2(XO_4)_3$, where M and M′ are normally di- and tripositive cations and X includes Si, Al, Ga, and Ge, has eight-coordinate sites that can be occupied by lanthanoid ions, (Fig. 23.7). Yttrium aluminium garnet (YAG) is the host material for neodymium ions in the laser material Nd:YAG. Here we see a result of the similarities in the chemistry of yttrium from Group 3 and many lanthanoids. The contraction in the ionic radius of the lanthanoids with increasing atomic number means that many Ln^{3+} ions have an ionic radius similar to that of Y^{3+} and there-fore partial replacement of this ion by a lanthanide ion (for instance, Nd^{3+} in Nd:YAG) is easily achieved.

23.7 Coordination compounds

Key point: High coordination numbers with ligands adopting geometries that minimize interligand repulsions are the norm for the lanthanoids.

The adoption by the relatively large, hard Lewis acid Ln^{3+} ions of structures with high coordination numbers and with a variety of coordination environments in the solid state is repeated in solution. The variation in structure adopted is consistent with the view that the spatially buried f electrons have no significant stereochemical influence, and consequently ligands adopt positions that minimize interligand repulsions. In addition, polydentate lig-ands must satisfy their own stereochemical constraints, much as for the s-block ions and Al^{3+} complexes. For example, many lanthanoid complexes have been formed with crown ether and β-diketonate ligands. The coordination numbers for $[Ln(OH_2)_n]^{3+}$ in aqueous solution are thought to be 9 for the early lanthanoids and 8 for the later, smaller members of the series, but these ions are highly labile and the measurements are subject to consider-able uncertainty. Similarly, a striking variation is observed for the coordination numbers and structures of lanthanoid salts and complexes. For example, the small ytterbium cation, Yb^{3+}, forms the seven-coordinate complex $[Yb(acac)_3(OH_2)]$, and the larger La^{3+} is eight-coordinate in $[La(acac)_3(OH_2)_2]$. The structures of these two complexes are approximately a capped trigonal prism (**2**) and a square antiprism (**3**), respectively.

The partially fluorinated β-diketonate ligand $[C_3F_7COCHCOC(CH_3)_3]^-$, nicknamed fod, produces complexes with Ln^{3+} that are volatile and soluble in organic solvents. On account of their volatility, these complexes can be used as precursors for the synthesis of lanthanoid-containing materials by vapour deposition, such as the high temperature su-perconductors (Section 25.5).

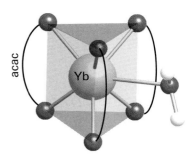

2 $[Yb(acac)_3OH_2]$

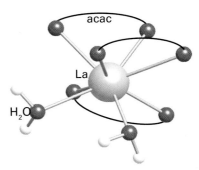

3 $[La(acac)_3(OH_2)_2]$

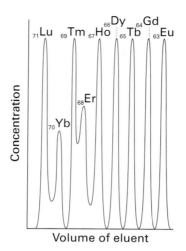

Figure 23.8 Elution of heavy lanthanoid ions from a cation exchange column using ammonium 2-hydroxyisobutyrate as the eluent. Note that the higher atomic number lanthanoids elute first because they have smaller radii and are more strongly complexed by the eluent.

Charged ligands generally have the highest affinity for the smallest Ln^{3+} ion, and the resulting increase in formation constants from large, lighter Ln^{3+} (on the left of the series) to small, heavier Ln^{3+} (on the right of the series) provides a convenient method for the chromatographic separation of these ions (Fig. 23.8). In the early days of lanthanoid chemistry, before ion-exchange chromatography was developed, tedious repetitive crystallizations were used to separate the elements.

Lanthanoid complexes have found applications as **shift reagents** in NMR spectroscopy. The chemical shift of a proton (Section 8.5) is markedly shifted when it is in the vicinity of a paramagnetic centre and the δ-range considerably expanded. Hence addition of a small amount of a paramagnetic lanthanoid complex to a solution of a complex organic molecule results in marked changes to the spectrum as the two molecules are in close proximity in the solvent. This technique is of considerable use when NMR spectra are collected on low- and mid-field instruments as the resonances are spread out over a greater range of δ values, effectively increasing the resolution of the instrument by separating otherwise overlapping resonances. Europium and Yb complexes generally induce a downfield shift whereas Pr and Dy produce upfield shifts. Commonly used shift reagents include $Ln(fod)_3$ with Ln = Eu, Pr, and Yb. Chiral lanthanide shift reagents can be used to undertake an assay of enantiomeric composition as the NMR resonances of the two enantiomers are different and their separation improved in the presence of the shift reagent, Section 27.20.

23.8 Organometallic compounds

Key points: The organometallic compounds of the lanthanoids are dominated by good donor ligands, with complexes of acceptor ligands being rare; the bonding in complexes is best treated on the basis of ionic interactions; there are similarities between the organometallic compounds of the lanthanoids and those of the early d metals.

In line with the picture of the lanthanoid ions as having no directional frontier orbitals, it should not be surprising that they do not exhibit a rich organometallic chemistry. In particular, the lack of any orbitals that can backbond to organic fragments (because the 5d orbitals are empty and the 4f orbitals are buried) restricts the number of bonding modes that are available. Most of the ligands discussed in Chapter 22 cannot bond in the fashion described there. In addition, the strongly electropositive nature of the lanthanoids means that they need good donor, not good acceptor, ligands. Thus alkoxide, amide, and halide ligands, which are both σ and π donors, are common, whereas CO and phosphine ligands, which are both σ donors and π acceptors, and which do such a good job in stabilizing low oxidation state d-metal complexes, are rarely seen in lanthanoid chemistry. Indeed, under normal laboratory conditions, neutral metal carbonyl compounds are unknown for the f-block elements.

As an example of the contrasting behaviour of organometallic complexes of the d and f blocks it should be noted that the first η^2-alkene complex of a lanthanoid was characterized only in 1987, a century and a half after the first d-metal alkene complex, Zeise's salt, was isolated. This lanthanoid alkene complex, $[(Cp^*)_2Yb(C_2H_4)Pt(PPh_3)_2]$ (4), has a particularly electron-rich alkene as it is already bound to an electron-rich Pt(0) centre, and it is thought that the alkene therefore needs no backbonding from the Yb atom to stabilize it.

The bonding in the organometallic lanthanoid complexes that does form is predominantly ionic and is governed by electrostatic factors and steric requirements. Consequently there is no need for the 18-electron rule to be obeyed. Although there is a chemical uniformity across the f block, there is also a gradation of change with steric factors tending to dominate: a small change in ligand size can make a significant difference to reactivity. Owing to the weakness of the bonds, the lanthanoid complexes remain strong Lewis acids and the complexes are therefore very sensitive to air and moisture.

The first organometallic compounds of the lanthanoids were cyclopentadienyl compounds: G. Wilkinson made a large number of $Ln(Cp)_3$ compounds with a wide variety of electron counts in 1954. The large lanthanoid ions can easily accommodate three cyclopentadienyl ligands, and they even tend to oligomerize, indicating that there is yet more space for additional ligands. Compounds of substituted cyclopentadienyl ligands are possible but the limit seems to have been reached with the sterically demanding pentamethylcyclopentadienyl (Cp^*) ligand. It was, however, nearly 40 years after the original $Ln(Cp)_3$

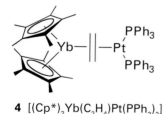

4 $[(Cp^*)_2Yb(C_2H_4)Pt(PPh_3)_2]$

compounds were isolated that $Ln(Cp^*)_3$ compounds were obtained, and even then they exist in equilibrium:

$$Sm(\eta^5\text{-}Cp^*)_3 \ (5) \rightleftarrows Sm(\eta^1\text{-}Cp^*)(\eta^5\text{-}Cp^*)_2 \ (6)$$

The great majority of lanthanoid organometallic compounds formally contain Ln(III) with a limited number of Ln(II) compounds: no other oxidation states are known. σ-Bonded alkyl groups are common, with compounds containing the cyclopentadienyl ligands tending to dominate. It is best to consider cyclopentadienyl compounds as containing Cp^- groups electrostatically bound to a central Ln^{3+} (or Ln^{2+}) cation; this view is supported by the observation that the La compounds are diamagnetic. Compounds containing η^8-cyclooctatetraene ligands are known, such as $Ce(C_8H_8)_2$ (7), and, as noted in Section 22.11, it is best to consider these to be complexes of the electron-rich $C_8H_8^{2-}$ ion as ligand.

Current research into lanthanoid organometallic compounds typically involves compounds of the type $[(Cp)_2LnR]_2$, $[(Cp)_2LnR(sol)]$, $[(Cp^*)LnRX(sol)]$, and $[(Cp^*)LnR_2(sol)]$. A number of arene complexes are also known, such as $[(C_6Me_6)Sm(AlCl_4)_3]$ (8), where it is thought that the bonding is largely the result of an electrostatically induced dipole between the Sm^{3+} ion and the electron-rich ring.

As well as an extensive organometallic chemistry of Ln(III) there are a number of Ln(II) molecular complexes that have been synthesized, including several for lanthanoids that rarely exist in this oxidation state. The species Er(II), Sm(II), and Yb(II) have extensive chemistries of this type, for example the compounds $Ln(Cp^*)_2$, but compounds of Tm(II), Dy(II), and Tm(II) have also been obtained, including $TmI_2(DME)_3$. No lanthanoid has two stable states with oxidation numbers differing by 2, so there is no possibility of oxidative addition or reductive elimination reactions: σ-bond metathesis-type reactivity dominates (Section 22.23).

In addition to the handling problems caused by the extreme sensitivity to air and moisture of the lanthanoid organometallic compounds, their study by NMR has been inhibited by the fact that they are all paramagnetic. Useful comparisons can be drawn through the study of compounds of the 4d-metal yttrium because Y^{3+} has the same charge as and a size similar to a typical Ln^{3+} ion, but is diamagnetic. In addition, yttrium is 100 per cent ^{89}Y with $I = \frac{1}{2}$, so it is possible to measure Y−C and Y−H coupling constants quite easily and thus get additional structural information.

There are strong similarities between the chemical properties of the early d-block organometallic compounds (those of Groups 3 to 5) and those of the f block. These similarities are to be expected because the early d metals are also strongly electropositive, have a limited number of d electrons to backbond with ligands, and have a limited number of accessible oxidation states. An example of a lanthanoid dinitrogen complex is $(\mu\text{-}N_2)$ $[(C_5Me_4H)_2La(THF)]_2$.

Although the organometallic chemistry of the lanthanoids is less rich than that of the d metals, there are some striking examples of unusual reactions. For instance, a lanthanoid organometallic compound can be used for the activation of the C−H bond in methane. This discovery was based on the observation that $^{13}CH_4$ exchanges ^{13}C with the CH_3 group attached to Lu:

$$Lu(Cp^*)_2CH_3 + {}^{13}CH_4 \rightarrow Lu(Cp^*)_2({}^{13}CH_3) + CH_4$$

This reaction can be carried out in deuterated cyclohexane with no evidence for activation of the cyclohexane C−D bond, presumably because cyclohexane is too bulky to gain access to the Lu atom. A mechanism involving a four-centre σ-bond metathesis-type intermediate has been proposed (Section 22.23):

$$\begin{array}{ccccc}
Me\text{—}H & & Me\text{- - -}H & & Me \quad H \\
| & \longrightarrow & \vdots \quad \vdots & \longrightarrow & | \quad | \\
Lu\text{—}Me & & Lu\text{- -}Me & & Lu \quad Me
\end{array}$$

Lanthanoid complexes are used in the Ziegler−Natta polymerization of alkenes (Section 26.16). Dinitrogen complexes of lanthanoids were first reported in 1988.

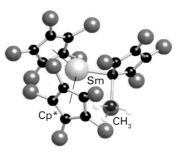

5 $[Sm(\eta^5\text{-}Cp^*)_2(\eta_1\text{-}Cp^*)]$

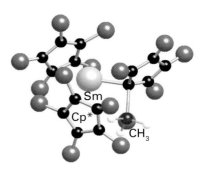

6 $[Sm(\eta^5\text{-}Cp^*)_2(\eta_1\text{-}Cp^*)]$

Dy	Dysprosium
Er	Erbium
Sm	Samarium
Tm	Thulium
Yb	Ytterbium

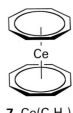

7 $Ce(C_8H_8)_2$

8 $[Sm(C_6Me_6)(AlCl_4)_3]$

9 [(Cp)$_2$Ln(μ-H)$_2$Ln(Cp)$_2$]

EXAMPLE 23.2 Accounting for the organometallic reactivity of a lanthanoid

Suggest a likely reaction pathway for the following transformation:

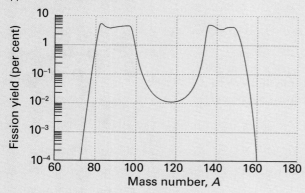

Answer The reaction with dihydrogen cannot proceed through a dihydrogen complex, oxidative addition to a dihydride, and then reductive elimination of butane because the lanthanoid is incapable of oxidative addition reactions. A possible scenario involves a σ-bond metathesis reaction through an intermediate such as

$$
\begin{array}{ccc}
\text{H} & \text{- - -} & \text{H} \\
\vdots & & \vdots \\
\text{Ln} & \text{- - - -} & \text{Bu}
\end{array}
$$

Self-test 23.2 The product of the reaction above is in fact a hydride bridged dimer (**9**). Suggest a strategy to ensure that the hydride is monomeric.

Actinoid chemistry

The chemical properties of the actinoids show less uniformity across the series than those of the lanthanoids. However, the radioactivity associated with most of the actinoids has hindered their study. Because the later actinoids are available in such tiny amounts, little is known about their reactions. The early actinoids, particularly uranium and plutonium, are of great importance in the generation of power through nuclear fission and their chemical properties have been investigated thoroughly (Box 23.2).

BOX 23.2 Nuclear fission and nuclear waste management

The fission of heavy elements, such as ^{235}U, can be induced by bombardment by neutrons. Thermal neutrons (neutrons with low velocities) bring about the fission of ^{235}U to produce two nuclides of medium mass, and a large amount of energy is released because the binding energy per nucleon decreases steadily for atomic numbers beyond about 26 (Fe; see Fig. 1.2). That unsymmetrical fission of the uranium nucleus occurs is shown by the double-humped distribution of fission products (Fig. B23.2) with maxima close to mass numbers 95 (isotopes of Mo) and 135 (isotopes of Ba). Almost all the fission products are unstable nuclides. The most troublesome are those with half-lives in the range of years to centuries: these nuclides decay fast enough to be highly radioactive but not sufficiently fast to disappear in a convenient time.

Figure B23.2 The double-humped distribution of fission products of uranium.

The first nuclear power plants relied on the fission of uranium to generate heat. The heat was used to produce steam, which can drive turbines in much the same way as conventional power plants use the burning of fuels to produce heat. However, the energy produced by the fission of a heavy element is huge in comparison with the burning of conventional fuels: for instance, the complete combustion of 1 kg of octane produces approximately 50 MJ, whereas the energy liberated by the fission of 1 kg of ^{235}U is approximately 2 TJ (1 TJ = 10^{12} J), 40 000 times as much. Later designs of nuclear power plants used plutonium, normally mixed with uranium.

Nuclear power offers the potential of enormous quantities of energy at low cost and with little contribution to carbon dioxide emissions. In some countries, such as France, nuclear energy is by far the largest contributor to their national power production. Improvements in nuclear power plant safety are being developed. A great concern, the disposal of the radioactive waste products, is a highly active area for inorganic chemistry research. At present, the best methods involve the immobilization of radioactive ions through various reactions that include vitrification (trapping the radionuclides in a glass), ion-exchange reactions with zeolites and similar porous materials, and the formation of synthetic rock materials (Synroc) which can then be buried. In the last process, high-level nuclear waste containing actinoids is mixed with several titanium-based complex oxides, such as perovskite, CaTiO$_3$, and zirconolite, CaZrTi$_2$O$_7$. This mixture is then heated to produce a hard rock-like material where the actinoids have become immobilized in the complex oxides. The Synroc product can then be stored for long periods, often underground, with minimal likelihood of the radionuclides entering the environment.

23.9 General trends

Key point: The early actinoids do not exhibit the chemical uniformity of the lanthanoids; they exist in diverse oxidation states with An(III) progressively more stable across the series.

The 14 elements from thorium (Th, $Z = 90$, $5f^1$) to lawrencium (Lr, $Z=103$, $5f^{14}$) involve the progressive completion of the 5f subshell, and in this sense are analogues of the lanthanoids. However, the actinoids do not exhibit the chemical uniformity of the lanthanoids. Whereas, like the lanthanoids, a common oxidation state of the actinoids is An(III), unlike the lanthanoids the early members of the series occur in a rich variety of other oxidation states. The Frost diagrams in Fig. 23.9 show that oxidation numbers higher than +3 are easily accessible and often preferred for the early elements of the block (Th, Pa, U, Np, Pu) whereas +3 becomes predominant in Am and beyond. Linear or nearly linear AnO_2^+ and AnO_2^{2+} ions are the dominant aqua species for oxidation numbers +5 and +6 exhibited by U and Np. Unlike in the lanthanoids, the f orbitals of the early actinoids extend into the bonding region, so the spectra of their complexes are strongly affected by ligands. The 5f and 6d orbitals are less compact than the 4f and 5d orbitals and the electrons in them are more available for bonding. Thus the outermost electron configuration of U is $5f^36d^17s^2$ with all six electrons available for bonding. At around Am and Cm the 5f and 6d orbital energies are lower due to their poor screening by the other electrons, so that the later actinoids behave like the lanthanoids with only the outermost three electrons being readily ionized.

The striking differences between the chemical properties of the lanthanoids and early actinoids led to controversy about the most appropriate placement of the actinoids in the periodic table. For example, before 1945, periodic tables usually showed U below W because both elements have a maximum oxidation number of +6. The emergence of oxidation state An(III) for the later actinoids was a key point in determining their current placement. The similarity of the heavy actinoids and the lanthanoids is illustrated by their similar elution behaviour in ion-exchange separation (compare Figs 23.8 and 23.10).

EXAMPLE 23.3 Assessing the redox stability of actinoid ions

Use the Frost diagram for thorium (Fig. 23.9) to describe the relative stability of Th(II) and Th(III).

Answer We need to use the interpretation of Frost diagrams described in Section 5.13. The initial slope in the Frost diagrams indicates that the Th^{2+} ion might be readily attained with a mild oxidant. However, Th^{2+} lies above the lines connecting Th(0) with the higher oxidation states, so it is susceptible to disproportionation. Thorium(III) is readily oxidized to Th(IV) and the steep negative slope indicates that it is susceptible to oxidation by water:

$$Th^{3+}(aq) + H^+(aq) \rightarrow Th^{4+}(aq) + \tfrac{1}{2}H_2(g)$$

We can confirm from *Resource section 3* that because $E^{\ominus} = +3.8\,V$, this reaction is highly favoured. Thus, Th(IV) will dominate in aqueous solution.

Self-test 23.3 Use the Frost diagrams and data in *Resource section 3* to determine the most stable uranium ion in acid aqueous solution in the presence of air and give its formula.

Bk	Berkelium
Cf	Californium
Cm	Curium
Es	Einsteinium

Like the lanthanoids, the actinoids have large atomic and ionic radii (the radius of an An^{3+} ion is typically about 5 pm larger than its Ln^{3+} congener) and as a result often have high coordination numbers. For example, uranium in solid UCl_4 is eight-coordinate and in solid UBr_4 it is seven-coordinate in a pentagonal-bipyramidal array. Solid-state structures with coordination numbers up to 12 have been observed.

Because of the small quantities of material available in most cases and their intense radioactivity, most of the chemical properties of the transamericium elements (the elements following americium, $Z = 95$) have been established by experiments carried out on a microgram scale or even on just a few hundred atoms. For example, the transamericium ion complexes have been adsorbed on and eluted from a single bead of ion-exchange material of diameter 0.2 mm. For the heaviest and most unstable post-actinoids, such as hassium (Hs, $Z = 108$), the lifetimes are too short for chemical separation and the identification of the element is based exclusively on the properties of the radiation it emits.

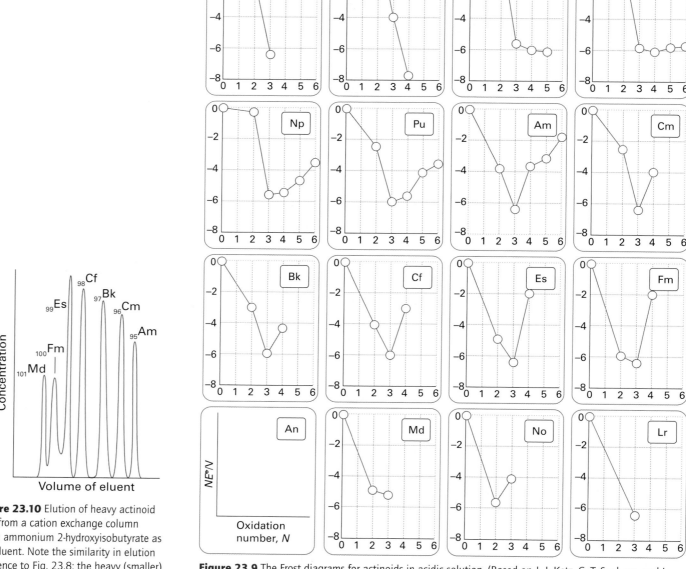

Figure 23.9 The Frost diagrams for actinoids in acidic solution. (Based on J. J. Katz, G. T. Seaborg, and L. Morss, *Chemistry of the actinide elements.* Chapman and Hall, London (1986).)

Figure 23.10 Elution of heavy actinoid ions from a cation exchange column using ammonium 2-hydroxyisobutyrate as the eluent. Note the similarity in elution sequence to Fig. 23.8: the heavy (smaller) An³⁺ ions elute first.

23.10 Electronic spectra

Key points: The electronic spectra of the early actinoids have contributions from ligand-to-metal charge transfer, 5f → 6d, and 5f → 5f transitions. The uranyl ion fluoresces strongly.

Transitions between electronic states involving only the f orbitals, the 5f and 6d orbitals, and ligand-to-metal charge transfer (LMCT) are all possible for the actinoid ions. The f−f transitions are broader and more intense than for the lanthanoids because the 5f orbitals interact more strongly with the ligands. Their molar absorption coefficients typically lie in the range $10-100$ dm³ mol⁻¹ cm⁻¹. The most intense absorptions are associated with LMCT transitions. For instance, LMCT transitions result in the intense yellow colour of the uranyl ion, UO_2^{2+}, in solution and its compounds. In species such as U^{3+} (f^3) transitions such as $5f^26d^1 \leftarrow 5f^3$ occur at wavenumbers between 20 000 and

33 000 cm^{-1} (500–300 nm) giving solutions and compounds of this ion a deep orange–red colour. For Np^{3+} and Pu^{3+}, with increasing effective nuclear charge, the separation of the 5f and 6d levels increases and the corresponding transitions move into the UV region of the spectrum; Np^{3+} solutions are violet and those of Pu^{3+} are light violet–blue due mainly to f–f transitions.

The uranyl ion, UO_2^{2+}, is also strongly fluorescent, with a strong emission between 500 and 550 nm on excitation with UV radiation (Fig. 23.11). This property has in the past been used for colouring glass, where the addition of 0.5–2 per cent of uranyl salts produces a bright golden yellow colour. The glass has a bright, green–yellow fluorescence in sunlight which adds to its attractiveness; when viewed under UV radiation it glows an intense green. However as this glass is radioactive its commercial production has been largely phased out in recent years.

23.11 Thorium and uranium

Key points: The common nuclides of thorium and uranium exhibit only low levels of radioactivity, so their chemical properties have been extensively developed; the uranyl cation is found in complexes with many different ligand donor atoms; the organometallic compounds of the elements are dominated by pentamethylcyclo-pentadienyl complexes.

Because of their ready availability and low level of radioactivity, the chemical manipulation of Th and U can be carried out with ordinary laboratory techniques. As indicated in Fig. 23.9, the most stable oxidation state of Th in aqueous solution is Th(IV). This oxidation state also dominates the solid-state chemistry of the element. Eight-coordination is common in simple Th(IV) compounds. For example, ThO_2 has the fluorite structure (in which a Th atom is surrounded by a cubic array of O^{2-} ions) and in $ThCl_4$ and ThF_4 the coordination numbers are also 8 with dodecahedral and square antiprism symmetry, respectively. The coordination number of Th in $[Th(NO_3)_4(OPPh_3)_2]$ is 10 (**10**), with the NO_3^- ions and triphenylphosphine oxide groups arranged in a capped cubic array around the Th atom. The very unusual coordination number of 11 is exhibited by Th in its hydrated nitrate $Th(NO_3)_4.5H_2O$ with the Th^{4+} ion coordinated to four NO_3^- ions in a bidentate fashion and with three H_2O molecules (**11**).

The chemical properties of U are more varied than those of Th because the element has access to oxidation states from U(III) to U(VI), with U(IV) and U(VI) the most common. Uranium halides are known for the full range of oxidation states U(III) to U(VI), with a trend towards decreasing coordination number with increasing oxidation number. The U atom is nine-coordinate in solid UCl_3, eight-coordinate in UCl_4, and six-coordinate for the U(V) and U(VI) chlorides U_2Cl_{10} (Fig 23.12) and UCl_6, both of which are molecular compounds. The high volatility of UF_6 (it sublimes at 57°C) together with the occurrence of fluorine in a single isotopic form account for the use of this compound in the separation of the uranium isotopes by gaseous diffusion or centrifugation. In gas diffusion, the lighter $^{235}UF_6$ molecules travel at greater average speeds and strike the walls of a container

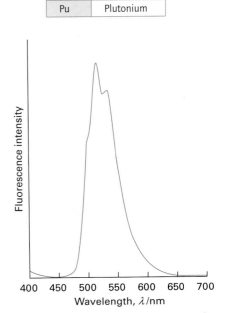

| Np | Neptunium |
| Pu | Plutonium |

Figure 23.11 The fluorescence spectrum of the UO_2^{2+} ion in aqueous solution.

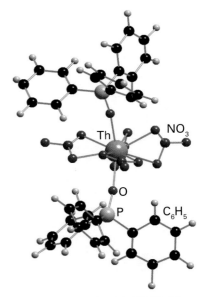

10 $[Th(NO_3)_4(OPPh_3)_2]$

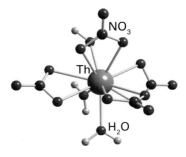

11 $[Th(NO_3)_4(OH_2)_3]$

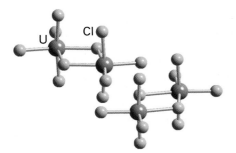

Figure 23.12 The crystal structure of U_2Cl_{10} consists of discrete molecules formed from pairs of edge-sharing UCl_6 octahedra.

12 $M(Cp)_4$, M = Th, U

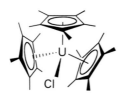

13 $U(Cp^*)_3Cl$

14 $M(C_8H_8)_2$, M = Th, U

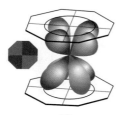

15

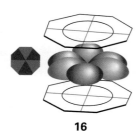

16

more frequently than their heavier isotopologues such as $^{238}UF_6$. If the container-wall is permeable, $^{235}UF_6$ will diffuse through it slightly more rapidly. A series of such 'diffusers', sometimes numbering thousands, enables enrichment of the gas in $^{235}UF_6$.

Uranium metal does not form a passivating oxide coating, so it becomes corroded on prolonged exposure to air to give a complex mixture of oxides. These oxides include UO_2, U_3O_8, and several polymorphs of the stoichiometry UO_3. The dioxide UO_2 adopts the fluorite structure but also takes up interstitial O atoms to form the non-stoichiometric series UO_{2+x}, $0 < x < 0.25$. One form of uranium trioxide, δ-UO_3, adopts the ReO_3 structure type (Section 24.8b).

The most important oxide is UO_3, which dissolves in acid to give the uranyl ion, UO_2^{2+} $[O=U=O]^{2+}$. In water, this ion forms complexes with many anions, such as NO_3^- and SO_4^{2-}. In contrast to the angular shape of the VO_2^+ ion and similar d^0 complexes, the AnO_2^{n+} unit, with An = U, Np, Pu, and Am, maintains its linearity in all complexes. Both f-orbital bonding and relativistic effects have been invoked to explain this linearity. Compounds containing the UO_2^{2+} ion show $(2+n)$-coordination, $n = 4$, 5, and 6, with the four to six additional ligands forming a planar or near planar arrangement around the uranyl unit. Thus the structure of UO_2F_2 has a slightly puckered ring of six F^- ions around the UO_2^{2+} unit. The stability of the UO_2^{2+} dication means that in most reactions it remains inert; recently, however, it has been shown that it is possible to reduce this species, when trapped within a rigid framework, to produce $[O = U(V) - OR]^+$.

The separation of U from most other metals is accomplished by the extraction of the neutral uranyl nitrato complex $[UO_2(NO_3)_2(OH_2)_4]$ from the aqueous phase into a polar organic phase, such as a solution of tributylphosphate dissolved in a hydrocarbon solvent. This kind of solvent-extraction process is used to separate actinoids from other fission products in spent nuclear fuel.

The organometallic chemistry of U and Th is reasonably well developed and shows many similarities to that of the lanthanoids, except that Th and U occur in a number of oxidation states and are larger than the typical Ln ion. Thus, compounds are dominated by those containing good donor ligands, such as σ-bonded alkyl and cyclopentadienyl groups. The increased size of Th and U compared with typical lanthanoids means that the tetrahedral species $Th(Cp)_4$ and $U(Cp)_4$ (**12**) can be isolated as monomers, and not only can $U(Cp^*)_3$ be isolated but so too can $U(Cp^*)_3Cl$ (**13**). As with lanthanoid organometallic compounds, actinoid organometallic compounds do not obey the 18-electron rule (Section 22.1).

Sandwich compounds are possible with the η^8-cyclooctatetraene ligand, and both thorocene, $Th(C_8H_8)_2$ and uranocene, $U(C_8H_8)_2$ (**14**) are known and are sufficiently stable not to react with water. Compared with lanthanoid complexes of cyclooctatetraene, the bonding in uranocene and thorocene is complicated by the extension of the f orbitals beyond the core of the atom. In theory, therefore, not only is δ bonding possible (**15**), but so is ϕ bonding (**16**). Quite how much of a role ϕ bonds play in the bonding in uranocene or thorocene is the subject of considerable debate, but it is interesting to note that actinoid cyclooctatetraene compounds are the only 'real' compounds (as opposed to metal dimers in the gas phase) that might contain any contribution from ϕ bonding.

23.12 Neptunium, plutonium, and americium

Key points: Oxidation states higher than $+3$ become increasing less accessible between Np and Am, although all these actinoids can form AnO_2^{n+} species in aqueous solution.

The three elements Np, Pu, and Am form compounds containing similar species, although there are significant differences in the stabilities of the main oxidation states. The Frost diagrams in Fig. 23.9 summarize their behaviour. Neptunium dissolves in dilute acids to produce Np^{3+}, which is readily oxidized by air to produce Np^{4+}. Increasingly strong oxidizing agents produce NpO_2^+ (Np(V)) and NpO_2^{2+} (Np(VI)). The four oxidation states of plutonium, Pu(III), Pu(IV), Pu(V), and Pu(VI), are separated from each other by less than 1 V and solutions of Pu often contain a mixture of the species Pu^{3+}, Pu^{4+}, and PuO_2^{2+} (PuO_2^+ has a tendency to disproportionate to Pu^{4+} and PuO_2^{2+}). The ion Am^{3+} is the most stable species in solution, reflecting the tendency for the An(III) to dominate actinoid chemistry for the high atomic number elements. Under strongly

Am	Americium
Np	Neptunium
Pu	Plutonium

oxidizing conditions AmO_2^+ and AmO_2^{2+} can be formed; Am(IV) disproportionates in acidic solutions.

The An(IV) oxides NpO_2, PuO_2, and AmO_2, which are formed by heating the elements or their salts in air, all adopt the fluorite structure. Lower oxides include Np_3O_8, Pu_2O_3, and Am_2O_3. The trichlorides, $AnCl_3$, can be obtained by direct reaction of the elements at 450°C and have structures analogous to that of $LnCl_3$ with a nine-coordinate An atom. Tetrafluorides are known for all three actinoids though only Np and Pu form tetrachlorides, further demonstrating the difficulty in raising americium to Am(IV). Both Np and Pu form hexafluorides which, like UF_6, are volatile solids.

All three metals have species analogous to the uranyl ion, forming NpO_2^{2+}, PuO_2^{2+}, and AmO_2^{2+}, which can be extracted from aqueous solution by tributylphosphate as $AnO_2(NO_3)_2(OP(OBu)_3)_2$. The tetrahalides are Lewis acids and form adducts with electron pair donors such as DMSO, as in $AnCl_4(Me_2SO)_7$. Neptunium forms a number of organometallic compounds that are analogues of those of uranium, such as $Np(Cp)_4$.

Although these elements are highly toxic due to their radioactivity, this property is put to good use in many smoke detectors, which contain a minute amount (0.2 µg) of ^{241}Am. This isotope is an α-emitter with a half-life of 432 years and the α-particles enter a space between two electrodes where they ionize the air, so permitting a small current to flow between the electrodes. However, when smoke particles enter the space between the electrodes they absorb the α-particles (and any ions formed by the α-particles adhere to the smoke particle surfaces) and the resulting reduction in current is detected, setting off the alarm. The isotope ^{241}Am is used in this application as it is a strong source of α-particles; however these do not escape from the smoke detector due to their short path-lengths in solids and do not present a hazard.

FURTHER READING

S.A. Cotton, *Lanthanides and actinides*. Macmillan, London (1991).

N. Kaltsoyannis and P. Scott, *The f elements*. Oxford University Press (1999).

G. Seaborg, J. Katz, and L.R. Morss, *Chemistry of the actinide elements*. Chapman and Hall, London (1986).

D.M.P. Mingos and R.H. Crabtree (ed.). *Comprehensive organometallic chemistry III*. Elsevier, Oxford (2006). Volume 4 (ed. M. Bochmann) deals with Groups 3 and 4 and the lanthanoids and actinoids.

W.J. Evans, *Adv. Organomet. Chem.*, 1985, **24**, 131, C.J. Schaverien, *Adv. Organomet. Chem.*, 1994, **36**, 283, and W.J. Evans and B.L. Davis, *Chem. Rev.*, 2002, **102**, 2119. Reviews of organolanthanoid chemistry.

T.J. Marks and A. Streitwieser, p. 1547, and T.J. Marks, p. 1588, in G. Seaborg, J. Katz, and L.R. Morss, *Chemistry of the actinide elements*, Vol. 2. Chapman and Hall, London (1986). For a survey of organoactinoid chemistry.

EXERCISES

23.1 (a) Give a balanced equation for the reaction of any of the lanthanoids with aqueous acid. (b) Justify your answer with reduction potentials and with a generalization on the most stable positive oxidation states for the lanthanoids. (c) Name the two lanthanoids that have the greatest tendency to deviate from the usual positive oxidation state and correlate this deviation with electronic structure.

23.2 Explain the variation in the ionic radii between La^{3+} and Lu^{3+}.

23.3 From a knowledge of their chemical properties, speculate on why Ce and Eu were the easiest lanthanoids to isolate before the development of ion-exchange chromatography.

23.4 How would you expect the first and second ionization energies of the lanthanoids to vary across the series? Sketch the graph that you would get if you plotted the third ionization energy of the lanthanoids against atomic number. Identify elements at any peaks or troughs and suggest a reason for their occurrence.

23.5 Derive the ground state term symbol for the Tb^{3+} ion.

23.6 Predict the magnetic moment of a compound containing the Tb^{3+} ion.

23.7 Explain why stable and readily isolable carbonyl complexes are unknown for the lanthanoids.

23.8 Suggest a synthesis of neptunocene from $NpCl_4$.

23.9 Account for the similar electronic spectra of Eu^{3+} complexes with various ligands and the variation of the electronic spectra of Am^{3+} complexes as the ligand is varied.

23.10 Predict a structure type for BkN based on the ionic radii $r(Bk^{3+}) = 96$ pm and $r(N^{3-}) = 146$ pm.

23.11 Describe the general nature of the distribution of the elements formed in the thermal neutron fission of ^{235}U, and decide which of the following highly radioactive nuclides are likely to present the greatest radiation hazard in the spent fuel from nuclear power reactors: (a) ^{39}Ar, (b) ^{228}Th, (c) ^{90}Sr, (d) ^{144}Ce.

PROBLEMS

23.1 Lanthanoid coordination compounds rarely exhibit isomerism in solution. Suggest two factors that might cause this phenomenon, explaining your reasoning. (See D. Parker, R.S. Dickins, H. Puschmann, C. Crossland, and J.A.K. Howard, *Chem. Rev.*, 2002, **102**, 1977.)

23.2 Neither lanthanoid nor actinoid organometallic compounds obey the 18-electron rule. Discuss the reasons, using the structures of the tris(Cp) and tris(Cp*) Ln and An complexes as examples. (See W.J. Evans and B.L. Davis, *Chem. Rev.*, 2002, **102**, 2119.)

23.3 The bonding in the linear uranyl ion, OUO^{2+}, is often explained in terms of significant π bonding using 5f orbitals on the metal. Using the f orbitals illustrated in Fig. 1.16, construct a reasonable molecular orbital diagram for π bonding with the appropriate oxygen p orbitals.

23.4 The existence of a maximum oxidation number of +6 for both U ($Z = 92$) and W ($Z = 74$) prompted the placement of U under W in early periodic tables. When the element after uranium, Np ($Z = 93$), was discovered in 1940 its properties did not correspond to those of Re ($Z = 75$), and this cast doubt on the original placement of U. (See G.T. Seaborg and W.D. Loveland, *The elements beyond uranium*. Wiley-Interscience, New York (1990), p. 9 *et seq.*) Using standard potential data from *Resource section 3*, discuss the differences in oxidation state stability between Np and Re.

23.5 The processing of spent nuclear fuel and separation of the lighter actinoid elements, U, Np, and Pu, is an important industrial process. Discuss the chemistry involved in the various methods used to extract and separate these elements.

23.6 Summarize the arguments related to the placement of the lanthanoids and actinoids in the periodic table. See *J. Chem. Ed.*, 2008, **85**, 1482 and related papers.

PART 3
Frontiers

Inorganic chemistry is advancing rapidly at its frontiers, especially where research impinges on other disciplines such as the life sciences, condensed-matter physics, materials science, and environmental chemistry. These swiftly developing fields also represent many areas of inorganic chemistry where novel types of compounds are used in catalysis, electronics, and pharmaceuticals. The aim of this section of the book is to demonstrate the vigorous nature of contemporary inorganic chemistry by building on the introductory and descriptive material in Parts 1 and 2.

These *Frontiers* chapters open with a discussion in Chapter 24 of materials chemistry, focusing on solid-state compounds, their synthesis, structure, and electronic, magnetic, and optical properties. One area that has developed enormously in the past decade has been that of nanomaterials, and in Chapter 25 we provide a comprehensive introduction to inorganic nanochemistry. Chapter 26 covers catalysis involving inorganic compounds and discusses the basic concepts relating to catalytic reactions at metal centres.

Finally, we turn to the frontier where inorganic chemistry meets life. Chapter 27 discusses the function of different elements in cells and intracellular compartments and the various and extraordinarily subtle ways in which they are exploited. It also describes the structures and functions of complexes and materials that are formed in the biological environment and how inorganic elements are used in medical treatments.

Solid-state and materials chemistry

24

The area of 'materials chemistry' is developing rapidly and there has been a great increase in interest in the synthesis and properties of novel inorganic solids. In this chapter we discuss a number of areas of current importance and investigation. Initially, we describe how inorganic materials are synthesized as bulk solids and on substrates. The role of defects in controlling structure and ion migration in solids is then discussed with examples from materials used in sensors and in energy generation and storage. We then extend these concepts to key classes of inorganic materials, with sections on intercalation compounds, complex electronic oxides (such as high-temperature superconductors), magnetic compounds (including materials exhibiting giant magnetoresistance), framework structures (such as zeolites and their analogues), and molecular materials. Throughout the chapter we shall see how the synthesis and properties of these solids correlate with their crystal and electronic structures.

The chemistry of the solid state is a vigorous and exciting area of inorganic research, partly on account of the technological applications of materials but also because their properties are challenging to understand. We shall draw on some of the concepts developed in Chapter 3 for discussing solids, such as lattice energetics and band structure. We also introduce some concepts that are needed to discuss the dynamics of events that occur in the interior of solids. We shall need to go beyond the views that solids have a well-defined stoichiometry and that their atoms are in fixed locations, for interesting phenomena arise from nonstoichiometry and ion mobility. The fact that in solid materials atoms and ions can interact in a cooperative manner gives rise to many of the fascinating and useful aspects of their chemical properties.

Much of the current research in solid-state chemistry is motivated by the search for commercially useful materials, such as components of batteries and fuel cells, catalysts for hydrocarbon interconversion, and improved electronic and photonic devices for information processing and storage. The scope for the synthesis of new inorganic solids is enormous. For example, although it is known that 100 structural types account for 95 per cent of the known binary (A_aB_b, such as brass, CuZn) or ternary ($A_aB_bC_c$) intermetallic compounds, there are plenty of opportunities for extending these studies to the synthesis and characterization of four-, five-, and six-component systems.

Among the many other areas of materials chemistry that are currently being explored are new microporous solids for use in molecular separations and heterogeneous catalysis; these materials were discussed in Section 14.15, but recent advances in this area are developed here and in Chapter 26. Furthermore, because the properties and uses of a material are dependent on its physical size and form, there has been an explosion of work on nanomaterials, in which the focus is on inorganic solids with controlled sub-micrometre dimensions; this topic is treated in Chapter 25.

Computer modelling is used to develop numerical models of the structures and properties of materials. It has been applied to a wide range of inorganic materials to understand and predict how atomic and electronic structure control their physical properties. Some of the most successful applications have been in areas related to defect structures, ionic mobility, and catalysis on surfaces.

Synthesis of materials

Much of synthetic inorganic chemistry, including the coordination chemistry of the metals and organometallic chemistry, makes use of the conversion of molecules by the replacement of one ligand by another in a solution-based reaction. Processes of this type generally have relatively small activation energies and can be undertaken at low temperatures, typically between 0°C and 150°C, and in solvents that permit the migration of reacting species. Rapid molecular migration in solvents results in fairly short reaction times. The formation of solid materials by reaction of solids, however, involves rather different reactions as the high lattice energies of their extended structures need to be overcome and ion migration in the solid state is normally slow except at very elevated temperatures. Some inorganic materials can be prepared from solutions at lower temperatures, where the building blocks are condensed together to form the extended structure.

24.1 The formation of bulk material

New materials can be obtained by two main methods. One is the direct reaction of two or more solids. The other is the linking of polyhedral building units from solution and deposition of the newly formed solid.

(a) Methods of direct synthesis

Key point: Many complex solids can be obtained by direct reaction of the components at high temperatures.

The most widely used method for the synthesis of bulk inorganic solids involves heating the solid reactants together at a high temperature, typically between 500 and 1500°C, for an extended period. Normally a complex oxide may be obtained by heating a mixture of all the oxides of the various metals present; alternatively, simple compounds that decompose to give the oxides may be used instead of the oxide itself. Thus, ternary oxides, such as $BaTiO_3$, and quaternary oxides, such as $YBa_2Cu_3O_7$, are synthesized by heating together the following mixtures for several days:

$$BaCO_3(s) + TiO_2(s) \xrightarrow{1000°C} BaTiO_3(s) + CO_2(g)$$

$$\tfrac{1}{2}Y_2O_3(s) + 2BaCO_3(s) + 3CuO(s) + \tfrac{1}{4}O_2 \xrightarrow{930°C/air\,and\,450°C/O_2} YBa_2Cu_3O_7(s) + 2CO_2(g)$$

High temperatures are used in these syntheses to accelerate the slow diffusion of ions in solids and to overcome the high Coulombic attractions between the ions. The direct method is applicable to many other inorganic material types, such as the syntheses of complex chlorides and dense, anhydrous metal aluminosilicates:

$$3CsCl(s) + 2ScCl_3(s) \rightarrow Cs_3Sc_2Cl_9(s)$$
$$NaAlO_2(s) + SiO_2(s) \rightarrow NaAlSiO_4(s)$$

Most simple binary oxides are available commercially as pure, polycrystalline powders with typical particle dimensions of a few micrometres. Alternatively, the decomposition of a simple metal salt precursor, either prior to or during reaction, leads to a finely divided oxide. Such precursors include metal carbonates, hydroxides, oxalates, and nitrates. An additional advantage of precursors is that they are normally stable in air whereas many oxides are hygroscopic and pick up carbon dioxide from the air. Thus, in the synthesis of $BaTiO_3$, barium carbonate, $BaCO_3$, which starts to decompose to BaO above 900°C, would be ground together with TiO_2 in the correct stoichiometric proportions using a pestle and mortar. The mixture is then transferred to a crucible, normally constructed of an inert material such as vitreous silica, recrystallized alumina, or platinum, and placed in a furnace. Even at high temperatures the reaction is slow and typically takes several days.

A variety of methods can be used to improve reaction rates, including pelletizing the reaction mixture under high pressure to increase the contact between the reactant particles, regrinding the mixture periodically to introduce virgin reactant interfaces, and the use of 'fluxes', low-melting solids that aid the ion diffusion processes. The size of the reactant particles is a major factor in controlling the time it takes a reaction to proceed to completion. The larger the particles the lower their total surface area, and therefore the smaller the area at which the reaction can take place. Furthermore, the distances over which diffusion of ions must occur are much greater for larger particles, which are typically several microns in size for a polycrystalline material. In order to increase the rate of reaction and allow solid state reactions to occur at lower temperatures, reactants having small particle sizes, between 10 nm and 1 μm, and large surface areas are often deliberately employed. These submicron or nanoparticles of metal oxides can be prepared by using some of the methods described in Chapter 25 such as spray pyrolysis, where an aqueous solution of the metal salt is sprayed on to a hot surface to evaporate the solvent and decompose the salt to a fine-particle oxide. In the synthesis of $LiNiO_2$, by reaction of LiOH and NiO in air

$$2\,LiOH\,(s) + 2\,NiO(s) + \tfrac{1}{2}\,O_2 \rightarrow 2\,LiNiO_2(s) + H_2O(g)$$

the use of highly reactive nickel oxide particles produced by spray pyrolysis allows the reaction to occur to completion at under 700°C and in only 3 hours, which is at a lower temperature and much faster than would be needed with polycrystalline NiO.

The reaction environment may need to be controlled if a particular oxidation state is required or one of the reactants is volatile. Solid-state reactions can be carried out in a controlled atmosphere, using a tube furnace in which a gas can be passed over the reaction mixture while it is being heated. An example of such a reaction is the use of an inert gas to prevent oxidation, as in the preparation of $TlTaO_3$:

$$Tl_2O(s) + Ta_2O_5(s) \xrightarrow{\text{N}_2\,/600°C} 2\,TlTaO_3(s)$$

High gas pressures may also be used to control the composition of the reaction product. For example, Fe(III) is normally obtained in oxygen at or near normal pressures but at high pressures Fe(IV) may be formed, as in the production of Sr_2FeO_4 from mixtures of SrO and Fe_2O_3 under several hundred atmospheres of oxygen. For volatile reactants the reaction mixture is normally sealed in a glass tube, under vacuum, prior to heating. Examples of such reactions are

$$Ta(s) + S_2(l) \xrightarrow{500°C} TaS_2(s)$$

$$Tl_2O_3(l) + 2\,BaO(s) + 3\,CaO(s) + 4\,CuO(s) \xrightarrow{860°C} Tl_2Ba_2Ca_3Cu_4O_{12}(s)$$

The sulfur and thallium(III) oxide are volatile at the reaction temperatures and would be lost from the reaction mixture in an open vessel, leading to products of the incorrect stoichiometry.

High pressures can also be used to affect the outcome of a solid-state chemical reaction. Specialised apparatus, typically based on large presses, allows reactions between solids to take place at pressures of up to about 100 GPa (1 Mbar) at temperatures close to 1500°C. Reactions carried out under such conditions promote the formation of dense, higher co-ordination number structures. An example is the production of $MgSiO_3$, with a perovskite-like structure and six-coordinate Si in an octahedral SiO_6 unit, rather than the normal tetrahedral SiO_4 unit. Small-scale reactions can be undertaken at very high pressures in 'diamond anvil cells' in which the faces of two opposed diamonds are pushed together in a vice-like apparatus to generate pressures of up to 100 GPa.

(b) Solution methods

Key point: Frameworks formed from polyhedral species can often be obtained by condensation reactions in solution.

Many inorganic materials, especially framework structures, can be synthesized by crystallization from solution. Although the methods used are very diverse, the following are typical reactions that occur in water:

$$ZrO_2(s) + 2\,H_3PO_4(l) \rightarrow Zr(HPO_4)_2.H_2O(s) + H_2O(l)$$

$$12\,NaAlO_2(s) + 12\,Na_2SiO_3(s) + (12 + n)\,H_2O$$
$$\xrightarrow{90°C} Na_{12}[Si_{12}Al_{12}O_{48}].nH_2O \text{ (Zeolite LTA)(s)} + 24\,NaOH(aq)$$

Solution methods are extended by using **hydrothermal techniques**, in which the reacting solution is heated above its normal boiling point in a sealed vessel. Such reactions are important for the synthesis of open-structure aluminosilicates (zeolites), analogous porous structures based on linked oxo-polyhedra (Section 24.13a), and related **metal-organic frameworks** (MOF) in which metal ions are linked by coordinating organic species, such as carboxylates (Section 24.13b). These porous structures are often thermodynamically metastable with respect as conversion to denser structure types so they cannot be made by direct high-temperature reactions. For example, the sodium aluminosilicate zeolite $Na_{12}[Si_{12}Al_{12}O_{48}].nH_2O$ formed in solution converts on heating above 800°C to the dense aluminosilicate $NaSiAlO_4$. More recently, other solvents such as liquid ammonia, super-critical CO_2, and organic amines have been used in so-called **solvothermal reactions**.

A reaction in solution can also be used as an initial stage in the synthesis of many materials, particularly oxides, normally obtained through direct high-temperature reaction. The advantages of starting with solutions is that the reactants are mixed at the atomic level, so overcoming the problems associated with the direct reaction of two or more solid phases consisting of micrometre-sized particles. In the simplest reaction of this type, a solution of metal ions (for example, solutions of metal nitrates) is converted to a solid through a variety of methods such as evaporation of the solvent, precipitation as a simple mixed metal salt, or formation of a gel. This solid is then heated to produce the target material. Two examples are

$$2\,La^{3+}(aq) + Cu^{2+}(aq) \xrightarrow{\;OH^-(aq)\;} 2\,La(OH)_3{\cdot}Cu(OH)_2(s)$$
$$\xrightarrow{\;600°C\;} La_2CuO_4(s) + 4\,H_2O(g)$$

$$Zn^{2+}(aq) + 2\,Fe^{2+}(aq) + 3C_2O_4^{2-}(aq) \rightarrow ZnFe_2(C_2O_4)_3(s)$$
$$\xrightarrow{\;700°C\;} ZnFe_2O_4(s) + 4\,CO(g) + 2\,CO_2(g)$$

As well as the advantages of reduced reaction times, a result of the intimate mixing of the reactants, the final decomposition temperature is somewhat lower than that needed for the direct reaction of the oxides. The use of a lower temperature can also have the effect of reducing the size of the particles formed in the reaction. Further discussion of routes involving gel formation, or so-called 'sol–gel processes', are included in Sections 24.8 and 25.4.

Although high-temperature direct-combination methods and solvothermal techniques are the most commonly used methods in materials chemistry, some reactions involving solids can occur at low temperatures if there is no major change in structure. These so-called 'intercalation reactions' are discussed in Section 24.10.

24.2 Chemical deposition

Key point: Thermal decomposition of volatile inorganic compounds can be used to deposit films of materials, particularly those with uses in electronic devices, on substrates.

The technological applications of the materials described in this chapter often require the inorganic material to be generated as a thin layer or film, for example on a silicon substrate, and it has become necessary to develop techniques to deposit films of many inorganic materials. The principal method is **chemical vapour deposition** (CVD), in which a volatile inorganic compound is decomposed above the substrate. When the compound is a metallo-organic complex (that is, a complex of a metal atom with organic ligands), this route is known as **metallo-organic chemical vapour deposition** (MOCVD). A large area of research has grown up around the design and synthesis of volatile inorganic compounds that can be used for CVD.

For many electronic compounds, simple metal alkyls provide a route to depositing the metal or, through reaction with other gas molecules, its compounds. Thus Me_2Zn, which has a vapour pressure of 0.3 bar at room temperature, can react with H_2S above a substrate to generate the Group 12/16 (II/VI) semiconductor ZnS (and methane). For others, such as In, the metal alkyl is a solid with a low vapour pressure at room temperature (for Me_3In, less than 2 mbar) and it cannot easily be used. In such cases other volatile compounds must be found. This search focuses on other organometallic compounds such as carbonyls, cyclopentadienyl complexes, acetoacetonates, and thiocarbamates, although it should be noted than many of these volatile metal compounds are highly toxic. For complex oxides, such as the superconducting cuprate $YBa_2Cu_3O_7$, suitable precursors are required for each of the metals; the challenge is to find volatile molecules of the electropositive elements Ba and Y, which normally form ionic compounds. Metal β-diketonates have been found to be useful in this respect, as compounds such as the 2,2,6,6-tetramethyl-3,5-heptanedione complex of yttrium sublime at about 150°C.

Another approach to molecules that can be used for CVD involves incorporating into a single-molecule precursor more than one of the atom types to be deposited. This procedure has the potential advantage of improving the control of the product stoichiometry. Thus, zinc sulfide can be deposited from a variety of zinc thiocomplexes, such as $Zn(S_2PMe_2)_2$. A future goal is to make complex volatile molecules containing, for example, several different metal atoms that can be deposited simultaneously to make a complex oxide.

Other methods of deposition of thin films, at the nanometre scale, of inorganic materials on substrates include sputtering and laser ablation; these techniques are dealt with in greater detail in Section 25.5.

EXAMPLE 24.1 Synthesizing complex oxides

How would you synthesize a sample of the high-temperature superconductor $NdBa_2Cu_3O_{7-x}$?

Answer We need to think of an analogous compound and adapt its preparation to this compound. The same method as used for preparing $YBa_2Cu_3O_7$ can be used but with the appropriate lanthanoid oxide. That is, use the reaction of neodymium oxide, barium carbonate, and copper(II) oxide at 940°C followed by annealing under pure oxygen at 450°C.

Self-test 24.1 How would you prepare a sample of Sr_2MoO_4?

Defects and ion transport

As discussed in Section 3.16, all solids above $T = 0$ contain defects, imperfections of structure or composition. It is also possible to introduce defects (extrinsic defects) deliberately into a material through mechanisms such as doping. These defects, which are mainly interstitials (Frenkel-type) or vacancies (Schottky-type), are important because they influence properties such as electrical conductivity and chemical reactivity. Electrical conduction can arise from the motion of ions through the solid, and this motion is often enhanced by the presence of defects. Materials with high ionic conductivity have important applications in sensors and fuel cells of various kinds.

24.3 Extended defects

Key point: Wadsley defects are shear planes that collect defects along certain crystallographic directions.

The defects discussed in Chapter 3 were point defects. Such defects entail a significant local distortion of the structure and in some instances localized charge imbalances too. Therefore, it should not be surprising that defects may cluster together and sometimes form lines and planes.

Tungsten oxides illustrate the formation of planes of defects. As illustrated in Fig. 24.1, the idealized structure of WO_3 (which is usually referred to as the 'ReO$_3$ structure'; see below) consists of WO_6 octahedra sharing all vertices. To picture the formation of the defect plane, we imagine the removal of shared O atoms along a diagonal. Then adjacent slabs slip past each other in a motion that results in the completion of the vacant coordination sites around each W atom. This shearing motion creates edge-shared octahedra along a diagonal. The resulting structure was named a **crystallographic shear plane** by A.D. Wadsley, who first devised this way of describing extended planar defects.

Crystallographic shear planes randomly distributed in the solid are called **Wadsley defects**. Such defects lead to a continuous range of compositions, as in tungsten oxide, which ranges from WO_3 to $WO_{2.93}$ (made by heating and reducing WO_3 with tungsten metal). If, however, the crystallographic shear planes are distributed in a nonrandom, periodic manner, so giving rise to a new unit cell, then we should regard the material as a new stoichiometric phase. Thus, when even more O^{2-} ions are removed from tungsten oxide, a series

(a)

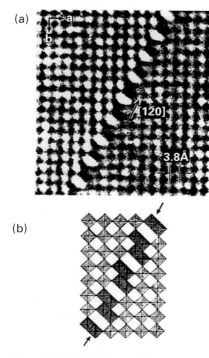

(b)

Figure 24.2 (a) High resolution electron micrograph lattice image of a crystallographic shear plane in WO_{3-x} (b) The oxygen octahedral polyhedra that surround the W atoms imaged in the electron micrograph. Note the edge-shared octahedra along the crystallographic shear plane. [Reproduced by permission from S. Iijima, *J. Solid State Chem.* 1975, **14**, 52].

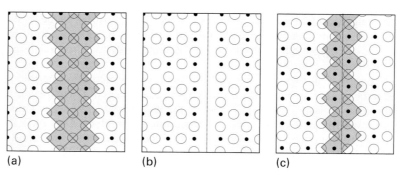

Figure 24.1 The concept of a crystallographic shear plane illustrated by the (100) plane of the ReO_3 structure. (a) A plane of metal, Re, and oxygen, O, atoms. The octahedron around each metal atom is completed by a plane of oxygen atoms above and below the plane illustrated here. Some of the octahedra are shaded to clarify the processes that follow. (b) Oxygen atoms in the plane perpendicular to the page are removed, leaving two planes of metal atoms that lack their sixth oxygen ligand. (c) The octahedral coordination of the two planes of metal atoms is restored by translating the right slab as shown. This creates a plane (labelled a shear plane) vertical to the paper in which the MO_6 octahedra share edges.

of discrete phases having ordered crystallographic shear planes and compositions W_nO_{3n-2} ($n = 20, 24, 25,$ and 40) are observed. Compounds with closely spaced compositions that contain shear planes are known for oxides of W, Mo, Ti, and V and some of their complex oxides, for example the tungsten bronzes $M_8W_9O_{47}$, M = Nb, Ta and the 'Magnéli phases', V_nO_{2n-1} ($n = 3-9$). Electron microscopy (Section 25.3) provides an excellent method of observing these defects experimentally because it reveals both ordered and random arrays of shear planes (Fig. 24.2).

24.4 Atom and ion diffusion

Key point: The diffusion of ions in solids is strongly dependent on the presence of defects.

One reason why diffusion in solids is much less familiar than diffusion in gases and liquids is that at room temperature it is generally very much slower. This slowness is why most solid-state reactions are undertaken at high temperatures (Section 24.1). However, there are some striking exceptions to this generalization. Diffusion of atoms or ions in solids is in fact very important in many areas of solid-state technology, such as semiconductor manufacture, the synthesis of new solids, fuel cells, sensors, metallurgy, and heterogeneous catalysis.

The rates at which ions move through a solid can often be understood in terms of the mechanism for their migration and the activation barriers the ions encounter as they move. The lowest energy pathway generally involves defect sites with the roles summarized in Fig. 24.3. Materials that show high rates of diffusion at moderate temperatures have the following characteristics:

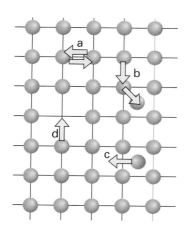

Figure 24.3 Some diffusion mechanisms for ions or atoms in a solid: (a) two atoms or ions exchange positions, (b) an ion hops from a normally occupied site in the structure to an interstitial site, which produces a vacancy which can then be filled by movement of an ion from another site, (c) an ion hops between two different interstitial sites and (d) an ion or atom moves from a normally occupied site to a vacancy, so producing a new vacant site.

- Low-energy barriers: so temperatures at (or a little above) 300 K are sufficient to permit ions to jump from site to site.

- Low charges and small radii: so, for example, the most mobile cation (other than the proton) and anion are Li^+ and F^-. Reasonable mobilities are also found for Na^+ and O^{2-}. More highly charged ions develop stronger electrostatic interactions and are less mobile.

- High concentrations of intrinsic or extrinsic defects: defects typically provide a low-energy pathway for diffusion through a structure that does not involve the energy penalties associated with continuously displacing ions from normal, favourable ion sites. These defects should not be ordered, as for crystallographic shear planes (Section 24.3), because such ordering removes the diffusion pathway.

- Mobile ions are present as a significant proportion of the total number of ions.

Figure 24.4 shows the temperature dependence of the diffusion coefficients, which are a measure of mobility, for specified ions in a selection of solids at high temperatures. The slopes of the lines are proportional to the activation energy for migration. Thus Na^+ is highly mobile and has a low activation energy for motion through β-alumina, whereas Ca^{2+} in CaO is much less mobile and has a high activation energy for hopping through the rock-salt structure.

As with the mechanisms of most chemical reactions, the evidence for the mechanism of a particular diffusion process is circumstantial. The individual events are never directly observed but are inferred from the influence of experimental conditions on diffusion rates. For example, a detailed analysis of the thermal motion and distributions of ions in crystals based on X-ray and neutron diffraction provides strong hints about the ability of ions to move through the crystal, including their most likely paths. Computer modelling, which is often an elaboration of the ionic model (Chapter 3), also gives very useful guidance to the feasibility of these migration mechanisms.

24.5 Solid electrolytes

Any electrochemical cell, such as a battery, fuel cell, electrochromic display, or electrochemical sensor, requires an electrolyte. In many applications, an ionic solution (for example, dilute sulfuric acid in a lead–acid battery) is an acceptable electrolyte, but because it is often desirable to avoid a liquid phase due to possible spillage, there is considerable interest in the development of solid electrolytes. Two important and thoroughly studied solid electrolytes with mobile cations are silver tetraiodomercurate(II), Ag_2HgI_4, and sodium β-alumina with the composition $Na_{1+x}Al_{11}O_{17+x/2}$. Other recently developed fast-cationic conductors include NASICON (a name formed from the letters in sodium, <u>Na</u>, <u>S</u>uper<u>i</u>onic <u>Con</u>ductor) of composition $Na_{1+x}Zr_2P_{3-x}Si_xO_{12}$ and a number of proton conductors that operate at or a little above room temperature, such as $CsHSO_4$ above 160°C.

Solids exhibiting high anion mobility are rarer than cationic conductors and generally show high conductivity only at elevated temperatures: anions are typically larger than cations and so the energy barrier for diffusion through the solid is high. As a consequence, fast anion conduction in solids is limited to F^- and O^{2-} (with ionic radii 133 pm and 140 pm, respectively). Despite these limitations, anionic conductors play an important role in sensors and fuel cells, where a typical material is 'yttrium-stabilized zirconia' (YSZ), of composition $Y_xZr_{1-x}O_{2-x/2}$). Table 24.1 summarizes some typical ionic conductivity values of solid electrolytes and other ionically conducting media.

(a) Solid cationic electrolytes

Key points: Solid inorganic electrolytes often have a low-temperature form in which the ions are ordered on a subset of sites in the structure; at higher temperatures the ions become disordered over the sites and the ionic conductivity increases.

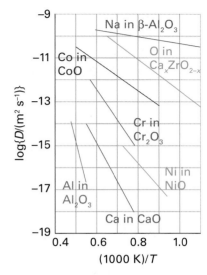

Figure 24.4 The diffusion coefficients (on a logarithmic scale) as a function of inverse temperature for the mobile ion in selected solids.

Table 24.1 Comparative values of ionic and electronic conductivity

Material	Conductivity/(S m^{-1})[†]
Ionic conductors	
Ionic crystals	$<10^{-16}-10^{-2}$
Example: LiI at 298°C	10^{-4}
Solid electrolytes	$10^{-1}-10^{3}$
Example: YSZ at 600°C	1
AgI at 500°C	10^{2}
Strong (liquid) electrolytes	$10^{-1}-10^{3}$
Example: 1 M NaCl(aq)	10^{2}
Electronic conductors	
Metals	$10^{3}-10^{6}$
Semiconductors	$10^{-3}-10^{2}$
Insulators	$<10^{-7}$

[†] The symbol S denotes siemens; 1 S = 1 Ω^{-1}, where 1 Ω (ohm) = 1 V A^{-1}

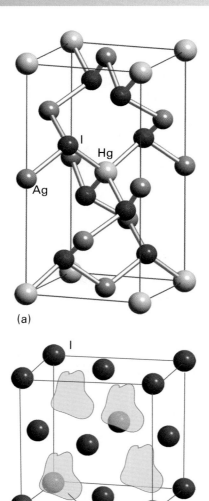

(a)

(b)

Ag, Hg, or vacancy

Figure 24.5 (a) Low-temperature ordered structure of Ag_2HgI_4. (b) High-temperature disordered structure showing the cation disorder. Ag_2HgI_4 is an Ag^+ ion conductor in the high temperature form.

Below 50°C, Ag_2HgI_4 has an ordered crystal structure in which Ag^+ and Hg^+ ions are tetrahedrally coordinated by I^- ions and there are unoccupied tetrahedral holes (Fig. 24.5a). At this temperature its ionic conductivity is low. Above 50°C, however, the Ag^+ and Hg^{2+} ions are randomly distributed over the tetrahedral sites (Fig. 24.5b) and as a result there are many more sites that Ag^+ ions can occupy within the structure than there are Ag^+ ions present. At this temperature the material is a good ionic conductor, largely on account of the mobility of the Ag^+ ions between the different sites available for it. The close-packed array of polarizable I^- ions is easily deformed and results in a low activation energy for the migration of an Ag^+ ion from one ion site to the next. There are many related solid electrolytes having similar structures containing soft anions, such as AgI and $RbAg_4I_5$, both of which have highly mobile Ag^+ ions so that the conductivity of $RbAg_4I_5$ at room temperature is greater than that of aqueous sodium chloride.

Sodium β-alumina is an example of a mechanically hard material that is a good ionic conductor. In this case, the rigid and dense Al_2O_3 slabs are bridged by a sparse array of O^{2-} ions (Fig. 24.6). The plane containing these bridging ions also contains Na^+ ions, which can move from site to site because there are no major bottlenecks to hinder their motion. Many similar rigid materials having planes or channels through which ions can move are known; they are called **framework electrolytes**. Another closely related material, sodium β″-alumina, has even less restricted motion of ions than β-alumina, and it has been found possible to substitute doubly charged cations such as Mg^{2+} or Ni^{2+} for Na^+. Even the large lanthanoid cation Eu^{2+} can be introduced into β″-alumina, although the diffusion of such ions is slower than that of their smaller counterparts. The material NASICON, mentioned earlier, is a nonstoichiometric, solid-solution system with a framework constructed from ZrO_6 octahedra and PO_4 tetrahedra, corresponding to the parent phase of composition $NaZr_2P_3O_{12}$ (Fig. 24.7). A solid solution can be obtained by partially replacing P by Si to give $Na_{1+x}Zr_2P_{3-x}Si_xO_{12}$ with an increase in the number of Na^+ ions for charge balance. In this material, the full set of possible Na sites is only partially filled and these sites lie within a three-dimensional network of channels that allow rapid migration of the remaining Na^+ ions. Other classes of materials currently being investigated as fast cation conductors include Li_4GeO_4 doped with V on the Ge sites ($Li_{4-x}(Ge_{1-x}V_x)O_4$, a lithium-ion conductor with vacancies on the Li^+ ion sub-lattice), the perovskite $La_{0.6}Li_{0.2}TiO_3$, and sodium yttrium silicate, $Na_5YSi_4O_{12}$ (a sodium-ion conductor).

EXAMPLE 24.2 Correlating conductivity and ion size in a framework electrolyte

Conductivity data on β-alumina containing monopositive ions of various radii show that Ag^+ and Na^+ ions, both of which have radii close to 100 pm, have activation energies for conductivity close to 17 kJ mol^{-1} whereas that for Tl^+ (radius 149 pm) is about 35 kJ mol^{-1}. Suggest an explanation of the difference.

Answer One way of approaching this problem is to think of the constrictions to migration that are related to the sizes of the ions. In sodium β-alumina and related β-aluminas, a fairly rigid framework provides a two-dimensional network of passages that permit ion migration. Judging from the experimental results, the bottlenecks for ion motion appear to be large enough to allow Na^+ or Ag^+ (ionic radii close to 100 pm) to pass quite readily (with a low activation energy) but too small to let the larger Tl^+ (ionic radius 149 pm) pass through as readily.

Self-test 24.2 Why does increased pressure reduce the conductivity of K^+ in β-alumina more than that of Na^+ in β-alumina?

Because the reactivity of a bulk material is related to the presence of crystal defects and to the processes of atom and ion diffusion, by modelling an ion diffusion process information can be obtained on both ion conduction and the reaction mechanisms of a solid. Such a study has been undertaken on Li_3N with the aim of determining the energy barriers for Li^+ ions moving through the structure. Comparison of the values for the various barriers between all the different possible sites that Li^+ could occupy as it migrates through the solid in turn allows proposals to be made for the conduction pathway and a value for ionic conductivity to be calculated. Diffusion of Li^+ ions in the Li_2N plane (Fig. 24.8) was found to have a much lower energy barrier than for motion perpendicular to the plane. The effect of replacing the Li^+ ions between the layers, as in Li_2MN (where M is a 3d-series metal such as Ni) and thereby reducing the energy barrier can also be studied; this kind of investigation is important for understanding potential

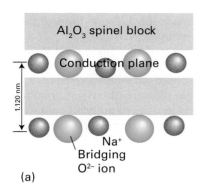

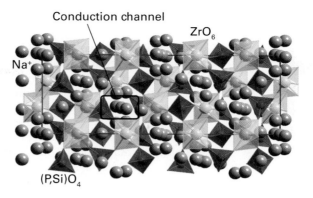

Figure 24.7 The $Na_{1+x}Zr_2P_{3-x}Si_xO_{12}$ (NASICON) structure shown as linked $(P,Si)O_4$ tetrahedra and ZrO_6.

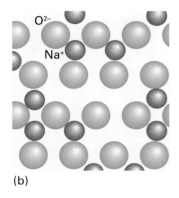

Figure 24.6 (a) Schematic side view of β-alumina showing the Na_2O conduction planes between Al_2O_3 slabs. The O atoms in these planes bridge the two slabs. (b) A view of the conduction plane. Note the abundance of mobile ions and vacancies in which they can move.

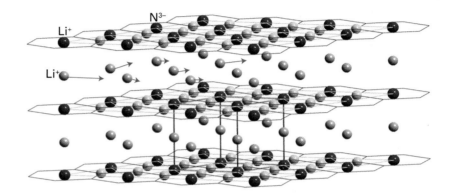

Figure 24.8 The Li_3N structure showing possible pathways for Li^+ ion diffusion between the Li_2N planes.

applications of these materials as the electrolyte in rechargeable lithium-ion batteries (Section 24.7h).

(b) Solid anionic electrolytes

Key point: Anion mobility can occur at high temperatures in certain structures that contain high levels of anion vacancies.

Michael Faraday reported in 1834 that red-hot solid PbF_2 is a good conductor of electricity. Much later it was recognized that the conductivity arises from the mobility of F^- ions through the solid. The property of anion conductivity is shared by other crystals having the fluorite structure. Ion transport in these solids is thought to be by an **interstitial mechanism** in which an F^- ion first migrates from its normal position into an interstitial site (a Frenkel-type defect, Section 3.16) and then moves to a vacant F^- site.

Structures that have large numbers of vacant sites generally show the highest ionic conductivities because they provide a path for ion motion (although at very high levels of defects, clustering of the defects or the vacancies can lower the conductivity). These vacancies, which are equivalent to extrinsic defects, can be introduced in fairly high numbers into many simple oxides and fluorides by doping with appropriately chosen metal ions in different oxidation states. Zirconia, ZrO_2, at high temperature has a fluorite structure, but on cooling the pure material to room temperature it distorts to a monoclinic polymorph. The cubic fluorite structure may be stabilized at room temperature by replacing some Zr^{4+} with other ions, such as the similarly sized Ca^{2+} and Y^{3+} ions. Doping with these ions of lower oxidation number results in the introduction of vacancies on the anion sites to preserve the charge neutrality of the material and produces, for example, $Y_xZr_{1-x}O_{2-x/2}$, the material mentioned previously as 'yttrium stabilized zirconia' (YSZ). This material has completely occupied cation sites in the fluorite structure but high levels of anion vacancies, with

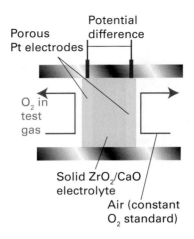

Porous Pt electrodes

Potential difference

O_2 in test gas

Solid ZrO_2/CaO electrolyte

Air (constant O_2 standard)

Figure 24.9 An oxygen sensor based on the solid electrolyte $Zr_{1-x}Ca_xO_{2-x}$.

$0 \leq x \leq 0.15$. These vacant sites provide a path for oxide-ion diffusion through the structure so that a typical electrical conductivity, in, for example, $Ca_{0.15}Zr_{0.85}O_{1.85}$ is 5 S cm^{-1} at 1000°C;[1] note that this conductivity is much lower than typical solid-state cation conductivities—even at these very high temperatures—due to the large anion size.

The high oxide-ion conductivity of calcium-oxide doped zirconia is exploited in a solid-state electrochemical sensor for measuring the partial pressure of oxygen in automobile exhaust systems (Fig. 24.9).[2] The platinum electrodes in this cell adsorb O atoms and, if the partial pressures of oxygen are different between the sample and reference side, there is a thermodynamic tendency for oxygen to migrate through the electrolyte as the O^{2-} ion. The thermodynamically favoured processes are:

High $p(O_2)$ side:

$$\tfrac{1}{2}O_2(g) + Pt(s) \rightarrow O\ (Pt,\ surface)$$

$$O\ (Pt,\ surface) + 2\ e^- \rightarrow O^{2-}\ (ZrO_2)$$

Low $p(O_2)$ side:

$$O^{2-}\ (ZrO_2) \rightarrow O\ (Pt,\ surface) + 2\ e^-$$

$$O\ (Pt,\ surface) \rightarrow \tfrac{1}{2}O_2(g) + Pt(s)$$

The cell potential is related to the two oxygen partial pressures (p_1 and p_2) by the Nernst equation (Section 5.5), for the half-cell reaction $O_2 + 4\,e^- \rightarrow 2\,O^{2-}$, which occurs at both electrodes

$$E_{cell} = \frac{RT}{4F}\ln\frac{p_1}{p_2}$$

so a simple measurement of the potential difference provides a measure of the oxygen partial pressure in the exhaust gases.

■ **A brief illustration.** According to this equation, the potential difference produced by an oxygen sensor operating at 1000 K in an exhaust system, with air on one side ($p(O_2) = 0.2$ atm) and a burnt fuel/air mixture ($p(O_2) = 0.001$ atm) on the other, is about 0.1 V. ■

Behaviour similar to that of YSZ is encountered for other compounds that adopt the fluorite structure, such as PbF_2 as discovered by Faraday. As noted previously, the anionic conductivities remain low even at high temperatures, so many other complex metal oxides are currently being investigated with the aim of achieving high mobilities at low temperatures. Some compounds that show promising behaviour include $La_2Mo_2O_9$, barium indate ($Ba_2In_2O_5$), BIMEVOX (a d-metal doped bismuth vanadium oxide), the apatite structure of $La_{9.33}Si_6O_{26}$, and strontium- and magnesium-doped lanthanum gallate (Sr,Mg-doped $LaGaO_3$, or LSGM). As well as uses in sensor devices, materials possessing oxide- and proton-ion conductivity are important in a number of fuel cell types (Box 24.1).

(c) Mixed ionic-electronic conductors

Key point: Solid materials can exhibit both ionic and electronic conductivity.

Most ionic conductors, such as sodium β″-alumina and YSZ, have low electronic conductivity (that is, conduction by electron rather than ion motion). Their application as solid electrolytes, in sensors for instance, requires this feature to avoid shorting-out the cell. In some cases a combination of electronic and ionic conductivity is desirable, and this type of behaviour can be found in some d-metal compounds where defects allow O^{2-} conduction and the metal d orbitals provide an electronic conduction band. Many such materials are perovskite-based structures with mixed oxidation states at the B cation sites (Section 3.9). Two examples are $La_{1-x}Sr_xCoO_{3-y}$ and $La_{1-x}Sr_xFeO_{3-y}$. These oxide systems are good electronic conductors with partially filled bands as a result of the nonintegral d-metal oxidation number and can

---

[1]The symbol S denotes siemens; $1\ S = 1\ \Omega^{-1}$, where $1\ \Omega$ (ohm) $= 1\ V\ A^{-1}$.

[2]The signal from this sensor is used to adjust the air/fuel ratio and thereby the composition of the exhaust gas being fed to the catalytic converter.

BOX 24.1 Solid oxide fuel cells

A fuel cell consists of an electrolyte sandwiched between two electrodes; oxygen passes over one electrode and the fuel over the other, generating electricity, water, and heat. The general operation and construction of a fuel cell that converts a fuel, such as hydrogen, methane, or methanol, by reaction with oxygen into electrical energy (and combustion products H_2O and CO_2) was described in Box 10.1. A variety of materials can be used as the electrolyte in such cells, including phosphoric acid, proton-exchange membranes, and, in *solid oxide fuel cells* (SOFCs), oxide-ion conductors (Fig. B24.1).

SOFCs operate at high temperatures and use an oxide-ion conductor as the electrolyte. The design of a typical SOFC is shown in the illustration. Each cell generates a limited potential difference but, as with the cells of a battery, a connected stack may be constructed in series to increase the potential difference and power supplied. Cells are connected electrically through an 'interconnect' that can also be used to isolate the fuel and air supplies for each cell.

The attraction of SOFCs is based on several aspects, including the clean conversion of fuel to electricity, low levels of noise pollution, the ability to cope with different fuels, but most significantly a high efficiency. Their high

efficiency is a result of their high operating temperatures which are typically between 500 and 1000°C. In high-temperature SOFCs the interconnect may be a ceramic such as lanthanum chromite (the perovskite $LaCrO_3$) or, if the temperature is below 1000°C, an alloy such as Y/Cr may be used. The oxide-ion conductor used as the electrolyte in these very high-temperature SOFCs is normally yttrium-stabilized zirconia (YSZ).

Intermediate-temperature SOFCs, which typically operate between 500 and 700°C, have a number of advantages over very high temperature devices in that there is reduced corrosion, a simpler design, and the time to heat the system to the operating temperature is much reduced. However, such devices require a material with excellent oxide-ion conductivities at lower temperatures to act as the electrolyte. The best developed intermediate-temperature (less than about 600°C) SOFC consists of an anode of Gd-doped CeO_2 (CGO)/Ni, an electrolyte of Gd-doped cerium oxide, and a cathode of the perovskite LSCF $(La,Sr)(Fe,Co)O_3$. The electrolyte material CGO possesses a much higher ionic conductivity than YSZ at these lower temperatures. Unfortunately, however, its electronic conductivity is also higher and the use of a CGO electrolyte can result in reduced efficiency as energy is wasted due to electrons flowing through the electrolyte. For this reason new and better oxide-ion conductors are being sought, such as those mentioned in the text.

One of the main advantages of SOFCs over other fuel cell types is their ability to handle more convenient hydrocarbon fuels: other types of fuel cells have to rely on a clean supply of hydrogen for their operation. Because SOFCs operate at high temperature there is the opportunity to convert hydrocarbons catalytically to hydrogen and carbon oxides within the system (Section 26.15). Because of their size and the requirement to be heated to, and operate at, high temperatures, the applications of SOFCs focus on medium- to large-scale static systems, including small residential systems producing about 2 kW.

Oxygen or air $O_2 + 4e^- \rightarrow 2O^{2-}$

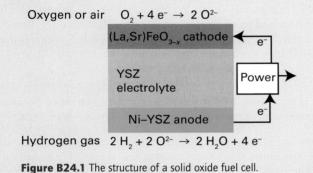

Hydrogen gas $2H_2 + 2O^{2-} \rightarrow 2H_2O + 4e^-$

Figure B24.1 The structure of a solid oxide fuel cell.

conduct by O^{2-} migration through the perovskite O^{2-} ion sites. This type of material is of use in solid oxide fuel cells (SOFC, Box 24.1), one type of fuel cell mentioned in Box 5.1, in which one electrode has to allow diffusion of ions through a conducting electrode.

Metal oxides, nitrides, and fluorides

In this section, we explore the binary compounds of N, O, and F with metals. These compounds, particularly oxides, are central to much solid-state chemistry on account of their stability, ease of synthesis, and variety in composition and structure. These attributes lead to the vast number of compounds that have been synthesized and the ability to tune the properties of a compound for a specific application based on their electronic or magnetic characteristics. As we shall see, a discussion of the chemical properties of these compounds also provides insight into defects, nonstoichiometry, and ion diffusion, and into the influence of these characteristics on physical properties.

The chemical properties of metal fluorides parallel much of that of metal oxides, but the lower charge of the F^- ion means that equivalent stoichiometries are produced with cations having lower charges, as in $KMn^{(II)}F_3$ compared with $SrMn^{(IV)}O_3$. Compounds containing the nitride ion, N^{3-}, in combination with one or more metal ion have only recently been developed to a significant extent; the area of metal nitride chemistry is one where rapid advances are being made in terms of new materials and structure types.

24.6 Monoxides of the 3d metals

The monoxides of most of the 3d metals adopt the rock-salt structure, so it might at first appear that there is little to say about them and that their properties should be simple.

Table 24.2 Monoxides of the 3d-series metals

Compound	Structure	Composition, x	Electrical character
CaO_x	Rock-salt	1	Insulator
TiO_x	Rock-salt	0.65 – 1.25	Metallic
VO_x	Rock-salt	0.79 – 1.29	Metallic
MnO_x	Rock-salt	1 – 1.15	Semiconductor
FeO_x	Rock-salt	1.04 – 1.17	Semiconductor
CoO_x	Rock-salt	1 – 1.01	Semiconductor
NiO_x	Rock-salt	1 – 1.001	Insulator
CuO_x	PtS (linked CuO_4 square-planes)	1	Semiconductor
ZnO_x	Wurtzite	Slight Zn excess	Wide band gap n-type semiconductor

In fact, the actual stoichiometries formed around the ideal composition MO and the structures and properties of these oxides are more interesting, and as a result they have been repeatedly investigated (Table 24.2). In particular, the compounds provide examples of how mixed oxidation states and defects lead to nonstoichiometry for solids that are nominally TiO, VO, FeO, CoO, and NiO. All these compounds are usually obtained with significant deviations from the nominal MO stoichiometry.

(a) Defects and nonstoichiometry

Key point: The nonstoichiometry of $Fe_{1-x}O$ arises from the creation of vacancies on the Fe^{2+} octahedral sites, with each vacancy charge-compensated by the conversion of two Fe^{2+} ions to two Fe^{3+} ions.

The origin of nonstoichiometry in FeO has been studied in more detail than that in most other MO compounds. In fact, it is found that stoichiometric FeO does not exist but rather a range of iron-deficient compounds in the range $Fe_{1-x}O$, $0.13 < x < 0.04$, can be obtained by quenching (cooling very rapidly) iron(II) oxide from high temperatures. The compound $Fe_{1-x}O$ is in fact metastable at room temperature: it is thermodynamically unstable with respect to disproportionation into iron metal and Fe_3O_4 but does not convert for kinetic reasons. The general consensus is that the structure of $Fe_{1-x}O$ is derived from the rock-salt FeO structure by the presence of vacancies on the Fe^{2+} octahedral sites, and that each vacancy is charge-compensated by the conversion of two adjacent Fe^{2+} ions to two Fe^{3+} ions. The relative ease of oxidizing Fe(II) to Fe(III) accounts for the fairly broad range of compositions of $Fe_{1-x}O$. At high temperatures, the interstitial Fe^{3+} ions associate with the Fe^{2+} vacancies (or defects) to form clusters distributed throughout the structure (Fig. 24.10).

Similar defects and the clustering of defects appear to occur with all other 3d-metal monoxides, with the possible exception of NiO. The range of nonstoichiometry in $Ni_{1-x}O$ is extremely narrow, but conductivity and the rate of atom diffusion vary with oxygen partial pressure in a manner that suggests the presence of isolated point defects. Both CoO and NiO occur in a metal-deficient state, although their range of compositions is again not as broad as that of FeO. As indicated by standard potentials in aqueous solution, Fe(II) is more easily oxidized than either Co(II) or Ni(II); this solution redox chemistry correlates well with the much smaller range of oxygen deficiency in NiO and CoO. Chromium(II) oxide, like FeO, spontaneously disproportionates:

$$3\, Cr^{(II)}O(s) \rightarrow Cr_2^{(III)}O_3(s) + Cr^{(0)}(s)$$

However, the material can be stabilized by crystallization in a copper(II) oxide matrix.

Both CrO and TiO have structures that show high levels of defects on both the cation and anion sites forming metal-rich or metal-deficient stoichiometries ($Ti_{1-x}O$ and TiO_{1-x}). In fact, TiO has large numbers of vacancies, in equal amounts, on both cation and anion sublattices, rather than the expected perfect, defect-free structure. Note that significant deviations from the stoichiometry MO are out of the question for Group 2 metal oxides, such as CaO, because for these elements M^{3+} ions are chemically inaccessible.

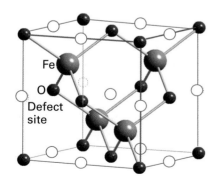

Figure 24.10 Defect sites proposed for $Fe_{1-x}O$. Note that the tetrahedral Fe^{3+} interstitials (grey spheres) and octahedral Fe^{2+} vacancies (circles) are clustered together.

(b) Electronic properties

Key points: The 3d-metal monoxides MnO, FeO, CoO, and NiO are semiconductors; TiO and VO are metallic conductors.

The 3d-metal monoxides MnO, $Fe_{1-x}O$, CoO, and NiO have low electrical conductivities that increase with temperature (corresponding to semiconducting behaviour) or have such large band gaps that they are insulators. The electron or hole migration in these oxide semiconductors is attributed to a hopping mechanism. In this model, the electron or hole hops from one localized metal atom site to the next. When it lands on a new site it causes the surrounding ions to adjust their locations and the electron or hole is trapped temporarily in the potential well produced by this distortion. The electron resides at its new site until it is thermally activated to migrate into another nearby site. Another aspect of this charge-hopping mechanism is that the electron or hole tends to associate with local defects, so the activation energy for charge transport may also include the energy of freeing the hole from its position next to a defect.

Hopping contrasts with the band model for semiconductivity discussed in Section 3.20, where the conduction and valence electrons occupy orbitals that spread through the whole crystal. The difference stems from the less diffuse d orbitals in the monoxides of the mid-to-late 3d metals, which are too compact to form the broad bands necessary for metallic conduction. When NiO is doped with Li_2O in an O_2 atmosphere a solid solution $Li_x(Ni^{2+})_{1-2x}(Ni^{3+})_xO$ is obtained, which has greatly increased conductivity for reasons similar to the increase in conductivity of Si when doped with B. The characteristic pronounced increase in electronic conductivity with increasing temperature of metal oxide semiconductors is used in 'thermistors' to measure temperature.

In contrast to the semiconductivity of the monoxides in the centre and right of the 3d series, TiO and VO have high electronic conductivities that decrease with increasing temperature. This metallic conductivity persists over a broad composition range from highly oxygen-rich $Ti_{1-x}O$ to metal-rich TiO_{1-x}. In these compounds, a conduction band is formed by the overlap of the t_{2g} orbitals of metal ions in neighbouring octahedral sites that are oriented towards each other (Fig. 24.11). The radial extension of the d orbitals of these early d-block elements is greater than for elements later in the period, and a band results from their overlap (Fig. 24.12); this band is only partly filled. The widely varying compositions of these monoxides also appear to be associated with the electronic delocalization: the conduction band serves as a rapidly accessible source and sink of electrons that can readily compensate for the formation of vacancies.

(c) Magnetic properties

Key point: The 3d-metal monoxides MnO, FeO, CoO, and NiO order antiferromagnetically with Néel temperatures that increase from Mn to Ni.

As well as having electronic properties that are the result of interactions between the d electrons, d-metal monoxides have magnetic properties that derive from cooperative interaction of the individual atomic magnetic moments (Section 20.8). The overall magnetic structure of MnO and the other 3d-series metal monoxides is shown in Fig. 24.13. The Néel temperatures (T_N, the temperature of the paramagnetic/antiferromagnetic transition, Section 20.8) of the series of d-metal oxides are as follows:

MnO	FeO	CoO	NiO
122 K	198 K	271 K	523 K

These values reflect the strength of the superexchange spin interactions along the $M-O-M$ directions, which in the rock-salt type structure propagate in all three unit cell directions. As the size of the M^{2+} ion decreases from Mn to Ni, the superexchange mechanism becomes stronger due to the increased metal–oxygen orbital overlap and T_N increases.

24.7 Higher oxides and complex oxides

Binary metal oxides that do not have a 1:1 metal:oxygen ratio are known as **higher oxides**. Compounds containing ions of more than one metal are often termed **complex oxides** or **mixed oxides** and include compounds containing three elements, ternary oxides (for instance,

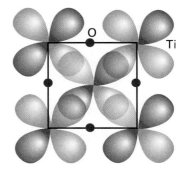

Figure 24.11 Overlap of the d_{zx} orbitals in TiO to give a t_{2g} band. In the perpendicular directions the d_{yx} and d_{zy} orbitals overlap in an identical manner.

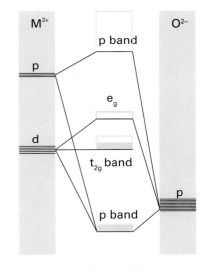

Figure 24.12 Molecular orbital energy level diagram for early d-metal monoxides. The t_{2g} band is only partly filled and metallic conduction results.

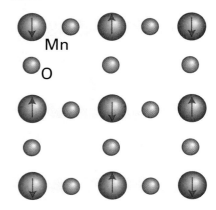

Figure 24.13 The arrangement of electron spins on the Mn^{2+} ions in the antiferromagnetic state of MnO.

LaFeO$_3$), and four, quaternary (for instance, YBa$_2$Cu$_3$O$_7$), or more elements. This section describes the structures and properties of some of the more important complex, mixed metal oxides.

(a) The M$_2$O$_3$ corundum structure

Key point: The corundum structure is adopted by many oxides of the stoichiometry M$_2$O$_3$, including Cr-doped aluminium oxide (ruby).

α-Aluminium oxide (the mineral *corundum*) adopts a structure that can be modelled as a hexagonal close-packed array of O^{2-} ions with the cations in two-thirds of the octahedral holes (Fig. 24.14). The corundum structure is also adopted by the oxides of Ti, V, Cr, Rh, Fe, and Ga in their +3 oxidation states. Two of these oxides, Ti$_2$O$_3$ and V$_2$O$_3$, exhibit metallic-to-semiconducting transitions below 410 and 150 K, respectively (Fig. 24.15). In V$_2$O$_3$, the transition is accompanied by antiferromagnetic ordering of the spins. The two insulators Cr$_2$O$_3$ and Fe$_2$O$_3$ also display antiferromagnetic ordering.

Another interesting aspect of the M$_2$O$_3$ compounds is the formation of solid solutions of dark-green Cr$_2$O$_3$ and colourless Al$_2$O$_3$ to form brilliant red ruby. As remarked in Section 20.7, this shift in the ligand-field transitions of Cr^{3+} stems from the compression of the O^{2-} ions around Cr^{3+} in the Al$_2$O$_3$ host structure. (In Al$_2$O$_3$, a = 475 pm and c = 1300 pm; in Cr$_2$O$_3$ the lattice constants are 493 pm and 1356 pm, respectively.) The compression shifts the absorption towards the blue as the strength of the ligand field increases, and the solid appears red in white light. The responsiveness of the absorption (and fluorescence) spectrum of Cr^{3+} ions to compression is sometimes used to measure pressure in high-pressure experiments. In this application, a tiny crystal of ruby in one segment of the sample can be interrogated by visible light and the shift in its fluorescence spectrum provides an indication of the pressure inside the cell.

(b) Rhenium trioxide

Key point: The rhenium trioxide structure can be constructed from ReO$_6$ octahedra sharing all vertices in three dimensions.

The rhenium trioxide structure type is very simple, consisting of a cubic unit cell with Re atoms at the corners and O atoms at the mid-point of each edge (Fig. 24.16). Alternatively, the structure can be considered to be derived from ReO$_6$ octahedra sharing all vertices. The structure is also closely related to a perovskite structure (Section 3.9) in which the A-type cation has been removed from the unit cell. Materials adopting the rhenium trioxide structure are relatively rare. This rarity is in part due to the requirement of the oxidation state M(VI) when M is in combination with oxygen. Rhenium(VI) oxide, ReO$_3$, itself and one form of UO$_3$ (δ-UO$_3$) have this structure type and WO$_3$ exists in a slightly distorted version of it. In WO$_3$, the WO$_6$ octahedra are slightly distorted and tilted relative to each other so that the W−O−W bond angle is not 180°.

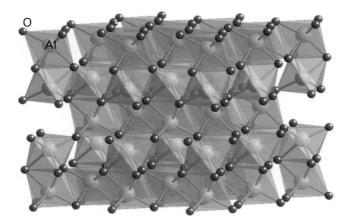

Figure 24.14 The corundum structure, as adopted by Al$_2$O$_3$, with cations occupying two-thirds of the octahedral holes between layers of close-packed oxide ions.

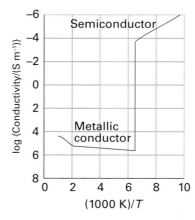

Figure 24.15 The temperature dependence of the electrical conductivity of V$_2$O$_3$ showing the metal-to-semiconductor transition.

Rhenium trioxide itself is a bright red lustrous solid. Its electrical conductivity at room temperature is similar to that of copper metal. The band structure for this compound contains a band derived from the Re t_{2g} orbitals and the O2p orbitals (Fig. 24.17). This band can contain up to six electrons per Re atom but is only partially filled for the Re^{6+} d^1 configuration, so producing the observed metallic properties.

(c) Spinels

Key point: The observation that many d-metal spinels do not have the normal spinel structure is related to the effect of ligand-field stabilization energies on the site preferences of the ions.

The d-block higher oxides Fe_3O_4, Co_3O_4, and Mn_3O_4, and many related mixed-metal compounds, such as $ZnFe_2O_4$, have very useful magnetic properties. They all adopt the structural type of the mineral spinel, $MgAl_2O_4$, and have the general formula AB_2O_4. Most oxide spinels are formed with a combination of A^{2+} and B^{3+} cations (that is, as $A^{2+}B_2^{3+}O_4$ in $Mg^{2+}[Al^{3+}]_2O_4$), although there are a number of spinels that can be formulated with A^{4+} and B^{2+} cations (as $A^{4+}B_2^{2+}O_4$ as in $Ge^{4+}[Co^{2+}]_2O_4$. The spinel structure was described briefly in Section 3.9, where we saw that it consists of an fcc array of O^{2-} ions in which the A ions reside in one-eighth of the tetrahedral holes and the B ions inhabit half the octahedral holes (Fig. 24.18); this structure is commonly denoted $A[B_2]O_4$, where the atom type in the square bracket represents that occupying the octahedral sites. In the inverse spinel structure, the cation distribution is $B[AB]O_4$, with the more abundant B-type cation distributed over both coordination geometries. Lattice enthalpy calculations based on a simple ionic model indicate that, for A^{2+} and B^{3+}, the normal spinel structure, $A[B_2]O_4$, should be the more stable. The observation that many d-metal spinels do not conform to this expectation has been traced to the effect of ligand-field stabilization energies on the site preferences of the ions.

The **occupation factor**, λ, of a spinel is the fraction of B atoms in the tetrahedral sites: $\lambda = 0$ for a normal spinel and $\lambda = \frac{1}{2}$ for an inverse spinel, $B[AB]O_4$; intermediate λ values indicate a level of disorder in the distribution, where B-type cations occupy that portion of the tetrahedral sites. The distribution of cations in (A^{2+},B^{3+}) spinels (Table 24.3) illustrates that for d^0 A and B ions the normal structure is preferred as predicted by electrostatic considerations. Table 24.3 shows that, when A^{2+} is a d^6, d^7, d^8, or d^9 ion and B^{3+} is Fe^{3+}, the inverse structure is generally favoured. This preference can be traced to the lack of ligand-field stabilization (Section 20.1 and Fig. 20.13) of the high-spin d^5 Fe^{3+} ion in either the octahedral or the tetrahedral site and the ligand-field stabilization of the other d^n ions in the octahedral site. For other combinations of d-metal ions on the A and B sites the relative ligand-field stabilization energies of the different arrangements of the two ions on the octahedral and tetrahedral sites need to be calculated. It is also important to note that simple ligand-field stabilization appears to work over this limited range of cations. More detailed analysis is necessary when cations of different radii are present or any ions that are present do not adopt the high-spin configuration typical of most metals in spinels (for instance, Co^{3+} in Co_3O_4, which is low-spin d^6). Moreover, because λ is often found to depend on the temperature, care has to be taken in the synthesis of a spinel with a specific

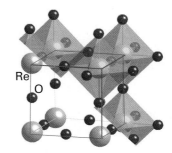

Figure 24.16 The ReO_3 structure shown as the unit cell and the ReO_6 octahedra forming the unit cell.

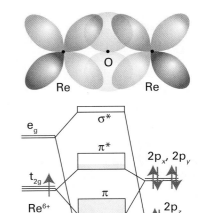

Figure 24.17 The band structure of ReO_3.

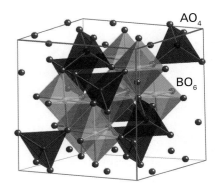

Figure 24.18 A segment of the spinel (AB_2O_4) unit cell showing the tetrahedral environment of A ions and the octahedral environments of B ions. (Compare with Fig. 3.44.)

Table 24.3 Occupation factor, λ, in some spinels*

	A	Mg^{2+}	Mn^{2+}	Fe^{2+}	Co^{2+}	Ni^{2+}	Cu^{2+}	Zn^{2+}
B		d^0	d^5	d^6	d^7	d^8	d^9	d^{10}
Al^{3+}	d^0	0	0	0	0	0.38	0	
Cr^{3+}	d^3	0	0	0	0	0	0	0
Mn^{3+}	d^4	0						0
Fe^{3+}	d^5	0.45	0.1	0.5	0.5	0.5	0.5	0
Co^{3+}	d^6					0		0

* $\lambda = 0$ corresponds to a normal spinel; $\lambda = 0.5$ corresponds to an inverse spinel.

distribution of cations because slow cooling or quenching of a sample from a high reaction temperature can produce quite different cation distributions.

> **EXAMPLE 24.3** Predicting the structures of spinel compounds
>
> Is $MnCr_2O_4$ likely to have a normal or inverse spinel structure?
>
> **Answer** We need to consider whether there is a ligand-field stabilization. Because Cr^{3+} (d^3) has a large ligand-field stabilization energy ($1.2\Delta_O$ from Table 20.2) in the octahedral site (but a much smaller one in a tetrahedral field) whereas the high spin d^5 Mn^{2+} ion does not have any LFSE, a normal spinel structure is expected. Table 24.3 shows that this prediction is verified experimentally.
>
> **Self-test 24.3** Table 24.3 indicates that $FeCr_2O_4$ is a normal spinel. Rationalize this observation.

The inverse spinels of formula AFe_2O_4 are sometimes classified as **ferrites** (the same term also applies in different circumstances to other iron oxides). When $RT > J$, where J is the energy of interaction of the spins on different ions, ferrites are paramagnetic. However, when $RT < J$, a ferrite may be either ferrimagnetic or antiferromagnetic. The antiparallel alignment of spins characteristic of antiferromagnetism is illustrated by $ZnFe_2O_4$, which has the cation distribution $Fe[ZnFe]O_4$. In this compound the Fe^{3+} ions (with $S = \frac{5}{2}$) in the tetrahedral and octahedral sites are antiferromagnetically coupled, through a superexchange mechanism (Section 20.8), below 9.5 K to give nearly zero net magnetic moment to the solid as a whole; note that Zn^{2+} as a d^{10} ion makes no contribution to the magnetic moment of the material.

The compound $CoAl_2O_4$ is among the normal spinels in Table 24.3 with $\lambda = 0$ and thus has the Co^{2+} ions at the tetrahedral sites. The colour of $CoAl_2O_4$ (an intense blue) is that expected of tetrahedral Co^{2+}. This property, coupled with the ease of synthesis and stability of the spinel structure, has led to cobalt aluminate being used as a pigment ('cobalt blue'). Other mixed d-metal spinels that exhibit strong colours, for example $CoCr_2O_4$ (green), $CuCr_2O_4$ (black), and $(Zn,Fe)Fe_2O_4$ (orange/brown), are also used as pigments, with applications that include colouring various construction materials, such as concrete.

(d) Perovskites and related phases

Key points: The perovskites have the general formula ABX_3, in which the 12-coordinate hole of a ReO_3-type BX_3 structure is occupied by a large A ion; the perovskite barium titanate, $BaTiO_3$, exhibits ferroelectric and piezoelectric properties associated with cooperative displacements of the ions.

The perovskites have the general formula ABX_3, in which the 12-coordinate hole of BX_3 (as in ReO_3) is occupied by a large A ion (Fig. 24.19; a different view of this structure is given in Fig. 3.42) The X ion is most frequently O^{2-} or F^- (as in $NaFeF_3$) although nitride- and hydride-containing perovskites can also be synthesized, such as in $LiSrH_3$. Perovskite itself is named after the naturally occurring oxide mineral $CaTiO_3$ and the largest class of perovskites are those with the anion as oxide. This breadth of perovskites is widened by the observation that solid solutions and nonstoichiometry are also common features of the perovskite structure, as in $Ba_{1-x}Sr_xTiO_3$ and $SrFeO_{3-y}$. Some metal-rich materials adopt the perovskite structure with the normal distribution of cations and anions partially inverted, for instance $SnNCo_3$.

The perovskite structure is often observed to be distorted such that the unit cell is no longer centrosymmetric and the crystal acquires an overall permanent electric polarization as a result of ion displacements. Some polar crystals are **ferroelectric** in the sense that they resemble ferromagnets, but instead of the electron spins being aligned over a region of the crystal, the electric dipole moments of many unit cells are aligned. As a result, the relative permittivity, which reflects the polarity of a compound, for a ferroelectric material often exceeds 1×10^3 and can be as high as 1.5×10^4; for comparison, the relative permittivity of liquid water is about 80 at room temperature. Barium titanate, $BaTiO_3$, is the most extensively studied example of such a material. At temperatures above 120°C this compound has the perfect cubic perovskite structure. At room temperature, it adopts a lower symmetry, tetragonal unit cell in which the various ions can be considered as having been displaced from their normal high symmetry sites (Fig. 24.20). This displacement results in a spontaneous polarization of the unit cell and formation of an electric dipole; coupling between these ion displacements and therefore the induced dipoles is very weak. Application

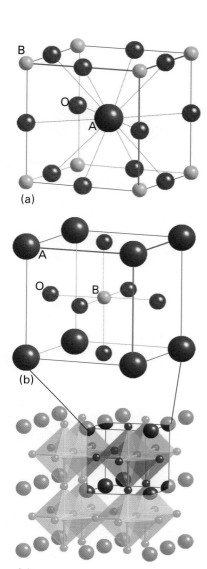

Figure 24.19 Views of the perovskite (ABO_3) structure (a) emphasizing the 12-fold coordination of the larger A cation and showing the relationship with the ReO_3 structure of Fig. 23.22b, (b) highlighting the octahedral coordination of the B cation. (c) A polyhedral representation accentuating the BO_6 octahedra.

of an external electric field aligns these dipoles throughout the material, resulting in a bulk polarization in a particular direction, which can persist after removal of the electric field. The temperature below which this spontaneous polarization can occur and the material behaves as a ferroelectric is called the **Curie temperature** (T_C)(Section 20.8). For $BaTiO_3$ $T_C = 120°C$. The high relative permittivity of barium titanate leads to its use in capacitors, where its presence allows up to 1000 times the charge to be stored in comparison with a capacitor with air between the plates. The introduction of dopants into the barium titanate structure, forming solid solutions, allows various properties of the compound to be tuned. For example, the replacement of Ba by Sr or of Ti by Zr causes a sharp lowering of T_C.

Another characteristic of many crystals, including a number of perovskites that lack a centre of symmetry, is **piezoelectricity**, the generation of an electrical field when the crystal is under stress or the change in dimensions of the crystal when an electrical field is applied. Piezoelectric materials are used for a variety of applications, such as pressure transducers, ultramicromanipulators (where very small movements can be controlled), sound detectors, and as the probe support in scanning tunnelling microscopy. Some important examples are $BaTiO_3$, $NaNbO_3$, $NaTaO_3$, and $KTaO_3$.

Although a noncentrosymmetric structure is required for both ferroelectric and piezo-electric behaviour, the two phenomena do not necessarily occur for the same crystal. For example, quartz, which does not have the perovskite structure, is permanently polarized; although quartz is piezoelectric it is not ferroelectric because an external electric field cannot reverse the polarization. Quartz is widely used to set the clock rate of microprocessors and watches because a thin sliver oscillates at a specific frequency to produce a small oscillating electrical field. This frequency is very insensitive to temperature.

Another prototypical structure, that of potassium tetrafluoridonickelate(II), K_2NiF_4 (Fig. 24.21), is related to perovskite. The compound can be thought of as containing individual slices from the perovskite structure that share the four F atoms from the octahedra within the layer and have terminal F atoms above and below the layer. These layers are displaced relative to each other and separated by the K^+ ions (which are nine-coordinate, to eight F atoms of one layer and one terminal F atom from the next).

Compounds with the K_2NiF_4 structure have come under renewed investigation because some high-temperature superconductors, such as $La_{1.85}Sr_{0.15}CuO_4$, crystallize with this structure. Apart from their importance in superconductivity, compounds with the K_2NiF_4 structure also provide an opportunity to investigate two-dimensional magnetic domains as coupling between electron spins is much stronger within the layers of linked octahedra than between the layers.

The K_2NiF_4 structure has been introduced as being derived from a single slice of the perovskite structure; other related structures are possible where two or more perovskite layers are displaced horizontally relative to each other. Structures with K_2NiF_4 at one end of the range (a single perovskite layer) and perovskite itself at the other (an infinite number of such layers) are known as **Ruddlesden–Popper phases**. They include $Sr_3Fe_2O_7$ with double layers and $Ca_4Mn_3O_{10}$ with triple layers (Fig. 24.22).

(e) High-temperature superconductors

Key point: High-temperature cuprate superconductors have structures related to perovskite.

The versatility of the perovskites extends to superconductivity because most of the high-temperature superconductors (which were first reported in 1986) can be viewed as variants of the perovskite structure. Superconductors have two striking characteristics. Below a critical temperature, T_c (not to be confused with the Curie temperature of a ferroelectric, T_C), they enter the superconducting state and have zero electrical resistance. In this super-conducting state they also exhibit the **Meissner effect**, the exclusion of a magnetic field. The Meissner effect is the basis of the common demonstration of superconductivity in which a pellet of superconductor levitates above a magnet. It is also the basis for a number of potential applications of superconductors that include magnetic levitation, as in MAGLEV trains.

Following the discovery in 1911 that mercury is a superconductor below 4.2 K, physicists and chemists made slow but steady progress in the discovery of superconductors with higher values of T_c at a rate of about 3 K per decade. After 75 years, T_c had been edged up to 23 K in Nb_3Ge. Most of these superconducting materials were metal alloys, although superconductivity had been found in many oxides and sulfides (Table 24.4); magnesium diboride is superconducting below 39 K (see Box 13.4). Then, in 1986, the

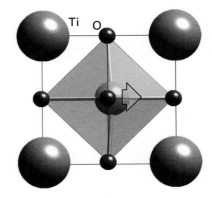

Figure 24.20 The tetragonal $BaTiO_3$ structure showing the local Ti^{4+} ion displacement that leads to the ferroelectric behaviour of this material.

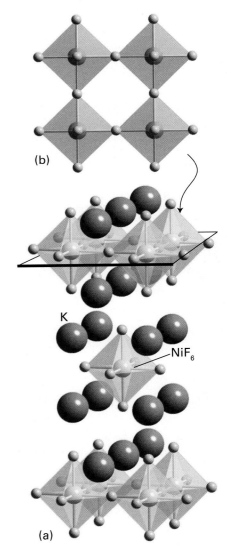

Figure 24.21 The K_2NiF_4 structure. (a) The displaced layers of NiF_6 octahedra interspersed with K^+ ions and (b) a view of one layer of composition NiF_4 showing the corner sharing octahedra linked through F.

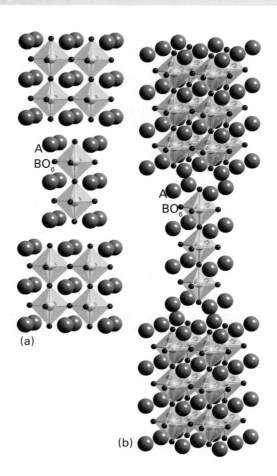

Figure 24.22 The Ruddlesden–Popper phases of stoichiometry (a) $A_3B_2O_7$ and (b) $A_4B_3O_{10}$ formed respectively from two and three perovskite layers of linked BO_6 octahedra separated by A-type cations.

Table 24.4 Some materials that exhibit superconductivity below the critical temperature, T_c

Element	T_c/K	Compound	T_c/K
Zn	0.88	Nb_3Ge	23.2
Cd	0.56	Nb_3Sn	18.0
Hg	4.15	$LiTi_2O_4$	13.7
Pb	7.19	$K_{0.4}Na_{0.6}BiO_3$	29.8
Nb	9.50	$YBa_2Cu_3O_{10}$	93
		$Tl_2Ba_3Ca_3Cu_4O_{12}$	134
		MgB_2	40
		K_3C_{60}	39
		$PbMo_6S_8$	15.2
		NbPS	12

first **high-temperature superconductor** (HTSC) was discovered. Several materials are now known with T_c well above 77 K, the boiling point of the relatively inexpensive refrigerant liquid nitrogen, and in a few years the maximum T_c was increased by more than a factor of five to around 134 K.

Two types of superconductors are known:

- **Type I** show abrupt loss of superconductivity when an applied magnetic field exceeds a value characteristic of the material.

- **Type II** superconductors, which include high-temperature materials, show a gradual loss of superconductivity above a critical field denoted H_c.[3]

Figure 24.23 shows that there is a degree of periodicity in the elements that exhibit superconductivity. Note in particular that the ferromagnetic metals Fe, Co, and Ni do not display superconductivity; nor do the alkali metals and the coinage metals Cu, Ag, and Au. For simple metals, ferromagnetism and superconductivity never coexist, but in some of the oxide superconductors ferromagnetism and superconductivity appear to coexist on different regions of the structure of the same solid.

The first HTSC reported was $La_{1.8}Ba_{0.2}CuO_4$ (T_c = 35 K), which is a member of the solid-solution series $La_{2-x}Ba_xCuO_4$ in which Ba replaces a proportion of the La sites in La_2CuO_4. This material has the K_2NiF_4 structure type with layers of edge-sharing CuO_6 octahedra separated by the La^{3+} and Ba^{2+} cations, although the octahedra are axially elongated by a Jahn–Teller distortion (Section 20.1g). A similar compound with Sr replacing Ba in this structure type, as in $La_{1.8}Sr_{0.2}CuO_4$ (T_c = 38 K), is also known.

One of the most widely studied HTSC oxide materials, $YBa_2Cu_3O_{7-x}$ (T_c = 93 K; informally this compound is called '123', from the proportions of metal atoms in the compound, or YBCO, pronounced 'ib-co'), has a structure similar to perovskite but with missing O atoms. In terms of the structure shown in Fig. 24.11, the stoichiometric $YBa_2Cu_3O_7$ unit cell consists of three simple perovskite cubes stacked vertically with Y and Ba in the A sites

[3]The Chevrel phases discussed in Section 24.11 have the highest observed values of H_c.

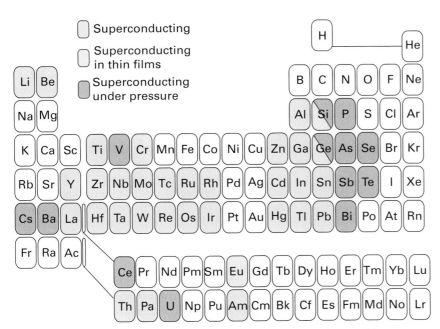

Figure 24.23 Elements that show superconductivity under the specified conditions.

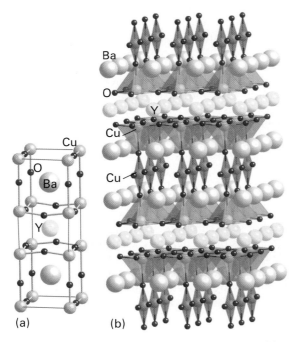

(a) (b)

Figure 24.24 Structure of the $YBa_2Cu_3O_7$ superconductor. (a) The unit cell and (b) oxygen polyhedra around the copper ions showing the layers formed from linked CuO_5 square pyramids and chains formed from corner-linked CuO_4 square planes.

of the original perovskite and Cu atoms in the B sites. However, unlike in a true perovskite structure, the B sites are not surrounded by an octahedron of O atoms: the 123 structure has a large number of sites that would normally be occupied by O but are in fact vacant. As a result, some Cu atoms have five O atom neighbours in a square-pyramidal arrangement and others have only four, as square-planar CuO_4 units. Similarly, the Y and Ba in the A sites have less than 12-coordination. The compound $YBa_2Cu_3O_7$ readily loses oxygen from some sites within the CuO_4 square planes, forming $YBa_2Cu_3O_{7-x}$ ($0 < x < 1$), but as x increases above 0.1 the critical temperature drops rapidly from 93 K. A sample of the 123 material made in the laboratory and heated under pure oxygen at 450°C as the final stage of its preparation is typically oxygen deficient with $x \approx 0.1$.

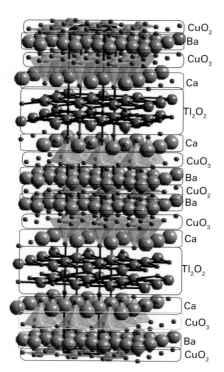

Figure 24.25 $Tl_2Ba_3Ca_2Cu_3O_{10}$ structure formed from three oxygen-deficient perovskite layers produced from linked CuO_4 square planes and square-based pyramids separated by Ca on the A type cation position. Double layers of stoichiometry Tl_2O_2 with rock-salt type arrangements of the Tl and O atoms are interleaved between the multiple perovskite layers.

1 fod

2 Y(thd)₃

If we assign the usual oxidation numbers $N_{ox}(Y) = +3$, $N_{ox}(Ba) = +2$, and $N_{ox}(O) = -2$, then the average oxidation number of copper turns out to be $+2.33$, so it is inferred that $YBa_2Cu_3O_{7-x}$ is a mixed oxidation state material that contains Cu^{2+} and Cu^{3+}. Note that in $YBa_2Cu_3O_{7-x}$ the material formally contains some Cu^{3+} until x increases above 0.5. An alternative view is that the number of electrons in $YBa_2Cu_3O_{7-x}$ is such that a partially filled band is present: this view is consistent with the high electrical conductivity and metallic behaviour of this oxide at room temperature (Section 3.19). If this band can be considered as being constructed from Cu3d orbitals, then the partial filling is a result of holes in this level (corresponding to Cu^{3+}); another possible description is that the band also involves O2p orbitals, suggesting that the material can be considered to contain Cu^{2+} and O^-.

The square-planar CuO_4 units in $YBa_2Cu_3O_{7-x}$ are arranged in chains and the CuO_5 units link together to form infinite sheets. The stoichiometry of an infinite sheet of vertex-sharing CuO_4 square planes is CuO_2. The addition of one or two additional apical O atoms in the cases where the layers are constructed from, respectively, linked square-based pyramids or octahedra maintains the CuO_2 sheet. This structural feature is also seen in all other oxocuprate HTSCs. It is thought that it is an important component of the mechanism of superconduction in these materials.

Some HTSC and other superconducting materials are listed in Table 24.4. All of them may be considered to have at least part of their structure derived from that of perovskite, as a layer of linked CuO_n ($n = 4, 5, 6$) polyhedra is a section of that structural type. Lying between these cuprate layers (which may include up to six such perovskite-derived CuO_2 sheets) can be a variety of other simple structural units, containing s- and p-block metals in combination with oxygen, such as rock-salt and fluorite structures. Thus $Tl_2Ba_2Ca_2Cu_3O_{10}$ can be considered as having three perovskite layers based on Cu, O, and Ca separated by double layers of a rock-salt structure built from Tl and O; the Ba lie between the rock-salt and perovskite layers (Fig. 24.25).

The synthesis of high-temperature superconductors has been guided by a variety of qualitative considerations, such as the demonstrated success of the layered structures and of mixed-oxidation-state Cu in combination with heavy p-block elements. Additional considerations are the radii of ions and their preference for certain coordination environments. Many of these materials are prepared simply by heating an intimate mixture of the metal oxides to 800–900°C in an open alumina crucible. Others, such as mercury- and thallium-containing complex copper oxides, require reactions involving the volatile and toxic oxides Tl_2O and HgO; in such cases the reactions are normally carried out in sealed gold or silver tubes.

Thin films are needed if superconductors are to be used in electronic devices. Their preparation is an active area of research and a promising strategy is chemical vapour deposition. The general strategy is to form a thin film by decomposing a thermally unstable compound on a hot solid substrate material (Fig. 24.26). Fluorinated acetylacetonato complexes of the metals (using ligands such as fod, **1**) are sometimes used because they are more volatile than simple acetylacetonato complexes. These complexes, such as $Cu(acac)_2$, $Y(thd)_3$ (**2**), and $Ba(fod)_2$, are swept into the reaction chamber by slightly moist oxygen gas. When conditions are properly controlled, this gaseous mixture reacts on the hot substrate to produce the desired $YBa_2Cu_3O_{7-x}$ film. The film may be amorphous and require subsequent heating to form a crystalline product.

There is, as yet, no settled explanation of high-temperature superconductivity. It is believed that the movement of pairs of electrons, known as 'Cooper pairs' and responsible for conventional superconductivity, is also important in the high-temperature materials, but the mechanism for pairing is hotly debated.

(f) Other superconducting oxides

The observation of superconductivity in the complex cuprates is unusual in terms of the high critical temperatures reached but many other oxides and oxide phases demonstrate a transition to zero electrical resistance, albeit normally at considerably lower temperatures. Some compositionally simple examples include phases from the solid solution $Li_{1+x}Ti_{2-x}O_4$, (which adopts the spinel structure and has $T_c = 13.7$ K for $x = 0$) and $Na_{0.35}CoO_2.H_2O$ with $T_c \approx 5$ K (see also Table 24.4).

The complex bismuth oxides of composition $(K_{0.87}Bi_{0.13})BiO_3$ (for which $T_c = 10.2$ K) and $(Ba_{0.6}K_{0.4})BiO_3$ ($T_c = 30$ K), which adopt perovskite structures with Bi as the B-type cation, are among a number of similar bismuthates that exhibit superconductivity.

Several complex oxides that adopt the pyrochlore structure (Fig 24.27) and have composition $M_{2-x}B_2O_{7-x}$, where M is a Group I, Group II, or post-transition metal cation, such as Cs, Ca, or Cd and where B is a heavy d metal, show superconductivity. For example, $Cd_2Re_2O_7$ is superconducting below 1.4 K and KOs_2O_6 is superconducting below 10 K. Very recently, interest has centred on a new family of superconductors with critical temperatures approaching the best cuprates. These new lanthanoid iron arsenic oxides of composition $LnFeAs(O,F)_{1-x}$ were first reported for Ln = La in $LaFeAsO_{1-x}F_x$ with $T_c = 26$ K, and for compositionally similar Pr and Sm compounds critical temperatures of 52 K and 55 K, respectively, have been achieved. The structures of these compounds are based on alternating layers of composition LnO and FeAs (Fig 24.28).

(g) Colossal magnetoresistance

Key point: Perovskites with Mn on the B cation sites can show very large changes in resistance on application of a magnetic field, known as colossal magnetoresistance.

Manganites, which are Mn(III) and Mn(IV) complex oxides, with the generic solid solution formulation $Ln_{1-x}A_xMnO_3$ (A = Ca, Sr, Pb, Ba; Ln = typically La, Pr, or Nd), order ferromagnetically on cooling below room temperature, with Curie temperatures typically between 100 K and 250 K, and simultaneously transform from insulators (at the higher temperature) to poor metallic conductors. These materials also exhibit **magnetoresistance**, a marked decrease of their resistance on the application of a magnetic field near and just above their Curie temperatures (Fig. 24.29). Recent investigations have shown that, for these manganites, the decrease in resistance can be by as much as 11 orders of magnitude, and for this reason these compounds have been named **colossal magnetoresistance manganites**.[4]

Colossal magnetoresistance (CMR) manganites have perovskite-type structures with the A cation sites occupied by a mixture of Ln^{3+} and A^{2+} cations and the B site occupied by Mn. The oxidation number of Mn in these solid solutions varies between +3 and +4 as the proportion of A^{2+} is changed. Pure $LaMnO_3$ orders antiferromagnetically below its Néel temperature ($T_N = 150$ K), but with an increase in x in $Ln_{1-x}A_xMnO_3$, corresponding to an increase in Mn^{4+} content, the manganites order ferromagnetically on cooling.

The observation that the transition to conducting electron behaviour and ferromagnetism occurs simultaneously on cooling the $Ln_{1-x}A_xMnO_3$ manganites and the origin of the CMR effect are not completely understood, but it is known that they are based on a so-called **double exchange** (DE) mechanism between the Mn(III) and Mn(IV) species present in these materials. The basic process of this mechanism is the transfer of an electron from

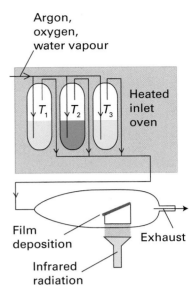

Argon, oxygen, water vapour

Heated inlet oven

T_1 T_2 T_3

Film deposition

Exhaust

Infrared radiation

Figure 24.26 Schematic diagram of the chemical vapour deposition of superconductor thin films. A reactive carrier gas mixture (Ar, O_2, and H_2O) is passed through traps of the volatile metal precursors held at temperatures T_1, T_2, and T_3 which provide the desired vapour pressure of each reactant. Deposition occurs on the wedge-shaped block, which is heated using an IR lamp. After the deposition, the resulting film is annealed at high temperature to improve the crystallinity of the film.

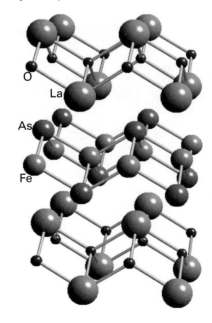

O

La

As

Fe

Figure 24.28 The structure of the superconductor LaFeAsO delineating the two types of layer.

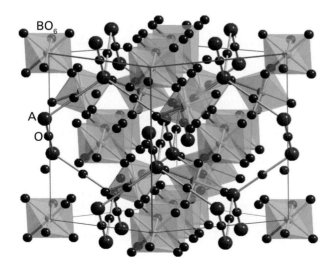

BO_6

A

O

Figure 24.27 The pyrochlore structure adopted by many compounds of stoichiometry $A_2B_2O_7$. The BO_6 octahedra are shown which form channels containing the A-type cations and oxide ions.

[4] Albert Fert and Peter Grünberg were awarded the 2007 Nobel Prize in Physics for their discovery of this phenomenon.

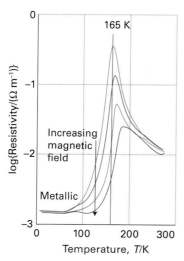

Figure 24.29 The resistivity as a function of temperature at various magnetic fields of a material displaying colossal magnetoresistance. At 165 K application of a magnetic field will cause a change in resistivity of about two orders of magnitude.

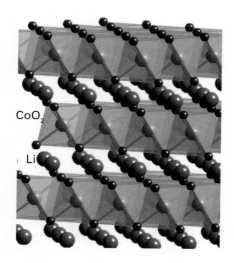

Figure 24.30 The structure of LiCoO$_2$ shown as layers of linked CoO$_6$ octahedra separated by Li$^+$ ions; lithium may be de-intercalated electrochemically from between the layers.

Mn^{3+}(t$_{2g}^3$e$_g^1$) to Mn^{4+}(t$_{2g}^3$) through the O atom, so that the Mn(III) and Mn(IV) positions change places. In manganites at high temperatures, the electrons in these systems effectively become trapped on a specific site, leading to ordering of the Mn(III) and Mn(IV) species, that is the trapping results in charge ordering. The charge-ordered state is generally associated with insulating and paramagnetic behaviour whereas the charge-disordered state, in which the electron can move between sites, is associated with metallic behaviour and ferromagnetism. The high-temperature charge-ordered state can be transformed into a metallic, spin-ordered (ferromagnetic) state just by the application of a magnetic field, therefore application of a magnetic field to a manganite just above a critical temperature causes the transformation of charge ordering into electron delocalization and hence a massive decrease in resistance.

The CMR effect is being developed for magnetic data-storage devices, such as computer hard drives. Further work using these compounds is aimed at **spintronics**, where, rather than use electron movements to transmit information, as is the basis of electronics and the functioning of the silicon chip, the movement of spin through materials could be used in a similar way. As spin-transfer is much faster than electron motion and does not develop heat through resistive effects, computing devices based on spintronics should have much higher processing powers and not require cooling; the latter is an increasing problem with semiconductor technology as transistors become ever more densely packed on computer processors.

(h) Rechargeable battery materials

Key point: The redox chemistry associated with the extraction and insertion of metal ions into oxide structures is exploited in rechargeable batteries.

The existence of complex oxide phases that demonstrate good ionic conductivity associated with the ability to vary the oxidation state of a d-metal ion has led to the development of materials for use as the cathode in rechargeable batteries (see Box 11.1). Examples include LiCoO$_2$, with a layer-type structure based on sheets of edge-linked CoO$_6$ octahedra separated by Li$^+$ ions (Fig. 24.30), and various lithium manganese spinels, such as LiMn$_2$O$_4$. In each of these compounds the battery is charged by removing the mobile Li$^+$ ions from the complex metal oxide, as in

$$LiCoO_2 \rightarrow CoO_2 + Li^+ + e^-$$

The battery is discharged through the reverse electrochemical reaction. Lithium cobalt oxide, which is used in many commercial lithium-ion batteries, has many of the characteristics required for this type of application. The specific energy (the stored energy divided by the mass) of LiCoO$_2$ (140 W h kg^{-1}) is maximized by using light elements such as Li and Co; the 3d metals are almost invariably used in such applications because they are the lowest density elements with variable oxidation states. The high mobility of the Li$^+$ ion and good reversibility of electrochemical charging and discharging stem from the lithium ion's small ionic radius and the layer-like structure of LiCoO$_2$, which allows the Li$^+$ to be extracted without major disruption of the structure. High capacities are obtained from the large amount of Li (one Li$^+$ ion for each LiCoO$_2$ formula unit) that may be reversibly extracted (about 500 discharge/recharge cycles) from the compound, and the current is delivered at a constant and high potential difference (of between 3.5 and 4 V). The high potential difference is partly due to the high oxidation states of cobalt (+3 and +4) that are involved.

Because cobalt is expensive and fairly toxic, the search continues for even better oxide materials than LiCoO$_2$. New materials will be required to demonstrate the high levels of reversibility found for lithium cobaltate and considerable effort is being directed at doped forms of LiCoO$_2$ and of LiMn$_2$O$_4$ spinels and at nanostructured complex oxides (Section 25.4), which, because of their small particle size, can offer excellent reversibility. There is also a high level of interest in LiFePO$_4$, which shows good characteristics for a cathode material and contains cheap and nontoxic iron.

The other electrode in a rechargeable lithium ion battery can simply be Li metal, which completes the overall cell reaction through the process Li(s) $\rightarrow$ Li$^+$ + e$^-$. Lithium ions then migrate to the cathode through an electrolyte, which is typically an anhydrous lithium salt, such as LiPF$_4$ or LiC(SO$_2$CF$_3$)$_3$ dissolved in a polymer, such as poly(propene carbonate).

However, the use of Li metal has a number of problems associated with its reactivity and volume changes that occur in the cell. Therefore, an alternative anode material that is frequently used in rechargeable batteries is graphitic carbon, which can intercalate, electrochemically, large quantities of Li to form LiC_6 (Section 14.5). As the cell is discharged, Li is transferred from between the carbon layers at the anode and intercalated into the metal oxide at the cathode (and vice versa on charging) with the following overall processes :

$$Li_yC_6 + Li_{1-x}CoO_2 \underset{\text{discharge (reaction proceeds to the left)}}{\overset{\text{charge (reaction proceeds to the right)}}{\rightleftharpoons}} C_6 + Li_{1-x+y}CoO_2$$

24.8 Oxide glasses

The term **ceramic** is often applied to all inorganic nonmetallic, nonmolecular materials, including both amorphous and crystalline materials, but the term is commonly reserved for compounds or mixtures that have undergone heat treatment. The term **glass** is used in a variety of contexts but for our present purposes it implies an amorphous ceramic with a viscosity so high that it that can be considered to be rigid. A substance in its glassy form is said to be in its **vitreous** state. Although ceramics and glasses have been utilized since antiquity, their development is currently an area of rapid scientific and technological progress. This enthusiasm stems from interest in the scientific basis of their properties and the development of novel synthetic routes to new high-performance materials. We confine our attention here to glasses. The most familiar glasses are alkali-metal or alkaline-earth-metal silicates and borosilicates.

(a) Glass formation

Key points: Silicon dioxide readily forms a glass because the three-dimensional network of strong covalent Si–O bonds in the melt does not easily break and reform on cooling; the Zachariasen rules summarize the properties likely to lead to glass formation.

A glass is prepared by cooling a melt more quickly than it can crystallize. Cooling molten silica, for instance, gives vitreous quartz. Under these conditions the solid has no long-range order as judged by the lack of X-ray diffraction peaks, but spectroscopic and other data indicate that each Si atom is surrounded by a tetrahedral array of O atoms. The lack of long-range order results from variations of the Si–O–Si angles. Figure 24.31a illustrates in two dimensions how a local coordination environment can be preserved but long-range order lost by variation of the bond angles around O. This loss of long-range order is readily apparent when X-rays are scattered from a glass (Fig. 24.31b); in contrast to a long-range, periodically ordered crystalline material, where diffraction gives rise to a series of diffraction maxima (Section 8.1), the X-ray diffraction pattern obtained from a glass shows only broad features as the long-range order is lost. Silicon dioxide readily forms a glass because the three-dimensional network of strong covalent Si–O bonds in the melt does not readily break and reform on cooling. The lack of strong directional bonds in metals and simple ionic substances makes it much more difficult to form glasses from these materials. Recently, however, techniques have been developed for ultrafast cooling and, as a result, a wide variety of metals and simple inorganic materials can now be frozen into a vitreous state.

The concept that the local coordination sphere of the glass-forming element is preserved but that bond angles around O are variable was originally proposed by W.H. Zachariasen in 1932. He reasoned that these conditions would lead to similar molar Gibbs energies and molar volumes for the glass and its crystalline counterpart. Zachariasen also proposed that the vitreous state is favoured by polyhedral corner-sharing O atoms, rather than edge- or face-shared, which would enforce greater order. These and other **Zachariasen rules** hold for common glass-forming oxides, but exceptions are known.

An instructive comparison between vitreous and crystalline materials is seen in their change in volume with temperature (Fig. 24.32). When a molten material crystallizes, an abrupt change in volume (usually a decrease) occurs. By contrast, a glass-forming material that is cooled sufficiently rapidly persists as a metastable supercooled liquid. When cooled below the **glass transition temperature**, T_g, the supercooled liquid becomes rigid, and this change is accompanied by only an inflection in the cooling curve rather than an abrupt change in slope. The rates of crystallization are very slow for many complex metal silicates, phosphates, and borates, and it is these compounds that often form glasses.

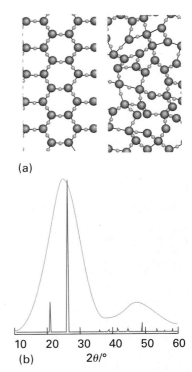

(a)

(b)

Figure 24.31 (a) Schematic representation of a two-dimensional crystal, left, compared with a two-dimensional glass, right. (b) The powder X-ray diffraction pattern of a glass (SiO$_2$, orange) contrasted with that of a crystalline solid (quartz SiO$_2$, blue). The long-range order, which gives rise to the sharp diffraction maxima in quartz, is no longer present in amorphous SiO$_2$ and only broad features are seen.

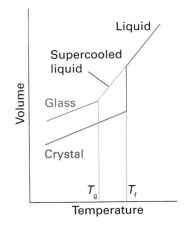

Figure 24.32 Comparison of the volume change for supercooled liquids and glasses with that for a crystalline material. The glass transition temperature is T_g.

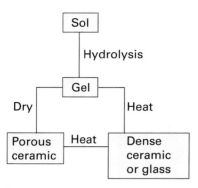

Figure 24.33 Schematic diagram of the sol-gel process. When the gel is dried at high temperatures dense ceramics or glasses are formed. Drying at low temperatures above the critical pressure of water produces porous solids known as xerogels or aerogels.

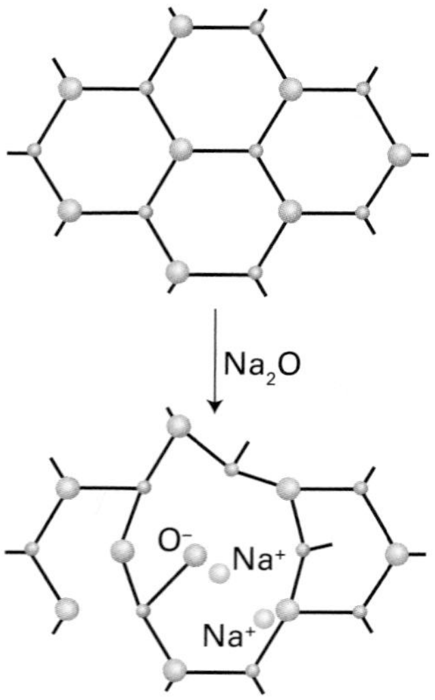

Figure 24.34 The role of a modifier is to introduce O^{2-} ions and cations that disrupt the lattice.

Another route to glasses is the **sol-gel process**, which is described schematically in Fig. 24.33 and discussed more fully in Section 25.4. As described there, the sol-gel process is also used to produce crystalline ceramic materials and high-surface-area compounds such as silica gel. A typical process involves the addition of a metal alkoxide precursor to an alcohol, followed by the addition of water to hydrolyse the reactants. This hydrolysis leads to a thick gel that can be dehydrated and **sintered** (heated below its melting point to produce a compact solid). For example, ceramics containing TiO_2 and Al_2O_3 can be prepared in this way at much lower temperatures than required to produce the ceramic from the simple oxides. Often a special shape can be fashioned at the gel stage. Thus, the gel may be shaped into a fibre and then heated to expel water; this process produces a glass or ceramic fibre at much lower temperatures than would be necessary if the fibre were made from the melt of the components.

(b) Glass composition, production, and application

Key points: Low-valence metal oxides, such as Na_2O and CaO, are often added to silica to reduce its softening temperature by disrupting the silicon–oxygen framework. Other cations may be incorporated into glasses, giving applications as diverse as lasers and nuclear waste containment.

Although vitreous silica is a strong glass that can withstand rapid cooling or heating without cracking, it has a high glass transition temperature and therefore must be worked at inconveniently high temperatures. Therefore, a **modifier**, such as Na_2O or CaO, is commonly added to SiO_2. A modifier disrupts some of the $Si-O-Si$ linkages and replaces them with terminal $Si-O^-$ links that associate with the cation (Fig. 24.34). The consequent partial disruption of the $Si-O$ network leads to glasses that have lower softening points. The common glass used in bottles and windows is called 'sodalime glass' and contains Na_2O and CaO as modifiers. When B_2O_3 is used as a modifier, the resulting 'borosilicate glasses' have lower thermal expansion coefficients than sodalime glass and are less likely to crack when heated. Borosilicate glass (such as Pyrex®) is therefore widely used for ovenware and laboratory glassware.

Glass formation is a property of many oxides, and practical glasses have been made from sulfides, fluorides, and other anionic constituents. Some of the best glass formers are the oxides of elements near silicon in the periodic table (B_2O_3, GeO_2, and P_2O_5), but the solubility in water of most borate and phosphate glasses and the high cost of germanium limit their usefulness.

The development of transparent crystalline and vitreous materials for light transmission and processing has led to a revolution in signal transmission. For example, optical fibres are currently being produced with a composition gradient from the interior to the surface. This composition gradient modifies the refractive index and thereby decreases light loss. Fluoride glasses are also being investigated as possible substitutes for oxide glasses because oxide glasses contain small numbers of OH groups, which absorb near-infrared radiation and attenuate the signal. Doping lanthanoid ions into the glass fibres produces materials that can be used to amplify the signals by using lasing effects. Optical circuit elements are being developed that may eventually replace all the components in an electronic integrated circuit and lead to very fast optical computers.

As modifier cations effectively become trapped in a glass, which is chemically inert and thermodynamically very stable with respect to transformation to a soluble, crystalline phase, such glassy materials offer a potential method for containing and storing nuclear waste. Thus vitrification of metal oxides containing a radioactive species with glass-forming oxides produces a stable glass that can be stored for long periods, allowing the radionucleide to decay (Section 23.1). Examples of such materials are borosilicate glasses and combinations of such glasses with crystalline oxide phases ('Synroc').

The incorporation of functional inorganic compounds into glasses has led to the development of so-called 'smart glasses' that have switchable properties, such as **electrochromism**, the ability to change colour or light transmission properties in response to application of a potential difference, and reversible **photochromism**, the ability to change colour under certain light conditions. An electrochromic glass generally consists of a glass coated with colourless, tungsten trioxide, WO_3, or a sandwich-like layer structure consisting of these two components. When a potential difference is applied to the WO_3 layer, cations are inserted and the W is partially reduced to form $M_xW(VI,V)O_3$, which is dark blue. The coating maintains its dark colour until the potential difference is reversed, when the glass becomes colourless. Application of electrochromic glasses include privacy glass, auto-dimming rear view mirrors in vehicles, and aircraft windows. A similar coating is used in self-cleaning glasses where a metal oxide coating on the glass acts as a photocatalyst for the breakdown of organic dirt on their surfaces. A photochromic glass incorporates a small amount of a colourless silver halide, normally AgCl, within the glass. When exposed to UV radiation, such as that in sunlight, the AgCl dissociates to form small clusters of Ag atoms, which absorb light across the visible region of the spectrum, imparting a grey colour to the glass. When the UV radiation is removed, the AgCl reforms and the glass returns to its optically transparent state. Photochromic glasses are widely used in the lenses of spectacles and sunglasses.

24.9 Nitrides and fluorides

The solid-state chemistry of the metals in combination with anions other than O^{2-} is not as highly developed or extensive as that of the complex oxides described so far. However, complex nitrides and fluorides and mixed anion compounds are of growing importance.

(a) Nitrides

Key points: Complex metal nitrides and oxide nitrides are materials containing the N^{3-} anion; many new compounds of this type have recently been synthesized.

Simple metal nitrides of main-group elements, such as AlN, GaN, and Li_3N, have been known for decades. Many of the recent advances in nitride chemistry have centred on d-metal compounds and complex nitrides. That nitrides are less common than oxides stems, in part, from the high enthalpy of formation of N^{3-} compared with that of O^{2-}. Furthermore, because many nitrides are sensitive to oxygen and water, their synthesis and handling are problematic. Some simple metal nitrides can be obtained by the direct reaction of the elements, for example Li_3N is obtained by heating lithium in a stream of nitrogen at 400°C. The instability of sodium nitride allows sodium azide to be used as a nitriding agent:

$$2\,NaN_3(s) + 9\,Sr(s) + 6\,Ge(s) \xrightarrow{750°C,\,sealed\,Nb\,tube} 3\,Sr_3Ge_2N_2(s) + 2\,Na(g)$$

The ammonolysis of oxides (the dehydrogenation of NH_3 by oxides with the formation of water as a byproduct) provides a convenient route to some nitrides. For instance, tantalum nitride can be obtained by heating tantalum pentoxide in a fast-flowing stream of ammonia:

$$3\,Ta_2O_5(s) + 10\,NH_3(l) \xrightarrow{700°C} 2\,Ta_3N_5(s) + 15\,H_2O(g)$$

In such reactions, the equilibrium is driven towards the products by removal of the steam in the gas flow. Similar reactions may be used for the preparation of complex nitrides from complex oxides, although competing reactions involving partial reduction of the metal oxide by ammonia can also occur. In all these reactions the complete elimination of O^{2-} ions from the product can be troublesome, so the reactions give products that contain both the oxide and nitride ions:

$$Ca_2Ta_2O_7(s) + 2\,NH_3(g) \xrightarrow{800°C} 2\,CaTaO_2N(s) + 3\,H_2O(g)$$

In comparison with the O^{2-} ion, the higher charge of the N^{3-} ion results in a greater degree of covalence in its bonding and therefore nitrides, particularly those of less electropositive elements such as d metals, should not be described in purely ionic terms. There is also a tendency with nitrides for the formation of compounds in which the metallic element is in a lower oxidation state because nitrogen, on account of its high bond energy, is not as potent an oxidant as oxygen or fluorine. Thus, whereas heating titanium in oxygen readily produces TiO_2, Ti_2N and TiN are known but Ti_3N_4 is difficult to prepare and poorly characterized. Likewise, V_3N_5 is unknown whereas V_2O_5 is readily obtained from the decomposition of many vanadium salts in air.

Many of the early d-metal nitrides are interstitial compounds and are used as high-temperature refractory ceramics. Similarly, the nitrides of Si and Al, such as Si_3N_4 (Fig. 24.35), are stable at very high temperatures, particularly under nonoxidizing

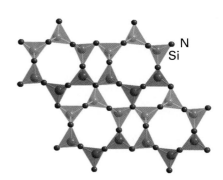

Figure 24.35 The structure of Si_3N_4 shown as linked SiN_4 tetrahedra.

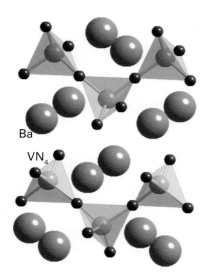

Ba

VN_4

Figure 24.36 Ba_2VN_3, containing one-dimensional chains of corner-sharing VN_4 tetrahedra separated by Ba^{2+} ions.

conditions, and are used for crucibles and furnace elements. Recently, GaN, which can exist in both wurtzite and sphalerite structural types, has been the focus of considerable research on account of its semiconducting properties. The nitride Li_3N has an unusual structure based on hexagonal Li_2N^- layers separated by Li^+ ions (Fig. 24.8). These Li^+ ions are highly mobile, as is expected from the existence of free space between the layers, and this compound and other structurally related materials are being studied for possible use in rechargeable batteries. Among the many complex nitrides that have been synthesized are materials of stoichiometry AMN_2, such as $SrZrN_2$ and $CaTaN_2$, with structures based on sheets formed from MN_6 octahedra sharing edges (see the discussion of $LiCoO_2$ in Section 24.7h) and A_2MN_3, such as Ba_2VN_3, containing one-dimensional chains of corner-sharing MN_4 tetrahedra (Fig. 24.36). Analogues of the zeolites containing nitride bridges rather than oxide bridges have also been prepared.

(b) Fluorides and other halides

Key point: Because fluorine and oxygen have similar ionic radii, fluoride solid-state chemistry parallels much of oxide chemistry.

The ionic radii of F^- and O^{2-} are very similar (at between 130 and 140 pm), and as a result metal fluorides show many stoichiometric and structural analogies with the complex oxides but with lower charge on the metal ion to reflect the lower charge on the F^- ion. Many binary metal fluorides adopt the simple structural types expected on the basis of the radius-ratio rule (Section 3.10). For example, FeF_2 and PdF_2 have a rutile structure and AgF has a rock-salt structure; similarly, NbF_3 adopts the ReO_3 structure. For complex fluorides, analogues of typical oxide structural types are well known, including perovskites (such as $KMnF_3$), Ruddlesden–Popper phases (for example $K_3Co_2F_7$), and spinels (Li_2NiF_4). Synthetic routes to complex fluorides also parallel those for oxides. For instance, the direct reaction of two metal fluorides yields the complex fluoride, as in

$$2\,LiF(s) + NiF_2(s) \rightarrow Li_2NiF_4(s)$$

Like some complex oxides, some complex fluorides may be precipitated from solution

$$MnBr_2(aq) + 3\,KF(aq) \rightarrow KMnF_3(s) + 2\,KBr(aq)$$

As with the ammonolysis of oxides to produce nitrides and oxide-nitrides, the formation of oxide-fluorides is possible by the appropriate treatment of a complex oxide, as in

$$Sr_2CuO_3(s) \xrightarrow{\ F_2/200°C\ } Sr_2CuO_2F_{2+x}(s)$$

$Sr_2CuO_2F_{2+x}$ is a superconductor with $T_c = 45$ K.

Fluoride analogues of the silicate glasses, which are based on linked SiO_4 tetrahedra, exist for small cations that form tetrahedral units in combination with F^-; an example is $LiBF_4$, which contains linked BF_4 tetrahedra. Lithium borofluoride glasses are used to contain samples for X-ray work because they are highly transparent to X-rays on account of their low electron densities. Framework and layer structures based on linked MF_4 (M = Li, Be) tetrahedra have also been described. Some metal fluorides are used as fluorination agents in organic chemistry. However, few of the solid complex metal fluorides are technologically important in comparison with the wealth of applications associated with analogous complex oxides.

Metal chloride structures reflect the greater covalence associated with bonding to chloride in comparison with fluoride: the chlorides are less ionic and have structures with lower coordination numbers than the corresponding fluorides. Thus, simple metal chlorides, bromides, and iodides normally adopt the cadmium-chloride or cadmium-iodide structures based on sheets formed from edge-sharing MX_6 octahedra. Complex chlorides often contain the same structural unit, for example $CsNiCl_3$ has chains of edge-sharing $NiCl_6$ octahedra separated by Cs^+ ions. Many analogues of oxide structures also occur among the complex chlorides, such as $KMnCl_3$, K_2MnCl_4, and Li_2MnCl_4, which have the perovskite, K_2NiF_4, and spinel structures, respectively.

Chalcogenides, intercalation compounds, and metal-rich phases

The soft chalcogens S, Se, and Te form binary compounds with metals that commonly have quite different structures from the corresponding oxides, nitrides, and fluorides. As we saw in Sections 3.9, 16.11, and 19.9, this difference is consistent with the greater covalence of the compounds of sulfur and its heavier congeners. For example, we noted there that MO compounds generally adopt the rock-salt structure whereas ZnS and CdS can crystallize with either of the sphalerite or the wurtzite structures in which the lower coordination numbers indicate the presence of directional bonding. Similarly, the d-block monosulfides generally adopt the more characteristically covalent nickel-arsenide structure rather than the rock-salt structure of alkaline-earth oxides such as MgO. Even more striking are the layered MS_2 compounds formed by many d-block elements in contrast to the fluorite or rutile structures of many d-block dioxides.

Before we discuss these compounds, we should note that there are many metal-rich compounds that disobey simple valence rules and do not conform to an ionic model. Some examples are Ti_2S, Pd_4S, V_2O, and Fe_3N, and even the alkali metal suboxides, such as Cs_3O (Section 11.8). The occurrence of metal-rich phases is generally associated with $M-M$ interactions. Many other intermetallic compounds display stoichiometries that cannot be understood in terms of conventional valence rules. Included among them are the important permanent-magnet materials $Nd_2Fe_{17}B$ and $SmCo_5$.

24.10 Layered MS_2 compounds and intercalation

The layered metal sulfides and their intercalation compounds were introduced in Section 19.9. Here we develop a broader picture of their structures and properties.

(a) Synthesis and crystal growth

Key point: d-Metal disulfides are synthesized by the direct reaction of the elements in a sealed tube and purified by using chemical vapour transport with iodine.

Compounds of the chalcogens with d metals are prepared by heating mixtures in a sealed tube (to prevent the loss of the volatile elements). The products obtained in this manner can have a variety of compositions. The preparation of crystalline dichalcogenides suitable for chemical and structural studies is often performed by **chemical vapour transport** (CVT), as described below. It is possible in some cases simply to sublime a compound, but the CVT technique can also be applied to a wide variety of nonvolatile compounds in solid-state chemistry.

In a typical procedure, the crude material is loaded into one end of a borosilicate or fused quartz tube. After evacuation, a small amount of a CVT agent is introduced and the tube is sealed and placed in a furnace with a temperature gradient. The polycrystalline and possibly impure metal chalcogenide is vaporized at one end and redeposited as pure crystals at the other (Fig. 24.37). The technique is called chemical vapour *transport* rather than *sublimation* because the CVT agent, which is often a halogen, produces an intermediate volatile species, such as a metal halide. Generally, only a small amount of transport agent is needed because on crystal formation it is released and diffuses back to pick up more reactant. For example, TaS_2 can be transported with I_2 in a temperature gradient. The reaction with I_2 to produce gaseous products

$$TaS_2(s) + 2I_2(g) \rightarrow TaI_4(g) + S_2(g)$$

is endothermic, so the equilibrium lies further to the right at 850°C than at 750°C. Consequently, although TaI_4 is formed at 850°C, at 750°C the mixture deposits TaS_2. If, as occasionally is the case, the transport reaction is exothermic, the solid is carried from the cooler to the hotter end of the tube.

(b) Structure

Key points: Elements on the left of the d block form sulfides consisting of sandwich-like layers of the metal coordinated to six S ions; the bonding between the layers is very weak.

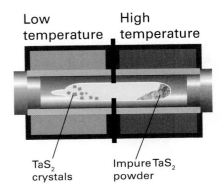

Figure 24.37 Vapour transport crystal growth and purification of TaS_2. A small quantity of I_2 is present to serve as a transport agent.

As we saw in Section 19.9, the d-block disulfides fall into two classes: layered materials are formed by metals on the left of the d block, and compounds containing formal S_2^{2-} ions are formed by metals in the middle and towards the right of the block (such as pyrite, FeS_2). We concentrate here on the layered materials.

In TaS_2 and many other layered disulfides, the d-metal ions are located in octahedral holes between close-packed AB layers (Fig. 24.38a). The Ta ions form a close-packed layer denoted X, so the metal and adjoining sulfide layers can be portrayed as an AXB sandwich. These sandwich-like slabs form a three-dimensional crystal by stacking in sequences such as AXBAXBAXB..., where the strongly bound AXB slabs are held to their neighbours by weak dispersion forces. An alternative view of these MS_2 structures, which have the metal ions in octahedral holes, is as MS_6 octahedra sharing edges (Fig. 24.39a), which reinforces the idea of the greater degree of covalent bonding that occurs in these materials than in, for instance, Li_2S, which has the antifluorite structure.

The Nb atoms in NbS_2 reside in the trigonal-prismatic holes between sulfide layers that are in register with one another (AA, Fig. 24.38b and Fig 24.39b). The Nb atoms, which are strongly bonded to the adjacent sulfide layers, form a close-packed array denoted m, so we can represent each slab as AmA or CmC. These slabs form a three-dimensional crystal by stacking in a pattern such as ...AmACmCAmACmC.... Weak dispersion forces also contribute to holding these AmA and CmC slabs together. Polytypes (versions that differ only in the stacking arrangement along a direction perpendicular to the plane of the slabs) can occur. Thus, NbS_2 and MoS_2 form several polytypes, including one with the sequence CmCAmABmB. Molybdenum sulfide, MoS_2, is used as a high performance lubricant in, for example, racing cars and machining, as it can be used at much higher temperatures and pressures than oils. Lubrication occurs with a dry coating of the material as the MoS_2 layers are able to slip over each other easily due to the weak interlayer interactions.

It has been convenient to describe—somewhat simplistically—the layered structures in terms of cations and anions. However, to account for the significant covalence, the electrical conductivities, and the chemical properties of these layers we need to invoke the more sophisticated band model of their electronic structure. Some approximate band structures derived primarily from molecular orbital calculations and photoelectron spectra are shown in Fig. 24.40. They show that the dichalcogenides with octahedral and trigonal-prismatic metal sites have low-lying bands composed primarily of chalcogen s and p orbitals, higher-energy bands derived primarily from metal d orbitals, and still higher in energy a variety of metal and chalcogen bands.

When the metal atom occupies an octahedral site, its d orbitals are split into a lower t_{2g} set and a higher e_g set, just as in localized complexes. These atomic orbitals combine to give a t_{2g} band and an e_g band. The broad t_{2g} band may accommodate up to six electrons per metal atom. Thus TaS_2, with one d electron per metal atom, has an only partly filled t_{2g} band, and is a metallic conductor. A trigonal-prismatic ligand field leads to a low-energy a_1' level and two doubly-degenerate higher-energy orbitals designated e' and e'', respectively. In this case, the lowest band needs only two electrons per atom to fill it.

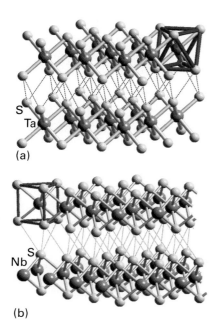

S
Ta
(a)

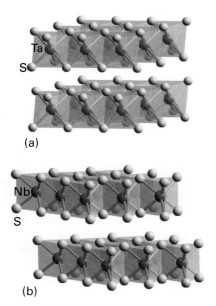

S
Nb
(b)

Figure 24.38 (a) The structure of TaS_2 (CdI_2-type). The Ta atoms reside in octahedral sites between the AB layers of S atoms. (b) The NbS_2 structure; the Nb atoms reside in trigonal-prismatic sites, between the sulfide layers.

Ta
S
(a)

Nb
S
(b)

Figure 24.39 The metal disulfide structures of Figure 24.38 drawn as layers of MS_6 polyhedra sharing edges: (a) octahedra in TaS_2, (b) trigonal prisms in NbS_2.

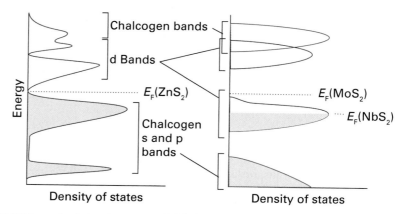

Chalcogen bands

d Bands

$E_F(ZnS_2)$

$E_F(MoS_2)$

$E_F(NbS_2)$

Energy

Chalcogen s and p bands

Density of states

Density of states

Figure 24.40 Approximate band structures of dichalcogenides. (a) Octahedral MS_2 compounds; for ZrS_2 (d^0) only the sulfur s and p bands are full and the compound is a semiconductor. (b) Trigonal prismatic MS_2 compounds. NbS_2 (d^1) is metallic whereas MoS_2 (d^2) is a semiconductor.

Thus MoS_2, with trigonal-prismatic Mo sites and two d electrons, has a filled $a_1{'}$ band and is an insulator (more precisely, a large band-gap semiconductor).

(c) Intercalation and insertion

Key points: Insertion compounds can be formed from the d-metal disulfides either by direct reaction or electrochemically; insertion compounds can also be formed with molecular guests.

We have already introduced the idea that alkali metal ions may insert between graphite sheets (Section 14.5), metal disulfide slabs (Section 19.9), and metal oxide layers (as in Li_xCoO_2, Section 24.7h) to form intercalation compounds. For a reaction to qualify as an intercalation, or as an **insertion reaction**, the basic structure of the host should not be altered when it occurs. Reactions in which the structure of one of the solid starting materials is not radically altered are called **topotactic reactions**. They are not limited to the type of insertion chemistry we are discussing here. For example, hydration, dehydration, and ion exchange reactions may also be topotactic.

The π conduction and valence bands of graphite are contiguous in energy (we have seen in fact that graphite is formally a semimetal, Section 3.19) and the favourable Gibbs energy for intercalation arises from the transfer of an electron from the alkali metal atom to the graphite conduction band. The insertion of an alkali metal atom into a dichalcogenide involves a similar process: the electron is accepted into the d band and the charge-compensating alkali metal ion diffuses to positions between the slabs. Some representative alkali metal insertion compounds are listed in Table 24.5.

The insertion of alkali metal ions into host structures can be achieved by direct combination of the alkali metal and the disulfide:

$$TaS_2(s) + x\,Na(g) \xrightarrow{800°C} Na_x TaS_2(s)$$

with $0.4 < x < 0.7$. Insertion may also be achieved by using a highly reducing alkali metal compound, such as butyllithium, or the electrochemical technique of **electrointercalation** (Fig. 24.41). One advantage of electrointercalation is that it is possible to measure the amount of alkali metal incorporated by monitoring the current (I) passed during the synthesis (using $n_e = It/F$). It also is possible to distinguish solid-solution formation from discrete-phase formation. As illustrated in Fig. 24.42, the formation of a solid solution is characterized by a gradual change in potential as intercalation proceeds. In contrast, the formation of a new discrete phase yields a steady potential over the range in which one solid phase is being converted into the other, followed by an abrupt change in potential when that reaction is complete.

Insertion compounds are examples of mixed ionic and electronic conductors. In general, the insertion process can be reversed either chemically or electrochemically. This reversibility makes it possible to recharge a lithium cell by removal of Li from the compound. In a clever synthetic application of these concepts, the previously unknown layered disulfide VS_2 can be prepared by first making the known layered compound $LiVS_2$ in a high-temperature process. The Li is then removed by reaction with I_2 to produce the metastable layered VS_2, which has the TiS_2 structure:

$$2\,LiVS_2(s) + I_2(s) \rightarrow 2\,LiI(s) + 2\,VS_2(s)$$

Insertion compounds also can be formed with molecular guests. Perhaps the most interesting guest is the metallocene $Co(\eta^5\text{-}Cp)_2$, where $Cp = C_5H_5$ (Section 22.14), which can be incorporated into a variety of hosts with layered structures, such as TiS_2, $TiSe_2$, and TaS_2, to the extent of about 0.25 $Co(\eta^5\text{-}Cp)_2$ per MS_2 or MSe_2. This limit appears to correspond to the space available for forming a complete layer of $Co(\eta^5\text{-}Cp)_2^+$ ions. The organometallic compound appears to undergo oxidation upon intercalation, so the favourable Gibbs energy in these reactions arises in the same way as in alkali metal intercalation. In agreement with this interpretation, $Fe(\eta^5\text{-}Cp)_2$, which is more difficult to oxidize than its Co analogue, does not intercalate (Section 22.19).

We can imagine the insertion of ions into one-dimensional channels, between two-dimensional planes of the type we have been discussing, or into channels that intersect to form three-dimensional networks (Fig. 24.43). Aside from the availability of a site for a guest to enter, the host must provide a conduction band of suitable energy to take up

Table 24.5 Some alkali metal intercalation compounds of chalcogenides

Compound	Δ/pm*
$K_{1.0}ZrS_2$	160
$Na_{1.0}TaS_2$	117
$K_{1.0}TiS_2$	192
$Na_{0.6}MoS_2$	135
$K_{0.4}MoS_2$	214
$Rb_{0.3}MoS_2$	245
$Cs_{0.3}MoS_2$	366

* The change in interlayer spacing as compared with the parent MS_2 phase.

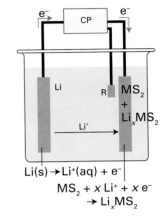

Figure 24.41 Schematic experimental arrangement for electrointercalation. A polar organic solvent (such as polypropene carbonate) containing an anhydrous lithium salt is used as an electrolyte. R is a reference electrode and CP is a coulometer (to measure the charge passed) and a potential controller.

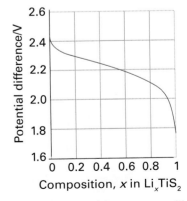

Figure 24.42 Potential versus composition diagram for the electrointercalation of lithium into titanium disulfide. The composition, x, in Li_xTiS_2 is calculated from the charge passed in the course of electrointercalation.

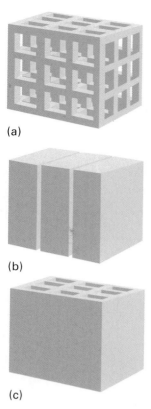

(a)

(b)

(c)

Figure 24.43 Schematic representation of host materials for intercalation reactions. (a) A three-dimensional host with intersecting channels. (b) a two-dimensional layered compound, and (c) a host containing one-dimensional channels.

Table 24.6 Some three-dimensional intercalation compounds

Phase	Composition, x
$Li_x[Mo_6S_8]$	$0.65-2.4$
$Na_x[Mo_6S_8]$	3.6
$Ni_x[Mo_6Se_8]$	1.8
H_xWO_3	$0-0.6$
H_xReO_3	$0-1.36$

electrons reversibly (or, in some cases, be able to donate electrons to the host). Table 24.6 illustrates that a wide variety of hosts are possible, including metal oxides and various ternary and quaternary compounds. We see that intercalation chemistry is by no means limited to graphite and layered disulfides.

24.11 Chevrel phases and chalcogenide thermoelectrics

Key point: A Chevrel phase has a formula such as Mo_6X_8 or $A_xMo_6S_8$, where Se or Te may take the place of S and the intercalated A atom may be a variety of metals such as Li, Mn, Fe, Cd, and Pb.

We close this section on sulfide materials with a brief discussion of an interesting class of ternary compounds first reported by R. Chevrel in 1971. These compounds, which illustrate three-dimensional intercalation, have formulas such as Mo_6X_8 and $A_xMo_6S_8$; Se or Te may take the place of S and the intercalated A atom may be a variety of metals such as Li, Mn, Fe, Cd, or Pb. The parent compounds Mo_6Se_8 and Mo_6Te_8 are prepared by heating the elements at about 1000°C. A structural unit common to this series is M_6S_8, which may be viewed as an octahedron of M atoms face-bridged by S atoms, or alternatively as an octahedron of M atoms in a cube of S atoms (Fig. 24.44). This type of cluster is also observed for some halides of the Periods 4 and 5 early d-block elements, such as the $[M_6X_8]^{4-}$ cluster found in Mo and W dichlorides, bromides, and iodides.

Figure 24.45 shows that in the three-dimensional solid the Mo_6S_8 clusters are tilted relative to each other and relative to the sites occupied by intercalated ions. This tilting allows a secondary donor–acceptor interaction between vacant Mo $4d_{z^2}$ orbitals (which project outward from the faces of the Mo_6S_8 cube) and a filled donor orbital on the S atoms of adjacent clusters.

One of the physical properties that has drawn attention to the Chevrel phases is their superconductivity. Superconductivity persists up to 14 K in $PbMo_6S_8$, and it also persists to very high magnetic fields, which is of considerable practical interest because many applications involve high fields (over 25 T), for example, for the next generation of NMR instruments. In this respect the Chevrel phases appear to be significantly superior to the newer oxocuprate high-temperature superconductors.

Further possible application of Chevrel phases is in the thermoelectric devices used to convert heat into electrical energy or in devices that use electrical energy directly for cooling purposes. The ideal thermoelectric materials have good electrical conductivity and with low thermal conductivities. This requirement often involves designing a material with a combination of structural elements: one that allows for rapid electron transport, a property associated with crystalline solids where there is little electron scattering by the regular placement of atoms, and a second disordered or glassy structural feature, which scatters the vibrational modes responsible for heat transport, thus producing low thermal conductivity. In Chevrel phases the ability to incorporate various cations between the M_6X_8 blocks that 'rattle' around on their sites reduces the material's thermal conductivity but not at the expense of electronic conductivity.

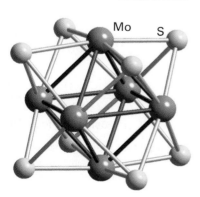

Figure 24.44 The Mo_6S_8 unit present in a Chevrel phase $Pb_xMo_6S_8$.

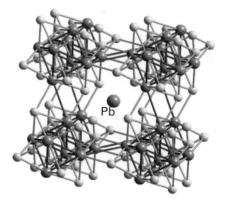

Figure 24.45 The structure of a Chevrel phase showing the canted Mo_6S_8 units forming a slightly distorted cube around a Pb atom. An Mo atom in one cube can act as the acceptor for an electron pair donated by an S atom in a neighbouring cage.

Many other metal chalcogenide phases are being investigated for thermoelectric device applications. The compounds Bi_2Te_3 and Bi_2Se_3, in common with many metal chalcogenides, have layer-like structures similar to those of the MS_2 phases described in Section 24.10b, and devices built on these materials can be designed to produce good electrical conductivity within the layers but poor thermal conductivity perpendicular to them, leading to useful thermoelectric efficiencies. Another important family of thermoelectric materials are the so-called 'skutterudites' (named after the mineral *skutterudite*, $CoAs_3$) with the general formula $M_x(Co,Fe,Ni)(P,As,Sb)_3$ with various inserted cations, M, such as Ln or Na, as in $Na_{0.25}FeSb_3$. The skutterudite structure is similar to that of ReO_3 (Section 24.7) although the $CoAs_6$ octahedra are tilted to produce some large cavities within the structure into which the cations may be inserted. These cations reduce the thermal conductivity of the structure by absorbing heat to rattle around within their cavities, while the high electronic conductivity of the $(Co,Fe,Ni)(P,As,Sb)_3$ network is maintained.

Framework structures

Much of this chapter has concerned structures derived from close-packed anions accompanying d-metal ions typically with coordination number 6. Many of these structures (for example, ReO_3, perovskites, and MS_2) can also be described in terms of linked polyhedra in which MX_6 octahedra connect through their vertices or edges to form a variety of arrays. Coordination number 6 is preferred for most d metals in their typical oxidation states but for smaller metal species (for example, the later 3d series and the lighter p-block metals and metalloids such as Al and Si), fourfold tetrahedral coordination to O is commonplace and leads to structures that are best described as based on linked MO_4 tetrahedra. These tetrahedral units may be the only building block present, as in zeolites, or may link together with metal–oxygen octahedra to produce new structural types (Fig. 24.46). Many of these linked polyhedral structures are known as **framework structures**.

24.12 Structures based on tetrahedral oxoanions

As remarked above, the elements able to form very stable tetrahedral MO_4 species that can link together into framework structures are the later 3d series and the lighter p-block

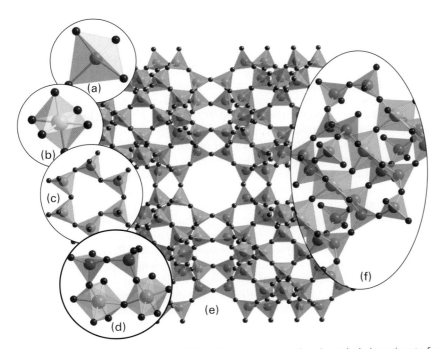

Figure 24.46 Tetrahedral (a) and octahedral (b) units are linked together through their vertices to form larger units, called secondary building units such as (c) and (d), which in turn bridge through their vertices to form framework structures such as (e) and (f).

metals and metalloids. These ions are so small that they coordinate strongly to four O atoms in preference to higher coordination numbers; the principal examples are SiO_4, AlO_4, and PO_4, although GaO_4, GeO_4, AsO_4, BO_4, BeO_4, LiO_4, $Co(II)O_4$, and ZnO_4 are all well known in these structural types. Other tetrahedral units that have been found only rarely in framework structures include $Ni(II)O_4$, $Cu(II)O_4$, and InO_4. We concentrate on the structures derived from the most commonly found units in the zeolites (framework aluminosilicates with large pores), aluminophosphates, and phosphates.

In zeolites synthesized from solution, where all the vertices of the SiO_4 and AlO_4 building units are shared between two tetrahedra, **Lowenstein's rule** states that no O atom is shared between two AlO_4 tetrahedra. Lowenstein's rule applies only to frameworks synthesized from solution, as directly linked AlO_4 tetrahedra are readily obtained in compounds generated from reactions at high temperatures. This indicates that in aqueous solutions the bridging oxygen in a $[O_3Al-O-SiO_3]$ unit is stabilized relative to one in a $[O_3Al-O-AlO_3]$ unit, presumably due to the higher formal charge on Si.

> ■ **A brief illustration.** To determine the maximum Al:Si ratio in a zeolite we note that the maximum Al content is reached when alternate tetrahedra in the structure are those of AlO_4 and each vertex is linked to four SiO_4 units (and, similarly, each SiO_4 tetrahedron is surrounded by four AlO_4 units). Thus the highest ratio achievable is 1:1. ■

(a) Contemporary zeolite chemistry

Key points: New zeolite framework structures are synthesized by using complex template molecules; important applications of zeolites include gas absorption and ion exchange.

The role of structure-directing agents in the synthesis of zeolitic materials was described in Section 14.15. Following the discovery of a large number of new microporous structures from the 1950s onwards, many of which were prepared in the laboratory using organic templates, more recent work on zeolites has been directed towards the systematic study of template–framework relationships. These studies can be divided into two major categories. One is to understand the interaction between the framework and template through computer modelling and experiment (Section 26.14). The other is to design templates with specific geometries to direct the formation of zeolites with particular pore sizes and connectivity.

One particular area that has become the focus of much attention is the use of bulky organic and organometallic molecules as templates in the quest for new, very large pore structures. This approach has been used to make the first zeolitic materials containing 14-ring channels.[5] Thus, the microporous silica UTD-1 (Fig. 24.47a) has been prepared by using a permethylated bis-cyclopentadienyl cobalt metallocene and the siliceous CIT-5 (CFI structure type) was prepared by using a polycyclic amine and lithium (Fig. 24.47b). Other more complex amines have been synthesized with the aim of using them as templates in zeolite synthesis and to obtain desirable pore geometries. The addition of fluorides into the zeolite precursor gel improves reaction rates and acts as a template for some of the smaller cage units, for example where linked TO_4 (T = Si, Al, P, etc.) tetrahedra arranged at the corners of a cube surround a central F^- ion.

As part of the overall growth in the number of known synthetic zeolites, there has been a significant increase in the proportion that can be made in (essentially) pure silica form so that over 20 structural types of zeolitic silica polymorphs are now known. The use of low H_2O/SiO_2 ratios is a key factor for producing these materials and the new 'silica' phases so produced are of unusually low density, for example a purely siliceous framework with the same topology as the naturally occurring mineral chabazite, $Ca_{1.85}(Al_{3.7}Si_{8.3}O_{24})$, is the least dense silica polymorph known, with—according to the normal atomic radii—only 46 per cent of the unit cell volume occupied. These pure silica zeolites have no overall framework charge and no extra-framework cations; as a result they are hydrophobic, leading to specific applications in molecular absorption of low polarity molecules and catalysis (Section 26.14).

The experimental determination of the location of template molecules by X-ray and neutron diffraction has proved important for establishing template–framework relationships. Examples of such work include establishing the location of the templating ions in

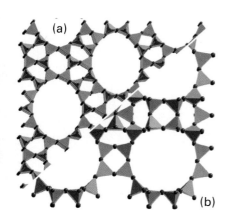

Figure 24.47 Representations as linked tetrahedra of synthesized zeolites showing the main channels in (a) UTD-1 and (b) CIT-5.

[5] An n-ring channel (in this case $n = 14$) refers to the number of tetrahedral units (MO_4) linked together to define the circumference of the channel: the larger n is, the greater is the diameter of the channel.

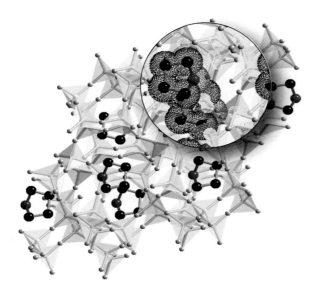

Figure 24.48 Calculated and experimental position of a pyrrolidinium cation in a zeolite cavity. The inset shows how the shape of the cavities replicates the shape of the templating amine cation.

fluoride–silicalite at the channel intersections (Fig. 24.48). These approaches to the determination of structure are often used in association with computer modelling (Chapter 8).

Much of the attention directed at new zeolites focuses on obtaining ever larger pore sizes or a particular pore geometry, with the eventual aim of improving their catalytic properties (Section 26.14). For example, larger pore sizes would allow larger, more complex organic molecules to undergo transformations inside the zeolite cavity. However, we should also consider the other main applications of zeolites, which are as absorbents for small molecules and as ion exchangers. Zeolites are excellent absorbents for most small molecules such as H_2O, NH_3, H_2S, NO_2, SO_2, and CO_2, linear and branched hydrocarbons, aromatic hydrocarbons, alcohols, and ketones in the gas or liquid phase. Zeolites with different sized pores may be used to separate mixtures of molecules based on size, and this application has led to their description as **molecular sieves**. Through the correct selection of zeolite pore, it is possible to control the rates of diffusion of various molecules with different effective diameters, leading to separation and purification. Figure 24.49 illustrates this application schematically.

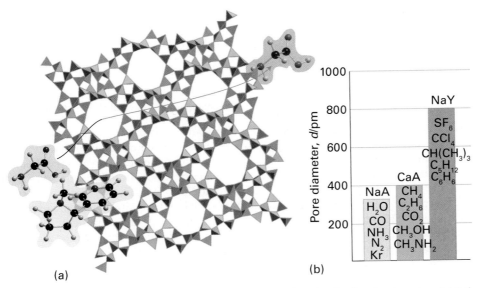

(a)

(b)

Figure 24.49 The use of porous zeolite structures for the separation of molecules of various sizes. (a) Only the smaller molecule can diffuse into and then out of the zeolite pore so a membrane of this material may be used to separate this mixture. (b) The maximum molecular size that can be absorbed into various zeolite channel diameters. NaY is faujasite, a moderately large pore zeolite.

Industrial applications of zeolites for separation and purification include petroleum refining processes, where they are used to remove water, CO_2, chlorides, and mercury, the desulfurization of natural gas (where the removal of H_2S and other sulfurous compounds protects transmission pipelines and removes the undesirable smell from home supplies), the removal of H_2O and CO_2 from air before liquefaction and separation by cryogenic distillation, and the drying and the removal of odours from pharmaceutical products. One area of growing importance is the separation of air into its main components other than by cryogenic means. Many dehydrated zeolites adsorb N_2 more strongly into their pores than O_2 and this is believed to be due to a stronger interaction of the quadrupolar nitrogen (^{14}N, $I = 1$) compared to oxygen (^{16}O, $I = 0$) with the cations in the zeolite pores. By passing air over a bed of zeolite at controlled pressure it is possible to produce oxygen of over 95 per cent purity. Zeolites such as NaX (the sodium-exchanged form of an aluminium-rich zeolite of structure type X) and CaA, a calcium-exchanged form of zeolite framework structure type A, also known as LTA, were originally developed for this purpose. The selectivity of the process can be much improved by exchanging selected cations into the zeolite pores and thus changing the dimensions of the sites onto which the nitrogen and oxygen molecules adsorb. Thus lithium-, calcium-, strontium-, and magnesium-exchanged forms of the faujasite (FAU, type X) and Linde Type A (LTA) structures are now used very effectively in this process.

The excellent ion-exchange properties of zeolites result from their open structures and ability to trap significant quantities of cations selectively within these pores. High capacities for ion exchange are derived from the large numbers of exchangeable cations. This is especially true of those zeolites with a high proportion of Al in the framework, which includes the zeolites with the LTA and gismondine (GIS) topologies that have the highest attainable Si:Al ratio of 1:1. The ion-exchange selectivity has led to a major application of zeolites as a 'builder' in laundry detergents, where they are used to remove the 'hard' ions Ca^{2+} and Mg^{2+} and replace them with 'soft' ions such as Na^+. Phosphates, which may also be used as detergent builders, have been the subject of environmental concerns linked to rapid algal growth and eutrophication of natural waters (Section 5.15). Spherical Na-form LTA zeolite particles, a few micrometres in diameter, are small enough to pass through the openings in the weave of clothing and so may be added to detergents to remove the hard cations in natural waters, and are then washed away harmlessly into the environment. A different zeolite framework, zeolite P with the GIS framework topology, has been developed as an alternative builder, with improved properties associated with its extremely high selectivity for Ca^{2+} ions.

Another important area where the ion-exchange properties of zeolites are exploited is in the trapping and removal of radionuclides from nuclear waste. Several zeolites, including the widely used clinoptilolite, have high selectivities for the larger alkali metal and alkaline earth metal cations, which in nuclear waste include ^{137}Cs and ^{90}Sr (Fig. 24.50). These zeolites may be vitrified by further reaction with glass-forming oxides, as discussed in Section 24.8.

(b) Aluminophosphates

Key point: The structures and physical properties of aluminophosphates parallel those of zeolites.

The structural and electronic equivalence of two silicate tetrahedra $(SiO_4)_2$ and the aluminophosphate unit (AlO_4PO_4) can be recognized in the simple compounds SiO_2 and $AlPO_4$, both of which adopt a similar range of dense polymorphs, including the quartz structure. The development of zeolites with very high Si content, which are effectively silica polymorphs, in turn led to the discovery of the aluminophosphate (ALPO) framework structures based on a 1:1 mixture of AlO_4 and PO_4 tetrahedra; $AlPO_4$ itself has the same structure as quartz (SiO_2) but with an arrangement of alternating Al and Si atoms at the centres of the tetrahedra. A wide range of ALPOs has been developed that parallel the zeolites in, for example, their synthesis under hydrothermal conditions (although in acid conditions rather than the basic ones used for zeolites) and their adsorption and catalytic properties. Again, organic template molecules have been designed and used to prepare many different ALPO structures, such as those with the structure codes VPI (with a large channel formed from 18 AlO_4/PO_4 tetrahedra), DAF, CIT, and STA (Fig. 24.51). Although aluminophosphate frameworks are neutral, substitution of either Al(III) or P(V) with metal ions of lower charge leads to the formation of 'solid acid catalysts' that can, for example, convert

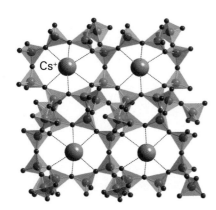

Figure 24.50 The clinoptilolite structure highlighting the relationship between the trapped Cs^+ ions, shown as spheres with ionic radius 180 pm, and the framework.

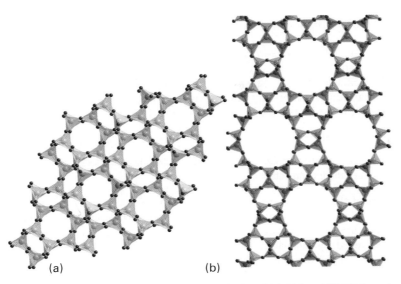

Figure 24.51 The frameworks, formed from linked oxotetrahedra, of (a) DAF and (b) CIT. In each case the main channels present are emphasized by viewing the structure along them.

methanol into hydrocarbons selectively, although the aluminosilicate zeolites, particularly ZSM-5, remain the best materials for this application (Section 26.14). The incorporation of Co and Mn, both of which have redox properties, into ALPO frameworks yields materials that can be used for the oxidation of alkanes.

(c) Phosphates and silicates

Key point: Calcium hydrogenphosphates are inorganic materials used in bone formation.

Another tetrahedral oxoanion that is frequently incorporated into framework materials is the phosphate group, PO_4^{3-}, although, as we shall see in the next section, many other tetrahedral units also form such structures.

Simple phosphate structures are described in Section 15.15 and are generally formed from linked PO_4 tetrahedra in chains, cross-linked chains, and cyclic units. We consider just one metal phosphate material in more detail here, namely calcium hydrogenphosphate, and closely related materials. The principal mineral present in bone and teeth is hydroxyapatite, $Ca_5(OH)(PO_4)_3$, the structure of which consists of Ca^{2+} ions coordinated by PO_4^{3-} and OH^- groups to produce a rigid three-dimensional structure (Fig. 24.52). The mineral *apatite* is the partially fluoride-substituted $Ca_5(OH,F)(PO_4)_3$. Related biominerals are $Ca_8H_2(PO_4)_6$ and amorphous forms of calcium phosphate itself. Biominerals are discussed in more detail in Sections 25.10–12 and Chapter 27.

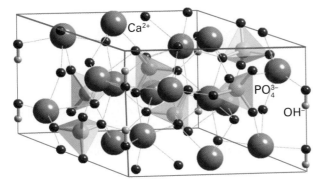

Figure 24.52 The structure of hydroxyapatite, $Ca_5(OH)(PO_4)_3$, shown as Ca^{2+} ions coordinated by phosphate and OH^- ions into a strong three-dimensional structure. The separate O atoms are actually OH^- ions with the H lying outside the unit cell.

24.13 Structures based on octahedra and tetrahedra

Many metals adopt MO_6 octahedral coordination in their oxo compounds. This polyhedral structural unit can be regarded as being formed by locating metal ions in the octahedral holes of a close-packed O^{2-} ion array in, for example, MgO with the rock-salt structure. The MO_6 polyhedral building unit may also be incorporated into framework-type structures, often in combination with tetrahedral oxo species.

(a) Clays, pillared clays, and layered double hydroxides

Key point: Sheet-like structures, found in many metal hydroxides and clays, can be constructed from linked metal oxo tetrahedra and octahedra.

The diameters of the largest pores found in synthetic zeolites are of the order of 1.2 nm. In an attempt to increase this diameter and allow larger molecules to be absorbed into inorganic structures, chemists have turned to mesoporous materials (Section 25.9) and structures produced by 'pillaring' (that is, stacking and connecting together two-dimensional materials). The two-dimensional nature of many d-metal disulfides and some of their intercalation compounds are discussed in Sections 19.9 and 24.10. Similar intercalation reactions when applied to aluminosilicates from the clay family allow the synthesis of large-pore materials.

Naturally occurring clays such as kaolinite, hectorite and montmorillonite have layer structures like those shown in Fig. 24.53. The layers are constructed from vertex- and edge-sharing octahedra, MO_6, and tetrahedra, TO_4, and exist as double layer systems (two layers, one formed from octahedra and one from tetrahedra, Fig. 24.53), as in kaolinite, and triple layer (a central layer based on octahedra sandwiched between two tetrahedra-based layers) systems, as in bentonite. The metal atoms, M and T, contained within the layers, which have an overall negative charge, are typically Si (on tetrahedral sites) and Al (occupying octahedral and tetrahedral sites). Small singly and doubly charged ions, such as Li^+ and Mg^{2+}, occupy sites between the layers. These interlayer cations are often hydrated and can readily be replaced by ion exchange. Other materials with similar structures are the layered double hydroxides with structures similar to that of $Mg(OH)_2$, the naturally occurring mineral *brucite*.

In the pillaring of clays, the species exchanged into the interlayer region is selected for size. Ions such as alkylammonium ions and polynuclear hydroxometal ions may replace the alkali metal, as shown schematically in Fig. 24.54. The most widely used pillaring species are of the polynuclear hydroxide type and include $Al_{13}O_4(OH)_{28}^{3+}$, $Zr_4(OH)_{16-n}^{n+}$, and $Si_8O_{12}(OH)_8$, the first consists of a central AlO_4 tetrahedron surrounded by octahedrally coordinated Al^{3+} ions as $Al(O,OH)_6$ species. The pillaring process can be followed by powder X-ray diffraction because it leads to expansion of the interlayer spacing, corresponding to an increase in the c lattice parameter.

Once an ion such as $Al_{13}O_4(OH)_{28}^{3+}$ has been incorporated between the layers, heating the modified clay results in its dehydration and the linking of the ion to the layers (Fig. 24.54). The resulting product is a pillared clay with excellent thermal stability to at least 500°C. The expanded interlayer region can now absorb large molecules in the same way as zeolites. However, because the distribution of pillaring ions between the layers is difficult to control, the pillared clay structures are less regular than zeolites. Despite this lack of uniformity, pillared clays have been widely studied for their potential as catalysts because they act in a similar way to zeolites, as acid catalysts promoting isomerization and dehydration.

(b) Advances in inorganic framework chemistry

Key point: Enormous structural diversity can be obtained from linked polyhedra and, with the use of templates, can lead to remarkable porous frameworks.

The extensive development of aluminosilicates and zeolites has motivated synthetic inorganic chemists to seek similar structural types built from other tetrahedral and octahedral polyhedra for use in ion exchange, absorption, and catalysis. The use of different and larger polyhedra provides more flexibility in the framework topologies, and there is also the potential to incorporate d-metal ions with their associated properties of colour,

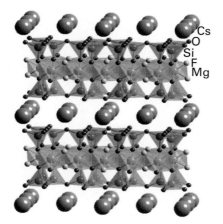

Cs
O
Si
F
Mg

Figure 24.53 Sheet-like structures of the clay hectorite which consist of layers of linked octahedra and tetrahedra centred on, typically, Al, Si, or Mg and separated by cations such as K^+ or Cs^+.

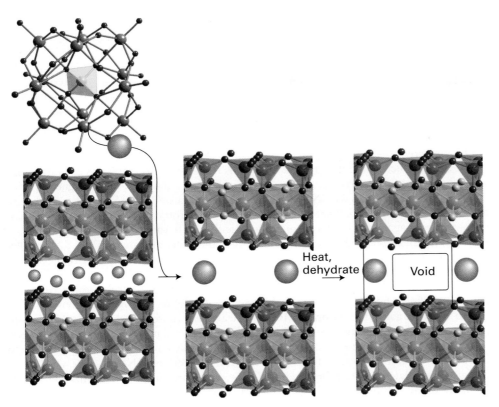

Figure 24.54 Schematic representation of pillaring of a clay by ion exchange of a simple monatomic interlayer cation with large polynuclear hydroxometallate followed by dehydration and cross-linking of the layers to form the cavities.

redox properties, and magnetism. Tetrahedral species that have been incorporated into such frameworks, as well as the aluminate, silicate, and phosphate groups mentioned previously, include ZnO_4, AsO_4, CoO_4, GaO_4, and GeO_4. Octahedral units are mainly based on d metals and the heavier and larger metals from Groups 13 and 14. Other polyhedral units such as five-coordinate square pyramids also occur, but less commonly.

Zeolite analogues are called **zeotypes**. Ring sizes larger than 12 tetrahedrally coordinated atoms were first seen in metallophosphate systems, and structures with 20 and 24 linked tetrahedra have been made. Aluminophosphates were the first microporous frameworks synthesized that contain polyhedra with coordination numbers greater than 4. Other so-called **hypertetrahedral frameworks** are now well established, such as the titanosilicate families (which have four-coordinate Si and five- and six-coordinate Ti sites) and a series of octahedral molecular sieves based on linked MnO_6 units.

Good examples of this structural family are the titanosilicate zeotypes built from SiO_4 tetrahedra and various TiO_n polyhedra with $n = 4$–6. These compounds are made under the hydrothermal conditions similar to those used for synthesizing many zeolites but by using a source of Ti, such as $TiCl_4$ or $Ti(OC_2H_5)_4$, that hydrolyses under the basic conditions in the autoclave. Templates may also be used in such media and act as structure-directing units giving rise to particular pore sizes and geometries. In a typical reaction the titanosilicate ETS-10 (Engelhard TitanoSilicate 10) is prepared by the reaction of $TiCl_4$, sodium silicate, sodium hydroxide, and sometimes a template such as tetraethylammonium bromide, in a sealed polytetrafluorethylene-lined autoclave at between 150 and 230°C. The number of titanosilicates continues to grow, but two materials, ETS-10 and $Na_2Ti_2O_3SiO_4\cdot 2H_2O$, are worthy of further consideration here as specific examples of this type of material.

ETS-10 is a microporous material built from TiO_6 octahedra and SiO_4 tetrahedra, with the TiO_6 groups linked together in chains (Fig. 24.55a). The structure has 12-membered rings (that is, pores formed from 12-unit polyhedra) in all three directions.

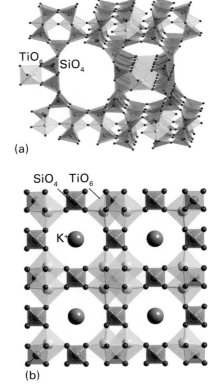

(a)

(b)

Figure 24.55 The structures of (a) the titanosilicate ETS-10, which is built from chains of TiO_6 octahedra linked by SiO_4 tetrahedra, and (b) $K_3H(TiO)_4(SiO_4)_3\cdot 4H_2O$, a synthetic analogue of a natural mineral pharmacosiderite.

NiO$_6$ —— PO$_4$

Figure 24.56 The framework structure of VSB-1 consists of linked NiO$_6$ octahedra and PO$_4$ tetrahedra. The main channel is bounded by 24 such units and has a diameter such that molecules up to 0.88 nm in diameter may pass through it.

Na$_2$Ti$_2$O$_3$SiO$_4$.2H$_2$O also has TiO$_6$ octahedra, but in this structure these octahedra form clusters with four TiO$_6$ units linked by the tetrahedral SiO$_4$ units (as in Fig. 24.55a). This connectivity gives rise to large octagonal pores containing hydrated Na$^+$ ions. This compound, in common with several other titanosilicates such as K$_3$H(TiO)$_4$(SiO$_4$)$_3$.4H$_2$O, a synthetic analogue of a natural mineral pharmacosiderite (Fig. 24.55b), shows excellent ion-exchange properties, particularly with large cations. These large ions replace Na$^+$ with very high selectivity, so that, for example, Cs$^+$ and Sr^{2+} ions may be extracted into the titanosilicate structure from their dilute solutions. This ability has led to the development of these materials for removal of radionuclides from nuclear waste, in which ^{137}Cs and ^{90}Sr are highly active.

The aim of much of the recent work on porous framework structures has been to incorporate d-metal ions. Materials that have been synthesized include antiferromagnetic porous iron(III) fluorophosphates with Néel temperatures in the range 10–40 K. These temperatures are relatively high for iron clusters linked by phosphate groups and indicate the presence of moderately strong magnetic interactions. Zeotypic cobalt(II), vanadium(V), titanium(IV), and nickel(II) phosphates have been prepared, including the material known as VSB-1 (Versailles-Santa Barbara), which was the first microporous solid with 24-membered ring tunnels, and is simultaneously porous, magnetic, and an ion-exchange medium (Fig. 24.56).

Two other rapidly growing areas of research into framework structures are materials based on sulfides and on the use of small organic molecules to link the metal ions, so giving rise to **metal–organic frameworks** (MOFs). Tetrahedral coordination is common in simple metal sulfide chemistry, as we saw in the discussion of ZnS in wurtzite and sphalerite (Section 3.9), and some compounds may be considered as containing linked fragments of these structures known as **supertetrahedral clusters**. For example, [N(CH$_3$)$_4$]$_4$[Zn$_{10}$S$_4$(SPh)$_{16}$] contains the isolated supertetrahedral unit Zn$_{10}$S$_{20}$ terminated by phenyl groups (Fig. 24.57).

Metal–organic frameworks have structures that are based on bidentate or polydentate organic ligands lying between the metal atoms. Examples of simple ligands used to build these often porous frameworks are CN$^-$, nitriles, amines, and carboxylates. Many hundreds of these compounds have now been made, and include positively charged frameworks with balancing anions in the cavities, for instance Ag(4,4'-bpy)NO$_3$, and electrically neutral frameworks, such as Zn$_2$(1,3,5-benzenetricarboxylate)NO$_3$.H$_2$O. C$_2$H$_5$OH, from which the H$_2$O molecules may be removed reversibly. The compounds have properties analogous to those of zeolites and of note is the framework of a chromium terephthalate (1,4-benzenedicarboxylate) which has pore diameters of about 3 nm and a specific internal surface area for N$_2$ of over 5000 m^2 g^{-1} (Fig. 24.58).

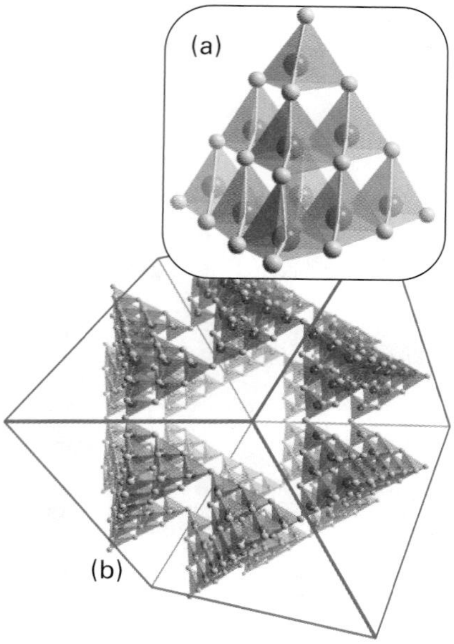

(a)

(b)

Figure 24.57 (a) An isolated supertetrahedral unit of stoichiometry Zn$_{10}$S$_{20}$ formed from individual ZnS$_4$ tetrahedra. (b) These supertetrahedral units may be linked together as building blocks to form large porous three-dimensional structures.

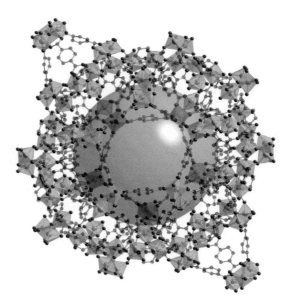

Figure 24.58 A metal-organic framework (MOF) formed from chromium terephthalate units showing CrO$_6$ octahedra (gold) linked by the organic anions, The large central sphere shows the extent of the pore in this material, which can be filled with solvent or gas molecules.

These materials have potential applications in gas storage, for example of H_2 and CO_2. The very large pores allow large, complex molecules such as many pharmaceutically active compounds to be inserted into them and a further application is the use of these MOFs as drug delivery agents. Here an active compound, for example the painkiller ibuprofen, is adsorbed within the MOF pore from which it can then be released slowly, where required, in the human body.

Hydrides and hydrogen-storage materials

The development of materials for use in a future hydrogen-based energy economy is a key challenge facing inorganic chemists. One of the main areas where new materials are needed is in hydrogen storage. High-pressure gas cylinders and liquification routes are unlikely to meet the requirements of some hydrogen-storage applications in, for example, transport or distributed power supplies because of their weight and safety issues. New materials for storing hydrogen will be required to have high capacities, both volumetrically and gravimetrically, while also being relatively cheap. Technical targets that have been proposed for transportation applications are materials that have 6–9 per cent of the system mass as hydrogen and cost a few dollars per kilowatt hour of stored energy. A further requirement is that the hydrogen should become available from the storage system at between 60 and 120°C. Two main approaches to these new materials are being pursued: chemically bound hydrogen in, for example, metal hydrides and new porous or high-surface-area compounds that physisorb hydrogen. In this section the inorganic chemistry behind these two approaches is discussed, together with a description of the latest advances in the materials that have been produced.

24.14 Metal hydrides

Key point: Many metal hydrides and complex metal hydrides such as alanates (hydridoaluminates), amides, and borohydrides release H_2 when heated and are potential materials for hydrogen storage.

Hydrogen forms metal hydrides by reaction with many metals and metal alloys (Section 10.6). In many cases this process can be reversed, liberating H_2 and regenerating the metal or alloy, by heating the (complex) metal hydride. These metal hydrides have an important safety advantage over pressurized or liquefied hydrogen (and many physisorbed hydrogen systems), that are also being considered for hydrogen storage, where rapid uncontrolled release of hydrogen could be a concern (Boxes 10.4, 13.3). For hydrogen storage applications compounds that contain hydrogen in combination with the light elements (such as Li, Be, Na, Mg, B, and Al) offer some of the most promising materials, particularly as H:M ratios can reach 2 or more. Simple metal hydride systems have been dealt with in the appropriate descriptive chemistry section of Chapters 10, 11 and 12, including discussion of the Group 1 and Group 2 hydrides. Here we expand on the chemistry of these compounds and the related complex metal hydrides.

(a) Magnesium-based metal hydrides

Magnesium hydride, MgH_2 (rutile structure type, Fig. 24.59) potentially offers a high capacity of 7.7 per cent hydrogen by mass combined with the low cost of magnesium and good reversibility for hydrogen uptake and evolution at high temperatures. However, its decomposition temperature of 300°C under 1 atm of hydrogen gas is uneconomically high and considerable effort has been expended to develop new related materials that have lower hydrogen release temperatures. Doping of MgH_2 with a variety of metals has been undertaken to produce the solid solutions $Mg_{1-x}M_xH_{2\pm y}$ for M = Al, V, Ni, Co, Ti, Ge, and La/Ni mixtures, some of which have slightly lower decomposition temperatures. More promising are changes in the microstructure of MgH_2 which are caused by ball-milling (where the solid is shaken vigorously with small hard spheres inside a canister), particularly when done in combination with other materials, such as metal (Ni, Pd) and metal oxides (V_2O_5, Cr_2O_3). Ball-milling of MgH_2 increases the surface area of the solid by about a factor of 10 and the number of defects in the crystallites, promoting both the adsorption and de-adsorption of hydrogen.

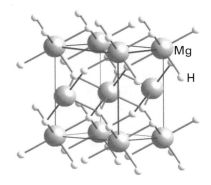

Figure 24.59 The structure of MgH_2.

(b) Complex hydrides

Complex hydrides include the tetrahydridoaluminates (containing AlH_4^-), amides, NH_2^-, and boranates (hydridoborates containing BH_4^-), which can contain very high levels of hydrogen, for example $LiBH_4$ is 18 per cent hydrogen by mass. For these systems their decomposition to liberate hydrogen is a problem: tetrahydridoborates (borohydrides), for instance, decompose only near 500°C. For some applications that do not require reversibility, the so-called 'one-pass hydrogen storage systems', the hydrogen can be liberated by treatment with water.

> ■ **A brief illustration.** The molar mass of Li_2BeH_4 is $(2 \times 6.94) + 9.01 + (4 \times 1.01)$ g mol^{-1} = 26.93 g mol^{-1} of which 4.04 g mol^{-1} is H, so the hydrogen content is (4.04 g mol^{-1}/26.93 g mol^{-1}) $\times$ 100% = 15.0%. ■

Both $NaAlH_4$ and Na_3AlH_6 have good theoretical hydrogen storage capacities, totalling 7.4 and 5.9 per cent by mass, respectively, and low cost. Their structures are based on the tetrahedral AlH_4^- and octahedral AlH_6^{3-} complex anions (Fig. 24.60). These systems exhibit poor reversibility in their decomposition reactions and proceed in stages, which is not ideal for applications, particularly as the final amounts of hydrogen are evolved only above 400 °C.

$$3\,NaAlH_4(s) \xrightleftharpoons{200°C} Na_3AlH_6(s) + 2\,Al(s) + 3\,H_2(g)$$

yield: 3.6 per cent H_2 by mass

$$2\,Na_3AlH_6(s) \xrightleftharpoons{260°C} 6\,NaH(s) + 2\,Al(s) + 3\,H_2(g)$$

yield: 1.8 per cent H_2 by mass

$$2\,NaH(s) \xrightleftharpoons{425°C} 2\,Na(s) + H_2(g)$$

yield: 2.0 per cent H_2 by mass

Various additives such as Ti and Zr have been used to prepare doped materials which, in some cases, have enhanced the rates of hydrogenation and dehydrogenation. Ball-milling to produce smaller particle sizes and strained materials also seems to improve rates of hydrogen evolution.

The lithium tetrahydridoaluminates $LiAlH_4$ and Li_3AlH_6 have higher mass percentages of hydrogen than the corresponding sodium compounds and would be attractive for hydrogen storage except for their chemical instability; $LiAlH_4$ initially decomposes easily but the resulting products cannot be rehydrogenated back to $LiAlH_4$. Furthermore, LiH, one of the initial products, loses hydrogen only above 680°C.

Lithium nitride, Li_3N, reacts with hydrogen to form $LiNH_2$ and LiH:

$$Li_3N(s) + 2\,H_2(g) \rightarrow LiNH_2(s) + 2\,LiH(s)$$

This mixture evolves hydrogen above 230°C, also producing Li_2NH, and has a theoretical hydrogen storage capacity of 6 per cent H_2 by mass. One problem with nitrides is that partial decomposition to produce ammonia can occur.

Lithium tetrahydridoborate, $LiBH_4$, is theoretically a very promising material for hydrogen storage with 18 per cent H_2 by mass (Box 10.4), but it does not undergo a reversible reaction involving hydrogen evolution and uptake. The compound $Li_3Be_2H_7$ has excellent reversibility but only above 150°C. Other complex lithium amide-boranates, such as $Li_3(NH_2)_2BH_4$ (Fig 24.61) and $Li_4(NH_2)_2(BH_4)_2$, have been proposed as hydrogen-storage materials and seem to offer reduced evolution of ammonia compared with $LiNH_2$ but with only limited reversibility. Related metal amine complexes, for example $Mg(NH_3)_6Cl_2$, have also been proposed as 'indirect' hydrogen storage materials (where evolved ammonia might be converted in a secondary reaction to hydrogen) as they have high hydrogen contents.

(c) Intermetallic compounds

Several types of intermetallic compound are being investigated as potential hydrogen-storage systems as, in general, they have excellent reversible hydrogen uptake at low pressures (1–20 bar) at just above room temperature (Table 24.7). However, the mass percentage of

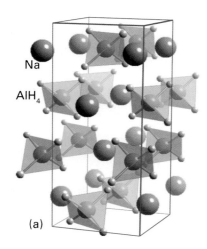

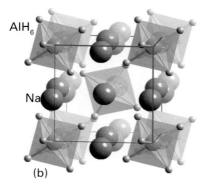

Figure 24.60 The structures of (a) $NaAlH_4$ and (b) Na_3AlH_6 depicting the hydride coordination around aluminium.

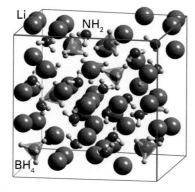

Figure 24.61 The structure of $Li_3(NH_2)_2BH_4$.

Table 24.7 Intermetallic structure types for hydrogen storage*

Type	Metals	Typical hydride composition	Mass % H_2	p_{eq}, T
Element	Pd	$PdH_{0.6}$	0.56	0.020 bar, 298 K
AB_5	$LaNi_5$	$LaNi_5H_6$	1.5	2 bar, 298 K
AB_2 (Laves)	ZrV_2	$ZrV_2H_{5.5}$	3.0	10^{-8} bar, 323 K
AB	FeTi	$FeTiH_2$	1.9	5 bar, 303 K
A_2B	Mg_2Ni	Mg_2NiH_4	3.6	1 bar, 555 K
BCC	TiV_2	TiV_2H_4	2.6	10 bar, 313 K

* The values of p_{eq} and T are the pressure and temperature conditions for phase formation/decomposition.

hydrogen that can be adsorbed by these materials is relatively low due to the high molar masses of the metals. One system, the AB_5 type phases, is typified by $LaNi_5$, which absorbs hydrogen to give $LaNi_5H_6$, but this corresponds to only 1.5 per cent hydrogen by mass.

A series of alloys termed **Laves phases** are adopted by some intermetallics of the stoichiometry AB_2 where A = Ti, Zr, or Ln and B is a 3d metal such as V, Cr, Mn, or Fe. These materials have high capacities and good kinetics for hydrogen adsorption, forming compounds such as $ZrFe_2H_{3.5}$ and $ErFe_2H_5$ with up to 2 per cent hydrogen by mass. However, the hydrides of these Laves phases are thermodynamically very stable at room temperature, restricting the reverse desorption of hydrogen.

The compound Mg_2NiH_4, which contains 3.6 per cent hydrogen by mass, is formed by heating the alloy Mg_2Ni to 300°C under 25 kbar of hydrogen

$$Mg_2Ni(s) + 2H_2(g) \rightarrow Mg_2NiH_4$$

The structure of Mg_2NiH_4 consists, at low temperature, of an ordered array of Mg, Ni, and H but transforms, at high temperature or extended ball-milling, to a cubic phase with H^- ions randomly distributed throughout the arrangement of Mg and Ni atoms (Fig. 24.62).

A third class of intermetallic compounds being studied is the so-called 'Ti-based BCC alloys' such as FeTi and alloys of similar composition with various quantities of other d metals, such as Ti/V/Cr/Mn. Hydrogen capacities of near 2.5 per cent by mass can be achieved for these alloys but high temperatures and pressures are needed to reach these values.

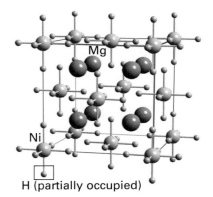

Figure 24.62 The idealized structure of cubic Mg_2NiH_4; hydrogen sites are only partially occupied.

24.15 Other inorganic hydrogen-storage materials

Key point: High surface area and porous inorganic compounds can adsorb high levels of hydrogen gas.

Physisorption on to the surface of very high surface area materials, including any internal pores, offers high hydrogen-storage capacities. However, these high values may be reached only by cooling the system or use of very high pressures, neither of which may be applicable to eventual applications. Some inorganic systems currently being studied include metal alloys that have been templated by using zeolites, inorganic clathrates, and metal-organo frameworks (MOFs, Section 24.13b), all of which have highly porous structures formed from relatively light elements.

Carbon in its various forms is also being studied as a hydrogen-storage material. Graphite itself in a nanostructure form (Section 25.7) adsorbs 7.4 mass per cent hydrogen under 1 MPa (10 bar) of hydrogen, although more typically activated carbon/graphite materials develop surface monolayers equivalent to 1.5 to 2 per cent hydrogen by mass. Carbon nanotubes (Section 25.7), with their curved surfaces, tend to show enhanced adsorption of hydrogen in comparison with graphite sheets, with reported values of up to 8 per cent by mass at 77 K.

Zeolites (Section 24.12) have also been proposed as possible hydrogen-storage materials and many of the different topologies have been studied. The best framework types include faujasites (FAU, zeolites X and Y), zeolite A (LTA), and chabazite, with maximum capacities of between 2.5 and 2.0 per cent hydrogen by mass.

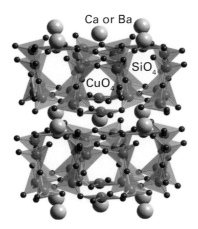

Figure 24.63 The square-planar copper/oxygen environment formed by Si_4O_{10} groups in Egyptian blue ($CaCuSi_4O_{10}$) and Chinese blue ($BaCuSi_4O_{10}$).

Inorganic pigments

Many inorganic solids are intensely coloured and are used as pigments in colouring inks, plastics, glasses, and glazes. Whereas many insoluble organic compounds (for example the C.I. Pigment Red 48, which is calcium 4-((5-chloro-4-methyl-2-sulfophenyl) azo)-3-hydroxy-2-naphthalenecarboxylic acid) are also used as pigments, inorganic materials often have advantages in terms of applications associated with their chemical, light, and thermal stability. Pigments were originally developed from naturally occurring compounds such as hydrated iron oxides, manganese oxides, lead carbonate, vermilion (HgS), orpiment (As_2S_3), and copper carbonates. These compounds were even used in prehistoric cave paintings. Synthetic pigments, which are often analogues of naturally occurring compounds, were developed by some of the earliest chemists and alchemists, and the first synthetic chemists were probably those involved in making pigments. Thus, the pigment Egyptian blue ($CaCuSi_4O_{10}$) was made from sand, calcium carbonate, and copper ores as long as 3000 years ago. This compound and a structural analogue, Chinese blue ($BaCuSi_4O_{10}$), which was first made about 2500 years ago, have a structure containing square-planar copper(II) ions surrounded by Si_4O_{10} groups (Fig. 24.63). Inorganic pigments continue to be important commercial materials and this section summarizes some of the recent advances in this field.

As well as producing the colours of inorganic pigments as a result of the absorption and reflection of visible light, some solids are able to absorb energy of other wavelengths (or types, for example electron beams) and emit light in the visible region. This **luminescence** is responsible for the properties of inorganic phosphors (Box 24.2).

24.16 Coloured solids

Key point: Intense colour in inorganic solids can arise through d−d transitions, charge transfer (and the analogous interband electron transfer), or intervalence charge transfer.

BOX 24.2 Inorganic phosphors

Luminescence is the emission of light by materials that have absorbed energy in some form. Photoluminescence occurs when photons, usually in the ultraviolet region of the electromagnetic spectrum, are the source of energy and the output usually is visible light. **Cathodoluminescence** uses electron beams as a source of energy and **electroluminescence** uses electrical energy as the energy source. Two types of photoluminescence can be distinguished: **fluorescence**, which has a period of less than 10^{-8} s between photon absorption and emission, and **phosphorescence**, for which there are much longer delay times (Section 20.7).

Photoluminescent materials, frequently referred to as **phosphors**, generally consist of a host structure, such as ZnS (wurtzite structure; Section 3.9), $CaWO_4$ (scheelite structure type with discrete WO_4^{2-} tetrahedra separated by Ca^{2+} ions), or Zn_2SiO_4, into which an **activator ion** is introduced by doping. These activator ions are certain d-metal or lanthanoid ions, such as Mn^{2+}, Cu^{2+}, and Eu^{2+}, that have the ability to absorb and emit light of the desired wavelengths. In some cases a second dopant is added as a **sensitizer** to aid the absorption of light of the desired wavelength. Well-known applications of such materials include fluorescent lamps and television screens, where there is a need for materials that fluoresce in specific regions of the visible spectrum.

Many host–activator combinations have been studied as potential phosphor materials with the aim of producing a material that efficiently converts UV radiation or cathode rays (electron beams) to pure emission of a desirable colour. The modification of the host structure and the nature and environment of the activator ion allows these properties to be tuned (see table).

In many fluorescent lamps, a mercury discharge produces UV radiation at 254 and 185 nm. Then a coating of ZnS that has been doped with various activators produces fluorescence at several wavelengths that combine together to give an effective white light.

Phosphor host	Activator	Colour
Zn_2SiO_4	Mn^{2+}	Green
$CaMg(SiO_3)_2$ diopside	Ti	Blue
$CaSiO_3$	Mn	Yellow/orange
$Ca_5(PO_4)_3(F,Cl)$	Mn	Orange
ZnS	Ag^+, Cu^{2+}, Mn^{2+}	Blue, green, yellow

Colour television requires three primary-colour, cathodoluminescent materials to produce a picture. These phosphors are typically ZnS:Ag^+ (blue), ZnS:Cu^+ (green), and YVO_4:Eu^{3+} (red). In YVO_4:Eu^{3+} the vanadate group absorbs the incident electron energy and the activator is Eu^{3+}. The emission mechanism involves electron transfer between the vanadate group and Eu^{3+} and the efficiency of this process depends on the M−O−M bond angle in a similar way to the superexchange process in antiferromagnets (Section 20.8). The nearer this angle is to 180°, the faster and more efficient is the electron transfer process. In YVO_4:Eu this angle is 170° and this material is a highly efficient phosphor.

Anti-Stokes phosphors convert two or more photons of lower energy to one of higher energy (such as infrared radiation to visible light). They act by absorbing two or more photons in the excitation process before emitting one photon. The best anti-Stokes phosphors have ionic host structures such as YF_3, $NaLa(WO_4)_2$, and $NaYF_4$ doped with Yb^{3+} as a sensitizer ion (to absorb the IR radiation) and Er^{3+} as an activator (emitting visible light). Applications include night-vision binoculars.

The blue colour of $CoAl_2O_4$ and $CaCuSi_4O_{10}$ stems from the presence of d−d transitions in the visible region of the electromagnetic spectrum. The characteristic intense colour of cobalt aluminate is a result of having a noncentrosymmetric tetrahedral site for the metal ion, which removes the constraint of the Laporte selection rule of octahedral environments (Section 20.6). The chemical and thermal stabilities are due to the location of the Co^{2+} ion in the close-packed oxide arrangement. Other inorganic pigments with colours based on d−d transitions include Ni-doped TiO_2 (yellow) and $Cr_2O_3.nH_2O$ (green).

Colour also arises in many inorganic compounds from charge transfer (Section 20.5) or what is often electronically an equivalent process in solids, the promotion of an electron from a valence band (derived mainly from anion orbitals) into a conduction band (derived mainly from metal orbitals). Charge-transfer pigments include compounds such as lead chromate ($PbCrO_4$), containing the yellow–orange chromate(VI) anion, and $BiVO_4$ with the yellow vanadate(V) anion. The compounds CdS (yellow) and CdSe (red) both adopt the wurtzite structure and their colour arises from transitions from the filled valence band (which is mainly derived from chalcogenide p orbitals) to orbitals based mainly on Cd. For a material with a band gap of 2.4 eV (as for CdS at 300 K) these transitions occur as a broad absorption corresponding to wavelengths shorter than 515 nm, thus CdS is bright yellow as the blue part of the visible spectrum is fully absorbed. For CdSe the band gap is smaller on account of the higher energies of the Se 4p orbitals and the absorption edge shifts to lower energies. As a result, only red light is not absorbed by the material. In some mixed-valence compounds, electron transfer between differently charged metal centres can also occur in the visible region, and as these transfers are often fully allowed they give rise to intense colour. Prussian blue, $[Fe^{(III)}]_4[Fe^{(II)}(CN)_6]_3$ (Fig. 24.64), is one such compound and its dark blue colour has resulted in its widespread use in inks. Intensely coloured Ru compounds, such as the tris(carboxyl)-terpyridine complex $[Ru(2,2',2''-(COOH)_3-terpy)(NCS)_3]$, absorb efficiently right across the visible and near IR regions of the spectrum and are used as photosensitizers in Gräztel-type solar cells.

Inorganic radicals often have fairly low-energy electronic transitions that can occur in the visible region. Two examples are NO_2 (brown) and ClO_2 (yellow). One inorganic pigment is based on an inorganic radical, but because of the high reactivity normally associated with main-group compounds containing unpaired electrons, this species is trapped inside a zeolite cage. Thus the royal-blue pigment ultramarine, a synthetic analogue of the naturally occurring semi-precious stone lapis lazuli, has the idealized formula $Na_8[SiAlO_4]_6.(S_3)_2$ and contains the S_3^- polysulfide radical anion occupying a sodalite cage formed by the aluminosilicate framework (Fig. 24.65).

Current developments in inorganic pigment chemistry are focused on finding replacements for some of the yellow and red materials that contain heavy metals, such as Cd and Pb. Although these materials themselves are not toxic, as the compounds are very stable and the metal is difficult to leach into the environment, their synthesis and disposal can be problematic. Compounds that have been investigated to replace cadmium chalcogenides and lead-based pigments are the lanthanoid sulfides, such as Ce_2S_3 (red), and early d-metal oxide-nitrides, such as $Ca_{0.5}La_{0.5}Ta(O_{1.5}N_{1.5})$ (orange). In both cases, the replacement of O^{2-} by S^{2-} or N^{3-} ion has the effect of narrowing the band gap in these solids (compared to the colourless solids CeO_2 and $Ca_2Ta_2O_7$, which have large band gaps and absorb only in the UV region of the spectrum) and bringing the electron excitation energy into the visible. However, neither of these materials has the stability of the cadmium- and lead-based pigments.

24.17 White and black pigments

Some of the most important compounds used to modify the visual characteristics of polymers and paints have visible-region absorption spectra that result in them appearing either white (ideally no absorption in the visible region) or black (complete absorption between 380 and 800 nm).

(a) White pigments

Key point: Titanium dioxide is used almost universally as a white pigment.

White inorganic materials can also be classified as pigments and vast quantities of these compounds are synthesized for applications such as the production of white plastics and paints. Important commercial compounds of this class that have been used extensively

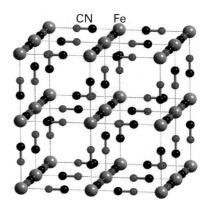

Figure 24.64 Prussian blue, $[Fe^{(III)}]_4[Fe^{(II)}(CN)_6]_3$. The iron ions are linked through cyanide, forming a cubic unit cell.

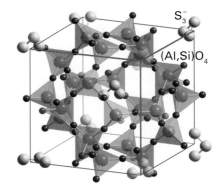

Figure 24.65 One of the sodalite cages present in ultramarine, $Na_8[SiAlO_4]_6.(S_3)_2$. The framework consists of linked SiO_4 and AlO_4 tetrahedra surrounding a cavity containing the polysulfide radical ion S_3^- and Na^+ ions (the latter omitted for clarity).

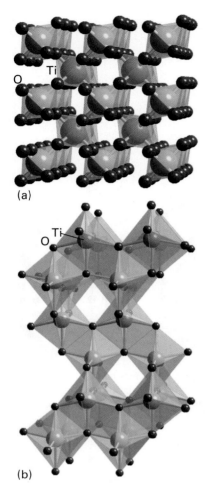

(a)

(b)

Figure 24.66 TiO_2 exists as several polymorphs that can be described in terms of linked TiO_6 octahedra, including the (a) rutile and (b) anatase forms.

historically are TiO_2, ZnO, and ZnS, lead(II) carbonate, and lithopone (a mixture of ZnO and $BaSO_4$); note that none of the metals in these materials has an incomplete d-electron shell that might otherwise induce colour through d−d transitions. Titanium dioxide, TiO_2, in either its rutile or anatase forms (Fig. 24.66), is produced from titanium ores, often ilmenite, $FeTiO_3$, by the *sulfate process* (which involves dissolution in concentrated H_2SO_4 and subsequent precipitation through hydrolysis) or the *chloride process* (which is based on the reaction of mixed complex titanium oxides with chlorine to produce $TiCl_4$, which is then combusted with oxygen at over 1000°C). These routes produce very high quality TiO_2 free from impurities (which is essential for a bright white pigment) of controlled particle size. The desirable qualities of TiO_2 as a white pigment derive from its excellent light scattering power, which in turn is a result of its high refractive index ($n_r = 2.70$), the ability to produce very pure materials of a desired particle size, and its good light-fastness and weather resistance. Uses of titanium dioxide, which nowadays dominates the white-pigment market, include paints, coatings, and printing ink (where it is often used in combination with coloured pigments to increase their brightness and hiding power), plastics, fibres, paper, white cements, and even foodstuffs (where it can be added to icing sugar, sweets, and flour to improve their brightness).

(b) Black, absorbing, and specialist pigments

Key points: Special colour, light absorbing, and interference effects can be induced in inorganic materials used as pigments.

The most important black pigment is *carbon black*, which is a better defined, industrially manufactured form of soot. Carbon black is obtained by partial combustion or pyrolysis (heating in the absence of air) of hydrocarbons. The material has excellent absorption properties right across the visible region of the spectrum and applications include printing inks, paints, plastics, and rubber. Copper(II) chromite, $CuCr_2O_4$, with the spinel structure (Fig. 24.18), is used less frequently as a black pigment. These black pigments also absorb light outside the visible region, including the infrared, which means that they heat up readily on exposure to sunlight. Because this heating can have drawbacks in a number of applications, there is interest in the development of new materials that absorb in the visible region but reflect infrared wavelengths; $Bi_2Mn_4O_{10}$ is one compound that exhibits these properties.

Examples of more specialist inorganic pigments are magnetic pigments based on coloured ferromagnetic compounds such as Fe_3O_4 and CrO_2, and anticorrosive pigments such as zinc phosphates. The deposition of inorganic pigments as thin layers on to surfaces can produce additional optical effects beyond light absorption. Thus deposition of TiO_2 or Fe_3O_4, as thin layers a few hundred nanometres thick, on flakes of mica produces lustrous or pearlescent pigments where interference effects between light scattered from the various surfaces and layers produces shimmering and iridescent colours.

Semiconductor chemistry

The basic inorganic chemistry of semiconducting materials, particularly their electronic band structures, was covered in Chapter 3. The aim of this section is to discuss the inorganic semiconducting compounds themselves in more detail and to describe some of the applications that result from using chemistry to control their electronic properties.

Semiconductors are classified on the basis of their composition. To produce materials with band gaps typical of a semiconductor (a few electronvolts, corresponding to 100–200 kJ mol^{-1}), compounds usually contain the p-block metals and Group 13/14 metalloids, often in combination with heavier chalcogenides and pnictides. For these combinations, the atomic orbitals form into bands with energies such that the valence and conduction band separation is in the desired range of 0.2–4 eV. Materials based on more electropositive and electronegative elements (for example, alkaline earth metal oxides) are much more ionic in character with much wider separations of the valence and conduction bands, and are insulators. A further factor that influences the band gap in semiconductors is particle size (Chapter 25).

24.18 **Group 14 semiconductors**

Key point: Crystalline and amorphous silicon are cheap semiconducting materials and are widely used in electronic devices.

The most important semiconducting material is Si, which in its pure crystalline form (with a diamond structure) has a band gap of 1.1 eV. As would be expected from considerations of atomic radii, orbital energies, and the extent of orbital overlap, Ge has a smaller band gap, 0.66 eV, and C as diamond has a band gap of 5.47 eV. When doped with a Group 13 or 15 element, Si—and indeed C and Ge—are extrinsic semiconductors. The conductivity of pure Si, an intrinsic semiconductor, is around 10^{-2} S cm^{-1} at room temperature but increases by several orders of magnitude on doping with either a Group 13 element (to give a p-type semiconductor) or a Group 15 element (to give an n-type semiconductor) and thus the properties of doped Si can be tuned for a particular semiconductor application.

Amorphous Si can be obtained by chemical vapour deposition or by heavy-ion bombardment of crystalline Si. The deposited material, which is often obtained by thermal decomposition of SiH$_4$, contains a small proportion of H that is present in Si—H groups in a three-dimensional glass-like structure with many Si—Si links. The lack of regular structure in this material and the presence of Si—H groups alters the semiconducting properties of the material considerably. One of the main applications of amorphous Si is in silicon solar cells. Thin films of p- and n-type amorphous Si forming a p–n junction generate current when illuminated (Box 24.3). The electron and hole pairs produced from the energy supplied by the incident photon separate rather than recombine because of the normal p–n junction bias, with the tendency for the electrons to travel towards the p-type Si and holes towards the n-type. If a load is connected across the junction then a current can flow and electrical energy is generated from the electromagnetic illumination. The efficiency of such devices depends on a number of factors. For example, amorphous Si absorbs solar radiation 40 times more efficiently than does single-crystal Si, so a film only about 1 µm thick can absorb 90 per cent of the usable solar energy. Also, the lifetimes and mobility of the electrons and holes are longer in amorphous Si, which results in high photoelectric efficiencies (the proportion of radiant energy converted into electrical energy) of the order of 10 per cent. Other economic advantages are that amorphous Si can be produced at a lower temperature and can be deposited on low-cost substrates. Amorphous Si solar cells are widely used in pocket calculators but, as production costs diminish, are likely to find much wider applications as renewable energy devices.

24.19 **Semiconductor systems isoelectronic with silicon**

Key points: Semiconductors formed from equal amounts of Group 13/15 or Group 12/16 elements are isoelectronic with silicon and can have enhanced properties based on changes in the electronic structure and electron motion.

Gallium arsenide, GaAs, is one of a number of so-called Group 13/15 (or, still more commonly, III/V semiconductors), which also include GaP, InP, AlAs, and GaN, formed by combination of equal amounts of a Group 13 and a Group 15 element. Ternary and quaternary Group 13/15 compounds, such as Al$_x$Ga$_{1-x}$As, InAs$_{1-y}$P$_y$, and In$_x$Ga$_{1-x}$As$_{1-y}$P$_y$, can also be formed and many of them also have valuable semiconducting properties. Note that these compositions are isoelectronic with pure Group 14 elements, but the changes in the element electronegativity and thus bonding type (for instance, pure Si can be considered as having purely covalent bonding whereas GaAs has a small degree of ionic character due to the difference in the electronegativities of Ga and As) leads to changes in the band structures and fundamental properties associated with electron motion through the structures.

One of the advantageous properties of GaAs is that semiconductor devices based on it respond more rapidly to electrical signals than those based on silicon. This responsiveness makes GaAs better than silicon for a number of tasks, such as amplifying the high-frequency (1–10 GHz) signals of satellite TV. Gallium arsenide can be used with signal frequencies up to about 100 GHz. At even higher frequencies materials such as indium phosphide (InP) may be used. At present, frequencies above about 50 GHz are rarely used commercially, so most of the electronics in the world tend to be based on silicon, with some GaAs, and only a few InP devices. Gallium arsenide is also far more expensive than Si, both in terms of the cost of raw materials and the chemical processes required to produce the pure material.

BOX 24.3 p–n Junctions and LEDs

A p–n junction can be constructed from two pieces of silicon, one of which is n-type and the other p-type (Fig B24.2), where the illustration shows the band structure of the junction. The Fermi levels in the differently doped materials are different but when they are placed in contact electrons will flow from the n-type (high potential) to the p-type (low potential) region across the junction so as to reach an equilibrium distribution in which the Fermi levels are equal. If a potential difference is applied in the right direction across the junction this process will continue, with electrons able to move from n-type to p-type and a current flows in this direction. However, current can flow only in this direction, from n-type to p-type. The p–n junction thus forms the basis of a rectifier, allowing current to pass in only one direction.

A light-emitting diode (LED) is a p–n junction semiconductor diode that emits light when current is passed. The transfer of the electron from the conduction band of the n-type material to the valence band of the p-type semiconductor is accompanied by the emission of light. LEDs are highly monochromatic, emitting a pure colour in a narrow frequency range. The colour is controlled by the band gap, with small band gaps producing radiation in the infrared and red regions of the electromagnetic spectrum and larger band gaps resulting in emission in the blue and ultraviolet regions (see table). It is possible to produce white light with a single LED by using a phosphor layer (yttrium aluminium garnet) on the surface of a blue, gallium nitride, LED. These white light LEDs are highly efficient at converting electricity into light, much more so than incandescent lamps and even 'low-energy' fluorescent lights (Box 24.2). At present they are more expensive to produce than fluorescent light bulbs and their use is limited to small-scale battery driven devices, but they are likely to form the basis of many lighting products of the future.

Another advantage of being able to produce blue LEDs is that they can be used as lasers in high-capacity optoelectronic storage devices. The wavelength of blue light produced by these LEDs (405 nm) is shorter than that used in DVD format devices (red light, 650 nm), which allows data to be written in smaller bits on an optical disc.

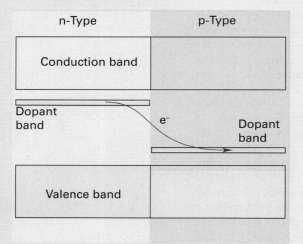

Figure B24.2 The structure of a p–n junction.

LED colours

LED colour	Chip material	
	Low brightness	High brightness
Red	GaAsP/GaP	AlInGaP
Orange	GaAsP/GaP	AlInGaP
Amber	GaAsP/GaP	AlInGaP
Yellow	GaP	–
Green	GaP	GaN
Turquoise	–	GaN
Blue	–	GaN

In some Group 13/15 semiconductors, such as GaN, the cubic, diamond-like, sphalerite structure is only metastable, the stable polymorph being the hexagonal, wurtzite structure. Both structures can be grown by altering the synthetic routes and conditions. Because of their large and direct energy band gaps, these semiconductors allow the fabrication of luminescent devices that produce blue light at high intensity (Box 24.3), and their stabilities to high temperatures and good thermal conductivities also make them valuable for the fabrication of high-power transistors.

The Group 12/16 (II/VI) semiconductors comprise the compounds containing Zn, Cd, and Hg as cations and O, S, Se, and Te as anions. These semiconductor materials can crystallize in either the cubic sphalerite phase or the hexagonal wurtzite phase and the form synthesized has characteristic semiconducting properties. For example, the band gap is 3.64 eV in cubic ZnS but 3.74 eV in hexagonal ZnS. These Group 12/16 compounds are more ionic in nature than the Group 13/15 semiconductors and Group 14 elements, particularly for the lighter elements, and the band gap is around 3–4 eV for ZnO and ZnS but 1.475 eV for CdTe. Although amorphous Si is the leading thin-film photovoltaic (PV) material, cadmium telluride (CdTe) is also being studied for similar applications.

Some other semiconducting oxides and sulfides were mentioned in Section 3.20, and research continues to seek other complex metal oxides and chalcogenides with improved characteristics. For example, the current world record thin-film solar cell efficiency of 17.7 per cent is held by a device based on copper indium diselenide (CuInSe$_2$; CIS), which has a similar structure to cubic ZnS but with an ordered distribution of Cu and In atoms in the tetrahedral holes.

Molecular materials and fullerides

The majority of compounds discussed so far in this chapter have been materials with extended structures in which ionic or covalent interactions link all the atoms and ions together into a three-dimensional structure. Examples include infinite structures based on ionic, as in NaCl, or covalent interactions, as in SiO_2. These materials are widely used in applications such as heterogeneous catalysis, rechargeable batteries, and electronic devices due to the chemical and thermal stability that derives from their linked structures. It is often possible to tune the properties of many solids exactly, for example by doping, introducing defects, or forming solid solutions. However, control of the arrangement of atoms into a particular structure cannot be achieved to the same degree for solids as for molecular systems. Thus a coordination or organometallic chemist can modify a molecule by introducing a wide range of ligands, often through simple substitution reactions. The desire to combine the synthetic and chemical flexibility of molecular chemistry with the properties of classical solid-state materials has led to the rapid emergence of the area of **molecular materials chemistry**, where functional solids are produced from linked and interacting molecules or molecular ions.

24.20 Fullerides

Key points: Solid C_{60} can be considered as a close-packed array of fullerene molecules interacting only weakly through van der Waals forces; holes in arrays of C_{60} molecules may be filled by simple and solvated cations and small inorganic molecules.

The chemical properties of C_{60} span many of the conventional borders of chemistry and include the chemistry of C_{60} as a ligand (Section 14.6). In this section we describe the solid-state chemistry of solid fullerene, $C_{60}(s)$, and the M_nC_{60} fulleride derivatives that contain discrete C_{60}^{n-} molecular anions. The synthesis and chemistry of the more complex carbon nanotubes are discussed in Section 25.7.

Crystals of C_{60} grown from solution may contain included solvent molecules, but with the correct crystallization and purification methods—for example, using sublimation to eliminate the solvent molecules—pure C_{60} crystals may be grown. The solid structure has a face-centred cubic array of the C_{60} molecules as shown in Fig. 24.67, as would be expected on the basis of efficient packing of these almost spherical molecules. At room temperature the molecules can rotate freely on their lattice positions and powder X-ray diffraction data collected from crystalline C_{60} are typical of an fcc lattice with a lattice parameter of 1417 pm. The molecules are separated by a distance of 296 pm, which is similar to the value found for the interlayer separation in graphite (335 pm). On cooling the solid, the rotation halts and adjacent molecules align relative to each other such that an electron-rich region of one C_{60} molecule is close to an electron-poor region in its neighbour.

Exposure of solid C_{60} to alkali metal vapour results in the formation of a series of compounds of formula M_xC_{60}, the precise stoichiometry of the product depending on the composition of the reactant mixture. With excess alkali metal, compounds of composition M_6C_{60}, M = K, Rb, Cs, and sometimes Na and Li, are formed. The structure of K_6C_{60} is body-centred cubic; C_{60}^{6-} molecular ions occupy sites at the cell corners and body centre and the K^+ ions fill a portion of the sites, with approximately tetrahedral coordination to four C_{60} molecular ions, near the centre of each of the faces (Fig. 24.68). Of most interest are the compounds with the stoichiometries M_3C_{60}, which become superconducting in the temperature range 10–40 K depending on the type of metal. The stoichiometry K_3C_{60} is obtained by filling all the tetrahedral and all the octahedral holes in the cubic close-packed C_{60}^{3-} arrangement (Fig. 24.69). K_3C_{60} becomes superconducting on cooling to 18 K, although gradual replacement of K by the larger alkali metal ions raises T_c, so that for Rb_3C_{60} T_c = 29 K and for $CsRb_2C_{60}$ T_c = 33 K. Note that Cs_3C_{60} does not form the same fcc structure as the other M_3C_{60} phases (in fact it has a structure based on a body-centred arrangement of C_{60}^{3-} anions) and is not superconducting at normal pressures; however, it can be made superconducting, with a critical temperature of 40 K at 12 kbar.

Other more complex species can be incorporated into a matrix of C_{60} units. Molecular species such as iodine (I_2) or phosphorus molecules (P_4 tetrahedra) can fill spaces between the C_{60} molecules in close-packed arrays. Solvated cations can also occupy the tetrahedral and octahedral holes in a similar way to the simple alkali metal cations.

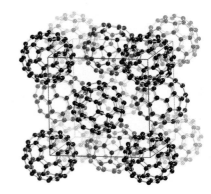

Figure 24.67 The arrangement of C_{60} molecules in a face-centred cubic lattice in the crystalline material.

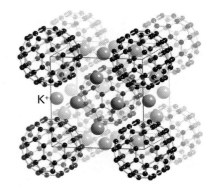

Figure 24.68 The structure of K_6C_{60} with a body-centred cubic unit cell with C_{60}^{6-} molecular ions at the cell corners and body centre, and K^+ ions occupying half of the sites in the faces of the cell which have approximate tetrahedral coordination to four C_{60} molecular ions.

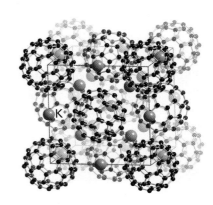

Figure 24.69 The structure of K_3C_{60} is obtained by filling all the tetrahedral and all the octahedral holes in the close-packed C_{60}^{3-} lattice with K^+ ions.

Thus, $Na(NH_3)_4CsNaC_{60}$, which is obtained by the ammoniation of Na_2CsC_{60}, contains Na^+ ions solvated with ammonia molecules on the octahedral site and uncoordinated Cs^+ and Na^+ ions on the (twice as abundant) tetrahedral sites in an fcc arrangement of C_{60}^{6-} molecular ions.

24.21 Molecular materials chemistry

The ability to modify the shapes, and thus the packing and arrangements of inorganic molecules in the solid state, is one valuable aspect of molecular materials chemistry. This capability, when associated with some of the specific properties of inorganic compounds such as the unpaired d electrons of d metals, can allow control of magnetic and electronic properties. This section considers a number of such inorganic molecular materials being developed at the frontiers of the subject.

(a) One-dimensional metals

Key points: A stack of molecules that interact with each other along one dimension, as occurs in a number of crystalline platinum complexes, can show conductivity in that direction; a Peierls distortion ensures that no one-dimensional solid is a metallic conductor below a critical temperature.

A one-dimensional metal is a material that exhibits metallic properties along one direction in the crystal and nonmetallic properties orthogonal to that direction. Such properties arise when the orbital overlap occurs along a single direction in the crystal (as in VO_2). Several classes of one-dimensional metals are known and include $(SN)_x$ and organic polymers, such as doped polyacetylenes $[(CH)I_{0.25}]_n$, but this section is concerned specifically with chains of interacting d metals, particularly Pt.

In such materials, the structural requirements for a one-dimensional metal are satisfied by the presence of square-planar complexes that stack one above another (Fig. 24.70). The ligands surrounding the metal atom ensure large interchain separations, of at least 900 pm, while the average intrachain metal–metal distance is less than 300 pm. Square-planar complexes are commonly found for metal ions with d^8 configurations, and the overlap of orbitals between d^8 species is greatest for the heavy d metals of Period 6 (which use 5d orbitals). Hence the compounds of interest are mainly associated with Pt(II) and Ir(I), where a band is formed from overlapping d_{z^2} and p_z orbitals. The d_{z^2} band is full for Pt(II), and a partially full level is achieved by oxidation of the platinum. Many d^8 integral oxidation number tetracyanoplatinate(II) complexes are semiconductors with $d_{Pt-Pt} > 310$ pm, and the partial oxidation of these salts results in Pt–Pt distances of less than 290 pm and metallic behaviour. Typical means of oxidizing the chain are the incorporation of extra anions into the structure or the removal of cations. The first one-dimensional metal Pt complex was made in 1846 by oxidation of a solution of $K_2Pt(CN)_4.3H_2O$ with bromine, which on evaporation gave crystals of $K_2Pt(CN)_4Br_{0.3}.3H_2O$, known as KCP (as in Fig. 24.70).

The electronic properties of one-dimensional solids are not quite as simple as has been implied by the discussion so far, as a theorem due to Rudolph Peierls states that, at $T = 0$, *no* one-dimensional solid is a metal! The origin of Peierls' theorem can be traced to a hidden assumption in the discussion so far: we have supposed that the atoms lie in a line with a regular separation. However, the actual spacing in a one-dimensional solid (and any solid) is determined by the distribution of the electrons, not vice versa, and there is no guarantee that the state of lowest energy is a solid with a regular lattice spacing. In fact, in a one-dimensional solid at $T = 0$, there always exists a distortion, a **Peierls distortion**, which leads to a lower energy than in the perfectly regular solid.

An idea of the origin and effect of a Peierls distortion can be obtained by considering a one-dimensional solid of N atoms and N valence electrons (Fig. 24.71). Such a line of atoms distorts to one that has long and short alternating bonds. Although the longer bond is energetically unfavourable, the strength of the short bond more than compensates for the weakness of the long bond and the net effect is a lowering of energy below that of the regular solid. Now, instead of the electrons near the Fermi surface being free to move through the solid, they are trapped between the longer-bonded atoms (these electrons have antibonding character, and so are found outside the internuclear region between strongly bonded atoms). The Peierls distortion introduces a band gap in the centre of the original conduction band, and the filled orbitals are separated from the empty orbitals. Hence, the distortion results in a semiconductor or insulator, not a metallic conductor.

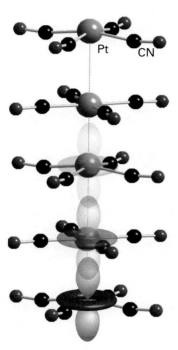

Figure 24.70 A representation of the infinite chain structure of KCP $(K_2Pt(CN)_4Br_{0.3}.3H_2O)$ and a schematic illustration of its d band.

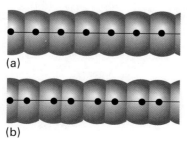

(a)

(b)

Figure 24.71 The formation of a Peierls distortion; the energy of the line of atoms with alternating bond lengths (b) is lower than that of the uniformly spaced atoms (a).

The conduction band in KCP is a d band formed principally by overlap of $Pt5d_{z^2}$ orbitals. The small proportion of Br in the compound, which is present as Br^-, removes a small number of electrons from this otherwise full d band, so turning it into a conduction band. Indeed, at room temperature, doped KCP is a lustrous bronze colour with its highest conductivity along the axis of the Pt chain. However, below 150 K the conductivity drops sharply on account of the onset of a Peierls distortion. At higher temperatures, the motion of the atoms averages the distortion to zero, the separation is regular (on average), the gap is absent, and the solid is metallic. Mixed-valence platinum chain complexes, such as $[Pt(en)_2][PtCl_2(en)_2](ClO_4)_4$, may be isolated as insulated, one-dimensional wires by surrounding the Pt chain with a sheath of electronically inert molecules such as anionic lipids. Nanowires of this type are discussed more fully in Section 25.7.

The observation of metallic behaviour in one-dimensional solids formed from interacting molecules has in turn led to a search for superconductivity in this class of materials. One type of material in which superconductivity has been found is a series of metal complexes derived from the organic metal and superconductors based on stacked molecules with interacting π systems. Thus salts of TCNQ (7,7,8,8-tetracyano-*p*-quinodimethane (**3**) with TTF (tetrathiafulvalene, **4**) show metallic properties, and salts of tetramethyltetraselenafulvalene (TMTSF, **5**) such as $(TMTSF)_2ClO_4$ show superconductivity, albeit at less than 10 K and often only under pressure. Molecular metal complexes involving types of sulfur-containing ligand, such as dmit ($dmit^{2-}$ = 1,3-dithiol-2-thione-4,5-dithiolato), can also show superconductivity. The compound $[TTF][Ni(dmit)_2]_2$, which consists of stacks of both TTF and $Ni(dmit)_2$, shows superconductivity at 10 kbar and below 2 K.

(b) Molecular magnets

Key point: Molecular solids containing individual molecules, clusters, or linked chains of molecules can show bulk magnetic effects such as ferromagnetism.

Molecular inorganic magnetic materials, in which individual molecules, or units constructed from such molecules, contain d-metal atoms with unpaired electrons, is a class of compounds of growing interest. Generally the phenomena associated with long-range interaction of electron spins, such as ferromagnetism and antiferromagnetism, are much weaker as the short, superexchange-type pathways that are found in metal oxides do not exist. However, as with all molecular systems the opportunity exists to tune interactions between metal centres by tailoring the ligand properties.

Examples of ferromagnetic molecular inorganic compounds are decamethylferrocene tetracyanoethenide (TCNE), $[Fe(\eta^5\text{-}Cp^*)_2(C_2(CN)_4)]$ (**6**, $Cp^* = C_5Me_5$) and the analogous manganese compound. These materials, which have structures based on chains of alternating $[M(\eta^5\text{-}Cp^*)_2]^+$ and $TCNE^-$ ions, show ferromagnetism along the chain direction below $T_C = 4.8$ K (for M = Fe) and 6.2 K (for M = Mn). An alternative approach to molecular-based compounds exhibiting magnetic ordering consists of assembling chains of magnetically interacting centres. For example, MnCu(2-hydroxy-1,3-propenebisoxamato).$3H_2O$ (**7**) consists of chains of alternating Mn(II) and Cu(II) ions bridged by the ligand. The magnetic moments on the metal ions in the chains order ferromagnetically below 115 K. This ferromagnetic ordering occurs initially only along the chains, that is in one dimension, due to the stronger interactions through the ligands and shorter Mn—Cu distances, but this material orders fully in three dimensions at 4.6 K once the individual chains interact with each other magnetically.

The incorporation of several d-metal ions into a single complex provides an opportunity to produce a molecule that acts as a tiny magnet. Such compounds have been termed **single-molecule magnets** (SMMs). One example is the complex manganese acetate $Mn_{12}O_{12}$ $(O_2CMe)_{16}(H_2O)_4.2MeCO_2H.4H_2O$, which contains a cluster of 12 Mn(III) and Mn(IV) ions linked through O atoms with the metal oxide unit terminated by the acetate groups (Fig. 24.72). Another is $[Mn_{84}O_{72}(O_2CMe)_{78}(OMe)_{24}(MeOH)_{12}(H_2O)_{42}(OH)_6].xH_2O.$ $yCHCl_3$, which contains 84 Mn(III) ions in a large doughnut-shaped molecule 4 nm in diameter. The ability to magnetize such individual SMMs potentially provides a route to storing information at extremely high densities because their dimensions, a few nanometres, are much less than that of the typical domain of magnetic material used in a conventional magnetic data storage medium.

3 TCNQ

4 TTF

5 TMTSF

6 $[(\eta^5\text{-}Cp^*)_2Fe(TCNE)]$

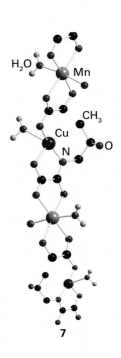

7

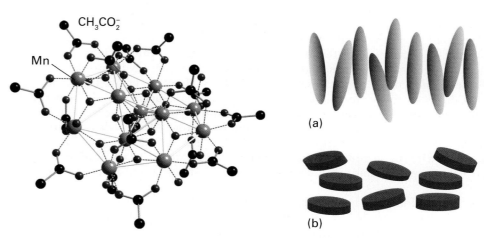

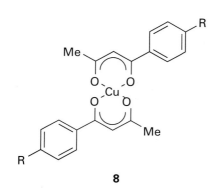

Figure 24.72 The core of the single molecule magnetic compound $Mn_{12}O_{12}(O_2CMe)_{16}(H_2O)_4.2MeCO_2H$ $.4H_2O$ that contains 12 Mn(III) and Mn(IV) ions.

Figure 24.73 Schematic diagram of liquid crystalline materials based on (a) calamitic (rod-like) or (b) discotic (disc-like) molecules.

(c) Inorganic liquid crystals

Key point: Inorganic metal complexes with disc- or rod-like geometries can show liquid crystalline properties.

Liquid crystalline, or **mesogenic**, compounds possess properties that lie between those of solids and liquids and include both. For instance, they are fluid, but with positional order in at least one dimension. Such materials have become widely used in displays. The molecules that form liquid crystalline materials are generally **calamitic** (rod-like) or **discotic** (disc-like), and these shapes lead to the ordered liquid-type structures in which the molecules align in a particular direction (Fig. 24.73). Although most liquid crystalline materials are totally organic there is a growing number of inorganic liquid crystals based on the coordination compounds of metals and on organometallic compounds. These metal-containing liquid crystals show similar properties to the purely organic systems but offer additional properties associated with a d-metal centre, such as redox and magnetic effects.

As the requirement for liquid crystalline behaviour is a rod- or disc-shaped molecule, many of the metal-containing systems are based around the low coordination geometries of the later d metals, particularly the square-planar complexes found for Groups 10 and 11. Thus the β-diketone complex (8) has a square-planar Cu^{2+} ion coordinated to four O atoms from two β-diketones with long pendant alkyl groups. This copper(II) material is paramagnetic but also forms a **nematic phase** in which the rod-shaped molecules align predominantly in one direction (as in Fig. 24.73).

8

FURTHER READING

A.R. West, *Basic solid state chemistry*. Wiley, New York (1999). A good comprehensive guide to the fundamentals of the solid state from an inorganic chemist's perspective.

A.K. Cheetham and P. Day (ed.), *Solid state chemistry: compounds*. Oxford University Press (1992). A useful collection of chapters covering the key compound types in materials chemistry.

R.M. Hazen, *The breakthrough: the race for the superconductor*. Summit Books, New York (1988). A readable narrative of the discovery of high-temperature superconductors.

R.C. Mehrotra, Present status and future potential of the sol–gel process. *Struct. Bonding*, 1992, 77, 1. A good review of sol–gel chemistry.

A.K. Cheetham, G. Férey, and T. Loiseau, Open-framework inorganic materials. *Angew. Chem., Int. Ed. Engl.*, 1999, 38, 3268. An excellent review of progress in, and the structural chemistry of, framework solids.

D.W. Bruce and D. O'Hare, *Inorganic materials*. Wiley, New York (1997). Collection of reviews on various topics of materials chemistry, highlighting molecular systems.

M.T. Weller, *Inorganic materials chemistry*. Oxford Chemistry Primers 23. Oxford University Press (1994). Introductory text covering some aspects of solid-state chemistry and materials characterization.

C.N.R. Rao and J. Golalakrishnan, *New directions in solid state chemistry*. Cambridge University Press (1997). Review of current hot topics and research in solid-state chemistry.

L.V. Interrante, L.A. Casper, and A.B. Ellis (ed.), *Materials chemistry: an emerging discipline*. Advances in Chemistry Series no. 245. American Chemical Society, Washington, DC (1995). A series of chapters on a broad range of inorganic and organic solids.

S.E. Dann, *Reactions and characterization of solids*. Royal Society of Chemistry, Cambridge (2000). A good introductory text on solid-state chemistry.

A.F. Wells, *Structural inorganic chemistry*. Oxford University Press (1985). A comprehensive and systematic volume on structural solid-state chemistry.

U. Müller, *Inorganic structural chemistry*. Wiley, New York (1993). A useful text on structural solid-state chemistry with numerous illustrations.

B.D. Fahlman *Materials chemistry*. Springer, Dordrecht (2007). Good coverage of semiconductors, metals and alloys and characterization methods.

P. Day. *Molecules into materials: case studies in materials chemistry-mixed valency, magnetism and superconductivity*. World Scientific Publishing, Singapore (2007). A collection of papers demonstrating how this important area of materials chemistry developed.

P. Ball *Made to measure: new materials for the 21st century*. Princeton University Press, Princeton, (1997). A very readable overview of materials from the perspective of applications, covering fuel cells, ultra-hard materials, and smart materials.

J.N. Lalena and D.A. Cleary. *Principles of inorganic materials design*. John Wiley and Sons, New Jersey (2005) Good overview of inorganic materials with a strong theoretical perspective.

A Züttel, A Borgschulte, and L. Schlapbach. *Hydrogen as a future energy carrier*. Wiley VCH, Weinheim (2008). The latest developments and thinking on the hydrogen energy economy.

M.D. Hampton, D.V. Schur, S.Yu Zaginaichenko, and V.I. Trefilov. *Hydrogen materials science and chemistry of metal hydrides*. Kluwer Academic Publishers, Dordrecht (2002). A comprehensive review of hydrogen storage materials.

EXERCISES

24.1 When NiO is doped with small quantities of Li_2O the electronic conductivity of the solid increases. Provide a plausible chemical explanation for this observation. (*Hint*: Li^+ occurs on Ni^{2+} sites.)

24.2 Is a crystallographic shear plane a defect, a way of describing a new structure, or both? Explain your answer.

24.3 How might you distinguish experimentally the existence of a solid solution from a series of crystallographic shear plane structures for a material that appears to have variable composition?

24.4 Sketch the wurtzite crystal structure and locate on the sketch the interstitial sites to which a cation might migrate. What is the nature of the bottleneck between the normal and interstitial sites?

24.5 Outline how you could prepare samples of (a) $MgCr_2O_4$, (b) $SrFeO_3Cl$, (c) Ta_3N_5.

24.6 Identify the likely products of the reactions

(a) $Li_2CO_3 + CoO \xrightarrow{800°C, O_2}$

(b) $2 Sr(OH)_2 + WO_3 + MnO \xrightarrow{900°C, O_2}$

24.7 Draw one unit cell in the ReO_3 structure showing the M and O atoms. Does this structure appear to be sufficiently open to undergo Na^+ ion intercalation? If so, where might the Na^+ ions reside?

24.8 Explain the variation of the magnetic susceptibility of a material that orders antiferromagnetically with $T_N = 240$ K, with temperature between 4 K and 600 K.

24.9 Magnetic measurements on the ferrite $CoFe_2O_4$ indicate 3.4 spins per formula unit. Suggest a distribution of cations between octahedral and tetrahedral sites that would satisfy this observation.

24.10 Given ligand-field Δ_O and Δ_T values of 1400 and 620 cm^{-1}, respectively, for Fe^{3+} and 860 and 360 cm^{-1}, respectively, for Ni^{2+}, determine the site preference for A = Ni^{2+} and B = Fe^{3+} in normal compared with inverse spinel, assuming that ligand-field stabilization is dominant.

24.11 Which of the materials (a) $YBa_2Cu_4O_8$, (b) $Ca_{1.8}Na_{0.2}CuO_2Cl_2$, (c) $Gd_2Ba_2Ti_2Cu_2O_{11}$, (d) $SrCuO_{2.12}$, which all contain layers of the stoichiometry CuO_2, might be expected to be high-temperature superconductors?

24.12 Classify the oxides (a) BeO, (b) TiO_2, (c) La_2O_3, (d) B_2O_3, (e) GeO_2 into glass forming and nonglass forming.

24.13 Which metal sulfides might be glass forming?

24.14 Write possible formulas for a sulfide and a fluoride that might adopt the spinel structure.

24.15 Describe two methods that could be used to prepare the intercalation compound $LiTiS_2$.

24.16 Which of the elements Be, Ca, Ga, Zn, P, and Cl might form structures in which the element is incorporated into a framework as an oxotetrahedral species?

24.17 Propose formulas for structures that would be isomorphous with SiO_2 and zeolites of the same stoichiometry involving Al, P, B, and Zn, or mixtures thereof, replacing Si.

24.18 Calculate the mass percentage of hydrogen in $NaBH_4$ and state whether or not this material might be suitable for hydrogen storage.

24.19 Substitution of Mg by small amounts of Li and Al into MgH_2 improves its hydrogen-storage properties. Write a formula for this lithium aluminium magnesium dihydride and explain how Li and Al would be incorporated into the structures.

24.20 Egyptian blue, $CaCuSi_4O_{10}$, is pale blue and the spinel $CuAl_2O_4$ is an intense blue–green. Explain the difference.

24.21 Place the semiconductors AlP, BN, InSb, and C(diamond) in order of the expected band gap.

24.22 Describe the structures of Na_2C_{60} and Na_3C_{60} in terms of hole filling in a close-packed array of fulleride molecular ions.

PROBLEMS

24.1 The compound Fe_xO generally has $x < 1$. Describe the probable metal ion defect that leads to x being less than 1.

24.2 The compound Ag_2HgI_4 is a good electrical conductor above 50°C. Speculate on why Ag^+ rather than Hg^{2+} is the mobile species.

24.3 Solid PbF_2 and ZrO_2 at elevated temperatures are good anion conductors. Describe their crystal structure and why it may be conducive to ion transport.

24.4 Describe the inorganic chemistry involved in the operation of a lambda sensor (an exhaust gas oxygen sensor) in a vehicle engine.

24.5 To obtain high oxidation states for d metals in complex oxides, compounds are normally prepared at as low a temperature as possible commensurate with the reaction. Discuss the thermodynamic reasons for the use of these conditions and explain why the optimum temperature for producing YBCO involves a final annealing stage at about 450°C under pure oxygen.

24.6 Heating a complex oxide under ammonia can lead to nitridation or reduction. Describe possible products that might be formed when Sr_2WO_4 is heated in ammonia.

24.7 What are the advantages of sol–gel routes, in comparison with direct high-temperature reaction methods, in the synthesis of complex metal oxides?

24.8 Describe the issue of occupation factors in Fe_3O_4, Cr_3O_4, and Mn_3O_4 and how the occupation factor is related to ligand-field stablization factors.

24.9 The perovskites, such as ABO_3, often have the dipositive ion in the A site and an ion of higher oxidation state in the B site. What are the factors that lead to this site preference?

24.10 Superconductors are often classified as type I or II. Describe the physical characteristic that determines the classification of a superconductor into one or other of these two types.

24.11 The superconducting compound $YBa_2Cu_3O_7$ is described as having a perovskite-like structure. Ignoring problems of charge on the ions, describe the difference between this structure and that of perovskite.

24.12 Discuss how various chemical substitutions led to the discovery, and subsequent optimization of their properties, of the superconducting materials with the LnFeOAs structure type.

24.13 State Zachariasen's two generalizations that favour glass formation and apply them to the observation that cooling molten CaF_2 leads to a crystalline solid whereas cooling molten SiO_2 at a similar rate produces a glass.

24.14 Starting with $TiCl_4$ and $AlCl_3$, give a plausible procedure for the formation of a ceramic containing polycrystalline Al_2O_3 and TiO_2.

24.15 Reaction of ZrS_2 (c lattice parameter 583 pm) with $Co(\eta^5-C_5H_5)_2$ gives a compound with a c lattice parameter of 1164 pm and reaction with $Co(\eta^5-C_5Me_5)_2$ gives a product with a corresponding lattice parameter of 1161 pm: ZrS_2 does not react with $Fe(\eta^5-C_5H_5)_2$. Explain these observations.

24.16 Describe the interactions between Mo_6S_8 units in a Chevrel phase.

24.17 Discuss differences in the properties of zeolites, zeotypes (frameworks built from oxotetrahedral species other than AlO_4 and SiO_4), and metal organic frameworks.

24.18 Which class of material is the most promising for hydrogen storage in transport applications? What are the different requirements of a static system that might be used to store hydrogen produced from renewable energy?

24.19 What are the properties of an ideal inorganic pigment?

24.20 Compare and contrast the chemistries of graphite and C_{60} with respect to their compounds in association with the alkali metals.

24.21 'The increased reactivity that allows the design and synthesis of materials based on molecular units also means that these compounds are unsuitable for many applications that currently use inorganic materials.' Discuss this remark.

Nanomaterials, nanoscience, and nanotechnology

25

Inorganic chemistry often focuses on entities with atomic and molecular dimensions ranging from 0.1 to 10 nm, whereas solid-state chemistry and materials chemistry have traditionally been concerned with solid materials with dimensions greater than 100 nm. There is currently a great deal of interest in materials that exist between these two scales because they can exhibit unique properties. This chapter focuses on materials that have a critical dimension between 1 and 100 nm. It introduces the fundamental physical and chemical principles that illustrate why these nanomaterials have generated such intense and broad interest, and describes the technology used to generate and exploit them. In the first part of this chapter we introduce nanomaterials, nanoscience, and nanotechnology from a historical perspective, together with definitions and examples. In the second part we discuss the advances in characterization and fabrication methods that allow us to make high-quality nanomaterials. The bulk of the chapter is the third part, where we use examples of nanomaterials to introduce how chemical principles and materials properties are different or superior in the nanoscale regime.

All materials exhibit a hierarchy of structural levels. Chemists have long been familiar with the atomic, molecular, unit-cell, and electronic structural levels because atomic-scale interactions play a primary role in many of the bulk properties exhibited by real materials. Moreover, deviations on an atomic level from the basic structural motif are of great interest because they can affect the bulk properties of materials, as we have seen in connection with the electrical conductivities of semiconductors and solids that contain point defects (Chapter 3).

Materials scientists have long been interested in the structure of materials from the micro (10^{-6} m, 1 μm, '1 micron') to macro (bulk) scales, as well as in the properties of materials that are controlled by structures that exist on and interact over these scales. They have studied how the physical properties at these scales deviate from those predicted by extrapolation from longer scales and deviate from the basic structural motif, such as the presence of line, plane, and volume defects. For example, crystals often contain an enormous number of regions where one part of the crystal is displaced relative to the other part. These dislocations affect strength, crystal growth, electrical conductivity, and other physical properties.

This chapter is concerned with **nanomaterials**, which are materials that have some critical dimension between 1 and 100 nm. Over this intermediate scale, effects from the sub-micrometre- and micrometre-length scales can play equal roles and quantum effects can intervene to give rise to fascinating properties.

Fundamentals

It will prove helpful to understand nanomaterials from historical, empirical, and chemical perspectives.

25.1 Terminology and history

Key points: A nanomaterial is any material that has a critical dimension on the scale of 1 to 100 nm; a more exclusive definition is that a nanomaterial is a substance that exhibits properties absent in both the molecular and bulk solid state on account of it having a critical dimension in this range.

The prefix 'nano' will occur in many combinations throughout this chapter. **Nanoscience** is the study of the properties of matter that have length scales between 1 and 100 nm. **Nanotechnology** is the collection of procedures for manipulating matter on this scale in order to build nanosized entities for useful purposes. Some definitions, however, are more restrictive. Thus, a 'nanomaterial' is taken to be a solid material that exists over the scale of 1 to 100 nm *and* exhibits novel properties that are related to its scale. Likewise, 'nanoscience' is also sometimes restricted to the study of *new* effects that arise only in materials that exist on the nanoscale, and 'nanotechnology' is similarly restricted to the procedures for creating new functionalities that are possible only by manipulating matter on the nanometre-scale.

The original version of nanotechnology occurred in nature, where organisms developed an ability to manipulate light and matter on an atomic scale to build devices that perform specific functions, such as storing information, reproducing themselves, and moving about. In this sense, DNA is the ultimate nanomaterial, as it stores information as the sequence of base pairs that are spaced about 0.3 nm apart. Folded DNA molecules have an information density of more than about 1 Tb cm^{-2} (1 Tb = 10^{12} bits). Photosynthesis is another example of biological nanotechnology in which nanostructures are exploited to absorb light, separate electric charge, shuttle protons around, and ultimately convert solar energy into biologically useful chemical energy. Our quest to harness solar energy has also turned to the nanoscale, where nanocatalysts are improving our ability to convert light to electrical energy by using photovoltaic materials.

Humans have practised nanotechnology for centuries, although it was more of an art than a science or engineering discipline. For example, gold and silver salts have long been used to colour glass: gold has been used to produce red stained glass and silver used to produce yellow. In stained glass, the metal atoms form nanoparticles (known previously as 'colloidal particles') with optical properties that depend strongly on their size. Metallic nanopigments are now becoming a focal point of biomedical nanotechnology because they can be used to tag DNA and other nanoparticles. Other traditional examples of nanotechnology include the photosensitive nanosized particles in silver halide emulsions used in photography and the nanosized carbon granules in the 'carbon black' used for reinforcing tyres and in printer's ink.

The science and engineering of nanotechnology began to take shape in the latter half of the twentieth century. One significant leap forward in nanotechnology occurred when Gerd Binnig and Heinrich Rohrer developed the scanning tunnelling microscope, which was the forerunner to all other forms of scanning probe microscopy (Section 25.3). Later, a scanning probe tip was used to rearrange atoms on a surface to spell out words, so demonstrating an ability to manipulate and characterize nanoscale structures. Work in the areas of nanoscience and nanotechnology has thus been multidisciplinary as well as diverse in scope. This chapter aims to introduce this broad field, highlighting the importance of inorganic chemistry in the multidisciplinary research directed towards understanding nanomaterials.

25.2 Novel optical properties of nanomaterials

Confinement effects lead to some of the most fundamental manifestations of nanoscale phenomena in materials and are frequently used as a point of departure for the study of nanoscience. Novel optical properties appear in nanoparticles as a result of such effects and are being exploited for information, biological, sensing, and energy technologies.

(a) Semiconducting nanoparticles

Key points: The colour of quantum dots is dictated by quantum confinement phenomena and particle localization; quantization of the HOMO–LUMO bands leads to novel optical effects involving interband and intraband transitions.

Semiconducting nanoparticles have been investigated intensively for their optical properties. These particles are often called **quantum dots** (QD) because quantum effects become important in these three-dimensionally confined particles (dots). Two important effects occur in semiconductors when electrons are confined to tiny regions. First, the HOMO–LUMO energy gap increases from the value observed in bulk crystals. Second, the energy levels of electrons in the LUMOs (and holes—the absence of electrons—in the HOMOs) are quantized, like those of a particle in a box. Both effects play an important role in determining the optical properties of QDs.

Quantum confinement, the trapping of electrons and holes in tiny regions, provides a method of tailoring or engineering the band gap of materials. The crucial feature is that as the critical dimension of a material decreases, the band gap increases. Transitions of electrons between states in the valence band (the so-called HOMO states) and the conduction band (the LUMO states) are called **interband transitions**, and the minimum energy for these transitions is increased in QDs relative to those in bulk semiconductors. The wavelengths of interband transitions depend on the size of the dots and it is possible to tailor their luminescence simply by changing their size. A key example of these QD materials is CdSe. By varying the size of the CdSe nanoparticle it is possible to tune emission over the entire visible spectrum, making them ideal for LED and fluorescent display technologies (Box 25.1). Another exciting use of QDs is as chromophores for 'biotags', where dots of different sizes are functionalized to detect different biological analytes. It turns out that, although they emit at specific tuneable wavelengths, QDs exhibit broadband absorption for energies above the band gap. The intriguing point for bioapplications is that it is possible to excite an array of distinct QD chromophores with a single broadband excitation and to simultaneously detect multiple analytes by their distinct optical emissions. These materials have been used to image breast cancer cells and live nerve cells to track small molecule transport to specific organelles.

The second manifestation of quantum confinement is that the available quantized energy levels inside a QD have no net linear momentum and therefore transitions between them do not require any momentum transfer. As a result, the transition probabilities between any two states are high. This lack of momentum dependence also explains the broadband absorption nature of QDs because probabilities are high for most transitions from the occupied valence band states to unoccupied conduction band states. Probabilities are also high for **intraband transitions**, transitions of electrons between states in the LUMO band or of holes in the HOMO band. These relatively intense intraband transitions are typically in the infrared region of the spectrum and are currently being exploited to make devices such as infrared photodetectors, sensors, and lasers.

(b) Metallic nanoparticles

Key points: The colours of metallic nanoparticles dispersed in a dielectric medium are dominated by localized surface plasmon absorption, the collective oscillation of electrons at the metal–dielectric interface.

The optical properties of metallic nanoparticles arise from a complex electrodynamic effect that is strongly influenced by the surrounding dielectric medium. Light impinging on metallic particles causes optical excitations of their electrons. The principal type of optical excitation that occurs is the collective oscillation of electrons in the valence band of the metal. Such coherent oscillations occur at the interface of a metal with a dielectric medium and are called **surface plasmons**.

In bulk particles, the surface plasmons are travelling waves and are characterized by a linear momentum. To excite plasmons using photons in bulk metals, the momenta of the plasmon and the photon must match. This matching is possible only for very specific geometries of the interaction between light and matter, and is a weak contributor to the optical properties of the metal. In nanoparticles, however, the surface plasmons are localized and have no characteristic momentum. As a result, the momenta of the plasmon and the photon do not need to match and plasmon excitation occurs with a greater intensity. The peak intensity of the surface plasmon absorption for gold and silver occurs in the optical region of the spectrum, and so these metallic nanoparticles are useful as pigments.

The characteristics of plasmon absorption depend strongly on the metal and the dielectric surroundings as well as on the size and shape of the nanoparticle. To control the dielectric surroundings, so-called **core–shell composite nanoparticles** have been designed in which metallic shells of nanometre-scale thickness encapsulate a dielectric nanoparticle. Metallic nanoparticles and metallic nanoshells are used as dielectric sensors because their optical properties change when they come into contact with different dielectric materials. In particular, biological sensing is of interest because biological analytes can bind to the surface of the nanoparticle, causing a detectable shift in the plasmon absorption band.

Gold nanoparticles are common examples of metallic nanoparticles and have found practical applications as biological and chemical sensors, 'smart bombs' for cancer therapy, and optical switching and fluorescent display materials. Many of these applications are possible owing to the development of techniques to bind photoresponsive chromophores

BOX 25.1   CdSe Nanocrystals for LEDs

Cadmium selenide (CdSe) nanocrystals have found application in a wide variety of commercial venues including LEDs (Fig. B25.1), solar cells, fluorescent displays, and *in vivo* cellular imaging of cancer cells. Their high photoluminescence efficiencies and emission colour tunability based on nanocrystal size make them attractive for full-colour displays. Close-packed QD monolayers have been self-assembled using soft nanolithography contact printing to achieve light-emitting devices. In soft nanolithography, photolithography or electron beam lithography is used to produce a pattern in a layer of photoresist on the surface of a silicon wafer. This process generates a bas-relief master that contains islands of photoresist that stand out from the Si surface. A chemical precursor to polydimethylsiloxane (PDMS), a free-flowing liquid, is then poured over the bas-relief master and cured into the rubbery solid. After curing, a PDMS stamp that matches the original pattern reproduces features from the master as small as a few nanometres. Whereas making a master is expensive, copying the pattern on PDMS stamps is inexpensive and easy. One stamp can be used in various inexpensive ways, including microcontact printing and micromoulding to make nanostructures. These techniques can be employed to produce subwavelength optical devices, waveguides, and optical polarizers used in optical fibre networks and eventually perhaps in all-optical computers.

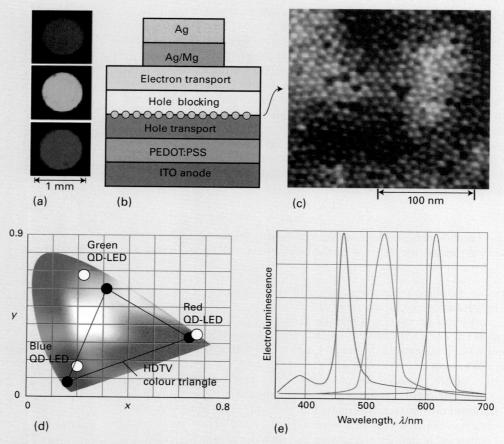

Figure B25.1 (a) Electroluminescent (EL) red, green, and blue QD-LED pixels with the device structure shown in (b). (b) Schematic diagram shows the cross section of a typical QD-LED. (c) A high-resolution AFM micrograph shows a close-packed monolayer of QDs deposited on top of the hole-transporting polymer layer prior to deposition of hole-blocking and electron-transporting layers. (d) Chromaticity diagram shows the positions of red, green, and blue QD-LED colours, an HDTV colour triangle is shown for comparison. (e) Normalized EL spectra of fabricated QD-LEDs corresponding to the colour coordinates in (d). QD-LED images and EL spectra are taken at video brightness (100 cd m^{-2}), which corresponds to the applied current density of 10 mA cm^{-2} for red QD-LEDs, 20 mA cm^{-2} for green QD-LEDs, and 100 mA cm^{-2} for blue QD-LEDs. (Based on L. Kim, *et al.*, *Nano Lett.*, 2008, **8**, 4513.)

to the surface of the nanoparticles. Gold nanoparticles can be used tagged with biomolecules that afford delivery to specific cells and have found applications as immunoprobes for early detection of disease. The nanomaterials have also been used as key electron-transfer agents in a new generation of solar materials that couples photoresponsive π-conjugated molecules such as pyrene to gold nanoparticle surfaces. Chromophore-functionalized gold nanoparticles afford unique device architectures and flexibility in design as these hybrid materials can be coupled covalently to conductive glass substrates that serve as electrodes, so leading to improved charge transport and photoefficiencies.

Silver nanoplates with sizes in the range 40–300 nm have been synthesized by a simple room-temperature solution-phase chemical reduction method in the presence of diluted cetyltrimethylammonium bromide (CTAB, $(C_{16}H_{33})(CH_3)_3NBr$). These plates are single crystals having as their basal plane the (111) plane of face-centred cubic silver. The stronger adsorption of CTAB on the (111) basal plane than on the (100) side plane of these plates may account for the anisotropic growth of nanoplates. As discussed earlier, metal nanoparticles have interesting optical properties related to their surface plasmon excitations. The optical (in-plane dipole) plasmon resonance peaks can be shifted to wavelengths of 1000 nm in the near-IR when the aspect ratio (the ratio of long-axis to short-axis, or width to thickness) of the nanoplates reaches 9. Such control over the optical properties of simple metals opens up new possibilities for various near-IR applications, including remote sensing (Fig. 25.1).

Characterization and fabrication

The burgeoning activities during the late twentieth century in the areas of nanomaterials, nanoscience, and nanotechnology were intimately tied to advances in characterization and fabrication. This section introduces the basic concepts of nanoscale characterization and fabrication methods that underly the design of nanomaterials and the development of nanotechnologies.

25.3 Characterization methods

The great advances made in nanosciences and nanotechnology would not have occurred without the ability to characterize the nanoscale structural, chemical, and physical properties of materials. Moreover, the direct observation of nanostructure allows meaningful relationships between processing and properties to be made.

(a) Scanning probe microscopy

Key points: Scanning tunnelling microscopy uses tunnelling current from a sharp conductive tip to image and characterize a conductive surface; atomic force microscopy responds to intermolecular forces to image the surface.

Scanning tunnelling microscopy (STM) and **atomic force microscopy** (AFM) were the first of a series of **scanning probe microscopies** (SPM), which are techniques that allow three-dimensional imaging of the surface of materials by using a sharply pointed probe brought into close proximity (or in contact) with the specimen and constructing an image by scanning the probe over the surface of the specimen and monitoring the spatial variation in the value of a physical parameter, such as potential difference, electric current, magnetic field, or mechanical force.

In STM, an atomically sharp conductive tip is scanned at about 0.3–10 nm above the surface of the sample. Electrons in the conductive tip, which is held either at a constant potential or at a constant height above the sample, can tunnel through the gap with a probability that is related exponentially to the distance from the surface. As a result, the electron tunnelling current reflects the distance between the sample and the tip. If the tip is scanned over the surface at a constant height, then the current represents the change in the topography of the surface. The movement of the tip at a constant height with great precision is accomplished by using piezoelectric ceramics for the displacement of the tip.

Typically, STM probes are sharpened to have only a few atoms at the end. The simplest method of creating such sharp tips is through electrochemical etching. In one technique, a thin tungsten wire is placed in a solution of potassium hydroxide and an electric current is passed through it. STM probes have been used to manipulate individual atoms and molecules on cooled surfaces to achieve patterning on the atomic scale. Figure 25.2 shows an STM study of trimesic acid (TMA) molecules adsorbed on a graphite substrate. The TMA molecules act as a host for the incorporation of C_{60} as a molecular guest. The STM tip was used to nudge these adsorbed C_{60} molecules from one cavity of the host structure to the next, giving rise to what is jocularly referred to as 'nanosoccer'.

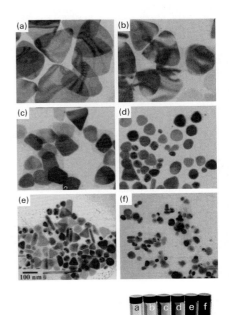

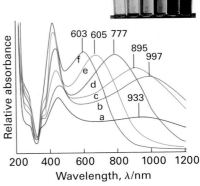

Figure 25.1 (Top) TEM images of the silver nanoplates obtained with different seed amounts: (a) 0.1 cm³, (b) 0.25 cm³, (c) 0.5 cm³, (d) 1.0 cm³, (e) 2.0 cm³, and (f) 3.0 cm³. (Bottom) Absorption spectra of the corresponding silver nanoplate solutions. (S. Chen and D.L. Carroll, *J. Phys. Chem. B*, 2004, **108**, 5500.)

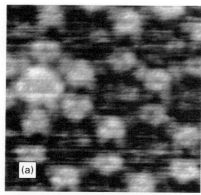

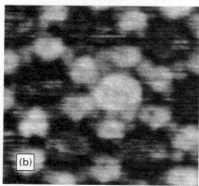

Figure 25.2 Playing nanosoccer: (a) STM topograph images of the starting situation with a single C_{60} guest molecule inside the trimesic acid host network. (b) Final result of the lateral manipulation with the whole molecule imaged in the adjacent target cavity. (S.J.H. Griessl, *et al.*, *J. Phys. Chem. B* 2004, **108**, 1155.)

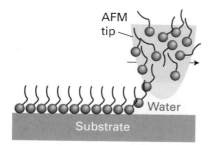

Figure 25.3 The process of dip-pen nanolithography. Organothiols move from an AFM tip through a water meniscus forming a self-assembled monolayer on a gold substrate. (Adapted from C. Mirkin, Nanoscience Boot Camp at Northwestern University 2001.)

In AFM, the atoms at the tip of the probe interact with the surface atoms of the sample through intermolecular forces (such as van der Waals interactions). The cantilever holding the probe bends up and down in response to the forces and the extent of deflection is monitored with a reflected laser beam. Variations on AFM include:

frictional force microscopy, which measures variations in the lateral forces on the tip that are based on chemical variations on the surface

magnetic force microscopy, which uses a magnetic tip to image magnetic structure

electrostatic force microscopy, which uses the tip to sense electric fields

scanning capacitance microscopy, in which the tip is used as an electrode in a capacitor.

Molecular recognition AFM is also now being carried out in which the tip is functionalized with specific ligands and the interaction between the tip and surface is measured. Such microscopes can provide resolution of the chemical properties on the surface. AFM can also act as a patterning tool. In **dip-pen nanolithography** (DPN, Fig. 25.3) the tip of an AFM is used as an ink pen. By using an ink containing molecular entities, **self-assembled monolayers** (SAMs) can be formed through molecular transport of the nanoink to the solid substrate surface. These monolayers often involve specific covalent interactions between the S atom of organothiols and Au surfaces.

Scanning near-field optical microscopy (SNFOM) combines the local interaction of a scanning probe and a specimen with well-established methods of optical spectroscopy. Conventional optical spectroscopy is limited in resolution by the wavelength of the light source, but when the light source is brought to within a few atomic diameters of the surface and a nanometre-scale aperture is used, the local confinement of the electromagnetic field allows for much improved spatial resolution (of about 20 nm) to investigate local chemical properties of nanoscale structures.

(b) Scanning electron microscopy techniques

Key point: Transmission and scanning electron microscopes use electrons to image the sample in a similar fashion to optical microscopes.

An imaging electron microscope operates like a conventional optical microscope but instead of imaging photons, as in the visible microscope, electron microscopes use electrons. In these instruments, electron beams are accelerated through 1–200 kV and electric and magnetic fields are used to focus the electrons. In **transmission electron microscopy** (TEM) the electron beam passes through the thin sample being examined and is imaged on a phosphorescent screen. In **scanning electron microscopy** (SEM) the beam is scanned over the object and the reflected (scattered) beam is then imaged by the detector. The ultimate resolution of SEM depends on how sharply the incident beam is focused on the sample, how it is moved over the sample, and how much the beam spreads out into the sample before reflecting. In both microscopes, the electron probes cause the production of X-rays with energies characteristic of the elemental composition of the material. As such, **energy-dispersive spectroscopy** of these characteristic X-rays is used to quantify the chemical make-up of materials by using electron microscopes.

The primary advantage of SEM over TEM is that it can form images of electron-opaque samples without the need for difficult specimen preparations. It is therefore the electron microscopy method of choice for the straightforward characterization of materials. However, SEM samples need to be conductive because otherwise electrons can collect on the sample and interact with the electron beam itself, resulting in blurring. Nonconductive samples must therefore be coated with a thin layer of metal, usually gold or aluminium. The use of **focused ion beams** (FIBs; high energy streams of Ga^{3+} ions) to cut slices of samples has revolutionized TEM sample preparation, shortening times and allowing for *in situ* studies of materials' architectures. The FIB technique can be used in concert with SEM to examine internal morphologies. Damaged or dysfunctional integrated circuits can be sliced open to investigate the origin of failure and examine links between materials' defects and their conductive behaviour.

25.4 Top-down and bottom-up fabrication

Key points: Top-down fabrication methods carve out or add on nanoscale features to a bulk material by using physical methods; bottom-up fabrication methods assemble atoms or molecules in a controlled manner to build nanomaterials piece by piece.

There are two basic techniques for the fabrication of nanoscale entities. The first is to take a macroscale (or microscale) object and carve out nanoscale patterns. Methods of this sort are called **top-down approaches**. In top-down approaches, patterns are first designed on a large scale, and their lateral dimensions are reduced and then used to transfer the nanoscaled features into or on to the bulk material. Physical interactions are used in top-down fabrication approaches, such as photolithography, e-beam lithography, and soft lithography. The most common and well-known approach is **photolithography**, the technique used to fabricate very large-scale integrated circuits having feature dimensions on the 100 nm scale (Fig. 25.4). The second technique is to build larger objects by controlling the arrangement of their component smaller-scale objects. Methods of this sort are called **bottom-up approaches** and start with control over the arrangements of atoms and molecules. This chapter emphasizes the bottom-up approach to nanoscale fabrication because of its focus on the interactions of atoms and molecules and their arrangement into larger functional structures.

The two basic approaches most widely used to prepare nanomaterials, solution methods and vapour-phase methods, are bottom-up methods because control over the arrangement of individual atoms is exerted to achieve larger-scale structures.

(a) Solution-based synthesis of nanoparticles

Key points: Solution-based synthetic methods are the main techniques for nanoparticle synthesis because they have atomically mixed and highly mobile reagents, allow for the incorporation of stabilizing molecules, and have been widely successful in practice; the two stages of crystallization from solution are nucleation and growth.

As discussed in Chapter 24, the two basic techniques used to generate inorganic solids are direct-combination methods and solution-based methods. The former does not lend itself well to the synthesis of nanoparticles because the reactants tend to be larger than nanoparticles and therefore require long times to reach equilibrium. Moreover, the use of elevated temperatures leads to particle growth and coarsening during the reaction period, resulting in large crystallite sizes. There are, however, some examples of using mechanical ball-milling at low temperature to break macroscopic powders into nanoparticles. However, solution-based methods permit excellent control over the crystallization of inorganic materials and are widely used. The techniques used to make nanoparticles are similar to those described in Chapter 24 to make solid-state compounds. Special care must be taken to control the size and shape of the particles during nanoparticle synthesis, as well as their uniformity in size and shape. By fine-tuning the crystallization process from solution, highly monodisperse, uniformly shaped nanoparticles of a wide range of compositions can be prepared from combinations of elements from throughout the periodic table.

Because the reactants in solution-based methods are mixed on an atomic scale and solvated in a liquid medium, diffusion is fast and diffusion distances are typically small. Therefore, reactions can be carried out at low temperature, which minimizes the thermally driven particle growth that is problematic in direct combination methods. Although the specifics of each reaction differ greatly, the basic stages in solution chemistry are:

1. Solvate the reactant species and additives.
2. Form stable solid nuclei from solution.
3. Grow the solid particles by addition of material until the reactant species are consumed.

The basic aim in solution synthesis is to generate in a controlled manner the simultaneous formation of large numbers of stable nuclei that undergo little further growth. If growth is to occur, it should occur independently of the nucleation step because then all particles have a chance to grow to similar sizes. If performed successfully, the particles will be monodisperse and in the nanometre range. The drawback to the solution method is that the particles can undergo **Ostwald ripening**, in which smaller particles in the distribution redissolve and their solvated species reprecipitate on to larger particles, so increasing the size distribution and decreasing the total particle count. To prevent this unwanted ripening, **stabilizers**, surfactant molecules that help to stabilize the particles against growth and dissolution, are added.

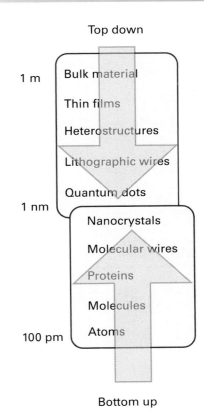

Figure 25.4 The two techniques for making nanoscale structures. The top-down technique starts with larger objects that are whittled down into nanoscale objects; the bottom-up technique starts with smaller objects that are combined into nanoscale objects.

EXAMPLE 25.1 Controlling nucleation during nanoparticle synthesis

Use thermodynamic arguments related to the crystallization of a phase to describe how to generate a short burst of homogeneous nucleation in a solution.

Answer We need to be aware that the formation of stable solid nuclei is controlled through the thermodynamic driving force of the reaction, which can be controlled by adjusting the composition of the different solution species and the temperature (as well as the pressure for hydrothermal techniques). As with most chemical reactions, we need to note that there is an activation barrier to the crystallization reaction, which prevents the instantaneous formation of many stable nuclei. To overcome this barrier in a short single burst, we convert the solution quickly to a highly nonequilibrium state so that the tendency to nucleate becomes very strong. Widespread homogeneous nucleation of many particles then occurs. This nucleation quickly moves the system back towards equilibrium and suppresses further nucleation.

Self-test 25.1 Use a similar argument to describe the synthesis of core-shell nanoparticles.

As there are many methods to synthesize nanoparticles, we limit this discussion to a few well-known examples. The first example is metallic nanoparticles, such as gold. In 1857, Michael Faraday found that reduction of an aqueous solution of $[AuCl_4]^-$ with phosphorus in CS_2 produced a deep-red suspension that contained nanoparticles of gold. Because sulfur forms chemical bonds to gold, sulfur-containing species are good stabilizing agents and the most widely used stabilizer for gold nanoparticles contain a thiol group ($-SH$). An approach has been developed in the same spirit as Faraday's to control the size and dispersity of gold nanoparticles by using $[AuCl_4]^-$ and thiol stabilizers. The reaction is relatively simple and the procedure produces air-stable gold nanoparticles with diameters between 1.5 and 5.2 nm. In the so-called **Brust-Schiffrin method**, $[AuCl_4]^-$ is first transferred from water to methylbenzene (toluene) by using tetraoctylammonium bromide as a phase-transfer agent. The methylbenzene contains dodecanethiol as a stabilizer and, after transfer, $NaBH_4$ is used as a reducing agent to precipitate Au nanoparticles with dodecanethiol surface groups:

Transfer:

$$[AuCl_4]^-(aq) + N(C_8H_{17})_4^+(sol) \rightarrow N(C_8H_{17})_4^+(sol) + AuCl_4^-(sol)$$

Precipitation (of the $[Au_m(C_{12}H_{25}SH)_n]$ nanoparticle):

$$m\,[AuCl_4]^-(sol) + n\,C_{12}H_{25}SH(sol) + 3m\,e^- \rightarrow 4m\,Cl^-(sol) + [Au_m(Cl_{12}H_{25}SH)_n](sol)$$

where sol is methylbenzene. The ratio of stabilizer ($C_{12}H_{25}SH$) to metal (Au) controls the particle size in the sense that higher stabilizer:metal ratios lead to the smaller metal core sizes. By adding the $NaBH_4$ reductant quickly and cooling the system as soon as possible after the reaction terminates, smaller and more monodisperse nanoparticles are formed. The rapid addition of reductant improves the probability of simultaneous formation of all nuclei. By cooling the solution quickly, both post-nucleation growth and dissolution of particles are minimized. Similar approaches can be used for other metal nanoparticles.

Quantum dots of materials such as GaN, GaP, GaAs, InP, InAs, ZnO, ZnS, ZnSe, CdS, and CdSe have been investigated for their optical properties (Section 25.2) because their interband absorption and fluorescence occur in the visible spectrum. As an early example of their preparation, dimethylcadmium is dissolved in mixture of trioctylphosphine (TOP) and trioctylphosphine oxide (TOPO), and the selenium source, which is often Se dissolved in TOP or TOPO, is added at room temperature. The solution is injected into a reactant vessel containing vigorously stirred, hot TOPO, which permits widespread nucleation of TOPO-stabilized CdSe QDs. The addition of the room-temperature liquid lowers the temperature of the solution and prevents further nucleation or growth (because the activation barriers are so high). The solution is then reheated to a temperature that allows for slow growth but no further nucleation. This step leads to nanoparticles with narrow size distributions and sizes in the range 2–12 nm. Alternative, less hazardous synthetic methods have been developed and are being pursued.

Cadmium sulfide can be grown in pH-controlled aqueous solutions of Cd(II) salts with polyphosphate stabilizers by the addition of a sulfur source. For example, at pH = 10.3 the addition of Na_2S causes the precipitation of CdS nanoparticles from aqueous solutions containing $Cd(NO_3)_2$ and sodium polyphosphate. These QDs range in size from 1 to 10 nm, and the size is controllable through the reactant concentrations and the rate of addition of the reactant.

Oxide nanoparticles can also be grown by solution methods. Many applications use colloidal particles of oxides such as SiO_2 and TiO_2 for food, ink, paints, coatings, and so on. Many of the efforts to achieve controlled oxide nanoparticle growth stem from earlier work in traditional ceramic and colloidal applications, where particle sizes from 1 nm to 1 μm are used. Silica, SiO_2, and titania, TiO_2, are probably the best-known oxides grown from solution. Typical schemes involve the controlled hydrolysis of metal alkoxides. All successful hydrolysis reactions of metal alkoxides aim to follow the same basic rules as described above: the controlled nucleation stage and slow growth stage are performed independently. In all cases, strict monitoring of the pH, precursor chemistry, reactant concentration, rate of addition of reactant, and temperature is required to control the final size and shape of the particles.

An important example of nanoparticle oxide use is in the photoelectrochemical solar cell known as the **Grätzel cell**. This cell uses nanocrystalline TiO_2 as a medium to transfer electrons from an organoruthenium dye to a conductor. Nucleation occurs in the hydrolysis of titanium isopropoxide that is added dropwise into vigorously stirred 0.1 M $HNO_3(aq)$. The filtered nanoparticles are then placed in an autoclave and allowed to grow by the hydrothermal addition of material. The size, shape, and state of agglomeration are controlled by adjusting the conditions of either the nucleation or the growth stage.

(b) Vapour-phase synthesis of nanoparticles

Key points: Vapour-phase synthetic methods are alternative techniques for nanoparticle synthesis because they have atomically mixed and highly mobile reagents, can be controlled by varying the conditions, and have been widely successful in practice.

The same fundamentals concerning nucleation and growth that are relevant to solution synthesis apply to vapour-phase synthesis. The vapour needs to be supersaturated to the point at which a high density of homogeneous nucleation events produces solid particles in one short burst, and growth must be limited and controlled in a subsequent step, if it is to occur at all. Commercially, vapour-phase synthesis is carried out to produce nanoscale carbon black and fumed silica in large quantities. Metals, oxides, nitrides, carbides, and chalcogenides can also easily be formed by using vapour-phase techniques.

There are, however, several differences between vapour-phase and solution-based techniques. In the latter, stabilizers can be added in a straightforward and controllable fashion, and particles remain dispersed and independent of one another. In vapour-phase techniques, however, surfactants or stabilizers are not easily added (although solid-state stabilizers are available). Without surface stabilizers, nanoparticles tend to agglomerate into larger particles. If temperatures during the process are high enough, the particles sinter and coalesce into one larger particle. Although the 'soft agglomerates' formed under the influence of weak intermolecular forces can usually be redispersed, the 'hard agglomerates' (partially sintered-formed at intermediate temperature) and coalesced particles (formed at high temperatures) cannot be. Therefore, techniques that lead to soft agglomerates are generally best suited to nanoparticle production. Because of these differences, the size dispersion of nanoparticles tends to be better for solution-based techniques than for vapour-phase techniques.

Vapour-phase techniques are attractive synthesis methods for particles when continuous operation is required or when solution methods do not produce good quality nanoparticles. They can also produce complex materials in a straightforward manner, such as doped binary and ternary compounds or composite particles. Vapour-phase techniques are classified by the physical state of the precursor used as a reagent and by the reaction method, such as 'plasma synthesis' or 'flame pyrolysis'. In each case, the reagent is converted to a supersaturated or superheated vapour that is allowed to react or to cool to force nucleation. Solid reagents are vaporized and then recondensed in gas condensation methods (thermal evaporation to produce vapour), laser ablation, sputtering, and spark discharge methods. Liquid or vapour precursors are used in spray pyrolysis, flame synthesis, laser pyrolysis, plasma synthesis, and chemical vapour deposition. In chemical vapour deposition, the vapour is transported to a substrate where reaction and solid nucleation occurs (Fig. 25.5). In a variation of this procedure, gaseous precursors are delivered into a hot-walled

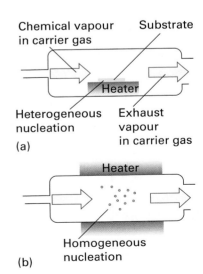

Figure 25.5 Chemical vapour methods to achieve (a) thin film growth and (b) nanoparticle production.

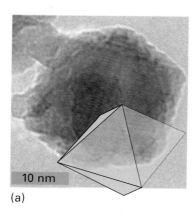

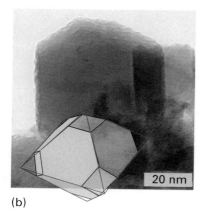

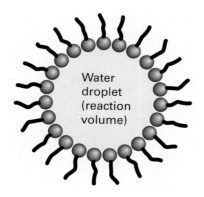

Figure 25.6 TEM images of nanoparticles of $Ni_{0.5}Zn_{0.5}Fe_2O_4$ synthesized by plasma torch synthesis. The particle shapes are superimposed: (a) an octahedron and (b) a truncated octahedron. (Based on images provided by M. McHenry and R. Swaminatham.)

reactor and allowed to react homogeneously in the vapour phase to nucleate solids, and particles are collected downstream. Particle sizes are controlled by the flow rates, precursor chemistry, concentrations, and residence times in the reactor. Examples of materials synthesized by this technique include metals, oxides, nitrides, and carbides such as SiC, SiO_2, Si_3N_4, $SiC_xO_yN_z$, TiO_2, TiN, ZrO_2, and ZrN.

Both solution and vapour-phase methods can be used to make composite nanoparticles, such as core–shell nanocomposites. In both techniques the approach is to grow a second phase on an initial nucleus or nanoparticle. Reaction design to produce core–shell nanoparticles by solution is straightforward provided the solution characteristics of both materials are similar. In practice, however, it is difficult to find materials where the synthesis conditions overlap. Vapour-phase techniques offer another approach to the design of core–shell particles, by injecting a second vapour into a reactor at the growth stage.

Another straightforward approach to vapour-phase nanoparticle production is **plasma synthesis**, which can be used to synthesize elemental solids, alloys, and oxides, as well as core–shell nanoparticles. In this method, gas or solid particles are fed into a plasma where they vaporize and ionize to highly energetic charged species. Inside a plasma, the temperatures can exceed 10 kK. On leaving the plasma, the temperature falls rapidly and crystallization occurs under conditions far from equilibrium. The nanoparticles are collected downstream from the plasma zone. Depending on the carrier gas (that is, the oxidation conditions), elemental solids, core–shell, or compound particles can be formed. Figure 25.6 shows two examples of nickel/zinc ferrite particles.

25.5 Templated synthesis using frameworks, supports, and substrates

Heterogeneous nucleation, nucleation that occurs on an existing surface, can be used to generate important kinds of nanostructures, including zero-, one-, and two-dimensional materials. The methods are similar to those already described: they are either physical or chemical and involve crystallization from either liquids or vapours. The principal difference is that an outside agent is also involved and permits the direct control of the nanoparticle formation. The outside agent can be a framework or a support structure that limits the size of the reaction volume, as described in Section 25.5a for nanoparticle synthesis using inverse micelle frameworks and embossed supports. During thin film growth, the outside agent is a substrate. Because of its importance in nanomaterials synthesis, thin film growth on substrates is described in Sections 25.5b and c.

(a) Nanosized reaction vessels

Key points: By carrying out reactions in nanoscale reaction vessels, the ultimate dimensions of solid products are confined to the vessel size; a reverse micelle has an aqueous core in which reactions can occur.

When the synthesis of a particle is carried out in a nano-vessel, the ultimate particle size is limited by the vessel size. One popular route is the **inverse micelle synthesis** approach. An inverse micelle consists of a two-phase dispersion of immiscible liquids, such as water and a nonpolar oil. By including amphipathic surfactant molecules (molecules having a polar and a nonpolar end), the aqueous phase can be stabilized as dispersed spheres with a size dictated by the water:surfactant ratio (Fig. 25.7). The size of the crystalline particles is limited by the micelle volume, which can be controlled on the nanoscale. Examples of nanoparticles formed by this technique are Cu, Fe, Au, Co, CdS, CdSe, ZrO_2, ferrites, and core–shell particles such as Fe/Au.

An alternative approach to nanopatterning is **laser-assisted embossing**, which has been used to generate close-packed arrays of hemispherical nanowells having volumes of the order of zeptolitres ($1\ zL = 10^{-21}\ L = 10^{-25}\ m^3 = 10^3\ nm^3$). The volumes of the nanowells are manipulated by changing the intensity of the laser, by altering the pressure between the mould and the substrate, and by growing an amorphous oxide on the patterned silicon prior to nanocrystallization. Nanowells with diameters as small as 50 nm have been used as reaction chambers for the preparation of semiconducting nanocrystals. Under most growth conditions, independent nanocrystals are formed in individual nanowells (Fig. 25.8).

Figure 25.7 An inverse micelle. The hydrophilic heads of molecules surround the water droplet and the hydrophobic tails contact the nonpolar solvent, which acts as the dispersed medium. Reactants are solvated in the water droplet and then caused to react in these spatially confined vessels.

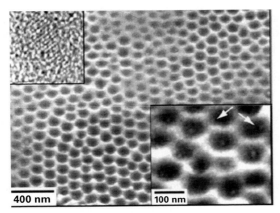

Figure 25.8 An SEM image of an array of nanoscale wells that contain one CdS nanocrystal per well. The lower inset magnifies a small area of the array; the white arrows indicate crystals formed at the edges of the wells. The upper inset is a TEM image of an individual CdS nanocrystal removed from the array of nanowells by sonication. (J.E. Barton and T.W. Odom, *Nano Lett.*, 2004, **4**, 1525.)

EXAMPLE 25.2 **Assessing the role of the volume of nano-reaction vessels**

The use of nano-reaction vessels combines solution-based synthesis techniques directly with size-limited growth. Assume that a surface is embossed with 100 nm-wide hemispherical imprints that are then used as reaction vessels. (a) What is the volume of the reaction vessel? (b) What will be the size of an NaCl particle that is crystallized in this vessel to produce a single spherical particle? Assume a solution of 1.00 M NaCl(aq) and that the crystallization goes to completion. The density of NaCl is 2170 kg m^{-3} and its molar mass is 58.44 g mol^{-1}.

Answer (a) The volume of each hemispherical imprint is half the volume of a sphere of radius 50 nm:

$$V_{solution} = \tfrac{1}{2}(\tfrac{4}{3}\pi r^3) = \tfrac{2}{3}\pi(5.0\times10^{-8}\,\text{m})^3 = 2.62 \times 10^{-22}\,\text{m}^3$$

(b) Because the concentration of the solution is 1.00 mol dm^{-3}, the amount of NaCl in the hemispherical imprint is

$$n = \left(1.00 \times 10^3\,\text{mol m}^{-3}\right) \times \left(2.62 \times 10^{-22}\,\text{m}^3\right) = 2.62 \times 10^{-19}\,\text{mol}$$

The mass of this amount of NaCl is

$$m = \left(2.62 \times 10^{-19}\,\text{mol}\right) \times \left(58.44\,\text{g mol}^{-1}\right) = 1.53 \times 10^{-17}\,\text{g}$$

or 1.53×10^{-20} kg of NaCl in the solution. The volume of this NaCl is therefore

$$V_{NaCl} = \frac{1.53 \times 10^{-20}\,\text{kg}}{2170\,\text{kg m}^{-3}} = 7.06 \times 10^{-24}\,\text{m}^3$$

To determine the size of the particle, we use the definition of volume and solve for the diameter (twice the radius):

$$d = 2r = 2\left(\tfrac{3}{4\pi}V\right)^{1/3} = 2\left(\tfrac{3}{4\pi}\times 7.06 \times 10^{-24}\,\text{m}^3\right)^{1/3} = 2.38 \times 10^{-8}\,\text{m}$$

Hence, the diameter of the particle is 23.8 nm.

Self-test 25.2 For the same solution, how small would the hemispherical imprint need to be to make a particle 2 nm in diameter?

As discussed earlier, high-temperature (150–250°C) solution synthesis is one of the most successful methods for achieving size and shape control of nanocrystalline materials through nucleation and growth in the presence of selective surfactant molecules, including TOPO. However, inhomogeneity in the injection of precursors, mixing of the reactants, and temperature gradients in the reaction flask can lead to problems with control over the size distribution of nanoparticles. Small reaction vessels, such as reverse micelles and laser-embossed surfaces, provide an appealing alternative for controlling chemical and thermal homogeneity. The use of nanofabricated structures made from these materials enables the growth and assembly of nanocrystals that are potentially more interesting than inorganic salts and organic molecular crystals. This approach to the growth of nanocrystals has several advantages, including facile syntheses based on bulk reactions, flexible size control achieved by using different concentrations of precursor materials, realization of isolated nanocrystals in ordered arrays, and reusable templates.

(b) Physical vapour deposition

Key points: In the physical vapour deposition methods, a vapour of atoms, ions, or clusters physically adsorb to the surface and combine with other species to create a solid; molecular beam epitaxy is a technique where evaporated species from elemental charges are directed as a beam at a substrate where growth occurs.

In **physical vapour deposition** (PVD) methods, vapours are delivered from their source to a solid substrate on which they crystallize. The arriving gaseous species are typically atoms, ions, or clusters of elements. There are several general forms of PVD that are widely used: **molecular beam epitaxy** (MBE), sputtering, and **pulsed laser deposition** (PLD). The gas-phase species can have either relatively low kinetic energies on arrival (as in MBE) or relatively high kinetic energies (as in sputtering and PLD). The most important feature that all PVD methods have in common is the ability to achieve complex film stoichiometries, either through the transfer of the different species between the source and the film or by using multisource systems to compensate for deviations from the ideal film composition. Because vapour deposition methods allow single atomic layers to be deposited in a controlled fashion on a support or substrate, nanoscale architectures can be built from the bottom up. In this section we describe some fundamental principles for these processes in systems that allow for real-time, *in situ* control over the nanostructure of materials.

Molecular beam epitaxy is an ultrahigh vacuum technique for growing thin **epitaxial films**, films that have a definite crystallographic relationship with the underlying substrate. In MBE, molecular beams are formed by heating elemental sources until atoms evaporate and are transported ballistically to the substrate surface with relatively low kinetic energies (of the order of 1 eV). The overall pressure is kept very low to minimize collisions within the beam. To prevent species from bouncing off other surfaces of the system and back towards the substrate, much of the system is cooled with liquid nitrogen. In that way, stray elemental species adsorb onto surfaces and remain there. This cooling also helps to maintain the vacuum. Film stoichiometry and film growth rates are highly dependent on the beam fluxes, which can be controlled by adjusting the temperatures of the elemental sources.

Homoepitaxy is the epitaxial growth of a thin film of a material on a substrate of the same material. **Heteroepitaxy** is the epitaxial growth of a thin film of a material on a substrate of a different material. Heteroepitaxy introduces a strain between the growing material and the substrate caused by their crystallographic mismatch. This strain can lead to a self-assembly process where nano-islands form in ordered arrays on the surface to minimize strain energies. By alternating the deposition of two materials by MBE, a multilayer structure of ordered QDs can be created (Fig. 25.9). By tailoring the structure of the substrate appropriately, nanowires and nanowells can also be created, as well as the superlattices and artificially layered crystals we describe later.

Pulsed laser deposition is a versatile PVD technique that can be used to synthesize a wide variety of high-quality thin films (Fig. 25.10). In PLD, a pulsed laser is used to ablate a target, which releases a plume of atomized and ionized particles from its surface that condenses onto a nearby target. The PLD process usually produces films of composition identical to that of the substrate, which is a major simplification compared with techniques that require fine-tuning or expensive control equipment to achieve a specific stoichiometry. This simplicity makes PLD attractive for exploratory materials chemistry because the cation stoichiometry is controlled externally and independently of the deposition parameters. The technique also allows control over a number of deposition conditions that alter the thermodynamics and kinetics of growth, including substrate temperature, laser energy, and atmospheric conditions (partial pressures of O_2, N_2, or Ar); it can be carried out at pressures at or below 500 mTorr (about 100 Pa). Pulsed laser deposition can achieve average growth

(a)

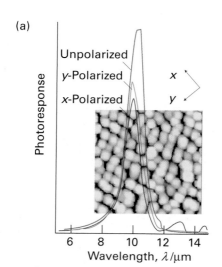

(b)

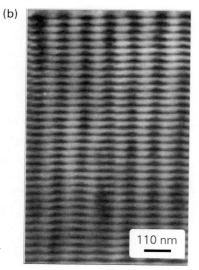

110 nm

Figure 25.9 (a) Photoresponse (in arbitrary units) of $(In_{0.2}Ga_{0.8})As$ quantum dots formed using MBE. The IR absorption band arises from intraband absorption. The inset is an AFM image of the quantum dots. The quantum dots vary in lateral dimensions between 15 and 20 nm and in vertical height in the range 3–7 nm. (b) A cross-sectional TEM micrograph of an ordered superlattice of $(In_{0.2}Ga_{0.8})$As quantum dots in a GaAs matrix. (Courtesy of E. Towe.)

rates similar to those of MBE (for example, 100 pm s^{-1}) but the instantaneous growth is much higher because growth occurs only in the short intervals (of about 1 μs) immediately after the 10−20 ns laser pulse ablates the target. Therefore, for a very short period there is a very high arrival rate and a very considerable supersaturation at the substrate surface, leading to very high nucleation rates. An additional difference from MBE is that in PLD the energies of the particles arriving at the substrate are usually 10–100 eV, which is large enough that there is the potential that arriving particles could damage the growing film. Nevertheless, PLD has been used to grow a variety of high-quality superlattices, such as the SrMnO$_3$ and PrMnO$_3$ superlattice film shown in Fig. 25.11 and described later.

(c) Chemical vapour deposition

Key point: In chemical vapour deposition methods, a vapour of molecules chemically interact or decompose at or near the substrate, where they adsorb on the surface and combine with other species to create a solid and residual gaseous products.

The control over complex stoichiometries, the ability to achieve monolayer-by-monolayer growth, and the attainment of high-quality films is not limited to physical vapour techniques. Chemical techniques, such as **metal/organic chemical vapour deposition** (MOCVD) and **atomic layer deposition** (ALD), also provide these levels of control. In contrast to physical methods, in which species condense directly onto a substrate and react with one another, chemical techniques require that a precursor decomposes chemically on or near the substrate in order to deliver the reactant species to the growing film. Therefore, in chemical vapour methods, the decomposition thermodynamics of the selected precursors must be considered because the vapour often contains elements that must not be incorporated in the growing films. Typically, the kinetic energy of the arriving species is relatively low in chemical techniques, and the temperature of the substrate can be controlled externally.

The layout of a typical chemical vapour deposition (CVD) system was shown Fig. 25.5. Such systems normally operate at moderate vacuum or even at atmospheric pressure (in the region of 0.1–100 kPa). Their growth rates can be quite high, more than 10 times that of MBE or PLD. In CVD techniques, chemical decomposition of the feed molecules proceeds upstream of a substrate surface on which the desired product will be grown. The decomposition of the gaseous reactants is activated by high temperatures, lasers, or plasmas. Large numbers of different materials have been grown using CVD methods. For Group 13/15 (III/V) semiconductors (such as GaAs), the typical sources are organometallic precursors for the Group 13 element (such as Ga(CH$_3$)$_3$), and hydrides or chlorides for the Group 15 element (such as AsH$_3$). A typical reaction is

$$Ga(CH_3)_3(g) + AsH_3(g) \xrightarrow{550-650°C} GaAs(s) + 3CH_4(g)$$

carried out in a hydrogen environment. The CVD technique can be fine-tuned to produce very high-quality films.

The drawbacks of CVD include the use of toxic chemicals, turbulent flow in the reaction chamber, the incorporation of unwanted chemical species owing to incomplete decomposition or exhaust, and the use of high pressure (which prevents the use of *in situ* diagnostics). This same technique, however, can be used without a substrate to create nanoparticles by pyrolysis of the vapour species. Furthermore, MOCVD and MBE have been combined to allow for *in situ* monitoring of growth. This technique has been called **chemical beam epitaxy** (CBE). These low pressures decrease the interactions in the reactant stream and lead to more ballistic transport of the chemical vapour species to the substrate, where they decompose and react to form a film.

A final chemical approach aims to control the precise chemical interactions that occur at a surface. In the process called **atomic layer deposition** (ALD), chemical species are delivered sequentially to a substrate on which a single monolayer deposits. The excess reactant is removed. Repetition of this monolayer coverage, subsequent reaction, and removal of excess reactants allows for precise control over the growth of complex materials. In this process, it is necessary to control the chemical species and their interactions to ensure monolayer coverage only and facile subsequent reaction. To do so means that the vapour of each reagent must interact in the proper manner with the film layer deposited previously. The technique produces flat, homogeneously coated layers. Some examples of ALD-grown nanomaterials are Al$_2$O$_3$, ZrO$_2$, HfO$_2$, CuS, and BaTiO$_3$.

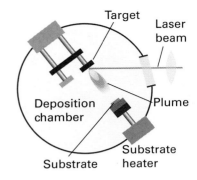

Figure 25.10 A pulsed laser deposition chamber used to produce nanostructured super lattices and artificially layered thin films.

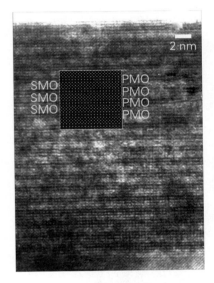

Figure 25.11 Cross-sectional TEM image of a 2 × 2 superlattice of SrMnO$_3$ and PrMnO$_3$ prepared using laser-MBE. The inset shows a calculated image confirming the (SrMnO$_3$)$_2$(PrMnO$_3$)$_2$ structure. (Reprinted with permission from B. Mercey, *et al.*, In situ monitoring of the growth and characterization of (PrMnO$_3$)$_n$ (SrMnO$_3$)$_n$ superlattices. *J. Appl. Phys.*, 2003, **94**, 2716, American Institute of Physics.)

Self-assembled nanostructures

One of the frontiers of nanoscience lies at methods that bridge the bottom-up and top-down approaches to achieve nanoscale structures. This frontier is the focus of the section that follows. The specific examples that are discussed reflect the key advantages of, and challenges facing, bottom-up methods.

25.6 Control of nanoarchitecture

In order to achieve specific commercial applications, it is important to be able to control the architecture of nanomaterials. Self-assembly, supramolecular chemistry, and morphosynthesis all offer routes to the control of the dimensionality of self-assembled nanostructures.

(a) Self-assembly

Key point: Components that can self-assemble fall between the sizes that can be controlled chemically and those that can be manipulated by conventional manufacturing; self-assembly offers the crucial technique to bridge top-down and bottom-up methods.

Various definitions of self-assembly have been proposed. They include the noncovalent interaction of two or more molecular subunits to form an aggregate with novel structure and properties that are determined by the nature and positioning of the components, the spontaneous assembly of molecules into structured, stable, noncovalently joined aggregates, the spontaneous formation of higher-ordered structures, and the process by which specific components spontaneously assemble in a highly selective fashion into a well-defined, discrete supramolecular architecture. Because the components that assemble fall between the sizes that can be controlled chemically and those that can be manipulated by conventional manufacturing, self-assembly may offer the crucial technique to bridge top-down and bottom-up methods. An example of the application of methods that bridge top-down and bottom-up assembly has been the development of microdevices capable of heterogeneous catalysis, acting as lasers, and as the basis of gas sensing.

Potential self-assembly components must be mobile, therefore self-assembly usually takes place in fluid phases or on smooth surfaces. **Static self-assembly** occurs when the system is at a global or local equilibrium, such as in liquid crystals. **Dynamic self-assembly** occurs when the system is dissipating energy, such as in an oscillating chemical reaction. **Templated self-assembly** occurs when systems are organized based on inter-actions bet-ween the components and regular features in the environment. The QD self-assembly that occurs during strained heteroepitaxy in MBE and the use of laser-embossed surfaces for small reaction vessel synthesis in ordered arrays (Section 25.5) are examples of templated self-assembly. Finally, **biological self-assembly** occurs in systems involving life, such as cells and tissues: entire organisms are elaborate examples of biological self-assembly.

(b) Morphosynthesis

Key point: Morphosynthesis is the control of architecture and morphology and the patterning of inorganic materials with nanoscale to macroscopic-scale dimensions through changes in synthesis parameters.

Morphosynthesis is the control of architecture and morphology and the patterning of inorganic materials with nanoscale to macroscale dimensions through changes in synthesis parameters. As an example, sodium polyacrylate has been used as a structure-directing or templating agent that, when coupled with controls over temperature, pH, and the concentration of reagents, can be used to tune the nanoarchitectures of $BaSO_4$ (Fig. 25.12).

Another intriguing example of morphosynthesis is the use of small molecules, such as ethylenediamine, to control crystal face growth to achieve ZnS nanowires. Zinc sulfide is a large band-gap semiconductor that has been widely used commercially as a phosphor in luminescent devices because of its emission in the visible range. Ethylenediamine molecules can act as templates in a solvothermal route to direct the formation of nanorods. The structures

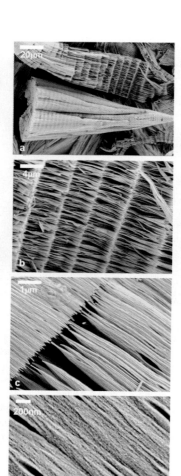

Figure 25.12 SEM illustrating the multiple levels of order in $BaSO_4$ nanostructures controlled through morphosynthesis: (a) funnel-like structure with well-aligned cones; (b) magnified section showing the alignment in the superstructure; (c) magnified section showing the alignment of the nanostructured bundles; (d) magnified section showing the surface of the bundles. (Reproduced with permission from S.-H. Yu, *et al.*, *Nano Lett.*, 2003, **3**, 379.)

BOX 25.2 Cu₂S/CNT: Novel solar cell and glucose sensor materials

Nanocrystals (NCs) and carbon nanotubes (CNTs) have found practical applications as solar cells and sensing devices. Semiconducting nanocrystalline oxides have been used to generate efficient photocurrent by an interaction between the excited NCs and the conductive CNTs, thus demonstrating their importance as foundations to build light-harvesting assemblies. These third-generation solar cells could push the efficiency of solar energy conversion beyond what traditional silicon-based solar cells can offer. Silicon-based solar cells range in 15–20 per cent conversion of solar power to usable electricity, whereas new third-generation solar cells can reach an efficiency of 35–40 per cent.

The size and morphology of the Cu₂S NCs depend strongly on the concentration of the precursors, Cu(acac)₂ and elemental S (Fig. B25.2). When the concentration of Cu(acac)₂ is near 0.05 M, the Cu₂S NCs are all spherical with diameters in the range 3–5 nm, with an average of 4 nm. On increasing the concentration of Cu(acac)₂ to 0.10 M, the spherical NCs increase in size, averaging about 8 ± 1 nm in diameter. As the concentration of Cu(acac)₂ is increased to 0.15 M, the NCs become triangular in shape, with an edge size of about 12 nm. The density of the NCs on the MWCNTs

decreases with this change of shape. The lattice structure of the MWCNTs acts as a substrate and template for the growth of the Cu₂S nanocrystals. This triangular shape of the Cu₂S NCs has been observed only in the presence of the MWCNTs. It is thought that the formation of triangular NCs at higher precursor concentration can be explained by kinetic arguments, assisted by the lattice matching with the graphite layers.

These NC–CNT structures can also serve as biosensor platforms owing to their ability to promote electron-transfer reactions with enzymes and other biomolecules. NCs deposited on CNTs have been shown to be excellent amperometric sensors for glucose over a range of concentrations. The key to these devices is to have the active Cu₂S NCs in electrical contact with the backing electrodes with the MWCNTs acting as a bridge for electrical conductivity. The redox reaction provides mobile electrons and the electrode coated with these NC–CNT hybrid materials acts as an electrical conductor. Amperometric measurements can then be made to monitor the extent of the redox reaction to detect glucose levels. Glucose detection using these nanostructures was very good, with a 10^{-5} M selectivity and a 5×10^{-8} M detection limit.

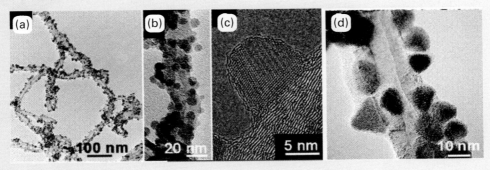

Figure B25.2 (a, b) TEM and HRTEM images of Cu₂S-MWCNT nanostructures. (c) TEM image of higher-density spherical Cu₂S NCs using 0.10 M Cu(acac)₂(aq). (d) HRTEM image of the 0.10 M Cu(acac)₂ Cu₂S NCs with average size of 8 ± 1nm. (H. Lee, *et al. Nano Lett.*, 2007, **7**, 778.)

of the nanoparticles for solar cell and glucose sensor applications can be modified by changing the concentration of the copper(II) acac complexes that are used as templates for the growth of copper(I) sulfide nanoparticles on carbon nanotubes (Box 25.2).

(c) Supramolecular chemistry and dimensonality

Key point: Supramolecular chemistry is regarded as a most promising method for designing and controlling the bottom-up assembly of nanometre-scale objects with rationally designed dimensionality.

A general concept for generating ordered structures is based on the recognition-derived spontaneous assembly of complementary subunits. In the field of supramolecular chemistry, one of the key challenges is the characterization of the supramolecules themselves, in particular the characterization of discrete, highly symmetrical, large, self-assembled entities.

A great number of biological systems are formed as a result of the formation of many weak hydrogen bonds and van der Waals interactions. Synthetic systems can take advantage not only of these bonds but also of metal–ligand bonding interactions, the latter being much stronger and generally highly directional. There are some important examples of biological systems using coordination chemistry for self-assembly, such as the metal-ion activated proteins, including Ca²⁺ signalling proteins and Zn fingers (Sections 27.4 and 27.5). As a result of various d-metal coordination geometries, metal complexes provide a pool of different acceptor subunits that can be linked together by donor building blocks to form rigid frameworks. The final shape of the self-assembled entity is defined by both the metal atom coordination geometry and the orientation of the interaction sites in a given ligand. This methodology has been successfully used to synthesize the cuboctahedron shown in Fig. 25.13.

Many self-assembled nanostructures use similar approaches to take advantage of encoded information within the components of the nanostructure building block. Molecular simulations have been performed to study the self-assembly of nanoparticles functionalized with oligomeric tethers attached to specific locations on the nanoparticle surface (Fig. 25.14). These results suggest that tethered nanobuilding blocks constitute a unique class of macromolecule that can be used to assemble novel nanomaterials.

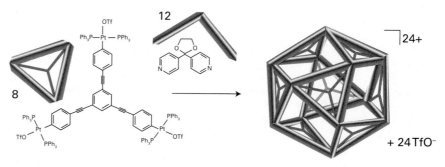

Figure 25.13 A cuboctahedron synthesized through self-assembly of two distinct building blocks. (Adapted from S. Leininger, *et al.*, *Chem. Rev.*, 2000, **100**, 853.)

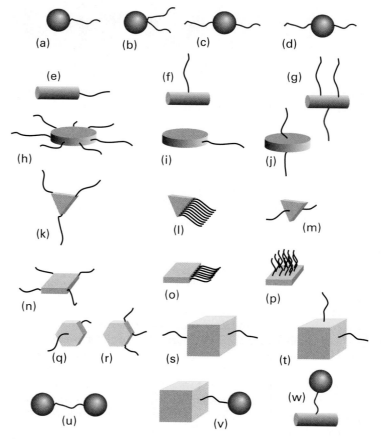

Figure 25.14 Representative tethered nanobuilding blocks: (a–d) tethered nanospheres; (e–g) tethered nanorods; (h–p) tethered nanoplates, including circular, triangular, and rectangular plates; (q, r) tethered nanowheels; (s, t) tethered nanocubes; (u–w) triblock nanobuilding blocks created by joining two nano-objects with a polymer tether. (Adapted from Z. Zhang, *et al.*, *Nano Lett.*, 2003, **3**, 1341.)

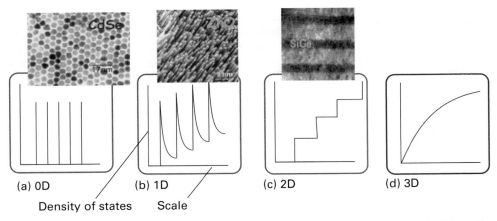

(a) 0D (b) 1D (c) 2D (d) 3D

Density of states   Scale

Figure 25.15 Effects of confinement of charge carriers on the density of states for nanomaterials with specific dimensionality. (Adapted from A. Mrumjadar, Nonoscience Boot Camp at Northwestern University 2001.)

Specific tethering groups influence the nature of the self-assembly of the nanobuilding blocks into larger ordered nanostructures.

Control over dimensionality in materials can yield unique control over their physical properties. For example, dimensionality has a noted influence on the density of electronic states (Fig. 25.15). In the next few sections we examine specific examples of how that control has been achieved and the novel properties that have been reported.

25.7 One-dimensonal control: carbon nanotubes and inorganic nanowires

Key point: Dimensionality plays a crucial role in determining the properties of materials.

The elongated one-dimensional morphology of nanorods, nanowires, nanofibres, nanowhiskers, nanobelts, and nanotubes has been studied extensively because one-dimensional systems are the lowest dimensional structures that can be used for efficient transport of electrons and optical excitation. They are therefore expected to be crucial to the function and integration of nanoscale devices. Not much is known, however, about the nature of localization that could preclude transport through one-dimensional systems. Such systems should possess discrete molecular-like states extending over large distances and exhibit some exotic phenomena, such as the effective separation of the spin and charge of an electron. There are also many applications where one-dimensional nanostructures could be exploited, including nanoelectronics, very strong and tough composites, functional nanostructured materials, and novel probe microscopy tips.

To address these fascinating fundamental scientific issues and potential applications, two important questions pose key challenges to the fields of condensed matter chemistry and physics research. First, how can atoms or other building blocks be assembled rationally into structures with nanometre-sized diameters but much greater lengths? Second, what are the intrinsic properties of these quantum wires and how do these properties depend, for example, on their diameter and their structure? A key class of nanomaterials that offers potential answers to these questions is **carbon nanotubes** (CNTs). Carbon nanotubes are perhaps the best example of novel nanostructures fabricated through bottom-up chemical synthesis approaches. They have very simple chemical composition and atomic bonding configuration but exhibit remarkably diverse structures and unparalleled physical properties. These novel materials have found application as chemical sensors, fuel cells, field-effect transistors, electrical interconnects, and mechanical reinforcers.

The structure and properties of buckminsterfullerene, C_{60}, and related species have produced much activity in the chemistry and materials sciences communities. The now familiar soccer-ball structure of C_{60} itself was discussed in Chapter 14. Carbon nanotubes were discovered in the early 1990s by electron microscopy. The bonding, local coordination, and general structure of CNTs are similar to those of buckminsterfullerene, but CNTs can have a greatly extended length, leading to a tube rather than a ball structure (Fig. 25.16).

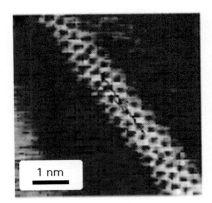

1 nm

Figure 25.16 STM image of a single-walled carbon nanotube. (J. Hu, *et al.*, *Acc. Chem. Res.*, 1999, **32**, 435.) (Reproduced with permission from the American Chemical Society.)

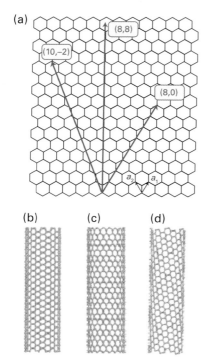

Figure 25.17 (a) The honeycomb structure of a graphene sheet. Single-walled carbon nanotubes can be formed by folding the sheet along lattice vectors, two of which are shown as a_1 and a_2. Folding along the (8,8), (8,0), and (10,–2) vectors leads to armchair (b), zigzag (c), and chiral (d) tubes, respectively. (Based on H. Dai, *Acc. Chem. Res.*, 2002, **35**, 1035.) (Reproduced with permission from the American Chemical Society.)

Carbon nanotubes are cylindrical shells formed conceptually by rolling graphene (graphite-like) sheets into closed tubular nanostructures with diameters matching that of C_{60} (0.5 nm) but lengths up to micrometres. A single-walled nanotube (SWNT) is formed by rolling a sheet of graphene into a cylinder along an (*m,n*) lattice vector in the graphene plane (Fig. 25.17). The (*m,n*) indices determine the diameter and chirality of the CNT, which in turn control its physical properties. Most CNTs have closed ends where hemispherical units cap the hollow tubes.

Carbon nanotubes self-assemble into two distinct classes, SWNTs and multiwalled carbon nanotubes (MWNTs). In MWNTs, the tube wall is composed of multiple graphene sheets. **Morphosynthetic control** provides routes to tune the details of these self-assembled nanostructures. This self-assembly occurs during synthesis using several of the fabrication techniques we have already described, including pulsed laser ablation, laser-assisted catalytic growth, chemical vapour deposition (CVD) based on hydrocarbon gases, and carbon arc discharge. All these strategies rely on vaporizing carbon and condensing some fraction into extended nanostructures. **Laser vaporization methods** typically make relatively small amounts of nanocarbons; specialized CVD techniques have been developed to synthesize CNTs in quantities in excess of a few milligrams. In the **CVD approach**, a hydrocarbon gas such as methane is decomposed at elevated temperatures and C atoms are condensed on to a cooled substrate that may contain various catalysts, such as Fe. This CVD method is attractive because it produces **open-ended tubes** (which are not produced in the other methods), allows continuous fabrication, and can easily be scaled up to large-scale production. Because the tubes are open, the method also allows for the use of the nanotube as a templating agent.

In the **arc discharge method**, extremely high temperatures are obtained by shorting two carbon rods together, which causes a plasma discharge. Such plasmas easily achieve temperatures in excess of where carbon vaporizes (at about 4500 K). Low potential differences and moderately high currents are needed to produce this arc. The typical CNT formed by either the arc method or CVD is multiwalled. To encourage SWNT formation, it is necessary to add a metal catalyst such as Co, Fe, or Ni to the carbon source. These metal catalyst particles block the end-cap of each carbon hemisphere and thus promote SWNT growth. In addition, the growth directions of the nanotubes can be controlled by van der Waals forces, applied electric fields, and patterning of the metal catalyst onto different substrates. The patterned growth approach is feasible with discrete catalytic nanoparticles and scalable on large wafers to achieve large arrays of nanowires (Fig. 25.18).

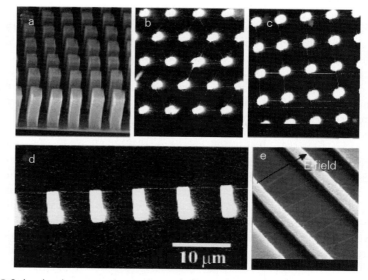

Figure 25.18 Ordered carbon nanotube structures obtained by direct chemical vapour deposition synthesis. (a) An SEM image of self-oriented MWNT arrays. Each tower-like structure is formed by many closely packed multiwalled nanotubes. Nanotubes in each tower are oriented perpendicular to the substrate. (b) SEM top view of a hexagonal network of SWNTs (the line-like structures) suspended on top of silicon posts (the bright dots). (c) SEM top view of a square network of suspended SWNTs. (d) Side view of a suspended SWNT power line on silicon posts (the bright lines). (e) SWNTs suspended by silicon structures (the bright regions). The nanotubes are aligned along the electric field direction. (H. Dai, *Acc. Chem. Res.*, 2002, **35**, 1035.) (Reproduced with permission from the American Chemical Society.)

The repeating axial hexagonal patterns of CNTs are graphitic structures; however, the electrical properties of nanotubes depend on the relative orientation of the repeating hexagons. The nanotubes can be either semiconductors or metallic conductors. When oriented in the chair configuration, CNTs exhibit remarkably high electrical conductivity. Electrons can travel through the micrometre-length nanowire with zero scattering and zero heat dissipation. CNTs also have very high thermal conductivity, comparable to the best thermal conductors known (such as their counterparts, diamond and graphite). Therefore, they are being heralded as the ideal nanomaterial for the development of interconnects for integrated circuits. They may solve two key challenges in the computer industry: heat dissipation and increased processing speeds.

Open-ended CNTs can also be used for nanorobotic spot welding to deposit metal at specific sites for connections in integrated circuits (Fig. 25.19). Copper is a good choice for these applications because it has high thermal and electrical conductivity. It also has a very low bonding energy to C (0.10–0.14 eV/atom), so can be deposited easily on surfaces and structures from the encapsulating CNT. The deposition is accomplished by making physical contact between a STM probe and a copper-filled CNT tip, and then applying a potential difference to create a circuit with the probe as the anode. This procedure generates sufficient heat for the Cu atoms to be able to migrate along the tube and be deposited where they are required. The nanowelding process is of interest because its advantages include efficiency (a very low current can induce melting and flow), three-dimensional manipulation of welding site by using piezoelectric distortion, rapid melting, precise control of small mass delivery, and the compatibility of Cu with conventional semiconductors.

In addition to CNTs, parallel methods have been discovered for making nanotubes out of materials that share bonding characteristics with C, including semiconductors and metal oxides. More specifically, BN, ZnO, ZnSe, ZnS, InP, GaAs, InAs, and GaN have all been made into nanotubes. The novel electronic properties and small sizes of these nanotubes make them appealing as inorganic nanowires.

Core–sheath wires similar to macroscopic coaxial wires are of great interest. One method to prepare them uses a colloid-templating strategy that involves a layer-by-layer assembly of polyelectrolytes and inorganic nanoparticles on submicrometre- and micrometre-sized polystyrene (PS) latex particles. A number of methods, including laser ablation, carbothermal reduction, and several solution-based methods, have been developed to generate one-dimensional nanostructures with coaxial structures. For instance, by using a novel nanowire-templating technique based on the layer-by-layer approach and calcining (heating in air to drive off volatile templating agents), ordered Au/TiO$_2$ core–sheath nanowire arrays have been fabricated. A template-grown gold nanowire array is used as a positive template, and then a cationic polyelectrolyte and an inorganic precursor are assembled on gold nanowires by the layer-by-layer technique. Calcination then converts the inorganic precursor to titanium dioxide (Fig. 25.20).

Figure 25.19 FESEM image of Cu-filled CNTs; all have sharp tips filled with metal nanoneedles. These CNTs are up to 5 μm long, with outer diameters in a range of 40–80 nm. (L. X. Dong, *et al.*, *Nano Lett.* 2007, **7**, 58.)

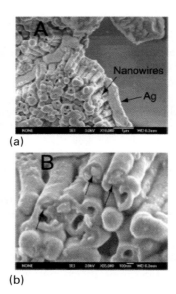

Figure 25.20 (a) Low-magnification and (b) high-magnification SEM images of Au/TiO$_2$ core–sheath nanowire arrays. (Y.-G. Guo, *et al.*, *J. Phys. Chem. B*, 2003, **107**, 5441.)

An advantage of the layer-by-layer nanowire-templating approach over other methods is that nanoscale control can be exerted over the sheath thickness by varying the number of deposition cycles. Titanium dioxide has been one of the most investigated oxide materials because of its technological importance. It has been widely used for photocatalysis and environmental clean-up applications, on account of its strong oxidizing power, chemical inertness, and nontoxicity. These core–sheath nanowires have well-defined diameters and lengths, largely determined by the dimensions of the templates, and porous sheaths with thicknesses controlled by the number of layers deposited. This method opens a door to the use of a wide variety of nanowires as the positive templates for the fabrication of core–sheath nanostructured materials for chemical sensors, photocatalysis, light energy conversion devices, and nanoscale electronic and optoelectronic devices. More specifically, Au and Ag nanowires have been targeted as scaffolds for optical sensing of environmentally and biologically important analytes with very high selectivity and sensitivity.

25.8 Two-dimensional control: quantum wells and solid-state superlattices

In this section, we describe the design of so-called *two-dimensional materials*, which are materials that have macroscopic length scales in two dimensions and a nanoscale length scale along the third dimension.

Thin-film processing methods permit the deposition of films only one atomic (or one unit cell) layer thick. By varying sequentially the types of atomic (or unit cell) layers being deposited, it is possible to control the material architecture along the growth direction at a sub-nanometre scale, thereby allowing the bottom-up development of artificially layered nanostructures. A **quantum well** (QW) is a thin layer of one material sandwiched between two thick layers of another material and is the two-dimensional equivalent of a zero-dimensional QD. In a **superlattice**, two (or more) materials alternate with an artificially induced periodicity along the growth direction. Superlattices often have repeat periods of about 1.5–20 nm or greater, and sublayer thicknesses that range from two unit cells to many tens of unit cells. Artificial crystal structures usually have repeat distances similar to bulk crystals (about 0.3–2.0 nm) and have sublayer thicknesses that range from an atomic layer to two unit cells (about 1 nm).

These structures have found broad commercial application as key device elements in computer chip manufacturing, including hard disk read heads. To control a physical property such as absorption and emission, a quantum well can be used in a similar fashion to the QDs described earlier. The optical properties of quantum wells can be enhanced in quantum well superlattices (often called *multiple quantum well structures*), which are used in lasers. To control a physical property such as hardness, which is tied to dislocation formation and motion, it is known that many interfaces between dissimilar materials are required. Hardness can be increased substantially in such superlattices, as compared to monolithic films, ranking them

among the hardest known materials and making them successful coatings in the tooling industry. To control a property such as ferroelectricity or superconductivity, which are intimately
tied to the crystal structures, it is important to be able to engineer artificial crystal structures.

(a) Quantum wells

**Key points: Quantum wells consist of a thin small band-gap material sandwiched between thick lay-
ers of large-gap materials; multiple quantum wells can enhance the effects of quantum wells when the
wells do not interact.**

A quantum well is typically composed of two semiconductor materials with different band
gaps, such as $Al_{1-x}Ga_xAs$ and GaAs. The smaller band-gap material (GaAs) is sandwiched
between layers of the larger band-gap material ($Al_{1-x}Ga_xAs$), and the thickness of the layer
of the small band-gap material is confined to the nanometre scale (Fig. 25.21). Electrons are
confined in the layer of the small band-gap material along the growth direction; the barrier
height is the difference in the energy levels of the band edges for the two materials. Therefore, the energy states in the smaller gap material are quantized along the growth direction,
similar to the quantization of energy in a QD. The electronic wave functions are spatially extended in the direction parallel to the layer and therefore the energy levels are not quantized.

As for a QD, the optical properties of quantum wells can be tailored and both interband
(between the valence and conduction band) and intraband (between the quantized
sub-bands) absorption and emission can be controlled. Both $In_{1-x}Ga_xAs/GaAs$ and
$Al_{1-x}Ga_xAs/GaAs$ quantum wells have been widely studied, and the optical transition has
been observed to move to higher energies compared to the bulk when the thickness drops
down to about 20 nm. The main use of quantum wells is in semiconductor lasers, where
the small-gap quantum well is the active layer in the device. Typical QW lasers use the interband transitions, but intraband lasers are also being developed and are discussed below.

Many of the effects that occur in quantum wells can be enhanced by using **superlattice
structures**, the periodic repetition of quantum wells along the special direction (the z-axis).
In the context of semiconductors these superlattices are called **multiple quantum well**
(MQW) structures. If the active layers (the small band-gap layers) do not interact, then the
electrons are confined to a given layer and are unable to tunnel between them. The use of
MQW structures in this case increases the absorption or emission from a given device, as
there are multiple levels. For example, an MQW laser has much higher power output than
the corresponding single quantum-well laser.

If the wide band-gap layers are thin enough, one QW interacts with the adjacent QW
and electrons can tunnel between them. This phenomenon is used in **quantum cascade** (QC)
lasers, which operate at high powers in the IR region. The laser characteristics of these materials are fundamentally different from those of semiconductor diodes and MQW lasers in a
variety of ways. In a QC laser, only one type of carrier (electrons) is necessary to achieve laser
action; in the other two, both electrons and holes are required. In addition, in the QC laser,
the transitions are intraband transitions that arise from the quantization of the valence band.

The QC lasers are constructed from 13/15 (III/V) semiconducting materials such as GaAs,
InAs, and AlAs. All these superlattice systems have been grown by solid-source molecular
beam epitaxy on single-crystal substrates. The heterostructures that comprise an individual
layer in such a superlattice are far more complex than the simple unit-cell periodicities found
in most artificially layered structures or superlattices discussed below. For example, a single
functional unit of an AlAs/GaAs superlattice QC laser has a layer sequence (in nanometres) of

2.4 4.3 1.8 4.3 1.4 4.4 1.3 4.5 1.2 4.5 1.1 4.8 0.8 5.2 0.6 5.8 0.6 6.2 0.5 6.6 0.5   7.2   0.5   8.0

with GaAs layers in red, AlAs in blue, and Si-doped layers in green. This sequence leads
to a functional layer thickness of 72.8 nm, which is much larger than most superlattice
periods. The superlattice itself consists of 40 such layers.

(b) Solid-state superlattices

**Key points: Artificially layered materials have a periodic repeat along the growth direction of a thin
film; the periodic repeat is controlled by the number and type of sublayers deposited in sequence.**

The periodic repeat along the growth direction of a thin film in a superlattice is controlled
by the number and type of sublayers deposited in sequence, whereas the lateral periodicity is determined by the coherency—the matching of lattice characteristics—between the

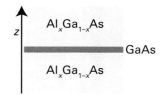

Figure 25.21 An $(Al_xGa_{1-x}As)-(GaAs)-$
$(Al_xGa_{1-x}As)$ quantum well; the thickness of
the GaAs layer is on the nanoscale.

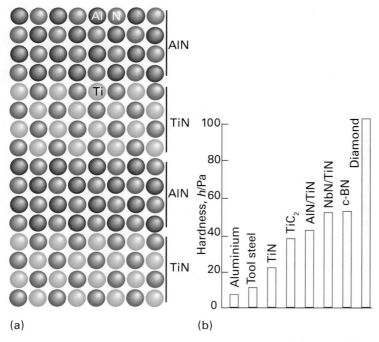

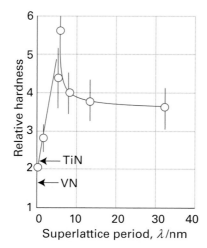

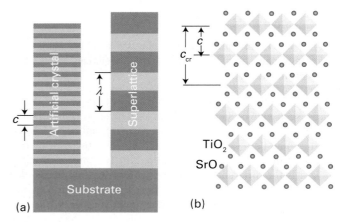

Figure 25.22 (a) The structure of an ultrahard AlN/TiN superlattice and (b) the hardness of a nitride superlattice and commonly used hard materials. (Adapted from S.A. Barnett and A. Madan, *Physics World*, 1998, **11**, 45.)

Figure 25.23 (a) The structure of an AB artificial crystal and an AB superlattice; c and λ represent the repeat period during growth of the artificial structure and the superlattice period, respectively. (b) The structure of Sr_2TiO_4 as an artificially layered oxide. The polyhedra represent Ti-centred, corner-sharing TiO_6 octahedra and the spheres represent Sr^{2+} cations. The lateral coherency between the SrO and TiO_2 sublayers is excellent in this layered structure and the crystallographic repeat parameter, c_{cr}, is twice the growth repeat period, c.

Figure 25.24 The dependence of hardness on the superlattice period for $(TiN)_m(VN)_m$ superlattices. (Adapted from U. Helmersson, *et al.*, *J. Appl. Phys.*, 1987, **62**, 481.)

sublayers (Figures 25.22 and 25.23). It should be kept in mind that the superlattice is built bottom-up with periods in the nanometre range and total thicknesses in the micrometre range. The goal of materials chemists and engineers is to use these synthesis skills to control the properties and functionality of materials of this kind.

Superlattice nitrides are ranked among the hardest known materials. The superlattice period and the chemical composition play important roles in determining the mechanical properties of such compounds, as the interfaces between the two nitride layers are responsible for the enhanced hardness. Figure 25.24 shows, for instance, that the maximum hardness of a typical superlattice occurs for periodicities in the range 5–10 nm. These superlattices have been successfully deposited using sputtering, pulsed laser deposition, and molecular beam epitaxy. Such superlattices are used on cutting tools and are made economically by using sputtering to form them.

Perovskite oxides (Section 24.7) are used in numerous applications on account of their ferroelectric, acoustic, microwave, electronic, magnetic, and optical properties. The most widely used perovskite is $BaTiO_3$, which is a very important dielectric material. It has been discovered that layering ferroelectric $BaTiO_3$ with the isostructural perovskite $SrTiO_3$ can lead to enhanced dielectric properties arising from lattice strain. The techniques of PLD, MBE, and CVD have all been used to create $SrTiO_3/BaTiO_3$ superlattice thin films (Fig. 25.25). Although the misfit between the two crystal structures is small enough to allow for epitaxial growth and the formation of coherent interfaces between each bilayer, the stress introduced at the interface is sufficient to improve the dielectric response, specifically the remanent polarization (the polarization in the absence of an applied field) of the superlattices, compared to undoped $BaTiO_3$.

It is important to control layer thickness precisely to ensure the formation of coherently strained interfaces that lead to improved dielectric response. The technique of PLD has been used to deposit alternating blocks of $SrTiO_3$ and $BaTiO_3$ having unequal layer thicknesses of $SrTiO_3/BaTiO_3$ = 15/3, 10/3, 3/10, and 3/15, and the net dipole moment density of superlattices has been reported to be in the order of 46 μC cm^{-2}, whereas that of bulk films of the parent phases is about 14 μC cm^{-2}.

Perovskite-based structures can also exhibit interesting magnetic effects. In particular, manganese-based perovskite films such as $(La,Sr)MnO_3$ possess useful ferromagnetic and magnetoresistive properties. In general, these characteristics arise from electron exchange interactions occurring between Mn and O that are mediated by the various cations in the structure, and can be altered significantly when these cations are spatially ordered. The PLD procedure has been used to deposit superlattices of $LaMnO_3/SrMnO_3$; the Mn^{3+} cations are found in the $LaMnO_3$ layers and the Mn^{4+} are found in the $SrMnO_3$ layers. Thus, the thin-film superlattice technique allows for precise ordering of A-site cations (La, Sr), which in turn causes ordering of the Mn in its different oxidation states. In superlattices of $(LaMnO_3)_m(SrMnO_3)_m$, samples with $m \leq 4$ possess magnetic properties just like solid solutions of $La_{0.5}Sr_{0.5}MnO_3$, whereas superlattices with larger periods have significantly higher resistivities and lower Curie temperatures. Similar effects are observed in superlattices with unequal layer thicknesses, such as $(LaMnO_3)_m(SrMnO_3)_n$ and $(PrMnO_3)_m(SrMnO_3)_m$ systems (Fig. 25.11).

The magnetic properties of thin-film superlattices in the family $La(Sr)MnO_3/LaMO_3$ (M = Fe, Cr, Co, and Ni) exhibit a wide range of properties depending on both superlattice repeat period and the identity of M. For instance, when M = Co or Ni the Curie temperature is increased but when M = Fe or Cr it is decreased.

Interesting electronic effects are observed in superlattice thin films consisting of spinel Fe_3O_4 and rock-salt NiO layers. Superlattices of these two structures are well matched because both have the same O sublattice and differ only in the locations of the cations. Superlattices of composition Fe_3O_4/NiO have been grown by MBE for a variety of layer thicknesses. The electronic conduction mechanism in many d-metal oxides such as these is **electron hopping**, a process that has a short mean free path. If this distance is approximately the same as the crystal lattice spacings it can lead to novel electrical properties for artificially layered materials with short lattice spacings. Thus, Fe_3O_4/NiO superlattices with bilayer repeat periods of 6.8 nm have electrical conductivities that differ by factors in the range 10^6–10^8 in different directions, which is among the highest known electrical anisotropy for any material. This result effectively demonstrates both the high degree of chemical modulation available by superlattice deposition and the extent of electrical conduction localization that can be achieved as a consequence.

25.9 Three-dimensional control

The design and synthesis of three-dimensional (3D) supramolecular architectures with tunable, nanoporous, open channel structures have attracted considerable attention because of their potential applications as molecular sieves, sensors, size-selective separators, and catalysts.

(a) Mesoporous materials

Key points: Mesoporous materials and metal–organic frameworks have ordered pore structures that are defined over the nanoscale; the synthesis of these materials relies upon control of self-assembly; guests can be incorporated into the inorganic host framework for novel applications.

An important class of three-dimensionally ordered nanomaterials is **mesostructured nanomaterials**. Mesoporous materials are well known in heterogeneous catalysis (Chapter 26) and are of great interest because of our ability to tune their pore sizes from 1.5 to 10 nm (Fig. 25.26). Mesoporous inorganic nanomaterials are synthesized in a multistep process based on the initial self-assembly of surfactant molecules and block copolymers that self-organize into supramolecular structures (which are liquid crystalline assemblies of cylindrical, spherical, or lamellar micelles; Fig. 25.27). These supramolecular frameworks serve as structure-directing templates for the growth of mesostructured inorganic materials (often silica or titania). During a solvothermal reaction step, oxide particles (silica, Fig. 25.27) form at the surfaces of the hexagonal rods, assembling around the supramolecular structures. The templating agent can be removed through an acid wash or by calcination to yield inorganic materials having hexagonal pores with uniform and controllable dimensions. This porosity offers unique capabilities for both catalysis and inclusion chemistry.

A wide variety of mesostructured and mesoporous inorganic materials have been obtained by varying the choice of templating agent and reaction conditions (Fig. 25.28). For instance, the family known as M41S has silica or alumina-silica inorganic phases and various cationic surfactants leading to three distinct types of structures: hexagonal lamellar (MCM-50), cubic (MCM-48), and hexagonal (MCM-41) (Fig. 25.28).

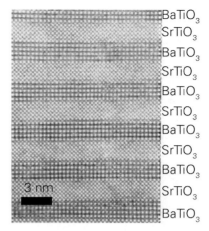

Figure 25.25 TEM image of a $SrTiO_3$/$BaTiO_3$ superlattice. (Reprinted from D. G. Schlom, *et al.*, Oxide nano-engineering using MBE. *Mater. Sci. Eng. B*, 2001, **87**, 282, with permission from Elsevier.)

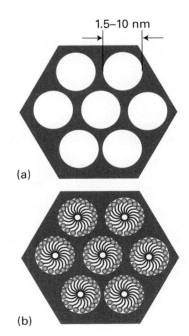

Figure 25.26 A hexagonal mesoporous structure with (a) controlled nanoporosity and (b) functionalized pores.

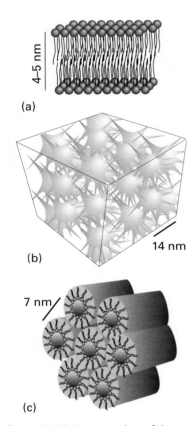

(a)

(b)

14 nm

7 nm

(c)

Figure 25.28 Representations of three types of ordered mesoporous solids. (a) Lamellar (layered materials), (b) cubic (complex arrangements), and (c) hexagonal (honeycomb). (Adapted from A. Mueller and D.F. O'Brien, *Chem. Rev.*, 2002, **102**, 729.)

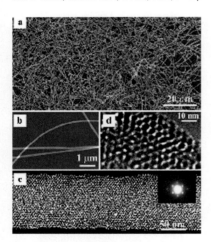

Figure 25.29 (a) SEM image of mesoporous silica nanofibres. (b) Low-magnification TEM image of the nanofibres. (c) High magnification TEM image of one nanofibre. The inset is a selected-area electron diffraction pattern of the nanofibre. (d) High-resolution TEM image recorded at the edge of one nanofibre. (J. Wang, *et al.*, *Chem. Mater.*, 2004, **16**, 5169.)

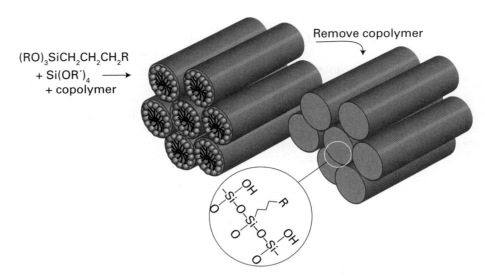

Remove copolymer

$(RO)_3SiCH_2CH_2CH_2R$
$+ Si(OR')_4$
$+ copolymer$

Figure 25.27 Block-copolymer structure-directing agents self-assemble into micellar rods that form hexagonal arrays that can be removed to yield nanoporous silica matrices for inclusion chemistry and catalysis. (Adapted from M.E. Davis, *Chem. Rev.*, 2002, **102**, 3601.)

The shape of the surfactant, and the water content used during synthesis controls the resulting nanoarchitecture, as can be seen from the illustration. Surfactants can also be used to tailor the structure. For example, surfactant cetyltrimethylammonium cations (C_{16}TMAC) have been used to fabricate silica nanofibres with hexagonal pores.

Mesoporous nanomaterials also offer routes to functionalize pores to increase catalytic activity and selectivity. They have received much attention as host materials for the inclusion of numerous guests such as organometallic complexes, polymers, d-metal complexes, macromolecules, and optical laser dyes. The silica nanofibres shown in Fig. 25.29 can even be used as hosts to grow nanowires of various other oxide materials.

EXAMPLE 25.3 **Controlling nanoporosity in molecular sieves**

(a) Compare the nanoarchitecture of ZSM-5 zeolite and MCM-41; include in your comparison descriptions of the dimensionality of tunnels and relative pore sizes. (b) Cetyltrimethylammonium $[C_{16}H_{33}N(CH_3)_3]^+$ cation and tetrapropylammonium $[N(CH_2CH_2CH_3)_4]^+$ cation are surfactants used in the syntheses of these materials. Which surfactant is a better choice for the synthesis of MCM-41?

Answer We need to consider the relative size of the pores in these two catalytic materials and the chain length and size of the surfactant used for templating their porosity. (a) ZSM-5 is a microporous zeolite with pore sizes of approximately 0.5 nm. It has three-dimensional intersecting pores similar to the cubic mesoporous phases but with smaller dimensions. MCM-41 is a mesoporous solid with hexagonal, one-dimensional pores with tunable dimensions of 2 nm up to 10 nm. (b) The choice of surfactant must match the pore size of the solid. The longer chain surfactant (cetyltrimethylammonium cation) is a better choice for the self-assembly of the MCM-41 material owing to its longer hydrocarbon tails and larger spontaneous curvature. This increased tail size encourages larger pore dimensions that are a direct consequence of the packing of the surfactants within the micellar rods that lead to the hexagonal mesophase.

Self-test 25.3 Which of these materials, MCM-41 or ZSM-5, is more likely to be used as a host material for the entrapment of QDs?

(b) Metal–organic polyhedra and frameworks

Key points: Metal–organic polyhedra and frameworks offer tunable catalytic surfaces and high surface areas; they are assembled using supramolecular chemistry.

A second class of three-dimensionally controlled nanostructures is **metal–organic frameworks** (MOFs). These frameworks are self-assembled from a careful choice of metal ions, bridging organic ligands, and/or **metal–organic polyhedra** (MOP). They offer an alternative approach to the synthesis of open porous materials using **supramolecular assembly** and are being called the 'new zeolites'. These macroscopic materials have some of the highest reported surface areas and, as such, have found practical application for methane and hydrogen storage, catalysis, and drug delivery.

As discussed earlier, supramolecular chemistry relies on bridging together nanobuilding blocks (NBBs) to assemble macroscopic materials. One new strategy in design is the incorporation of **nanoporosity** (pores with dimensions of 1 to 100 nm) within the NBB or MOP. The synthesis of a nanoporous cuboctahedron NBB (Fig. 25.30a) was achieved by linking rigid square molecules of $Cu_2(CO_2)_4$ with m-BDC (1,3-benzenedicarboxylate). The overall NBB formula is $Cu_{25}(m\text{-BDC})_{24}(DMF)_{14}(H_2O)_{10}(H_2O)_{50}(DMF)_6(C_2H_5OH)_6$ and is termed more conveniently MOP-1. This compound is constructed from 12 paddle-wheel units (themselves NBBs) bridged by m-BDC to yield a large metal-carboxylate polyhedron. The simplest way to view the overall structure is by considering its relationship to the cuboctahedron (shown in Fig. 25.13), where each square and link has been replaced by the paddle-wheel NBB and the m-BDC (two-connector) units, respectively, to give an *expanded-augmented cuboctahedron* (or a truncated cuboctahedron). The ability to synthsize the cuboctahedron illustrates the feasibility of obtaining crystals of large porous metal–organic polyhedra in which rigid NBBs are an integral part of a well-defined structure. More recently, a zeolitic imidazolate framework compound, ZIF-69, has been reported that can store 83 dm^3 of carbon dioxide per litre (1 dm^3) of material at 0°C and 1 atm.

MOPs have also been designed for selective catalysis that mimic active sites in enzymes. Enzymatic catalysis often relies on the highly constrained environment of substrates within the active site. Recent examples in supramolecular organometallic chemistry mimic this approach by using host–guest interactions. A supramolecular M_4L_6 tetrahedral assembly has been developed to provide a host that mediates a variety of organic and organometallic transformations, including the isomerization of allylic alcohols. The $M_4L_6{}^{12-}$ host is formed from the self-assembly of four octahedral metal centres (M = Fe(III) or Ga(III)) at the vertices and six bis-catecholamide naphthalene ligands (L^{4-}) that span the edges of a tetrahedron (Fig. 25.30b). The host has been found to encapsulate monocationic guests

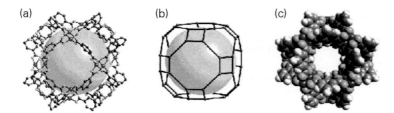

Figure 25.30a The crystal structure of MOP-1 (drawn using coordinates obtained for c-MOP-1) showing (a) 12 paddle-wheel units (Cu, red; O, blue; C, grey) linked by m-BDC to form (b) a large truncated cuboctahedron of 1.5 nm diameter void (yellow sphere); the grey spheres (carbon atoms) represent the polyhedron constructed by linking together only the carboxylate C atoms in (a) to form linked square NBBs, an arrangement that provides for (c) a large porous polyhedron with triangular and square windows. Hydrogen atoms in light grey; otherwise same colouring scheme as in (a). (M. Eddaoudi, *et al.*, *J. Am. Chem. Soc.*, 2001, **123**, 4368.) (Reproduced with permission from the American Chemical Society.)

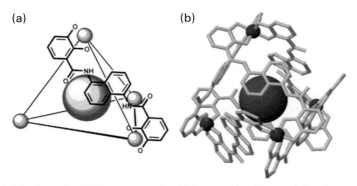

Figure 25.30b (a) Schematic of MOP structure of an M_4L_6 assembly with one of the edge-spanning ligands displayed containing an organometallic guest. (b) Crystal structure of Fe_4L_6 assembly with an organometallic guest. (D. H. Leung, *et al.*, *J. Am. Chem. Soc.*, 2007, **129**, 2746.) (Reproduced with permission from the American Chemical Society.)

ranging from NH_4^+ cations to cationic organometallic species containing Rh metal centres. The catalytic activity is controlled by the size and shape of the host cavity.

Some of the key challenges in the area of mesostructured materials are achieving control over pore size and shape, the presence of counterions and solvents within the channels, the interpenetration of different networks, the framework structural instability in the absence of guest molecules, and the often low thermal stability of the host framework. In some cases, the incorporation of solvent molecules can be used to achieve desired structures. For example, a robust three-dimensional porous structure of formula $[Ln_2(PDC)_3(DMF)_2]_\infty$ has been constructed from lanthanoid (Ln) cations (Er^{3+} or Y^{3+}) and the non-linear anionic bridging ligand, pyridine-3,5-dicarboxylate (PDC^{2-}) in dimethylformamide (DMF). The solvated framework polymers undergo a solid-state, crystal-to-crystal reaction on heating. Through loss of both sorbed and coordinated solvent and rearrangement of the framework core a porous MOF structure is formed with retention of structural integrity (Fig. 25.31). These mesoporous materials have proven to be effective absorbants for H_2, N_2, and benzene.

(c) Inorganic–organic nanocomposites

Key points: Class I inorganic–organic materials have noncovalent interactions and class II have some covalent interactions; sol–gel and self-assembly methods are key chemical routes to hybrid nanocomposite design and synthesis; the properties of polymer nanocomposites are controlled through the nature of the polymeric and inorganic phases, as well as through their dispersions and interactions.

Inorganic–organic nanocomposites are a third class of three-dimensional ordered materials that possess chemical and physical properties that can be tuned by using the synergistic association of organic and inorganic components at nanoscales. These hybrid materials originated in the paint and polymer industries where inorganic fillers and pigments were dispersed in organic materials (including solvents, surfactants, and polymers) to fabricate commercial products with improved materials performance. Hybrid nanomaterials are also of interest because their mechanical properties occupy a niche between glasses and polymers, so achieving enhanced strength and robustness. Hybrid nanomaterials have been reported with excellent laser efficiencies, photostability, and ultrafast photochromic response. They also can have very large second-order and third-order nonlinear optical response, which is important in frequency conversion and optical switching for telecommunications.

Nanocomposites have already entered the market-place in sunscreens, fire-retardant fabrics, stain-resistant clothing, thermoplastics, water filters, and automobile parts. Examples include the use of nylon-6/montmorillonite clay nanocomposites for timing-belt covers, television screens that are coated with indigo dyes embedded in a silica/zirconia matrix, organically doped sol–gel glassware, and sol–gel-entrapped enzymes.

Hybrid nanomaterials offer the materials chemist novel routes that use rational materials design to optimize structure–property relationships. Their nanoarchitectures and resulting properties depend on the chemical nature of the components and the synergy between them. A key part of the design of these hybrids is the selective tuning of the nature, extent, and accessibility of the interfaces between the inorganic and the organic building blocks.

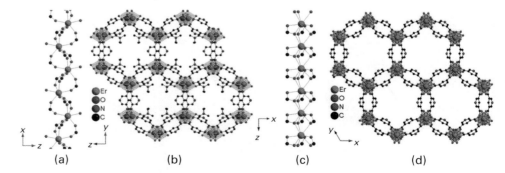

(a)   (b)   (c)   (d)

Fig 25.31 Views of the MOF structures of $[Er_2(PDC)_3(DMF)_2]_\infty$: (a) zigzag connection between adjacent erbium centres; (b) channels with a view along the orthorhombic x axis showing coordinated DMF molecules protruding into the channel space. Views of the structure after DMF solvent removal and rearrangement: (c) the shorter linear connection between adjacent Er^{3+} centres; (d) view of the open channels along the hexagonal z axis. (J. Jia, *et al.*, *Inorg. Chem.*, 2006, **45**, 8838.) (Reproduced with permission from the American Chemical Society.)

The interfaces fall into two main classes. **Class I** corresponds to hybrid materials where no covalent or ionic bonds are present between the organic and inorganic phases. In Class I materials, the various components interface through weak interactions such as hydrogen bonding, van der Waals contacts, and electrostatic forces. In **Class II** materials, at least some of the organic and inorganic components are linked through strong chemical bonds (covalent, ionic, or Lewis acid–base bonds).

An important feature in the tailoring of hybrid networks is the chemical pathway used to design a given material. The main chemical routes are summarized in Fig. 25.32. It is possible to control the nature of the inorganic precursors (by using the methods described in Section 25.5), the nature of the organic material, and the method of assembling the composite material. The templated-growth fabrication processes often rely on organic molecules and macromolecules as structure-directing agents to achieve the construction of complex hierarchical architectures. It turns out that molecular and supramolecular interactions between template molecules (surfactants, amphiphilic block copolymers, organogelators, etc.) and hybrid or metal–oxo-based networks and NBBs allow a rich variety of nanocomposite materials to be prepared. Figure 25.33 illustrates two approaches to the assembly of NBBs into larger, ordered structures.

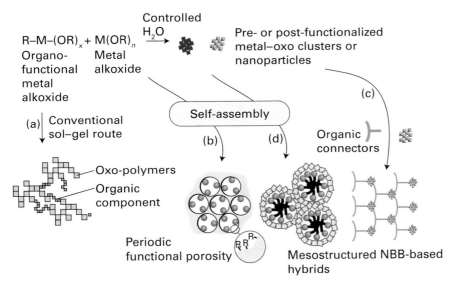

Figure 25.32 Different routes to obtain hybrid nanomaterials. Route (a) is a conventional sol–gel route that leads to 'ordinary' hybrids. Routes (b) and (d) involve the use of templates capable of self-assembly, giving rise to mesoorganized phases. Routes (c) and (d) involve the assembly of nanobuilding blocks (NBBs). (Adapted from C. Sanchez, *et al.*, *Chem. Mater.*, 2001, **13**, 3061.)

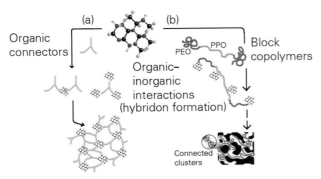

Figure 25.33 Different approaches to the assembly of nanobuilding blocks (NBBs) involving clusters and different connectors. (a) Ordered cluster dispersions based on the covalent linkage of complementary organic and inorganic NBBs. (b) Creation of a bicontinuous mesostructured hybrid arrangement. Interactions between the template and the inorganic NBB include covalent bonding and van der Waals forces (hydrophobic contacts). (Adapted from C. Sanchez, *et al.*, *Chem. Mater.*, 2001, **13**, 3061.)

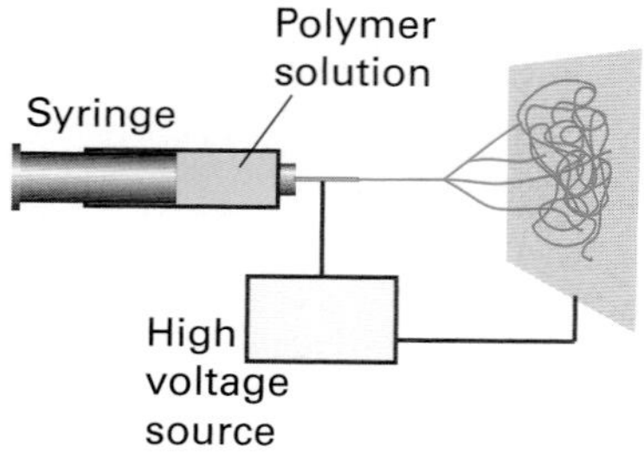

Important examples of Class I are **polymer nanocomposites** (PNCs). These materials are composed of inorganic nanoparticles dispersed in a polymeric matrix. Early commercial PNCs used two-dimensional ordered or lamellar clays, such as sodium montmorillonite (Na-MMT), dispersed in the polymer matrix. Such dispersants (or fillers) have sandwich-type structures (with channels between layers), a total thickness of 0.3–1 nm, and a length of 50–100 nm for each layer; these sandwich structures typically agglomerate into micrometre-sized agglomerates. The dispersion of the inorganic phase within the organic polymeric matrix in a PNC is obtained by intercalation (inserting the polymer between the layered sheets of the clay). Intercalation involves the expansion of the interlamellar spacing as a consequence of ion exchange by using organic amines or quaternary ammonium salts. Exfoliation is achieved through reactive chemical compounding or by intensive melt mixing of the clay and polymer phases. Without proper dispersion, the nanosized filler particles aggregate into larger clusters, causing degradation in the properties of the composite.

Many techniques can be used to control the state of dispersion as well as the nature of the bond between nanoparticle and matrix, including the use of silanes, grafting, and CVD. Silanes are being produced that perform both tasks simultaneously: they both tailor the surface properties of the filler, which promotes the coupling of the nanoparticle to the polymer matrix, and act as a surfactant, reducing the surface energy of the filler to prevent agglomeration and promote dispersion. Radiation copolymer grafting using γ-rays, electron beams, or X-rays is very effective in modifying polymer surfaces. In most cases the polymer chains that are created strengthen the polymer–filler bonds. This strengthening is also evident in CVD as the gaseous molecules that form a film on the substrate typically increase the interactions between filler and polymer.

The nature of the filler (dispersant) and any cavities (voids) present in a PNC greatly influences the mechanical properties of the composites because they can control the distribution of stress through the composite matrix. Both the yield strength (a measure of the material's ability to resist permanent deformation) and toughness (a measure of the energy absorbed prior to fracture) of nanocomposites are controlled by the nanoparticle size and dispersion and nanoparticle-to-polymer contact interactions. Therefore, control of the nanoparticle dispersion and alignment of these fillers offers a way to tailor the mechanical properties of the nanocomposites.

One important class of PNCs uses nanocarbons as the dispersants. Single-walled nanotubes (SWNTs) have exceptional mechanical properties; their very high Young's modulus (their resistance to mechanical stress), low densities, and high aspect ratios make them very attractive as fillers to enhance polymer strength. SWNTs have high tensile strength, almost 100 times that of steel, whereas their density is only about one-sixth that of steel. SWNT-reinforced composites have been developed by the physical mixing of SWNTs in solutions of preformed polymers, *in situ* polymerization in the presence of SWNTs, surfactant-assisted processing of SWNT–polymer composites, and chemical modification of the incorporated SWNTs. Melt processing can be carried out easily by compression moulding at high temperatures and pressures followed by rapid quenching. The so-called **electrospinning methods** use electrostatic forces to distort a droplet of polymer solution into a fine filament, which is then deposited on a substrate (Fig. 25.34). The nanofibres created from the electrospinning can be configured into a variety of forms including membranes, coatings, and films, and can be deposited on targets of different shapes. Fibres can be prepared with diameters smaller than 3 nm, with high surface-to-volume and length-to-diameter ratios, and with controlled pore sizes. Electrospun polymer nanocomposites have been made with homogeneously dispersed SWNTs and have exhibited significantly improved mechanical strength.

Different organo-functional groups covalently attached to the nanotubes through oxidation reactions have been used to improve their chemical compatibility with specific polymers (Fig. 25.35). The use of SWNTs with different functionalities allows the study of the importance of the interfacial interactions between the filler and the polymer matrix. This chemical functionalization is an effective approach towards improving the processability of the nanocomposite and the chemical compatibility of the components. As described before, specific functionalization can be used to separate the SWNT bundles and prevent their agglomeration.

Metal oxide fillers have also been used to optimize polymer strengths and thermal properties. A particularly interesting example involves the control over the mechanical properties of alumina/polymethylmethacrylate (PMMA) nanocomposites by engineering the distribution of weak particle-to-polymer interactions. The filler nanoparticles

Polymer solution

Syringe

High voltage source

Figure 25.34 The electrospinning apparatus. (Adapted from R. Sen, *et al.*, *Nano Lett.*, 2004, **4**, 459.)

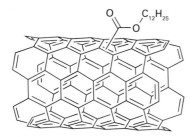

Figure 25.35 Ester-functionalized SWNTs (SWNT-COO(CH$_2$)$_{11}$CH$_3$, EST-SWNTs). (Based on R. Sen *et al.*, *Nano Lett.*, 2004, **4**, 459.)

(alumina) are dispersed in the composite matrix in the form of 'net-like' structures as opposed to 'isolated island' structures formed when microparticles are dispersed in the polymer matrix. In contrast to microparticle–polymer composites, these net-like structures inhibit crazing, the propagation of microscopic cracks during tensional loading. Alumina/ PMMA composites also display higher yield strengths and a shift from brittle to ductile behaviour when nanoparticulate alumina is used (Fig. 25.36). Ductility is advantageous as the composite can then be drawn into thin wires and can withstand sudden impact, thus making the composite more resilient mechanically. The addition of nano-alumina with the anti-agglomerant methacrylic acid lowers the glass transition temperature enough to change the stress–strain curve of the composite and shift from brittle to ductile under tensile loading.

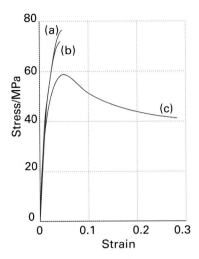

Figure 25.36 Typical stress–strain curves for (a) neat PMMA, (b) 2 per cent by mass as-received micrometre-sized alumina-filled PMMA composite, and (c) 2.2 per cent by mass 38 nm (MAA) alumina/PMMA nanocomposite. Although the strength decreases slightly for (c) (related to the decreased stress at the curve maximum), the overall ductility (related to the total strain) and toughness (related to the area under the curve) are greatly improved. (Adapted from B.J. Ash, *et al., Macromolecules*, 2004, **37**, 1358.) (Reproduced with permission from the American Chemical Society.)

Bioinorganic nanomaterials

Biological phenomena, such as DNA condensation, intercellular transport, tissue assembly, respiration, photosynthesis, and reproduction, originate and operate at the nanoscale. Because biological processes use and manipulate materials at this length scale, it is not surprising that scientists and engineers are looking to natural systems for inspiration and to reach a better understanding of the design of robust and useful nanomaterials. This interest has spawned an extensive research effort termed **biomimetics**, the mimicking of biological systems. Of particular interest to us are materials that bridge solid-state inorganic materials and living cells, the so-called **bioinorganic materials** (Chapter 27). Their nano counterparts, the bioinorganic nanomaterials, promise to have a profound impact on the way we live; key applications include drug delivery, medical diagnostics, cancer therapy, and environmental pollution control, as well as chemical and biological sensing.

Many biological structures (people, for instance) are produced from self-assembly of functional building blocks, which have complex morphological architectures themselves. Despite success in understanding the basic principles of the biological assembly process, as well as in making inorganic materials through biological templating, it remains a key challenge to mimic natural pathways as efficient routes for fabricating artificial bionanomaterials. In this section we discuss some recent advances and offer a taste of the complexity, diversity, and architectures of exquisite bionanomaterials by using key examples of these types of materials.

25.10 DNA and nanomaterials

Key point: Interactions between gold nanoparticles and DNA can cause the self-assembly of DNA into condensates and the ordering of nanoparticles into regular arrays that can be used for biosensors.

DNA condensates are self-assembled nanostructures. The architecture of these condensates is driven by the components that become woven into the nanoassembly. Electrostatic interactions drive the formation of DNA condensates. DNA is a negatively charged polyelectrolyte that interacts with positively charged ions, molecules, or modified nanoparticles to form ordered structures and convert from their random coils to a more compact form. This compact form, called a **condensate**, is essential for the organization of DNA into chromatin, a component of genes. Mimicking or controlling the DNA condensation process is the basis of the design of nonviral gene delivery vectors that have practical application in the treatment of cancer, drug delivery, and biosensing.

Studies of DNA condensation *in vitro* have focused on the use of complex cations, including hexaamminecobalt(III) or polyamines as the positive charge centre in the condensate. Complex formation in these systems can be expected to mimic DNA wrapping around charged particles in living cells, such as positively charged histone proteins. Because the cationic particles can be fine-tuned with respect to size and charge density, the effect of those parameters on the complex formation patterns with DNA have begun to be understood. Gold nanoparticles have been suggested as an effective transfection agent (a device for introducing exogenous DNA into a cell) and, as such, it is important to understand the interactions of nanoparticles with DNA.

Figure 25.37 shows AFM images of complexes of lysine-modified gold nanoparticles and DNA at different nanoparticle/DNA ratios. These results indicate the possibility of developing

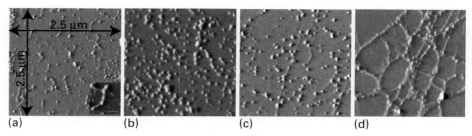

Figure 25.37 AFM images of complexes of lysine-modified gold nanoparticles and DNA at different nanoparticle/DNA ratios (a) 10:1, (b) 20:1, (c) 50:1, and (d) 100:1. Several wrapped molecules are seen along with free nanoparticles in (a) and (b). Network formation is seen in (c) and (d). The inset in (a) shows a single DNA molecule wrapping around six nanoparticles. (Reproduced by permission from M. Ganguli, *et al.*, *Langmuir*, 2004, **20**, 5165.)

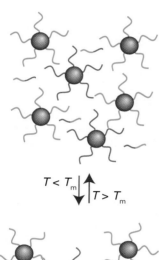

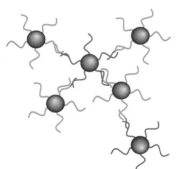

Figure 25.38 DNA templating of inorganic semiconducting nanoparticles. (N.L. Rosi and C.A. Mirkin, *Chem. Rev.*, 2005, **105**, 1547.)

functionalized gold nanoparticle-DNA into a model system to study DNA condensation *in vivo*. These smart nanomaterials have far-reaching consequence for the design of nanoscale self-assembled materials that can be used as building blocks for active nanodevices.

Another exciting accomplishment in the synthesis of bionanomaterials is the use of DNA to drive the assembly of inorganic nanoparticles (Fig. 25.38). First, nanoparticles are functionalized with single strands of DNA. Then, the complementary strand of DNA is introduced from an analyte or bioagent, leading to dimerization with the DNA on the nanoparticles and the self-assembly of the nanoparticles into ordered arrays. Detectable differences in optical properties occur between the disordered and ordered states and result in colour changes. This work has helped to make sophisticated optical-based biosensor arrays with remarkable sensitivity and selectivity for different analytes and bioagents.

25.11 Natural and artificial nanomaterials: biomimetics

Key point: Biological materials can be used as templates in the design of nanoinorganics having specific architectures that mimic the structure of the natural material.

The formation mechanisms in fossilization, including those for siliceous woods, offer efficient methods to reproduce morphological hierarchies of original plant matter through the replacement of the organic components by silica. Artificial fossilization processes can be realized by carefully lining the morphologically complex surfaces of the biological structure with inorganic layers followed by removal of the organic template. Natural materials such as wood and eggshell membrane have been used as templates for the preparation of macroporous silica, zeolites, and titanium dioxide from both precursor sol–gel solutions and suspensions of nanocrystals. Unfortunately, morphological replication has been achieved only over the micrometre scale and the nanoscale details of the biological templates have not yet been reproduced.

An artificial fossilization process has been developed by taking advantage of a surface sol–gel process that can replicate nanoscale features of biological templates (Box 25.3). The surface sol–gel process consists of two steps. First, metal alkoxides are adsorbed from the solution on to hydroxylated substrate surfaces. Then the adsorbed species are hydrolysed to yield nanometre-thick oxide films. Natural cellulose fibres possess surface –OH groups and provide a template for using the surface sol–gel process. The outer diameter of the TiO_2 nanotube varies from 30 to 100 nm, and the thickness of the tube is uniform along its length, with a wall thickness of about 10 nm. The nanotube assembly exhibits the original morphology of interwoven cellulose fibres. The 'titania paper' produced in this way records the morphological information of the original paper at the nanoscale and offers a remarkable example of successful biotemplating of metal oxide nanomaterials.

25.12 Bionanocomposites

Key points: Bionanomaterials synthesis offers new routes to realize smart materials with advanced mechanical strengths and improved performance over natural materials; protein engineering offers an opportunity to insert species that attach selectively to inorganic materials.

BOX 25.3 Artificial fossils

An artificial 'titania fossil' of paper has been created. A titania gel film was deposited on the morphologically complex surface of paper, and the resultant paper/titania composite was calcined to remove the original filter paper. In a typical procedure, a piece of commercial filter paper was placed in a suction filtering unit, washed by suction filtration with ethanol, and then dried by air flow. The 10 cm³ of titanium butoxide solution was passed through the filter paper for 2 minutes. Two 20 cm³ portions of ethanol were filtered immediately to remove the unreacted metal alkoxide and 20 cm³ of water was passed through to promote hydrolysis and condensation. Finally, the filter paper was dried in flowing air. By repeating this filtration/deposition cycle, thin titania gel layers covered the surface of the cellulose fibres. The resultant paper/titania composite was calcined in air at 723 K for 6 hours to remove the filter paper.

The resulting titania fossils possessed, except for some shrinkage in size owing to the calcination step, the morphological characteristics of the original filter paper, as shown in Fig. B25.3. The titania sheet is self-supporting and 0.22 mm thick. In this case, deposition of titania thin films was repeated 20 times. Besides paper, other morphologies have been produced using a similar approach (Fig. B25.2d,e). The sheet size and thickness depend on the original filter paper used. The original morphology of the filter paper was found to be replicated by titania films, and the cellulose fibres were precise copies of irregular titania nanotubes, which are clearly seen in the illustrations.

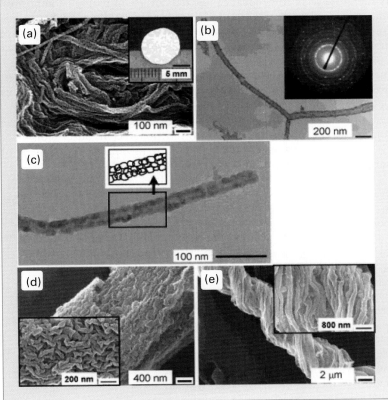

Figure B25.3 (a) A field emission scanning electron micrograph (FE-SEM) of titania paper. The inset shows the photograph of a sheet of titania paper. (b, c) Transmission electron micrographs of individual titania nanotubes, isolated from the assembly. The inset in (c) gives a schematic illustration of the boxed area, showing that the titania nanotube wall is composed of fine anatase particles. (d) An SEM image of titania cloth and (e) an SEM image of titania thread. (J. Huang and T. Kunitake, *J. Am. Chem. Soc.*, 2003, **125**, 11834.)

Millions of bone fractures resulting in hospitalization occur worldwide every year. Modern treatments for severe bone injuries replace a nonuniform defect with a permanent biomaterial that may corrode, wear, and ultimately cause severe infection. In addition, these biomaterials typically exhibit mechanical strength far greater than that of bone, resulting in stress shielding and eventual bone resorption around the implant. There is a social need for a bone tissue engineering scaffold that has mechanical properties similar to bone and that will facilitate bone growth into the defect, degrade slowly and naturally, and eventually be replaced by natural bone tissue without threat of infection. The primary obstacle towards this goal is matching the unique mechanical properties of natural bone tissue with a degradable, synthetic nanomaterial.

The versatile and important structural properties of bone are derived from the nanoscale interactions between its inorganic and organic components. There are two types of bone, *trabecular bone*, the weaker and porous central part, and *cortical bone*, the stronger and denser outer shell of bone. Hydroxyapatite nanocrystals provide compressive strength as they precipitate onto bundles of collagen fibres that impart tensile strength. Degradable biomaterials composed of inorganic and organic components can have mechanical properties closely matching natural bone tissue, and they may be suitable for load-bearing bone tissue engineering applications. Here we describe one illustrative example of these approaches that combines polymeric components with inorganic components.

Poly(propene fumarate) (PPF) cross-linked with poly(propene fumarate)-diacrylate (PPF-DA) is an injectable, biodegradable material. Although PPF shows great potential as a material for bone tissue engineering where mechanical properties such as tensile strength are not crucial, such as the trabecular bone, its mechanical properties are inferior to those required for the more demanding human bone applications, such as cortical bone. A promising strategy to improve the mechanical properties of a polymer is to mimic the architecture of bone or to incorporate inorganic particles into the polymer matrix. For example, significant increases have been observed in the compressive mechanical properties of PPF by the incorporation of calcium phosphate. The extent to which inorganic particles modify the polymer properties is associated with their size, shape, and uniformity of dispersion, as well as with the degree of interaction between the inorganic and organic components (similar to the red abalone shell, Box 25.4). Chemically modified inorganic materials that allow for optimal dispersion and interaction with the polymer matrix therefore seem to be ideal candidates for use in bionanocomposites. A hybrid alumoxane nanoparticle, or a partially hydrolysed aluminium oxyhydroxide with surrounding organic material (Fig. 25.39), with a long carbon chain and a reactive double bond dispersed in PPF/PPF-DA, shows over a threefold increase in strength over polymer resin alone.

Several approaches have been successful for attaching nucleic acids to nanoparticles, which is then followed by DNA hybridization like that mentioned in Section 25.10. Various functional inorganic nanomaterials, such as semiconductors, fullerenes, metals, and oxides, have been considered for use in bottom-up approaches to the design of systems that bridge inorganic and biological materials. One approach is to functionalize the surface of nanoparticles for selective molecular attachment; such materials are used in the development of biosensors, nanoprobes, drug transporters, and other smart devices (Box 25.5).

BOX 25.4   A natural nanocomposite: the shell of red abalone

Nacre (mother-of-pearl) is an outstanding example of a natural nanocomposite that achieves enhanced structural integrity by combining (on the nanoscale) individual components that are either brittle or structurally weak. Red abalone shell is composed of about 95 per cent by mass inorganic aragonite (a polymorph of $CaCO_3$), with only a small percentage of an organic biopolymer. This ceramic/polymer composite material exhibits natural beauty in conjunction with a nanostructure that affords high-quality mechanical properties. Compared to its constituent materials, the laminated structure of red abalone achieves approximately a twofold increase in strength and a thousandfold increase in toughness, owing to the interfaces between the ceramic matrix and the biopolymer assemblies. These remarkable increases in material strength and toughness have inspired chemists and materials scientists to develop synthetic, biomimetic nanocomposites that attempt to reproduce Nature's composite systems.

The abalone shell is a highly ordered nanocomposite material (Fig. B25.4) that is extraordinarily tough, hard, and strong. Nanoscale asperities (spikes) exist on the ceramic platelet surfaces that serve to lock together neighbouring platelets, thus resisting fracture and enhancing material strength. The formation mechanisms of these nanoscale asperities are still unknown but they play an essential role in stabilizing the material against fracture. In the nacreous layer, cobble-like polygonal nanograins are the basic building blocks that construct individual aragonite platelets. The nanograin-structured aragonite platelets are somewhat ductile and the organic biopolymer serves as an adhesive to hold the aragonite platelets together. The nacreous region of the abalone shell has such a great increase in toughness over its constituent materials because this nanocomposite has excellent crack-tip deflection properties (or the ability to prevent catastrophic failure by crack propagation), the ability to deform by nanograin or platelet slip, and its possession of a well-dispersed organic adhesive holding everything together. By contrast, the outer prismatic layer does not show these crack diversion mechanisms and serves mainly as a brittle outer shield for the abalone.

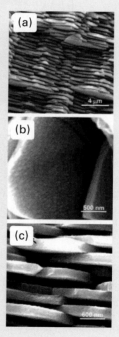

Figure B25.4 SEM images of the nacreous fracture shell section taken at increasing magnification. The brick-and-mortar architecture is shown in (a) and (c). Nanoscale asperities on the aragonite platelet surface are readily observed in (b). (X. Li, *et al.*, *Nano Lett.*, 2004, **4**, 613.)

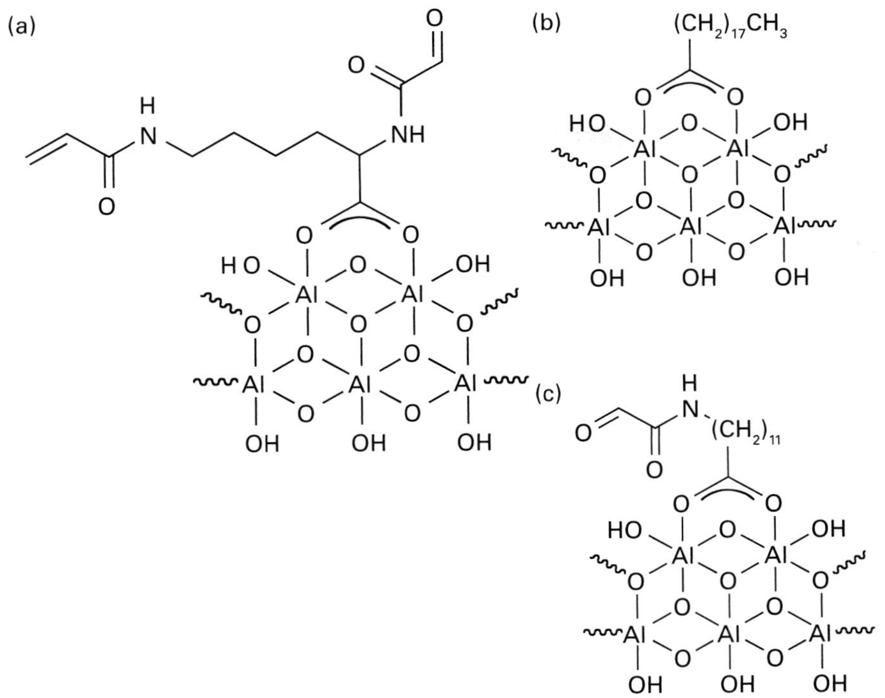

(a)

(b) $(CH_2)_{17}CH_3$

(c)

Figure 25.39 Chemical structures of modified alumoxanes: (a) diacryloyl lysine–alumoxane (activated), (b) stearic acid–alumoxane (surfactant), and (c) acryloyl undecanoic amino acid–alumoxane (hybrid). (R.A. Horch, *et al.*, *Biomacromolecules*, 2004, **5**, 1990.) (Reproduced with permission from the American Chemical Society.)

BOX 25.5 Clay/DNA bioinorganic nanostructures

Biomolecules have been used frequently in the design and assembly of functional nanostructured materials and as templates for the directed growth of nanoparticles and nanowires. The advantage of using biomolecules to develop the self-assembly of extended nanostructures into a functional circuit is that molecular recognition already exists for these materials. Recent studies have led to the push for the development of DNA-based nanostructures as nonviral vectors for cellular delivery and transfection and as active elements in nanocircuitry and biosensors. Clay minerals are attractive for a large number of applications owing to their versatility, ease of synthetic control, wide range of compositions, low toxicity and low cost. Clays may be classified into two main types: cationic clays with negatively charged alumino silicate layers and anionic clays with positively charged hydroxide layers.

Bottom-up self-assembly of DNA/organoclay nanocomposites occurs by two methods which incorporate aminopropyl (AMP) groups, with approximate unit cell composition $[H_2N(CH_2)_3]_8[Si_8Mg_6O_{16}(OH)_4]$, as the structural organoclay building block. The AMP clay exercises electrostatic induction and capture of DNA to form both intercalative mesolamellar assemblies and wrapped single molecules (Fig. B25.5). This method for developing nanocomposites is facile and results in ordered mesolamellar structures with biofunctionality. The overall positive charge of AMP clay layers shield negatively charged DNA, reducing the electrostatic repulsion interaction between cell membrane and anionic DNA to facilitate the potential for cell uptake by endocytosis.

The development of intercalated mesolamellar structures has widespread applications, ranging from pharmaceuticals and cosmetics to agricultural and environmental uses. Protonation of aminopropyl side chains in AMP leads to exfoliation (delamination or separation of the clay nanoparticles into dispersed nanosheets). Charge transport within the clay is enhanced by this exfoliation process and leads to reassembly with negatively charged drug molecules and biomolecules. The denaturation of biomolecules occurs easily under a number of conditions, including sensitivity to pH, temperature, solvent/buffer, and ionic effects. Organoclay wrapping has been utilized as

an effective method for increasing protein stability. Similarly, organoclay wrapping has been employed as a method for increasing DNA stability. The increased thermal stability for intercalated DNA nanocomposites affords new applications in drug delivery and transfection studies for gene therapy.

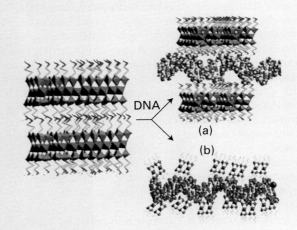

Figure B25.5 Preparation of DNA/organoclay nanostructures. (a) Protonation of aminopropyl side chains of as-synthesized AMP clay in water results in exfoliation and formation of dispersed nanosheets. Addition of stoichiometric quantities of DNA gives rise to electrostatically induced reassembly of the organoclay layers in association with biomolecule intercalation to produce an ordered mesolamellar nanocomposite. (b) Exfoliation and fractionation by gel chromatography results in organoclay polycationic clusters that bind and condense to produce an ultrathin organoclay sheath on individual DNA molecules, leading to molecular-scale isolation of the double-helical strands. (A.J. Patil, *et al.*, *Nano Lett.* 2007, **7**, 2660.)

(a)

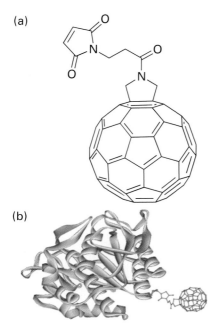

(b)

Figure 25.40 (a) Structure of *N*-(3-maleimidopropionyl)-3,4-fulleropyrrolidine. (b) The protein conjugated to *N*-(3-maleimidopropionyl)-3,4-fulleropyrrolidine (shown to scale). (P. Nednoor, *et al., Biocon. Chem.*, 2004, **15**, 12.)

Proteins have been bound to nanoparticles by using a single site on the protein to yield hybrid nanomaterials. An important example of this type of material consists of proteins attached to fullerenes, which take advantage of the latter's rich chemical and electronic properties, small dimensions, and potential applications. Alterations to the amino acid sequence of a protein are carried out by site-directed mutagenesis (SDM). By using SDM, a single cysteine residue, which selectively attaches to the fullerene, has been introduced in the protein sequence away from the active site to avoid deactivation on binding to the nanoparticles. Figure 25.40 illustrates one approach to create a covalent linkage between nanosized fullerenes and an active protein by using a functional tether (a bridging molecule). Many of these functionalized fullerenes have found biological applications as inhibitors of HIV protease, as light-sensitive biochemical probes, and in photodynamic tumour therapies. Engineering covalent linkages between nanomaterials and proteins has outstanding potential for future applications in drug delivery and enhanced drug efficacy.

There is a new world of inorganic chemistry, as we have sought to show, waiting in the wings of our subject: this intermediate domain between the atomic and the bulk is waiting for discovery and exploitation. With exploitation comes the responsibility of careful consideration of the effects that nanomaterials will have on our environment both as novel materials for water filtration and remediation of biohazards and as sources of toxic pollutants (Box 25.6). It is essential to have a reliable and extensive grounding in the conventions of inorganic chemistry, to note what Nature has already achieved, and—above all—to build on it with imagination.

BOX 25.6 Environmental ceria nanocomposites for water treatment and CO removal

Clean water and air are crucial for human survival. Toxic substances like As, Cr, and CO can contaminate air and drinking water, causing injury and death for those who consume them. Arsenic has been used for years as a poison and is a potential carcinogen. Ingested Cr(VI) can cause liver and kidney damage and is a potential carcinogen; inhaled Cr(VI) is known to cause lung cancer. Several researchers have explored the use of ceria (CeO_2) for the removal of these and other pollutants from water and air. Nanoscale ceria has a higher surface area and catalytic activity, making it even more suitable for catalysis and adsorption. Ceria nanocrystals with included Au nanoparticles can oxidize CO at lower temperatures compared to bulk ceria. Ceria nanoparticles supported on carbon nanotubes are able to adsorb high amounts of As and Cr. However, nanoparticles are too small to be removed by conventional microfiltration, making the nanoparticles difficult to recover for re-use and potentially creating new health risks as these nanomaterials may easily enter drinking water streams and have shown high toxicity when inhaled.

One possible solution to this dilemma is to create ceria microparticles with nanoscale features. This approach preserves enhanced nanoscale properties such as surface area while making the composite easier to handle and filter. Ceria micro/nanocomposites for use in water treatment and CO removal have been prepared by dissolving cerium(III) chloride hydrate, urea, and tetrabutylammonium bromide (TBAB) in ethylene glycol and stirring the solution for 30 minutes at 180°C. The resulting ceria precursor is then calcined in air at 450°C for 2 hours to remove organic components of the precursor, leaving only ceria (CeO_2). In order to fabricate the ceria-supported gold catalyst, $HAuCl_4$ and NaOH are combined in water and added to a suspension of the ceria nanocomposite in water.

Ceria nanocomposites with a self-assembled flowerlike structure have been fabricated by using TBAB as a templating agent. The self-assembly

process of the ceria nanocomposite was studied by collecting samples of the ceria precursor at various times after the appearance of precipitate in the ethylene glycol solution (Fig B25.6). After the precipitate appears the ceria precursor consists of nanoparticles about 100 nm in diameter; 11 minutes later there are some microscale petals among the nanoparticles, after 19 minutes, flowerlike microstructures are visible, and at 30 minutes the precursor consists entirely of flowerlike microparticles. SEM imaging reveals microparticles 4–6 μm in diameter consisting of multiple microscale petals both before and after calcination. The urea was necessary to

Figure B25.6 SEM images of the as-prepared ceria precursors collected at different time intervals after precipitation of ceria: (a) 3 min, (b) 11 min, (c) 19 min, and (d) 30 min. (L.-S. Zhong, *et al., Chem. Mater.* 2007, **19**, 1648.)

maintain the proper pH for precipitate formation and the TBAB is necessary for the assembly of the petals into the flowerlike microstructure. This is an interesting example of morphosynthetic control of structure by careful control of pH, temperature, and templating agent.

These ceria nanocomposites can be used to remove As(V) and Cr(VI) from water. The maximum adsorption capacity of the nanocomposite is 14.4 mg g^{-1} for As(V) and 5.9 mg g^{-1} for Cr(VI). The microcrystalline ceria nanocomposite can be removed from the water after adsorption by centrifugation, so demonstrating the benefit of an overall microscale size. Commercial ceria powder was used to remove the pollutants under the same conditions and found to be much less effective. This is probably due to the higher surface area of the ceria nanocomposite (34.1 m^2 g^{-1}) compared to that of the commercial ceria (2.0 m^2 g^{-1}).

Ceria-supported gold catalysts have also been used to remove CO by oxidation. The ceria-supported gold catalyst was capable of converting 100 per cent of the CO at a temperature of only 125°C. Commerical ceria powders currently in use require 500°C to convert all of the CO. It is likely that the high catalytic ability of the ceria-supported gold catalyst is due to the CO adsorbing on the Au nanoparticles and reacting with the oxygen stored in the ceria.

FURTHER READING

G.A. Ozin and A.C. Arsenault, *Nanochemistry: A chemical approach to nanomaterials*. Springer-Verlag, New York (2005). An invaluable reference book for undergraduate and graduate students looking for an easy way to educate themselves with the up-to-date advances made in chemical patterning, self-assembly, and nanomaterial synthesis.

C.P. Poole and F.J. Owens, *Introduction to nanotechnology*. Wiley-Interscience, Hoboken (2003). This book provides key chapters on a variety of nanomaterials systems, including quantum structures, magnetic nanomaterials, nanoelectro-mechanical systems (NEMS), carbon nanotubes, and nanocomposites, with emphasis on characterization and synthesis strategies for each system.

The rise of nanotech. Reprinted by *Scientific American*, 2007. This collection of easy-to-understand articles that originally appeared in *Scientific American* in 2001 offers an outstanding overview of the key issues of nanoscience, including coverage of top-down versus bottom-up methods of synthesis and applications of nanotechnology in medicine, computers, and telecommunications.

M. Ratner and D. Ratner, *Nanotechnology: the next big idea*. Prentice Hall, Upper Saddle River (2003). A wonderful general approach to nanoscience with broad coverage of key issues in the field, including molecular electronics and smart materials.

M. Wilson, K. Kannangara, G. Smith, M. Simmons, and B. Raguse (ed.), *Nanotechnology: basic science and emerging technologies*. CRC Press, Boca Raton (2002). This is another good introductory book on many areas of nanotechnology.

B. Bhushan (ed.), *Handbook of nanotechnology*. Springer, Berlin (2004). An overarching technical introduction to nanotechnology.

P.N. Prasad, *Nanophotonics*. Wiley-Interscience, Hoboken (2003). A good introduction to optical properties and processes in nanostructured materials.

Supramolecular chemistry and self-assembly. Special issue of *Science* (March 2002). This collection of articles describes key research on the overlap between supramolecular chemistry and self-assembled nanostructures; it includes surveys of chemical approaches to link nanobuilding blocks to generate extended framework structures.

Inorganic–organic nanocomposites. Special issue of *Chemistry of Materials* (October 2001). This collection of articles describes nanocomposites with broad applications in photonics, electronics, and chemical sensing.

J. Hu, T.W. Odom, and C.M. Leiber, Chemistry and physics in one dimension: synthesis and properties of nanowires and nanotubes. *Acc. Chem. Res.*, 1999, **32**, 435. An excellent review of issues of controlled dimensionality in nanomaterials, including important pioneering work on the transport properties of carbon nanotubes.

T. Masciangioli and W.-X. Zhang, Environmental technologies at the nanoscale. *Environ. Sci. Technol., A-Pages*, 2003, **37**, 102A. A short survey of the future impacts of nanoscience on our environment.

C.J. Murphy, Optical sensing with quantum dots. *Anal. Chem.*, 2002, **74**, 520A. A well-constructed survey of the key advantages and limitations of using QDs for optical sensing.

K.G. Thomas and P.V. Kamat, Chromophore-functionalized gold nanoparticles. *Acc. Chem. Res.*, 2003, **36**, 888. A thorough review of Au nanoparticles and the chemistry of surface functionalization with various chromophores to achieve metal hybrid nanomaterials for optoelectronic, light harvesting, and sensing applications.

EXERCISES

25.1 (a) Compare the surface area for two spherical objects: one has a diameter of 10 nm and the other has a diameter of 1000 nm. (b) Describe whether these two objects are considered nanoparticles by using the size-based definition of a nanomaterial. (c) Using a surface-area-related property, describe what must hold true for either of these objects to be considered nanoparticles using the size/properties-based definition of a nanomaterial.

25.2 Quantum confinement bestows unique properties on semiconductor nanocrystals compared to the bulk semiconductor. For this to happen, some characteristic length that describes an electron in a crystal and the size of the crystal must be similar. Describe a characteristic length for an electron and how it becomes comparable to the particle size in quantum confinement.

25.3 Explain why quantum dots might be superior to organic fluorophores for bio-imaging applications.

25.4 Compare and contrast the band energies for a quantum dot nanocrystal and a bulk semiconductor.

25.5 (a) Explain the difference between the top-down and bottom-up methods of fabrication of materials. Be specific and provide one example of each. (b) Give one advantage and one disadvantage for each synthesis method.

25.6 (a) Give a definition for scanning probe microscopy that makes clear both what it is and why it is called scanning probe microscopy. (b) Using a material of interest to you, pick any scanning probe microscopy method and describe how you might use it to characterize an important aspect of your material.

25.7 Distinguish between SEM and TEM. What is the principal difference in sample preparation and detection?

25.8 (a) Describe the three basic steps in nanoparticle formation from solution. (b) Explain why two steps should occur independently to achieve a uniform size distribution. (c) What are stabilizer molecules used for in nanoparticle synthesis?

25.9 Explain whether vapour-phase or solution-based techniques typically lead to (a) larger size distributions in nanoparticle synthesis, (b) agglomerated particles that are strongly bonded to one another in so-called hard agglomerates.

25.10 (a) Draw a schematic diagram of a core–shell nanoparticle. (b) Briefly describe how core–shell nanoparticles could be made using either vapour-phase or solution-based techniques. (c) For what purpose would core–shell nanoparticles be used?

25.11 (a) Discuss the difference between homogeneous and heterogeneous nucleation from the vapour phase. (b) Which type of nucleation is preferred for the growth of a thin film in this process? (c) Which type of nucleation is preferred for the growth of nanoparticles in this process?

25.12 Describe the difference between a physical vapour and a chemical vapour with respect to the type and stability of the vapour species.

25.13 (a) Superlattices of ordered quantum dots (QDs) and an overlaying semiconductor can be considered core–shell QD arrays. What is the purpose of building multiple layers on the QD? (b) What are the limitations on the order and types of semiconductors that can be put together?

25.14 (a) Give two examples of applications of quantum wells. (b) Describe why quantum wells are used and if either molecular materials or traditional solid-state materials can exhibit similar properties. (c) How are quantum wells made?

25.15 Use any of the solid-state superlattices described in the chapter to explain why the use of nanostructured materials led to enhanced properties.

25.16 (a) What is the relevance of self-assembly to the fabrication of nanomaterials? (b) What role will it play in nanotechnology?

25.17 Discuss briefly the features common to self-assembly processes.

25.18 Distinguish between static and dynamic self-assembly. Give an example of each type.

25.19 How is a self-assembled monolayer (SAM) comprising gold-organothiol linkages related to the cell membrane within our bodies? Use a sketch to make your comparison, define the term surfactant, and distinguish between the hydrophilic and hydrophobic regions.

25.20 Define morphosynthesis. Give an example of how this approach can be used to control nanoarchitecture.

25.21 (a) Describe the two classes of inorganic–organic nanocomposites based on their bonding types. (b) Give one example of a nanocomposite in each class.

25.22 (a) Explain why the state of dispersion of inorganic nanoparticles is important in inorganic–organic nanocomposites. (b) Use the concept of oil and water dispersions to explain why highly dispersed inorganic nanoparticles might be difficult to achieve in these nanocomposites.

25.23 Give an example of a bionanomaterial and its application in nanotechnology.

25.24 (a) Define biomimetics. (b) Describe biomimetics with respect to how artificial fossilization is used to create titania paper.

25.25 Describe why mechanical properties are a key metric of the quality of artificial bone materials. Use the example described in the text to explain how chemistry plays a major role in enhancing the mechanical properties of bionanocomposite bone material.

PROBLEMS

25.1 The synthesis method described in the chapter to generate CdSe quantum dots involves the use of rather toxic compounds. Explore the chemical literature to find a more recent example that uses less toxic substances. Describe the solvation step, the nucleation step, and the growth step in each case. Comment on the size dispersions in each case. (See the following articles as a start: G.C. Lisensky and E.M. Boatman, *J. Chem. Educ.*, 2005, **82**, 1360; W. William Yu and X.-G. Peng, *Angew. Chem., Int. Ed. Engl.*, **2002**, 41, 2368.)

25.2 The Grätzel cell has been described as a useful photoelectrochemical cell. Describe how the photoelectrochemical cell differs from the photovoltaic cell. Describe why the nanostructure TiO_2 is important for the improved performance. What other inorganic species play an important role in functionalizing the Grätzel cell? (See M. Grätzel, *Nature*, 2001, **414**, 338.)

25.3 In Fig. 25.11, a TEM micrograph is shown of a superlattice of $(PrMnO_3)_2(SrMnO_3)_2$ deposited on $SrTiO_3$ epitaxially. (a) What is meant by 'epitaxy'? (b) Describe whether the diagram illustrates that there is atomic order between all of the layers (the contrast represents columns

of atoms). (c) The material was grown layer-by-layer by using a physical vapour method. Of the following combinations of growth, which refer to homoepitaxial situations and which refer to heteroepitaxial combinations: $PrMnO_3$ on $SrTiO_3$; $PrMnO_3$ on $PrMnO_3$; $SrMnO_3$ on $PrMnO_3$?

25.4 Compare and contrast the atomic-layer deposition (based on chemical interactions) and the MBE deposition (based on physical interactions of atomic beams) with respect to system cost, speed of process, ambient atmospheric conditions, *in situ* monitoring abilities, and precision ability to design artificial structures and superlattices. Describe which has been used to produce more materials and describe the general quality of the materials.

25.5 One of the key challenges of nanotechnology is achieving mechanical devices that function on the nanoscale. One such device is a Brownian ratchet. Describe how the Brownian ratchet mechanism has been used to sort DNA. (See J.S. Bader *et al.*, *Proc. Natl. Acad. Sci. USA*, 1999, **96**, 13165.)

25.6 Carbon nanotubes have been suggested for their use as wires in molecular electronics. Describe the challenges of using nanotubes as wires in terms of connecting two functional electronic devices. Describe a possible technique to overcome some of the problems.

25.7 Show that folded DNA has an information density (density of base-pairs) equal to 1 Tb cm^{-2}. IBM has proposed a scanning probe data storage device called the Millipede, in which an array of scanning probe tips reads and writes nanoscale marks on a substrate. What would the footprint (the mark plus the space between it and one of its neighbours) of a square mark need to be to attain an information density of 1 Tb cm^{-2}? Describe the method proposed to make such marks and to read them.

25.8 Nanotubes are widely known for carbon. Find an example of an inorganic nanotube not based on carbon and describe its synthesis and properties as compared to the corresponding bulk material. Contrast its structure compared to the carbon nanotubes discussed in this chapter.

26 Catalysis

In this chapter we apply the concepts of organometallic chemistry, coordination chemistry, and materials chemistry to catalysis. We emphasize general principles, such as the nature of catalytic cycles, in which a catalytic species or surface is regenerated in a reaction, and the delicate balance of reactions required for a successful cycle. We see that there are numerous requirements for a successful catalytic process: the reaction being catalysed must be thermodynamically favourable and fast enough when catalysed; the catalyst must have an appropriate selectivity towards the desired product and a lifetime long enough to be economical. We then survey homogeneously catalysed reactions and show how proposals about mechanisms are invoked. The final part of the chapter develops a similar theme in heterogeneous catalysis, and we shall see that many parallels exist between homogeneous and heterogeneous catalysis. In neither type of catalysis are mechanisms necessarily finally settled and there is still considerable scope for making new discoveries.

A **catalyst** is a substance that increases the rate of a reaction but is not itself consumed. Catalysts are widely used in nature, in industry, and in the laboratory, and it is estimated that they contribute to one-sixth of the value of all manufactured goods in industrialized countries. As shown in Table 26.1, 16 of the top 20 synthetic chemicals in the USA are produced directly or indirectly by catalysis. For example, a key step in the production of a dominant industrial chemical, sulfuric acid, is the catalytic oxidation of SO_2 to SO_3. Ammonia, another chemical essential for industry and agriculture, is produced by the catalytic reduction of N_2 by H_2. Inorganic catalysts are also used for the production of the major organic chemicals and petroleum products, such as fuels, petrochemicals, and polyalkene plastics. Catalysts play a steadily increasing role in achieving a cleaner environment, through, for example, the destruction of pollutants (as with the catalytic converters found on the exhaust systems of vehicles), the development of better industrial processes that are more efficient with higher product yields and fewer unwanted byproducts and in clean energy generation in fuel cells. Industrially important catalysts are almost invariably inorganic (which justifies their discussion in this book). Enzymes, a class of biochemical catalysts, often with a metal ion at the centre of a complex molecule, are discussed in Chapter 27.

In addition to their economic importance and contribution to the quality of life, catalysts are interesting in their own right: the subtle influence a catalyst has on reagents can completely change the outcome of a reaction. The understanding of the mechanisms of catalytic reactions has improved considerably in recent years with the greater availability of isotopically labelled molecules, improved methods for determining reaction rates, improved spectroscopic and diffraction techniques, and much more reliable molecular orbital calculations.

General principles

A catalysed reaction is faster than an uncatalysed version of the same reaction because the catalyst provides a different reaction pathway with a lower activation energy. The term *negative catalyst* is sometimes applied to substances that retard reactions. Substances that block one or more elementary steps in a catalytic reaction are called **catalyst poisons**.

Table 26.1 The top 20 synthetic chemicals in the USA in 2008 (based on mass)

Rank	Chemical	Catalytic process	Rank	Chemical	Catalytic process
1	Sulfuric acid	SO_2 oxidation, heterogeneous	12	Ammonium nitrate	Precursors catalytic
2	Ethene	Hydrocarbon cracking, heterogeneous	13	Urea	NH_3 precursor catalytic
3	Propene	Hydrocarbon cracking, heterogeneous	14	Ethylbenzene	Alkylation of benzene, homogeneous
3	Polyethene	Polymerization, heterogeneous	15	Styrene	Dehydrogenation of ethylbenzene,
5	Chlorine	Electrolysis, not catalytic	16	HCl	heterogeneous
6	Ammonia	$N_2 + H_2$, heterogeneous	17	Cumene	Precursors catalytic
7	Phosphoric acid	Not catalytic			Alkylation of benzene,
8	1,2-Dichloroethane	Ethene + Cl_2, heterogeneous	18	Ethylene oxide	heterogeneous
9	Polypropene	Polymerization, heterogeneous	19	Ammonium sulfate	Ethene + O_2, heterogeneous
10	Nitric acid	$NH_3 + O_2$, heterogeneous	20	Sodium carbonate	Precursors catalytic
11	Sodium hydroxide	Electrolysis, not catalytic			Not catalytic

Source: Facts & Figures for the Chemical Industry, *Chem. Eng. News*, 2009, **87**, 33.

26.1 The language of catalysis

Before we discuss the mechanism of catalytic reactions, we need to introduce some of the terminology used to describe the rate of a catalytic reaction and its mechanism.

(a) Energetics

Key point: A catalyst increases the rates of processes by introducing new pathways with lower Gibbs energies of activation; the reaction profile contains no high peaks and no deep troughs.

A catalyst increases the rates of processes by introducing new pathways with lower Gibbs energies of activation, $\Delta^{\ddagger}G$. We need to focus on the Gibbs energy profile of a catalytic reaction, not just the enthalpy or energy profile, because the new elementary steps that occur in the catalysed process are likely to have quite different entropies of activation. A catalyst does not affect the Gibbs energy of the overall reaction, $\Delta_r G^{\ominus}$, because G is a state function.[1] The difference is illustrated in Fig. 26.1, where the overall reaction Gibbs energy is the same in both energy profiles. Reactions that are thermodynamically unfavourable cannot be made favourable by a catalyst.

Figure 26.1 also shows that the Gibbs energy profile of a catalysed reaction contains no high peaks and no deep troughs. The new pathway introduced by the catalyst changes the mechanism of the reaction to one with a very different shape and with lower maxima. However, an equally important point is that stable or nonlabile catalytic intermediates do not occur in the cycle. Similarly, the product must be released in a thermodynamically favourable step. If, as shown by the blue line in Fig. 26.1, a stable complex were formed with the catalyst, it would turn out to be the product of the reaction and the cycle would terminate. Similarly, impurities may suppress catalysis by coordinating strongly to catalytically active sites and act as catalyst poisons.

(b) Catalytic cycles

Key point: A catalytic cycle is a sequence of reactions that consumes the reactants and forms products, with the catalytic species being regenerated after the cycle.

The essence of catalysis is a cycle of reactions in which the reactants are consumed, the products are formed, and the catalytic species is regenerated. A simple example of a catalytic

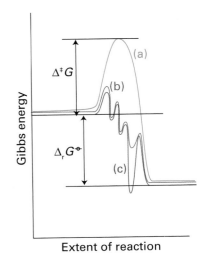

Figure 26.1 Schematic representation of the energetics of a catalytic cycle. The uncatalysed reaction (a) has a higher $\Delta^{\ddagger}G$ than a step in the catalysed reaction (b). The Gibbs energy of the overall reaction, $\Delta_r G^{\ominus}$, is the same for routes (a) and (b). The curve (c) shows the profile for a reaction mechanism with an intermediate that is more stable than the product.

[1] That is, G depends only on the current state of the system and not on the path that led to the state.

cycle involving a homogeneous catalyst is the isomerization of prop-2-en-1-ol (allyl alcohol (CH_2=$CHCH_2OH$)) to prop-1-en-1-ol (CH_3CH=$CHOH$) with the catalyst [$Co(CO)_3H$]. The first step is the coordination of the reactant to the catalyst. That complex isomerizes in the coordination sphere of the catalyst and goes on to release the product and reform the catalyst (Fig. 26.2). Once released the prop-1-en-1-ol tautomerises to propanal (CH_3CH_2CHO). As with all mechanisms, this cycle has been proposed on the basis of a range of information like that summarized in Fig. 26.3. Many of the components shown in the diagram were encountered in Chapter 21 in connection with the determination of mechanisms of substitution reactions. However, the elucidation of catalytic mechanisms is complicated by the occurrence of several delicately balanced reactions, which often cannot be studied in isolation.

Two stringent tests of any proposed mechanism are the determination of rate laws and the elucidation of stereochemistry. If intermediates are postulated, their detection by magnetic resonance and IR spectroscopy also provides support. If specific atom-transfer steps

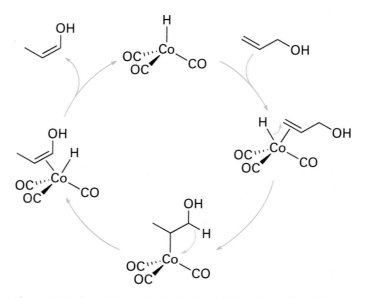

Figure 26.2 The catalytic cycle for the isomerization of prop-2-en-l-ol to prop-1-en-l-ol.

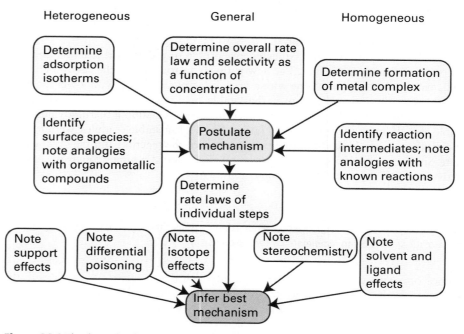

Figure 26.3 The determination of catalytic mechanisms.

are proposed, then isotopic tracer studies may serve as a test. The influences of different ligands and different substrates are also sometimes informative. Although rate data and the corresponding laws have been determined for many overall catalytic cycles, it is also necessary to determine rate laws for the individual steps in order to have reasonable confidence in the mechanism. However, because of experimental complications, it is rare that catalytic cycles are studied in this detail.

(c) Catalytic efficiency and lifetime

Key points: A highly active catalyst, one that results in a fast reaction even in low concentrations, has a large turnover frequency. A catalyst must be able to survive a large number of catalytic cycles if it is to be of use.

The **turnover frequency**, f, is often used to express the efficiency of a catalyst. For the conversion of A to B catalysed by Q and with a rate v,

$$A \xrightarrow{\ Q\ } B \qquad v = \frac{d[B]}{dt} \tag{26.1}$$

then provided the rate of the uncatalysed reaction is negligible, the turnover frequency is

$$f = \frac{v}{[Q]} \tag{26.2}$$

A highly active catalyst, one that results in a fast reaction even in low concentrations, has a high turnover frequency.

In heterogeneous catalysis, the reaction rate is expressed in terms of the rate of change in the amount of product (in place of concentration) and the concentration of catalyst is replaced by the amount present. The determination of the number of active sites in a heterogeneous catalyst is particularly challenging, and often the denominator [Q] in eqn 26.2 is replaced by the surface area of the catalyst.

The **turnover number** is the number of cycles for which a catalyst survives. If it is to be economically viable, a catalyst must have a large turnover number. However, it may be destroyed by side reactions to the main catalytic cycle or by the presence of small amounts of impurities in the starting materials (the feedstock). For example, many alkene polymerization catalysts are destroyed by O_2, so in the synthesis of polyethene (polyethylene) and polypropene (polypropylene) the concentration of O_2 in the ethene or propene feedstock should be no more than a few parts per billion.

Some catalysts can be regenerated quite readily. For example, the supported metal catalysts used in the reforming reactions that convert hydrocarbons to high-octane gasoline become covered with carbon because the catalytic reaction is accompanied by a small amount of dehydrogenation. These supported metal particles can be cleaned by interrupting the catalytic process periodically and burning off the accumulated carbon.

(d) Selectivity

Key point: A selective catalyst yields a high proportion of the desired product with minimum amounts of side products.

A **selective catalyst** yields a high proportion of the desired product with minimum amounts of side products. In industry, there is considerable economic incentive to develop **selective catalysts**. For example, when metallic silver is used to catalyse the oxidation of ethene with oxygen to produce oxirane (ethylene oxide, **1**), the reaction is accompanied by the more thermodynamically favoured but undesirable formation of CO_2 and H_2O. This lack of selectivity increases the consumption of ethene, so chemists are constantly trying to devise a more selective catalyst for oxirane synthesis. Selectivity can be ignored in only a very few simple inorganic reactions, where there is essentially only one thermodynamically favourable product, as in the formation of NH_3 from H_2 and N_2. One area where selectivity is of considerable and growing importance is asymmetric synthesis, where only one enantiomer of a particular compound is required and catalysts may be designed to produce one chiral form in preference to any others.

1 Oxirane (ethylene oxide)

26.2 Homogeneous and heterogeneous catalysts

Key points: Homogeneous catalysts are present in the same phase as the reagents, and are often well defined; heterogeneous catalysts are present in a different phase from the reagents.

Catalysts are classified as **homogeneous** if they are present in the same phase as the reagents; this normally means that they are present as solutes in liquid reaction mixtures. Catalysts are **heterogeneous** if they are present in a different phase from that of the reactants; this normally means that they are present as solids with the reactants present either as gases or in solution. Both types of catalysis are discussed in this chapter and will be seen to be fundamentally similar.

From a practical standpoint, homogeneous catalysis is attractive because it is often highly selective towards the formation of a desired product. In large-scale industrial processes, homogeneous catalysts are preferred for exothermic reactions because it is easier to dissipate heat from a solution than from the solid bed of a heterogeneous catalyst. In principle, every homogeneous catalyst molecule in solution is accessible to reagents, potentially leading to very high activities. It should also be borne in mind that the mechanism of homogeneous catalysis is more accessible to detailed investigation than that of heterogeneous catalysis as species in solution are often easier to characterize than those on a surface and because the interpretation of rate data is frequently easier. The major disadvantage of homogeneous catalysts is that a separation step is required.

Heterogeneous catalysts are used very extensively in industry and have a much greater economic impact than homogeneous catalysts. One attractive feature is that many of these solid catalysts are robust at high temperatures and therefore tolerate a wide range of operating conditions. Reactions are faster at high temperatures, so at high temperatures solid catalysts generally produce higher outputs for a given amount of catalyst and reaction time than homogeneous catalysts operating at lower temperatures in solutions. Another reason for their widespread use is that extra steps are not needed to separate the product from the catalyst, resulting in efficient and more environmentally friendly processes. Typically, gaseous or liquid reactants enter a tubular reactor at one end, pass over a bed of the catalyst, and products are collected at the other end. This same simplicity of design applies to the catalytic converter used to oxidize CO and hydrocarbons and reduce nitrogen oxides in automobile exhausts (Fig. 26.4), see also Box 26.1.

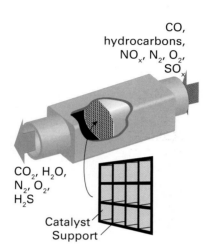

Figure 26.4 A heterogeneous catalyst in action. The automobile catalytic converter oxidizes CO and hydrocarbons, and reduces nitrogen and sulfur oxides. The particles of a metal catalyst are supported on a robust, ceramic honeycomb.

Homogeneous catalysis

Here we concentrate on some important homogeneous catalytic reactions based on organometallic compounds and coordination complexes. We describe their currently favoured mechanisms, but it should be noted that, as with nearly all mechanistic proposals,

---

BOX 26.1 Catalytic converters

Catalytic convertors are used to reduce toxic emissions from internal combustion engines, which include nitrogen oxides (NO_x), carbon monoxide, and unburnt hydrocarbons (HC). By 2009, when the Euro V emission standards came into use in Europe, the levels of these compounds in exhaust gases were restricted to 0.50 (CO), 0.23 (NO_x + HC) and 0.18 (HC) g km^{-1}, respectively; these values are similar to those demanded in California. The catalytic converter consists of a honeycomb stainless steel or ceramic structure on to which first silica and alumina are deposited followed by a mixture of platinum, rhodium, and palladium as nanoparticles, with diameters typically between 10 and 50 nm. A three-way catalytic converter, used with petrol (gasoline) engines undertakes the following three reactions

$$2\,NO_x(g) \rightarrow x\,O_2(g) + N_2(g)$$

$$2\,CO(g) + O_2(g) \rightarrow 2\,CO_2(g)$$

$$C_xH_{2x+2}(g) + 2x\,O_2(g) \rightarrow x\,CO_2(g) + 2x\,H_2O(g)$$

The first stage of the catalytic converter involves reduction of the NO_x on a reduction catalyst, which consists of a mixture of platinum and rhodium; rhodium is highly reactive towards NO. The second stage involves catalytic oxidation, which removes the unburnt hydrocarbons and carbon monoxide by oxidizing them on the mixed metal platinum/palladium catalyst. A two-way catalytic converter used in conjunction with most diesel engines undertakes only the oxidation reactions (the second and third of those above).

In order to facilitate the near complete conversion of the gases emerging from the engine the correct initial fuel/air mixture is used. The ideal stoichiometric air:fuel mixture entering the engine is 14.7:1 so that after combustion the gases entering the catalytic converter are about 0.5 per cent oxygen. If richer or leaner air:fuel mixtures are used (that is, with lower and higher air contents, respectively), then the level of oxygen entering the exhaust stream may be too high or low for effective operation of the catalytic converter. For these reasons, various metal oxides, particularly Ce_2O_3 and CeO_2, are incorporated into the catalytic coating to store and release oxygen as the oxygen content of the gases emerging from the engine changes.

catalytic mechanisms are subject to refinement or change as more detailed experimental information becomes available. Unlike simple reactions, a catalytic process frequently contains many steps over which the experimentalist has little control. Moreover, highly reactive intermediates are often present in concentrations too low to be detected spectroscopically. The best attitude to adopt towards these catalytic mechanisms is to learn the pattern of transformations and appreciate their implications but be prepared to accept new mechanisms that might be indicated by future work.

The scope of homogeneous catalysis ranges across hydrogenation, oxidation, and a host of other processes. Often the complexes of all metal atoms in a group will exhibit catalytic activity in a particular reaction, but the 4d-metal complexes are often superior as catalysts to the complexes of their lighter and heavier congeners. In some cases the difference may be associated with the greater substitutional lability of 4d organometallic compounds in comparison with their 3d and 5d analogues. It is often the case that the complexes of costly metals must be used on account of their superior performance compared with the complexes of cheaper metals.

Chapters 21 and 22 described reactions that take place at metal centres: these reactions lie at the heart of the catalytic processes we now consider. In general, we need to invoke a variety of the processes described in those chapters, and it will be useful to review the following sections:

ligand substitution reactions: Sections 21.0–9 and Section 22.21

redox reactions: Sections 21.10–12

oxidative addition and reductive elimination reactions: Section 22.22

migratory insertion reactions: Section 22.24

1,2-insertions and β-hydride eliminations: Section 22.25

Together with the direct attack on coordinated ligands, these reaction types (and in some cases their reverse), often in combination, account for the mechanisms of most of the homogeneous catalytic cycles that have been proposed for organic transformations. Other reactions yet to be fully investigated will undoubtedly use other reaction steps.

26.3 Alkene metathesis

Key points: Alkene metathesis reactions are catalysed by homogeneous organometallic complexes that allow considerable control over product distribution; a key step in the reaction mechanism is the dissociation of a ligand from a metal centre to allow an alkene to coordinate.

In an **alkene metathesis** reaction carbon–carbon double bonds are redistributed, as in the cross metathesis reaction:

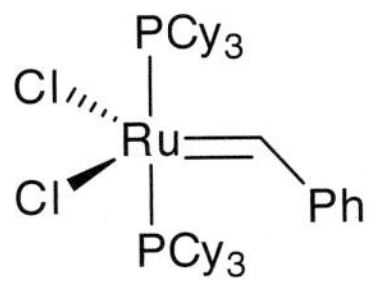

Alkene metathesis was first reported in the 1950s with poorly defined mixtures of reagents, such as WCl_6/Bu_4Sn and MoO_3/SiO_2, being used to bring about a number of different reactions (Table 26.2). In recent years, a number of newer catalysts have been introduced, and the development of the well-defined ruthenium alkylidene compound (**2**) by Grubbs in 1992, now known as **Grubbs' catalyst**, was of seminal importance.[2]

Alkene metathesis reactions proceed through a metallacyclobutane intermediate:

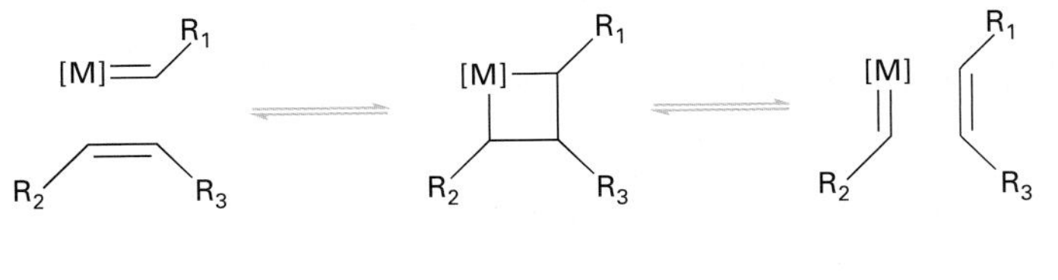

2

[2]The 2005 Nobel prize was awarded to Robert Grubbs, Yves Chauvin, and Richard Schrock for their work on developing metathesis catalysts.

Table 26.2 The scope of the alkene metathesis reaction

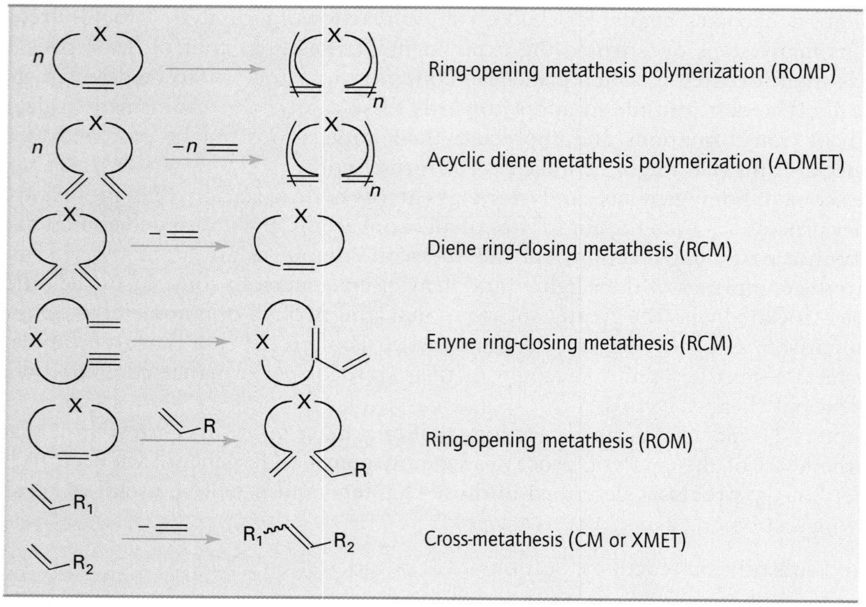

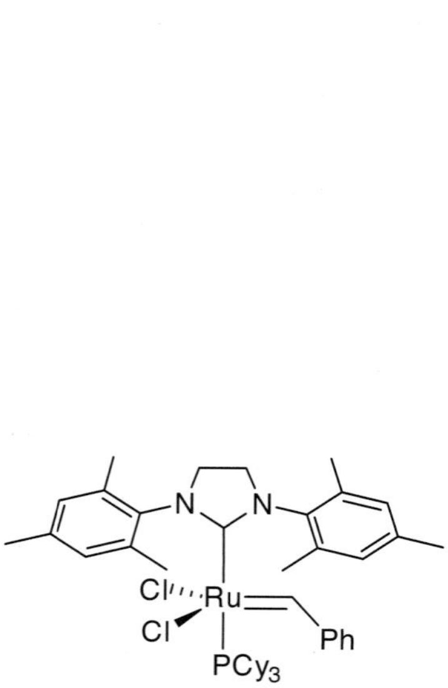

In the case of Grubbs' catalyst, it is known that the dissociation of a PCy_3 ligand from the Ru metal centre is crucial in allowing the alkene molecule to coordinate prior to metalla-cyclobutane formation.

The identification of this mechanism led Grubbs to replace one of the PCy_3 ligands with a bis(mesityl) N-heterocyclic carbene (NHC) ligand, reasoning that the stronger σ-donor and poorer π-acceptor ability of the NHC ligand would both encourage PCy_3 dissociation and stabilize the alkene complex. In a triumph of rational design, the so-called **second-generation Grubbs' catalyst** (3) proved to be more active than the original bisphosphine complex. The second-generation Grubbs' catalyst is active in the presence of a large number of different functional groups on substrates and can be used in many solvent systems. It is commercially available and has been widely used, including in the total synthesis of a number of natural products.

The driving force for alkene metathesis reactions varies. For ROM and ROMP (see Table 26.2) it is the release of ring strain from a strained starting material that provides the energy to drive the reaction. For metathesis reactions that result in the generation of ethene (such as RCM or CM) it is the removal of the liberated ethene that can be used to encourage the formation of the desired products. Where there is no clearly identifiable thermodynamically favourable product possible, mixtures of alkenes result, with their relative proportions being determined by the statistical likelihood of their formation.

26.4 Hydrogenation of alkenes

Key points: Wilkinson's catalyst, $[RhCl(PPh_3)_3]$, and related complexes are used for the hydrogenation of a wide variety of alkenes at pressures of hydrogen close to 1 atm or less; suitable chiral ligands can lead to enantioselective hydrogenations.

The addition of hydrogen to an alkene to form an alkane is favoured thermodynamically ($\Delta_r G^\ominus = -101$ kJ mol^{-1} for the conversion of ethene to ethane). However, the reaction rate is negligible at ordinary conditions in the absence of a catalyst. Efficient homogeneous and heterogeneous catalysts are known for the hydrogenation of alkenes and are used in such diverse areas as the manufacture of margarine, pharmaceuticals, and petrochemicals.

One of the most studied catalytic systems is the Rh(I) complex $[RhCl(PPh_3)_3]$, which is often referred to as **Wilkinson's catalyst**. This useful catalyst hydrogenates a wide variety of alkenes and alkynes at pressures of hydrogen close to 1 atm or less at room temperature. The dominant cycle for the hydrogenation of terminal alkenes by Wilkinson's catalyst is shown in Fig. 26.5. It involves the oxidative addition of H_2 to the 16-electron complex $[RhCl(PPh_3)_3]$ (A), to form the 18-electron dihydrido complex (B). The dissociation of a

3

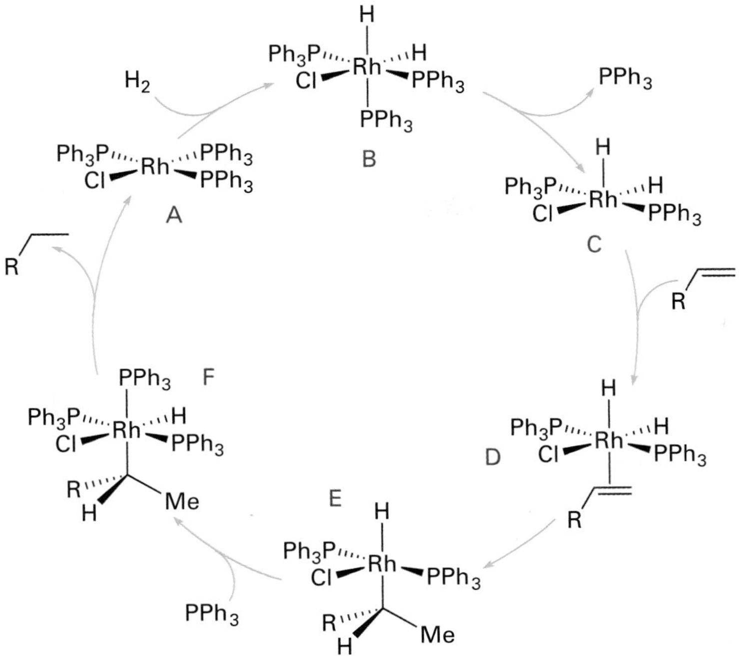

Figure 26.5 The catalytic cycle for the hydrogenation of terminal alkenes by Wilkinson's catalyst.

phosphine ligand from (B) results in the formation of the coordinatively unsaturated complex (C), which then forms the alkene complex (D). Hydrogen transfer from the Rh atom in (D) to the coordinated alkene yields a transient 16-electron alkyl complex (E). This complex takes on a phosphine ligand to produce (F), and hydrogen migration to carbon results in the reductive elimination of the alkane and the reformation of (A), which is set to repeat the cycle. A parallel but slower cycle (which is not shown) is known in which the order of H_2 and alkene addition is reversed. Another cycle is known, based around the 14-electron intermediate $[RhCl(PPh_3)_2]$. Even though there is very little of this species present, it reacts much faster with hydrogen than $[RhCl(PPh_3)_3]$ and makes a significant contribution to the catalytic cycle. In this cycle, (E) would eliminate alkane directly, regenerating $[RhCl(PPh_3)_2]$, which rapidly adds H_2 to give (C).

Wilkinson's catalyst is highly sensitive to the nature of the phosphine ligand and the alkene substrate. Analogous complexes with alkylphosphine ligands are inactive, presumably because they are more strongly bound to the metal atom and do not readily dissociate. Similarly, the alkene must be just the right size: highly hindered alkenes or the sterically unencumbered ethene are not hydrogenated by the catalyst, presumably because the sterically crowded alkenes do not coordinate and ethene forms a strong complex that does not react further. These observations emphasize the point made earlier that a catalytic cycle is usually a delicately poised sequence of reactions, and anything that upsets its flow may block catalysis or alter the mechanism.

Wilkinson's catalyst is used in laboratory-scale organic synthesis and in the production of fine chemicals. Related Rh(I) phosphine catalysts that contain a chiral phosphine ligand have been developed to synthesize optically active products in **enantioselective reactions** (reactions that produce a particular chiral product). The alkene to be hydrogenated must be **prochiral**, which means that it must have a structure that leads to R or S chirality when complexed to the metal. The resulting complex will have two diastereomeric forms depending on which face of the alkene coordinates to the metal atom. In general, diastereomers have different stabilities and labilities, and in favourable cases one or the other of these effects leads to product enantioselectivity. Enantioselectivities are normally measured in terms of the **enantiomeric excess** (ee), which is defined as the percentage yield of the major enantiomeric product minus the percentage yield of the minor enantiomeric product.

■ **A brief illustration.** A reaction that gives 51 per cent of one enantiomer and 49 per cent of another would be described as having an enantiomeric excess of 2 per cent; a reaction that gave 99 per cent of one enantiomer and 1 per cent of another would have ee = 98 per cent. ■

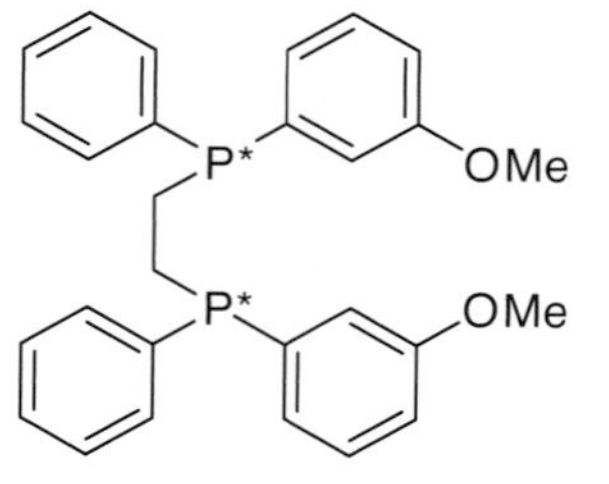

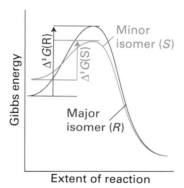

4 DiPAMP

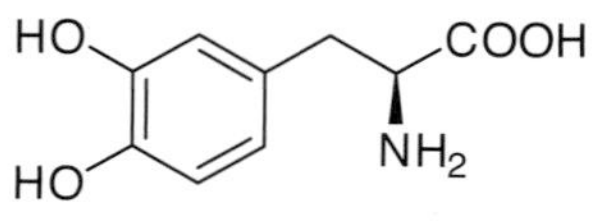

5 L-Dopa

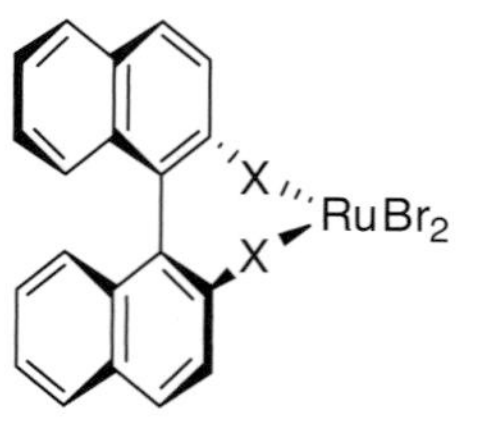

6 [Ru(BINAP)Br₂], X = PPh₂

Gibbs energy

Minor isomer (*S*)

$\Delta^{\ddagger}G(R)$ $\Delta^{\ddagger}G(S)$

Major isomer (*R*)

Extent of reaction

Figure 26.6 Kinetically controlled stereoselectivity. Note that $\Delta^{\ddagger}G_S < \Delta^{\ddagger}G_{R'}$ so the minor isomer reacts faster than the major isomer.

An enantioselective hydrogenation catalyst containing a chiral phosphine ligand referred to as DiPAMP (**4**) is used to synthesize L-dopa (**5**), a chiral amino acid used to treat Parkinson's disease. An interesting detail of the process is that the minor diastereomer in solution leads to the major product. The explanation of the greater turnover frequency of the minor isomer lies in the difference in activation Gibbs energies (Fig. 26.6). Spurred on by clever ligand design and using a variety of metals, this field has grown rapidly and provides many clinically useful compounds; of particular note are systems derived from ruthenium(II) BINAP (**6**).[3]

26.5 Hydroformylation

Key point: The mechanism of hydrocarbonylation is thought to involve a pre-equilibrium in which octacarbonyldicobalt combines with hydrogen at high pressure to give a monometallic species that brings about the actual hydrocarbonylation reaction.

In a **hydroformylation reaction**, an alkene, CO, and H₂ react to form an aldehyde containing one more C atom than in the original alkene:

$$RCH=CH_2 + CO + H_2 \rightarrow RCH_2CH_2CHO$$

The term 'hydroformylation' derived from the idea that the product resulted from the addition of methanal (formaldehyde, HCHO) to the alkene, and the name has stuck even though experimental data indicate a different mechanism. A less common but more appropriate name is **hydrocarbonylation**. Both cobalt and rhodium complexes are used as catalysts. Aldehydes produced by hydroformylation are normally reduced to alcohols that are used as solvents and plasticizers, and in the synthesis of detergents. The scale of production is enormous, amounting to millions of tonnes annually.

The general mechanism of cobalt-carbonyl-catalysed hydroformylation was proposed in 1961 by Heck and Breslow by analogy with reactions familiar from organometallic chemistry (Fig. 26.7). Their general mechanism is still invoked, but has proved difficult to verify in detail. In the proposed mechanism, a pre-equilibrium is established in which octacarbonyldicobalt combines with hydrogen at high pressure to yield the known tetracarbonylhydridocobalt complex (A):

$$[Co_2(CO)_8] + H_2 \rightarrow 2\,[Co(CO)_4H]$$

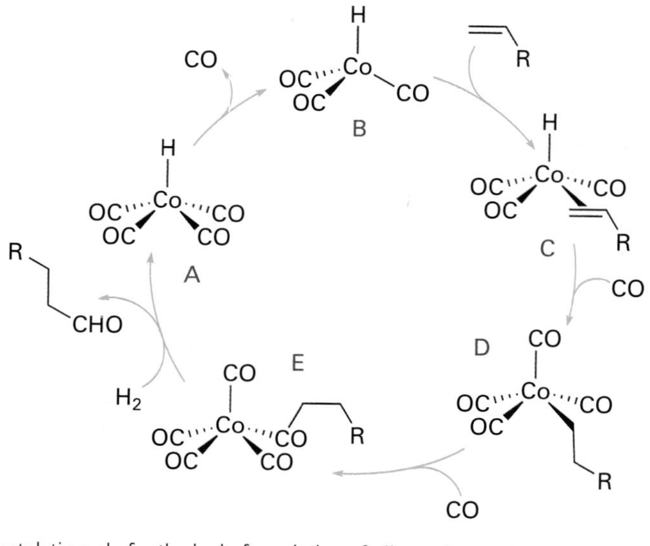

Figure 26.7 The catalytic cycle for the hydroformylation of alkenes by a cobalt carbonyl catalyst.

[3]Ryoji Noyori and William Knowles were jointly awarded the 2001 Nobel prize for their work on asymmetric hydrogenation. The prize was shared with Barry Sharpless for his work on asymmetric oxidations (Section 26.7).

This complex, it is proposed, loses CO to produce the coordinatively unsaturated complex [Co(CO)$_3$H] (B):

$$[Co(CO)_4H] \rightarrow [Co(CO)_3H] + CO$$

It is thought that [Co(CO)$_3$H] then coordinates an alkene, producing (C) in Fig. 26.7, whereupon the coordinated hydrido ligand migrates onto the alkene, and CO recoordinates. The product at this stage is a normal alkyl complex (D). In the presence of CO at high pressure, (D) undergoes migratory insertion and coordinates another CO, yielding the acyl complex (E), which has been observed by IR spectroscopy under catalytic reaction conditions. The formation of the aldehyde product is thought to occur by attack of either H$_2$ (as depicted in Fig. 26.7) or the strongly acidic complex [Co(CO)$_4$H] to yield an aldehyde and generate [Co(CO)$_4$H] or [Co$_2$(CO)$_8$], respectively. Either of these complexes will regenerate the coordinatively unsaturated [Co(CO)$_3$H].

A significant portion of branched aldehyde is also formed in the cobalt-catalysed hydroformylation. This product may result from a 2-alkylcobalt intermediate formed when reaction of (C) leads to an isomer of (D), with hydrogenation then yielding a branched aldehyde as set out in Fig. 26.8. When the linear aldehyde is required, such as for the synthesis of biodegradable detergents, the isomerization can be suppressed by the addition of an alkylphosphine to the reaction mixture. One plausible explanation is that the replacement of CO by a bulky ligand disfavours the formation of complexes of sterically crowded 2-alkenes:

$$K \ll 1$$

Here again we see an example of the powerful influence of ancillary ligands on catalysis.

Another effective hydroformylation catalyst precursor is [Rh(CO)H(PPh$_3$)$_3$] (7), which loses a phosphine ligand to form the coordinatively unsaturated 16-electron complex [Rh(CO)H(PPh$_3$)$_2$], which promotes hydroformylation at moderate temperatures and 1 atm. This behaviour contrasts with the cobalt carbonyl catalyst, which typically requires 150°C and 250 atm. The rhodium catalyst is useful in the laboratory as it is effective under convenient conditions. Because it favours linear aldehyde products, it competes with the phosphine-modified cobalt catalyst in industry.

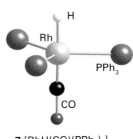

7 [RhH(CO)(PPh$_3$)$_3$]

Figure 26.8 The formation of branched aldehydes in hydroformylation reactions occurs when the alkyl group is not terminally bound.

EXAMPLE 26.1 Interpreting the influence of chemical variables on a catalytic cycle

An increase in CO partial pressure above a certain threshold decreases the rate of the cobalt-catalysed hydroformylation of 1-pentene. Suggest an interpretation of this observation.

Answer The decrease in rate with increasing partial pressure suggests that CO suppresses the concentration of one of the catalytic species. An increase in CO pressure will lower the concentration of $[Co(CO)_3H]$ in the equilibrium

$$[Co(CO)_4H] \rightleftharpoons [Co(CO)_3H] + CO$$

This type of evidence was used as the basis for postulating the existence of $[Co(CO)_3H]$ as an important intermediate, even though it is not detected spectroscopically in the reaction mixture.

Self-test 26.1 Predict the influence of added triphenylphosphine on the rate of hydroformylation catalysed by $[Rh(CO)H(PPh_3)_3]$.

26.6 Wacker oxidation of alkenes

Key points: The Wacker process is used to produce acetaldehyde from ethene and oxygen; the most successful system uses a palladium catalyst to oxidize the alkene, with the palladium being reoxidized via a secondary copper catalyst.

The **Wacker process** is used primarily to produce ethanal (acetaldehyde) from ethene and oxygen:

$$C_2H_4 + \tfrac{1}{2}O_2 \rightarrow CH_3CHO \qquad \Delta_r G^\ominus = -197 \text{ kJ mol}^{-1}$$

Its invention at the Wacker Consortium für Elektrochemische Industrie in the late 1950s marked the beginning of an era of production of chemicals from petroleum feedstock. Although the Wacker process is no longer of major industrial concern, it has some interesting mechanistic features that are worth noting.

The actual oxidation of ethene is known to be caused by a palladium(II) salt:

$$C_2H_4 + PdCl_2 + H_2O \rightarrow CH_3CHO + Pd(0) + 2\,HCl$$

The exact nature of the Pd(0) species is unknown, but it probably is present as a mixture of compounds. The slow oxidation of Pd(0) back to Pd(II) by oxygen is catalysed by the addition of Cu(II), which shuttles back and forth to Cu(I):

$$Pd(0) + 2\,[CuCl_4]^{2-} \rightarrow Pd^{2+} + 2\,[CuCl_2]^- + 4\,Cl^-$$

$$2\,[CuCl_2]^- + \tfrac{1}{2}O_2 + 2\,H^+ + 4\,Cl^- \rightarrow 2\,[CuCl_4]^{2-} + H_2O$$

The overall catalytic cycle is shown in Fig. 26.9. Detailed stereochemical studies on related systems indicate that the hydration of the alkene–Pd(II) complex (B) occurs by the attack of H_2O from the solution on the coordinated ethene rather than the insertion of coordinated OH. Hydration, to form (C), is followed by two steps that isomerize the coordinated alcohol. First, β-hydrogen elimination occurs with the formation of (D), and then migration of a hydride results in the formation of (E). Elimination of the ethanal and an H^+ ion then leaves Pd(0), which is converted back to Pd(II) by the auxiliary copper(II)-catalysed air oxidation cycle.

One important observation that the mechanism must account for is that, when the reaction is carried out in the presence of D_2O, no deuterium is incorporated into the final product. This observation suggests that either intermediate (D) is very short lived and does not exchange the Pd–H for a Pd–D, or that intermediate (C) rearranges directly to (E).

Alkene ligands coordinated to Pt(II) are also susceptible to nucleophilic attack, but only palladium leads to a successful catalytic system. The principal reason for palladium's unique behaviour appears to be the greater lability of the 4d Pd(II) complexes in comparison with their 5d Pt(II) counterparts. Furthermore, the potential for the oxidation of Pd(0) to Pd(II) is more favourable than for the corresponding Pt couple.

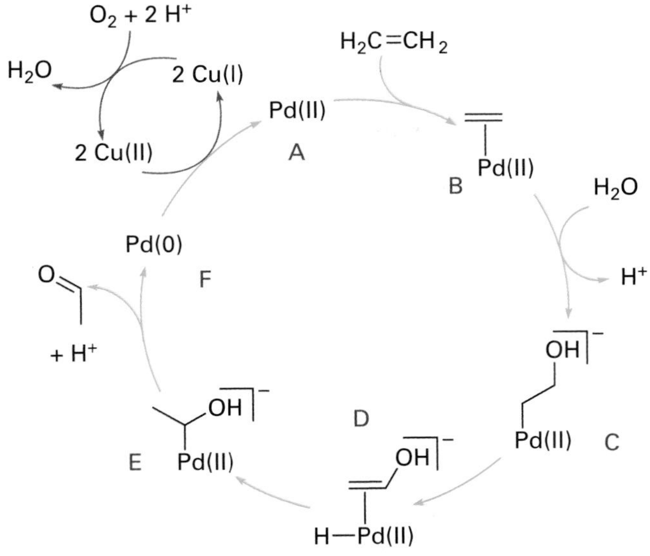

Figure 26.9 The catalytic cycle for the palladium-catalysed oxidation of alkenes to aldehydes.

26.7 Asymmetric oxidations

Key point: Appropriate chiral ligands can be used in conjunction with d-metal catalysts to induce chirality into oxidation products of organic substrates.

In addition to catalysing reductions, d-metal complexes are also active in oxidations. For example, in the **Sharpless epoxidation**, prop-2-en-1-ol (allyl alcohol) or a derivative is oxidized with *tert*-butylhydroperoxide, in the presence of a Ti catalyst with diethyl tartrate as a chiral ligand, producing an epoxide:

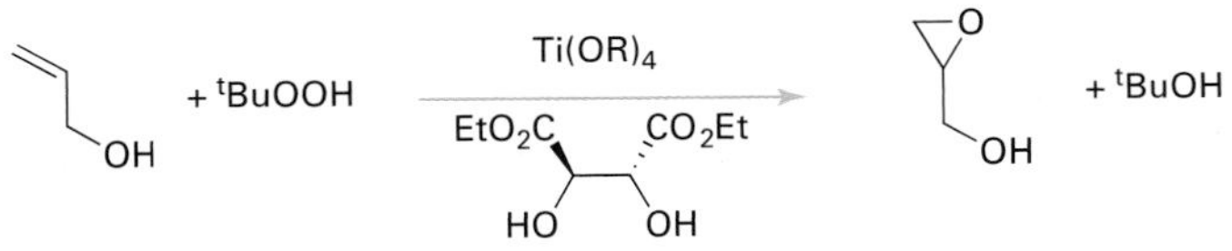

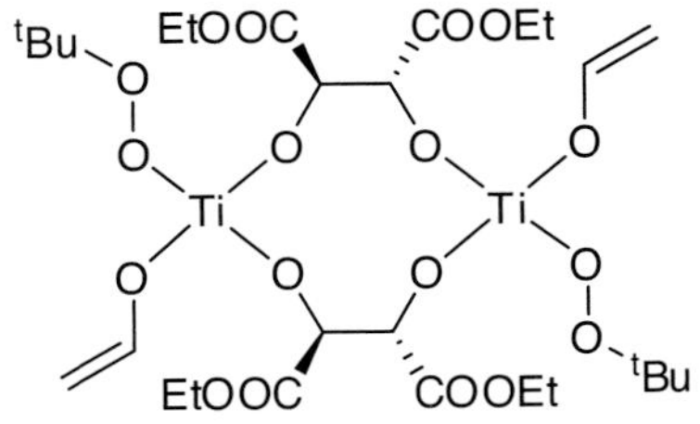

The reaction is thought to go through a transition state in which both the peroxide and the allyl alcohol are coordinated to a Ti atom through their O atoms. Each Ti atom is known to have one diethyl tartrate attached to it, and the chiral environment that the diethyl tartrate produces around the Ti atom is sufficient to differentiate the two prochiral faces of the allyl alcohol. Additional experimental evidence points to a dimeric intermediate (**8**). Enantiomeric excesses of greater than 98 per cent have been reported for Sharpless epoxidations.

A **Jacobsen oxidation** is a reaction in which the catalyst is a Mn complex of the mixed 2 × N + 2 × O donor ligand known as salen (Fig. 26.10). Hypochlorite ions (ClO⁻) are used to oxidize the Mn(III) complex to a Mn(V) oxide, which is then able to deliver its O atoms to an alkene to generate an epoxide. This oxidation has been used with a wide variety of substrates and routinely delivers enantiomeric excesses greater than 95 per cent. The mechanism of the reaction has not been defined precisely, but proposals include the existence of a dimeric form of the catalyst or a radical oxygen transfer step.

26.8 Palladium-catalysed C—C bond-forming reactions

Key points: A number of palladium-catalysed coupling reactions are known; they all proceed through oxidative addition of reagents at the metal centre followed by the reductive elimination of the two fragments.

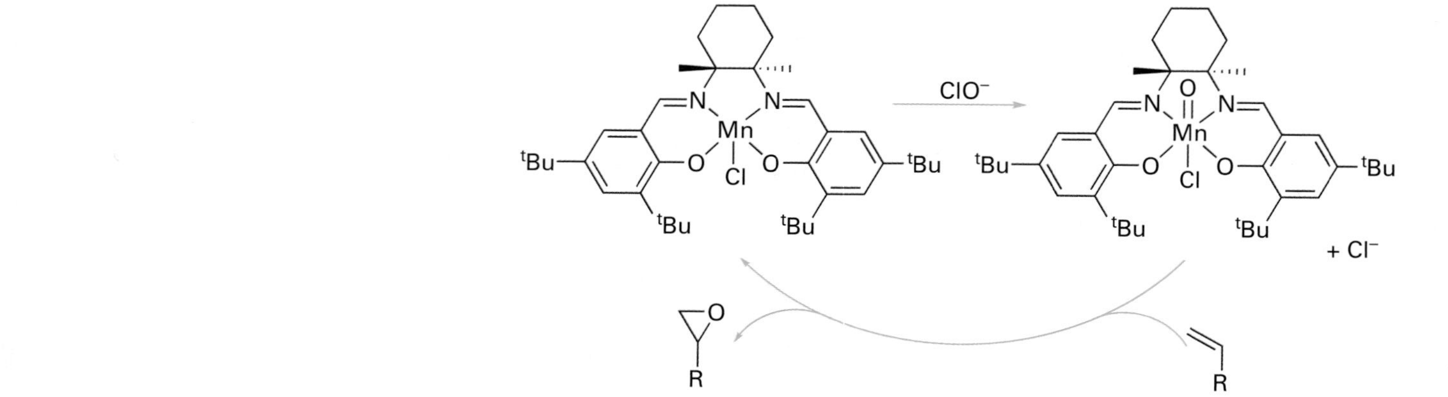

Figure 26.10 The Jacobsen epoxidation relies on a manganese complex of a salen-based ligand and chlorate(I).

A large number of palladium-catalysed carbon–carbon bond-forming ('coupling') reactions are known. They include the coupling of a Grignard with an aryl halide and the Heck, Stille, and Suzuki coupling reactions:

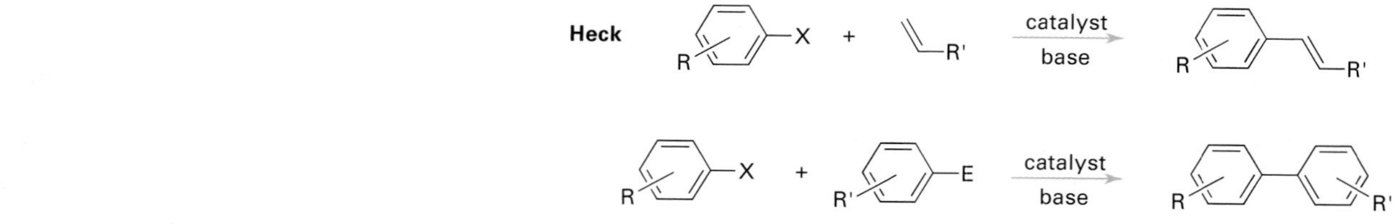

Suzuki: E = $B(OH)_2$
Stille: E = SnR_3

Normally, either a Pd(II) complex, such as $[PdCl_2(PPh_3)_2]$, in the presence of additional phosphine or a Pd(0) compound, such as $[Pd(PPh_3)_4]$, is used as the catalyst, although many other Pd/ligand combinations are active. The precise reaction pathway is unclear (and probably differs with each Pd/ligand/substrate combination) but it is apparent that all these reactions follow the same general sequence. Figure 26.11 shows an idealized catalytic cycle for the coupling of an ethenyl group with an aryl halide. An initial oxidative

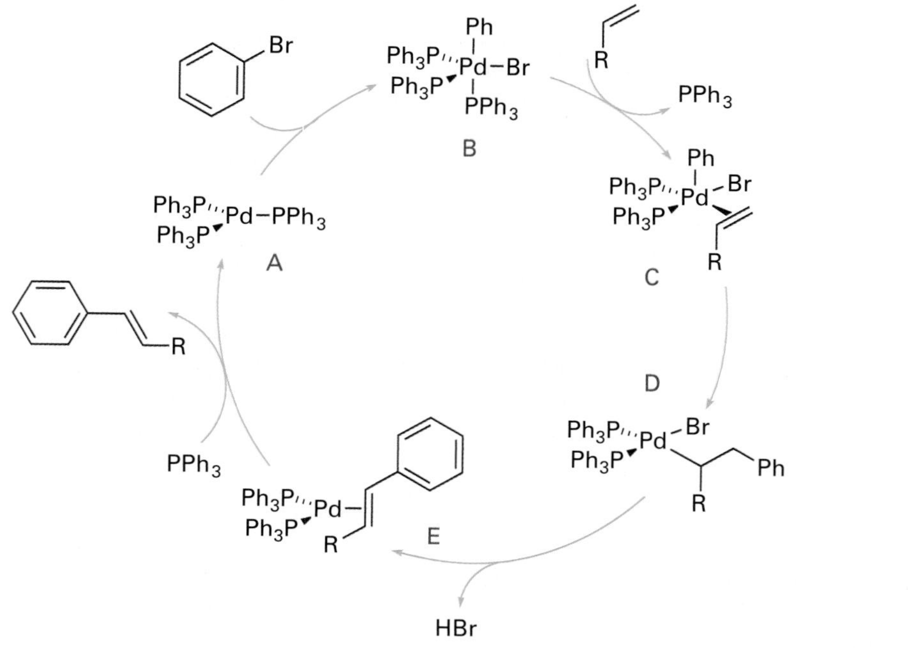

Figure 26.11 An idealized catalytic cycle for the coupling of a substituted prop-1-ene to an aryl halide in the Heck reaction.

addition of an aryl–halogen bond to an unsaturated Pd(0) complex (A) results in a Pd(II) species (B). Coordination of an alkene results in complex (C); 1,2-insertion results in an alkyl complex (D), which can be deprotonated with the loss of the halide to give the organic product attached to the palladium atom (E).

In other palladium-catalysed coupling reactions, such as that of a Grignard reagent with an aryl halide, initial oxidative addition proceeds as in Fig. 26.11. The second organic group is thought to be introduced with the Grignard reagent behaving as the nucleophilic R^- group displacing the halide at the metal centre in (B), to give two organic fragments attached to the Pd atom, as indicated in Fig. 26.12. These two adjacent fragments can then couple and reductively eliminate to regenerate the starting Pd(0) species (A).

In all palladium-catalysed coupling reactions, it is necessary for the two fragments that are coupling to be *cis* to each other at the metal centre before insertion or reductive elimination can take place; this requirement has led to the use of chelating diphosphines such as dppe (**9**) and the ferrocene derivative (**10**). Palladium-catalysed coupling reactions are tolerant to a wide range of substitution on both fragments and, with suitable substrates and appropriate chiral ligands, asymmetric reactions are possible too. The reaction

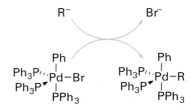

Figure 26.12 The exchange of a halide for an organic fragment at a Pd centre can be thought of as nucleophilic displacement.

was the first published example of an **asymmetric Suzuki reaction**.

26.9 Methanol carbonylation: ethanoic acid synthesis

Key point: Rhodium and iridium complexes are highly active and selective in the carbonylation of methanol to form acetic acid.

The time-honoured method for synthesizing ethanoic (acetic) acid is by aerobic bacterial action on dilute aqueous ethanol, which produces vinegar. However, this process is uneconomical as a source of concentrated ethanoic acid for industry. A highly successful commercial process is based on the carbonylation of methanol:

$$CH_3OH + CO \rightarrow CH_3COOH$$

The reaction is catalysed by all three members of Group 9 (Co, Rh, and Ir). Originally a Co complex was used, but then a Rh catalyst developed at Monsanto greatly reduced the cost of the process by allowing lower pressures to be used. As a result, the rhodium-based **Monsanto process** was used throughout the world. Subsequently, British Petroleum developed the **Cativa process**, which uses a promoted Ir catalyst. Both processes are highly selective and generate ethanoic acid of sufficient purity that it can be used in human food.

The Monsanto and Cativa processes follow essentially the same reaction sequence, so the rhodium-based cycle described here captures the principal features of the iridium-based process too (Fig. 26.13). Under the conditions used, I^- ions that are present react with methanol to set up an appreciable concentration of iodomethane in the first step of the reaction. Starting with the four-coordinate, 16-electron complex $[Rh(CO)_2I_2]^-$ (A), the next step is the oxidative addition of iodomethane to produce the six-coordinate 18-electron complex $[Rh(Me)(CO)_2I_3]^-$ (B). This step is followed by methyl migration, yielding a 16-electron acyl complex (C). Coordination of CO restores an 18-electron complex (D), which is then set to undergo reductive elimination of acetyl iodide with the regeneration of $[Rh(CO)_2I_2]^-$. Water then hydrolyses the acetyl iodide to acetic acid and regenerates HI. Under normal operating conditions, the rate-determining step for the rhodium-based system is the oxidative addition of iodomethane, whereas for the iridium-based system it is the migration of the methyl group. An important feature is that methyl migration on iridium is favoured by formation of a neutral intermediate, and iodide-accepting promoters help facilitate substitution of an iodide ligand by CO in the Ir analogue of complex (B).

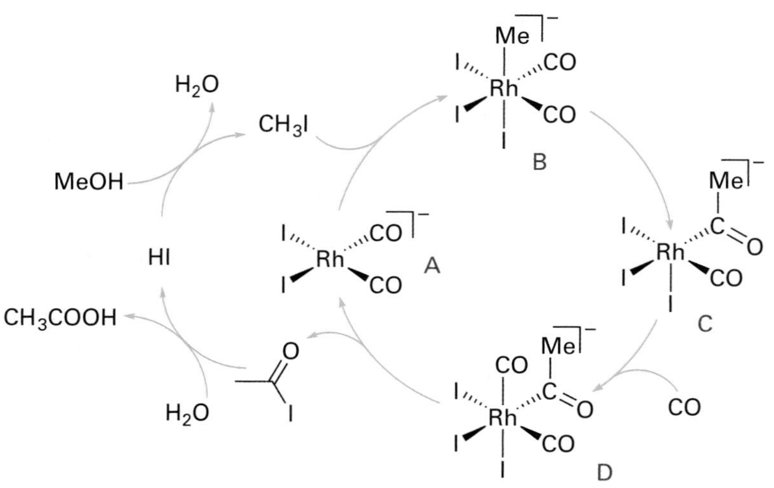

Figure 26.13 The catalytic cycle for the formation of ethanoic (acetic) acid with a rhodium-based catalyst. The oxidative addition step (A → B) is rate determining.

Heterogeneous catalysis

Numerous industrial processes are facilitated by heterogeneous catalysis. Practical heterogeneous catalysts are high-surface-area materials that may contain several different phases and operate at pressures of 1 atm and higher. In some cases the bulk of a high-surface-area material serves as the catalyst, and such a material is called a **uniform catalyst.** A simple example is a very finely divided metal as in skeletal nickel. Another example is the catalytic zeolite ZSM-5, which contains channels or pores through which molecules diffuse, producing a high internal surface area for the reaction. More often, **multiphasic catalysts** are used, which consist of a high-surface-area material that serves as a support on to which an active catalyst is deposited (Fig. 26.14). Heterogeneous catalysts generally fall into two categories in terms of the location of the active surfaces. Many heterogeneous catalysts are finely divided solids where the active sites lie on the particle surfaces; others, particularly the microporous zeolite family and mesoporous materials, have pore-like structures and the active sites are the internal surfaces, such as pores and cavities, within the individual crystallites. We shall discuss examples that illustrate some of the range of heterogeneous catalysts, but first we need to describe some of the unique mechanistic features they exhibit. We concentrate on the inorganic chemistry involved in reactions on surfaces, not the physical chemical aspects of adsorption and reaction.

26.10 The nature of heterogeneous catalysts

There are many parallels between the individual reaction steps encountered in heterogeneous and homogeneous catalysis, but we need to consider some additional points.

(a) Surface area and porosity

Key point: Heterogeneous catalysts are high-surface-area materials formed as either finely divided substrates or crystallites with accessible internal pores.

An ordinary dense solid is unsuitable as a catalyst because its surface area is quite low. Thus α-alumina, which is a dense material with a low specific surface area, is used much less as a catalyst support than the microcrystalline solid γ-alumina, which can be prepared with small particle size and therefore a high specific surface area (the surface area divided by the mass of the sample). The high surface area results from the many small but connected particles like those shown in Fig. 26.14, and a gram or so of a typical catalyst support has a surface area equal to that of a tennis court. Similarly, polycrystalline quartz is not used as a catalyst support but the high-surface-area versions of SiO_2 are widely used. In a typical heterogeneous catalyst this substrate surface is coated with active sites or particles, such as metals or metal oxides, producing a large number of active sites.

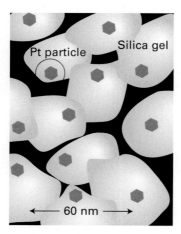

Figure 26.14 Schematic diagram of metal particles supported on a finely divided silica such as silica gel.

Both γ-alumina and high-surface-area silica are metastable materials, but under ordinary conditions they do not convert to their more stable phases (α-alumina and polycrystalline quartz, respectively). The preparation of γ-alumina involves the dehydration of an aluminium oxide hydroxide:

$$2\,AlO(OH) \xrightarrow{\ \Delta\ } \gamma\text{-}Al_2O_3 + H_2O$$

Similarly, high-surface-area silica is prepared from the acidification of silicates to produce $Si(OH)_4$, which rapidly forms a hydrated silica gel from which much of the adsorbed water can be removed by heating. When viewed with an electron microscope, the texture of the silica or alumina appears to be that of a rough gravel bed with irregularly shaped voids between the interconnecting particles (as in Fig. 26.14). Other high-surface-area materials used as supports in heterogeneous catalysts include TiO_2, Cr_2O_3, ZnO, MgO, and carbon.

Zeolites (Section 24.13) are examples of uniform catalysts. They are prepared as very fine crystals that contain large regular channels and cages defined by the crystal structure (Fig. 26.15). The openings in these channels vary from one crystalline form of the zeolite to the next, but are typically between 0.3 and 2 nm. The zeolite absorbs molecules small enough to enter the channels and excludes larger molecules. This selectivity, in combination with catalytic sites inside the cages, provides a degree of control over catalytic reactions that is unattainable with silica gel or γ-alumina. The synthesis of new zeolites and similar shape-selective solids and the introduction of catalytic sites into them is a vigorous area of research (Section 24.12).

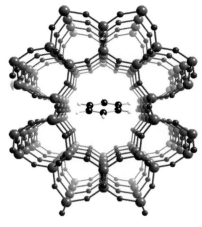

Figure 26.15 A view into the channels of zeolite theta-1 with an absorbed benzene molecule in the large central channel. (Based on A. Dyer, *An introduction to molecular sieves*. Wiley, Chichester (1988).)

(b) Surface acidic and basic sites

Key point: Surface acids and bases are highly active for catalytic reactions such as the dehydration of alcohols and isomerization of alkenes.

When exposed to atmospheric moisture, the surface of γ-alumina is covered with adsorbed water molecules. Dehydration at 100 to 150°C leads to the desorption of water, but surface OH groups remain and act as weak Brønsted acids:

OH   OH   OH   OH
 | | | |
Al   Al   Al ⟶ Al   Al   Al $+ H_2O$
                         \\ O /

At even higher temperatures, adjacent OH groups condense to liberate more H_2O and generate exposed Al^{3+} Lewis acid sites as well as O^{2-} Lewis base sites (**11**). The rigidity of the surface permits the coexistence of these strong Lewis acid and base sites, which would otherwise immediately combine to form Lewis acid–base complexes. Surface acids and bases are highly active for catalytic reactions such as the dehydration of alcohols and isomerization of alkenes. Similar Brønsted and Lewis acid sites exist on the interior of certain zeolites. Different oxides and their mixtures show variations in surface acidity; thus a SiO_2/TiO_2 mixture is more acidic than SiO_2/Al_2O_3, and promotes different catalytic reactions.

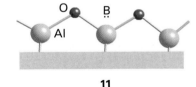

11

EXAMPLE 26.2 Using IR spectra to probe molecular interaction with surfaces

Infrared spectra of hydrogen-bonded pyridine complexes (X–H....py) show bands near 1540 cm^{-1}, and Lewis acid complexes of pyridine (py), such as $Cl_3Al(py)$, display bands near 1465 cm^{-1}, due to Al—py Lewis acid–base interactions. A sample of γ-alumina that had been pretreated by heating to 200°C and then cooled and exposed to pyridine vapour had absorption bands near 1540 cm^{-1} and none near 1465 cm^{-1}. Another sample that was heated to 500°C, cooled, and then exposed to pyridine had bands near 1540 and 1465 cm^{-1}. Correlate these results with the statements made in the text concerning the effect of heating γ-alumina. (Much of the evidence for the chemical nature of the γ-alumina surface comes from experiments like these.)

Answer The positions of the absorption bands seen in the IR spectrum are characteristic of the various functional groups of the molecules present. The types of species present at each stage of the reaction can be surmised by assigning the bands observed in the spectrum to these groups. In the text above it is stated that when heated above 150°C the surface H_2O is lost but OH$^-$ bound to Al^{3+} remains. These groups, which appear to be mildly acidic as judged by colour indicators, interact with pyridine to produce absorption bands at about 1540 cm^{-1} that indicate the presence of hydrogen-bonded pyridine generated in the reaction

OH + py   →   O—H- - -py
| |
Al   Al

When heated to 500°C, much but not all the OH is lost as H_2O, leaving behind O^{2-} and exposed Al^{3+}. The evidence is the appearance of the 1465 cm^{-1} absorption bands, which are indicative of $Al^{3+}-NC_5H_5$, as well as the 1540 cm^{-1} band from the residual O–H···py interactions.

Self-test 26.2 What intensities for the diagnostic IR bands would be expected in the spectrum obtained from a sample of γ-alumina heated at 900°C, cooled in the absence of water, and then exposed to pyridine vapour?

Highly active acidic and basic surfaces act as useful substrates for depositing other catalytic centres, particularly metal particles. Treatment of γ-alumina with H_2PtCl_6 followed by heating in a reducing environment produces Pt particles of dimensions 1–50 nm distributed over the alumina surface.

(c) Surface metal sites

Key points: Very small metal particles on ceramic oxide substrates are very active catalysts for a range of reactions.

Metal particles are often deposited on supports to provide a catalyst. For example, finely divided Pt/Re alloys distributed on the surface of γ-alumina particles are used to interconvert hydrocarbons, and finely divided Pt/Rh rhodium alloy particles supported on γ-alumina are used in the catalytic converters of vehicles to promote the combination of O_2 with CO and hydrocarbons to form CO_2 and reduction of nitrogen oxides to nitrogen (Box 26.1). A supported metal particle about 2.5 nm in diameter has about 40 per cent of its atoms on the surface, and the particles are protected from fusing together into bulk metal by their separation. The high proportion of exposed atoms is a great advantage for these small supported particles, particularly for metals such as platinum and the even more expensive rhodium.

The metal atoms on the surface of metal clusters are capable of forming bonds such as M–CO, M–CH$_2$R, M–H, and M–O (Table 26.3). Often the nature of surface ligands is inferred by comparison of IR spectra with those of organometallic or inorganic complexes. Thus, both terminal and bridging CO groups can be identified on surfaces by IR spectroscopy, and the IR spectra of many hydrocarbon ligands on surfaces are similar to those of discrete organometallic complexes. The case of the N_2 ligand is an interesting contrast because coordinated N_2 was identified by IR spectroscopy on metal surfaces before dinitrogen complexes had been prepared.

The development of new techniques for studying single-crystal surfaces has greatly expanded our knowledge of the surface species that may be present in catalysis. For example, desorption of molecules from surfaces (thermally or by ion or atom impact) combined with mass spectrometric analysis of the desorbed substance provides insight into the chemical identity of surface species. Similarly, Auger and X-ray photoelectron spectroscopy (XPS, Section 8.8) provide information on the elemental composition of surfaces. Low-energy electron diffraction (LEED) provides information about the structure of single-crystal surfaces and, when adsorbate molecules are present, their arrangement on the surface. One important finding from LEED is that the adsorption of small molecules on a surface may bring about a structural modification of the surface. This surface reconstruction is often observed to reverse when desorption occurs. Scanning tunnelling microscopy (STM, Section 25.3a) provides an unrivalled method for locating adsorbates on surfaces. This striking technique provides a contour map of single-crystal surfaces at or close to atomic resolution.

Although most of these modern surface techniques cannot be applied to the study of supported multiphasic catalysts, they are very helpful for revealing the range of probable surface species and circumscribing the structures that may plausibly be invoked in a mechanism of heterogeneous catalysis. The application of these techniques to heterogeneous catalysis is similar to the use of X-ray diffraction and spectroscopy for the characterization of organometallic homogeneous catalyst precursors and model compounds.

Table 26.3 Chemisorbed ligands on surfaces

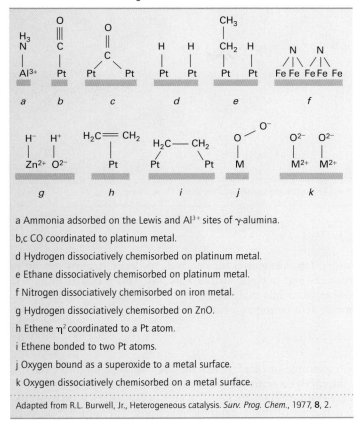

a Ammonia adsorbed on the Lewis and Al^{3+} sites of γ-alumina.

b,c CO coordinated to platinum metal.

d Hydrogen dissociatively chemisorbed on platinum metal.

e Ethane dissociatively chemisorbed on platinum metal.

f Nitrogen dissociatively chemisorbed on iron metal.

g Hydrogen dissociatively chemisorbed on ZnO.

h Ethene η^2 coordinated to a Pt atom.

i Ethene bonded to two Pt atoms.

j Oxygen bound as a superoxide to a metal surface.

k Oxygen dissociatively chemisorbed on a metal surface.

Adapted from R.L. Burwell, Jr., Heterogeneous catalysis. *Surv. Prog. Chem.*, 1977, **8**, 2.

(d) Chemisorption and desorption

Key point: Adsorption is essential for heterogeneous catalysis to occur but must not be so strong that it blocks the catalytic sites and prevents further reaction.

The adsorption of molecules on surfaces often activates molecules just as coordination activates molecules in complexes. The desorption of product molecules that is necessary to refresh the active sites in heterogeneous catalysis is analogous to the dissociation of a complex in homogeneous catalysis.

Before a heterogeneous catalyst is used it is usually 'activated'. Activation is a catch-all term. In some instances it refers to the desorption of adsorbed molecules such as water from the surface, as in the dehydration of γ-alumina. In other cases it refers to the preparation of the active site by a chemical reaction, such as by reduction of metal oxide particles to produce active metal particles.

An activated surface can be characterized by the adsorption of various inert and reactive gases. The adsorption may be either **physisorption**, when no new chemical bond is formed, or **chemisorption**, when surface–adsorbate bonds are formed (Fig. 26.16). Low-temperature physisorption of a gas such as nitrogen is useful for the determination of the total surface area of a solid, whereas chemisorption is used to determine the number of exposed reactive sites. For example, the dissociative chemisorption of H_2 on supported platinum particles reveals the number of exposed surface Pt atoms.

The interaction of small molecules with metal surfaces is similar to their interaction with low-oxidation-state metal complexes. Table 26.4 shows that a wide range of metals chemisorb CO, and that fewer are capable of chemisorbing N_2, just as there is a much wider variety of metals that form carbonyls than form dinitrogen complexes. Furthermore, just as with metal carbonyl complexes, both bridging and terminal CO surface species have been identified by IR spectroscopy. The dissociative chemisorption of H_2 is analogous to the oxidative addition of H_2 to metal complexes (Sections 10.5 and 22.7).

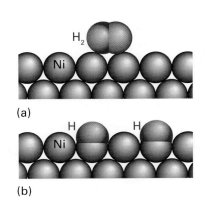

Figure 26.16 Schematic representation of (a) physisorption and (b) chemisorption of hydrogen and a nickel metal surface.

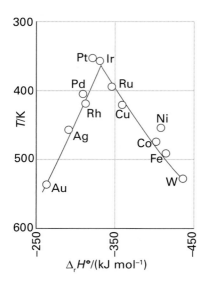

Figure 26.17 A volcano diagram. In this case the reaction temperature for a set rate of methanoic (formic) acid decomposition plotted against the stability of the corresponding metal methanoate as judged by its enthalpy of formation. (Based on W.J.M. Rootsaert and W.M.H. Sachtler, *Z. Phyzik. Chem.* 1960, **26** 16.)

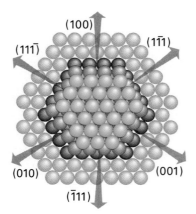

Figure 26.18 Some possible metal crystal planes that might be exposed on a metal surface to a reactive gas. The planes labelled [$\bar{1}$11], [1$\bar{1}$1], etc. are hexagonally close-packed and the planes represented by [100], [010], etc. have square arrays of atoms.

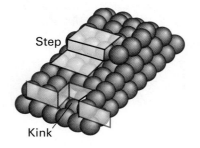

Figure 26.19 Schematic representation of surface irregularities, steps, and kinks.

Table 26.4 The abilities of metals to chemisorb simple gas molecules

	O_2	C_2H_2	C_2H_4	CO	H_2	CO_2	N_2
Ti, Zr, Hf, V, Ta, Cr, Mo, W, Fe, Ru, Os	+	+	+	+	+	+	+
Ni, Co	+	+	+	+	+	+	+
Rh, Pd, Pt, Ir	+	+	+	+	+	+	+
Mn, Cu	+	+	+	+	±	+	+
Al, Au	+	+	+	+	−	−	−
Na, K	+	+	−	−	−	−	−
Ag, Zn, Cd, In, Si, Ge, Sn, Pb, As, Sb, Bi	+	−	−	−	−	−	−

+ Strong chemisorption, ± weak, − unobservable.
Adapted from G.C. Bond, *Heterogeneous catalysis*, Oxford University Press (1987).

Although adsorption is essential for catalysis to occur, it must not be so strong as to block the active sites and prevent further reaction. This factor is in part responsible for the limited number of metals that are effective catalysts. The catalytic decomposition of methanoic (formic) acid on metal surfaces

$$HCOOH \xrightarrow{\quad M \quad} CO + H_2O$$

provides a good example of this balance between adsorption and catalytic activity. It is observed that the catalysis is most effective using metals for which the metal methanoate is of intermediate stability (Fig. 26.17). The plot in Fig. 26.17 is an example of a 'volcano diagram', and is typical of many catalytic reactions. The implication is that the earlier d-block metals form very stable surface compounds whereas the later noble metals such as silver and gold form very weak surface compounds, both of which are detrimental to a catalytic process. Between these extremes the metals in Groups 8 to 10 have high catalytic activity, especially the platinum metals (Group 10). In Section 26.4 we saw a similar high activity of these metal complexes in the homogeneous catalysis of hydrocarbon transformations.

The active sites of heterogeneous catalysts are not uniform and many diverse sites are exposed on the surface of a poorly crystalline solid such as γ-alumina or a noncrystalline solid such as silica gel. However, even highly crystalline metal particles are not uniform. A crystalline solid has typically more than one type of exposed plane, each with its characteristic pattern of surface atoms (Fig. 26.18). In addition, single-crystal metal surfaces have irregularities such as steps that expose metal atoms with low coordination numbers (Fig. 26.19). These highly exposed, coordinatively unsaturated sites appear to be particularly reactive. As a result, the different sites on the surface may serve different functions in catalytic reactions. The variety of sites also accounts for the lower selectivity of many heterogeneous catalysts in comparison with their homogeneous analogues.

(e) Surface migration

Key point: Adsorbed atoms and molecules migrate over metal surfaces.

The surface analogue of fluxional mobility in clusters is diffusion, and there is abundant evidence for the diffusion of chemisorbed molecules or atoms on metal surfaces. For example, adsorbed H atoms and CO molecules are known to move over the surface of a metal particle. These diffusion pathways generally involve the adsorbed molecules moving through a variety of different coordination sites on the metal surface. So, for example, CO migration can result from a molecule moving between sites interacting with one (terminal CO) and between two and four (bridging CO) metal atoms on the surface. The energy barrier to this process is relatively low, a few tens of kilojoules per mole, and thus migration rates are very high under typical catalytic reaction conditions. This mobility is important in catalytic reactions as it allows atoms or molecules to find and approach one another rapidly.

26.11 Hydrogenation catalysts

Key point: Alkenes are hydrogenated on supported metal particles by a process that involves H_2 dissociation and migration of $H^·$ to an adsorbed ethene molecule. Skeletal nickels can be used to reduce alkanals to alkanols.

A milestone in heterogeneous catalysis was Paul Sabatier's observation in 1890 that nickel catalyses the hydrogenation of alkenes. He was in fact attempting to synthesize $Ni(C_2H_4)_4$ in response to Mond, Langer, and Quinke's synthesis of $Ni(CO)_4$ (Section 22.18). However, when he passed ethene over heated nickel he detected ethane. His curiosity was sparked, so he included hydrogen with the ethene, whereupon he observed a good yield of ethane.

The hydrogenation of alkenes on supported metal particles is thought to proceed in a manner very similar to that in metal complexes. As pictured in Fig. 26.20, H_2, which is dissociatively chemisorbed on the surface, is thought to migrate to an adsorbed ethene molecule, giving first a surface alkyl and then the saturated hydrocarbon. When ethene is hydrogenated with D_2 over platinum, the simple mechanism depicted in Fig. 26.20 indicates that CH_2DCH_2D should be the product. In fact, a complete range of $C_2H_nD_{6-n}$ ethane isotopologues is observed. It is for this reason that the central step is written as reversible; the rate of the reverse reaction must be greater than the rate at which the ethane molecule is formed and desorbed in the final step.

One of the most important classes of heterogeneous hydrogenation catalysts are the 'skeletal nickels', which are used for various processes such as conversion of alkanals (aldehydes) to alkanols (alcohols), as in

$$CH_3CH_2CH_2CHO + H_2 \xrightarrow{85°C, 2.5 \, atm \, H_2} CH_3CH_2CH_2CH_2OH$$

and reduction of alkylchloronitroanilines to the corresponding amines. Skeletal nickels and similar catalytically active metal alloys are produced by preparing a metal alloy such as NiAl at high temperature and then selectively dissolving most of the aluminium by treatment with sodium hydroxide. Other metals, such as molybdenum and chromium, can be added to the original alloy and may act as promoters that affect the catalyst's reactivity and selectivity for certain reactions. The resulting spongy or porous metals are rich in nickel (>90 per cent) and their high surface areas lead to very high catalytic activities. One further application of these catalysts is to the conversion of naturally occurring polyunsaturated fats, which are liquids, to solid polyhydrogenated fats, as in margarine.

26.12 Ammonia synthesis

Key point: Catalysts based on iron metal are used for the synthesis of ammonia from nitrogen and hydrogen.

The synthesis of ammonia has already been discussed from several different viewpoints (Section 15.6). Here we concentrate on details of the catalytic steps. The formation of ammonia is exergonic and exothermic at 25°C, the relevant thermodynamic data being $\Delta_f G^\ominus = -116.5 \, kJ \, mol^{-1}$, $\Delta_f H^\ominus = -146.1 \, kJ \, mol^{-1}$, and $\Delta_f S^\ominus = -199.4 \, J \, K^{-1} \, mol^{-1}$. The negative entropy of formation reflects the fact that two NH_3 molecules form in place of four reactant molecules.

The great inertness of N_2 (and to a lesser extent H_2) requires that a catalyst be used for the reaction. Iron metal, together with small quantities of alumina and potassium salts and other promoters, is used as the catalyst. Extensive studies on the mechanism of ammonia synthesis indicate that the rate-determining step under normal operating conditions is the dissociation of N_2 coordinated to the catalyst surface. The other reactant, H_2, undergoes much more facile dissociation on the metal surface and a series of insertion reactions between adsorbed species leads to the production of NH_3:

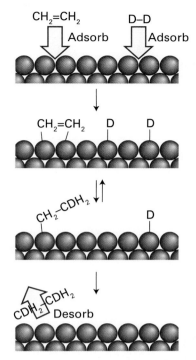

Figure 26.20 Schematic diagram of the stages involved in the hydrogenation of ethene by deuterium on a metal surface.

Because of the slowness of the N_2 dissociation, it is necessary to run the ammonia synthesis at high temperatures, typically 400°C. However, because the reaction is exothermic, high temperature reduces the equilibrium constant of the reaction. To recover some of this reduced yield, pressures in the order of 100 atm are used to favour the formation products. A catalyst operating at room temperature that could give good equilibrium yields of NH_3, such as the enzyme nitrogenase (Section 27.13), has long been sought.

In the course of developing the original ammonia synthesis process, Haber, Bosch, and their co-workers investigated the catalytic activity of most of the metals in the periodic table and found that the best are Fe, Ru, and U promoted by small amounts of alumina and potassium salts. Cost and toxicity considerations led to the choice of iron as the basis of the commercial catalyst. The role of the various promoters, particularly K, in the Fe metal catalyst had been the subject of much scientific research. G. Ertl[4] found that, in the presence of potassium, N_2 molecules adsorb more readily on the metal surface and the adsorption enthalpy is made more exothermic by about 12 kJ mol^{-1}, probably as a result of the increased electron-donating abilities of the Fe/K surface. The more strongly adsorbed N_2 molecule is then cleaved more easily in the rate-determining step in the process.

26.13 Sulfur dioxide oxidation

Key point: The most widely used catalyst for the oxidation of SO_2 to SO_3 is molten potassium vanadate supported on a high-surface-area silica.

The oxidation of SO_2 to SO_3 is a key step in the production of sulfuric acid (Section 16.13). The reaction of sulfur with oxygen to produce SO_3 gas is exergonic ($\Delta_r G^{\ominus} = -371$ kJ mol^{-1}) but very slow, and the principal product of combustion is SO_2:

$$S(s) + O_2(g) \rightarrow SO_2(g)$$

The combustion is followed by the catalytic oxidation of SO_2:

$$SO_2(g) + \tfrac{1}{2}O_2(g) \rightarrow SO_3(g)$$

This step is also exothermic and thus, as with ammonia synthesis, has a less favourable equilibrium constant at elevated temperatures. The process is therefore generally run in stages. In the first stage, the combustion of sulfur raises the temperature to about 600°C, but by cooling and pressurizing before the catalytic stage the equilibrium is driven to the right and high conversion of SO_2 to SO_3 is achieved.

Several quite different catalytic systems have been used to catalyse the combination of SO_2 with O_2. The most widely used catalyst at present is potassium or caesium vanadate molten salt covering a high-surface-area silica. The current view of the mechanism of the reaction is that the rate-determining step is the oxidation of V(IV) to V(V) by O_2 (Fig. 26.21). In the melt, the vanadium and oxide ions are part of a polyvanadate complex (Section 19.8), but little is known about the evolution of the oxo species.

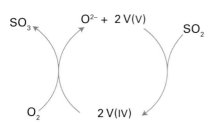

Figure 26.21 Cycle showing the key elements involved in the oxidation of SO_2 by V(V) compounds.

26.14 Catalytic cracking and the interconversion of aromatics by zeolites

Key points: Zeolite catalysts have strongly acidic sites that promote reactions such as isomerization via carbonium ions; shape selectivity may arise at various stages of the reaction due to the relative dimensions of the zeolite channels and reactant, intermediate, and product molecules.

The zeolite-based (Sections 14.15 and 24.12) heterogeneous catalysts play an important role in the interconversion of hydrocarbons and the alkylation of aromatics as well as in oxidation and reduction. Two important zeolites used for such reactions are faujasite (Fig 26.22), also known as zeolite X or zeolite Y (the X or Y terminology is defined by the Si:Al ratio of the material, X has a higher Al content) and zeolite ZSM-5, an aluminosilicate

---

[4] Gerhard Ertl was awarded the 2007 Nobel Prize for chemistry.

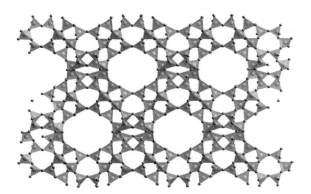

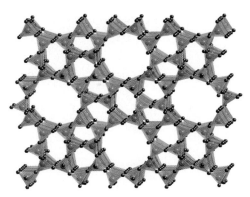

Figure 26.22 The zeolite faujasite (also known as zeolite X or Y) framework structure showing the large pore in which catalytic cracking occurs. Tetrahedra are SiO_4 or AlO_4.

Figure 26.23 The zeolite ZSM-5 structure highlighting the channels along which small molecules may diffuse. Tetrahedra are SiO_4 or AlO_4.

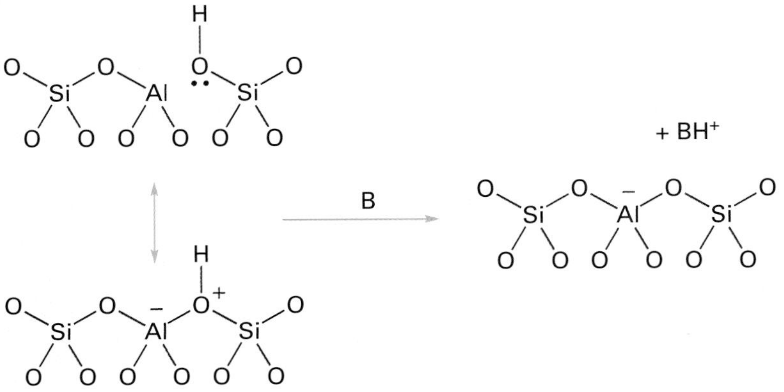

Figure 26.24 The Brønsted acid site in H-ZSM5 and its interaction with a base, typically an organic molecule. (Based on W.O. Haag, R.M. Lago, and P.B. Weisz, *Nature*, 1984, **309**, 589.)

zeolite with a high Si content.[5] The channels of ZSM-5 consist of a three-dimensional maze of intersecting tunnels (Fig. 26.23). As with other aluminosilicate catalysts, the Al sites are strongly acidic. The charge imbalance of Al^{3+} in place of tetrahedrally coordinated Si(IV) requires the presence of an added positive ion. When this ion is H^+ (Fig. 26.24) the Brønsted acidity of the aluminosilicate can be higher than that of concentrated H_2SO_4 and is termed a superacid (Section 4.15); the turnover frequency for hydrocarbon reactions at these sites can be very high.

Natural petroleum consists of only about 20 per cent of alkanes suitable for use in petrol (gasoline) and diesel with chain lengths ranging from C_5H_{12}, pentane, to $C_{12}H_{26}$. Conversion of the higher molar mass hydrocarbons to the valuable lighter ones involves breaking the C−C bonds but also structural rearrangement of the hydrocarbons through dehydrogenation, isomerization, and aromatization reactions. All these processes are catalysed by a solid acid catalyst based on alumina, silica, and zeolites. Acidic clays and mixed Al_2O_3/SiO_2 were originally used for this process in the 1940s but since the 1960s these catalysts have been largely superseded by zeolites. The principal zeolite used for catalytic cracking is zeolite Y, in which the extra-framework cations have been replaced with lanthanoid ions, typically a mixture of La, Ce, and Nd. The mechanism of the catalytic cracking initially involves protonation of the alkane or alkene chain by the Brønsted acid sites in the zeolite pores followed by cleavage of the C−C bond in the β-position to the C atom carrying the positive charge. For example

$$RCH_2C^+HCH_2(CH_2)_2R' \rightarrow RCH_2CH{=}CH_2 + {}^+CH_2CH_2R'$$

[5] The catalyst was developed in the research laboratories of Mobil Oil; the initials stand for Zeolite Socony-Mobil.

Acidic zeolite catalysts also promote rearrangement reactions via carbonium ions. For example, the isomerization of 1,3-dimethylbenzene to 1,4-dimethylbenzene probably occurs by the following steps:

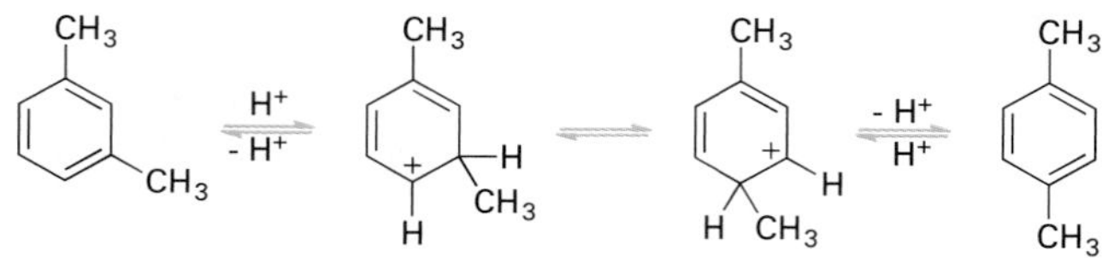

Reactions such as dimethylbenzene (xylene) isomerization and methylbenzene (toluene) disproportionation illustrate the selectivity that can be achieved with acidic zeolite catalysis. The shape selectivity of these zeolite catalysts has been attributed to a variety of processes.

In reactant selectivity the molecular sieving abilities of zeolites are important as only molecules of appropriate size and shape can enter the zeolite pores and undergo a reaction. In product selectivity, a reaction product molecule that has dimensions compatible with the channels will diffuse faster, allowing it to escape; molecules that do not fit the channels diffuse slowly and, on account of their long residence in the zeolite, have ample opportunity to be converted to the more mobile isomers that can escape rapidly. A currently more favoured view of zeolite selectivity is based on transition state selectivity, where the orientation of reactive intermediates within the zeolite channels favours specific products. In the case of dimethylbenzene (xylene) isomerization the narrower intermediates formed during the generation of 1,4-dialkylbenzene molecules fit better within the pores.

Another common reaction in zeolites is the alkylation of aromatics with alkenes.

EXAMPLE 26.3 Proposing a mechanism for the alkylation of benzene

In its protonated form, ZSM-5 catalyses the reaction of ethene with benzene to produce ethylbenzene. Write a plausible mechanism for the reaction.

Answer We should recall that protonated forms of zeolite catalysts are very strong acids. Therefore, a mechanism involving protonation of the organic species present in the system is the likely pathway. The acidic form of ZSM-5 is strong enough to generate carbocations from aliphatic hydrocarbons so the initial stage would be:

$$CH_2CH_2 + H^+ \rightarrow CH_2CH_3^+$$

As we saw in Section 4.10, the carbocation can attack benzene as a strong electrophile. Subsequent deprotonation of the intermediate yields ethylbenzene:

$$CH_2CH_3^+ + C_6H_6 \rightarrow C_6H_5CH_2CH_3 + H^+$$

Self-test 26.3 A pure silica analogue of ZSM-5 can be prepared. Would you expect this compound to be an active catalyst for benzene alkylation? Explain your reasoning.

Mesoporous silicates discovered in the 1990s (Sections 24.13 and 25.9) have large ordered arrays of pores in the range $1-20$ nm and generate very high specific surface areas (of over 1000 $m^2\,g^{-1}$). The large pores allow larger molecules to undergo catalytic processes, although their acidity is weak compared with the zeolites. More importantly other catalytic centres, such as nanoparticles of metals, alloys, and metal oxides such as platinum or Pt/Sn, may be deposited within the mesostructured channels. As one example, Co deposited on the mesoporous silica support MCM-41[6] promotes the cycloaddition of alkynes with alkenes and carbon monoxide to produce cyclopentenones in the Pauson–Khand reaction:

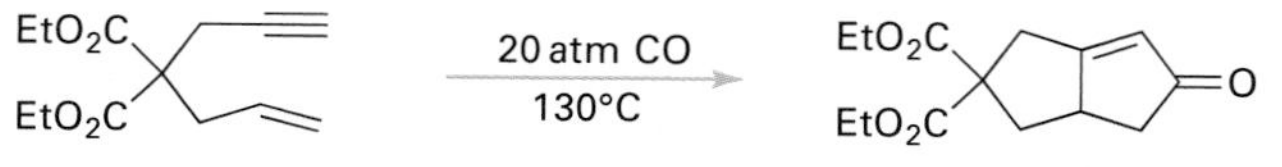

Other porous inorganic framework materials (Section 24.14) are also being investigated as potential catalysts, either in their own right or as hosts for active species. For example,

---

[6] MCM-41 stands for Mobile Crystalline Material.

the d-metal constituents of metal-organic frameworks can take part in redox reactions and also have very large pores that can incorporate nanoparticles of metals, such as Pt.

26.15 Fischer–Tropsch synthesis

Key point: Hydrogen and carbon monoxide can be converted to hydrocarbons and water by reaction over iron or cobalt catalysts.

The conversion of **syngas**, a mixture of H_2 and CO, to hydrocarbons over metal catalysts was first discovered by Franz Fischer and Hans Tropsch at the Kaiser Wilhelm Institute for Coal Research in Müllheim in 1923. In the Fischer–Tropsch reaction, CO reacts with H_2 to produce hydrocarbons, which can be written symbolically as the formation of the chain extender $(-CH_2-)$, and water:

$$CO + 2H_2 \rightarrow -CH_2- + H_2O$$

The process is exothermic, with $\Delta_r H^\ominus = -165$ kJ mol^{-1}. The product range consists of aliphatic straight-chain hydrocarbons that include methane (CH_4) and ethane, LPG $(C_3$ to $C_4)$, gasoline $(C_5$ to $C_{12})$, diesel $(C_{13}$ to $C_{22})$, and light and heavy waxes $(C_{23}$ to C_{32} and $>C_{33}$, respectively). Side reactions include the formation of alcohols and other oxygenated products. The distribution of the products depends on the catalyst and the temperature, pressure, and residence time. Typical conditions for the Fischer–Tropsch synthesis are a temperature range of 200–350°C and pressures of 15–40 atm.

There is general agreement that the first stages of the hydrocarbon synthesis involve the adsorption of CO on the metal, followed by its cleavage to give a surface carbide (and water), and the successive hydrogenation of such species to surface methyne (CH), methylene (CH_2), and methyl (CH_3) species, but there is still debate on what happens next and how chain growth occurs. One proposal suggests a polymerization of bridging surface $-CH_2-$ groups initiated by a surface $-CH_3$ group. However, the fact that many such species have been isolated and are stable as metal complexes suggests that the mechanism is probably not so simple. Other mechanistic investigations of these processes have suggested an alternative possibility for chain growth in the hydrocarbon synthesis, namely that it proceeds by combination of surface bridging $-CH_2-$ groups and alkenyl chains $(M-CH=CHR)$, rather than by the combination of alkyl chains $(M-CH_2CH_2R)$ with methylene groups in the surface.

Several catalysts have been used for the Fischer–Tropsch synthesis; the most important are based on Fe and Co. Cobalt catalysts have the advantage of a higher conversion rate and a longer life (of over 5 years). The cobalt catalysts are in general more reactive for hydrogenation and produce fewer unsaturated hydrocarbons and alcohols than iron catalysts. Iron catalysts have a higher tolerance for sulfur, are cheaper, and produce more alkene products and alcohols. The lifetime of the iron catalysts is short, however, and in commercial installations generally limited to about 8 weeks.

26.16 Alkene polymerization

Key points: Heterogeneous Ziegler–Natta catalysts are used in alkene polymerization; the Cossee-Arlman mechanism describes their function; low molar mass homogeneous catalysts also catalyse the alkene polymerization reaction; considerable control over polymer tacticity is possible with judicious ligand design.

Polyalkenes, which are among the most common and useful class of synthetic polymers, are most often prepared by use of organometallic catalysts, either in solution or supported on a solid surface. The development of alkene polymerization catalysts in the second half of the twentieth century, producing polymers such as polypropene and polystyrene, ushered in a revolution in packaging materials, fabrics, and constructional materials.

In the 1950s J.P. Hogan and R.L. Banks discovered that chromium oxides supported on silica, a so-called **Philips catalyst**, polymerized alkenes to polyenes. Also in the 1950s K. Ziegler, working in Germany, developed a catalyst for ethene polymerization based on a catalyst formed from $TiCl_4$ and $Al(C_2H_5)_3$, and soon thereafter G. Natta in Italy used this type of catalyst for the stereospecific polymerization of propene (see below). Both the **Ziegler–Natta catalysts** and the chromium-based polymerization catalysts are widely used today.

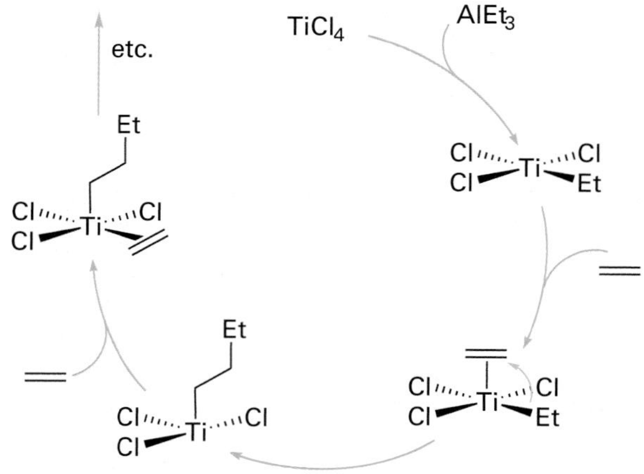

Figure 26.25 The Cossee-Arlman mechanism for the catalytic polymerization of ethene. Note that the Ti atoms are not discrete but are part of an extended structure containing bridging chlorides.

The full details of the mechanism of Ziegler–Natta catalysts are still uncertain, but the **Cossee–Arlman mechanism** is regarded as highly plausible (Fig. 26.25). The catalyst is prepared from $TiCl_4$ and $Al(C_2H_5)_3$, which react to give polymeric $TiCl_3$ mixed with $AlCl_3$ in the form of a fine powder. The alkylaluminium alkylates a Ti atom on the surface of the solid and an ethene molecule coordinates to the neighbouring vacant site. In the propagation steps for the polymerization, the coordinated alkene undergoes a migratory insertion reaction. This migration opens up another neighbouring vacancy, and so the reaction can continue and the polymer chain can grow. The release of the polymer from the metal atom occurs by β-hydrogen elimination, and the chain is terminated. Some catalyst remains in the polymer, but the process is so efficient that the amount is negligible.

The proposed mechanism of alkene polymerization on a Philips catalyst involves the initial coordination of one or more alkene molecules to a surface Cr(II) site followed by rearrangement to metallocycloalkanes on a formally Cr(IV) site. Unlike Ziegler–Natta catalysts, the solid-phase catalyst does not need an alkylating agent to initiate the polymerization reaction; instead, this species is thought to be generated by the metallocycloalkane directly or by formation of an ethenylhydride by cleavage of a C−H bond at the chromium site.

Homogeneous catalysts related to the Philips and Ziegler–Natta catalysts provide additional insight into the course of the reaction and are of considerable industrial significance in their own right, being used commercially for the synthesis of specialized polymers. Most examples are from Group 4 (Ti, Zr, Hf) and are based on a bis(cyclopentadienyl) metal system: the tilted ring complex $[Zr(\eta^5\text{-}Cp)_2(CH_3)L]^+$ (**12**) is a good example. These Group 4 metallocene complexes catalyse alkene polymerization by successive insertion steps that involve prior coordination of the alkene to the electrophilic metal centre. Catalysts of this type are used in the presence of the so-called methyl aluminoxane (MAO), a poorly defined compound of approximate formula $(MeAlO)_n$, which, among other functions, serves to methylate a starting chloride complex.

Additional complications arise with alkenes other than ethene. We shall discuss only terminal alkenes such as propene and styrene, as these are relatively simple. The first complication to consider arises because the two ends of the alkene molecule are different. In principle, it is possible for the polymer to form with the different ends head-to-head (**13**), head-to-tail (**14**), or randomly. Studies on catalysts such as (**12**) show that the growing chain migrates preferentially to the more highly substituted C atom of the alkene, thus giving a polymer chain that contains only head-to-tail orientations:

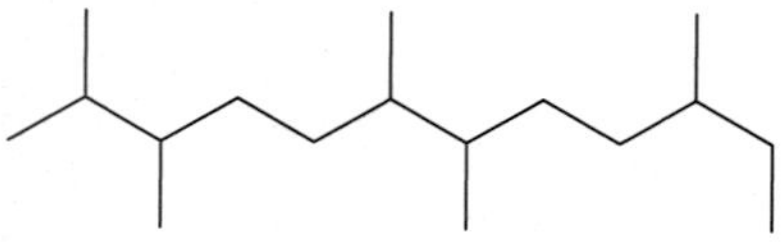

12 $[Zr(\eta^5\text{-}Cp)_2(CH_3)L]^+$

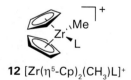

13

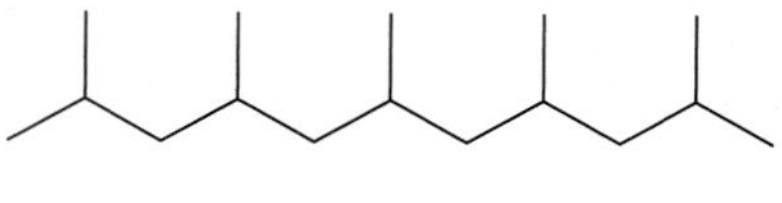

14

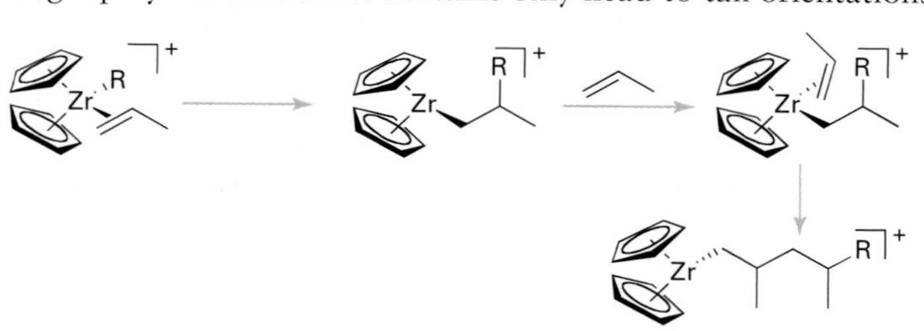

If we consider propene, we can see that the coordinated alkene induces less steric strain if its smaller CH$_2$ end is pointing into the cleft of the (Cp)$_2$Zr catalyst (**15**), rather than the larger methyl substituted end, (**16**). The migrating polymer chain is thus adjacent to the methyl-substituted end of the propene molecule, and it is to this methyl-substituted C atom that the chain attaches, giving a head-to-tail sequence to the whole polymer chain.

The second structural modification of polypropene is its **tacticity**, the relative orientations of neighbouring groups in the polymer. In a regular **isotactic** polypropene, all the methyl groups are on the same side of the polymer backbone (**17**). In regular **syndiotactic** polypropene, the orientation of the methyl groups alternates along the polymer chain (**18**). In an **atactic** polypropene, the orientation of neighbouring methyl groups is random (**19**). Control of the tacticity of a polymer is equivalent to controlling the stereospecificity of the reaction steps. The orientation of neighbouring groups is not simply of academic interest because the orientation has a significant effect on the properties of the bulk polymer. For example, the melting points of isotactic, syndiotactic, and atactic polypropene are 165°C, 130°C, and below 0°C, respectively.

It is not possible to control the tacticity of polypropene with a Zr catalyst such as (**15**), and an atactic polymer results. However, with other catalysts it is possible to control the tacticity. The type of catalyst normally used to control the tacticity has a metal atom bonded to two indenyl groups that are linked by a CH$_2$CH$_2$ bridge. Reaction of the bis(indenyl) fragment with a metal salt gives rise to three compounds: two enantiomers (**20**) and (**21**), which have C$_2$ symmetry, and a nonchiral compound (**22**). These compounds are called *ansa*-metallocenes (the name is derived from the Latin for handle and used to indicate a bridge). It is possible to separate the two enantiomers from the nonchiral compounds and both enantiomers of these *ansa*-metallocenes can catalyse the stereoregular polymerization of propene.

If we now consider the coordination of propene to one of the enantiomeric compounds (**20**) or (**21**) there is a second constraint in addition to the steric factor mentioned above (the CH$_2$ group pointing towards the cleft). Of the two potential arrangements of the methyl group shown in (**23**) and (**24**), the latter is disfavoured by a steric interaction with the phenyl ring of the indenyl group. During the polymerization reaction, the R group migrates preferentially to one side of the propene molecule; coordination of another alkene is then followed by migration, and so on. Figure 26.26 shows how an isotactic polypropene then results.

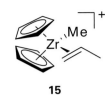

15

16

17

18

19

EXAMPLE 26.4 **Controlling the tacticity of polypropene**

Show that polymerization of propene with a catalyst containing CH$_2$CH$_2$-linked fluorenyl and cyclopentadienyl groups (**25**) should result in syndiotactic polypropene.

Answer We need to consider how the propene reactant will coordinate to the catalyst: in complex (**25**) a coordinated propene will always coordinate with the methyl group pointing away from the fluorenyl and towards the cyclopentadienyl ring. A series of sequential alkene insertions, as outlined in Fig. 26.27, will therefore lead to a product that should be syndiotactic.

Self-test 26.4 Demonstrate that polymerization of propene with a simple Zr(Cp)$_2$Cl$_2$ catalyst would give rise to atactic polypropene.

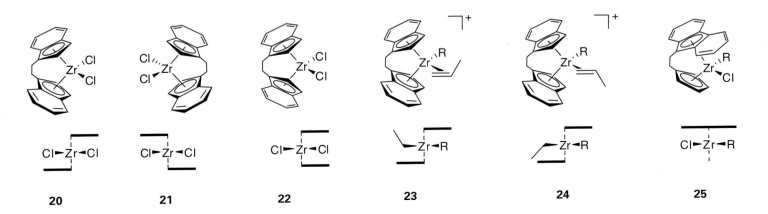

20   **21**   **22**   **23**   **24**   **25**

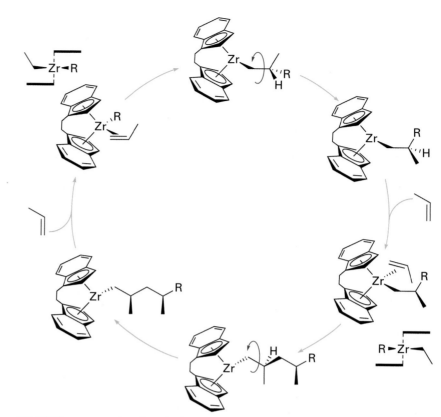

Figure 26.26 When propene is polymerized with an indenyl metallocene catalyst, isotactic polypropene results. The zirconium species all have a single positive charge; this has been omitted for clarity.

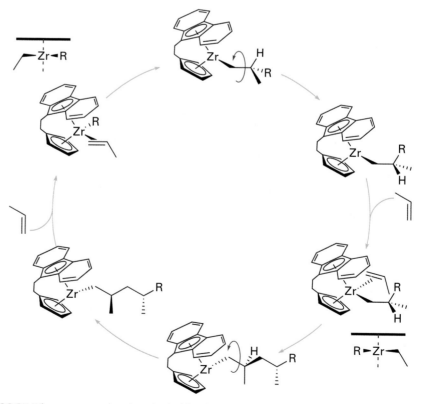

Figure 26.27 When a propene is polymerized with a fluorenyl metallocene catalyst, syndiotactic polypropene results. The zirconium species all have a single positive charge; this charge has been omitted for clarity.

26.17 **Electrocatalysis**

Key points: Overpotentials represent the kinetic barrier for electrochemical reactions and electrocatalysts may be used to increase current densities in such processes.

Kinetic barriers are quite common for electrochemical reactions at the interface between a solution and an electrode and, as mentioned in Section 5.18, it is common to express these barriers as overpotentials η (eta), the potential in addition to the zero-current cell potential (the emf) that must be applied to bring about an otherwise slow reaction within the cell. The overpotential is related to the current density, j (the current divided by the area of the electrode), that passes through the cell by[7]

$$j = j_0 e^{a\eta} \tag{26.3}$$

where j_0 and a are best regarded for our purposes as empirical constants. The constant j_0, the **exchange current density**, is a measure of the rates of the forward and reverse electrode reactions at dynamic equilibrium. For systems obeying these relations, the reaction rate (as measured by the current density) increases rapidly with increasing applied potential difference when $a\eta > 1$. If the exchange current density is high, an appreciable reaction rate may be achieved with only a small overpotential. If the exchange current density is low, a high overpotential is necessary. There is therefore considerable interest in increasing the exchange current density. In an industrial process an overpotential in a synthetic step is very costly because it represents wasted energy.

A catalytic electrode surface can increase the exchange current density and hence dramatically decrease the overpotential required for sluggish electrochemical reactions, such as H_2, O_2, or Cl_2 evolution and consumption. For example, 'platinum black', a finely divided form of platinum, is very effective at increasing the exchange current density and hence decreasing the overpotential of reactions involving the consumption or evolution of H_2. The role of platinum is to dissociate the strong $H-H$ bond and thereby reduce the large barrier that this bond strength imposes on reactions involving H_2. Palladium also has a high exchange current density (and hence requires only a low overpotential) for H_2 evolution or consumption.

The effectiveness of metals can be judged from Fig. 26.28, which also gives insight into the process. The volcano-like plot of exchange current density against $M-H$ bond enthalpy suggests that $M-H$ bond formation and cleavage are both important in the catalytic process. It appears that an intermediate $M-H$ bond energy leads to the proper balance for the existence of a catalytic cycle and the most effective metals for electrocatalysis are clustered around Group 10.

Ruthenium dioxide is an effective catalyst for both O_2 and Cl_2 evolution and it also is a good electrical conductor. It turns out that at high current densities RuO_2 is more effective for the catalysis of Cl_2 evolution than for O_2 evolution. RuO_2 is therefore extensively used as an electrode material in the commercial production of chorine. The electrode processes that contribute to this subtle catalytic effect do not appear to be well understood.

There is great interest in devising new catalytic electrodes, particularly ones that decrease the O_2 overpotential on surfaces such as graphite. Thus, tetrakis(4-N-methylpyridyl) porphyriniron(II), [Fe(TMPyP)]$^{4+}$ (**26**), has been deposited on the exposed edges of graphite electrodes (on which O_2 reduction requires a high overpotential) and the resulting electrode surface was found to catalyse the electrochemical reduction of O_2. A plausible explanation for this catalysis is that [Fe^{III}(TMPyP)] attached to the electrode (indicated below by an asterisk) is first reduced electrochemically:

$$[Fe^{III}(TMPyP)]^* + e^- \rightarrow [Fe^{II}(TMPyP)]^*$$

The resulting [Fe^{II}(TMPyP)] forms an O_2 complex:

$$[Fe^{II}(TMPyP)]^* + O_2 \rightarrow [Fe^{II}(O_2)(TMPyP)]^*$$

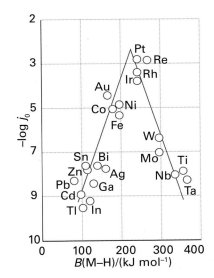

Figure 26.28 The rate of H_2 evolution expressed as the logarithm of the exchange current density plotted against $M-H$ bond energy.

[7]The exponential relation between the current and the overpotential is accounted for by the Butler–Volmer equation, which is derived by applying transition-state theory to dynamical processes at electrodes. See P. Atkins and J. de Paula, *Physical chemistry*. Oxford University Press and W.H. Freeman & Co (2010).

This iron(II) porphyrin oxygen complex then undergoes reduction to water and hydrogen peroxide:

$$[Fe^{II}(O_2)(TMPyP)]^* + ne^- + nH^+ \rightarrow [Fe^{II}(TMPyP)]^* + (H_2O_2, H_2O)_n$$

Although details of the mechanism are still elusive, this general set of reactions is in harmony with electrochemical measurements and the known properties of iron porphyrins. The investigation of porphyrin iron complexes as catalysts was motivated by Nature's use of metalloporphyrins for oxygen activation (Section 27.10).

One application where cheap and efficient electrocatalysts are much needed is in PEM fuel cells (PEM stands for 'proton exchange membrane' and sometimes 'polymer electrolyte membrane'; Box 5.1). As these systems operate at quite low temperatures, 50–100°C, they are suitable for applications that involve transport and mobile power, such as mobile telephones (cell phones). The electrolyte, a polymer that conducts protons but not electrons, separates an anode in contact with hydrogen gas and a cathode over which flows oxygen gas. As noted above hydrogen gas is readily dissociated at the anode using a platinum metal catalyst but the oxygen reduction reaction at the cathode presents more of a problem. There is a substantial overpotential associated with this reduction, which considerably reduces the efficiency of the cell, giving operating voltages near 0.7 V as compared with the theoretical value of 1.23 V. Platinum may again be used to catalyse the reaction but even so the reaction is inefficient and use of large quantities of platinum costly. Considerable research effort is being applied to finding better electrocatalysts and recently the alloy Pt_3Ni (111 surface) has proved to have greatly enhanced properties.

26.18 New directions in heterogeneous catalysis

Key point: Continuing developments in heterogeneous catalysis include research on new compositions for the controlled partial oxidation of hydrocarbons.

The development of solid-phase catalysts is very much a frontier subject for inorganic chemistry with the continuing discovery of compositions for promoting reactions, particularly for petrochemicals. One very active area is the investigation of selective heterogeneous oxidation catalysts, which allow partial oxidation of hydrocarbons to useful intermediates in, for example, the polymer and pharmaceutical industries. Examples of such reactions include alkene epoxidation, aromatic hydroxylation, and ammoxidation (an oxidation in the presence of ammonia that generates nitriles) of alkanes, alkenes, and alkyl aromatics. In all these cases it is desirable to produce the products without complete oxidation of the hydrocarbon to carbon dioxide.

One example where new catalysts are being investigated is in the partial oxidation of benzene to phenol. At present, the three-step **cumene process** produces about 95 per cent of the phenol used in the world and gives propanone (acetone) as a byproduct, although the market for acetone is oversupplied from other industrial processes. The cumene process involves three stages, namely alkylation of benzene with propene to form cumene (a process catalysed by phosphoric acid or aluminium chloride), direct oxidation of cumene to cumene hydroperoxide using molecular oxygen, and finally cleavage of cumene hydroperoxide to phenol and acetone, which is catalysed by sulfuric acid. Far better would be a single-stage process that accomplishes the reaction

$$C_6H_6 + \tfrac{1}{2}O_2 \rightarrow C_6H_5OH$$

Examples of catalysts investigated for this and similar processes include iron-containing zeolites with the silicalite framework type (Section 24.13), mixtures of $FeCl_3/SiO_2$, photocatalysts based on $Pt/H_2SO_4/TiO_2$, and various vanadium salts.

Hybrid catalysis

Chemists have begun to look at catalytic systems that cannot simply be defined as either homogeneous or heterogeneous. Research has centred on trying to achieve the best of both systems: the high selectivity of homogeneous catalysts coupled with the ease of separation of heterogeneous catalysts. This approach is sometimes referred to as 'heterogenizing' homogeneous catalysts.

26.19 Tethered catalysts

Key point: The tethering of a catalyst to a solid support allows easy separation of the catalyst with little loss in catalyst activity.

One popular technique has been the **tethering** of a homogeneous catalyst to a solid support. Thus, a hydrogenation catalyst such as Wilkinson's catalyst can be attached to a silica surface by means of a long hydrocarbon chain:

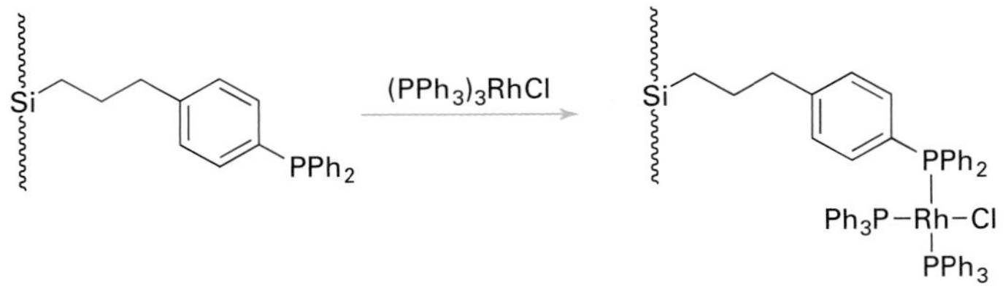

When the silica monolith is immersed in a solvent, the rhodium-based catalytic site behaves as though it is in solution, and reactivity is largely unaffected. Separation of the products from the catalyst simply requires the decanting of the solvent. Commercially available functionalized silica precursors include amino-, acrylate-, allyl-, benzyl-, bromo-, chloro-, cyano-, hydroxy-, iodo-, phenyl-, styryl-, and vinyl-substituted reagents. Relatively simple reactions can result in the synthesis of a whole host of further reagents (such as the phosphine compound in the scheme above). In addition to silica, polystyrene, polyethene, polypropene, and various clays have been used as the solid support and have led to reports of the successful heterogenization of most reactions that rely on soluble metal complexes.

In some cases the activity of a supported catalyst is greater than that of its unsupported analogue. This improvement normally takes the form of enhanced selectivity brought about by the steric demands of approaching a catalyst constrained to a surface or an increase in catalyst turnover frequencies brought about by protection from the support. Often, however, supported catalysts suffer from catalyst leaching and reduced activity.

26.20 Biphasic systems

Key point: Biphasic systems offer another way of combining the selectivity of homogeneous catalysts with the ease of separation of heterogeneous catalysts.

Another popular method of attempting to combine the best of both homogeneous and heterogeneous catalysts has been the use of two liquid phases that are not miscible at room temperature but are miscible at higher temperatures. The existence of immiscible aqueous/organic phases together with a compound that facilitates the transfer of reagents between the two phases will be familiar; two other systems worthy of note are ionic liquids and the 'fluorous biphase' system.

Ionic liquids are typically derived from 1,3-dialkylimidazolium cations (**27**) with counterions such as PF_6^-, BF_4^-, and $CF_3SO_3^-$. These systems have melting points of less than (and often much less than) 100°C, a very low viscosity, and an effectively zero vapour pressure. If it can be arranged that the catalysts are preferentially soluble in the ionic phase (for instance, by making them ionic), immiscible organic solvents can be used to extract organic products. As an example, consider the hydroformylation of alkenes by a Rh catalyst with phosphine ligands. When the ligand used is triphenylphosphine, the catalyst is extracted from the ionic liquid together with the products. However, when an ionic sulfonated triphenylphosphine is used as the ligand, the catalyst remains in the ionic liquid and product separation from the catalyst is complete. It should be borne in mind that the ionic liquid phase is not always unreactive, and may induce alternative reactions.

The fluorous biphase system, which typically consists of a fluorinated hydrocarbon and a 'normal' organic solvent such as methylbenzene, offers two principal advantages over aqueous/organic phases: the fluorous phase is unreactive, so sensitive groups (such as hydrolytically unstable groups) are stable in it, and polyfluorinated and hydrocarbon solvents, which are not miscible at room temperatures, become so on heating, giving rise to a genuinely homogeneous system. A catalyst that contains polyfluorinated groups is preferentially

27

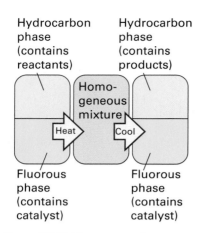

Figure 26.29 The sequence of processes that occurs during the use of a fluorous biphase catalyst.

retained in the fluorous solvent, with the reactants (and products) preferentially soluble in the hydrocarbon phase. Figure 26.29 indicates the type of sequence that is used. Separation of the products from the catalyst becomes a trivial matter of decanting one liquid from another. A number of ligand systems that confer fluorous solubility on a catalyst have been developed; typically they are phosphine based such as (**28**), (**29**), and (**30**). Rhodium- and Pd-based catalysts of these ligands have then been prepared. With a fluorous solvent such as perfluoro-1,3-dimethylcyclohexane (**31**), miscibility with an organic phase can be achieved at 70°C, and catalysts have been used in hydrogenation (Section 26.4), hydroformylation (Section 26.5), and hydroboration reactions.

FURTHER READING

R. Whyman, *Applied organometallic chemistry and catalysis*. Oxford University Press (2001).

G.W. Parshall and S.D. Ittle, *Homogeneous catalysis*. Wiley, New York (1992).

See H.H. Brintzinger, D. Fischer, R. Mülhaupt, B. Rieger, and R.M. Waymouth, *Angew. Chem., Int. Ed. Engl.*, 1995, **34**, 1143 for a review of the area of control of polymer tacticity.

For a good review of the Stille reaction that touches on the mechanism of all palladium-catalysed coupling reactions, see P. Espinet and A.M. Echavarren, *Angew. Chem., Int. Ed. Engl.*, 2004, **43**, 4704.

A whole issue of a major journal has been devoted to the subject of supported homogeneous catalysts. See *Chem. Rev.*, 2002, **102**, 3215.

Fluorous biphase systems are discussed in greater detail by E.G. Hope and A.M Stuart, *J. Fluorine Chem.*, 1999, **100**, 75.

For a review of the development of alkene metathesis catalysts, see T.M. Trnka and R.H. Grubbs, *Acc. Chem. Res.*, 2001, **34**, 18.

V. Ponec and G.C. Bond, *Catalysis by metals and alloys*. Elsevier, Amsterdam (1995). Comprehensive discussion of the basis of chemisorption and catalysis by metals.

R.D. Srivtava, *Heterogeneous catalytic science*. CRC Press, Boca Raton (1988). A survey of experimental methods and several major heterogeneous catalytic processes.

M. Bowker, *The basis and applications of heterogeneous catalysis*. Oxford Chemistry Primers 53. Oxford University Press (1998). Concise coverage of heterogeneous catalysis.

J.M. Thomas and W.J. Thomas, *Principles and practice of heterogeneous catalysis*. VCH, Weinheim (1997). Readable introduction to the fundamental principles of heterogeneous catalysis written by world-renowned experts.

K.M. Neyman and F. Illas, Theoretical aspects of heterogeneous catalysis: applications of density functional methods. *Catalysis Today*, 2005, **105**, 15. Modelling methods applied to heterogeneous catalysis.

M.A. Keane, Ceramics for catalysis. *J. Mater. Sci.*, 2003, **38**, 4661. Overview of heterogeneous catalysis illustrated with three established methods: (i) catalysis using zeolites; (ii) catalytic converters; (iii) solid oxide fuel cells.

F.S. Stone, Research perspectives during 40 years of the Journal of Catalysis. *J. Catal.*, 2003, **216**, 2. A historical perspective on developments in catalysis.

G. Rothenberg, *Catalysis: concepts and green applications*. Wiley VCH (2008). Catalysis and sustainability.

D.K. Chakrabarty and B. Viswanathan, *Heterogeneous catalysis*. New Age Science Ltd (2008).

EXERCISES

26.1 Which of the following constitute genuine examples of catalysis and which do not? Present your reasoning. (a) The addition of H_2 to C_2H_4 when the mixture is brought into contact with finely divided platinum. (b) The reaction of an H_2/O_2 gas mixture when an electrical arc is struck. (c) The combination of N_2 gas with lithium metal to produce Li_3N, which then reacts with H_2O to produce NH_3 and LiOH.

26.2 Define the terms (a) turnover frequency, (b) selectivity, (c) catalyst, (d) catalytic cycle, (e) catalyst support.

26.3 Classify the following as homogeneous or heterogeneous catalysis and present your reasoning. (a) The increased rate in the presence of NO(g) of SO_2(g) oxidation by O_2(g) to SO_3(g). (b) The hydrogenation of liquid vegetable oil using a finely divided nickel catalyst. (c) The conversion of an aqueous solution of D-glucose to a D,L mixture catalysed by HCl(aq).

26.4 You are approached by an industrialist with the proposition that you develop catalysts for the following processes at 80°C with no input of electrical energy or electromagnetic radiation:

(a) The splitting of water into H_2 and O_2.

(b) The decomposition of CO_2 into C and O_2.

(c) The combination of N_2 with H_2 to produce NH_3.

(d) The hydrogenation of the double bonds in vegetable oil.

The industrialist's company will build the plant to carry out the process and the two of you will share equally in the profits. Which of these would be easy to do, which are plausible candidates for investigation, and which are unreasonable? Describe the chemical basis for the decision in each case.

26.5 Addition of PPh_3 to a solution of Wilkinson's catalyst, $[RhCl(PPh_3)_3]$, reduces the turnover frequency for the hydrogenation of propene. Give a plausible mechanistic explanation for this observation.

26.6 The rates of H_2 gas absorption (in $dm^3 \, mol^{-1} \, s^{-1}$) by alkenes catalysed by $[RhCl(PPh_3)_3]$ in benzene at 25°C are: hexene, 2910; *cis*-4-methyl-2-pentene, 990; cyclohexene, 3160; 1-methylcyclo-hexene, 60. Suggest the origin of the trends and identify the affected reaction step in the proposed mechanism (Fig. 26.5).

26.7 Infrared spectroscopic investigation of a mixture of CO, H_2, and 1-butene under conditions that bring about hydroformylation indicate the presence of compound (E) in Fig. 26.7 in the reaction mixture. The same reacting mixture in the presence of added tributylphosphine was studied by infrared spectroscopy and neither (E) nor an analogous phosphine-substituted complex was observed. What does the first observation suggest as the rate-limiting reaction in the absence of

phosphine? Assuming the sequence of reactions remains unchanged, what are the possible rate-limiting reactions in the presence of tributylphosphine?

26.8 Show how reaction of MeCOOMe with CO under conditions of the Monsanto ethanoic acid process can lead to ethanoic anhydride.

26.9 Suggest reasons why (a) ring-opening alkene metathesis polymerization and (b) ring-closing metathesis reactions proceed.

26.10 (a) Starting with the alkene complex shown in Fig. 26.9 with *trans*-DHC=CHD in place of C_2H_4, assume dissolved OH^- attacks from the side opposite the metal. Give a stereochemical drawing of the resulting compound. (b) Assume attack on the coordinated *trans*-DHC=CHD by an OH^- ligand coordinated to Pd, and draw the stereochemistry of the resulting compound. (c) Does the stereochemistry differentiate these proposed steps in the Wacker process?

26.11 Aluminosilicate surfaces in zeolites act as strong Brønsted acids, whereas silica gel is a very weak acid. (a) Give an explanation for the enhancement of acidity by the presence of Al^{3+} in a silica lattice. (b) Name three other ions that might enhance the acidity of silica.

26.12 Why is the platinum/rhodium catalyst in automobile catalytic converters dispersed on the surface of a ceramic rather than used in the form of a thin metal foil?

26.13 Alkanes are observed to exchange hydrogen atoms with deuterium gas over some platinum metal catalysts. When 3,3-dimethylpentane in the presence of D_2 is exposed to a platinum catalyst and the gases are observed before the reaction has proceeded very far, the main product is $CH_3CH_2C(CH_3)_2CD_2CD_3$ plus unreacted 3,3-dimethylpentane. Devise a plausible mechanism to explain this observation.

26.14 The effectiveness of platinum in catalysing the reaction $2 H^+(aq) + 2 e^- \rightarrow H_2(g)$ is greatly decreased in the presence of CO. Suggest an explanation.

26.15 Describe the role of electrocatalysts in reducing the overpotential in the oxygen reduction reaction in fuel cells.

PROBLEMS

26.1 Consider the validity of each of the following statements and provide corrections where required. (a) A catalyst introduces a new reaction pathway with lower enthalpy of activation. (b) As the Gibbs energy is more favourable for a catalytic reaction, yields of the product are increased by catalysis. (c) An example of a homogeneous catalyst is the Ziegler−Natta catalyst made from $TiCl_4$(l) and $Al(C_2H_5)_3$(l). (d) Highly favourable Gibbs energies for the attachment of reactants and products to a homogeneous or heterogeneous catalyst are the key to high catalytic activity.

26.2 A catalyst might not just lower the enthalpy of activation, but might make a significant change to the entropy of activation. Discuss this phenomenon. (See A. Haim, *J. Chem., Educ.*, 1989, **66**, 935.)

26.3 The addition of promoters can further enhance the rate of a catalysed reaction. Describe how the promoters allowed the iridium-based Cativa process to compete with the rhodium-based process in the carbonylation of methanol. (See A. Haynes, P.M. Maitlis, G.E. Morris, G.J. Sunley, H. Adams, and P.W. Badger, *J. Am. Chem. Soc.*, 2004, **126**, 2847.)

26.4 When direct evidence for a mechanism is not available, chemists frequently invoke analogies with similar systems. Describe how J.E. Bäckvall, B. Åkermark, and S.O. Ljunggren (*J. Am. Chem. Soc.*, 1979, **101**, 2411) inferred the attack of uncoordinated water on η^2-C_2H_4 in the Wacker process.

26.5 Whereas many enantioselective catalysts require the precoordination of a substrate, this is not always the case. Use the example of asymmetric epoxidation to demonstrate the validity of this statement, and indicate the advantages that such catalysts might have over catalysts that require precoordination of a substrate. (See M. Palucki, N.S. Finney, P.J. Pospisil, M.L. Güler, T. Ishida, and E.N. Jacobsen, *J. Am. Chem. Soc.*, 1998, **120**, 948.)

26.6 Discuss shape selectivity with respect to catalytic processes involving zeolites.

26.7 Summarize the potential impact of heterogeneous oxidation catalysts in chemistry. (See J.M. Thomas and R. Raja, Innovations in oxidation catalysis leading to a sustainable society. *Catalysis Today*, 2006, **117**, 22.

26.8 Discuss the applications and mechanisms of oxidation and ammoxidation catalysts such as bismuth molybdate. (See, for example, R.K. Grasselli, *J. Chem. Educ.*, 1986, **63**, 216.)

26.9 Discuss the advantages of a solid support in catalysis by reference to the use of $(Ni(POEt)_3)_4$ in alkene isomerization. (See A.J. Seen, *J. Chem. Educ.*, 2004, **81**, 383 and K.R. Birdwhistell and J. Lanza, *J. Chem. Educ.*, 1997, **74**, 579.)

27 Biological inorganic chemistry

Organisms have exploited the chemical properties of the elements in remarkable ways, providing examples of coordination specificities that are far higher than observed in simple compounds. This chapter describes how different elements are taken up selectively by different cells and intracellular compartments and the various ways they are exploited. We discuss the structures and functions of complexes and materials that are formed in the biological environment in the context of the chemistry covered earlier in the text.

Biological inorganic chemistry ('bioinorganic chemistry') is the study of the 'inorganic' elements as they are utilized in biology. The main focus is on metal ions, where we are interested in their interaction with biological ligands and the important chemical properties they are able to exhibit and impart to an organism. These properties include ligand binding, catalysis, signalling, regulation, sensing, defence, and structural support.

The organization of cells

To appreciate the role of the elements (other than C, H, O, and N) in the structure and function of organisms we need to know a little about the organization of the 'atom' of biology, the cell, and its 'fundamental particles', the cell's constituent organelles.

27.1 The physical structure of cells

Key points: Living cells and organelles are enclosed by membranes; the concentrations of specific elements may vary greatly between different compartments due to the actions of ion pumps and gated channels.

Cells, the basic unit of any living organism, range in complexity from the simplest types found in prokaryotes (bacteria and bacteria-like organisms now classified as archaea) and the much larger and more complex examples found in eukaryotes (which include animals and plants). The main features of these cells are illustrated in the generic model shown in Fig. 27.1. Crucial to all cells are membranes, which act as barriers to water and ions and make possible the management of all mobile species and of electrical currents. Membranes are lipid bilayers, approximately 4 nm thick, in which are embedded protein molecules and other components. Bilayer membranes have great lateral strength but they are easy to bend. The long hydrocarbon chains of lipids make the membrane interior very hydrophobic and impermeable to ions, which must instead travel through specific channels, pumps, and other receptors provided by special membrane proteins. The structure of a cell also depends on osmotic pressure, which is maintained by high concentrations of solutes, including ions, imported during active transport by pumps.

Prokaryotic cells consist of an enclosed aqueous phase, the **cytoplasm**, which contains the DNA and most of the materials used and transformed in the biochemical reactions. Bacteria are classified according to whether they are enclosed by a single membrane or have an additional intermediate aqueous space, the **periplasm**, between the outer membrane and the cytoplasmic membrane, and are known as 'Gram-positive' or 'Gram-negative', respectively, depending on their response to a staining test with the dye crystal violet. The much more extensive cytoplasm of eukaryotic cells contains subcompartments (also enclosed within lipid bilayers) known as **organelles**, which have highly specialized functions. Organelles include the **nucleus** (which houses DNA), **mitochondria** (the 'fuel cells' that carry out respiration), **chloroplasts** (the 'photocells' that harness light energy), the **endoplasmic reticulum** (for protein synthesis), **Golgi** (vesicles containing proteins for export), **lysosomes** (which contain degradative enzymes and help rid the cell of waste), **peroxisomes** (which remove harmful hydrogen peroxide), and other specialized processing zones.

27.2 The inorganic composition of cells

Key points: The major biological elements are oxygen, hydrogen, carbon, nitrogen, phosphorus, sulfur, sodium, magnesium, calcium, and potassium. The trace elements include many d metals, as well as selenium, iodine, silicon, and boron.

Table 27.1 lists many of the elements known to be used in living systems, although not necessarily by higher life forms. All the second- and third-period elements except Be, Al, and the noble gases are used, as are most of the 3d elements, whereas Cd, Br, I, Mo, and W are the only heavier elements so far confirmed to have a biological function. Several others, such as Li, Ga, Tc, Ru, Gd, Pt, and Au, have important and increasingly well-understood applications in medicine.

The biologically essential elements can be classified as either 'major' or 'trace'. Although a good idea of the biological abundances of different elements is given in Table 27.1, the levels vary considerably among organisms and different components of organisms. For example, Ca has little role in microorganisms but is abundant in higher life forms, whereas the use of Co by higher organisms depends on it being incorporated into a special cofactor (cobalamin) by microorganisms. There is probably a universal requirement for K, Mg, Fe, and Mo. Vanadium is used by lower animals and plants as well as some bacteria. Nickel is essential for most microorganisms, and is used by plants, but there is no evidence for any direct role in animals. Nature's use of different elements is largely based on their availability. For example, Zn has widespread use (and, together with Fe, ranks among the

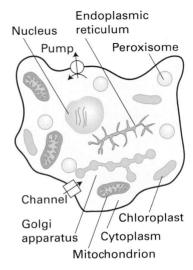

Figure 27.1 The layout of a generic eukaryotic cell showing the cell membrane, various kinds of compartments (organelles), and the membrane-bound pumps and channels that control the flow of ions between compartments.

Table 27.1 The approximate concentrations, log ([J]/mol dm⁻³), where known, of elements (apart from C, H, O, N, P, S, and Se) in different biological zones

Element	External fluids (sea water)	Free ions in external fluids (blood plasma)	Cytoplasm (free ions)	Comments on status in cell
Na	$>10^{-1}$	10^{-1}	$<10^{-2}$	Not bound
K	10^{-2}	4×10^{-3}	$\leq 3 \times 10^{-1}$	Not bound
Mg	$>10^{-2}$	10^{-3}	$c.\ 10^{-3}$	Weakly bound as ATP complex
Ca	$>10^{-3}$	10^{-3}	$c.\ 10^{-7}$	Concentrated in some vesicles
Cl	10^{-1}	10^{-1}	10^{-2}	Not bound
Fe	10^{-17} (Fe(III))	10^{-16} (Fe(III))	$<10^{-7}$ (Fe(II))	Too much unbound Fe is toxic (Fenton chemistry) in and out of cells
Zn	$<10^{-8}$	10^{-9}	$<10^{-11}$	Totally bound, but may be exchangeable
Cu	$<10^{-10}$ (Cu(II))	10^{-12}	$<10^{-15}$ (Cu(I))	Totally bound, not mobile. Mostly outside cytoplasm
Mn	10^{-9}		$c.\ 10^{-6}$	Higher in chloroplasts and vesicles
Co	10^{-11}		$<10^{-9}$	Totally bound (cobalamin)
Ni	10^{-9}		$<10^{-10}$	Totally bound
Mo	10^{-7}		$<10^{-7}$	Mostly bound

most abundant biological trace elements) whereas Co (a comparatively rare element) is essentially restricted to cobalamin. The early atmosphere (over 2.3 Ga ago[1]), being highly reducing, enabled Fe to be freely available as soluble Fe(II) salts, whereas Cu was trapped as insoluble sulfides (as was Zn). Indeed, Cu is not found in the archaea (which are believed to have evolved in pre-oxygenic times), including the hyperthermophiles, organisms that are able to survive at temperatures in excess of 100°C. These organisms are found in deep sea hydrothermal vents and terrestrial hot springs and are good sources of enzymes that contain W, the heaviest element known to be essential to life. The finding that W, Co, and for the most part Ni are used only by more primitive life forms probably reflects their special role in the early stages of evolution.

(a) Compartmentalization

Key point: Different elements are strongly segregated inside and outside a cell and among different internal compartments.

Compartmentalization is the distribution of elements inside and outside a cell and between different internal compartments. The maintenance of constant ion levels in different biological zones is an example of 'homeostasis' and it is achieved as a result of membranes being barriers to passive ion flow. An example is the large difference in concentration of K^+ and Na^+ ions across cell membranes. In the cytoplasm, the K^+ concentration may be as high as 0.3 M whereas outside it is usually less than 5×10^{-3} M. By contrast, Na^+ is abundant outside a cell but scarce inside; indeed, the low intracellular concentration of Na^+, which has characteristically weak binding to ligands, means that it has few specific roles in biochemistry. Another important example is Ca^{2+}, which is almost absent from the cytoplasm (its free concentration is below 1×10^{-7} M) yet is a common cation in the extracellular environment and is concentrated in certain organelles, such as mitochondria. That pH may also vary greatly between different compartments has particularly important implications because sustaining a transmembrane proton gradient is a key feature in photosynthesis and respiration.

The distributions of Cu and Fe provide another example: Cu enzymes are often **extracellular**, that is they are synthesized in the cell and then secreted outside the cell, where they catalyse reactions involving O_2. By contrast, Fe enzymes are contained inside the cell. This difference can be rationalized on the basis that the inactive trapped states of these elements are Fe(III) and Cu(I) (or even metallic Cu) and organisms have stumbled on the expediency of keeping Fe in a relatively reducing environment and Cu in a relatively oxidizing environment.

The selective uptake of metal ions has potential industrial applications, for many organisms and organs are known to concentrate particular elements. Thus, liver cells are a good source of cobalamin[2] (Co) and milk is rich in Ca. Certain bacteria accumulate Au and thus provide an unusual way for procuring this precious metal. Compartmentalization is an important factor in the design of metal complexes that are used in medicine (Sections 27.17–20).

The very small size of bacteria and organelles raises an interesting point about scale, as species present at very low concentrations in very small regions may be represented by only a few individual atoms or molecules. For example, the cytoplasm in a bacterial cell of volume 10^{-15} dm³ at pH = 6 will contain less than 1000 'free' H^+ ions. Indeed, any element nominally present at less than 1 nmol dm⁻³ may be completely absent in individual cases. The word 'free' is significant, particularly for metal ions such as Zn^{2+} that are high in the Irving–Williams series; even a eukaryotic cell with a total Zn concentration of 0.1 mmol dm⁻³ may contain very few uncomplexed Zn^{2+} ions.

Two important issues arise in the context of compartmentalization. First, the process requires energy because ions must be pumped against an adverse gradient of chemical potential. However, once a concentration difference has been established, there is a difference in electrical potential across the membrane dividing the two regions. For instance, if the concentrations of K^+ ions on either side of a membrane are $[K^+]_{in}$ and $[K^+]_{out}$, then the

---

[1] Current geological and geochemical evidence date the advent of atmospheric O_2 at between 2.2 and 2.4 Ga ago (1 Ga = 10⁹ a). It is likely that this gas arose by the earliest catalytic actions of the photosynthetic Mn cluster described in Section 27.10.

[2] In nutrition, the common complexes of cobalamin that are ingested are known as vitamin B_{12}.

contribution to the potential difference $\Delta\phi$ across the membrane is

$$\Delta\phi = \frac{RT}{F} \ln \frac{[K^+]_{in}}{[K^+]_{out}} \qquad (27.1)$$

This difference in electrical potential is a way of storing energy, which is released when the ions flood back to their natural concentrations. Second, the selective transport of ions must occur through ion channels built from membrane-spanning proteins, some of which release ions on receipt of an electrical or chemical signal whereas others, the **transporters** and **pumps**, transfer ions against the concentration gradient by using energy provided by adenosine triphosphate (ATP) hydrolysis. The selectivity of these channels is exemplified by the highly discriminatory transport of K^+ as distinct from Na^+ (Section 27.3).

Proteins, the most important sites for metal ion coordination, are not permanent species but are ceaselessly degraded by enzymes (proteases), releasing both amino acids and metal ions to provide materials for new molecules.

EXAMPLE 27.1 **Assessing the role of phosphate ions**

Phosphate is the most abundant small anion in the cytoplasm. What implications does this abundance have for the biochemistry of Ca^{2+}?

Answer We can approach this problem by considering how Ca^{2+} is compartmentalized. In a eukaryotic cell Ca^{2+} is pumped out of the cytoplasm (to the exterior or into organelles such as mitochondria) using energy derived from ATP hydrolysis. Spontaneous influx of Ca^{2+} occurs under the action of special channels or if the cell boundary is damaged. The solubility product of $Ca_3(PO_4)_2$ is very low and it could precipitate inside the cell if the Ca^{2+} concentration rises above a critical value.

Self-test 27.1 Is Fe(II) expected to be present in the cell as uncomplexed ions?

(b) Biological metal-coordination sites

Key points: The major binding sites for metal ions are provided by the amino acids that make up protein molecules; the ligation sites range from backbone peptide carbonyls to the side chains that provide more specific complexation; nucleic acids and lipid head groups are usually coordinated to major metal ions.

Metal ions coordinate to proteins, nucleic acids, lipids, and a variety of other molecules. For instance, ATP is a tetraprotic acid and is always found as its Mg^{2+} complex (**1**); DNA is stabilized by weak coordination of K^+ and Mg^{2+} to its phosphate groups but destabilized by binding of soft metal ions such as Cu(I) to the bases. Macromolecules known as ribozymes may represent an important stage in the early evolution of life forms and are catalytic molecules composed of RNA and Mg^{2+}. The binding of Mg^{2+} to phospholipid head groups is important for stabilizing membranes. There are a number of important small ligands, apart from water and free amino acids, which include sulfide, sulfate, carbonate, cyanide, carbon monoxide, and nitrogen monoxide, as well as organic acids such as citrate that form reasonably strong polydentate complexes with Fe(III).

As will be familiar from introductory chemistry, a protein is a polymer with a specific sequence of amino acids linked by peptide bonds (**2**). A 'small' protein is generally regarded as one with molar mass below $20\ kg\ mol^{-1}$, whereas a 'large' protein is one having a molar mass above $100\ kg\ mol^{-1}$. The principal amino acids are listed in Table 27.2. Proteins are

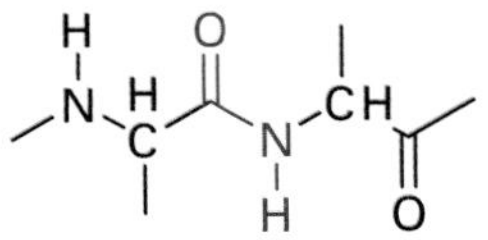

2 Peptide bond

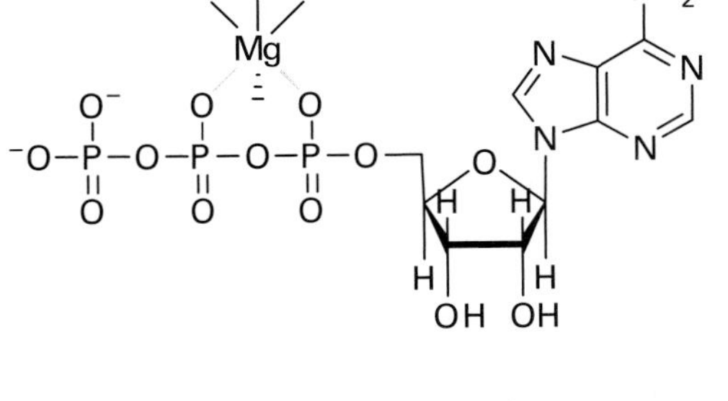

1 Mg–ATP complex

synthesized, a process called **translation** (of the genetic code carried by DNA), on a special assembly called a *ribosome*. A protein may be processed further by **post-translational modification**, a change made to the protein structure, which includes the binding of **cofactors** such as metal ions.

Metalloproteins, proteins containing one or more metal ions, perform a wide range of specific functions. These functions include oxidation and reduction (for which the most

Table 27.2 The amino acids and their codes

Amino acid	Structure in peptide chain (side chain shown in blue)	Three-letter abbreviation	One-letter abbreviation
Alanine	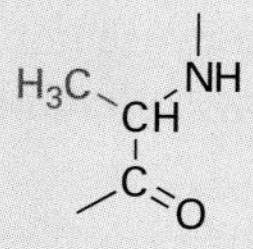	Ala	A
Arginine	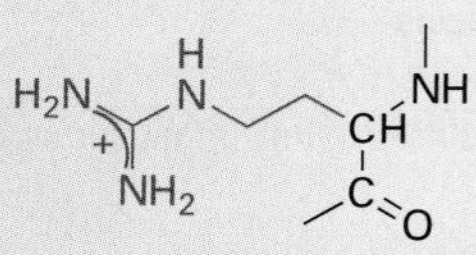	Arg	R
Asparagine	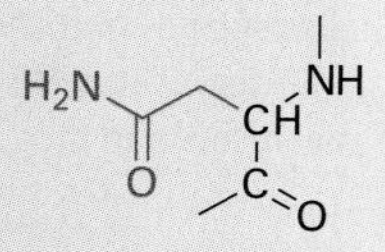	Asn	N
Aspartic acid	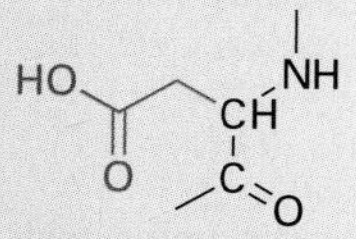	Asp	D
Cysteine	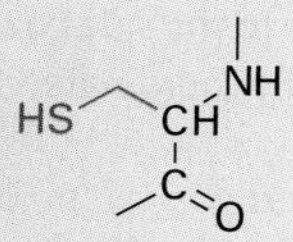	Cys	C
Glutamic acid	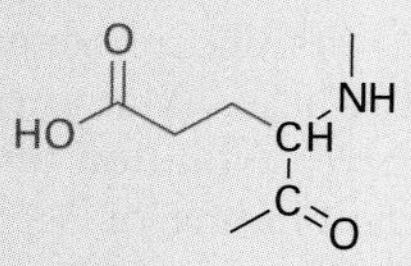	Glu	E
Glutamine	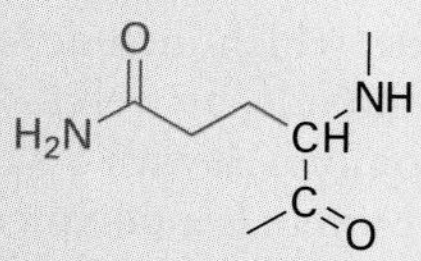	Gln	Q
Glycine	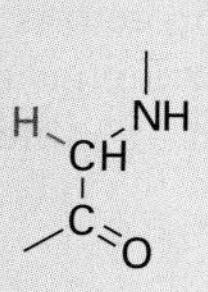	Gly	G
Histidine	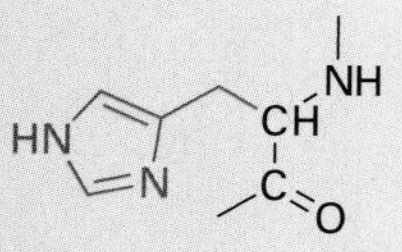	His	H
Isoleucine	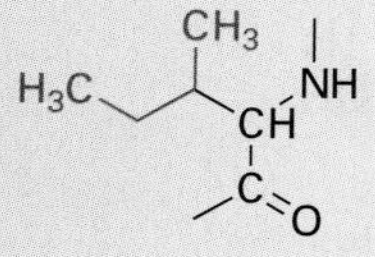	Ile	I

Amino acid	Structure	3-letter	1-letter
Leucine	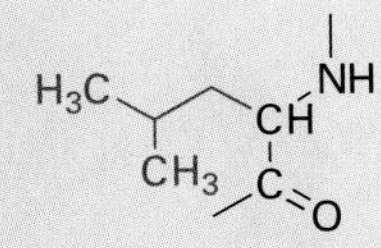 (structure)	Leu	L
Lysine	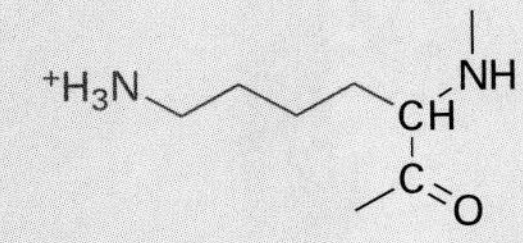 (structure)	Lys	K
Methionine	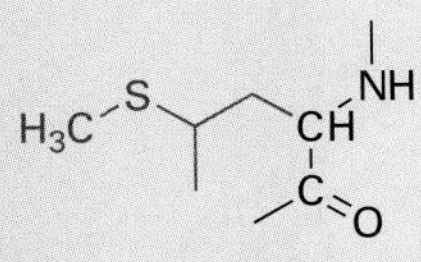 (structure)	Met	M
Phenylalanine	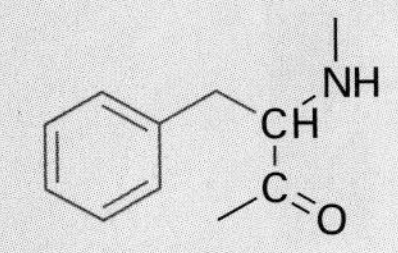 (structure)	Phe	F
Proline	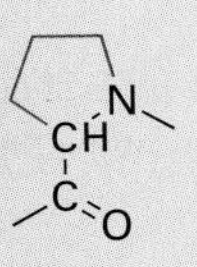 (structure)	Pro	P
Serine	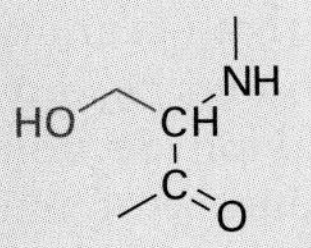 (structure)	Ser	S
Threonine	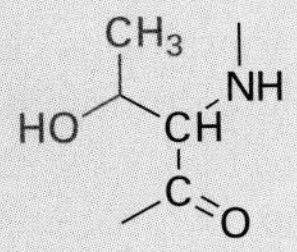 (structure)	Thr	T
Tryptophan	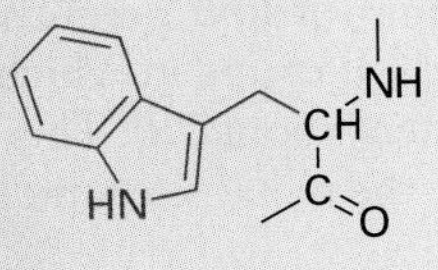 (structure)	Trp	W
Tyrosine	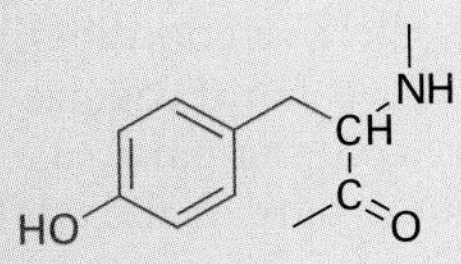 (structure)	Tyr	Y
Valine	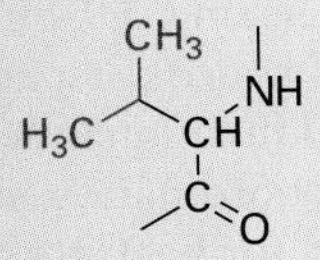 (structure)	Val	V

important elements are Fe, Mn, Cu, and Mo), radical-based rearrangement reactions and methyl-group transfer (Co), hydrolysis (Zn, Fe, Mg, Mn, and Ni), and DNA process-ing (Zn). Special proteins are required for transporting and storing different metal at-oms. The action of Ca^{2+} is to alter the conformation of a protein (its shape) as a step in cell signalling (a term used to describe the transfer of information between and within cells). Such proteins are often known as **metal ion-activated proteins**. Hydrogen bonding between main-chain −NH and CO groups of different amino acids results in **second-ary structure** (Fig. 27.2). The -helix regions of a polypeptide provide flexible mobility

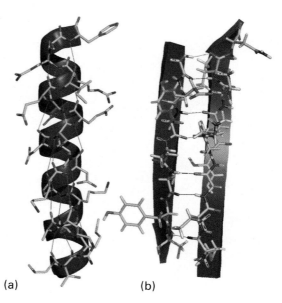

Figure 27.2 The most important regions of secondary structure, (a) α helix, (b) β sheet, showing hydrogen bonding between main-chain amide and carbonyl groups and their corresponding representations.

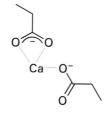

3 Ca²⁺ coordination

4 Cu–imidazole coordination

5 Zn–cysteine coordination

6 Fe–methionine coordination

7 Fe–tyrosine coordination

(like springs) and are important in converting processes that occur at the metal site into conformational changes; by contrast, a **χ-sheet** region confers rigidity to support a pre-organized coordination sphere suited to a particular metal ion (Sections 7.14 and 11.16). The secondary structure is largely determined by the sequence of amino acids: thus the α helix is favoured by chains containing alanine and lysine but is destabilized by glycine and proline. A protein that lacks its cofactor (such as the metal ions required for normal activity) is called an **apoprotein**; an enzyme with a complete complement of cofactors is known as a **holoenzyme**.

An important factor influencing metal-ion coordination in proteins is the energy required to locate an electrical charge inside a medium of low permittivity. To a first approximation, protein molecules may be regarded as oil drops in which the interior has a much lower relative permittivity (about 4) than water (about 78). This difference leads to a strong tendency to preserve electrical neutrality at the metal site, and hence influence the redox chemistry and Brønsted acidity of its ligands.

All amino acid residues can use their peptide carbonyl (or amide-N) as a donor group, but it is the side chain that usually provides more selective coordination. By referring to Table 27.2 and from the discussion in Section 4.12, we can recognize donor groups that are either chemically hard or soft and that therefore confer a particular affinity for specific metal ions. Aspartate and glutamate each provide a hard carboxylate group, and may use one or both O atoms as donors (**3**). The ability of Ca²⁺ to have a high coordination number and its preference for hard donors are such that certain Ca²⁺-binding proteins also contain the unusual amino acids γ-carboxyglutamate and hydroxyaspartate (generated by post-translational modification), which provide additional functionalities to enhance binding. Histidine, which has an imidazole group with two coordination sites, the ε-N atom (more common) and the δ-N atom, is an important ligand for Fe, Cu(**4**), and Zn. Cysteine has a thiol S atom that is expected to be unprotonated (thiolate) when involved in metal coordination. It is a good ligand for Fe, Cu, and Zn (**5**), as well as for toxic metals such as Cd and Hg. Methionine contains a soft thioether S donor that stabilizes Fe(II)(**6**) and Cu(I). Tyrosine can be deprotonated to provide a phenolate O donor atom that is a good ligand for Fe(III) (**7**). Selenocysteine (a specially coded amino acid in which Se replaces S) has also been identified as a ligand, for example it is found as a ligand to Ni in some hydrogenases (Section 27.14). A modified form of lysine, in which the side-chain –NH₂ has reacted with a molecule of CO₂ to produce a carbamate, is found as a ligand to Mg in the crucial photosynthetic enzyme known as rubisco (Section 27.9) and in other enzymes such as urease, where it is a ligand for Ni(II).

The primary and secondary structures of a polypeptide molecule can enforce unusual metal coordination geometries that are rarely encountered in small complexes. Protein-induced strain is an important possibility, for example the protein may impose a coordination geometry on the metal ion that resembles the transition state for the particular process being executed.

(c) Special ligands

Key point: Metal ions may be bound in proteins by special organic ligands such as porphyrins and pterin-dithiolenes.

The porphyrin group (**8**) was first identified in haemoglobin (Fe) and a similar macrocycle is found in chlorophyll (Mg). There are several classes of this hydrophobic macrocycle, each differing in the nature of the side chains. The corrin ligand (**9**) has a slightly smaller ring size and coordinates Co in cobalamin (Section 27.11). Rather than show these macrocycles in full, we shall use shorthand symbols such as (**10**) to show the complexes they form with metals. Almost all Mo and W enzymes have the metal coordinated by a special ligand known as *molybdopterin* (**11**). The donors to the metal are a pair of S atoms from a dithiolene group that is covalently attached to a pterin. The phosphate group is often joined to a nucleoside base X, such as guanosine 5′-phosphate (GMP), resulting in the formation of a diphosphate bond. Why Mo and W are coordinated by this complex ligand is unknown, but the pterin group could provide a good electron conduit and facilitate redox reactions.

(d) The structures of metal coordination sites

Key point: The likelihood that a protein will coordinate a particular kind of metal centre can be inferred from the amino acid sequence and ultimately from the gene itself.

The structures of metal coordination sites have been determined mainly by X-ray diffraction (now mostly by using a synchrotron, Section 8.1) and sometimes by NMR spectroscopy.[3] The basic structure of the protein can be determined even if the resolution is too low to reveal details of the coordination at the metal site. The packing of amino acids in a protein is far denser than is commonly conveyed by simple representations, as may be seen by comparing the representations of the structure of the K^+ channel in Fig. 27.3. Thus, even the substitution of an amino acid that is far from a metal centre may result in significant structural changes to its coordination shell and immediate environment. Of special interest are channels or clefts that allow a substrate selective access to the active site, pathways for long-range electron transfer (metal centres positioned less than 1.5 nm apart), pathways for long-range proton transfer (comprising chains of basic groups such as carboxylates and water molecules in close proximity, usually less than 0.3 nm apart), and tunnels for small gaseous molecules (which can be revealed by placing the crystal under Xe, an electron-rich gas).

8 Porphyrin^{2-}

9 Corrin$^-$

10

11 Molybdopterin as ligand

EXAMPLE 27.2 Interpreting the coordination environments of metal ions

Simple Cu(II) complexes have four to six ligands with trigonal-bipyramidal or tetragonal geometries, whereas simple complexes of Cu(I) have four or fewer ligands, and geometries that range between tetrahedral and linear. Predict how a Cu-binding protein will have evolved so that the Cu can act as an efficient electron-transfer site.

Answer Here we are guided by Marcus theory (Section 21.12). An efficient electron-transfer reaction is one that is fast despite having a small driving force. The Marcus equation tells us that an efficient electron-transfer site is one for which the reorganization energy is small. The protein enforces on the Cu atom a coordination sphere that is unable to alter much between Cu(II) and Cu(I) states (see Section 27.8).

Self-test 27.2 In certain chlorophyll cofactors the Mg is axially coordinated by a methionine-S ligand. Why is this unusual coordination choice achieved in a protein?

---

[3] The atomic coordinates of proteins and other large biological molecules are stored in a public repository known as the Protein Data Bank located at http://www.rcsb.org/pdb/home/home.do. Each set of coordinates corresponding to a particular structure determination is identified by its 'pdb code'. A variety of software packages are available to construct and examine protein structures generated from these coordinates.

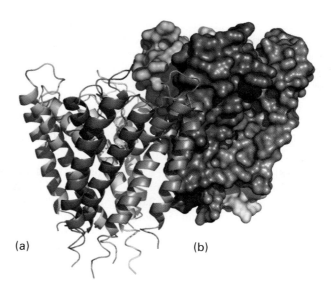

(a)   (b)

Figure 27.3 Illustrations of how protein structures are represented to reveal either (a) secondary structure or (b) the filling of space by nonhydrogen atoms. The example shows the four subunits of the K⁺ channel, which is found mainly embedded in the cell membrane.

Other physical methods described in Chapter 8 provide less information on the overall structure but are useful for identifying ligands. Thus, EPR spectroscopy is very important for studying d-block metals, especially those engaged in redox chemistry, because at least one oxidation state usually has an unpaired electron. The use of NMR is restricted to proteins smaller than 20–30 kg mol⁻¹ because tumbling rates for larger proteins are too slow and ¹H resonances are too broad to observe unless shifted away from the normal region ($\delta \approx 1-10$) by a paramagnetic metal centre. Extended X-ray absorption fine-structure spectroscopy (EXAFS, Section 8.9) can provide structural information on metal sites in amorphous solid samples, including frozen solutions. Vibrational spectroscopy (Section 8.4) is increasingly being used: IR spectroscopy is particularly useful for ligands such as CO and CN⁻, and resonance Raman spectroscopy is very helpful when the metal centre has strong electronic transitions, such as occur with Fe porphyrins. Mössbauer spectroscopy (Section 8.7) plays a special role in studies of Fe sites. Perhaps the greatest challenge is presented by Zn^{2+}, which has a d^{10} configuration that provides no useful magnetic or electronic signatures.

Metal ion binding sites can often be predicted from a gene sequence. **Bioinformatics**, the development and use of software to analyse and compare DNA sequences, is a powerful tool because many proteins that bind metal ions or have a metal-containing cofactor occur at cellular levels below that normally detectable directly by analysis and isolation. A particularly common sequence of the human genome encodes the so-called **Zn finger domain**, thereby identifying proteins that are involved in DNA binding (Section 27.5). Likewise, it can be predicted whether the protein that is encoded is likely to bind Cu, Ca, an Fe-porphyrin, or different types of Fe−S clusters. The gene can be cloned and the protein for which it encodes can be produced in sufficiently large quantities by 'overexpression' in suitable hosts, such as the common gut bacterium *Escherichia coli* or yeast, to enable it to be characterized. Furthermore, the use of genetic engineering to alter the amino acids in a protein, the technique of **site-directed mutagenesis**, is a powerful principle in biological inorganic chemistry. This technique often permits identification of the ligands to particular metal ions and the participation of other residues essential to functions such as substrate binding or proton transfer.

Although structural and spectroscopic studies give a good idea of the basic coordination environment of a metal centre, it is by no means certain that the same structure is retained in key stages of a catalytic cycle, in which unstable states are formed as intermediates. The most stable state of an enzyme, in which form it is usually isolated, is called the 'resting state'. Many enzymes are catalytically inactive on isolation and must be subjected to an activation procedure that may involve reinsertion of a metal ion or other cofactor or removal of an inhibitory ligand.

Intense efforts have been made to model the active sites of metalloproteins by synthesizing analogues. The models may be divided into two classes: those designed to mimic the structure and spectroscopic properties of the real site, and those synthesized with the intention of mimicking a functional activity, most obviously catalysis. Synthetic models not only illuminate the chemical principles underlying biological activity but also generate new directions for coordination chemistry. As we shall see throughout this chapter, the difficulty is that an enzyme not only imposes some strain on the coordination sphere of a metal atom (even a porphyrin ring is puckered in most cases) but also provides, at fixed distances, functional groups that provide additional coulombic and hydrogen-bonding interactions essential for binding and activating substrates. Indeed, the active site of a metalloenzyme is arguably the ultimate example of supramolecular chemistry.

Transport, transfer, and transcription

In this section we turn to three related aspects of the function of biological molecules containing metal ions, and see their role in the transport of ions through membranes, the transport and distribution of molecules through organisms, and the transfer of electrons. Metal ions also play an important role in the transcription of genes.

27.3 Sodium and potassium transport

Key points: Transport across a membrane is active (energized) or passive (spontaneous); the flow of ions is achieved by proteins known as ion pumps (active) and channels (passive).

In Chapter 11 we saw that differentiating between Na^+ and K^+, two ions that are very similar except for their radii (*Resource section 1*), is achieved through their selective complexation by special ligands, such as crown ethers and cryptands, with dimensions appropriate for coordination to one particular kind of ion. Organisms use this principle in the molecules known as **ionophores**, which have hydrophobic exteriors that make them soluble in lipids. The antibiotic valinomycin (Section 11.16) is an ionophore that has a high selectivity for K^+, which is coordinated by six carbonyl groups. It enables K^+ to pass through a bacterial cell membrane and thereby dissipate the electrical potential difference, so causing the bacterium's death.

At the higher end of the complexity scale are the ion channels, which are large membrane-spanning proteins that allow selective transport of K^+ and Na^+ (as well as Ca^{2+} and Cl^-) and are responsible for electrical conduction in nervous systems as well as in coupled transport of solutes.[4] Figure 27.4 shows the important structural aspects of the potential-gated K^+ channel. Moving from the inside surface of the membrane, the enzyme has a pore (which can open and close on receipt of a signal) leading into a central cavity about 1 nm in diameter; up to this stage K^+ ions can remain hydrated. Polypeptide helices pointing at this cavity have their partial charges directed in such a way as to favour population by cations, resulting in a local K^+ concentration of approximately 2 M. Above the central cavity the tunnel contracts into a **selectivity filter** consisting of helical ladders of closely spaced peptide carbonyl-O donors that form a sequence of four cubic eightfold coordination sites. During operation of the channel these sites are occupied, at any one time, by a queue of two K^+ ions and two H_2O molecules in alternate fashion, as in $\cdots K^+ \cdots H_2O \cdots K^+ \cdots H_2O \cdots$. The rate of passage of K^+ ions through the selectivity filter is close to the limit for diffusion control. A plausible mechanism for selective K^+ transport (Fig. 27.5) involves concerted displacement of K^+ ions between adjacent cubic carbonyl-O sites through intermediate, unstable octahedral states in which the K^+ ions are coordinated equatorially by four carbonyl-O donors and axially by the two intervening H_2O molecules. This mechanism is not

---

[4] Roderick MacKinnon shared the 2003 Nobel Prize for Chemistry for his elucidation of the structures and mechanisms of ion channels.

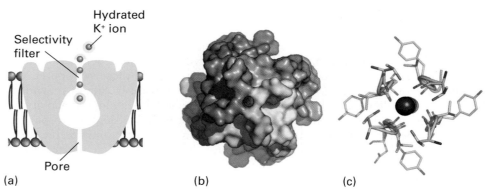

Figure 27.4 (a) Schematic structure of the K⁺ channel showing the different components and the transport of K⁺ ions: the blue halo represents hydration. (b) View of the enzyme from inside the cell showing the entrance pore that admits hydrated ions. (c) View looking up the selectivity filter showing how mobile dehydrated K⁺ ions are coordinated by peptide carbonyl-O atoms provided by each of the four subunits. Note the almost fourfold symmetry axis.

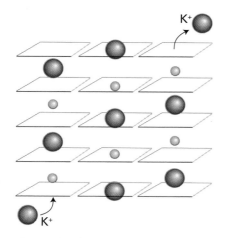

Figure 27.5 Mechanism of transport of K⁺ ions through the selectivity filter of the K⁺ channel. Green spheres represent water molecules.

effective for Na⁺ because the cavity is too large, which accounts for the 10^4-fold selectivity of the channel for K⁺ over Na⁺. The binding is weak and fast because it is important to convey but not to trap K⁺.

Overall, the K⁺ channel acts by the mechanism illustrated in Fig. 27.6. Charged groups on the molecule move in response to a change in the membrane potential and cause the intracellular pore to open, so allowing entry of hydrated K⁺ ions. Selective binding of dehydrated K⁺ ions occurs in the filter region, a potential drop across the membrane is sensed, and the cavity closes. At this point the filter opens up to the external surface, where the K⁺ concentration is low and the K⁺ ions are hydrated and released. This release causes the protein to switch back to the original conformation and K⁺ ions again enter the filter.

The Na⁺/K⁺ pump (Na⁺/K⁺-ATPase), the enzyme that maintains the concentration differential of Na⁺ and K⁺ inside and outside a cell, is another example of the high discrimination between alkali metal ions that has evolved with biological ligands. The ions are pumped against their concentration gradients by coupling the process to ATP hydrolysis. The mechanism, which is outlined in Fig. 27.7, involves conformational changes induced by ATP-driven protein phosphorylation.

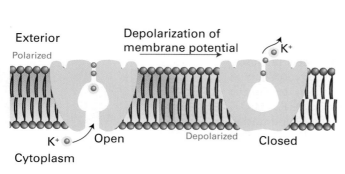

Figure 27.6 Proposed mechanism of action of the K⁺ channel. The potential difference across the membrane is sensed by the protein, which causes the pore to open, allowing hydrated ions to enter the cavity. After shedding their hydration sphere, K⁺ ions pass up the selectivity filter at rates close to diffusion control.

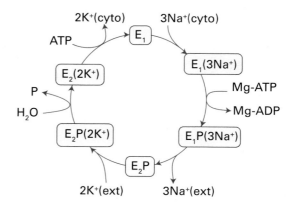

Figure 27.7 General principle of the Na⁺,K⁺-ATPase (the Na pump). Release of two K⁺ ions into the cytoplasm is accompanied by binding of ATP (from the cytoplasm) and conversion of the enzyme into state 1, which binds three Na⁺ ions from the cytoplasm. A phosphate group (P) is transferred to the enzyme, which opens to the external side, expels three Na⁺ ions, then binds two K⁺ ions. Release of the phosphate group causes release of K⁺ into the cytoplasm and the cycle begins again.

EXAMPLE 27.3 Assessing the role of ions in active and passive transport

The toxic species Tl⁺ (radius 150 pm) is used as an NMR probe for K⁺ binding in proteins. Explain why Tl⁺ is suited for this purpose and account for its high toxicity.

Answer To address this question we need to recall from Chapter 13 that Tl, in common with other heavy, post-d-block elements, displays the inert-pair effect, a preference for forming compounds in which its oxidation number is 2 less than the group oxidation number. Thallium (Group 13) thus resembles the heavy Group 1 elements (in fact, TlOH is a strong base) and Tl⁺ can replace K⁺ in complexes, with the advantage that it can be studied by NMR spectroscopy (^{203}Tl and ^{205}Tl have $I = \frac{1}{2}$). The similarity with K⁺ allows Tl⁺, a toxic element, free entry into a cell because it is 'recognized' by the Na⁺/K⁺-ATPase. But once inside, more subtle differences in chemical properties, such as the tendency of Tl to form more stable complexes with soft ligands, are manifested and become lethal.

Self-test 27.3 Explain why the intravenous fluid used in hospital procedures contains NaCl, not KCl.

27.4 Calcium signalling proteins

Key point: Calcium ions are suitable for signalling because they exhibit fast ligand exchange and a large, flexible coordination geometry.

Calcium ions play a crucial role in higher organisms as an intracellular messenger, providing a remarkable demonstration of how organisms have exploited the otherwise rather limited chemistry of this element. Fluxes of Ca²⁺ trigger enzyme action in cells in response to receiving a hormonal or electrical signal from elsewhere in the organism. Calcium is particularly suited for signalling because it has fast ligand-exchange rates, intermediate binding constants, and a large, flexible coordination sphere.

Calcium signalling proteins are small proteins that change their conformation depending on the binding of Ca²⁺ at one or more sites; they are thus examples of the metal ion-activated proteins mentioned earlier. Every muscle movement we make is stimulated by Ca²⁺ binding to a protein known as troponin C. The best-studied Ca²⁺-regulatory protein is calmodulin (17 kg mol⁻¹, Fig. 27.8): its roles include activating protein kinases that catalyse phosphorylation of proteins and activating NO-synthase, a Fe-containing enzyme responsible for generating the intercellular signalling molecule nitric oxide. Calmodulin has four Ca²⁺-binding sites (one is shown as **12**) with dissociation constants lying close to 10⁻⁶. The binding of Ca²⁺ to the four sites alters the protein conformation and it is then recognized by a target enzyme.[5]

Calcium signalling requires special Ca²⁺ pumps, which are large, membrane-spanning enzymes that pump Ca²⁺ out of the cytoplasm, either out of the cell altogether or into Ca-storing organelles such as the endoplasmic reticulum or the mitochondria. As with Na⁺/K⁺-ATPases, the energy for Ca²⁺ pumping comes from ATP hydrolysis. Hormones or electrical stimuli open specific channels (analogous to K⁺ channels) that release Ca²⁺ into the cell. Because the level in the cytoplasm before the pulse is low, the influx easily raises the Ca²⁺ concentration above that needed for Ca²⁺-binding proteins such as calmodulin (or troponin C in muscle). The action can be short-lived, so that after a pulse of Ca²⁺ the cell is quickly evacuated by the calcium pump.

Although Ca²⁺ is invisible to most spectroscopic methods, some Ca proteins, such as calmodulin or troponin C, are small enough to be studied by NMR. Because of their preference for large multicarboxylate ligands, the lanthanoid ions (Section 23.7) have been used as probes for Ca binding, exploiting their properties of paramagnetism (as chemical shift reagents in NMR spectroscopy) and fluorescence. Intracellular concentrations of Ca are monitored by using special fluorescent polycarboxylate ligands (**13**) that are introduced to the cell as their esters, which are hydrophobic and able to cross the membrane lipid barrier. Once in the cell, enzymes known as esterases hydrolyse the esters and release the ligands, which respond to changes in Ca²⁺ concentration in the range 10⁻⁷–10⁻⁹ M.

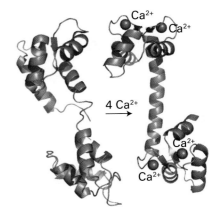

Figure 27.8 The binding of four Ca²⁺ to apocalmodulin causes a change in the protein conformation, converting it to a form that is recognized by many enzymes. The high proportion of α-helix is typical of proteins that are activated by metal-ion binding.

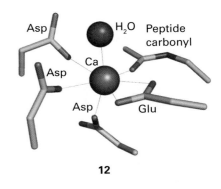

12

13 FURA-2

---

[5] The occupation of a binding site is 50 per cent when the concentration (in mol dm⁻³) of species is equal to the dissociation constant (the reciprocal of the association constant).

27.5 Zinc in transcription

Key points: Zinc fingers are protein structural features produced by coordination of Zn to specific histidine and cysteine residues; a sequence of these fingers enables the protein to recognize and bind to precise sequences of DNA base pairs and plays a crucial role in transferring information from the gene.

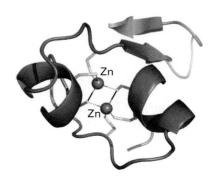

The roles of Zn are either catalytic, which we will deal with later, or structural and regulatory. Unlike Ca and Mg, Zn forms more stable complexes with softer donors, so it is not surprising that it is usually found coordinated in proteins through histidine and cysteine residues. Typical catalytic sites (**14**) commonly have three permanent protein ligands and an exchangeable ligand (H_2O), whereas structural Zn sites (**15**) are coordinated by four 'permanent' protein ligands.

Transcription factors are proteins that recognize certain regions of DNA and control how the genetic code is interpreted as RNA. It has been known since the 1980s that many DNA-binding proteins contain repeating domains that are folded in place by the binding of Zn and form characteristic folds known as 'zinc fingers' (Fig. 27.9). In a typical case, one side of the finger provides two cysteine-S donors and the other side provides two histidine-N donors and folds as an α helix. Each 'finger' makes recognitory contacts with specific DNA bases. As shown in Fig. 27.10, the zinc fingers wrap around sequences of DNA that they are able to recognize by acting collectively. The high fidelity of transcription factors is the result of a number of such contacts being made along the DNA chain at the beginning of the sequence that is transcribed.

The characteristic residue sequence for a Zn-finger motif is

$$-(\text{Tyr},\text{Phe})-X-\text{Cys}-X_{2-4}-\text{Cys}-X_3-\text{Phe}-X_5-\text{Leu}-X_2-\text{His}-X_{3-5}-\text{His}-$$

where the amino acid X is variable. Aside from the 'classical' $(\text{Cys})_2(\text{His})_2$ zinc finger, others have been discovered that have $(\text{Cys})_3\text{His}$ or $(\text{Cys})_4$ coordination, together with more elaborate examples having 'Zn-thiolate clusters', such as the so-called GAL4 transcription factor in which two Zn atoms are linked by bridging cysteine-S ligands (**16**). Various protein folds are produced, with faintly jocular names, such as 'zinc knuckles', joining an

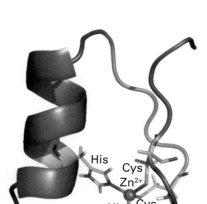

Figure 27.9 Zinc fingers are protein folds that form a sequence able to bind to DNA. A typical finger is formed by the coordination of Zn(II) to two pairs of amino acid side chains located either side of the 'fingertip'.

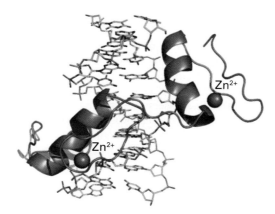

Figure 27.10 A pair of zinc fingers interacting with a section of DNA.

16 $Zn_2(\text{Cys})_6$

increasingly large family. Higher order Zn-thiolate clusters are found in proteins known as metallothioneins and some Zn-sensor proteins (see Section 27.16).

Zinc is particularly suited for binding to proteins to hold them in a particular conformation: Zn^{2+} is high in the Irving–Williams series (Section 20.1) and thus forms stable complexes, particularly to S and N donors. It is also redox inactive, which is an important factor because it is crucial to avoid oxidative damage to DNA. Other examples of structural zinc include insulin and alcohol dehydrogenase. The lack of good spectroscopic probes for Zn, however, has meant that even though it is tightly bound in a protein, it is difficult to confirm its binding or deduce its coordination geometry in the absence of direct structural information from X-ray diffraction or NMR. However, some elegant measurements have exploited the ability of Co^{2+}, which is coloured and paramagnetic, or Cd^{2+}, which has useful NMR properties, to substitute for and report on the Zn site. These substitutions depend on strong similarities between the metal ions: like Zn, Co^{2+} readily forms tetrahedral complexes, whereas Cd lies directly below Zn in the periodic table. For many Zn enzymes, it is found that the surrogate metals are as active as Zn itself (Section 27.9).

27.6 Selective transport and storage of iron

Key points: The uptake of Fe into organisms involves special ligands known as siderophores; transport in the circulating fluids of higher organisms requires a protein called transferrin; Fe is stored as ferritin.

Iron is essential for almost all life forms; however, Fe is also difficult to obtain, yet any excess presents a serious toxic risk. Nature has at least two problems in dealing with this element. The first is the insolubility of Fe(III), which is the stable oxidation state found in most minerals. As the pH increases, hydrolysis, polymerization, and precipitation of hydrated forms of the oxide occur. Polymeric oxide-bridged Fe(III) is the thermodynamic sink of aerobic Fe chemistry (as seen in a Pourbaix diagram, Section 5.14). The insolubility of rust renders the straightforward uptake by a cell very difficult. The second problem is the toxicity of 'free-Fe' species, particularly through the generation of OH radicals. To prevent Fe from reacting with oxygen species in an uncontrolled manner, a protective coordination environment is required. Nature has evolved sophisticated chemical systems to execute and regulate all aspects, from the primary acquisition of Fe, to its subsequent transport, storage, and utilization in tissue. The 'Fe cycle' as it affects a human is summarized in Fig. 27.11.

(a) Siderophores

Siderophores are small polydentate ligands that have a very high affinity for Fe(III). They are secreted from many bacterial cells into the external medium, where they sequester Fe to give a soluble complex that re-enters the organism at a specific receptor. Once inside the cell, the Fe is released.

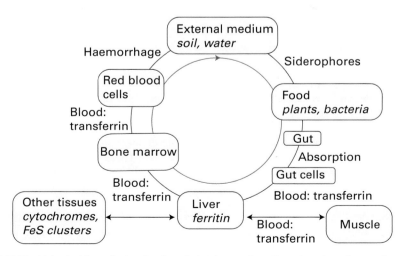

Figure 27.11 The biological Fe cycle showing how Fe is taken up from the external medium and guarded carefully in its travels through organisms.

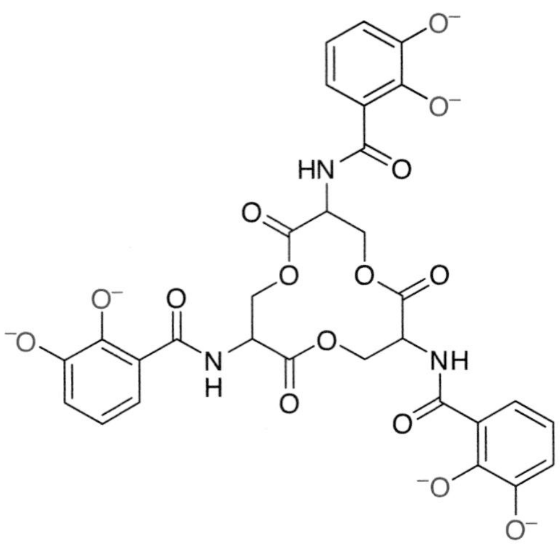

17 Enterobactin

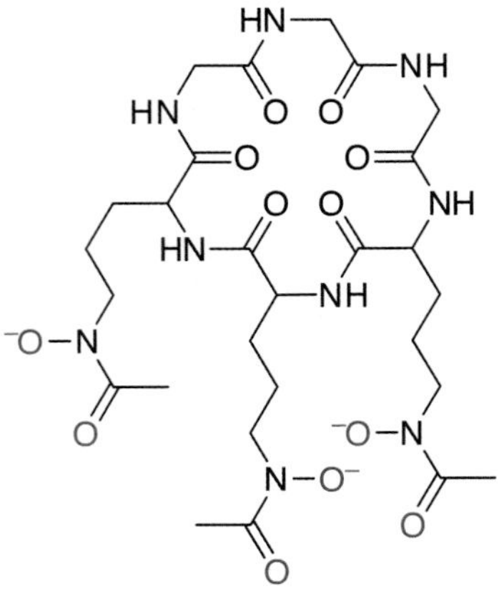

18 Ferrichrome

Aside from citrate (the Fe(III) citrate complex is the simplest Fe transport species in biology) there are two main types of siderophore. The first type is based on phenolate or catecholate ligands, and is exemplified by enterobactin (**17**), for which the value of the association constant for Fe(III) is 10^{52}, an affinity so great that enterobactin enables bacteria to erode steel bridges. The second type of siderophore is based on hydroxamate ligands and is exemplified by ferrichrome (**18**), a cyclic hexapeptide consisting of three glycine and three N-hydroxyl-l-ornithines.

All Fe(III) siderophore complexes are octahedral and high spin. Because the donor atoms are hard O or N atoms and negatively charged, they have a relatively low affinity for Fe(II). Synthetic siderophores are proving to be very useful agents for the control of 'iron overload', a serious condition affecting large populations of the world, particularly South-East Asia (Section 27.17).

(b) Iron-transport proteins in higher organisms

There are several important, structurally similar Fe-transport proteins known collectively as **transferrins**. The best characterized examples are serum transferrin (in blood plasma), ovotransferrin (in egg-white), and lactoferrin (in milk). The apoproteins are potent antibacterial agents as they deprive microbes of their iron. Transferrins are also present in tears, serving to cleanse eyes after irritation. All these transferrins are glycoproteins (protein molecules modified by covalently bound carbohydrate) with molar masses of about 80 kg mol^{-1} and containing two separated and essentially equivalent binding sites for Fe. Complexation of Fe(III) at each site involves simultaneous binding of HCO_3^- or CO_3^{2-} and release of H$^+$:

$$apo\text{-}TF + Fe(III) + HCO_3^- \rightarrow TF\text{-}Fe(III)\text{-}CO_3^{2-} + H^+$$

where TF denotes transferrin. For each site, the association constant under physiological conditions (pH = 7) is in the range 10^{22}–10^{26}. However, its value depends strongly on the pH and this dependence is the main factor controlling Fe uptake and release.

Transferrin consists of two very similar parts, termed the **N-lobe** and the **C-lobe** (Fig. 27.12). The protein is a product of gene duplication because the structure of the first half of the molecule can almost be overlaid on the second half. Each half consists of two domains, 1 and 2, which together form a cleft with a binding site for Fe(III). There is a considerable proportion of α helix, resulting in flexibility. Complexation with Fe(III) causes a conformational change consisting of a hinge motion involving domains 1 and 2 at each lobe. Binding of Fe(III) causes the domains to come together.

In each active site (**19**), a single Fe atom is coordinated by widely dispersed amino acid side chains from both domains and the connecting region, hence the change in conformation that occurs. The protein ligands are carboxylate-O (Asp), two phenolate-O (Tyr), and an imidazole-N (His). Only one of the aspartate carboxylate-O atoms is coordinated. The protein ligands form part of a distorted octahedral coordination sphere. The coordination

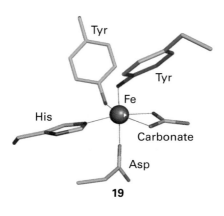

19

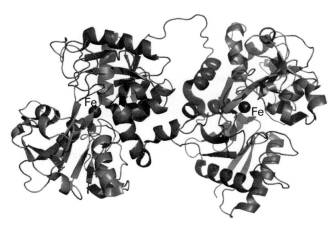

Figure 27.12 Structure of the Fe-transport protein transferrin: the identical halves of the molecule each coordinate to a single Fe(III) atom (the black spheres) between two lobes. This coordination causes a conformational change that allows transferrin to be recognized by the transferrin receptor.

is completed by bidentate binding to the exogenous carbonate, which is referred to as a **synergistic ligand** because Fe binding depends on its presence. In certain cases phosphate is bound instead of carbonate. As expected from the predominantly anionic ligand set, Fe(III) binds much more tightly than Fe(II). However, ions similar to Fe(III), particularly Ga(III) and Al(III), also bind tightly, so that these metals can use the same transport system to gain access to tissues.

(c) Release of iron from transferrin

Cells in need of Fe produce large amounts of a protein called the **transferrin receptor** ($180 \, \text{kg mol}^{-1}$), which is incorporated within their plasma membrane. This protein binds Fe-loaded transferrin. The most favoured mechanism for Fe uptake involves the Fe-loaded transferrin receptor complex entering the cell by a process known as **endocytosis**. In endocytosis, a section of the cell membrane is engulfed by the wall, along with its component membrane-bound proteins, to form a vesicle. The pH within this vesicle is then lowered by a membrane-bound H^+-pumping enzyme that is also swallowed by the cell. The subsequent release of Fe(III) is probably linked to the coordination of carbonate, which is synergistic in the sense that it is necessary for the binding of Fe but is unstable at low pH. Indeed, from *in vitro* studies it is known that Fe is released by lowering the pH to about 5 for serum transferrin and to 2–3 for lactoferrin. The vesicle then splits and the TF-receptor complex is returned to the plasma membrane by **exocytosis**, and Fe(III), probably now complexed by citrate, is released to the cytoplasm.

(d) Ferritin, the cellular Fe store

Ferritin is the principal store of non-haem Fe in animals (most Fe is occupied in haemoglobin and myoglobin) and, when fully loaded, contains 20 per cent Fe by mass. It occurs in all types of organism, from mammals to prokaryotes. In mammals, it is found particularly in the spleen and in blood. Ferritins have two components, a 'mineral' core that contains up to 4500 Fe atoms (mammalian ferritin) and a protein shell. Apoferritin (the protein shell devoid of Fe) can be prepared by treatment of ferritin with reducing agents and an Fe(II) chelating ligand (such as 1,10-phenanthroline or 2,2′-bipyridyl). Dialysis then yields the intact shell.

Apoferritins have average molar masses in the range 460 to 550 kg mol^{-1}. The protein shell (Fig. 27.13) consists of 24 subunits that link together to form a hollow sphere with twofold, threefold (as shown in the illustration), and fourfold symmetry axes. Each subunit consists of a bundle of four long and one short α helices, with a loop that forms a section of β sheet with a neighbouring subunit. The mineral core is composed of hydrated Fe(III) oxide with varying amounts of phosphate, which helps anchor it to the internal surface. The structure as revealed by X-ray or electron diffraction resembles that of the mineral ferrihydrite, $5\text{Fe}_2\text{O}_3.9\,\text{H}_2\text{O}$, which is based on an hcp array of O^{2-} and OH^- ions, with Fe(III) layered in both the octahedral and tetrahedral sites (20).

The threefold and fourfold symmetry axes of apoferritin are, respectively, hydrophilic and hydrophobic pores. The threefold-axis pores are suited for the passage of ions. However, the

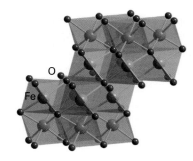

Figure 27.13 The structure of ferritin, showing the arrangement of subunits that make up the protein shell.

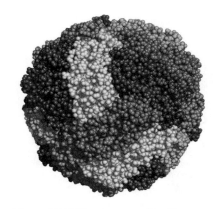

20

ferrihydrite core is insoluble and Fe must be mobilized. The most feasible mechanism so far proposed for the reversible incorporation of Fe in ferritin involves its transport in and out as Fe(II), perhaps as the Fe^{2+} ion, which is soluble at neutral pH, but more likely some type of 'chaperone' complex. Oxidation to Fe(III) is thought to occur at specific di-iron binding sites known as **ferroxidase centres,** present in each of the subunits. Oxidation to Fe(III) involves the coordination of O_2 and inner-sphere electron transfer:

$$2\,Fe(II) + O_2 + 2\,H^+ \rightarrow 2\,Fe(III) + H_2O_2$$

The mechanism by which Fe is released almost certainly involves its reduction back to the more mobile Fe(II).

27.7 Oxygen transport and storage

Dioxygen, O_2, is a special molecule that has not always been available to biology; in fact, to many life forms it is highly toxic. As the waste product of oxygenic photosynthesis that began with cyanobacteria more than 2 Ga ago, O_2 is a biogenic substance, one that owes it existence to solar energy capture by living organisms. As we shall see in Section 27.10, the great thermodynamic advantage of having such a powerful oxidant available undoubtedly led to the evolution of higher organisms that now dominate Earth. Indeed, the requirement for O_2 became so important as to necessitate special systems for transporting and storing it. Apart from the difficulty in supplying O_2 to buried tissue, there is the problem of achieving a sufficiently high concentration in aqueous environments. This problem is overcome by special metalloproteins known as O_2 **carriers.** In mammals and most other animals and plants, these special proteins (myoglobin and haemoglobin) contain an Fe porphyrin cofactor. Animals such as molluscs and arthropods use a Cu protein called haemocyanin, and some lower invertebrates use an alternative type of Fe protein, haemerythrin, which contains a dinuclear Fe site.

(a) Myoglobin

Key points: The deoxy form containing high-spin five-coordinate Fe(II) reacts rapidly and reversibly with O_2 to produce low-spin six-coordinate Fe(II); a slow autoxidation reaction releases superoxide and produces Fe(III), which is inactive in binding O_2.

Myoglobin[6] is an Fe protein (17 kg mol^{-1}, Fig. 27.14) that coordinates O_2 reversibly and controls its concentration in tissue. The molecule contains several regions of α helix, implying mobility, with the single Fe porphyrin group located in a cleft between helices E and F. Two propionate substituents on the porphyrin interact with solvent H_2O molecules on the surface of the protein. The fifth ligand to the Fe is provided by a histidine-N from helix F, and the sixth position is the site at which O_2 is coordinated. In common terminology, the side of the haem plane at which exchangeable ligands are bound is known as the **distal region**, while that below the haem plane is known as the **proximal region**. The histidine on helix F is one of two that are present in all species. Such 'highly conserved' amino acids are a strong indication that evolution has determined that they are essential for function. The other conserved histidine is located on helix E.

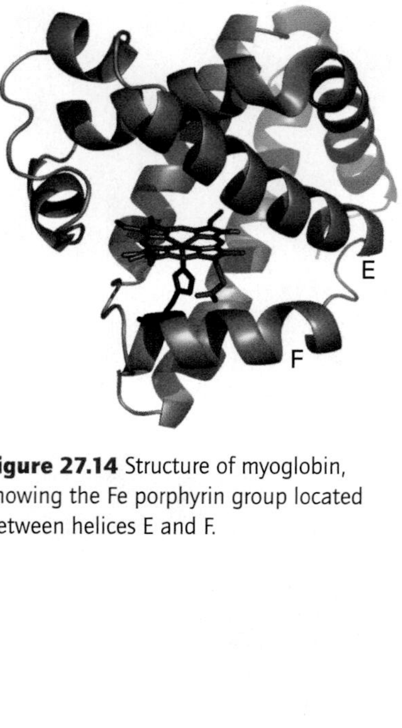

Deoxymyoglobin (Mb) is bluish red and contains Fe(II); this is the oxidation state that binds O_2 to give the familiar bright red oxymyoglobin (oxyMb). In some instances deoxymyoglobin becomes oxidized to Fe(III), which is called metmyoglobin (metMb) and is unable to bind O_2. This oxidation may occur by a ligand substitution-induced redox reaction in which Cl$^-$ ions displace bound O_2 as superoxide:

$$Fe^{(II)}O_2 + Cl^- \rightarrow Fe^{(III)}Cl + O_2^-$$

In healthy tissue, an enzyme (methaemoglobin reductase) is available to reduce the met form back to the Fe(II) form.

Figure 27.14 Structure of myoglobin, showing the Fe porphyrin group located between helices E and F.

---

[6] Myoglobin was the first protein for which the three-dimensional structure was determined by X-ray diffraction. For this achievement, John Kendrew shared the 1962 Nobel Prize for Chemistry with Max Perutz, who solved the structure of haemoglobin.

The Fe in deoxymyoglobin is five-coordinate, high-spin, and lies above the plane of the ring. When O_2 binds it is coordinated end-on to the Fe atom, the electronic structure of which is tuned by the F helix histidine ligand (Fig. 27.15). The unbound end of the O_2 molecule is fastened by a hydrogen bond to the imidazole-NH of the histidine in helix E. The coordination of O_2 (a strong-field π-acceptor ligand) causes the Fe(II) to switch from high-spin (equivalent to $t_{2g}^4 e_g^2$) to low-spin (t_{2g}^6) and, with no d electrons in antibonding orbitals, to shrink slightly and move into the plane of the ring. The bonding is often expressed in terms of Fe(II) coordination by singlet O_2, in which the doubly occupied antibonding $2\pi_g$ orbital of O_2 acts as a σ donor and the empty $2\pi_g$ orbital of O_2 accepts an electron pair from the Fe (Fig. 27.16). An alternative description is often considered, in which the bonding is expressed in terms of low-spin Fe(III) coordinated by superoxide, O_2^-. With this model, the formation of metmyoglobin by reaction with anions is a simple ligand displacement.

(b) Haemoglobin

Key point: Haemoglobin consists of a tetramer of myoglobin-like subunits, with four Fe sites that bind O_2 cooperatively.

Haemoglobin (Hb, 68 kg mol^{-1}, Fig. 27.17) is the O_2 transport protein found in special cells known as *erythrocytes* (red blood cells): a litre of human blood contains about 150 g of Hb. Simplistically, Hb can be thought of as a tetramer of myoglobin-like units with a cavity in the middle. There are in fact two types of Mb-like subunits, which differ slightly in their structures, and Hb is referred to as an $\beta_2\chi_2$ tetramer.

The O_2 binding curves for Mb and Hb are shown in Fig. 27.18: it is highly significant that the curve for Hb is sigmoidal, which indicates that uptake and release of successive O_2 molecules is cooperative. At low O_2 partial pressure and greater acidity (as in venous blood and muscle tissue following aggressive exercise) Hb has low affinity for O_2. This low affinity enables Hb to transfer its O_2 to Mb. As the pressure increases, so does the affinity of Hb for O_2 and as a result Hb can pick up O_2 in the lungs. This change in affinity can be attributed to there being two conformations. The **tensed state** (T) has a low affinity and the **relaxed state** (R) has a high affinity. Deoxy-Hb is T and fully loaded oxy-Hb is R.

To understand the molecular basis of cooperativity we need to refer to Fig. 27.15. Binding of the first O_2 molecule to the T-state molecule is weak, but the decrease in the size of the Fe allows it to move into the plane of the porphyrin ring. This motion is particularly important for Hb because it pulls on the proximal histidine ligand and helix F moves. This movement is transmitted to the other O_2 binding sites, the effect being to move the other Fe atoms closer to their respective ring planes and thereby convert the protein into the R state. The way is thereby opened for them to bind O_2, which they now do with greater ease, although the statistical probability decreases as saturation is approached.

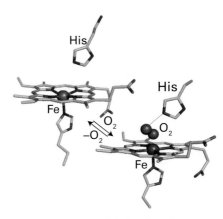

Figure 27.15 Reversible binding of O_2 to myoglobin: coordination by O_2 causes the Fe to become low spin and move into the plane of the porphyrin ring.

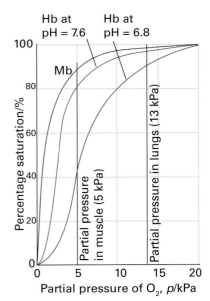

Figure 27.18 Oxygen binding curves for myoglobin and haemoglobin showing how cooperativity between the four sites in haemoglobin gives rise to a sigmoidal curve. The binding of the first O_2 molecule to haemoglobin is unfavourable, but it results in a greatly enhanced affinity for subsequent O_2 molecules.

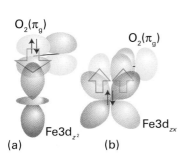

Figure 27.16 The orbitals used to form the Fe−O_2 adduct of myoglobin and haemoglobin. This model considers the O_2 ligand to be in a singlet state, in which the full $2\pi_g$ orbital donates an electron pair and the other $2\pi_g$ orbital acts as a p-electron pair acceptor.

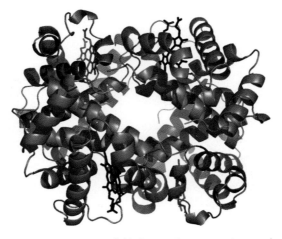

Figure 27.17 Haemoglobin is an $\alpha_2\beta_2$ tetramer. Its α and β subunits are very similar to myoglobin.

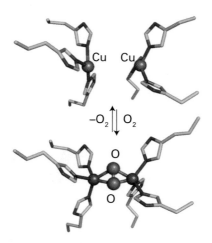

Figure 27.19 Binding of O_2 at the active site of haemocyanin causes the two Cu atoms to be brought closer together. The O_2 complex is regarded as a binuclear Cu(II) centre in which the two Cu atoms are bridged by an η^2,η^2-peroxide.

(c) Other oxygen transport systems

Key point: Arthropods and molluscs use haemocyanin and certain marine worms use haemerythrin.

In many organisms, such as arthropods and molluscs, O_2 is transported by the Cu protein haemocyanin, which, unlike haemoglobin, is extracellular, as is common for Cu proteins. Haemocyanin is oligomeric, with each monomer containing a pair of Cu atoms in close proximity. Deoxyhaemocyanin (Cu(I)) is colourless but it becomes bright blue when O_2 binds.

The active site is shown in Fig. 27.19. In the deoxy state, each Cu atom is three-coordinate and bound in a pyramidal array by three histidine residues. The two Cu atoms are so far apart (460 pm) that there is no direct interaction between them. The low coordination number is typical of Cu(I), which is normally two- to four-coordinate. Rapid and reversible coordination of O_2 occurs between the two Cu atoms in a bridging dihapto manner (μ-$\eta^2\eta^2$) and the low vibrational wavenumber of the coordinated O_2 molecule (750 cm^{-1}) shows it has been reduced to peroxide O_2^{2-}, with an accompanying lowering of the bond order from 2 to 1. To accommodate the binding of O_2, the protein adjusts its conformation to bring the two Cu atoms closer together. The Cu sites become five-coordinate, which is typical of Cu(II).

Haemerythrin is an example of a special class of dinuclear Fe centres that are found in a number of proteins with diverse functions, such as methane monooxygenase, and some ribonucleotide reductases and acid phosphatases. The two Fe atoms in the active site of haemerythrin (**21**) are each coordinated by amino acid side chains but are also linked by two bridging carboxylate groups and a small ligand. In the reduced form, which binds O_2 reversibly, this small ligand is an OH$^-$ ion. Coordination of O_2 occurs at only one of the Fe atoms and the distal O atom forms a hydrogen bond to the H atom of the bridging hydroxide.

(d) Reversible O_2 binding by small-molecule analogues

Key point: Proteins binding O_2 reversibly do so by preventing its reduction and eventual O—O bond cleavage. This protection is difficult to achieve with small molecules. Certain elaborate macrocyclic Fe(II) complexes exhibit reversible O_2 binding by providing steric hindrance to attack on the coordinated O_2.

Much effort has been spent on synthesizing simple complexes that coordinate O_2 reversibly and could be used as blood substitutes in special circumstances, such as emergency surgery. The problem is that although O_2 reacts with d-block metal ions to form complexes in which the O—O bond is retained (as in superoxo and peroxo species), these products tend to undergo irreversible decomposition involving rapid O—O bond cleavage and formation of water or oxides. Overcoming this problem requires complexes designed to protect the coordinated O—O ligand, preventing it from reacting further. Sterically hindered Fe(II) complexes such as the 'basket' porphyrin (**22**) achieve this protection by preventing a second Fe(II) complex from attacking the distal O atom of the superoxo species to form a bridged peroxo intermediate. As we shall see in Section 27.10, peroxo complexes of Fe(III) tend to undergo rapid O—O bond proteolysis, resulting in formation of H_2O and Fe(IV)=O.

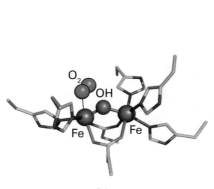

21

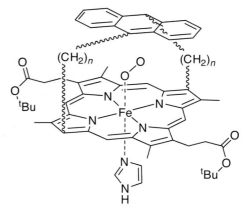

22

Simple Cu complexes that can coordinate O_2 reversibly are also rare, but studies have revealed interesting chemistry that is particularly relevant for developing catalysts for oxygenation reactions. Analogues of the dinuclear Cu(I) centre of haemocyanin react with O_2 but have a strong tendency to undergo further reactions that involve cleavage of the $O-O$ bond, an example being the rapid equilibrium between μ-η^2:η^2 peroxo-dicopper(II) and bis(μ-oxo)copper(III) complexes shown in Fig. 27.20.

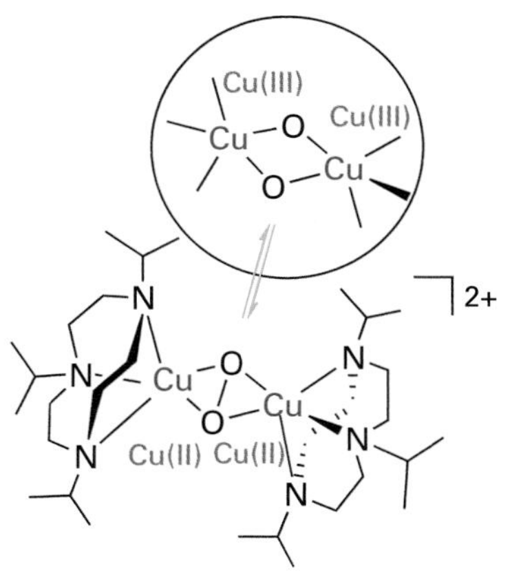

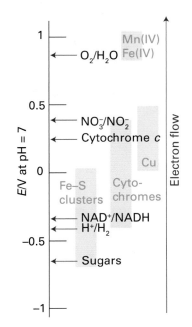

Figure 27.20 Rapid equilibrium between μ-$\eta^2\eta^2$ peroxo-dicopper(II) and bis (μ-oxo) copper(III) in a model complex for the active site of haemocyanin.

EXAMPLE 27.5 Identifying how biology compensates for strong competition by CO

Carbon monoxide is well known to be a strong inhibitor of O_2 binding by myoglobin and haemoglobin, yet relative to O_2 its binding is much weaker in the protein compared to a simple Fe-porphyrin complex. This suppression of CO binding is important as even trace levels of CO would otherwise have serious consequences for aerobes. Suggest an explanation.

Answer We need to consider how CO and O_2, both of which are π-acceptor ligands, differ in terms of the orbitals they use for bonding to a metal atom. The binding of O_2 is nonlinear (see Figs 27.15 and 27.16) and the distal O atom is well positioned to form a hydrogen bond to the distal imidazole. By contrast, CO adopts a linear FeCO arrangement and does not participate in the additional bonding.

Self-test 27.5 Suggest a reaction sequence accounting for why simple Fe-porphyrin complexes are unable to bind O_2 reversibly, but give products that include oxo-bridged dinuclear Fe(III) porphyrin species.

27.8 Electron transfer

In all but a few interesting cases, the energy for life stems ultimately from the Sun, either directly in photosynthesis or indirectly by acquiring energy-rich compounds (fuel) from photosynthesizing organisms. Energy can be acquired as a flow of electrons from fuel to oxidant. Important fuels include fats, sugars, and H_2, and important biological oxidants include O_2, nitrate, and even H^+. As estimated from Fig. 27.21, oxidation of sugars by O_2 provides a lot of energy (over 4 eV per O_2 molecule), and is the reason for the success of aerobic organisms over the anaerobic ones that once dominated the Earth.

(a) General considerations

Key points: Electron flow along electron-transport chains is coupled to chemical processes such as ion (particularly H^+) transfer; the simplest electron-transfer centres have evolved to optimize fast electron transfer.

In organisms, electrons are abstracted from food (fuel) and flow to an oxidant, down the potential gradient formed by the sequence of acceptors and donors known as a **respiratory chain** (Fig. 27.22).[7] Apart from flavins and quinones, which are redox-active organic cofactors, these acceptors and donors are metal-containing electron transfer (ET) centres, which fall into three main classes, namely FeS clusters, cytochromes, and Cu sites. These enzymes are generally bound in a membrane, across which the energy from ET is used to sustain a transmembrane proton gradient: this is the basis of the **chemiosmotic theory**. The counterflow of H^+, through a rotating enzyme known as ATP synthase, drives the phosphorylation of ADP to ATP. Many membrane-bound redox enzymes are **electrogenic proton pumps**, which means they directly couple long-range ET to proton transfer through specific internal channels.

We shall examine the properties of the three main types of ET centre. The same rules concerning outer-sphere electron-transfer that were discussed in Section 21.12 apply to metal centres in proteins, and we should note that organisms have optimized the structures and properties of these centres to achieve efficient long-range electron transfer.

In the following discussion it will be useful to keep in mind that reduction potentials depend on several factors (Section 5.10). Besides ionization energy and ligand environment (strong donors stabilize high oxidation states and lower the reduction potential; weak donors, π acceptors, and protons stabilize low oxidation states and raise the reduction potential), an active site in a protein is also influenced by the relative permittivity (which

Figure 27.21 The 'redox spectrum' of life.

---

[7] An overall current of about 80 A flows through the mitochondrial respiratory chains in an average human.

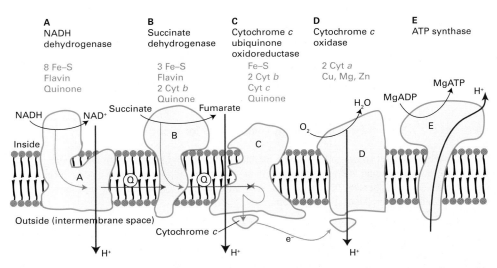

A NADH dehydrogenase	**B** Succinate dehydrogenase	**C** Cytochrome *c* ubiquinone oxidoreductase	**D** Cytochrome *c* oxidase	**E** ATP synthase
8 Fe–S Flavin Quinone	3 Fe–S Flavin 2 Cyt *b* Quinone	Fe–S 2 Cyt *b* Cyt *c* Quinone	2 Cyt *a* Cu, Mg, Zn	

Figure 27.22 The mitochondrial respiratory electron transfer (ET) chain consists of several metalloenzyme molecules that use the energy of electron transport to transport protons across a membrane. The proton gradient is used to drive ATP synthesis.

stabilizes centres with low overall charge), the presence of neighbouring charges, including those provided by other bound metal ions, and the availability of hydrogen-bonding interactions that will also stabilize reduced states.

When we consider the kinetics of electron transfer, it will similarly be useful to keep in mind that 'efficiency' means that electron transfer is fast even when the reaction Gibbs energy is low and therefore that the reorganization energy λ of Marcus theory (Section 21.12) is low. This requirement is met by providing a ligand environment that does not alter significantly when an electron is added and by burying the site so that water molecules are excluded. Intersite distances are generally less than 1.4 nm in order to facilitate electron tunnelling, although it is still debated whether electron transfer in proteins depends mainly on distance alone or whether the protein can provide special pathways.

(b) Cytochromes

Key points: Cytochromes operate in the potential region -0.3 to $+0.4$ V; they have a combination of low reorganization energy and extended electron coupling through delocalized orbitals.

Cytochromes were identified many years ago as cell pigments (hence the name). They contain an Fe porphyrin group and the term 'cytochrome' can refer to both an individual protein and a subunit of a larger enzyme that contains the cofactor. Cytochromes use the Fe^{3+}/Fe^{2+} couple and are generally six-coordinate (**23**), with two stable axial bonds to amino acid donors, and the Fe is usually low spin in both oxidation states. This contrasts with ligand-binding Fe-porphyrin proteins such as haemoglobin, for which the sixth coordination site is either empty or occupied by an H_2O molecule.

A good way to consider the capability of cytochromes for fast electron transfer is to treat the d orbitals of Fe(III) and Fe(II) in terms of an octahedral ligand field and to consider the overlap between the electron-rich but nearly nonbonding t_{2g} orbitals (the configurations are t_{2g}^5 and t_{2g}^6 in Fe(III) and Fe(II), respectively) and the orbitals of the porphyrin. The electron enters or leaves an orbital having π overlap with the π^* antibonding molecular orbital on the ring system. This arrangement provides enhanced electron transfer because the d orbitals of the Fe atom are effectively extended out to the edge of the porphyrin ring, so decreasing the distance over which an electron must transfer between redox partners (Fig. 27.23).

The paradigm of cytochromes is **mitochondrial cytochrome *c*** (12 kg mol^{-1}, Fig. 27.24). This cytochrome is found in the mitochondrial intramembrane space, where it supplies electrons to cytochrome *c* oxidase, the enzyme responsible for reducing O_2 to H_2O at the end of the energy-transducing respiratory chain (Section 27.10). The fifth and sixth ligands to Fe in cytochrome *c* are histidine (imidazole-N) and methionine (thioether-S, **23**). Methionine is not a common ligand in metalloproteins but because it is a neutral, soft

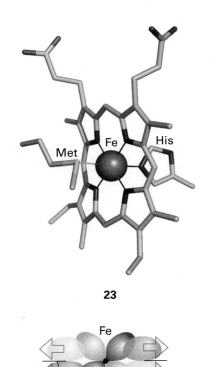

23

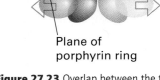

Figure 27.23 Overlap between the t_{2g} orbitals of the Fe and low-lying empty π^* orbitals on the porphyrin effectively extends the Fe orbitals out to the periphery of the ring.

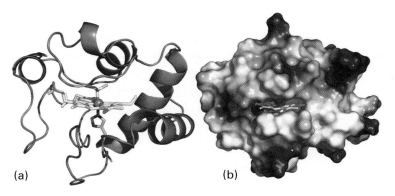

(a) (b)

Figure 27.24 Different views (but from the same viewpoint) of mitochondrial cytochrome *c*. (a) The secondary structure and the position of the haem cofactor. (b) The surface charge distribution that guides the docking with its natural redox partners (red and blue areas represent patches of negative and positive charge, respectively).

donor it is expected to stabilize Fe(II) rather than Fe(III). The reduction potential of cytochrome *c* is +0.26 V, at the higher end of values for cytochromes in general. Cytochromes vary in the identity of the axial ligands as well as the structure of the porphyrin ligand (the notation *a*, *b*, *c*, *d*,... defines positions of absorption maxima in the visible region, but also refers to variations in the substituents on the porphyrin ring). Many cytochromes, in particular those sandwiched between membrane-spanning helices, have bis(histidine) axial ligation (**24**).

In cytochrome *c*, the edge of the porphyrin ring is exposed to solvent and is the most likely site for electrons to enter or leave. Specific protein–protein interactions are important for obtaining efficient electron transfer and the region around the exposed edge of the porphyrin ring in cytochrome *c* provides a pattern of charges that are recognized by cytochrome *c* oxidase and other redox partners. One example in particular has been well studied, that of cytochrome *c* with yeast cytochrome *c* peroxidase. The driving force for electron transfer from either of the two catalytic intermediates of peroxidase (Section 27.10) to each reduced cytochrome *c* is approximately 0.5 V. Figure 27.25 shows the structure of a bimolecular complex formed between cytochrome and cytochrome *c* peroxidase. Electrostatic interactions guide the two proteins together within a distance that is favourable for fast electron tunnelling between cytochrome *c* and two redox centres on the peroxidase, the haem cofactor and tryptophan-191 (Section 27.10).

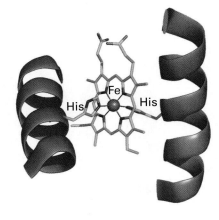

24

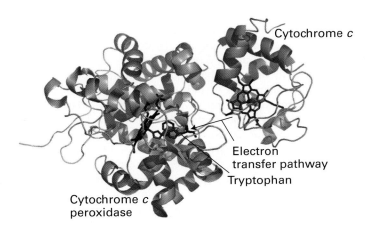

Cytochrome *c*

Electron transfer pathway

Tryptophan

Cytochrome *c* peroxidase

Figure 27.25 The bimolecular ET complex between cytochrome *c* and cytochrome *c* peroxidase produced by co-crystallization of cytochrome *c* with the Zn derivative of cytochrome *c* peroxidase. The orientation suggests an electron transfer pathway between the haem groups of cytochrome *c* and cytochrome *c* peroxidase that includes tryptophan.

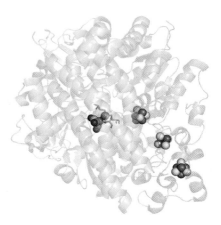

Figure 27.26 A series of three Fe–S clusters provides a long-range electron transfer pathway to the buried active site in hydrogenases.

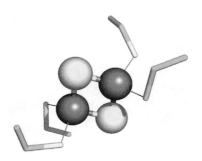

25 [2Fe–2S]

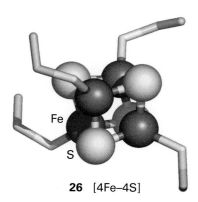

Fe

S

26 [4Fe–4S]

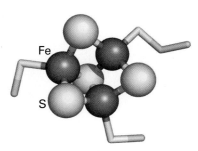

Fe

S

27 [3Fe–4S]

(c) Iron–sulfur clusters

Key points: Iron–sulfur clusters generally operate at more negative potentials than cytochromes; they are composed of high-spin Fe(III) or Fe(II) with sulfur ligands in a mainly tetrahedral environment.

Iron–sulfur clusters are widespread in biology, although their importance was not established as early as cytochromes on account of their lack of distinctive optical characteristics. By convention, FeS clusters are represented by square brackets showing how many Fe and non-protein S atoms are present, as in [2Fe–2S] (**25**), [4Fe–4S] (**26**), and [3Fe–4S] (**27**). The efficacy of FeS clusters as fast ET centres is largely due to their being able to delocalize the added electron to varying degrees, which minimizes bond length changes and decreases the reorganization energy. The presence of sulfur ligands to provide good lead-in groups is also important. Small ET proteins containing FeS clusters are known as **ferredoxins**, whereas in many large enzymes, FeS clusters are arranged in a relay, less than 1.5 nm apart, to link remote redox sites in the same molecule. The relay concept is illustrated in Fig. 27.26 with a class of enzymes known as **hydrogenases**, which we discuss further in Section 27.14.

In nearly all cases, the Fe atoms are tetrahedrally coordinated by cysteine thiolate (RS$^-$) groups as the protein ligands. The overall assembly, including the protein ligands, is known as an 'FeS centre'. Examples are known in which one or more of the Fe atoms is coordinated by non-thiolate amino acid ligands, such as carboxylate, imidazole, and alkoxyl (serine), or by an exogenous ligand such as H_2O or OH$^-$, and the coordination number about the Fe subsite may be increased to six. The cubane [4Fe–4S] (**26**) and cuboidal [3Fe–4S] (**27**) clusters are obviously closely related, and may even interconvert within a protein by the addition or removal of Fe from one subsite. Larger clusters also occur, such as the 'super clusters' [8Fe–7S] and [Mo–7Fe–8S–X] found in nitrogenase (Section 27.13).

Despite the presence of more than one Fe atom, FeS clusters generally carry out single electron transfers and are good examples of mixed-valence systems comprising Fe(III) and Fe(II). The redox state of a cluster is commonly represented by summing the charges due to Fe (3+ or 2+, respectively) and S atoms (2−) and the resultant overall charge, which is referred to as the **oxidation level**, is written as a superscript.

$$[2Fe-2S]^{2+} + e^- \rightarrow [2Fe-2S]^+ \qquad E^\ominus = 0 \text{ to } -0.4 \text{ V}$$

2 Fe(III)   {Fe(III):Fe(II)}

$S = 0$ $S = \frac{1}{2}$

$$[3Fe-4S]^+ + e^- \rightarrow [3Fe-4S] \qquad E^\ominus = +0.1 \text{ to } -0.4 \text{ V}$$

3Fe(III)   {2Fe(III):Fe(II)}

$S = \frac{1}{2}$ $S = 2$

$$[4Fe-4S]^{2+} + e^- \rightarrow [4Fe-4S]^+ \qquad E^\ominus = -0.2 \text{ to } -0.7 \text{ V}$$

{2Fe(III):2Fe(II)}   {Fe(III):3Fe(II)}

$S = 0$ $S = \frac{1}{2}$

Most FeS centres have negative reduction potentials (usually more negative than −0.2 V) so the reduced forms are good reducing agents: exceptions are [4Fe–4S] clusters that operate instead between the +3 and +2 oxidation states (these are called 'HiPIP' centres because they were originally discovered in a protein called *high-potential iron protein* for which the reduction potential is 0.35 V) and so-called **Rieske centres**, which are [2Fe–2S] clusters having one Fe subsite coordinated by two neutral imidazole ligands rather than cysteine (**28**). Coordination by histidine stabilizes the iron as Fe(II) and usually raises the reduction potential to a much more positive value (above +0.2 V). The half-reactions have been written to include the spin states of FeS clusters: individual Fe atoms are high spin, as expected for tetrahedral coordination by S^{2-}, and different magnetic states arise from ferromagnetic and antiferromagnetic coupling (Section 20.8). These magnetic properties are very important, as they allow the centres to be investigated by EPR (Section 8.6).

A major question is how FeS centres are synthesized and inserted into the protein. This process has been studied mostly in prokaryotes, from which it is known that specific proteins are involved in the supply and transport of Fe and S atoms, their assembly into clusters, and their transfer to target proteins. Free sulfide (H_2S, HS$^-$, or S^{2-}) in a cell is highly poisonous, so it is produced only when required by an enzyme called cysteine desulfurase, which breaks down cysteine to yield S^{2-} ions and alanine.

(d) Copper electron-transfer centres

Key point: The protein overcomes the large inherent difference in preferred geometries for Cu(II) and Cu(I) by constraining Cu in a coordination environment that does not change upon electron transfer.

The so-called 'blue' Cu centre is the active site of a number of small electron-transfer proteins as well as larger enzymes (the blue Cu oxidases) that contain, in addition, other Cu sites. Blue Cu centres have reduction potentials for the Cu(II)/Cu(I) redox couple that lie in the range $0.15-0.8$ V and so they are generally more oxidizing than cytochromes. The name stems from the intense blue colour of pure samples in the oxidized state, which arises from ligand(thiolate)-to-metal charge transfer. In all cases, the Cu is shielded from solvent water and coordinated by a minimum of two imidazole-N and one cysteine-S in a nearly trigonal planar manner, with one or two longer bonds to axial ligands. The most studied examples are plastocyanin (Fig. 27.27), a small electron carrier protein in chloroplasts, and azurin, a bacterial electron carrier. These small proteins have a β-barrel structure, which holds the Cu coordination sphere in a rigid geometry. Indeed, the crystal structures of oxidized, reduced (**29**, numbers refer to bond distances in pm), and apo forms reveal that the ligands remain in essentially the same position in all cases. As a result, the blue Cu centre is well suited to undergo fast and efficient electron transfer because the reorganization energy is small.

The dinuclear Cu centre known as Cu$_A$ is present in cytochrome c oxidase and N$_2$O reductase. The two Cu atoms (**30**) are each coordinated by two imidazole groups and a pair of cysteine thiolate ligands act as bridging ligands. In the reduced form, both Cu atoms are Cu(I). This form undergoes one-electron oxidation to give a purple, paramagnetic species in which the unpaired electron is shared between the two Cu atoms. Once again, we see how delocalization assists electron transfer because the reorganization energy is lowered.

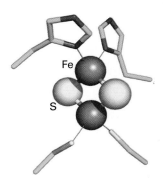

28 Rieske [2Fe–2S] centre

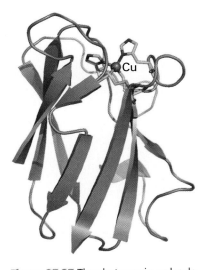

Figure 27.27 The plastocyanin molecule.

EXAMPLE 27.6 Explaining the function of ET centres

The reduction potential of Rieske FeS centres is very pH dependent, unlike the standard FeS centres that have only thiolate ligation. Suggest an explanation.

Answer We need to refer back to Sections 5.6 and 5.14 to see how protonation equilibria influence reduction potentials. Each of the two imidazole ligands that coordinate one of the Fe atoms in the Rieske [2Fe−2S] cluster is electrically neutral at pH = 7 and the proton located on the noncoordinating N atom is easily removed. The pK_a depends on the oxidation level of the cluster, and there is a large region of pH in which the imidazole ligands are protonated in the reduced form but not in the oxidized form. As a result, the reduction potential depends on pH.

Self-test 27.6 Simple Cu(II) compounds show a large EPR hyperfine coupling to the Cu nucleus ($I = \frac{3}{2}$ for ^{65}Cu and ^{63}Cu), whereas the EPR spectra of blue Cu proteins show a much smaller hyperfine coupling. What does this suggest about the nature of the ligand coordination at blue Cu centres?

Catalytic processes

The classic role of enzymes is as highly selective catalysts for the myriad chemical reactions that take place in organisms and sustain the activities of life. In this section we view some of the most important examples in terms of the suitability of certain elements for their roles.

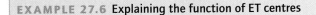

29a Oxidized plastocyanin   **29b** Reduced plastocyanin

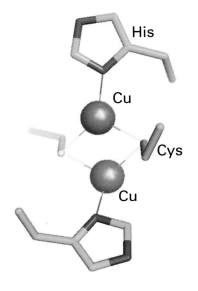

30

27.9 Acid–base catalysis

Biological systems rarely have the extreme pH conditions under which catalysis by free H^+ or OH^- can occur; indeed, the result would be indiscriminate because all hydrolysable bonds would be targets. One way that organisms have solved this problem has been to harness properties of certain metal ions and build them into protein structures designed to accomplish specific (Brønsted) acid–base reactions (Section 4.1).

Organisms make extensive use of Zn for achieving acid–base catalysis, but not to the exclusion of other metals. For example, in addition to the numerous enzymes that feature Fe(II) and Fe(III), Mg(II) serves as the catalyst in pyruvate kinase (phosphate ester hydrolysis) and ribulose bisphosphate carboxylase (CO_2 incorporation into organic molecules), Mn is the catalyst in arginase (for the hydrolysis of arginine, yielding urea and L-ornithine), and Ni(II) is the active metal in urease (for the hydrolysis of urea, yielding ammonia and, ultimately, carbon dioxide). Because of their importance to industry and medicine, many of these enzymes have been studied in great detail and model systems have been synthesized in efforts to reproduce catalytic properties and understand the mode of action of inhibitors. Many of these sites (including arginase-[Mn,Mn] and urease-[Ni,Ni]) contain two or more metal ions in an arrangement unique to the protein and difficult to model with simple ligands.

(a) Zinc enzymes

Key point: Zinc is well suited for catalysing acid–base reactions as it is abundant, redox inactive, forms strong bonds to donor groups of amino acid residues, and exogenous ligands such as H_2O are exchanged rapidly.

A Zn^{2+} ion has high rates of ligand exchange and its polarizing power means that the pK_a of a coordinated H_2O molecule is quite low. Combined with strong binding to protein ligands, rapid ligand exchange (coordinated H_2O or substrate molecules), a reasonably high electron affinity, flexibility of coordination geometry, and no complicating redox chemistry, Zn is well suited to its role in catalysing specific acid–base reactions. The large family of Zn enzymes include carbonic anhydrase, carboxypeptidases, alkaline phosphatase, β-lactamase (responsible for penicillin resistance in bacteria), and alcohol dehydrogenase. Typically, the Zn is coordinated by three amino acid ligands (in contrast to zinc fingers, which have four) and one exchangeable H_2O molecule (**14**).

The mechanisms of Zn enzymes are normally discussed in terms of two limiting cases. In the **Zn-hydroxide mechanism**, the Zn functions by promoting deprotonation of a bound water molecule, so creating an optimally positioned OH^- nucleophile that can go on to attack the carbonyl C atom:

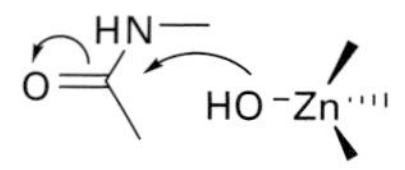

In the **Zn-carbonyl mechanism**, the Zn ion acts directly as a Lewis acid to accept an electron pair from the carbonyl O atom, and its role is therefore analogous to H^+ in acid catalysis:

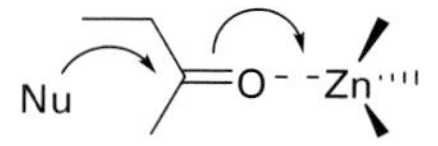

Similar reactions occur with other X=O groups, particularly the P=O of phosphate esters. There is an obvious advantage of achieving such catalysis at Zn or some other acid species that is anchored in a stereoselective environment.

The formation and transport of CO_2 is a fundamental process in biology. The solubility of CO_2 in water depends on its hydration and deprotonation to form HCO_3^-. However, the uncatalysed reaction at pH = 7 is very slow, the forward process occurring with a rate constant of less than 10^{-3} s^{-1}. Because turnover of CO_2 by biological systems is very high, such a rate is far too slow to sustain vigorous aerobic life in a complex organism. In photosynthesis, only CO_2 can be used by the enzyme known as 'rubisco' (Section 27.9(b)) so rapid dehydration of HCO_3^- is essential and is one of the first steps in the production of biomass. The CO_2/HCO_3^- equilibrium is also important (in addition to its role in CO_2 transport) because it provides a way of regulating tissue pH.

In 1932 an enzyme, carbonic anhydrase (CA, or carbon dioxide dehydratase) was identified that catalyses this reaction with a remarkable rate enhancement, and it was found to contain one Zn atom per molecule. It is now known that there are several forms of CA, all of which are monomers with molar mass close to 30 kg mol^{-1}. The best-studied enzyme is CA II from red blood cells, which has a turnover frequency for CO_2 hydration of about 10^6 s^{-1}, making it one of the most active of all enzymes. The crystal structure of human CA II shows that the Zn atom is located in a conical cavity about 1.6 nm deep, which is lined with several histidine residues. The Zn is coordinated by three His-N ligands and one H_2O molecule in a tetrahedral arrangement (Fig. 27.28). The neutral N-ligands lower the pK_a of the bound H_2O by about 3 units (compared to the aqua ion), so creating a high local concentration of OH$^-$ as attacking nucleophile. Other groups in the active site pocket, including noncoordinating histidines and ordered water molecules, are important for mediating proton transfer (which is the rate-limiting factor) and for binding the CO_2 substrate (which does not coordinate to Zn). Carbonic anhydrases show some variation in their Zn coordination environment, with some CAs from higher plants having two cysteines and one histidine.

The mechanism of action of CA (Fig. 27.29) is best described in terms of a Zn-hydroxide mechanism. Proton transfers are very fast, aided by a hydrogen-bonding network that extends from the protein surface to the active site. The key feature is the acidity of the H_2O molecule coordinated to Zn, as the coordinated HO$^-$ ion that is produced after deprotonation is sufficiently nucleophilic to attack a nearby CO_2 molecule bound noncovalently. This attack results in a coordinated HCO$_3^-$ ion, which is then released. Small analogues of CA that have been studied, such as (**31**), reproduce the substrate binding and acid–base properties of the enzyme, but their turnover frequencies are orders of magnitude lower.

Carboxypeptidase (CPD, 34.6 kg mol^{-1}) is an exopeptidase, an enzyme that catalyses the hydrolysis of C-terminal amino acids containing an aromatic or bulky aliphatic side chain. There are two types of Zn-containing enzyme, and both are synthesized as inactive precursors in the pancreas for secretion into the digestive tract. The better studied is CPD A, which acts on terminal aromatic residues; whereas CPD B acts on basic residues. The X-ray structure of CPD shows that the Zn is located to one side of a groove in which the substrate is bound. It is coordinated by two histidine-N ligands, one glutamate-CO$_2^-$ (bidentate), and one weakly bound H_2O molecule (Fig. 27.30). Structures of the enzyme

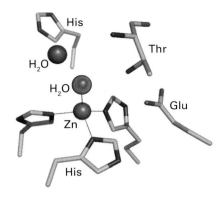

Figure 27.28 The active site of carbonic anhydrase.

31

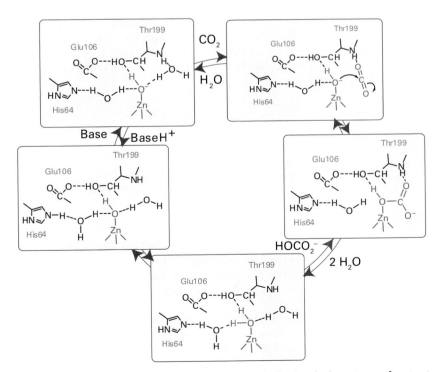

Figure 27.29 The mechanism of action of carbonic anhydrase indicating the importance of proton transfer in this very fast reaction.

Tyr248

His94

Figure 27.30 Structure of the active site of carboxypeptidase with a peptide inhibitor (red) bound to it.

obtained in the presence of glycyl inhibitors show that the H_2O molecule has moved away from the Zn atom, which has become coordinated instead to the carbonyl-O of the glycine, suggesting a Zn-carbonyl mechanism. The guanidinium group of a nearby arginine binds the terminal carboxyl group, while the tyrosine provides aromatic/hydrophobic recognition.

Alkaline phosphatase (AP) introduces us to catalytic Zn centres that contain more than one metal atom. The enzyme catalyses the hydrolysis of phosphate monoesters:

$$R - OPO_3^{2-} + H_2O \rightarrow ROH + HOPO_3^{2-}$$

It occurs in tissues as diverse as the intestine and bone, where it is found in the membranes of osteoblasts, cells that form the sites of nucleation of hydroxyapatite crystals. AP catalyses the general breakdown of organic phosphates, including ATP, to provide the phosphate required for bone growth. As its name implies, its optimum pH is in the mild alkali region. The active site of AP contains two Zn atoms located only about 0.4 nm apart, with a Mg ion nearby. The crystal structure of the enzyme–phosphate complex (**32**) reveals that the phosphate ion (the product of the normal reaction) bridges the two Zn atoms.

Alcohol dehydrogenase (ADH) is discussed here even though it is classed as a redox enzyme because the role of Zn once again is as a Lewis acid. The reaction catalysed is the reduction of NAD^+ by alcohol:

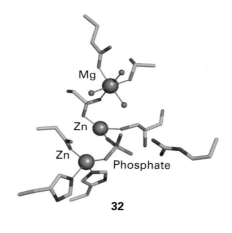

32

NADH

aldehyde

NAD^+

alcohol

The Zn activates the C−OH group towards transfer of the H as a hydride entity to a molecule of NAD^+. It is easy to visualize this reaction in the opposite direction, in which the Zn atom polarizes the carbonyl group and induces attack by the nucleophilic hydridic H atom from NADH. Alcohol dehydrogenase is an α_2 dimer that contains both catalytic and structural Zn sites.

Cadmium, the element below Zn in Group 12 and normally regarded as highly toxic, is now recognized as being an essential nutrient for certain organisms. In 2005, a carbonic anhydrase isolated from the marine phytoplankton *Thalassiosira weissflogii* was discovered to contain Cd at its active site. In contrast to cases in which Cd is simply able to substitute for Zn, this enzyme is specific for Cd^{2+}. The surface waters in which *Thalassiosira weissflogii* grows are extremely low in Zn^{2+} and its growth in the laboratory is stimulated by adding Cd^{2+}.

(b) Magnesium enzymes

Key points: The major direct catalytic function of magnesium is as the catalytic centre of ribulose bisphosphate carboxylase.

The Mg^{2+} cation confers less polarization of coordinated ligands than Zn^{2+} (we often refer to Mg^{2+} as being a 'weaker acid' than Zn^{2+}); however, compared to Zn it is much more mobile and cells contain high concentrations of uncomplexed Mg^{2+} ions. Its major role in enzyme catalysis is as the Mg–ATP complex (**1**), which is the substrate in **kinases**, the enzymes that transfer phosphate groups thereby activating the target compound or causing it to change its conformation. Kinases are controlled by calmodulin (Section 27.4) and other proteins, so they are part of the signalling mechanism in higher organisms.

An important example of an Mg enzyme in which Mg acts separately from ATP is ribulose 1,5-bisphosphate carboxylase, commonly known as 'rubisco'. This enzyme, the most abundant in the biosphere, is responsible for the production of biomass by oxygenic photosynthetic organisms and removal of CO_2 from the atmosphere (to the extent, globally, of over 10^{11} t of CO_2 per year). Rubisco is an enzyme of the **Calvin cycle**, the stages of photosynthesis that can occur in the dark, in which it catalyses the incorporation of CO_2 into a molecule of ribulose 1,5-bisphosphate (Fig. 27.31). The Mg^{2+} ion is octahedrally coordinated by carboxylate groups from glutamate and aspartate residues, three coordinated H_2O molecules, and a carbamate derived from a lysine residue. The carbamate is formed by a reaction between CO_2 and the side-chain $-NH_2$, in an activation process that is necessary for Mg^{2+} to bind. In the catalytic cycle, the binding of ribulose 1,5-bisphosphate displaces two H_2O molecules, and proton abstraction assisted by the carbamate results in a coordinated enolate. This intermediate reacts with CO_2, forming a new C–C bond, then the product is cleaved to yield two new three-carbon species and the cycle continues. The reactive enolate will also react with O_2, in which case the result is an oxidative degradation of substrate: for this reason the enzyme is often called ribulose 1,5-bisphosphate carboxylase-oxygenase. We note the difference from Zn, which would favour ligation by softer ligands and a lower coordination number. Rubisco requires a metal ion that combines good Lewis acidity with weak binding and high abundance.

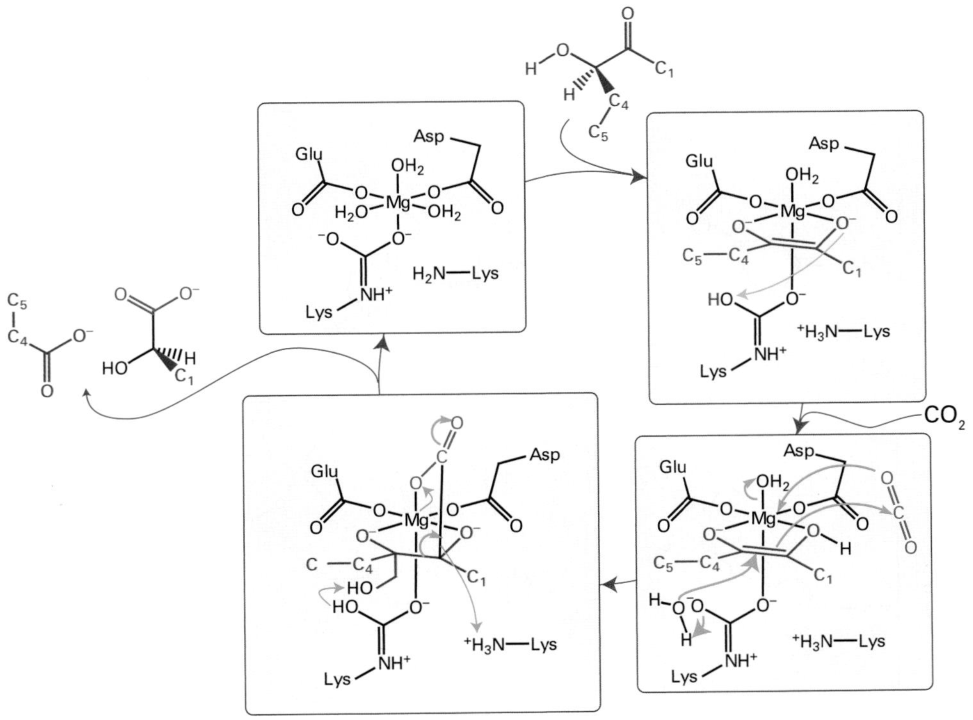

Figure 27.31 Mechanism of action of ribulose 1,5-bisphosphate carboxylase, the enzyme responsible for removing CO_2 from the atmosphere and 'fixing' it in organic molecules in plants.

(c) Iron enzymes

Key points: Acid phosphatases contain a dinuclear metal site containing Fe(III) in conjunction with Fe, Zn, or Mn; aconitase contains a [4Fe-4S] cluster, one subsite of which is modified to manipulate the substrates.

Acid phosphatases, sometimes known as 'purple' acid phosphatases (PAPs) on account of their intense colour, occur in various mammalian organs, particularly the bovine spleen and porcine uterus. Acid phosphatases catalyse hydrolysis of phosphate esters, with optimal activity under mild acid conditions. They are involved in bone maintenance and hydrolysis of phosphorylated proteins (therefore they are important in signalling). They may also have other functions, such as Fe transport. The pink or purple colours of acid phosphatases are due to a tyrosinate → Fe(III) charge-transfer transition at 510–550 nm ($\varepsilon = 4000$ dm³ mol⁻¹ cm⁻¹). The active site contains two Fe atoms linked by ligands, similar to haemerythrin (**21**). Acid phosphatases are inactive in the oxidized {Fe(III)Fe(III)} state in which they are often isolated. In the active state, one Fe is reduced to Fe(II). Both Fe atoms are high spin and remain so throughout the various stages of reactions.

Acid phosphatases also occur in plants, and in these enzymes the reducible Fe is replaced by Zn or Mn. The active site of an acid phosphatase from sweet potato (**33**) shows how phosphate becomes coordinated to both Fe(III) and Mn(II) ions. In the mechanism shown in Fig. 27.32, rapid binding of the phosphate group of the ester occurs to the M(II) subsite, then the P atom is attacked by an OH⁻ ion that is formed at the more acidic Fe(III) subsite. The FeZn centre is also found in an important enzyme called *calcineurin*, which catalyses the phosphorylation of serine or threonine residues on certain protein surfaces, in particular a transcription factor involved in controlling the immune response. Calcineurin is activated by Ca²⁺ binding, both directly and through calmodulin.

Aconitase is an essential enzyme of the **tricarboxylic acid cycle** (also known as the Krebs cycle, or citric acid cycle), the main source of energy production in higher organisms, where it catalyses the interconversion of citrate and isocitrate in a reaction that formally involves dehydration and rehydration, and proceeds through an intermediate, aconitate, which is released in small amounts:

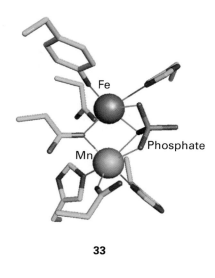

33

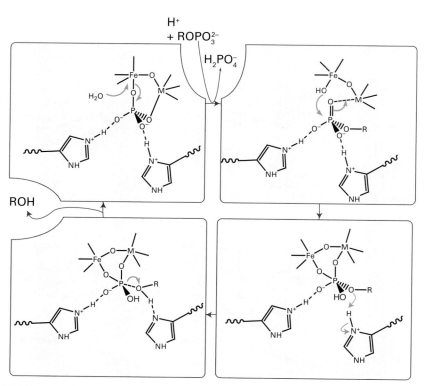

Figure 27.32 Proposed mechanism of action of acid phosphatase. The metal site M(II) is occupied by Fe (most common in animals) or by Mn or Zn (in plants).

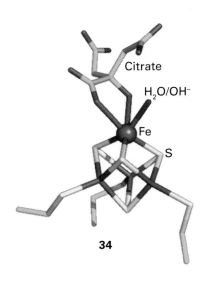

34

The active form of the enzyme contains a [4Fe–4S] cluster, which degrades to [3Fe–4S] when the enzyme is exposed to air. The site of catalysis is the Fe atom that is lost on oxidation. The structure shows that this unique subsite is not coordinated by a protein ligand but by an H_2O molecule, which explains why this Fe is more readily removed.

A plausible mechanism for the action of aconitase, based on structural, kinetic, and spectroscopic evidence, involves the binding of citrate to the active Fe subsite, which increases its coordination number to 6. An intermediate in the catalytic cycle is 'captured' for X-ray diffraction investigation, using a site-directed mutant that can bind the citrate but cannot complete the reaction (**34**). The Fe atom polarizes a C–O bond and OH is abstracted, while a nearby base accepts a proton. The substrate now swings round, and the OH and H are reinserted onto different positions. A form of aconitase that is found in cytoplasm has another intriguing role, that of an Fe sensor (Section 27.15).

27.10 Enzymes dealing with H_2O_2 and O_2

In Section 27.7 we saw how organisms have evolved systems that transport O_2 reversibly and deliver it unchanged to where it is required. In this section we describe how O_2 is reduced catalytically, either for production of energy or synthesis of oxygenated organic molecules. We start by considering a simpler case, that of the reduction of hydrogen peroxide, as this discussion introduces Fe(IV) as a key intermediate in so many biological processes. We end by completing a remarkable cycle, the production of O_2 from H_2O, catalysed by a unique Mn/Ca cluster.

(a) Peroxidases

Key points: Peroxidases catalyse reduction of hydrogen peroxide; they provide important examples of Fe(IV) intermediates that can be isolated and characterized.

Haem-containing peroxidases, as exemplified by horseradish peroxidase (HRP) and cytochrome *c* peroxidase (CcP), catalyse the reduction of hydrogen peroxide:

$$H_2O_2(aq) + 2e^- + 2H^+(aq) \rightarrow 2H_2O(l)$$

The intense chemical interest in these enzymes lies in the fact that they are the best examples of Fe(IV) in chemistry. Iron(IV) is an important catalytic intermediate in numerous biological processes involving oxygen. Catalase, which catalyses the thermodynamically favourable disproportionation of H_2O_2 and is one of the most active enzymes known, is also a peroxidase. The active site of yeast cytochrome *c* peroxidase shown in Fig. 27.33 indicates how the substrate is manipulated during the catalytic cycle. The proximal ligand is the imidazole side chain of a histidine and the distal pocket, like myoglobin, also contains an imidazole side chain, but there is also a guanidinium group from arginine.

The catalytic cycle shown in Fig. 27.34 starts from the Fe(III) form. A molecule of H_2O_2 coordinates to Fe(III) and the distal histidine mediates proton transfer so that both H atoms are placed on the remote O atom. The simultaneous bond polarization by the guanidinium side chain results in heterolytic cleavage of the O–O bond: one half leaves as H_2O and the other remains bound to the Fe atom to produce a highly oxidizing intermediate. Although it is instructive to regard this system as a trapped O atom (or an O^{2-} ion bound to Fe(V)), detailed measurements by EPR and Mössbauer spectroscopy show that this highly oxidizing intermediate (which is known historically as 'Compound I') is in fact Fe(IV) and an organic cation radical. In HRP the radical is located on the porphyrin ring whereas in cytochrome peroxidase it is located on nearby peptide residue tryptophan-191. Descriptions of the Fe−O bonding range from Fe(IV)=O ('ferryl') to Fe(IV)−O··H, in which the O atom is either protonated or linked by a hydrogen bond to a donor group.

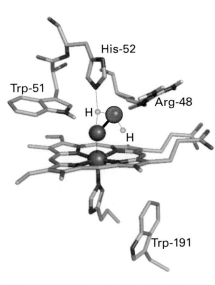

Figure 27.33 The active site of yeast cytochrome *c* peroxidase showing amino acids essential for activity and indicating how peroxide is bound in the distal pocket.

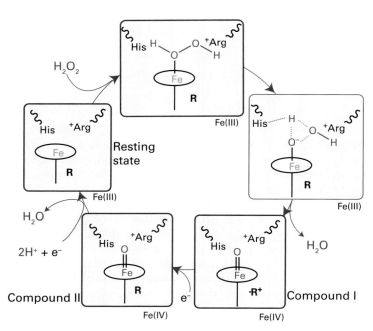

Figure 27.34 The catalytic cycle of haem-containing peroxidases.

Compound I is reduced back to the resting Fe(III) state by two one-electron transfers from either organic substrates or cytochrome c (Fig. 27.25).

(b) Oxidases

Key points: Oxidases are enzymes that catalyse the reduction of O_2 to water or hydrogen peroxide without incorporation of O atoms into the oxidizable substrate; they include cytochrome c oxidase, the enzyme that is a basis for all higher life forms.

Cytochrome c oxidase is a membrane-bound enzyme that catalyses the four-electron reduction of O_2 to water, using cytochrome c as the electron donor. The potential difference between the two half-cell reactions is over 0.5 V but this value does not reflect the true thermodynamics because the actual reaction catalysed by cytochrome c oxidase is:

$$O_2(g) + 4e^- + 8H^+(\text{inside}) \rightarrow 2H_2O(l) + 4H^+(\text{outside})$$

This reaction includes four H^+ that are not consumed chemically but are 'pumped' across the membrane against a concentration gradient. Such an enzyme is called an **electrogenic ion pump** (or *proton pump*). In eukaryotes, cytochrome oxidase is located in the inner membrane of mitochondria and has many subunits (Fig. 27.35), although a simpler enzyme is produced by some bacteria. It contains three Cu atoms and two haem-Fe atoms, as well as a Mg atom and a Zn atom. The Cu and Fe atoms are arranged in three main sites. The active site for O_2 reduction consists of a myoglobin-like Fe-porphyrin (haem-a_3) that is situated close to a 'semi-haemocyanin-like' Cu (known as Cu_B) coordinated by three histidine ligands (**35**). One of the histidine imidazole ligands to the Cu is modified by formation of a covalent bond to an adjacent tyrosine. Electrons are supplied to the dinuclear site by a second Fe porphyrin (haem-a) that is six-coordinate, as expected for an electron-transfer centre. These centres are located in subunit 1. Subunit 2 contains the dinuclear Cu_A centre that was described in Section 27.8, which is believed to be the immediate acceptor of the electron arriving from cytochrome c. The electron transfer sequence is therefore

$$\text{Cytochrome } c \rightarrow Cu_A \rightarrow \text{haem } a \rightarrow \text{binuclear site}$$

The enzyme contains two proton-transfer channels, one of which is used to supply the protons needed for H_2O production while the other is used for protons that are being pumped across the membrane. Figure 27.36 shows the proposed catalytic cycle.

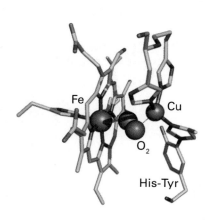

35

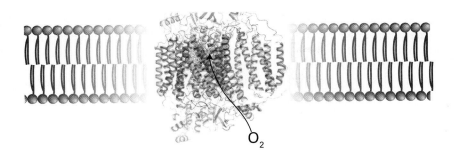

Figure 27.35 The structure of cytochrome *c* oxidase as it occurs in the membrane, showing the locations of the redox centres and the sites for reaction with O$_2$ and cytochrome *c*. See (**33**) for a more detailed view of the active site.

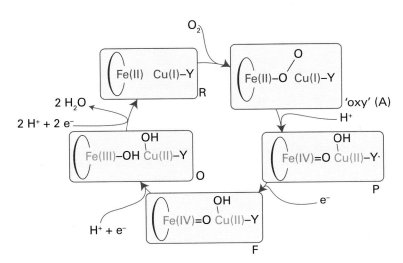

Figure 27.36 The catalytic cycle of cytochrome *c* oxidase. The intermediates are labelled according to current convention. Electrons are provided from the other haem and Cu$_A$. During the cycle, an additional four H$^+$ are pumped across the membrane.

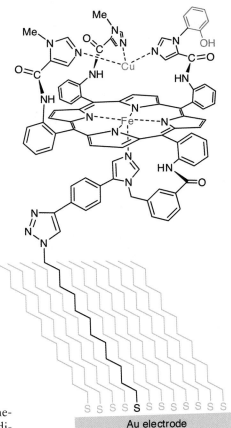

36

Starting from the state in which the active site is Fe(II)−Cu(I), O$_2$ binds to give an intermediate (oxy) that resembles oxymyoglobin. However, unlike oxymyoglobin, this intermediate takes up the other electron that is immediately available, producing a peroxy species that quickly breaks down to give an intermediate known as P. Species P has been trapped and studied by optical and EPR spectroscopy, which show that it contains Fe(IV) and an organic radical that may be located on the unusual His−Tyr pair (Y·). The oxido-Fe(IV) (ferryl) group is formed by heterolytic cleavage of O$_2$, producing a water molecule. The role of a cation radical is again noted: without this radical, the Fe would have to be assigned as Fe(V).

It is vital that intermediates such as peroxide are not released during conversion of O$_2$ to water. Studies with an elaborate model complex (**36**) which can be attached to an electrode show that the presence of the phenol is crucial because it allows all four electrons necessary for the reduction of the O$_2$ to be provided rapidly without relying on long-range electron transfer, which is slow through the long-chain aliphatic linker. If the phenolic −OH group is replaced by −OCH$_3$, hydrogen peroxide is released during O$_2$ reduction because the methoxy derivative is unable to form an oxidized radical.

The blue Cu oxidases contain a blue Cu centre that removes an electron from a substrate and passes it to a trinuclear Cu site that catalyses the reduction of O$_2$ to H$_2$O. Two examples, ascorbate oxidase and a larger class known as laccases, are well characterized, whereas another protein, ceruloplasmin, occurs in mammalian tissue and is the least well understood. Ascorbate oxidase occurs in the skins of fruit such as squash. Its role may be twofold: to protect the flesh of the fruit from O$_2$ and to oxidize phenolic substrates to intermediates that will form the skin of the fruit. Laccases are widely distributed, particularly

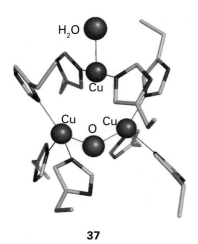

37

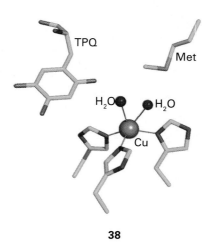

38

in plants and fungi, from which they are secreted to catalyse the oxidation of phenolic substrates. The active site at which O_2 is reduced (**37**) is well buried. It contains a pair of Cu atoms linked in the oxidized form by a bridging O atom, with a third Cu atom situated very close by, completing an almost triangular arrangement.

Amine oxidases catalyse the oxidation of amines to aldehydes by using just a single Cu atom that shuttles between Cu(II) and Cu(I), yet the enzyme carries out a two-electron reduction of O_2, producing a molecule of H_2O_2. The problem is overcome because, like cytochrome *c* peroxidase and cytochrome *c* oxidase, amine oxidases have an additional oxidizing source located near to the metal, in this case a special cofactor called topaquinone (TPQ), which is formed by post-translational oxidation of tyrosine (**38**).

EXAMPLE 27.7 Interpreting reduction potentials

The four-electron reduction potential for O_2 is +0.82 V at pH = 7. Cytochrome *c*, the electron donor to cytochrome *c* oxidase, has a reduction potential of +0.26 V, whereas the organic substrates of fungal laccases often have values as high as +0.7 V. What is the significance of these data in terms of energy conservation?

Answer Although cytochrome *c* oxidase and laccase both catalyse the efficient four-electron reduction of O_2, we need to consider their different biological functions. Laccases are efficient catalysts of phenol oxidation, the driving force being small. Cytochrome oxidase is a proton pump and approximately 2 eV (4 × 0.56 eV) of Gibbs energy is available from oxidation of cytochrome *c* to drive proton transfer across the mitochondrial inner membrane.

Self-test 27.7 Before the discovery of the unusual active site structures in amine oxidase and another Cu enzyme called galactose oxidase, Cu(III) was proposed as a catalytic intermediate. What properties would be expected of this state ?

(c) Oxygenases

Key points: Oxygenases catalyse the insertion of one or both O atoms derived from O_2 into an organic substrate; monooxygenases catalyse insertion of one O atom while the other O atom is reduced to H_2O; dioxygenases catalyse the incorporation of both O atoms.

Oxygenases catalyse the insertion of one or both O atoms of O_2 into substrates, whereas with the Fe^- and Cu^- containing oxidases both O atoms end up as H_2O. Oxygenases are often referred to as *hydroxylases* when the O atom is inserted into a C–H bond. Most oxygenases contain Fe, the rest contain Cu or flavin, an organic cofactor. There are many variations. Monooxygenases catalyse reactions of the type

$$R-H+O_2+2H^++2e^-\rightarrow R-O-H+H_2O$$

in which electrons are supplied by an electron donor such as an FeS protein. Monoxygenases can also catalyse the epoxidation of alkenes. Dioxygenases catalyse the insertion of both atoms of O_2 into substrates, and no additional electron donor is required. Two C–H bonds on the same molecule may be oxygenated:

$$H-R-R'-H + O_2 \rightarrow H-O-R-R'-O-H$$

The Fe enzymes are divided into two main classes, haem and non-haem. We discuss the haem enzymes first, the most important type being cytochrome P450.

Cytochrome P450 (or just P450) refers to an important and widely distributed group of haem-containing monooxygenases. In eukaryotes, they are localized particularly in mitochondria, and in higher animals they are concentrated in liver tissue. They play an essential role in biosynthesis (for example steroid transformations), such as the production of progesterone. The designation 'P450' arises from the intense absorption band that appears at 450 nm when solutions containing the enzyme or even crude tissue extracts are treated with a reducing agent and carbon monoxide, which produces the Fe(II)–CO complex. Most P450s are complex membrane-bound enzymes that are difficult to isolate. Much of what we know about them stems from studies carried out with an enzyme P450$_{cam}$, which is isolated from the bacterium *Pseudomonas putida*.

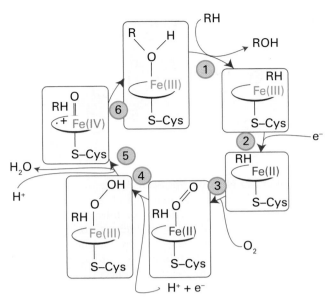

Figure 27.37 The catalytic cycle of cytochrome P450.

This organism uses camphor as its sole source of carbon, and the first stage is oxygenation of the 5-position:

$$O_2, 2H^+, 2e^- \xrightarrow{\text{cytochrome P450}} + H_2O$$

The catalytic cycle has been studied using a combination of kinetic and spectroscopic methods (Fig. 27.37). Starting from the resting enzyme, which is Fe(III), the binding of the substrate in the active site pocket (1) induces release of the coordinated H_2O molecule. This step is detected as a change in spin state from low spin ($S = \frac{1}{2}$) to high spin ($S = \frac{5}{2}$) and the reduction potential increases, causing an electron to be transferred (2) from a small [2Fe−2S]-containing protein known as putidaredoxin. The five-coordinate Fe(II) that is formed resembles deoxymyoglobin and binds O_2 (3). Unlike in myoglobin, addition of a second electron is both thermodynamically and kinetically favourable. The subsequent reactions (4−6) are very fast, but it is thought that an Fe(III) peroxide intermediate is formed that undergoes rapid heterolytic O−O cleavage to produce a species similar to Compound I of peroxidases. In what is known as the **oxygen rebound mechanism**, the Fe(IV)=O group abstracts an H atom from the substrate and then inserts it back as an OH radical. This process is remarkable, as it amounts to the 'taming' of an O atom or OH radical by its attachment to Fe.

Other P450s are thought to operate by similar mechanisms but differ in the architecture of the active site pocket. That site, unlike the site in peroxidases, is predominantly hydrophobic, with specific polar groups present to orient the organic substrate so the correct R−H bond is brought close to the Fe=O entity.

Non-haem oxygenases are widely distributed and are usually dioxygenases. Most contain a single Fe atom at the active site and are classified according to whether the active species in the protein is Fe(III) or Fe(II). In the Fe(III) enzymes, which are also (historically) known as **intradiol oxygenases**, the Fe atom functions as a Lewis acid catalyst and activates the organic substrate towards attack by noncoordinating O_2:

By contrast, in the Fe(II) enzymes, which are known historically as **extradiol oxygenases**, the Fe binds O_2 directly and activates it to attack the organic substrate:

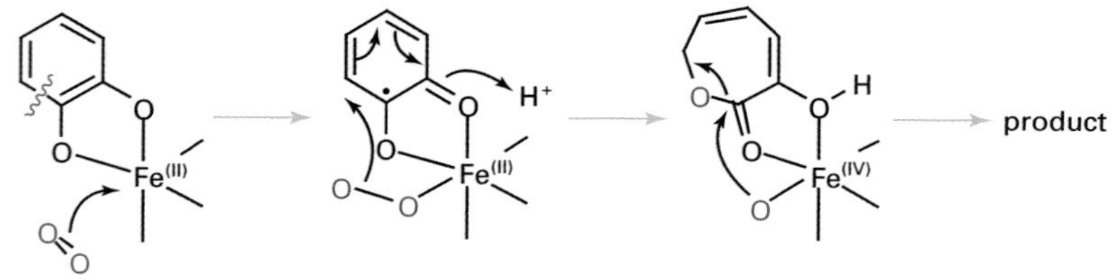

The Fe(III) enzymes are exemplified by protocatechuate 3,4-dioxygenase: the Fe is high spin and tightly coordinated by a set of protein ligands that includes two His-N and two Tyr-O, the latter hard donors being particularly suitable for stabilizing Fe(III) relative to Fe(II). The Fe(III) enzymes are deep red due to an intense tyrosinate-to-Fe(III) charge transfer transition. The Fe(II) enzymes are exemplified by catechol 2,3-dioxygenase: the Fe is high spin and coordinated within the protein by a set of ligands that includes two His-N and one carboxylate group. The binding is weak, reflecting the low position of Fe(II) in the Irving–Williams series (Section 20.1). This weak binding, together with the difficulty of observing useful spectroscopic features (such as EPR spectra), has made these enzymes much more difficult to study than the Fe(III) enzymes.

A particularly important class of Fe(II) oxygenases use a molecule of 2-oxoglutarate as a second substrate:

$$RH + O_2 + {}^-O \cdots \cdots O^- \longrightarrow ROH + {}^-O \cdots \cdots O^- + CO_2$$

The principle of **oxo-glutarate-dependent oxygenases** is that the transfer of one O atom of O_2 to 2-oxoglutarate (also known as α-ketoglutarate) results in its irreversible decarboxylation, thus driving insertion of the other O atom into the primary substrate. Examples include enzymes that serve in cell signalling by modifying an amino acid in certain transcription factors (Section 27.15).

Oxygenases play a crucial role in the metabolism of methane, a greenhouse gas. Of all hydrocarbons, methane contains the strongest C–H bonds and is the most difficult to activate. Methane-metabolizing bacteria produce two types of enzyme that catalyse the conversion of methane to methanol (a more useful chemical and fuel) and thus attract much industrial interest. One is a membrane-bound enzyme that contains Cu atoms. This enzyme, known as 'particulate' methane monooxygenase (p-mmo), is expressed when high levels of Cu are available. The other enzyme, soluble methane monooxygenase (s-mmo), contains a dinuclear Fe active site (**39**) that is related to haemerythrin (**21**) and acid phosphatase (**33**). The mechanisms are not established but Fig. 27.38 shows a plausible catalytic cycle for s-mmo. The intermediate Fe(IV) species that is proposed differs from those we have encountered up to now, as the O_2-derived oxido ligands are bridging rather than terminal.

Despite the importance of Fe(IV) as an enzyme intermediate, small Fe(IV) complexes that could provide important models for understanding the enzymes have been elusive. The easiest to prepare are haem analogues in which Fe is equatorially ligated by porphyrin and which can be formed by reacting the Fe(II) or Fe(III) forms with a peroxo acid. Small models of non-haem Fe(IV) species have now been prepared. The mononuclear complex (**40**) containing the pentadentate pentaaza ligand *N*,*N*-bis(2-pyridylmethyl)-*N*-bis(2-pyridyl)methylamine is formed by treating the Fe(II) complex with the oxo-transfer agent iodosylbenzene. It is fairly stable at room temperature and has been structurally characterized by X-ray diffraction. A powerful oxidizing agent, it can also be generated in acetonitrile solution by bulk electrolysis of the Fe(II) complex in the presence of water, and the standard potential for Fe(IV)/Fe(III) is estimated to be 0.9 V relative to the ferrocinium/ferrocene couple. Complex (**40**) and similar species are paramagnetic ($S = 1$) and show characteristic absorption bands in the near-infrared. The Fe–O bond length in (**40**) is 164 pm, which is fully consistent with a multiple bond in which the O atom is acting as a π donor. Complex (**40**) is able to oxygenate C–H bonds in a variety of hydrocarbons,

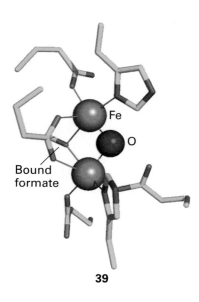

39

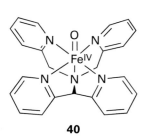

40

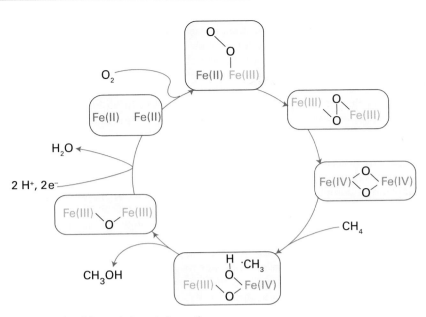

Figure 27.38 A plausible catalytic cycle for methane monooxygenase.

including cyclohexane. The bis(µ-oxo)Fe(IV) complex (**41**) has been proposed as a structural analogue of the reactive intermediate formed in s-mmo.

Tyrosinase and catechol oxidase, two enzymes responsible for producing melanin-type pigments, each contain a strongly coupled dinuclear Cu centre that coordinates O_2 in a manner similar to haemocyanin. However, unlike in haemocyanin, the ligands σ*-to-Cu charge transfer is enhanced sufficiently to activate the coordinated O_2 for electrophilic attack at a phenolic ring of the substrate. The structure of the active site of catechol oxidase complexed with the inhibitor phenolthiourea (**42**) shows how the phenol ring of the substrate can be oriented in close proximity to a bridging O_2. Copper enzymes are also responsible for the production of important neurotransmitters and hormones, such as dopamine and noradrenaline. These enzymes contain two Cu atoms that are well separated in space and uncoupled magnetically.

(d) Photosynthetic O_2 production

Key points: Biological solar energy capture by photoactive centres results in the generation of species with sufficiently negative reduction potentials to reduce CO_2 to produce organic molecules; in higher plants and cyanobacteria, the electrons are derived from water, which is converted to O_2 by a complex catalytic centre containing four Mn atoms and one Ca atom.

Photosynthesis is the production of organic molecules using solar energy. It is conveniently divided into the **light reactions** (the processes by which electromagnetic energy is trapped) and the **dark reactions** (in which the energy acquired in the light reactions is used to convert CO_2

41

42

and H_2O into carbohydrates). We have already mentioned the most important of the dark reactions, the incorporation of CO_2 into organic molecules, which is catalysed by rubisco. In this section, we describe some of the roles that metals play in the light reactions.

The basic principle of photochemical energy capture, applied in a technology to produce H_2 from water, was described in Chapter 10 (Box 10.3). We can view photosynthesis in an analogous way in that H_2 is 'stored' by reaction with CO_2. In biology, photons from the Sun excite pigments present in giant membrane-bound proteins known as **photosystems**. The most important pigment, chlorophyll, is a Mg complex that is very similar to a porphyrin (8). Most chlorophyll is located in giant proteins known as **light-harvesting antennae**, the name perfectly describing their function, which is to collect photons and funnel their energy to enzymes that convert it into electrochemical energy. This energy conversion uses further chlorophyll complexes that become powerful reducing agents when excited by light. Each electron released by excited chlorophyll travels rapidly down a sequence of protein-bound acceptors, including FeS clusters, and (through the agency of ferredoxin and other redox enzymes) is eventually used to reduce CO_2 to carbohydrate. Immediately after releasing an electron, the chlorophyll cation, a powerful oxidant, must be rapidly reduced by using an electron from another site to avoid wasting the energy by recombination (simple reversal of electron flow). In 'oxygenic' photosynthesis, which occurs in green algae, cyanobacteria, and most importantly in green plants, each such 'restoring' electron is provided from a water molecule, resulting in production of O_2.

In green plants, photosynthesis occurs in special organelles known as *chloroplasts*. Plant chloroplasts have two photosystems, I and II, operating in series, that allow low-energy light (approximately 680–700 nm, >1 eV) to span the large potential range (>1 V) within which water is stable. The arrangement of proteins is depicted in Fig. 27.39. Some of the energy of the photosynthetic electron transfer chain is used to generate a transmembrane proton gradient which in turn drives the synthesis of ATP, as in mitochondria. Photosystem I lies at the low-potential end, its electron donor is the blue Cu protein plastocyanin that has been reduced using the electrons generated by photosystem II; in turn, the electron donor to photosystem II is H_2O. Thus green plants dispose of the oxidizing power by converting H_2O into O_2. This four-electron reaction is remarkable because no intermediates are released. The catalyst, called the 'oxygen evolving centre' (OEC) also has a special significance because its action, commencing over 2 Ga ago, has provided essentially all the O_2 we have in the atmosphere. The OEC is the only enzyme active site known to produce an

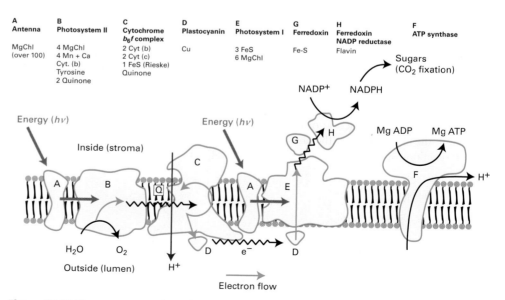

Figure 27.39 The arrangement of proteins in the photosynthetic electron-transport chain (The Mg–chlorophyll complex is represented 'MgChl'.) **A.** Antenna ('light harvesting') complex. **B.** Photosystem II. **C.** The 'cytochrome b_6f complex' (this is similar to **complex III** in the mitochondrial ET chain). **D.** Plastocyanin (soluble). **E.** Photosystem I. **F.** ATPase. **G.** Ferredoxin (Fe–S). **H.** Ferredoxin–NADP+ reductase (flavin). Blue arrows show transfer of energy. Note how the overall transfer of electrons is from Mn (high potential) to FeS (low potential): this apparently 'uphill' flow reflects the crucial input of energy at each photosystem.

O–O bond from two H_2O molecules, and there is much interest in producing functional models of this catalyst for photochemical water splitting (Box 10.3).

The OEC is a metal oxide cluster, containing four Mn atoms and one Ca atom, that is located in subunit D1 of photosystem II. Subunit D1 has long attracted interest because the cell replaces it at frequent intervals as it quickly becomes worn out by oxidative damage. X-ray diffraction data indicate that the metal atoms are arranged as a [3MnCa–4O] cubane connected to a fourth 'dangling' Mn (**43**). The OEC exploits the oxidizing abilities of Mn(IV) and Mn(V), coupled with that of a nearby tyrosine residue, to oxidize H_2O to O_2. Successive photons received by photosystem II result in the OEC being progressively oxidized (the acceptor, an oxidized chlorophyll known as P680$^+$, has a reduction potential of approximately 1.3 V) through a series of states designated S0 to S4, as shown in Fig. 27.40. Apart from S4, which rapidly releases O_2 and has not been isolated, these states are identified in kinetic studies by their characteristic spectroscopic properties, for example S2 shows a complex multi-line EPR spectrum. Note that the Mn ligands are hard O-atom donors and Mn(III) (d^4), Mn(IV) (d^3), and Mn(V) (d^2) are hard metal ions.

Based on the available structural evidence, different models have been proposed for the mechanism of O_2 evolution from two H_2O molecules. The overriding barrier to O_2 formation lies in forming the weak peroxidic O–O bond, following which the formation of O=O is energetically easy (Section 16.1 and *Resource section 3*). First, as the Mn sites are progressively oxidized, coordinated H_2O molecules become increasingly acidic and lose protons, progressing from H_2O through OH$^-$ to O^{2-}. Second, computational studies suggest that Mn(V)=O is best regarded as Mn(IV)–O•, in which the oxido ligand is electron deficient and has appreciable radical character. An H_2O or OH$^-$ ligand coordinated at one Mn subsite (or the Ca) could attack such an electron-deficient O-ligand, perhaps formed at the 'dangling' Mn that is not part of the cubane. Such an attack by coordinated H_2O or OH$^-$ on Mn(IV)–O• would result in a Mn(III)-peroxide species that is easily converted to O_2 by using the reservoir of oxidizing power that has accumulated on the cluster. The presence of the Ca^{2+} is essential, and the only metal ion that can be substituted is Sr^{2+}. A possible role for the Ca is that it provides a site that will remain permanently in the +2 oxidation state and provide a rapid and stable binding site for incoming H_2O, whereas if this subsite were occupied by a fifth Mn atom, the latter would certainly become oxidized and the advantage would be lost.

27.11 The reactions of cobalt-containing enzymes

Key points: Nature uses cobalt in the form of complexes with a macrocyclic ligand, known as corrin. Complexes in which the fifth ligand is a benzimidazole that is covalently linked to the corrin ring are known as cobalamins. Cobalamin enzymes catalyse methyl transfer and dehalogenation.

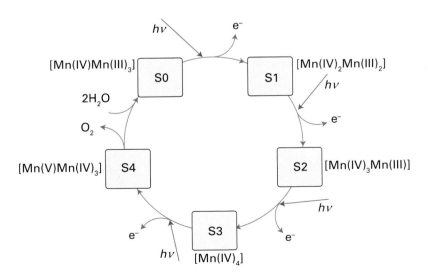

Figure 27.40 The S-cycle for evolution of O_2 by successive one-electron oxidations of the [4MnCa–4O] cluster of photosystem II. The formal oxidation numbers of the Mn components are indicated, but H$^+$ transfers are omitted. In chloroplasts that have become adapted to dark conditions, the cycle 'rests' in the S1 state.

In Coenzyme B$_{12}$ the sixth ligand is deoxyadenosine, which is coordinated through a Co—C bond; enzymes containing coenzyme B$_{12}$ catalyse radical-based rearrangements.

Cobalt macrocycle complexes are cofactors in enzymes that catalyse methyl transfer reactions and they are also important for dehalogenation and radical-based rearrangements (for example, isomerizations). The macrocycle is a corrin ring (**9**), which is similar to porphyrin (**8**), except there is less conjugation and it has a smaller ring (15-membered instead of 16-membered). The five-coordinate complex known as **cobalamin** includes a fifth nitrogen donor in one of the axial positions: usually this ligand is a dimethylbenzimidazole that is covalently linked to the corrin ring through a nucleotide, but a histidine residue is also commonly encountered. The more elaborate structure known as **coenzyme B$_{12}$** (**44**) is an important enzyme cofactor for radical rearrangements: the sixth ligand, R, is 5′-deoxyadenosine, which is bonded to the Co atom through the —CH$_2$— group, making coenzyme B$_{12}$ a rare example of a naturally occurring organometallic compound.[8] The sixth ligand is exchangeable and the complex is ingested in the form of species such as aquacobalamin, hydroxocobalamin, or cyanocobalamin, known generally as vitamin B$_{12}$. Cobalamin is essential for higher organisms (the human requirement is only a few milligrams per day) but it is synthesized only by microorganisms. Like Fe porphyrins, the Co corrins are enzyme cofactors and exert their activities when bound within a protein.

The Co atom can exist in three oxidation states under physiological conditions, Co(III), Co(II), and Co(I), all of which are low spin. The electronic structure of Co is crucial to its biological activity. As expected, the Co(III) form (d^6) is an 18-electron, six-coordinate species (**45**). The Co(II) form (**46**) is 17-electron, five-coordinate and has its unpaired electron in the d$_{z^2}$ orbital. These species are termed 'base-on' forms because the fifth nitrogen ligand is coordinated. The Co(I) form (**47**) is a classic 16-electron, four-coordinate square-planar species, due to dissociation of both axial ligands. The square-planar structure is a 'base-off' form.

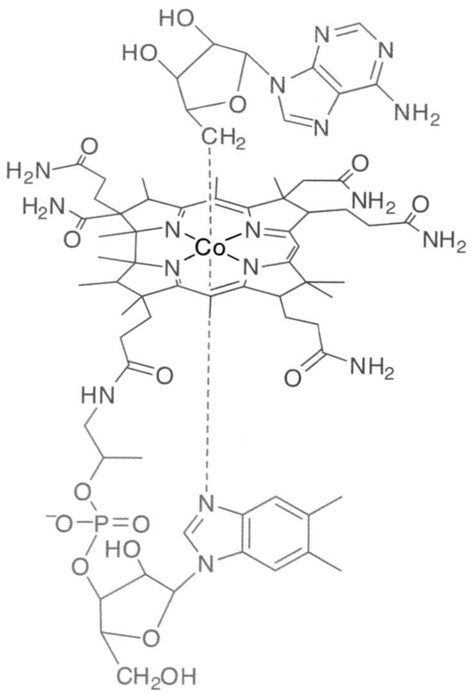

44 Coenzyme B$_{12}$

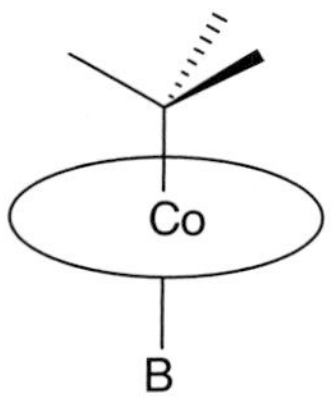

45

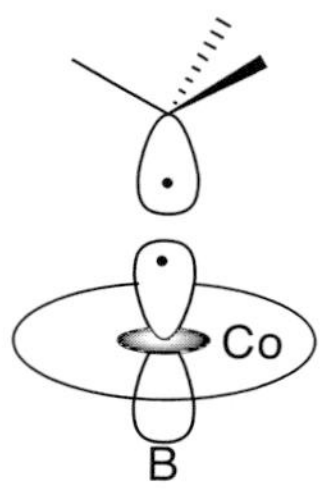

46

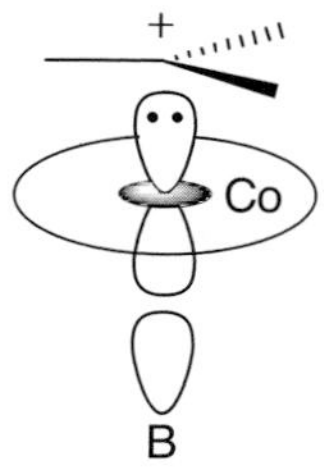

47

[8] Coenzyme B$_{12}$ was one of the earliest molecules to be structurally characterised by X-ray diffraction methods (Dorothy Crowfoot Hodgkin, Nobel Prize for Chemistry, 1964). In 1973, Robert Woodward and Albert Eschenmoser published the total synthesis of B$_{12}$, the most complex natural product to be synthesized at that time, involving almost 100 steps.

Methyl transfer reactions of cobalamins exploit the high nucleophilicity of square-planar Co(I). A particularly important example is methionine synthase, which is responsible for the biosynthesis of methionine. Methionine is produced by transferring a CH_3 group, derived from the methyl carrier methyl hydrofolate, to homocysteine. Not only is methionine an essential amino acid, but also accumulation of homocysteine (which occurs if activity is impaired) is associated with serious medical problems. The mechanism involves a 'base-on/base-off' cycle in which Co(I) abstracts an electrophilic $-CH_3$ group (effectively CH_3^+) from a quaternary N atom on N^5-tetrahydrofolate to produce methylcobalamin, which then transfers $-CH_3$ to homocysteine (Fig. 27.41). Methylcobalamin is the methyl-transferring cofactor for a wide variety of biosynthetic pathways, including the production of antibiotics. In anaerobic microbes, methylcobalamin is involved in the synthesis of acetyl coenzyme A, an essential metabolite, and in production of methane by methanogens.

Radical-based rearrangements catalysed by coenzyme B_{12} (but see Box 27.1) include isomerizations (mutases) and dehydration or deamination (lyases). The generic reaction is

Dehydration and deamination occur after two $-OH$ or $-OH$ and $-NH_2$ become placed on the same carbon atom and hence are triggered by isomerization:

Radical-based rearrangements occur by a mechanism involving initiation of radical formation that begins with enzyme-induced weakening of the Co−C (adenosine) bond. In the free state, the Co−C bond dissociation energy is about 130 kJ mol^{-1}, but when bound in the

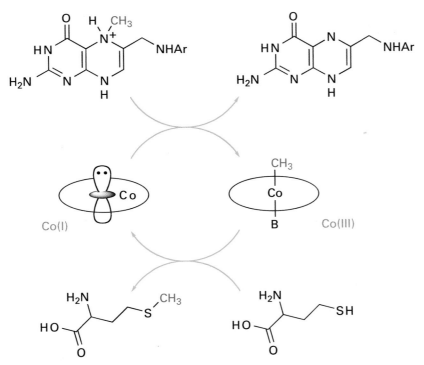

Figure 27.41 The mechanism of methionine synthase. Co(I) is a strong nucleophile and attacks the electrophilic quaternary-CH_3 group on the methyl carrier tetrahydrofolate. The resulting Co(III) methyl complex transfers CH_3^+ to homocysteine.

BOX 27.1   Iron–sulfur enzymes in radical reactions

The discovery in 1970 of an enzyme, lysine 2,3-aminomutase, which catalyses the radical-based rearrangement of an amino acid without any involvement of coenzyme B$_{12}$, initiated development of a whole new area of biochemistry that has led to B$_{12}$ being relegated to second place in regard to catalyzing these rearrangements. Lysine 2,3-aminomutase belongs to a large class of FeS enzymes now known as the radical S-adenosylmethionine (SAM) superfamily. Radical SAM enzymes include those responsible for synthesis of essential vitamins, such as vitamin H (biotin), vitamin B$_1$ (thiamine), haem, and molybdopterin (Section 27.12), as well as those that undertake routine repair of DNA.

Interconversion of L-lysine and L-β-lysine involves migration of the α-amino group to the β-carbon atom. L-β-lysine is required by certain bacteria for antibiotic synthesis.

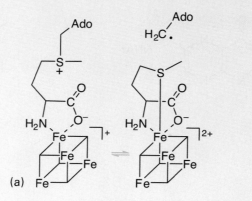

L-lysine

L-β-lysine

As with B$_{12}$ enzymes, this reaction involves a 5'-deoxyadenosyl radical, but in radical SAM enzymes the radical is generated by reductive cleavage of the S-adenosylmethionine cation using a special [4Fe–4S] cluster. The reaction sequence begins with reduction of the [4Fe–4S] cluster: [4Fe–4S]$^+$ is a powerful reductant and SAM is reductively cleaved at the tertiary S$^+$ to produce methionine, which remains coordinated to the special Fe of the cluster, and the deoxyadenosyl radical, which now abstracts a hydrogen atom from lysine and induces rearrangement. The X-ray structure of the precursor (Fig. B27.1) reveals how the different groups are arranged in space. Based on various lines of spectroscopic evidence, likely mechanisms for forming a 5'-deoxyadenosyl radical (Fig. B27.2) involve the tertiary S

attacking the unique Fe subsite (a) or a cluster S atom (b). In either case this attack is followed by rapid electron transfer and bond cleavage.

The amino acid part sequence $-$CxxxCxxC$-$ is the characteristic coordination motif for the [4Fe–4S] cluster in a radical SAM protein. More than 2000 proteins have now been identified by searching for the equivalent base sequence occurring in genes.

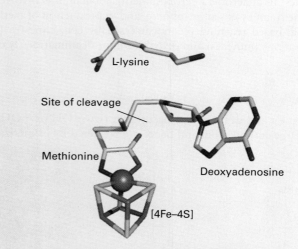

Figure B27.1 Structure of the active site of lysine 2,3-aminomutase, showing the arrangement of the precursor state in which S-adenosylmethionine is coordinated to the [4Fe–4S] cluster. Lysine, the substrate, is held close by.

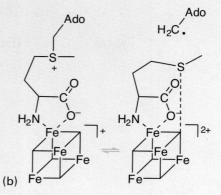

Figure B27.2 Two possible mechanisms by which the reactive 5'-deoxyadenosyl radical is generated by the [4Fe–4S]$^{2+}$ cluster in radical SAM enzymes.

enzyme the bond is substantially weakened, resulting in homolytic cleavage of the Co$-$CH$_2$R bond. This step results in five-coordinate low-spin Co(II) and a CH$_2$R radical, which gives rise to controlled radical chemistry in the enzyme active site pocket (Fig. 27.42). Important examples are methylmalonyl CoA mutase and diol dehydratases.

In 1970, an alternative system for catalyzing radical-based rearrangements was discovered that does not depend on Co. As described in Box 27.1, the enzymes involved use instead a [4Fe$-$4S] cluster to generate the active deoxyadenosyl radical by reductive cleavage of its methionyl derivative.

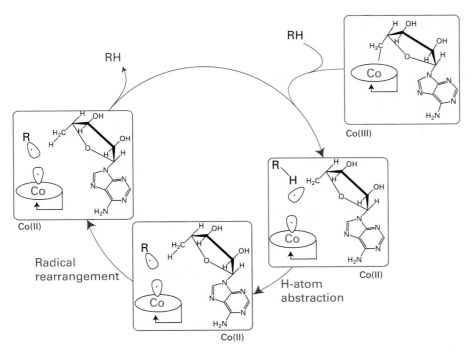

Figure 27.42 The principle of radical-based rearrangements by coenzyme B_{12}. Homolytic cleavage of the Co−C bond results in low-spin Co(II) ($d_{z^2}^1$) and a carbon radical that abstracts an H atom from the substrate RH. The substrate radical is retained in the active site and undergoes rearrangement before the hydrogen atom transfers back.

EXAMPLE 27.8 **Identifying the significance of the d-electron configuration of cobalamin**

Why is a Co-based macrocyclic complex (rather than an Fe complex like haem) well suited for radical-based rearrangements?

Answer To answer this question, we need to consider the electron configuration of Co(II) complexes in which there is a strong equatorial ligand field. Radical-based rearrangements depend on homolytic cleavage of the Co−C bond, which generates the adenosine radical and leaves an electron in the d_{z^2} orbital of the Co. This is the stable configuration for a low-spin Co(II) (d^7) complex, but for Fe, this configuration would require the oxidation state Fe(I), which is not normally encountered in coordination complexes.

Self-test 27.8 Provide an explanation for why the toxicity of mercury is greatly increased by the action of enzymes containing cobalamin.

27.12 Oxygen atom transfer by molybdenum and tungsten enzymes

Key points: Mo is used to catalyse O atom transfer in which the O atom is provided by a water molecule; related chemistry, but in more reducing environments, is displayed by W.

Molybdenum and tungsten are the only heavier elements known so far to have specific functions in biology. Molybdenum is widespread across all life forms, and this section deals with its presence in enzymes other than nitrogenase (Section 27.13). By contrast, W has so far only been found in prokaryotes. Molybdenum enzymes catalyse the oxidation and reduction of small molecules, particularly inorganic species. Reactions include oxidation of sulfite, arsenite, xanthine, aldehydes, and carbon monoxide, and reduction of nitrate and dimethyl sulfoxide (DMSO).

Both Mo and W are found in combination with an unusual class of cofactors (**11**), at which the metal is coordinated by a dithiolene group. In higher organisms, Mo is coordinated by one pterin dithiolene along with other ligands that often include cysteine. This is illustrated by the active site of sulfite oxidase, where we see also how the pterin ligand is twisted (**48**). In prokaryotes, Mo enzymes have two pterin cofactors coordinated to the metal atom and although the role of such an elaborate ligand is not entirely clear, it is potentially redox active and may mediate long-range electron transfer. Coordination of the

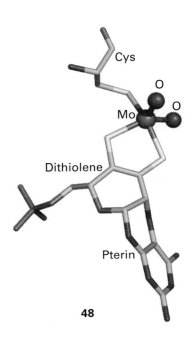

48

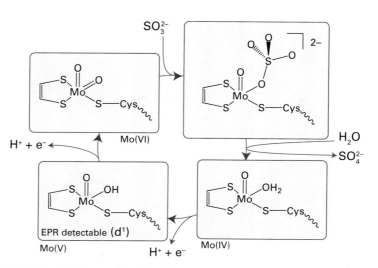

Figure 27.43 Oxidation of sulfite to sulfate by sulfite oxidase, illustrating the direct O-atom transfer mechanism for Mo enzymes.

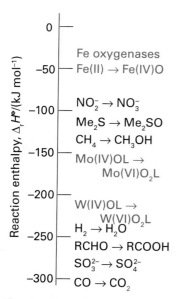

Figure 27.44 Structure of sulfite oxidase showing the Mo and haem domains. The region of polypeptide linking the two domains is highly mobile and is not resolved by crystallography.

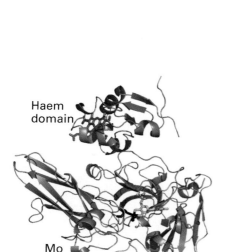

Figure 27.45 Scale showing relative enthalpies for O-atom transfer. Fe(IV) oxido species are powerful O-atom donors, whereas Mo(IV) and W(IV) are good O-atom acceptors.

Mo is usually completed by ligands derived from H_2O, specifically H_2O itself, OH^-, and O^{2-}. Molybdenum is suited for its role because it provides a series of three stable oxidation states, Mo(IV), Mo(V), and Mo(VI), related by one-electron transfers that are coupled to proton transfer. Typically, Mo(IV) and Mo(VI) differ in the number of oxido groups they contain, and Mo enzymes are commonly considered to couple one-electron transfer reactions with O-atom transfer. In humans and other mammals, an inability to synthesize molybdenum cofactor has serious consequences. Sulfite oxidase deficiency is a rare, inherited defect (sulfite ions are very toxic) and it is often fatal.

A mechanism often considered for Mo enzymes is direct O-atom transfer, which is illustrated in Fig. 27.43 for sulfite oxidase. The S atom of the sulfite ion attacks an electron-deficient O atom coordinated to Mo(VI), leading to Mo–O bond cleavage, formation of Mo(IV), and dissociation of SO_4^{2-}. Reoxidation back to Mo(VI), during which a transferable O atom is regained, occurs by two one-electron transfers from an Fe-porphyrin that is located on a mobile 'cytochrome' domain of the enzyme (Fig. 27.44). The intermediate state containing Mo(V) (d^1) is detectable by EPR spectroscopy.

This kind of oxygenation reaction can be distinguished from that of the Fe and Cu enzymes described previously, because with Mo enzymes the oxido group that is transferred is not derived from molecular O_2 but from water. The Mo(VI)=O unit can transfer an O atom, either directly (inner sphere) or indirectly to reducing (oxophilic) substrates, such as SO_3^{2-} or AsO_3^{2-}, but cannot oxygenate C–H bonds. Figure 27.45 shows the reaction enthalpies for O-atom transfer: we see that the highly oxidizing Fe species formed by reaction with O_2 are able to oxygenate all substrates, whereas Mo(VI) oxo species are limited to more reducing substrates and Mo(IV) is able to extract an O atom from nitrate.

As expected from its position below Mo in Group 6, the lower oxidation states of W are less stable than those of Mo, so W(IV) species are usually potent reducing agents. This potency is illustrated by the W-containing formate dehydrogenases present in certain primitive organisms, which catalyse the reduction of CO_2 to formate, the first stage in non-photosynthetic carbon assimilation. This reaction does not involve O-atom insertion but rather the formation of a C–H bond. One mechanism that has been proposed is

Microbial oxidation of CO to CO_2 is important for removing more than 100 Mt of this toxic gas from the atmosphere every year. As shown in Box 27.2, this reaction is carried out by enzymes that contain either a Mo-pterin/Cu cofactor or an air-sensitive [Ni–4Fe–5S] cluster.

BOX 27.2 **Life on carbon monoxide**

Contrary to what is expected from its well-known toxicity, carbon monoxide is one of Nature's most essential small molecules. Even at atmospheric levels (0.05−0.35 ppm) CO is scavenged by a diverse range of microbial organisms for which it provides a source of carbon for growth and a 'fuel' for energy (CO is a stronger reducing agent than H_2). Two unusual enzymes, known as carbon monoxide dehydrogenases, catalyse the rapid oxidation of CO to CO_2. Aerobes use an enzyme containing an unusual Mo-pterin group (Fig. B27.3). A sulfido ligand on the Mo atom is shared with a Cu atom that is coordinated to one other ligand, a cysteine-S, completing the linear arrangement that is so common for Cu(I). A possible mechanism for CO_2 formation involves one of the oxo-groups of Mo(VI) attacking the C-atom of a CO that is coordinated to Cu(I). In contrast, certain anaerobes with the unique ability to live on CO as sole energy and carbon source use an enzyme that contains an unusual [Ni4Fe–5S] cluster. X-ray diffraction studies on crystals of the Ni enzyme incubated in the presence of HCO_3^- at different potentials have revealed the structure of an intermediate showing CO_2 coordinated by Ni and Fe (Fig. B27.4). This intermediate supports a mechanism in which, during the conversion of CO to CO_2, CO binds to the Ni site (which is square-planar Ni(II)) and the C-atom is attacked by a OH^- ion that was coordinated to the pendant Fe atom.

Carbon monoxide is carried in mammals by haemoglobin, and it is estimated that about 0.6 per cent of the total haemoglobin of an average healthy human is in the carbonylated form. Free CO is produced by the action of haem oxygenase, a P450-type enzyme that catalyses the first step in haem degradation. In addition to releasing Fe and CO, the breakdown of haem produces biliverdin and bilirubin, the familiar green and yellow pigments responsible for the appearance of a bruise. Like NO, CO appears to be a cell signalling agent and, in low amounts, it has important therapeutic effects, including suppression of hypertension (high blood pressure) and protection against tissue rejection following an organ transplant. There is therefore considerable interest in developing pharmaceutical agents, such as the water-soluble complex $[Ru(CO)_3Cl(glycinate)]$, to release CO slowly and supplement the action of haem oxygenase.

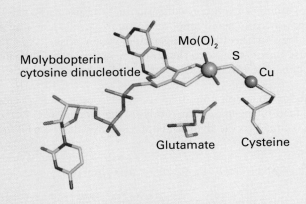

Figure B27.3 The Mo-pterin cofactor of CO-dehydrogenases found in aerobes.

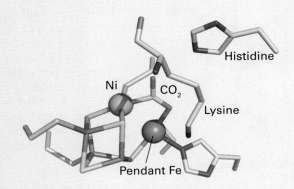

Figure B27.4 X-ray crystallographic observation of an intermediate in the interconversion between CO and CO_2, catalysed by the [Ni4Fe−5S] cluster in CO-dehydrogenases from anaerobes.

Biological cycles

Nature is extraordinarily economical and maximizes its use of elements that have been taken up from the non-biological, geological world, often with great difficulty. We have already seen how iron is assimilated by using special ligands, and how this hard-earned resource is stored in ferritin. Thus useful species are recycled rather than returned to the environment. An important example is nitrogen, which is so hard to assimilate from its unreactive gaseous source, N_2, and the elusive gas H_2, rapid cycling of which is achieved by microbes in processes analogous to electrolytic and fuel cells.

27.13 **The nitrogen cycle**

Key points: The nitrogen cycle involves enzymes containing Fe, Cu, and Mo, often in cofactors having very unusual structures; nitrogenase contains three different kinds of FeS cluster, one of which also contains Mo and a small interstitial atom.

The global biological nitrogen cycle involves organisms of all types and a diverse variety of metalloenzymes (Fig. 27.46 and Box 15.2). The cycle can be divided into uptake of usable nitrogen (assimilation) from nitrate or N_2 and denitrification (dissimilation). The nitrogen cycle involves many different organisms and a variety of metal-containing enzymes.

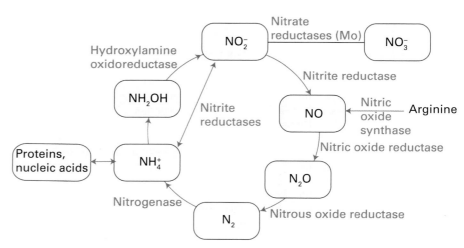

Figure 27.46 The biological nitrogen cycle.

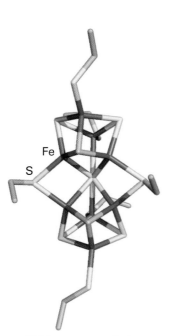

49 Nitrogenase P-cluster
[8Fe-7S]

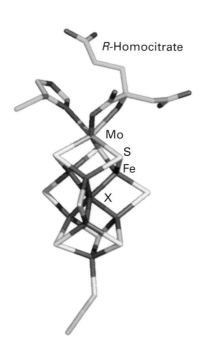

50 Nitrogenase FeMoco
[Mo7Fe–8S,X]

Many of the compounds are toxic or environmentally challenging. Ammonia is a crucial compound for the biosynthesis of amino acids and NO_3^- is used as an oxidant. Molecules such as NO are produced in small amounts to serve as cell signalling agents that play a crucial in physiology and health. Nitrous oxide, which is isoelectronic with CO_2, is a potential greenhouse gas: its release to the atmosphere depends on the balance between activities and abundancies of NO reductase and N_2O reductase across the biological world.

The so-called 'nitrogen-fixing' bacteria found in soil and root nodules of certain plants contain an enzyme called nitrogenase that catalyses the reduction of N_2 to ammonia in a reaction that is coupled to the hydrolysis of 16 molecules of ATP and the production of H_2:

$$N_2 + 8\,H^+ + 8\,e^- + 16\,ATP \rightarrow 2\,NH_3 + H_2 + 16\,ADP + 16\,PO_3(OH)^{2-}$$

'Fixed' nitrogen is essential for the synthesis of amino acids and nucleic acids, so it is central to agricultural production. Industrial production of ammonia by the Haber process (Section 26.12) involves reaction of N_2 and H_2 at high pressures and high temperatures; by contrast, nitrogenase produces NH_3 under normal conditions, and it is small wonder that it has attracted so much attention. Indeed, the mechanism of activation of the N_2 molecule by nitrogenase has inspired coordination chemists for several decades. The process is very costly in terms of energy for a reaction that is not very unfavourable thermodynamically. However, as we saw in Section 15.6, N_2 is an unreactive molecule and energy is required to overcome the high activation barrier for its reduction.

Nitrogenase is a complex enzyme that consists of two types of protein: the larger of the two is called the 'MoFe-protein' and the smaller is the 'Fe-protein' (Fig. 27.47). The Fe-protein contains a single [4Fe–4S] cluster that is coordinated by two cysteine residues from each of its two subunits. The role of the Fe-protein is to transfer electrons to the MoFe-protein in a reaction that is far from understood: in particular, it is unclear why each electron transfer is accompanied by the hydrolysis of two ATP molecules, which are bound to the Fe-protein.

The MoFe protein is $\alpha_2\beta_2$, each $\alpha\beta$ pair of which contains two types of supercluster. The [8Fe−7S] cluster (**49**) is known as the 'P-cluster' and is thought to be an electron transfer centre, whereas the other cluster (**50**), formulated as [Mo7Fe−8S,X] and known as 'FeMoco' (FeMo cofactor), is thought to be the site at which N_2 is reduced to NH_3. The Mo is coordinated also by an imidazole-N from histidine and two O atoms from an exogenous molecule R-homocitrate. The mechanism of N_2 reduction is still unresolved, the main question being whether N_2 is bound and reduced at the Mo atom or at some other site. Here it is significant that nitrogenases are known in which the Mo atom is replaced by a V or Fe atom, arguing against a specific role for Mo. The cage-like cluster provided by the six central Fe atoms also coordinates a small central atom ('X'),

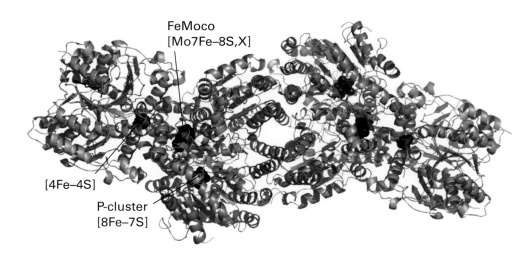

FeMoco
[Mo7Fe–8S,X]

[4Fe–4S]

P-cluster
[8Fe–7S]

Figure 27.47 The structure of nitrogenase showing the Fe protein and the MoFe protein complexed with each other. Positions of the metal centres (black) are indicated. The MoFe protein is an $\alpha_2\beta_2$ of different subunits (red and blue) and it contains two P-clusters and two MoFe cofactors. The Fe protein (green) has a [4Fe−4S] cluster and it is also the site of binding and hydrolysis of Mg-ATP.

which is proposed to be C, N, or O. It was nearly ten years after the determination of the structure of the MoFeS cage that this small central atom was discovered, after improvements in resolution and detailed consideration of the X-ray interference characteristics. The six Fe atoms would otherwise be three-coordinate with a flattened trigonal-pyramidal geometry.

Nitrate reductase is another example of a Mo enzyme involved in the transfer of an O atom, in this case catalysing a reduction reaction (the standard potential for the NO_3^-/NO_2^- couple corrected to pH = 7 is +0.4 V, thus NO_3^- is quite strongly oxidizing). The other enzymes in the nitrogen cycle contain either haem or Cu as their active sites. There are two distinct classes of nitrite reductase. One is a multi-haem enzyme that can reduce nitrite all the way to NH_3. The other class contains Cu and carries out one-electron transfer, producing NO: it is a trimer of identical subunits, each of which contains one 'blue' Cu (mediating long-range electron transfer to the electron donor, usually a small 'blue' Cu protein) and a Cu centre with more conventional tetragonal geometry that is thought to be the site of nitrite binding.

The nitrogen cycle is notable for using some of the most unusual redox centres yet encountered as well as some of the strangest reactions. Another unusual cofactor is a [4Cu−S] cluster, named Cu_Z, that is found in N_2O reductase and has the structure shown in (**51**). It is puzzling how this centre is able to bind and activate N_2O, which is a poor ligand. Long-range electron transfer in N_2O reductase is carried out by a Cu_A centre, the same as found for cytochrome *c* oxidase.

Of particular importance for humans are two enzymes that manipulate NO. One of these, NO synthase, is a haem enzyme responsible for producing NO, by oxidation of L-arginine, on receipt of a signal. Its activity is controlled by calmodulin (Section 27.4). The other enzyme is guanylyl cyclase, which catalyses the formation of the important regulator cyclic guanosine monophosphate (cGMP) from guanosine triphosphate. Nitric oxide binds to the haem Fe of guanylyl cyclase, displacing a histidine ligand and activating the enzyme. Another interesting NO-binding protein is nitrophorin, which is found in some bloodsucking parasites, notably 'kissing bugs' (predacious bugs of the family *Reduviidae*). Nitrophorin binds NO tightly until it is injected into a victim, where a change in pH causes its release. The free NO causes dilation of the surrounding blood vessels, rendering the victim a more effective blood donor.

EXAMPLE 27.9 Identifying intermediates formed during the reduction of N_2 to NH_3

Suggest likely intermediates formed during the six-electron reduction of an N_2 molecule to NH_3.

Answer To identify possible intermediates we need to recall from Chapter 5 that p-block elements normally undergo two-electron transfers that are accompanied by proton transfers. Because N_2 has a triple bond, the

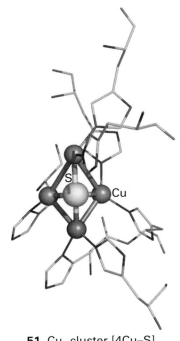

51 Cu_Z cluster [4Cu–S]

two N atoms will remain bonded together throughout most of the six-electron reduction. We would propose diazene (N_2H_2) and hydrazine (N_2H_4) along with their deprotonated conjugate bases.

Self-test 27.9 The MoFe cofactor can be extracted from nitrogenase by using dimethylformamide (DMF), although it is catalytically inactive in this state. Suggest experiments that could establish if the structure of the species in DMF solution is the same as present in the enzyme.

27.14 The hydrogen cycle

Key point: The active sites of hydrogenases contain Fe or Ni, along with CO and CN ligands.

It has been estimated that 99 per cent of all organisms utilize H_2. Even if these species are almost entirely microbes, the fact remains that almost all bacteria and archaea possess extremely active metalloenzymes, known as hydrogenases, that catalyse the interconversion of H_2 and H^+ (as water). The elusive molecule H_2 is produced by some organisms (it is a waste product) and used by others as a fuel, helping to explain why so little H_2 is in fact detected in the atmosphere (Box 10.1). Human breath contains measurable amounts of H_2 due to the action of bacteria in the gut. Hydrogenases are very active enzymes, with turnover frequencies (molecules of substrate transformed per second per molecule of enzyme) exceeding 10 000 s^{-1}. They are therefore attracting much attention for the insight they can provide regarding clean production of H_2 (Section 10.4, Box 10.3) and oxidation of H_2 in fuel cells—technology that currently depends greatly on Pt (Box 5.1).

There are three classes of hydrogenase, based on the structure of the active site. All contain Fe and some also contain Ni. The two best-characterized types are known as [NiFe]-hydrogenases and [FeFe]-hydrogenases, and structures of the active sites of two representative enzymes are shown as (**52**) and (**53**). The active sites contain at least one CO ligand, and provide further examples of biological organometallic active sites (in addition to coenzyme B_{12}). Further ligation is provided by CN^-, cysteine (and sometimes selenocysteine), and the active site of [FeFe]-hydrogenases (the site is commonly referred to as an 'H-cluster') contains a [4Fe-4S] cluster linked by a bridging cysteine thiolate, and an unusual bridging bidentate ligand that is thought to be either dithiomethylamine or dithiomethylether. These fragile active sites are buried deeply within the enzyme, thus necessitating special pores and pathways to convey H_2 and H^+, and a relay of FeS clusters for long-range electron transfer (as shown in Fig. 27.26).

The [FeFe]-enzymes tend to operate in the direction of H_2 production; they are usually found in strictly anaerobic organisms and are very sensitive to O_2. The mechanism of catalysis is uncertain, but it involves the participation of unusual Fe(I) species: the form assigned as Fe(I)Fe(I) binds H^+ and that assigned as Fe(II)Fe(I) binds H_2, probably in an analogous manner to the dihydrogen complex shown (**11**) in Section 10.6. The [NiFe]-enzymes are noted for H_2 oxidation and a catalytically active form assigned by EPR spectroscopy as Ni(III)–H⁻ (a hydrido complex) has been identified (**54**). Single-crystal EPR studies show that the unpaired electron is coupled strongly to an H-atom nucleus lying along an axis that points towards the Fe (that is, the H is in a bridging position). The active sites of [NiFe]- and [FeFe]-hydrogenases react with O_2, often irreversibly, and this sensitivity is a limiting factor in developing and exploiting renewable H_2 production by microorganisms (Section 10.4 and Box 10.1).

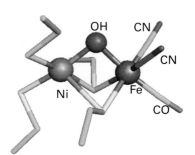

52 [NiFe]-hydrogenase

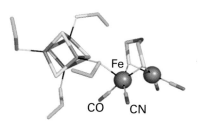

53 [FeFe]-hydrogenase

54

Sensors

A number of metalloproteins are used to detect and quantify the presence of small molecules, particularly O_2, NO, and CO. These proteins therefore act as sensors, alerting an organism to an excess or deficit of particular species, and triggering some kind of remedial action. Special proteins are also used to sense the levels of metals such as Cu and Zn that are otherwise always strongly complexed in a cell.

27.15 **Iron proteins as sensors**

Key point: Organisms use sophisticated regulatory systems based on Fe-containing proteins to adapt quickly to changes in cellular concentrations of Fe and O_2.

We have already seen how FeS clusters are used in electron transfer and catalysis (Sections 27.8 and 27.9). The coordination of an FeS cluster ties together different parts of a protein and thus controls its tertiary structure. The sensitivity of the cluster to oxygen, electrochemical potential, or Fe and S concentrations makes it able to be an important sensory device. In the presence of O_2 or other potent oxidizing agents, [4Fe−4S] clusters have a tendency (controlled by the protein) to degrade, producing [3Fe−4S] and [2Fe−2S] species. The cluster may be removed completely under some conditions (Fig. 27.48). The principle behind an organism's exploitation of FeS clusters as sensors is that the presence or absence of a particular cluster (the structure of which is very sensitive to Fe or oxygen) alters the conformation of the protein and determines its ability to bind to nucleic acids.

In higher organisms, the protein responsible for regulating Fe uptake (transferrin) and storage (ferritin) is an FeS protein known as the **iron regulatory protein** (IRP), which is closely related to aconitase (Section 27.9) but is found in the cytoplasm rather than in mitochondria. It acts by binding to specific regions of messenger RNA (mRNA) that carry the genetic command (transcribed from DNA) to synthesize transferrin receptor or ferritin. A specific interaction region on the RNA is known as the **iron-responsive element** (IRE). The principle is outlined in Fig. 27.49. When Fe levels are high, a [4Fe−4S] cluster is present, and the protein does not bind to the IRE that controls translation of ferritin. In this case binding would be a 'stop' command, and the cell will respond by synthesizing ferritin. Simultaneously, binding of the [4Fe−4S]-loaded protein to the transferrin receptor IRE destabilizes the RNA, so transferrin receptor is not made. When Fe levels are high, the opposite actions occur: a [4Fe−4S] cluster is formed, ferritin synthesis is activated, and transferrin receptor synthesis is switched off (repressed).

The common gut bacterium *E. coli* derives energy either by aerobic respiration (using a terminal oxidase related to cytochrome *c* oxidase, Section 27.10) or by anaerobic respiration with an oxidant such as fumaric acid, or nitrate, using the Mo enzyme nitrate

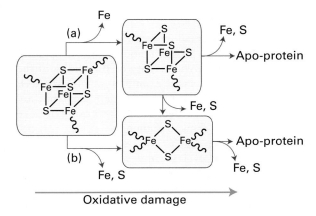

Figure 27.48 The degradation of Fe−S clusters forms the basis for a sensory system The [4Fe−4S] cluster cannot support a state in which all Fe are Fe(III); thus severe oxidizing conditions, including exposure to O_2, causes their breakdown to [3Fe−4S] or [2Fe−2S] and eventually complete destruction. Degradation to [3Fe−4S] (a) requires only removal of an Fe subsite whereas degradation to [2Fe−2S] (b) may require rearrangement of the ligands (cysteine) and produce a significant protein conformational change. These processes link cluster status to availability of Fe as well as O_2 and other oxidants and provide the basis for sensors and feedback control.

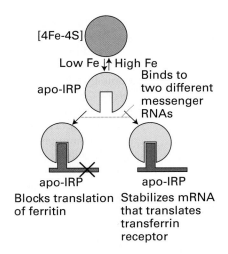

Figure 27.49 Interactions of iron-regulatory protein with iron-responsive elements on the RNAs responsible for synthesizing ferritin or transferrin receptor depend on whether an Fe−S cluster is present and form the basis for regulation of cellular Fe levels.

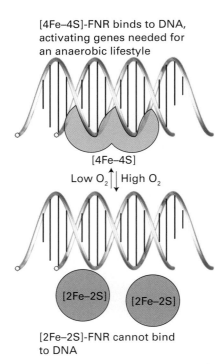

[4Fe–4S]-FNR binds to DNA, activating genes needed for an anaerobic lifestyle

[4Fe–4S]

Low O_2 ⇅ High O_2

[2Fe–2S] [2Fe–2S]

[2Fe–2S]-FNR cannot bind to DNA

Figure 27.50 The principle of operation of the fumarate nitrate regulatory system that controls aerobic versus anaerobic respiration in bacteria.

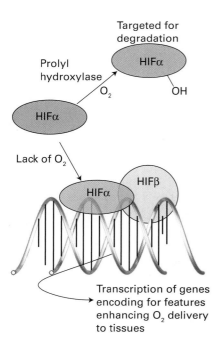

Targeted for degradation

HIFα

Prolyl hydroxylase

O_2 OH

HIFα

Lack of O_2

HIFα HIFβ

Transcription of genes encoding for features enhancing O_2 delivery to tissues

Figure 27.51 The principle of O_2 sensing by prolyl oxygenases.

reductase (Sections 27.12 and 27.13). The problem the organism faces is how to sense whether O_2 is present at a sufficiently low level to warrant inactivating the genes functioning in aerobic respiration and activate instead the genes producing enzymes necessary for the less efficient anaerobic respiration. This detection is achieved by an FeS protein called fumarate nitrate regulator (FNR). The principle is outlined in Fig. 27.50. In the absence of O_2, FNR is a dimeric protein with one [4Fe–4S] cluster per subunit. In this form it binds to specific regions of DNA repressing transcription of the aerobic enzymes and activating transcription of enzymes such as nitrate reductase. When O_2 is present, the [4Fe–4S] cluster is degraded to a [2Fe–2S] cluster and the dimer breaks up so that it cannot bind to DNA. The genes encoding aerobic respiratory enzymes are thus able to be transcribed, whereas those for anaerobic respiration are repressed.

In higher animals, the system that regulates the ability of cells to cope with O_2 shortage involves an Fe oxygenase. Prolyl hydroxylases catalyse the hydroxylation of specific proline residues in proteins, thus altering their properties. In higher animals, one such target protein is a transcription factor called hypoxia inducible factor (HIF), which mediates the expression of genes responsible for adapting cells to low-O_2 conditions (hypoxia). We should bear in mind here that the internal environment of cells and cell compartments is usually quite reducing, equivalent to an electrode potential below –0.2 V, and even though we regard O_2 as essential for higher organisms, its actual levels may be fairly low. When O_2 levels are above a safe threshold, prolyl hydroxylases catalyse oxygenation of two conserved proline residues of HIF, causing the transcription factor to be recognized by a protein that induces its degradation by proteases. Hence genes such as those ultimately responsible for producing more red blood cells (which will help an individual to cope better when O_2 supply is a problem) are not activated. The principle is outlined in Fig. 27.51.

Although essential for higher organisms, O_2 requires stringent control of its four-electron reduction, and increasing amounts of research are being carried out to prevent and cure malfunctions of normal O_2 consumption. The term 'oxidative stress' is used to describe conditions in which the normal function of an organism is threatened by a build up of partially reduced O_2 intermediates, such as superoxides, peroxides, and hydroxyl radicals known collectively as **reactive oxygen species** (ROS). Prolonged exposure to ROS is associated with premature aging and certain cancers. To avoid or minimize oxidative stress cells must first sense ROS and then produce agents to destroy them. Both sensory and attack agents are proteins with active groups such as metal ions (particularly Fe, Cu) and exposed, redox-active cysteine thiols.

An underlying principle of haem sensors is that the small molecules being sensed are π acceptors that can bind strongly to the Fe and displace an indigenous ligand. This binding results in a change in conformation that alters catalytic activity or the ability of the protein to bind to DNA. Of the indigenous ligands, two classic examples are NO and CO; although we have long thought of these molecules as toxic to higher life forms, they are becoming well established as hormones. Indeed, there is evidence that the sensing of trace levels of CO is important for controlling circadian rhythms in mammals.

The enzyme guanyl cyclase senses NO, a molecule that is now well established as a hormone that delivers messages between cells. Guanyl cyclase catalyses the conversion of guanidine monophosphate (GMP) to cyclic GMP (cGMP), which is important for activating many cellular processes. The catalytic activity of guanyl cyclase increases greatly (by a factor of 200) when NO binds to the haem, but (as is obviously important) by a factor of only 4 when CO is bound.

An excellent, atomically defined example of CO sensing is provided by a haem-containing a transcription factor known as CooA. This protein is found in some bacteria that are able to grow on CO as their sole energy source under anaerobic conditions. Whether growth on CO takes place depends on the ambient CO level, as an organism will not waste its resources synthesizing the necessary enzymes when the essential substrate is not present. CooA is a dimer, each subunit of which contains a single b-type cytochrome (the sensor) and a 'helix-turn-helix' protein fold that binds to DNA (Fig. 27.52). In the absence of CO, each Fe(II) is six-coordinate and both its axial ligands are amino acids of the protein, a histidine imidazole and, unusually, the main-chain $-NH_2$ group of a proline that is also the N-terminal residue of the other subunit. In this form, CooA cannot bind to the specific DNA sequence to transcribe the genes for synthesizing the CO-oxidizing enzymes necessary for existing on CO. When CO is present it binds to the Fe, displacing the distal proline residue and causing CooA to adopt a conformation that will bind to the DNA. The likelihood of NO binding in place of CO to cause a false transcriptional response is prevented because NO not only displaces proline but also results in dissociation of the proximal histidine; the NO complex is thus not recognized.

27.16 Proteins that sense Cu and Zn levels

Key point: Cu and Zn are sensed by proteins with binding sites specially tailored to meet the specific coordination preferences of each metal atom.

The levels of Cu in cells are so strictly controlled that almost no uncomplexed Cu is present. An imbalance in Cu levels is associated with serious health problems such as Menkes disease (Cu deficiency) and Wilson's disease (Cu accumulation). Most of what we know about how Cu levels are sensed and converted to cell signals stems from studies on the *E. coli* system, which involves a transcription factor called CueR (Fig. 27.53). This protein binds Cu(I) with high selectivity, although it also binds Ag(I) and Au(I). Metal coordination causes a conformational change that enables CueR to bind to DNA at a receptor site that controls transcription of an enzyme known as CopA, which is an ATP-driven Cu pump. CopA is located in the cytoplasmic membrane and exports Cu into the periplasm. In CueR, the Cu(I) is coordinated by two cysteine-S atoms arranged in a linear coordination geometry. Titrations using CN^- as a buffer show that Cu^+ is bound with a dissociation constant of approximately 10^{-21}. As may be understood by reference to Section 7.3, this ligand environment leads to remarkably selective binding for d^{10} ions, and measurements with Ag and Au show that these ions are taken up with similar affinities.

Most of our current insight about Zn sensing, as for Cu, is provided by studies of bacterial systems. The major difference with respect to Cu is that although Zn is also coordinated (mainly) by cysteine thiolates, the geometry is tetrahedral rather than linear. *E. coli* contains a Zn^{2+}-sensing transcription factor known as ZntR that is closely related to CueR. The factor ZntR contains two Zn-binding domains each of which coordinates a pair of Zns using cysteine and histidine ligands. The surrounding protein fold is shown in Fig. 27.53, for comparison with CueR. The extent to which these dynamic Zn-binding sites can be identified with zinc fingers remains unclear.

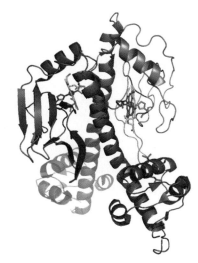

Figure 27.52 The structure of CooA, a bacterial CO sensor and transcription factor. The molecule, a dimer of two identical subunits (represented as red and blue), has two haem-binding domains and two 'helix-turn-helix' domains that recognize a section of DNA. The protein ligands to the Fe atoms are a histidine from one subunit and an N-terminal proline from the other. The binding of CO and displacement of proline disrupts the assembly and allows CooA to bind to DNA.

EXAMPLE 27.10 Identifying links between redox chemistry and metal ion sensing

Suggest a way in which the binding of Cu or Zn to their respective sensor proteins might be linked to the level of cellular O_2.

Answer To address this problem we recall from Chapter 16 that strong S–S bonds arise from the combination of S atoms or radicals. A pair of cysteines that are coordinating a metal ion or are able to approach each other at close range can undergo oxidation by O_2 or other oxidants, resulting in the formation of a disulfide bond (cystine). This reaction prevents the cysteine S atoms from acting as ligands and provides a way in which even redox inactive metals such as Zn may be involved in sensing O_2.

Self-test 27.10 Why might Cu sensors function to bind Cu(I) rather than Cu(II)?

Biomineralization

Key points: Calcium compounds are used in exoskeletons, bones, teeth, and other devices; some organisms use crystals of magnetite, Fe_3O_4, as a compass; plants produce silica-based protective devices.

Biominerals can be either infinite covalent networks or ionic. The former include the silicates, which occur extensively in the plant world. Leaves, even whole plants, are often covered with silica hairs or spines that offer protection against predatory herbivores. Ionic biominerals are mainly based on calcium salts, and exploit the high lattice energy and low solubilities of these compounds. Calcium carbonate (calcite or aragonite, Section 12.10) is the material present in sea shells and eggshells. These minerals persist long after the organism has died, indeed chalk is a biogenic mineral, a result of the process of

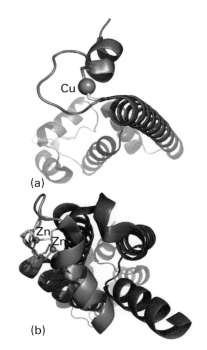

Figure 27.53 Comparison of (a) the Cu- and (b) the Zn-binding sites in the respective transcription factors CueR and ZntR. Note how Cu(I) is recognized by a linear binding site (to two cysteines) whereas Zn is recognized by an arrangement of Cys and His ligands that binds two Zn(II) atoms together with a bridging phosphate group.

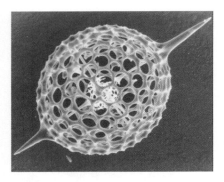

Figure 27.54 The porous silica structure of a radiolarian microskeleton showing the large radial spines. (Photograph supplied by Professor S. Mann, University of Bristol.)

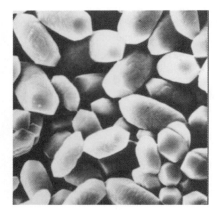

Figure 27.55 Our gravity sensor. Crystals of biologically formed calcite that are found in the inner ear. (Photograph supplied by Professor S. Mann, University of Bristol.)

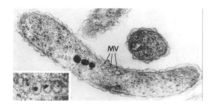

Figure 27.56 Magnetite crystals in magnetotactic bacteria. These are tiny compasses that guide these organisms to move vertically in river bed sludge. MV indicates empty vesicles (Photograph supplied by Professor S. Mann, University of Bristol.)

calciferation of prehistoric organisms. Calcium phosphate (hydroxyapatite, Section 15.4) is the mineral component of bones and teeth, which are particularly good examples of how organisms fabricate 'living' composite materials. Indeed, the different properties of bone found among species (such as stiffness) are produced by varying the amount of organic component, mostly the fibrous protein collagen, with which hydroxyapatite is associated. High hydroxyapatite/collagen ratios are found for large marine animals, whereas low ratios are found for animals requiring agility and elasticity.

Biomineralization is crystallization under biological control. There are some striking examples that have no counterparts in the laboratory. Perhaps the most familiar example is the exoskeleton of the sea urchin, which comprises large sponge-like plates containing continuous macropores 15 μm in diameter: each of these plates is a single crystal of Mg-rich calcite. Large single crystals support other organisms, such as diatoms and radiolarians, with their skeletons of silica cages (Fig. 27.54).

Biominerals produce some intriguing gadgetry. Figure 27.55 shows crystals of calcite that are part of the gravity sensor device in the inner ear. These crystals are located on a membrane above sensory cells and any acceleration or change of posture that causes them to move results in an electrical signal to the brain. The crystals are uniform in size and spindle-shaped so that they move evenly without becoming hooked together. The same property is seen for crystals of magnetite (Fe_3O_4) that are found in a variety of magnetotactic bacteria (Fig. 27.56). There are considerable variations in size and shape across different species, but within any one species, the crystals, formed in magnetosome vesicles, are uniform. Magnetotactic bacteria live in fluid sediment suspensions in marine and freshwater environments, and it is thought that their microcompasses allow them to swim always in a downwards direction to maintain their chemical environment during turbulent conditions.

There is great interest in how biomaterials are formed, not least because of the inspiration they provide for nanotechnology (Chapter 25). The formation of biominerals involves the following hierarchy of control mechanisms:

1. Chemical control (solubility, supersaturation, nucleation).
2. Spatial control (confinement of crystal growth by boundaries such as cells, subcompartments, and even proteins in the case of ferritin, Section 27. 6).
3. Structural control (nucleation is favoured on a specific crystal face).
4. Morphological control (growth of the crystal is limited by boundaries imposed by organic material that grows with time).
5. Constructional control (interweaving inorganic and organic materials to form a higher-order structure, such as bone).

Bone is continually being dissolved and reformed; indeed it functions not only as a structural support but also as the central Ca store. Thus, during pregnancy, bones tend to be raided for their Ca in a process called **demineralization**, which occurs in special cells called *osteoclasts*. Depleted or damaged bones are restored by mineralization, which occurs in cells called *osteoblasts*. These processes involve phosphatases (Section 27.9).

The chemistry of elements in medicine

Serendipity has played an important role in drug discovery, with many effective treatments arising from chance discoveries. There appears to be a special role for compounds containing metals that are not otherwise present in biological systems, for example Pt, Au, Ru, and Bi. A major challenge in pharmacology is to determine the mechanism of action at the molecular level, bearing in mind that the drug that is administered is unlikely to be the molecule that reacts at the target site. This is particularly true for metal complexes, which are usually more susceptible to hydrolysis than organic molecules. In general, the mechanism of action is proposed by extrapolation of *in vitro* studies. Orally administered drugs are highly desirable because they avoid the trauma and potential hazards of injection; however, they may not pass through the gut wall or survive hydrolytic enzyme action. Inorganic compounds are also used in the diagnosis of disease or damage, a particularly interesting example being the use of radioactive technetium.

27.17 Chelation therapy

Key points: The treatment of Fe overload involves sequestration of Fe by ligands based on or inspired by siderophores.

'Iron overload' is the name given to several serious conditions that affect a large proportion of the world's population. Here we recall that, despite its great importance, Fe is potentially a highly toxic element, particularly in its ability to produce harmful radicals by reaction with O_2, and its levels are normally strictly controlled by regulatory systems. In many groups of people, a genetic disorder results in breakdown of this regulation. One kind of iron overload is caused by an inability for the body to produce sufficient porphyrin. Other problems are caused by faults in the regulation of Fe levels by ferritin or transferrin production. These disorders are treated by **chelation therapy**, the administration of a ligand to sequester Fe and allow it to be excreted. Desferrioxamine ('Desferral', **55**) is a ligand that is similar to the siderophores described in Section 27.6. It is a very successful agent for iron overload, apart from the trauma of its introduction into the body, which involves it being plumbed into an intravenous supply.

A special case of chelation therapy is the treatment of individuals who have been contaminated with Pu following exposure to nuclear weapons. In its common oxidation states, Pu(IV) and Pu(III) have similar charge densities to Fe(III) and Fe(II). Siderophore-like chelating ligands have been developed, such as 3,4,3-LIMACC (**56**), which contains four catechol groups.

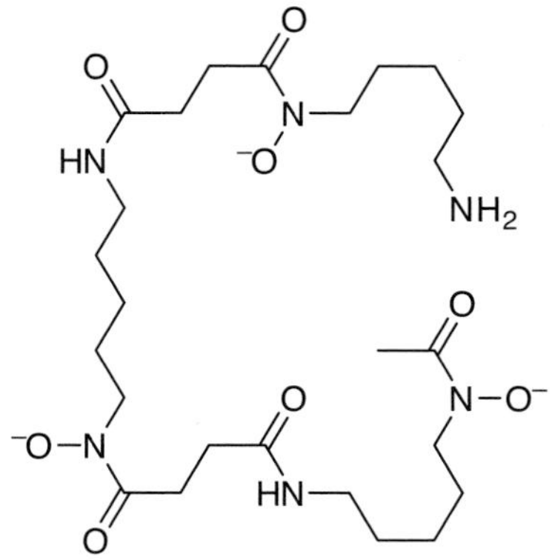

55 Desferral

27.18 Cancer treatment

Key points: The complex *cis*-[PtCl$_2$(NH$_3$)$_2$] results in the inhibition of DNA replication and prevention of cell division; other drugs cause DNA to be degraded by oxygenation.

'Cancer' is a term that covers a large number of different types of the disease, all characterized by the uncontrolled replication of transformed cells that overwhelm the normal operation of the body. The principle of treatment is to apply drugs that destroy these malignant cells selectively.

The remarkable action of the complex *cis*-[PtCl$_2$(NH$_3$)$_2$] (**57**, known as cisplatin) was discovered in 1964 while examining the effect of an electric field on the growth of bacteria. The behaviour of a colony of bacteria suspended in solution between two platinum electrodes was observed, and it was noted that the cells continued to grow in size, forming long filaments, but stopped replicating. The effect was traced to a complex that was

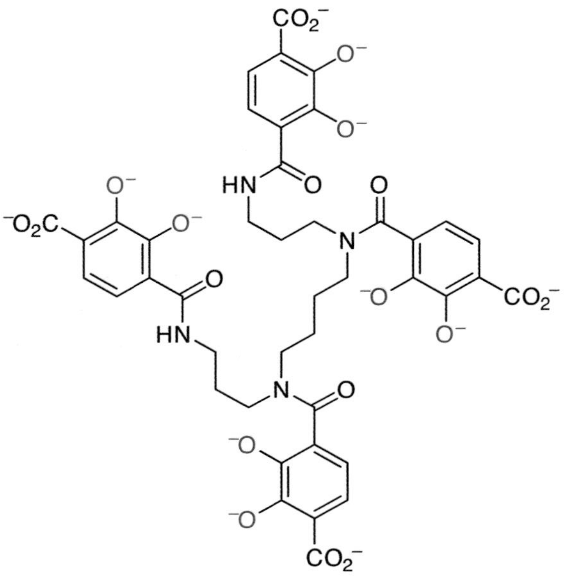

56

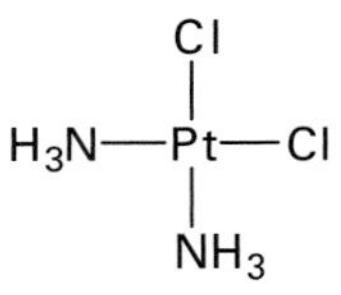

57 Cisplatin

formed electrochemically by dissolution of Pt into the electrolyte, which contained NH_4Cl. Since then, cisplatin has been a successful drug for the treatment of many forms of cancer, particularly testicular cancer, for which the success rate approaches 100 per cent. The other geometric isomer, *trans*-[$PtCl_2(NH_3)_2$], is inactive.

The ultimate molecular basis of the chemotherapeutic action of cisplatin and related drugs is thought to be the formation of a stable complex between Pt(II) and DNA. Cisplatin is administered into the bloodstream of the patient, where, because the plasma contains high concentrations of Cl⁻, it tends to remain as the neutral dichlorido species. The electrical neutrality of the dichlorido complex facilitates its passage through the cell and nuclear membranes. Once it is subjected to the lower Cl⁻ concentrations inside the cell (Table 27.1), the Cl⁻ ligands are replaced by H_2O, and the resulting cationic species (with charges +1 or +2) are attracted electrostatically to DNA and form inner-sphere complexes in which the $-Pt(NH_3)_2$ fragment becomes coordinated to the N atoms of the nucleotide bases. Some classic studies have shown that the preferred target is a pair of N atoms on consecutive guanine bases in the same strand. Complexes of the $-Pt(NH_3)_2$ fragment with oligonucleotides have been studied by X-ray crystallography and ^{195}Pt-NMR (Fig. 27.57). Complexation with Pt causes the helix to bend and partially unwind. It is thought that this distortion renders the DNA incapable of replication or repair. The distortion also makes the DNA recognizable by 'high mobility group' proteins that bind to bent DNA; the cell may thus be targeted for its own death.

Despite its efficacy, cisplatin has highly undesirable side effects, in particular it causes serious damage to the kidneys before it is eventually excreted. Great efforts have been made to find Pt complexes that are effective with fewer side effects. One example in clinical use is carboplatin (**58**). Effective drugs may also include trinuclear Pt(II) (**59**) as well as Pt(IV) complexes such as satraplatin (**60**), which can be administered orally.

Other metal complexes are being discovered that bind by intercalation within the DNA interior and offer improved efficacy over Pt drugs. They include Ru(III) complexes such as *fac*-[$RuCl_3(NH_3)_3$], which are believed to function by providing a source of Ru(III) that is carried to cancer cells by transferrin, and the complex (**61**), which may be activated by reduction to Ru(II) *in vivo*. There is increasing interest in organometallic compounds. The Ru(II) arene complex (**62**), which possesses high anti-cancer activity, is believed to coordinate to guanine-N in a similar way to Pt complexes but the interaction is supplemented by intercalation of the biphenyl group within the hydrophobic DNA core as well as hydrogen bonding between guanine and the $-NH_2$ groups of the en ligand. Even common Ti compounds such as $TiCp_2Cl_2$ have undergone clinical trials. Metallo-supramolecular 'cylinders' formed by placing a metal cation at either end of a bundle of ligands (**63**) have much larger dimensions that mimic those of Zn fingers. The cylinders bind in the major groove of DNA, causing it to form small coils.

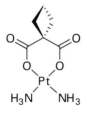

58 Carboplatin

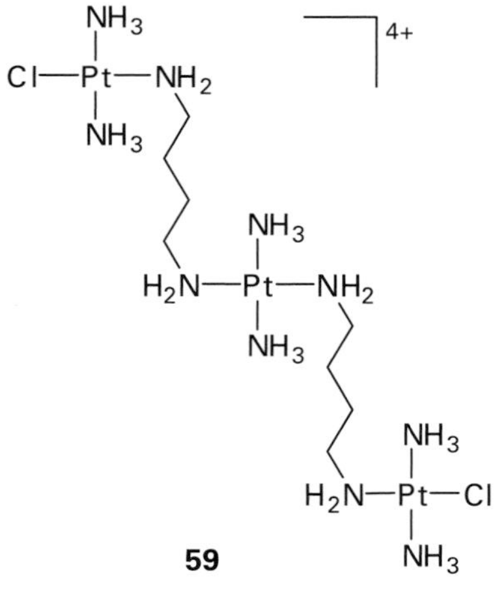

59

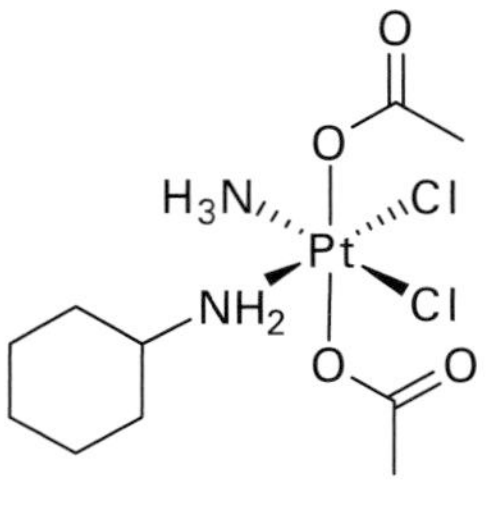

60

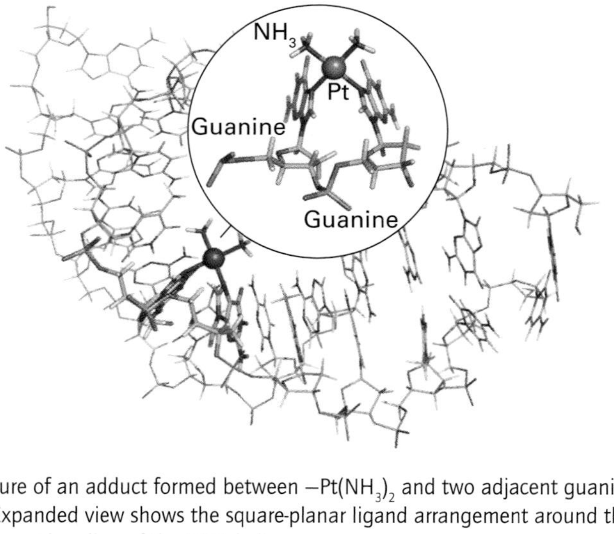

Figure 27.57 Structure of an adduct formed between $-Pt(NH_3)_2$ and two adjacent guanine bases on an oligonucleotide. Expanded view shows the square-planar ligand arrangement around the Pt atom. Coordination of Pt causes bending of the DNA helix.

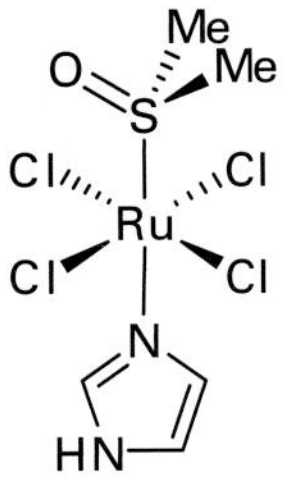

61 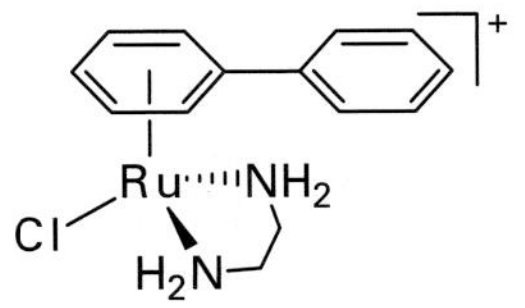 **62** 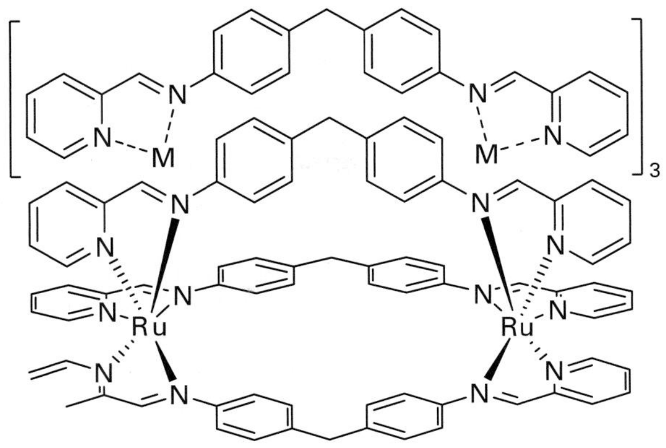 **63**

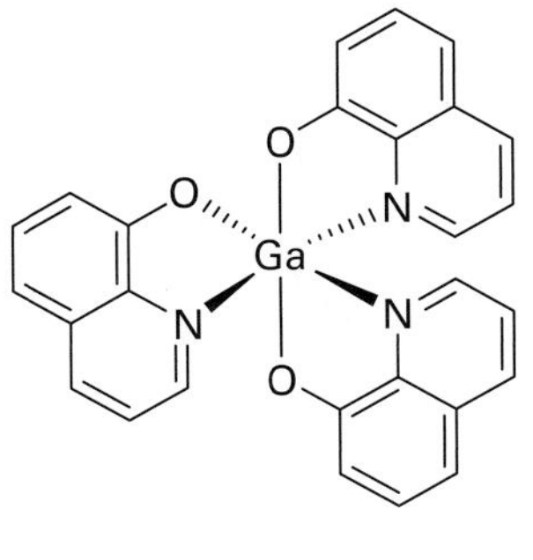

Compounds of Ga(III) are under investigation as anti-cancer drugs. Like Fe(III), Ga(III) is a hard Lewis acid and the two metal ions have similar radii, however Ga(III) is not easily reduced to Ga(II) and any redox or O_2 binding proteins that have incorporated Ga in place of Fe will be inactive. It is thought that Ga(III) enters cells using the same transport systems as Fe. The target for Ga is the Fe-containing enzyme ribonucleotide reductase, which is essential for producing the bases used in DNA. Compounds undergoing trials range from simple salts like gallium nitrate to complexes such as (**64**) that can pass through the intestinal wall.

The challenge with cancer chemotherapy is to identify complexes that select malignant cells and ignore healthy cells. Some Ru complexes undergo selective interaction with DNA, and are activated on irradiation, becoming potent oxidizing agents capable of carrying out cleavage of phosphodiester linkages. This method of treating cancers is known as **phototherapy**. Bleomycin (**65**) is representative of a class of drug that appears to function by binding to DNA and generating, on reaction with O_2, an Fe(IV) (ferryl) species that oxygenates particular sites and leads to degradation.

27.19 Anti-arthritis drugs

Key point: Complexes of Au are effective against rheumatoid arthritis.

Gold drugs are used in the treatment of rheumatoid arthritis, an inflammatory disease that affects the tissue around joints. The inflammation arises by the action of hydrolytic enzymes in cell compartments known as lysosomes that are associated with the Golgi apparatus (see Fig. 27.1). Although the mechanism of action is not established, it is known that Au accumulates in the lysosomes, and it is therefore possible that Au inhibits these hydrolytic enzymes. Another hypothesis is that Au(I) compounds deactivate singlet O_2, a harmful species that can be formed by oxidation of superoxide. The mechanism for this reaction might involve promotion of intersystem crossing by the high spin-orbit coupling constant of the heavy element.

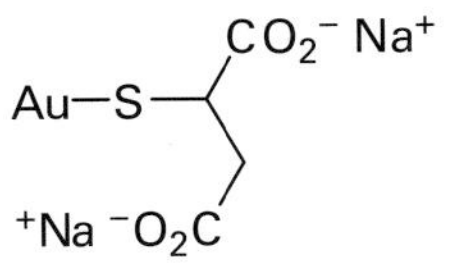

Commonly administered drugs include sodium aurothiomalate ('myochrisin', **66**), sodium aurothioglucose ('solganol', **67**; the linkage between units is uncertain), and others, all of which feature Au(I) with linear coordination. Many are water-soluble polymers, but they cannot be administered orally because they undergo acid hydrolysis in the stomach. By contrast, the compound known as auranofin (**68**) can be given orally. Because Au(I) is chemically soft it is likely that it targets sulfur groups such as cysteine side chains in proteins. It is much more likely to survive in biological environments than Au(III), which is highly oxidizing. As expected, Au compounds lead to side effects, which include skin allergies as well as kidney and gastrointestinal problems.

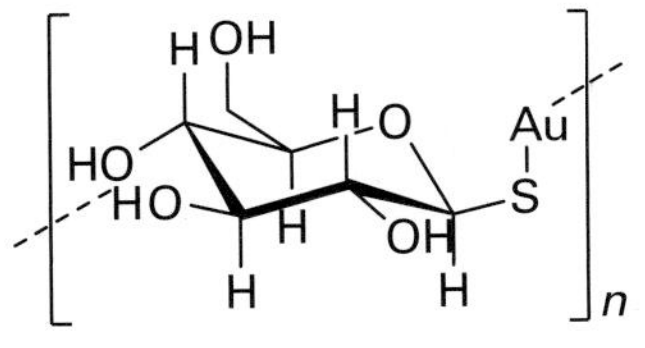

27.20 Imaging agents

Key point: Particular organs and tissues are targeted according to the ligands that are present.

64

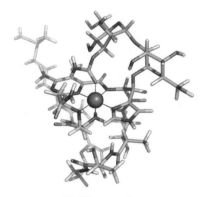

65 Bleomycin

66 Myochrisin

67 Solganol

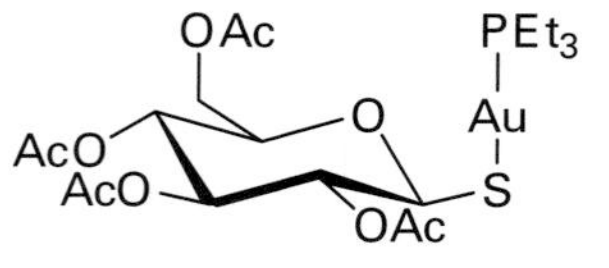

68 Auranofin

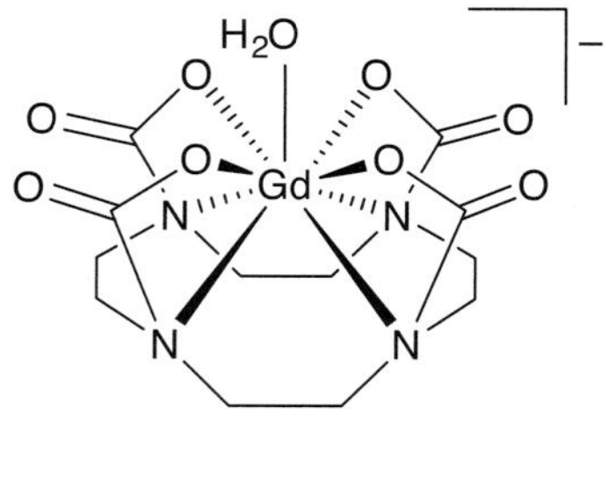

69 Dotarem

70 Cardiolyte

71 Tc-MAG-3

72

73 Ceretec

Complexes of gadolinium(III) (f^7) are used in magnetic resonance imaging (MRI), which has become an important technique in medical diagnosis. Through their effect on the relaxation time of ^{1}H-NMR spectroscopic resonances, Gd(III) complexes are able to enhance the contrasts between different tissues and highlight details such as the abnormalities of the blood–brain barrier. A number of Gd(III) complexes are approved for clinical use, each exhibiting different degrees of rejection or retention by certain tissues, as well as stability, rates of water exchange, and magnitude of relaxation parameters. All are based on chelating ligands, particularly those having multiple carboxylate groups. One example is the complex (**69**) formed with the macrocyclic aminocarboxylate ligand DOTA, which is known as dotarem.

Technetium is an artificial element that is produced by a nuclear reaction, but it has found an important use as an imaging agent. The active radionuclide is ^{99m}Tc (m for metastable), which decays by γ-emission and has a half-life of 6 h. Production of ^{99m}Tc involves bombarding ^{98}Mo with neutrons and separating it as it is formed from the unstable product ^{99}Mo:

$$^{98}\text{Mo} \xrightarrow{\text{neutron capture}} {}^{99}\text{Mo} \xrightarrow{\text{β-decay, 90h}} {}^{99m}\text{Tc} \xrightarrow{\text{γ-emission, 6h}}$$
$$^{99}\text{Tc} \xrightarrow{\text{β-decay, 200ka}} {}^{99}\text{Ru}$$

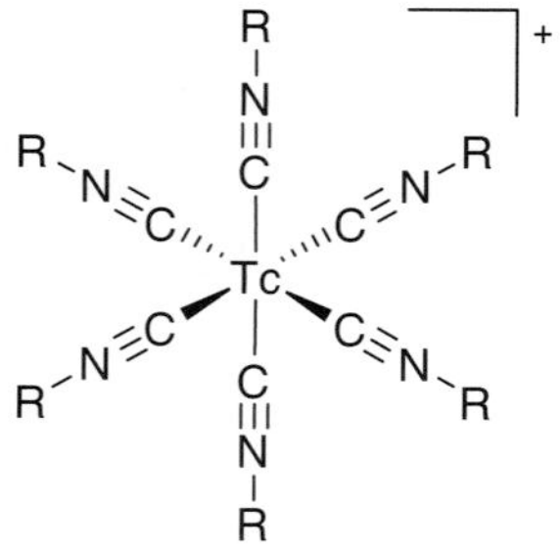

High-energy γ-rays are less harmful to tissue than α- or β-particles. The chemistry of technetium resembles manganese except that higher oxidation states are much less oxidizing.

A variety of substitution-inert Tc complexes can be made that, when injected into the patient, target particular tissues and report on their status. Complexes have been developed that target specific organs such as the heart (revealing tissue damage due to a heart attack), kidney (imaging renal function), or bone (revealing abnormalities and fracture lines). A good basis for organ targeting appears to be the charge on the complex: cationic complexes target the heart, neutral complexes target the brain, and anionic complexes target the bone and kidney. Of the different imaging agents, $[\text{Tc(CNR)}_6]^+$ (**70**) is the best established: known as cardiolyte, it is widely used as a heart imaging agent. The compound of Tc(V) with mercaptoacetyltriglycine (**71**), known as Tc-MAG-3, is used to image kidneys because of its rapid excretion. For imaging bone, complexes of Tc(VII) with hard diphosphonate ligands (**72**) are effective. Brain imaging is carried out with compounds such as ceretec (**73**).

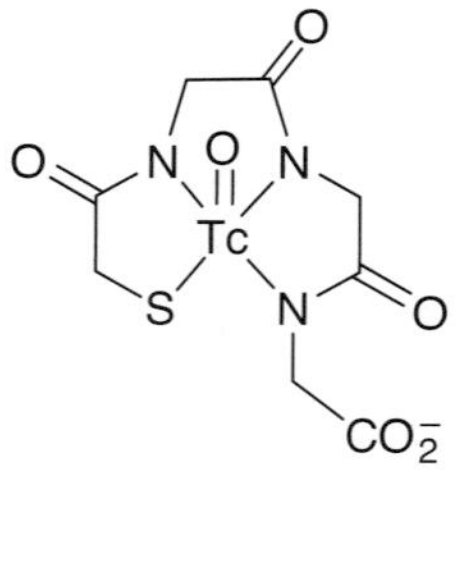

To produce Tc tracers, radioactive $^{99}\text{MoO}_4^{2-}$ is passed onto an anion exchange column, where it binds tightly until it decays to the pertechnate ion $^{99m}\text{TcO}_4^-$ and the lower charge causes it to be eluted. The eluate is treated with a reducing agent, usually Sn(II), and the ligands required to convert it into the desired imaging agent. The resulting compound is then administered to the patient.

Tracers are being developed that are much more specific for their targets. These tracers contain the metal in a stable coordination sphere that is covalently linked to a biologically active fragment. An example is the Gd contrast agent EP-210R (**74**), which contains four Gd^{3+} complexes linked to a peptide that recognizes and binds to fibrin, a molecule produced by thrombi (blood clots).

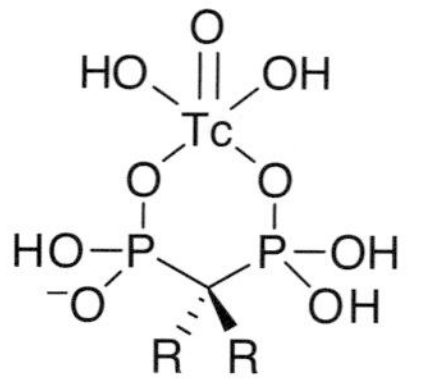

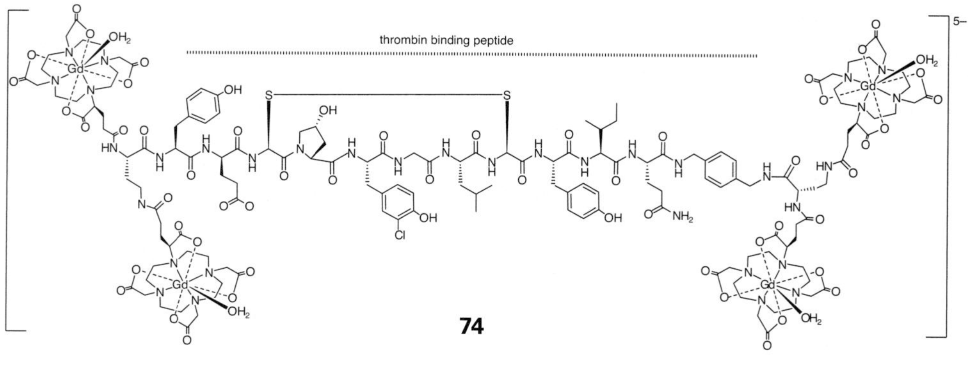

74

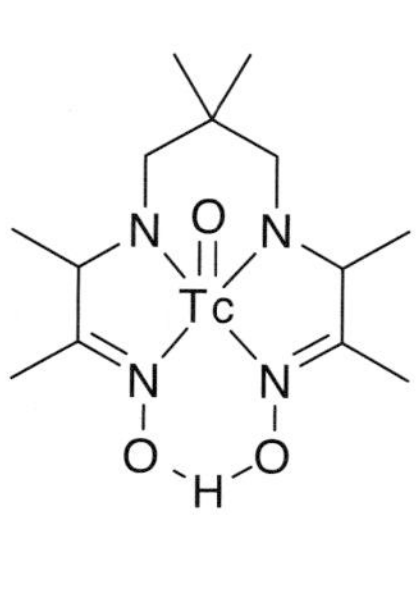

Perspectives

In this final section, we stand back from the material in the chapter and review it from a variety of different perspectives, from the point of view of individual elements and from the point of view of the contribution of bioinorganic chemistry to urgent social problems.

27.21 The contributions of individual elements

Key point: Elements are selected by Nature for their inherent useful properties and their availability.

In this section we summarize the major roles of each element and correlate what we have discussed with emphasis on the element rather than the type of reaction that is involved.

Na, K, and Li The ions of these elements are characterized by weak binding to hard ligands and their specificity is based on size and hydrophobicity that arises from a lower charge density. Compared to Na^+, K^+ is more likely to be found coordinated within a protein and is more easily dehydrated. Both Na^+ and K^+ are important agents in controlling cell structure through osmotic pressure, but whereas Na^+ is ejected from cells, K^+ is accumulated, contributing to a sizeable potential difference across the cell membrane. This differential is maintained by ion pumps, in particular the Na^+,K^+-ATPase, also known as the Na-pump. The electrical energy is released by specific gated ion channels, of which the K^+ channel (Section 27.3) has been studied the most.

An important related issue is the widespread use of simple Li compounds (particularly Li_2CO_3) as psychotherapeutic agents in the treatment of mental disorders, notably bipolar disorder (manic depression). One possibility is that hydrated Li^+ binds tightly in the Na^+ or K^+ channels in place of the dehydrated ions that are normally transported selectively in the filter region of these proteins. As Li^+ ions are highly labile, we can be confident that the aqua ion itself, or a complex with an abundant ligand, is the active species. As a simple aqua ion, Li^+ is strongly solvated, in fact the solvated radius is greater than that of Na^+.

Mg Magnesium ions are the dominant $2+$ ion in cytoplasm and the only ones to occur above millimolar levels in the free, uncomplexed state. The energy currency for enzyme catalysis, ATP, is always present as its Mg^{2+} complex. Magnesium has a special role in the light-harvesting molecule chlorophyll because it is a small $2+$ cation that is able to adopt octahedral geometry and can stabilize a structure without promoting energy loss by fluorescence. The Mg^{2+} ion is a weak acid catalyst and is the active metal ion in rubisco, the highly abundant enzyme responsible for removing from the atmosphere some 100 Gt of CO_2 per year. Rubisco is activated by weak binding of Mg^{2+} to two carboxylates and a special carbamate ligand, leaving three exchangeable water molecules.

Ca Calcium ions are important only in eukaryotes. The bulk of biological Ca is used for structural support and devices such as teeth. The selection of Ca for this function is due to the insolubility of Ca carbonate and phosphate salts. However, a tiny amount of Ca is used as the basis of a sophisticated intracellular signalling system. The principle of this process is that Ca is suited for rapid coordination to hard acid ligands, especially carboxylates from protein side chains, and has no preference for any particular coordination geometry.

Mn Manganese has several oxidation states, most of which are very oxidizing. It is well suited as a redox catalyst for reactions involving positive reduction potentials. One reaction in particular, in which H_2O is used as the electron donor in photosynthesis, is responsible for producing almost all the O_2 in the Earth's atmosphere. This reaction involves a special Mn_4Ca cluster. Manganese(II) is also used as a weak acid–base catalyst in some enzymes. Spectroscopic detectability varies depending on oxidation state: EPR has been useful for Mn(II) and for particular states of the Mn cluster that constitutes the catalyst for the evolution of O_2.

Fe Versatile Fe is probably essential to all organisms and was certainly a very early element in biology. Three oxidation states are important, namely Fe(II), Fe(III), and Fe(IV). Active sites based on Fe catalyse a great variety of redox reactions ranging from electron transfer to oxygenation, as well as acid–base reactions that include reversible O_2 binding, dehydration/hydration, and ester hydrolysis. Iron-containing active sites feature ligands ranging from soft donors such as sulfide (as in FeS clusters) to hard donors such as carboxylate. The porphyrin macrocycle is particularly important as a ligand. Iron(II) in various coordination environments is used to bind O_2, either reversibly or as a prerequisite for activation. Iron(III) is a good Lewis acid, whereas the $Fe(IV)=O$ (ferryl) group may be considered as Nature's way of managing a reactive O atom for insertion into C–H bonds. Cells contain very little uncomplexed Fe(II) and extremely low levels of Fe(III). These ions are toxic, particularly in terms of their reaction with peroxides, which generates the hydroxyl radical. Primary uptake into organisms from minerals poses problems because Fe is found predominantly as Fe(III), salts of which are insoluble at neutral pH (see Fig. 5.12). Iron uptake, delivery, and storage are controlled by sophisticated transport systems, including a special storage protein known as ferritin. Iron porphyrins (as found in cytochromes) show intense UV–visible absorption bands and most active sites with unpaired electrons give rise to characteristic EPR spectra.

Co Cobalt and nickel are among the most ancient biocatalysts. Cobalt is processed only by microorganisms and higher organisms must ingest it as vitamin B_{12} in which Co is complexed by a special macrocycle called corrin. Complexes in which the fifth ligand is a benzimidazole that is covalently linked to the corrin ring are known as cobalamins. Cobalamins are cofactors in enzymes that catalyse alkyl transfer reactions and many radical-based rearrangements. Alkyl transfer reactions exploit the high nucleophilicity of Co(I). In the special cofactor known as coenzyme B_{12}, the sixth ligand to Co(III) is a carbanion donor atom from deoxyadenosine. Radical-based rearrangements involve the ability of coenzyme B_{12} to undergo facile homolytic cleavage of the Co−C bond, producing stable low-spin Co(II) and a carbon radical that can abstract a hydrogen atom from substrates. Cobalamin-containing enzymes show strong UV–visible absorption bands; EPR spectra are observed for Co(II).

Ni Nickel is important in bacterial enzymes, notably hydrogenases, where it also uses the +3 and +1 oxidation states, which are rare in conventional chemistry. A particularly remarkable enzyme, coenzyme A synthase, uses Ni to produce CO and then react it with $CH_3−$ (provided by a cobalamin enzyme) to produce a C−C bond in the form of an acetyl ester. Nickel is also found in plants as the active site of urease. Urease was the first enzyme to be crystallized (in 1926), yet it was not until 1976 that it was discovered to contain Ni.

Cu Unlike Fe, copper probably became important only after O_2 had become established in the Earth's atmosphere and it became available as soluble Cu(II) salts rather than insoluble sulfides (Cu_2S). The main role of Cu is in electron transfer reactions at the higher end of the potential scale and catalysis of redox reactions involving O_2. It is also used for reversible O_2 binding. Both Cu(II) and Cu(I) are strongly bound to biological ligands, particularly soft bases. Free Cu ions are highly toxic and almost absent from cells.

Zn Zinc is an excellent Lewis acid, forming stable complexes with ligands such as N and S donors, and catalysing reactions such as ester and peptide hydrolysis. The biological importance of Zn stems largely from its lack of redox chemistry, although its common adoption of di-, tri-, and tetrathiolate ligation provides a link to the redox chemistry of cysteine/cystine interconversions. Zinc is used as a structure former in enzymes and proteins that bind to DNA. A major problem has been the lack of good spectroscopic methods for studying this d^{10} ion. In some cases, Zn enzymes have been studied by EPR, after substituting the Zn by Co(II).

Mo and W Molybdenum is an abundant element that is probably used by all organisms as a redox catalyst for the transfer of O atoms derived from H_2O. In these oxo-transfer enzymes the Mo is always part of a larger pterin-containing cofactor in which it is coordinated by a special dithiolene ligand. Interconversion between Mo(IV) and Mo(VI) usually results in a change in the number of terminal oxo ligands, and recovery of the starting material occurs by single-electron transfer reactions with Mo(V) as an intermediate. Aside from oxo-transfer and related reactions, Mo has another intriguing role, that of nitrogen fixation, in which it is part of a special FeS cluster. Use of W is confined to prokaryotes, where it is also used as a redox catalyst, but in reactions where a stronger reducing agent is required.

Si Silicon is often neglected among biological elements, yet its turnover in some organisms is comparable to that of carbon. Silica is an important material for the fabrication of the exoskeleton and of prickly defensive armour in plants.

Pt, Au, Bi, and Ru These elements have no known deliberate biological functions and are foreign agents to biological systems, acting under normal conditions as poisons. However, used in controlled procedures, and dressed up by complexation to target a particular site, they are potent drugs, active against a range of diseases and disorders.

27.22 **Future directions**

Key points: Biological metals and metalloproteins have important futures in medicine, energy production, green synthesis, and nanotechnology.

The pioneering studies of the structures and mechanism of ion channels mentioned in Section 27.3 are providing important new leads in neurophysiology, including the rational design of drugs that can block or modify their action in some way. New functions for Ca are continually emerging, and one intriguing aspect is its role in determining the left–right asymmetry of higher organisms, a prime example being the specific placements of heart and liver in the body cavity. The so-called **Notch signalling pathway** in embryonic cells depends on transient extracellular bursts of Ca^{2+} that are dependent in some way on the activity of an H^+/K^+-ATPase. There is also a growing awareness of the role of Zn and Zn transport proteins in control of cellular activity, and also of neural transmission. Indeed, the term **metalloneurochemistry** has been coined to describe the study of metal-ion function in the brain and nervous system at the molecular level. An important challenge is to map out the distribution and flow of Zn in tissue such as brain, and advances are being made in the design of fluorescent ligands that will bind Zn selectively at cellular levels and report on its transport across different zones, for example the synaptic junctions. Metal ions are involved in protein folding, and it is believed that Cu, in particular, may have an important role in fatal neurodegenerative disorders. These roles include controlling the behaviour of prions involved in transmittable diseases such as spongiform encephalopathy (Creuzfeldt–Jakob disease, the human form of 'mad cow' disease) as well as amyloid peptides that are implicated in Alzheimer's disease.

In many regions of the world, rice is the staple food but this commodity is low in Fe. Thus transgenic techniques are being used to improve Fe content. The object is to produce better plant siderophores and improve Fe storage (by enhanced expression of the ferritin gene).

Enzymes tend to show much higher catalytic rates and far higher selectivity than synthetic catalysts, leading naturally to greater efficiency and lower energy costs. The principal disadvantages of using enzymes as industrial catalysts are their lower thermal stability, limitations on solvent and pH conditions, and a large mass per active unit. There is much interest in achieving enzyme-like catalytic performance with small synthetic molecules, a concept that is known as 'bioinspired catalysis'. The idea is to reproduce, using all the tools of synthetic chemistry, the properties of an enzyme trimmed down to its smallest fully functional component. Examples of bioinspired catalysts have already been described: areas of particular interest for industrial production are the conversion of methane to methanol (Section 27.10), activation of N_2 to produce cheap fertilizers, and production of hydrogen.

In the not so distant future, when fossil fuels have been depleted, H_2 will become an important energy carrier, used either directly or indirectly (after conversion into fuels such as alcohol) to power vehicles of all kinds. One of the scientific challenges is how to obtain efficient electrolytic production of H_2 from water, given that electricity will be widely available from a variety of sources. This process requires demanding conditions of temperature and overpotential (Section 10.4), or catalysts that are currently based on Pt and other precious metals. However, Nature has already shown us that rapid hydrogen cycling is possible under mild conditions by using just the common metals Fe and Ni. A related challenge is the synthesis of efficient electrocatalysts that can convert water to O_2 without requiring a large overpotential, not because there is a need for O_2 itself, but because it is an essential byproduct of electrolytic or photolytic H_2 production (see Box 10.3). Once again, we can turn to the biosphere for inspiration because by elucidating the mechanism of the Mn catalyst, we might synthesize new catalysts that are both cheap and durable.

We have seen the exquisite structures of materials that are produced by organisms. This understanding is now leading to new directions in nanotechnology (Chapter 25). For example, sponge-like single crystals of calcite, having intricate morphological features, have been produced on polymer membranes formed by templating the skeletal plates of the sea urchin. Another recent development is the production of Pd nanoclusters by hydrogen-oxidizing bacteria, the action of hydrogenases making available controlled electron flow to effect the electroplating of Pd on to microscopic sites.

FURTHER READING

J.J.R. Frausto da Silva and R.J.P. Williams, *The biological chemistry of the elements*. Oxford University Press (2001). An excellent, detailed book that looks at the broader picture of the relationship between elements and life.

L. Que Jr. and W.B. Tolman, *Bio-coordination chemistry. Comprehensive coordination chemistry, Vol. 8*. Elsevier (2004). A text providing particularly detailed insight into model compounds.

R.R. Crichton, F. Lallemand, I.S.M. Psalti, and R.J. Ward, *Biological inorganic chemistry*. Elsevier (2007). A modern introduction to biological inorganic chemistry.

E. Gouaux and R. MacKinnon, Principles of selective ion transport in channels and pumps. *Science*, 2005, **310**, 1461. An article linking detailed three-dimensional structural data of giant proteins with physiological function and the chemistry of Group 1 and 2 metal ions and Cl^-.

R.K.O. Sigel and A.M. Pyle, Alternative roles for metal ions in enzyme catalysis and the implications for ribozyme chemistry. *Chem. Rev.*, 2007, **107**, 97. A review describing the role of metal ions, particularly Mg, as active centres in catalysts based on RNA instead of proteins.

E. Kimura, Model studies for molecular recognition of carbonic anhydrase and carboxypeptidase. *Acc. Chem. Res.*, 2001, **34**, 171. This review describes the acid–base and catalytic properties of small Zn complexes in an effort to understand how Zn enzymes function.

K.N. Ferreira, T.M. Iverson, K. Maghlaoui, J. Barber, and S. Iwata, Architecture of the photosynthetic oxygen-evolving center. *Science*, 2004, **303**, 1831. A seminal article on the structure of the catalyst responsible for O_2 in the atmosphere.

L. Que, Jr., The road to non-heme oxoferryls and beyond. *Acc. Chem. Res.*, 2007, **40**, 493. A stimulating account of efforts to understand the chemistry of Fe(IV) species and produce new catalysts for organic oxygenation reactions.

E.A. Lewis and W.B. Tolman, Reactivity of dioxygen–copper systems. *Chem. Rev.*, 2004, **104**, 1047. A review of small molecule analogues of Cu-containing enzymes.

S.C. Wang and P.A. Frey, S-adenosylmethionine as an oxidant: the radical SAM superfamily. *Trends in Biochemical Sciences*, 2007, **32**, 101. An authoritative account of the discovery of radical SAM enzymes. It categorises the different classes of enzymes and describes the mechanisms by which the [4Fe−4S] cluster cleaves SAM to initiate radical reactions.

H.B. Gray, B.G. Malmström, and R.J.P. Williams, Copper coordination in blue proteins. *J. Biol. Inorg. Chem.*, 2000, **5**, 551. An article that draws together and summarizes the theories spanning more than 30 years of research on blue Cu centres.

P.J. Kiley and H. Beinert, The role of Fe−S proteins in sensing and regulation in bacteria. *Curr. Opin. Chem. Biol.*, 2003, **6**, 182. A review describing how Fe−S clusters are involved in sensing.

J. Green and M.S. Paget, Bacterial redox centres. *Nature Reviews*, 2004, **2**, 954. A general review on the mechanisms by which reactive oxygen species are sensed in cells.

S. Mann, *Biomineralisation: principles and concepts in bioinorganic materials chemistry*. Oxford University Press (2001).

M.D. Archer and J. Barber (ed.), *Molecular to global photosynthesis*. Imperial College Press (2004).

C.W. Cady, R.H. Crabtree, and G.W. Brudvig, Functional models for the oxygen-evolving complex of photosystem II. *Coord. Chem. Rev.*, 2008, **252**, 444. A review of recent efforts to understand and mimic the chemistry of the Mn_4Ca cluster that converts water to O_2.

M.J. Hannon, Supramolecular DNA recognition. *Chem. Soc. Rev.*, 2007, **36**, 280. An account of efforts to make large metal complexes that can recognise certain DNA sequences.

M.A. Jakupec, M. Galanski, V.B. Arion, C.G. Hartinger, and B. Keppler, Antitumour metal compounds: more than theme and variations. *Dalton Transactions*, 2008, **2**,183.

P.C.A. Bruijnincx and P. J Sadler, New trends for metal complexes with anticancer activity. *Curr. Opin. Chem. Biol.*, 2008, **12**, 197. Two complementary reviews of recent developments in anti-cancer drugs based on metal complexes.

P. Caravan, Strategies for increasing the sensitivity of gadolinium-based MRI contrast agents. *Chem. Soc. Rev.*, 2006, **35**, 512. A review of developments in producing selective Gd-based magnetic resonance imaging reagents.

B.E. Mann and R. Motterlini, CO and NO in medicine. *Chem. Commun.*, 2007, 4197. A review of the roles of NO and CO in biology and medicine.

EXERCISES

27.1 Calcium-binding proteins can be studied by using lanthanoid ions (Ln^{3+}). Compare and contrast the coordination preferences of the two types of metal ion and suggest techniques in which lanthanoid ions would be useful.

27.2 In zinc enzymes, 'spectroscopically silent' Zn(II) can often be replaced by Co(II) with high retention of activity. Explain the principles by which this substitution can be exploited to obtain structural and mechanistic information.

27.3 Compare and contrast the acid–base catalytic activities of Zn(II), Fe(III), and Mg(II).

27.4 Propose physical methods that would allow you to determine whether a reactive intermediate isolated by rapid freeze quenching contains Fe(V).

27.5 Figure 27.58 shows Mössbauer spectra of a sample of ferredoxin from chloroplasts at 77 K. With reference to Section 8.7, interpret the data with regard to the oxidation states and spin states of the two Fe atoms and comment on the electron delocalization at this temperature.

27.6 The structure of the P-cluster in nitrogenase differs significantly between oxidized and reduced states. Comment on this observation in the light of proposals that it participates in long-range electron transfer.

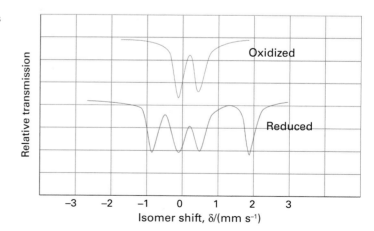

Figure 27.58 Mössbauer spectra of a sample of ferredoxin from chloroplasts at 77 K.

27.7 Microorganisms can synthesize the acetyl group ($CH_3CO−$) by direct combination of methyl groups with CO. Make some predictions about the metals that are involved.

PROBLEMS

27.1 With reference to details that have been discussed in Section 27.3 for the K⁺ channel, predict the properties of Na⁺, Ca²⁺, and Cl⁻ binding sites that would be important for providing selectivity in their respective transmembrane ion transporters.

27.2 Comment on the implications of discovering, by microwave detection, substantial levels of O_2 on a planet in another solar system.

27.3 Apart from direct O-atom transfer (Fig. 27.43), another mechanism proposed for Mo enzymes is indirect O-atom transfer, also known as *coupled electron–proton transfer*. In this mechanism, as shown in Fig. 27.59 for sulfite oxidase, the O atom that is transferred originates instead from an uncoordinated H_2O molecule. Propose a way to distinguish between direct and indirect O-atom transfer mechanisms.

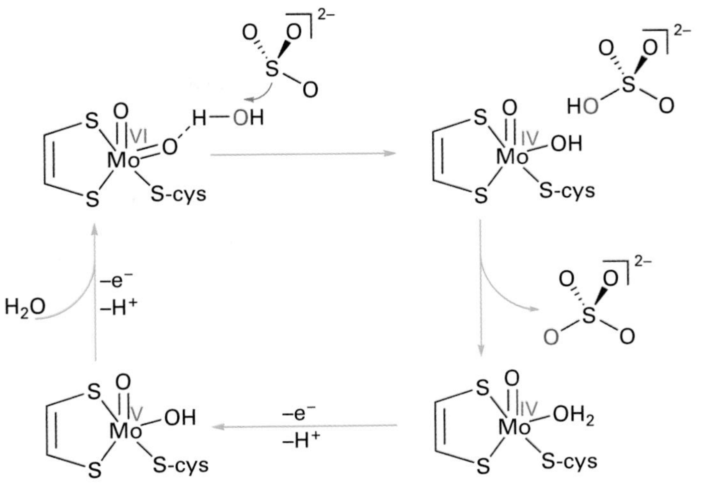

27.4 In justifying research into small molecule catalysts for producing NH_3 from N_2, it is sometimes stated that nitrogenase is an 'efficient' enzyme: how true is this statement? Comment critically on the argument that 'knowing the three-dimensional structure of nitrogenase has not enlightened us as to its mechanism of action' and discuss how this view might be valid more generally for enzymes for which a structure is known.

27.5 'In Mo enzymes, the bond between a terminal oxo anion and Mo(VI) is usually written as a double bond, whereas it is more correctly assigned as a triple bond.' Discuss this statement. Suggest how a terminal oxido ligand influences the reactivity of other

Figure 27.59 The mechanism referred to in Problem 27.3.

coordination sites on the Mo atom and explain how a terminal sulfido ligand (as occurs in xanthine oxidase) would alter the properties of the active site.

Resource section 1
Selected ionic radii

Ionic radii are given (in picometres, pm) for the most common oxidation states and coordination geometries. The coordination number is given in parentheses. All d-block species are low-spin unless labelled with[†], in which case values for high-spin are quoted. Most data are taken from R.D. Shannon, *Acta Cryst.*, 1976, **A32**, 751, where values for other coordination geometries can be found. Where Shannon values are not available, Pauling ionic radii are quoted and are indicated by *.

1	2	3	4	5	6	7	8	9	10	11	12	13	14	15	16	17	18
Li^+ 59 (4) 76 (6) 92 (8)	Be^{2+} 27 (4) 45 (6)											B^{3+} 11 (4) 27 (6)	C^{4+} 15 (4) 16 (6)	N^{3-} 146 (4)	O^{2-} 138 (4) 140 (6) 142 (8)	F^- 131 (4) 133 (6)	Ne^+ 112*
														N^{3+} 16 (6)			
Na^+ 99 (4) 102 (6) 32 (8)	Mg^{2+} 49 (4) 72 (6) 103 (8)											Al^{3+} 39 (4) 53 (6)	Si^{4+} 26 (4) 40 (6)	P^{5+} 29 (4) 38 (6)	S^{2-} 184 (6)	Cl^- 181 (6)	Ar^+ 154*
														P^{3+} 44 (6)	S^{6+} 12 (4) 29 (6)	Cl^{7+} 8 (4) 27 (6)	
															S^{4+} 37 (6)		
K^+ 137 (4) 138 (6) 151 (8)	Ca^{2+} 100 (6) 112 (8)	Sc^{3+} 75 (6) 87 (8)	Ti^{4+} 42 (4) 61 (6) 74 (8)	V^{5+} 36 (4) 54 (6)	Cr^{6+} 26 (4) 44 (6)	Mn^{7+} 25 (4) 46 (6)	Fe^{6+} 25 (4)	Co^{4+} 40 (4) 53 (6)[†]	Ni^{4+} 48 (6)	Cu^{3+} 54 (6)	Zn^{2+} 60 (4) 74 (6) 90 (8)	Ga^{3+} 47 (4) 62 (6)	Ge^{4+} 39 (4) 53 (6)	As^{5+} 34 (4) 46 (6)	Se^{2-} 198 (6)	Br^- 196 (6)	Kr^+ 169*
			Ti^{3+} 67 (6)	V^{4+} 58 (6) 72 (8)	Cr^{5+} 49 (6)	Mn^{6+} 26 (4)	Fe^{4+} 58 (6)	Co^{3+} 55 (6)	Ni^{3+} 56 (6)	Cu^{2+} 57 (4) 73 (6)		Ge^{2+} 73 (6)	As^{3+} 58 (6)	Se^{6+} 28 (4) 42 (6)	Br^{7+} 39 (6)		
			Ti^{2+} 86 (6)	V^{3+} 64 (6)	Cr^{4+} 41 (4) 55 (6)	Mn^{5+} 33 (4) 63 (6)	Fe^{3+} 49 (4)[†] 55 (6) 78 (8)[†]	Co^{2+} 58 (4)[†] 65 (6) 90 (8)	Ni^{2+} 55 (4) 69 (8)	Cu^+ 60 (4) 77 (6)					Se^{4+} 50 (6)		
				V^{2+} 79 (6)	Cr^{3+} 62 (6)	Mn^{4+} 37 (4) 53 (6)	Fe^{2+} 63 (4)[†] 61 (6) 92 (8)[†]										
					Cr^{2+} 73 (6)	Mn^{3+} 65 (6)											
						Mn^{2+} 67 (6) 96 (8)											
Rb^+ 148 (6) 160 (8)	Sr^{2+} 118 (6) 126 (8)	Y^{3+} 90 (6) 102 (8)	Zr^{4+} 59 (4) 72 (6) 84 (8)	Nb^{5+} 48 (4) 64 (6) 74 (8)	Mo^{6+} 41 (4) 59 (6)	Tc^{7+} 37 (4) 56 (6)	Ru^{8+} 36 (4)	Rh^{5+} 55 (6)	Pd^{4+} 62 (6)	Ag^{3+} 67 (4) 75 (6)	Cd^{2+} 78 (4) 95 (6) 110 (8)	In^{3-} 62 (4) 80 (6) 92 (8)	Sn^{4+} 55 (4) 69 (6) 81 (8)	Sb^{5+} 60 (6)	Te^{6+} 43 (4) 56 (6)	I^- 220 (6)	Xe^+ 190*
				Nb^{4+} 68 (6) 79 (8)	Mo^{5+} 46 (4) 61 (6)	Tc^{5+} 60 (6)	Ru^{7+} 38 (4)	Rh^{4+} 60 (6)	Pd^{3+} 76 (6)	Ag^{2+} 79 (4) 94 (6)			Sn^{2+} 102 (6)	Sb^{3+} 76 (6)	Te^{4+} 66 (4) 97 (6)	I^{7+} 42 (4) 53 (6)	Xe^{8+} 40 (4) 48 (6)
				Nb^{3+} 72 (6)	Mo^{4+} 65 (6)	Tc^{4+} 66 (6) 95 (8)	Ru^{5+} 71 (6)	Rh^{3+} 67 (6)	Pd^{2+} 64 (4) 86 (6)	Ag^+ 67 (2) 100 (4) 115 (6)							
					Mo^{3+} 69 (6)		Ru^{4+} 62 (6)		Pd^+ 59 (2)								
							Ru^{3+} 68 (6)										

1	2	3	4	5	6	7	8	9	10	11	12	13	14	15	16	17	18
Cs^+ 167 (6) 174 (8)	Ba^{2+} 135 (6) 142 (8)	La^{3+} 103 (6) 116 (8)	Hf^{4+} 58 (4) 71 (6) 83 (8)	Ta^{5+} 64 (6) 74 (8)	W^{6+} 42 (4) 60 (6)	Re^{7+} 38 (4) 53 (6)	Os^{8+} 39 (4)	Ir^{5+} 57 (6)	Pt^{5+} 57 (6)	Au^{5+} 57 (6)	Hg^{2+} 96 (4) 102 (6) 114 (8)	Tl^{3+} 75 (4) 89 (6) 98 (8)	Pb^{4+} 65 (4) 78 (6) 94 (8)	Bi^{5+} 76 (6)	Po^{6+} 67 (6)	At^{7-} 62 (6)	
				Ta^{4+} 68 (6)	W^{5+} 62 (6)	Re^{6+} 55 (6)	Os^{7+} 53 (6)	Ir^{4+} 63 (6)	Pt^{4+} 63 (6)	Au^{3+} 68 (4) 85 (6)	Hg^+ 119 (6)	Tl^- 150 (6) 159 (8)	Pb^{2+} 119 (6) 129 (8)	Bi^{3+} 103 (6) 117 (8)	Po^{4+} 94 (6) 108 (8)		
				Ta^{3+} 72 (6)	W^{4+} 66 (6)	Re^{5+} 58 (6)	Os^{6+} 55 (6)	Ir^{3+} 68 (6)	Pt^{2+} 60 (4) 80 (6)	Au^+ 137 (6)							
						Re^{4+} 63 (6)	Os^{5+} 58 (6)										
							Os^{4+} 63 (6)										

1	2
Fr^+ 196 (6)	Ra^{2+} 170 (8)

Lanthanoids

1	2	3	4	5	6	7	8	9	10	11	12	13	14
Ce^{4+} 87 (6) 97 (8)	Pr^{4+} 85 (6) 96 (8)	Nd^{3+} 98 (6) 111 (8)	Pm^{3+} 97 (6) 109 (8)	Sm^{3+} 96 (6) 108 (8)	Eu^{3+} 95 (6) 107 (8)	Gd^{3+} 94 (6) 105 (8)	Tb^{4+} 76 (6) 88 (8)	Dy^{3+} 91 (6) 103 (8)	Ho^{3+} 90 (6) 102 (8)	Er^{3+} 89 (6) 100 (8)	Tm^{3+} 88 (6) 99 (8)	Yb^{3+} 87 (6) 99 (8)	Lu^{3+} 86 (6) 98 (8)
Ce^{3+} 101 (6) 114 (8)	Pr^{3+} 99 (6) 113 (8)	Nd^{2+} 129 (8)		Sm^{2+} 127 (8)	Eu^{2+} 117 (6) 125 (8)		Tb^{3+} 92 (6) 104 (8)	Dy^{2+} 107 (6) 119 (8)			Tm^{2+} 103 (6) 109 (8)	Yb^{2+} 102 (6) 114 (8)	

Actinoids

1	2	3	4	5	6	7	8	9	10	11	12	13	14
Th^{4+} 94 (6) 110 (8)	Pa^{5+} 78 (6) 95 (8)	U^{6+} 52 (4) 73 (6) 100 (8)	Np^{7+} 72 (6)	Pu^{6+} 71 (6)	Am^{4+} 85 (6) 95 (8)	Cm^{4+} 85 (6) 95 (8)	Bk^{4+} 63 (6) 93 (8)	Cf^{4+} 82 (6) 92 (8)	Es	Fm	Md	No^{2+} 110 (6)	Lr
	Pa^{4+} 90 (6) 101 (8)	U^{5+} 76 (6)	Np^{6+} 72 (6)	Pu^{5+} 74 (6)	Am^{3+} 98 (6) 123 (8)	Cm^{3+} 97 (6)	Bk^{3+} 96 (6)	Cf^{3+} 95 (6)					
	Pa^{3+} 104 (6)	U^{4+} 89 (6) 100 (8)	Np^{5+} 75 (6)	Pu^{4+} 86 (6) 96 (8)	Am^{2+} 126 (8)								
		U^{3+} 103 (6)	Np^{4+} 87 (4) 98 (6)	Pu^{3+} 100 (6)									
			Np^{3+} 101 (6)										
			Np^{2+} 110 (6)										

Resource section 2
Electronic properties of the elements

Ground-state electron configurations of atoms are determined experimentally from spectroscopic and magnetic measurements. The results of these determinations are listed below. They can be rationalized in terms of the building-up principle, in which electrons are added to the available orbitals in a specific order in accord with the Pauli exclusion principle. Some variation in order is encountered in the d and f blocks to accommodate the effects of electron–electron interaction more faithfully. The closed-shell configuration $1s^2$ characteristic of helium is denoted [He] and likewise for the other noble-gas element configurations. The ground-state electron configurations and term symbols listed below have been taken from S. Fraga, J. Karwowski, and K.M.S. Saxena, *Handbook of atomic data*. Elsevier, Amsterdam (1976).

The first three ionization energies of an element E are the energies required for the following processes:

$$I_1: E(g) \rightarrow E^+(g) + e^-(g)$$

$$I_2: E^+(g) \rightarrow E^{2+}(g) + e^-(g)$$

$$I_3: E^{2+}(g) \rightarrow E^{3+}(g) + e^-(g)$$

The electron affinity E_{ea} is the energy *released* when an electron attaches to a gas-phase atom:

$$E_{ea}: E(g) + e^-(g) \rightarrow E^-(g)$$

The values given here are taken from various sources, particularly C.E. Moore, *Atomic energy levels*, NBS Circular 467, Washington (1970) and W.C. Martin, L. Hagan, J. Reader, and J. Sugar, *J. Phys. Chem. Ref. Data*, 1974, **3**, 771. Values for the actinoids are taken from J.J. Katz, G.T. Seaborg, and L.R. Morss (ed.), *The chemistry of the actinide elements*. Chapman & Hall, London (1986). Electron affinities are from H. Hotop and W.C. Lineberger, *J. Phys. Chem. Ref. Data*, 1985, **14**, 731.

For conversions to kilojoules per mole and reciprocal centimetres, see the back inside cover.

Atom			Ionization energy (eV)			Electron affinity E_{ea} (eV)
			I_1	I_2	I_3	
1	H	$1s^1$	13.60			+0.754
2	He	$1s^2$	24.59	54.51		-0.5
3	Li	[He]$2s^1$	5.320	75.63	122.4	+0.618
4	Be	[He]$2s^2$	9.321	18.21	153.85	≤0
5	B	[He]$2s^2 2p^1$	8.297	25.15	37.93	+0.277
6	C	[He]$2s^2 2p^2$	11.257	24.38	47.88	+1.263
7	N	[He]$2s^2 2p^3$	14.53	29.60	47.44	-0.07
8	O	[He]$2s^2 2p^4$	13.62	35.11	54.93	+1.461
9	F	[He]$2s^2 2p^5$	17.42	34.97	62.70	+3.399
10	Ne	[He]$2s^2 2p^6$	21.56	40.96	63.45	-1.2
11	Na	[Ne]$3s^1$	5.138	47.28	71.63	+0.548
12	Mg	[Ne]$3s^2$	7.642	15.03	80.14	≤0
13	Al	[Ne]$3s^2 3p^1$	5.984	18.83	28.44	+0.441
14	Si	[Ne]$3s^2 3p^2$	8.151	16.34	33.49	+1.385
15	P	[Ne]$3s^2 3p^3$	10.485	19.72	30.18	+0.747
16	S	[Ne]$3s^2 3p^4$	10.360	23.33	34.83	+2.077
17	Cl	[Ne]$3s^2 3p^5$	12.966	23.80	39.65	+3.617
18	Ar	[Ne]$3s^2 3p^6$	15.76	27.62	40.71	-1.0
19	K	[Ar]$4s^1$	4.340	31.62	45.71	+0.502
20	Ca	[Ar]$4s^2$	6.111	11.87	50.89	+0.02
21	Sc	[Ar]$3d^1 4s^2$	6.54	12.80	24.76	
22	Ti	[Ar]$3d^2 4s^2$	6.82	13.58	27.48	
23	V	[Ar]$3d^3 4s^2$	6.74	14.65	29.31	
24	Cr	[Ar]$3d^5 4s^1$	6.764	16.50	30.96	
25	Mn	[Ar]$3d^5 4s^2$	7.435	15.64	33.67	
26	Fe	[Ar]$3d^6 4s^2$	7.869	16.18	30.65	
27	Co	[Ar]$3d^7 4s^2$	7.876	17.06	33.50	
28	Ni	[Ar]$3d^8 4s^2$	7.635	18.17	35.16	
29	Cu	[Ar]$3d^{10} 4s^1$	7.725	20.29	36.84	
30	Zn	[Ar]$3d^{10} 4s^2$	9.393	17.96	39.72	
31	Ga	[Ar]$3d^{10} 4s^2 4p^1$	5.998	20.51	30.71	+0.30
32	Ge	[Ar]$3d^{10} 4s^2 4p^2$	7.898	15.93	34.22	+1.2
33	As	[Ar]$3d^{10} 4s^2 4p^3$	9.814	18.63	28.34	+0.81
34	Se	[Ar]$3d^{10} 4s^2 4p^4$	9.751	21.18	30.82	+2.021
35	Br	[Ar]$3d^{10} 4s^2 4p^5$	11.814	21.80	36.27	+3.365
36	Kr	[Ar]$3d^{10} 4s^2 4p^6$	13.998	24.35	36.95	-1.0
37	Rb	[Kr]$5s^1$	4.177	27.28	40.42	+0.486
38	Sr	[Kr]$5s^2$	5.695	11.03	43.63	+0.05
39	Y	[Kr]$4d^1 5s^2$	6.38	12.24	20.52	
40	Zr	[Kr]$4d^1 5s^2$	6.84	13.13	22.99	

Atom		Ionization energy (eV)			Electron affinity E_{ea} (eV)
		I_1	I_2	I_3	
41 Nb	[Kr]$4d^45s^1$	6.88	14.32	25.04	
42 Mo	[Kr]$4d^55s^1$	7.099	16.15	27.16	
43 Tc	[Kr]$4d^55s^2$	7.28	15.25	29.54	
44 Ru	[Kr]$4d^75s^1$	7.37	16.76	28.47	
45 Rh	[Kr]$4d^85s^1$	7.46	18.07	31.06	
46 Pd	[Kr]$4d^{10}$	8.34	19.43	32.92	
47 Ag	[Kr]$4d^{10}5s^1$	7.576	21.48	34.83	
48 Cd	[Kr]$4d^{10}5s^2$	8.992	16.90	37.47	
49 In	[Kr]$4d^{10}5s^25p^1$	5.786	18.87	28.02	+0.3
50 Sn	[Kr]$4d^{10}5s^25p^2$	7.344	14.63	30.50	+1.2
51 Sb	[Kr]$4d^{10}5s^25p^3$	8.640	18.59	25.32	+1.07
52 Te	[Kr]$4d^{10}5s^25p^4$	9.008	18.60	27.96	+1.971
53 I	[Kr]$4d^{10}5s^25p^5$	10.45	19.13	33.16	+3.059
54 Xe	[Kr]$4d^{10}5s^25p^6$	12.130	21.20	32.10	-0.8
55 Cs	[Xe]$6s^1$	3.894	25.08	35.24	
56 Ba	[Xe]$6s^2$	5.211	10.00	37.51	
57 La	[Xe]$5d^16s^2$	5.577	11.06	19.17	
58 Ce	[Xe]$4f^15d^16s^2$	5.466	10.85	20.20	
59 Pr	[Xe]$4f^36s^2$	5.421	10.55	21.62	
60 Nd	[Xe]$4f^46s^2$	5.489	10.73	20.07	
61 Pm	[Xe]$4f^56s^2$	5.554	10.90	22.28	
62 Sm	[Xe]$4f^66s^2$	5.631	11.07	23.42	
63 Eu	[Xe]$4f^76s^2$	5.666	11.24	24.91	
64 Gd	[Xe]$4f^75d^16s^2$	6.140	12.09	20.62	
65 Tb	[Xe]$4f^96s^2$	5.851	11.52	21.91	
66 Dy	[Xe]$4f^{10}6s^2$	5.927	11.67	22.80	
67 Ho	[Xe]$4f^{11}6s^2$	6.018	11.80	22.84	
68 Er	[Xe]$4f^{12}6s^2$	6.101	11.93	22.74	
69 Tm	[Xe]$4f^{13}6s^2$	6.184	12.05	23.68	
70 Yb	[Xe]$4f^{14}6s^2$	6.254	12.19	25.03	
71 Lu	[Xe]$4f^{14}5d^16s^2$	5.425	13.89	20.96	
72 Hf	[Xe]$4f^{14}5d^26s^2$	6.65	14.92	23.32	

Atom		Ionization energy (eV)			Electron affinity E_{ea} (eV)
		I_1	I_2	I_3	
73 Ta	[Xe]$4f^{14}5d^36s^2$	7.89	15.55	21.76	
74 W	[Xe]$4f^{14}5d^46s^2$	7.89	17.62	23.84	
75 Re	[Xe]$4f^{14}5d^56s^2$	7.88	13.06	26.01	
76 Os	[Xe]$4f^{14}5d^66s^2$	8.71	16.58	24.87	
77 Ir	[Xe]$4f^{14}5d^76s^2$	9.12	17.41	26.95	
78 Pt	[Xe]$4f^{14}5d^96s^1$	9.02	18.56	29.02	
79 Au	[Xe]$4f^{14}5d^{10}6s^1$	9.22	20.52	30.05	
80 Hg	[Xe]$4f^{14}5d^{10}6s^2$	10.44	18.76	34.20	
81 Tl	[Xe]$4f^{14}5d^{10}6s^26p^1$	6.107	20.43	29.83	
82 Pb	[Xe]$4f^{14}5d^{10}6s^26p^2$	7.415	15.03	31.94	
83 Bi	[Xe]$4f^{14}5d^{10}6s^26p^3$	7.289	16.69	25.56	
84 Po	[Xe]$4f^{14}5d^{10}6s^26p^4$	8.42	18.66	27.98	
85 At	[Xe]$4f^{14}5d^{10}6s^26p^5$	9.64	16.58	30.06	
86 Rn	[Xe]$4f^{14}5d^{10}6s^26p^6$	10.75			
87 Fr	[Rn]$7s^1$	4.15	21.76	32.13	
88 Ra	[Rn]$7s^2$	5.278	10.15	34.20	
89 Ac	[Rn]$6d^17s^2$	5.17	11.87	19.69	
90 Th	[Rn]$6d^27s^2$	6.08	11.89	20.50	
91 Pa	[Rn]$5f^26d^17s^2$	5.89	11.7	18.8	
92 U	[Rn]$5f^36d^17s^2$	6.19	14.9	19.1	
93 Np	[Rn]$5f^46d^17s^2$	6.27	11.7	19.4	
94 Pu	[Rn]$5f^67s^2$	6.06	11.7	21.8	
95 Am	[Rn]$5f^77s^2$	5.99	12.0	22.4	
96 Cm	[Rn]$5f^76d^17s^2$	6.02	12.4	21.2	
97 Bk	[Rn]$5f^97s^2$	6.23	12.3	22.3	
98 Cf	[Rn]$5f^{10}7s^2$	6.30	12.5	23.6	
99 Es	[Rn]$5f^{11}7s^2$	6.42	12.6	24.1	
100 Fm	[Rn]$5f^{12}7s^2$	6.50	12.7	24.4	
101 Md	[Rn]$5f^{13}7s^2$	6.58	12.8	25.4	
102 No	[Rn]$5f^{14}7s^2$	6.65	13.0	27.0	
103 Lr	[Rn]$5f^{14}6d^17s^2$	4.6	14.8	23.0	

Resource section 3
Standard potentials

The standard potentials quoted here are presented in the form of Latimer diagrams (Section 5.12) and are arranged according to the blocks of the periodic table in the order s, p, d, f. Data and species in parentheses are uncertain. Most of the data, together with occasional corrections, come from A.J. Bard, R. Parsons, and J. Jordan (eds.), *Standard potentials in aqueous solution*. Marcel Dekker, New York (1985). Data for the actinoids are from L.R. Morss, *The chemistry of the actinide elements*, Vol. 2 (eds. J.J. Katz, G.T. Seaborg, and L.R. Morss). Chapman & Hall, London (1986). The value for $[Ru(bpy)_3]^{3+/2+}$ is from B. Durham, J.L. Walsh, C.L. Carter, and T.J. Meyer, *Inorg. Chem.*, 1980, **19**, 860. Potentials for carbon species and some d-block elements are taken from S.G. Bratsch, *J. Phys. Chem. Ref. Data*, 1989, **18**, 1. For further information on standard potentials of unstable radical species see D.M. Stanbury, *Adv. Inorg. Chem.*, 1989, **33**, 69. Potentials in the literature are occasionally reported relative to the standard calomel electrode (SCE) and may be converted to the H^+/H_2 scale by adding 0.2412 V. For a detailed discussion of other reference electrodes, see D.J.G. Ives and G.J. Janz, *Reference electrodes*. Academic Press, New York (1961).

s Block • Group 1

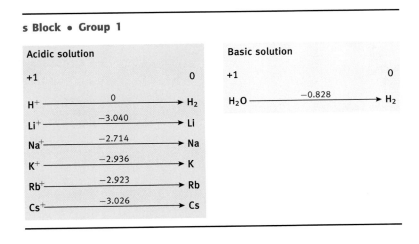

s Block • Group 2

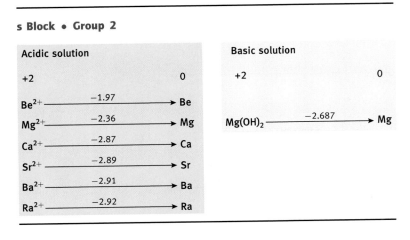

p Block • Group 13

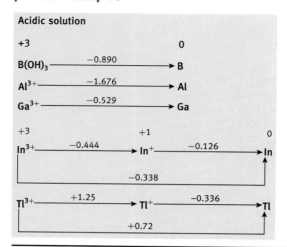

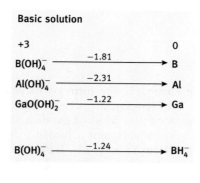

p Block • Group 14

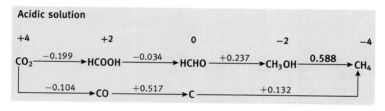

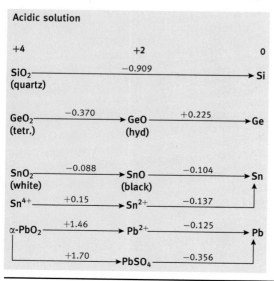

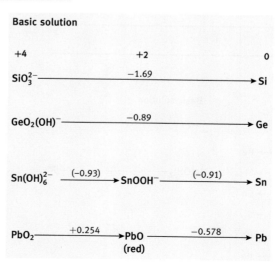

p Block • Group 15

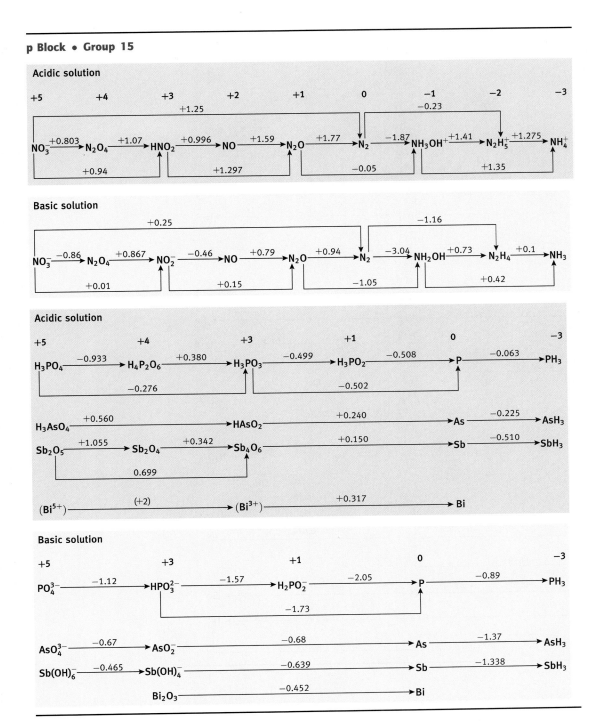

Acidic solution

+5	+4	+3	+2	+1	0	−1	−2	−3

NO_3^- →(+0.803) N_2O_4 →(+1.07) HNO_2 →(+0.996) NO →(+1.59) N_2O →(+1.77) N_2 →(−1.87) NH_3OH^+ →(+1.41) $N_2H_5^+$ →(+1.275) NH_4^+

+1.25
−0.23
+0.94
+1.297
−0.05
+1.35

Basic solution

NO_3^- →(−0.86) N_2O_4 →(+0.867) NO_2^- →(−0.46) NO →(+0.79) N_2O →(+0.94) N_2 →(−3.04) NH_2OH →(+0.73) N_2H_4 →(+0.1) NH_3

+0.25
−1.16
+0.01
+0.15
−1.05
+0.42

Acidic solution

+5	+4	+3	+1	0	−3

H_3PO_4 →(−0.933) $H_4P_2O_6$ →(+0.380) H_3PO_3 →(−0.499) H_3PO_2 →(−0.508) P →(−0.063) PH_3

−0.276
−0.502

H_3AsO_4 →(+0.560) $HAsO_2$ →(+0.240) As →(−0.225) AsH_3

Sb_2O_5 →(+1.055) Sb_2O_4 →(+0.342) Sb_4O_6 →(+0.150) Sb →(−0.510) SbH_3

0.699

(Bi^{5+}) →(+2) (Bi^{3+}) →(+0.317) Bi

Basic solution

+5	+3	+1	0	−3

PO_4^{3-} →(−1.12) HPO_3^{2-} →(−1.57) $H_2PO_2^-$ →(−2.05) P →(−0.89) PH_3

−1.73

AsO_4^{3-} →(−0.67) AsO_2^- →(−0.68) As →(−1.37) AsH_3

$Sb(OH)_6^-$ →(−0.465) $Sb(OH)_4^-$ →(−0.639) Sb →(−1.338) SbH_3

Bi_2O_3 →(−0.452) Bi

p Block • Group 16

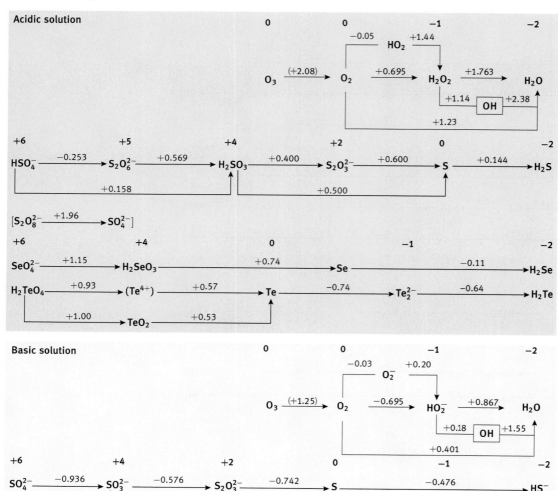

p Block • Group 17

Acidic solution

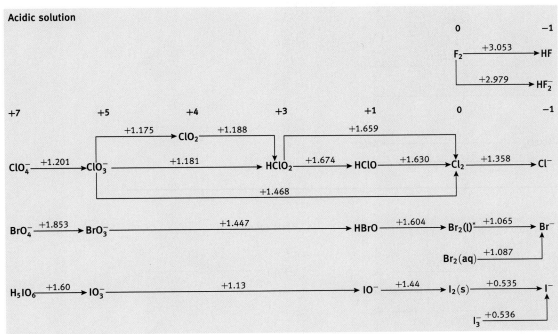

*Bromine is not sufficiently solube in water at room temperature to achieve unit activity. Therefore, the value for a saturated solution in contact with $Br_2(l)$ should be used in all practical calculations.

Basic solution

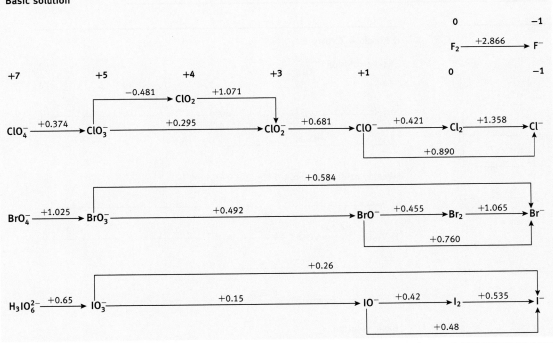

p Block • Group 18

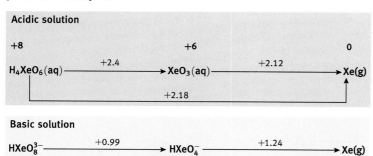

Acidic solution

+8		+6		0
$H_4XeO_6(aq)$	$\xrightarrow{+2.4}$	$XeO_3(aq)$	$\xrightarrow{+2.12}$	$Xe(g)$

$+2.18$

Basic solution

$HXeO_6^{3-} \xrightarrow{+0.99} HXeO_4^- \xrightarrow{+1.24} Xe(g)$

d Block • Group 3

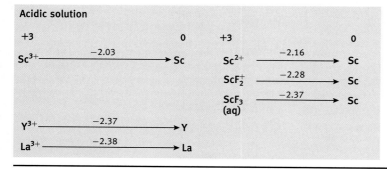

Acidic solution

+3		0
Sc^{3+}	$\xrightarrow{-2.03}$	Sc

+3		0
Sc^{2+}	$\xrightarrow{-2.16}$	Sc
ScF_2^+	$\xrightarrow{-2.28}$	Sc
ScF_3 (aq)	$\xrightarrow{-2.37}$	Sc

Basic solution

+3		0
$Sc(OH)_3$	$\xrightarrow{-2.60}$	Sc

$Y^{3+} \xrightarrow{-2.37} Y$

$La^{3+} \xrightarrow{-2.38} La$

d Block • Group 4

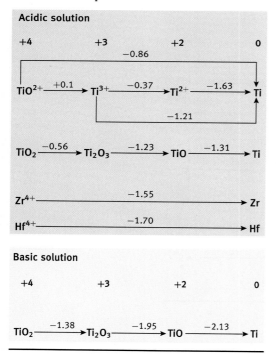

Acidic solution

+4	+3	+2	0
	-0.86		
$TiO^{2+} \xrightarrow{+0.1}$	$Ti^{3+} \xrightarrow{-0.37}$	$Ti^{2+} \xrightarrow{-1.63}$	Ti
		-1.21	

$TiO_2 \xrightarrow{-0.56} Ti_2O_3 \xrightarrow{-1.23} TiO \xrightarrow{-1.31} Ti$

$Zr^{4+} \xrightarrow{-1.55} Zr$

$Hf^{4+} \xrightarrow{-1.70} Hf$

Basic solution

+4	+3	+2	0
$TiO_2 \xrightarrow{-1.38}$	$Ti_2O_3 \xrightarrow{-1.95}$	$TiO \xrightarrow{-2.13}$	Ti

d Block • Group 5

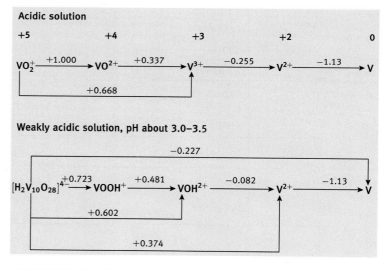

Acidic solution

+5	+4	+3	+2	0
$VO_2^+ \xrightarrow{+1.000}$	$VO^{2+} \xrightarrow{+0.337}$	$V^{3+} \xrightarrow{-0.255}$	$V^{2+} \xrightarrow{-1.13}$	V
	$+0.668$			

Weakly acidic solution, pH about 3.0–3.5

		-0.227		
$[H_2V_{10}O_{28}]^{4-} \xrightarrow{+0.723}$	$VOOH^+ \xrightarrow{+0.481}$	$VOH^{2+} \xrightarrow{-0.082}$	$V^{2+} \xrightarrow{-1.13}$	V
	$+0.602$			
	$+0.374$			

Basic solution

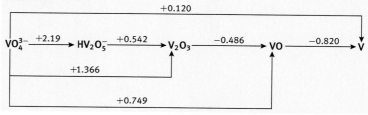

		$+0.120$		
$VO_4^{3-} \xrightarrow{+2.19}$	$HV_2O_5^- \xrightarrow{+0.542}$	$V_2O_3 \xrightarrow{-0.486}$	$VO \xrightarrow{-0.820}$	V
	$+1.366$			
	$+0.749$			

d Block • Group 5 (Continued)

Acidic solution

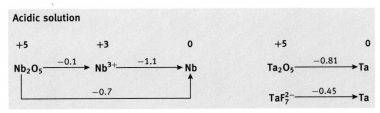

d Block • Group 6

Acidic solution

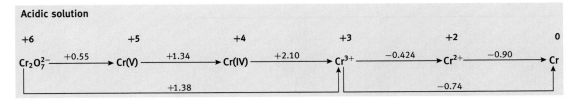

Neutral solution

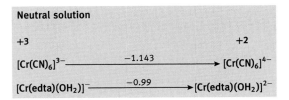

Basic solution

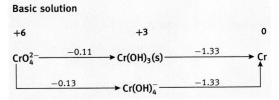

Acidic solution

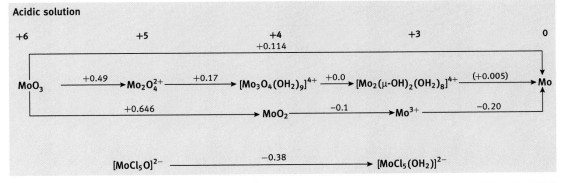

Neutral solution

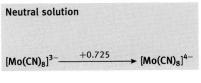

Basic solution

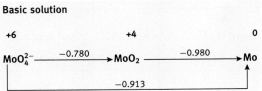

Acidic solution

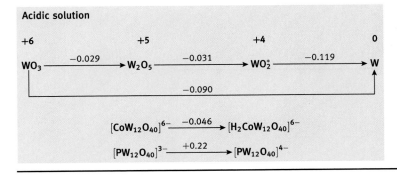

d Block • Group 6 *(Continued)*

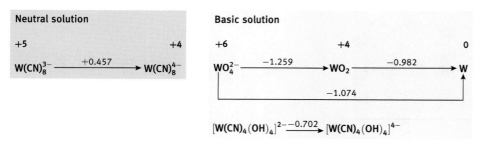

Neutral solution

+5   +4

$W(CN)_8^{3-}$ —— $+0.457$ —→ $W(CN)_8^{4-}$

Basic solution

+6   +4   0

WO_4^{2-} —— -1.259 —→ WO_2 —— -0.982 —→ W

—— -1.074 ——

$[W(CN)_4(OH)_4]^{2-}$ —-0.702→ $[W(CN)_4(OH)_4]^{4-}$

*Probably $[W_3(\mu_3\text{-O})(\mu\text{-O})_3(OH_2)_9]^{4+}$. See S.P. Gosh and E.S. Gould, *Inorg. Chem.*, 1991, **30**, 3662.

d Block • Group 7

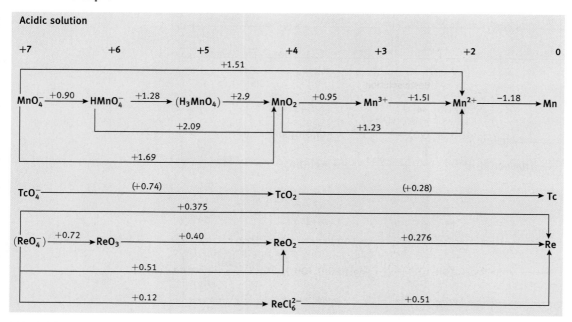

Acidic solution

+7   +6   +5   +4   +3   +2   0

MnO_4^- —$+0.90$→ $HMnO_4^-$ —$+1.28$→ (H_3MnO_4) —$+2.9$→ MnO_2 —$+0.95$→ Mn^{3+} —$+1.5I$→ Mn^{2+} —-1.18→ Mn

+1.51
+2.09
+1.69
+1.23

TcO_4^- —$(+0.74)$→ TcO_2 —$(+0.28)$→ Tc

+0.375
(ReO_4^-) —$+0.72$→ ReO_3 —$+0.40$→ ReO_2 —$+0.276$→ Re
+0.51
+0.12 —→ $ReCl_6^{2-}$ —$+0.51$→

Basic solution

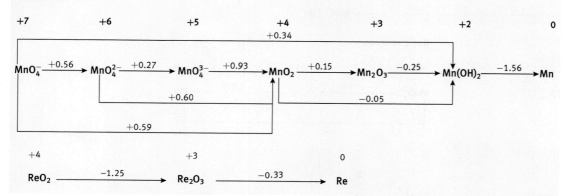

+7   +6   +5   +4   +3   +2   0

+0.34
MnO_4^- —$+0.56$→ MnO_4^{2-} —$+0.27$→ MnO_4^{3-} —$+0.93$→ MnO_2 —$+0.15$→ Mn_2O_3 —-0.25→ $Mn(OH)_2$ —-1.56→ Mn

+0.60
+0.59
−0.05

+4   +3   0

ReO_2 —-1.25→ Re_2O_3 —-0.33→ Re

d Block • Group 8

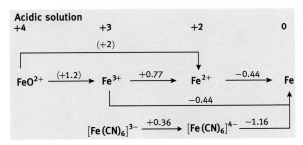

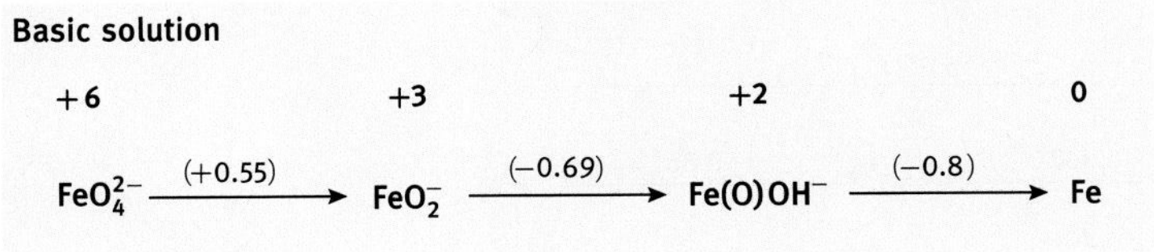

Acidic solution

+4	+3	+2	0

FeO^{2+} $\xrightarrow{(+1.2)}$ Fe^{3+} $\xrightarrow{+0.77}$ Fe^{2+} $\xrightarrow{-0.44}$ Fe

(+2) over $FeO^{2+} \to Fe^{2+}$

-0.44 ($Fe^{3+} \to Fe$)

$[Fe(CN)_6]^{3-}$ $\xrightarrow{+0.36}$ $[Fe(CN)_6]^{4-}$ $\xrightarrow{-1.16}$

Basic solution

+6	+3	+2	0

FeO_4^{2-} $\xrightarrow{(+0.55)}$ FeO_2^- $\xrightarrow{(-0.69)}$ $Fe(O)OH^-$ $\xrightarrow{(-0.8)}$ Fe

Acidic solution

+8	+7	+6	+4	+3	+2	0

+1.04

RuO_4 $\xrightarrow{+0.99}$ RuO_4^- $\xrightarrow{+1.6}$ RuO_2^+ $\xrightarrow{+1.5}$ $(Ru(OH)_2^{2+})^*$ $\xrightarrow{+0.86}$ Ru^{3+} $\xrightarrow{+0.25}$ Ru^{2+} $\xrightarrow{+0.8}$ Ru

+1.4

+0.68

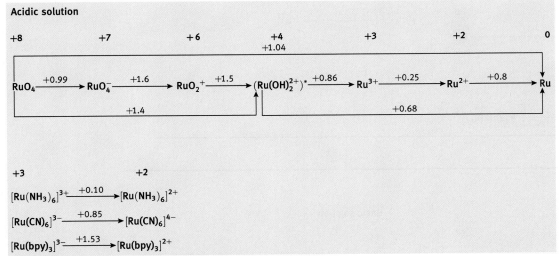

+3	+2

$[Ru(NH_3)_6]^{3+}$ $\xrightarrow{+0.10}$ $[Ru(NH_3)_6]^{2+}$

$[Ru(CN)_6]^{3-}$ $\xrightarrow{+0.85}$ $[Ru(CN)_6]^{4-}$

$[Ru(bpy)_3]^{3-}$ $\xrightarrow{+1.53}$ $[Ru(bpy)_3]^{2+}$

*Likely to be $H_n[Ru_4O_6(OH_2)_{12}]^{(4+n)+}$. See A. Patel and D.T. Richen, *Inorg. Chem.*, 1991, **30**, 3792.

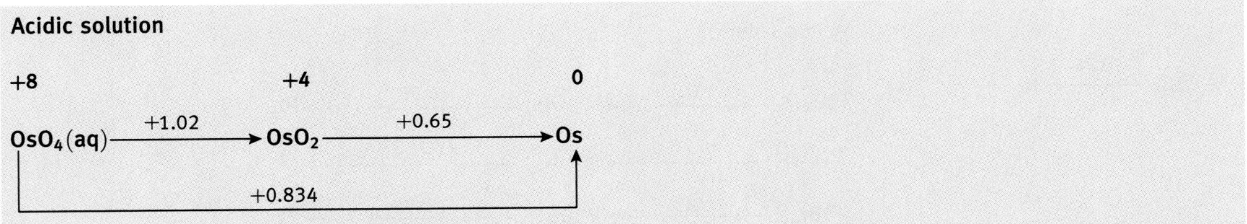

Acidic solution

+8	+4	0

$OsO_4(aq)$ $\xrightarrow{+1.02}$ OsO_2 $\xrightarrow{+0.65}$ Os

+0.834

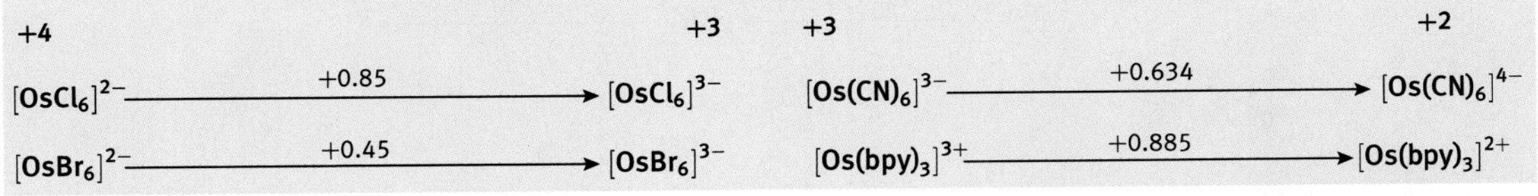

+4	+3	+3	+2

$[OsCl_6]^{2-}$ $\xrightarrow{+0.85}$ $[OsCl_6]^{3-}$ $[Os(CN)_6]^{3-}$ $\xrightarrow{+0.634}$ $[Os(CN)_6]^{4-}$

$[OsBr_6]^{2-}$ $\xrightarrow{+0.45}$ $[OsBr_6]^{3-}$ $[Os(bpy)_3]^{3+}$ $\xrightarrow{+0.885}$ $[Os(bpy)_3]^{2+}$

d Block • Group 9

Acidic solution

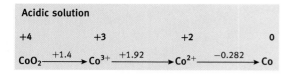

+4   +3   +2   0

$CoO_2 \xrightarrow{+1.4} Co^{3+} \xrightarrow{+1.92} Co^{2+} \xrightarrow{-0.282} Co$

Basic solution

+4   +3   +2   0

$CoO_2 \xrightarrow{(+0.7)} Co(O)OH \xrightarrow{(-0.22)} Co(OH)_2 \xrightarrow{-0.873} Co$

Neutral solution

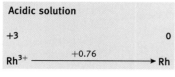

+3   +2

$[Co(NH_3)_6]^{3+} \xrightarrow{+0.058} [Co(NH_3)_6]^{2+}$

$[Co(phen)_3]^{3+} \xrightarrow{+0.33} [Co(phen)_3]^{2+}$

$[Co(ox)_3]^{3-} \xrightarrow{+0.57} [Co(ox)_3]^{4-}$

Acidic solution

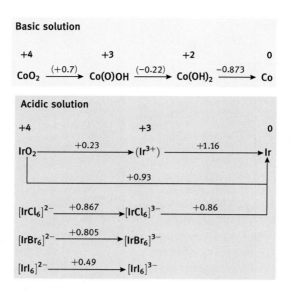

+4   +3   0

$IrO_2 \xrightarrow{+0.23} (Ir^{3+}) \xrightarrow{+1.16} Ir$

$\xrightarrow{+0.93}$

$[IrCl_6]^{2-} \xrightarrow{+0.867} [IrCl_6]^{3-} \xrightarrow{+0.86}$

$[IrBr_6]^{2-} \xrightarrow{+0.805} [IrBr_6]^{3-}$

$[IrI_6]^{2-} \xrightarrow{+0.49} [IrI_6]^{3-}$

Acidic solution

+3   0

$Rh^{3+} \xrightarrow{+0.76} Rh$

Neutral solution

+3   +2

$[Rh(CN)_6]^{3-} \xrightarrow{+0.9} [Rh(CN)_6]^{4-}$

d Block • Group 10

Acidic solution

+4   +3   +2   0

$NiO_2 \xrightarrow{+1.59} Ni^{2+} \xrightarrow{-0.257} Ni$

Basic solution

$NiO_2^* \xrightarrow{+0.7} NiOOH \xrightarrow{+0.52} Ni(OH)_2 \xrightarrow{-0.72} Ni$

Neutral solution

$[Ni(NH_3)_6]^{2+} \xrightarrow{-0.49} Ni$

Acidic solution

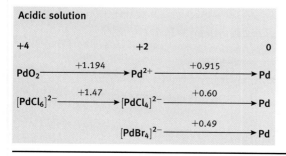

+4   +2   0

$PdO_2 \xrightarrow{+1.194} Pd^{2+} \xrightarrow{+0.915} Pd$

$[PdCl_6]^{2-} \xrightarrow{+1.47} [PdCl_4]^{2-} \xrightarrow{+0.60} Pd$

$[PdBr_4]^{2-} \xrightarrow{+0.49} Pd$

Basic solution

+4   +2   0

$PdO_2 \xrightarrow{+1.47} PdO \xrightarrow{+0.897} Pd$

Acidic solution

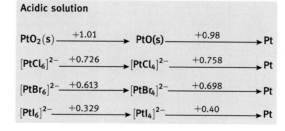

$PtO_2(s) \xrightarrow{+1.01} PtO(s) \xrightarrow{+0.98} Pt$

$[PtCl_6]^{2-} \xrightarrow{+0.726} [PtCl_4]^{2-} \xrightarrow{+0.758} Pt$

$[PtBr_6]^{2-} \xrightarrow{+0.613} [PtBr_4]^{2-} \xrightarrow{+0.698} Pt$

$[PtI_6]^{2-} \xrightarrow{+0.329} [PtI_4]^{2-} \xrightarrow{+0.40} Pt$

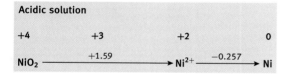

d Block • Group 11

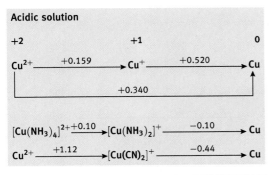

Acidic solution

+2   +1   0

Cu^{2+} $\xrightarrow{+0.159}$ Cu^+ $\xrightarrow{+0.520}$ Cu

$\xrightarrow{+0.340}$

$[Cu(NH_3)_4]^{2+}$ $\xrightarrow{+0.10}$ $[Cu(NH_3)_2]^+$ $\xrightarrow{-0.10}$ Cu

Cu^{2+} $\xrightarrow{+1.12}$ $[Cu(CN)_2]^+$ $\xrightarrow{-0.44}$ Cu

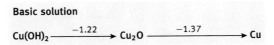

Basic solution

$Cu(OH)_2$ $\xrightarrow{-1.22}$ Cu_2O $\xrightarrow{-1.37}$ Cu

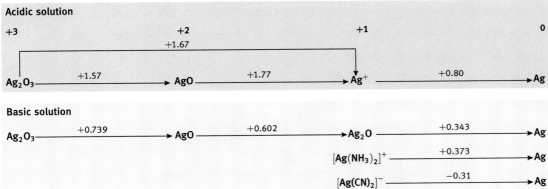

Acidic solution

+3   +2   +1   0

$\xrightarrow{+1.67}$

Ag_2O_3 $\xrightarrow{+1.57}$ AgO $\xrightarrow{+1.77}$ Ag^+ $\xrightarrow{+0.80}$ Ag

Basic solution

Ag_2O_3 $\xrightarrow{+0.739}$ AgO $\xrightarrow{+0.602}$ Ag_2O $\xrightarrow{+0.343}$ Ag

$[Ag(NH_3)_2]^+$ $\xrightarrow{+0.373}$ Ag

$[Ag(CN)_2]^-$ $\xrightarrow{-0.31}$ Ag

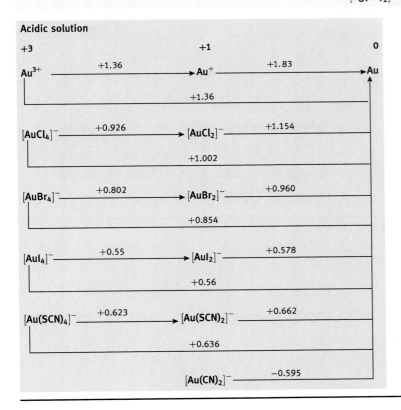

Acidic solution

+3   +1   0

Au^{3+} $\xrightarrow{+1.36}$ Au^+ $\xrightarrow{+1.83}$ Au

$\xrightarrow{+1.36}$

$[AuCl_4]^-$ $\xrightarrow{+0.926}$ $[AuCl_2]^-$ $\xrightarrow{+1.154}$

$\xrightarrow{+1.002}$

$[AuBr_4]^-$ $\xrightarrow{+0.802}$ $[AuBr_2]^-$ $\xrightarrow{+0.960}$

$\xrightarrow{+0.854}$

$[AuI_4]^-$ $\xrightarrow{+0.55}$ $[AuI_2]^-$ $\xrightarrow{+0.578}$

$\xrightarrow{+0.56}$

$[Au(SCN)_4]^-$ $\xrightarrow{+0.623}$ $[Au(SCN)_2]^-$ $\xrightarrow{+0.662}$

$\xrightarrow{+0.636}$

$[Au(CN)_2]^-$ $\xrightarrow{-0.595}$

d Block • Group 12

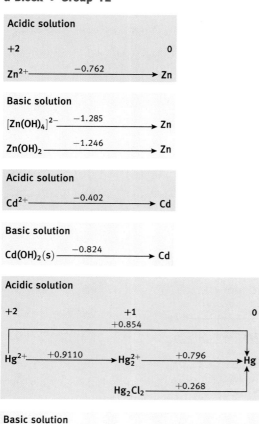

Acidic solution

+2   0

Zn^{2+} ——————-0.762——————→ Zn

Basic solution

$[Zn(OH)_4]^{2-}$ ———-1.285———→ Zn

$Zn(OH)_2$ ———-1.246———→ Zn

Acidic solution

Cd^{2+} ——————-0.402——————→ Cd

Basic solution

$Cd(OH)_2(s)$ ———-0.824———→ Cd

Acidic solution

+2   +1   0

                    $+0.854$

Hg^{2+} —$+0.9110$→ Hg_2^{2+} —$+0.796$→ Hg

                    Hg_2Cl_2 —$+0.268$→

Basic solution

HgO ——————$+0.0977$——————→ Hg

f Block • Lanthanoids

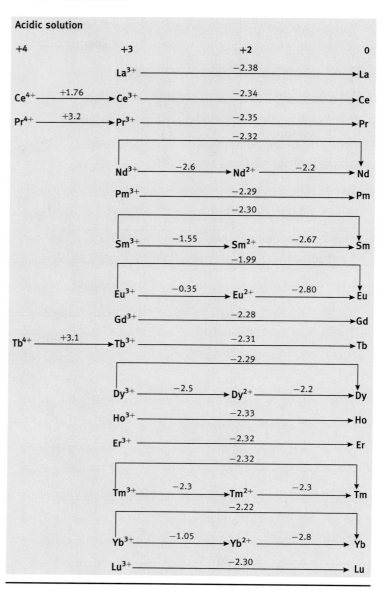

Acidic solution

+4   +3   +2   0

La^{3+} ——————-2.38——————→ La

Ce^{4+} —$+1.76$→ Ce^{3+} ——-2.34——→ Ce

Pr^{4+} —$+3.2$→ Pr^{3+} ——-2.35——→ Pr

                    -2.32

Nd^{3+} ——-2.6——→ Nd^{2+} ——-2.2——→ Nd

Pm^{3+} ——————-2.29——————→ Pm

                    -2.30

Sm^{3+} ——-1.55——→ Sm^{2+} ——-2.67——→ Sm

                    -1.99

Eu^{3+} ——-0.35——→ Eu^{2+} ——-2.80——→ Eu

Gd^{3+} ——————-2.28——————→ Gd

Tb^{4+} —$+3.1$→ Tb^{3+} ——-2.31——→ Tb

                    -2.29

Dy^{3+} ——-2.5——→ Dy^{2+} ——-2.2——→ Dy

Ho^{3+} ——————-2.33——————→ Ho

Er^{3+} ——————-2.32——————→ Er

                    -2.32

Tm^{3+} ——-2.3——→ Tm^{2+} ——-2.3——→ Tm

                    -2.22

Yb^{3+} ——-1.05——→ Yb^{2+} ——-2.8——→ Yb

Lu^{3+} ——————-2.30——————→ Lu

f Block • Actinoids

Acidic solution

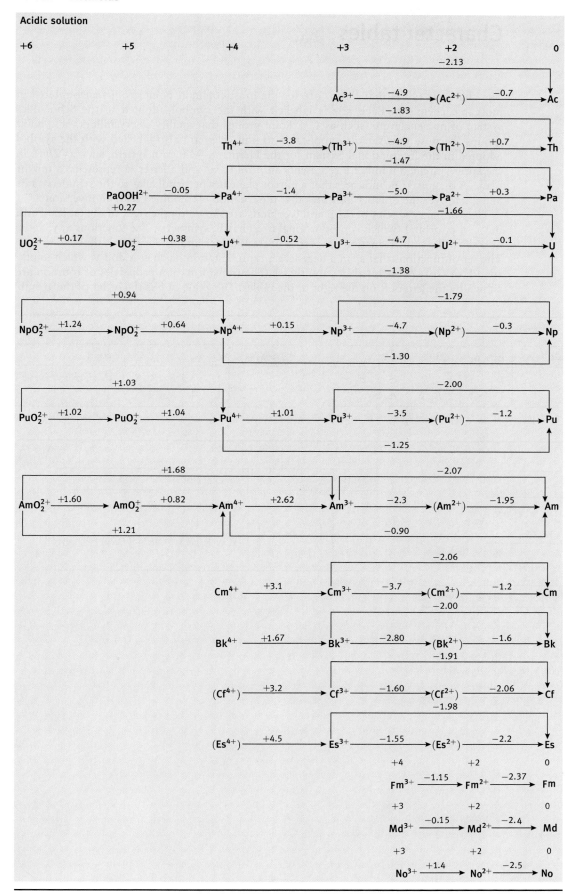

Resource section 4
Character tables

The character tables that follow are for the most common point groups encountered in inorganic chemistry. Each one is labelled with the symbol adopted in the Schoenflies system of nomenclature (such as C_{3v}). Point groups that qualify as crystallographic point groups (because they are also applicable to unit cells) are also labelled with the symbol adopted in the International System (or the Hermann–Mauguin system, such as $2/m$). In the latter system, a number n represents an n-fold axis and a letter m represents a mirror plane. A diagonal line indicates that a mirror plane lies perpendicular to the symmetry axis and a bar over the number indicates that the rotation is combined with an inversion.

The symmetry species of the p and d orbitals are shown on the right of the tables. Thus, in C_{2v}, a p_x orbital (which is proportional to x) has B_1 symmetry. The functions x, y, and z also show the transformation properties of translations and of the electric dipole moment. The set of functions that span a degenerate representation (such as x and y, which jointly span E in C_{3v}) are enclosed in parentheses. The transformation properties of rotation are shown by the letters R on the right of the tables. The value of h is the order of the group.

The groups C_1, C_s, C_i

C_1 (1)	E	$h = 1$
A	1	

$C_s = C_h$ (m)	E	σ_h	$h = 2$	
A′	1	1	x, y, R_z	x^2, y^2, z^2, xy
A″	1	−1	z, R_x, R_y	yz, zx

$C_i = S_2$ (1)	E	i	$h = 2$	
A_g	1	1	R_x, R_y, R_z	$x^2, y^2, z^2, xy, zx, yz$
A_u	1	−1	x, y, z	

The groups C_n

C_2 (2)	E	C_2	$h = 2$	
A	1	1	z, R_z	x^2, y^2, z^2, xy
B	1	−1	x, y, R_x, R_y	yz, zx

C_3 (3)	E	C_3	C_3^2	$\varepsilon = \exp(2\pi i/3)$	$h = 3$	
A	1	1	1	z, R_z	$x^2 + y^2, z^2$	
E	$\begin{Bmatrix} 1 & \varepsilon & \varepsilon^* \\ 1 & \varepsilon^* & \varepsilon \end{Bmatrix}$			$(x, y)(R_x, R_y)$	$(x^2 - y^2, xy)$ (yz, zx)	

C_4 (4)	E	C_4	C_2	C_4^3	$h = 4$	
A	1	1	1	1	z, R_z	$x^2 + y^2, z^2$
B	1	−1	1	−1		$x^2 - y^2, xy$
E	$\begin{Bmatrix} 1 & i & -1 & -i \\ 1 & -i & 1 & i \end{Bmatrix}$				$(x, y)(R_x, R_y)$	(yz, zx)

The groups C_{nv}

C_{2v} (2mm)	E	C_2	σ_v (xz)	σ_v' (yz)		$h = 4$
A_1	1	1	1	1	z	x^2, y^2, z^2
A_2	1	1	−1	−1	R_z	xy
B_1	1	−1	1	−1	x, R_y	zx
B_2	1	−1	−1	1	y, R_x	yz

C_{3v} (3m)	E	$2C_3$	$3\sigma_v$		$h = 6$
A_1	1	1	1	z	$x^2 + y^2, z^2$
A_2	1	1	−1	R_z	
E	2	−1	0	$(x, y)\,(R_x, R_y)$	$(x^2 - y^2, xy)(zx, yz)$

C_{4v} (4mm)	E	$2C_4$	C_2	$2\sigma_v$	$2\sigma_d$		$h = 8$
A_1	1	1	1	1	1	z	$x^2 + y^2, z^2$
A_2	1	1	1	−1	−1	R_z	
B_1	1	−1	1	1	−1		$x^2 - y^2$
B_2	1	−1	1	−1	1		xy
E	2	0	−2	0	0	$(x, y)\,(R_x, R_y)$	(zx, yz)

C_{5v}	E	$2C_5$	$2C_5^2$	$5\sigma_v$		$h = 10, \alpha = 72°$
A_1	1	1	1	1	z	$x^2 + y^2, z^2$
A_2	1	1	1	−1	R_z	
E_1	2	$2\cos\alpha$	$2\cos 2\alpha$	0	$(x, y)\,(R_x, R_y)$	(zx, yz)
E_2	2	$2\cos 2\alpha$	$2\cos\alpha$	0		$(x^2 - y^2, xy)$

C_{6v} (6mm)	E	$2C_6$	$2C_3$	C_2	$3\sigma_v$	$3\sigma_d$		$h = 12$
A_1	1	1	1	1	1	1	z	$x^2 + y^2, z^2$
A_2	1	1	1	1	−1	−1	R_z	
B_1	1	−1	1	−1	1	−1		
B_2	1	−1	1	−1	−1	1		
E_1	2	1	−1	−2	0	0	$(x, y)\,(R_x, R_y)$	(zx, yz)
E_2	2	−1	−1	2	0	0		$(x^2 - y^2, xy)$

$C_{\infty v}$	E	$2C_\phi$	$\infty\sigma_v$		$h = \infty$
$A_1\ (\Sigma^+)$	1	1	1	z	$x^2 + y^2, z^2$
$A_2\ (\Sigma^-)$	1	1	−1	R_z	
$E_1\ (\Pi)$	2	$2\cos\phi$	0	$(x, y)\,(R_x, R_y)$	(zx, yz)
$E_2\ (\Delta)$	2	$2\cos 2\phi$	0		$(xy, x^2 - y^2)$

The groups D_n

D_2 (222)	E	$C_2(z)$	$C_2(y)$	$C_2(x)$		$h = 4$
A	1	1	1	1		x^2, y^2, z^2
B_1	1	1	−1	−1	z, R_z	xy
B_2	1	−1	1	−1	y, R_y	zx
B_3	1	−1	−1	1	x, R_x	yz

D_3 (32)	E	$2C_3$	$3C_2$		$h = 6$
A_1	1	1	1		$x^2 + y^2, z^2$
A_2	1	1	1	z, R_z	
E	2	−1	0	$(x, y)\,(R_x, R_y)$	$(x^2 - y^2, xy)(zx, yz)$

The groups D_{nh}

D_{2h} (mmm)	E	$C_2(z)$	$C_2(y)$	$C_2(x)$	i	$\sigma(xy)$	$\sigma(xz)$	$\sigma(yz)$	$h = 8$	
A_g	1	1	1	1	1	1	1	1		x^2, y^2, z^2
B_{1g}	1	1	−1	−1	1	1	−1	−1	R_z	xy
B_{2g}	1	−1	1	−1	1	−1	1	−1	R_y	zx
B_{3g}	1	−1	−1	1	1	−1	−1	1	R_x	yz
A_u	1	1	1	1	−1	−1	−1	−1		
B_{1u}	1	1	−1	−1	−1	−1	1	1	z	
B_{2u}	1	−1	1	−1	−1	1	−1	1	y	
B_{3u}	1	−1	−1	1	−1	1	1	−1	x	

D_{3h} (6m2)	E	$2C_3$	$3C_2$	σ_h	$2S_3$	$3\sigma_v$	$h = 12$	
A_1'	1	1	1	1	1	1		$x^2 + y^2, z^2$
A_2'	1	1	−1	1	1	−1	R_z	
E'	2	−1	0	2	−1	0	(x, y)	$(x^2 - y^2, xy)$
A_1''	1	1	1	−1	−1	−1		
A_2''	1	1	−1	−1	−1	1	z	
E''	2	−1	0	−2	1	0	(R_x, R_y)	(zx, yz)

D_{4h} (4/mmm)	E	$2C_4$	C_2	$2C_2'$	$2C_2''$	i	$2S_4$	σ_h	$2\sigma_v$	$2\sigma_d$	$h = 16$	
A_{1g}	1	1	1	1	1	1	1	1	1	1		$x^2 + y^2, z^2$
A_{2g}	1	1	1	−1	−1	1	1	1	−1	−1	R_z	
B_{1g}	1	−1	1	1	−1	1	−1	1	1	−1		$x^2 - y^2$
B_{2g}	1	−1	1	−1	1	1	−1	1	−1	1		xy
E_g	2	0	−2	0	0	2	0	−2	0	0	(R_x, R_y)	(zx, yz)
A_{1u}	1	1	1	1	1	−1	−1	−1	−1	−1		
A_{2u}	1	1	1	−1	−1	−1	−1	−1	1	1	z	
B_{1u}	1	−1	1	1	−1	−1	1	−1	−1	1		
B_{2u}	1	−1	1	−1	1	−1	1	−1	1	−1		
E_u	2	0	−2	0	0	−2	0	2	0	0	(x, y)	

D_{5h}	E	$2C_5$	$2C_5^2$	$5C_2$	σ_h	$2S_5$	$2S_5^2$	$5\sigma_v$	$h = 20, \alpha = 72°$	
A_1'	1	1	1	1	1	1	1	1		$x^2 + y^2, z^2$
A_2''	1	1	1	−1	1	1	1	−1	R_z	
E_1'	2	$2\cos\alpha$	$2\cos 2\alpha$	0	2	$2\cos\alpha$	$2\cos 2\alpha$	0	(x, y)	
E_2'	2	$2\cos 2\alpha$	$2\cos\alpha$	0	2	$2\cos 2\alpha$	$2\cos\alpha$	0		$(x - y^2, xy)$
A_1''	1	1	1	1	−1	−1	−1	−1		
A_2''	1	1	1	−1	−1	−1	−1	1	z	
E_1''	2	$2\cos\alpha$	$2\cos 2\alpha$	0	−2	$-2\cos\alpha$	$-2\cos 2\alpha$	0	(R_x, R_y)	(zx, yz)
E_2''	2	$2\cos 2\alpha$	$2\cos\alpha$	0	−2	$-2\cos 2\alpha$	$-2\cos\alpha$	0		

The groups D_{nh} (continued)

D_{6h} (6/mmm)	E	$2C_6$	$2C_3$	C_2	$3C_2'$	$3C_2''$	i	$2S_3$	$2S_6$	σ_h	$3\sigma_d$	$3\sigma_v$	$h = 24$	
A_{1g}	1	1	1	1	1	1	1	1	1	1	1	1		$x^2 + y^2, z^2$
A_{2g}	1	1	1	1	−1	−1	1	1	1	1	−1	−1	R_z	
B_{1g}	1	−1	1	−1	1	−1	1	−1	1	−1	1	−1		
B_{2g}	1	−1	1	−1	−1	1	1	−1	1	−1	−1	1		
E_{1g}	2	1	−1	−2	0	0	2	1	−1	−2	0	0	(R_x, R_y)	(zx, yz)
E_{2g}	2	−1	−1	2	0	0	2	−1	−1	2	0	0		$(x^2 - y^2, xy)$
A_{1u}	1	1	1	1	1	1	−1	−1	−1	−1	−1	−1		
A_{2u}	1	1	1	1	−1	−1	−1	−1	−1	−1	1	1	z	
B_{1u}	1	−1	1	−1	1	−1	−1	1	−1	1	−1	1		
B_{2u}	1	−1	1	−1	−1	1	−1	1	−1	1	1	−1		
E_{1u}	2	1	−1	−2	0	0	−2	−1	1	2	0	0	(x, y)	
E_{2u}	2	−1	−1	2	0	0	−2	1	1	−2	0	0		

$D_{\infty h}$	E	$\infty C_2'$	$2C_\phi$	i	$\infty\sigma_v$	$2S_\phi$	$h = \infty$	
$A_{1g}(\Sigma_g^+)$	1	1	1	1	1	1		$z^2, x^2 + y^2$
$A_{1u}(\Sigma_u^+)$	1	−1	1	−1	1	−1	z	
$A_{2g}(\Sigma_g^-)$	1	−1	1	1	−1	1	R_z	
$A_{2u}(\Sigma_u^-)$	1	1	1	−1	−1	−1		
$E_{1g}(\Pi_g)$	2	0	$2\cos\phi$	2	0	$-2\cos\phi$	(R_x, R_y)	(zx, yz)
$E_{1u}(\Pi_u)$	2	0	$2\cos\phi$	−2	0	$2\cos\phi$	(x, y)	
$E_{2g}(\Delta_g)$	2	0	$2\cos 2\phi$	2	0	$2\cos 2\phi$		$(xy, x^2 - y^2)$
$E_{2u}(\Delta_u)$	2	0	$2\cos 2\phi$	−2	0	$-2\cos 2\phi$		
$\vdots$	$\vdots$	$\vdots$	$\vdots$	$\vdots$	$\vdots$	$\vdots$		

The groups D_{nd}

$D_{2d} = V_d$ (42m)	E	$2S_4$	C_2	$2C_2'$	$2\sigma_d$	$h = 8$	
A_1	1	1	1	1	1		$x^2 + y^2, z^2$
A_2	1	1	1	−1	−1	R_z	
B_1	1	−1	1	1	1		$x^2 - y^2$
B_2	1	−1	1	−1	1	z	xy
E	2	0	−2	0	0	$(x, y) (R_x, R_y)$	(zx, yz)

D_{3d} (3m)	E	$2C_3$	$3C_2$	i	$2S_6$	$3\sigma_d$	$h = 12$	
A_{1g}	1	1	1	1	1	1		$x^2 + y^2, z^2$
A_{2g}	1	1	−1	1	1	−1	R_z	
E_g	2	−1	0	2	−1	0	(R_x, R_y)	$(x^2 - y^2, xy) (zx, yz)$
A_{1u}	1	1	1	−1	−1	−1		
A_{2u}	1	1	−1	−1	−1	1	z	
E_u	2	−1	0	−2	1	0	(x, y)	

The groups D_{nd} (continued)

D_{4d}	E	$2S_8$	$2C_4$	$2S_8^3$	C_2	$4C_2'$	$4\sigma_d$		$h = 16$
A_1	1	1	1	1	1	1	1		$x^2 + y^2, z^2$
A_2	1	1	1	1	1	–1	–1	R_z	
B_1	1	–1	1	–1	1	1	–1		
B_2	1	–1	1	–1	1	–1	1	z	
E_1	2	$\sqrt{2}$	0	$-\sqrt{2}$	–2	0	0	(x, y)	
E_2	2	0	–2	0	2	0	0		$(x^2 - y^2, xy)$
E_3	2	$-\sqrt{2}$	0	$\sqrt{2}$	–2	0	0	(R_x, R_y)	(zx, yz)

The cubic groups

T_d ($43m$)	E	$8C_3$	$3C_2$	$6S_4$	$6\sigma_d$		$h = 24$
A_1	1	1	1	1	1		$x^2 + y^2 + z^2$
A_2	1	1	1	–1	–1		
E	2	–1	2	0	0		$(2z^2 - x^2 - y^2, x^2 - y^2)$
T_1	3	0	–1	1	–1	(R_x, R_y, R_z)	
T_2	3	0	–1	–1	1	(x, y, z)	(xy, yz, zx)

O_h ($m3m$)	E	$8C_3$	$6C_2$	$6C_4$	$3C_2(=C_4^2)$	i	$6S_4$	$8S_6$	$3\sigma_h$	$6\sigma_d$		$h = 48$
A_{1g}	1	1	1	1	1	1	1	1	1	1		$x^2 + y^2 + z^2$
A_{2g}	1	1	–1	–1	1	1	–1	1	1	–1		
E_g	2	–1	0	0	2	2	0	–1	2	0		$(2z^2 - x^2 - y^2, x^2 - y^2)$
T_{1g}	3	0	–1	1	–1	3	1	0	–1	–1	(R_x, R_y, R_z)	
T_{2g}	3	0	1	–1	–1	3	–1	0	–1	1		(xy, yz, zx)
A_{1u}	1	1	1	1	1	–1	–1	–1	–1	–1		
A_{2u}	1	1	–1	–1	1	–1	1	–1	–1	1		
E_u	2	–1	0	0	2	–2	0	1	–2	0		
T_{1u}	3	0	–1	1	–1	–3	–1	0	1	1	(x, y, z)	
T_{2u}	3	0	1	–1	–1	–3	1	0	1	–1		

The icosahedral group

I	E	$12C_5$	$12C_5^2$	$20C_3$	$15C_2$		$h = 60$
A_1	1	1	1	1	1		$x^2 + y^2 + z^2$
T_1	3	$\frac{1}{2}(1+\sqrt{5})$	$\frac{1}{2}(1-\sqrt{5})$	0	–1	(x, y, z) (R_x, R_y, R_z)	
T_2	3	$\frac{1}{2}(1-\sqrt{5})$	$\frac{1}{2}(1+\sqrt{5})$	0	–1		
G	4	–1	–1	1	0		
H	5	0	0	–1	1		$(2z^2 - x^2 - y^2, x^2 - y^2, xy, yz, zx)$

Further information: www.oxfordtextbooks.co.uk/orc/ichem5e

Resource section 5
Symmetry-adapted orbitals

Table RS5.1 gives the symmetry classes of the s, p, and d orbitals of the central atom of an AB_n molecule of the specified point group. In most cases, the z-axis is the principal axis of the molecule; in C_{2v} the x-axis lies perpendicular to the molecular plane.

The orbital diagrams that follow show the linear combinations of atomic orbitals on the peripheral atoms of AB_n molecules of the specified point groups. Where a view from above is shown, the dot representing the central atom is either in the plane of the paper (for the D groups) or above the plane (for the corresponding C groups). Different phases of the atomic orbitals ($+1$ or -1; amplitudes) are shown by different colours. Where there is a large difference in the magnitudes of the orbital coefficients in a particular combination, the atomic orbitals have been drawn large or small to represent their relative contributions to the linear combination. In the case of degenerate linear combinations (those labelled E or T), any linearly independent combination of the degenerate pair is also of suitable symmetry. In practice, these different linear combinations look like the ones shown here, but their nodes are rotated by an arbitrary axis around the z-axis.

Molecular orbitals are formed by combining an orbital of the central atom (as in Table RS5.1) with a linear combination of the same symmetry.

Table RS5.1 Symmetry species of orbitals on the central atom

	$D_{\infty h}$	C_{2v}	D_{3h}	C_{3v}	D_{4h}	C_{4v}	D_{5h}	C_{5v}	D_{6h}	C_{6v}	T_d	O_h
s	Σ	A_1	A_1'	A_1	A_{1g}	A_1	A_1'	A_1	A_{1g}	A_1	A_1	A_{1g}
p_x	Π	B_1	E'	E	E_u	E	E_1'	E_1	E_{1u}	E_1	T_2	T_{1u}
p_y	Π	B_2	E'	E	E_u	E	E_1'	E_1	E_{1u}	E_1	T_2	T_{1u}
p_z	Σ	A_1	A_2''	A_1	A_{2u}	A_1	A_2''	A_1	A_{2u}	A_1	T_2	T_{1u}
d_{z^2}	Σ	A_1	A_1'	A_1	A_{1g}	A_1	A_1'	A_1	A_{1g}	A_1	E	E_g
$d_{x^2-y^2}$	Δ	A_1	E'	E	B_{1g}	B_1	E_2'	E_2	E_{2g}	E_2	E	E_g
d_{xy}	Δ	A_2	E'	E	B_{2g}	B_2	E_2'	E_2	E_{2g}	E_2	T_2	T_{2g}
d_{yz}	Π	B_2	E''	E	E_g	E	E_1''	E_1	E_{1g}	E_1	T_2	T_{2g}
d_{zx}	Π	B_1	E''	E	E_g	E	E_1''	E_1	E_{1g}	E_1	T_2	T_{2g}

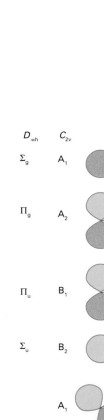

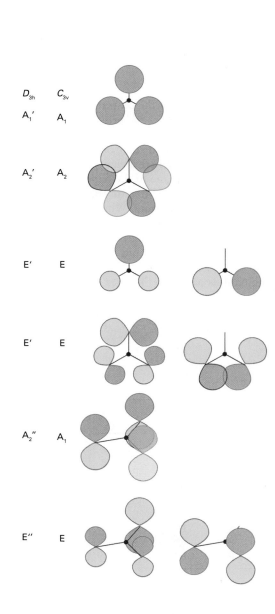

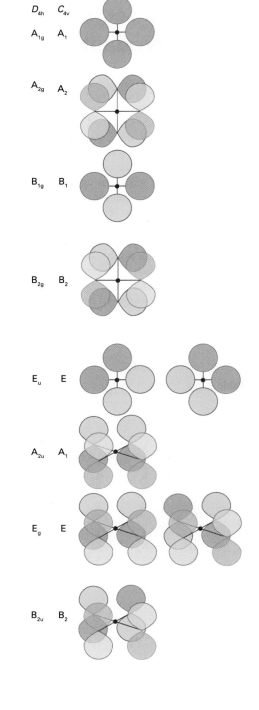

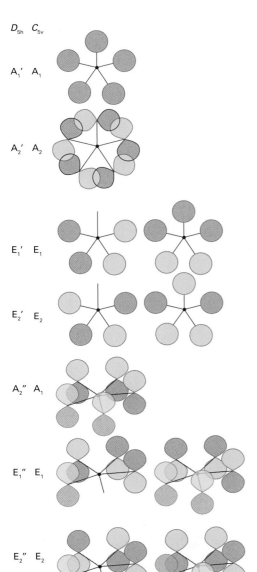

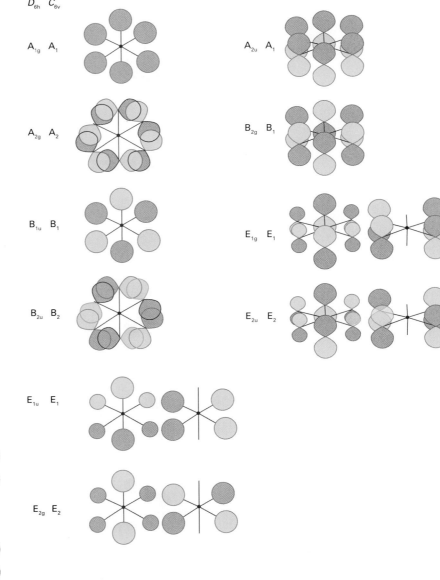

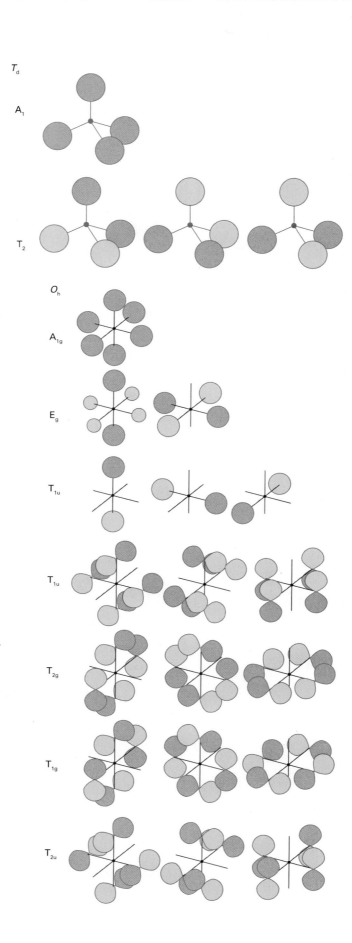

Resource section 6
Tanabe–Sugano diagrams

This section collects together the Tanabe–Sugano diagrams for octahedral complexes with electron configurations d^2 to d^8. The diagrams, which were introduced in Section 20.4, show the dependence of the term energies on ligand-field strength. The term energies E are expressed as the ratio E/B, where B is a Racah parameter, and the ligand-field splitting Δ_O is expressed as Δ_O/B. Terms of different multiplicity are included in the same diagram by making specific, plausible choices about the value of the Racah parameter C, and these choices are given for each diagram. The term energy is always measured from the lowest energy term, and so there are discontinuities of slope where a low-spin term displaces a high-spin term at sufficiently high ligand-field strengths for d^4 to d^8 configurations. Moreover, the noncrossing rule requires terms of the same symmetry to mix rather than to cross, and this mixing accounts for the curved rather than the straight lines in a number of cases. The term labels are those of the point group O_h.

The diagrams were first introduced by Y. Tanabe and S. Sugano, *J. Phys. Soc. Japan*, 1954, **9**, 753. They may be used to find the parameters Δ_O and B by fitting the ratios of the energies of observed transitions to the lines. Alternatively, if the ligand-field parameters are known, then the ligand-field spectra may be predicted.

---

1. d^2 with $C = 4.428B$

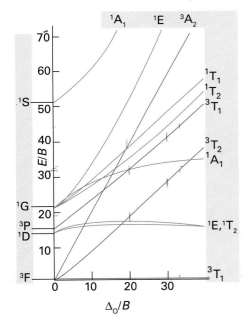

2. d^3 with $C = 4.502B$

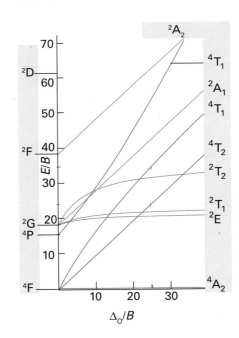

3. d⁴ with C = 4.611B

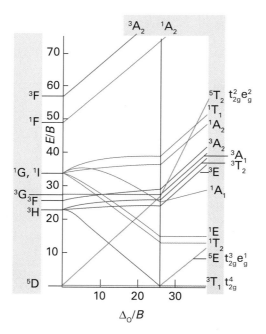

4. d⁵ with C = 4.477B

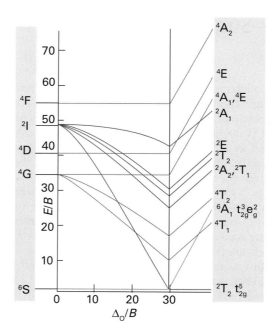

5. d⁶ with C = 4.808B

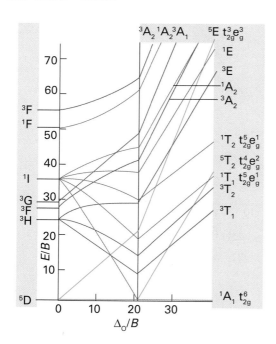

6. d⁷ with C = 4.633B

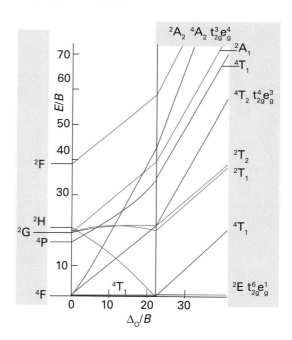

7. d⁸ with $C = 4.709B$

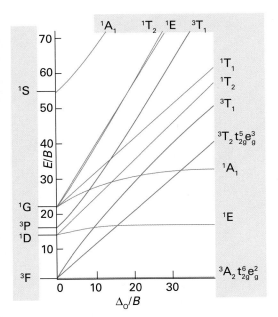

Index